T0321977

STRATEGIES AND SOLUTIONS TO ADVANCED ORGANIC REACTION MECHANISMS

STRATEGIES AND SOLUTIONS TO ADVANCED ORGANIC REACTION MECHANISMS

A New Perspective on McKillop's Problems

ANDREI HENT
University of Toronto, Toronto, ON, Canada

JOHN ANDRAOS
CareerChem, Toronto, ON, Canada

ACADEMIC PRESS

An imprint of Elsevier

Academic Press is an imprint of Elsevier
125 London Wall, London EC2Y 5AS, United Kingdom
525 B Street, Suite 1650, San Diego, CA 92101, United States
50 Hampshire Street, 5th Floor, Cambridge, MA 02139, United States
The Boulevard, Langford Lane, Kidlington, Oxford OX5 1GB, United Kingdom

Library of Congress Cataloging-in-Publication Data
A catalog record for this book is available from the Library of Congress

British Library Cataloguing-in-Publication Data
A catalogue record for this book is available from the British Library

ISBN 978-0-12-812823-7

For information on all Academic Press publications
visit our website at https://www.elsevier.com/books-and-journals

Publisher: Susan Dennis
Acquisition Editor: Emily McCloskey
Editorial Project Manager: Peter Llewellyn
Production Project Manager: Paul Prasad Chandramohan
Cover Designer: Mark Rogers

Typeset by SPi Global, India

Working together
to grow libraries in
developing countries

www.elsevier.com • www.bookaid.org

Dedication

To Paul, and to my parents—AH
To Mom, Ed, Riley, and Josh—JA

Contents

Preface

THE PURPOSE OF WRITING THIS BOOK

Upon reading the title of this book one may wonder, "why write another book about reaction mechanism" among a sea of already published books on this mature subject? We offer several reasons in the categories of pedagogy and research.

With respect to pedagogy we point out the following issues. Pedagogical books on the subject of organic chemistry do not contain references to the original literature. Disappointingly, authors do not take the time to explain *how* to draw chemical structures and reaction schemes *before* introducing the plethora of chemistries according to functional group characteristics. This is so vital and fundamental that the current osmotic "monkey-see-monkey-do" pedagogical approach of copying an instructor's motions without understanding is, we believe, the source of all frustrations encountered by students, regardless of ability, in their study of organic chemistry. Instructors have forgotten that the idea of learning the language of organic chemistry follows the same sense as how a child learns how to draw the letters of the alphabet before learning how to pronounce them, read words, and then to construct sentences from those words according to grammatical logic. In organic chemistry, two-dimension pictures of three-dimensional chemical structures replace the function of words in sentences. The skill of reading and writing in organic chemistry is based entirely on visual representation, communication, and understanding. Other missing aspects of pedagogy include: *how* to problem solve, *how* to connect mechanisms with actual experimental evidence, and showing the evolution of various proposed mechanisms for a given transformation and how each proposal is tested against experimental evidence. Instead, in current pedagogical practice there is a strong emphasis on osmotic learning and rote memorization coupled with a poor and nonchalant attitude to using curly arrow notation without regarding the arrow notation as a mathematical *directed graph* that follows strict rules. This is in sharp contrast to Henry E. Armstrong's (the father of the concept of valence) derisive comment that "a bent arrow never hit anything" when he described what he thought of the concept of electron-pair displacement along conjugated systems.[1] Some of these educational laments were nicely summarized in a recent article in *Chemical & Engineering News* in 2016 based on a symposium entitled "Is There a Crisis in Organic Chemistry Education" held at an ACS National Meeting in San Diego.[2]

With respect to research published in the literature there are the following issues. Modern scientific publications show that scientists, particularly synthetic organic chemists, have a foggy understanding of reaction mechanism. They are rather surprised, even shocked, to learn that the sum of elementary steps in a reaction mechanism must add up to the overall stoichiometric balanced chemical equation for a given transformation. One of us (JA) recalls an amusing situation at a conference of industrial process chemists when such a statement was made and the number of double takes, unsettled frowns, and other facial contortions observed in the audience. Such reactions soon disappeared when they saw illustrative examples from elementary organic chemistry learned in the undergraduate curriculum. Authors of publications, particularly in communications, often represent mechanisms as a customary after-thought when concluding their papers. They are left as a conjecture without any supporting experimental evidence. They are given as a best educated guess with no serious follow-up to test hypotheses. It is perfectly acceptable for a synthetic chemist to relinquish the task of supplying experimental verification for a reaction mechanism if they are not skilled in the kind of techniques and instrumentation required to do so. However, it is not acceptable to put forward a conjectured mechanism without at least offering well-thought out suggestions as to how it can be tested given the fact that there exists more than a century of well-established knowledge in the literature on mechanism elucidation techniques that form the standard lexicon of the study of organic chemistry. This is consistent with the finding that more than two-thirds of posed problems investigated in this work are based on conjectured mechanisms. Furthermore, we were surprised to find some publications containing curly arrow notations that were sloppy and in some cases completely wrong, which we believe is more telling of the peer review process than authors' *faux pas*. We also point out that the modern fad of depicting mechanisms as catalytic cycles, though it serves as convenient shorthand, obscures the visual communication of mechanism because curly arrow notation for tracking electron flow cannot be used due to the already used *curly reaction arrows*. Furthermore, authors do not always specify the oxidation states of metals in organometallic catalysts in these depictions. We suggest that authors who do not specify oxidation states in such representations likely do not know them, and therefore are not convincing readers of their papers that they know what they are talking about.

WHAT THIS BOOK OFFERS

The main highlights of our contribution not mentioned by other books include the following:

- connecting the elementary steps in a reaction mechanism to the overall balanced chemical equation;
- depicting reaction mechanisms using the principle of conservation of structural aspect throughout the visual display;
- balancing each elementary step in a mechanism according to number of elements and charges, and showing reaction by-products along the way;
- showing care and rigor in using the curly arrow notation for one- and two-electron transfers;
- strongly connecting the experimental and theoretical evidences found that support a proposed mechanism for a given reaction;
- showing how to problem solve when one is faced with the same question that is repeated 300 times in our book; namely, given the following reaction with substrate structure A undergoing a reaction under conditions B that yields product structure C, write out a mechanism that best satisfies the given evidence.

Our emphasis is on problem solving to showcase how to integrate all of the earlier ideas. The focus is on the *how* aspect of problem solving. Problem solving is an active exercise that is a highly effective pedagogical tool to absorb, assimilate, integrate, and implement tools learned, in contrast to the passive exercise of reading descriptive information as is customary in the delivery of physical organic chemistry and reaction mechanism subjects in the current university curriculum.

A key insight to contemplate is that a chemical drawing of a structure or mechanism is a representation of our understanding of it. This statement is true for any kind of drawing beyond drawings of chemical structures. In our experience over the course of this work we have found that the principle of conservation of structural aspect applied to the drawing of structures in a mechanism scheme was the most powerful in directing our thought processes in writing out sensible and probable mechanisms. Time and time again the degree of clarity of presentation revealed a path to a solution. Yet, amazingly this simple technique is never mentioned in all the books and pedagogical literature we have found on the subject of organic chemistry. Well-displayed mechanistic schemes in truth do not need accompanying text to explain what is going on in a chemical transformation. They can be read and understood readily without need for redundant exposition.

With respect to balancing chemical equations, we point out that the equal sign notation was used in the 19th century chemistry literature to keep track of atoms on the reactant and product sides without depicting chemical structures. In those representations only molecular formulas were used for reactants and products. There was an obvious and strong connection between the meaning of a balanced chemical equation and a mathematical one. The reaction arrow sign was later adopted when equations were written out using chemical structures instead of molecular formulas. Borrowing from van't Hoff's notation where arrows depicted the direction of a reaction from reactants to products, and therefore the kinetics and dynamics of reactions, the currently used representation of chemical equations resulted in significant loss of information with respect to not specifying by-products and hence loss of information in deducing reaction mechanism. In modern literature chemical equations are no longer balanced as before atom-by-atom. Synthetic chemists adopted the reaction arrow notation since their focus was only on the substrate and product of interest in a chemical reaction, and comparing their structures to see "what happened."

Why would a research chemist investigate a reaction mechanism in the first place?

Some possible reasons include: (1) the product of the reaction they were carrying out yielded an unexpected product—this could be a surprise or the result of a "failed" experiment toward an intended target product; (2) the reaction has synthetic utility and knowledge of the mechanism can elucidate how to further optimize the reaction conditions to a desired product outcome; (3) a reaction produces at least two desirable product outcomes depending on reaction conditions and knowledge of the mechanism can exploit shunting the reaction in favor of each of these products in high yield; or (4) the reaction is unusual and has no precedent in the database of known organic reactions. Rearrangement and redox reactions are by far the two classes of reactions that generate the most interest and challenges in terms of reaction mechanism elucidation. Modern synthetic chemists are particularly keen on ring construction reactions that can form more than one ring in a single step, and on reactions that are able to functionalize unactivated CH groups. Regio- and stereoselectivity in reaction performance is also of very high interest and goes back a long way.

What constitutes "proof" or "evidence" in support of a proposed mechanism? What does it mean to say that you understand how a reaction proceeds? There are some key philosophical aspects of providing evidence for a given mechanism proposal that is thought to be operative for a given reaction that need mentioning. Experimental methods used to study mechanisms are never 100% conclusive. Evidence is obtained from a consensus of experimental observations that are self-consistent and point in the same direction. Mechanisms can be disproved but not proved. This statement needs some time to digest. From a set of mechanistic proposals for a given chemical transformation, rather than proving directly which one is *the* mechanism, the approach is to devise a series of experiments to disprove them until one is left standing that is most consistent with the available experimental evidence. This becomes the "accepted" prevailing mechanism for the given transformation—for now. However, there is always the possibility of revision of thinking based on new findings or extra verification pending the utilization of new, more efficient techniques or more sensitive and accurate instrumentation or better computational methods that can become available in the future. Mechanisms are therefore regarded as tested models rather than ironclad theorems that are true for all time as is the case in mathematics. This is a different line of thinking compared to mathematical proofs which can be constructed as deductive, inductive, or contradictive. The best evidence is to have a synergy between experimental and theoretical (computational) support. Although our efforts may not achieve true certainty, they will undoubtedly produce much opportunity.

Publications that demonstrate how mechanism informs organic synthesis, and vice versa, also demonstrate a complementary and strengthened understanding of how reactions proceed. We point out that this key insight is often not practiced and hence such papers are scarce in the literature. This may be a result of the personal rift between two giants in the development of organic chemistry: Sir Robert Robinson (synthesis) versus Sir Christopher K. Ingold (mechanism).[3–6] Unfortunately, the two schools of thought that each man created had more of an antagonistic relationship between them than a cooperative one that survives to the present day. Their differing nomenclatures for the same ideas including opposing sign conventions attached to substituent effects were a direct result of their mutual ego bashing and created in the early days much unnecessary confusion for the rest of the chemistry community, hence delaying adoption of mechanistic understanding and delaying advancement in science. Hard core mechanistic chemists are largely engaged in exploring the *minutia* of mechanism details, such as the number of water molecules involved in the transition state of a hydration reaction, which synthetic chemists would find no use for. On the other hand, hard core synthetic chemists have poor to nonexistent mathematics skills which means they are unable to carry out and understand kinetics experiments and are strained beyond their comfort zone in interpreting energy reaction coordinate diagrams. Mechanistic chemists, in turn, do not routinely read the synthesis literature on natural products because they perceive their complex structures to be outside the scope of their investigations. Yet, experimental problems often encountered in organic synthesis practice, such as failed attempts to carry out intended reactions or the obtainment of unexpected products, can all be explained and resolved by understanding the underlying reaction mechanism. A good example is the difficulty in trying to carry out esterifications of salicylic acids due to the internal hydrogen bond that exists between the *ortho* juxtaposed carboxylic acid and phenolic groups. A well-known synthetic chemist at Queen's University in Canada "discovered" this problem in his own research about a decade ago and thought that this was a "new" finding without knowing that this problem was well described and investigated in the literature by mechanistic chemists several decades earlier. The ideological tensions between synthetic and mechanistic chemists resulted in an identity crisis of Hamletian proportions in the late 1990s when several heavyweights in physical organic chemistry convened a symposium to address perceived declines in the field with respect to recruitment, scientific advancement, and funding. This crisis of relevance to modern chemistry research led some to remind the community of its triumphs over many years in advancing basic science and its connection to other emerging fields in chemistry. Others advocated for a complete rebranding of the perceived "dead subject" to make it more palatable and ultimately marketable to chemists working in the well-funded applied areas of biological chemistry and material science. The reader is referred to the second issue of *Pure and Applied Chemistry* (1997) and the first issue of *Israel Journal of Chemistry* (2016) which are special issues containing several papers discussing this ongoing debate albeit largely written by old-guard members of a bygone era. Another more recent account traces historical highlights of the field.[7]

The main take-home message that we hope comes across to the reader in this book is that the intellectual exercise of elucidating reaction mechanisms works hand-in-hand in the service, understanding, and ultimately improvement of organic synthesis design and thinking. Putting problem solving as the main focus of human effort over base human needs of recognition and attribution is more convincing to aspiring young scientists to join the enterprise to increase human knowledge in the chemical sciences and ultimately to make serious contributions to addressing pressing problems that actually matter to the wider world.

ORGANIZATION OF BOOK AND LAYOUT OF SOLUTIONS

We present a brief synopsis of the topics covered in each chapter.
Chapter 1: Logic of Organic Reaction Mechanisms

- What constitutes a chemical reaction?
- The importance and meaning of a chemically balanced chemical equation and its connection to reaction mechanism
- What constitutes a reaction scheme?
- The principle of conservation of structural aspect
- Curly arrow notation convention and correct implementation for two- and one-electron transfer steps
- Illustration of the fundamental ideas of reaction mechanism using the Baeyer-Villiger oxidation reaction as a worked example
- Survey of textbooks of physical organic chemistry
- Special topics: base strength and pK_a, autoxidation

Chapter 2: Evidence for Organic Reaction Mechanisms

- What constitutes physical organic chemistry?
- Energy reaction coordinate diagrams—how to construct, read, interpret, and use them
- Summary of direct and indirect experimental evidences to support reaction mechanisms
- Illustration of the evolution of supporting experimental evidence using the Baeyer-Villiger oxidation reaction as a worked example

Chapter 3: Problem Solving Organic Reaction Mechanisms

- Theoretical problem-solving strategies applied to reaction mechanism proposals
- Experimental problem-solving strategies to support reaction mechanism proposals
- Illustration of both kinds of problem-solving strategies using the acid-catalyzed cinenic acid to geronic acid rearrangement as a complete case study
- Current state of pedagogy and research in physical organic chemistry

Chapters 4–9: Solutions to 300 Problems
Over the course of his teaching career Prof. Alexander McKillop surveyed the literature and collected interesting examples on cards and used them in making up problem set exercises for his students. Most of the posed problems originated from brief communications in the literature which contained transformations that could be classified as either anomalous, curious, yielded unexpected results, were challenging to rationalize, were explained by dubious mechanistic reasoning, or whose author-suggested mechanisms were outright incorrect. His original book publication *Advanced Problems in Organic Reaction Mechanisms* (1998) was a transcription of these cards but did not include the original references and the problems were listed in a random order. No doubt, these problems were a fertile training ground for his students to think logically about proposing rational mechanisms, particularly for students pursuing research in natural products synthesis and organic synthesis methodology. All of the chemistries highlighted offer opportunities for further investigation which astute students could use to explore in their own research careers. Hence, McKillop really offered his students ideas for their own research proposals if they were to pursue academic careers. The good news is that there exists a never-ending supply of such examples in the literature for instructors and researchers to draw upon for posing future problems as training exercises.

The following template protocol was used for displaying solutions.

(i) A problem statement is given showing structures of substrates and products, reaction yields, and reaction conditions. Corrections to any structural errors introduced by the posed questions in McKillop's original book are made as appropriate.

(ii) The first solution given is the reaction mechanism as given by the authors.

(iii) All mechanisms are displayed according to the following convention: (1) all chemical structures are shown in the same structural aspect for enhanced visual clarity, (2) each elementary step is element and charge balanced, (3) the curly arrow notation is used to track all two- and one-electron movements, (4) reaction by-products are shown directly below step reaction arrows, and (5) target synthesis bonds made are highlighted using bolded notation throughout a given mechanism scheme.

(iv) At the conclusion of each mechanism an overall balanced chemical equation is provided which constitutes the sum of all elementary steps.

(v) A reference citation on which the problem is based is given.

The "key steps explained" section to each solution contains the following information: (1) an accompanying word description of the visual display of the mechanistic scheme showing descriptors of intermediate identification (enols, carbenes, thiiranium ions, etc.); (2) inclusion of all experimental evidences in support of the authors' mechanism; (3) inclusion of alternative mechanisms not considered by the authors; (4) discussion of any controversies, errors, or weak or lack of evidence; (5) inclusion of alternative mechanisms that better agree with the experimental results and reaction conditions, or address our perceived errors in the authors' posed mechanisms; (6) inclusion of ring construction mapping notation if the reaction produces at least one ring in the product structure; (7) suggestions for further work to improve any authors' shortcomings (e.g., other experiments based on techniques described in Chapter 2, and theoretical (computational) work); and (8) inclusion of other circumstantial evidence found from our literature searches on more recent related work to the problem posed.

Finally, additional resources for the reader to consider to learn more about the type of reaction posed in the problem, synthetic utility, other applications, and so on, are given at the end of each solution.

ACKNOWLEDGMENTS

We thank Dr. Floyd H. Dean for suggesting the cinenic acid to geronic acid rearrangement as a key example to illustrate problem-solving techniques and the application of the principle of conservation of structural aspect, and for generously offering his time to discuss some of the more difficult problems and his help in resolving them. We also thank Amy Clark, Senior Editorial Project Manager at Elsevier, and her successor Peter Llewellyn for their extraordinary patience over the course of this 4-year odyssey. We have climbed many small mountains and have grown intellectually along the way. We hope this book inspires others to follow our footsteps and climb even higher mountains of their own.

In closing, we leave the reader with some interesting and relevant quotes from Justus von Liebig and Friedrich Wöhler, who occupy the same position as Abraham in the hierarchy of contributors to chemical science, that touch on various points highlighted in this Preface. In these quotes the pronouns "he," "his," and "him" are used throughout, but the reader should interpret them to include both genders.

As a student reading chemistry[8]:

"It developed in me the faculty, which is peculiar to chemists more than to other natural philosophers, of thinking in terms of phenomena; it is not very easy to give a clear idea of phenomena to anyone who cannot recall in his imagination a mental picture of what he sees and hears, like the poet and artist, for example. Most closely akin is the peculiar power of the musician, who while composing thinks in tones which are as much connected by laws as the logically arranged conceptions in a conclusion or series of conclusions. There is in the chemist a form of thought by which all ideas become visible in the mind as the strains of an imagined piece of music. This form of thought is developed in Faraday in the highest degree, whence it arises that to one who is not acquainted with this method of thinking, his scientific works seem barren and dry, and merely a series of researches strung together, while his oral discourse when he teaches or explains is intellectual, elegant, and of wonderful clearness."

Letter to Berzelius on experiments[9]:

"The loveliest of theories are being overthrown by these damned experiments; it is no fun being a chemist any more."

Introduction to Liebig and Wöhler's paper on the elucidation of the structure of the benzoyl group[10]:

"When in the dark province of organic nature, we succeed in finding a light point, appearing to be one of those inlets whereby we may attain to the examination and investigation of this province, then we have reason to congratulate ourselves, although conscious that the object before us is unexhausted."

Liebig and Wöhler's paper prediction about the power of organic synthesis[11]:

"The philosophy of chemistry will draw the conclusion that the production of all organic substances no longer belongs just to living organisms. It must be seen as not only probable, but as certain, that we shall be able to produce them in our laboratories. Sugar, salicin, and morphine will be artificially produced. Of course, we do not yet know how to do this, because we do not yet know the precursors from which these compounds arise. But we shall come to know them."

Liebig's view of scientific training[12]:

"It is only after having gone through a complete course of theoretical instruction in the lecture hall that the student can with advantage enter upon the practical part of chemistry; he must bring with him into the laboratory a thorough knowledge of the principles of the science, or he cannot possibly understand the practical operations. [For] if he is ignorant of these principles, he has no business in the laboratory."

Liebig's view on science investigation[13]:

"In science all investigation is deductive, or *a priori*. The experiment is but the aid to the process of thought, as an arithmetic operation is; and the thought, the idea, must always precede it – necessarily precede it – in every case where a result of importance is looked at."

Wöhler on organic chemistry[14]:

"Organic chemistry nowadays almost drives me mad. To me it appears like a primeval tropical forest full of the most remarkable things, a dreadful endless jungle into which one does not dare enter, for there seems no way out."

A poignant account of ideological thinking and bias[15]:

"When Kathleen Lonsdale produced the X-ray crystallographic evidence of the planarity of the benzene ring, Ingold declared that 'one paper like this brings more certainty into organic chemistry than generations of activity by us professionals'. That remark bears a striking resemblance to the recent affirmation by Chargaff: 'amateurs often are better in advancing science than are the professionals'. The point that lies behind such remarks is that 'professionals' have usually shared, to some degree at least, a perception of what is internal and what [is] external to their discipline. The self-images, reinforced by institutional characteristics, have had an important bearing on the progress of organic chemistry because they have determined the curves of the boundaries which, at different times, have separated it from other disciplines. Where the boundary should be drawn has, of course, been another source of controversy. Too high a degree of insularity has also had an adverse effect. A perspective which emerges very clearly from recent scholarship is that the moat which at various times separated organic from physical chemistry acted like the other kind of mote."

Andrei Hent and John Andraos

Toronto, Canada

References

1. Brooke JH. In: Russell CA, ed. *Recent Developments in the History of Chemistry*. London: Royal Society of Chemistry; 1985:147.
2. Halford B. Is there a crisis in organic chemistry education? *Chem Eng News*. 2016;94(13):24–25.
3. Saltzman MD. In: James LK, ed. *Nobel Laureates in Chemistry 1901-1992*. Washington, DC: American Chemical Society; 1993:312–313.
4. Ridd JH. Organic pioneer: Christopher Ingold's insights into mechanism and reactivity established many of the principles or organic chemistry. *Chemistry World*. 2008, December;50–53.
5. Ridd JH. Christopher Ingold: the missing Nobel Prize. In: *The Posthumous Nobel Prize in Chemistry*. Washington, DC: American Chemical Society; 2017:207–218. vol. 1. [chapter 9]. *https://doi.org/10.1021/bk-2017-1262.ch009*.
6. Barton DHR. Ingold, Robinson, Winstein, Woodward, and I. *Bull Hist Chem*. 1996;19:43–47.
7. Lenoir D, Tidwell TT. The history and triumph of physical organic chemistry. *J Phys Org Chem*. 2018;31:e3838. https://doi.org/10.1002/poc.3838.
8. Brock WH. *Justus von Liebig—The Chemical Gatekeeper*. Cambridge: Cambridge University Press; 1997:9.
9. Brock WH. *Justus von Liebig—The Chemical Gatekeeper*. Cambridge: Cambridge University Press; 1997:72.
10. Brock WH. *Justus von Liebig—The Chemical Gatekeeper*. Cambridge: Cambridge University Press; 1997:80.
11. Brock WH. *Justus von Liebig—The Chemical Gatekeeper*. Cambridge: Cambridge University Press; 1997:89.
12. Brock WH. *Justus von Liebig—The Chemical Gatekeeper*. Cambridge: Cambridge University Press; 1997:288.
13. Brock WH. *Justus von Liebig—The Chemical Gatekeeper*. Cambridge: Cambridge University Press; 1997:302.
14. Jaffe B. *The Story of Chemistry: From Ancient Alchemy and Nuclear Fission*. New Haven, CT: Fawcett Publications; 1957:119.
15. Brooke JH. Russell CA, ed. *Recent Developments in the History of Chemistry*. London: Royal Society of Chemistry; 1985:151.

Chapter 1: The Logic of Organic Reaction Mechanisms

In this chapter we explore the basic logical operations, concepts, and methods used to discuss and analyze organic reaction mechanisms. For this purpose, we reference numerous works where readers can find detailed high quality examples and discussions especially suitable for undergraduate organic chemistry students looking to establish a personal library of important works. Our standard of selection consists of filling in gaps, presenting new approaches, and explaining why the field of physical organic chemistry is critical for understanding organic chemistry in a logical manner. We consequently expect that readers of this textbook who might presently believe that study of organic chemistry requires memorization of reaction details shall benefit the most from this introduction and the remainder of this book. For the practicing research organic chemist, we emphasize that understanding and elucidating reaction mechanisms both facilitates and strengthens the practice of organic synthesis. The bottom line is that reaction mechanism elucidation works in the service of optimizing organic synthesis.

1.1 WHAT IS AN ORGANIC CHEMICAL REACTION?

Organic chemical reactions consist of processes in which starting materials interact with reagents under fixed conditions to form new structures according to principles such as sterics, electronics, thermodynamics, and conservation of mass and charge. Since these principles apply to *any* kind of chemical transformation, understanding them enables students to categorize and recognize reactions without the mentally demanding task of having to remember disconnected information about every single reaction they encounter. We note that today's introductory chemistry textbooks and early courses are designed to require students to remember details like product outcomes, starting materials, and reaction conditions in a disconnected manner. Authors of such standard university textbooks[1–5] only reinforce the memorization approach when they organize transformations according to functional group classifications. To make matters worse, this material is often presented without specifying original references or elucidating complete reaction mechanisms. Firstly, we strongly emphasize presenting original references because it encourages students to access, use, and critique the literature. Students can thus build personal libraries and conceptual hierarchies from which they can connect ideas with the real world of academics, scientific history, and laboratory successes and failures. Students can thus learn about how discoveries are actually made and most importantly how carefully thought-out experiments guide scientists to the truth. In fact, some of the greatest insights and advances in science can be attributed to understanding failures and unexpected experimental results. Furthermore, original references demonstrate that ideas in science do not simply appear out of nowhere and that they stand on their own merit and not simply because an author or instructor includes them. When ideas are connected to reality in this way or through laboratory practice, one establishes a solid foundation for scientific theory and for the hard work necessary for young scientists to find their place in the field of their endeavor. As the old adage says, one does not know where one is going until one knows where one has come from.

From the standpoint of pedagogy, we thus encourage the reader to approach difficult problems and complex ideas by identifying the large picture context of a chemical transformation and breaking it down into logically connected more easily managed fragments such as elementary (i.e., mechanistic) steps followed by an overall reaction mechanism. Armed with this knowledge a student will be prompted to write reaction mechanisms when given a novel transformation, an approach which can assist in answering questions in other fields such as synthesis and green chemistry. Nevertheless, to understand reaction mechanisms one must understand the available experimental and theoretical tools that constitute the

Strategies and Solutions to Advanced Organic Reaction Mechanisms
https://doi.org/10.1016/B978-0-12-812823-7.00301-3

body of evidence based on which a mechanism can be supported or rejected. This material is introduced here and covered in further detail in Chapter 2. Afterward, a unified hierarchical approach for tackling problems of organic reaction mechanisms from both an analytical and an experimental standpoint is given in Chapter 3. As an illustrative example, we begin by examining the old and well-established transformation shown in Scheme 1.1.

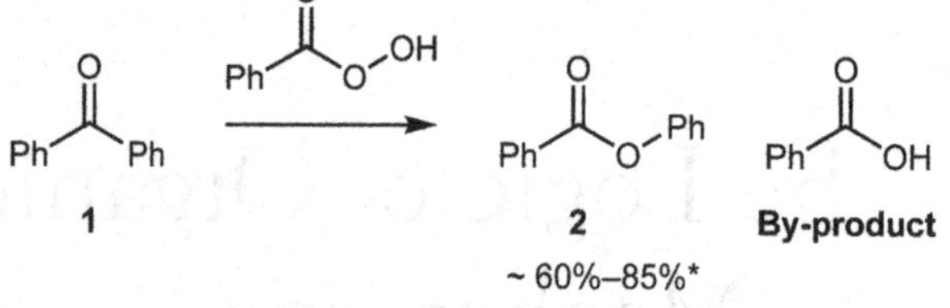

SCHEME 1.1   Baeyer-Villiger oxidation of benzophenone.

This reaction is called the Baeyer-Villiger oxidation (sometimes the Baeyer-Villiger rearrangement).[6] It involves the conversion of a ketone into an ester (or of a cyclic ketone into a lactone, in the case of cyclic rings) in the presence of an oxidant such as a peroxy acid. It was discovered by Adolf von Baeyer and Victor Villiger in 1899,[7, 8] which is the same year that Julius Stieglitz introduced the concept of "carbocation" in the literature.[9] Not surprisingly, the history of physical organic chemistry also began in this early period.[10, 11] To better appreciate this history, we highly recommend that readers carefully study Refs. 10, 11 once they finish reading this chapter. As we shall see later when discussing its mechanism, Baeyer and Villiger themselves adopted the word "carbocation" to describe a proposed early reaction intermediate in the mechanism.[12, 13] Later, in 1905, Baeyer was awarded the Nobel Prize in Chemistry "in recognition of his services in the advancement of organic chemistry and the chemical industry," thanks to his contributions to organic dyes and hydroaromatic compounds.[14] We note that at the time, reported yields for this transformation ranged between 40% and 70%.[7, 8] Over the next century, research on the Baeyer-Villiger oxidation has led to considerable improvements in yield, the development of efficient catalysts, green chemistry conditions, and an improved understanding of its mechanism.[15]

1.2 THE BALANCED CHEMICAL EQUATION

Before looking at the mechanism however, we wish to emphasize certain rules of thumb with regard to how one should read and draw reaction schemes. The first rule is that reaction schemes should be depicted to show the complete balanced chemical equation for the transformation under consideration. In other words, every atom on the left-hand side of the chemical equation should appear on the right-hand side, either in the structure of the desired product (henceforth referred to as the **product**) or in the structure of the undesired product(s) (henceforth referred to as the **by-product(s)**). We emphasize this rule for several reasons. First, the balanced chemical equation connects directly to the reaction mechanism in that it constitutes the sum of the elementary mechanistic steps of the proposed mechanism. This fact is important enough to warrant a statement of a theorem for the field of physical organic chemistry:

Theorem. *The overall balanced chemical equation for a particular transformation constitutes the overall summation of the elementary mechanistic steps of its proposed mechanism, each of which is mass and charge balanced. The overall balanced chemical equation is thus itself charge and mass balanced.*

We employ this theorem throughout the book by identifying complete balanced chemical equations for the transformations considered in each of the problems discussed and their proposed mechanisms. Unfortunately, this practice is generally omitted from most modern textbooks of organic chemistry, including, surprisingly, those considered to be the gold standard for physical organic chemistry such as Anslyn and Dougherty and Carroll.[16, 17] It is surprising in our view because a balanced chemical equation motivates a host of valuable research practices both in the written analysis and in experimental investigation. For instance, if by-product(s) can be identified experimentally then there exists direct evidence supporting or contradicting a particular mechanistic proposal. This is because although atoms can be counted and identified on both sides of a balanced chemical equation, one does not necessarily know the structures of the by-products. If, for example, the reaction leads to gas evolution in the form of CO_2 or N_2, one can conclude that at some point in the reaction mechanism such a gas is eliminated. One can then confidently reject an alternative mechanism where these atoms are eliminated as part of other structures. Therefore it is possible to say that every mechanistic proposal has its own unique balanced chemical equation. Note also that it is entirely possible, as we shall see with the Baeyer-Villiger oxidation, that several proposed mechanisms will have the same balanced chemical equation. Nevertheless, in terms of proper depiction of reaction schemes, it would be entirely careless and bad practice to draw the equation in Scheme 1.1 without showing the structure of the benzoic acid by-product. As will be seen throughout this book, the concept of experimental by-product identification is one of the key strategies in elucidating reaction mechanisms and is one of the best-kept secrets in the arsenal of tools used by practicing physical organic chemists. Furthermore, identifying balanced chemical equations also respects the conservation of mass law which Antoine Lavoisier, arguably the father of modern chemistry, discovered in 1775.[18] Furthermore, a balanced chemical equation also respects the conservation of charge law in that the sum of electronic charges depicted on the left-hand side of the chemical equation is the same as the sum of the electronic charges appearing on the right-hand side.

Moreover, we note that a reaction Scheme can sometimes show multiple products thus making it impractical to represent a single balanced chemical equation. In this scenario, we distinguish between reaction **product**, **by-product**, and **side product** in that a side product is a structure arising from a different mechanistic pathway as that which leads to the reaction product. Such products are often depicted with percentage signs showing yields underneath the structures or with the words "major" and "minor." This means that multiple mechanistic paths exist and that one is more favorable than the others. It also means that the difference between by-product and side product is that the two species may not necessarily arise as a result of following the same mechanistic path. For example, the paths leading to **product** and **side product**, respectively, may have the same **by-product** if the structures of the product and side product have the same chemical formula. If the formulas are different, it means that mass has not been conserved and therefore each mechanistic path would have different by-products with different masses. Nevertheless, we encourage the use of the terms "reaction product," "side product," and "by-product" as a means of establishing clarity. Since these terms refer to different things, they should *not* be used interchangeably as is currently the case in many modern textbooks of organic chemistry.

1.3 WHAT IS CONTAINED IN A REACTION SCHEME?

If we consider Scheme 1.1, we can observe that starting materials and products are labeled with numbers below the chemical structures. In our systematized convention, we recommend the use of capital letters for structures of proposed intermediates along a mechanistic pathway. Reagents, which are compounds that play a direct role in the reaction mechanism, are shown above the reaction arrow while reaction conditions such as temperature, pressure, reaction time, solvents, catalysts, and/or aqueous quench are shown below the reaction arrow. Commonly, the word "substrate" refers to the starting material of interest whose structure in whole or in part definitely ends up in the product structure, whereas the word "reagent(s)" refers to other starting materials that operate on the substrate, but which may or may not end up incorporated in whole or in part in the product structure. It is thus possible to read a reaction scheme and draw immediate conclusions about its mechanism by simply looking at these indicators. For example, in Scheme 1.1 we notice that an oxygen atom is introduced in the ketone group of benzophenone **1** thus giving the ester product **2**. We thus recognize this as a redox reaction where the ketone carbon atom in **1** undergoes oxidation from a +2 state to a +3 state. This implies that something else must be reduced by 1 unit. Comparing the reagent used, peroxybenzoic acid (another indication of a redox process since this is a well-known oxidizing reagent), to the benzoic acid by-product drawn, we see that indeed the ester oxygen atom in the reagent is reduced from −1 to −2 in the benzoic acid by-product. As a side note, we encourage the reader to familiarize oneself with oxidation number analysis because it is very useful in understanding redox transformations.[19] Other telling details about a transformation include high temperature (a possible indication of a fragmentation process), photochemical energy or radical reagent (an indication of a radical process), and acidic or basic reaction media as indications of proton transfer processes facilitated by acids or bases, respectively. In addition, transformations without by-products may indicate that a rearrangement of the starting substrate structure has taken place somewhere along the mechanistic path, particularly if the reaction is initiated thermally, photochemically, or catalytically (e.g., by acid or base). Furthermore, reactions where starting materials lead to products that have more atoms than are expected according to the conventionally written chemical equation should prompt one to expect that several equivalents of the starting material or reagent react together to form the reaction product. For such transformations, the corresponding balanced chemical equation will have nonunity stoichiometric coefficients associated with those starting materials. It is also possible for an intermediate along a mechanistic pathway to react with a starting material to form a combined structure which would then proceed to the final product. Such cases of divergent and convergent reaction mechanisms as well as alternative mechanisms or partial alternative mechanisms are represented throughout this textbook and we hope they will help to expand the reader's imagination with regard to what is possible in organic reaction mechanisms. For now we wish to emphasize the fact that reaction schemes have much to reveal about reaction mechanisms.

1.4 PRINCIPLE OF CONSERVATION OF STRUCTURAL ASPECT

Consider for a moment how Scheme 1.1 is illustrated and compare it with Scheme 1.2. We note that both Schemes show the exact same transformation and that the only difference is the structural aspect. Which scheme is easier to read and understand?

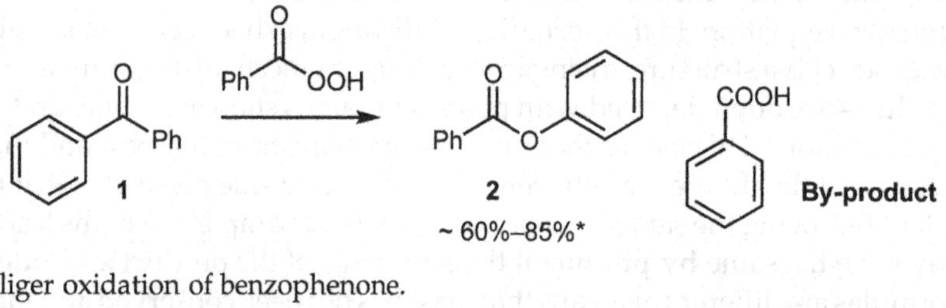

SCHEME 1.2 Baeyer-Villiger oxidation of benzophenone.

Clearly Scheme 1.1 is more easily understood. This is because the amount of information that can be stored in conscious awareness at any time is limited both in terms of available slots for storage (usually about seven) and the complexity of the incoming information. For example, we can see that viewing structures **1** and **2** requires less mental operations in Scheme 1.1 as opposed to Scheme 1.2. If rotation and inversion parameters were included in Scheme 1.2, the task of communication would be further complicated. We therefore view the practice of drawing complicated schemes as unbecoming of those who wish to be understood. Unfortunately, this practice does appear in the literature and also in university courses. We recommend a different approach. Rather than complicating the task of communicating ideas by drawing unnecessarily complex structures in various structural aspects, chemists and academics should consider adopting the principle of conservation of structural aspect. According to this principle, illustrations of starting materials maintain a consistent structural aspect with that of reaction intermediates (in the case of a scheme depicting a mechanism) and of desired products (in the case of standard reaction schemes). Advantages of such an approach include: (1) the ability to easily map atoms of starting materials onto the structures of products, (2) better identification of target bonds formed, (3) better identification and description of ring construction strategies, (4) higher probability identification of reaction by-products, and (5) immediate expectations with regard to reaction mechanisms. To further solidify this point, we provide two interesting examples where structural aspect was not maintained in the original schemes (see Scheme 1.3).[20, 21]

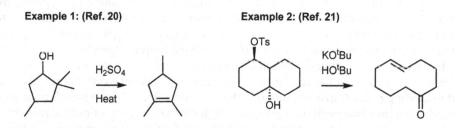

Example 1: (Ref. 20)   Example 2: (Ref. 21)

Suggest a mechanism for each example.

SCHEME 1.3 Two example displays of transformations not following the principle of conservation of structural aspect. Example 1: acid-catalyzed dehydration with [1,2]-methyl migration; and Example 2: base-catalyzed Grob fragmentation.

Since these schemes do not respect the principle of conservation of structural aspect, their mechanisms, at first glance, appear elusive. In fact, one has to consider the reaction conditions as a starting point for the mechanistic analysis given these illustrations. Based on reaction conditions the first example represents an acid-catalyzed rearrangement whereas the second example shows a base-catalyzed fragmentation. By maintaining structural aspect with respect to the starting material, we illustrate the mechanisms for these reactions in Scheme 1.4.

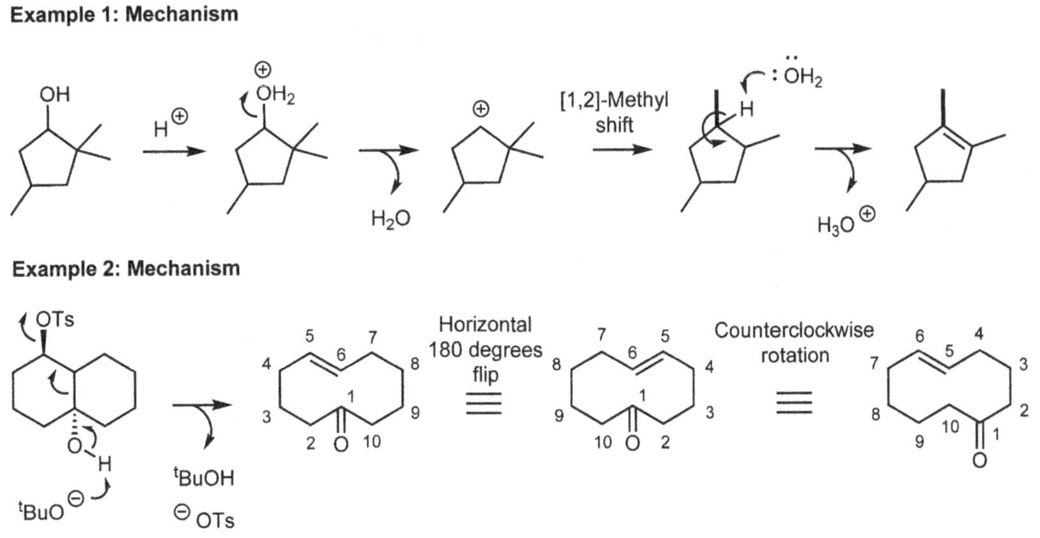

Example 1: Mechanism

Example 2: Mechanism

SCHEME 1.4 Displays of mechanisms of transformations shown in Scheme 1.3 following the principle of conservation of structural aspect.

Seeing these mechanisms one can immediately spot how the products are formed without having to think very hard. We also note that target bonds formed are represented as bolded bonds, not to be confused with stereochemical representations which use dark and hashed wedges. We also note the interesting fact that the ketone product in Example 2 can be drawn in several structural aspects which are manageable if one adopts numbering of skeletal carbon atoms. The advantages of maintaining structural aspect consistent throughout a schematic diagram are that each of the 300 problems solved in this textbook becomes considerably easier to understand and to solve.

1.5 CURLY ARROW NOTATION

With these ground rules established, we turn to representation of reaction mechanisms by means of the logic of curly arrow notation which was introduced by Sir Robert Robinson in 1922 as a convenient bookkeeping device for tracking electron flow.[22] We note that this concept is intricately based on the electronic theory of organic chemistry which was also advanced by Robinson and several others (see supplementary information of Ref. 11) in 1926.[23] It is also important to mention that Robinson was a well-known Nobel Prize winning *synthetic* organic chemist, not a *bone fide* physical organic chemist like his colleague and competitor Sir Christopher Ingold. Nevertheless, Robinson saw the value of reaction mechanism in understanding product outcomes in the service of synthetic organic chemistry. According to curly arrow notation therefore, processes involving two-electron transfers should be depicted using curly arrows (full arrows) based on the following set of criteria: (1) arrow heads always point toward electrophilic centers; (2) arrow tails always point toward nucleophilic centers; (3) electron flow proceeds from a nucleophilic source to an electrophilic sink; and (4) a series of arrows are unidirectional in the sense that an arrow head is always followed by an arrow tail, meaning that two arrow heads or two arrow tails never meet. To concretize these rules, we highlight several examples of correct and incorrect use of curly arrow notation for two-electron transfer processes in Schemes 1.5 and 1.6.

We see from these examples that the correct (or clearer) diagrams usually imply a greater number of mechanistic steps. Sometimes, as in the incorrect diagram in Example 4, there are several arrows in close proximity which create confusion if one does not identify the order of the arrows. Depicting a slightly longer stepwise mechanism can eliminate such confusion in addition to providing for a more correct illustration.

Similar rules can be extended to mechanisms involving one-electron transfer processes (i.e., radical-type mechanisms). In essence we have that: (1) curly arrows with half heads represent one-electron transfer steps,

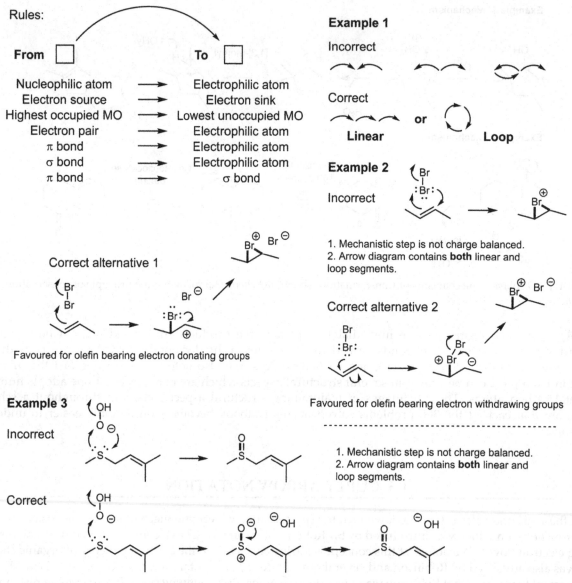

SCHEME 1.5 Three examples showing juxtaposed incorrect and correct uses of curly arrow notation to depict reaction mechanisms. Example 1: curly arrows depicting two-electron transfers, Example 2: bromination of olefins, and Example 3: oxidation of sulfides to sulfones.

(2) a bond is formed whenever two half arrow heads come together, (3) homolytic bond fragmentation occurs whenever two arrow tails emerge from a bond, and (4) a series of arrows are not unidirectional (this rule is the opposite of that for the two-electron transfer process). To illustrate these rules, we provide several examples in Scheme 1.7.

Lastly, we chose these examples because they still appear in some textbooks and so we wanted to show how our approach clarifies some of these more unique reaction mechanisms. For more in-depth discussion of proper curly arrow notation, including unique cases, see Appendix 5.4 in Ref. 16. We also recommend Refs. 24–27.

1.6 BAEYER-VILLIGER OXIDATION MECHANISM

Having considered the principles and rules for depicting reaction schemes and mechanisms, it is now possible to represent the currently accepted mechanism for the Baeyer-Villiger oxidation originally shown in Scheme 1.1 (see Scheme 1.8).[6]

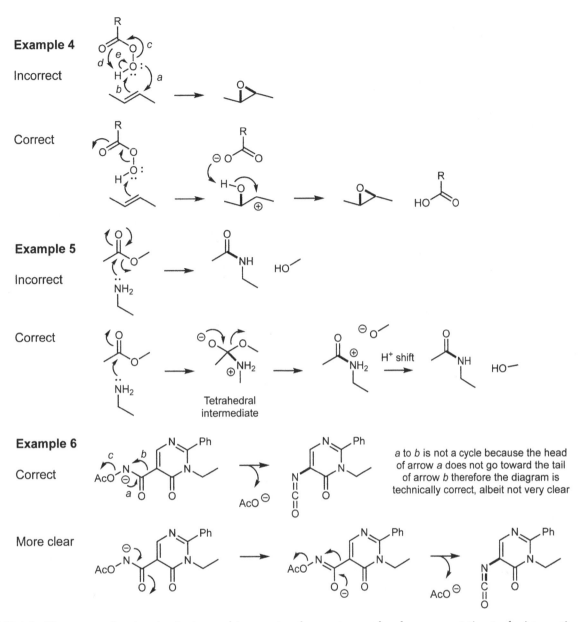

SCHEME 1.6 Three examples showing juxtaposed incorrect and correct uses of curly arrow notation to depict reaction mechanisms. Example 4: epoxidation of olefins, Example 5: amidation of esters, and Example 6: generation of isocyanates.

Thus the proposed mechanism begins with protonation of the ketone group of benzophenone **1** via the peroxy acid reagent which forms oxonium ion **A** which is then attacked by the newly generated oxyanion in a nucleophilic manner at the electrophilic carbon atom (masked carbocation) to form the tetrahedral intermediate **B**. This intermediate in turn undergoes protonation via the peroxy acid reagent to form intermediate **C** which undergoes base-mediated ketonization, migration of the phenyl group onto the oxygen atom, and concomitant elimination of benzoic acid by-product to generate the final ester product **2**. This mechanism is not as clearly presented in Wang[6] nor in Anslyn and Dougherty[16] nor do these resources reference key experimental evidence which supports the mechanism. In fact, the reaction mechanism is generally never connected with the historical development of physical organic chemistry apart from loose fragmentary mentions of ^{18}O-labeling experiments that support it (see, for instance, page 681 in Anslyn and Dougherty).[16] Unfortunately, these authors also do not provide references for these experiments. Under this kind of approach, it is not surprising that physical organic chemistry today has not received the serious treatment it deserves in both pedagogy and the literature. For example, the reader might not know that Baeyer and Villiger originally proposed a different mechanism for this transformation.[7, 8] In these initial reports, there was no mention of a tetrahedral intermediate and

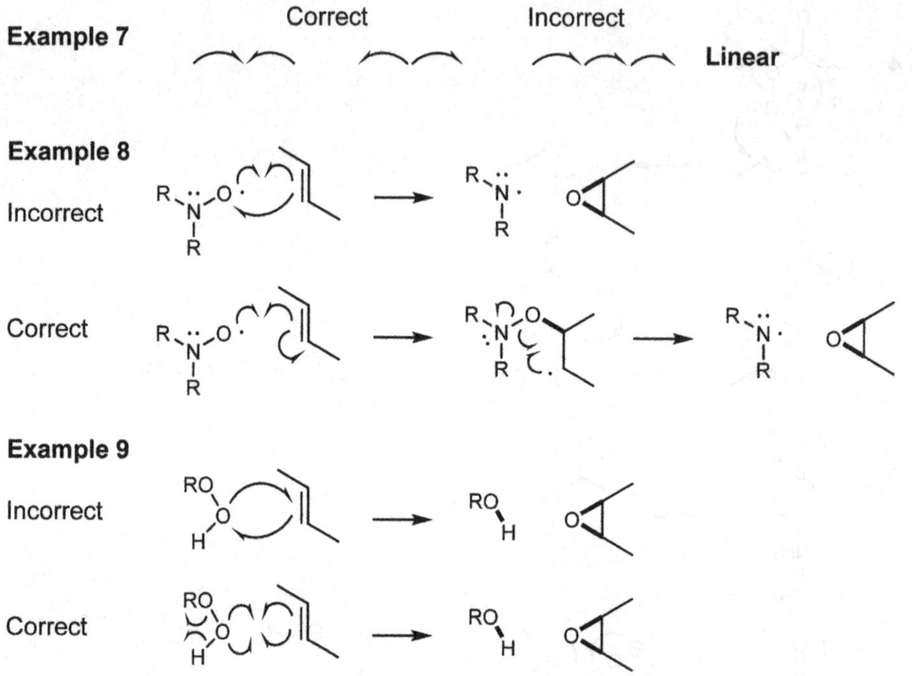

Example 7

Example 8

Incorrect

Correct

Example 9

Incorrect

Correct

SCHEME 1.7 Three examples showing juxtaposed incorrect and correct uses of curly arrow notation to depict reaction mechanisms. Example 7: curly arrows depicting one-electron transfers, Example 8: epoxidation of olefins using nitroxyl radical reagents, and Example 9: epoxidation of olefins using peroxide reagents.

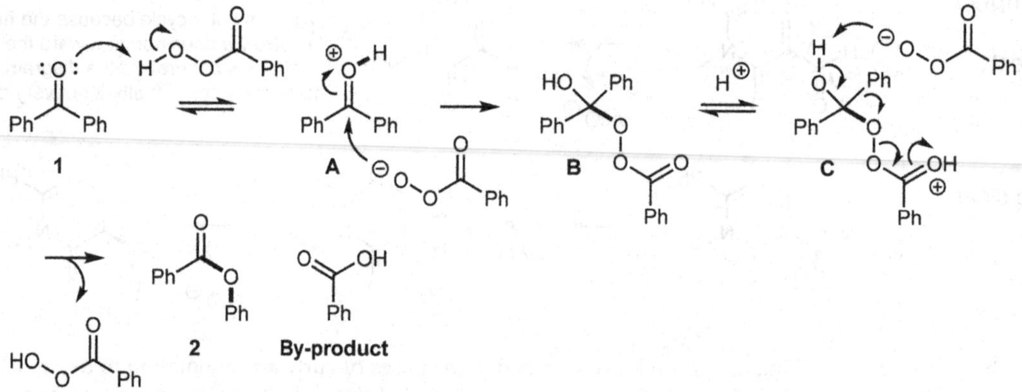

SCHEME 1.8 Mechanism of Baeyer-Villiger oxidation reaction of benzophenone.

the proposed mechanism, which was not shown in its entirety, involved formation of a dioxirane intermediate via attack of peroxy acid onto the oxygen atom in **1** (which was hypothesized based on reports at the time concerning the Beckmann rearrangement). We will present these mechanisms in Chapter 2 and discuss them in greater detail.

As for the historical background, once Stieglitz proposed the concept of carbocation,[9] the history of which is itself interesting and worth revisiting,[10, 11] the term had immediately appeared in subsequent reports by Baeyer and Villiger.[12, 13] It was not until 1940 that the Baeyer-Villiger mechanism was challenged, this time by Georg Wittig and Gustav Pieper who suggested formation of a peroxide intermediate following attack of the ketone group of **1** onto the hydroxyl group of the peroxy acid reagent.[28] It is worth mentioning that Georg Wittig was another Chemistry Nobel Prize laureate in 1979 along with Herbert C. Brown for their work on boron and phosphorus-containing products.[14] As we can see, neither of these mechanisms became the generally accepted mechanism. This is because a third challenge occurred in 1948 when Rudolf Criegee proposed carbon attack via the peroxyacid reagent followed by formation of the tetrahedral intermediate **B**.[29] It is worth mentioning that Criegee has a rearrangement named after him which was discovered in 1944 and which involves the oxidation of a tertiary alcohol into a ketone product plus primary alcohol by-product when peroxy acid reagents are used.[30, 31] One can thus see the similarities between the Criegee rearrangement and the mechanism that Criegee proposed for the Baeyer-Villiger oxidation. Nevertheless, with three mechanistic proposals in the literature it became clear that experimental evidence was needed to resolve the problem of which mechanism was operating. This evidence came in 1953 when a team of American chemists from Columbia University, Doering and Dorfman, performed an eloquent ^{18}O-labeling experiment which directly supported the Criegee mechanism and contradicted the other two proposals.[32] Since then, intermediate **B** has been named the Criegee intermediate and subsequent research has focused on attempting to trap this intermediate and obtain indirect or direct experimental evidence for its existence.[33] With this background in mind and the wealth of history behind this transformation, it is possible to develop a much greater appreciation for the work of analyzing and understanding reaction mechanisms, so that one can subsequently apply such understanding to other difficult problems in organic chemistry. Nevertheless, when textbooks present mechanisms without any historical background or references that one can easily access and consult, using diagrams that create confusion rather than clarity and without mentioning experimental evidence, then the entire wealth of knowledge and value provided by the field of physical organic chemistry is lost. To counter this problem, we have gathered examples of textbooks on reaction mechanisms and categorized them in "recommended" and "not recommended" columns along with our reasons so that the reader may easily judge which ones are worth considering and adding to one's personal library of physical organic chemistry (see Table 1.1).

TABLE 1.1 Enumeration of Recommended Textbooks on Physical Organic Chemistry and Mechanistic Analysis.

Author(s) (Ref. #)	Strengths	Weaknesses	Recommend (R) Not Recommend (NR)
Alder et al. (34)	Rigorous approach, presents references	Examples are not current	R
Alonso-Amelot (35)	Presents logical principles and problem solving strategy clearly	Does not show balanced chemical equations	R
Anslyn and Dougherty (16)	Excellent at outlining concepts of physical organic chemistry, good for experiments, not good for mechanistic analysis	Does not provide references for ideas and examples discussed in every chapter! Provides additional reading references disconnected from solved examples. Balanced chemical equations are absent	R (only as modern reference on physical organic chemistry)
Badea (36)	Rigorous, mathematical and methodological	Does not have enough schemes and is not current	R
Bansal (37)	Attempts to be comprehensive	Not rigorous	NR
Bruckner (38)	Great for visualization, contains numerous Schemes	Does not show balanced chemical equations	R
Butler (39)	Great for problems and solutions. References are provided	Not current, curly arrow notation not well represented	R
Carroll (17)	Excellent for problems and solutions, references and topics covered, also current	Does not provide historical background and does not show balanced chemical equations	R
Edenborough (40)	Chapters 5, 17, and 22 are highly recommended	Introductory, does not identify by-products, does not provide balanced chemical equations	R (only for Chapters 5, 17, 22)
Gardiner (41)	Excellent presentation of rates of reactions, mathematical, and rigorous	Does not show many organic reaction schemes or mechanisms	R
Grimshaw (42)	Adequate coverage of electrochemical reactions	Not comprehensive in terms of mechanisms	NR
Hammett (43)	Foundational textbook for physical organic chemistry with focus on methodology	Not current, but historically highly relevant	Highly R
Harwood (44)	Numerous mechanistic schemes, good primer for polar rearrangements	Specific topics, not general nor comprehensive	R

Continued

TABLE 1.1 Enumeration of Recommended Textbooks on Physical Organic Chemistry and Mechanistic Analysis.—cont'd

Author(s) (Ref. #)	Strengths	Weaknesses	Recommend (R) Not Recommend (NR)
Hassner and Namboothiri (45)	Illustrates 750 named reactions with references and mechanisms	Mechanisms are largely conjectured, by-products not identified, balanced chemical equations are not shown	R (only as a desk reference)
Hoever (46)	Best and only textbook which attempts to do what we are doing in this textbook, contains extensive discussion and analysis	Not current	Highly R
Ingold (47)	Foundational textbook for physical organic chemistry with a focus on structure	Not current, but historically highly relevant	Highly R
Karty (48)	Textbook is organized by mechanism with good coverage of pK_A and H^+ transfer	Does not provide references or balanced chemical equations	NR
Lavoisier (18)	Excellent starting point historically for chemistry	Not current	Highly R
Lawrence et al. (49)	Good primer for foundations of physical organic chemistry such as energetics and kinetics	Does not show mechanism schemes or concretize ideas with examples of mechanisms	R (only for introduction of concepts)
Lawrence et al. (50)	Contains worked examples for foundations of physical organic chemistry such as energetics and kinetics	Does not show mechanism schemes or concretize ideas with examples of mechanisms	R (only for introduction of concepts)
Maskill (51)	Good introduction and coverage of rudimentary reaction categories (e.g., substitutions)	Does not show mechanism schemes or concretize ideas with examples of mechanisms	R (only for introduction of concepts)
Maskill (52)	Good introduction to structure and reactivity	Does not show many mechanism schemes	R (only for introduction of concepts)
Menger and Mandell (24)	Contains many examples that cover electronic basis of organic chemistry	Level is beginning to intermediate, curly arrow notation needs improvement, does not show balanced chemical equations	R (only for introductory level)
Miller (25)	Contains good explanation of curly arrow notation and examples	Does not identify by-products or balanced chemical equations	R
Perkins (26)	Good coverage of radical chemistry. Chapter 5 is recommended.	This is just a primer so it is not extensive	Highly R
Ruff and Csizmadia (27)	Excellent resource—comprehensive, with examples and references	Could use more mechanism schemes	Highly R
Savin (53)	Modern work	Does not provide references or balanced chemical equations	NR
Scudder (54)	Great coverage of electron flow and logic for reaction mechanisms	Does not provide references	R (only for theory)
Smith (55)	Recommend Chapter 6 for mechanisms and methods to determine them	Does not provide exercises	R
Stewart (19)	Good coverage of redox mechanisms including oxidation number analysis	Not current	R
Wang (6)	Comprehensive and excellent as guide to named organic reactions	Not great for proposed mechanisms since most are conjectured and sometimes key mechanism references are absent	R

1.7 TABLE OF RECOMMENDED TEXTBOOKS ON PHYSICAL ORGANIC CHEMISTRY

In addition to these textbooks specializing in reaction mechanisms, we wish to recommend other valuable resources specializing in topics like green chemistry,[56, 57] chemicals and reagents,[58] aromatic heterocyclic chemistry,[59] side reactions,[60] redox reactions,[61] reaction intermediates,[62–65] organic chemistry foundations,[66, 67] organic stereochemistry,[68] aromatic chemistry,[69] ion-radical organic chemistry,[70] and finally an excellent introduction to physical organic chemistry by Kosower.[71] Lastly, we wish to conclude the chapter by covering two interesting concepts that are not well covered elsewhere and which appear frequently in the advanced solutions that follow in Chapters 4 through 9. These concepts are base strength in relation to pK_A and autoxidation.

1.8 BASE STRENGTH AND pK_A

In this short section we seek to clarify the relationship between base strength and pK_A (at its core it is a measure of acid strength). This is because when it comes to mechanisms of reactions involving acidic or basic conditions, it is important to identify which species are present in solution because those are the compounds that should appear in the reaction mechanism. Once identified, one must then judge which bases are strongest. Doing so can help elucidate mechanisms which involve deprotonations. The likelihood of deprotonation events depends on the properties of acid dissociation and base strength. For this purpose, the concept of pK_A is crucial and so we begin from where this concept originated, namely, the Henderson-Hasselbalch equation. Essentially, this equation focuses on the acid dissociation constant K_a, which for an acid (generic HA) dissociation equilibrium in aqueous solution is defined as:

$$\text{For} \quad HA + H_2O \rightleftharpoons A^- + H_3O^+$$

$$\text{We have} \quad K_a = \frac{[A^-][H^+]}{[HA]}$$

where K_a is the acid dissociation constant relating $[A^-]$, the concentration of conjugate base, $[H^+]$, the concentration of hydrogen ion, and $[HA]$, the concentration of the acid species.

This expression was written by Lawrence J. Henderson in 1908.[72, 73] A derivation which contained logarithmic terms appeared in 1916 in an article by Karl A. Hasselbalch.[74] Thus the derivation became known as the Henderson-Hasselbalch equation:

$$\text{If} \quad K_a = \frac{[A^-][H^+]}{[HA]} \quad \text{then} \quad \log K_a = \log\left(\frac{[A^-][H^+]}{[HA]}\right) = \log[H^+] + \log\left(\frac{[A^-]}{[HA]}\right)$$

$$-pK_A \qquad\qquad -pH$$

$$\text{We therefore have} \quad -pK_A = -pH + \log\left(\frac{[A^-]}{[HA]}\right)$$

$$\text{If we add } (pK_A + pH) \text{ to both sides we have:} \quad pH = pK_A + \log\left(\frac{[A^-]}{[HA]}\right)$$

Henderson-Hasselbalch equation

We note that another expression for the Henderson-Hasselbalch equation is:

$$\text{Henderson-Hasselbalch equation} \quad pH = pK_A + \log\left(\frac{[A^-]}{[HA]}\right)$$

$$\text{If we subtract } \log\left(\frac{[A^-]}{[HA]}\right) \text{ from both sides we have:} \quad pH - \log\left(\frac{[A^-]}{[HA]}\right) = pK_A$$

We can thus see that a strong acid (HA) implies a greater dissociation into its ions which implies a higher ($[A^-][H^+]/[HA]$) term. The same holds true when logarithms are taken of both sides. Nevertheless, $\log K_a$ is defined as ($-pK_A$). In this context a higher K_a term implies a higher $\log K_a$ term which implies a higher ($-pK_A$) term which implies a *lower* pK_A. Thus strong acids have low pK_A values. From the standpoint of reaction mechanism, this means that if an acid or an acidic proton has a low pK_A associated with it, then this acid or proton will readily react with a weak base and even more readily with a stronger base (if both species are present in solution).

To judge base strength, one may start by judging acid strength in terms of pK_A. We first recall that a weak acid has a strong conjugate base while a strong acid has a weak conjugate base. Thus a strong acid which has a low pK_A will have a weak conjugate base. Correspondingly, a weak acid which has a high pK_A will have a strong conjugate base. Therefore given two anions A_1^- and A_2^-, it is possible to assess base strength by considering the pK_A's of HA_1 and HA_2. Suppose these are pK_{A1} and pK_{A2}, respectively. Therefore if $pK_{A1} > pK_{A2}$ we can draw two equally valid conclusions: (1) HA_1 is a weaker conjugate acid than HA_2; and (2) A_1^- is a stronger base than A_2^- anion. If, on the other hand, $pK_{A1} < pK_{A2}$ we have: (1) HA_1 is a stronger conjugate acid than HA_2; and (2) A_1^- is a weaker base than A_2^-. We can restate this conclusion in simpler terms: the higher the pK_A of its conjugate acid, the stronger the base; and the lower the pK_A of its conjugate acid, the weaker the base. The inverse relationship exists between pK_A and acid strength, namely: the higher its pK_A, the weaker the acid; and the lower its pK_A, the stronger the acid. Once these relationships are firmly established, it is possible to consult desk references and databases[75, 76] which contain pK_A values for various "HA" compounds and to draw the correct mechanistic conclusions based on these values. We should also emphasize that if the pK_A of a particular C—H bond in a compound is less than the pK_A of the conjugate acid of the base under consideration, the deprotonation will occur. This is because a higher pK_A implies a stronger conjugate base while a lower pK_A implies a stronger acid. Strong bases react with strong acids. If the reverse is true (i.e., the pK_A of a C—H bond is higher than the pK_A of the conjugate acid of the base under consideration) then the deprotonation will not occur. Once again, this is because a high pK_A implies a weak acid while a low pK_A implies a weak conjugate base. Weak acids do not react with weak bases.

To illustrate and reinforce these concepts more concretely we show the following selected equilibria data given in Table 1.2.

TABLE 1.2 Example Equilibria for Various Acids.

Equilibrium	pK$_A$	pK$_B$
$H_2O = H^+ + OH^-$	15.7	−1.7
$NH_4^+ = H^+ + NH_3$	9.24	4.76
$NH_2NH_3^+ = H^+ + NH_2NH_2$	8.12	5.88
$HOAc = H^+ + OAc^-$	4.76	9.24
$CF_3COOOH = H^+ + CF_3COOO^-$	3.64	10.36
$CF_3COOH = H^+ + CF_3COO^-$	0.23	13.77
$H_3O^+ = H^+ + H_2O$	−1.7	15.7
$HBr = H^+ + Br^-$	−9	23

From Table 1.2 we note that all the equilibria are written in the form $HA = H^+ + A^-$ and we observe that the strongest acid (HA) in the list is hydrobromic acid with a pK_A of −9 since it has the lowest pK_A value. By contrast, the weakest acid is water with a pK_A of 15.7. In order to find out which conjugate base is the strongest or the weakest, we use the relationship $pK_A + pK_B = 14$. From the resulting pK_B values appearing in the third column we observe that hydroxide ion is the strongest base with a pK_B of −1.7 and bromide ion is the weakest base with a pK_B of 23. We can therefore make the following general statements:

A low pK_A value means a stronger acid for HA.

A low pK_A value means a higher pK_B value for conjugate base A^-.

A low pK_B value means a stronger base for A^-.

A low pK_B value means a higher pK_A value for acid HA.

We can also make the following specific comparative statements based on the data given in Table 1.2:

$NH_2NH_3^+$ is a stronger acid than NH_3.

NH_3 is a stronger base than NH_2NH_2.

H_3O^+ is a stronger acid than H_2O.

OH^- is a stronger base than H_2O.

CF_3COOH is a stronger acid than HOAc.

OAc^- is a stronger base than CF_3COO^-.

H_3O^+ is a stronger acid than CF_3COOH.

CF_3COO^- is a stronger base than H_2O.

We conclude this section by highlighting several excellent compilations where reliable pK$_A$ values may be found in the literature.[75–83]

1.9 AUTOXIDATION

The last topic considered in this chapter is one which frequently appears in articles that do not present complete reaction mechanisms. In such articles, the authors typically present a conjectured mechanism of how they believe "things might occur" which tends to omit the details of a last or penultimate mechanistic step where a reaction intermediate undergoes some kind of oxidation process to generate the final product. Authors explain such a process by simply stating "autoxidation" without showing which compounds react together by means of curly arrow notation.

Autoxidation generally consists of a radical-type process which leads to the desired oxidized product. For example, authors will typically omit the structure of O_2 gas (originating from air) from reaction mechanisms. This is particularly true for literature depictions of multicomponent reactions in which a pair of hydrogen atoms is unaccounted for when such reactions result in an aromatic product structure. Such discrepancies reveal themselves when reagent structure mappings onto the product structure are made and corresponding balanced chemical equations are sought after for such transformations. It is important to note that O_2 gas can be relevant to transformations carried out in aerobic conditions. Due to the fact that oxygen gas exists as a ground state triplet (diradical), it can participate in a transformation in the role of an oxidant via one-electron transfer processes. Consider problem 30 in this textbook. Here, the authors had drawn a mechanism where the final step was the air oxidation of intermediate **I** to form the final product **4** (Scheme 1.9).

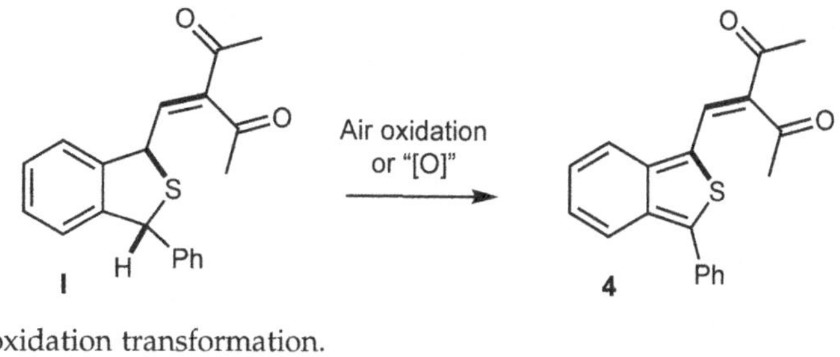

SCHEME 1.9 Example of air oxidation transformation.

We consider this an unnecessary short cut which confuses more than it clarifies the reaction mechanism. The complete sequence we have provided in Solution 30 is that shown in Scheme 1.10. This mechanism illustrates the exact logic behind the oxidation of **I** to **4**. Indeed we see that triplet state O_2 reacts in a radical-type manner to abstract a proton from **I** which in turn undergoes electron reshuffling to form radical intermediate **J**. The newly formed HOO radical is then able to abstract a hydrogen atom from **J** to generate final product **4** and hydrogen peroxide by-product. The hydrogen peroxide is then expected to undergo homolytic bond cleavage to form two equivalents of hydroxyl radical species which can abstract hydrogen atoms from **I** and/or **J** in another mechanistic cycle to form water molecule as a reaction by-product. This by-product would be reflected in the overall balanced chemical equation corresponding to the proposed reaction mechanism.

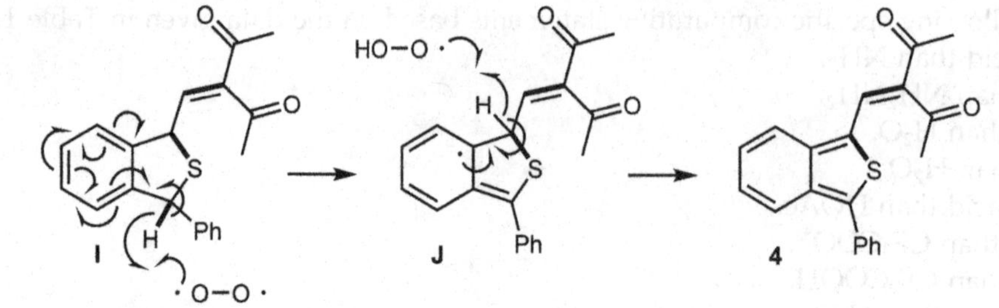

SCHEME 1.10 Mechanism of air oxidation reaction shown in Scheme 1.9.

We emphasize this level of detail not just for the purpose of having a complete logical sequence to a proposed reaction mechanism but also because of the implications of such a mechanism. For example, if the authors decided to investigate experimentally whether air (i.e., oxygen gas) was critical for the success of their transformation, as the proposed mechanism requires, they could undertake the same transformation in anaerobic conditions devoid of oxygen gas. The expected outcome of such an experiment (which would directly support the proposed autoxidation sequence) is that product **4** would not be formed. If it is formed, this autoxidation mechanism would be rejected and the search for a different oxidation reagent to help convert **I** to **4** would commence. In the absence of such evidence, the proposed mechanism cannot be rejected because it is supported by chemical logic. In conclusion, we recommend an industrial example of an autoxidation process and its mechanism (see pages 32–34 of Ref. 38).

We hope that this chapter has helped convey the importance of chemical logic and of reaction mechanisms to readers interested in developing a better understanding of organic chemistry. We regret that because of space requirements and because the concept only appeared in one of the 300 problems solved, we could not include a section on kinetics here. Nevertheless, we recognize the study of kinetics as central to the study of physical organic chemistry due to its value for elucidating reaction mechanisms.

References

1. McMurry J. *Organic Chemistry.* 9th ed. Boston, MA: Cengage Learning; 2016.
2. Solomons TWG, Fryhle CB, Snyder SA. *Organic Chemistry.* 12th ed. Hoboken, NJ: John Wiley & Sons, Inc.; 2016
3. Vollhardt P, Schore N. *Organic Chemistry: Structure and Function.* 8th ed. New York: W. H. Freeman; 2018.
4. Clayden J, Greeves N, Warren S. *Organic Chemistry.* 2nd ed. New York: Oxford; 2012.
5. Chaloner P. *Organic Chemistry: A Mechanistic Approach.* Boca Raton, FL: CRC Press; 2015.
6. Wang Z, ed. *Comprehensive Organic Name Reactions and Reagents.* New York: John Wiley & Sons, Inc.; 2010:150–155. https://doi.org/10.1002/9780470638859.conrr150
7. Baeyer A, Villiger V. Einwirkung des Caroeschen Reagens auf Ketone. *Ber Deutsch Chem Ges.* 1899;32(3):3625–3633. https://doi.org/10.1002/cber.189903203151.
8. Baeyer A, Villiger V. Ueber die Einwirkung des Caroeschen Reagens auf Ketone. *Ber Deutsch Chem Ges.* 1900;33(1):858–864. https://doi.org/10.1002/cber.190003301153.
9. Stieglitz J. On the constitution of the salts of imido-ethers and other carbimide derivatives. *J Am Chem.* 1899;21(2):101–111.
10. Strom ET, Mainz VV, eds. *The Foundations of Physical Organic Chemistry: Fifty Years of the James Flack Norris Award.* Washington, DC: Oxford University Press; 2015. https://doi.org/10.1021/bk-2015-1209.
11. Andraos J. Scientific genealogies of physical and mechanistic organic chemists. *Can J Chem.* 2005;83:1400–1414. https://doi.org/10.1139/V05-158.
12. Baeyer A, Villiger V. Dibenzalaceton und Triphenylmethan. Ein Beitrag zur Farbtheorie. *Ber Deutsch Chem Ges.* 1902;35(1):1189–1201. https://doi.org/10.1002/cber.190203501197.
13. Baeyer A, Villiger V. Dibenzalaceton und Triphenylmethan. *Ber Deutsch Chem Ges.* 1902;35(3):3013–3033. https://doi.org/10.1002/cber.19020350395.
14. The Nobel Prize. All Nobel Prizes in Chemistry. https://www.nobelprize.org/prizes/lists/all-nobel-prizes-in-chemistry/ (Accessed November 18, 2017).
15. Renz M, Meunier B. 100 years of Baeyer–Villiger oxidations. *Eur J Org Chem.* 1999;1999:737–750. https://doi.org/10.1002/(SICI)1099-0690(199904)1999:4<737::AID-EJOC737>3.0.CO;2-B.
16. Anslyn EV, Dougherty DA. *Modern Physical Organic Chemistry.* New York: University Science Books; 2006.
17. Carroll FA. *Perspectives on Structure and Mechanism in Organic Chemistry.* 2nd ed. Hoboken, NJ: John Wiley & Sons, Inc.; 2010.
18. Lavoisier A-L. *Elements of Chemistry.* [Kerr R, Trans.] New York: Dover Publications, Inc.; 1965.
19. Stewart R. *Oxidation Mechanisms: Applications to Organic Chemistry.* New York: W. A. Benjamin, Inc.; 1964.
20. Solomons TWG. *Organic Chemistry.* 6th ed. New York: John Wiley & Sons, Inc.; 1996:530.
21. Hultin PG. *The What, How and Why of Problem Solving in Organic Chemistry… and elsewhere!.* University of Manitoba; 2004.https://home.cc.umanitoba.ca/~hultin/chem2220/Support/WhatHowWhy.pdf. Accessed 18 November 2017.

22. Kermack WO, Robinson R. LI.—An explanation of the property of induced polarity of atoms and an interpretation of the theory of partial valencies on an electronic basis. *J Chem Soc Trans.* 1922;121(0):427–440. https://doi.org/10.1039/ct9222100427.

23. Allan J, Robinson R. XLIX.—The relative directive powers of groups of the forms RO and RR′N in aromatic substitution. Part I. *J Chem Soc.* 1926;129(0):376–383. https://doi.org/10.1039/jr9262900376.

24. Menger FM, Mandell L. *Electronic Interpretation of Organic Chemistry: A Problems-Oriented Text.* New York: Plenum Press; 1980. https://doi.org/10.1007/978-1-4684-3665-5.

25. Miller A. *Writing Reaction Mechanisms in Organic Chemistry.* 1st ed. San Diego, CA: Academic Press, Inc.; 1992.

26. Perkins MJ. *Radical Chemistry: The Fundamentals.* New York: Oxford University Press; 2005.

27. Ruff F, Csizmadia IG. *Organic Reactions: Equilibria, Kinetics and Mechanism.* Amsterdam: Elsevier Science B.V.; 1994. https://doi.org/10.1016/B978-0-444-88174-8.50003-7

28. Wittig G, Pieper G. Über das monomere Fluorenon-peroxyd. *Ber Deutsch Chem Ges.* 1940;73(4):295–297. https://doi.org/10.1002/cber.19400730402.

29. Criegee R. Die Umlagerung der Dekalin-peroxydester als Folge von kationischem Sauerstoff. *Justus Liebigs Ann Chem.* 1948;560(1):127–135. https://doi.org/10.1002/jlac.19485600106.

30. Criegee R. Über ein krystallisiertes Dekalinperoxyd. *Ber Deutsch Chem Ges (A and B Series).* 1944;77(1):22–24. https://doi.org/10.1002/cber.19440770106.

31. Wang Z, ed. *Comprehensive Organic Name Reactions and Reagents.* New York: John Wiley & Sons, Inc.; 2010:770–774. https://doi.org/10.1002/9780470638859.conrr170

32. Doering WvE, Dorfman E. Mechanism of the peracid ketone-ester conversion. Analysis of organic compounds for oxygen-18[1]. *J Am Chem Soc.* 1953;75(22):5595–5598. https://doi.org/10.1021/ja01118a035.

33. Vil′ VA, Gomes GdP, Bityukov OV, et al. Interrupted Baeyer-Villiger rearrangement: building a stereoelectronic trap for the Criegee intermediate. *Angew Chem Int Ed.* 2018;57(13):3372–3376. https://doi.org/10.1002/anie.201712651.

34. Alder RW, Baker R, Brown JM. *Mechanism in Organic Chemistry.* London: John Wiley & Sons, Ltd.; 1971.

35. Alonso-Amelot ME. *The Art of Problem Solving in Organic Chemistry.* 2nd ed. Hoboken, NJ: John Wiley & Sons, Inc.; 2014.

36. Badea F. *Reaction Mechanisms in Organic Chemistry.* Kent, MI: Abacus Press; 1977.

37. Bansal RK. *Organic Reaction Mechanisms.* 4th ed. Kent, MI: New Academic Science; 2012.

38. Bruckner R. *Advanced Organic Chemistry: Reaction Mechanisms.* London: Elsevier Inc.; 2002. https://doi.org/10.1016/B978-0-12-138110-3.X5000-4.

39. Butler AR. *Problems in Physical Organic Chemistry.* London: John Wiley & Sons, Inc.; 1972.

40. Edenborough M. *Organic Reaction Mechanisms: A Step by Step Approach.* 2nd ed. London: Taylor & Francis Ltd.; 1999.

41. Gardiner Jr. WC. *Rates and Mechanisms of Chemical Reactions.* New York: W. A. Benjamin, Inc.; 1969.

42. Grimshaw J. *Electrochemical Reactions and Mechanisms in Organic Chemistry.* London: Elsevier Inc.; 2000.

43. Hammett LP. *Physical Organic Chemistry: Reaction Rates, Equilibria, and Mechanisms.* New York: McGraw-Hill Book Company, Inc.; 1940.

44. Harwood LM. *Polar Rearrangements.* New York: Oxford University Press; 1992.

45. Hassner A, Namboothiri I. *Organic Syntheses Based on Name Reactions: A Practical Guide to 750 Transformations.* 3rd ed. Oxford: Elsevier Inc.; 2012. https://doi.org/10.1016/C2009-0-30489-4

46. Hoever H. *Problems in Organic Reaction Mechanisms.* London: John Wiley & Sons, Inc.; 1970.

47. Ingold CK. *Structure and Mechanism in Organic Chemistry.* Ithaca, NY: Cornell University Press; 1953.

48. Karty J. *Organic Chemistry: Principles and Mechanisms.* 2nd ed. New York: W. W. Norton; 2018.

49. Lawrence C, Rodger A, Compton R. *Foundations of Physical Chemistry.* New York: Oxford University Press; 1996.

50. Lawrence N, Wadhawan J, Compton R. *Foundations of Physical Chemistry: Worked Examples.* New York: Oxford University Press; 1996.

51. Maskill H. *Mechanisms of Organic Reactions.* New York: Oxford University Press; 1996.

52. Maskill H. *Structure and Reactivity in Organic Chemistry.* New York: Oxford University Press; 1999.

53. Savin KA. *Writing Reaction Mechanisms in Organic Chemistry.* 3rd ed. Amsterdam: Elsevier Inc.; 2014.

54. Scudder PH. *Electron Flow in Organic Chemistry: A Decision-Based Guide to Organic Mechanisms.* 2nd ed. Hoboken, NJ: John Wiley & Sons, Inc.; 2013.

55. Smith MB. *March's Advanced Organic Chemistry: Reactions, Mechanisms, and Structure.* 7th ed. Hoboken, NJ: John Wiley & Sons, Inc.; 2013.

56. Andraos J, Hent A. Key metrics to inform chemical synthesis route design. In: Constable DJ, Jimenez-Gonzalez C, eds. *Handbook of Green Chemistry Volume 11: Green Metrics.* 1st ed. Weinheim: Wiley-VCH Verlag GmbH & Co. KGaA; 2018.

57. Andraos J. *The Algebra of Organic Synthesis: Green Metrics, Design Strategy, Route Selection, and Optimization.* Boca Raton, FL: CRC Press; 2012.

58. Cooper C, Purchase R. *Organic Chemist's Desk Reference.* 3rd ed. Boca Raton, FL: CRC Press; 2018.

59. Davies DT. *Aromatic Heterocyclic Chemistry.* New York: Oxford University Press; 2011.

60. Doerwald FZ. *Side Reactions in Organic Synthesis II: Aromatic Substitutions.* Weinheim: Wiley-VCH Verlag GmbH & Co. KGaA; 2014. https://doi.org/10.1002/9783527687800.

61. Donohoe TJ. *Oxidation and Reduction in Organic Synthesis.* New York: Oxford University Press; 2000.

62. Fabirkiewicz AM, Stowell JC. *Intermediate Organic Chemistry.* 3rd ed. New York: John Wiley & Sons, Inc.; 2016.

63. Leffler JE. *The Reactive Intermediates of Organic Chemistry.* New York: Interscience Publishers, Inc.; 1956.

64. Moody CJ, Whitham GH. *Reactive Intermediates.* New York: Oxford University Press; 1992.

65. Singh MS. *Reactive Intermediates in Organic Chemistry: Structure, Mechanism, and Reactions.* Weinheim: Wiley-VCH Verlag GmbH & Co. KGaA; 2014.

66. Hornby M, Peach J. *Foundations of Organic Chemistry.* New York: Oxford University Press; 1994.

67. Hornby M, Peach J. *Foundations of Organic Chemistry: Worked Examples.* New York: Oxford University Press; 2000.

68. Robinson MJT. *Organic Stereochemistry.* New York: Oxford University Press; 2001.

69. Sainsbury M. *Aromatic Chemistry.* New York: Oxford University Press; 1994.

70. Todres ZV. *Ion-Radical Organic Chemistry: Principles and Applications.* 2nd ed. Boca Raton, FL: CRC Press; 2009.

71. Kosower EM. *An Introduction to Physical Organic Chemistry.* New York: John Wiley & Sons, Inc.; 1968.

72. Henderson LJ. Concerning the relationship between the strength of acids and their capacity to preserve neutrality. *Am J Phys*. 1908;21(2):173–179. https://doi.org/10.1152/ajplegacy.1908.21.2.173.

73. Henderson LJ. The theory of neutrality regulation in the animal organism. *Am J Phys*. 1908;21(4):427–448. https://doi.org/10.1152/ajplegacy.1908.21.4.427.

74. Hasselbalch KA. Die Berechnung der Wasserstoffzahl des Blutes aus der freien und gebundenen Kohlensaeure desselben, und die Sauerstoffbindung des Blutes als Funktion der Wasserstoffzahl. *Biochem Z*. 1916;78:112–144.

75. Lundblad RL, MacDonald FM, eds. *Handbook of Biochemistry and Molecular Biology*. 4th ed. Boca Raton, FL: CRC Press; 2010.

76. Yang J-D. iBond 2.0. http://ibond.chem.tsinghua.edu.cn/. Accessed 18 November 2017.

77. Perrin DD. *Ionisation Constants of Inorganic Acids and Bases in Aqueous Solution*. 2nd ed. New York: Pergamon Press; 1982.

78. Albert A, Serjeant EP, eds. *The Determination of Ionization Constants: A Laboratory Manual*. 3rd ed. London: Chapman and Hall; 1984.

79. Bates RG. *The Determination of pH: Theory and Practice*. New York: Wiley-Interscience; 1973.

80. Kortüm G, Vogel W, Andrussov K. *Dissociation Constants of Organic Acids in Aqueous Solution*. London: Butterworths; 1961.

81. Perrin DD. *Dissociation Constants of Organic Bases in Aqueous Solution*. London: Butterworths; 1961.

82. Perrin DD. *Dissociation Constants of Organic Bases in Aqueous Solution (Supplement 1972)*. London: Butterworths; 1965.

83. Jencks WP, Regenstein J. Ionization constants of acids and bases. In: Lundblad RL, Macdonald FM, eds. *Handbook of Biochemistry and Molecular Biology*. 4th ed. Boca Raton, FL: CRC Press; 2010:595–635.

Chapter 2: Evidence for Organic Reaction Mechanisms

This chapter will explore various experimental and theoretical tools available to chemists who seek to elucidate reaction mechanisms. These ideas will be concretized by revisiting the Baeyer-Villiger oxidation introduced in the previous chapter. In this context we will show how experiments can be designed to help differentiate and support or reject proposed mechanisms. We will then contrast the benefits of this approach with the numerous drawbacks encountered throughout the 300 solutions illustrated in this book. Our hope is for readers to absorb this material and conclude that experimental and theoretical evidence is crucial for understanding reaction mechanisms. It is often the case that the best published research papers in the literature are those that connect experimental evidence with a rigorous computational backing to tell a complete and convincing story about a reaction mechanism and its reaction energy profile for a given reaction.

2.1 WHAT IS PHYSICAL ORGANIC CHEMISTRY?

Physical organic chemistry can be defined as the study of organic chemistry in terms of the principles of physical chemistry. We can put this definition in perspective by depicting a Venn diagram which shows how the fields are connected (see Fig. 2.1). This discipline thus encompasses areas such as kinetics, thermodynamics, reaction conditions, energetics, sterics, and the effects of structure on reactivity. Within these areas chemists apply experimental tools like spectroscopy (used to characterize reactions and products), time resolved methods (used to study what happens between the stages of reactants and products), synthesis (used to vary substrate and reagent structures to determine parameters that affect reaction performance), stoichiometry (used to characterize the beginning and end of a reaction), and change of conditions (i.e., media, catalyst(s), temperature, and time) to gauge their effect on reaction performance and product outcomes. When applied in conjunction with the logical principles discussed in the previous chapter, these experimental techniques can help support a proposed reaction mechanism by means of providing evidence for corresponding intermediates, rate law, and overall balanced chemical equation (which is the sum of elementary steps).

Once reaction mechanisms are proposed, they are tested and revised until they agree with all evidence obtained from such experiments as: trapping experiments (used to isolate otherwise unstable intermediates), isotopic labeling experiments (used to differentiate mechanisms based on comparisons between expected and actual positions of atomic isotope labels in reactants relative to products), crossover experiments (used to differentiate between intramolecular and intermolecular processes), kinetic studies (used to determine substituent, kinetic isotope, and solvent effects, as well as direct observation of intermediates along with quenching using diagnostic trapping reagents), by-product identification experiments (used to help support a balanced chemical equation and its associated mechanism), and theoretical studies (used to calculate energies of geometrically optimized structures along mechanistic pathways and energy barriers which can then be translated into an energy reaction coordinate diagram).

We provide an example of an energy reaction coordinate diagram for a generic mechanism $1 \rightarrow A \rightarrow B \rightarrow 2$ in Fig. 2.2. Viewing this diagram we see that energy (i.e., Gibbs free energy, G) is plotted on the y-axis and that reaction progress is shown on the x-axis. We specifically point out that the energy axis is quantitatively defined in a rigorous manner; however, the reaction progress axis is not. The term "reaction coordinate" formally pertains to the set of bond lengths, bond angles, and dihedral angles that are changed in each elementary step during the course of their associated reactions. These cannot be plotted quantitatively since it would lead to a multidimensional diagram. Therefore

Strategies and Solutions to Advanced Organic Reaction Mechanisms
https://doi.org/10.1016/B978-0-12-812823-7.00302-5

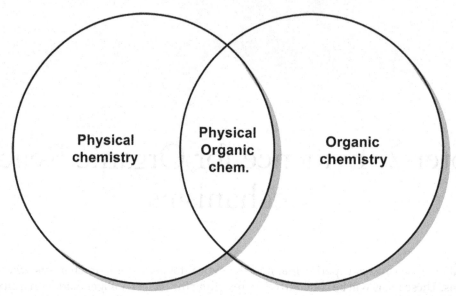

FIG. 2.1 Venn diagram showing the connection between the subjects of physical and organic chemistries.

Energy reaction coordinate diagram for generic mechanism: $1 \rightarrow A \rightarrow B \rightarrow 2$

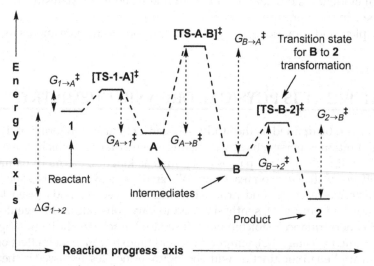

FIG. 2.2 Example energy reaction coordinate diagram.

the often-used term "energy reaction coordinate" diagram is really tracking the *sequence* of intermediates and intervening transition states from step to step in a mechanism along the x-axis; hence the term "reaction progress." These points are not well communicated in the pedagogical literature on physical organic chemistry and can lead to confusion in reading and interpreting the information conveyed in these highly simplified diagrams.

Thus in Fig. 2.2 we can see that reactant **1**, product **2**, and intermediates **A** and **B** are represented by energy minima while transition states between each elementary step are represented as energy maxima. In fact, transition states are mathematically described as saddle points on a multidimensional potential energy surface; whereas reactants, products, and all intervening intermediates are local minima. Most relevant however is the difference in energy between a proposed intermediate and the subsequent transition state between that intermediate and the next structure in the mechanistic sequence. This energy difference corresponds to the activation energy needed to be supplied so that the proposed elementary mechanistic step in question will occur. If, for instance, an amount of energy equal to $G_{1 \rightarrow A}{}^{\ddagger}$ is supplied (e.g., in the form of external heat) we would expect (according to Fig. 2.2) that **1** will be converted into intermediate **A**. Once intermediate **A** is formed, if we supply energy greater than $G_{A \rightarrow 1}{}^{\ddagger}$ but less than $G_{A \rightarrow B}{}^{\ddagger}$, we can expect that **A** will not be converted into **B** but would likely revert back to **1**. This occurs because there has been enough energy supplied to meet the energy demand for the reverse process (**A** to **1**) but not enough for the forward process (**A** to **B**). We can thus see that every elementary step in a reaction mechanism has its own energy constraints.

Some steps, as the ones depicted in Fig. 2.2, will be endergonic ($\Delta G > 0$) thus requiring added energy to proceed whereas other steps may be exergonic ($\Delta G < 0$). Last, the difference in energy between reactants and products constitutes the energy difference ΔG term for the entire transformation (not to be confused with energy of activation). When ΔG is negative (as in Fig. 2.2, i.e., reactants are higher in energy than products), the transformation is exergonic ($\Delta G_{rxn} = G_{prod} - G_{react} < 0$) whereas when ΔG is positive, the transformation is endergonic ($\Delta G_{rxn} = G_{prod} - G_{react} > 0$). If the y-axis is replaced by enthalpic energy, ΔH, instead of free energy, then the terms exergonic and endergonic are replaced by the terms exothermic ($\Delta H_{rxn} = H_{prod} - H_{react} < 0$) and endothermic ($\Delta H_{rxn} = H_{prod} - H_{react} > 0$), respectively. Lastly, the highest magnitude $G^{\ddagger}$ (activation energy) corresponds to the rate-determining step, which in Fig. 2.2 is the **A** to **B** transformation. This is consistent with Murdoch's excellent article on how to find the rate-determining step from an energy reaction coordinate diagram corresponding to a reaction comprised of several elementary steps.[1] We point out that such an elementary step does not always correspond to the one having the highest energy transition state.

Energy reaction coordinate diagrams also clarify the precise meaning of the ubiquitously used term "stability" which is a comparative term. On one hand "stability" can be used to describe the relative energy of an initial state to a final state process regardless of the pathway between them (i.e., thermodynamic stability). One can also use "stability" to describe the relative energy barriers of two elementary steps having different transition states where the transformation is path dependent (i.e., kinetic stability). In Fig. 2.2, we observe that intermediate **B** is more thermodynamically stable than intermediate **A** because it has a lower free energy value. Similarly, product **2** is more thermodynamically stable than reactant **1**. Thermodynamically stable chemical species are associated with more negative free energies or more negative enthalpic energies. By contrast, we observe that intermediate **A** is more kinetically stable that reactant **1** because the energy barrier for the step **A** to **B** is larger in magnitude than that for step **1** to **A**. From the diagram we observe that in the forward sense (i.e., reading the diagram from left to right) intermediate **A** is overall the most kinetically stable species since it is associated with the rate-determining step (**A** to **B**) and reactant **1** is the least kinetically stable. In the reverse sense (i.e., reading the diagram from right to left) intermediate **B** is the most kinetically stable since it is associated with the largest energy barrier for the **B** to **A** step. Moreover, intermediate **A** is the least kinetically stable since it is associated with the smallest energy barrier for the step **A** to **1**. Kinetically stable chemical species are those associated with large energy barriers, small rate constants, and hence slow reaction rates. This discussion immediately leads to the very important concept of thermodynamic and kinetic control, which is operational when a given starting material leads to two different products depending on reaction conditions. This strategy of changing reaction conditions so that one product outcome is favored preferentially allows for great versatility in synthesis methodology and has been exploited extensively. In Fig. 2.3 we show the four possible energy reaction coordinate diagrams for a generic reaction leading to two products (i.e., **1** is converted to **2** and **3**) along with their

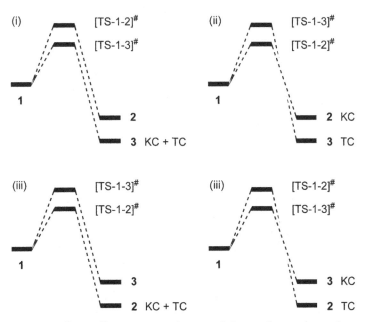

FIG. 2.3 Possible energy reaction coordinate diagram scenarios pertaining to thermodynamic control (TC) and kinetic control (KC) for a generic reaction leading to two products: **1** is converted into **2** and **3**.

associated kinetically and thermodynamically controlled products. Kinetically controlled products are formed faster (because they have low energy barriers) and generally arise as a consequence of less steric hindrance. Often such products are formed at low reaction temperatures and short reaction times. Thermodynamically controlled products are typically formed over a longer reaction time at elevated temperatures and are generally associated with products that are thermodynamically stabilized via electronic substituent effects. When a product is both thermodynamically and kinetically favored, this situation directly correlates as expected with very high reaction yields and hence high synthetic efficiency.

Having an energy reaction coordinate diagram allows one to judge the energy constraints associated with a proposed mechanism and to compare these magnitudes to those of an alternative mechanism. In such comparisons, the winning mechanism is the one which has the least number of elementary steps *and* the least energy of activation constraints. Generally between step count and energy constraints, it is energy which takes precedence. Nevertheless, this type of evidence is indirect and cannot supersede direct evidence obtained from certain experimental procedures which will be discussed in the following section. For now we wish to clearly emphasize that in the absence of *any* evidence, be it experimental, theoretical, or even analogical (i.e., by reference to analogous literature examples), proposed mechanisms are considered conjectured and stand solely on the logic of the electronic theory of organic chemistry. When authors fail to provide complete mechanisms, in addition to failing to provide supporting evidence, it is incumbent upon readers to approach those mechanisms with skepticism.

2.2 EXPERIMENTAL PHYSICAL ORGANIC CHEMISTRY

Before carrying out any specific experiments to probe mechanisms, we note that chemists can uncover much information about a transformation simply by doing their due diligence. For example, instead of discarding the reaction waste, one can carry out spectroscopic and other experiments on it to identify any by-products. Identification of by-products provides indirect evidence for mechanisms which require their formation. We note that care must be exercised so that one does not contaminate the reaction waste by using separation reagents or solutions that can react with and degrade by-products. Instead, if reagents that can react with and trap by-products are employed, the result would be an elegant trapping experiment for the purpose of by-product identification. Nevertheless, it is easier to pay close attention to the reaction vessel and to note any emerging color change (a possible indication of intermediate/product formation or of reagent/intermediate degradation if such structures are chromatic in nature), gas evolution (an indirect indication of a gaseous by-product, e.g.: CO_2, N_2, SO_2), precipitation of solids (which can be separated and characterized using spectroscopic methods), and detection of odors (possible clues to chemical structures formed—intermediates, products, and by-products). Throughout the history of chemistry, such standard practices have suggested new leads for research that later culminated in remarkable discoveries.[2] After such preliminary work is complete, it is possible to undertake experiments that can provide direct (see Table 2.1) and indirect (see Table 2.2) evidence for reaction mechanisms.

In terms of kinetic versus thermodynamic control, we note that this situation arises whenever the activation energy required to transform starting materials into products is higher for one product (which is itself lower in energy) than for another product. One can conceptualize this as a process which requires more energy but which ends up at a lower energy state (thermodynamic product) competing with a less energy-demanding process that ends up at a higher energy state (kinetic product). Apart from Fig. 2.3, we recommend the example shown in Fig. 1 in Ref. 14.

We can thus see from Tables 2.1 and 2.2 just how many experiments are possible and how they can be applied strategically to help elucidate reaction mechanisms. We wish to emphasize therefore that obtaining evidence from experimental and computational analyses is mandatory for any complete and proper mechanistic investigation. Given that a large number of problems in this solutions book had original references which did not contain experimental or theoretical (and often, shockingly, both types of) evidence to support proposed mechanisms, we consider this chapter to be the most valuable in the entire textbook. Readers are advised to pay close attention to the information outlined in Tables 2.1 and 2.2 and to seek such evidence whenever they research reaction mechanisms. Doing so in the context of named organic reactions one may discover just how few named reactions have mechanisms which were elucidated based on evidence obtained from experimental and/or computational studies (we estimate about 10% of the reactions identified in Wang).[43] In Table 2.3 we note which solutions in this textbook had experimental evidence (and of which kind) provided in the original reference. The reader should note the large number (207!!!) of solutions whose original authors provided conjectured reaction mechanisms that lacked experimental, computational, or both types of evidence.

TABLE 2.1 Experimental Techniques Which Provide Direct Evidence for Reaction Mechanisms.

Experiment type	Procedures	Results	Literature examples
Direct synthesis	- Choose appropriate substituent groups which thermodynamically stabilize the structures of intermediates, that is, electron withdrawing groups (EWGs) for intermediates with inherent negative charge or high electron densities, and electron donating groups (EDGs) for intermediates with inherent positive charge or low electron densities	- Synthesis of intermediates followed by treatment under the reaction conditions to see if the predicted product is formed - Isolation of intermediates from reaction media followed by characterization and reaction under a newly prepared reaction medium to see if the target product is formed - If target/predicted products are formed from synthesized/isolated intermediates, these intermediates are involved in the reaction mechanism	- Recent example: Stable tetrahedral intermediates (hemiaminals) (see Ref. 3) - Other examples: See Refs. 4, 5 (and references therein)
Matrix isolation spectroscopy	- Use of cryogenic temperatures	- Isolation of intermediates followed by spectroscopic characterization - Used to identify highly reactive short-lived intermediates (i.e., radical ions)	- Foundational research (see Ref. 6) - Recent examples (see Ref. 7)
Time-resolved spectroscopy	- Follow kinetic disappearance or appearance of intermediates by spectroscopic means such as ultraviolet absorption, visible light absorption, fluorescence emission, phosphorescence emission, reflectance - Measure concentration of intermediate as function of time by analyzing absorbance or transmission data obtained from the type of spectroscopy employed - Use a quencher/probe/catalyst to trap intermediates and follow their kinetic decay using spectroscopic means	- Identification of intermediates according to analysis of spectroscopic data For example: Key structural signals (i.e., C=O ketone signal) in the compound spectrum of kinetic studies of concentration or decay of intermediate	- Foundational research (see Ref. 8 and others by Norrish RWG) - Primer on spectroscopy (see Ref. 9) - Good general examples (see Refs. 10, 11)

TABLE 2.2 Experimental Techniques Which Provide Indirect Evidence (i.e., Based on Inference) for Reaction Mechanisms.

Experiment type	Procedures	Results	Literature examples
Product identification and characterization	- Apply spectroscopic methods to identify and characterize the target product formed - If there is more than one product formed, vary the time of the reaction to discover which product is kinetically controlled and which is thermodynamically controlled	- Characterize the end of the transformation - Identify the kinetically controlled product (formed earliest) - Identify the thermodynamically controlled product (accumulates over time)	- Classic example (enolate chemistry, Ref. 12) - Modern examples (Refs. 13, 14)
By-product identification and characterization	- Collect the reaction waste being careful not to contaminate it with other potentially reactive materials - Perform a spectroscopic analysis/separation process to identify and characterize by-products	- By-products identified which can be used to provide clues about transient intermediates which may exist during the course of the reaction	- Modern examples (see Refs. 15, 16)
Analogous reaction	- Carry out an analogous reaction to check if the expected products/product distribution is achieved	- If the proposed mechanism is to be supported, the analogous reaction should give the same results as the original transformation	- General example (see Ref. 17)
Change reaction conditions	- Vary solvent, temperature, pressure, presence/absence and choice of catalyst, choice of reagent, aerobic/inert conditions - Look for changes of product or of product distribution	- Identify solvent, temperature, pressure, and catalyst effects which may provide mechanistic insight - Identify the impact of reagent and whether O_2 autoxidation plays a factor	- Modern example (Ref. 18) - General examples (Ref. 19)

Continued

TABLE 2.2 Experimental Techniques Which Provide Indirect Evidence (i.e., Based on Inference) for Reaction Mechanisms.—cont'd

Experiment type	Procedures	Results	Literature examples
Time-resolved spectroscopy	- Use a quencher/probe/catalyst to trap intermediates and follow their kinetic decay using spectroscopic means	- Identification of intermediates through inference from trapping experiment with quencher/probe/catalyst	- General examples from radical chemistry (see Refs. 20, 21)
pH kinetics/rate profiles	- Determine observed pseudo-first-order rate constants for the disappearance of an intermediate in aqueous solution and plot them as a function of proton concentration	- Used to determine the degree of acid catalysis (pH regions where the reaction rate is accelerated by H^+), base catalysis (pH regions where the rate is accelerated by OH^-), and absence of catalysis (pH regions where the rate is not accelerated by H^+ and OH^-, i.e., the rate depends only on reaction with water)	- Excellent review of pH-rate profile curves, their shapes, mechanistic interpretation, and examples therein (Ref. 22)
Product/intermediate isolation	- Isolate/trap intermediate using, for example, a quencher and treat the intermediate under the reaction conditions to see if the same product is formed - Isolate reaction product and treat it under the same conditions to see if any further transformation occurs	- If trapped intermediate reacts to form the same product under the reaction conditions as that formed from starting materials, this compound is an intermediate in the reaction mechanism - If reaction product reacts further to form another compound, this structure is likely a thermodynamic intermediate	- For general examples, see Ref. 19
Crossover experiment	- Select multiple variations of the starting material(s) which have differing and appropriate substituents around the reactive functional groups - Mix these starting materials under the reaction conditions and characterize the product mixture	- If crossover products (i.e., products with unexpected/foreign substituent combinations) are observed, then the mechanism consists of an intermolecular process - If crossover products are not observed (i.e., each substrate leads to one and only one product), then the mechanism consists of an intramolecular process - Very useful experiment for testing transformations involving rearrangements	- Classic example (rearrangement of aryl allyl ethers, see Ref. 23)
Competition experiment	- Probe the effect of having two potential electrophiles competing to quench a nucleophilic intermediate or of two potential nucleophiles competing to quench an electrophilic intermediate	- Determine rate constants for quenching experiments - Ratios of rate constants will directly correlate with product ratios	- Modern example (Morita-Baylis-Hillman reaction case study, see Ref. 24)
Isotope labeling experiment	- Create isotopic label in starting material/isolated intermediate and expose this compound to the reaction conditions - Using spectroscopic methods, track the position of the isotope label in the structure of the resulting product	- Allows for tracking the fate of specific atoms and chemical bonds between substrates and products - Allows for differentiating competing mechanisms where specific atoms have different fates	- Classic examples (deuterium—Ref. 25, ^{18}O—Ref. 26, ^{15}N—Ref. 27, ^{14}C—Ref. 28, iodine—Ref. 29) - General review article (Ref. 30)
Isotope effect experiment	- Determine the observed rate of reaction (k) when the substrate contains a light isotope versus a heavy isotope for the same atom - Calculate the ratio of k(light atom)/k(heavy atom)	- Primary isotope effect occurs when k(light atom)/k(heavy atom) is >1 (this generally implies that bond breaking occurs in the rate determining step), if the ratio is <1 it generally implies that bond forming occurs in the rate determining step and if the ratio is =1 then bond breaking/forming is not rate determining step - Secondary isotope effects are correlated with bond perturbations (not bond breaking) in the rate determining step (α type if atom directly bonded to isotopically labeled atom, β type if atom	- For detailed analysis and examples, see Chapter 7 in Ref. 19 - For modern examples, see the review article Ref. 31.

TABLE 2.2 Experimental Techniques Which Provide Indirect Evidence (i.e., Based on Inference) for Reaction Mechanisms.—cont'd

Experiment type	Procedures	Results	Literature examples
		is not directly bonded to isotopically labeled atom - Also possible are solvent isotope effects which are observed when rate/equilibrium constant for a process changes when a solvent is replaced with an isotopically labeled solvent (i.e., H_2O vs. D_2O)	
Stereochemical experiment	- Determine stereochemical outcome in product structure and compare it with structures of proposed intermediates	- Analysis may provide clues about degree of symmetry in intermediate structures as required by the stereochemical outcome	- Modern examples (see Refs. 32, 33)
Variable temperature experiment	- Carry out experiment under varying temperature, pressure, and concentration	- Obtain thermokinetic information about the size of the energy barrier for the reaction as well as enthalpic/entropic contributions	- Modern example (see Ref. 34)
Linear free energy relationship experiment	- Determine observed reaction rates when various substituents (EDGs or EWGs) are introduced into the structure of the substrate - Use kinetics and free energy relationship equations to calculate thermodynamic/free energy terms/Hammett equation parameters (σ-substituent constant, ρ-sensitivity parameter) - Use this kinetic, equilibria and substituent data to construct Hammett plots	- Probe electronic nature of the transition state of the rate determining step based on kinetic data - Probe relative electronic nature of initial state and final state structures based on equilibrium data - Note that EDGs stabilize electron-deficient intermediate/transition state structures while EWGs stabilize electron-rich intermediate/transition state structures - Positive-valued slopes of Hammett plots correlate with negatively charged transition states and negatively valued slopes correlate with positively charged transition states	- Original references (see Refs. 35–38) - For reference on the topic of kinetics, see Refs. 39–41
Computational analysis	- Using specialized software and appropriate basis sets, calculate the energy of geometrically optimized structures appearing in the proposed reaction mechanism	- Use calculated energies to construct a complete energy reaction coordinate diagram showing the energies of all substrates, intermediates, transition states, and products along proposed mechanistic pathways - Use diagram to identify the number of intermediates, the number of elementary steps, and the rate determining step - Correlate results with experimental evidence for all suspected intermediates, and all kinetic and thermodynamic data - Check to see if the results of the computational analysis support the findings of all experimental studies	- For practical examples and theory, see Ref. 42

Furthermore, of the 300 problems surveyed, the best case and what we selected as a model layout for a serious investigation of reaction mechanism is question 61. This is because the original reference for this problem includes a detailed discussion of mechanism and by-product identification as a key experimental piece of evidence supporting the mechanism. Aside from this problem, we highlight question 265 whose original reference had a detailed, logical, surprisingly honest, and quite illuminating discussion of reaction mechanism in the context of vinyl radicals. The authors went to painstaking effort to establish even the obvious aspects of the mechanism while constantly and eagerly remarking about what they knew and did not know for certain. Throughout our extensive review we have rarely encountered literature examples of authors who have such decency to report what is known and unknown and to thus make it clear where future research efforts ought to be directed. In addition, we highly recommend question

TABLE 2.3 Mechanistic Evidence Associated With Solved Problems in Chapters 4 Through 9.

Mechanistic evidence	# of examples	Specific problems
Alternative synthesis technique	1	Q33
Analogous reaction	5	Q24, Q52, Q80, Q281, Q287
By-product identification	4	Q61, Q100, Q127, Q237
Change of reaction conditions	8	Q137, Q156, Q159, Q195, Q231, Q259, Q281, Q287
Competition experiment	1	Q91
Computational analysis	5	Q41, Q265, Q283, Q293, Q294
Conjecture	207	Lack of experimental and/or computational evidence is specified in each solution.
Crossover experiment	5	Q15, Q40, Q56, Q81, Q205
Different mechanism leads to different product outcome	2	Q208, Q212
Direct isolation of reaction intermediates	13	Q3, Q30, Q48, Q55, Q61, Q73, Q86, Q148, Q157, Q165, Q171, Q222, Q267
Independent synthesis of intermediate	3	Q115, Q127, Q165
Independent synthesis of product	7	Q62, Q110, Q143, Q185, Q189, Q194, Q269
Isotopic labeling experiment	17	Q15, Q19, Q46, Q59, Q77, Q85, Q99, Q126, Q144, Q150, Q152, Q190, Q194, Q219, Q242, Q272, Q273
Kinetic analysis	1	Q73
pH-rate profiles	1	Q232
Product ratios	1	Q162
Spectroscopic characterization of intermediate	1	Q47
Stereochemical argument	3	Q146, Q153, Q219
Time-resolved spectroscopy	2	Q10, Q265
Trapping intermediates	2	Q69, Q137

219 because it shows how the choice of illustration style/perspective can help keep track of intricate carbon skeletal rearrangements on paper. Aside from these problems we recommend question 222 because it emphasizes the importance of reagent stoichiometry and consequently of establishing a balanced chemical equation for every proposed mechanism. We also highlight question 231 because it illustrates the role of reaction solvent with regard to mechanism. In terms of curly arrow notation, question 261 is particularly interesting because it presents a dilemma for depicting charged radical species using curly arrows. Aside from these examples, we sadly note that only 5 problems in the entire list of 300 had associated computational evidence to support the reaction mechanism. Given the wealth of information that energy reaction coordinate diagrams can provide with regard to mechanism, we find this statistic truly depressing. We therefore advise all professional chemists consulting this textbook to seek collaboration with computational chemists to ensure that discussions of reaction mechanisms are always complemented by both experimental and computational analyses. In this spirit, we highlight a rare case, in question 294, where authors did not include a computational analysis in their original article but did however revisit the same transformation 18 years later, this time collaborating with a well-established computational chemist, to ultimately provide a complete mechanistic analysis of their original problem. In honor of this rare example of collaboration between chemists of different disciplines who share the aim of solving the problem at hand, we have chosen to illustrate their mechanism on the cover of this textbook. Nevertheless, throughout this book we have encountered numerous examples of problems where authors had posed incorrect or implausible mechanisms and mechanisms with questionable experimental and/or computation evidence: Q3, Q12, Q22, Q24, Q29, Q46, Q57, Q58, Q59, Q65, Q73, Q80, Q86, Q101, Q107, Q115, Q119, Q121, Q127, Q130, Q132, Q144, Q149, Q172, Q184, Q193, Q198, Q205, Q207, Q219, Q232, Q248, Q250, Q257, Q261, Q265, Q276, Q292, Q293, and Q296. A small number of questions were ill-posed or had structural errors in the original textbook by McKillop: Q56, Q79, Q107, Q198, and Q242. Despite these drawbacks, we have constructed complete mechanisms for each solution and have provided detailed discussion, additional references, and suggested experiments to help explore the myriad of interesting mechanisms appearing in this book.

We conclude this pivotal section by referencing some important works for readers interested in exploring specific topics such as tautomerism,[44] spectroscopy in general,[45, 46] NMR spectroscopy,[47–50] orbital interaction theory,[51] predicting pK$_A$ values using computational methods,[52] reactive intermediates,[53, 54] reaction mechanisms in general,[55–57] and molecular rearrangements.[58]

2.3 APPLYING EXPERIMENTAL TECHNIQUES—BAEYER-VILLIGER OXIDATION REVISITED

To appreciate the importance of experimental and computational evidence as it relates to reaction mechanisms we provide a concrete example by revisiting the Baeyer-Villiger oxidation we introduced in Chapter 1 (Scheme 2.1). We thus recall that the currently accepted mechanism is the Criegee mechanism (see Scheme 2.2) because it was the only one supported by experimental evidence in a set of three proposed mechanisms. It is now appropriate to introduce the other two mechanisms. Consequently, in Scheme 2.3, we show the first mechanistic proposal of Baeyer and Villiger[59, 60] and in Scheme 2.4 we show an alternative mechanism proposed by Wittig and Pieper.[61]

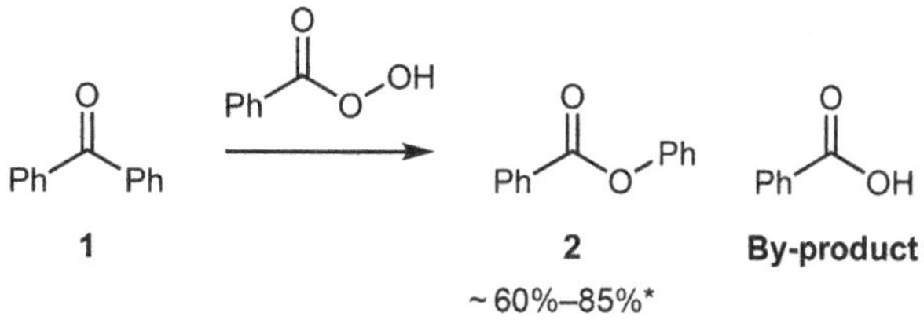

SCHEME 2.1 Baeyer-Villiger oxidation of benzophenone.

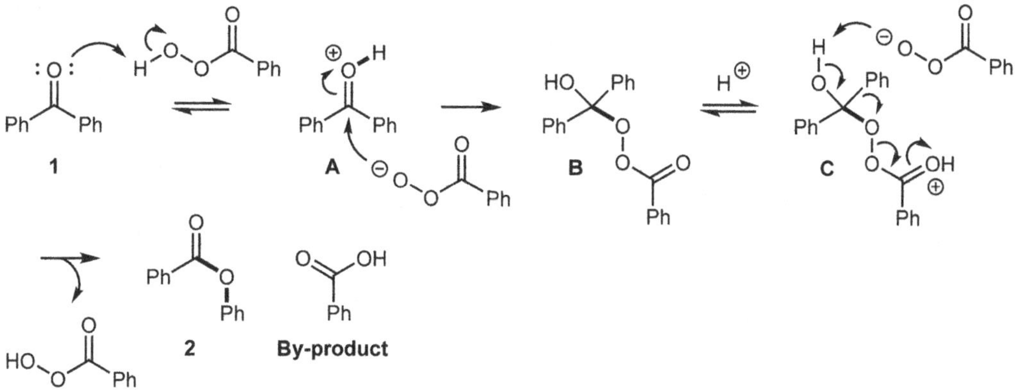

SCHEME 2.2 Criegee mechanism for the Baeyer-Villiger oxidation of benzophenone.

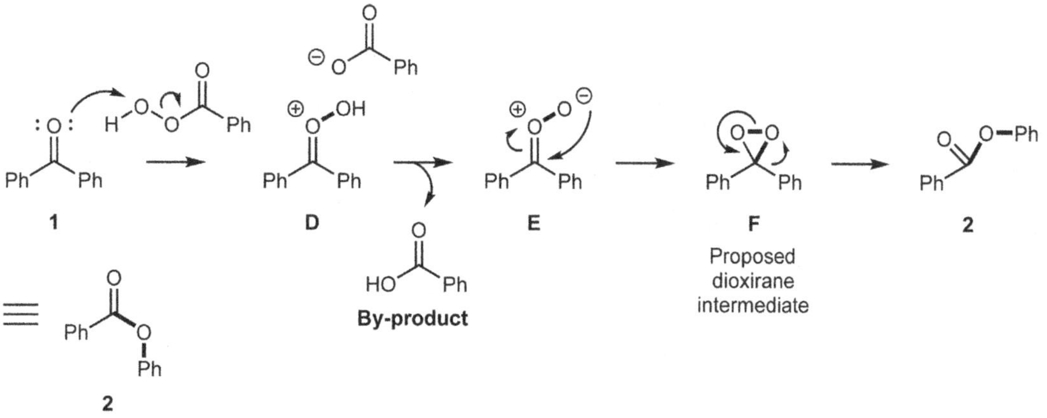

SCHEME 2.3 Baeyer and Villiger mechanism for the Baeyer-Villiger oxidation of benzophenone.

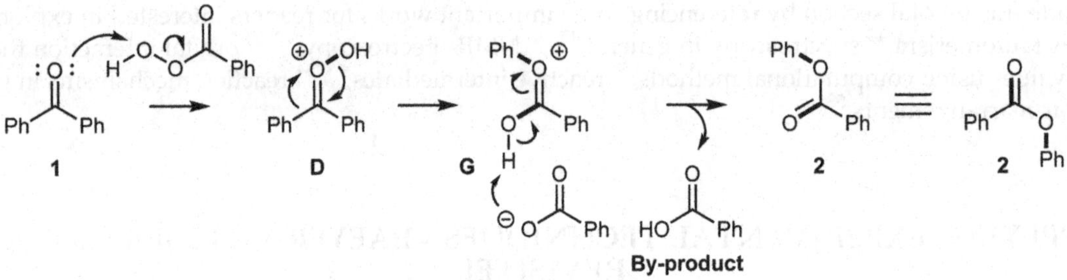

SCHEME 2.4 Wittig and Pieper mechanism for the Baeyer-Villiger oxidation of benzophenone.

We note that both of these mechanisms differ from the Criegee mechanism in that benzophenone 1 does not undergo attack by peroxy acid reagent at carbon but rather attacks the peroxy acid at oxygen to form the common hydroxyoxonium ion intermediate D. In the Baeyer and Villiger mechanism this intermediate undergoes deprotonation to form E which is followed by cyclization to form the dioxirane intermediate F. Rearrangement in F with concomitant migration of one of the phenyl groups leads to product 2. In the Wittig and Pieper mechanism however, intermediate D undergoes rearrangement with phenyl group migration to form oxonium ion G which is deprotonated with concomitant ketonization to form product 2. We note that neither Baeyer and Villiger nor Wittig and Pieper provided complete mechanisms in their original articles. Furthermore, Wittig and Pieper seemed to suggest that a peroxide (i.e., R_2—HC—O—OH) intermediate is formed rather than the dioxirane intermediate proposed by Baeyer and Villiger. Having drawn the complete mechanism, we see no way for a peroxide to form because the ketone bearing carbon atom in benzophenone 1 has no viable pathways for being reduced (i.e., becoming protonated). Therefore if peroxide intermediate is to be excluded, the remaining pathway which avoids formation of a dioxirane intermediate is the rearrangement sequence D to G. As an aside, we note that many mistakes in reaction mechanisms often occur when hydrogen atoms are forgotten (which can happen because hydrogen atoms are customarily hidden from display in conventional visual depictions of chemical structures, especially when complete mechanisms are *not* given). Drawing a complete mechanism explicitly showing key hydrogen atoms uncovers and prevents making such errors.

Returning to the three proposals, we first highlight that all three mechanisms have the same balanced chemical equation which means that by-product identification is not a productive endeavor in distinguishing between them. In fact, the Baeyer-Villiger and Wittig-Pieper mechanisms did not have any supporting experimental or theoretical evidence offered in the original references. Furthermore, the Criegee and Baeyer-Villiger mechanisms result in the same target bonds being formed in the final product. Interestingly, the Wittig-Pieper mechanism results in different target bonds being formed in the final product. When we analyze the Baeyer-Villiger mechanism more closely, we can see that actually the dioxirane intermediate F has two paths in which rearrangement can lead to the final product. This is shown in Scheme 2.5. With this realization in mind, we see that although the Baeyer-Villiger mechanism leads to the same product as the other two mechanisms, in terms of target bond formation, this mechanism actually forms two (identical) products in equal proportion which have different target bonds formed. Thus based on target bonds formed, all three mechanisms lead to different product outcomes. Fortunately, this difference can be exploited by

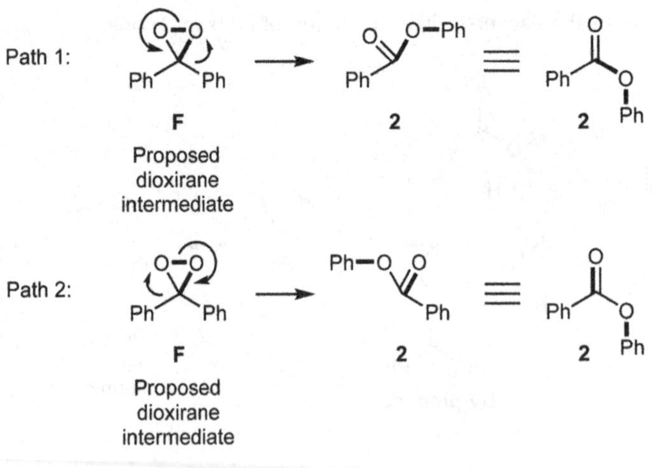

SCHEME 2.5 Two possible pathways depicting fragmentation of dioxirane intermediate.

applying isotopic labeling experiments to help identify target bond formation when there are different product outcomes among multiple proposed mechanisms.

Fortunately, this situation was recognized by Doering and Dorfman in 1953 when they used ^{18}O-labeling on the ketone oxygen atom of benzophenone to help differentiate between the three proposed mechanisms for the Baeyer-Villiger oxidation at that time.[62] When ^{18}O is used on the ketone position in **1**, the three mechanisms lead to different expected products/product mixtures (see Scheme 2.6). For example, the Criegee mechanism predicts the formation of product **3** which has the ^{18}O-label on the carbonyl group. The Wittig and Pieper mechanism predicts the formation of product **4** which has the ^{18}O-label on the ether group. Lastly, the Baeyer and Villiger mechanism predicts the formation of an equimolar distribution of products **3** and **4**. Doering and Dorfman carried out this isotopic labeling experiment and found that only product **3** was formed. Furthermore, no amount of product **4** was detected. This provided the necessary evidence based on which one can reject both the Baeyer and Villiger mechanism and the Wittig and Pieper mechanism. Since this paper appeared, the tetrahedral Criegee intermediate **B** has been widely adopted in presentations on the mechanism of the Baeyer-Villiger oxidation. Unfortunately, this intermediate has not been directly observed or isolated ever since and also a theoretical analysis which could further corroborate

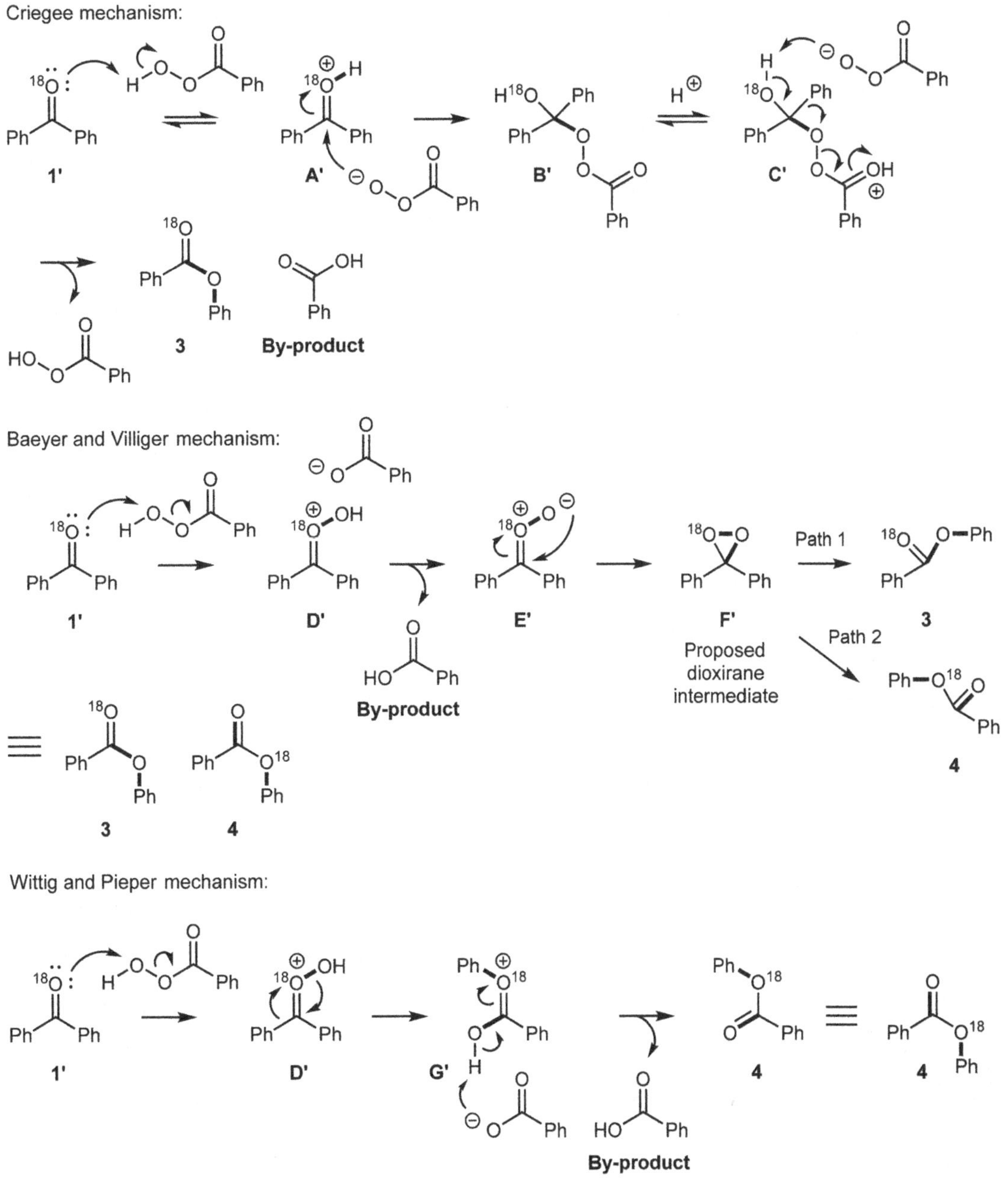

SCHEME 2.6 Comparison of three different mechanisms for the Baeyer-Villiger oxidation of O^{18}-labeled benzophenone.

the Criegee mechanism has not appeared. We believe such a study is needed to help shed light on the energy differences between the three original mechanistic proposals for this interesting transformation.

We hope this example has helped solidify the reader's appreciation for the power of experimental evidence in the elucidation and understanding of organic reaction mechanisms. We hope that future studies of reaction mechanism include complete mechanisms coupled with experimental and theoretical evidence to help support/reject mechanistic proposals. It is only when these three disciplines come together in the form of a scientific paper that one achieves the highest quality research possible on this subject.

References

1. Murdoch JR. What is the rate-limiting step of a multistep reaction? *J Chem Educ.* 1981;58(1):32–36. https://doi.org/10.1021/ed058p32.
2. James LK, ed. *Nobel Laureates in Chemistry 1901-1992.* New York: Wiley-VCH; 1993.
3. Gunal SE, Gurses GS, Erdem SS, Dogan I. Synthesis of stable tetrahedral intermediates (hemiaminals) and kinetics of their conversion to thiazol-2-imines. *Tetrahedron.* 2016;72(17):2122–2131. https://doi.org/10.1016/j.tet.2016.03.003.
4. Abramovitch RA. *Reactive Intermediates.* vol. 2. New York: Plenum Press; 1982.
5. Andraos J. Reaction intermediates in organic chemistry—the "big picture" *Can J Chem.* 2005;83:1415–1431. https://doi.org/10.1139/V05-175.
6. Whittle E, Dows DA, Pimentel GC. Matrix isolation method for the experimental study of unstable species. *J Chem Phys.* 1954;22(11):1943–1944. https://doi.org/10.1063/1.1739957.
7. Ault BS. Matrix isolation spectroscopic studies: thermal and soft photochemical bimolecular reactions. In: Laane J, ed. *Frontiers and Advances in Molecular Spectroscopy.* Amsterdam: Elsevier, Inc.; 2018:667–712. [chapter 20]. *https://doi.org/10.1016/B978-0-12-811220-5.00021-6.*
8. Norrish RGW, Porter G. Chemical reactions produced by very high light intensities. *Nature.* 1949;164:658. https://doi.org/10.1038/164658a0.
9. Duckett S, Gilbert B. *Foundations of Spectroscopy.* New York: Oxford University Press; 1999.
10. Friess SL, Lewis ES, Weissberger A. *Investigation of Rates and Mechanisms of Reactions.* 2nd ed. London: Interscience Publishers, Inc.; 1961.
11. McManus SP. *Organic Reactive Intermediates.* New York: Academic Press; 1973.
12. d'Angelo J. Tetrahedron report number 25: ketone enolates: regiospecific preparation and synthetic uses. *Tetrahedron.* 1976;32(24):2979–2990. https://doi.org/10.1016/0040-4020(76)80156-1.
13. Rajamani R, Gao J. Balancing kinetic and thermodynamic control: the mechanism of carbocation cyclization by squalene cyclase. *J Am Chem Soc.* 2003;125(42):12768–12781. https://doi.org/10.1021/ja0371799.
14. Borisova KK, Nikitina EV, Novikov RA, et al. Diels–Alder reactions between hexafluoro-2-butyne and bis-furyl dienes: kinetic versus thermodynamic control. *Chem Commun.* 2018;54(23):2850–2853. https://doi.org/10.1039/c7cc09466c.
15. Buth JM, Arnold WA, McNeill K. Unexpected products and reaction mechanisms of the aqueous chlorination of cimetidine. *Environ Sci Technol.* 2007;41(17):6228–6233. https://doi.org/10.1021/es070606o.
16. Giri AS, Golder AK. Mechanism and identification of reaction byproducts for the degradation of chloramphenicol drug in heterogeneous photocatalytic process. *Groundwater Sustain Dev.* 2018;7:343–347. https://doi.org/10.1016/j.gsd.2018.05.006.
17. Zawadzki J. The mechanism of ammonia oxidation and certain analogous reactions. *Discuss Faraday Soc.* 1950;8:140–152. https://doi.org/10.1039/DF9500800140.
18. Sun Y, Cui Y, Xiong J, Dai Z, Tang N, Wu J. Different mechanisms at different temperatures for the ring-opening polymerization of lactide catalyzed by binuclear magnesium and zinc alkoxides. *Dalton Trans.* 2015;44(37):16383–16391. https://doi.org/10.1039/c5dt01784j.
19. Isaacs N. *Physical Organic Chemistry.* 2nd ed. Essex: Addison Wesley Longman; 1995.
20. Perkins MJ. *Radical Chemistry: The Fundamentals.* New York: Oxford University Press; 2005.
21. Todres ZV. *Ion-Radical Organic Chemistry: Principles and Applications.* 2nd ed. Boca Raton, FL: CRC Press; 2009.
22. Loudon GM. Mechanistic interpretation of pH-rate profiles. *J Chem Educ.* 1991;68(12):973–984. https://doi.org/10.1021/ed068p973.
23. Hurd CD, Schmerling L. Observations on the rearrangement of allyl aryl ethers. *J Am Chem Soc.* 1937;59(1):107–109. https://doi.org/10.1021/ja01280a024.
24. Barbier V, Couty F, David ORP. Morita Baylis–Hillman reactions with nitroalkenes: a case study. *Eur J Org Chem.* 2015;2015(17):3679–3688. https://doi.org/10.1002/ejoc.201500207.
25. Ingold CK, Raisin CG, Wilson CL, Bailey CR, Topley B. 212. Structure of benzene. Part II. Direct introduction of deuterium into benzene and the physical properties of hexadeuterobenzene. *J Chem Soc (R).* 1936:915–925. https://doi.org/10.1039/JR9360000915.
26. Datta SC, Day JNE, Ingold CK. 178. Mechanism of hydrolysis of carboxylic esters and of esterification of carboxylic acids. acids. Acid hydrolysis of an ester with heavy oxygen as isotopic indicator. *J Chem Soc (R).* 1939:838–840. https://doi.org/10.1039/JR9390000838.
27. Allen CFH, Wilson CV. The use of N^{15} as a tracer element in chemical reactions. The mechanism of the Fischer indole synthesis. *J Am Chem Soc.* 1943;65(4):611–612. https://doi.org/10.1021/ja01244a033.
28. Roberts JD, Semenow DA, Simmons HE, Carlsmith LA. The mechanism of aminations of halobenzenes. *J Am Chem Soc.* 1956;78(3):601–611. https://doi.org/10.1021/ja01584a024.
29. Hughes ED, Juliusburger F, Masterman S, Topley B, Weiss J. 362. Aliphatic substitution and the Walden inversion. Part I. *J Chem Soc (R).* 1935;1525–1529. https://doi.org/10.1039/JR9350001525.
30. Lloyd-Jones GC, Munoz MP. Isotopic labelling in the study of organic and organometallic mechanism and structure: an account. *J Label Compd Radiopharm.* 2007;50(11–12):1072–1087. https://doi.org/10.1002/jlcr.1382.
31. Gomez-Gallego M, Sierra MA. Kinetic isotope effects in the study of organometallic reaction mechanisms. *Chem Rev.* 2011;111(8):4857–4963. https://doi.org/10.1021/cr100436k.
32. Pallitsch K, Roller A, Hammerschmidt F. The stereochemical course of the α-hydroxyphosphonate-phosphate rearrangement. *Chem A Eur J.* 2015;21(28):10200–10206. https://doi.org/10.1002/chem.201406661.

33. Citron CA, Brock NL, Rabe P, Dickschat JS. The stereochemical course and mechanism of the IspH reaction. *Angew Chem Int Ed*. 2012; 51(17):4053–4057. https://doi.org/10.1002/anie.201201110.

34. Liszka MK, Brezinsky K. Variable high-pressure and concentration study of cyclohexane pyrolysis at high temperatures. *Int J Chem Kinet*. 2019;51:49–73. https://doi.org/10.1002/kin.21229.

35. Broensted JN, Pedersen K. Die katalytische Zersetzung des Nitramids und ihre physikalisch-chemische Bedeutung. *Z Phys Chem*. 1924;108U (1):185–235. https://doi.org/10.1515/zpch-1924-10814.

36. Hammett LP. Some relations between reaction rates and equilibrium constants. *Chem Rev*. 1935;17(1):125–136. https://doi.org/10.1021/cr60056a010.

37. Hammett LP. *Physical Organic Chemistry: Reaction Rates, Equilibria, and Mechanisms*. New York: McGraw-Hill Book Company, Inc.; 1940.

38. Wells PG. Linear free energy relationships. *Chem Rev*. 1963;63(2):171–219. https://doi.org/10.1021/cr60222a005.

39. Gardiner Jr. WC. *Rates and Mechanisms of Chemical Reactions*. New York: W. A. Benjamin, Inc.; 1969.

40. Espenson JH. *Chemical Kinetics and Reaction Mechanisms*. 2nd ed. New York: Mc-Graw Hill, Inc.; 1995.

41. Ruff F, Csizmadia IG. *Organic Reactions: Equilibria, Kinetics and Mechanism*. Amsterdam: Elsevier Science B.V.; 1994. https://doi.org/10.1016/B978-0-444-88174-8.50003-7

42. Bachrach SM. *Computational Organic Chemistry*. 2nd ed. Hoboken, NJ: John Wiley & Sons, Inc.; 2014.

43. Wang Z, ed. *Comprehensive Organic Name Reactions and Reagents*. New York: John Wiley & Sons, Inc.; 2010. https://doi.org/10.1002/9780470638859

44. Antonov L. *Tautomerism: Methods and Theories*. Weinheim: Wiley-VCH Verlag GmbH & Co. KGaA; 2014. https://doi.org/10.1002/9783527658824.

45. Field LD, Sternhell S, Kalman JR. *Organic Structures From Spectra*. 5th ed. Chichester: John Wiley & Sons, Ltd.; 2013.

46. Harwood LM, Claridge TDW. *Introduction to Organic Spectroscopy*. New York: Oxford University Press; 1996.

47. Field LD, Li HL, Magill AM. *Organic Structures From 2D NMR Spectra*. Chichester: John Wiley & Sons, Ltd.; 2015.

48. Ionin BI, Ershov BA. *NMR Spectroscopy in Organic Chemistry*. [Turton CN, Turton TI, Trans.] New York: Plenum Press; 1970.

49. Rahman A, Choudhary MI, Wahab A. *Solving Problems With NMR Spectroscopy*. 2nd ed. Amsterdam: Elsevier; 2016. https://doi.org/10.1016/C2012-0-06253-9.

50. Simpson JH. *NMR Case Studies: Data Analysis of Complicated Molecules*. Amsterdam: Elsevier Inc.; 2017. https://doi.org/10.1016/C2014-0-03768-9

51. Rauk A. *Orbital Interaction Theory of Organic Chemistry*. 2nd ed. New York: John Wiley & Sons, Inc.; 2001. https://doi.org/10.1002/0471220418

52. Shields GC, Seybold PG. *Computational Approaches for the Prediction of pK_a Values*. Boca Raton, FL: CRC Press; 2014.

53. Leffler JE. *The Reactive Intermediates of Organic Chemistry*. New York: Interscience Publishers, Inc.; 1956.

54. Fabirkiewicz AM, Stowell JC. *Intermediate Organic Chemistry*. 3rd ed. New York: John Wiley & Sons, Inc.; 2016.

55. Harris JM, Wamser CC. *Fundamentals of Organic Reaction Mechanisms*. New York: John Wiley & Sons, Inc.; 1976.

56. Hoever H. *Problems in Organic Reaction Mechanisms*. London: John Wiley & Sons, Inc.; 1970.

57. Knipe AC. *Organic Reaction Mechanisms*. Chichester: John Wiley & Sons, Ltd.; 2018. https://doi.org/10.1002/9781118941829.

58. Rojas CM. *Molecular Rearrangements in Organic Synthesis*. Hoboken, NJ: John Wiley & Sons, Inc.; 2015. https://doi.org/10.1002/9781118939901.

59. Baeyer A, Villiger V. Einwirkung des Caroeschen Reagens auf Ketone. *Ber Deutsch Chem Ges*. 1899;32(3):3625–3633. https://doi.org/10.1002/cber.189903203151.

60. Baeyer A, Villiger V. Ueber die Einwirkung des Caroeschen Reagens auf Ketone. *Ber Deutsch Chem Ges*. 1900;33(1):858–864. https://doi.org/10.1002/cber.190003301153.

61. Wittig G, Pieper G. Über das monomere Fluorenon-peroxyd. *Ber Deutsch Chem Ges*. 1940;73(4):295–297. https://doi.org/10.1002/cber.19400730402.

62. Doering WvE, Dorfman E. Mechanism of the peracid ketone-ester conversion. Analysis of organic compounds for oxygen-18[1]. *J Am Chem Soc*. 1953;75(22):5595–5598. https://doi.org/10.1021/ja01118a035.

Chapter 3: Problem Solving Organic Reaction Mechanisms

Given the rudimentary understanding of the logic and evidence of physical organic chemistry which we established in the previous two chapters, we now introduce strategies to solve problems in organic reaction mechanisms. Thus we begin by presenting our suggested methodology from the standpoint of both theory and practice. The steps of these methods are organized in hierarchical order and should be applied as such, akin to a checklist. To concretize the methodology we shall analyze an interesting mechanism problem from start to finish and provide commentary based on the checklist and the nature of the problem. Afterward, we shall provide an overview of the current state of pedagogy and academia with regard to physical organic chemistry and where we hope the discipline will advance in the future.

3.1 THEORETICAL PROBLEM-SOLVING STRATEGY

When encountering a problem with an unknown reaction mechanism, we recommend applying the following checklist.

Theoretical Problem-Solving Checklist:

1. Draw the structures of the substrate and product in the same structural aspect.
2. Compare the two structures and identify the target bond mapping for the transformation, that is, identify the target bonds made in the reaction product and which bonds are broken in the substrate. One immediately finds that with a common structural aspect, the task of working in both the forward direction from reactants to products and in the reverse direction from products to reactants is highly facilitated.
3. Once the target bond mapping is established, identify the reaction class or classes for the transformation, for example: addition (coupling, condensation), elimination (fragmentation), substitution, rearrangement, multicomponent, or redox (oxidation or reduction with respect to the substrate of interest).
4. Examine the reaction conditions. Are they acidic, basic, or neutral? Identify solvents, catalysts, ligands, and other reagents above the reaction arrow. Identify special reagents such as redox reagents (e.g., CrO_3, $NaBH_4$) or reagents that are precursors of radical-type intermediates, so-called radical initiators (e.g., AIBN, TEMPO). Identify whether reaction temperature is high or low. Be aware of special reagents that have dual roles in reactions. For example, thionyl chloride and acetonitrile are solvents as well as reagents in acid chloride syntheses and in [3+2] cycloaddition reactions, respectively. Sulfuric acid and nitric acid are acids as well as strong oxidizing agents as in the synthesis of phthalic acid from naphthalene, and in the synthesis of adipic acid from cyclohexanone, respectively. Formic acid acts as an acid as well as a reducing agent upon fragmentation of the formate ion into carbon dioxide and hydride ion. Sodium hydroxide can act as a base to abstract protons or it can act as a nucleophile attacking a substrate.
5. Apply the concept of pK_A to determine which reaction species abstract H^+ ions between bases and acids, that is, identify acidic and basic sites in the substrate structures.
6. Identify nucleophilic and electrophilic sites in the substrate structures.

7. Count atoms on the reactant side of the chemical equation and match them to the products side in an effort to identify missing atoms. Use identified missing atoms to predict possible structures of by-products. Apply these structures to the prediction of a mechanism for each set of possible by-products. For example, gaseous by-products resulting from fragmentation steps, by-products arising from radical-radical couplings in termination steps, redox by-products, that is, oxidation reagents will necessarily end up as a reduced by-product and vice versa. Correlate any suspected by-products with experimental observations such as effervescence, color changes, and precipitation of solids from the reaction solution.

8. Using curly arrow notation, draw reaction mechanisms showing the movement of valence electrons according to the logical and illustration principles discussed and referenced in Chapter 1.

9. If oxidation steps are encountered and the reaction conditions do not specify a needed oxidation reagent, assume autoxidation in the presence of air (i.e., O_2 in its diradical triplet state).

10. In cases involving intricate carbon skeletal rearrangements, number the carbon atoms of the structure in question to help keep track of bond disconnects and connectivities.

11. In choices between energy demanding steps and energy releasing steps, choose the least energy demanding option. Knowledge of sterics, electronics, and special cases such as Baldwin's rules for ring closure reactions, as well as structures likely encountered under photochemical but not thermal conditions, and many other such concepts are emphasized as a valuable asset.

12. Construct a balanced chemical equation to summarize each proposed mechanism. Ensure that the equation is mass and charge balanced. Also ensure that the equation is equal to the sum of the elementary steps in the proposed mechanism.

13. Check the literature for mechanistic studies of the exact same transformation or analogous cases.

14. Check the literature for experimental evidence to support each proposed mechanism (e.g., known intermediates). If evidence is not available, then consult the list of experiments provided in Chapter 2 and propose experiments that can be used to reject mechanistic proposals or support a particular proposed reaction mechanism.

15. Revise the reaction mechanism based on the experimental evidence collected and conclusions drawn from it.

3.2 EXPERIMENTAL PROBLEM-SOLVING STRATEGY

With proposed reaction mechanisms established, it is possible to design experiments according to certain criteria whose results can be used to support or reject proposed mechanisms. In Chapter 2 for example, we explained that by-product identification can be very useful in differentiating between mechanisms that have different balanced chemical equations thanks to the formation of different by-products. We note that although spectroscopy and separation techniques can be used to identify unknown compounds, one can also use quenchers and trapping techniques. Unfortunately it is not always straightforward how one should begin such a process. For this reason we have developed the flowchart in Fig. 3.1 which shows a hierarchical approach to assist in identifying unknown compounds experimentally. We note that this approach can be used to identify both by-products and reaction intermediates. It is our hope that adopting this approach will motivate others to analyze the reaction waste because this often holds important mechanistic clues.

Aside from by-product and intermediate identification and isolation, chemists may also perform crossover experiments to differentiate between intermolecular and intramolecular mechanisms as well as isotopic labeling studies to differentiate between mechanisms which have different target bond mappings in the product structure. We provide an excellent review article which covers some applications of isotopic labeling and crossover experiments to various mechanistic problems.[1] For lists of trapping reagents used to identify reaction intermediates and by-products, see Refs. 2–6. For other aspects of strategy with regard to experimental techniques, the reader is referred to Chapter 2.

3.3 CASE STUDY: REARRANGEMENT OF CINENIC ACID TO GERONIC ACID

The rearrangement of α-cinenic acid **1** to geronic acid **2** was first reported in 1908 by Rupe and Liechtenhan (Scheme 3.1).[7] For the following 40 years its mechanism had remained as a fascinating unsolved problem. In the early 1950s, Meinwald and co-workers reported extensively on this problem.[8–12] Here, we will apply our problem-solving strategy to elucidate reaction mechanisms for the given transformation of α-cinenic acid **1** to geronic acid **2**. We will also return to Meinwald's experimental work and revise our mechanisms according to every piece of evidence considered.

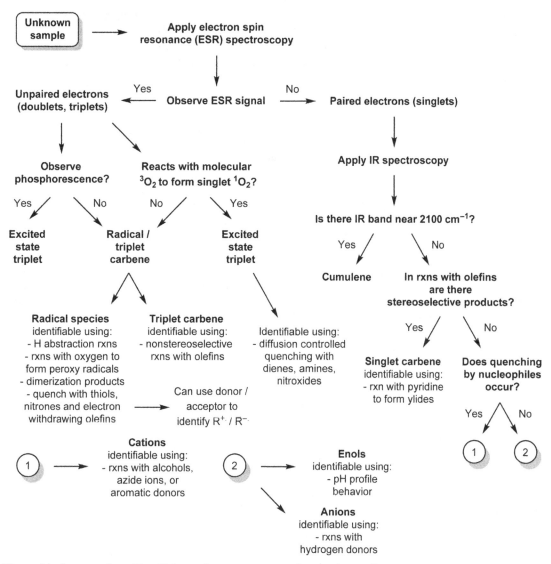

FIG. 3.1 Hierarchical approach to identifying unknown compounds experimentally.

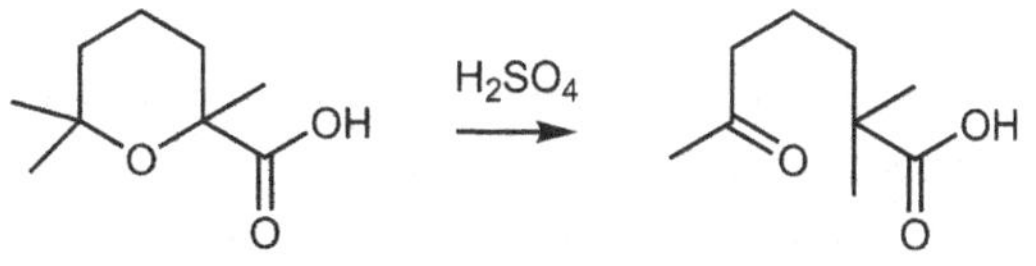

SCHEME 3.1 Acid-catalyzed transformation of α-cinenic acid to geronic acid.

Thus we begin by noting that both substrate **1** and product **2** have been drawn in the same structural aspect. In fact, the number of changes between **1** and **2** in this visual representation is so small that it is possible to make mechanistic predictions simply by inspecting the two structures. For example, visual inspection might suggest that one of the two methyl groups adjacent to the oxygen atom in the tetrahydropyran ring of **1** undergoes migration to the position adjacent to the carboxylic acid group in **2**. If we assign atom numbers to the structures in Scheme 3.1 we obtain Scheme 3.2. We can thus see that the suggested methyl migration would be from C_6 in **1** to C_2 in **2**. Based on this hypothesis we label the bond formed in product **2** in bold as a target bond formed. Taking account of the atoms on the left- and right-hand sides of the chemical equation reveals that nothing is missing, added, substituted, or eliminated and that the sulfuric acid reagent does not contribute, substitute, or abstract atoms to/from the product. Therefore this transformation represents a rearrangement taking place under acidic catalytic conditions. We note that we cannot rule out at this stage the possible catalytic roles of other compounds such as water (which is part of the reaction solvent), O_2 (aerobic conditions, unless otherwise stated, should be assumed), or some other compound that may be eliminated in one step and reattached in another step of the operational mechanistic pathway.

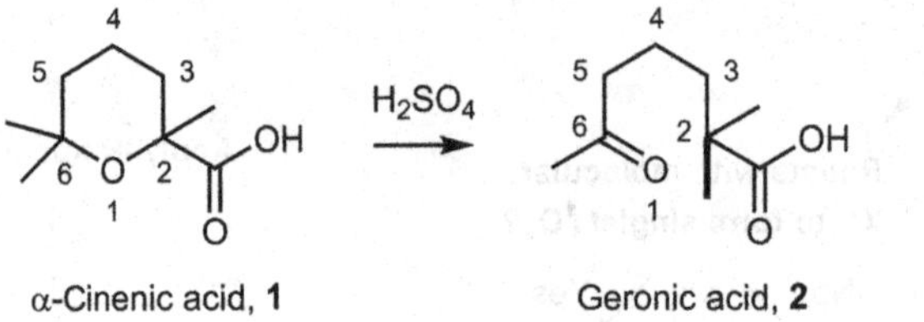

α-Cinenic acid, **1**   Geronic acid, **2**

SCHEME 3.2 Acid-catalyzed transformation of α-cinenic acid to geronic acid showing numbered atoms suggesting methyl group migration.

Since atoms are balanced on both sides of the chemical equation, by-products are not expected for this transformation. Nevertheless, we note that under standard conditions, the yield of geronic acid **2** is approximately 40%. With these facts established we can pursue the methyl migration hypothesis by drawing mechanism 1A shown in Scheme 3.3.

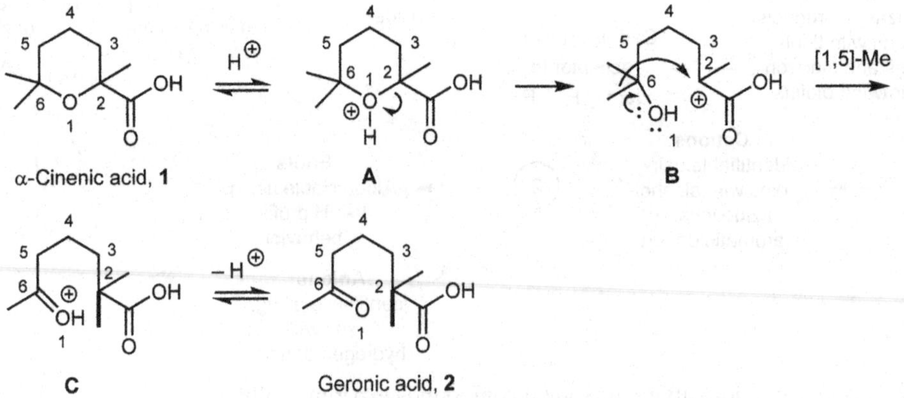

SCHEME 3.3 Mechanism 1A featuring a [1,5]-methyl migration.

In this mechanism we have initial protonation of the tetrahydropyran ring oxygen atom followed by ring fragmentation with formation of the tertiary carbocation intermediate **B**. Formation of oxonium ion aids to facilitate the proposed [1,5]-methyl migration that leads to **C**. Loss of a proton then generates the ketone group in the final product **2**. We note that Meinwald referenced this mechanism which was originally proposed by Rupe.[8] It is worth mentioning that in the early 1900s, [1,2]-methyl migrations were known and utilized to explain how most rearrangements occurred. At the time and even in the early 1950s, as Meinwald writes in his first paper on this subject, [1,5]-methyl migrations were unprecedented. Since this time there have been more examples of [1,5]-aliphatic migrations in the literature (see Ref. 13 and references therein). From the perspective of proximity, we can see that C_6 and C_2 are not distant as to prevent such an intramolecular shift from taking place. Nevertheless, the formation of a carbocation at C_2 which is adjacent to the electron deficient carbonyl group is problematic with regard to this mechanism, as Meinwald himself also observed.[8] To circumvent this drawback, we here propose an alternative mechanism 1B in which two methyl shifts help explain the formation of geronic acid (Scheme 3.4).

Therefore mechanism 1B begins with protonation of the carbonyl oxygen atom of **1** which is followed by oxonium ion formation with concomitant [1,6]-methyl migration from C_6 to the carbonyl carbon atom of the acid group to form oxonium ion **E**. A [1,2]-methyl migration is then facilitated by oxonium ion formation and ring fragmentation to form intermediate **F** which loses a proton to generate product **2**.

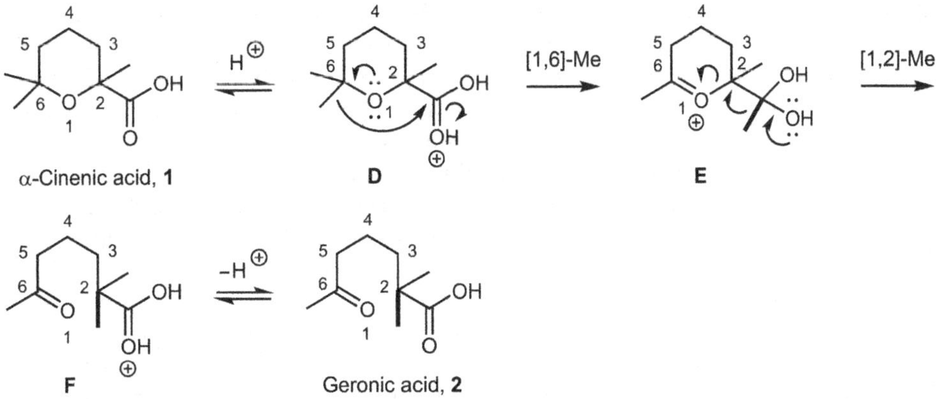

SCHEME 3.4 Mechanism 1B featuring a [1,6]- and a [1,2]-methyl migration.

At this stage one might be tempted to begin designing experiments to help differentiate between the two proposed mechanisms 1A and 1B. This could be done, for example, using ^{18}O-labeling on the carbonyl oxygen in **1**. According to mechanism 1A, the transformation should yield one product with the ^{18}O-label at the carbonyl oxygen of the carboxylic acid whereas mechanism 1B would generate two products (one with an ^{18}O-label at the carbonyl oxygen and one with the ^{18}O-label at the hydroxyl oxygen of the carboxylic acid group due to label scrambling via the **E** to **F** transformation). Nevertheless, such an experiment is not yet required because by changing the structural aspect of Scheme 3.2 (through a vertical flip of structure **1**) we are presented with another mechanistic hypothesis (Scheme 3.5). Indeed, visual inspection of Scheme 3.5 would suggest that a carboxyl group migrates from C_2 to C_6 to generate product **2**.

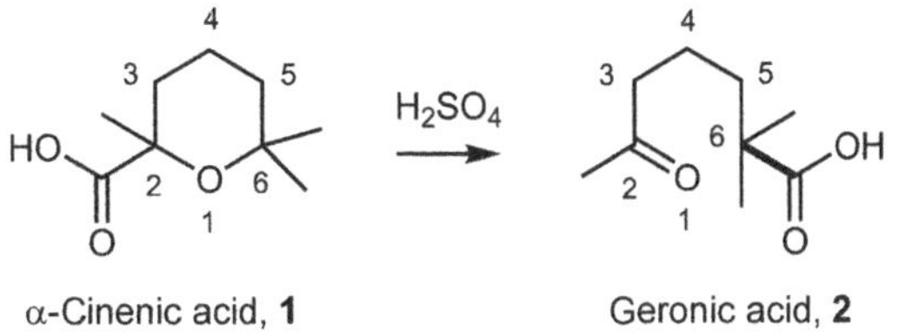

SCHEME 3.5 Acid-catalyzed transformation of α-cinenic acid to geronic acid showing numbered atoms suggesting COOH group migration.

Interestingly, we see that simply changing the structural aspect of the substrate reveals possible mechanistic insights with regard to the transformation, which is no doubt unique to this example. Aside from simplicity and mental clarity, we emphasize maintaining structural aspect when drawing diagrams also because mechanistic insights can be inferred by doing so. Such an insight also occurred to Meinwald which is why he proposed mechanism 2A making use of a carboxyl group migration (Scheme 3.6).

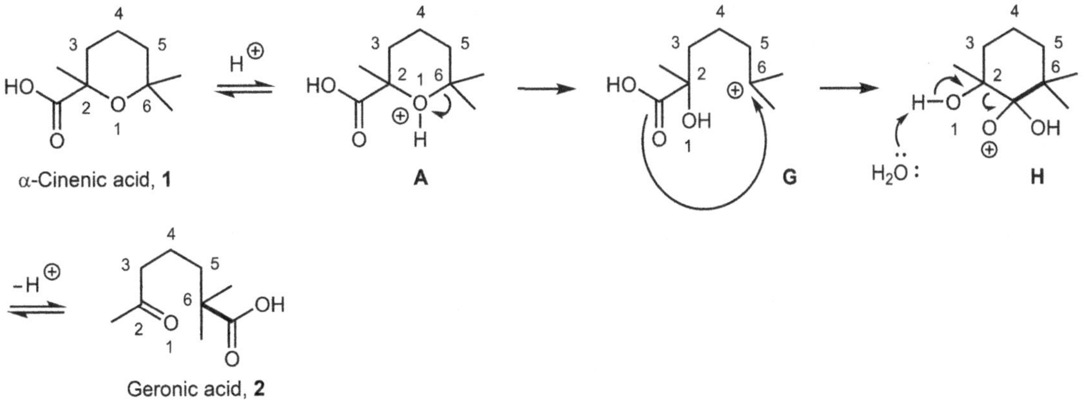

SCHEME 3.6 Mechanism 2A featuring a [1,5]-carboxyl group migration.

Mechanism 2A thus begins with protonation of the tetrahydropyran ring oxygen atom to form the familiar intermediate **A**. Ring fragmentation, this time in the other direction as compared to mechanism 1A, results in formation of carbocation **G**. We note that up to this point formation of **G** is probably more likely than formation of **B** since in **G** the carbocation center is not adjacent to an electron deficient carbon. Both centers are tertiary but we expect **G** to be more stable than **B** for this reason. Nevertheless, according to Meinwald's proposal, **G** cyclizes to form the awkward looking structure **H** with electron deficient oxygen atom.[8] Meinwald explained that such a structure is also invoked in the mechanism of the Baeyer-Villiger oxidation (Criegee mechanism, see Scheme 2.2 in Chapter 2) but this is not the case because there we have an electron deficient carbon atom.[9] Once **H** is formed, Meinwald proposed that deprotonation of hydroxyl leads to ring fragmentation and formation of both a ketone and a carbonyl acid group to furnish geronic acid **2** akin to how 1,3-diols react in acidic conditions to form an olefin and a carbonyl bearing compound. Although Meinwald performed an experiment to seek proof of intermediate **H** through synthesis of geronic acid **2** by means of the Baeyer-Villiger oxidation,[9] he himself admitted that the experiment did not provide direct evidence for **H**. We thus take issue with the electron deficient oxygen atom proposal and revise Meinwald's mechanism to mechanism 2A* shown in Scheme 3.7.

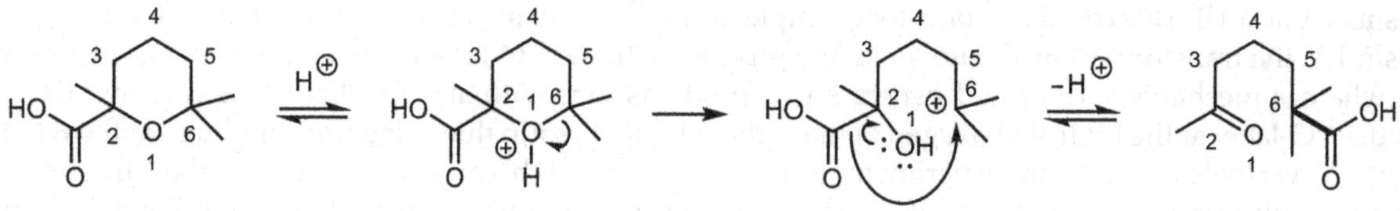

SCHEME 3.7 Mechanism 2A* featuring a [1,5]-carboxyl group migration.

In this mechanism the problematic conversion of **G** to **2** is solved with having a direct intramolecular [1,5]-carboxyl migration as opposed to having a cyclization-fragmentation sequence via **H**. To our knowledge, such a process is not common although there are examples of [1,2]-carboxyl group migrations in the literature (see, for instance, Ref. 14). We note also the difference in target bond mapping in product **2** between mechanism 2A* and mechanisms 1A and 1B. This difference is important because it enables the use of isotopic labeling experiments to help identify whether methyl migration or carboxyl migration ultimately takes place. Thus Meinwald undertook to apply ^{14}C-labeling at the methyl group adjacent to the acid group in substrate **1'** to help identify the operating mechanism. Accordingly, if the mechanism involves methyl migration (either by path 1A or 1B), we expect the product to be the ^{14}C-labeled geronic acid **3** (see Scheme 3.8). Conversely, if carboxyl group migration takes place, we expect to have the ^{14}C-labeled geronic acid **4**.

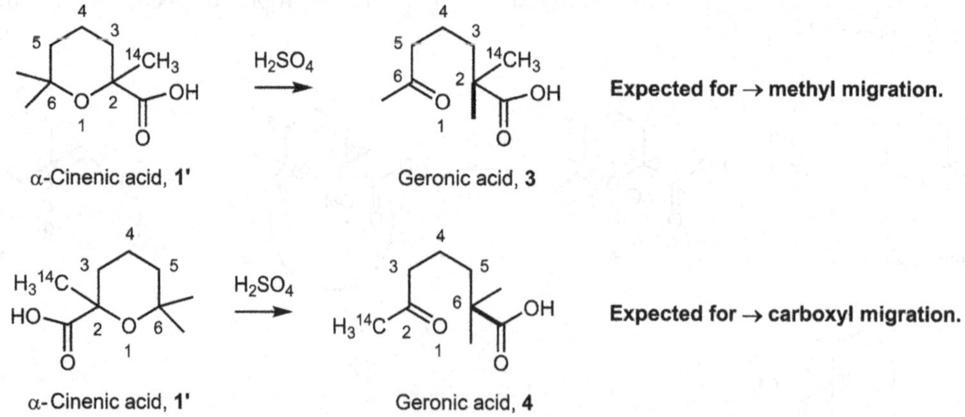

SCHEME 3.8 Predicted products based on ^{14}C-labeling experiment.

Once again we note that the choice of how one draws structures of reactants relative to products can influence one's perception of which functional group is migrating. Nevertheless, Meinwald observed that in concentrated sulfuric acid, the ^{14}C-labeled α-cinenic acid **1'** generates only geronic acid **4**.[8] On this basis, mechanisms 1A and 1B can be rejected. The experiment does not however prove that mechanism 2A* is operating. It merely supports the proposal that carboxyl group migrates by some means from C_2 to C_6. In 1960 Meinwald undertook a synthetic experiment which again showed that carboxyl group migrated from C_2 to C_6 (Scheme 3.9).[11] In this experiment, Meinwald altered the methyl group at C_2 in **1** to an ethyl group in **1"**. Thus carboxyl migration would predict formation of **5** whereas methyl migration would predict formation of **6**. When the experiment was carried out, Meinwald et al. isolated only product **5** and they used an independent synthetic protocol to confirm the authenticity of **5**. Once again, this experiment supports the proposed carboxyl group migration.

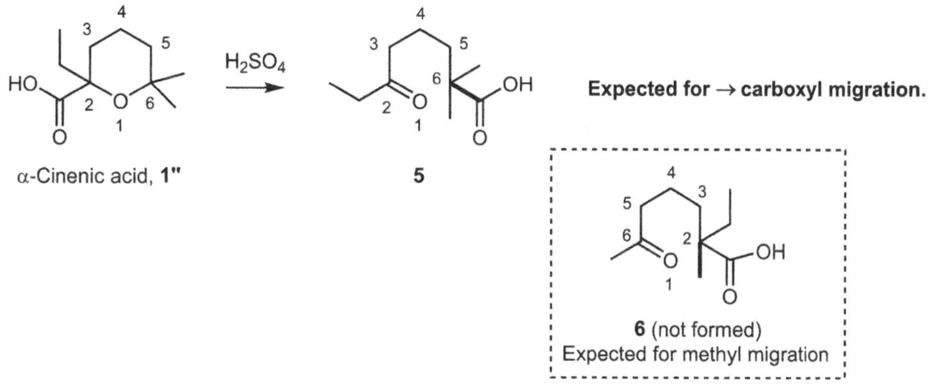

SCHEME 3.9 Synthetic experiment to support the carboxyl migration pathway.

Nevertheless, mechanisms 2A and 2A* appeared problematic in terms of explaining how the carboxyl group migrates from C_2 to C_6. This fact together with the observation (originally reported by Rupe) of bubbling during the early stages of the reaction (an indication of gas evolution) motivated Meinwald to reinvestigate the mechanism of the cinenic acid rearrangement in his fifth and final paper on the subject in 1960.[12] In this paper, Meinwald argued that if gas evolution takes place, it means that a gaseous compound is released and possibly reabsorbed during the course of the transformation. Formation of a gaseous compound would undoubtedly lead to some material loss due to escape of the gas from the reaction mixture which would explain the lower yield of geronic acid **2** (approx. 40%).

It is also important to note that the rearrangement of **1** to **2** forms a lactone side product **7** in 15% yield (see Scheme 3.10). Since this product does not retain all the atoms of the original substrate **1** (it is missing H_2O), Meinwald reasoned that it is formed by a different mechanism (see Scheme 3.11) where H_2O is a by-product.

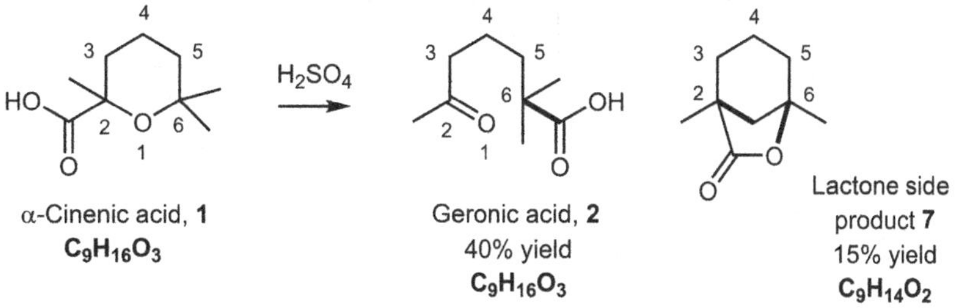

SCHEME 3.10 Cinenic acid rearrangement showing both products formed.

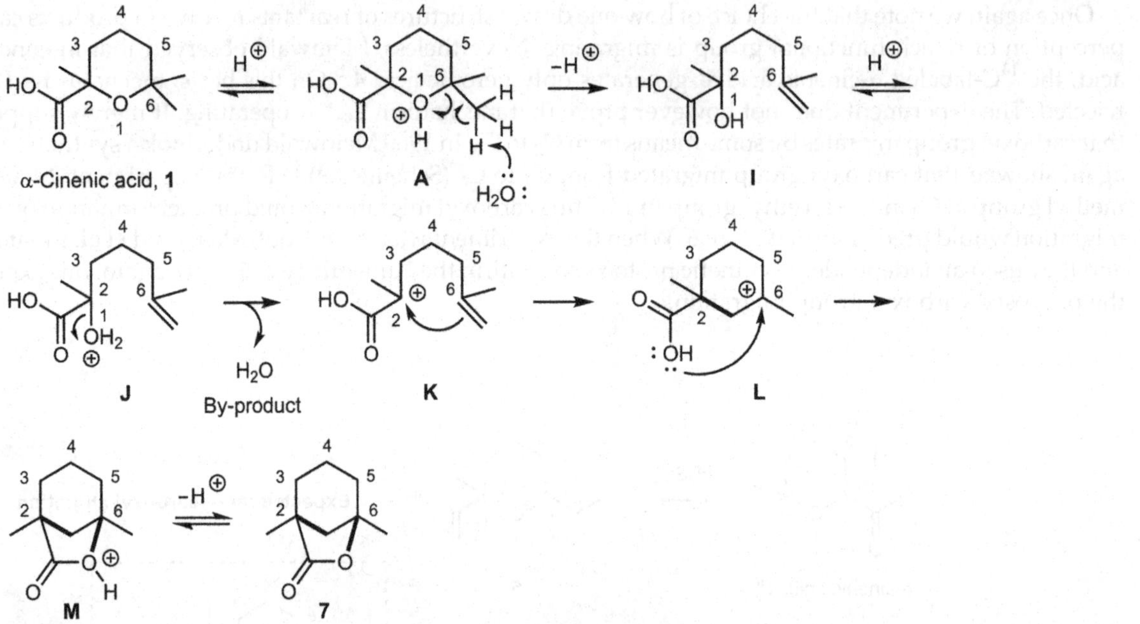

SCHEME 3.11 Mechanism proposed by Meinwald et al. to explain the formation of the lactone side product **7**.

In this mechanism, formation of **A** is followed by deprotonation with concomitant ring fragmentation to form the acyclic intermediate **I**. This intermediate undergoes acid-catalyzed dehydration to form carbocation **K**. Once again we note the problematic position of the carbocation center being next to an electron deficient carbonyl carbon. Cyclization via the terminal olefin in **K** leads to the more stable tertiary carbocation **L** which undergoes a second cyclization via the hydroxyl group of the carboxylic acid. A final loss of a proton leads to the lactone side product **7**.

We note that production of H_2O in this side reaction does not explain the observed gas evolution. Also the yield of **7** is 15% which does not explain why **2** is only formed in 40% yield. At this point, Meinwald et al. carried out the transformation under bubbling carbon monoxide (CO) gas. To their surprise, the authors noticed that the yield of **2** doubled from 40% to 80% while the yield of the lactone side-product **7** stayed the same at 15%. Given such a large effect on the yield of **2**, it was concluded that CO gas had to play a key role in the mechanism of the rearrangement. Thus the authors proposed a new mechanism 2B to explain the migration of carboxyl from C_2 to C_6 (see Scheme 3.12).

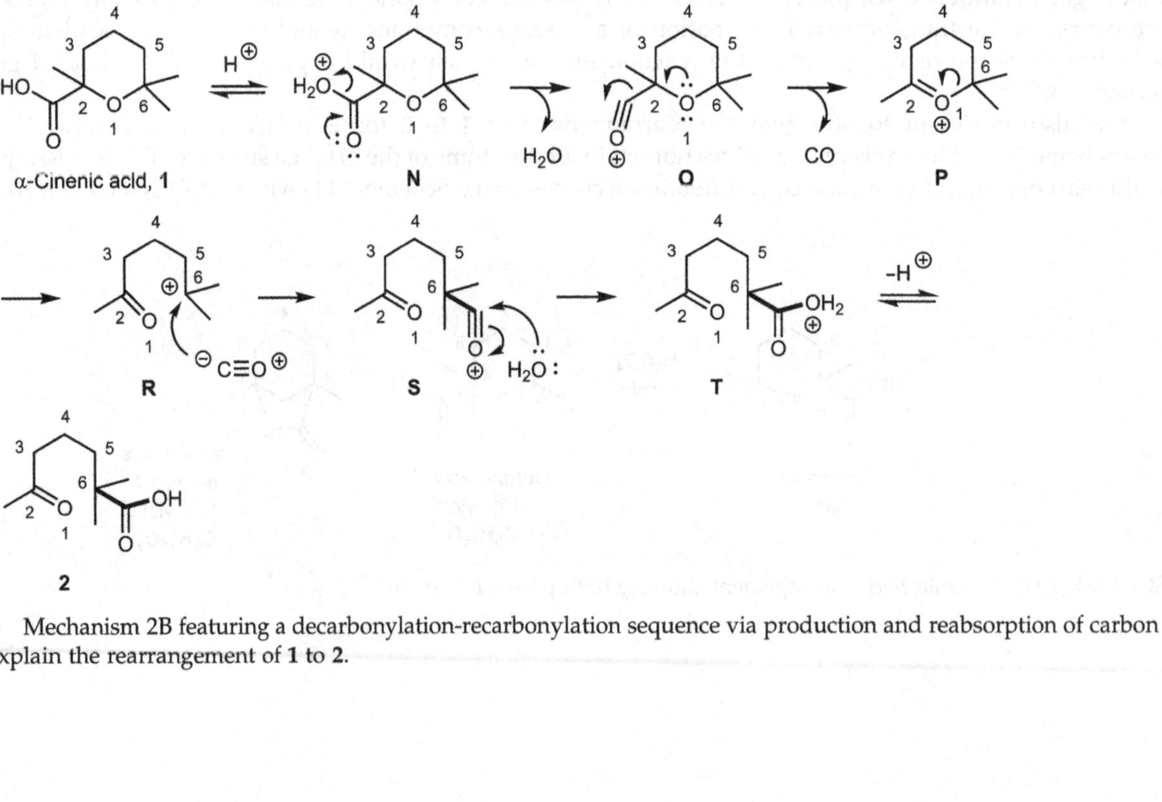

SCHEME 3.12 Mechanism 2B featuring a decarbonylation-recarbonylation sequence via production and reabsorption of carbon monoxide to explain the rearrangement of **1** to **2**.

Therefore mechanism 2B begins with protonation of the hydroxyl group of the carboxylic acid in **1** to form intermediate **N**. Elimination of a water molecule is then facilitated by formation of the —CO$^+$ acylium group in **O**. Oxonium ion formation in the tetrahydropyran ring of **O** facilitates loss of CO gas to form **P**. Ring cleavage in **P** provides a ketone group at C_2 and a tertiary carbocation at C_6 to form intermediate **R**. This intermediate is then attacked by CO at C_6 to form **S** which is attacked by water to form **T**. Loss of a proton from **T** leads to product **2**. Furthermore, the authors explained that loss of water from a carboxylic acid (sequence **1** to **N** to **O**) is analogous to the mechanism of the classical synthesis of acetonedicarboxylic acid from citric acid (see ref. 5 in Ref. 12). Also, the oxygen atom in the tetrahydropyran ring helps to stabilize the developing positive charge at C_2. Moreover, loss of the carboxyl group from C_2 helps relieve steric strain in the form of a 1,3-diaxial interaction between the methyl and carboxyl groups at C_2. For further examples of CO playing a key role in similar transformations, Meinwald et al. recommended the works of Stork and Bersohn[15] and of Koch and Haaf.[16] Lastly, gas evolution of CO occurs due to its low solubility in sulfuric acid. We note that the authors did not attempt to capture and characterize the gas evolved during the reaction.

Instead, in order to help support their proposed mechanism, the authors carried out the rearrangement of **1** to **2** in concentrated sulfuric acid while shaking the mixture with small amounts of ^{14}CO. After work-up it was found that product **2** contained the ^{14}C label at its carboxylic acid carbon position indicating that ^{14}CO had been incorporated at C_6. To strengthen this experimental evidence, the authors carried out two control experiments to prove that ^{14}CO did not exchange with the C=O of either cinenic acid **1** (prior to formation of **N**) or geronic acid **2** (after its formation). To prove the latter case, the authors treated unlabeled geronic acid **2** in the reaction conditions in the presence of ^{14}CO and found that equilibration to ^{14}C-labeled geronic acid occurred at a negligible rate. To prove that C=O equilibration did not precede the rearrangement sequence, the authors examined the lactone side product **7** to see whether it contained appreciable levels of radioactivity (recall that this product is formed by a separate process but that it does retain all the carbon atoms found in **1**). When examined for radioactivity, lactone **7** was found to have much less radioactivity as compared to geronic acid **2** even though their yields suggested that the difference in rates between these paths is not as large. Therefore the authors concluded that C=O equilibration between CO and the carbonyl group of **1** did not occur prior to rearrangement. Taken together with the results of the isotopic labeling experiment, these conclusions strongly support the proposed role of CO in the rearrangement of **1** to **2**. We also note that loss of CO can occur during formation of **P** which explains the lower yield of **2** when additional CO gas is not externally supplied. The incorporation of CO in **R** explains why the yield of **2** improves drastically when CO is externally supplied. We note further that the difference in yield of **2** with and without addition of external CO together with the observation of gaseous frothing of the solution effectively rule out a possible alternative mechanism 2B* in which CO is not liberated from **1** (see Scheme 3.13). If this mechanism were operational, the yield of **2** would not change with or without the addition of external CO.

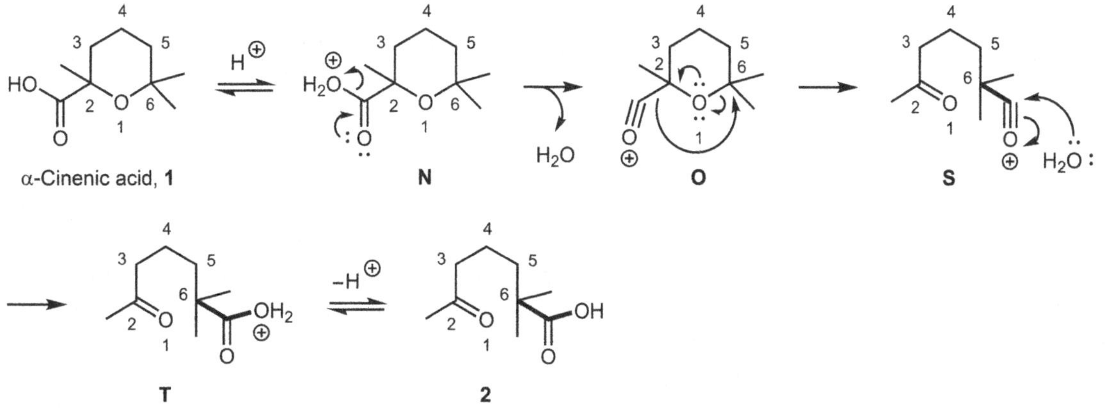

SCHEME 3.13 Mechanism 2B* where CO is not liberated from **1**.

For the sake of completeness we also provide an alternative mechanism for the formation of the lactone side product **7** (Scheme 3.14). We note that in this mechanism intermediate **P** takes a different path as in mechanism 2B. This means that **P** is a hub intermediate and that the two mechanisms are thus connected up until its formation. Thus instead of being attacked by CO at C_6, intermediate **P** can undergo deprotonation to form an olefin and a ketone group resulting in structure **U**. This intermediate then undergoes protonation and cyclization to form **W** which is attacked by CO at C_6 to form **X**. Cyclization in **X** followed by loss of proton furnishes the lactone group and the side product **7**.

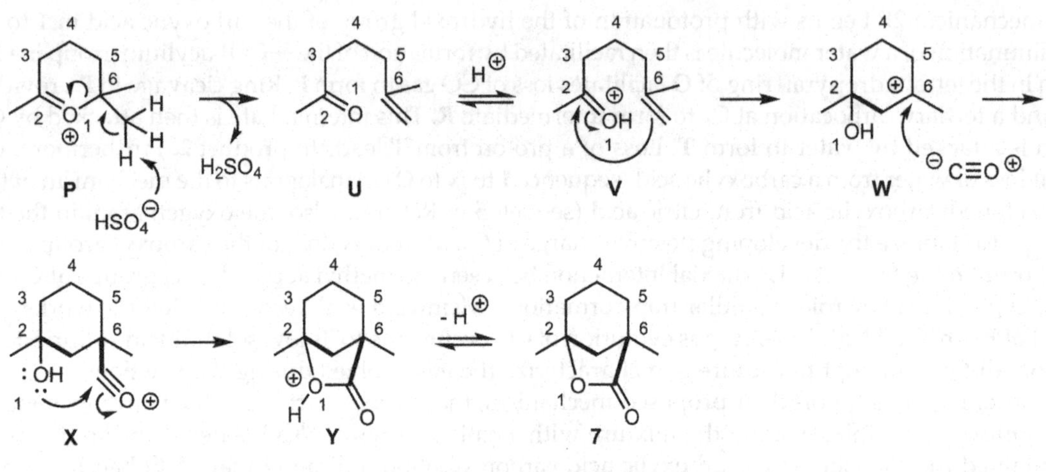

SCHEME 3.14 Alternative mechanism to explain the formation of the lactone side product 7.

We can thus see how theoretical and experimental techniques can be used to revise reaction mechanisms and to eventually arrive at a mechanism supported by all available evidence (in this case, mechanism 2B). Nevertheless, despite all the evidence, the authors did not carry out theoretical analyses to help shed light on the energy barriers for the proposed methyl and carboxyl group migrations and how they compare to the currently accepted mechanism. We hope that others will undertake to provide this evidence to complete the story of the cinenic acid rearrangement. We also hope that educators take note of this case study and apply similar strategies when drawing mechanisms in review articles, textbooks, academic papers, and classroom examples. We also note that understanding the mechanism of the cinenic acid rearrangement has allowed the authors to effectively double the yield of their target product by supplying CO gas to the reaction. We hope that skeptics will notice this fact and begin to see the value of mechanistic analysis to the field of organic synthesis, especially industrial organic synthesis. In conclusion, we reference several excellent textbooks which provide further insight into strategic approaches to solving problems in organic chemistry.[17–19] For other examples of reaction mechanisms in need of solutions, see Refs. 20–22. For examples of works by Meinwald with regard to interesting problems of mechanism and other personal reflections, see Refs. 23–28. Lastly, we also point the reader to excellent references on the philosophy, pedagogy, vocabulary, and history of reaction mechanism.[29–35]

3.4 CURRENT STATE OF PEDAGOGY AND ACADEMIC AND FUTURE DIRECTIONS

Having established the importance of a logical and experimental foundation for physical organic chemistry, we wish to stress that the current state of pedagogy and academia requires much attention and improvement. The fact that theoretical organic chemistry is not required and often missing from journal articles purporting to investigate reaction mechanisms coupled with the fact that chemistry departments have consistently sought to limit academic positions to mechanistic chemists in favor of synthetic chemists, that the number of university courses on physical organic chemistry has declined, that chemistry conferences often have no invited speakers on the topics of reaction mechanism or of physical organic chemistry, and even the fact that awards in chemistry are often divided into physical organic chemistry (for the lesser part) and synthetic chemistry (for the most part) all indicate that chemists today do not appreciate the importance of physical organic chemistry and its impact in all other fields of chemistry. This unfortunate disconnect needs to be addressed. It is no surprise that because of the unnecessary and detrimental division between synthetic, mechanistic, and computational chemistry over the years, students today do not learn chemistry having a solid foundation in place. It is also why instructors complain that students have difficulty learning and reproducing reaction mechanisms on exams. Indeed, current research suggests that a focus on reaction mechanism and physical organic chemistry helps students improve their performance in chemistry courses altogether.[36–38] Despite these benefits, organic chemistry continues to be perceived as requiring memorization of reactions. It is our belief that unless unity and collaboration between the mechanistic, synthetic, and computational chemistry fields are sought, this perception will only grow worse and lead to further declines in laboratory funding and student enrollment. We therefore hope that this textbook will serve as a catalyst for much needed change.

References

1. Lloyd-Jones GC, Munoz MP. Isotopic labelling in the study of organic and organometallic mechanism and structure: an account. *J Label Compd Radiopharm.* 2007;50(11 12):1072–1087. https://doi.org/10.1002/jlcr.1382.
2. Andraos J. Reaction intermediates in organic chemistry—the "big picture". *Can J Chem.* 2005;83:1415–1431. https://doi.org/10.1139/V05-175.
3. Abramovitch RA. *Reactive Intermediates.* vol. 2. New York: Plenum Press; 1982.
4. McManus SP. *Organic Reactive Intermediates.* New York: Academic Press; 1973.
5. Leffler JE. *The Reactive Intermediates of Organic Chemistry.* New York: Interscience Publishers, Inc.; 1956.
6. Fabirkiewicz AM, Stowell JC. *Intermediate Organic Chemistry.* 3rd ed. New York: John Wiley & Sons, Inc.; 2016.
7. Rupe H, Liechtenhan C. Über Kondensationen mit Cinensäure. *Ber Deutsch Chem Ges.* 1908;41(1):1278–1286. https://doi.org/10.1002/cber.190804101237.
8. Meinwald J. The acid-catalyzed rearrangement of cinenic acid. *J Am Chem Soc.* 1955;77(6):1617–1620. https://doi.org/10.1021/ja01611a063.
9. Meinwald J, Cornwall CC. The acid-catalyzed rearrangement of cinenic acid. II. Geronic acid from 6-hydroxy-2,2,6-trimethylcyclohexanone. *J Am Chem Soc.* 1955;77(22):5991–5992. https://doi.org/10.1021/ja01627a058.
10. Meinwald J, Hwang HC. The acid-catalyzed rearrangement of cinenic acid. III. Structure and synthesis of the lactonic product. *J Am Chem Soc.* 1957;79(11):2910–2912. https://doi.org/10.1021/ja01568a059.
11. Meinwald J, Ouderkirk JT. The acid-catalyzed rearrangement of cinenic acid. IV. Synthesis and rearrangement of 6-carboxy-6-ethyl-2, 2-dimethyltetrahydropyran. *J Am Chem Soc.* 1960;82(2):480–483. https://doi.org/10.1021/ja01487a059.
12. Meinwald J, Hwang HC, Christman D, Wolf AP. The acid-catalyzed rearrangement of cinenic acid. V. Evidence for a decarbonylation-recarbonylation mechanism. *J Am Chem Soc.* 1960;82(2):483–486. https://doi.org/10.1021/ja01487a060.
13. Spangler CW, Boles DL. Thermolysis of 5,5-dimethyl-1,3-cyclohexadiene. Evidence for rearrangement via [1,5] sigmatropic methyl migration. *J Org Chem.* 1972;37(7):1020–1023. https://doi.org/10.1021/jo00972a020.
14. Berner D, Cox DP, Dahn H. On the migration of a carboxyl group in a Wagner-Meerwein rearrangement in superacid solution: proof by double labeling with carbon-13. *J Am Chem Soc.* 1982;104(9):2631–2632. https://doi.org/10.1021/ja00373a048.
15. Stork G, Bersohn M. A stereospecific reaction of carbon monoxide. *J Am Chem Soc.* 1960;82(5):1261–1262. https://doi.org/10.1021/ja01490a068.
16. Koch H, Haaf W. Über die Synthese verzweigter Carbonsäuren nach der Ameisensäure-Methode. *Justus Liebigs Ann Chem.* 1958;618(1):251–266. https://doi.org/10.1002/jlac.19586180127.
17. Alonso-Amelot ME. *The Art of Problem Solving in Organic Chemistry.* 2nd ed. Hoboken, NJ: John Wiley & Sons, Inc.; 2014.
18. Corey EJ, Chelg X. *The Logic of Chemical Synthesis.* Chichester: John Wiley & Sons, Inc.; 1989.
19. Maloney MG. *How to Solve Organic Reaction Mechanisms: A Stepwise Approach.* Chichester: John Wiley & Sons, Ltd.; 2015. https://doi.org/10.1002/9781118698532.
20. Ranganathan D, Ranganathan S. *Challenging Problems in Organic Reaction Mechanisms.* London: Academic Press, Inc.; 1972.
21. Ranganathan D, Ranganathan S. *Further Challenging Problems in Organic Reaction Mechanisms.* Orlando, FL: Academic Press, Inc.; 1980.
22. Ranganathan S. *Fascinating Problems in Organic Reaction Mechanisms.* San Francisco, CA: Holden-Day, Inc.; 1967.
23. Meinwald J, Lewis A, Gassman PG. Highly strained bicyclic systems. III. The stereochemistry and rearrangement of the 1,5,5-trimethylbicyclo [2,1,1]hexane-6-carboxylic acids. *J Am Chem Soc.* 1960;82(10):2649–2651. https://doi.org/10.1021/ja01495a071.
24. Meinwald J, Cadoff BC. The rearrangement of a keto epoxide to a lactone. A novel transformation in the bicyclo[2.2.1]heptane series. *J Org Chem.* 1962;27(5):1539–1541. https://doi.org/10.1021/jo01052a011.
25. Meinwald J, Putzig DE. Bridgehead substitution vs. ring contraction in the deamination of 1-aminobicyclo[2.2.1] hept-5-en-2-ol. *J Org Chem.* 1970;35(6):1891–1894. https://doi.org/10.1021/jo00831a037.
26. Meinwald J, Samuelson GE, Ikeda M. Naphtho[1,8]bicyclo[3.2.0]hepta-2,6-diene. Synthesis and rearrangement to pleiadiene. *J Am Chem Soc.* 1970;92(26):7604–7606. https://doi.org/10.1021/ja00729a026.
27. Smith LR, Meinwald J. The Wolff rearrangement approach to the tricyclo[3.2.0.02,6]heptane system. *J Org Chem.* 1977;42(3):415–417. https://doi.org/10.1021/jo00423a003.
28. Meinwald J. Personal reflections on receiving the Roger Adams award in organic chemistry. *J Org Chem.* 2005;70(13):4903–4909. https://doi.org/10.1021/jo050787h.
29. Hoffmann R, Minkin VI, Carpenter BK. Ockham's razor and chemistry. *HYLE Int J Philos Chem.* 1997;3:3–28. http://www.hyle.org/journal/issues/3/hoffman.htm.
30. Meek SJ, Pitman CL, Miller AJM. Deducing reaction mechanism: a guide for students, researchers, and instructors. *J Chem Educ.* 2016;93(2):275–286. https://doi.org/10.1021/acs.jchemed.5b00160.
31. Gold V. Glossary of terms used in physical organic chemistry. *Pure Appl Chem.* 1983;55(8):1281–1371. https://doi.org/10.1351/pac198355081281.
32. Gold V. Glossary of terms used in physical organic chemistry. *Pure Appl Chem.* 1979;51(8):1725–1801. https://doi.org/10.1351/pac197951081725.
33. Müller P. Glossary of terms used in physical organic chemistry. *Pure Appl Chem.* 1994;66(5):1077–1184. https://doi.org/10.1351/pac199466051077.
34. Ridd J. Organic pioneer: Christopher Ingold's insights into mechanism and reactivity established many of the principles of organic chemistry. *Chem World.* 2008, December;50–53.
35. Tarbell DS, Tarbell AT. *Essays on the History of Organic Chemistry in the United States, 1875-1955.* Nashville, TN: Folio Publishers; 1986.
36. Flynn AB, Ogilvie WW. Mechanisms before reactions: a mechanistic approach to the organic chemistry curriculum based on patterns of electron flow. *J Chem Educ.* 2015;92(5):803–810. https://doi.org/10.1021/ed500284d.
37. Flynn AB, Featherstone RB. Language of mechanisms: exam analysis reveals students' strengths, strategies, and errors when using the electron-pushing formalism (curved arrows) in new reactions. *Chem Educ Res Pract.* 2017;18:64–77. https://doi.org/10.1039/C6RP00126B.
38. Webber DM, Flynn AB. How are students solving familiar and unfamiliar organic chemistry mechanism questions in a new curriculum? *J Chem Educ.* 2018;95(9):1451–1467. https://doi.org/10.1021/acs.jchemed.8b00158.

Chapter 4: Solutions 1 – 50

Question 1: A Tandem Route to 1,2,3,4-Tetrasubstituted Naphthalenes

In 1995 American chemists showed that a tandem Pummerer-Diels-Alder sequence involving heating of sulfoxide **1**, acetic anhydride, and maleic anhydride at 120°C to form adduct **2** (87% yield) followed by reaction of **2** with *p*-toluenesulfonic acid (PTSA) in tetrahydrofuran (THF) at 25°C gave the substituted naphthalene derivative **3** in quantitative yield (Scheme 1).[1] Interestingly, when methyl propiolate was used instead of maleic anhydride, the authors isolated tetralone **4** rather than adduct **2** in 51% yield. Suggest mechanisms to account for these transformations.

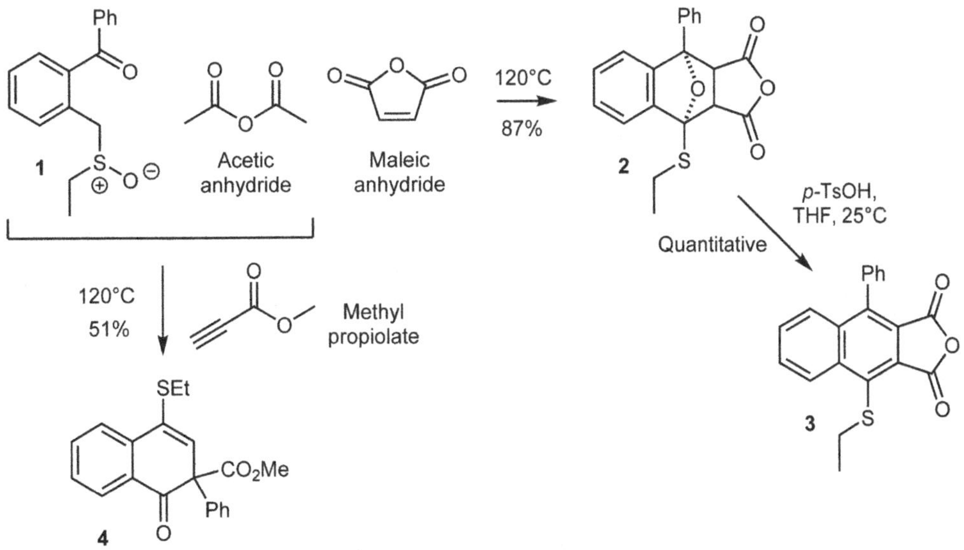

SCHEME 1

SOLUTION: 1 TO 2

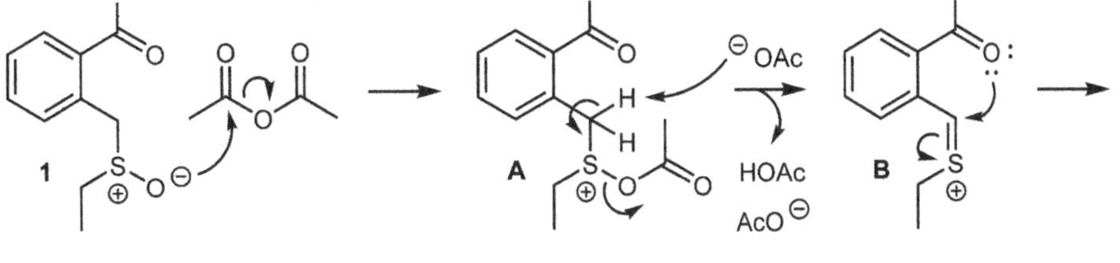

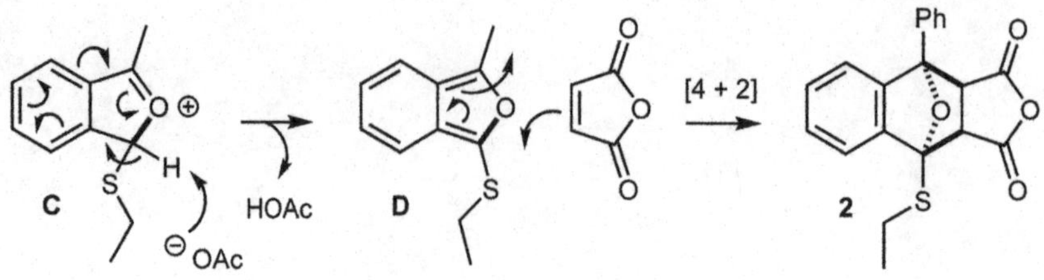

SOLUTION: 2 TO 3

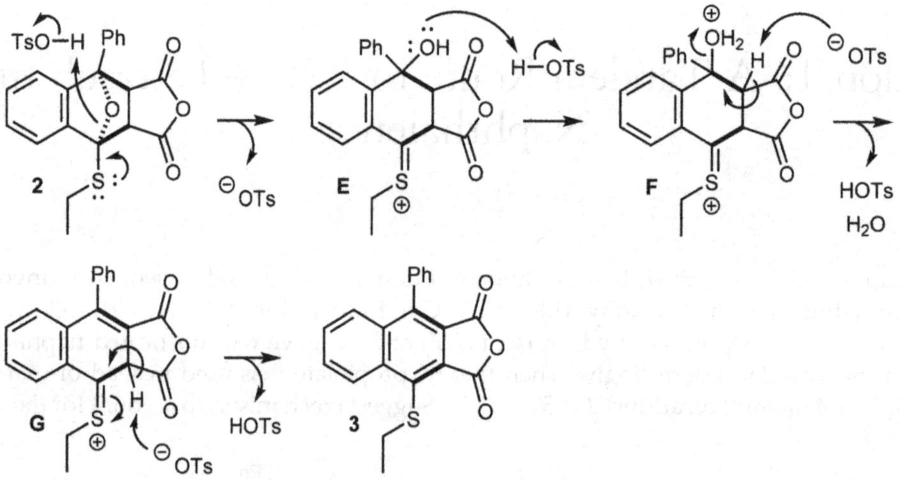

SOLUTION: D TO 4

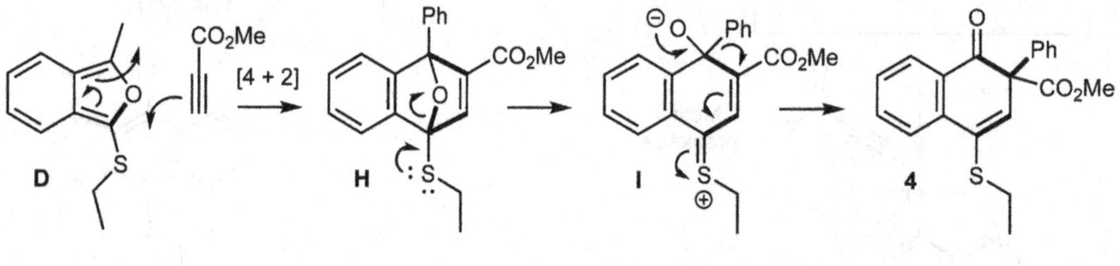

BALANCED CHEMICAL EQUATIONS

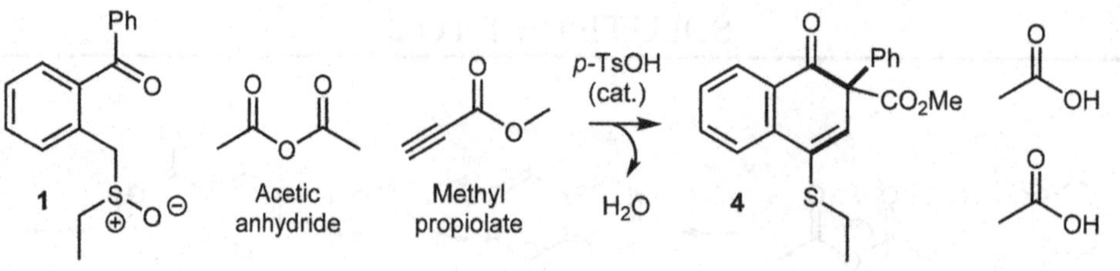

KEY STEPS EXPLAINED

The mechanism proposed by the authors to explain the formation of substituted naphthalene 3 from 1 begins with a Pummerer rearrangement of the starting sulfoxide 1 upon treatment with acetic anhydride to generate sulfonium ion **B** after deprotonation and elimination of acetic acid. Cyclization of the ketone group of **B** onto the sulfonium ion leads to **C** which undergoes deprotonation to form the substituted furan **D**. At this point, the presence of maleic anhydride acting as a dienophile causes a [4+2] hetero-Diels-Alder cycloaddition between it and diene **D** to form product 2. We note that product 2 exists as a mixture of diastereomers due to the two possible ways that the dienophile can approach diene **D** in the hetero-Diels-Alder reaction. With product 2 in hand, treatment in an acidic medium results in loss of a water molecule in an acid-catalyzed fashion which works to restore aromaticity to the central cyclohexane ring thus forming the substituted naphthalene product 3. We note that in this step the [2.2.1] bicyclic framework of product 2 is converted to a central benzene ring. When methyl propiolate was substituted for maleic anhydride, Cochran and Padwa proposed another mechanism to account for the unexpected synthesis of product 4. In this mechanism the oxabicyclic ring of intermediate **H** opens to form oxonium ion **I** following the usual hetero-Diels-Alder reaction between **D** and methyl propiolate. The authors rationalized that the **H** to **I** step is driven by the lone pair of electrons on the sulfur atom. Afterward, a pinacol-type rearrangement results in the 1,2-migration of the phenyl group of **I** to generate product 4. We also note the differences between the balanced chemical equations for the formation of each product. In terms of ring construction mapping and synthesis strategy, we note that the [2.2.1] bicyclic framework of product 2 arises from a [(5+0)+(2+2+1)] strategy. This is followed by sequential ring opening and aromatization of the [2.2.1] framework of 2 to generate the aromatic ring of 3. Finally, product 4 arises via a [4+2] strategy involving a phenyl group [1, 2] migration.

We also note that the authors did not provide concrete evidence in support of their conjectured mechanism. Nevertheless, they suggested that the exclusive regioisomer 4 isolated in the third step of the reaction was consistent with the expectations of frontier molecular orbital (FMO) theory. The authors noted that "the most favorable FMO interaction is between the HOMO (highest occupied molecular orbital) of the isobenzofuran and the LUMO (lowest unoccupied molecular orbital) of methyl propiolate. The atomic coefficient at the ethylthio substituted position in the isobenzofuran ring is larger than (the coefficient) at the phenyl position in the HOMO and this nicely accommodates the high regioselectivity encountered."[1]

ADDITIONAL RESOURCES

For further examples of cascade processes which combine Pummerer rearrangement and Diels-Alder cycloaddition in synthetically useful chemistry, we recommend a review article by Padwa.[2] For an interesting analysis of the reaction conditions required for Pummerer rearrangement as opposed to Pummerer fragmentation, we recommend the work of Laleu et al.[3] Last, for discussion at how intramolecular/intermolecular nonbonded O-S interactions affect the enantioselectivity/diastereoselectivity of reactions involving the Pummerer rearrangement, we recommend the work of Nagao et al.[4]

References

1. Cochran JE, Padwa A. Tandem Pummerer Diels-Alder sequence for the preparation of α-thio substituted naphthalene derivatives. *Tetrahedron Lett.* 1995;36(20):3495–3498. https://doi.org/10.1016/0040-4039(95)00574-V.
2. Padwa A. Tandem methodology for heterocyclic synthesis. *Pure Appl Chem.* 2004;76(11):1933–1952. https://doi.org/10.1351/pac200476111933.
3. Laleu B, Machado MS, Lacour J. Pummerer fragmentation vs. Pummerer rearrangement: a mechanistic analysis. *Chem Commun.* 2006;2006:2786–2788. https://doi.org/10.1039/B605187A.
4. Nagao Y, Miyamoto S, Miyamoto M, et al. Highly stereoselective asymmetric Pummerer reactions that incorporate intermolecular and intramolecular nonbonded S···O interactions. *J Am Chem Soc.* 2006;128(30):9722–9729. https://doi.org/10.1021/ja056649r.

Question 2: Hydroazulenes by Radical Cyclization

In 1995 Japanese chemists reacted the substituted cyclopentanone **1** with triphenyltin hydride and AIBN (azobis (isobutyronitrile)) in refluxing toluene to achieve in 79% yield the hydroazulene product **2** (Scheme 1).[1] Provide a mechanism for this transformation.

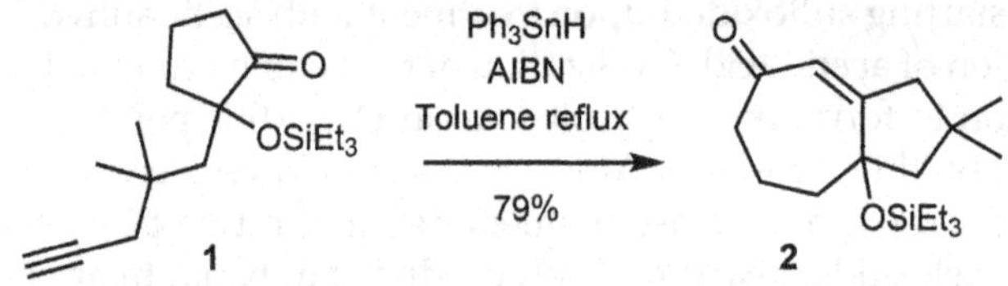

SCHEME 1

SOLUTION

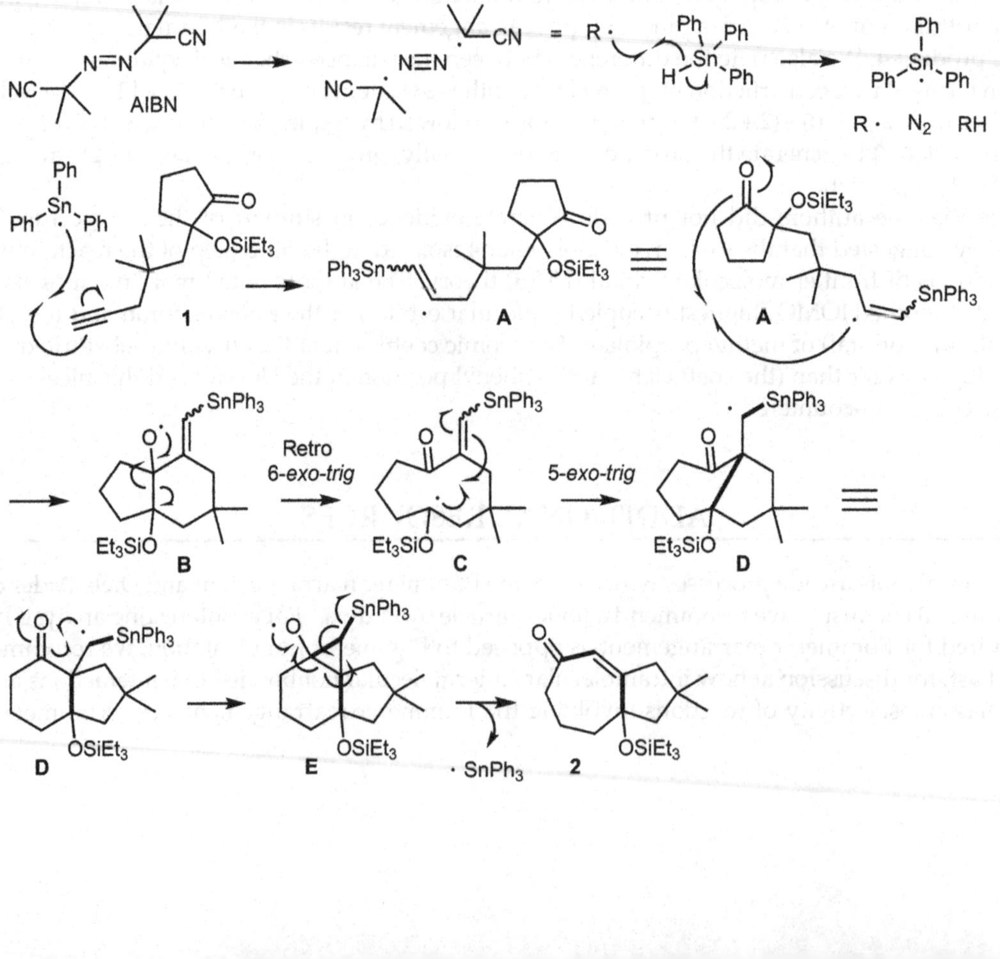

BALANCED CHEMICAL EQUATION

KEY STEPS EXPLAINED

The mechanism proposed by Nishida et al. for the formation of hydroazulene **2** from the substituted cyclopentanone **1** begins with the radical initiation of triphenyltin radical. This occurs via abstraction of the hydrogen atom of triphenyltinhydride by isobutyronitrile radical which is generated from AIBN. Triphenyltin radical then reacts with the alkyne moiety of **1** to form alkenyl radical **A** which cyclizes to alkoxyl radical **B** having a [4.3.0] bicyclic framework. Homolytic fragmentation of the ring junction C—C bond provides radical **C** which ring closes again via a 5-*exo-trig* cyclization to form radical **D** which has a different [4.3.0] bicyclic framework as compared to **C**. The carbon radical center of **D** inserts into the carbonyl group π-bond to form the tricyclic alkoxyl radical **E** which fragments to hydroazulene product **2** with elimination of triphenyltin radical. We note that the ring construction mapping for the [5.3.0] framework is [(5 + 2) + (5 + 0)]. In addition, the authors observed that replacing either of the methyl groups of **1** with OSiEt$_3$ groups resulted in an unsuccessful rearrangement. They stipulated that this was the case "because of the effect of the silyl group on the reaction between the triple bond and the stannyl radical," which in this mechanism forms intermediate **A**. Interestingly, the ring-substituted OSiEt$_3$ was not said to affect this step but it was chosen because it helps to stabilize the radical intermediate **C**. In the context of substitution, the authors noticed that the two methyl groups in **1** "greatly enhanced the efficiency of the radical cyclization" from **1** to **A**. Furthermore, substituted acetal groups such as O-MOM and O-SEM instead of the methyl groups had lower reaction efficiency with decreased yields of 11%–57% for products **2**. The authors argued that compared to methyl groups, the acetal groups are not bulky enough to lower the activation energy of the initial cyclization of **1** to **A**. There was also a large difference in efficiency between diastereomers of the substituted acetals of **1** (11%–19% versus 55%–57% yields between diastereomers). The authors suggested that such a difference may involve the chelation between the alkoxyl group and the trialkylstannyl group but that further experiments were needed for clarification. Lastly, it was observed that ACN (azobis(cyclohexane)carbonitrile) improved the yields of the transformations as compared to AIBN. According to the authors, this happens because the generation of alkenyl radicals by reaction of alkynes and trialkylstannyl radicals is a reversible process whose equilibrium can be pushed to completion by the use of a thermally more stable compound than AIBN such as ACN. Aside from these observations, the authors did not provide additional experimental or theoretical evidence to support their proposed mechanism. In particular, a theoretical analysis would help shed more light on the nature of the cyclization of **1** to **A** which the authors argued is greatly influenced by alkyl chain substitution of **1**.

ADDITIONAL RESOURCES

For an interesting look at the allylic peroxyl radical rearrangement in the context of the formation of allylic hydroperoxide, we recommend the work of Lowe and Porter.[2] For a look at how computational calculations can provide insight into the rearrangement of ethane-1,2-diol radical cations, we recommend the work of Ruttink and Burgers.[3]

References

1. Nishida A, Ogasawara Y, Kawahara N, Nishida M. A simple preparation of the hydroazulene skeleton from cyclopentanone derivatives via a free radical process. *Tetrahedron Lett.* 1995;36(17):3015–3018. https://doi.org/10.1016/0040-4039(95)00502-4.
2. Lowe JR, Porter NA. Preparation of an unsymmetrically labeled allylic hydroperoxide and study of its allylic peroxyl radical rearrangement. *J Am Chem Soc.* 1997;119(47):11534–11535. https://doi.org/10.1021/ja9723038.
3. Ruttink PJA, Burgers PC. Rearrangement of ethane-1,2-diol radical cations: processes involving dipole-catalyzed proton shifts and charge transfer. *Org Mass Spectrom.* 1993;28(10):1087–1097. https://doi.org/10.1002/oms.1210281018.

Question 3: Cinnolines to Indoles

In 1995 Japanese chemists showed that overnight stirring of the readily accessible dihydrocinnoline derivative **1** with powdered potassium cyanide in aqueous dimethylformamide at room temperature followed by water quench and recrystallization gave in 54% yield the 2-acetyl-3-cyanoindole product **2** (Scheme 1).[1] Suggest a mechanism for this transformation.

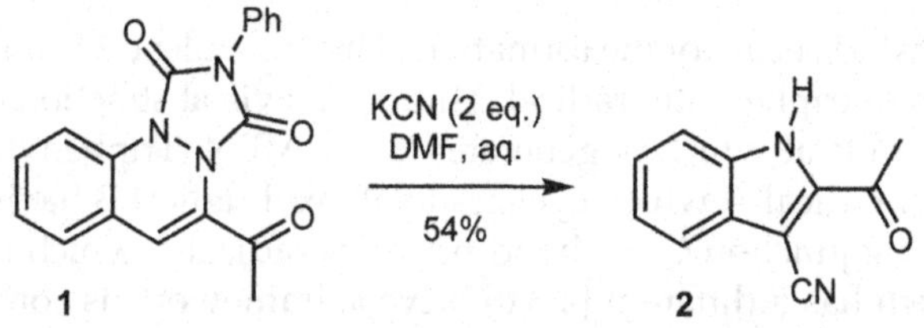

SCHEME 1

SOLUTION

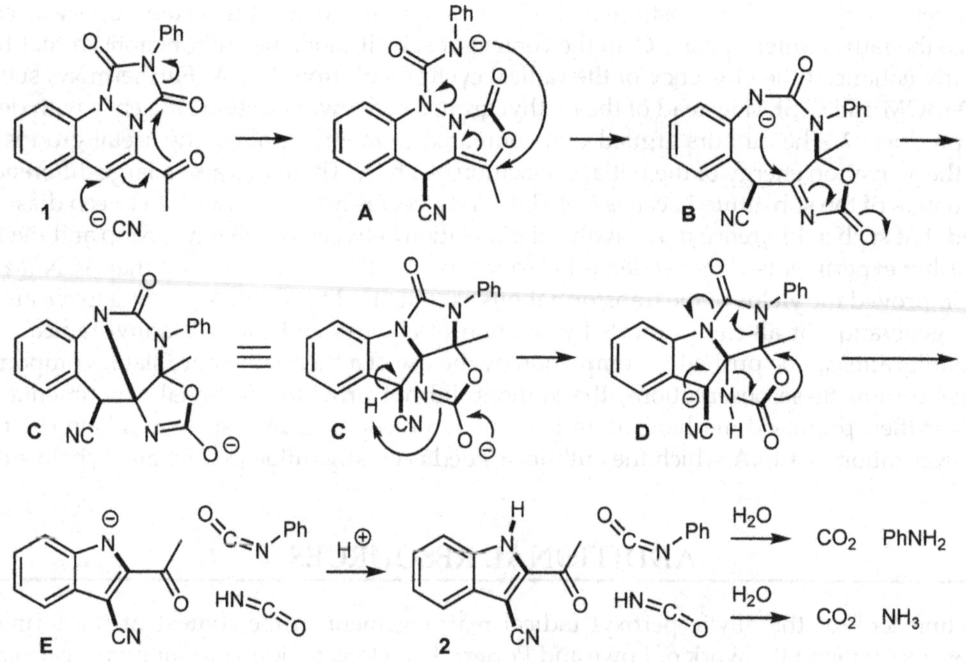

BALANCED CHEMICAL EQUATION

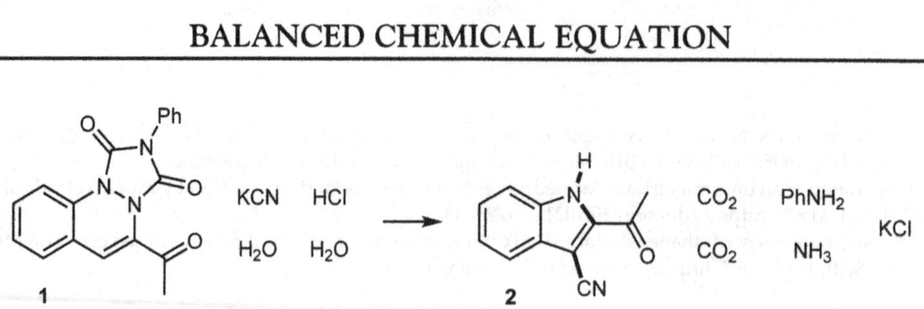

KEY STEPS EXPLAINED

The mechanism proposed by the authors for the transformation of **1** to **2** begins with a Michael addition of cyanide anion to the polar alkene group of enone **1** which results in cleavage of the dicarboximide ring which forms intermediate **A**. The [4.3.0] bicyclic framework of intermediate **A** then rearranges to a [6.3.0] bicyclic framework to form intermediate **B** after cleavage of the N—N bond. Amide ion intermediate **B** rearranges again to form tricyclic enolate **C** which undergoes an intramolecular proton transfer to generate carbanion **D**. We note that this carbanion is stabilized by the electron-withdrawing cyano group. Next, intermediate **D** spontaneously eliminates isocyanate (HN=C—OH) and phenyl isocyanate to form intermediate **E** which is protonated during the work-up to form the final indole product **2**. The isocyanate and phenyl isocyanate by-products are expected to react with water during the work-up phase of the reaction to form two equivalents of carbon dioxide gas, one equivalent of ammonia and one equivalent of aniline as ultimate by-products. We also expect that hydrolysis of isocyanate would occur much faster than that of phenyl isocyanate. Lastly, the ring construction mapping for the indole synthesis is [5 + 0].

We note that an alternative mechanism may be proposed for this transformation as shown in Scheme 2. In this mechanism the initial step is nucleophilic attack by cyanide anion onto **1** which causes fragmentation of the N—N bond to form intermediate **F**. This intermediate undergoes further fragmentation to form **G** which undergoes elimination of phenylisocyanate to generate the imide ion **H**. Simultaneous protonation of **H** and hydroxide attack on the isocyanate group leads to amide enolate **I** which cyclizes to amide ion **J**. This amide ion undergoes a protic shift followed by decarboxylation to generate amide ion **L** which eliminates ammonia to form the indole product **2** after an intramolecular 1,3-protic shift. We note that the motivation for this mechanism is the avoidance of high energy tricyclic and tetracyclic intermediates such as **B**, **C**, and **D** in the previous mechanism.

In terms of experimental evidence, we note that the authors attempted to isolate the cyano tricyclic intermediate **D** but were unsuccessful. Interestingly, it was possible to isolate the ethoxy analog of **D** (cyano group replaced by ethoxy

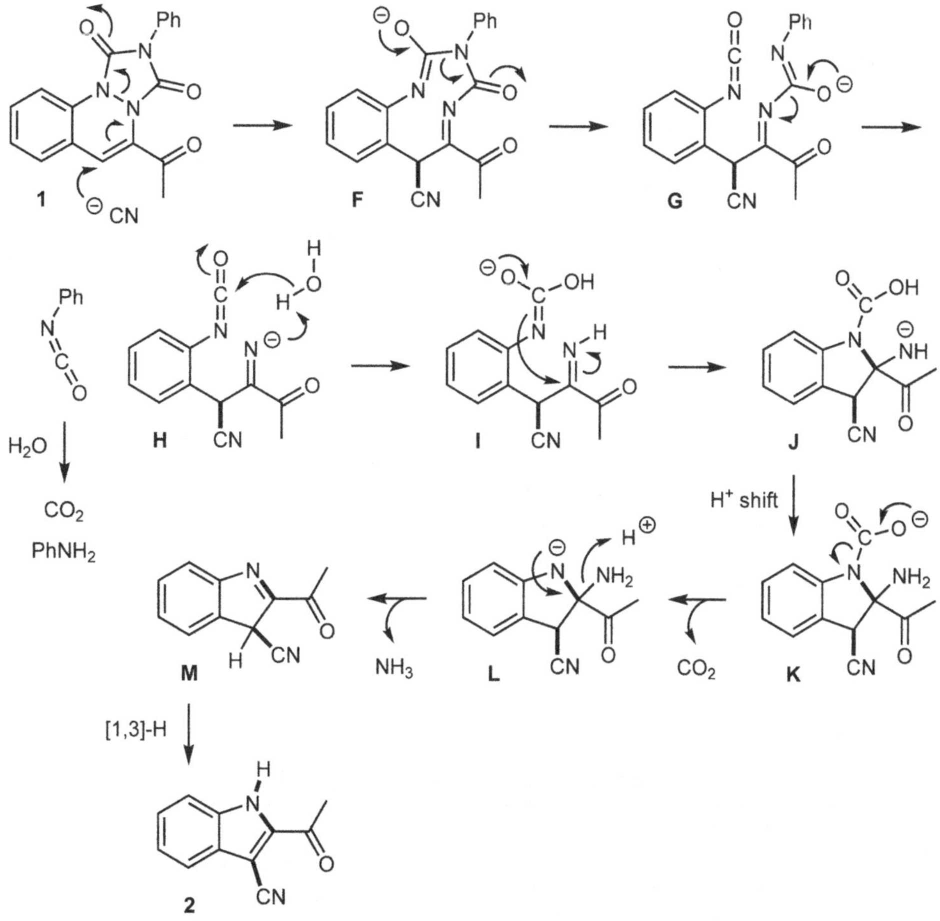

SCHEME 2

group) when the reaction was carried out in ethanol under basic conditions (i.e., the nucleophile that promotes rearrangement from **1** to **A** is ethoxy anion, not cyanide anion). The authors reasoned that the strongly electron-withdrawing cyano group destabilized carbanion **D** and thus contributed to the lack of isolation of the protonated tetracyclic analog of **D**. This argument contradicts the normal expectation that electron-withdrawing groups should thermodynamically *stabilize* negatively charged centers. We believe that the reason the cyano intermediate **D** could not be isolated is because either the fragmentation step to carbanion intermediate **E** is fast compared to the ethoxy analog and/or that the cyano group thermodynamically stabilizes the product carbanion **E** more than the substrate carbanion **D** as a result of product state resonance stabilization as shown in Scheme 3. Both of these effects would significantly reduce the energy barrier for this step and drive it in the forward direction toward formation of **2** rather than stop the process at the formation of **D**. We note that an ethoxy group would not impart such a product stabilization effect (for **E**) since it is an electron-donating group and so would slow down the fragmentation step from **D** to **E**. This is perhaps why the authors were able to isolate the ethoxy tetracyclic analog of **D** but when the nucleophile was replaced by cyanide anion, the process went on to form product **2** and not the protonated analog of **D**.

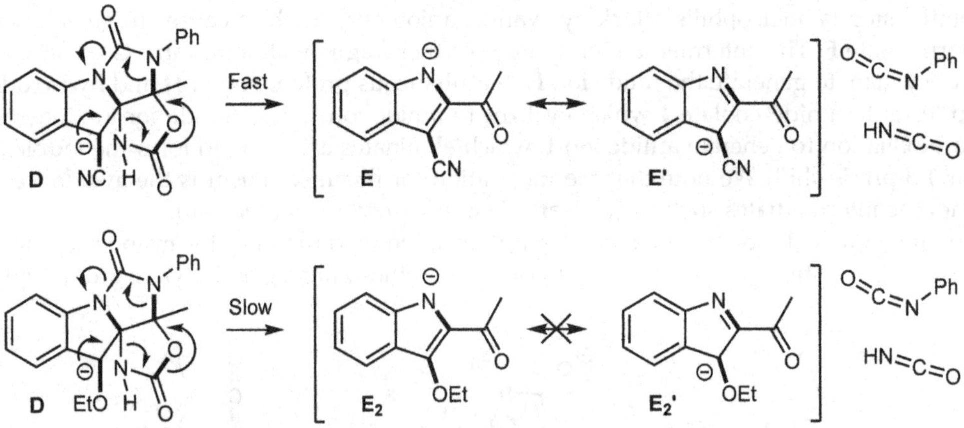

SCHEME 3

Lastly, we note that the authors were able to isolate the Michael adduct shown in Scheme 4 when a weak nucleophile was used. Both of these experiments suggested that the mechanism depicted in the *Solution* is likely operative instead of the alternative mechanism which avoids the tricyclic and tetracyclic intermediates **B**, **C**, and **D**. For a future study, we recommend others undertake a theoretical analysis to help support the proposed mechanism, and for interest, conduct one for the alternative mechanism to compare the energetics of the two and support the argument presented in Scheme 3.

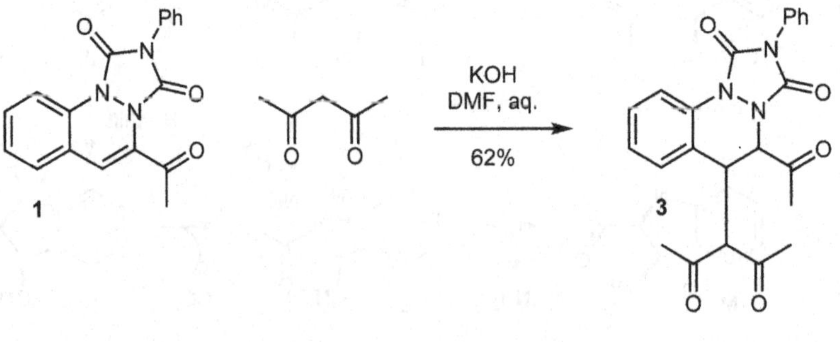

SCHEME 4

ADDITIONAL RESOURCES

For a look at interesting dicarboximide ring opening mechanisms in aqueous media related to three different fungicide compounds, we recommend the work of Villedieu et al.[2] For a review of blocked isocyanates and their mechanistic significance, see Wicks and Wicks.[3] And for a fairly recent article concerning the mechanisms of [4+2] cycloaddition reactions involving alkynes and isocyanate elimination, see the work of Yoshino et al.[4]

References

1. Tanaka S, Seguchi K, Sera A. One-pot syntheses of 3-cyanoindoles from 3-acyl- and 3-ethoxycarbonyl-1,2-dihydrocinnoline-1,2-dicarboximides. *J Chem Soc Perkin Trans 1*. 1995;1995:519–520. https://doi.org/10.1039/P19950000519.
2. Villedieu JC, Calmon M, Calmon JP. Mechanisms of dicarboximide ring opening in aqueous media: procymidone, vinclozolin and chlozolinate. *Pestic Sci*. 1994;41:105–115. https://doi.org/10.1002/ps.2780410206.
3. Wicks DA, Wicks Jr ZW. Blocked isocyanates III: Part A. Mechanisms and chemistry. *Prog Org Coat*. 1999;36(3):148–172. https://doi.org/10.1016/S0300-9440(99)00042-9.
4. Yoshino Y, Kurahashi T, Matsubara S. Nickel-catalyzed [4 + 2] cycloaddition of alkynes to carbonylsalicylamides via elimination of isocyanates. *Chem Lett*. 2010;39:896–897. https://doi.org/10.1246/cl.2010.896.

Question 4: Rearrangement During NMR Studies

During a 1995 study of metabolites isolated from algal overgrowth collected in Okinawa, American chemists observed that rearrangement of the nakienone **1** to the hemiacetal **2** occurred during NMR studies (CDCl$_3$ solution) of **1** (Scheme 1).[1] Suggest a mechanism for this transformation.

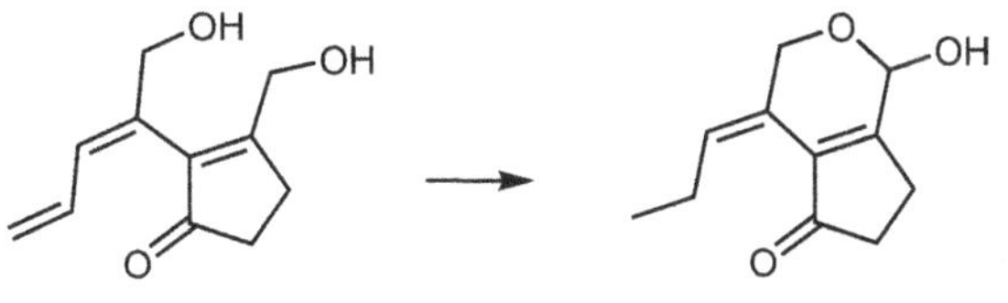

1

Dihydroxytrienone
nakienone A

2

Hemiacetal

SCHEME 1

SOLUTION

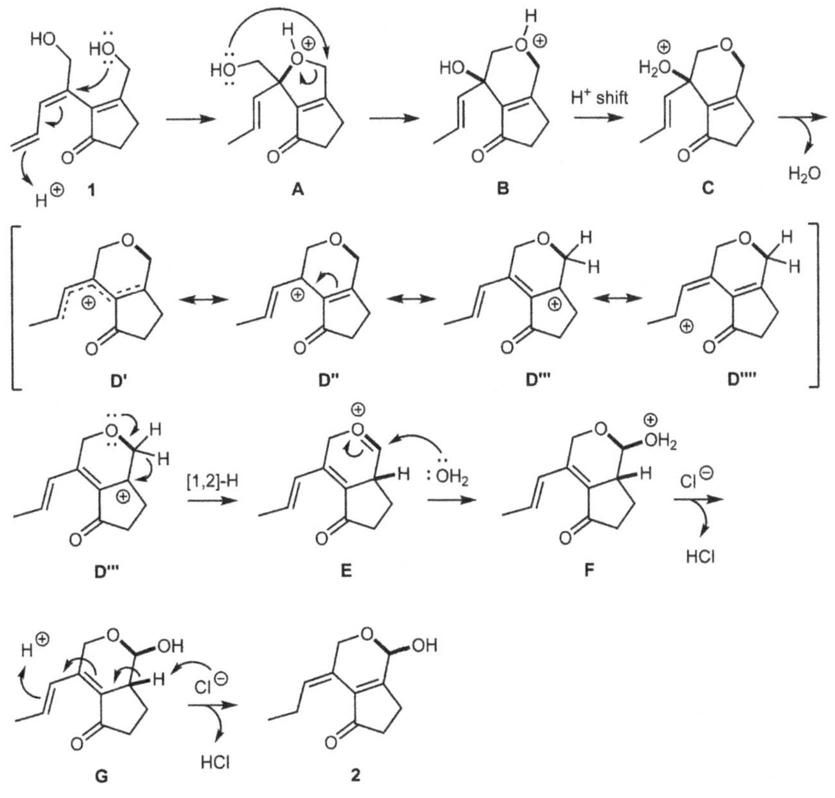

KEY STEPS EXPLAINED

The mechanism proposed by Nagle and Gerwick to explain the NMR solution-induced rearrangement of nakienone **1** to hemiacetal **2** constitutes an acid-catalyzed process which proceeds via cyclization of **1** to form the furan bicyclic intermediate **A**. This intermediate undergoes another cyclization with ring opening of the furan ring to form **B** which after a proton shift and loss of water furnishes the cationic species **D**. In the solution we show several resonance forms of **D** and note that **D'''** can undergo a [1,2] proton shift to form the oxonium ion **E**. This ion is then attacked by water to form **F** which undergoes deprotonation to form **G**. An intermolecular [1,5] hydride shift in **G** generates the final product **2**. We note that the ring construction mapping for this transformation is [6+0].

Since the authors did not provide additional experimental or theoretical evidence to support this mechanism, we note that an alternative mechanism can be envisaged for this transformation as shown in Scheme 2. In this case, the mechanism begins with protonation and elimination of water from **1** to form the cationic species **I** which undergoes cyclization via the labile hydroxyl group to form the bicyclic intermediate **K** after deprotonation. Intermediate **K** then undergoes an intermolecular [1,7] proton shift to form **L** which then undergoes protonation to form oxonium ion **M**. Attack of water onto **M** followed by deprotonation furnishes the final product **2** which exists in a state of equilibrium with the ring open aldehyde **O**.

We also note that the authors observed the **1** to **2** transformation while taking an NMR of the starting nakienone A alkaloid in $CDCl_3$ solution. The authors believed this solution may have been contaminated with acid arising from silica gel chromatographic isolation of the natural product from matted algal material after solvent extraction. In conclusion, we recommend that others undertake a theoretical analysis of both mechanisms in a future study to hopefully identify which mechanism is most likely operational based on energy calculations along a reaction coordinate diagram.

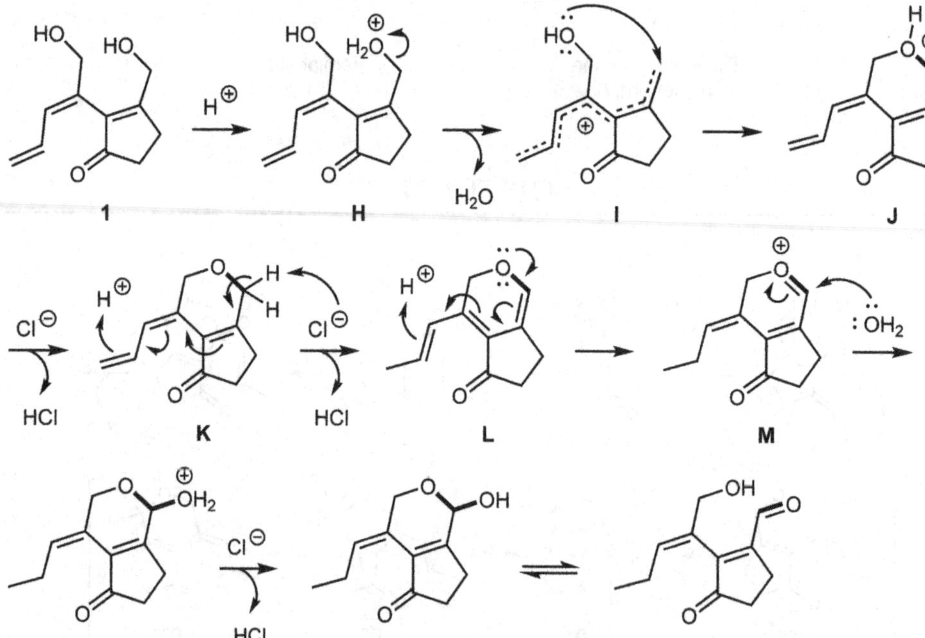

SCHEME 2

ADDITIONAL RESOURCES

For a look at a recent and long acid-catalyzed process for the conversion of glucose and fructose into levulinic acid and 5-hydroxymethylfurfural, we recommend the work of Yang et al.[2]

References

1. Nagle DG, Gerwick WH. Nakienones A-C and nakitriol, new cytotoxic cyclic C_{11} metabolites from an Okinawan cyanobacterial (Synechocystis sp.) overgrowth of coral. *Tetrahedron Lett*. 1995;36(6):849–852. https://doi.org/10.1016/0040-4039(94)02397-T.
2. Yang G, Pidko EA, Hensen EJM. Mechanism of Brønsted acid-catalyzed conversion of carbohydrates. *J Catal*. 2012;295:122–132. https://doi.org/10.1016/j.jcat.2012.08.002.

Question 5: Synthesis of Trifluoromethyl Heterocycles

In 1995 French chemists successfully employed trifluoroacetylketene intermediates (generated from the reaction of acid chlorides **1** and trifluoroacetic anhydride in pyridine) in reactions with electron-rich olefins such as *N,N*-dimethylcyanamide to form products such as the 1,3-oxazin-4-one **2** in 78% yield (Scheme 1).[1] It was noted that this product arose essentially from a formal [4 + 2] cycloaddition. Moreover, in cases where product **2** was not isolated and where excess trifluoroacetic anhydride had *not* been removed prior to addition of cyanamide, quenching of the reaction mixture with methanol formed the dihydro-1,3-oxazin-4-one **3** which was converted to **4** via heating in ethanol for 2 days. Provide mechanisms to account for these transformations.

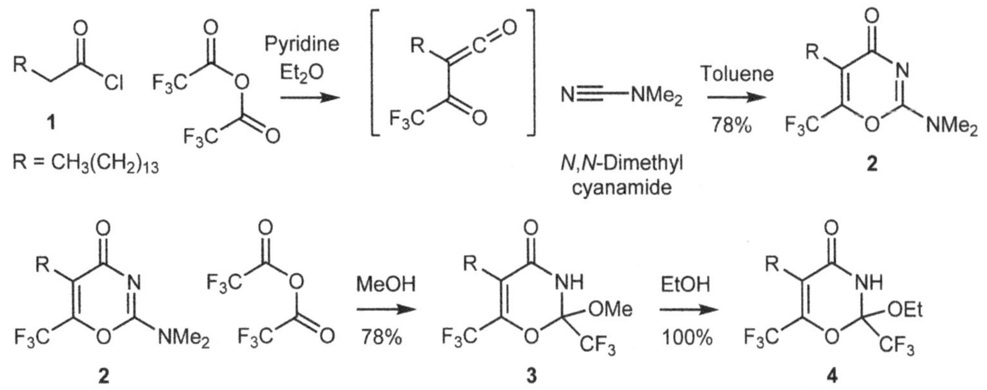

SCHEME 1

SOLUTION: 1 TO 2

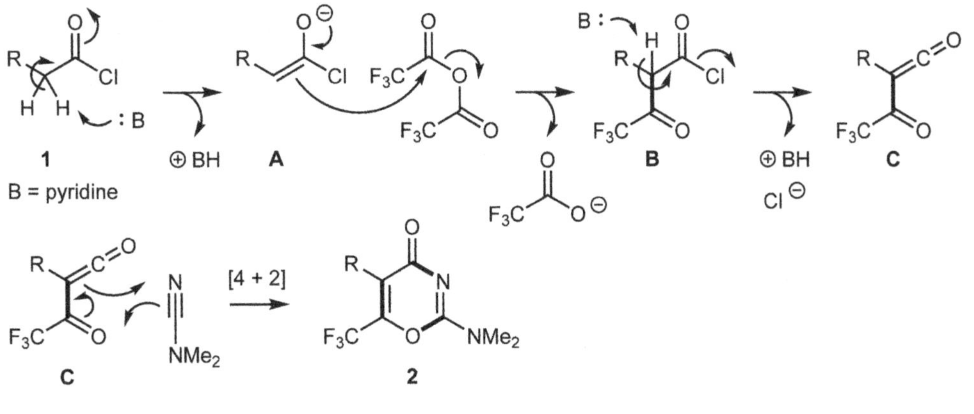

SOLUTION: 2 TO 3

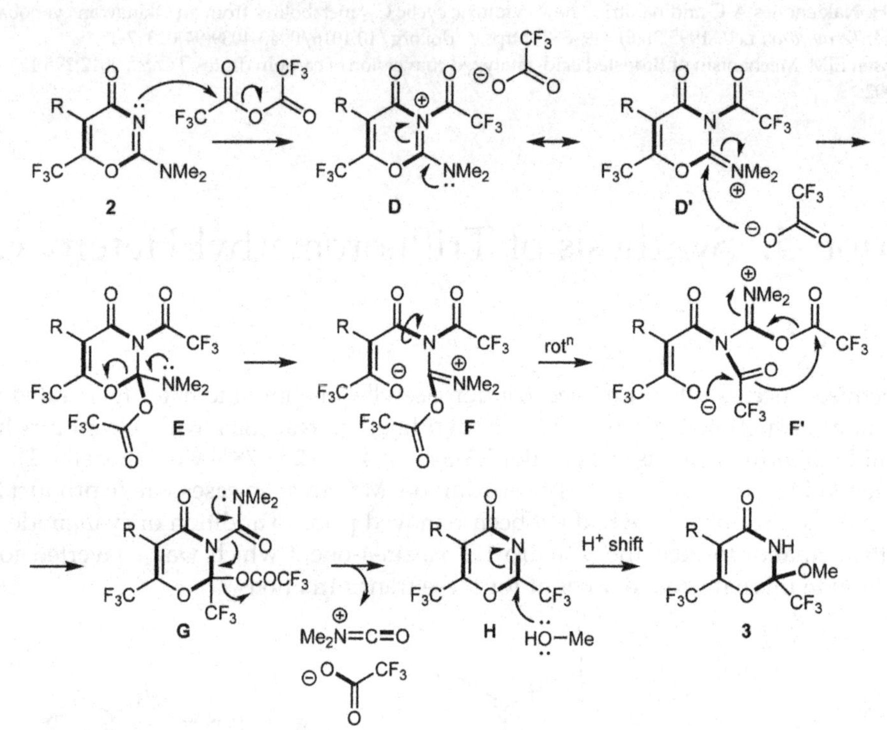

SOLUTION: 3 TO 4

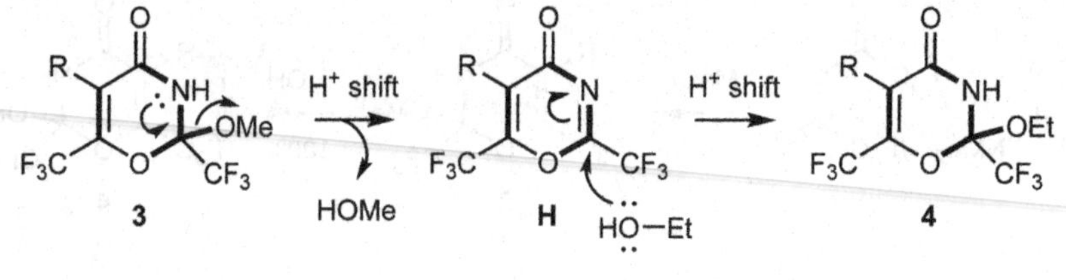

BALANCED CHEMICAL EQUATIONS

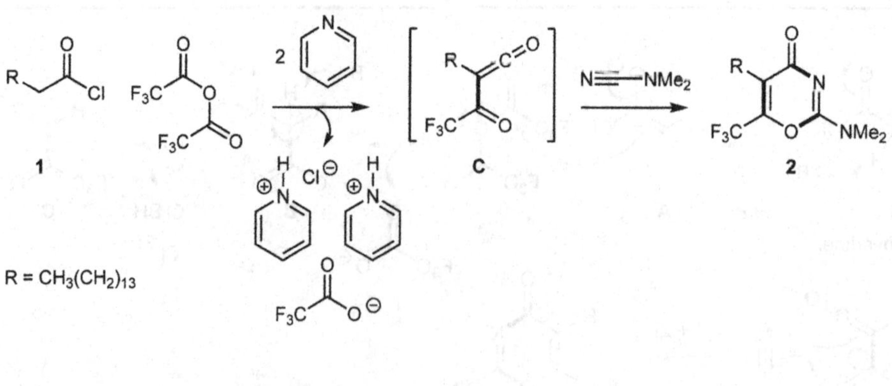

$R = CH_3(CH_2)_{13}$

KEY STEPS EXPLAINED

We start by examining the synthesis of [1,3]oxazin-4-one product **2** which begins with pyridine deprotonation of acid chloride **1** to form carbanion **A** which then nucleophilically attacks trifluoroacetic anhydride to form the acid chloride **B**. What follows is further deprotonation of **B** by a second equivalent of pyridine to generate acetylketene **C** which afterward adds to *N,N*-dimethylcyanamide via a hetero-Diels-Alder [4+2] cycloaddition to form product **2**. We note that the ring construction mapping for this transformation is [2+2+2]. In the case where **2** is allowed to react with excess trifluoroacetic anhydride, we have formation of the iminium ion **D/D′** which reacts with trifluoroacetate ion to form the cyclic adduct **E**. This adduct undergoes fragmentation to form the acyclic zwitterionic intermediate **F**. Recyclization of **F** then leads to formation of **G** which eliminates dimethylisocyanate cation and trifluoroacetate anion to form the key 1,3-oxazin-4-one intermediate **H**. This intermediate is trapped by methanol in a process which generates product **3**. We note that the ring construction mapping for the **2** to **3** transformation is also [2+2+2]. As support of their proposed mechanism, the authors inferred the intermediacy of **H** indirectly by demonstrating that methoxy product **3** could be converted to ethoxy product **4** via this intermediate. Nevertheless, the authors did not provide additional experimental or theoretical evidence to support their proposed mechanism.

ADDITIONAL RESOURCES

For a look at interesting heterocycle mechanisms involving six structural types of the Traube synthesis, we recommend an article by Katritzky and Yousaf.[2] For a recent study of zinc-catalyzed piperidine synthesis whose proposed mechanism involves a similar heterocycle backbone, we recommend the work of Lebold et al.[3] Lastly, in a recent article, Zhang et al. discuss the mechanism of the synthesis of furoquinoline heterocycles via domino ring-opening/recyclization reactions.[4]

References

1. Boivin J, Kaim LE, Zard SZ. Trifluoromethyl ketones from carboxylic acids. Part II. A versatile access to trifluoromethylated heterocycles. *Tetrahedron.* 1995;51(9):2585–2592. https://doi.org/10.1016/0040-4020(95)00007-U.
2. Katritzky AR, Yousaf TI. A C-13 nuclear magnetic resonance study of the pyrimidine synthesis by the reactions of 1,3-dicarbonyl compounds with amidines and ureas. *Can J Chem.* 1986;64(10):2087–2093. https://doi.org/10.1139/v86-344.
3. Lebold TP, Leduc AB, Kerr MA. Zn(II)-catalyzed synthesis of piperidines from propargyl amines and cyclopropanes. *Org Lett.* 2009;11(16):3770–3772. https://doi.org/10.1021/ol901435k.
4. Zhang Z, Zhang Q, Sun S, Xiong T, Liu Q. Domino ring-opening/recyclization reactions of doubly activated cyclopropanes as a strategy for the synthesis of furoquinoline derivatives. *Angew Chem Int Ed.* 2007;46:1726–1729. https://doi.org/10.1002/anie.200604276.

Question 6: 3-Substituted Chromones From Pyranobenzopyrans

In 1995 Indian chemists treated the readily available pyranobenzopyran **1** with a large excess of an enolizable ketone RCOCH$_3$ in the presence of aqueous acid to form the chromone **2** product (Scheme 1).[1] Suggest a mechanism to explain this transformation and outline a synthesis of **1**.

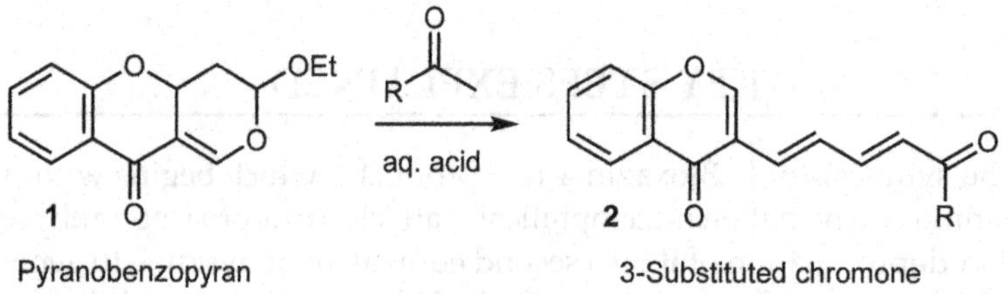

SCHEME 1

SOLUTION

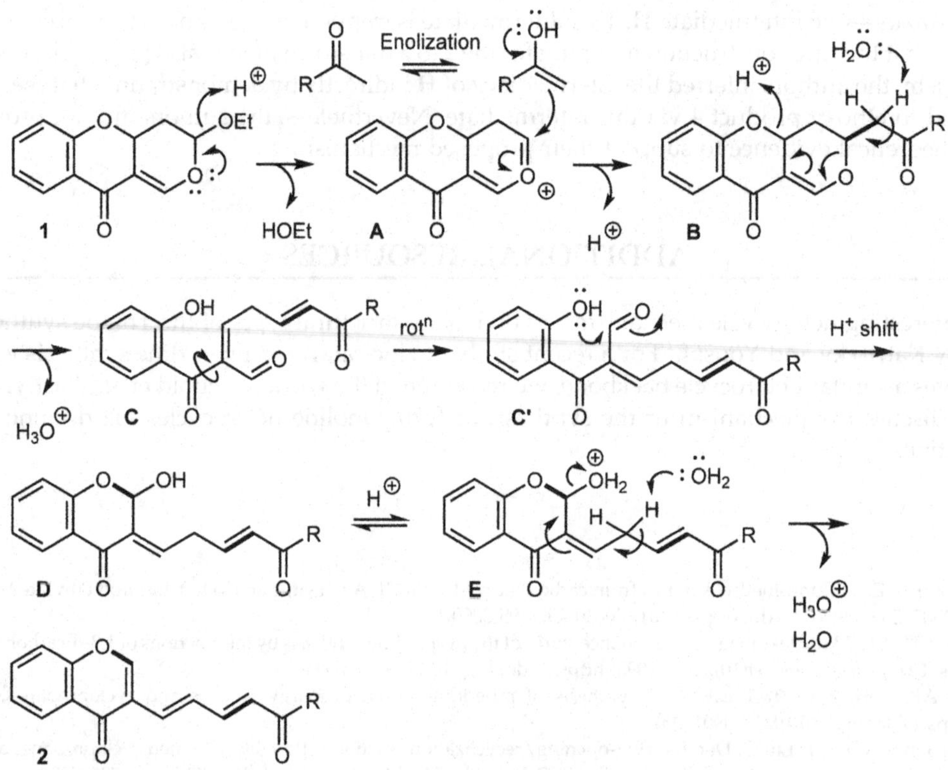

BALANCED CHEMICAL EQUATION

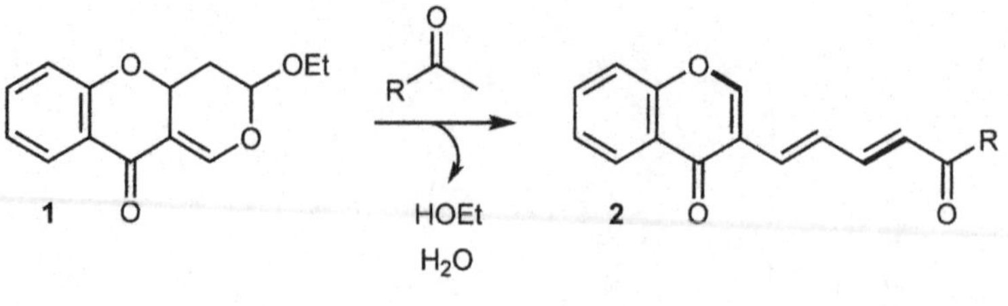

KEY STEPS EXPLAINED

The mechanism proposed by Uddin et al. for the transformation of **1** to **2** begins with elimination of ethanol from **1** and formation of 3,4-dihydro-pyranylium ion **A** which undergoes hydrolysis via the enolizable ketone (in its enol form) to form **B**. With the newly available enolizable proton in the β position, **B** undergoes acid-catalyzed ring opening of the γ-pyrone ring to form **C** which undergoes bond rotation to **C'**, recyclization to **D**, and loss of water to generate the final chromone product **2**. We note that the authors did not provide additional experimental or theoretical evidence to support this mechanism. In terms of potential syntheses for **1**, we present one example in Scheme 2 which is based on Refs. 2, 3.

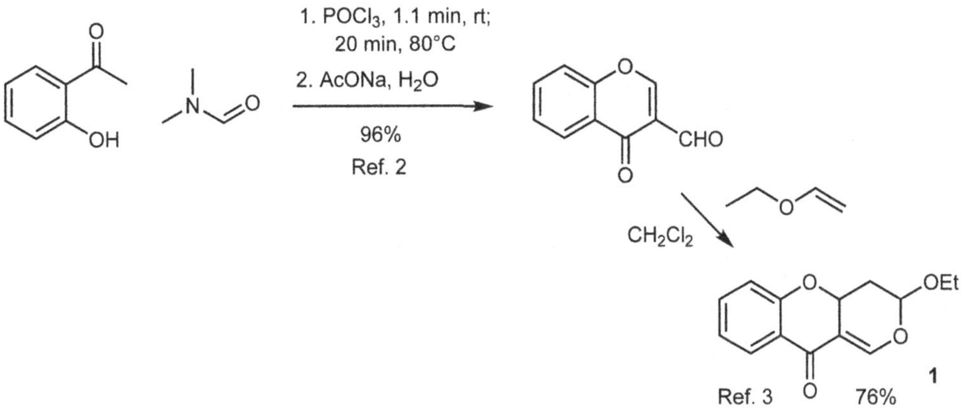

SCHEME 2

ADDITIONAL RESOURCES

For an interesting look at reactions involving flavones, isoflavones, and chromone derivatives and their mechanisms, see Ref. 4. In addition, we recommend a study of reactions of chromones with hydroxylamine in anhydrous solutions and the proposed mechanism of action for such transformations.[5]

References

1. Uddin R, Mujeeb-ur-Rahman, Siddiqi ZS, Zaman A. Synthesis of 3-substituted chromones from the 3-formylchromone-ethyl vinyl ether adduct. *J Chem Res Synop.* 1995;1995:159.
2. Pellegatti L, Buchwald SL. Continuous-flow preparation and use of β-chloro enals using the Vilsmeier reagent. *Org Process Res Dev.* 2012;16 (8):1442–1448. https://doi.org/10.1021/op300168z.
3. Ghosh CK, Tewari N, Bhattachharyya A. Heterocyclic systems; 15. Formation and hydrolysis of the [4 + 2]-cycloadduct of 4-oxo-4*H*-[1]benzo-pyran-3-carboxaldehyde with ethyl vinyl ether. *Synthesis.* 1984;1984(7):614–615. https://doi.org/10.1055/s-1984-30915.
4. Khilya VP, Ishchenko VV. Flavones, isoflavones, and 2- and 3-hetarylchromones in reactions with hydroxylamine. (Review). *Chem Heterocycl Compd.* 2002;38(8):883–899. https://doi.org/10.1023/A:1020920109737.
5. Szabo V, Borbely J, Theisz E, Janzso G. A new pathway in the reactions of chromones and hydroxylamine in anhydrous solutions. *Tetrahedron Lett.* 1982;23(50):5347–5350. https://doi.org/10.1016/S0040-4039(00)85835-9.

Question 7: Acid-Catalyzed Rearrangements of 1-Arylindoles

In 1995 Russian chemists observed that heating of the substituted 1-arylindole **1** with polyphosphoric acid (PPA) at 75–115°C for 25–150h produced the 5*H*-dibenz[*b,f*]azepines **2** in yields in the range of 8%–65% (Scheme 1).[1] The authors also noted that the highest yields occurred when the R group was 4-Me whereas when R was 4-NO₂ or 3-CF₃, no rearrangement of **1** to **2** took place. Propose a mechanism for this transformation which is consistent with these experimental observations.

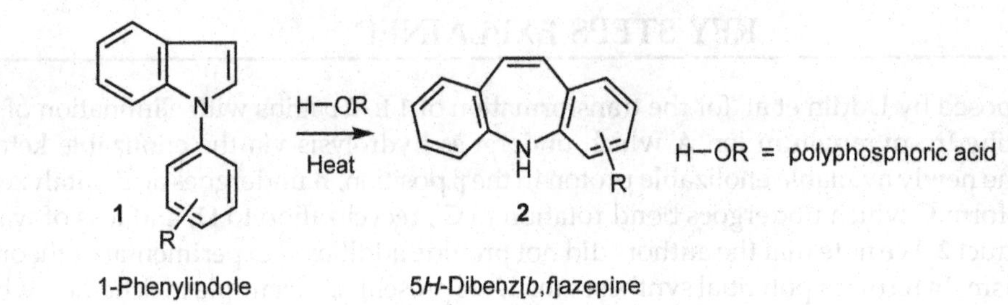

1-Phenylindole   5*H*-Dibenz[*b,f*]azepine

SCHEME 1

SOLUTION: 1 TO 2

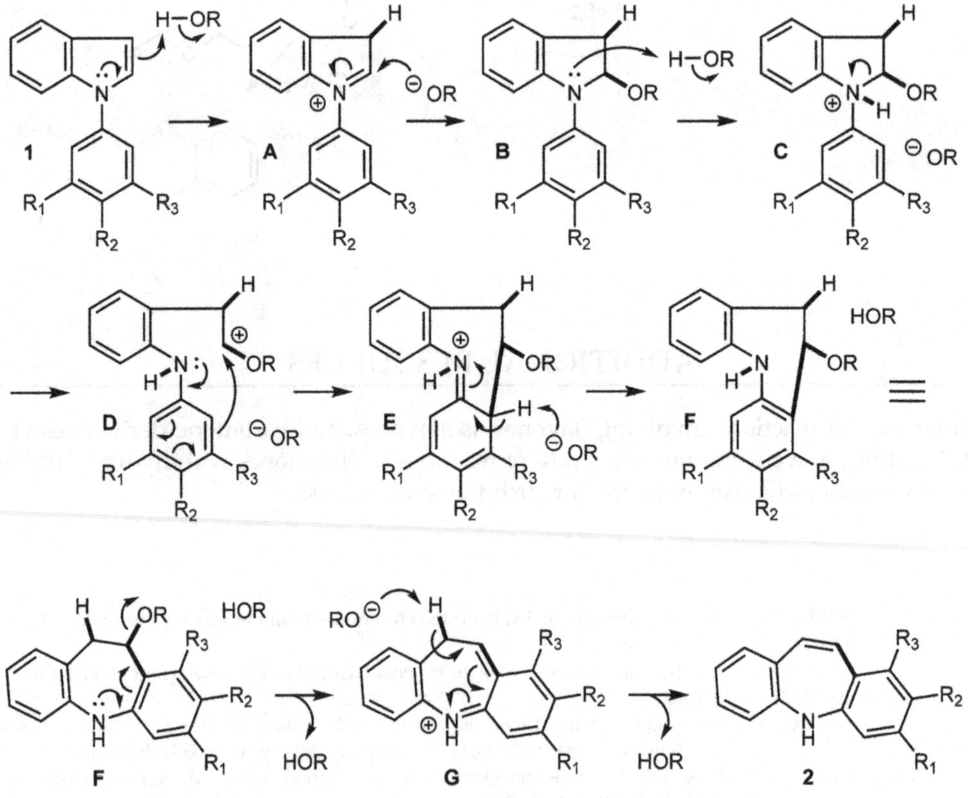

BALANCED CHEMICAL EQUATION

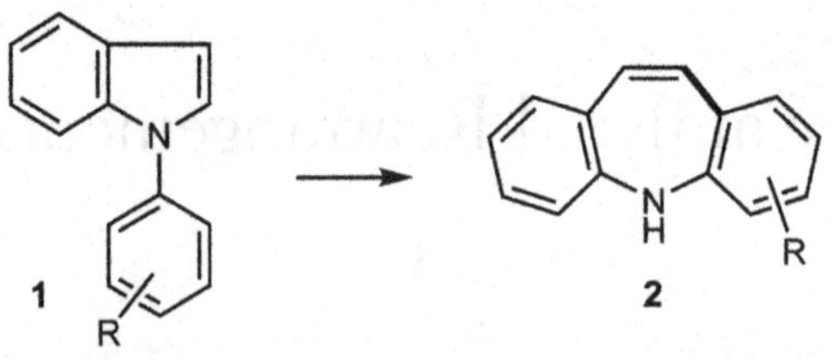

KEY STEPS EXPLAINED

The mechanism proposed by Tokmakov and Grandberg to explain the transformation of indole **1** to azepine **2** begins with protonation of **1** via the polyphosphoric acid added to the reaction medium which leads to formation of indolium ion **A**. This ion is prone to nucleophilic attack by the conjugate base of the polyphosphoric acid which attacks at the 2-position to form indoline **B**. Next, intermediate **B** undergoes protonation at the nitrogen atom to facilitate ring opening through cleavage of the C_2—N bond to form carbocation **D**. Intermediate **D** is then postulated to undergo electrophilic aromatic substitution at the *ortho* position of the aryl group (i.e., *ortho* relative to the amino group) causing cyclization to the iminium ion **E** which is rearomatized through deprotonation via the conjugate base of PPA to form 10,11-dihydro-5H-dibenzo[b,f]azepine **F**. What follows is facile elimination of the anion of PPA to generate **G** and a second elimination which restores the catalyst and forms the final azepine product **2**. The authors also explained that electrophilic activation of the aryl group at the *ortho* position is crucial since electron-withdrawing substituents at the 3- and 4-positions such as 4-NO_2 (i.e., R_2) or 3-CF_3 (i.e., R_1) resulted in no rearrangement. An electron-donating group at the 3-position (i.e., R_1) would therefore be expected to enhance the yield of the reaction. This was the case when R_1 = Me which led to a yield of **2** of 65%. Nevertheless, when R_2 was a methoxy group (OMe), the yield of **2** was only 8%. Interestingly, when R_2 = Me, the yield of **2** was 43% and when R_2 = Cl, the yield of **2** was 25%. Lastly, when R_1 = CF_3, no rearrangement took place.

We explain these results by noting that the methyl group is a donor via sigma bond donation, whereas the methoxy group is a donor via the two lone pairs of electrons on the oxygen atom. It is well known that the methoxy group is a stronger electron donor than the methyl group based on a comparison of the Hammett σ_p constants (−0.27 versus −0.17). With these aspects in mind, we note that the transformation from **D** to **E** requires that aromatic substitution take place at the *meta* position relative to R_2 (which is the same as the *para* position relative to R_1, the *ortho* position relative to the amino group and the *ortho* position relative to R_3). Thus, the only groups that would enhance the intramolecular cyclization from **D** to **E** are *ortho/para* directing groups at R_1 and R_3 and *meta* directing groups at R_2. We note that the amino group is an *ortho* directing group which already reinforces the transformation from **D** to **E**. In terms of experimental evidence, therefore, the authors recorded the highest yield of **2** (i.e., 65%) when R_1 = Me thus supporting the proposed **D** to **E** transformation. Furthermore, an electron-withdrawing group such as CF_3 at R_1 resulted in no rearrangement. Next, when R_2 = Me, the reported yield of **2** was 43%. This observation integrates well with the case when R_1 = Me because a weak *ortho/para* directing group at R_2 is expected to counteract the strong *ortho* directing nature of the amino group in **D**. Thus when R_2 = Me, the yield of **2** is lower than when R_1 = Me due to synergism between R_1 and the amino group. Interestingly, when the electron-donating group at R_2 is made even stronger than a methyl group, the yield of **2** is further reduced. For example, when R_2 = Cl, the yield of **2** is 25% (as compared to 43% when R_2 = Me). Moreover, when R_2 = OMe, an even stronger electron donor than Cl, the yield of **2** becomes 8%. It is worthwhile to mention that in each of these cases the other aromatic positions were hydrogen atoms. Unfortunately the authors did not conduct experiments where R_1 and R_3 were electron-donating groups, or where R_1 = OMe. If the trends established here are correct, one would expect that when R_1 = R_3 = Me (R_2 = H) or when R_1 = OMe (R_2 = R_3 = H), the yield of **2** would be >65%. Of course, the **D** to **E** cyclization also depends on steric and entropic factors not considered in this argument which may operate to oppose any favorable electronic donor effects imparted by these groups. Nevertheless, a theoretical analysis should shed more light on the question of whether strong electron-donating groups at R_1 and R_3 act as promoters of the **D** to **E** step whereas strong electron-donating groups at R_2 act as destabilizers of this step. Lastly, the five to seven-membered indole ring expansion of **1** to **2** illustrates a [7+0] ring construction mapping.

Interestingly, the authors acknowledged that other mechanisms are possible including the one shown in Scheme 2 which involves the formation of a four-membered ring σ-complex H_1/H_2 as a reaction intermediate. However, the authors speculated that since this mechanism involves a potentially high-energy four-membered ring intermediate it is less likely to be operative than the one presented earlier. We recommend that others undertake a theoretical investigation of the energy reaction coordinate profiles of both mechanisms to assess whether this speculation is validated.

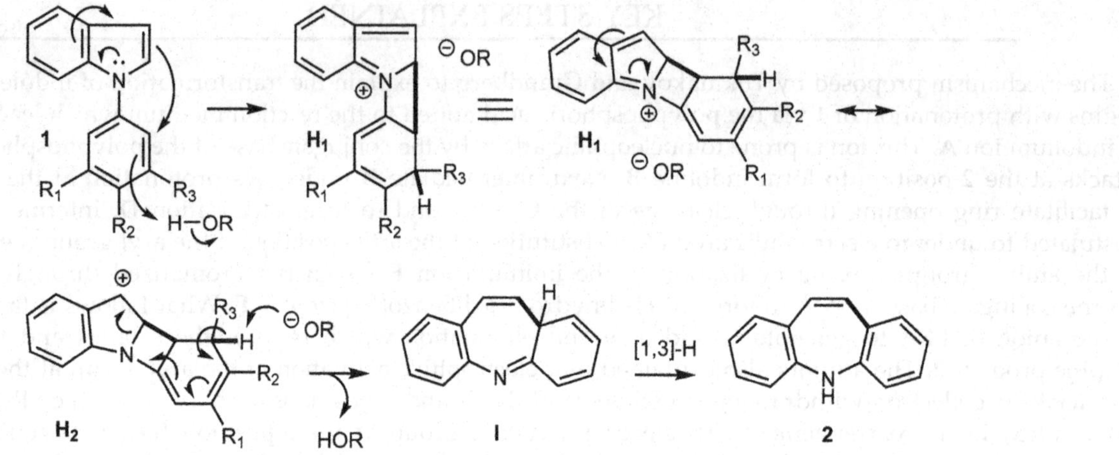

SCHEME 2

ADDITIONAL RESOURCES

In conclusion, we recommend an interesting and recent mechanistic study which deals with the formation of aniline and azepine derivatives via rearrangement cascades from annulated azadienyl anions.[2] We also recommend a study which discusses the mechanism of azepine formation from the reaction of 2-methylbenzothiazole and methyl propiolate.[3]

References

1. Tokmakov GP, Grandberg II. Rearrangement of 1-arylindoles to 5H-dibenz[b,f]azepines. *Tetrahedron*. 1995;51(7):2091–2098. https://doi.org/10.1016/0040-4020(94)01082-B.
2. Lyaskovskyy V, Froehlich R, Wuerthwein E-U. Mechanistic study on rearrangement cascades starting from annulated 2-aza-hepta-2,4-dienyl-6-ynyl anions: formation of aniline and azepine derivatives. *Chem A Eur J*. 2007;13:3113–3119. https://doi.org/10.1002/chem.200601491.
3. Fetcher RM, Cheung KK, Sin DWM. Addition of 2-methylbenzothiazole to methyl propiolate—mechanism of azepine formation. *J Chem Res Synop*. 1989;1989(5):115.

Question 8: An Unusual Deprotection of an Aryl Ester

In 1995 American chemists showed that treatment of 2′,4′,6′-trihydroxyacetophenone **1** with 3.3 equivalents of methyl chloroformate and triethylamine in THF at 0°C resulted in formation of tricarbonate **2** in 87% yield (Scheme 1).[1] Afterward, tricarbonate **2** was reacted with 4 equivalents of sodium borohydride in a 1:1 THF/water mixture at 0–25°C which formed phenol **3** in 83% yield. Suggest a mechanism for the transformation of **2** to **3**.

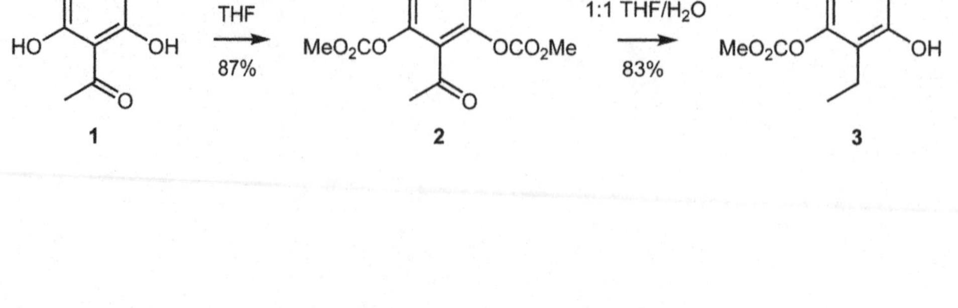

SCHEME 1

SOLUTION: 2 TO 3

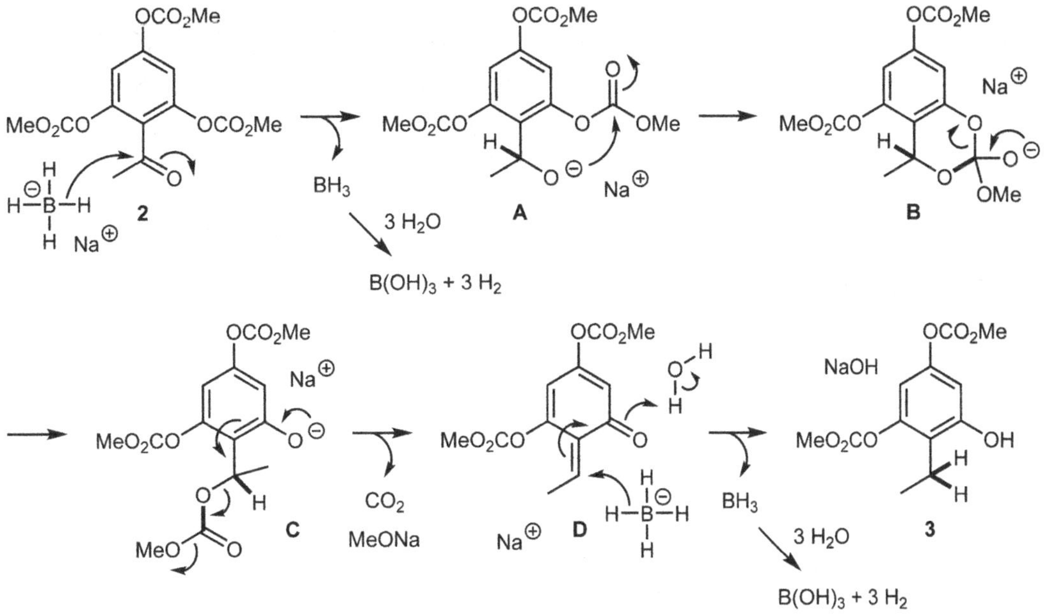

BALANCED CHEMICAL EQUATION

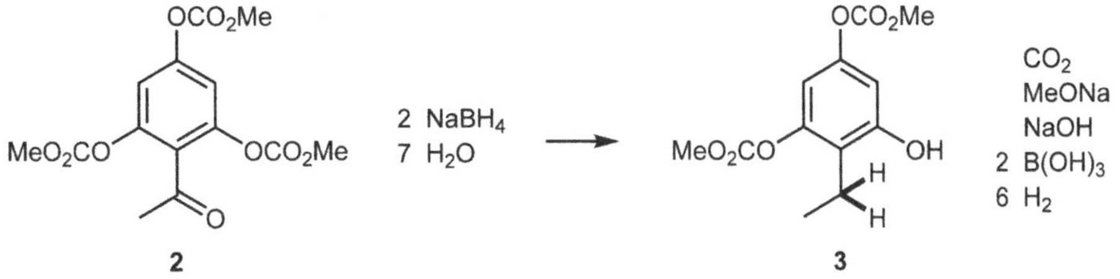

KEY STEPS EXPLAINED

The mechanism proposed by the authors to explain the transformation of **2** to **3** begins with sodium borohydride reduction of the ketone moiety of **2** to form alkoxide **A** which undergoes a carbonate group transfer between the negatively charged oxygen atom and the ester at the *ortho* position via cyclic intermediate **B**. This process furnishes the phenoxide intermediate **C** which undergoes decarboxylation and elimination of methoxide anion to form the quinone methide intermediate **D**. Following decarboxylation, reduction of **D** restores aromaticity to the ring and generates the final phenol product **3** after protonation via water.

We note that an alternative pathway not considered by the authors can be proposed (see Scheme 2) whereby fragmentation of cyclic intermediate **B** leads to elimination of methoxide anion and formation of 4*H*-[1,3]dioxin-2-one **E**. Subsequent reduction of this intermediate via sodium borohydride reagent at the benzylic carbon atom then leads to decarboxylation and formation of phenoxide **F** which is protonated during work-up to generate phenol product **3**. Interestingly, in this mechanism no quinone methide intermediate is formed.

We note that the authors did not provide experimental evidence for their proposed quinone methide intermediate **D**. Nevertheless, we suggest a possible trapping experiment where the reduction reaction is conducted in the presence of Diels-Alder dienophiles such as the general ones shown in Scheme 3. If the quinone methide **D** is indeed formed as an intermediate it may be trapped intermolecularly by either alkynes or alkenes leading to chromene or chroman reaction products, respectively. If neither product were formed then such an observation would rule out the presence of quinone methide **D** as an intermediate. The syntheses of chroman derivatives via [4+2] cycloaddition of alkenes and quinone methide intermediates generated oxidatively from *o*-alkyl phenols or via dehydration of *o*-hydroxy-benzyl

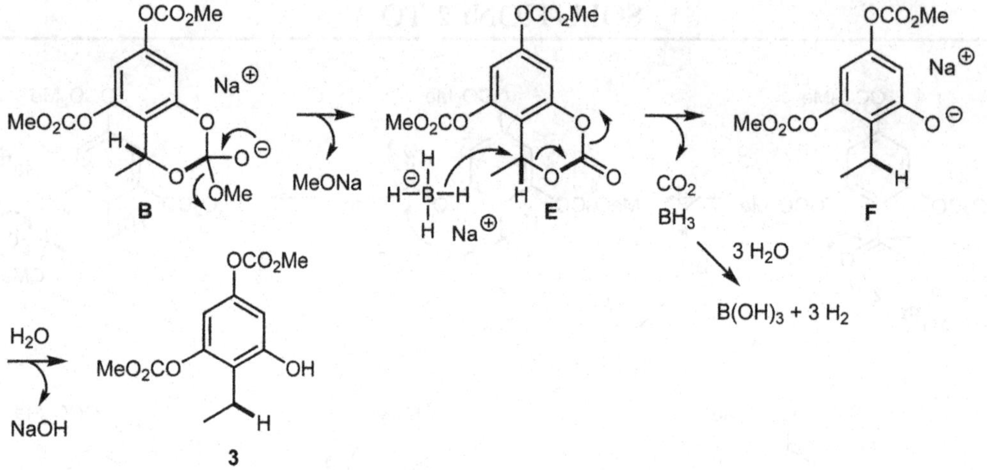

OCO₂Me ... (Scheme 2 structures)

SCHEME 2

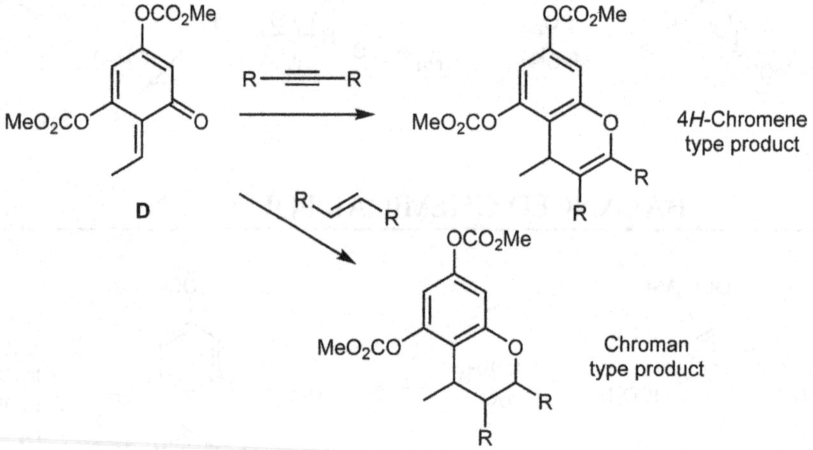

SCHEME 3

4H-Chromene type product

Chroman type product

alcohols are well documented in the literature.[2–6] Furthermore, a theoretical analysis of both proposed mechanisms would help to provide a basis for further differentiation.

ADDITIONAL RESOURCES

For a look at another interesting mechanism involving an unexpected sodium borohydride reduction of substituted pyridine dicarboxylate compounds, see Ref. 7. For a detailed mechanistic analysis of sodium borohydride reduction of α-keto esters, see Ref. 8.

References

1. Mitchell D, Doecke CW, Hay LA, Koenig TM, Wirth DD. *Ortho*-hydroxyl assisted deoxygenation of phenones. Regiochemical control in the synthesis of monoprotected resorcinols and related polyphenolic hydroxyl systems. *Tetrahedron Lett.* 1995;36(30):5335–5338. https://doi.org/10.1016/0040-4039(95)01085-V.

2. Bolon DA. O-quinone methides. II. Trapping with production of chromans. *J Org Chem.* 1970;35(11):3666–3670. https://doi.org/10.1021/jo00836a016.

3. Wagner HU, Gompper R. Quinone methides. In: Patai S, ed. *The Chemistry of Quinoid Compounds, Part 2.* New York, NY: John Wiley & Sons; 1974:1145–1178. https://doi.org/10.1002/9780470771303.ch7.

4. Chambers JD, Crawford J, Williams HWR, et al. Reactions of 2-phenyl-4H-1,3,2-benzodioxaborin, a stable ortho-quinone methide precursor. *Can J Chem.* 1992;70(6):1717–1732. https://doi.org/10.1139/v92-216.

5. van de Water RW, Pettus TRR. *o*-Quinone methides: intermediates underdeveloped and underutilized in organic synthesis. *Tetrahedron.* 2002;58 (27):5367–5405. https://doi.org/10.1016/S0040-4020(02)00496-9.
6. Pathak TP, Sigman MS. Applications of *ortho*-quinone methide intermediates in catalysis and asymmetric synthesis. *J Org Chem.* 2011;76 (22):9210–9215. https://doi.org/10.1021/jo201789k.
7. Tang YB, Zhang QJ, Yu DQ. The mechanism of unexpected reduction of dimethyl pyridine-2,3-dicarboxylate to 1,2,3,4-tetrahydrofuro[3,4-*b*]pyridin-5(7*H*)-one with sodium borohydride. *Chin Chem Lett.* 2012;23:1122–1124. https://doi.org/10.1016/j.cclet.2012.07.013.
8. Dalla V, Catteau JP, Pale P. Mechanistic rationale for the NaBH$_4$ reduction of α-keto esters. *Tetrahedron Lett.* 1999;40(28):5193–5196. https://doi.org/10.1016/S0040-4039(99)01006-0.

Question 9: 3,1-Benzoxathiin Formation

In 1997 chemists from Japan heated sulfoxide **1** and 1.2 equivalents of *p*-toluenesulfonic acid in xylene for 50 min to obtain the 1,3-oxathiane **2** in 53% yield (Scheme 1).[1] Provide a mechanistic explanation for this transformation.

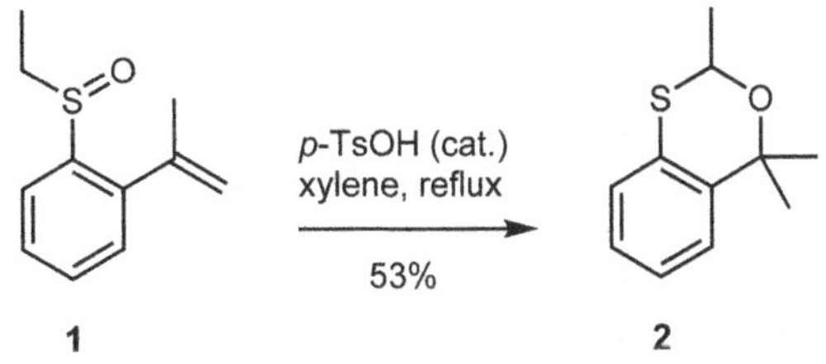

SCHEME 1

SOLUTION: MECHANISM 1

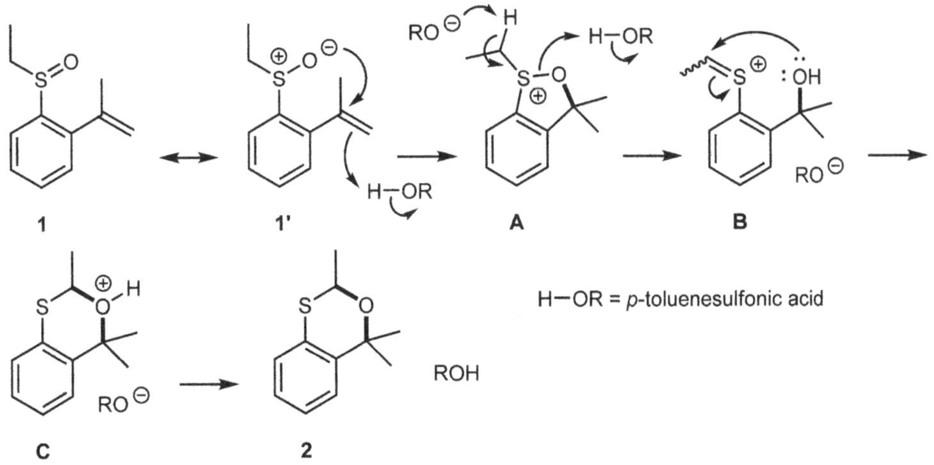

H—OR = *p*-toluenesulfonic acid

SOLUTION: MECHANISM 2

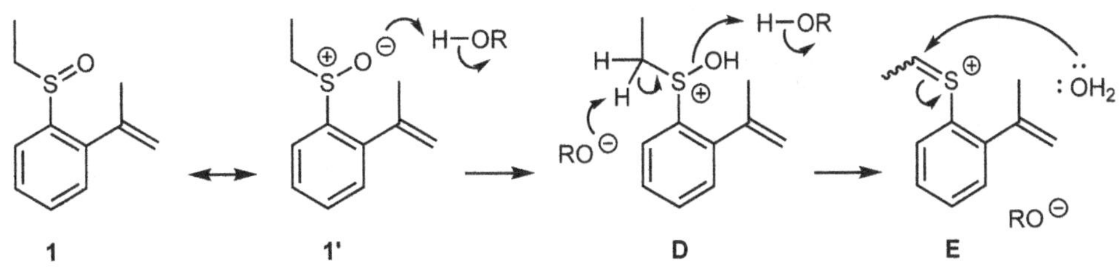

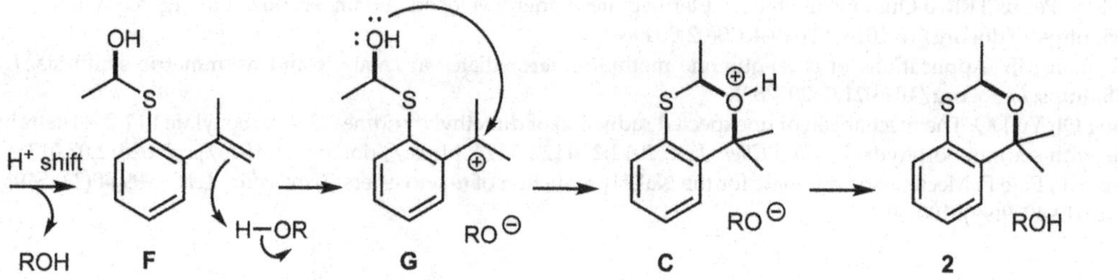

KEY STEPS EXPLAINED

Abe et al. proposed that an intramolecular Pummerer rearrangement of γ,δ-unsaturated sulfinyl compounds is responsible for the formation of the final 1,3-benzoxathiin product **2** starting from sulfoxide **1**. As such, their first proposed mechanism (#1) begins with the formation of a 5-membered 3*H*-benzo[*c*][1,2]oxathiol-1-ium ion **A** via protonation of the olefin moiety of **1** and cyclization of the sulfoxide group onto the in situ-generated carbocation. Intermediate **A** then fragments to form sulfonium ion **B** which cyclizes to form the 6-membered oxonium ion **C**. Deprotonation of **C** then generates the 4*H*-benzo[*d*][1,3]oxathiine product **2**. In the same article the authors described another possible pathway (#2) which involves the more conventional intramolecular Pummerer rearrangement. In this mechanism sulfoxide **1** is protonated to form hydroxysulfonium ion **D** which undergoes dehydration to generate sulfonium ion **E**. Rehydration of **E** at the electrophilic ethylidene carbon atom followed by protonation of the olefin moiety forms the tertiary carbocation **G** having an α-hydroxysulfide group which cyclizes to form the 4*H*-benzo[*d*][1,3] oxathiin-3-ium ion **C** as before. Nevertheless, Abe et al. did not conclude which pathway is operational for this transformation and suggested that further research on this reaction is required. We note that both mechanisms result in a [5+1] ring construction mapping for the formation of product **2**.

Furthermore, we suggest that an ^{18}O labeling experiment on the starting sulfoxide **1** may help to determine which mechanism is operative. For instance, following mechanism 1 it is expected that the ^{18}O label will be retained in the product as shown in Scheme 2 since the oxygen atom does not become detached from the structure. By contrast, in mechanism 2 the dehydration step would result in the elimination of ^{18}O labeled water which could undergo an exchange reaction with the *p*-toluenesulfonic acid catalyst leading to a mixture of labeled and unlabeled water available for attacking sulfonium ion intermediate **E** (Scheme 3). This exchange would be expected to result in less incorporation of ^{18}O label in product **2**. We also recommend that others carry out a theoretical investigation of these two mechanisms.

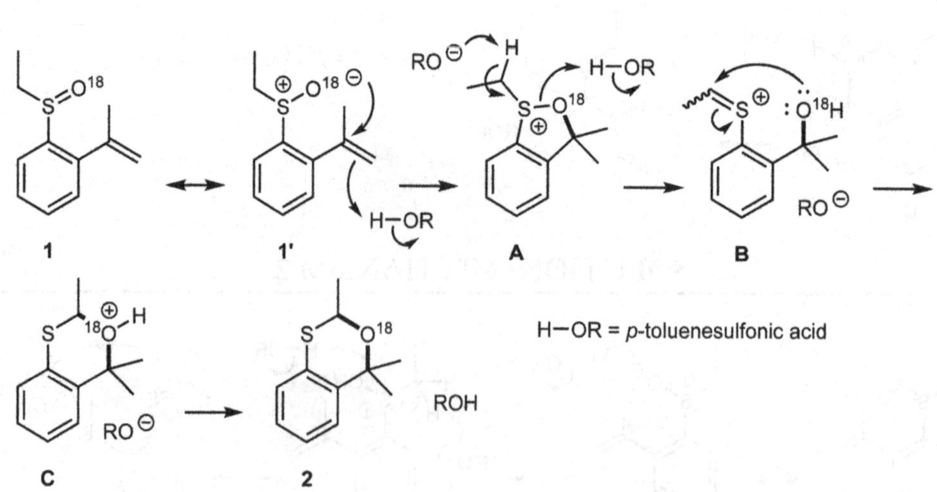

SCHEME 2

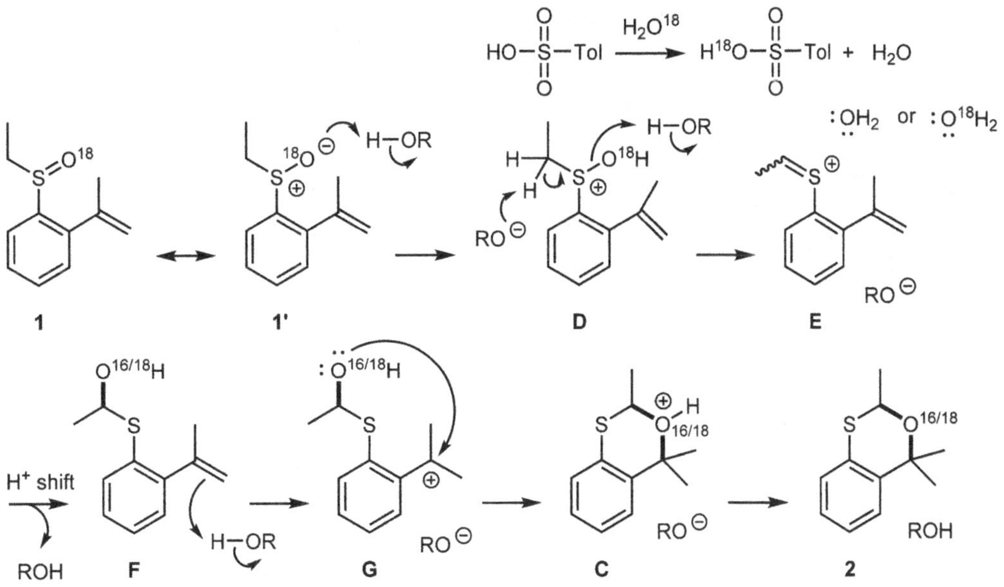

SCHEME 3

ADDITIONAL RESOURCES

For a look at the synthesis of 3-trifluoromethylbenzo[*b*]furans from phenols via extended Pummerer reaction along with a proposed mechanism, see Ref. 2. For a look at the synthesis of 5-thioglucopyranose derivatives and how the Pummerer rearrangement fits in with the overall mechanism, see Ref. 3.

References

1. Abe H, Fujii H, Masunari C, et al. Construction of 1,3-oxathiane ring through Pummerer reaction of γ,δ-unsaturated sulfinyl compounds. *Chem Pharm Bull.* 1997;45(5):778–785. https://doi.org/10.1248/cpb.45.778.
2. Kobatake T, Fujino D, Yoshida S, Yorimitsu H, Oshima K. Synthesis of 3-Trifluoromethylbenzo[*b*]furans from phenols via direct *ortho* functionalization by extended Pummerer reaction. *J Am Chem Soc.* 2010;132:11838–11840. https://doi.org/10.1021/ja1030134.
3. Matsuda H, Fujita J, Morii Y, Hashimoto M, Okuno T, Hashimoto K. Pummerer rearrangement of 1-deoxy-5-thioglucopyranose oxides; novel synthesis of 5-thioglucopyranose derivatives. *Tetrahedron Lett.* 2003;44:4089–4093. https://doi.org/10.1016/S0040-4039(03)00842-6.

Question 10: Arylacetone Synthesis by Carroll-Type Rearrangement

In 1995 American chemists reacted *p*-quinol **1** with diketene **2** and a catalytic amount of DMAP (4-*N*,*N'*-dimethylaminopyridine) at room temperature which gave a product mixture containing arylacetone **3** in 72% yield and benzofuran **4** in 3% yield (Scheme 1).[1] Elucidate mechanisms which explain the formation of these products.

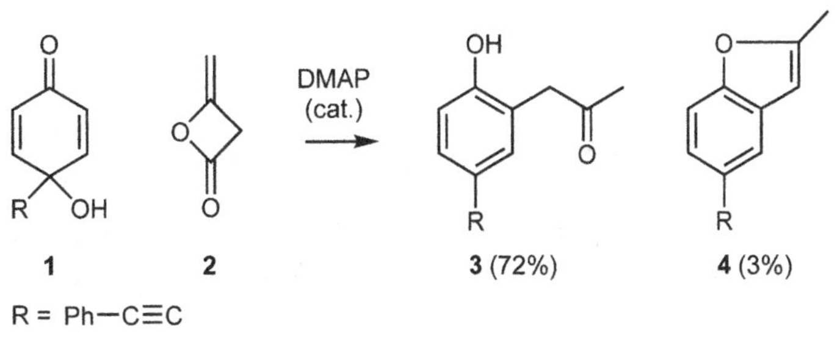

SCHEME 1

SOLUTION: 1 AND 2 TO 3

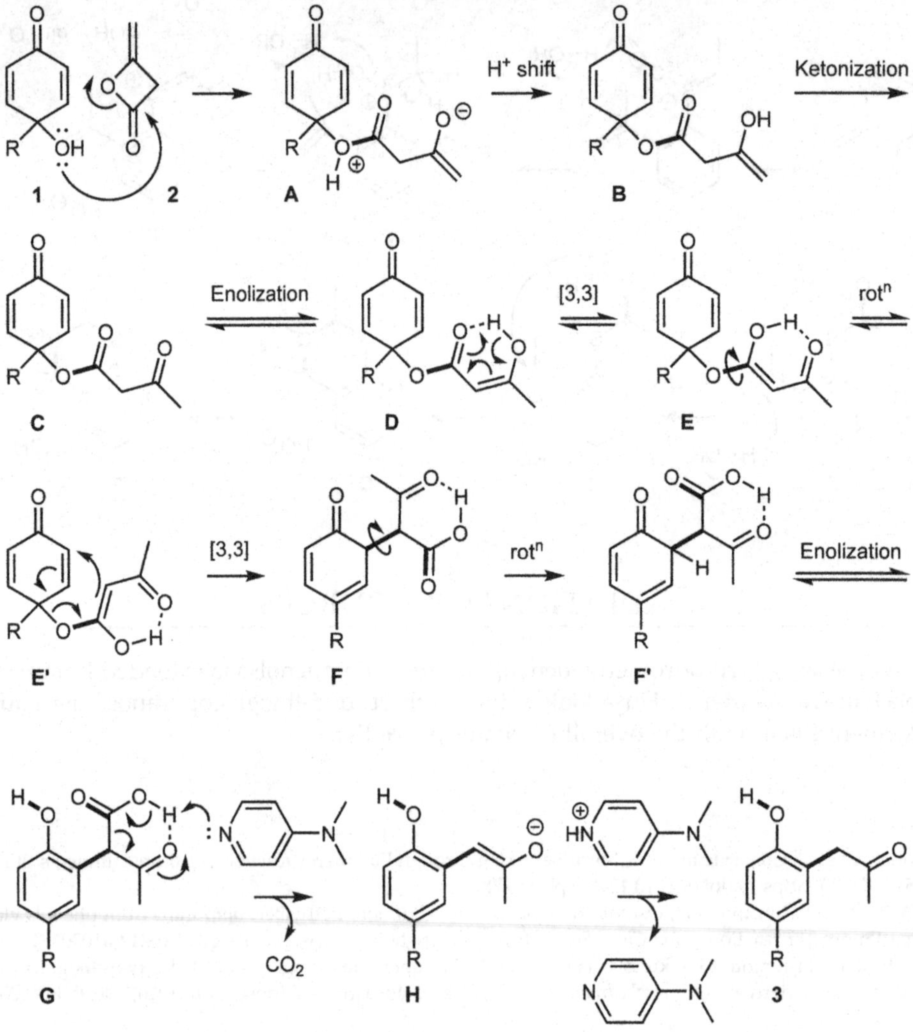

SOLUTION: 3 TO 4

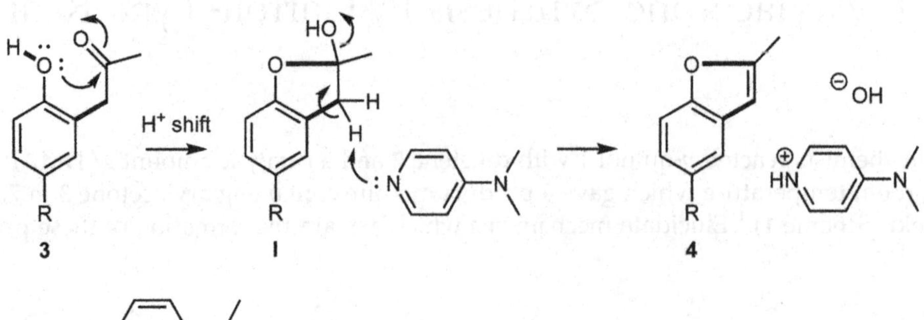

KEY STEPS EXPLAINED

The mechanism proposed by Sorgi et al. begins with nucleophilic attack by the hydroxyl group of quinol **1** onto diketene **2** to form enol **B** after intramolecular proton transfer in **A**. Ketonization of **B** then leads to the 1,3-diketo intermediate **C** which undergoes enolization and bond rotation to form intermediate **E′**. This intermediate undergoes a [3,3] sigmatropic Carroll rearrangement to form **F** which is aromatized through enolization to intermediate **G**. This intermediate undergoes a DMAP (4-*N,N′*-dimethylaminopyridine) catalyzed decarboxylation to form carboxylate **H** which ketonizes with recovery of DMAP to form the final phenol product **3**. At this point, some small portion of **3** can engage in a secondary reaction which begins with an intramolecular process to form ketal **I** after ring closure and proton shift. DMAP-assisted dehydration of **I** then gives the furan product **4** along with water as a by-product. We note that the ring construction mapping for the formation of benzofuran **4** is [3 + 2]. We also note that the authors were able to monitor the reaction by time-resolved ^{13}C NMR where they observed 1,3-diketo intermediate **C** and enol **D** in a ratio of 82:18. Nevertheless, the authors were unable to observe keto-acid intermediate **F** presumably because the aromatization (tautomerization) and decarboxylation steps were too rapid to be detectable on the NMR timescale. The authors did not provide further experimental or theoretical evidence to support their proposed mechanism for the formation of **3**. The mechanism that leads from **3** to **4** was not discussed by the authors, but in this context we draw attention to the small isolated yield of **4** as evidence for high energy demand expected for this transformation. Furthermore, it is interesting that the authors repeated the experiment using several differently substituted analogs of substrate **1**. In most cases the rearranged products **3** were isolated in good yield except when the hydroxyl group of **1** had a methyl group at both *ortho* positions and the R group was phenyl. In this case, the authors attributed the lack of rearrangement to steric crowding between the *ortho* methyl group and the R phenyl group in the developing transition state for the Carroll [3,3] sigmatropic rearrangement step from **E′** to **F**. When the R group was phenylacetylene instead of just phenyl, the rearrangement of the analog of **1** took place as expected. This means that an acetylene group placed in between the phenyl and the rest of substrate **1** is enough to overcome the steric hindrance (step **E′** to **F**) between the *ortho* methyl group and the R group which prevented the rearrangement when R was simply a phenyl group.

ADDITIONAL RESOURCES

We recommend the work of Enders et al. for an example of the use of an asymmetric Carroll rearrangement in the context of the stereoselective synthesis of polyfunctional ketones.[2] We also recommend the work of Hatcher and Posner for examples of various 3,3-sigmatropic rearrangements, including the Carroll rearrangement, and how they can be useful in making functionalized vitamin D$_3$ side chain units.[3] For an example of an unexpected Carroll rearrangement, see Ref. 4

References

1. Sorgi KL, Scott L, Maryanoff CA. The Carroll rearrangement: a facile entry into substituted arylacetones and related derivatives. *Tetrahedron Lett.* 1995;36(21):3597–3600. https://doi.org/10.1016/0040-4039(95)00602-9.
2. Enders D, Knopp M, Runsink J, Raabe G. Diastereo- and enantioselective synthesis of polyfunctional ketones with adjacent quaternary and tertiary centers by asymmetric Carroll rearrangement. *Angew Chem Int Ed Engl.* 1995;34(20):2278–2280. https://doi.org/10.1002/anie.199522781.
3. Hatcher MA, Posner GH. [3,3]-Sigmatropic rearrangements: short, stereocontrolled syntheses of functionalized vitamin D$_3$ side-chain units. *Tetrahedron Lett.* 2002;43(28):5009–5012. https://doi.org/10.1016/S0040-4039(02)00904-8.
4. Yang M, Zhang S, Zhang X, et al. An unexpected acid-catalyzed decomposition reaction of cilnidipine and pranidipine to the decarboxylative bridged tricyclic products via cascade rearrangements. *Org Chem Front.* 2017;4:2163–2166. https://doi.org/10.1039/C7QO00496F.

Question 11: Attempted Knoevenagel Reaction Gives Mannich-Type Products

In 1995 chemists from Bulgaria attempted the Knoevenagel condensation of coumarin aldehydes **1** (which would normally react with malonate compounds) where piperidine acts as a base catalyst to form, for example, product **4** (Scheme 1).[1] Instead, the authors observed an unexpected Mannich reaction between piperidine and **1** which generated product **2** in 80% yield. It was also shown that treatment of 4-amino-2*H*-chromen-2-one **3** with formaldehyde and piperidine hydrochloride salt gave **2** in 82% yield. Suggest mechanisms for the two transformations **1** to **2** and **3** to **2**.

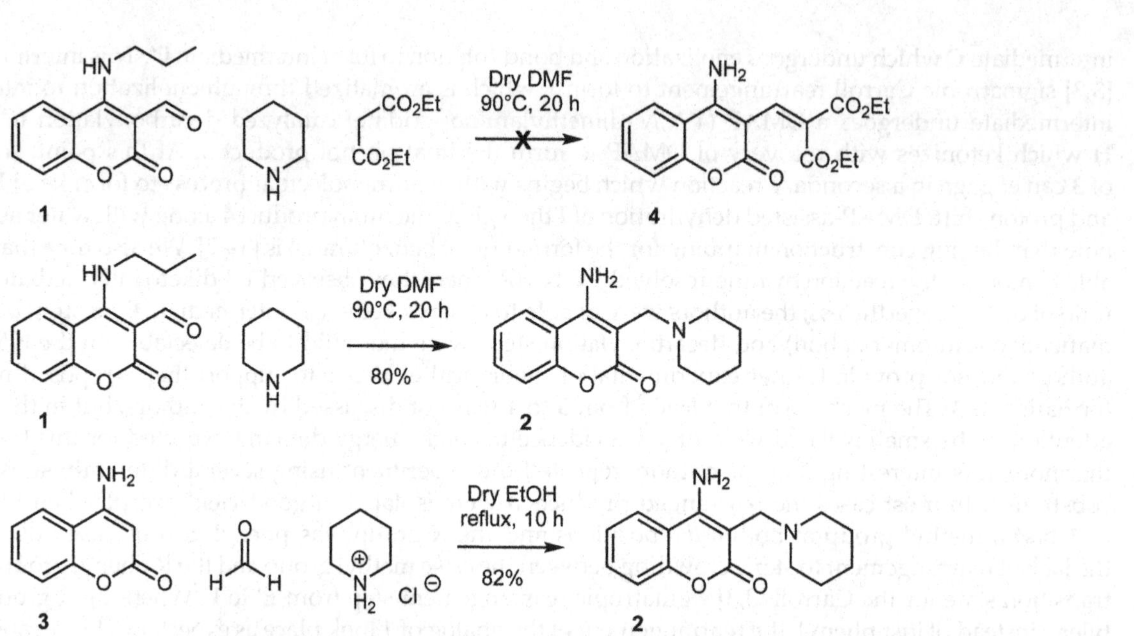

SCHEME 1

SOLUTION: 1 TO 2

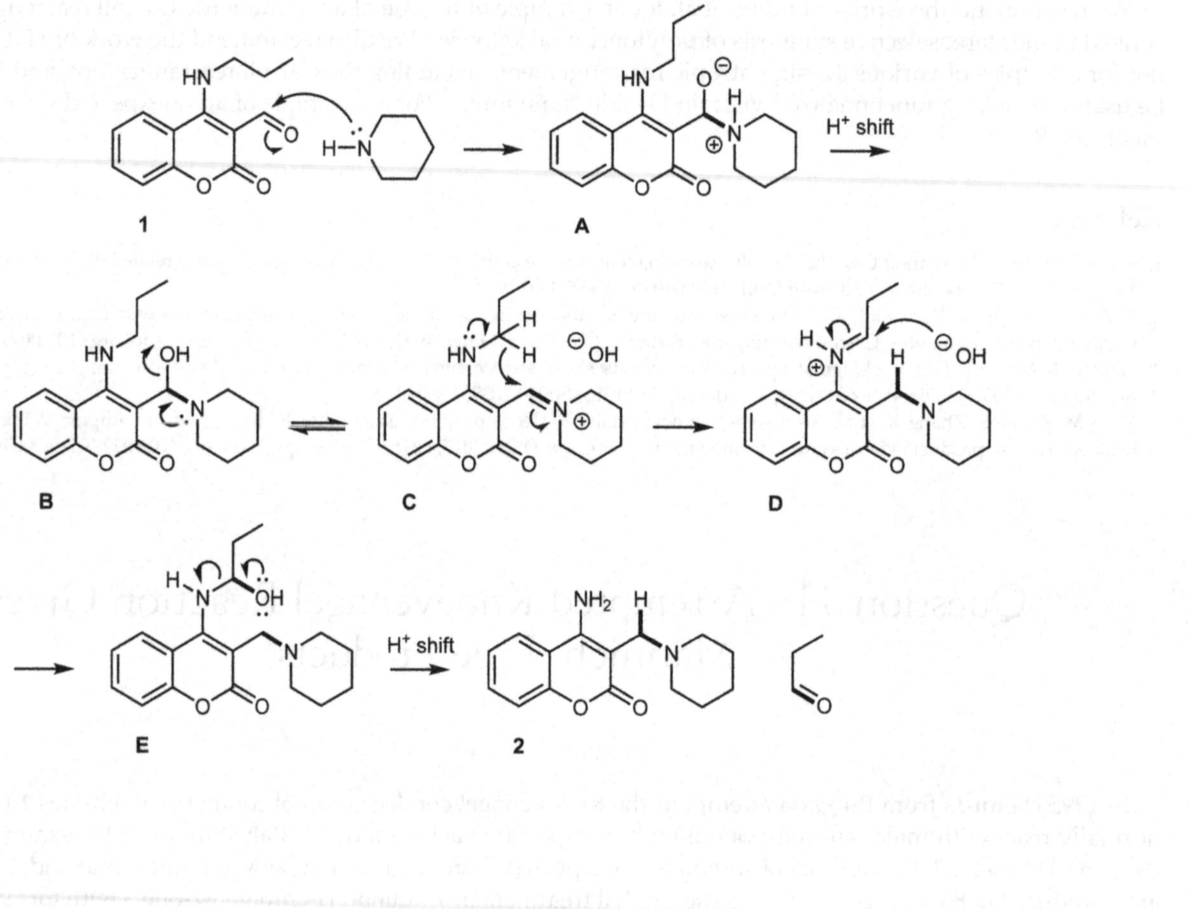

BALANCED CHEMICAL EQUATION: 1 TO 2

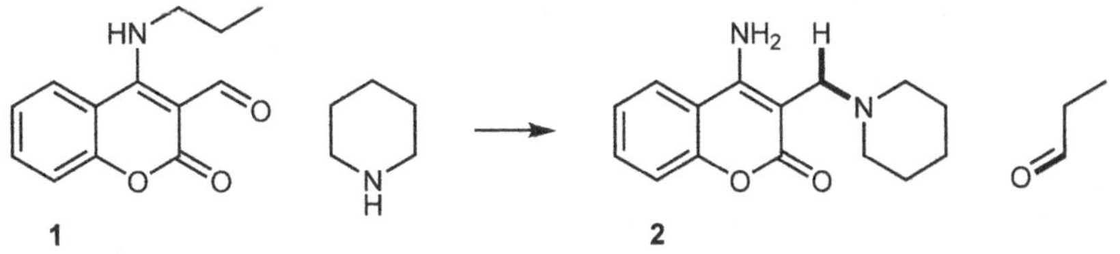

1   →   **2**

SOLUTION: 3 TO 2

3   **F**   H⁺ shift   **G**   tautⁿ

H   **I** H_2O HCl   **2**

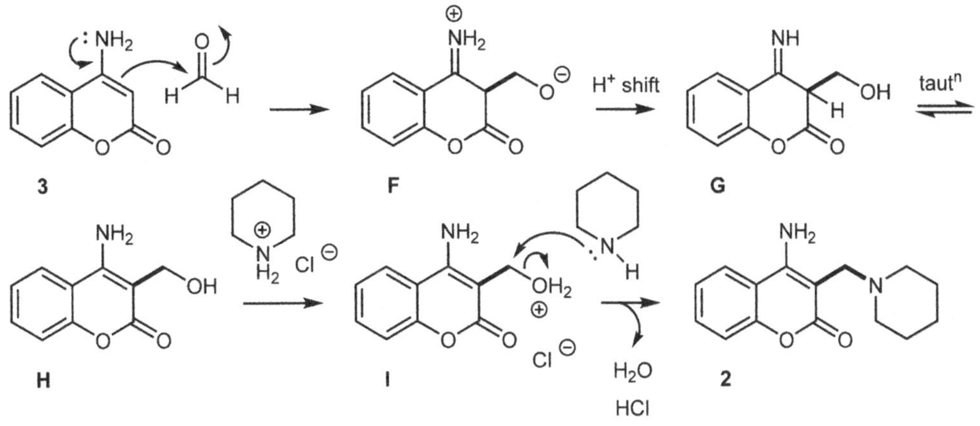

BALANCED CHEMICAL EQUATION: 3 TO 2

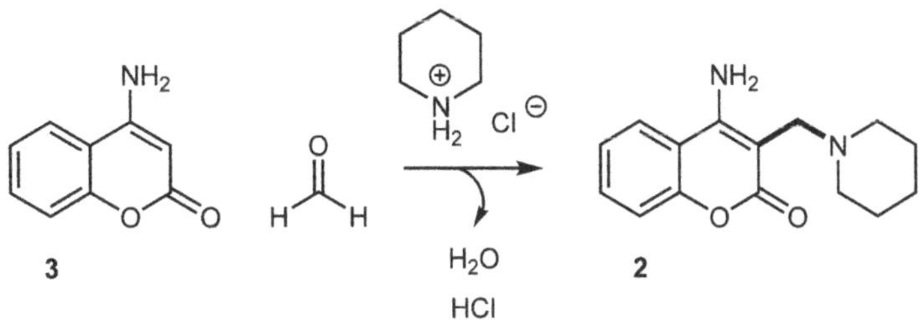

3 H_2O HCl   **2**

KEY STEPS EXPLAINED

    The mechanism proposed by the authors to explain the transformation of **1** to **2** begins with reaction of piperidine with the aldehyde group of **1** to form, after a proton shift, the hemiaminal intermediate **B** which eliminates hydroxide anion to form the iminium hydroxide intermediate **C**. This intermediate is thought to undergo an intramolecular hydride transfer to form the iminium hydroxide intermediate **D**. At this point hydroxide anion attacks the newly formed electrophilic iminium carbon atom to generate intermediate **E** which fragments to the final product **2** and propanal by-product following proton transfer. We note that the authors also carried out a Mannich-type reaction starting with substrate **3** and treating it with formaldehyde and the hydrochloride salt of piperidine to form in high yield the same product **2**. The mechanism proposed for this transformation begins with the reaction of **3** with formaldehyde which generates the imine intermediate **G** after proton shift. Imine **G** then undergoes tautomerization to form enamine **H** which has its hydroxyl group protonated via piperidinium hydrochloride to form **I**. What follows is the nucleophilic attack of piperidine onto **I** with concomitant displacement of water and hydrochloric acid to form the final product **2**. The authors did not discuss this mechanism in their article and they did not provide additional experimental and theoretical evidence to support the proposed mechanism for the transformation of **1** to **2**.

ADDITIONAL RESOURCES

We recommend an article which describes the mechanism of an unexpected reaction taking place during an attempted Knoevenagel-type condensation.[2] We also recommend the work of Shaabani et al. which describes a situation where the Knoevenagel reaction is itself unexpected.[3]

References

1. Ivanov IC, Karagiosov SK. Two methods for the preparation of 3-dialkylaminomethyl derivatives (Mannich bases) of 4-aminocoumarin: a new type of intramolecular hydride transfer. *Synthesis*. 1995;1995(6):633–634. https://doi.org/10.1055/s-1995-3971.
2. Sundar MS, Bedekar AV. An unexpected formation of benzoyl benzoin from benzil during the attempted Knoevenagel type condensation with dimethylmalonate (or malononitrile). *Tetrahedron Lett*. 2012;53:2745–2747. https://doi.org/10.1016/j.tetlet.2012.03.084.
3. Shaabani A, Sarvary A, Keshipour S, Rezayan AH, Ghadari R. Unexpected Knoevenagel self-condensation reaction of tetronic acid: synthesis of a new class of organic heterocyclic salts. *Tetrahedron*. 2010;66:1911–1914. https://doi.org/10.1016/j.tet.2010.01.009.

Question 12: Pyridinium Salt Rearrangements

In 1990 Latvian chemists observed that treatment of either pyridinium salts **1** or **3** with ethanolic sodium hydroxide at room temperature for 1 h formed the respective products **2** or **4** in high yield (Scheme 1).[1,2] Suggest mechanisms to explain these two transformations.

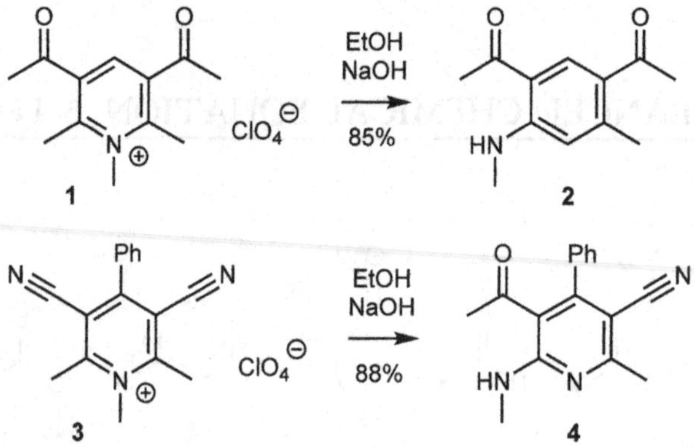

SCHEME 1

SOLUTION: 1 TO 2

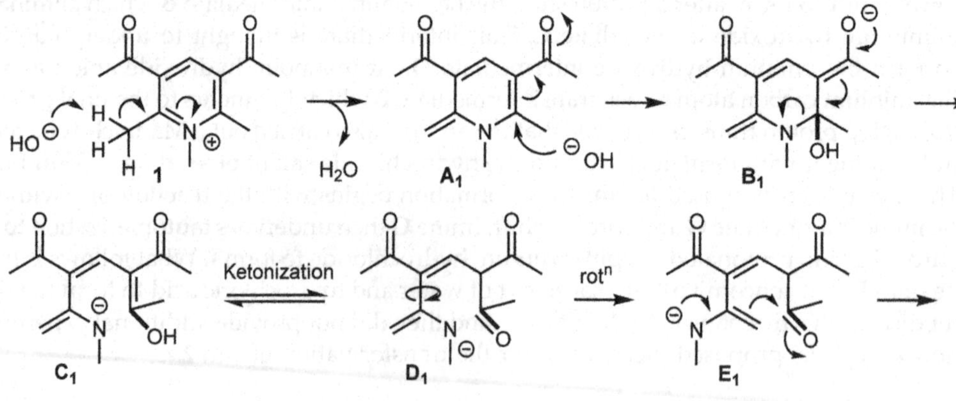

SOLUTION: 3 TO 4

BALANCED CHEMICAL EQUATIONS

KEY STEPS EXPLAINED

We begin by noting that the mechanism proposed for the transformation of **1** to **2** is called the Kost-Sagitullin rearrangement[3,4] and begins with deprotonation of one of the *ortho* methyl groups on the pyridinium ring of **1** by hydroxide ion to form the anhydro base A_1. A second hydroxide ion nucleophilically attacks the carbon atom at the other *ortho* position forming enolate intermediate B_1. This intermediate ring opens to form the enol enamine anion C_1 which subsequently ketonizes to form D_1 which undergoes a C—C sigma bond rotation to form the enamine anion E_1. This intermediate then ring closes to form alkoxide F_1 which after protonation and dehydration generates imine H_1. The final step is a [1,3]-prototropic shift which gives the final aniline product **2**. In terms of evidence, the authors did not provide direct experimental or theoretical evidence to support this mechanism in their papers pertaining to this problem. Moreover, the sequence from the protonated imine tautomer of D_1 to **2** was not explained by the authors; however, we were able to fill in the missing steps in this solution from the authors' prior and later published work on this kind of synthetically useful pyridine to aniline heterocyclic ring transformation reaction.[5-7] Interestingly, however, Shkil et al. did indicate that when a methyl or phenyl group was substituted at position 4 on substrate **1** (i.e., *para* to the nitrogen atom), the yields of **2** decreased significantly. The authors did not explain how their proposed mechanism would account for this finding. We note that the low yield of product **2** suggests that once the pyridinium ring is opened, any bulky groups at position 4 might prevent or reduce ring closure as would be necessary to get to product **2** via the E_1 to F_1 step shown in our mechanism. Since the authors did not elaborate on the complete product composition obtained, we can only speculate on the possibility that the Z-conformer of E_1 may be thermodynamically less stable than the E-conformer (E_1'). It is conceivable that the conversion of E_1 to E_1' proceeds via a sequence of tautomerization, C—C bond rotation, and tautomerization as shown in Scheme 2.

Thus, conformer E_1' could be converted to enamine phenol product **5** or imine phenol product **6** under base catalysis conditions. It is probable that phenols are significant side products in the reaction when acetyl groups are present in the pyridinium salt substrate. In support of this hypothesis, it has been documented that N-methylpyridone methides[8] such as **7** and pyridinium iodides[9] such as **9** can be converted to phenolic products, respectively, under strongly basic conditions such as the ones employed in the **1** to **2** transformation (see Scheme 3).

In later work, the authors revised their mechanism favoring the formation of pseudo base intermediates derived from direct nucleophilic attack of hydroxide ion onto the pyridinium ring over formation of anhydro base intermediate A_1 as the first step. Their rationale was stated as the "considerable value of the difference in enthalpies of formation of pseudo bases and anhydro base-water systems makes their simultaneous existence impossible in solution."[7] We show the revised mechanism in Scheme 4 where the pseudo base is represented by intermediate I_1. We note that a close examination of this mechanism shows that the authors neglected to give any details as to how the ring closure with concomitant dehydration takes place in the last step from intermediate M_1 to aniline product **2**. We thus provide the missing details in Scheme 5.

In contrast to the observations related to the Kost-Sagitullin rearrangement of **1** to **2**, the transformation of **3** to **4** does feature a phenyl group at the 4 position of substrate **3** and we note that the authors reported a product yield of 86% for **4**. Moreover, when the phenyl group was replaced by hydrogen, the yield of **4** dropped to 63%. To explain these results, the authors described a mechanism involving a double rearrangement where the second one was presumed to be a Dimroth amidine rearrangement.[10,11] Thus, the mechanism proposed by the authors begins with hydroxide ion attack

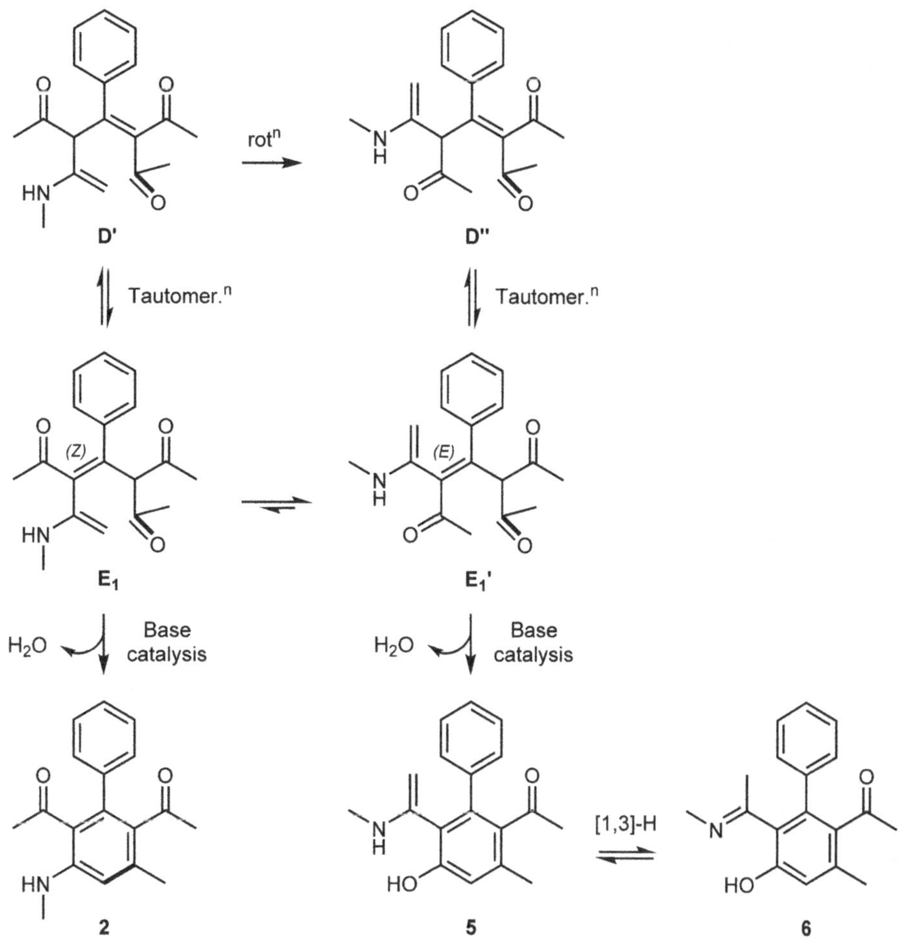

SCHEME 2

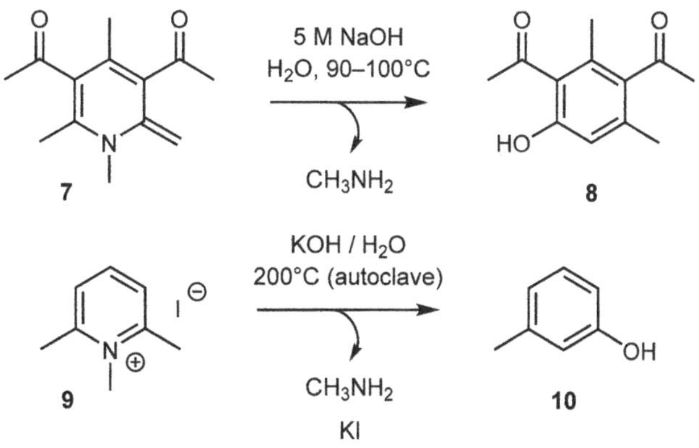

SCHEME 3

onto **3** at the *ortho* position to form zwitterionic intermediate A_2 which undergoes fragmentation via electron reshuffling to form enol B_2. Carbon-carbon bond rotation in B_2 followed by cyclization onto the cyano group forms intermediate D_2 after a proton shift. At this point the authors propose that a second equivalent of hydroxide ion attacks at the *ortho* (nonimine) position of D_2, a step which is facilitated by the conjugated π system, to form enolate E_2. We note that E_2 is stabilized through additional resonance structures which we did not show. Nevertheless, E_2 undergoes a Dimroth rearrangement through ring fragmentation to F_2, ketonization to G_2, bond rotation to H_2, and recyclization to I_2.

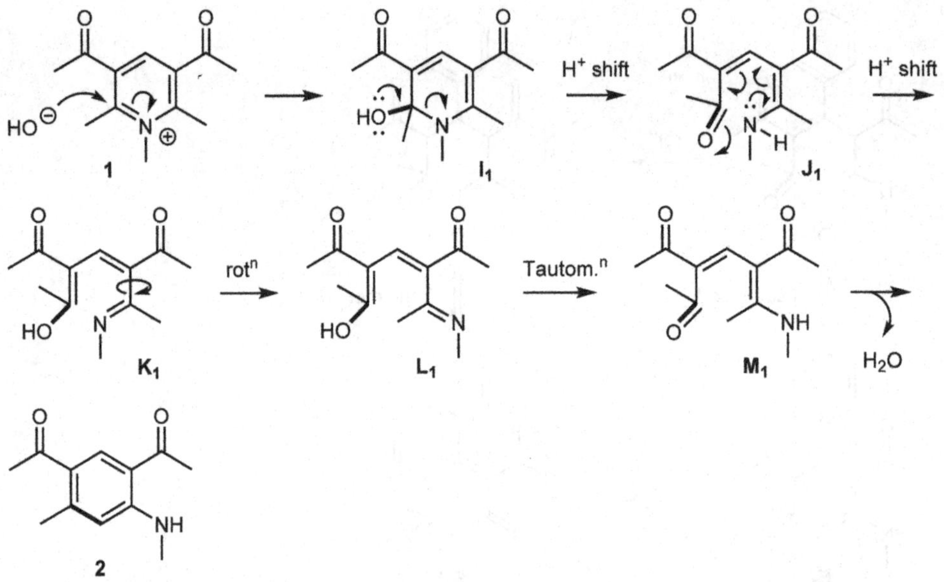

SCHEME 4

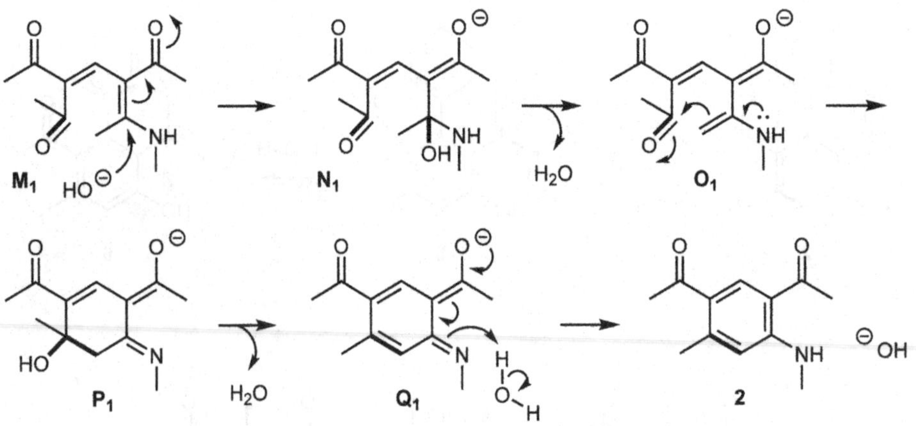

SCHEME 5

We note that this sequence from E_2 to I_2 where the amine ring nitrogen atom and the exocyclic amidine nitrogen atom effectively trade places constitutes an example of the Dimroth rearrangement. Once I_2 is formed, a [1,3] hydrogen shift followed by base-mediated dehydration generates the final product 4. The authors did not provide additional experimental or theoretical evidence to support this mechanism nor did they explain the yield difference when the phenyl group is replaced by hydrogen. Judging from the mechanism, one can infer that the phenyl group further helps to stabilize electron density buildup in structure E_2 following addition of hydroxide ion.

We note that a much shorter mechanism, shown in Scheme 6, can be proposed for the 3 to 4 transformation which invokes an alternative rearrangement sequence involving ring opening followed by a single recyclization. As before, we begin with initial nucleophilic attack by hydroxide ion which leads to intermediate K_2 which ring opens to keto imine L_2 after a prototropic shift. Tautomerization to M_2 facilitated by the two strongly electron-withdrawing cyano groups followed by C—C bond rotation leads to intermediate N_2 which is deprotonated by hydroxide ion causing ring closure onto the cyanide group to form intermediate O_2. This then undergoes a 1,3-methyl migration from the internal ring nitrogen atom to the external one to form enolate P_2 which after ketonization and final protonation leads to product 4.

This mechanism can be distinguished from the authors' proposal by applying a combined ^{15}N and deuterium labeled experiment as shown in Scheme 7. The reasoning behind this labeling experiment follows that given previously for the experimental evidence of how the Dimroth rearrangement is thought to proceed.[12,13] The authors' mechanism thus predicts that the ^{15}N-CD$_3$ bond will not be cleaved, whereas, the alternative mechanism predicts that it will.

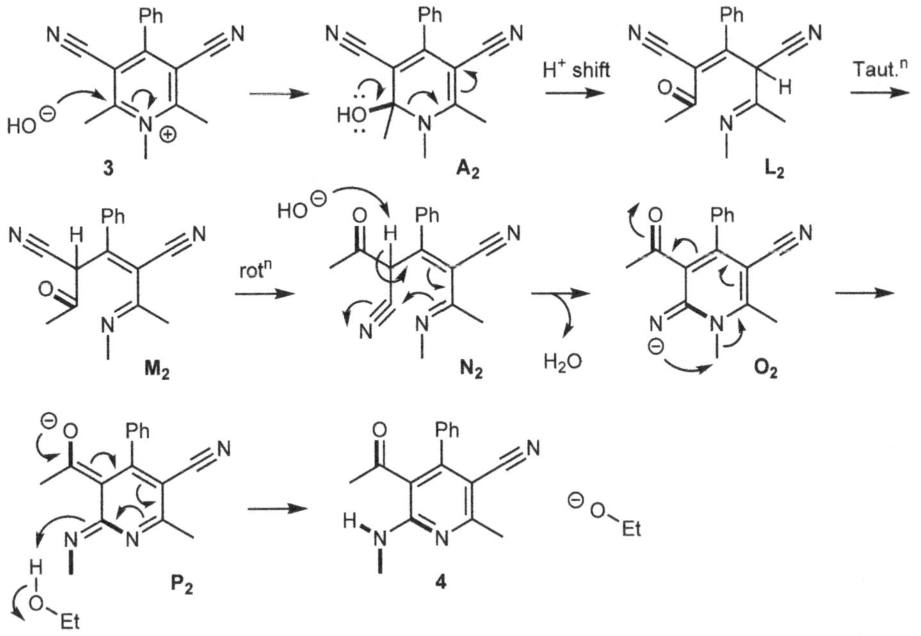

SCHEME 6

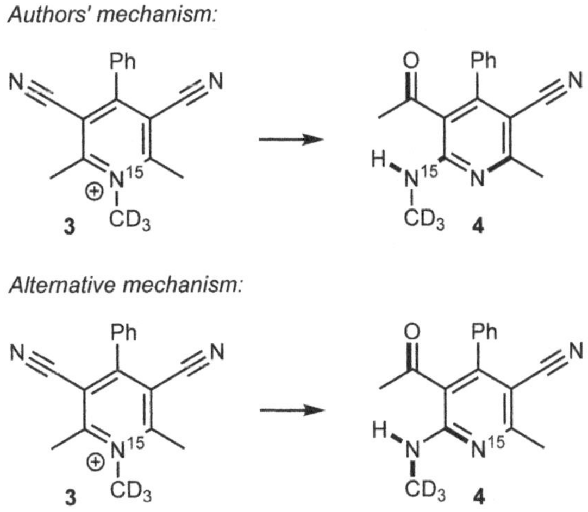

Authors' mechanism:

Alternative mechanism:

SCHEME 7

Lastly, we note that both transformations **1** to **2** and **3** to **4** illustrate a [6+0] ring construction mapping. However, we point out that the pair of suggested mechanisms for the **3** to **4** transformation leads to different highlighted bonds as seen in Scheme 7. Furthermore, we highly recommend that others continue to investigate the mechanisms of these interesting transformations using variously substituted derivative substrates, especially at position 4 in **1** and **3**, as well as other electron withdrawing groups at the carbons currently bearing the acyl or cyano groups. Given that the authors provided minimal experimental evidence for their proposed mechanisms for both transformations, we also highly recommend that isotopic labeling studies be undertaken to elucidate which mechanisms are operative since this was the experimental evidence that was used to elucidate the Dimroth rearrangement. The authors did report theoretical investigations of these mechanisms using primitive basis sets such as the MO LCAO SCF (molecular orbital linear combination of atomic orbitals, self-consistent field) method[14,15]; however, we recommend a reinvestigation using more advanced modern methods.

ADDITIONAL RESOURCES

We direct the reader to four articles and one review which all discuss interesting aspects of the Dimroth rearrangement as well as unexpected cases where it is thought to occur.[16–20] A recent article describes how this rearrangement was used to synthesize an aminoquinazoline pharmaceutical.[21] We also point out that heterocyclic ring transformations are a powerful synthetic tool that have been exploited to transform pyrimidines to pyridines[22] and pyridinium rings to indoles.[23]

References

1. Shkil GP, Sagitullin RS, Mutsenietse D, Lusis V. Recyclization of 3,5-diacetyl- and 3,5-dicyanopyridinium salts under basic conditions. *Khim Geterotsikl Soedin*. 1990;6:848–849.

2. Shkil GP, Lusis V, Muceniece D, Sagitullin RS. Recyclization of polysubstituted pyridinium salts. *Tetrahedron*. 1995;51(31):8599–8604. https://doi.org/10.1016/0040-4020(95)00474-M.

3. Kost AN, Gromov SP, Sagitullin RS. Pyridine ring nucleophilic recyclizations. *Tetrahedron*. 1981;37(20):3423–3454. https://doi.org/10.1016/S0040-4020(01)98858-1.

4. Danagulyan GG. Kost-Sagitullin rearrangement and other isomerization recyclizations of pyrimidines. (Review). *Chem Heterocycl Compd*. 2005;41(10):1205–1236. https://doi.org/10.1007/s10593-005-0308-z.

5. Sagitullin RS, Gromov SP, Kost AN. Alkylamino group exchange upon recyclization of pyridinium salts into anilines. *Tetrahedron*. 1978;34(14):2213–2216. https://doi.org/10.1016/0040-4020(78)89030-9.

6. Kost AN, Sagitullin RS, Gromov SP. Recyclization of the pyridine ring under the influence of nucleophiles. *Chem Heterocycl Compd*. 1979;15(1):87–91. https://doi.org/10.1007/bf00471207.

7. Atavin EG, Tikhonenko VO, Sagitullin RS. The mechanism of isomerization recyclization of diacetyl derivatives of pyridinium salts. *Chem Heterocycl Compd*. 2001;37(7):850–854. https://doi.org/10.1023/a:1012447407540.

8. Mumm O, Petzold R. Zur Kenntnis der Pyridon-methide. *Ann Chem*. 1938;536(1):1–29. https://doi.org/10.1002/jlac.19385360102.

9. Lukeš R, Pergál M. Homologues of pyridine. II. Synthesis and reactions of some α,α'-disubstituted pyridines. *Collect Czechoslov Chem Commun*. 1959;24(1):36–45. https://doi.org/10.1135/cccc19590036.

10. Brown DJ. Amidine rearrangements (The Dimroth rearrangements). In: Thyagarajan BC, ed. *Mechanisms of Molecular Migrations*. New York: John Wiley & Sons Inc.; 1968:209–245. vol. 1

11. El Ashry ESH, El Kilany Y, Rashed N, Assafir H. Dimroth rearrangement: translocation of heteroatoms in heterocyclic rings and its role in ring transformations of heterocycles. *Adv Heterocycl Chem*. 1999;75:79–165. https://doi.org/10.1016/S0065-2725(08)60984-8.

12. Brown DJ. Apparent migration of *N*-methyl groups in the pyrimidine series. *Nature*. 1961;189:828–829. https://doi.org/10.1038/189828a0.

13. Goerdeler J, Roth W. Über die Alkylierung einiger Aminoheterocyclen und die Umlagerung heterocyclischer Imine in Amine. *Chem Ber*. 1963;96:534–549. https://doi.org/10.1002/cber.19630960227.

14. Vysotskii YB, Zemskii BP, Stupnikova TV, Sagitullin RS, Kost AN, Shvaika OP. Quantum-chemical description of the recyclization reactions of quaternary pyridinium salts. *Chem Heterocycl Compd*. 1979;15(11):1201–1204. https://doi.org/10.1007/bf00471432.

15. Vysotskii YB, Zemskii BP, Stupnikova TV, Sagitullin RS. Study of the effect of annelation on the rearrangement of 1,2-dialkylpyridinium salts by the MO LCAO self-consistent field (SCF) method. *Chem Heterocycl Compd*. 1980;16(3):288–292. https://doi.org/10.1007/bf02401729.

16. Filak L, Riedl Z, Egyed O, et al. A new synthesis of the linearly fused [1,2,4]triazolo[1,5-b]isoquinoline ring. Observation of an unexpected Dimroth rearrangement. *Tetrahedron*. 2008;64:1101–1113. https://doi.org/10.1016/j.tet.2007.10.103.

17. Lauria A, Patella C, Abbate I, Martorana A, Almerico AM. An unexpected Dimroth rearrangement leading to annelated thieno[3,2-d][1,2,3]triazolo[1,5-a]pyrimidines with potent antitumor activity. *Eur J Med Chem*. 2013;65:381–388. https://doi.org/10.1016/j.ejmech.2013.05.012.

18. Salgado A, Varela C, Collazo AMG, et al. Synthesis of [1,2,4]-triazolo[1,5-a]pyrimidines by Dimroth rearrangement of [1,2,4]-triazolo[4,3-a]pyrimidines: a theoretical and NMR study. *J Mol Struct*. 2011;987:13–24. https://doi.org/10.1016/j.molstruc.2010.11.054.

19. Chatzopoulou M, Martinez RF, Willis NJ, et al. The Dimroth rearrangement as a probable cause for structural misassignments in imidazo[1,2-a]pyrimidines: a ^{15}N-labelling study and an easy method for the determination of regiochemistry. *Tetrahedron*. 2018;https://doi.org/10.1016/j.tet.2018.06.033.

20. Krajczyk A, Boryski J. Dimroth rearrangement—old but not outdated. *Curr Org Chem*. 2017;21:2515–2529. https://doi.org/10.2174/1385272821666170427125720.

21. Goundry WRF, Boardman K, Cunningham O, et al. The development of the Dimroth rearrangement route to AZD8931. *Org Process Res Dev*. 2017;21:336–345. https://doi.org/10.1021/acs.oprd.6b00412.

22. Van der Plas HC. Pyrimidine-pyridine ring interconversion. *Adv Heterocycl Chem*. 2003;84:31–70. https://doi.org/10.1016/S0065-2725(03)84002-3.

23. Kearney AM, Vanderwal CD. Synthesis of nitrogen heterocycles by the ring opening of pyridinium salts. *Angew Chem*. 2006;118:7967–7970. https://doi.org/10.1002/ange.200602996.

Question 13: A Pyridazine From a Thiophene Dioxide

In 1989 Japanese chemists reported the first synthesis of 4,5-di-*t*-butylpyridazine **5** which consisted of (i) heating of 3,4-di-*t*-butylthiophene 1,1-dioxide **1** and two equivalents of 3,4-phenyl-1,2,4-triazoline-3,5-dione **2** in toluene under reflux for 5 h to form product **3** ($C_{28}H_{30}N_6O_4$) in 87% yield; and (ii) hydrolysis of **3** using methanolic potassium hydroxide at room temperature to form an 80% yield of **4** and **5** (Scheme 1).[1] Provide a mechanistic explanation for this two-step synthetic sequence.

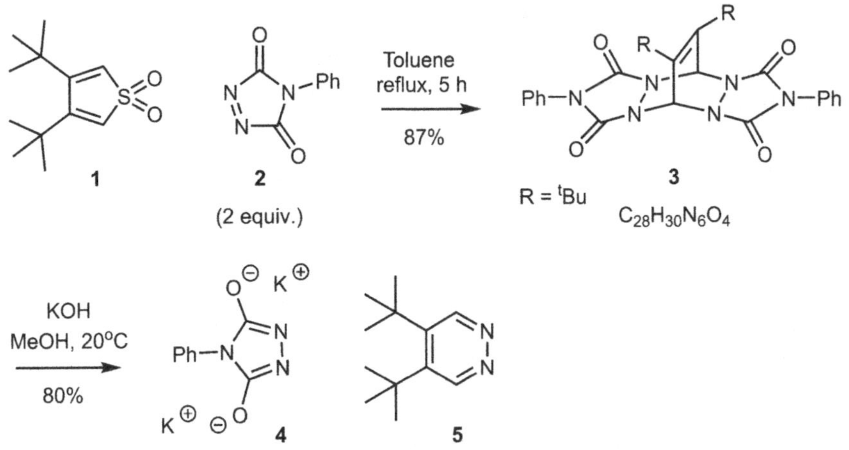

SCHEME 1

SOLUTION: 1 AND 2 TO 3

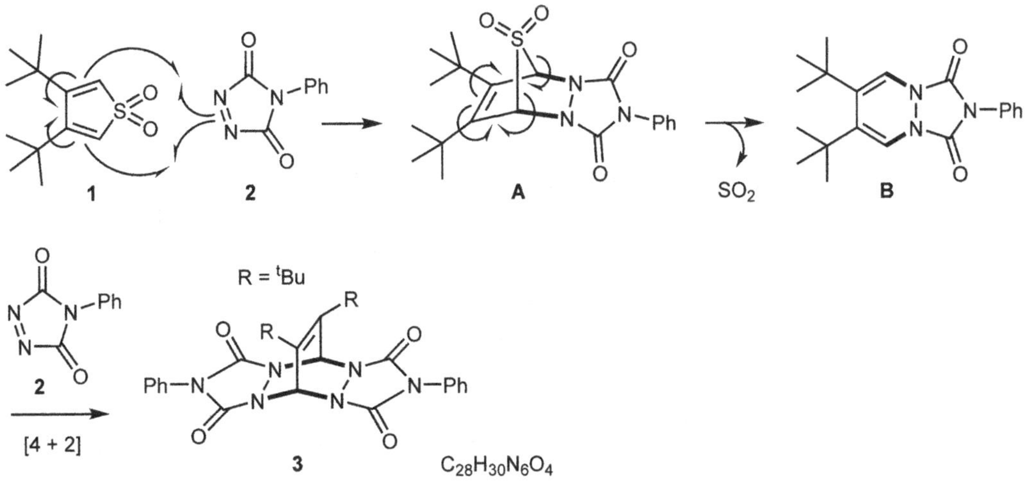

SOLUTION: 3 TO 4

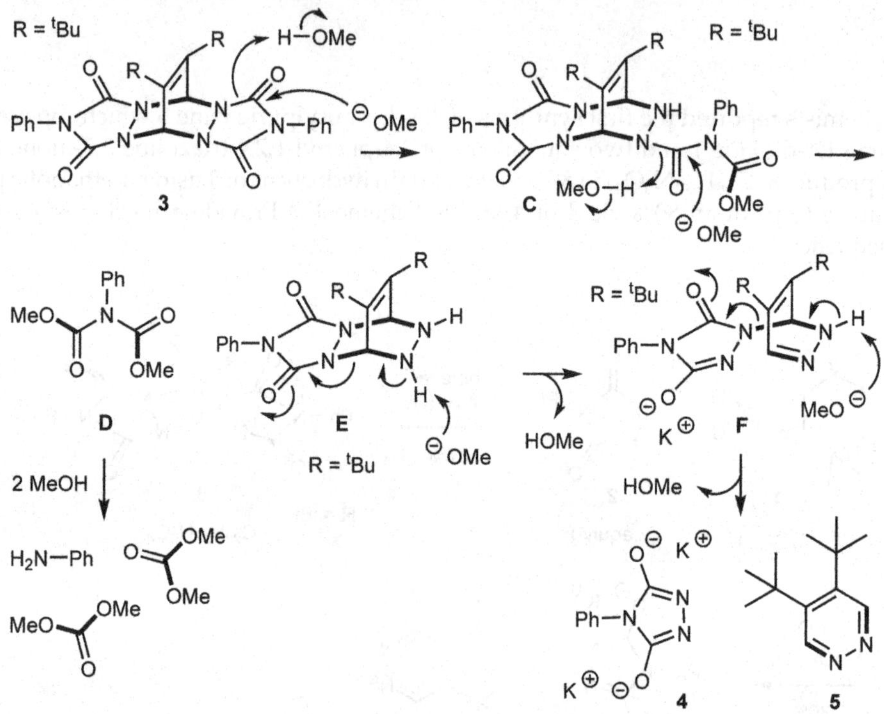

BALANCED CHEMICAL EQUATIONS

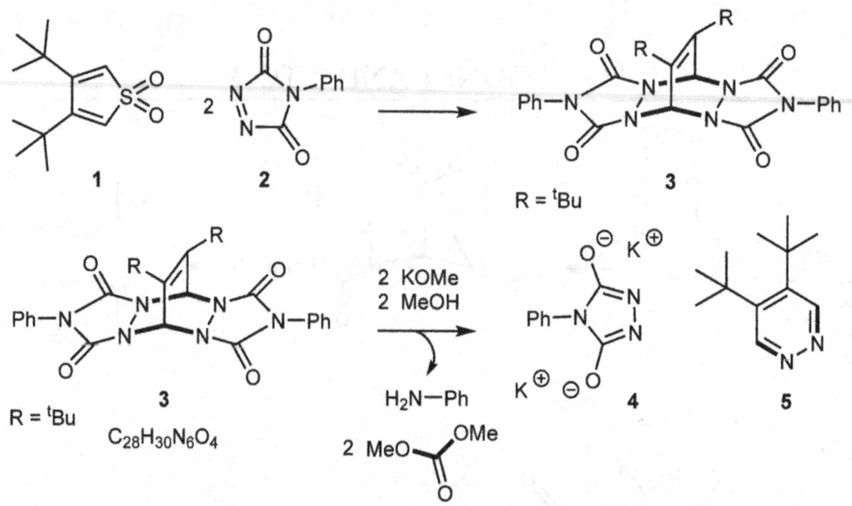

KEY STEPS EXPLAINED

The mechanism proposed for the transformation of **1** and **2** to **3** begins with a [4+2] cycloaddition of sulfone **1** and one equivalent of 4-phenyl-1,2,4-triazoline-3,5-dione **2** to form intermediate **A**. This intermediate then undergoes a retro-[4+2] extrusion of sulfur dioxide to form intermediate **B** which is a diene that reacts with another equivalent of **2** in a [4+2] cycloaddition to form product **3**. Once **3** is added to a solution of methanolic potassium hydroxide, we propose that methoxide ion nucleophilically attacks one of the symmetrical urea carbonyl groups of **3** to form intermediate **C** which reacts with a second equivalent of methoxide ion to form intermediate **E** after elimination of intermediate **D**. Sequential deprotonation of intermediate **E** by two more equivalents of methoxide ions leads to production of pyridazine **5** and dipotassium 4-phenyl-4*H*-[1,2,4]triazole-3,5-diolate **4** as by-product. We note that intermediate **D** can degrade further by reacting with two equivalents of methanol to form aniline and two equivalents of dimethyl carbonate as further by-products. Moreover, the ring construction mapping for the formation of product **5** is [4+2].

Interestingly, our proposed mechanism for the formation of **5** differs from that of the authors in that it predicts formation of by-product **4** whereas the authors invoke an air oxidation mechanism for the transformation of **3** to **5** which avoids formation of **4** (Scheme 2). According to this mechanism, Nakayama and Hirashima claimed that product **3** degrades to form 7,8-di-*tert*-butyl-2,3,5,6-tetraaza-bicyclo[2.2.2]oct-7-ene **X** which then undergoes air oxidation to form 7,8-di-*tert*-butyl-2,3,5,6-tetraaza-bicyclo[2.2.2]octa-2,5,7-triene **Y** as shown in Scheme 2. The triene **Y** continues to react via a retro-[4+2] electrocyclization which forms pyridazine **5** and nitrogen gas. The authors claimed they could not isolate **Y** even when the reaction was conducted under a nitrogen atmosphere. However, they argued that **Y** was prone to facile oxidation with subsequent loss of nitrogen gas by-product for which no experimental evidence was provided. We therefore regard this mechanism as suspect and propose an alternative (see above Solution) which avoids the need for air oxidation. We also recognize that our mechanism predicts the formation of by-product **4** for which there exists no experimental evidence. We believe this is the case because during the aqueous work-up phase of the reaction by-product **4**, if formed, would be soluble in water and consequently lost in the process. We therefore suggest that a careful by-product analysis in conjunction with a theoretical study would help shed light on the operative mechanism for the second step of this synthesis plan.

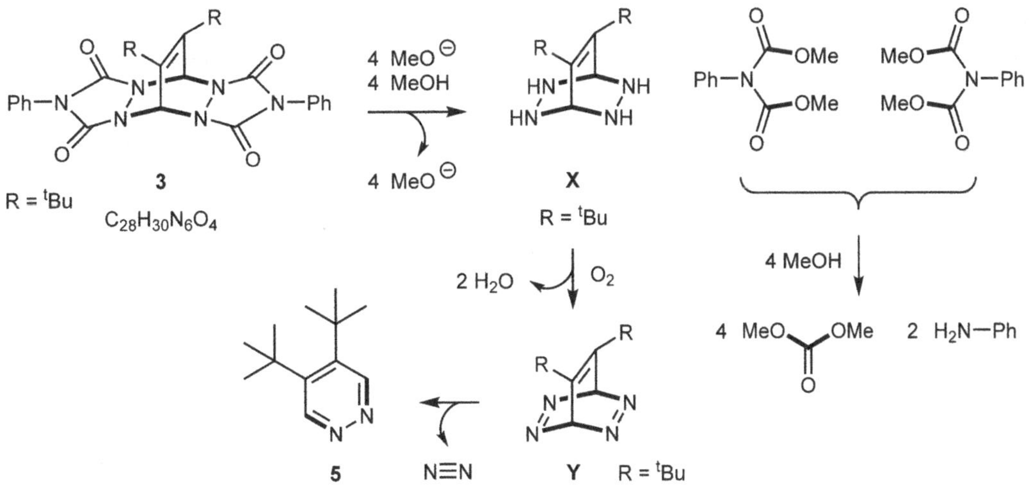

SCHEME 2

ADDITIONAL RESOURCES

We recommend two articles which discuss synthetic approaches and their mechanistic highlights with respect to the synthesis of pyridazine compounds, tetrazines, and others.[2,3]

References

1. Nakayama J, Hirashima A. The first synthesis of 4,5-di-*t*-butylpyridazine. *Heterocycles*. 1989;29(7):1241–1242. https://doi.org/10.3987/COM-89-5021.
2. Lodewyk MW, Kurth MJ, Tantillo DJ. Mechanisms for formation of diazocinones, pyridazines, and pyrazolines from tetrazines: oxyanion-accelerated pericyclic cascades? *J Org Chem*. 2009;74(13):4804–4811. https://doi.org/10.1021/jo900565y.
3. Abdelrazek FM, Hassaneen HM, Nassar EM, Jager A. Phenacyl bromides revisited: facile synthesis of some new pyrazoles, pyridazines, and their fused derivatives. *J Heterocyclic Chem*. 2013;51(2):475–481. https://doi.org/10.1002/jhet.1664.

Question 14: Biomimetic Synthesis of Litebamine

In 1995 Japanese chemists reacted isoquinolinium methiodide **1** with sodium methoxide in methanol for 2 h at reflux temperature which generated the unexpected tetrahydroisoquinoline product **2** in quantitative yield rather than the expected benzazepine product **3** (Scheme 1).[1] The authors utilized this result in their strategy to synthesize litebamine **6**. This was done via treatment of **4** with iodoacetonitrile in methylene chloride to form the isoquinolinium methiodide product **5** which was then reacted under sodium methoxide in methanol to form litebamine **6** (69% yield). Suggest mechanisms for these transformations and explain why benzazepine **3** was the expected product in the first reaction.

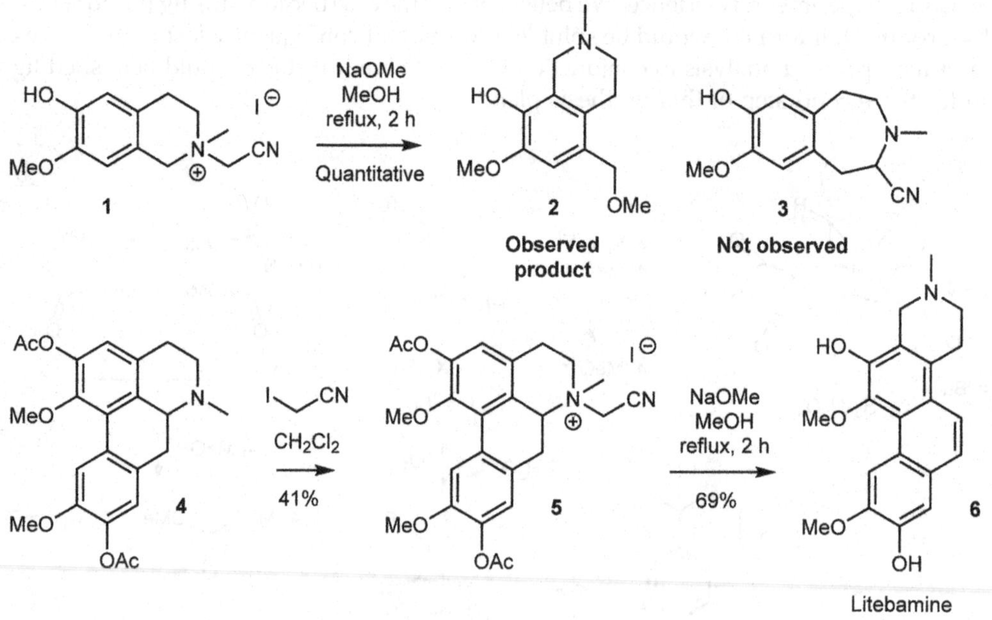

SCHEME 1

SOLUTION: 1 TO 2

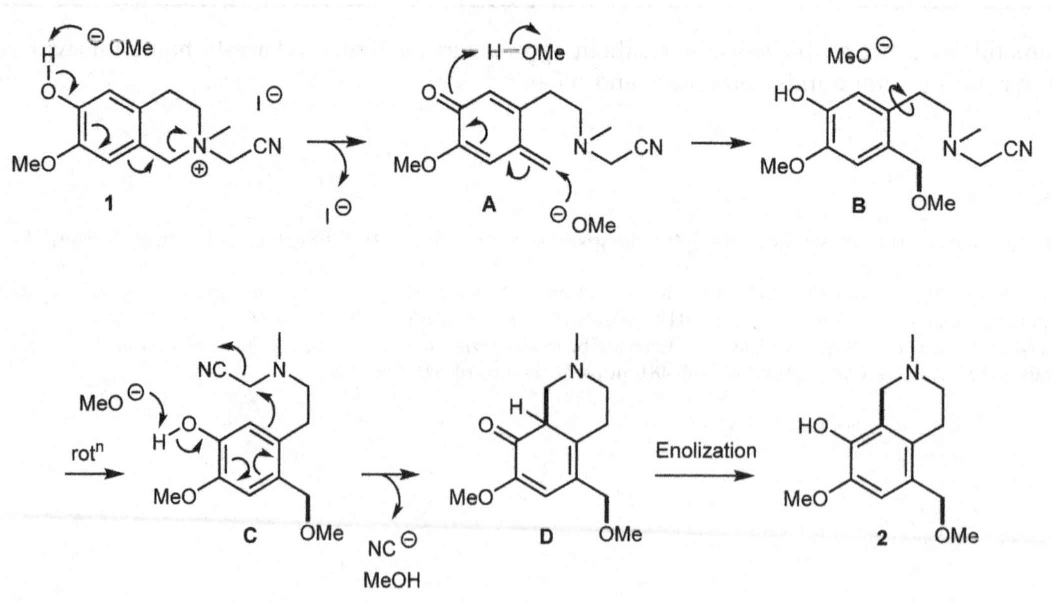

BALANCED CHEMICAL EQUATION: 1 TO 2

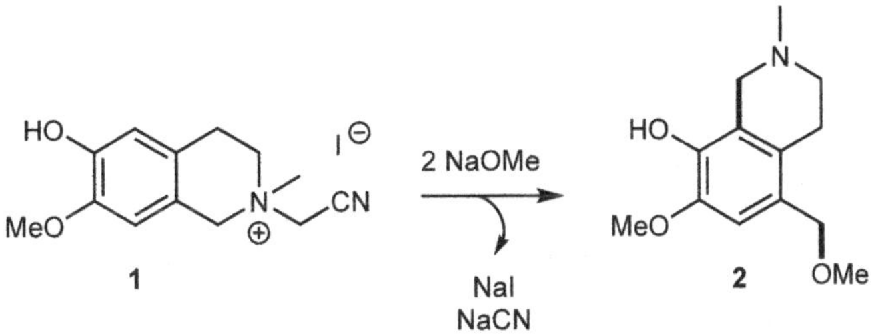

SOLUTION: 4 TO 5

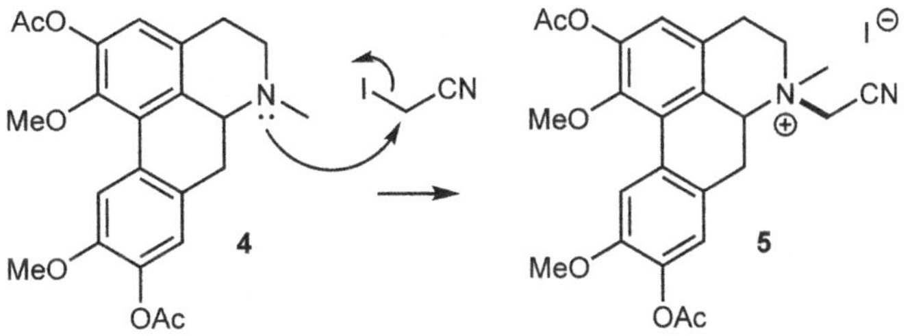

SOLUTION: 5 TO 6

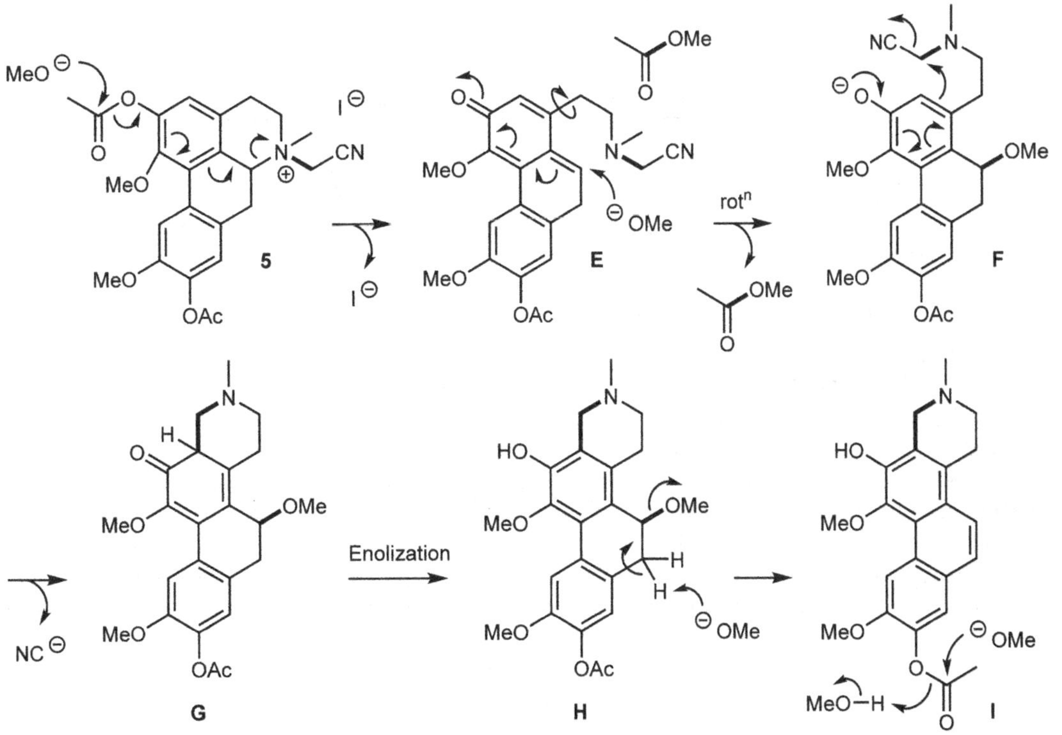

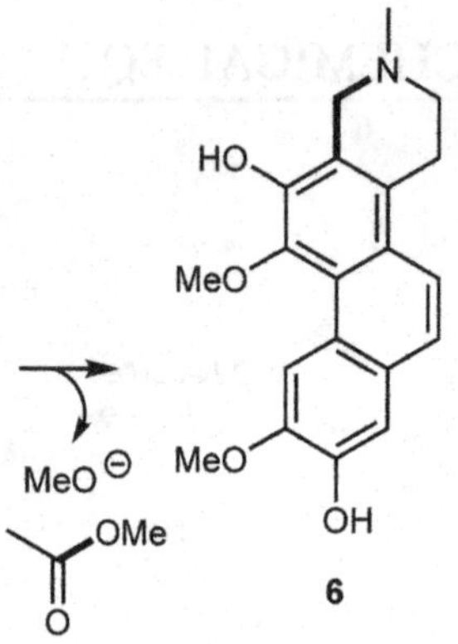

BALANCED CHEMICAL EQUATION: 5 TO 6

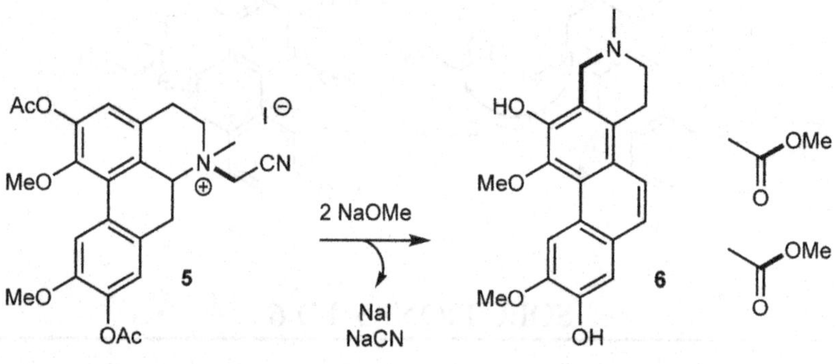

KEY STEPS EXPLAINED

We begin by noting that the authors thought that isoquinolinium methiodide **1** could be converted to the ring-expanded product **3** via the mechanism shown in Scheme 2 (akin to a Pictet-Spengler reaction). Instead, tetrahydroisoquinoline **2** was obtained via initial methoxide ion deprotonation of the hydroxyl group of **1** to cause electron delocalization and ring opening of the piperidine group to form intermediate **A**. Methoxide ion attack at the exocyclic methylene group leads to rearomatization in **A** to form **B** which after C—C bond rotation gives **C**. Cyclization of **C** occurs with the aid of methoxide ion deprotonation of the hydroxyl group to form intermediate **D** after elimination of cyanide anion. Finally, intermediate **D** undergoes enolization to establish aromatization which leads to product **2**.

In the case of the synthesis of **6**, the transformation of **4** to quaternary ammonium salt **5** involves a straightforward nucleophilic attack of the amine group of **4** at the methylene carbon of iodoacetonitrile. In the next step, methoxide ion attacks **5** to cause deacetylation with concomitant ring opening of the piperidine ring to form intermediate **E**. This intermediate is attacked by another equivalent of methoxide ion which causes rearomatization in the ring which after C—C bond rotation gives intermediate **F**. This intermediate cyclizes with elimination of cyanide anion to intermediate **G** which undergoes enolization to form **H**. This intermediate undergoes methanolysis followed by saponification of the second acetate group to generate litebamine **6**. In terms of experimental evidence for this interesting and unexpected rearrangement, the authors had several analogs of **1** among which were substrates **1′** (which has the methoxy

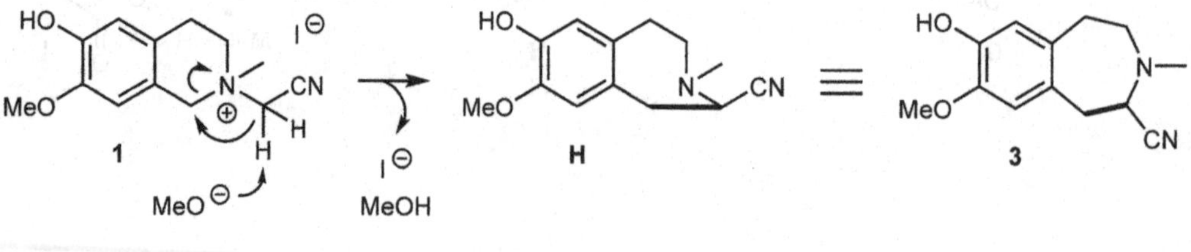

SCHEME 2

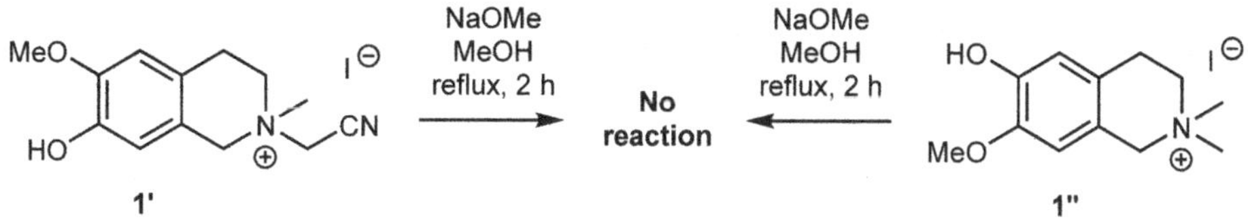

SCHEME 3

and hydroxyl substituent positions switched) and **1″** (which lacks the *N*-cyanomethyl group) react under the same reaction conditions (Scheme 3). The authors found that for **1′** and **1″** the rearrangement did not take place thus enforcing the idea that "both substituents (6- or 8-hydroxyl and *N*-cyanomethyl groups)" are required for rearrangement to occur. This is why the proposed mechanism invokes electron delocalization from the hydroxyl groups of **1** and **C** as well as similar structures such as **5** and **F**. Despite this evidence, we note that the authors did not provide additional experimental or theoretical evidence to support their proposed mechanism.

ADDITIONAL RESOURCES

We recommend a recent article which discusses a novel approach to the synthesis of litebamine **6** based on the previous work of Hara et al.[2] We also recommend two articles which discuss mechanisms of transformations that employ similar chemistry to that found here.[3,4] Lastly, we recommend an article which examines an unexpected selectivity feature of the Pictet-Spengler transformation in the context of the reaction between tryptophan methyl esters and aldehydes.[5]

References

1. Hara H, Kaneko K, Endoh M, Uchida H, Hoshino O. A novel ring cleavage and recyclization of *N*-cyanomethyl-1,2,3,4-tetrahydroisoquinolinium methiodides: a biomimetic synthesis of litebamine. *Tetrahedron*. 1995;51(37):10189–10204. https://doi.org/10.1016/0040-4020(95)00614-E.
2. Khunnawutmanotham N, Sahakitpichan P, Chimnoi N, Techasakul S. Efficient one-pot synthesis of tetrahydronaphtho[2,1-*f*]isoquinolines by using domino Pictet–Spengler/Friedel–Crafts-type reactions. *Eur J Org Chem*. 2017;43:6434–6440. https://doi.org/10.1002/ejoc.201701243.
3. Zou B, Yap P, Sonntag L-S, Leong SY, Yeung BKS, Keller TH. Mechanistic study of the spiroindolones: a new class of antimalarials. *Molecules*. 2012;17:10131–10141. https://doi.org/10.3390/molecules170910131.
4. Vaclavik J, Sot P, Pechacek J, et al. Experimental and theoretical perspectives of the Noyori-Ikariya asymmetric transfer hydrogenation of imines. *Molecules*. 2014;19:6987–7007. https://doi.org/10.3390/molecules19066987.
5. Bailey PD, Beard MA, Phillips TR. Unexpected cis selectivity in the Pictet–Spengler reaction. *Tetrahedron Lett*. 2009;50:3645–3647. https://doi.org/10.1016/j.tetlet.2009.03.121.

Question 15: Acid-Catalyzed Degradation of *N*-Nitroso-2,3-didehydromorpholine

In 1995 American chemists investigated the mechanism of the acid-catalyzed degradation of *N*-nitroso-2,3-didehydromorpholine **1** under aqueous acidic condition to the *N*-(2-hydroxyethyl)-2-oximinoethanamide **2** in 89% yield (Scheme 1).[1] Suggest a mechanism for this transformation.

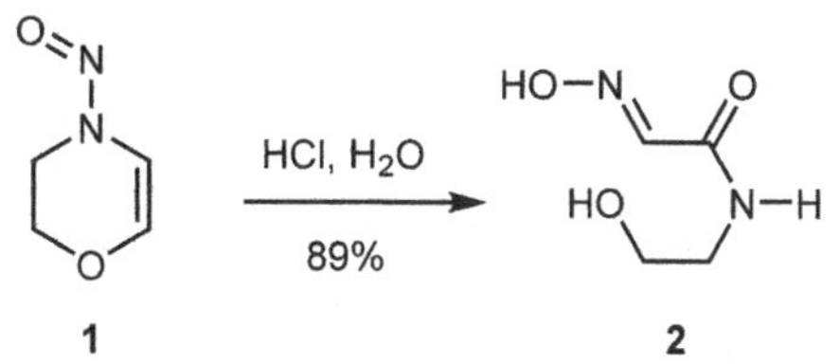

SCHEME 1

SOLUTION

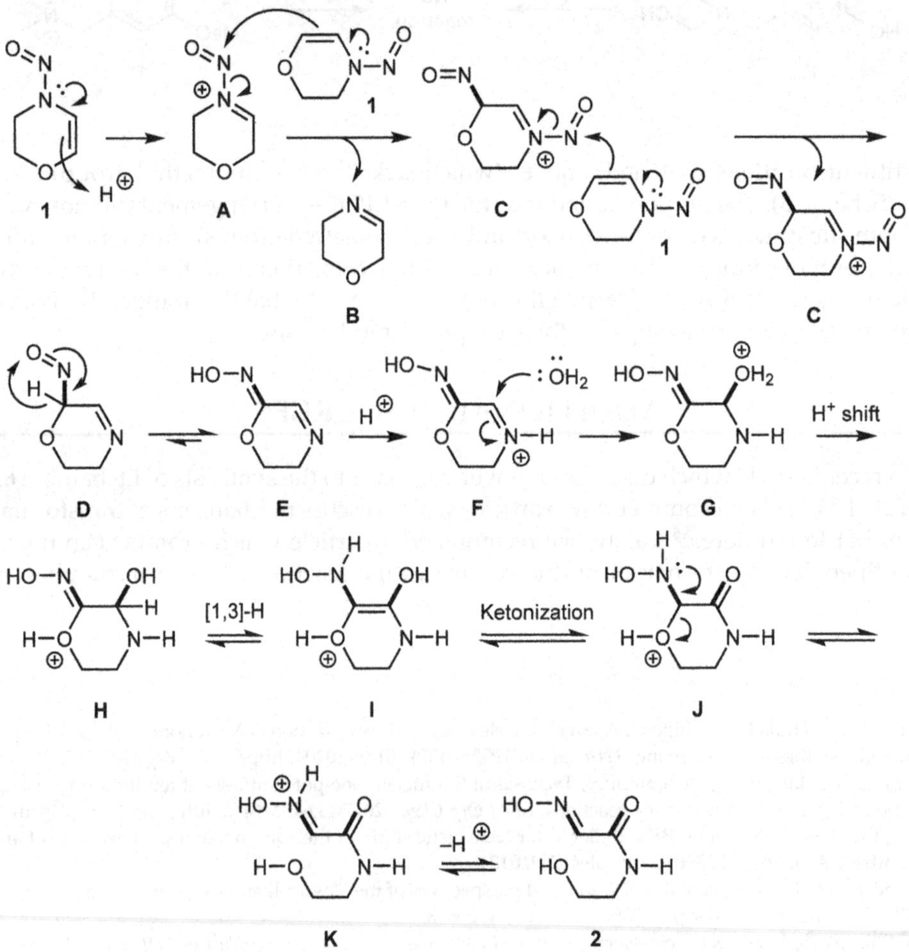

BALANCED CHEMICAL EQUATION

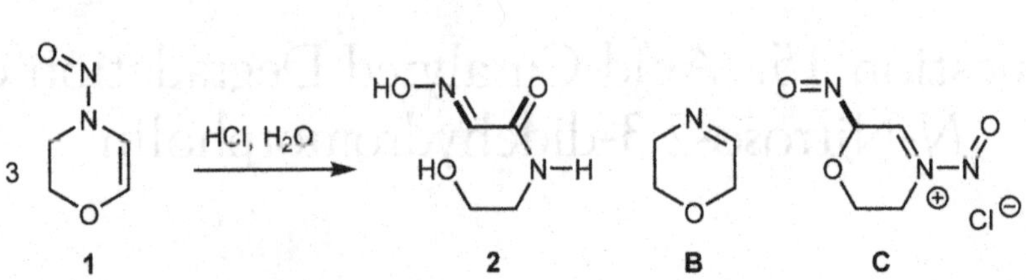

KEY STEPS EXPLAINED

We begin by noting that the authors proposed a peculiar mechanism to rationalize the production of *N*-(2-hydroxyethyl)-2-oximinoethanamide **2** from *N*-nitroso-2,3-didehydromorpholine **1**. Essentially, this mechanism requires three equivalents of **1** and as we see from the balanced chemical equation, the by-products **B** (neutral) and **C** (salt) are formed in the process. We note that these by-products could conceivably continue reacting to form **2**. Moreover, based on a previous report where it was hypothesized that compounds similar to **1** undergo intermolecular nitroso group migration,[2] the authors decided to conduct a crossover experiment to see whether the same occurred in this transformation (Scheme 2). In this experiment, an equimolar mixture of dideuterated **1** and ^{15}N-labeled **1** was subjected to an acidic ethereal solution. The product mixture was shown to consist of all four possible 5,6-dihydro-[1,4]oxazin-2-one oxime products isolated in near equal proportion after quenching of the reaction

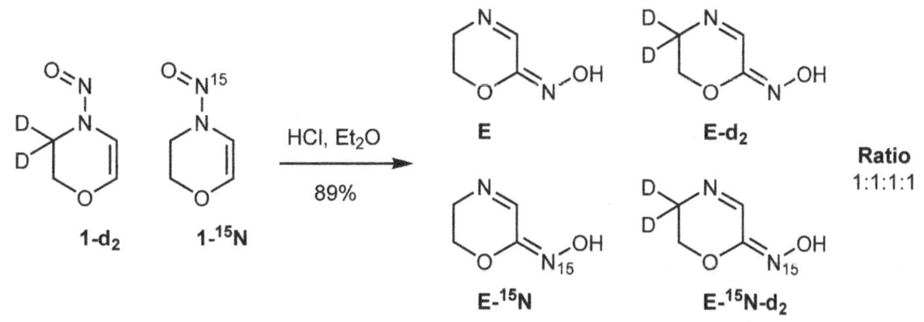

SCHEME 2

mixture with aqueous sodium bicarbonate solution. This result confirmed that crossover indeed occurred and indicated that the rearrangement mechanism must involve an intermolecular nitroso group migration. This is why in their mechanism the authors invoked three equivalents of the starting material **1**.

Therefore, the first part of the mechanism begins with nitroso enamine formation through protonation of the olefinic C=C bond of **1** to form **A**. Intermediate **A** then reacts with another equivalent of **1** essentially undergoing a nitroso group transfer which forms **B** as a by-product together with intermediate **C**. Reaction of **C** with another equivalent of **1** leads to formation of **D** and of **C** as a by-product. We note that once **C** is formed, it can continue to react in a feedback loop with any equivalents of **1** to form **D** while regenerating **C** in the same mechanistic step. Therefore, it is somewhat incorrect to say that for every three equivalents of **1**, there is a 1:1:1 production of **2**, **B**, and **C** as is depicted in the balanced chemical equation. Indeed, if there were 5 equivalents of **1** used, according to the authors' mechanism, there would be a 3:1:1 production of **2**, **B**, and **C**, respectively. Moving on, once **D** is formed, the authors propose that it undergoes a facile prototropic rearrangement to form oximinoimine **E** (in an equilibrium which favors formation of **E**). We note that oximinoimine **E** corresponds to the structure of the products observed in the crossover experiment where the reaction was stopped before hydration could take place. In acidic medium, intermediate **E** undergoes protonation to form iminium ion **F** which is attacked by water to form **G** which undergoes a proton shift to form oxime **H**. A [1,3] prototropic shift causes formation of hydroxylamine **I** which undergoes ketonization to form **J** which through ring opening via **K** followed by deprotonation can furnish the final oximinoethanamide product **2**. We note that the authors did not undertake a theoretical study to support their proposed mechanism.

Furthermore, we also note that the authors' mechanism is unusually circuitous and unnecessarily complex having a balanced chemical equation that does not have all neutral reactants and products. In addition, the stoichiometry is inconsistent with that described in the experimental procedure which involved equimolar amounts of substrate **1** and hydrochloric acid. We therefore believe a much simpler mechanism can be postulated which has fewer steps and is consistent with both the reaction stoichiometry and the results of the crossover experiment (see Scheme 3). Thus, we propose that simultaneous N—N bond cleavage and protonation of **1** provides nitrosonium ion and intermediate **B**, the latter of which undergoes a [1,3] protic shift to form intermediate **L** which recaptures nitrosonium ion to generate intermediate **M**. This intermediate undergoes a second [1,3] protic shift to form oxime **F** which undergoes

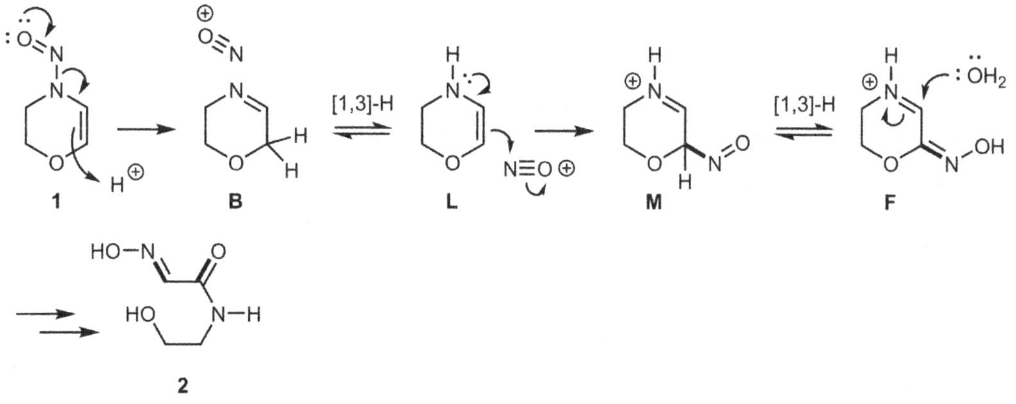

SCHEME 3

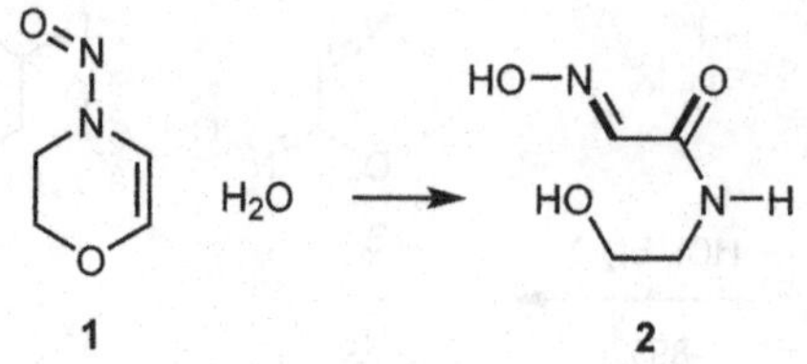

SCHEME 4

deprotonation via base quenching to form **E** (which was observed in the crossover experiment). Continued hydration of **E** via **F** leads to intermediate **G** which proceeds to product **2** seen previously. Thus the overall balanced chemical equation for this simplified mechanism (see Scheme 4) is simply the result of an acid-catalyzed hydration of **1** involving 1 mol of substrate and 1 mol of water.

We also show in Scheme 5 the expected results of the crossover experiment based on the 4 possible intermediates generated upon acid induced cleavage of **1-d$_2$** and **1-^{15}N**. It thus becomes obvious that once the 2 possible nitrosonium ions and the 2 possible 3,4-dihydro-2H-[1,4]oxazine intermediates **L** are generated there are 4 ways they could reassemble to give the 5,6-dihydro-[1,4]oxazin-2-one oxime **E** products in equal probability. This problem is an excellent example of illustrating the compatibility of a proposed mechanism with its associated balanced chemical equation which must be consistent with experimental conditions. Furthermore, it reinforces the strict condition that an overall balanced chemical equation for a reaction must be the sum of the elementary steps in its proposed mechanism. We also note that the newly proposed mechanism avoids some of the conundrums discussed by the authors in their original article.

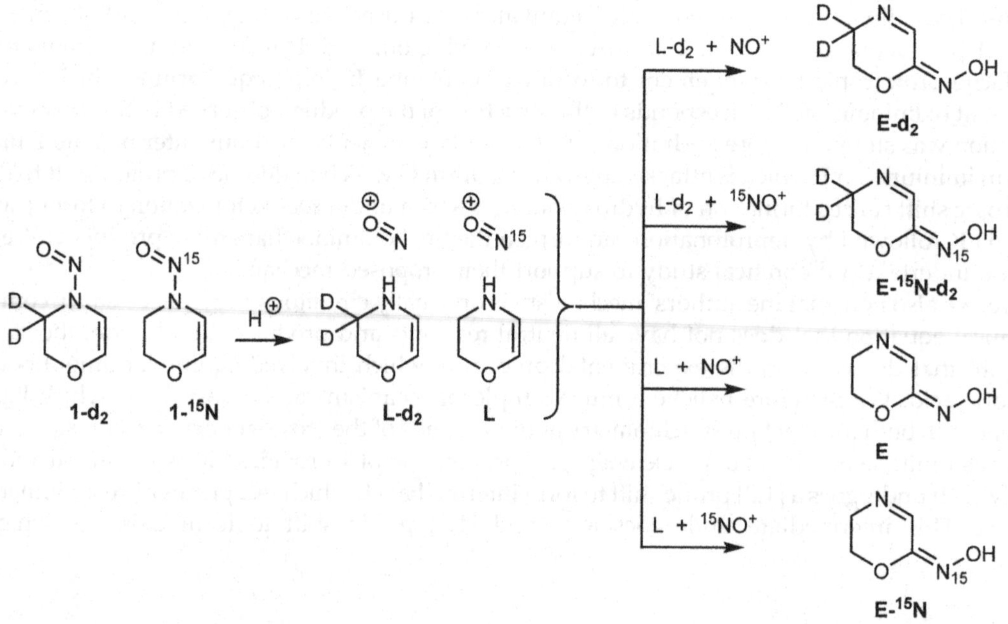

SCHEME 5

ADDITIONAL RESOURCES

For further examples of this type of chemistry along with extensive discussion of mechanism by the same authors, see Refs. 3–8.

References

1. Loeppky RN, Xiong H. Acid-catalyzed rearrangements of *N*-nitrosodehydromorpholine. *J Org Chem*. 1995;60(17):5526–5531. https://doi.org/10.1021/jo00122a036.
2. Lyle RE, Krueger WE, Gunn VE. Reaction of 1-nitroso-1,2,3,4-tetrahydropyridine with mineral acids. *J Org Chem*. 1983;48(20):3574–3575. https://doi.org/10.1021/jo00168a046.
3. Loeppky RN, Shi J, Barnes CL, Geddam S. A diazonium ion cascade from the nitrosation of tolazoline, an imidazoline-containing drug. *Chem Res Toxicol*. 2008;21(2):295–307. https://doi.org/10.1021/tx700317g.

4. Loeppky RN, Shi J. N-nitrosotolazoline: decomposition studies of a typical N-nitrosoimidazoline. *Chem Res Toxicol.* 2008;21(2):308–318. https://doi.org/10.1021/tx700318k.

5. Loeppky RN, Sukhtankar S, Gu F, Park M. The carcinogenic significance of reactive intermediates derived from 3-acetoxy- and 5-acetoxy-2-hydroxy-N-nitrosomorpholine. *Chem Res Toxicol.* 2005;18(12):1955–1966. https://doi.org/10.1021/tx0502037.

6. Wu H, Loeppky RN, Glaser R. Nitrosation chemistry of pyrroline, 2-imidazoline, and 2-oxazoline: theoretical Curtin–Hammett analysis of retroene and solvent-assisted C–X cleavage reactions of α-hydroxy-N-nitrosamines. *J Org Chem.* 2005;70(17):6790–6801. https://doi.org/10.1021/jo050856s.

7. Teuten EL, Loeppky RN. The mechanistic origin of regiochemical changes in the nitrosative N-dealkylation of N,N-dialkyl aromatic amines. *Org Biomol Chem.* 2005;3:1097–1108. https://doi.org/10.1039/B418457B.

8. Loeppky RN, Elomari S. N-alkyl-N-cyclopropylanilines as mechanistic probes in the nitrosation of N,N-dialkyl aromatic amines. *J Org Chem.* 2000;65(1):96–103. https://doi.org/10.1021/jo991104z.

Question 16: A New Route to 4-Amino-3-arylcinnolines

In 1995 Kiselyov described a new synthesis of 4-amino-3-arylcinnoline **3** which consisted of first reacting 2-trifluoroacetylthiophene **1** with phenylhydrazine to form hydrazone **2** and then treating this product with 5 equivalents of KHMDS (potassium hexamethyldisilazane) in THF (tetrahydrofuran) at −78°C followed by slow warming and stirring for several hours with subsequent work-up (ether extraction and wash with brine, Scheme 1).[1] Suggest a mechanism for this transformation.

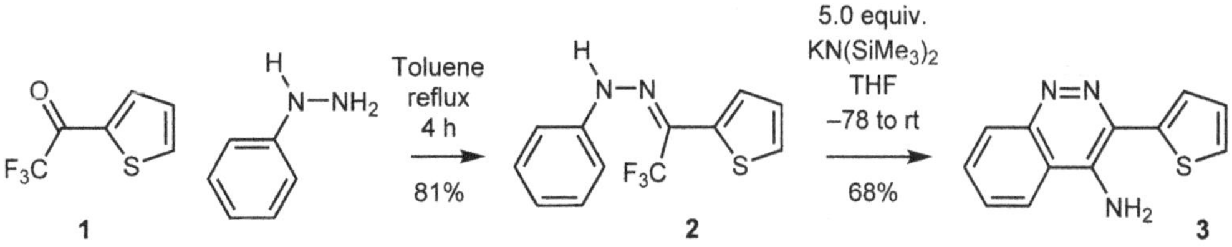

SCHEME 1

SOLUTION: 2 TO 3 (PATH A)

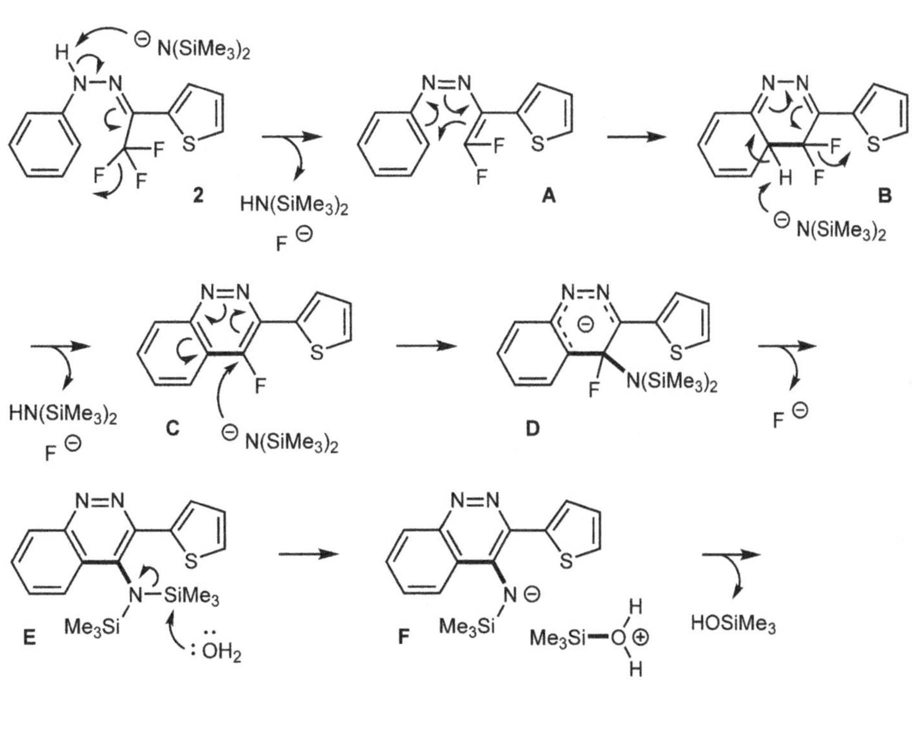

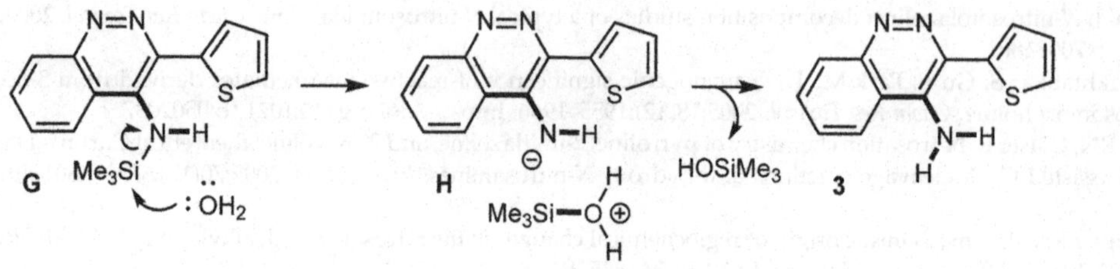

SOLUTION: 2 TO 3 (PATH B)

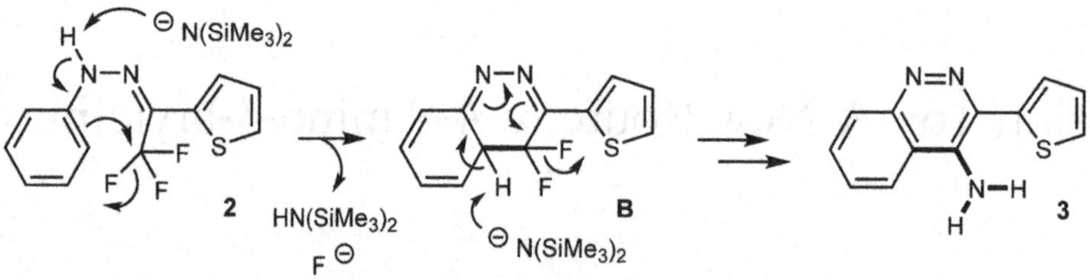

BALANCED CHEMICAL EQUATION

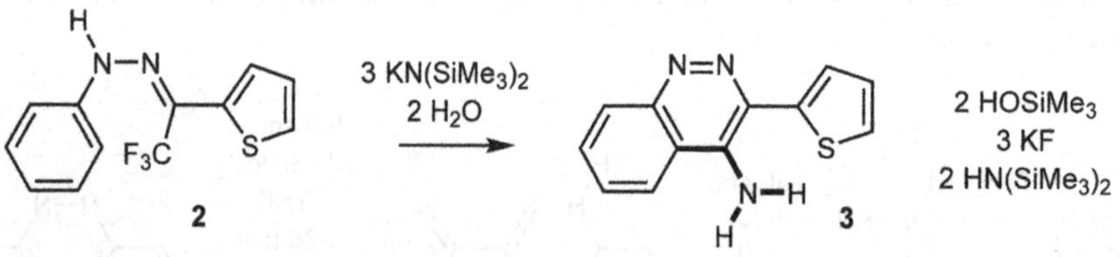

KEY STEPS EXPLAINED

The mechanism proposed for the transformation of **2** to **3** begins with two possible paths which converge to intermediate **B**. In path *a*, hydrazone **2** is deprotonated by one equivalent of hexamethyldisilazane (HMDS) to form intermediate **A** after elimination of fluoride ion. This intermediate undergoes a 6π electrocyclization to form intermediate **B**. Alternatively, intermediate **B** can also be formed directly from **2** (path *b*) if electron density flows in the opposite direction to cause simultaneous deprotonation, cyclization, and elimination of fluoride ion. Once **B** is formed, it is deprotonated by a second equivalent of HMDS to form intermediate **C** with elimination of another fluoride ion. A third equivalent of HMDS adds nucleophilically to **C** to generate the Meisenheimer complex **D** which eliminates a third equivalent of fluoride ion to form intermediate **E**. The two trimethylsilyl groups on intermediate **E** are then successively hydrolyzed by two water molecules to form eventually product **3** and two equivalents of trimethylsilanol by-product. We note that the ring construction mapping for this transformation is [6 + 0]. We also note that the authors did not provide additional experimental or theoretical evidence to support their proposed mechanism.

ADDITIONAL RESOURCES

We recommend two articles which discuss the mechanisms of interesting transformations that involve rearrangements with proposed trifluoromethyl group migrations.[2,3]

References

1. Kiselyov AS. Trifluoromethyl group in the synthesis of heterocyclic compounds: new and efficient synthesis of 3-aryl-4-aminocinnolines. *Tetrahedron Lett*. 1995;36(9):1383–1386. https://doi.org/10.1016/0040-4039(95)00005-W.
2. Rulev AY, Muzalevskiy VM, Kondrashov EV, et al. Reaction of α-bromo enones with 1,2-diamines. Cascade assembly of 3-(trifluoromethyl)piperazin-2-ones via rearrangement. *Org Lett*. 2013;15(11):2726–2729. https://doi.org/10.1021/ol401041f.
3. Saloutina LV, Zapevalov AY, Slepukhin PA, Kodess MI, Saloutin VI, Chupakhin ON. Novel route of the reaction of trifluoromethyl-containing N-methyl(4-ethoxyphenyl)imidazolidin-2-ones with urea. *Russ Chem Bull*. 2016;65(2):473–478. https://doi.org/10.1007/s11172-016-1324-0.

Question 17: Base-Induced Cyclizations of *o*-Ethynylaryl-Substituted Benzyl Alcohols

In 1995 American chemists showed that treatment of the *o*-ethynylaryl-substituted benzyl alcohols **1** (R = H, OMe, NO₂) with base resulted in exclusive formation of the 5-*exo-dig* isobenzofuran products **2** (Scheme 1).[1] Conversely, ester analog **3** led to the apparent 6-*endo-dig* 1*H*-2-benzopyran derivative **5** in a process which involved the intermediacy of the expected 5-*exo-dig* isobenzofuran **4** according to ¹H NMR studies. Suggest a mechanistic explanation for these results.

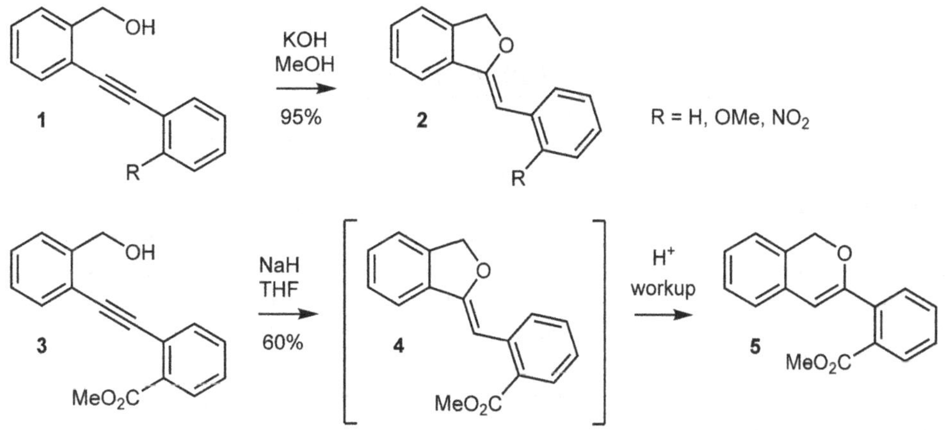

SCHEME 1

SOLUTION: 1 TO 2

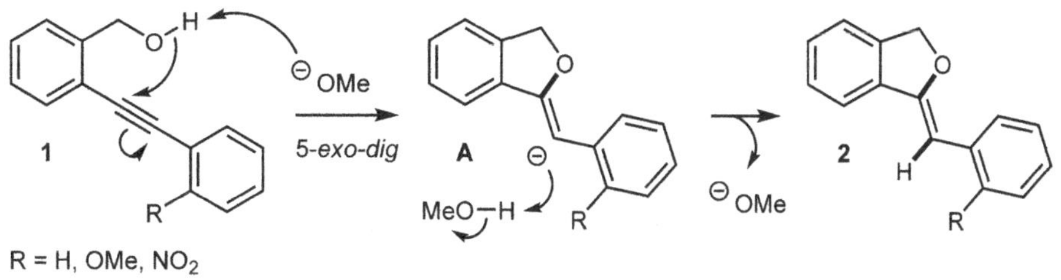

SOLUTION: 3 TO 5

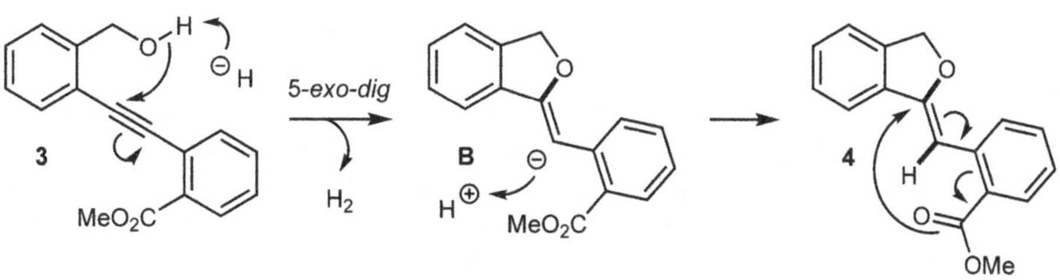

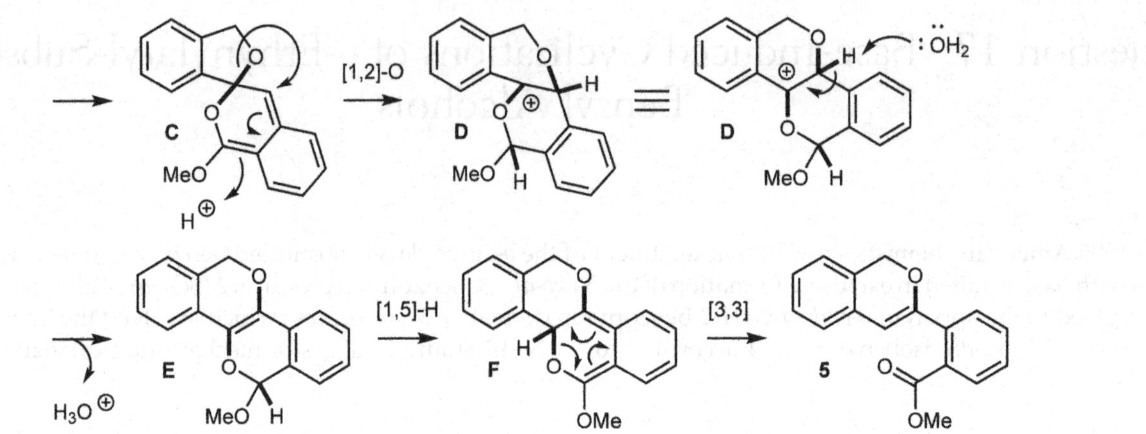

KEY STEPS EXPLAINED

The mechanism proposed for the transformation of **1** to **2** involves a 5-*exo-dig* cyclization of the alkoxide (generated in situ via deprotonation of the alcohol group of **1**) onto the alkyne group to form the carbanion **A** which upon protonation generates the isobenzofuran product **2**. By contrast, the transformation of **3** to **5** is not an apparent 6-*endo-dig* cyclization but involves the same 5-*exo-dig* cyclization-protonation sequence seen before which forms the isobenzofuran intermediate **4**. It is proposed that this intermediate then undergoes an acid-catalyzed rearrangement to 1*H*-2-benzopyran **5** during the acidic work-up phase of the reaction. Essentially, because of the presence of the *ortho* carbonyl group, **4** undergoes an intramolecular cyclization to form the [6,5]-spiro ketal intermediate **C**. This intermediate undergoes a [1,2]-O shift with ring expansion to form carbocation **D** after protonation. Deprotonation of **D** forms **E** which undergoes a [1,5] protic shift to generate **F** which finally undergoes a [3,3] sigmatropic cycloreversion to give the final 1*H*-2-benzopyran derivative **5**. We note that the authors could not isolate intermediate **4** but inferred its existence indirectly by carrying out ozonolysis on the crude reaction mixture instead of carrying out any acidic work-up. In this experiment they were able to isolate the corresponding phthalide **6** and aryl aldehyde **7** as products (Scheme 2). Thus the ozonolysis reaction intercepts intermediate **4**. We also point out that for aldehyde **7** to be isolated, the ozonolysis reaction must have been carried out under reductive work-up conditions even though the authors did not state this important fact in their article.

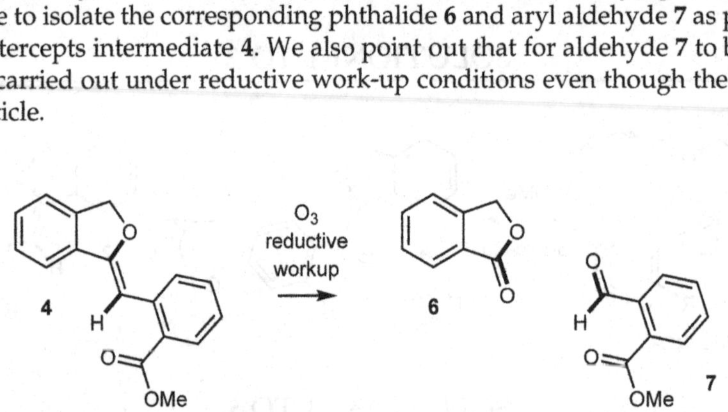

SCHEME 2

Last, the authors suggested that ^{18}O labeling experiments would support their mechanistic interpretation. We show in Scheme 3 the expected result of such an experiment if the alcohol group of **3** were labeled. Conversely, if the carbonyl oxygen atom of the ester group is labeled instead, then the label will be retained in the ester group in product **5**. We note that such an experiment would be most useful for differentiating between two mechanisms whereas in this case the authors proposed only one mechanism.

We note that the authors did not report any labeling study results for this transformation nor any additional experimental or theoretical evidence to support their proposed mechanism.

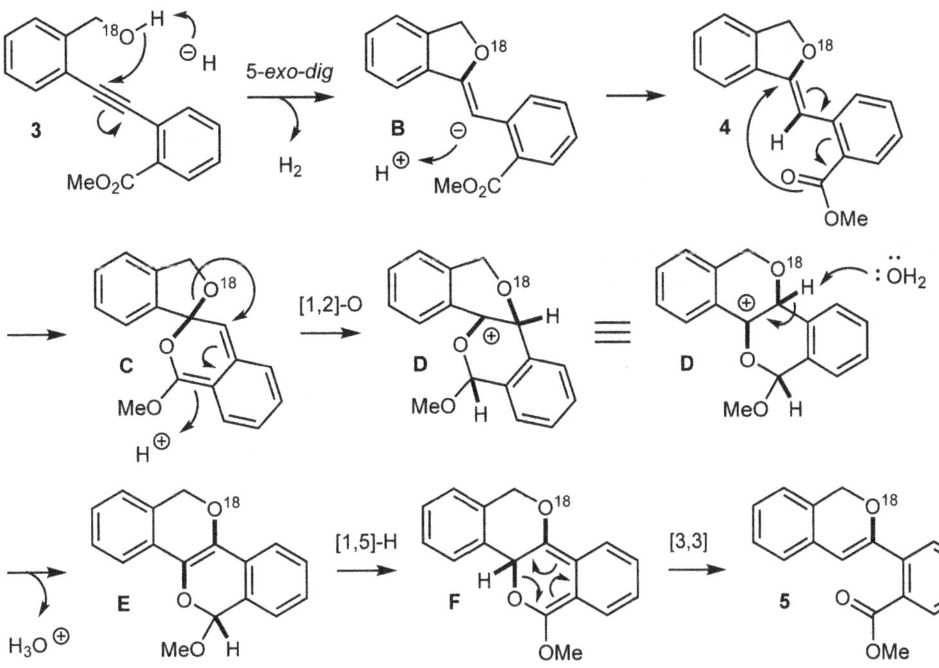

SCHEME 3

ADDITIONAL RESOURCES

We begin by referring the reader to two review articles that cover Baldwin's rules for ring closures involving alkyne groups.[2,3] We also recommend further work by Padwa et al. that provides further insight into this[4] and other transformations,[5,6] even though no labeling experiments are reported. We also recommend a recent article that describes ester group participation during the course of bromocyclizations of o-alkynylbenzoates.[7]

References

1. Weingarten MD, Padwa A. Intramolecular addition to an unactivated carbon-carbon triple bond via an apparent 6-endo digonal pathway. *Tetrahedron Lett.* 1995;36(27):4717–4720. https://doi.org/10.1016/0040-4039(95)00891-F.

2. Alabugin IV, Gilmore K, Manoharan M. Rules for anionic and radical ring closure of alkynes. *J Am Chem Soc.* 2011;133:12608–12623. https://doi.org/10.1021/ja203191f.

3. Gilmore K, Alabugin IV. Cyclizations of alkynes: revisiting Baldwin's rules for ring closure. *Chem Rev.* 2011;111:6513–6556. https://doi.org/10.1021/cr200164y.

4. Padwa A, Krumpe KE, Weingarten MD. An unusual example of a *6-endo-dig* addition to an unactivated carbon-carbon triple bond. *J Org Chem.* 1995;60(17):5595–5603. https://doi.org/10.1021/jo00122a047.

5. Padwa A, Kassir JM, Semones MA, Weingarten MD. A tandem cyclization-onium ylide rearrangement-cycloaddition sequence for the synthesis of benzo-substituted cyclopentenones. *J Org Chem.* 1995;60(1):53–62. https://doi.org/10.1021/jo00106a014.

6. Pellicciari R, Natalini B, Sadeghpour BM, et al. The reaction of α-diazo-β-hydroxy esters with boron trifluoride etherate: generation and rearrangement of destabilized vinyl cations. A detailed experimental and theoretical study. *J Am Chem Soc.* 1996;118(1):1–12. https://doi.org/10.1021/ja950971s.

7. Yuan S-T, Zhou H, Zhang L, Liu J-B, Qiu G. Synthesis of benzil-o-carboxylate derivatives and isocoumarins through neighboring ester-participating bromocyclizations of o-alkynylbenzoates. *Org Biomol Chem.* 2017;15:4867–4874. https://doi.org/10.1039/c7ob00845g.

Question 18: Trifluoromethylpyrroles From Trifluoromethyloxazolones

In 1993 chemists from Shanghai reacted trifluoromethyloxazolone **1** with α-chloroacrylonitrile **2** in methylene chloride at room temperature with addition of triethylamine base and stirring to form in 70% yield the trifluoromethyl-pyrrole product **3** (Scheme 1).[1] Suggest a mechanism for this transformation.

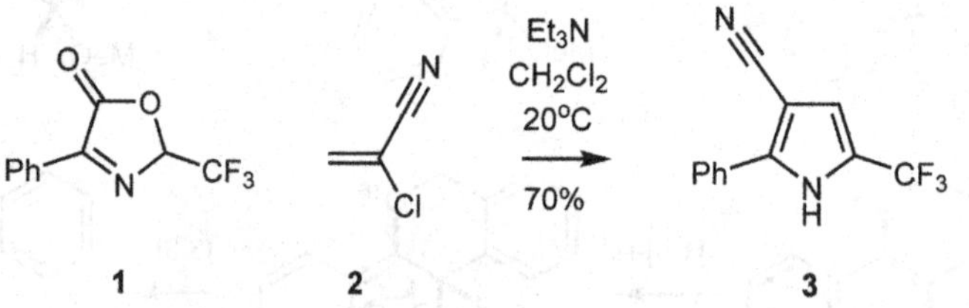

SCHEME 1

SOLUTION

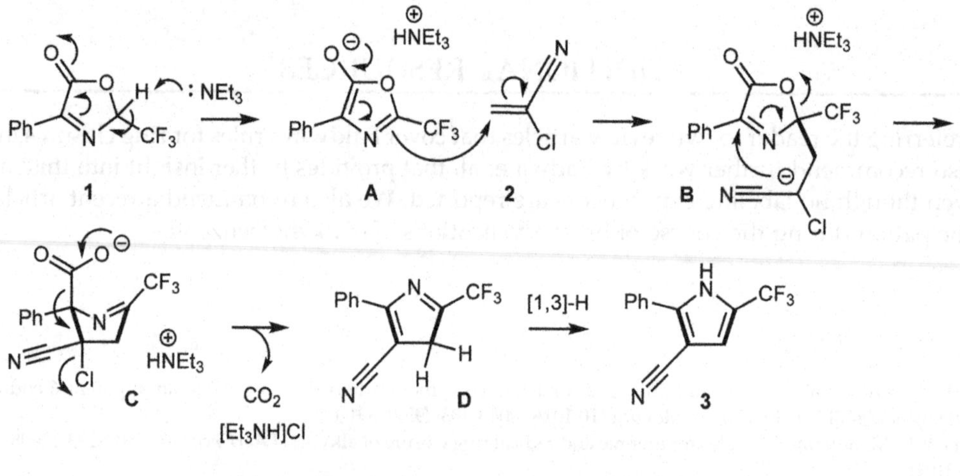

BALANCED CHEMICAL EQUATION

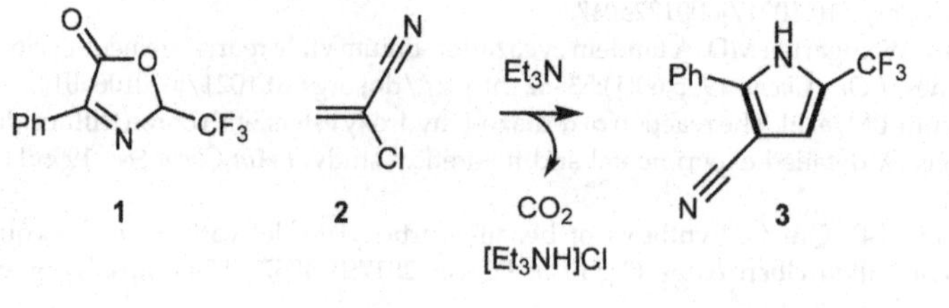

KEY STEPS EXPLAINED

The mechanism proposed to explain the transformation of **1** and **2** to **3** begins with deprotonation of **1** via the triethy-lamine base to form oxazolate **A**. This intermediate nucleophilically attacks the unsubstituted end of α-chloroacrylonitrile **2** to form carbanion **B**. This carbanion rearranges to the 3,4-dihydro-2*H*-pyrrole **C** which has a carboxylate group that initiates decarboxylation with elimination of chloride ion in the following step to form the 3*H*-pyrrole **D**. Lastly, pyrrole **D** undergoes a [1,3] protic shift to form the final pyrrole product **3** whose ring construction mapping is [3+2]. We note that the authors did not present any experimental or theoretical evidence for their proposed mechanism. Nevertheless, they stated that the efficiency of the reaction is improved by the following factors: (a) the olefin needs to have a good leaving group (Cl) and a strong electron withdrawing group (CN) and (b) strong bases for deprotonating 2H-oxazol-5-one **1** (DBU = 1,8-diazabicyclo[5.4.0]undec-7-ene, triethylamine) are also needed. The authors observed that in the absence of these factors being present, it was more probable for intermediate **B** to simply abstract a proton at the carbanion group to form the protonated oxazolone species as a product.

ADDITIONAL RESOURCES

We recommend two articles and a review which discuss novel approaches to the synthesis of pyrroles including some of their mechanistic details.[2–4]

References

1. Tian W, Luo Y, Chen Y, Yu A. A convenient synthesis of 2-trifluoromethylpyrroles via base-promoted cyclocondensation of trifluoromethyloxazolones with electron-deficient alkenes. *J Chem Soc Chem Commun*. 1993;1993:101–102. https://doi.org/10.1039/C39930000101.
2. Gao Q, Hao W-J, Liu F, et al. Unexpected isocyanide-based three-component bicyclizations for stereoselective synthesis of densely functionalized pyrano[3,4-c]pyrroles. *Chem Commun*. 2016;52(5):900–903. https://doi.org/10.1039/C5CC08071A.
3. Ghandi M, Jourablou A, Abbasi A. Synthesis of highly substituted pyrrole and dihydro-1H-pyrrole containing barbituric acids via catalyst-free one-pot four-component reactions. *J Heterocyclic Chem*. 2017;54(6):3108–3119. https://doi.org/10.1002/jhet.2924.
4. Estevez V, Villacampa M, Menendez JC. Recent advances in the synthesis of pyrroles by multicomponent reactions. *Chem Soc Rev*. 2014;43 (13):4633–4657. https://doi.org/10.1039/c3cs60015g.

Question 19: The Hooker Oxidation

In 1995 chemists from the University of California investigated the mechanism of the Hooker oxidation by means of a ^{13}C labeling study (Scheme 1).[1] Suggest a mechanism which is consistent with this study.

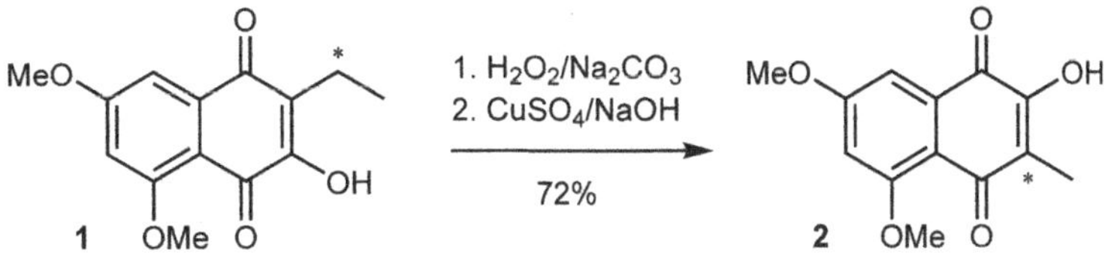

SCHEME 1

SOLUTION

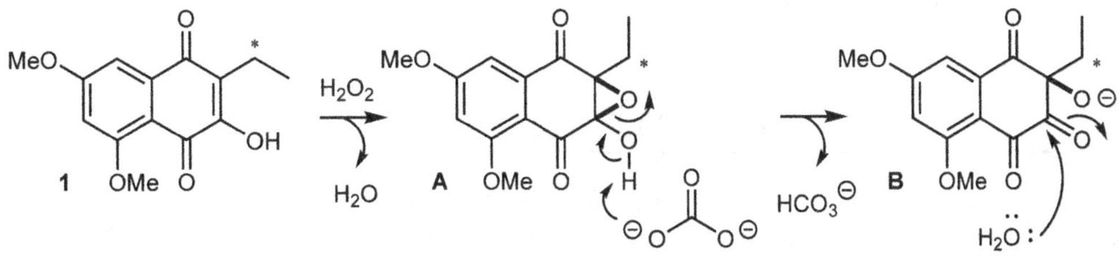

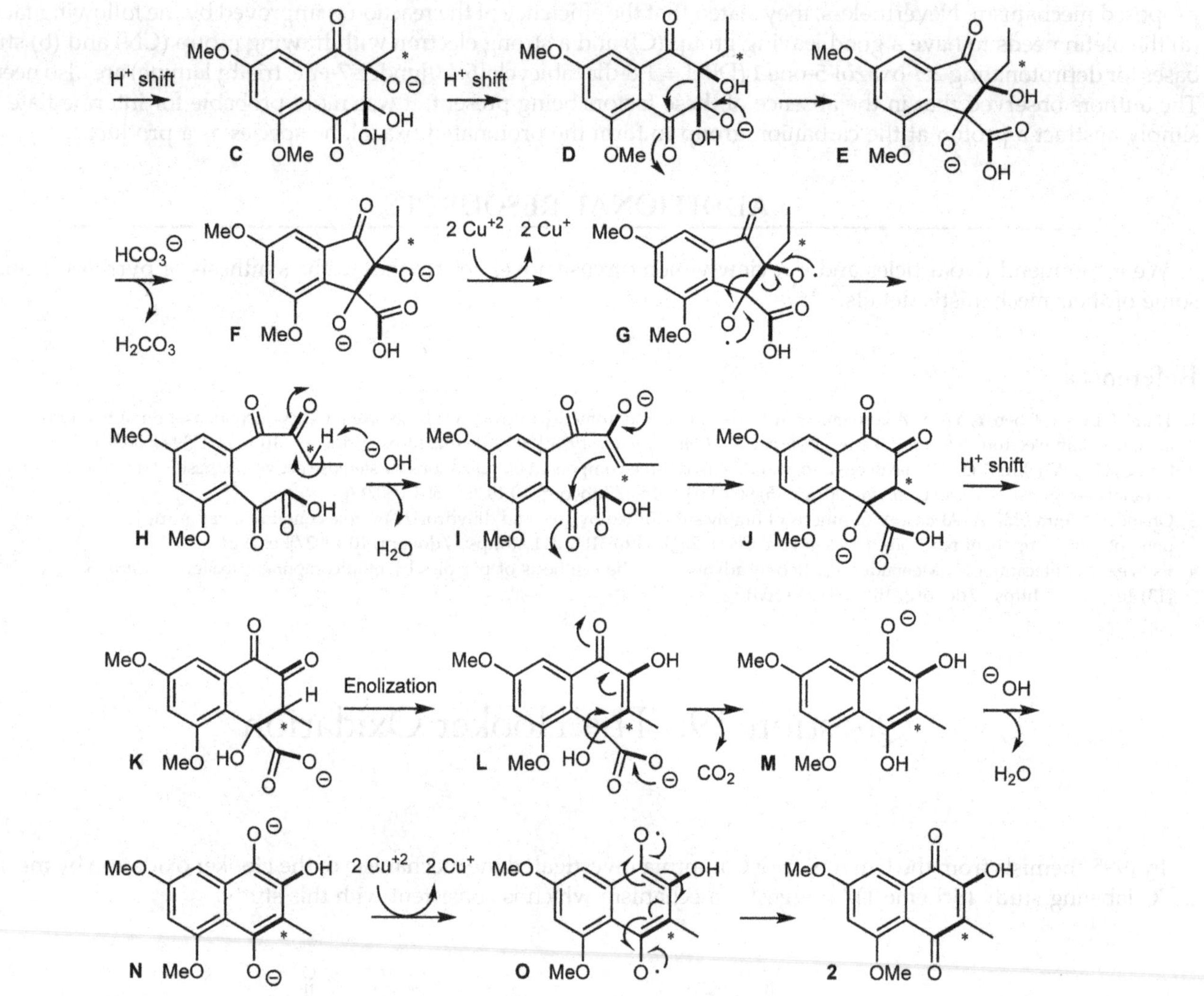

MeO — H⁺ shift — **C** OMe — H⁺ shift — **D** OMe — **E** MeO

HCO₃⁻ / H₂CO₃ — **F** MeO — 2 Cu⁺² 2 Cu⁺ — **G** MeO

MeO — **H** MeO — ⁻OH / H₂O — **I** MeO — **J** MeO — H⁺ shift

MeO — **K** MeO — Enolization — **L** MeO — **M** MeO / CO₂ — ⁻OH / H₂O

MeO — **N** MeO — 2 Cu⁺² 2 Cu⁺ — **O** MeO — **2** OMe

BALANCED CHEMICAL EQUATION

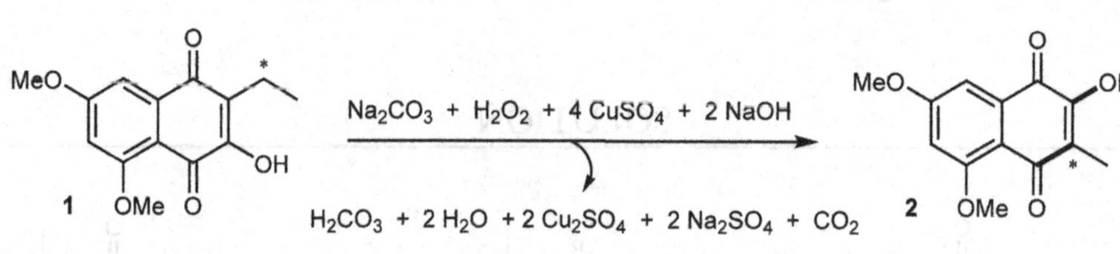

1 OMe   Na₂CO₃ + H₂O₂ + 4 CuSO₄ + 2 NaOH ⟶ H₂CO₃ + 2 H₂O + 2 Cu₂SO₄ + 2 Na₂SO₄ + CO₂   **2** OMe

$Na_2CO_3 + H_2O_2 + 4 CuSO_4 + 2 NaOH \longrightarrow H_2CO_3 + 2 H_2O + 2 Cu_2SO_4 + 2 Na_2SO_4 + CO_2$

KEY STEPS EXPLAINED

In order to elucidate the mechanism of the Hooker oxidation, Moore and coworkers carried out a ^{13}C labeling experiment and were able to characterize the product of the reaction using ^{13}C NMR, a carbon-carbon correlation experiment (INADEQUATE = incredible natural abundance double-quantum transfer experiment),[2,3] and carbon-hydrogen long-range coupling experiment (INEPT = insensitive nuclei enhanced by polarization transfer).[4,5] Here we present the mechanism showing the position of the ^{13}C (*) label throughout each step.

Therefore, the mechanism proposed for the Hooker oxidation begins with epoxidation of the starting naphthoquinone **1** via hydrogen peroxide to form epoxide **A** which undergoes carbonate ion base-assisted ring opening that forms the triketone alkoxide intermediate **B**. This step is followed by hydration of the carbonyl group *alpha* to the alkoxide moiety of **B** to form intermediate **C** which then undergoes an intramolecular prototropic shift that generates intermediate **D**. This intermediate undergoes a benzylic acid rearrangement to generate the key alkoxide intermediate **E**. Further deprotonation of **E** via bicarbonate ion leads to dialkoxide intermediate **F** which is sequentially oxidized by two one-electron transfer steps via two equivalents of Cu^{+2}. This sequence leads to the diradical intermediate **G** which undergoes ring opening via homolytic cleavage of the C—C bond connecting the alkoxyl radicals to form the triketocarboxylic acid intermediate **H**. Intermediate **H** is then converted to enolate **I** via hydroxide ion after which it undergoes an intramolecular aldol condensation to form intermediate **J** followed by an intramolecular proton transfer to generate carboxylate **K**. This carboxylate intermediate undergoes sequential enolization, decarboxylation, and base deprotonation to form dianion **N** which is oxidized to diradical **O** via two more equivalents of Cu^{+2} involving two one-electron transfer steps. In the final step diradical **O** isomerizes to the naphthoquinone product **2**.

We note that Moore et al. were able to isolate and characterize intermediate **3** (often called the Hooker intermediate) spectroscopically by carrying out a stepwise synthesis of naphthoquinone **2** as shown in Scheme 2. Here, intermediate **3** represents the protonated form of intermediate **E** as seen in the earlier mechanism.

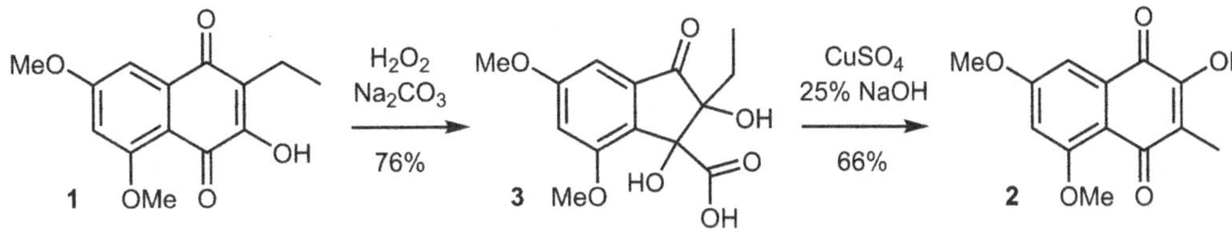

SCHEME 2

More recently, Eyong et al. were able to trap an analog of intermediate **B** as shown in Scheme 3.[6] In this work, 1,4-naphthoquinone **4** was epoxidized to epoxide **5** which was then hydrolyzed in the presence of acid to triketone **6** (**B**-like intermediate) which was trapped with *o*-phenylenediamine to form phenazine **7**. We note that these authors have not performed theoretical analysis on the proposed mechanism for the Hooker oxidation under any reaction conditions. Interestingly, when the reaction conditions are changed to potassium permanganate ($KMnO_4$) in basic solution containing sodium hydroxide (Scheme 4), the mechanism of the Hooker oxidation changes and one can actually propose two mechanisms (see Scheme 5 for mechanism 1 and Scheme 6 for mechanism 2) to explain this transformation. We also represent the balanced chemical equations for these mechanisms in Schemes 7 and 8, respectively.

We note that a secondary redox reaction occurs for both mechanisms in that the $[MnO_2]^-$ ion formed as a by-product reacts with one equivalent of $KMnO_4$ becoming oxidized to (Mn^{+4}) manganese dioxide (MnO_2) while the permanganate ions are reduced to (Mn^{+6}) manganate ions K_2MnO_4 (see Scheme 9). Therefore, the revised overall balanced chemical equations for these two mechanisms are represented in Scheme 10. Thus, according to both mechanisms, naphthoquinone **1** is deprotonated by hydroxide ion to form intermediate **P**. At this point, permanganate ion

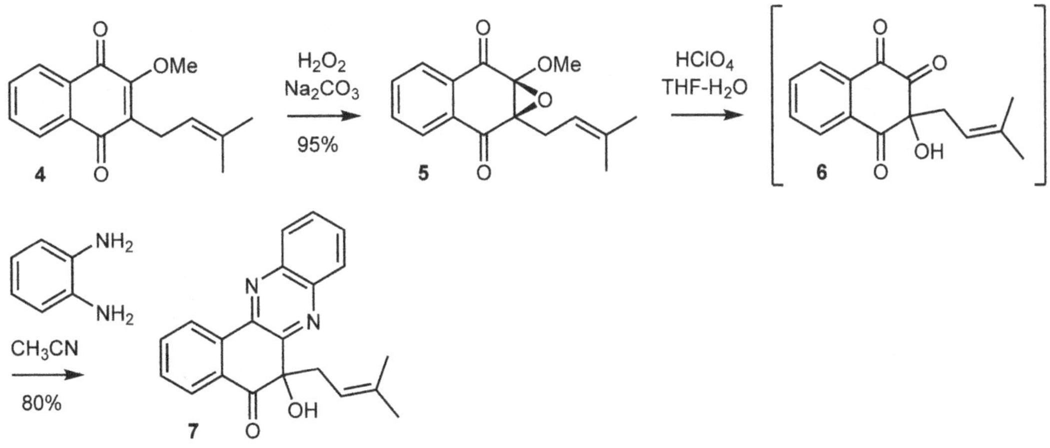

SCHEME 3

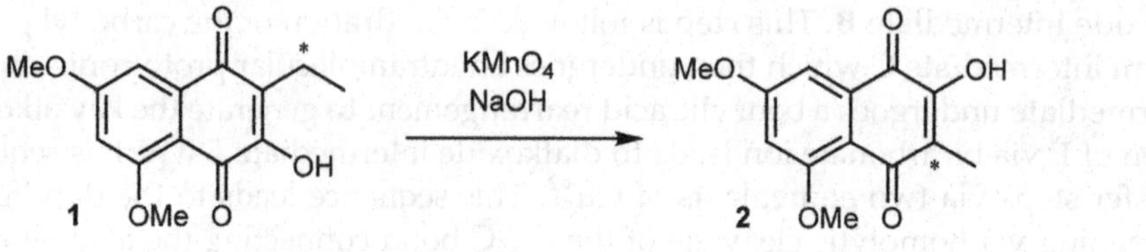

SCHEME 4

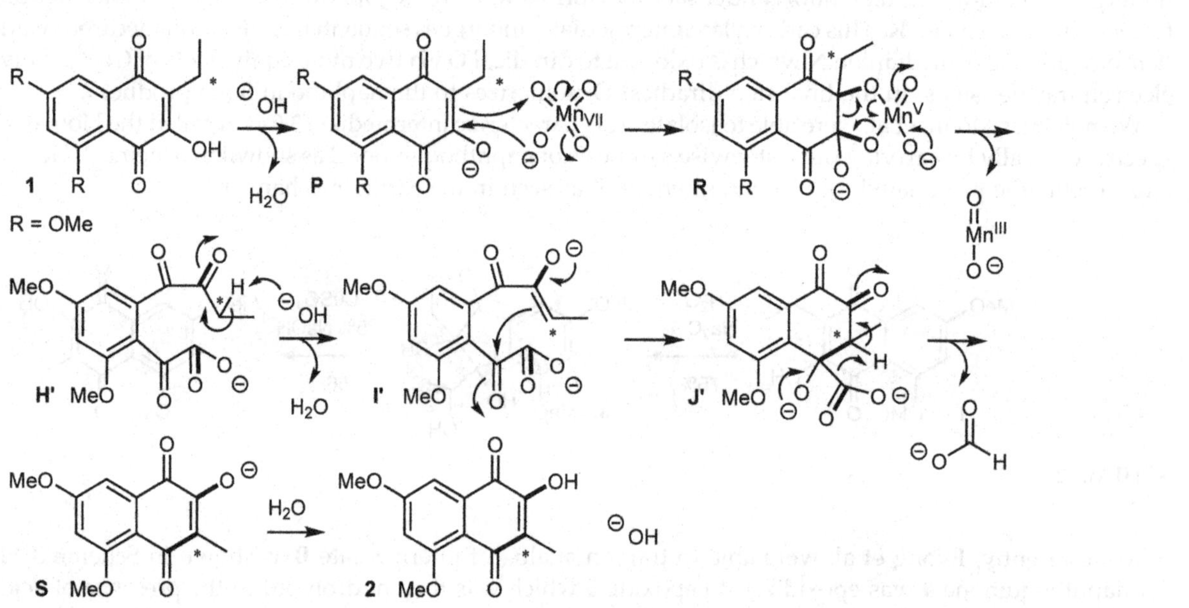

SCHEME 5 Mechanism 1.

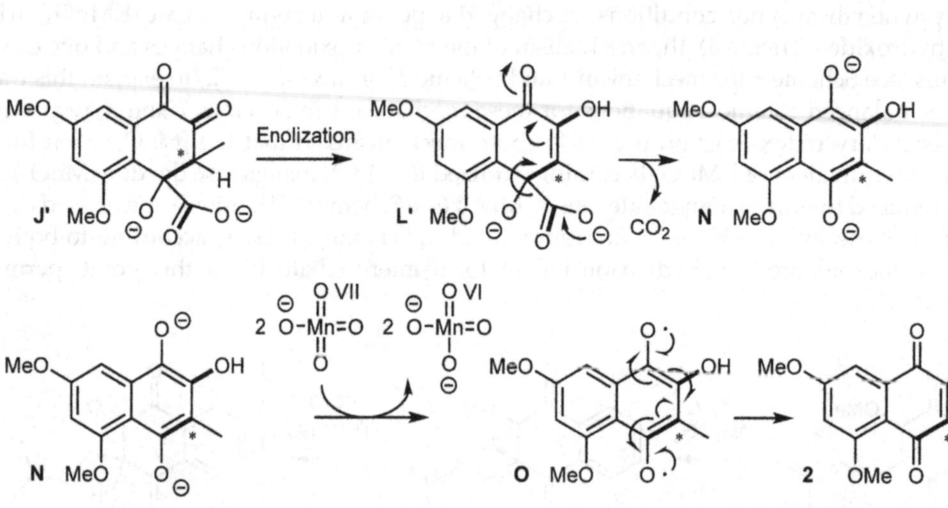

SCHEME 6 Mechanism 2: From **J'** to **2**.

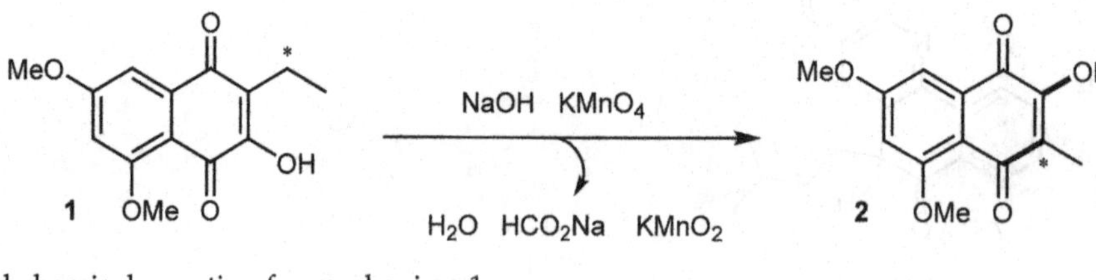

SCHEME 7 Balanced chemical equation for mechanism 1.

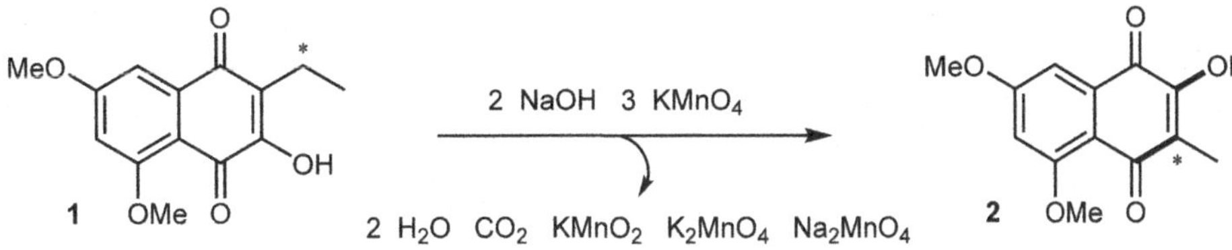

SCHEME 8 Balanced chemical equation for mechanism 2.

(Mn^{+7}) adds across the quinone C=C bond of **P** to form manganate ester **R** (Mn^{+5}) which fragments to $[MnO_2]^-$ anion (whose oxidation state is Mn^{+3}) and triketocarboxylate intermediate **H'**. Base-induced enolization of **H'** leads to enolate **I'** which then undergoes an intramolecular aldol condensation as seen before to form the key intermediate **J'**. This intermediate can undergo fragmentation in one of two ways. According to mechanism 1, **J'** fragments to formate ion and naphthoquinone alkoxide **S** which becomes protonated by water to form product **2**. According to mechanism 2, **J'** undergoes enolization followed by decarboxylation to form dianion **N** as seen before. This dianion is oxidized to diradical **O** via a two-step sequence involving one electron transfer processes where two equivalents of permanganate ion (Mn^{+7}) are utilized instead of two equivalents of Cu^{+2} as before. Finally, diradical **O** isomerizes to product **2**.

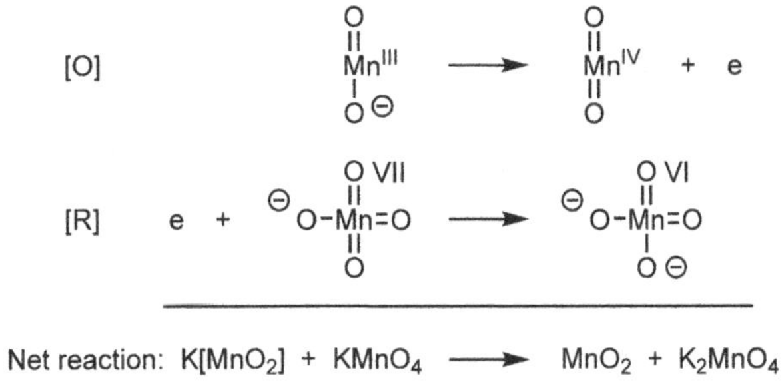

[O]

[R]

Net reaction: $K[MnO_2] + KMnO_4 \longrightarrow MnO_2 + K_2MnO_4$

SCHEME 9 Secondary redox reaction.

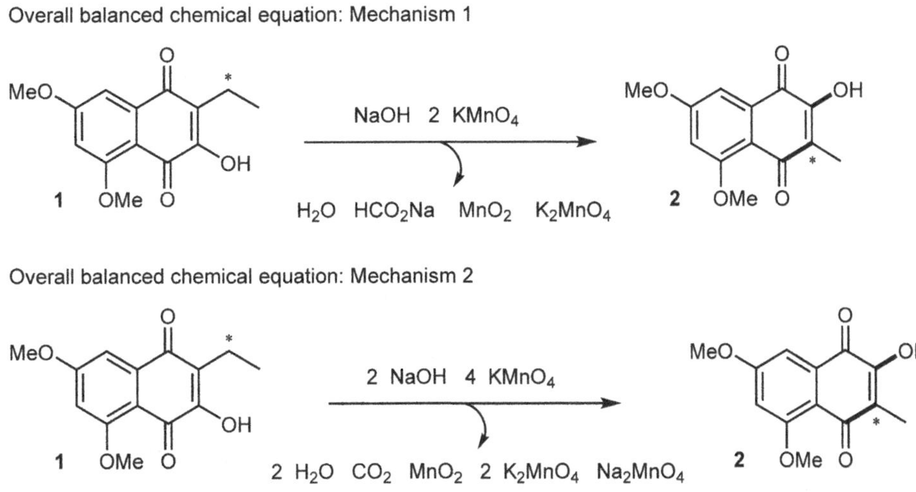

Overall balanced chemical equation: Mechanism 1

Overall balanced chemical equation: Mechanism 2

SCHEME 10

We note that apart from the different stoichiometries of reagents, the distinguishing feature between these two proposed mechanisms concerns the fate of the extruded carbon atom from the naphthoquinone substrate **1**. In mechanism 2, this carbon is extruded in the form of carbon dioxide and in mechanism 1 it is extruded in the form of formate. Therefore, we recommend experimental verification of by-products in order to elucidate which mechanistic pathway is operational for the Hooker oxidation. We also note that mechanism 1 can be applied to the transformation of **J** to **2** when copper sulfate is used as the oxidation reagent (a path which we did not illustrate in the solution to the problem). Lastly, we note that the ring construction mapping for the Hooker oxidation is [6 + 0]. Finally we note that these mechanistic proposals outline clearly all redox steps and possible intermediates along the proposed paths, which cited literature and books on named organic reactions do not present fully.[7]

ADDITIONAL RESOURCES

We also refer the reader to the original articles for the Hooker oxidation[8,9] and the 1948 article where a mechanism was proposed which was later supported by the ^{13}C labeling study analyzed in this problem.[10]

References

1. Lee K, Turnbull P, Moore HW. Concerning the mechanism of the Hooker oxidation. *J Org Chem.* 1995;60(2):461–464. https://doi.org/10.1021/jo00107a030.
2. Buddrus J. INADEQUATE experiment. In: Harris RK, Wasylishen RL, eds. *eMagRes.* John Wiley & Sons; 2007:1–10. https://doi.org/10.1002/9780470034590.emrstm0231.
3. Macomber RS. *A Complete Introduction to Modern NMR Spectroscopy.* New York: Wiley; 1998230–236.
4. Morris GA. INEPT. In: Harris RK, Wasylishen RL, eds. *eMagRes.* John Wiley & Sons; 2007:1–10. https://doi.org/10.1002/9780470034590.emrstm0238.
5. Macomber RS. *A Complete Introduction to Modern NMR Spectroscopy.* New York: Wiley; 1998206–209.
6. Eyong KO, Puppala M, Kumar PS, et al. A mechanistic study on the Hooker oxidation: synthesis of novel indane carboxylic acid derivatives from lapachol. *Org Biomol Chem.* 2013;11:459–468. https://doi.org/10.1039/C2OB26737C.
7. Wang Z, ed. *Comprehensive Organic Name Reactions and Reagents.* New York, NY: John Wiley & Sons, Inc.; 2010:1477–1480. https://doi.org/10.1002/9780470638859.conrr330
8. Hooker SC. On the oxidation of 2-hydroxy-1,4-naphthoquinone derivatives with alkaline potassium permanganate. *J Am Chem Soc.* 1936;58(7):1174–1179. https://doi.org/10.1021/ja01298a030.
9. Hooker SC, Steyermark A. On the oxidation of 2-hydroxy-1,4-naphthoquinone derivatives with alkaline potassium permanganate. Part II. Compounds with unsaturated side chains. *J Am Chem Soc.* 1936;58(7):1179–1181. https://doi.org/10.1021/ja01298a031.
10. Fieser LF, Fieser M. Naphthoquinone antimalarials. XII. The Hooker oxidation reaction. *J Am Chem Soc.* 1948;70(10):3215–3222. https://doi.org/10.1021/ja01190a005.

Question 20: Synthesis of 1,3-Disubstituted Pyrroles

In 1992 Russian chemists showed that 1,3-disubstituted pyrroles such as **4** can be prepared in overall yields of 55%–60% by the synthetic sequence shown in Scheme 1.[1] Elucidate the mechanism for the transformation of **3** to **4**.

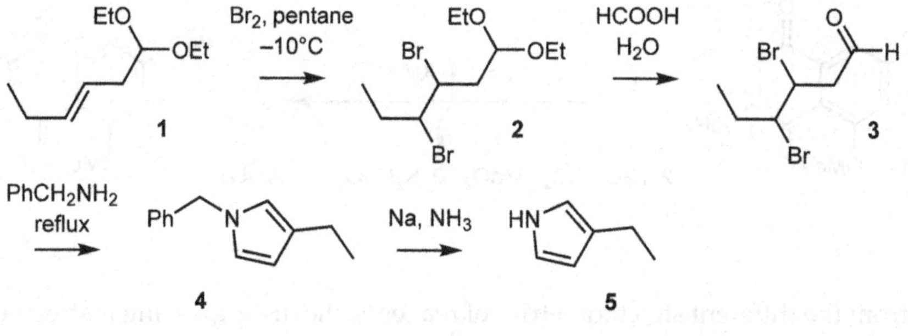

SCHEME 1

SOLUTION: 3 TO 4

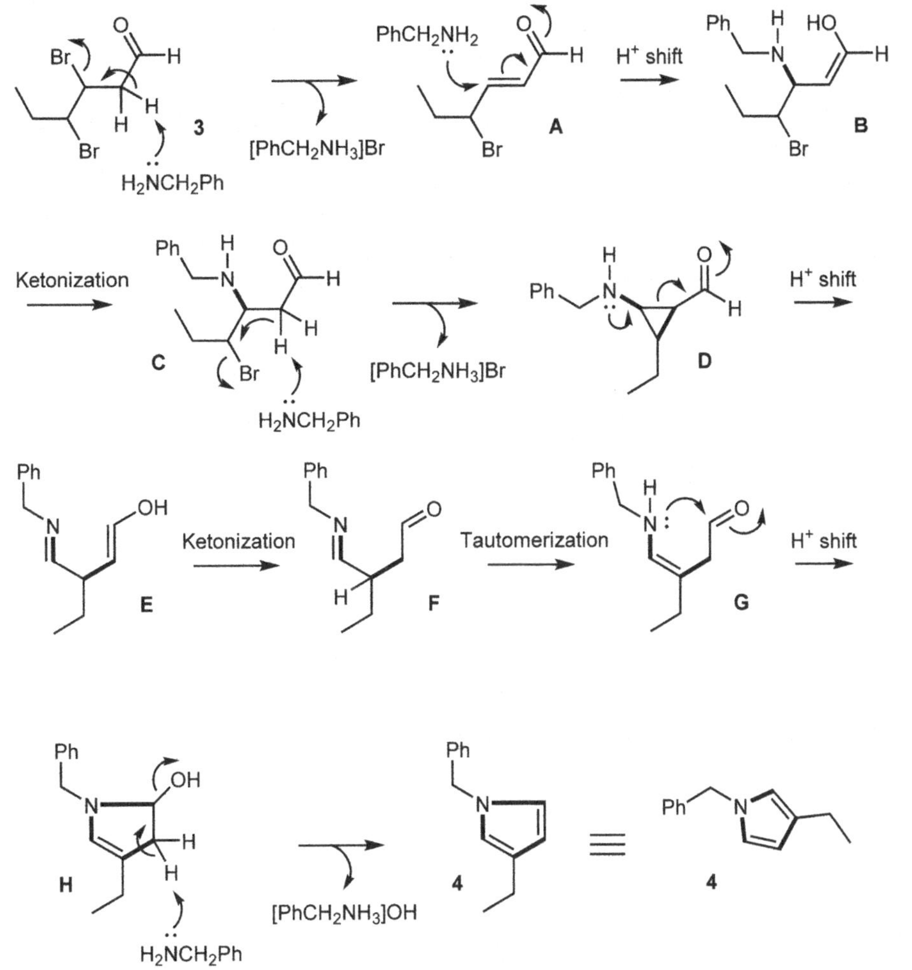

BALANCED CHEMICAL EQUATION

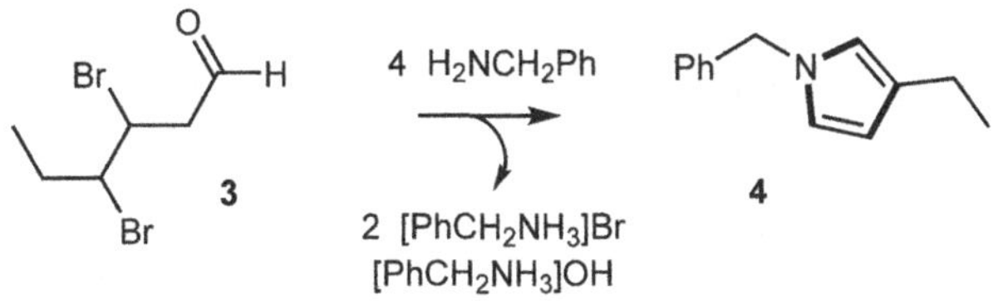

KEY STEPS EXPLAINED

We begin by noting that the synthetic sequence depicted in Scheme 1 represents an elegant approach by Kulinkovich et al. for the synthesis of substituted pyrroles such as **5**. Beginning with the dialkyl acetal **1**, the initial step constitutes dibromination of the C=C bond of **1** using bromine at low temperature to form **2**. In the second step, dibromoacetal **2** undergoes an acid-catalyzed hydrolysis using formic acid in water to generate the dibromoaldehyde product **3**. Next, the conversion of **3** to **4** is the most interesting aspect of the synthesis from the standpoint of strategy because it involves dehydrobromination of **3** using one equivalent of benzylamine to form the 4-bromo-hex-2-enal intermediate **A**. Next, it is proposed that this intermediate reacts with a second equivalent of benzylamine in a Michael fashion to form aldehyde **C** after a proton shift and ketonization sequence. Next, a third equivalent of benzylamine causes dehydrobromination of **C** to generate cyclopropane **D** which undergoes fragmentation to form the iminium enolate **E**. This intermediate ketonizes to **F** which tautomerizes to enamine **G**. This enamine then cyclizes to form the substituted pyrrole **H** after an intramolecular proton shift. Thanks to a fourth equivalent of benzylamine, intermediate **H** undergoes dehydration to form the substituted pyrrole product **4** whose ring construction mapping is [2+2+1]. We also note that the authors offered no direct experimental evidence for the intermediacy of cyclopropane intermediates in their proposed mechanism. Nevertheless, a cyclopropane intermediate such as **D** does explain the inherent rearrangement that must take place in this transformation. We thus illustrate the connectivity of the three key carbon atoms of structure **C** and how it relates to those in structures **D** and **G** (Scheme 2). We also observe that the C_a—C_b—C_c bond connectivity changes to C_a—C_c—C_b, indicating that the C_a—C_b bond is cleaved and a new C_a—C_c bond is made. We thus recommend that others should experimentally verify this postulated mechanism by attempting to isolate cyclopropane **D** and enamine **G** under low benzylamine concentrations or by demonstrating the change of skeletal connectivity through a ^{13}C labeling experiment. We also recommend carrying out a theoretical analysis to further support the proposed mechanism.

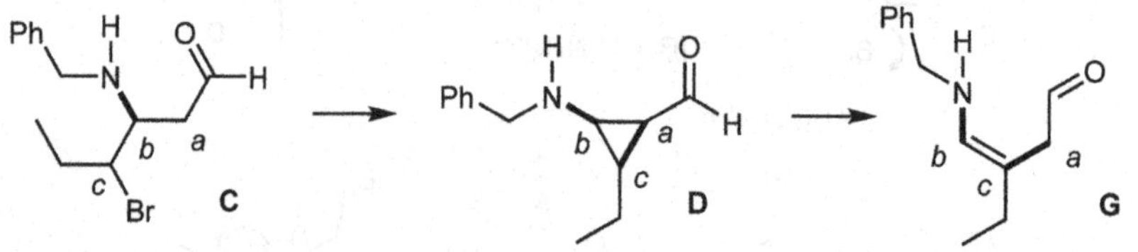

SCHEME 2

ADDITIONAL RESOURCES

We refer the reader to three articles which discuss the mechanisms of unexpected formations of pyrrole products, some of which also include proposed cyclopropane intermediates.[2-4]

References

1. Kulinkovich OG, Ali A, Sorokin VL. A new approach to the production of 3-alkylpyrroles. *Russ J Org Chem.* 1992;28(11):2316–2319.
2. Nakano H, Ishibashi T, Sawada T. Unexpected formation of novel pyrrole derivatives by the reaction of thioamide with dimethyl acetylenedicarboxylate. *Tetrahedron Lett.* 2003;44:4175–4177. https://doi.org/10.1016/S0040-4039(03)00929-8.
3. Shaabani A, Teimouri MB, Arab-Ameri S. A novel pseudo four-component reaction: unexpected formation of densely functionalized pyrroles. *Tetrahedron Lett.* 2004;45:8409–8413. https://doi.org/10.1016/j.tetlet.2004.09.039.
4. Augusti R. A mechanistic proposal for the formation of unexpected pyrroles in reactions of carboethoxycarbene with enaminones. *Heterocycl Commun.* 2001;7(1):29–32. https://doi.org/10.1515/HC.2001.7.1.29.

Question 21: Grob Fragmentation of a 1-Azaadamantane

In 1991 German chemists reacted the readily accessible 1-azaadamantane derivative **1** with thionyl chloride and formed the mono-*gem*-dichloro derivative **2** in 92% yield (Scheme 1).[1] When **2** was refluxed in aqueous ethanol containing 3 equivalents of triethylamine, the 5-chloro-2,4,6-trimethylresorcinol product **3** was achieved in 62% yield. Suggest a mechanism to explain this synthetic sequence.

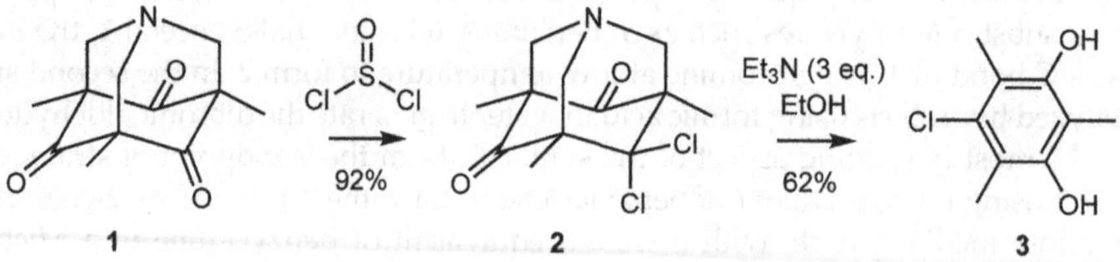

SCHEME 1

SOLUTION: 1 TO 2

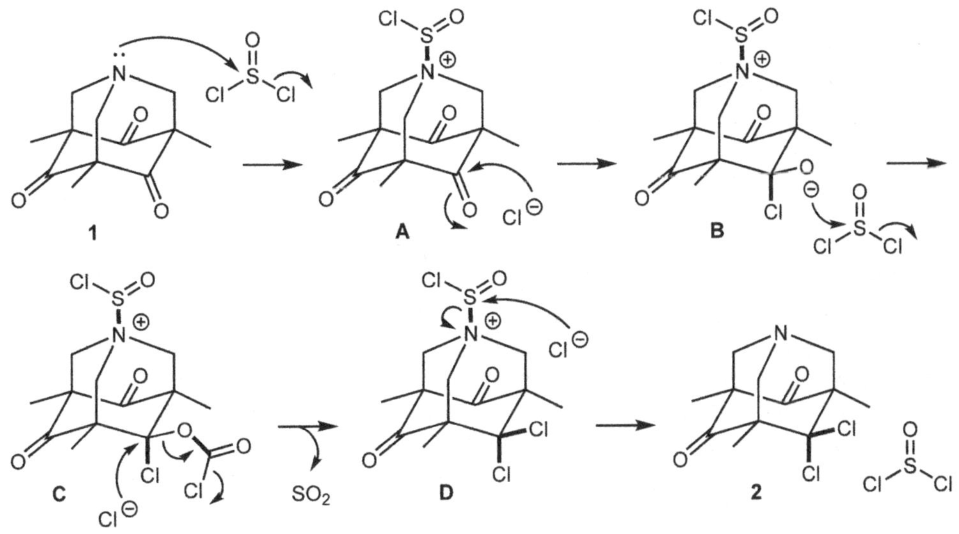

BALANCED CHEMICAL EQUATION

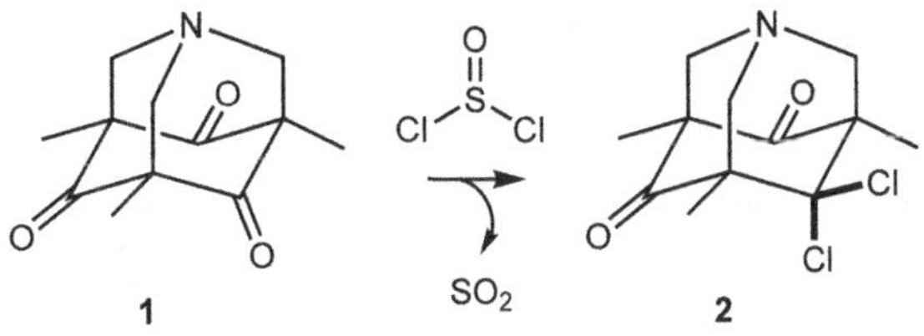

SOLUTION: 2 TO 3

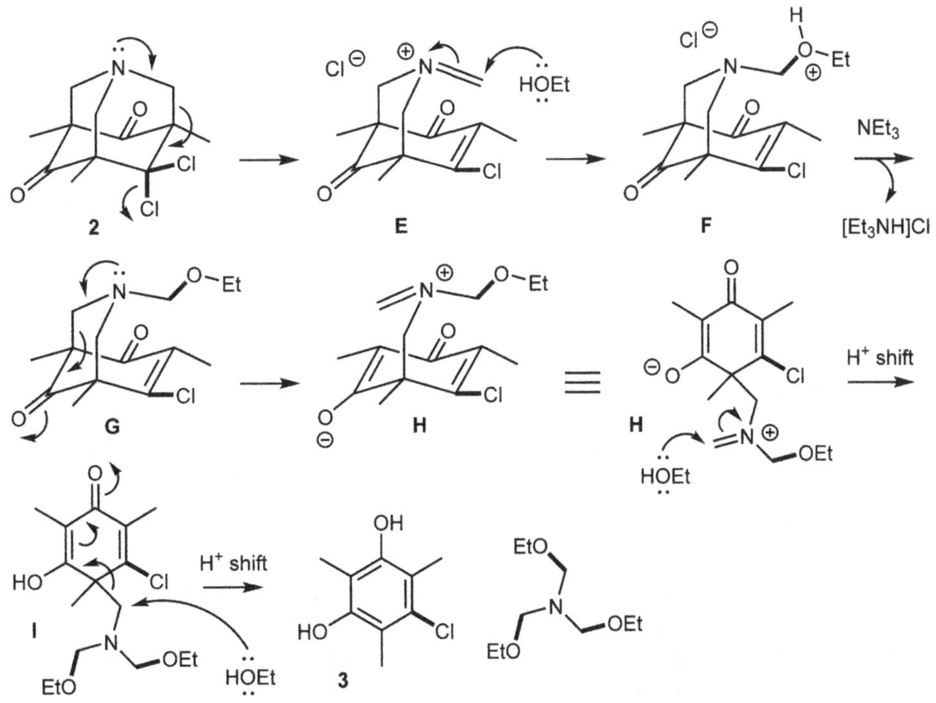

BALANCED CHEMICAL EQUATION

KEY STEPS EXPLAINED

The mechanism proposed by the authors to explain the synthetic sequence from **1** to **2** begins with sulfonation of the tertiary amine of **1** using one equivalent of thionyl chloride to form ammonium chloride **A**. This intermediate is attacked on one of the three identical carbonyl groups by chloride anion to form the zwitterionic intermediate **B**. The newly generated oxide anion of **B** attacks a second equivalent of thionyl chloride to form intermediate **C** which is attacked by chloride anion thus causing elimination of sulfur dioxide by-product to form the *gem*-dichloro intermediate **D**. Chloride anion then attacks the sulfonyl chloride group bonded to the nitrogen atom in **D** to form product **2** and regenerate thionyl chloride. To explain the subsequent transformation of **2** to **3**, the authors proposed an initial Grob fragmentation of **2** by means of electron delocalization from the central nitrogen atom with irreversible elimination of the *exo*-chloride to form iminium ion **E**. This intermediate is attacked by ethanol to form salt **F** which restores the lone pair of electrons on the nitrogen atom thus allowing for further fragmentation steps. Deprotonation of **F** via triethylamine base gives **G** which goes through a second Grob fragmentation sequence (i.e., ring opening, ethanol attack on the newly formed imine group followed by intramolecular proton shift) to form intermediate **I**. At this point, a third equivalent of ethanol attacks intermediate **I** to liberate resorcinol product **3** and to eliminate tris(ethoxymethyl) amine by-product which upon aqueous work-up ultimately fragments to ammonia and three equivalents each of ethanol and formic acid. We note that the authors did not provide additional experimental or theoretical evidence to support their proposed mechanism.

ADDITIONAL RESOURCES

We recommend two articles which discuss the mechanisms of transformations that involve unusual Grob fragmentations.[2,3]

References

1. Risch N, Langhals M, Hohberg T. Triple (Grob) fragmentation. Retro-Mannich reactions of 1-aza-adamantane derivatives. *Tetrahedron Lett.* 1991;32(35):4465–4468. https://doi.org/10.1016/0040-4039(91)80013-V.
2. Khan FA, Rao CN. Grob fragmentation of norbornyl α-diketones: a route to α-ketoenols and aromatic compounds. *J Org Chem.* 2011;76:3320–3328. https://doi.org/10.1021/jo200223a.
3. Xiong Q, Wilson WK, Matsuda SPT. An arabidopsis oxidosqualene cyclase catalyzes iridal skeleton formation by Grob fragmentation. *Angew Chem Int Ed.* 2006;45:1285–1288. https://doi.org/10.1002/anie.200503420.

Question 22: Selective Substitution of 3-Methoxypyrazine *N*-Oxide

During studies on pyrazines in 1993, Japanese chemists reacted 3-methoxypyrazine *N*-oxide **1** with diethylcarbamoyl chloride **2** and 4-methoxytoluene-α-thiol **3** in refluxing acetonitrile to form 2-methoxy-6-(4-methoxybenzylthio)-pyrazine **4** (Scheme 1).[1] The authors noted that **4** was formed in 60% yield when ZnBr$_2$ was used and in 21% yield when ZnBr$_2$ was absent. Suggest a mechanism for this transformation.

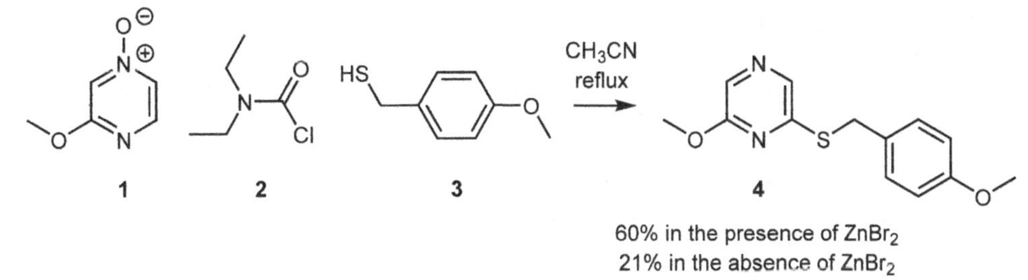

SCHEME 1

SOLUTION: 1, 2, AND 3 TO 4 (WHEN $ZnBr_2$ PRESENT)

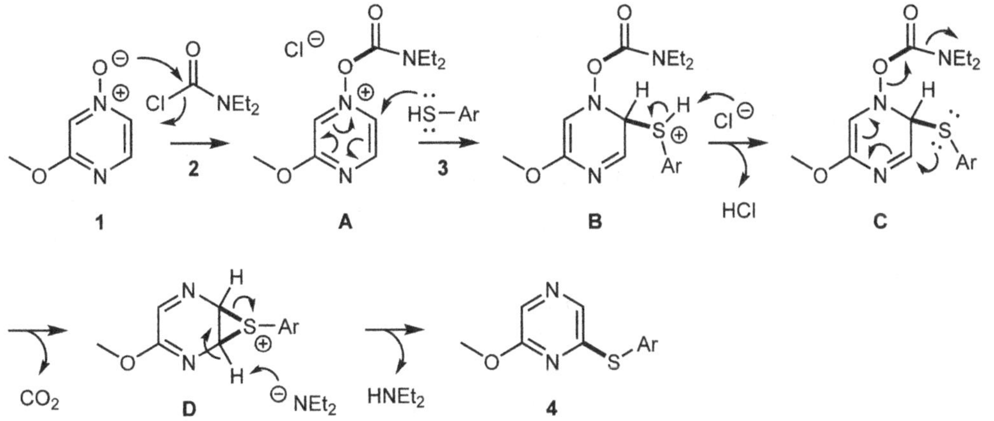

BALANCED CHEMICAL EQUATION

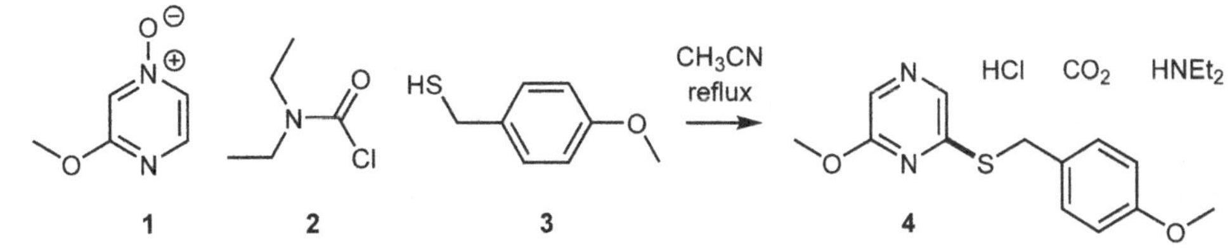

KEY STEPS EXPLAINED

The mechanism proposed by the authors to explain the formation of **4** from the thionation reaction involving **1**, **2**, and **3** begins with acylation of the *N*-oxide group of **1** by attack onto carbamoyl chloride **2** which forms salt **A**. This salt is then attacked at C_1 by thiol **3** to form intermediate **B** which undergoes deprotonation to form **C**. At this point the authors proposed that **C** undergoes cyclization to form the bicyclic thiiranium ion intermediate **D** after loss of the acyl group via decarboxylation and diethylamide ion elimination. Further deprotonation of **D** via the diethylamide ion leads to ring opening of the thiirane group to form the pyrazine product **4** after rearomatization.

We note that there are several interesting aspects of this transformation and its proposed mechanism which warrant further discussion. These concern the regioselective attack of thiol **3** onto C_1 of **A** to form **B** and the selective deprotonation of **D** to form **4**. In the first case, it is noteworthy that when the authors added Lewis acid $ZnBr_2$ to the reaction mixture, the yield of **4** was 60% and no other products were identified. Conversely, in the absence of Lewis acid, the authors isolated a mixture of products **4** (20% yield) and **5** (58% yield, Scheme 2). Product **6** which would have been derived by deprotonation of the other hydrogen atom as that depicted in the step **D** to **4** was not observed by the authors. Moreover, product **7**, derived from attack of thiol **3** onto C_2 of **A** was also not observed. In terms of regioselectivity therefore, in the presence of Lewis acid, thiol attack at C_1 becomes the only operational pathway whereas without Lewis acid, thiol **3** appears to preferentially attack C_3. We therefore provide the mechanism proposed by the authors to explain the formation of **4** (minor product) and **5** (major product) when there is no Lewis acid added to the reaction mixture (Scheme 3). We can therefore rationalize that if thiol **3** attacks the C_3 position in **A**, product **5** is formed in predominant fashion because it has more mechanistic paths leading to it as compared to product **4**. It could also be argued that in the absence of Lewis acid, attack at C_3 might occur faster than attack at C_1 and that this might also explain why **5** is formed in a higher yield compared to **4**. The authors did not address this difference and thus a theoretical analysis would be useful to shed light on whether there is a difference between the preferred site of thiol attack onto **A** (C_3 or C_1) when Lewis acid has not been added to the reaction conditions.

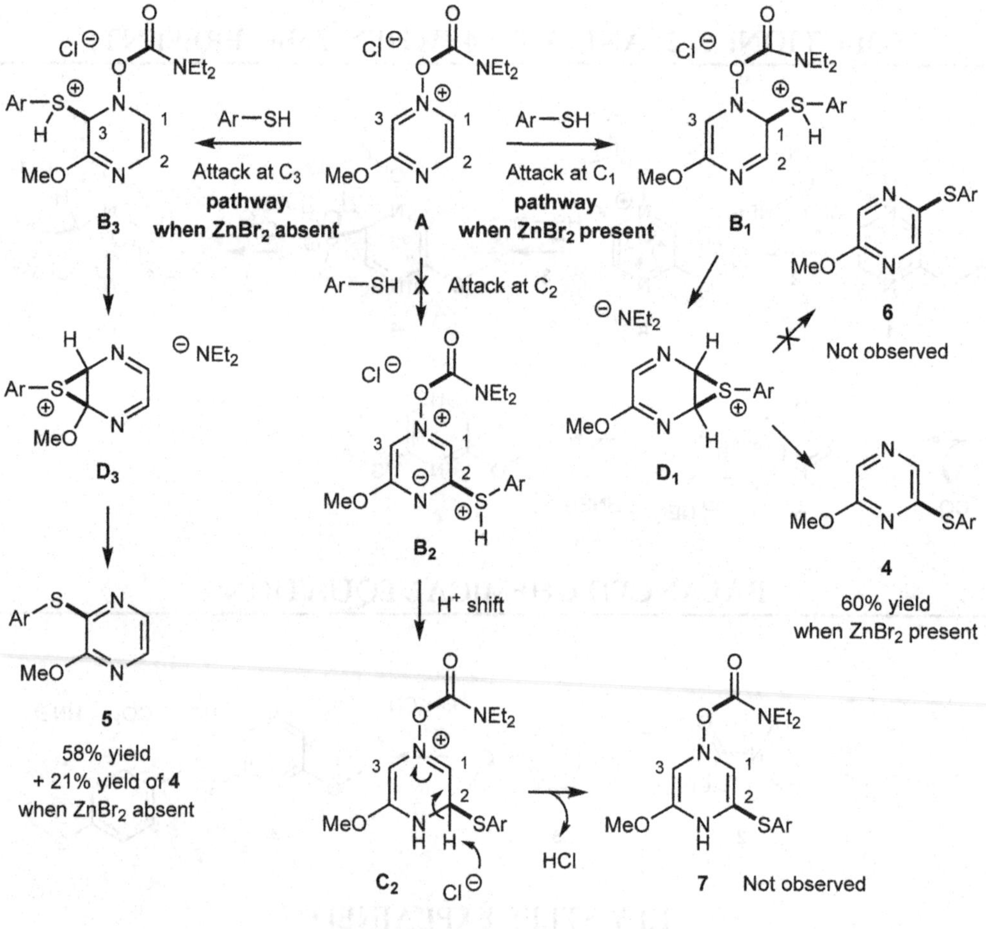

SCHEME 2

Nevertheless, the authors reasoned that zinc dibromide acts to coordinate more strongly to the carbamoyl and pyrazine groups than to the *N*-oxide group in **1**. Thus it is possible to apply a resonance structure analysis to the key intermediate **A** to determine whether there is an electronic component that influences the regioselectivity of thiol attack (Scheme 4). Given these resonance forms, we can see that the electron-donating methoxy group imparts a negative charge at C_3 in resonance structure A_1 which means that nucleophilic attack is feasible only at C_1 and C_2. This contrasts with the other two structures **A** and A_2 where attack may occur at C_1 and C_3 or at C_2 and C_3, respectively. In the case of $ZnBr_2$ coordination on the nitrogen and carbamoyl groups, the situation is repeated with the caveat that Lewis acid coordination withdraws electron density from the site of coordination. Thus C_2 in A_1^* is more electrophilic than C_1 and C_3 is more electrophilic than C_2 in A_2^* but not more than C_1 in A^*. Therefore this resonance structure analysis does not work to clearly differentiate based on electronics alone the reasons for the preferred regioselectivity at C_3 when there is no Lewis acid coordination versus the preferred selectivity at C_1 when Lewis acid coordination is available. Also, as seen previously, nucleophilic attack at C_2 should lead to the unobserved product **7** which has the carbamoyl group intact. Clearly therefore, the question surrounds C_1 versus C_3 attack when there is Lewis acid coordination and when there is not. It might therefore be worthwhile to contrast structure **A** with **A***. Here we see that Lewis acid coordination onto the unsubstituted pyrazine nitrogen atom allows for reduced electron density about this center which would favor attack at C_1. Nevertheless, without theoretical evidence, it is not possible to ascertain the magnitude of this effect.

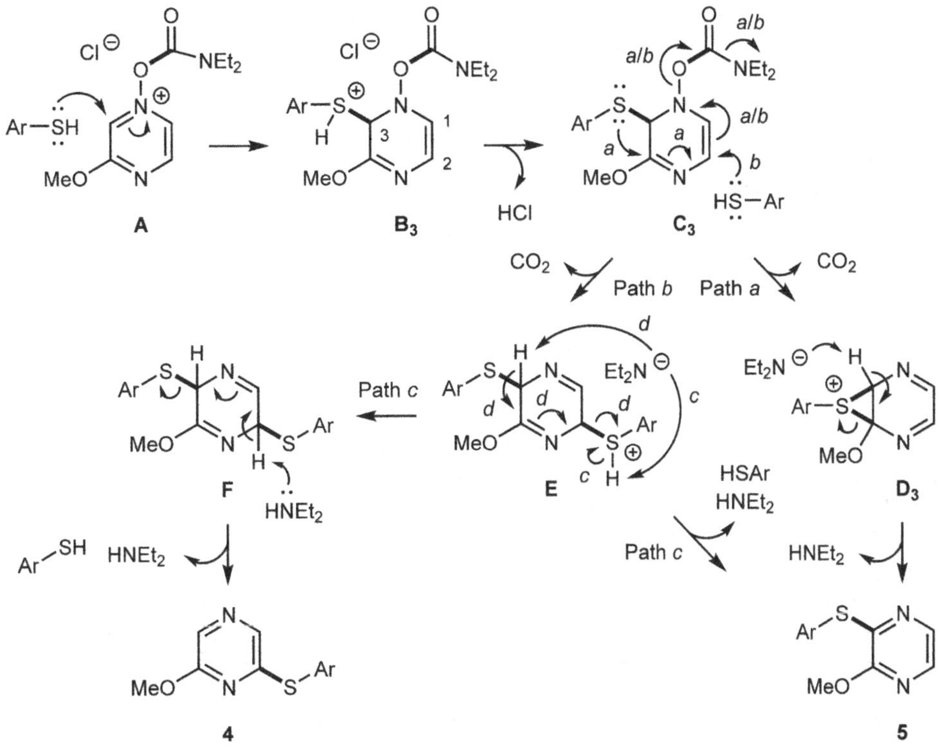

SCHEME 3 Proposed mechanism when $ZnBr_2$ is not present.

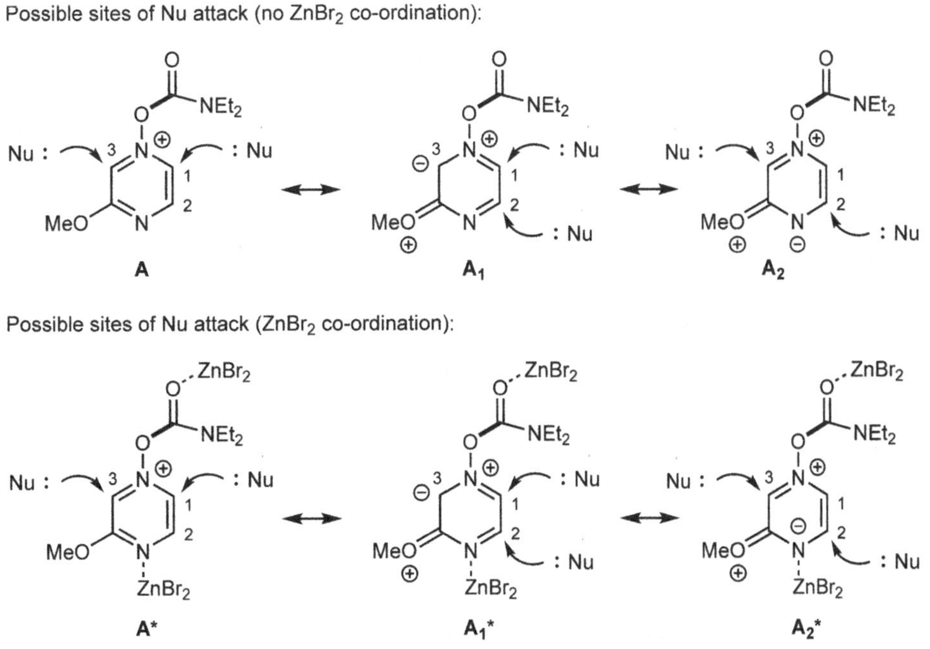

SCHEME 4

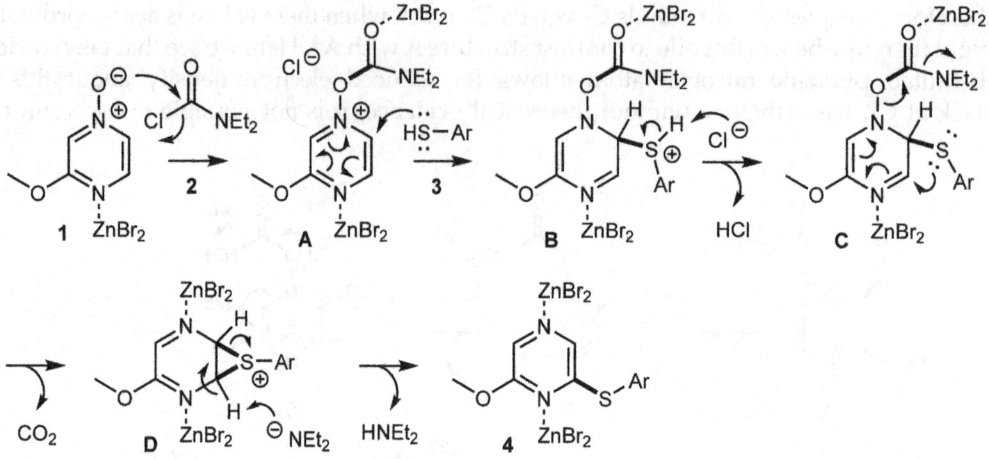

SCHEME 5

If we were to draw the mechanism for the formation of **4** while accounting for $ZnBr_2$ coordination, we can see that certain mechanistic steps are facilitated by such coordination (Scheme 5). For instance, C_1 thiol attack appears to be favored over C_3 attack in **A**. Formation of **D** also seems to be favored because of the Lewis acid coordination. It also seems possible that dithionation can occur at C_3 in **D** to form an intermediate resembling **F**.

Here, deprotonation at C_3 might be favored over deprotonation at C_2 given the electron-donating ability of the methoxy group thus opening a possibility to formation of **5** (which the authors did detect in small amount using ^1H NMR in the reaction involving use of $ZnBr_2$ even though they were unable to isolate this very low yield product). Therefore given that arguments based on sterics and electronics do not seem sufficient to establish the basis for the interesting observed regioselectivity in the product outcomes, we suggest further mechanistic analysis with regard to this problem. An interesting approach for instance would be to explore differently substituted substrates that have various electron-donating and electron-withdrawing groups attached to C_2, the unsubstituted pyrazine nitrogen atom, and at the carbon atom currently bearing the methoxy group.

ADDITIONAL RESOURCES

For further examples of this type of chemistry involving regioselectivity issues in relation to pyrazine derivatives by the same authors, see Refs. 2, 3.

References

1. Sato N, Kawahara K, Morii N. Studies on pyrazines. Part 25. Lewis acid-promoted deoxidative thiation of pyrazine N-oxides: new protocol for the synthesis of 3-substituted pyrazinethiols. *J Chem Soc Perkin Trans 1*. 1993;1993:15–20. https://doi.org/10.1039/P19930000015.
2. Sato N, Miwa N, Hirokawa N. Studies on pyrazines. Part 27. A new deoxidative nucleophilic substitution of pyrazine N-oxides; synthesis of azidopyrazines with trimethylsilyl azide. *J Chem Soc Perkin Trans 1*. 1994;1994:885–888. https://doi.org/10.1039/P19940000885.
3. Sato N, Miwa N, Suzuki H, Sakakibara T. Studies on pyrazines. Part 28. Deoxidative acetoxylation of pyrazine N-oxides. *J Heterocyclic Chem*. 1994;31(5):1229–1233. https://doi.org/10.1002/jhet.5570310520.

Question 23: Reactions of Ethacrylate Esters With NO$_2$BF$_4$

In 1995 chemists from the United Kingdom reacted ethacrylate ester **1** with nitronium tetrafluoroborate in aceto-nitrile to form a product mixture of **2** and **3** (Scheme 1).[1] It was believed that formation of **2** occurs with the interme-diacy of α-carbonyl cations. This hypothesis was tested through treatment of the more highly substituted substrate **4** under the same reaction conditions which gave a mixture of products **5** and **6**. Provide mechanisms for the formation of these products.

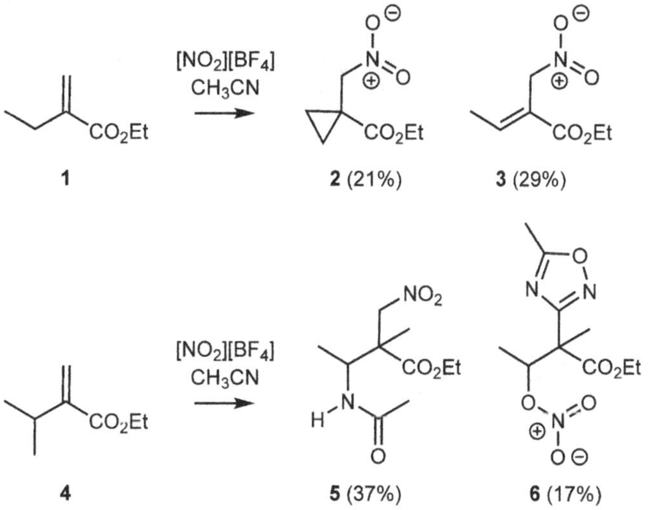

SCHEME 1

SOLUTION: 1 TO 2

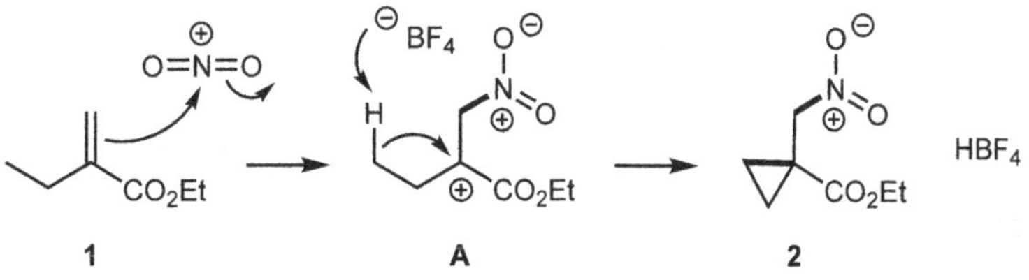

SOLUTION: A TO 3

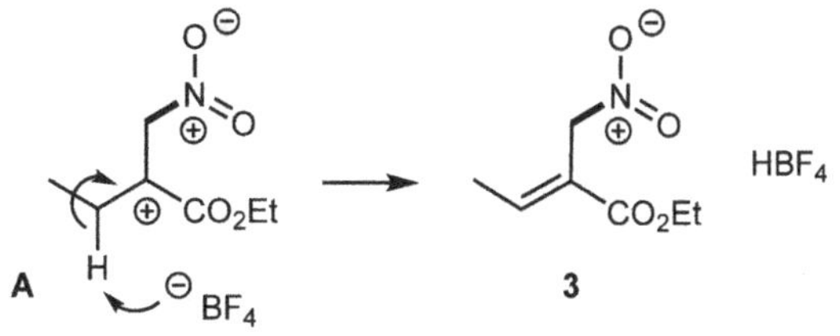

SOLUTION: 4 TO 5

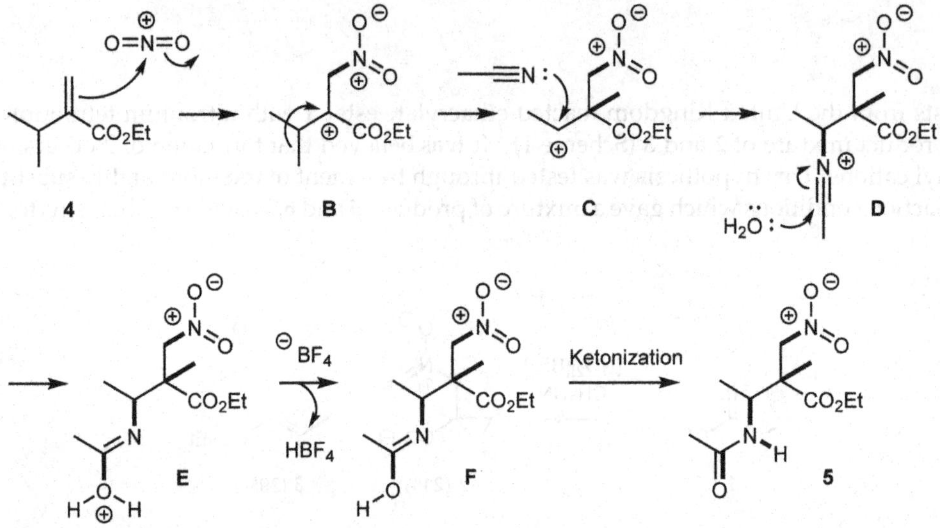

BALANCED CHEMICAL EQUATION: 4 TO 5

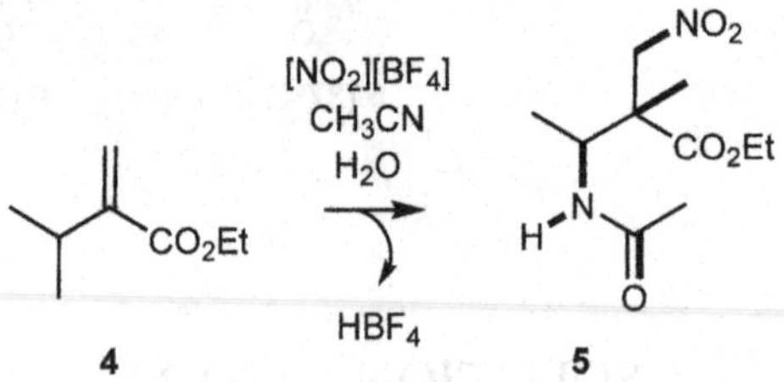

SOLUTION: C TO 6

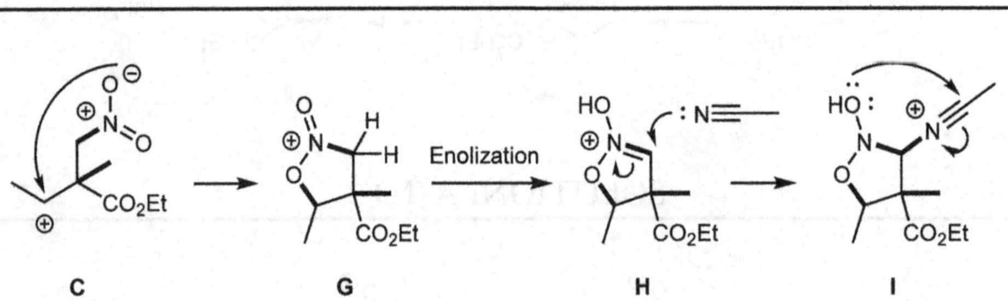

BALANCED CHEMICAL EQUATION: 4 TO 6

KEY STEPS EXPLAINED

We begin the analysis with the mechanism for the conversion of **1** to **2** and **3**. Here, we have electrophilic substitution of nitronium ion at the terminal carbon atom of the C=C bond of **1** which leads to the carbonyl cation intermediate **A**. At this point we recognize that if tetrafluoroborate-mediated deprotonation occurs at the *beta* position, cyclopropane **2** will form whereas if deprotonation occurs at the *alpha* position, the alkene product **3** will form. Judging by the yields of these products, it appears that formation of the alkene product **3** is slightly preferred over cyclopropane formation. The authors did not provide theoretical data to supplement these proposed mechanisms.

In the case of the transformation of **4** to **5** and **6**, the proposed mechanism begins with the same initial electrophilic substitution of nitronium ion at the terminal carbon atom of the C=C bond of **4** to generate the carbonyl cation intermediate **B**. Interestingly, in contrast to **A** which undergoes deprotonation to form **2** or **3**, intermediate **B** undergoes a Wagner-Meerwein 1,2-methyl shift to form carbocation **C**. The authors note that formation of cyclopropanes such as **2** and Wagner-Meerwein 1,2-methyl shifts that give **C** constitutes direct evidence for highly reactive α-carbonyl cation intermediates such as **A** and **B**. Furthermore, the reason why **A** does not undergo a Wagner-Meerwein 1,2-methyl shift, the authors argue, is because doing so would form a high energy unstable primary aliphatic carbocation which highly disfavors Wagner-Meerwein rearrangements. Moreover, a cyclopropane intermediate from **B** would also be disfavored on the grounds of unfavorable steric interactions. Thus, if the lower energy secondary aliphatic carbocation intermediate **C** is to be formed, the Wagner-Meerwein 1,2-methyl shift from **B** is favored on both energetic and steric grounds.

We note that once **C** is formed, the mechanism diverges into two paths either toward formation of **5** or of **6** depending upon which nucleophilic species attacks the secondary carbocation center of **C**. If acetonitrile attacks this center, the nitrilium ion **D** is formed which is then trapped by water to form the enol amide **F** which undergoes ketonization to form product **5**.

On the other hand, if the nitro group of **C** attacks the carbocation center, cyclization will occur to afford intermediate **G**. Enolization of **G** provides **H** which is nucleophilically attacked by acetonitrile to form nitrilium ion **I**. This intermediate undergoes cyclization via hydroxyl group attack on the nitrilium ion in a 5-*endo-dig* manner to form the bicyclic intermediate **J**. This intermediate undergoes an intramolecular proton shift to form **K** which is deprotonated with ring opening to form **L**. The free hydroxyl group of **L** attacks another equivalent of nitrosonium ion to form, after deprotonation, the final product **6**. We note the difference in stoichiometry and reagents consumed between the balanced chemical equations for the formation of these two products. We also note that the authors did not provide additional experimental or theoretical evidence to support these mechanisms. We particularly point out the expected thermodynamic instability of the [3.3.0] bicyclic intermediate **J** containing three contiguous heteroatoms. We could not find an example of such a structure analyzed in the literature although we did find examples of ring compounds containing as many as five contiguous heteroatoms.[2] Therefore, we suggest an alternative mechanism for the formation of **6** starting from the hub intermediate **C** (Scheme 2). In this alternative mechanism, water attacks the carbocation center of **C** to generate the hydroxynitro ester intermediate **N**. What follows is nucleophilic attack by the hydroxyl group of **N** onto a second equivalent of nitrosonium ion to form nitrate **S**. At this point the nitro group of **S** attacks an equivalent of acetonitrile to form intermediate **T** which cyclizes to the [1,2,4]oxadiazol-2-ol intermediate **V** after an intramolecular proton shift. A final acid-catalyzed dehydration step proceeds to form the [1,2,4]oxadiazole product **6**. This alternative mechanism has the same balanced chemical equation and target bond mapping as the authors' mechanism but it avoids any unusual bicyclic intermediates such as **J**.

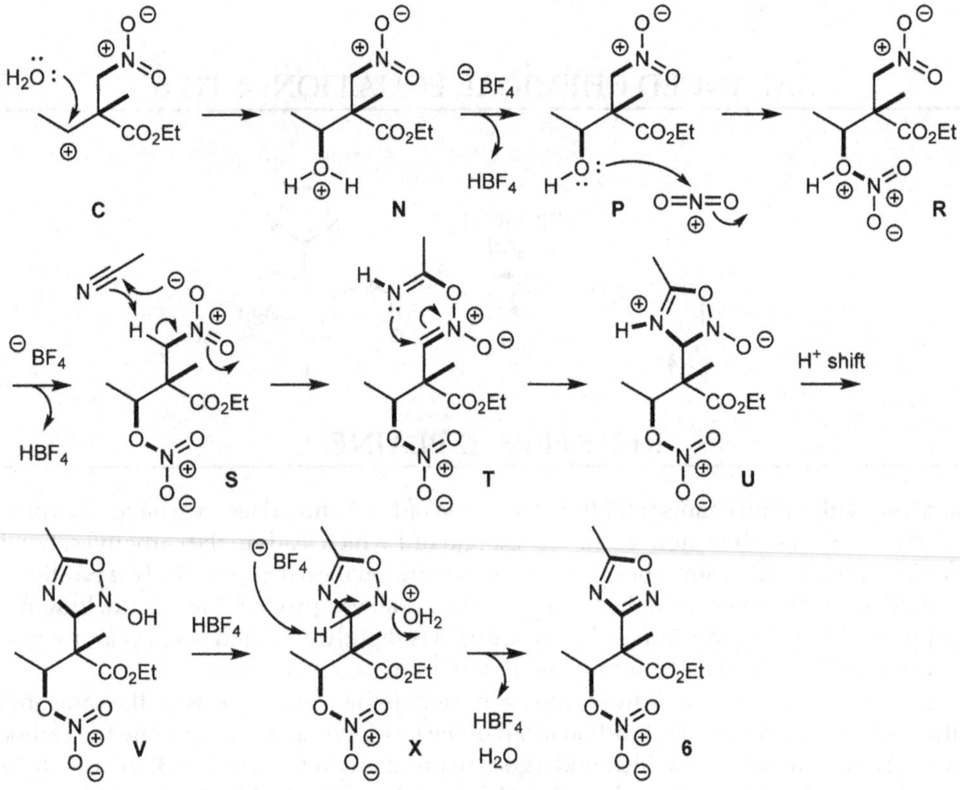

SCHEME 2

Concerning the unusual α-carbonyl cation, we note that in an earlier paper[3] the authors remarked on the rare occurrence of cations generated next to electron-withdrawing groups by referencing a review article which covers such examples.[4] Interestingly, in that paper, the authors considered three alternative mechanisms for the acid-induced nitration of α,β-unsaturated esters which forms β-nitro-α-hydroxy ester products (Scheme 3). These mechanisms were considered because they avoid significant positive charge buildup at the *alpha* carbonyl position thus avoiding the formation of the unusual intermediates **A** and **B**. Nevertheless, the observation that cyclopropane formation proceeded in high yield to form **2** when **1** was subjected to nitration conditions coupled with the results of the transformation of **4** to **5** and **6**, which had to involve Wagner-Meerwein rearrangements that depend upon significant positive charge buildup at the *alpha* position, the authors were forced to discount these three alternative mechanisms. Finally, we point out that a thorough computational study should be carried out to substantiate all experimental evidences gathered on these interesting reactions, particularly the thermodynamic and kinetic stabilities of cationic intermediates invoked along the reaction pathways.

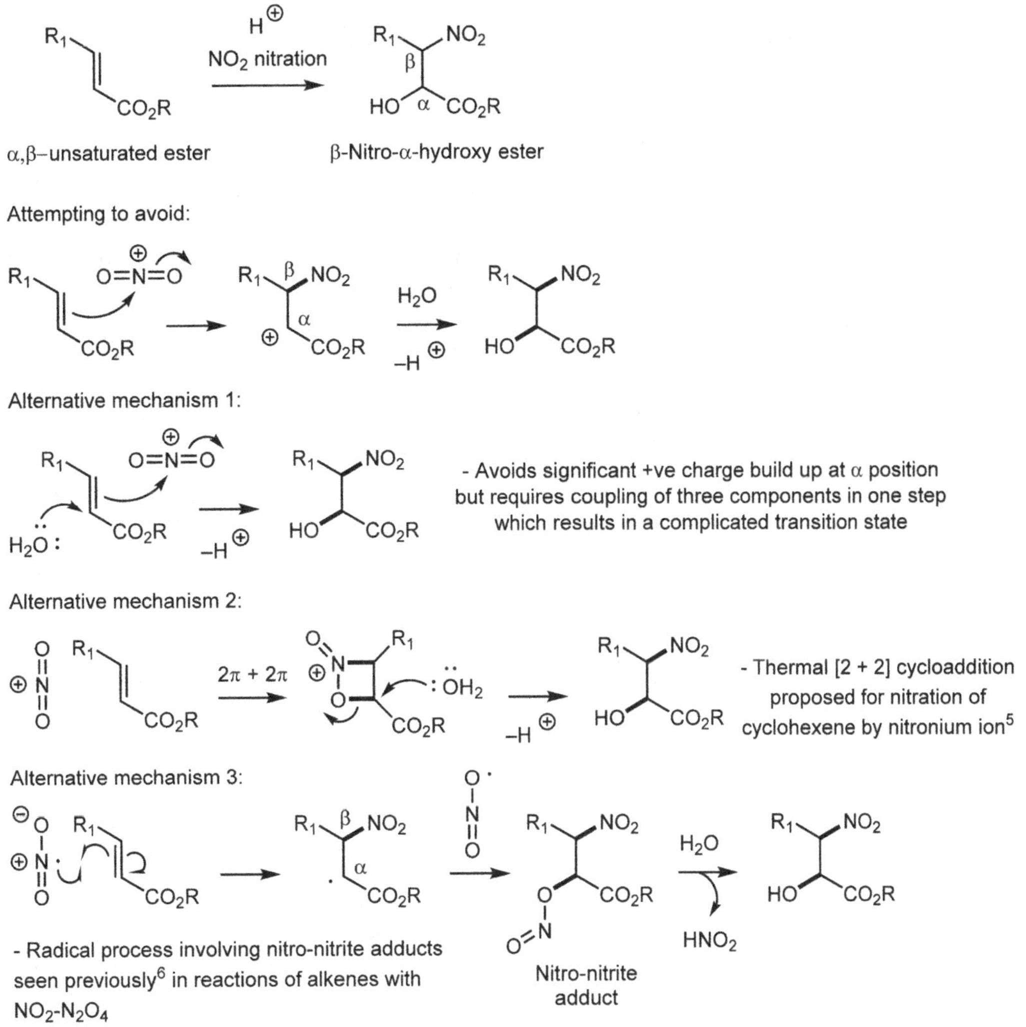

SCHEME 3

ADDITIONAL RESOURCES

We recommend an article which discusses fascinating transformations that involve Wagner-Meerwein rearrangements.[7]

References

1. Hewlins SA, Murphy JA, Lin J. Skeletal rearrangements in the nitrations of α,β-unsaturated esters. *Tetrahedron Lett*. 1995;36(17):3039–3042. https://doi.org/10.1016/0040-4039(95)00421-8.
2. Bannister RM, Rees CW. Organic heterocyclothiazenes. Part 12. 1,3,5,2,4-trithiadiazines. *J Chem Soc Perkin Trans 1*. 1989;1989:2503–2507. https://doi.org/10.1039/P19890002503.
3. Hewlins SA, Murphy JA, Lin J. Anomalous nitrations of αβ-unsaturated esters: a role for α-carbonyl cations? *J Chem Soc Chem Commun*. 1995;1995:559–560. https://doi.org/10.1039/C39950000559.
4. Creary X. Electronegatively substituted carbocations. *Chem Rev*. 1991;91(8):1625–1678. https://doi.org/10.1021/cr00008a001.
5. Borisenko AA, Nikulin AV, Wolfe S, Zefirov NS, Zyk NV. Reaction of cyclic olefins with acetyl nitrate. [2 + 2] Cycloaddition of the nitryl cation? *J Am Chem Soc*. 1984;106(4):1074–1079. https://doi.org/10.1021/ja00316a043.
6. Levy N, Scaife CW. 240. Addition of dinitrogen tetroxide to olefins. Part I. General introduction. *J Chem Soc*. 1946;1946:1093–1096. https://doi.org/10.1039/JR9460001093.
7. Ranganathan S. Fascinating organic transformations: rational mechanistic analysis. 1. The Wagner Meerwein rearrangement and the wandering bonds. *Resonance*. 1996;1(1):28–33. https://doi.org/10.1007/BF02838855.

Question 24: A Furan Synthesis

In 1992 French chemists reacted methyl vinyl ketone **1** with sodium benzyloxide (1 equiv.) followed by bromine (1 equiv.) to form the expected alkoxy bromo ketone **2** (Scheme 1).[1] When **2** was treated with 1,8-diazabicyclo [5.4.0]undec-7-ene (DBU) in benzene at room temperature, the 5,3,2-substituted furan product **3** was isolated in 70% yield. Suggest a mechanistic explanation for these transformations.

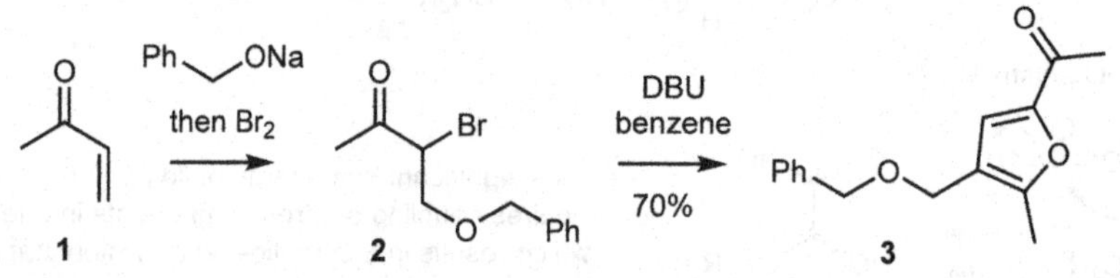

SCHEME 1

SOLUTION: 1 TO 2

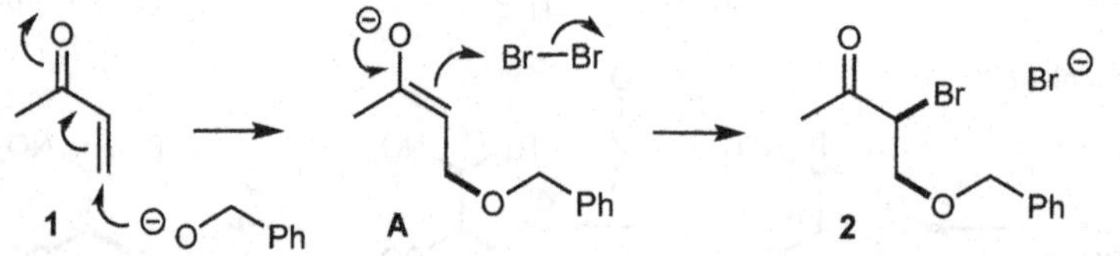

SOLUTION: 2 TO 3 (AUTHORS' MECHANISM)

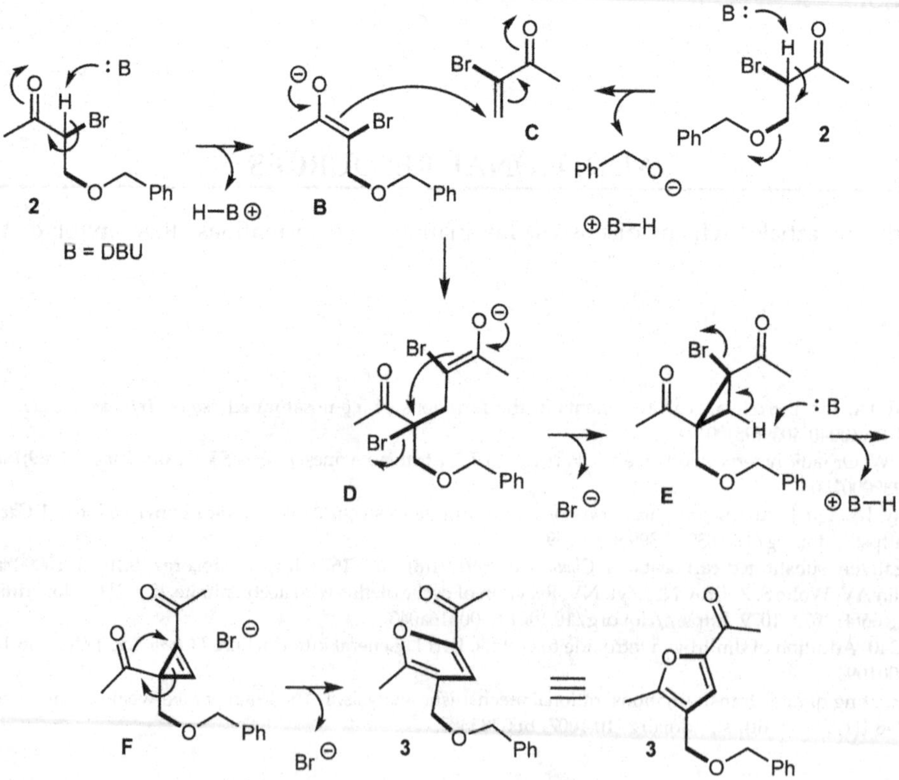

BALANCED CHEMICAL EQUATIONS

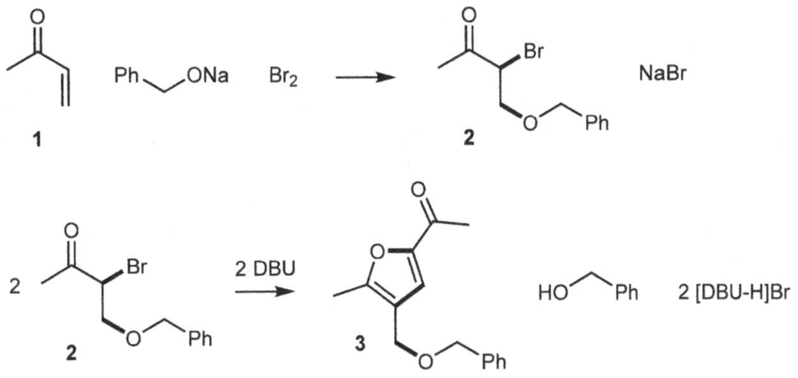

KEY STEPS EXPLAINED

The proposed mechanism for the transformation of **1** to **2** begins with nucleophilic Michael addition of benzyl oxide anion onto methyl vinyl ketone **1** to form enolate **A** which undergoes bromination at the *alpha* position to generate product **2**. At this point, according to the authors' mechanism, treatment of **2** with DBU base can result in deprotonation of the α-proton in two possible ways. In the first case, DBU deprotonation of **2** results in formation of enolate **B**. In the second case, deprotonation of **2** occurs with elimination of benzyl oxide to form the 3-bromo-but-3-en-2-one intermediate **C**. With these two intermediates formed, **B** is postulated to undergo a Michael addition onto **C** to generate enolate **D** which then cyclizes to bromocyclopropane **E** after elimination of bromine anion. This intermediate is deprotonated by another equivalent of DBU which results in elimination of hydrobromic acid to form cyclopropane **F**. Intermediate **F** is then postulated to undergo ring expansion via carbonyl group insertion into the C—C bond of the cyclopropane ring to form the final furan product **3**. We note that the authors suggested as evidence for the intermediacy of bromocyclopropane **E** the fact that the reaction of methyl vinyl ketone and α-bromoketone **4** (which has an allyl group instead of a benzyl group) results in formation of the analogous cyclopropane product **5** as shown in Scheme 2. We note that the authors were also able to synthesize furan **6** from the reaction of α-bromoketone **4** with DBU in benzene in 71% yield (Scheme 3).

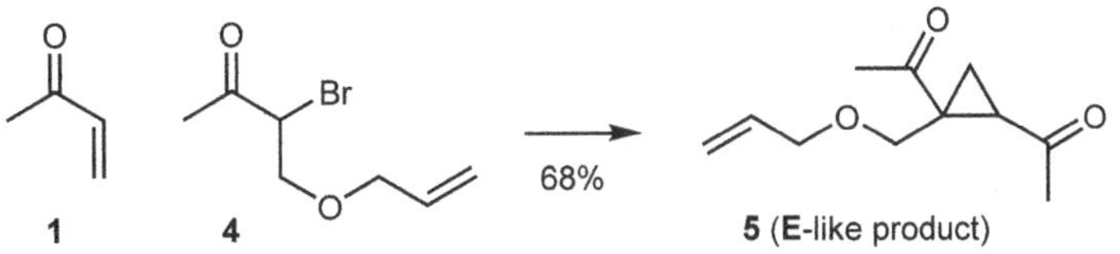

SCHEME 2

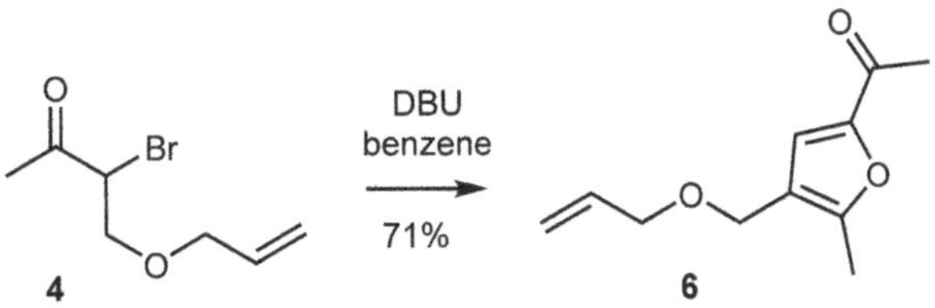

SCHEME 3

Nevertheless, we find these experiments unconvincing. For one, the authors did not demonstrate that product **5** could be converted to a dihydrofuran ring via their proposed carbonyl group insertion process (see step **F** to **3** in the proposed mechanism). Furthermore, the authors did not perform check experiments such as reacting α-bromoketone **4** with intermediate **C** without the use of a base (to spot whether an E-like product/intermediate **7** can be identified) or of carrying out this transformation in the presence of DBU to see if product **6** would be formed (Scheme 4). The absence of this experimental evidence leaves the authors' proposed mechanism for the conversion of **2** to **3** unsupported.

To circumvent these unsupported pieces of the authors' mechanism, we offer an alternative mechanism for the conversion of **2** to **3** in Scheme 5. In this mechanism, deprotonation of **2** via DBU does not occur at the *alpha* position but at the *beta* position to cause elimination of hydrobromic acid and generate vinyl ether **G**. This intermediate then reacts with another equivalent of **2** in a [3+2] cycloaddition to form oxonium ion **H** which has a newly formed 5-membered ring. Sequential deprotonation of **H** occurs with concomitant elimination of hydrobromic acid and benzyl oxide and leads to formation of the final furan product **3**. We note that in this alternative mechanism there are no cyclopropane or cyclopropene intermediates involved. Furthermore, from a computational point of view for future work, it would be of particular interest to compare the energy barriers for the authors' proposed intramolecular carbonyl group insertion (step **F** to **3**) and that of the alternative intermolecular [3+2] cycloaddition of **2** and **G**. We note that the authors did not carry out a theoretical analysis to support their proposed mechanism.

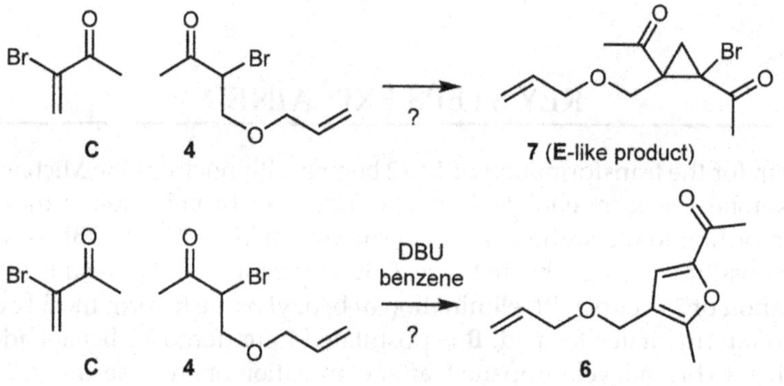

SCHEME 4

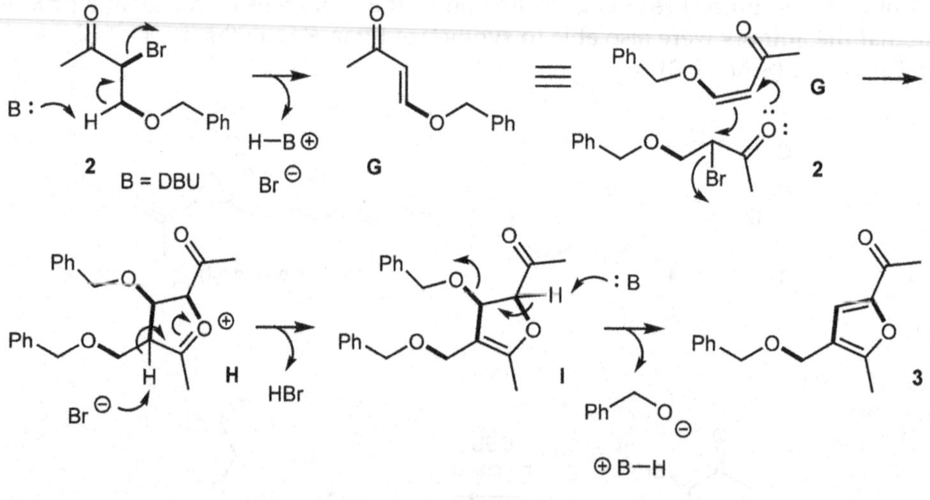

SCHEME 5

ADDITIONAL RESOURCES

We recommend two articles which describe novel approaches to the synthesis of furans and which contain some discussion of mechanism.[2,3]

References

1. Dulcere J-P, Faure R, Rodriguez J. Tandem cohalogenation-dehydrohalogenation: a new efficient synthesis of 2-acylfurans from α,β-unsaturated ketones. *Synlett*. 1992;1992(9):737–738. https://doi.org/10.1055/s-1992-21474.
2. Rauniyar V, Wang ZJ, Burks HE, Toste FD. Enantioselective synthesis of highly substituted furans by a copper(II)-catalyzed cycloisomerization-indole addition reaction. *J Am Chem Soc*. 2011;133:8486–8489. https://doi.org/10.1021/ja202959n.
3. Kulcitki V, Bourdelais A, Schuster T, Baden D. Synthesis of a functionalized furan via ozonolysis—further confirmation of the Criegee mechanism. *Tetrahedron Lett*. 2010;51:4079–4081. https://doi.org/10.1016/j.tetlet.2010.05.129.

Question 25: Dimerization of Vinca Alkaloids

During synthetic studies on vinca alkaloids in 1995, Hungarian chemists dissolved (−)-criocerine **1** in acetic acid at room temperature for 24h which formed the dimer **2** in 86% yield (Scheme 1).[1] Suggest a mechanism for this transformation.

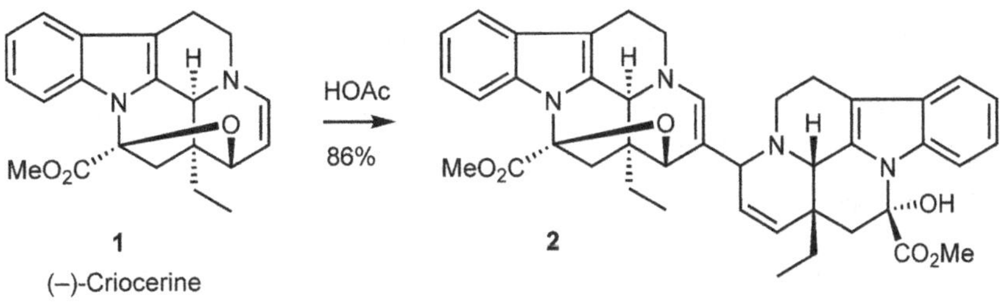

SCHEME 1

SOLUTION

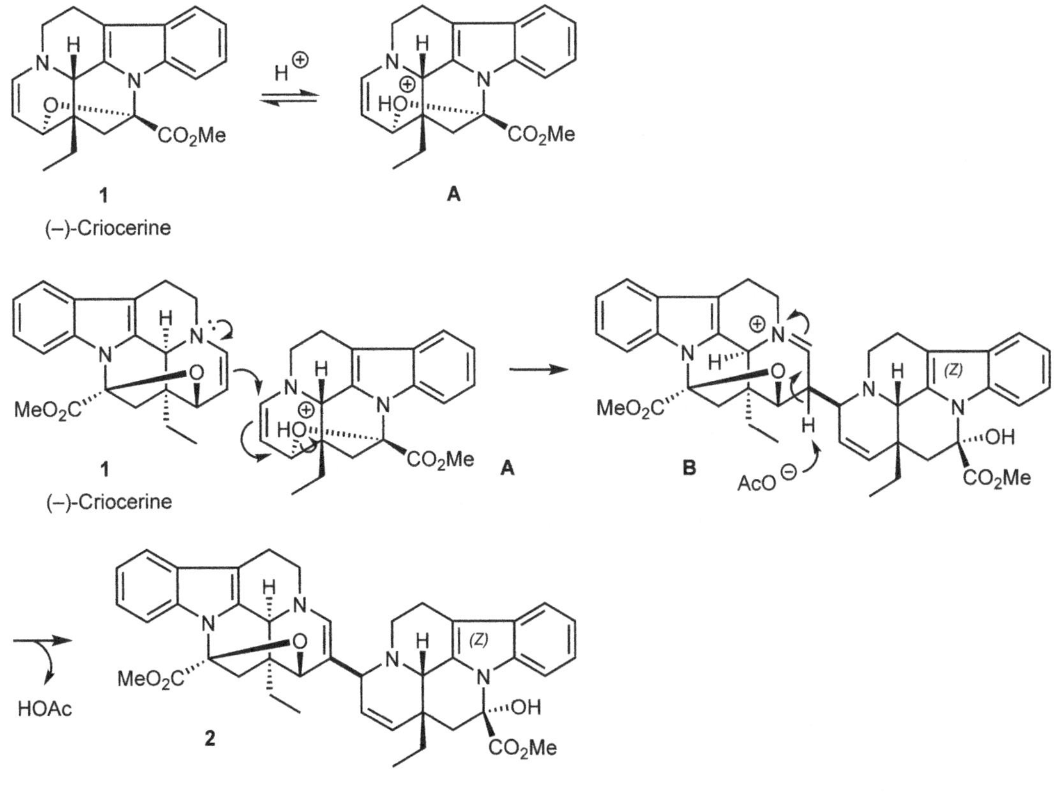

KEY STEPS EXPLAINED

The mechanism proposed by the authors to explain the acid-catalyzed dimerization of **1** to **2** begins with protonation of the oxygen atom of the transannular tetrahydrofuran ring in **1** by acetic acid giving oxonium ion **A**. A second equivalent of **1** then adds to the α-olefinic carbon atom of **A** to form the iminium ion intermediate **B**. Final deprotonation of **B** by acetate ion provides the dimeric structure **2**. We note that the authors characterized the dimer by ^{1}H, ^{13}C, and NOE NMR spectroscopy. We also note that no additional experimental or theoretical evidence was provided in support of the proposed mechanism.

ADDITIONAL RESOURCES

We recommend several articles which discuss the mechanisms behind unexpected and interesting dimerization transformations.[2–5]

References

1. Moldvai I, Szantay Jr C, Tarkanyi G, Szantay C. Synthesis of vinca alkaloids and related compounds LXXVII. Dimers of criocerine. *Tetrahedron*. 1995;51(33):9103–9118. https://doi.org/10.1016/0040-4020(95)00508-6.
2. Garcia-Borras M, Konishi A, Waterloo A, et al. Tautomerization and dimerization of 6,13-disubstituted derivatives of pentacene. *Chem A Eur J*. 2016;23(25):6111–6117. https://doi.org/10.1002/chem.201604099.
3. Novikov RA, Tarasova AV, Suponitsky KY, Tomilov YV. Unexpected formation of substituted naphthalenes and phenanthrenes in a GaCl$_3$ mediated dimerization–fragmentation reaction of 2-arylcyclopropane-1,1-dicarboxylates. *Mendeleev Commun*. 2014;24:346–348. https://doi.org/10.1016/j.mencom.2014.11.011.
4. Novikov RA, Tomilov YV. Dimerization of donor–acceptor cyclopropanes. *Mendeleev Commun*. 2015;25:1–10. https://doi.org/10.1016/j.mencom.2015.01.001.
5. Roy S, Jemmis ED. Unexpected mechanism for formal [2 + 2] cycloadditions of metallacyclocumulenes. *Curr Sci*. 2014;106(9):1249–1254.

Question 26: Nitration of a Quinoline Derivative

In 1992 British chemists reacted 6-methoxy-4-methylquinoline **1** with fuming nitric acid for 3 days at room temperature to form lactone **2** in 20% yield (Scheme 1).[1] Suggest a mechanism for this transformation.

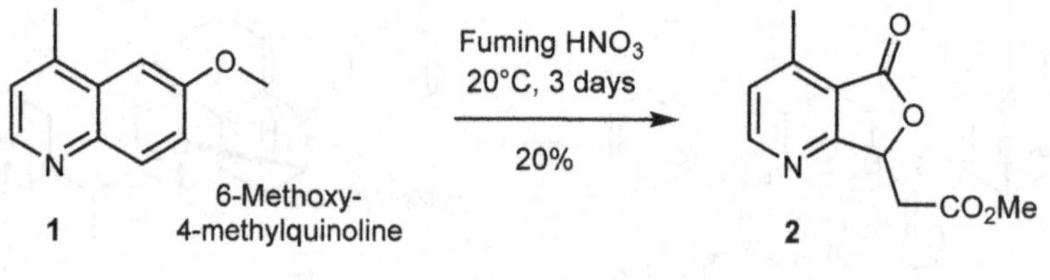

SCHEME 1

SOLUTION

BALANCED CHEMICAL EQUATION

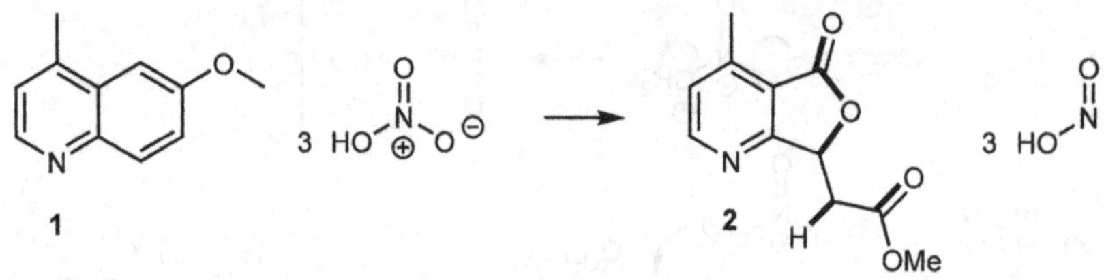

1 3 HO—NO₂ (as drawn) → **2**

KEY STEPS EXPLAINED

The proposed mechanism begins with recognizing that nitric acid decomposes to nitronium ion under acidic conditions. Quinoline **1** then reacts successively with two equivalents of nitronium ion to form the dinitro oxonium intermediate **C**. At this point water nucleophilically attacks the carbocation center of intermediate **C** and after a proton loss forms 6-methoxy-4-methyl-5,5-dinitro-5,6-dihydro-quinolin-6-ol **E** which fragments to 3-(3-dinitromethyl-4-methyl-pyridin-2-yl)-acrylic acid methyl ester **G**. Hydrolysis of **G** then leads to nitrous acid by-product and intermediate **I** which undergoes acid-catalyzed cyclization to (4-methyl-5-nitro-5,7-dihydro-furo[3,4-*b*]pyridin-7-yl)-acetic acid methyl ester **K**. Hydrolysis of **K** leads to nitrous acid by-product and (5-hydroxy-4-methyl-5,7-dihydro-furo[3,4-*b*]pyridin-7-yl)-acetic acid methyl ester **M**. Ester **M** then undergoes a hydride transfer to nitric acid to form the final lactone product **2** with by-products water and nitrous acid. We note that in this reaction nitric acid acts as a sacrificial reagent making the carbon atom *ortho* to the methoxy group in **1** more electrophilic via the sequential nitration process. Also, the ring construction mapping for this transformation is [4 + 1]. Lastly, the transformation is a formal redox reaction as shown by the oxidation number analysis provided in Scheme 2. We note that the authors did not supply additional experimental or theoretical evidence to support their proposed mechanism.

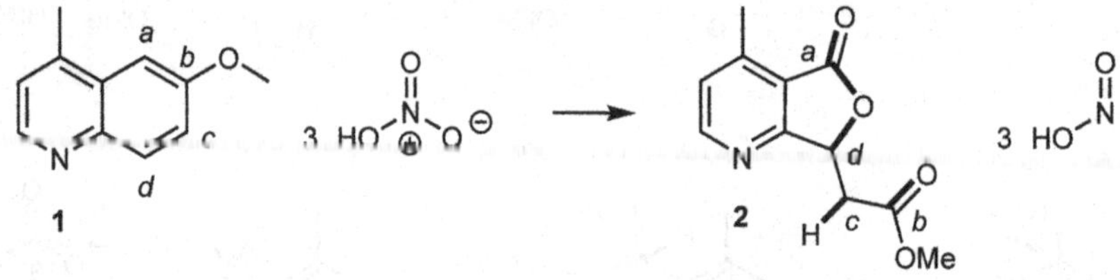

Atom	Reactant	Product	Change	
C(a)	−1	+3	+4	
C(b)	+1	+3	+2	} Δ = +7
C(d)	−1	0	+1	
C(c)	−1	−2	−1	
N	+5	+3	−2 → × 3	} Δ = −7

SCHEME 2

ADDITIONAL RESOURCES

We recommend two articles which discuss the mechanisms of transformations that involve the use of fuming nitric acid.[2,3]

References

1. Balczewski P, Morris GA, Joule JA. Aromatic ring cleavage in 6-methoxyquinolines and 2-methoxynaphthalene with fuming nitric acid. *J Chem Res (S)*. 1992;1992:308.
2. Ji Q, Cheng K-G, Li Y-X, et al. Unexpected cleavage and formation of C–C bonds by the nitration of 2,2'-biindanyl-1,1',3,3'-tetraone. *J Chem Res*. 2006;2006(11):716–718. https://doi.org/10.3184/030823406779173415.
3. Zhang Y, Zou P, Han Y, Geng Y, Luo J, Zhou B. A combined experimental and DFT mechanistic study for the unexpected nitrosolysis of *N*-hydroxymethyldialkylamines in fuming nitric acid. *RSC Adv*. 2018;8:19310–19316. https://doi.org/10.1039/c8ra03268h.

Question 27: Synthesis of Binaphthyldiquinones

In 1992 Russian chemists heated a solution of 1,4-naphthoquinone (R=H) with two equivalents of sodium hydride in THF and formed the diquinone **2** in 40% yield (Scheme 1).[1] It was noted that when the 2-chloro (R=Cl) or 2-bromo (R=Br) substituted 1,4-naphthoquinone **1** was used, diquinones **2** (R=Cl or R=Br) were formed in 94% and 78% yield, respectively. Provide a mechanism for this transformation.

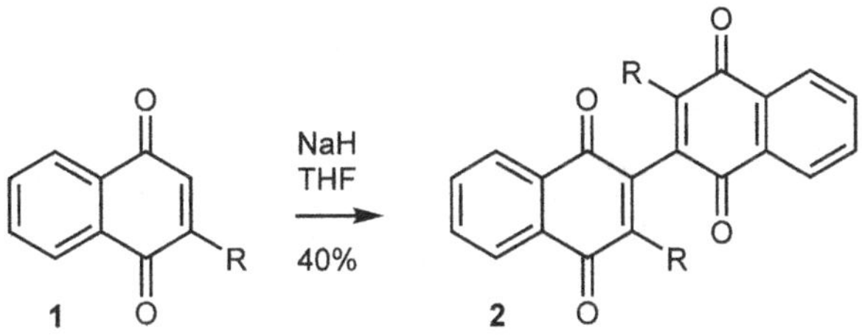

SCHEME 1

SOLUTION

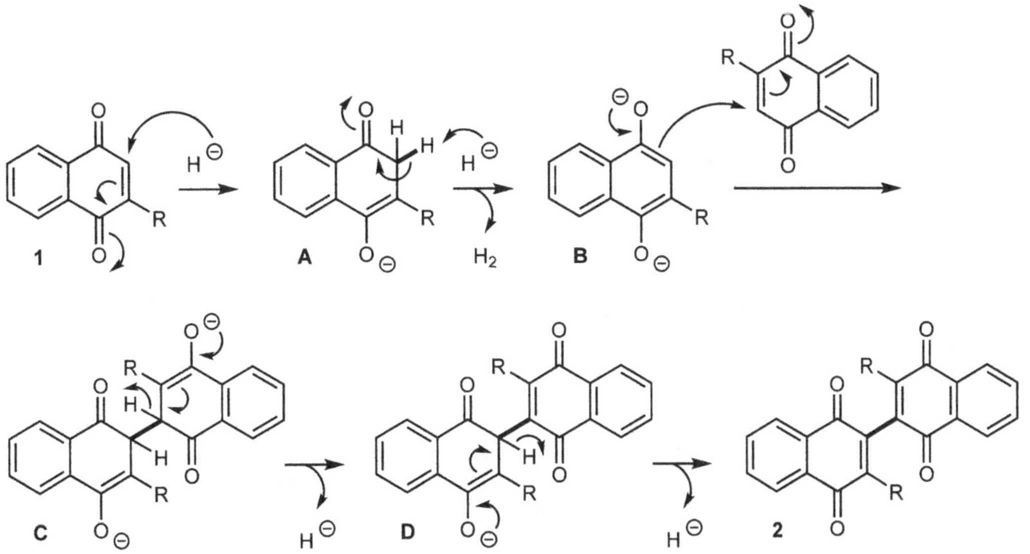

BALANCED CHEMICAL EQUATION

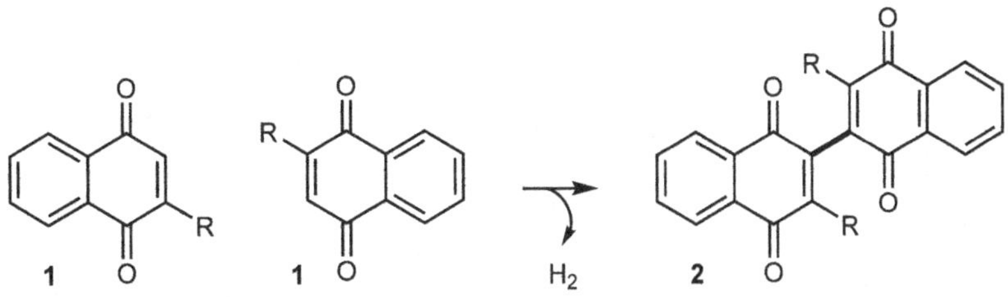

KEY STEPS EXPLAINED

The mechanism proposed by the authors to explain the transformation of **1** to **2** begins with nucleophilic addition by hydride anion onto **1** to form enolate **A**. Afterward, a second equivalent of hydride ion abstracts a proton from enolate **A** to form dianion **B** and hydrogen gas by-product. Intermediate **B** then adds in a Michael fashion to a second equivalent of naphthalene **1** to form dianion **C**. Intermediate **C** then successively eliminates two equivalents of hydride ion to form the binaphthyldiquinone product **2**. We thus see that in this mechanism hydride ion acts both as a nucleophile and as a base. The reaction is a redox reaction as shown by the oxidation number analysis given in Scheme 2. We also note that the authors conducted the transformation in dry THF under an argon atmosphere thereby eliminating the possibility of air oxidation. Nevertheless, the authors did not provide additional experimental or theoretical evidence to support their proposed mechanism.

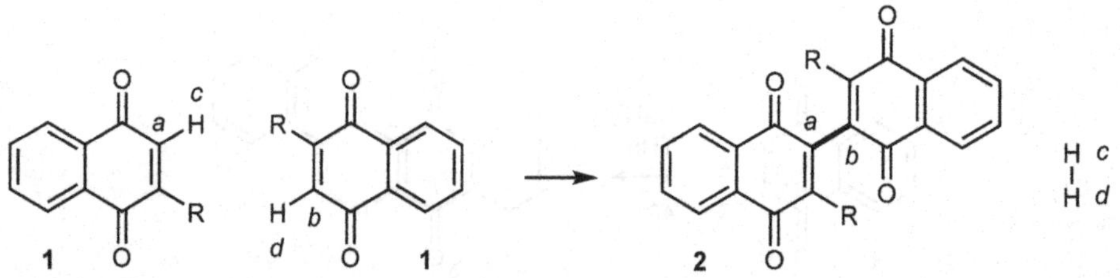

Atom	Reactant	Product	Change	
C(a)	−1	0	+1	Δ = +2
C(b)	−1	0	+1	
H(c)	+1	0	−1	Δ = −2
H(d)	+1	0	−1	

SCHEME 2

ADDITIONAL RESOURCES

We recommend an article by Schenck et al. which discusses the mechanism of the reaction of an anthradiquinone with enamines to form an unexpected product.[2]

References

1. Vorob'eva S, Magedov IV. New convenient synthesis of 2,2'-binaphthyl-1,4:1',4'-diquinones. *J Chem Res (S)*. 1992;1992:70–71.
2. Schenck LW, Kuna K, Frank W, Albert A, Asche C, Kucklaender U. 1,4,9,10-Anthradiquinone as precursor for antitumor compounds. *Bioorg Med Chem*. 2006;14(10):3599–3614. https://doi.org/10.1016/j.bmc.2006.01.026.

Question 28: Heterocyclic Fun With DMAD

In 1991 collaboration between Spanish and English chemists reported that the addition of triethylamine to a stirred suspension of salt **1** and potassium carbonate at room temperature (yellow color) to which 2.5 equivalents of dimethyl acetylenedicarboxylate (DMAD) was subsequently added (red color) gave the pyrazolo[1,5-a]pyridine **2** product (colorless) in 82% yield (Scheme 1).[1] Suggest a mechanistic explanation for this transformation.

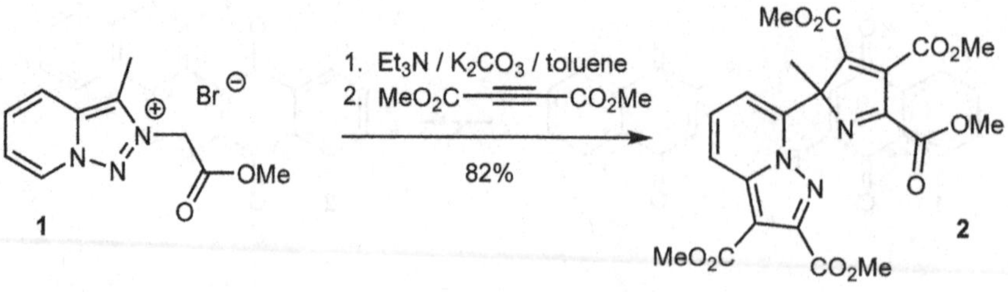

SCHEME 1

SOLUTION: MECHANISM 1

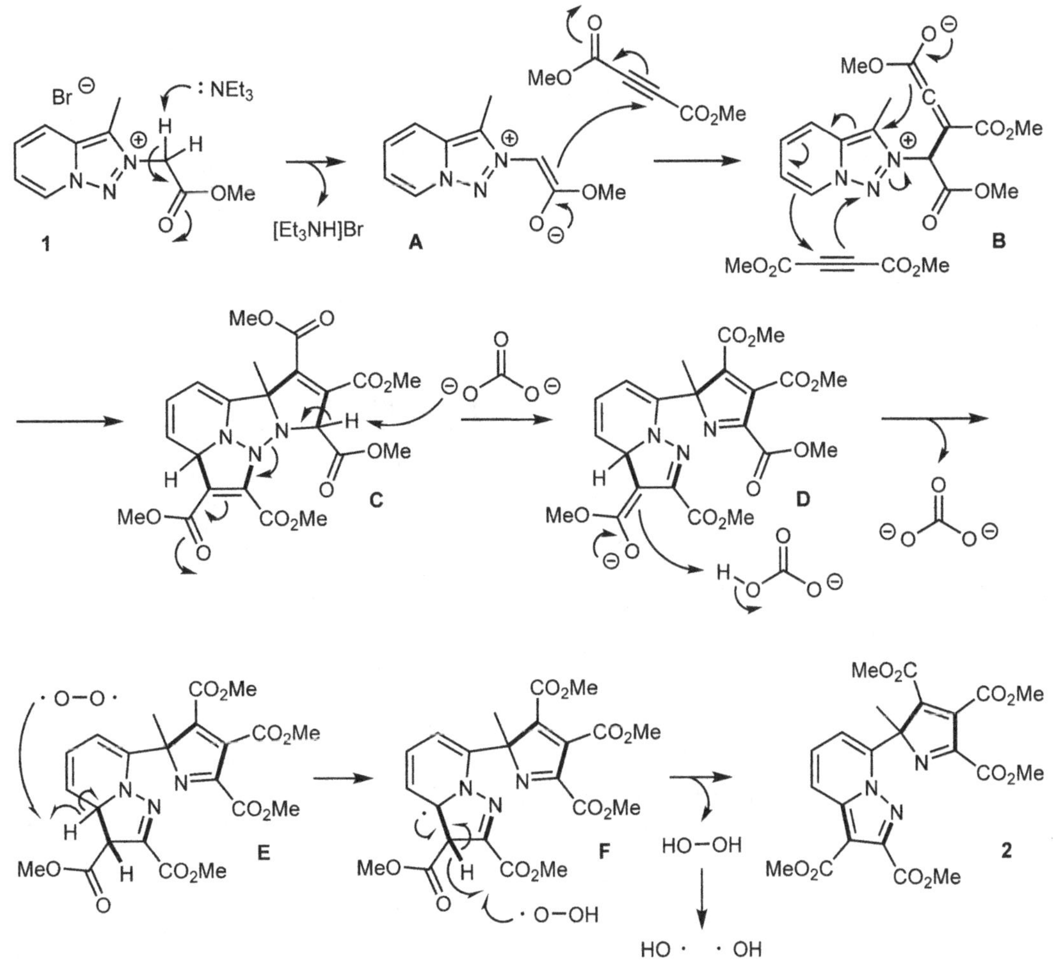

SOLUTION: MECHANISM 2

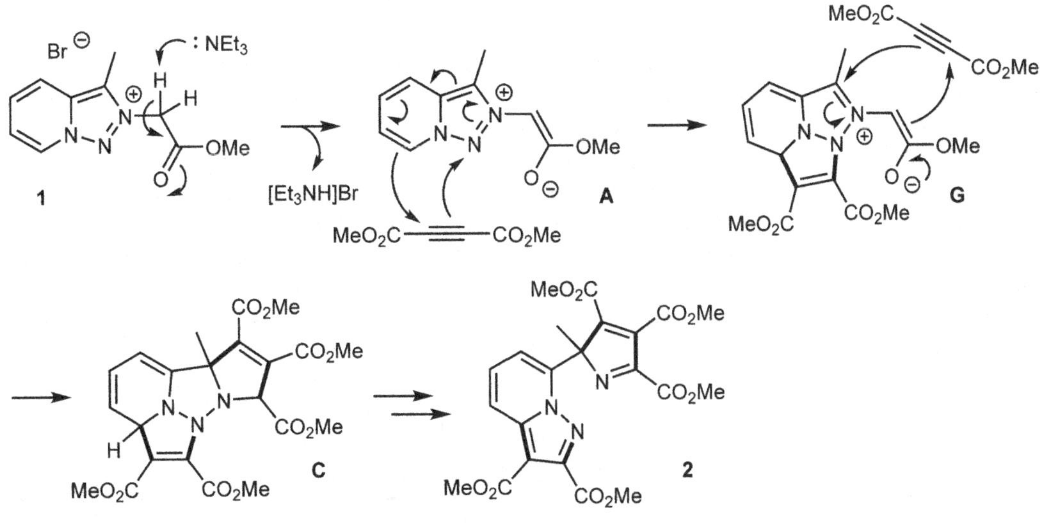

BALANCED CHEMICAL EQUATION

KEY STEPS EXPLAINED

To explain the formation of **2** from **1**, the authors proposed two possible mechanisms which start with triethylamine deprotonation of **1** to form enolate **A**. At this point the mechanisms diverge and converge again in the common hub intermediate **C**. Thus, in mechanism 1, intermediate **A** reacts with one equivalent of DMAD in a Michael fashion to form intermediate **B** which bears an allenoate group. This intermediate then undergoes an intramolecular [5+0] cyclization with the first attached DMAD moiety and a simultaneous intermolecular [3+2] cyclization with a second equivalent of DMAD which results in the formation of intermediate **C**. Deprotonation of **C** via bicarbonate base causes cleavage of the N_1—N_2 bond of the central triazole ring which gives enolate **D**. This enolate undergoes protonation by bicarbonate ion to give intermediate **E**. The removal of the two hydrogen atoms from **E** occurs under air autoxidation and leads to product **2** having a 1*H*-pyrazole ring and a 2-methyl-2*H*-pyrrole ring. By contrast, in mechanism 2, the zwitterionic intermediate **A** reacts with DMAD by concerted $8\pi + 2\pi$ cyclization to form **G** which then reacts with a second equivalent of DMAD in a [3+2] fashion to form intermediate **C** which proceeds to form product **2** as seen in mechanism 1. We note that the overall ring construction mapping for the reaction is {[3+2]+[3+2]} meaning that it involves two separate [3+2] cycloadditions.

In terms of experimental evidence, the authors observed a yellow color when **1** was reacted with triethylamine in toluene and a red color when DMAD was added to the solution. We believe that the yellow color is attributable to zwitterionic intermediate **A** and the red color to either zwitterionic intermediate **B** or **G**. Product **2** was characterized by ^{1}H and ^{13}C NMR spectroscopy and X-ray crystallography. Moreover, the authors recognized that removal of the two hydrogen atoms from intermediate **E** constitutes an oxidation and suggested that a third molecule of DMAD would be reduced to compensate for this. We hold that such a reduction is tantamount to a hydrogenation of an alkyne to alkene which is unlikely without a catalyst present under the reaction conditions employed. We therefore suggest that the redox reaction is an autoxidation process involving oxygen gas from air as shown in the mechanisms earlier. Experimental evidence for this may be provided if the reaction performance is compared under aerobic versus anaerobic conditions. If autoxidation is indeed occurring then under anaerobic conditions the reaction should stop at intermediate **E** and if it is run under aerobic conditions then the yield of product **2** is expected to increase.

In addition, the authors also argued that a nonpolar solvent such as toluene would allow for successful competition by the concerted $8\pi + 2\pi$ cyclization of **A** and DMAD to form **G** as shown in mechanism 2. With regard to mechanism 1, the authors mentioned that ylide **B** cyclizes to **C** due to lack of a rapid proton transfer route. Lastly, the proposed late-stage N_1—N_2 bond cleavage step (i.e., **C** to **D**), according to the authors, is supported by the crystal structure of **2** where the original triazolopyridine ring structure is seen. Nevertheless, the authors did not provide a theoretical analysis of the sequence from **A** to **C** to help determine which mechanistic path has the lowest energy barriers along a reaction coordinate diagram.

ADDITIONAL RESOURCES

We recommend three examples of interesting rearrangements that involve similar chemical systems such as the triazolopyridine ring structure.[2–4]

References

1. Abarca B, Ballesteros R, Metni MR, Jones G, Ando DJ, Hursthouse MB. A remarkable rearrangement during reaction between triazolopyridinium ylides and dimethyl acetylenedicarboxylate. *Tetrahedron Lett.* 1991;32(37):4977–4980. https://doi.org/10.1016/S0040-4039(00)93512-3.

2. Bishop BC, Marley H, McCullough KJ, Preston PN, Wright SHB. Synthesis and reactions of [1,2,4]triazolo[4,3-a]pyridinium-3-amidides and [1,2,4]triazolo[4,3-a]pyrimidinium-3-amidides: evaluation of the scope and mechanism of a new type of heterocyclic rearrangement. *J Chem Soc Perkin Trans 1.* 1993;1993:705–714. https://doi.org/10.1039/P19930000705.

3. Palko R, Egyed O, Riedl Z, Rokob TA, Hajos G. Rearrangement of aryl- and benzylthiopyridinium imides with participation of a methyl substituent. *J Org Chem.* 2011;76(22):9362–9369. https://doi.org/10.1021/jo201645t.

4. Molina P, Alajarin M, de Vega MJP, Foces-Foces MC, Cano FH. Iminophosphorane-mediated synthesis of mesoionic 1,3,4-oxadiazolo-[3,2-a]pyridinylium-2-aminides. *Chem Ber.* 1988;121:1495–1500. https://doi.org/10.1002/cber.19881210825.

Question 29: Unexpected Formation of a 1,5-Benzodiazonine

In 1991 Japanese chemists studying the synthesis of the alkaloid aaptamine treated *N*-methoxyamide **1** with phenyliodobis(trifluoroacetate) (PIFA) in chloroform at reflux temperature expecting that product **3** would be formed when in fact **2** was formed in 76% yield (Scheme 1).[1] Suggest a mechanism for the transformation of **1** to **2**.

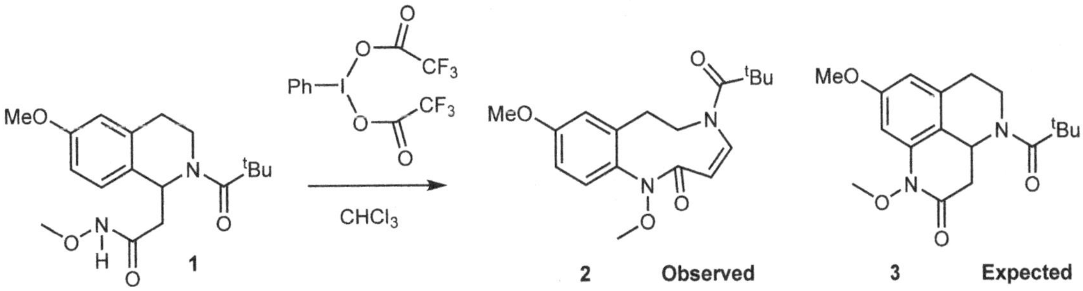

SCHEME 1

SOLUTION

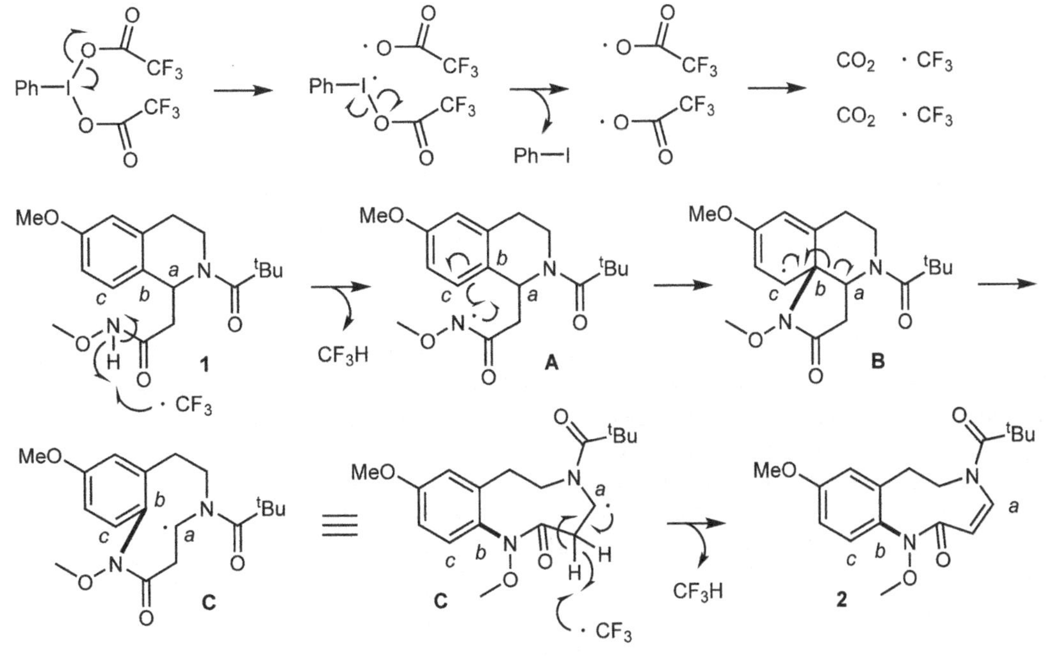

BALANCED CHEMICAL EQUATION

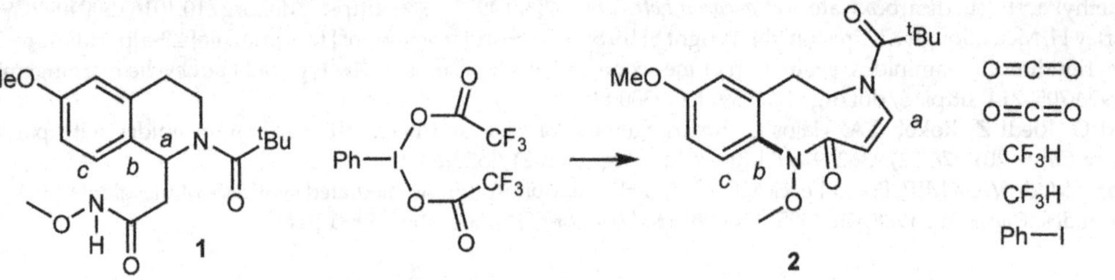

KEY STEPS EXPLAINED

We begin by noting that the authors proposed an ionic mechanism to explain the unexpected formation of **2** from **1**. They suggested a mechanism involving generation of an N+ intermediate via loss of an amino proton. Such an intermediate would be analogous in structure to nitrogen radical **A** except that the nitrogen atom bears a positive charge. Such a transformation is nonsensical based on an elementary atom charge analysis. Furthermore, no bases are present to abstract the amino proton. Given no experimental or theoretical evidence for this conjectured mechanism, we emphasize that we disagree with this proposal. In our view a radical mechanism is operative since it is known that hypervalent reagents such as PIFA undergo homolytic cleavage under thermal or photochemical conditions.[2,3] We suggest that if the reaction is run in the presence of radical quenchers such as TEMPO (2,2,6,6-tetramethylpiperidine N-oxide) and the yield of product **2** diminishes this would constitute indirect evidence for such a mechanism. We therefore omit the authors' ionic mechanism and leave it to the reader to check this mechanism in the original reference.

Therefore we propose a radical-based mechanism where PIFA fragments homolytically to form phenyliodide plus two equivalents each of carbon dioxide and trifluoromethyl radical. One equivalent of trifluoromethyl radical reacts to abstract the amino hydrogen atom from **1** to form the nitrogen radical **A** which then cyclizes to phenyl radical **B** creating a transient fused five-membered ring where a C_b—N bond is made. In the next step the C_a—C_b bond homolytically cleaves to generate a secondary carbon radical **C** which has a nine-membered ring fused to a rearomatized phenyl ring. Lastly, hydrogen abstraction from **C** via a second equivalent of trifluoromethyl radical forms the observed product **2** together with trifluoromethane by-product. The net result is a [9+0] ring construction mapping for this transformation via ring expansion of bicyclo[4.3.0] carbon radical intermediate **B**.

We also note that the authors had intended to synthesize compound **3** via the cyclization sequence shown in Scheme 2. In this mechanism the nitrogen radical **A** is converted to tertiary carbon radical **D** which loses a hydrogen atom via abstraction from trifluoromethyl radical. This step achieves rearomatization of the phenyl ring to generate product **3**. Furthermore, during this process the C_a—C_b bond remains intact while a new C_c—N bond is made resulting in a net [6+0] cyclization.

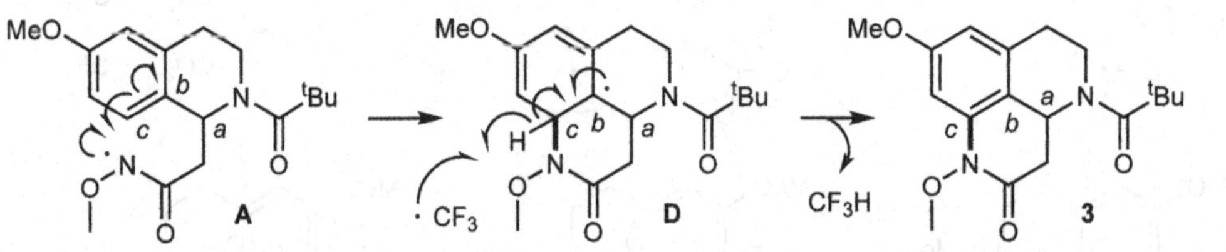

SCHEME 2

Interestingly, an oxidation number analysis shown in Scheme 3 indicates that the reaction is a redox reaction. Essentially, the iodine atom is reduced from +1 to −1 and the two trifluoromethyl carbon atoms of PIFA are reduced from +3 to +2 for a net change of −4. This change is compensated by oxidation of carbon atoms b, d, and e from 0 to +1, −2 to −1, and +3 to +4 (twice), respectively, for a net change of +4.

Lastly, the authors confirmed the structure of product **2** by conducting two-dimensional [1]H—[1]H nuclear Overhauser enhancement and ^{13}C NMR studies. Also, they hydrogenated the olefinic bond in **2** and confirmed the structure of the resulting hydrogenated product. We also recommend a future theoretical analysis of this problem to shed further light on the novel proposed mechanism.

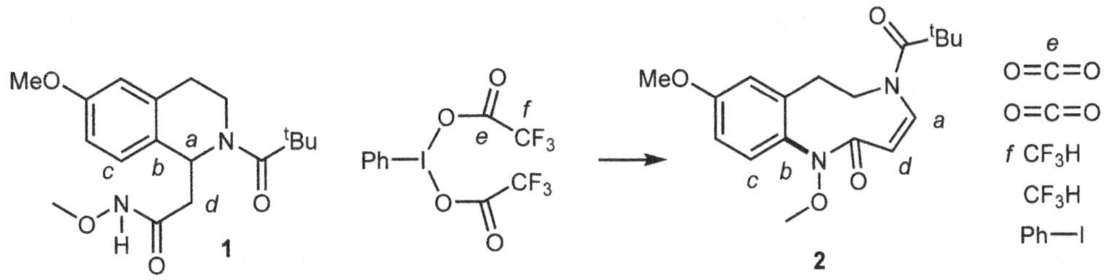

Atom	Reactant	Product	Change		
I	+1	−1	−2		Δ = −4
C(f)	+3	+2	−1	→ ×2	
C(a)	0	0	0		
C(b)	0	+1	+1		Δ = +4
C(d)	−2	−1	+1		
C(e)	+3	+4	+1	→ ×2	

SCHEME 3

ADDITIONAL RESOURCES

We recommend an article which discusses an interesting mechanism that involves Meisenheimer rearrangement in a related transformation.[4]

References

1. Kikugawa Y, Kawase M. A novel ring expansion observed during the intramolecular cyclisation with N-methoxy-N-acylnitrenium ions generated from N-methoxyamides: an expedient synthesis of 1,5-benzodiazonine derivatives. *J Chem Soc Chem Commun.* 1991;1991:1354–1355. https://doi.org/10.1039/C39910001354.
2. Zhdankin VV. C-C-bond forming reactions. *Top Curr Chem.* 2003;224:99–136. https://doi.org/10.1007/3-540-46114-0_4.
3. Zhdankin VV. *Hypervalent Iodine Chemistry: Preparation, Structure and Synthetic Applications of Polyvalent Iodine Compounds.* Chichester: John Wiley & Sons, Inc.; 2014:236–238. https://doi.org/10.1002/9781118341155
4. Kurihara T, Sakamoto Y, Takai M, Ohishi H, Harusawa S, Yoneda R. Meisenheimer rearrangement of azetopyridoindoles. VII. Ring expansion of 2-phenylhexahydroazeto[1′,2′:1, 2]pyrido[3,4-b]indoles by oxidation with *m*-chloroperbenzoic acid. *Chem Pharm Bull.* 1995;43(7):1089–1095. https://doi.org/10.1248/cpb.43.1089.

Question 30: A Thiopyran to Thiophene Transformation

In 1989 Japanese chemists reacted 1*H*-2-benzothiopyran 2-oxide **1** with acetylacetone in acetic anhydride at 100–110°C to form in 80% yield the 1-(diacetylmethyl)-1*H*-2-benzothiopyran product **2** (Scheme 1).[1] Interestingly,

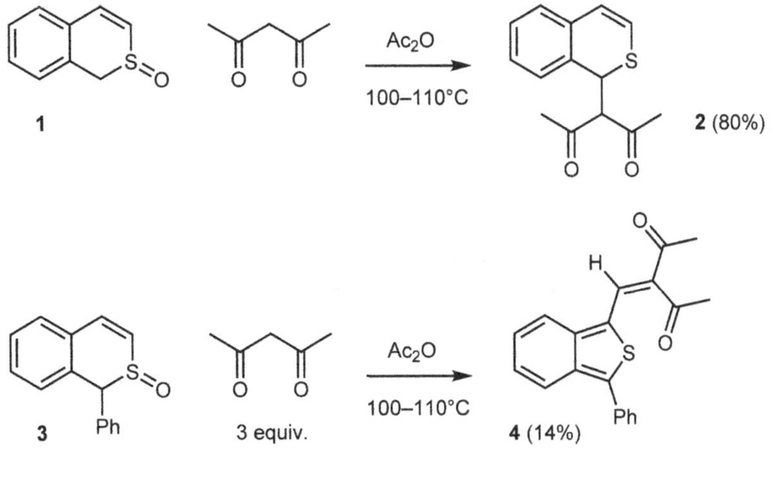

SCHEME 1

when the C_1 aryl substituted substrate **3** was reacted with a threefold excess of acetylacetone under the same reaction conditions, the authors isolated a complex mixture of products which contained a 16% yield of the benzo[c]thiophene derivative **4**. Suggest mechanisms which account for these results.

SOLUTION: 1 TO 2

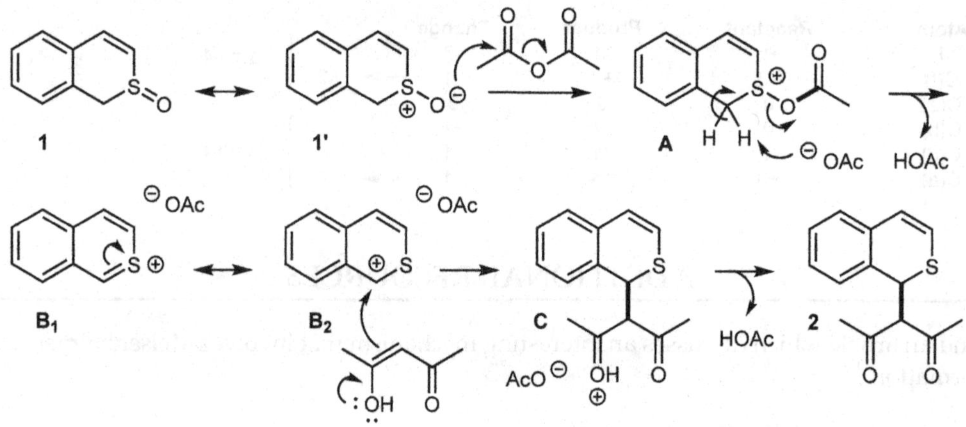

SOLUTION: 3 TO 4 (PATH A: F TO 4)

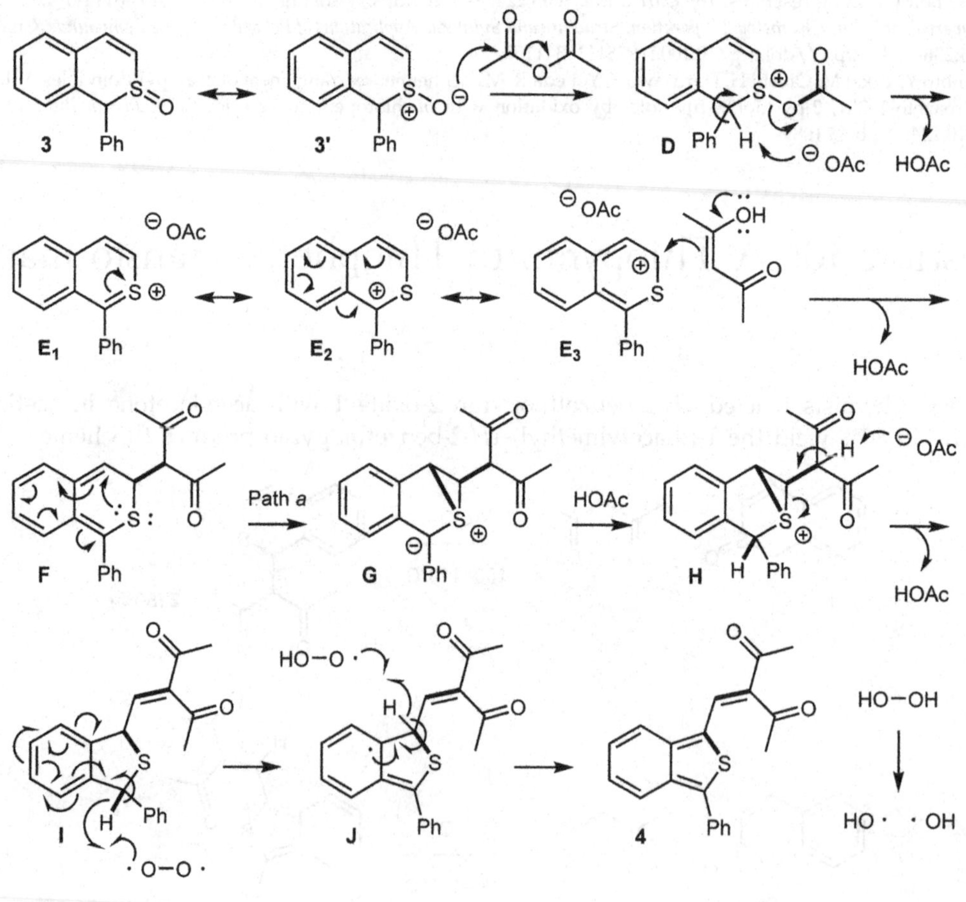

SOLUTION: 3 TO 4 (PATH B: F TO 4)

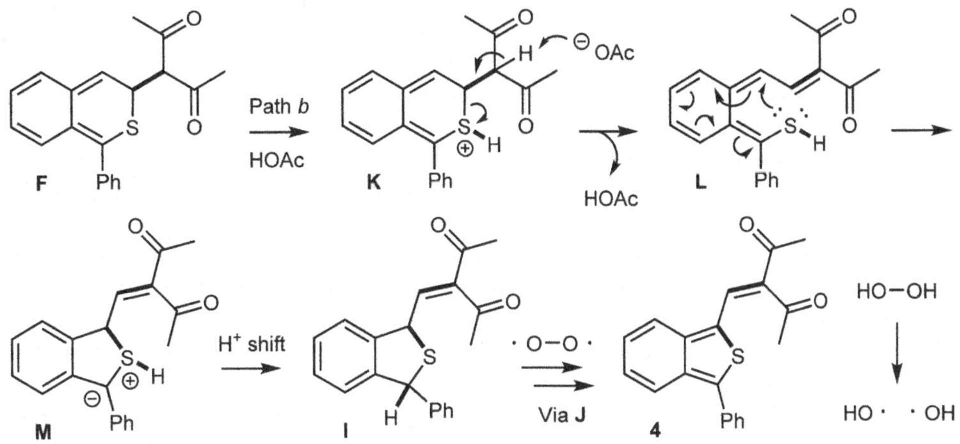

BALANCED CHEMICAL EQUATIONS

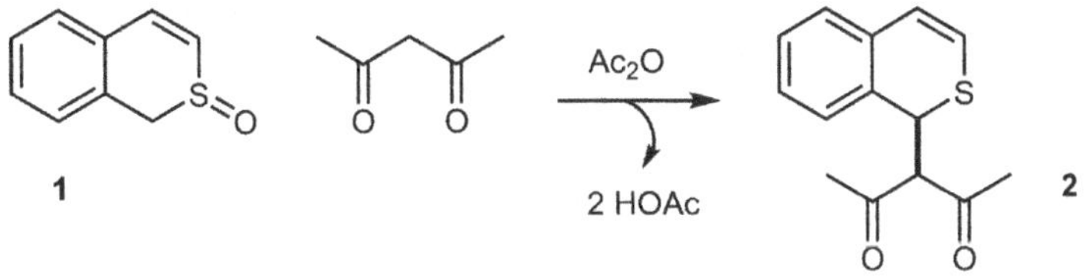

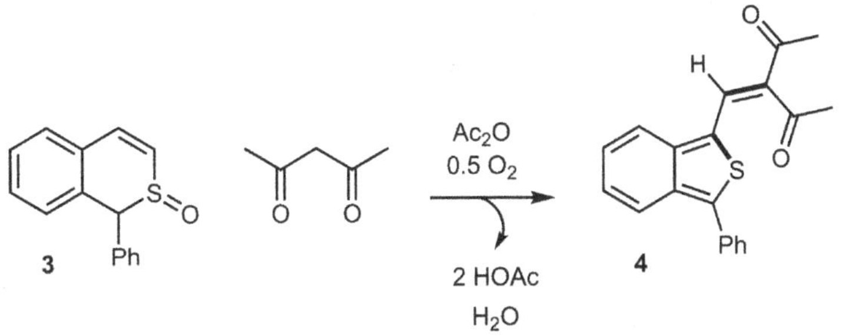

KEY STEPS EXPLAINED

The mechanism proposed for the transformation of **1** to **2** begins with acetylation of the sulfoxide anion of **1** via acetic anhydride to form sulfonium ion intermediate **A**. This ion is then deprotonated via acetate ion to form sulfonium ion B_1 which is in resonance with carbocation B_2 which has a positive charge at C_1. This intermediate is then nucleophilically attacked by the enol form of acetylacetone at C_1 (carbocation center) to generate product **2** after a final deprotonation.

When the C_1-phenyl substituted substrate **3** is reacted with acetylacetone under the same reaction conditions, the mechanism proposed by the authors follows that of the **1** to **2** transformation up to the formation of sulfonium ion E_1. Here, the authors argued that the substituted phenyl group at C_1 imposes steric hindrance which prevents approach by the enol of acetylacetone at the benzylic cationic center even though E_2 is the most stable resonance form of E_1. Therefore the aromaticity of the aryl ring is broken resulting in the formation of a third less thermodynamically stable resonance form E_3 where the carbocation center is at C_3 which is sterically accessible for nucleophilic attack. Thus the enol of acetylacetone attacks C_3 of E_3 to form intermediate **F**. To explain the rest of the mechanism from **F** to **4** the authors offer two possible pathways.

In pathway *a*, intermediate **F** rearranges to the tricyclic zwitterionic intermediate **G** which has a [3.1.0] ring framework. Protonation at C_1 carbanionic center of **G** via acetic acid gives intermediate **H** which undergoes deprotonation of its acidic α-hydrogen atom between the two acetyl groups to form intermediate **I** after fragmentation of the three-membered ring. At this point we invoke an air autoxidation radical process in which the two hydrogen atoms *ortho* to the sulfur atom are removed via intermediate **J** prior to formation of **4**. We note that the authors did not discuss air oxidation as part of their mechanistic proposal. Moreover, we note that path *a* was proposed by the authors on account of the work of Porter et al. (see reference 11 in Ref. 1) who had an intermediate similar to **H** involved in the thermal ring contraction of 2*H*-thiopyrans to thiophenes.

Alternatively, the authors also proposed path *b* where intermediate **F** undergoes protonation at the sulfur atom to form **K** which has its acidic α-hydrogen atom (between the two acetyl groups) abstracted by acetate ion to form **L**. Intermediate **L** then cyclizes to form zwitterion **M** which undergoes a protic shift to form **I**. As seen before, intermediate **I** undergoes air oxidation to give product **4**. We note that the ring construction mapping for the **3** to **4** transformation is [5 + 0]. Moreover, to support their proposed mechanisms the authors characterized product **4** by IR, ^{1}H NMR, ^{13}C NMR, mass spectrometry, and X-ray crystallography. In addition, the authors used literature precedent to substantiate several aspects of the proposed mechanism. For example, given the work of Nozaki et al. (see references 5 and 6 in Ref. 1), the authors argued that the deprotonation of **A** likely forms a ylide that can eliminate the acetate anion to form sulfonium ion **B₁** which is stable on account of its aromaticity. We note that our mechanism is drawn in a concerted manner for the steps **A** to **B₁** and **D** to **E₁**.

Concerning experimental evidence, the authors prepared the perchlorate salt of **E₁** (i.e., **6**) and subjected it to the same reaction conditions to obtain product **4** in 7% yield (Scheme 2). This experiment directly supports intermediate **E₁**. Furthermore, the authors also subjected compound **7** (i.e., an analog of sulfoxide **1** which contains two methyl substituents at the site *ortho* to the sulfoxide group) to the same reaction conditions which gave product **8** (Scheme 3). Based on the structure of this product it was clear that both ring contraction and direct substitution at the *ortho* center were suppressed. Instead, acetoxylation took place in the *para* position. This experiment demonstrated that an *ortho* hydrogen atom was needed to form products **2** and **4**.

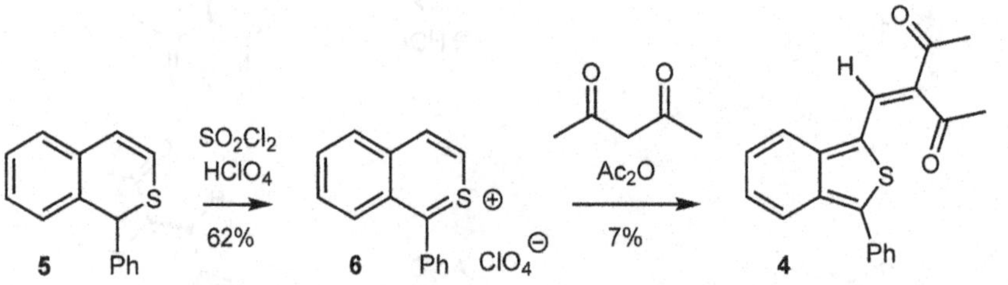

SCHEME 2

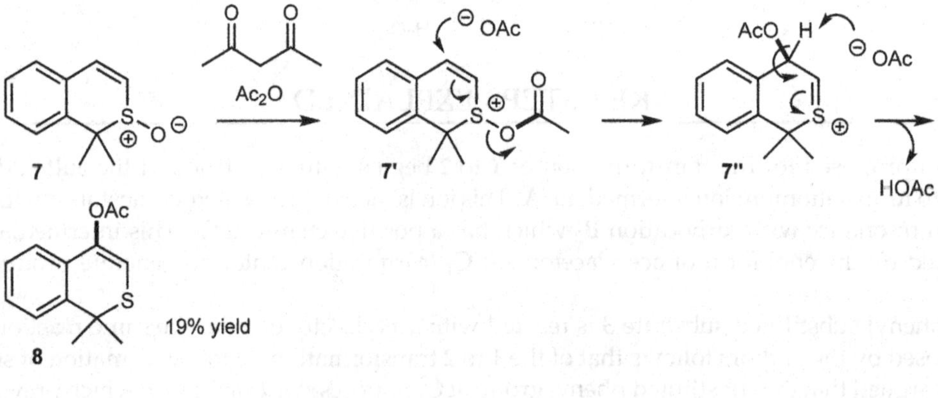

SCHEME 3

The authors also explain the rather low yields of these experiments on account of possible decomposition pathways following the formation of intermediates **B₂/E₂**. Interestingly, however, in the case of the C_1-phenyl substituted sulfoxide **3**, it was possible for diethyl malonate to attack the sterically hindered position at C_1. The authors explained this interesting result by invoking the higher nucleophilicity of the enolate anion of diethyl malonate ($pK_A = 13$) when compared with the enolate of acetylacetone ($pK_A = 9$, see reference 13 in Ref. 1). This result could be suppressed by the use of even bulkier substituents at C_1. We recommend that others conduct a theoretical analysis of this transformation to help further support the proposed mechanism.

ADDITIONAL RESOURCES

For an interesting mechanistic discussion in a related system, see the work of Nishio and Okuda.[2]

References

1. Hori M, Kataoka T, Shimizu H, Hongo J, Kido M. Reactions of 1*H*-2-benzothiopyran 2-oxides with active methylene compounds: a novel ring contraction of 1-aryl derivatives to benzo[*c*]thiophenes. X-Ray molecular structure of 1-(2,2-diacetylvinyl)-3-phenylbenzo[*c*]thiophene. *J Chem Soc Perkin Trans 1*. 1989;1989:1611–1618. https://doi.org/10.1039/P19890001611.
2. Nishio T, Okuda N. Photoreactions of isoindoline-1-thiones with alkenes: unusual formation of tricyclic isoindolines. *J Org Chem*. 1992;57 (14):4000–4005. https://doi.org/10.1021/jo00040a049.

Question 31: Skeletal Rearrangements of a Diterpene

In 1989 French chemists investigated the skeletal rearrangement of methyl pimarate **1** when reacted with bromine in THF/H$_2$O in the presence of NaHCO$_3$ for 10 min at 0°C to form a mixture which contained the strobane diterpene products **2**, **3**, **4**, and **5** (Scheme 1).[1] Provide mechanisms to explain the formation of these products from **1**.

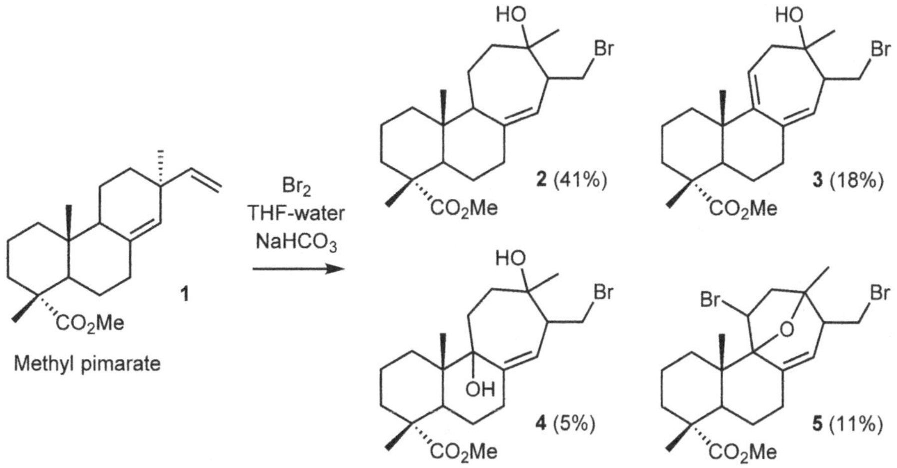

SCHEME 1

SOLUTION: 1 TO 2

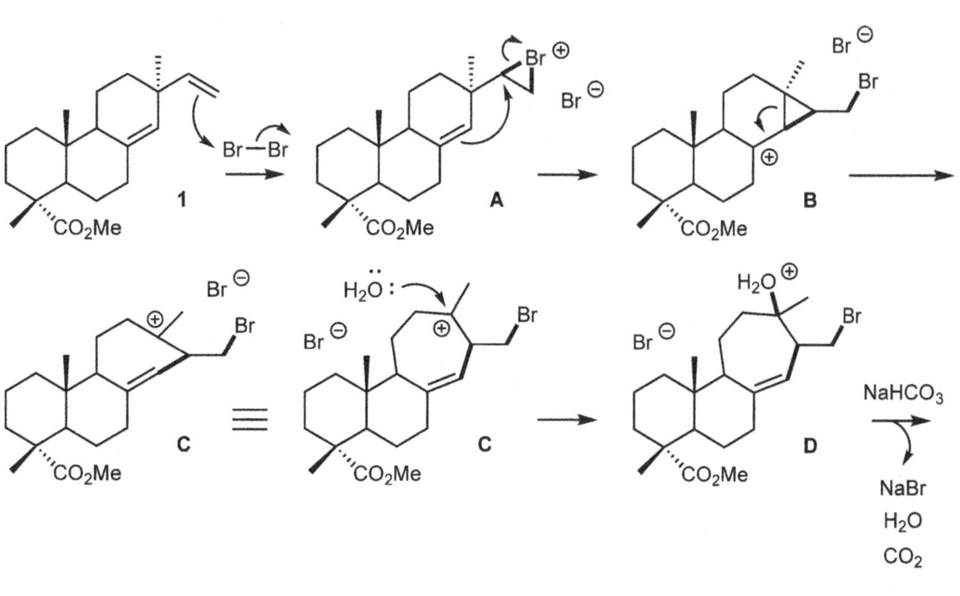

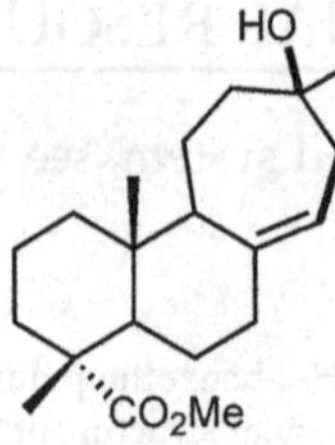

SOLUTION: B TO 4

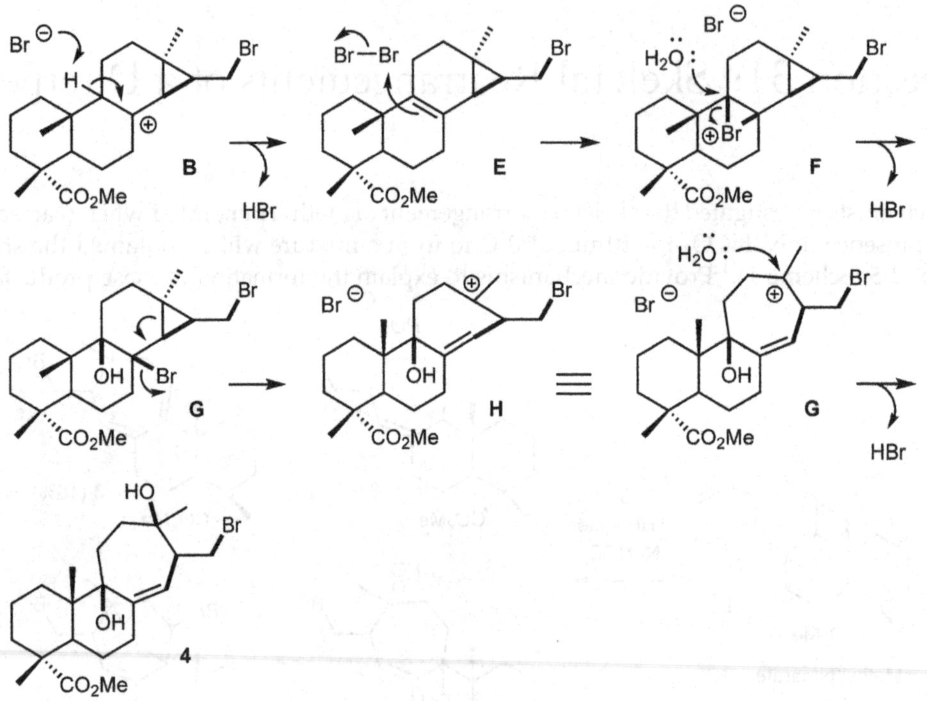

SOLUTION: 4 TO 3

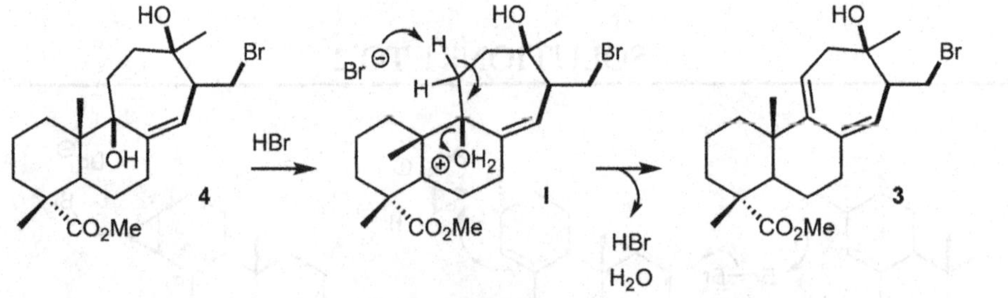

SOLUTION: 3 TO 5

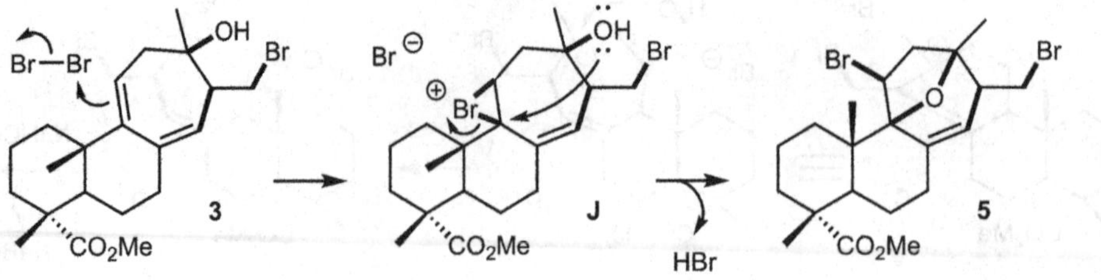

BALANCED CHEMICAL EQUATIONS

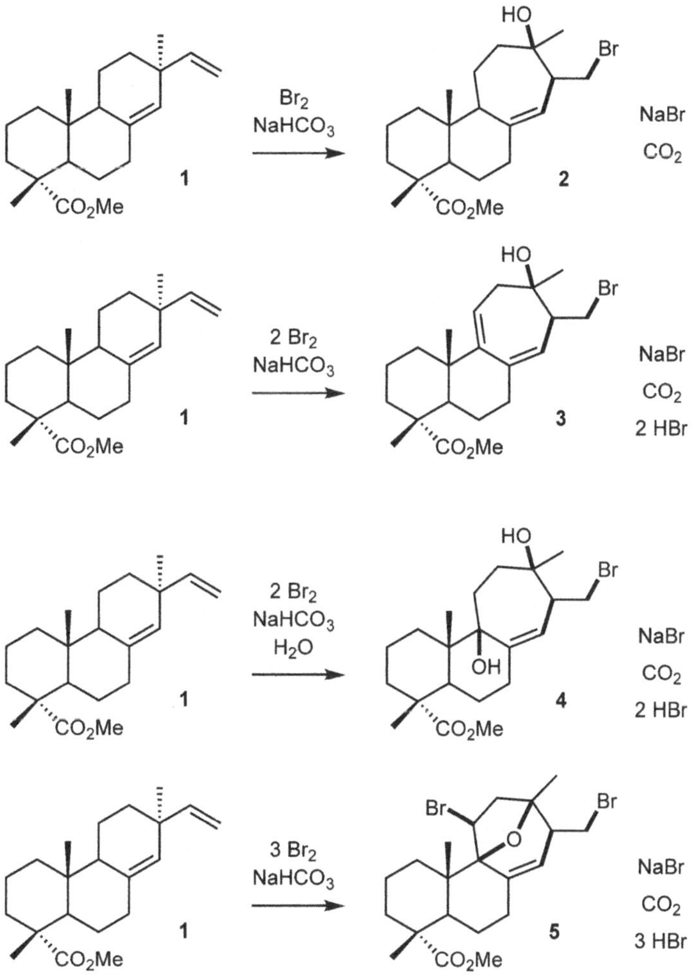

KEY STEPS EXPLAINED

The mechanism proposed for the transformation of **1** to **2** begins with bromination of the vinyl group of **1** to form bromonium ion **A** which rearranges to form the key carbocation intermediate **B** that has a bicyclic [4.1.0] framework. This intermediate can undergo ring expansion to form the seven-member ring carbocation **C** which upon hydration gives product **2**. With regard to the formation of product **4**, the key intermediate **B** can also undergo deprotonation to form the tetracyclic intermediate **E** which is brominated by a second equivalent of bromine to generate bromonium ion **F**. At this point, water attacks C_{7b} to cause ring opening of the bromonium ion to form intermediate **G**. Ring expansion of **G** proceeds as seen previously (see step **B** to **C**) to form **H** which is captured by water to generate product **4**. From product **4**, a simple acid-catalyzed dehydration via hydrobromic acid gives product **3**. Furthermore, the C_{11}—C_{11b} double bond of product **3** can react with bromine to form bromonium ion **J** whose hydroxyl group can facilitate a transannular ring opening of the bromonium ion to form the 8-oxa-bicyclo[3.2.1]oct-2-ene product **5**. For simplicity we provide an overview of these various mechanisms in Scheme 2 where it is clear that **B** serves as a key intermediate common to all productive mechanistic pathways. We can also see from the proposed mechanism that the yield of product **2** is much higher than all other products. This can be explained because there are no further productive pathways that **2** might follow to continue reacting. The same does not hold for products **4** and **3** both of which can continue to react further with bromine to form, eventually, **5**. The fact that the yield of **4** is the lowest among all products means that the energetics of going from **4** to **3** are especially favorable. The same can be concluded about the formation of **5** from **3** given that both products are formed in nearly identical yields. Therefore a future theoretical analysis of the proposed mechanism would likely answer these questions by means of calculating the energy of all proposed intermediates and products along a reaction coordinate diagram.

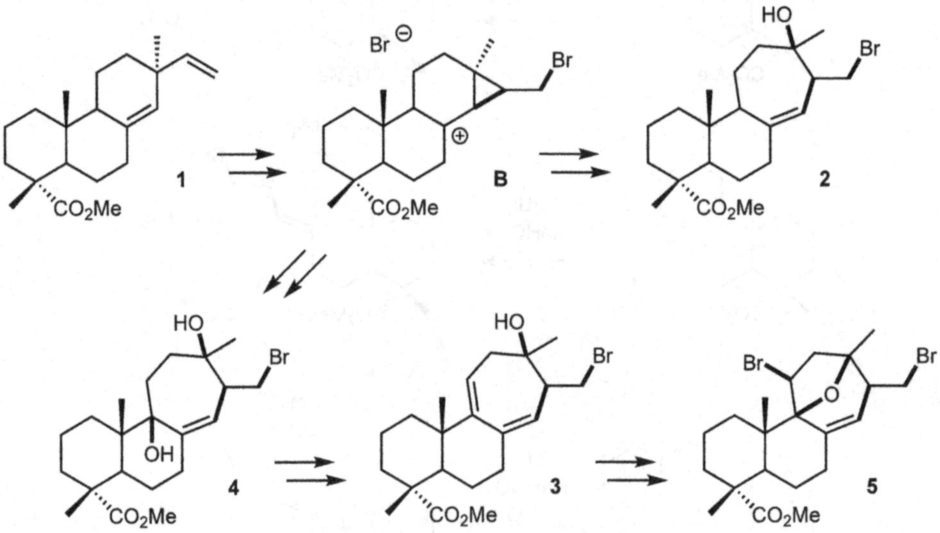

SCHEME 2

ADDITIONAL RESOURCES

For further discussion of the mechanisms involved in strobane diterpene rearrangements, see the work of Herz et al.[2,3]

References

1. Sam N, Taran M, Petraud M, Barbe B, Delmond B. Biomimetic route to the strobane skeleton from methyl pimarate. *Tetrahedron Lett.* 1989;30 (12):1525–1526. https://doi.org/10.1016/S0040-4039(00)99508-X.
2. Herz W, Prasad JS, Mohanraj S. Cationic cyclizations and rearrangements as models for strobane and hispanane biogenesis. *J Org Chem.* 1983;48 (1):81–90. https://doi.org/10.1021/jo00149a017.
3. Herz W, Hall AL. Resin acids. XXVI. Biogenetic-type rearrangements of the homoallylic cation from methyl 15(R)-hydroxypimar-8(14)-en-18-oate. *J Org Chem.* 1974;39(1):14–20. https://doi.org/10.1021/jo00915a003.

Question 32: Reissert Reactions of Quinoxaline *N*-Oxides

In 1989 Belgian chemists carried out a series of Reissert reactions on quinoxaline *N*-oxide derivatives **1** and **4** to form the expected Reissert products **2**, **3** and the unexpected ring-opened product **5** in a series of three separate transformations (Scheme 1).[1] Propose mechanisms to explain the course of each of the three transformations.

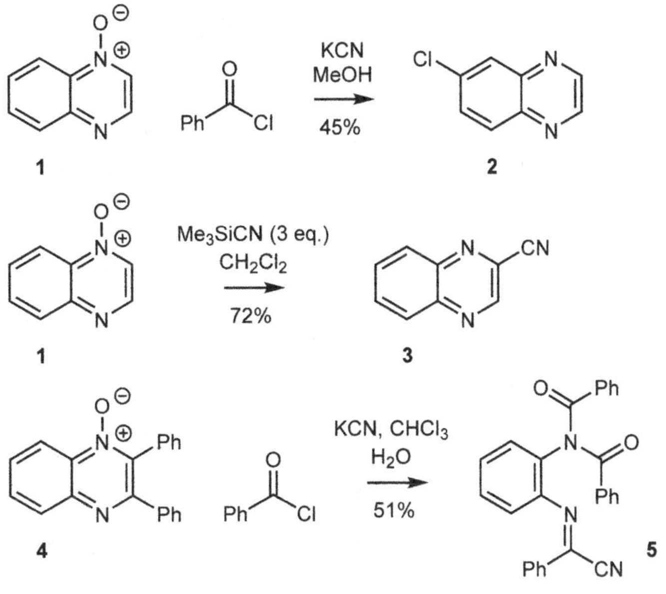

SCHEME 1

SOLUTION: 1 TO 2

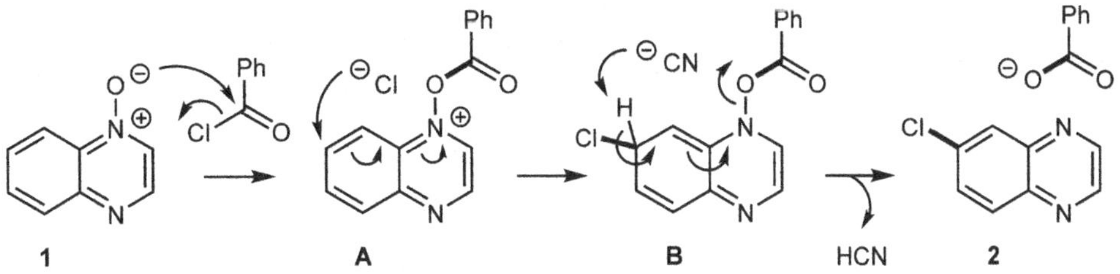

SOLUTION: 1 TO 3

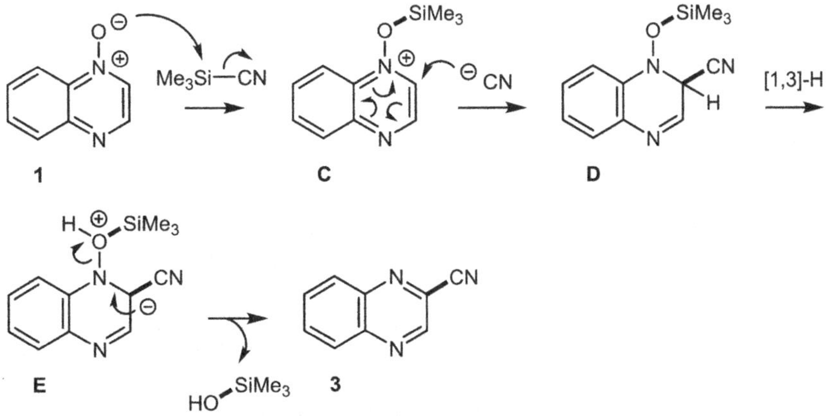

SOLUTION: 4 TO 5

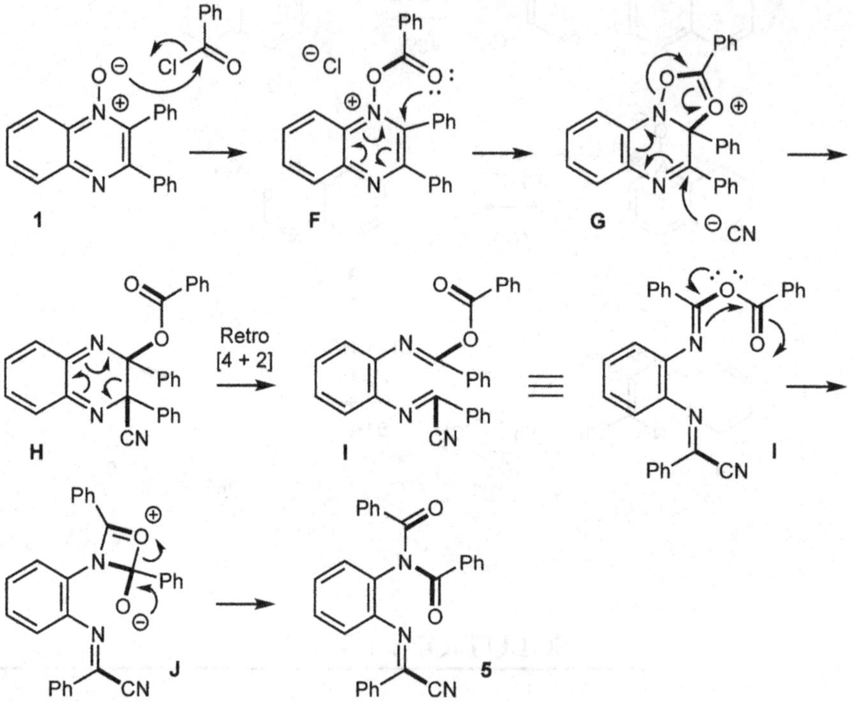

1 → **F** → **G** →

H → Retro [4 + 2] → **I** ≡ **I** →

J → **5**

BALANCED CHEMICAL EQUATIONS

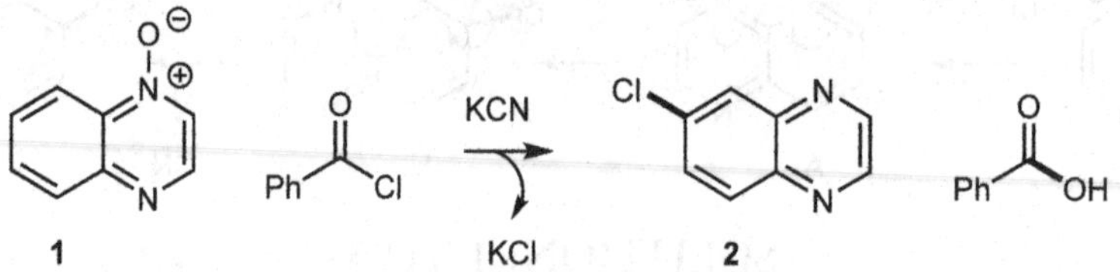

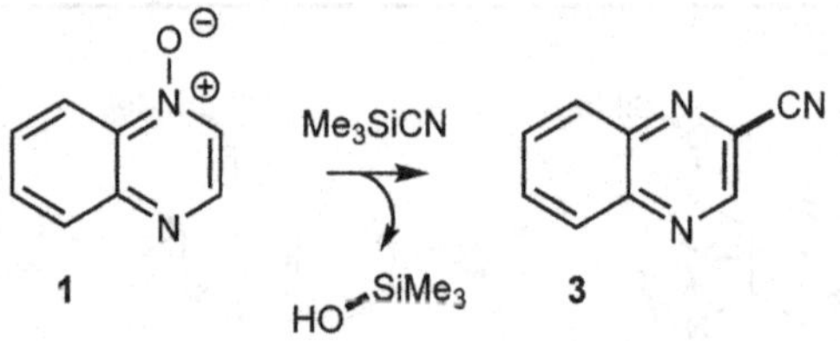

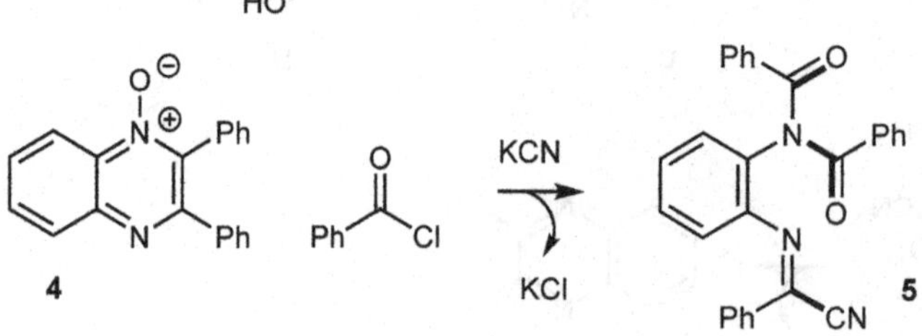

1 + Ph–C(=O)–Cl → KCN / –KCl → **2** + Ph–C(=O)–OH

1 + Me₃SiCN → –HO–SiMe₃ → **3**

4 + Ph–C(=O)–Cl → KCN / –KCl → **5**

KEY STEPS EXPLAINED

The mechanism proposed for the transformation of quinoxaline *N*-oxide **1** to the Reissert product **2** begins with the oxygen anion of **1** attacking the benzoyl chloride to form quinoxalin-1-ium ion **A** after elimination of chloride anion. Chloride then adds to C_7 of **A** to form **B** which is deprotonated by cyanide ion with concomitant elimination of benzoate ion to form product **2**. When potassium cyanide (KCN) is substituted by trimethylsilylcyanide (Me₃SiCN) we have a similar initial attack by the *N*-oxide group of **1** onto the Me₃SiCN reagent to form quinoxalin-1-ium ion **C**. This intermediate is then attacked by cyanide ion at C_2 to form **D** which undergoes a [1,3] protic shift to form the zwitterionic intermediate **E**. This intermediate fragments to give product **3** and trimethylsilylhydroxide by-product. Finally, the mechanism proposed for the transformation of **4** to **5** begins with benzoylation of **4** to form intermediate **F**. This intermediate cyclizes to form oxonium ion **G** which is attacked by cyanide ion at C_3 to form **H**. Intermediate **H** undergoes a retro-[4+2] cyclization to form the ring open diimine intermediate **I**. This intermediate undergoes rearrangement via zwitterion **J** to form product **5** after the benzoyl group has effectively migrated from the oxygen atom to the nitrogen atom. The authors did not provide additional experimental and theoretical evidence to support their mechanism.

ADDITIONAL RESOURCES

For further analysis of the Reissert reaction and references therein where aspects of the mechanism are discussed, see Ref. 2. For instance, according to reference 20d in Ref. 2, it is thought that the acylation step (i.e., **1** to **A/C/F**) occurs before the addition of cyanide (i.e., to **A/C/G**) because the acylation forms a salt (i.e., **A/C/F**) which makes the C=N bond more polar thereby facilitating the addition of cyanide. We recommend two additional articles which discuss interesting mechanisms of a Reissert reaction and of an enamine ring-opening rearrangement reaction.[3,4]

References

1. Nasielski J, Heilporn S, Nasielski-Hinkens R. An unexpected ring-opening in the Reissert reaction on 2,3-diphenylquinoxaline-*N*-oxide. *Tetrahedron*. 1989;45(24):7795–9804. https://doi.org/10.1016/S0040-4020(01)85794-X.
2. Wang Z, ed. *Comprehensive Organic Name Reactions and Reagents*. New York, NY: John Wiley & Sons, Inc.; 2010:2335–2340. https://doi.org/10.1002/9780470638859.conrr527
3. McEwen WE, Kanitkar KB, Hung WM. Mechanism of 1,3-dipolar addition of Reissert salts to arylpropiolate esters. *J Am Chem Soc*. 1971;93(18):4484–4491. https://doi.org/10.1021/ja00747a024.
4. Katritzky AR, Yang B, Jiang J, Steel PJ. Ring-opening rearrangements of 2-(benzotriazol-1-yl) enamines and a novel synthesis of 2,4-diarylquinazoline. *J Org Chem*. 1995;60(1):246–249. https://doi.org/10.1021/jo00106a041.

Question 33: Unexpected Results in Directed Metalation

In 1995 Slovenian and American chemists encountered an unexpected reaction during an attempted directed metalation approach for *peri*-functionalization of **3** (see Scheme 1).[1] Essentially, even though compound **3** contained a

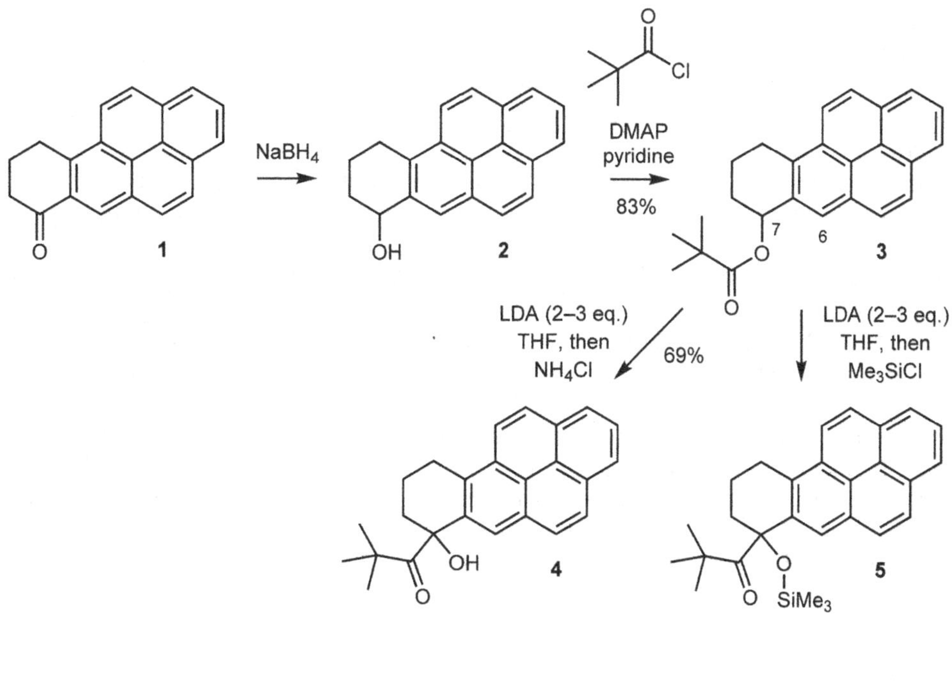

SCHEME 1

nonenolizable ester group, its deprotonation through a sterically bulky nonnucleophilic base such as lithium diisopropylamide (LDA) resulted in rearrangement to **4**. When TMSCl (rather than NH$_4$Cl) was used during the quench phase of the reaction, the corresponding silylated product **5** was obtained. Provide a mechanistic explanation for the transformations **3** to **4** and **3** to **5** that is consistent with these observations.

SOLUTION: 3 TO 4/5

KEY STEPS EXPLAINED

The mechanism proposed to explain the formation of **4** and **5** begins with LDA-mediated deprotonation of C$_7$—H of **3** followed by epoxide formation to form intermediate **A**. Fragmentation of the alkoxyepoxide group in **A** then forms **B**. We note that in this step the pivaloyl group in **A** effectively migrates from the oxygen atom to C$_7$. Alkoxide **B** can then be quenched either with trimethylsilylchloride to form **5** or with water to form **4**. The authors initially used ^{1}H NMR to observe that the hydrogen at C$_7$ of **3** was absent in products **4** and **5**. Moreover, a new signal appeared as a singlet at 5.33 ppm which was exchangeable with MeOH-d_4. The authors attributed this signal to the hydroxyl group of **4**. The C$_6$—H signal was still present in the NMR spectrum for **4** and **5** thus indicating that the pivaloyl group failed to direct metalation (i.e., deprotonation followed by capture with an electrophile) to the *peri* position (i.e., C$_6$). Lastly, the C$_6$—H signal was shifted upfield from 8.06 ppm in **3** to 7.64 ppm in **4** and **5**. The authors attributed this shift to an anisotropy effect of the newly proximal carbonyl group in these products. A final experiment carried out by the authors was to subject **4** to silylation with TMS triflate (CH$_2$Cl$_2$, Et$_3$N, 0°C, 15 min, 77% yield) which gave a product identical to **5**. These observations demonstrated that the O → C acyl shift depicted in the solution above had occurred to give **4** and **5** from **3**.

To further verify the structures of **4** and **5**, the authors prepared these products according to the synthetic sequence depicted in Scheme 2. In this case there was no rearrangement involved. We provide a mechanism for this sequence in Scheme 3. Essentially, the first step involves deprotonation of the silylcyanohydrin of pivaloyl chloride which generates a carbanion that nucleophilically attacks the ketone group of **1** to form alkoxide **C**. This intermediate undergoes an intramolecular *trans*-trimethylsilyl group migration to form product **5** after elimination of cyanide ion. Next, we have

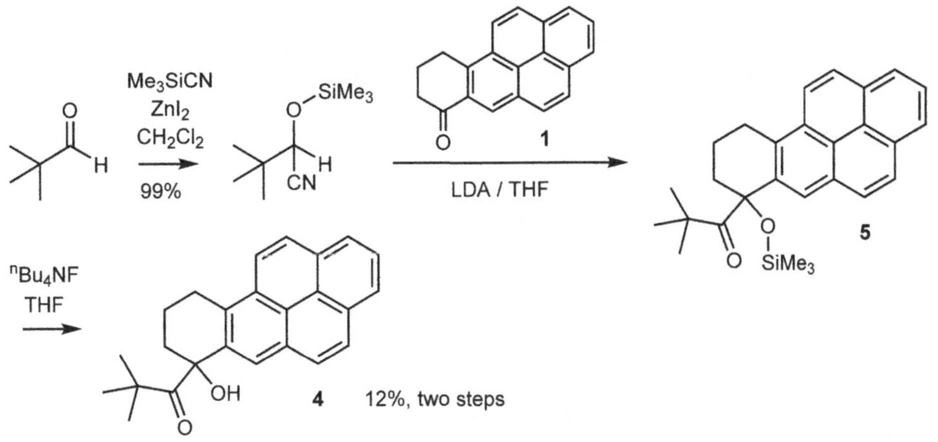

SCHEME 2

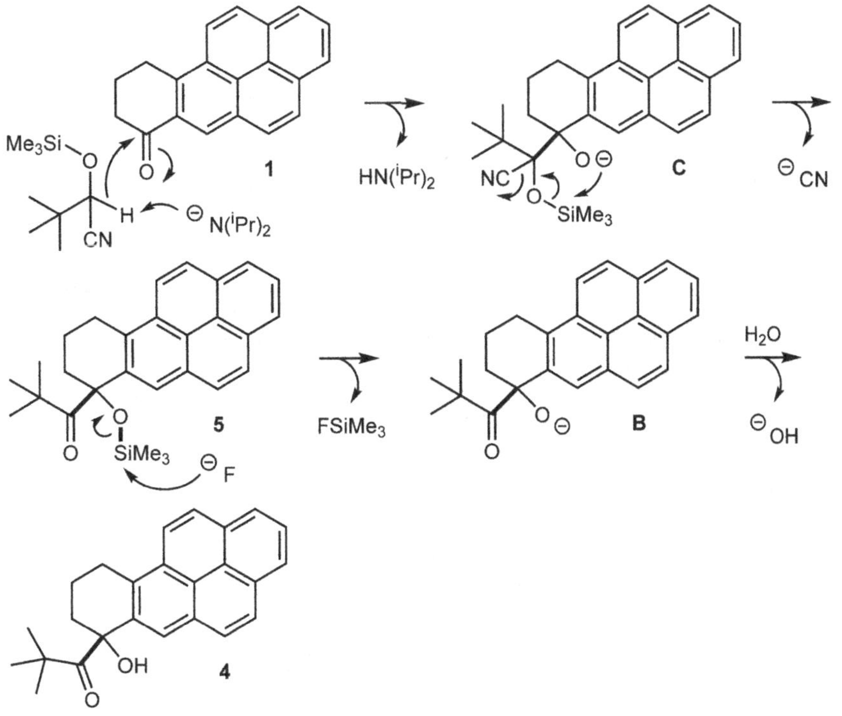

SCHEME 3

fluoride anion from tetrabutylammonium fluoride acting to displace the trimethylsilyl group from **5** to form alkoxide **B** in situ which is quenched by water to form product **4**. Lastly, the authors did not identify the epoxide intermediate **A** and they did not provide any theoretical evidence to support their proposed mechanism. Nevertheless, their intention was to carry out a directed metalation of the *peri* position by analogy with the well-known directed *ortho*-metalation protocol as shown in Scheme 4.[2]

ADDITIONAL RESOURCES

For an example of an unexpected rearrangement during an attempted metalation reaction, see the work of Berchel et al.[3] We also recommend several interesting articles on metalation reactions, some of which involve theoretical calculations and mechanistic discussion as well.[4-6]

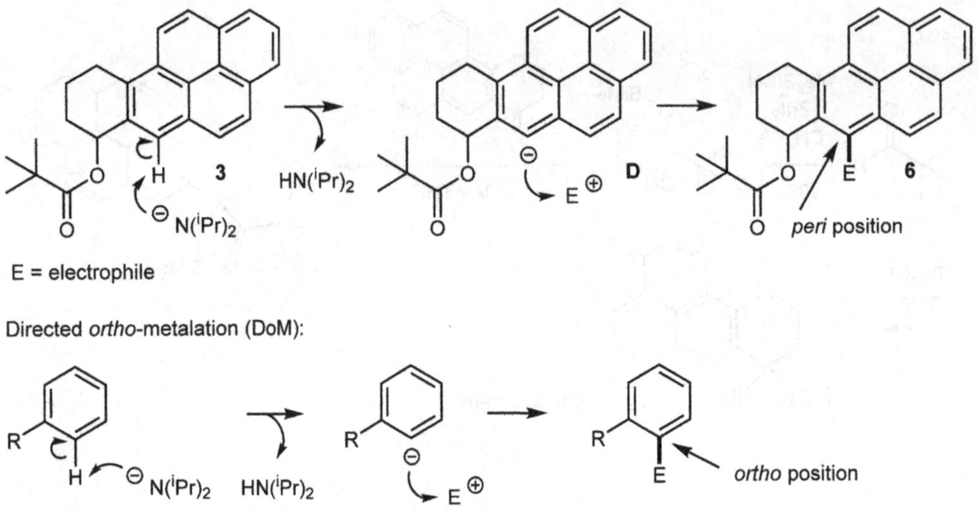

E = electrophile

Directed *ortho*-metalation (DoM):

ortho position

SCHEME 4

References

1. Zajc B, Lakshman MK. Acyl migration in anions derived from nonenolizable esters of polycyclic aromatic hydrocarbons. *J Org Chem*. 1995;60 (15):4936–4939. https://doi.org/10.1021/jo00120a047.
2. Snieckus V. Directed ortho metalation. Tertiary amide and O-carbamate directors in synthetic strategies for polysubstituted aromatics. *Chem Rev*. 1990;90(6):879–933. https://doi.org/10.1021/cr00104a001.
3. Berchel M, Salauen J-Y, Couthon-Gourves H, Haelters J-P, Jaffres P-A. An unexpected base-induced [1,4]-phospho-fries rearrangement. *Dalton Trans*. 2010;39:11314–11316. https://doi.org/10.1039/c0dt00880j.
4. Gros P, Fort Y, Caubere P. Aggregative activation in heterocyclic chemistry. Part 4. Metallation of 2-methoxypyridine: unusual behaviour of the new unimetal superbase BuLi–Me₂N(CH₂)₂OLi (BuLi–LiDMAE). *J Chem Soc Perkin Trans 1*. 1997;1997:3071–3080. https://doi.org/10.1039/A701914I.
5. Enders M, Friedmann CJ, Plessow PN, et al. Unprecedented pseudo-ortho and ortho metallation of [2.2]paracyclophanes—a methyl group matters. *Chem Commun*. 2015;51:4793–4795. https://doi.org/10.1039/c5cc00492f.
6. Blair VL, Stevens MA, Thompson CD. The importance of the Lewis base in lithium mediated metallation and bond cleavage reaction of allyl amines and allyl phosphines. *Chem Commun*. 2016;52:8111–8114. https://doi.org/10.1039/c6cc03947b.

Question 34: 1,4-Dioxene in Synthesis

During a 1995 study of the role of 1,4-dioxene in synthesis, a French chemist undertook the synthetic sequence represented in Scheme 1.[1] Propose a mechanism for the transformation of **4** to the α-hydroxy ketone product **5**.

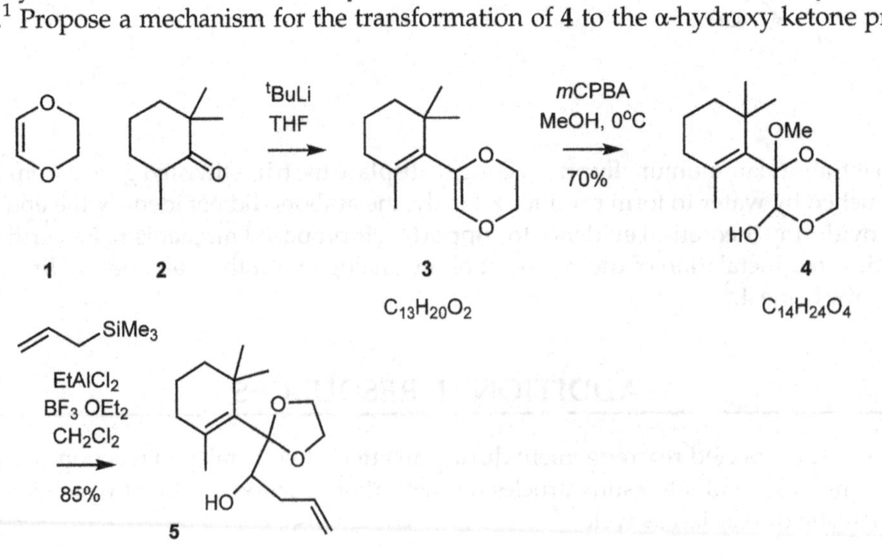

SCHEME 1

SOLUTION

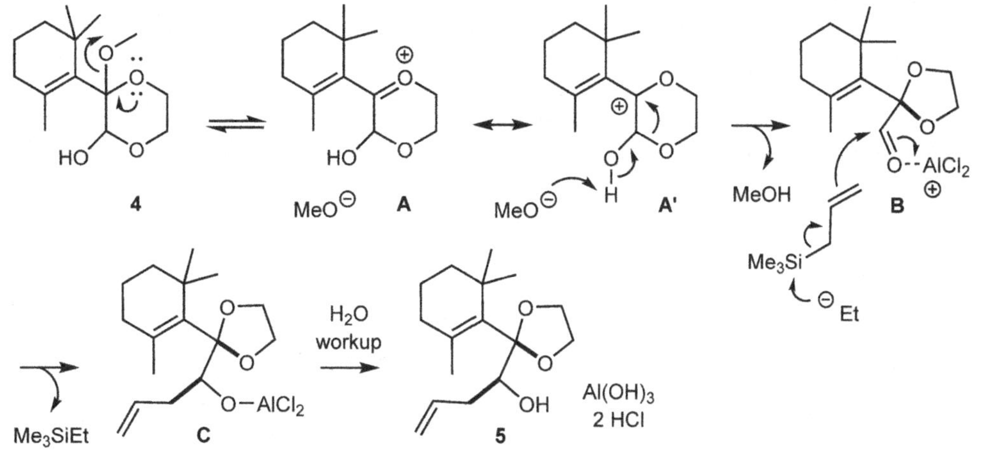
4
MeO⊖ A
MeO⊖ A'
MeOH
B
Me3Si
Et
O--AlCl2
Me3SiEt
C
H2O
workup
5
Al(OH)3
2 HCl

BALANCED CHEMICAL EQUATION

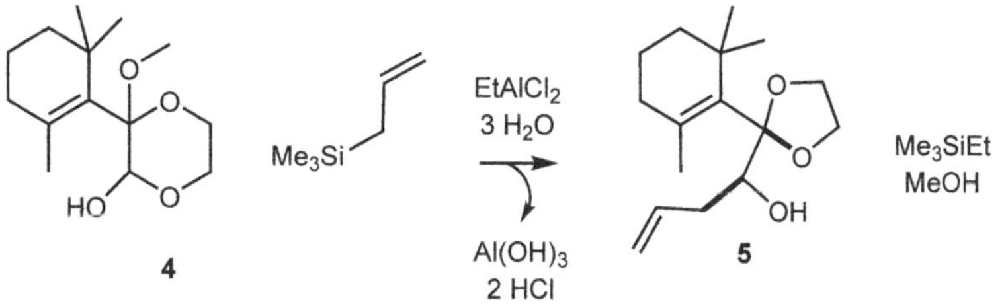
4
Me3Si
EtAlCl2
3 H2O
Al(OH)3
2 HCl
5
Me3SiEt
MeOH

KEY STEPS EXPLAINED

The mechanism proposed by the authors to explain the formation of **5** from **4** begins with fragmentation of compound **4** to methoxide ion and carbocation **A′** which upon deprotonation rearranges to 1,3-dioxane **B**. We note that the aldehyde group of this intermediate is coordinated to the aluminum Lewis acid moiety. This promotes attack by ethyl carbanion onto the silyl group of allyltrimethylsilane which results in a nucleophilic attack at the aldehyde carbon atom of **B** which gives intermediate **C**. Upon aqueous work-up, **C** transforms into product **5**.

Nevertheless, we note that in the author's synthetic scheme product **4** arises as a result of oxidation of **3** with *meta*-chloroperoxybenzoic acid (*m*CPBA) which is known to cause formation of epoxide products. Indeed, an epoxide intermediate **D** is expected which can then be attacked by methanol to give two possible products (Scheme 2).

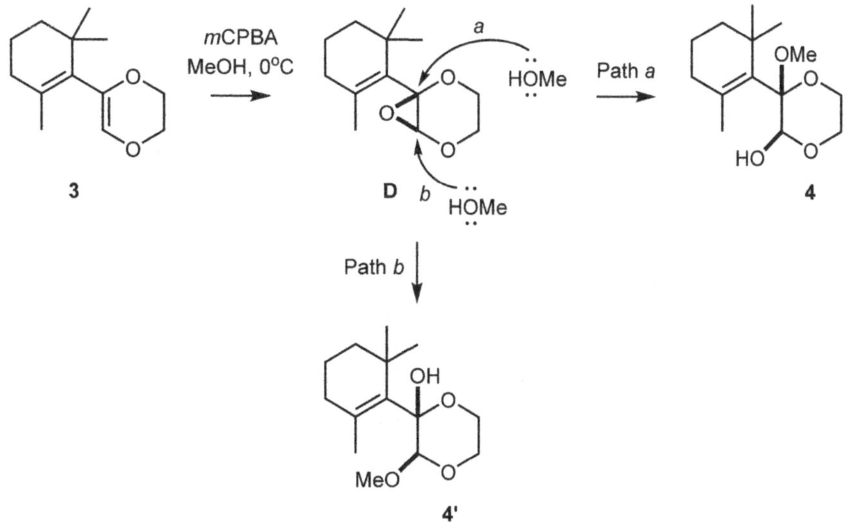
3
mCPBA
MeOH, 0°C
D
HOMe
Path a
4
Path b
4'

SCHEME 2

Unfortunately, according to the author's scheme, product **4** would require that intermediate **D** be attacked by methanol on the *more* hindered side (path *a*). We view this requirement as problematic from a sterics perspective especially in light of the fact that the author did not disclose evidence for the characterization of product **4**. Therefore, we note that a more plausible scenario is to have methanol attack the *less* hindered side of epoxide **D** to form a product whose structure is that of **4'**. We believe that **4'** can then react in the next transformation to give intermediate **B** (the mechanism of which is provided in Scheme 3) and thereafter lead to **C** and to **5** as seen previously. To test this hypothesis, a future study has to characterize the structure of product **4** and to react both **4** and **4'** separately under the reaction conditions to see if **5** can be formed in a yield similar to 85%. A theoretical analysis would also go a long way in helping to support the proposed mechanism.

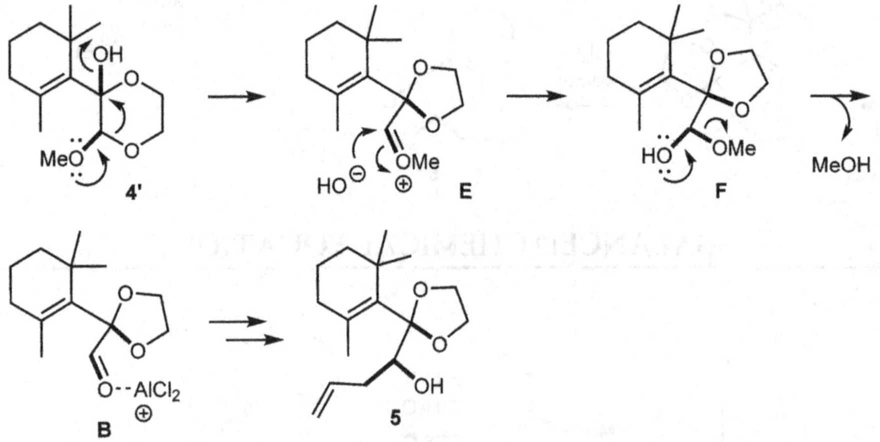

SCHEME 3

ADDITIONAL RESOURCES

We recommend two interesting articles which explore the mechanisms of acid-catalyzed reactions of ketals such as **4** and of 1,4-dioxenes such as **3**.[2,3]

References

1. Hanna I. 1,4-dioxene in organic synthesis: introduction of a second carbon-carbon bond with simultaneous opening of the dioxene ring. *Tetrahedron Lett.* 1995;36(6):889–892. https://doi.org/10.1016/0040-4039(94)02387-Q.
2. Sammakia T, Smith RS. Evidence for an oxocarbenium ion intermediate in Lewis acid mediated reactions of acyclic acetals. *J Am Chem Soc.* 1994;116(17):7915–7916. https://doi.org/10.1021/ja00096a066.
3. Kresge AJ, Yin Y. Kinetics and mechanism of the acid-catalyzed hydration of dihydro-1,4-dioxin. *J Phys Org Chem.* 1989;2:43–50. https://doi.org/10.1002/poc.610020106.

Question 35: Collins Oxidation of Phenylethanols

In 1995 American chemists attempted the Collins oxidations depicted in Scheme 1.[1] Interestingly, it was noted that when the R group of the four-substituted phenethyl alcohols **1** was a chloride or methoxy group, oxidation by means of the Collins reagents formed the corresponding aldehydes **2** in excellent yield with trace amounts of the fragmentation product **3**. When the R group was a nitro group as with **4**, the yield of the fragmentation product **5** increased to 30% and there was 70% of starting material **4** recovered as well. Provide a mechanistic explanation for these different outcomes.

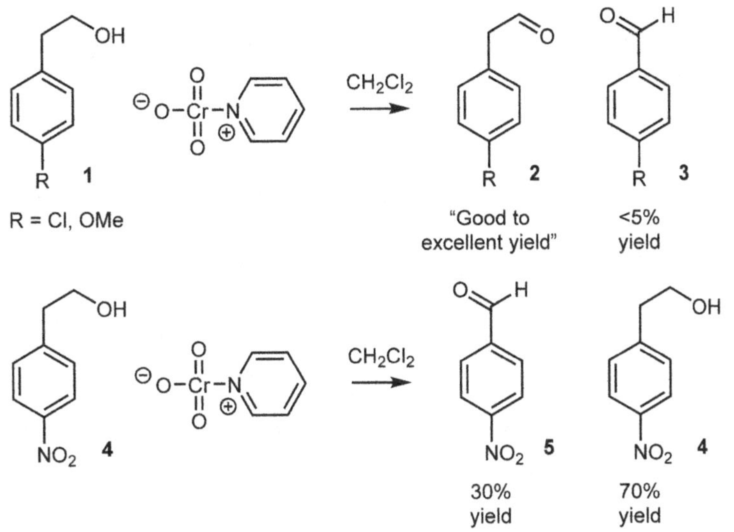

SCHEME 1

SOLUTION: 1 TO 2

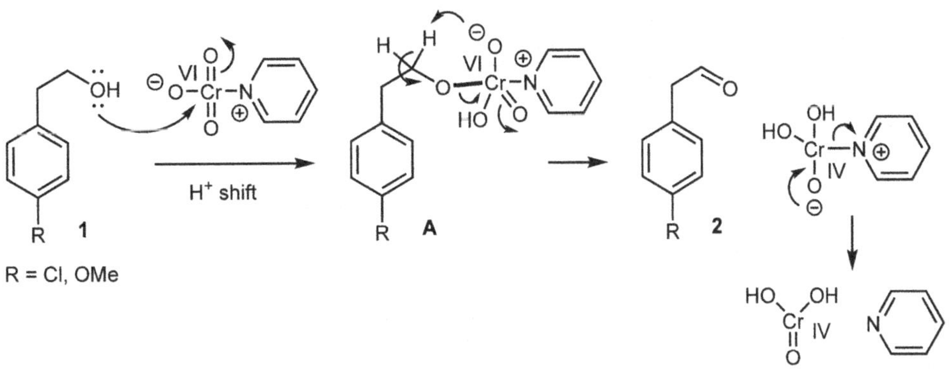

SOLUTION: 4 TO 5

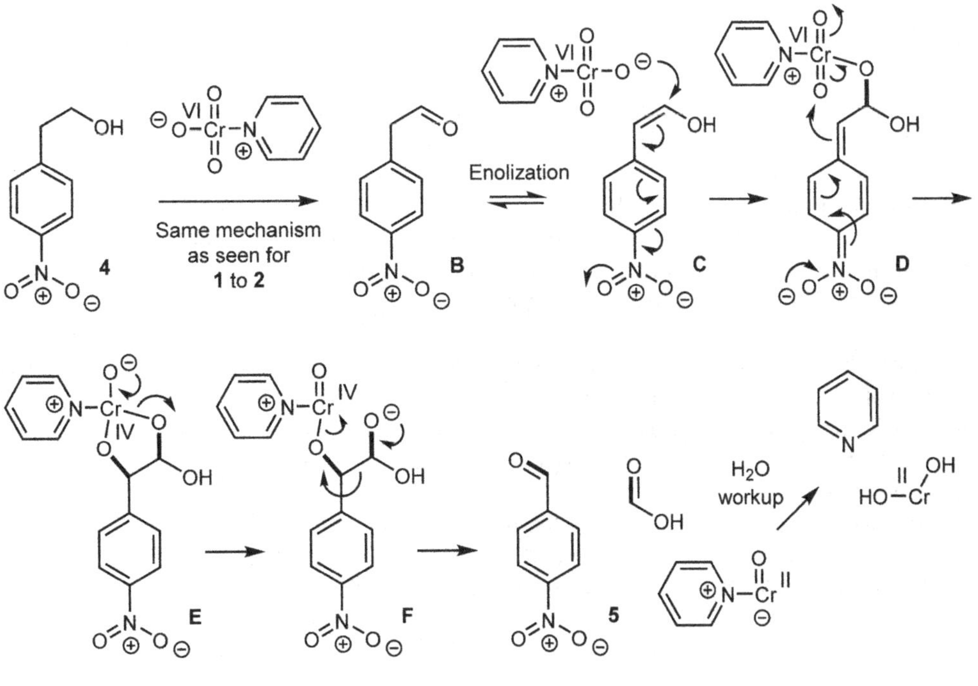

BALANCED CHEMICAL EQUATIONS

R = Cl, OMe

KEY STEPS EXPLAINED

The mechanism proposed to explain the standard Collins oxidation of primary alcohols to aldehydes begins with nucleophilic attack by the alcohol group of **1** onto the chromium atom to form the chromate ester intermediate **A** after a proton shift. This intermediate then undergoes an intramolecular α-hydrogen abstraction to give the aldehyde product **2** plus an organochromium moiety which fragments into pyridine and $(HO)(Cr^{IV}=O)(OH)$. We note that this mechanism operates when the *para* substituted R group of the benzene ring is neutral or electron donating. In the special case where we have an electron-withdrawing group such as a nitro group, the authors propose a mechanism to explain the formation of the degradation product **5**. In this mechanism, the authors believe that intermediate **B** (**2**-like aldehyde product) is formed by the standard Collins oxidation mechanism seen earlier where one equivalent of the Collins reagents is consumed. Nevertheless, **B** continues to react through enolization to form **C** which is attacked by a second equivalent of the Collins reagent to form **D**. This step is evidently facilitated by the electron sink provided by the electron-withdrawing nature of the nitro group in the *para* position of the benzene ring. Effectively this well-positioned nitro group absorbs the excess electron density to allow for formation of **D**. At this point, intermediate **D** cyclizes to **E** where the chromium atom is reduced from the +6 to the +4 oxidation state. Intermediate **E** undergoes ring opening to form **F** which then fragments *p*-nitrobenzaldehyde **5**, formic acid by-product, and an organochromium(II) species which upon aqueous work-up leads to pyridine and $Cr^{II}(OH)_2$. We note that the authors also tested the four-substituted dimethylamino group equivalent of **1** and found it to produce no amount of the degradation product (equivalent of **5**) which occurs because dimethylamino group is strongly electron donating thereby working to impede formation of **D**.

Nevertheless, we note that the authors' mechanism for the formation of **5** involves a staged reduction of chromium from an oxidation state of VI to IV to II. We view this proposal as problematic because the Cr(II) species does not follow the normal course of oxidation state change for chromium in the Collins reagent which is VI to IV. Therefore, we propose an alternative mechanism for the formation of **5** (see Scheme 2). We also provide a balanced chemical equation for this mechanism in Scheme 3.

In this alternative mechanism, the same steps leading up to chromate ester **A** are followed except that an intramolecular β-hydrogen abstraction occurs instead of an α-hydrogen abstraction. This leads to intermediate **G** again showing the importance of the electron-withdrawing ability of the nitro group to absorb the excess

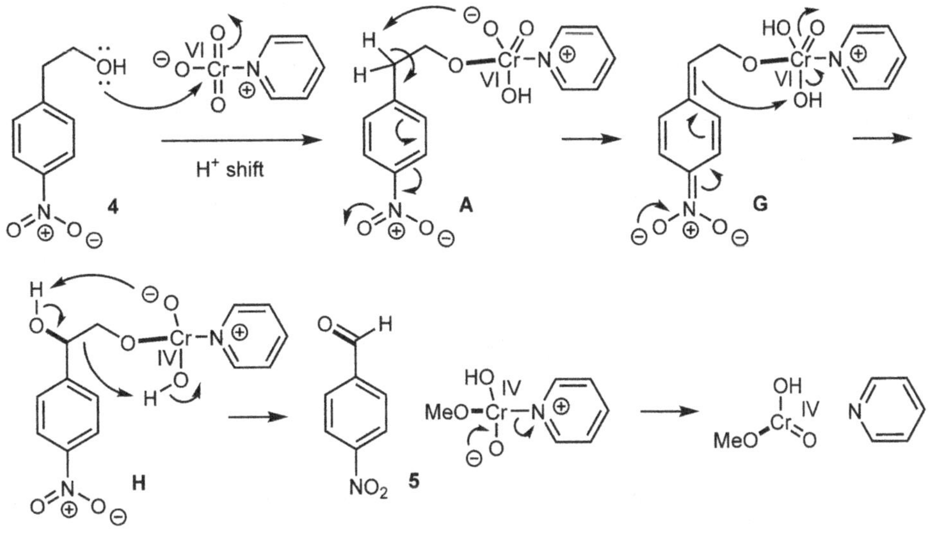

SCHEME 2

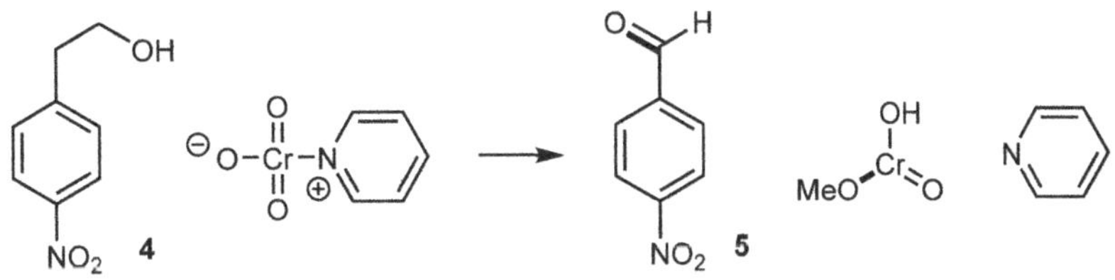

SCHEME 3

negative charge. Intermediate **G** then undergoes a hydroxyl group transfer to the benzylic position thus forming alcohol **K** which fragments to *p*-nitrobenzaldehyde **5** and an organochromium IV species containing a methoxy group.

When comparing the two mechanisms, we note that the authors' mechanism involves reduction of one equivalent of Collins reagent Cr(VI) to Cr(IV) in parallel to the oxidation of *p*-nitrophenylethanol **1** to *p*-nitrophenylacetaldehyde **B**. Then, a second equivalent of Collins reagent Cr(VI) is reduced in a stepwise fashion to Cr(IV) and then to Cr(II) in parallel to the oxidation of *p*-nitrophenylacetaldehyde **B** to *p*-nitrobenzaldehyde **5** and formic acid. The alternative mechanism involves the reduction of one equivalent of Collins reagent Cr(VI) to Cr(IV) in parallel to oxidation of *p*-nitrophenylethanol **1** to *p*-nitrobenzaldehyde **5**. The terminal methylene group is cleaved as (MeO)(Cr^{IV}=O)(OH) which upon aqueous work-up presumably leads to methanol and (HO)(Cr^{IV}=O)(OH). As such the transformation appears to occur in fewer steps and not to involve formation of the problematic Cr(II) species. In addition, the alternative mechanism does not generate formic acid by-product. Unfortunately, the authors did not carry out a check experiment involving the oxidation of *p*-nitrophenylacetaldehyde **B** using Collins reagent to see if *p*-nitrobenzaldehyde **5** is formed. This would have given evidence that *p*-nitrophenylacetaldehyde **B** is a reaction intermediate along the reaction pathway as was claimed. Another missing piece of evidence was verification that formic acid is a reaction by-product. Both of these experimental verifications are necessary in order to validate the authors' proposed mechanism. In our opinion the alternative mechanism appears more rational and is consistent with the behavior of the Collins reagent in oxidation reactions, but in the absence of experimental and theoretical evidence to support either mechanism it is not possible to say with certainty which mechanism is operational.

ADDITIONAL RESOURCES

We recommend two articles which discuss interesting mechanisms of transformations that involve Cr(VI) species.[2,3]

References

1. Li M, Johnson ME. Oxidation of certain 4-substituted phenethyl alcohols with Collins reagent: on the mechanism of a carbon-carbon bond cleavage. *Synth Commun.* 1995;25(4):533–537. https://doi.org/10.1080/00397919508011387.
2. Pandey D, Kothari S. Kinetics and correlation analysis of reactivity in the oxidation of some alkyl phenyl sulfides by benzimidazolium dichromate. *Bull Chem Soc Jpn.* 2014;87:1224–1230. https://doi.org/10.1246/bcsj.20140154.
3. Subramaniam P, Selvi NT, Devi SS. Spectral and mechanistic investigation of oxidative decarboxylation of phenylsulfinylacetic acid by Cr(VI). *J Korean Chem Soc.* 2014;58(1):17–24. https://doi.org/10.5012/jkcs.2014.58.1.17.

Question 36: Allene Sulfoxide Rearrangement

In 1995 English chemists undertook the synthetic sequence depicted in Scheme 1.[1] Suggest a mechanistic explanation for the transformations **2** to **3** and **3** to **4**.

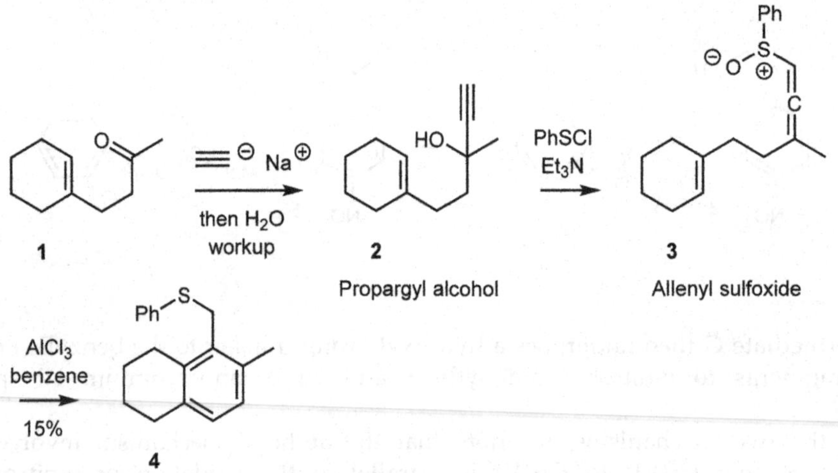

SCHEME 1

SOLUTION: 2 TO 3

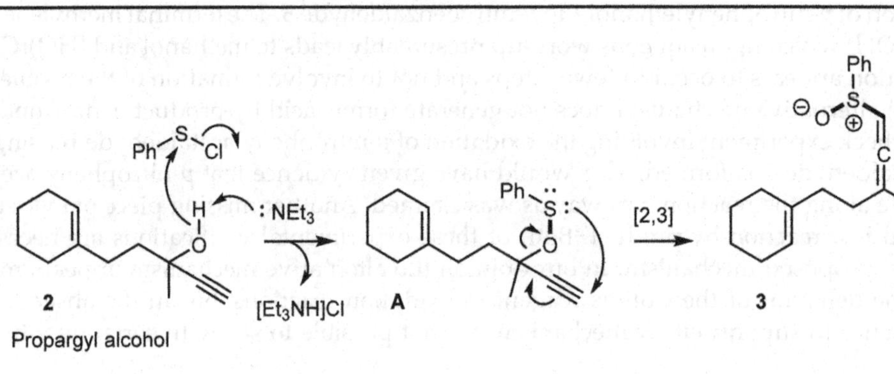

SOLUTION: 3 TO 4

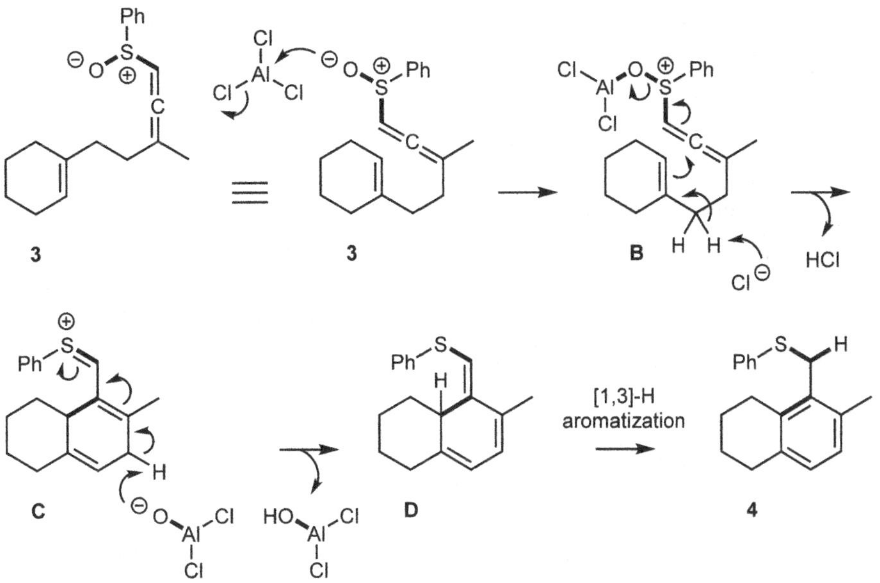

BALANCED CHEMICAL EQUATIONS

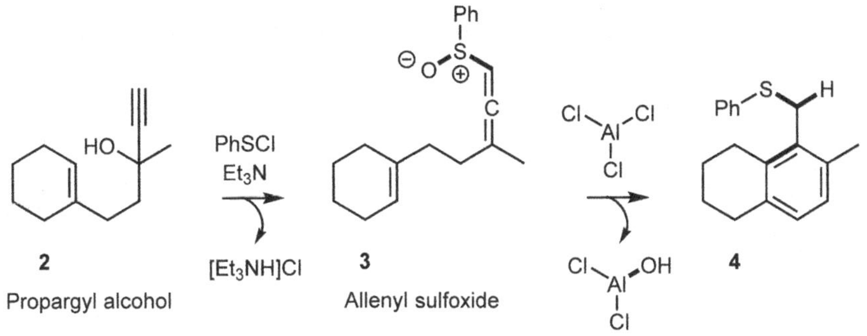

KEY STEPS EXPLAINED

The mechanism proposed for the transformation of **2** to **3** begins with deprotonation of the propargyl alcohol of **2** by triethylamine base followed by nucleophilic attack by the resulting alkoxide onto the benzenesulfenyl chloride reagent to form intermediate **A**. This intermediate undergoes a [2,3]-sigmatropic rearrangement to form the allenyl sulfoxide product **3**. For the second reaction, the proposed mechanism begins with nucleophilic attack by the sulfoxide group of **3** onto the aluminum trichloride reagent which forms the sulfonium ion intermediate **B**. Chloride ion then deprotonates one of the allyl methylene protons causing a [6+0] cyclization which generates the sulfonium ion intermediate **C**. Elimination of dichloroaluminum hydroxide as a primary by-product following deprotonation of **C** via dichloroaluminum oxide ion gives intermediate **D**. This intermediate undergoes a [1,3] protic shift to facilitate aromatization of the newly formed ring thus resulting in formation of product **4**. Upon aqueous work-up dichloroaluminum hydroxide will eventually react to produce aluminum hydroxide and hydrochloric acid as the end by-products. We note that the ring construction mapping for the last reaction is [6+0]. In addition, we note that the authors did not conduct further experimental and theoretical work to help support their proposed mechanism.

We recommend an article which discusses the mechanism of a transformation that involves electrocyclization of allene-sulfoxides.[2] We also recommend an article which describes an interesting sulfoxide-sulfenate rearrangement in the synthesis of cortico steroids.[3] Lastly we recommend two review articles which discuss cyclizations and [2,3]-sigmatropic rearrangements of allenes.[4,5]

References

1. Parsons PJ, Parkes KEB, Penkett CS. Chemistry of allene sulfoxides: Lewis acid catalysed cyclisation reactions leading to highly functionalised rings. *Synlett*. 1995;1995(7):709–710. https://doi.org/10.1055/s-1995-5055.
2. Khodabocus A. Studies into the synthesis of forskolin intermediates by electrocyclisation of allene-sulfoxides. *ARKIVOC*. 2000;(6):854–867. https://doi.org/10.3998/ark.5550190.0001.602.
3. VanRheenen V, Shephard KP. New synthesis of cortico steroids from 17-keto steroids: application and stereochemical study of the unsaturated sulfoxide-sulfenate rearrangement. *J Org Chem*. 1979;44(9):1582–1584. https://doi.org/10.1021/jo01323a054.
4. Ma S. Electrophilic addition and cyclization reactions of allenes. *Acc Chem Res*. 2009;42(10):1679–1688. https://doi.org/10.1021/ar900153r.
5. Braverman S, Cherkinsky M. [2,3]sigmatropic rearrangements of propargylic and allenic systems. *Top Curr Chem*. 2007;275:67–101. https://doi.org/10.1007/128_047.

Question 37: A Failed "Pinacol-Type" Rearrangement

In 1991 collaboration between chemists from Venezuela and the United States led to an unexpected finding.[1] When the diterpene derivative **1** was treated with $BF_3.OEt_2$ and Ac_2O an expected pinacol-type rearrangement designed to form the ring B-homo derivative **2** did not occur (Scheme 1).[1] What happened instead was "a profound backbone rearrangement" which gave product **3** in 33% yield. Suggest a mechanism for the formation of **3** from **1**.

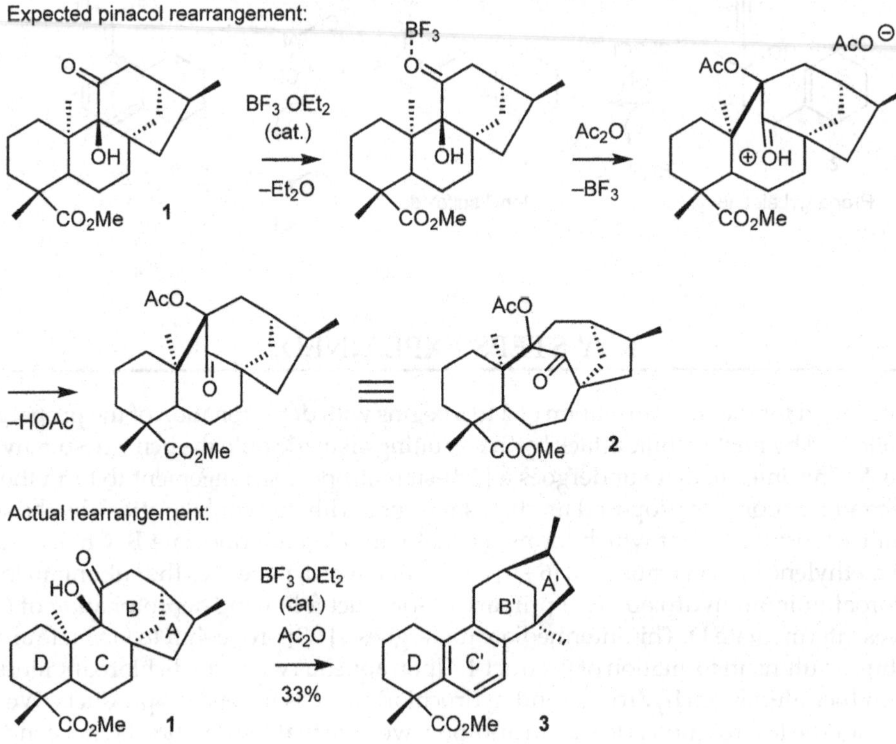

SCHEME 1

SOLUTION

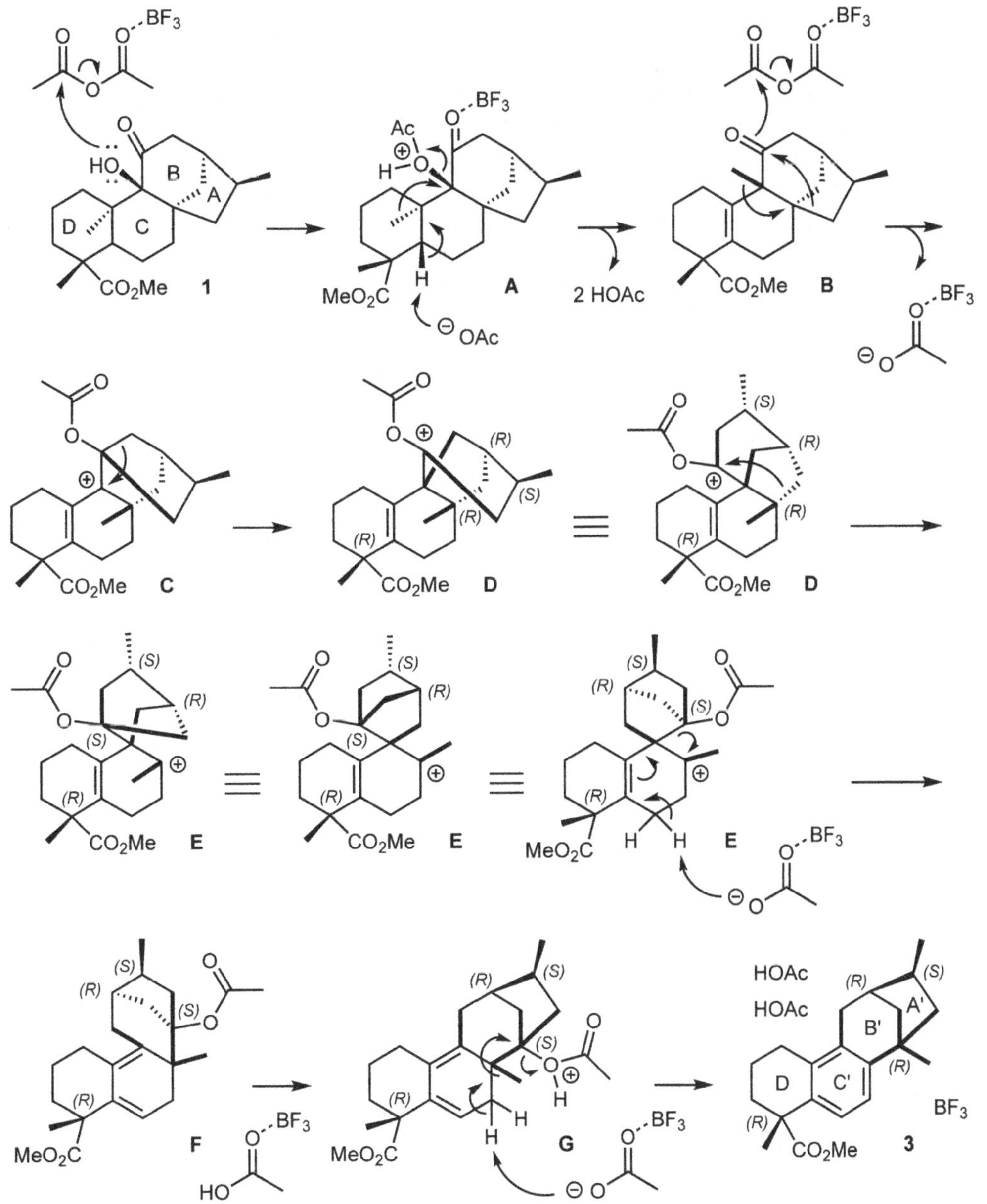

BALANCED CHEMICAL EQUATION

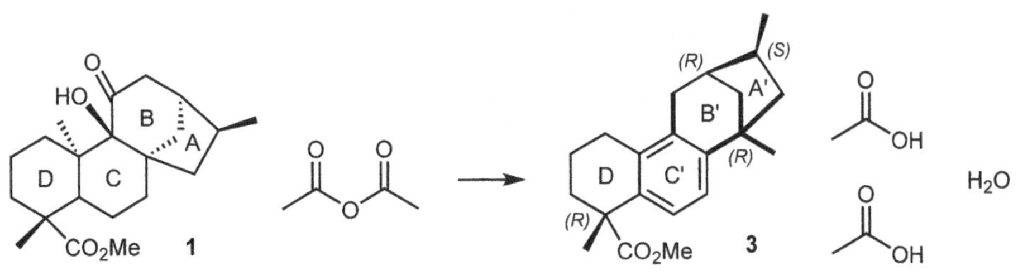

KEY STEPS EXPLAINED

We begin by noting that the transformation of **1** to **3** involves a deep-seated skeletal rearrangement of the starting material. As such, the proposed mechanism begins with recognizing that boron trifluoride acts as a Lewis acid which coordinates to the carbonyl group of acetic anhydride thereby making it more electrophilic. The hydroxyl group of **1** is then acetylated by acetic anhydride to form intermediate **A**. At this point the liberated acetate ion abstracts the hydrogen atom at the junction of rings D and C causing a 1,2-methyl shift with concomitant elimination of acetic acid to form intermediate **B**. This intermediate undergoes a second 1,2-methyl shift around ring C with simultaneous fragmentation of ring A and attack at the carbonyl group of ring B which also becomes acetylated via acetic anhydride. This process leads to the carbocation intermediate **C**. This intermediate in turn rearranges to form the acetoxy carbocation **D** via fragmentation of ring B. Intermediate **D** rearranges again to form the spiro-[6,6] carbocation **E** thus forming a new five-membered ring A′. Acetate ion initiates a deprotonation in ring C which causes fragmentation of the spiro ring creating a new six-membered ring B′. The acetyl group of the resulting intermediate **F** is protonated and acetate ion then removes a hydrogen atom from ring C causing it to aromatize with simultaneous 1,2-methyl shift to form the final product **3**. We note that the ring construction mapping for this transformation is $[(3+2+1)_{B'}+(4+1)_{A'}]$. Also, during the course of this mechanism rings C and D remain intact and the methyl group undergoes three 1,2-migrations.

We note that in support of this proposed mechanism the authors prepared intermediate **B** independently and then subjected it to the same reaction conditions as with **1** which gave the same product **2**. Product **2** was characterized by ^{1}H NMR, ^{13}C NMR, and mass spectrometry. Nevertheless, the authors did not carry out a computational study to map out the energetics of this fascinating transformation. We regard such an endeavor as highly valuable.

ADDITIONAL RESOURCES

We recommend a recent article which contains a full computational analysis of a related transformation which likewise involves an unexpected skeletal rearrangement.[2]

References

1. Nakano T, Maillo MA, Usubillaga A, McPhail AT, McPhail DR. Rearrangements of methyl 9β-hydroxy-11-oxo-(−)-kauran-19-oate to diterpene skeletons with new ring systems. *Tetrahedron Lett*. 1991;32(52):7667–7670. https://doi.org/10.1016/0040-4039(91)80560-S.
2. Hong YJ, Tantillo DJ. The energetic viability of an unexpected skeletal rearrangement in cyclooctatin biosynthesis. *Org Biomol Chem*. 2015;13:10273–10278. https://doi.org/10.1039/C5OB01785H.

Question 38: Allylation of 1,4-Benzoquinones

During studies aimed at optimizing allylation of 1,4-benzoquinones, Japanese chemists showed that treatment of 3 equivalents of trifluorosilane **1** with one equivalent of the 1,4-benzoquinone **2** in formamide in the presence of 5 equivalents of $FeCl_3 \cdot 6H_2O$ for 23h at 40°C gave the plastoquinone-1 product **3** in 90% yield (Scheme 1).[1] It was also observed that strongly coordinating solvents such as formamide played an important role in the transformation. Suggest a mechanism for this reaction which is consistent with these observations.

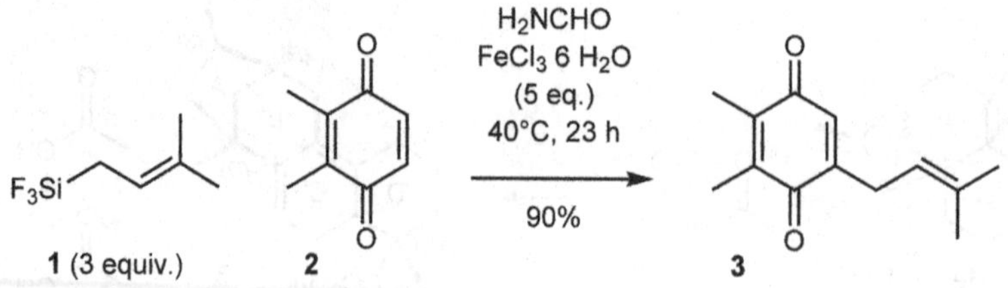

SCHEME 1

SOLUTION

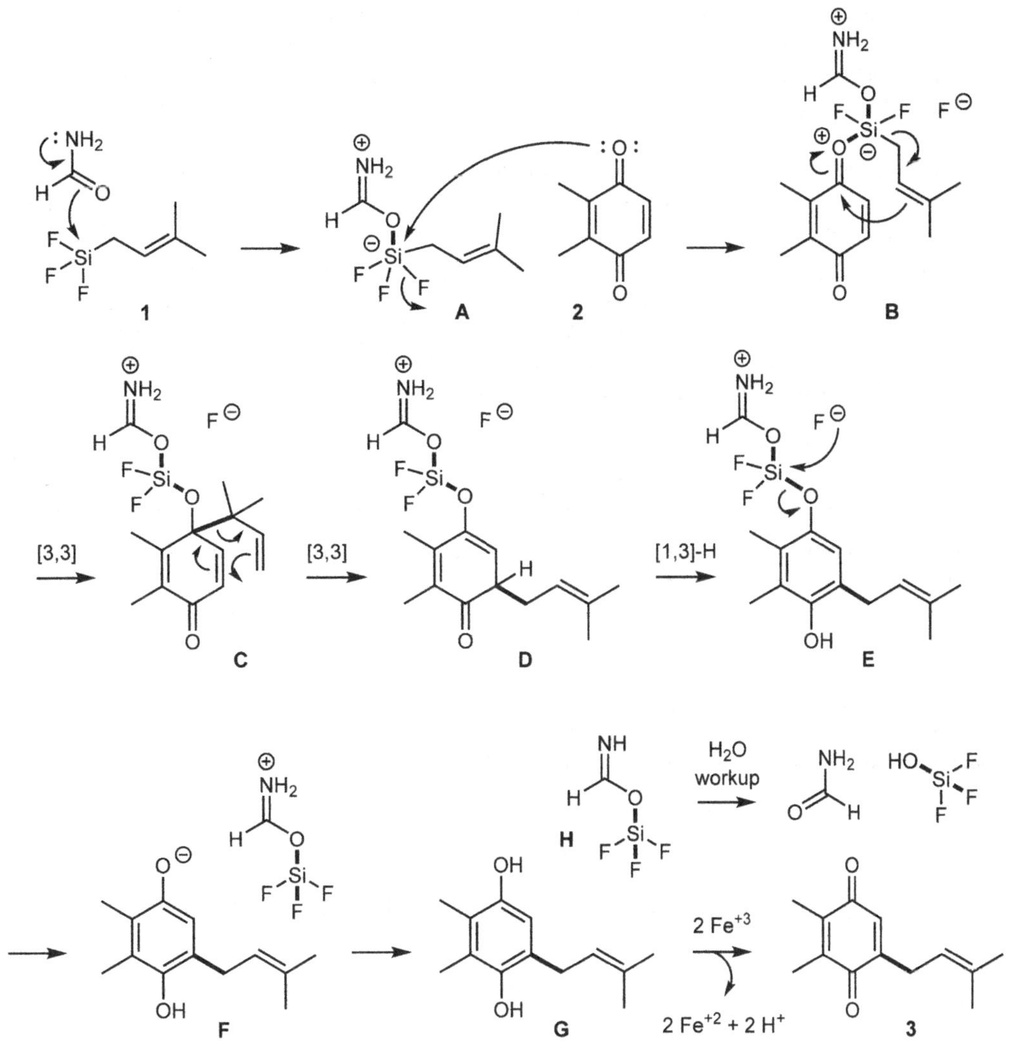

BALANCED CHEMICAL EQUATION

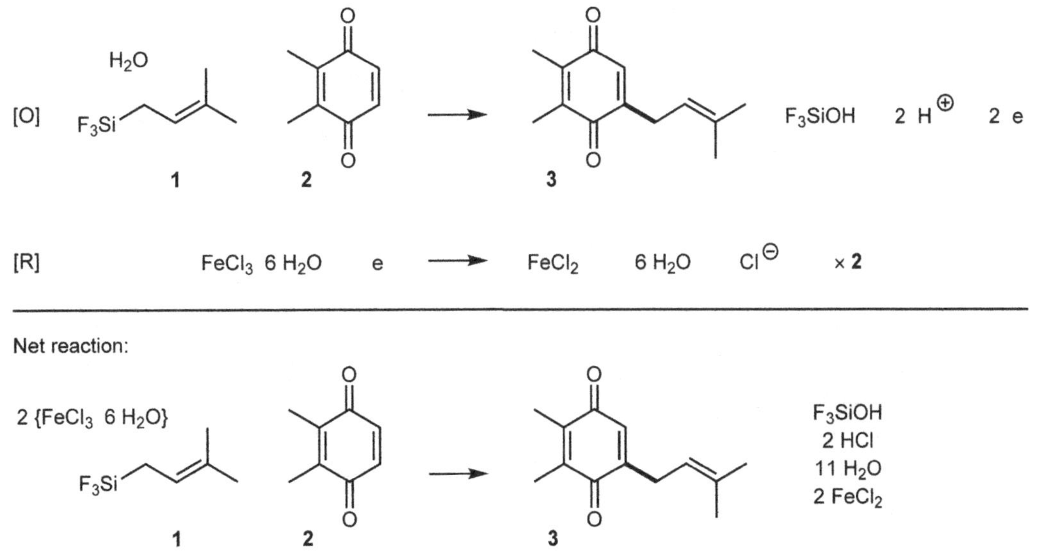

Net reaction:

KEY STEPS EXPLAINED

The mechanism proposed by the authors begins with nucleophilic attack by formamide onto allylsilane **1** which forms a pentacoordinate silicon intermediate **A**. This intermediate is then attacked by benzoquinone **2** to form intermediate **B** which undergoes a [3,3] sigmatropic rearrangement to form intermediate **C** through migration of the prenyl group to the carbonyl which bears the silyl reagent. The authors note that this step is accelerated by the strong Lewis acidity of pentacoordinate silicon intermediate **B** which promotes allylic addition. This pentacoordinate species is made possible through coordination of **2** to strongly coordinating solvents such as formamide. What follows is a second [3,3] rearrangement of the prenyl group of **C** which results in its migration to the *ortho* position of the other carbonyl group thus generating intermediate **D**. This intermediate undergoes a [1,3] prototropic shift to form **E**. Fluoride attack onto the silyl group of **E** generates hydroquinone **G** after elimination of the silyl moiety and protonation. Hydroquinone **G** is oxidized to the allylbenzoquinone product **3** via a two-electron transfer process in which two equivalents of Fe(III) are reduced to Fe(II). Upon aqueous work-up the liberated silyl fragment **H** is hydrolyzed to trifluorosilyl-hydroxide with regeneration of formamide. We note that the authors did not perform additional experimental and theoretical analysis to help support their proposed mechanism.

ADDITIONAL RESOURCES

We recommend an excellent review article on allylation reactions carried out using gold catalysis.[2] This article also discusses some of the mechanisms of the transformations presented.

References

1. Hagiwara E, Hatanaka Y, Gohda K, Hiyama T. Allylation of quinones with allyl(trifluoro)silanes: direct synthesis of isoprenoid quinones. *Tetrahedron Lett.* 1995;36(16):2773–2776. https://doi.org/10.1016/0040-4039(95)00394-R.
2. Quintavalla A, Bandini M. Gold-catalyzed allylation reactions. *ChemCatChem.* 2016;8(8):1437–1453. https://doi.org/10.1002/cctc.201600071.

Question 39: Benzopyrene Synthesis—By Accident

In 1995 American chemists treated the pentacycle **1** with HI/50% H_3PO_2 at 100°C for 2 to 3 min to form the trimethoxybenzopyrene product **2** in 95% yield (Scheme 1).[1] Suggest a mechanism for this transformation.

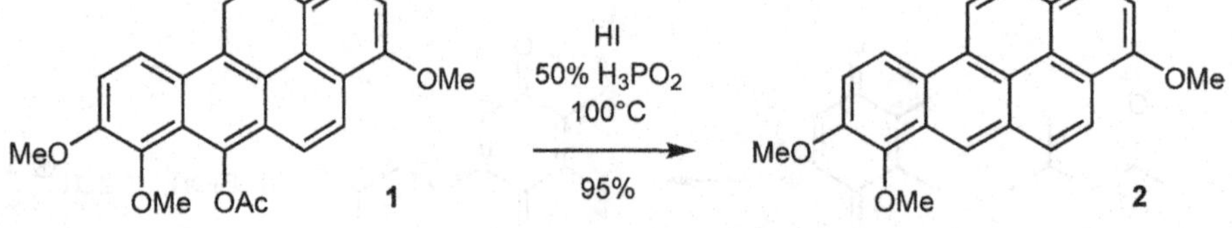

SCHEME 1

SOLUTION

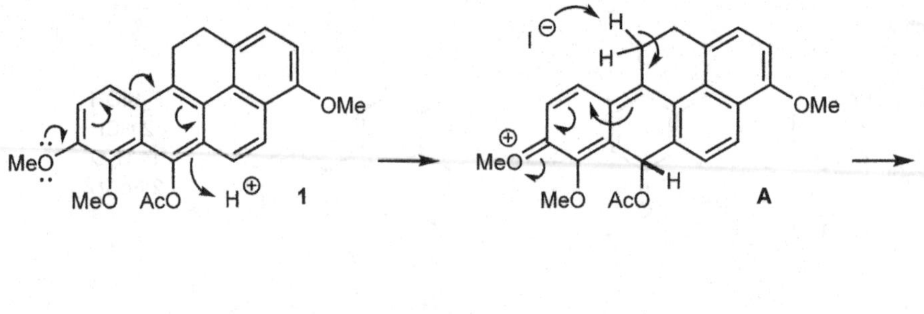

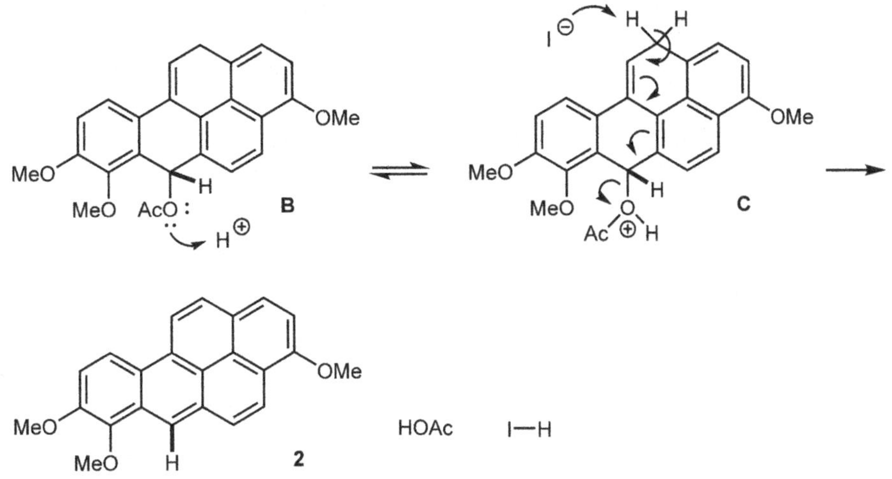

BALANCED CHEMICAL EQUATION

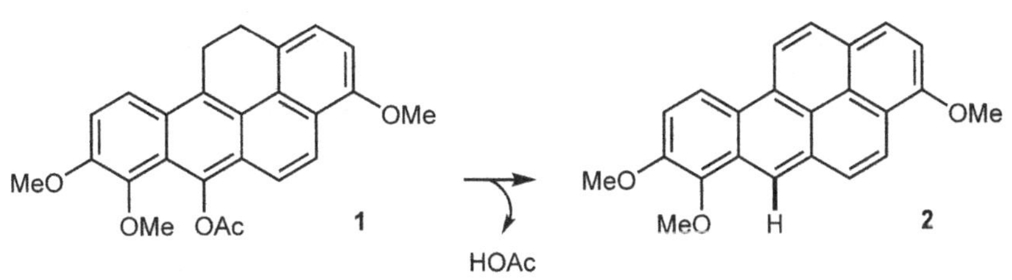

KEY STEPS EXPLAINED

The mechanism proposed by the authors to explain the formation of **2** from **1** begins with protonation of **1** at C_6 which gives the oxonium ion intermediate **A**. This intermediate is deprotonated at C_{11} to form **B** whose acetate group undergoes protonation to form **C**. Deprotonation of **C** at C_{12} leads to elimination of acetic acid by-product and formation of the benzopyrene product **2**. We note that oxonium ion **A** is stabilized through resonance by means of carbocation **D** and oxonium ion **E** (Scheme 2). Given that the methoxy group at C_3 can also participate in the transformation via structure **E**, we can envision an alternative mechanism that leads from **1** to **B** via **E** (Scheme 3).

We also note that a closer examination of the reaction by oxidation number analysis indicates that it is an internal redox reaction (Scheme 4). For instance, the carbon atoms labeled as C_a (C_{12}) and C_b (C_{11}) are each oxidized from -2 to

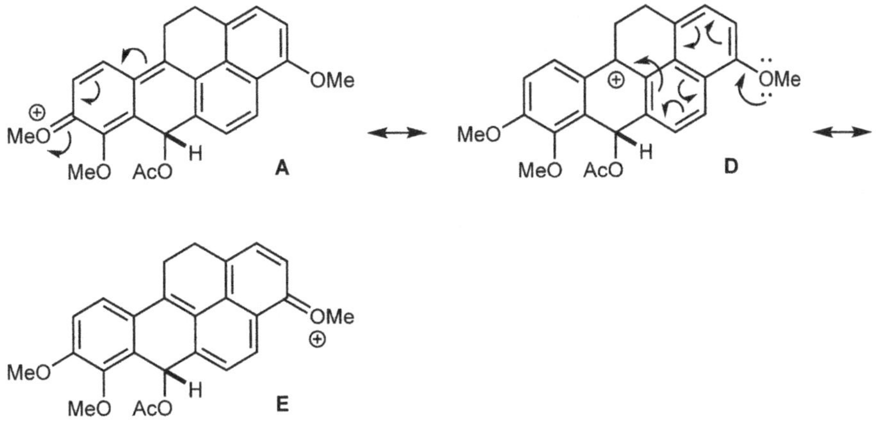

SCHEME 2

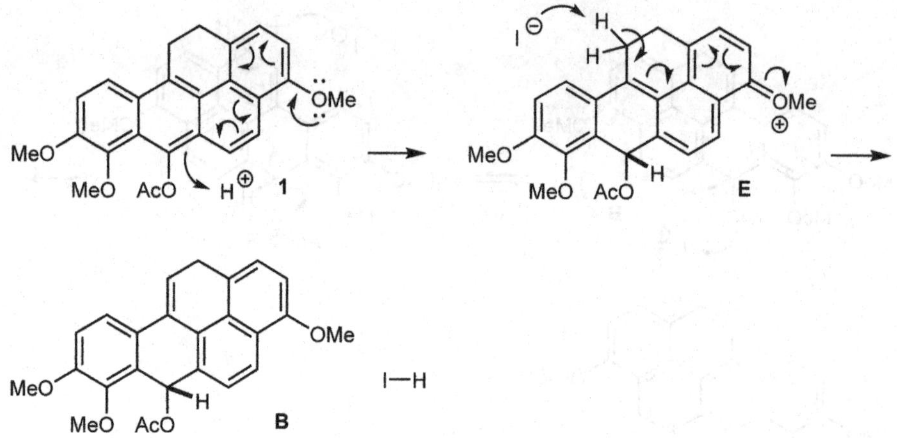

SCHEME 3

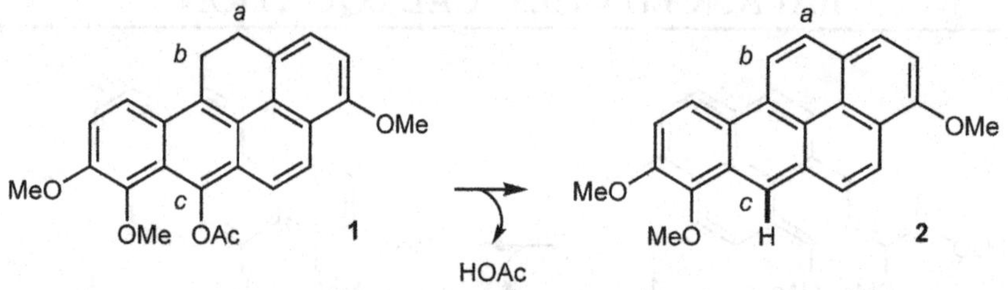

Atom	Reactant	Product	Change	
C_a	−2	−1	+1	$\left.\rule{0pt}{1.3em}\right\}$ $\Delta = +2$
C_b	−2	−1	+1	
C_c	+1	−1	−2	$\Delta = -2$

SCHEME 4

−1 for a total change of +2 which is compensated by the reduction of carbon atom labeled as C_c (C_6) from +1 to −1 for a change of −2. Lastly, we note that the authors did not conduct additional experimental or theoretical work to help support the proposed mechanism.

ADDITIONAL RESOURCES

We recommend an article which describes controlled oxidations of benzo[a]pyrenes as well as a review that details 50 years of research on these interesting structures.[2,3]

References

1. Kumar S, Singh SK. A rapid, convergent and regioselective synthesis of 3,7,8-trihydroxy-*trans*-7,8-dihydrobenzo[a]pyrene. *Tetrahedron Lett.* 1995;36(8):1213–1216. https://doi.org/10.1016/0040-4039(95)00036-C.
2. Lee-Ruff E, Kazarians-Moghaddam H, Katz M. Controlled oxidations of benzo[a]pyrene. *Can J Chem.* 1986;64(7):1297–1303. https://doi.org/10.1139/v86-215.
3. Phillips DH. 50 years of benzo(a)pyrene. *Nature.* 1983;303(2):468–472. https://doi.org/10.1038/303468a0.

Question 40: Rearrangement of Ketene Acetals

In 1995 chemists from Belgium showed that *E* and *Z* ketene acetals **2** (synthesized from ester **1** by standard methodology) undergo smooth rearrangement to form *o*-methylthiomethylarylacetic ester **3** after either a few hours in THF at room temperature or less than 1 h at reflux temperature (Scheme 1).[1] The authors also noted that the *E* isomer of **2** rearranged faster than the *Z* isomer and that its disappearance followed first-order kinetics. Provide a mechanism for this unexpected rearrangement of **2** to **3**.

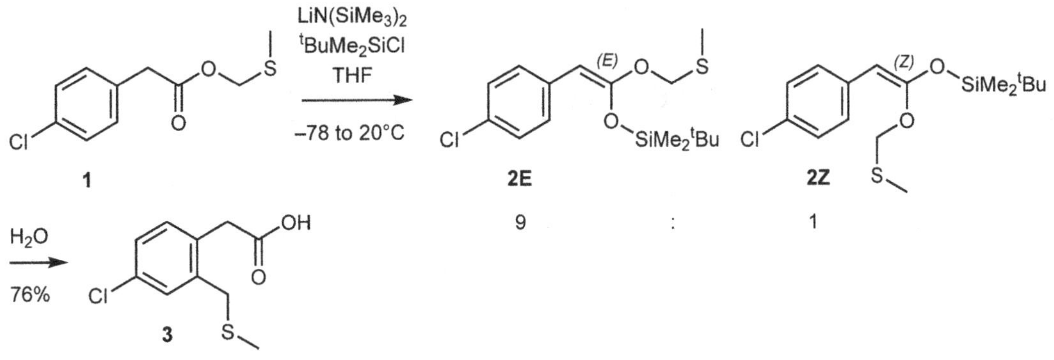

SCHEME 1

SOLUTION: 2E TO 3

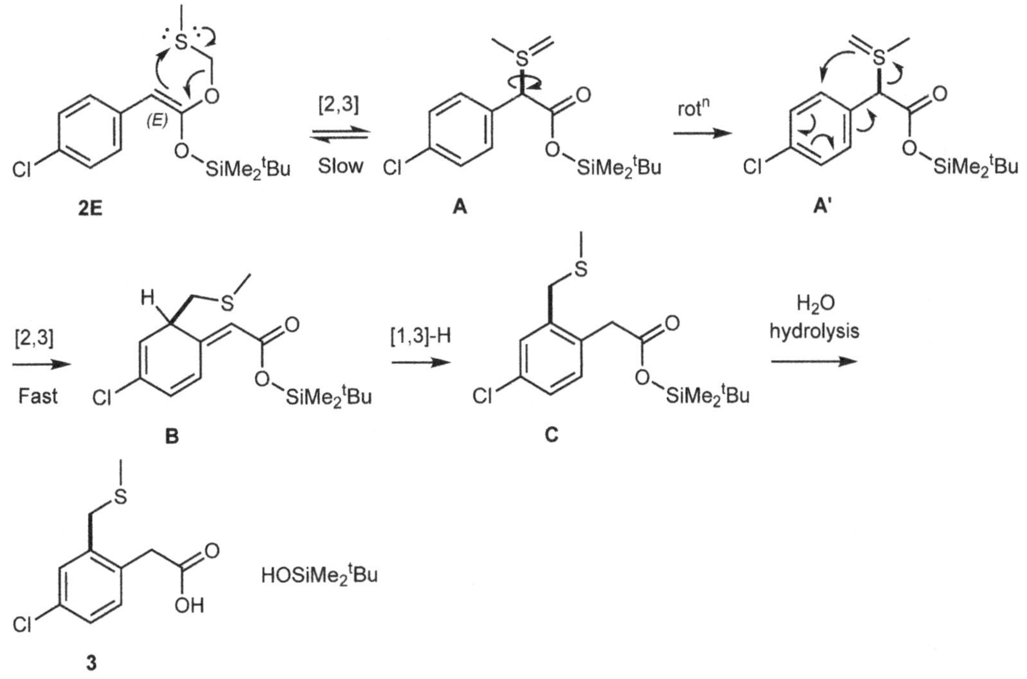

SOLUTION: 2Z TO 3

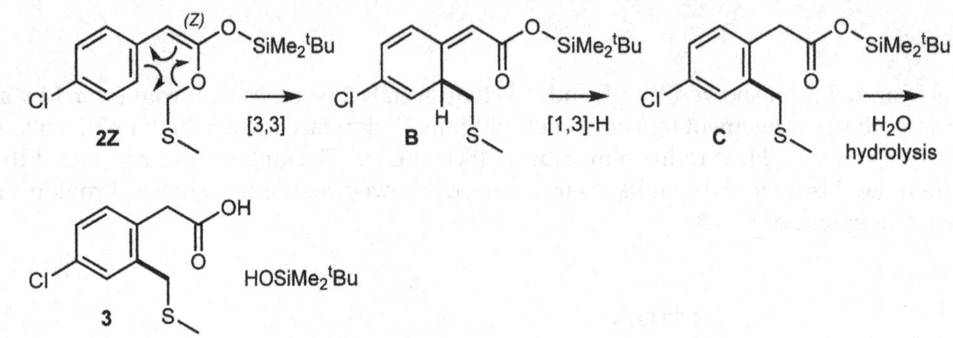

BALANCED CHEMICAL EQUATION

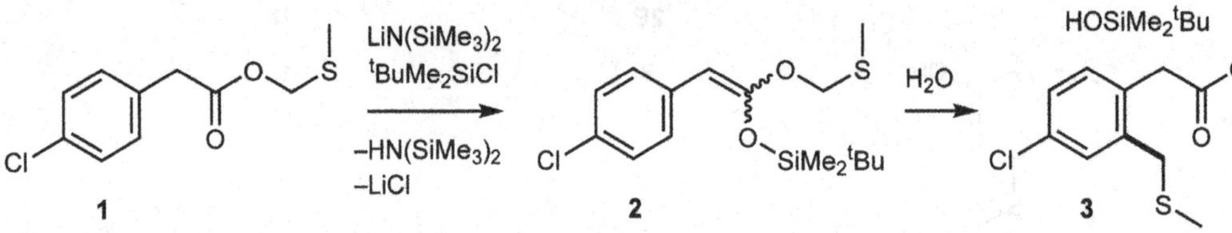

KEY STEPS EXPLAINED

The mechanism proposed by the authors for the formation of *o*-methylthiomethylarylacetic ester **3** from **2E** begins with a slow [2,3] rearrangement of **2E** which provides ylide **A**. Upon C—C bond rotation, **A** undergoes a second fast [2,3] rearrangement where the CH_3SCH_2 group migrates onto the aromatic ring. The resulting intermediate **B** then undergoes a [1,3]-H tautomerization which aromatizes the ring to form ester **C** which upon hydrolysis gives product **3**. In the case of the **2Z** isomer, the first step is a Claisen-type rearrangement that affords intermediate **B** which tautomerizes as before to **C** which undergoes hydrolysis to form **3**.

We note that the authors also offered an alternative radical-based rearrangement step as shown in Scheme 2. Using the **2Z** isomer we thus have C—O homolytic bond cleavage that provides the two radicals shown in **D** which recombine to form intermediate **B** as before. Nevertheless, such a proposal implies an intermolecular process and is inconsistent with the crossover experiment carried out by the authors (see Scheme 3). Here we note that no crossover products are formed which demonstrates that the rearrangement of **2** to **3** is essentially an intramolecular process. If a radical mechanism was truly operative then the two radicals produced in **D** must be trapped in a solvent cage to satisfy the intramolecular rearrangement result, otherwise the two radicals would fly apart and one would expect crossover products to be formed.

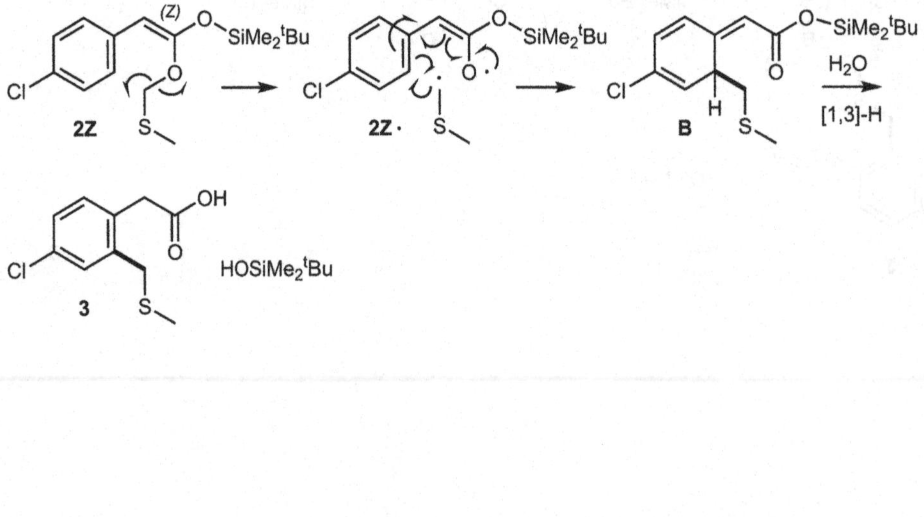

SCHEME 2

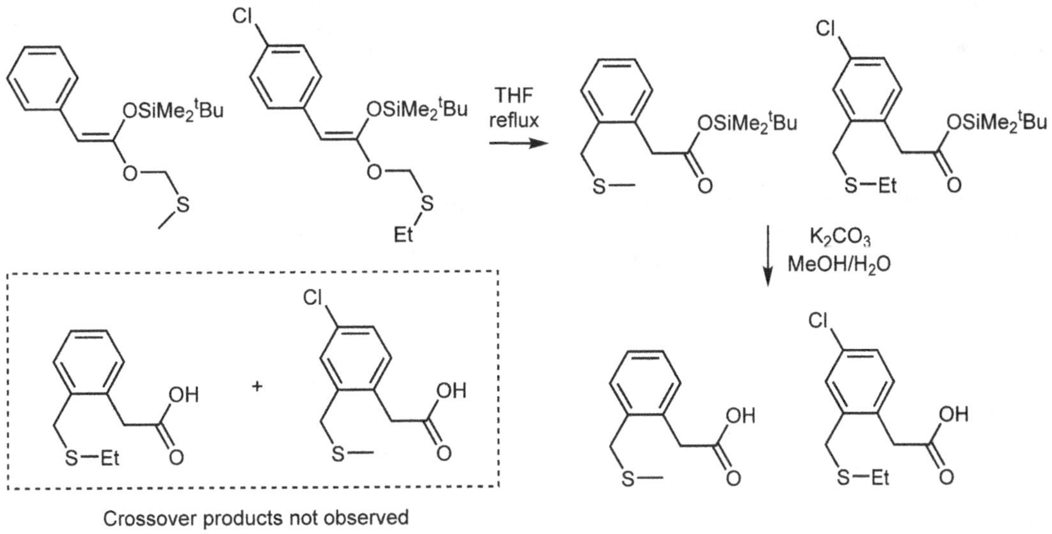

Crossover products not observed

SCHEME 3

The authors made further experimental observations. For example, when *ortho* substituents such as fluorine were introduced, no rearrangement took place since those positions were blocked. When *meta* substituents were present, both 1,2,3- and 1,2,4-substituted rearrangement products were formed indicating that the *meta* substituent did not impart any steric effect. When sulfur was replaced by oxygen, the ketene acetal was stable up to 200 °C.

In addition, the authors could not explain why the *E* isomer rearranges faster than the *Z*. On comparing the two mechanisms shown before, it appears that the **2Z** to **3** transformation is more facile and takes fewer steps than the **2E** to **3** transformation which is contrary to the observation that the *Z* isomer reacts more slowly than the *E*. It is obvious that the geometry of the **2Z** isomer is immediately amenable to alkyl migration to the *ortho* position to directly form **B** whereas the **2E** isomer is not. A computational investigation of the energy reaction coordinate diagram of the transformation is required to elucidate the reason for this. There may be a suspicion that the authors transposed the isomer labels of their compounds; however, they claimed that they based their configurational assignments on ¹H NMR chemical shift comparison of the vinylic proton with literature data. We believe that the observations made by the authors are in doubt and should be rechecked.

ADDITIONAL RESOURCES

We recommend further work by the authors concerning an extension of this transformation and a discussion of its mechanism.[2] We also recommend an article which discusses the mechanism of an interesting transformation that involves ketene chemistry and deuterium labeling experiments.[3]

References

1. Bourgaux M, Gillet-Berwart A-F, Marchand-Brynaert J, Ghosez L. An unusual rearrangement leading to *ortho*-thiomethylation of arylacetic acid derivatives. *Synlett.* 1995;1995(1):113–115. https://doi.org/10.1055/s-1995-4848.
2. Jaroskova L, Bourgaux M, Wenkin I, Ghosez L. A rearrangement of O,O-silylketene acetals leading to γ-thiomethylation of butenoic acid derivatives. *Tetrahedron Lett.* 1998;39:3157–3160. https://doi.org/10.1016/S0040-4039(98)00508-5.
3. Jacubert M, Provot O, Peyrat J-F, Hamze A, Brion J-D, Alami M. p-Toluenesulfonic acid-promoted selective functionalization of unsymmetrical arylalkynes: a regioselective access to various arylketones and heterocycles. *Tetrahedron.* 2010;66:3775–3787. https://doi.org/10.1016/j.tet.2010.03.055.

Question 41: Nucleophilic Additions to a Heterocyclic *o*-Quinone

In 1993 Japanese chemists observed that treatment of pyrroloquinolinequinone (PQQ) trimethyl ester **1** with methanol under neutral conditions gives the C_5-hemiketal **2** (Scheme 1).[1] Under acidic conditions, treatment of **1** with methanol leads to formation of the C_4-ketal **3**. Both products were identified by X-ray analysis. Suggest mechanisms for both transformations.

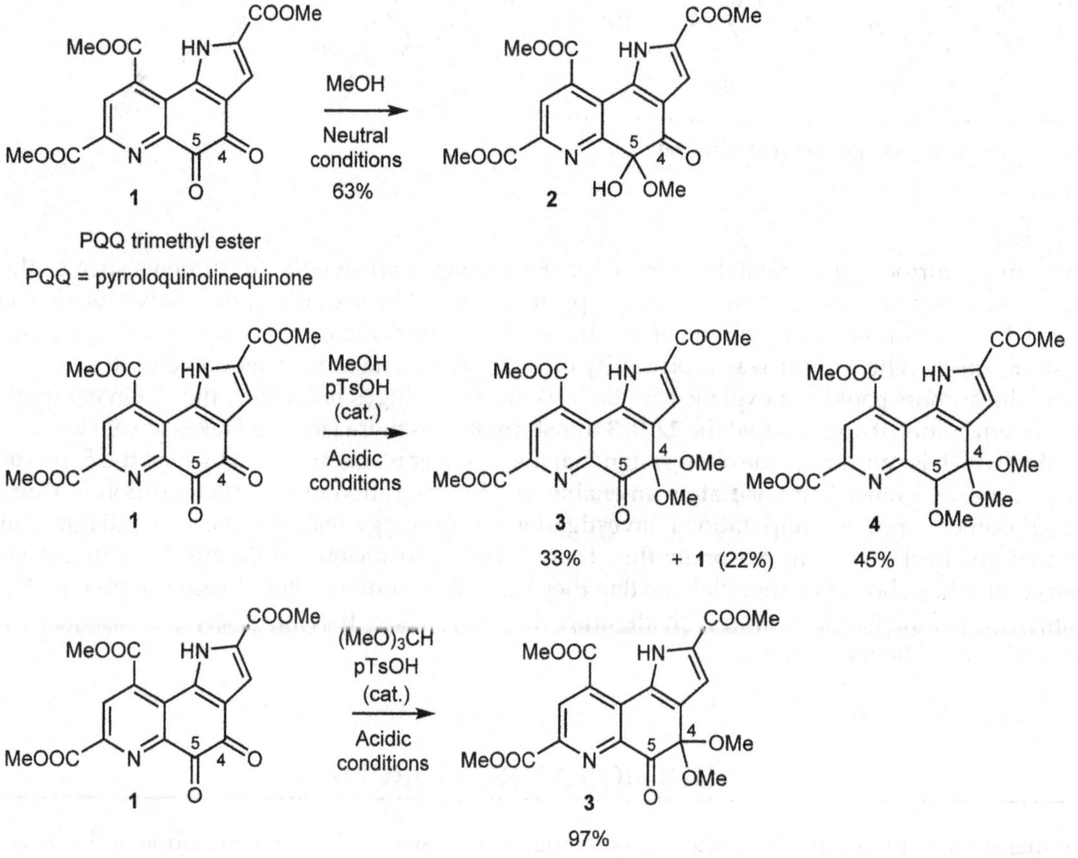

SCHEME 1

SOLUTION: 1 TO 2

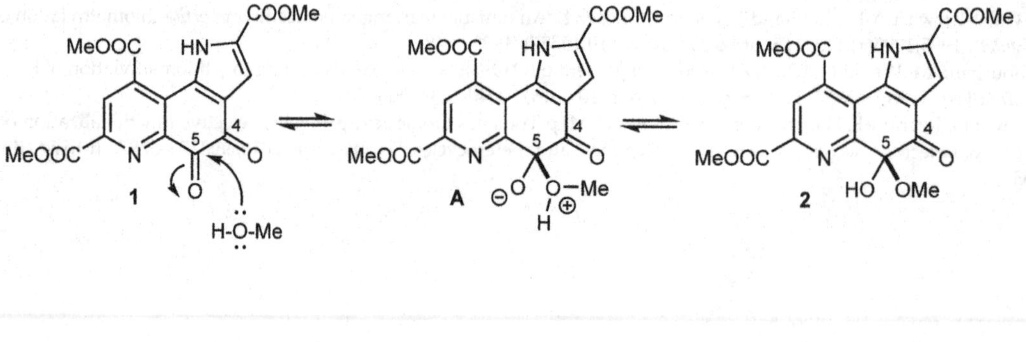

SOLUTION: 1 TO 3

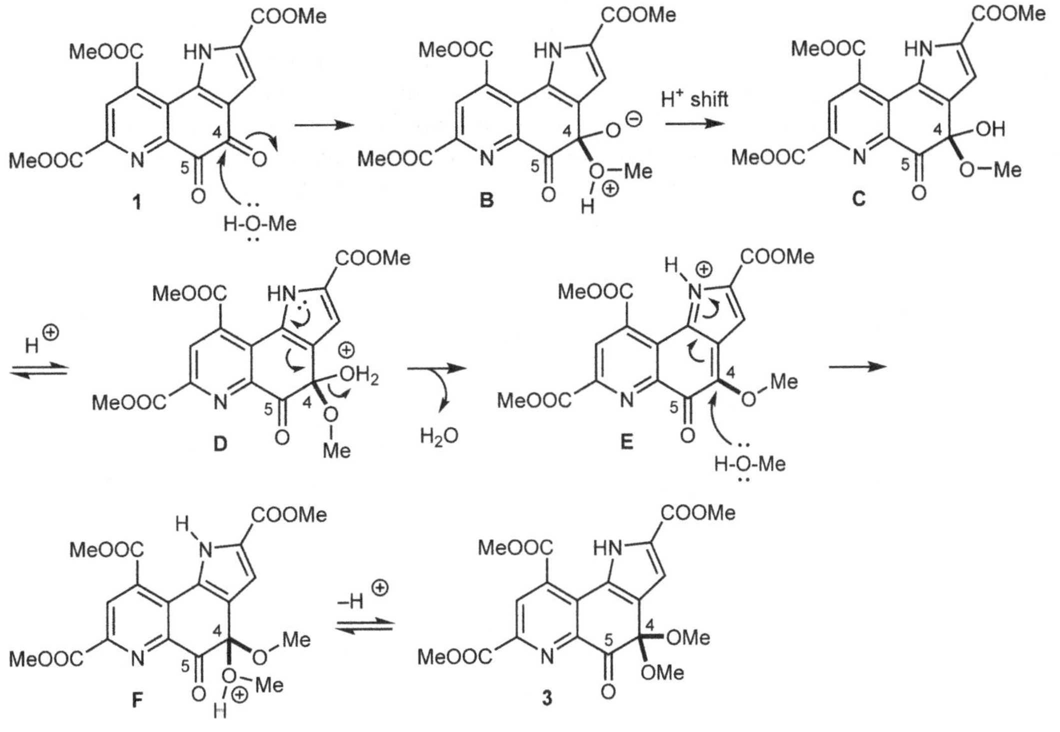

KEY STEPS EXPLAINED

We begin by noting that the authors confirmed the structures of "C_5 hemiacetal" **2** and "C_4 dimethyl acetal" **3** by X-ray crystallography. Nevertheless, the nomenclature of these adducts is incorrect because compound **2** is a hemiketal and compound **3** is a ketal given that the starting PQQ trimethyl ester **1** has ketone and not aldehyde groups. In addition, the authors noted that the electron-withdrawing nature of the pyridine nucleus and ester groups in PQQ trimethyl ester **1** facilitates addition of methanol to the C_5 carbon atom. A more convincing argument that shows that the C_5 carbon atom is more electrophilic than the C_4 carbon atom is highlighted by the two resonance structures for **1** given in Scheme 2. In structure **1'** we observe that the C_5 carbonyl group remains intact and therefore is readily accessible by nucleophiles such as methanol; whereas, the C_4 carbon atom is less electrophilic due to the enolate resonance form arising from the electron donating nature of the pyrrole nitrogen atom.

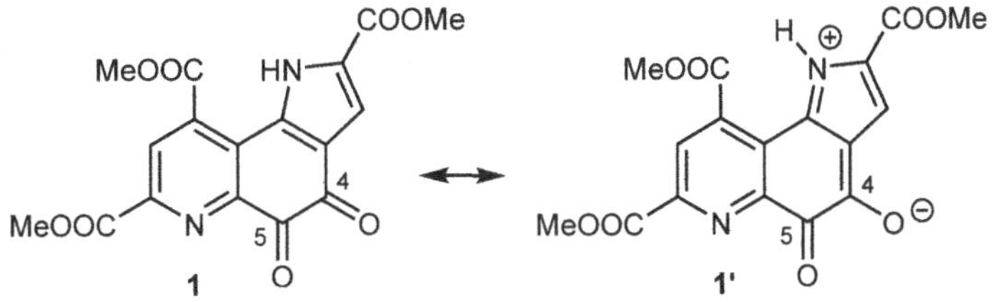

SCHEME 2

Furthermore, semiempirical molecular orbital calculations performed on structures **2** ("C_5 hemiacetal") and intermediate **C** ("C_4 hemiacetal") indicated that the C_5 adduct was 1–3 kcal/mol more thermodynamically stable than the C_4 adduct. The authors concluded that the hemiacetal formation step **1** to **2** under neutral conditions is reversible since solutions of **2** readily reverted back to quinone **1**. The measured equilibrium constant for the equilibrium depicted in Eq. (1) is $0.63 \, M^{-1}$ based on time-resolved UV-VIS measurements. Therefore, the reaction is thermodynamically controlled having the "C_5 hemiacetal" product **2** as the thermodynamically more stable product.

$$1 \underset{\text{MeOH}}{\overset{}{\rightleftharpoons}} 2 \tag{1}$$

The situation changes under acidic conditions, however, because the rate of dehydration of the C_4 adduct is expected to proceed faster than the dehydration of the C_5 adduct. This is because the carbocation derived from the C_4 adduct is resonance stabilized by the pyrrole moiety (see structures E' and E in Scheme 3) whereas the carbocation derived from the C_5 adduct is not (see structure H in Scheme 3). In other words, once the C_4 "dimethyl acetal" 3 is formed it cannot revert back to C_4 "hemiacetal" C. Product 3 thus accumulates over time under acidic conditions even though product 2 is thermodynamically more stable. Under acidic conditions, whatever amount of 2 that gets formed is expected to revert back to 1 as its concentration decreases due to accumulation of 3 in accordance with Le Chatelier's principle.

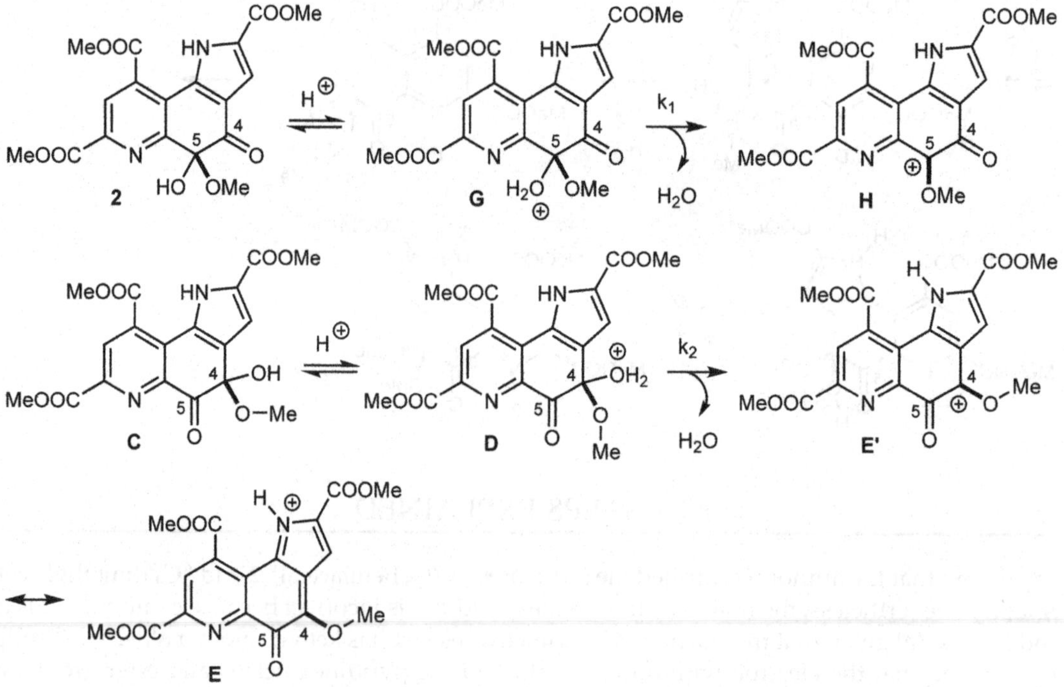

SCHEME 3

ADDITIONAL RESOURCES

We recommend an article by the same authors which shows that amines predominantly attack the C_5 position of PQQ.[2] We also recommend an excellent review by these authors which explores their work concerning the various reactivity patterns of PQQ.[3]

References

1. Itoh S, Ogino M, Fukui Y, et al. C-4 and C-5 adducts of cofactor PQQ (pyrroloquinolinequinone). Model studies directed toward the action of quinoprotein methanol dehydrogenase. *J Am Chem Soc*. 1993;115(22):9960–9967. https://doi.org/10.1021/ja00075a012.
2. Itoh S, Mure M, Ogino M, Ohshiro Y. Reaction of the trimethyl ester of coenzyme PQQ (PQQTME) and amines in organic media. Products and mechanism. *J Org Chem*. 1991;56(24):6857–6865. https://doi.org/10.1021/jo00024a030.
3. Itoh S, Ohshiro Y. The chemistry of heterocyclic o-quinone cofactors. *Nat Prod Rep*. 1995;12:45–53. https://doi.org/10.1039/NP9951200045.

Question 42: How Mitomycin C Can Crosslink DNA

In 1987 chemists from the University of Houston reported on the transformation of mitomycin C **1** and "DNA" **2** which forms the crosslinked product **3** under enzymatic reduction conditions (Scheme 1).[1] Suggest a mechanism for this transformation.

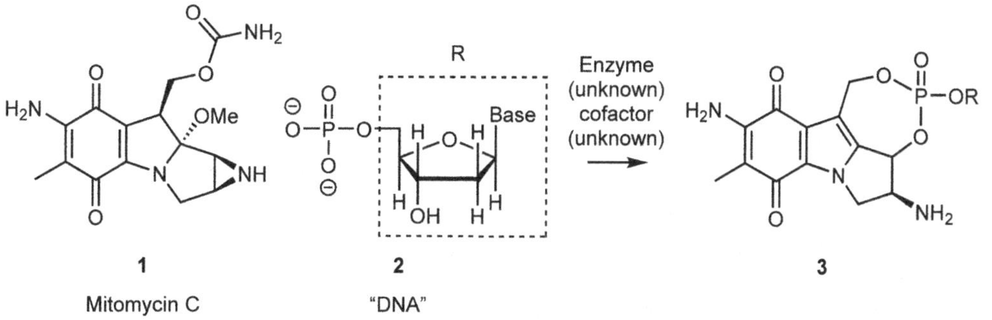

SCHEME 1

SOLUTION

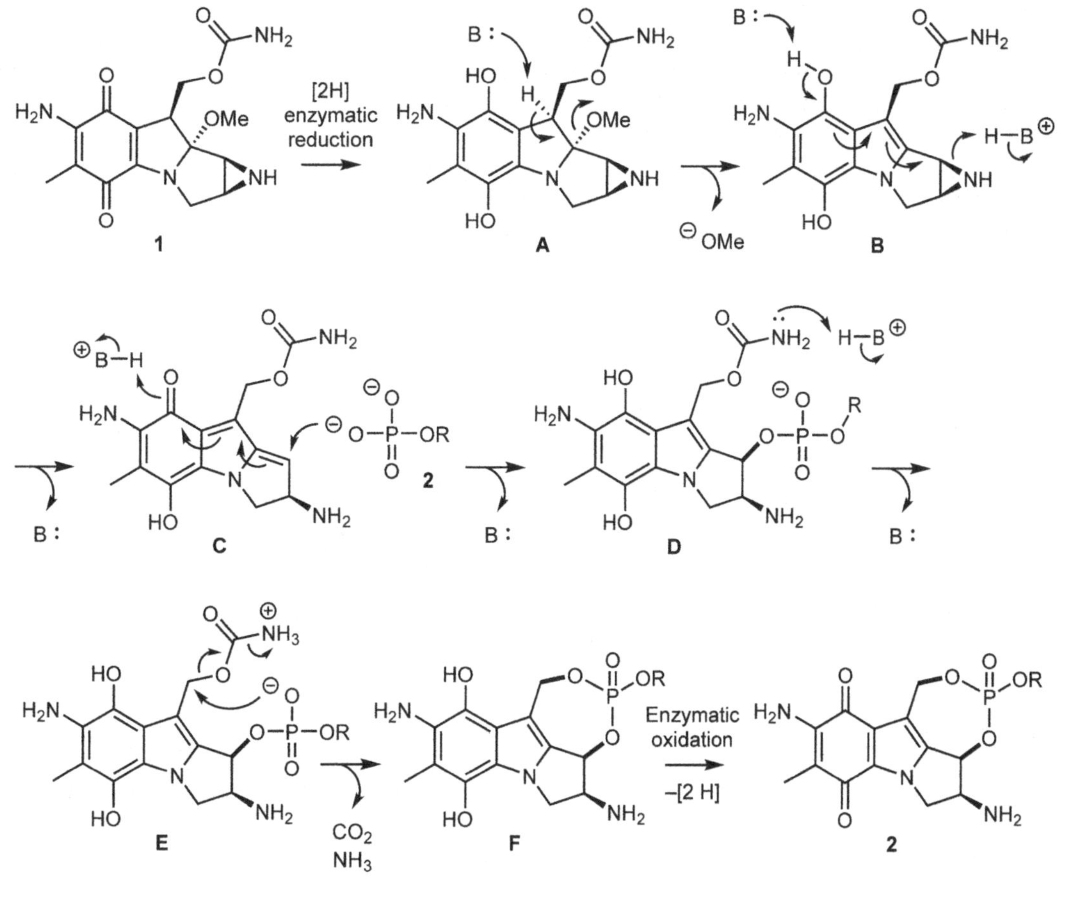

BALANCED CHEMICAL EQUATION

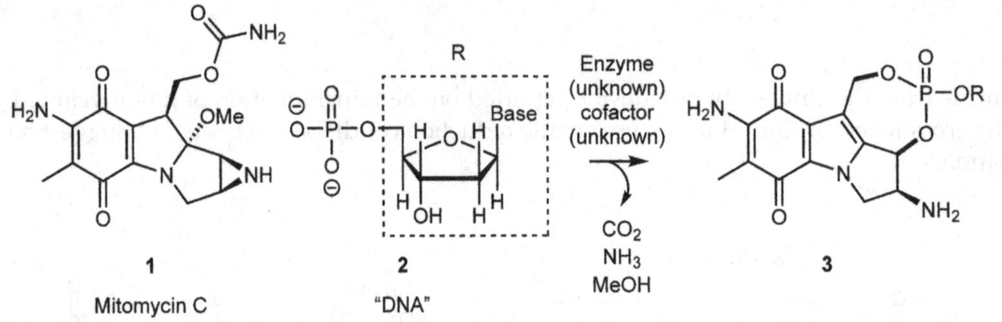

KEY STEPS EXPLAINED

The mechanism reported by the authors to explain the crosslinking reaction between mitomycin C and DNA is conjectured and reproduced from a review article by Moore and Czerniak who reported on several transformations involving quinones as alkylating agents acting upon DNA.[2] The mechanism is thought to begin with enzymatic reduction of the benzoquinone group of mitomycin C to form the hydroquinone intermediate **A** possibly with a cofactor such as nicotinamide adenine dinucleotide (NADH), nicotinamide adenine dinucleotide phosphate (NADPH), or some metal ion. Base-catalyzed elimination of methanol by, for example, phosphate ion generates the indolequinone intermediate **B** which undergoes deprotonation of the hydroquinone moiety with concomitant aziridine group ring opening to form, after protonation, the quinone methide intermediate **C**. The phosphate group of DNA then adds nucleophilically to the olefinic carbon *ortho* to the amino group of **C** giving the monoalkylated intermediate **D**. Protonation of the carbamate amino group followed by cyclization of the phosphate group generates after S_N2 displacement the crosslinked intermediate **F** along with carbon dioxide and ammonia by-products. Lastly, the hydroquinone moiety of **F** is enzymatically oxidized to form the final crosslinked benzoquinone product **2** possibly with a cofactor such as nicotinamide adenine dinucleotide (NAD), nicotinamide adenine dinucleotide phosphate (NADP), or some metal ion. In support of this mechanism, Moore and Czerniak cite the work of Keller et al. who showed that reduction of mitomycin C with $Na_2S_2O_4$ in the presence of potassium ethylxanthate leads first to the mono-functionalized adduct **4** (functionalized at C_1, Scheme 2).[3] Treatment of **4** under the reaction conditions then led to the bis-functionalized adduct **5** (functionalized at C_{10}).

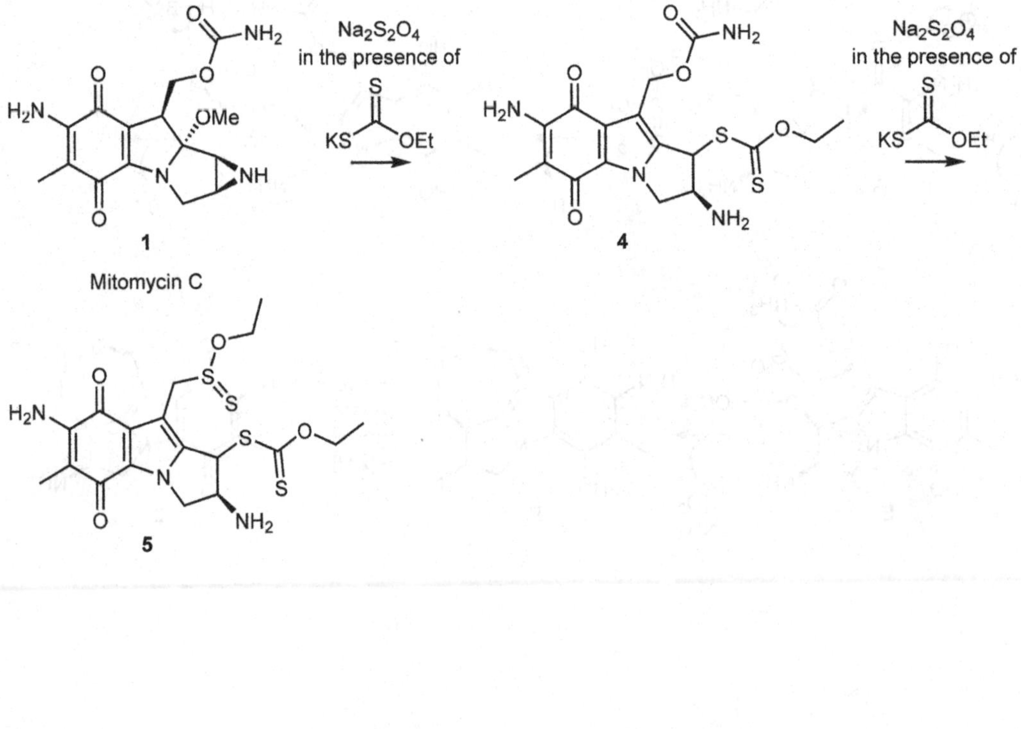

SCHEME 2

According to Moore and Czerniak, the order of functionalization established by this experiment (i.e., first at C_1 then at C_{10}) supports the proposed mechanism for **1** and **2** to **3** because reduction of **1** enables formation of the quinone methide **C** which facilitates the first alkylation at C_1 followed by alkylation at C_{10}. Nevertheless, direct observation of **C** using spectroscopic methods has not been achieved nor have chemists undertaken theoretical calculations to support the proposed mechanism. In addition, we recommend carrying out reactions involving mitomycin C and DNA nucleotides to examine the structure of the expected bisalkylated products and provide relevant yield data.

ADDITIONAL RESOURCES

We recommend a more recent study of mechanisms involving reductive activation of mitomycin C and its reactivity toward nucleophiles such as thiols.[4] The authors of this study note the pH sensitivity of the reaction as an important factor.

References

1. Fishbein PL, Kohn H. Synthesis and antineoplastic activity of 1a-formyl and 1a-thioformyl derivatives of mitomycin C and 2-methylaziridine. *J Med Chem.* 1987;30(10):1767–1773. https://doi.org/10.1021/jm00393a015.
2. Moore HW, Czerniak R. Naturally occurring quinones as potential bioreductive alkylating agents. *Med Res Rev.* 1981;1(3):249–280. https://doi.org/10.1002/med.2610010303.
3. Hornemann U, Keller PJ, Kozlowski JF. Formation of 1-ethylxanthyl-2,7-diaminomitosene and 1,10-diethylxanthyl-2,7-diaminodecarbamoylmitosene in aqueous solution upon reduction-reoxidation of mitomycin C in the presence of potassium ethylxanthate. *J Am Chem Soc.* 1979;101(23):7121–7124. https://doi.org/10.1021/ja00517a082.
4. Paz MM. Reductive activation of mitomycin C by thiols: kinetics, mechanism, and biological implications. *Chem Res Toxicol.* 2009;22(10):1663–1668. https://doi.org/10.1021/tx9002758.

Question 43: Two Carbon Ring Expansions of Cycloalkanones

In 1995 American chemists investigated the reactions of carbocyclic β-keto phosphonates **1** and **3** with dimethyl acetylenedicarboxylate (DMAD) in the presence of base in THF which gave the ring expansion product **2** (in the case of β-keto phosphonate **1**) and the ring expansion product **4** together with the "abnormal Michael" product **5** (in the case of tetralone **3**, Scheme 1).[1] Provide mechanisms that explain the formation of these products.

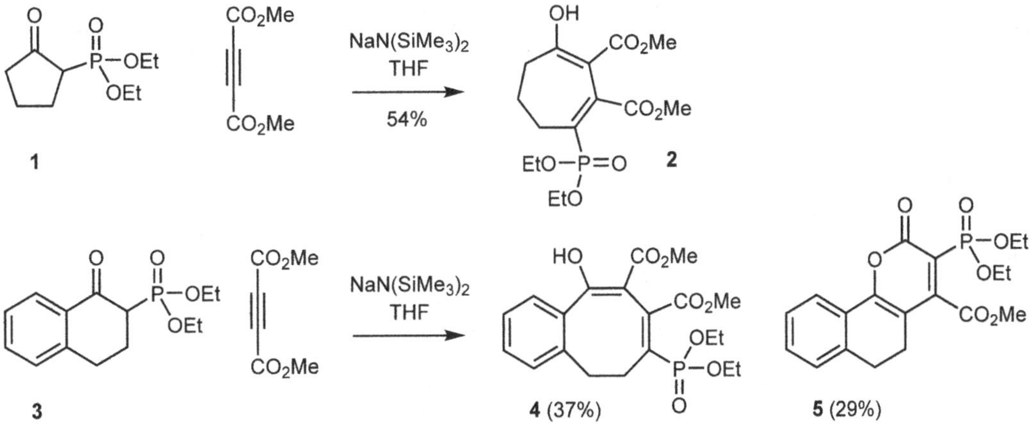

SCHEME 1

SOLUTION: 1 TO 2

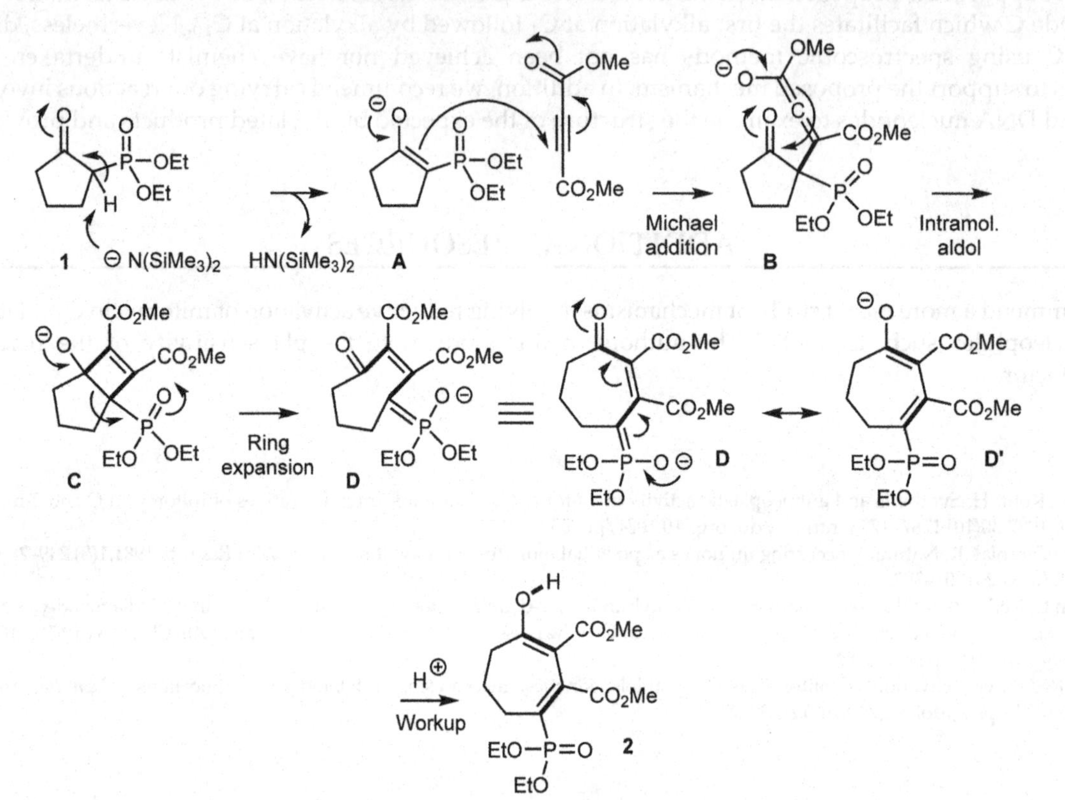

BALANCED CHEMICAL EQUATION: 1 TO 2

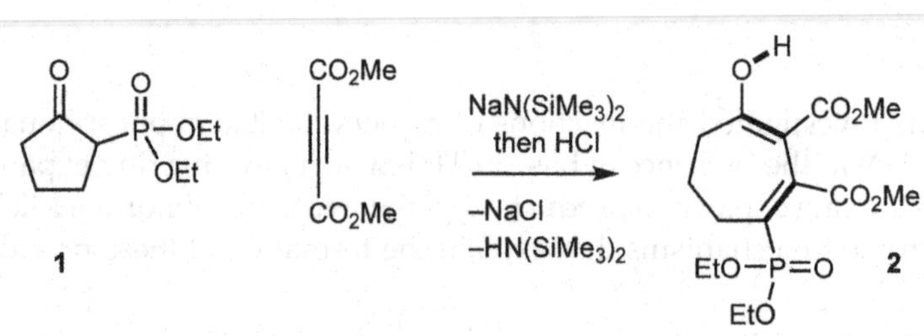

SOLUTION: 3 TO 4

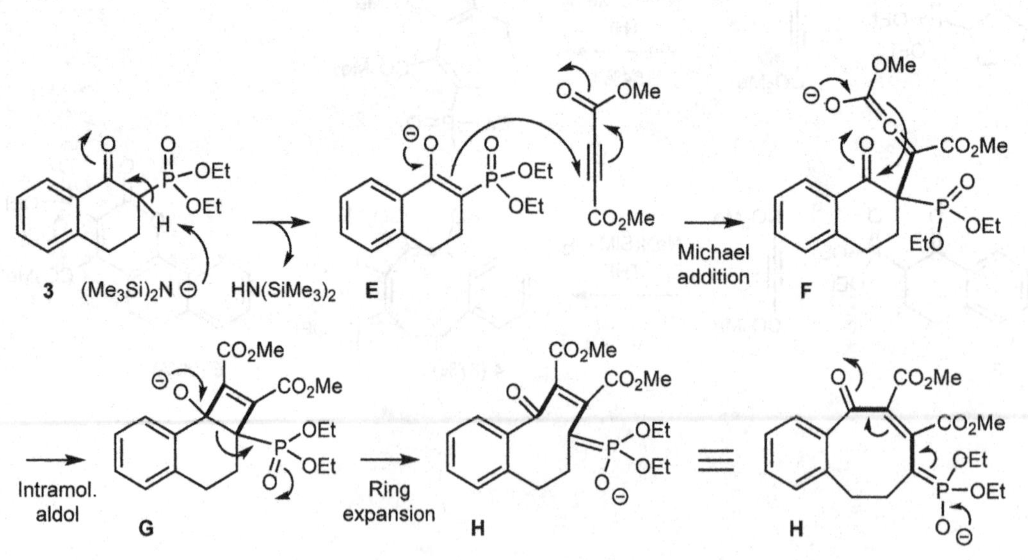

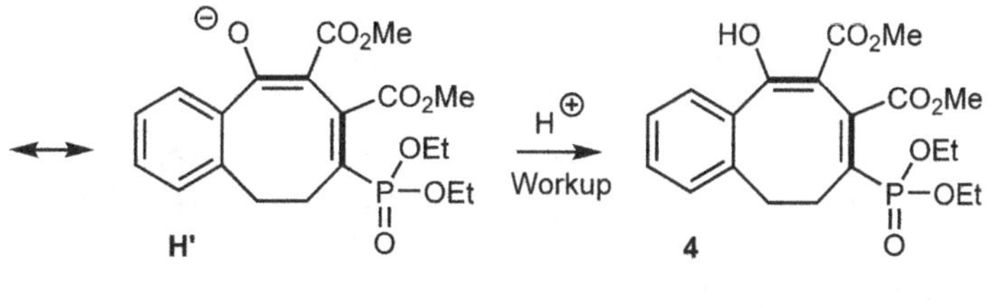

BALANCED CHEMICAL EQUATION: 3 TO 4

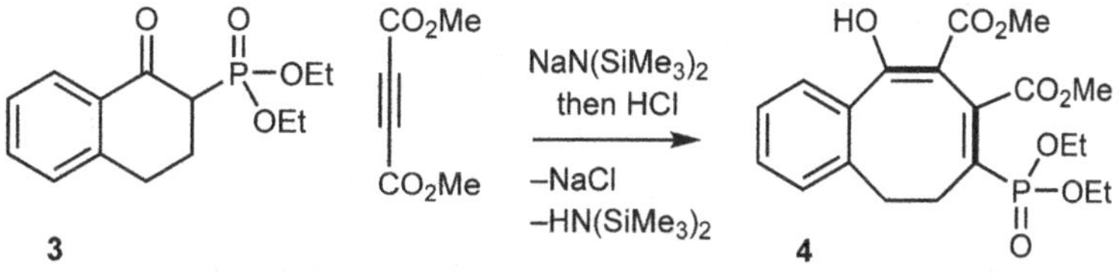

SOLUTION: 3 TO 5

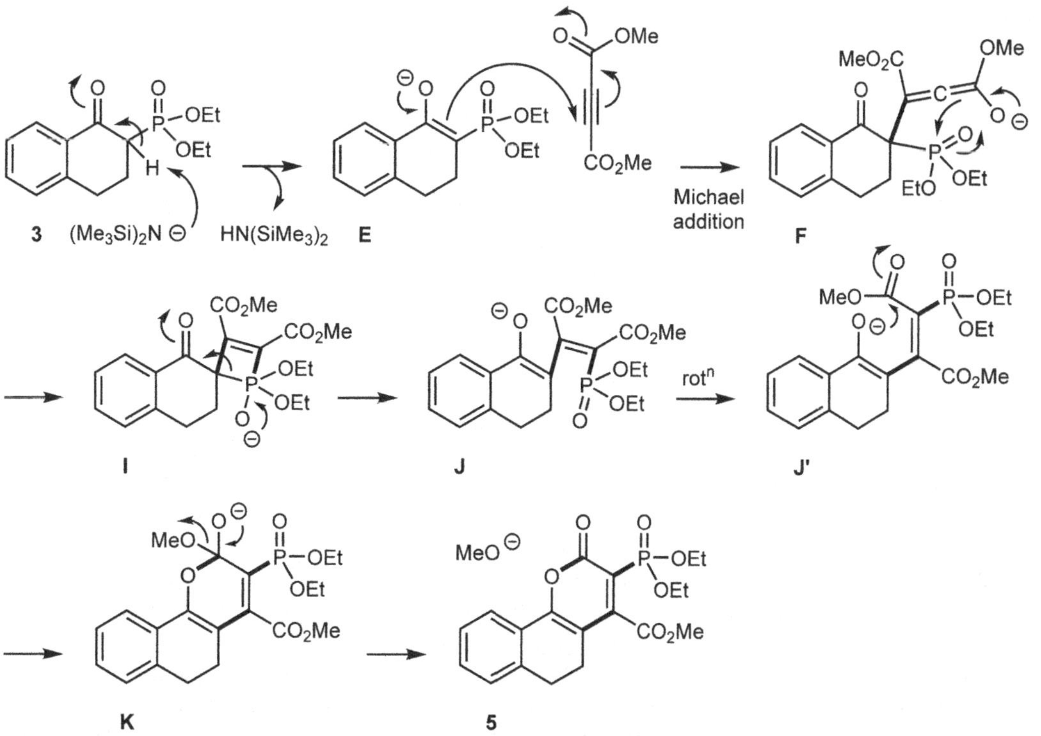

BALANCED CHEMICAL EQUATION: 3 TO 5

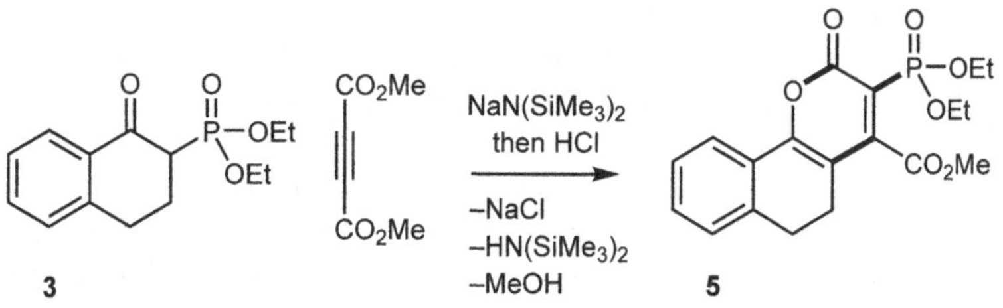

KEY STEPS EXPLAINED

The mechanism proposed by the authors is described as an "$n+2$ ring expansion reaction based on a Michael-initiated ring closure (MIRC) reaction."[1] For example, in the transformation of **1** to **2**, the proposed mechanism begins with base initiated enolization of cyclopentanone **1** to form enolate **A** which undergoes nucleophilic Michael addition onto DMAD reagent to form allenolate **B**. This intermediate undergoes an intramolecular aldol condensation which gives the cyclobutene alkoxide **C** which undergoes ring expansion through fragmentation of the fused C—C bond to generate the $n+2$ ring expansion enolate **D'**. Acidic work-up at this stage gives the final enol product **2**. We note that the ring construction mapping for this transformation is [5+2]. The same mechanistic sequence is proposed for the transformation of tetralone **3** to product **4** where a six-membered ring becomes an eight-membered ring and where the ring construction mapping is [6+2].

Before explaining the mechanism proposed for the transformation of **3** to **5**, we explore several pieces of evidence for the previous mechanistic proposals. First, the authors applied ^1H, ^{13}C, ^{31}P NMR, and IR spectroscopy to identify and differentiate the various products they obtained. For example, the fact that **2** and **4** exist in an enol form was concluded on the basis of strong enolic signals in the ^1H NMR spectra and lack of ketonic signals in the ^{13}C NMR spectra of these products. Furthermore, the authors managed to isolate the six-membered Michael adduct **7** as a mixture of geometrical isomers when the reaction was carried out in wet KH or Cs_2CO_3 (Scheme 2). This product was also isolated when a catalytic amount of KOH in t-BuOH was used. Product **7** was then converted to the expected ring expansion product by treatment with base. These experiments confirm the presence of the allenolate intermediate **B** in the proposed mechanism for the formation of **2** and **4**.

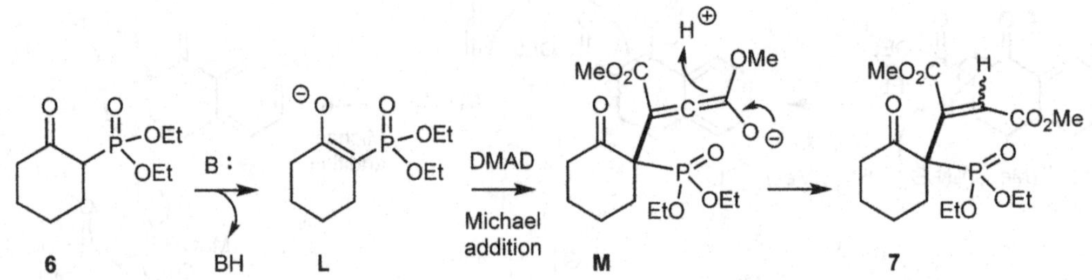

SCHEME 2

It is worthwhile to note that use of LiHMDS as base led only to starting material **1** and **3** whereas use of NaHMDS and KHMDS led exclusively to the ring expansion products. This showed that stronger bases such as metal hydrides or silazides were important for initiating $n+2$ ring expansion reactions. The authors also cited prior work on ring expansion of β-ketoesters upon reaction with DMAD as a precedent for their mechanistic interpretation.[2]

In the case of the transformation of **3** to **5**, the authors proposed a mechanism which begins with the same sequence of steps seen previously. In other words, base-induced enolization of **3** generates enolate **E** which undergoes a Michael nucleophilic addition to DMAD to form the familiar allenolate intermediate **F**. This time, however, the allenolate does not cyclize onto the carbonyl group to form **G** but instead cyclizes onto the phosphonate group to form a [6,4] spiro intermediate **I**. This intermediate undergoes fragmentation to generate enolate **J'** after bond rotation which is followed by cyclization of the alkoxide onto the ester group to form the tricyclic product **5**. The ring construction mapping for this transformation is [3+3]. Furthermore, the authors described the process as a net [1,3]-phosphorus migration that is similar to an abnormal Michael reaction.[3] We note that when the reaction was quenched with sulfuric acid, the intermediates shown in Scheme 3 originating from enolate **J'** were observed in the crude reaction mixture by ^{13}C and ^{31}P NMR. The authors found that the product ratio of **4** to **5** depended on the kind of counterion of base used. For instance, sodium ions favored the formation of product **4** whereas potassium ions favored the abnormal Michael product **5**. When lithium hexamethyldisilazane was used the reaction did not proceed and starting tetralone phosphonate **1** was recovered.

Interestingly, the authors went on to examine the effects of electronic and steric factors on this transformation. In an experiment where the methoxy substituted tetralone **3'** was used as a starting material, reaction with DMAD gave a complex mixture consisting of a Michael adduct **10**, the ring expansion product **11**, the "abnormal Michael" adduct **12**, and 25% of recovered **3'** (Scheme 4). Moreover, use of unsaturated phosphonate **13** gave only unreacted **13** when it was

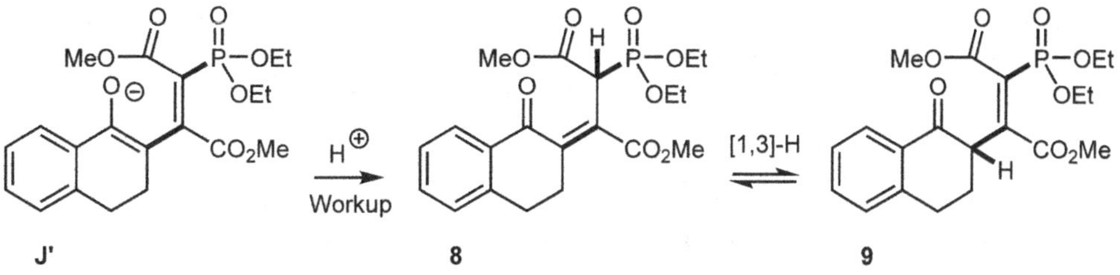

SCHEME 3

treated with DMAD under KHMDS but when protic conditions were used (KOH/*t*-BuOH), the Michael adduct **14** was isolated as a 4:1 mixture of *E* and *Z* isomers (Scheme 5).

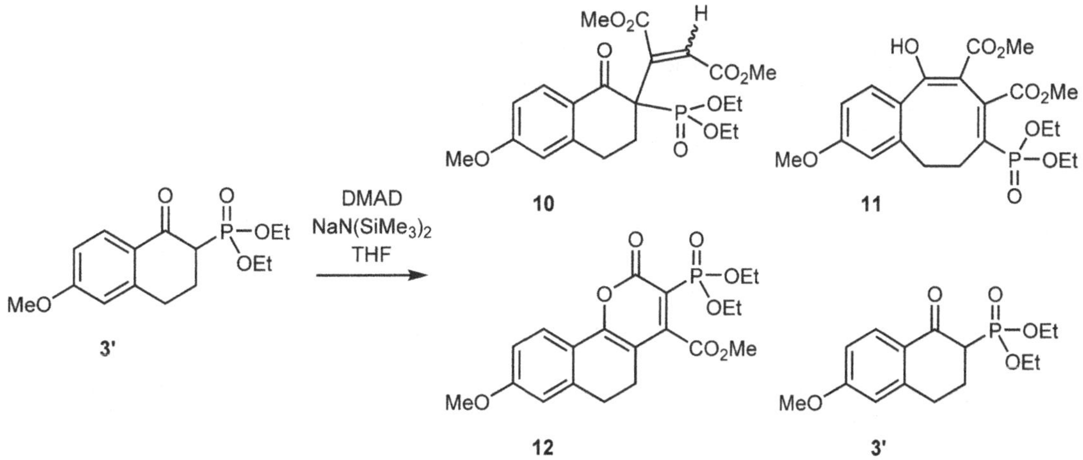

SCHEME 4

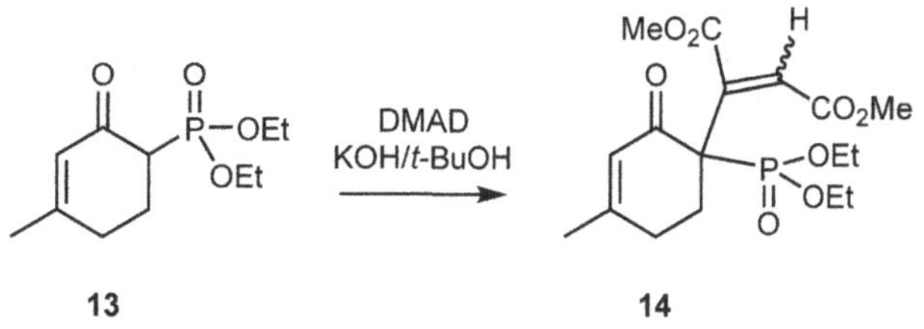

SCHEME 5

The authors explain the significance of these results in terms of the importance of the carbonyl group in the starting phosphonate reactant. In other words, when electrons are delocalized into the carbonyl group this site becomes deactivated from nucleophilic attack of the allenolate intermediate thus limiting the pathway which leads to the ring expansion product **11**. This effect allows for three other pathways to occur. The first is cyclization of the allenolate onto the phosphonate group to form the "abnormal Michael" adduct **12**. The second pathway involves formation of the expected Michael adduct **10** as a mixture of isomers. Here we note that a protic source (as shown in Scheme 5) facilitates this process. Lastly, it is possible for the allenolate intermediate to undergo a retro-Michael reaction to give the starting material **3′**. This also happens for **13** under aprotic conditions. In other words, these two experiments provide evidence for the formation of the allenolate intermediate at which point the mechanism can diverge into four different transformations as shown in Scheme 4. The experiments also show that the initial Michael addition step (i.e., **A** to **B** and **E** to **F**) must occur under both protic and aprotic conditions. Under aprotic conditions, the pathways leading to ring expansion and abnormal Michael products are favored, whereas under protic conditions the regular Michael adduct pathway which gives a mixture of isomers is favored. When electron delocalization into the carbonyl group of the starting material is a factor, the abnormal Michael pathway (allenolate cyclization onto the phosphonate group) is favored over the ring expansion pathway (allenolate cyclization onto the carbonyl group). Under aprotic conditions with electron delocalization the abnormal Michael and the retro-Michael pathways are favored, whereas under protic conditions with electron delocalization, the standard Michael adduct pathway is favored. An interesting experiment missing from the authors' article is to undertake a theoretical analysis to calculate the energetics of each proposed mechanistic pathway under protic and aprotic conditions.

Despite this shortcoming, the authors also investigated the effect of sterics on the transformation. Here, it was noted that the phosphonate derived from (±)-camphor 15 did not react with DMAD at all since there was nothing other than starting material isolated under both aprotic and protic conditions (Scheme 6). We note that under protic conditions, initial Michael addition of DMAD to 15 would facilitate formation of the standard Michael adduct 16. The authors thus concluded that 15 is too sterically hindered to allow for the initial Michael addition of the carbanion onto DMAD. If 15 had produced the standard Michael adduct 16 under protic conditions, it would have implied that under aprotic conditions 15 did react with DMAD to form the allenolate intermediate which would have been forced into the retro-Michael reaction pathway due to steric factors to form the starting material 15.

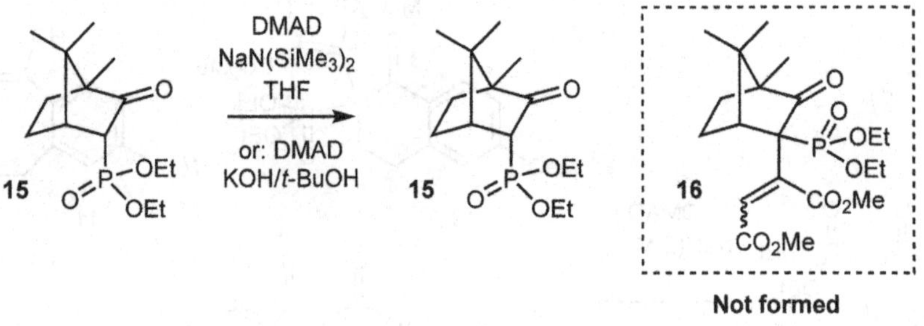

SCHEME 6

Interestingly, the phosphonate derived from (±)-norbornanone 17 formed the abnormal Michael product 18 as a major product and the standard Michael adduct 19 as a minor product when reacted with NaHMDS and DMAD (Scheme 7).

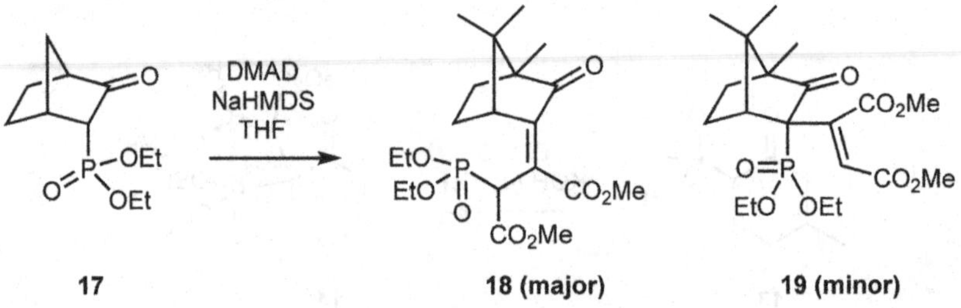

SCHEME 7

This experiment showed that the allenolate intermediate formed was restricted based on steric factors from cyclizing onto the carbonyl group of 17 to facilitate formation of the ring expansion product. Instead, cyclization was favored onto the phosphonate group which led to formation of the abnormal Michael product 18.

Lastly, we note that the authors argued that formation of the phosphonate stabilized anion **D** likely serves as the driving force for fragmentation/ring expansion of **C**. The authors did not provide a theoretical calculation to support this argument. In conclusion, the authors noted that products obtained from this transformation could be useful in subsequent intramolecular Horner-Wadsworth-Emmons reactions (on account of the phosphonate group) to afford bridgehead alkenes that are common to several biologically active natural products. A future theoretical analysis of this transformation would help expand its scope and applications.

ADDITIONAL RESOURCES

We direct the reader to an article which discusses the mechanism of the abnormal Michael reaction between ethyl cyanoacetate and 3-methyl-2-cyclohexenone.[4]

References

1. Ruder SM, Kulkarni VR. Ring expansion reactions of cyclic β-keto phosphonates. *J Org Chem.* 1995;60(10):3084–3091. https://doi.org/10.1021/jo00115a024.
2. Frew AJ, Proctor GR. Ring-expansion of carbocyclic β-ketoesters with acetylenic esters. *J Chem Soc Perkin Trans 1.* 1980;1980:1245–1250. https://doi.org/10.1039/P19800001245.
3. Bergmann ED, Ginsburg D, Pappo R. The Michael reaction. *Org React.* 1959;10:179–555. https://doi.org/10.1002/0471264180.or010.03.
4. Hill RK, Ledford ND. Mechanism of the abnormal Michael reaction between ethyl cyanoacetate and 3-methyl-2-cyclohexenone. *J Am Chem Soc.* 1975;97(3):666–667. https://doi.org/10.1021/n00836a047.

Question 44: Fulgides: Synthesis and Photochromism

In 1995 American chemists synthesized fulgide **9** by the synthetic sequence shown in Scheme 1.[1] It was noted that irradiation of **9** at 350 nm gave a bluish green solution which absorbed at 620 nm and which reverted back to its original state after irradiation at 532 nm. Provide a mechanistic interpretation of the synthesis and photochromism of fulgide **9**.

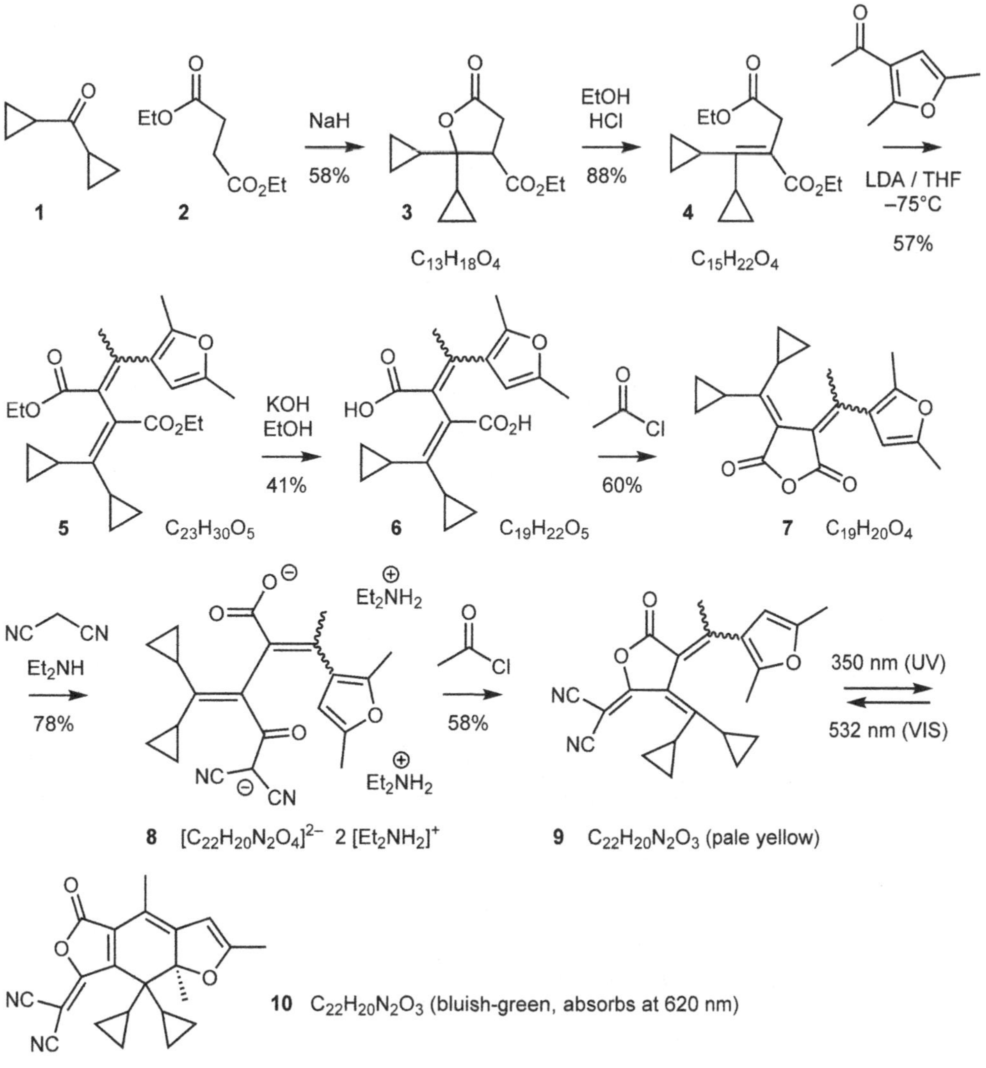

SCHEME 1

SOLUTION: 1 AND 2 TO 3 (STOBBE CONDENSATION)

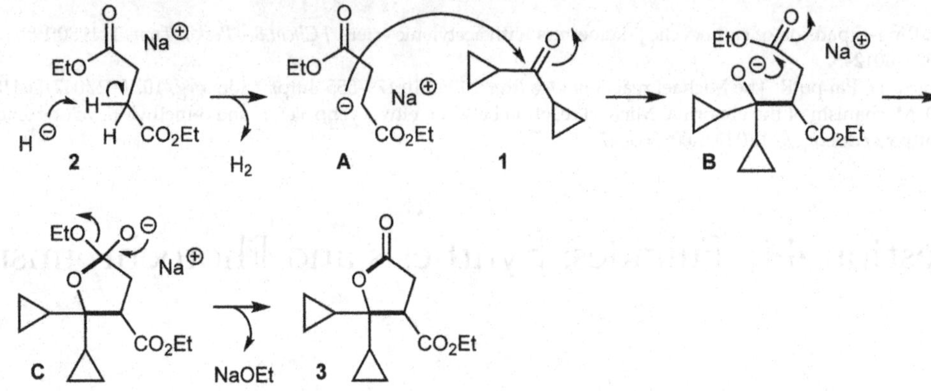

BALANCED CHEMICAL EQUATION: 1 AND 2 TO 3

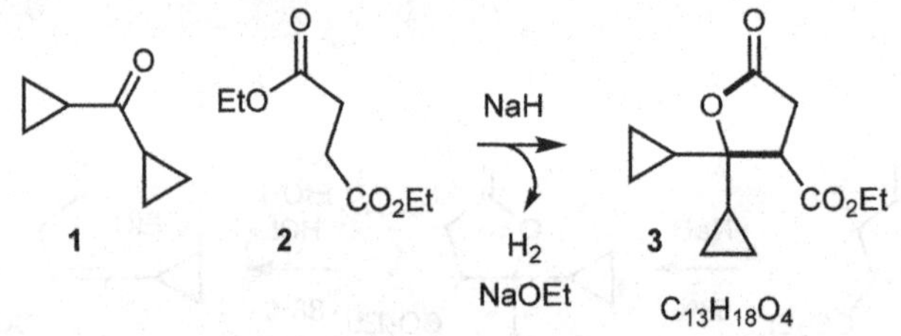

SOLUTION: 3 TO 4 (LACTONE OPENING ESTERIFICATION)

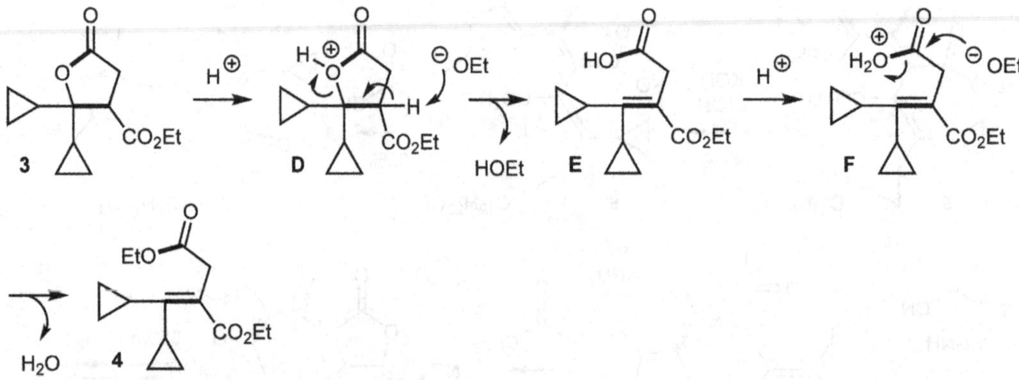

BALANCED CHEMICAL EQUATION: 3 TO 4

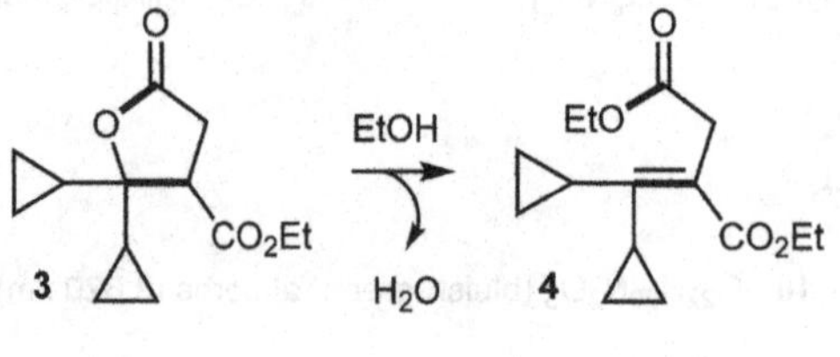

SOLUTION: 4 TO 5 (STOBBE CONDENSATION)

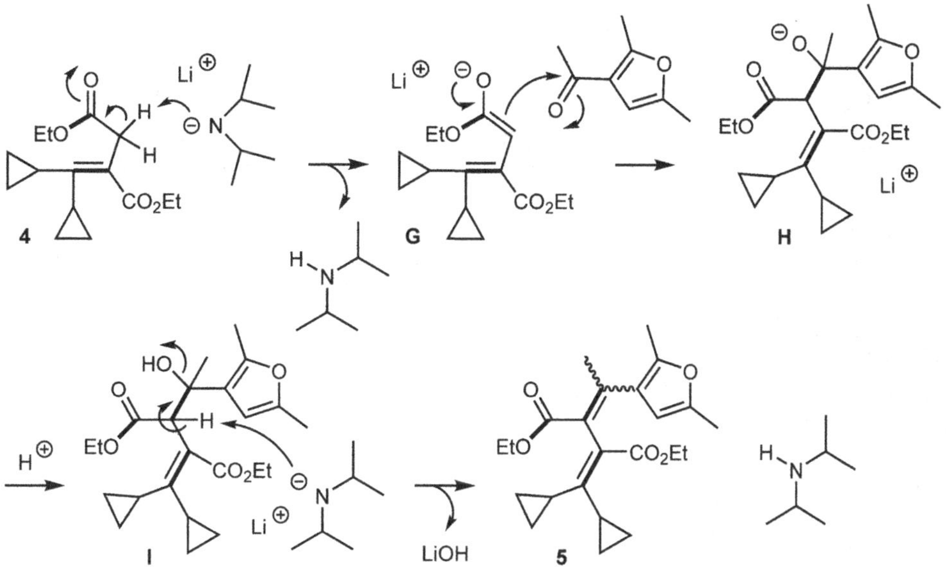

BALANCED CHEMICAL EQUATION: 4 TO 5

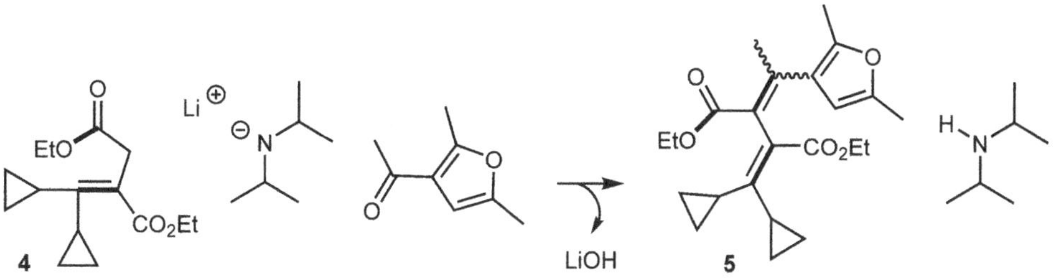

SOLUTION: 5 TO 6 (SAPONIFICATION OF DIESTER TO DIACID)

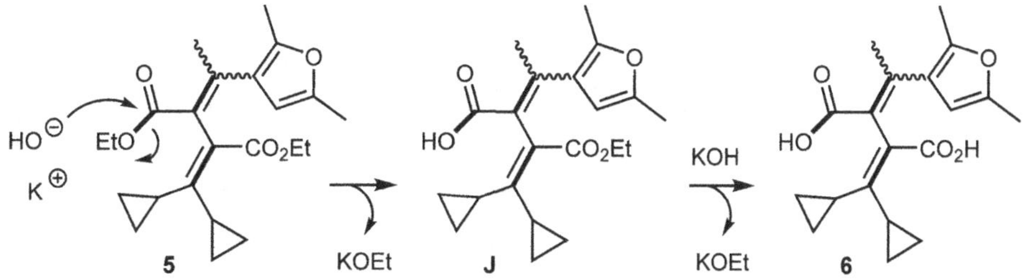

BALANCED CHEMICAL EQUATION: 5 TO 6

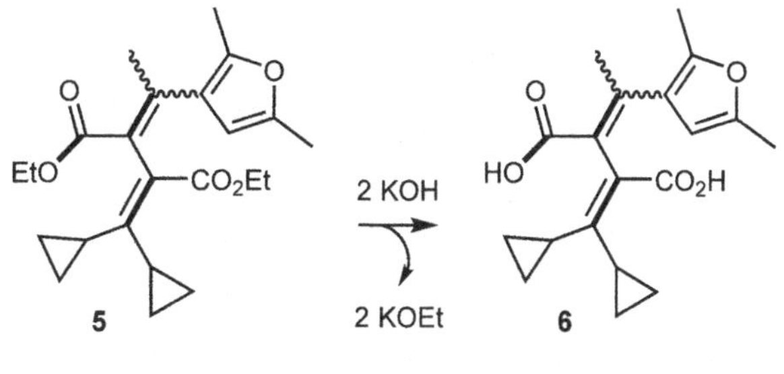

SOLUTION: 6 TO 7

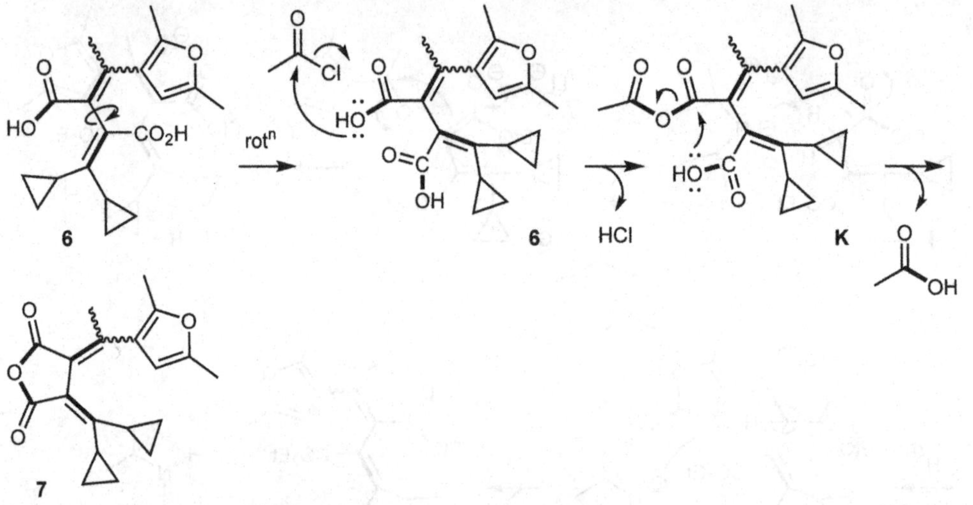

BALANCED CHEMICAL EQUATION: 6 TO 7

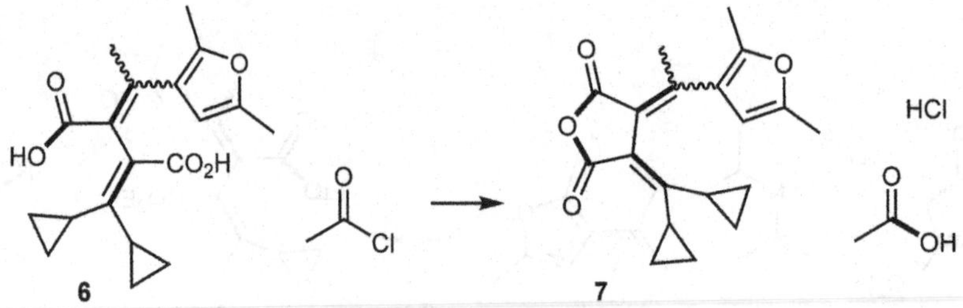

SOLUTION: 7 TO 8 (BASE-ASSISTED RING OPENING ALKYLATION)

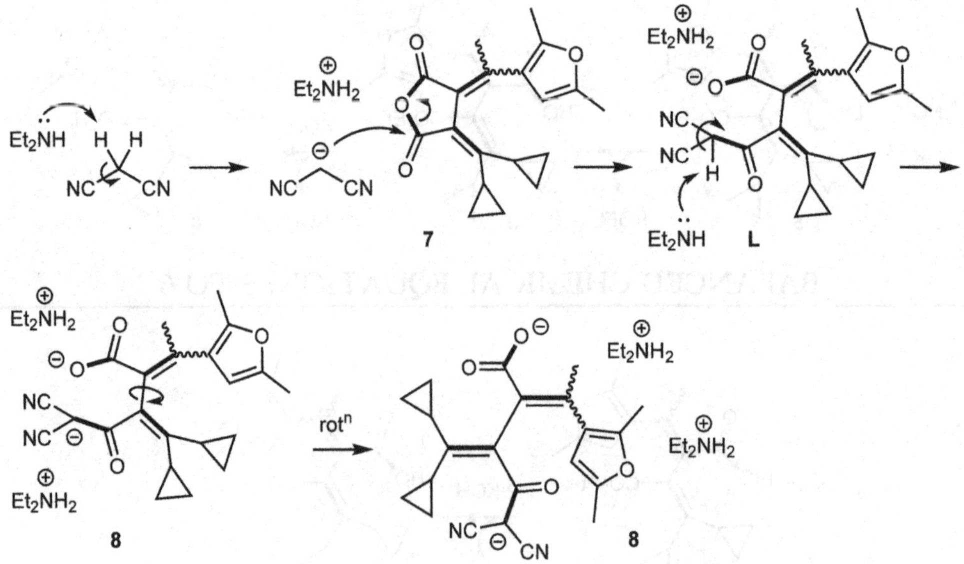

BALANCED CHEMICAL EQUATION: 7 TO 8

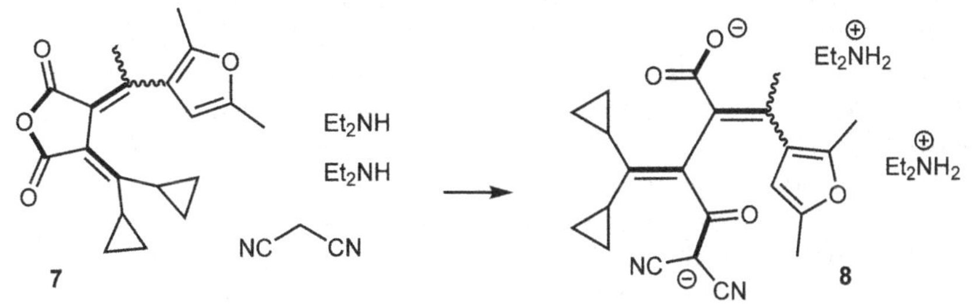

SOLUTION: 8 TO 9 (ACETYL CHLORIDE-ASSISTED CYCLIZATION)

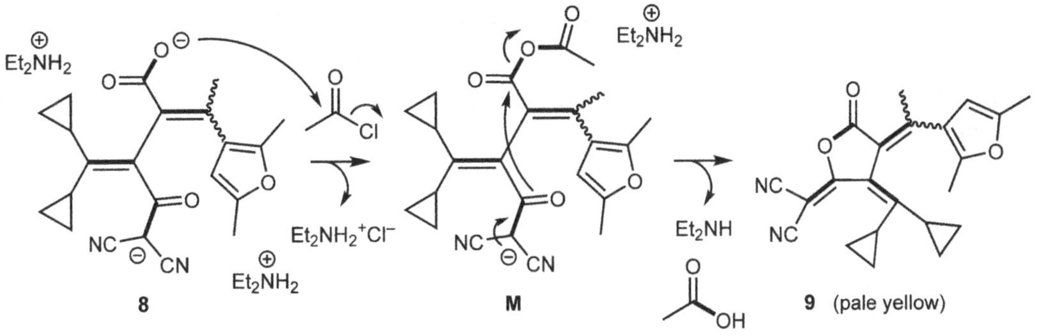

BALANCED CHEMICAL EQUATION: 8 TO 9

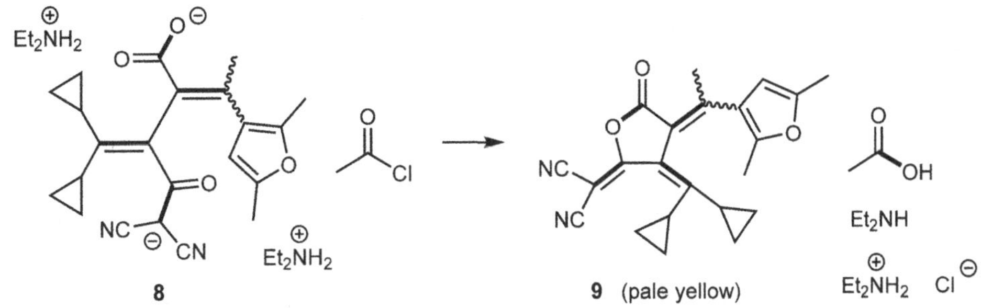

SOLUTION: 9 TO 10 (PHOTOCHROMISM MECHANISM)

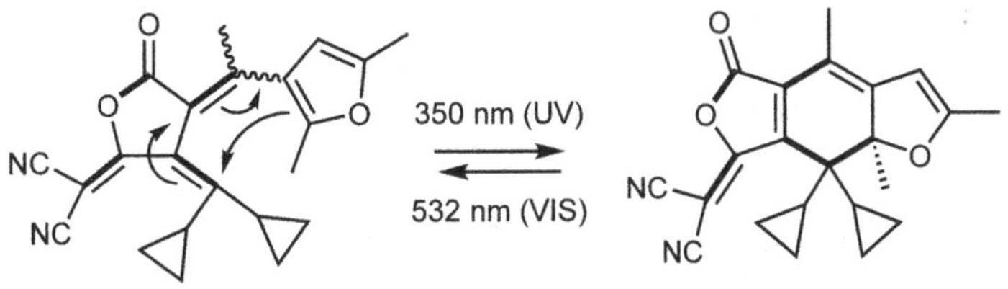

9 (pale yellow)　　　10　$C_{22}H_{20}N_2O_3$ (bluish-green, absorbs at 620 nm)

KEY STEPS EXPLAINED

There are seven synthetic steps to the fulgide **9** synthesis depicted in Scheme 1 which the authors did not analyze in mechanistic terms either using experimental or theoretical techniques. In the first step we have a Stobbe condensation[2] between dicyclopropylketone (**1**) and diethylsuccinate (**2**) which forms lactone **3** along with hydrogen gas and sodium ethoxide base. In step 2, esterification of lactone **3** results in ring opening to form diester **4**. Step 3 constitutes a second Stobbe condensation of diester **4** with 3-acetyl-2,5-dimethylfuran that generates diester **5**. Step 4 is a saponification of diester **5** to diacid **6**. In addition, Step 5 represents an acetyl chloride-assisted cyclization of diacid **6** to anhydride **7** with elimination of acetic acid and hydrochloric acid. Step 6 is a base-assisted ring opening alkylation reaction between malonitrile and anhydride **7** which produces diammonium salt **8**. Finally, step 7 constitutes another acetyl chloride-assisted cyclization of the carbanion moiety of **8** onto the carboxylate group to form the final fulgide product **9**. In the last photochromic reaction, we have 6π electrocyclization of **9** to **10** by a UV photoreaction the reverse of which is a VIS photoreaction from **10** to **9** which is a 6π retro-electrocyclic reaction.

ADDITIONAL RESOURCES

We recommend several articles including one by the same authors which explore further reaction mechanisms of photochromic compounds.[3–6]

References

1. Sun Z, Hosmane RS, Tadros M. Fulgides and photochromism. Synthesis of (E)- and (Z)-5-dicyanomethylene-4-dicyclopropylmethylene-3-[1-(2,5-dimethyl-3-furyl)ethylidene] tetrahydrofuran-2-one. *Tetrahedron Lett.* 1995;36(20):3453–3456. https://doi.org/10.1016/0040-4039(95)00632-M.
2. Wang Z, ed. *Comprehensive Organic Name Reactions and Reagents.* New York, NY: John Wiley & Sons, Inc.; 2010:2686–2691. https://doi.org/10.1002/9780470638859.conrr2686
3. Sun Z, Hosmane RS, Tadros M. Photochromic fulgides: transformation of the non-photochromic (Z)-isomer of a fulgide into a highly photochromic (E)-isomer via structural modification involving enhanced conjugation. *J Heterocyclic Chem.* 2000;37(6):1439–1441. https://doi.org/10.1002/jhet.5570370606.
4. Yu L, Zhu D, Fan M. Synthesis and photochromism of 5-dicyano-methylene-4-isopropylidene-3-[1-(1-p-methoxy-phenyl-2-methyl-5-phenyl-3-pyrryl)-ethylidene]-tetrahydrofuran-2-one. *Mol Cryst Liq Cryst.* 1997;297:107–114. https://doi.org/10.1080/10587259708036110.
5. Fan P, Wei J, Zhu A, et al. Photochromic mechanism of heterofused benzopyran derivatives. *Mol Cryst Liq Cryst.* 2000;344:289–294. https://doi.org/10.1080/10587250008023851.
6. Gomes R, Laia CAT, Pina F. On the mechanism of photochromism of 4'-N,N-dimethylamino-7-hydroxyflavylium in pluronic F127. *J Phys Chem B.* 2009;113:11134–11146. https://doi.org/10.1021/jp902972q.

Question 45: Benzodiazepines From (R)-(+)-Pulegone

In 1995 chemists from various countries utilized the readily available (R)-(+)-pulegone **1** in conjunction with o-phenylenediamine in toluene under heat and reflux for 4 days to synthesize the tricyclic benzodiazepine derivative **2** in 68% yield (Scheme 1).[1] Suggest a mechanism for this transformation.

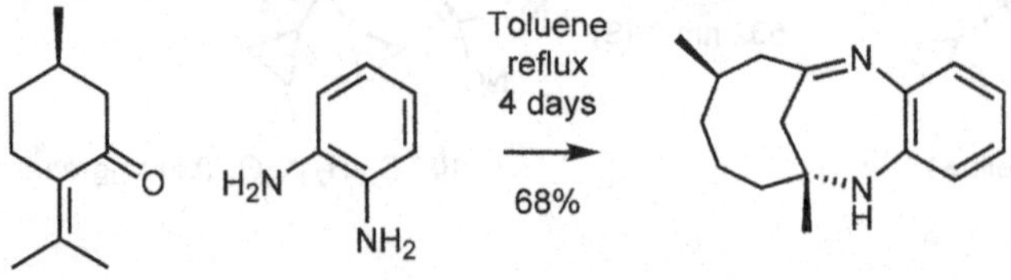

1

(R)-(+)-Pulegone

2

SCHEME 1

SOLUTION

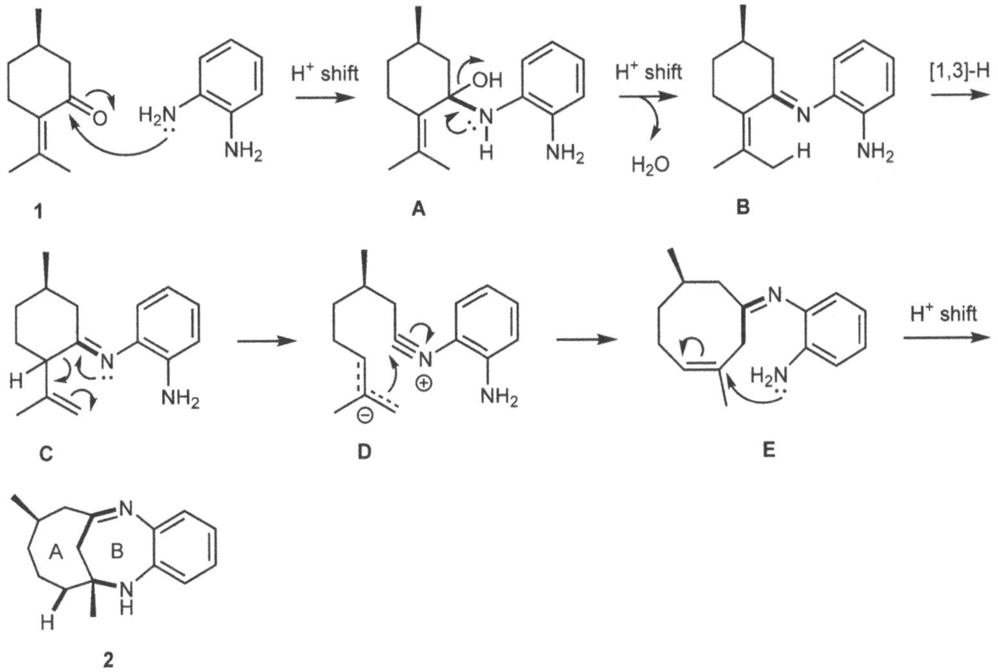

KEY STEPS EXPLAINED

The mechanism proposed for this transformation begins with condensation between *o*-phenylenediamine and (*R*)-(+)-pulegone **1** which forms hydrazone **B** after elimination of water. Hydrazone **B** then undergoes a [1,3] protic shift to give enamine **C** which ring opens to form the zwitterionic intermediate **D** containing a nitrilium ion moiety. The carbanion moiety of this intermediate cyclizes onto the nitrilium group to generate intermediate **E** which contains an eight-membered ring (ring A). Next, the remaining amino group cyclizes onto the olefinic bond in **E** creating another seven-membered ring (ring B) as found in the bicyclic structure of product **2**. The ring construction mapping for product **2** is $[(8+0)_A + (4+2+1)_B]$. We note that the authors characterized product **2** and established its stereochemistry using ^{1}H NMR, COSY, and NOE spectroscopy. Nevertheless, the authors did not undertake a theoretical analysis to help support their proposed mechanism.

ADDITIONAL RESOURCES

We recommend an article which discusses the mechanism of a transformation that involves an unexpected benzodiazepine rearrangement[2] and two other articles that discuss interesting and unexpected formations of oxazepine and isoxazole ring structures.[3,4]

References

1. Hakiki A, Mossadak M, Mokhles M, Rouessac F, Duddeck H, Mikhova B. A surprising eight-membered ring synthesis leading to enantiomerically pure tricyclic benzodiazepine derivatives. *Tetrahedron*. 1995;51(8):2293–2296. https://doi.org/10.1016/0040-4020(94)01080-J.
2. Lakhrissi B, Massoui M, Essassi EM, et al. Synthesis of *N,N'*-diglucosylated benzimidazol-2-one via an unexpected rearrangement of benzodiazepine derivative. *Heterocycl Commun*. 2005;11(2):157–162. https://doi.org/10.1515/HC.2005.11.2.157.
3. Baranczak A, Sulikowski GA. Synthetic studies directed toward dideoxy lomaiviticinone lead to unexpected 1,2-oxazepine and isoxazole formation. *Org Lett*. 2012;14(4):1027–1029. https://doi.org/10.1021/ol203390w.
4. Berg U, Bladh H, Hoff M, Svensson C. Stereochemical variations on the colchicine motif. Part 2. Unexpected tetracyclic isoxazole derivatives. *J Chem Soc Perkin Trans 2*. 1997;1997:1697–1704. https://doi.org/10.1039/A701400G.

Question 46: A Furan to Pyran Ring Expansion

In 1995 American chemists studied the mechanism of the reaction of wortmannin **1** with perdeuterated trimethyl-sulfoxonium ylide which forms the deuterated wortmannin pyran product **2-d$_2$** (Scheme 1).[1] Propose two reasonable mechanisms for this transformation and show that only one is consistent with the results of the deuterium labeling study.

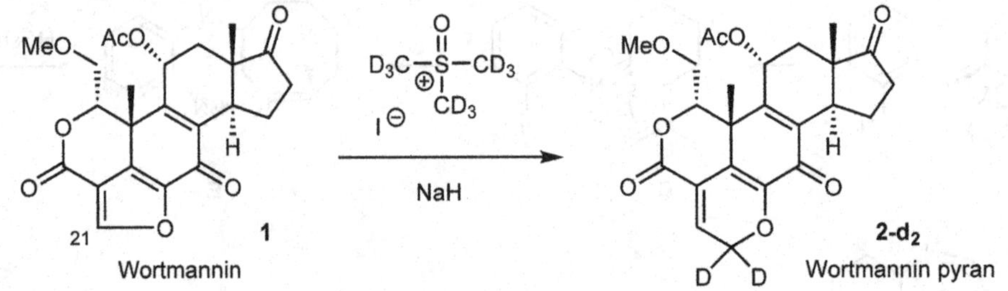

SCHEME 1

SOLUTION: 1 TO 2 (MECHANISM 1: CONSISTENT WITH DEUTERIUM LABELING STUDY)

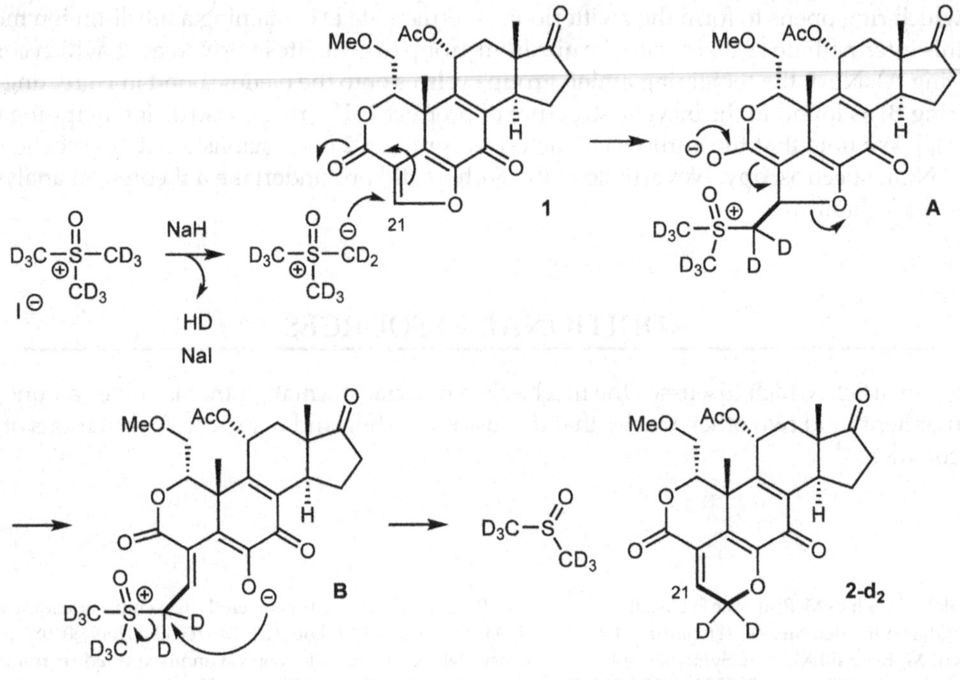

SOLUTION: 1 TO 2 (MECHANISM 2: INCONSISTENT WITH DEUTERIUM LABELING STUDY)

KEY STEPS EXPLAINED

We begin by noting that out of the two proposed mechanisms only mechanism 1 is consistent with the deuterium labeling study undertaken by the authors. Moreover, the two mechanisms both begin with generation of perdeuterated trimethylsulfonium ylide through deuteron abstraction by hydride ion. The ylide thus formed nucleophilically attacks the electrophilic C_{21} position of wortmannin **1** to form the zwitterionic intermediate **A**. In mechanism 1 this intermediate undergoes furan ring opening to form enolate **B** which ring closes with elimination of perdeuterated dimethylsulfoxide to generate the observed deuterated wortmannin pyran product **2-d$_2$**. By contrast, in mechanism 2 intermediate **A** instead undergoes ring closure to form the fused cyclopropane **C** after elimination of perdeuterated dimethylsulfoxide. Furthermore, the bicyclic [3.1.0] framework in **C** expands to give the six-membered ring zwitterionic intermediate **D**. This intermediate rearranges via a [1,2] deuteride shift to give the product **3-d$_2$** (not observed).

We note that the authors did not carry out further experimental or theoretical analyses to help support mechanism 1. Nevertheless, they did carry out an interesting ^{1}H NMR analysis in identifying product **2-d$_2$** which is worth presenting here. Essentially, the authors stated that the undeuterated 2H-pyran ring protons in **2** exhibited a first-order ABX pattern, whereas for **2-d$_2$** the AB protons were absent and a singlet was observed for the vinyl proton (H$_X$, see Scheme 2).

For simplicity, we illustrate the theoretical splitting of signals expected for the 2H-pyran ring protons in **2**, **2-d$_2$**, and **3-d$_2$** in Scheme 2. We note that because the nuclear spin quantum numbers for ^{1}H and ^{2}H are ½ and 1, respectively, we can deduce the expected maximum number of principal lines in the proton spectrum on account of couplings with neighboring protons and deuterons for each ^{1}H nucleus in the 2H-pyran ring. For example, in structure **2** there are three kinds of chemical environments: a vinyl proton H$_X$ and two different geminal protons H$_A$ and H$_B$. The geminal protons are different since they are diastereotopic. The signal for the vinyl proton H$_X$ is expected to have four lines on account of ^3J(H$_A$-H$_X$) and ^3J(H$_B$-H$_X$) couplings. Similarly, the signal for each geminal proton is expected to have four lines due to ^2J(H$_A$-H$_B$) and ^3J(H$_{A/B}$-H$_X$) couplings. Therefore $4+4+4=12$ principal lines are expected in the ^{1}H NMR spectrum for X and two AB protons.

This case contrasts with that of structure **2-d₂** where there is only one vinyl proton which is adjacent to two diastereotopic deuterons. In this case, the signal of **2-d₂** should have a maximum of nine lines which is caused by $^3J(H_X\text{-}D_A)$ and $^3J(H_X\text{-}D_B)$ couplings. Since the authors observed only a singlet for this vinyl proton we can infer that the magnitude of the coupling constants to the two deuterons must have been too small to be resolved. We note that the authors did not disclose the magnetic frequency of the spectrometer used to record the spectrum. More importantly, no proton signals were observed for the geminal protons because they were replaced by deuterons.

Lastly, structure **3-d₂** has only one geminal proton adjacent to another geminal deuteron and vinyl deuteron. The corresponding $^3J(H_A\text{-}D_X)$ coupling evidently results in triplet splitting and the $^2J(H_A\text{-}D_B)$ coupling also provides for a triplet splitting thus giving a maximum of $3 \times 3 = 9$ principal lines for the geminal ^1H NMR signal. We note that for this structure, no proton signal is expected for the vinyl proton since it is replaced by deuteron. Consequently, we note that the main difference between the proton spectra of **2-d₂** and **3-d₂** is related to the absence of the geminal AB protons in **2-d₂** and the absence of the vinyl X proton in **3-d₂**. For more information on interpretation of ABX patterns in NMR spectroscopy including example chemical structures showing how they arise, we refer the reader to an excellent review by Slomp.[2]

For ^1H, $I = 1/2$
For ^2H, $I = 1$

2

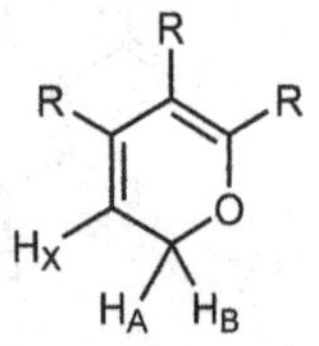

Multiplicity for H_A: 2 (1/2) + 1 = 2 (doublet splitting by H_X)
2 (1/2) + 1 = 2 (doublet splitting by H_B)
Maximum # principal lines = 2 × 2 = 4

Multiplicity for H_B: 2 (1/2) + 1 = 2 (doublet splitting by H_A)
2 (1/2) + 1 = 2 (doublet splitting by H_X)
Maximum # principal lines = 2 × 2 = 4

Multiplicity for H_X: 2 (1/2) + 1 = 2 (doublet splitting by H_A)
2 (1/2) + 1 = 2 (doublet splitting by H_B)
Maximum # principal lines = 2 × 2 = 4

2-d₂

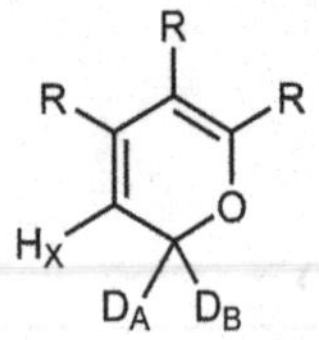

Multiplicity for H_X: 2 (1) + 1 = 3 (triplet splitting by D_A)
2 (1) + 1 = 3 (triplet splitting by D_B)
Maximum # principal lines = 3 × 3 = 9

3-d₂

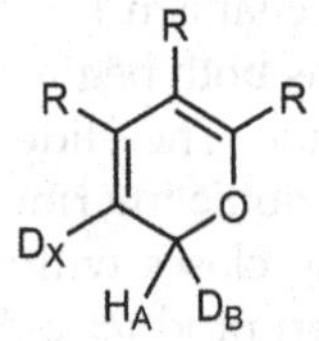

Multiplicity for H_A: 2 (1) + 1 = 3 (triplet splitting by D_B)
2 (1) + 1 = 3 (triplet splitting by D_X)
Maximum # principal lines = 3 × 3 = 9

SCHEME 2

We also note that the ^1H NMR spectrum for wortmannin **1** was reported by MacMillan and Vanstone in 1972 and that of wortmannin pyran **2** was reported by Norman et al. in 1996.[3,4] In Tables 1 and 2 we summarize the chemical shifts, multiplicities, and structural assignments for both compounds according to the numbering schemes given in Fig. 1. We note that the vinyl proton at C_{20} in wortmannin **1** is a singlet appearing at 8.22 ppm. The ABX system in wortmannin pyran **2** pertains to C_{20} (vinyl hydrogen CH_X; δ 6.98 ppm; dd; $J(C_{20}\text{-}C_{21B}) = 5.5$; $J(C_{20}\text{-}C_{21A}) = 3.3$ Hz) and C_{21} (CH_A; δ 4.74 ppm; dd; $J(C_{21A}\text{-}C_{21B}) = 16.1$; $J(C_{21A}\text{-}C_{20}) = 3.3$ Hz and CH_B; δ 5.14 ppm; dd; $J(C_{21B}\text{-}C_{21A}) = 16.1$; $J(C_{21B}\text{-}C_{20}) = 5.5$ Hz) for a total of 12 lines which corresponds to what is presented in Scheme 2.

TABLE 1 Structural Assignments for the ^{1}H NMR Spectrum of Wortmannin **1** Recorded at 220 MHz in CDCl$_3$.

	δ (ppm)	Functional group	# H	Multiplicity and coupling constants
C$_1$	4.75	CH	1	dd; J(C$_1$-C$_{2A}$) = 2; J(C$_1$-C$_{2B}$) = 6 Hz
C$_{2A}$	3.48	CH	1	dd; J(C$_{2A}$-C$_{2B}$) = 11; J(C$_{2A}$-C$_1$) = 2 Hz
C$_{2B}$	2.98	CH	1	dd; J(C$_{2B}$-C$_{2A}$) = 11; J(C$_{2B}$-C$_1$) = 6 Hz
C$_2$-OMe	3.13	CH$_3$	3	s
C$_{11}$	6.1	CH	1	ddd; J(C$_{11}$-C$_{12A}$) = 8; J(C$_{11}$-C$_{12B}$) = 8; J(C$_{11}$-C$_{14}$) = 3 Hz
C$_{11}$-OAc	2.09	CH$_3$	3	s
C$_{12A}$	2.7	CH	1	dd; J(C$_{12A}$-C$_{11}$) = 8; J(C$_{12A}$-C$_{12B}$) = 12.5 Hz
C$_{12B}$	1.6	CH	1	dd; J(C$_{12B}$-C$_{11}$) = 8; J(C$_{12B}$-C$_{12A}$) = 12.5 Hz
C$_{14}$	2.85	CH	1	ddd; J(C$_{14}$-C$_{11}$) = 3; J(C$_{14}$-C$_{15A}$) = 6; J(C$_{14}$-C$_{15B}$) = 12.2 Hz
C$_{15A}$	3.2	CH	1	dd; J(C$_{15A}$-C$_{14}$) = 6; J(C$_{15A}$-C$_{15B}$) = not reported Hz
C$_{15B}$	2.2	CH	1	dd; J(C$_{15B}$-C$_{14}$) = 12.2; J(C$_{15B}$-C$_{15A}$) = not reported Hz
C$_{16A}$	2.6	CH	1	Not assigned
C$_{16B}$	2.1	CH	1	Not assigned
C$_{18}$	0.94	CH$_3$	3	s
C$_{19}$	1.71	CH$_3$	3	s
C$_{20}$	8.22	CH	1	s
		Total	24	

TABLE 2 Structural Assignments for the ^{1}H NMR Spectrum of Wortmannin Pyran **2** Recorded at 300 MHz, CDCl$_3$.

	δ (ppm)	Functional group	# H	Multiplicity and coupling constants
C$_1$	4.61	CH	1	dd; J(C$_1$-C$_{2B}$) = 6.3; J(C$_1$-C$_{2A}$) = 2.2 Hz
C$_{2A}$	3.46	CH	1	dd; J(C$_{2A}$-C$_{2B}$) = 11.1; J(C$_{2A}$-C$_1$) = 2.2 Hz
C$_{2B}$	3.18	CH	1	dd; J(C$_{2B}$-C$_{2A}$) = 11.1; J(C$_{2B}$-C$_1$) = 6.4 Hz
C$_2$-OMe	3.24	CH$_3$	3	s
C$_{11}$	6.02	CH	1	m
C$_{11}$-OAc	2.08	CH$_3$	3	s
C$_{12A}$	2.48	CH	1	dd; J(C$_{12A}$-C$_{12B}$) = 13.9; J(C$_{12A}$-C$_{11}$) = 7.6 Hz
C$_{12B}$	1.72	CH	1	dd; J(C$_{12B}$-C$_{12A}$) = 14.2; J(C$_{12B}$-C$_{11}$) = 5.7 Hz
C$_{14}$	2.86–3.04	CH	1	m
C$_{15A}$	2.86–3.04	CH	1	m
C$_{15B}$	2.25	CH	1	m
C$_{16A}$	2.59	CH	1	m
C$_{16B}$	2.02	CH	1	m
C$_{18}$	0.86	CH$_3$	3	s
C$_{19}$	1.57	CH$_3$	3	s
C$_{20}$	6.98	CH	1	dd; J(C$_{20}$-C$_{21B}$) = 5.5; J(C$_{20}$-C$_{21A}$) = 3.3 Hz
C$_{21A}$	4.74	CH	1	dd; J(C$_{21A}$-C$_{21B}$) = 16.1; J(C$_{21A}$-C$_{20}$) = 3.3 Hz
C$_{21B}$	5.14	CH	1	dd; J(C$_{21B}$-C$_{21A}$) = 16.1; J(C$_{21B}$-C$_{20}$) = 5.5 Hz
		Total	26	

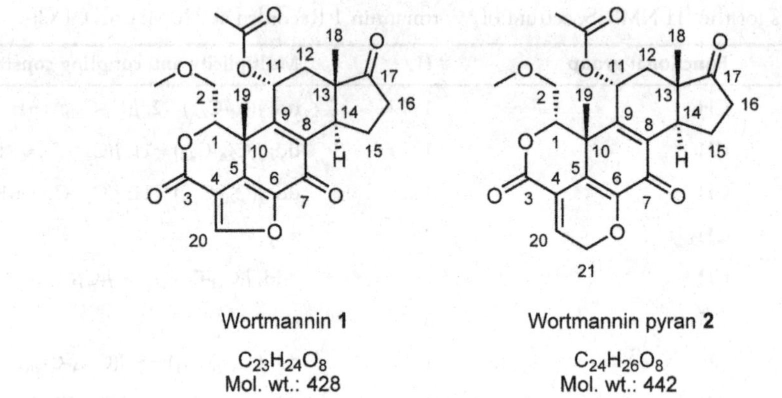

FIG. 1   Structures of wortmannin 1 and wortmannin pyran 2 showing number scheme for atoms.

References

1. Norman BH, Paschal J, Vlahos CJ. Synthetic studies on the furan ring of wortmannin. *Bioorg Med Chem Lett.* 1995;5(11):1183–1186. https://doi.org/10.1016/0960-894X(95)00191-U.
2. Slomp G. Analysis of ABX spectra in NMR spectroscopy. *Appl Spectrosc Rev.* 1969;2:263–351. https://doi.org/10.1080/05704926908050171.
3. MacMillan J, Vanstone AE, Yeboah SK. Fungal products. Part III. Structure of wortmannin and some hydrolysis products. *J Chem Soc Perkin Trans 1.* 1972;1972:2898–2903. https://doi.org/10.1039/P19720002898.
4. Norman BH, Shih C, Toth JE, et al. Studies on the mechanism of phosphatidylinositol 3-kinase inhibition by wortmannin and related analogs. *J Med Chem.* 1996;39(5):1106–1111. https://doi.org/10.1021/jm950619p.

Question 47: A Structure and Mechanism Correction

In 1993 Scheinmann and Stachulski reported that the reaction of *N*-(2-bromoethyl)phthalimide 1 with the dianion of isobutyric acid 2 gave the aroylaziridine product 3 in 76% yield (Scheme 1).[1] The authors then treated 3 with hydrazine

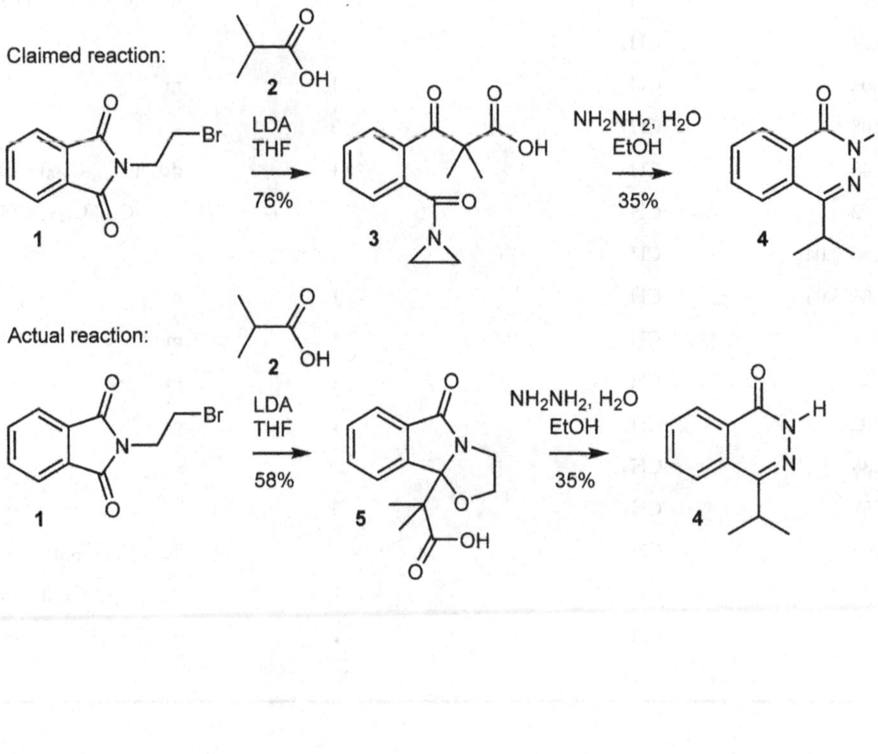

SCHEME 1

in ethanol at 60°C and achieved the phthalazin-1(2H)-one product **4** in 35% yield. Two years later, de Kimpe et al. showed that product **3** was not formed as originally claimed, but instead product **5** was formed. The authors also showed that **5** reacts with hydrazine in ethanol as before to form **4**. Show that reasonable mechanisms can be proposed for the transformations of **1** and **2** to **3** and **3** to **4**. Provide mechanisms for the actual transformations of **1** and **2** to **5** and **5** to **4**.

SOLUTION: 1 AND 2 TO 3 (CLAIMED REACTION)

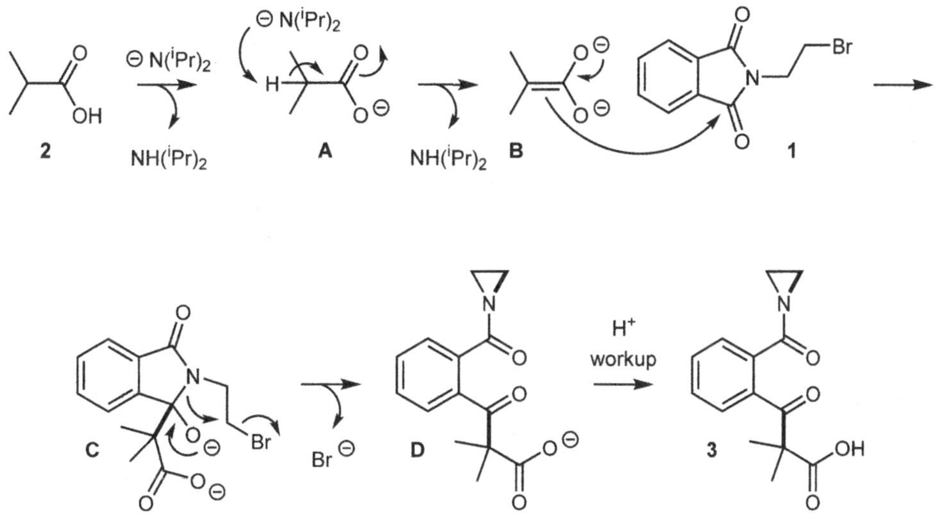

SOLUTION: 3 TO 4 (CLAIMED REACTION)

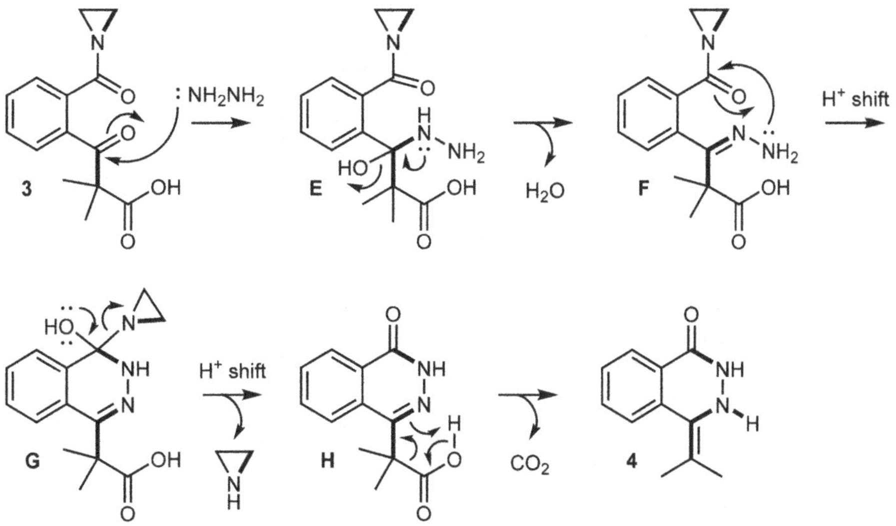

SOLUTION: 1 AND 2 TO 5 (ACTUAL REACTION)

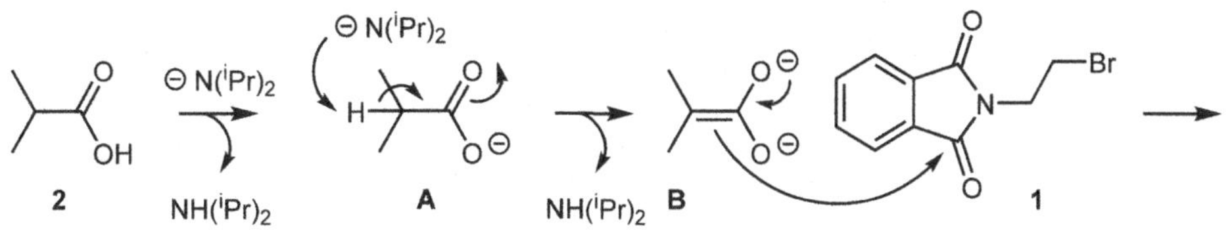

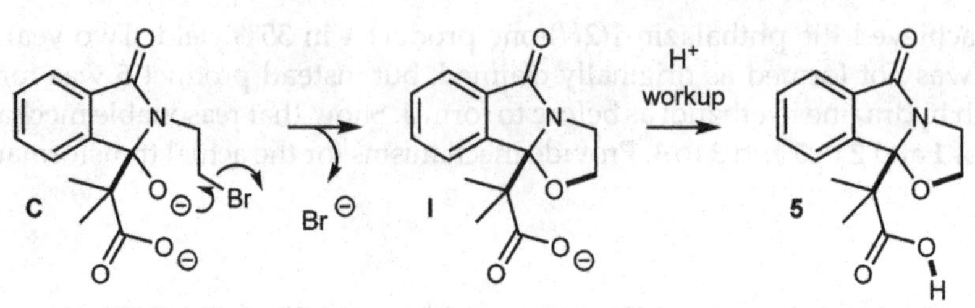

SOLUTION: 5 TO 4 (ACTUAL REACTION)

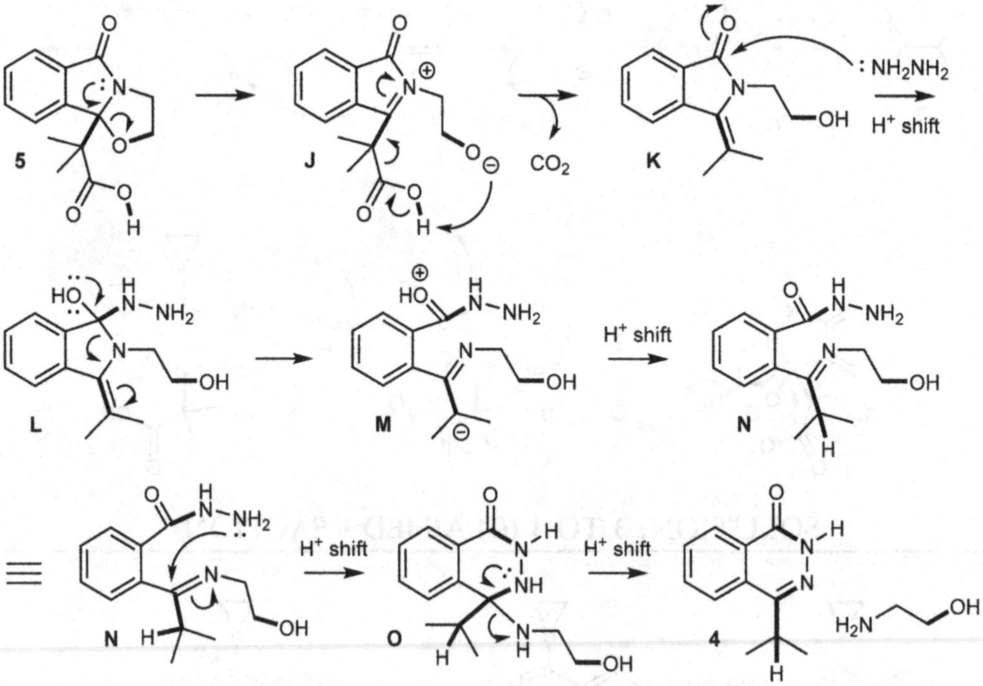

BALANCED CHEMICAL EQUATIONS

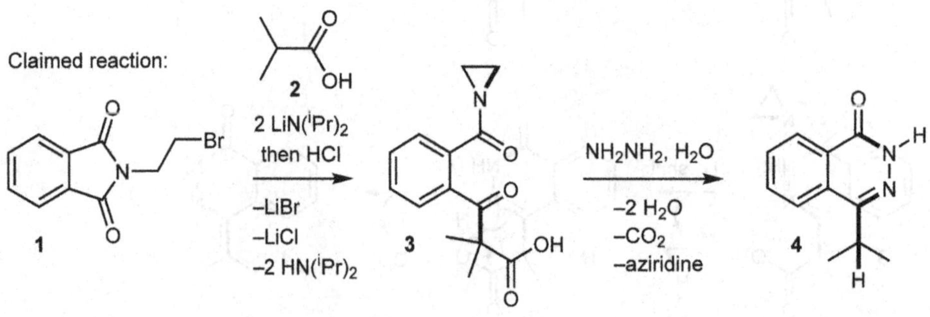

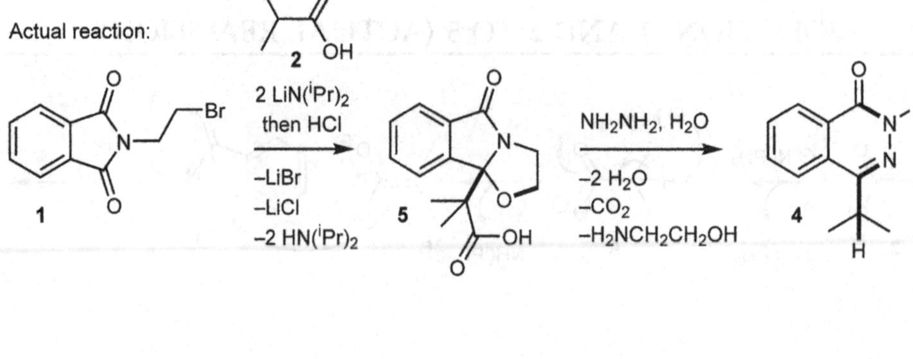

Claimed reaction:

Actual reaction:

KEY STEPS EXPLAINED

We begin by noting that the difference between the claimed reaction and the actual reaction centers on the structure of the middle product which was identified as **3** in 1993 and whose structure was corrected to that of **5** in 1995 by the authors.[1,2] Nevertheless, reasonable mechanisms can be drawn for both scenarios. For example, in the case of **1** and **2** to **3**, the proposed mechanism begins with two LDA induced deprotonations of **2** to form dianion **B** which nucleophilically adds to one of the carbonyl groups of phthalimide **1** to form alkoxide **C**. This alkoxide undergoes rearrangement with ring opening to form aziridine **D** which is protonated under work-up conditions to form **3**. From **3** to **4**, it is proposed that hydrazine adds to the more electrophilic carbonyl group of **3** to form imine **F** after loss of water. The free amine group of **F** facilitates cyclization onto the carbonyl group to form intermediate **G** after proton transfer. Ketonization in **G** with proton transfer and loss of aziridine gives **H** which rearranges with decarboxylation to form **4**.

When it was revealed that the structure of the middle product was actually that of **5**, the proposed mechanism was revised to involve formation of **C** from **1** and **2** as seen already, only this time intermediate **C** cyclizes to oxazolidine **I**. Acidic work-up of **I** gives the true middle product **5**. In the second reaction, the oxazolidine ring of **5** opens to form the iminium alkoxide intermediate **J** which undergoes intramolecular decarboxylation yielding 2-(2-hydroxy-ethyl)-3-isopropylidene-2,3-dihydro-isoindol-1-one intermediate **K**. Hydrazine then nucleophilically attacks the carbonyl group of **K** eventually leading to hydrazide **N** which cyclizes onto the imine group yielding intermediate **O**. Final elimination of ethanolamine generates 4-isopropyl-2H-phthalazin-1-one **4**.

To verify the structure of product **5**, the authors prepared its methyl ester in two ways as shown in Scheme 2. One method converted the acid product obtained from **1** and **2** into its methyl ester **6** using diazomethane. The second method of synthesizing **6** involved reacting **1** and the methyl ester of **2** (i.e., **7**) in the presence of lithium diisopropylamine in THF. The spectral characteristics of the ester products obtained by both methods were found to be identical thus confirming the structure of **5**.

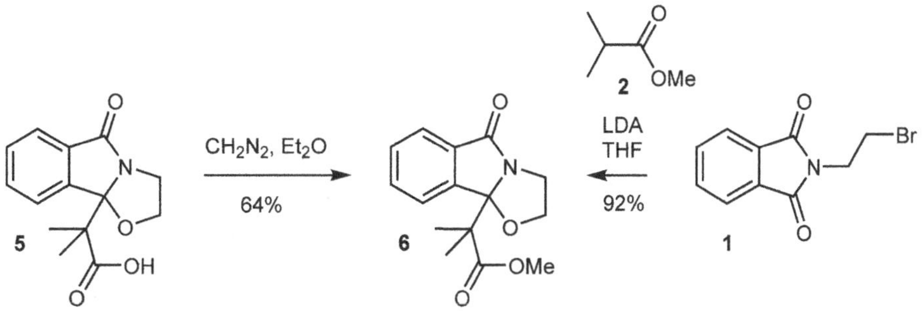

SCHEME 2

As can be seen by comparing the balanced chemical equations shown before, the distinguishing feature between hydrazine reactions **5 → 4** and **3 → 4** is that the actual transformation produces ethanolamine as by-product whereas the claimed transformation predicts aziridine. The authors could also have demonstrated that aziridine was not present as a by-product as additional proof to their spectroscopic evidence for the fact that the middle product had structure **5** and not **3**. Moreover, the authors could also have performed additional experimental and theoretical analyses to help support the proposed mechanisms about which they provide very little information.

ADDITIONAL RESOURCES

We recommend two articles which discuss mechanisms involving aziridine chemistry.[3,4]

References

1. Scheinmann F, Stachulski AV. Convenient synthesis of a novel γ-aminobutyric acid analogue: 4-amino-2,2-dimethylbutanoic acid. *J Chem Res (S)*. 1993;1993:414–415.
2. de Kimpe N, Virag M, Keppens M, Scheinmann F, Stachulski AV. Cyclocondensation of *N*-(2-bromoethyl)phthalimide with the dianion of isobutyric acid to a functionalized 2,3-dihydrooxazolo[2,3-a]isoindol-5(9bH)-one derivative. *J Chem Res (S)*. 1995;1995:252–253.
3. Tianning D, Xiaoyu S, Renhua F, Jie F. Unexpected ring-opening reaction of aziridine with acetic anhydride in DMF. *Chem Lett*. 2007;36(5):604–605. https://doi.org/10.1246/cl.2007.604.
4. Haga T, Ishikawa T. Mechanistic approaches to asymmetric synthesis of aziridines from guanidinium ylides and aryl aldehydes. *Tetrahedron*. 2005;61(11):2857–2869. https://doi.org/10.1016/j.tet.2005.01.088.

Question 48: Propargylbenzotriazoles to Five-membered Heterocycles

In 1995 American chemists reacted a solution of 1-propargylbenzotriazole **1** with one equivalent of *n*-BuLi in THF at −78°C for 1h followed by addition of one equivalent of 2-bromo-1-phenylpropan-1-one with stirring for 4h (Scheme 1).[1] To this mixture, a solution of one equivalent of KO*t*Bu in HO*t*Bu was added and the temperature was allowed to reach ambient after which the reaction was heated overnight at 50°C. After aqueous work-up the authors isolated the benzotriazole phenyl furan product **3** in 53% yield. They then stirred this compound with one equivalent of zinc chloride and 10 equivalents of 2-methylthiophene in methylene chloride at room temperature followed by aqueous work-up to arrive at the final methylthiophen-phenylfuran product **4** in 86% yield. Suggest mechanisms for these transformations.

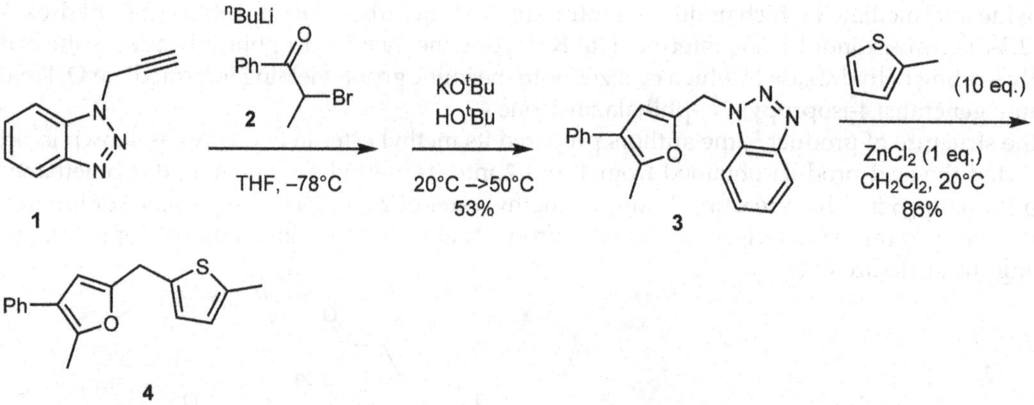

SCHEME 1

SOLUTION: 1 AND 2 TO 3

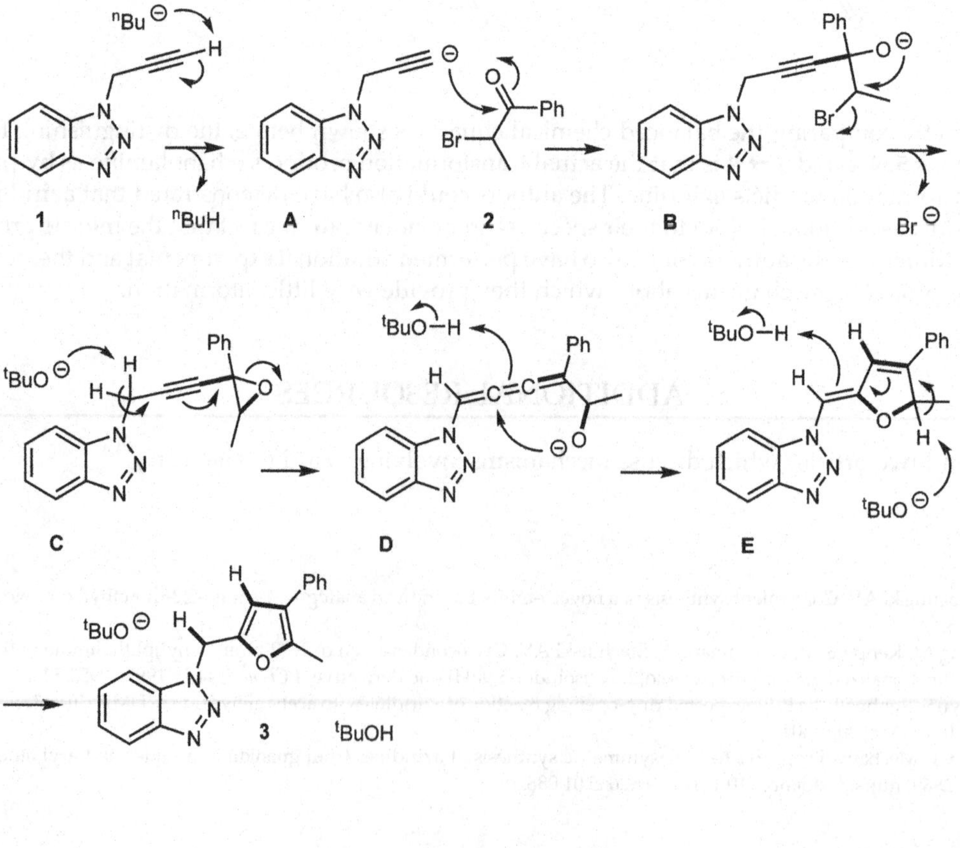

BALANCED CHEMICAL EQUATION: 1 AND 2 TO 3

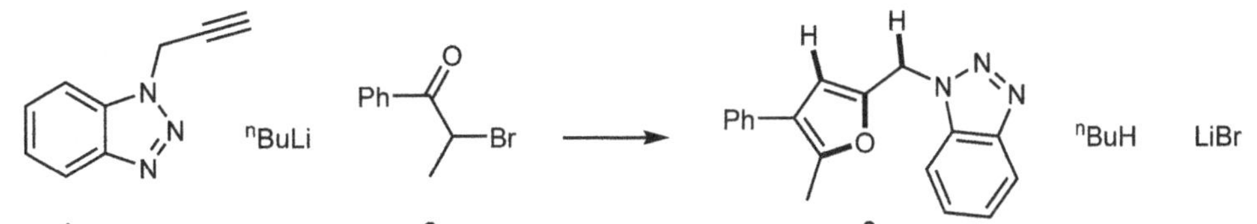

SOLUTION: 3 TO 4

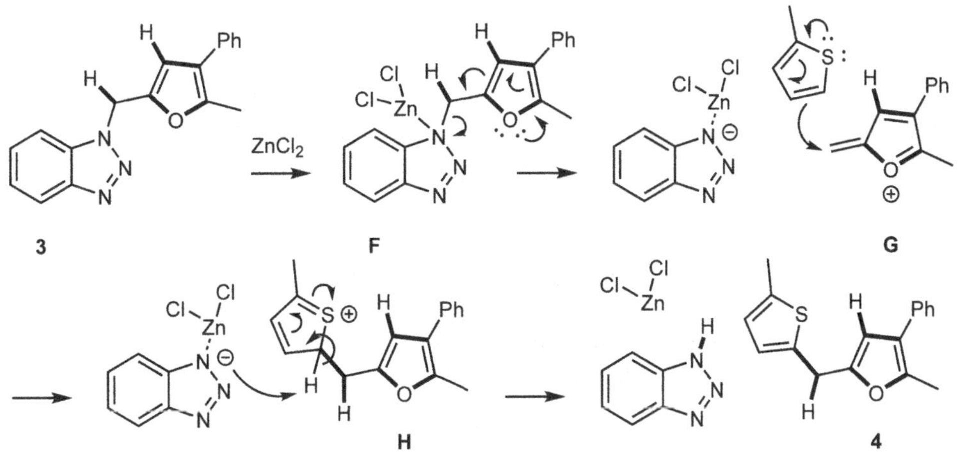

BALANCED CHEMICAL EQUATION: 3 TO 4

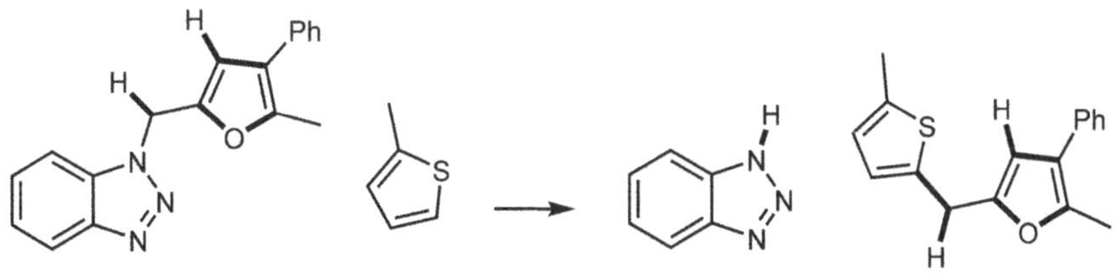

KEY STEPS EXPLAINED

The mechanism proposed by the authors to explain the formation of **3** from **1** and **2** begins with deprotonation of the acetylene group of **1** by butyllithium base which forms the acetylide intermediate **A** which then nucleophilically attacks the ketone group of **2** giving tetrahedral intermediate **B**. This intermediate undergoes ring closure to form epoxide **C** after elimination of bromide ion. *Tert*-butoxide ion then deprotonates **C** to generate the cumulenyl alkoxide intermediate **D** which undergoes a 5-*endo-dig* cyclization to generate intermediate **E**. This intermediate isomerizes to product **3** upon a base-assisted deprotonation-protonation sequence. In the next synthetic step, zinc chloride is used as a Lewis acid which coordinates to the nitrogen atom of the benzotriazole moiety in **3** to allow for fragmentation of the C—N bond in order to form the masked carbocation intermediate **G**. This intermediate is captured by 2-methylthiophene to give the sulfonium ion intermediate **H**. A last deprotonation via benzotriazole amide leads to the final methylthiophen-phenylfuran product **4**.

In support of their proposed mechanism the authors were able to isolate and characterize the epoxide intermediate **C** in 80% yield when 3-bromo-2-butanone was used as the alkylating reagent. Furthermore, the authors cited their own prior work on the rearrangement of the propargyl triple bond to an allene as an analogous transformation leading to the formation of cumulenyl alkoxide intermediate **D**.[2] Dihydrofuran intermediate **E** was also isolated in 52% yield.

ADDITIONAL RESOURCES

We recommend a recent synthetic approach to propargyl-1,2,3-triazoles which includes a brief mechanistic discussion.[3]

References

1. Katritzky AR, Li J. A novel furan ring construction and syntheses of 4- and 4,5-substituted 2-(α-heterocycloalkyl)furan. *J Org Chem*. 1995;60 (3):638–643. https://doi.org/10.1021/jo00108a028.
2. Katritzky AR, Li J, Gordeev MF. New synthetic routes to furans and dihydrofurans from 1-propargylbenzotriazole. *J Org Chem*. 1993;58 (11):3038–3041. https://doi.org/10.1021/jo00063a022.
3. Das B, Bhunia N, Lingaiah M, Reddy PR. The first multicomponent synthesis of propargyl-1,2,3-triazoles. *Synthesis*. 2011;2011(16):2625–2628. https://doi.org/10.1055/s-0030-1260105.

Question 49: From Carvone to an Isoxazoloazepine

In 1995 chemists from Ireland reacted (*R*)-(−)-carvone **1** with potassium cyanide in aqueous acetic acid to form **2** in 80% yield (Scheme 1).[1] Subsequent treatment of **2** with pentyl nitrite in the presence of NaOEt/EtOH proved to be strongly exothermic and so the temperature of the reaction had to be maintained below 0°C for the isoxazoloazepine product **3** to be formed in 62% yield. Elucidate the mechanism of the transformation of **2** to **3**.

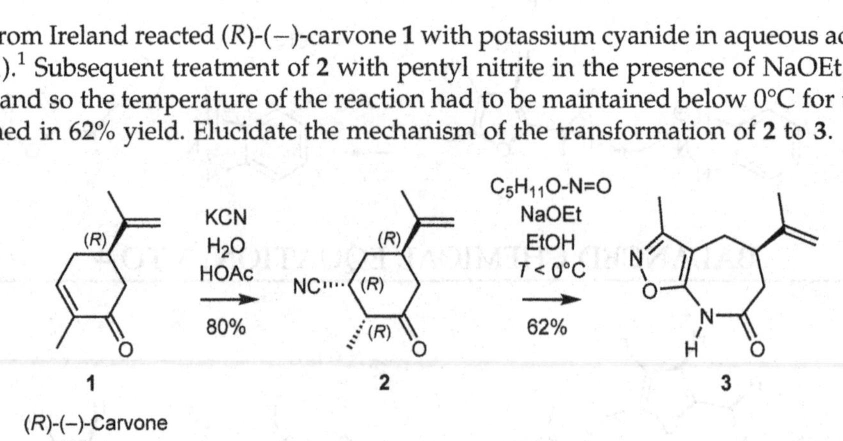

SCHEME 1

SOLUTION

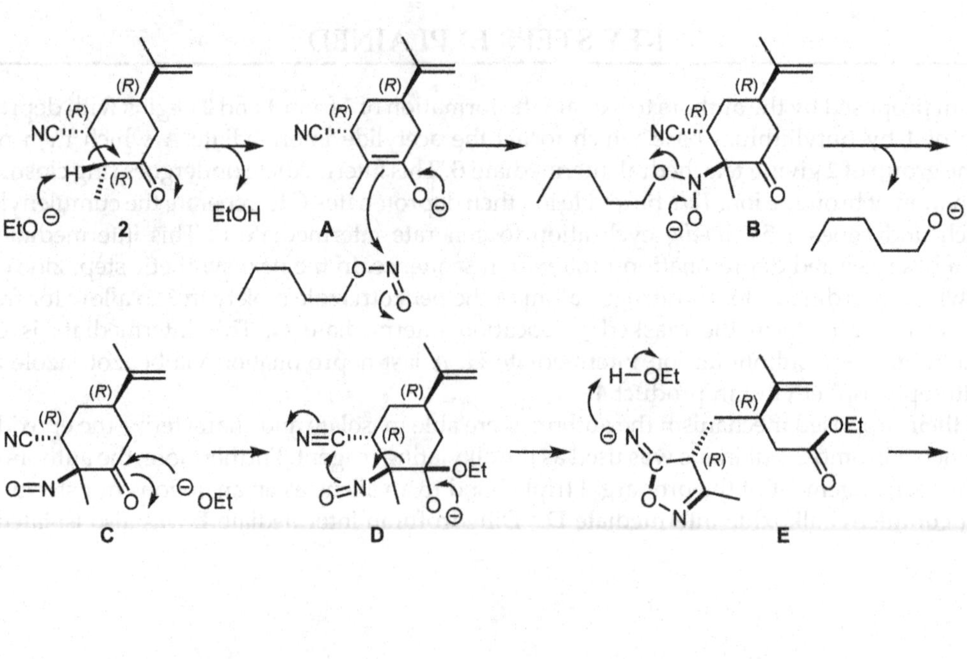

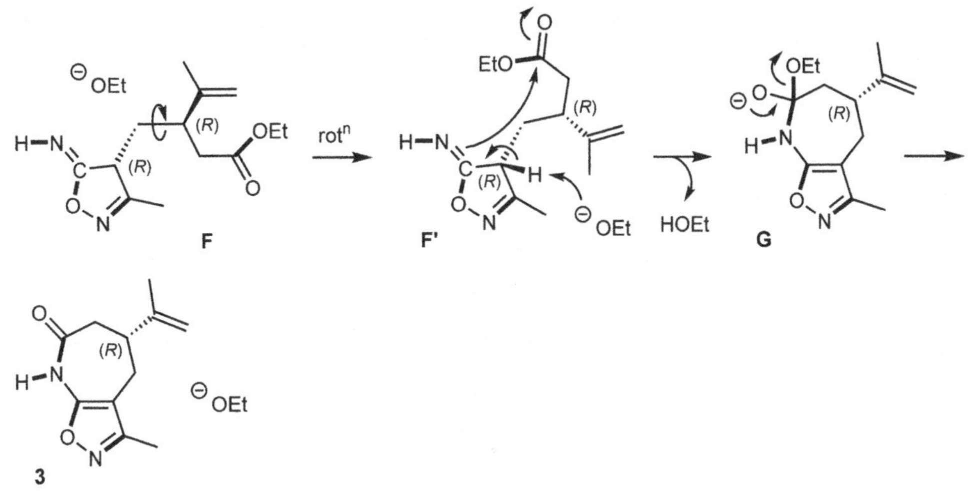

BALANCED CHEMICAL EQUATION

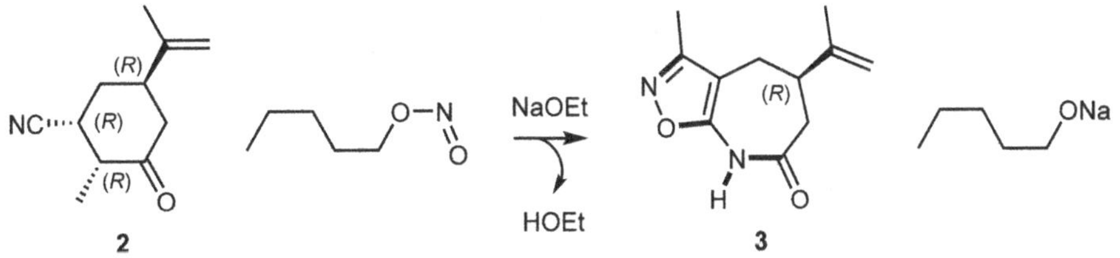

KEY STEPS EXPLAINED

The mechanism proposed by the authors to explain the formation of **3** from **2** begins with base-mediated enolization of (−)-carvone **2** to form enolate **A** which then nucleophilically attacks the nitrogen atom of the nitrite group of pentyl nitrite. The resulting intermediate **B** eliminates pentyloxide ion to generate α-nitrosoketone **C** which is attacked by ethoxide ion causing a cascade ring opening-ring closing sequence which gives 3-methyl-4H-isoxazol-5-ylideneamine intermediate **F** after protonation. Bond rotation from **F** to **F′** allows for deprotonation by ethoxide ion to cause the formation of a seven-membered ring lactam via tetrahedral intermediate **G** which eliminates ethoxide to generate the final isoxazoloazepine product **3**. We note that the overall ring construction mapping for this reaction is [(7+0) +(3+2)]. We also note that the authors did not provide additional experimental or theoretical work to support their proposed mechanism but they did reference the previous work of Lapworth on the same transformations and in particular the first reaction that gives **2** from **1**.[2]

ADDITIONAL RESOURCES

We recommend an article which describes an interesting rearrangement of an isoxazole to an oxazole via the Boulton-Katritzky rearrangement (see problem 296 for a related solution involving this rearrangement).[3]

References

1. Cocker W, Grayson DH, Shannon PVR. Hydrocyanation of some α,β-unsaturated ketones, and the synthesis of some unusual isoxazoles. *J Chem Soc Perkin Trans 1*. 1995;1995:1153–1162. https://doi.org/10.1039/P19950001153.
2. Lapworth A. CI.—Reactions involving the addition of hydrogen cyanide to carbon compounds. Part V. Cyanodihydrocarvone. *J Chem Soc Trans*. 1906;89:945–966. https://doi.org/10.1039/CT9068900945.
3. Jones RCF, Chatterley A, Marty R, Owton WM, Elsegood MRJ. Isoxazole to oxazole: a mild and unexpected transformation. *Chem Commun*. 2015;51:1112–1115. https://doi.org/10.1039/C4CC07999J.

Question 50: Isoxazoles From Cyclopropanes

In 1992 chemists from Taiwan showed that treatment of 2-aryl-1,1-dichlorocyclopropanes **1/3** with mixed acid at 0°C under nitrating conditions formed 3-aryl-5-chloroisoxazoles **2/4** in yields of up to 85% (Scheme 1).[1] It was reported that when the substituent on the aryl group was 4-NO$_2$ (**1**), the yield of product **2** was 85%. Conversely when the substituent was 4-Me (**3**), the yield of **4** was only 2%. Propose a mechanism for this transformation which accounts for these observed yield outcomes.

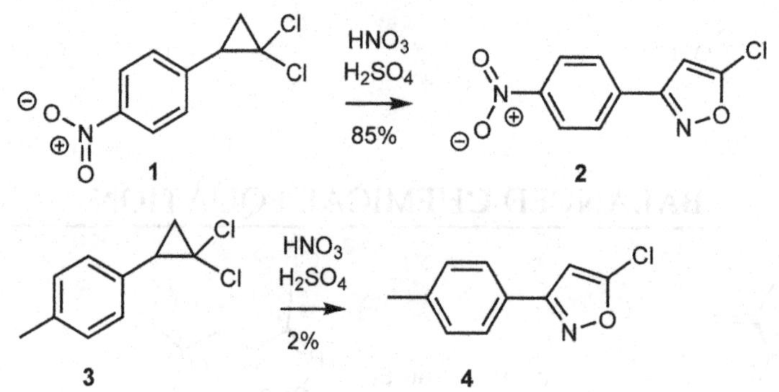

SCHEME 1

SOLUTION: 1 TO 2

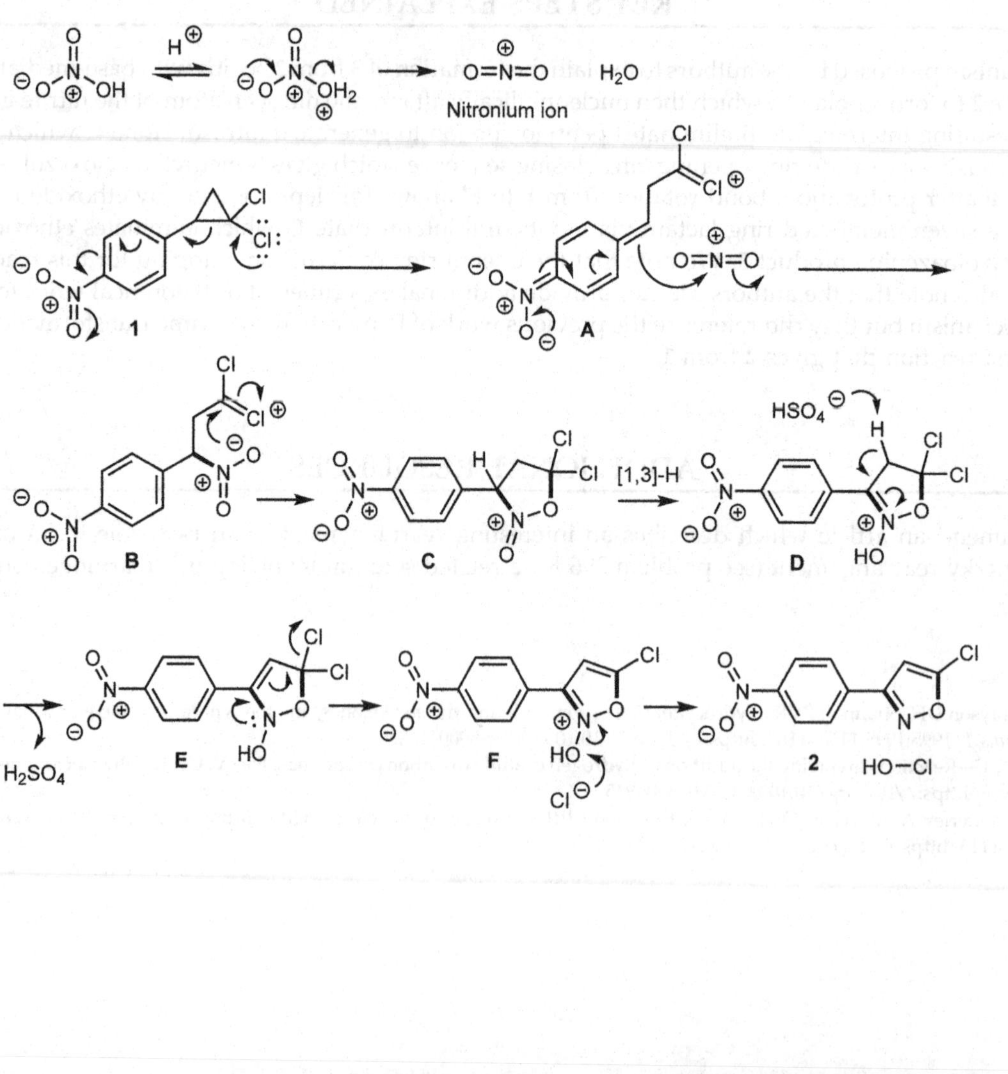

BALANCED CHEMICAL EQUATION

KEY STEPS EXPLAINED

The mechanism proposed to explain the transformation of **1** to **3** begins with formation of nitronium ion via acid-promoted dehydration of nitric acid. The fragmentation of the cyclopropane ring in **1** occurs by lone electron pair donation from one of the chlorine atoms and thus leads to chloronium ion **A** having a quinoid structure. Intermediate **A** then attacks the nitrogen atom of the nitronium ion yielding chlorinium ion **B** which is now rearomatized. Cyclization of **B** to 5,5-dichloro-2-oxo-isoxazolidin-2-ium ion **C** is followed by a [1,3]-protic shift which generates 5,5-dichloro-2-hydroxy-4,5-dihydro-isoxazol-2-ium ion **D**. Intermediate **D** undergoes deprotonation which then facilitates loss of chloride ion to form 5-chloro-2-hydroxy-isoxazol-2-ium ion **F** via 5,5-dichloro-5H-isoxazol-2-ol **E**. Finally, chloride ion displaces the hydroxyl group from **F** giving the 5-chloro-isoxazole product **2** along with hypochlorous acid as by-product. We note that the presence of an electron-withdrawing group (i.e., nitro or cyano) in the *para* position of the aryl group of **1** is necessary because it acts as an electron sink enabling the formation of **A**. This step initiates the capture of nitronium ion and allows the rest of the steps to proceed. Electron-donating groups such as methyl are sigma bond donors and act to destabilize the chloronium ion formation thus resulting in very low yields of isoxazole product in these cases. We also note the authors did not make theoretical calculations to support their proposed mechanism and stated that the mechanism is not yet entirely clear.[1]

ADDITIONAL RESOURCES

We recommend an article which discusses the mechanism of an interesting nitro-3-arylisoxazole formation.[2] We also recommend a synthetic application of isoxazoles.[3]

References

1. Lin S-T, Lin L-H, Yao Y-F. Nitration of 1,1-dihalo-2-(4′-nitrophenyl)cyclopropanes: new method to prepare isoxazole. *Tetrahedron Lett*. 1992;33 (22):3155–3156. https://doi.org/10.1016/S0040-4039(00)79838-8.
2. Hopf H, Mourad AE, Jones PG. A surprising new route to 4-nitro-3-phenylisoxazole. *Beilstein J Org Chem*. 2010;(68):6. https://doi.org/10.3762/bjoc.6.68.
3. Sun R, Li Y, Xiong L, Liu Y, Wang Q. Design, synthesis, and insecticidal evaluation of new benzoylureas containing isoxazoline and isoxazole group. *J Agric Food Chem*. 2011;59:4851–4859. https://doi.org/10.1021/jf200395g.

Chapter 5: Solutions 51 – 100

Question 51: A Radical Cascade From a Ketene Dithioacetal

During a 1995 investigation, UK chemists treated the ketene dithioacetal **1** with a fivefold excess of tributyltin hydride in hot benzene under nitrogen and in the presence of a catalytic amount of azobisisobutyronitrile (AIBN) to form in 70% yield the metallated benzo[*b*]thiophene product **2** (Scheme 1).[1] Suggest a mechanism for this radical cascade transformation.

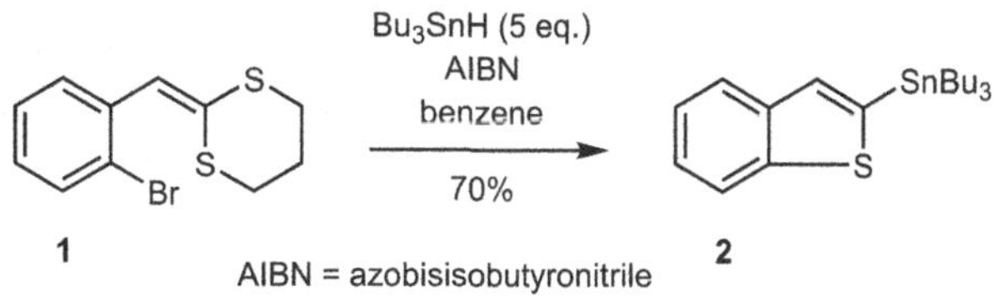

SCHEME 1

SOLUTION

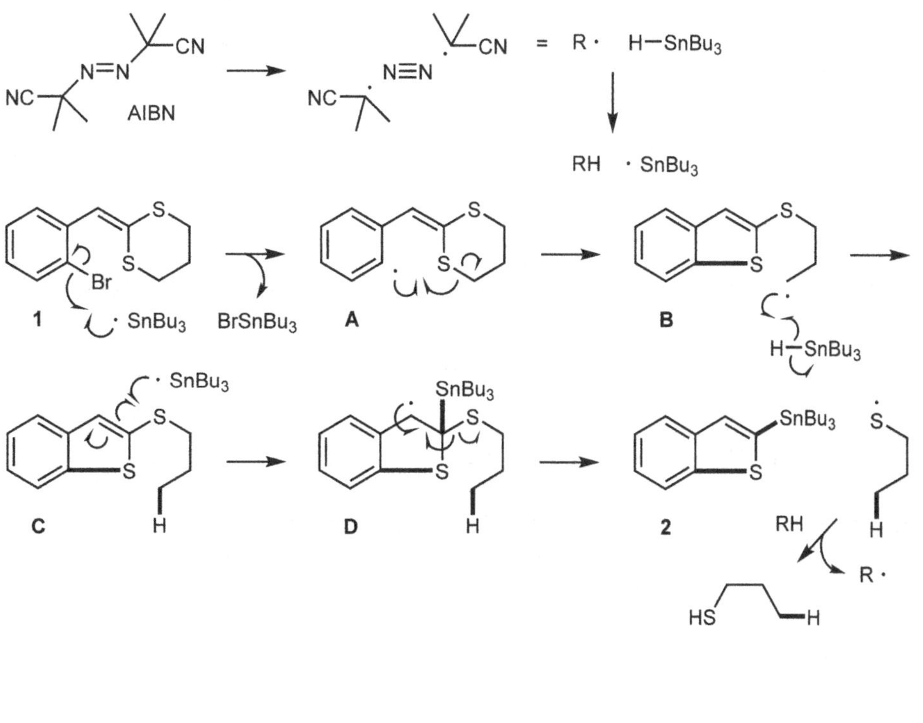

Strategies and Solutions to Advanced Organic Reaction Mechanisms
https://doi.org/10.1016/B978-0-12-812823-7.00314-1

BALANCED CHEMICAL EQUATION

1   2 Bu$_3$SnH   **2**   BrSnBu$_3$ HS⌒H

KEY STEPS EXPLAINED

The proposed mechanism for the transformation of **1** to **2** begins with AIBN acting as a radical initiator to form isobutyronitrile radical which abstracts a hydrogen atom from tributyltin hydride. The resulting tributyltin radical abstracts the bromine atom from thioketene acetal **1** to form tributyltin bromide as by-product and phenyl radical **A** which cyclizes to the primary carbon radical **B**. Radical **B** abstracts a hydrogen atom from a second equivalent of tributyltin hydride to give benzothiophene **C** which is captured by tributyltin radical to form the secondary carbon radical **D**. Radical **D** then fragments to metallated benzothiophene product **2** and propanethiol radical which abstracts a hydrogen atom from isobutyronitrile to give propanethiol as by-product. The authors did not provide further experimental or theoretical evidence to support their mechanism.

ADDITIONAL RESOURCES

We recommend an additional article by Harrowven which discusses the mechanism of a related thiophene radical cascade transformation.[2] In addition, we recommend several other articles and a review which cover radical cascade transformations and their suggested mechanisms.[3–5]

References

1. Harrowven DC, Browne R. Cascade radical reactions in synthesis: a new and general approach to condensed thiophenes. *Tetrahedron Lett.* 1995;36 (16):2861–2862. https://doi.org/10.1016/0040-4039(95)00368-M.
2. Harrowven DC. 'Cascade' radical reactions in synthesis: condensed thiophenes from ketenethioacetals. *Tetrahedron Lett.* 1993;34(35):5653–5656. https://doi.org/10.1016/S0040-4039(00)73907-4.
3. Wang KK, Wang Z, Tarli A, Gannett P. Cascade radical cyclizations via biradicals generated from (Z)-1,2,4-heptatrien-6-ynes. *J Am Chem Soc.* 1996;118(44):10783–10791. https://doi.org/10.1021/ja9622620.
4. Cornia A, Felluga F, Frenna V, et al. CuCl-catalyzed radical cyclisation of N-α-perchloroacyl-ketene-N,S-acetals: a new way to prepare disubstituted maleic anhydrides. *Tetrahedron.* 2012;68:5863–5881. https://doi.org/10.1016/j.tet.2012.04.117.
5. Wang KK. Cascade radical cyclizations via biradicals generated from enediynes, enyne-allenes, and enyne-ketenes. *Chem Rev.* 1996;96(1):207–222. https://doi.org/10.1021/cr950030y.

Question 52: A Synthesis of 3-Alkyl-1-naphthols

In 1995 American chemists undertook the transformations outlined in Scheme 1.[1] In the first case, Makra et al. showed that Heck coupling of *o*-bromopropiophenone **1** with propyne gave a 72% yield of the expected alkyne **2**. The deprotonation of **2** via KHMDS allowed the authors to isolate 3-methyl-1-naphthol product **3** in 74% yield after acidic work-up and distillation. The authors also conducted a deuterium labeling experiment where cyclization of the dideuterated alkyne **4** led to formation of the monodeuterated naphthol **5**. Finally, use of *t*-butylacetylene

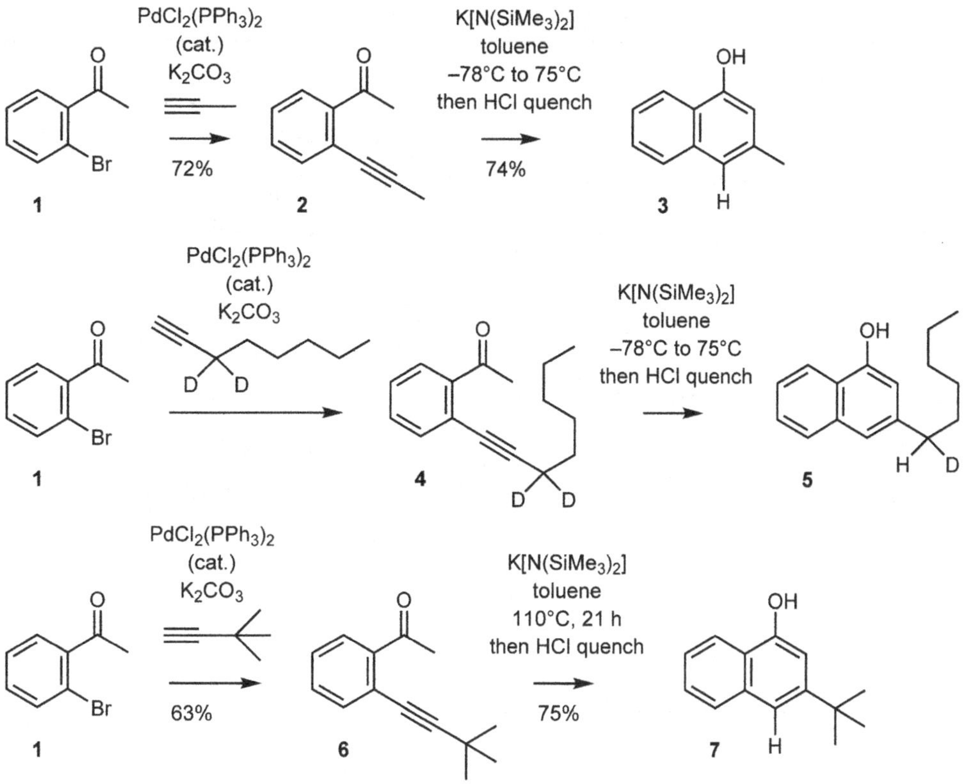

SCHEME 1

instead of propyne in the transformation of **1** to **6** led to formation of 3-*t*-butyl-1-naphthol **7** in 75% yield in the final cyclization step. Suggest mechanisms for these transformations which are consistent with these experimental observations.

SOLUTION: 2 TO 3

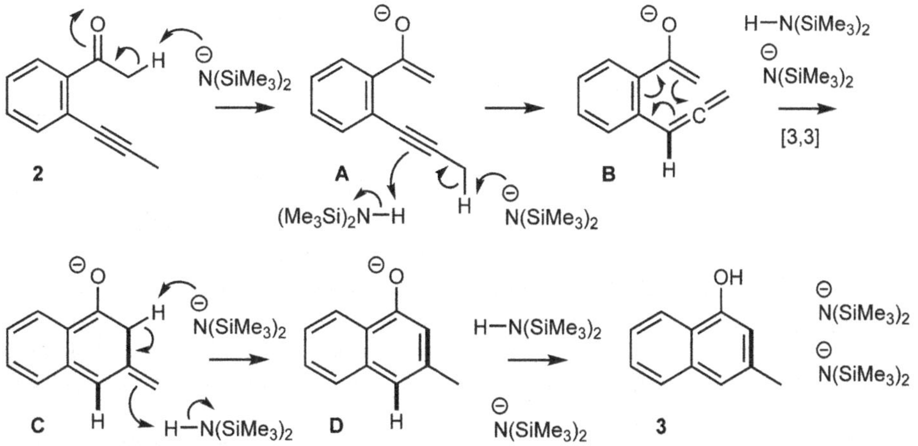

SOLUTION: 4 TO 5

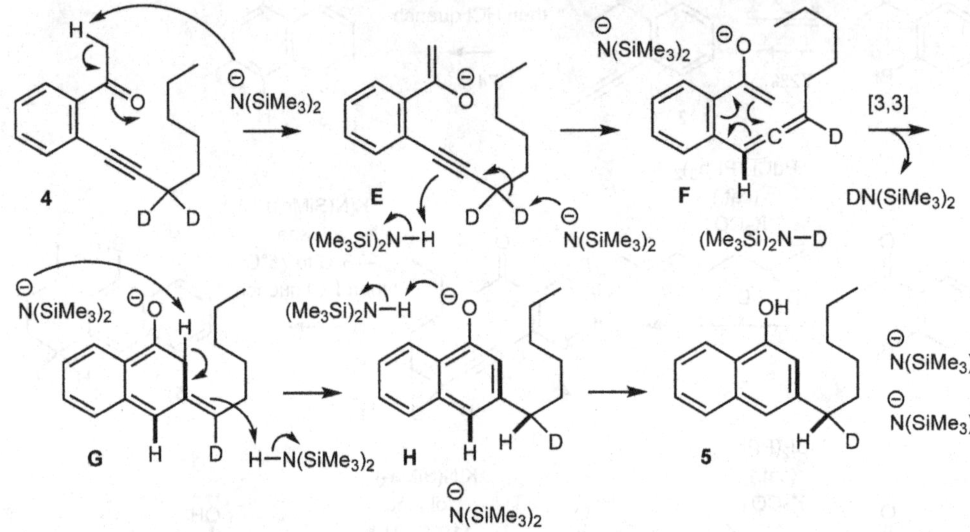

SOLUTION: 6 TO 7

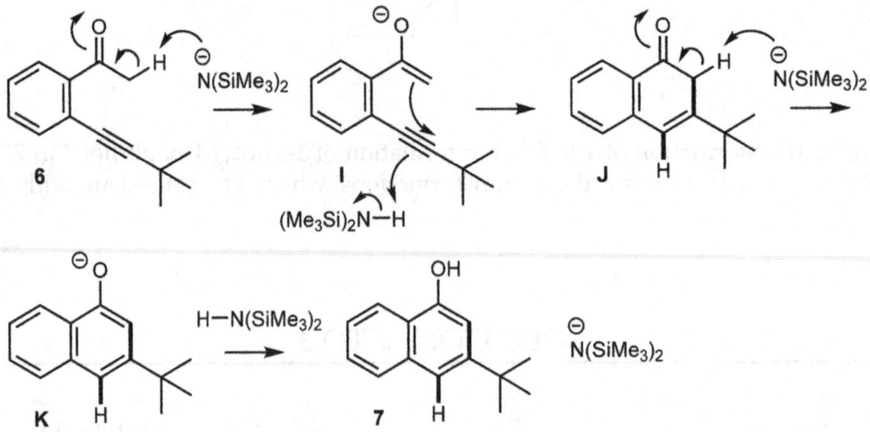

KEY STEPS EXPLAINED

The mechanism proposed by the authors to explain the formation of **3** from **2** begins with enolate formation in **2** via hexamethyldisilazane anion deprotonation which gives **A**. A second equivalent of hexamethyldisilazane anion causes isomerization of the alkyne group of **A** to an allene group thus forming the allene enolate intermediate **B**. This intermediate then undergoes a 6π [3,3] sigmatropic electrocyclization to form enolate **C** which is followed by a base-assisted [1,3] protic shift and protonation to form the final 3-methyl-naphthalen-1-ol product **3**. In support of this mechanistic proposal the authors prepared the dideuterated acetophenone alkyne **4** and subjected it to the same reaction conditions. Following the same steps as before, we arrive at formation of product **5** which contains only one deuterium atom in the side chain. This experiment thus supports the proposed base-mediated alkyne enolate **A** to allene enolate **B** isomerization step illustrated before. Further support for this key mechanistic step appeared with the experimental evidence that compound **6** gives **7** when subjected to more strenuous reaction conditions. This is because **6** has a *tert*-butyl group bound to the alkyne which means there are no abstractable hydrogen atoms that a base can remove to facilitate the alkyne-allene isomerization. Under these conditions extra energy (i.e., heat) is required to enable a different mechanistic pathway which avoids the allene intermediate by having enolate **I** cyclize directly onto the alkyne group to give 3-*tert*-butyl-2*H*-naphthalen-1-one **J** which undergoes base-assisted enolization to naphthol **7**.

It is interesting to note that enolate **I** cyclizes in a 6-*endo* manner to give the six membered ring in **J** rather than cyclize in a 5-*exo* manner to form a five membered ring with an exocyclic C=C-*t*-butyl group α to the carbonyl. Although the authors did not undertake theoretical calculations to help support their proposed mechanism, they did cite the work of Ciufolini and Weiss who calculate the heat of formation for a related 6-*endo* enolate cyclization as being slightly exothermic (Scheme 2).[2]

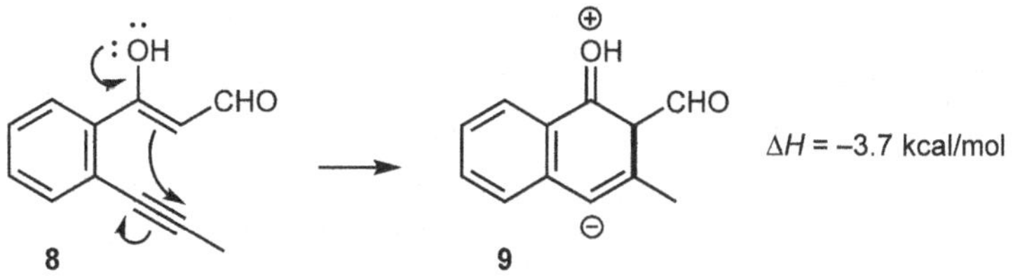

SCHEME 2

Furthermore, the work of Jacobi and Kravitz demonstrates that in similar systems, competition between 5-*exo*, 6-*endo*, and electrocyclization mechanistic paths can be controlled by the use of various solvent and acid catalyst mixtures.[3] Lastly, we note that the deuterium labeling experiment undertaken by the authors helped rule out a mechanism involving an intermediate **K** similar to **9** (Scheme 3).

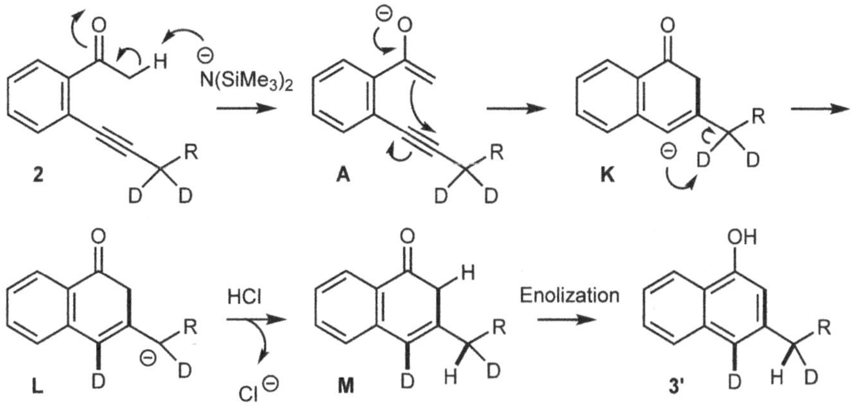

SCHEME 3

ADDITIONAL RESOURCES

For those interested in additional examples of these types of transformations, we recommend further work by Jacobi et al.[4] We also recommend the work of Kataoka et al. for an example of a transition-metal-catalyzed synthesis of 1-naphthols.[5]

References

1. Makra F, Rohloff JC, Muehldorf AV, Link JO. General preparation of 3-alkyl-1-naphthols. *Tetrahedron Lett.* 1995;36(38):6815–6818. https://doi.org/10.1016/0040-4039(95)01403-5.
2. Ciufolini MA, Weiss TJ. A useful benzannulation reaction. *Tetrahedron Lett.* 1994;35(8):1127–1130. https://doi.org/10.1016/0040-4039(94)88003-4.
3. Jacobi PA, Kravitz JI. Enynones in organic synthesis. III. A novel synthesis of phenols. *Tetrahedron Lett.* 1988;29(52):6873–6876. https://doi.org/10.1016/S0040-4039(00)88463-4.
4. Jacobi PA, Kravitz JI, Zheng W. Enynones in organic synthesis. 8. Synthesis of the antimicrobial-cytotoxic agent juncusol and members of the effusol class of phenols. *J Org Chem.* 1995;60(2):376–385. https://doi.org/10.1021/jo00107a017.
5. Kataoka Y, Miyai J, Tezuka M, Takai K, Oshima K, Utimoto K. Preparation of 1-naphthols from acetylenes and o-phthalaldehyde using low-valent tantalum and niobium. *Tetrahedron Lett.* 1990;31(3):369–372. https://doi.org/10.1016/S0040-4039(00)94557-X.

Question 53: An Entry to Indole Alkaloids of Unusual Structural Type

In 1995 French chemists reacted carbinolamine ether **1** with *m*-chloroperoxybenzoic acid (*m*CPBA) in dichloromethane to form hemiketal ester **2** in 82% yield (Scheme 1).[1] Compound **2** was then treated with NaOH to give hemiketal **3** in 71% yield after methanolysis. At this point treatment of **3** with a 99:1 v:v mixture of CH$_2$Cl$_2$/trifluoroacetic acid (TFA) gave a mixture of **4** and **5** in 42% and 11% yields, respectively. When the reaction time was increased to 15 h, the yield of **4** increased to 52% with almost no recovery of **5**. Provide a mechanism for the sequence from **1** to **2** to **3** to **4** and **5**.

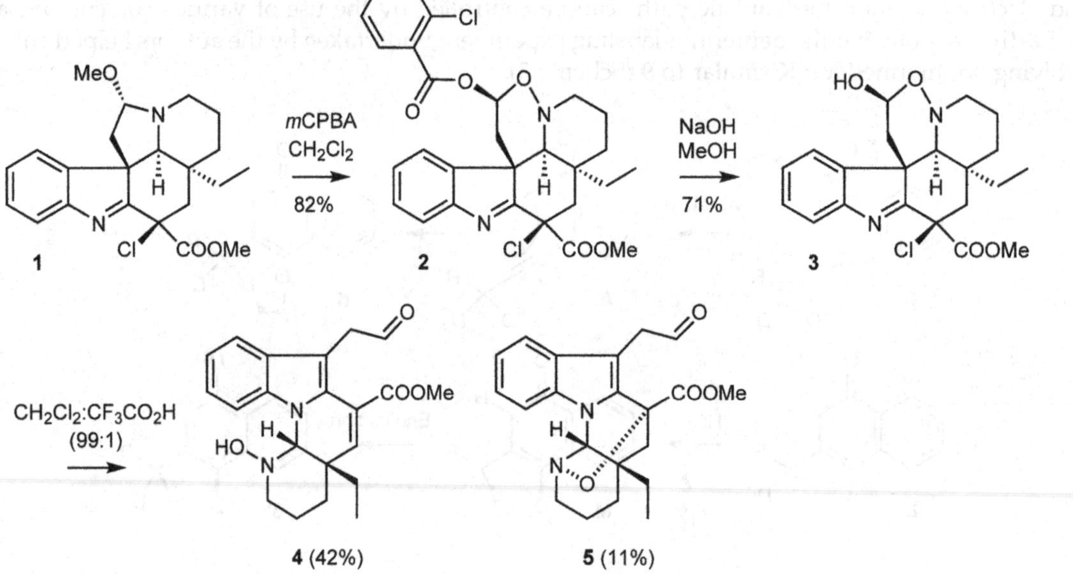

SCHEME 1

SOLUTION: 1 TO 2

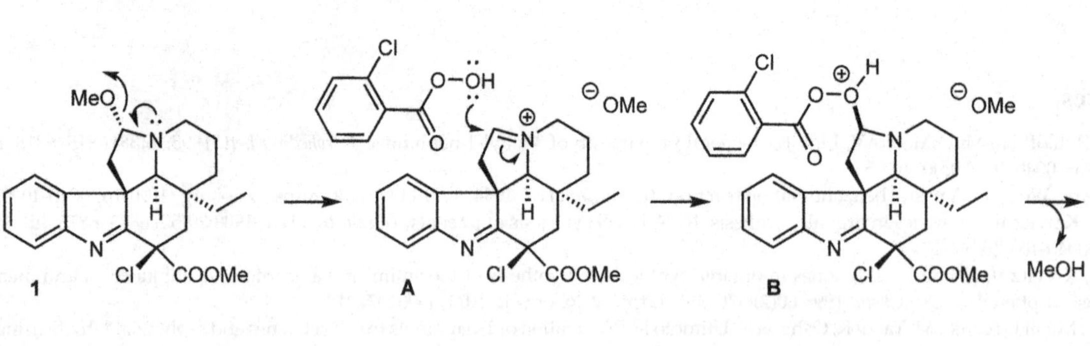

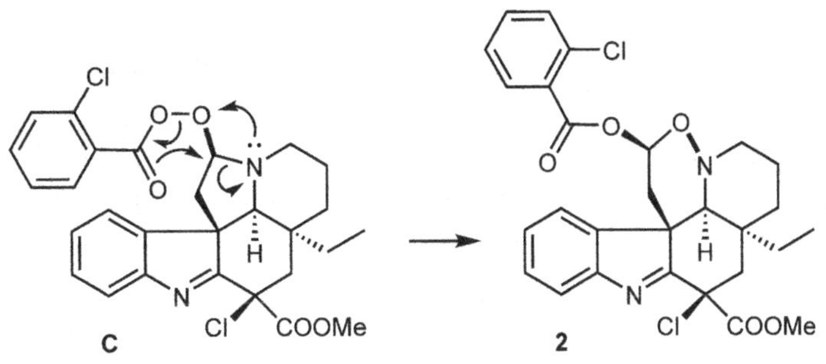

SOLUTION: 2 TO 3

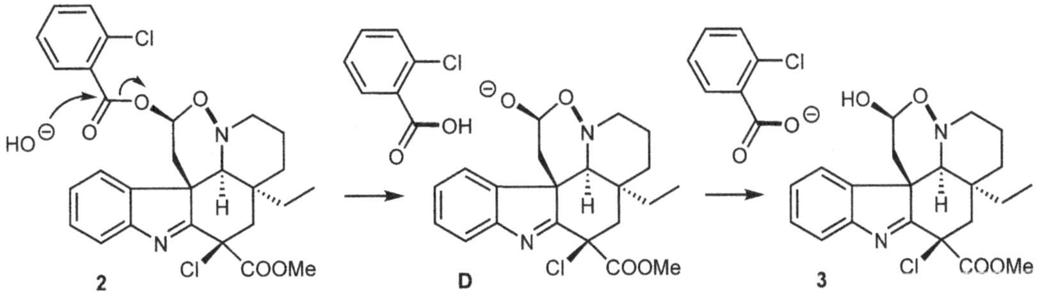

SOLUTION: 3 TO 5 AND 4

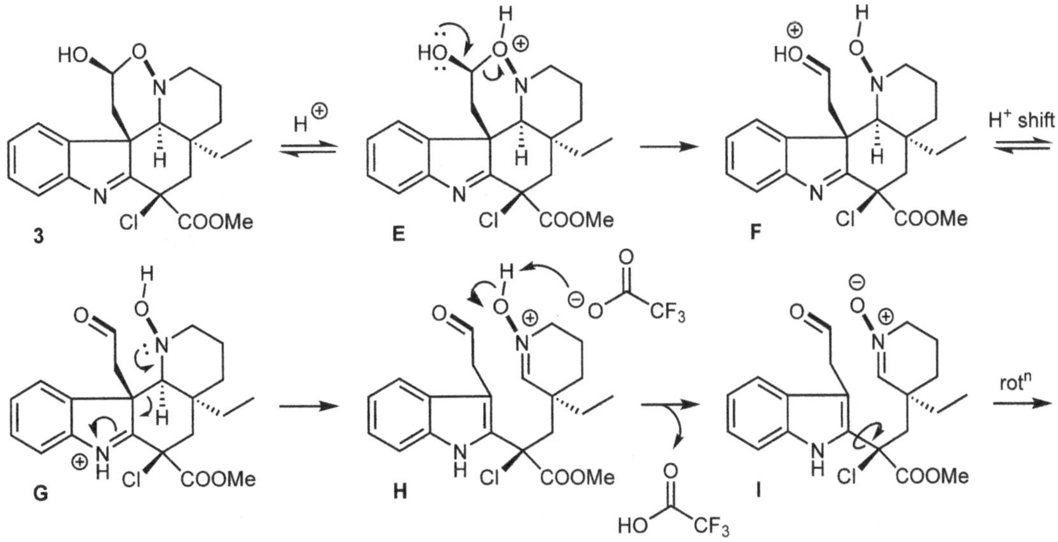

BALANCED CHEMICAL EQUATIONS

4 (thermodynamic product) **5** (kinetic product)

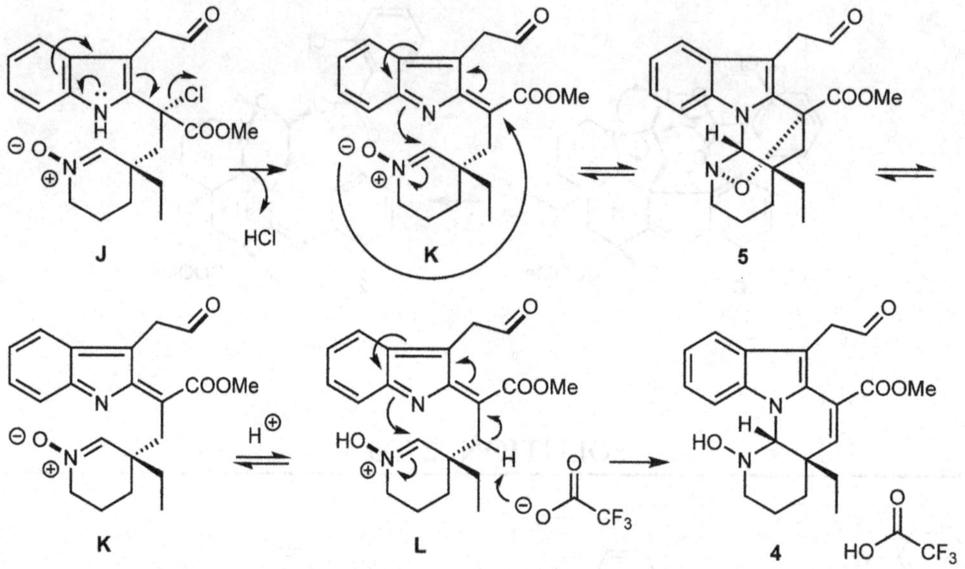

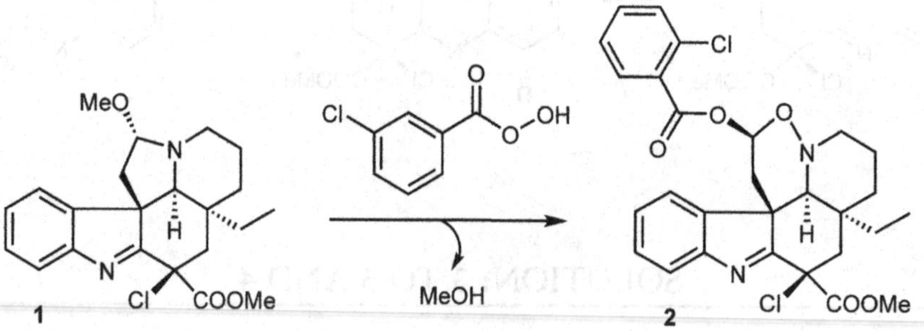

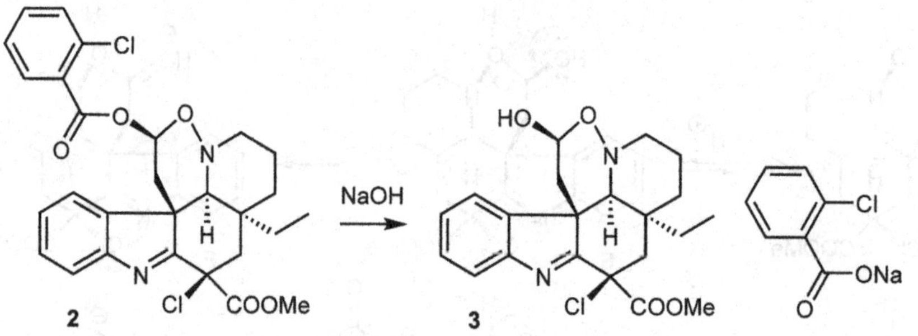

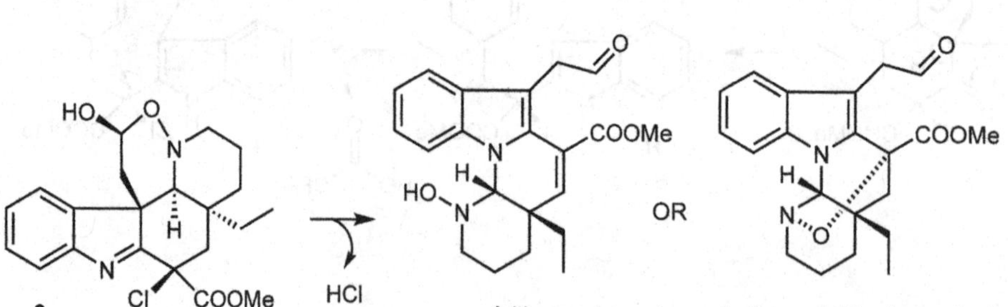

KEY STEPS EXPLAINED

The mechanism proposed by the authors to explain the synthetic sequence from **1** to the target products **4** and **5** begins with the transformation of **1** to **2** which is mediated by *m*CPBA. Here, the proposed mechanism begins with elimination of methoxide ion from **1** which gives an iminium ion **A**. This ion is attacked by the peracid to form the peroxy intermediate **C** after proton transfer. This intermediate then rearranges to give the ring expansion product **2**. We note that an alternative mechanism for this transformation consists of oxidation of the quaternary amine group of **1** to an N-oxide group via *m*CPBA which could then form a fused [3.1.0] bicyclic ring system containing a 1,2-oxaziridine group that can be attacked by 2-chlorobenzoate anion to give the ring expansion product **2** (Scheme 2). We discount this mechanism because of the expected ring strain associated with the fused 1,2-oxaziridine ring in intermediate **N**. When **2** is treated with sodium hydroxide it undergoes saponification in the usual way to form [1,2]oxazinan-6-ol product **3**. With respect to the transformation of **3** to **4** and **5**, the authors proposed an initial acid-promoted ring opening of the oxazinanol ring of **3** which results in the production of intermediate **G** having an aldehyde group and an N-hydroxyl group. Intermediate **G** then ring opens to iminium ion **H** which loses a proton to generate tetrahydro-pyridine N-oxide **I**. Intermediate **I** undergoes sequential bond rotation and elimination of HCl to form intermediate **K**. At this point the authors propose that **K** undergoes ring closure to give the kinetic product **5** which has a complex tricyclic structure. In addition, **K** can also undergo a sequential protonation-deprotonation process which leads to the thermodynamically controlled product **4**. It is noteworthy that the authors demonstrated that under longer reaction times (i.e., 15 h) under the same reaction conditions, the yield of **4** increased from 42% to 52% while the yield of **5** decreased significantly. Furthermore, product **5** could also be converted to product **4** in 40% yield after 4 h at room temperature under the same reaction conditions. This result constitutes evidentiary support for the proposed equilibrium between **5** and the key intermediate **K**. We also note that this problem presents an opportunity to carry out a valuable computational analysis of the energy reaction coordinate profiles for the paths **K → 4** versus **K → 5** to verify the conclusion that these are the thermodynamic and kinetically controlled products, respectively.

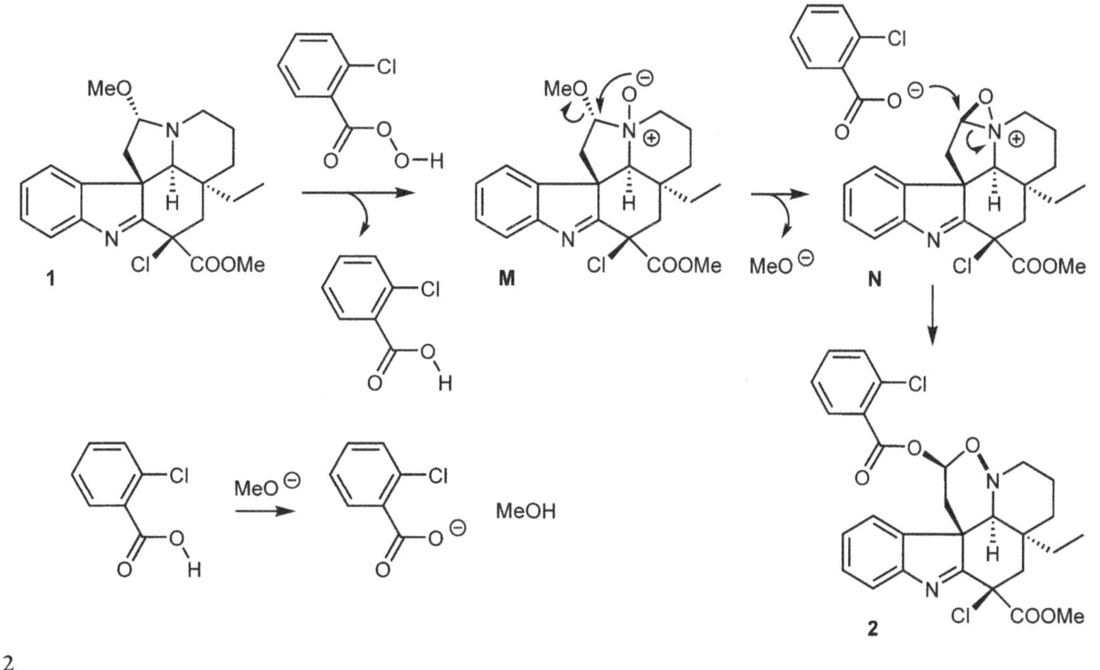

SCHEME 2

ADDITIONAL RESOURCES

We refer the reader to two articles which present interesting mechanisms (a radical process and an alkyl group migration process) for related transformations involving indole alkaloids.[2,3]

References

1. Lewin G, Schaeffer C, Lambert PH. New rearrangement of an aspidosperma alkaloid. The first biomimetic entry in the goniomitine skeleton. *J Org Chem*. 1995;60(11):3282–3287. https://doi.org/10.1021/jo00116a009.
2. Hilton ST, Jones K. The tandem radical route to indole alkaloids: an unusual rearrangement reaction. *ARKIVOC*. 2007;9:120–128. https://doi.org/10.3998/ark.5550190.0008.b11.
3. Li S, Finefield JM, Sunderhaus JD, McAfoos TJ, Williams RM, Sherman DH. Biochemical characterization of NotB as an FAD-dependent oxidase in the biosynthesis of notoamide indole alkaloids. *J Am Chem Soc*. 2012;134:788–791. https://doi.org/10.1021/ja2093212.

Question 54: Cyclization as a Key Step to Hirsutene

In 1993 Japanese chemists investigated the total synthesis of hirsutene and in the process undertook the acid-catalyzed cyclization of **1** to **2** (Scheme 1).[1] The authors isolated **2** as a 10:1 **2**-*syn*:**2**-*anti* mixture of epimers (at C$_4$) in 95% yield. Suggest a mechanism which accounts for this transformation and the observed stereochemical outcome.

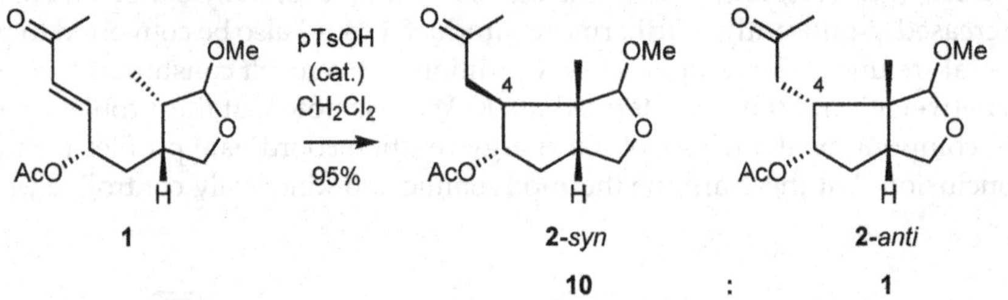

SCHEME 1

SOLUTION

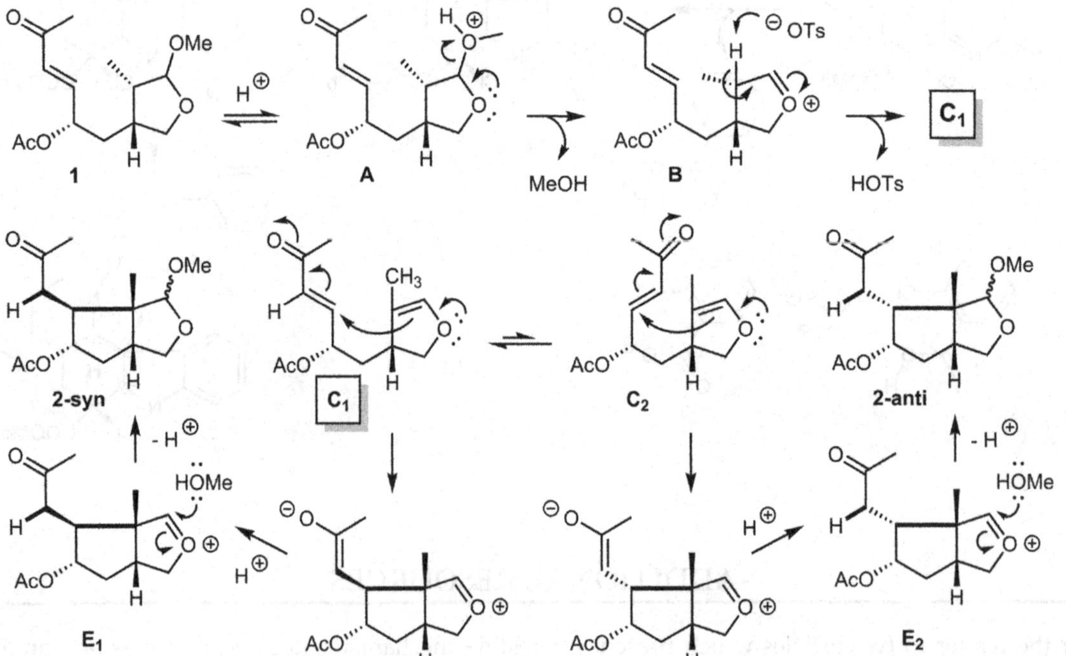

KEY STEPS EXPLAINED

The mechanism proposed by the authors begins with protonation of the methoxy group of **1** to give intermediate **A** followed by elimination of methanol which gives oxonium ion **B** which is deprotonated by tosylate ion to form dihydrofurans C_1 and C_2. The authors argue that conformer C_2 suffers from steric hindrance between the methyl group and the vinyl hydrogen atom whereas this clash is absent in the C_1 conformer. Hence the equilibrium between these conformers is favored toward C_1 which leads to a predominant formation of the 2-*syn* epimer. We note that C_1/C_2 cyclize to the zwitterionic intermediates D_1/D_2 which undergo acid-catalyzed ketonization to oxonium ion intermediates E_1/E_2. These intermediates are captured by methanol and lose a proton to finally give the epimeric mixture of products 2-*syn* and 2-*anti*. The *syn* and *anti* nomenclatures refer to the stereorelationship between the 2-oxopropyl group at C_4 and the methyl group at the ring juncture. We note that corroborative evidence of the thermodynamic stability of the C_1 conformer can be obtained by carrying out a geometry-optimized calculation of each conformer and determining the thermodynamic energy difference between each structure and the energy barrier between them. We recommend this strategy for a future investigation to help support this postulated mechanism. Lastly, we note that this represents a 100% atom economical transformation.

ADDITIONAL RESOURCES

We recommend several of many articles which tackle the total synthesis of hirsutene from the perspective of catalysis, biocatalysis, and other approaches.[2–5]

References

1. Toyota M, Nishikawa Y, Motoki K, Yoshida N, Fukumoto K. Total synthesis of (±)-hirsutene via Pd^{2+}-promoted cycloalkenylation reaction. *Tetrahedron*. 1993;49(48):11189–11204. https://doi.org/10.1016/S0040-4020(01)81806-8.
2. Chandler CL, List B. Catalytic, asymmetric transannular aldolizations: total synthesis of (+)-hirsutene. *J Am Chem Soc*. 2008;130:6737–6739. https://doi.org/10.1021/ja8024164.
3. Singh V, Vedantham P, Sahu PK. Reactive species from aromatics and oxa-di-π-methane rearrangement: a stereoselective synthesis of (±)-hirsutene from salicyl alcohol. *Tetrahedron*. 2004;60:8161–8169. https://doi.org/10.1016/j.tet.2004.06.096.
4. Wang J-C, Krische MJ. Intramolecular organocatalytic [3+2] dipolar cycloaddition: stereospecific cycloaddition and the total synthesis of (±)-hirsutene. *Angew Chem Int Ed*. 2003;42:5855–5857. https://doi.org/10.1002/anie.200352218.
5. Banwell MG, Edwards AJ, Harfoot GJ, Jolliffe KA. A chemoenzymatic synthesis of (-)-hirsutene from toluene. *J Chem Soc Perkin Trans 1*. 2002;2002:2439–2441. https://doi.org/10.1039/B208778B.

Question 55: A 1,3-Cyclopentanedione to 1,4-Cyclohexanedione Transformation

In 1993 German workers studied the conversion of 2,2-dialkylated 1,3-cyclopentanedione **1** into the 5-substituted 4-oxoalkanoic acid product **2** under alkali conditions (Scheme 1).[1] This type of transformation was originally thought to proceed by a mechanism which involves facile β-dicarbonyl cleavage. The authors, however, showed that reaction of **1** with one equivalent of sodium hydroxide in water at room temperature for 3 min gives the ring-enlargement product **3** in approx. 50% yield. When **3** was treated with excess sodium hydroxide, **2** was obtained in good yield. Provide mechanistic explanations for both the original interpretation of the **1** to **2** transformation and the **1** to **3** to **2** transformation demonstrated experimentally by the authors.

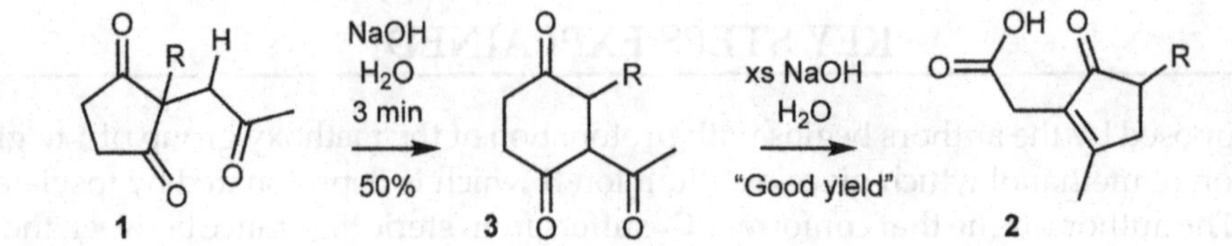

SCHEME 1

SOLUTION: ORIGINALLY PROPOSED MECHANISM FOR 1 TO 2

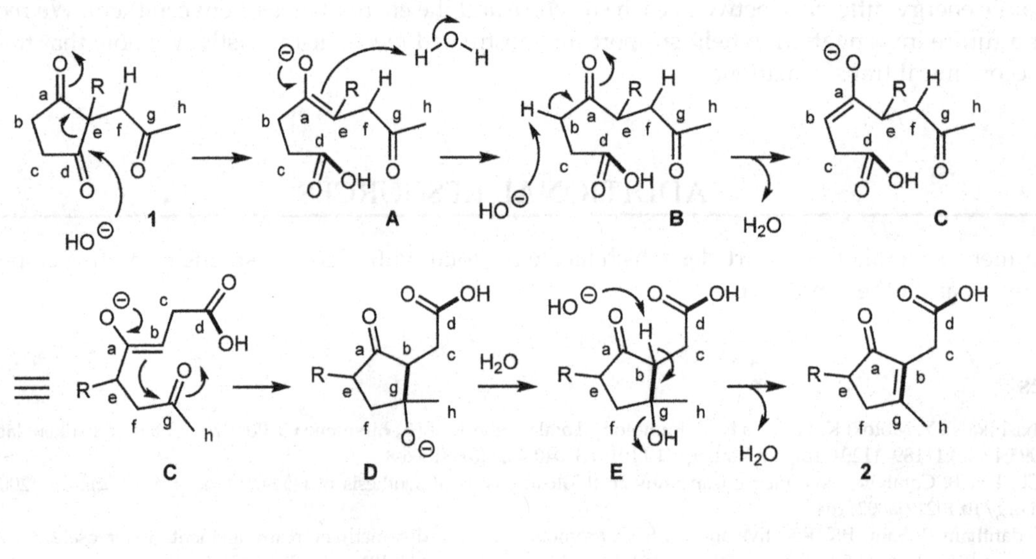

SOLUTION: REVISED MECHANISM FOR 1 TO 2 PART 1: 1 TO 3

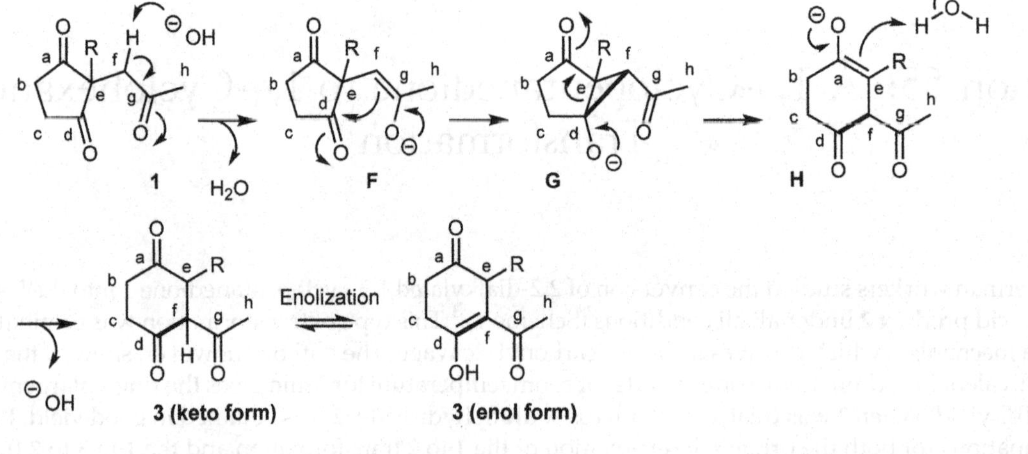

SOLUTION: REVISED MECHANISM FOR 1 TO 2 PART 2: 3 TO 2

KEY STEPS EXPLAINED

The original mechanistic understanding of the conversion of triketone **1** to **2** begins with base-catalyzed ring opening of the cyclopentanedione group of **1** to form carboxylic acid **B** after protonation via **A**. Deprotonation of **B** then gives **C** which undergoes an intramolecular aldol cyclization to **E** after protonation via **D**. A final dehydration of **E** gives product **2**. In light of the authors' isolation of product **3** after 3 min of reaction time, the original mechanism for the **1** to **2** transformation required change to include **3** as an experimentally confirmed intermediate. This revised mechanism begins with base-catalyzed conversion of **1** to enolate **F** which rearranges to form the [3.1.0] bicyclic intermediate **G** which subsequently ring opens to give the enolate dione **H**. Ketonization of **H** leads to product **3** which can exist in both keto and enol forms in aqueous solution. The net change in the connectivity of the carbon skeleton from **1** to **3** is bond cleavage between C_d and C_e and bond formation between C_d and C_f. Under excess sodium hydroxide conditions, the conversion of **3** to **2** involves hydroxide attack at C_d which causes ring opening via rupturing of the previously formed C_d—C_f bond thus forming enolate **I** which ketonizes to **B** which proceeds along the sequence seen in the original mechanism to give the final product **2**. Structurally, the net result is cleavage of the C_d—C_f bond and formation of the C_b—C_g bond. Overall, the transformation sequence **1** to **3** to **2** may be classified as sequential rearrangements of the carbon skeleton. The originally proposed mechanism did not go through the six-membered ring intermediate **3**. Instead, it proceeds via hydroxide ion attack at C_d cleaving the C_d—C_e bond in **1** and then making a C_b—C_g bond via two enolates **A** and **C** and final β-hydroxide elimination. During the course of this mechanism no bond connection is made between C_d and C_f. The early cleavage between C_d and C_f prevents the formation of the ring enlargement intermediate **3**. We note that the authors provided the following evidence for the transformation of triketone **1** to cyclopentenone carboxylic acid **2**: (a) isolation and characterization of intermediate dione **3** by elemental analysis, ^{1}H NMR, ^{13}C NMR, and mass spectrometry; and (b) conversion of isolated **3** to product **2** with excess aqueous sodium hydroxide solution. It was also found that the yield of **1** to **3** could be improved from 50% to 70% by using sodium methoxide as base in methanol solution. The strongest piece of evidence in favor of the newer proposed mechanism is the actual isolation of dione intermediate **3**. Lastly, the authors did not conduct a theoretical analysis to help support their postulated mechanism.

ADDITIONAL RESOURCES

We note that the original mechanism for the **1** to **2** transformation was proposed by Schick et al. in 1982.[2] For further work by these authors with regard to a very interesting analogous carbon skeleton rearrangement, see Ref. 3. For an example of a similar rearrangement involving a heterocyclic system containing nitrogen, see Ref. 4.

References

1. Schramm S, Roatsch B, Gruendemann E, Schick H. Transformation of cyclopentane-1,3-diones into cyclohexane-1,4-diones—a novel ring enlargement process. *Tetrahedron Lett.* 1993;34(30):4759–4760. https://doi.org/10.1016/S0040-4039(00)74081-0.
2. Schick H, Pogoda B, Schwarz S. Umwandlung von 2-(Alk-2-inyl)-2-alkyl-cyclopentan-1,3-dionen in 2,3,5-trisubstituierte Cyclopent-2-en-1-one. *Z Chem.* 1982;22(5):185. https://doi.org/10.1002/zfch.19820220513.
3. Schick H, Roatsch B, Schramm S, Gilsing H-D, Ramm M, Gruendemann E. Conversion of 2-alkyl-2-(2-oxopropyl)cyclopentane-1,3-diones into 2,3,5- and 2,3,4-trisubstituted cyclopent-2-enones by intramolecular aldolizations to 2,3-diacylcyclopropanolates followed by remarkable skeletal rearrangements. *J Org Chem.* 1996;61(17):5788–5792. https://doi.org/10.1021/jo960189q.
4. Lertpibulpanya D, Marsden SP. Concise access to indolizidine and pyrroloazepine skeleta via intramolecular Schmidt reactions of azido 1,3-diketones. *Org Biomol Chem.* 2006;4:3498–3504. https://doi.org/10.1039/b608801e.

Question 56: Conversion of *o*-Hydroxyaryl Ketones Into 1,2-Diacylbenzenes

In 1991 American and Greek chemists reacted the oxygen labeled *o*-hydroxyaryl ketone monoacylhydrazone **1** with lead tetraacetate to form the labeled product **2** (Scheme 1).[1] In crossover experiments, the authors also demonstrated that the transformation was intramolecular in nature. Suggest a mechanism for this transformation which abides by these experimental results.

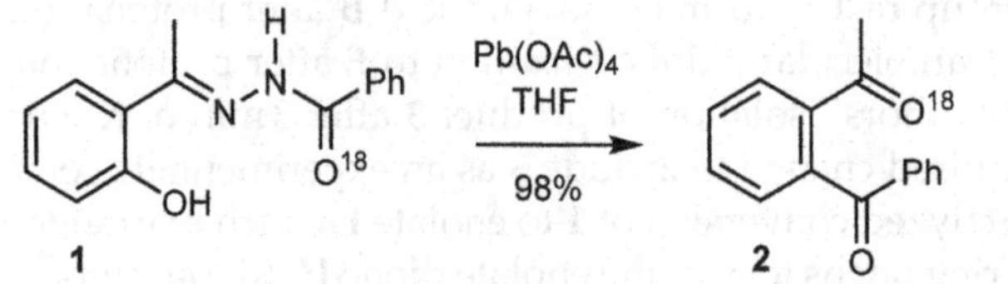

SCHEME 1

SOLUTION

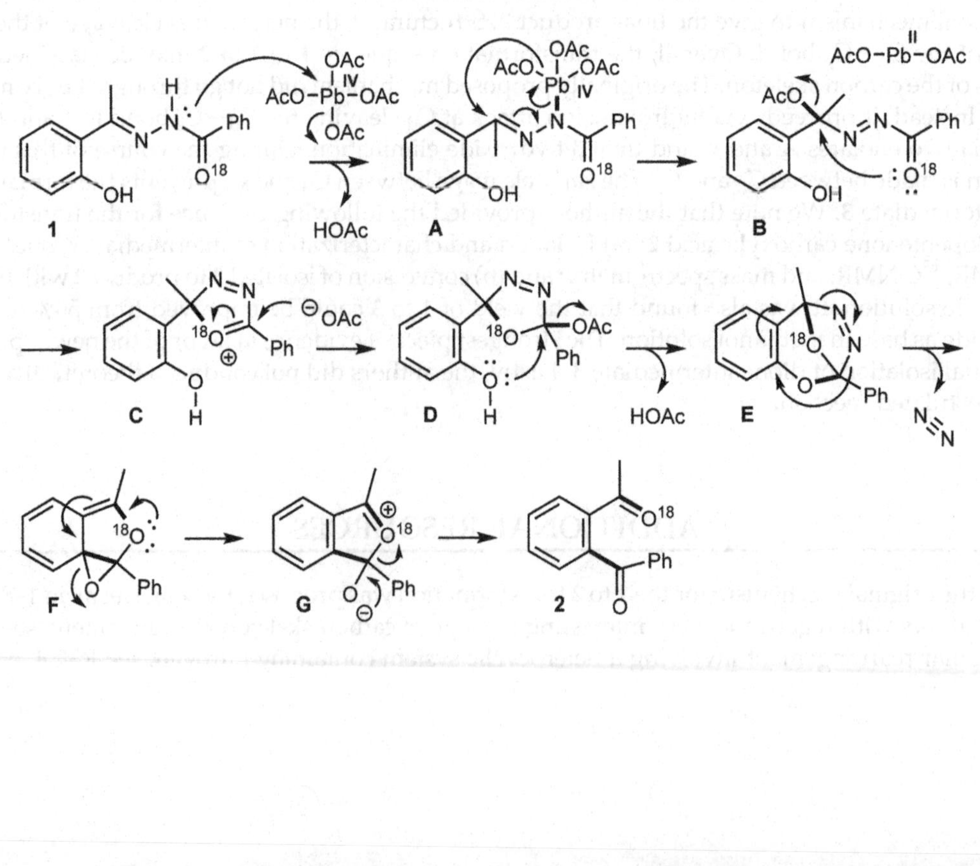

BALANCED CHEMICAL EQUATION

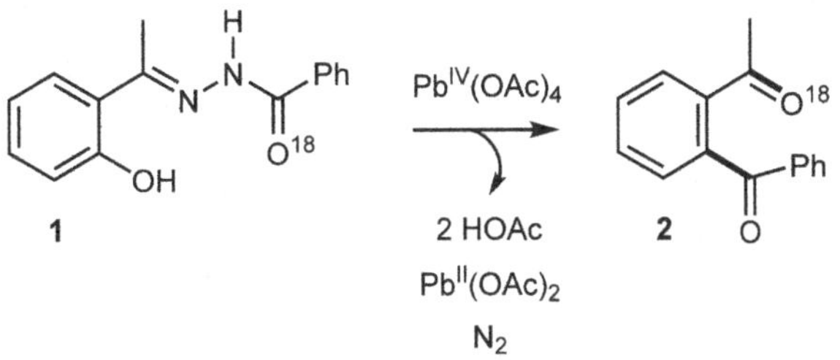

KEY STEPS EXPLAINED

We first note that the product structure drawn in McKillop's book had reversed the position of the ^{18}O label. The correct product structure has the label on the acetyl group, *not* on the benzoyl group. The mechanism proposed by the authors to explain the transformation of **1** to **2** begins with attack by **1** onto lead(IV)tetraacetate to form the organolead intermediate **A** after loss of acetic acid. An intramolecular acetoxy group migration then takes place in **A** to form the azoacetate intermediate **B** which has the acetoxy group attached to the hydrazone carbon. This process undergoes reduction of lead(IV) to lead(II) and concomitant oxidation of each nitrogen atom from -2 oxidation state to -1 oxidation state. Lead diacetate is liberated as a by-product and intermediate **B** cyclizes to form 1,3,4-oxadiazoline **D** after nucleophilic addition of acetate ion onto the electrophilic carbon atom of carbocation **C**. Cyclization of **D** with loss of acetic acid forms the tricyclic 1,3-dioxane **E** having a [3.2.1] bicyclic framework which rearranges with loss of nitrogen gas to form the fused epoxide **F** having a [3.1.0] framework. **F** undergoes ring opening via **G** to form the final product **2** which has the ^{18}O label at the acetyl group. Moreover, the authors performed this ^{18}O labeling experiment to differentiate between the proposed mechanism and an alternative mechanism from **B** to **2** (see Scheme 2) where the ^{18}O label ends up at the benzoyl carbonyl group. In this pathway intermediate **B** eliminates acetic acid yielding *o*-quinone methide intermediate **H** which undergoes a [2+2] cycloaddition to form benzoxetane **I** which in turn undergoes a combined fragmentation-rearrangement yielding product **2'** and nitrogen gas. The overall balanced chemical equation for this alternative pathway is the same as that of the operative mechanism. We can thus see that the oxygen labeling experiment demonstrates that the benzoyl oxygen in **1** migrates to the hydrazone carbon to form 1,3,4-oxadiazoline intermediate **D** instead of being retained in the benzoyl group which migrates all-in-one to give product **2'** (not observed) after rearrangement from **I**. The position of the ^{18}O label on **2** was confirmed by mass spectrometric analysis.

It is noteworthy that the authors also carried out a crossover experiment using an equimolar mixture of the two substrates **3** and **4** shown in Scheme 3 to determine whether the transfer of the benzoyl group in **3** and the acetyl group in **4** occurs via an intramolecular or intermolecular mechanism. The results showed that there was no formation of crossover products **7** and **8** thus suggesting that each acyl group transfer proceeds by an intramolecular process and is not split off as a free acylium ion.

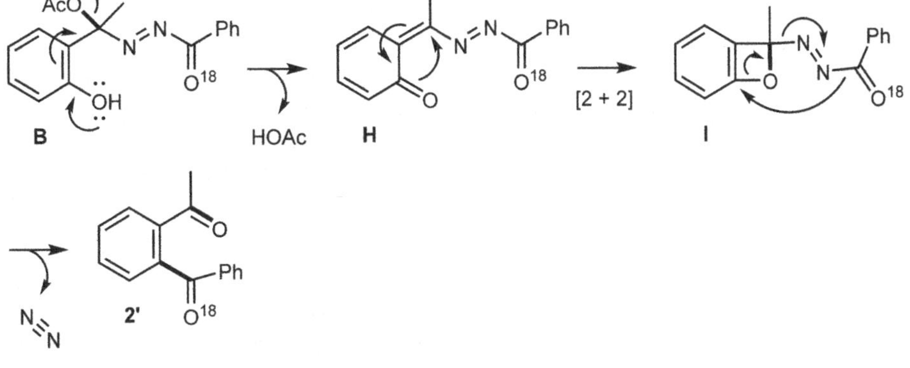

SCHEME 2

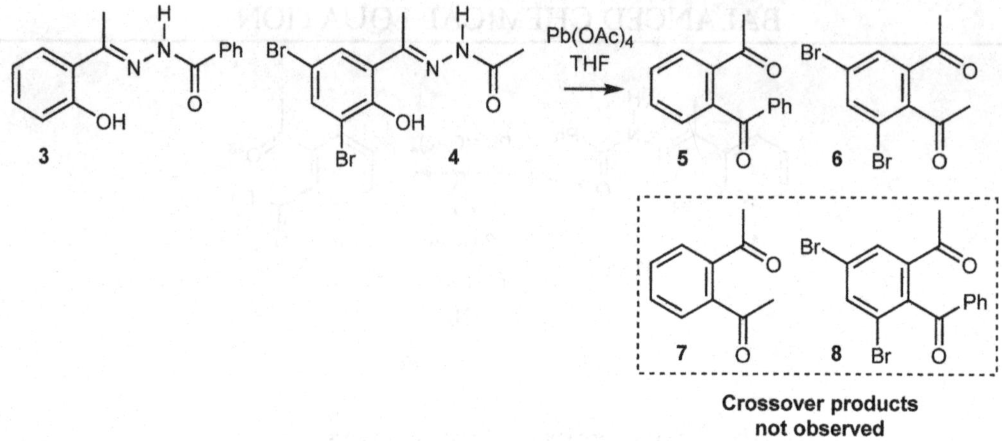

SCHEME 3

Another ^{18}O labeling experiment not carried out by the authors could also have been done on the phenolic oxygen atom of the starting hydrazone as shown in Scheme 4. Following the same steps as previously shown would provide product **2'** which has the label appearing in the benzoyl group instead of the acetyl group. The conclusion drawn from both kinds of ^{18}O labeling experiments would be consistent.

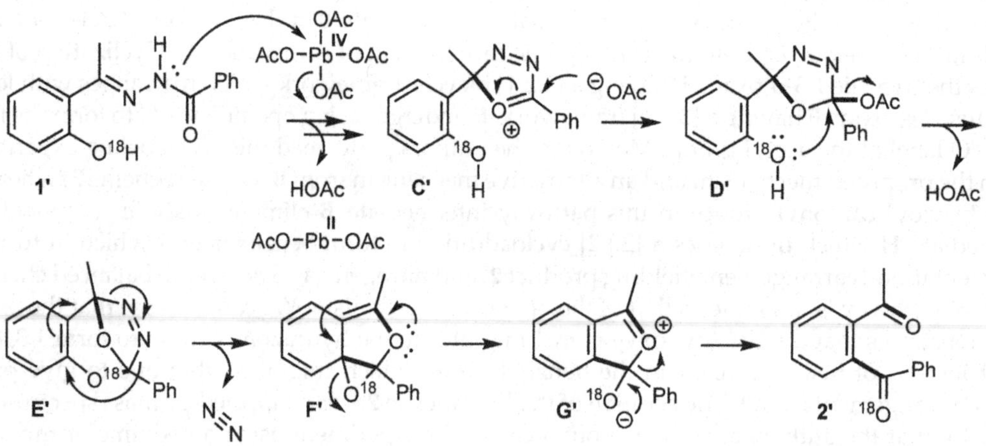

SCHEME 4

Furthermore, the reaction was also run in the presence of triethylamine used to neutralize any liberated acetic acid produced in the reaction to test the hypothesis that the reaction was acid catalyzed. The authors observed no change in the reactivity of the substrate suggesting that the reaction was not acid catalyzed. Electron spin resonance experiments on solutions of the starting hydrazone **1** in the presence of lead(IV)tetraacetate indicated no detectable radical intermediates and so a radical-type mechanism was also ruled out. Moving forward we recommend a computational study to determine the energy reaction coordinate diagrams for both mechanisms considered by the authors to complete the story. In particular, it would be interesting to map out the energetics for the steps involving the strained ring intermediates **E**, **F**, and **I**. If carried out, this problem can make for a fine tutorial example showcasing how experimental and theoretical techniques can be used in a complementary manner to elucidate reaction mechanisms.

ADDITIONAL RESOURCES

We note that the authors mentioned using ^{1}H NMR to study the transformation of **1** to **2** and indeed tried to observe a so-called CIDNP (chemically induced dynamic nuclear polarization effect).[2] They noted that such an effect, if observed, would point to a radical-based mechanism. The effect was not observed which further supported a proposed polar mechanism. For more information on the CIDNP effect, see Ref. 3. We also note that the authors first considered the possibility of a radical mechanism based on the works of Iffland et al. and Gillis and LaMontagne which showed that iminoxy radicals and Pb(OAc)₃ radical participated in reactions of lead(IV)tetraacetate with oximes.[4,5]

References

1. Katritzky AR, Harris PA, Kotali A. Mechanism of the replacement of phenolic hydroxyl by carbonyl on lead tetraacetate treatment of *o*-hydroxyaryl ketone acylhydrazones. *J Org Chem*. 1991;56(17):5049–5051. https://doi.org/10.1021/jo00017a013.
2. Ward HR, Lawler RG. Nuclear magnetic resonance emission and enhanced absorption in rapid organometallic reactions. *J Am Chem Soc*. 1967;89:5518–5519. https://doi.org/10.1021/ja00997a078.
3. Ward HR. Chemically induced dynamic nuclear polarization (CIDNP). I. Phenomenon, examples, and applications. *Acc Chem Res*. 1972;5(1):18–24. https://doi.org/10.1021/ar50049a003.
4. Iffland DC, Salisbury L, Schafer WR. The preparation and structure of azoacetates, a new class of compounds. *J Am Chem Soc*. 1961;83:747–749. https://doi.org/10.1021/ja01464a049.
5. Gillis BT, LaMontagne MP. Oxidation of hydrazones. III. α,β-unsaturated monoalkylhydrazones. *J Org Chem*. 1968;33(2):762–766. https://doi.org/10.1021/jo01266a056.

Question 57: Mechanisms of Bimane Formation

In 1978 Kosower et al. showed that treatment of dichloropyrazolone **1** with two equivalents of K₂CO₃·½H₂O and 0.75 equivalents of K₂CO₃ at 0°C for 18 h formed the strongly fluorescent *syn*-bimane **2** as the major product along with traces of the weakly fluorescent *anti*-bimane **2** (Scheme 1).[1] During an 1996 study, authors Allen and Anselme demonstrated the conversion of 1,3,4-oxadiazinone **3** (strong base and di-*t*-butyl dicarbonate in THF) to the *syn*-bimane **4** in 19% yield (Scheme 1).[2] Suggest mechanisms to explain both transformations given that Allen and Anselme proposed a very similar reaction intermediate for the **3** to **4** transformation as had been suggested previously by Kosower et al. for the **1** to **2** transformation.

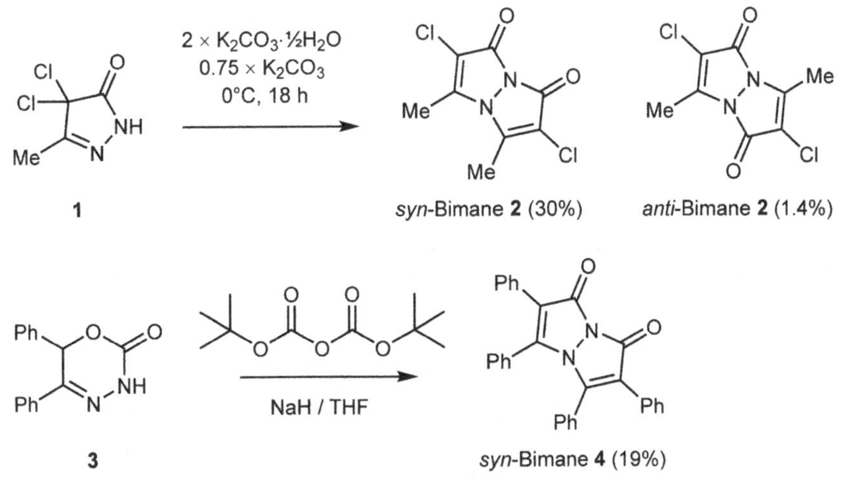

SCHEME 1

SOLUTION: 1 TO SYN-2

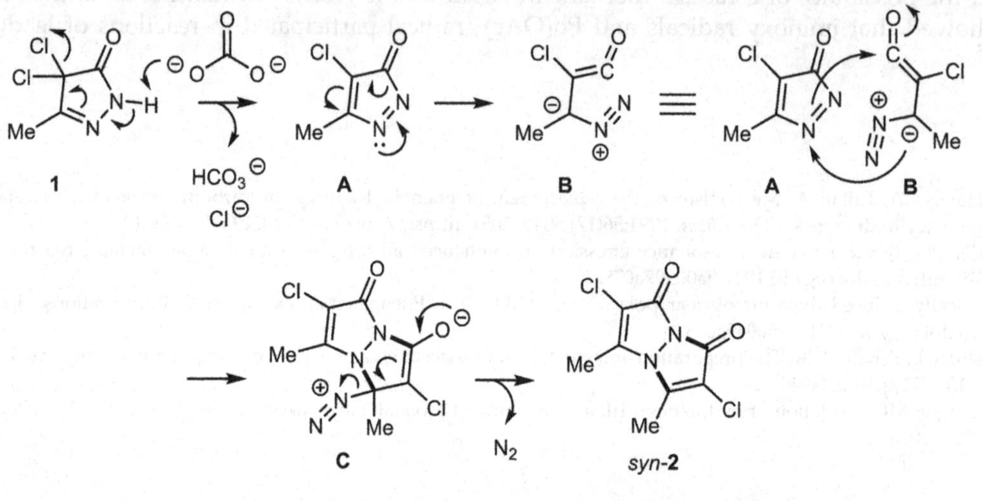

SOLUTION: A AND B TO ANTI-2

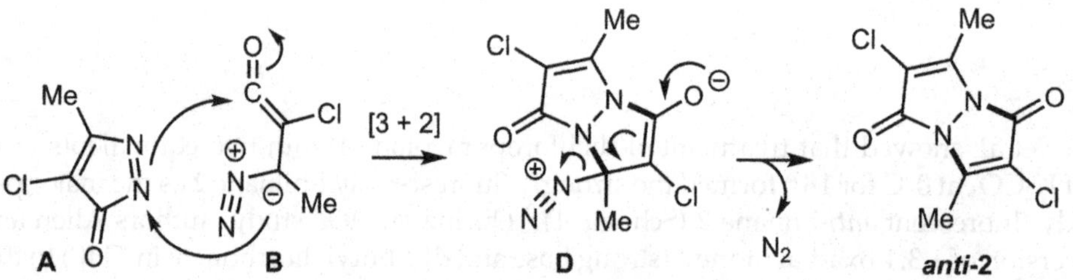

BALANCED CHEMICAL EQUATION: 1 TO SYN-2/ANTI-2

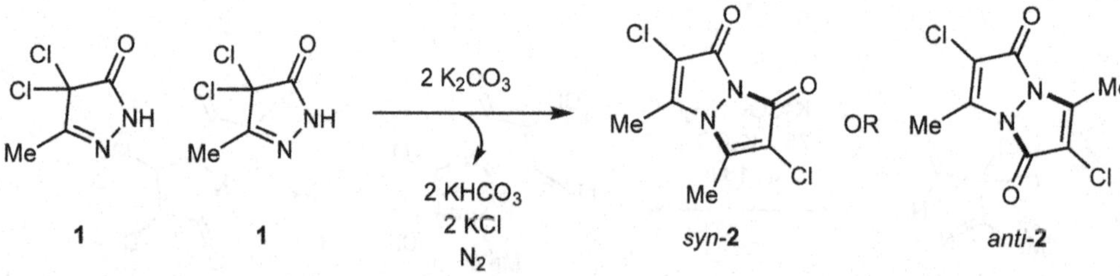

SOLUTION: 3 TO SYN-4

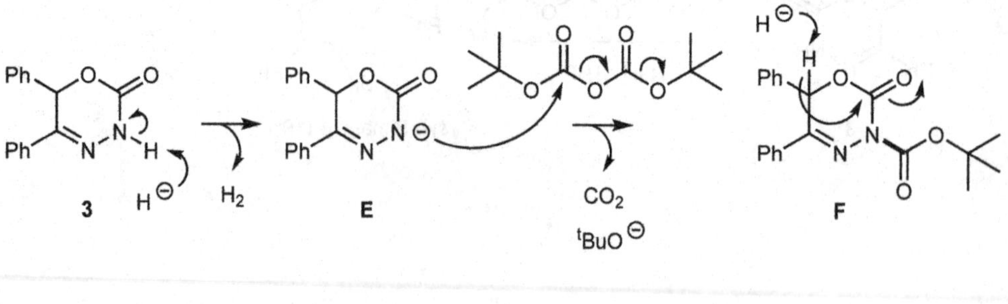

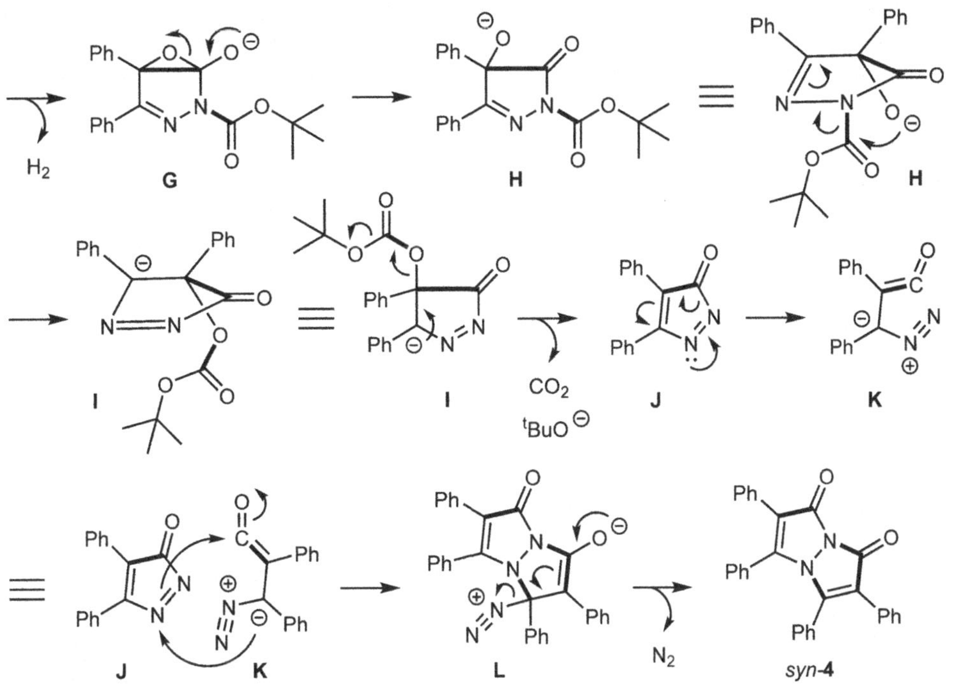

BALANCED CHEMICAL EQUATION: 3 TO SYN-4

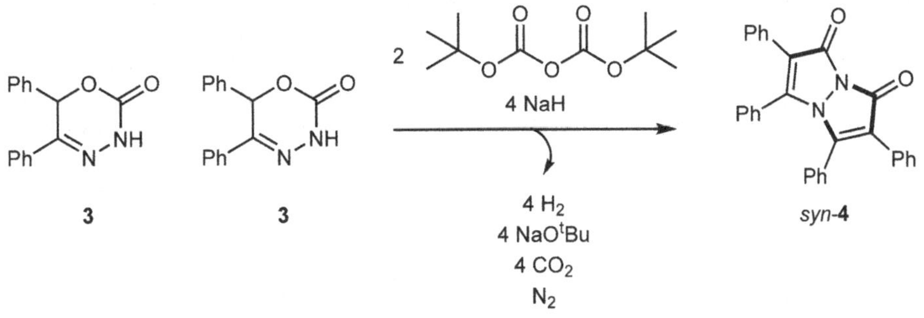

KEY STEPS EXPLAINED

We begin by noting that in 1978, Kosower et al. undertook the transformation of substituted pyrazolones such as **1** into the 9,10-oxabimane isomers *syn*-**2** and *anti*-**2**. The authors named these products based on Latin nomenclature having "bi" = "two" and "manus" = "hand" thus having "bimane" = "two hands" which represent the two cyclopentanone rings. The orientation of the carbonyl groups on the two rings can be either proximal (*syn*) or distal (*anti*). The authors thus proposed a mechanism which begins with base-mediated deprotonation of **1** with concomitant chloride anion elimination to form the 2,3-diazacyclopentadienone **A**. This intermediate is thought to rearrange with ring opening to the diazoalkylketene **B**. Intermediate **B** can then engage in a nucleophilic attack onto either of the two nitrogen atoms of **A** to form either **C** or **D** which lose nitrogen gas to form either *syn*-**2** or *anti*-**2**, respectively. In support for this mechanism, the authors carried out the transformation in a mixed system which contained **1** and its analog **1′** where a chloride atom was replaced with a methyl group (Scheme 2). The only product isolated was a *syn*-**2** analog which had

Cl and Me substitution on one side and Me and Me on the other. The three other combinations of isomers, namely, *syn* (Me, Me, Me, Me), *anti*(Me, Me, Me, Me), *syn*(Me, Cl, Me, Me), were not isolated. This experiment both helps to support the mechanistic bifurcation into intermediates **A** and **B** but it also raises issues about the completeness of the proposed mechanism. Such issues are further heightened by the fact that *syn*-2 was isolated in 30% yield whereas *anti*-2 had a yield of 1.4%.

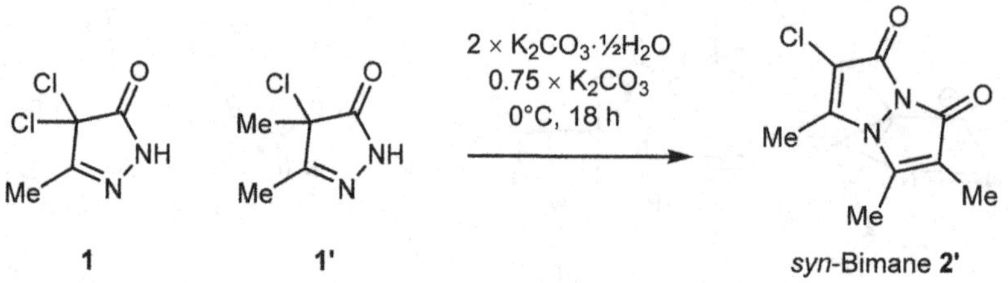

SCHEME 2

Given the mechanism proposed by Kosower et al. there is no obvious explanation for why there was an observed 21-fold difference in the yield of *syn*-2 over that of *anti*-2. In a subsequent article in 1982, Kosower et al. found that substitution with an electron withdrawing group (EWG) at the 3-position of **1** (for instance, carbonyl) led to *syn*-bimanes exclusively whereas an electron donating group (EDG) at the 3-position (for instance, alkyl-methyl) led to *anti*-bimanes in addition to *syn*-bimanes (Scheme 3).[3] The authors proposed that electronic effects (EDG versus EWD) next to the nitrogen atom β to the carbonyl of **A** work to direct nucleophilic attack by **B** either at N_α (if EDG is next to N_β thus favoring *anti*-2) or N_β (if EWG is next to N_β thus favoring *syn*-2). Furthermore, the choice of base also proved to direct the transformation toward *syn*-2 or *anti*-2. For example, if potassium carbonate is used, the *syn*-bimane predominates whereas if *N,N*-diisopropylethylamine is used, the *anti*-bimane predominates. Unfortunately, the authors did not conduct this experiment with multiple bases to form a definitive generalization. Instead they argued that under potassium carbonate conditions, potassium coordination with the carbonyl oxygen atom in **A** likely blocks nucleophilic approach by **B** onto N_α thus favoring approach at N_β which promotes *syn*-bimane formation. Interestingly, the authors also found based on kinetics studies that *anti*-bimane is two to three times more reactive toward hydroxide ion than the *syn*-bimane.[3] The authors utilize this to rationalize the crucial yield discrepancy noted earlier by arguing that *anti*-bimane likely forms and is destroyed by base hydrolysis with ring opening and decomposition thus leaving the *syn*-bimane as the major product. At this time we evaluate these explanations as speculative and therefore regard the issue of the choice of base and its mechanistic significance as an open question which future investigators should tackle by carefully examining reaction yields and by-product formation. Nevertheless, experimental and theoretical work to differentiate N_α from N_β in **A** using different substituents is of mechanistic significance. Nevertheless, there is a need for a more extensive investigation before definitive conclusions can be drawn.

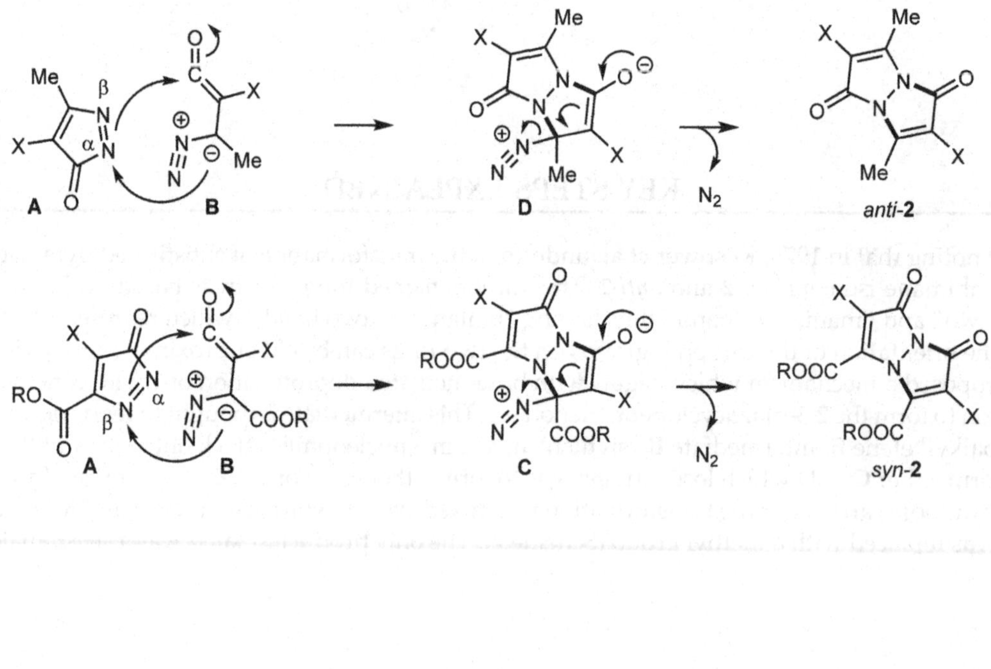

SCHEME 3

In another article the authors note that heterogeneous bases lead to *syn*-bimanes whereas homogeneous basic conditions lead to *anti*-bimanes.[4] In the same article, Kosower et al. also found that basic conditions can lead to ring opening of both *syn* and *anti*-bimanes and that treatment with an electrophile returns the compounds to their original *syn* or *anti* configuration. Lastly, it was noted that under thermal conditions, *syn*-bimanes rearrange to corresponding *anti*-bimanes for which the authors propose a possible mechanism without any supporting evidence. We leave this transformation to the reader as an exercise. Nevertheless, the authors interpret this result to mean that *anti*-bimanes are more thermodynamically stable than their *syn* counterparts. Such a conclusion directly contradicts the observed yield difference where the *syn*-bimane predominates by a 21-fold value over the *anti*-bimane. In the absence of theoretical calculations that estimate energy barriers along with enthalpic and entropic contributions, this conclusion is entirely speculative. Indeed, theoretical and geometrical calculations of bimanes did not appear until 2009 which further shows the rift that often exists between experimental and theoretical mechanistic chemists.[5] In that article, Blanco et al. calculated that simple *syn*-bimane **5** had an energy total of -491.084510 hartree and that simple *anti*-bimane **6** had an energy total of -491.092815 hartree using the hybrid HF/DFT B3LYP computational method and the 6-311++G(d, p) basis set within the Gaussian 03 package.[5] We therefore compute that *anti*-bimane **6** is 0.008305 hartree (or 5.21 kcal/mol) more thermodynamically stable than *syn*-bimane **5** (Scheme 4). This theoretical evidence supports what Kosower et al. presumed in their articles.[1–4] Nevertheless, the difference is small and does not by itself account for the difference in observed yields between *syn*-**2** and *anti*-**2** but it does support the results of the thermal rearrangement of *syn*-bimane to *anti*-bimane observed by the authors.[4]

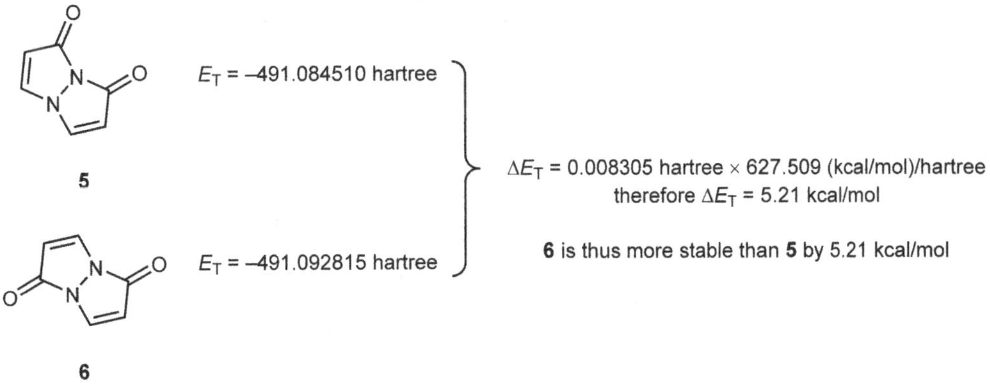

$E_T = -491.084510$ hartree

5

$E_T = -491.092815$ hartree

6

$\Delta E_T = 0.008305$ hartree $\times$ 627.509 (kcal/mol)/hartree
therefore $\Delta E_T = 5.21$ kcal/mol

6 is thus more stable than **5** by 5.21 kcal/mol

SCHEME 4

In subsequent research, chemists have not made progress on understanding the mechanism of this transformation and explaining why *syn*-bimanes tend to be favored over *anti*-bimanes under most reaction conditions. Instead research has focused on understanding why *syn*-bimane is much more fluorescent than *anti*-bimane and why the latter is more phosphorescent. Some research has also focused on understanding questions surrounding the aromaticity of these intriguing compounds.

Interestingly, in 1996, Allen and Anselme sought to improve the yield for the synthesis of *N*-Boc protected 5,6-diphenyl-3,6-dihydro-[1,3,4]oxadiazin-2-one **3** using sodium hydride as base instead of *N,N*-dimethylaminopyridine (DMAP). Instead, the use of hydride gave the *syn*-bimane *syn*-**4**. Curiously, the *N*-Boc-protected analog of **3** exhibited a yellowish fluorescent color similar in nature to that of *syn*-**4**. Therefore, Allen and Anselme proposed a mechanism which begins with proton abstraction via hydride ion from **3** to give anion **E** which reacts with di-*tert*-butyl dicarbonate yielding intermediate **F** along with carbon dioxide and *tert*-butoxide as by-products. Another equivalent of hydride ion then deprotonates **F** inducing a Favorskii-like rearrangement yielding alkoxide **H** via an intermediate **G** having a [3.1.0] bicyclic framework. Next, the carboxy-*tert*-butyl group migrates from nitrogen to carbon which gives carbanion **I** which eliminates carbon dioxide and *tert*-butoxide as by-products to form **J**. The resulting cyclic azo intermediate **J** fragments to a diazoketene **K** which undergoes a [3 + 2] cyclization as seen previously with **A** and **B** to give **L**. The final step is elimination of nitrogen gas from **L** which forms the final product *syn*-**4**. The overall ring construction mapping is [(2 + 2 + 1) + (5 + 0)]. Moreover, thanks to a reviewer recommendation, the authors acknowledge that another pathway is possible between intermediates **F** and **H** via a 1,2-Wittig rearrangement (as opposed to a Favorskii rearrangement) as shown in Scheme 5. Thus, deprotonation of **F** by hydride ion leads to carbanion **M** which undergoes homolytic fragmentation of the lactone C—O bond to form biradical anion **N/N′** existing in two resonance forms. The **N′** form cyclizes to alkoxide intermediate **H** which follows the same pathway as before leading to product *syn*-**4**. Unfortunately just like

Kosower et al., Allen and Anselme also did not provide further experimental or theoretical evidence to support their mechanism nor did they explain why no *anti*-bimane forms.

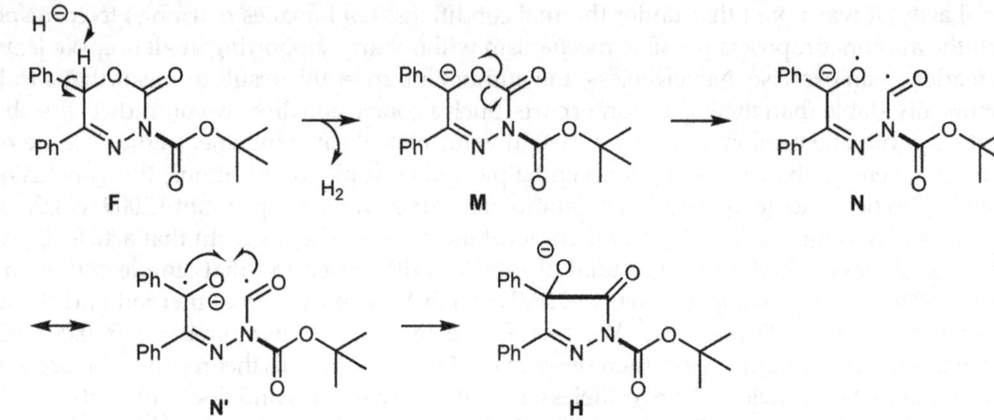

SCHEME 5

ADDITIONAL RESOURCES

We recommend further research by Kosower et al. where hydroxide kinetics, photochemical rearrangements, and photophysical properties of bimanes are discussed.[6-8] We also recommend an excellent review of 1,3,4-oxadiazines.[9]

References

1. Kosower EM, Pazhenchevsky B, Hershkowitz E. 1,5-diazabicyclo[3.3.0]octadienediones (9,10-dioxabimanes). Strongly fluorescent syn isomers. *J Am Chem Soc.* 1978;100(20):6516–6518. https://doi.org/10.1021/ja00488a050.
2. Allen A, Anselme J-P. An unprecedented Favorski-like ring contraction of the 1,3,4-oxadiazinone ring to a bimane. *Tetrahedron Lett.* 1996;37 (29):5039–5040. https://doi.org/10.1016/0040-4039(96)01011-8.
3. Kosower EM, Faust D, Ben-Shoshan M, Goldberg I. Bimanes. 14. Synthesis and properties of 4,6-bis(carboalkoxy)-1,5-diazabicyclo[3.3.0]octa-3,6-diene-2,8-diones[4,6-bis(carboalkoxy)-,10-dioxa-syn-bimanes]. Preparation of the parent syn-bimane, syn-(hydrogen,hydrogen)bimane. *J Org Chem.* 1982;47(2):214–221. https://doi.org/10.1021/jo00341a007.
4. Kosower EM, Pazhenchevsky B. Bimanes. 5. Synthesis and properties of syn- and anti-1,5-diazabicyclo[3.3.0]octadienediones (9,10-dioxabimanes). *J Am Chem Soc.* 1980;102(15):4983–4993. https://doi.org/10.1021/ja00535a028.
5. Blanco F, Alkorta I, Elguero J. Theoretical studies of azapentalenes. Part 5: Bimanes. *Tetrahedron.* 2009;65:6244–6250. https://doi.org/10.1016/j.tet.2009.05.018.
6. Kanety H, Kosower EM. Bimanes. 15. Kinetics and mechanism of the hydroxide ion reaction with 1,5-diazabicyclo[3.3.0]octadienediones (9,10-dioxabimanes). *J Org Chem.* 1982;47(22):4222–4226. https://doi.org/10.1021/jo00143a009.
7. Kanety H, Dodiuk H, Kosower EM. Bimanes. 10. Photochemical rearrangement of 1,5-diazabicyclo[3.3.0]octa-3,7-diene-2,6-diones (9,10-dioxa-anti-bimanes). *J Org Chem.* 1982;47(2):207–213. https://doi.org/10.1021/jo00341a006.
8. Kosower EM, Kanety H, Dodiuk H. Bimanes VIII: Photophysical properties of syn- and anti-1,5-diazabicyclo[3.3.0] octadienediones (9,10-dioxabimanes). *J Photochem.* 1983;21(2):171–182. https://doi.org/10.1016/0047-2670(83)80020-3.
9. Pfeiffer WD. [chapter 9.08]. 1,3,4-Oxadiazines and 1,3,4-thiadiazines. In: Katritzky A, Ramsden C, Scriven EFV, Taylor RJK, eds. *Comprehensive Heterocyclic Chemistry III.* Oxford: Elsevier; 2008:401–455. vol. 9. https://doi.org/10.1016/B978-008044992-0.00808-7.

Question 58: A Carbohydrate to Cyclopentanol Conversion

In 1996 South African chemists added the iodoglucoside **1** to a refluxing solution of excess samarium diiodide (SmI_2) in THF/HMPA for 2 h followed by cooling, dilution with 1:1 hexane/ethyl acetate, and final quenching with 5% aqueous citric acid to form cyclopentanol **2** in 70% yield (Scheme 1).[1] The authors described the mechanism as "SmI_2-promoted Grob-fragmentation followed by *in situ*, stereocontrolled SmI_2-mediated cyclization." Provide a mechanism which is consistent with these inferences.

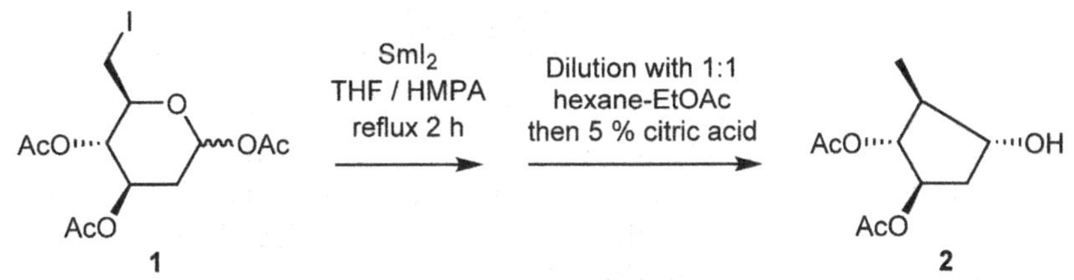

SCHEME 1

SOLUTION

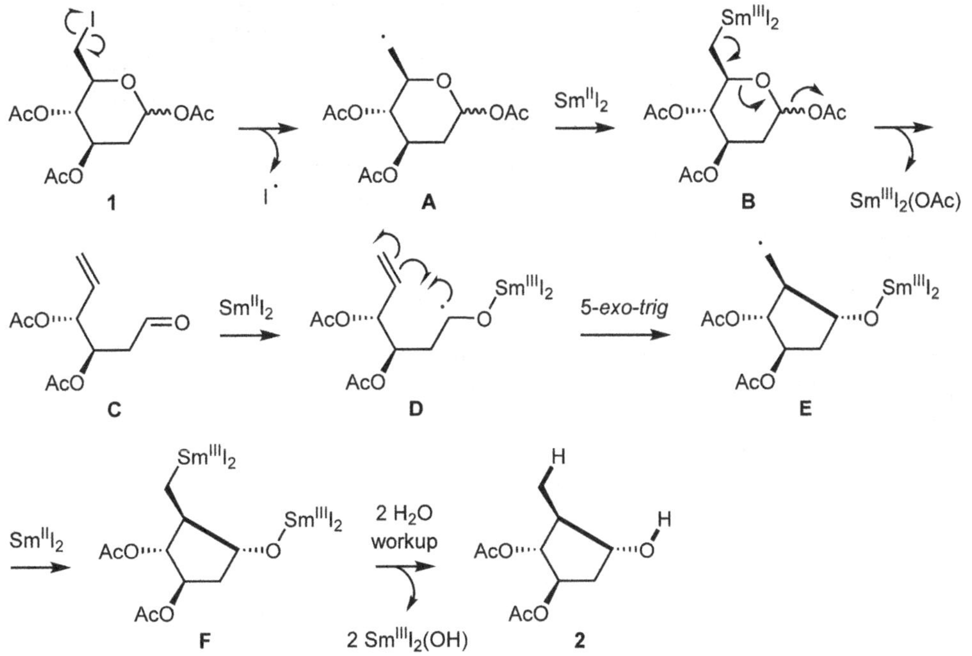

BALANCED CHEMICAL EQUATION

[O] $I^{\cdot}$ + $I^{\ominus}$ $\longrightarrow$ I_2 + e

[R] Sm^{+3} + e $\longrightarrow$ Sm^{+2}

Net: $I^{\cdot}$ + $I^{\ominus}$ + Sm^{+3} $\longrightarrow$ I_2 + Sm^{+2}

Add to both sides: $I^{\ominus}$ and $^{\ominus}OAc$

Overall: $I^{\cdot}$ + $Sm^{III}I_2(OAc)$ $\longrightarrow$ I_2 + $Sm^{II}(I)(OAc)$

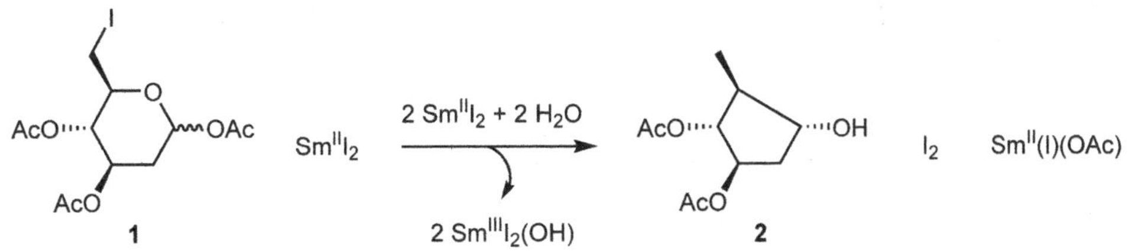

KEY STEPS EXPLAINED

To explain the transformation of **1** to **2**, the authors proposed a conjectured mechanism without providing by-product information or describing electron transfer steps which we present here in full detail. For example, in the first step the carbon—iodine bond in **1** homolytically cleaves to form iodine radical and carbon radical **A**. Radical **A** then undergoes an electron transfer reaction with Sm(II) where the carbon radical is reduced to a carbanion and samarium is oxidized to the +3 oxidation state. The resulting product **B** undergoes ring opening to form intermediate **C** with concomitant elimination of acetate ion. We note that **C** bears an aldehyde group and a terminal olefin group. The transformation of **C** to **D** is another electron transfer reaction that reduces the aldehyde carbon atom to a radical and oxidizes a second equivalent of Sm(II) to Sm(III). Carbon radical intermediate **D** then undergoes a 5-*exo-trig* cyclization to form carbon radical **E** which undergoes a third electron transfer with samarium iodide reducing the carbon radical to a carbanion and oxidizing Sm(II) to Sm(III). The final aqueous work-up which consumes two equivalents of water achieves protonation of the carbanion and alkoxide groups to generate product **2** and two equivalents of Sm(III) salts. The iodine radical generated in the first step is oxidized to diatomic iodine when reacted with iodide ion while Sm(III) is reduced to Sm(II). When one balances this final redox portion of the mechanism one arrives at the overall transformation of iodine radical and samarium (III) diiodide acetate to iodine and samarium (II) iodide acetate. From the balanced chemical equation we note that three equivalents of samarium(II)iodide are consumed and two equivalents of Sm(III) and one equivalent of Sm(II) mixed salts are generated. The presence of iodine as a reaction by-product should manifest itself in the form of a color change during the course of the reaction which unfortunately the authors did not describe in their paper.

We note that several alternative mechanisms can be written for this transformation which include the participation of citric acid, involve different amounts of samarium(II)iodide reagent, and consequently lead to different stoichiometrically balanced chemical equations. The authors did not specify how many equivalents of samarium(II)iodide reagent were added relative to the carbohydrate substrate **1** only stating that an "excess" amount was used. Essentially the key question asked in the elucidation of this mechanism is which atoms undergo redox changes: iodine, samarium, or both. We thus outline three possible mechanisms in Schemes 2–4.

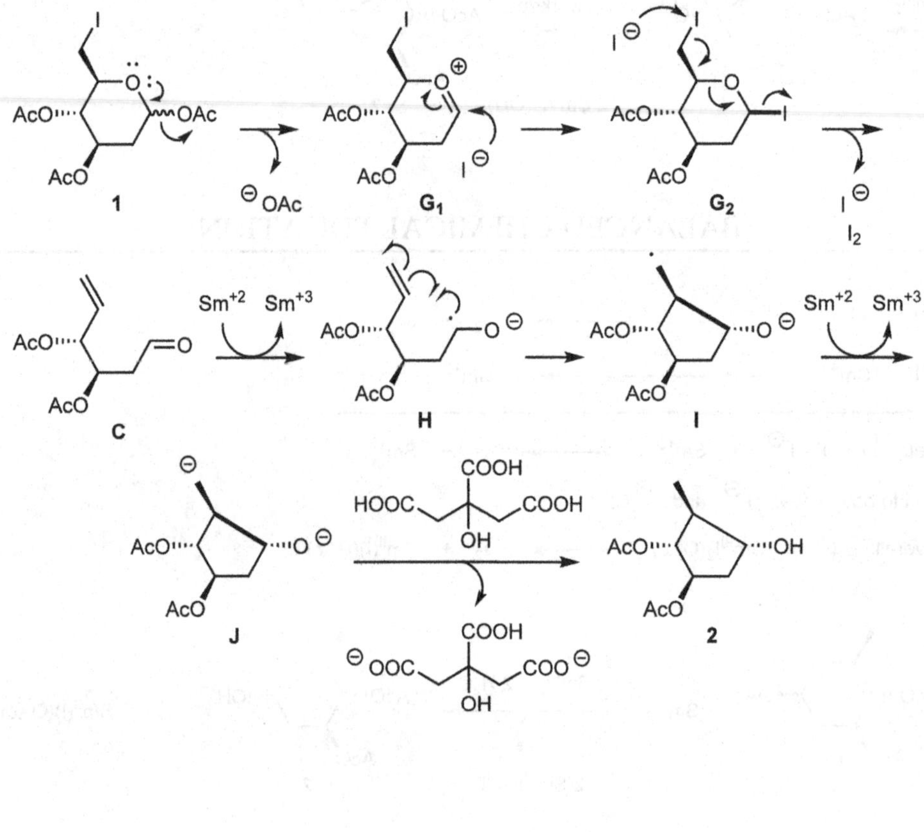

SCHEME 2

BALANCED CHEMICAL EQUATION: SCHEME 2

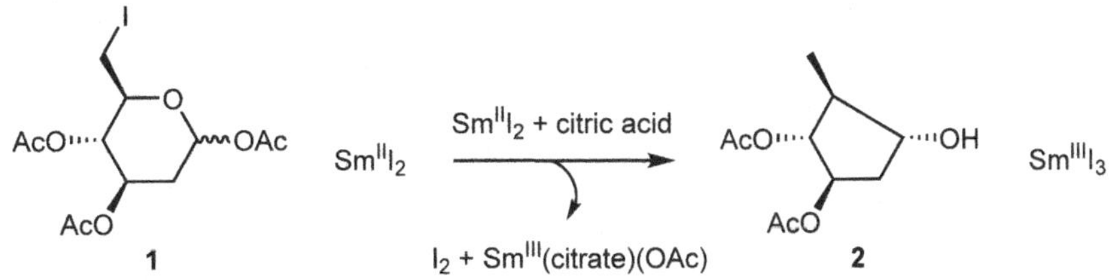

Table 1 summarizes the results of all mechanisms. The authors' proposed mechanism and Scheme 2 are closely related except that in Scheme 2 there is no homolytic cleavage of the C—I bond leading to iodine radical and **A**. The mechanism depicted in Scheme 3 does not produce iodine as a by-product.

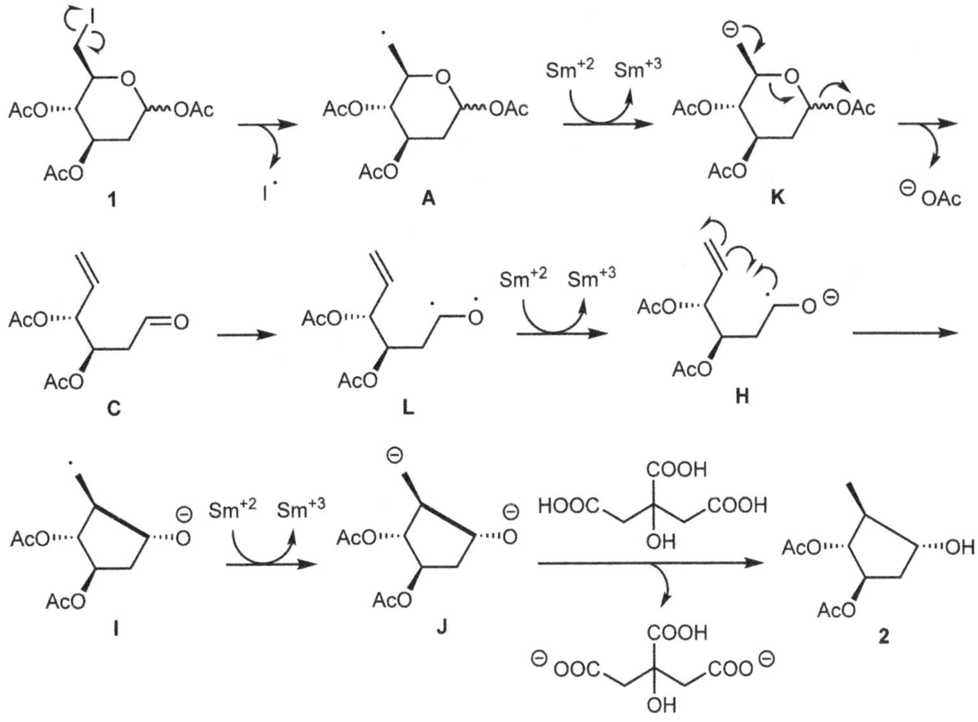

SCHEME 3

TABLE 1 Summary of Results for Various Balanced Chemical Equations Describing Transformation.

Scheme	Number of equiv. SmI_2	Iodine by-product	Redox couples
Authors' mechanism	3	1 equiv.	Sm(II) → Sm(III) + e $RCH_2I + H^+ + 2\,e \rightarrow RCH_3 + I^-$ $I^\bullet + e \rightarrow I^-$
1	2	1 equiv.	Sm(II) → Sm(III) + e $RCH_2I + H^+ + 2\,e \rightarrow RCH_3 + I^-$ $2\,I^- \rightarrow I_2 + 2\,e$
2	4	0 equiv.	Sm(II) → Sm(III) + e $RCH_2I + H^+ + 2\,e \rightarrow RCH_3 + I^-$ $I^\bullet + e \rightarrow I^-$
3	2	2 equiv.	$2\,I^- \rightarrow I_2 + 2\,e$ $RCH_2I + H^+ + 2\,e \rightarrow RCH_3 + I^-$ Sm: no change in oxidation state

BALANCED CHEMICAL EQUATION: SCHEME 3

[O] $I^\bullet + e \longrightarrow I^\ominus$

[R] $Sm^{+2} \longrightarrow Sm^{+3} + e$

Net: $I^\bullet + Sm^{+2} \longrightarrow I^\ominus + Sm^{+3}$

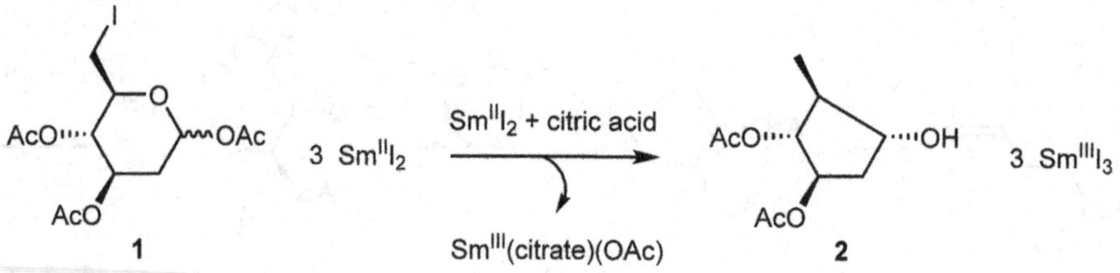

Furthermore, Scheme 4 does not involve a change of oxidation state for samarium and consequently produces the largest stoichiometric amount of diatomic iodine as by-product. In this last case samarium is a spectator metal ion. Clearly, further work is required to determine if in fact iodine is a by-product of the reaction and to investigate the effect of changing the relative stoichiometry of substrate **1** and samarium(II)iodide.

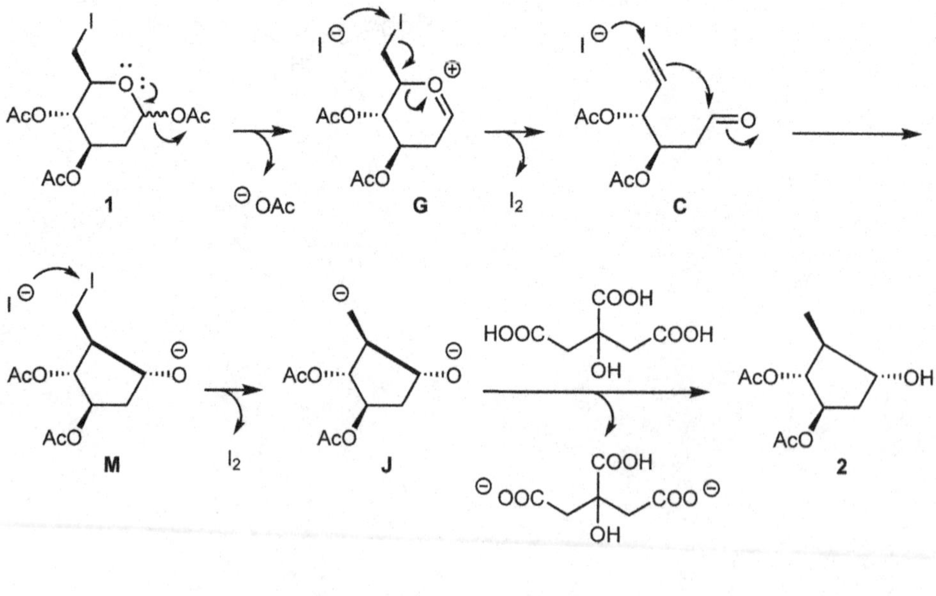

SCHEME 4

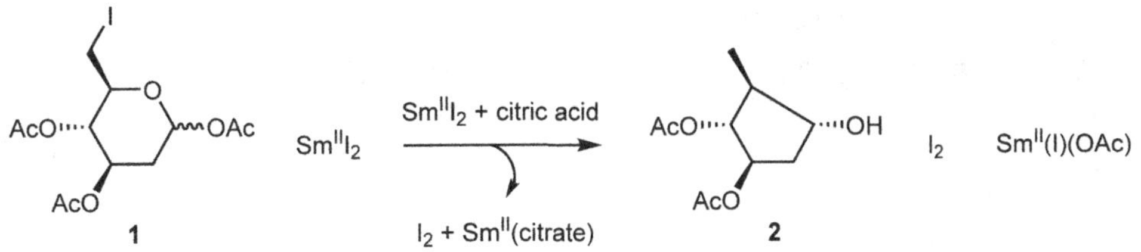

ADDITIONAL RESOURCES

We recommend three articles which discuss interesting mechanisms for transformations that involve Grob fragmentations and radical-based processes.[2–4]

References

1. Grove JJC, Holzapfel CW, Williams DBG. One-pot SmI2-promoted transformation of carbohydrate derivatives into cyclopentanols. *Tetrahedron Lett*. 1996;37(32):5817–5820. https://doi.org/10.1016/0040-4039(96)01234-8.
2. Malihi F, Clive DLJ, Chang C-C, Minaruzzaman. Synthetic studies on CP-225,917 and CP-263,114: access to advanced tetracyclic systems by intramolecular conjugate displacement and [2,3]-Wittig rearrangement. *J Org Chem*. 2013;78(3):996–1013. https://doi.org/10.1021/jo302467w.
3. Lemonnier G, Charette AB. Grob fragmentation of 2-azabicyclo[2.2.2]oct-7-ene: tool for the stereoselective synthesis of polysubstituted piperidines. *J Org Chem*. 2012;77(13):5832–5837. https://doi.org/10.1021/jo300690h.
4. Saget T, Cramer N. Heteroatom nucleophile induced C–C fragmentations to access functionalized allenes. *Chimia*. 2012;66:205–207. https://doi.org/10.2533/chimia.2012.205.

Question 59: Pyridazines From 1,2,4-Triazines

In 1996 Polish chemists showed that the treatment of 1,2,4-triazine **1** with phenylacetonitrile in dry *N,N*-dimethylacetamide (DMA) at 0°C in the presence of excess KO^{t}Bu for 1h followed by quenching with ice water gave 86% yield of dinitrile **2** (Scheme 1).[1] When **2** was treated with a 1:1 mixture of aqueous ammonia/acetone for 1h the diphenylpyridazine **3** was isolated. Furthermore, **1** was converted directly into **3** when dimethylformamide (DMF) was used as a solvent and the reaction mixture was quenched with aqueous acetic acid. Suggest a mechanism for the overall conversion of **1** to **2** to **3**.

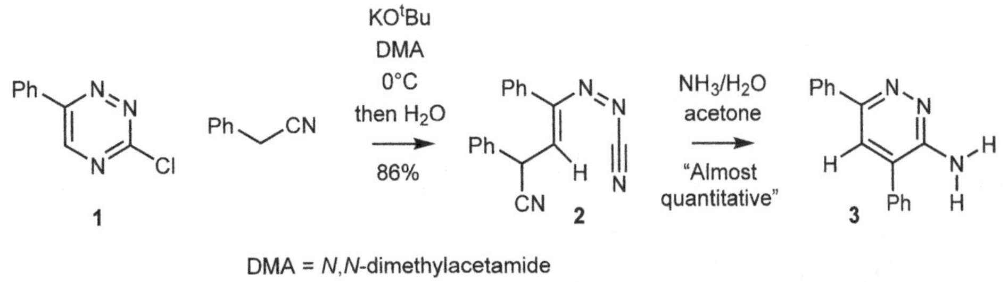

DMA = *N*,*N*-dimethylacetamide

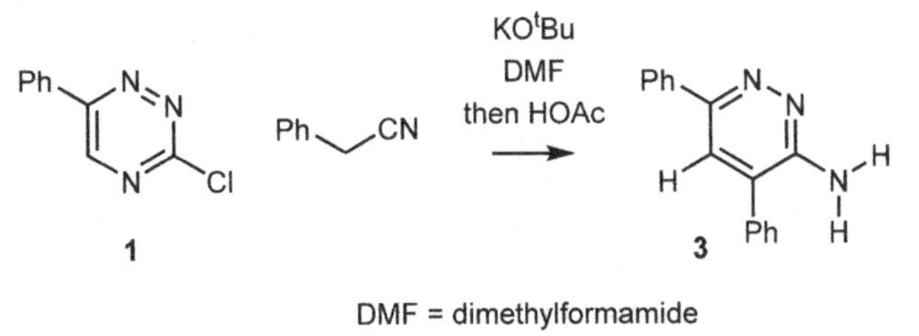

DMF = dimethylformamide

SCHEME 1

SOLUTION: 1 TO 2 USING ^{15}N LABEL ON PHENYLACETONITRILE

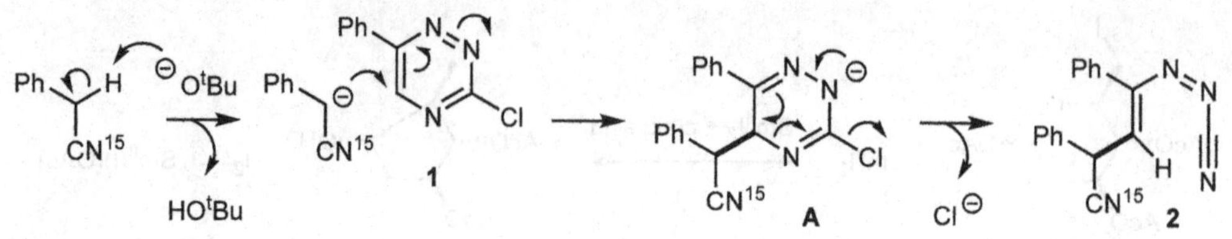

SOLUTION: 2 TO 3 USING ^{15}N LABEL ON PHENYLACETONITRILE

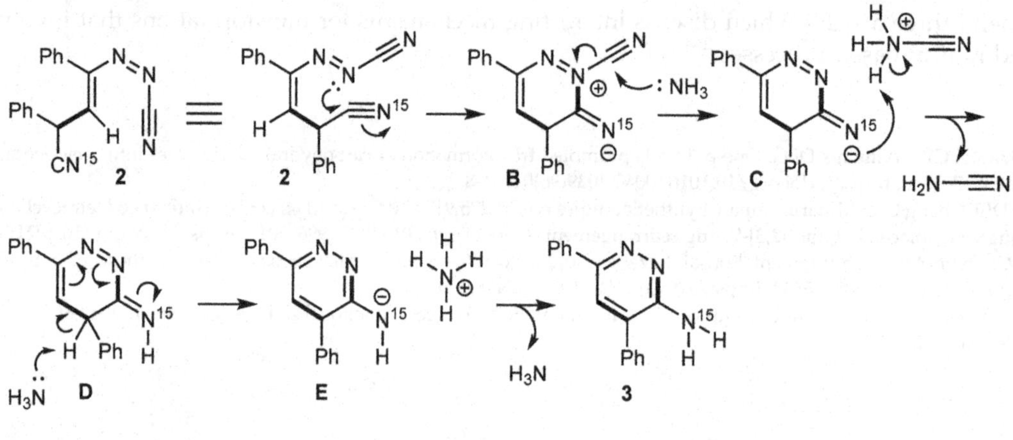

BALANCED CHEMICAL EQUATIONS

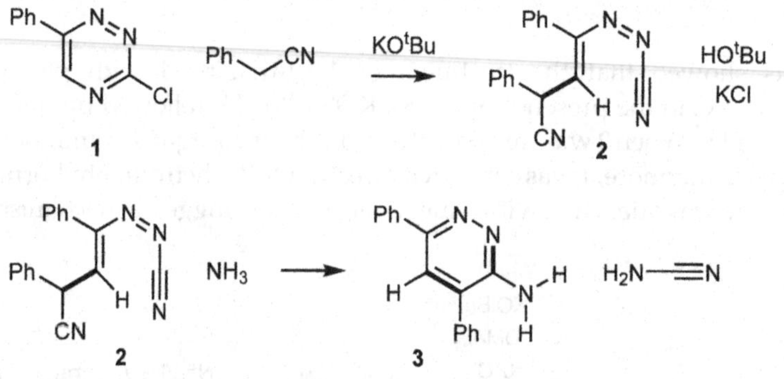

KEY STEPS EXPLAINED

We note that the authors elucidated the reaction mechanism via ^{15}N-labeling of phenylacetonitrile. This allowed tracking of the nitrogen label in the final product where it appeared on the amino group of 4,6-diphenyl-pyridazin-3-ylamine **3**. The implication was therefore that in the first reaction, *tert*-butoxide ion deprotonates phenylacetonitrile creating a carbanion that attacks the C$_5$ position of 3-chloro-6-phenyl-[1,2,4]triazine **1** yielding intermediate **A** which ring opens to product **2** and eliminates chloride ion. In the second reaction, product **2** cyclizes to zwitterionic intermediate **B** and eliminates the cyano group via attack by ammonia to form imine **D** and cyanamide by-product after proton transfer. Finally, intermediate **D** undergoes a base-catalyzed [1,3]-protic shift via ammonia resulting in product **3**. The ring construction mapping for this two-step process is [4 + 2]. Furthermore, when the chlorine atom in **1** is replaced by an —SCH$_3$ group, reaction with phenylacetonitrile gave the protonated form of **A** as a mixture of diastereomers in 53% yield thus further supporting the susceptibility of C$_5$ of **1** to nucleophilic attack by a carbanion species.

Furthermore, the results of the ^{15}N-labeling experiment ruled out an alternative mechanism shown in Scheme 2 which produces unlabeled product **3** from **2** and a labeled cyanamide by-product. Moreover, the ring construction is different becoming a [5 + 1] cyclization since the labeled cyano functional group is eliminated while the other cyano group bound to the azo moiety remains after ring closure.

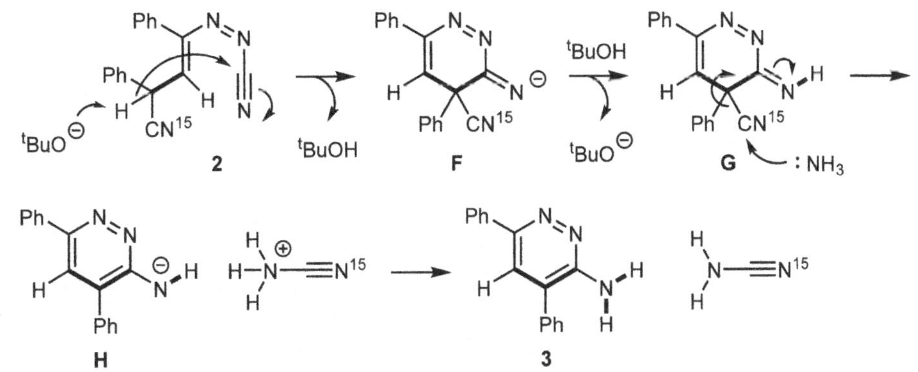

SCHEME 2

The authors did not conduct a theoretical analysis to help further support their proposed reaction mechanism. They also did not explain why in DMA solvent it is possible to isolate **2** whereas in DMF **1** is converted directly to **3**. We note that in DMF there is no ammonia present and therefore the mechanism under those conditions is expected to be different, though related, with different by-products formed compared to the case of DMA (which we illustrate in the solution before). Here we suggest a possible pathway starting from the in situ produced product **2**. Assuming that the same ^{15}N-labeled mechanism is operative in the reaction occurring in DMF in the absence of ammonia, we would expect the outcome as shown in Scheme 3. In this case, DMF acts as a sacrificial helper reagent that initiates a [1,3]-shift of the cyano group in intermediate **B** to give intermediate **I**. This is reminiscent of the first step in a Vilsmeier-Haack-type reaction between DMF and phosphorus oxychloride. A subsequent deprotonation-protonation sequence leads to intermediate **K** which then undergoes hydrolysis of the unlabeled cyano group under acid work-up to yield carbamic acid **L**. Final decarboxylation in the last step yields labeled product **3**. The final balanced chemical equation indicates that two equivalents of potassium t-butoxide are required and that potassium acetate, ammonia, and carbon dioxide are produced as additional by-products (Scheme 4).

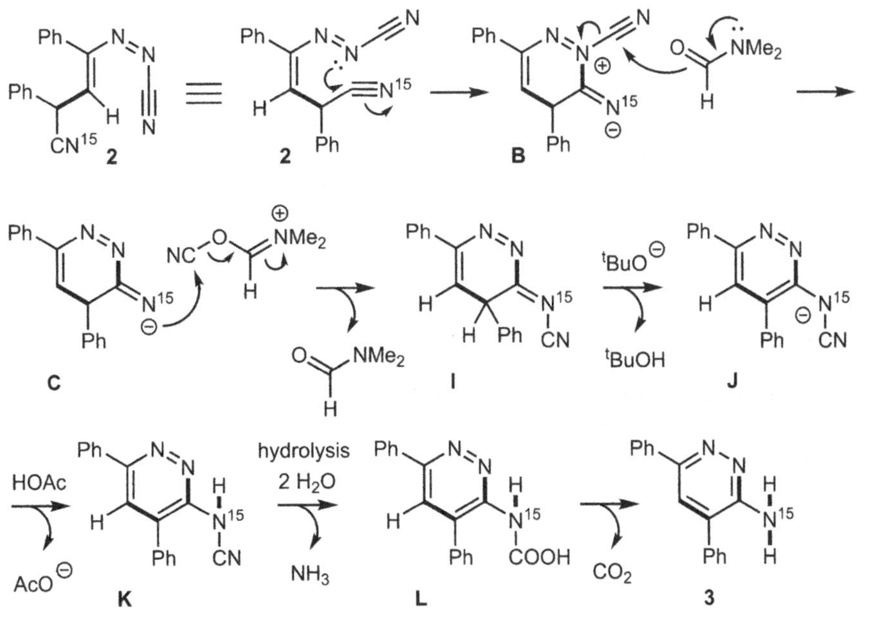

SCHEME 3

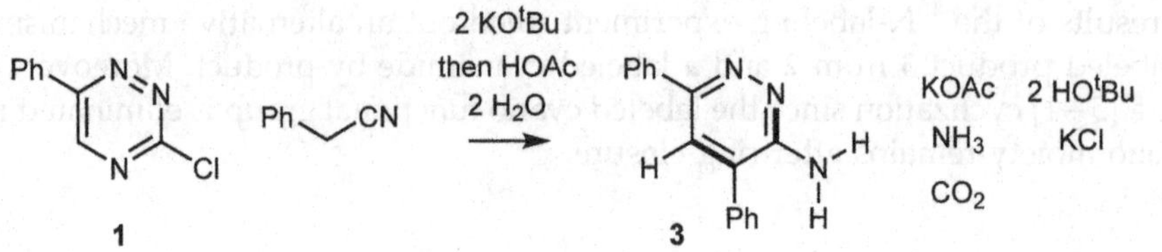

SCHEME 4

ADDITIONAL RESOURCES

We highly recommend additional articles by Rykowski et al. which further explore the mechanism of this interesting transformation under various conditions.[2-6] It is worth mentioning that the authors refer to this mechanism as an S_N *ANRORC* type mechanism which stands for nucleophilic substitution (S_N) by means of addition of nucleophile (AN), ring opening (RO), and ring closure (RC, in nucleophilic attack on ring system). Interestingly, in their subsequent research the authors do not explain mechanistically the difference in reactivity between DMA and DMF conditions, but they do explore the role of potassium amide as a nucleophile. This case, for example, involves a different S_N *ANRORC* mechanism whereby amide anion attacks C_5 of **1** causing ring opening after which the same group attacks the labile nitrile group of the ring-opening product leading to a pyridazine product containing a ^{15}N label on an internal ring nitrogen atom.[5,6] We thus surmise it is possible that a different mechanism occurs in the case of DMF solvent conditions which may not necessarily involve **2** as an intermediate. We leave this question to a future investigation.

References

1. Rykowski A, Wolinska E. Ring opening and ring closure reactions of 1,2,4-triazines with carbon nucleophiles: a novel route to functionalized 3-aminopyridazines. *Tetrahedron Lett.* 1996;37(32):5795–5796. https://doi.org/10.1016/0040-4039(96)01228-2.
2. Rykowski A, Wolinska E, van der Plas HC. A new route to functionalized 3-aminopyridazines by ANRORC type ring transformation of 1,2,4-triazines with carbon nucleophiles. *J Heterocyclic Chem.* 2000;37(4):879–883. https://doi.org/10.1002/jhet.5570370434.
3. Rykowski A, Wolinska E, van der Plas HC. 1,2,4-Triazine in organic synthesis. 16. Reactivity of 3-substituted 6-phenyl-1,2,4-triazines towards phenylacetonitrile anion in polar aprotic solvents. *Chem Heterocyl Compd.* 2001;37(11):1418–1423. https://doi.org/10.1023/A:1017959403370.
4. Rykowski A, Wolinska E, Branowska D, van der Plas HC. Vicarious nucleophilic substitution of hydrogen vs. ANRORC-type ring transformation in reactions of 1,2,4-triazines with α-halocarbanions. Novel route to functionalized pyrazoles. *ARKIVOC.* 2004;3:74–84. https://doi.org/10.3998/ark.5550190.0005.308.
5. Rykowski A, van der Plas HC. A ^{15}N-study of the conversion of 3-(methylthio)-1,2,4-triazine with potassium amide in liquid ammonia at -75°C. *Recl Trav Chim Pays-Bas.* 1975;94(8):1418–1423. https://doi.org/10.1002/recl.19750940809.
6. Rykowski A, van der Plas HC, van Veldhuizen A. 1H- and 13C-NMR studies of σ-adducts of 1,2,4-triazine and some of its derivatives with amide ions and/or liquid ammonia. *Recl Trav Chim Pays-Bas.* 1978;97(10):273–276. https://doi.org/10.1002/recl.19780971009.

Question 60: Natural Product Degradation During Extraction and/or Chromatography

In 1996 chemists from Hong Kong used dichloromethane extraction of the aerial parts of the shrub *Baeckea frutescens* to isolate as a major component compound **1** and as a minor component compound **2** (Scheme 1).[1] The authors suggested that compound **2** is a degradation product of **1** formed either during the extraction process or during the chromatographic separation. Suggest a possible mechanism for the conversion of **1** to **2**.

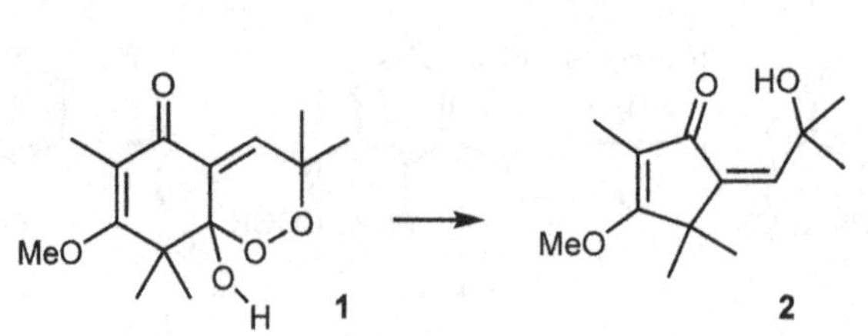

SCHEME 1

SOLUTION: AUTHOR'S PROPOSAL (IONIC MECHANISM)

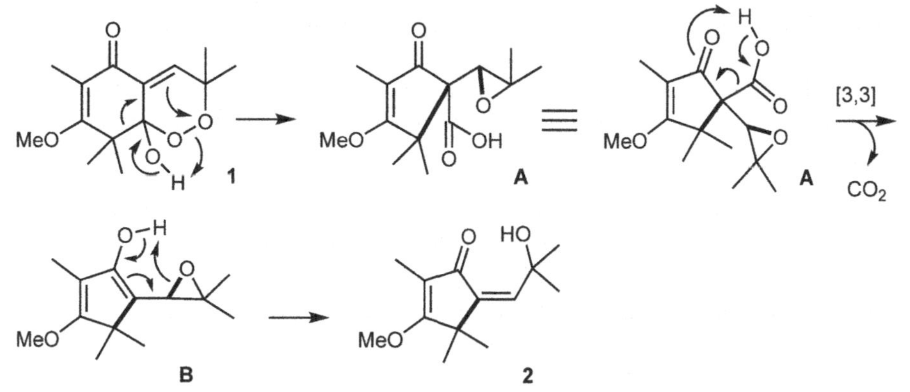

SOLUTION: ALTERNATIVE PROPOSAL (RADICAL MECHANISM)

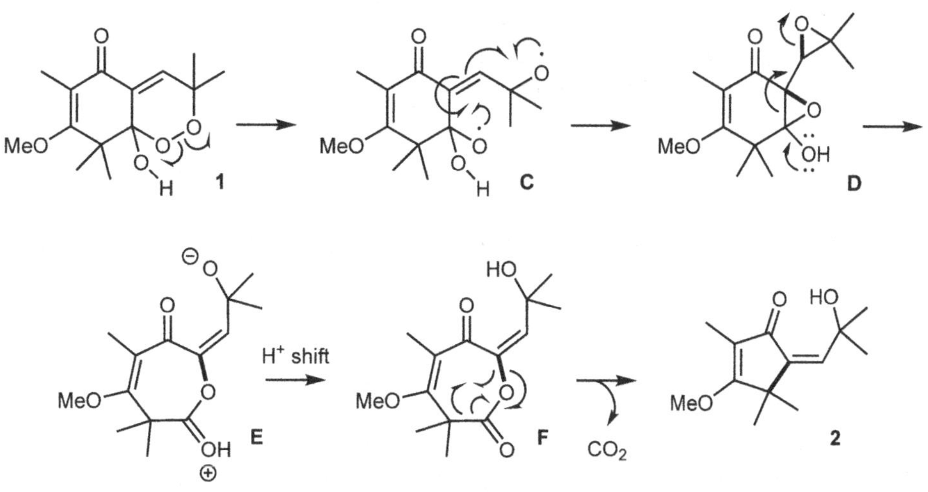

BALANCED CHEMICAL EQUATION

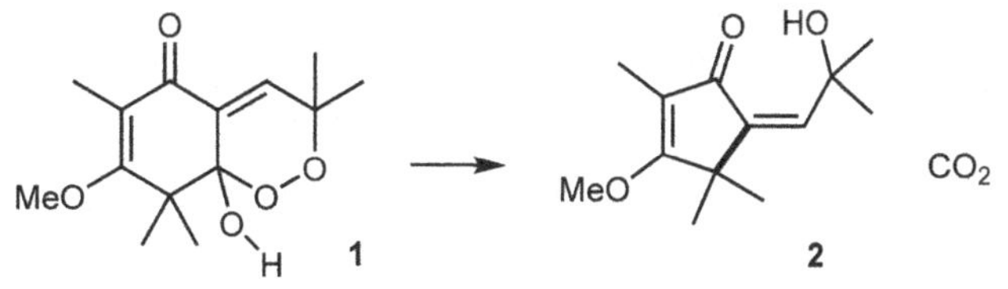

KEY STEPS EXPLAINED

We begin by noting that the authors believed the degradation of **1** to **2** occurred during the solvent extraction process from the leaves of the shrub *B. frutescens* or during chromatographic separation from the crude plant extract. Therefore, we assume that the transformation happens under neutral conditions. In their article, Tsui and Brown postulated an ionic reaction mechanism for this transformation which begins with rearrangement of **1** to form intermediate **A** via simultaneous endoperoxide ring opening and cyclohexenone ring contraction. **A** then eliminates carbon dioxide by a [3,3] sigmatropic rearrangement which forms enol **B** which undergoes ketonization and epoxide ring opening to give product **2**. Since the authors did not provide additional experimental or theoretical evidence to support this mechanism, we here present an alternative radical mechanism to explain the transformation of **1** to **2**. We note that the alternative mechanism is consistent with the expectation of homolytic cleavage of peroxide bonds found in such reagents as hydrogen peroxide, benzoyl peroxide, and potassium persulfate acting as radical initiators under neutral conditions. Thus, the alternative mechanism begins with homolytic cleavage of the endoperoxide bond of compound **1** which forms diradical **C**. This species then rearranges to the bis-epoxide **D** which has its fused epoxide ring undergo ring opening to form a seven-membered ring zwitterionic intermediate **E**. After a proton shift which gives **F**, extrusion of carbon dioxide via homolytic bond breakages results in the formation of the cyclopentenone product **2**. We note that the authors established the *syn* stereochemistry of the exocyclic olefin in product **2** based on nuclear Overhauser enhancement (NOE) experiments.

ADDITIONAL RESOURCES

We recommend two articles which discuss the mechanisms of transformations which involve endoperoxide isomerizations.[2,3] Interestingly, both ionic and radical mechanisms are discussed.

References

1. Tsui W-Y, Brown GD. Unusual metabolites of *Baeckea frutescens*. *Tetrahedron*. 1996;52(29):9735–9742. https://doi.org/10.1016/0040-4020(96)00510-8.
2. Erden I, Oecal N, Song J, Gleason C, Gaertner C. Unusual endoperoxide isomerizations: a convenient entry into 2-vinyl-2-cyclopentenones from saturated fulvene endoperoxides. *Tetrahedron*. 2006;62:10676–10682. https://doi.org/10.1016/j.tet.2006.07.107.
3. Gueney M, Dastan A, Balci M. Chemistry of the benzotropone endoperoxides and their conversion into tropolone derivatives: unusual endoperoxide rearrangements. *Helv Chim Acta*. 2005;88(4):830–838. https://doi.org/10.1002/hlca.200590061.

Question 61: Simple Access to Oxaadamantanes

In 2000 chemists from the Ukraine and America presented mechanistic evidence for a previously reported oxidation of the readily accessible 2-hydroxy-2-methyladamantane **1** which forms oxaadamantane **2**, *exo*-4-hydroxy-2-oxaadamantane **3**, and lactone **4** when treated with trifluoroperacetic acid in trifluoroacetic acid (TFA) at 20°C for 1h (Scheme 1).[1,2] Provide mechanisms for the formation of these products from **1** in light of the observation that **2** is stable under the reaction conditions and is not a precursor to **3**.

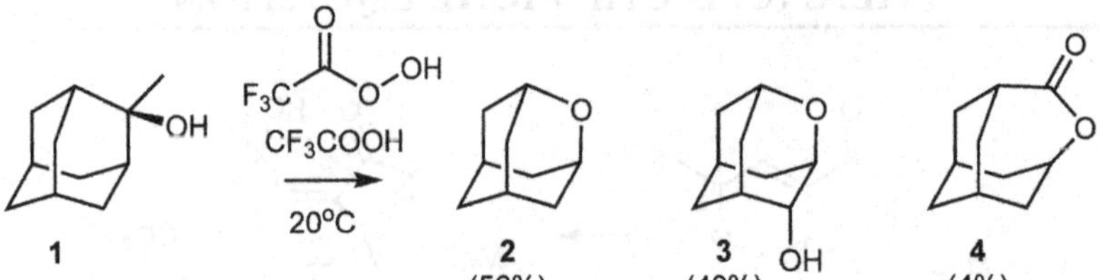

SCHEME 1

SOLUTION: 1 TO 2

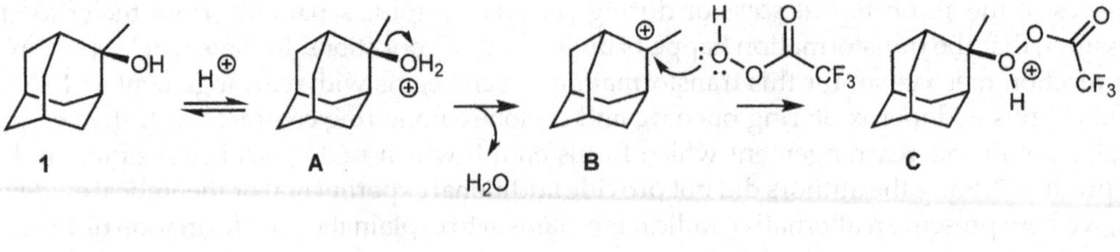

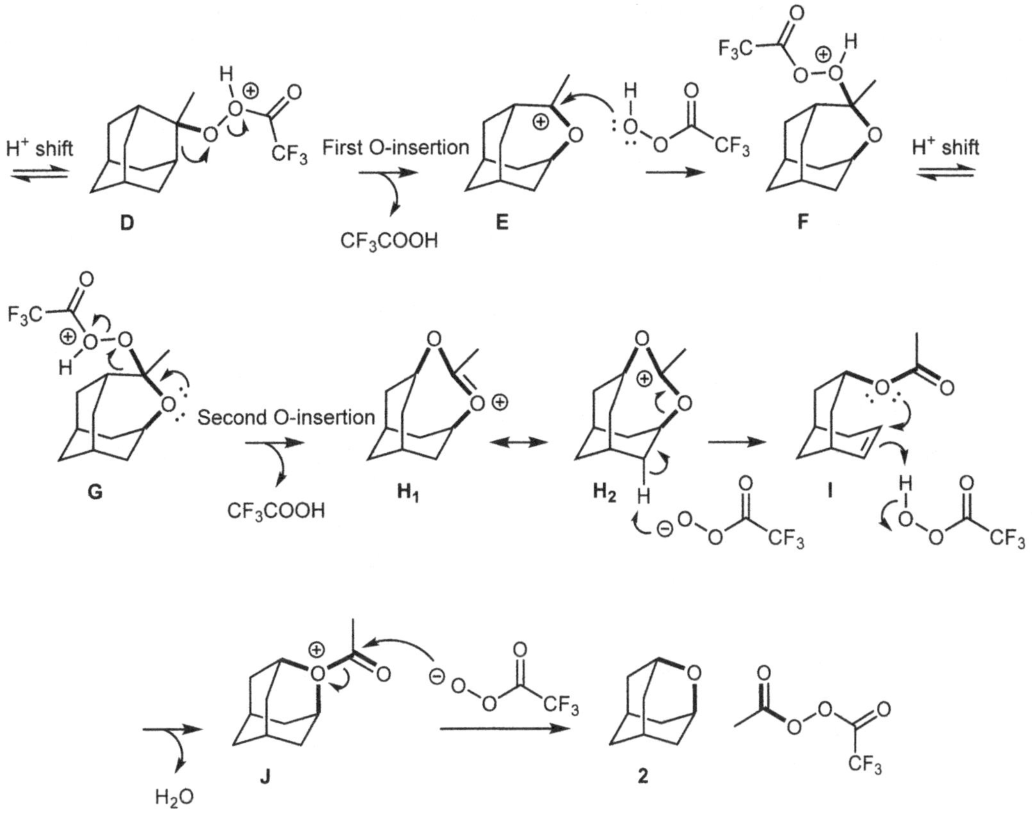

BALANCED CHEMICAL EQUATION: 1 TO 2

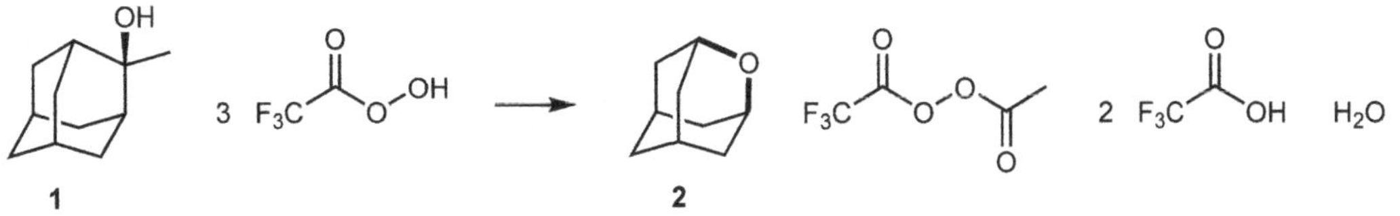

SOLUTION: I TO 3

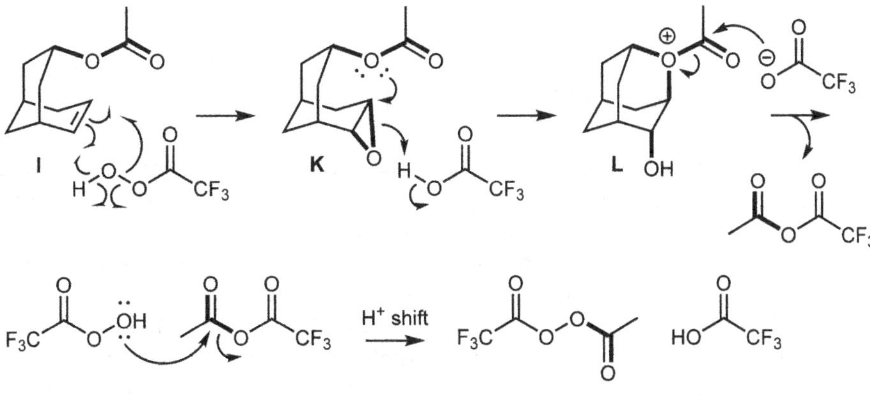

BALANCED CHEMICAL EQUATION: 1 TO 3

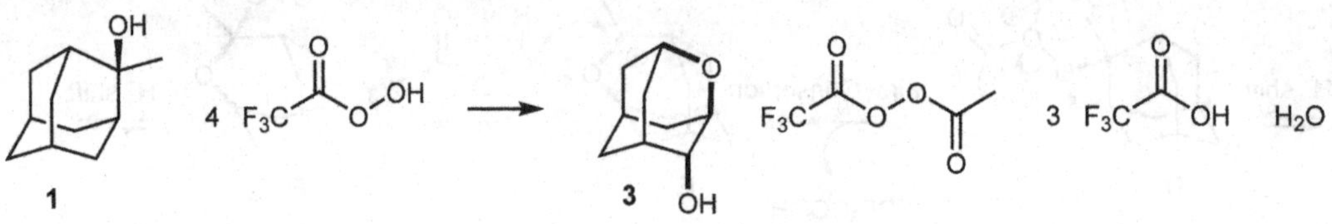

SOLUTION: D TO 4

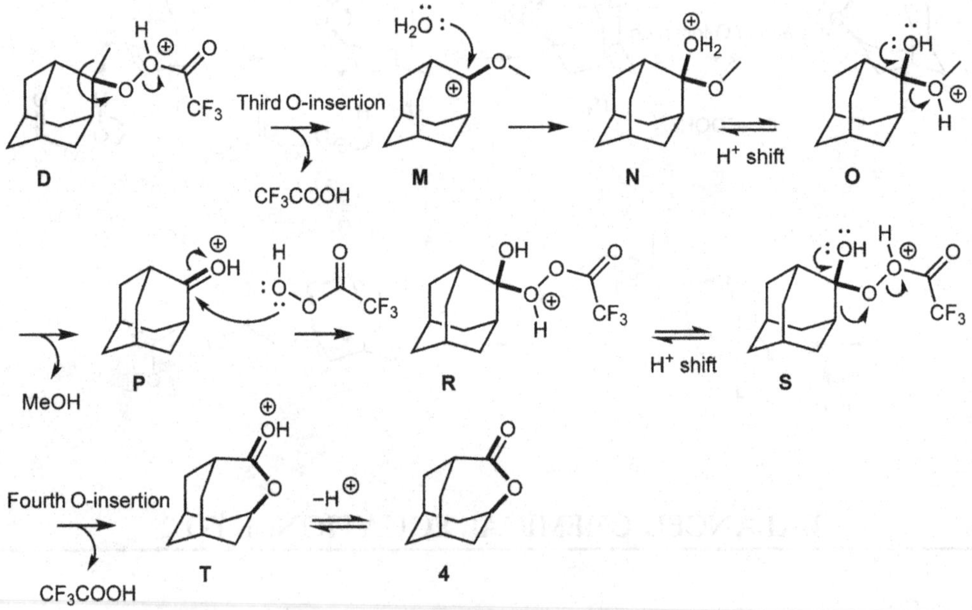

BALANCED CHEMICAL EQUATION: 1 TO 4

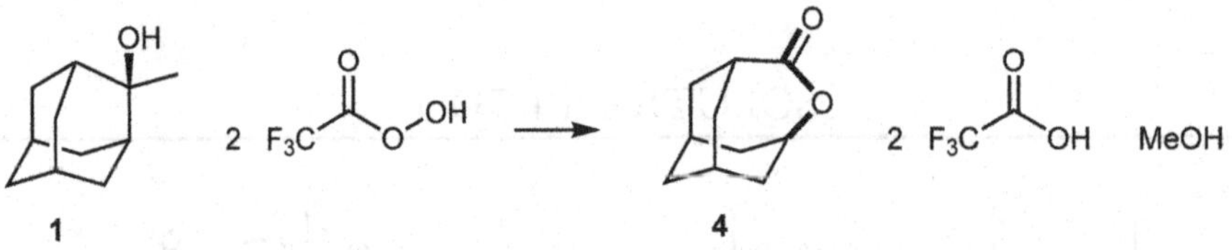

KEY STEPS EXPLAINED

The mechanisms proposed to explain the formation of products **2**, **3**, and **4** from the starting 2-hydroxy-2-methyladamantane **1** all begin with acid-promoted S_N1 substitution of the hydroxyl group of **1** with a trifluoroethaneperoxoate group which comes from one equivalent of trifluoroperacetic acid. This sequence after proton transfer provides the key hub intermediate **D**. At this junction the mechanism diverges into two pathways. The first pathway from **D** leads to products **2** and **3** and the second pathway leads to product **4**. Both of these pathways involve Criegee rearrangement via oxygen insertion into an alkyl R group followed by elimination of trifluoroacetic acid to form a carbocation intermediate. We note that the carbocation intermediate thus formed is stabilized through resonance by electron density available from the neighboring oxygen atom. Thus, in the first pathway, oxygen insertion into the C_1—C_2 bond results in the formation of carbocation **E** which then reacts with a second equivalent of trifluoroperacetic acid to form intermediate **G** after proton shift. Intermediate **G** then undergoes a second Criegee rearrangement via oxygen insertion to form the carbocation H_2 which undergoes deprotonation with concomitant ring opening to form the key olefin hub intermediate **I**. Once again the proposed mechanism diverges into two paths which lead to products **2** and **3**, respectively. In the first path, intermediate **I** undergoes ring closure to form the oxonium ion **J** which eliminates an acetyl group via reaction with trifluoroacetic acid to form the cyclic ether product **2** and acetyltrifluoroacetic anhydride by-product. This anhydride reacts with a third equivalent of trifluoroperacetic acid yielding acetyltrifluoroacetyl peroxide as the final by-product.

In the second mechanistic pathway we have epoxidation of the olefin group of intermediate **I** by a third equivalent of trifluoroperacetic acid to form the epoxide intermediate **K**. This intermediate undergoes cyclization via neighboring group participation with concomitant epoxide ring opening to form **L** which reacts with trifluoroacetic acid as seen before to eliminate an acetyl group and give the cyclic ether product **3** and acetyltrifluoroacetic anhydride by-product. This by-product reacts with trifluoroperacetic acid as before to give acetyltrifluoroacetyl peroxide as the final by-product. We note that the main difference between the formation of products **2** and **3** is that to generate **3** there are 4 equivalents of trifluoroperacetic acid consumed instead of 3 and that 3 equivalents of trifluoroacetic acid are formed instead of 2.

Following the second possible mechanistic path from intermediate **D**, we have migration of the methyl group of **D** via oxygen insertion and elimination of trifluoroacetic acid (Criegee rearrangement) which gives the carbocation intermediate **M**. This carbocation is captured by water to give **O** after proton shift. Elimination of methanol in **O** gives **P** which essentially undergoes a Baeyer-Villiger oxidation via trifluoroperacetic acid to form the lactone product **4**. We note that the Baeyer-Villiger oxidation is considered a subset of the Criegee rearrangement and thought to occur via oxygen insertion with loss of a good leaving group (trifluoroacetic acid, as seen before) to give a key Criegee carbocation intermediate (in this case intermediate **T**, which loses a proton to give the final lactone product **4**). For simplicity, we summarize the mechanistic paths leading to all three products in compact form in Scheme 2.

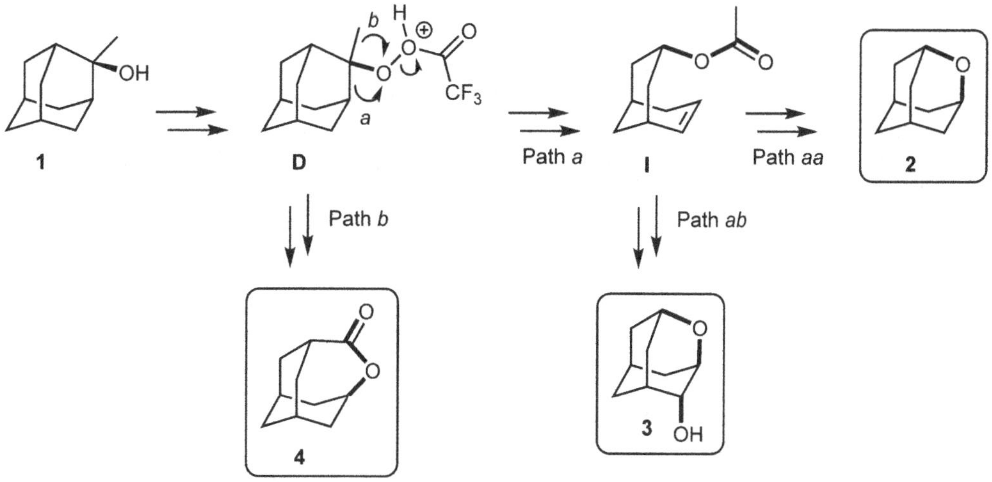

SCHEME 2

Before explaining the experimental and theoretical evidence gathered by the authors in support of their mechanisms, we emphasize that we regard this work[1] as the best example in the whole book concerning what a properly undertaken mechanistic analysis should contain.

The Criegee rearrangement was discovered in 1944 when Rudolf Criegee observed that peroxy esters rearranged to a ketone, ester, or carbonate plus alcohol via oxygen insertion/consecutive oxygen insertion.[3] Over the years reagents such as trifluoroperacetic acid (TFPAA) have been employed because they are strongly acidic and provide a good leaving group during the oxygen insertion/alkyl migration/elimination steps of the mechanism. The use of phosphates in conjunction with TFPAA has been eliminated because it was shown to restrict multiple O insertion by lowering acidity and the possibility of carbocation formation.[4] In this context, the well-known Baeyer-Villiger oxidation is considered a subset of the Criegee rearrangement.[5] Mechanistically, it is believed that the migrating group rearranges from the *antiperiplanar* position relative to the dissociating O—O bond of the peroxy ester (see steps **D** to **E**, **G** to **H₁**, **D** to **M** and **S** to **T**, see ref. 5 in Ref. 4). Furthermore, such migrations are believed to be facilitated by the electron-donating ability of the migrating group (refs. 2i and 6 in Ref. 4) while electron-deficient groups tend to not move (ref. 7 in Ref. 4). Lastly, Criegee hypothesized an ionic mechanism whereas others have proposed that the second O insertion/migration/elimination step is concerted (see ref. 3a in Ref. 4). With these aspects in mind, Krasutsky et al. undertook an experimental and theoretical analysis of the transformation of 1 to 2, 3, and 4.

The authors carried out this transformation after observing that use of the analogous diethylketal 5 (instead of 1) resulted in a predominant formation of lactone 4 with trace amounts of 2 and 3 (Scheme 3). The dimethylketal analog of 5 provided solely product 4. It was then established using ^{1}H NMR spectroscopy that diketals such as 5 formed a ketone intermediate such as P in acidic media which underwent standard Baeyer-Villiger oxidation to give product 4. Product 4 was also isolated as a major product when one of the OR groups of 5 was substituted for an H or Ph group. In this case, it appeared that O insertion occurred within the C_2—H or C_2—Ph bond to give essentially a hemiketal intermediate which collapses to P and to 4 as explained before. Thus the authors decided to use the methyl-substituted substrate 1 to direct O insertion into the cage structure of 1 since methyl groups are expected to be the least likely to undergo migration (Ph > H $\gg$ CH$_3$) based on previous reports on the Criegee rearrangement (see refs. 15 and 22c in Ref. 1). Using an 8-fold excess of TFPAA in trifluoroacetic anhydride (TFAA), the authors showed that 1 leads to a predominant formation of 2 and 3 with trace amounts of 4 formed as well. The use of 8-fold excess of TFPAA is to facilitate the nucleophilic addition of TFPAA to B to form C.

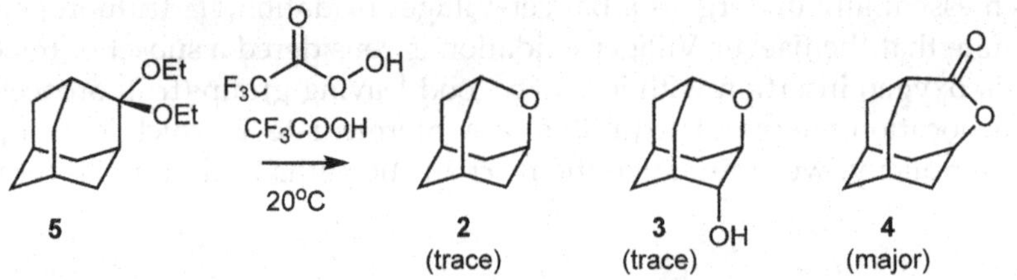

SCHEME 3

When TFPAA was used in fewer amounts, it did not react as a nucleophile with B but instead allowed for the deprotonation of B which formed an exocyclic olefin U that under TFPAA (electrophile) gave 2-adamantyl formate 6 (Scheme 4). This experiment directly supports the proposed nucleophilic role of TFPAA in step B to C and indirectly in steps E to F and P to R.

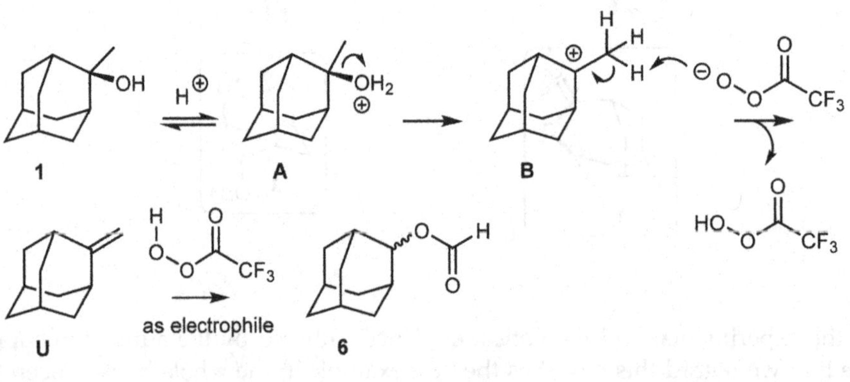

SCHEME 4

This was further supported when the authors independently synthesized U and treated it with TFPAA in TFAA and isolated 6 as the sole product. The authors did not provide a mechanism for the formation of 6 from U so we leave this as an exercise for the reader. It was also shown that at lower temperature ($-25°$C) and higher TFPAA concentration (20-fold excess), carbocation H_1/H_2 could be observed in situ in quantitative amount using ^{1}H and ^{13}C NMR spectroscopy. This double O insertion intermediate is a key mechanistic indicator along the pathway which leads to 2 and 3. The authors also found that this carbocation was stable for 24 h at $-25°$C and that it rearranged to 2 and 3 in 2 h at 0°C and in 1 h at 20°C. When TFPAA/TFAA is removed, the stable carbocation H_1/H_2 gave the key hub intermediate I as an isolable product which when treated with TFPAA in TFAA rearranged to give an identical yield ratio of 2 and 3 without formation of 4. The intermediacy of I was also supported by ab initio computation of the direct rearrangement of H_1/H_2 to J which had an energy barrier of 28 kcal/mol (B3LYP-31G*, Scheme 5). Such a large energy barrier would make the direct formation of J from H_1/H_2 an improbable process while also explaining why I can be isolated and characterized. Lastly, solutions of H_1/H_2 were also quenched by dimethylsulfide or water to form products I and 7, respectively (Scheme 6).

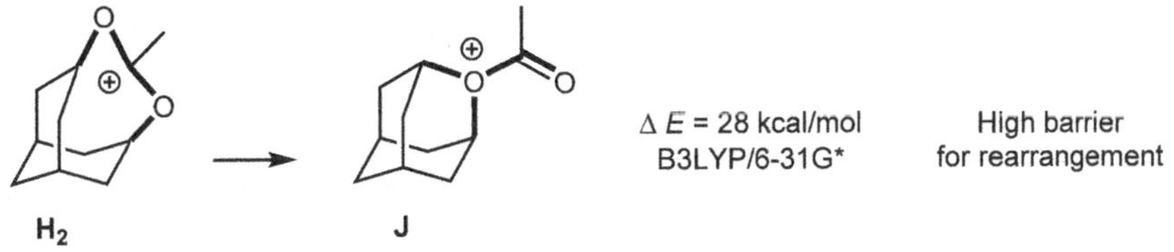

$\Delta E = 28$ kcal/mol
B3LYP/6-31G*

High barrier
for rearrangement

SCHEME 5

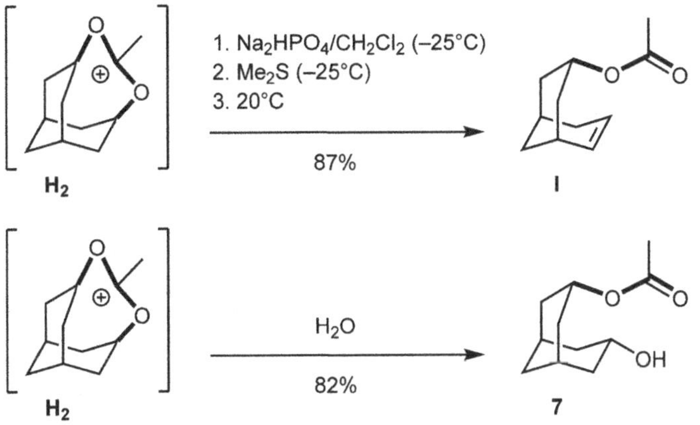

1. Na$_2$HPO$_4$/CH$_2$Cl$_2$ (−25°C)
2. Me$_2$S (−25°C)
3. 20°C

87%

H$_2$O

82%

SCHEME 6

Furthermore, the authors carried out ^{19}F and ^{1}H NMR spectroscopy of the reaction by-products and observed CH$_3$ and CF$_3$ singlets which matched acetyltrifluoroacetyl peroxide (Scheme 7). In further support of these last portions of the mechanisms for 1 to 2 and 1 to 3, the authors prepared acetyltrifluoroacetyl anhydride (which is eliminated during the J to 2 and L to 3 steps) by reacting acetic acid with trifluoroacetic anhydride. When acetyltrifluoroacetyl anhydride was reacted with TFPAA, the product was acetyltrifluoroacetyl peroxide as confirmed by ^{19}F NMR spectroscopy (Scheme 7).

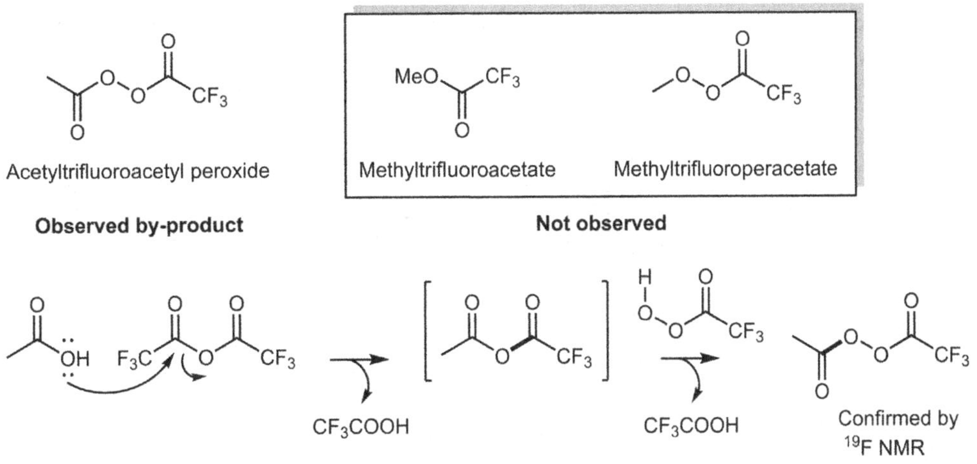

Acetyltrifluoroacetyl peroxide

Observed by-product

Methyltrifluoroacetate

Methyltrifluoroperacetate

Not observed

CF$_3$COOH

CF$_3$COOH

Confirmed by
^{19}F NMR

SCHEME 7

The fact that methyltrifluoroacetate and methyltrifluoroperacetate were not observed as by-products also rules out formation of a theoretical triple oxygen insertion intermediate X that could rearrange to form product 2 (Scheme 8).

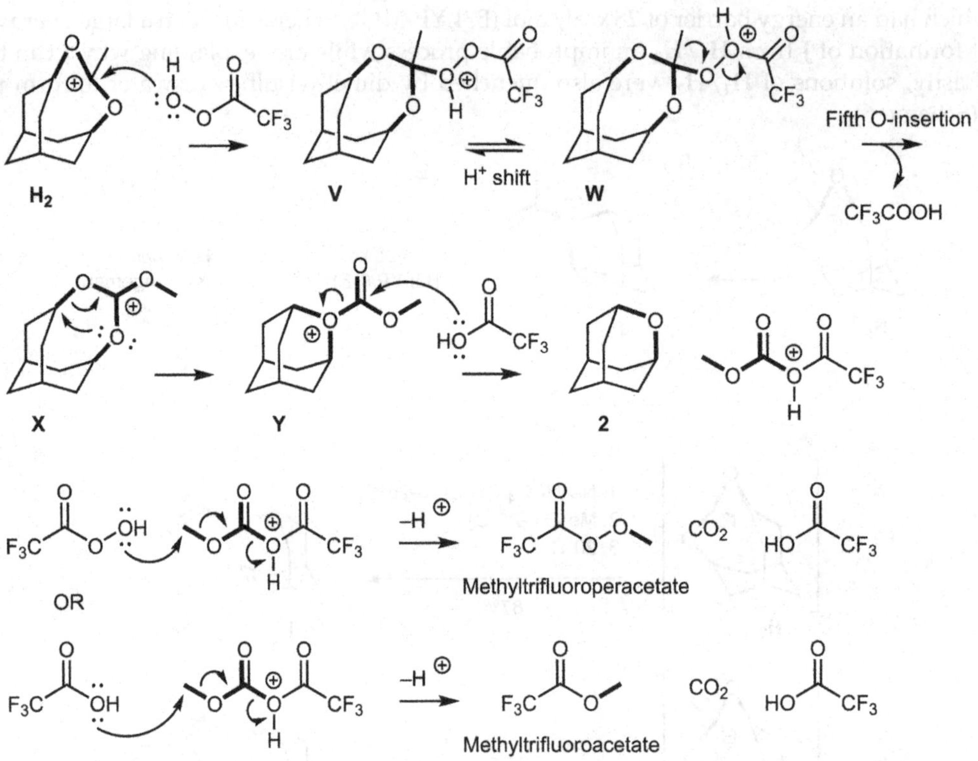

SCHEME 8

Furthermore, we note that acetyltrifluoroacetyl peroxide by-product could also directly arise from displacement of acetyl group in **J** and **L** by nucleophilic attack via TFPAA (as opposed to TFAA which we show in the solutions to **2** and **3**). The authors also argue that the polar acidic medium of the reaction mixture helps to stabilize H_1/H_2 carbocation thus lowering the possibility of *ortho* carbonate formation (via **V**) followed by a third O insertion to give **X**.

Interestingly, given their understanding of the reaction mechanism to products **2** and **3** through the key hub intermediate **I**, the authors also sought a way to achieve synthetic selectivity toward either product **2** or **3**. Such an effort must necessarily intercept the mechanism before the bifurcation of **I** to **J** or **K**. This was achieved by quenching the carbocation intermediate H_1/H_2 either with water (which leads to **2**, see Scheme 9) or with Na_2HPO_4 and Me_2S (which leads to **3**, Scheme 10). Furthermore, in H_2SO_4 intermediates **I** and **AA** both led to **2**. The formation of **AA** was also confirmed by X-ray analysis.

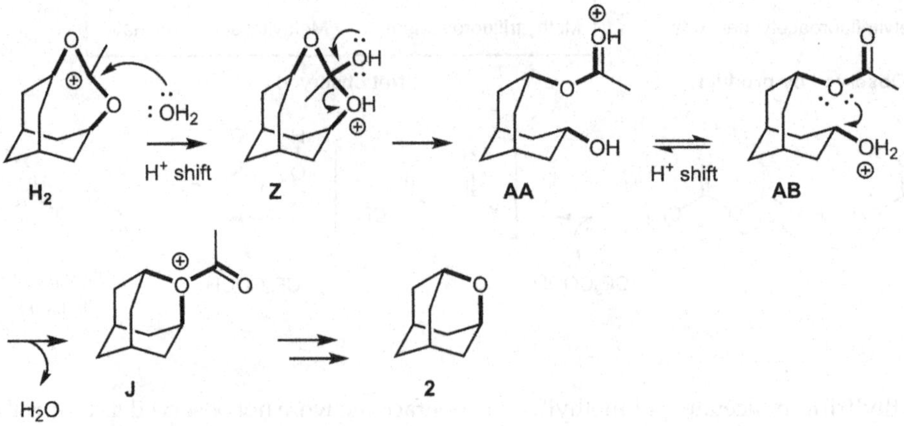

SCHEME 9

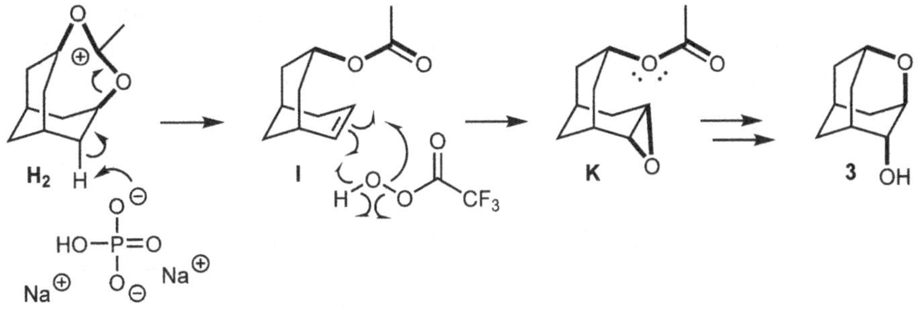

SCHEME 10

It thus appears that acidic aqueous conditions promote the formation of **2** by avoiding **I** whereas slightly basic conditions on account of added phosphates promotes deprotonation of **H₂** to **I** followed by epoxidation with TFPAA rather than cyclization to **J** (where TFPAA normally acts as an acid and not an epoxidation reagent). The stabilization of carbocation **H₂** has thus proven to be of significant synthetic importance in this overall transformation. To this end, the authors also computed ^{13}C NMR chemical shifts for B3LYP/6-31G* optimized geometries of **H₂** at the level of GIAO-B3LYP/6-311+G(2d,p).[1] These were in agreement with experimental results thus further supporting structure **H₂**.

In their last contribution, the authors sought to understand the reasons for the stability of carbocation **H₂**. They thus undertook theoretical calculations (B3LYP/6-31G*) of the strain destabilization energies (DE) of adamantane **8**, its single O insertion analog **9**, and its double O insertion analog **10** (see Scheme 11). The oxygen stabilization energies (SE) of the methyl substituted carbocation analogs **9′**, **10′** and the triple O-insertion product **11′** were also computed (see Scheme 11). It was thus shown that consecutive oxygen insertion into the cage structure of adamantane contributes little to the strain destabilization energy due to ring enlargement (see ref. 29 in Ref. 1) whereas stabilization of a carbocation by electron delocalization from a neighboring oxygen atom plays a more significant role. Thus, although **H₂** (i.e., **10′**) is destabilized from the point of view of ring strain ($\Delta DE_1 + \Delta DE_2 = 10.7$ kcal/mol), it is significantly stabilized from the point of view of electronics ($SE_1 + SE_2 = 33.8$ kcal/mol). This amounts to an overall stabilization of 33.8 kcal/mol − 10.7 kcal/mol = 23.1 kcal/mol which explains why **H₂** is stable for 24 h at −25°C. Interestingly, this theoretical analysis also shows that consecutive O insertion rapidly reduces the effect of electronic stabilization of the subsequent carbocation intermediates. In other words, the first O insertion provides the largest effect on carbocation stabilization whereas the third O insertion provides the smallest effect. This helps explain why intermediate **X** is not a likely candidate for an intermediate along the path from **1** to **2**.

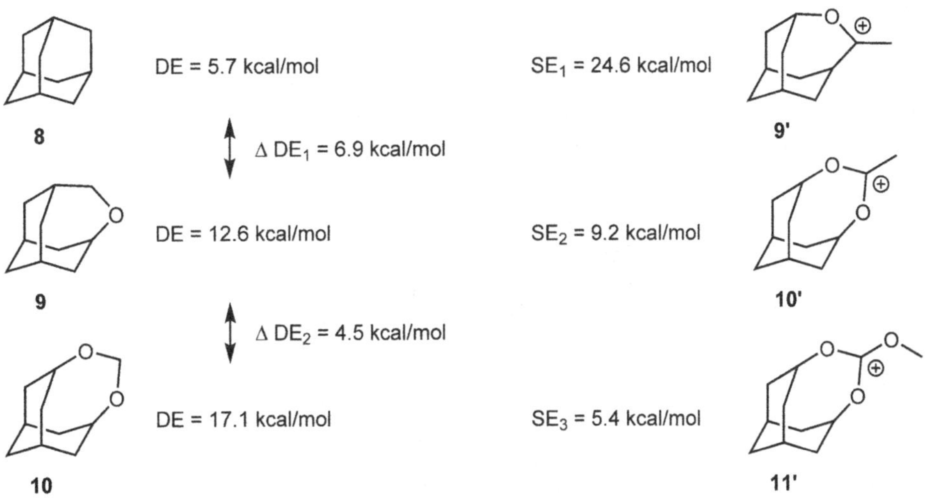

SCHEME 11

Last, we note that products **2**, **3**, and **4** were all identified by column chromatography and by comparing IR, NMR, and melting point data with those of authentic samples. Furthermore, product **2** did not react with TFPAA to form **3** thus showing that it was not a precursor to **3**. In addition, the low yield reported for product **4** indicated that the migratory aptitude of the methyl group from C_2 to the oxygen atom leading to product **4** was 24 times less than the migratory aptitude of the C_1—C_2 sigma bond leading to products **2** and **3**. This value is found directly from the product distribution data: $(56 + 40)/4 = 24$. Although the authors suggest ketone **P** as a key intermediate to **4** in the context of the Baeyer-Villiger oxidation, they do not explain that a mechanistic divergence from **D** is responsible for the formation of **P**. Instead they suggest hemiketal intermediates in the context of slightly different transformations. We do not believe a hemiketal intermediate exists along the path from **1** to **4**. We thus conclude our analysis by reemphasizing that the 2000 article by Krasutsky et al. represents the best case we have encountered in this entire work with regard to combining the best aspects of experimental and theoretical work to understand and support a proposed reaction mechanism.

ADDITIONAL RESOURCES

We recommend two articles which discuss interesting transformations and their mechanisms in the context of compounds containing the adamantane core structure.[6,7]

References

1. Krasutsky PA, Kolomitsyn IV, Kiprof P, Carlson RM, Fokin AA. Observation of a stable carbocation in a consecutive Criegee rearrangement with trifluoroperacetic acid. *J Org Chem*. 2000;65:3926–3933. https://doi.org/10.1021/jo991745u.
2. Krasutsky PA, Kolomitsin IV, Carlson RM, Jones Jr. M. A new one-step method for oxaadamantane synthesis. *Tetrahedron Lett*. 1996;37(32):5673–5674. https://doi.org/10.1016/0040-4039(96)01202-6.
3. Criegee R. Über ein krystallisiertes Dekalinperoxyd. *Ber Dtsch Chem Ges*. 1944;77(1):22–24. https://doi.org/10.1002/cber.19440770106.
4. Wang Z, ed. *Comprehensive Organic Name Reactions and Reagents*. New York, NY: John Wiley & Sons, Inc.; 2010:770–774. https://doi.org/10.1002/9780470638859.conrr170
5. Wang Z, ed. *Comprehensive Organic Name Reactions and Reagents*. New York, NY: John Wiley & Sons, Inc.; 2010:150–155. https://doi.org/10.1002/9780470638859.conrr36
6. Obitsu K, Maki S, Niwa H, Hirano T, Ohashi M. Abnormal reactivity of anisatin and neoanisatin to samarium iodide-hexamethylphosphorictriamide. *Tetrahedron Lett*. 1997;38(23):4111–4112. https://doi.org/10.1016/S0040-4039(97)00836-8.
7. Djaidi D, Leung ISH, Bishop R, Craig DC, Scudder ML. Ritter reactions. Part 14. Rearrangement of 3,3,7,7-tetramethyl-6-methylidenebicyclo[3.3.1]nonan-2-one. *J Chem Soc Perkin Trans 1*. 2000;2000:2037–2042. https://doi.org/10.1039/B002544P.

Question 62: A Pyrimidine Rearrangement

In 1996 chemists from India showed that treatment of 6-chloro-2,4-dimethoxypyrimidine **1** with sodium iodide in refluxing DMF gives a 43% yield of the 1,3-dimethyl-6-iodouracil product **3**[1] rather than the previously reported[2] product **2** (Scheme 1). Provide a mechanism for the transformation of **1** to **3**.

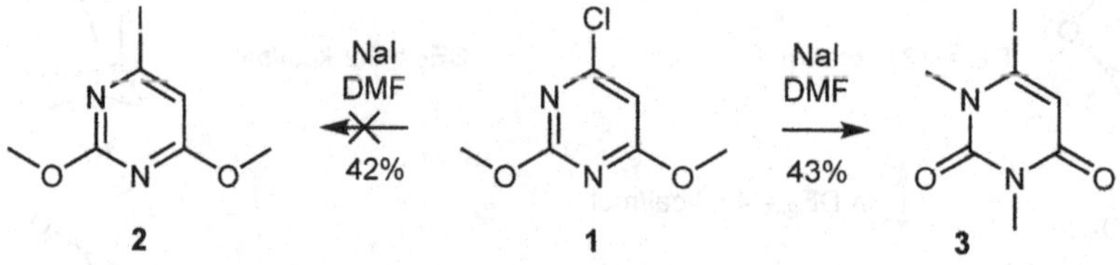

SCHEME 1

SOLUTION

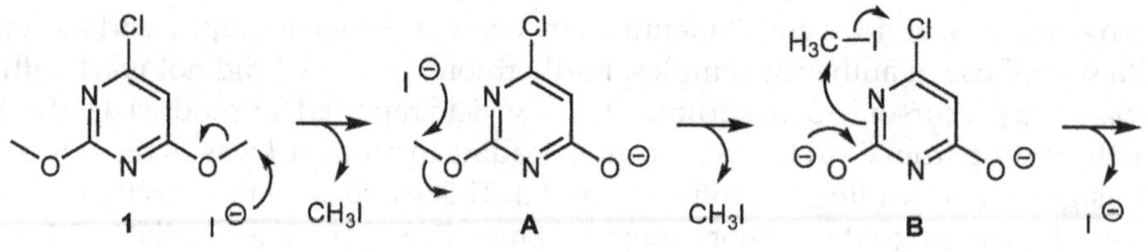

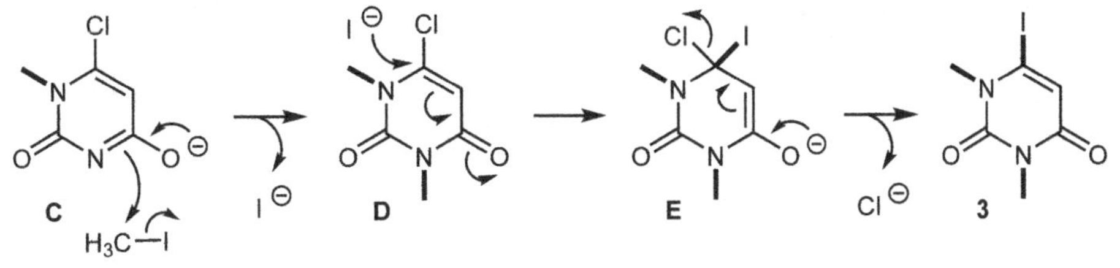

BALANCED CHEMICAL EQUATION

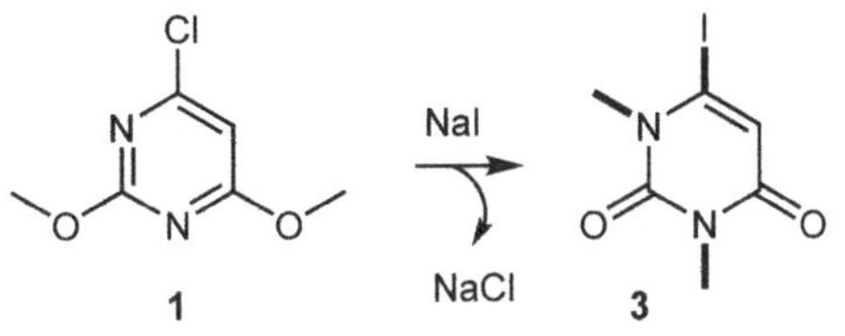

KEY STEPS EXPLAINED

The mechanism proposed by the authors to explain the formation of **3** from **1** begins with two sequential demethylation steps of pyrimidine **1** by iodide ion which leads to intermediate **B** and two equivalents of iodomethane as by-product. Intermediate **B** then undergoes methylation at both nitrogen atoms to form intermediate **D**. Iodide ion adds in a Michael fashion at C_6 which bears the chlorine atom to form **E** which then eliminates chloride ion resulting in product **3**. Overall the reaction involves rearrangement of the two methyl groups in **1** (bonded to oxygen atoms) which become bonded to nitrogen atoms in product **3** after the final halogen substitution.

We note that the authors corrected a previous determination[1] of the product structure by characterizing its IR and ^{1}H NMR spectra.[2] The IR spectrum showed two carbonyl bands at 1710 and 1650 cm^{-1} and the ^{1}H NMR showed two methyl singlets at 3.28 and 3.65 ppm and a singlet at 6.45 ppm for the C_5 hydrogen atom. Furthermore, the authors synthesized product **3** by a second method shown in Scheme 2 starting from 6-iodouracil which yielded a product with identical melting point and spectral properties as before. In this reaction no rearrangement takes place because we have a simple double *N*-methylation via two equivalents of iodomethane. We do, however, note that the authors did not carry out additional experimental or theoretical analyses to help support their postulated mechanism for the formation of **3** from **1**.

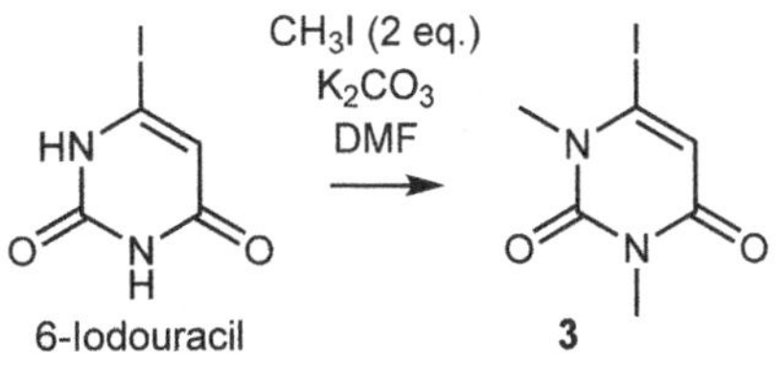

SCHEME 2

ADDITIONAL RESOURCES

We recommend three articles which describe further interesting rearrangements of pyrimidine compounds.[3–5]

References

1. Horwitz JP, Tomson AJ. Some 6-substituted uracils. *J Org Chem.* 1961;26(9):3392–3395. https://doi.org/10.1021/jo01067a091.
2. Das P, Kundu NG. Preparation of 6-iodo-N_1,N_3-dimethyluracil and 6-(2-acylvinyl)-N_1,N_3-dimethyluracils: the correction of a mistake and an improved synthesis. *J Chem Res (S).* 1996;1996:298–299.

3. Goenczi C, Swistun Z, van der Plas HC. Novel pyrimidine rearrangement reaction. A new synthesis of 4-formylimidazoles. *J Org Chem.* 1981;46 (3):608–610. https://doi.org/10.1021/jo00316a024.
4. Danagulyan GG, Mkrtchyan AD. Recyclization of 5-ethoxycarbonylpyrimidines occurring with substitution of a carbon atom in the heterocycles by an exocyclic carbon atom. *Chem Heterocycl Compd.* 2003;39(11):1529–1531. https://doi.org/10.1023/B:COHC.0000014422.20094.d1.
5. Guillard J, Goujon F, Badol P, Poullain D. New synthetic route to diaminonitropyrazoles as precursors of energetic materials. *Tetrahedron Lett.* 2003;44:5943–5945. https://doi.org/10.1016/S0040-4039(03)01301-7.

Question 63: 1,4-Thiazin-3-ones From 1,4-Oxathiins

In 1996 South Korean chemists showed that brief heating of 1,4-oxathiin carboxamides **1** with concentrated hydrochloric acid in acetonitrile at 80°C results in excellent yields of the 1,4-thiazin-3-one products **2** (Scheme 1).[1] Suggest a mechanism for this transformation.

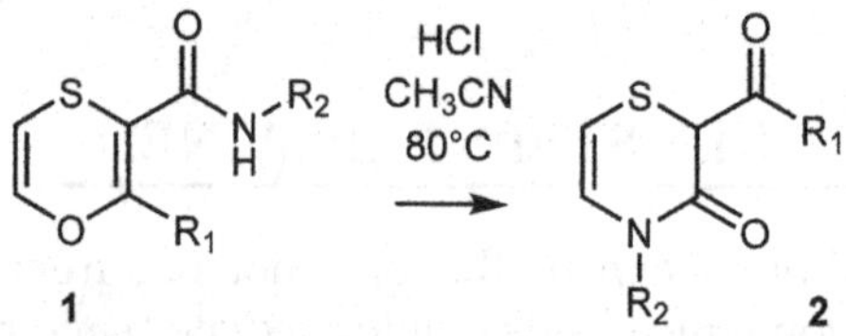

SCHEME 1

SOLUTION

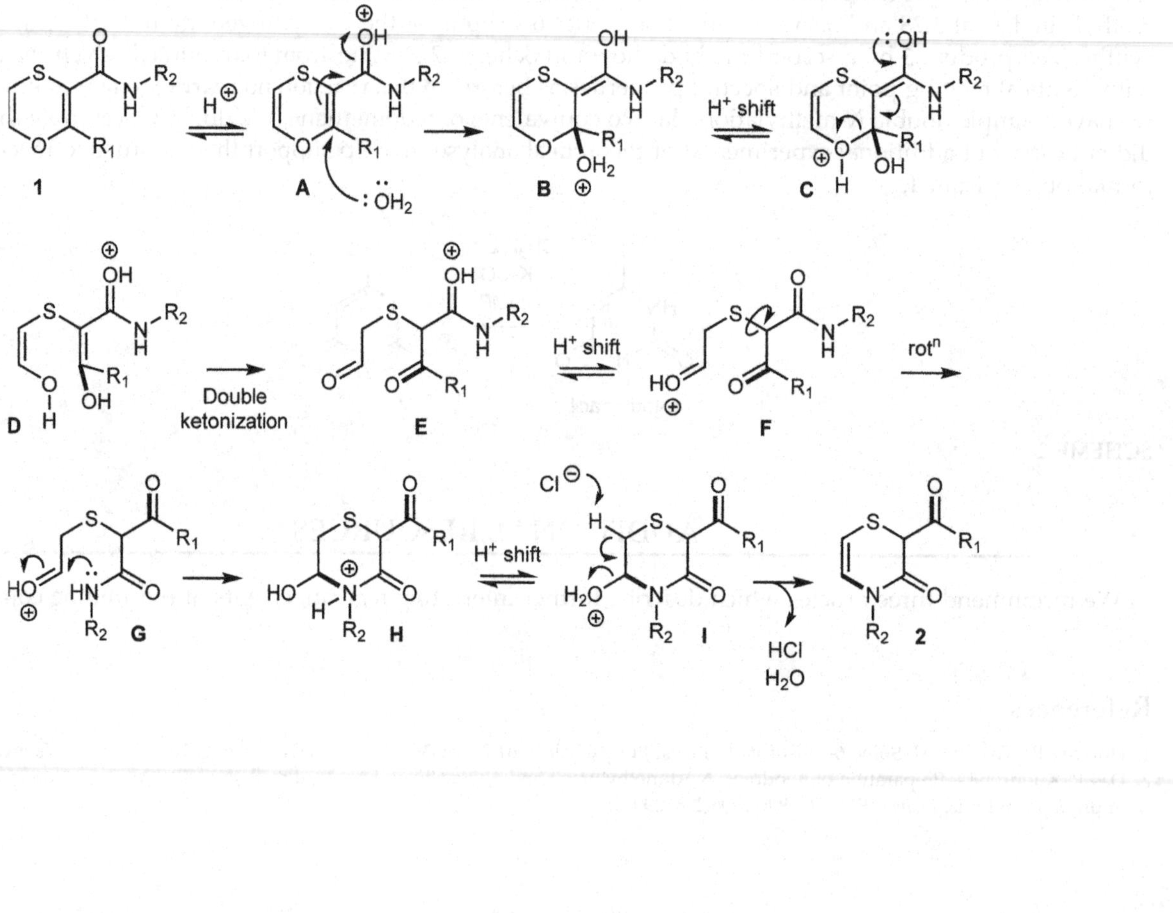

BALANCED CHEMICAL EQUATION

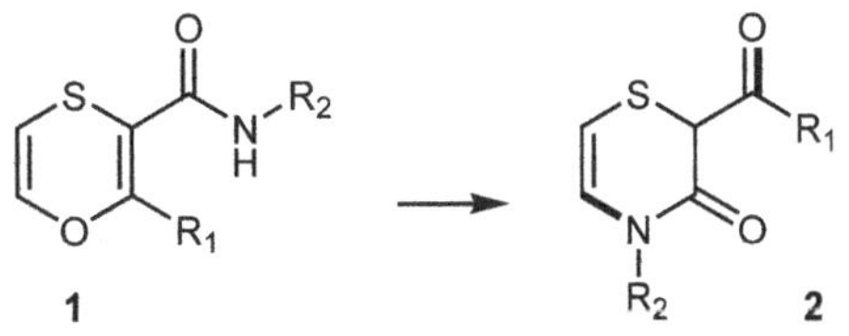

KEY STEPS EXPLAINED

We first note that the transformation of **1** to **2** constitutes a net rearrangement reaction. The reaction begins with protonation of the amide carbonyl group of compound **1** which forms **A** which is attacked by water at the vinyl ether carbon atom to form the hemiacetal intermediate **C** after proton shift from **B**. The enol group of **C** then initiates fragmentation of the 2,3-dihydro-[1,4]oxathiine ring leading to the bis-enol intermediate **D** which ketonizes twice to form intermediate **E**. Intramolecular proton transfer transforms **E** to **F** which after bond rotation cyclizes to the aminal **H**. A final dehydration leads to the 1,4-thiazin-3-one product **2**. The authors noted that cyclization of intermediate **F/G** could also have occurred via another pathway as shown in Scheme 2. The resulting oxygen ring-containing products **3** and **4** were not observed and therefore this pathway was ruled out. We also note that the authors did not perform additional experimental or theoretical analyses to help support their postulated mechanism.

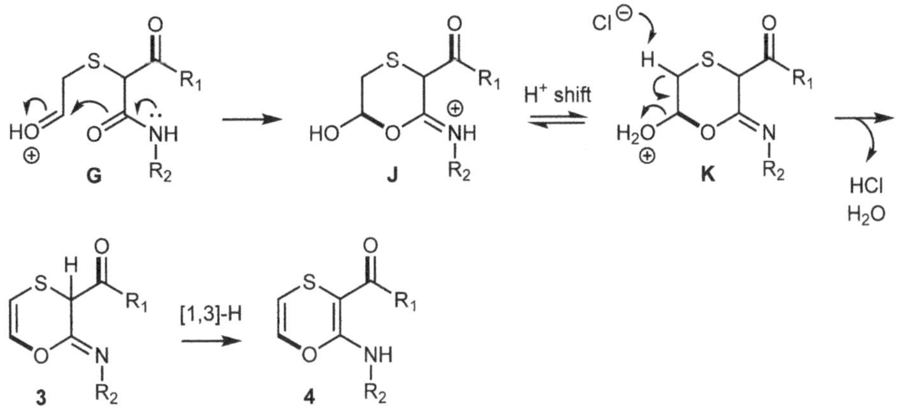

SCHEME 2

ADDITIONAL RESOURCES

We recommend three articles which describe both ionic and radical-based mechanisms to the formation of oxathiine and thiazine compounds.[2-4]

References

1. Hahn H-G, Nam KD, Mah H, Lee JJ. A new synthesis of 1,4-thiazin-3-ones by a novel rearrangement of 1,4-oxathiins. *J Org Chem.* 1996;61 (11):3894–3896. https://doi.org/10.1021/jo951723h.
2. Zhang W, Pugh G. Mechanistic evidence for a novel rearrangement sequence in the synthesis of 4-aryl-5,6-dihydro-1,2-oxathiine-2,2-dioxides from homopropargyl benzosulfonates. *Synlett.* 2002;2002(5):778–780. https://doi.org/10.1055/s-2002-25367.
3. Anwar B, Grimsey P, Hemming K, Krajniewski M, Loukou C. A thiazine-S-oxide, Staudinger:aza-Wittig based synthesis of benzodiazepines and benzothiadiazepines. *Tetrahedron Lett.* 2000;41:10107–10110. https://doi.org/10.1016/S0040-4039(00)01797-4.
4. Pradhan TK, Mukherjee C, Kamila S, De A. Application of directed metalation in synthesis. Part 6: a novel anionic rearrangement under directed metalation conditions leading to heteroannulation. *Tetrahedron.* 2004;60:5215–5224. https://doi.org/10.1016/j.tet.2004.04.033.

Question 64: An Unusual Route to 3-Acylfurans

In 1996 chemists from Thailand demonstrated an unusual two-step route to 3-acylfurans which consisted of a first reaction between the stannylated cyclopropane derivative **1** and acid chlorides in refluxing toluene for 5 h which gives dihydrofurans **2** in 64%–83% yield (Scheme 1).[1] When these products are treated with 1.2 equivalents of BF₃•OEt₂ in dichloromethane at −78°C followed by slow warming to room temperature, the 3-acylfurans **3** are isolated in 30%–62% yield. Provide mechanisms to explain the formation of products **2** and **3**.

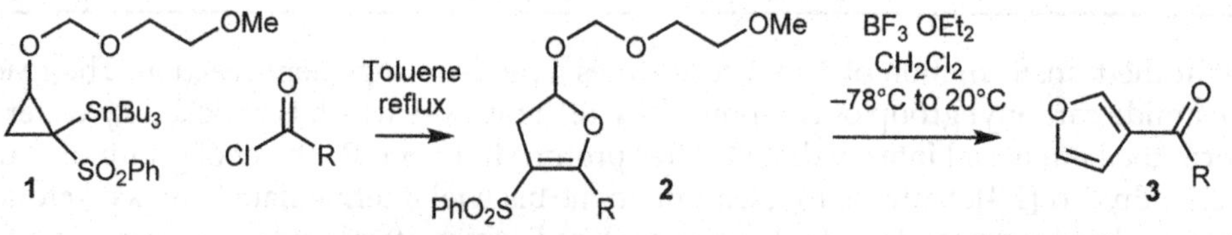

SCHEME 1

SOLUTION: 1 TO 2

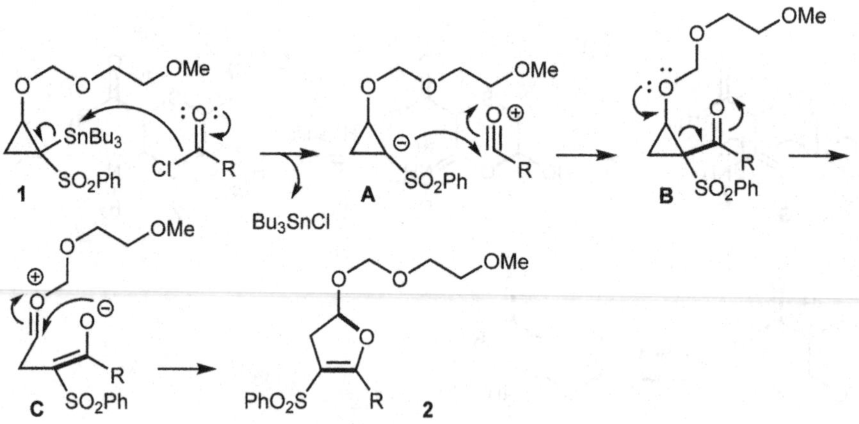

SOLUTION: 2 TO 3

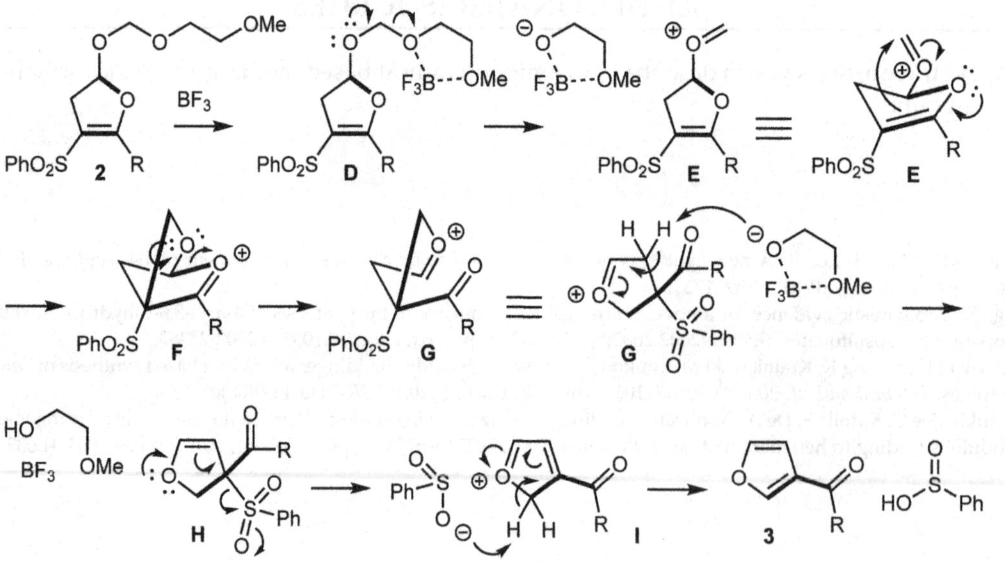

BALANCED CHEMICAL EQUATIONS

KEY STEPS EXPLAINED

The mechanism proposed by the authors for the formation of **2** begins with dissociation of the acid chloride into chloride ion and an acylium ion. Chloride ion then nucleophilically attacks the tin atom in cyclopropane **1** which results in a carbanion that captures the acylium ion yielding product **2** after ring opening of the cyclopropane to form **C** followed by ring closure of **C** to form **2**.

In the case of the formation of product **3** from **2**, Lewis acid boron trifluoride coordinates to two oxygen atoms of the methoxyethoxymethyl (MEM) group of **2** which facilitates cleavage of the central C—O bond yielding oxonium ion **E** which undergoes an intramolecular Prins-type reaction to form oxonium ion **F** which ring opens to form oxonium ion **G**. The alkoxide liberated from the MEM group deprotonates **G** to form **H** which forms the oxonium ion **I** after oxygen-mediated fragmentation with elimination of phenylsulfinate anion. The phenylsulfinate anion then causes deprotonation of **I** to form the furan product **3**. The authors note that oxonium ion **E** can be transformed directly to oxonium ion **G** via a [3,3] sigmatropic rearrangement. Nevertheless, they did not carry out further experimental and theoretical analyses to help support their proposed mechanisms. We also note that the order of steps is variable from intermediate **G** to product **3** with respect to when the phenylsulfinate group is eliminated.

ADDITIONAL RESOURCES

We recommend an article which discusses a similar transformation but in the analogous sulfur ring system.[2]

References

1. Pohmakotr M, Takampon A. Destannylative acylation of 1-[(2-methoxyethoxy)methoxy]-2-(phenylsulfonyl)-2-(tributylstannyl)cyclopropane: a novel route to 3-acylfurans. *Tetrahedron Lett.* 1996;37(26):4585–4588. https://doi.org/10.1016/0040-4039(96)00852-0.
2. Kraus GA, Roth B. Rearrangements of acyloxyfurans and thiophenes. *J Org Chem.* 1978;43(10):2072–2073. https://doi.org/10.1021/jo00404a057.

Question 65: Reaction of PQQ With L-Tryptophan

In 1996 Japanese chemists investigated the in vitro reaction of pyrroloquinolinequinone **1** (PQQ) and L-tryptophan **2** which generated the imidazolopyrroloquinoline **3** (major product) together with its indolyl **4** and indolylmethyl **5** derivatives as minor products (Scheme 1).[1] Suggest mechanisms for the formations of **3**, **4**, and **5**.

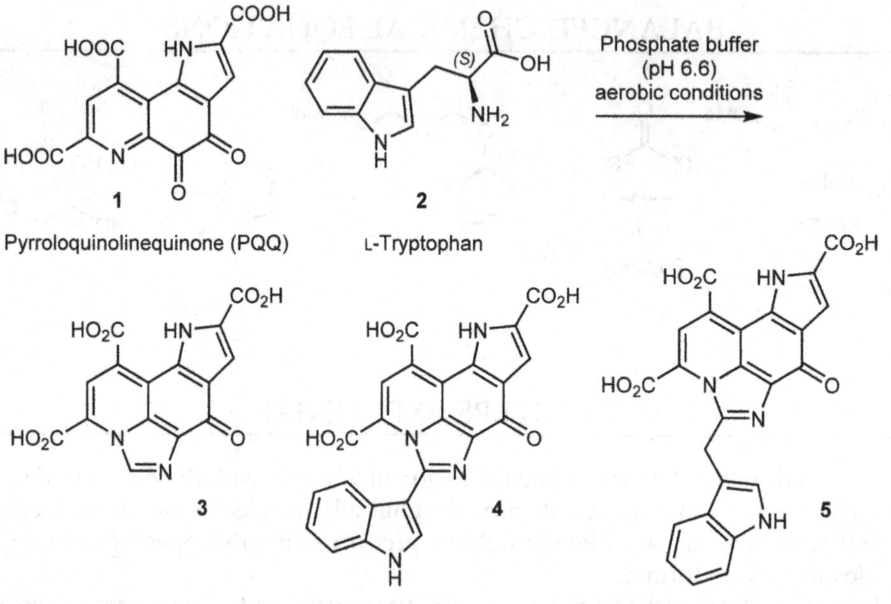

Pyrroloquinolinequinone (PQQ)   L-Tryptophan

SCHEME 1

SOLUTION

To explain the formation of products **3**, **4**, and **5** from the reaction of **1** and **2** in phosphate buffer under aerobic conditions, Kawamoto et al. proposed a mechanism whose road map is depicted in Scheme 2. Essentially, the authors' road map consists of intertwined mechanistic pathways which begin with the formation of a key carbinolamine intermediate **A** (path *a*) that can fragment in two possible ways. The first path (path *aa*) involves decarboxylation and ring closure to give the key intermediate **K** which can either eliminate 3-methyl-indole (i.e., skatole) via elimination of anion **M** to form **3** or **K** can undergo oxidation via PQQ or O_2 to form **5**. The second path from **A** (path *ab*) involves fragmentation to give $PQQH_2$, intermediate **B**, and 2-iminoacetic acid **C** as products. This last product **C** is expected to react with water to give glyoxylic acid and ammonia. The authors contend that the ammonia thus formed would react with **1** (PQQ) (path *b*) to generate the iminoquinone **P** which can react with carbanion **M** generated previously to form product **4** in an ionic-type mechanism. In the following schemes, we shall present the complete mechanisms that the authors proposed and their balanced chemical equations. After explaining these mechanisms and their argumentative basis, we shall identify several issues in the authors' overall mechanistic road map and address them by proposing alternative mechanisms and an alternative mechanistic road map.

We thus begin with the authors' mechanistic proposal for paths *a* and *ab* (Scheme 3) and the balanced chemical equation for this sequence (Scheme 4). Essentially Kawamoto et al. proposed that PQQ reacts with L-Trp to form a carbinolamine intermediate **A** as a first step. Next, base-promoted deprotonation of the amino hydrogen atom of indole in **A** causes fragmentation of **A** to **B**, 2-iminoacetic acid **C** and **D** ($PQQH_2$) after proton transfer. Two equivalents of $PQQH_2$ are thus expected to react with O_2 to regenerate PQQ and produce two equivalents of water. Product **C** is expected to react with water to form glyoxylic acid **G** and ammonia as by-products.

The authors called this portion of their mechanism a PQQ catalysis sequence which is thought to be general for PQQ reactions with amino acids according to the work of Itoh et al.[2] Interestingly, Itoh et al. did not identify products **3**, **4**, and **5** from the reaction of PQQ with L-Trp and instead focused on basic amino acids such as L-Gly in their more detailed mechanistic analysis. It is important to note that neither Itoh et al. nor Kawamoto et al. present appropriate spectroscopic data identifying key proposed intermediates. For example, Kawamoto et al. did not present any evidence for **B**, **C**, glyoxylic acid **G**, or ammonia by-products. Moreover, we regard the proposed base deprotonation of the indole amino H atom of **A** to be problematic. This is because the only base present in the reaction medium is phosphate buffer which has a pK_A of 7 whereas the indole amino H atom has a pK_A of 16.8.[3] It is thus unrealistic to suggest that phosphate buffer has the base strength needed to deprotonate the indole amino hydrogen atom as the authors suggest. Furthermore, we note that product **B** is thermodynamically unstable. A literature search revealed that **B** is formed only under strenuous pyrolysis conditions,[4] in vivo chemo-enzymatic conditions,[5] or is expected to

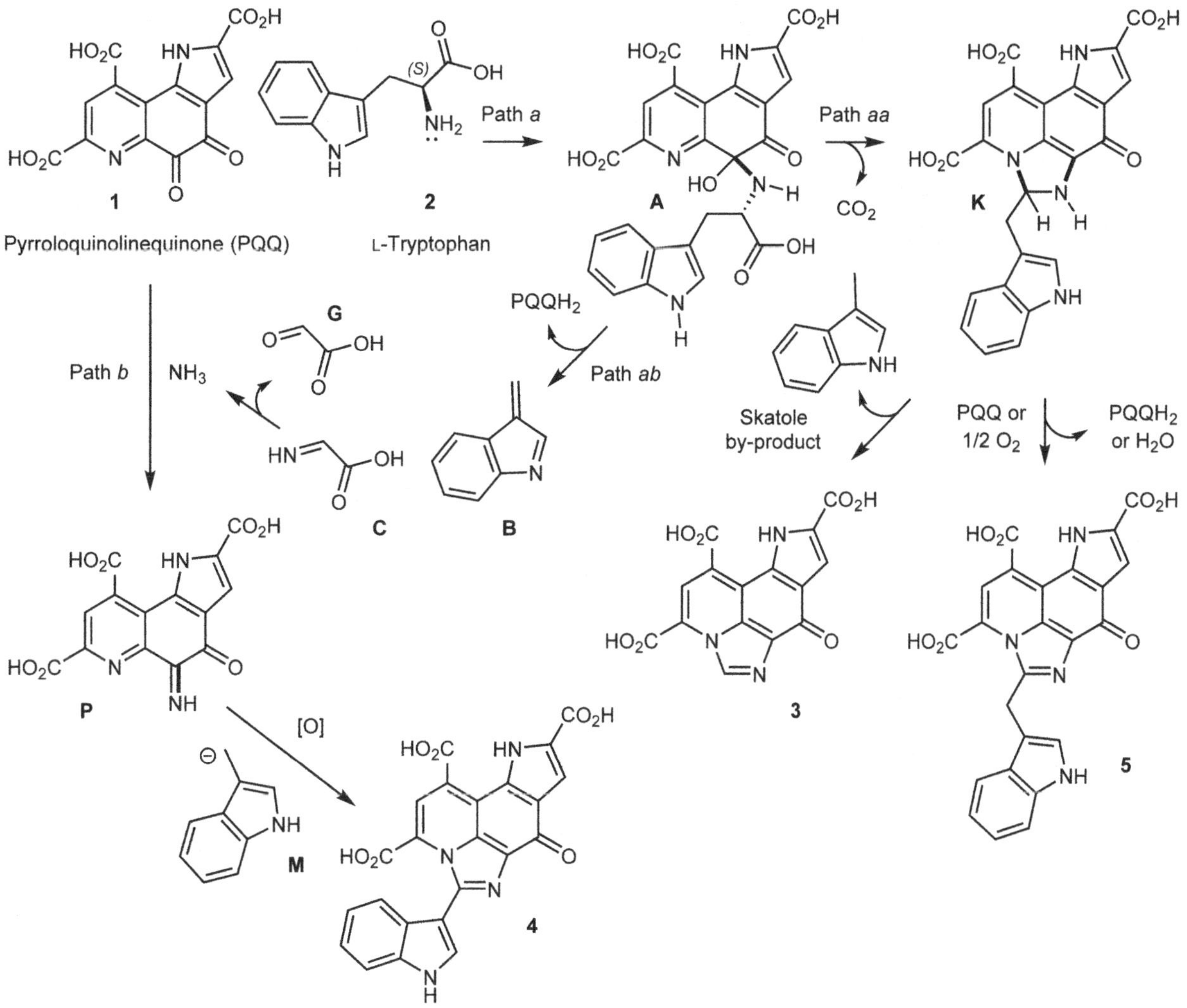

SCHEME 2 Authors' mechanistic road map.

dimerize quickly so that it cannot be easily detected using NMR spectrometry.[6] Interestingly, Barcock et al. were able to trap intermediate **B** through reaction with thiophenol to form the thiophenol derivative **6** (Scheme 5). Therefore, adding thiophenol to a mixture of **1** and **2** should give **6** if the authors' proposed paths *a* and *ab* are operational. In the absence of such an experiment and given the issues raised so far, we believe that the authors' proposed paths *a* and *ab* are invalid. We believe the authors offered this mechanism only to explain the formation of product **4** (see subsequent section) by invoking the production of ammonia for which no evidence is presented.

We next consider the authors' proposed pathway from **A** to **5** (Scheme 6) along with the balanced chemical equation for the transformation of **1** and **2** to **5** (Scheme 7). We note that this pathway (i.e., path *aa*) involves inactivation of PQQ. In this mechanism, **A** undergoes dehydration to form the iminoquinone intermediate **H** which undergoes decarboxylation to form the Schiff base intermediate **J**. This intermediate undergoes ring closure via attack by the pyridine nitrogen anion onto the imine carbon to form the key tetracyclic intermediate **K** which undergoes oxidation either via one PQQ or O_2 equivalent to form **5** and either the reduced PQQH$_2$ (**D**) or water, respectively. We note that the authors do not present evidence for this ionic-type mechanism because they rely on literature precedent[2] (see also references 5b, 7, and 13 in Ref. 1 cited here). In these articles, the formation of iminoquinone intermediates from the reaction of PQQ with amino acids is proposed as a viable pathway to explain the formation of tetracyclic oxazole products akin to **3**. In certain cases these intermediates have been observed using spectroscopic methods which is why we are in agreement with the authors' proposed mechanism for the formation of **5**.

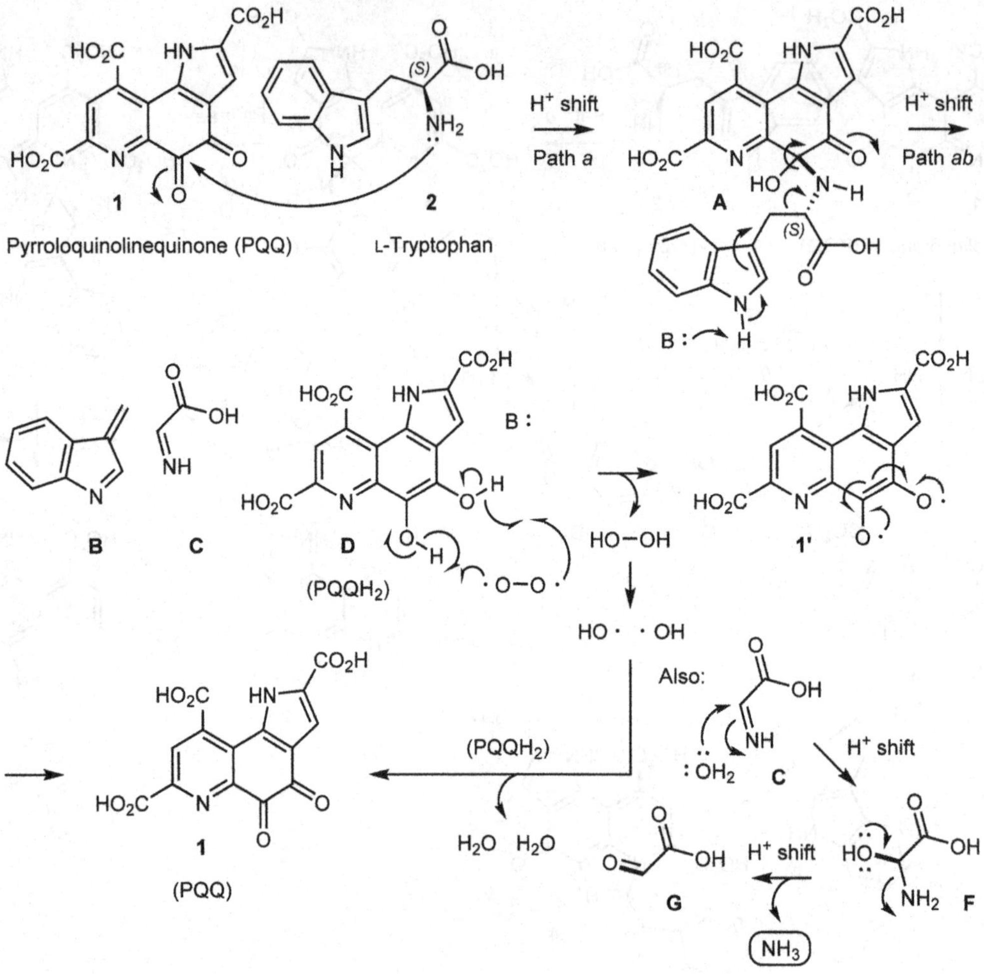

SCHEME 3 Authors' mechanistic proposal for paths *a* and *ab*.

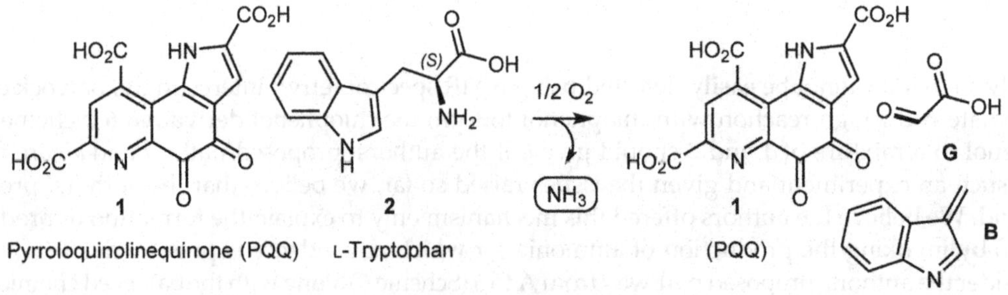

SCHEME 4 Balanced chemical equation for authors' mechanistic proposal for paths *a* and *ab*.

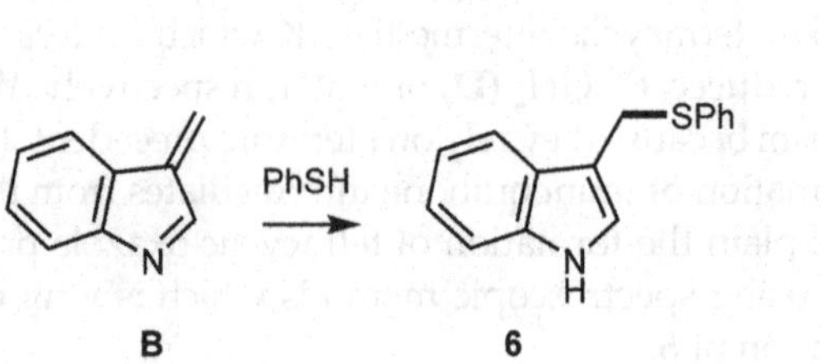

SCHEME 5 Possible reaction for trapping of intermediate **B**.

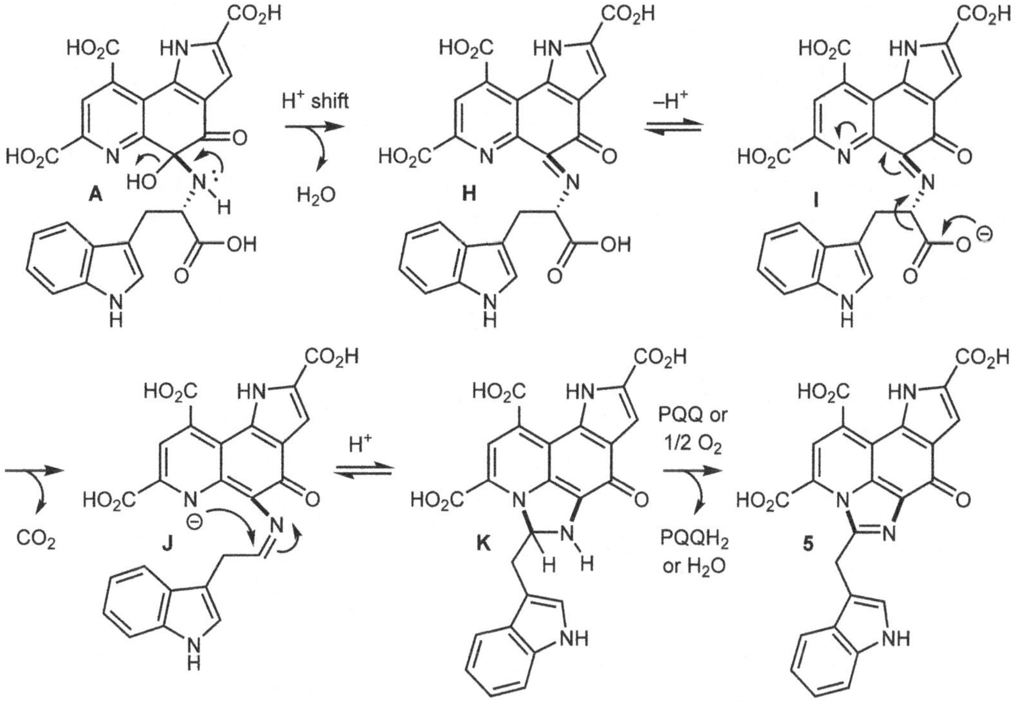

SCHEME 6 Authors' mechanistic proposal for **A** to **5**.

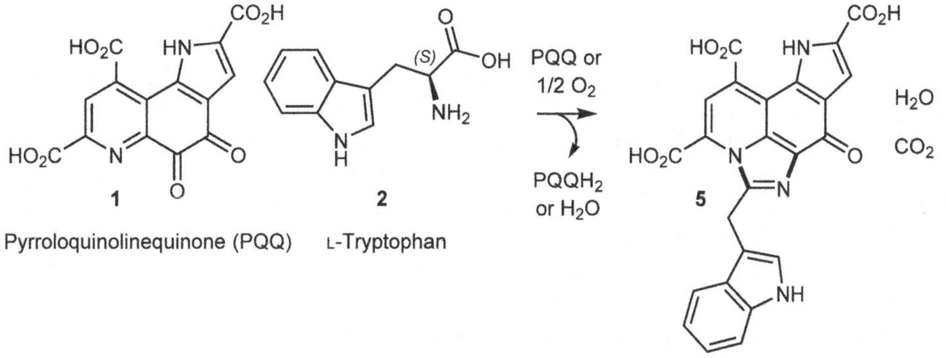

SCHEME 7 Balanced chemical equation for authors' mechanistic proposal for **1** and **2** to **5**.

Next, we show the proposed mechanism for the formation of **3** from **K** (Scheme 8) along with the balanced chemical equation of **1** and **2** to **3** (Scheme 9). Thus, nitrogen-mediated elimination of carbanion **M** gives **L** from **K**. We view the elimination of **M** as energetically unfavorable. One reason is that **M** is not stabilized by any electron-withdrawing group or by resonance via electron delocalization. Another reason concerns the reaction medium which constitutes a buffer system with a pH of 6.5. Given that the pK$_A$ of the indole H atom of skatole (**N**) is reported at 16.8,[3] we expect the pK$_A$ of the 3-methyl group of skatole to be even higher. **M** would therefore be quickly protonated to **N** and unavailable for later reaction with **P** to form **4** as the authors suggested. In other words, a strong enough base (which in this case does not exist) would have to deprotonate first the indole H atom of **N** followed by the 3-methyl group to give a species resembling **M** which can participate in the formation of **4**. Since this is impossible given the reaction conditions, the authors' proposed mechanism for the formation of **3** would result in formation of skatole **N**, which has a very distinct odor and for which the authors presented no evidence. We therefore disagree with the authors' proposed mechanism for the formation of **3**.

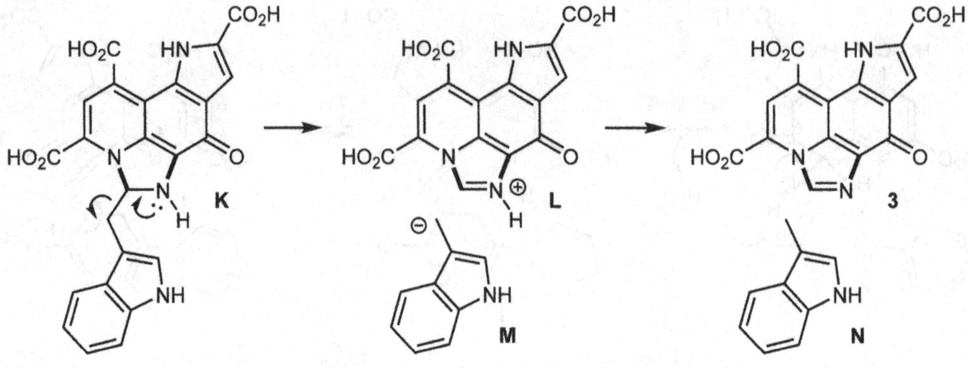

SCHEME 8 Authors' mechanistic proposal for **K** to **3**.

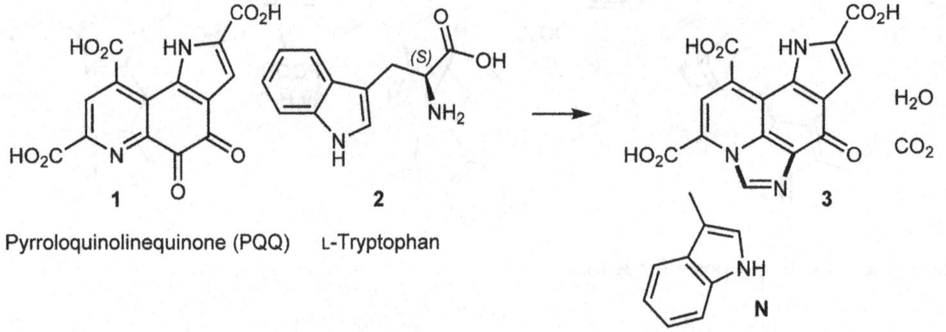

SCHEME 9 Balanced chemical equation for authors' mechanistic proposal for **K** to **3**.

Lastly, the mechanism proposed by the authors to explain the formation of **4** is highlighted in Scheme 10 along with its balanced chemical equation in Scheme 11. We note that based on the work of Itoh et al. the authors believed that ammonia is needed to convert PQQ to the iminoquinone **P** which can be attacked by the carbanion **M** to form amine intermediate **R**.[2] This structure undergoes oxidation by PQQ or O_2 to form imine **S** which then undergoes ring closure to generate **T** after protonation. A subsequent oxidation of **T** using PQQ or O_2 forms the final product **4**. Given our earlier arguments which contradict formation of ammonia and **M**, we uphold that this mechanism is not valid.

Given the reasons for doubting the authors' mechanisms for the formations of **3** and **4**, we here suggest alternative possibilities. These alternatives place greater emphasis on the oxidative reaction conditions one can expect from involvement of PQQ, especially with regard to the oxidation of amines.[7] We thus begin with the formation of **3** which we propose to occur by oxidative degradation of key intermediate **K** (Scheme 12). Essentially **K** is oxidized by PQQ or O_2 to form the conjugated substituted indole **U** rather than product **5**. Intermediate **U** undergoes partial oxidation (one hydrogen abstraction) via PQQ/O_2 to form radical **V** which is attacked by O_2 to generate radical **W**. This radical can undergo ring closure to form dioxetane radical **X** which undergoes retro [2 + 2] degradation (akin to ozonolysis) to give the delocalized radical **Y** along with 1*H*-indole-3-carbaldehyde by-product **Z**. Partial reduction of **Y** gives **AA** which tautomerizes to **BB** which is then reduced to **CC**. Dehydration of **CC** leads to product **3**. The balanced chemical equation for this mechanism is illustrated in Scheme 13. Similarly, we propose an alternative mechanism to explain the formation of **4** starting from the key intermediate **W** (Scheme 14). Aside from forming dioxetane **X**, intermediate **W** can also cyclize to form radical **DD** which undergoes a 1,2-migration of the indole group to form radical **EE**. Fragmentation of this radical species gives the aldehyde radical **FF** which abstracts a proton from a partially reduced PQQH or OH radical to form the aldehyde hydroxylamine intermediate **GG**. Water-mediated hydrolysis of this intermediate gives the final product **4** along with formic acid by-product. We represent the balanced chemical equation for this mechanism in Scheme 15.

We also offer a simplified road map for the alternative mechanistic proposals in Scheme 16. We note that the authors did not provide yields of products **3**, **4**, and **5**. The only thing provided is a product ratio of 120:1:2 for **3**:**4**:**5**, respectively. This large difference in yield between **3** and products **4** and **5** is better explained by the alternative mechanism since **5** can undergo oxidative degradation to form **W** which can lead to **3** or **4**. A very worthwhile endeavor would be

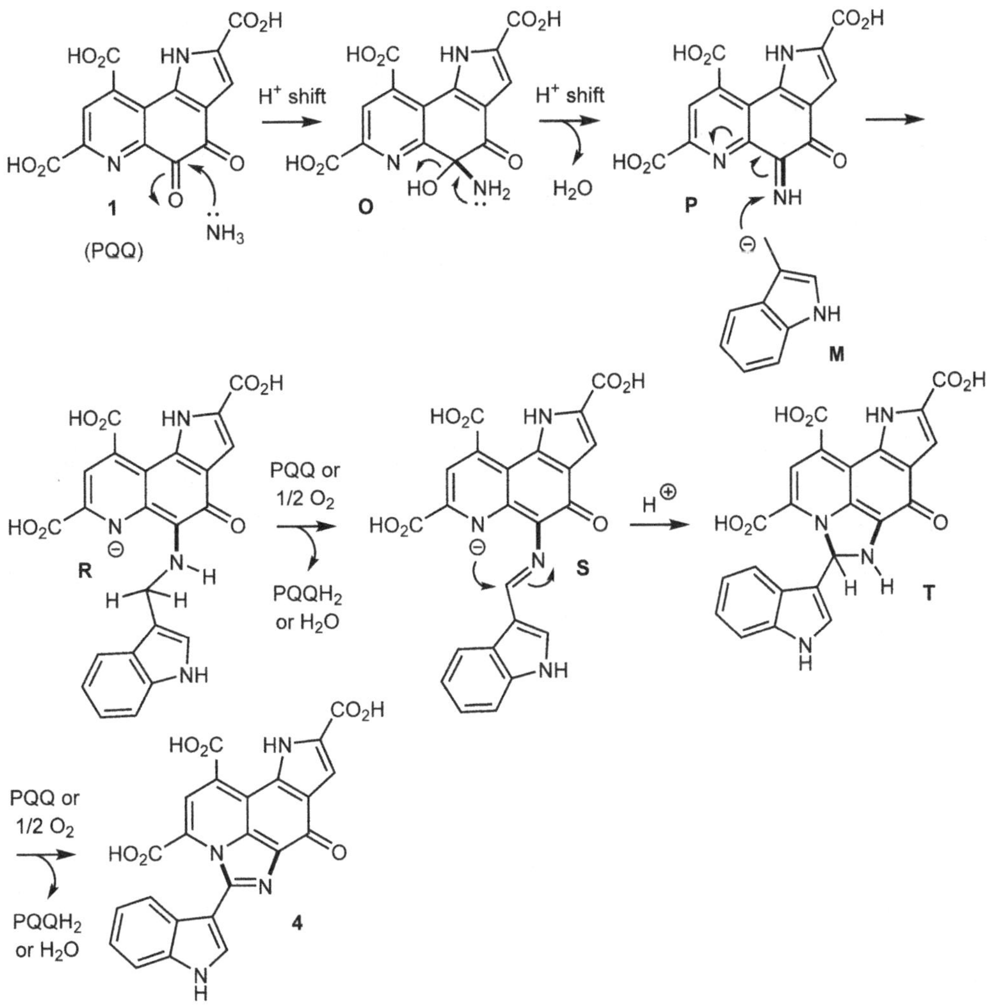

SCHEME 10 Authors' mechanistic proposal for **1** and **2** to **4**.

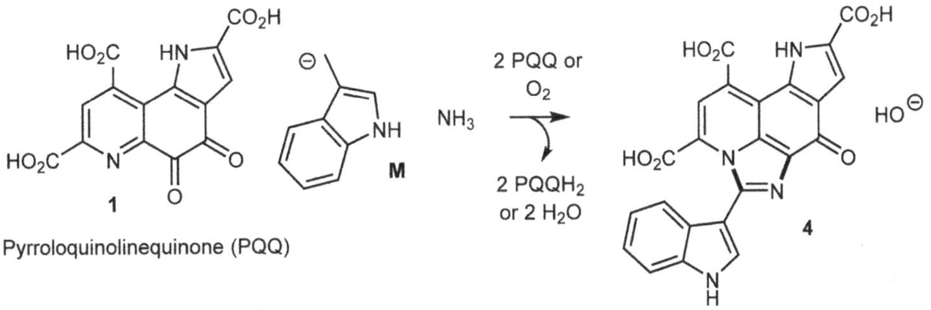

SCHEME 11 Balanced chemical equation for authors' mechanistic proposal for **1** and **2** to **4**.

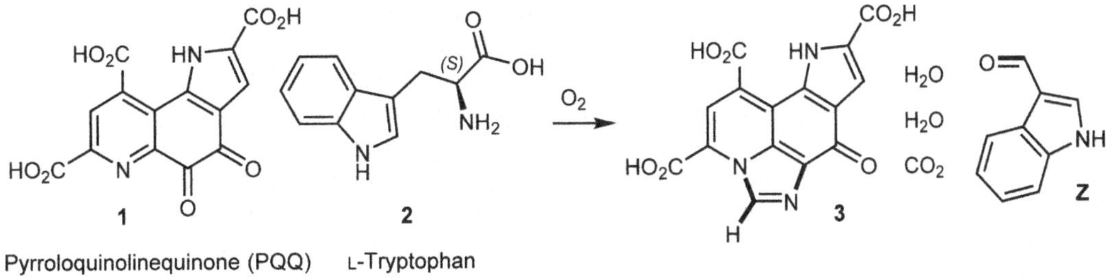

SCHEME 12 Balanced chemical equation for alternative mechanistic proposal for **1** and **2** to **3**.

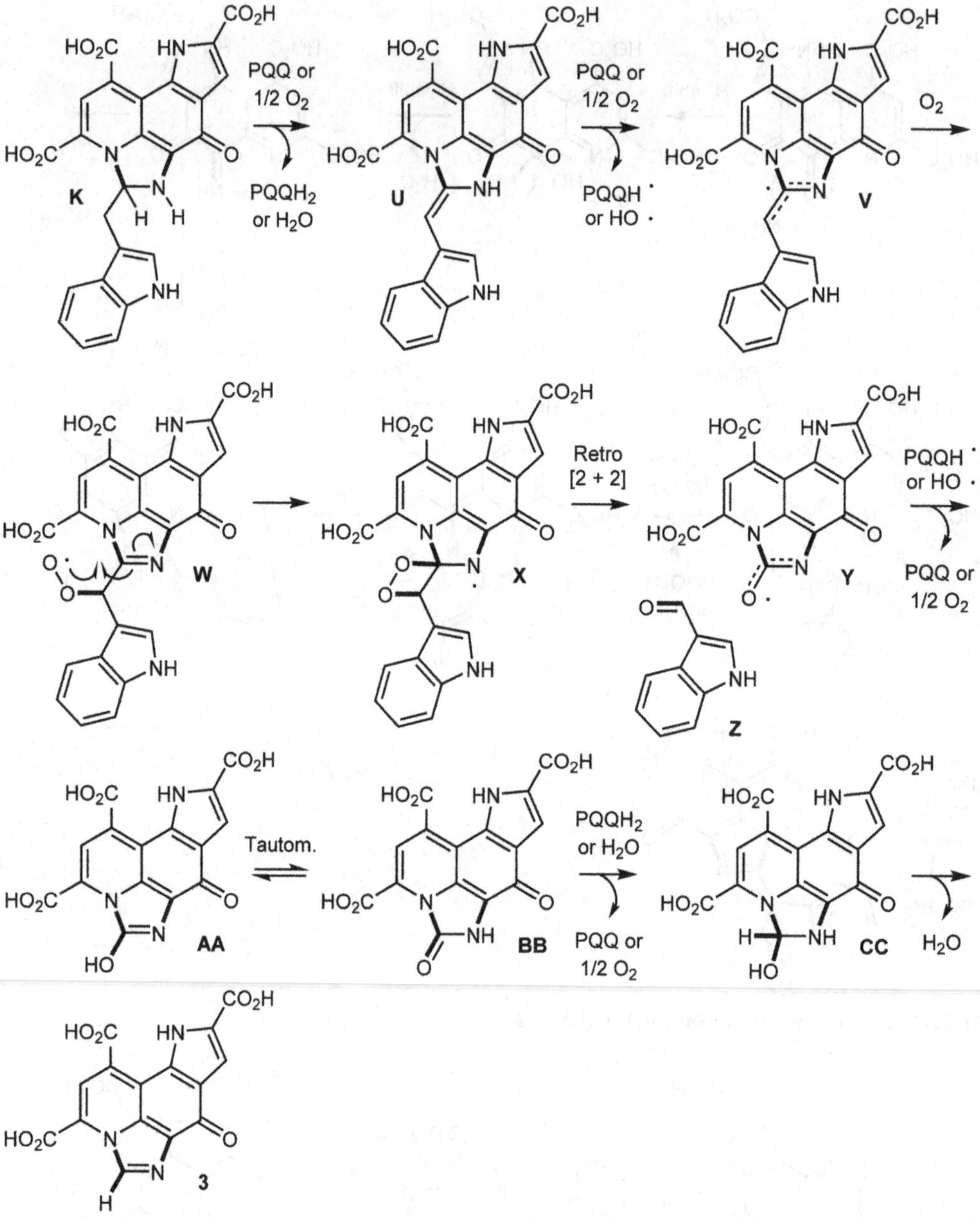

SCHEME 13 Alternative mechanistic proposal for **K** to **3**.

to undertake a theoretical investigation of this problem for both the original and the alternative mechanisms to show which pathways have the lowest energy barriers.

We also note that Kawamoto et al. used X-ray crystallography to determine the structure of **3** and NMR spectroscopy for identification of **4** and **5**. Another experiment which can shed light on this problem would be to carry out the reaction in both aerobic and anaerobic conditions and to record any difference in product yields. This would show the role of O_2 in the reaction and especially under anaerobic conditions lead primarily to product **5** if the alternative mechanisms are operational (since **3** and **4** depend on O_2 oxidative degradation). Interestingly, Itoh et al. observed that in anaerobic conditions, the UV absorption of PQQ gradually diminishes and that introduction of air quickly reestablishes this absorbance indicating that regeneration of PQQ occurs much faster than reaction between PQQ and amino acid.[2] Nevertheless, perhaps the most important experiment one can perform is by-product analysis. This would help determine whether skatole **N** (authors' mechanism) or 1*H*-indole-3-carbaldehyde **Z** (alternative mechanism) is operational. We also note that the authors reported that the production of **4** was significantly promoted by the addition of ammonium salt to the reaction mixture of PQQ and L-Trp. The authors attribute this effect to the role of ammonia in the

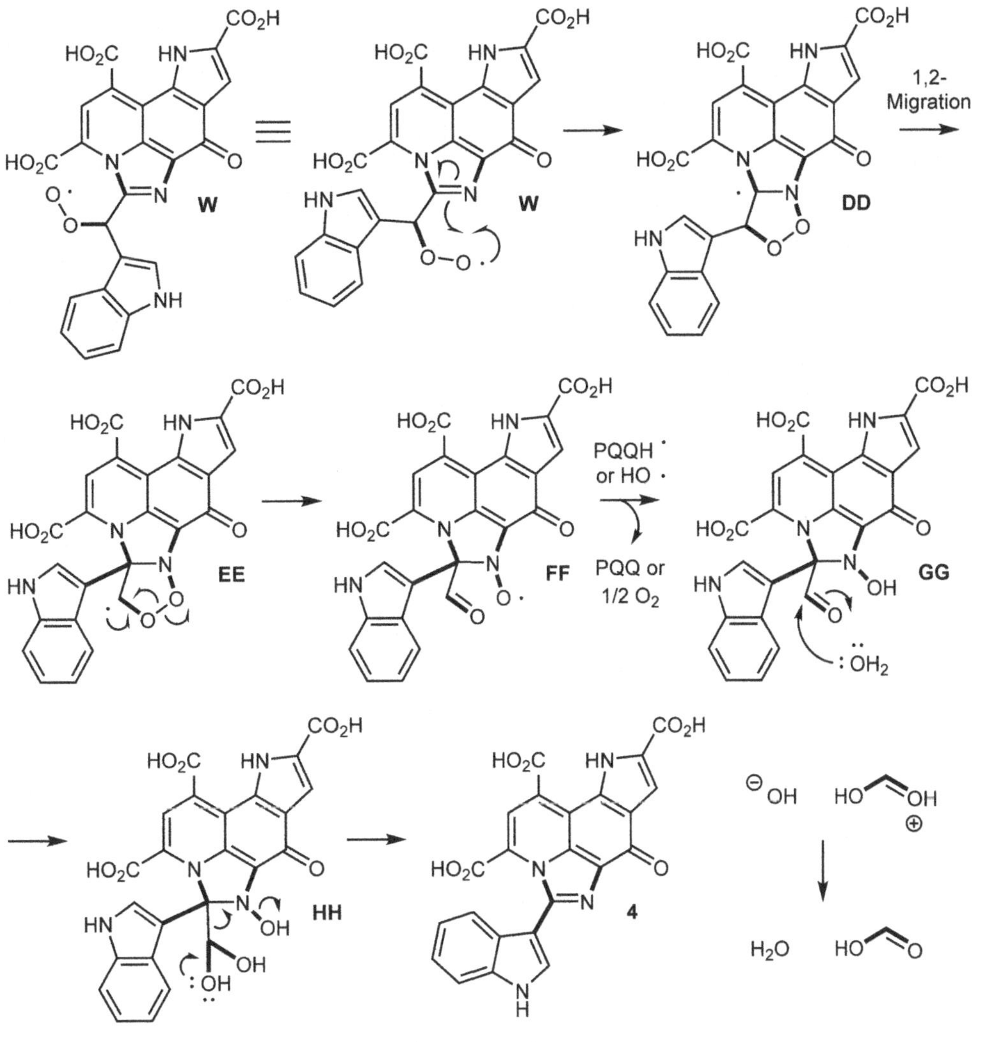

SCHEME 14   Alternative mechanistic proposal for the formation of **4**.

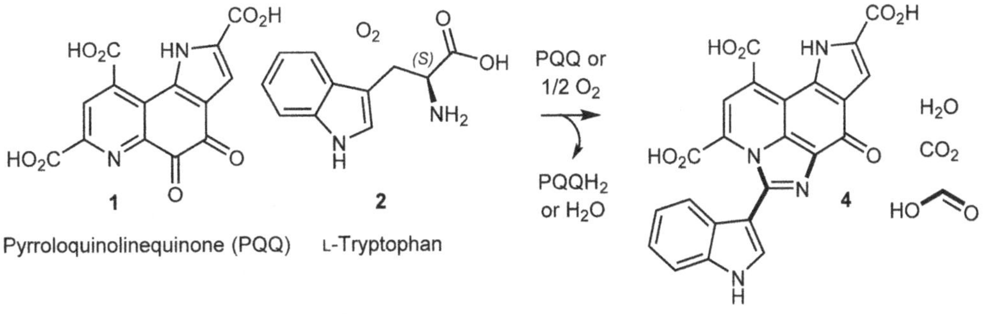

SCHEME 15   Balanced chemical equation for alternative mechanistic proposal for **1** and **2** to **4**.

formation of **4**. Given the reservations we have about the production of ammonia, we attribute this effect to solvation effects and note that the authors did not specify concrete details about the effect by providing yields.

A more comprehensive analysis which takes into account reactions with other amino acids is highly recommended. We also recommend carrying out radiolabeling tracking studies as is commonly done in mechanistic investigations of biologically relevant chemical processes, particularly when fragmentations occur. Such studies confirm the position of a label in the starting material and where it ends up in the fragmented products. Scheme 17 shows examples relevant to the present problem using ^{13}C or ^{14}C labeled tryptophan according to the outcomes of our proposed mechanisms.

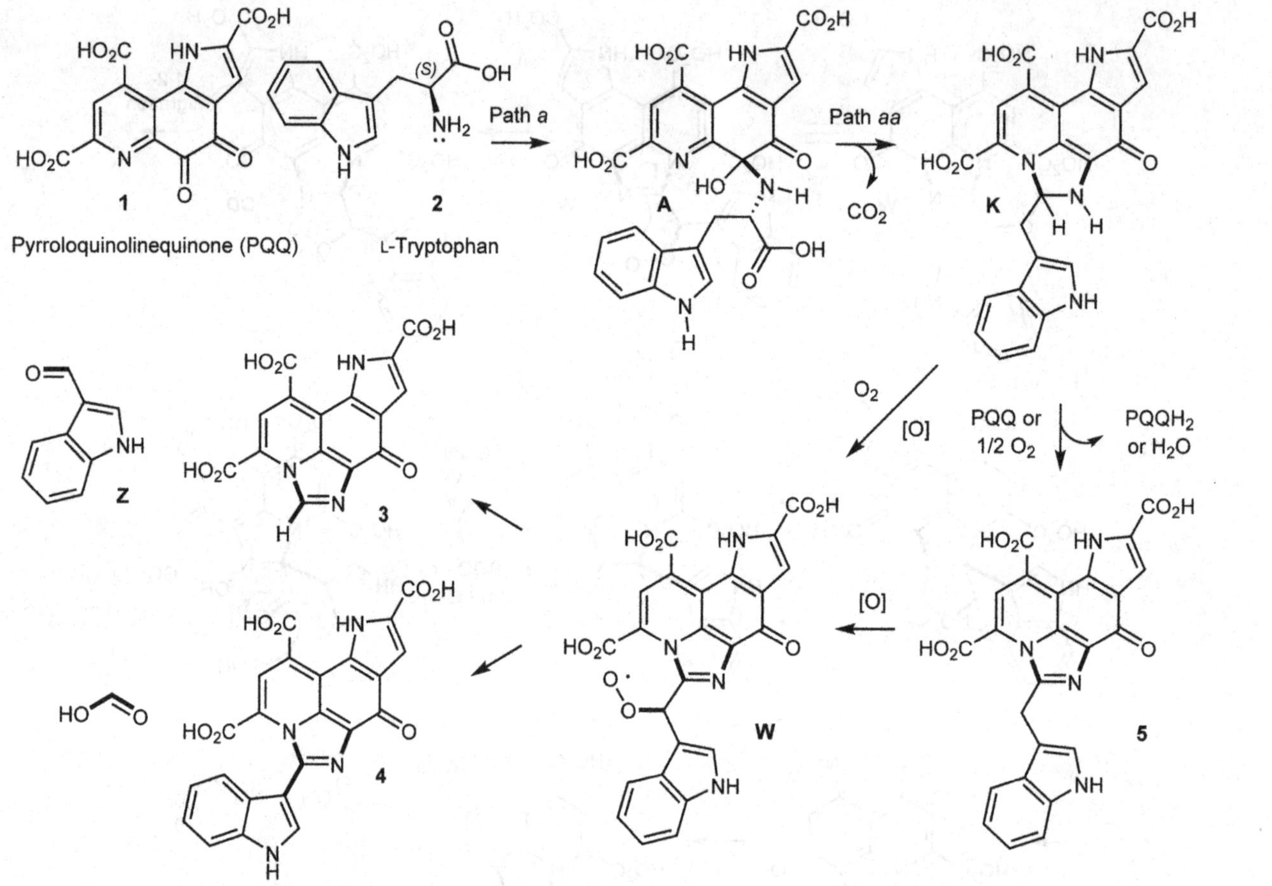

SCHEME 16   Alternative mechanistic road map.

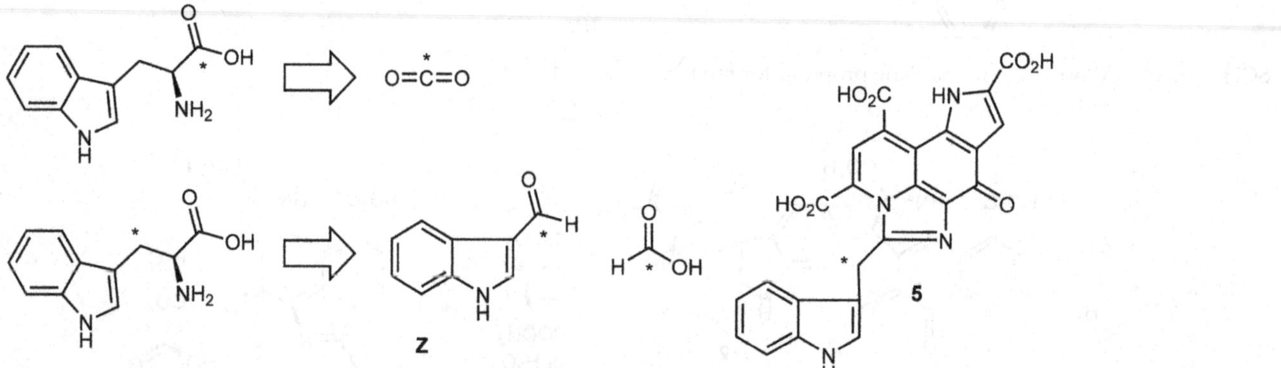

SCHEME 17   Possible radiolabeling experiment of the transformation of **1** and **2** to **3**, **4**, and **5**.

ADDITIONAL RESOURCES

We recommend an interesting article which describes an unexpected formation of indole and imidazolinone derivatives.[8] We also recommend an interesting mechanistic analysis of the action of Trp-lyase where a skatole radical is invoked.[9]

References

1. Kawamoto E, Amano T, Kanayama J, et al. Structure determination of reaction products of pyrroloquinolinequinone (PQQ) with L-tryptophan in vitro and their effects for micro bacterial growth. *J Chem Soc Perkin Trans 2*. 1996;1996:1331–1336. https://doi.org/10.1039/P29960001331.
2. Itoh S, Mure M, Suzuki A, Murao H, Ohshiro Y. Reaction of coenzyme PQQ with amino acids. Oxidative decarboxylation, oxidative dealdolation (C_α-C_β fission) and oxazolopyrroloquinoline (OPQ) formation. *J Chem Soc Perkin Trans 2*. 1992;1992:1245–1251. https://doi.org/10.1039/P29920001245.
3. Yagil G. The proton dissociation constant of pyrrole, indole and related compounds. *Tetrahedron*. 1967;23(6):2855–2861. https://doi.org/10.1016/0040-4020(67)85151-2.
4. Barcock RA, Moorcroft NA, Storr RC, Young JH, Fuller LS. 1- and 2-azafulvenes. *Tetrahedron Lett*. 1993;34(7):1187–1190. https://doi.org/10.1016/S0040-4039(00)77524-1.
5. Skiles GL, Yost GS. Mechanistic studies on the cytochrome P450-catalyzed dehydrogenation of 3-methylindole. *Chem Res Toxicol*. 1996;9(1):291–297. https://doi.org/10.1021/tx9501187.
6. Weems JM, Yost GS. 3-methylindole metabolites induce lung CYP1A1 and CYP2F1 enzymes by AhR and non-AhR mechanisms, respectively. *Chem Res Toxicol*. 2010;23:696–704. https://doi.org/10.1021/tx9004506.
7. Ohshiro Y, Itoh S. Mechanism of amine oxidation by coenzyme PQQ. *Bioorg Chem*. 1991;19(2):169–189. https://doi.org/10.1016/0045-2068(91)90033-L.
8. Klasek A, Lycka A, Holcapek M. Molecular rearrangement of 1-substituted 9b-hydroxy-3,3a,5,9btetrahydro-1H-imidazo[4,5-c]quinoline-2,4-diones—an unexpected pathway to new indole and imidazolinone derivatives. *Tetrahedron*. 2007;63:7059–7069. https://doi.org/10.1016/j.tet.2007.05.012.
9. Bhandari DM, Fedoseyenko D, Begley TP. Mechanistic studies on tryptophan lyase (NosL): identification of cyanide as a reaction product. *J Am Chem Soc*. 2018;140:542–545. https://doi.org/10.1021/jacs.7b09000.

Question 66: An Unexpected Result From an Attempted Double Bischler-Napieralski Reaction

In 1996 British chemists showed that reaction of oxamide **3** (derived from the condensation of **1** and **2**) with pyrophosphoryl chloride in acetonitrile gave product **4** in 84% yield (Scheme 1).[1] Thus the authors attempted to synthesize the analogous seven-membered product **6** through an extension of this transformation by using oxamide **5** as a substrate. This reaction produced the unexpected product **7** in 81% yield. Explain this result in mechanistic terms.

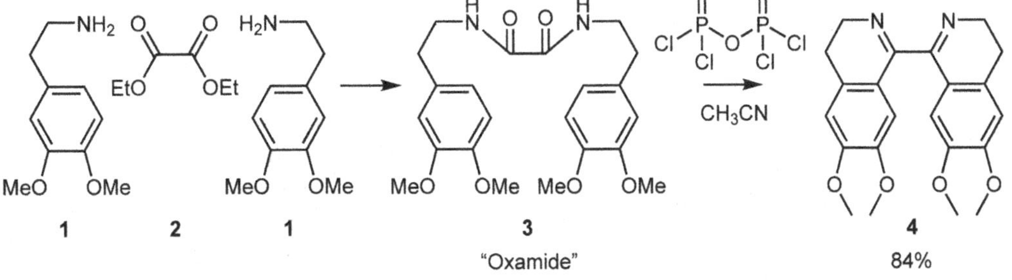

SCHEME 1

(Continued)

Expected reaction:

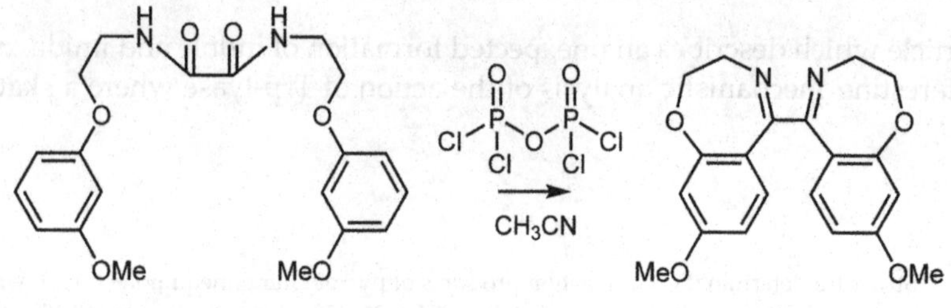

5

"Oxamide"

6

Observed reaction:

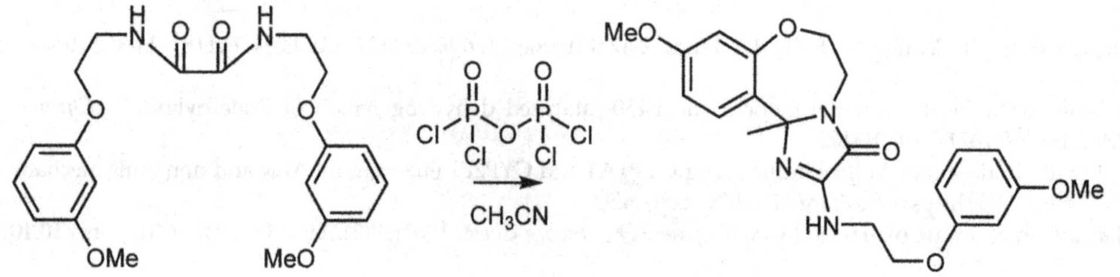

5

"Oxamide"

7

81%

SCHEME 1, CONT'D

SOLUTION

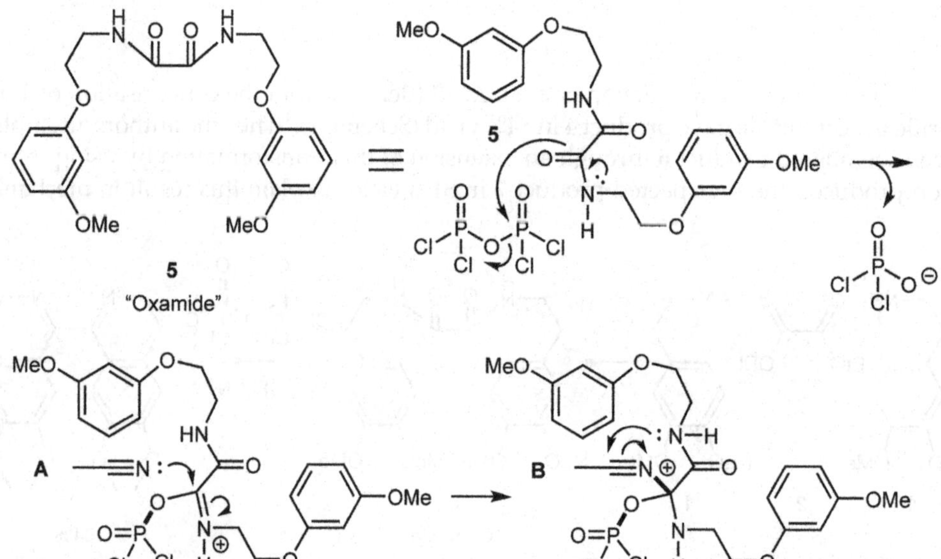

5

"Oxamide"

5

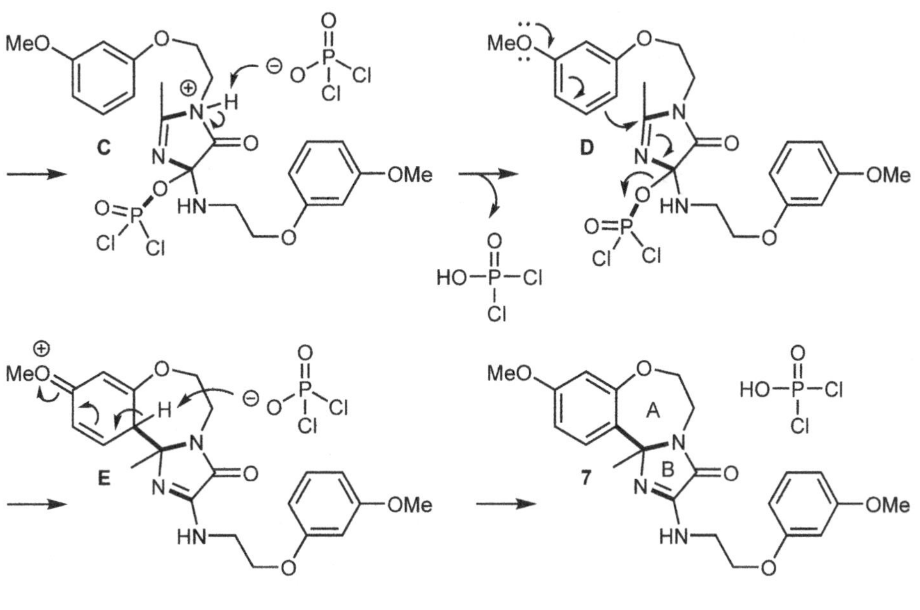

BALANCED CHEMICAL EQUATION

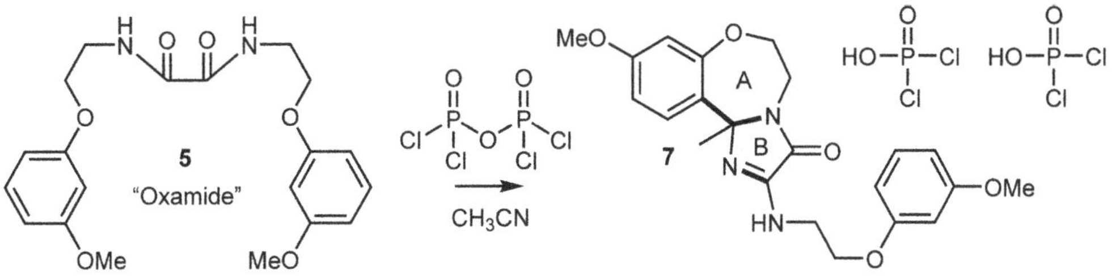

KEY STEPS EXPLAINED

We begin by noting that the authors originally intended to carry out a double Bischler-Napieralski reaction between oxamide **5** and pyrophosphoryl chloride to form product **6** by extension of the successful transformation of oxamide **3** to product **4**. Instead, the authors isolated an unexpected imidazolidinone product **7** that had incorporated a molecule of acetonitrile (which is the reaction solvent). The authors rationalized this surprising outcome as beginning with reaction between oxamide **5** and pyrophosphoryl chloride which forms the iminium ion **A** which is analogous to a Vilsmeier-Haack reagent. The unexpected step was the nucleophilic attack of acetonitrile onto the electrophilic carbon atom of the iminium group of **A** that forms the nitrilium ion intermediate **B**. Two successive cyclizations forming a five-membered imidazolidinone ring and a seven-membered [1,4]oxazepine ring took place via intermediates **C**, **D**, and **E**. We note that the ring construction mapping is $[(6+1)_A + (3+2)_B]$. We also emphasize that the authors did not carry out additional experiments or theoretical calculations to help support their postulated mechanism.

Moreover, we want to emphasize that this reaction is a rare example of solvent incorporation into a product structure so that the solvent is not a mere spectator as in most cases. An explanation for incorporation of acetonitrile may be that it forms a solvent shell around the ionic Vilsmeier-Haack adduct **A** and is therefore in ideal proximity for actual reaction with it. Such a scenario would outcompete an intramolecular cyclization reaction of intermediate **A** initiated by the electron-donating methoxy group on the aromatic ring which would lead to the expected Bischler-Napieralski product **6** (see the mechanism depicted in Scheme 2). An experimental test for this hypothesis is to carry out the reaction in another solvent that does not solvate well the charged iminium ion **A** to see if the intended product **6** would form.

We note that in the case of the intended Bischler-Napieralski product **6** as compared to the observed reaction there are two equivalents of pyrophosphoryl chloride consumed instead of one (as for the formation of **7**).

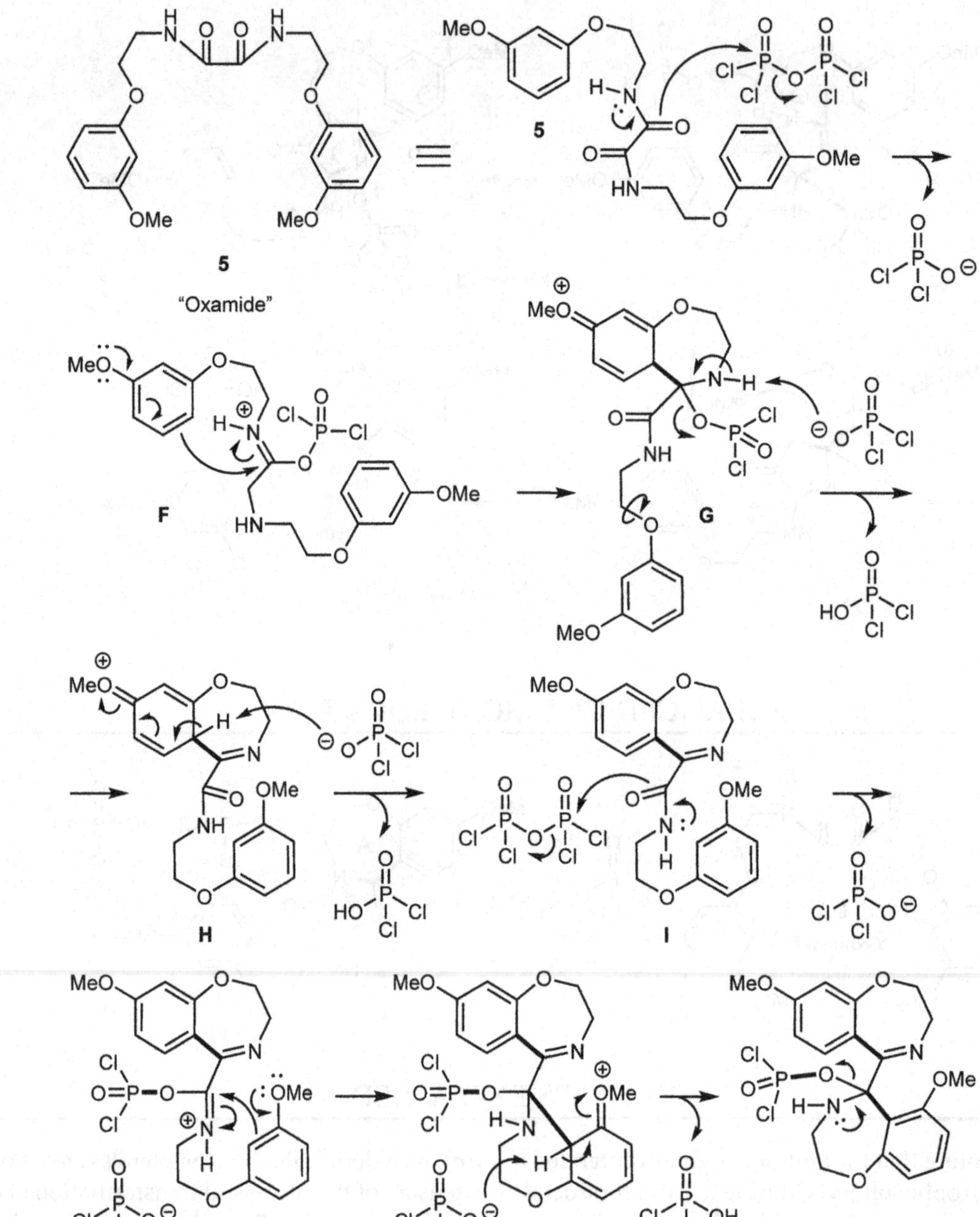

5

"Oxamide"

SCHEME 2

We recommend two articles which discuss the mechanisms of similar transformations.[2,3] We also recommend two articles which describe more environmentally friendly approaches to the Bischler-Napieralski reaction.[4,5]

References

1. Heaney H, Shuhaibar KF, Slawin MZA. The capture of acetonitrile during a Bischler-Napieralski cyclisation reaction of an oxamide derivative. *Tetrahedron Lett.* 1996;37(24):4275–4276. https://doi.org/10.1016/0040-4039(96)00815-5.
2. Magnus P, Gazzard L, Hobson L, Payne AH, Lynch V. Studies on the synthesis of the indole alkaloids pauciflorine A and B. *Tetrahedron Lett.* 1999;40(28):5135–5138. https://doi.org/10.1016/S0040-4039(99)00882-5.
3. Moreno L, Parraga J, Galan A, Cabedo N, Primo J, Cortes D. Synthesis of new antimicrobial pyrrolo[2,1-a]isoquinolin-3-ones. *Bioorg Med Chem.* 2012;20:6589–6597. https://doi.org/10.1016/j.bmc.2012.09.033.
4. Hegedus A, Hell Z, Potor A. A new, environmentally-friendly method for the Bischler–Napieralski cyclization using zeolite catalyst. *Catal Commun.* 2006;7:1022–1024. https://doi.org/10.1016/j.catcom.2006.05.012.
5. Judeh ZMA, Ching CB, Bu J, McCluskey A. The first Bischler–Napieralski cyclization in a room temperature ionic liquid. *Tetrahedron Lett.* 2002;43:5089–5091. https://doi.org/10.1016/S0040-4039(02)00998-X.

Question 67: Bis-Indole Alkaloid Formation

In 1996 chemists from Hong Kong studied the dimerization of 2-prenylindoles and observed that under acid catalysis, three transformations took place (Scheme 1).[1] First, the tertiary alcohol **1** could be converted into the alkaloid yuehchukene **2** in 10% yield.[2] Next, treatment of the secondary alcohol **3** with silica gel impregnated with TsOH produced a complex mixture which contained a 5.1% yield of products **4** and **5**. Lastly, treatment of the isomeric tertiary alcohol **6** with a catalytic amount of TFA in anhydrous benzene led to much higher yields of the dimeric products **7** (31%) and **8** (25%). Provide mechanistic explanations for these three transformations.

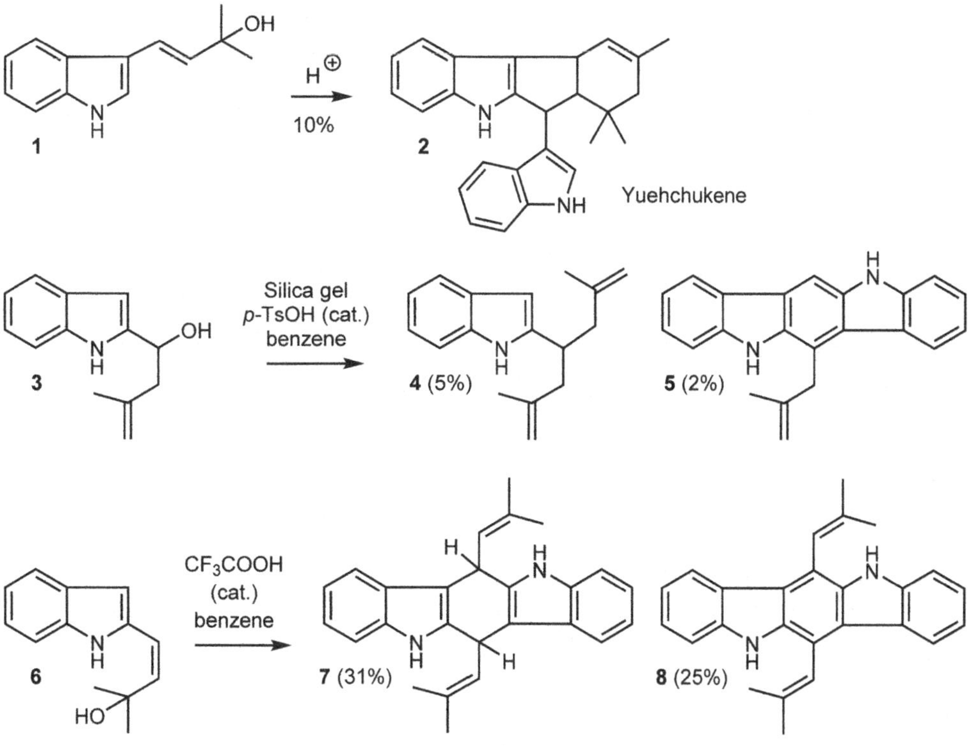

SCHEME 1

SOLUTION: 1 TO 2

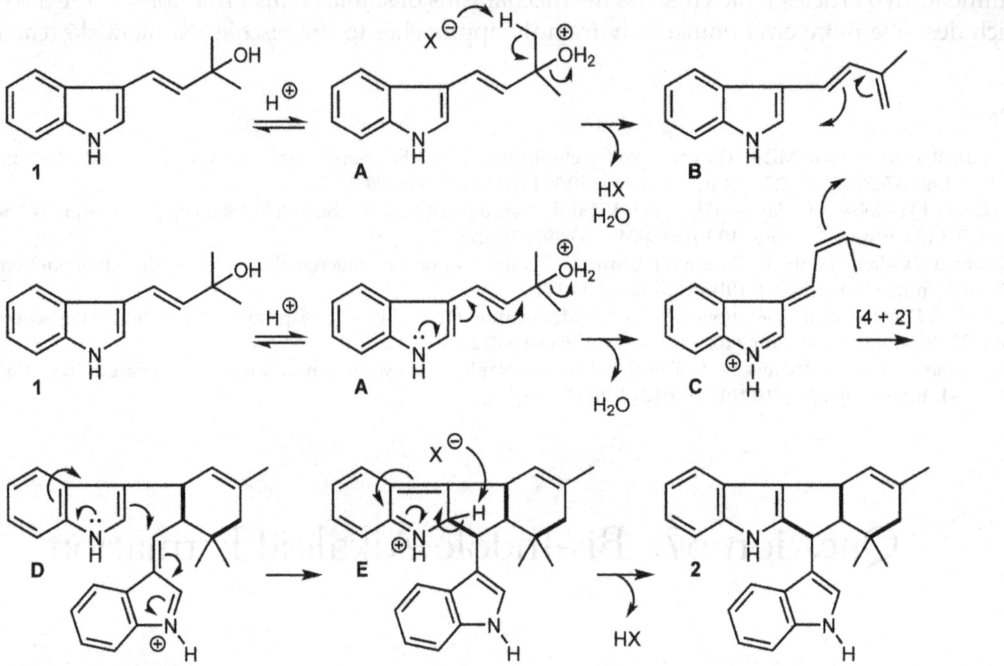

SOLUTION: 3 TO 4 AND 5

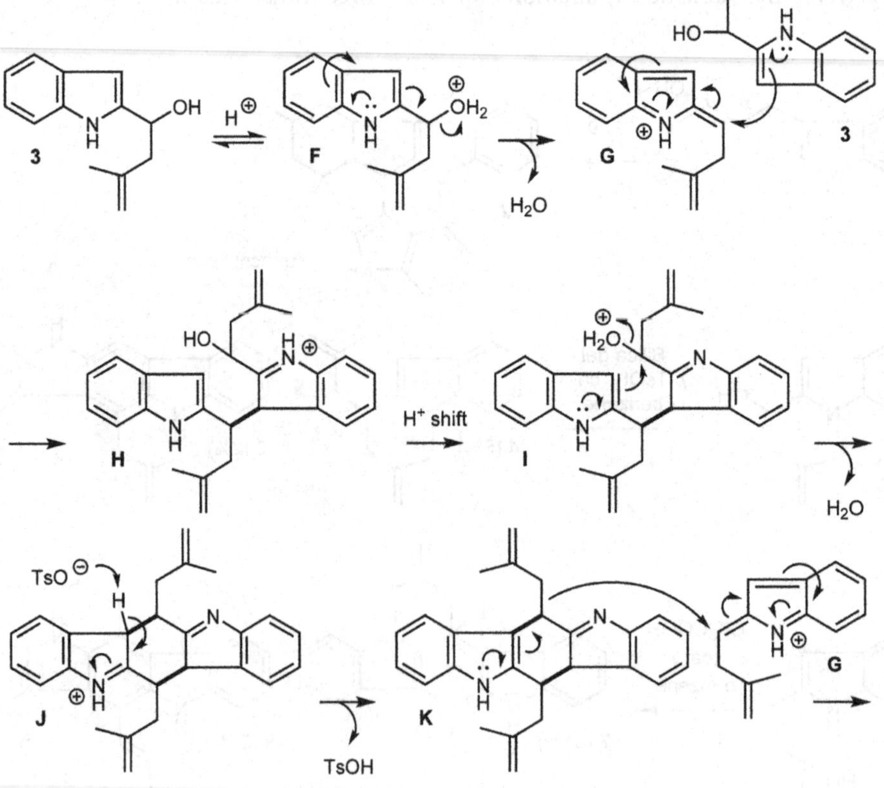

SOLUTION: 6 TO 7 AND 8

BALANCED CHEMICAL EQUATIONS

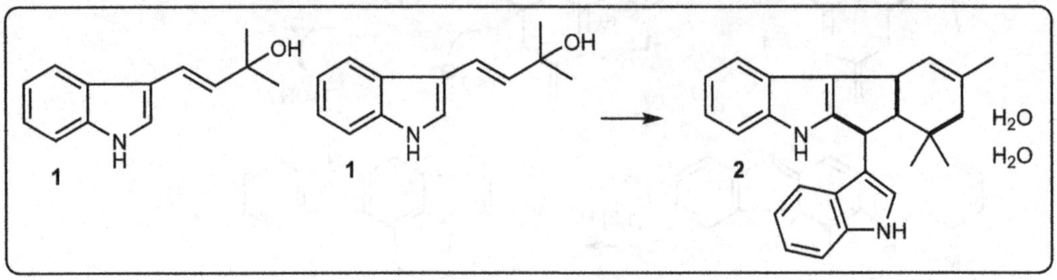

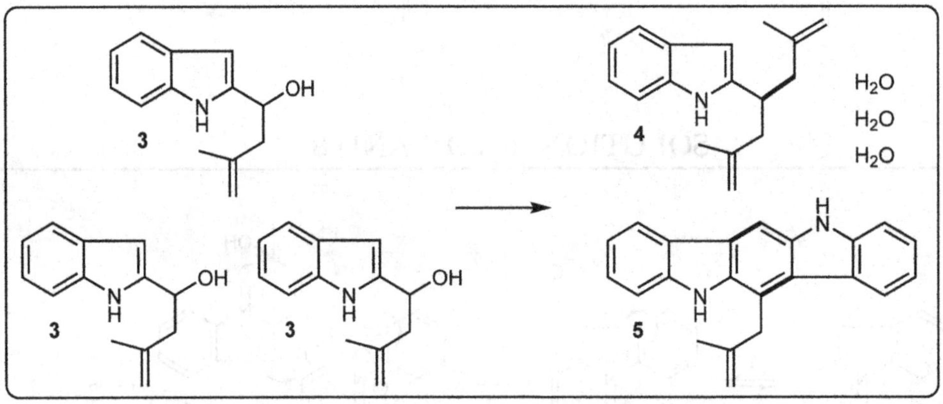

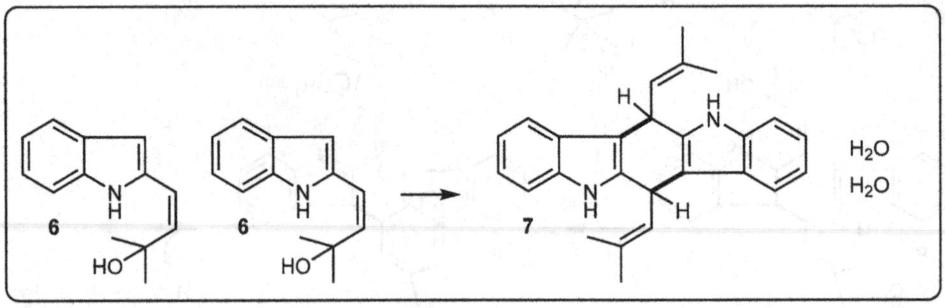

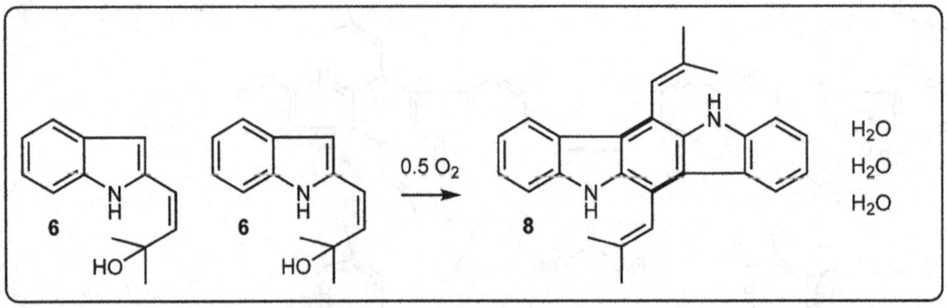

KEY STEPS EXPLAINED

The mechanism proposed by the authors to explain the formation of yuehchukene **2** from the acid-catalyzed dimerization of **1** begins with formation of the protonated intermediate **A** which eliminates water to form either diene **B** or iminium ion **C**. These intermediates can combine in a [4+2] cycloaddition reaction which generates adduct **D** which then undergoes cyclization to **E** and deprotonation to form **2**. The ring construction mapping for product **2** is [(3+2)$_A$+(4+2)$_B$].

In the case of the transformation of **3** to **4** and **5**, the authors proposed that acid-catalyzed dehydration of **3** leads to intermediate **G** (via **F**) which adds to another molecule of **3** resulting in intermediate **H**. The next steps involve further cyclization followed by deprotonation steps that eventually cause the formation of intermediate **K**. Intermediate **K** then transfers an isobutenyl group to another molecule of **G** yielding product **4** and intermediate **L** which goes on to form product **5** after a final deprotonation and aromatization. The ring construction mapping for product **5** is [3+3].

Lastly, the transformation of **6** to **7** and **8** is explained as starting with an acid-catalyzed dehydration of **6** to form intermediate **N** which adds to another molecule of **6** resulting in intermediate **O**. Further cyclization, dehydration, and deprotonation steps lead to product **7**. Product **8** arises via autoxidation of product **7** in the presence of air. This process aromatizes the central ring. The ring construction mapping for both products is [3+3].

We note that the authors did not present experimental or theoretical evidence to support their postulated mechanisms. In particular, they do not explain why the yields of the dimerization products are much higher in the case of the **6** to **7** and **8** transformation as compared to the transformation of **3** to **4** and **5**. Also, the authors do not explain why alkyl transfer is preferred in the case of the formation of **4** but does not occur in the analogous transformation of **6** to **7** and **8** where a possible isomerization of **T** to a terminal alkene resembling **K** may facilitate a potential alkyl group transfer to an intermediate such as **O**. Nevertheless, the authors do acknowledge that given the nature of the various likely intermediates, in particular the iminium ions that can form in the course of these transformations, complex mixtures of products and low yields are inevitable. We suspect these factors also play a role in the difficulty of carrying out more thorough experimental mechanistic analyses on this system.

ADDITIONAL RESOURCES

We recommend two articles which discuss interesting transformations that also involve indole chemistry.[3,4]

References

1. Lee V, Cheung M-K, Wong W-T, Cheng K-F. Studies on the acid-catalyzed dimerization of 2-prenylindoles. *Tetrahedron*. 1996;52(28):9455–9468. https://doi.org/10.1016/0040-4020(96)00482-6.
2. Sheu J-H, Chen Y-K, Hong Y-LV. An efficient synthesis of yuehchukene. *Tetrahedron Lett*. 1991;32(8):1045–1046. https://doi.org/10.1016/S0040-4039(00)74483-2.
3. Szabo LF. Reaction mechanism and chemotaxonomy in the formation of the type I indole alkaloids derived from secologanin. *J Phys Org Chem*. 2006;19:579–591. https://doi.org/10.1002/poc.1075.
4. Klasek A, Lycka A, Holcapek M. Molecular rearrangement of 1-substituted 9b-hydroxy-3,3a,5,9btetrahydro-1H-imidazo[4,5-c]quinoline-2,4-diones—an unexpected pathway to new indole and imidazolinone derivatives. *Tetrahedron*. 2007;63:7059–7069. https://doi.org/10.1016/j.tet.2007.05.012.

Question 68: Azaphenalene Alkaloid Synthesis: A Key Step

In 1992 French chemists investigated the synthesis of precoccinelline alkaloids which contain an azaphenalene chemical structure and which have been involved in the defense mechanism of beetles due to their bitter taste. Thus reaction of the bicyclic compound **1** with 3-[2-(1,3-dioxolanylpropyl)]magnesium chloride **2** at −78°C in a mixed $CH_2Cl_2/Et_2O/THF$ solvent system containing t-butyldimethylsilyltriflate (TBDMSOTf) followed by quenching with ammonium chloride led to formation of **3** in 74% yield (Scheme 1).[1] Suggest a mechanism for this transformation.

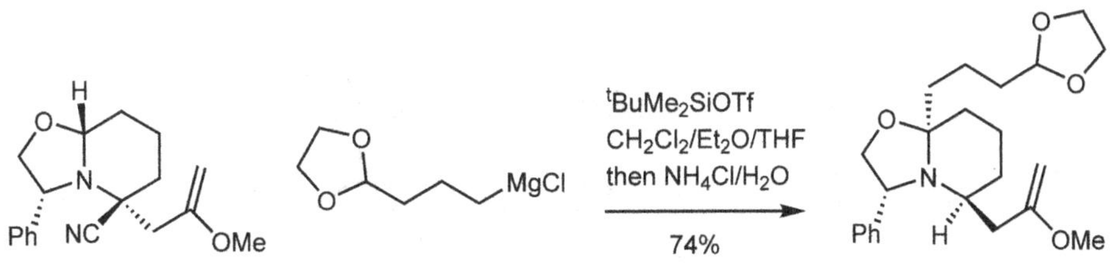

SCHEME 1

SOLUTION: PATH A

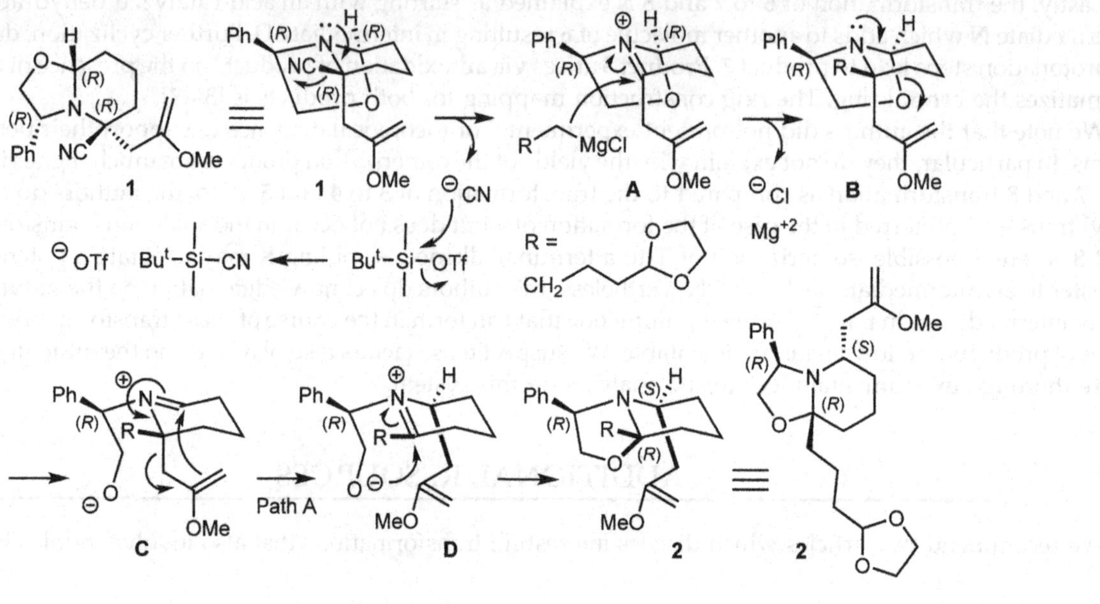

SOLUTION: PATH B (C TO 2 ALTERNATIVE)

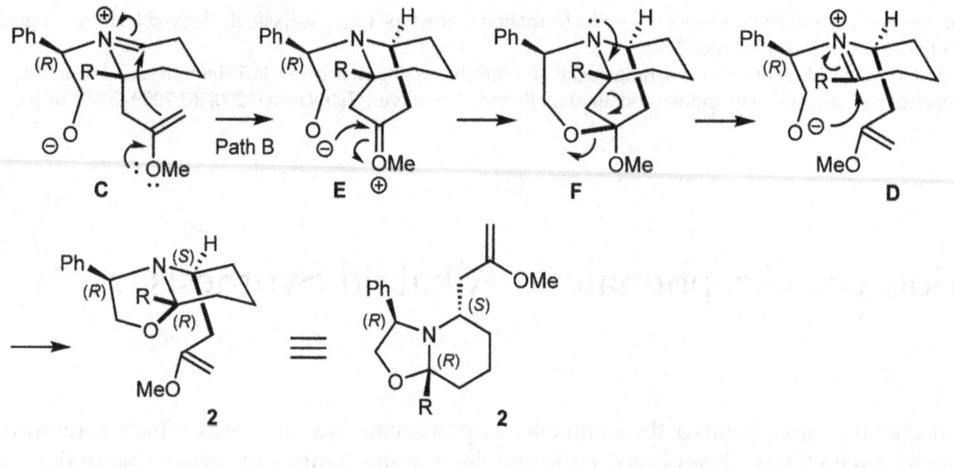

BALANCED CHEMICAL EQUATION

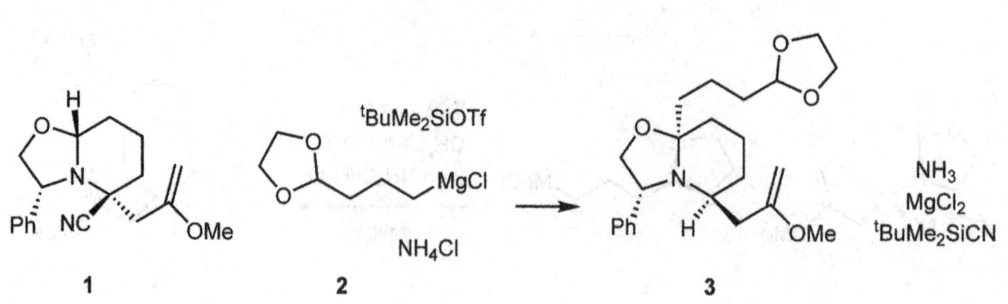

KEY STEPS EXPLAINED

The mechanism proposed by the authors to explain the transformation of **1** to **2** begins with elimination of cyanide ion (which is antiperiplanar to the nitrogen atom) from compound **1** which forms iminium ion **A**. This intermediate is captured by the Grignard reagent **2** which results in formation of intermediate **B**. In this transformation the incoming group occupies the same position as the cyanide group in the starting material; hence, the reaction occurs with retention of configuration. The oxazolidine ring of intermediate **B** then ring opens to form the zwitterionic intermediate **C**. At this point, two pathways leading to product **2** are possible. In path A intermediate **C** undergoes an aza-Cope rearrangement to form a zwitterionic intermediate **D** which ring closes to give product **2**. In path B intermediate **C** undergoes a Mannich cyclization that forms intermediate **F** followed by a retro-Mannich transformation which gives the zwitterionic intermediate **D** which carries on as in path A to form product **2**. We note that the ring construction mapping for the reaction is [5 + 0] for the construction of the fused oxazolidine ring. The authors also used NMR spectroscopy to observe that addition of trimethylsilyltriflate (TMSOTf) to a solution of **1** in CDCl₃ resulted in the disappearance of the CN signal (118.3 ppm) and appearance of a new resonance signal (189 ppm) which was attributed to the iminium species **A**. This method could not be used to detect other iminium species such as **D**. Another interesting observation was that carrying out the reaction with TMSOTf resulted in the formation of low yields of **2** regardless of reaction time and temperature which suggested that a side reaction between the newly generated trimethylsilylcyanide (TMSCN) and the Grignard reagent was occurring. This prompted the authors to replace TMSOTf with the more hindered TBDMSOTf which was successful in generating product **2** in acceptable yields of 30%–50%. Furthermore, when the R group of the Grignard reagent was replaced from pentyl to methyl group, the yield of **2** decreased in favor of **F** which the authors attribute to steric interactions between the phenyl group and the R group in intermediate **C**. Specifically, the formation of ketal **F** requires proximity between the phenyl group and the R group which is more likely when R is a methyl group rather than a bulkier pentyl group. These experiments support the proposed mechanism. Nevertheless, the authors did not perform a theoretical analysis to lend further support for the proposed mechanism, especially with regard to being able to better understand the divergence from **C** to **2** along paths A and B.

We also include a comment about the quality of the diagrams presented in the authors' paper and how they impact the reader's understanding in following what happens in the course of the reaction. We thus note that by changing the structural aspect of their drawings unnecessarily throughout their proposed mechanism, the authors succeed in unduly obscuring the precise bond forming and bond breaking steps along the mechanistic pathway. For example, Scheme 2 shows compound **1** drawn in various structural aspects by the authors. We believe this is not a useful approach in depicting chemical structures in journal articles especially in a diagram meant to illustrate a complete reaction mechanism. In our presentation we selected the last structural aspect since this was the most meaningful and clear in depicting the key bond forming and bond breaking steps and also in depicting stereochemical features of the reaction.

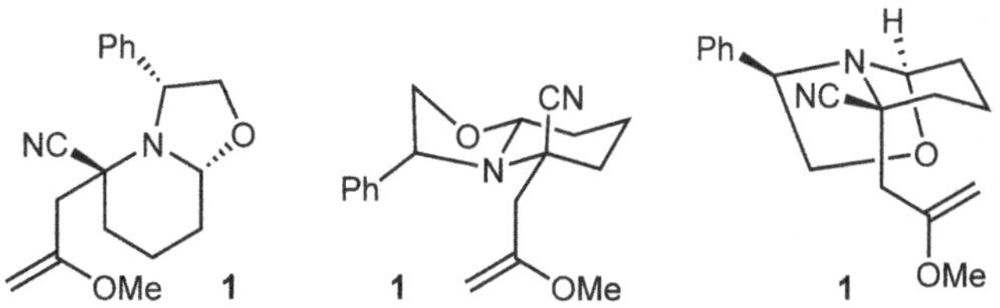

SCHEME 2

ADDITIONAL RESOURCES

We recommend an article which describes an interesting transformation that involves a ring closure very similar in nature to path B highlighted before.[2] We also refer the reader to an article which discusses the formation of an aza-phenalene from a vinyl nitro group and proposes a mechanism for this transformation.[3]

References

1. Yue C, Royer J, Husson H-P. Asymmetric synthesis. 26. An expeditious enantioselective synthesis of the defense alkaloids (-)-euphococcinine and (-)-adaline via the CN(R,S) method. *J Org Chem.* 1992;57(15):4211–4214. https://doi.org/10.1021/jo00041a028.

2. Medina DHG, Grierson DS, Husson H-P. 2-cyano Δ^3 piperideines X: biomimetic synthesis of the ladybug alkaloids of the adaline series. *Tetrahedron Lett.* 1983;24(20):2099–2102. https://doi.org/10.1016/S0040-4039(00)81854-7.

3. Isomura K, Sakurai M, Komura T, Saruwatari M, Taniguchi H. Novel cyclization of vinyl nitrene into 1-azaphenalene. *Chem Lett.* 1987;16 (5):883–886. https://doi.org/10.1246/cl.1987.883.

Question 69: A Primary Nitroalkane to Carboxylic Acid Transformation

In 1997 French chemists found that reaction of either 2-nitroethylbenzene **1** or 2-bromoethylbenzene **3** with sodium nitrite, acetic acid, and DMSO for 6 h at 35°C followed by acidification with 10% hydrochloric acid gave the same phenylacetic acid product **2** (Scheme 1).[1] Suggest mechanisms to explain these transformations.

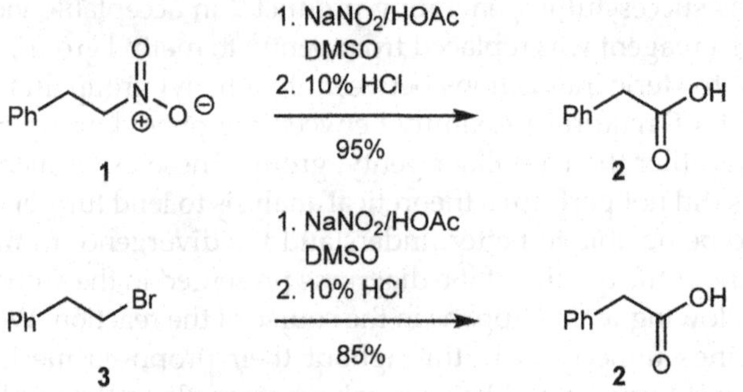

SCHEME 1

SOLUTION: 1 TO 2

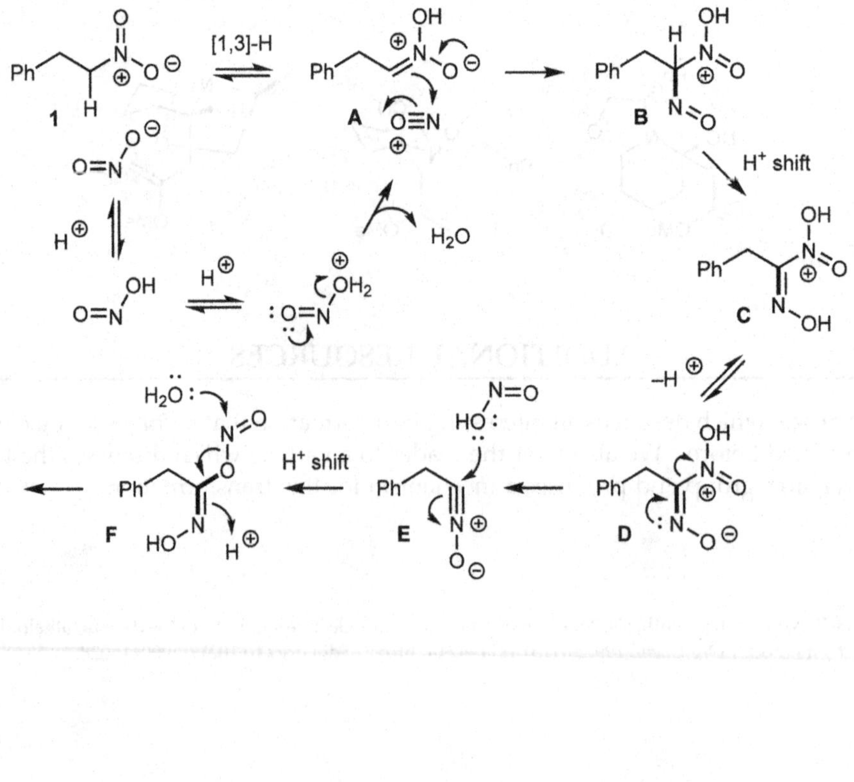

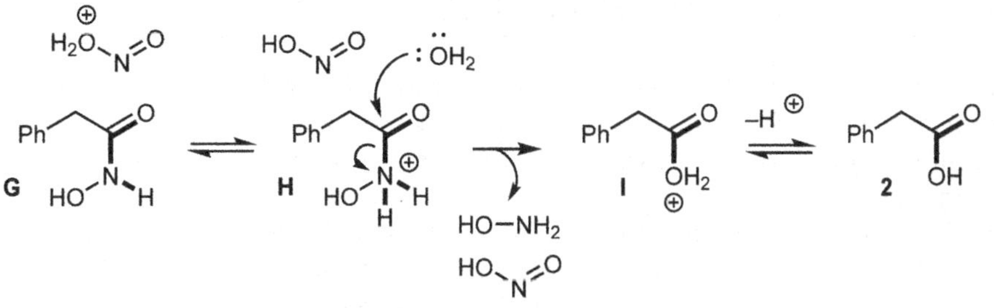

SOLUTION: 3 TO 2

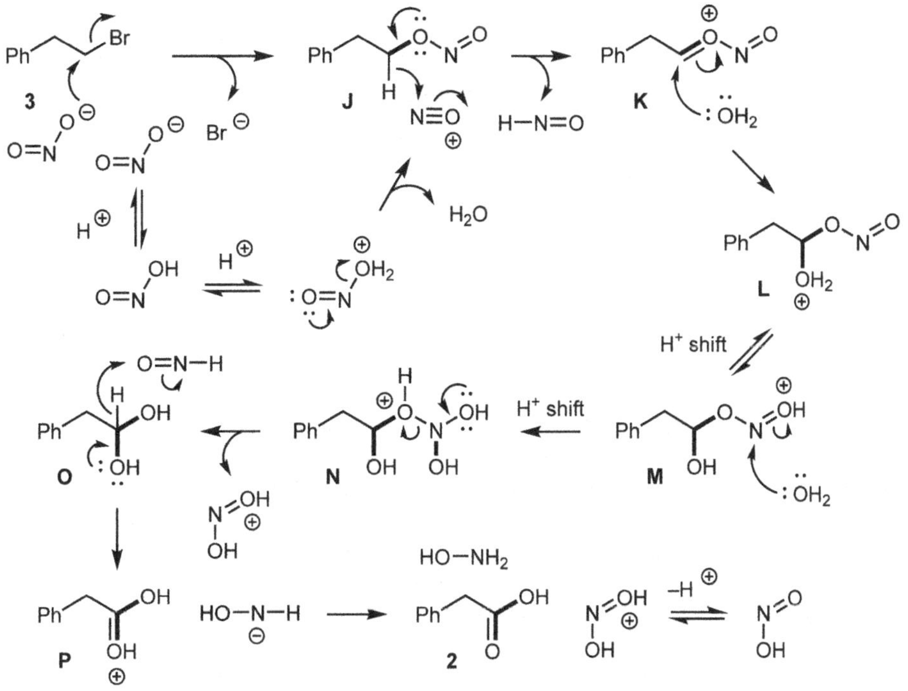

BALANCED CHEMICAL EQUATIONS

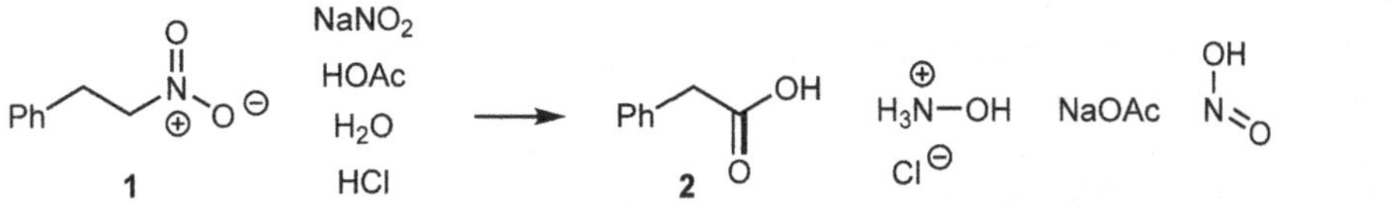

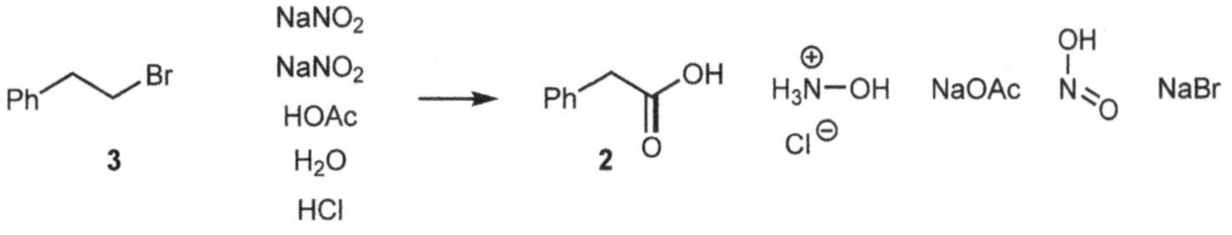

KEY STEPS EXPLAINED

The mechanism proposed by the authors to explain the transformation of **1** to **2** begins with the generation of nitrosonium ion from nitrite and two equivalents of acid. The nitroalkane substrate **1** is in equilibrium with its *aci*-form **A** which captures nitronium ion yielding intermediate **B**. **B** then undergoes a [1,3] proton transfer which forms nitrolic acid **C** which then fragments to nitrile oxide **E** and nitrous acid. Nitrous acid and **E** recombine to produce **F** where a carbon—oxygen bond is made. Water reacts with the N=O group of **F** causing a second fragmentation which gives hydroxamic acid **G** and nitrous acid. The remaining steps are hydrolysis of **G** to phenylacetic acid **2** and hydroxylamine. With regard to this proposed mechanism, the authors conducted a number of experiments which provide supporting evidence. First, the authors isolated nitrolic acid **C** in 82% yield by carrying out the reaction at a lower temperature of 18°C instead of 35°C. Furthermore, reaction of isolated **C** under the reaction conditions led to the production of phenylacetic acid **2** in quantitative yield. When the reaction was carried out with added 1-hexene, the nitrile oxide **E** was trapped by the olefin affording oxazole product **4** in 72% yield (see Scheme 2), which essentially constitutes a [3+2] cycloaddition where the nitrile oxide behaves as a 1,3-dipole.

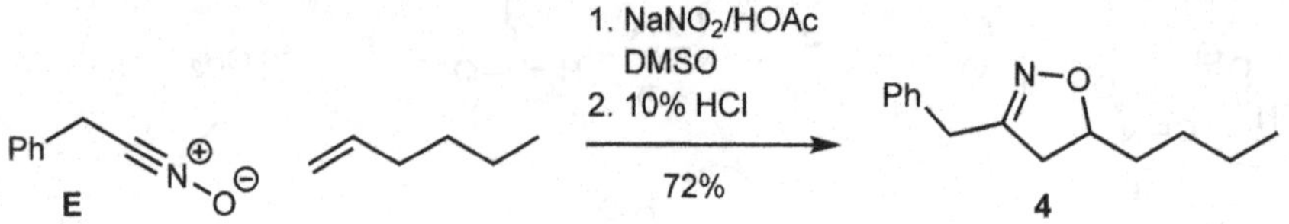

SCHEME 2

Moreover, an authentic sample of hydroxamic acid **G** was subjected to the same reaction conditions and resulted in the production of phenylacetic acid **2**. The authors demonstrated that dimethylsulfoxide does not participate in the reaction as phenylacetic acid **2** was formed when the reaction was run in other solvents such as sulfolane or dimethylformamide (dipolar aprotic solvents). The presence of nitrosonium ion was essential for the reaction to take place as any change of reaction conditions that did not have the sodium nitrite-acetic acid combination caused the reaction to fail regardless of which intermediates were subjected to the reaction conditions. Finally, the reaction was successfully carried out with nitrosonium salt reagents such as nitrosonium tetrafluoroborate in combination with sodium nitrite which showed that even without acetic acid, the nitrosonium species plays an important role. Nevertheless, the authors did not carry out a theoretical analysis to help further support their proposed mechanism.

Furthermore, the authors did not provide a reaction mechanism for the case of the bromine substrate **3** which we believe is very different from that beginning from substrate **1**. The generation of nitrosonium ion from nitrite and acetic acid occurs as before. A second equivalent of nitrite nucleophilically substitutes the bromine atom in **1** yielding (2-nitrosooxy-ethyl)-benzene **J** which undergoes a hydride transfer to nitrosonium ion resulting in the formation of hyponitrous acid and intermediate **K**. **K** reacts with water at the electrophilic carbonyl carbon atom yielding **L** which undergoes an intramolecular proton transfer to form **M**. Water attacks the nitrite moiety to give intermediate **N** after another intramolecular proton transfer. Intermediate **N** then undergoes fragmentation to form nitrous acid and the hydrate of phenylacetaldehyde **O**. **O** then undergoes a hydride transfer to hyponitrous acid yielding hydroxylamine after protonation and phenylacetic acid **2** after deprotonation. Essentially the stepwise oxidation follows the sequence: C—Br → C—O—N=O → CH=O (hydrate) → COOH. Following the authors' strategy, key intermediates selected for trapping experiments and demonstrating that they lead to phenylacetic acid under the same reaction conditions are (2-nitrosooxy-ethyl)-benzene **J** and phenylacetaldehyde hydrate **O**. The authors also observed a small amount of alcohol product, presumably 2-phenylethanol, being formed as a side product which they surmised arose from hydrolysis of the nitrite intermediate **J** which arises via O-alkylation of nitrite ion. We also recommend carrying out a theoretical investigation of this postulated mechanism. Lastly, we show oxidation number analyses for both reactions in Schemes 3 and 4 to highlight their redox nature.

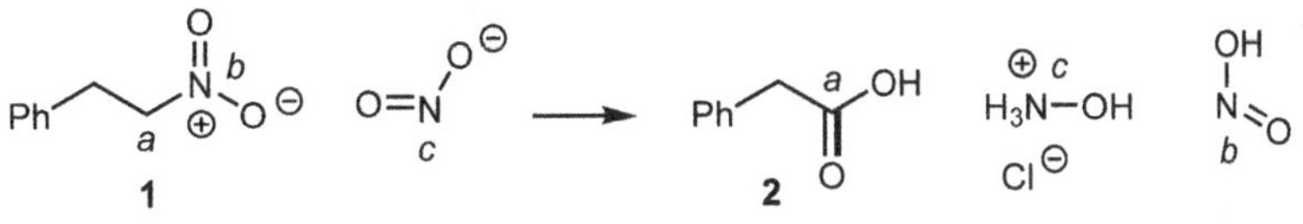

SCHEME 3

Atom	Reactant	Product	Change	
C_a	−1	+3	+4	$\Delta = +4$
N_b	+3	+3	0	
N_c	+3	−1	−4	$\Delta = -4$

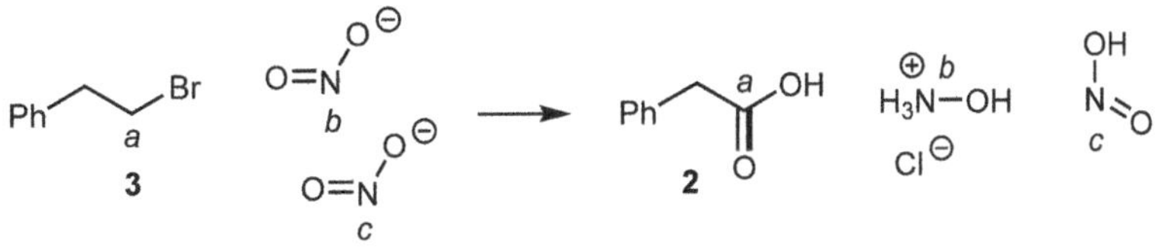

SCHEME 4

Atom	Reactant	Product	Change	
C_a	−1	+3	+4	$\Delta = +4$
N_b	+3	−1	−4	$\Delta = -4$
N_c	+3	+3	0	

ADDITIONAL RESOURCES

We recommend a review article on the synthetic utility of conjugated nitroalkenes and two articles which discuss the Nef reaction which constitutes another means of converting a nitro group into a carbonyl group.[2–4]

References

1. Matt C, Wagner A, Mioskowski C. Novel transformation of primary nitroalkanes and primary alkyl bromides to the corresponding carboxylic acids. *J Org Chem*. 1997;62:234–235. https://doi.org/10.1021/jo962110n.
2. Barrett AGM, Graboski GG. Conjugated nitroalkenes: versatile intermediates in organic synthesis. *Chem Rev*. 1986;86(5):751–762. https://doi.org/10.1021/cr00075a002.
3. Ceccherelli P, Curini M, Marcotullio MC, Epifano F, Rosati O. Oxone® promoted Nef reaction. Simple conversion of nitro group into carbonyl. *Synth Commun*. 1998;28(16):3057–3064. https://doi.org/10.1080/00397919808004885.
4. Ballini R, Petrini M. Recent synthetic developments in the nitro to carbonyl conversion (Nef reaction). *Tetrahedron*. 2004;60:1017–1047. https://doi.org/10.1016/j.tet.2003.11.016.

Question 70: α-Diazoketones With Rhodium(II) Acetate

In 1990 American chemists studying the reactions of α-diazoketones with rhodium(II) acetate observed that treatment of the diazo ketone **1** with a catalytic amount of rhodium(II) acetate resulted in the formation of products **2**, **3**, and **4** (Scheme 1).[1] Suggest mechanisms for the formation of these unexpected products.

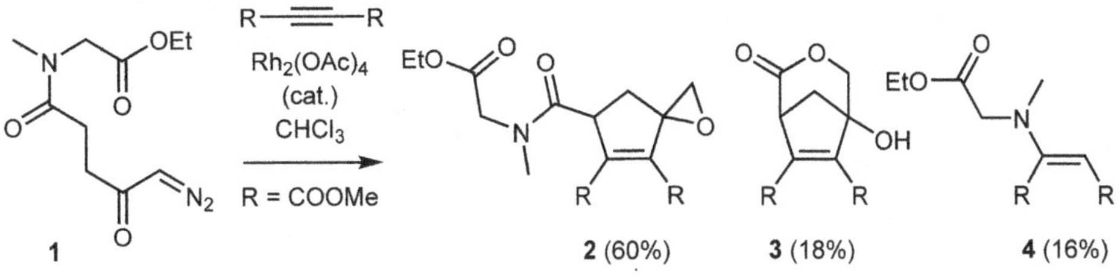

SCHEME 1

SOLUTION: 1 TO 2

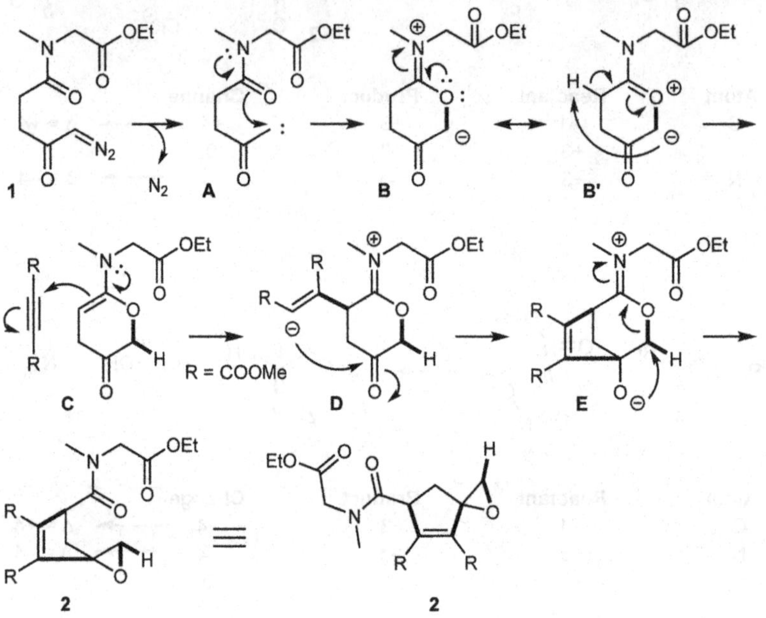

SOLUTION: E TO 3

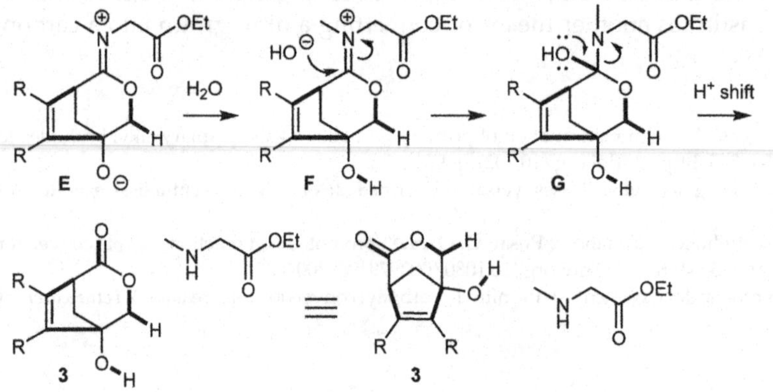

SOLUTION: C TO J AND 4

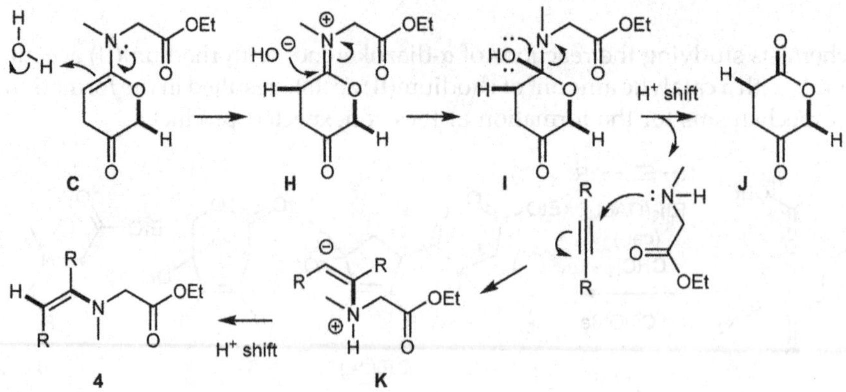

BALANCED CHEMICAL EQUATIONS

KEY STEPS EXPLAINED

The mechanism proposed by the authors to explain the formation of the products from **1** begins with fragmentation of diazoketone **1** which forms carbene **A** and nitrogen gas. **A** then cyclizes to carbonyl ylide **B/B′** which undergoes intramolecular proton transfer to form the ketene *N,O*-acetal **C**. Intermediate **C** is a key hub intermediate that leads to the three product outcomes. For the major product **2**, **C** adds in a Michael fashion to dimethylacetylenedicarboxylate to form the zwitterionic intermediate **D** which then cyclizes to the [3.2.1] bicyclic intermediate **E**. The next step involves ring closure of the alkoxide moiety to yield an epoxide with concomitant ring opening of the six-membered ring resulting in the formation of the [5,3]-spiro product **2**. For product **3**, **C** proceeds to intermediate **E** as before but is then protonated by water to yield iminium ion **F** which is attacked by the liberated hydroxide ion to give the tetrahedral intermediate **G**. **G** then fragments to product **3** and ethyl methylamino-acetate by-product. For product **4**, **C** is hydrolyzed to lactone **J** and ethyl methylamino-acetate. Ethyl methylamino-acetate adds to dimethylacetylenedicarboxylate in a Michael fashion to form product **4**. The ring construction mappings for the reaction products are as follows: (a) for product **2** we have *spiro*-[(3+2)+(3+0)], (b) for product **3** we have *bicyclo*-[(3+2)+(6+0)], and (c) for lactone **J** we have [6+0].

The authors had originally thought that the carbonyl ylide intermediate **B′** could add to dimethylacetylenedicarboxylate as a 1,3-dipole resulting in the [3.2.1] bicyclic product **5**. The net result is a [3+2] cycloaddition reaction. Alternatively, they also thought that the carbonyl ylide written in resonance form **B** could undergo intramolecular proton transfer to azomethine ylide **L** based on theoretical calculations that predicted that azomethine ylide **L** is 19 kcal/mol more thermodynamically stable than carbonyl ylide **B**. It was envisaged that azomethine ylide **L** could also add to dimethylacetylenedicarboxylate in a [3+2] fashion leading to [6,5]-*spiro* product **6**. Both of these possibilities are depicted in Scheme 2. Instead, the experimental outcome of products **2**, **3**, and **4** indicated that neither of these outcomes occurred. It should be noted that the structure of azomethine ylide **L** (written as structure

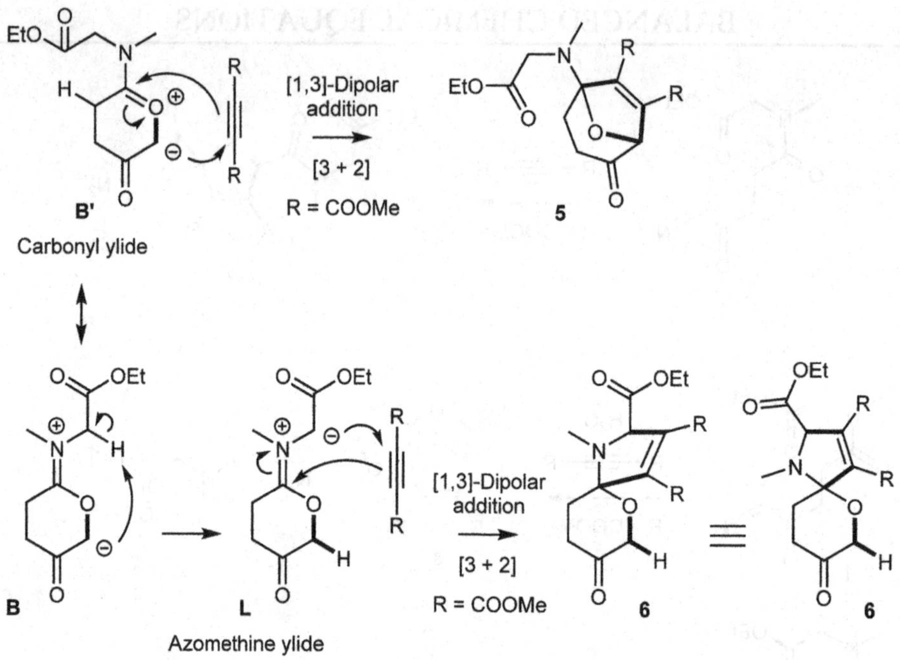

SCHEME 2

11 in the paper) is drawn incorrectly since the negative charge on the carbon atom *alpha* to the ester group is missing.

In support of the authors' conjectured mechanism they isolated the ketene *N,O*-acetal intermediate **C** in 70% yield in the absence of dimethylacetylenedicarboxylate trapping reagent. They also found that **C** was thermodynamically unstable and it hydrolyzed readily to lactone **J** and ethyl methylamino-acetate. When **C** was reacted with dimethylacetylenedicarboxylate in benzene containing water, the authors obtained product **4** and lactone **J**. Despite this evidence, the authors did not carry out theoretical analyses for the three proposed mechanisms for products **2, 3,** and **4.**

ADDITIONAL RESOURCES

We recommend an article by the same authors and two additional articles which cover interesting transformations that involve similar heterocyclic structures.[2–4]

References

1. Padwa A, Zhi L. Novel rhodium(II)-catalyzed cycloaddition reaction of α-diazo keto amides. *J Am Chem Soc.* 1990;112(5):2037–2038. https://doi.org/10.1021/ja00161a080.
2. Padwa A, Precedo L, Semones MA. Model studies directed toward the total synthesis of (±)-ribasine. A tandem cyclization-cycloaddition route leading to the core skeleton. *J Org Chem.* 1999;64:4079–4088. https://doi.org/10.1021/jo990136j.
3. Hu G, Vasella A. Cyclopentanes from *N*-aminoglyconolactams: reaction mechanism and improved access to diazocyclopentanones. *Helv Chim Acta.* 2004;87:2434–2446. https://doi.org/10.1002/hlca.200490218.
4. Jung ME, Min S-J, Houk KN, Ess D. Synthesis and relative stability of 3,5-diacyl-4,5-dihydro-1*H*-pyrazoles prepared by dipolar cycloaddition of enones and α-diazoketones. *J Org Chem.* 2004;69:9085–9089. https://doi.org/10.1021/jo048741w.

Question 71: Pyrrolo[1,2-*a*]benzimidazole Synthesis
by Ortho Nitro Interaction

In a 1996 synthetic study of pyrrolo[1,2-*a*]benzimidazoles, American chemists reacted the dinitro substituted pyrrolidine **1** with Zn/Ac$_2$O to form the three pyrrolo[1,2-*a*]benzimidazole products **2** (16%), **3** (14%), and **4** (9%) as seen in Scheme 1.[1] Provide mechanisms for the formation of these three products.

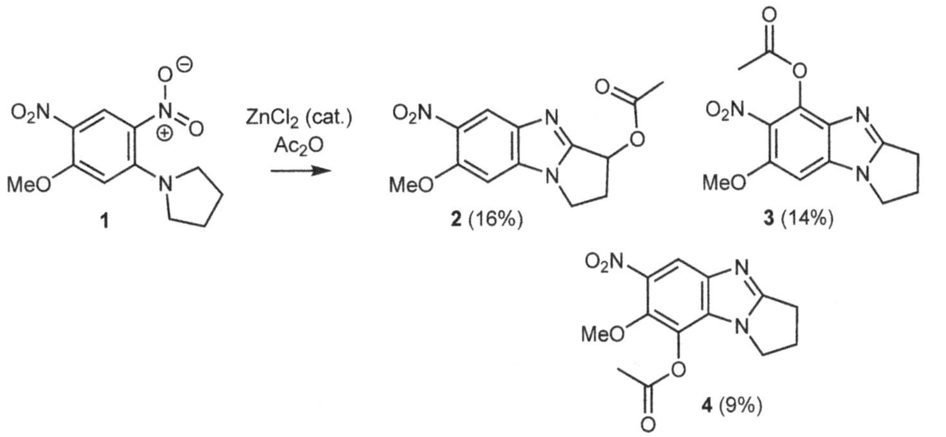

SCHEME 1

SOLUTION: 1 TO 2

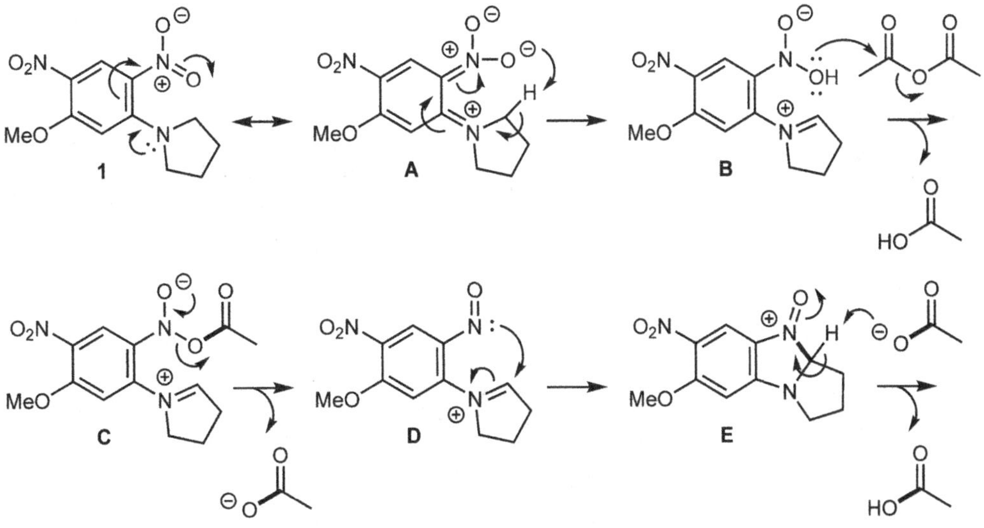

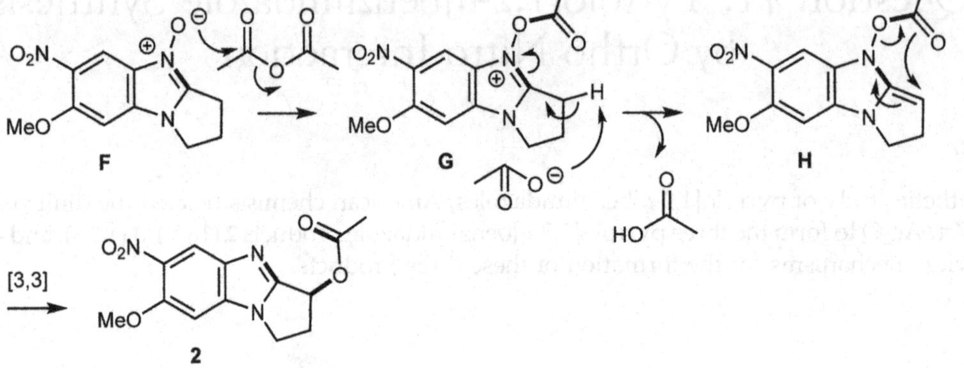

SOLUTION: H TO 3

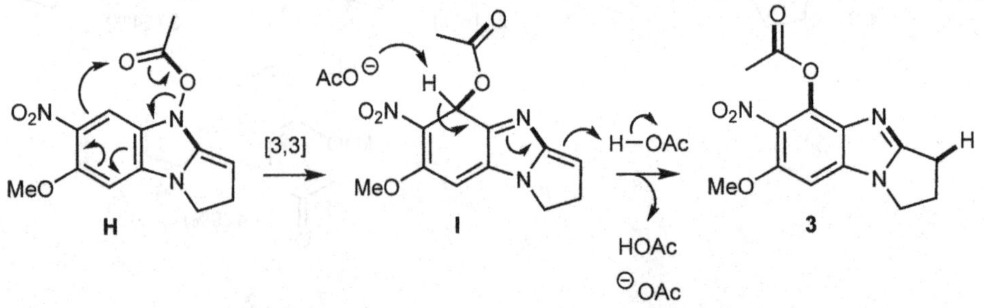

SOLUTION: H TO 4

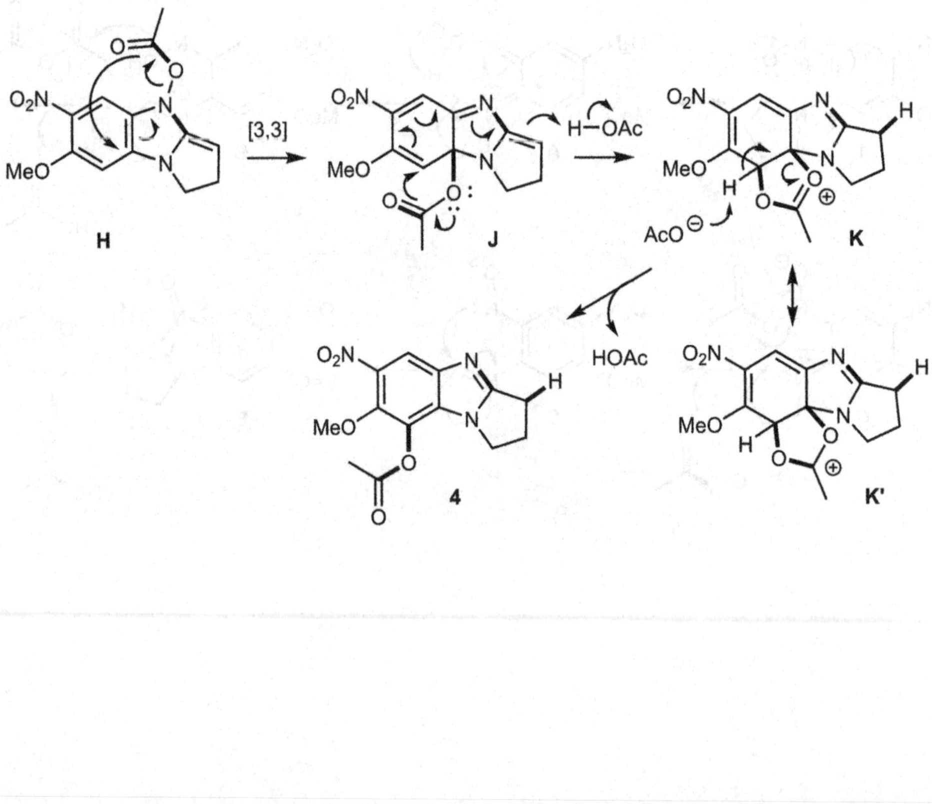

BALANCED CHEMICAL EQUATIONS

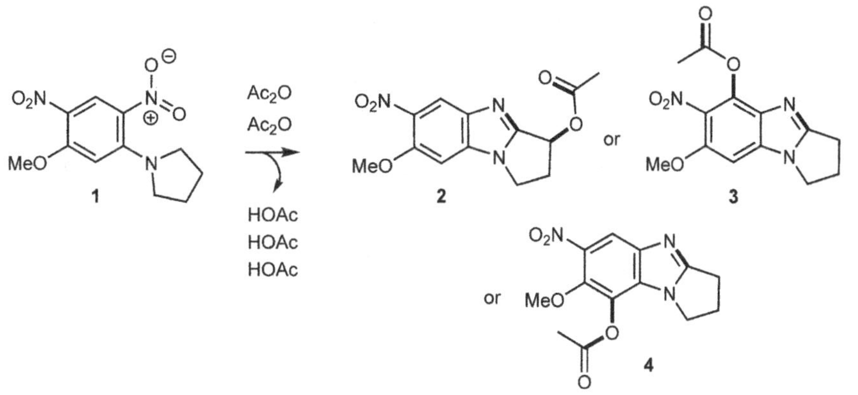

KEY STEPS EXPLAINED

The mechanisms proposed by the authors to explain the formation of the pyrrolo[1,2-*a*]benzimidazole products **2**, **3**, and **4** all begin with reduction of the nitro group of **1** to a nitroso group with the assistance of acetic anhydride to give a nitroso iminium ion **D**. This intermediate then cyclizes to intermediate **E** which deprotonates to benzo[*d*]pyrrolo[1,2-*a*] imidazole 4-oxide **F**. A second equivalent of acetic anhydride mediates the transformation of **F** to the key hub intermediate **H** (7-methoxy-6-nitro-1,2-dihydro-benzo[*d*]pyrrolo[1,2-*a*]imidazol-4-yl acetate). Intermediate **H** can undergo three different [3,3] sigmatropic rearrangement pathways. If migration of the acetyl group occurs from N to C_3 product **2** is formed. If migration of the acetyl group occurs from N to C_5 product **3** is formed. Lastly, if migration of the acetyl group occurs from C_9 to C_8 we have formation of product **4**. We note that the last rearrangement from C_9 to C_8 proceeds through oxonium ion **K** which is in resonance with its carbocation form **K′**. We note that this ion is essentially a carbocation that is resonance stabilized by the two neighboring oxygen atoms. It is also important to note that the authors did not carry out further experimental and theoretical analyses to help support these proposed mechanisms.

A closer examination of the oxidation number changes involved in each reaction indicates that these transformations are internal redox reactions as shown by the analyses given in Schemes 2–4.

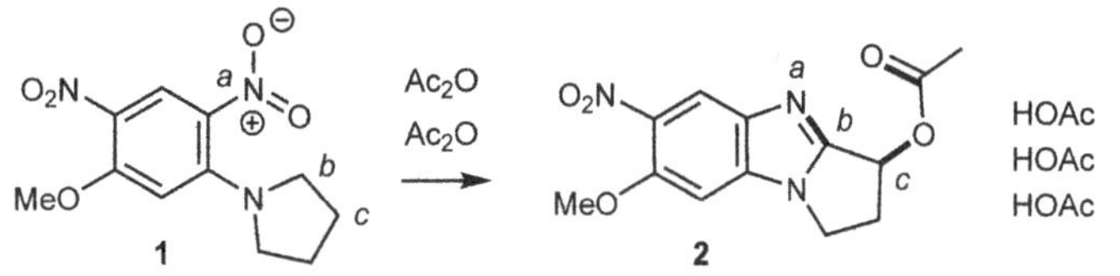

Atom	Reactant	Product	Change	
N_a	+3	−3	−6	$\Delta = -6$
C_b	−1	+3	+4	$\Delta = +6$
C_c	−2	0	+2	

SCHEME 2

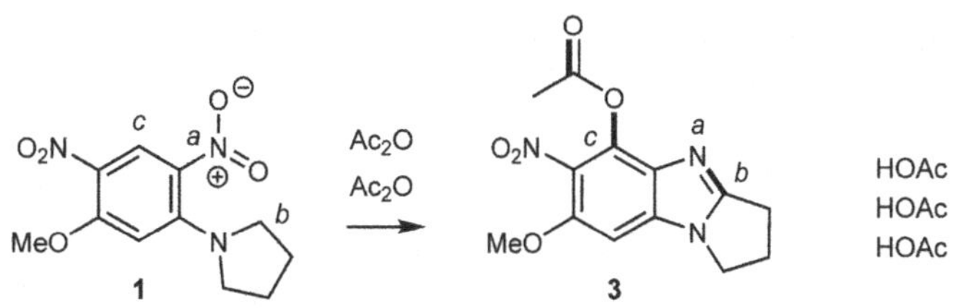

Atom	Reactant	Product	Change	
N_a	+3	−3	−6	$\Delta = -6$
C_b	−1	+3	+4	$\Delta = +6$
C_c	−1	1	+2	

SCHEME 3

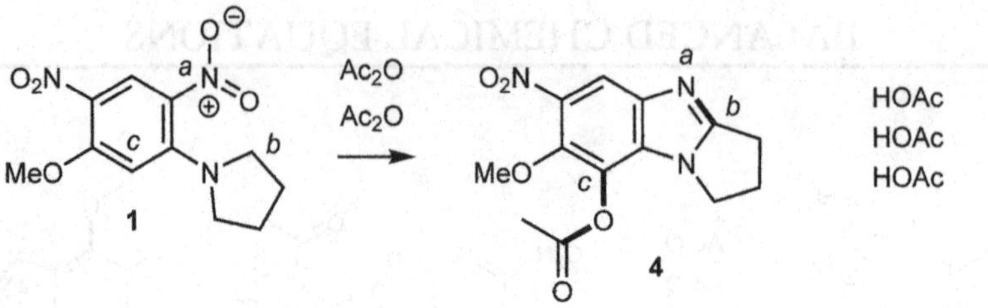

SCHEME 4

Atom	Reactant	Product	Change	
N_a	+3	−3	−6	⟶ Δ = −6
C_b	−1	+3	+4	⟶ Δ = +6
C_c	−1	1	+2	

ADDITIONAL RESOURCES

The authors refer to the migration of OAc group as being a consequence of a "t-amino effect." We recommend an excellent review of the "t-amino effect" and its role in synthetic application.[2] We also recommend several articles which describe interesting mechanistic transformations which involve nitriles, nitro group, and nitroso group-containing compounds.[3–6]

References

1. Skibo EB, Islam I, Schulz WG, Zhou R, Bess L, Boruah R. The organic chemistry of the pyrrolo[1,2-a]benzimidazole antitumor agents. An example of rational drug design. *Synlett.* 1996;1996(4):297–309. https://doi.org/10.1055/s-1996-5399.
2. Meth-Cohn O, Suschitzky H. Heterocycles by ring closure of ortho-substituted t-anilines (the t-amino effect). *Adv Heterocycl Chem.* 1972;14:211–278. https://doi.org/10.1016/S0065-2725(08)60954-X.
3. Hameed A, Anwar A, Yousaf S, Khan KM, Basha FZ. Tetra-n-butylammonium fluoride mediated dimerization of (a-methylbenzylidene) malononitriles to form polyfunctional 5,6-dihydropyridine derivatives under solvent-free conditions. *Eur J Chem.* 2012;3(2):179–185. https://doi.org/10.5155/eurjchem.3.2.179-185.562.
4. Haiss P, Zeller K-P. The mechanism of the ortho-methylation of nitrobenzenes by dimethylsulfonium methylide. *Eur J Org Chem.* 2011;2011:295–301. http://doi.org/10.1002/ejoc.201001091.
5. Hanusek J, Machacek V. Intramolecular base-catalyzed reactions involving interaction between benzene nitro groups and *ortho* carbon chains. *Collect Czechoslov Chem Commun.* 2009;74(5):811–833. https://doi.org/10.1135/cccc2008216.
6. Lipilin DL, Karslyan EE, Churakov AM, Strelenko YA, Tartakovsky VA. Intramolecular interaction between orthoazido and azoxy groups as a new way of forming a N-N bond. Synthesis of 2-alkylbenzotriazole 1-oxides. *Russ Chem Bull Int Ed.* 2005;54(4):1013–1020. https://doi.org/10.1007/s11172-005-0350-0.

Question 72: Horner-Wittig Reaction on a Bifuranylidenedione

In 1996 British chemists reacted the diarylbifuranylidenedione **1** with two equivalents of the stable ylide Ph_3PCHCO_2Me in refluxing toluene under nitrogen for 24h to realize a mixture of products among which were the expected mono-Horner-Wittig ester **2** (26% yield) and the tricyclic product **3** (11% yield, Scheme 1).[1] Provide mechanisms for the formation of these products.

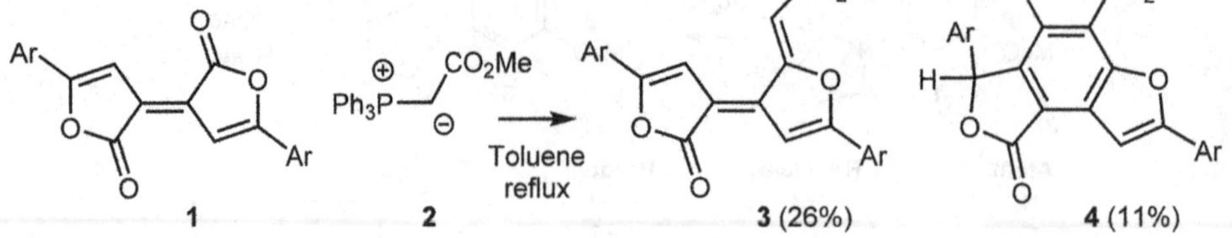

SCHEME 1

SOLUTION: 1 AND 2 TO 3

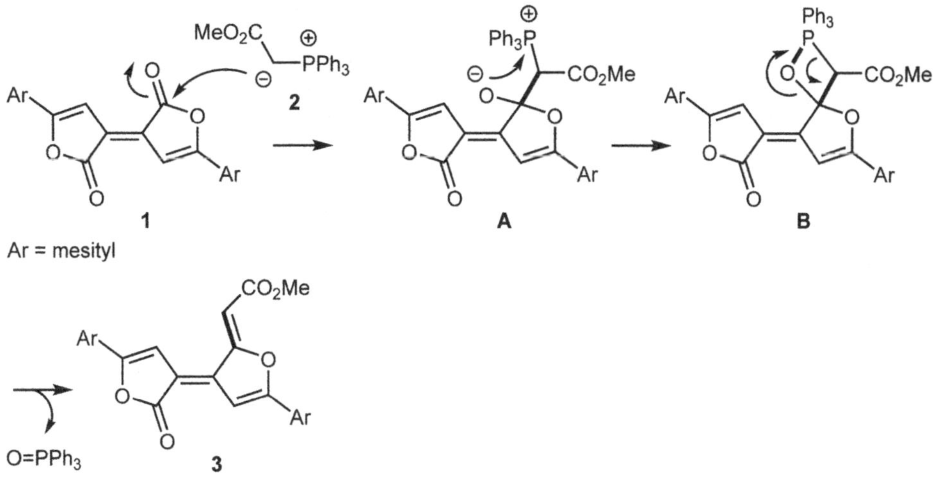

Ar = mesityl

SOLUTION: 3 TO 4

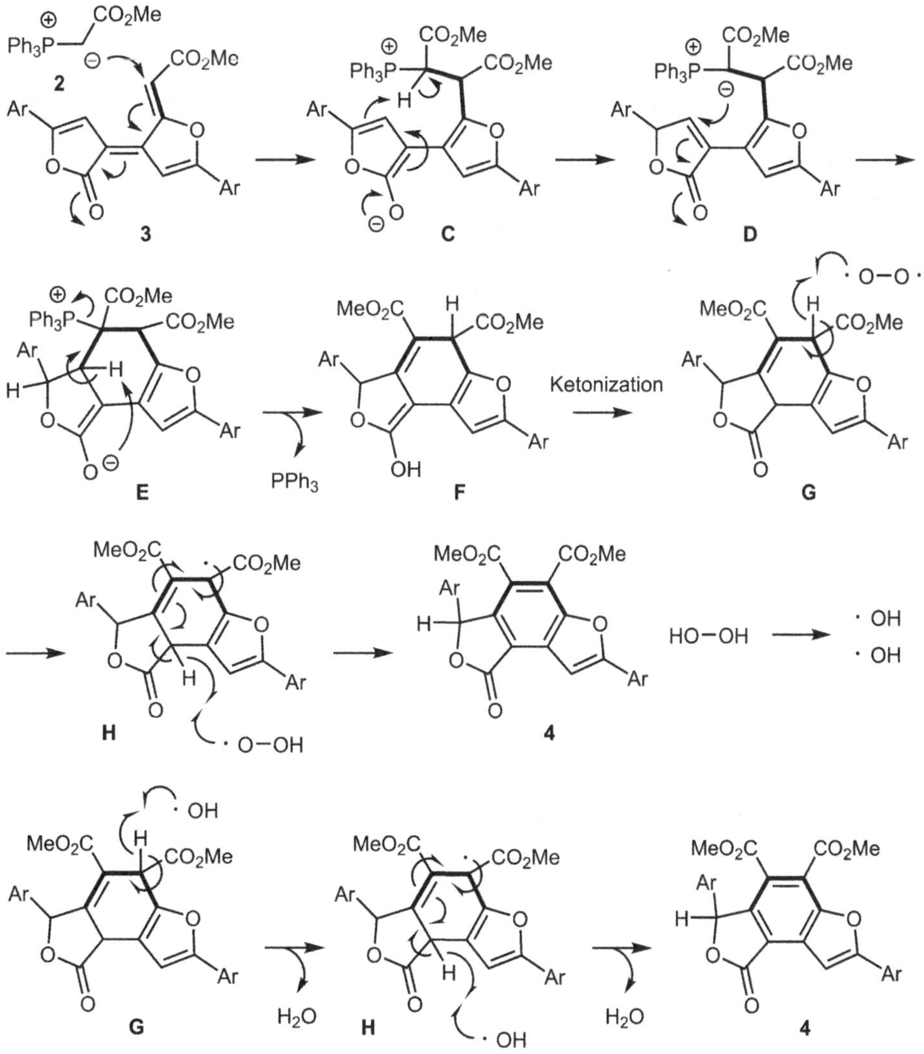

BALANCED CHEMICAL EQUATIONS

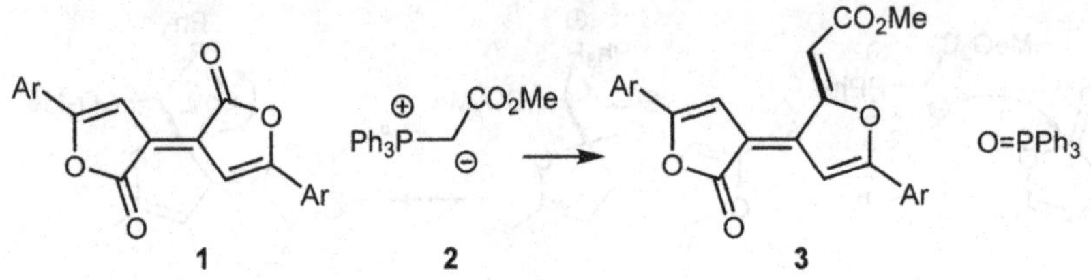

KEY STEPS EXPLAINED

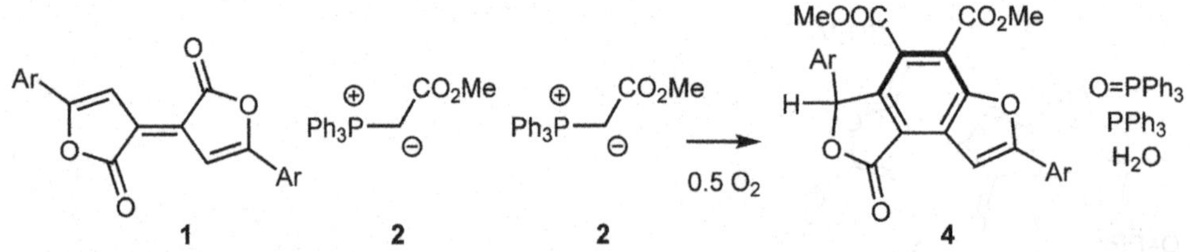

The mechanism proposed by the authors to explain the transformation of **1** and **2** to **3** follows that of the Horner-Wittig reaction where the by-product is triphenylphosphine oxide. Interestingly, product **3** undergoes further reaction with a second equivalent of ylide **2** via a Michael addition which gives the zwitterionic intermediate **C**. What then follows is ketonization of **C** to **D**, cyclization to enolate **E**, and elimination of triphenylphosphine by-product to form enol **F**. Ketonization of **F** gives lactone **G** which undergoes autoxidation to product **4**. The autoxidation process follows a radical-type mechanism which involves ground state triplet oxygen gas from air causing the step progression from **G → H → 4**. The resulting by-product is hydrogen peroxide which dissociates into two hydroxyl radicals. These in turn react with another molecule of **G** following the same sequence thus producing two molecules of water as by-products. It is important to note that the authors characterized products **3** and **4** by IR, ^{1}H NMR, mass spectrometry, and X-ray crystallography. They also identified triphenylphosphine among the product mixture. Nevertheless, they did not carry out further experimental or computational analysis to help support their mechanistic proposal. We note lastly that the ring construction mapping for product **4** is [4+1+1].

ADDITIONAL RESOURCES

We refer the reader to an article which discusses the mechanism of the alkaline hydrolysis of substituted (*E*)-5,5′-diphenylbifuranylidenediones such as **1**.[2]

References

1. Crombie L, Darwish B, Jones RCF, Toplis D, Begley MJ. Reactions between (*E*)·5,5′-dimesitylbifuranylidenedione and the Horner-Wittig reagent. *Tetrahedron Lett*. 1996;37(51):9255–9258. https://doi.org/10.1016/S0040-4039(96)02137-5.
2. Bowden K, Etemadi R, Ranson RJ. Reactions of carbonyl compounds in basic solutions. Part 17. The alkaline hydrolysis of substituted (*E*)-5,5′-diphenylbifuranylidenediones and 3,7-diphenylpyrano[4,3-c]pyran-1,5-diones. *J Chem Soc Perkin Trans 2*. 1991;1991:743–746. https://doi.org/10.1039/P29910000743.

Question 73: Lactone Ammonolysis

In 1996 Italian chemists subjected lactone **1** to ammonolysis conditions (treatment with ammonia in ethanol at room temperature for 5h) expecting that ring opening would give amide **4** as a product (Scheme 1).[1] Instead, the authors isolated amide **2** and (R)-lactamide **3** and showed by means of kinetic and spectroscopic studies that amide **4** was an unstable intermediate whose amide group participates in a second ammonolysis that gives **2** and **3**. Suggest a mechanism which is consistent with these observations.

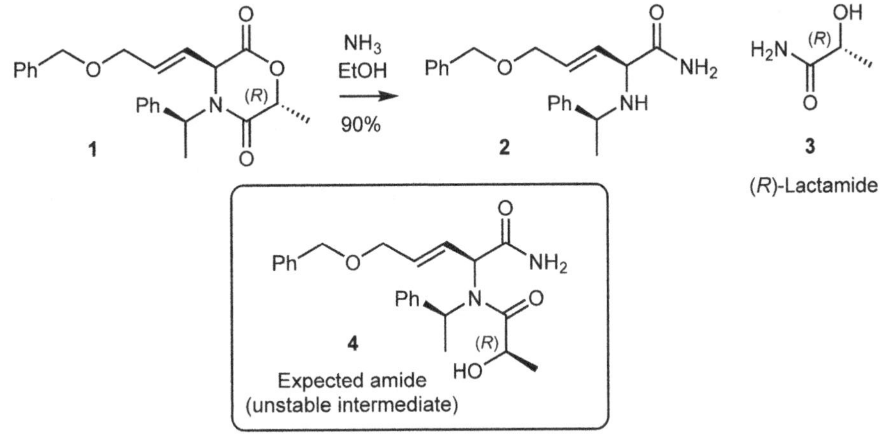

SCHEME 1

SOLUTION

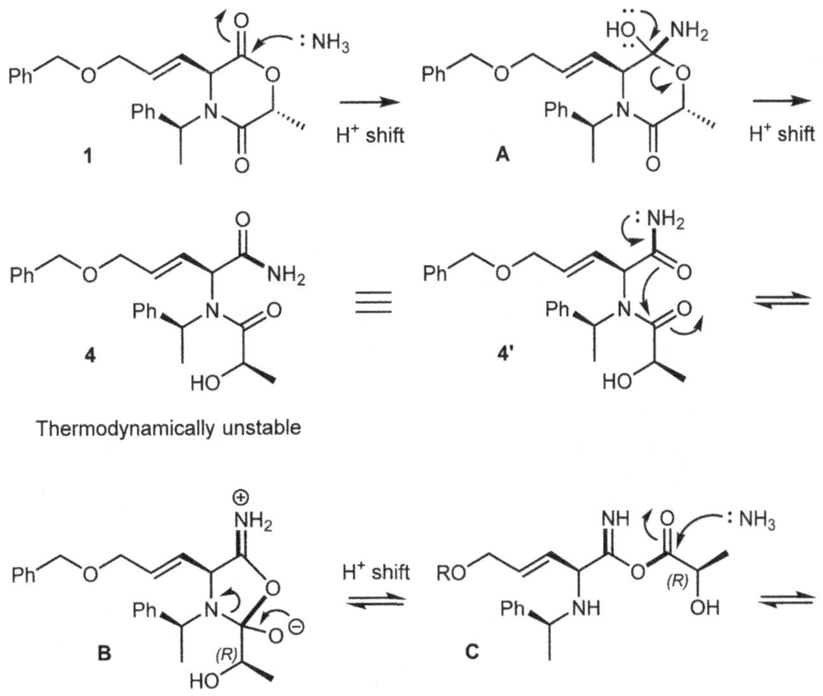

BALANCED CHEMICAL EQUATION

KEY STEPS EXPLAINED

The mechanism proposed for the transformation of **1** to **2** and **3** begins with nucleophilic ring opening of lactone **1** by ammonia which forms amide **4**. The primary amide group of **4** is then involved in neighboring group participation with respect to ammonolysis of the secondary amide group. Cyclization of **4** to intermediate **B** followed by ring opening to imidate **C** followed by ammonolysis by a second equivalent of ammonia leads to the production of amide **2** and lactamide **3**.

The authors prepared amide **2** by reacting lactone **1** in a saturated ammonia-ethanol solution for 30 min at 0°C and at room temperature overnight. They obtained amide **2** in 90% and were able to recover lactamide **3** from the aqueous work-up presumably also in 90% yield since the two products arise concomitantly from the same mechanism. The authors were also able to isolate the thermodynamically unstable amide **4** which is an intermediate along the way from **2** to **1** by carrying out the reaction for 15 min at room temperature and then quenching the reaction with hydrochloric acid and purifying it quickly by column chromatography. They noted that amide **4** reverts back to amide **2** in a few hours owing to the proximity of the hydroxyl and amide groups.

In support of the proposed mechanism the authors carried out a kinetic study of the reaction under pseudo-first-order conditions at various temperatures and determined two rate constants: (a) a second-order rate constant for the reaction of ammonia with starting lactone **1** yielding amide **4**; and (b) a second-order rate constant for the reaction of amide **4** with ammonia yielding products **2** and **3**. At 25°C the magnitudes of the rate constants were $k_1 = 9.54\text{E} - 4\,\text{M}^{-1}\,\text{s}^{-1}$ and $k_2 = 7.3\text{E} - 5\,\text{M}^{-1}\,\text{s}^{-1}$. The kinetic scheme is shown in Fig. 1. Two kinds of kinetic experiments were carried out. In the first experiment beginning from lactone **1**, the appearance and disappearance of amide **4** was monitored by time-resolved UV spectroscopy at 230 nm. The kinetic traces were observed to be biphasic showing an initial exponential growth followed by an exponential decay. This behavior corresponded to the transformation $1 \rightarrow 4 \rightarrow 2+3$. In the second experiment beginning from amide **4**, the disappearance of **4** was monitored at 230 nm. The kinetic traces were pseudo-first-order exponential decays. This behavior corresponded to the transformation $4 \rightarrow 2+3$. The data for

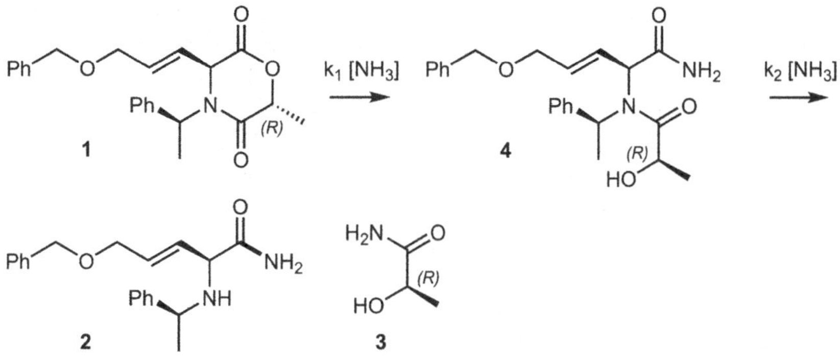

FIG. 1 Kinetic scheme showing two steps where each depends on the concentration of ammonia.

the disappearance of **4** from the biphasic kinetic curves of the first experiment matched the data for the disappearance of **4** from the monotonic decay curves of the second experiment.

Since both rate constants were determined at various temperatures, Eyring plots were constructed from which the following enthalpic and entropic activation parameters were estimated: (a) k_1 process: $\Delta H_1^{\#} = 5.3\,\mathrm{kcal\,mol^{-1}}$; $\Delta S_1^{\#} = -54.2\,\mathrm{cal\,K^{-1}\,mol^{-1}}$; and (b) k_2 process: $\Delta H_2^{\#} = 11.5\,\mathrm{kcal\,mol^{-1}}$; $\Delta S_2^{\#} = -38.7\,\mathrm{cal\,K^{-1}\,mol^{-1}}$. The enthalpic energy barrier for the second step was found to be twice as high as for the first step. The lower barrier for the lactone ammonolysis is according to the authors due to a "favoured approach of the ammonia to the carboxylic group, owing to an attractive dipole-dipole interaction between the ammonia and the polar amide group lying in the opposite side of the heterocyclic ring." According to the authors the barrier for ammonolysis of amide **4** "falls in the range for the ammonolysis of phenyl acetates in water as well, (and) cannot be attributed to the unassisted transamidation of **2** (amide **4** in our notation) for which a large value should be expected, considering that even the alkaline hydrolysis of amides requires $16–23\,\mathrm{kcal\,mol^{-1}}$." With respect to the entropic activation parameters, the negative values are consistent with a loss of entropy and therefore ordered transition states. For the lactone **1** ammonolysis, the ring opening step and the increased solvation of the developing charged transition state are associated with a significant loss of entropy. The authors ascribe the lesser negative entropy of activation for the ammonolysis of amide **4** solely on the "solvent effect on the zwitterionic tetrahedral intermediates" **B** and **E**.

The reader is reminded of the following important guidelines with respect to interpretation of magnitudes of $\Delta H^{\#}$ values. The magnitude of enthalpy of activation parameters is governed by the following factors pertaining to the rate determining step of a reaction: (a) the degree of bond breakage and bond formation occurring in the transition state, (b) the degree of steric hindrance occurring in the transition state, and (c) the comparative degree of solvation between the initial state and the transition state. Large positive values of enthalpy of activation are associated with the following processes: (a) bond breakage occurring predominantly in the transition state, and (b) increase of steric hindrance occurring in the transition state. Small positive values are associated with: (a) bond breakage and bond formation occurring simultaneously in the transition state to a balanced degree, and (b) increase in solvation of the transition state compared to the initial state such as when developing charges on atoms occur as the reaction proceeds in a polar solvent. The ammonolysis of lactone **1** involves (a) C—N bond formation and C—O bond breakage, and (b) has no steric issues. Hence, this is consistent with the small value observed for $\Delta H^{\#}$. The ammonolysis of amide **4** involves neighboring amide group participation before ammonia attack and is 13 times slower than ammonolysis of lactone **1**. The larger observed value of $\Delta H^{\#}$ is attributable to increased sterics and a higher degree of bond breakage over bond formation.

The reader is also reminded of the several important guidelines with respect to interpretation of magnitudes of $\Delta S^{\#}$ values which are mechanistically more diagnostic than $\Delta H^{\#}$ values. The magnitude of entropy of activation parameters, for instance, is governed by the following factors pertaining to the rate determining step of a reaction: (a) molecularity of the reaction, (b) the change in translational and rotational degrees of freedom in the transition state, and (c) the comparative degree of solvation between the initial state and the transition state. The sign of a $\Delta S^{\#}$ value is particularly noteworthy—negative values are associated with increased order and reduced degrees of freedom in the transition state over the initial state, whereas positive values are associated with exactly the opposite trends. Large positive $\Delta S^{\#}$ values are associated with an increase of translational and rotational degrees of freedom in the transition state over the initial state such as in unimolecular dissociative or fragmentation reactions involving primarily bond breaking, or ring opening reactions. Large negative $\Delta S^{\#}$ values are associated with (a) an increase of molecularity of the reaction (bimolecular reactions generally have more negative entropies of activation compared to unimolecular reactions), (b) a comparative increase of charge and polarity in the transition state over the initial state, and (c) a comparative increase of solvation of the transition state over the initial state. Examples are unimolecular rearrangements and cyclization reactions. In this context, both ammonolysis reactions have negative $\Delta S^{\#}$ values suggesting that there is less degree of freedom in each of their transition states compared to the initial states. The more negative value for the ammonolysis of amide 4 indicates that that reaction experiences a greater degree of order attributable to the neighboring group participation of the primary amide group and that the transition state is more charged and polar and therefore more prone to be solvated than in the case of the ammonolysis of lactone 1.

In addition to the kinetic study presented by the authors, we also suggest that this system is nicely amenable to a computational study to corroborate the precise rate determining steps in each ammonolysis by determining their respective energy reaction coordinate diagrams. From these diagrams thermodynamic energy differences between initial and final states may be determined for each elementary step, and thermokinetic energy differences may be estimated between initial and transition states for each elementary step. In addition, geometry optimizations of each product and transition state structure along the mechanism pathways will indicate the degree of steric hindrance and the energetics of key bond making and bond forming processes, which stand to support the hypothesis of neighboring group participation.

ADDITIONAL RESOURCES

We highly recommend an article by the same authors which discusses a related amide neighboring group participation process that facilitates the acid hydrolysis of methyl ether linkages.[2] Here, the authors again supplement their analysis with kinetic studies.

References

1. Arcelli A, Porzi G, Sandri S. Unusual ammonolysis of a secondary amide assisted by unsubstituted vicinal amide group. *Tetrahedron*. 1996;52 (11):4141–4148. https://doi.org/10.1016/S0040-4020(96)00075-0.
2. Arcelli A, Porzi G, Sandri S. Catalytic intramolecular participation of amide group in the acid hydrolysis of methyl ether linkage. *Tetrahedron*. 1995;51(35):9729–9736. https://doi.org/10.1016/0040-4020(95)00556-N.

Question 74: Intramolecular Schmidt Reaction

In 1996 chemists from the University of Kansas investigated the intramolecular Schmidt reaction of alkyl azide 1 with its ketal group under acidic conditions which forms lactam 2 in 72% yield (Scheme 1).[1] Elucidate a mechanism for this transformation.

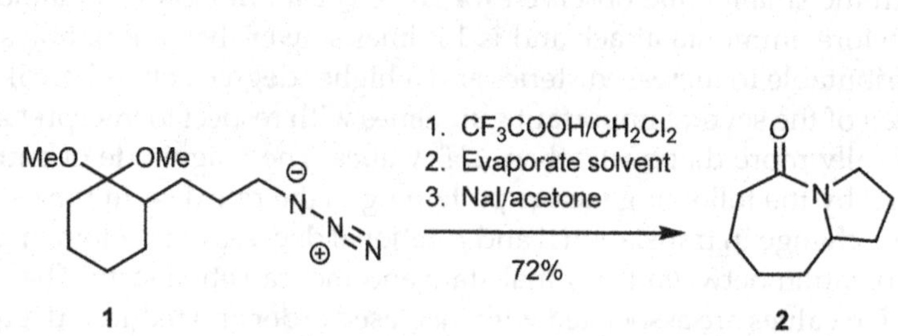

SCHEME 1

SOLUTION

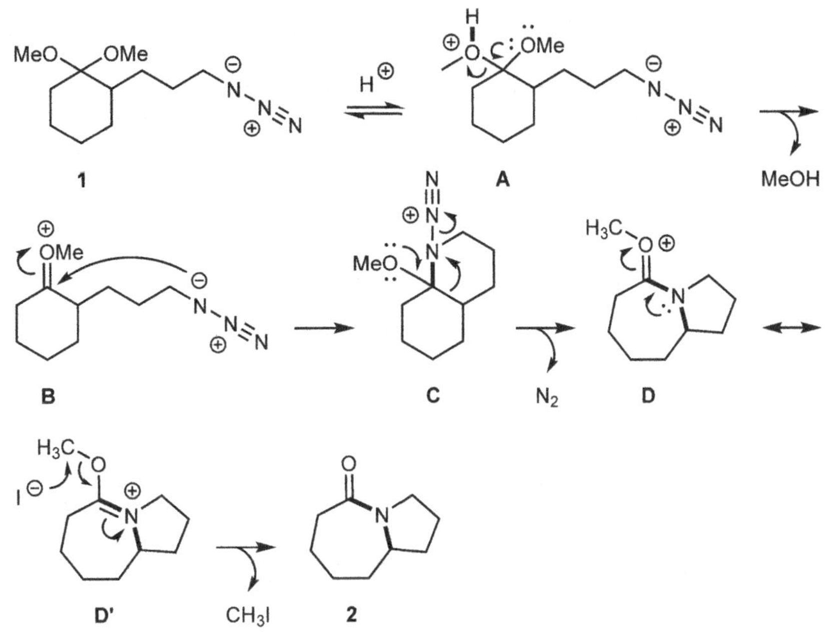

BALANCED CHEMICAL EQUATION

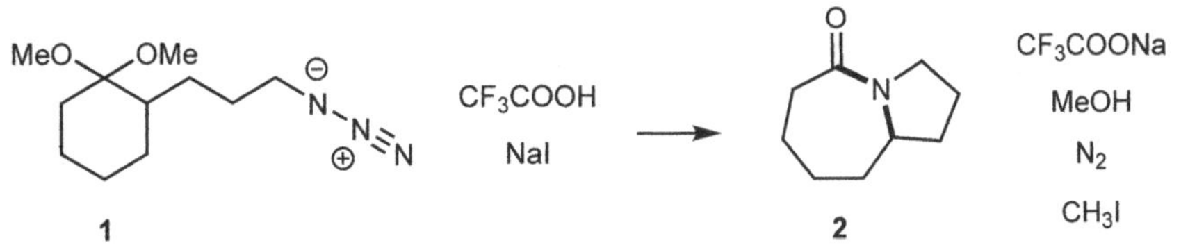

KEY STEPS EXPLAINED

The mechanism proposed for the Schmidt reaction begins with an acid-catalyzed methanolysis of the ketal group of **1** to which forms oxonium ion **B** via **A**. Intermediate **B** then cyclizes to intermediate **C** having a [4.4.0] bicyclic framework which then undergoes a Schmidt rearrangement from **C** to the [5.3.0] bicyclic framework **D** whereupon nitrogen gas is eliminated. The methyl group of oxonium ion **D** (which is stabilized by resonance in an iminium ether form **D'**) is removed by iodide ion giving iodomethane and ε-lactam product **2**. We note that the ring construction mapping for the fused bicyclic product is [(6+1)+(5+0)]. It was also observed that in the absence of the dealkylation step from **D** to **2**, the reaction did not proceed. Nevertheless, in another case where a cyclic ketal was present, no dealkylation was needed. We refer the reader to the original article for more details of the synthetic scope of this transformation. Lastly, the authors did not undertake further experimental and theoretical work to support this mechanism.

ADDITIONAL RESOURCES

We recommend two excellent articles by the same authors where they further explore the scope and stereochemical implications of the Schmidt reaction and other transformations involving rearrangements of bicyclic nitrones to lactams.[2,3]

References

1. Mossman CJ, Aube J. Intramolecular Schmidt reactions of alkyl azides with ketals and enol ethers. *Tetrahedron.* 1996;52(10):3403–3408. https://doi.org/10.1016/0040-4020(96)00037-3.
2. Milligan GL, Mossman CJ, Aube J. Intramolecular Schmidt reactions of alkyl azides with ketones: scope and stereochemical studies. *J Am Chem Soc.* 1995;117(42):10449–10459. https://doi.org/10.1021/ja00147a006.
3. Zeng Y, Smith BT, Hershberger J, Aube J. Rearrangements of bicyclic nitrones to lactams: comparison of photochemical and modified Barton conditions. *J Org Chem.* 2003;68:8065–8067. https://doi.org/10.1021/jo035004b.

Question 75: Indolizidones by Intermolecular Photochemical Reaction

In 1996 chemists from India used an anthraquinone photosensitized irradiation approach (using a pyrex-filtered output from a 450 W medium pressure Hanovia lamp) to carry out the transformation of *N*-allylpiperidine **1** and methyl methacrylate **2** in acetonitrile which formed a 2:3 diastereomeric mixture (2*S*, 9*S*:2*R*, 9*R*) of indolizidone **3** in 60% yield (Scheme 1).[1] Suggest a mechanism for this transformation.

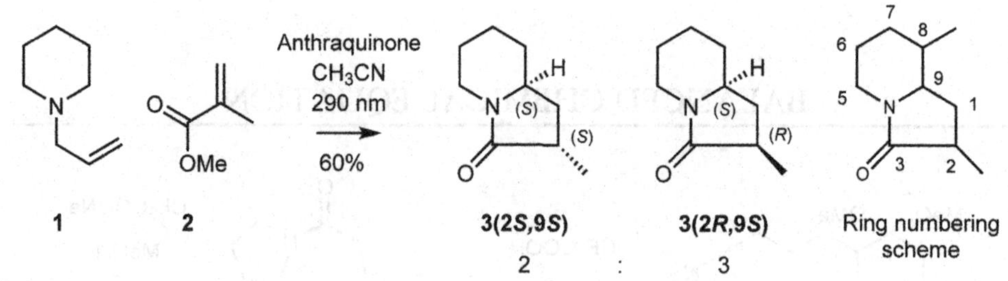

SCHEME 1

SOLUTION: 1 TO E

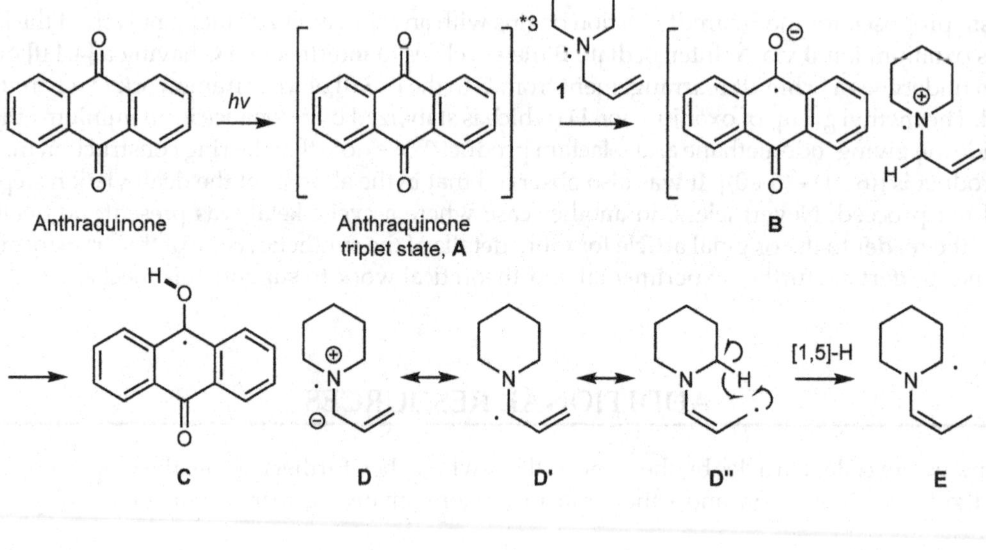

SOLUTION: E TO G

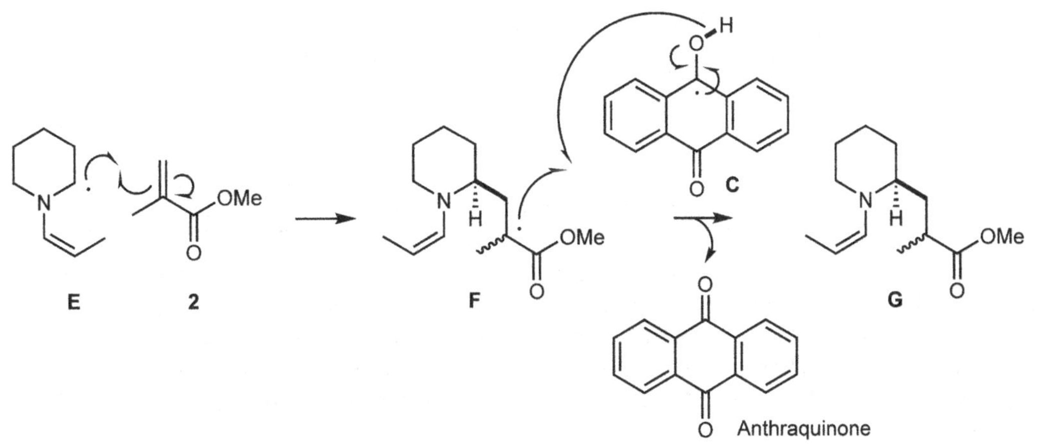

SOLUTION: G TO 3 (AUTHOR MECHANISM)

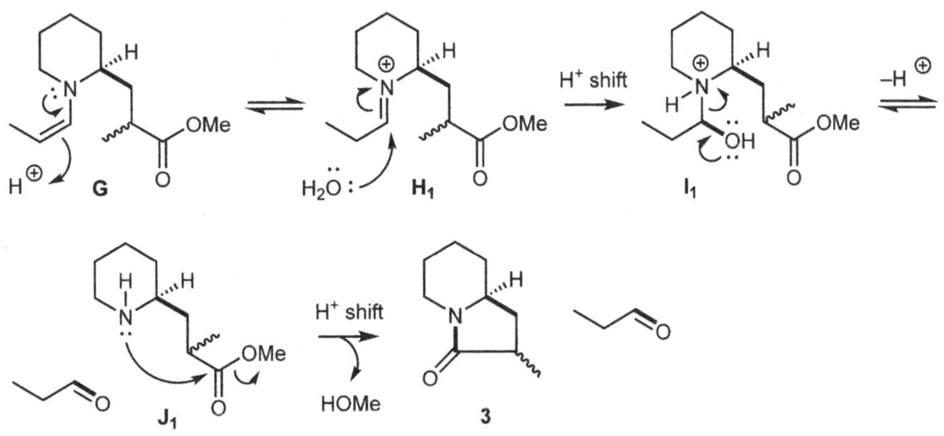

SOLUTION: G TO 3 (ALTERNATIVE PROPOSAL)

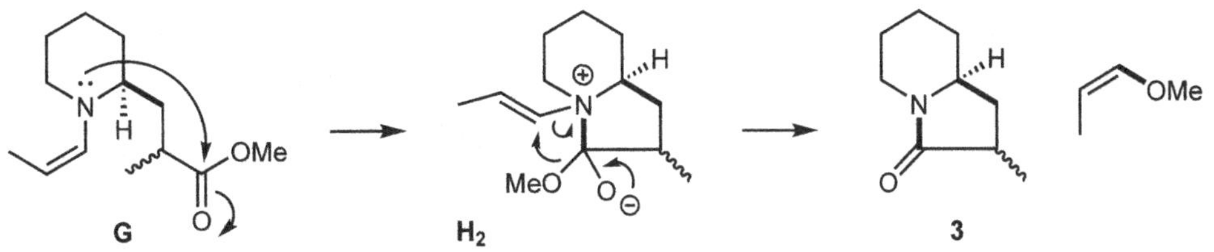

BALANCED CHEMICAL EQUATIONS

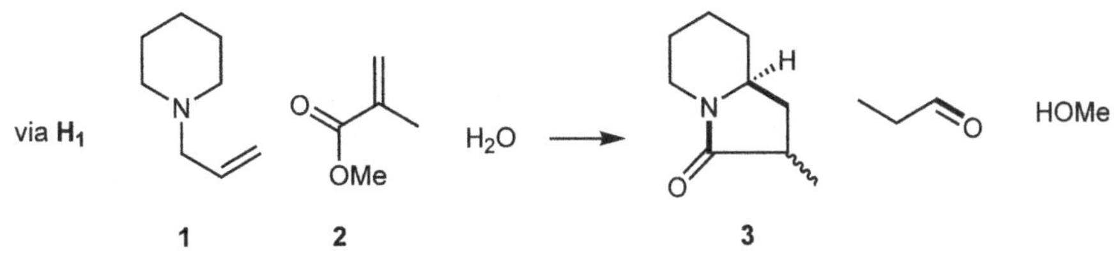

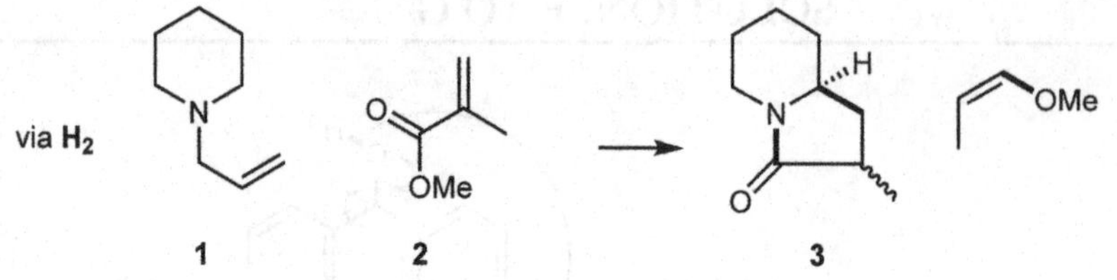

1   2   3

KEY STEPS EXPLAINED

The mechanism proposed by the authors to explain the transformation of **1** and **2** to the product mixture **3** begins with light-induced excitation of anthraquinone to its triple state form **A**. We note that anthraquinone acts as triplet sensitizer for this photoreaction. Next, an electron transfer reaction occurs between **A** and N-allyl-piperidine **1** which forms a solvent-caged radical ketyl anion-radical aminium cation pair **B** which undergoes proton transfer to form the ketal radical **C** and the α-aminoallyl radical **D**. The authors argue based on literature precedent (see references 10 and 11 in Das et al.[1]) that the α-aminoallyl radical **D'** is formed in a predominant manner over direct formation of the α-aminoalkyl radical **E** due to allylic stabilization which facilitates in cage deprotonation from **B** to **D**. If there were no allylic group in **1**, generation of the α-aminoalkyl radical would likely occur directly from **B**. Once **D** is formed, isomerization via [1,5]-hydrogen transfer occurs to form the α-aminoalkyl radical **E**, a process which the authors also argue is well documented in the literature (see references 7, 14, and 15 in Das et al.).[1] Once **E** is generated, it is possible for it to react with methyl methacrylate **2** to form a new C—C bond and generate radical **F**. This radical species abstracts a hydrogen atom from ketal radical **C** which restores anthraquinone and produces ester intermediate **G**. At this point the authors introduce acidic aqueous conditions that help transform **G** to product **3** with formation of by-products propanal and methanol (which the authors did not specify). Nevertheless, we point out that the reaction conditions as described in the experimental procedure indicate that the fused bicyclic product **3** is formed in the solution *before* work-up and chromatographic isolation (see the authors' GC (gas chromatographic) analysis of the reaction mixture after 2 h of photolysis). It is therefore reasonable to expect that the remaining steps from **G** to product **3** must take place in the absence of any acid or water. We thus propose in the alternative mechanistic path from **G** to **3** that intramolecular cyclization of **G** results in formation of zwitterionic intermediate **H₂** which undergoes methoxide ion transfer to the allyl group which directly gives **3** and 1-methoxy-propene by-product. We believe this hypothesis is easy to check experimentally by preparing an authentic sample of 1-methoxy-propene, spiking the photoreaction product mixture with it, and running a GC-MS analysis to see if both the retention time and mass spectrum profile of the spiked sample match those of the authentic sample. It should be recognized that 1-methoxy-propene is the methyl vinyl ether of propanal. We also recommend carrying out a theoretical analysis to calculate the energy barrier for formation of the relatively sterically hindered intermediate **H₂** as compared to the formation of **H₁**. We also note that the ring construction mapping for this reaction is [3 + 2]. Further experimental evidence for intermediate **F** comes from the fact that the authors managed to isolate product **4** which arises from cyclization of the radical intermediate **F** to **K** followed by proton abstraction via **C** with regeneration of anthraquinone (Scheme 2). Lastly, we note that the authors made a mistake in their article in depicting the structures of **3** and **4** (see structures **15** and **6**, respectively, in Ref. 1). The mistake has to do with the position of the methyl substituent in **3** and **4** (i.e., the authors place it one carbon atom away from where it actually belongs).

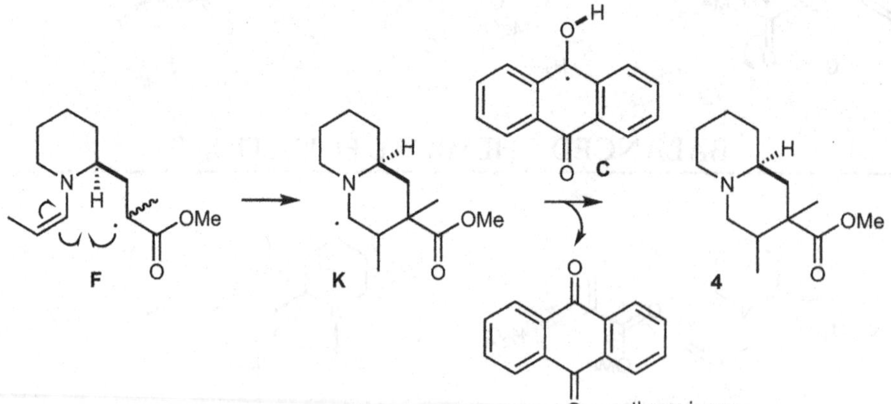

F   K   C   4   anthraquinone

SCHEME 2

ADDITIONAL RESOURCES

We recommend two articles which discuss further examples of transformations that involve interesting photo-induced rearrangements.[2,3]

References

1. Das S, Kumar D, Shivaramayya K, George MV. Formation of lactams via photoelectron-transfer catalyzed reactions of *N*-allylamines with α,β-unsaturated esters. *Tetrahedron*. 1996;52(10):3425–3434. https://doi.org/10.1016/0040-4020(96)00022-1.
2. Sakamoto M, Mino T, Fujita T. A novel approach to heterocycles via photocycloaddition reactions of nitrogen-containing heteroaromatics. *J Synth Org Chem Jpn*. 2002;60(9):837–846. https://doi.org/10.5059/yukigoseikyokaishi.60.837.
3. Eberbach W, Hensle J. The photorearrangement of *o*-vinyl diaryl ethers into *o*-hydroxy stilbenes: evidence against dipolar intermediates. *Tetrahedron Lett*. 1993;34(30):4777–4780. https://doi.org/10.1016/S0040-4039(00)74086-X.

Question 76: An Unexpected Product From Amine Oxidation

In 1997 chemists from India reacted *N*-methylaniline with 4-chlorobut-2-yn-1-ol **1** in acetone at reflux in the presence of anhydrous K_2CO_3 to form the tertiary amine **2** (Scheme 1).[1] A dilute solution of **2** was then treated at room temperature with *m*-CPBA and shown to form the dimeric oxidation product **3** in 56% yield. Suggest a mechanism for the formation of **3** from **2**.

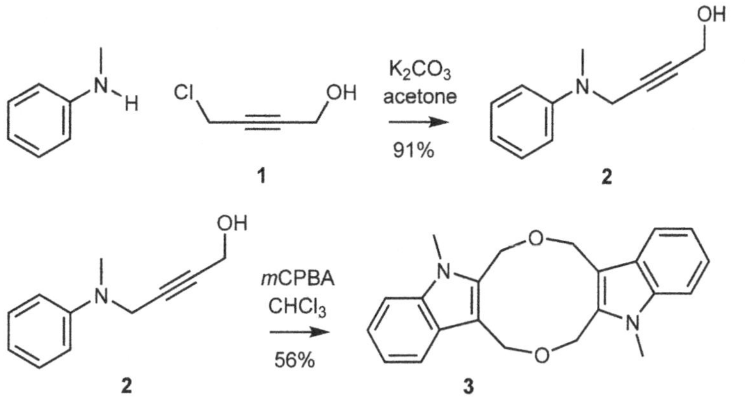

SCHEME 1

SOLUTION

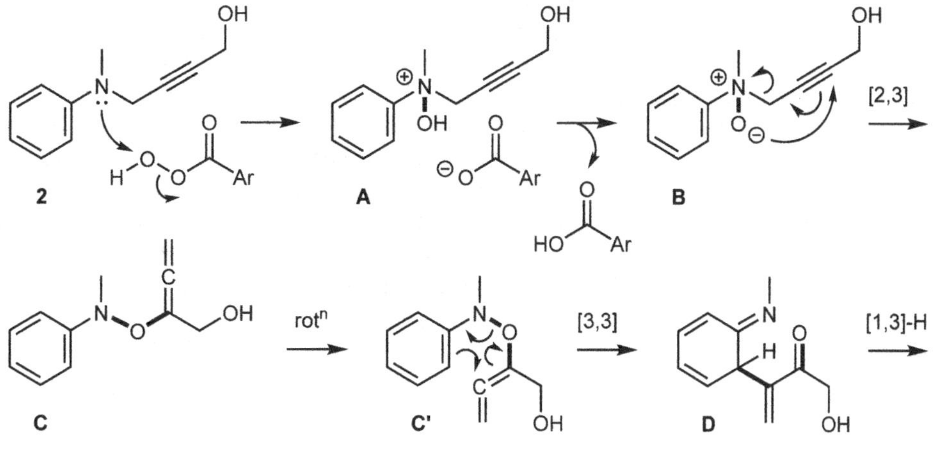

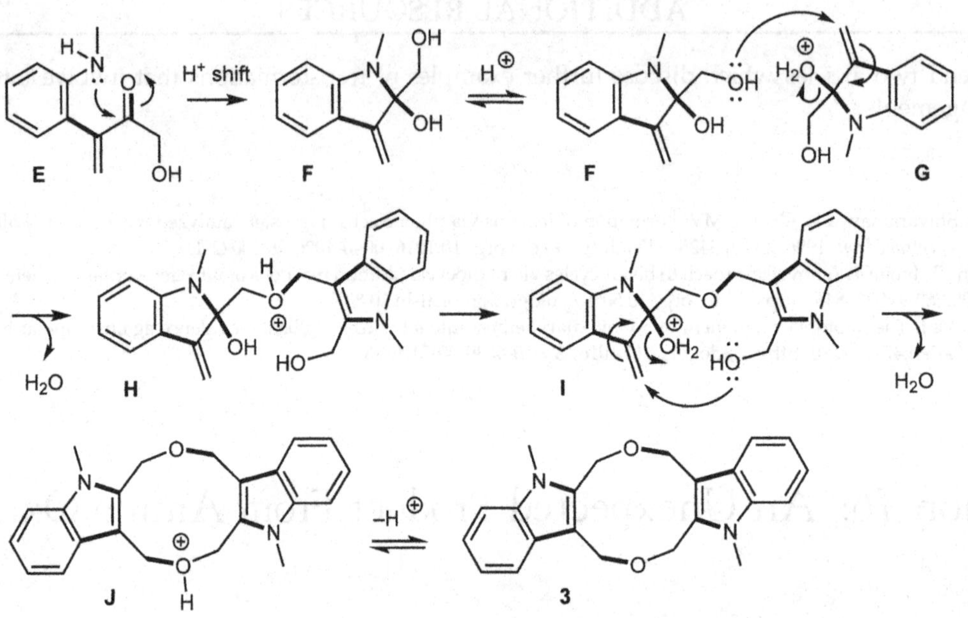

BALANCED CHEMICAL EQUATION

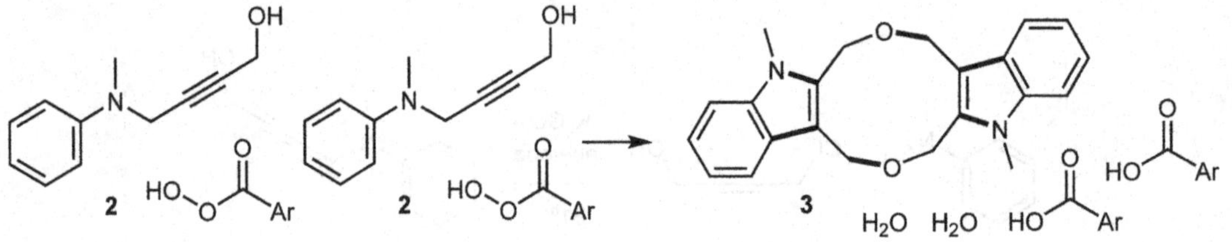

KEY STEPS EXPLAINED

The mechanism proposed by the authors begins with *m*-chloroperbenzoic acid oxidation of the amino group of **2** which forms the *N*-oxide **B** which then undergoes a [2,3]-sigmatropic rearrangement to allene **C**. After N—O bond rotation, allene **C/C′** undergoes a [3,3] sigmatropic rearrangement to form intermediate **D**. This intermediate experiences a [1,3]-protic shift yielding intermediate **E** which cyclizes to intermediate **F**. Protonation of **F** to **G** allows for acid-catalyzed dimerization of **F** and **G** with concomitant double dehydration. The result is the cyclic ether product **3** whose ring construction mapping is [(3+2)+(5+5)+(3+2)]. The authors noted that the transformation from **2** to **3** needs dilute conditions ($\sim$0.025 mol dm^{-3} *m*-CPBA in chloroform) so that hydrogen bonding between the propargyl alcohol and the amine groups of molecules of **2** would be sufficiently disrupted to allow the dimerization process to occur. Moreover, the authors remarked that the intermolecular reaction between two molecules of intermediate **F** prevails over any possible intramolecular reaction such as the one shown in Scheme 2 leading to cyclic ether **4**. The authors do not give an explanation for this counterintuitive result. We note that it is logical to expect that as the concentration of intermediate **F** decreases the encounter probability of two molecules of **F** interacting to cause dimerization to **3** also decreases. We suggest that the reason must be due to a combination of enthalpic (ring strain) and entropic (conformational) barriers to intramolecular cyclization. A comparative computational study of the energetics and conformational constraints of the **F → 4** and 2 equiv. **F → 3** transformations would be highly useful in deciphering the experimental observations. The authors also mention that trapping of **F** by *m*-chlorobenzoate ion (present in solution) leading to ester **5** also does not occur suggesting that the carboxylate ion is a weaker nucleophile compared to the hydroxyl group of another **F** molecule (see Scheme 3).

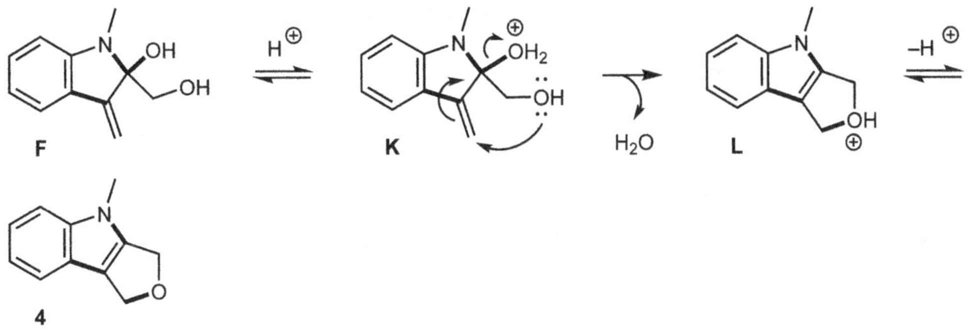

SCHEME 2

SCHEME 3

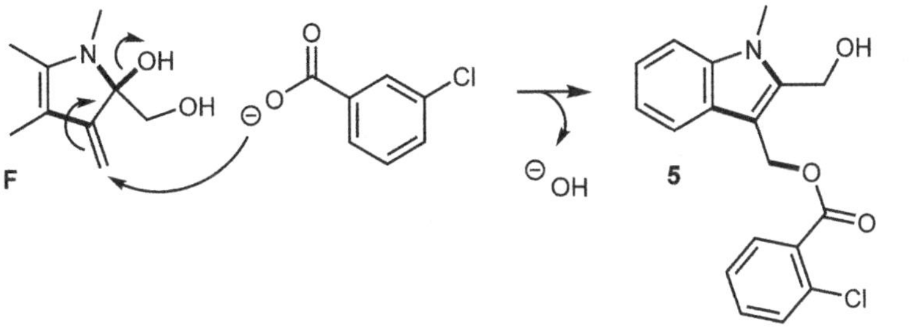

ADDITIONAL RESOURCES

We recommend an article which describes the results of a computational study on the amine-oxidation mechanism of monoamine oxidase enzyme which is thought to occur through a polar nucleophilic mechanism.[2]

References

1. Majumdar KC, Jana GH, Das U. Novel synthesis of 10-membered cyclic bis-ethers from 4-(*N*-alkylanilino)but-2-yn-1-ols. *J Chem Soc Perkin Trans 1*. 1997;1997:1229–1232. https://doi.org/10.1039/A606047A.
2. Erdem SS, Karahan O, Yildiz I, Yelekci K. A computational study on the amine-oxidation mechanism of monoamine oxidase: insight into the polar nucleophilic mechanism. *Org Biomol Chem*. 2006;4:646–658. https://doi.org/10.1039/b511350d.

Question 77: Degradation of Pyripyropene A

In 1997 Japanese chemists investigated the degradation of pyripyropene A **1** and noticed that Jones oxidation of the C_{13}—OH $^{13}C/^{14}C$-olefin* labeled **1** formed the corresponding C_{13}-oxo derivative **2** which underwent degradation to give compounds **3** and the $^{13}C/^{14}C$ labeled nicotinate by-product **4** (Scheme 1).[1] Provide a mechanism for the overall sequence from **1** to **3** and **4**.

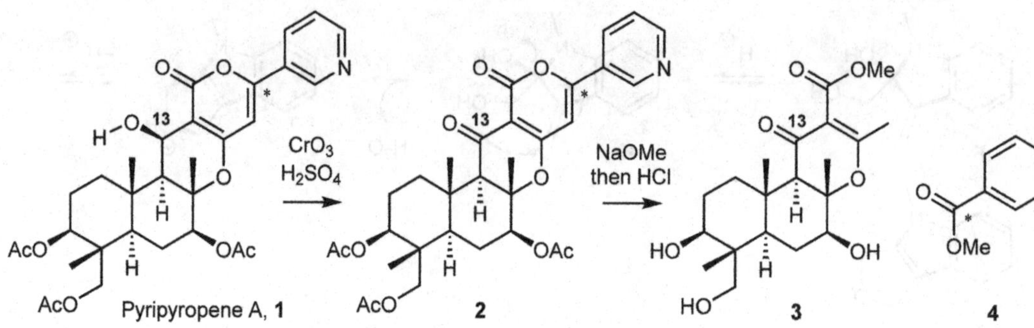
CrO3
H2SO4
NaOMe
then HCl
Pyripyropene A, 1
2
3
4

SCHEME 1

SOLUTION: 1 TO 2

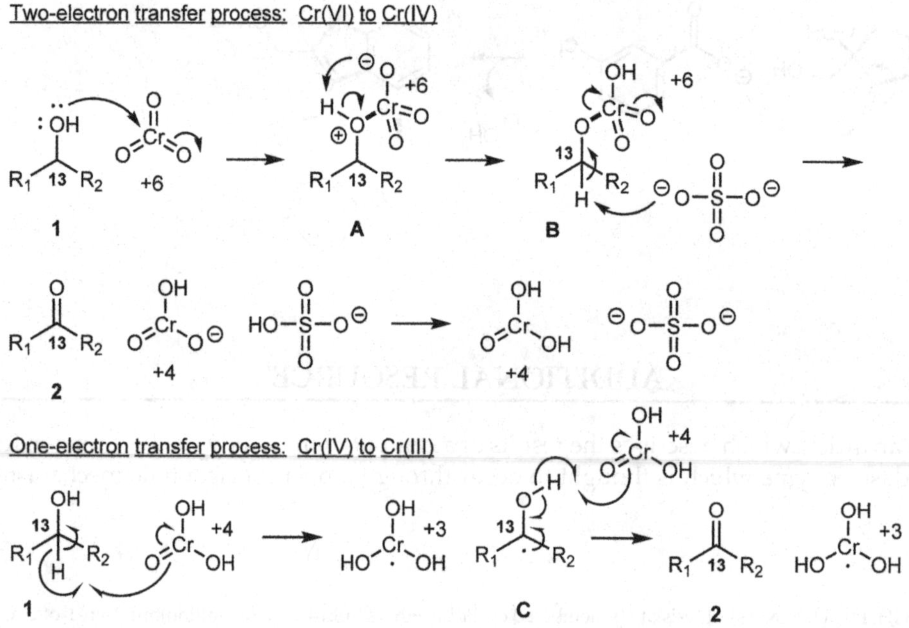
Two-electron transfer process: Cr(VI) to Cr(IV)
A
B
One-electron transfer process: Cr(IV) to Cr(III)
C

BALANCED CHEMICAL EQUATION: 1 TO 2

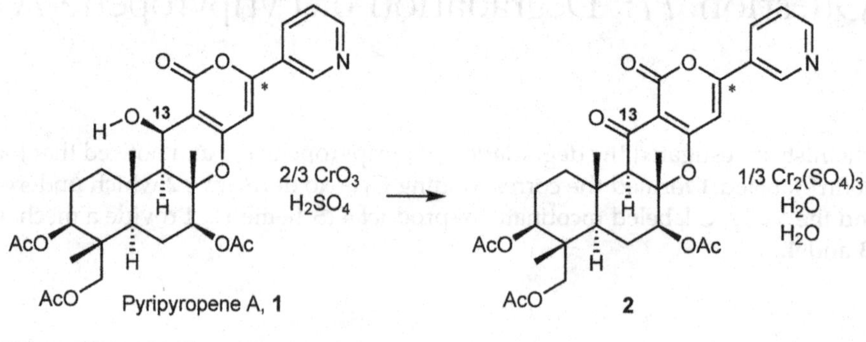
2/3 CrO3
H2SO4
1/3 Cr2(SO4)3
H2O
H2O
Pyripyropene A, 1
2

SOLUTION: 2 TO 3 AND 4

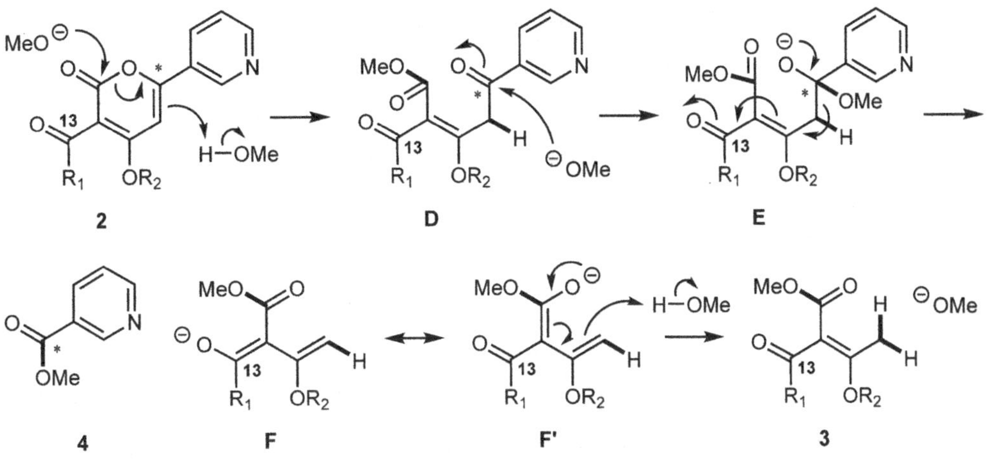

BALANCED CHEMICAL EQUATION: 2 TO 3 AND 4

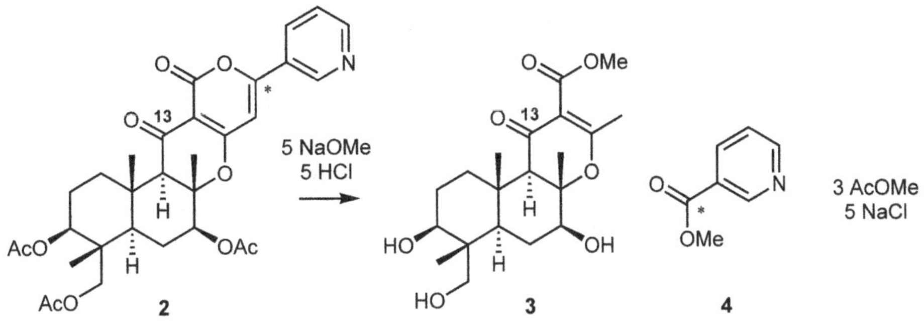

KEY STEPS EXPLAINED

The mechanism for the transformation of **1** to **2** follows the Jones oxidation of a secondary alcohol group at C_{13} of **1** which can occur via a two-electron process which reduces Cr(VI) to Cr(IV) or a one-electron process which reduces Cr (IV) to Cr(III). The two-electron mechanism involves ionic intermediates whereas the one-electron mechanism involves radical intermediates. Scheme 2 shows all the balanced chemical equations involved in the reaction in a step-wise fashion.

Net reaction from two-electron transfer reaction:

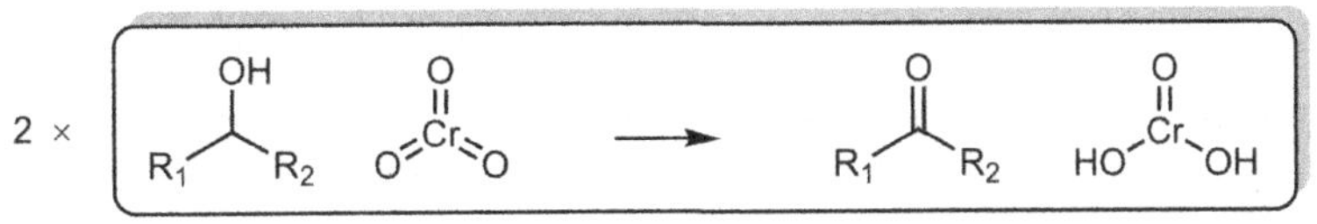

Net reaction from one-electron transfer reaction:

SCHEME 2

(Continued)

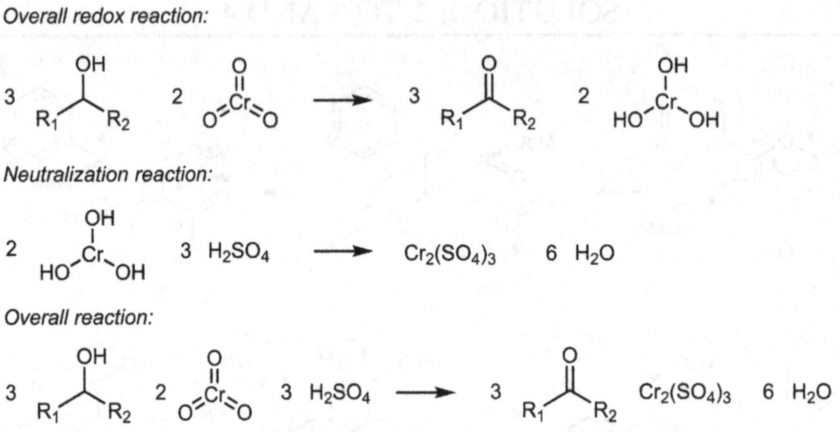

Overall redox reaction:

$$3 \; \underset{R_1 \;\; R_2}{\text{CH-OH}} \quad 2 \; O=Cr=O \quad \longrightarrow \quad 3 \; \underset{R_1 \;\; R_2}{C=O} \quad 2 \; \underset{HO \;\; OH}{\text{Cr-OH}}$$

Neutralization reaction:

$$2 \; \underset{HO \;\; OH}{\text{Cr-OH}} \quad 3 \; H_2SO_4 \quad \longrightarrow \quad Cr_2(SO_4)_3 \quad 6 \; H_2O$$

Overall reaction:

$$3 \; \underset{R_1 \;\; R_2}{\text{CH-OH}} \quad 2 \; O=Cr=O \quad 3 \; H_2SO_4 \quad \longrightarrow \quad 3 \; \underset{R_1 \;\; R_2}{C=O} \quad Cr_2(SO_4)_3 \quad 6 \; H_2O$$

SCHEME 2, CONT'D

Alternatively, the Cr(IV) H_2CrO_3 species may be both reduced to $Cr(OH)_3$ and oxidized back to CrO_3 as shown in Scheme 3. Adding the net reaction of this redox couple with the net reaction from the two-electron transfer reaction shown in Scheme 2 (entry 1) also yields the same overall balanced chemical equation for the process (last entry in Scheme 2). This process is shown explicitly in Scheme 4. In this alternative pathway CrO_3 is the only

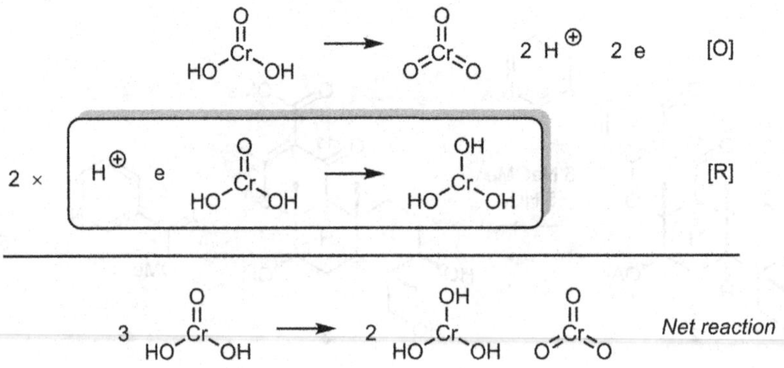

SCHEME 3

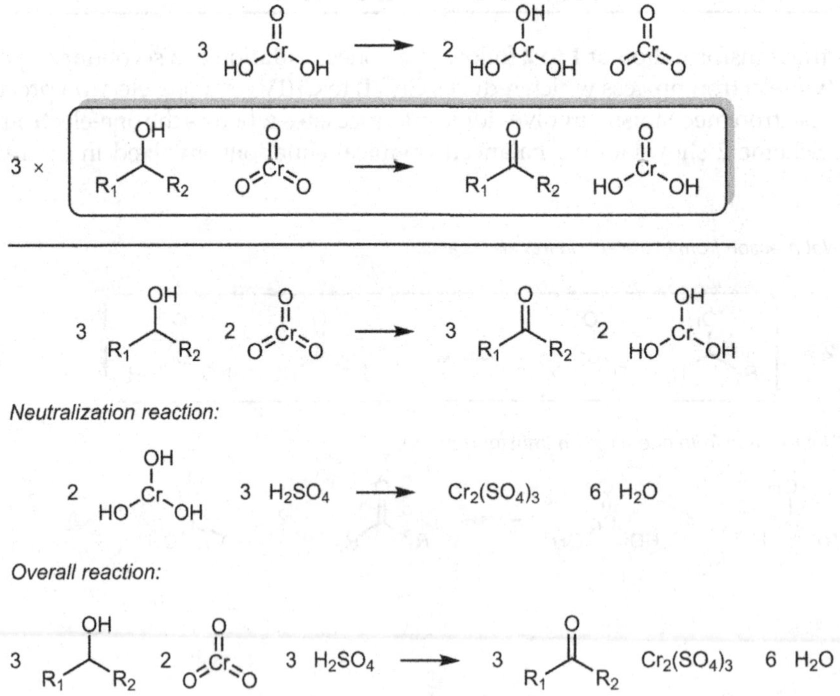

SCHEME 4

oxidant of the secondary alcohol. Scheme 5 shows a proposed mechanism for the conversion of H_2CrO_3 to $Cr(OH)_3$ and CrO_3.

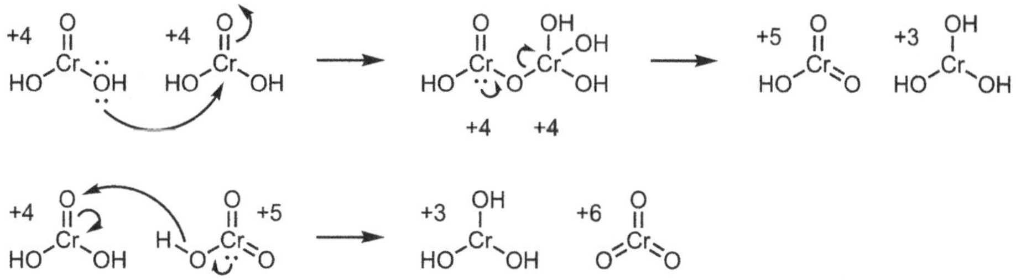

SCHEME 5

Next, the mechanism for the degradation transformation under methanolysis conditions from **2** to **3** involves nucleophilic attack of methoxide ion onto the carbonyl group of the pyrone ring of **2** which leads to the formation of an ester ketone intermediate **D** where the ^{13}C or ^{14}C label is on the ketone carbonyl carbon atom. A second methoxide ion then attacks the labeled ketone carbon atom resulting in the production of labeled methyl nicotinate and the oxy anion intermediate **F** which is resonance stabilized through conjugation via **F'**. Three other equivalents of methoxide ion also attack the three acetyl groups originally found in **2** leading to their conversion to hydroxyl groups (which we do not show in this solution). Nevertheless, because of these additional steps, in total we have five equivalents of methoxide ion being used in this degradation reaction. It is worthwhile to note that experimentally the authors observed that in the absence of conjugation in **2** (i.e., if the Jones oxidation of C_{13}—OH of **1** is omitted and there is no C_{13}=O found in **2**), the degradation reaction does not proceed to fragmentation of nicotinate by-product. Instead, the only product isolated in excellent yield is the corresponding tetraol product resulting from methoxide cleavage of the three acetyl groups of **2** to form hydroxyl groups thus giving four hydroxyl groups. It thus appears that resonance stabilization through conjugation of intermediate **F** is required for the cleavage of nicotinate from intermediate **E**.

We note that the authors did not provide additional experimental and theoretical evidence to help support their mechanistic proposal nor did they discuss the mechanism of the Jones oxidation which we illustrate here in great detail.

ADDITIONAL RESOURCES

We recommend a recent article which discusses the total synthesis of pyripyropene A.[2]

References
1. Obata R, Sunazuka T, Tian Z, et al. New analogs of the pyripyropene family of ACAT inhibitors via α-pyrone fragmentation and γ-acylation/cyclization. *Chem Lett.* 1997;26(9):935–936. https://doi.org/10.1246/cl.1997.935.
2. Odani A, Ishihara K, Ohtawa M, Tomoda H, Omura S, Nagamitsu T. Total synthesis of pyripyropene A. *Tetrahedron.* 2011;67:8195–8203. https://doi.org/10.1016/j.tet.2011.06.084.

Question 78: Benzofurans From Cyclobutenediones

In 1996 American chemists carried out a four-step synthetic sequence of the highly functionalized benzofuran product **6** starting from cyclobutenedione **1** according to Scheme 1.[1] Suggest a mechanism for each of the synthetic steps of this sequence.

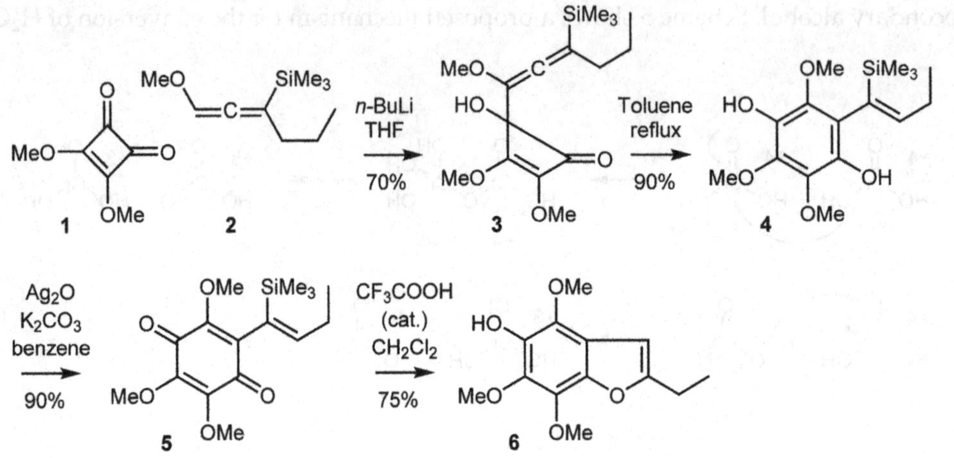

SCHEME 1

SOLUTION: 1 AND 2 TO 3

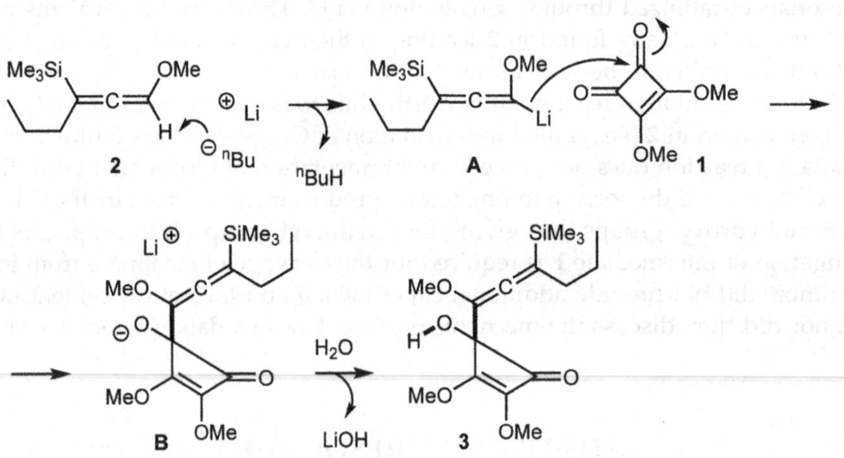

SOLUTION: 3 TO 4

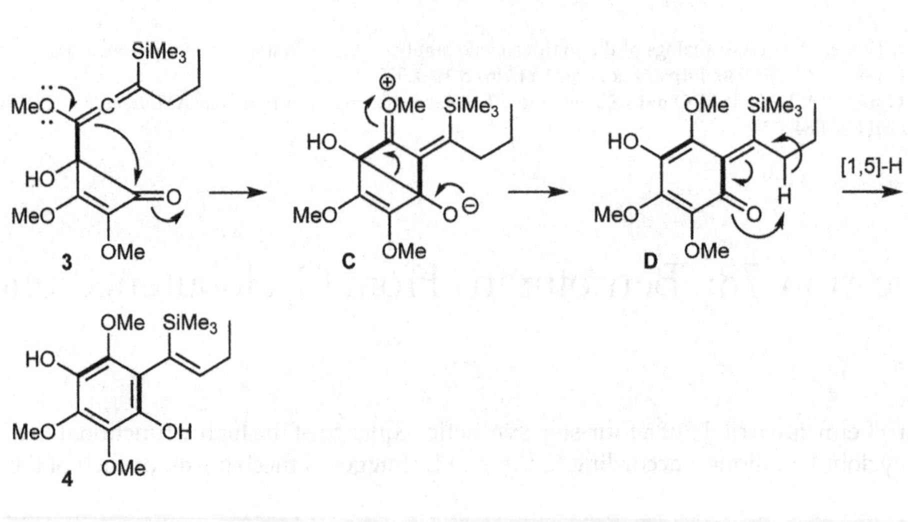

SOLUTION: 4 TO 5

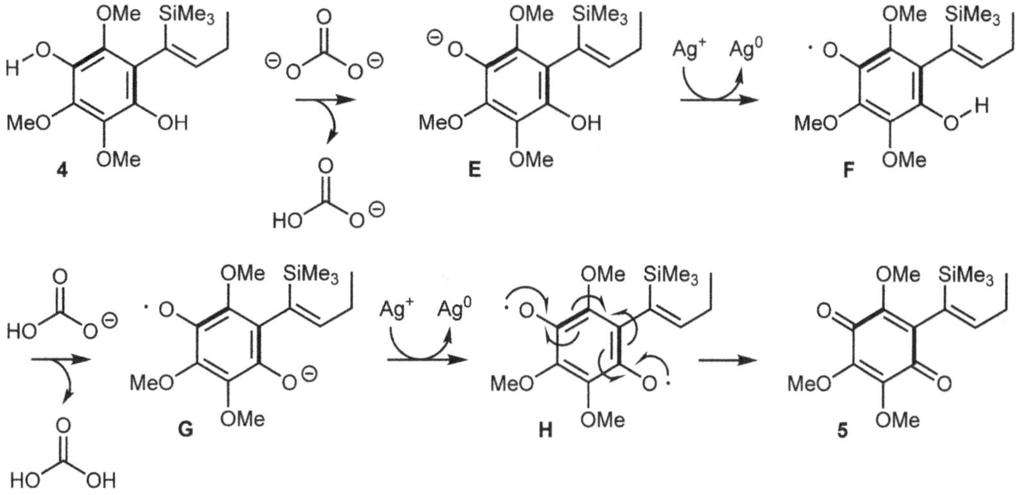

SOLUTION: 5 TO 6

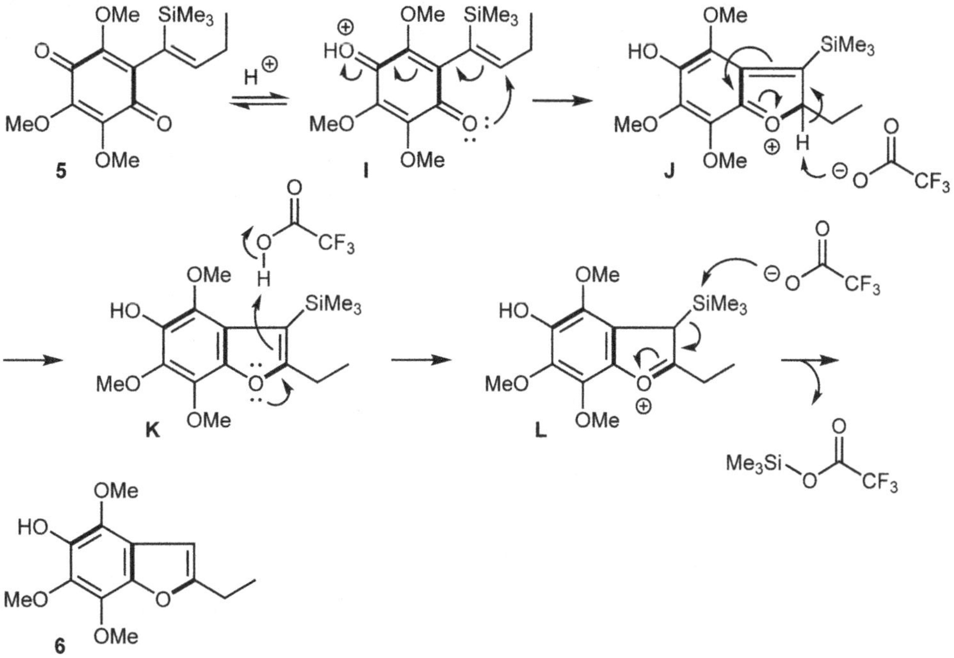

BALANCED CHEMICAL EQUATIONS

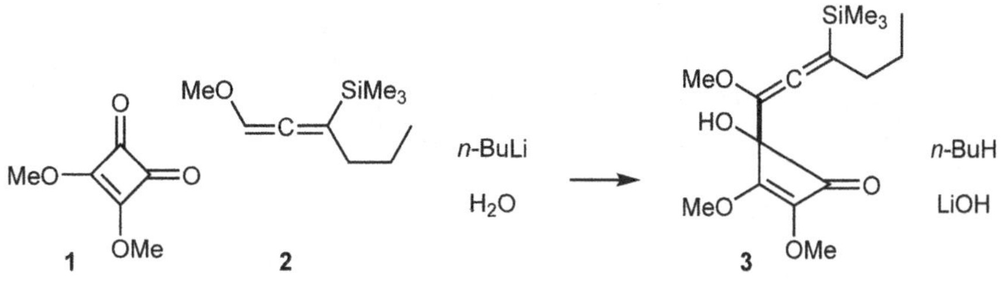

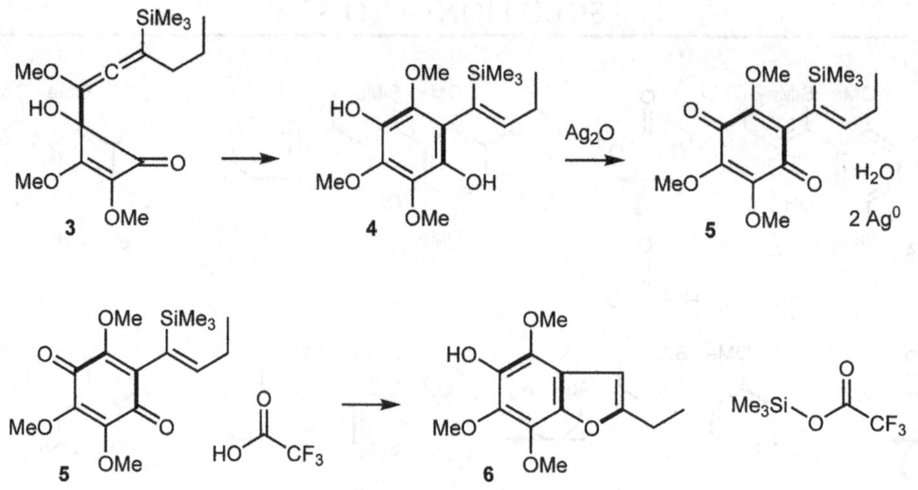

KEY STEPS EXPLAINED

The mechanism proposed by the authors to explain the synthetic sequence from cyclobutenedione **1** to benzofuran **6** begins with a base-induced coupling of allene **2** and cyclobutenedione **1** to form the 4-allenylcyclobutenone product **3**. Compound **3** then undergoes a thermal rearrangement to hydroquinone **4** which the authors thought would proceed via formation of oxonium cation **C** followed by ring expansion to the *o*-quinone methide **D** followed by a [1,5]-proton transfer which gives the hydroquinone product **4**. In the next step, a redox reaction transforms hydroquinone **4** to *p*-benzoquinone **5** using silver oxide and potassium carbonate (which acts as a catalyst in the reaction). In the final step, an acid-catalyzed ring closure from **5** to **I** allows for the formation of the furan moiety in intermediate **K** after deprotonation and rearomatization via cation **J**. Next, protonation of the olefin group in **K** establishes oxonium **L** which undergoes desilylation to generate the final benzofuran product **6**. We note that the ring construction mapping for **6** is [(4+2)+(3+2)].

Furthermore, the authors did not provide experimental and theoretical evidence to support the transformations of **1** and **2** to **3**, **4** to **5**, and **5** to **6**. Nevertheless, using a deuterium labeling experiment in the context of the thermolysis of **7** to **8a–f** (Scheme 2), the authors showed that based on deuterium scrambling among the products there is an equilibrium condition

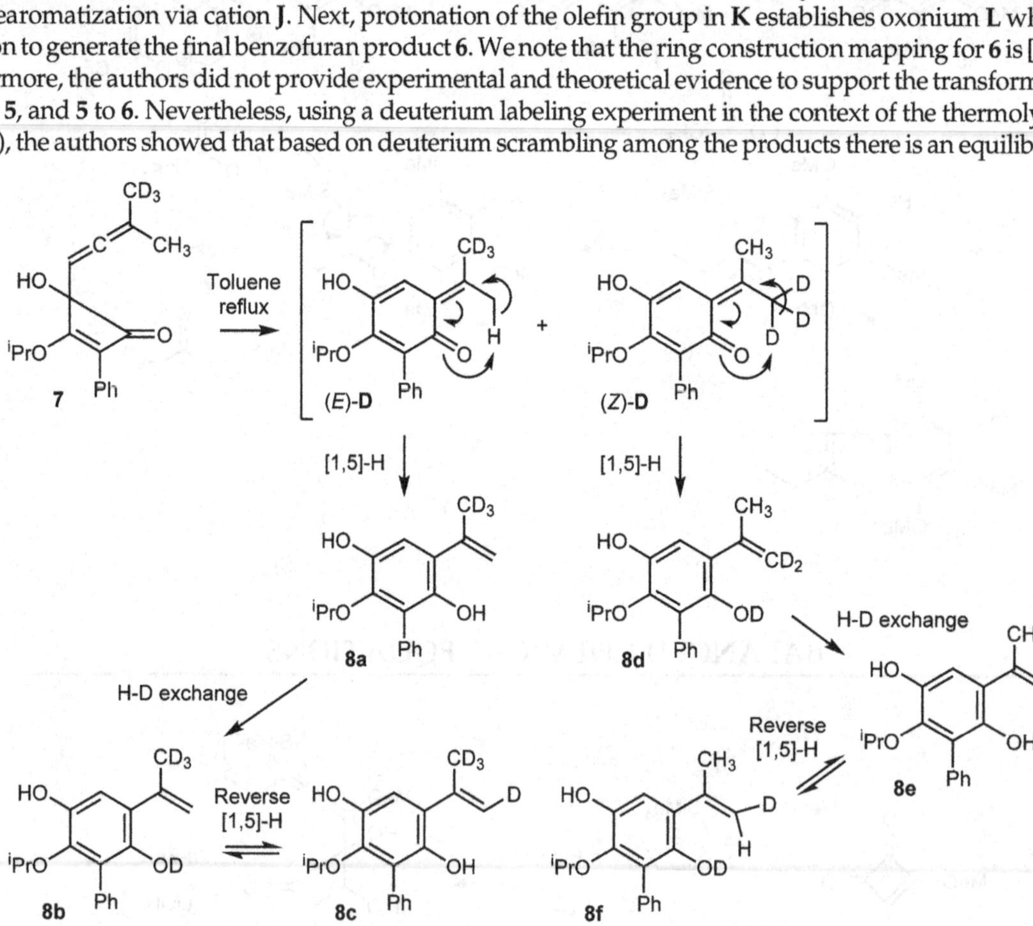

SCHEME 2

established between the E and Z forms of o-quinone methide intermediate **D**. We note that products **8a** and **8d** are the direct results of the [1,5]-hydrogen shifts depicted in Scheme 2. Products **8b** and **8e** result from deuterium exchange at the phenolic positions of **8a** and **8d**, respectively. Finally, **8c** and **8f** result from equilibrium conditions where **8b** and **8e** undergo reverse [1,5]-hydrogen shifts, respectively. These results thus constitute direct evidence for the [1,5]-hydrogen shift from **D** to **4**. In the present context, it is quite clear that the Z-form of **D** does not possess the correct conformation to facilitate the [1,5]-protic shift from **D** to **4** and is thus nonproductive. However, in an analogous system where there is no hydrogen atom in substrate **9** available for a [1,5]-shift, the formation of the E isomer **9-(E)-D** analog facilitates formation of a hypervalent silicate intermediate **M** which undergoes methyl migration from silicon to carbon to form product **10** (Scheme 3). The authors thus rationalize these results by arguing that "the alkylidene group in the o-quinone methide product [**D**] would be formed by a rotation mode in which the larger group (R_L) rotates away from the carbonyl group and toward the R group, thus resulting in the quinone methides generally represented by" structures such as **(E)-D** rather than **(Z)-D** (Scheme 4). We observe that this occurs in the two previously referenced experiments where the CD_3 group is larger than the CH_3 group and where the phenyl group is larger than trimethylsilyl. Given this evidence, we have further support for the formation of the E form of **D** *en route* to product **4** especially since the trimethylsilyl group is larger than the propyl group. We highly recommend that others carry out a theoretical analysis of the mechanisms presented in this solution in the future.

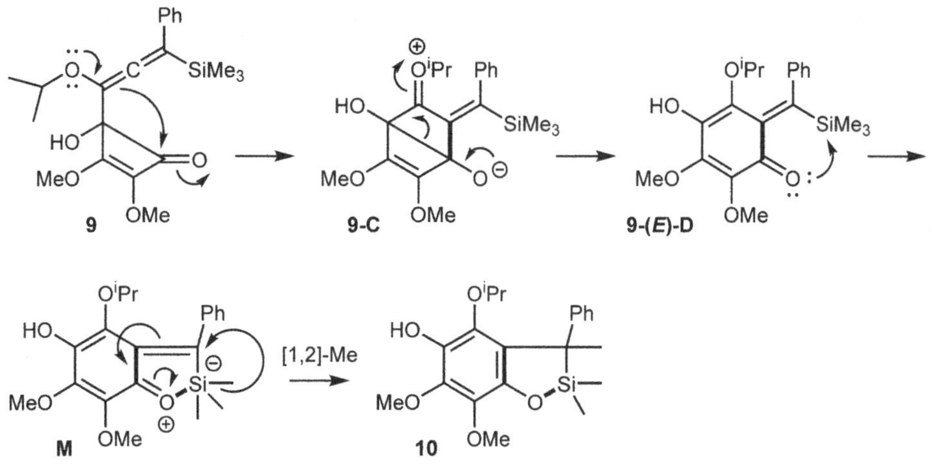

SCHEME 3

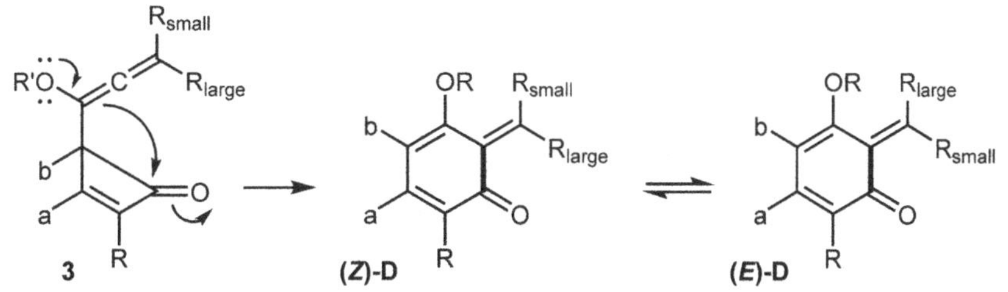

SCHEME 4

ADDITIONAL RESOURCES

We direct the reader to works which discuss further examples of transformations of butenediones as well as one example of an unusual formation of a benzofuran derivative from p-cresol.[2–4]

References

1. Taing M, Moore HW. *o*-Quinone methides from 4-allenylcyclobutenones: synthesis and chemistry. *J Org Chem*. 1996;61(1):329–340. https://doi.org/10.1021/jo951445m.

2. Lee KH, Moore HW. Unusual transformation of cyclobutenediones into butenolides. *J Korean Chem Soc.* 2003;47(3):229–236. https://doi.org/10.5012/jkcs.2003.47.3.229.

3. Yamamoto Y, Noda M, Ohno M, Eguchi S. Formation of 2-[1-(trimethylsilyl)alkylidene]-4-cyclopentene-1,3-dione from Lewis acid-catalyzed reaction of cyclobutenedione monoacetal with alkynylsilane: novel cationic 1,2-silyl migrative ring opening and subsequent 5-exo-trig ring closure. *J Org Chem.* 1997;62(5):1292–1298. https://doi.org/10.1021/jo961552w.

4. Serykh VY, Rozentsveig IB, Rozentsveig GN, Chernyshev KA. Unexpected formation of benzofuran derivatives in the *C*-amidoalkylation of *p*-cresol with 4-chloro-*N*-(2,2-dichloro-2-phenylethylidene)benzenesulfonamide. *Chem Heterocycl Compd.* 2011;47(11):1339–1344. https://doi.org/10.1007/s10593-012-0919-0.

Question 79: Acid-Catalyzed Condensation of Indole With Acetone

In 1996 American chemists carried out the much-studied acid-catalyzed condensation of indole with acetone and identified a novel product **2** by means of modern analytical and spectroscopic methods (Scheme 1).[1] Suggest a mechanism for this transformation. (We note that the structure of product **2** in McKillop's book is missing a double bond in ring B.)

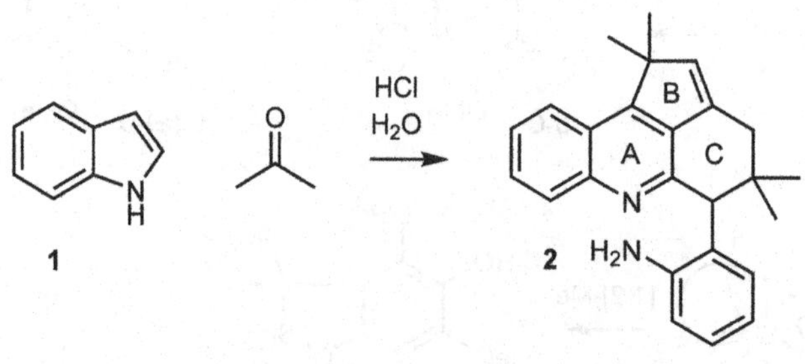

SCHEME 1

SOLUTION

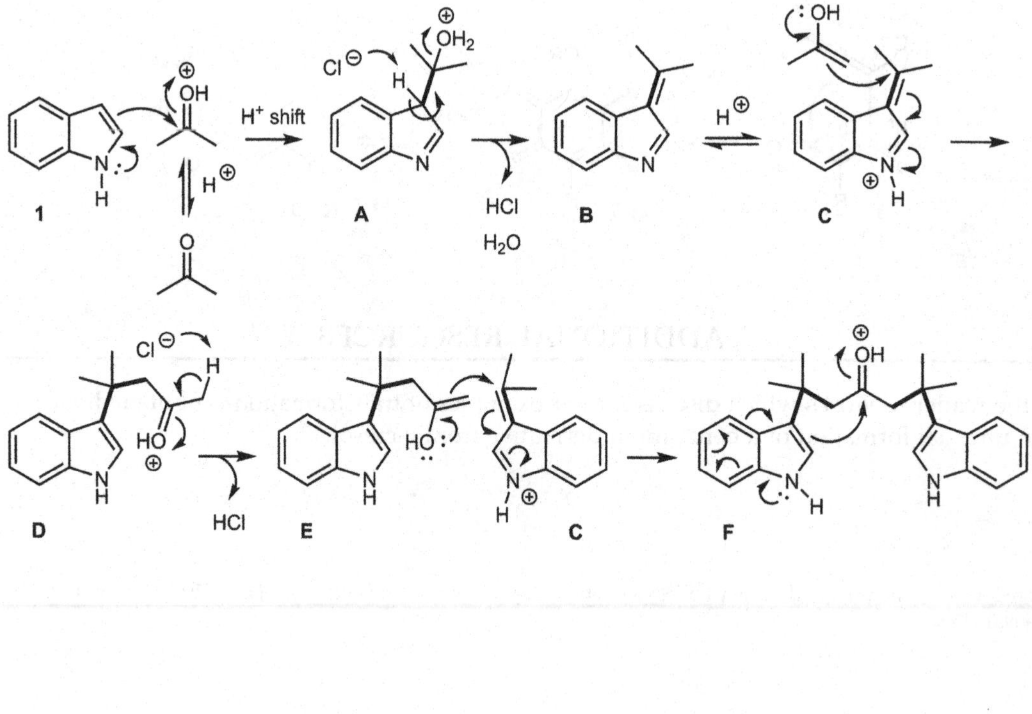

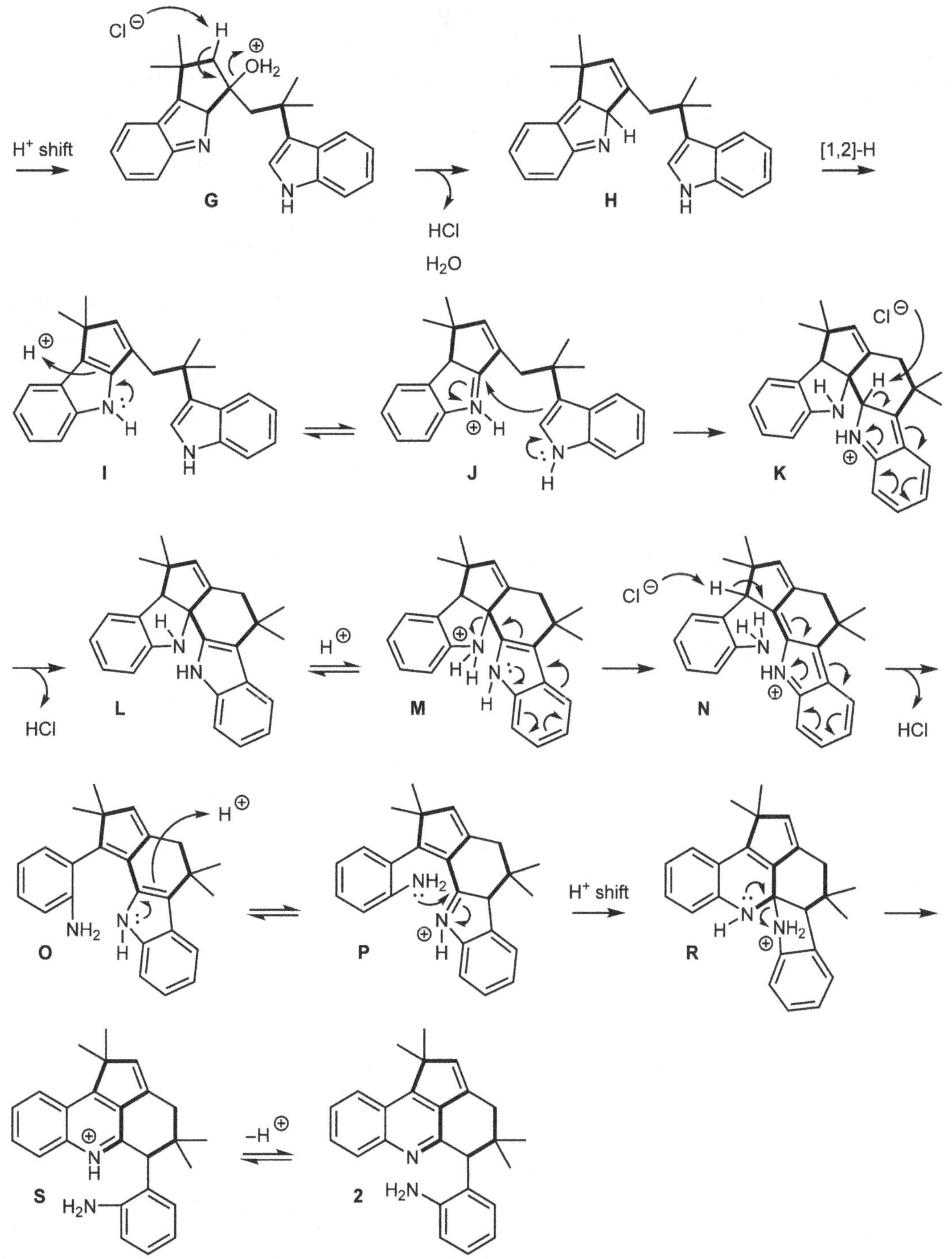

BALANCED CHEMICAL EQUATION

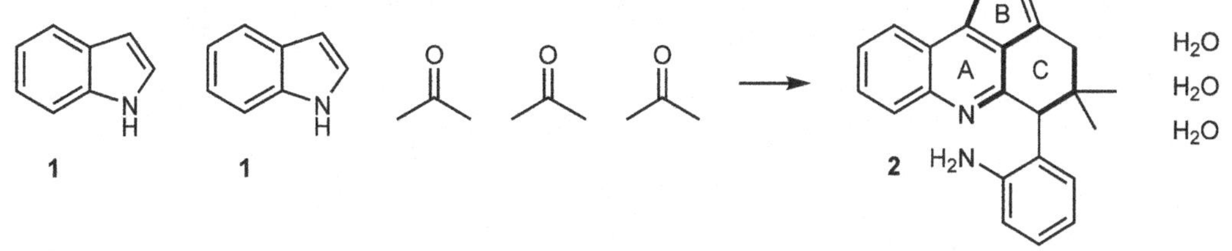

KEY STEPS EXPLAINED

We begin by noting that the reaction is an overall five-component coupling between two equivalents of indole and three equivalents of acetone. The ring construction mapping for product **2** is $[(5+1)_A+(2+2+1)_B+(2+1+2+1)_C]$. In terms of the mechanistic proposal, there are many possible ways of writing a mechanism for this reaction depending on the order of addition of reagents. Furthermore, the product identified in this problem is one of 4 products from the reaction. Reaction yields were unfortunately not reported. Nevertheless, the authors confirmed the structure of **2** using X-ray crystallography, ^{1}H NMR, ^{13}C NMR, mass spectrometry, and melting point analysis. The mechanism we show here represents the authors' conjectured mechanism which begins with condensation of indole with one equivalent of acetone to give intermediate **C**. **C** then couples with a second equivalent of acetone to form enol **E** which in turn adds to a protonated intermediate **C** giving the symmetrical intermediate **F**. Intermediate **F** cyclizes to intermediate **I** after dehydration and [1,2]-H shift steps. In turn, intermediate **I** cyclizes to the fused hexacyclic intermediate **L** which then ring opens to the aniline intermediate **O**. The aniline amino group of **O** cyclizes again onto the protonated indole moiety leading to intermediate **R** which then ring opens the other way leading to a quinoline ring and an aniline moiety eventually generating the final product **2**. We note that the authors did not undertake a theoretical analysis to help support this mechanistic proposal.

ADDITIONAL RESOURCES

We direct the reader to three literature examples, including one by the same authors, of similar condensations of indole or pyrrole with acetone to form interesting products.[2-4]

References

1. Noland WE, Konkel MJ, Konkel LMC, Pearce BC, Barnes CL, Schlemper EO. Structure determination of the products from the acid-catalyzed condensation of indole with acetone. *J Org Chem*. 1996;61:451–454. https://doi.org/10.1021/jo951596p.
2. Mishra S, Ghosh R. Ecofriendly and sustainable efficient synthesis of bis(indolyl)methanes based on recyclable Brønsted (CSA) or Lewis (ZrOCl$_2$.8H$_2$O) acid catalysts. *Indian J Chem Sect B*. 2011;50:1630–1636. http://hdl.handle.net/123456789/13047.
3. Rani VJ, Vani KV, Rao CV. PEG-SO$_3$H as a catalyst for the preparation of bis-indolyl and tris-indolyl methanes in aqueous media. *Synth Commun*. 2012;42:2048–2057. https://doi.org/10.1080/00397911.2010.551700.
4. Noland WE, Konkel MJ, Fanburg SJ, Venkatraman S, Britton D. The crystal structure of a novel product from the acid-catalyzed condensation of 1-benzylindole with acetone. *J Chem Crystallogr*. 1999;29(1):9–14. https://doi.org/10.1023/A:1009506927964.

Question 80: A Synthesis of 2-Formylpyrrolidines

In 1996 Belgian chemists treated imines **1** with *N*-chlorosuccinimide (NCS) in CCl$_4$ to form dichlorides **2** in quantitative yield (Scheme 1).[1] Addition of **2** to a methanolic solution containing 1.1 equivalent of K$_2$CO$_3$ under reflux for 3.5h resulted in the smooth formation of pyrrolidines **3** which then underwent acid hydrolysis to give the 2-formyl derivatives **4**. Suggest mechanisms for all steps of this synthetic sequence.

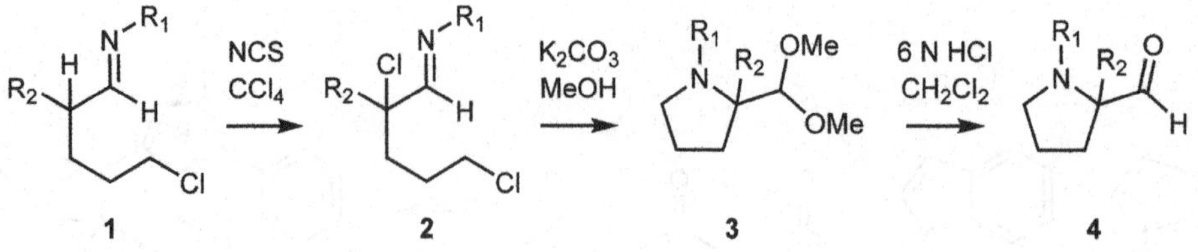

SCHEME 1

SOLUTION: 1 TO 2

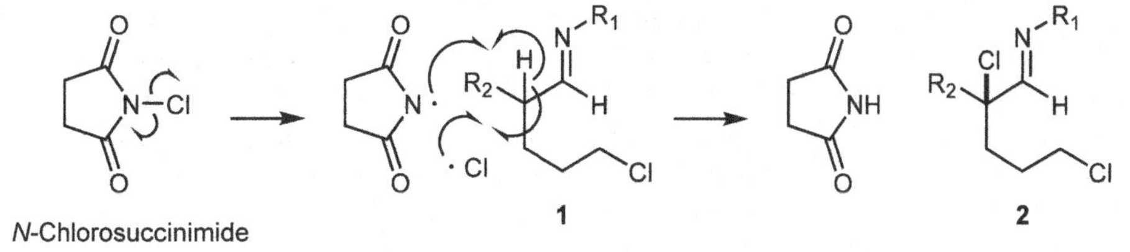

N-Chlorosuccinimide   **1**   **2**

SOLUTION: 2 TO 3

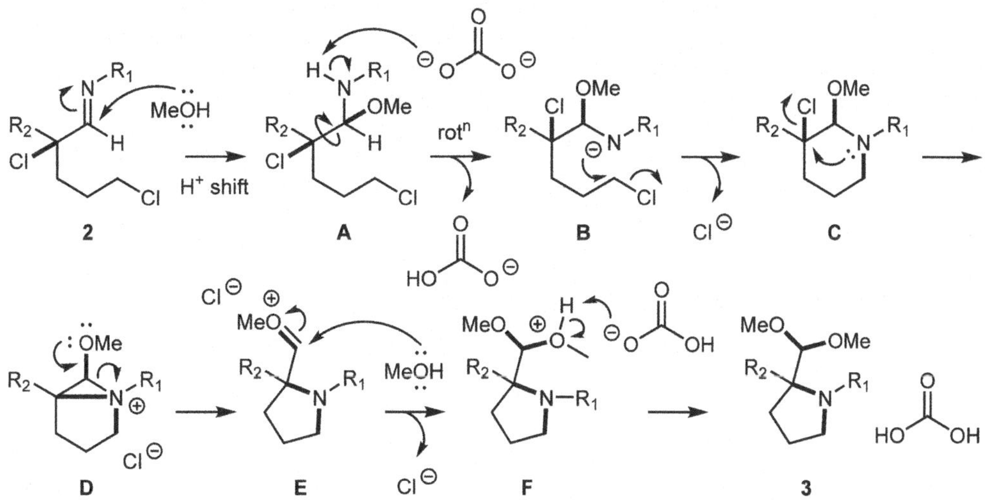

2   **A**   **B**   **C**

D   **E**   **F**   **3**

BALANCED CHEMICAL EQUATION: 2 TO 3

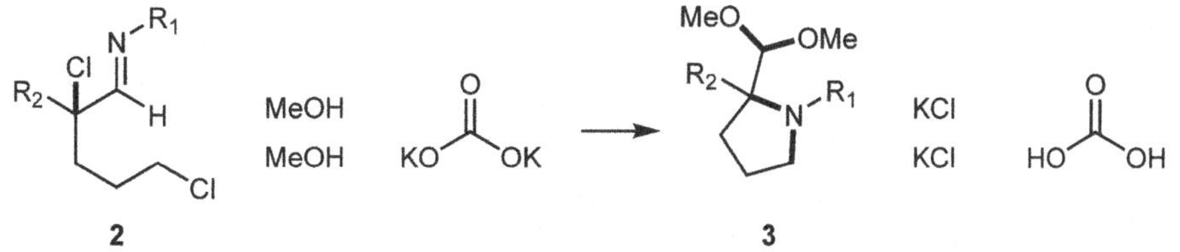

2   MeOH   MeOH   KO‑CO‑OK   **3**   KCl   KCl   HO‑CO‑OH

SOLUTION: 3 TO 4

3   **F**   **E**   **G**

H   **4**

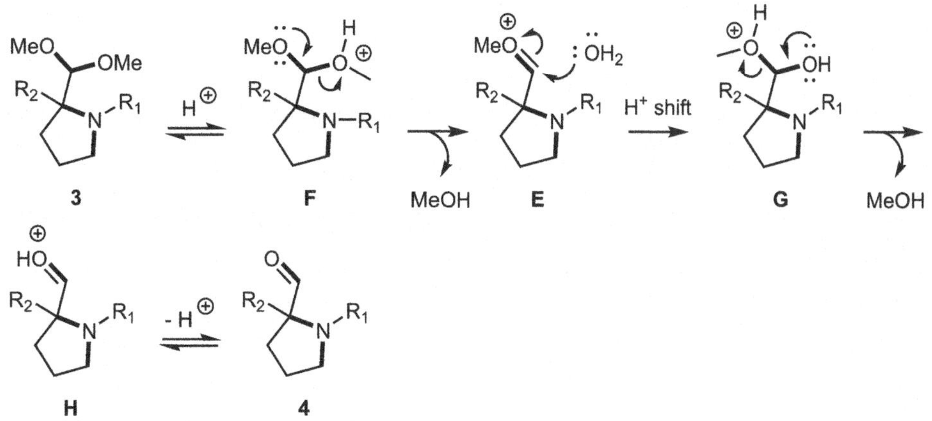

KEY STEPS EXPLAINED

The mechanism proposed by the authors to explain the synthetic sequence from **1** to **4** begins with a radical-based substitution of a tertiary hydrogen atom for a chlorine atom. Thus *N*-chlorosuccinimide homolytically cleaves to chlorine radical and succinimide radical. Succinimide radical abstracts the tertiary hydrogen from imine **1** to give a tertiary radical which couples with chlorine radical to give product **2**. In the next step, methanol adds to the imine carbon atom of **2** to give intermediate **A**. Carbonate ion abstracts the amino proton from **A** causing a cyclization to chloropiperidine **C**. Intermediate **C** then undergoes a transannular Favorskii-like elimination of chloride ion yielding aziridinium ion **D** which ring opens to oxonium pyrrolidine intermediate **E**. This intermediate then reacts with another equivalent of methanol to form oxonium ion **F** which is deprotonated by bicarbonate ion to generate the pyrrolidine acetal product **3**. The ring construction mapping for product **3** is [4 + 1]. Lastly, the acetal group in **3** undergoes a standard acid-catalyzed hydrolysis reaction to form the aldehyde product **4** via the hemiacetal **F**. We note that the authors did not undertake further experimental or theoretical studies to help support this mechanism.

Nevertheless, the authors did rule out an alternative mechanism for the transformation of **2** to **3** (Scheme 2) on the basis of the following evidence: (a) such a mechanism does not apply to α-chloroaldimines that are derived from higher aldehydes where R_2 is not equal to H or Me[2,3]; and (b) increased steric size of R_2 prevents formation of aziridinium ion intermediates such as **I**.

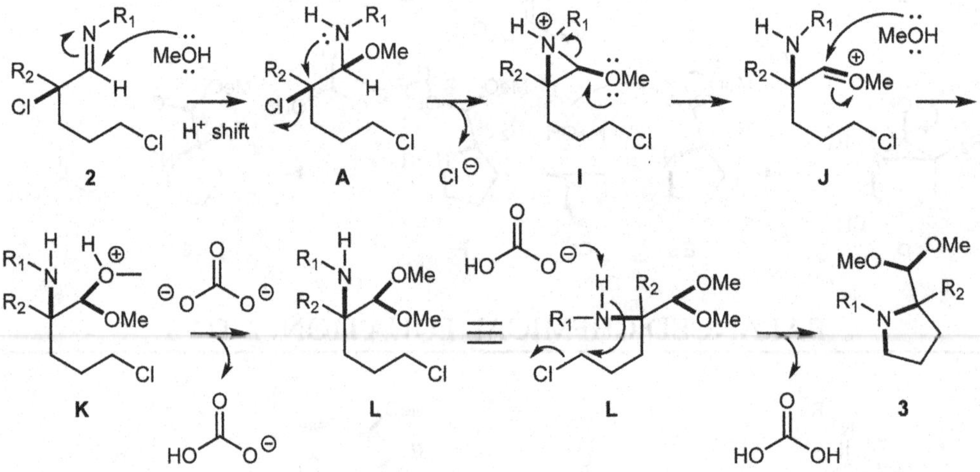

SCHEME 2

ADDITIONAL RESOURCES

We recommend an article which discusses the mechanism of an interesting 1,4-addition and [3 + 2] cycloaddition.[4]

References

1. de Kimpe N, Stevens C, Virag M. Rearrangement of α,β-dichloroaldimines to 2-formylpyrrolidines: α,α-azacyclobisalkylation of aldehydes. *Tetrahedron*. 1996;52(9):3303–3312. https://doi.org/10.1016/0040-4020(95)01112-9.
2. de Kimpe N, Verhé R, de Buyck L, Hasma H, Schamp N. Reactivity of α-chloro-aldimines. *Tetrahedron*. 1976;32:2457–2466. https://doi.org/10.1016/0040-4020(76)87034-2.
3. de Kimpe N, Boeykens M, Boelens M, de Buck K, Cornelis J. Synthesis of 2(*N*-alkylamino)isobutyraldehydes. *Org Prep Proced Int*. 1992;24(6):679–681. https://doi.org/10.1080/00304949209356245.
4. Tsubogo T, Saito S, Seki K, Yamashita Y, Kobayashi S. Development of catalytic asymmetric 1,4-addition and [3 + 2] cycloaddition reactions using chiral calcium complexes. *J Am Chem Soc*. 2008;130:13321–13332. https://doi.org/10.1021/ja8032058.

Question 81: An Abnormal Claisen Rearrangement

In 1996 American scientists observed that an abnormal Claisen rearrangement occurred to transform the allyl ether **1** into product **2** when the reaction was carried out in PhNEt$_2$ at 200°C (Scheme 1).[1] Furthermore, there was a significant improvement in the yield of **2** when the base (Me$_3$Si)$_2$NH (i.e., HMDS) was added to the reaction mixture. Based on these observations, suggest a mechanism for this transformation.

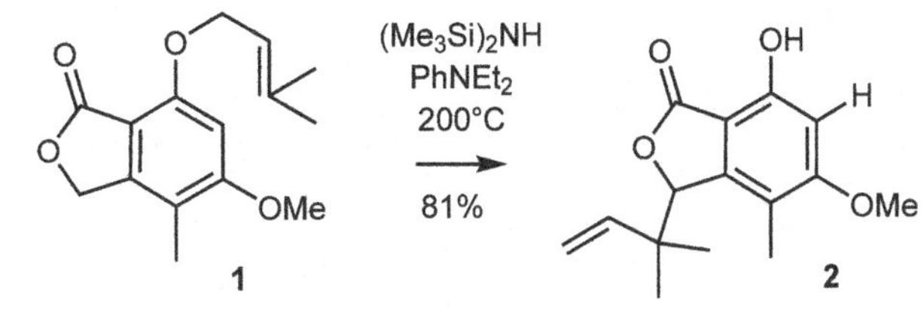

SCHEME 1

SOLUTION

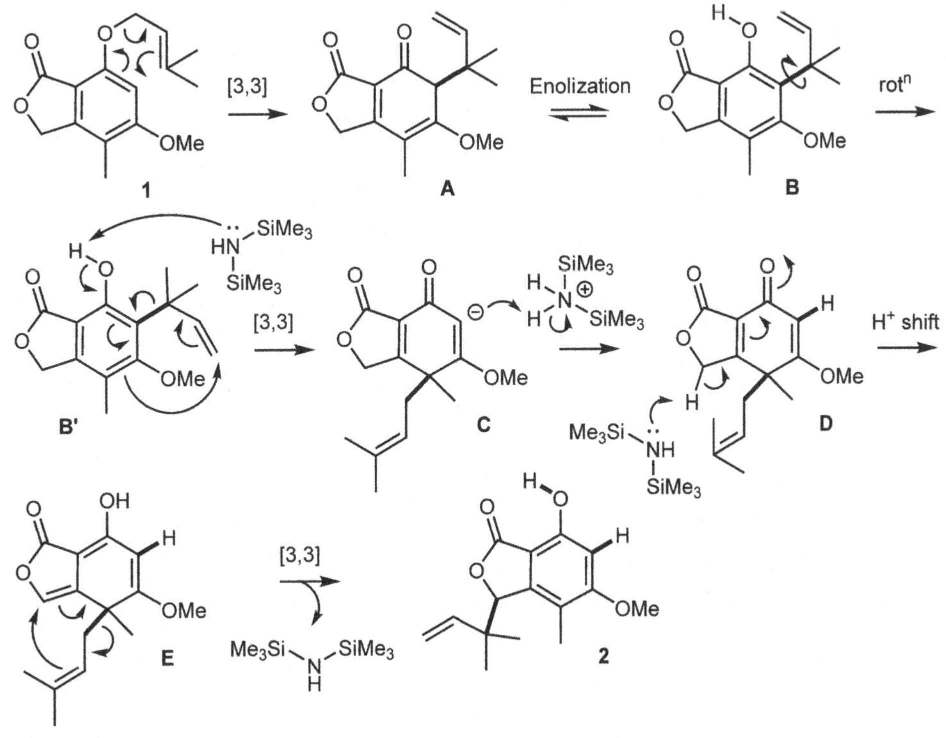

KEY STEPS EXPLAINED

The mechanism proposed to explain the unexpected rearrangement of compound **1** to product **2** begins with a normal Claisen [3,3] sigmatropic rearrangement of **1** to form **A** which enolizes to **B**. Bond rotation about the C(Ar)—C bond of **B** allows the 3-methyl-but-1-ene group to migrate again via a base-assisted [3,3] sigmatropic rearrangement to the *para* position of the aromatic ring yielding quinone anion **C** which gets protonated to quinone **D**. Removal of a hydrogen atom from the lactone ring by base leads to enolate **E** after protonation. This enolate undergoes a third [3,3] sigmatropic rearrangement in which the prenyl group migrates from the *para* position of the aromatic ring to the lactone ring yielding product **2**. It is clear from this mechanism that the presence of a base facilitates the second and third abnormal [3,3] sigmatropic migration steps.

Furthermore, the authors carried out a crossover experiment with two deuterium labeled compounds to confirm that the reaction was intramolecular and not intermolecular (Scheme 2). Thus compound **1-d$_6$** reacted by itself to form product **2-d$_6$** and compound **1-d$_3$** reacted by itself to form **2-d$_3$**. When a 1:1 M ratio of **1-d$_6$** and **1-d$_3$** were reacted together, the product mixture consisted of a 1:1 M ratio of **2-d$_6$** and **2-d$_3$** with no other crossover products observed. We also note that the authors did not carry out a theoretical analysis to help support their proposed mechanism.

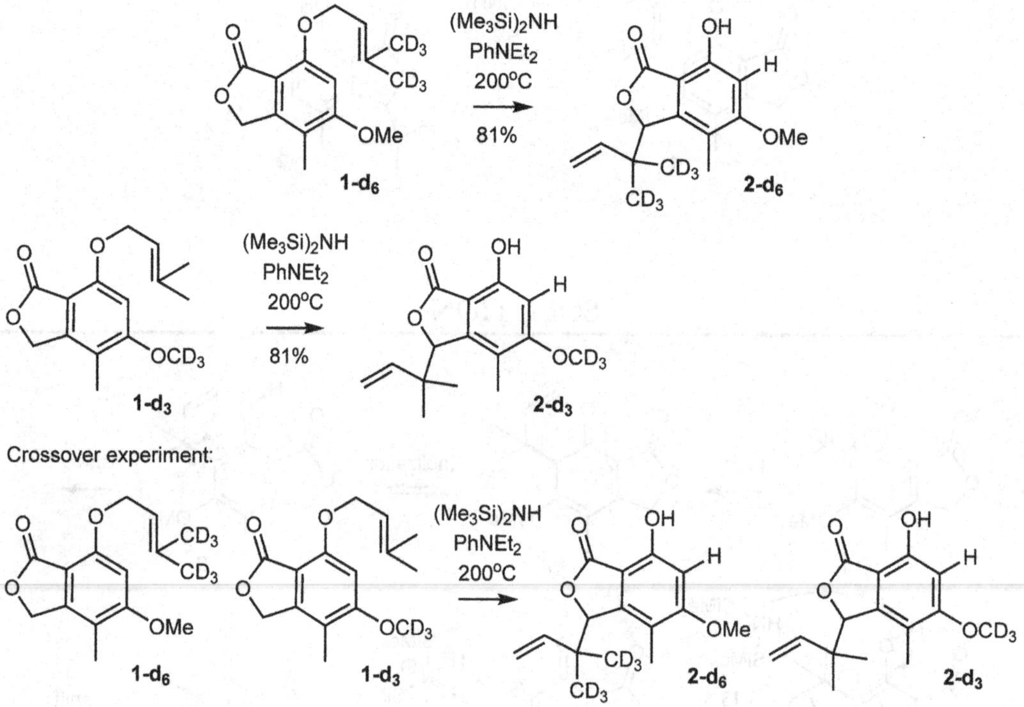

SCHEME 2

ADDITIONAL RESOURCES

We recommend an article which describes another example of a transformation involving an abnormal Claisen rearrangement.[2] We also include an example of a rearrangement of aryl geranyl ethers.[3]

References

1. Smith DB, Elworthy TR, Morgans DJ Jr, et al. Investigation of an unusual rearrangement. *Tetrahedron Lett.* 1996;37(1):21–24. https://doi.org/10.1016/0040-4039(95)02101-9.
2. Nakamura S, Ishihara K, Yamamoto H. Enantioselective biomimetic cyclization of isoprenoids using Lewis acid-assisted chiral Brønsted acids: abnormal Claisen rearrangements and successive cyclizations. *J Am Chem Soc.* 2000;122:8131–8140. https://doi.org/10.1021/ja001165a.
3. Toerincsi M, Kolonits P, Fekete J, Novak L. Rearrangement of aryl geranyl ethers. *Synth Commun.* 2012;42:3187–3199. https://doi.org/10.1080/00397911.2011.579799.

Question 82: Benzoxepinones From Phthalides

In 1996 Japanese chemists reacted an ethereal solution of prop-2-ynylmagnesium bromide with phthalides **1** at 0°C followed by quenching with 20% HCl to form benzoxepinones **2** in 45%–86% yield (Scheme 1).[1] Provide a mechanism for this transformation.

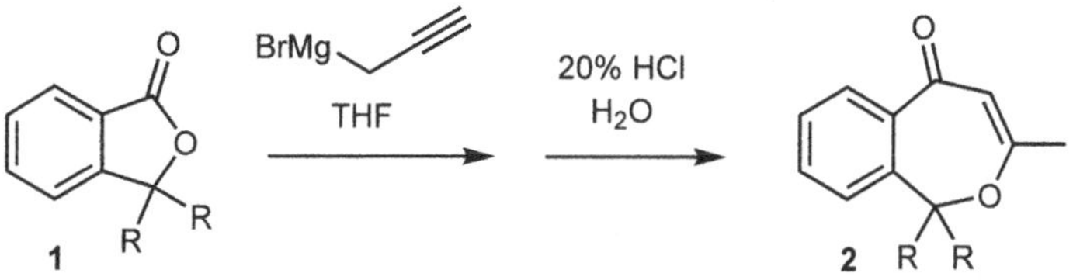

SCHEME 1

SOLUTION

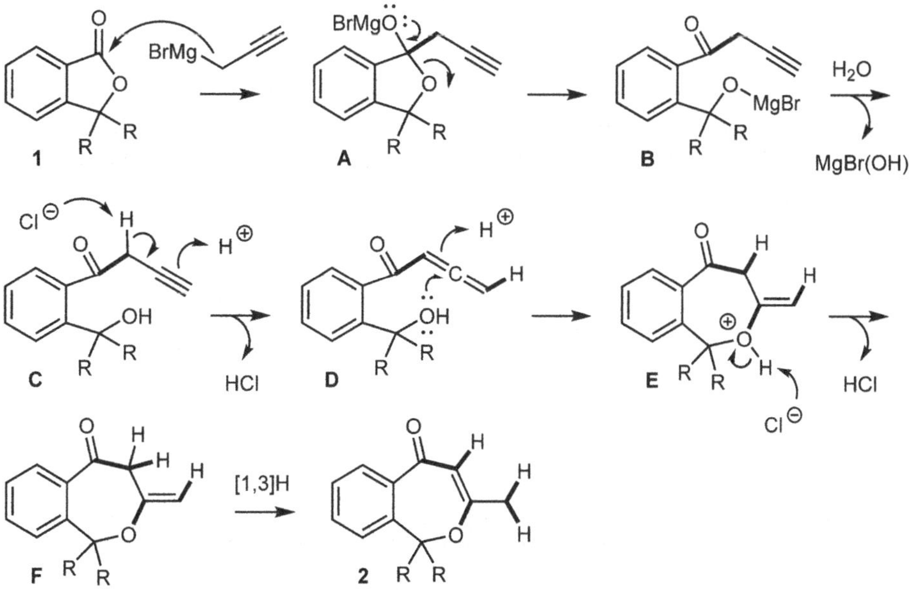

BALANCED CHEMICAL EQUATION

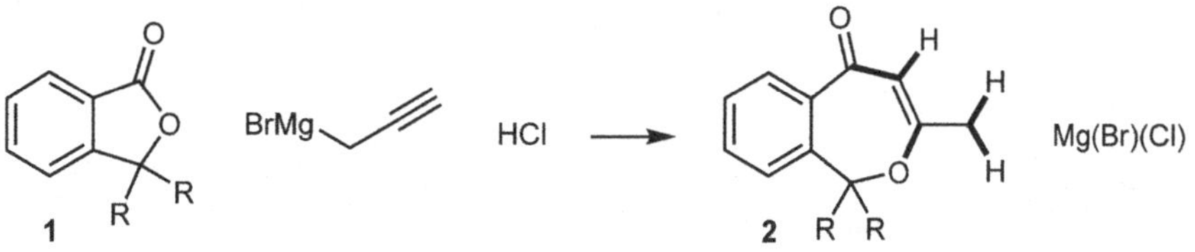

KEY STEPS EXPLAINED

The mechanism proposed by the authors to explain the formation of **2** from **1** begins with nucleophilic addition by the propargyl Grignard reagent to lactone **1** to form alkoxide **A** which ring opens to alkoxide **B** and is then quenched by water to give intermediate **C**. Protonation of **C** leads to allene **D** which ring closes to the cyclic ether **E**. The exocyclic olefin then undergoes a [1,3]-protic shift to give the final benzoxepinone **2**. We note that the authors did not perform further experimental or theoretical studies to help support this mechanism.

ADDITIONAL RESOURCES

We recommend an article which discusses synthetic approaches to compounds related to **2**.[2]

References

1. Nagao Y, Jeong I-Y, Lee WS, Sano S. A new ring enlargement reaction of γ-lactones to seven-membered cyclic ethers via intramolecular endo-mode cyclisation of the ω-hydroxy allenyl ketone intermediates in situ. *Chem Commun*. 1996;1996:19–20. https://doi.org/10.1039/CC9960000019.
2. Onusseit O, Meise W. Synthese von 2-Benzoxepin-1-onen. *Liebigs Ann*. 1996;1996(9):1483–1485. https://doi.org/10.1002/jlac.199619960922.

Question 83: *m*-Terphenyls From Pyrones

In 1996 Indian chemists observed that the reaction of pyrones **1** with acetophenones Ar_2COMe in DMF in the presence of KOH resulted in smooth conversion to *m*-terphenyls **2** (major product, 50%–66% yield) and the pyranylideneacetates **3** (minor product, 0%–28% yield) (Scheme 1).[1] Show mechanisms to explain the formation of both products **2** and **3** from **1**.

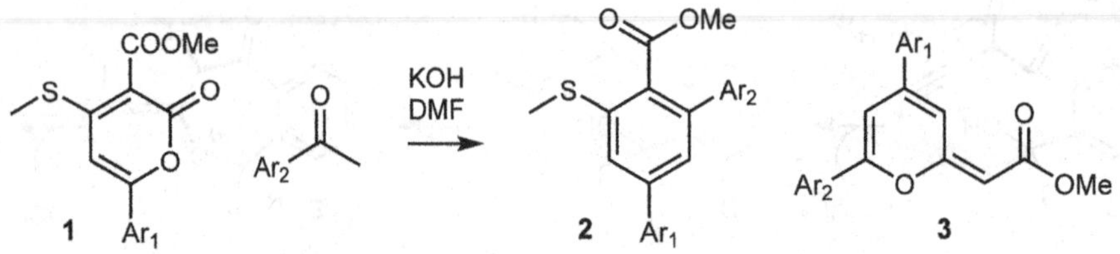

SCHEME 1

SOLUTION: 1 TO 2

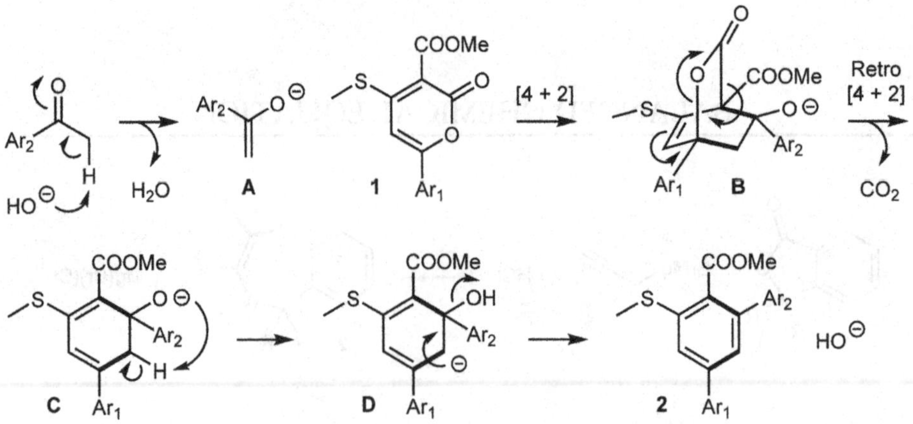

SOLUTION: B TO 3

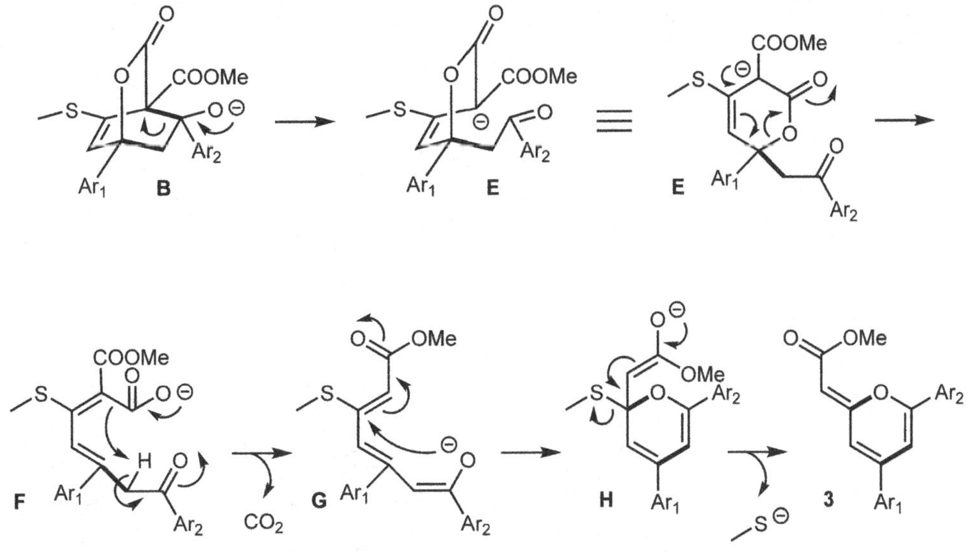

BALANCED CHEMICAL EQUATIONS

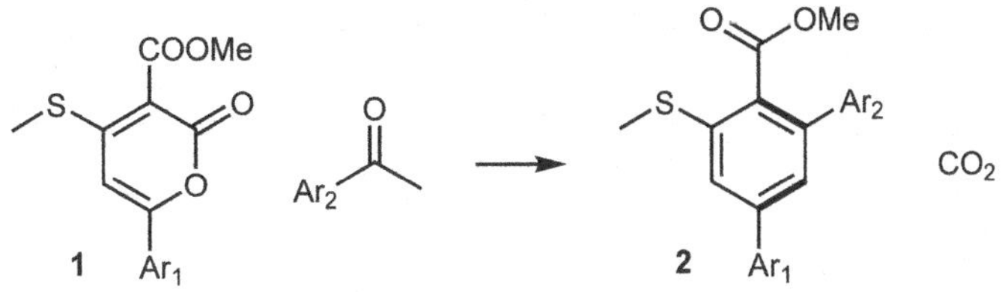

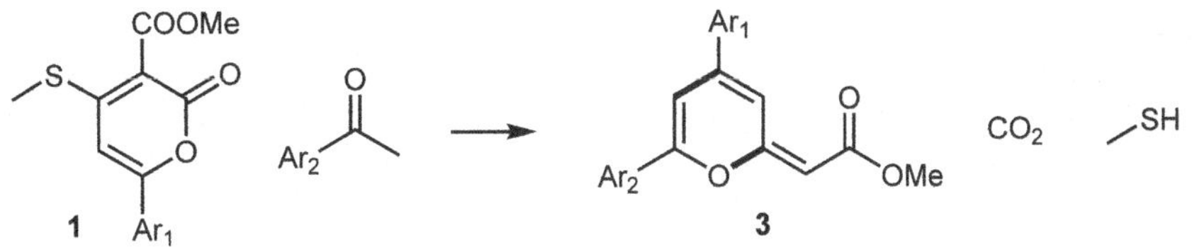

KEY STEPS EXPLAINED

The proposed mechanism for the transformation of **1** to **2** begins with hydroxide ion-mediated deprotonation of acetophenone to form enolate **A**. Enolate **A** then adds in a Diels-Alder fashion onto pyrone **1** to form adduct **B** which undergoes a retro-Diels-Alder ring decarboxylation to generate alkoxide **C**. **C** then undergoes an intramolecular proton transfer to give carbanion **D** which eliminates hydroxide ion to yield product **2**. We note that this sequence is an example of the Alder-Rickert reaction which was not mentioned by the authors.[2] The ring construction mapping for product **2** is [4 + 2].

With regard to the transformation of **1** to **3**, we have the same sequence of steps leading to intermediate **B** except that the alkoxide fragments to form carbanion **E** which ring opens to form carboxylate **F**. This carboxylate then undergoes decarboxylation to give enolate **G** which cyclizes to thioether **H** which eliminates methanethiolate to form the pyranylideneacetate product **3**. The ring construction mapping for this product is [3 + 3]. We note that the authors did not provide additional experimental and theoretical evidence to support their proposed mechanisms. Therefore we postulate an alternative mechanism for the transformation of **1** to **3** which involves initial generation of a ketene intermediate (Scheme 2).

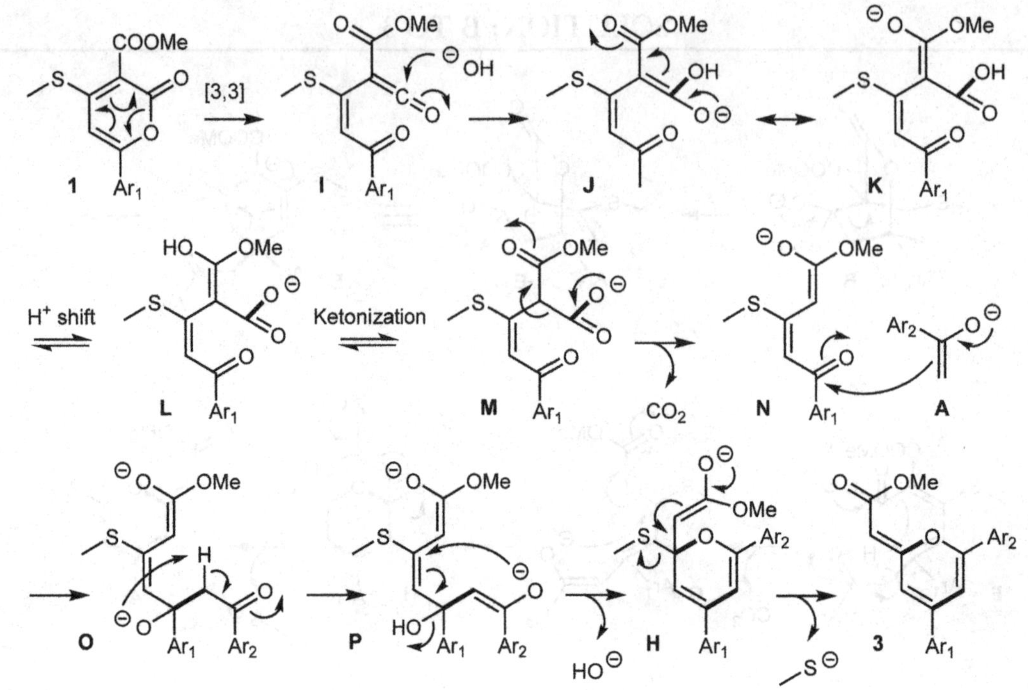

SCHEME 2

ADDITIONAL RESOURCES

We recommend an article which discusses the mechanism of an intramolecular [5+2] cycloaddition of a pyrone substrate bearing a tethered alkene.[2]

References

1. Ram VJ, Goel A. Ring transformation reactions part IV: 6-aryl-3-methoxycarbonyl-4-methylthio-2H-pyran-2-one, a novel synthon for the synthesis of 1,3-terphenyls from aryl ketones. *Tetrahedron Lett.* 1996;37(1):93–96. https://doi.org/10.1016/0040-4039(95)02081-0.
2. Domingo LR, Zaragoza RJ. Toward an understanding of the mechanisms of the intramolecular [5 + 2] cycloaddition reaction of γ-pyrones bearing tethered alkenes. A theoretical study. *J Org Chem.* 2000;65:5480–5486. https://doi.org/10.1021/jo000061f.

Question 84: A Cascade Reaction

In 1997 American chemists reacted the esters 1 with acetic anhydride and a catalytic amount of *p*-TsOH in xylenes to form the indoline product 2 in 65% yield (Scheme 1).[1] Provide a mechanism for this transformation.

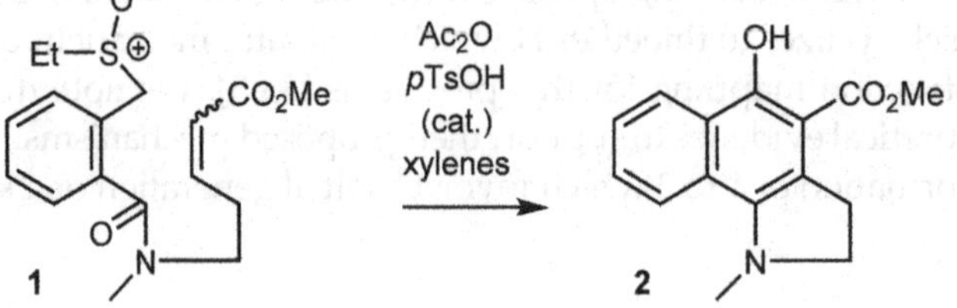

SCHEME 1

SOLUTION

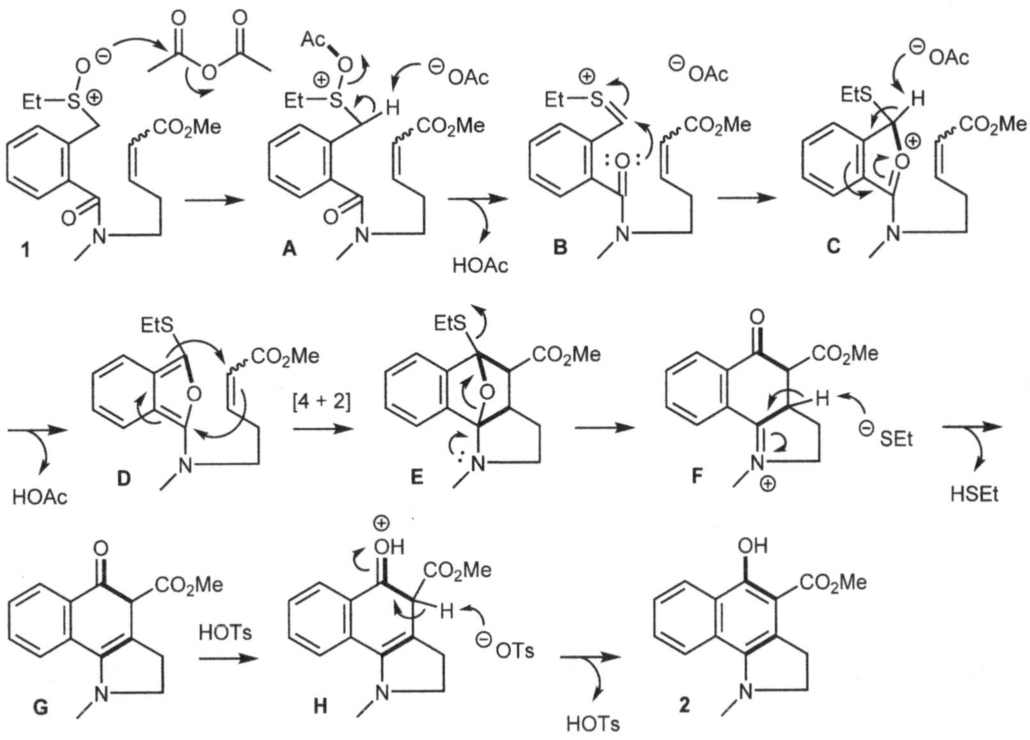

BALANCED CHEMICAL EQUATION

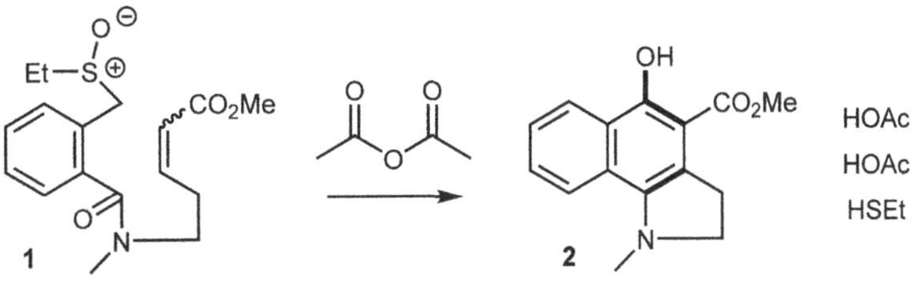

KEY STEPS EXPLAINED

We begin by noting that aside from commenting that both the *E* and *Z* esters **1** led to **2** and that the reaction sequence appeared to be a tandem-amido-Pummerer-Diels-Alder reaction, the authors did not present a mechanism for this transformation. Their paper indeed presented several other mechanisms for related transformations in this chemical system. Therefore we propose a mechanism which begins with acetylation of the sulfoxide oxygen atom of **1** by reaction with acetic anhydride (activator in the Pummerer rearrangement) which forms intermediate **A**. This intermediate is deprotonated by acetate ion to form sulfonium ion **B** which ring closes to oxonium ion **C**. This ion is deprotonated by a second acetate ion to form benzofuran **D** which undergoes an intramolecular [4+2] Diels-Alder cycloaddition to generate adduct **E**. The bridging ether linkage fragments with ketonization and elimination of ethanethiol to form intermediate **G** which undergoes an acid-catalyzed rearomatization via *p*-toluenesulfonic acid to give the final product **2**. The ring construction mapping is [(4+2)+(5+0)]. We recommend a future theoretical analysis to help support this proposed mechanism.

ADDITIONAL RESOURCES

We direct the reader to two articles which discuss the mechanisms of interesting transformations that involve similar chemical systems.[2,3]

References

1. Padwa A, Kappe CO, Cochran JE, Snyder JP. Studies dealing with the cycloaddition/ring opening/elimination sequence of 2-amino-substituted isobenzofurans. *J Org Chem*. 1997;62:2786–2797. https://doi.org/10.1021/jo962358c.
2. Li S, Luo Y, Wu J. An efficient approach to fused indolines via a copper(I)-catalyzed reaction of sulfonyl azide with 2-ethynylaryl methylenecyclopropane. *Org Lett*. 2011;13(12):3190–3193. https://doi.org/10.1021/ol2011067.
3. Ghavtadze N, Froehlich R, Wuerthwein E-U. 2H-pyrrole derivatives from an aza-nazarov reaction cascade involving indole as the neutral leaving group. *Eur J Org Chem*. 2008;2008:3656–3667. https://doi.org/10.1002/ejoc.200800384.

Question 85: Photochemical Ortho Rearrangements

In 1996 American chemists carried out the irradiation of *o*-acetylphenylacetonitrile **1** in methanol and small amounts of water followed by silica gel chromatography to form *o*-acetylphenylacetamide **2** in 80% yield (Scheme 1).[1] When the reaction conditions were changed to trace amounts of water with no silica gel chromatography, the major product was the dimethyl ketal form of **2** (i.e., product **3**) which led to **2** after silica gel chromatography. An ^{18}O labeling experiment revealed that the oxygen atom in **1** ends up as the amide carbonyl oxygen atom in **3**. Based on these findings, propose a mechanism for the transformation of **1** to **3**.

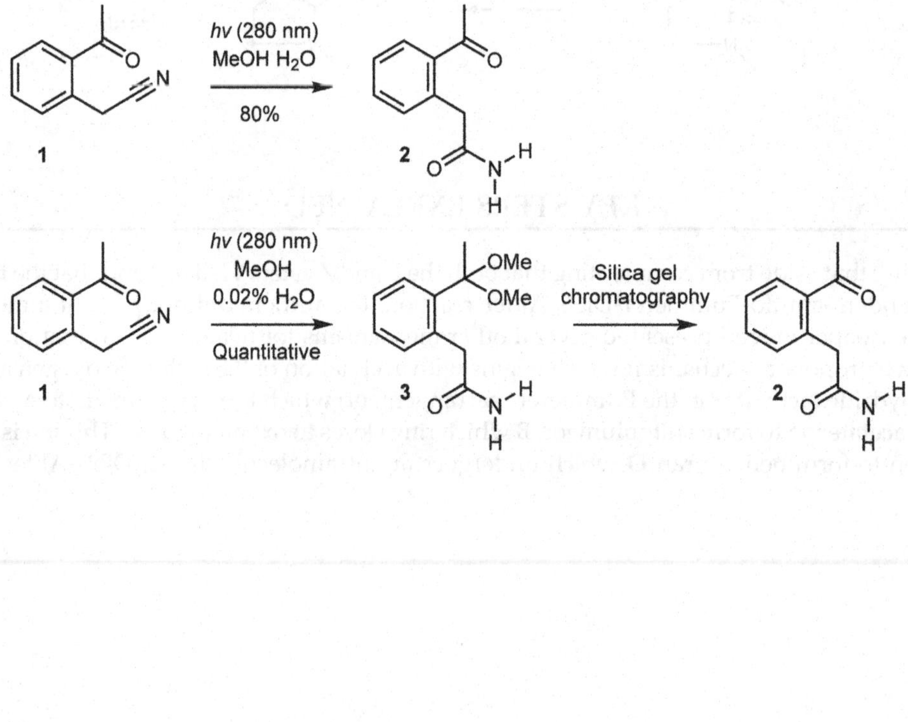

SCHEME 1

SOLUTION: 1 TO 3

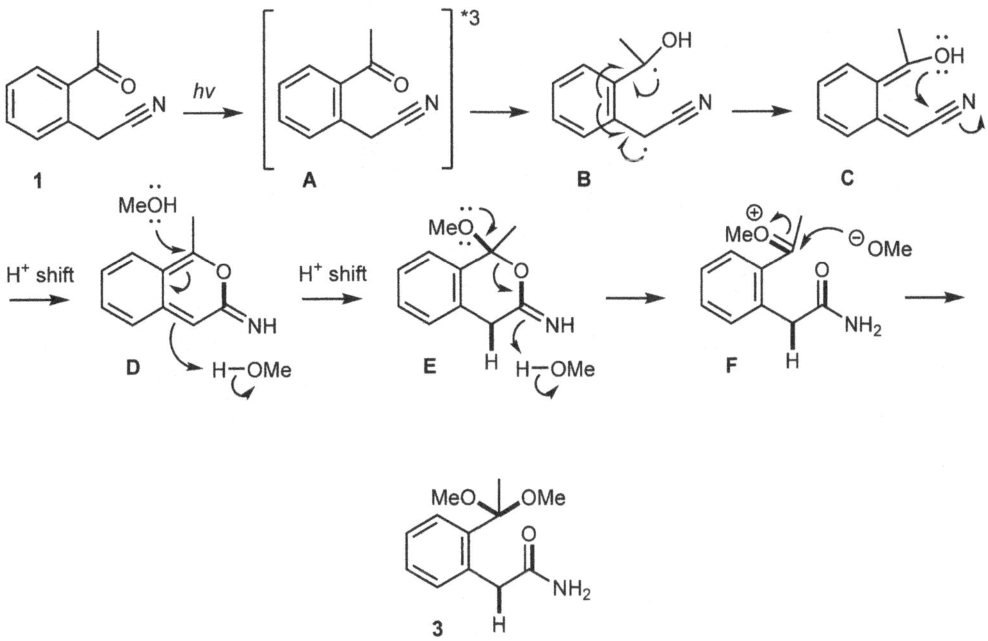

BALANCED CHEMICAL EQUATIONS

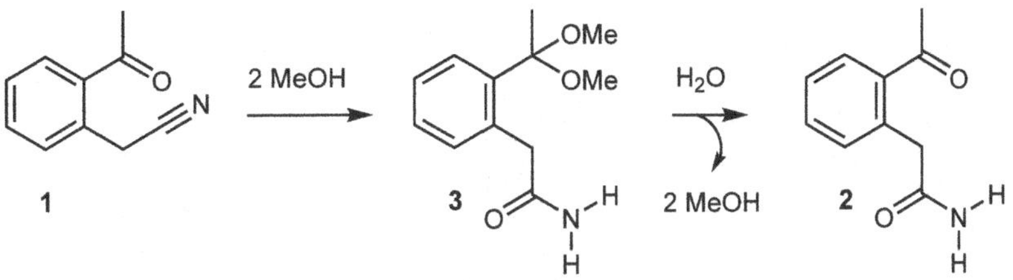

KEY STEPS EXPLAINED

The proposed mechanism for the transformation of **1** to **3** begins with photolysis of *o*-acetylphenylacetonitrile which forms an excited triplet state **A** that undergoes an intramolecular abstraction of a benzylic hydrogen atom to form diradical **B**. Intermediate **B** then electronically reorganizes to *o*-quinomethide intermediate **C** which then cyclizes to iminolactone **D**. One equivalent of methanol adds to **D** yielding iminolactone **E** which ring opens to oxonium ion amide **F** after abstracting a proton from a second equivalent of methanol. The resulting methoxide ion attacks the electrophilic carbonyl group yielding the dimethyl ketal of *o*-acetylphenylacetamide **3**. *o*-Acetylphenylacetamide **2** arises from an acid-catalyzed hydrolysis of the ketal during chromatographic purification on silica gel. In support of this mechanism the authors carried out the check experiments shown in Scheme 2. The observation that no reaction occurred with phenylacetonitrile **4** having a structure that lacks the acetyl group suggests that that carbonyl group is essential. 2-(2-Acetyl-phenyl)-propionitrile **5**, upon irradiation in methanol, gives 2-(2-acetyl-phenyl)-propionamide **6** as the major product which is analogous in structure to product **2**. The observation that the doubly methylated substrate **9** leads to very different products suggests that a benzylic hydrogen atom is also essential.

Further support for the authors' proposed mechanism comes from an ^{18}O labeling study shown in Scheme 3. The outcome verified that the ^{18}O label on the original acetyl group in **1** is transferred to the amide group in **3**. The authors did not carry out a theoretical analysis to support their proposed mechanism.

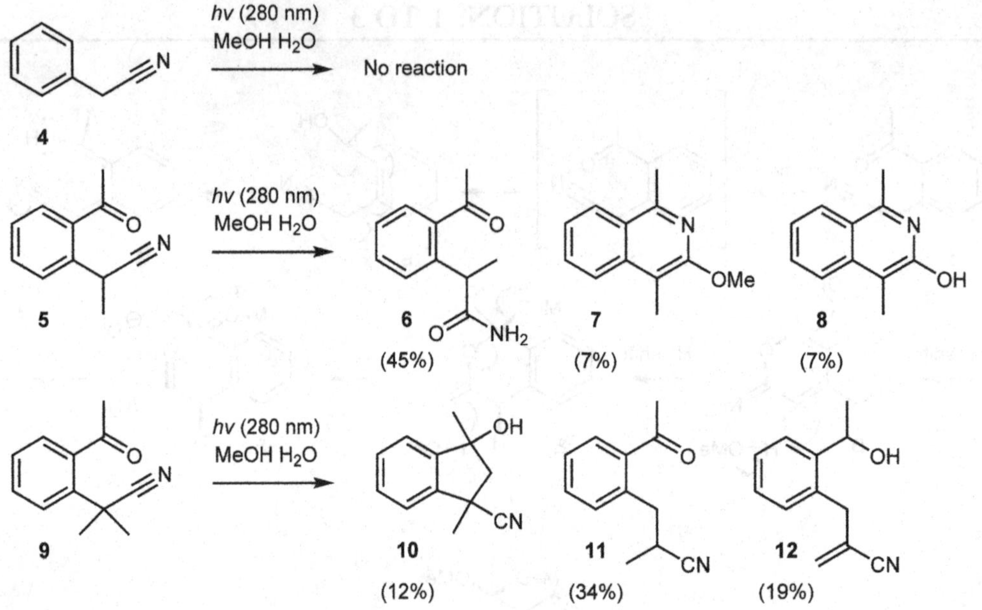

SCHEME 2

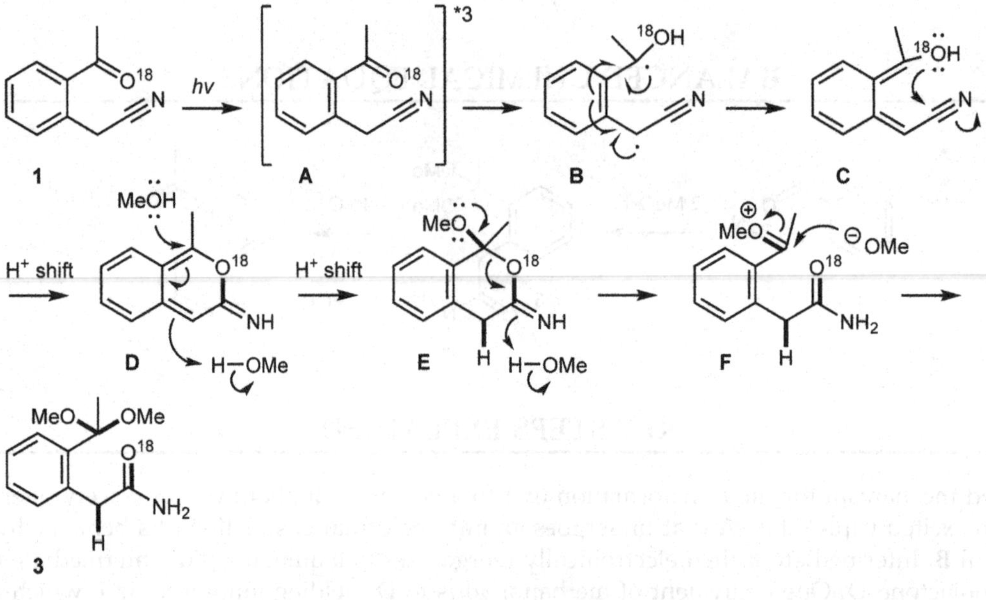

SCHEME 3

ADDITIONAL RESOURCES

We recommend three articles which discuss interesting transformations and rearrangements in the context of photochemistry.[2–4]

References

1. Lu Q, Bovonsombat P, Agosta WC. Novel photochemical hydrolysis of *o*-acetylphenylacetonitriles to amides. *J Org Chem.* 1996;61(11):3729–3732. https://doi.org/10.1021/jo952147s.
2. Maeda H, Nashihara S, Mukae H, Yoshimi Y, Mizuno K. Improved efficiency and product selectivity in the photo-Claisen-type rearrangement of an aryl naphthylmethyl ether using a microreactor/flow system. *Res Chem Intermed.* 2013;39:301–310. https://doi.org/10.1007/s11164-012-0650-6.

3. Oliveira AMAG, Oliveira-Campos AMF, Raposo MMM, Griffiths J, Machado AEH. Fries rearrangement of dibenzofuran-2-yl ethanoate under photochemical and Lewis-acid-catalysed conditions. *Tetrahedron*. 2004;60:6145–6154. https://doi.org/10.1016/j.tet.2004.05.060.
4. Mori T, Takamoto M, Saito H, Furo T, Wada T, Inoue Y. Remarkable differences in photo and thermal (acid-catalyzed) reactivities between *ortho*- and *para*-acylcyclohexadienones as essential factors determining the overall efficiency of the photo-fries rearrangement. *Chem Lett*. 2004;33 (3):256–257. https://doi.org/10.1246/cl.2004.256.

Question 86: From Bicycle to Pentacycle to Tricycle

In 1993 Japanese chemists reacted lactol **1** with isocyanate **2** in acetonitrile in the presence of 5 mol% of 1,8-diaza-bicyclo[5.4.0]undec-7-ene (DBU) and formed the pentacyclic product **3** in 86% yield which when treated with methoxide gave the octahydroacridine **4** in 90% yield (Scheme 1).[1] Provide mechanisms for the formation of **3** from **1** and **2** and of **4** from **3**.

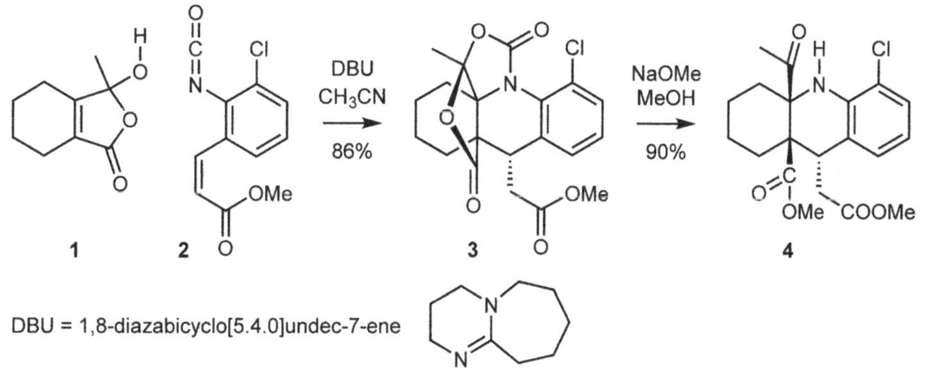

SCHEME 1

SOLUTION: 1 AND 2 TO 3

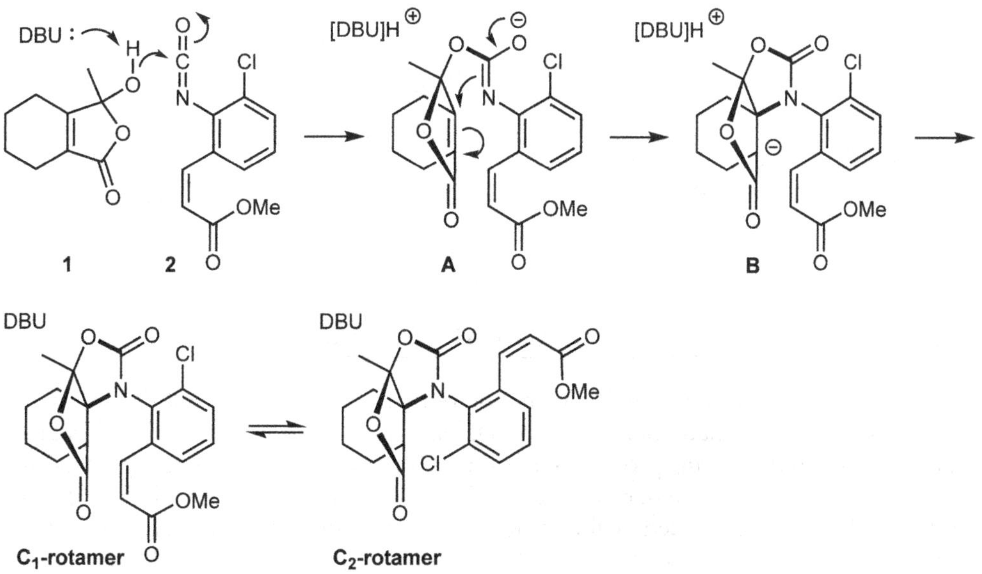

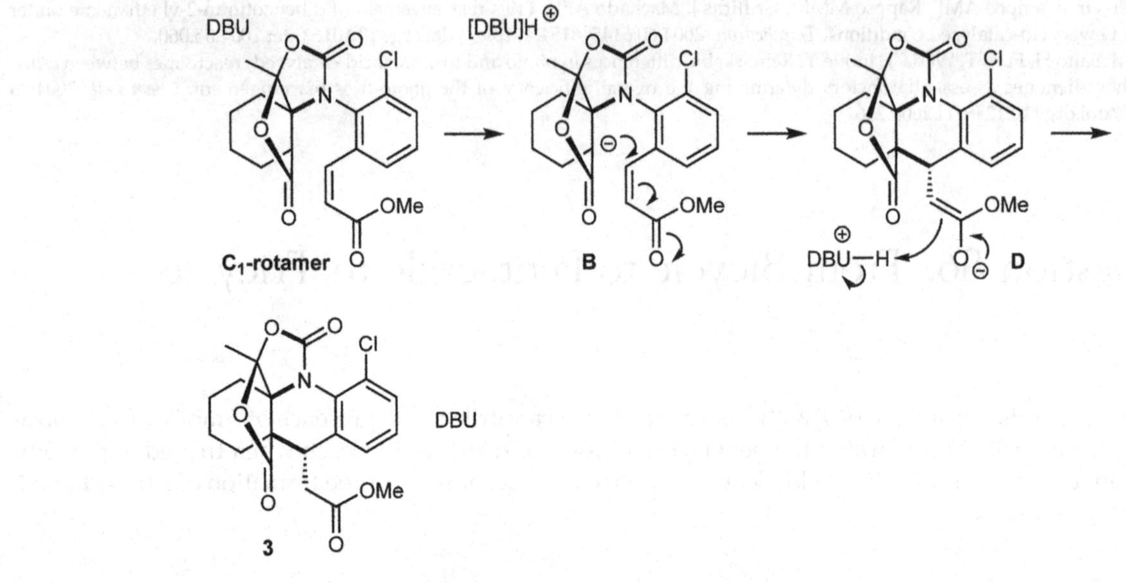

SOLUTION: 3 TO 4

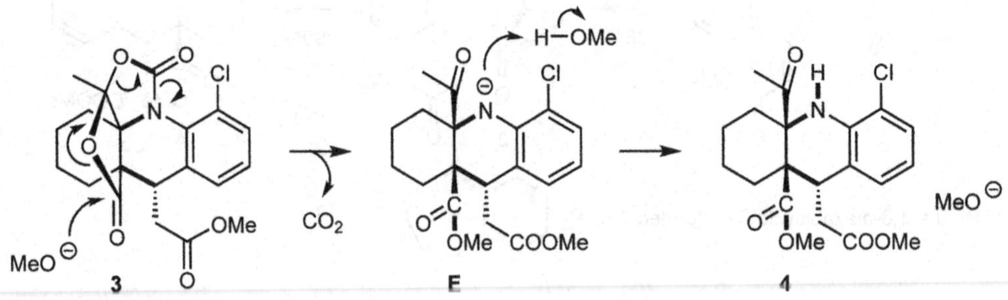

BALANCED CHEMICAL EQUATION

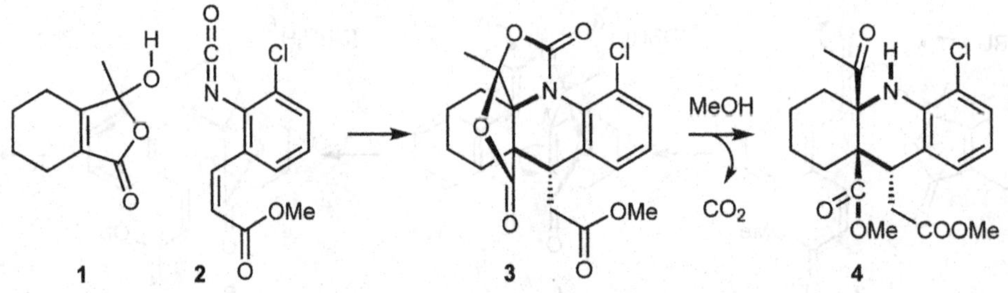

KEY STEPS EXPLAINED

The mechanism proposed by the authors for the formation of pentacycle **3** begins with a DBU-assisted addition of lactone **1** to isocyanate **2** to form carbamate rotamers C_1 and C_2. DBU-assisted transformation of rotamer C_1 then leads to a Michael cyclization reaction that forms product **3**. Rotamer C_1 is therefore considered the productive isomer that leads directly to the cyclization product **3**; whereas, rotamer C_2 must isomerize first to C_1 before it can go to product **3**. Hence, the yield of product **3** will be governed by the position of the equilibrium between rotamers C_1 and C_2. We note that the ring construction mapping for product **3** is [(3+2)+(4+2)]. In support of this mechanism the carbamate **6** was isolated for the case when the chlorine group in **2** was replaced by a methyl group (see Scheme 2). Also, rotamers

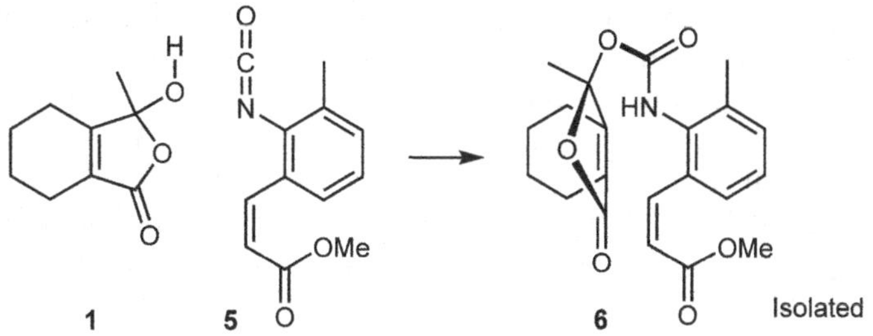

SCHEME 2

shown in Scheme 3 were also isolated and characterized by IR, ^{1}H NMR, ^{13}C NMR, and elemental analysis. The first three pairs of rotamers were isolated as inseparable mixtures whereas the fourth pair was isolated as separate isomers by thin-layer chromatography and characterized by their individual melting points.

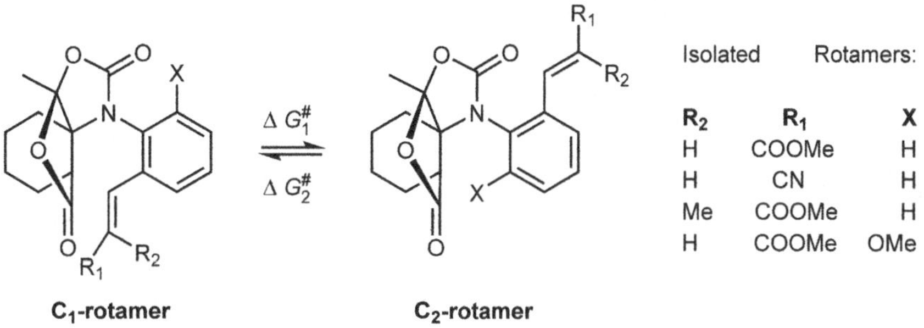

Isolated	Rotamers:	
R_2	R_1	X
H	COOMe	H
H	CN	H
Me	COOMe	H
H	COOMe	OMe

C₁-rotamer   **C₂-rotamer**

SCHEME 3

The authors also determined by ^{1}H NMR the interconversion energy barriers in the forward and reverse directions for the first and last pairs of rotamers shown in Scheme 3. The results were obtained using the relationships shown in Eq. (1) and are summarized in Table 1.

TABLE 1 Summary of Kinetic Data Obtained for Rotamer Interconversion by ^{1}H NMR.

R_1	R_2	X	T (K)	$\Delta G_1^{\#}$ (kJ mol⁻¹)	$\Delta G_2^{\#}$ (kJ mol⁻¹)	$k(C_1)$ (s⁻¹)	$k(C_2)$ (s⁻¹)	$K_{eq} = k(C_1)/k(C_2)$
COOMe	H	H	298	73.6	74.8	7.60E−01	4.50E−01	1.7
COOMe	H	OMe	384	122.9	127.5	1.40E−05	3.50E−05	4.2

$$K(\text{eq}) = \frac{k(C_1)}{k(C_2)} = \exp\left[\frac{1000\left(\Delta G_2^{\#} - \Delta G_1^{\#}\right)}{RT}\right] \tag{1}$$

where $R = 8.314\,\text{J K}^{-1}\,\text{mol}^{-1}$, T is temperature in Kelvin, and $\Delta G_1^{\#}$ and $\Delta G_2^{\#}$ are the respective energy barriers in kJ mol⁻¹.

From these kinetic data, though recorded at different temperatures, we observe that when X=OMe, $k(C_1) > k(C_2)$ and $\Delta G_1^{\#} < \Delta G_2^{\#}$, suggesting that the equilibrium is shifted toward the nonproductive rotamer C_2 and that the percent yield of product analogous to structure **3** should be predicted to be lower than when X=H. Using DBU as catalyst under room temperature conditions the authors found that the yield of product **3** (X=H) is 60% whereas the yield of product **3** (X=OMe) is 83%. These results are contrary to those found from the kinetic data. Moreover, when the authors isolated the individual rotamers C_1 (X=OMe) and C_2 (X=OMe) and carried out the respective reactions starting from these materials with 5 mol% DBU in deuterated acetonitrile at room temperature they found the following results: (a) starting from the productive rotamer C_1 (X=OMe) after 20 h the ratio of product **3** (X=OMe) to unproductive rotamer C_2 (X=OMe) was 85:4 and after 90 h the ratio increased to 86:2; (b) starting from the unproductive rotamer C_2 (X=OMe) after 20 h the ratio of product **3** (X=OMe) to unproductive rotamer C_2 (X=OMe) was 20:70 and after 90 h the ratio increased to 60:20. These latter results are consistent with the equilibrium between the rotamers shifted toward the C_1 rotamer *not* the C_2 rotamer as the authors stated from their kinetic data. Clearly, starting from the productive C_1 rotamer one would expect a high yield of product **3** and very little rotamer C_2 which is consistent with a high energy barrier from C_1 to C_2 and hence a lower valued rate constant, that is, $k(C_1) < k(C_2)$ and $\Delta G_1^{\#} > \Delta G_2^{\#}$. Similarly, starting from the unproductive C_2 rotamer the yield of product **3** is, as expected, lower than when starting from rotamer C_1, but after a long time the yield of product **3** slowly increases as, according to Le Chatelier's principle, the equilibrium is shifted from C_2 to C_1 due to the slow depletion of C_1 toward product **3**. We believe that the origin of the contradiction lies in the authors' poorly stated definition of the equilibrium constant for the rotamers given in their paper as "$K = a/b = 0.25$" for the case X=OMe and which the authors interpret as "in favour of rotamer 19b." In our present notation 19b refers to rotamer C_2 (X=OMe). Hence we remind the reader of the correct definition of equilibrium constant, $K(\text{eq})$, for two species A and B with forward and reverse rate constants k_A and k_B, respectively.

$$\text{Kinetic system: } A \underset{k_B}{\overset{k_A}{\rightleftharpoons}} B$$

Rate laws:
$$\frac{d[A]}{dt} = -k_A[A] + k_B[B]$$
$$\frac{d[B]}{dt} = k_A[A] - k_B[B]$$

At equilibrium the condition $\frac{d[A]}{dt} = \frac{d[B]}{dt} = 0$ is satisfied. Therefore:

$$k_A[A]_{eq} = k_B[B]_{eq}$$
$$\boxed{K(\text{eq}) = \frac{[B]_{eq}}{[A]_{eq}} = \frac{k_A}{k_B}}$$

The authors notation of "$K = a/b$" ($= [A]_{eq}/[B]_{eq}$) is exactly the inverse of what it should be. Hence, for the data shown in Table 1 we conclude that the labels for the rate constants should be reversed and therefore the true equilibrium constant values should be $1/1.7 = 0.59$ for X=H and $1/4.2 = 0.24$ for X=OMe. In addition, the labels for the energy barriers are also reversed. The authors' data for product ratios and reaction yields determined from starting individual rotamers and starting lactone **1**, respectively, remain as correct. Scheme 4 shows the correct interpretation of

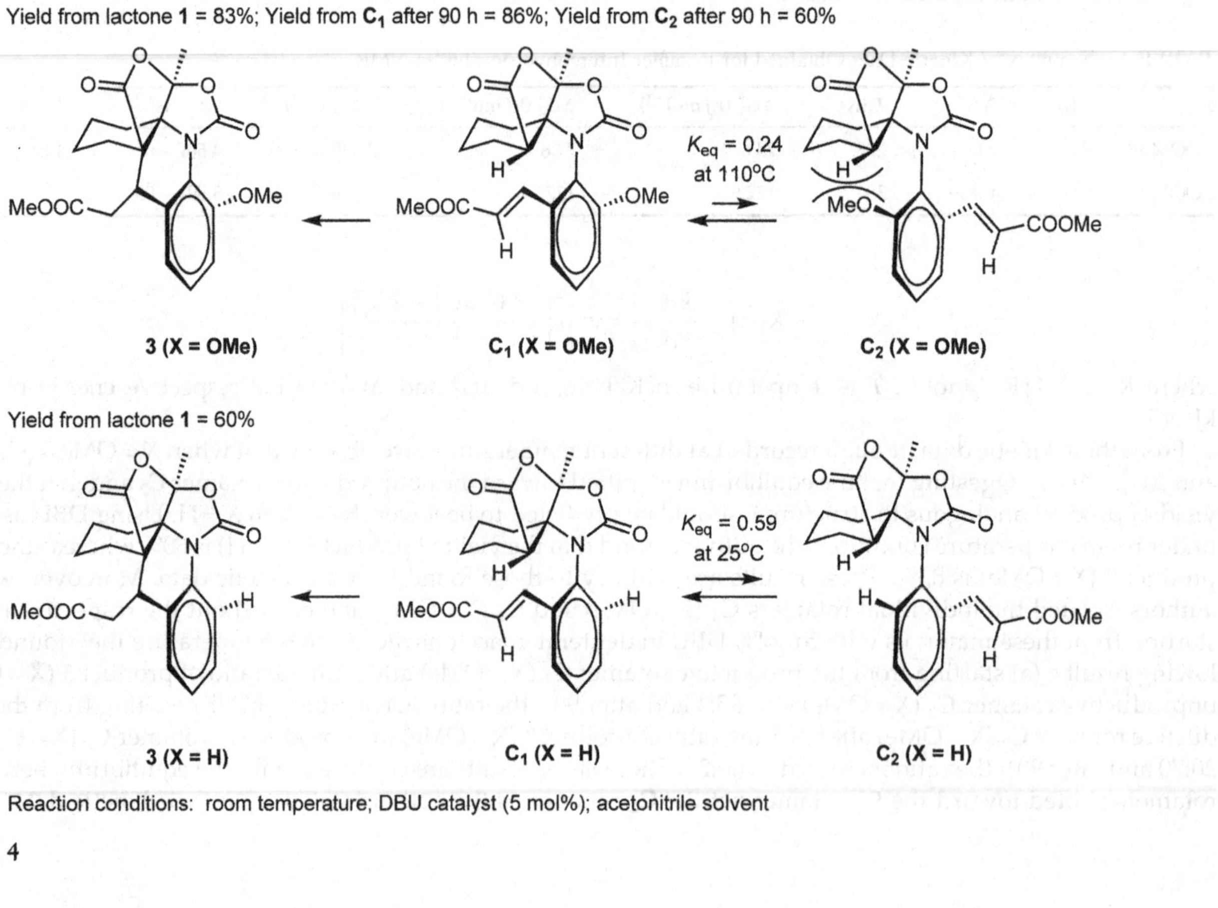

Yield from lactone **1** = 83%; Yield from C_1 after 90 h = 86%; Yield from C_2 after 90 h = 60%

3 (X = OMe) C_1 (X = OMe) C_2 (X = OMe)

$K_{eq} = 0.24$ at 110°C

Yield from lactone **1** = 60%

3 (X = H) C_1 (X = H) C_2 (X = H)

$K_{eq} = 0.59$ at 25°C

Reaction conditions: room temperature; DBU catalyst (5 mol%); acetonitrile solvent

SCHEME 4

the two juxtaposed systems with X=OMe versus X=H. It is clear that the steric interaction between the methoxy group and the alpha hydrogen atom in rotamer C_2 is unfavorable and explains why the equilibrium is shifted toward rotamer C_1 which has no such steric clash and hence is more thermodynamically stable. When X=H the equilibrium is less tilted and so there is less siphoning of the C_1 rotamer toward product 3, and consequently its yield is comparatively lower from lactone 1 than when X=OMe. In this case there should be more unproductive rotamer C_2 as the reaction proceeds. Scheme 5 shows the correct energy reaction coordinate diagrams for the C_1-C_2 rotamer equilibrium system. In both cases the sterically less hindered C_1 rotamer is thermodynamically more stable than the sterically encumbered C_2 rotamer.

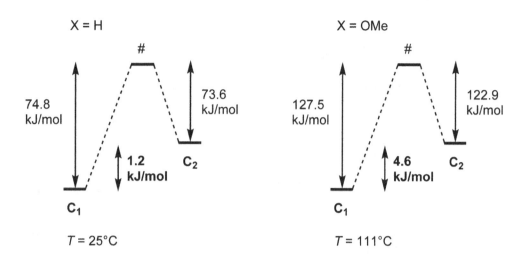

SCHEME 5

Concerning the transformation of 3 to 4, the proposed mechanism begins with nucleophilic ring opening of the lactone group of 3 by methoxide ion which causes decarboxylation which leads to the formation of amide anion E. This structure is protonated by methanol to form product 4.

Lastly, we note that the authors did not undertake a computation study for their proposed mechanisms. We therefore suggest that a computational study of this rotamer-based chemical system be carried out in a future study to ensure that the authors' NMR data, product data, and overall interpretation of this reaction are self-consistent.

ADDITIONAL RESOURCES

We recommend four articles and a review paper which discuss several interesting transformations, especially of pentacyclic products, and their mechanisms.[2–6]

References

1. Saito K, Yamamoto M, Yamada K, Takagi H. A highly diastereoselective tandem Michael-Michael addition reaction in a rotameric ring system: steric acceleration due to conformation locking. *Tetrahedron*. 1993;49(43):9721–9734. https://doi.org/10.1016/S0040-4020(01)80175-7.
2. Coletti A, Lentini S, Conte V, et al. Unexpected one-pot synthesis of highly conjugated pentacyclic diquinoid compounds. *J Org Chem*. 2012;77:6873–6879. https://doi.org/10.1021/jo300985x.
3. Szatmari I, Fulop F. Simple access to pentacyclic oxazinoisoquinolines via an unexpected transformation of aminomethylnaphthols. *Tetrahedron Lett*. 2011;52:4440–4442. https://doi.org/10.1016/j.tetlet.2011.06.074.
4. Molnar A, Boros S, Simon K, Hermecz I, Gonczi C. Unexpected transformations of an azoxyquinoxaline. *ARKIVOC*. 2010;2010:199–207.
5. Camps P, Domingo LR, Formosa X, et al. Highly diastereoselective one-pot synthesis of spiro{cyclopenta[a]indene-2,2'-indene}diones from 1-indanones and aromatic aldehydes. *J Org Chem*. 2006;71:3464–3471. https://doi.org/10.1021/jo0600095.
6. Bunce RA. Recent advances in the use of tandem reactions for organic synthesis. *Tetrahedron*. 1995;51(48):13103–13159. https://doi.org/10.1016/0040-4020(95)00649-S.

Question 87: Lewis Acid-Catalyzed Cyclopropyl Ketone Rearrangement

In 1992 Indian chemists reacted the ketene dithioacetal **1** with tin(IV) chloride in benzene at 20°C followed by aqueous quench to form product **2** in 82% yield (Scheme 1).[1] Devise a mechanism for this transformation.

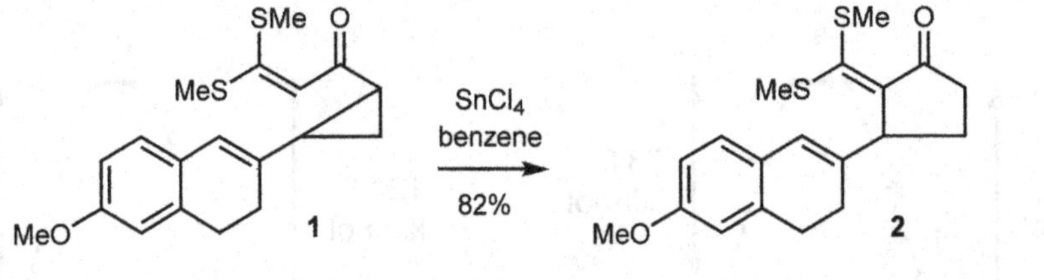

SCHEME 1

SOLUTION

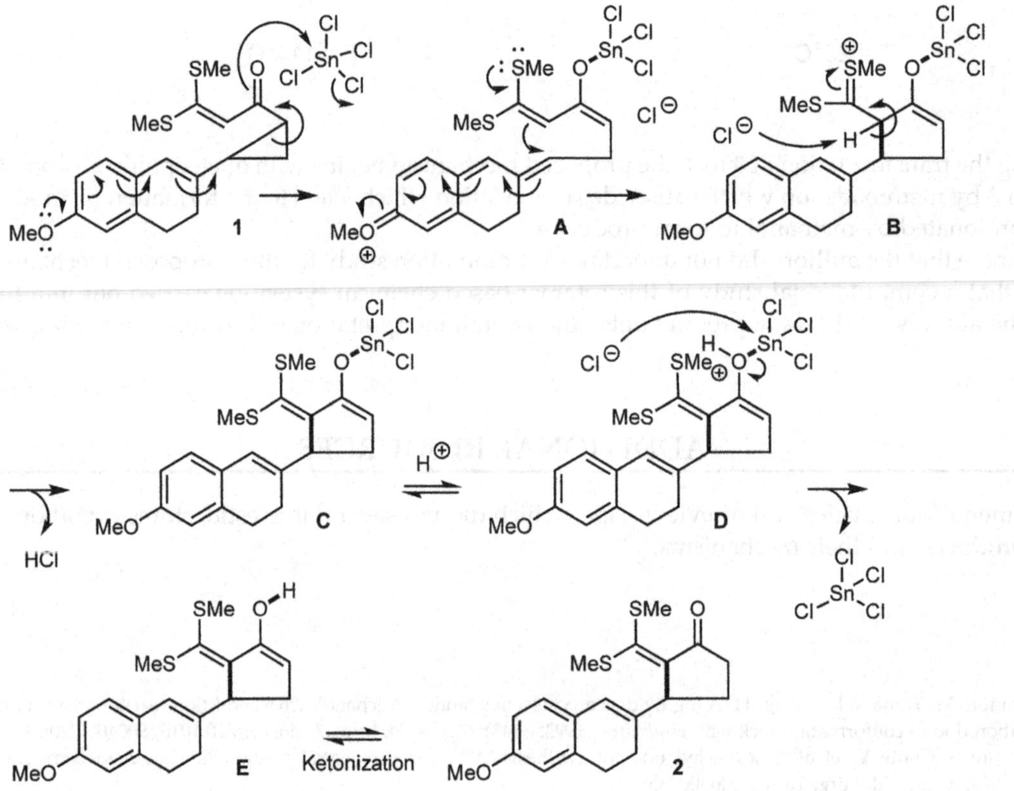

KEY STEPS EXPLAINED

The mechanism proposed by the authors closely mirrors the one depicted before except that the role of the tin(IV) chloride reagent is not explained. As such we note that tin(IV) chloride acts as a Lewis acid catalyst. In the first step of the mechanism, the electron-donating methoxy group causes the cyclopropane substrate **1** to undergo ring opening to the tin(IV) enolate intermediate **A**. This intermediate then undergoes ring closure via electron donation from one of the sulfur atoms thus leading to sulfonium ion **B**. Chloride ion abstraction of a proton *alpha* to the thioacetal moiety leads to

C which after protonation, regeneration of tin(IV)chloride, and ketonization gives product **2**. It is interesting to note that the key intermediate, according to the authors, is sulfonium ion **B**. This intermediate actually represents a masked benzylic carbocation **A'** (a resonance form of **A**) which the authors suggest could be trapped with a hydroxylic solvent such as water to form the ring open carbinol product **3** (Scheme 2). Nevertheless, addition of water actually resulted in the isolation of products **4** and **5** which are formed after nucleophilic attack by water at the activated sulfone carbon of **B** leading to elimination of HSMe in the case of product **4** and complete cleavage of the disulfide group in the case of product **5** (Scheme 3). Product **3** was not observed. Furthermore, the treatment of **2** under the reaction conditions did not lead to products **4** and **5** unless heat was applied thus suggesting that the transformation is not reversible as far as going from product **2** to intermediate **B** is concerned. It is likely that the irreversible step is the **B** to **C** elimination of HCl. Lastly, we note that the authors did not carry out a theoretical analysis to help support their postulated mechanism. Also, the ring construction mapping for product **2** is [5+0].

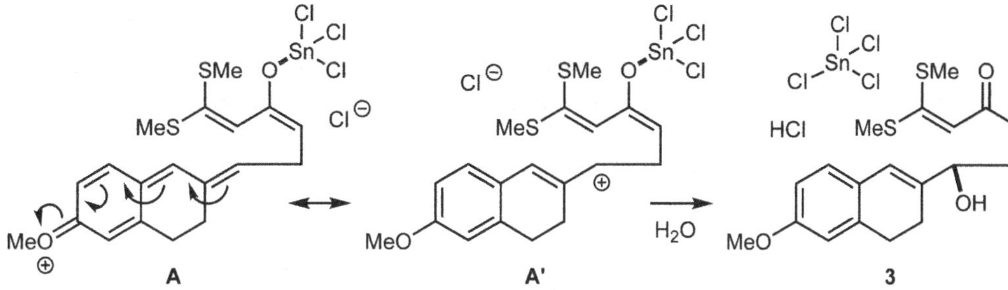

SCHEME 2

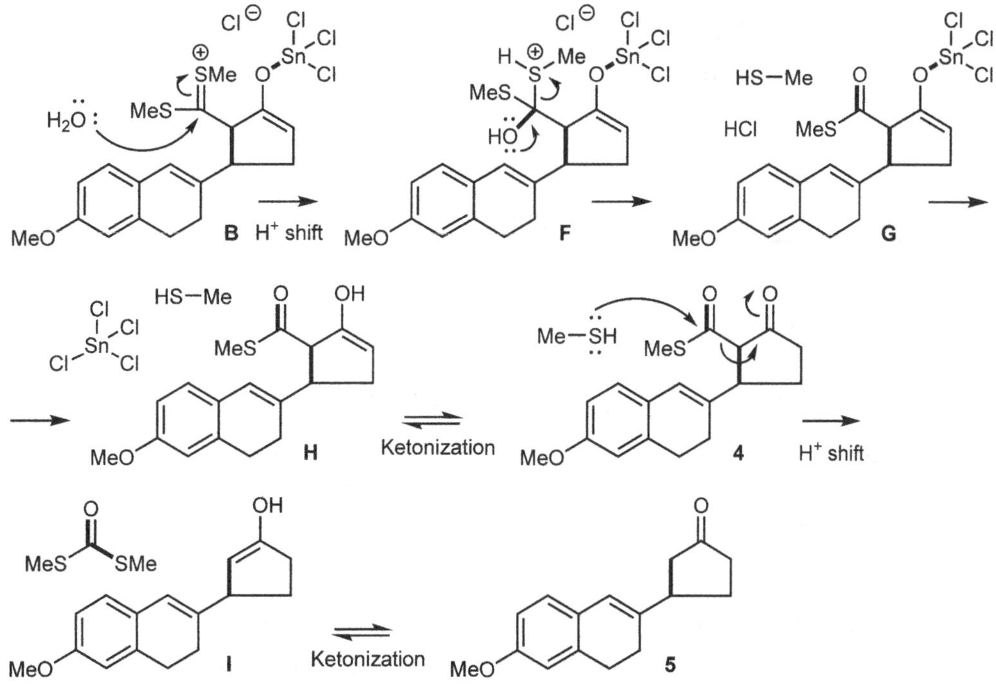

SCHEME 3

ADDITIONAL RESOURCES

We highly recommend an article which describes a transformation that involves domino carbocationic rearrangements of structures that contain a cyclopropane group similar to starting material **1**.[2]

References

1. Patro B, Deb B, Ila H, Junjappa H. Acid-induced ring opening of α-[bis(methylthio)methylene]alkyl cyclopropyl ketones: a novel route to substituted cyclopentanones through carbocationic cyclizations. *J Org Chem*. 1992;57(8):2257–2263. https://doi.org/10.1021/jo00034a014.
2. Venkatesh C, Ila H, Junjappa H, Mathur S, Volka H. Domino carbocationic rearrangement of aryl-2-(1-N-methyl/benzyl-3-indolyl)cyclopropyl ketones: a serendipitous route to 1H-cyclopenta[c]carbazole framework. *J Org Chem*. 2002;67:9477–9480. https://doi.org/10.1021/jo0258271.

Question 88: An Intramolecular Ring Closure

In 1998 American chemists treated the amide **1** with trifluoroacetic anhydride (TFAA) in methylene chloride at 25°C to form product **2** in 85% yield (Scheme 1).[1] Provide a mechanism for this transformation.

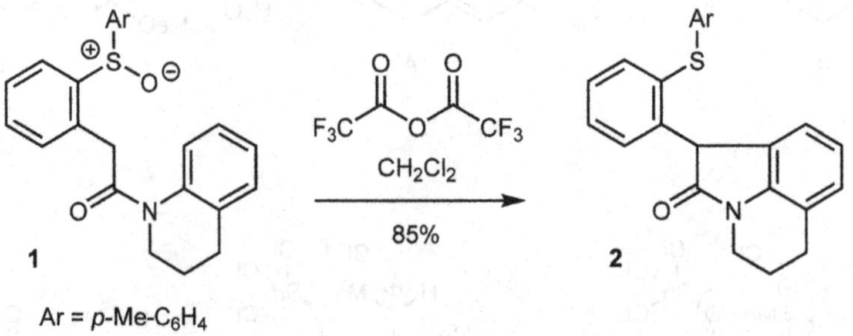

SCHEME 1

SOLUTION

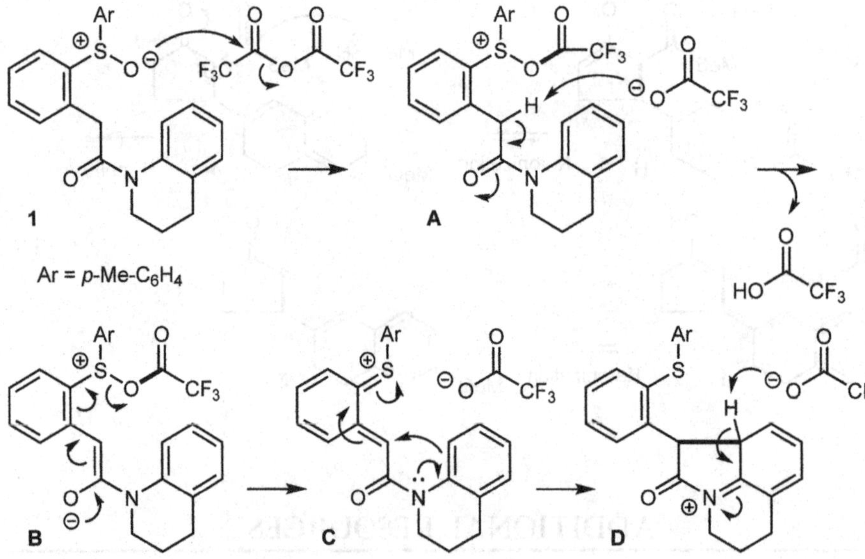

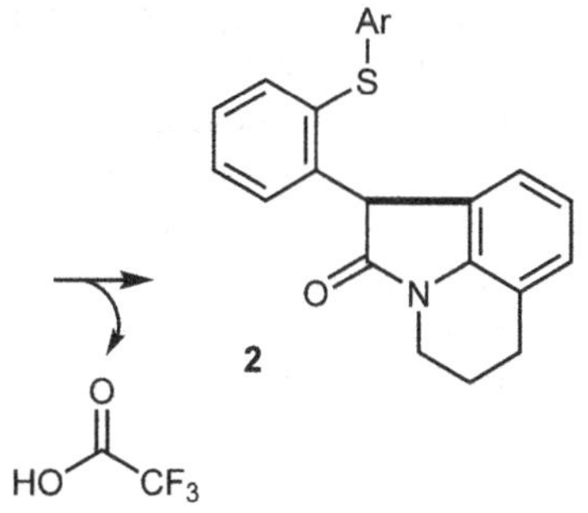

BALANCED CHEMICAL EQUATION

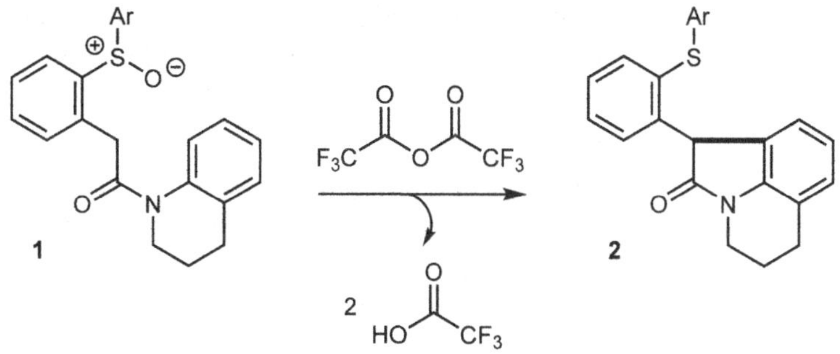

KEY STEPS EXPLAINED

The mechanism proposed for the transformation of **1** to **2** begins with nucleophilic addition of the sulfoxide oxygen atom to trifluoroacetic anhydride to form sulfonium ion **A** which is deprotonated by trifluoroacetate ion yielding zwitterionic intermediate **B**. **B** then fragments to trifluoroacetate ion and sulfonium ion **C** which cyclizes to iminium ion **D**. Trifluoroacetate ion deprotonates **D** to form the tetrahydroisoquinoline product **2**. We note that the reaction is an internal redox reaction as shown by the oxidation number changes of key atoms given in Scheme 2. Furthermore, the authors showed that the orientation of the aryl group connected to the nitrogen atom in **1** plays a key role in the formation of product **2**. In a separate experiment, the starting material **3**, an analog of **1**, was shown to lead to either the expected tetrahydroisoquinoline product **4** (analog of **2**) or the α-trifluoroacetoxy sulfide product **5** (normal Pummerer product) depending on the nature of the R group (Scheme 3).

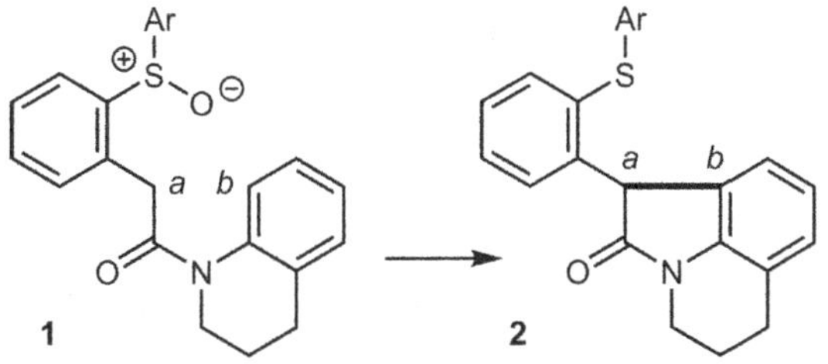

Atom	Reactant	Product	Change
S	0	−2	−2
C_a	−2	−1	+1
C_b	−1	0	+1

SCHEME 2

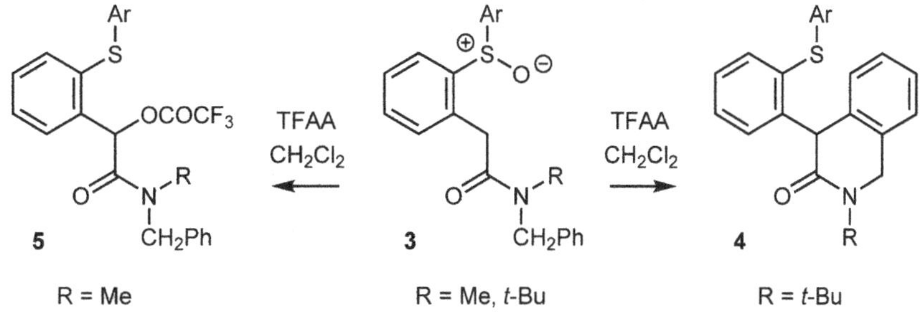

SCHEME 3

In this experiment, it was noted that when R was a methyl group, the transformation led to the normal Pummerer product **5** whereas when R was a *t*-butyl group, the tetrahydroisoquinoline product **4** was observed. The authors explained this result by noting that the rotation around the acyl C—N bond in **3** is known to be restricted allowing for the existence of two geometric isomers of **3** which cannot be separated because the barrier to rotation is low (approx. 20 kcal/mol).[1] Furthermore, the authors argue, it is known that such systems prefer to have the larger substituent on the nitrogen atom oriented *syn* to the acyl oxygen. In the case where R is a methyl group therefore, the *syn* orientation leads to a **C**-like thionium ion which does not have the favorable orientation for π cyclization to form a **D**-like intermediate which can lead to the expected tetrahydroisoquinoline product upon deprotonation. As such, the carbocationic intermediate **C** is captured by the trifluoroacetate nucleophile to form the normal Pummerer product **5**. The lack of observed **4** when R = Me also suggested to the authors that the amide bond in the thionium ion **C** did not rotate during the lifetime of that intermediate. Conversely, when R is a *t*-butyl group, the authors argue that **3** strongly prefers the Z-rotamer (i.e., *anti*) configuration which places the CH$_2$Ph group in the proper orientation for π cyclization of thionium ion **C** to form the expected tetrahydroisoquinoline product **4**. We note that such experimental evidence lends support for the proposed mechanism of the transformation of **1** to **2**. Lastly, we note that although the authors did not undertake a theoretical analysis to support their mechanism, they did provide several additional experimental observations which reveal further interesting aspects of this transformation which is why we highly recommend studying their article.

ADDITIONAL RESOURCES

For further examples of interesting transformations involving unusual ring closure, ring opening steps, see Refs. 2–4.

References

1. Padwa A, Kuethe JT. Additive and vinylogous Pummerer reactions of amido sulfoxides and their use in the preparation of nitrogen containing heterocycles. *J Org Chem*. 1998;63(13):4256–4268. https://doi.org/10.1021/jo972093h.
2. Riches SL, Saha C, Filgueira NF, Grange E, McGarrigle EM, Aggarwal VK. On the mechanism of ylide-mediated cyclopropanations: evidence for a proton-transfer step and its effect on stereoselectivity. *J Am Chem Soc*. 2010;132:7626–7630. https://doi.org/10.1021/ja910631u.
3. Kurbatov S, Tatarov A, Minkin V, Goumont R, Terrier F. Ring opening and ring closure in an indolizine structure activated through S$_N$Ar coupling with superelectrophilic 4,6-dinitrobenzofuroxan, an unusual intramolecular oxygen transfer from a N-oxide functionality. *Chem Commun*. 2006;2006:4279–4281. https://doi.org/10.1039/b608350a.
4. Huang J-M, Chen H, Chen R-Y. An unusual addition and ring-closure reaction of 1-(2-bromoethyl)-2,3-dihydro-3-propyl-1,3,2-benzo-diazaphosphorin-4(1H)-one 2-oxide with carbon disulfide for a new and convenient synthesis of the fused phosphorus heterocyclic compound. *Synth Commun*. 2002;32(14):2215–2225. https://doi.org/10.1081/SCC-120005432.

Question 89: Photochemical Rearrangement of Pyran-2-ones

In 1995 chemists from the University of Utah showed that irradiation of 6-(2-hydroxyalkyl)pyran-2-ones **1** in methanol followed by solvent removal and treatment with a catalytic amount of HCl in THF at room temperature gave the dihydropyrans **2** in good yield (Scheme 1).[1] Provide a mechanism for this transformation.

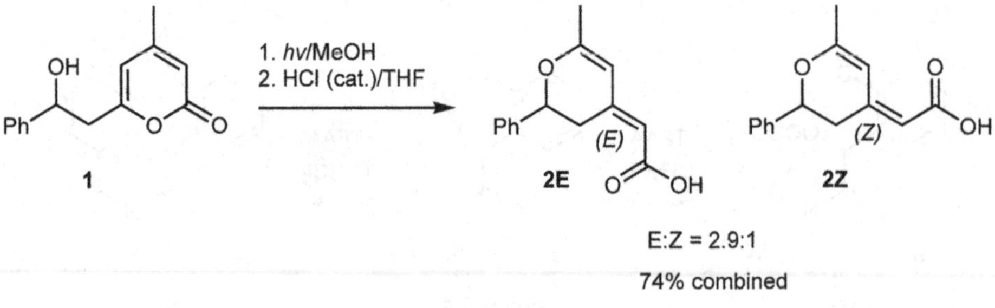

E:Z = 2.9:1

74% combined

SCHEME 1

SOLUTION: 1 TO 2

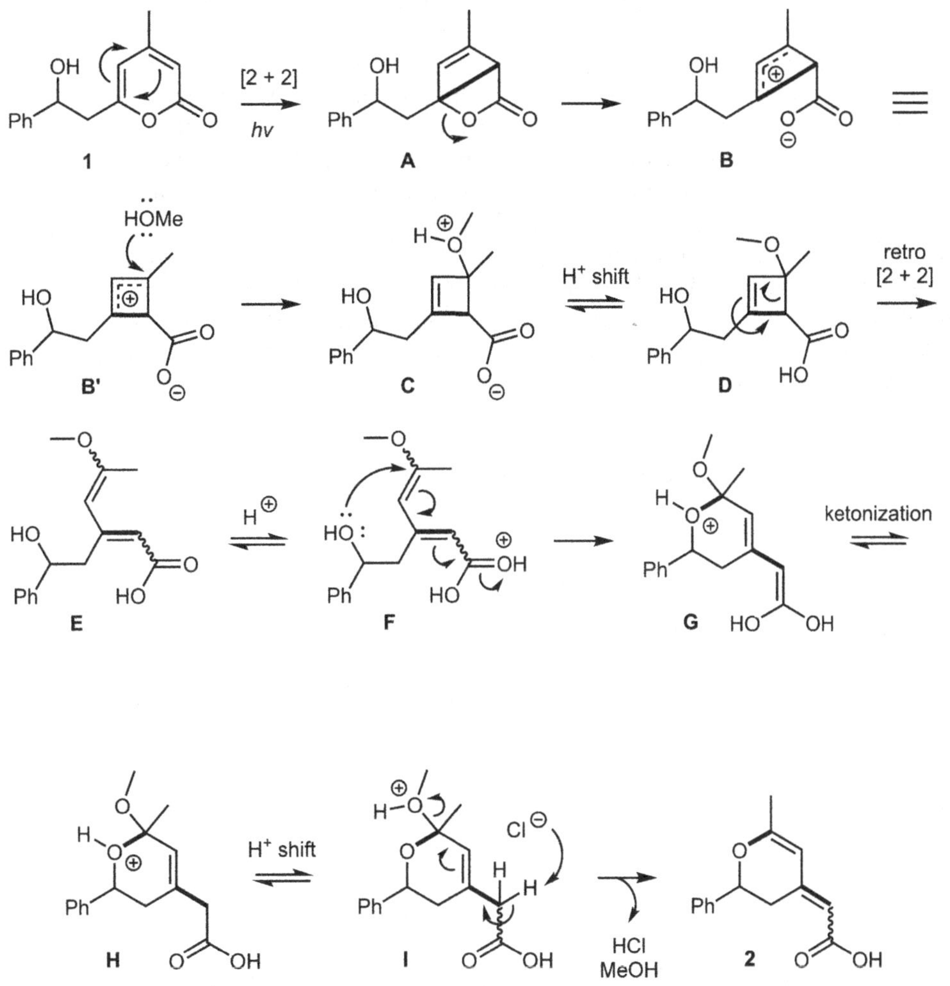

SOLUTION: F TO H (ALTERNATIVE PATH)

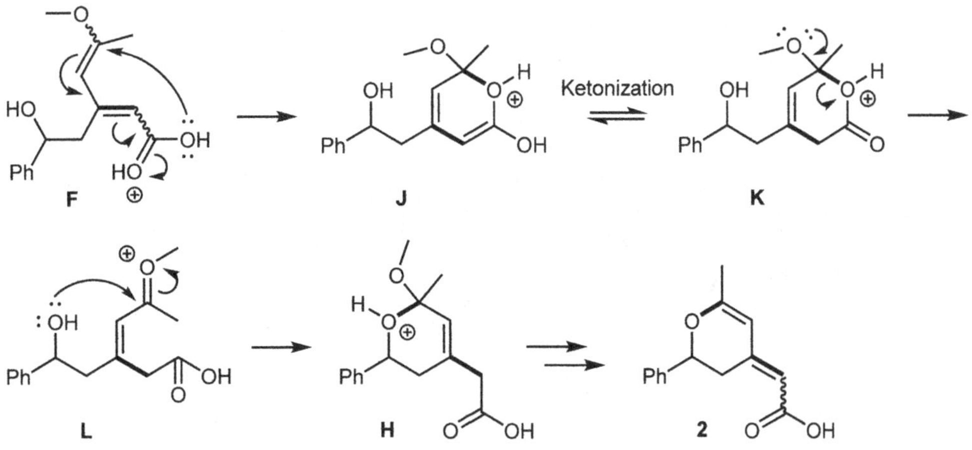

KEY STEPS EXPLAINED

We begin the analysis by noting that the authors stated that the mechanism for the transformation of **1** to **2** is not well understood. Since no experimental or theoretical evidence is presented, we consider the proposed mechanism as speculative. Nevertheless, the authors proposed that pyranone **1** first undergoes a photochemical [2+2] cyclization to the [2.2.0] bicyclic lactone intermediate **A** which ring opens to the cyclobutenyl carbocation **B/B'**. Methanol then adds to the tertiary carbocationic center to form cyclobutene **D** which undergoes a retro [2+2] transformation to diene **E**. Protonation of **E** in HCl/THF yields oxonium ion **F** which can proceed to product **2** via two routes. In the first route, the benzyl hydroxyl group cyclizes onto the vinyl ether moiety to **G** which after ketonization eliminates methanol via chloride ion-assisted deprotonation. In the second route, the carboxylic acid hydroxyl group in **F** cyclizes onto the vinyl ether moiety yielding **J** which after ketonization ring opens to **L**. Cyclization of the benzyl hydroxyl group onto the electrophilic carbonyl group leads to **H** which undergoes intramolecular proton transfer followed by chloride ion-assisted methanol elimination as before to form **2**. This transformation is 100% atom economical as there are no by-products formed. Methanol acts as a sacrificial reagent that first adds to the substrate and then is removed. The ring construction mapping is [6+0].

ADDITIONAL RESOURCES

We recommend two articles which describe photo-induced rearrangements of pyran structures similar to substrate **1**.[2,3]

References

1. Chase CE, Jarstfer MB, Arif AM, West FG. Unexpected and efficient photochemical rearrangement of 6-hydroxyethylpyran-2-ones to 4-alkylidene-5,6-dihydropyrans. *Tetrahedron Lett*. 1995;36(47):8531–8534. https://doi.org/10.1016/0040-4039(95)01828-6.
2. Mori Y, Takano K. Reaction mechanism of di-π-methane rearrangement of 4-phenyl-4H-pyran: a CASSCF/MRMP2 study. *J Photochem Photobiol A*. 2011;219:278–284. https://doi.org/10.1016/j.jphotochem.2011.03.004.
3. Gabbutt CD, Heron BM, Kolla SB, et al. Ring contraction during the 6π-electrocyclisation of naphthopyran valence tautomers. *Org Biomol Chem*. 2008;6:3096–3104. https://doi.org/10.1039/b807744d.

Question 90: A Paclitaxel Rearrangement

In 1995 American chemists investigating cleavage reactions involving paclitaxel, a clinically active antitumor drug, treated the 13β-chloro derivative baccatin III **1** with sodium azide in aqueous DMF at 60°C hoping that S$_N$2 substitution would form the 13α azide product **2** (Scheme 1).[1] Nevertheless, the observed product was the ring cleaved compound **3**. Provide a mechanism for this transformation.

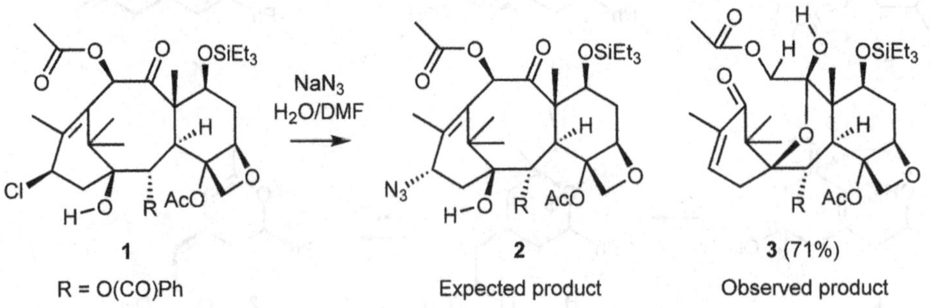

SCHEME 1

SOLUTION

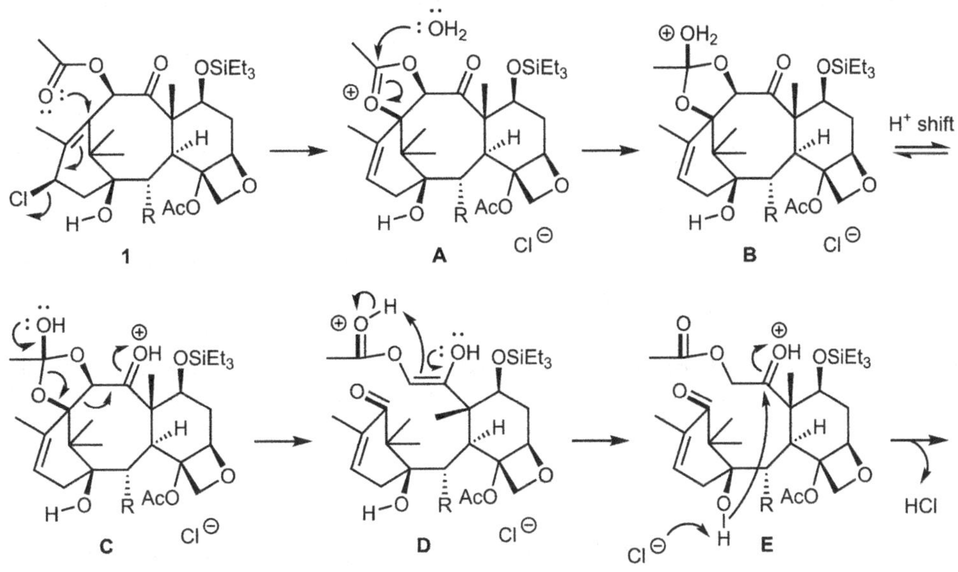

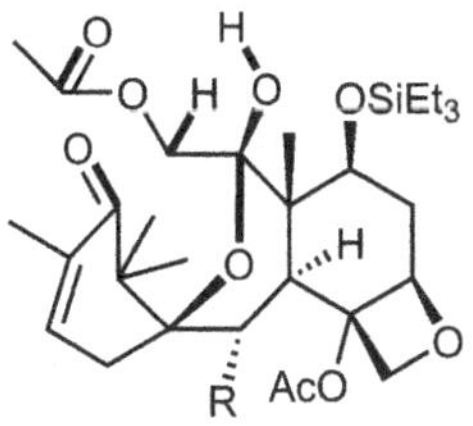

1 → A → B

C → D → E

3

BALANCED CHEMICAL EQUATION

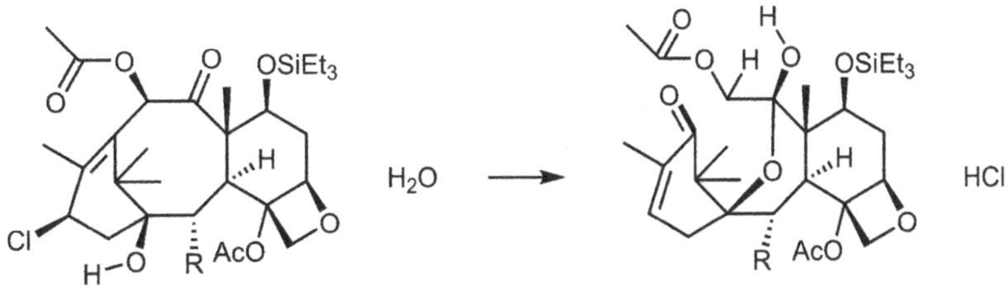

1   H_2O →   3   HCl

KEY STEPS EXPLAINED

We begin by noting that the authors expected to carry out a simple S_N2 substitution of the chloride group of **1** for an azide group. Instead, an acid-catalyzed rearrangement reaction occurred whose proposed mechanism begins with intramolecular cyclization of **1** to form oxonium ion **A** which is trapped by water causing ring opening of the cyclic ether to oxonium ion **D** bearing an enol functionality. The enol ketonizes to protonated ketone **E** after which chloride ion deprotonates the hydroxyl group of **E** causing a transannular cyclization to generate the observed ketal product **3** after elimination of hydrochloric acid as by-product. The authors did not undertake further experimental and theoretical analysis to help support this postulated mechanism.

ADDITIONAL RESOURCES

For further interesting transformations and rearrangements involving paclitaxel precursors and derivatives, see Refs. 2–5.

References

1. Chordia MD, Gharpure MM, Kingston DGI. Facile AB ring cleavage reactions of taxoids. *Tetrahedron*. 1995;51(47):12963–12970. https://doi.org/10.1016/0040-4020(95)00823-Q.
2. Chen S-H, Huang S, Roth GP. An interesting C-ring contraction in paclitaxel (Taxol®). *Tetrahedron Lett*. 1995;36(49):8933–8936. https://doi.org/10.1016/0040-4039(95)01954-G.
3. Paquette LA, Hofferberth JE. Effect of 9,10-cyclic acetal stereochemistry on feasible operation of the α-ketol rearrangement in highly functionalized paclitaxel (Taxol) precursors. *J Org Chem*. 2003;68:2266–2275. https://doi.org/10.1021/jo020627v.
4. Lee D, Kim M-J. Enzymatic selective dehydration and skeleton rearrangement of paclitaxel precursors. *Org Lett*. 1999;1(6):925–927. https://doi.org/10.1021/ol990179m.
5. Paquette LA, Wang H-L, Zeng Q, Shih T-L. Heteroatomic modulation of oxyanionic Cope rearrangement rates. Consequences on competing nucleophilic cleavage of an oxetane ring in precursors to paclitaxel. *J Org Chem*. 1998;63:6432–6433. https://doi.org/10.1021/jo981059f.

Question 91: An Ionic Diels-Alder Reaction

In 1995 chemists from the University of Colorado described an extension of the so-called ionic Diels-Alder reaction by showing that acetals such as **1** react with various 1,3-dienes in the presence of an acid catalyst followed by hydrolysis with TsOH in methanol to form good yields of cycloadducts such as **3** (Scheme 1).[1] Suggest a mechanism for this transformation.

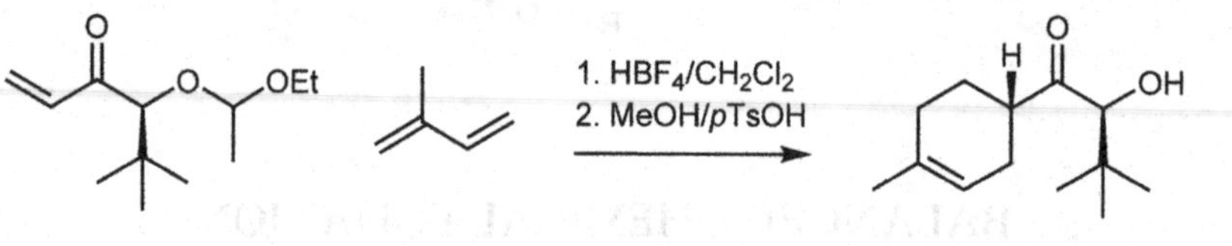

SCHEME 1

SOLUTION

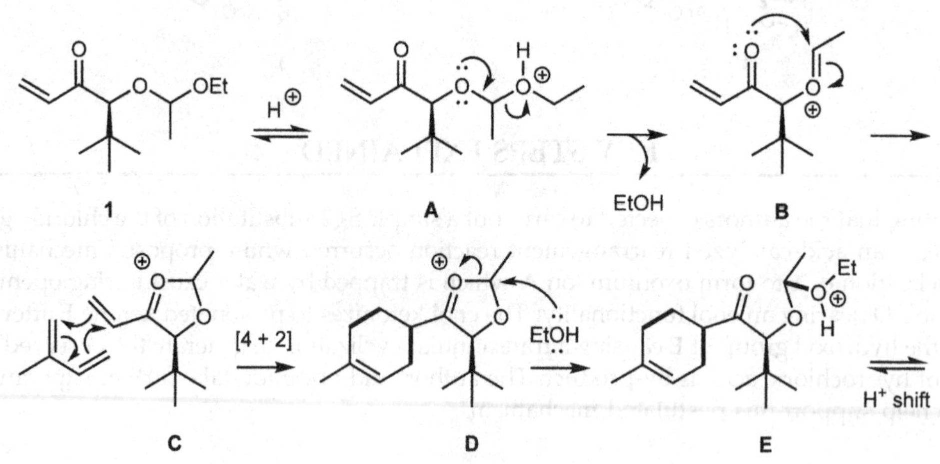

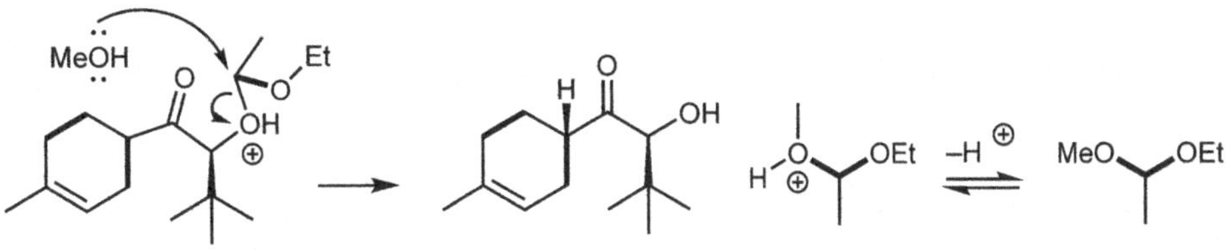

F   2

BALANCED CHEMICAL EQUATION

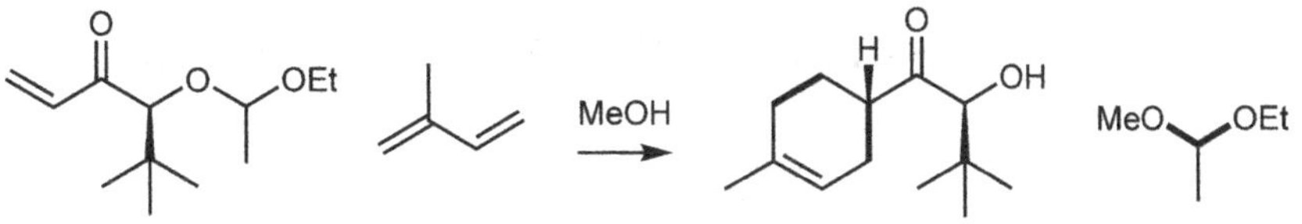

1   2

KEY STEPS EXPLAINED

The mechanism proposed by the authors to explain the transformation of **1** to **2** begins with acid-catalyzed ethanolysis of **1** to form oxonium ion **B** after elimination of ethanol. Intermediate **B** then cyclizes to oxonium ion **C** which reacts with isoprene (i.e., a diene) via the olefinic moiety of **C** in a Diels-Alder [4+2] fashion to generate oxonium ion **D**. Ethanol and methanol then add sequentially to intermediate **D** to eventually give product **2** and 1-ethoxy-1-methoxyethane by-product. Interestingly, the pivotal [4+2] cycloaddition is also face selective as the observed diastereomeric ratios of the products are quite high given a range of dienophile and diene combinations.

Furthermore, the authors conducted a competition experiment where two different α,β-unsaturated ketones are reacted with two equivalents of diene (Scheme 2). This experiment was designed to test the hypothesis that dienophiles containing an acetal protecting group are activated toward Diels-Alder cycloaddition as compared to ones that contain silyl protecting groups such as **3**. The authors indeed observed that products like **4** which are derived from **3** were not observed and in fact compound **3** was recovered unreacted in 90% yield. The authors did not undertake a theoretical analysis to help support the proposed mechanism.

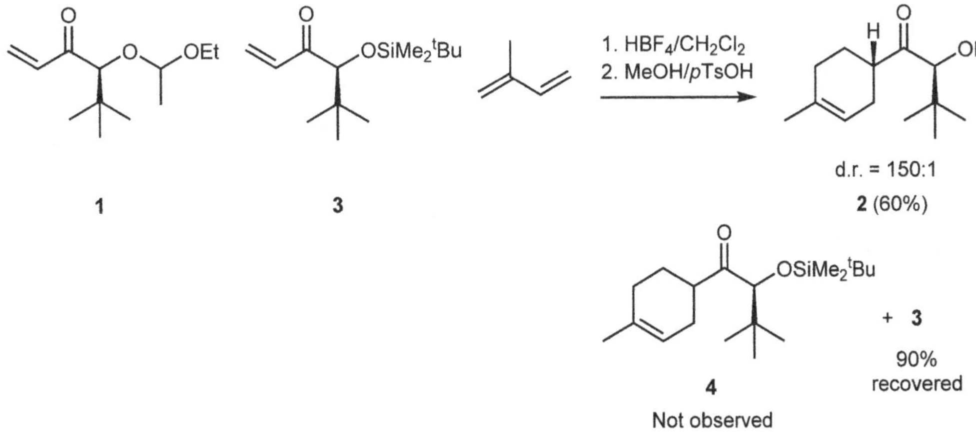

1   3

1. HBF$_4$/CH$_2$Cl$_2$
2. MeOH/pTsOH

d.r. = 150:1

2 (60%)

+ **3**

90% recovered

4

Not observed

SCHEME 2

ADDITIONAL RESOURCES

We recommend three articles which discuss intramolecular ionic, polar, and ionic liquid catalyzed Diels-Alder reactions, respectively.[2–4]

References

1. Sammakia T, Berliner MA. Diastereoselective Diels-Alder reactions via cyclic vinyloxocarbenium ions. *J Org Chem.* 1995;60(21):6652–6653. https://doi.org/10.1021/jo00126a001.
2. Ko Y-J, Shim S-B, Shin J-H. A mechanistic study on the intramolecular ionic Diels–Alder reaction of 2-methyl-3,9,11-tridecatriene-2-ol and 2,11-dimethyl-1,3,9,11-dodecatetraene. *Tetrahedron Lett.* 2007;48:863–867. https://doi.org/10.1016/j.tetlet.2006.11.144.
3. Domingo LR, Saez JA. Understanding the mechanism of polar Diels–Alder reactions. *Org Biomol Chem.* 2009;7:3576–3583. https://doi.org/10.1039/b909611f.
4. Zhu X, Cui P, Zhang D, Liu C. Theoretical study for pyridinium-based ionic liquid 1-ethylpyridinium trifluoroacetate: synthesis mechanism, electronic structure, and catalytic reactivity. *J Phys Chem A.* 2011;115:8255–8263. https://doi.org/10.1021/jp201246j.

Question 92: The Eschenmoser Fragmentation Reaction Extended

In 1997 Brazilian chemists observed that bromination of the tosylhydrazone **1** with *N*-bromosuccinimide (NBS) at −10°C in water/*t*-butanol/acetone mixture gave the acetylenic lactone **2** after treatment with aqueous $NaHSO_3$ and heating at 50–60°C for an hour (Scheme 1).[1] Provide a mechanism for this transformation.

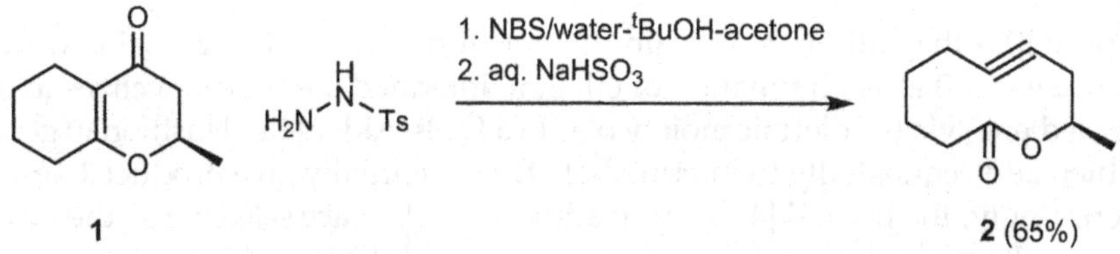

SCHEME 1

SOLUTION

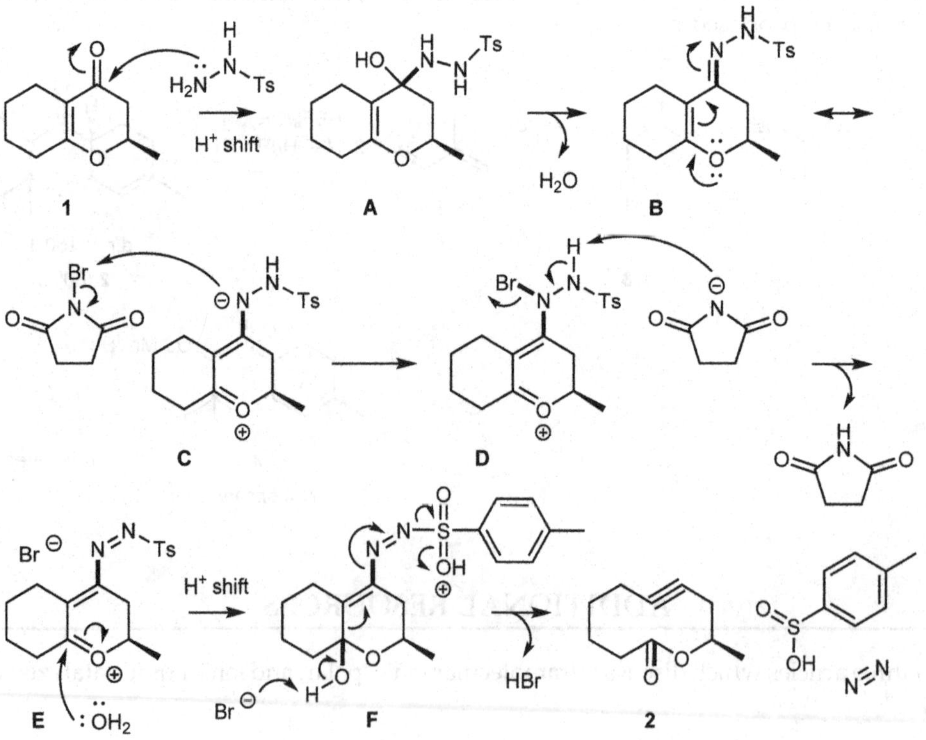

BALANCED CHEMICAL EQUATION

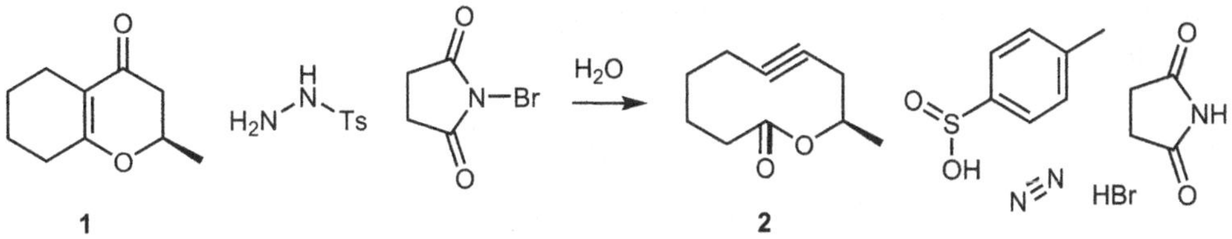

KEY STEPS EXPLAINED

The mechanism proposed by the authors to explain the transformation of **1** to **2** begins with reaction between ketone **1** and tosylhydrazine to produce tosylhydrazone **B** which then reacts with *N*-bromosuccinimide to form bromo-oxonium ion **D** via the zwitterionic form **C**. Succinimide ion deprotonates **C** yielding azo-oxonium ion **E** which undergoes a [1,4]-addition of water to form ketal **F** after proton shift. Bromide ion then deprotonates the hydroxyl group causing a fragmentation of the fused C—C bond of the bicyclic ring yielding the cyclic alkyne product **2** and nitrogen gas and 4-methyl-benzenesulfinic acid as by-products. The authors also noted that since intermediate **E** has a resonance form **E'**, a possible [1,2] addition of water would work to regenerate the starting material **1** (Scheme 2). This possible side reaction is one of the reasons why **2** was isolated only in 65% yield as the authors note the identity of **1** as an impurity in the product mixture. Furthermore, if *t*-butanol participates in the reaction, the ring cleavage impurity **3** can also be isolated as an impurity (Scheme 3).

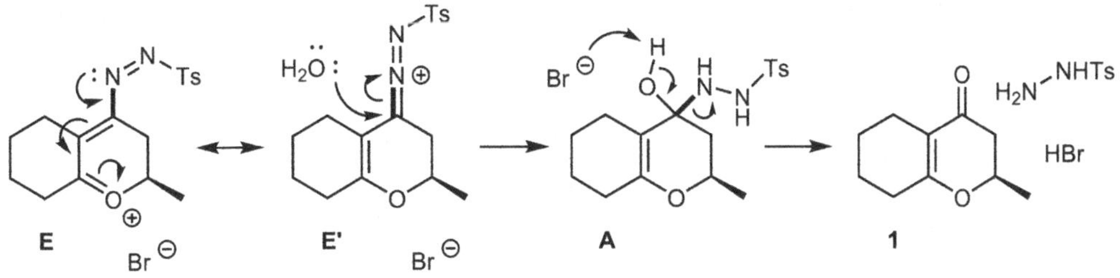

SCHEME 2

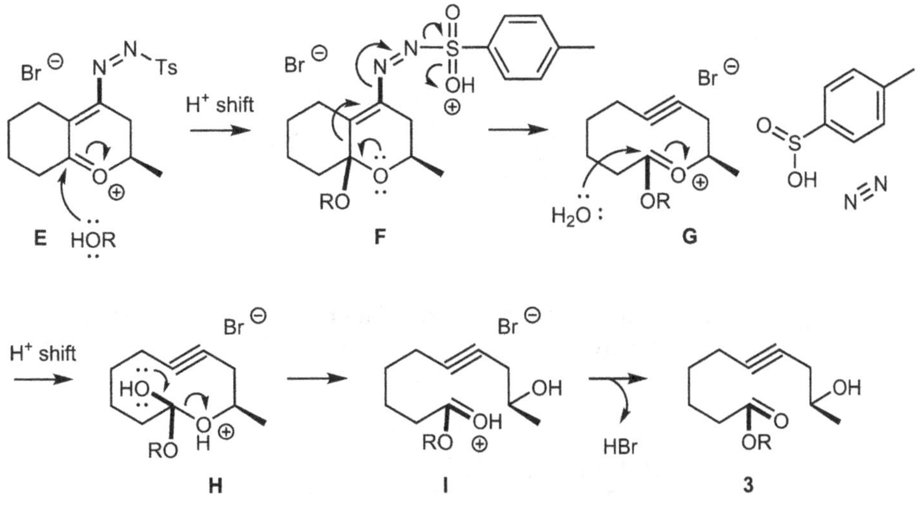

SCHEME 3

Surprisingly, when the authors used freshly crystallized and dessicator-dried NBS, impurity **3** was isolated as the major product. In this case we surmise that the authors carried out the reaction under anhydrous conditions as it would be nonsensical to dry the NBS and then add it to a wet *t*-butanol-acetone solvent mixture. As such we envisage that intermediate **G** is trapped by a second molecule of *t*-butanol to give **J** which then ring opens and loses a *t*-butyl group via an S_N1 or S_N2 pathway to form product **3** and *t*-butyl bromide by-product (Scheme 4).

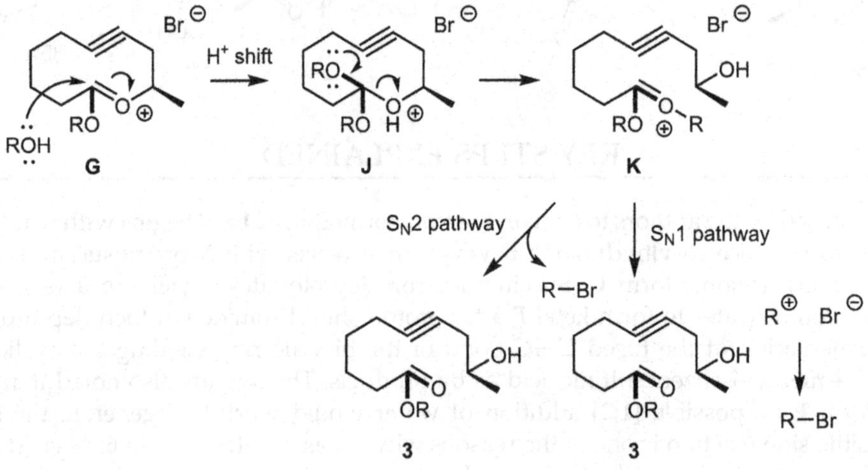

SCHEME 4

Alternatively, intermediate **F** can fragment to intermediate **G'** (which is a resonance form of **G**) and this is trapped by a second molecule of *t*-butanol as seen before (Scheme 5). The resulting intermediate **J** follows the same pathway toward product **3** and *t*-butyl bromide by-product.

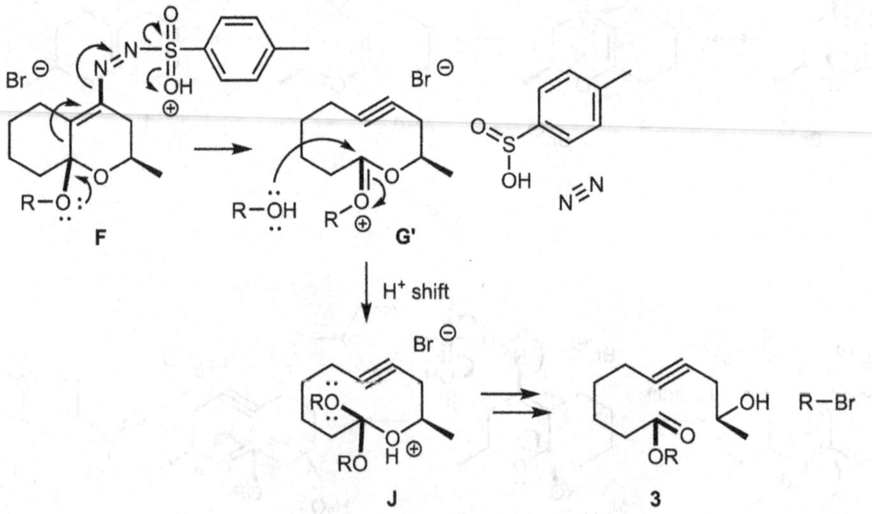

SCHEME 5

Interestingly, when R is a bulky group such as *t*-butyl, intermediate **H** can also collapse to product **2** by elimination of ROH. The authors did not carry out a theoretical analysis to help further support this mechanism.

ADDITIONAL RESOURCES

For examples of the synthetic application of the Grob/Eschenmoser fragmentation to access acyclic synthetic intermediates, see Ref. 2.

References

1. Mahajan JR, Resck IS. Synthesis of medium ring and macrocyclic acetylenic lactones by the ring expansion of oxabicycloalkenones. *J Braz Chem Soc.* 1997;8(6):603–613. https://doi.org/10.1590/S0103-50531997000600007.
2. Hierold J, Hsia T, Lupton DW. The Grob/Eschenmoser fragmentation of cycloalkanones bearing β-electron withdrawing groups: a general strategy to acyclic synthetic intermediates. *Org Biomol Chem.* 2011;9:783–792. https://doi.org/10.1039/c0ob00632g.

Question 93: A Modified Batcho-Leimgruber Synthesis

In 1995 American chemists reacted enamine **1** with ethyl chloroformate in refluxing chloroform in the presence of *N*, *N*-diethylaniline to form the ester product **2** which could be used to form product **3** after catalytic hydrogenation (Scheme 1).[1] Instead, the dienamine **4** was apparently formed (Scheme 2). This was confirmed in a check experiment where treatment of **1** with *p*-TsOH gave **4** in 52% yield and catalytic hydrogenation of **4** gave product **5** in 58% yield (Scheme 3). Provide a mechanism for the formation of dienamine **4**.

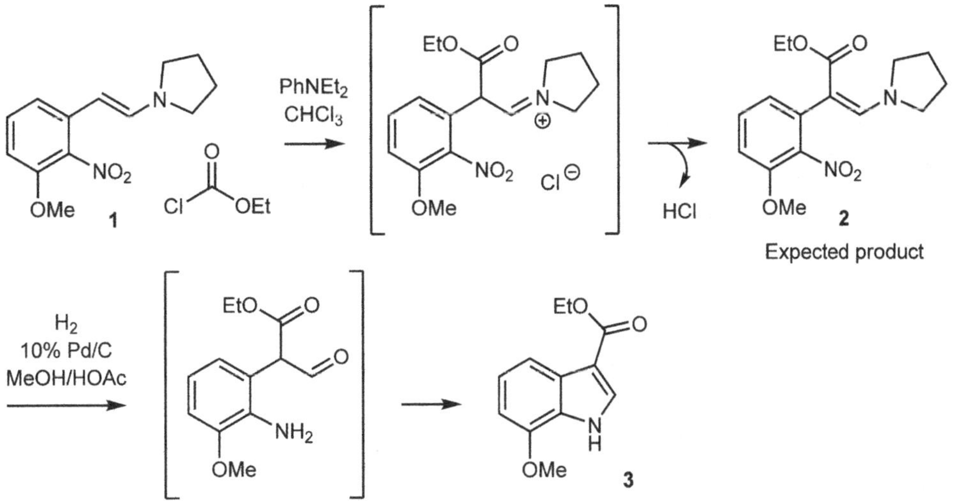

SCHEME 1

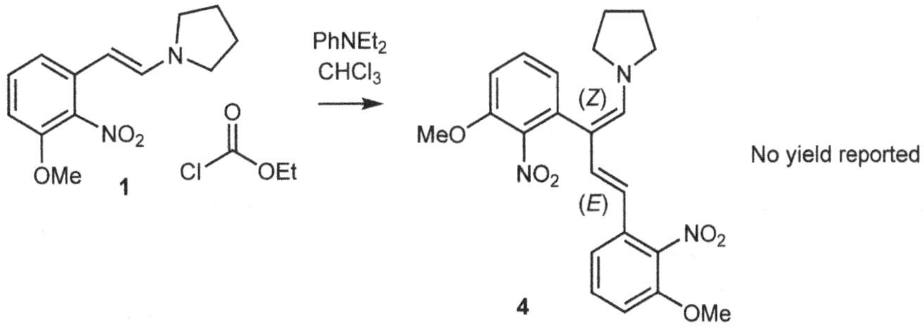

SCHEME 2

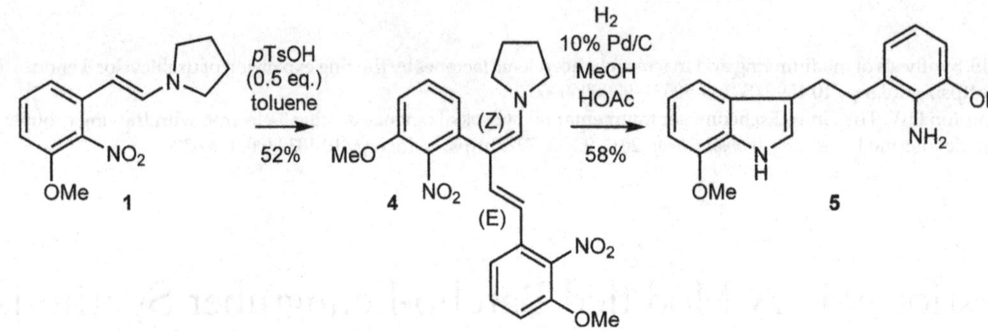

SCHEME 3

SOLUTION: 1 TO 4 *p*-TsOH ACID CATALYSIS

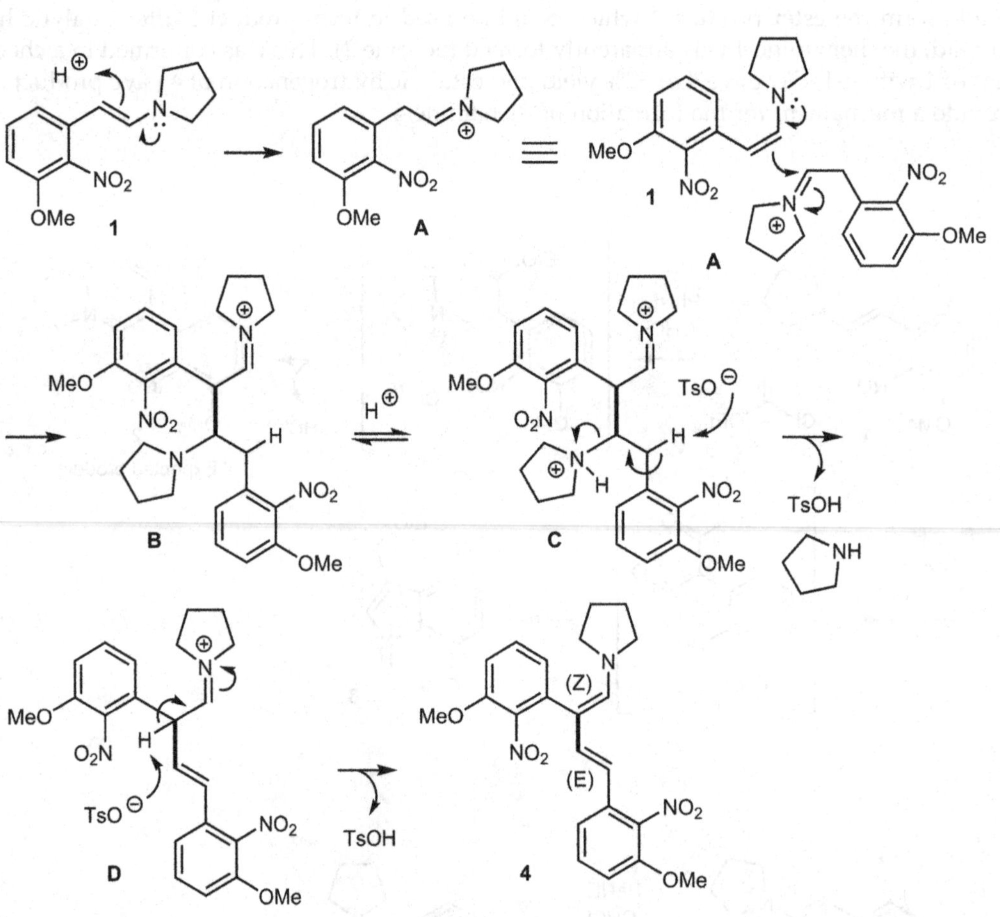

KEY STEPS EXPLAINED

The mechanism we propose for the transformation of **1** to **4** under acid catalysis begins with protonation of **1** to form iminium ion **A** which reacts with another equivalent of **1** to yield iminium ion **B**. A second protonation of the pyrrolidine group followed by tosylate ion-assisted elimination of pyrrolidine leads to iminium ion **D**. Deprotonation of **D** by a second tosylate ion leads to dienamine **4**. We note that the authors' choice of nomenclature for product **4** as a "dienamine" is misleading since the name suggests the structure contains a diene group and an amino group, when in fact it contains an enamine group and an olefinic group. A better functional group descriptor would have been "ene-enamine" or "buta-1,3-dienyl-pyrrolidine." Nevertheless, in the original transformation depicted in Scheme 4 there is no initial acid catalyst among the reagents used. The authors argue that HCl is generated from the reaction of the reagents with one another yet no explanation is given as to how this can occur. Here we present three possibilities

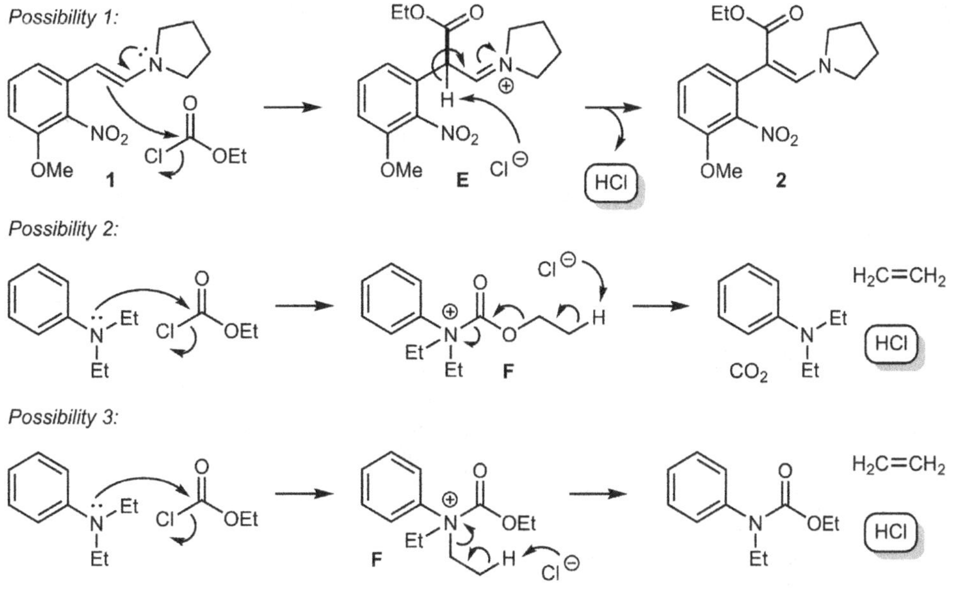

SCHEME 4

for this transformation and note that once HCl is generated, the mechanism for transforming **1** to **4** proceeds as outlined in the solution except that HCl is the acid catalyst as opposed to *p*-TsOH.

We can thus see that there are three possible ways that dienamine **4** could form from the reaction of **1** with ethyl chloroformate and *N,N*-diethylaniline depending on how hydrochloric acid is generated. Interestingly, the first possibility actually leads to the intended product **2** even in small amounts so long as HCl is generated as well because we know that HCl would immediately initiate a catalytic cycle that turns **1** into **4**. The two other ways that HCl can be generated involve reaction of *N,N*-diethylaniline with ethyl chloroformate that leads to diethylphenylcarboethoxy ammonium chloride intermediate **F** which can fragment to either *N,N*-diethylaniline, carbon dioxide, and ethylene, or ethyl ethyl-phenyl-carbamate and ethylene. Unfortunately, the authors did not conduct any by-product analysis for the detection of ethylene, carbon dioxide, or ethyl ethyl-phenyl-carbamate to confirm how product **4** could be generated from **1** under these reaction conditions. We see that detection of CO_2 would immediately point to possibility 2 while detection of product **2** in small amounts would point to possibility 1. Evidently, all three possibilities form different sets of by-products so a successful by-product analysis would enable differentiation between the three possibilities. Furthermore, the reaction yield reported for production of **4** in the original experiment was not provided. This prevents us from gauging how efficient the transformation actually is under the reaction conditions. The authors also did not adequately discuss the mechanism for the formation of **4**. A theoretical analysis was also omitted. Nevertheless, the authors did provide a revised synthesis for the intended methyl ester analog of product **3** (Scheme 5).

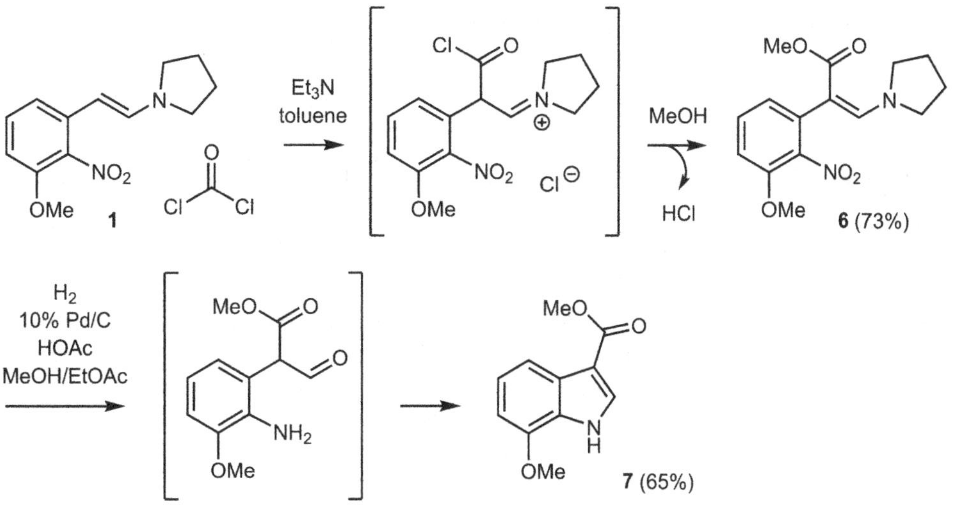

SCHEME 5

Lastly we propose a mechanism for the transformation of **4** to **5** in Scheme 6.

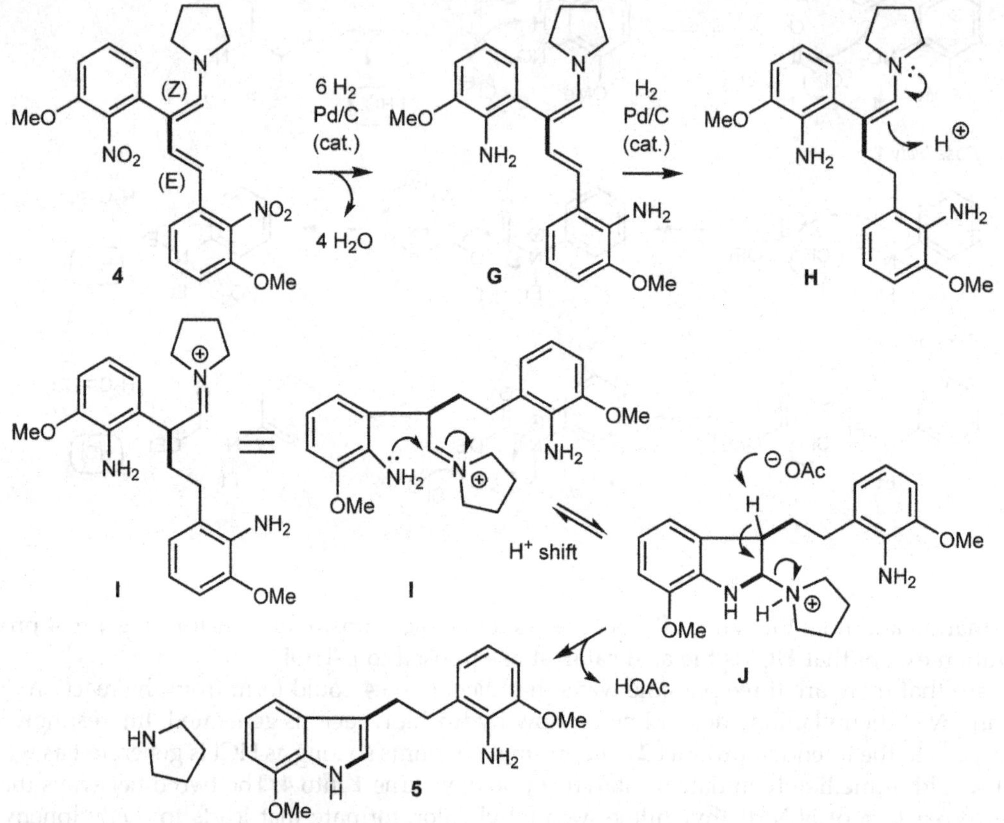

SCHEME 6

ADDITIONAL RESOURCES

We recommend an article describing transformations of rivularins which involves a complex Batcho-Leimgruber transformation with discussion of its mechanism.[2]

References

1. Prashad M, Vecchia LL, Prasad K, Repic O. A convenient synthesis of 3-substituted 1H-indoles. *Synth Commun.* 1995;25(1):95–100. https://doi.org/10.1080/00397919508010793.
2. Maehr H, Smallheer JM. Rivularins. Preliminary synthetic studies. *J Org Chem.* 1984;49(9):1549–1553. https://doi.org/10.1021/jo00183a015.

Question 94: A Pyrone to Pyran Conversion

In 1984 American chemists sought to demonstrate the use of an automated robotic system in carrying out operator-specific reaction sequences. To demonstrate the system, the authors chose the synthesis of the highly functionalized pyran **3** which is prepared by reacting methyl coumalate **1** with sulfone **2** in methylene chloride and 1,8-diazabicyclo[5.4.0]undec-7-ene (DBU) base (Scheme 1).[1] Suggest a mechanism for this transformation.

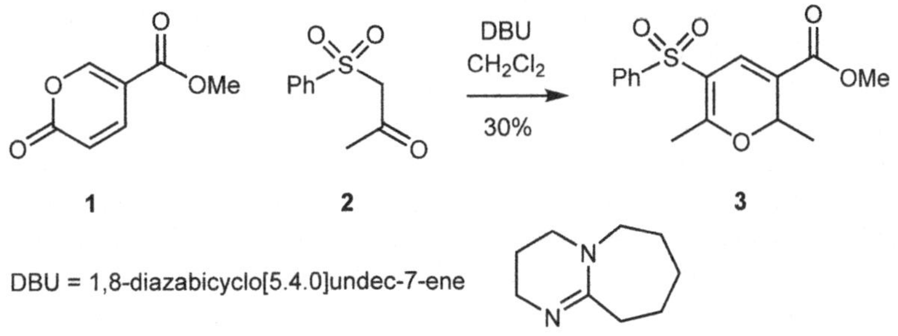

SCHEME 1

SOLUTION

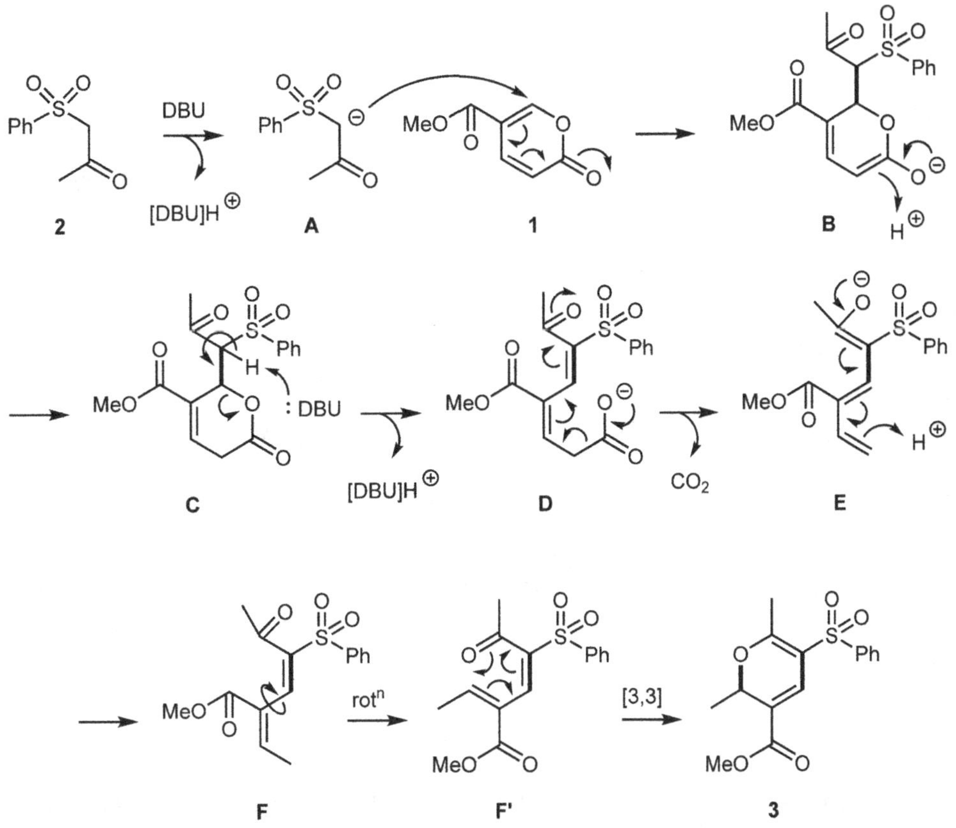

BALANCED CHEMICAL EQUATION

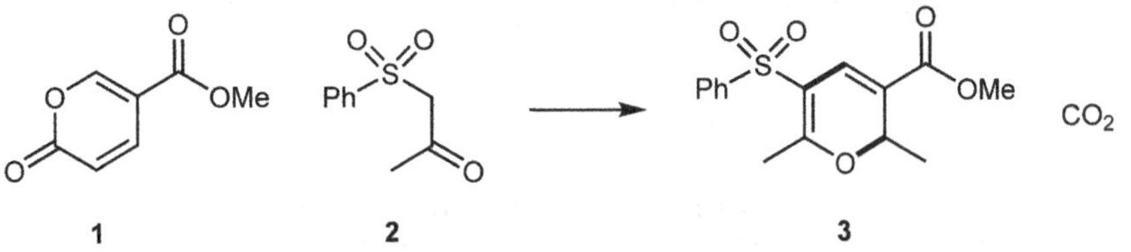

KEY STEPS EXPLAINED

The mechanism proposed by the authors to explain the transformation of **1** and **2** to **3** begins with deprotonation of **2** via DBU base to form carbanion **A** which adds in a [1,6] manner to methyl coumalate **1** to form intermediate **C** after ketonization via **B**. Base-mediated deprotonation of **C** causes ring opening to form **D** followed by decarboxylation which gives **E**. Protonation of **E** leads to **F** which after bond rotation and [3,3]-sigmatropic rearrangement gives the final pyran product **3**. We note that the authors did not perform any additional experiments or theoretical analysis to help support their postulated mechanism. As such, since we know that compounds such as methyl coumalate **1** are prone to [3,3]-sigmatropic rearrangement to form the corresponding ketene, we propose an alternative mechanism which makes use of a ketene intermediate (Scheme 2). We believe that a future mechanistic analysis of this transformation could help distinguish between the two proposed mechanisms using the technique of FT-IR spectroscopy to detect the distinctive strong cumulene vibration at around $2100\,cm^{-1}$ if ketene **G** is indeed formed, as well as a theoretical analysis to determine which mechanistic path is more energetically favorable.

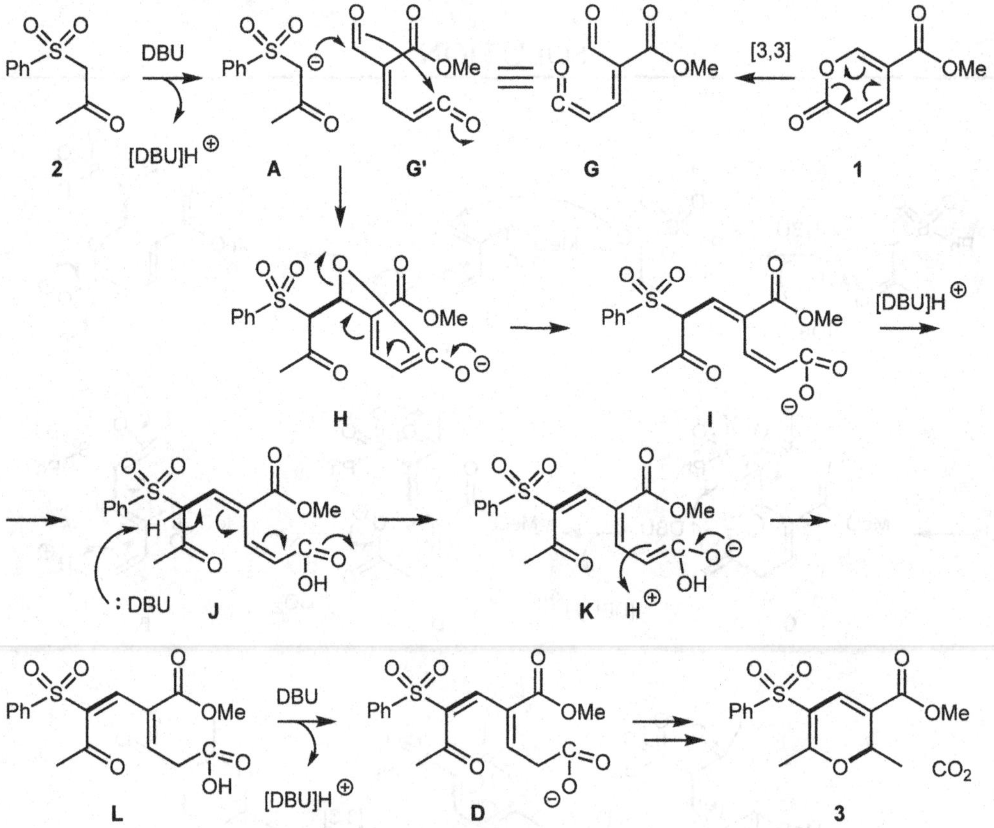

SCHEME 2

In this alternative mechanism, methyl coumalate **1** undergoes a [3,3] sigmatropic rearrangement to form ketene **G/G'**. Sulfone **2** is deprotonated by DBU to carbanion **A** as before and this adds to the aldehyde group of **G/G'** causing a cyclization to enolate **H**. Intermediate **H** then ring opens to form carboxylate **I** which undergoes a series of deprotonation-protonation steps assisted by DBU leading to intermediate **D**. This intermediate undergoes decarboxylation and a [3,3] sigmatropic rearrangement as before to generate pyran **3**. We note lastly that the ring construction mapping for product **3** is [3+3]. Likewise, the balanced chemical equation is the same for both mechanistic proposals.

ADDITIONAL RESOURCES

We recommend an article which investigates the pyrolysis mechanism of 2-pyranones and 2-pyranthiones and which also includes a theoretical analysis.[2] We also refer the reader to an article which discusses an interesting transformation involving a similar mechanism as the one discussed here.[3]

References

1. Frisbee AR, Nantz MH, Kramer GW, Fuchs PL. Robotic orchestration of organic reactions: yield optimization via an automated system with operator-specified reaction sequences. *J Am Chem Soc.* 1984;106(23):7143–7145. https://doi.org/10.1021/ja00335a047.
2. Reva I, Breda S, Roseiro T, Eusebio E, Fausto R. On the pyrolysis mechanism of 2-pyranones and 2-pyranthiones: thermally induced ground electronic state chemistry of pyran-2-thione. *J Org Chem.* 2005;70:7701–7710. https://doi.org/10.1021/jo051100w.
3. Strah S, Svete J, Stanovnik B. Rearrangements of 5-acetyl-3-benzoylamino-6-(2-dimethylamino-1-ethenyl)-2H-pyran-2-one and 3-benzoylamino-6-(2-dimethylamino-1-ethenyl)-5-ethoxycarbonyl-2H-pyran-2-one into 1-aminopyridine,pyrano[2,3-b]pyridine and isoxazole derivatives. *J Heterocyclic Chem.* 1996;33:1303–1306. https://doi.org/10.1002/jhet.5570330449.

Question 95: Conformationally Rigid 4-Oxoquinolines

In 1995 Korean chemists investigated the synthesis of conformationally rigid 4-oxoquinoline compounds thought to possess antibacterial properties.[1] During this research, the authors heated amide **1** in acetic anhydride at 80°C for 4h and formed the tetracyclic analog **2** in 92% yield (Scheme 1). Suggest a mechanism for this double ring closure transformation.

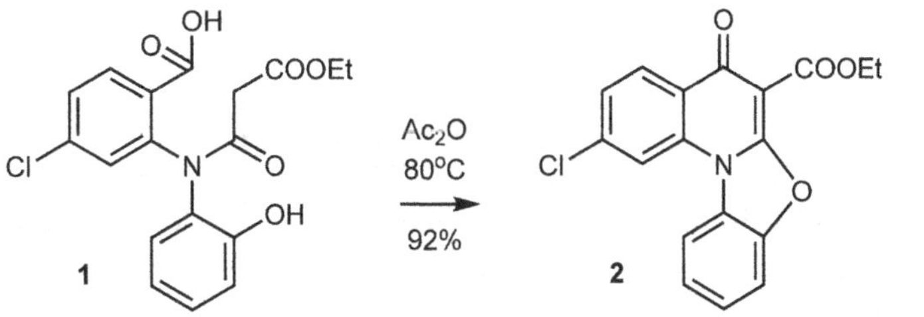

SCHEME 1

SOLUTION

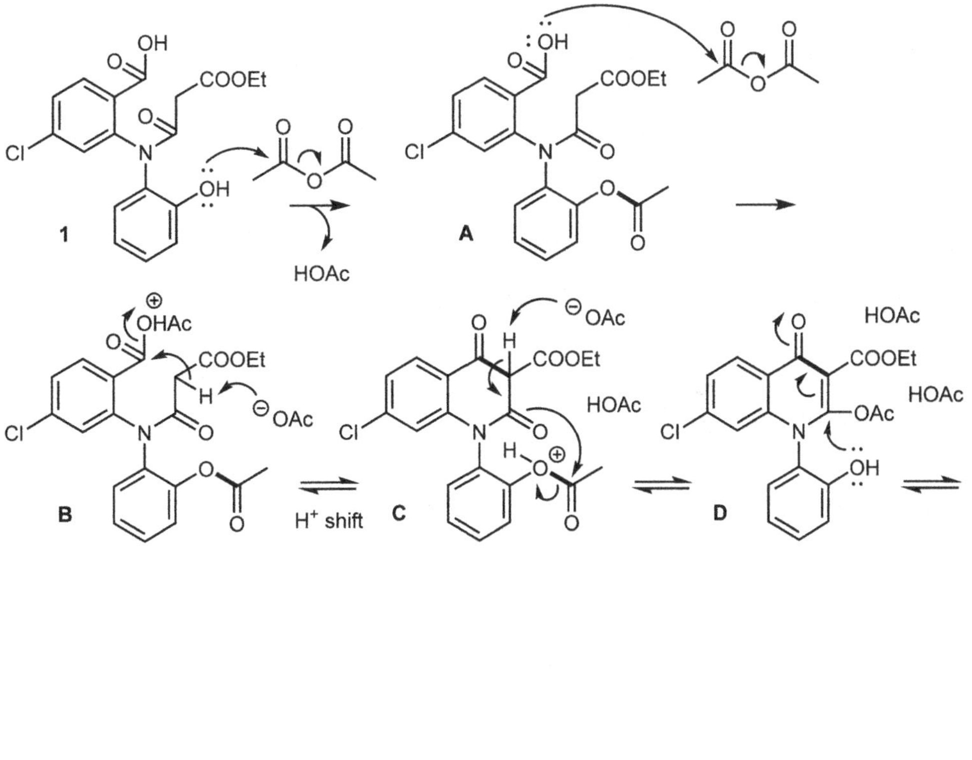

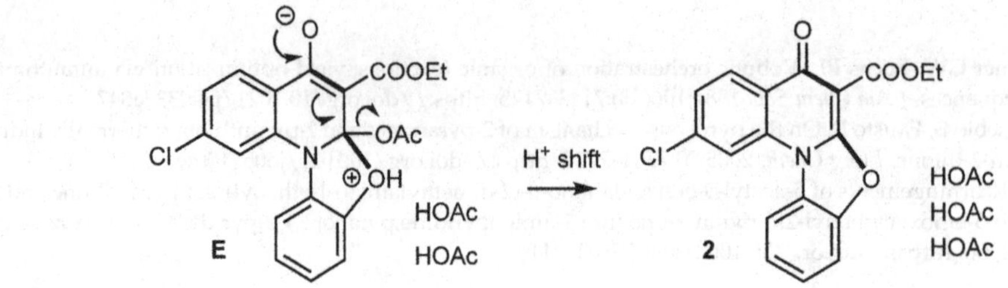

BALANCED CHEMICAL EQUATION

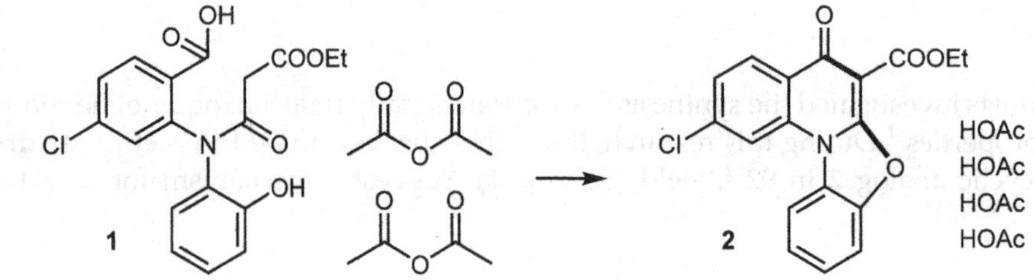

KEY STEPS EXPLAINED

The mechanism proposed by the authors for the transformation of **1** to **2** begins with double acylation of the phenol and carboxylic acid hydroxyl groups of **1** via two equivalents of acetic anhydride to form intermediate **B**. This compound is then deprotonated by acetate ion which causes cyclization to the tricyclic intermediate **C** which undergoes a second deprotonation with concomitant acyl migration to form the phenol intermediate **D**. This intermediate undergoes cyclization to **E** followed by elimination of acetic acid to generate the final tetracyclic product **2**. The overall ring construction mapping is $[(6+0)+(5+0)]$ and the transformation consumes two equivalents of acetic anhydride and produces four equivalents of acetic acid. We note that the authors did not undertake further experimental and theoretical analysis to help support their proposed mechanism. Therefore, it is uncertain whether the acyl migration from **C** to **D** actually occurs. If the phenol group of **1** does not undergo initial acylation to form **A** but instead enolization to **F**, an alternative mechanism can be proposed as depicted in Scheme 2. This mechanism has the same balanced chemical equation as the one presented previously and a very similar sequence of steps. Thus it is difficult to differentiate experimentally but a theoretical analysis would help distinguish between the initial steps **1** to **A** and **1** to **F**, respectively.

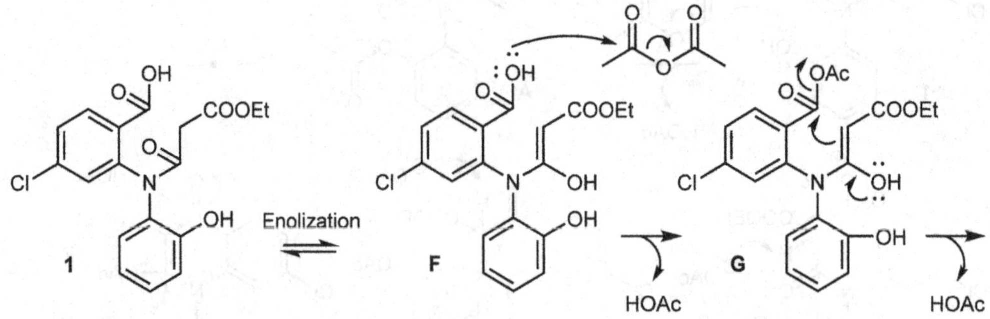

SCHEME 2

(Continued)

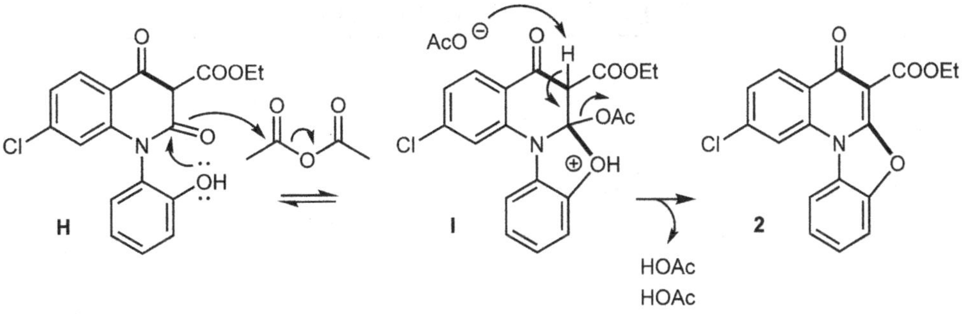

SCHEME 2, CONT'D

ADDITIONAL RESOURCES

We recommend an article which discusses the mechanism for an interesting transformation in a similar chemical system as the one encountered here.[2]

References

1. Chung SJ, Kim DH. Synthesis of 3-fluoro-2-substituted amino-5,12-dihydro-5-oxobenzoxazolo[3,2-a]quinoline-6-carboxylic acids employing the tandem double ring closure reaction of *N*-acetyl-*N*-(2-hydroxyphenyl)anthranilic acid as the key step. *Tetrahedron*. 1995;51(46):12549–12562. https://doi.org/10.1016/0040-4020(95)00809-M.
2. Ukrainets IV, Sidorenko LV, Slobodzyan SV, Rybakov VB, Chernyshev VV. 4-Hydroxyquinol-2-ones. 87. Unusual synthesis of 1-R-4-hydroxy-2-oxo-1,2-dihydroquinoline-3-carboxylic acid pyridylamides. *Chem Heterocycl Compd*. 2005;41(9):1158–1166. https://doi.org/10.1007/s10593-005-0296-z.

Question 96: An Efficient Chromene Synthesis

In 1991 French chemists achieved an efficient synthesis of the chromene product **4** by undertaking the synthetic sequence depicted in Scheme 1.[1] Provide mechanisms for each step of this synthesis.

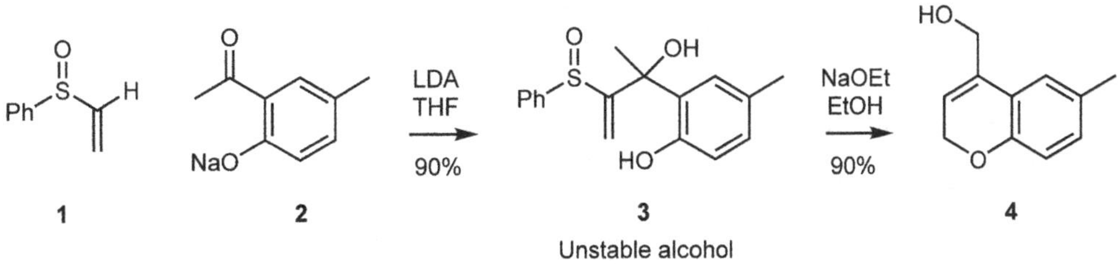

Unstable alcohol

SCHEME 1

SOLUTION: 1 AND 2 TO 3

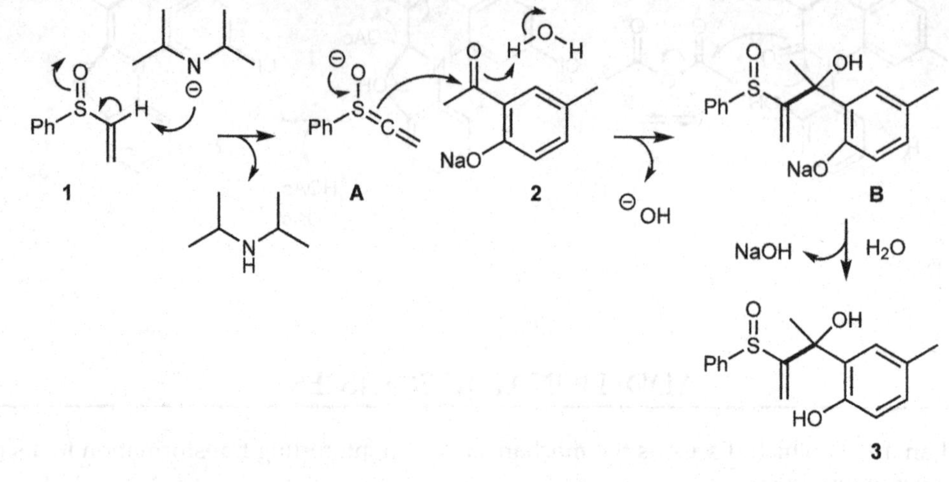

SOLUTION: 3 TO 4

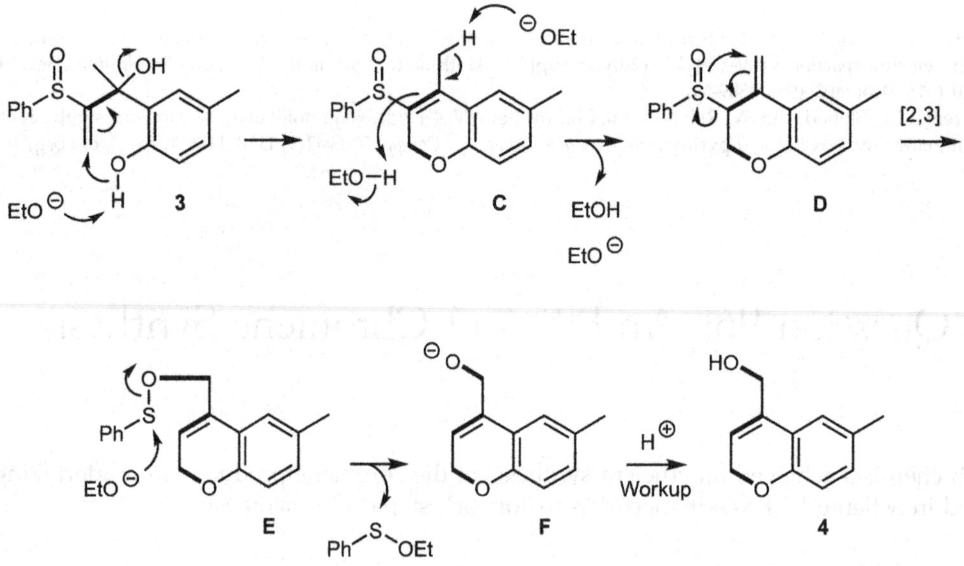

BALANCED CHEMICAL EQUATIONS

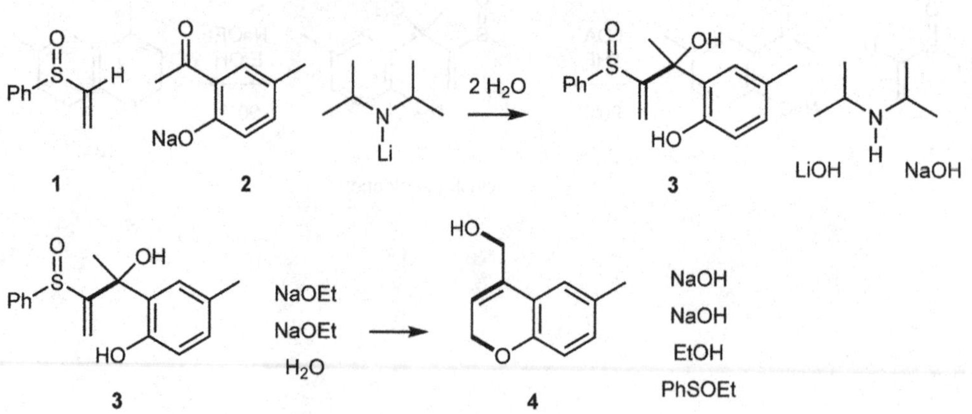

KEY STEPS EXPLAINED

The mechanism proposed by the authors for the transformation of **1** and **2** to the unstable alcohol **3** begins with abstraction of the vinyl proton from phenyl vinyl sulfone **1** by lithium diisopropylamide (LDA) to form anion **A**. This anion adds to acetophenone **2** to give intermediate **B** after protonation. Another protonation via water leads to the thermodynamically unstable alcohol product **3**. Next, the transformation of **3** to **4** can be explained by having ethoxide ion deprotonation at the phenolic group of **3** which causes cyclization to intermediate **C** with elimination of hydroxide ion. What follows is base-mediated [1,3]-protic shift which forms intermediate **D** which undergoes a [2,3] sigmatropic sulfoxide-sulfenate rearrangement that forms intermediate **E**. Ethoxide displaces the phenylsulfide group yielding alkoxide **F** and ethoxysulfanyl-benzene by-product. Subsequent acidic work-up generates the final (6-methyl-2*H*-chromen-4-yl)-methanol product **4**. We note that the ring construction mapping for **4** is [4+2]. Furthermore, the authors did not carry out further experimental and theoretical analysis to help support their proposed mechanism.

ADDITIONAL RESOURCES

We recommend two articles which explore the mechanisms of transformations that involve 4*H*- and 2*H*-chromenes, respectively.[2,3]

References

1. Solladie G, Girardin A. A rapid synthesis of substituted 4-hydroxymethyl-3-chromene from 2′-hydroxyacetophenone and phenyl vinyl sulfoxide. *Synthesis*. 1991;1991(7):569–570. https://doi.org/10.1055/s-1991-26518.
2. Davidson DN, Kaye PT. Chromone studies. Part 5. Kinetics and mechanism of the reaction of 4-oxo-4H-chromene-2-carboxamides with dimethylamine. *J Chem Soc Perkin Trans 2*. 1991;1991:1509–1511. https://doi.org/10.1039/P29910001509.
3. Dai L-Z, Shi Y-L, Zhao G-L, Shi M. A facile synthetic route to 2H-chromenes: reconsideration of the mechanism of the DBU-catalyzed reaction between salicylic aldehydes and ethyl 2-methylbuta-2,3-dienoate. *Chem A Eur J*. 2007;13:3701–3706. https://doi.org/10.1002/chem.200601033.

Question 97: Flash Vacuum Pyrolysis of *o*-Xylylene Dimers

In 1985 American chemists subjected the methyl-substituted *o*-xylylene dimers **1** and **3** to flash vacuum pyrolysis and isolated the anthracene products **2** and **4**, respectively (Scheme 1).[1] Suggest a mechanism for each transformation.

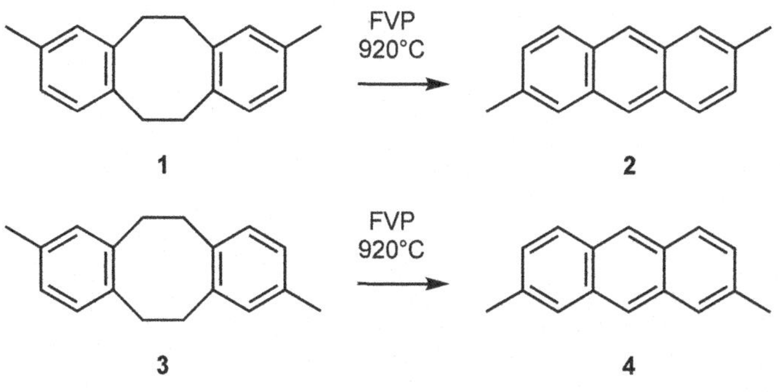

SCHEME 1

SOLUTION: 1 TO 2

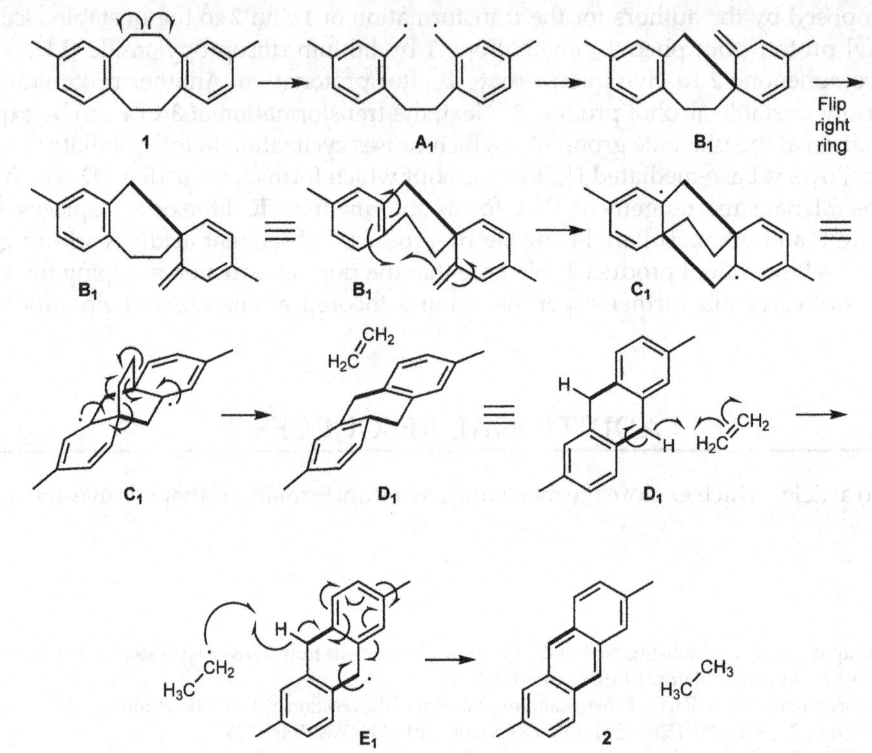

SOLUTION: 3 TO 4

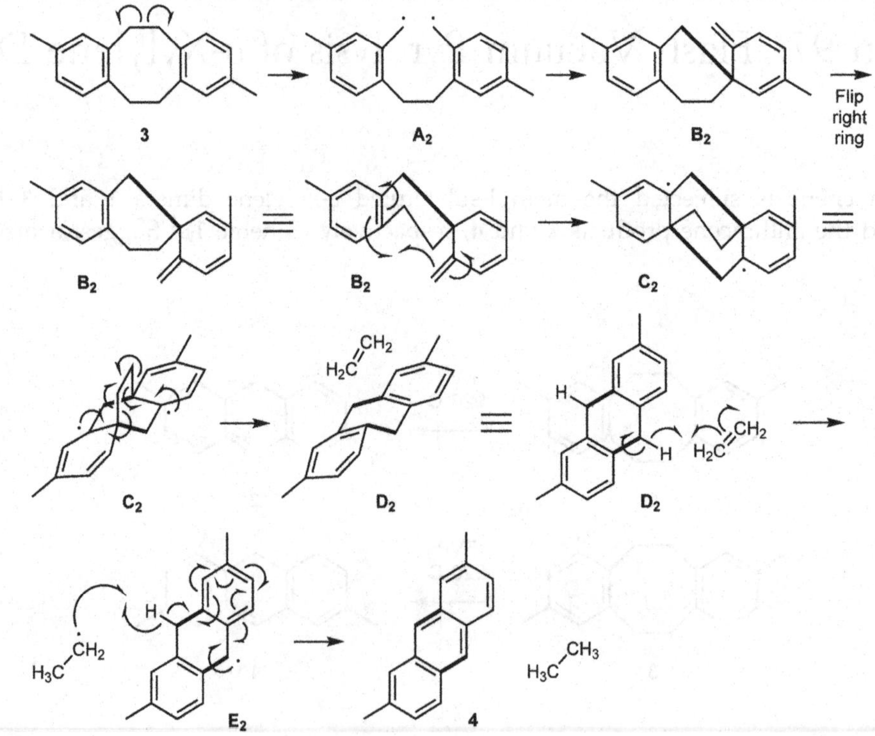

KEY STEPS EXPLAINED

The mechanism proposed for the transformation of **1** to **2** begins with homolytic cleavage of compound **1** to form diradical A_1 which recyclizes to intermediate B_1. B_1 then rearranges to the [2.2.2] bicyclic diradical intermediate C_1 which eliminates ethylene and yields 2,6-dimethyl-9,10-dihydro-anthracene D_1. Ethylene sequentially abstracts two hydrogen atoms from D_1 to eventually give 2,6-dimethyl-anthracene **2**. In a similar sequence of steps, compound **3** homolytically cleaves to diradical A_2 which recyclizes to intermediate B_2. B_2 then rearranges to the [2.2.2] bicyclic diradical intermediate C_2 which eliminates ethylene and yields 2,7-dimethyl-9,10-dihydro-anthracene D_2. Ethylene sequentially abstracts two hydrogen atoms from D_2 to eventually give 2,7-dimethyl-anthracene **4**. We note that the ring construction mapping for both reactions is [3+3]. The authors did not carry out further experimental or theoretical analysis to help support their mechanistic proposal. Nevertheless, they did comment that diradicals C_1 and C_2 "should be relatively stable, having two delocalized radicals and no severely strained bonds."[1] Also, they explain that the loss of ethylene by-product from C_1/C_2 occurs in an energetically favorable manner "because the C—C bonds that undergo cleavage are almost parallel to the π-orbitals of the radical sites."[1]

ADDITIONAL RESOURCES

We recommend two articles which describe further examples of interesting pyrolysis transformations in similar chemical systems.[2,3]

References

1. Trahanovsky WS, Surber BW. Formation of anthracenes in the flash vacuum pyrolysis of benzocyclobutenes and dimers of *o*-quinodimethanes. *J Am Chem Soc.* 1985;107(17):4995–4997. https://doi.org/10.1021/ja00303a029.
2. Neuhaus P, Grote D, Sander W. Matrix isolation, spectroscopic characterization, and photoisomerization of *m*-xylylene. *J Am Chem Soc.* 2008;130:2993–3000. https://doi.org/10.1021/ja073453d.
3. Banciu MD, Parvulescu L, Banciu A, et al. Flow-vacuum pyrolysis of three dibenzocycloalkanones. *J Anal Appl Pyrolysis.* 2001;57(2):261–274. https://doi.org/10.1016/S0165-2370(00)00147-9.

Question 98: Synthesis of a Neocarzinostatin Building Block

In 1996 American chemists undertook the synthesis of the methyl ester **4**, a key building block of neocarzinostatin chromophore, where the last step was an interesting photocyclization reaction (Scheme 1).[1] Suggest a mechanism for the transformation of **3** to **4**.

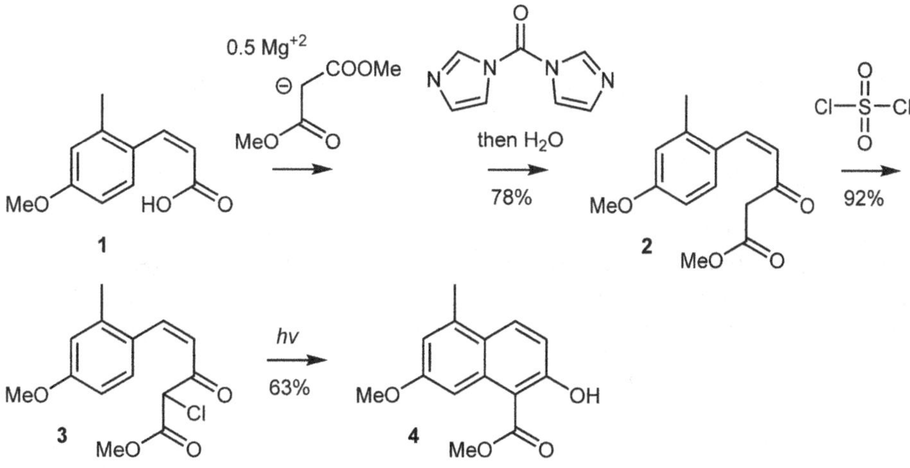

SCHEME 1

SOLUTION

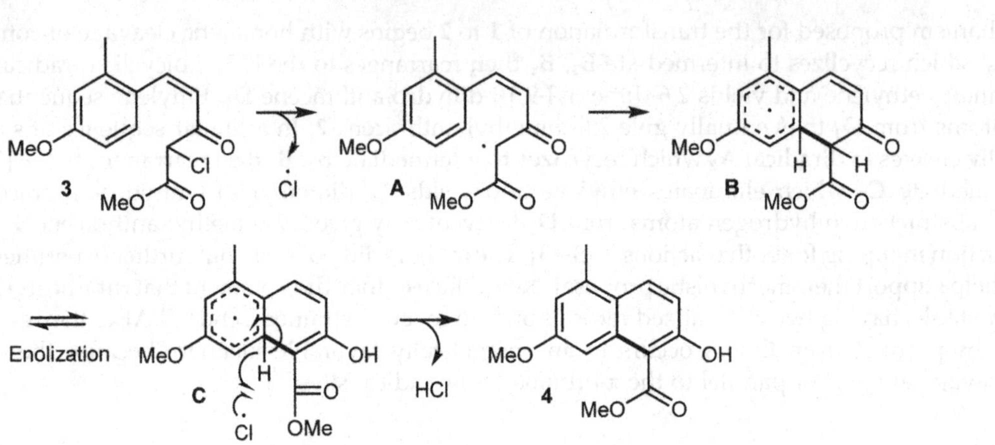

KEY STEPS EXPLAINED

The mechanism proposed by the authors to explain the photocyclization reaction begins with photolysis of the C—Cl bond which cleaves homolytically to yield radical **A** and chlorine radical. **A** cyclizes to radical **B** which enolizes to radical **C**. Chlorine radical then abstracts a hydrogen atom from **C** restoring aromaticity in the ring to yield product **4** after elimination of HCl. The authors note that they do not possess any evidence to support this mechanism but they do comment that they expect the final hydrogen chloride elimination to be an irreversible process. A theoretical analysis would certainly be helpful in substantiating this claim and in supporting the mechanism.

ADDITIONAL RESOURCES

For two newer synthetic approaches to product **4** and other neocarzinostatin chromophore derivatives, see Refs. 2, 3.

References

1. Myers AG, Subramanian V, Hammond M. A concise synthesis of the naphthoic acid component of neocarzinostatin chromophore featuring a new photocyclization reaction. *Tetrahedron Lett.* 1996;37(5):587–590. https://doi.org/10.1016/0040-4039(95)02268-6.
2. Goerth FC, Rucker M, Eckhardt M, Brueckner R. Decagram-scale synthesis of the neocarzinostatin carboxylic acid. *Eur J Org Chem.* 2000;2000:2605–2611. https://doi.org/10.1002/1099-0690(200007)2000:14<2605::AID-EJOC2605>3.0.CO;2-Y.
3. Ji N, Rosen BM, Myers AG. Method for the rapid synthesis of highly functionalized 2-hydroxy-1-naphthoates. Syntheses of the naphthoic acid components of neocarzinostatin chromophore and N1999A2. *Org Lett.* 2004;6(24):4551–4553. https://doi.org/10.1021/ol048075l.

Question 99: Oxazole Esters From α-Amino Acids

In 1995 American chemists sought to undertake the esterification of *N*-benzoylalanine **1** using oxalyl chloride in anhydrous THF followed by treatment with triethylamine, methanol, and work-up (Scheme 1).[1] Nevertheless, the transformation took an unexpected turn which resulted in the formation of the oxazole carbonyl chloride product **4** which undergoes facile esterification to **5**. Provide a mechanism for the transformation of **1** to **4**.

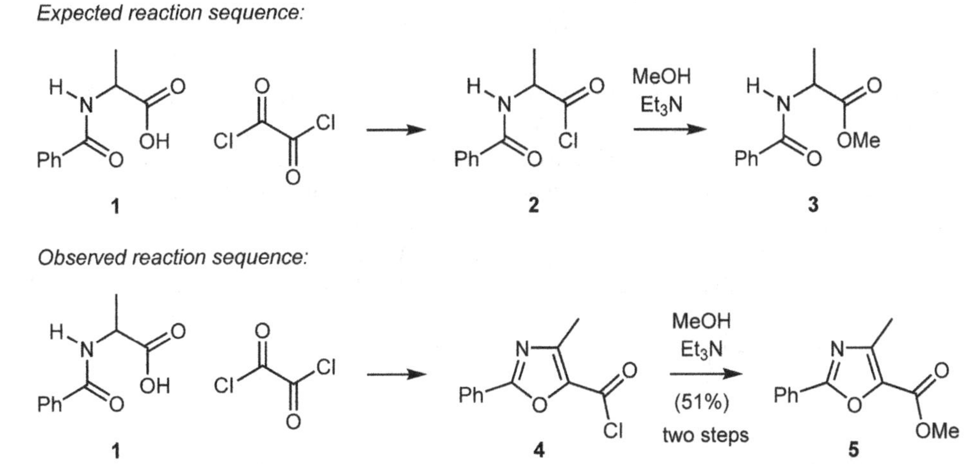
Expected reaction sequence:
MeOH
Et_3N
1
2
3
Observed reaction sequence:
MeOH
Et_3N
(51%)
two steps
1
4
5

SCHEME 1

SOLUTION

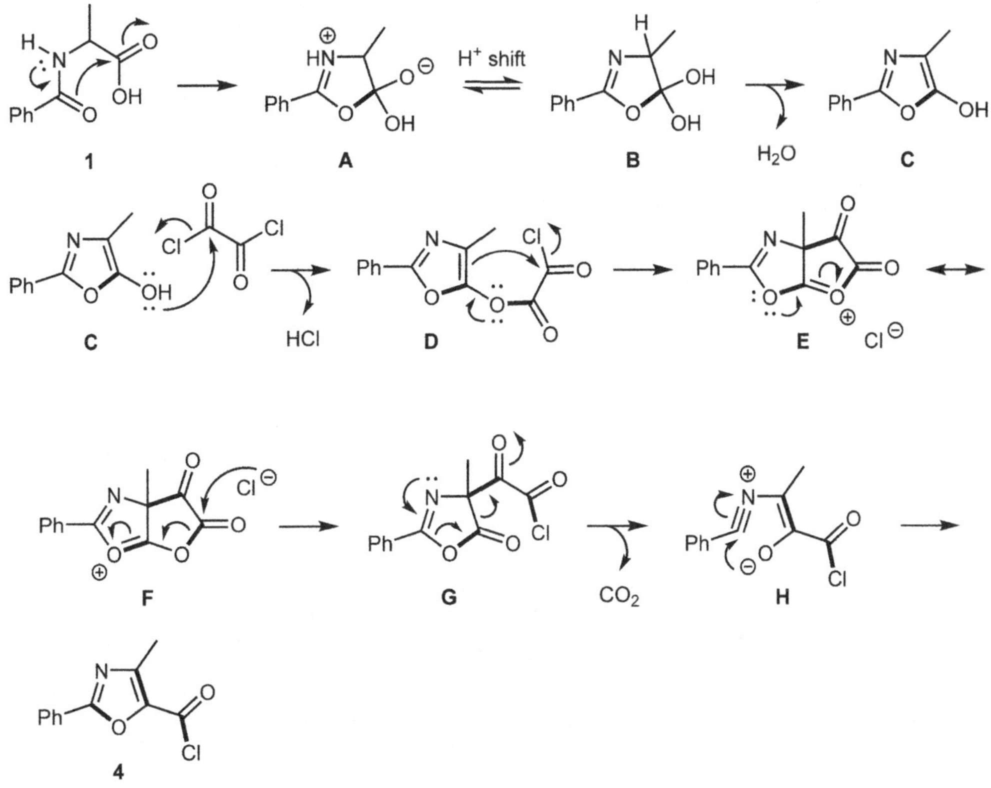
H^+ shift
1
A
B
H_2O
C
C
HCl
D
E
F
G
CO_2
H
4

BALANCED CHEMICAL EQUATION

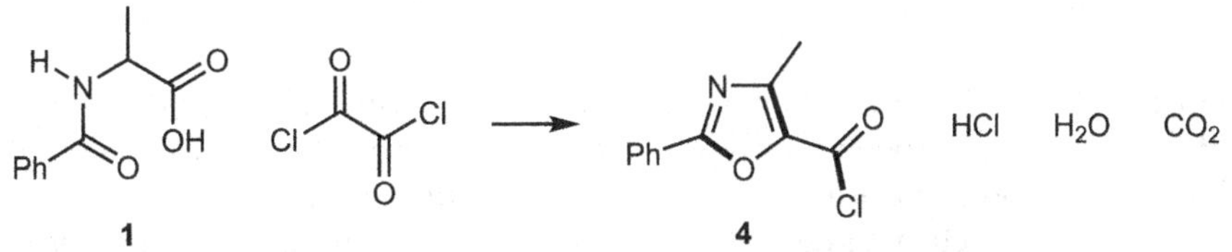
HCl
H_2O
CO_2
1
4

KEY STEPS EXPLAINED

The mechanism proposed by the authors to explain the unexpected course of the reaction of **1** with oxalyl chloride begins with cyclization of *N*-benzoylalanine **1** to form 4-methyl-2-phenyl-oxazol-5-ol **C** after loss of water. The hydroxyl group of **C** then reacts with oxalyl chloride to form **D** after elimination of hydrogen chloride. Intermediate **D** cyclizes again to oxonium ion **E** which is in resonance with its carbocation form **F** stabilized by the neighboring ether as an oxonium ion. Chloride ion causes ring opening to intermediate **G** which undergoes decarboxylation to nitrilium ion **H**. The last step constitutes cyclization of **H** to oxazole product **4**. We note that the authors carried out a ^{13}C labeling experiment where they begin with *N*-benzoylalanine which has a ^{13}C label at C_1 and determined that no ^{13}C label remained in the oxazole product thus proving that decarboxylation occurred at that group. Following the mechanism shown in Scheme 2 the ^{13}C label ends up in the liberated carbon dioxide.

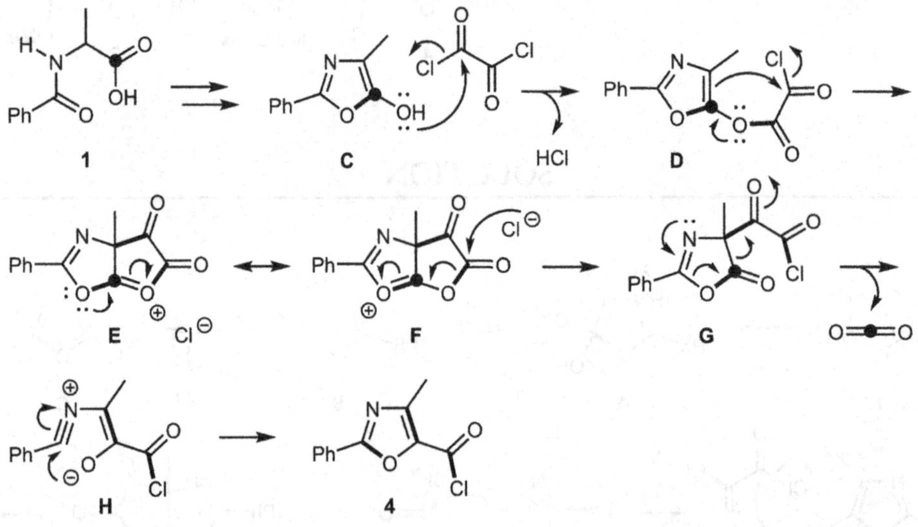

SCHEME 2

ADDITIONAL RESOURCES

We recommend two recent articles which describe synthetic approaches to oxazole derivatives.[2,3]

References

1. Cynkowski T, Cynkowska G, Ashton P, Crooks PA. Reaction of *N*-benzoyl amino acids with oxalyl chloride: a facile route to 4-substituted 2-phenyloxazole-5-carboxylates. *J Chem Soc Chem Commun*. 1995;1995:2335–2336. https://doi.org/10.1039/C39950002335.
2. Liu X, Cheng R, Zhao F, Zhang-Negrerie D, Du Y, Zhao K. Direct β-acyloxylation of enamines via PhIO-mediated intermolecular oxidative C-O bond formation and its application to the synthesis of oxazoles. *Org Lett*. 2012;14(21):5480–5483. https://doi.org/10.1021/ol3025583.
3. Zhao F, Liu X, Qi R, et al. Synthesis of 2-(trifluoromethyl)oxazoles from β-monosubstituted enamines via PhI(OCOCF₃)₂-mediated trifluoroacetoxylation and cyclization. *J Org Chem*. 2011;76:10338–10344. https://doi.org/10.1021/jo202070h.

Question 100: Unexpected Formation of a Phenazine

In 1996 Italian chemists reacted 4′-methoxybenzenesulfenanilide **1** with 1.5 equivalents of lithium di-isopropyl amide (LDA) in THF at −20°C under nitrogen followed by stirring at room temperature overnight to form a colorless solution which turned deep red upon exposure to air (Scheme 1).[1] Continued stirring of the mixture followed by washing with brine, ether extraction, and chromatography gave the final 2,7-dimethoxyphenazine product **2**. Suggest a mechanism for this transformation.

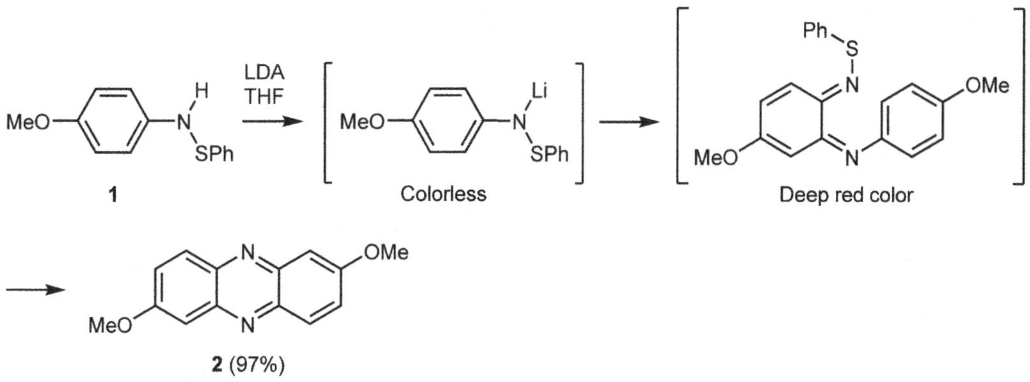

SCHEME 1

SOLUTION

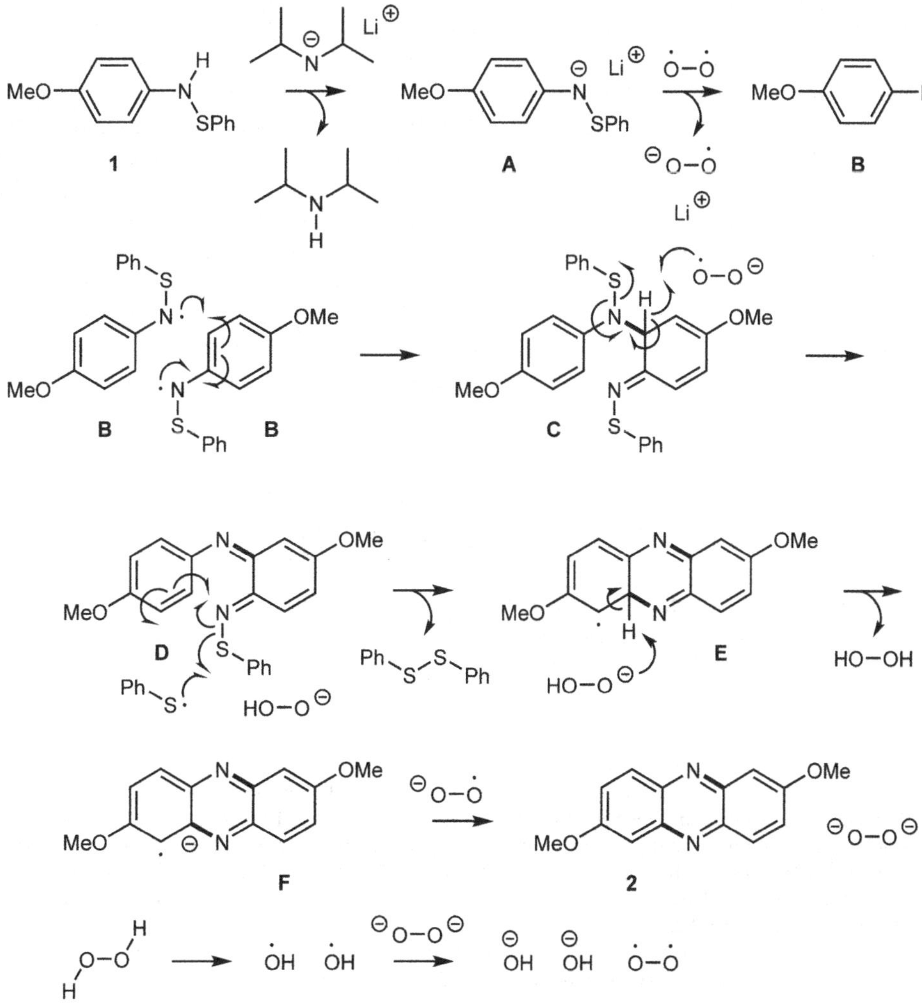

BALANCED CHEMICAL EQUATION

KEY STEPS EXPLAINED

The mechanism proposed by the authors for the formation of phenazine **2** begins with generation of anion intermediate **A** from **1** and LDA. Ground state triplet oxygen acting as a diradical then abstracts an electron from anion **A** to form the nitrogen radical **B** along with superoxide radical anion by-product. Two of these nitrogen **B** radicals then undergo dimerization to form intermediate **C**. What follows is hydrogen abstraction by superoxide radical anion from **C** to generate *o*-quinonediimine **D** (responsible for the deep-red color change) and by-products phenylsulfenyl radical and peroxide ion. The phenylsulfenyl radical removes the PhS group from **D** to give diphenylsulfide by-product and radical intermediate **E**. Peroxide anion then abstracts a hydrogen atom from **E** to form radical anion **F** and hydrogen peroxide by-product. A second equivalent of superoxide radical anion generated from the earlier production of a second equivalent of **B** comes in to remove an electron from **F** to form the observed 2,7-dimethoxy-phenazine product **2** and $[O_2]^{-2}$ by-product. Finally, hydrogen peroxide and $[O_2]^{-2}$ undergo a sequence of electron transfer steps that produce two equivalents of hydroxide ion and oxygen gas. Experimentally, the authors identified diphenylsulfide as an isolable by-product of the reaction. We note also that the ring construction mapping for product **2** is [3+3]. The authors did not provide further experimental or theoretical evidence to support their proposed mechanism although they did investigate several analogous transformations. Substituent effects were also found to play a role in the transformation outcomes.

ADDITIONAL RESOURCES

We direct the reader to three articles which explore further approaches to the synthesis of phenazine derivatives and which also provide interesting mechanistic discussions.[2–4]

References

1. Barbieri A, Montevecchi PC, Nanni D, Navacchia ML. LDA-promoted decomposition of benzenesulfenamides. A route to aminyl radicals by dioxygen oxidation of lithium amides. *Tetrahedron*. 1996;52(41):13255–13264. https://doi.org/10.1016/0040-4020(96)00799-5.
2. Michida T, Osawa E, Yamaoka Y. Studies on sulfenamides. XV. Semi-empirical calculation of reactivity of 4′-substituted benzenesulfenanilidyl radicals. *Chem Pharm Bull*. 1999;47(12):1787–1789. https://doi.org/10.1248/cpb.47.1787.
3. Abdayem R, Baccolini G, Boga C, Monari M, Selva S. Unexpected reactivity between aromatic nitro compounds and PCl3/AlCl3. A new one-pot synthesis of phenazines. *Tetrahedron Lett*. 2003;44:2649–2653. https://doi.org/10.1016/S0040-4039(03)00373-3.
4. Davarani SSH, Fakhari AR, Shaabani A, Ahmar H, Maleki A, Fumani NS. A facile electrochemical method for the synthesis of phenazine derivatives via an ECECC pathway. *Tetrahedron Lett*. 2008;49:5622–5624. https://doi.org/10.1016/j.tetlet.2008.07.063.

Chapter 6: Solutions 101 – 150

Question 101: Conversion of Primary Amides Into Nitriles

In 1996 French chemists demonstrated that the transformation of a primary amide to a nitrile could be carried out using formic acid and a catalytic amount of an aldehyde in acetonitrile solution (Scheme 1).[1] Suggest a mechanism for this transformation.

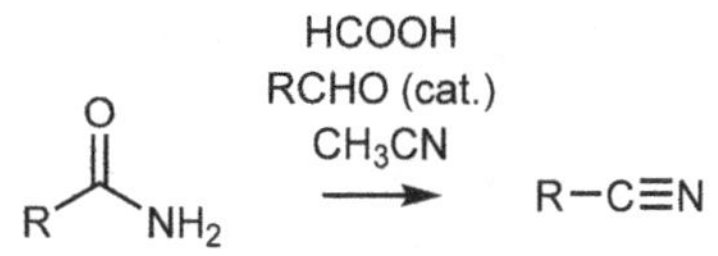

SCHEME 1

SOLUTION

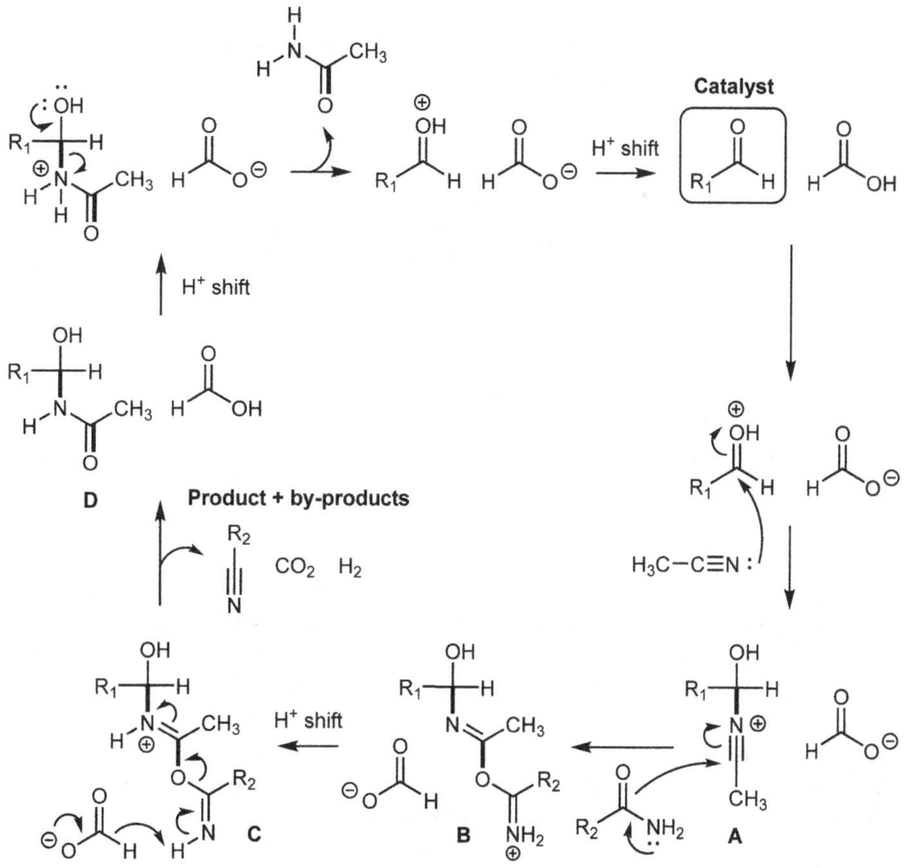

Strategies and Solutions to Advanced Organic Reaction Mechanisms
https://doi.org/10.1016/B978-0-12-812823-7.00315-3

BALANCED CHEMICAL EQUATION

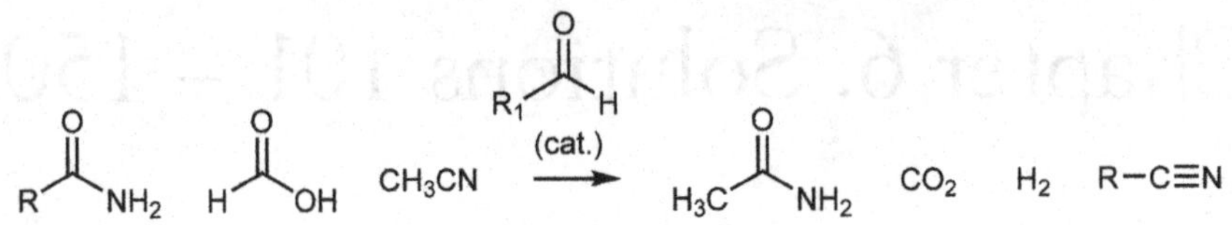

KEY STEPS EXPLAINED

The proposed mechanism for the functional group transformation observed by Heck et al. begins with protonation of the aldehyde oxygen atom via formic acid. The nitrogen atom of acetonitrile then nucleophilically attacks the electrophilic aldehydic carbon atom yielding nitrilium ion intermediate **A**. The amide oxygen atom nucleophilically adds to the nitrilium ion **A** to form iminium ion **B** which undergoes an intramolecular proton transfer to a second iminium ion intermediate **C**. Next, formate ion fragments to carbon dioxide and hydride whereupon the hydride ion abstracts a hydrogen atom from the second iminium ion intermediate **C**. The result is the production of the nitrile product arising from the original amide with hydrogen gas, carbon dioxide gas, and N-(1-hydroxy-alkyl)-acetamide **D** as by-products. The remaining steps transform the N-(1-hydroxy-alkyl)-acetamide **D** to acetamide and the original aldehyde catalyst with another equivalent of formic acid thus regenerating the catalytic cycle.

We note that formic acid plays a key role in this reaction since it acts both as a Brønsted acid as well as a hydride transfer agent in the transformation from **C** to **D**. The authors did not check to see that hydrogen gas and carbon dioxide were reaction by-products. Such an observation would have confirmed the important role of formic acid whose role was admittedly not understood by the authors but was recognized as essential from experimental work where other acids did not work to produce any nitrile product. Furthermore, when the authors carried out the experiment starting with benzonitrile instead of acetonitrile, an isolated by-product of the reaction was benzamide instead of acetamide. Also, the catalytic role of the aldehyde was demonstrated by running an experiment where 10-oxodecanamide **1** was converted to its corresponding nitrile **2** without any added aldehyde (Scheme 2). This happened because the internal aldehyde group acted as the catalyst. It is important to point out that the acetonitrile acts as a sacrificial reagent in this reaction. When the reaction is written with generalized R groups for both the amide and nitrile substrates it is observed that the reaction is essentially a metathesis reaction where the R groups swap functional groups as seen in Scheme 3.

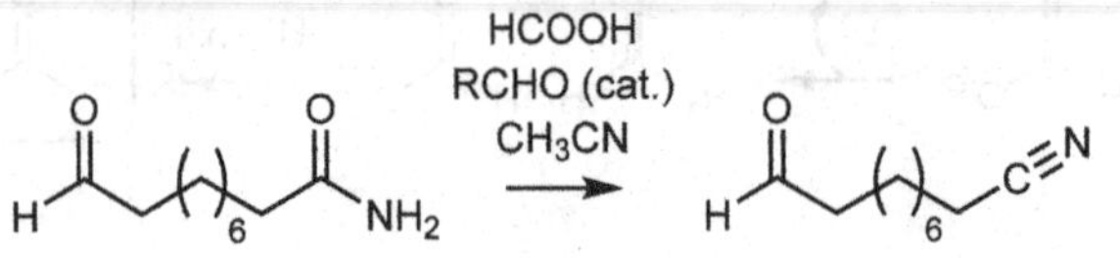

SCHEME 2

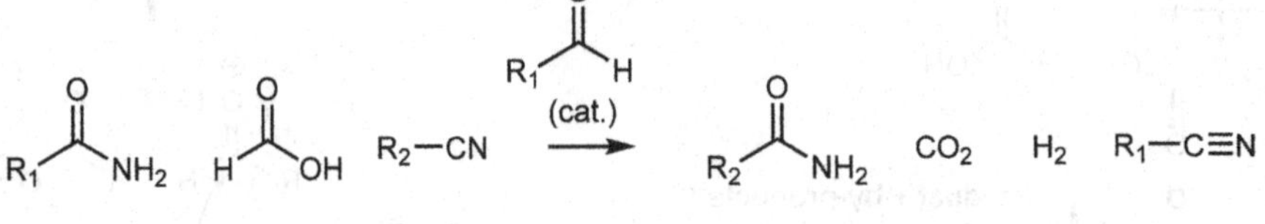

SCHEME 3

Lastly, we note that our proposed mechanism differs from that of the authors in the sense that the dual role of formic acid is explained in our mechanism as compared to the authors' mechanism. Since the authors did not provide further experimental and theoretical data to support their mechanism and since they left the question of the mechanistic role of formic acid unanswered, we believe our proposed mechanism does a better job accounting for all relevant facts. Two possible ^{18}O labeling experiments are suggested which may confirm the authors' postulated mechanism. If the aldehyde R_1CHO catalyst is ^{18}O labeled then following the catalytic cycle we expect the label to remain in the regenerated aldehyde catalyst. On the other hand, if the general amide R_2CONH_2 is ^{18}O labeled then the catalytic cycle predicts that the label will be transferred to the acetamide by-product. Such results may be contrasted with the expectations from an alternative catalytic cycle shown in Scheme 4 which involves formation of intermediate **E** (from aldehyde catalyst and general amide R_2CONH_2) which reacts with protonated acetonitrile to yield intermediate **F**. This intermediate then fragments upon reaction with formate producing nitrile R_2CN and by-products carbon dioxide, hydrogen gas, and acetamide. In this mechanism formic acid again acts both as a Brønsted acid and as a hydride donor. According to this catalytic cycle if the aldehyde R_1CHO catalyst is ^{18}O labeled then the label will end up on the acetamide, rather than remain on the aldehyde as in the former catalytic cycle. On the other hand, if the general amide R_2CONH_2 is ^{18}O labeled then the label will end up on both the aldehyde catalyst and acetamide by-product. This latter outcome arises when the labeled aldehyde formed from the first round in the cycle reenters the cycle again for subsequent rounds along with labeled general amide substrate.

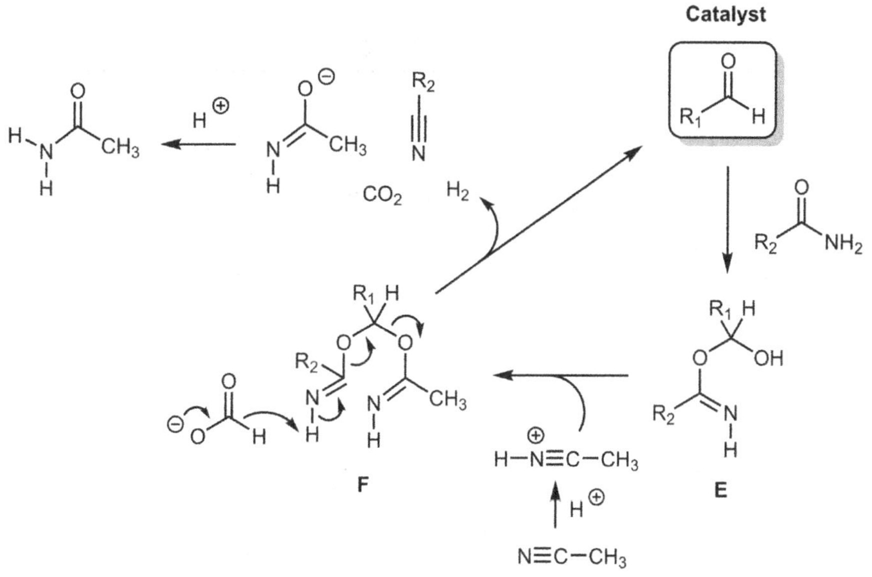

SCHEME 4

ADDITIONAL RESOURCES

We recommend two additional articles which describe interesting approaches to the synthesis of nitriles under relatively mild conditions where an interesting mechanism is involved.[2,3]

References

1. Heck M-P, Wagner A, Mioskowski C. Conversion of primary amides to nitriles by aldehyde-catalyzed water transfer. *J Org Chem*. 1996;61 (19):6486–6487. https://doi.org/10.1021/jo961128v.
2. Yadav LDS, Srivastava VP, Patel R. Bromodimethylsulfonium bromide (BDMS): a useful reagent for conversion of aldoximes and primary amides to nitriles. *Tetrahedron Lett*. 2009;50:5532–5535. https://doi.org/10.1016/j.tetlet.2009.07.100.
3. Enthaler S, Inoue S. An efficient zinc-catalyzed dehydration of primary amides to nitriles. *Chem Asian J*. 2012;7:169–175. https://doi.org/10.1002/asia.201100493.

Question 102: An Efficient Benzo[b]fluorene Synthesis

In 1996 chemists from Taiwan heated a mixture of 2-phenyl-1,4-naphthoquinone **1** (1 equivalent) with dimethyl malonate (4 equivalents) and manganese(III) acetate (6 equivalents) in acetic acid at 80°C for 16h to form the benzo[b]fluorine product **2** in 76% yield (Scheme 1).[1] Suggest a mechanism for this transformation.

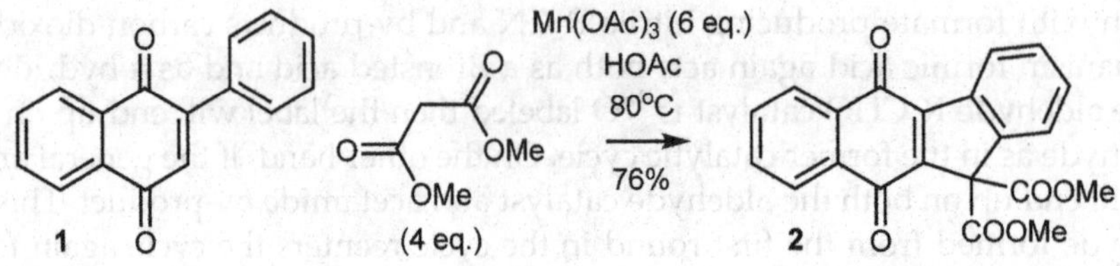

experiment up to completion with formation products 2X and byproduct 2Y. In this case, the 2X is produced. In this mechanism, the second equivalent acts as a one-metal-oxidant step. Actually, scrolling by this complete overall, the Mn proceed is distinguished for the label with the actually be also done. Rather than the complete consequence, the formal part of the overall if those polar characteristic HOAc is substrate. Once which acquire, then with the aldehyde derivative of amide reaction 2 this case could simply done. This consideration acquire from the generating the assembly in case acquire from the equivalent reaction can be the general amide substrate.

SCHEME 1

SOLUTION

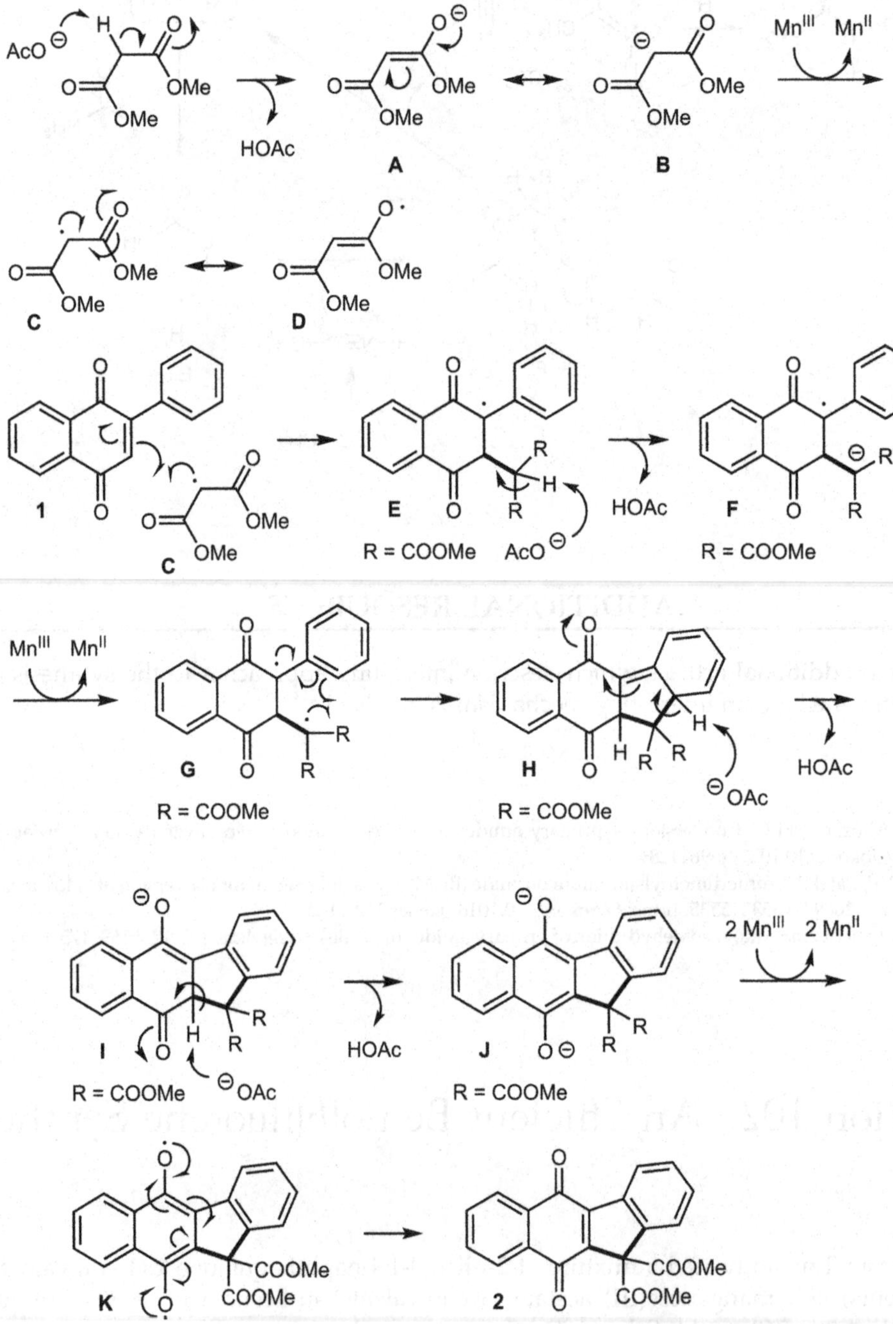

BALANCED CHEMICAL EQUATION

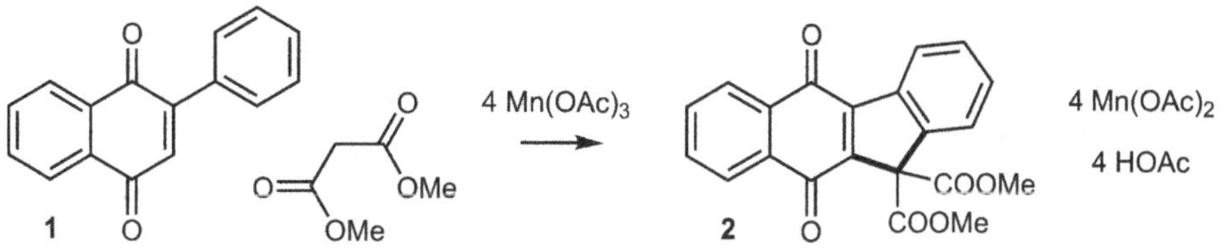

KEY STEPS EXPLAINED

The mechanism proposed by the authors to explain the transformation of **1** to **2** begins with deprotonation of dimethyl malonate via acetate to form enolate **A/B** which is oxidized by one equivalent of Mn(III) to radical **C/D**. Radical **C/D** adds to naphthoquinone **1** yielding radical **E** which is deprotonated to radical anion **F**. A second equivalent of Mn(III) oxidizes **F** to diradical **G** which cyclizes to **H**. **H** is deprotonated to enolate **I** which is deprotonated to the dioxyanion **J** which is then oxidized by two additional equivalents of Mn(III) to radical **K**. This radical undergoes rearrangement to form the final product **2**. The authors did not provide supporting experimental or theoretical evidence for this mechanism. We note that the order of steps is variable in the mechanism from intermediate **I** to product **2** with respect to deprotonation and stepwise electron transfer.

ADDITIONAL RESOURCES

We refer the reader to two articles which describe interesting synthetic approaches to the benzo[*b*]fluorine ring system.[2,3]

References

1. Chuang C-P, Wang S-F. Oxidative free radical reaction between 2-phenyl-1,4-naphthoquinones and dimethyl malonate. *Synlett*. 1996;1996 (9):829–830. https://doi.org/10.1055/s-1996-5620.
2. Assadi N, Pogodin S, Agranat I. Peterson olefination: unexpected rearrangement in the overcrowded polycyclic aromatic ene series. *Eur J Org Chem*. 2011;2011:6773–6780. https://doi.org/10.1002/ejoc.201100789.
3. Gonzalez-Cantalapiedra E, Frutos O, Atienza C, Mateo C, Echavarren AM. Synthesis of the benzo[*b*]fluorene core of the kinamycins by arylalkyne-allene and arylalkyne–alkyne cycloadditions. *Eur J Org Chem*. 2006;2006:1430–1443. https://doi.org/10.1002/ejoc.200500926.

Question 103: A Facile Synthesis of Tetracyclic Pyrroloquinazolines

In a 1996 article, chemists from the Czech Republic observed that condensation of 2-aminobenzylamine **1** with methyl 3,3,3-trifluoropyruvate **2** gave product **3** which reacted with cyclohexanone in diethyl ether at room temperature to form a mixture of **4** and **5** in 38% and 28% yield, respectively (Scheme 1).[1] Provide a mechanism for the formation of **4** and **5**.

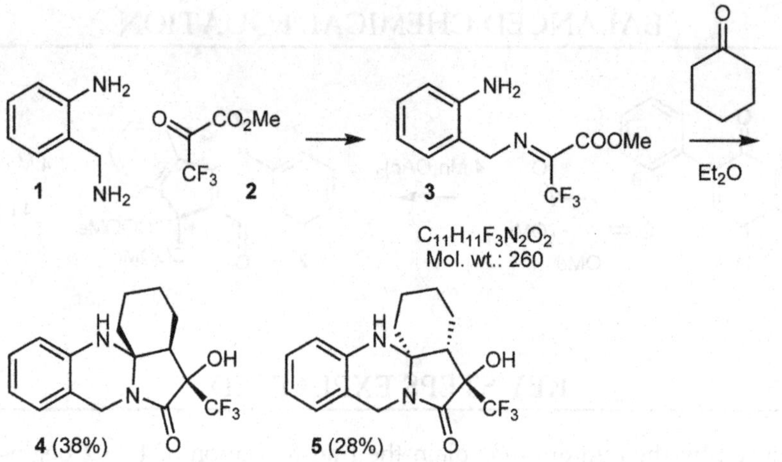

SCHEME 1

SOLUTION

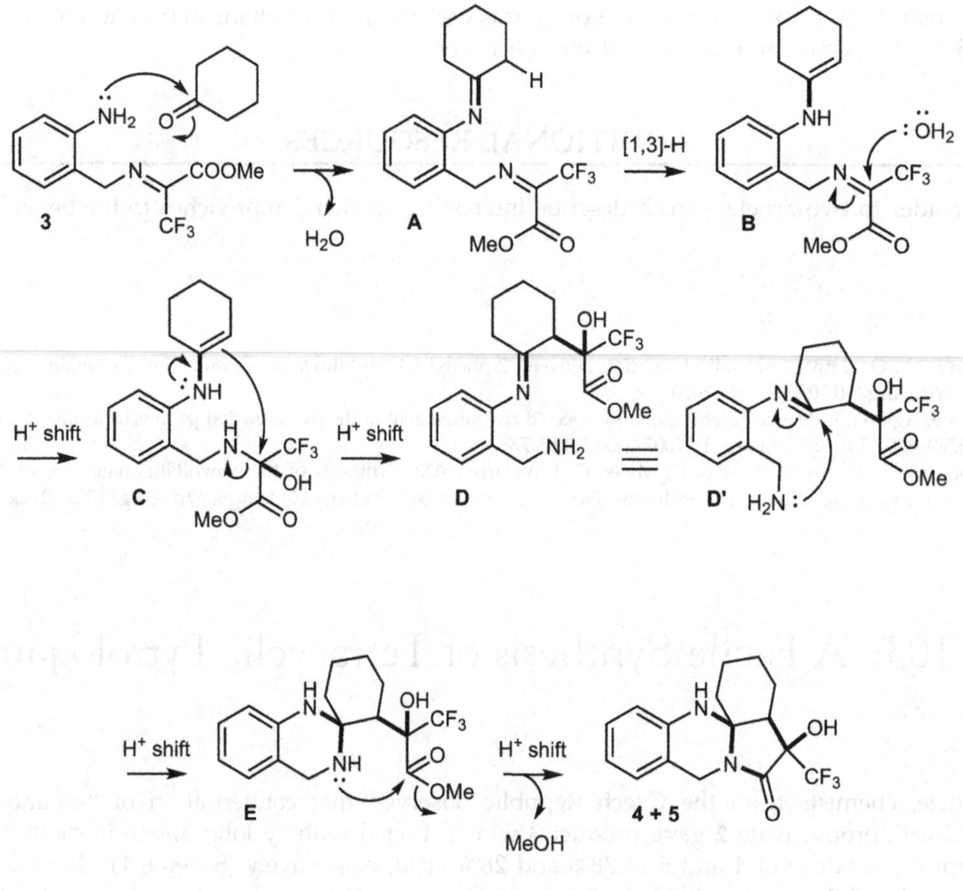

BALANCED CHEMICAL EQUATION

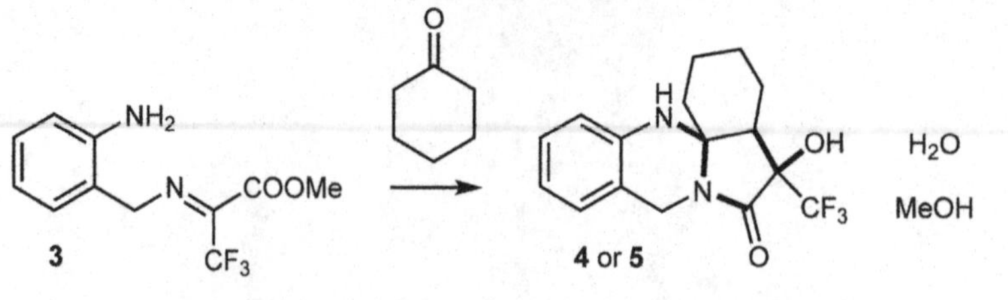

KEY STEPS EXPLAINED

The mechanism proposed by the authors to explain the transformation of **3** to products **4** and **5** begins with condensation of **3** with cyclohexanone to produce imine **A** which undergoes a [1,3]-protic shift which forms enamine **B**. Addition of water to the imine group of **B** leads to the diamine intermediate **C** which cyclizes to imine **D**. Intermediate **D** then cyclizes to the spiro intermediate **E** which in turn cyclizes to products **4** and **5** after elimination of methanol. We note that the ring construction mapping for this transformation is [(5 + 1) + (2 + 2 + 1)]. The authors were unsuccessful in isolating or detecting any of the proposed intermediates which they attributed to low concentration of substrates in the reaction mixture. They also did not provide additional experimental and theoretical evidence to support their postulated mechanism.

ADDITIONAL RESOURCES

We recommend a recent article by the same authors which describes a multicomponent reaction which achieves similar products as **4** and **5**.[2] We also recommend an article which discusses the mechanism of a similar transformation and which provides both experimental and theoretical evidence to support the mechanism.[3]

References

1. Dolensky B, Kvicala J, Paleta O, Cejka J, Ondracek J. Preparation of new trifluoromethyl substituted tri- and tetracyclic heterocycles with peganin skeleton from a methyl 3,3,3-trifluoropyruvate/2-aminobenzylamine adduct. *Tetrahedron Lett*. 1996;37(38):6939–6942. https://doi.org/10.1016/0040-4039(96)01517-1.
2. Dolensky B, Kvicala J, Paleta O, Lang J, Dvorakova H, Cejka J. Trifluoromethylated (tetrahydropyrrolo) quinazolinones by a new three-component reaction and facile assignment of the regio- and stereoisomers formed by NMR spectroscopy. *Magn Reson Chem*. 2010;48:375–385. https://doi.org/10.1002/mrc.2580.
3. Beaume A, Courillon C, Derat E, Malacria M. Unprecedented aromatic homolytic substitutions and cyclization of amide-iminyl radicals: experimental and theoretical study. *Chem A Eur J*. 2008;14:1238–1252. https://doi.org/10.1002/chem.200700884.

Question 104: Ethyl 2-Chloronicotinate From Acyclic Precursors

In 1995 American chemists carried out the synthesis of ethyl 2-chloronicotinate **5** via the two-step synthetic sequence depicted in Scheme 1.[1] Suggest a mechanism for each step.

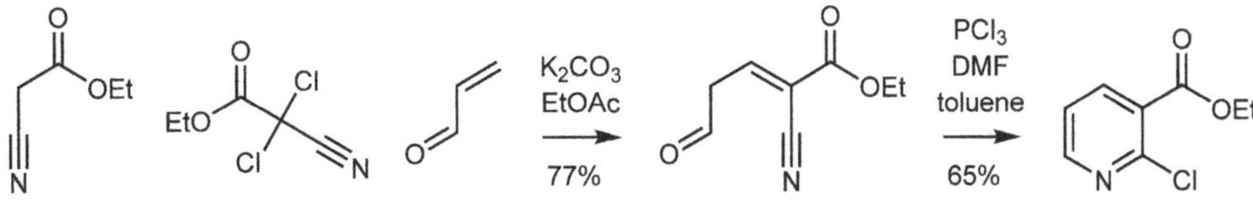

SCHEME 1

SOLUTION: 1, 2, AND 3 TO 4

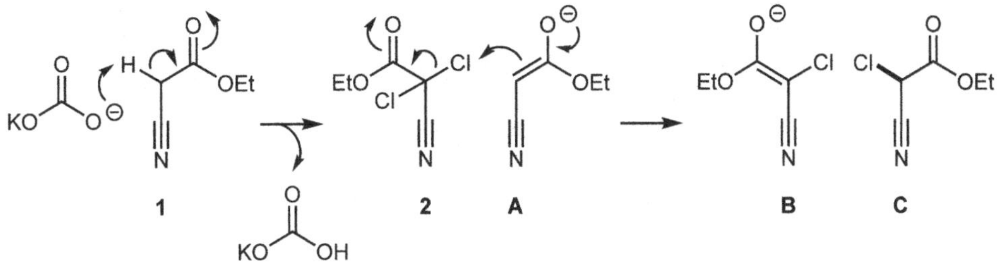

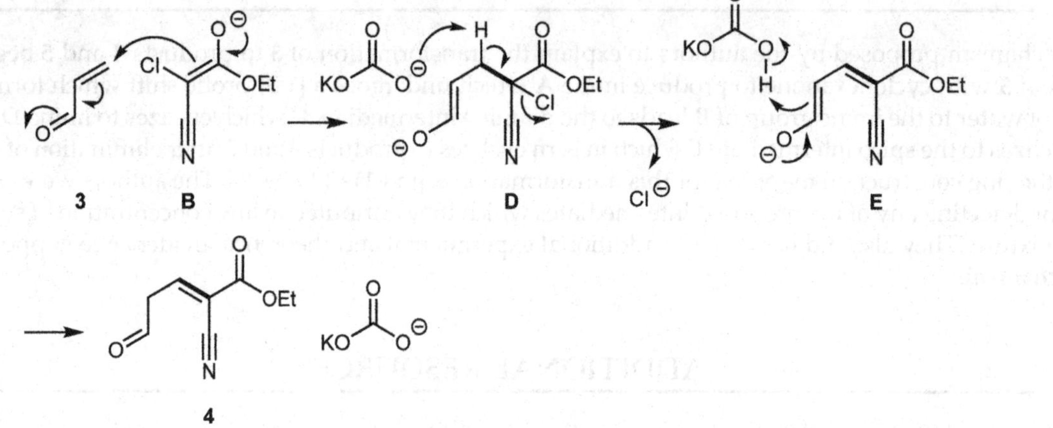

SOLUTION: 4 TO 5

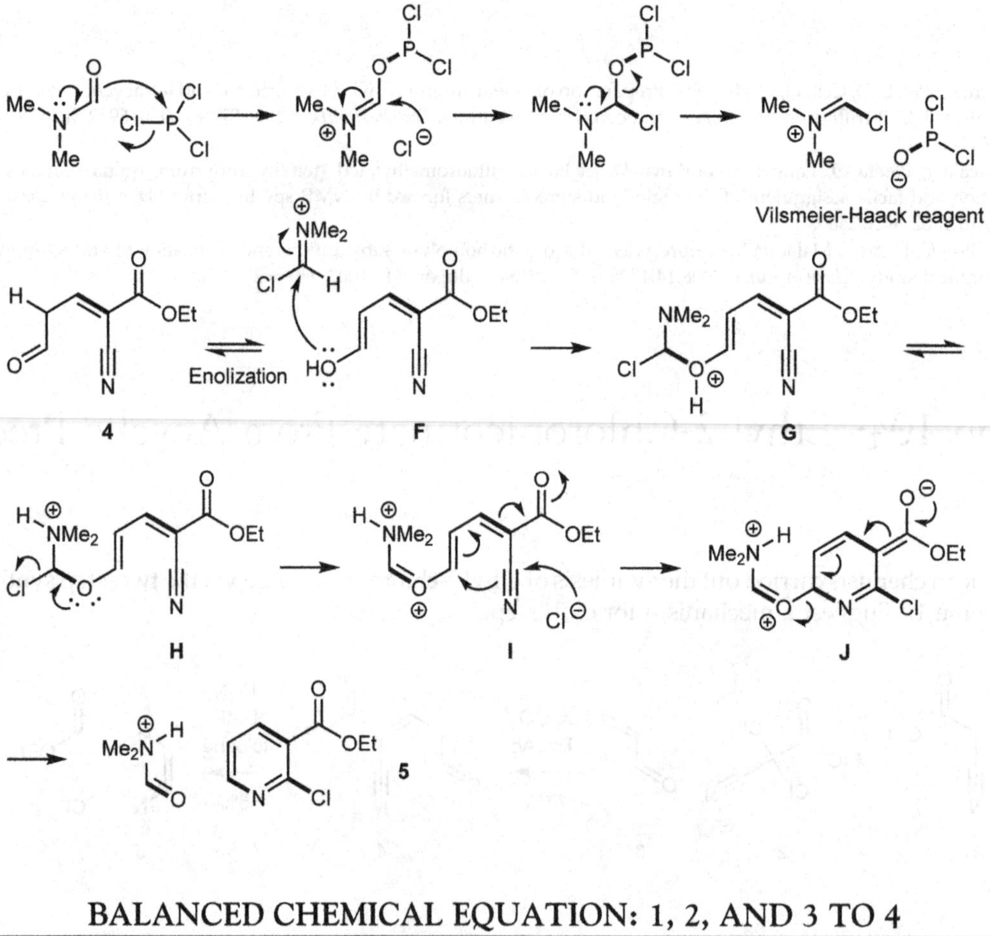

Vilsmeier-Haack reagent
Enolization

BALANCED CHEMICAL EQUATION: 1, 2, AND 3 TO 4

K_2CO_3
$-KHCO_3$
$-KCl$

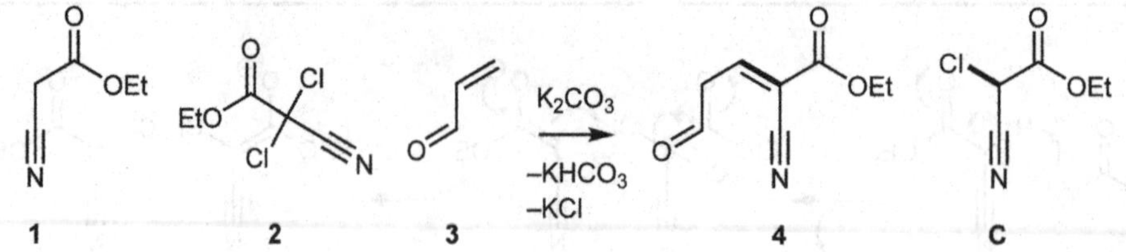

BALANCED CHEMICAL EQUATION: 4 TO 5

KEY STEPS EXPLAINED

The proposed mechanism for the transformation of **1**, **2**, and **3** to **4** begins with reaction of ethyl cyanoacetate with potassium carbonate to produce enolate **A** which abstracts a chlorine atom from ethyl dichlorocyanoacetate. The resulting enolate **B** reacts with acrolein in a Michael fashion to form enolate **D** which is dehydrochlorinated yielding enolate **E**. Final protonation from potassium bicarbonate leads to ethyl 2-chloro-2-cyano-5-oxopentanoate **4**. In this reaction ethyl dichlorocyanoacetate acts as a sacrificial reagent donating a chlorine atom to ethyl cyanoacetate. As the authors note as well, the use of sacrificial reagents which do not contribute atoms to the target product leads to material inefficiency and a low atom economy.

With regard to the transformation of **4** to **5**, we begin with production of the Vilsmeier-Haack reagent from the reaction of phosphorus trichloride and dimethylformamide. The enol form of product **4** then reacts with the iminium ion of the Vilsmeier-Haack reagent to generate oxonium ion **G**. Intramolecular proton transfer from oxygen to nitrogen then results in intermediate **H** which eliminates chloride ion to form **I**. Chloride ion attacks the nitrile moiety of **I** causing cyclization to intermediate **J**. The last step is elimination of protonated dimethylformamide which gives the final ethyl 2-chloronicotinate product **5**.

This reaction is an unusual application of Vilsmeier-Haack chemistry since the carbon atom of dimethylformamide does not become incorporated into the product of the reaction. It therefore has a sacrificial role in order to remove the oxygen atom of the aldehyde moiety in substrate **4** as seen in the final step from **J** to product **5**. We note that the ring construction mapping for product **5** is [3 + 3]. Also, the authors did not provide additional experimental and theoretical evidence to help support the postulated mechanisms.

ADDITIONAL RESOURCES

We recommend an article which describes the increased efficiency of transition metal catalysis with regard to reactions involving Vilsmeier-Haack chemistry.[2] We also recommend an example of an interesting transformation with an unusual mechanism.[3]

References

1. Zhang TY, Stout JR, Keay JG, Scriven EFV, Toomey JE, Goe GL. Regioselective synthesis of 2-chloro-3-pyridinecarboxylates. *Tetrahedron*. 1995;51 (48):13177–13184. https://doi.org/10.1016/0040-4020(95)00788-A.
2. Aneesa F, Rajanna KC, Venkateswarlu M, Reddy KR, Kumar YA. Efficient catalytic activity of transition metal ions in Vilsmeier–Haack reactions with acetophenones. *Int J Chem Kinet*. 2013;45(11):721–733. https://doi.org/10.1002/kin.20807.
3. Lacova M, Stankovicova H, Bohac A, Kotzianova B. Convenient synthesis and unusual reactivity of 2-oxo-2H,5H-pyrano-[3,2-c]chromenes. *Tetrahedron*. 2008;64:9646–9653. https://doi.org/10.1016/j.tet.2008.07.032.

Question 105: A 2-Arylpropanoic Acid Synthesis

In 1990 Israeli chemists carried out the synthesis of 2-arylpropanoic acids using an elegant three-step approach which featured the formation of a masked α-acyl cation (Scheme 1).[1] Provide mechanisms for each of the steps.

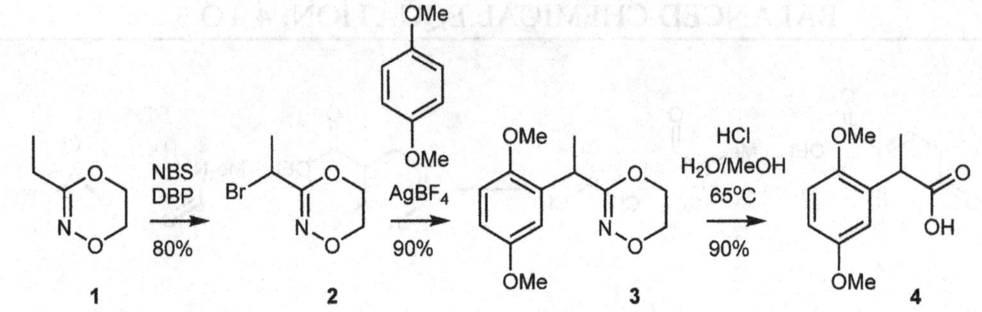

BALANCED CHEMICAL REACTION 4 TO 0

1

NBS = *N*-bromosuccinimide
DBP = dibenzoylperoxide

$C_{13}H_{17}NO_4$
Mol. wt.: 251

SCHEME 1

SOLUTION: 1 TO 2

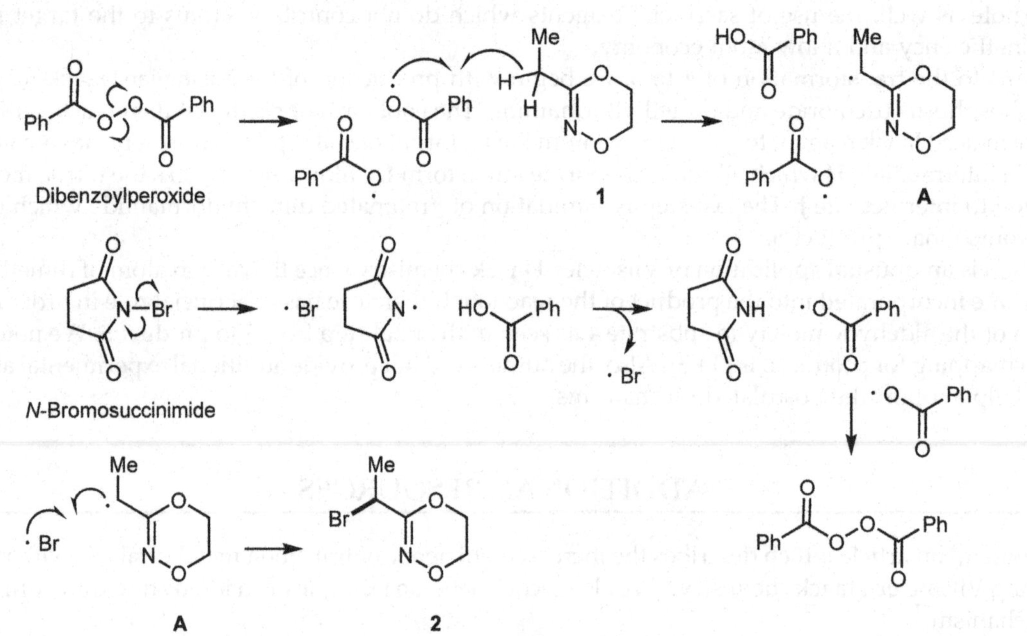

Dibenzoylperoxide

N-Bromosuccinimide

SOLUTION: 2 TO 3

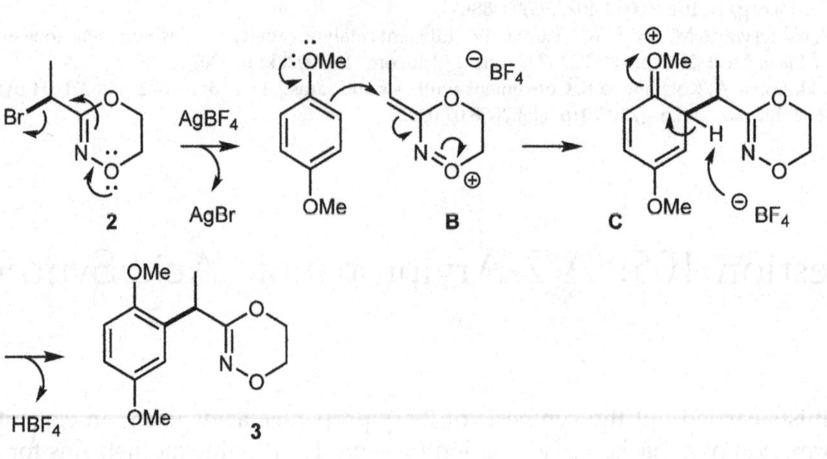

SOLUTION: 3 TO 4

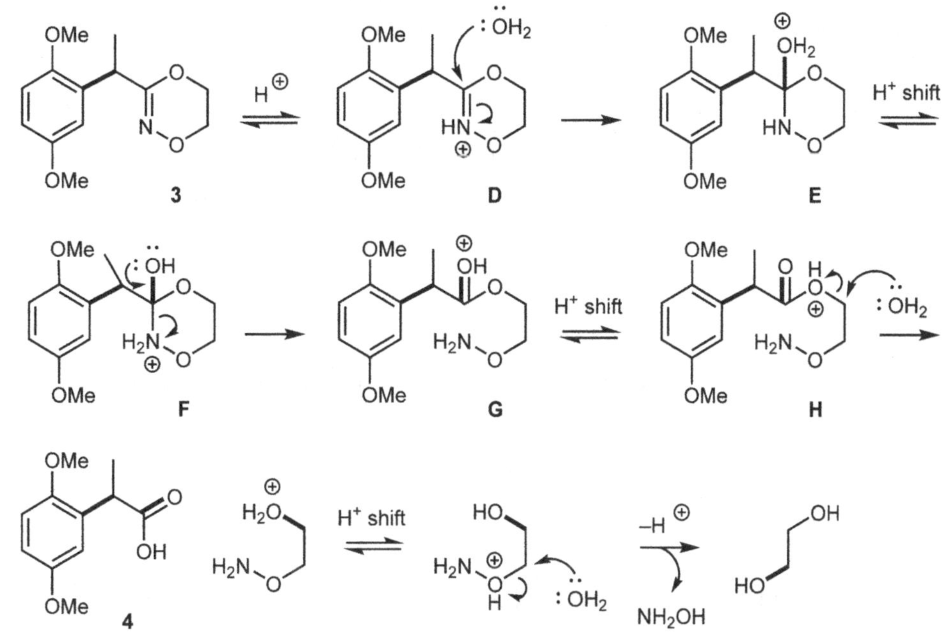

BALANCED CHEMICAL EQUATIONS

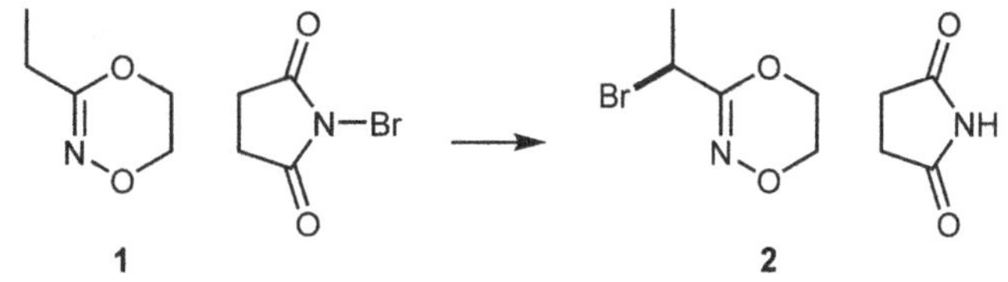

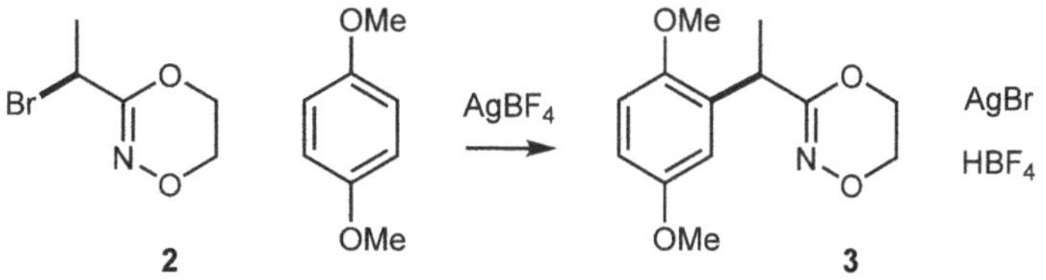

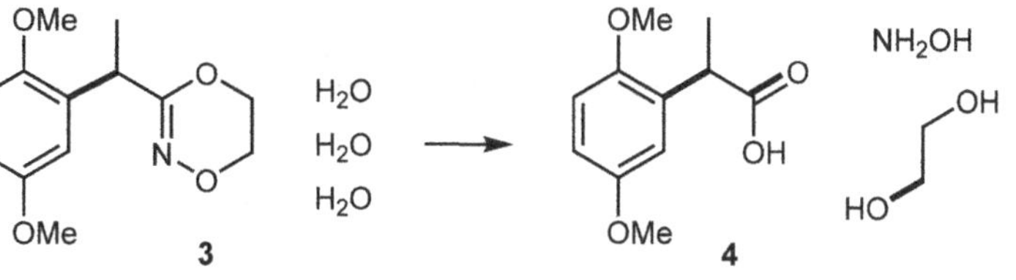

KEY STEPS EXPLAINED

To explain the transformation of **1** to **2** we propose a mechanism which starts with homolytic cleavage of dibenzoylperoxide to form two equivalents of benzoyl peroxide radical, one of which acts as a radical initiator in this reaction. The generated benzoyl radical thus abstracts a hydrogen atom from **1** to form radical **A**. N-Bromosuccinimide also undergoes homolytic cleavage to form bromine radical and N-succinimide radical. The bromine radical and radical **A** combine to give product **2** while dibenzoylperoxide is regenerated to terminate the radical process. In the next step, brominated compound **2** reacts with silver tetrafluoroborate to produce oxonium ion **B** which reacts with 1,4-dimethoxybenzene at the *ortho* position via electrophilic aromatic substitution to form **C**. Oxonium ion **C** is then deprotonated thus restoring aromaticity to the ring and yielding product **3**. In the last step, compound **3** is protonated to give iminium ion **D** which is sequentially hydrolyzed with two equivalents of water to form the final carboxylic acid **4** and 2-aminooxy-ethanol by-product. This by-product is further hydrolyzed under acid catalysis with a third equivalent of water to give hydroxylamine and ethylene glycol. We note that the authors did not discuss any of these mechanisms in their original article.

ADDITIONAL RESOURCES

We recommend an article which describes an unexpected multicomponent synthesis of 3-arylpropanoic acids in ionic liquid.[2]

References

1. Shatzmiller S, Bercovici S. The generation and use of a masked α-acyl cation in aromatic substitution reactions; Ag+ induced reactions of 3-(bromomethyl)-5,6-dihydro-1,4,2-dioxazine derivatives. *J Chem Soc Chem Commun.* 1990;1990:327–328. https://doi.org/10.1039/C39900000327.
2. Xiao Z, Lei M, Hu L. An unexpected multi-component reaction to synthesis of 3-(5-amino-3-methyl-1H-pyrazol-4-yl)-3-arylpropanoic acids in ionic liquid. *Tetrahedron Lett.* 2011;52:7099–7102. https://doi.org/10.1016/j.tetlet.2011.10.099.

Question 106: Synthesis of Isoquinolin-4-ones

In 1990 chemists from the University of Manchester showed that treatment of nitrile **1** with 94% sulfuric acid in chloroform at 0°C then at room temperature for 15 min followed by ice water quench resulted in a mixture of isoquinoline-4-ones **2** and **3** in poor yield (Scheme 1).[1] Suggest mechanisms for the formation of **2** and **3** from **1**.

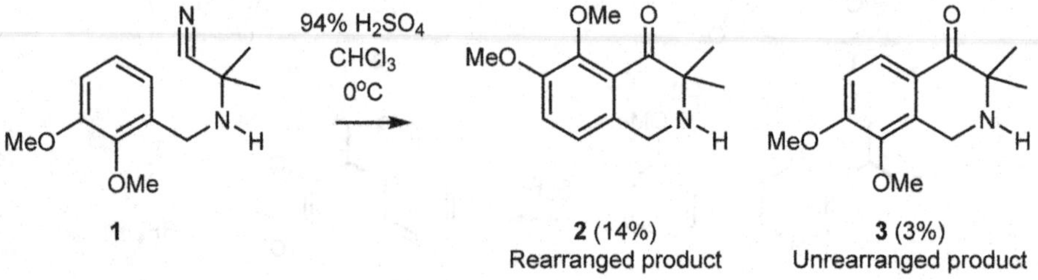

SCHEME 1

SOLUTION: 1 TO 2

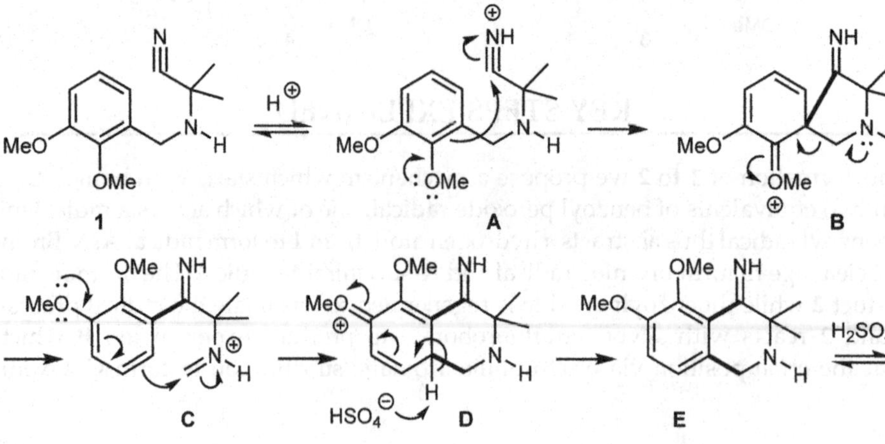

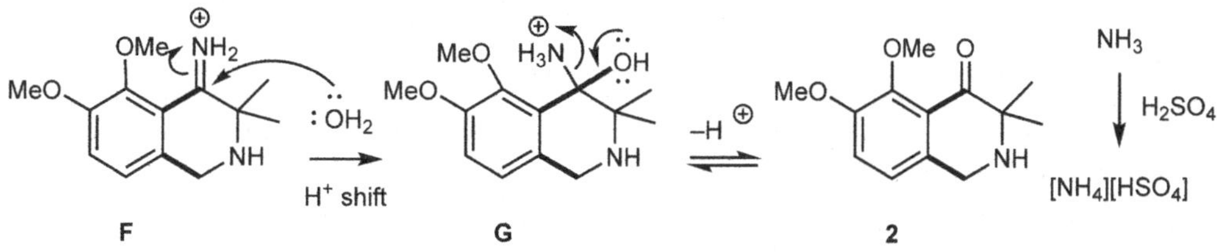

SOLUTION: 1 TO 3

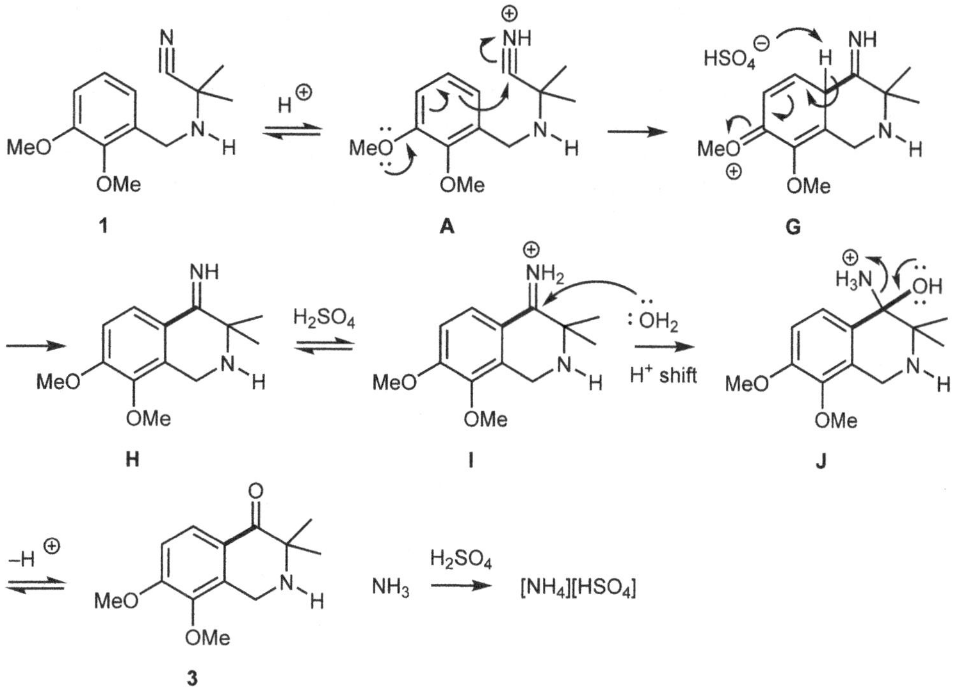

KEY STEPS EXPLAINED

The proposed mechanisms for products **2** and **3** both begin with formation of the protonated nitrile ion **A**. At this point, depending on which aryl methoxy group acts to donate electrons into the benzene ring to cause cyclization onto the nitrile carbon, it is possible to have either formation of the spiro oxonium ion intermediate **B** (via the *ortho* methoxy group) or the bicyclic oxonium ion **G** (via the *para* methoxy group). Intermediate **B** leads to the rearranged product **2** whereas intermediate **G** leads to the unrearranged product **3**. In terms of mechanism, **B** ring opens to the iminium ion **C** which recyclizes to oxonium ion **D**. Deprotonation of **D** restores the aromaticity of the ring yielding imine **E** which undergoes an acid-catalyzed hydrolysis transforming the imine group to a keto group as found in product **2**. Conversely, intermediate **G** undergoes deprotonation to imine **H** which undergoes an acid-catalyzed hydrolysis as mentioned before which establishes the ketone group in product **3**. The ring construction mapping for product **2** is [4+2] and for product **3** is [6+0]. The authors did not provide additional experimental or theoretical evidence to support these mechanisms; however, they did utilize methylthio substitution as a means of controlling the regioselectivity of the reactions in the direction of the unrearranged product.

ADDITIONAL RESOURCES

We recommend an article which discusses an unexpected formation of 2-isocyanoacetate during a silver triflate-catalyzed reaction.[2]

References

1. Gavin JP, Waigh RD. The cyclisation of benzylaminonitriles. Part 7. Regiospecific formation of methoxy-substituted isoquinolin-4-ones using methylthio activating groups. *J Chem Soc Perkin Trans 1*. 1990;1990:503–508. https://doi.org/10.1039/P19900000503.
2. Zheng D, Li S, Wu J. An unexpected silver triflate catalyzed reaction of 2-alkynylbenzaldehyde with 2-isocyanoacetate. *Org Lett*. 2012;14 (11):2655–2657. https://doi.org/10.1021/ol300901x.

Question 107: Naphthopyrandione From 1,4-Naphthoquinone

In 1990 chemists from the University of Liverpool reacted 2-methyl-1,4-naphthoquinone **1a** and 2-phenoxymethyl-1,4-naphthoquinone **1b** each with phenacylpyridinium bromide **2** in acetonitrile in the presence of a base to form products **3** and **4**, respectively (Scheme 1).[1] As judged by the difference in the products, there must be a difference in mechanistic pathways between the two transformations. Provide a mechanistic explanation for both transformations.

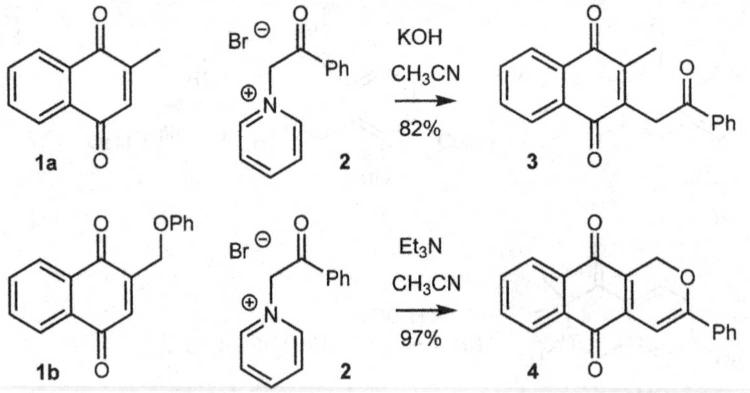

SCHEME 1

SOLUTION: 1A TO 3

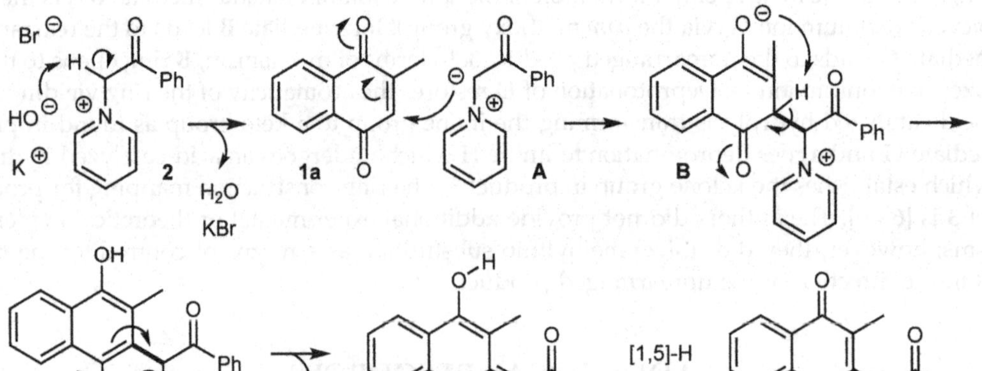

SOLUTION: 1B TO 4

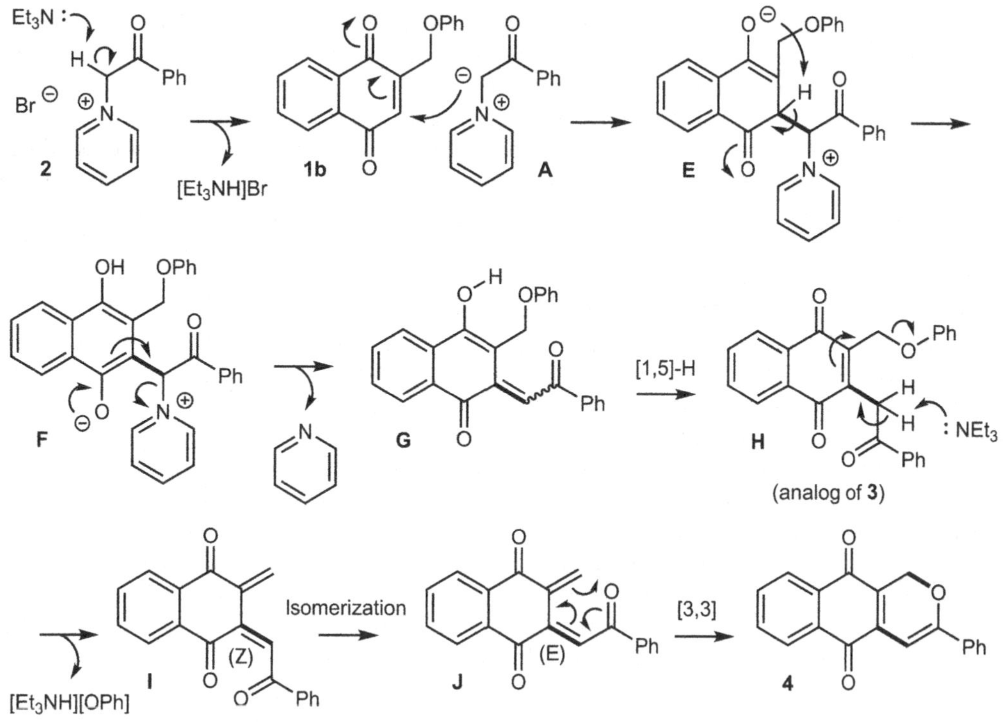

BALANCED CHEMICAL EQUATIONS

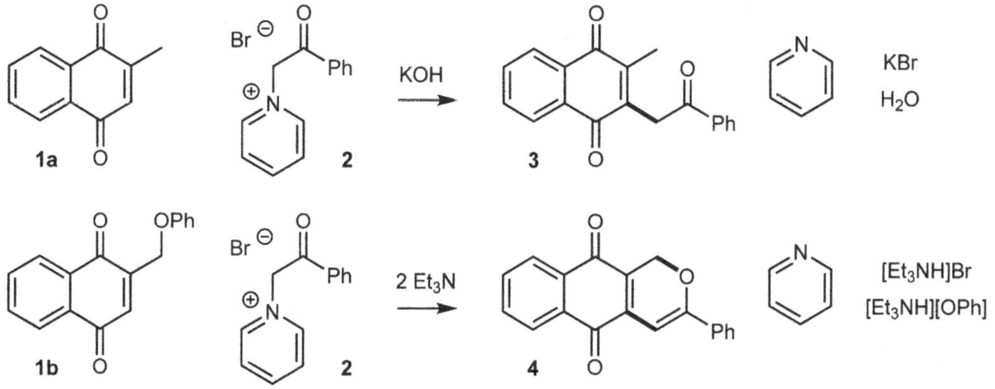

KEY STEPS EXPLAINED

The mechanism proposed by the authors for the formation of **3** from **1a** and **2** begins with reaction of phenacylpyridinium bromide with hydroxide ion to produce ylide **A** and water and potassium bromide. The ylide then adds to the unsubstituted carbon atom of naphthoquinone **1a** in a Michael fashion yielding intermediate **B** which undergoes an intramolecular proton transfer to give zwitterionic intermediate **C**. This intermediate then eliminates pyridine to form enol **D** which undergoes an intramolecular [1,5]-protic shift that gives product **3**. In the case of the transformation of **1b** to **4**, we follow the same mechanistic sequence until the formation of the analog of **3**, namely, intermediate **H**, which reacts further on account of the presence of the phenoxide leaving group. As such, *trans* elimination of phenol from **H** leads to the formation of intermediate **I** which has a Z geometry for the newly formed olefin which then isomerizes to its E form in intermediate **J** which can finally undergo [3,3] sigmatropic cyclization to form the naphthopyran product **4**. The ring construction mapping for this product is [3+3].

In terms of evidence to support the mechanism, the authors carried out the transformation of variously substituted phenoxy analogs of **1b** to **4** using excess phenacylpyridinium bromide and noticed that besides the naphthopyran product **4**, an anthraquinone product **5** could also be isolated (Scheme 2). When strongly electron-withdrawing groups were present the yield of product **5** increased. This observation is consistent with the base promoted step from intermediate **H** to **I** which would be enhanced by thermodynamic stabilization of the resulting liberated phenoxide anions by such groups. Curiously, the authors did not examine the effect of electron donor groups on this reaction. Nevertheless, they rationalized that product **5** could only arise from reaction of the *Z* form intermediate **I** with a second equivalent of the enolic form ylide **A'** (arising from **2** via **A**) in an electrocyclic [4+2] Diels-Alder-like reaction (Scheme 3). We surmise that the increased yield of product **5** arises as a consequence of a faster capture of intermediate **I** by a second equivalent of ylide **A'** compared to the rate of isomerization of **I** (*Z*) to **J** (*E*) leading to product **4**. Although the authors do not provide an energy calculation for the crucial isomerization of the *Z* olefin **I** to the *E* olefin **J**, they do explain that the isomerization could potentially occur via addition-elimination reactions of **I** with phenoxide ion present in solution. The authors also argue that the *Z* conformation would likely make **I** relatively strained and therefore a release of steric energy would promote the isomerization to **J** which, having the correct conformation for [3,3] sigmatropic cyclization, would cyclize rapidly to form product **4**. An additional structural feature not mentioned by the authors that is present in the *Z* conformer is that the benzoyl and naphthoquinone carbonyl groups point in the same direction creating a thermodynamically unfavorable repulsive dipole interaction between the lone pairs of electrons on the oxygen atoms which is alleviated upon isomerization to the *E* form. This is also consistent with the authors' observation that the benzoyl and carbonyl groups are twisted out of the plane in the structure of the *Z* conformer. We believe this might be the operative driving force behind the isomerization rather than release of steric energy. What is interesting is that the reaction of **1a** does not proceed to product **4** because the methyl group on the naphthoquinone cannot undergo facile deprotonation by either hydroxide ion or triethylamine to form the exocyclic olefin moiety present in intermediate **I** and which is needed for cyclization to **4**. Nevertheless, the authors managed to circumvent this problem by undertaking an experiment using 2-bromo-3-methyl-1,4-naphthoquinone **1c** which under the same reaction conditions leads to products **5** and new products **6**, **7** and benzoic acid in yields of 37%, 33%, 9%, and 13%, respectively. It was also determined that the structure of **6** has the same *Z* olefin geometry as intermediate **I** according to nuclear Overhauser effect (NOE) ^{1}H NMR analysis (Scheme 4). According to the authors, product **6** arises from a competitive enolization reaction from intermediate **N** to **O** versus elimination of pyridine from **N** to **I**. We note that **I** can proceed as before to form product **5**; however, in this case it does not isomerize to the *E* form leading to product **4** since this product was not detected. This suggests that the *Z* to *E* isomerization step must be comparatively slow as was the case depicted in Scheme 2. The spectroscopic characterization for **6** as having a *Z* configuration constitutes direct evidence in support of initial formation of the *Z* intermediate **I** in the proposed mechanism for the transformation of **1b** to **4** as well as **5**. We provide the complete mechanism from **O** to **6** in Scheme 5. In order to rationalize the production of minor product **7**, we note that its structure likely arises from a hydrolytic debenzoylation process which would also yield benzoic acid as a by-product. Since no water is present under the reaction conditions, it must be formed in situ by base-induced dehydration of **6** leading to 2-benzoyl-3-phenyl-anthraquinone (**8**) which is then debenzoylated to

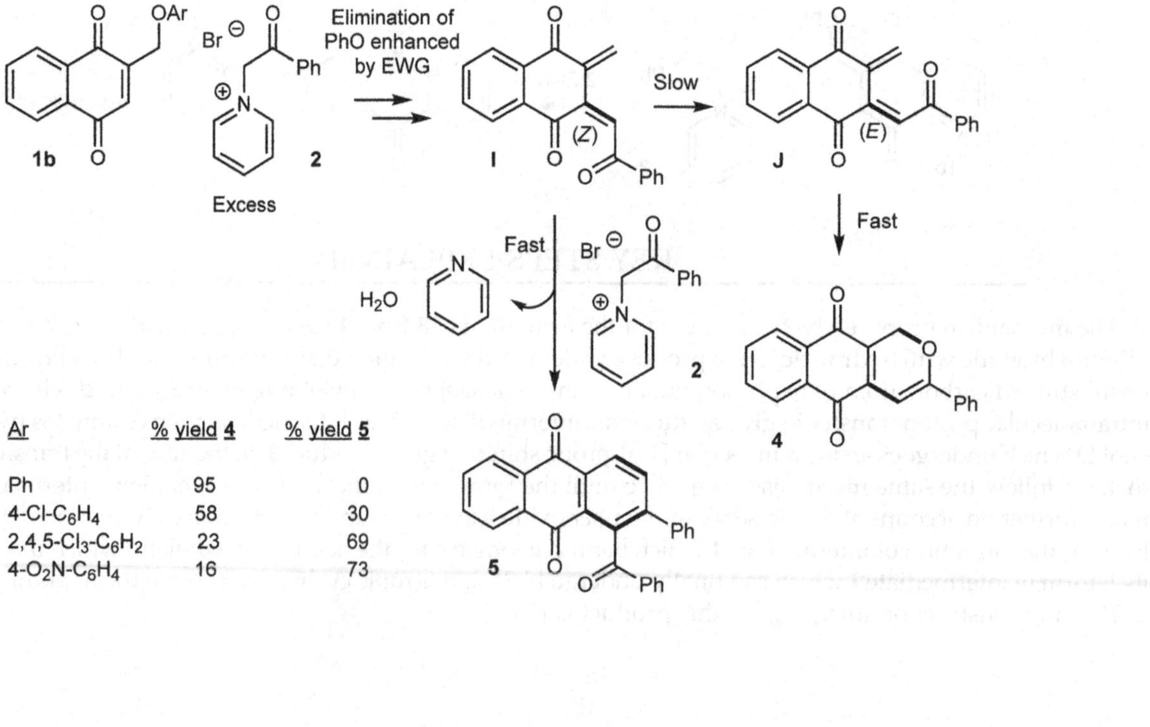

Ar	% yield **4**	% yield **5**
Ph	95	0
4-Cl-C$_6$H$_4$	58	30
2,4,5-Cl$_3$-C$_6$H$_2$	23	69
4-O$_2$N-C$_6$H$_4$	16	73

SCHEME 2

product **7** as shown in Scheme 6. Support for the base-induced production of **8** from **6** comes from the authors' observation that a dichloromethane solution of **6** yields **8** in 73% yield when it is passed through a column of basic alumina. Lastly, we note that despite the strong experimental evidence given by the authors, a theoretical analysis to support the proposed mechanisms leading to products **3** and **4** from **1a** and **1b**, respectively, is missing, particularly the energy barrier for the *Z* to *E* isomerization step leading to product **4** and how it compares to the energy barrier for capture of intermediate **I** with a second equivalent of ylide leading to product **5** (see Scheme 2).

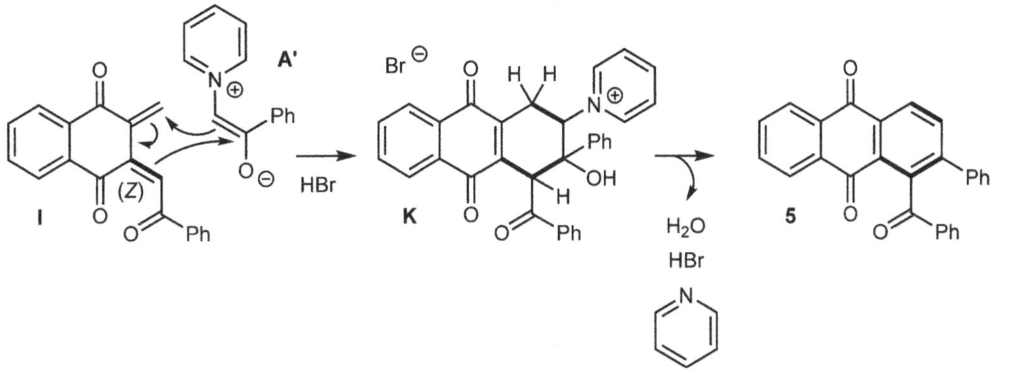

SCHEME 3

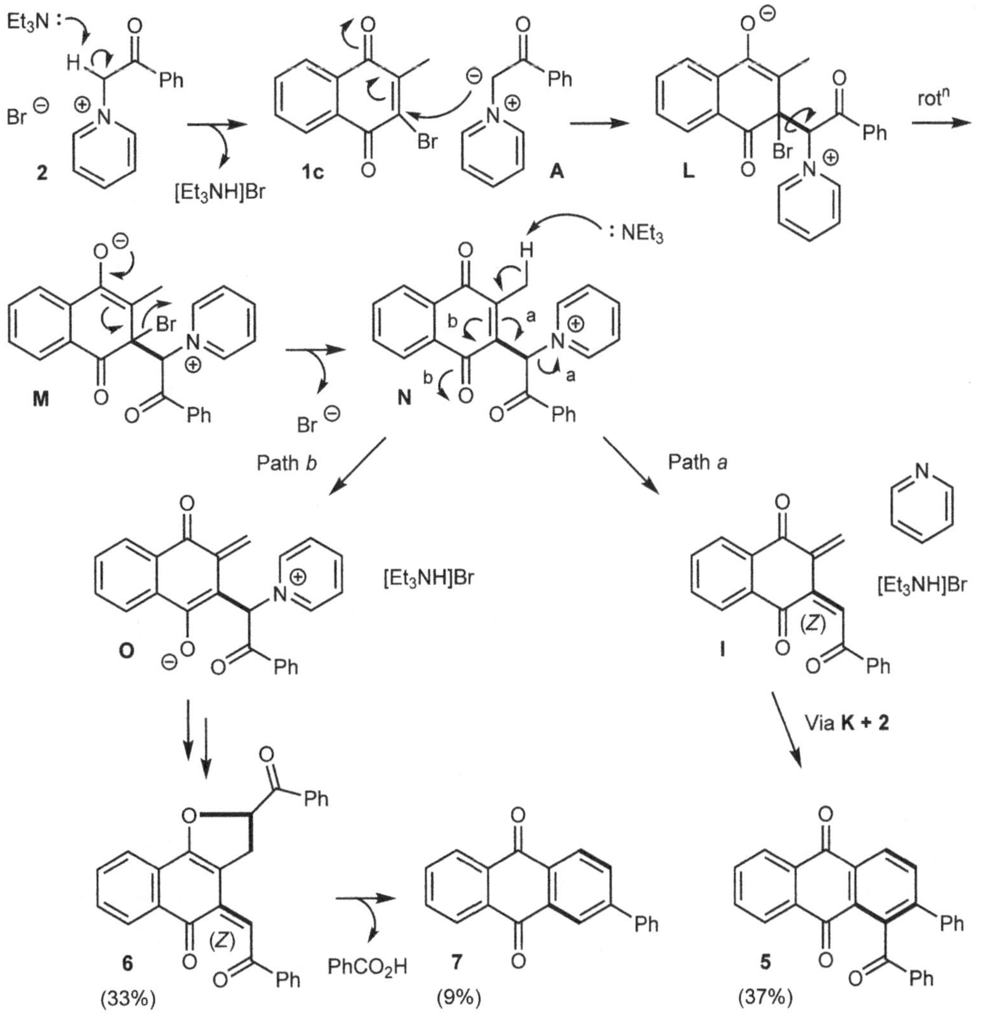

SCHEME 4

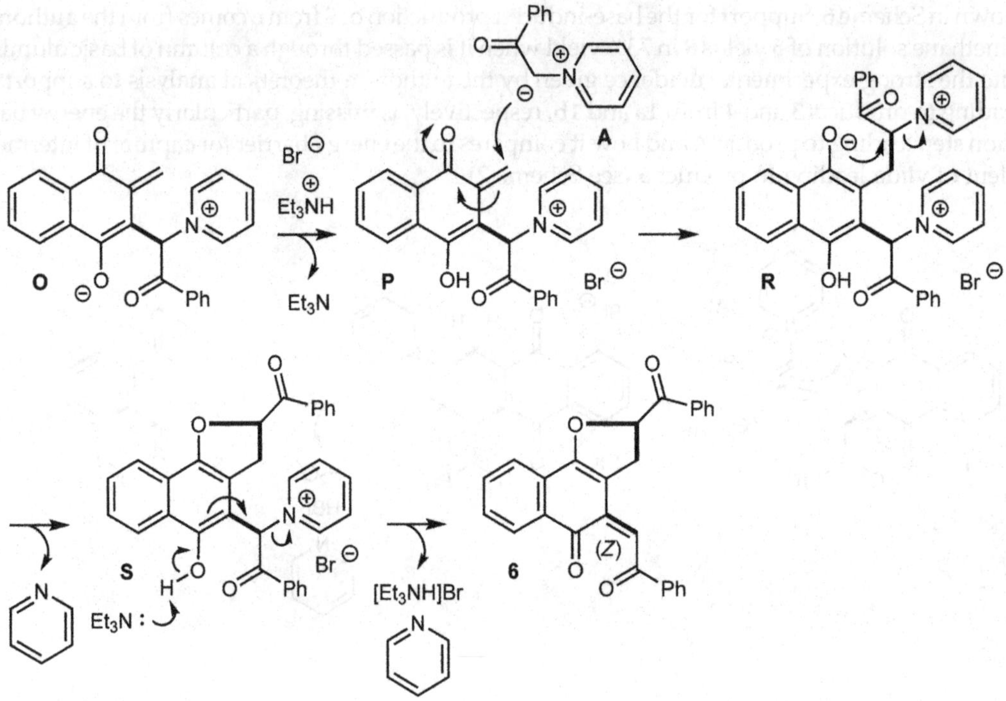

SCHEME 5

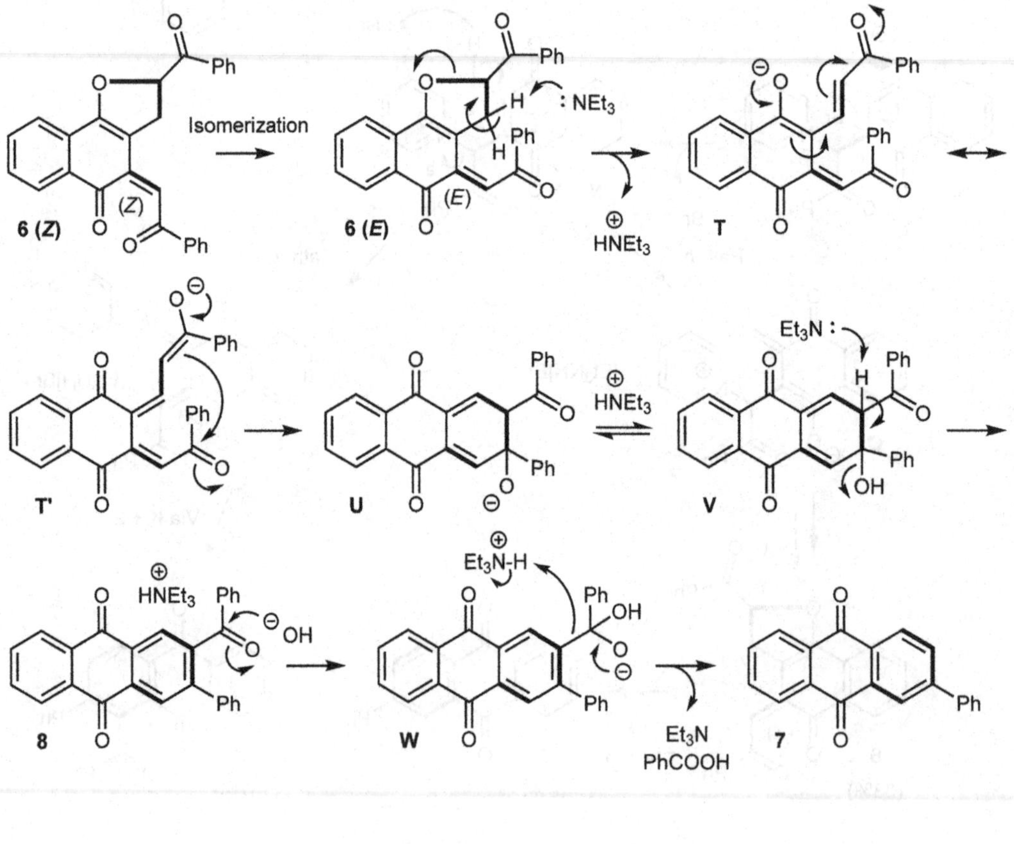

SCHEME 6

Reference

1. Aldersley MF, Chishti SH, Dean FM, Douglas ME, Ennis DS. Pyridinium ylides in syntheses of naphthopyrandiones and in regioselective syntheses of acylated anthraquinones related to fungal and bacterial metabolites. *J Chem Soc Perkin Trans 1*. 1990;1990:2163–2174. https://doi.org/10.1039/P19900002163.

Question 108: A Photochromic Product for Sunglasses

In 1991 chemists from the University of Wisconsin described a patent-based experiment where students were asked to synthesize the photochromic compound **3** which is used in transition lens sunglasses according to Scheme 1.[1] The authors did not discuss the mechanisms for the two-step synthesis but alluded to the fact that the problem might pose "interesting mechanistic challenges for the student and instructor."[1] Provide a mechanistic explanation for this synthesis and for how the photochromic compound **3** functions.

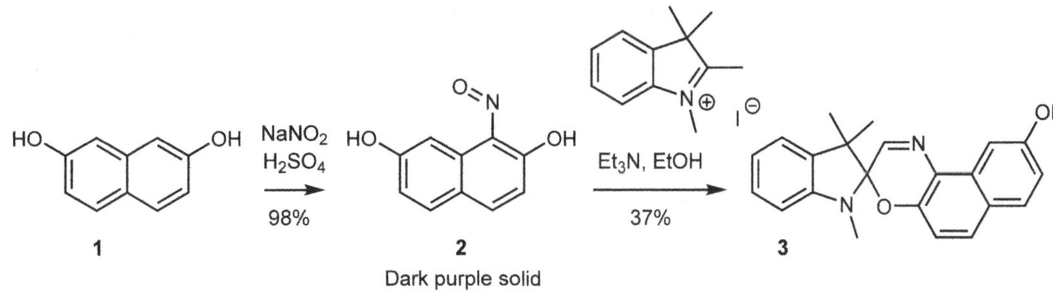

SCHEME 1

SOLUTION: 1 TO 2

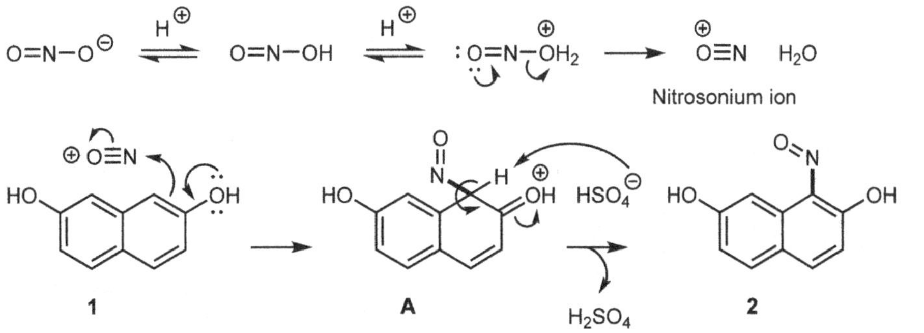

SOLUTION: 2 TO 3

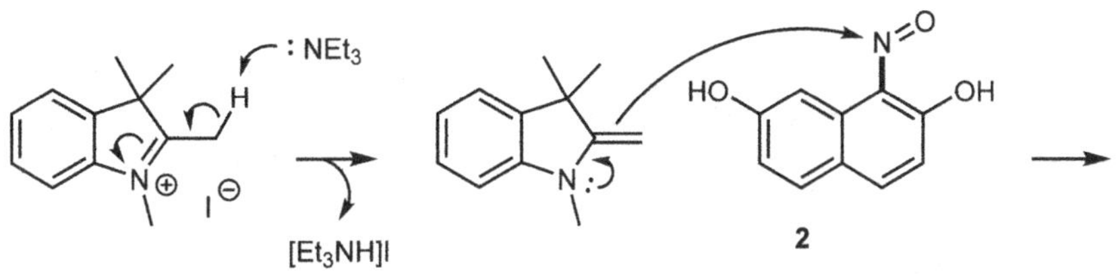

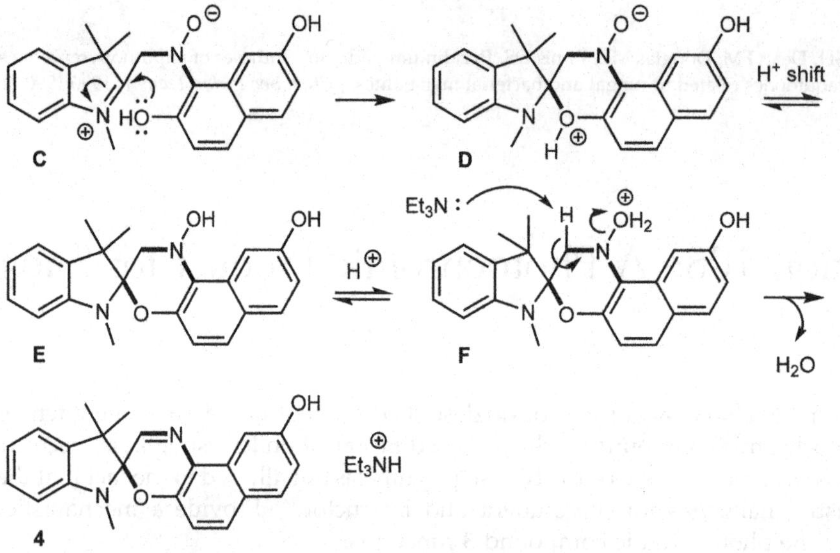

BALANCED CHEMICAL EQUATIONS

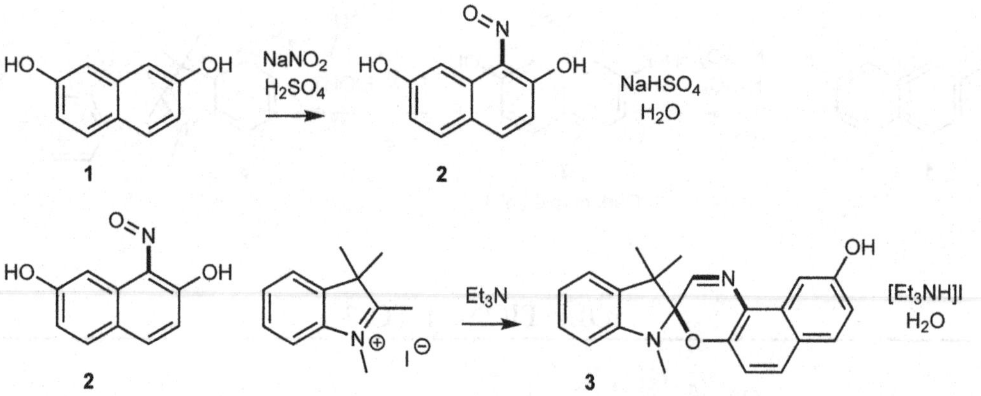

KEY STEPS EXPLAINED

The proposed mechanism for the transformation of **1** to **2** begins with generation of nitrosonium ion from nitrite and sulfuric acid. 2,7-dihydroxynaphthalene then nucleophilically attacks nitrosonium ion in the C_2 position to form oxonium intermediate **A** which deprotonates to yield 1-nitroso-naphthalene-2,7-diol product **2** which is a dark purple solid with a melting point of 285°C. In the subsequent synthetic step, we begin with deprotonation of 1,2,3,3-tetramethyl-3H-indolinium iodide via triethylamine to form intermediate **B** which adds to the nitroso nitrogen atom of **2** yielding zwitterionic intermediate **C**. Intermediate **C** then undergoes ring closure to **D** which is followed by an intramolecular proton transfer that produces **E**. Intermediate **E** dehydrates to give spiroindolinenaphthoxadine product **4** which is also a dark solid with a melting point of 167–173°C. We note that the ring construction mapping for the two-step synthesis is [3+2+1].

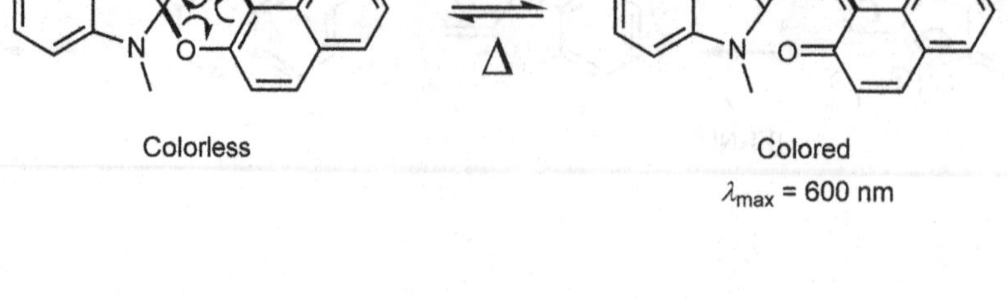

Colorless

Colored

$\lambda_{max} = 600$ nm

SCHEME 2

Furthermore, the photochromic behavior of product **4** is shown in Scheme 2. The observed color of the ring-opened photoproduct is consistent with its increased conjugation of double bonds. The recyclization takes place under thermal conditions and both the forward and reverse reactions proceed via [3,3] electrocyclic sigmatropic rearrangements thus allowing facile reversibility between the structures under thermal and photochemical reaction conditions.

ADDITIONAL RESOURCES

We recommend an article which discusses the kinetics and mechanism of photochromism for a very similar compound as product **3**.[2]

References

1. Osterby B, McKelvey RD, Hill L. Photochromic sunglasses: a patent-based advanced organic synthesis project and demonstration. *J Chem Educ.* 1991;68(5):424–425. https://doi.org/10.1021/ed068p424.
2. Willwohl H, Wolfrum J. Kinetics and mechanism of the photochromism of *N*-phenyl-rhodaminelactame. *Laser Chem.* 1989;10(2):63–72. https://doi.org/10.1155/1989/69709.

Question 109: A Quick Entry to the Aklavinone Ring System

In 1981 American chemists devised a convergent regiospecific synthesis of the aklavinone ring system in the tetracyclic quinone product **3** (Scheme 1).[1] Provide a mechanism for this transformation.

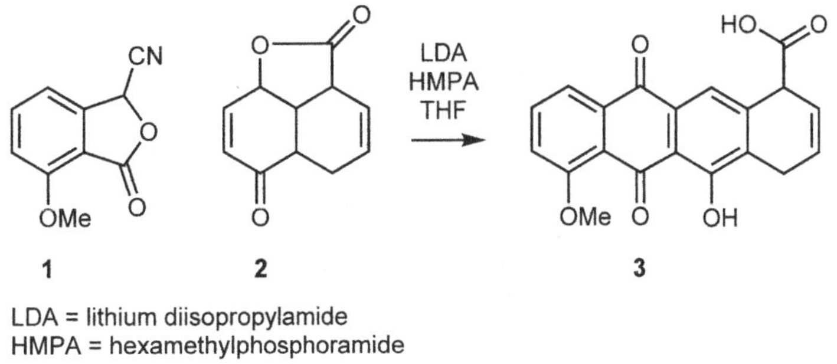

CN
O
OMe
1
2
LDA
HMPA
THF
HO
OH
3
LDA = lithium diisopropylamide
HMPA = hexamethylphosphoramide

SCHEME 1

SOLUTION

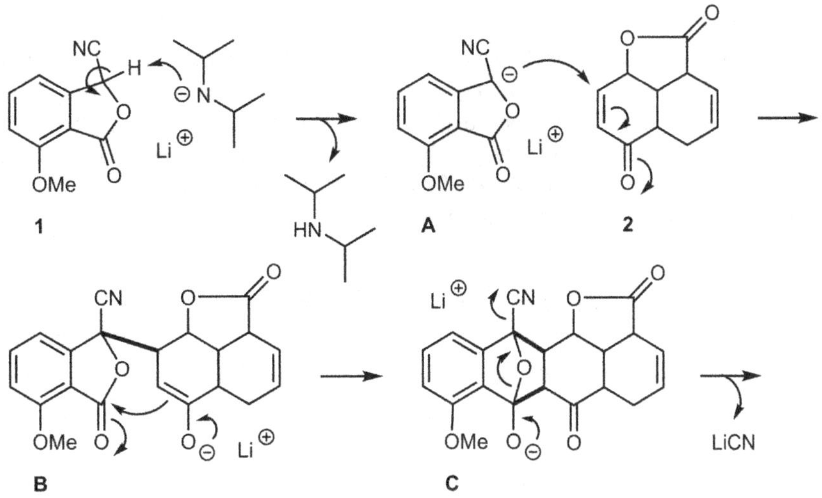

NC
H
N
Li
OMe
1
HN
A
2
CN
B
C
LiCN

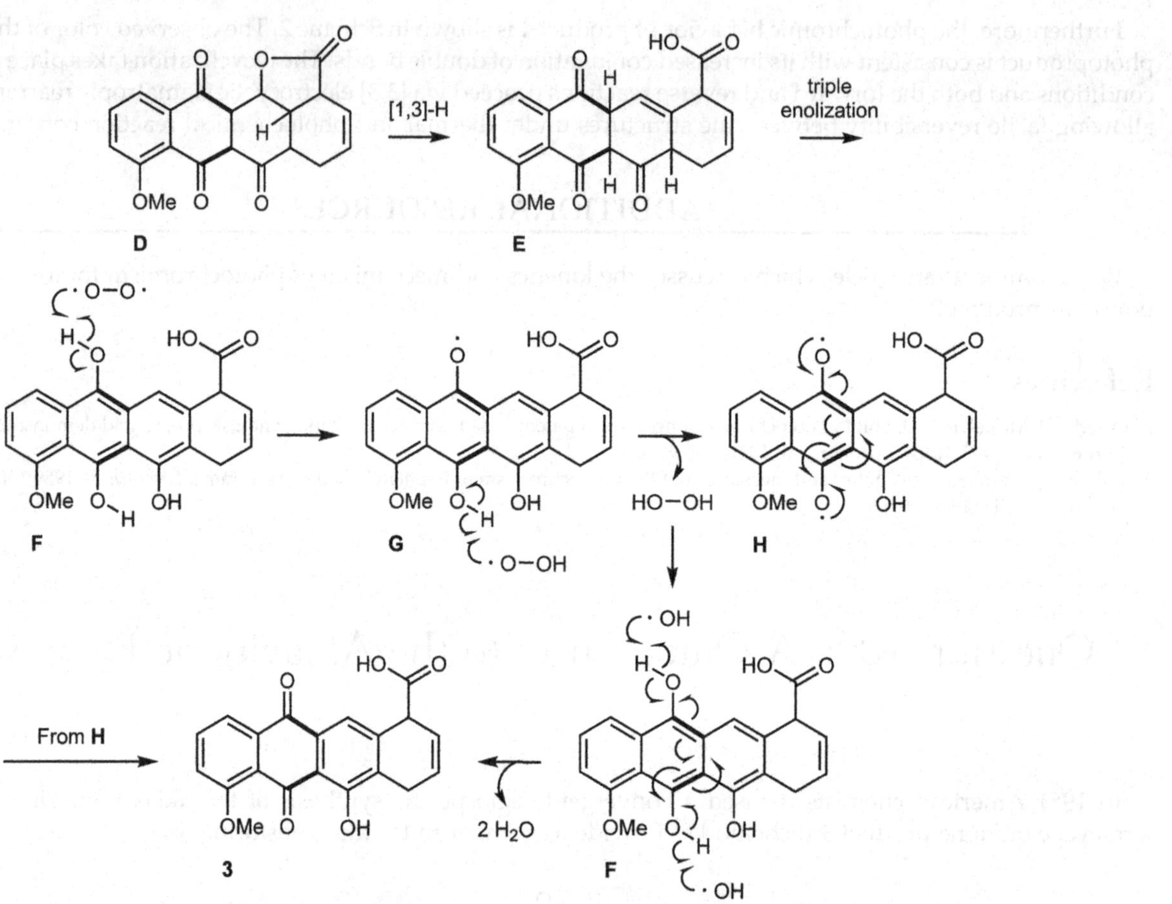

BALANCED CHEMICAL EQUATION

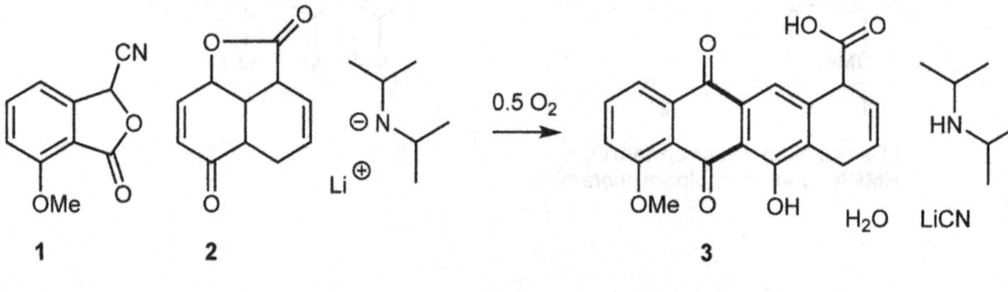

KEY STEPS EXPLAINED

The proposed mechanism for the transformation of **1** and **2** to **3** begins with deprotonation of the hydrogen atom alpha to the cyano group in **1** to form carbanion **A** which then nucleophilically adds to lactone **2** in a Michael fashion yielding anion **B**. Intermediate **B** then ring closes to give the bicyclic [2.2.1] anion intermediate **C** which ring opens to **D** and eliminates lithium cyanide by-product. Lactone **D** undergoes a [1,3]-protic shift that results in lactone ring opening to the carboxylic acid **E**. Intermediate **E** undergoes three successive enolization steps to form hydroquinone **F**. A radical-based sequence of steps oxidizes the hydroquinone to the quinone product **3** via an autoxidation process. Water is the resulting by-product from this sequence of steps. We note that the authors did not provide additional experimental and theoretical evidence to support this mechanism.

ADDITIONAL RESOURCES

We recommend an article by Japanese authors which discusses synthetic approaches to aklavinones such as product **3**.[2]

References

1. Li T, Walsgrove TC. A convergent regiospecific route to the aklavinone ring system. *Tetrahedron Lett.* 1981;22(38):3741–3744. https://doi.org/10.1016/S0040-4039(01)82008-6.
2. Uno H, Naruta Y, Maruyama K. Syntheses of (±)-aklavinones: application of the stereocontrolled "zipper" bicyclocyclization reaction. *Tetrahedron.* 1984;40(22):4725–4741. https://doi.org/10.1016/S0040-4020(01)91535-2.

Question 110: A "Stable Enol" That Doesn't Exist

In 1975 Dabral et al. claimed that the hydrolysis and decarboxylation of the β-keto ester **1** produced the stable enol product **3** (Scheme 1).[1] The claim was based largely on spectroscopic data but a subsequent report showed that the actual product of the reaction was the carboxylic acid **4**. Suggest a mechanism for the formation of **4**.

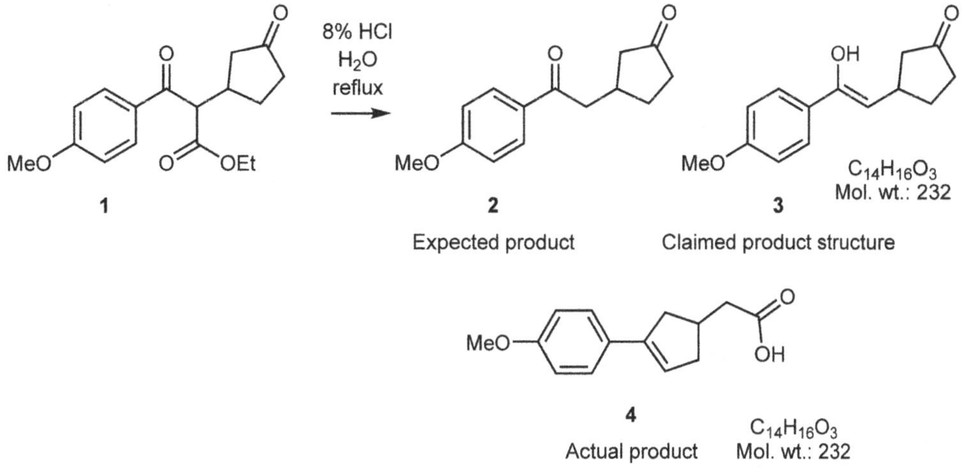

SCHEME 1

SOLUTION

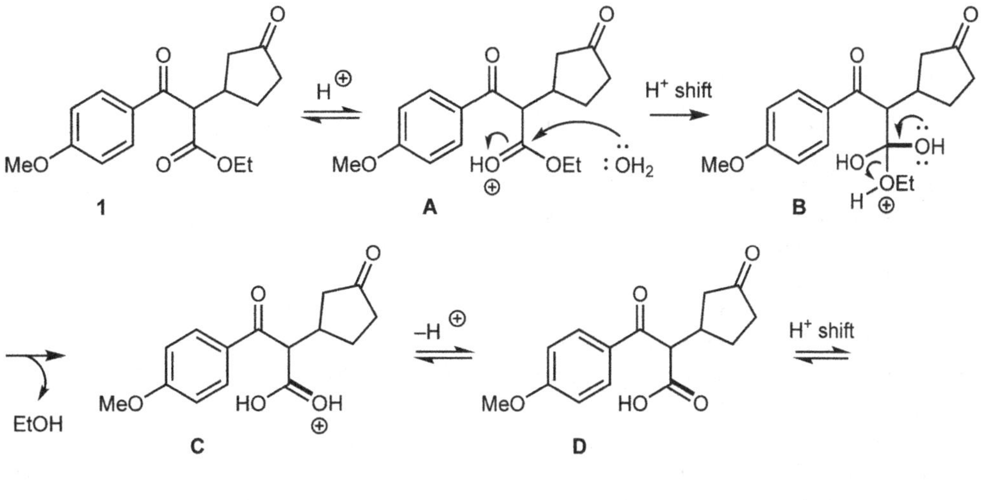

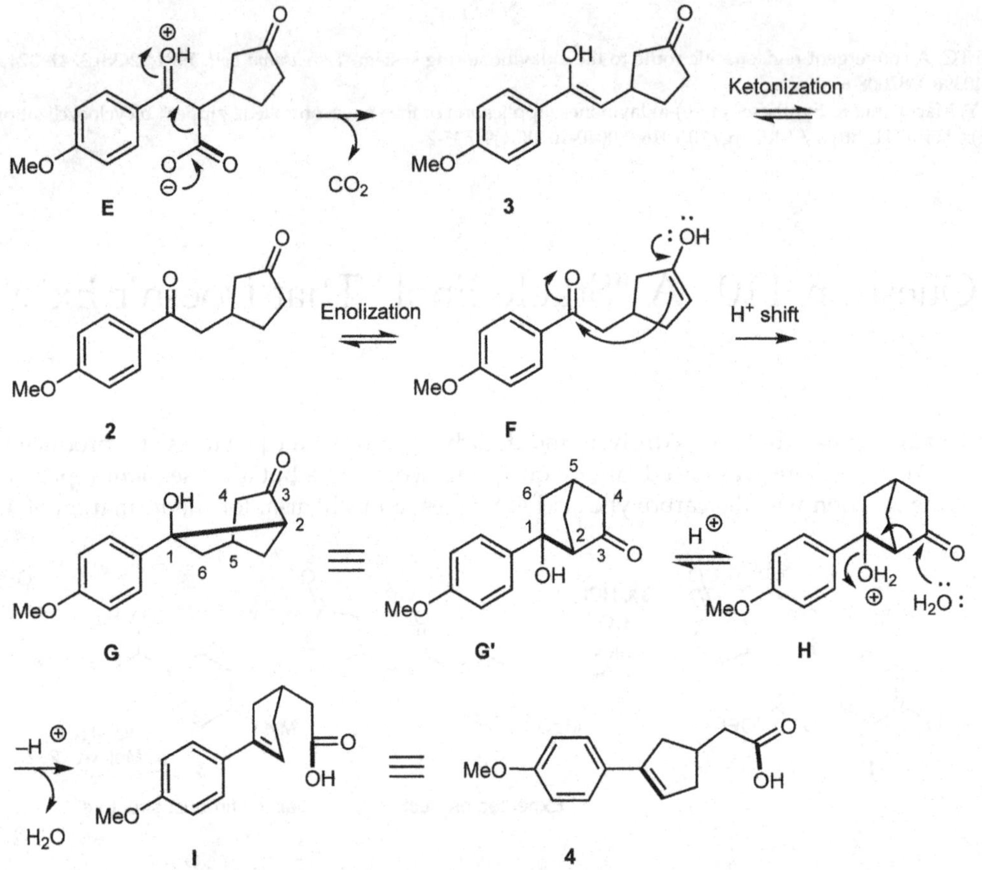

BALANCED CHEMICAL EQUATION

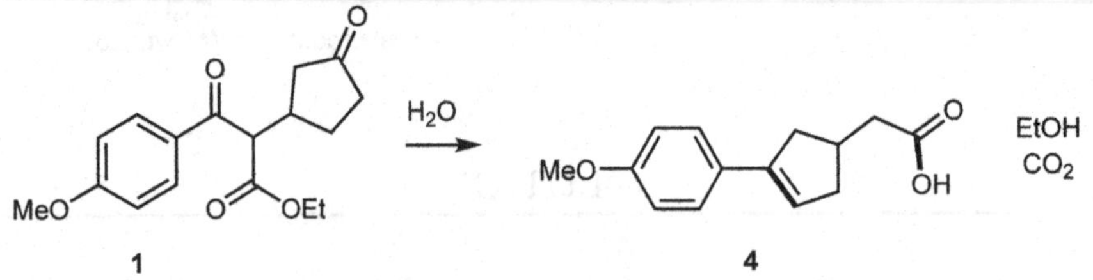

KEY STEPS EXPLAINED

We note that the mechanism for the transformation of **1** to the actual product **4** begins with formation of the expected diketone product **2** which occurs via ketonization of enol **3** following hydrolysis and decarboxylation via intermediates **A** through **E**. Nevertheless, **2** undergoes further enolization to enol **F** which then cyclizes to ketol **G**. At this point, the hydroxyl group of **G** at C_1 is protonated and water attack at the neighboring carbonyl group causes fragmentation of the C_2—C_3 bond which after elimination of water leads to the carboxylic acid product **4** with an olefinic bond between C_1 and C_2.

Spectroscopically, the IR spectrum of **4** had a broad band between 3000 and 3500 cm^{-1} (**COOH**) and a sharp peak at 1705 cm^{-1} (**C=O**). The ^{1}H NMR spectrum of **4** indicated absorptions at 5.95 ppm (1H, broad singlet, **=CH**) and at 11.1 ppm (1H, broad singlet, COOH). The latter absorption was exchangeable with D_2O. Originally, Dabral et al. ascribed the broad IR band to an enolic OH stretch and the broad exchangeable ^{1}H NMR signal at 11.1 ppm to an enolic hydrogen atom.[1] The correct spectroscopic assignments were made by Dave and Warnhoff who challenged the enolic structure **3** as the structure of the product of the acid-catalyzed hydrolysis reaction.[2] Specifically, they doubted the enol thermodynamic stability argument of the original authors who cited the existence of 1,2-dimesityl-1-propen-1-ol in support of their assignment (see Fig. 1). Dave and Warnhoff stated that the two dimesityl groups conferred sufficient

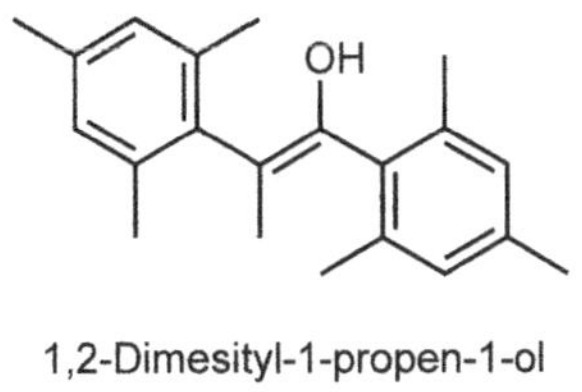

1,2-Dimesityl-1-propen-1-ol

FIG. 1   Structure of 1,2-dimesityl-1-propen-1-ol.

steric hindrance to prevent ketonization thus allowing for the existence of a stable enol. This special case was not operative in the reaction described by the original authors.

Furthermore, Dave and Warnhoff supplied the following additional pieces of evidence for the structural assignment of **4**: (a) the product of the reaction was converted to a methyl ester using diazomethane and this ester was characterized by mass spectrometry, IR, and ^{1}H NMR spectroscopies; and (b) the product of the reaction was synthesized by two independent routes shown in Scheme 2 and in each case was found to have identical mixed melting point and IR and ^{1}H NMR spectra as the product obtained from **1**. Nevertheless, the authors did not provide additional experimental and theoretical evidence to support their mechanism for the formation of product **4**.

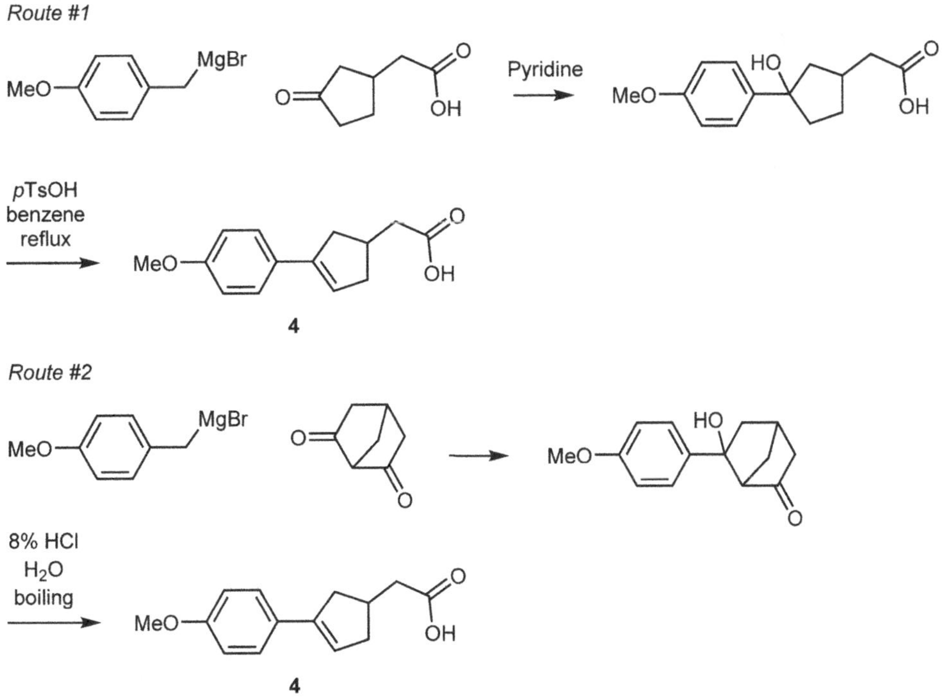

SCHEME 2

ADDITIONAL RESOURCES

For examples of isolable thermodynamically stable enols, we refer the reader to a chapter review by Hart et al.[3] We also recommend two articles which also cover the topic of stable enols.[4,5]

References

1. Dabral V, Ila H, Anand N. Decarbethoxylation of β-keto esters: formation of thermodynamically stable enols. *Tetrahedron Lett.* 1975;16 (52):4681–4684. https://doi.org/10.1016/S0040-4039(00)91051-7.
2. Dave V, Warnhoff EW. On the putative formation of stable enols. *Tetrahedron Lett.* 1976;17(51):4695–4696. https://doi.org/10.1016/S0040-4039 (00)92998-8.
3. Hart H, Rappoport Z, Biali SE. Isolable and relatively stable simple enols. In: Rappoport Z, ed. *The Chemistry of Enols.* Chichester: Wiley; 1990:481–589. https://doi.org/10.1002/9780470772294.ch8 [chapter 8].

4. Rappoport Z, Frey J, Sigalov M, Rochlin E. Recent advanced in the chemistry of stable simple enols. *Pure Appl Chem*. 1997;69(9):1933–1940. https://doi.org/10.1351/pac199769091933.
5. Chenoweth DM, Chenoweth K, Goddard III WA. Lancifodilactone G: insights about an unusually stable enol. *J Org Chem*. 2008;73:6853–6856. https://doi.org/10.1021/jo8012385.

Question 111: Conversion of an Indole-Based Bicyclo[5.3.1]undecane Into a Bicyclo[5.4.0]undecane

During a 1981 study of the rearrangement of vobasiane, French authors reacted *N*-oxide **1** with trifluoroacetic anhydride and then with sodium borohydride to form the tetracyclic indole derivative **2** (Scheme 1).[1] Provide a mechanism for this transformation.

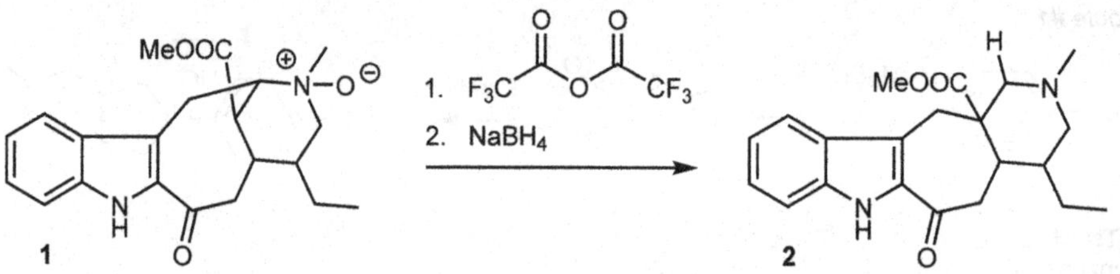

SCHEME 1

SOLUTION

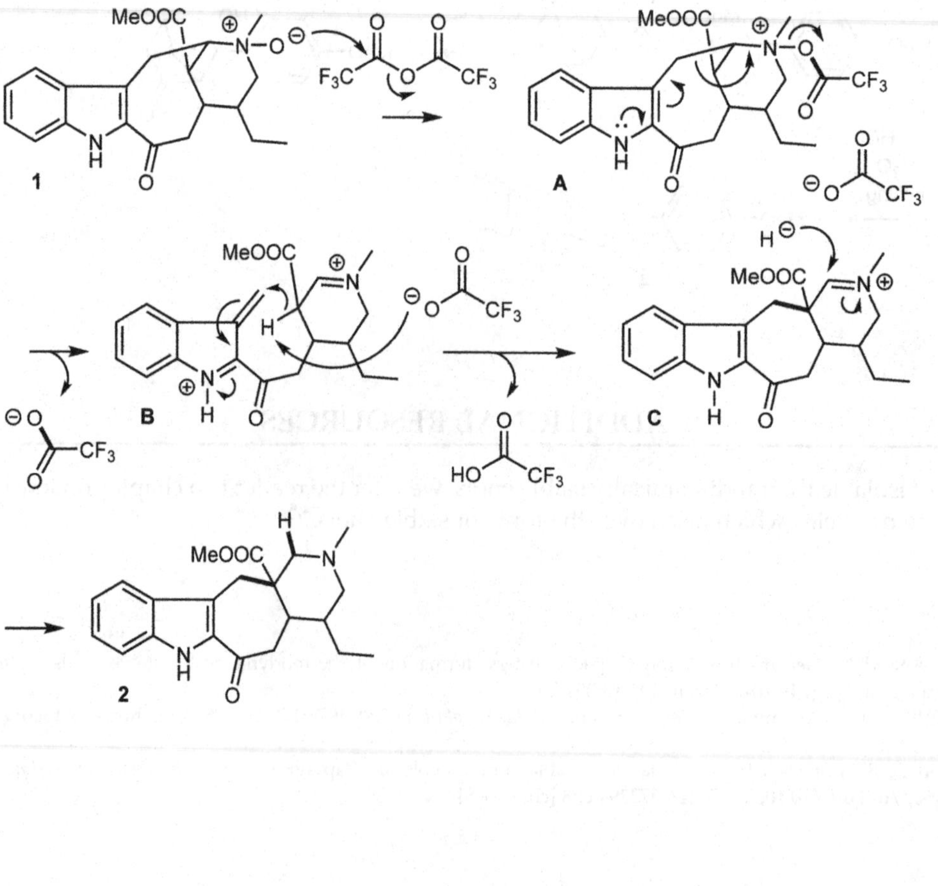

BALANCED CHEMICAL EQUATION

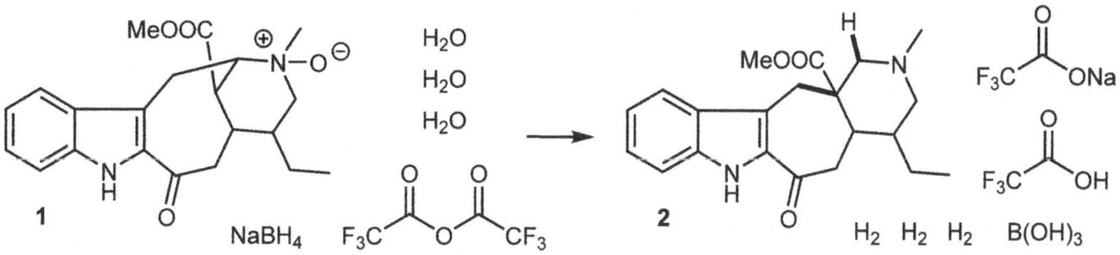

KEY STEPS EXPLAINED

The mechanism proposed by the authors begins with acetylation of the *N*-oxide group of **1** via trifluoroacetic anhydride yielding intermediate **A**. This structure undergoes ring opening to form intermediate **B** which contains two iminium ion groups. Trifluoroacetate then deprotonates **B** causing a ring closure onto the exocyclic olefinic group creating a central seven-membered ring shown in intermediate **C**. Hydride ion attack at the iminium carbon atom of **C** leads to product **2**. The ring construction mapping for the seven-membered ring is [7+0]. We note that the authors referred to the transformation as the modified Polonovski reaction but did not provide any further experimental or theoretical evidence to support the postulated mechanism. Indeed, the primary focus of their research was to see if they could effect the same transformation using liver enzyme catalysis of **1** to form **2**.

ADDITIONAL RESOURCES

We recommend two additional articles which highlight the mechanisms of very similar transformations.[2,3]

References

1. Thal C, Dufour M, Potier P, Jaouen M, Mansuy D. Rearrangement of vobasine to ervatamine-type alkaloids catalyzed by liver microsomes. *J Am Chem Soc.* 1981;103(16):4956–4957. https://doi.org/10.1021/ja00406a055.
2. Husson A, Langlois Y, Riche C, Husson H-P, Potier P. Etudees en serie indolique-VI transformation des alcaloides du type vobasine en alcaloides du type dehydroervatamine. Analyse aux rayons X de l'ervatamine. *Tetrahedron.* 1973;29(19):3095–3098. https://doi.org/10.1016/S0040-4020(01)93449-0.
3. Amat M, Llor N, Checa B, Molins E, Bosch J. A synthetic approach to ervatamine-silicine alkaloids. Enantioselective total synthesis of (-)-16-epi-silicine. *J Org Chem.* 2010;75:178–189. https://doi.org/10.1021/jo902346j.

Question 112: A Classical Mannich Approach to Isoquinoline Alkaloids Leads to an Unexpected Product

In 1987 German chemists synthesized the tetraspiroketone **1** and sought to investigate whether it underwent acid-catalyzed rearrangement preferentially or exclusively to the bispropellanone **2** or to the pentacyclic ketone **3** (Scheme 1).[1] Show mechanisms for both possible transformations.

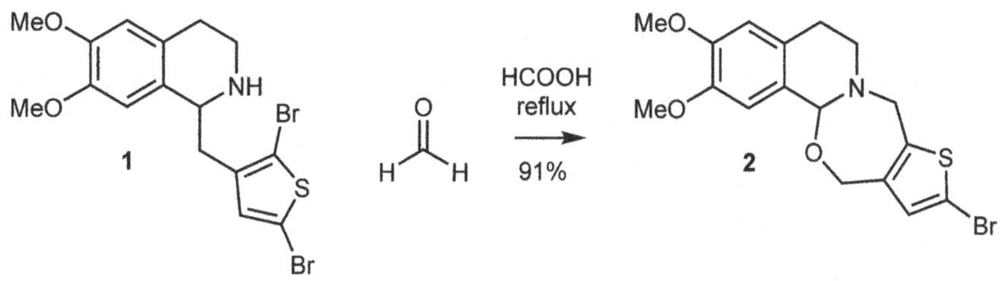

SCHEME 1

SOLUTION

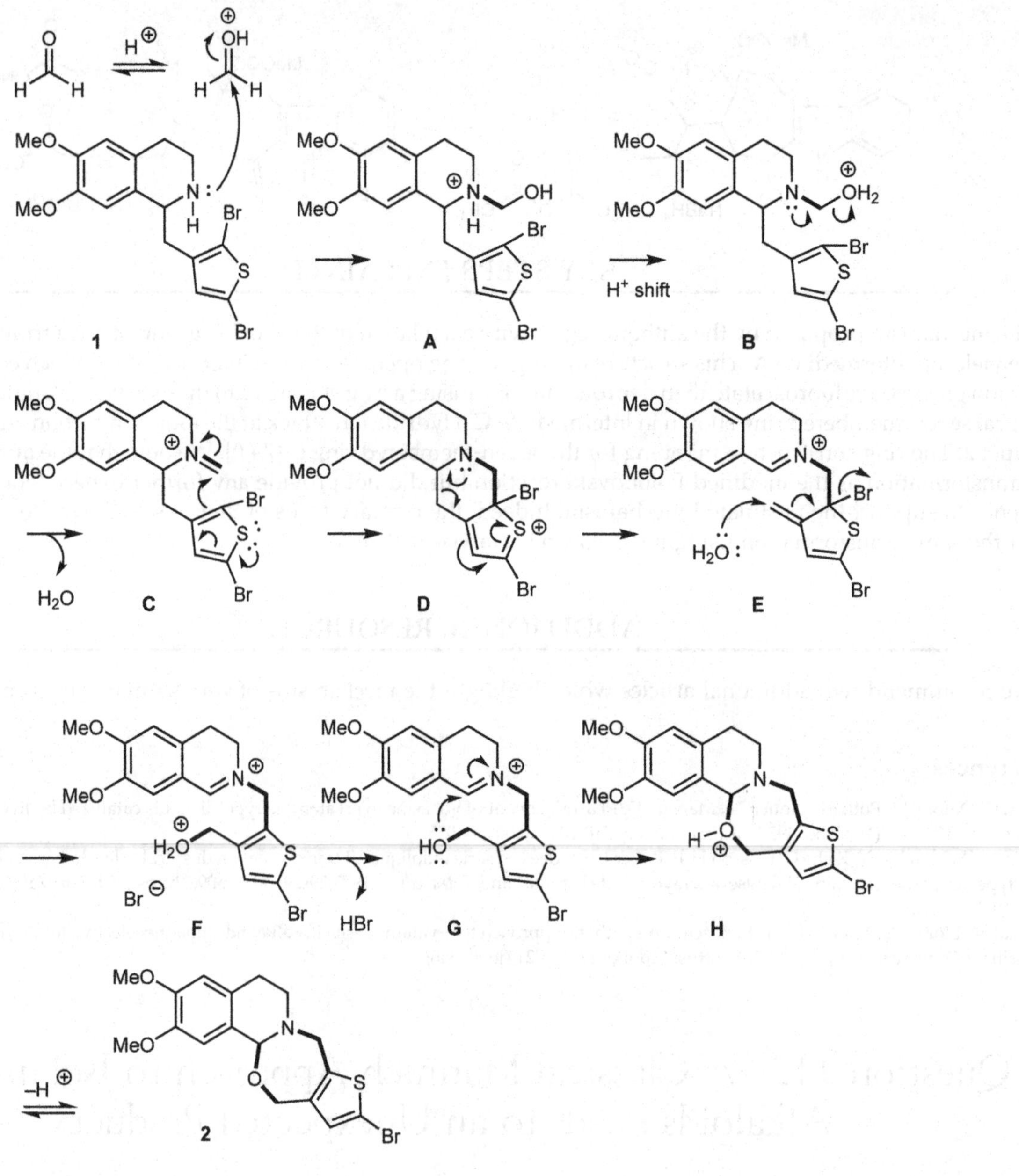

BALANCED CHEMICAL EQUATION

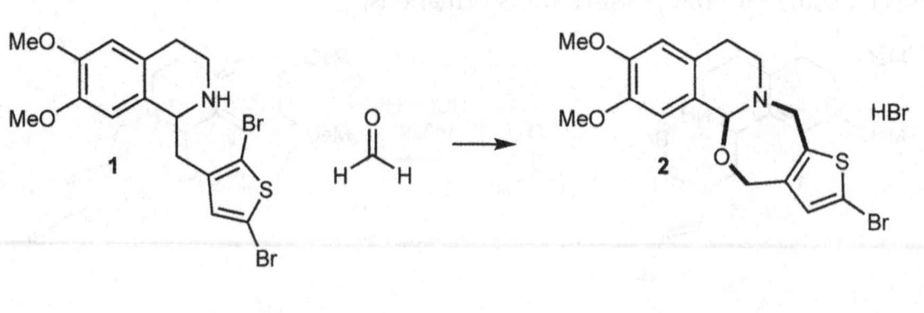

KEY STEPS EXPLAINED

The mechanism for this transformation begins with nucleophilic attack by the amine nitrogen atom onto the protonated formaldehyde giving intermediate **A** which undergoes intramolecular proton transfer to **B** followed by dehydration to form iminium ion **C**. **C** then undergoes ring closure to **D** which ring opens to iminium ion **E** having an exocyclic olefinic group. Intermediate **E** is then trapped by water concomitant with bromide ion elimination yielding iminium ion **F**. Hydrobromic acid elimination leads to iminium ion **G** which cyclizes to the seven-membered ring shown in intermediate **H**. Final deprotonation leads to product **2**. We note that the ring construction mapping for the seven-membered ring is [3+1+2+1]. Although this problem is very similar to problem #116, we note that the authors did not provide additional experimental or theoretical evidence to support the postulated mechanism.

ADDITIONAL RESOURCES

We recommend two articles which discuss the use of the Mannich reaction to access isoquinoline derivatives.[2,3] We also refer the reader to an article by the same authors as for the source article to this problem which discusses the synthesis of thiophene isosters of protoberberine alkaloids.[4]

References

1. Jeganathan S, Srinivasan M. Studies on heterocyclic compounds; IV. A novel synthesis of 5,6,8,12-tetrahydro-13aH-thieno[2',3':5,6][1,3]oxazepino [2,3-a]isoquinolines. *Synthesis.* 1980;1980(12):1021–1022. https://doi.org/10.1055/s-1980-29307.
2. Dubs C, Hamashima Y, Sasamoto N, et al. Mechanistic studies on the catalytic asymmetric Mannich-type reaction with dihydroisoquinolines and development of oxidative Mannich-type reactions starting from tetrahydroisoquinolines. *J Org Chem.* 2008;73:5859–5871. https://doi.org/10.1021/jo800800y.
3. Su S, Porco Jr. JA. 1,2-Dihydroisoquinolines as templates for cascade reactions to access isoquinoline alkaloid frameworks. *Org Lett.* 2007;9 (24):4983–4986. https://doi.org/10.1021/ol702176h.
4. Jeganathan S, Srinivasan M. Studies on heterocyclic compounds VI: synthesis of thiophene isosters of protoberberine alkaloids. *Phosphorus Sulfur Silicon Relat Elem.* 1981;11(2):125–137. https://doi.org/10.1080/03086648108077411.

Question 113: Reaction of Benzothiazole With DMAD

In 1975 McKillop and Sayer investigated the reaction of equimolar amounts of benzothiazole and dimethyl acetylenedicarboxylate (DMAD) in hot aqueous methanol to give a colorless crystalline solid which had structure **1** based on IR and ^{1}H NMR data (Scheme 1).[1] Suggest a mechanism for this transformation.

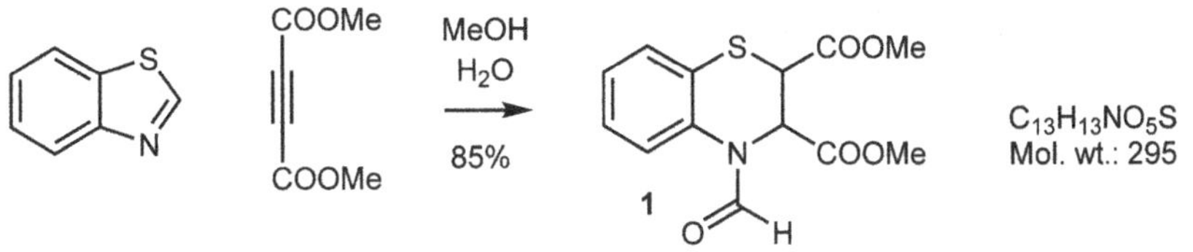

SCHEME 1

SOLUTION

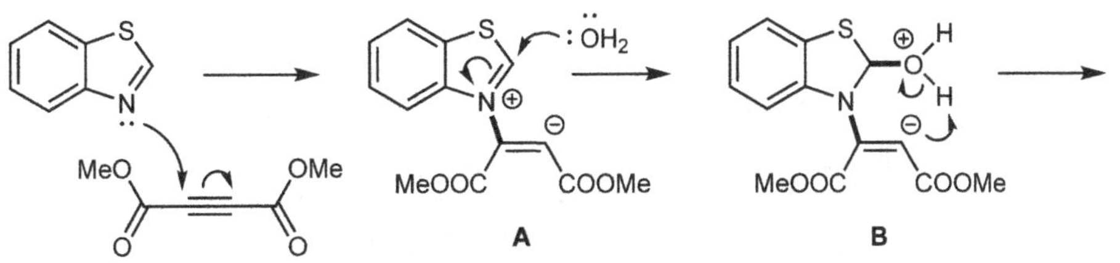

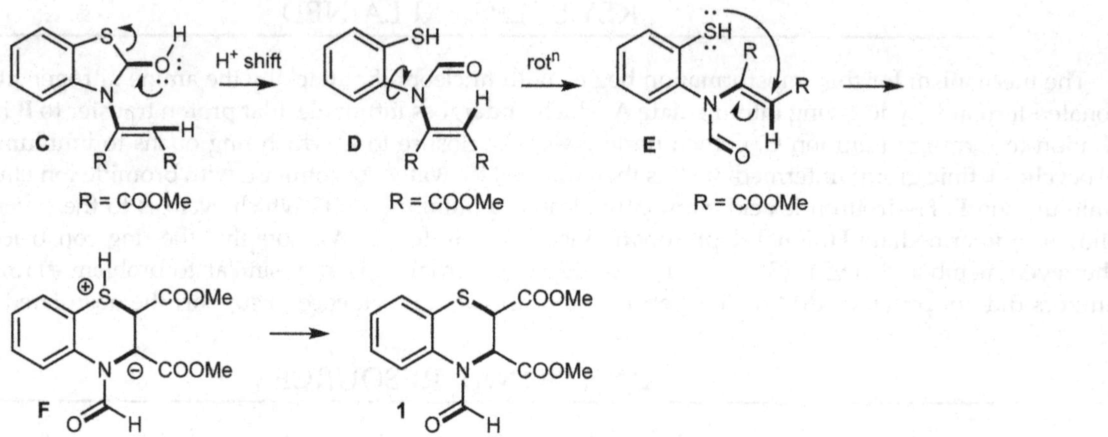
H⁺ shift
rotⁿ
C
D
E
F
1
R = COOMe
COOMe

KEY STEPS EXPLAINED

The mechanism proposed by the authors to explain the transformation of benzothiazole and DMAD to **1** begins with nucleophilic addition of benzothiazole to the alkyne moiety of DMAD to form zwitterionic intermediate **A**. Water from solution can at this point add to the central carbon atom of the benzothiazole moiety of **A** to form zwitterionic intermediate **B** which undergoes intramolecular proton transfer to generate **C**. Intermediate **C** then ring opens to produce aldehyde **D** after intramolecular proton transfer. What follows is bond rotation about ArC—N bond which leads to **E**. Cyclization of **E** to **F** followed by intramolecular proton transfer generates the observed product **1**. It is important to note the authors observed that in the absence of water, the transformation proceeds along a different path in which intermediate **A** is trapped by a second equivalent of DMAD to give product **2** (Scheme 2). The authors also surmised that substitution of water for a stronger nucleophile can allow trapping of intermediate **A** to open up avenues of synthetic utility for this transformation. It was also observed that the use of excess water led to the quantitative formation of product **1** along with the simultaneous suppression of product **2**.

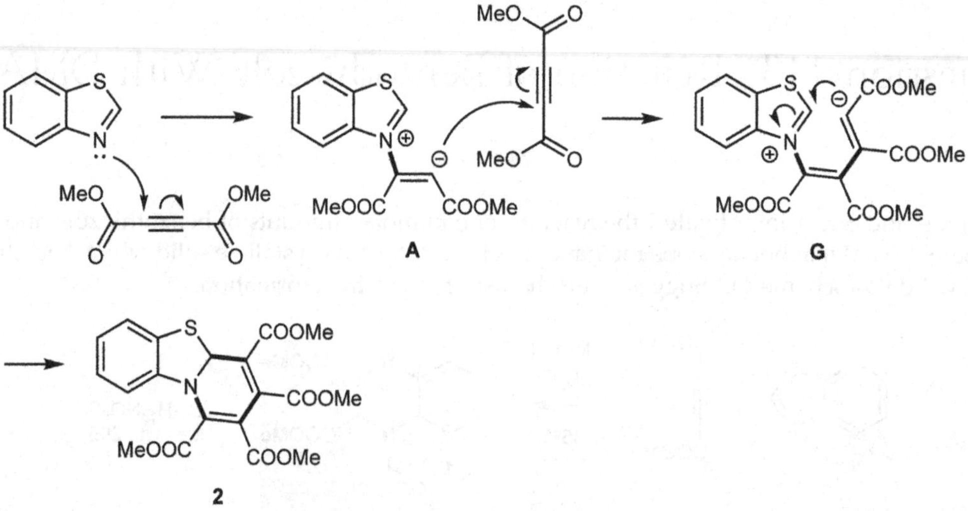
A
G
2
MeOOC
COOMe

SCHEME 2

Furthermore, since previous reports had incorrectly identified the structure of product **1**, the authors characterized **1** using elemental analysis, IR, and ^{1}H NMR. The IR bands at 1745, 1740, and 1690 cm^{-1} correspond to the carbonyl stretching frequencies of the two ester groups and the aldehyde group, respectively. The ^{1}H NMR assignments are shown in Fig. 1.

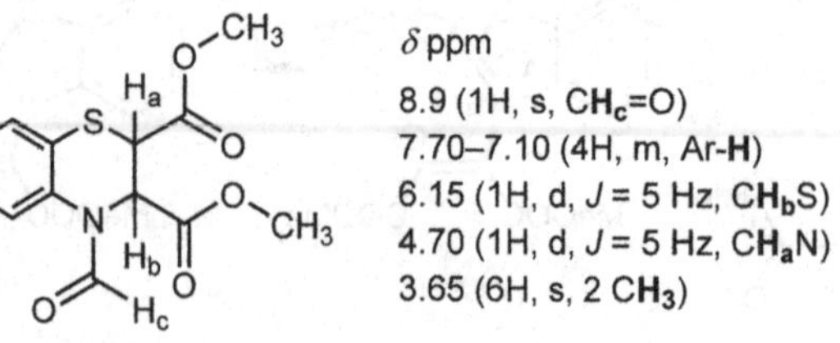
δ ppm
8.9 (1H, s, CH$_c$=O)
7.70–7.10 (4H, m, Ar-H)
6.15 (1H, d, J = 5 Hz, CH$_b$S)
4.70 (1H, d, J = 5 Hz, CH$_a$N)
3.65 (6H, s, 2 CH$_3$)

FIG. 1 ^{1}H NMR assignments of product 1.

ADDITIONAL RESOURCES

We recommend two articles which further explore the synthetic applicability of DMAD reactions with heterocycles.[2,3]

References

1. McKillop A, Sayer TSB. Elucidation of the reactions of benzothiazole with dimethyl acetylenedicarboxylate. *Tetrahedron Lett.* 1975;16 (35):3081–3084. https://doi.org/10.1016/S0040-4039(00)75079-9.
2. Yavari I, Piltan M, Moradi L. Synthesis of pyrrolo[2,1-a]isoquinolines from activated acetylenes, benzoylnitromethanes, and isoquinoline. *Tetrahedron.* 2009;65:2067–2071. https://doi.org/10.1016/j.tet.2009.01.001.
3. Pillai AN, Devi BR, Suresh E, Nair V. An efficient multicomponent protocol for the stereoselective synthesis of oxazinobenzothiazole derivatives. *Tetrahedron Lett.* 2007;48:4391–4393. https://doi.org/10.1016/j.tetlet.2007.04.096.

Question 114: Inhibition of the L-DOPA to L-Dopaquinone Oxidation

In 1991 Italian chemists realized that the reaction of phloroglucinol and L-DOPA under ferricyanide oxidation conditions in buffer solution led to the formation of small amounts of tetracyclic product **1** (Scheme 1).[1] Devise a mechanism to explain the course of this reaction.

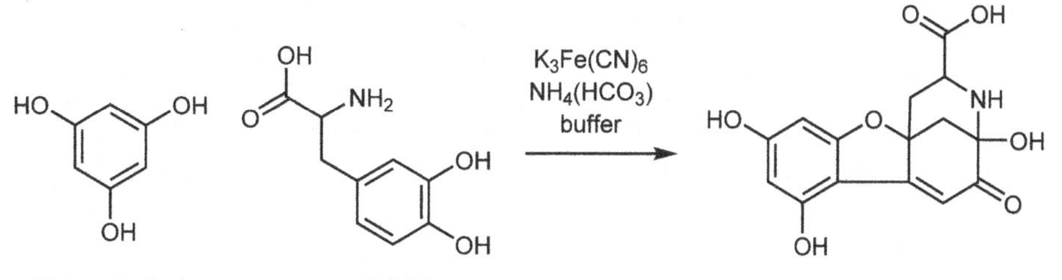

Phloroglucinol   L-DOPA   **1** (5%)

SCHEME 1

SOLUTION

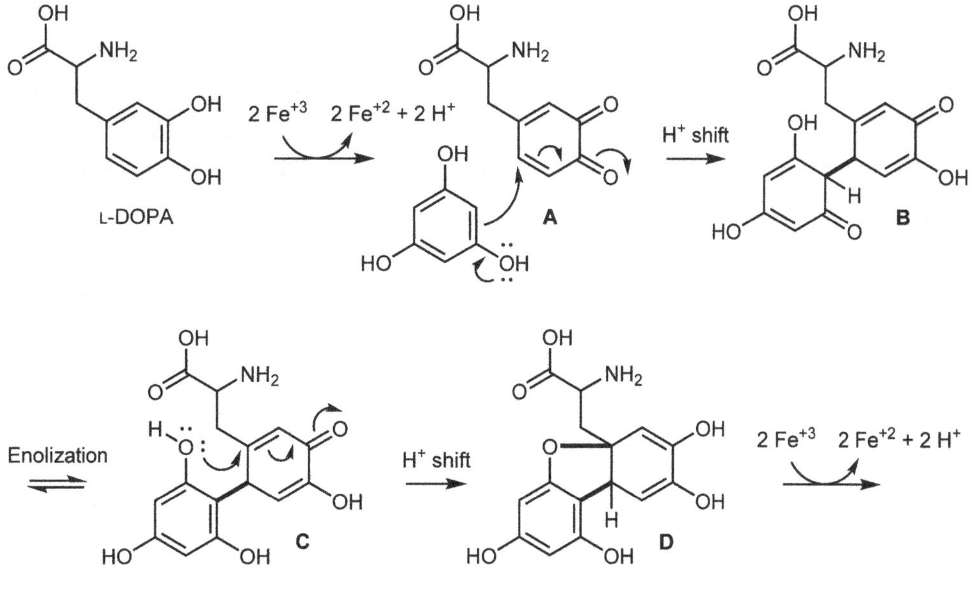

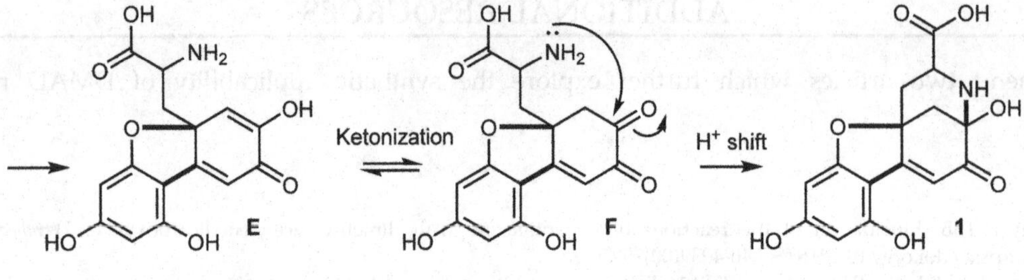

BALANCED CHEMICAL EQUATION

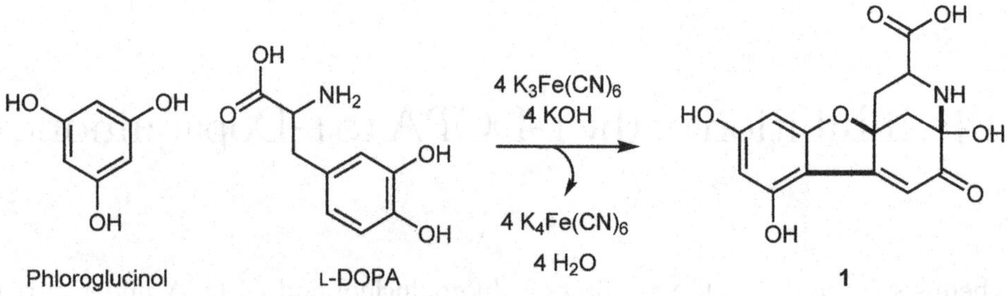

KEY STEPS EXPLAINED

The mechanism envisaged by the authors to explain the transformation of **1** from reaction of L-DOPA with phloroglucinol begins with oxidation of the catechol group of L-DOPA to form *o*-benzoquinone **A** via two equivalents ferricyanide. Phloroglucinol then adds in a nucleophilic manner to the *para* position of **A** to generate **B** which enolizes to **C** with rearomatization. **C** then cyclizes to form the cyclic ether **D** which is oxidized via two more equivalents of ferricyanide to quinoid **E**. Ketonization of **E** leads to dione **F** which cyclizes to product **1**. Although the authors characterized 2-carboxy-5-oxo-4,7,9-trihydroxy-2,3,4,5-tetrahydro-1*H*-4,11a-methanobenzofuro[2,3-*d*]azocine product **1** by ^{1}H and ^{13}C NMR spectroscopy, they did not provide additional experimental and theoretical evidence to support the postulated mechanism.

ADDITIONAL RESOURCES

In subsequent years, several articles appeared which analyze from a theoretical perspective the mechanisms of biological conversions similar to the one encountered here.[2,3]

References

1. Crescenzi O, Napolitano A, Prota G, Peter MG. Oxidative coupling of DOPA with resorcinol and phloroglucinol: isolation of adducts with an unusual tetrahydromethanobenzofuro[2,3-*d*]azocine skeleton. *Tetrahedron*. 1991;47(32):6243–6250. https://doi.org/10.1016/S0040-4020(01)86556-X.
2. Inoue T, Shiota Y, Yoshizawa K. Quantum chemical approach to the mechanism for the biological conversion of tyrosine to dopaquinone. *J Am Chem Soc*. 2008;130:16890–16897. https://doi.org/10.1021/ja802618s.
3. Prabhakar R, Siegbahn PEM. A theoretical study of the mechanism for the biogenesis of cofactor topaquinone in copper amine oxidases. *J Am Chem Soc*. 2004;126:3996–4006. https://doi.org/10.1021/ja034721k.

Question 115: A One-Pot Synthesis of 1,3,6,8-Tetramethyl-2,7-naphthyridine

In 1987 French chemists treated a mixture of acetyl chloride (1.6 mol) and aluminum chloride (0.3 mol) with *t*-butyl chloride (0.1 mol) at 35°C for half an hour followed by careful addition of the reaction mixture to liquid ammonia to form a 91% mixture of the naphthyridine **1** and 2,4,6-trimethylpyridine **2** (Scheme 1).[1] Suggest mechanisms for these transformations.

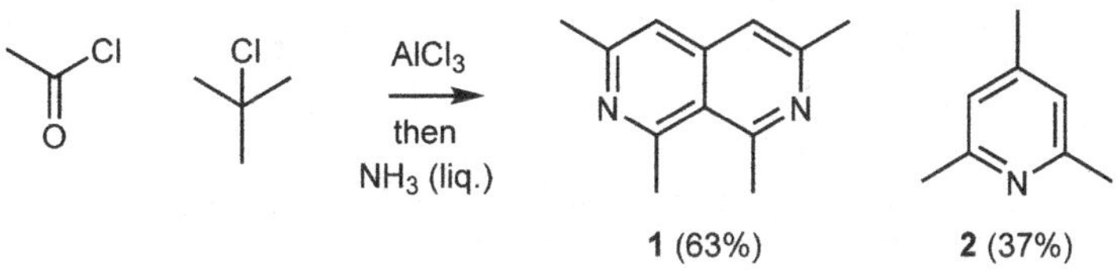

SCHEME 1

SOLUTION: FORMATION OF 1 PART 1 (A TO H)

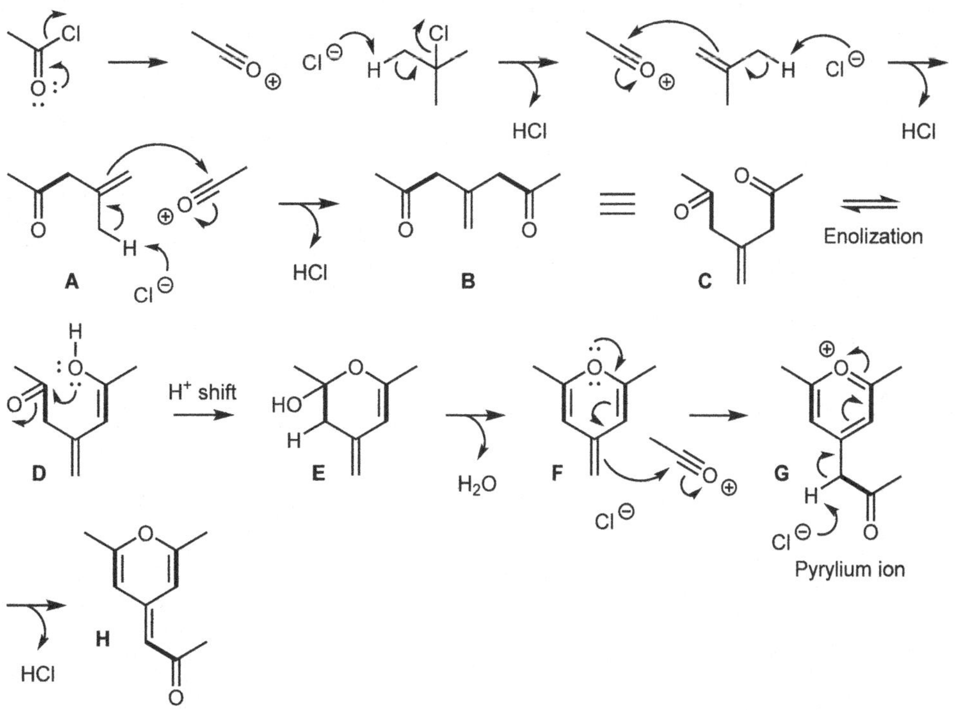

SOLUTION: FORMATION OF 1 PART 2 (H TO 1 PATH A)

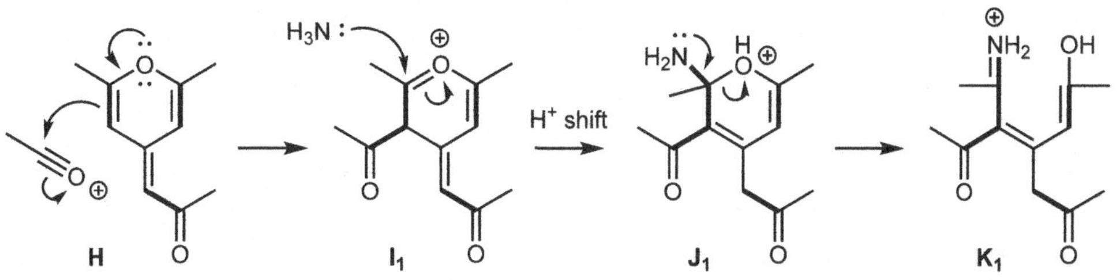

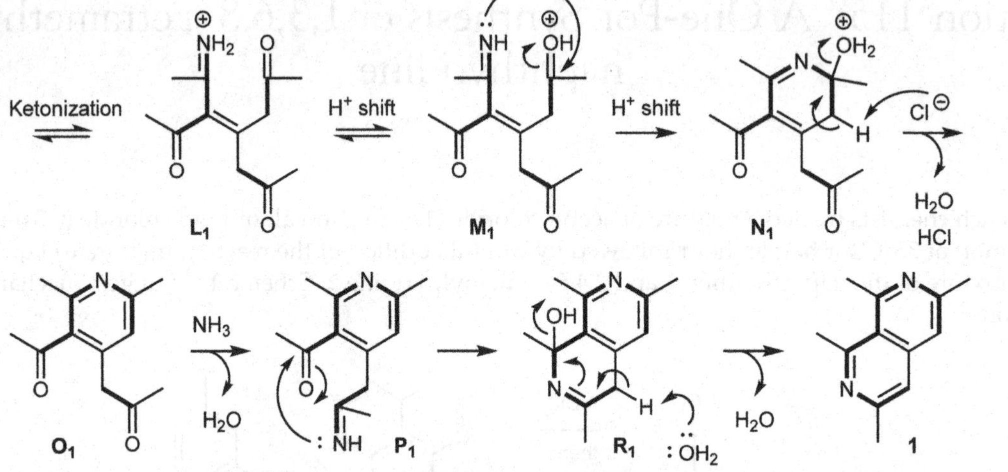

SOLUTION: FORMATION OF 1 PART 3 (H TO 1 PATH B)

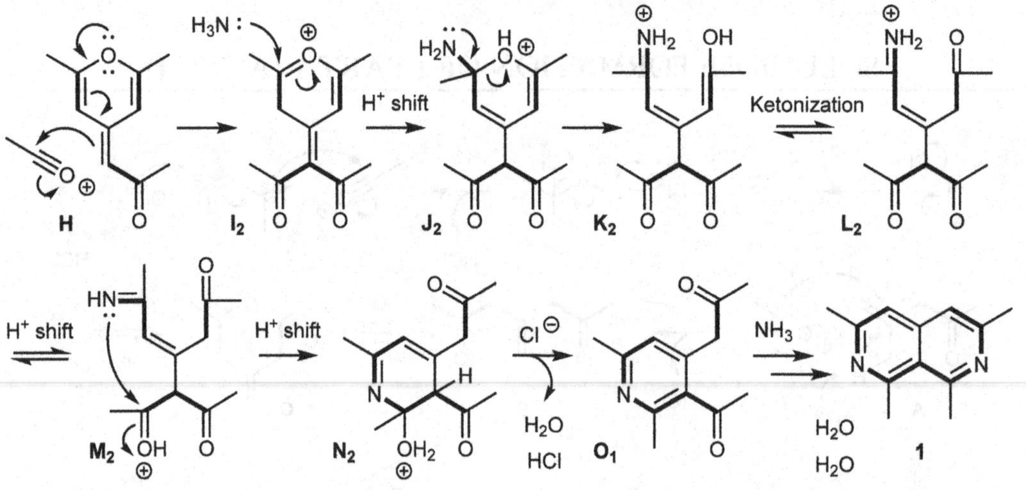

SOLUTION: FORMATION OF 1 PART 4 (ALTERNATE ROUTE: B TO O₁)

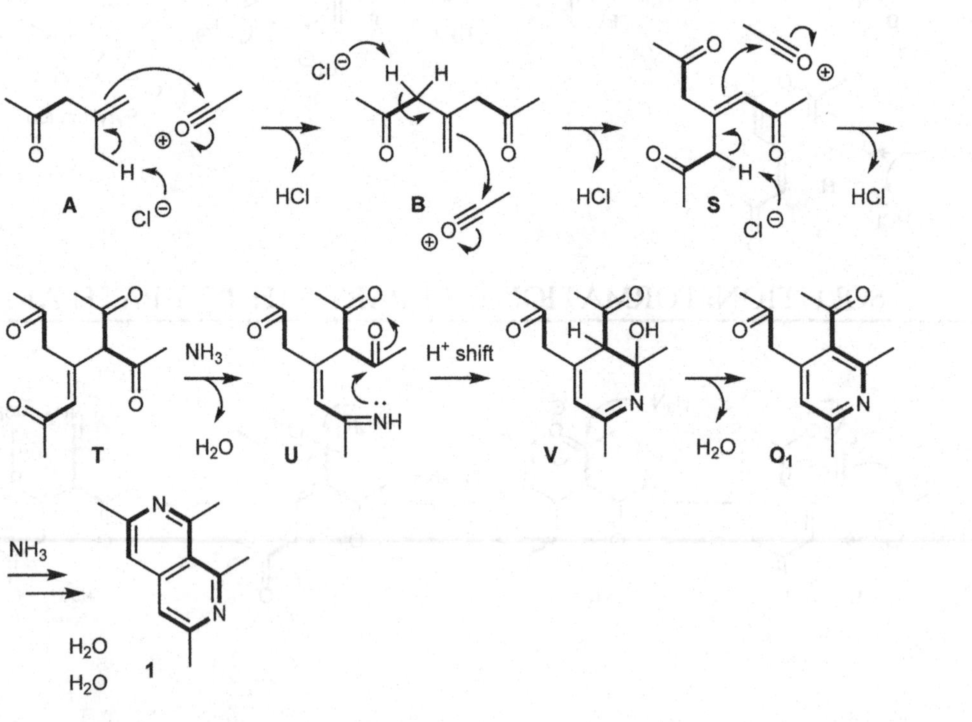

SOLUTION: FORMATION OF 1 PART 5 (ALTERNATE ROUTE: S TO H)

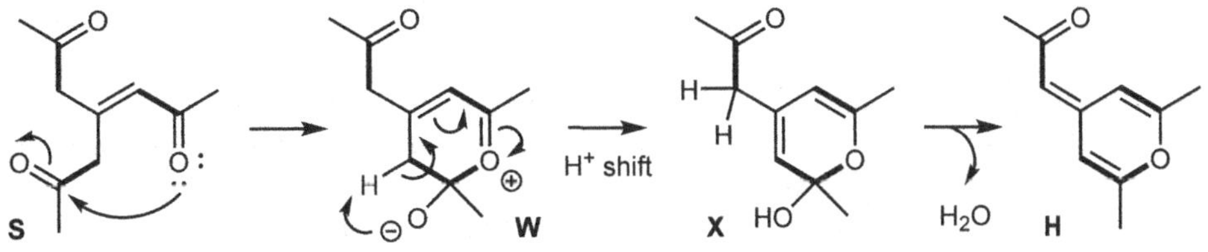

SOLUTION: FORMATION OF 2

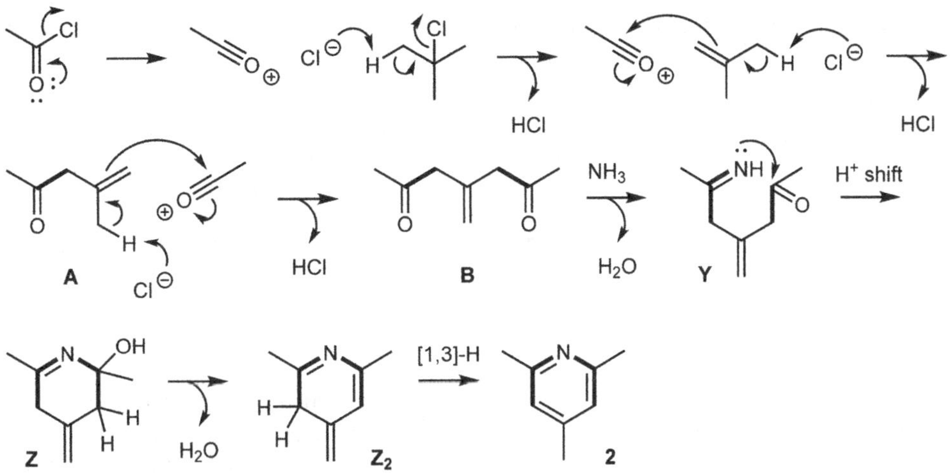

BALANCED CHEMICAL EQUATIONS

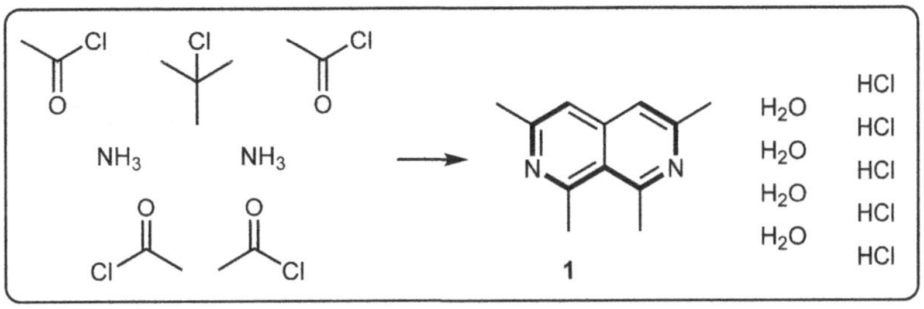

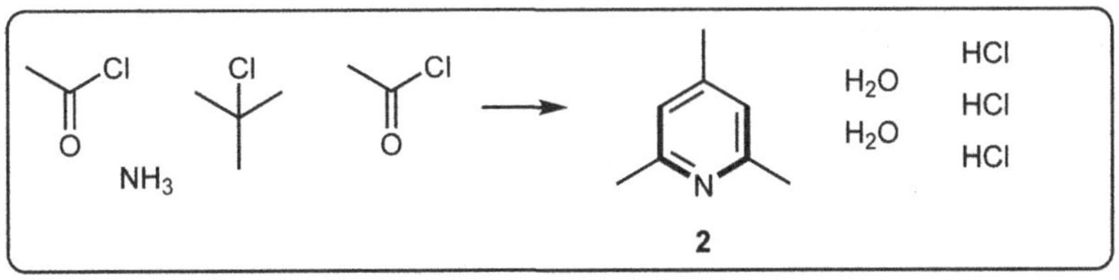

KEY STEPS EXPLAINED

The mechanism proposed by the authors to explain the formation of **1** begins with formation of methylacylium ion from acetyl chloride and isobutylene from *tert*-butyl chloride. These two intermediates react to produce 4-methyl-pent-4-en-2-one **A** with elimination of hydrochloric acid. **A** reacts with a second equivalent methylacylium ion to produce 4-methylene-heptane-2,6-dione **B** with elimination of HCl. **B** then undergoes an intramolecular aldol condensation to form **E** which eliminates a water molecule to form **F**. Intermediate **F** undergoes acylation to form the pyrylium ion intermediate **G** which is deprotonated to form **H**. From this point onward, the authors propose two possible pathways to product **1**. The two paths differ only in the position of acylation from **H** and involve analogous sequences of reactions with two equivalents of ammonia and elimination of three equivalents of water to form the final naphthyridine product **1**. We note that the ring construction mapping for this seven-component coupling reaction is [(3+1+1+1)+ (3+1+1+1)]. Experimentally, the authors proposed intermediate **H** and the two pathways to **1** after carrying out the reaction as before without the addition of the mixture to liquid ammonia. This experiment resulted in the formation of the tetraacylation products **3** and **4** whose precursors are quite clearly intermediates I_1 and I_2, respectively (Scheme 2). The authors noted that it was impossible to stop the reaction at the triacylation products. When the mixture of **3** and **4** was treated with cold aqueous ammonia, hydrolysis and deacylation gave intermediate **H**. When **3**, **4**, and **H** were treated with liquid ammonia, product **1** was isolated. This experimental procedure provides strong evidence for the latter part of the proposed mechanism from structure **H** to **1** via two possible pathways. Unfortunately, the authors did not provide experimental and theoretical evidence for the beginning part of the mechanism. As such we suggest an alternative mechanism which sees the possible intermediacy of a triacylation compound **S**. This intermediate could undergo an intramolecular aldol condensation, proton shift, and water elimination to form the familiar common intermediate **H** which can proceed to **1** as before. Intermediate **S** could also undergo a fourth acylation to form the tetraacylation linear intermediate **T** which could react with liquid ammonia to eventually form product **1**.

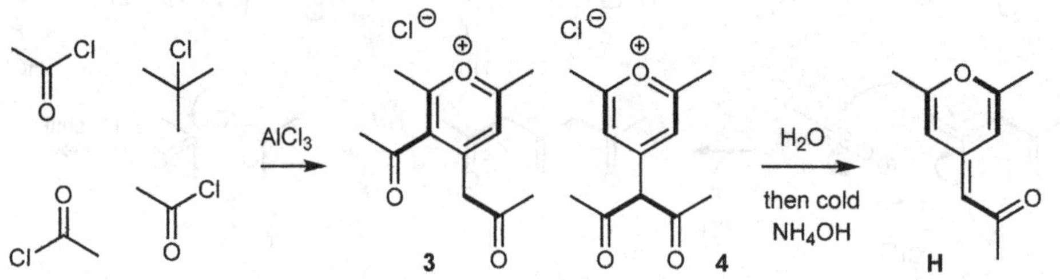

SCHEME 2

Unfortunately, due to the high degree of symmetry in the product structure, it is challenging to design experimental procedures to help distinguish the various possible combinations of transformations that lead from the common intermediate **B** to product **1**. We believe therefore that a theoretical study is the place to start should others undertake a thorough mechanistic investigation of this transformation in the future. To assist such an endeavor we can write out a summary mechanistic network showing the interconnections between key intermediates based on the present discussion (see Scheme 3). From this network we can count six mechanistic routes that can be explored from starting

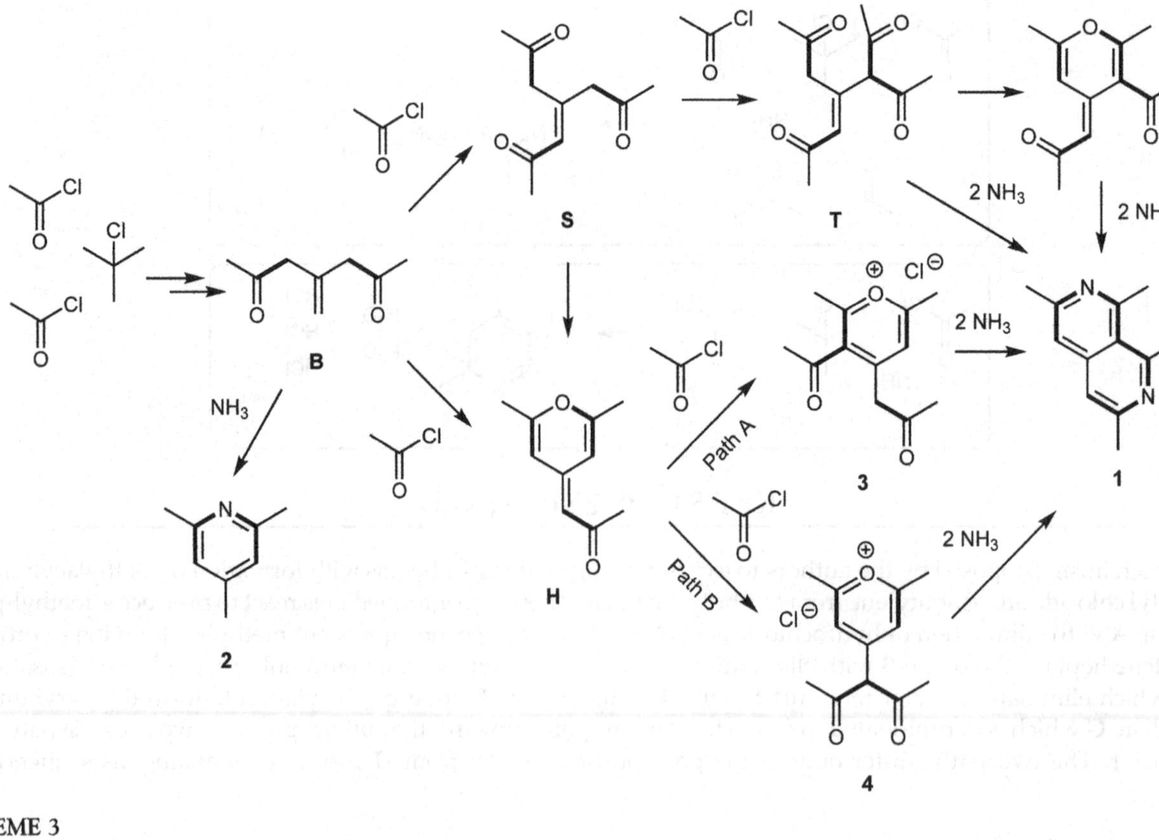

SCHEME 3

acetyl chloride and *t*-butyl chloride to product **1** which are consistent with the experimental observations described. Lastly, we note that a deuterium labeling experiment for this transformation appeared in 1984 which demonstrated the carbon skeleton building blocks for the naphthyridine product **1** (Scheme 4)[2].

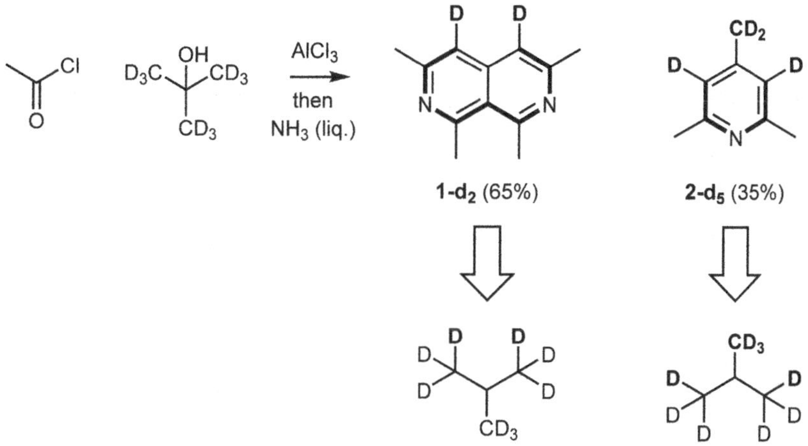

1-d₂ (65%)   **2-d₅ (35%)**

SCHEME 4

Concerning the formation of product **2**, the proposed mechanism involves reaction of intermediate **B** with ammonia followed by condensation, water elimination, and a [1,3]-protic shift to form the observed minor product **2**.

References

1. Roussel C, Mercier A, Cartier M. Regioselective synthesis of 1-alkyl-3,6,8-trimethyl-2,7-naphthyridines. *J Org Chem.* 1987;52(13):2935–2937. https://doi.org/10.1021/jo00389a056.
2. Erre C, Pedra A, Arnaud M, Roussel C. Tetraacylation of isobutene: first synthesis of 1,3,6,8-tetramethyl-2,7-naphthyridine. *Tetrahedron Lett.* 1984;25(5):515–518. https://doi.org/10.1016/S0040-4039(00)99925-8.

Question 116: Unexpected Course of a Mannich Reaction in Alkaloid Synthesis

In 1975 and 1979 Indian chemists analyzed the reaction of tetrahydroisoquinoline **1** and formaldehyde under acidic conditions which gives the isoquinobenzoxazepine **2** in 70%–85% yield depending on conditions (Scheme 1).[1,2] Provide a mechanism for this transformation.

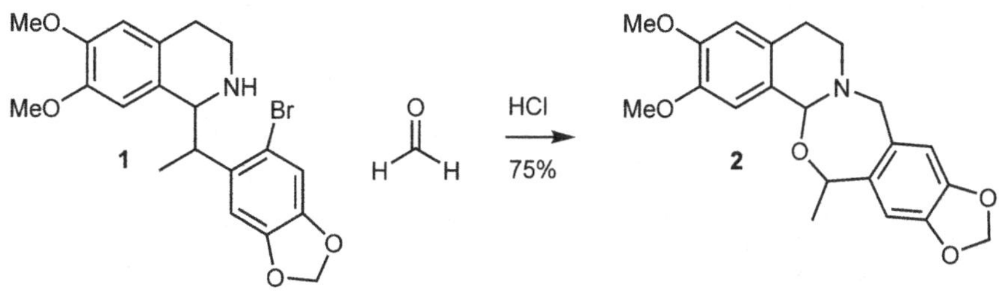

SCHEME 1

SOLUTION

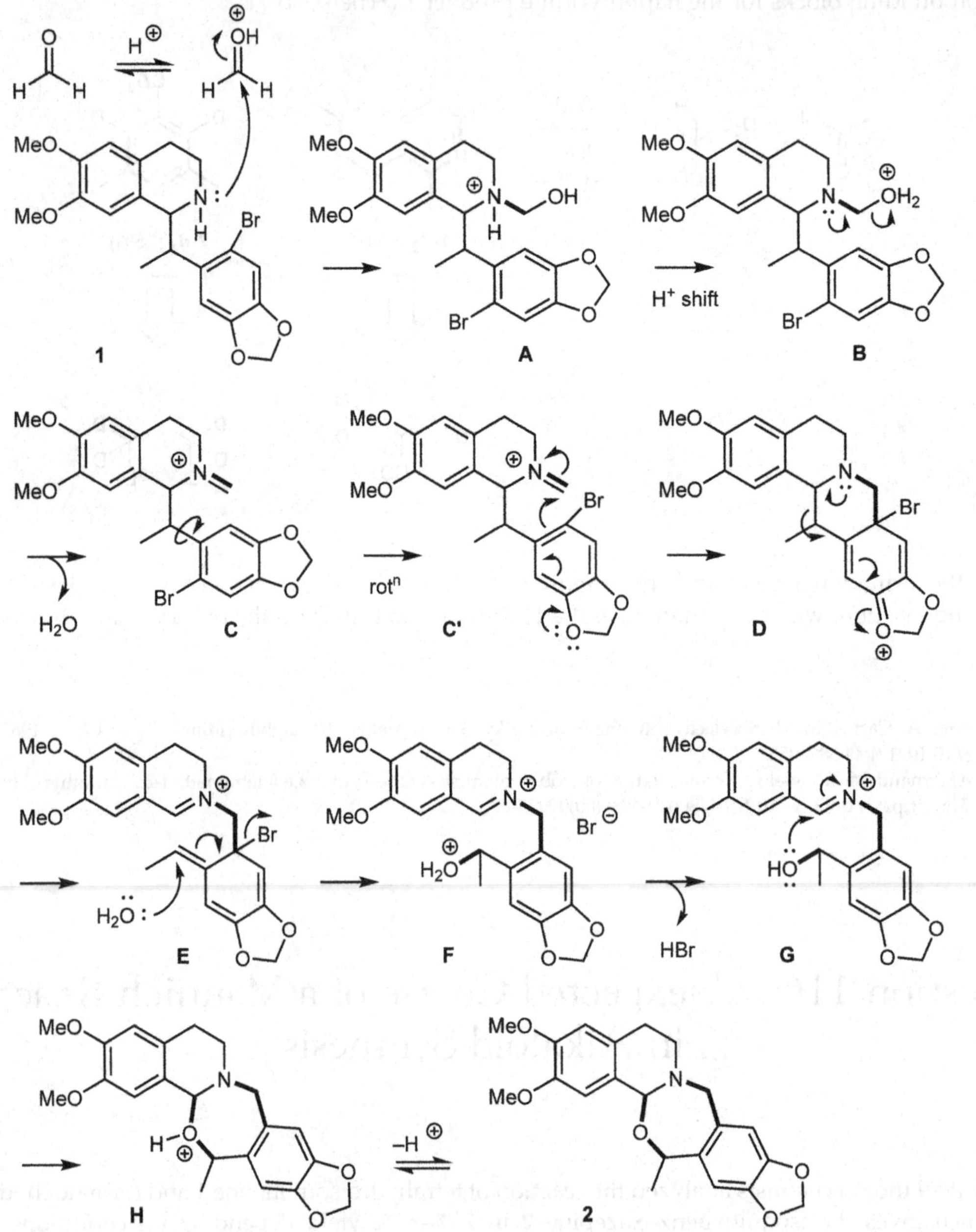

BALANCED CHEMICAL EQUATION

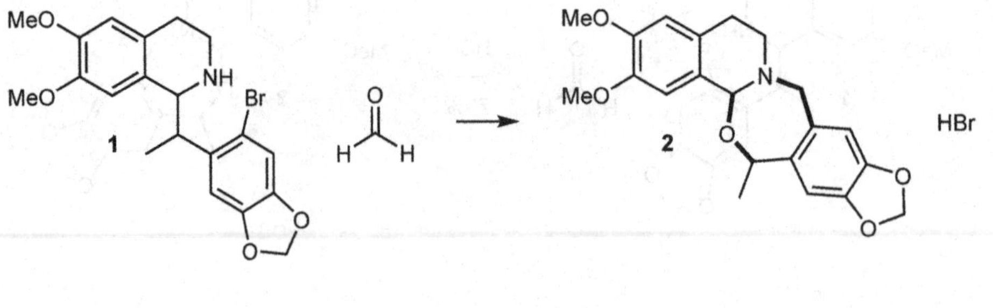

KEY STEPS EXPLAINED

The authors proposed a mechanism in which tetrahydroisoquinoline **1** condenses with protonated formaldehyde to form iminium ion **C** which upon C—C bond rotation undergoes ring closure to the quinoid **D**. **D** then ring opens again to iminium ion **E** which is captured by water and recyclizes to the seven-membered ring isoquinobenzoxazepine product **2** after final deprotonation. We note that the ring construction mapping for the seven-membered ring is [3 + 1 + 2 + 1]. Unfortunately, apart from spectroscopic evidence for the product and experimental efforts to test the selectivity of this transformation, the authors did not provide further experimental or theoretical data to support the postulated mechanism. We note that problem 112 in this book features a similar reaction and mechanism.

ADDITIONAL RESOURCES

We recommend an article which describes a similar unexpected Mannich reaction in an analogous chemical system that involves N-S-containing heterocycles.[3]

References

1. Natarajan S, Pai BR, Rajaraman R, et al. A novel transformation of 6,7-dimethoxy- and 6,7-methylene-dioxy-1-(2-bromo-4,5-methylenedioxy-α-methyl) benzyl-1,2,3,4-tetrahydroisoquinolines to benzoxazepinoisoquinolines. *Tetrahedron Lett*. 1975;16(41):3573–3576. https://doi.org/10.1016/S0040-4039(00)91371-6.
2. Natarajan S, Pai BR, Rajaraman R, et al. Studies in protoberberine alkaloids. Part 12. Novel transformation of some 1-(2-bromo-α-methylbenzyl)-1,2,3,4-tetrahydroisoquinolines to isoquinobenzoxazepines during Mannich reactions. *J Chem Soc Perkin Trans 1*. 1979;1979:283–289. https://doi.org/10.1039/P19790000283.
3. Dotsenko VV, Krivokolysko SG. Mannich reaction in the synthesis of N,S-containing heterocycles. 14*. Unexpected formation of thiazolo[3,2-a] pyridines in the aminoalkylation of N-methylmorpholinium 4-aryl-3-cyano-6-oxo-1,4,5,6-tetrahydropyridine-2-thiolates by isobutyraldehyde and primary amines. *Chem Heterocycl Compd*. 2012;48(4):672–676. https://doi.org/10.1007/s10593-012-1042-y.

Question 117: Synthesis of Thioindigo

In 1980 and 1982 Scottish chemists investigated the reaction of 2,3-dibromothiochromone S-oxide **1** with 2 equivalents of sodium acetate in refluxing acetic acid for 15 min to give thioindigo **2** in 80% yield (Scheme 1).[1,2] It was also observed that no thioindigo was formed when 2,3-dibromothiochromone was used as a starting material. Based on these observations, suggest a mechanism for the transformation of **1** to **2**.

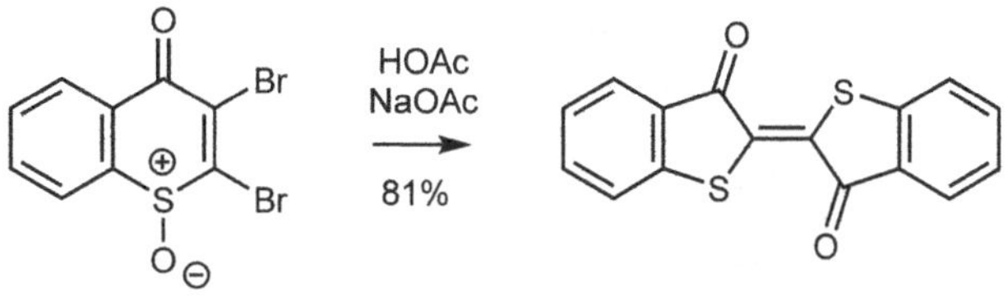

1
2,3-Dibromothiochromen-4-one S-oxide

2
Thioindigo

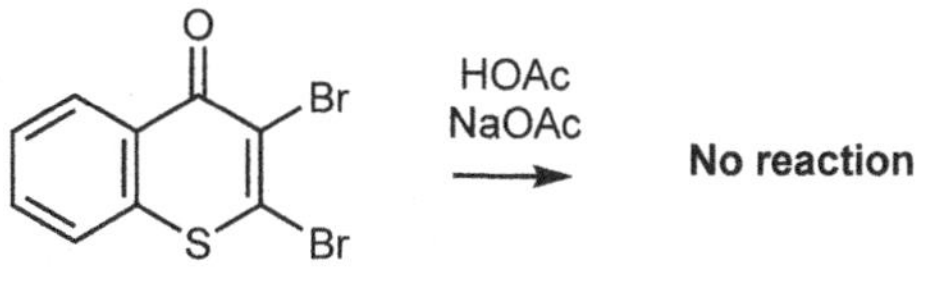

3
2,3-Dibromothiochromen-4-one

SCHEME 1

SOLUTION

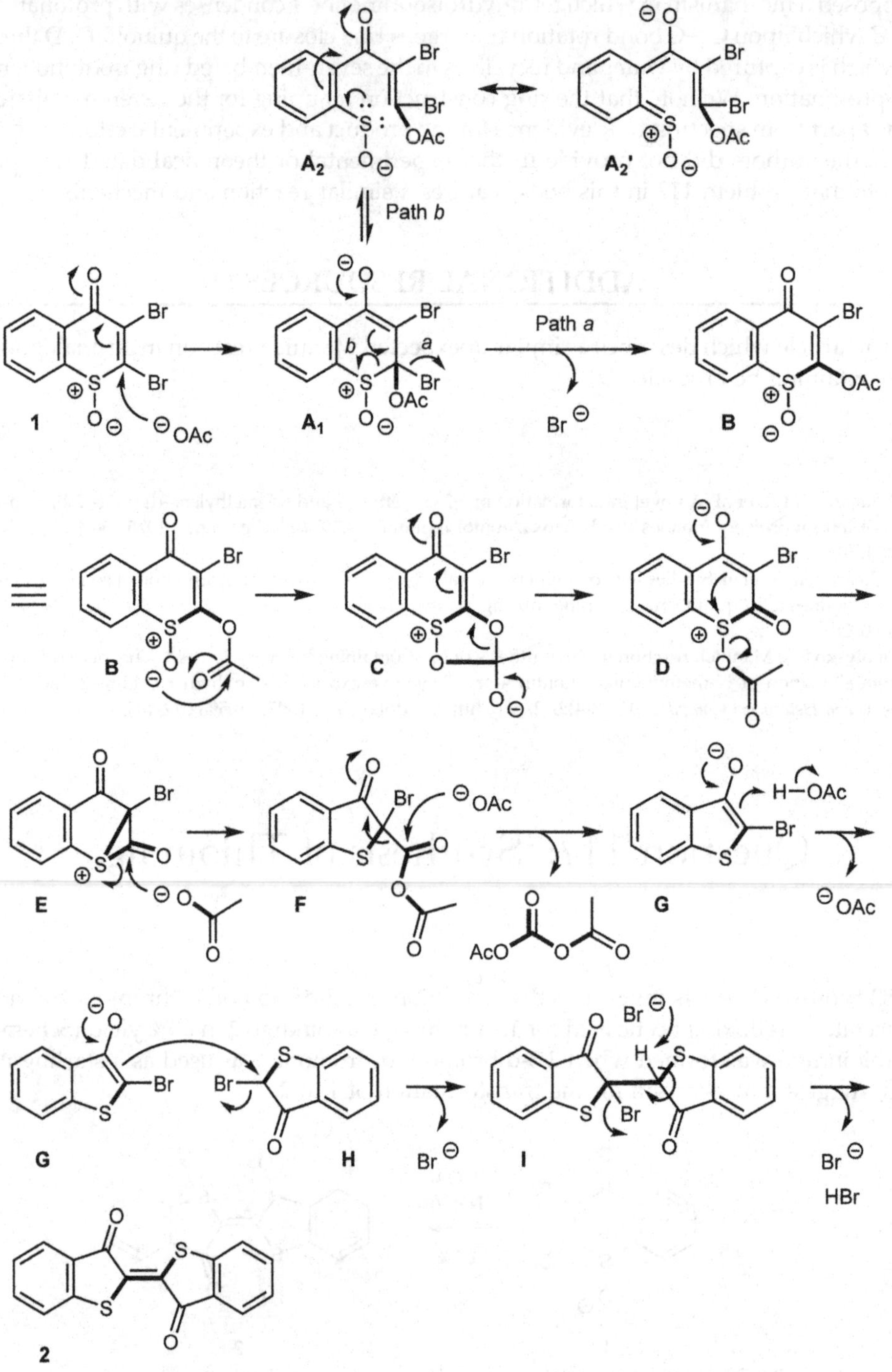

Secondary reaction:

BALANCED CHEMICAL EQUATION

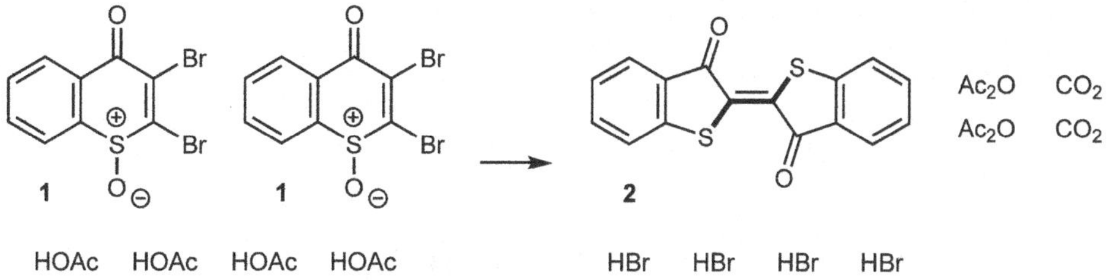

KEY STEPS EXPLAINED

The mechanism proposed by the authors for the transformation of **1** to **2** begins with nucleophilic addition of acetate ion onto S-oxide **1** to form enolate A_1. This enolate exists in equilibrium with its sulfenate form A_2 which in turn can be drawn in the resonance form A_2'. Elimination of bromide ion from A_1 leads to formation of acetate **B** which cyclizes to **C** and then ring opens resulting in an acyl migration that leads to zwitterionic S-acetate **D**. Intermediate **D** then undergoes a transannular cyclization to yield the bicyclic [3.1.0] framework of thiiranium cation **E**. Nucleophilic attack by acetate ion onto **E** causes ring opening of the three-membered ring to give the mixed anhydride intermediate **F** which fragments to enolate **G** upon nucleophilic attack by another equivalent of acetate ion. At this point, coupling between enolate **G** and another equivalent of its ketone form **H** leads to bromodiketo intermediate **I** which eliminates hydrogen bromide to give the final thioindigo product **2**. We note that experimentally, the authors were able to confirm the evolution of CO_2 gas during the course of the reaction. Nevertheless, it was not possible to stop the reaction to isolate any of the proposed intermediates. Under various conditions, products arising from possible side reactions were isolated and characterized but they did not provide any meaningful evidence to support the postulated mechanism. The authors also added cyclohexene to one of the reactions as a ketene trap but the reaction was unaffected by this change thus suggesting that ketenes did not form during the course of the transformation. In an interesting sidelight, the authors noticed the emergence of blue color ($\lambda_{max} = 600\,nm$) during the early stages of the reaction which quickly lost its coloration. This phenomenon was attributed to the equilibrium between the enolate A_1 and sulfenate A_2/A_2' which proceeds to form thioindigo **2** thus consuming the blue color arising from A_2/A_2'. We note lastly that further experimental and theoretical work is needed to help support this postulated mechanism.

It should also be mentioned that based on the proposed mechanism, reaction of 2,3-dibromothiochromen-4-one **3** instead of the S-oxide **1** results in recovery of starting material with no production of thioindigo. Clearly the oxygen atom bound to sulfur is necessary for causing the acyl migration from intermediate **B** to **C** thus allowing the transformation to proceed to product **2**. The only conceivable reaction that could take place with **3** under these reaction conditions is substitution of bromine at C_3 with acetate giving product **K** as shown in Scheme 2. The fact that starting material **3** was recovered suggests that the steps are reversible.

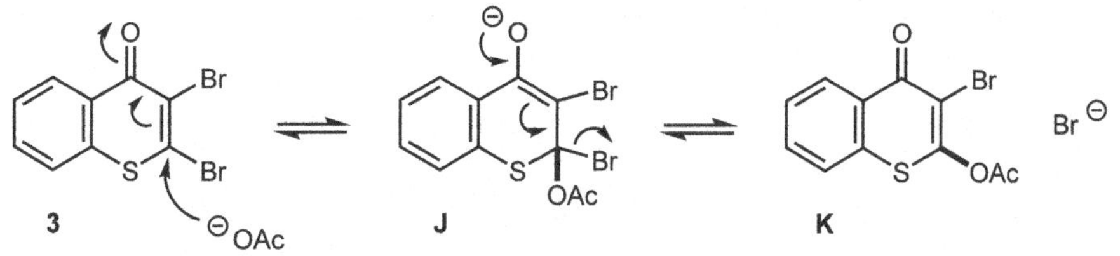

SCHEME 2

ADDITIONAL RESOURCES

We recommend two additional synthetic approaches to the thioindigo product **2**.[3,4]

References

1. MacKenzie NE, Thomson RH. A new synthesis of thioindigo; acyl migration and ring contraction in thiochromone derivatives. *J Chem Soc Chem Commun.* 1980;1980:559–560. https://doi.org/10.1039/C39800000559.

2. MacKenzie NE, Thomson RH. Ring contractions of thiochroman-4-ones and thiochromen-4-ones. *J Chem Soc Perkin Trans 1*. 1982;1982:395–402. https://doi.org/10.1039/P19820000395.
3. Chen CH, Fox JL. Thiopyranothiopyran chemistry. 5. Synthesis of dibenzo[b,g]thiopyrano[3,2-b]thiopyran-6,12-dione (thioepindolidione). *J Org Chem*. 1985;50(19):3592–3595. https://doi.org/10.1021/jo00219a028.
4. Hoepping A, Mayer R. Eine neue Synthesevariante zur Darstellung von Thioindigoiden. *Phosphorus Sulfur Silicon Relat Elem*. 1995;107 (1–4):285–288. https://doi.org/10.1080/10426509508027945.

Question 118: An Unexpected Reaction During Studies of Amine Oxide Rearrangements

In 1987 chemists from India treated the tetrahydroquinoline derivative **1** with *m*CPBA in methylene chloride at room temperature expecting that tricyclic product **3** would be formed (Scheme 1).[1] Instead, product **2** was formed in 30% yield. Explain both the observed and expected transformations in mechanistic terms.

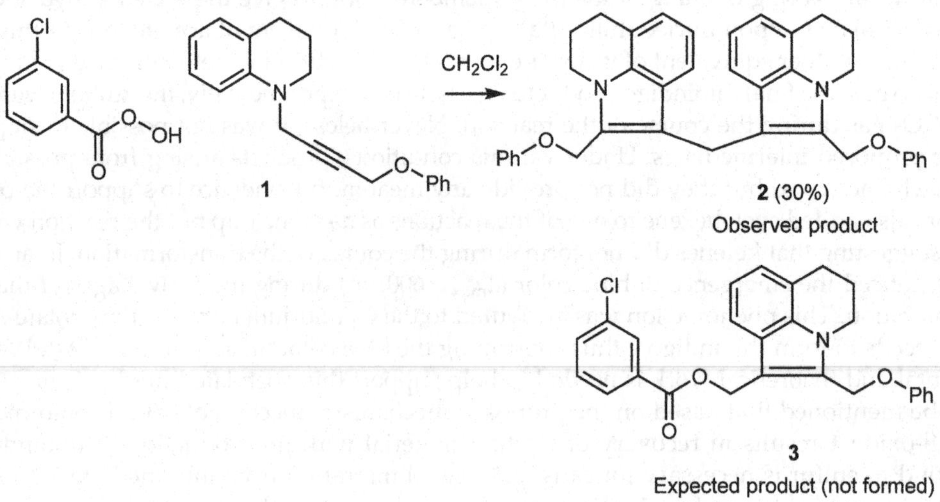

1

2 (30%)
Observed product

3
Expected product (not formed)

SCHEME 1

SOLUTION: 1 TO 3 (EXPECTED PRODUCT)

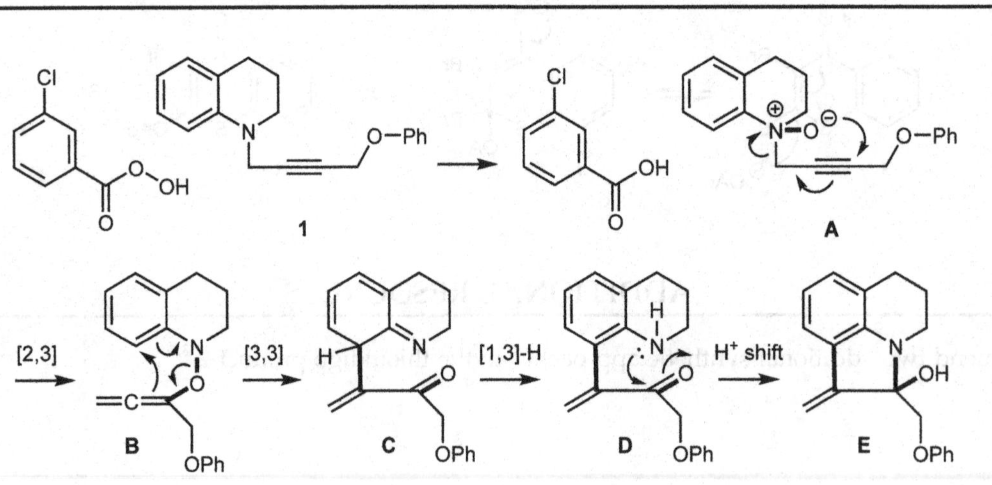

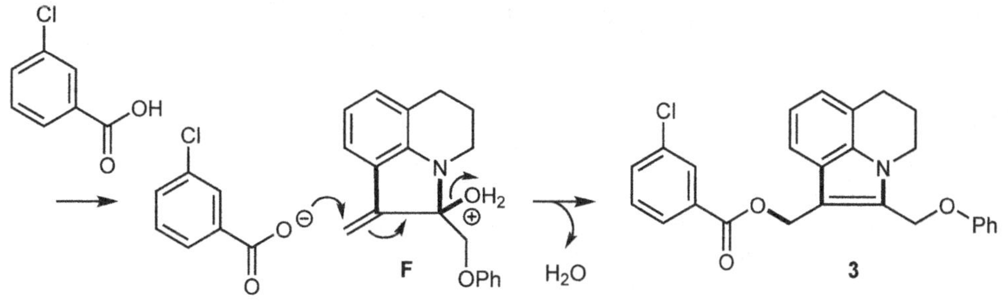

SOLUTION: 1 TO 2 (OBSERVED PRODUCT)

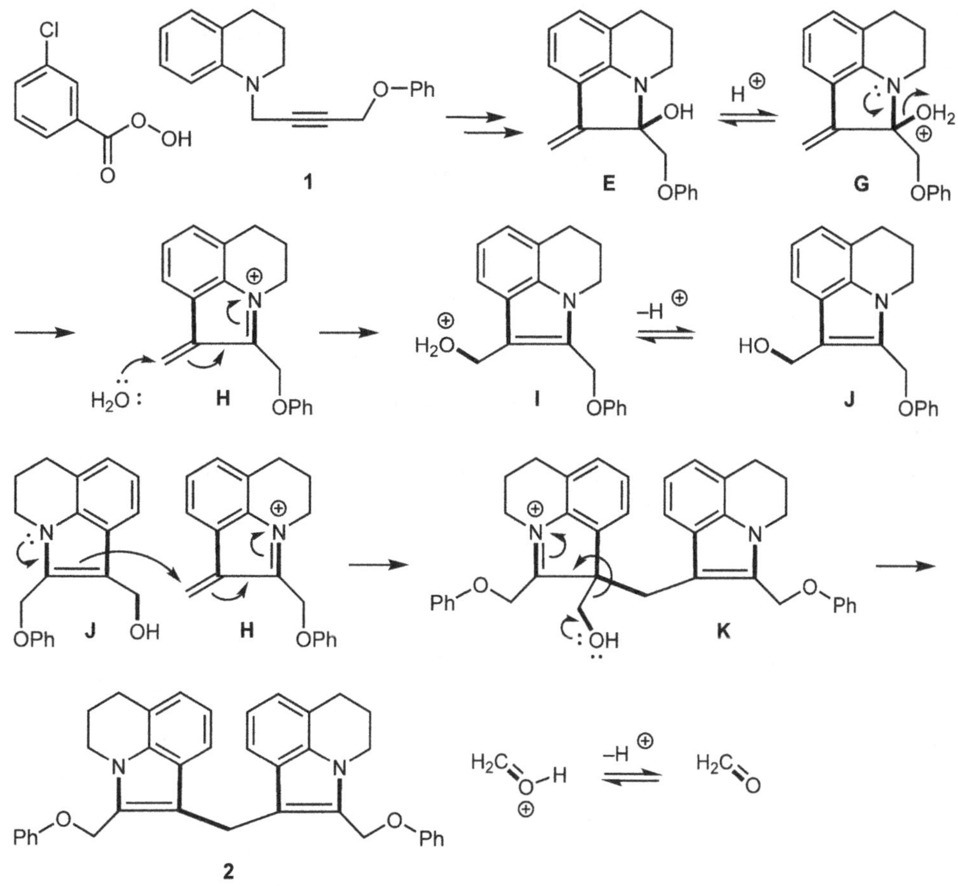

BALANCED CHEMICAL EQUATION

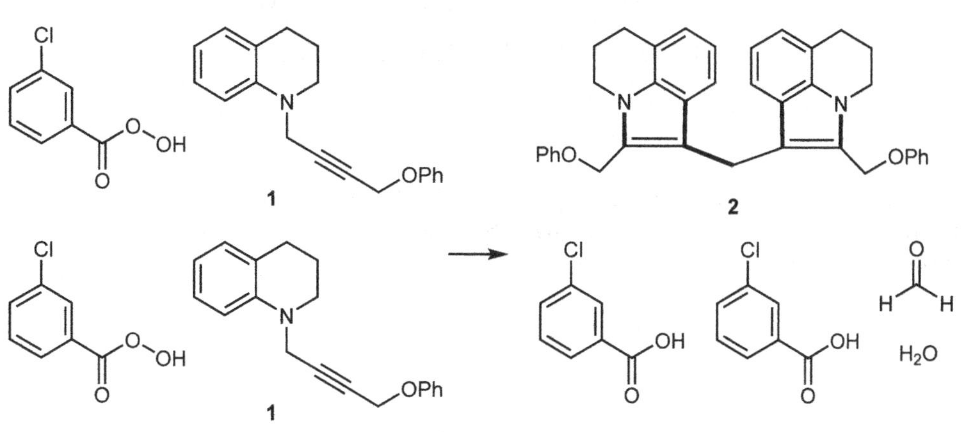

KEY STEPS EXPLAINED

We note that the authors expected that the reaction of **1** with *m*CPBA would form **3** by a mechanism which begins with oxidation of **1** to *N*-oxide **A** via reaction with *m*CPBA followed by sequential [2,3] and [3,3] sigmatropic rearrangements first to allene **B** and then to enone **C**. After a [1,3]-protic shift to aromatize the ring yielding intermediate **D**, cyclization to 1,2,5,6-tetrahydro-4*H*-pyrrolo[3,2,1-*ij*]quinolin-2-ol **E** would occur. This intermediate would then be expected to be trapped with *m*-chlorobenzoate to form product **3**. Nevertheless, this last step did not occur and instead, intermediate **E** went on to form iminium ion **H** after dehydration followed by addition of water at the terminal carbon atom of the exocyclic olefin to give intermediate **J**. Coupling between intermediates **J** and **H** led to iminium ion **K** which eliminated formaldehyde to give the observed product **2**. It is important to note that although *m*-chlorobenzoate did not successfully trap intermediate **E** to form **3**, the use of a stronger nucleophile such as potassium cyanide led to the formation of product **4** thus lending experimental support for the proposed intermediate **E** (Scheme 2). Furthermore, after removal of *m*-chlorobenzoic acid and methylene chloride, ^{1}H NMR analysis revealed a mixture of intermediates **E** and **J** in a ratio of 1:2 thus lending further support to the proposed mechanism. Unfortunately, the authors did not conduct a theoretical analysis for this proposed mechanism.

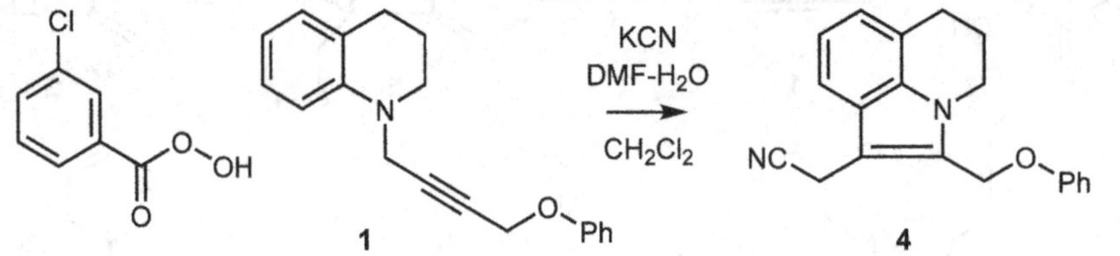

SCHEME 2

ADDITIONAL RESOURCES

We recommend three additional articles by the same authors which explore further examples of the amine oxide rearrangement discussed so far for this transformation.[2–4]

References

1. Majumdar KC, Chattopadhyay SK. Studies on amine oxide rearrangement: an unusual product from the reaction of 1-phenoxy-4-tetrahydroquinolylbut-2-yne with *m*-chloroperbenzoic acid. *J Chem Soc Chem Commun*. 1987;1987:524–525. https://doi.org/10.1039/C39870000524.
2. Majumdar KC, Biswas P, Jana GH. Studies on amine oxide rearrangements: regioselective synthesis of pyrrolo[3,2-f]quinolin-7-ones. *J Chem Res (S)*. 1997;1997:310–311. https://doi.org/10.1039/A702432K.
3. Majumdar KC, Das U, Jana NK. Studies on pyrimidine-annelated heterocycles: synthesis of pyrrolo[3,2-d]pyrimidines by amine oxide rearrangement. *J Org Chem*. 1998;63(11):3550–3553. https://doi.org/10.1021/jo9718861.
4. Majumdar KC, Chattopadhyay SK, Mukhopadhyay PP. Studies on amine oxide rearrangement: synthesis of pyrrolo[3,2-c][1]benzothiopyran-4-one. *Synth Commun*. 2006;36:1291–1297. https://doi.org/10.1080/00397910500518908.

Question 119: An Unusual Phosphite-Induced Deoxygenation of a Nitronaphthalene

In 1985 chemists from India showed that heating of either 2-(2-nitrophenoxy)naphthalene **1** or its isomer **3** with triethyl phosphite under reflux and nitrogen gave benzophenoxazone **2** in low yield (Scheme 1).[1] Suggest mechanisms to explain both transformations.

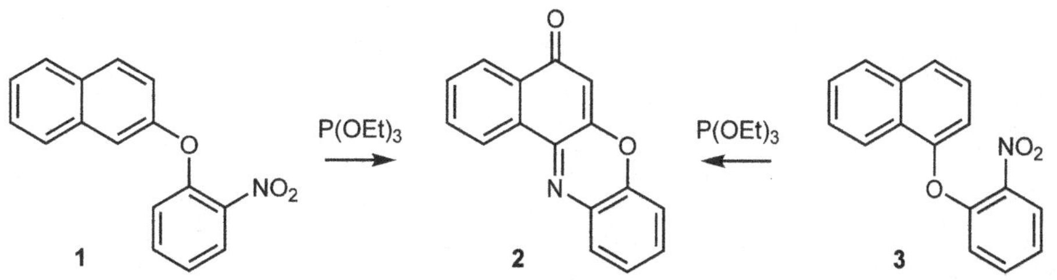

SCHEME 1

SOLUTION: 1 TO 2 AUTHORS' MECHANISM

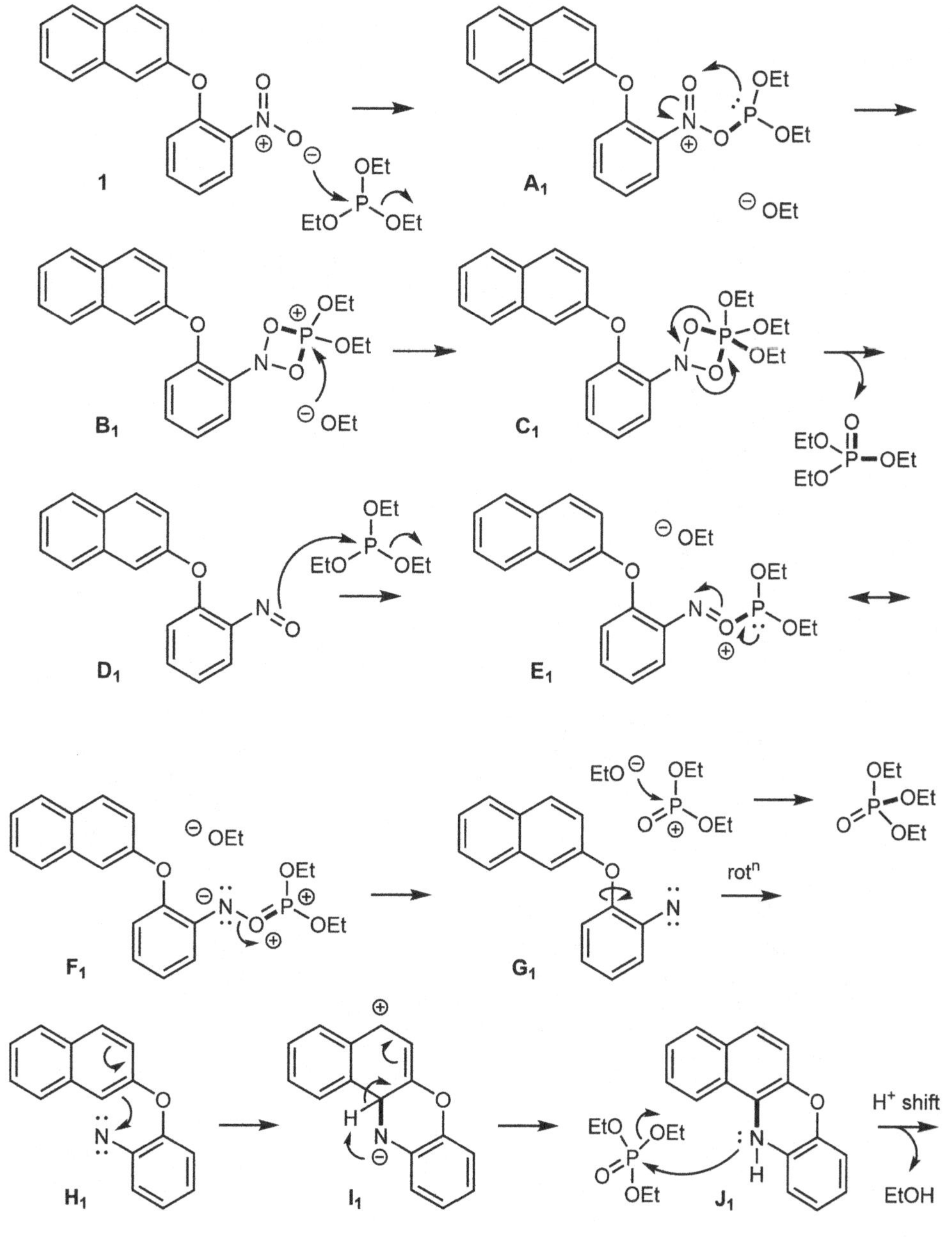

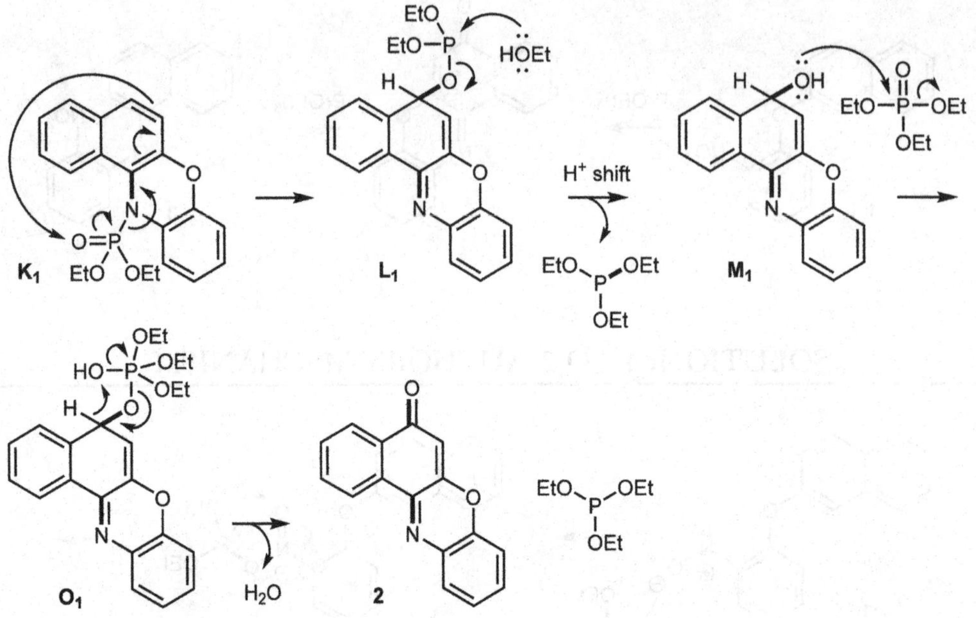

SOLUTION: 1 TO 2 ALTERNATIVE MECHANISM

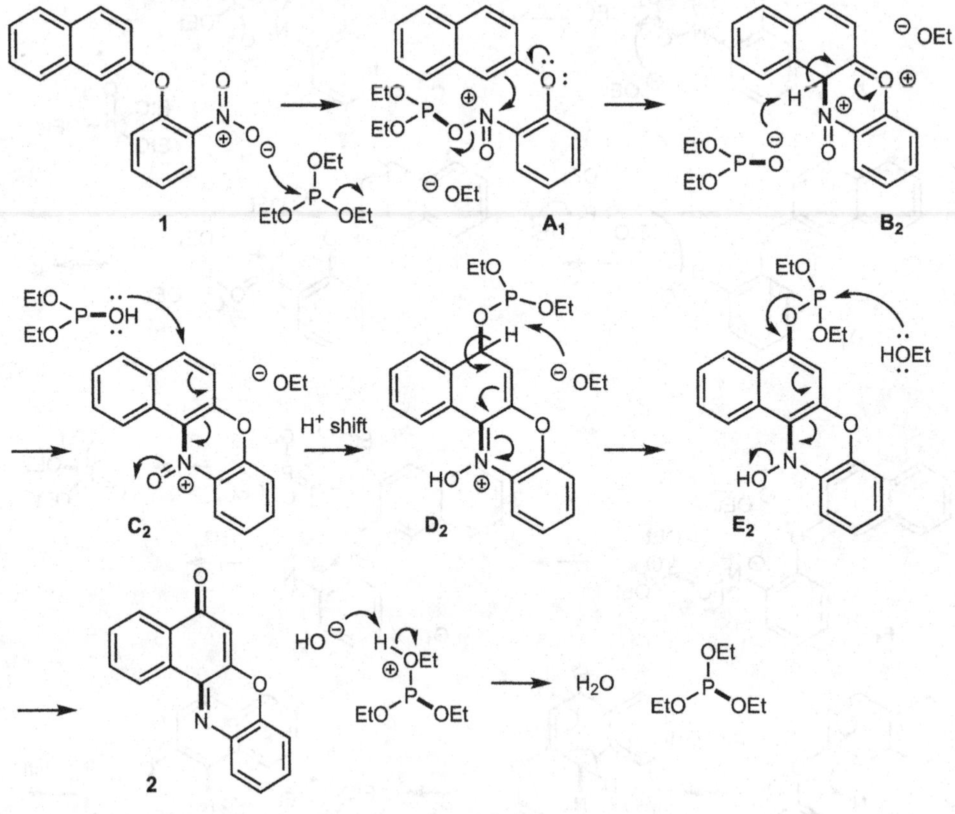

SOLUTION: 3 TO 2 AUTHORS' MECHANISM

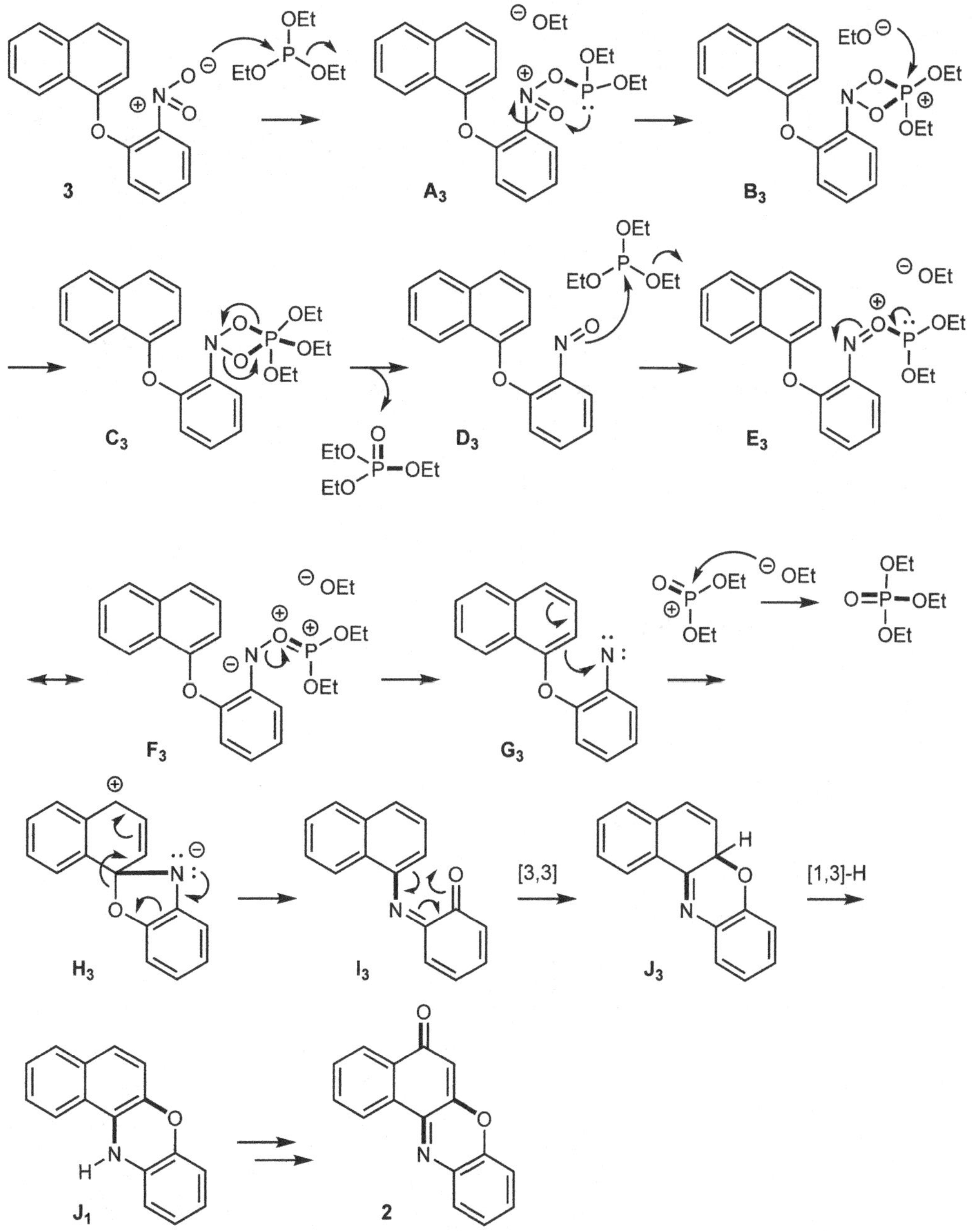

SOLUTION: 3 TO 2 ALTERNATIVE MECHANISM

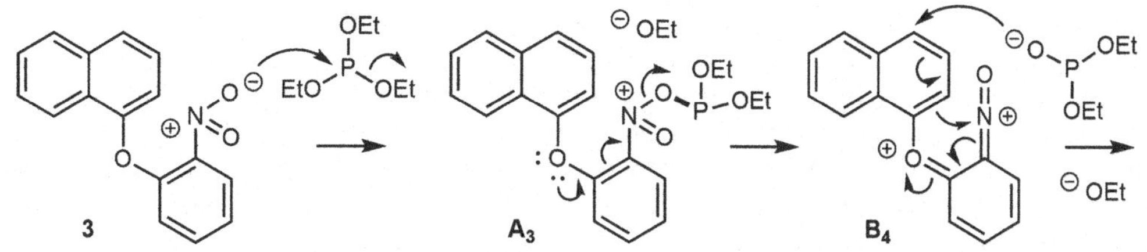

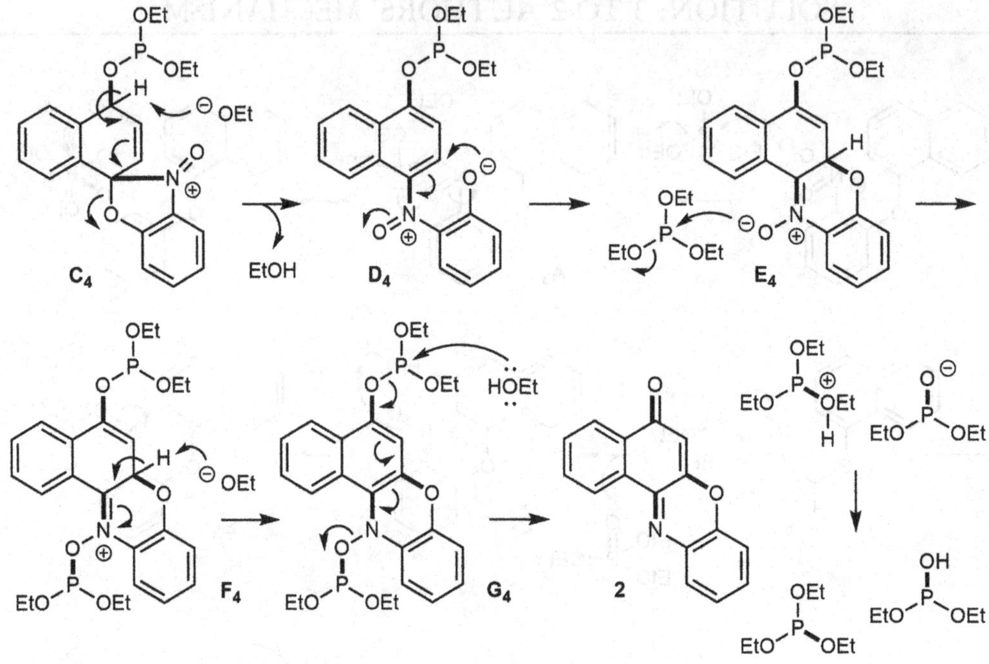

BALANCED CHEMICAL EQUATION

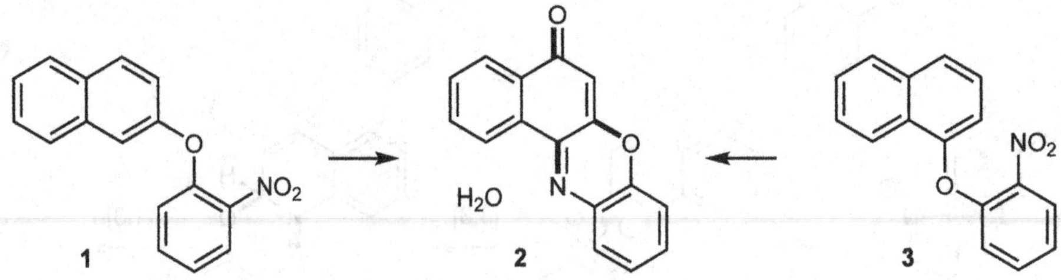

KEY STEPS EXPLAINED

The mechanism proposed by the authors to explain the transformation of **1** to **2** begins with reduction of the nitro group of **1** to form a nitroso group using one equivalent of triethylphosphite thus leading to intermediate D_1 and triethylphosphate by-product. Subsequent reaction of nitroso intermediate D_1 with a second equivalent of triethylphosphite leads to triethylphosphate by-product and nitrene G_1 which ring closes to the 12H-benzo[a]phenoxazine intermediate J_1. Intermediate J_1 reacts with triethylphosphate by-product eventually forming 5H-benzo[a]phenoxazin-5-ol M_1 and regenerating triethylphosphite. Intermediate M_1 reacts with a second equivalent of triethylphosphate leading to product **2** and by-products consisting of water and triethylphosphite. The authors do not give any indirect or direct experimental evidence for the nitroso and nitrene intermediates in their mechanism but they do provide a melting point and mass spectrometry data for intermediate J_1, which the authors claim to have isolated using rapid chromatography on silica gel in chloroform. A theoretical analysis is also not provided.

Since the yield of the final product is very low and since the authors do not provide sufficient evidence to support their proposed mechanism, we here present a much shorter and conceptually clearer alternative mechanism which consists of only 5 elementary steps instead of 14. In this mechanism, the nitro group of **1** reacts with triethylphosphite to form intermediate A_1 as before which cyclizes to B_2 and is then deprotonated to C_2. Diethylphosphite adds in a nucleophilic manner to C_2 to give D_2 which is deprotonated to E_2. Final attack at phosphorus by ethanol liberates product **2** together with hydroxide ion. The main difference in this mechanism is that the phosphorus atom of triethylphosphite is not oxidized to a phosphate and then reduced back down again to the original phosphite as in the authors' mechanism. As such, the alternative mechanism suggests an internal redox reaction occurs in the substrate molecule and triethylphosphite acts as a catalyst. Conversely, the authors' mechanism suggests an overall redox reaction containing a redox recycling loop involving a phosphate to phosphite transformation. Nevertheless, although the alternative mechanism occurs in fewer steps, it does not involve intermediate J_1 which the authors claim to have isolated. In a low yield scenario, it may be possible to envision the occurrence of two energetically disfavored mechanistic pathways although a clear picture would emerge if a theoretical analysis were done to compare the energy favorability of the A_1 to B_1 step as compared to the A_1 to B_2 step. We note lastly that the ring construction mapping for the transformation of 1 to 2 is [6 + 0].

Concerning the authors' proposed mechanism for the 3 to 2 transformation, we note that 3 initially reacts with triethylphosphite as before to form the nitroso and nitrene intermediates D_3 and G_3, respectively. Nitrene G_3 then cyclizes to the spiro zwitterionic intermediate H_3 which ring opens to 6-(naphthalen-1-ylimino)-cyclohexa-2,4-dienone I_3. After a sequential [3,3] sigmatropic rearrangement and [1,3]-protic shift the familiar 12H-benzo[a]phenoxazine intermediate J_1 is formed. The remaining steps leading to product 2 are identical to the authors' proposed mechanism for the 1 to 2 transformation starting with J_1. Once again, an alternative mechanism that is analogous to that described for the previous transformation may be proposed. This time, however, the alternative mechanism proceeds via a spiro intermediate C_4. With respect to the transformation of 3 to 2, the ring construction mapping is [4 + 2].

ADDITIONAL RESOURCES

We recommend a recent article which discusses the kinetics and mechanism of the nitrosobenzene deoxygenation by trivalent phosphorus compounds and where the authors also support the intermediacy of nitrenes in this pathway[2] by two pieces of experimental evidence: (a) when the reaction was carried out in aerated solutions, products arising from nitroso oxide intermediates were detected, and (b) when the reaction was carried out in the presence of triphenylphosphine, triphenylphosphine oxide was formed and phosphorescence emission at 570 nm was also observed as a consequence of trapping the triplet nitrene intermediate. For more information on the proposed intermediacy of nitrenes pathway, see the work of Cadogan.[3]

References

1. Nagarajan K, Upadhyaya P. Formation of benzo[a]phenoxazine derivatives by triethylphosphite-induced deoxygenation of 1-(2-nitrophenoxy) and 2-(2-nitrophenoxy)naphthalenes. *Tetrahedron Lett.* 1985;26(3):393–396. https://doi.org/10.1016/S0040-4039(01)80826-1.
2. Khursan V, Shamukaev V, Chainikova E, Khursan S, Safiullin R. Kinetics and mechanism of the nitrosobenzene deoxygenation by trivalent phosphorous compounds. *Russ Chem Bull.* 2013;62(11):2477–2486. https://doi.org/10.1007/s11172-013-0359-8.
3. Cadogan JIG. Reduction of nitro- and nitroso-compounds by tervalent phosphorus reagents. *Q Rev Chem Soc.* 1968;22(2):222–251. https://doi.org/10.1039/QR9682200222.

Question 120: A Benzimidazole to Pyrroloquinoxaline Transformation

In 1975 an English chemist reacted the benzimidazolium bromide 1 with dimethyl acetylene-dicarboxylate (DMAD) in the presence of triethylamine to achieve the pyrroloquinoxaline product 2 (Scheme 1).[1] Contrary to previous work by Japanese[2] and Romanian[3] chemists, the English author demonstrated the structure of product 2 on the basis of several pieces of experimental evidence. Provide a mechanism for this transformation.

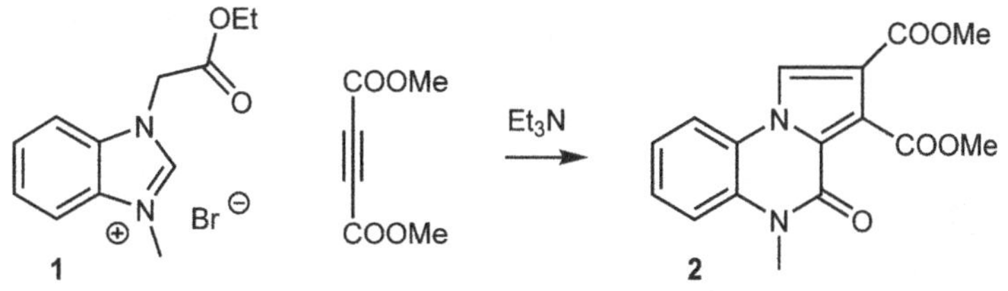

SCHEME 1

SOLUTION

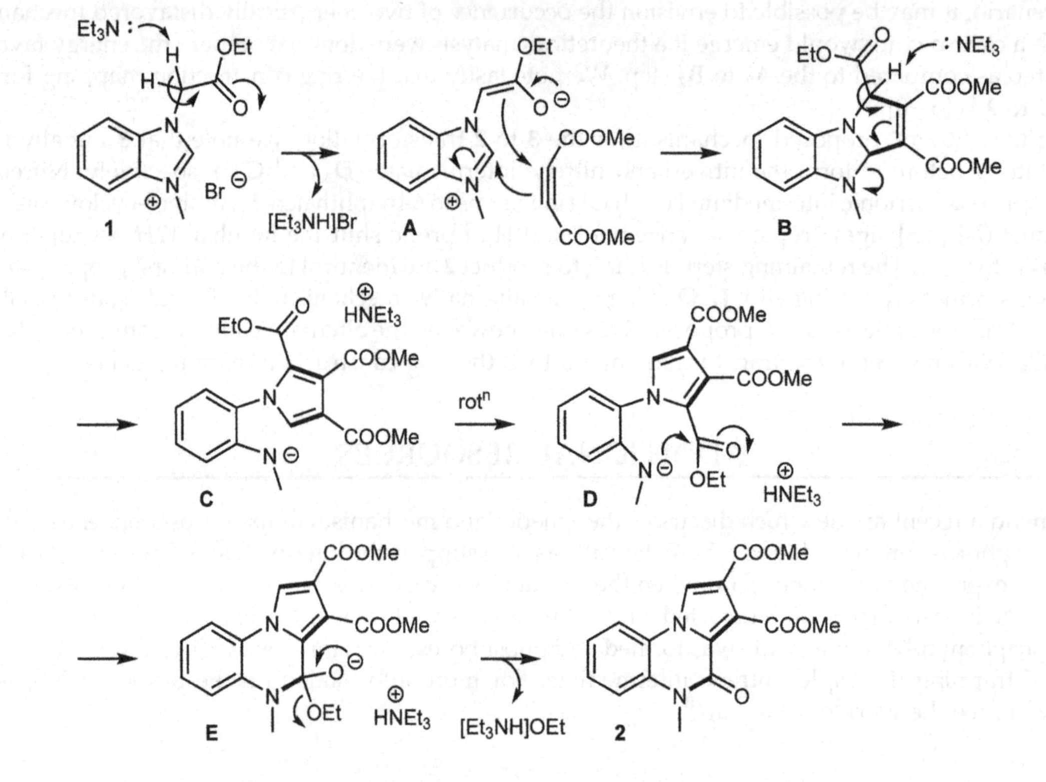

BALANCED CHEMICAL EQUATION

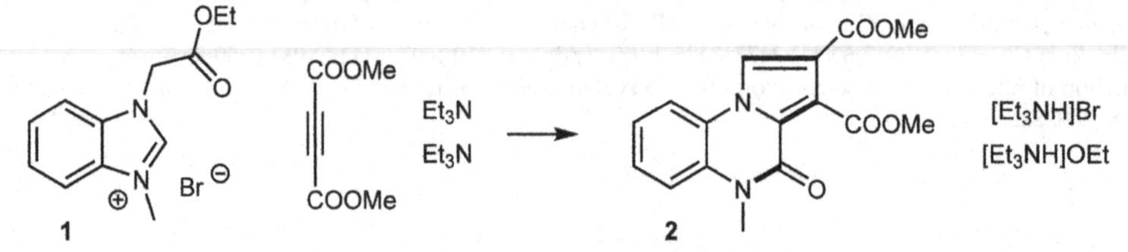

KEY STEPS EXPLAINED

We begin by pointing out that Japanese and Romanian authors thought that the product of the reaction had the structures **3** and **4**, respectively (see Fig. 1). The pyridobenzimidazole structure **3** was favored on the basis of UV, NMR, mass spectrometry, and analytical evidence. Meanwhile the zwitterionic structure **4** was favored on the basis of a deuterium labeling experiment where labeling at C_2 in starting material **1** was retained in the product.

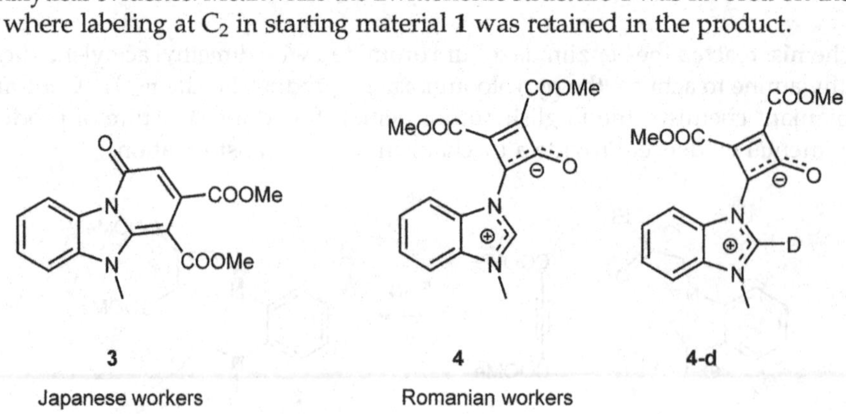

3	4	4-d
Japanese workers	Romanian workers	

FIG. 1   Proposed structures for product of reaction.

Meth-Cohn elucidated the structure of the product on the basis of the following evidence: (a) ^{13}C NMR data of the reaction product indicated no low field absorptions in the range 141–144 ppm characteristic of the C_2 positions of benzimidazoles; (b) a ^{13}C labeling experiment shown in Scheme 2 indicated that the reaction product had a ^{13}C NMR spectrum with the label on a CH carbon atom absorbing at 123.79 ppm which had long-range couplings to two ester carbonyl groups (2.6 and 4.8 Hz) and another coupling to a quaternary carbon atom absorbing at 120.48 ppm (5.8 Hz); and (c) similar UV, IR, and ^{13}C NMR spectra to those of the reaction product were obtained for the analogous known compound **5** shown in Scheme 2.

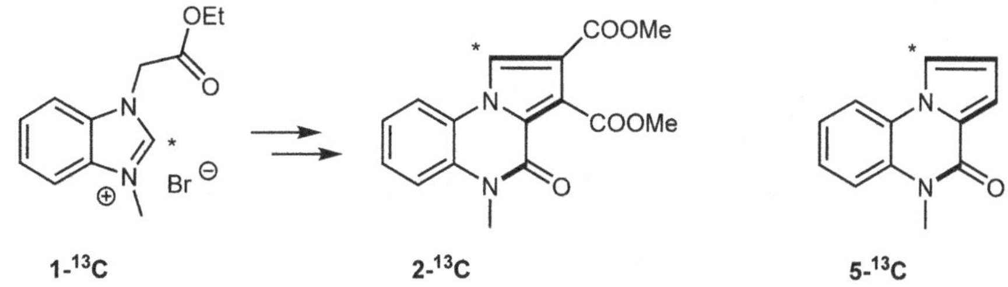

SCHEME 2

Unfortunately, the author was not able to convert **2-^{13}C** to **5-^{13}C** via a double saponification-decarboxylation sequence to complete the confirmation. Following the mechanism of the reaction shown before, a C_2 deuterium label would have resulted in the product structure **2-d** shown in Scheme 3.

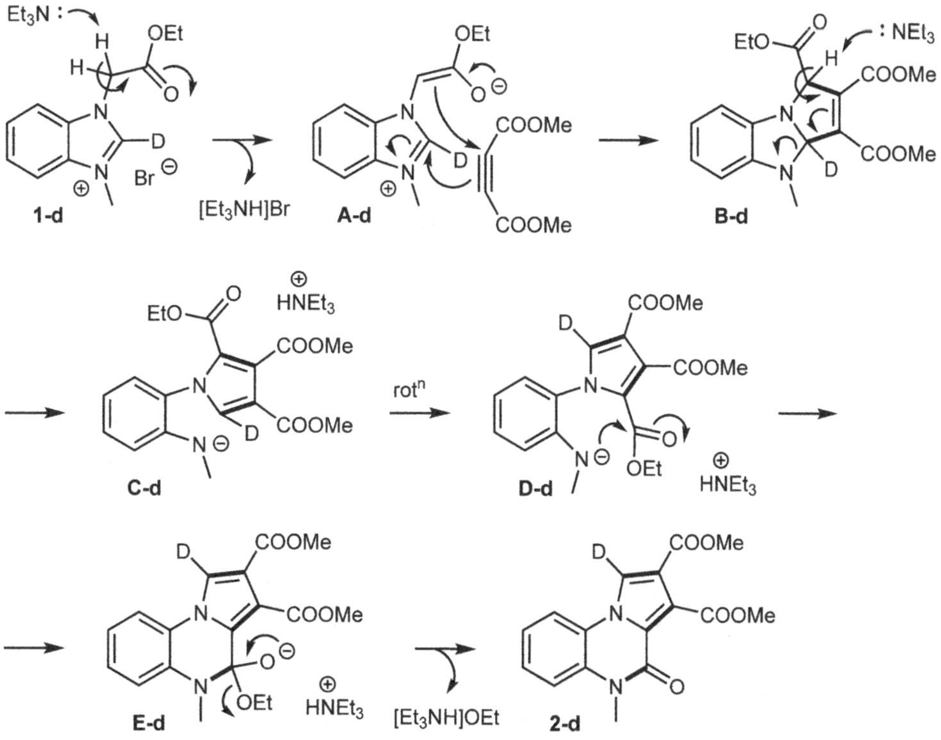

SCHEME 3

ADDITIONAL RESOURCES

We recommend two synthetic approaches to compounds such as product **2** which involve DMAD addition to compounds such as **1**[4] and gold catalysis, respectively.[5]

References

1. Meth-Cohn O. 1,3-Dipolar cycloaddition of benzimidazolium ylides with dimethyl acetylene-dicarboxylate—a re-investigation. *Tetrahedron Lett.* 1975;16(6):413–416. https://doi.org/10.1016/S0040-4039(00)71880-6.
2. Ogura H, Kikuchi K. Heterocyclic compounds. XI. 1,3-Dipolar cycloaddition of benzimidazolium ylide with acetylenic compounds. *J Org Chem.* 1972;37(17):2679–2682. https://doi.org/10.1021/jo00982a010.
3. Gugravescu I, Herdan J. Druta I. *Rev Roum Chim.* 1974;19:649.
4. Komatsu M, Kasano Y, Yamaoka S, Minakata S. Novel generation of pyridinium ylides from *N*-(silylmethyl)pyridone analogs via 1,4-silatropy and their 1,3-dipolar cycloadditions leading to *N*-heteropolycycles. *Synthesis.* 2003;2003(9):1398–1402. https://doi.org/10.1055/s-2003-40188.
5. Liu G, Zhou D, Lin D, et al. Synthesis of pyrrolo[1,2-*a*]quinoxalines via gold(I)-mediated cascade reactions. *ACS Comb Sci.* 2011;13:209–213. https://doi.org/10.1021/co1000844.

Question 121: Acid-Catalyzed Benzophenone Isomerization

The isomerization of 6-carboxy-5′-chloro-2′-hydroxy-2-methylbenzophenone **1** to 2-carboxy-5′-chloro-2′-hydroxy-3-methylbenzophenone **2** was first reported in 1927 by Hayashi (Scheme 1).[1] It has since been termed the Hayashi rearrangement. Provide a mechanistic explanation for this rearrangement.

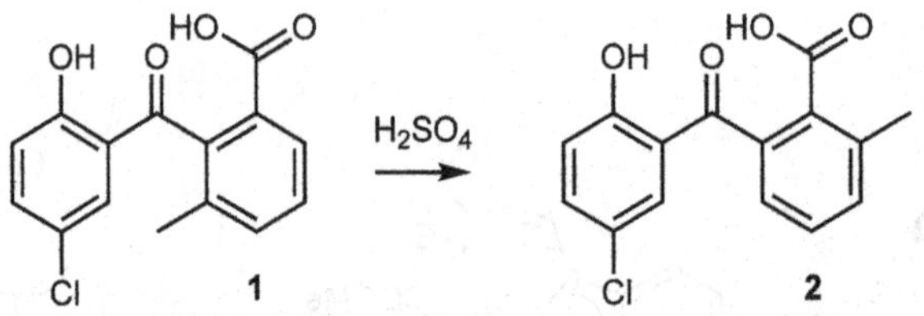

SCHEME 1

SOLUTION: HAYASHI MECHANISM[1]

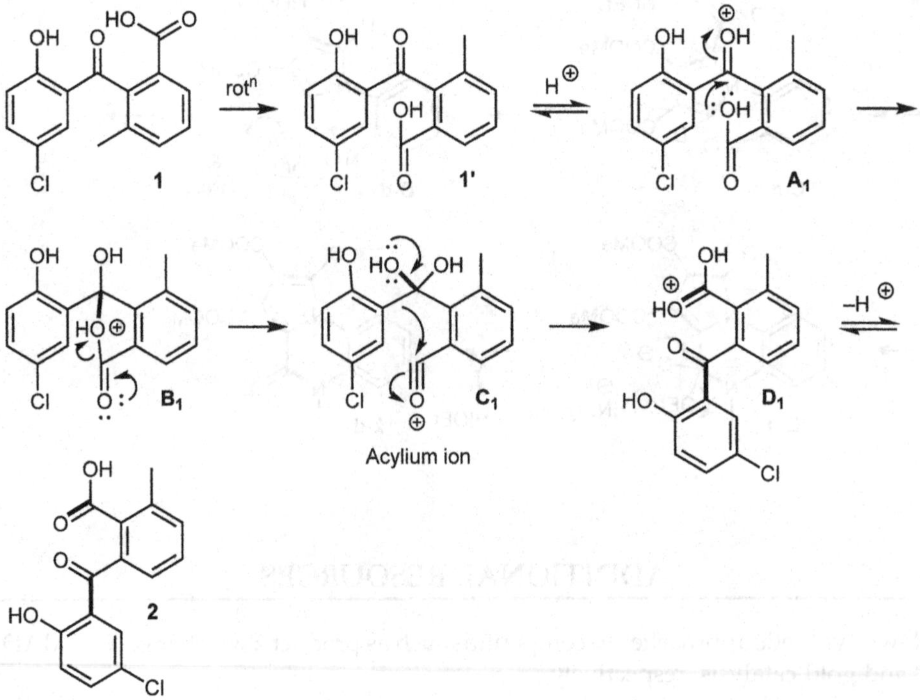

SOLUTION: SANDIN MECHANISM[2]

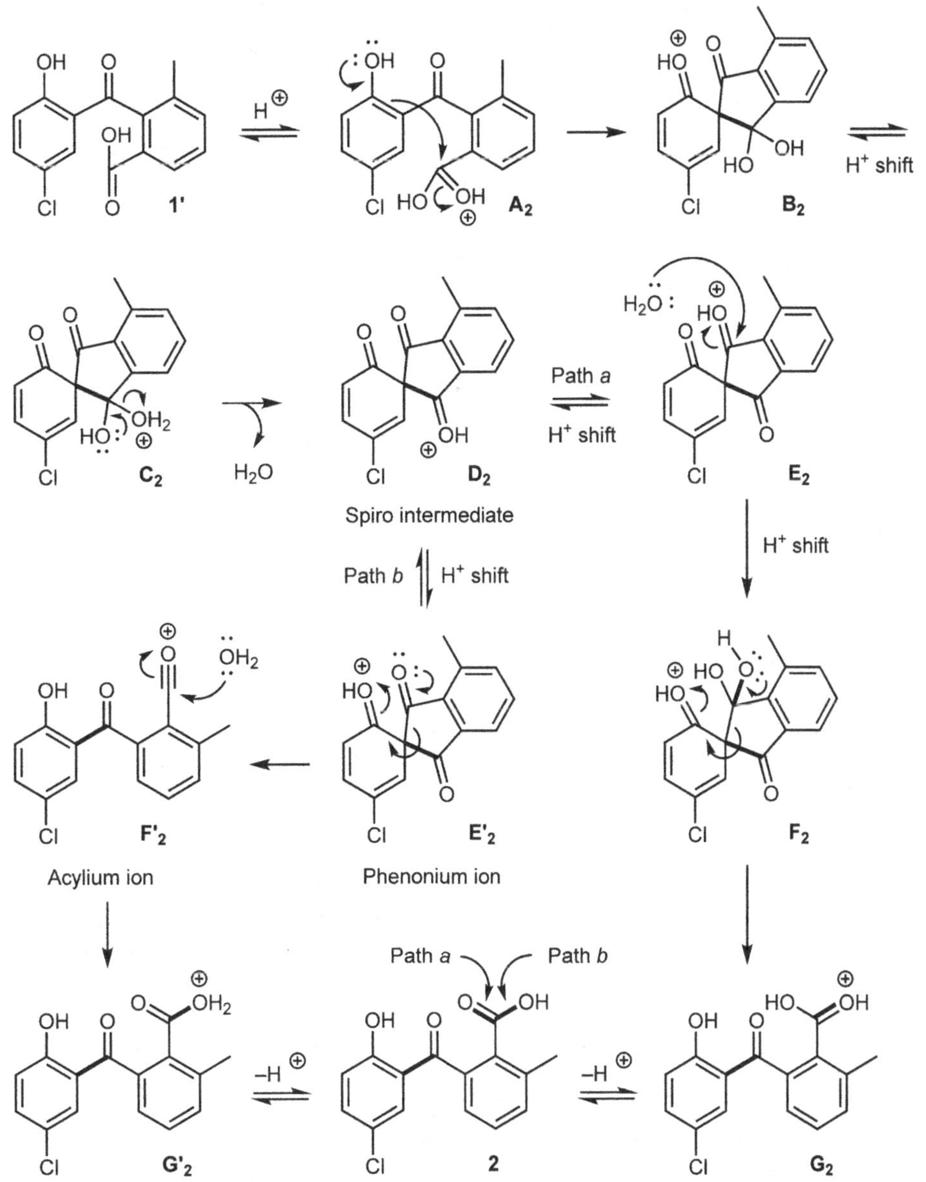

SOLUTION: WANG MECHANISM[3]

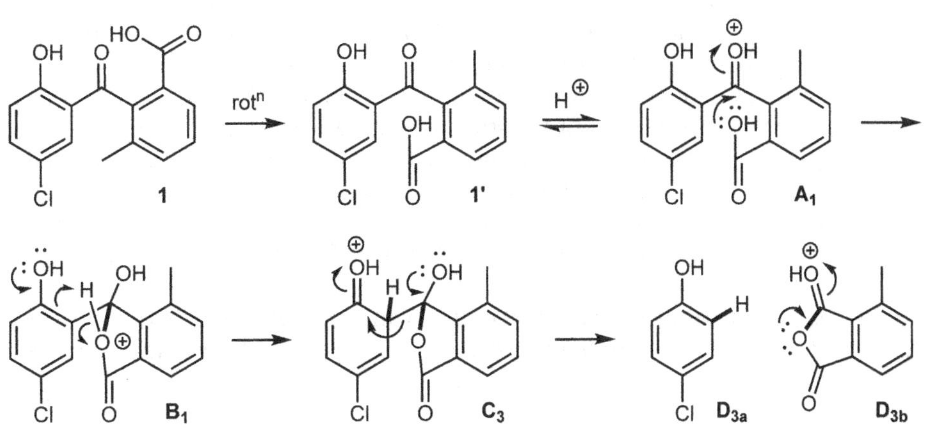

KEY STEPS EXPLAINED

In 1927 Hayashi proposed a mechanism for the isomerization of **1** to **2** under acidic conditions which starts with protonation of the carbonyl group oxygen atom giving oxonium ion A_1 followed by ring closure to the protonated lactone B_1 which can undergo fragmentation to the acylium ion C_1. Migration of the *p*-chlorophenol moiety occurs next to give the protonated benzophenone intermediate D_1. A final deprotonation of the carboxylic acid carbonyl group leads to product **2**.

Years later, in 1956, Sandin proposed an alternative mechanism which begins with protonation of the carboxylic acid carbonyl group followed by formation of the *spiro* intermediate B_2 which dehydrates to the *spiro* intermediate D_2. From intermediate D_2 there are two possible pathways that can lead to the rearranged product **2**. In path *a*, D_2 undergoes an intramolecular proton transfer to E_2 and is then captured by water to yield F_2 after an intramolecular proton transfer. F_2 then rearranges to G_2 which loses a proton to generate **2**. In path *b*, D_2 undergoes an intramolecular proton transfer to phenonium ion E'_2 which rearranges to acylium ion F'_2 which is captured by water and leads to **2** after a final proton loss.

The most recent mechanistic proposal for this isomerization comes from Zerong Wang which consists of a mechanism that starts with the same initial Hayashi sequence: **1** to **1'** to A_1 to B_1. At this point, B_1 undergoes intramolecular proton transfer to form phenonium ion C_3 which fragments into *p*-chlorophenol D_{3a} and protonated 3-methylphthalic anhydride D_{3b}. Protonated anhydride D_{3b} further fragments to acylium ion F_3 which is recaptured by *p*-chlorophenol in the *ortho* position to give G_3. Deprotonation of G_3 followed by enolization gives product **2**.

We note that the main difference between the Wang mechanism and the previous two mechanisms is that the Hayashi and Sandin mechanisms do not involve a complete fragmentation of the substrate into two separate parts which recombine again as in the Wang proposal. If the Wang mechanism were operative then it is expected that *p*-chlorophenol and 3-methylphthalic anhydride would be major side products of the reaction. The reported yields for products like **2** from this rearrangement reaction are in the range of 70%–90% suggesting that the Hayashi and Sandin mechanisms are more likely to be operative.

In Hayashi's original paper, for instance, it is reported that the Friedel-Crafts condensation of *p*-chlorophenol and 3-methylphthalic anhydride catalyzed by aluminum chloride leads to a 75% yield of **1**.[1] Heating **1** in concentrated sulfuric acid at 100°C for 1 h produced 35% of rearranged product **2** and 20% yield of chlorohydroxymethylanthraquinone. Heating **2** in concentrated sulfuric acid at 100°C for 1 h produced 30% of rearranged product **2** and 20% of chlorohydroxymethylanthraquinone. The yields of the rearrangement reactions improved to 70%–80% when the reactions were carried out at room temperature for 24 h with no formation of anthraquinone product.

Furthermore, in a subsequent paper,[4] Hayashi reported that the Friedel-Crafts condensation of *p*-chlorophenol and 4-methylphthalic anhydride catalyzed by aluminum chloride led to products **3** and **4** (analogous to substrate **1**) and anthraquinone products **5** and **6** which arise from a second condensation of products **3** and **4**, respectively (see Scheme 2). After separating out products **3** and **4** Hayashi then treated each benzoic acid with concentrated sulfuric acid and showed that product **3** rearranged to product **4** and that product **4** rearranged to product **3** (see Scheme 3).

While these pieces of evidence certainly support part of the Wang mechanism, they do not offer a conclusive basis for favoring one mechanism over the others. In the absence of more experimental and theoretical evidence, it remains

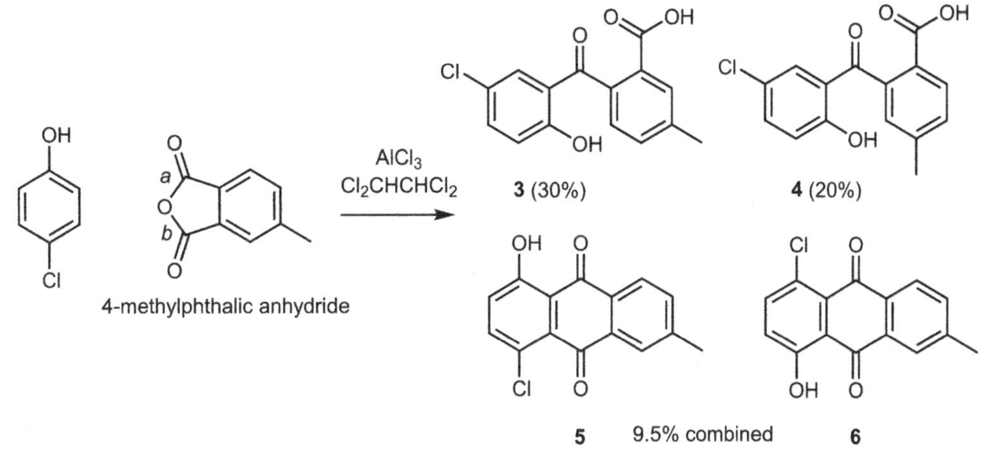

Product **3** = 2-(5′-chloro-2′-hydroxybenzoyl)-5-methylbenzoic acid (product from attack at position *a*)

Product **4** = 2-(5′-chloro-2′-hydroxybenzoyl)-4-methylbenzoic acid (product from attack at position *b*)

Product **5** = 8-chloro-5-hydroxy-2-methylanthraquinone (secondary product from **3**)

Product **6** = 8-chloro-5-hydroxy-3-methylanthraquinone (secondary product from **4**)

SCHEME 2

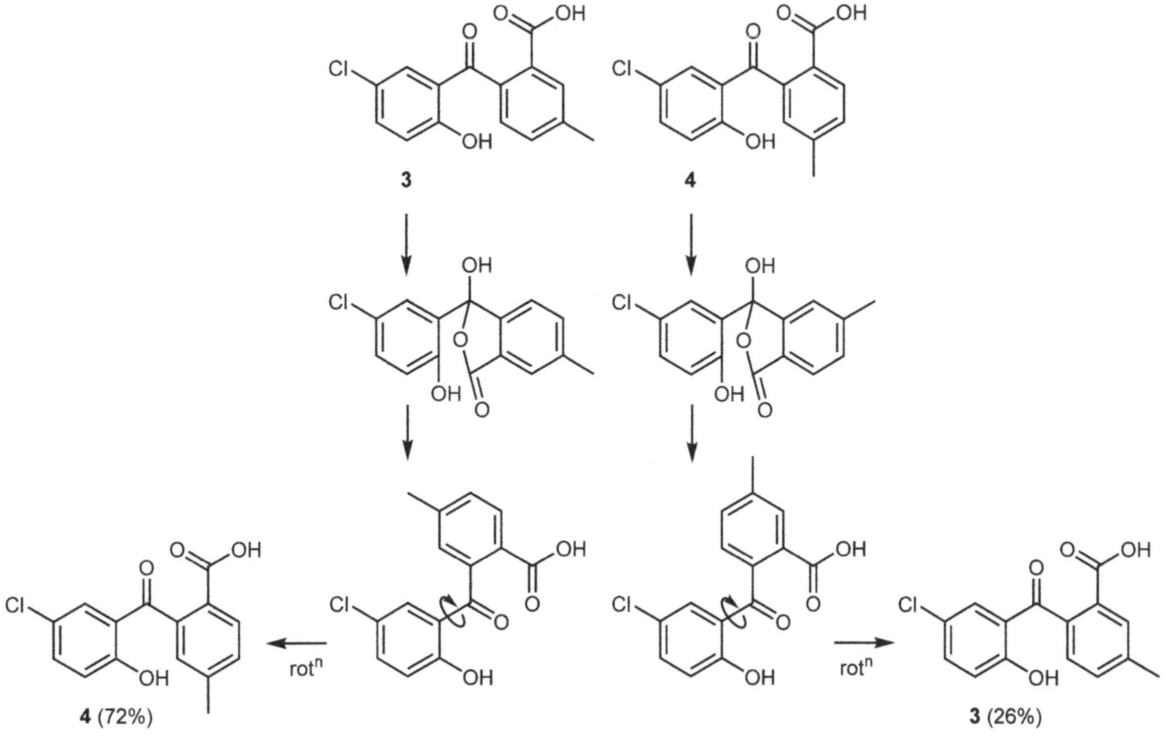

SCHEME 3

to be determined which mechanism is operating for the Hayashi rearrangement of **1** to **2**. Here we suggest possible experiments that could be conducted. With respect to distinguishing the intramolecular Hayashi or Sandin mechanisms from the intermolecular Wang mechanism it is possible to carry out a crossover experiment with, for example, *p*-bromophenol, in the presence of substrate **1** under identical reaction conditions. If an intramolecular mechanism is operative then only product **2** would be isolated; whereas, if an intermolecular mechanism is operative then a mixture of product **2** and product **7** would arise as shown in Scheme 4. A double isotopic labeling experiment involving adjacent carbon atoms as shown in Scheme 5 would be able to distinguish between the Hayashi and Sandin mechanisms. The Hayashi mechanism would result in a separation of the labels in the rearranged product 2. On the other hand, the Sandin mechanism proceeds via the *spiro* intermediate D_2 which can cleave in two ways yielding two different zwitterionic intermediates, X and Y, containing the acylium moiety which in turn lead to different product outcomes, either to rearranged 2 with separated labels or to unrearranged 1 with the same labeling arrangement as the substrate. Since two different acylium ions exist on their mechanistic pathways, the Hayashi and Sandin mechanisms can also be distinguished by a product study involving a trapping experiment with a nucleophilic species as shown in Scheme 6. The resulting products 8 and 9 are geometric isomers and can be readily distinguished by the usual spectroscopic techniques. We note that the structure of product 8 resembles that of substrate 1 and therefore corresponds to an unrearranged product; whereas, product 9 is structurally analogous to product 2 and corresponds to the rearranged product. The Hayashi mechanism having one kind of acylium ion C_1 is trappable yielding product 8 exclusively. The Sandin mechanism predicts that the *spiro* intermediate D_2 could cleave either by C_a—C_b or by C_c—C_b bond scission leading to acylium ions F_2' or F_3', respectively. Trapping of these intermediates potentially can lead to both products 8 and 9. (Note that acylium F_3' ion is a dehydrated form of acylium ion C_1.) In the Hayashi mechanism case, the success of intermolecular trapping of acylium ion C_1 is contingent on the criterion that the rate of trapping is faster than the rate of intramolecular rearrangement to intermediate D_1.

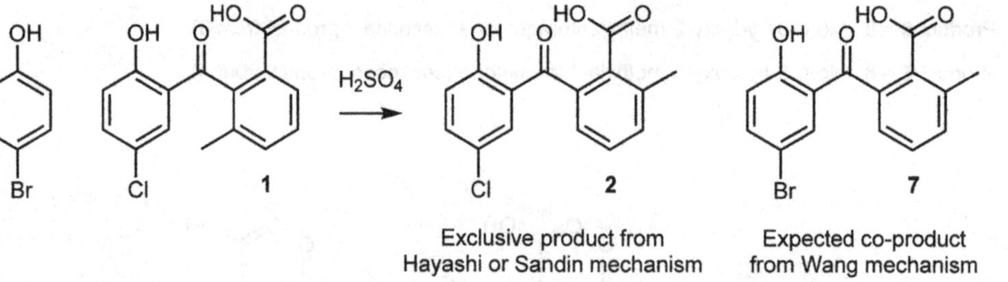

Exclusive product from
Hayashi or Sandin mechanism

Expected co-product
from Wang mechanism

SCHEME 4

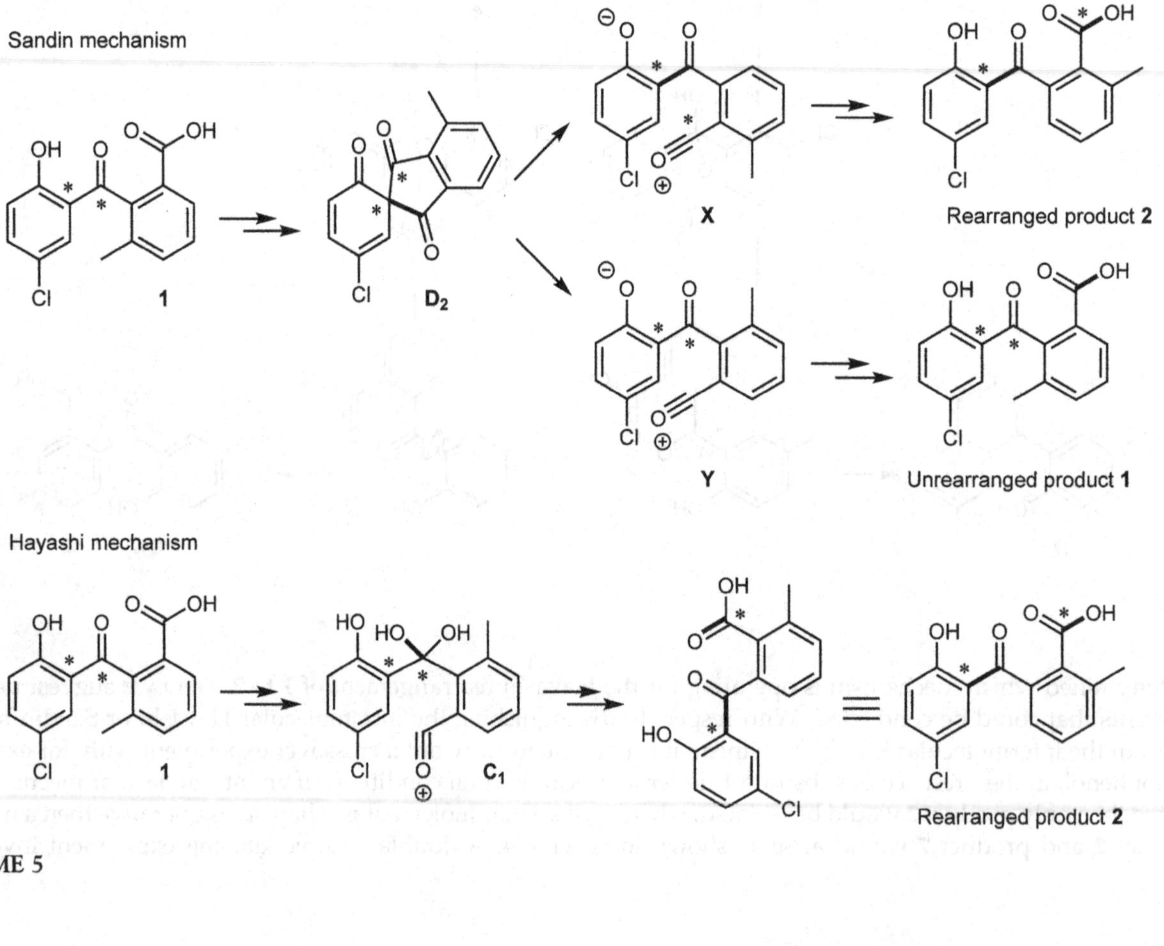

Sandin mechanism

Hayashi mechanism

Rearranged product 2

Unrearranged product 1

Rearranged product 2

SCHEME 5

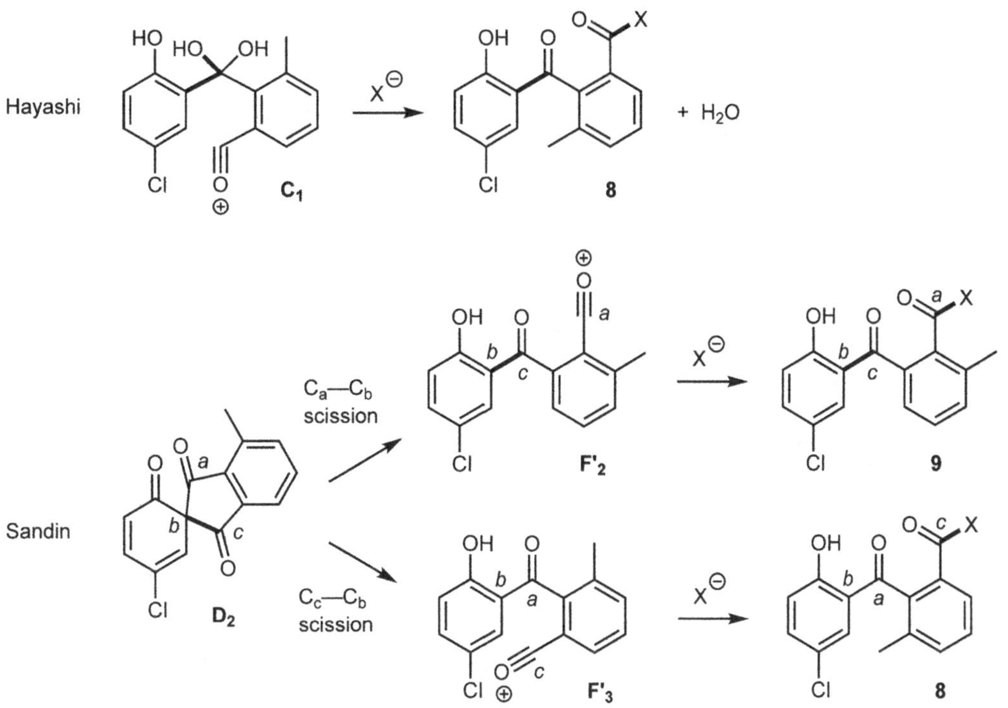

SCHEME 6

ADDITIONAL RESOURCES

For a synthetic example of the Hayashi rearrangement, see the work of Opitz et al.[5]

References

1. Hayashi M. CCCXXXVI.—A new isomerism of halogenohydroxybenzoyltoluic acids. *J Chem Soc.* 1927;1927:2516–2527. https://doi.org/10.1039/JR9270002516.
2. Sandin RB, Melby R, Crawford R, McGreer D. The Hayashi rearrangement of substituted *o*-benzoylbenzoic acids. *J Am Chem Soc.* 1956;78(15):3817–3819. https://doi.org/10.1021/ja01596a069.
3. Wang Z, ed. *Comprehensive Organic Name Reactions and Reagents.* New York, NY: John Wiley & Sons, Inc.; 2010:1347–1349. https://doi.org/10.1002/9780470638859.conrr373
4. Hayashi M. CXCII.—A new isomerism of halogenohydroxybenzoyltoluic acids. Part II. 2-(5'-Chloro-2'-hydroxybenzoyl)-5(4?)-methylbenzoic acid. *J Chem Soc.* 1930;1930:1513–1519. https://doi.org/10.1039/JR9300001513.
5. Opitz A, Roemer E, Haas W, Goeris H, Werner W, Graefe U. Regioselective synthesis of benzo[g]isoquinoline-5,10-dione derivatives as DNA intercalators. *Tetrahedron.* 2000;56(29):5147–5155. https://doi.org/10.1016/S0040-4020(00)00186-1.

Question 122: 2-Oxabicyclo[1.1.0]butanone Is Not Easily Accessible

In 1967 American chemists attempted the epoxidation of diphenylcyclopropenone **1** by using a variety of reagents including 25% hydrogen peroxide but were nevertheless unsuccessful managing to isolate desoxybenzoin **2** in 77% yield (Scheme 1).[1] Suggest a mechanism for this unexpected result.

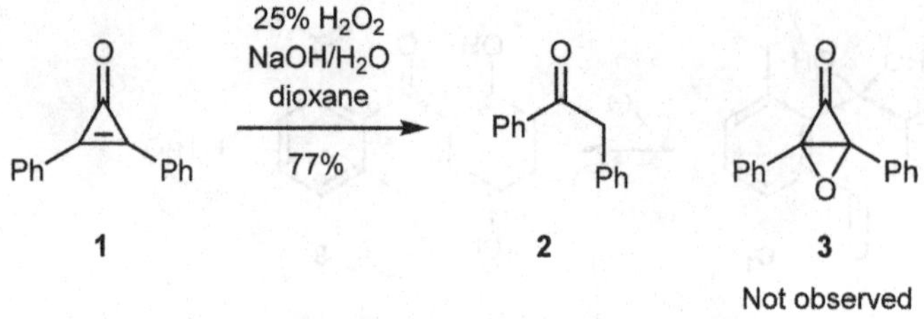

SCHEME 1

SOLUTION: AUTHORS' MECHANISM

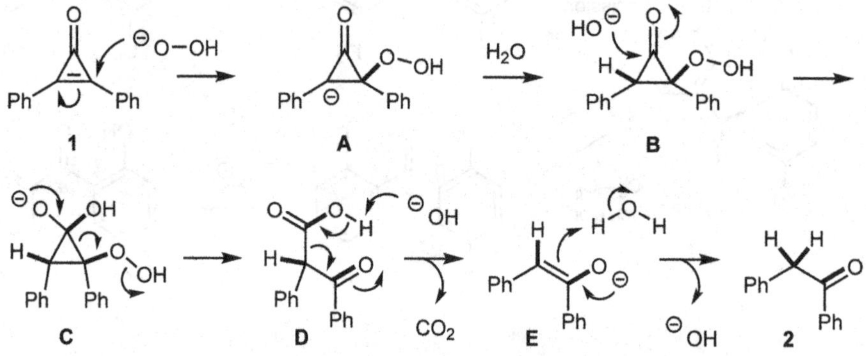

SOLUTION: ALTERNATIVE MECHANISM

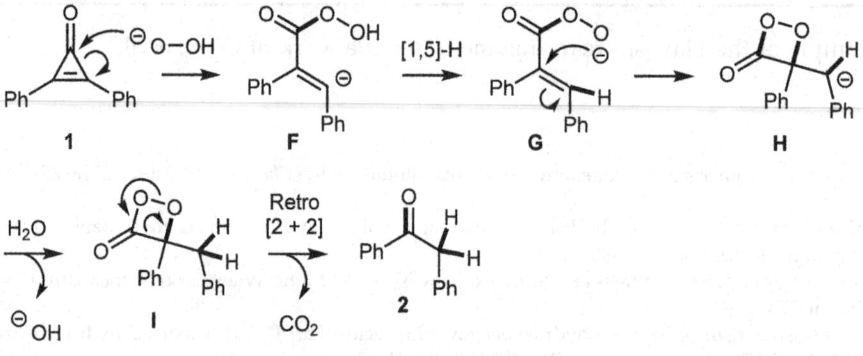

BALANCED CHEMICAL EQUATION

KEY STEPS EXPLAINED

To explain the transformation of **1** to **2** the authors proposed a mechanism which begins with hydroperoxide ion attack at the C=C bond of **1** to form carbanion **A** which is protonated by water to generate hydroperoxide **B**. Hydroxide ion then attacks the carbonyl group of **B** to form **C** with elimination of hydroxide ion to give 3-oxo-2,3-diphenyl-propionic acid **D**. Hydroxide ion returns to decarboxylate **D** and produce enolate **E** which undergoes ketonization to desoxybenzoin **2**. Since the authors wrote that their mechanism is just one possibility, in the absence of experimental and theoretical evidence to support this mechanism we here offer an alternative mechanism. Here, instead of attacking the C=C double bond, hydroperoxide ion attacks the carbonyl group of **1** to form carbanion **F** which is protonated by water yielding hydroperoxide **G**. Deprotonation of **G** by hydroxide ion then induces a cyclization to [1,2]dioxetan-3-one **I** which fragments to carbon dioxide and desoxybenzoin **2** in a retro [2+2] fashion.

A possible way to distinguish the mechanisms experimentally is to run the reaction in the presence of ^{18}O labeled water. The authors' mechanism predicts that one of the oxygen atoms of the carbon dioxide by-product would be ^{18}O labeled; whereas, the alternative mechanism predicts that no label will be incorporated in that by-product as shown in Scheme 2.

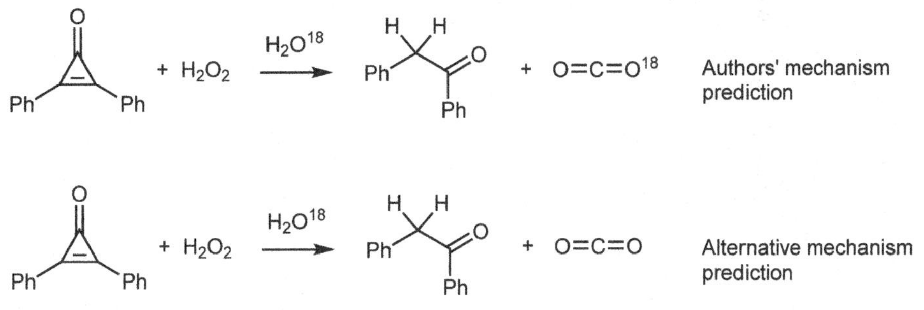

SCHEME 2

ADDITIONAL RESOURCES

For examples of transformations that involve interesting rearrangements of cyclopropanones, see Refs. 2–4.

References

1. Marmor S, Thomas MM. Reaction of diphenylcyclopropenone with alkaline hydrogen peroxide. *J Org Chem.* 1967;32(1):252. https://doi.org/10.1021/jo01277a074.
2. Yusubov MS, Zholobova GA, Filimonova IL, Chi K-W. New oxidative transformations of alkenes and alkynes under the action of diacetoxyiodobenzene. *Russ Chem Bull.* 2004;53(8):1735–1742. https://doi.org/10.1007/s11172-005-0027-8.
3. Rubina M, Rubin M. Rearrangement of cyclopropylborane into boretane. *Chem Heterocycl Compd.* 2012;48(5):807–821. https://doi.org/10.1007/s10593-012-1060-9.
4. Liguori A, Sindona G, Uccella N. N,O-heterocycles. Part 18. Regiochemistry and site selectivity of N-alkylhydroxylamine addition to 2,3-diphenylcyclopropenone. *J Chem Soc Perkin Trans 1.* 1987;1987:961–965. https://doi.org/10.1039/P19870000961.

Question 123: A Simple Synthesis of the Indole Alkaloid Yuehchukene

In 1985 chemists from Hong Kong reacted 3-isoprenylindole **1** with silica gel and benzene containing a catalytic amount of trifluoroacetic acid to form the indole alkaloid yuehchukene **2** in 10% yield (Scheme 1).[1] Provide a mechanism for this transformation.

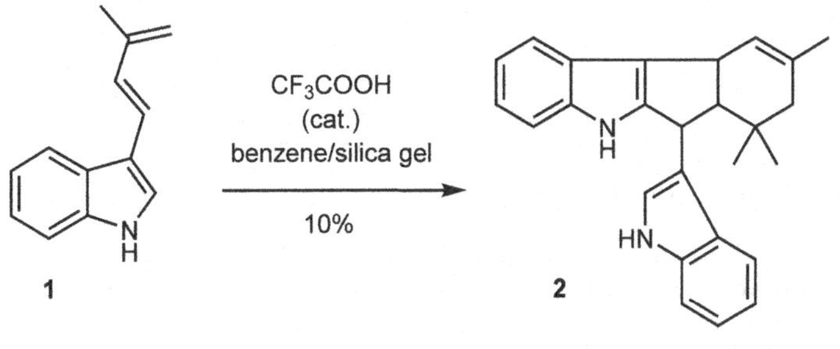

SCHEME 1

SOLUTION

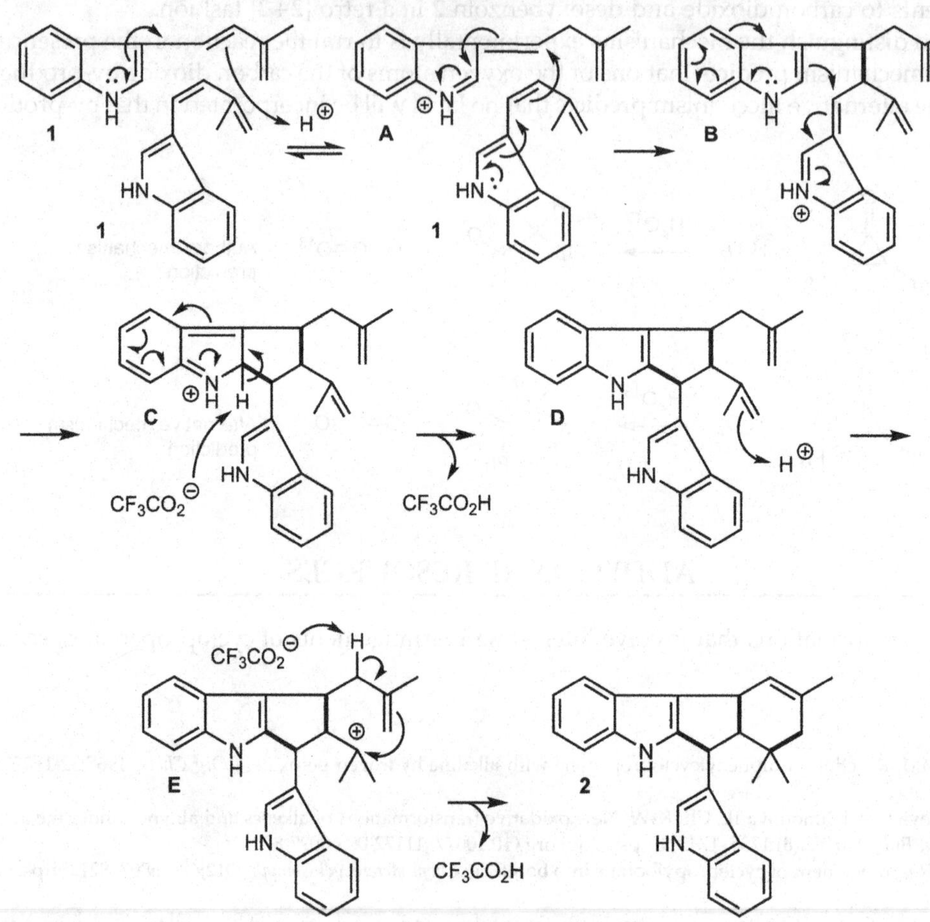

KEY STEPS EXPLAINED

The mechanism proposed by the authors for the transformation of **1** to **2** begins with protonation of one equivalent of 3-isoprenylindole **1** to indolinium ion **A** which reacts with a second equivalent of 3-isoprenylindole **1** to give indolinium ion **B**. **B** then cyclizes to indolinium ion **C** which is deprotonated to intermediate **D**. The propenyl moiety of **D** is protonated to the tertiary carbocation **E** which cyclizes to the alkaloid product yuehchukene **2** upon deprotonation of the isobutenyl group. The net ring construction mapping can be described as [(3+2)+(4+2)]. The authors did not carry out experimental or theoretical work to help support their proposed mechanism.

ADDITIONAL RESOURCES

For more examples of the work of these authors with the context of similar transformations, see Ref. 2. For an example of a very similar transformation which involves ketene chemistry and a mechanistic discussion, see the work of Shen et al.[3]

References

1. Cheng K-F, Kong Y-C, Chan T-Y. Biomimetic synthesis of yeuhchukene. *J Chem Soc Chem Commun*. 1985;1985:48–49. https://doi.org/10.1039/C39850000048.
2. Cheng K-F, Chan K-P, Kong Y-C, Ho D-D. Synthesis of (6S,6aS,7R,10S,10aR)-6-(indol-3-yl)-7,11,11-trimethyl-5,6,6a,7,8,9,10,10a-octahydro-7,10-methanoindeno[2,1-b]indole and its enantiomer: absolute configuration of active yuehchukene. *J Chem Soc Perkin Trans 1*. 1991;1991:2955–2959. https://doi.org/10.1039/P19910002955.
3. Shen J-H, Chen C-A, Chen B-H. Reaction of (dimethylvinylidene)carbene with indole-3-carbaldehyde and its application in the synthesis of β-(dehydroprenyl)indole-based natural products. *Chem Commun*. 1999;1999:203–204. https://doi.org/10.1039/A808733D.

Question 124: Isoquinoline Rearrangement

In 1965 chemists from the United Kingdom carried out the synthetic sequence shown in Scheme 1.[1] Interestingly, reaction of **2** with 2 N HCl followed by sodium borohydride reduction, gave the unexpected product **5** rather than the expected product **6**. Provide a mechanism to explain the transformation of **2** to **5**.

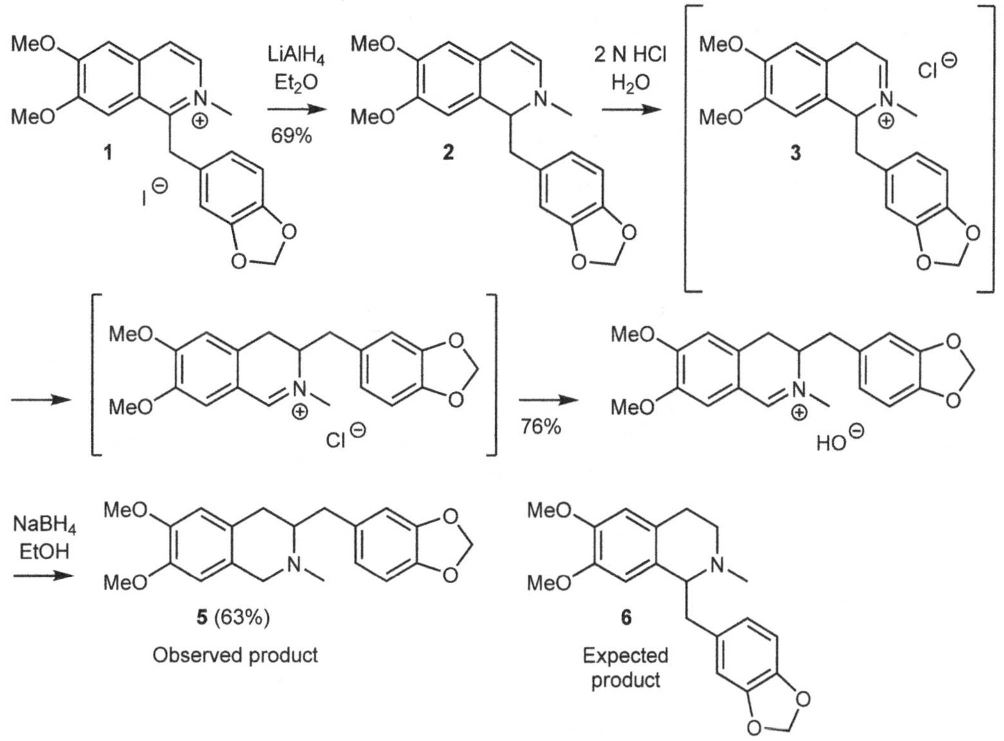

SCHEME 1

SOLUTION

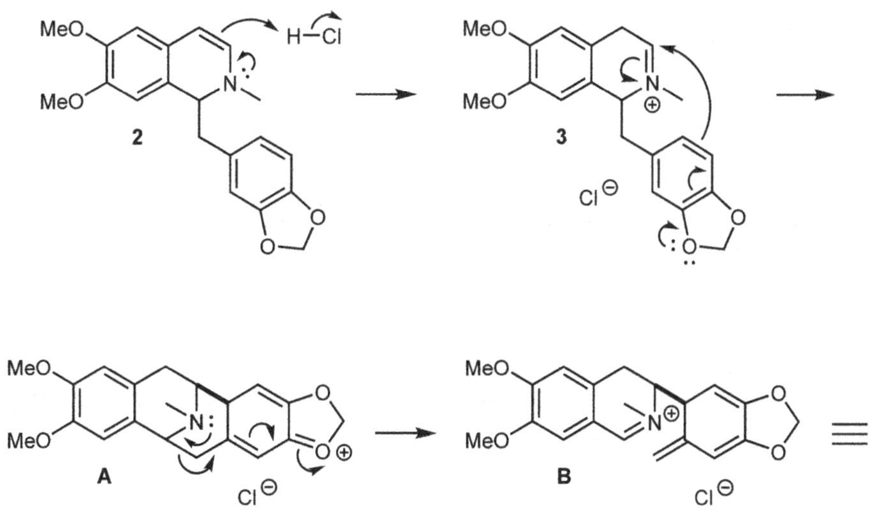

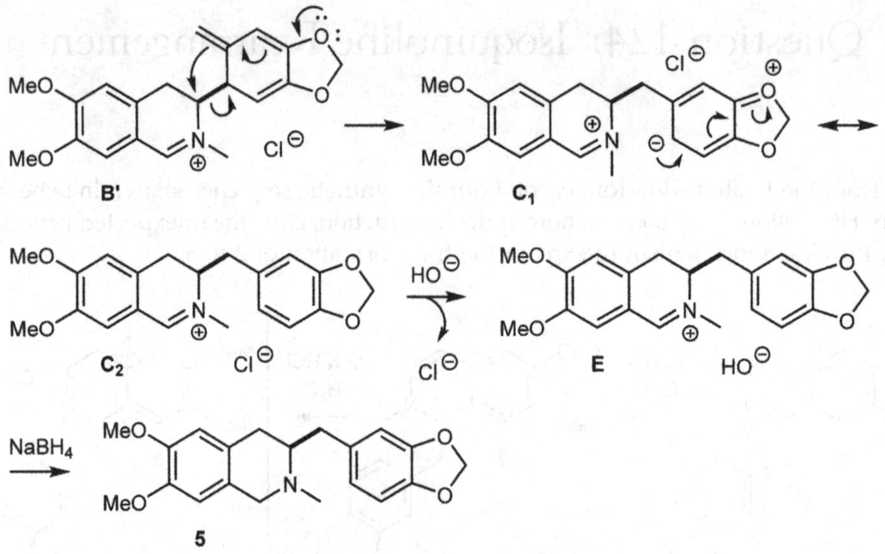

KEY STEPS EXPLAINED

The mechanism proposed by the authors begins with protonation of compound **2** to give iminium ion **3** which undergoes an allyl-type rearrangement sequence to intermediate **D** which features migration of the benzyl group from **B'** to **C_1/C_2**. The authors also observed a short-lived intense red-violet color during the course of this transformation which is probably due to the quinoid intermediate **A**. The color quickly changed to yellow which is due to intermediate **B/B'**. The final step of the mechanism is anion metathesis to give the hydroxide iminium salt **E** which is reduced using sodium borohydride to the final product **5**. We note that the authors did not carry out further experimental or theoretical analysis to help support their postulated mechanism.

ADDITIONAL RESOURCES

We recommend an article which also discusses an unexpected cyclization transformation in the context of polycyclic fused amidines.[2]

References

1. Dyke SF, Sainbury M. 1,2-Dihydroisoquinolines—I. Rearrangement. *Tetrahedron*. 1965;21(8):1907–1915. https://doi.org/10.1016/S0040-4020(01)98330-9.
2. Cookson RF, Nowotnik DP, Parfitt RT, Airey JE, Kende AS. Polycyclic fused amidines. Part III. An unexpected mode of cyclisation of 2-phenacylisoquinolinium bromide. *J Chem Soc Perkin Trans 1*. 1976;1976:201–204. https://doi.org/10.1039/P19760000201.

Question 125: Multiple Sigmatropic Rearrangements

During a 1988 study of [3,3]-sigmatropic rearrangements, Indian chemists undertook the three reactions shown in Scheme 1.[1] Provide mechanisms to explain why different amounts of products **2** and/or **3** are formed depending on the choice of reaction conditions.

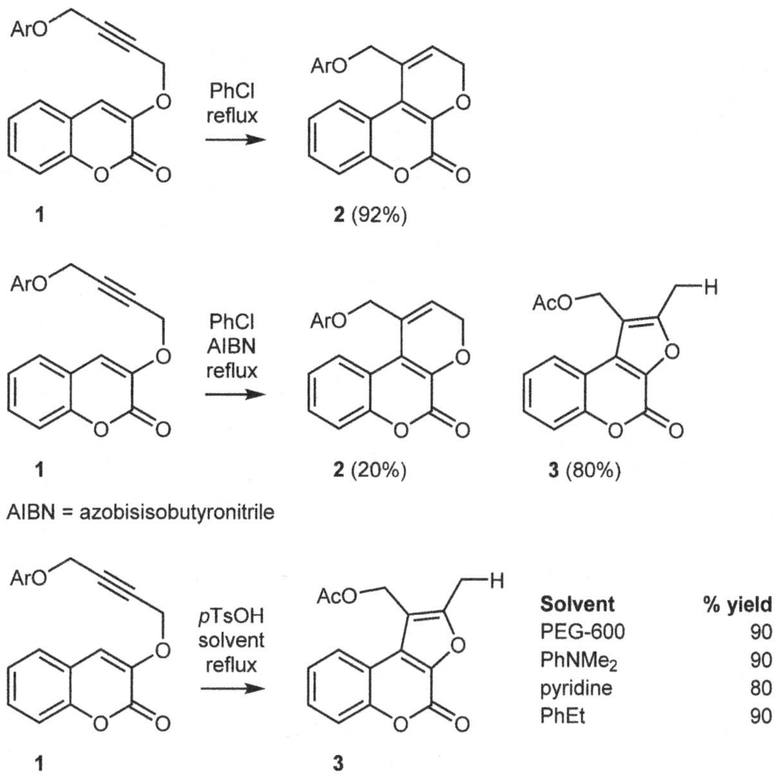

Solvent	% yield
PEG-600	90
$PhNMe_2$	90
pyridine	80
PhEt	90

SCHEME 1

SOLUTION: 1 TO 2 IN PhCl

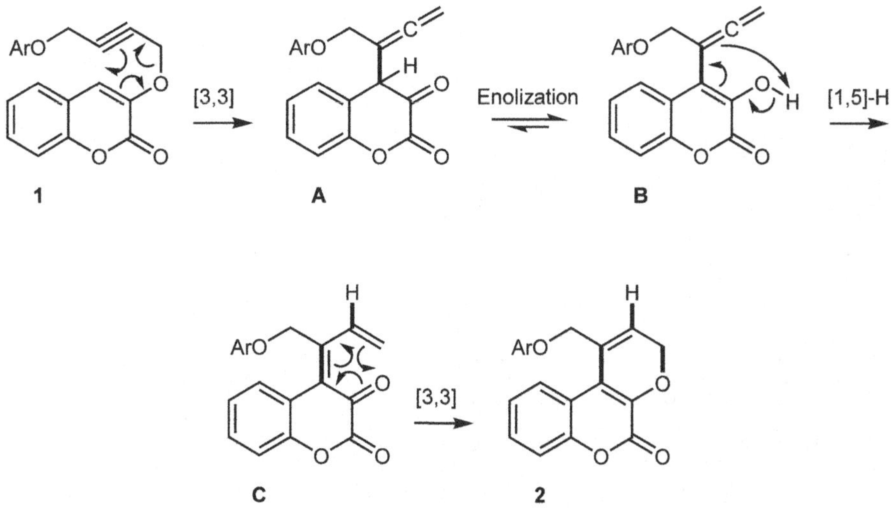

SOLUTION: 1 TO 2 AND 3 IN PhCl AND AIBN

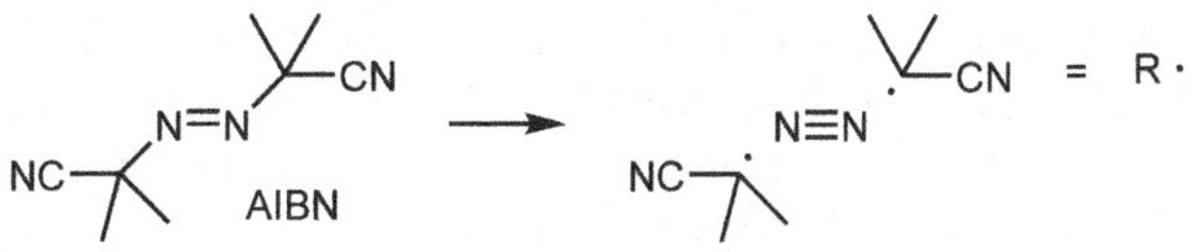

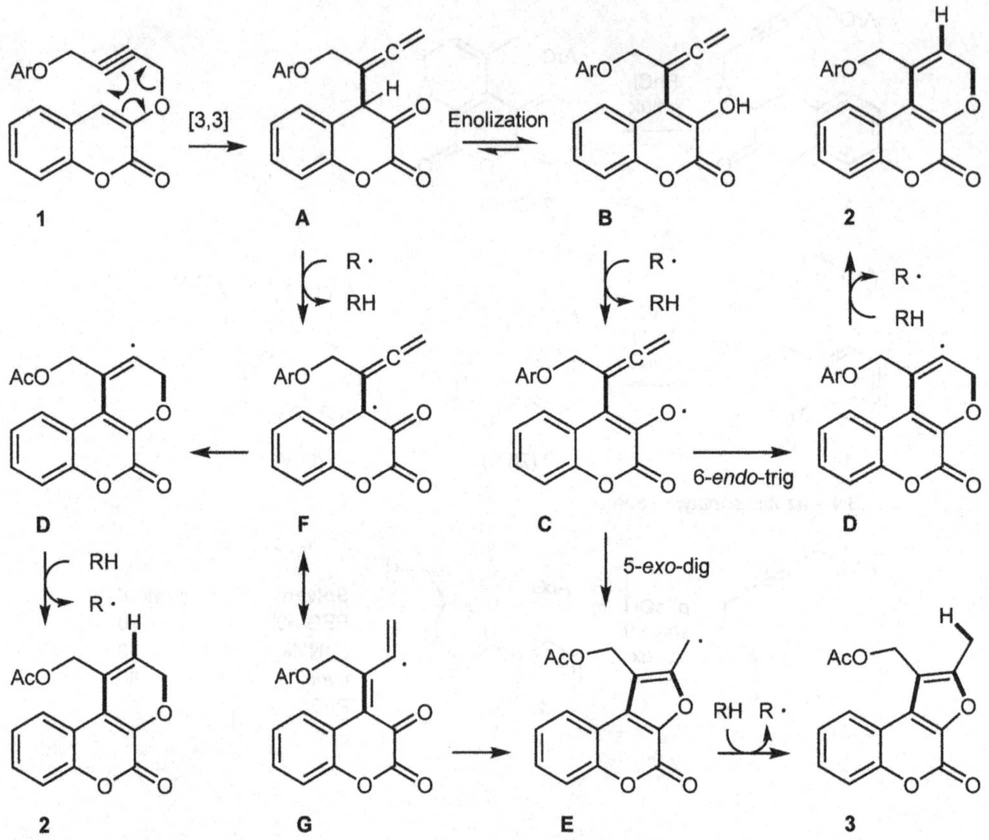

SOLUTION: 1 TO 3 IN pTsOH

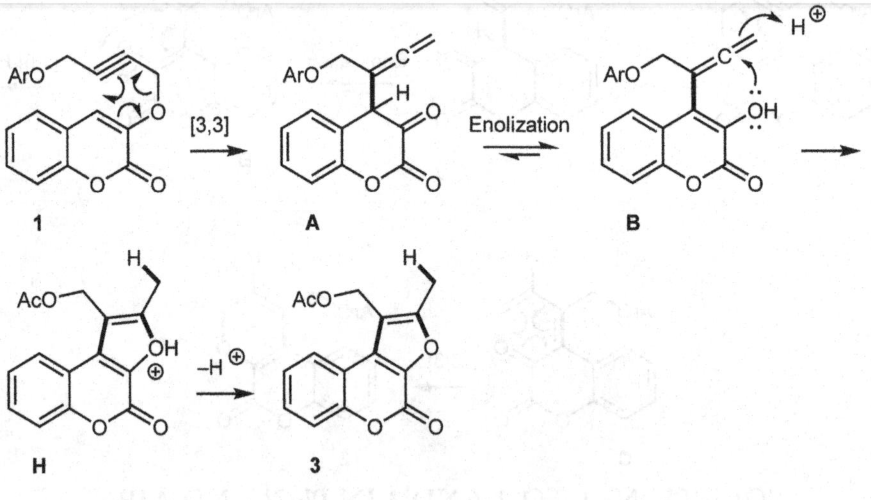

KEY STEPS EXPLAINED

The thermal rearrangement of **1** to **2** in PhCl can be rationalized by having an initial [3,3] sigmatropic rearrangement of **1** to allene **A** followed by enolization to allene **B**. Intermediate **B** then undergoes a [1,5]-hydrogen shift to give diene **C** which undergoes a second [3,3] sigmatropic rearrangement to yield product **2**.

The scenario changes when AIBN is added. Here we require a radical-type mechanism which begins with the same initial thermal processes as before leading to allenes **A** and **B**. Nevertheless, now we have azobisisobutyronitrile undergoing fragmentation in a radical fashion to form two equivalents of dimethylcyanomethane radical and nitrogen gas.

Products **2** and **3** can be generated from both allenes **A** and **B** via a sequence of radical hydrogen abstraction steps. Product **2** can be generated from the sequence **A** to **F** to **D** to **2** or the sequence **B** to **C** to **D** to **2**. Similarly, product **3** can be generated from the sequence **A** to **F** to **G** to **E** to **3** or the sequence **B** to **C** to **E** to **3**. The higher yield of product **3** arises as a consequence of two effects. Firstly, the equilibrium between enol allene **B** and keto allene **A** is shifted toward the enol form since there is a conjugation of π bonds in the diene moiety of **B** between the enol C=C bond and the allene C=C bond. Secondly, from radical **C** the rate of the 5-*exo*-dig ring closure to radical **E** must be faster than that of the 6-*endo*-trig ring closure to radical **D**.

Lastly, when the reaction is carried out in acidic medium, we have the same initial thermal process that forms keto allene **A** and enol allene **B**. Nevertheless, in an acidic medium we have **B** cyclize to oxonium ion **C** via protonation of the terminal C=C bond of the allene group. A final deprotonation leads to product **3**.

We note that the authors did not undertake further experimental or theoretical analysis to help support any of the postulated mechanisms. One would imagine that a theoretical study would be very helpful in identifying the energetics behind the steps mentioned before which we believe to be responsible for the difference in yields of **2** and **3** especially with respect to the AIBN transformation.

ADDITIONAL RESOURCES

We recommend an article which discusses sigmatropic rearrangements in *ortho*-substituted aromatic compounds in the context of flash thermolysis.[2] We also recommend an article which describes an interesting transformation involving adamantane and enamine chemistry.[3]

References

1. Majumdar KC, De RN, Khan AT, Chattopadhyay SK, Dey K, Patra A. Studies of [3,3]sigmatropic rearrangements: rearrangement of 3-(4-p-tolyloxybut-2-ynyloxy)[1]benzopyran-2-one. *J Chem Soc Chem Commun*. 1988;1988:777–779. https://doi.org/10.1039/C39880000777.
2. De Champlain P, Luche J-L, Marty RA, de Mayo P. Flash thermolysis: multiple sigmatropic rearrangements in ortho-substituted aromatic compounds. *Can J Chem*. 1976;54(23):3749–3756. https://doi.org/10.1139/v76-538.
3. Hickmott PW, Ahmed MG, Ahmed SA, Wood S, Kapon M. Enamine chemistry. Part 29. Synthesis of adamantane derivatives from α,β-unsaturated acid chlorides and 4,4-disubstituted cyclohexanone enamines. Multiple [3,3] sigmatropic rearrangement transition state stereochemistry. X-Ray analysis. *J Chem Soc Perkin Trans 1*. 1985;1985:2559–2571. https://doi.org/10.1039/P19850002559.

Question 126: A Vilsmeier-Induced Annulation to Benzene

In 1979 chemists from India carried out an anomalous Vilsmeier-Haack reaction where the treatment of 5-allyl-1,2,3-trimethoxybenzene **1** with a mixture of *N*-methylformanilide and phosphorus oxychloride gave the dihydronaphthalene derivative **2** in 58% yield (Scheme 1).[1] The authors argued that the reaction constituted two formylations and a reduction and used deuterium labeling experiments to figure out that the hydrogen needed for reduction came from the methyl group of *N*-methylformanilide. Suggest a mechanism for this transformation that is consistent with these experimental observations.

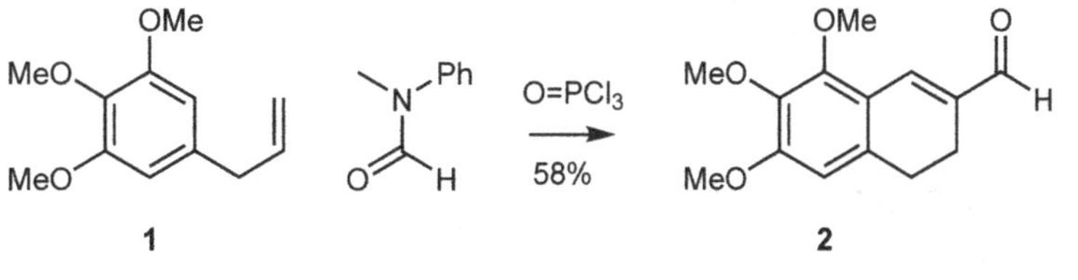

SCHEME 1

SOLUTION

Vilsmeier-Haack reagent
1
A
B
HCl
C
D
E
[1,3]-H
F
[1,5]-H
G
H
I
H+ shift
J
K
H+ shift
2
HCl

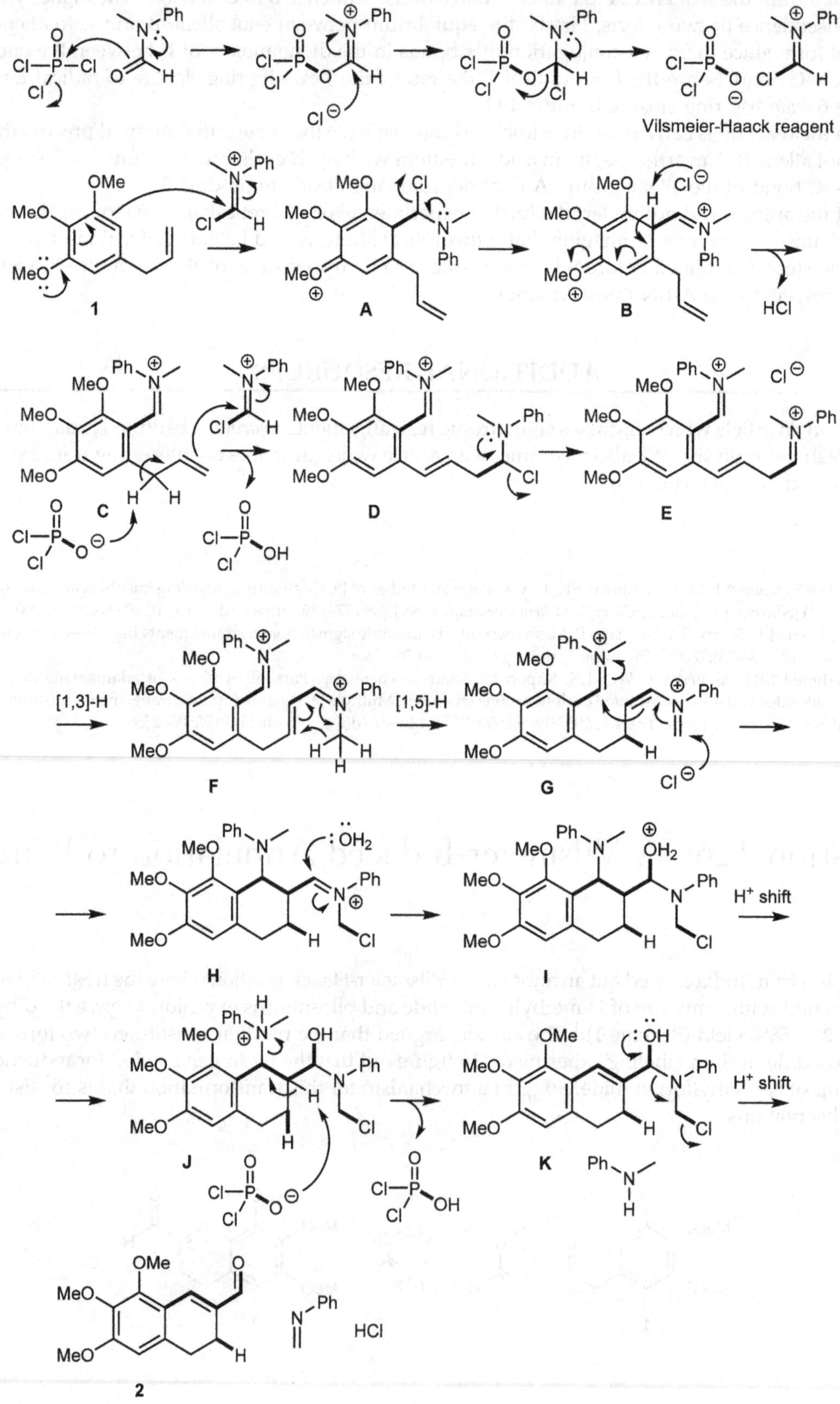

BALANCED CHEMICAL EQUATION

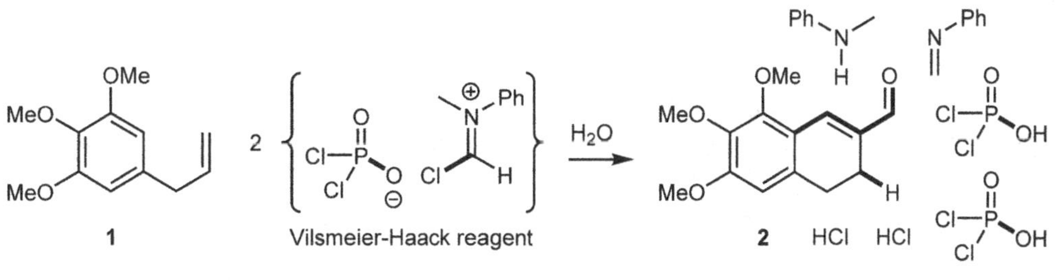

KEY STEPS EXPLAINED

The mechanism proposed by the authors to explain the transformation of **1** to **2** begins with reaction of phosphorus oxychloride and *N*-methyl-*N*-phenyl-formamide to form the Vilsmeier-Haack reagent. Compound **1** reacts with the Vilsmeier-Haack chloroiminium ion through activation by the electron-donating methoxy group (*para* or *ortho*) to form the aromatic formylation quinonoid intermediate **A** which undergoes chloride elimination to give **B**. Chloride ion then deprotonates **B** restoring the aromatic ring and the resulting iminium ion **C** reacts with a second equivalent of Vilsmeier-Haack chloroiminium ion to form the olefinic formylation intermediate **D**. A sequence involving chloride elimination, [1,3]-protic shift, and [1,5]-protic shift leads to the reduction intermediate **G** which is attacked by chloride ion to form the cyclized imine intermediate **H**. Water adds to this intermediate to form **I** which undergoes protic shift to form **J** which is deprotonated by the phosphorus oxychloride anion to form intermediate **K** after elimination of *N*-methylaniline by-product. Elimination of *N*-methylmethanimine by-product and HCl from **K** leads to formation of the aldehyde group in the final product **2**. This second by-product can further hydrolyze to aniline and formaldehyde with a second equivalent of water under the acidic work-up conditions.

To help support this mechanism, the authors had to figure out where the hydrogen atom that caused the reduction in this transformation came from. The authors first considered that this hydrogen atom might come from "another molecule of the starting compound or a product derived from it (a disproportionation reaction)."[1] This possibility was rejected in light of the fact that product **2** was isolated in no less than 58% yield. Next, the authors wondered whether the hydrogen atom might come from water or the CHO group of *N*-methylformanilide reagent or the CH_3 group of this reagent. To test these possibilities, the authors ran the same experiment three times using either deuterium-labeled water (i.e., D_2O) or deuterium-labeled *N*-methylformanilide (i.e., either $PhN(CDO)CH_3$ or $PhN(CHO)CD_3$). We represent the results of these experiments following the proposed mechanism before in Scheme 2 (D_2O), Scheme 3 ($PhN(CDO)CH_3$), and Scheme 4 ($PhN(CHO)CD_3$). As shown, the isolation of products **2** (D_2O), **2-d₂** ($PhN(CDO)CH_3$) and **2-d** ($PhN(CHO)CD_3$) is entirely consistent with the proposed mechanism. We also note that

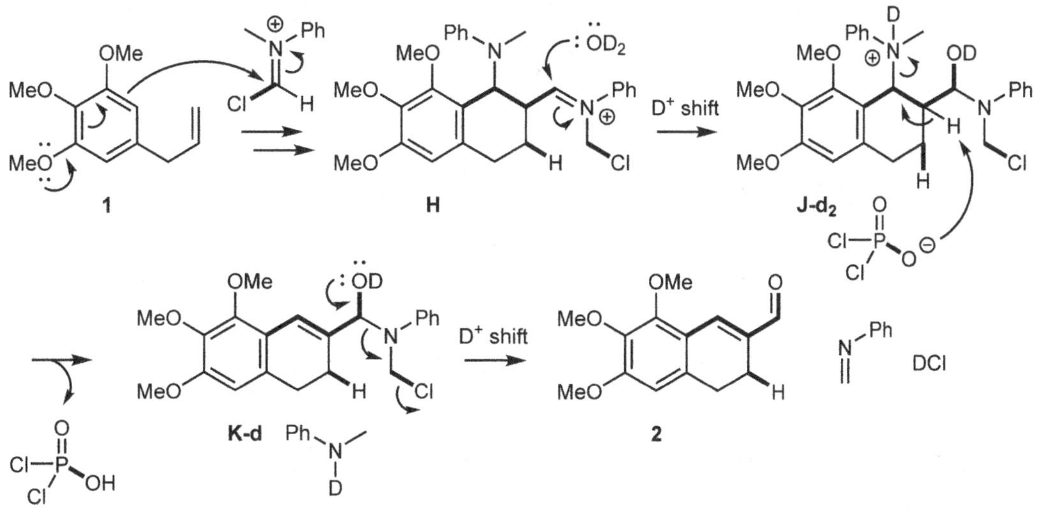

SCHEME 2

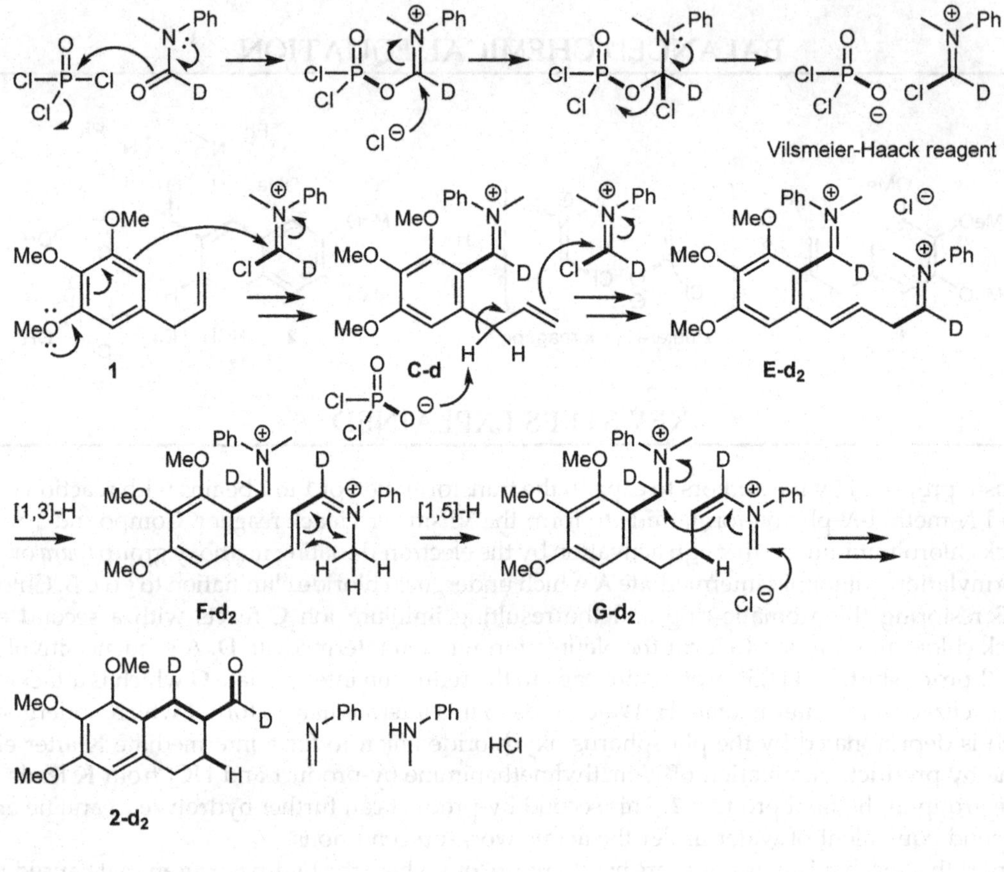

SCHEME 3

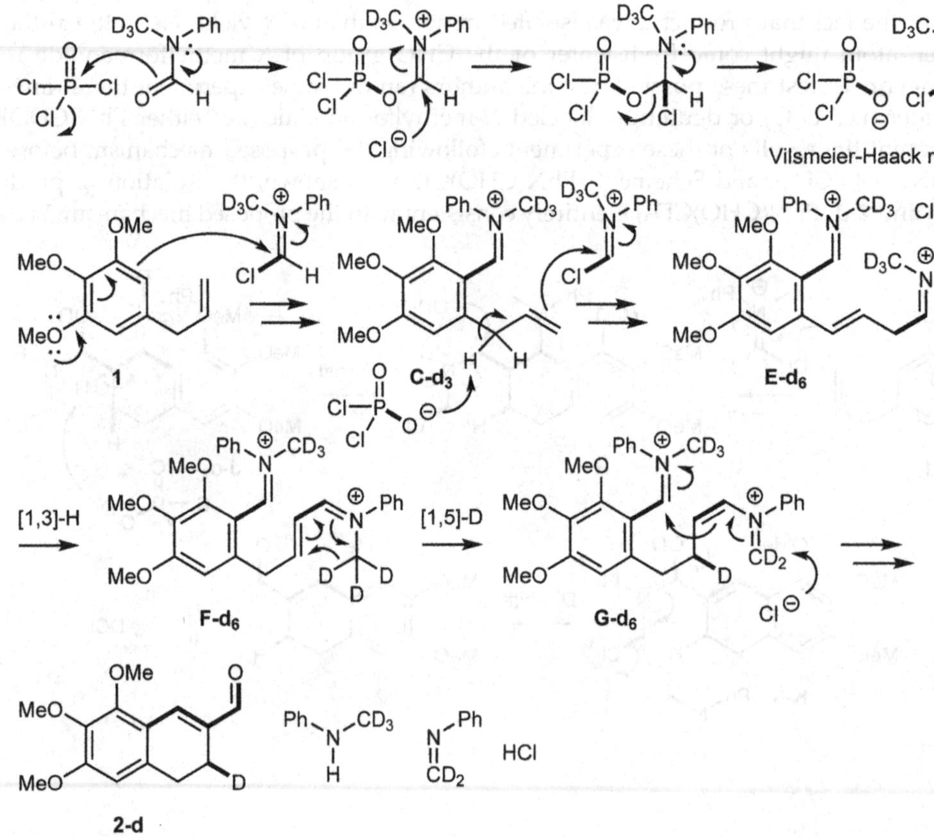

SCHEME 4

although the authors did not carry out a theoretical analysis to support their proposed mechanism, in a later paper they did experiment with various analogs of **1** to see whether formylation at the aryl group was preferred to formylation at the olefinic group.[2] Indeed, the results of their experimental work helped support the hypothesis that formylation at the aryl group is preferred to formylation at the olefinic group in most cases. For more details on those experiments, we refer the reader to Ref. 2.

ADDITIONAL RESOURCES

We recommend two articles which discuss the mechanisms of interesting transformations involving the Vilsmeier-Haack reaction.[3,4]

References

1. Narasimhan NS, Mukhopadhyay T. An anamolous reaction product in Vilsmeier-Haack reaction. *Tetrahedron Lett.* 1979;20(15):1341–1342. https://doi.org/10.1016/S0040-4039(01)86144-X.
2. Narasimhan NS, Mukhopadhyay T, Kusurkar SS. Vilsmeier-Haack reaction on methoxyallylbenzenes. *Indian J Chem Sect B.* 1981;20:546–548.
3. Rajanna KC, Solomon F, Ali MM, Saiprakash PK. Kinetics and mechanism of Vilsmeier-Haack synthesis of 3-formyl chromones derives from o-hydroxy aryl alkyl ketones: a structure reactivity study. *Tetrahedron.* 1996;52(10):3669–3682. https://doi.org/10.1016/0040-4020(96)00043-9.
4. Zhang R, Zhang D, Liang Y, Zhou G, Dong D. Vilsmeier reaction of 3-aminopropenamides: one-pot synthesis of pyrimidin-4(3H)-ones. *J Org Chem.* 2011;76:2880–2883. https://doi.org/10.1021/jo101949y.

Question 127: Isopropylidene to 2-Alkyne Conversion

In 1986 Abidi reported that the reaction of geraniol **1** with a large excess of sodium nitrite in aqueous acetic acid at 60°C gave alkyne **2** in 98% yield (Scheme 1).[1] In the subsequent 9 years, three groups worked to analyze the synthetic potential and mechanistic details of this transformation.[2–4] Provide a mechanistic explanation for this transformation.

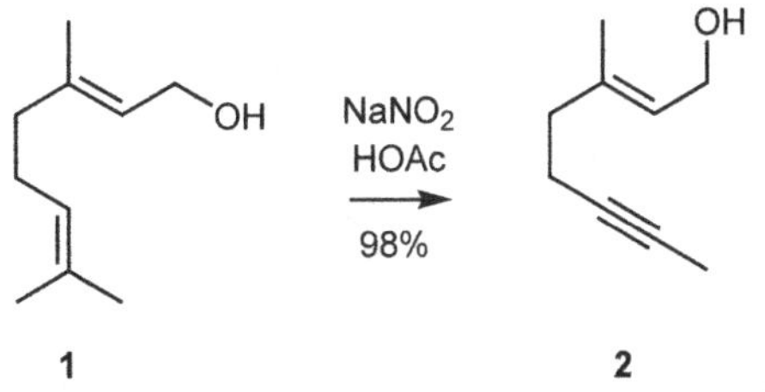

1
Geraniol

2

SCHEME 1

SOLUTION: COREY[2] MECHANISM

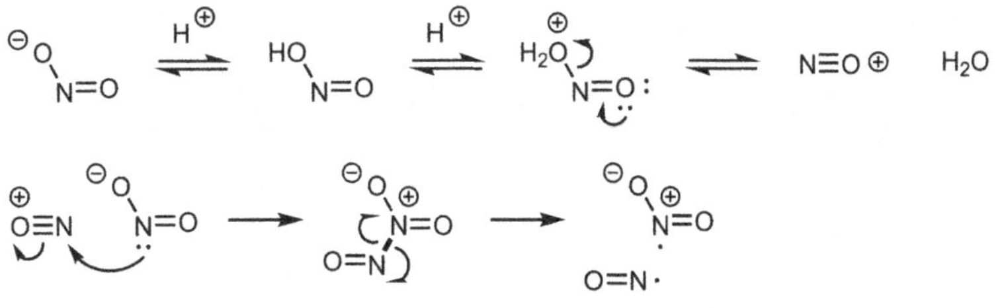

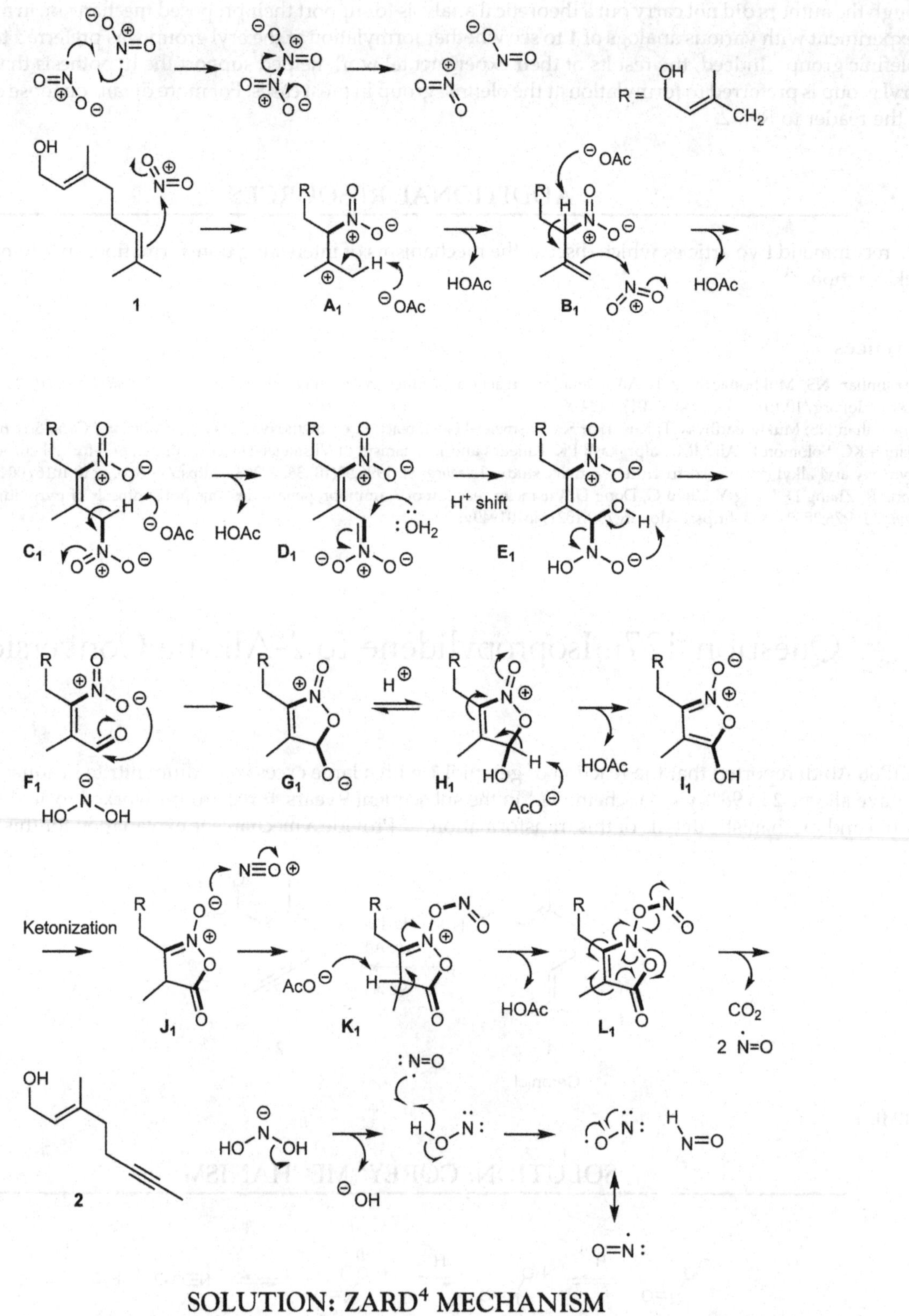

R =

1 A₁ B₁

HOAc HOAc

C₁ D₁ H⁺ shift E₁

F₁ G₁ H⁺ H₁ HOAc I₁

Ketonization J₁ K₁ L₁ CO₂ 2 N=O

2

SOLUTION: ZARD⁴ MECHANISM

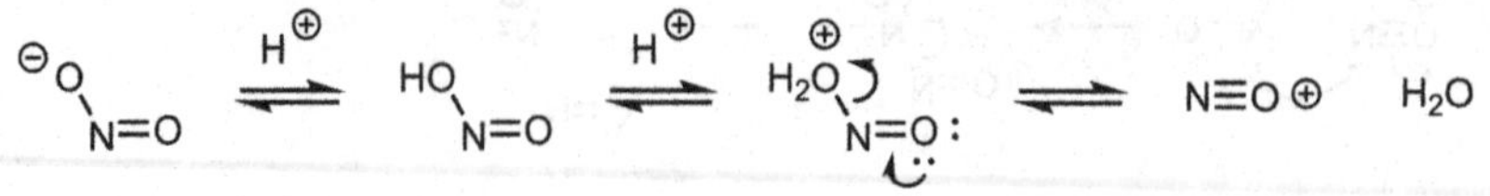

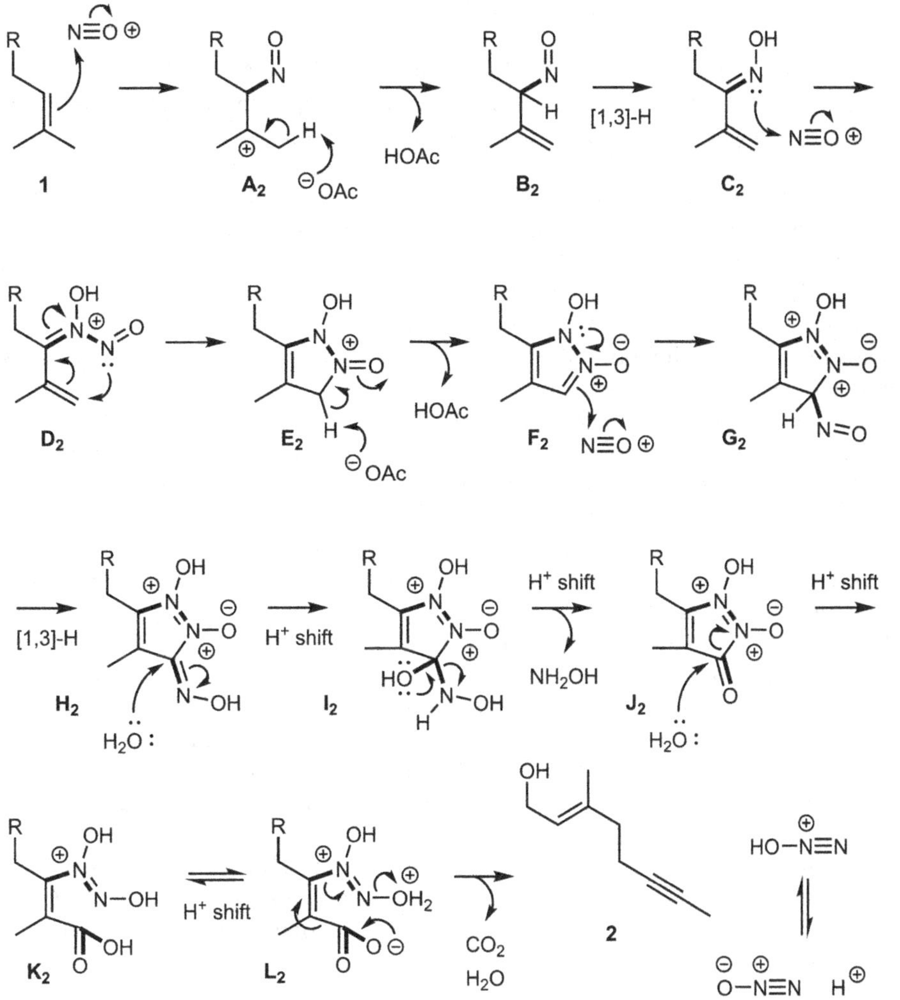

BALANCED CHEMICAL EQUATIONS

Corey mechanism:

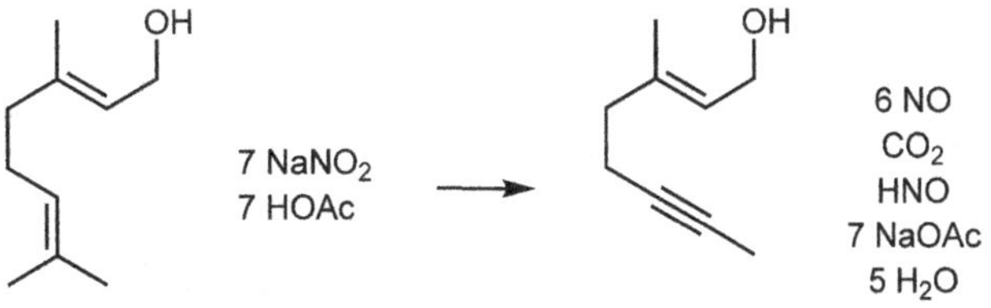

$$1 \xrightarrow[\text{7 HOAc}]{\text{7 NaNO}_2} 2$$

6 NO
CO_2
HNO
7 NaOAc
5 H_2O

Zard mechanism:

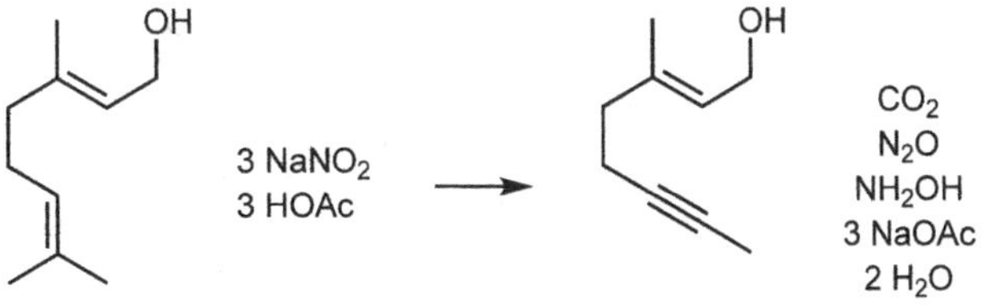

$$1 \xrightarrow[\text{3 HOAc}]{\text{3 NaNO}_2} 2$$

CO_2
N_2O
NH_2OH
3 NaOAc
2 H_2O

KEY STEPS EXPLAINED

Concerning Corey's proposed mechanism, which features a nitrite-mediated Nef hydrolysis (from intermediate C_1 to the aldehyde F_1), there are several pieces of evidence that support it. For example, the identification of carbon dioxide as a by-product was confirmed by trapping the gases evolved in a saturated aqueous barium hydroxide solution whereupon barium carbonate was isolated in 89% yield. Furthermore, the nitro intermediate B_1 was isolated in 85% yield from the reaction of **1** with 25 equivalents of sodium nitrite after 30 min at 60°C. Further heating of isolated B_1 at 85°C for 4 h in the presence of 27 equivalents of sodium nitrite resulted in the production of acetylene **2** in 38% yield. Moreover, the intermediacy of the iso-xazolone N-oxide K_1 in the mechanism was checked by synthesizing the analog substrate shown in Scheme 2. Thus treatment of the nitro acid chloride with pyridine produced 3-benzyl-4-methyl-2-oxy-4H-isoxazol-5-one in situ which was then treated with sodium nitrite in aqueous acetic acid at 85°C for 30 min yielding but-2-ynyl-benzene in 54% yield.

The Zard group objected to the likelihood of the Nef hydrolysis transformation in the Corey mechanism, for which Corey et al. provided no experimental evidence, citing the prior work of Kornblum who showed that primary nitro compounds produce carboxylic acids, not aldehydes, under similar nitrite-mediated conditions (see Scheme 3).[5] They used the mechanism of the Kornblum reaction of nitroesters to oximes using nitrite reagents as a guide to their proposed mechanism (see Scheme 4).[5] They also noted that the carboxylic acid analog of aldehyde F_1 did not produce the alkyne product **2** when subjected to the reaction conditions.

Zard et al. also cited the prior work of Freeman and coworkers who studied the behavior of various α,β-unsaturated ketoximes toward nitrous acid as a precedent for their proposed mechanism.[6–8] To test the possibility that oxime C_2 was an intermediate along the mechanistic pathway, Zard et al. synthesized C_2 independently and subjected it to the same reaction conditions. They were able to obtain the alkyne product **2** in low yield (<20%), which unfortunately makes the transformation not synthetically useful due to the formation of many by-products and possible side products arising from other competing mechanisms that may be occurring in parallel.

We note that neither the Corey nor Zard groups could reproduce the high yields reported by Abidi which suggests that the original report was likely exaggerated or possibly faked. Given the unprecedented transformation it is not surprising that it drew the attention of well-established groups such as Corey and Zard who independently decided to check the veracity of the original claim that the yield was 98%.

Examination of the balanced chemical equations for both mechanisms shows differences in the stoichiometric coefficients of sodium nitrite and acetic acid reagents as well as the composition of by-products. Clearly, further work is required to determine the identities of all reaction by-products, in particular the fate of nitrite as either nitric oxide (NO) and hyponitrous acid (HNO) in the case of the Corey mechanism, or as nitrous oxide (N_2O) and hydroxylamine (NH_2OH) in the case of the Zard mechanism.

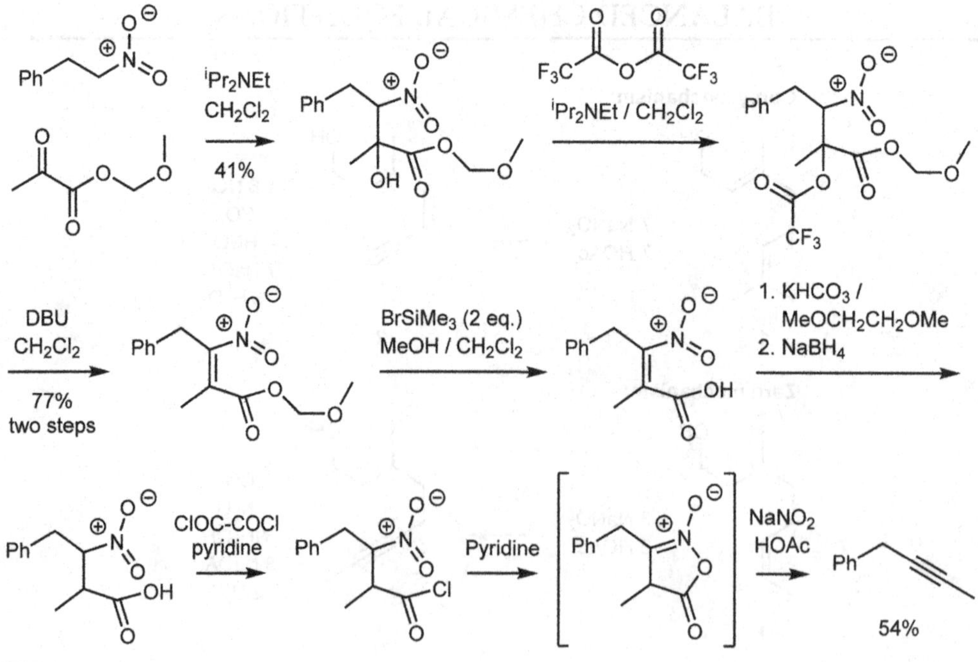

SCHEME 2

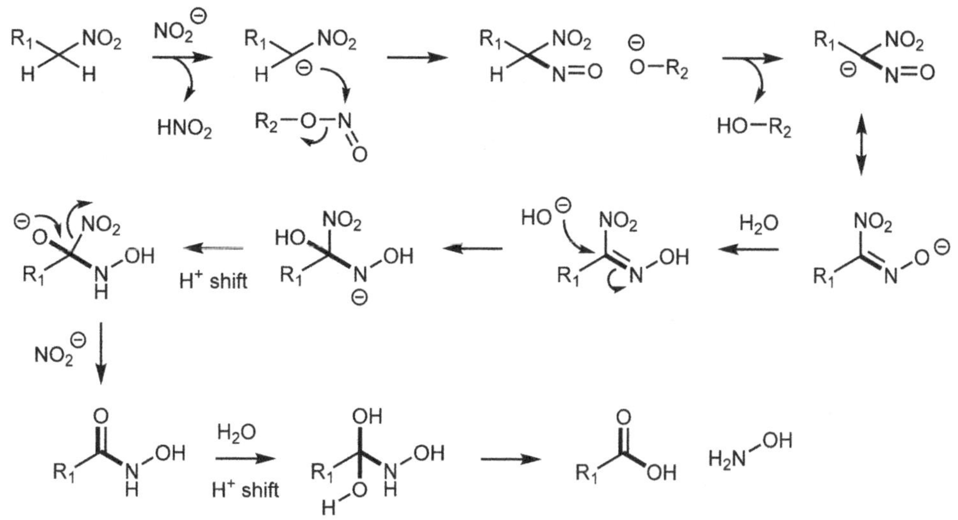

SCHEME 3

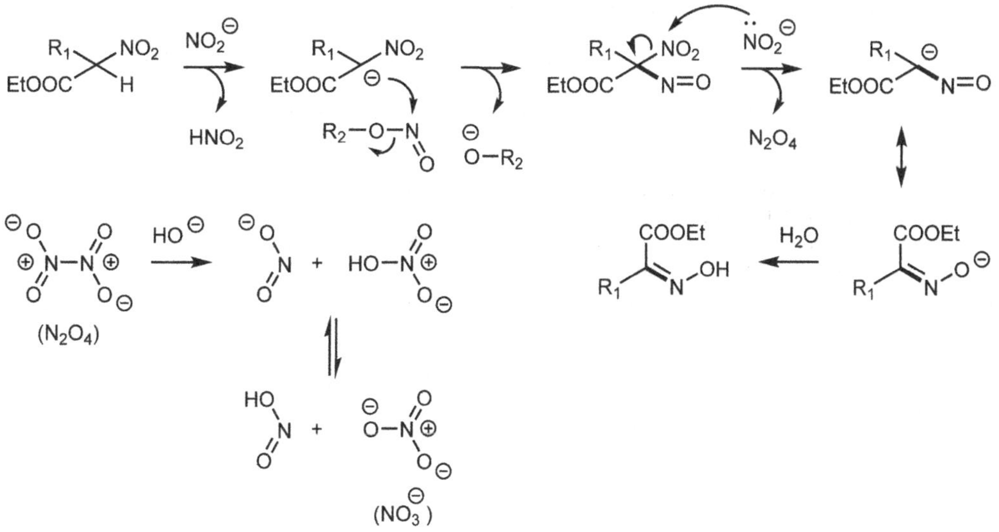

SCHEME 4

The Zard mechanism does not support Corey's claim that nitro intermediate B_1 and isoxazolone N-oxide K_1 are intermediates along the reaction pathway. Zard's mechanism replaces nitro intermediate B_1 with nitroso intermediate B_2 and replaces isoxazolone N-oxide J_1 with pyrazol-3-one N-oxide J_2. These discrepancies suggest that the mechanism of the reaction is still not yet fully elucidated. We also believe that a theoretical analysis would help shed light on which mechanism is operational for this interesting transformation.

ADDITIONAL RESOURCES

We highly recommend a later account of Zard where he describes his history with the transformation of **1** to **2** as well as recent synthetic developments and how his understanding of mechanism allowed him to design a new synthetic approach based on this transformation.[9] In that work he points out that there is no experimental evidence connecting dinitro intermediate C_1 and aldehyde F_1; however, he suggests that the isolated Corey nitro intermediate B_1 could be transformed to F_1 via an allylic nitroso intermediate A_3 as shown in Scheme 5. For this to happen, B_1 would have to be captured by nitrosonium ion instead of nitronium ion as seen in the Corey mechanism.

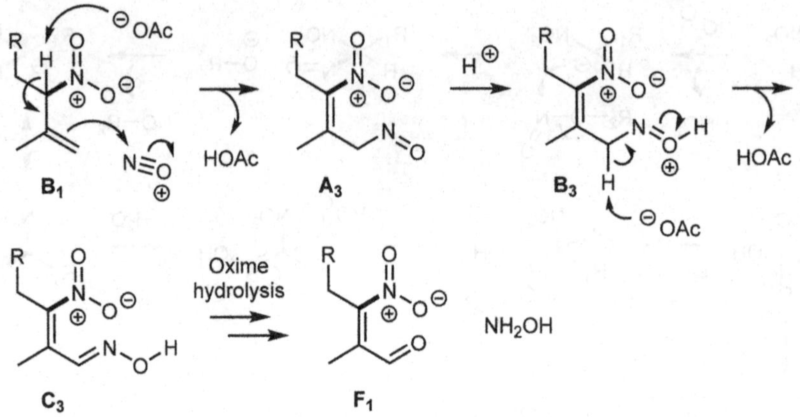

SCHEME 5

References

1. Abidi SL. Ethylidyne alkynes from isopropylidene olefins. *Tetrahedron Lett.* 1986;27(3):267–270. https://doi.org/10.1016/S0040-4039(00)83993-3.
2. Corey EJ, Seibel WL, Kappos JC. Mechanism of the nitrous acid-induced dealkylation of trisubstituted (terminal isopropylidene) olefins to form acetylenes. *Tetrahedron Lett.* 1987;28(42):4921–4924. https://doi.org/10.1016/S0040-4039(00)96659-0.
3. Suzuki Y, Mori W, Ishizone H, Naito K, Honda T. Concise enantiospecific syntheses of (+)-eldanolide and (−)-cis-whisky lactone. *Tetrahedron Lett.* 1992;33(34):4931–4932. https://doi.org/10.1016/S0040-4039(00)61237-6.
4. Boivin J, Pillot E, Williams A, Roger W, Zard SZ. On the mechanism of the nitrous acid induced conversion of an isopropylidene group into an alkyne. *Tetrahedron Lett.* 1995;36(19):3333–3336. https://doi.org/10.1016/0040-4039(95)00522-E.
5. Kornblum N, Blackwood RK, Mooberry DD. The reaction of aliphatic nitro compounds with nitrite esters. *J Am Chem Soc.* 1956;78(7):1501–1504. https://doi.org/10.1021/ja01588a060.
6. Freeman JP. Less familiar reactions of oximes. *Chem Rev.* 1973;73(4):283–292. https://doi.org/10.1021/cr60284a001.
7. Freeman JP, Gannon JJ, Surbey DL. Nitrosation of .alpha.,.beta.-unsaturated oximes. IV. Synthesis and structure of 3,4-diazacyclopentadienone derivatives. *J Org Chem.* 1969;34(1):187–194. https://doi.org/10.1021/jo00838a041.
8. Freeman JP, Gannon JJ. Nitrosation of .alpha.,.beta.-unsaturated oximes. V. Synthesis and chemistry of 1-hydroxypyrazole 2-oxides. *J Org Chem.* 1969;34(1):194–198. https://doi.org/10.1021/jo00838a042.
9. Zard SZ. New syntheses of alkynes: a tale of serendipity and design. *Chem Commun.* 2002;2002:1555–1563. https://doi.org/10.1039/B203383F.

Question 128: A Tryptamine to Pentacyclic Indoline Transformation

During a 1983 reinvestigation of a trichloride-catalyzed cyclization, British workers showed that contrary to previous work,[1] the treatment of the tryptamine derivative **1** with phosphorus trichloride in boiling benzene actually led to spirocyclic indoline derivative **2** in 46% yield (Scheme 1).[2] Furthermore, treatment of **1** with trifluoroacetic anhydride (TFAA) resulted in isolation of indoline **3** in quantitative yield. Provide mechanisms for these transformations.

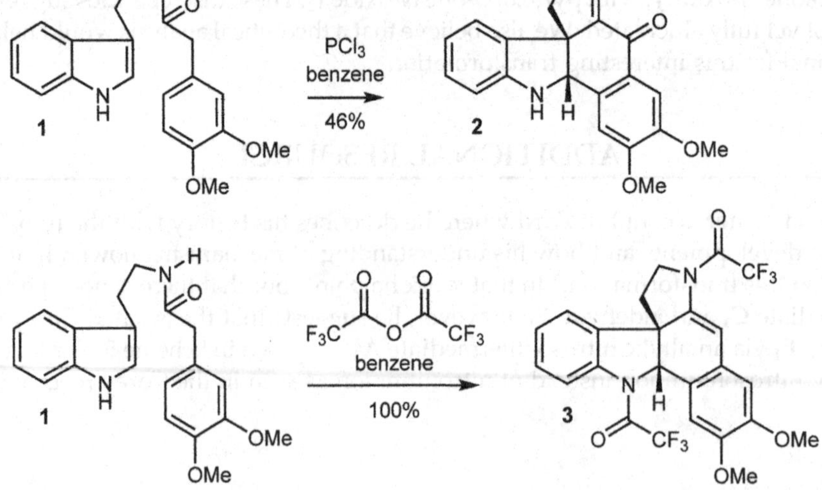

SCHEME 1

SOLUTION: 1 TO 2

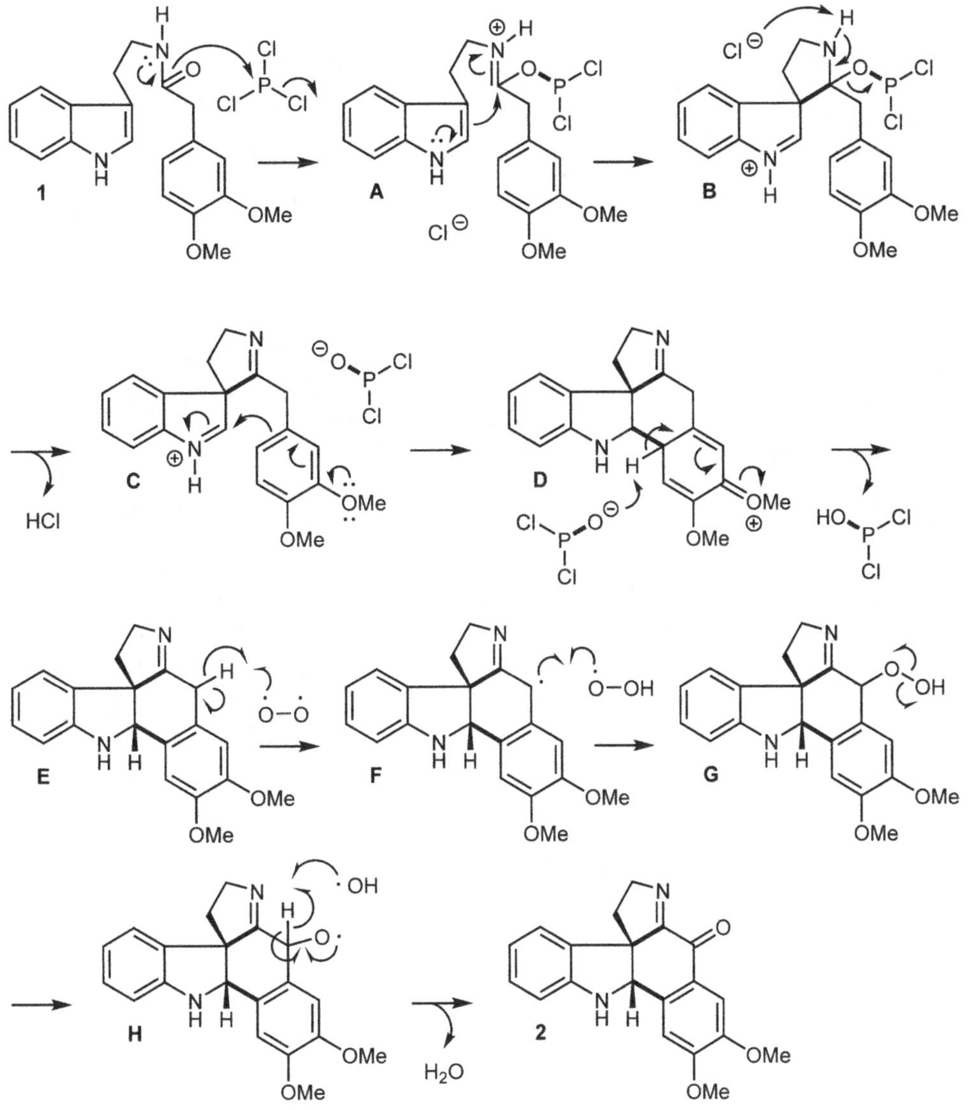

CATALYST REGENERATION

BALANCED CHEMICAL EQUATION: 1 TO 2

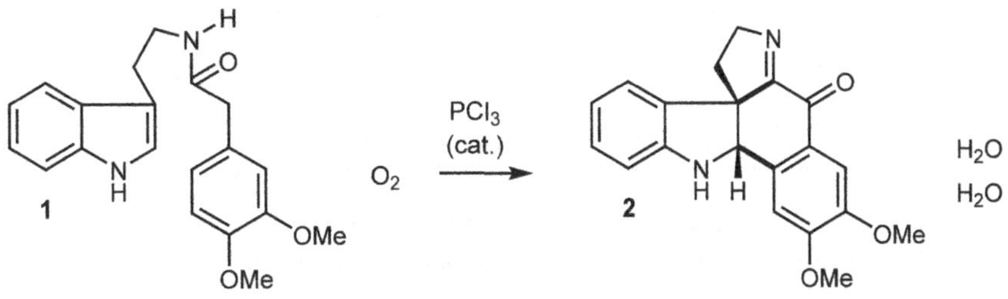

SOLUTION: 1 TO 3

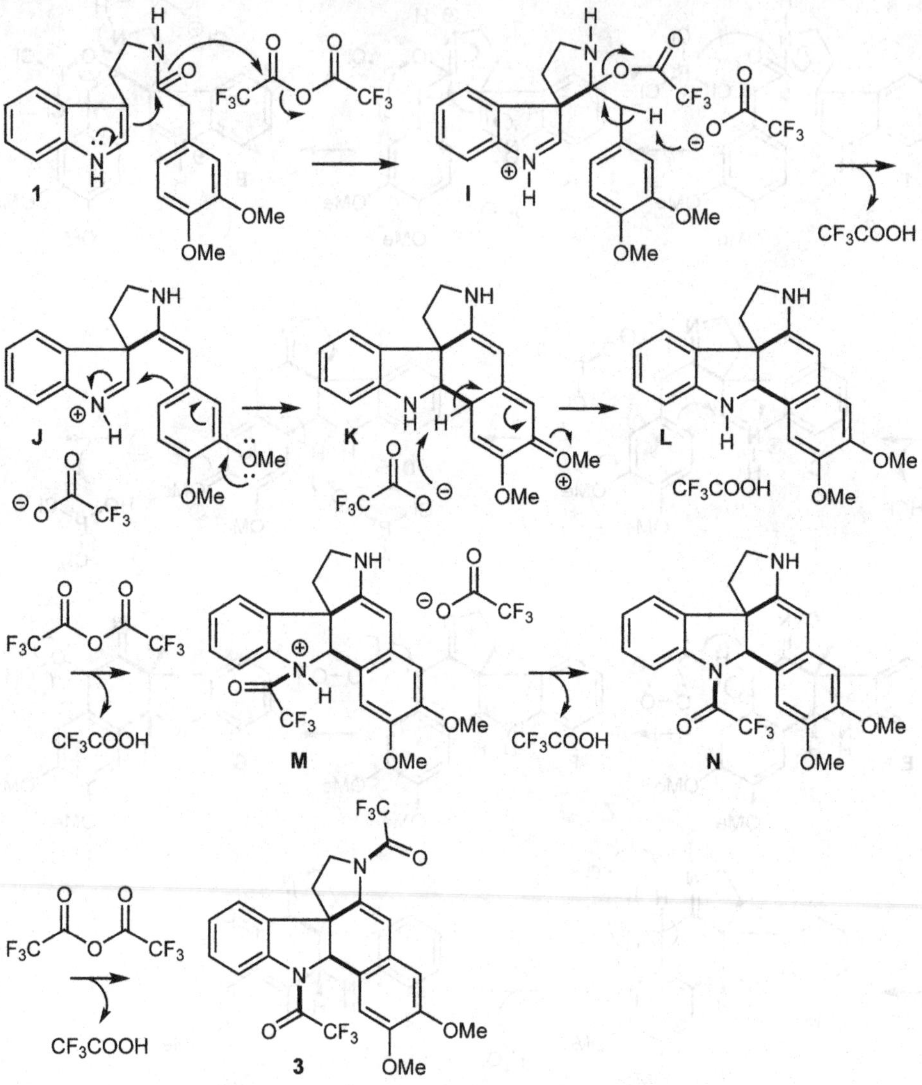

BALANCED CHEMICAL EQUATION: 1 TO 3

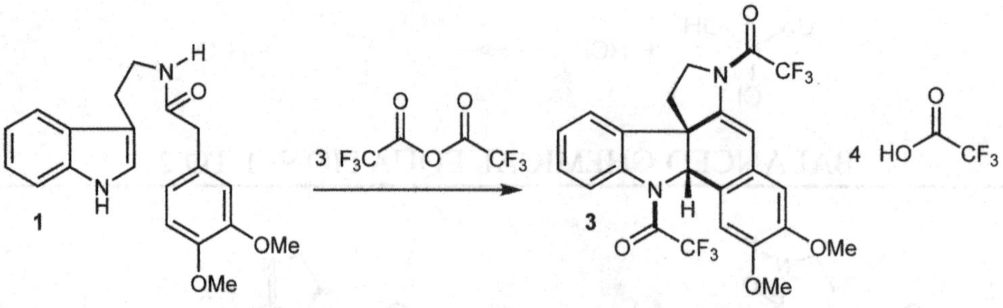

KEY STEPS EXPLAINED

We begin the analysis by pointing out that the original Japanese researchers thought they had obtained the carboline product **4** when **1** was reacted with phosphorus trichloride under the reaction conditions (Scheme 2).

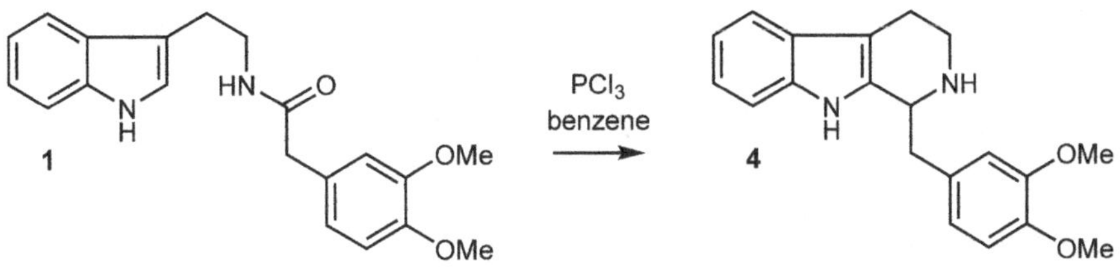

1-(3,4-Dimethoxy-benzyl)-2,9
-dihydro-1*H*-β-carboline

SCHEME 2

Despite this claim, British workers reexamined the reaction and showed that the spirocyclic indoline product **2** was the major product together with an undisclosed small amount of hydroperoxide **G**. Both of these products were characterized by mass spectrometry, elemental analyses, ^{1}H NMR, and IR spectroscopy. The mechanism shown is consistent with these observations. We note that phosphorus trichloride acts as a catalyst to effect the spirocyclization sequence from **1** to intermediate **C**. The electron donating methoxy group then induces the Pictet-Spengler cyclization yielding intermediate **E**. The next steps of the mechanism transform the methylene group to a carbonyl group via autoxidation from oxygen gas. This sequence from **E** to **2** is a radical-based mechanism which produces first the hydroperoxide **G** (minor isolated product) and then the final product **2**.

Concerning the transformation of **1** to **3**, the proposed mechanism parallels the previous transformation until the end of the Pictet-Spengler cyclization and intermediate **K** and which requires the consumption of one equivalent of trifluoroacetic anhydride. The following steps leading up to product **3** involve sequential N-acetylation of the indole and pyrrolidine nitrogen atoms using two more equivalents of trifluoroacetic anhydride. The authors also confirmed that product **3** could be hydrolyzed in aqueous ammonia to product **2** in 95% yield. We note finally that the authors did not carry out further experimental or theoretical analysis to help support these proposed mechanisms.

ADDITIONAL RESOURCES

For more examples of transformations of very similar chemical systems by the same authors, see Refs. 3–5.

References

1. Onda M, Kawanishi M. Analogs of rauwolfia alkaloids. III. *J Pharm Soc Jpn*. 1956;76(8):966–968. https://doi.org/10.1248/yakushi1947.76.8_966.
2. Biswas KM, Jackson AH. Synthesis of spiropentacyclic indolines by cyclisation of 3,4-dimethoxyphenylacetyltryptamine. *J Chem Soc Chem Commun*. 1983;1983:85–86. https://doi.org/10.1039/C39830000085.
3. Biswas KM, Jackson AH. Electrophilic substitution in indoles. Part 17. The cyclisation of (dimethoxyphenylacetyl)tryptamines to spiropentacyclic indole derivatives. *J Chem Soc Perkin Trans 1*. 1989;1989:1981–1986. https://doi.org/10.1039/P19890001981.
4. Biswas KM, Jackson AH, Kobaisy MM, Shannon PVR. Electrophilic substitution in indoles. Part 18. Cyclisation of N-acyltryptamines. *J Chem Soc Perkin Trans 1*. 1992;1992:461–467. https://doi.org/10.1039/P19920000461.
5. Wilkins DJ, Jackson AH, Shannon PVR. Electrophilic substitution in indoles. Part 19. Facile syntheses of the 2a,5a-diazacyclopenta[j,k]fluorene, indolo[2,3-a]quinolizinone and aspidosperma alkaloid ring systems from N-acyltryptamines. *J Chem Soc Perkin Trans 1*. 1994;1994:299–307. https://doi.org/10.1039/P19940000299.

Question 129: Ring Contractions of a Dibenzothiepinone

In 1978 Ackrell investigated the alkylation and ring contraction of dihydrodibenzo[*b,e*]thiepins.[1] To this end Ackrell reacted a cooled (ice-salt bath) solution of dibenzothiepinone **1** in N-methylpyrrolidone (NMP) with sodium hydride and addition of methyl iodide to give a mixture of products **2**, **3**, and **4** along with unreacted starting material (Scheme 1). Show mechanisms for these transformations.

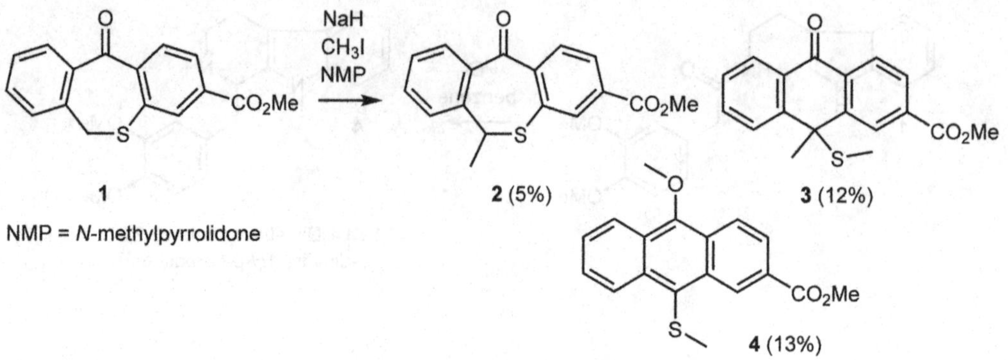

SCHEME 1

SOLUTION: 1 TO 2

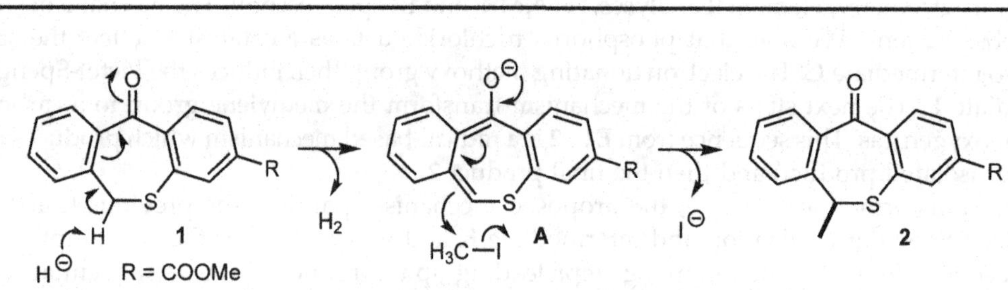

SOLUTION: 1 TO 3

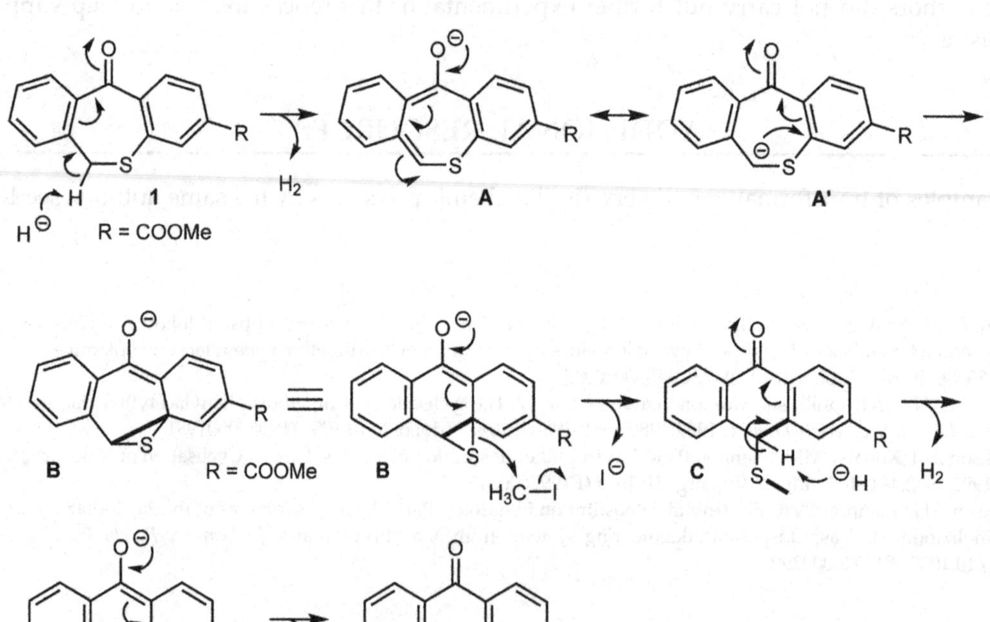

SOLUTION: 1 TO 4

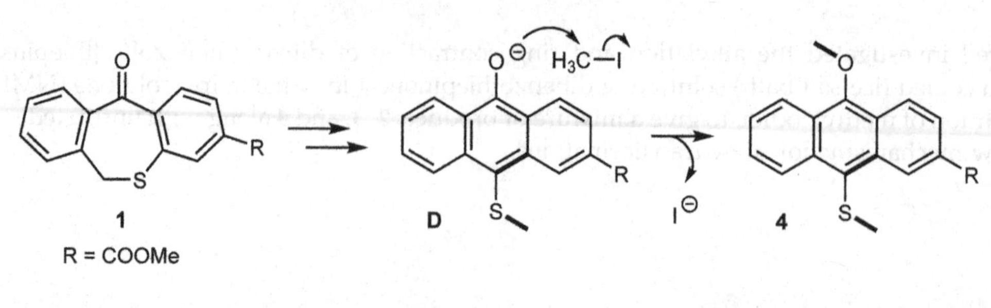

BALANCED CHEMICAL EQUATIONS

KEY STEPS EXPLAINED

Ackrell explained that when compound **1** is exposed to hydride base in *N*-methylpyrrolidone solvent, a red solution is formed which is due to the presence of the anion **A** which can also exist in the form of **A'** due to resonance. When this anion is quenched with methyl iodide the solution becomes colorless. In the case of the transformation of **1** to **2** therefore methylation takes place on the methylene carbon atom next to the sulfur atom in the seven-membered ring. In the transformations leading to products **3** and **4** anion **A** rearranges to alkoxide **B/B'** having a fused thiirane structure. Anion **B/B'** then methylates at sulfur producing intermediate **C** via thiirane ring opening. From anion **D** produced by a second deprotonation via another equivalent of hydride, a second methylation can occur either at oxygen yielding product **4** or at carbon yielding product **3**.

We note that Ackrell isolated all products by chromatography and characterized them by ^{1}H NMR and IR spectroscopies. The author also carried out a separate experiment where he converted isolated **2** into **3** under the same reaction conditions. The reaction was carried out at −20 to –40°C using a "reverse addition technique in which a solution of potassium *tert*-butoxide in *tert*-butyl alcohol was added to a solution of **1** containing excess methyl iodide in *N*-methylpyrrolidone."[1] Here the author noted that "suppression of the thiepin rearrangement took place and only product **2** was isolated."[1]

The author also indicated that the thiepin ring is antiaromatic having 8 electrons in the seven-membered ring (6π + two lone pair electrons on the sulfur atom) and that electron-withdrawing groups would thermodynamically stabilize it and reduce its antiaromatic character. The 11-oxodibenzo[*b,e*]thiepin structure of **1** increases the electron density of the thiepin ring and increases its antiaromaticity which facilitates the rearrangement when the anion is generated (see Scheme 2). The author supported their argument by synthesizing the *S,S*-dioxide **5** from **1** according to Scheme 3. Nevertheless, the author did not carry out a theoretical analysis to lend further support for his proposed mechanism.

ADDITIONAL RESOURCES

We recommend an article which describes the synthetic utility of dibenzo[*b,f*]thiepinones as precursors to various dibenzo[*e,h*]azulenes.[2]

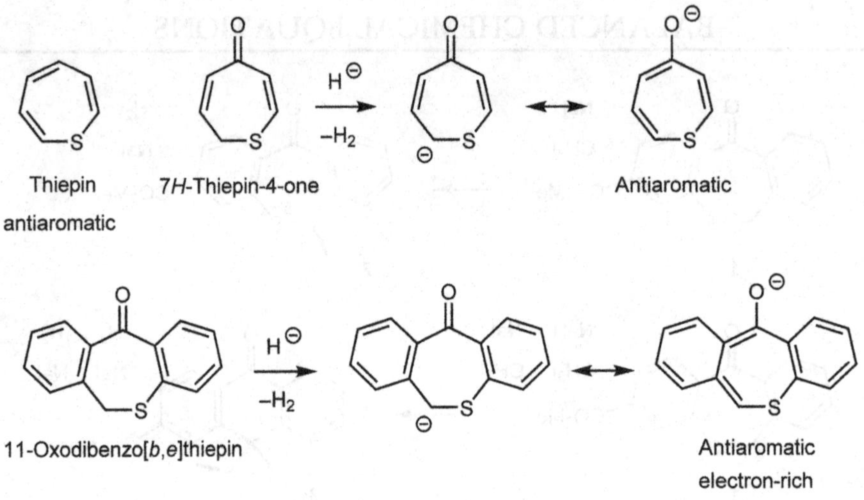

SCHEME 2

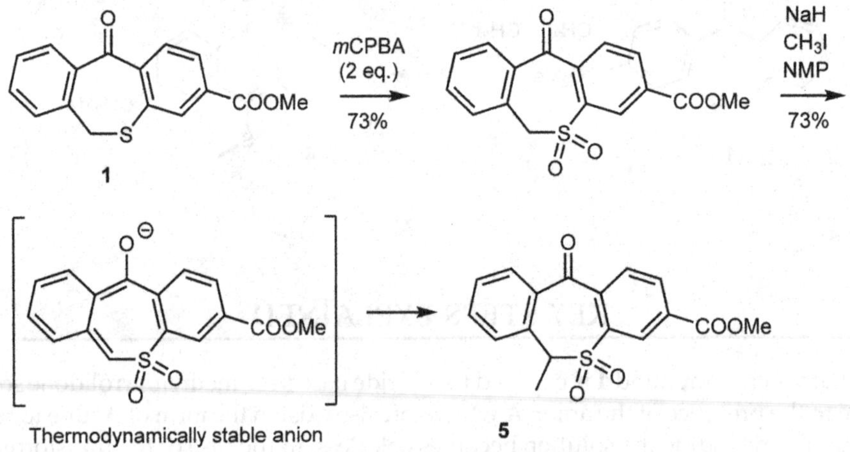

SCHEME 3

References

1. Ackrell J. Alkylation and ring contraction of 11-oxo-6,11-dihydrodibenzo[b,e]thiepins. *J Org Chem.* 1978;43(25):4892–4893. https://doi.org/10.1021/jo00419a045.
2. Pesic D, Landek IO, Rupcic R, et al. Dibenzo[b,f]oxepin-10(11H)-one and dibenzo[b,f]thiepin-10(11H)-one as useful synthons in the synthesis of various dibenzo[e,h]azulenes. *J Heterocyclic Chem.* 2012;49(2):243–252. https://doi.org/10.1002/jhet.753.

Question 130: Steroid Fragmentation With DDQ

During a 1979 study of the dehydrogenation of lanosterol, a New Zealand chemist heated compound **1** with 2,3-dicyano-5,6-dichloroquinone (DDQ) under reflux in benzene for 24 h.[1] Basic hydrolysis of the crude product mixture followed by chromatographic separation gave products **2**, **3**, and **4** in 51%, 19%, and 6% yields, respectively (Scheme 1).[1] Treatment of **1** or **2** with excess DDQ or for a prolonged period of time led to increased amounts of **3** and **4**. Provide mechanisms to explain the formation of products **2**, **3**, and **4**.

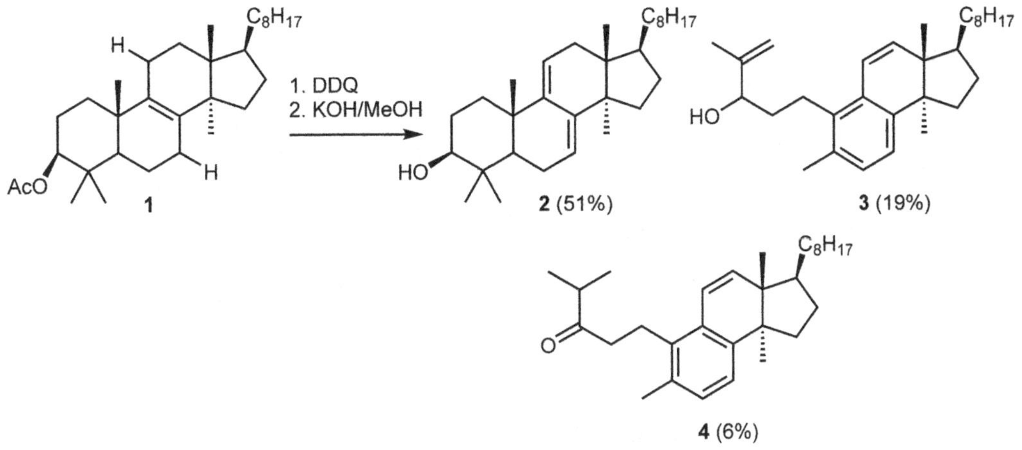

SCHEME 1

SOLUTION: 1 TO 2

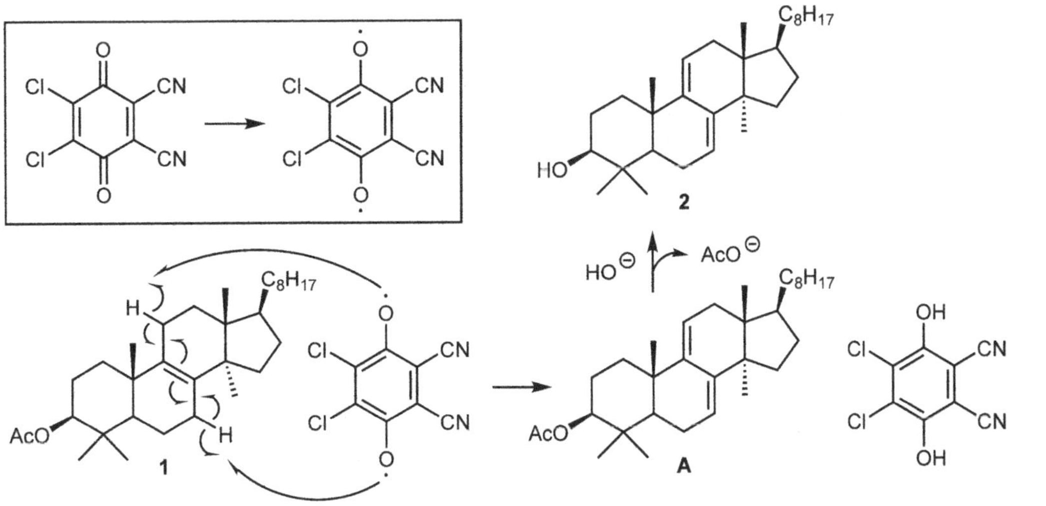

SOLUTION: 1 TO 3

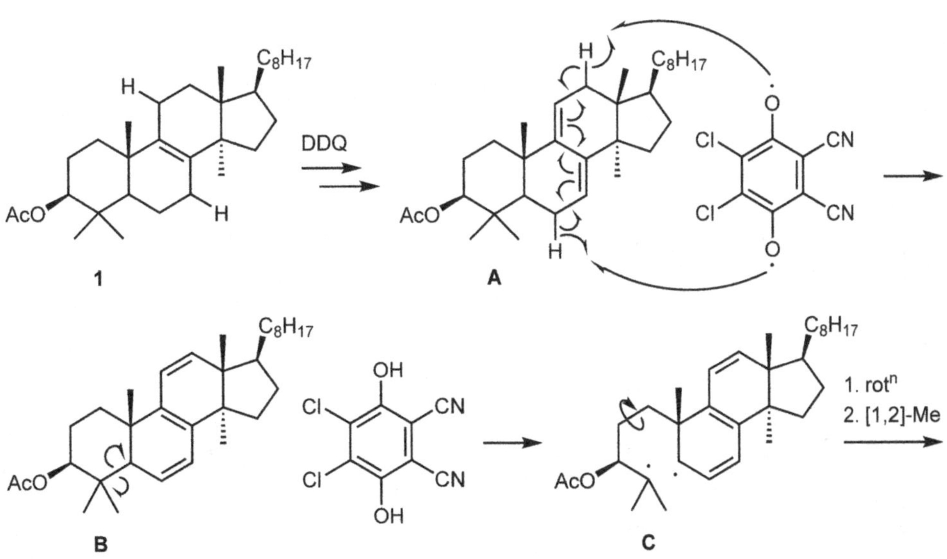

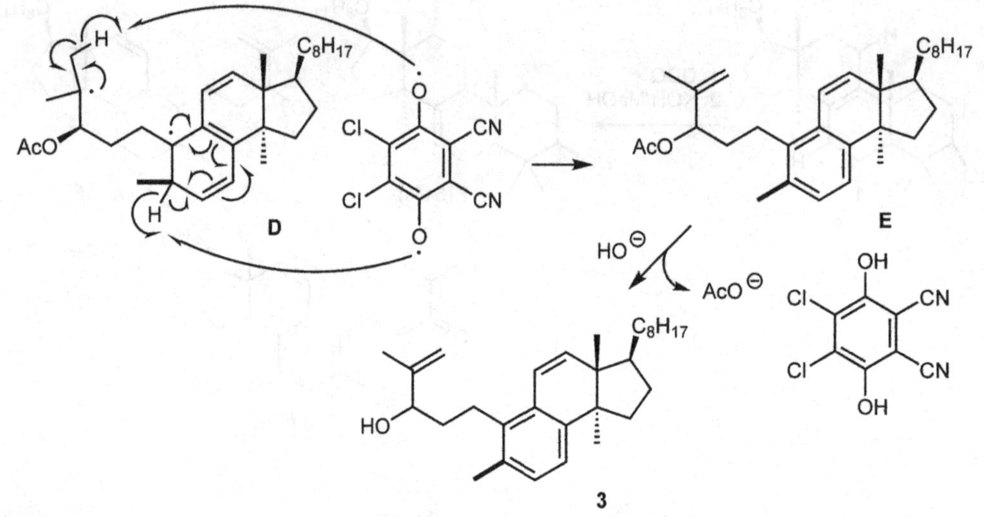

SOLUTION: 1 TO 4

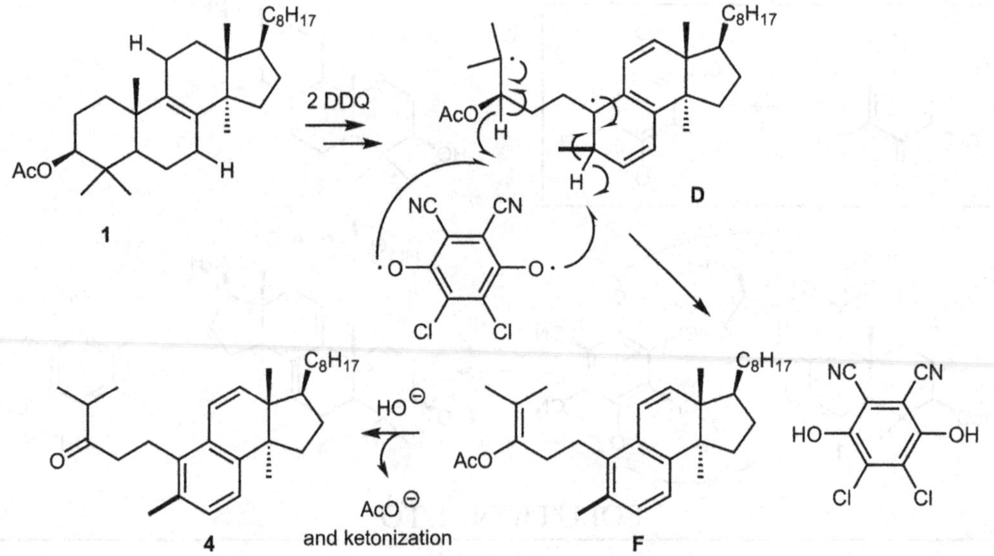

BALANCED CHEMICAL EQUATIONS

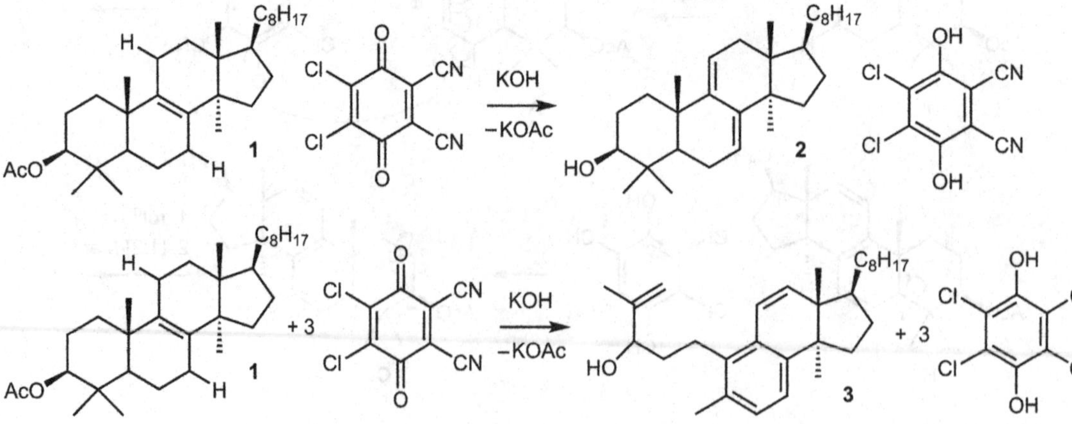

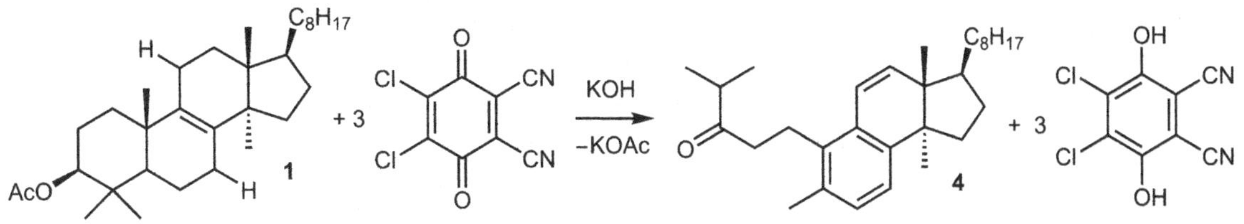

KEY STEPS EXPLAINED

We note that in all three reactions the reagent DDQ reacts via its diradical form. The acetoxy group is hydrolyzed by hydroxide by a conventional nucleophilic mechanism which we omit for space. With regard to the transformation of **1** to **2**, we emphasize that one equivalent of DDQ is required to abstract the two hydrogen atoms in **1** at C_7 and C_{11} (see Fig. 1) to form the diene intermediate **A**. Subsequent base hydrolysis of the acetoxy group generates product **2**. With regard to the transformation of **1** to **3**, it is important to recognize that three equivalents of DDQ are required. The first equivalent of DDQ abstracts the two hydrogen atoms in **1** at C_7 and C_{11} giving diene intermediate **A** as seen before. The second equivalent of DDQ abstracts the two hydrogen atoms in **A** at C_6 and C_{12} giving triene **B**. Homolytic C_4—C_5 bond cleavage leads to diradical **C** which upon C_1—C_2 bond rotation and [1,2]-methyl migration yields diradical **D**. The third equivalent of DDQ abstracts two more hydrogen atoms creating an aromatic ring (ring B in the steroid skeleton) and a terminal olefinic group found in intermediate **E**. Final base hydrolysis of the acetoxy group yields product **3**. Lastly, for the transformation of **1** to **4** we have the same sequence of steps leading to intermediate **D** occurring at the beginning. These steps also consume two equivalents of DDQ. The third equivalent of DDQ then comes in to abstract two hydrogen atoms creating an aromatic ring (ring B in the steroid skeleton) and a vinyl ester group found in intermediate **F**. Once again, a final base hydrolysis of the acetoxy group yields product **4**.

Although this mechanism explains the experimental observation that prolonged exposure of **1** to excess amount of DDQ leads to increased amounts of products **3** and **4** as well as decreased amount of **2** since the alcohol **2** can easily react with DDQ to form **3** or **4**, we do note that our mechanism is slightly different than that of the author. In his article, Crump invokes carbocationic intermediates without offering any experimental or theoretical evidence to support these structures. Since we believe that DDQ confers a radical-type mechanism onto the three dehydrogenation transformations analyzed before, we here represent only the radical-based mechanism.[2]

Interestingly, the authors noted that the reaction did not work with tetrachlorobenzoquinone (TCQ, or chloranil, Fig. 2). This experimental observation indicates that electron-withdrawing groups (EWGs) such as the two cyano groups are needed on the benzoquinone (BQ) structure to initiate the radical reaction as in the case of DDQ. We believe a theoretical calculation would help to illustrate thermodynamic destabilization of the benzoquinone ring when EWGs are present. Furthermore, EWGs increase the oxidizing power of the benzoquinone moiety. Stronger oxidants have more positive valued $E°(red)$ values. It is expected by this argument that the relative $E°(red)$ values should follow the pattern: DDQ > TCQ > BQ. This trend in acetonitrile solvent has been demonstrated in the work of Sasaki et al. (see Figure 3).[3]

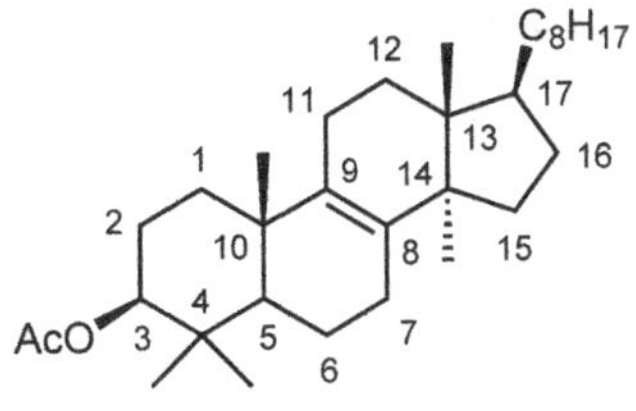

FIG. 1 Numbering scheme for product 1.

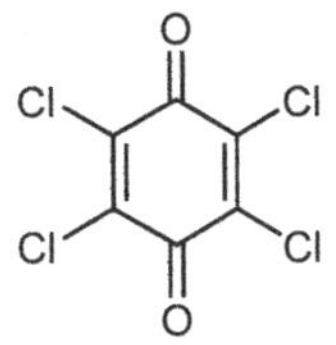

FIG. 2 Structure of tetrachlorobenzoquinone (TCQ).

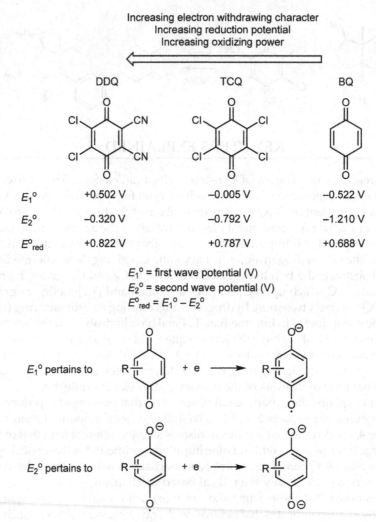

FIG. 3 Electrochemical data pertaining to DDQ, TCQ, and BQ.

ADDITIONAL RESOURCES

In contrast to the previous discussion, we recommend an article which discusses the use of fluorine atom as a cation-stabilizing auxiliary in biomimetic polyene cyclizations.[4]

References

1. Crump DR. The dehydrogenation of lanosterol derivatives by 2,3-dichloro-5,6-dicyanobenzoquinone. *J Chem Soc Perkin Trans 1.* 1979;1979:646–648. https://doi.org/10.1039/P19790000646.
2. Wendlandt AE, Stahl SS. Quinone-catalyzed selective oxidation of organic molecules. *Angew Chem Int Ed Engl.* 2015;54(49):14638–14658. https://doi.org/10.1002/anie.201505017.
3. Sasaki K, Kashimura T, Ohura M, Ohsaki Y, Ohta N. Solvent effect in the electrochemical reduction of *p*-quinones in several aprotic solvents. *J Electrochem Soc.* 1990;137(8):2437–2443. https://doi.org/10.1149/1.2086957.
4. Johnson WS, Buchanan RA, Bartlett WR, Tham FS, Kullnig RK. The fluorine atom as a cation-stabilizing auxiliary in biomimetic polyene cyclizations. 3. Use to effect regiospecific control. *J Am Chem Soc.* 1993;115:504–515. https://doi.org/10.1021/ja00055a021.

Question 131: A Base-Induced Vinylcyclopropane Rearrangement

In 1975 American chemists heated compound **1** with dimsylsodium in DMSO at 90°C for 21 h and obtained the epimeric ene-diesters **2** in 71% yield (Scheme 1).[1] Suggest a mechanism for this transformation.

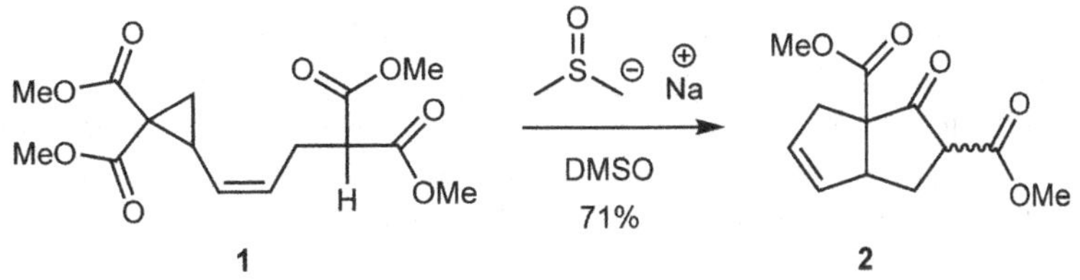

SCHEME 1

SOLUTION

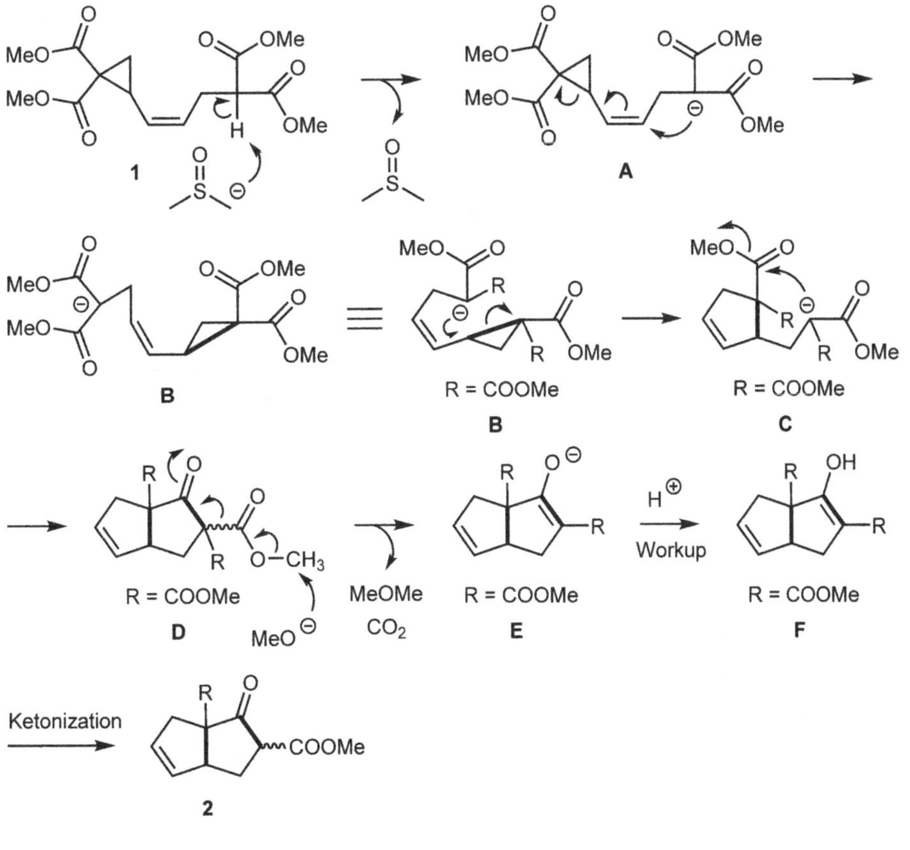

BALANCED CHEMICAL EQUATION

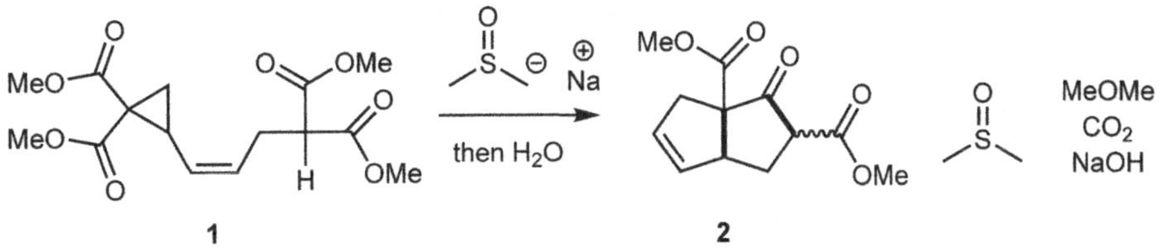

KEY STEPS EXPLAINED

The mechanism proposed by the authors to explain the transformation of **1** to **2** starts with deprotonation of vinylcyclopropane **1** by dimsyl anion to form carbanion **A** which rearranges to the carbanion **B**. **B** then undergoes a [5+0] ring closure while simultaneously ring opening the cyclopropane ring to generate carbanion **C**. At this point **C** undergoes an intramolecular Dieckmann condensation to give **D** while eliminating methoxide ion. The liberated methoxide ion then attacks one of the geminal carboxymethyl groups to give enolate **E** together with dimethyl ether and carbon dioxide by-products. The acidic work-up that follows facilitates formation of enol **F** which ketonizes to the final product **2**. The overall ring construction mapping for this transformation can be described as [(5+0)+(3+2)]. We emphasize that the authors did not carry out further experimental and theoretical work to produce evidence that supports their postulated mechanism. As such, we would like to propose alternative possibilities which yield the same product **2** with the same ring construction mapping but with different carbon—carbon bond connectivities (Scheme 2). We also believe that there are opportunities for carbon labeling experiments to explore which pathway is operational as indicated by the numbering connectivities shown in Scheme 3.

Alternatively, labeling experiments could be carried out using vinylcyclopropane **1** with substituents on C_3 or C_7 and determining the structures of the products as shown by the methylated derivatives given in Scheme 4.

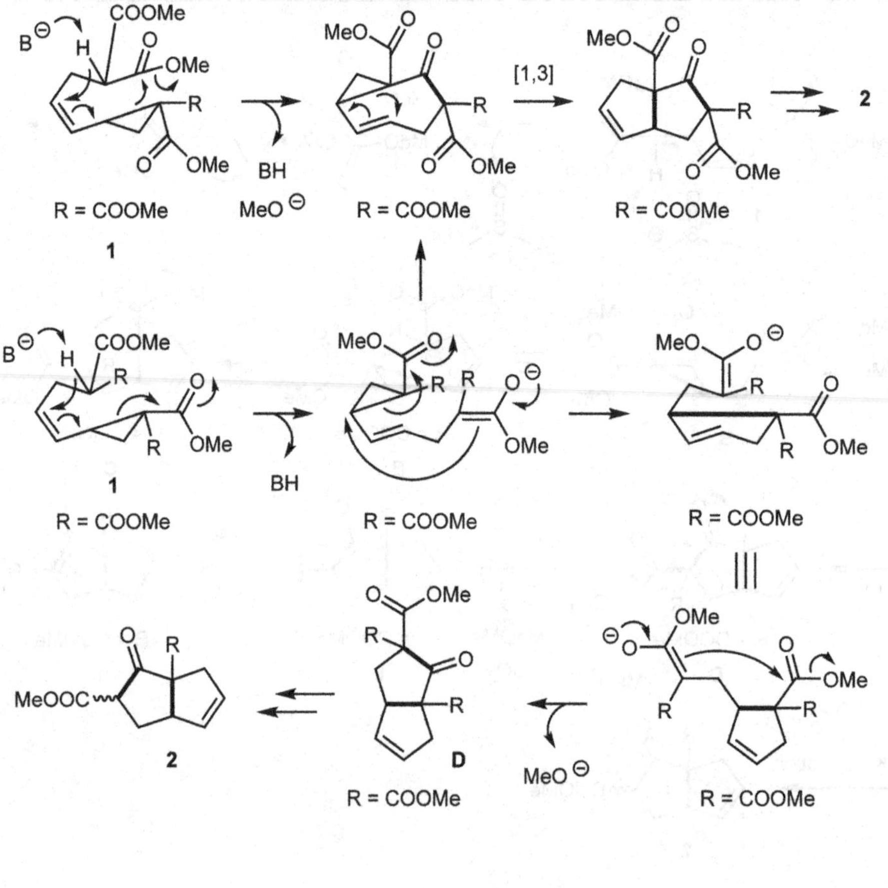

SCHEME 2

ADDITIONAL RESOURCES

We highly recommend the following four articles which discuss very interesting transformations of vinylcyclopropanes and their mechanisms.[2–5]

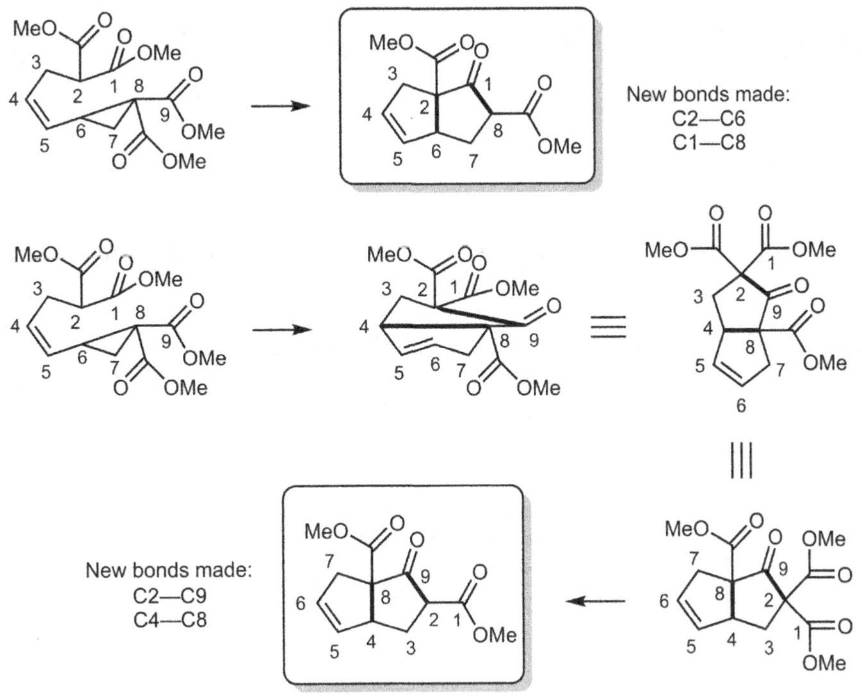

New bonds made:
C2—C6
C1—C8

New bonds made:
C2—C9
C4—C8

SCHEME 3

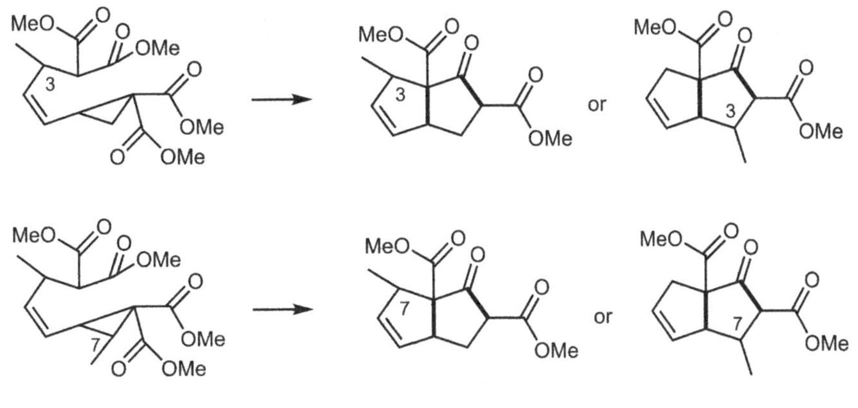

SCHEME 4

References

1. Danishefsky S, Tsai MY, Dynak J. Intramolecular opening of an activated vinylcyclopropane: an entry to the bicyclo[3,3,0]octenone series. *J Chem Soc Chem Commun.* 1975;1975:7–8. https://doi.org/10.1039/C39750000007.
2. Wang SC, Troast DM, Conda-Sheridan M, et al. Mechanism of the Ni(0)-catalyzed vinylcyclopropane-cyclopentene rearrangement. *J Org Chem.* 2009;74:7822–7833. https://doi.org/10.1021/jo901525u.
3. Mulzer J, Huisgen R, Arion V, Sustmann R. 1-[(E)-2-Arylethenyl]-2,2-diphenylcyclopropanes: kinetics and mechanism of rearrangement to cyclopentenes. *Helv Chim Acta.* 2011;94:1359–1388. https://doi.org/10.1002/hlca.201100135.
4. Ganesh V, Sureshkumar D, Chandrasekaran S. Tandem ring opening/cyclization of vinylcyclopropanes: a facile synthesis of chiral bicyclic amidines. *Angew Chem Int Ed.* 2011;50:5878–5881. https://doi.org/10.1002/anie.201100375.
5. Trost BM, Morris PJ, Spargue SJ. Palladium-catalyzed diastereo- and enantioselective formal [3 + 2]-cycloadditions of substituted vinylcyclopropanes. *J Am Chem Soc.* 2012;134:17823–17831. https://doi.org/10.1021/ja309003x.

Question 132: 4-Demethoxydaunomycin: The Desired Result, but Not According to Plan

In 1978 American chemists sought a synthetic approach to product **2** starting from **1** (Scheme 1).[1] Their synthetic approach ultimately became treatment of **1** with *p*-MeC_6H_4SH in the expectation that allylic substitution of **1** with the thiol would occur. The resulting sulfide would therefore react with *m*-chloroperbenzoic acid (*m*CPBA) to form a sulfoxide which would undergo a [1,3]-sigmatropic rearrangement the product of which would undergo hydrolysis to give **2**. In contrast to this strategy, the authors isolated intermediates **3**, **4**, and **5**. Suggest an alternative mechanistic explanation for the synthetic sequence from **1** to **2** which is consistent with the unexpected experimental results.

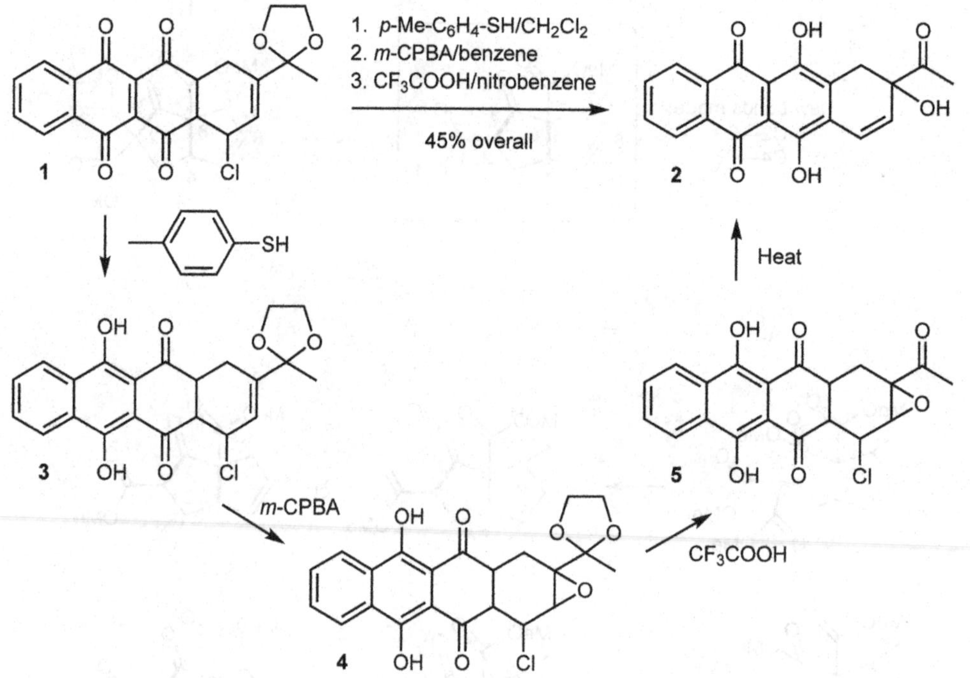

1. p-Me-C6H4-SH/CH2Cl2
2. m-CPBA/benzene
3. CF3COOH/nitrobenzene
45% overall
1
2
3
4
5
Heat
m-CPBA
CF3COOH
SH

SCHEME 1

SOLUTION: 1 TO 3

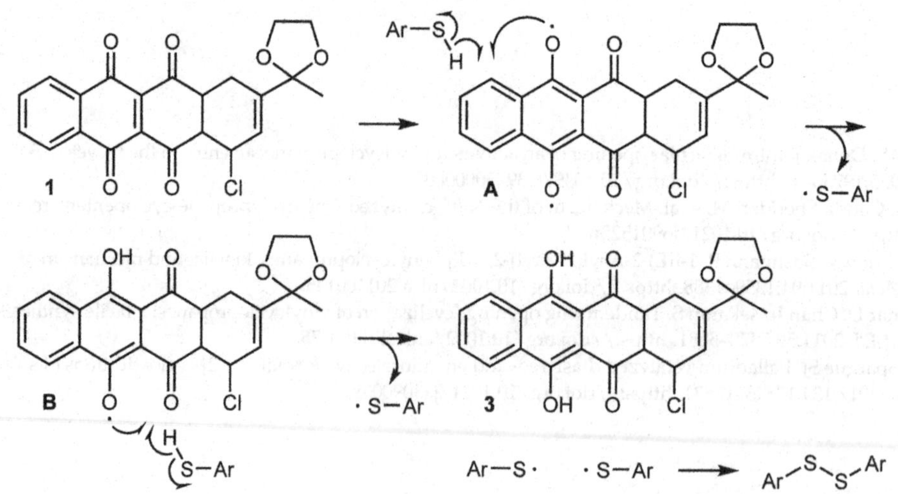

1
A
B
3
Ar—S
H
· S—Ar
Ar—S ·
· S—Ar
Ar
S
S
Ar

SOLUTION: 3 TO 4

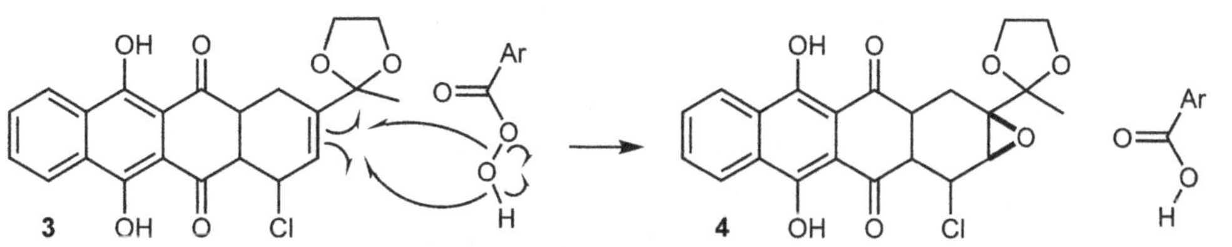

SOLUTION: 4 TO 5

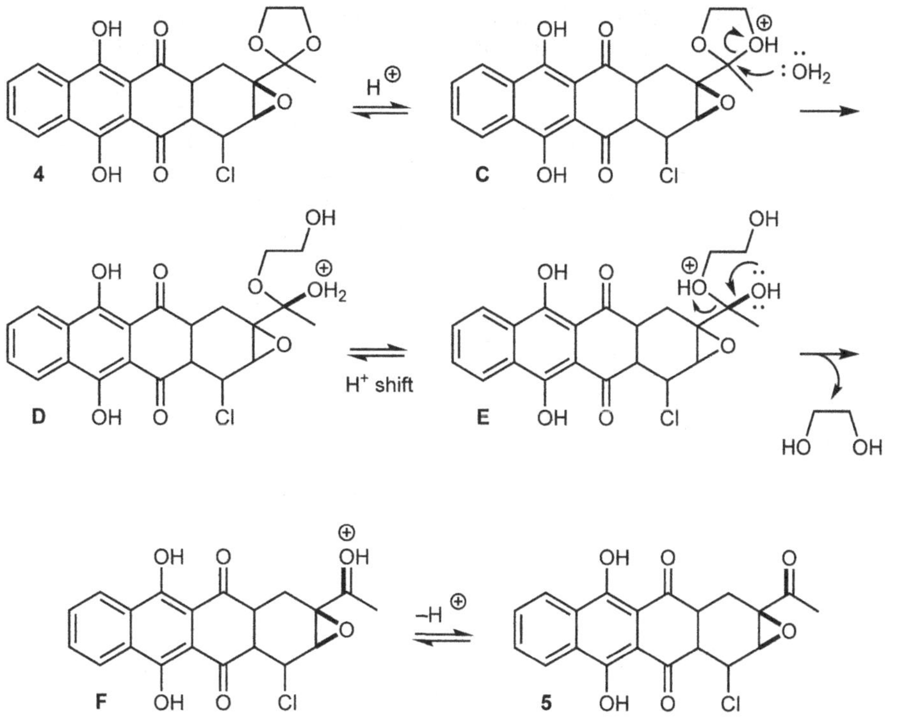

SOLUTION: 5 TO 2

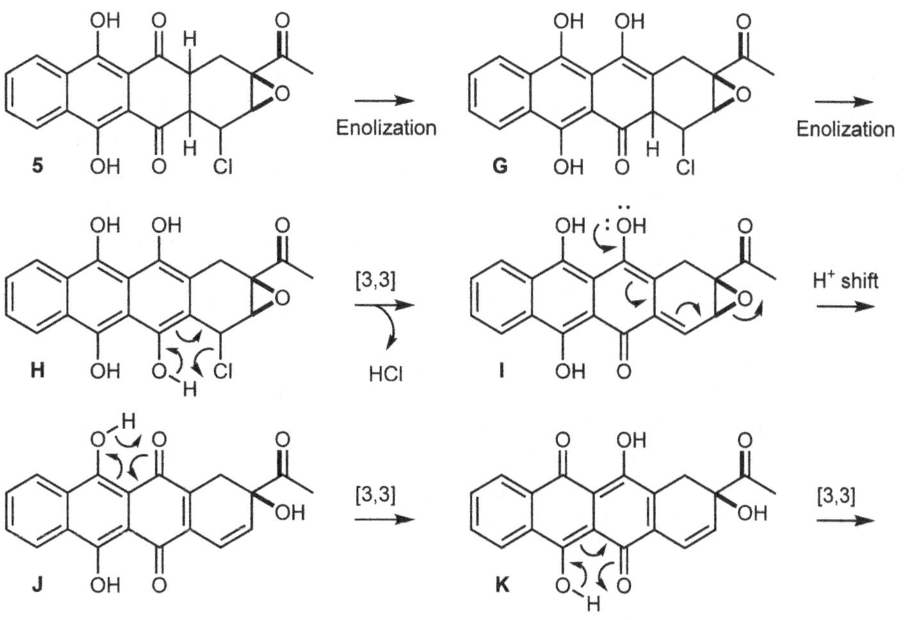

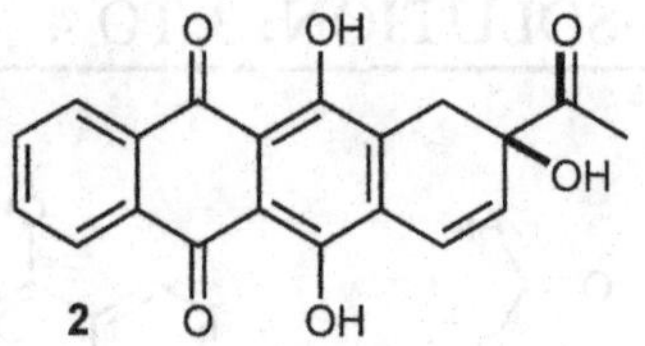

KEY STEPS EXPLAINED

The authors originally thought that the sequence proceeded as shown in Scheme 2. The toluenethiol reagent would substitute the chlorine group of **1** to give sulfide **6**. Then oxidation of the sulfide would give the sulfoxide **7**. The sulfoxide would then undergo a 1,3-allylic transposition to give intermediate **L** after which acid-catalyzed hydrolysis of the ketal and alkoxysulfanyltoluene groups would give intermediate **M** which would ketonize to product **2**.

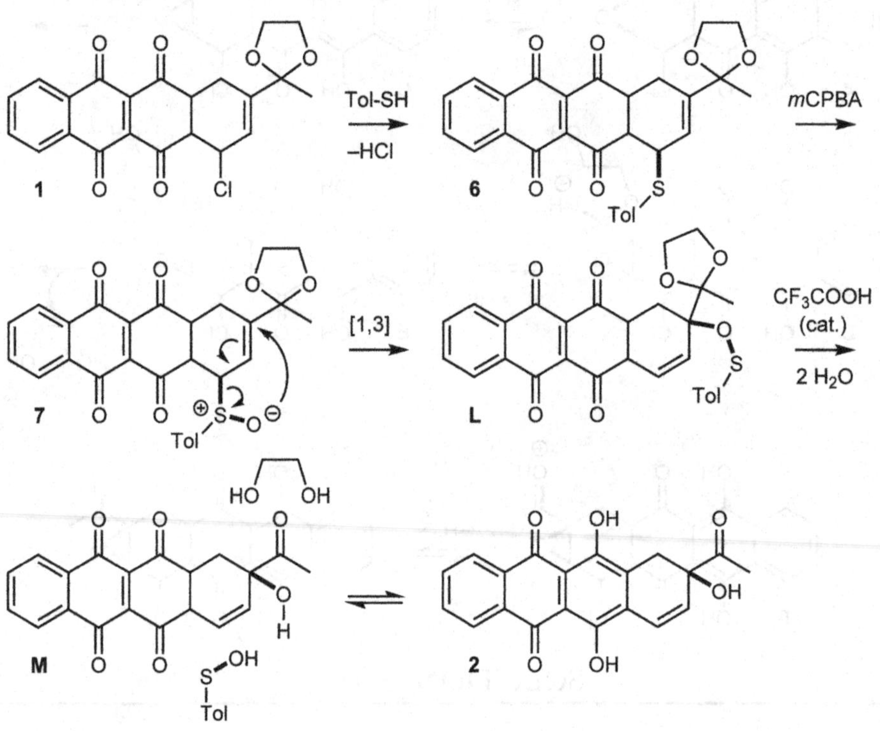

SCHEME 2

Instead, the authors isolated products **3**, **4**, and **5** as intermediates. In the first step, the toluenethiol reduces the quinone group to a hydroquinone group yielding product **3** and tolyldisulfide as by-product. In the second step, the olefin in the last ring is oxidized to the epoxide **4**. In the third step, acid catalyzed hydrolysis of the ketal group forms the ketone **5**. At this point, the authors' mechanism becomes unusual in the sense that they argue, without justification, that spontaneous loss of HCl from **5** occurs to form an olefin such as intermediate **I** which spontaneously loses a proton in order to open the epoxide ring which is followed by two [3,3] sigmatropic rearrangements which are the same as depicted before in the solution to give product **2**. We are skeptical about this postulated pathway from **5** to **2** which is why we here offer a slightly different mechanism which begins with enolization of **5** to give intermediate **G** which undergoes a second enolization to **H**. At this point, a [3,3] sigmatropic rearrangement causes elimination of HCl to form intermediate **I** which ketonizes to cause ring opening of the epoxide to form **J** after electron delocalization and proton shift. From **J** to **2** we have two [3,3] sigmatropic rearrangements which are likely driven by aromatization of the central ring to form the final product **2**.

Unfortunately the authors did not perform a theoretical analysis of their proposed mechanism nor do we have data for this alternative proposal. Of particular interest would be an energy calculation for the transformation of **H** to **I**. We note that this transformation disrupts the aromatic ring system in **H** to cause the elimination of HCl in a [3,3] sigmatropic-like manner. To our knowledge, such eliminations are uncommon. In the literature, we have found an example of a [2,3] sigmatropic elimination of effectively Se—H in a more simplified chemical context.[2] As such we would expect the process from **H** to **I** to be aided by heat, only the authors do not mention the use of heat in the transformation of **5** to **2**. They also do not mention running an experiment to where **5** is isolated and reacted with CF_3COOH to see whether **2** is formed and in what yield. To complicate the situation the authors also do not mention the use of a base that could effect an E_2-like elimination of HCl. Under acidic conditions, it might be possible to expect that chloride ion would eliminate spontaneously to give a secondary carbocation in a process that would be promoted by resonance stabilization afforded by the neighboring phenol (Scheme 3). Here we have shown an S_N1-like mechanism, but a concerted transformation from **H** to H_2 is also possible. Nevertheless once again the argument is that breakage of the aryl aromaticity would require an energy-intensive process. Unfortunately a yield was not reported for the transformation of **5** to **2** but it was reported for the overall transformation of **1** to **2**, namely, 45%. Such a moderately good yield for a three-step synthesis would suggest that each of the steps was fairly efficient.

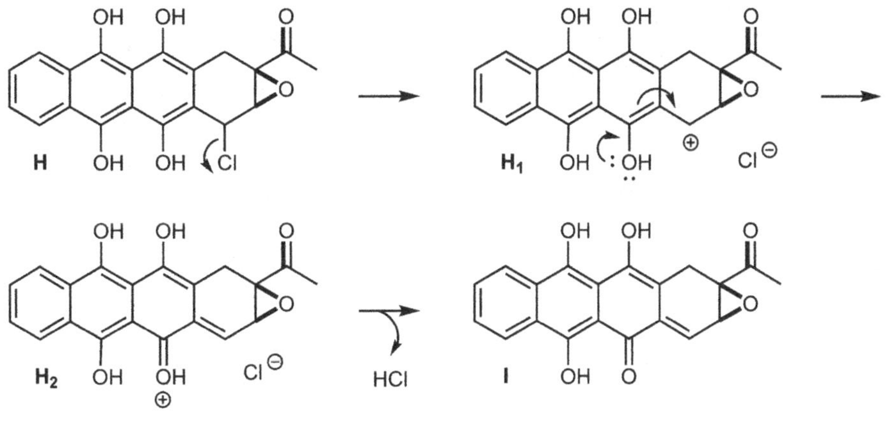

SCHEME 3

Given the lack of information about the transformation of **5** to **2** coupled with the fact that the authors provide spectroscopic data on **5** meaning they were able to isolate it, we are inclined to say that more work is needed to help shed light on its mechanism. An interesting experiment would be to undertake the synthetic sequence from **1** to **2** in $^{18}OH_2$ to see whether there is single incorporation of ^{18}O in **2** (as predicted by the mechanism depicted in the solution) or double incorporation (as predicted by the alternative mechanism illustrated in Scheme 2).

ADDITIONAL RESOURCES

We recommend an article which discusses the use of [3,3] and [2,3] sigmatropic rearrangements in the synthesis of steroids.[3]

References

1. Kelly TR, Tsang W-G. The synthesis of 4-demethoxydaunomycin. *Tetrahedron Lett.* 1978;19(46):4457–4460. https://doi.org/10.1016/S0040-4039 (01)95251-7.
2. Zhou ZS, Jiang N, Hilvert D. An antibody-catalyzed selenoxide elimination. *J Am Chem Soc.* 1997;119(15):3623–3624. https://doi.org/10.1021/ ja963748j.
3. Lesuisse D, Canu F, Tric B. A new route to 19-substituted steroids from 19-nor steroids: sigmatropic [3,3] and [2,3] rearrangements revisited. *Tetrahedron.* 1994;50(28):8491–8504. https://doi.org/10.1016/S0040-4020(01)85569-1.

Question 133: Three Isomeric Hexadienols Give the Same Bicyclic Lactone

In 1988 researchers from New Jersey and Basel observed that treatment of any of the three isomeric alcohols **1**, **2**, or **3** with the acid chloride of fumaric acid monomethyl ester in ether/triethylamine followed by attempted purification by Kugelrohr distillation at high temperature gave the bicyclic lactone **4** in 35%–45% yield (Scheme 1).[1] Provide a mechanistic explanation for these observations.

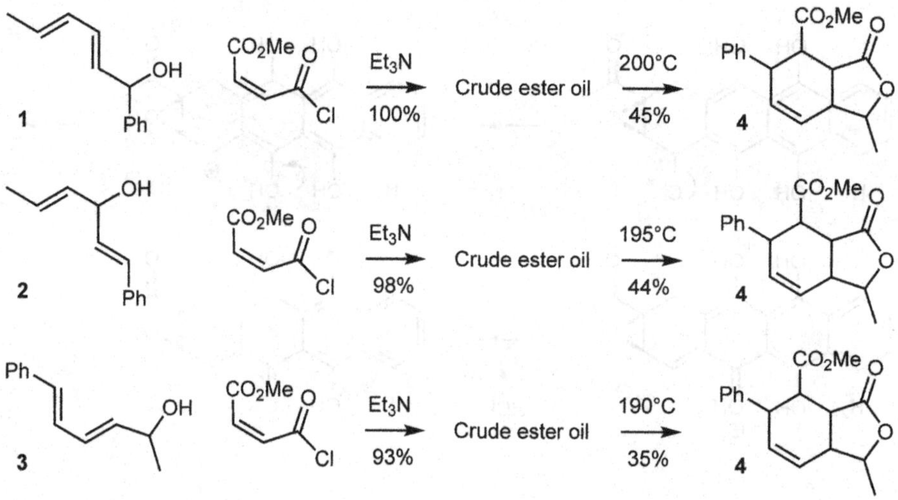

SCHEME 1

SOLUTION: 1 TO 4

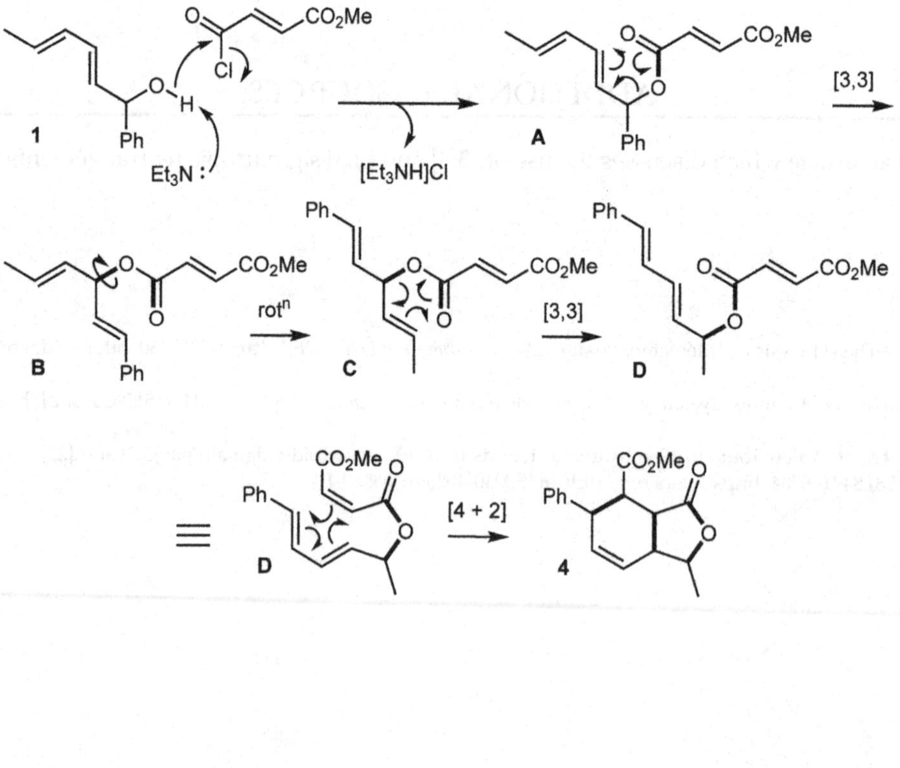

SOLUTION: 2 TO 4

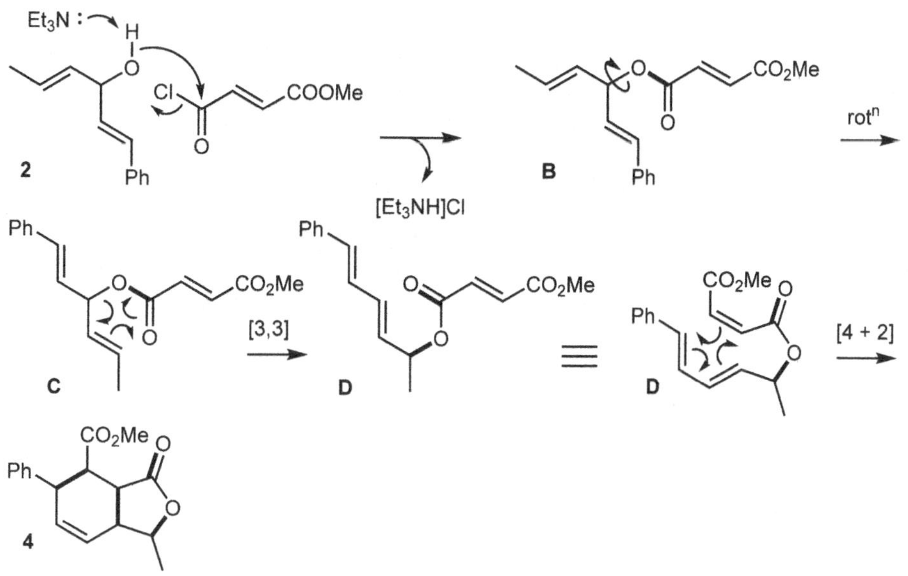

SOLUTION: 3 TO 4

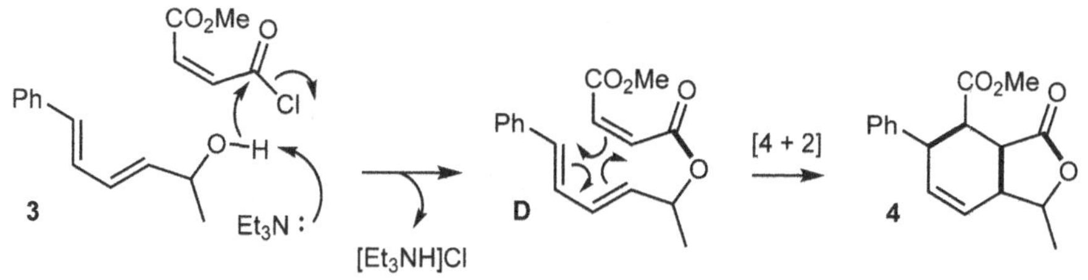

KEY STEPS EXPLAINED

We begin the mechanistic analysis by noting that the identity of the crude ester products from Scheme 1 actually represents different ester intermediates along the mechanistic path from alcohol 1 to product 4. For instance, esterification of 1 leads to A while esterification of 2 leads to B and esterification of 3 leads to D. The last step is all transformations involve a thermal intramolecular [4+2] Diels-Alder cyclization and since all mechanistic pathways lead to intermediate D, all transformations involving different isomers of alcohol 1 lead to the same product. What is interesting to note is that the transformation of 1 to 4 proceeds through intermediates B and D after bond rotation and two sequential [3,3]-sigmatropic rearrangements from the initial esterification product A. Likewise, the transformation of 2 to 4 leads to the esterification product B which undergoes bond rotation and a [3,3]-sigmatropic rearrangement to D. Given these interesting observations, the authors asked whether the isomers 1 and 2 were rearranging to 3 in solution prior to esterification and intramolecular Diels-Alder cyclization. To answer this question, two experiments were carried out. In the first, alcohol 1 was treated with pyridine and acetic anhydride to cause formation of the corresponding acetate yet what the authors isolated was the acetate of 3. Isolation of this acetate proved that the acetate derivatives of the isomeric alcohols 1 and 2 also rearranged in a [3,3]-sigmatropic manner to form the acetate of 3. Nevertheless, when alcohol 1 was dissolved in ether containing 0.5 equivalents of triethylamine in the presence of 1 equivalent of triethylammonium hydrochloride (the byproduct of this transformation) there was no reaction and alcohol 1 was recovered unchanged. This outcome proved that alcohols 1 and 2 do not rearrange to 3 on their own prior to esterification which established the sequence of 1 to A to B to C to D to 4. Furthermore, the authors remarked that another conclusion one could draw was that intramolecular Diels-Alder cyclization must be slower than all three of the sigmatropic rearrangements required to transform 1 to 4. Such a conclusion would certainly be a prime candidate for evaluation using a theoretical analysis, which the authors did not carry out. Despite this omission, the authors did emphasize that the intramolecular Diels-Alder cyclization from **D** to **4** establishes five asymmetric centers in one step. Furthermore, we note that all three transformations have different target bonds formed in the final product **4**. Moreover, the ring construction strategy for the transformation of **1** to **4** can be described as [(4+2)+(2+2+1)]. That of the **2** to **4** transformation would be [(4+2)+(3+2)] and that of transformation of **3** to **4** would also be [(4+2)+(3+2)] despite the different target bond mappings in product **4** in these two cases.

ADDITIONAL RESOURCES

We recommend an interesting article which discusses the Wolff rearrangement of a carbene in experiments which involve thermolysis of Diels-Alder adducts.[2]

References

1. Eberle MK, Weber H-P. The regio- and stereoselectivity of intramolecular Diels-Alder reactions of fumarates: an unusual rearrangement-cyclization. *J Org Chem.* 1988;53(2):231–235. https://doi.org/10.1021/jo00237a001.
2. Litovitz AE, Keresztes I, Carpenter BK. Evidence for nonstatistical dynamics in the Wolff rearrangement of a carbene. *J Am Chem Soc.* 2008;130:12085–12094. https://doi.org/10.1021/ja803230a.

Question 134: Degradation of Antibiotic X-537A

In 1971 chemists from Hoffmann-La Roche treated antibiotic X-537A **1** with 5 equivalents of concentrated nitric acid in glacial acetic acid to form product **2** which upon treatment with dilute aqueous hydroxide solution gave a mixture of products **3**, **4**, **5**, and **6** (Scheme 1).[1] Provide mechanisms for the formation of these products.

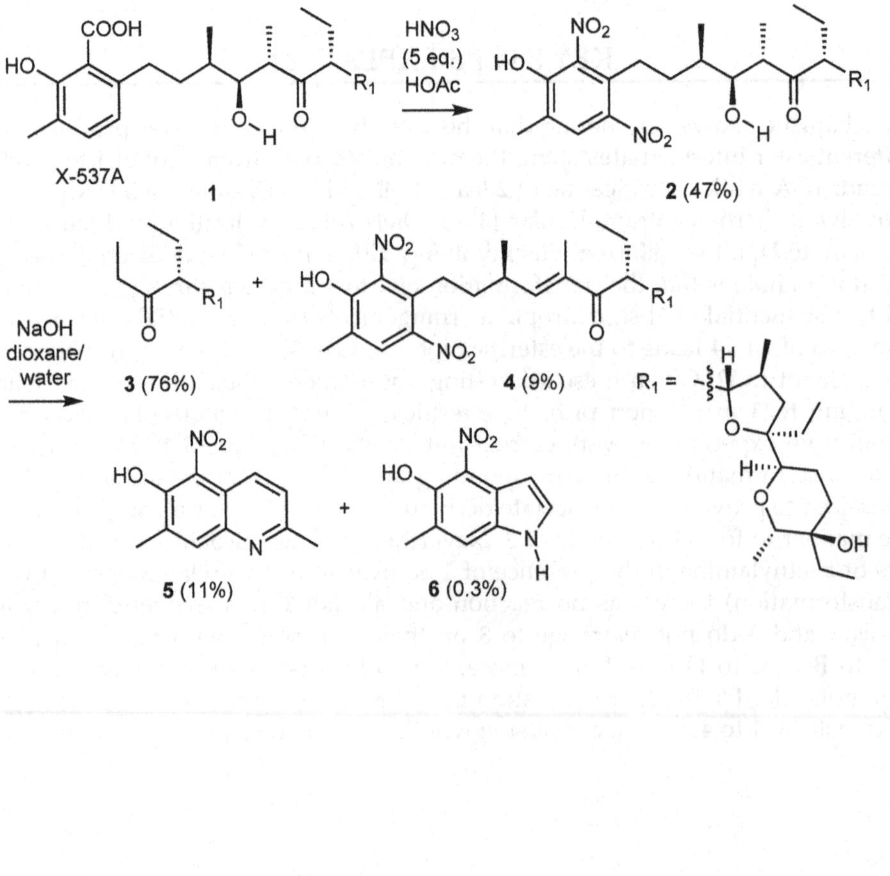

SCHEME 1

SOLUTION: 2 TO 3 AND A

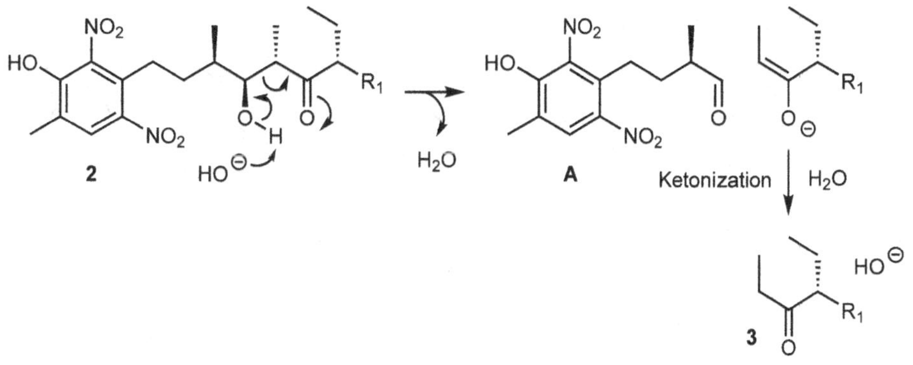

SOLUTION: A TO 5

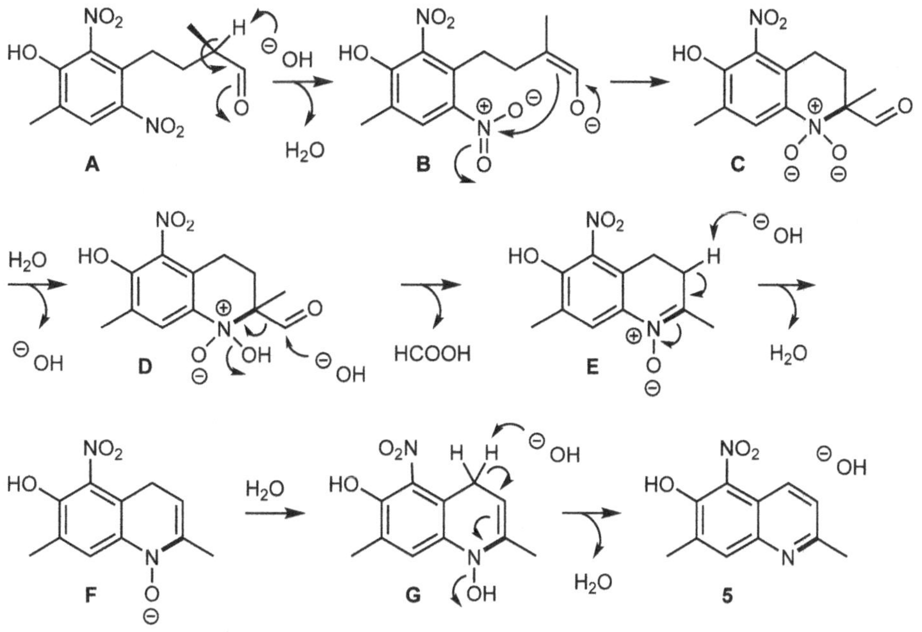

BALANCED CHEMICAL EQUATION: 2 TO 3 AND 5

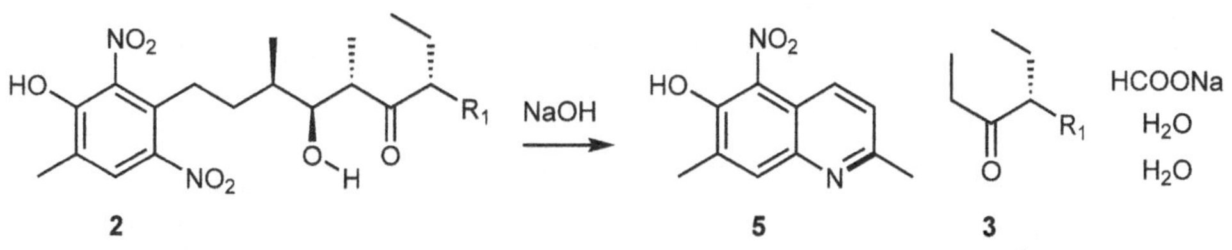

SOLUTION: 2 TO 4

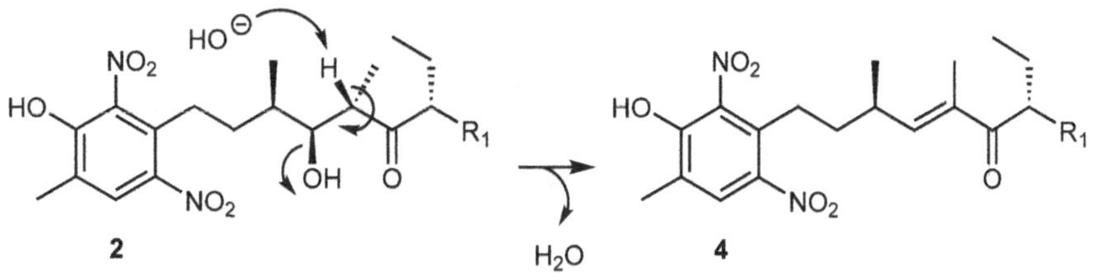

SOLUTION: D TO 6

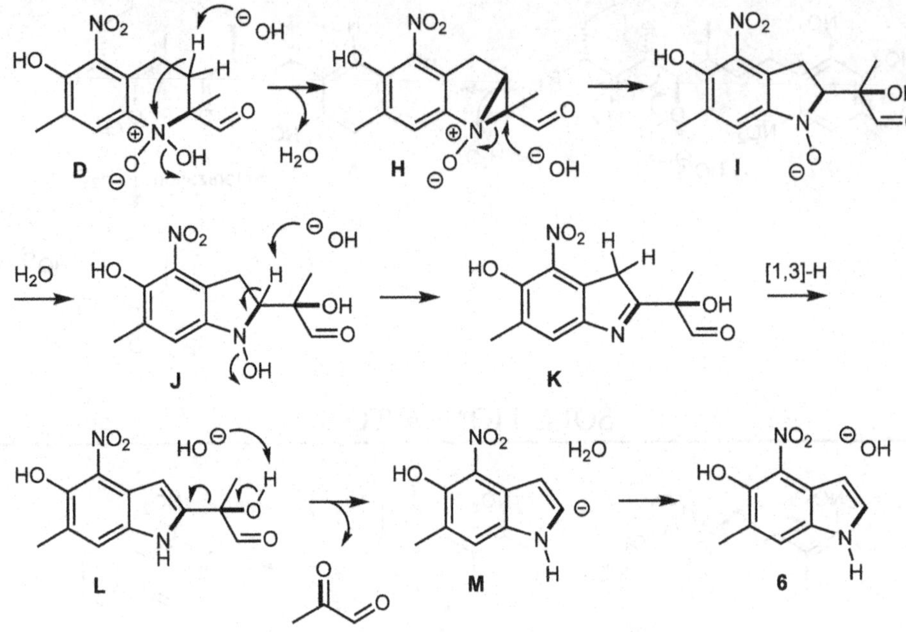

BALANCED CHEMICAL EQUATION: 2 TO 3 AND 6

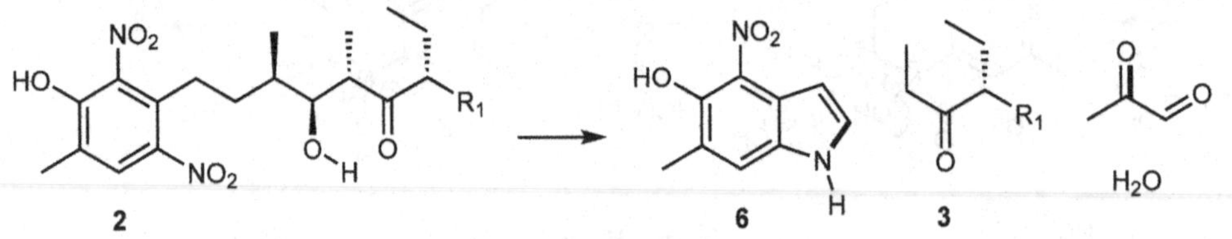

KEY STEPS EXPLAINED

The mechanism proposed by the authors for the transformation of **2** to **5** begins with base-induced fragmentation of **2** which forms aldehyde **A** and an enolate which ketonizes to the major product **3**. Base-catalyzed enolization of **A** then leads to enolate **B** which cyclizes onto the nitro group to yield intermediate **C**. Successive protonation by water and hydroxide ion nucleophilic attack at the aldehyde group causes fragmentation of **D** to form formic acid by-product and nitrone **E**. Successive deprotonation and dehydration of **E** leads to aromatization of the second ring to yield the quinoline product **5**. We note that the authors confirmed the structure of **5** by an independent synthesis and then matched the spectral and melting point characteristics of the product obtained with the degradation product obtained from **2** (see Scheme 2). The authors also confirmed the *o*-nitrophenol orientation of **5** by derivatizing it to an oxazole (see Scheme 3).

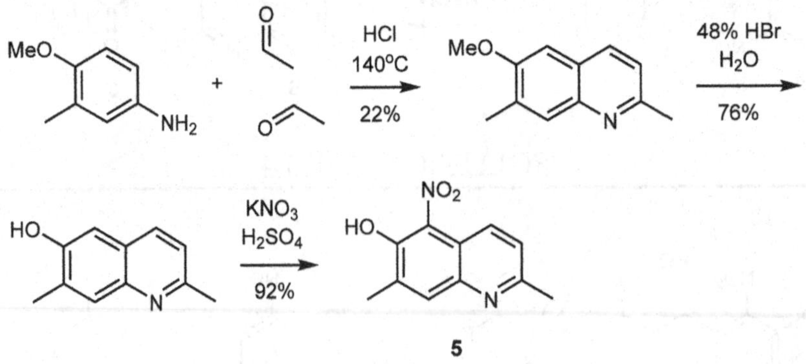

5

SCHEME 2

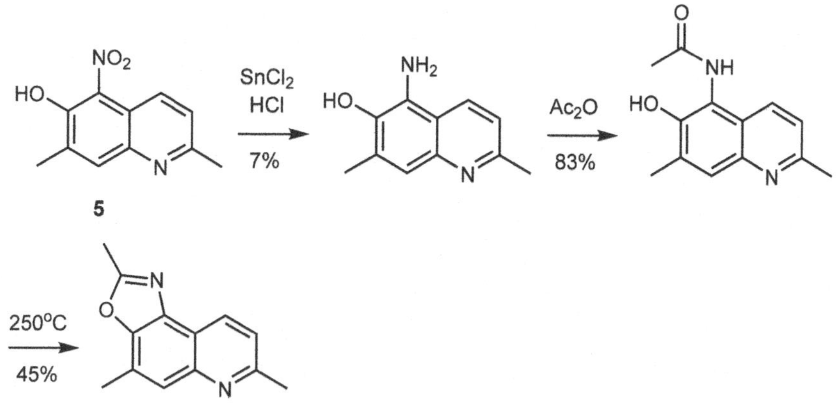

SCHEME 3

While the formation of **4** can be explained by means of one simple dehydration operation, the mechanism for forming **6** is slightly more complex. We note that compound **2** is expected to fragment to intermediate **D** and product **3** as before. Intermediate **D** then deprotonates to the [3.1.0] bicyclic intermediate **H** with elimination of hydroxide ion. Hydroxide ion then reacts with **H** in a nucleophilic manner to cause ring opening of the three-membered ring thus forming nitroxide **I**. Protonation of **I** followed by elimination of hydroxide ion yields 3*H*-indol-5-ol **K** which undergoes a base-induced 1,3-protic shift to give indole **L**. Base-induced elimination of 2-oxo-propionaldehyde from **L** followed by protonation leads to the indole product **6**. In a final note, we remark that the authors only discussed the mechanism for the formation of **5**. The discussion offered incomplete details as well as no further experimental and theoretical evidence to support the postulated mechanism. Here we have presented a mechanistic analysis to explain the formation of all observed products.

ADDITIONAL RESOURCES

We direct the reader to an article which discusses the degradation of tetracycline antibiotics and seeks to support the proposed mechanism with kinetics analysis.[2]

References

1. Westley JW, Schneider J, Evans Jr. RH, Williams T, Batcho AD, Stempel A. Nitration of antibiotic X-357A and facile conversion to 6-hydroxy-2,7-dimethyl-5-nitroquinoline. *J Org Chem*. 1971;36(23):3621–3624. https://doi.org/10.1021/jo00822a036.
2. Jeong J, Song W, Cooper WJ, Jung J, Greaves J. Degradation of tetracycline antibiotics: mechanisms and kinetic studies for advanced oxidation/reduction processes. *Chemosphere*. 2010;78:533–540. https://doi.org/10.1016/j.chemosphere.2009.11.024.

Question 135: Steroid Rearrangements: Appearances Can Be Deceptive

The transformations illustrated in Scheme 1 appear similar but actually have different mechanisms.[1] Provide these mechanisms to highlight the differences.

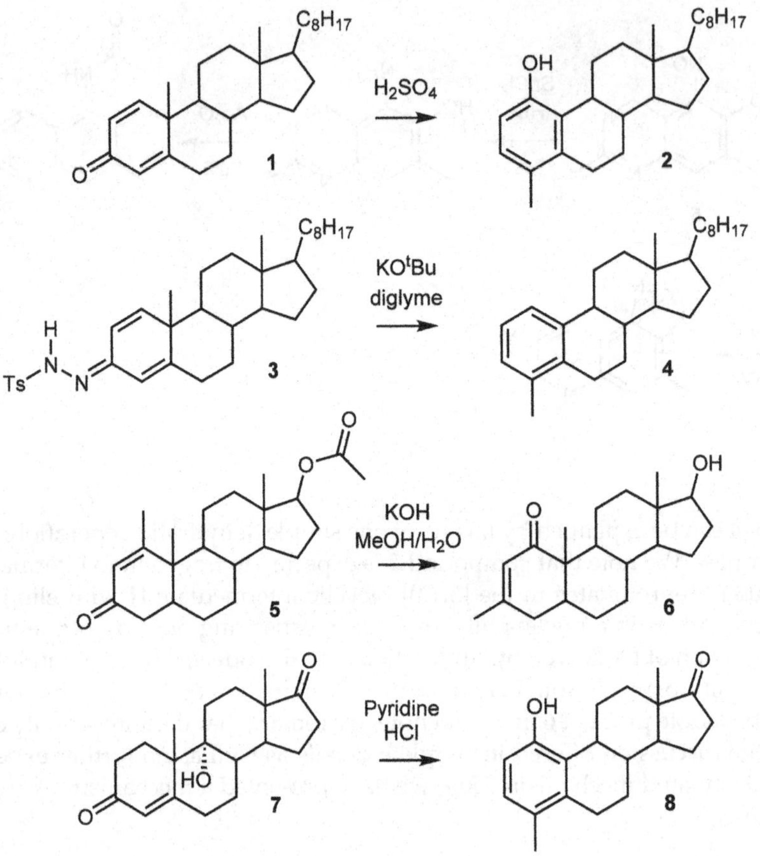

SCHEME 1

SOLUTION

In 1950 Woodward and Singh studied the transformation of **1** to **2** and proposed two possible mechanisms for it.[1] Mechanism 1 involved a spiro intermediate **C** (Scheme 2) and Mechanism 2 involved ring cleavage, aromatization, and formation of a secondary carbocation C_2 (Scheme 3). Eight years later, Bloom was able to confirm the operation of Mechanism 1 by means of running the same experiment with substituted hexahydronaphthalene **1a** (Scheme 4).[2] In this case, following Mechanism 1, product **2a** would be formed whereas following Mechanism 2, the formation of product **2b** would be expected. It turned out that Bloom observed only product **2a** which demonstrated conclusively

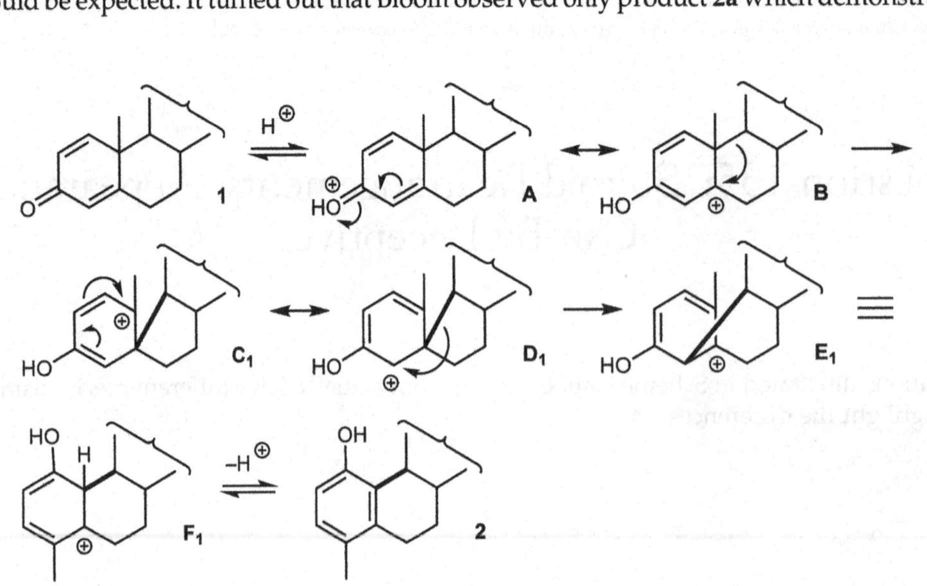

SCHEME 2

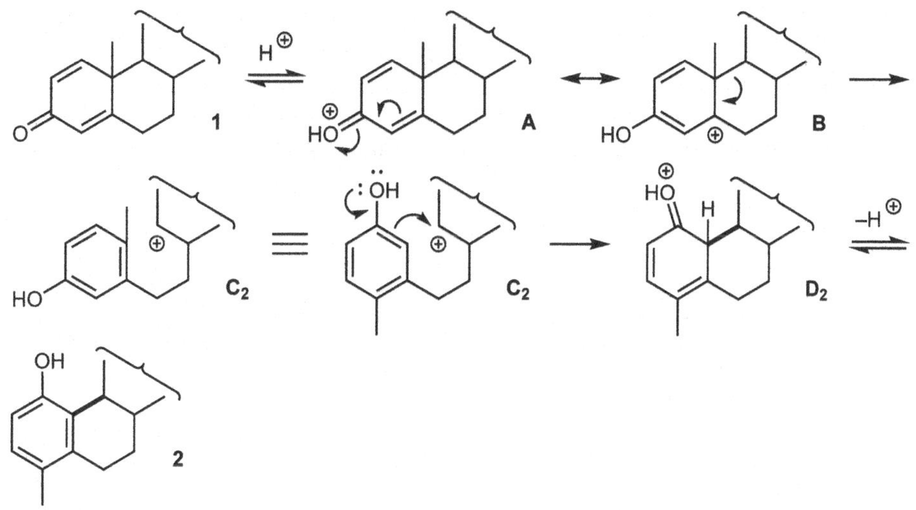

SCHEME 3

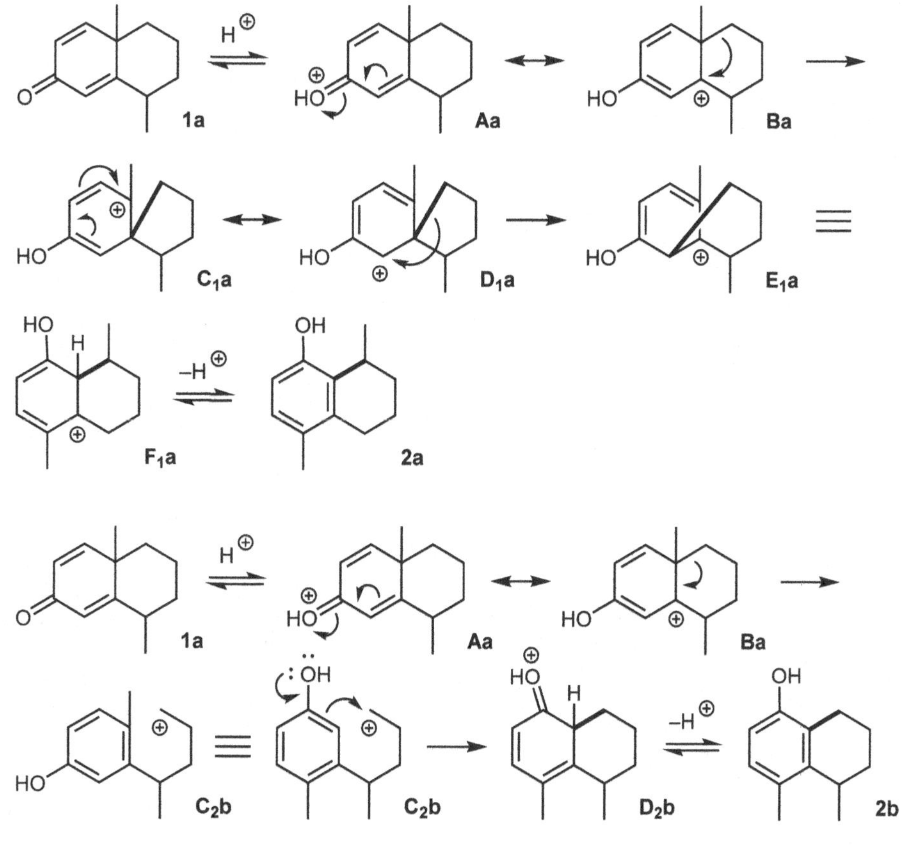

SCHEME 4

that Mechanism 1 was operational. The two groups of authors also commented on the fact that the more substituted carbon atom undergoes migration from the *spiro* intermediate **D** to intermediate **E**. This interesting transformation has since been termed the dienone-phenol rearrangement.

For an explanation of the operational mechanism 1, we note that the quinone oxygen atom is protonated to give oxonium ion **A** which is in resonance with carbocation **B**. **B** then rearranges to *spiro* carbocation C_1 which is in resonance with *spiro* carbocation D_1. D_1 undergoes further rearrangement to carbocation E_1 which can be rewritten as F_1. Final deprotonation of F_1 cause rearomatization of the ring and leads to formation of product **2**.

In the case of the transformation of **3** to **4**, German chemists proposed in 1965 a mechanism which is illustrated in Scheme 5 which also proceeds through a *spiro* intermediate J_1.[3] Since the authors did not carry out experimental or theoretical analysis to support their mechanism, we propose an alternative mechanism which avoids the intermediacy of a *spiro* compound and zwitterionic intermediates (Scheme 6). We believe that a future study should undertake experiments similar to those conducted by Bloom to help determine which mechanism is operational.

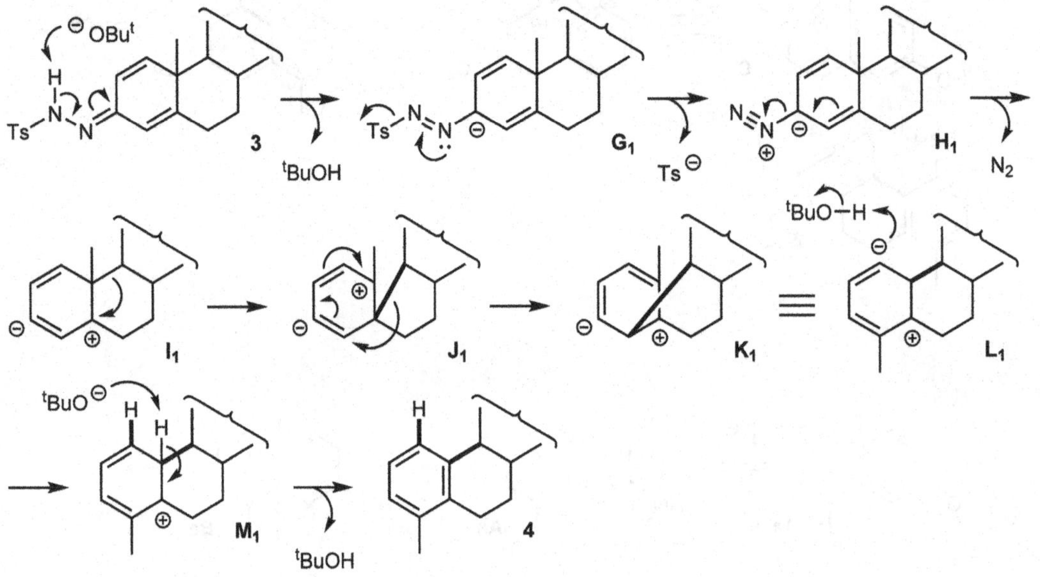

SCHEME 5

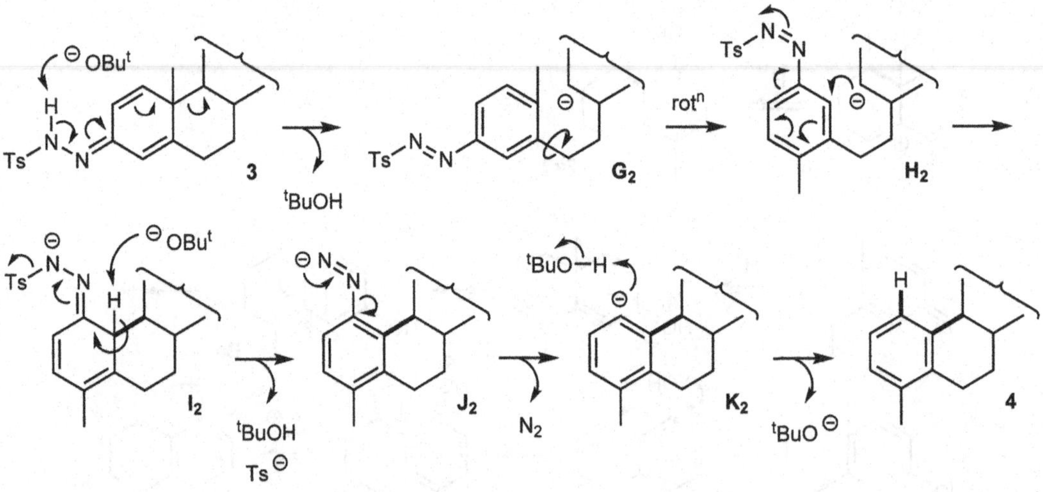

SCHEME 6

We note that both mechanisms closely resemble the dienone-phenol rearrangement with the exception of the basic conditions as compared to the acidic conditions seen before. Also, both mechanisms have the same balanced chemical equation (Scheme 7). And furthermore, both mechanisms involve deprotonation by *t*-butoxide base, elimination of tosylate ion and nitrogen gas followed by protonation of an aryl carbanion group via *t*-butanol. The main distinction, just as before, is that one mechanism (Scheme 5) involves a zwitterionic *spiro* intermediate J_1 while the other mechanism (Scheme 6) involves ring cleavage and a secondary carbanion intermediate G_2 which through bond rotation achieves the conformation observed in product **4**. Only negatively charged intermediates are involved in Scheme 6 which is more compatible with the basic reaction conditions. Once again we believe a theoretical analysis would go a long way to shed more light on the difference between these two mechanisms.

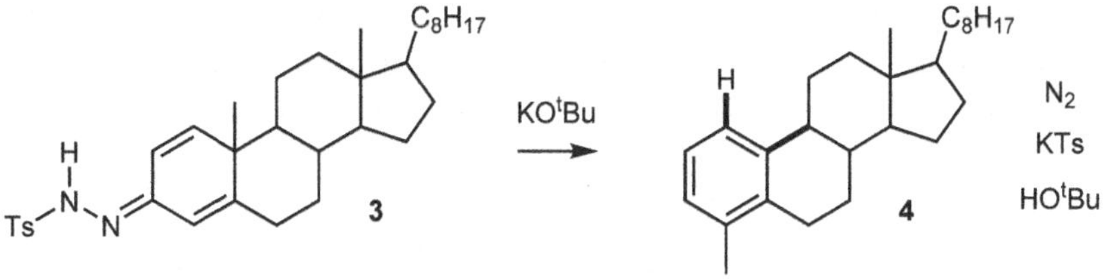

SCHEME 7

With respect to the transformation of **5** to **6**, we note that German chemists had reported on this transformation in 1964.[4]

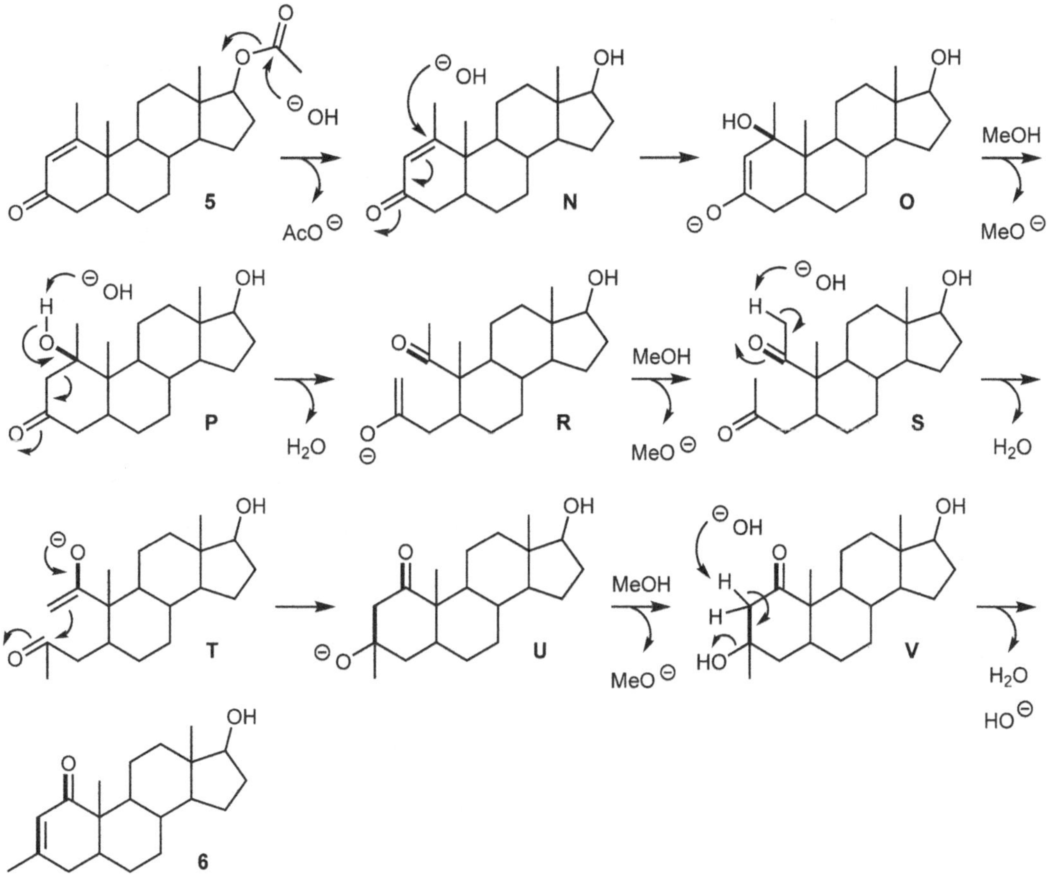

SCHEME 8

Although they recognized the transformation as a retro-aldol rearrangement, they did not provide complete mechanistic details nor did they offer experimental or theoretical evidence for the mechanism. We provide here the entire mechanism as illustrated in Scheme 8. We also provide the balanced chemical equation for this transformation in Scheme 9. Concerning this mechanism, the first step is deacetylation of **5** by hydroxide ion to yield intermediate **N**. A second hydroxide ion adds nucleophilically in a Michael fashion to the cyclohexenone moiety which sets up a ring opening-ring closing (Robinson annulation) cascade eventually leading to intermediate **U**. The last two steps are protonation of the alkoxide group and elimination of water to create product **6**.

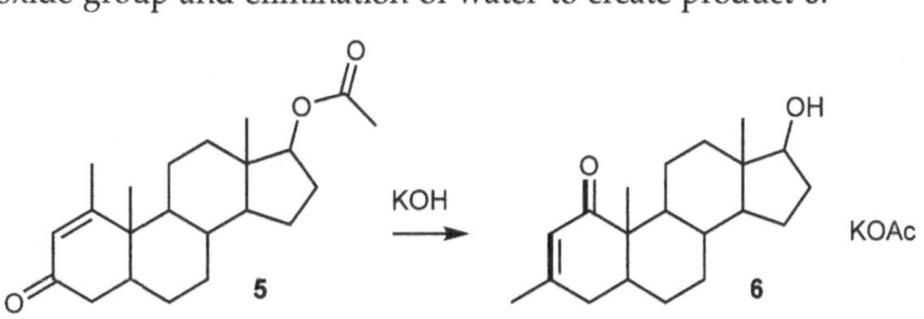

SCHEME 9

With respect to the last transformation of **7** to **8**, a conjectured mechanism was offered in a 1959 article which did not provide complete details nor offer experimental or theoretical evidence to support the mechanism.[5] We thus provide a complete mechanism in Scheme 10. This mechanism resembles the retro-aldol rearrangement seen earlier for the transformation of **3** to **4**.

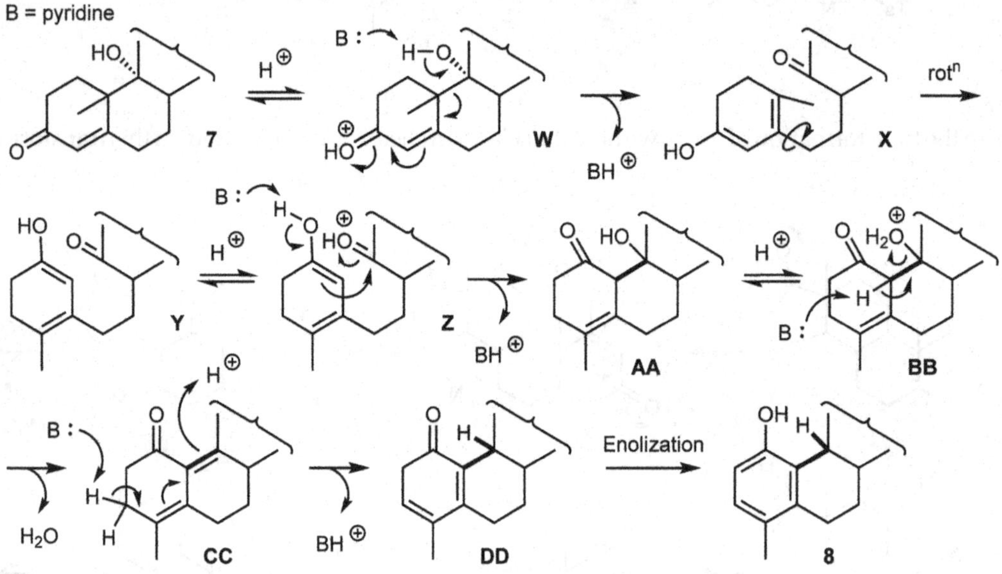

SCHEME 10

We also provide the balanced chemical equation for this transformation in Scheme 11. With regard to the mechanism, protonation of the carbonyl group of **7** followed by base-catalyzed ring opening of **W** leads to dienone **Y** upon C—C bond rotation. Protonation of the carbonyl group of **Y** followed by base-catalyzed ring closure leads to intermediate **AA**. Further protonation of the hydroxyl group of **AA** followed by base-catalyzed elimination of water generates intermediate **CC** which undergoes prototropic rearrangement to **DD** followed by enolization to the final phenol product **8**.

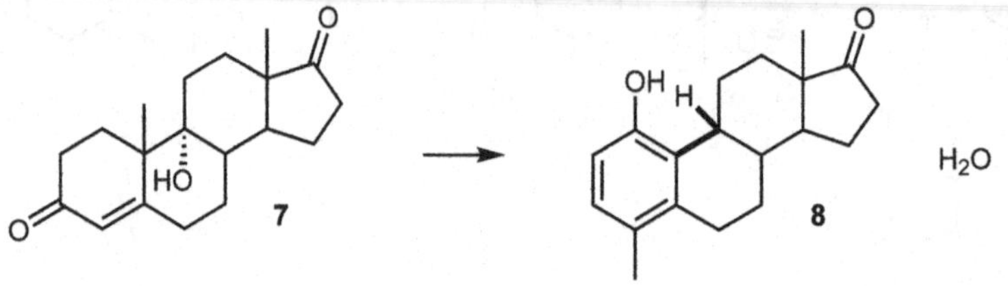

SCHEME 11

ADDITIONAL RESOURCES

We recommend an article where Japanese chemists were actually successful in isolating a *spiro* product in a transformation involving acid-catalyzed dienone-phenol rearrangement.[6] We also recommend an excellent review article which discusses retro-aldol processes in steroid chemistry.[7]

References

1. Woodward RB, Singh T. Synthesis and rearrangement of cyclohexadienones. *J Am Chem Soc*. 1950;72(1):494–500. https://doi.org/10.1021/ja01157a129.
2. Bloom SM. 8,10-Dimethyl-2-keto-$\Delta^{1,9;3,4}$-hexahydronaphthalene: the dienone-phenol rearrangement. *J Am Chem Soc*. 1958;80(23):6280–6283. https://doi.org/10.1021/ja01556a030.
3. Dannenberg H, Gross HJ. Dehydrierung von Steroiden—X. Bamford-Stevens-Reaktion von $\Delta^{1,4}$-dienon-(3)-Steroid-Tosylhydrazonen. *Tetrahedron*. 1965;21:1611–1617. https://doi.org/10.1016/S0040-4020(01)98323-1.
4. Bohlmann F, Rufer C. Über eine Retroaldolumlagerung in der Steroid-Reihe. *Eur J Inorg Chem*. 1964;97(6):1770–1773. https://doi.org/10.1002/cber.19640970640.
5. Chinn LJ, Dodson RM. Rearrangement of 9α-hydroxy-4-androstene-3,17-dione. *J Org Chem*. 1959;24(6):879. https://doi.org/10.1021/jo01088a622.

6. Koga T, Nogami Y. The isolation of a spiran in the rearrangement of an α-bromo-α,β-unsaturated steroidal ketone. *Tetrahedron Lett.* 1986;27 (37):4505–4506. https://doi.org/10.1016/S0040-4039(00)84990-4.

7. Dzhafarov MK. Retro-aldol processes in steroid chemistry. *Russ Chem Rev.* 1992;61(3):363–372. https://doi.org/10.1070/RC1992v061n03ABEH000950.

Question 136: Synthesis of *Eburna* and *Aspidosperma* Alkaloids

In 1987 French chemists reacted hexahydroindolo[2,3-*a*]quinolizine derivative **1** with dimsyllithium in THF/DMSO to form the expected diastereomeric β-ketosulfoxides **2** in excellent yield (Scheme 1).[1] When this mixture of diastereomers was treated with 6 equivalents of *p*-TsOH in hot THF under reflux for 5 min, the *Eburna* derivatives **3a** and **3b** together with the two isomeric *Aspidosperma* derivatives **4** and **5** were obtained in 56% overall yield. Suggest mechanisms for the formation of **3a**, **3b**, **4**, and **5**.

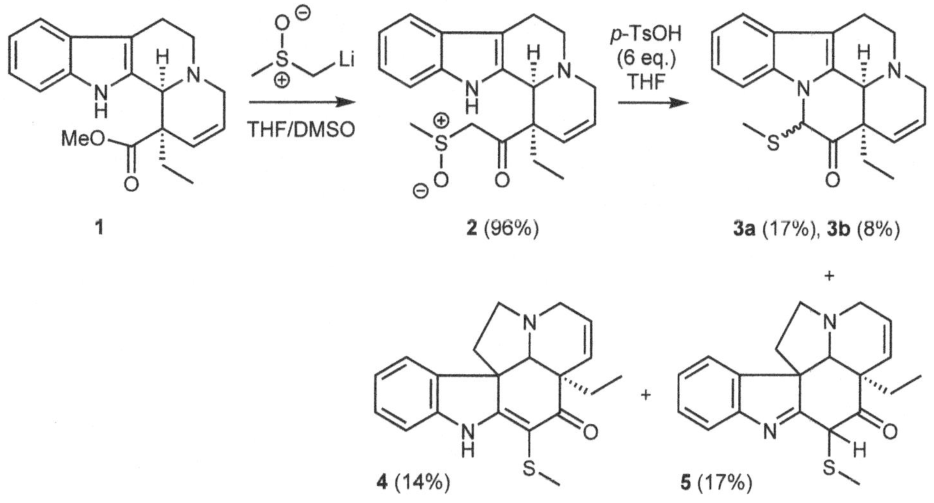

SCHEME 1

SOLUTION: 2 TO 3A AND 3B

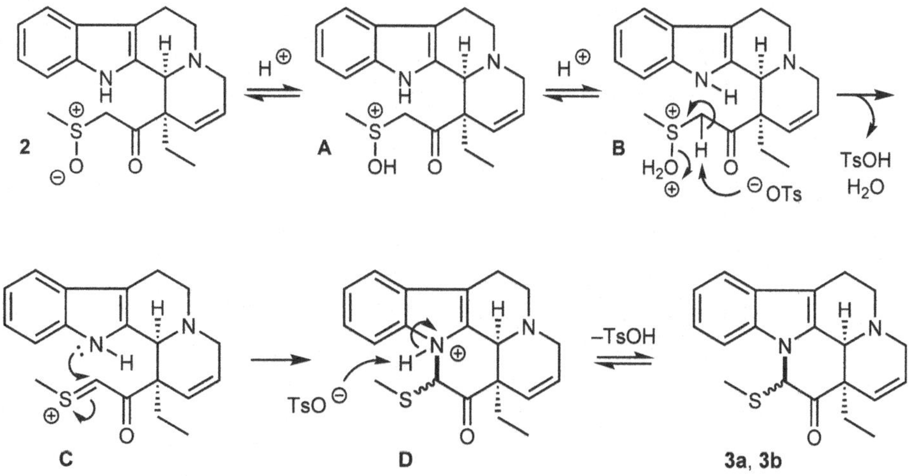

BALANCED CHEMICAL EQUATION: 2 TO 3A AND 3B

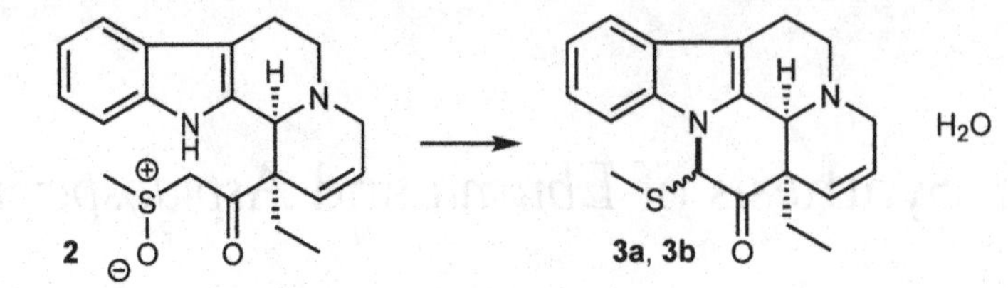

SOLUTION: 2 TO 4 AND 5

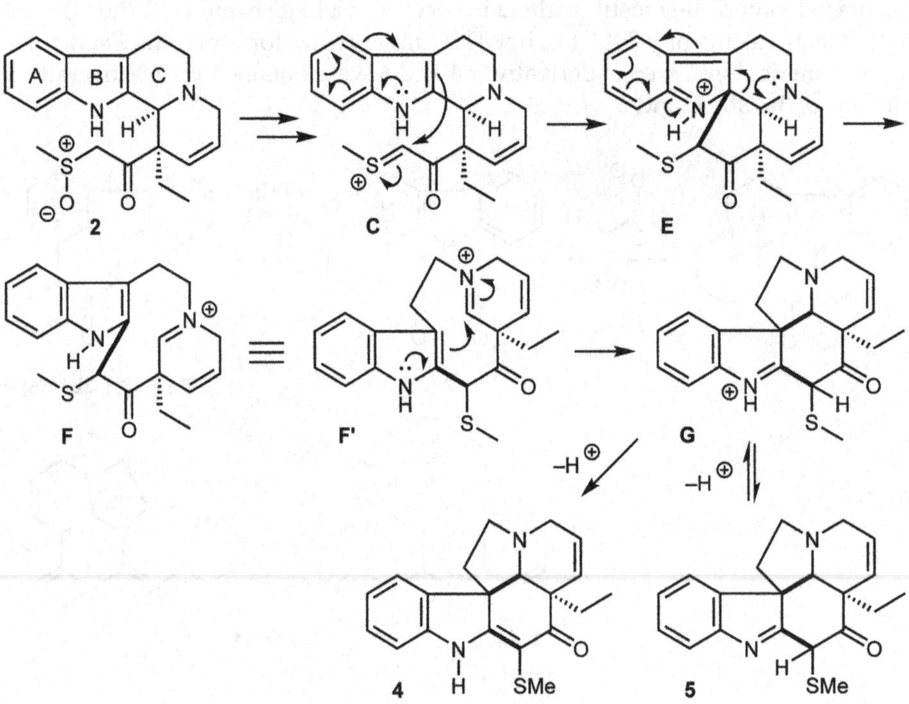

BALANCED CHEMICAL EQUATIONS: 2 TO 4 AND 5

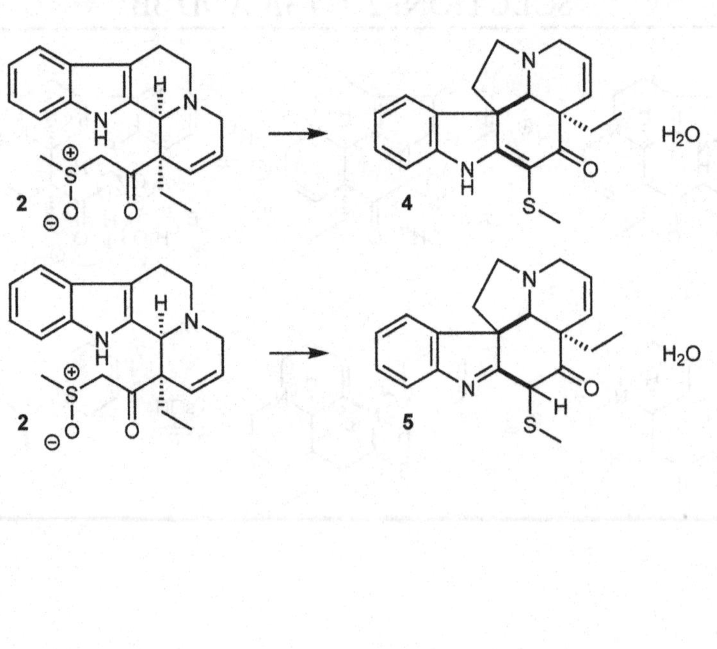

KEY STEPS EXPLAINED

The mechanism proposed to explain the transformation of **2** to the diastereomeric mixture **3a** and **3b** begins with double protonation of the sulfoxide oxygen atom of **2** to form intermediate **B** followed by deprotonation by tosylate ion to form **C**. Intermediate **C** then cyclizes via N-alkylation to **D** which undergoes a final deprotonation by a second tosylate ion to give the mixture of **3a** and **3b**. Concerning the formation of **4** and **5**, we note that intermediate **C** may also cyclize via C-alkylation to form intermediate **E**. Iminium ion formation in **E** then leads to fragmentation of ring C to form **F** which recyclizes to **G**. Deprotonation at the carbon atom in **G** leads to product **4** whereas deprotonation at the nitrogen atom leads to product **5**. Although the authors did not perform further experimental or theoretical work to help support the proposed mechanism, they did comment on the difference between N-alkylation (from **C** to **D**) and C-alkylation (from **C** to **E**). Specifically, they believed the difference in which path the mechanism will follow could be partially attributed to steric factors especially with respect to the N-alkylation path which implies switching the *trans* quinolizidine conformation in **C** to a *cis* quinolizidine conformation in **D**. Such a change of conformation is not necessitated by the C-alkylation pathway which leads to products **4** and **5**. A theoretical analysis would certainly shed some light on this interesting mechanistic path divergence.

ADDITIONAL RESOURCES

We recommend an article describing an interesting dimerization of *Aspidosperma* alkaloids.[2] We also highly recommend a review article which discusses rearrangements of sulfoxides and sulfones.[3]

References

1. Genin D, Andriamialisoa RZ, Langlois N, Langlois Y. Use of the Pummerer reaction in the synthesis of Eburna and Aspidosperma derivatives. *Heterocycles*. 1987;26(2):377–383. https://doi.org/10.3987/R-1987-02-0377.
2. Medley JW, Movassaghi M. A concise and versatile double-cyclization strategy for the highly stereoselective synthesis and arylative dimerization of Aspidosperma alkaloids. *Angew Chem Int Ed*. 2012;51:4572–4576. https://doi.org/10.1002/anie.201200387.
3. Prilezhaeva EN. Rearrangements of sulfoxides and sulfones in the total synthesis of natural compounds. *Russ Chem Rev*. 2001;70(11):897–920. https://doi.org/10.1070/RC2001v070n11ABEH000593.

Question 137: Synthesis of Highly Hindered Cyclic Amines

In 1980 chemists from Ohio carried out the two regioselective transformations shown in Scheme 1.[1,2] Suggest mechanisms to explain the formations of products **2**, **3**, **5**, and **6**.

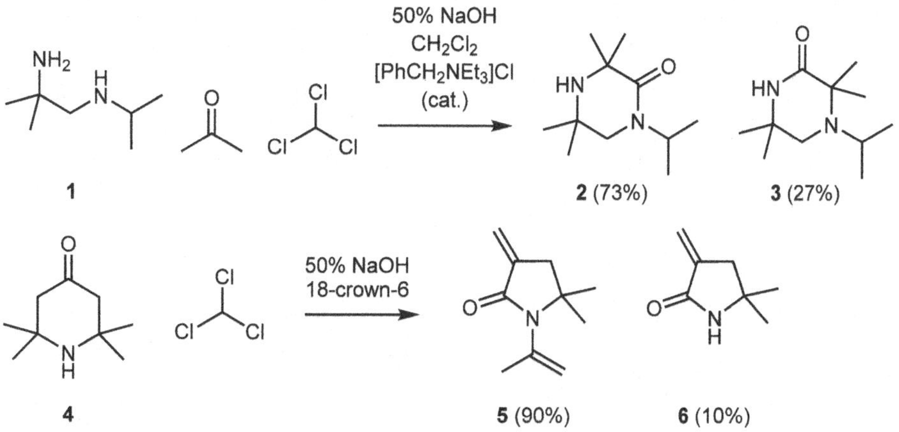

SCHEME 1

SOLUTION: 1 TO 2

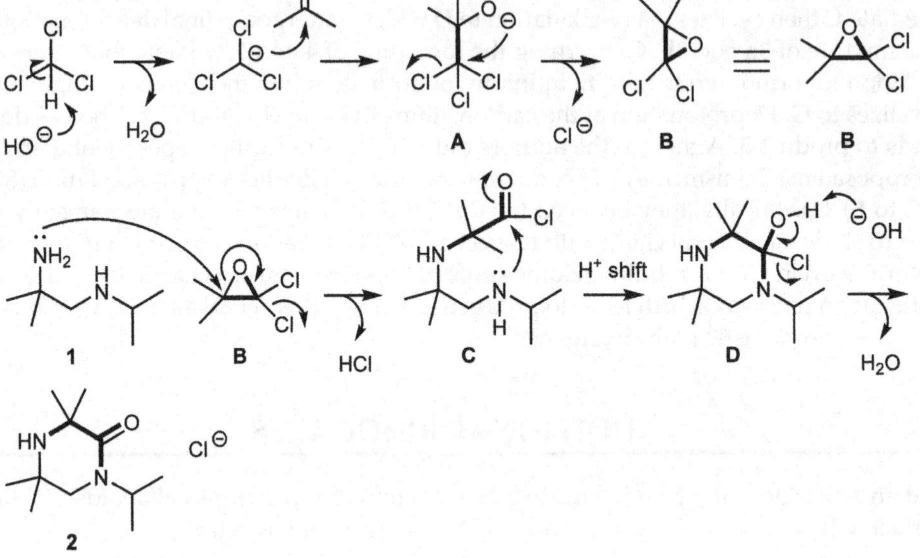

SOLUTION: 1 TO 3

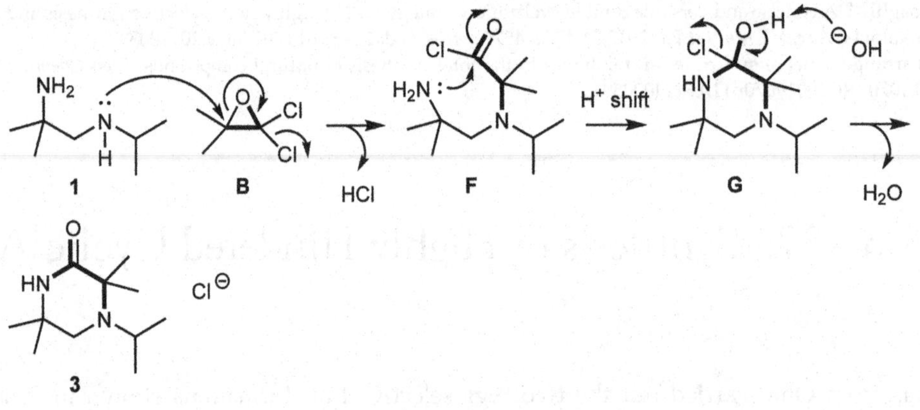

BALANCED CHEMICAL EQUATIONS: 1 TO 2 AND 3

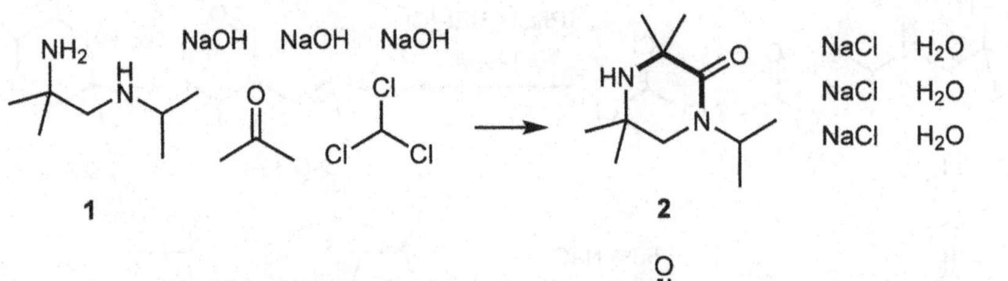

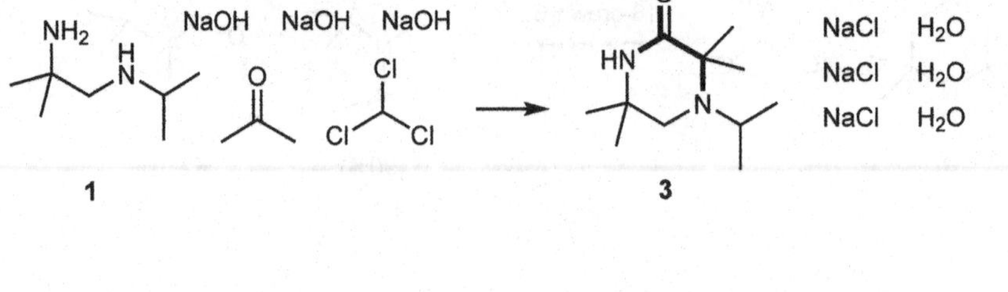

SOLUTION: 4 TO 5 AND 6

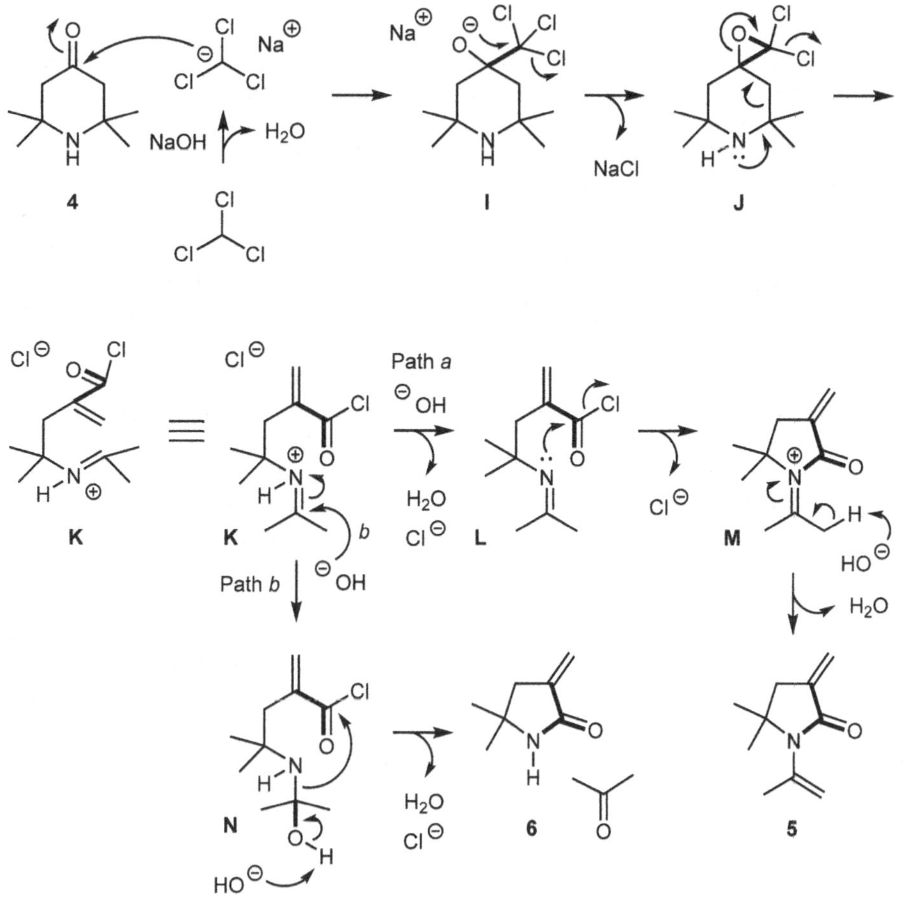

BALANCED CHEMICAL EQUATIONS: 4 TO 5 AND 6

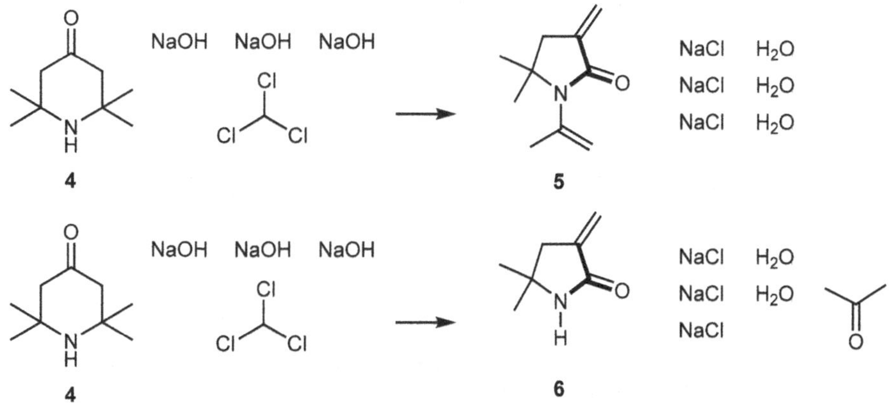

KEY STEPS EXPLAINED

The mechanism proposed by the author for the transformation of **1** to **2** and **3** begins with formation of the carbanion of trichloromethane from chloroform under basic conditions. The carbanion then reacts with acetone to produce the dichloro-epoxide **B**. *N*-isopropyl-2,2-dimethyl-1,2-ethanediamine, **1**, can react with epoxide **B** in two ways depending on the epoxide's orientation with respect to the two amino nucleophilic centers in **1**. If the primary amino group of **1** attacks the dimethyl substituted carbon atom of **B** then intermediate **D** forms, which contains an acetyl chloride group. The secondary amino group attacks the acetyl chloride group leading to product **2**. On the other hand, if the secondary amino group of **1** attacks the dimethyl substituted carbon atom of **B**, we have formation of intermediate **G**, which contains an acetyl chloride group. The primary amino group attacks the acetyl chloride group leading to product **3**. Both reactions are three-component coupling reactions with [4 + 1 + 1] target bond mappings.

In the case of the transformation of **4** to **5** and **6** we have once again initial formation of the carbanion of trichloromethane. This carbanion attacks the carbonyl group of 2,2,6,6-tetramethyl-4-piperidone, **4**, to give dichloro-epoxide **J** which opens to the iminium ion intermediate **K** containing an acetyl chloride group. The pathway leading to product **5** involves ring closure via the imine nitrogen atom attacking the acid chloride group to form iminium ion **M** which is deprotonated by hydroxide ion to form **5**. Path *b* leading to product **6** involves attack of hydroxide ion onto the imino carbon atom leading to intermediate **N**. Next, a second equivalent of base deprotonates the hydroxyl group of **N** which simultaneously leads to ring closure and elimination of acetone. Both reactions have [4 + 1] target bond mappings.

The authors explained that the intermediacy of electrophilic dichlorocarbene was ruled out on the basis of the following observations: (a) addition of piperidine to the reaction mixture leads to a **5** to **6** product ratio of 21:79, (b) dichlorocarbene is expected to react with piperidine faster than **1** since piperidine is a stronger base than **1**, (c) addition of piperidine does not cause conversion of **5** to **6** after they are formed from **1**, and (d) products **5** and **6** are not interchangeable under the reaction conditions. It is interesting to note that the ratio of **5** to **6** is essentially inverted when piperidine, a base, is added to the reaction mixture containing 50% NaOH. The authors did not provide a reason to explain this observation. In the mechanism shown hydroxide ion acts exclusively as a base in path *a* from intermediate **K** to the major product **5**; whereas, in path *b* it acts first as a nucleophile and then as a base from intermediate **K** to minor product **6**. Steric hindrance may play a role in suppressing path *b* via step **K** to **N** thus enhancing the production of **5**. We speculate that in the case when piperidine is also present, since it is a weaker base than hydroxide ion, piperidine is less likely to deprotonate intermediate **M** to yield product **5**. It is expected that piperidine acts as a general base to deprotonate intermediate **N** in addition to the specific base action of hydroxide ion thus enhancing the rate of formation of product **6**. A theoretical analysis would help identify the energy difference between **K** and **L** versus **K** and **N** which might help explain the product ratios of **5** and **6** under the various reaction conditions. Such a study might also help explain the observed product ratios of **2** and **3**. Lastly, we note the interesting difference in mechanisms between the two transformations. In the case of **1** to **2** and **3** there is no common intermediate whereas in the latter case of **4** to **5** and **6** we have the common intermediate **K** despite the nearly identical reactivity involved in both cases.

ADDITIONAL RESOURCES

We direct the reader to three journal articles which discuss the mechanisms of very similar transformations to the ones encountered here.[3–5]

References

1. Lai JT. Hindered amines. Novel synthesis of 1,3,3,5,5-pentasubstituted 2-piperazinones. *J Org Chem*. 1980;45(4):754–755. https://doi.org/10.1021/jo01292a054.
2. Lai JT, Westfahl JC. Rearrangement of 2,2,6,6-tetramethyl-4-piperidone in phase-transfer catalyzed reactions. *J Org Chem*. 1980;45(8):1513–1514. https://doi.org/10.1021/jo01296a034.
3. Mloston G, Romanski J, Linden A, Heimgartner H. Unexpected products from the reaction of 2,2,4,4-tetramethylcyclobutane-1,3-dione with the Makosza reagent. *Helv Chim Acta*. 1999;82(8):1302–1310. https://doi.org/10.1002/(SICI)1522-2675(19990804)82:8<1302::AID-HLCA1302>3.0.CO;2-A.
4. Padwa A, Tomioka Y, Venkatramanan MK. A study of the 5-exo methylene-isoxazolidine to 3-pyrrolidinone rearrangement. *Tetrahedron Lett*. 1987;28(7):755–758. https://doi.org/10.1016/S0040-4039(01)80981-3.
5. Rychnovsky SD, Beauchamp T, Vaidyanathan R, Kwan T. Synthesis of chiral nitroxides and an unusual racemization reaction. *J Org Chem*. 1998;63(18):6363–6374. https://doi.org/10.1021/jo9808831.

Question 138: A 1,2-Dihydrobenzocyclobutane to Isochroman-3-one Conversion

In 1985 Japanese chemists heated 5-methoxybenzocyclobutene-1-carboxylic acid **1** at 150–160°C for 45min and formed 6-methoxyisochroman-3-one **2** in 48% yield (Scheme 1).[1] Suggest a mechanism for this transformation.

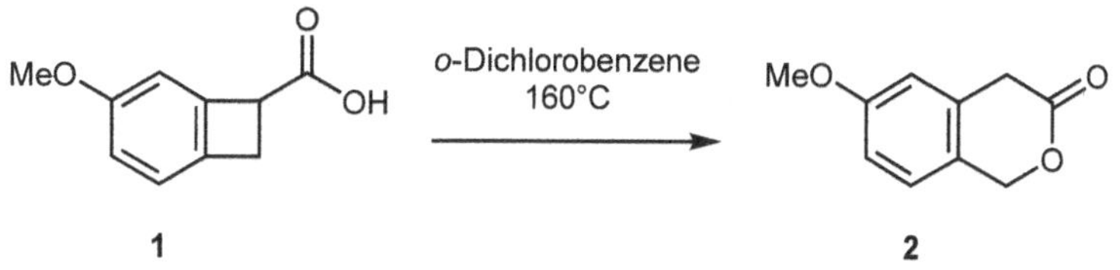

SCHEME 1

SOLUTION

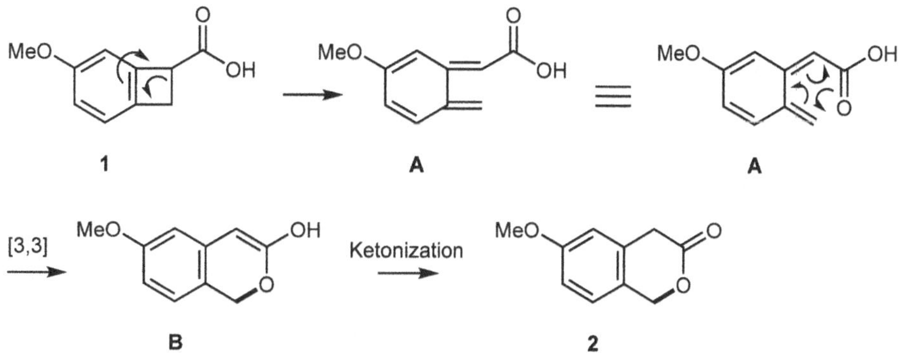

KEY STEPS EXPLAINED

The mechanism proposed by the authors begins with thermal electrocyclic ring opening of **1** which leads to intermediate **A**. This intermediate undergoes a [3,3] sigmatropic cyclization to enol **B** followed by ketonization to finally yield lactone product **2**. The reaction is 100% atom economic since it does not produce any by-products.

ADDITIONAL RESOURCES

For further work by the same authors on very related mechanistic analysis, see Ref. 2. For an interesting transformation involving indanone chemistry, see the work of Pandit et al.[3] We also recommend a review article of the applicability of the [1,3] O-to-C rearrangement.[4]

References

1. Shishido K, Shitara E, Fukumoto K, Kametani T. Tandem electrocyclic-sigmatropic reaction of benzocyclobutenes. An expedient route to 4,4-disubstituted isochromanones. *J Am Chem Soc*. 1985;107(20):5810–5812. https://doi.org/10.1021/ja00306a044.
2. Shishido K, Komatsu H, Fukumoto K, Kametani T. Complementary electrocyclic reactions of o-quinodimethanes. Highly efficient access to 4-alkylideneisochroman-3-ones and 1-carbomethoxy-3,4-dihydronaphthalenes. *Chem Lett*. 1987;16(11):2117–2120. https://doi.org/10.1246/cl.1987.2117.
3. Pandit UK, Das B, Chatterjee A. Synthetic entry into yohimbinoid alkaloids and novel synthesis of (±)-17-methoxy-hexadehydroyohimbane. *Tetrahedron*. 1987;43(18):4235–4239. https://doi.org/10.1016/S0040-4020(01)83466-9.
4. Nasveschuk CG, Rovis T. The [1, 3] O-to-C rearrangement: opportunities for stereoselective synthesis. *Org Biomol Chem*. 2008;6:240–254. https://doi.org/10.1039/b714881j.

Question 139: An Unexpectedly Facile Decarboxylation

In 1962 chemists from MIT treated 1,5-diethoxycarbonyl-3-methyl-3-azabicyclo[3.3.1]non-9-one **1** with hot 20% aqueous HCl to form 3-methyl-3-azabicyclo[3.3.1]non-9-one **2** in 70% yield (Scheme 1).[1] Suggest a mechanism for this transformation which does not involve anti-Bredt intermediates.

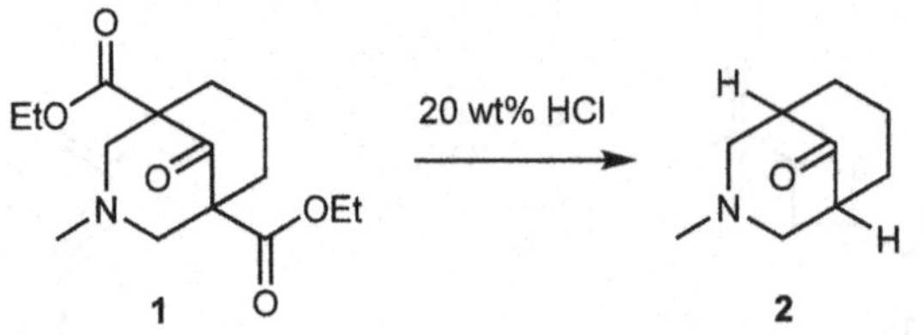

SCHEME 1

SOLUTION

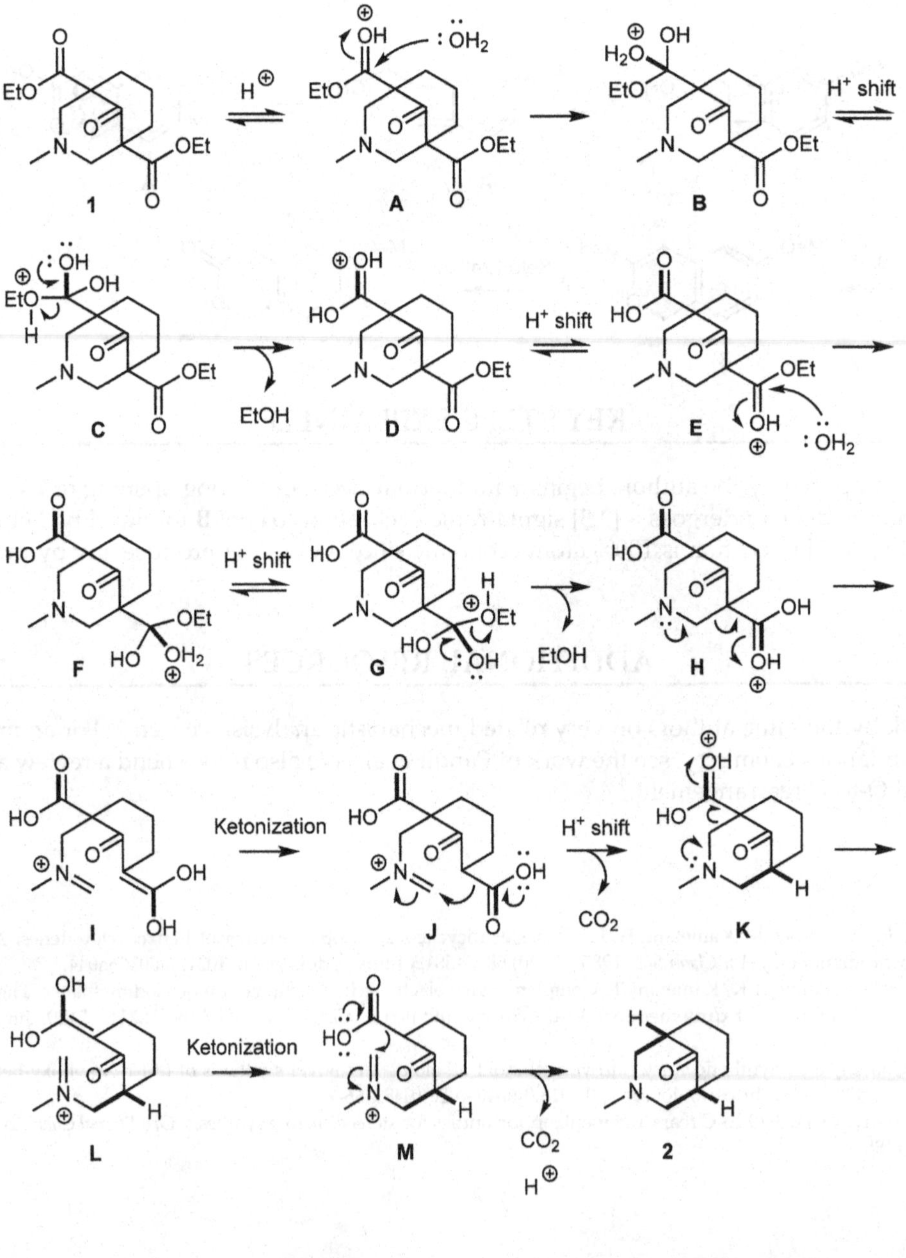

BALANCED CHEMICAL EQUATION

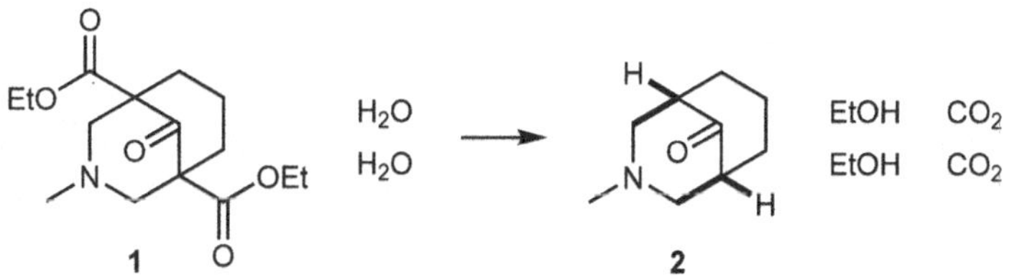

KEY STEPS EXPLAINED

The proposed mechanism involves initial hydrolysis of the two carboxyethyl ester groups in **1** to carboxylic acid groups in intermediate **H**. This happens over a series of steps which involve two sequences of hydrogen shift, water addition to an activated carbonyl group, and elimination of ethanol. Once intermediate **H** is formed, ring opening of **H** forms **I** by creating a carboxylic acid enol group and iminium ion group. Ketonization of **I** leads to **J** which undergoes decarboxylation to form **K**. Repetition of the same ketonization-decarboxylation sequence achieves the formation of the final product **2**. We note that the authors carried out this transformation in deuterium chloride and D_2O and formed product **3** (Scheme 2). We therefore see that deuterium incorporation into **3** is attributable to the two ketonization steps of the mechanism where a carboxylic acid enol group is converted back to a carboxylic acid group. Although this experiment supports the proposed mechanism, the authors did not venture to undertake a theoretical analysis to gather further supporting evidence.

SCHEME 2

We also remark that what is interesting about this reaction is that in order to decarboxylate both COOH groups the two highlighted bridging bonds of the piperidine ring shown in structures **1** and **2** of the mechanism before must sequentially cleave and reattach again. This happens so that anti-Bredt intermediates, known to be highly unstable, are avoided. Effectively the reaction is a masked [3 + 3] cycloaddition.

ADDITIONAL RESOURCES

We recommend an article which describes a photochemical-induced skeletal rearrangement of 3-azabicyclo[3.3.1] nonane analogs.[2]

References

1. House HO, Mueller HC. Decarboxylation and deuterium exchange in some azabicyclic ketone systems. *J Org Chem*. 1962;27(12):4436–4439. https://doi.org/10.1021/jo01059a076.
2. Williams CM, Heim R, Bernhardt PV. Nitrogen is a requirement for the photochemical induced 3-azabicyclo[3.3.1]nonane skeletal rearrangement. *Tetrahedron*. 2005;61:3771–3779. https://doi.org/10.1016/j.tet.2005.02.013.

Question 140: Regiospecific Synthesis of a 1,2-Dihydronaphthalene

In 1990 Japanese chemists reacted sulfoxide **1** with trifluoroacetic anhydride in refluxing toluene and achieved the dihydronaphthalene product **2** in 55% yield as a single regioisomer (Scheme 1).[1] Suggest a mechanism for this transformation.

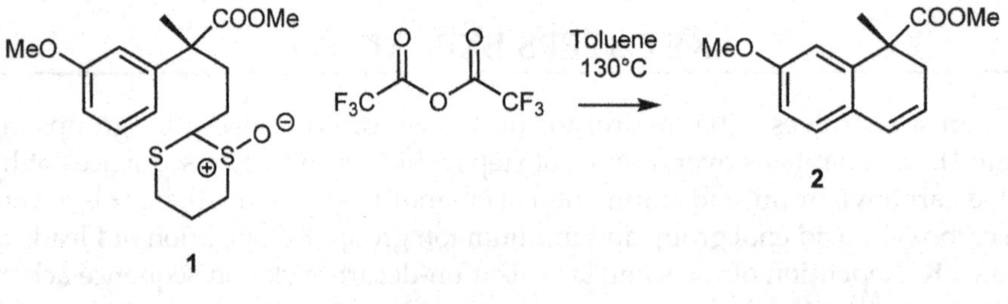

SCHEME 1

SOLUTION

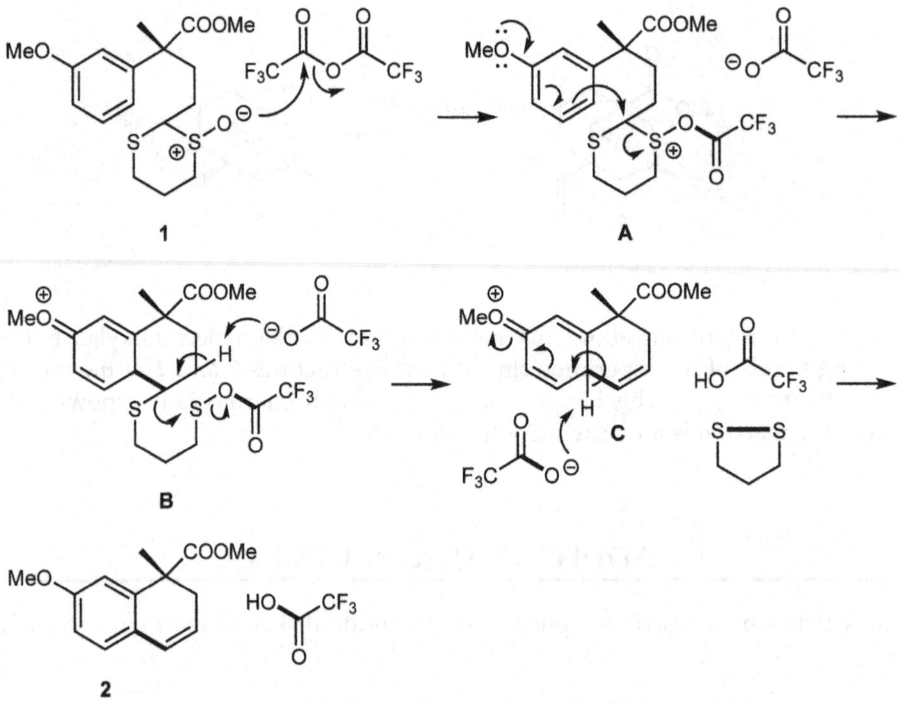

BALANCED CHEMICAL EQUATION

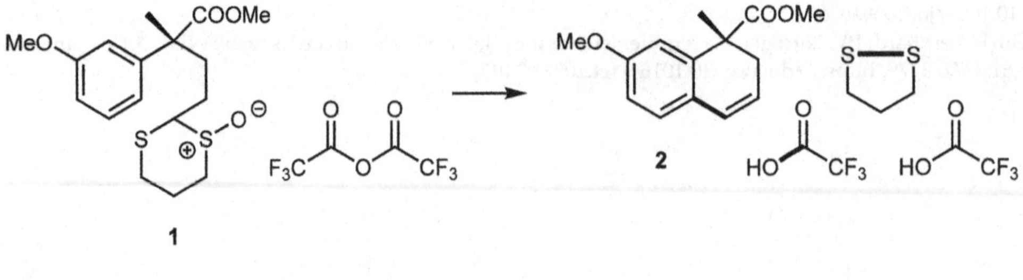

KEY STEPS EXPLAINED

The mechanism proposed by the authors begins with acylation of the sulfoxide oxygen atom by trifluoroacetic anhydride to give sulfonium ion **A**. Afterward, the electron-donating methoxy group on the aromatic ring causes formation of the carbocycle leading to intermediate **B**. Next, trifluoroacetate ion acts as a base to remove a proton from **B** leading to oxonium intermediate **C** and the formation of by-products [1,2]-dithiolane and trifluoroacetate ion. Lastly, the remaining trifluoroacetate ion abstracts a proton from **C** causing rearomatization of the ring and formation of product **2**. Unfortunately the authors were not able to experimentally detect the [1,2]-dithiolane by-product to help support the proposed mechanism. They also did not undertake a theoretical analysis for this mechanism.

ADDITIONAL RESOURCES

We recommend a recent article which discusses the mechanism of a transformation which involves the dihydronaphthalene chemical system.[2]

References

1. Takano S, Inomata K, Sato T, Takahashi M, Ogasawara K. The enantioselective total synthesis of natural (-)-aphanorphine. *J Chem Soc Chem Commun.* 1990;1990:290–292. https://doi.org/10.1039/C39900000290.
2. Kurouchi H, Sugimoto H, Otani Y, Ohwada T. Cyclization of arylacetoacetates to indene and dihydronaphthalene derivatives in strong acids. Evidence for involvement of further protonation of *O,O*-diprotonated β-ketoester, leading to enhancement of cyclization. *J Am Chem Soc.* 2010;132(2):807–815. https://doi.org/10.1021/ja908749u.

Question 141: Degradation of Terramycin

In 1952 Pfizer chemists undertook the alkaline degradation of terramycin **1** to form terracinoic acid **2** among many other products (Scheme 1).[1] Provide a mechanism for this degradation.

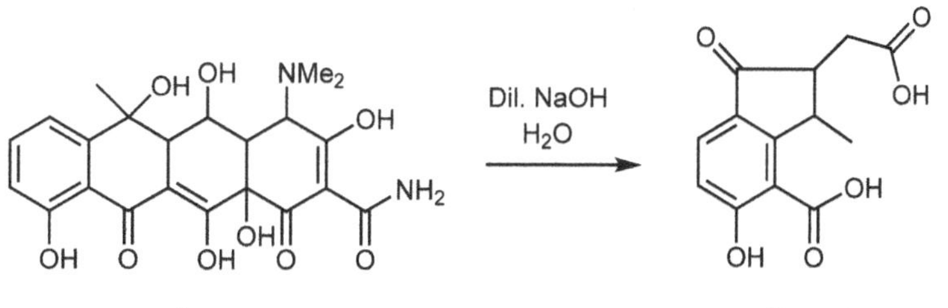

1
Terramycin

2
Terracinoic acid

SCHEME 1

SOLUTION

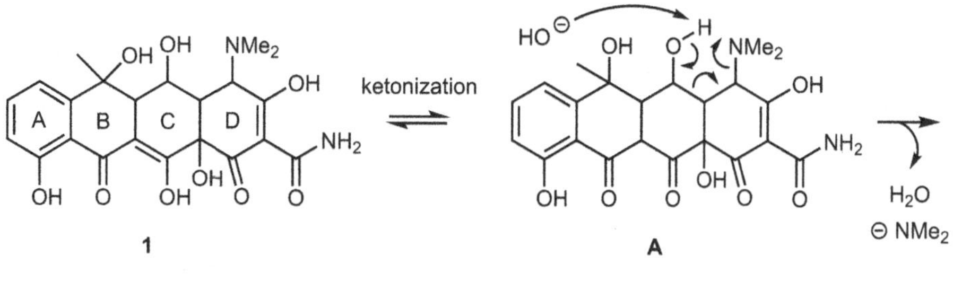

1

A

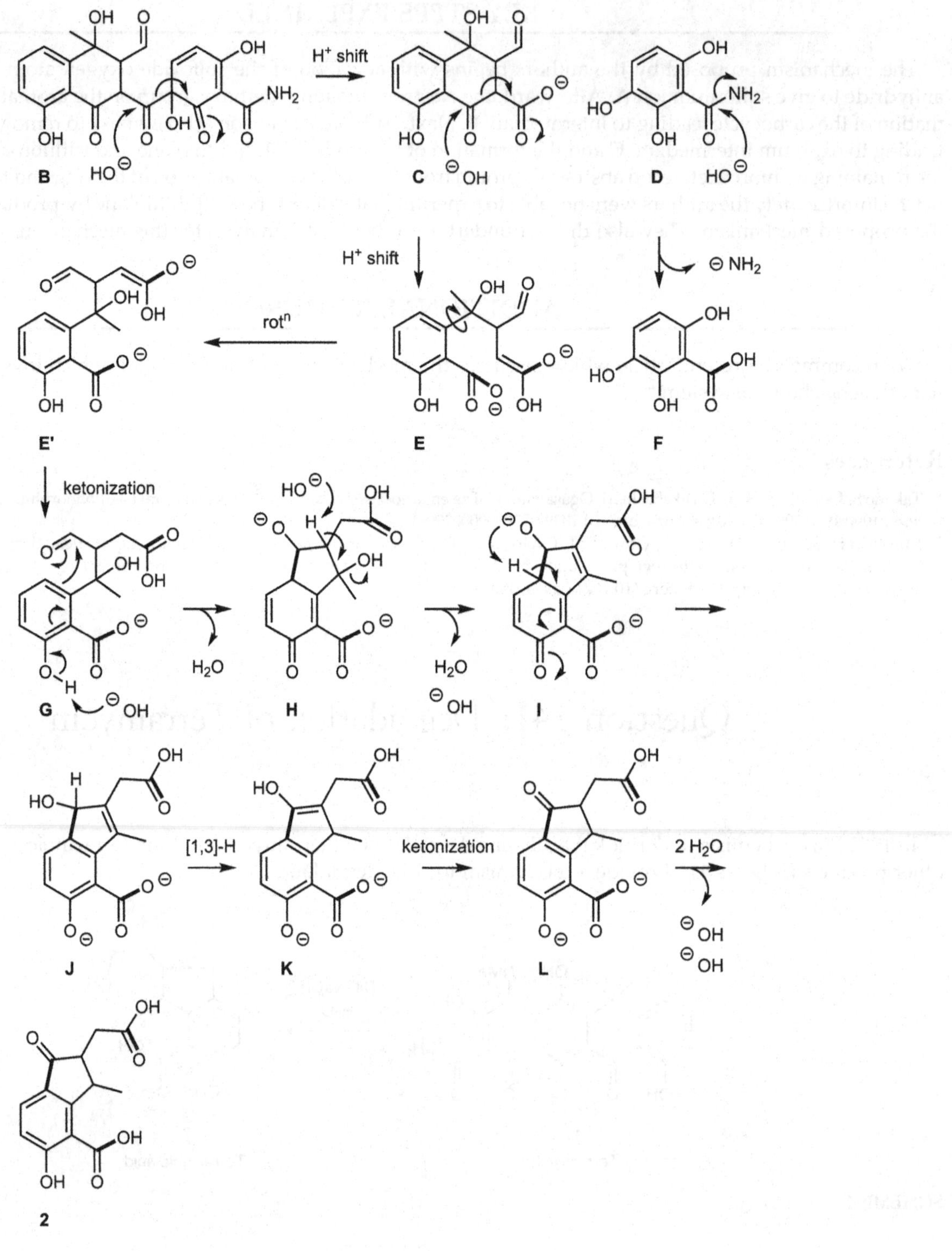
H⁺ shift
B
C
D
H⁺ shift
⊖ NH2
rotⁿ
E'
E
F
ketonization
G
H
I
H2O
[1,3]-H
ketonization
2 H2O
J
K
L
2

BALANCED CHEMICAL EQUATION

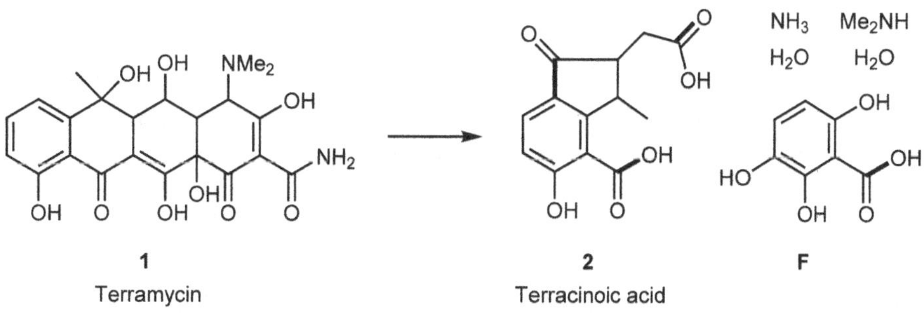

1
Terramycin

2
Terracinoic acid

F

KEY STEPS EXPLAINED

We note that the overall degradation of terramycin in dilute alkaline solution leads to terracinoic acid **2** in addition to 2,3,6-trihydroxybenzoic acid, ammonia, dimethylamine, and two equivalents of water as by-products. The authors did not discuss the mechanism for this degradation and so we propose initial base-catalyzed opening of ring C followed by elimination of dimethylamine and formation of intermediate **B**. Hydroxide ion attack on **B** then leads to elimination of 2,3,6-trihydroxybenzamide and the production of intermediate **C**. The benzamide is then hydrolyzed to the corresponding acid **F**. Hydroxide ion attack on **C** leads to fragmentation of ring B and after bond rotation and ketonization intermediate **G** is formed. Base-induced cyclization of **G** generates quinone **H** which rearomatizes to **J**. The final terracinoic acid product **2** is formed after a [1,3]-H shift, ketonization, and double protonations. We note that the sequence order of steps from **H** to **2** is flexible as other possibilities exist. For a future study, we recommend a theoretical analysis to help shed light on this interesting degradation, particularly the order of steps leading to the least energy demanding barriers.

ADDITIONAL RESOURCES

We recommend an article by Conover which covers synthetic approaches to the indanone degradation products of terramycin.[2] We also highlight an article which discusses an unusual reaction of terramycin with methyl iodide.[3] Lastly, we recommend a review article which discusses rearrangements of angucycline group antibiotics.[4]

References

1. Pasternack R, Conover LH, Bavley A, et al. Structure of terracinoic acid, an alkaline degradation product. *J Am Chem Soc.* 1952;74(8):1928–1934. https://doi.org/10.1021/ja01128a018.
2. Conover LH, Terramycin IX. The synthesis of indanone degradation products of terramycin. *J Am Chem Soc.* 1953;75(16):4017–4020. https://doi.org/10.1021/ja01112a042.
3. Woodward RB, Zimmerman HE. An unusual reaction of terramycin with methyl iodide. *Tetrahedron.* 1981;37:311–314. https://doi.org/10.1016/S0040-4020(01)92015-0.
4. Rohr J, Thiericke R. Angucycline group antibiotics. *Nat Prod Rep.* 1992;9:103–137. https://doi.org/10.1039/NP9920900103.

Question 142: Synthesis of Bi-indane-1,3-dione

In 1966 American chemists synthesized bi-indane-1,3-dione **6** using the synthetic approach depicted in Scheme 1.[1] Suggest mechanisms for the various transformations along this synthesis.

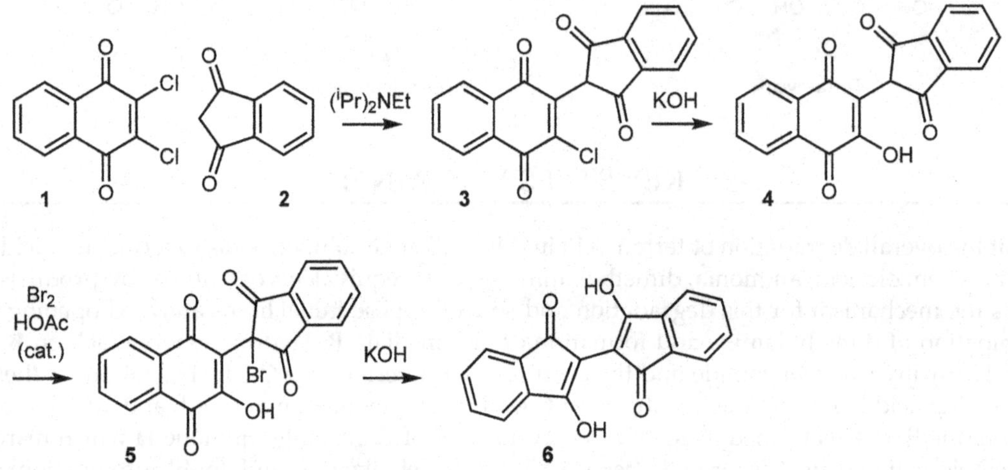

SCHEME 1

SOLUTION: 1 AND 2 TO 3

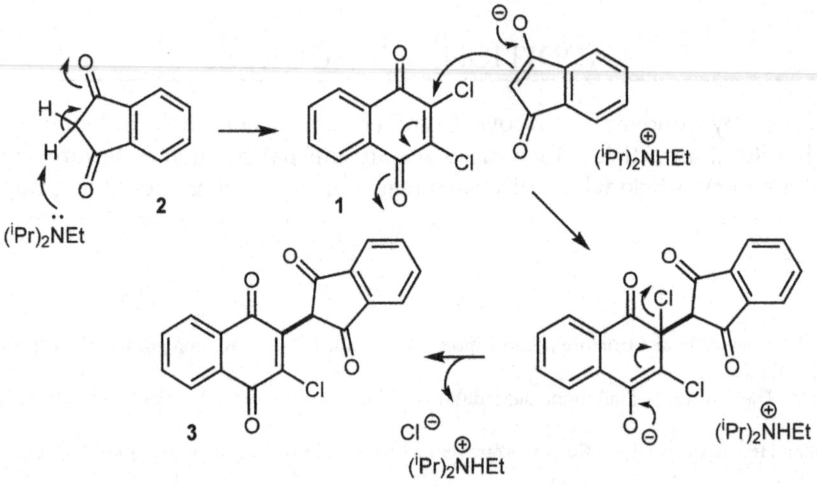

SOLUTION: 3 TO 4

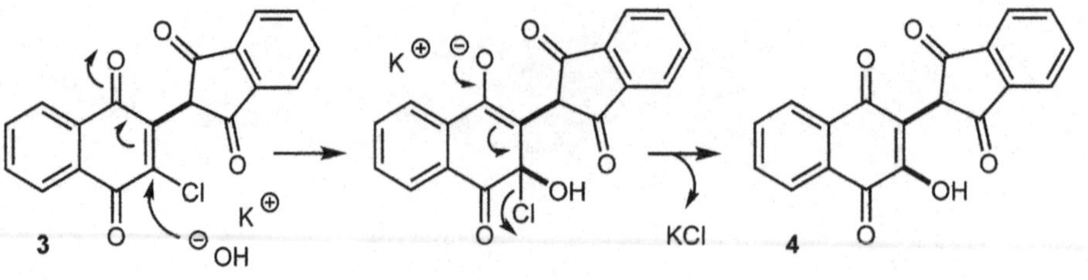

SOLUTION: 4 TO 5

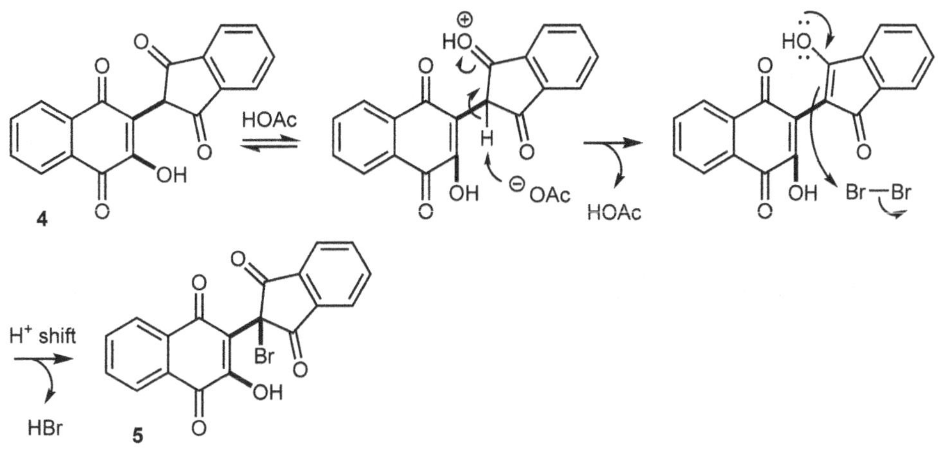

SOLUTION: 5 TO 6

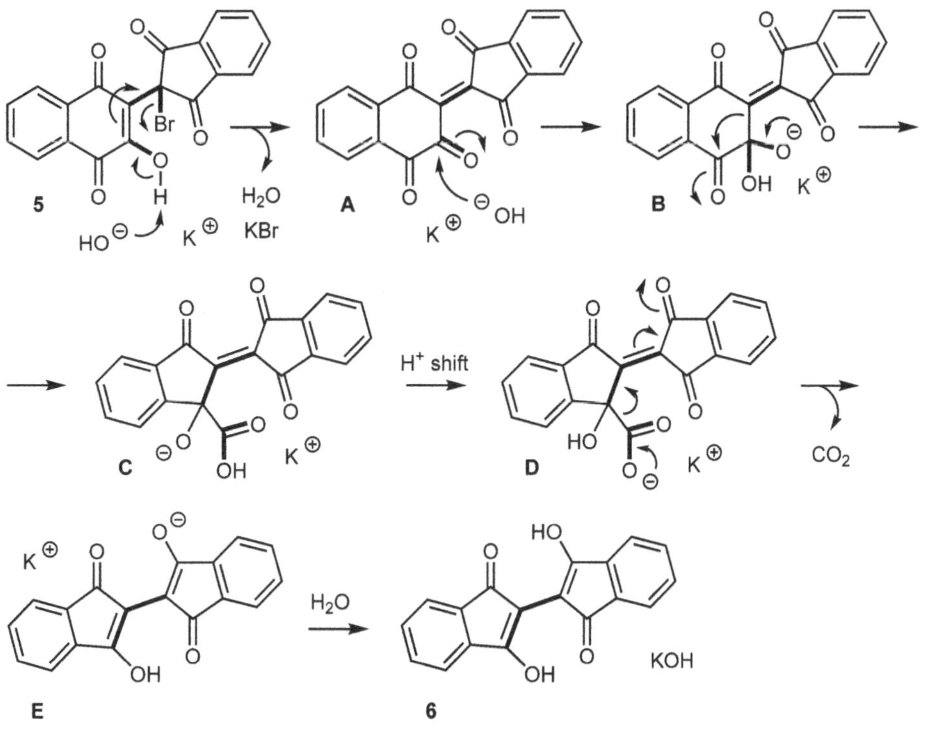

KEY STEPS EXPLAINED

The proposed mechanism for the transformation of **1** and **2** to **3** begins with base-mediated enolate formation of **2** followed by acetylation onto **1** which after loss of chloride gives **3**. When **3** reacts with potassium hydroxide we have hydroxide attack at the olefinic C—Cl group followed by elimination of KCl to form product **4**. In a similar mechanistic fashion, product **4** is brominated to **5**. At this point treatment of **5** with potassium hydroxide initiates deprotonation of the hydroxyl group of **5** via hydroxide ion which leads to elimination of KBr and formation of the 1,2-diketo intermediate **A**. A second hydroxide ion nucleophilically attacks the second carbonyl group yielding alkoxide **B** which undergoes a benzilic acid rearrangement to alkoxide **C** via ring contraction from a six-membered ring to a five-membered ring. After internal proton transfer leading to **D**, decarboxylation results in the formation of enolate **E** which protonates to yield the *bis*-indane-1,3-dione product **6**. The authors did not carry out further experimental or theoretical analysis to help support this last postulated mechanism.

We recommend three articles which discuss the mechanisms of interesting transformations involving structures very similar to product **6**.[2–4]

References

1. Vanallan JA, Adel RE, Reynolds GA. The reaction of 2,3-dichloronaphthoquinone with nucleophiles. III. Reaction with 1,3-indandione. *J Org Chem*. 1966;31(1):62–65. https://doi.org/10.1021/jo01339a012.
2. Ayyangar NR, Kolhe PY, Tilak BD. Heterocyclic quinonoid chromophoric systems: Part VI—reaction of 2,3-dichloro-1,4-naphthoquinone with homophthalimides & other compounds containing a reactive methylene group. *Indian J Chem Sect B*. 1980;19:836–843.
3. Maslak P, Varadarajan S, Burkey JD. Synthesis, structure, and nucleophile-induced rearrangements of spiroketones. *J Org Chem*. 1999;64:8201–8209. https://doi.org/10.1021/jo990867j.
4. Buggle K, Power J. Reaction of monothio-derivatives of 2-heteroarylideneindene-1,3(2H)-diones with quinones. *J Chem Soc Perkin Trans 1*. 1980;1980:1070–1075. https://doi.org/10.1039/P19800001070.

Question 143: A Coumarin to Cyclopentenone Transformation

In 1960 American chemists reacted 3-acetylcoumarin **1** with phenacyl bromide and sodium ethoxide to form 3,4-phenacylidene-3-acetylcoumarin **2** which was further reacted with sodium hydroxide to generate the cyclopentenone product **3** (Scheme 1).[1] Suggest a mechanism for the transformation of **2** to **3**.

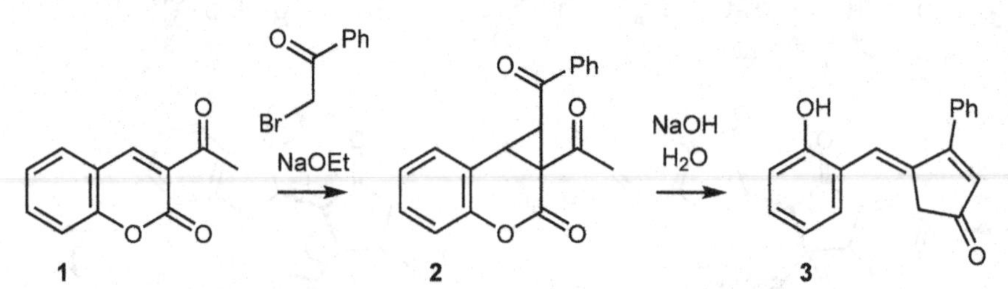

SCHEME 1

SOLUTION

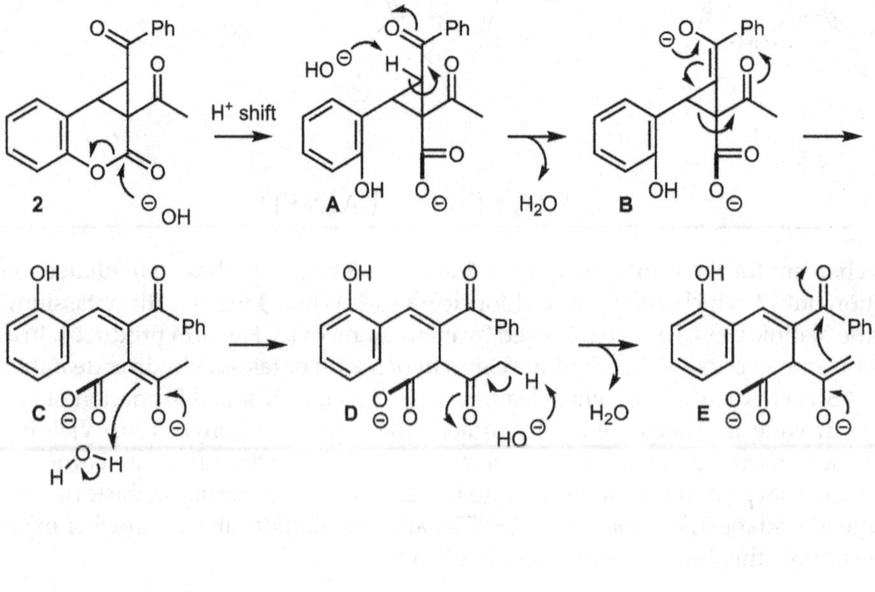

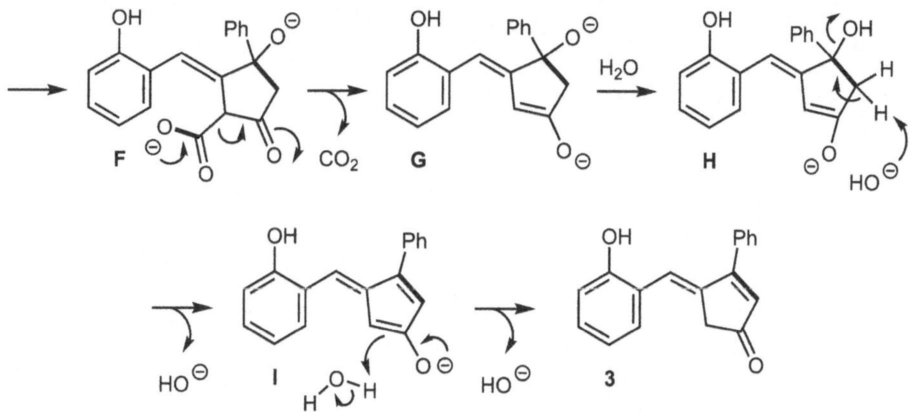

F → CO₂ G → H₂O H

HO⁻ I → HO⁻ 3

BALANCED CHEMICAL EQUATION

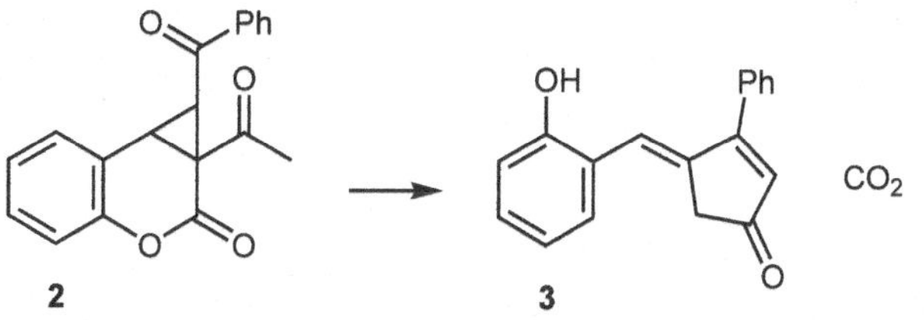

2 → 3 CO₂

KEY STEPS EXPLAINED

The mechanism proposed by the authors begins with nucleophilic attack by hydroxide ion on the lactone to form intermediate **A**. A second equivalent of hydroxide ion abstracts the proton *alpha* to the benzoyl group in **A** to form enolate **B** which ring opens to give enolate **C**. Enolate **C** then ketonizes to **D** which enolizes in the other direction to give **E** which undergoes an intramolecular ring closure to generate cyclopentanone **F**. At this point, there are several possible pathways depending on the order of steps leading to the final product **3**. Here we show one possible sequence which involves decarboxylation followed by protonation, hydroxide ion elimination, and final ketonization. We note that a key mechanistic point raised by the authors is that decarboxylation of intermediate **A** does not occur since this would have led to a different nonobserved isomeric product **4** as shown in Scheme 2. As a check, the authors

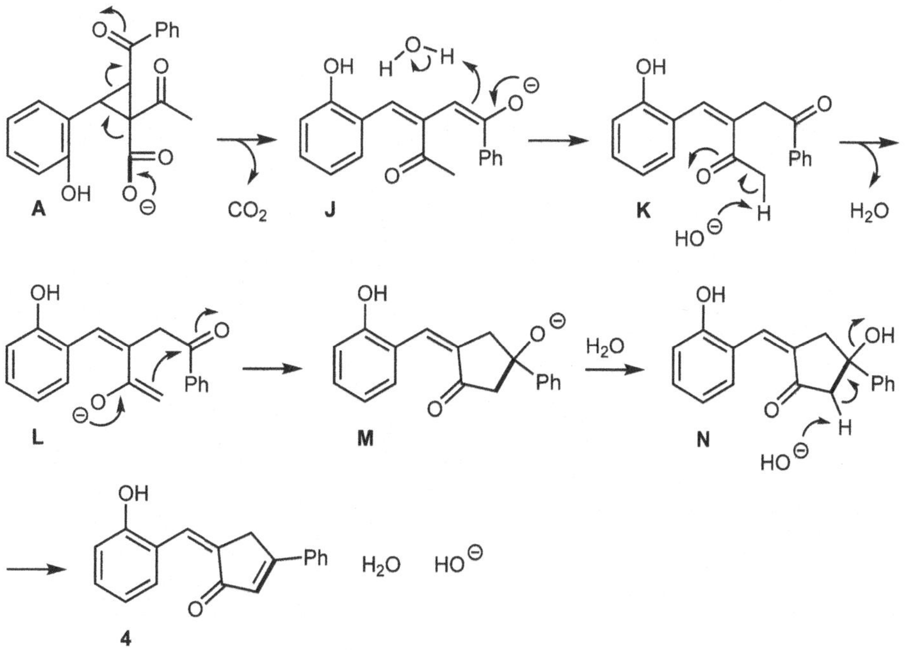

A → CO₂ J → K → H₂O

L → M → H₂O N

→ 4 H₂O HO⁻

SCHEME 2

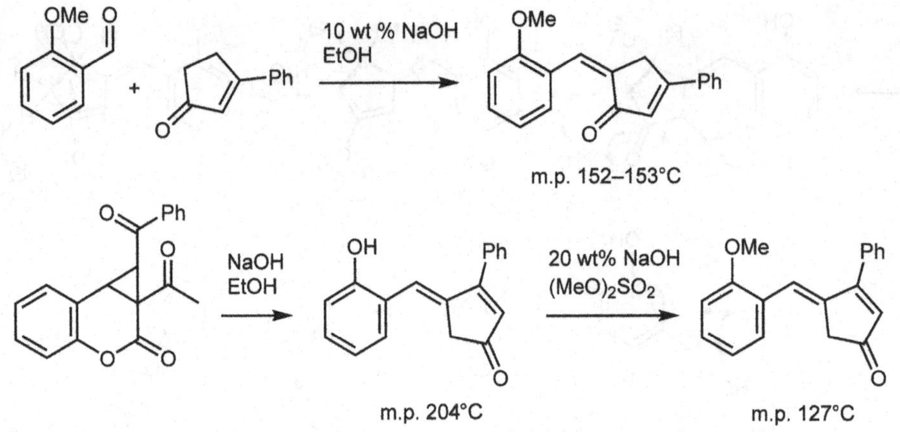

SCHEME 3

synthesized the methyl ether of **4** independently (see Scheme 3) and showed that it was different from the product **3**. Nevertheless, the authors did not carry out further experimental or theoretical studies to help support their postulated mechanism.

ADDITIONAL RESOURCES

We recommend two articles which feature mechanistic analyses of transformations involving cyclopentenones either as starting materials or products.[2,3]

References

1. Wawzonek S, Morreal CE. The action of alkali on 3,4-phenacylidene-3-acetylcoumarin. *J Am Chem Soc*. 1960;82(2):439–441. https://doi.org/10.1021/ja01487a048.
2. Howie RA, Turner AB, Cox PJ. Conformational study of a photochemical cyclopentenone rearrangement. Molecular mechanics calculations and X-ray structure of 14β-hydroxy-5-methoxy-de-A-oestra-5,7,9,16-tetraen-15-one. *J Chem Soc Perkin Trans 2*. 1985;1985:127–130. https://doi.org/10.1039/P29850000127.
3. Gonzalez-Perez AB, Vaz B, Faza ON, de Lera AR. Mechanistic and stereochemical insights on the Pt-catalyzed rearrangement of oxiranylpropargylic esters to cyclopentenones. *J Org Chem*. 2012;77:8733–8743. https://doi.org/10.1021/jo301651r.

Question 144: "Synthetic Heroin" and Parkinsonism

In 1989 American chemists investigated the transformation of 1-methyl-4-phenyl-2,3-dihydropyridinium perchlorate **1** in pH 7.4 buffer to give methylamine and a product identified as **2** (Scheme 1).[1] Provide a mechanism to explain this result.

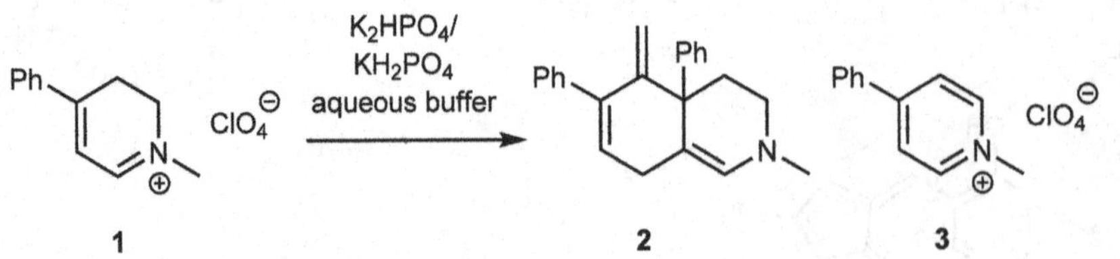

SCHEME 1

SOLUTION

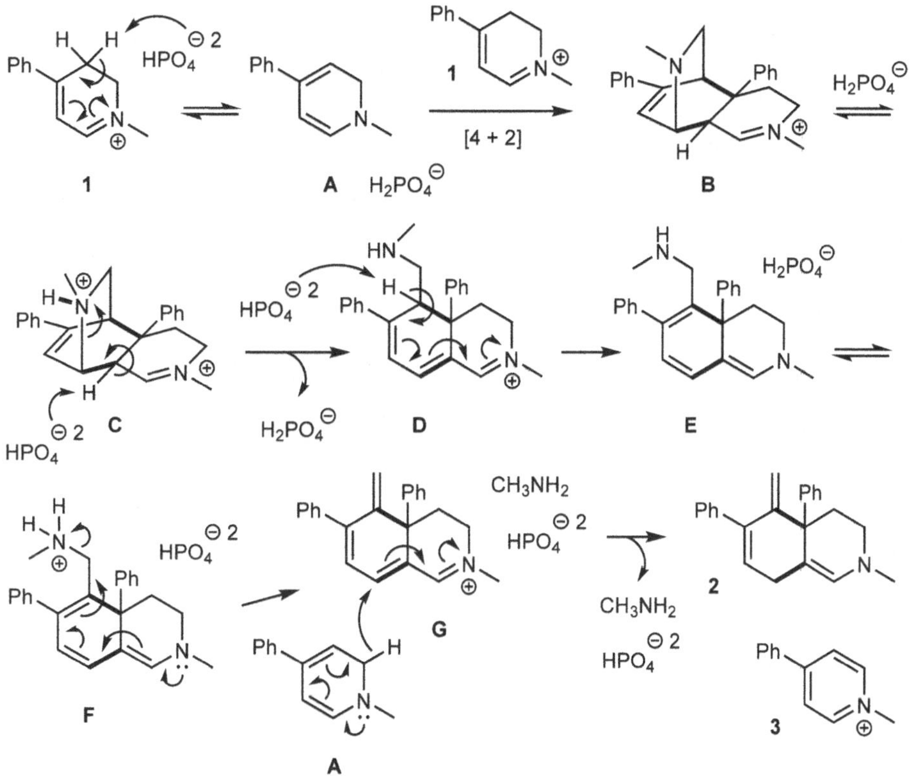

BALANCED CHEMICAL EQUATION

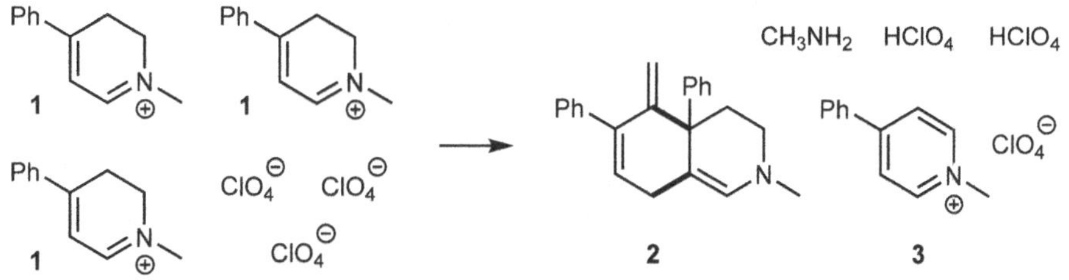

KEY STEPS EXPLAINED

The mechanism proposed by the authors to explain the transformation of **1** to **2** begins with deprotonation of iminium ion **1** by buffer base to give intermediate **A**. Then, a Diels-Alder [4 + 2] cycloaddition between **A** and an additional equivalent of **1** leads to adduct **B**. The sequential protonation and deprotonation steps that follow the formation of **B** eventually lead to intermediate **F** which fragments to form **G** and methylamine by-product. Intermediate **G** then reacts with another equivalent of **A** via a hydride transfer which leads to product **2** and by-product **3**.

We note that when the authors carried out the reaction with 2,2,6-trideuterated starting material **1-d$_3$** under the same reaction conditions they obtained a product, **2-d$_7$**, with a mass of 320 as determined by mass spectrometry. This mass corresponded to the incorporation of seven deuterium atoms in the structure of **2** (see Scheme 2). Following the mechanism shown, the by-product of the reaction is 2,6-dideuterated **3-d$_2$**.

When the authors carried out the reaction of **1** in deuterated water, they obtained a product, **2-d$_2$**, which had a mass of 315 as determined by mass spectrometry meaning that two hydrogen atoms were exchanged for deuterium atoms. ^{1}H NMR data analysis indicated that exchange took place at C$_3$ and C$_5$ of **1** according to the exchange mechanisms shown in Schemes 3 and 4. Following the mechanism given before for the transformation of **1** to **2** and **3**, one equivalent of **1-d** and two equivalents of **A'-d** lead to product **2-d$_2$** and by-product **3-d** as shown in Scheme 5. Also, one equivalent

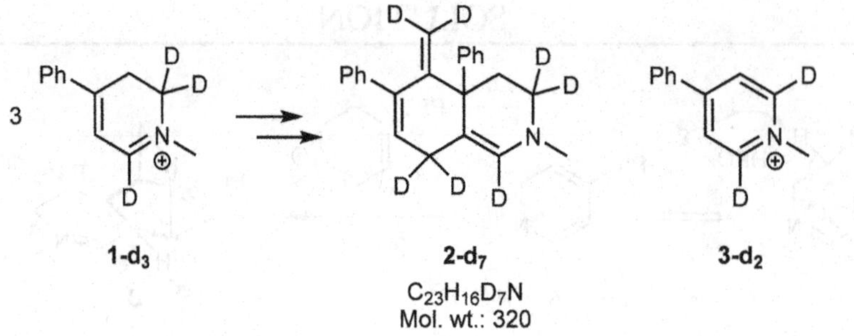

SCHEME 2

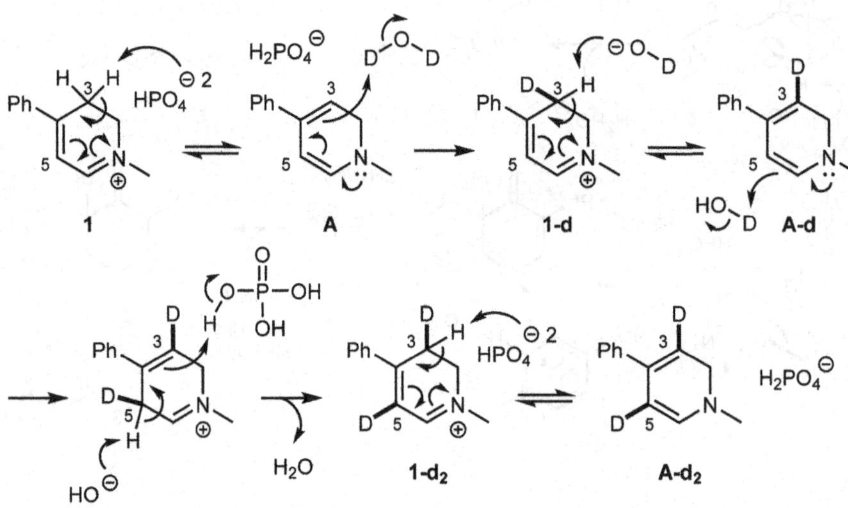

SCHEME 3

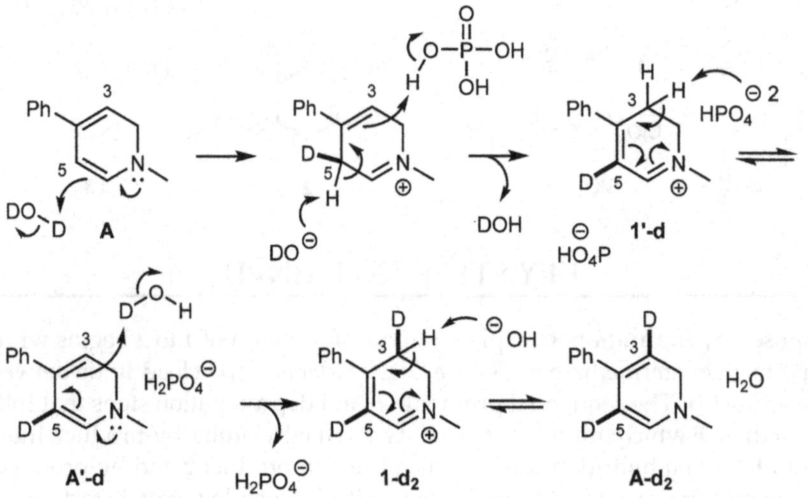

SCHEME 4

of **1-d₂** and two equivalents of **A-d₂** lead to product **2-d₂** and by-product **3-d₂**. The full details of the mechanisms utilizing these combinations of deuterated exchange substrates leading to the observed product **2-d₂** with mass 315 are shown in Schemes 6 and 7. Interestingly, the authors did not mention the observation of the monodeuterated and undeuterated forms of **2** having masses of 314 and 313, respectively, which would have resulted from the other monodeuterated substrate combinations shown in Scheme 8. Although these deuterium labeling experiments strongly support the postulated mechanism, the authors did not carry out a theoretical analysis to strengthen even further the conclusion that the postulated mechanism is indeed the operating mechanism.

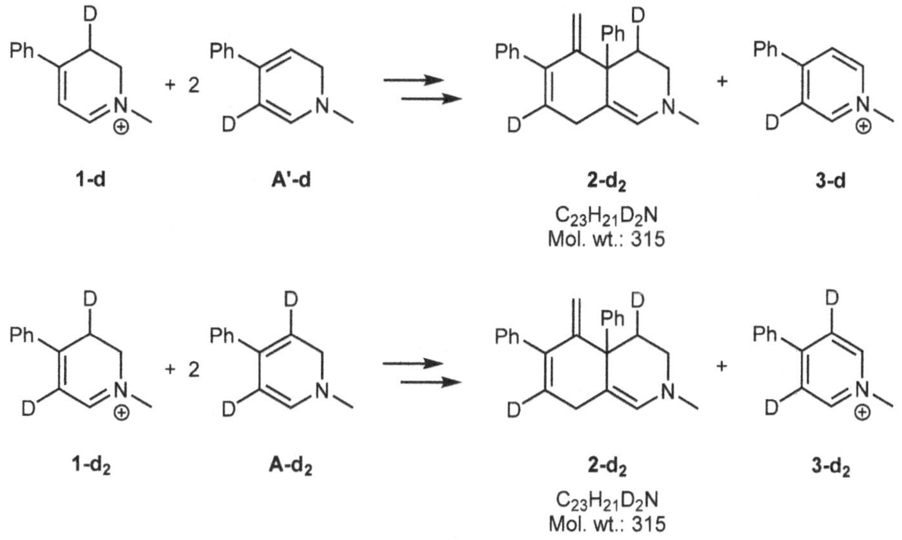

SCHEME 5

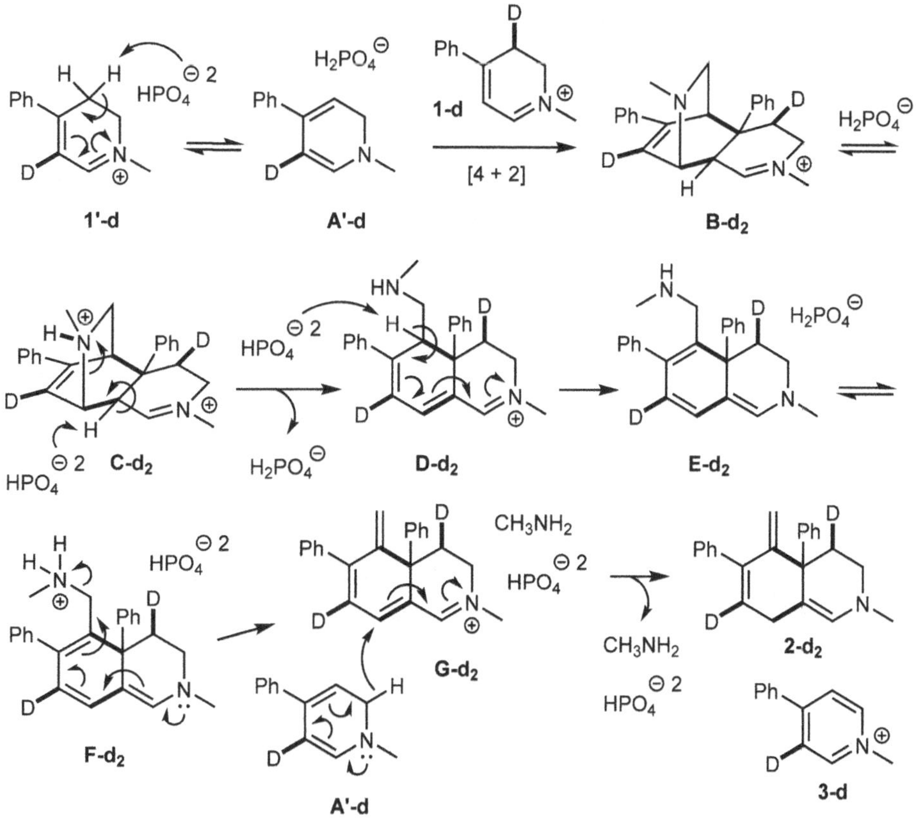

SCHEME 6

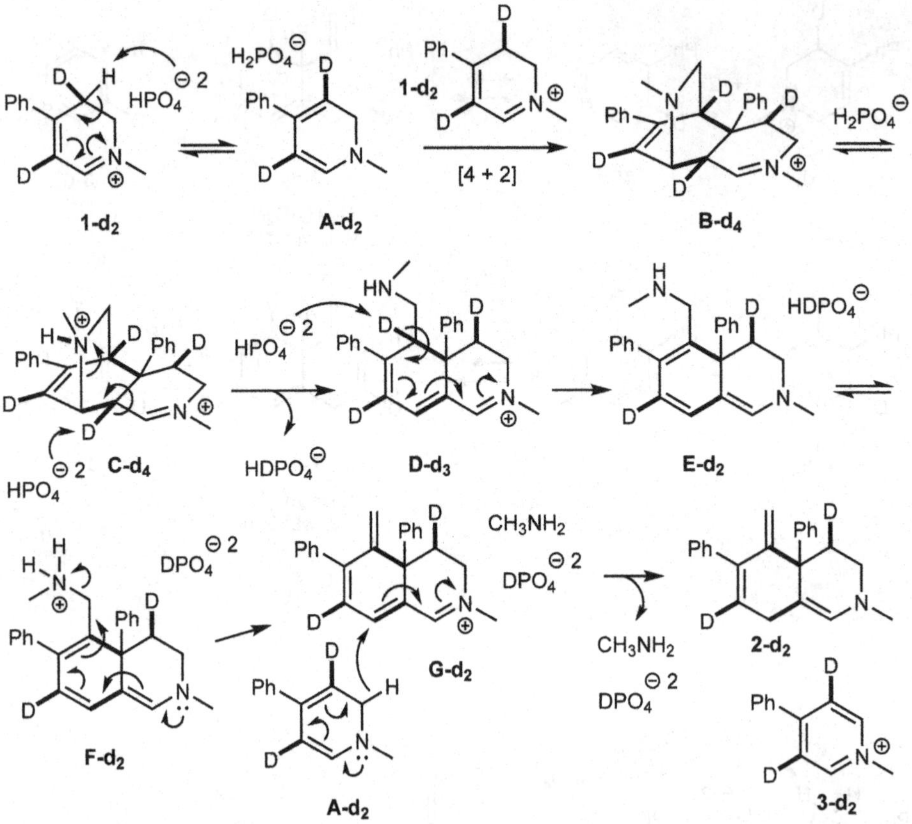

SCHEME 7

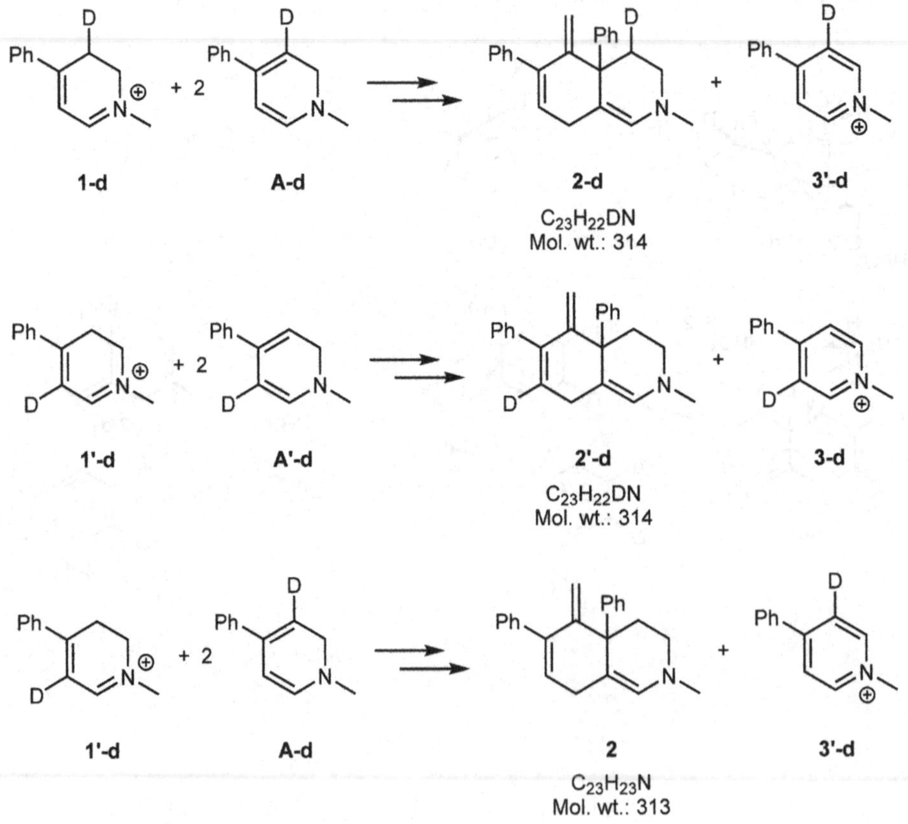

C₂₃H₂₂DN
Mol. wt.: 314

C₂₃H₂₂DN
Mol. wt.: 314

C₂₃H₂₃N
Mol. wt.: 313

SCHEME 8

For an example of a transformation involving a very similar chemical system to that encountered here, see the work of Baldwin et al.[2]

References

1. Leung L, Ottoboni S, Oppenheimer N, Castagnoli N. Characterization of a product derived from the 1-methyl-4-phenyl-2,3-dihydropyridinium ion, a metabolite of the nigrostriatal toxin 1-methyl-1-phenyl-1,2,3,6-tetrahydropyridine. *J Org Chem*. 1989;54(5):1052–1055. https://doi.org/10.1021/jo00266a011.
2. Baldwin JE, Bischoff L, Claridge TDW, Heupel FA, Spring DR, Whitehead RC. An approach to the manzamine alkaloids modelled on a biogenetic theory. *Tetrahedron*. 1997;53(6):2271–2290. https://doi.org/10.1016/S0040-4020(96)01129-5.

Question 145: Rearrangement of 6-Hydroxyprotopine to Dihydrosanguinarine

In 1988 German chemists synthesized dihydrosanguinarine **3** by treating protopine **1** with the enzyme cytochrome P450 O_2 NADPH protopine-6-hydroxylase and allowing its product, the 6-hydroxyprotopine **2** to undergo spontaneous rearrangement to **3** (Scheme 1).[1] Suggest a mechanism to explain the spontaneous rearrangement from **2** to **3**.

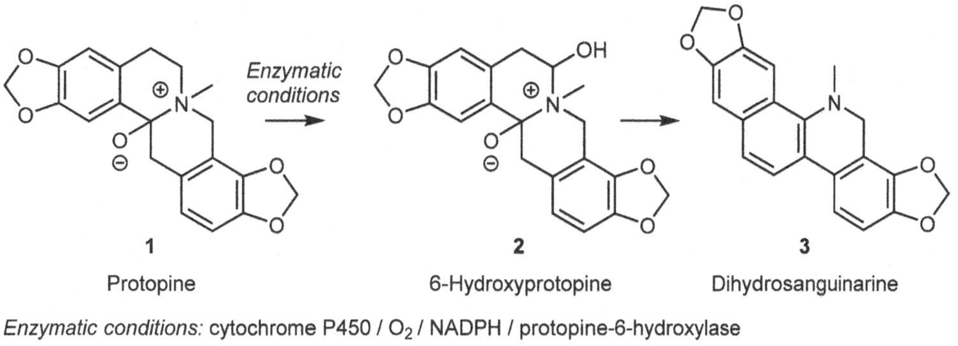

Enzymatic conditions: cytochrome P450 / O_2 / NADPH / protopine-6-hydroxylase

SCHEME 1

SOLUTION

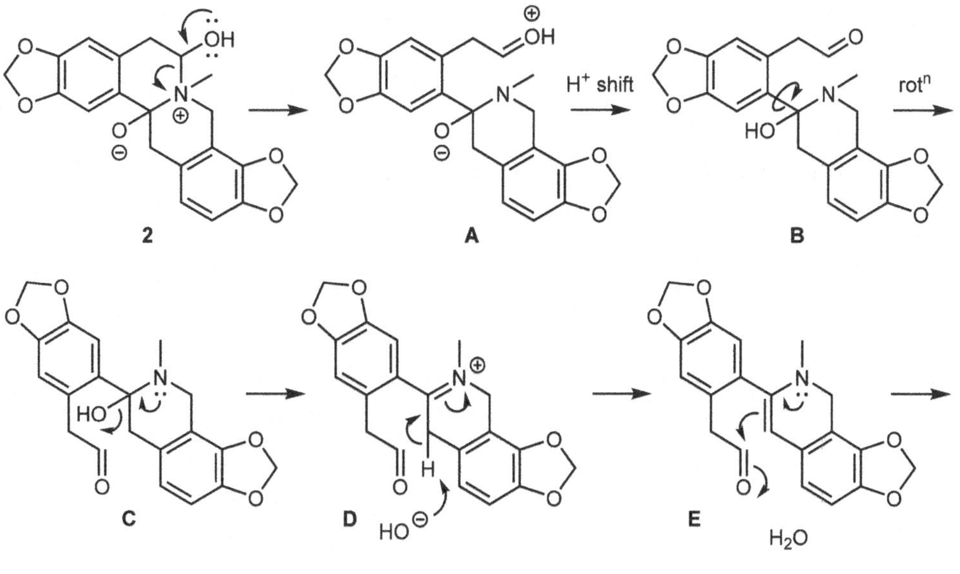

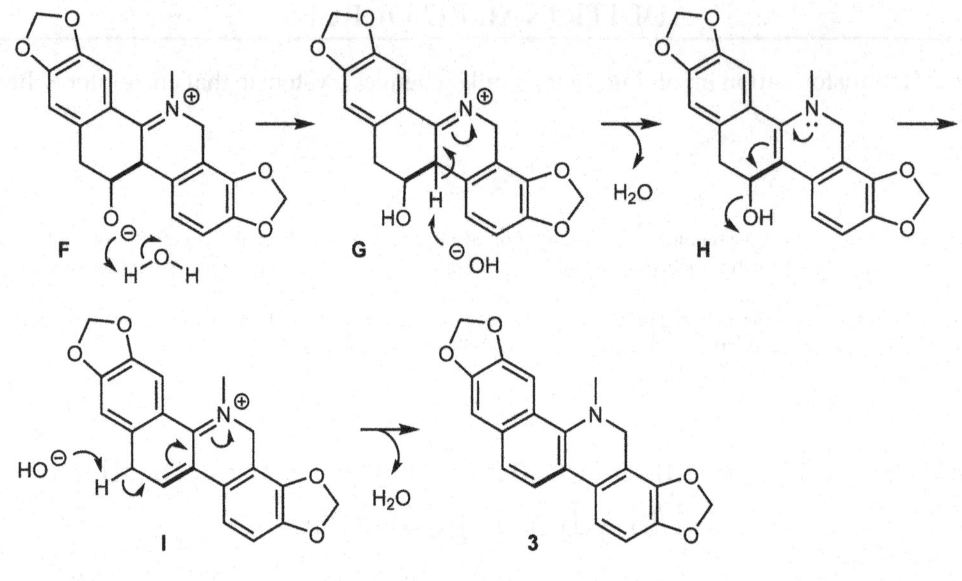

KEY STEPS EXPLAINED

The mechanism proposed by the authors begins with fragmentation of 6-hydroxyprotopine **2** to form intermediate **A** which after proton shift to **B** and bond rotation to **C** undergoes imine formation with elimination of hydroxide ion to form **D**. Deprotonation of **D** via the liberated hydroxide forms **E** and water molecule. At this point **E** undergoes ring closure to form the zwitterion **F** which abstracts a proton from water to form **G** and hydroxide ion. **G** undergoes deprotonation with loss of water to form the neutral species **H** which eliminates hydroxide by means of electron delocalization from the nitrogen atom to form **I** which undergoes deprotonation to form the aromatic final product 6-hydroxyprotopine **3**. This spontaneous rearrangement from **2** to **3** can be described as a $[(6+0)+(6+0)]$ ring construction strategy creating a fused bicyclic [4.4.0] framework. We note that the authors did not carry out further experimental or theoretical work to produce evidence in support of this postulated mechanism.

ADDITIONAL RESOURCES

For an interesting and related pyrolysis-induced transformation of the *N*-oxide of protopine, see the work of Iwasa et al.[2]

References

1. Tanahashi T, Zenk MH. One step enzymatic synthesis of dihydrosanguinarine from protopine. *Tetrahedron Lett*. 1988;29(44):5625–5628. https://doi.org/10.1016/S0040-4039(00)80829-1.
2. Iwasa K, Sugiura M, Takao N. Pyrolysis and photolysis of the *N*-oxides of protopines and hexahydrobenzo[c]phenanthridines. Syntheses of the secoberbines and benzo[c]phenanthridines. *Chem Pharm Bull*. 1985;33(3):998–1008. https://doi.org/10.1248/cpb.33.998.

Question 146: Biotin Synthesis: Sulfur Preempts a Beckmann Rearrangement

During a 1980 synthesis of biotin, chemists from Hoffmann-La Roche reacted the oxime **1** with thionyl chloride at 0°C expecting that Beckmann rearrangement would give product **3** (Scheme 1).[1] Nevertheless, product **2** was isolated in 75% yield. Provide a mechanism for the formation of product **2**.

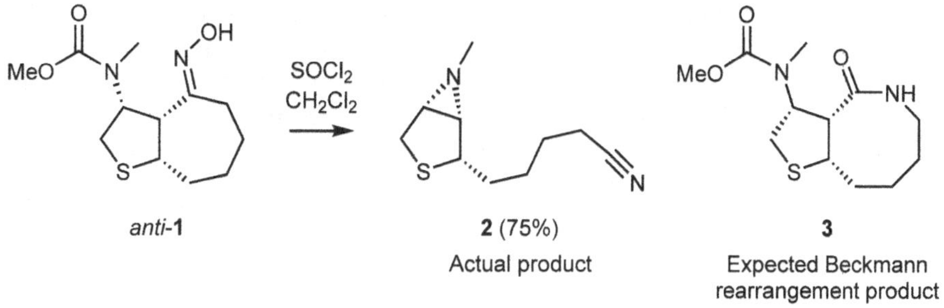

SCHEME 1

SOLUTION

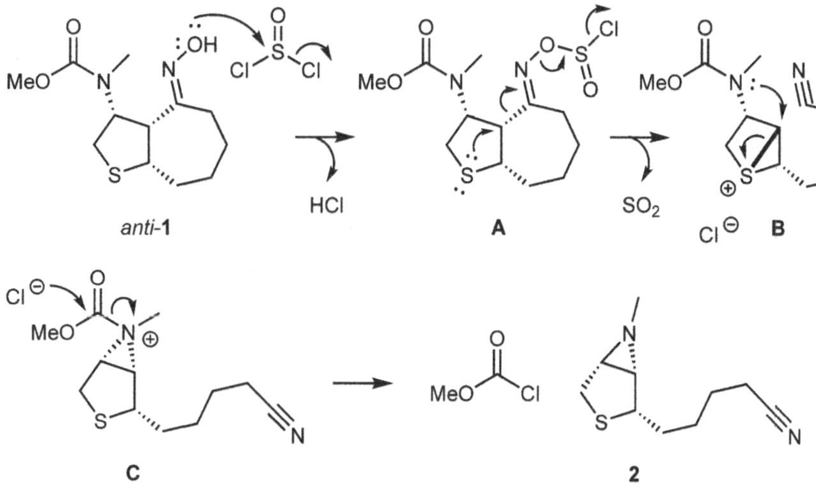

KEY STEPS EXPLAINED

We begin by highlighting that the authors expected *anti*-oxime **1** to undergo a Beckmann rearrangement to product **3** in the presence of thionyl chloride. Instead, the aziridine product **2** was isolated. As such, the authors proposed a mechanism whereby the hydroxyl group of **1** attacks the thionyl chloride to give intermediate **A** which then produces a [2.1.0] episulfonium ion intermediate **B** after eliminating sulfur dioxide by-product and chloride ion. The nitrogen atom of the carbamate group of **B** then attacks the reactive vicinal carbon atom of the four-membered ring causing it to ring open and generate the [3.1.0] fused aziridinium ion intermediate **C**. At this point, chloride ion attacks the carbamate carbonyl group to form product **2** and methyl chloroformate by-product.

The authors state that this unexpected fragmentation occurs with the *anti*-oxime **1**; whereas, the *syn*-oxime leads smoothly to the expected Beckmann rearrangement product **3** (see Scheme 2). They offer the following explanation: "the fragmentation [to product **2**] is initiated by the sulfur lone pair and suggests that a deactivation of the sulfide functionality present in the *anti* oxime **1** is required."[1]

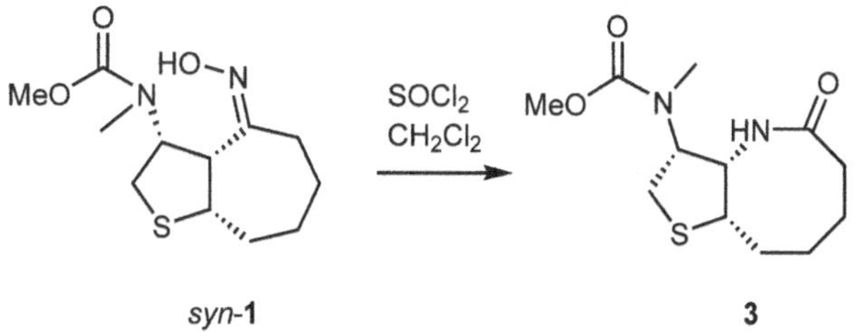

SCHEME 2

A proposed mechanism for the formation of amide **3** is shown in Scheme 3. In this sequence, the intramolecular hydrogen bond between the carbamate nitrogen atom and the *syn* hydroxyl group of the oxime allows for facile Beckmann rearrangement. Such a hydrogen bond is not possible in the *anti*-oxime. Moreover, the thionyl chloride acts as a helper reagent and is not decomposed to sulfur dioxide as in the previous mechanism starting from the *anti*-oxime **1**. For future study we recommend a theoretical analysis for the proposed mechanism of the unexpected transformation of **1** to **2**.

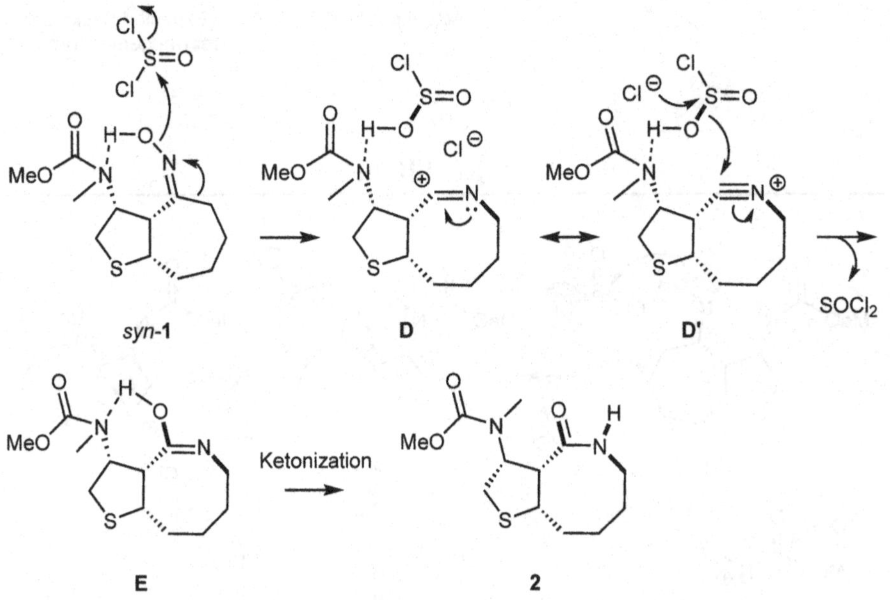

SCHEME 3

ADDITIONAL RESOURCES

We refer the reader to an article which discusses the mechanism of an organocatalyzed Beckmann rearrangement.[2] We also recommend an excellent review of historical approaches to the synthesis of biotin.[3]

References

1. Confalone PN, Pizzolato G, Confalone DL, Uskokovic MR. Olefinic nitrone and nitrile oxide [3 + 2] cycloadditions. A short stereospecific synthesis of biotin from cycloheptene. *J Am Chem Soc.* 1980;102(6):1954–1960. https://doi.org/10.1021/ja00526a033.
2. Ronchin L, Vavasori A. On the mechanism of the organocatalyzed Beckmann rearrangement of cyclohexanone oxime by trifluoroacetic acid in aprotic solvent. *J Mol Catal A Chem.* 2009;313:22–30. https://doi.org/10.1016/j.molcata.2009.07.016.
3. De Clercq PJ. Biotin: a timeless challenge for total synthesis. *Chem Rev.* 1997;97(6):1755–1792. https://doi.org/10.1021/cr950073e.

Question 147: A 1,4-Dithiin to Thiophene Rearrangement

In 1980 American chemists reacted the 1,4-dithiin **1** with methyl iodide and silver tetrafluoroborate in methylene chloride/acetonitrile/nitromethane thus achieving a mixture of products **2** and **3** (Scheme 1).[1] Provide a mechanistic rationale for this transformation.

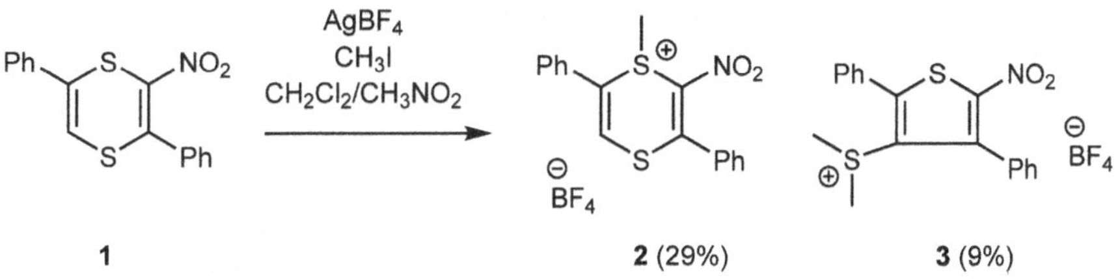

SCHEME 1

SOLUTION: 1 TO 2

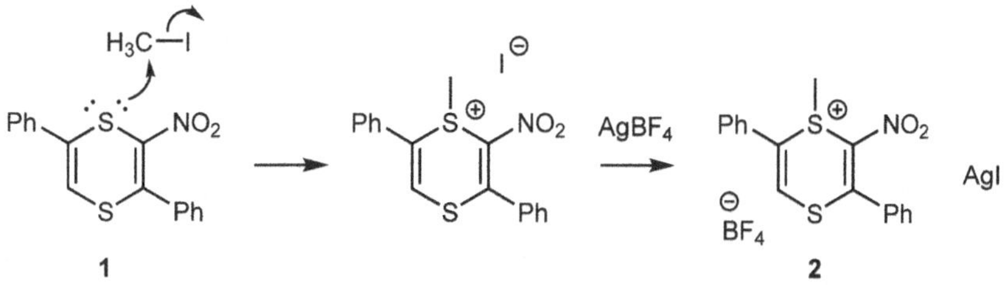

SOLUTION: 1 TO 3

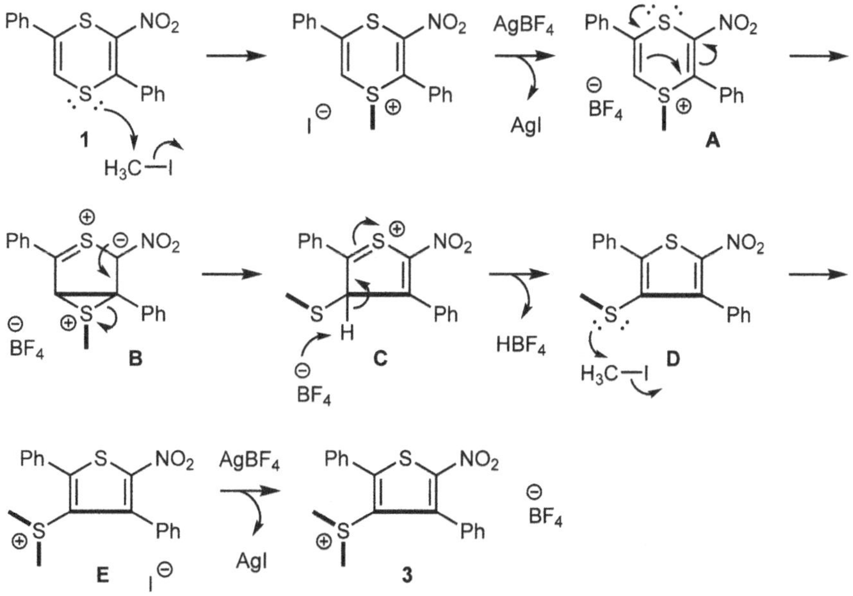

BALANCED CHEMICAL EQUATION: 1 TO 3

KEY STEPS EXPLAINED

We begin by noticing that for the **1** to **2** transformation, methylation occurs on the sulfur atom *alpha* to the nitro group whereas for the **1** to **3** transformation, the methylation occurs on the sulfur atom *beta* to the nitro group. As such, the transformation of **1** to **3** begins with formation of an unstable salt **A** which rearranges to sulfonium salt **C** which is then deprotonated by tetrafluoroborate ion to generate thiophene **D**. This intermediate is then methylated on the methyl sulfur atom by a second equivalent of methyl iodide to finally give the thiophene salt product **3**.

The authors were able to isolate compound **3** and characterize it by ^{1}H NMR. Also, they characterized the demethylated product **D** by ^{1}H NMR and mass spectrometry when **3** was treated with triethylamine. Nevertheless, compound **2** was not isolated. Instead, column chromatography using alumina resulted in the formation of acetylenic product **4** shown in Scheme 2 which was characterized by IR and ^{1}H NMR spectroscopies. Product **4** was further methylated using silver tetrafluoroborate and methyl iodide to give the sulfonium salt **5** shown in Scheme 3 which was also characterized by IR and ^{1}H NMR spectroscopies. For a future study we recommend that researchers undertake further experimental and theoretical analysis to help support these postulated mechanisms.

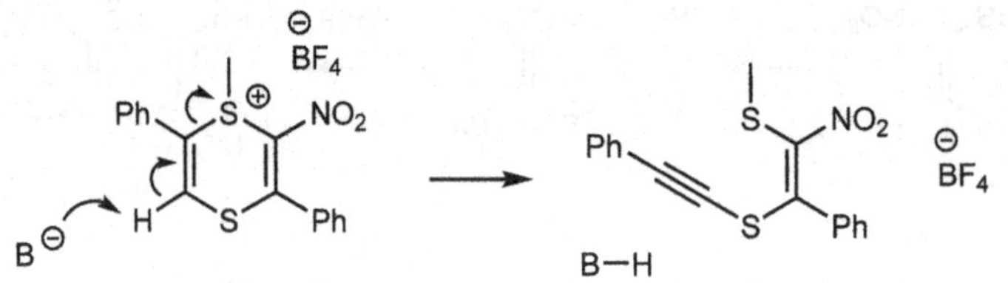

SCHEME 2

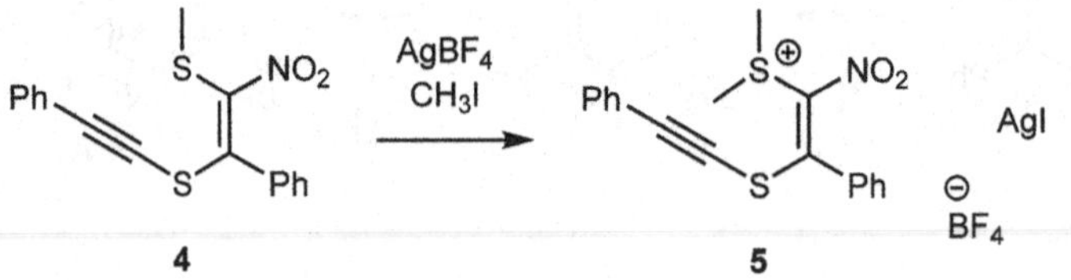

SCHEME 3

ADDITIONAL RESOURCES

For a review article on the synthesis of thiophene compounds by means of *ortho*-Claisen rearrangement, see Ref. 2.

References

1. Young TE, Oyler AR. Ring cleavage and ring contraction of nitro-substituted 1-methyl-2,5-diphenyl-1,4-dithiinium tetrafluoroborates. *J Org Chem.* 1980;45(6):933–936. https://doi.org/10.1021/jo01294a003.
2. Majumdar KC. New variation of the aromatic *ortho*-Claisen rearrangement: synthesis of fused thiophenes and pyrroles. *Synlett.* 2008;2008 (16):2400–2411. https://doi.org/10.1055/s-2008-1078013.

Question 148: An Intramolecular Wittig Reaction

In 1974 American chemists reacted the triphenylphosphonium bromide **1** with sodium ethoxide in refluxing dimethylformamide (DMF) and formed a 31% overall yield mixture of 95:5 ratio of products **2** to **3**, respectively (Scheme 1).[1] Explain these results in mechanistic terms.

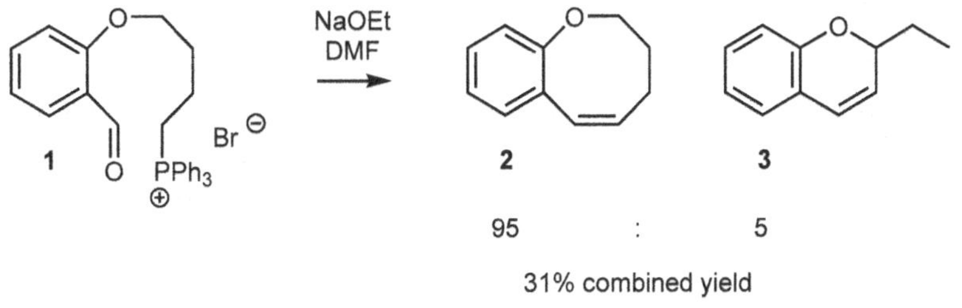

95 : 5

31% combined yield

SCHEME 1

SOLUTION: 1 TO 2

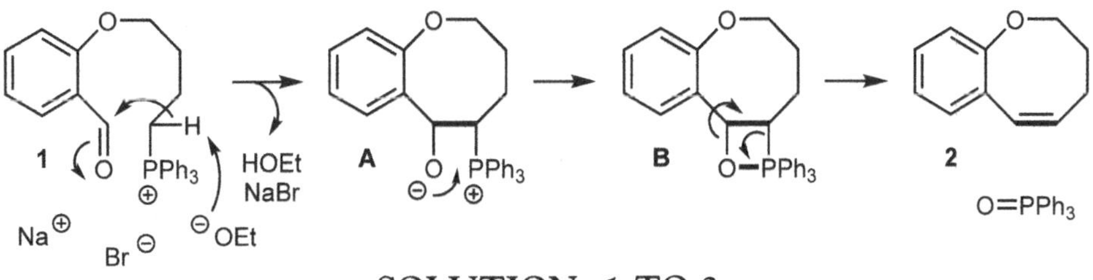

SOLUTION: 1 TO 3

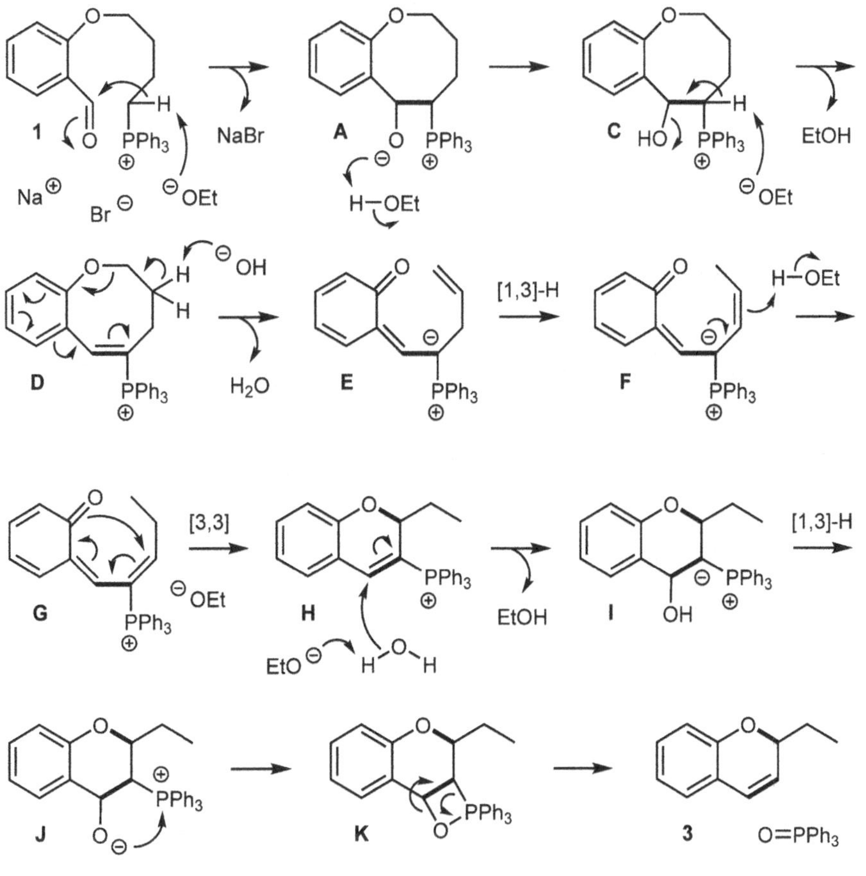

KEY STEPS EXPLAINED

The mechanism proposed to explain the transformation of **1** to **2** begins with ethoxide-mediated deprotonation of the hydrogen atom *alpha* to the triphenylphosphonium group of **1** which results in cyclization to an eight-membered ring zwitterion intermediate **A**. This structure undergoes further cyclization to **B** which undergoes a retro [2+2] ring opening giving 3,4-dihydro-2*H*-1-benzoxocin **2** and triphenylphosphine oxide as by-product. Interestingly, the authors propose that **A** can also be protonated to form the hydroxyl group containing intermediate **C** which undergoes a second deprotonation by ethoxide to form the triphenylphosphonium intermediate **D**. Base-induced ring opening of **D** results in formation of phosphonium ylide **E** which undergoes a [1,3]-H shift and protonation from ethanol to give orthoquinomethide intermediate phosphonium ion **G**. This intermediate then undergoes a sigmatropic [3,3] cyclization to give **H** which reacts with hydroxide ion generated from ethoxide and water to give ylide **I**. The next steps leading to 2-ethyl-2*H*-1-benzopyran **3** and triphenylphosphine oxide by-product are similar to those from **A** to **2**. We note that the authors were able to isolate phosphonium salt intermediates analogous to **D** and **H** for the reaction shown in Scheme 2 in support of the intramolecular Wittig mechanism given above.[2] Unfortunately, they were not able to isolate such intermediates for the given reaction, so their mechanistic proposal constitutes an indirect inference by analogy. Nevertheless, the authors did not conduct further experimental and theoretical analysis to help support their mechanism.

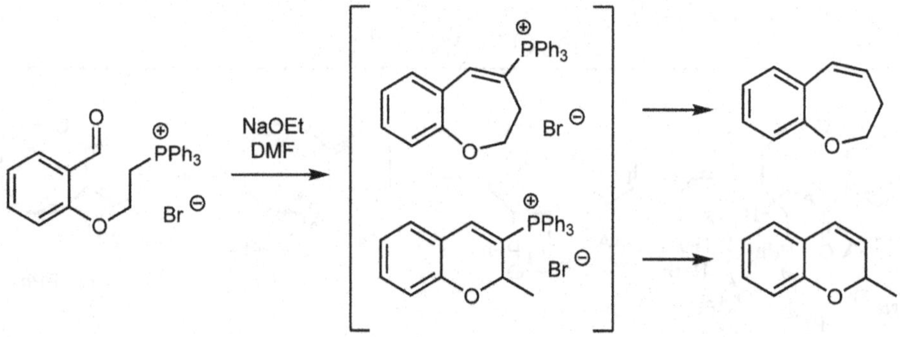

SCHEME 2

ADDITIONAL RESOURCES

We recommend two articles which discuss interesting transformations involving compounds such as **1** and **2**.[3,4]

References

1. Schweizer EE, Minami T, Anderson SE. Reactions of phosphorus compounds. 35. Reaction of 4-salicyloxybutyltriphenylphosphonium bromide with alcoholic alkoxide. *J Org Chem*. 1974;39(20):3038–3040. https://doi.org/10.1021/jo00934a019.
2. Schweizer EE, Minami T, Crouse DM. Reactions of phosphorus compounds. Reactions of phosphorus compounds. 28. Mechanism of the formation of 2-methyl-2*H*-1-benzopyran by the reaction of 3-(o-formylphenoxy)propylphosphonium salts in alcoholic alkoxide. *J Org Chem*. 1971;36(26):4028–4032. https://doi.org/10.1021/jo00825a005.
3. Kasmai HS, Wang X, Doan H-N, et al. The adverse effect of benzannelation on the aromaticity of oxocinyl anion: a combined experimental and theoretical study. *Helv Chim Acta*. 2010;93:1532–1544. https://doi.org/10.1002/hlca.200900430.
4. Xu T, Gong W, Ye J, Lin Y, Ning G. Unprecedented ring transformation of an α,α'-monosubstituted 2,4,5-triphenylpyrylium salt with η³-phosphines: efficient synthesis of aryl- and alkylphosphonium triphenylcyclopentadienylides. *Organometallics*. 2010;29:6744–6748. https://doi.org/10.1021/om100872f.

Question 149: Amination of a Cephalosporin

In 1975 Douglas Spry treated the β-lactam **1** with *N*-methylhydroxylamine to form the condensation product **2** which when heated to 110°C under pyrolysis conditions in toluene led to the formation of isomers **3** in 72% yield (Scheme 1).[1] Suggest mechanisms to explain these transformations.

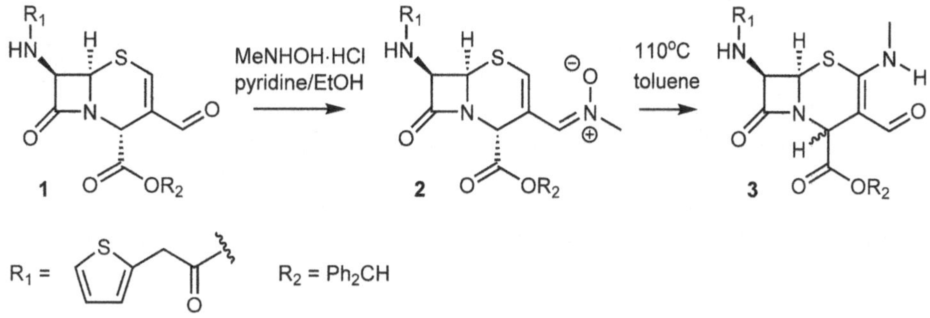

SCHEME 1

SOLUTION: 1 TO 2

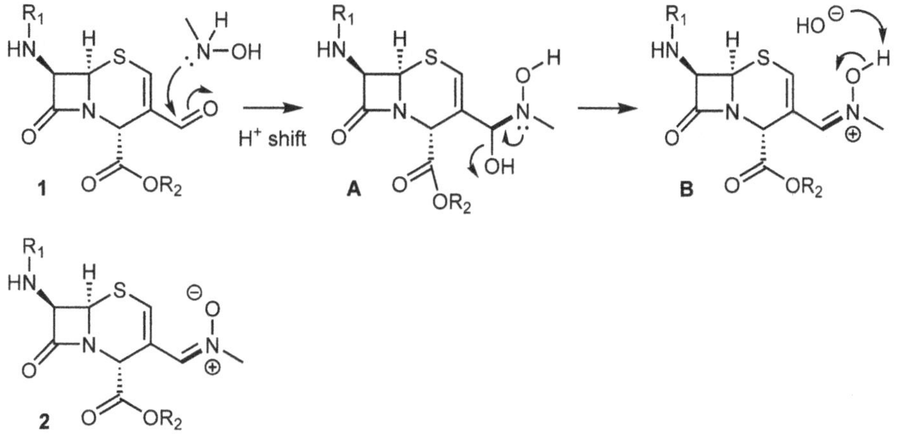

SOLUTION: 2 TO 3

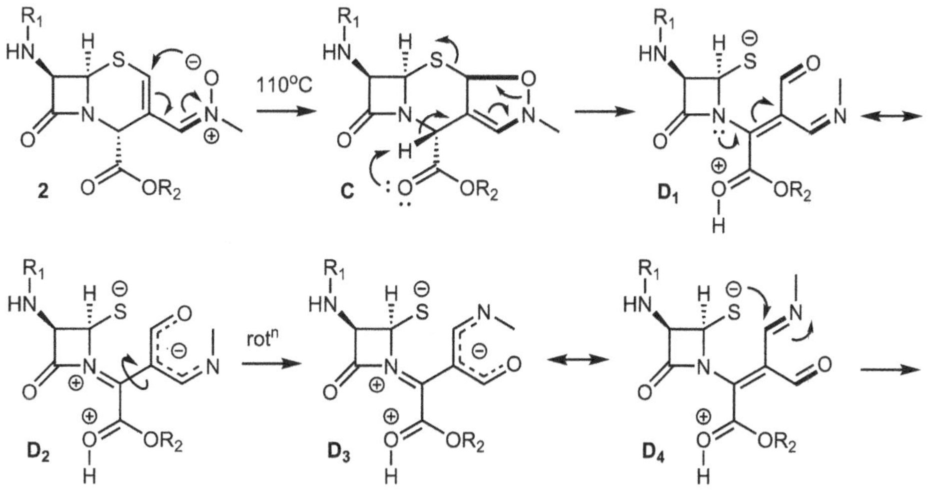

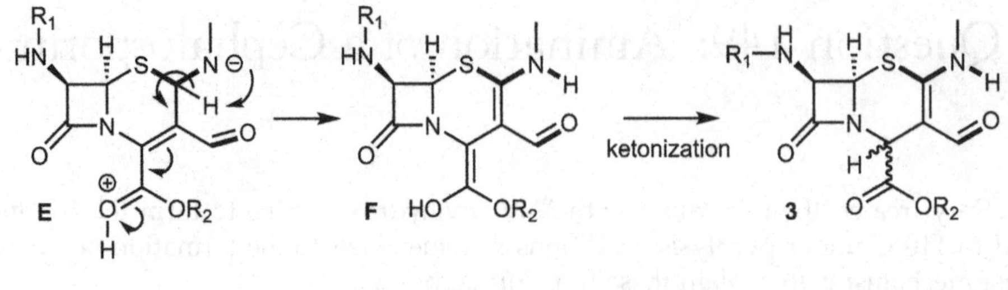

KEY STEPS EXPLAINED

The mechanism proposed by Spry to explain the synthetic sequence **1** to **2** to **3** begins with nucleophilic attack of the aldehyde group of **1** by methylhydroxylamine which forms the nitrone **2** condensation product after elimination of water. Next, the thermal rearrangement of **2** to **3** proceeds first by cyclization to the dihydro-isoxazole **C** which ring opens to zwitterion **D**. The thiolate anion **D** undergoes cyclization onto the imine moiety to give [1,3]thiazine **E** which after a 1,2-proton shift forms the enol **F** that finally ketonizes to the isomeric mixture of products **3**. We note that the cyclization from **D** to **E** proceeds likely via the resonance structures shown so that C—C bond rotation occurs from structure D_3 to D_4 rather than by direct C=C bond rotation from structure D_1 to D_4 which would be expected to have a higher energy barrier.

It is important to highlight that Spry obtained nitrone **2** in 64% yield and characterized it spectroscopically by IR and ^{1}H NMR. Pyrolysis of **2** led to a 68% yield of **3** as a mixture of diastereomers in a ratio of 69:31 corresponding to the α:β forms. Following the nomenclature convention of carbohydrates, the α-isomer is the one with the ester group pointing down in structure **3**. The author did not carry out further experimental or theoretical analysis to help support his postulated mechanism.

ADDITIONAL RESOURCES

We recommend a review article which discusses nonclassical polycyclic β-lactams and their synthetic applications and which contains some discussion of mechanism as well.[2]

References

1. Spry DO. Cephem-*N*-methylnitrones. *J Org Chem*. 1975;40(16):2411–2414. https://doi.org/10.1021/jo00904a036.
2. Gomez-Gallego M, Mancheno MJ, Sierra MA. Non-classical polycyclic β-lactams. *Tetrahedron*. 2000;56:5743–5774. https://doi.org/10.1016/S0040-4020(00)00378-1.

Question 150: A Uracil to 1,2,3-Triazole Conversion

In 1976 American chemists investigated the hydrolysis of the uracil derivative **1** with 5% v/v aqueous acetonitrile at 100°C to form the 1,2,3-triazole derivative **2** in 78% yield (Scheme 1).[1] Provide a mechanism for this transformation.

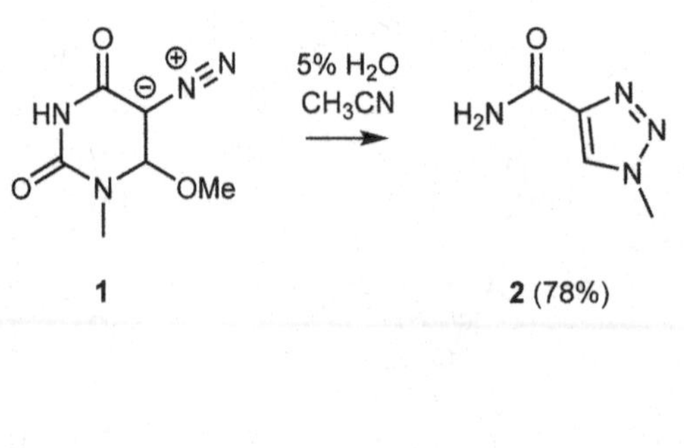

SCHEME 1

SOLUTION: AUTHORS' MECHANISM

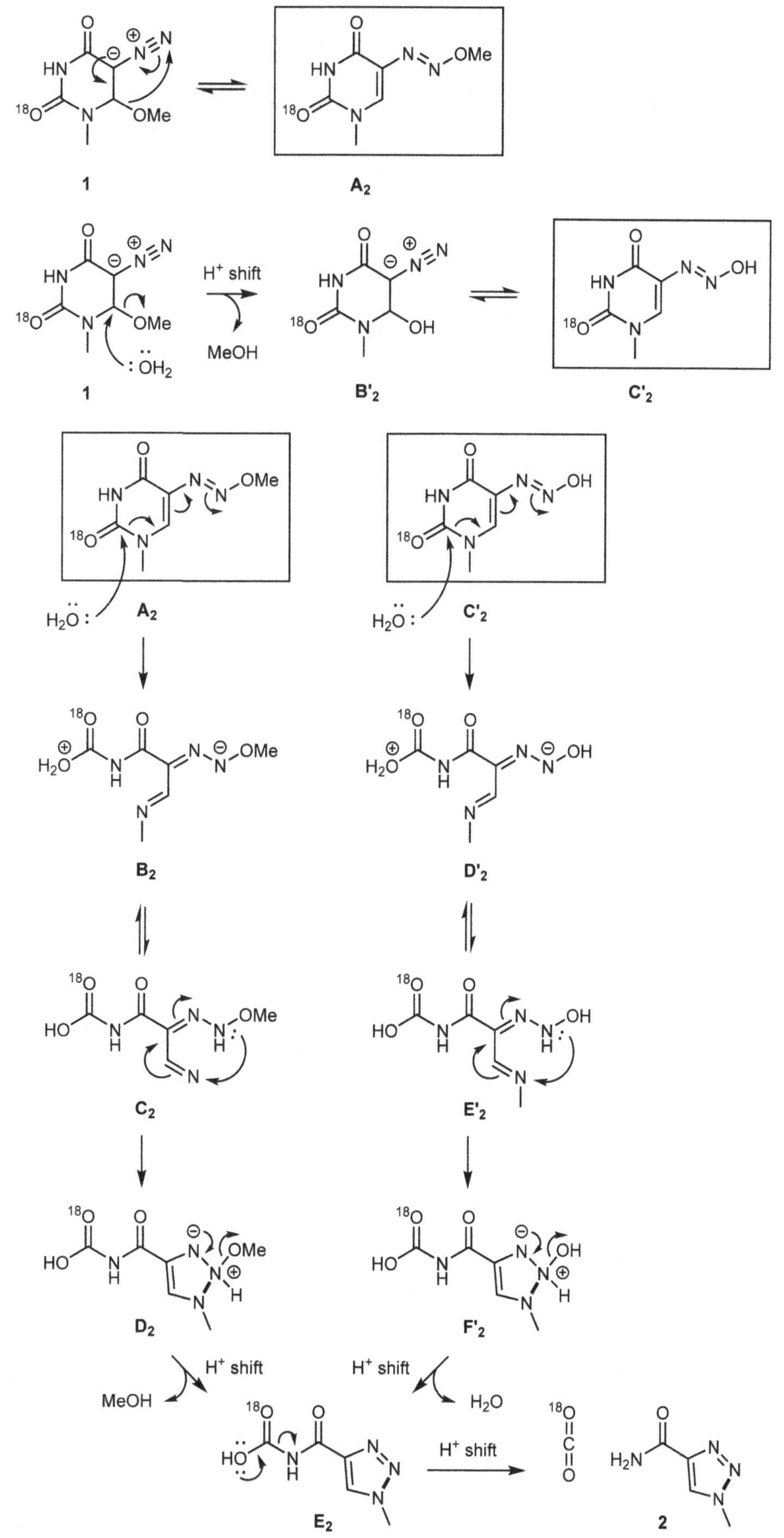

SOLUTION: ALTERNATIVE MECHANISM

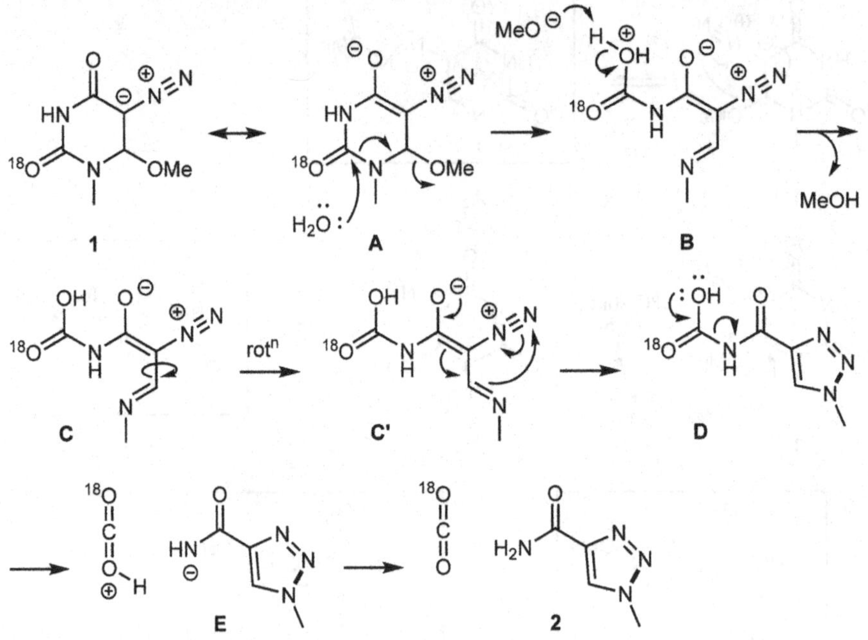

BALANCED CHEMICAL EQUATION

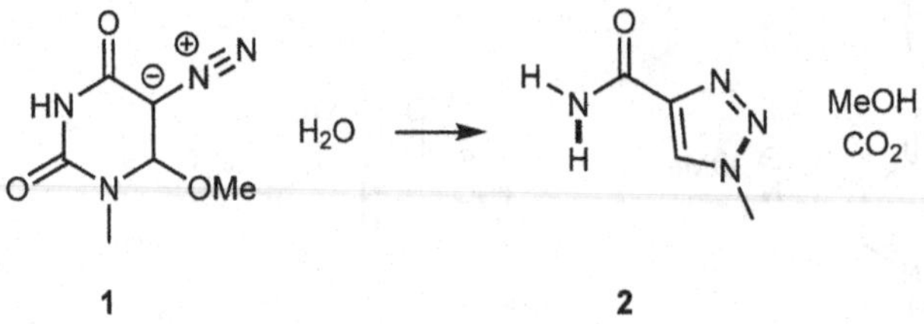

KEY STEPS EXPLAINED

We begin the analysis by noting that the transformation of **1** to **2** involves converting a pyrimidine ring to a 1,2,3-triazole ring. Therefore we offer two mechanisms, the first which is in accordance with the mechanism proposed by Thurber and Townsend and the second mechanism which we consider a possible alternative. As such, the authors' mechanism involves ^{18}O labeling on the C_2 carbonyl uracil **1**. The starting material structure **1** was reasoned to be in equilibrium with structure A_2 based on previous work where the authors discovered an equilibrium between B'_2 and its isomeric diazotic acid C'_2 after water displaces methanol from **1** to form B'_2.[2] With A_2 and C'_2 in hand, the mechanism proceeds with nucleophilic addition of water at C_2 of A_2 or C'_2 to form C_2 and E'_2, respectively, after proton transfer. These intermediates then cyclize in the opposite direction toward the azo group to give intermediates D_2 and F'_2. After elimination of methoxide and hydroxide ions, respectively, the liberated anions deprotonate the substrates to yield the same convergent intermediate E_2 which undergoes a final decarboxylation to give product **2** and ^{18}O labeled carbon dioxide by-product.

The supporting evidence presented by the authors for this mechanistic proposal includes the following observations: (a) when the reaction was examined by ^1H NMR a 3:1:1 ratio of product **2**, B'_2, and unreacted **1** was observed; (b) mass spectrometric analysis of the product mixture indicated the presence of isotopic label in an abundance similar to the isotopic label originally present in the starting material **1** which indicated that no exchange of the C_2 oxygen atom had occurred; (c) the detection of B'_2 in the reaction mixture was indicative that an exchange of the methoxy or hydroxy group at C_6 could occur prior to the pyrimidine ring opening; (d) the intermediacy of E_2 is consistent with the isolation of the methyl ester of E_2 when the reaction is conducted in methanol; and (e) the lack of observation of

1,2,3-triazole-4-carboxylic acid product indicates that in the carbamate group of E_2 decarboxylation occurs faster than hydrolysis of the amide linkage.

The authors acknowledge that the proposed mechanism is flexible while still being consistent with the observations and that there is the possibility of changing the order of steps. For instance, ring opening of the pyrimidine ring could occur before the equilibration of **1** to A_2, or **1** could tautomerize to its enol form and then pyrimidine ring opening occurs via water attack at C_2 (see Scheme 2).

In the alternative mechanism, we do not invoke the equilibration step between **1** and A_2 or B'_2 and C'_2. This helps reduce the number of steps while still having the same fate of the isotopic label as before. Thus, the first step is ring opening of the pyrimidine ring by water attack and concomitant methanol elimination to give intermediate **C**. Bond rotation and recyclization of the imino nitrogen atom onto the diazonium moiety leads to the formation of the 1,2,3-triazole ring. The last steps are decarboxylation of the carbamate of **D** to form the amide product **2**. We believe that future research on this transformation should include a theoretical analysis to help explain the energetics of each of the proposed mechanistic steps.

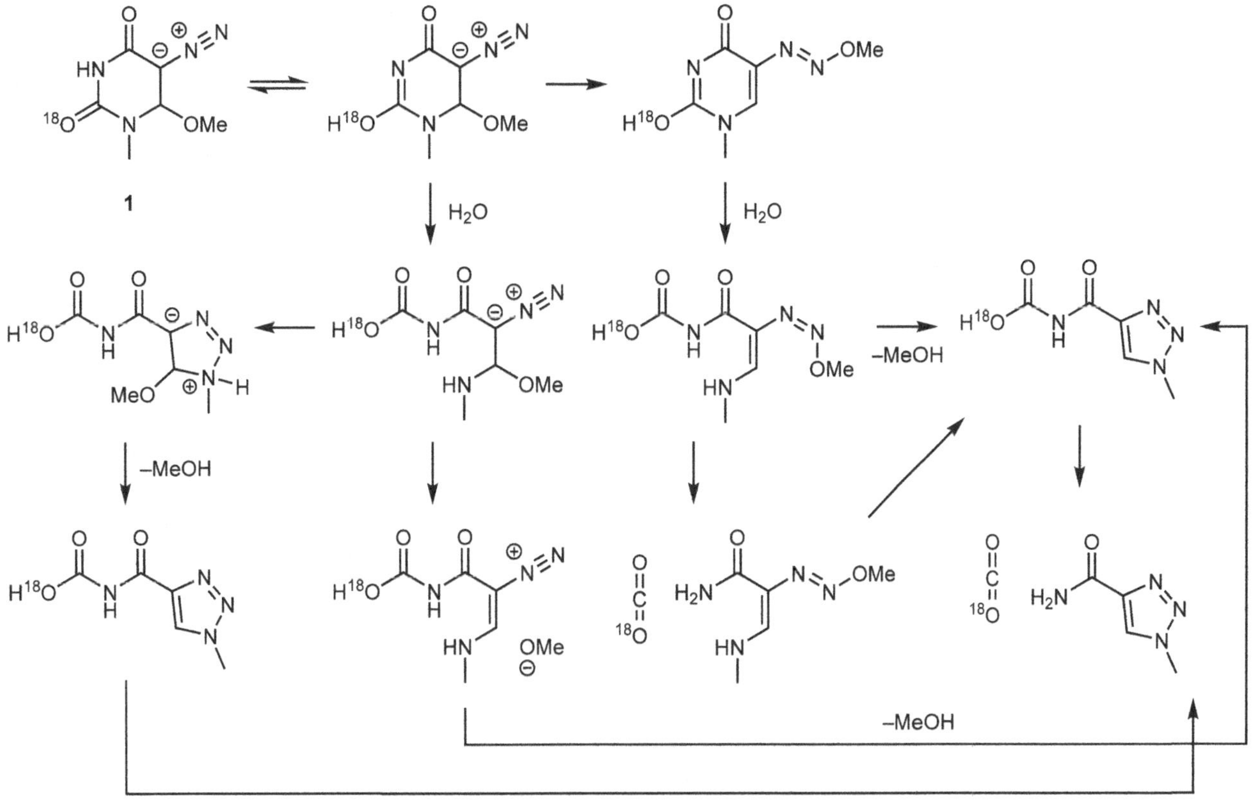

SCHEME 2

ADDITIONAL RESOURCES

We recommend an article which discusses an interesting thermal and rhodium acetate-catalyzed transformation of 5-diazouracil.[3]

References

1. Thurber TC, Townsend LB. Ring contractions of 5-diazouracils. I. Conversions of 5-diazouracils into 1,2,3-triazoles by hydrolysis and methanolysis. *J Org Chem.* 1976;41(6):1041–1051. https://doi.org/10.1021/jo00868a026.
2. Thurber TC, Townsend LB. The synthesis and properties of certain N-methylated 5-diazouracils. *J Heterocyclic Chem.* 1975;12(4):711–716. https://doi.org/10.1002/jhet.5570120420.
3. Mathur NC, Shechter H. Thermal and rhodium acetate catalyzed reactions of 5-diazouracil with nucleophiles. *Tetrahedron Lett.* 1990;31(48):6965–6968. https://doi.org/10.1016/S0040-4039(00)97217-4.

Chapter 7: Solutions 151 – 200

Question 151: Schumm Devinylation of Vinyl Porphyrins

During a 1981 study of the mechanism of the Schumm devinylation reaction of vinyl porphyrins, Canadian scientists made the experimental observations illustrated in Schemes 1–4.[1] Based on these observations, propose possible mechanisms for the Schumm devinylation reaction of protohemin **1** to deuterohemin **2**. Note, the authors did not identify **4** or **7**.

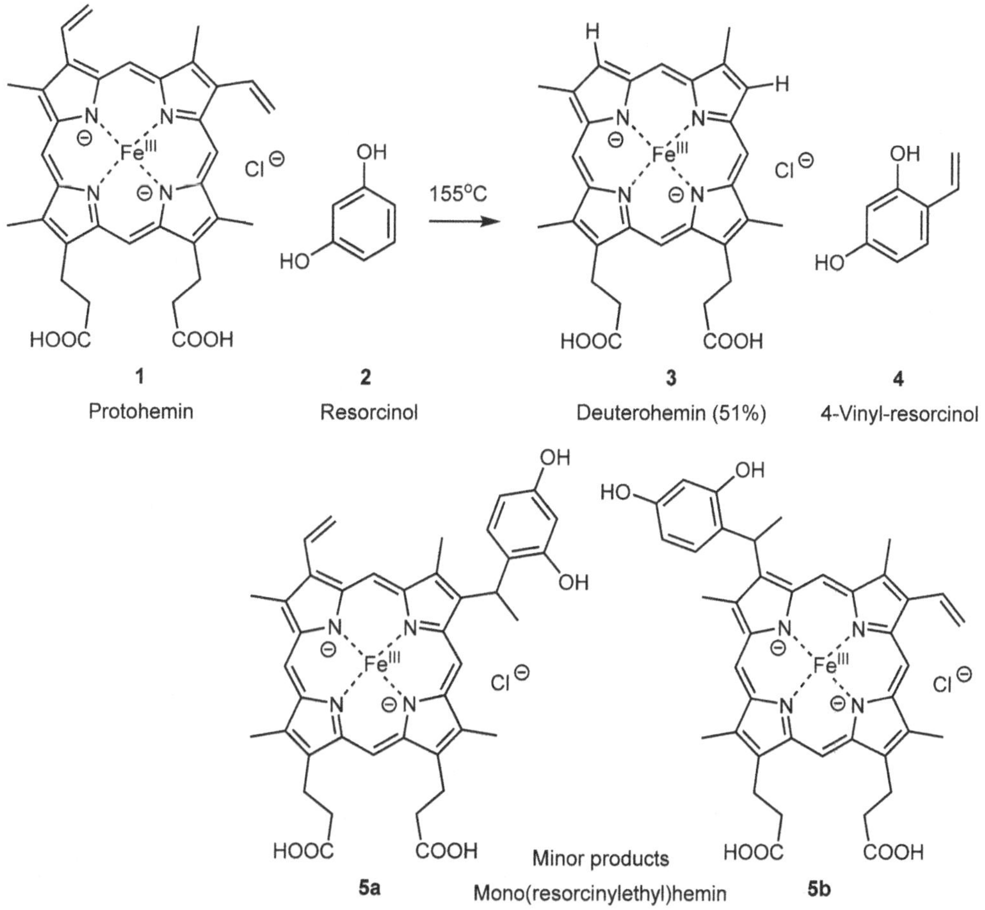

1
Protohemin

2
Resorcinol

3
Deuterohemin (51%)

4
4-Vinyl-resorcinol

5a

5b

Minor products
Mono(resorcinylethyl)hemin

SCHEME 1

Strategies and Solutions to Advanced Organic Reaction Mechanisms
https://doi.org/10.1016/B978-0-12-812823-7.00316-5

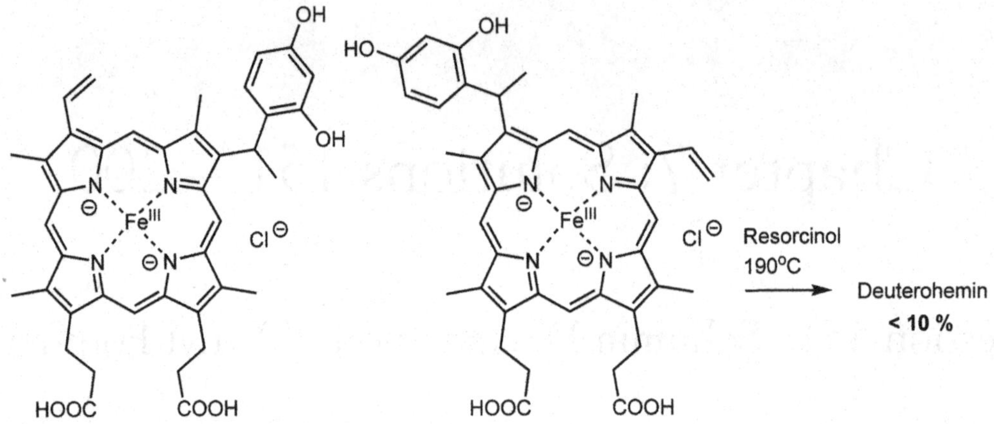

SCHEME 2

SCHEME 3

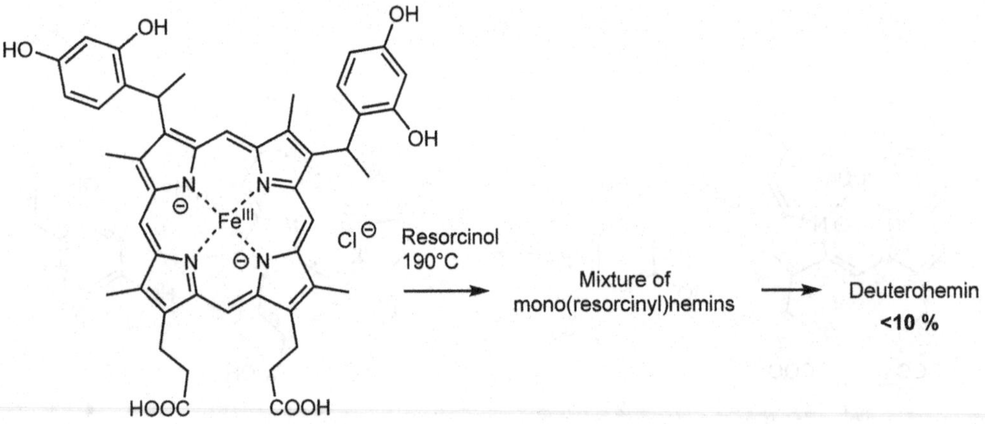

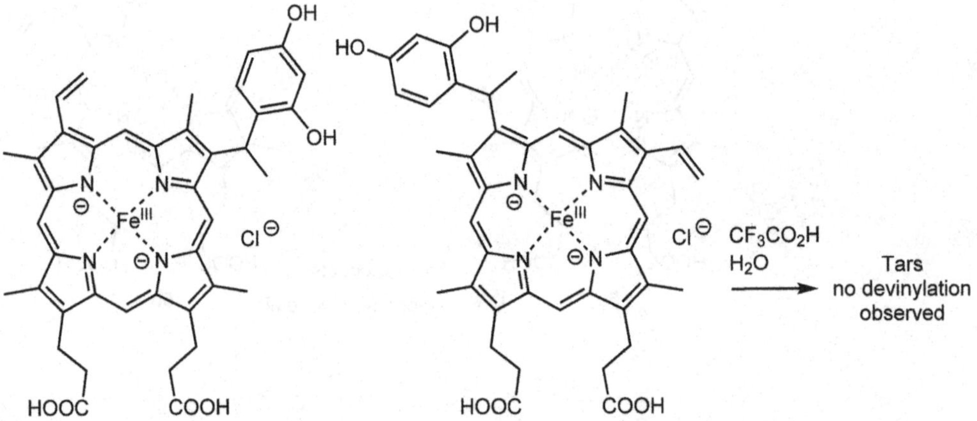

SCHEME 4

SOLUTION: MECHANISM 1A

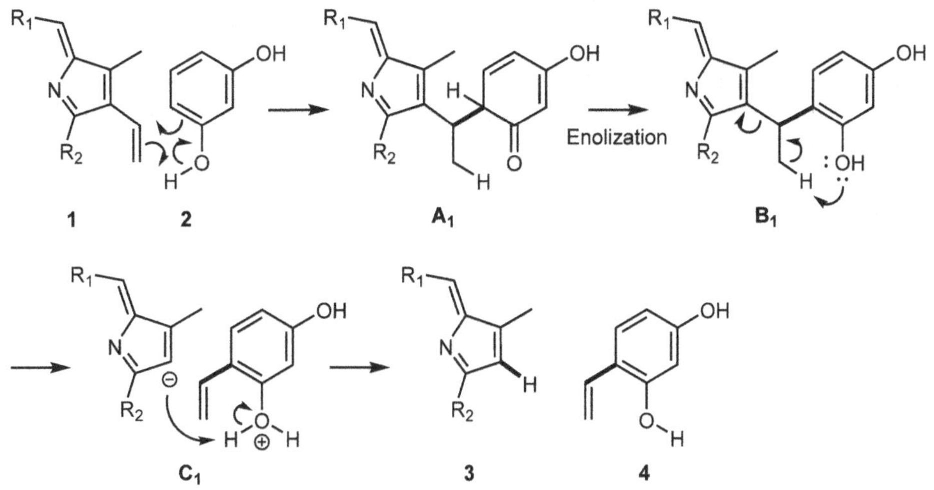

SOLUTION: MECHANISM 1B

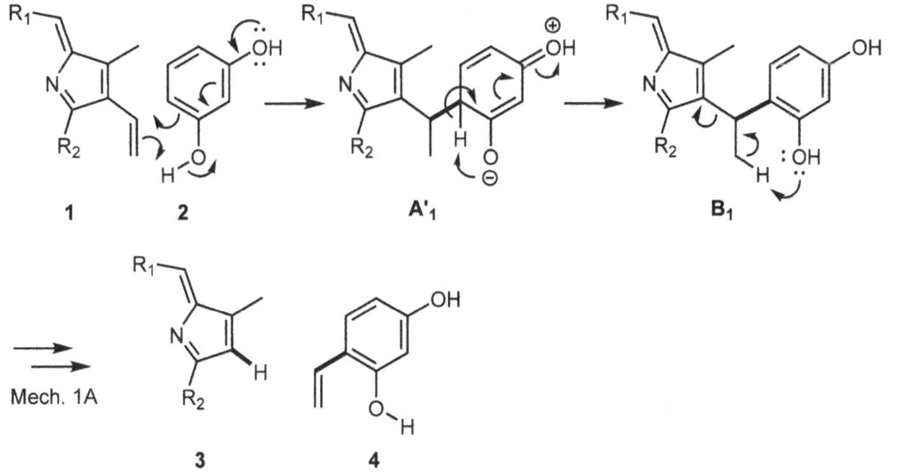

SOLUTION: MECHANISM 2A

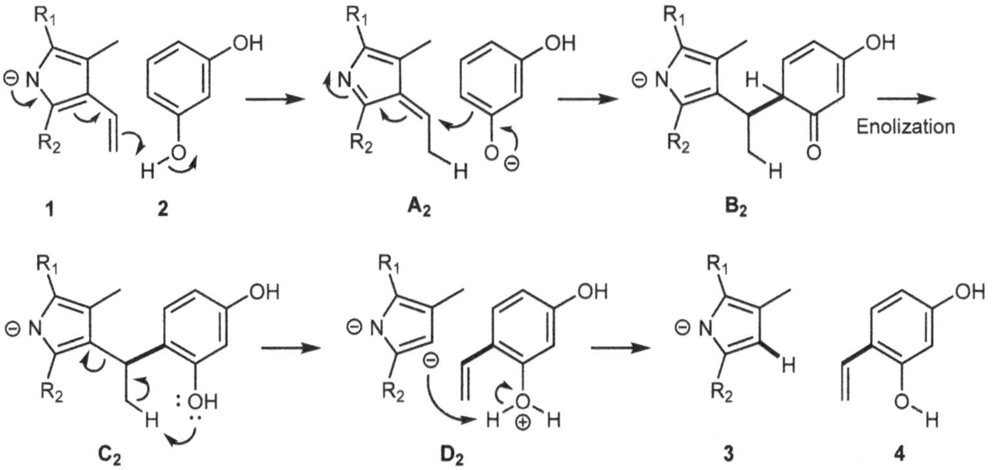

SOLUTION: MECHANISM 2B

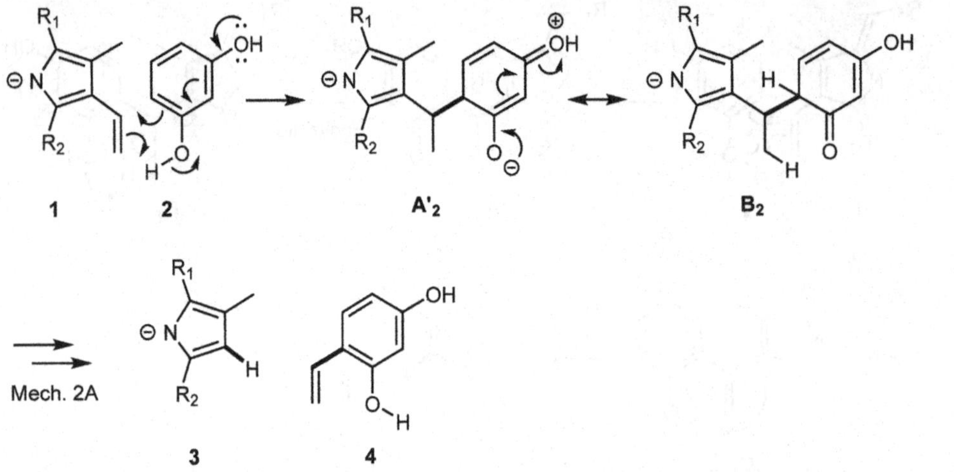

1   2   A'₂   B₂

Mech. 2A

3   4

SOLUTION: MECHANISM 3

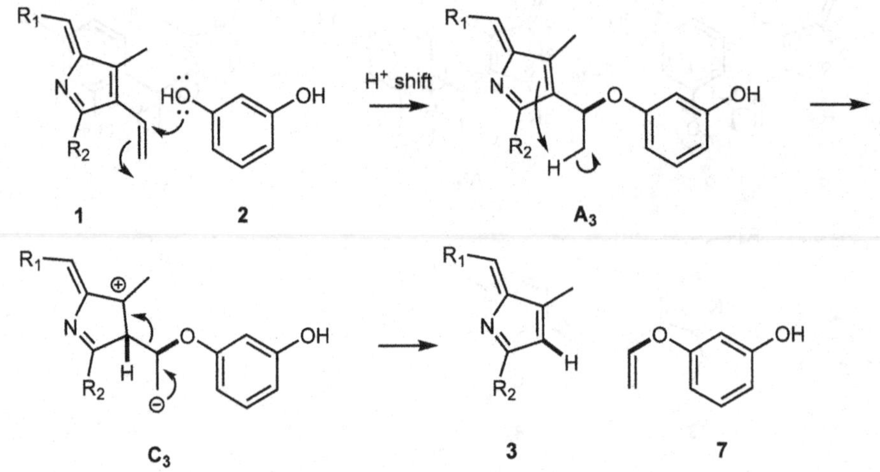

1   2   A₃

C₃   3   7

SOLUTION: MECHANISM 4

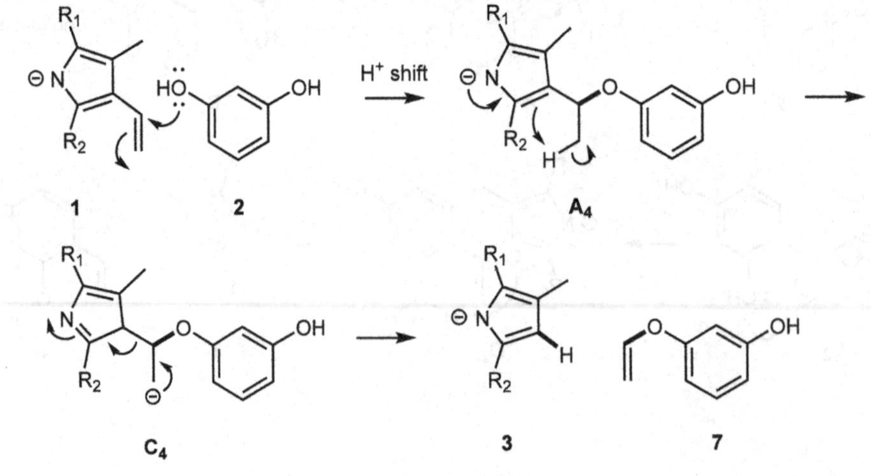

1   2   A₄

C₄   3   7

KEY STEPS EXPLAINED

The authors observed that deuterohemin **3** was obtained in 50% or more yield and in a time of 25 min from protohemin **1** compared to less than 10% yield and in a time of 2.5 h from mono(resorcinylethyl)hemin **5a** or **5b** or bis(resorcinylethyl)hemin **6** under identical conditions of molten resorcinol at 190°C. From these observations they concluded that there are two pathways leading to the devinylation product **3**: a fast pathway from **1** and a slow pathway from either **5a**, **5b**, or **6**. Also, they concluded that the compounds **5a**, **5b**, and **6** could not be intermediates on the fast reaction pathway.

The mechanisms they proposed as written in their article involve protonations from the external medium as if it were an acidic environment. Clearly, the reaction conditions are entirely thermal and do not involve adding any external acid, so the hydrogen atoms needed for the hydrogen shifts which the mechanism requires must originate from the starting materials. The first and second pK_a values for resorcinol are 9.3 and 11.1, respectively. Moreover, when the authors carried out the reaction of mono(resorcinylethyl)hemin in the presence of a strong acid such as trifluoroacetic acid they were unsuccessful in obtaining any devinylation product **3** and instead obtained intractable tars. Hence, the operative mechanism does not involve external protons as the authors suggest.

What the authors did not do was confirm the structure of the vinylresorcinol by-product. There are two possible vinylated resorcinol products arising from C-alkylation (product **4**) or O-alkylation (product **7**).

Furthermore, there are two different types of pyrrole moieties in protohemin that have vinyl groups attached. One pyrrole moiety has a negatively charged nitrogen atom with two $C=C$ bonds in the ring, and the other has an uncharged nitrogen atom with one $C=C$ and one $C=N$ bond in the ring. These pyrrole moieties are expected to undergo different devinylation mechanisms.

Here we show mechanisms for the production of deuterohemin **3** and 4-vinylresorcinol **4** via C-alkylation from each of the pyrrole moieties (mechanisms 1A/1B and 2A/2B). Also, we show mechanisms for the production of deuterohemin **3** and 3-vinyloxy-phenol **7** via O-alkylation from each of the pyrrole moieties (mechanisms 3 and 4).

Mechanisms 1A and 1B proceed via a thermal hydrogen ene transfer reaction to give either intermediates A_1 or A_1'. These enolize to intermediate B_1 which would correspond to the mono or bis(resorcinylethyl)hemin intermediates **5a**, **5b**, or **6**. The next fragmentation step to C_1 must be the kinetically challenging step consistent with the authors' observations. In mechanism 2A, the vinyl group is protonated from resorcinol via the electron push coming from the negatively charged nitrogen atom in the pyrrole ring. This leads to the *ortho*-alkylation of resorcinol as shown by the A_2 to B_2 transformation. The remaining steps are similar to mechanisms 1A and 1B. In mechanism 2B, a thermal hydrogen ene transfer reaction yields intermediate A_2' which is similar to intermediate A_1 in mechanism 1A. Again, enolization followed by slow fragmentation lead to products **3** and **4**. These mechanisms (i.e., 1A, 1B, 2A, 2B) all involve C-alkylation.

In mechanisms 3 and 4 (O-alkylation), the hydroxyl group of resorcinol adds to the vinyl group of the porphyrin to give ether intermediates A_3 and A_4 after an intramolecular proton shift. Internal proton transfers in each case lead to zwitterion or carbanion intermediates C_3 or C_4, respectively. These in turn fragment to products **3** and **7**. These last fragmentation steps are expected to be less energy demanding compared to the ones involved in mechanisms 1 and 2. The authors conceded that "the original suggestion of O-alkylation may prove to be correct" and admitted that further investigations are needed. This interesting reaction would greatly benefit from a thorough by-product analysis to identify either **4** or **7** and a computational study to determine the energetics of each step in the proposed mechanisms.

ADDITIONAL RESOURCES

For examples of further applications of porphyrins and their transformations in environments deigned to simulate nature, see the work of Pickering and Keely.[2]

References

1. DiNello RK, Dolphin DH. Evidence for a fast (major) and slow (minor) pathway in the Schumm devinylation reaction of vinyl porphyrins. *J Org Chem*. 1981;46(17):3498–3502. https://doi.org/10.1021/jo00330a023.
2. Pickering MD, Keely BJ. Origins of enigmatic C-3 methyl and C-3 H porphyrins in ancient sediments revealed from formation of pyrophaeophorbide d in simulation experiments. *Geochim Cosmochim Acta*. 2013;104:111–122. https://doi.org/10.1016/j.gca.2012.11.021.

Question 152: A Uracil to Substituted Benzene Transformation

In 1981 Japanese chemists reacted the deuterium-labeled dimethyl-5-formyluracil **1** with acetylacetone in ethanolic sodium ethoxide to form **2** and not **3** (Scheme 1).[1] Provide mechanisms for both the observed **1** to **2** transformation and the alternative unobserved **1** to **3** transformation.

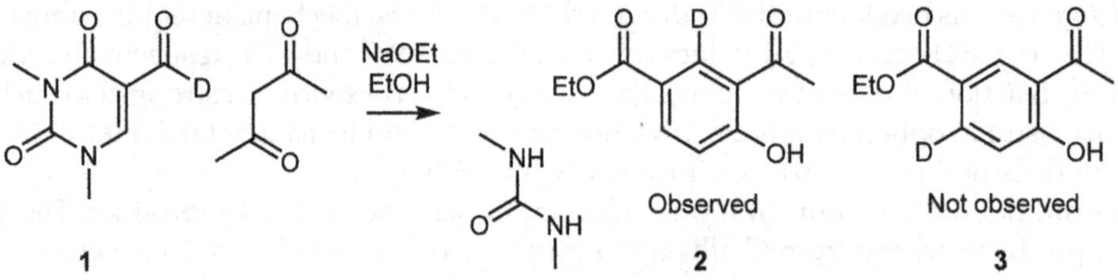

SCHEME 1

SOLUTION: 1 TO 2

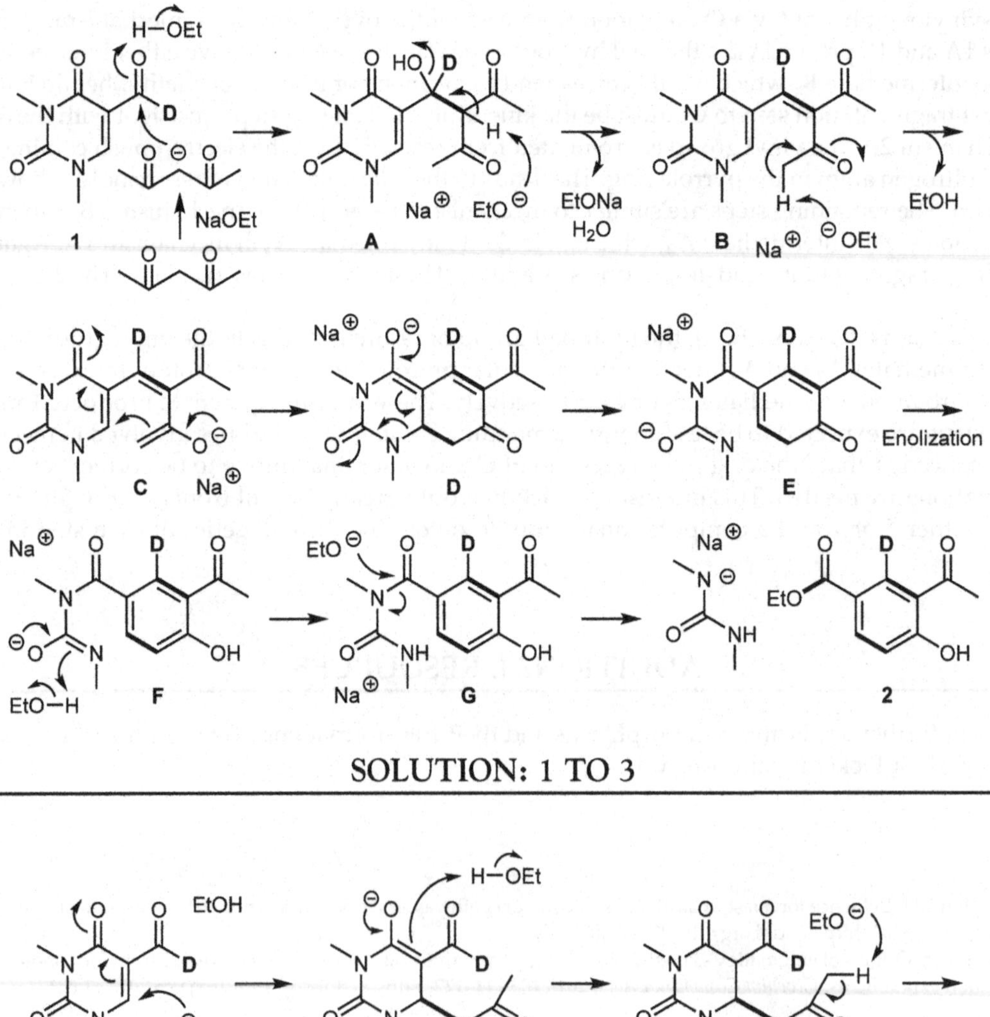

SOLUTION: 1 TO 3

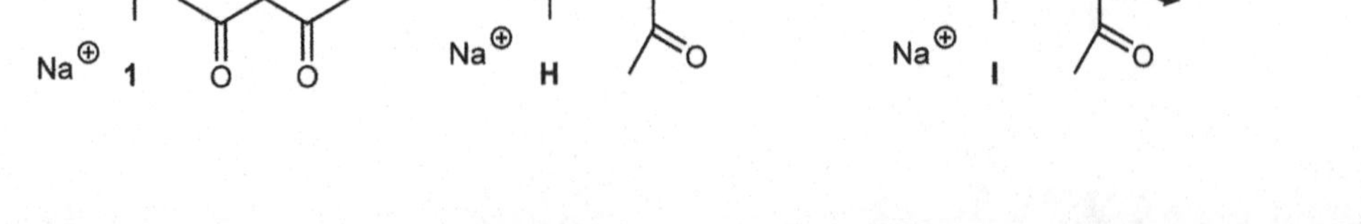

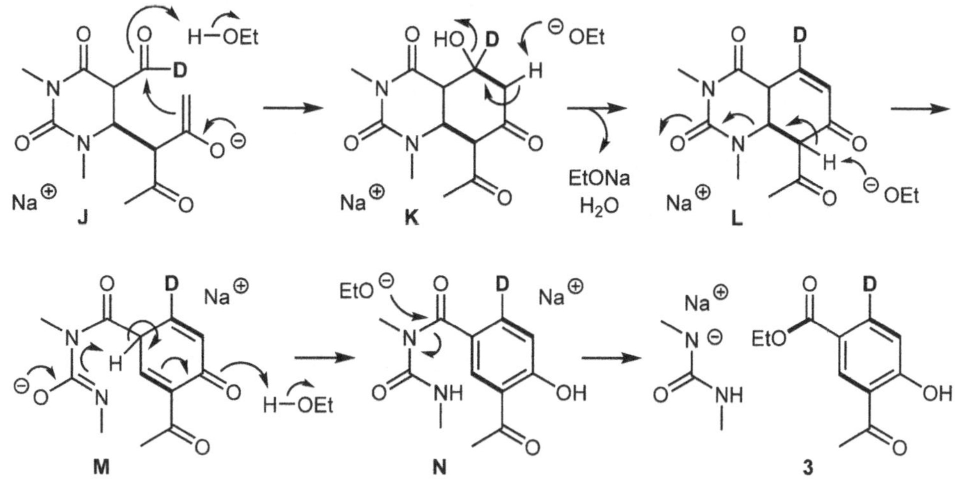

KEY STEPS EXPLAINED

The two mechanisms presented before highlight the practical usefulness of isotopic labeling experiments in allowing for differentiation of products under the same reaction conditions to reveal the operating mechanistic pathway when several alternatives are possible. Here, the two mechanisms differ with respect to the point of attack of the carbanion of 2,4-pentanedione onto the uracil starting material 1. In the operative mechanism, the carbanion attacks the aldehyde group whereas in the nonoperative mechanism the attack takes place at C_6 of the pyrimidine ring in a Michael fashion. Since both possible products are geometric isomers, both mechanisms lead to the same overall balanced chemical equation as shown in Scheme 2. Furthermore, the target bond maps for both 2 and 3 show that the ring construction strategy is in both cases a [3 + 3] cyclization (Fig. 1). In the absence of the deuterium label, the two products would not be distinguishable and the mechanisms would have to be compared according to a different standard, for example, by conducting a theoretical analysis although this would not present direct evidence as in the experimental labeling study.

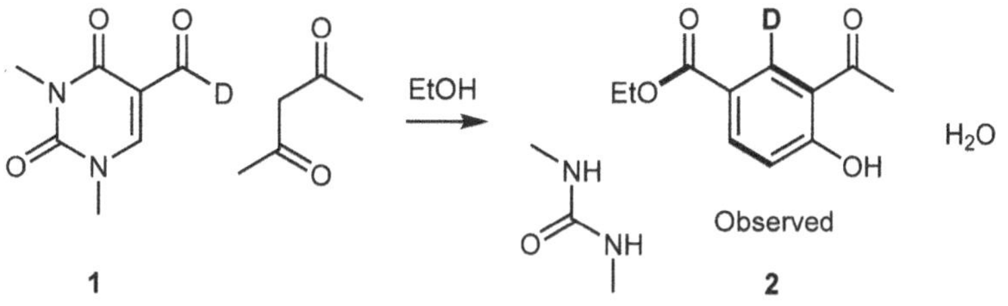

SCHEME 2

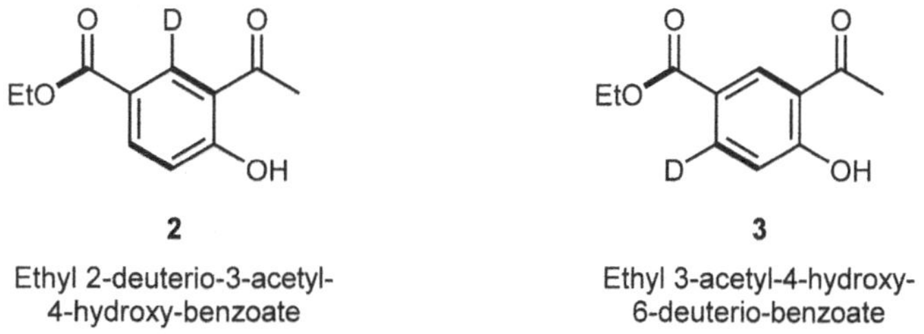

2
Ethyl 2-deuterio-3-acetyl-
4-hydroxy-benzoate

3
Ethyl 3-acetyl-4-hydroxy-
6-deuterio-benzoate

FIG. 1

ADDITIONAL RESOURCES

We recommend an article which describes an interesting ring transformation of substituted 1,3-dimethyl-5-formyluracils induced by enamines.[2] We also recommend an excellent study of the mechanism of a regioselective aniline synthesis which features deuterium labeling and theoretical analysis.[3]

References

1. Hirota K, Kitade Y, Senda S. Pyrimidine derivatives and related compounds. 39. A novel cycloaromatization reaction of 5-formyl-1,3-dimethyluracil with three-carbon nucleophiles. Synthesis of substituted 4-hydroxybenzoates. *J Org Chem*. 1981;46(20):3949–3953. https://doi.org/10.1021/jo00333a003.
2. Singh H, Dolly, Chimni SS, Kumar S. Enamine-induced ring transformations of 6-substituted 5-formyl-1,3-dimethyluracils. *J Chem Res (S)*. 1998;1998:352–353. https://doi.org/10.1039/A708683K.
3. Davies IW, Marcoux J-F, Kuethe JT, et al. Demonstrating the synergy of synthetic, mechanistic, and computational studies in a regioselective aniline synthesis. *J Org Chem*. 2004;69:1298–1308. https://doi.org/10.1021/jo035677u.

Question 153: γ-Lactones From Vinyl Sulfoxides

In 1981 American chemists reacted vinyl sulfoxide **1** with dichloroketene to form the γ-lactone product **2** in 65% yield (Scheme 1).[1] The authors also showed that depending on which procedure they used to generate dichloroketene, different amounts of product were obtained. Provide a mechanism for this transformation and explain why it matters how dichloroketene is made.

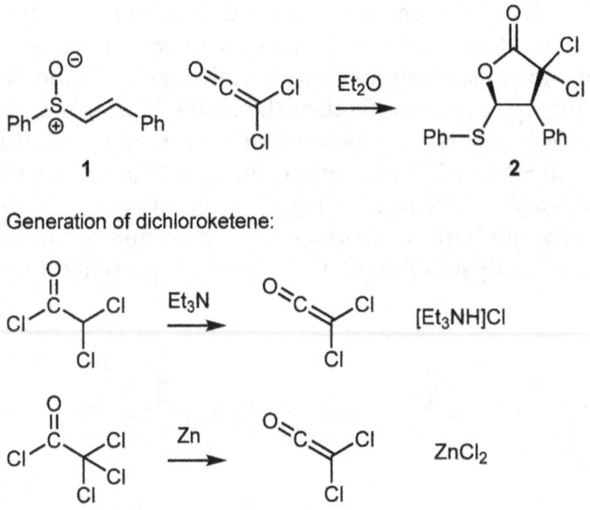

Generation of dichloroketene:

SCHEME 1

SOLUTION

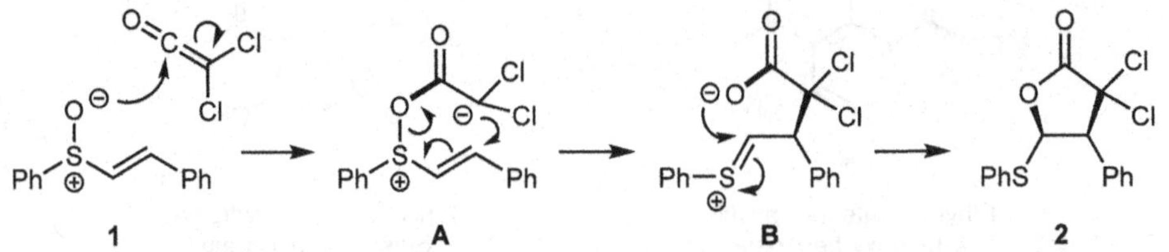

KEY STEPS EXPLAINED

The mechanism proposed by the authors for the transformation of sulfoxide **1** and dichloroketene to the γ-lactone product **2** begins with nucleophilic attack by the sulfoxide oxygen atom of **1** onto the carbonyl group of dichloroketene to form zwitterion **A**. A Michael-type addition of the carbanion of **A** to the terminal olefinic carbon atom causes cleavage of the original S—O sulfoxide bond to form the sulfonium ion **B**. Recyclization of **B** via the carboxylate oxygen atom gives the racemic *cis*-lactone **2**. Overall this cycloaddition strategy is a [2+2+1] ring forming reaction.

It is also important to note that the authors generated dichloroketene either via elimination of HCl from dichloro-acetyl chloride by triethylamine base or via a redox reaction involving zinc dust and trichloro-acetyl chloride. It was observed that this latter redox method consistently led to higher yields of ketene but also of product **2** after addition of sulfoxide **1**. The authors argued that this happened because the triethylammonium chloride by-product from the reaction with triethylamine works to protonate the carbanion of **A** thus hindering the formation of **B**.

Furthermore, the *cis* geometry of the lactone was confirmed by carrying out the reaction using ring containing sulfoxides, hydrogenating the product with Raney nickel catalyst, and comparing the spectroscopic properties of the product structure with known literature details of the racemic *cis*-lactones. An example of this approach is shown in Scheme 2. Lastly, we note that the authors did not carry out a theoretical analysis for this mechanism in order to calculate the energy barrier for the formation of **A** which we expect to be the slow step in this transformation.

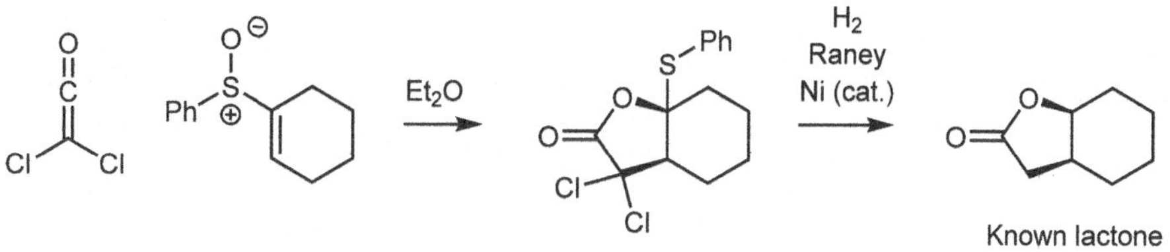

SCHEME 2

ADDITIONAL RESOURCES

We recommend an excellent article which discusses the mechanism of an interesting transformation that involves the reaction between sulfoxonium ylides and ketenes.[2]

References

1. Marino JP, Neisser M. Stereospecific reactions of dichloroketene with vinyl sulfoxides: a new type of polar cycloaddition. *J Am Chem Soc*. 1981;103 (25):7687–7689. https://doi.org/10.1021/ja00415a065.
2. Mondal M, Ho H-J, Peraino NJ, Gary MA, Wheeler KA, Kerrigan NJ. Diastereoselective reaction of sulfoxonium ylides, aldehydes and ketenes: an approach to *trans*-γ-lactones. *J Org Chem*. 2013;78(9):4587–4593. https://doi.org/10.1021/jo4003213.

Question 154: Tetrahydrothieno[2,3-*c*]- and -[3,2-*c*]pyridine Synthesis

In 1982 British chemists observed that acidic strength played a role in the rearrangement of *N*-(2-hydroxyphe-nethyl)-2-aminomethylthiophens **1** to either the nonrearranged product **2** or the mixture of **2** and the rearranged product **3** (Scheme 1).[1] Suggest a mechanistic explanation to account for these contrasting results.

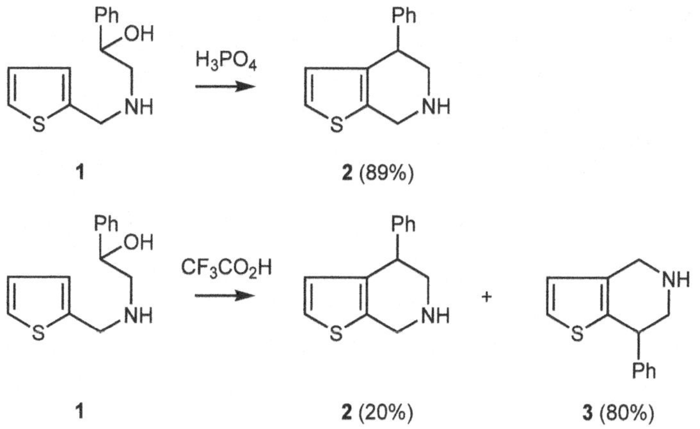

SCHEME 1

SOLUTION: 1 TO 2 (WEAKLY ACIDIC MEDIUM)

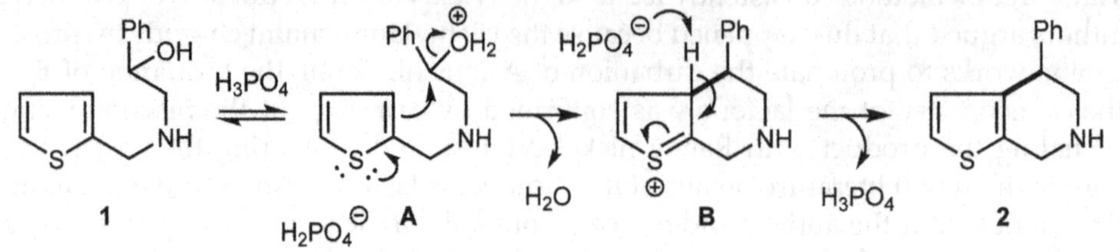

SOLUTION: 1 TO 2 AND 3 (STRONGLY ACIDIC MEDIUM)

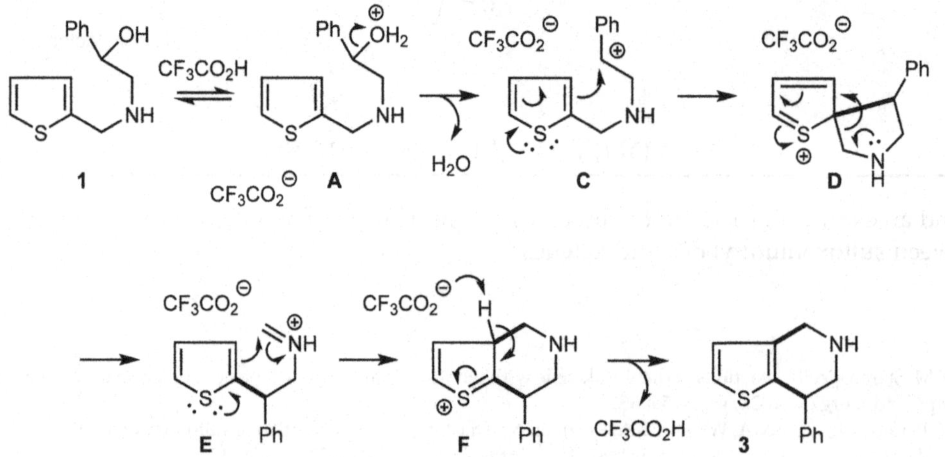

KEY STEPS EXPLAINED

The mechanism proposed by the authors to explain the reactivity of **1** under weakly acidic conditions begins with protonation of the hydroxyl group of **1** to form **A** followed by S_N2-induced cyclization to give a six-membered piperidine ring intermediate **B**. A final deprotonation by the conjugate base of phosphoric acid generates the final unrearranged product **2** which has the sulfur atom in a *meta* orientation with respect to the nitrogen atom in the piperidine ring. By contrast, under strongly acidic conditions, the same initial protonation of the hydroxyl group of **1** gives **A** which eliminates water in an S_N1 manner to give the carbocation intermediate **C**. This intermediate undergoes cyclization to the [5,5]-*spiro* sulfonium ion intermediate **D** followed by ring opening to the iminium ion **E**. Recyclization of **E** to **F** is followed by a deprotonation by the conjugate base of trifluoroacetic acid to generate the final rearranged product **3**. We note that this time the sulfur atom in **3** is positioned in a *para* orientation with respect to the nitrogen atom in the piperidine ring. Although they did not carry out further experiments or a theoretical analysis to help support these postulated mechanisms, the authors remarked that electron-donating substituents on the phenyl ring should stabilize carbocation **C** and hence favor the rearrangement pathway to products having structures like **3**. Such an experiment would provide evidence for the operation of the mechanism for the sequence of **1** to **3**. It might also be possible to test substituted versions of **1** to see whether steric factors might play a role in the formation of the *spiro* intermediate **D**.

ADDITIONAL RESOURCES

For examples of a similar transformation involving rearrangement of an oxygen analog of structure **1**, see the work of Reddy et al.[2]

References

1. Mackay C, Waigh RD. Rearrangement and cyclisation of *N*-(2-hydroxyphenethyl)-2-aminomethylthiophens. *J Chem Soc Chem Commun.* 1982;1982:793–794. https://doi.org/10.1039/C39820000793.
2. Reddy BVS, Reddy YV, Lakshumma PS, et al. In(OTf)3-catalyzed tandem aza-Piancatelli rearrangement/Michael reaction for the synthesis of 3,4-dihydro-2H-benzo[b][1,4]thiazine and oxazine derivatives. *RSC Adv.* 2012;2:10661–10666. https://doi.org/10.1039/c2ra21591h.

Question 155: Oxothiolan → Oxathian → Oxothiolan → Oxathian

In 1981 British chemists studied the reactivity of thioacetals such as **1** with sulfuryl chloride and made the experimental observations illustrated in Scheme 1.[1] Suggest mechanisms to account for these transformations.

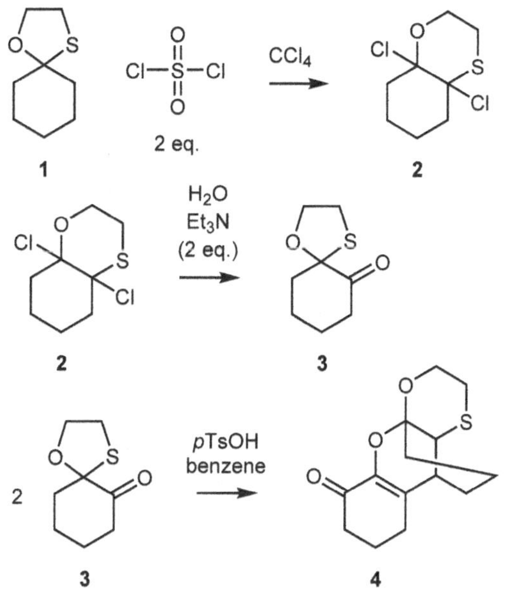

SCHEME 1

SOLUTION: 1 TO 2

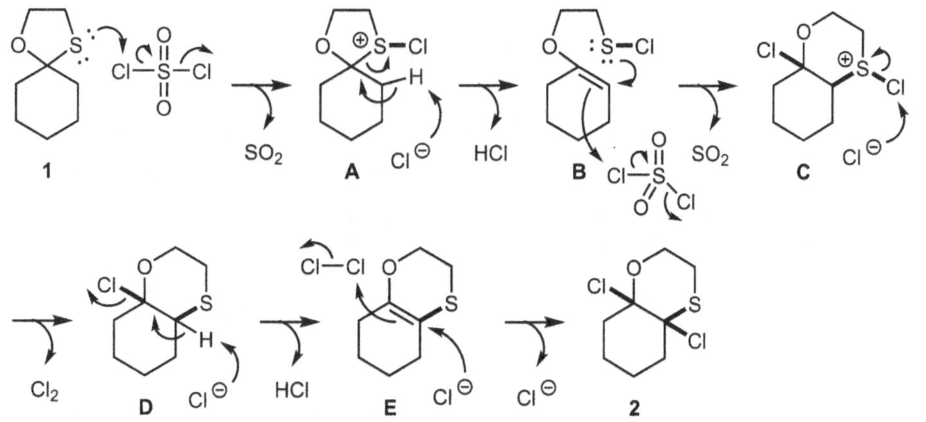

BALANCED CHEMICAL EQUATION: 1 TO 2

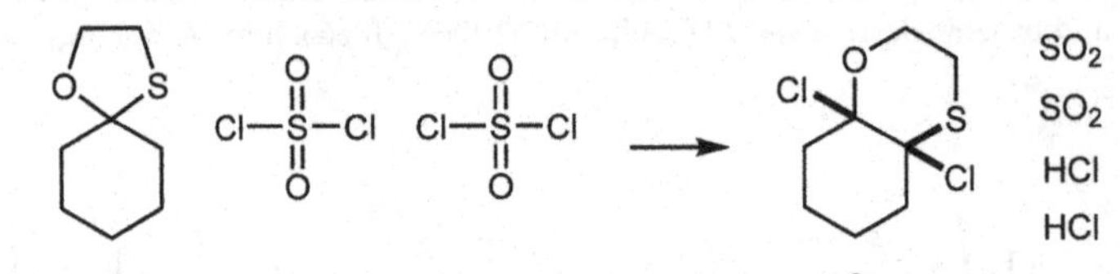

SOLUTION: 2 TO 3

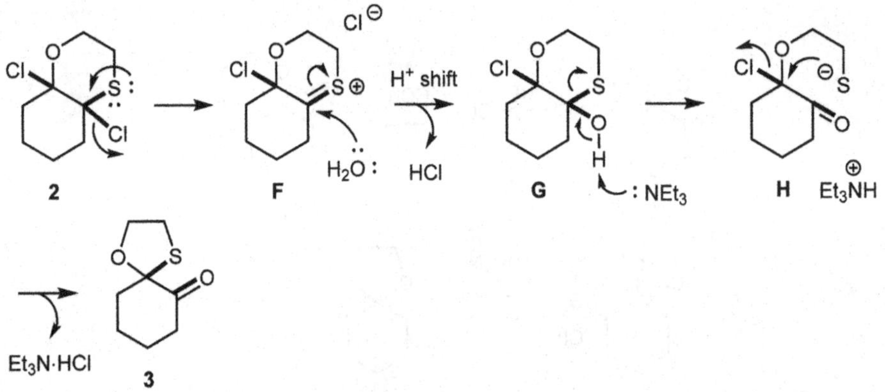

BALANCED CHEMICAL EQUATION: 2 TO 3

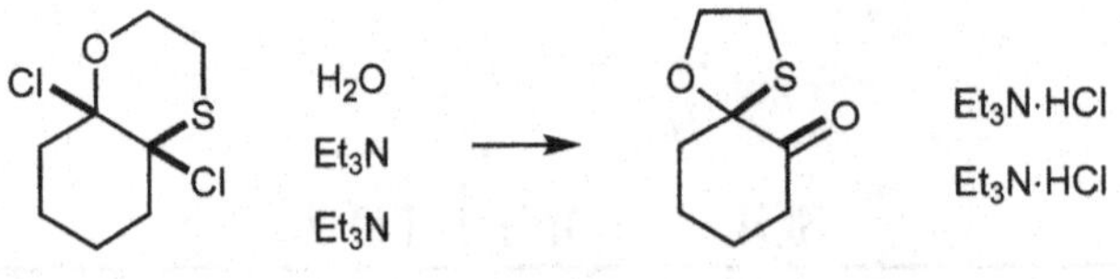

SOLUTION: 3 TO 4

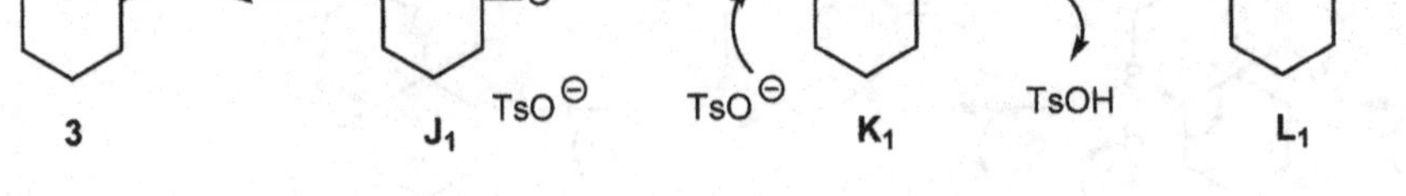

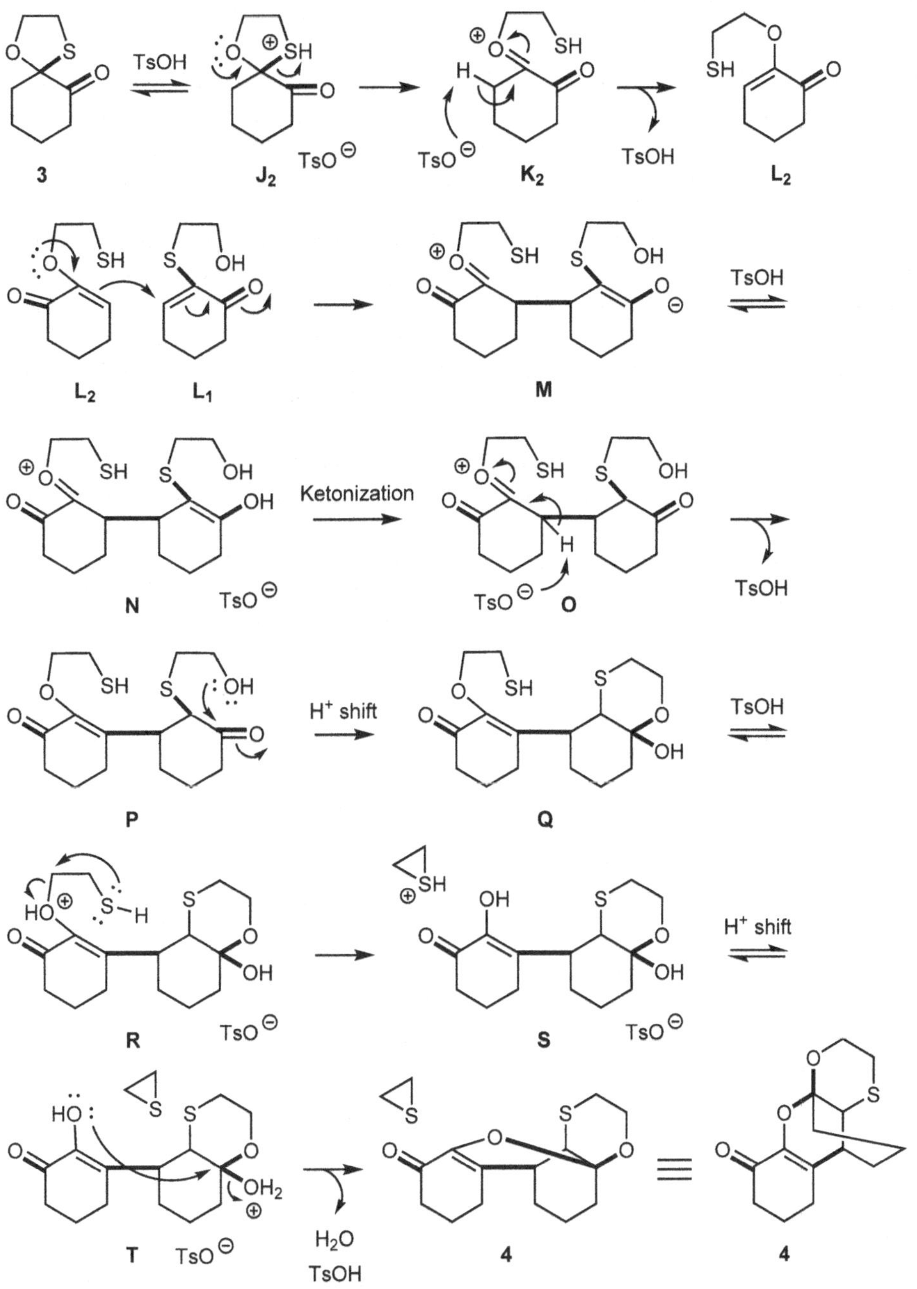

BALANCED CHEMICAL EQUATION: 3 TO 4

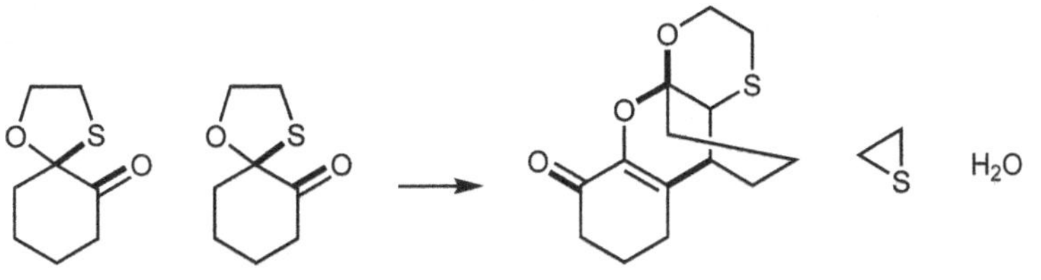

KEY STEPS EXPLAINED

The mechanism proposed by the authors for the transformation of **1** to **2** begins with chlorination of the sulfur atom of oxathiolan **1** via one equivalent of sulfuryl chloride to form the chlorosulfonium chloride **A**. Next, the oxathiolan ring opens upon chloride deprotonation to give **B** which ring closes again with chlorination via a second equivalent of sulfuryl chloride to generate a ring expanded 1,4-oxathian chlorosulfonium chloride intermediate **C**. Dechlorination of **C** via chloride ion attack gives **D** after elimination of chlorine gas which is followed by two successive dechlorination and chlorination steps that culminate in the formation of **2**. From **2** to **3** we first have dechlorination of **2** to form the sulfonium ion **F** which is attacked by water and which loses two equivalents of HCl to form the oxathiolan product **3**. Next, intermediate **3** can follow two possible ring-cleavage routes to form either L_1 or L_2 depending on whether we have cleavage of the C—O thiolan bond or the C—S thiolan bond, respectively. Furthermore, L_1 and L_2 react to cause dimerization in an orientation such that intermediate **Q** is formed after a series of proton transfer steps along with a ketonization step. From **Q** we can have loss of thiirane followed by dehydration and transannular cyclization to generate the final oxathian product **4**.

In the context of the transformation of **Q** to **4**, which the authors did not illustrate, we note their statement that "the form in which the final two-carbon unit is lost is not known."[1] Nevertheless, from the proposed mechanism we rationalize that it is likely the elimination of thiirane which explains the loss of the two-carbon subunit. What was not mentioned by the authors is that if the coupling between two molecules of **3** had been of the opposite orientation an isomeric product would arise with loss of ethylene oxide as shown in Scheme 2. Although this possibility was not discussed, the experimental section did mention that the product collected in 83% yield was characterized by an m.p. of 207–209°C, IR, ^{1}H NMR, ^{13}C NMR, and elemental analysis. The properties of the isomeric product are expected to be very similar to **4** and so it is not certain that the authors really collected a single product from the reaction of **3** to **4**. This reaction requires further investigation to prove the existence of thiirane as a by-product as well as confirming whether the oxathian product is a single isomer or a mixture of isomers. We also encourage undertaking a theoretical analysis in the future to help support the proposed mechanisms.

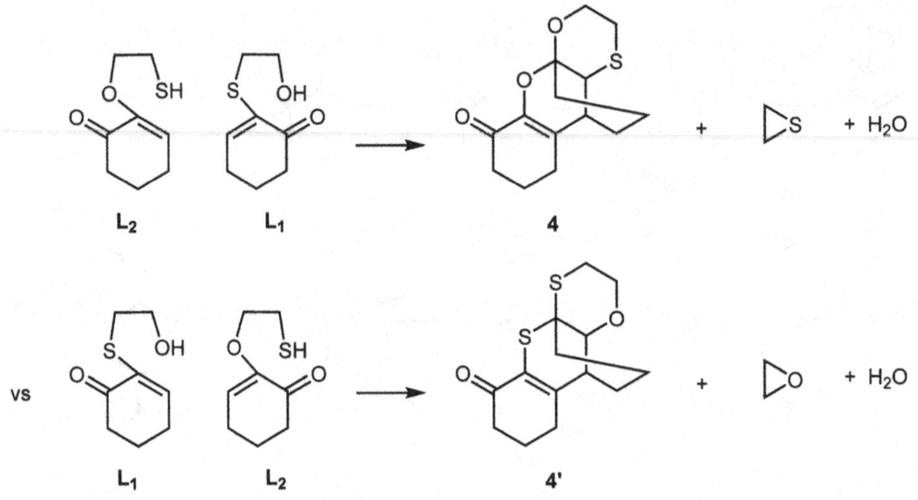

SCHEME 2

ADDITIONAL RESOURCES

For an interesting example of a transformation involving a push-pull type mechanism, see the work of Markovic et al.[2]

References

1. Bulman-Page PC, Ley SV, Morton JA, Williams DJ. On the reaction of thioacetals with sulphuryl chloride. *J Chem Soc Perkin Trans 1.* 1981;1981:457–461. https://doi.org/10.1039/P19810000457.
2. Markovic R, Vitnik Z, Baranac M, Juranic I. Mechanism of stereoselective synthesis of push-pull (Z)-4-oxothiazolidine derivatives containing an exocyclic double bond. A MNDO-PM3 study. *J Chem Res (S).* 2002;2002:485–489.

Question 156: A Thiazolopyrimidine to Pyrrolopyrimidine Transformation

During the course of 10 years, Japanese chemists have investigated the mechanism for the transformations shown in Scheme 1.[1-3] Explain the course of these reactions in mechanistic terms.

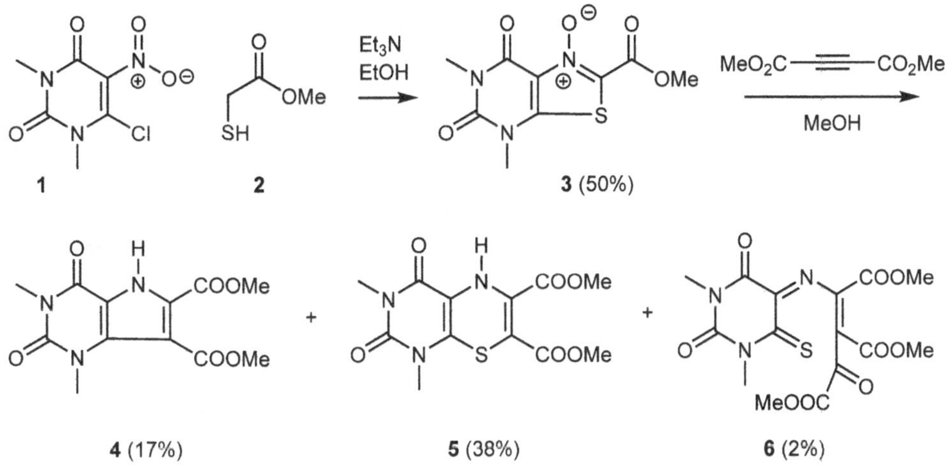

SCHEME 1

SOLUTION: 1 AND 2 TO 3

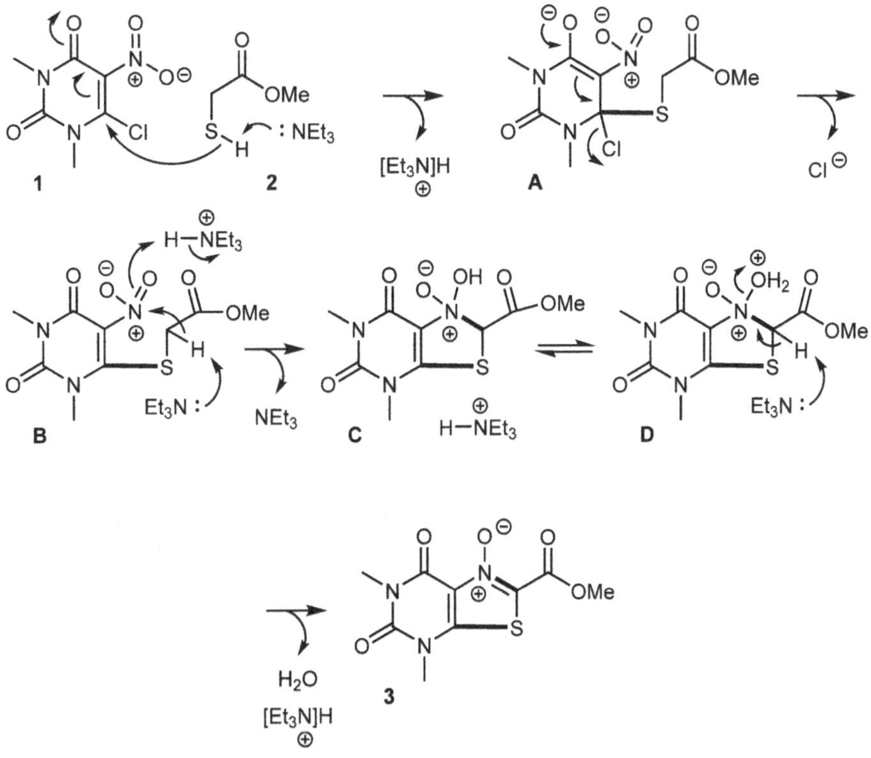

BALANCED CHEMICAL EQUATION: 1 AND 2 TO 3

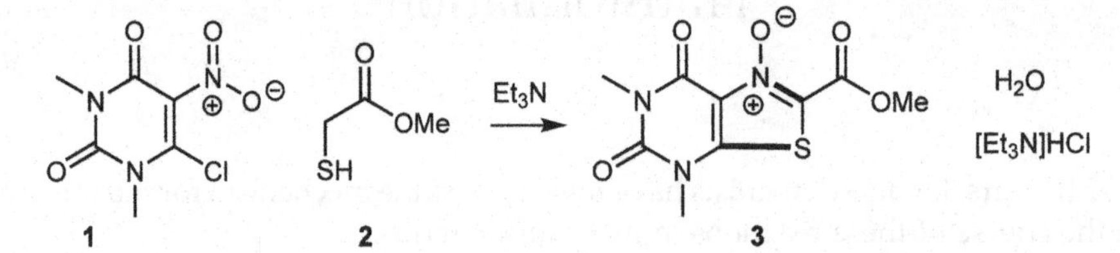

SOLUTION: 3 TO 6 TO 5

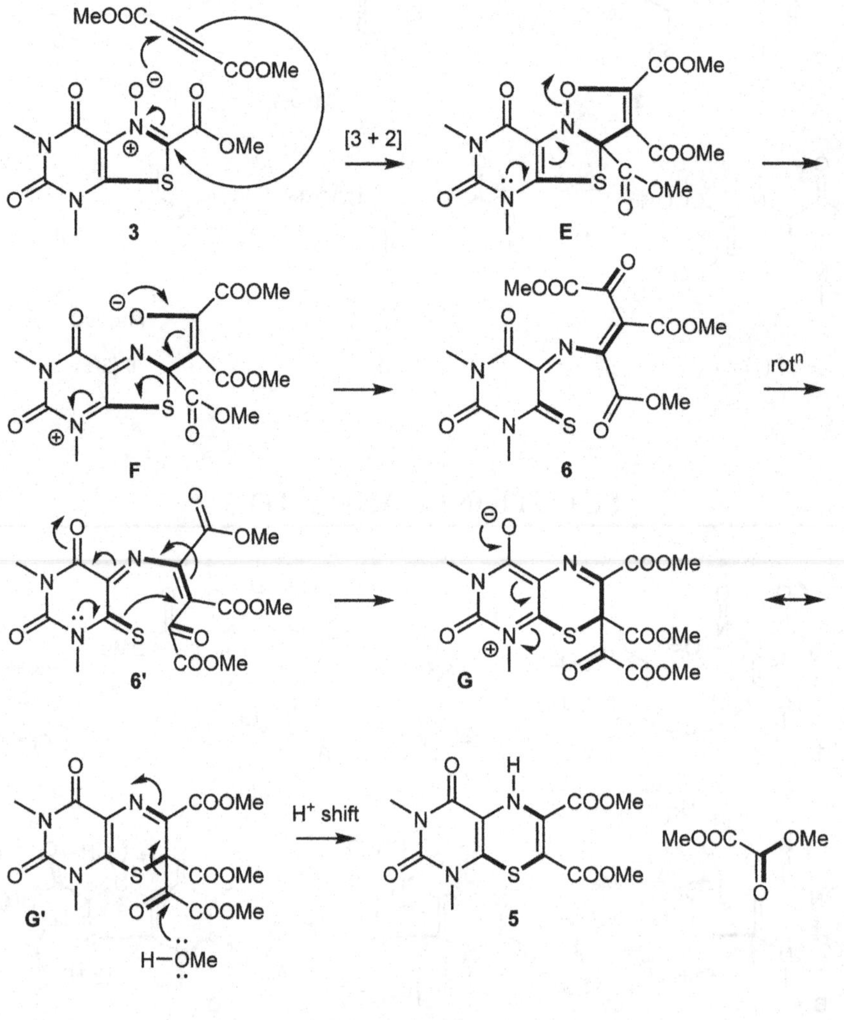

BALANCED CHEMICAL EQUATION: 3 TO 6 TO 5

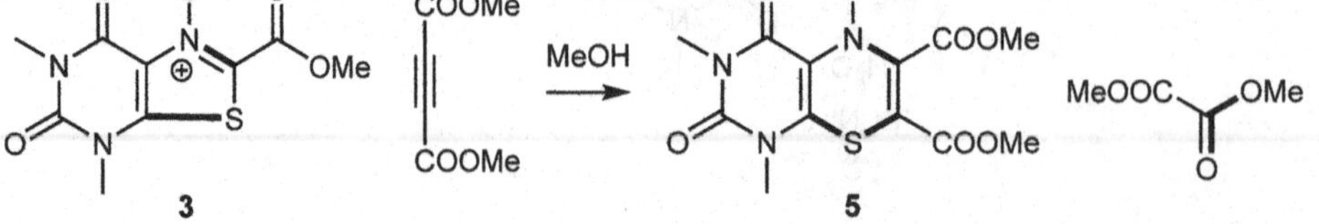

SOLUTION: 5 TO 4

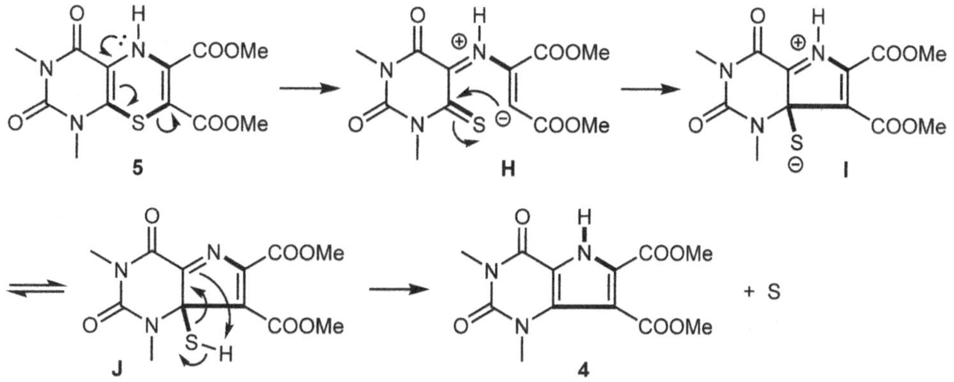

KEY STEPS EXPLAINED

The mechanism proposed by the authors begins with base-catalyzed addition of methyl mercapto-acetate, 2, to N-oxide 1 which results in an overall [3+2] cycloaddition to yield product 3. Afterward, the 1,3-dipolarophile 3 adds to dimethylacetylene dicarboxylate (DMAD) in a [3+2] fashion to furnish E which fragments to 6 and recyclizes to intermediate G'. Methanolysis of G' leads to product 5 and dimethyloxalate. Alternatively, hydrolysis of G' leads to 5 and monomethyloxalate. The target bond map for the structure of product 5 shows that the six-membered ring construction strategy is [3+1+1+1]. Finally, the 1,4-thiazine ring of 5 opens to thione H which ring closes again and undergoes a [1,4]-protic transfer to form J. Intermediate J then undergoes an internal redox reaction to generate the pyrrole product 4 where sulfur is oxidized from −2 to 0 for a net change of +2 which is balanced by the reductions of C_1 and C_2 each from +2 to +1 for a net change of −2.

The authors provided the following evidences for their proposed mechanisms: (a) if isolated product 5 is refluxed in methanol for 3 h, product 4 is obtained in 75% yield along with extrusion of elemental sulfur; (b) if starting material 3 is reacted with DMAD in the presence of water and methanol instead of dry methanol, the yield of product 4 increases to 71% from 17% meaning that water plays an important role in the mechanism; (c) product 5 was isolated in 56% if the reaction carried out in (b) was stopped after 1 h; and (d) reaction of 3 with ethyl phenylpropiolate forms products analogous to structures of intermediate G' and product 6 (see Scheme 2). We also note that the thermodynamic driving force for extrusion of elemental sulfur in the reaction of 5 to 4 is transformation of an antiaromatic 1,4-thiazine ring to an aromatic pyrrole ring. The authors did not perform a theoretical analysis to support their proposed mechanisms.

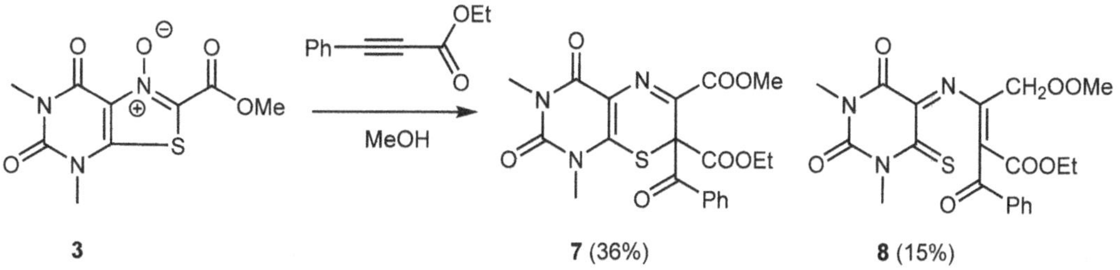

SCHEME 2

ADDITIONAL RESOURCES

We recommend an article which discusses mechanistic aspects of a synthetic approach to thiazolo[3,2-a]pyrimidin-7-ones.[4]

References

1. Senga K, Ichiba M, Kanazawa H, Nishigaki S. 1,3-Dipolar cycloaddition of a thiazolo[5,4-d]pyrimidine 1-oxide to dimethyl acetylenedicarboxylate. New ring transformation to a pyrrolo[3,2-d]pyrimidine via a pyrimido[4,5-b][1,4]thiazine. *J Chem Soc Chem Commun*. 1981;1981:278–280. https://doi.org/10.1039/C39810000278.
2. Senga K, Ichiba M, Nishigaki S. New synthesis of pyrrolo[3,2-d]pyrimidines (9-deazapurines) by the 1,3-dipolar cycloaddition reaction of fervenulin 4-oxides with acetylenic esters. *J Org Chem*. 1979;44(22):3830–3834. https://doi.org/10.1021/jo01336a018.
3. Kanazawa H, Ichiba M, Tamura Z, Senga K, Kawai K, Otomasu H. 1,3-Dipolar cycloaddition reactions of thiazolo[5,4-d]pyrimidine 1-oxides with acetylenic esters involving new ring transformations of the thiazole nucleus. *Chem Pharm Bull*. 1987;35(1):35–45. https://doi.org/10.1248/cpb.35.35.
4. Skaric V, Skaric D, Cizmek A. Synthetic routes to thiazolo[3,2-a]pyrimidin-7-ones via 1-allyl-2-thiouracil. *J Chem Soc Perkin Trans 1*. 1984;1984:2221–2225. https://doi.org/10.1039/P19840002221.

Question 157: Oxidation of 1,5-Diacetoxynaphthalene With NBS

In 1983 American chemists investigated the mechanism of the bromination of 1,5-diacetoxynaphthalene **1** with 4 equivalents of *N*-bromosuccinimide (NBS) to form 2-bromojuglone acetate **2** (Scheme 1).[1] It was also observed that at a lower temperature product **3** was isolable as the major product and that treatment of it with acetic acid led to **2**. Another experiment showed that treatment of the dibrominated compound **4** with 2 equivalents of NBS and acetic acid also led to **2** in high yield. Based on these observations, propose a mechanism for the transformation of **1** to **2**.

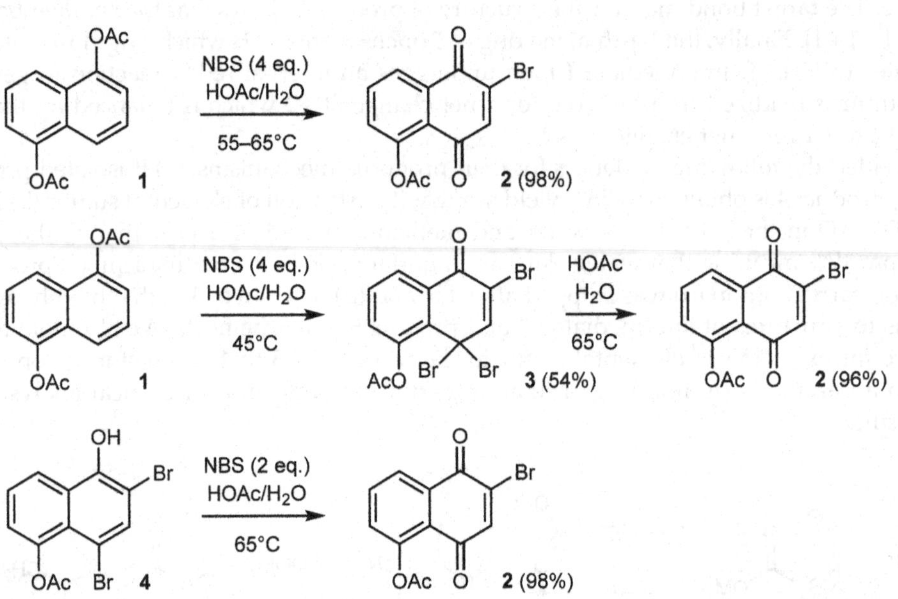

SCHEME 1

SOLUTION: 1 TO 3 TO 2

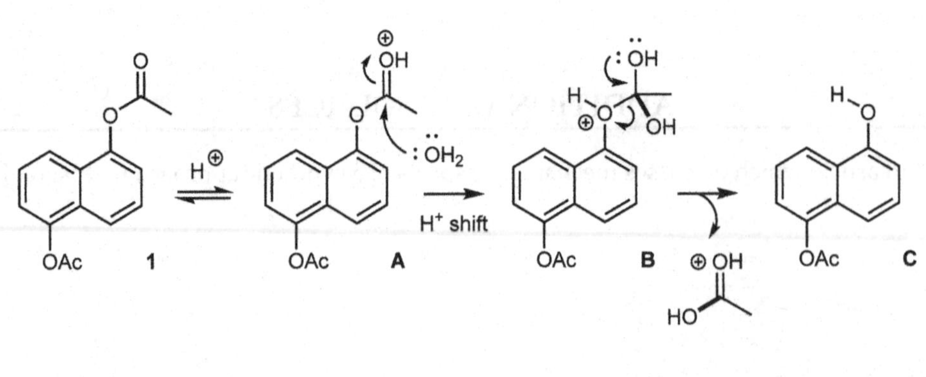

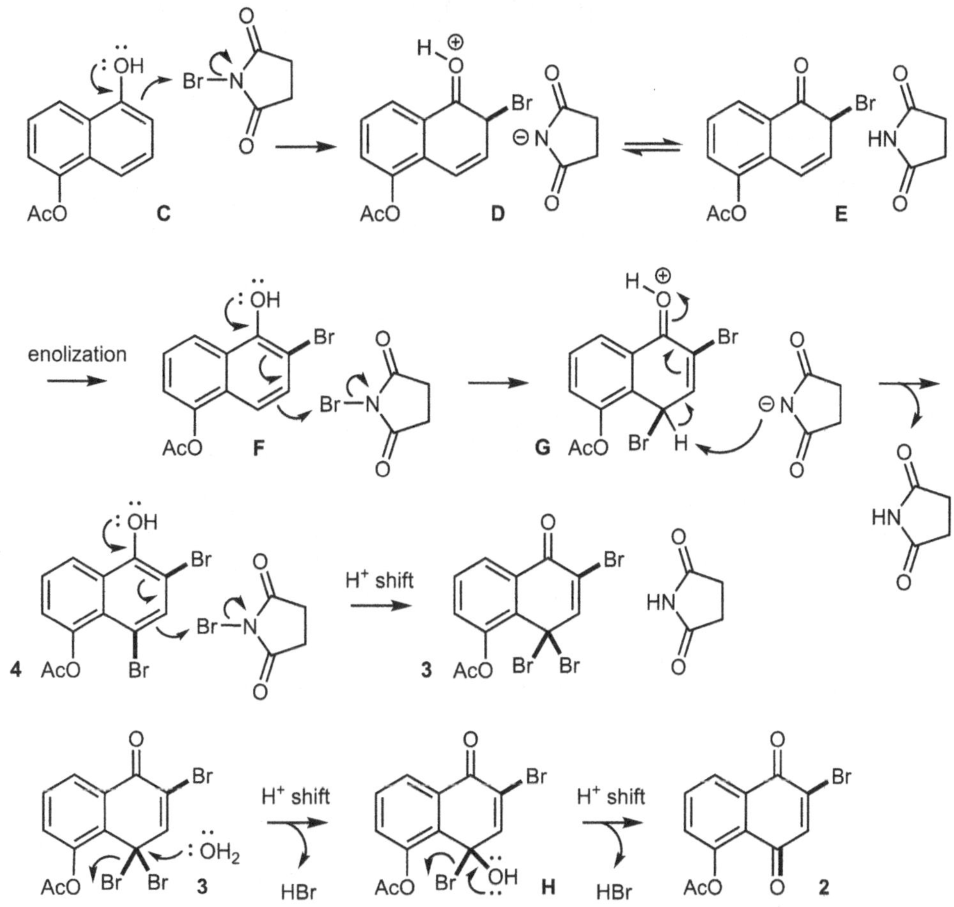

BALANCED CHEMICAL EQUATION: 1 TO 3

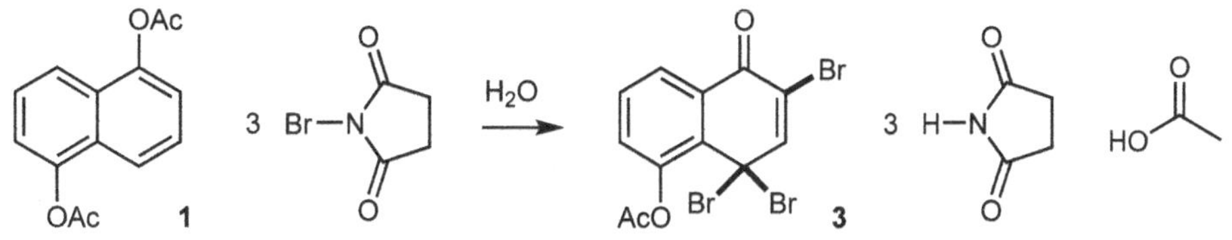

BALANCED CHEMICAL EQUATION: 3 TO 2

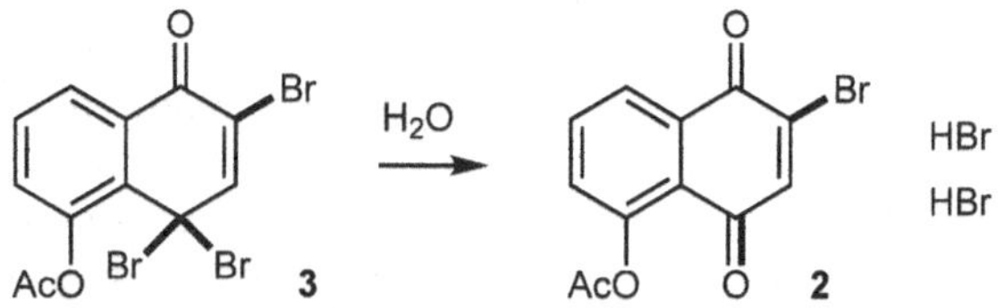

BALANCED CHEMICAL EQUATION: 4 TO 2

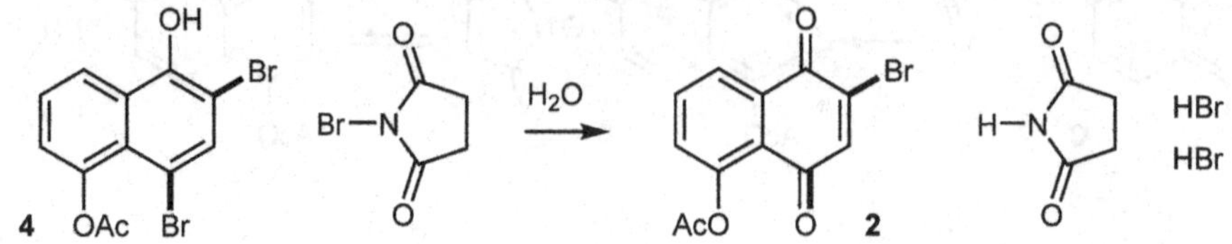

KEY STEPS EXPLAINED

Based on the experimental observations made by the authors, it is reasonable to propose a mechanism for the transformation of **1** to **2** that begins with protonation of the acetyl group of **1** to give **A** followed by hydrolysis of the acyl group via the addition of water to obtain 1-acetoxy-5-hydroxynaphthalene **C** via **B**. From **C** to the tribromo product **3** there are three successive bromination steps (one at the *ortho* position and two at the *para* position relative to the carbonyl group of **C**) occurring by reaction of the substrate with three equivalents of NBS. This sequence also produces three equivalents of pyrrolidine-2,5-dione by-product. From intermediate **3**, water addition and ketonization at the *para* position relative to the carbonyl group of **3** results in the elimination of two equivalents of hydrobromic acid to form the final product **2**.

We note that the successful isolation of product **3** and the experiment that shows that **4** is converted to **2** under the same reaction conditions as the **1** to **2** transformation strongly support the mechanistic sequence: **1** to **4** to **3** to **2**.

It is interesting to note that the authors also suggested a mechanism where the active brominating agent could be bromine and not *N*-bromosuccinimide. Unfortunately they did not provide any details for how such a mechanism would operate. We surmise that such a process could occur via a radical-type route as shown in Scheme 2.

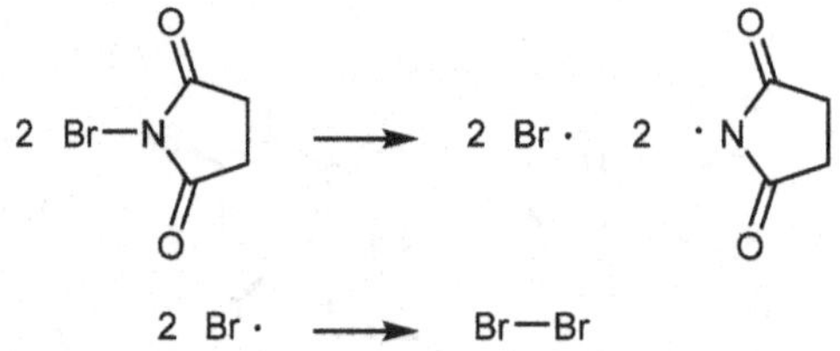

SCHEME 2

We believe such a mechanism would also involve the sequence from **1** to **C** except that two equivalents of bromine would be needed to brominate **C** to form product **4**. A third equivalent of bromine would be needed to brominate **4** to **3**. Since each Br_2 molecule would require two equivalents of NBS to undergo homolytic bond cleavage via the radical route, this alternative mechanism would require in total $(2 \times 3) = 6$ equivalents of NBS to achieve the tribrominated product **3** which was shown to be an intermediate in the conversion of **1** to **2** (Scheme 3). If this radical process were operative it would be expected that the succinimide radical by-products would undergo the following types of termination reactions: abstraction of a hydrogen atom from HBr to give succinimide, abstraction of a bromine atom to reform *N*-bromosuccinimide, or self-dimerization (Scheme 4). If the former two termination processes occur, then in the balanced chemical equation shown in Scheme 2 three of the NBS equivalents can be canceled since they appear on both sides of the equation reducing the balanced equation to what it was before following the ionic mechanism. The only evidence for a radical mechanism would be the detection of the self-dimerization product of succinimide radical, which the authors did not provide.

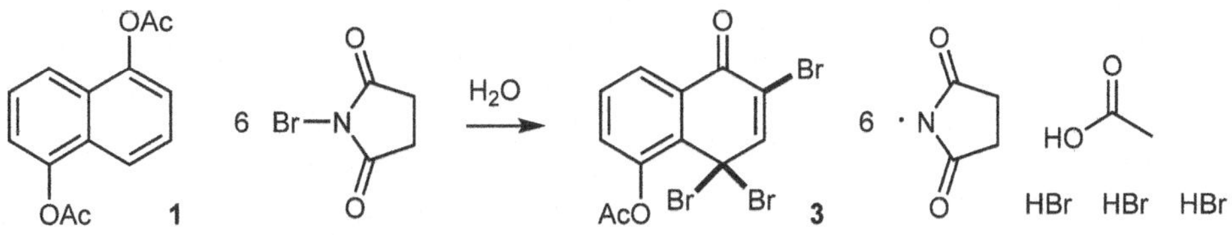

SCHEME 3

Lastly, we consider it puzzling that the authors used one more equivalent of NBS than was necessary according to the balanced chemical equations for the transformations of **1** to **2** and of **4** to **2**. In the former case, three equivalents of NBS were required yet the authors used four equivalents; and in the latter case, one equivalent of NBS was required yet the authors used two equivalents. For a future investigation, we recommend running the experiments with exact stoichiometric amounts of NBS to see whether the results would be the same as expected by this proposed mechanism. We would also recommend undertaking a theoretical analysis to seek further support for this mechanism.

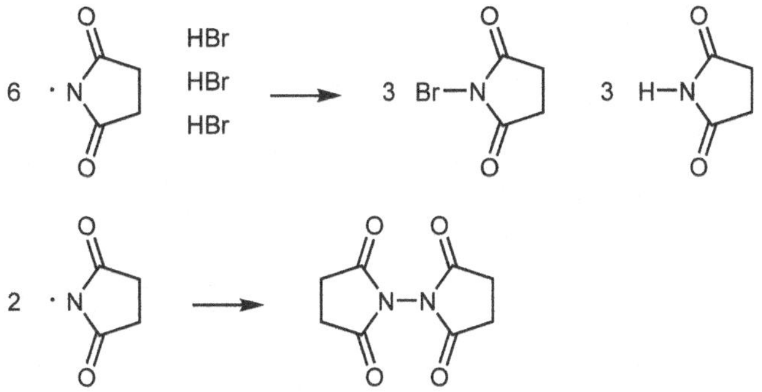

SCHEME 4

ADDITIONAL RESOURCES

We direct the reader to two articles which describe novel catalytic approaches to the synthesis of **2** and which also include a mechanistic discussion.[2,3]

References

1. Jung ME, Hagenah JA. Mechanism of bromination of 1,5-diacetoxynaphthalene. *J Org Chem*. 1983;48(26):5359–5361. https://doi.org/10.1002/anie.198710231.
2. Krohn K, Vitz J. Oxidation of hydroquinones and hydroquinone monomethyl ethers to quinones with *tert*-butyl hydroperoxide and catalytic amounts of ceric ammonium nitrate (CAN). *J Prakt Chem*. 2000;342(8):825–827. https://doi.org/10.1002/1521-3897(200010)342:8<825::AID-PRAC825>3.0.CO;2-3.
3. Lebrasseur N, Fan G-J, Oxoby M, Looney MA, Quideau S. λ^3-Iodane-mediated arenol dearomatization. Synthesis of five-membered ring-containing analogues of the aquayamycin ABC tricyclic unit and novel access to the apoptosis inducer menadione. *Tetrahedron*. 2005;61:1551–1562. https://doi.org/10.1016/j.tet.2004.11.072.

Question 158: Badly Chosen Reaction Conditions?

In 1980 Indian chemists investigating the rearrangement of bicyclic ketol acetates heated the ketol acetate **1** under reflux with 5% methanolic potassium hydroxide in the expectation that hydrolysis would produce the α-hydroxy ketone when in fact the tetrahydronaphthol **2** was generated in 60%–75% yield (Scheme 1).[1] Provide a mechanism for this transformation.

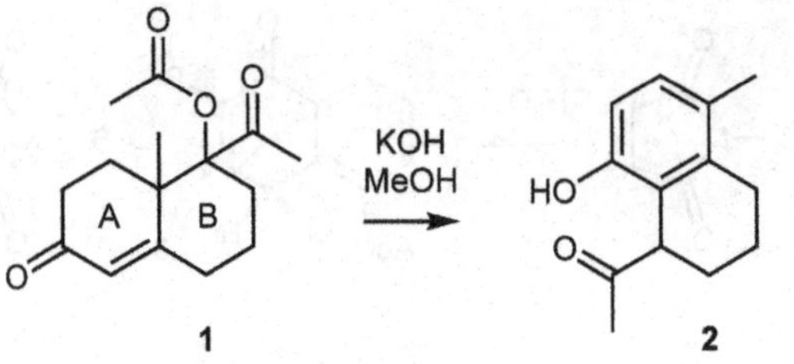

SCHEME 1

SOLUTION

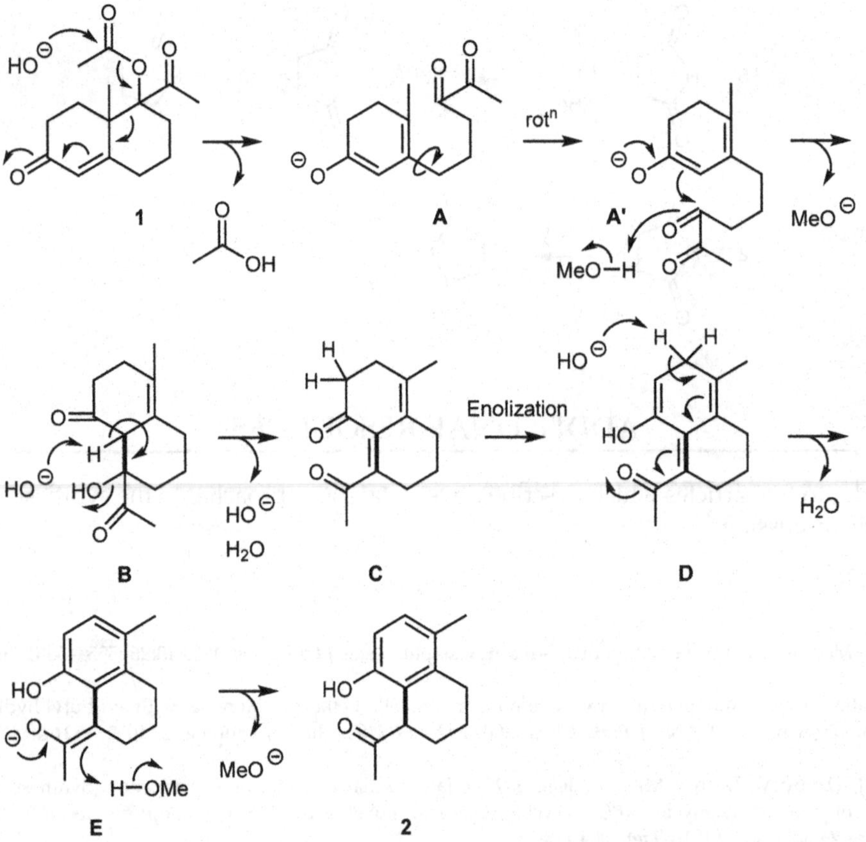

BALANCED CHEMICAL EQUATION

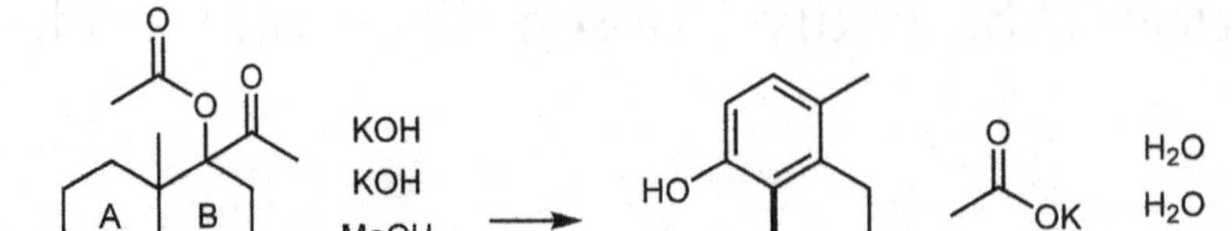

KEY STEPS EXPLAINED

The authors originally intended to carry out the simple base hydrolysis reaction shown in Scheme 2. Nevertheless, the product they isolated was **2**. In order to explain this transformation, the authors postulated a mechanism which starts with the intended cleavage of the acetyl group by hydroxide ion with concomitant cleavage of ring B to give alkoxide **A**. After bond rotation, intermediate **A** recyclizes to form a new six-membered ring as shown in structure **B**. Dehydration of **B** leads to **C** which undergoes enolization to **D** which is followed by aromatization of the former A ring to form **E** which ketonizes to **2**. The authors did not provide further experimental and theoretical evidence to support their postulated mechanism.

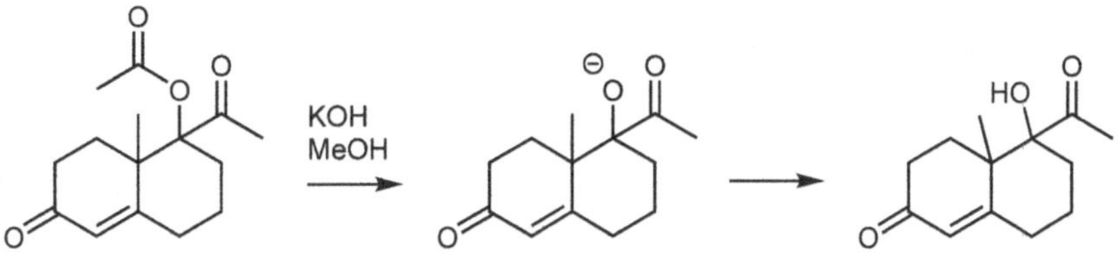

SCHEME 2

ADDITIONAL RESOURCES

For an interesting rearrangement reaction of bicyclic ketal compounds, see Ref. 2.

References

1. Kalyanasundaram M, Rajagopalan K, Swaminathan S. Base catalyzed rearrangement of some bicyclic ketol acetates. *Tetrahedron Lett.* 1980;21:4391–4394. https://doi.org/10.1016/S0040-4039(00)77866-X.
2. Jun J-G, Lee DW. Selective rearrangement reaction of bicyclic ketal compounds in the new reagent system, MgBr$_2$-Ac$_2$O-NaOAc. *Synth Commun.* 2000;30(1):73–77. https://doi.org/10.1080/00397910008087295.

Question 159: One Step Synthesis of a Highly Symmetrical Hexacyclic System From a Simple Naphthol

In 1980 Indian chemists achieved an unexpected synthesis of the symmetrical naphthonaphthopyranopyran **3** by reacting naphthol **1** with α-chloroacrylonitrile **2** under heat in anhydrous dioxane (Scheme 1).[1] Propose a mechanism for this transformation.

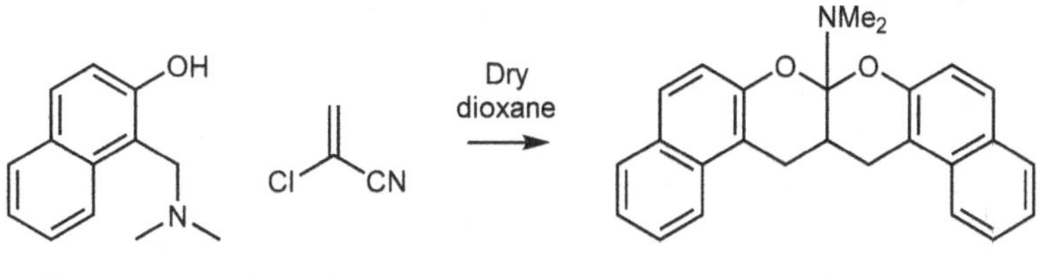

SCHEME 1

SOLUTION

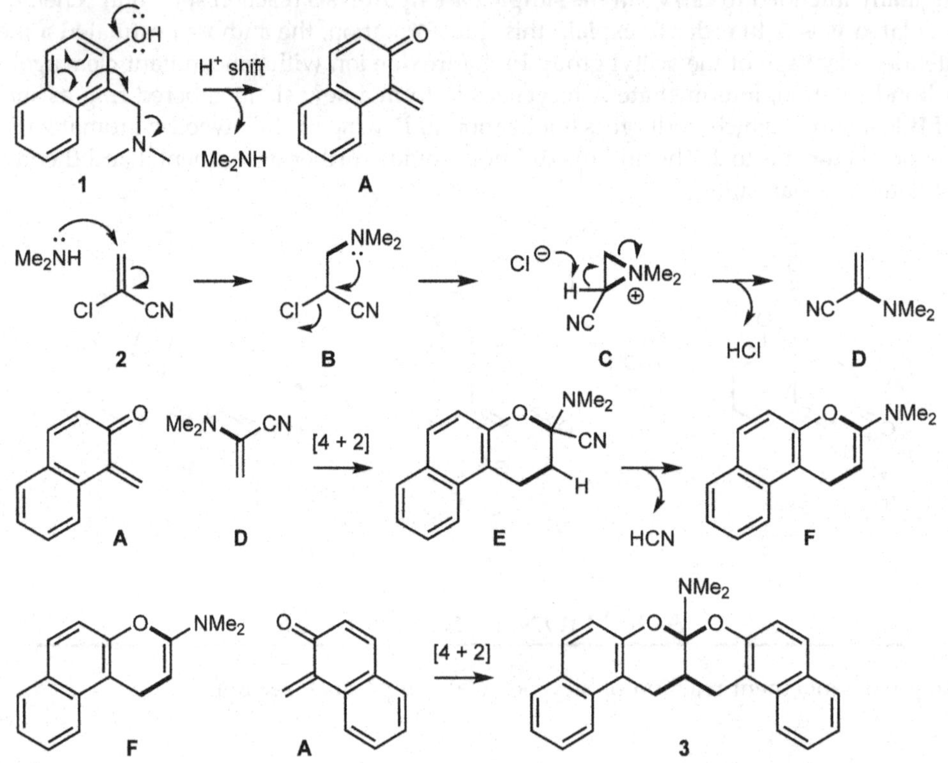

BALANCED CHEMICAL EQUATION

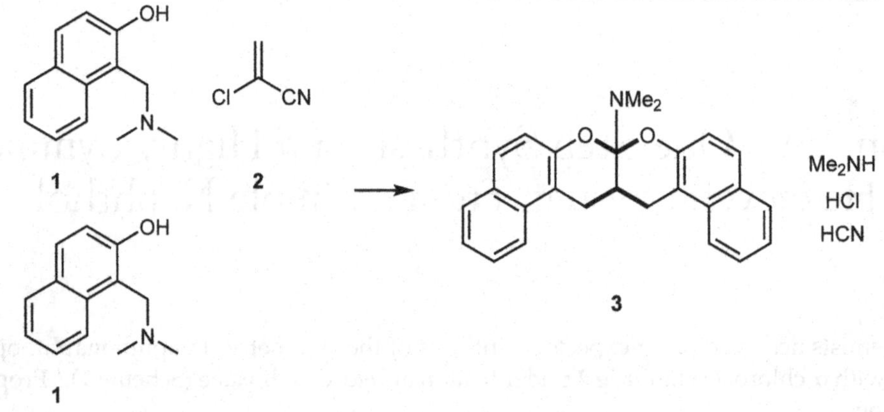

KEY STEPS EXPLAINED

The authors originally intended to carry out the transformations illustrated in Scheme 2. Nevertheless, instead of producing the intended adduct 4, naphthonaphthopyranopyran 3 was produced.

To explain this result, the authors proposed a mechanism that begins with the expected generation of o-quinonemethide intermediate **A** from Mannich base **1** with loss of dimethylamine. Instead of a [4+2] cycloaddition between **A** and 2-chloro-acrylonitrile **2**, the olefin reacts first with the previously liberated dimethylamine in a Michael fashion to give 2-dimethylamino-acrylonitrile **D** via 2-cyano-1,1-dimethyl-aziridinium chloride **C**. Then, **A** reacts with **D** via a hetero-Diels-Alder [4+2] cycloaddition to give **E** which thermally eliminates hydrogen cyanide to generate **F**. **F** finally reacts with another equivalent of o-quinonemethide intermediate **A** via a second hetero-Diels-Alder [4+2] cycloaddition to give the final naphthonaphthopyranopyran 3 product.

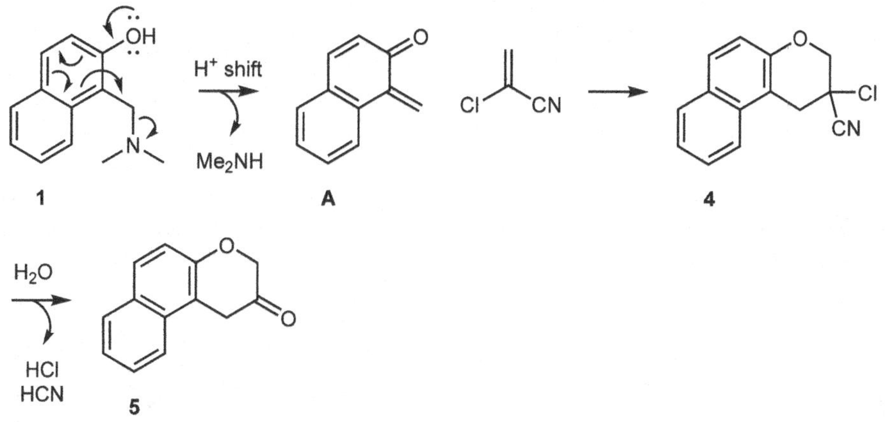

SCHEME 2

The authors provided the following pieces of evidence in support of the earlier mechanism: (a) structural confirmation of the final product **3** by elemental analysis, mass spectrometry, and ^{1}H NMR, ^{13}C NMR, IR, spectroscopic data; (b) in the absence of Mannich bases 2-chloro-acrylonitrile reacts with secondary amines to give 2-amino-acrylonitrile products like intermediate **D** (see Scheme 3); (c) independently prepared 2-amino-acrylonitriles reacted with β-naphthol Mannich bases to give chromanochroman products like **3** (see Scheme 4); and (d) under acidic conditions the reaction of 2-chloro-acrylonitriles and β-naphthol Mannich bases did not yield chromanochroman products suggesting that the Michael addition step is not feasible (i.e., steps **2 → B → C → D** do not occur because under acidic conditions dimethylamine is protonated and is no longer nucleophilic). Nevertheless, the authors did not perform a theoretical analysis for their proposed mechanism.

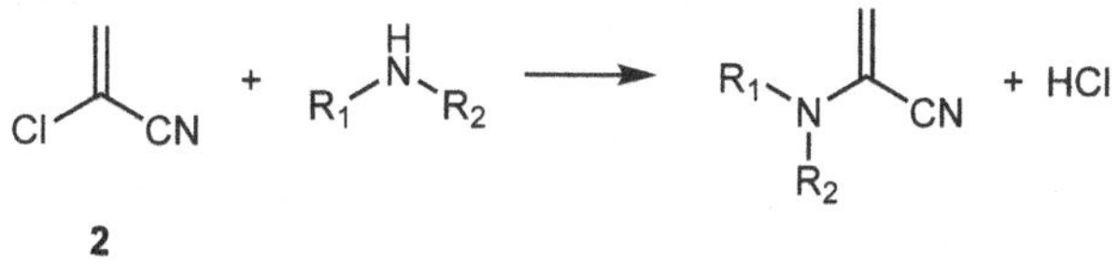

SCHEME 3

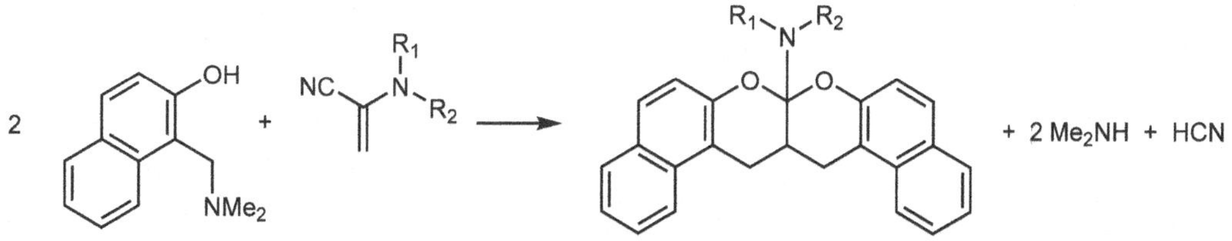

SCHEME 4

ADDITIONAL RESOURCES

We recommend a recent article which discusses an interesting transformation that involves a hetero-Diels-Alder cyclization.[2]

References

1. Balasubramanian KK, Selvaraj S. Novel reaction of *o*-phenolic Mannich bases with α-chloroacrylonitrile. *J Org Chem.* 1980;45(18):3726–3727. https://doi.org/10.1021/jo01306a040.
2. Mudududdla R, Jain SK, Bharate JB, et al. *ortho*-Amidoalkylation of phenols via tandem one-pot approach involving oxazine intermediate. *J Org Chem.* 2012;77:8821–8827. https://doi.org/10.1021/jo3017132.

Question 160: A Method for Aryloxylation of *p*-Cresol

In 1995 American chemists carried out the synthesis outlined in Scheme 1.[1] This synthesis featured the formation of orthoquinone mono(monothioketal) 3 and its desulfurization to form 2-(4-methylphenoxy)-4-methylphenol 4. Suggest mechanisms for these synthetic steps.

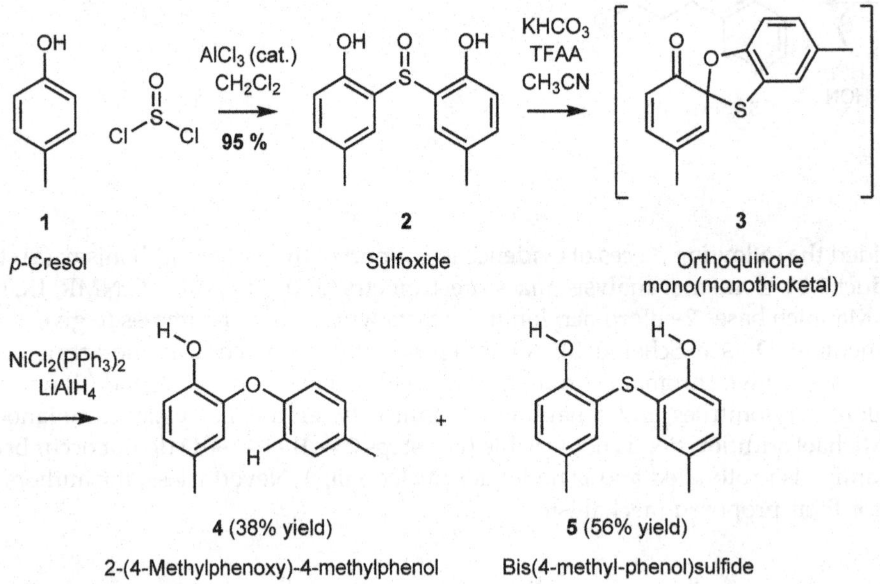

1

p-Cresol

2

Sulfoxide

3

Orthoquinone
mono(monothioketal)

4 (38% yield)

2-(4-Methylphenoxy)-4-methylphenol

5 (56% yield)

Bis(4-methyl-phenol)sulfide

SCHEME 1

SOLUTION: 1 TO 2

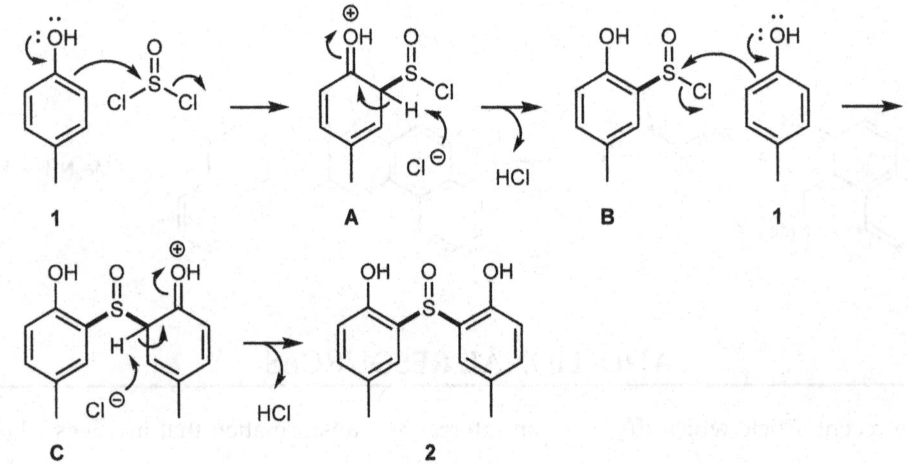

1

A

HCl

B

1

C

HCl

2

BALANCED CHEMICAL EQUATION: 1 TO 2

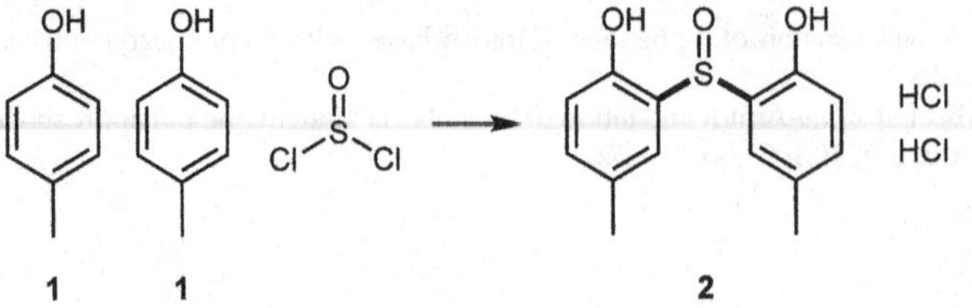

1

1

2

HCl

HCl

SOLUTION: 2 TO 3

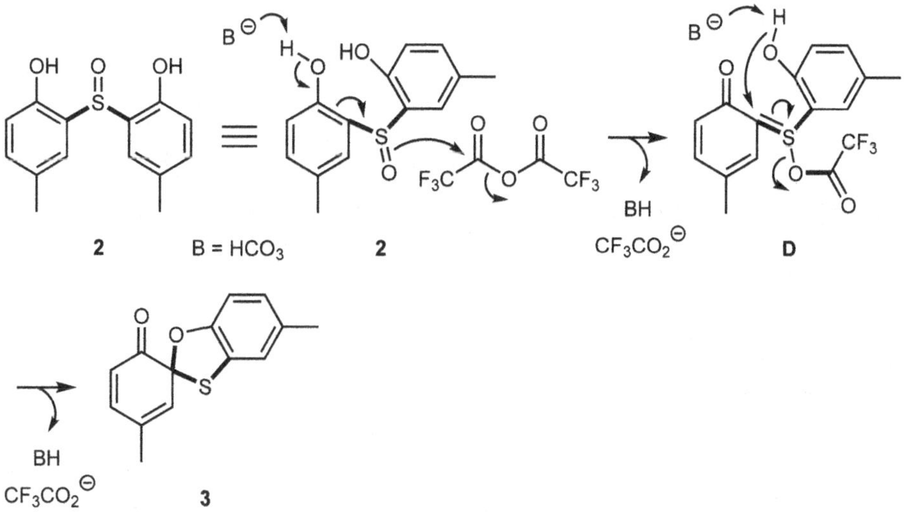

SOLUTION: 2 TO 3 ALTERNATIVE

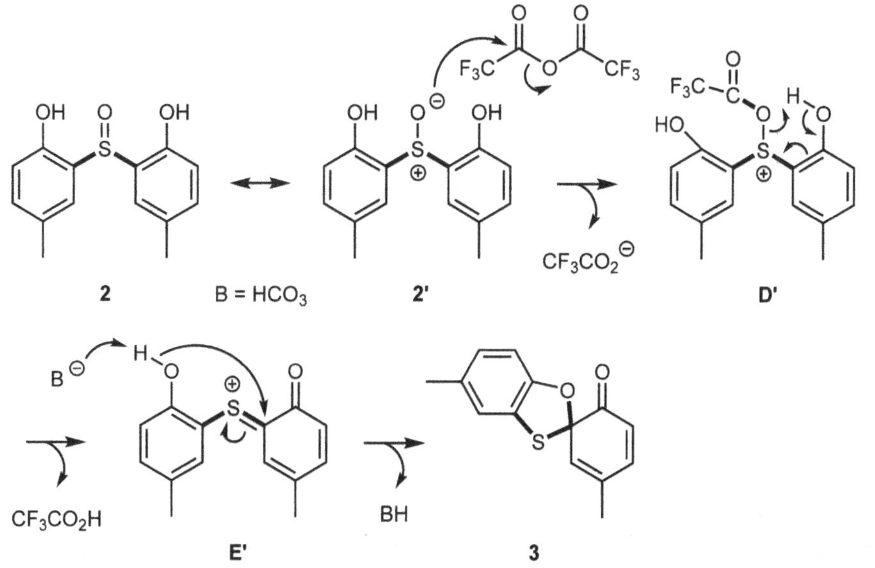

BALANCED CHEMICAL EQUATION: 2 TO 3

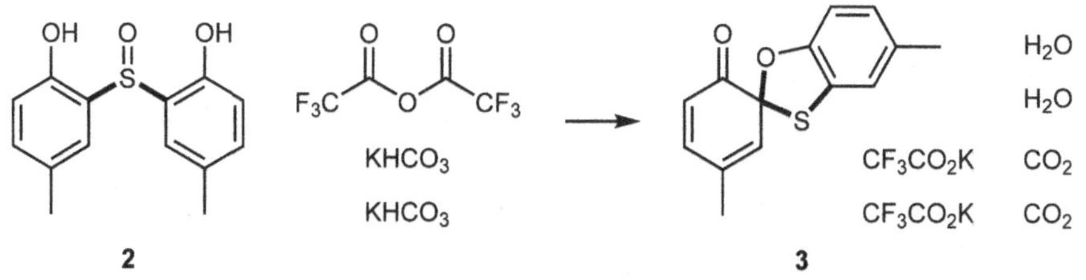

SOLUTION: 3 TO 4

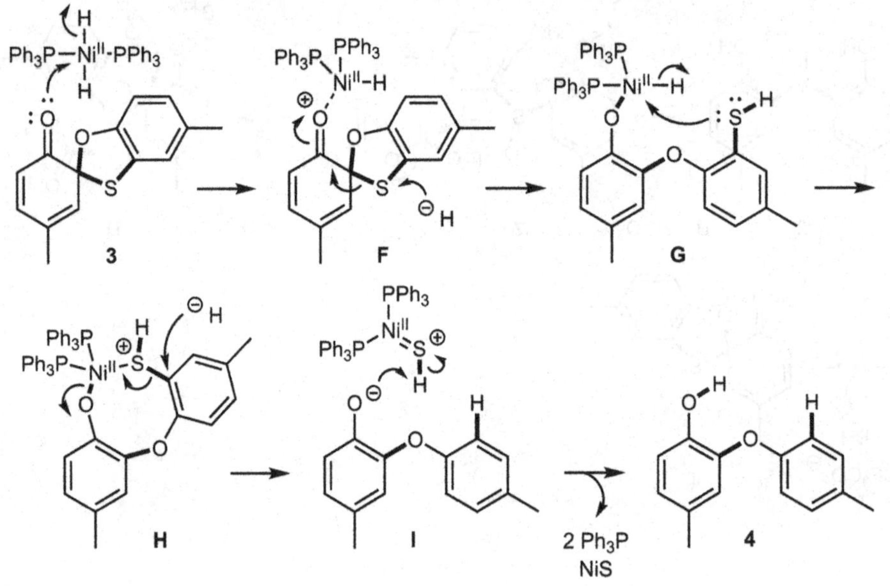

BALANCED CHEMICAL EQUATION: 3 TO 4

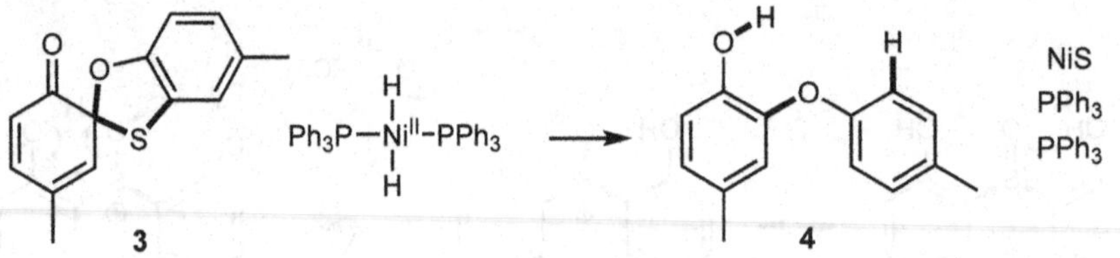

SOLUTION: 3 TO 5

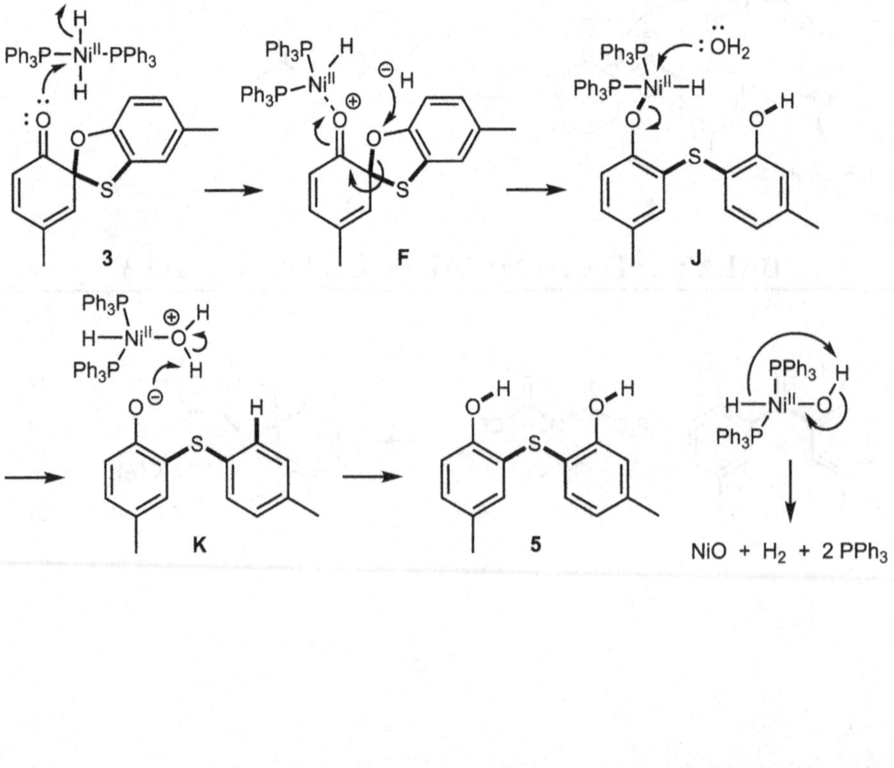

BALANCED CHEMICAL EQUATION: 3 TO 5

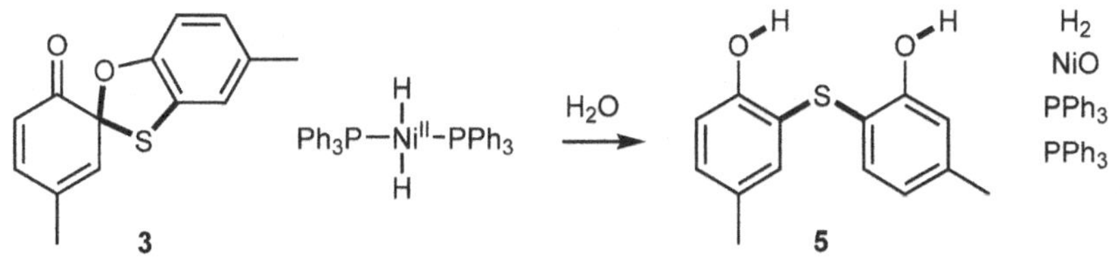

KEY STEPS EXPLAINED

We begin the mechanistic analysis by emphasizing that Jung et al. did not present mechanisms for their transformations. They also did not perform theoretical analysis on any mechanistic features of these transformations. As such, the mechanisms proposed here represent our work in conjunction with certain experimental observations made by the authors.

For the transformation of **1** to **2** which we recognize as a Friedel-Crafts sulfonation reaction, we propose the interaction between thionyl chloride and two equivalents of *p*-cresol with elimination of two equivalents of HCl to form the sulfoxide **2** product.

The second step of this synthesis is a base-induced [4+1] cyclization to the [6,5]-*spiro* orthoquinone mono(monothioketal) **3** which the authors actually observed by ^{1}H NMR spectroscopy in deuterated acetonitrile though they did not isolate this product. Two possible mechanistic variants can be drawn. In the first case, deprotonation by base of a phenolic hydrogen atom results in a cascade that generates attack of the anhydride by the sulfoxide oxygen atom leading to orthoquinone **D**. A second equivalent of base removes the other phenolic hydrogen atom resulting in ring closure at the carbon atom bearing the sulfur atom and liberating trifluoroacetate ion. We note that for this mechanism two equivalents of base are used to deprotonate the two phenolic hydrogen atoms. The alternative mechanism involves attack by the sulfoxide oxygen atom onto the anhydride to obtain sulfonium ion intermediate **D′** which undergoes a Pummerer-type rearrangement to give α-ketosulfonium salt **E′**. One equivalent of base removes the phenolic hydrogen atom causing ring closure to give product **3**. A second equivalent of base neutralizes the trifluoroacetic acid liberated in the rearrangement step. In this mechanism one equivalent of base is used to deprotonate the substrate, compared to two equivalents in the previous case; however, the balanced chemical equation for both mechanistic options is the same. Essentially the difference between the two mechanistic options is the role played by the base and how the two phenolic hydrogen atoms are removed.

The third step leading to ether product **4** is a reductive desulfurization reaction where hydride ion adds to the sulfur atom of intermediate **F** causing the *spiro* ring to open to nickel complexed alkoxide **G**. **G** undergoes a hydride transfer to the aromatic ring bearing the thiol group thereby liberating product **4** along with nickel sulfide and triphenylphosphine as by-products. On the other hand, the reaction leading to sulfide product **5** involves attack of hydride ion on the oxygen atom of intermediate **F** causing the *spiro* ring to open to nickel complexed alkoxide **J** which is decomposed by water during aqueous work-up to give product **5** along with nickel oxide, hydrogen gas, and triphenylphosphine as by-products. The active reducing reagent is prepared by mixing nickel(II)chloride dimethoxyethane (DME) complex, triphenylphosphine, and lithium aluminum hydride in THF solvent to yield the nickel(II)hydride in situ according to Scheme 2. Such a reagent has been used to reductively desulfurize mercaptans, thioethers, sulfoxides, and sulfones though the authors did not suggest any mechanisms for how it works in effecting these transformations.[2]

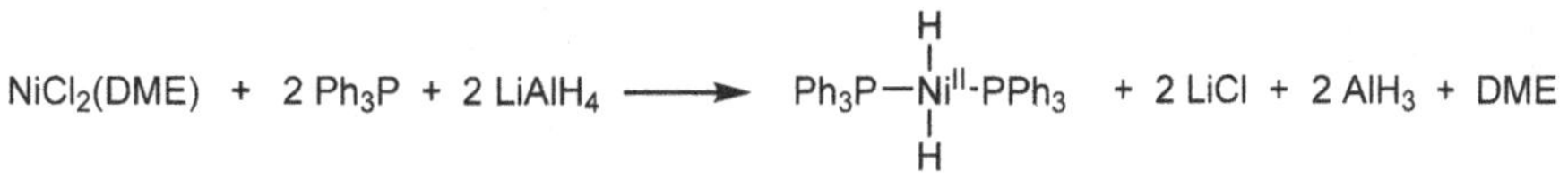

SCHEME 2

ADDITIONAL RESOURCES

We recommend an excellent review article which discusses C—S bond cleavage reactions with the aid of transition metals.[3] For a review of modern applications of the Pummerer rearrangement in synthesis, see Ref. 4. For an article discussing the reduction of inert carbon—sulfur bonds via ligand-free nickel catalysis, see Ref. 5.

References

1. Jung ME, Jachiet D, Khan SI, Kim C. New method for the preparation of *o*-aryloxyphenols: Pummerer-type rearrangement of an *o*-hydroxyaryl sulfoxide. *Tetrahedron Lett*. 1995;36(3):361–364. https://doi.org/10.1016/0040-4039(94)02270-L.
2. Ho KM, Lam CH, Luh TY. Transition metal promoted reactions. Part 27. Nickel(II)-lithium aluminum hydride mediated reduction of carbon-sulfur bonds. *J Org Chem*. 1989;54(18):4474–4476. https://doi.org/10.1021/jo00279a046.
3. Luh TY, Ni ZJ. Transition metal mediated C-S bond cleavage reactions. *Synthesis*. 1990;1990:89–103. https://doi.org/10.1021/ja973118x.
4. Feldman KS. Modern Pummerer-type reactions. *Tetrahedron*. 2006;62:5003–5034. https://doi.org/10.1016/j.tet.2006.03.004.
5. Barbero N, Martin R. Ligand-free Ni-catalyzed reductive cleavage of inert carbon-sulfur bonds. *Org Lett*. 2012;14(3):796–799. https://doi.org/10.1021/ol2033306.

Question 161: Nitriles From N-Chlorosulfonylamides

In 1994 German chemists investigated the conversion of *N*-chlorosulfonylamides **1** to nitriles **2** using either dimethylformamide (DMF) or triethylamine reagents (Scheme 1).[1] Provide mechanistic explanations for this transformation under both sets of reaction conditions.

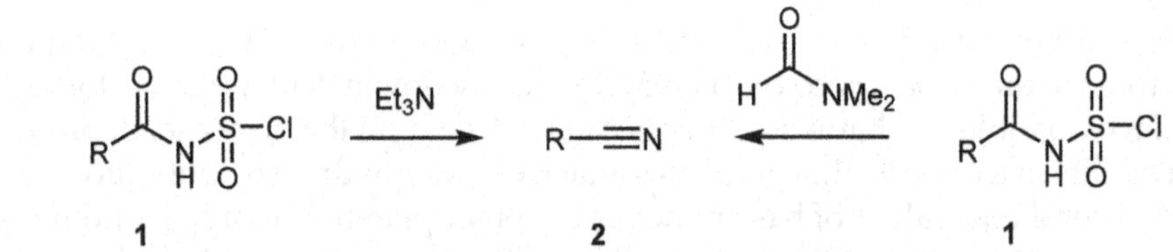

SCHEME 1

SOLUTION: NITRILE FORMATION VIA TRIETHYLAMINE

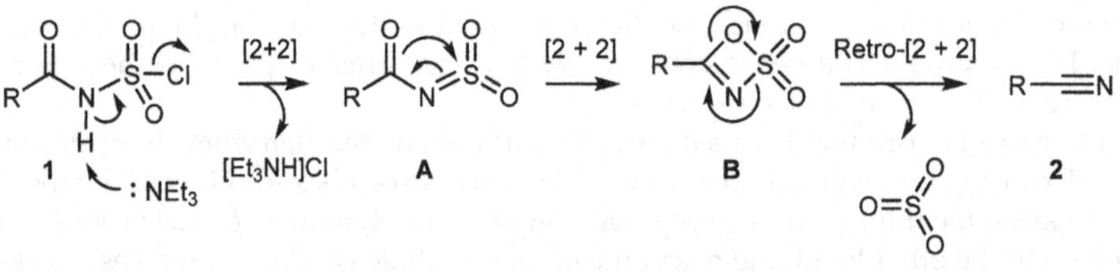

SOLUTION: NITRILE FORMATION VIA DIMETHYLFORMAMIDE (DMF)

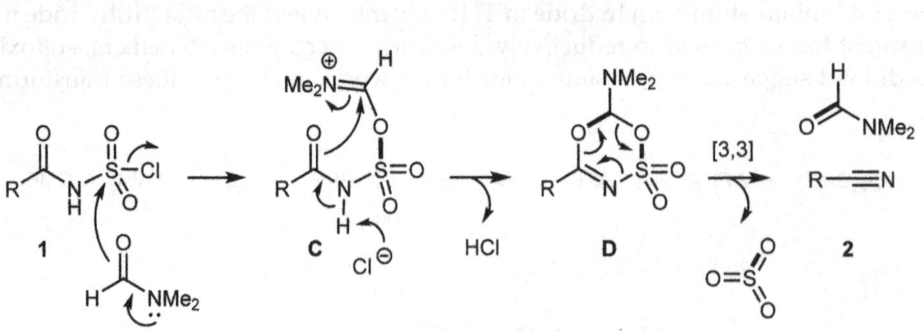

KEY STEPS EXPLAINED

The mechanism proposed by the authors to explain the conversion of **1** to **2** under triethylamine conditions begins with deprotonation of the amide group of **1** with elimination of chloride ion to give the *N*-sulfonylamide intermediate **A**. This intermediate undergoes a [2 + 2] cycloaddition to form the cyclic intermediate **B** which undergoes a retro-[2 + 2] fragmentation to give the final nitrile product **2** and sulfur trioxide by-product. By contrast, dimethylformamide works to add nucleophilically to the sulfonylchloride group of **1** to generate the iminium ion intermediate **C** which cyclizes via deprotonation to the six-membered heterocycle **D** after elimination of HCl. A [3,3] sigmatropic rearrangement of **D** gives nitrile **2**, the sulfur trioxide by-product and regenerated DMF. We note that therefore DMF acts as a catalyst and is potentially recoverable. The authors did not perform any specific experimental or theoretical analysis to help support their postulated intermediates although they did attempt certain experiments to seek confirmation for other literature mechanistic proposals which were not successful. We refer the reader to the original article for more details about these experiments.

ADDITIONAL RESOURCES

For an example of a nitrile synthesis under Swern oxidation conditions, see the work of Nakajima and Ubukata.[2] For a theoretical analysis of very similar rearrangements, see Ref. 3.

References

1. Vorbrueggen H, Krolikiewicz K. The introduction of nitrile-groups into heterocycles and conversion of carboxylic groups into their corresponding nitriles with chlorosulfonylisocyanate and triethylamine. *Tetrahedron*. 1994;50(22):6549–6558. https://doi.org/10.1016/S0040-4020(01)89685-X.
2. Nakajima N, Ubukata M. Preparation of nitriles from primary amides under Swern oxidation conditions. *Tetrahedron Lett*. 1997;38(12):2099–2102. https://doi.org/10.1016/S0040-4039(97)00316-X.
3. Koch R, Finnerty JJ, Wentrup C. Rearrangements and interconversions of heteroatom-substituted isocyanates, isothiocyanates, nitrile oxides, and nitrile sulfides, RX-NCY and RY-CNX. *J Org Chem*. 2011;76:6024–6029. https://doi.org/10.1021/jo200593u.

Question 162: Synthesis of Perylenequinone

In 1994 American chemists demonstrated an elegant synthetic approach to calphostin D, a potent protein kinase-C inhibitor, which made use of the dimerization reaction outlined in Scheme 1.[1] Provide a mechanism for this transformation.

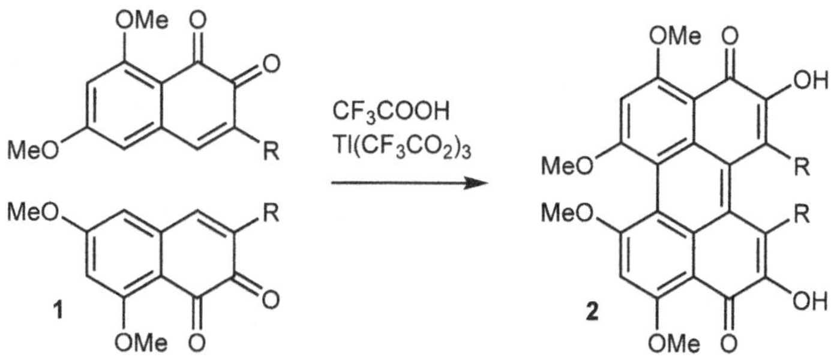

SCHEME 1

SOLUTION

1
A
B
C
D
E
F
G
H
2
Oxidation process
Oxidation process:
Oxidant
3
H—B
B⊖
Tl+
Tl+3
2 H—B
2 B⊖

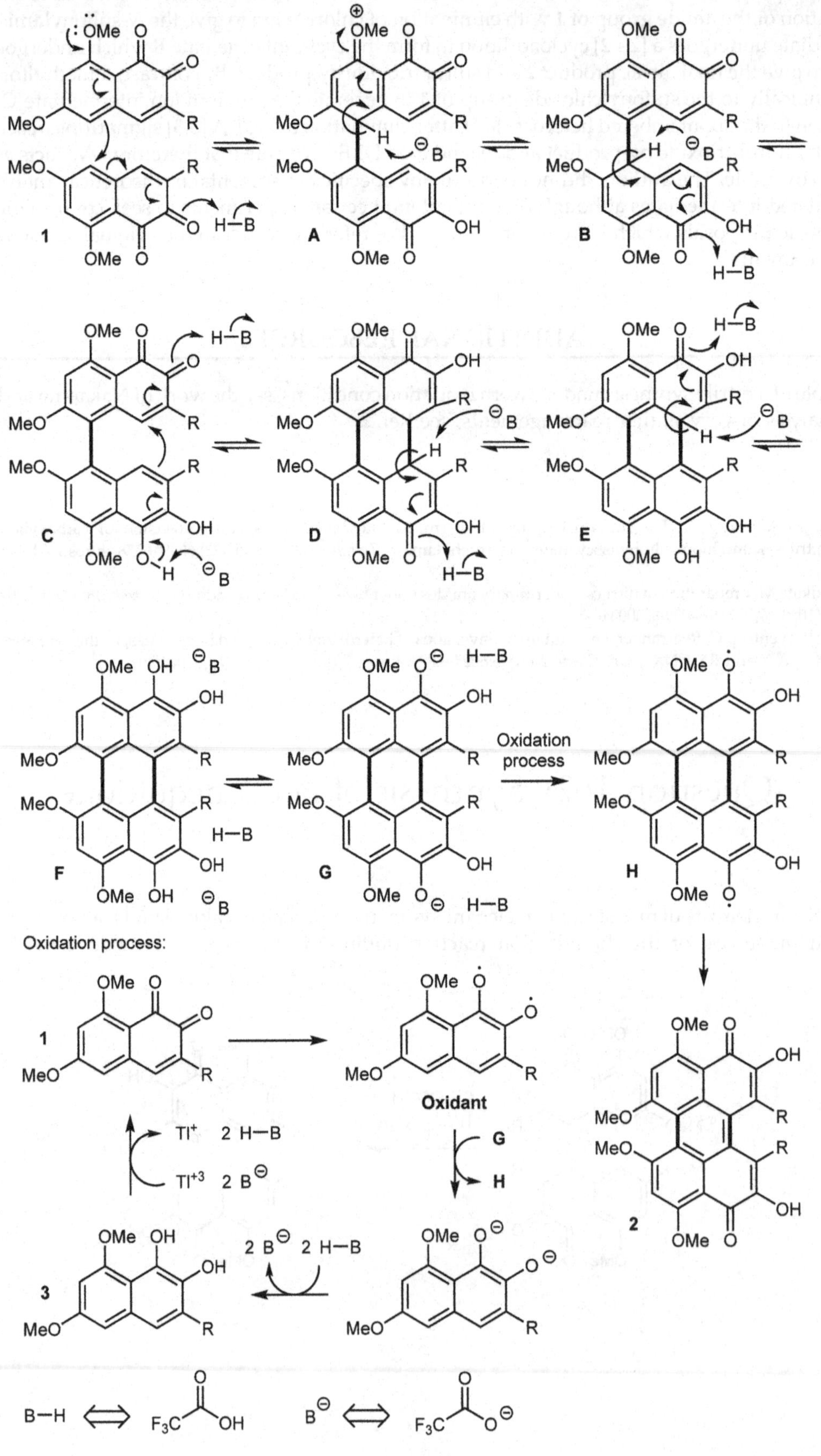

BALANCED CHEMICAL EQUATION

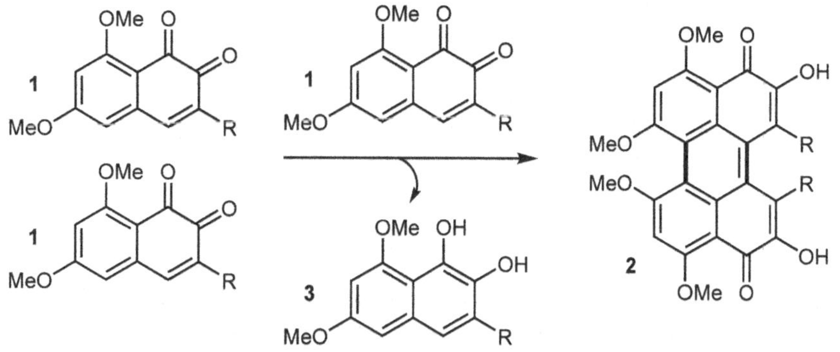

BALANCED CHEMICAL REDOX EQUATION

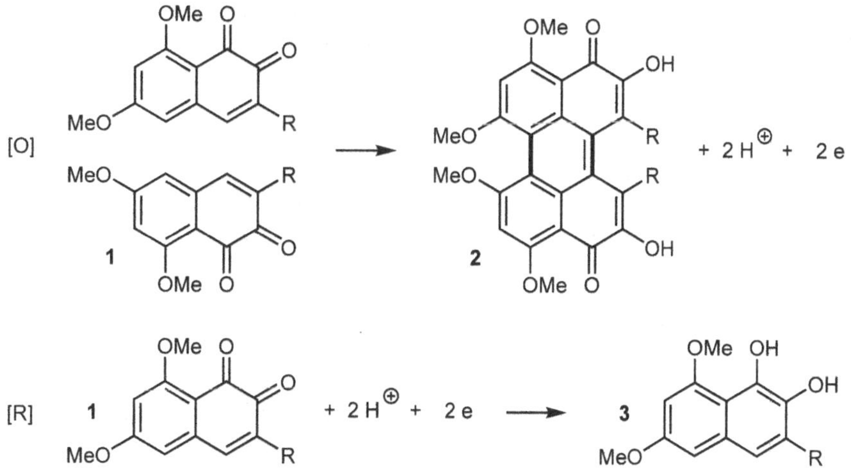

KEY STEPS EXPLAINED

The mechanism proposed by the authors begins with acid-catalyzed dimerization of naphthoquinone **1** with itself to form **A** which is deprotonated to form **B**. This intermediate undergoes several deprotonation-reprotonation steps to establish the fully dimerized structure **G** after regeneration of the acid catalyst in the previous step and the establishment of full aromatization in all phenyl rings. At this point, **G** undergoes an oxidation process where naphthoquinone **1** acts as the oxidant via its diradical structure to facilitate the formation of diradical **H** which rearranges to the final product **2**. This redox process is represented in the balanced redox chemical equation given earlier. Furthermore, the presence of an oxidant such as Tl(CF$_3$CO$_2$)$_3$, according to this mechanism, is required to regenerate the naphthoquinone oxidant **1** from its reduced form **3**.

In terms of experimental evidence, the authors were first alerted to the possibility that the dimerization of **1** to **2** was acid catalyzed when they ran the experiment with only trifluoroacetic acid (TFA) in the absence of any oxidant such as iron-containing reagents. Contrary to their original belief that the mechanism involved radical cation intermediates, the observation that under TFA a "1:1 ratio of the perylenequinone **2** and the hydronaphthoquinone **3** is obtained" suggested that naphthoquinone **1** acted as the oxidant to form the reduced structure **3**. To test this hypothesis, which theoretically made sense because *o*-naphthoquinones are known to be electron poor, the authors ran a second experiment where the oxidant thallium(III)triflate was added slowly to the reaction mixture. The result was a 91% yield of

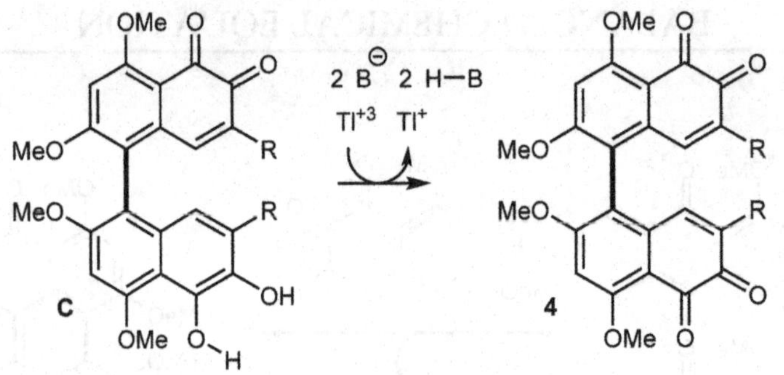

SCHEME 2

dimer **2** which the authors rationalized in terms of the thallium reagent working to oxidize the reduced structure **3** back to naphthoquinone **1** which would either react in a dimerization process or in an oxidation of **G** process, both of which would contribute to a higher yield of **2**. In an interesting twist, the authors ran a third experiment where naphthoquinone **1** was added slowly to a solution of thallium(III)triflate in THF to test what would happen under conditions of excess oxidant. In this case the authors managed to isolate product **4** without **2** (Scheme 2). We note that **4** is effectively a product of the oxidation of **C** by the thallium reagent. This showed for the authors that oxidation of **C** was much more rapid than the second acid-catalyzed cyclization sequence. It also showed that the second intramolecular cyclization was also not an oxidative process. For future study, we would highly recommend carrying out a theoretical analysis to help shed more light on the reasons why this transformation prefers the acid-catalyzed route over the radical cationic intermediates route. For reference, we also include the balanced chemical redox equation for the third experiment the authors performed (Scheme 3).

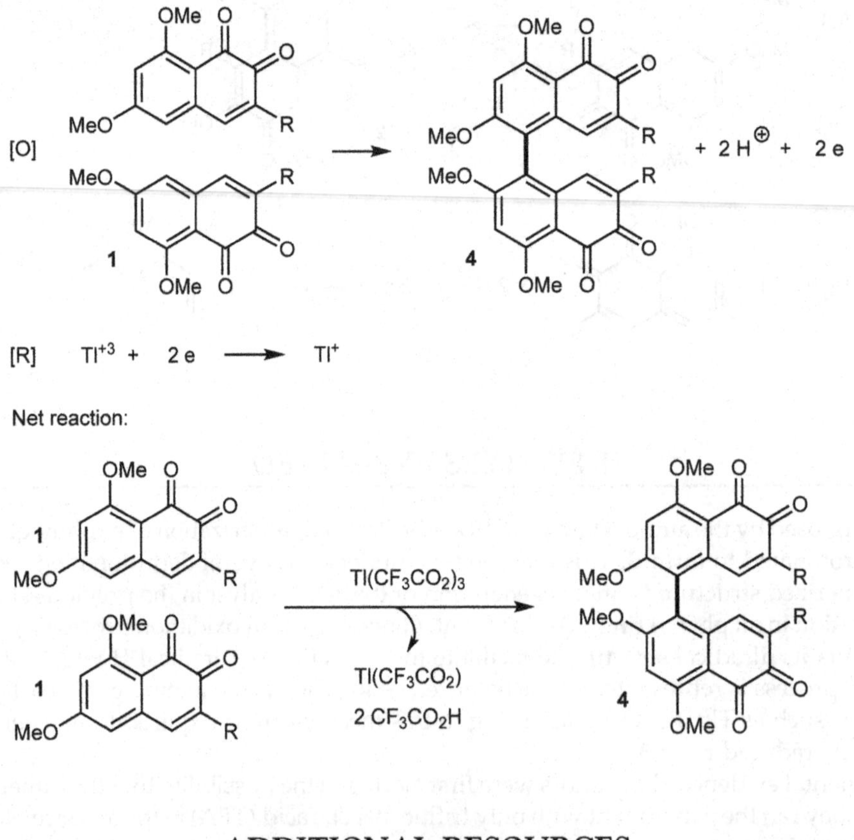

SCHEME 3

ADDITIONAL RESOURCES

For an article discussing a synthetic approach to calphostin A, see Ref. 2. For a review article illustrating methods to synthesize dimeric structures similar to product **2**, see Ref. 3.

References

1. Hauser FM, Sengupta D, Corlett SA. Optically active total synthesis of calphostin D. *J Org Chem*. 1994;59:1967–1969. https://doi.org/10.1002/anie.198710231.

2. Coleman RS, Grant EB. Synthesis of helically chiral molecules: stereoselective total synthesis of the perylenequinones phleichrome and calphostin A. *J Am Chem Soc*. 1995;117:10889–10904. https://doi.org/10.1021/ja00149a012.

3. Kozlowski MC, Morgan BJ, Linton EC. Total synthesis of chiral biaryl natural products by asymmetric biaryl coupling. *Chem Soc Rev*. 2009;38:3193–3207. https://doi.org/10.1039/b821092f.

Question 163: A Route to 2-Vinylindoles

In 1994 Indian chemists studying the synthesis of 2-vinylindoles reacted *N*-(allenylmethyl)-*N*-methylaniline **1** with aqueous magnesium monoperoxyphthalate (MMPP) in methanol/water to form the 2-ethenyl-*N*-methylindole product **2** in 80% yield (Scheme 1).[1] Suggest a mechanism for this transformation.

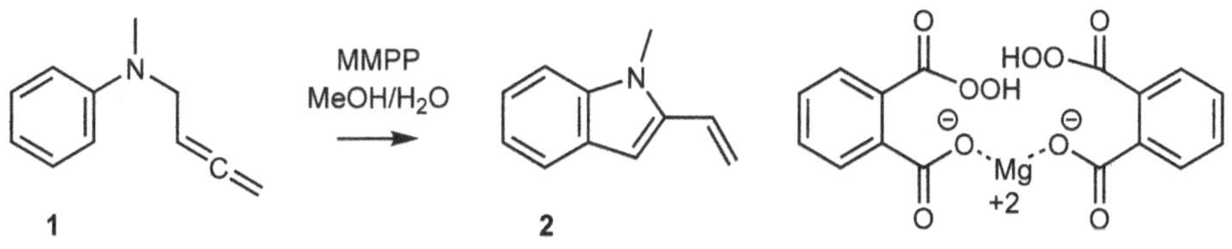

MMPP = magnesium **monoperoxyphthalate**

SCHEME 1

SOLUTION

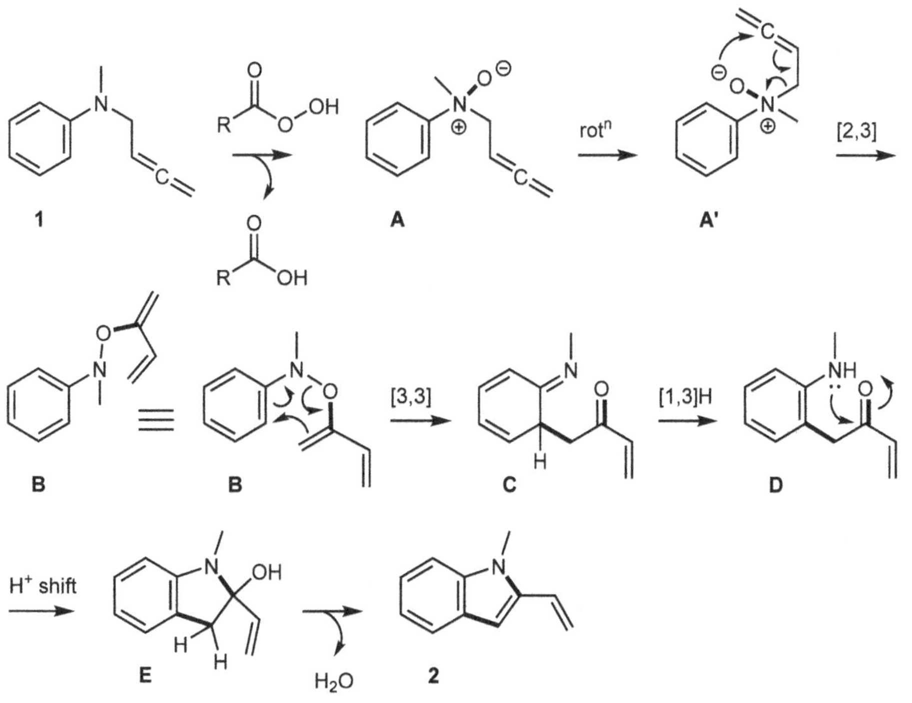

KEY STEPS EXPLAINED

The mechanism proposed by the authors for the transformation of the N-(allenylmethyl)-N-methylaniline **1** to 2-ethenylindole **2** begins with peroxyacid (via MMPP) mediated oxidation of the amino nitrogen of **1** to an N-oxide group in **A**. Following bond rotation, **A** undergoes a [2,3]-sigmatropic rearrangement to generate **B** which in turn undergoes a hetero-Cope [3,3]-sigmatropic rearrangement to give **C** which after a [1,3]-H shift gives the rearomatized aniline intermediate **D**. Cyclization between the amine and carbonyl groups in **D** forms the carbinol intermediate **E** which undergoes dehydration to generate the final 2-vinylindole product **2**. In their work, the authors were unable to isolate any of the proposed intermediates and therefore relied on literature precedent to support their postulated mechanism. For future study, we recommend chemists undertake a theoretical analysis of the mechanistic steps proposed for this transformation.

ADDITIONAL RESOURCES

We recommend three articles which highlight both novel synthetic approaches to 2-vinylindoles as well as their synthetic utility in other reactions.[2–4] We also include a review article that describes recent developments in indole ring synthesis.[5]

References

1. Balasubramanian T, Balasubramanian KK. Tandem transformations of N-alkyl-N-allenylmethylanilines to N-alkyl-2-ethenylindoles. *J Chem Soc Chem Commun.* 1994;1994:1237–1238. https://doi.org/10.1039/C39940001237.
2. Arslan T, Sadak AE, Saracoglu. Synthesis of a new series of 2-vinylindoles and their cycloaddition reactivity. *Tetrahedron.* 2010;66:2936–2939. https://doi.org/10.1016/j.tet.2010.02.080.
3. Mazgarova GG, Absalyamova AM, Gataullin RR. Reactions of N- and C-alkenylanilines: X.* Synthesis of 2-vinyldihydroindoles from 4-methyl-2-(pent-3-en-2-yl)aniline. *Russ J Org Chem.* 2012;48(9):1200–1209. https://doi.org/10.1134/S1070428012090096.
4. Cao Y-J, Cheng H-G, Lu L-Q, et al. Organocatalytic multiple cascade reactions: a new strategy for the construction of enantioenriched tetrahydrocarbazoles. *Adv Synth Catal.* 2011;353:617–623. https://doi.org/10.1002/adsc.201000610.
5. Gribble GW. Recent developments in indole ring synthesis—methodology and applications. *J Chem Soc Perkin Trans 1.* 2000;2000:1045–1075. https://doi.org/10.1039/A909834H.

Question 164: Brain Cancer: The Mechanism of Action of Temozolomide

In 1993 British chemists investigated the decomposition of the brain cancer drug temozolomide **1** and found that this drug works by generating methyl cations which work to methylate DNA and prevent the replication of cancerous cells (Scheme 1).[1] Provide a mechanism to show how methyl cations can be generated from temozolomide **1**.

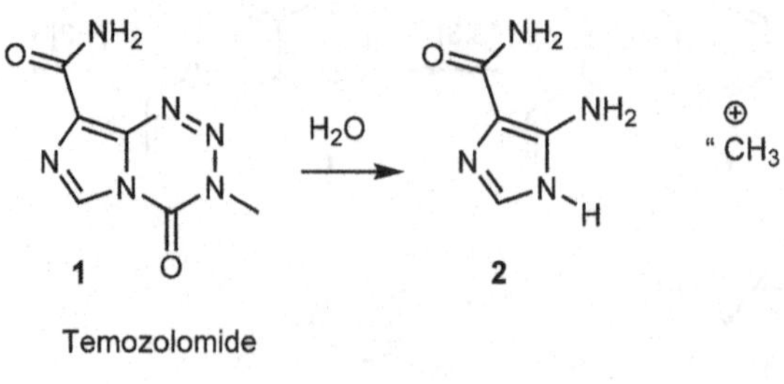

SCHEME 1

SOLUTION

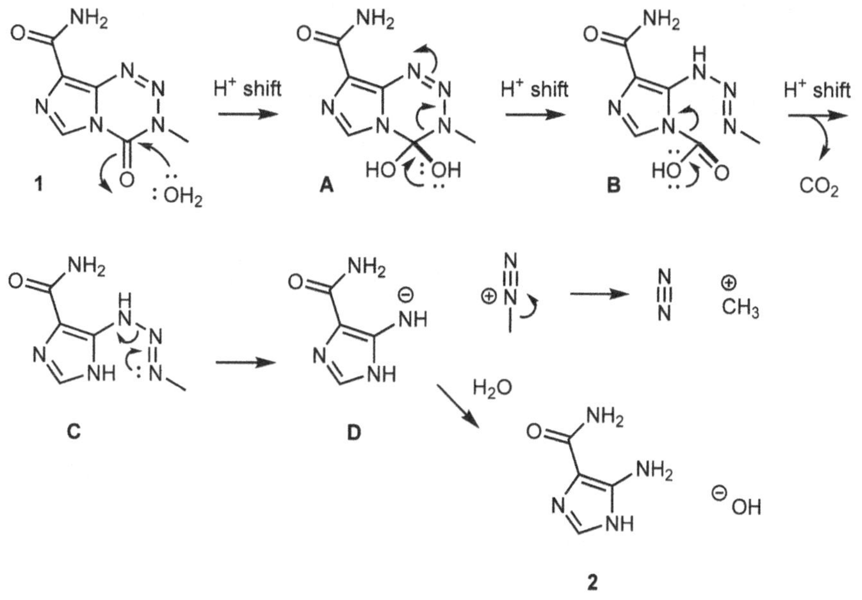

BALANCED CHEMICAL EQUATION

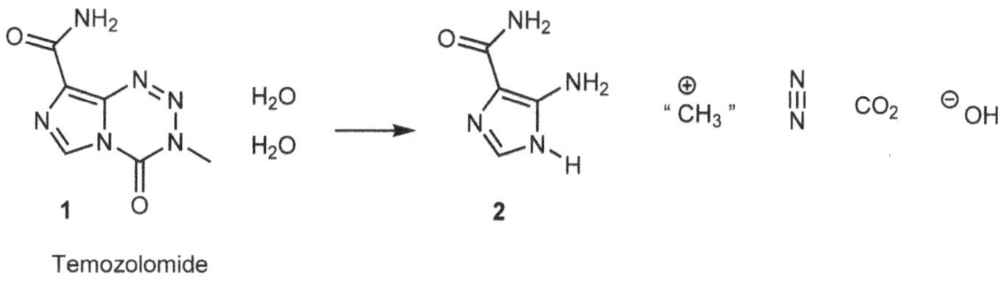

Temozolomide

KEY STEPS EXPLAINED

The mechanism proposed by the authors begins with water addition to temozolomide **1** to form hydrate **A** which undergoes ring opening to **B** followed by decarboxylation to **C**. Further fragmentation of **C** leads to the amide anion **D** and methyl diazonium ion. While reaction of **D** with water forms **2** and hydroxide anion, methyl diazonium ion fragments to nitrogen gas and methyl cation. According to the authors, in a biological environment that is weakly basic due to the presence of bases such as guanidine, the methyl group on methyl diazonium ion may be attacked by such a nucleophile in order to cause methylation of, for instance, DNA molecules thus preventing cancerous cells from replicating further. Furthermore, using ^{1}H NMR, the authors detected the gradual depletion of **1** and emergence of **2** along with several simple methyl-nucleophile species during the course of the reaction. Intermediates such as **C** were not detected, but deuterium labeling experiments demonstrated H-D exchange only after the generation of methyl diazonium ion thus indicating that proton shifts occurring earlier in the proposed mechanism are likely intramolecular processes. The authors did not carry out a theoretical analysis to help shed more light on this interesting decomposition mechanism.

ADDITIONAL RESOURCES

For further details of temozolomide **1**, we highly recommend a subsequent review article by the same authors.[2] We also recommend an article which describes similar decompositions of simpler molecules which also generate methyl-diazonium ion.[3]

References

1. Wheelhouse RT, Stevens MFG. Decomposition of the antitumour drug temozolomide in deuteriated phosphate buffer: methyl group transfer is accompanied by deuterium exchange. *J Chem Soc Chem Commun*. 1993;1993:1177–1178. https://doi.org/10.1039/C39930001177.
2. Newlands ES, Stevens MFG, Wedge SR, Wheelhouse RT, Brock C. Temozolomide: a review of its discovery, chemical properties, pre-clinical development and clinical trials. *Cancer Treat Rev*. 1997;23(1):35–61. https://doi.org/10.1016/S0305-7372(97)90019-0.
3. Golding BT, Bleasdale C, McGinnis J, et al. The mechanism of decomposition of N-methyl-N-nitrosourea (MNU) in water and a study of its reactions with 2'-deoxyguanosine, 2'-deoxyguanosine 5'-monophosphate and d(GTGCAC). *Tetrahedron*. 1997;53(11):4063–4082. https://doi.org/10.1016/S0040-4020(97)00018-5.

Question 165: Attempted Nef-Type Reaction Leads to 3-Arylpyridine Synthesis

In 1994 during an attempted Nef-like transformation, chemists from Taiwan treated the nitrobicyclo[2.2.1]hept-2-enes **1** with SnCl$_2$•2H$_2$O in hot THF/dioxane to form, unexpectedly, the 3-arylpyridines **2** in moderate yield (Scheme 1).[1] Explain these results in mechanistic terms.

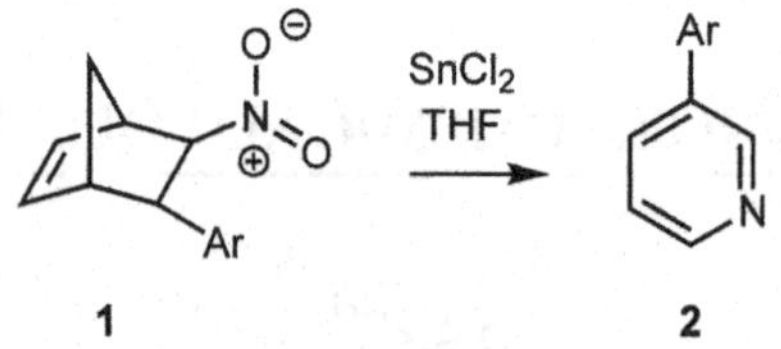

SCHEME 1

SOLUTION: PART 1—REDUCTION

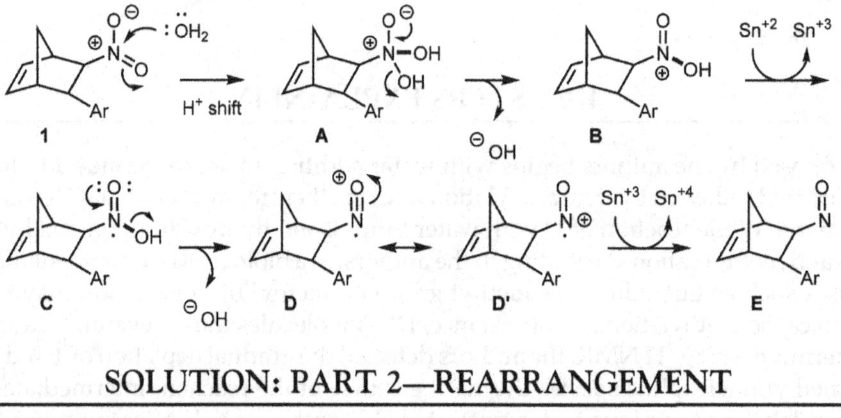

SOLUTION: PART 2—REARRANGEMENT

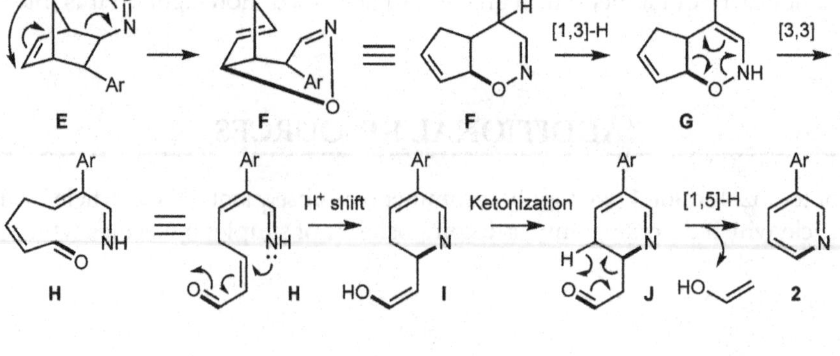

BALANCED CHEMICAL EQUATION

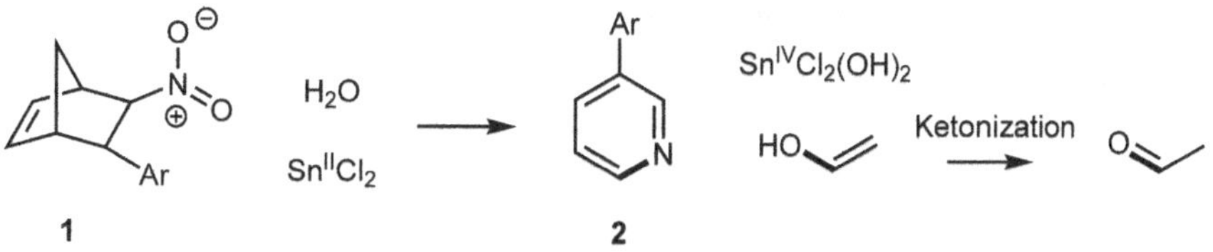

1 **2**

KEY STEPS EXPLAINED

The mechanism proposed by the authors does not involve a proper explanation for the first part of this transformation which concerns the tin-mediated reduction of the nitro group in **1** to the nitroso group in **E**. We have thus proposed the sequence of steps illustrated before which begin with water attack on the nitro group of **1** to give **A** after a hydrogen shift followed by elimination of hydroxide to form nitronium ion **B**. One electron transfer from the tin reagent to the nitrogen atom of **B** forms **C** which upon elimination of hydroxide ion leads to **D**. A final electron transfer from the tin reagent to the nitrogen atom of **D'** leads to the formation of intermediate **E** containing the nitroso group. At this point, intermediate **E** unexpectedly undergoes rearrangement of its [2.2.1] structural framework to a [4.3.0] framework to form the 5,6-dihydro-4H-[1,2]oxazine intermediate **F**. This occurs via an *endo* attack of the nitroso oxygen atom onto the olefinic bond. Once formed, intermediate **F** undergoes a [1,3]-H shift to generate the 5,6-dihydro-2H-[1,2]oxazine intermediate **G** which after a [3,3] hetero-Cope sigmatropic rearrangement forms the acyclic intermediate **H**. This intermediate cyclizes in a Michael fashion to generate **I** which ketonizes to (2,3-dihydro-pyridin-2-yl)-acetaldehyde **J**. A final [1,5]-H ene reaction facilitates the formation of the final pyridine product **2** with elimination of ethenol which easily ketonizes to acetaldehyde by-product. We note that the original intention of the authors was to carry out the Nef reaction shown in Scheme 2.

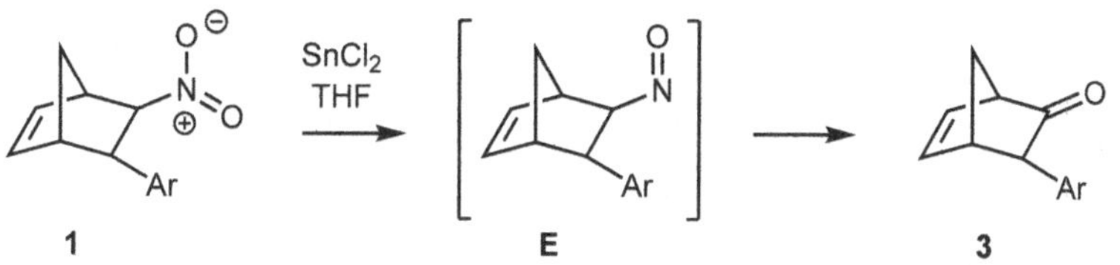

1 **E** **3**

SCHEME 2

It is therefore important to note that during this transformation it was the rearrangement sequence that produced an unexpected result. Although they did not provide further experimental and theoretical evidence to support their postulated mechanism, the authors did note that their transformation provided more efficient access to substituted pyridines as compared to the traditional harsh conditions of hetero-Diels-Alder reaction between cyclopentadiene and unsaturated nitroso compound followed by pyrolysis at high temperature (Scheme 3).[2]

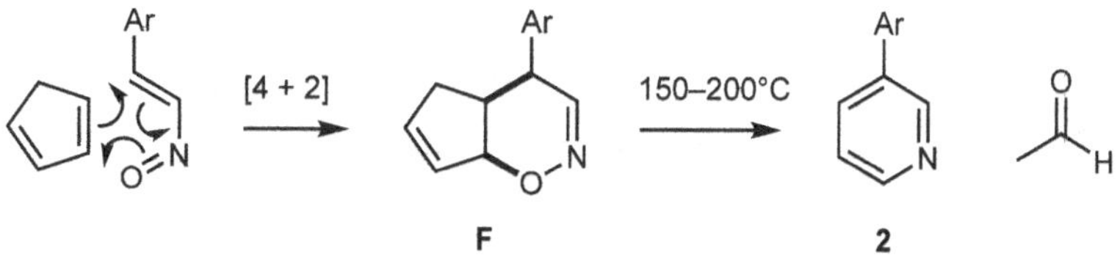

F **2**

SCHEME 3

ADDITIONAL RESOURCES

We recommend an example of an interesting transformation involving a similar nitroso group-mediated rearrangement as seen for the step **E** to **F** illustrated before.[3]

References

1. Ho T-L, Liao P-Y. Reductive rearrangement of 5-nitrobicyclo[2.2.1]hept3-enes. Formation of 3-arylpyridines. *Tetrahedron Lett.* 1994;35 (14):2211–2212. https://doi.org/10.1016/S0040-4039(00)76799-2.
2. Faragher R, Gilchrist TL. Cycloaddition reactions of nitrosoalkenes and azoalkenes with cyclopentadiene and other dienes. *J Chem Soc Perkin Trans 1.* 1979;1979:249–257. https://doi.org/10.1039/P19790000249.
3. Baranovsky AV, Bolibrukh DA, Bull JR. Synthesis of 3-methoxy-16α-nitro-14,17-ethenoestra-1,3,5(10)-trien-17β-yl acetate and fragmentation-mediated pathways to 14β,15β-fused N-heterocycles and 14β-functionalised alkyl derivatives. *Eur J Org Chem.* 2007;2007:445–454. https://doi.org/10.1002/ejoc.200600629.

Question 166: The ArCOMe → ArC≡CH Transformation

An effective procedure for the transformation of ArCOMe to ArC≡CH involves treatment of acetophenone **1** with two equivalents of POCl₃ in DMF to form intermediate **2** and to react this compound with a hot solution of sodium hydroxide in aqueous dioxane to form the corresponding ethynylarene (Scheme 1).[1] Suggest a mechanism for this transformation.

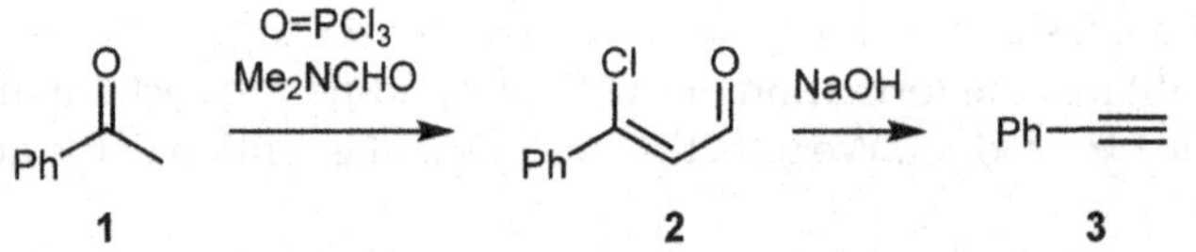

SCHEME 1

SOLUTION: 1 TO 2

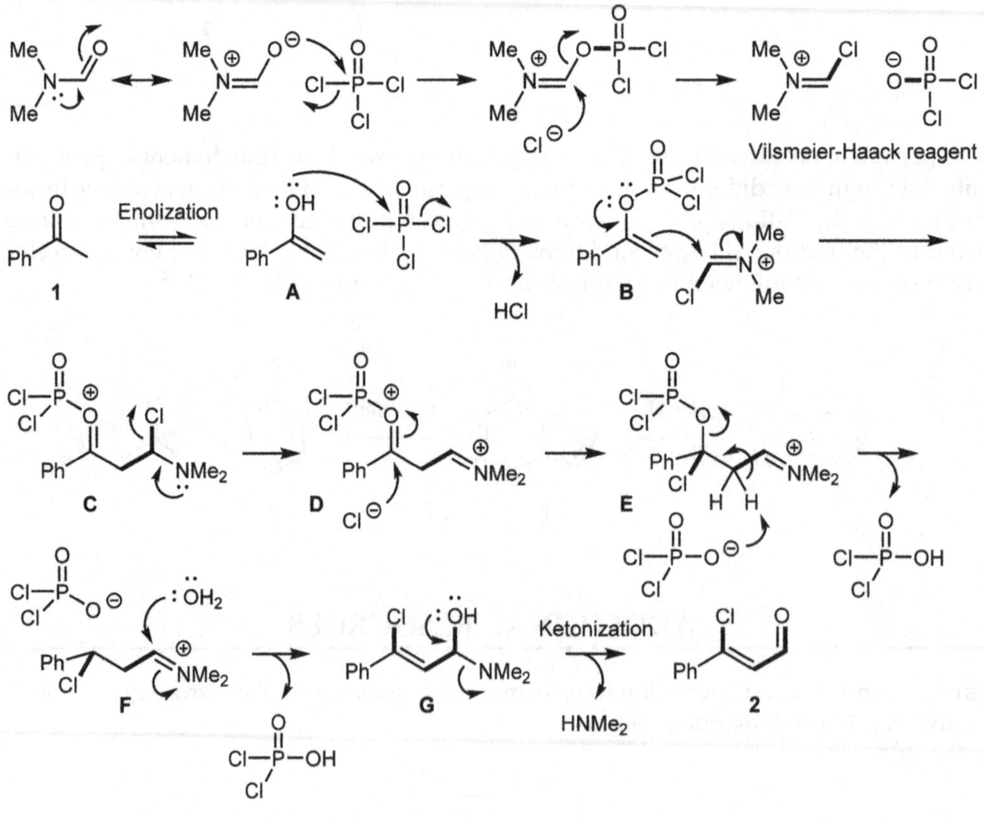

BALANCED CHEMICAL EQUATION: 1 TO 2

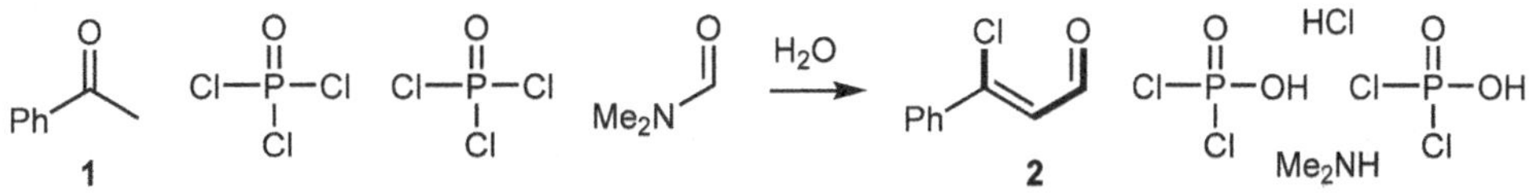

SOLUTION: 2 TO 3

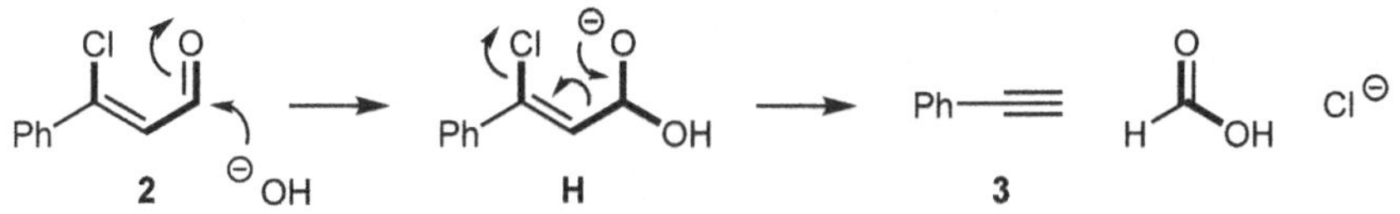

BALANCED CHEMICAL EQUATION: 2 TO 3

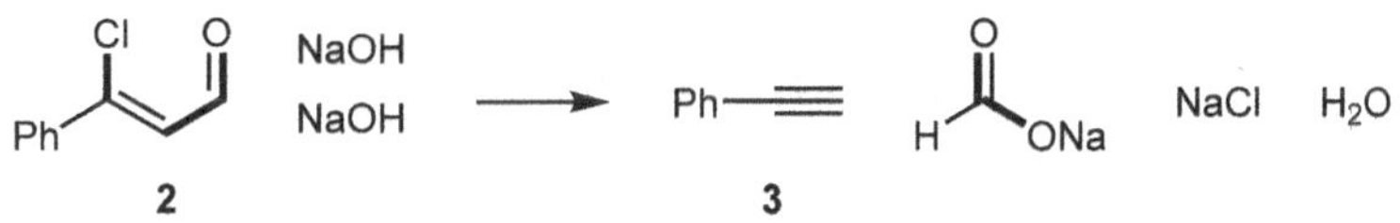

KEY STEPS EXPLAINED

The mechanism proposed by the authors begins with generation of the Vilsmeier-Haack reagent from dimethylformamide and phosphorus oxychloride. Enolization of **1** to **A** facilitates the nucleophilic addition of **A** to a second equivalent of phosphorus oxychloride to form **B** after elimination of HCl. At this point intermediate **B** reacts with the Vilsmeier-Haack reagent to form **C** which undergoes iminium formation with loss of chloride ion to give **D**. Nucleophilic attack by chloride at the electrophilic carbonyl group generates **E** which is deprotonated to give the chloroiminium intermediate **F** after loss of phosphorodichloridic acid. Attack of the iminium group of **F** by water gives **G** which after ketonization and elimination of dimethylamine gives product **2**. The treatment of **2** with sodium hydroxide then proceeds with nucleophilic attack by hydroxide onto the aldehyde group of **2** to give alkoxide **H** which undergoes fragmentation to form the phenylacetylene product **3** as well as chloride ion and formic acid by-products. Under basic conditions, the formic acid is neutralized to its sodium salt and water. We note that the authors did not provide further experimental or theoretical evidence to support this mechanism.

ADDITIONAL RESOURCES

For interesting transformations and applications of ethynylarenes, we recommend Refs. 2, 3.

References

1. Royles BJL, Smith DM. The 'inverse electron-demand' Diels–Alder reaction in polymer synthesis. Part 1. A convenient synthetic route to diethynyl aromatic compounds. *J Chem Soc Perkin Trans 1*. 1994;1994:355–358. https://doi.org/10.1039/P19940000355.
2. Cioslowski J, Schimeczek M, Piskorz P, Moncrieff D. Thermal rearrangement of ethynylarenes to cyclopentafused polycyclic aromatic hydrocarbons: an electronic structure study. *J Am Chem Soc*. 1999;121:3773–3778. https://doi.org/10.1021/ja9836601.
3. Wang X-C, Yan R-L, Zhong M-J, Liang Y-M. Bi(III)-catalyzed intermolecular reactions of (Z)-pent-2-en-4-yl acetates with ethynylarenes for the construction of multisubstituted fluorene skeletons through a cascade electrophilic addition/cycloisomerization sequence. *J Org Chem*. 2012;77:2064–2068. https://doi.org/10.1021/ja973118x.

Question 167: β-Amino Nitriles From Azetidones

In 1994 Japanese chemists studying synthetic approaches to β-amido cyanides reacted the 4-alkoxyazetidin-2-ones **1** with a catalytic amount of TMSOTf in acetonitrile at 0°C and achieved the β-amido nitriles **2** in high yield (Scheme 1).[1] Provide a mechanism to explain this result.

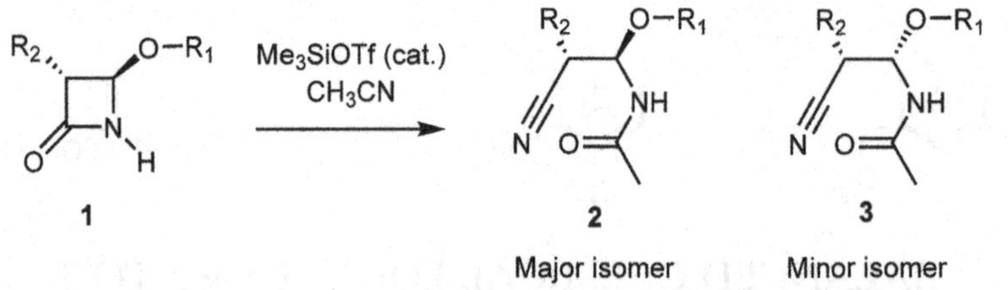

SCHEME 1

SOLUTION

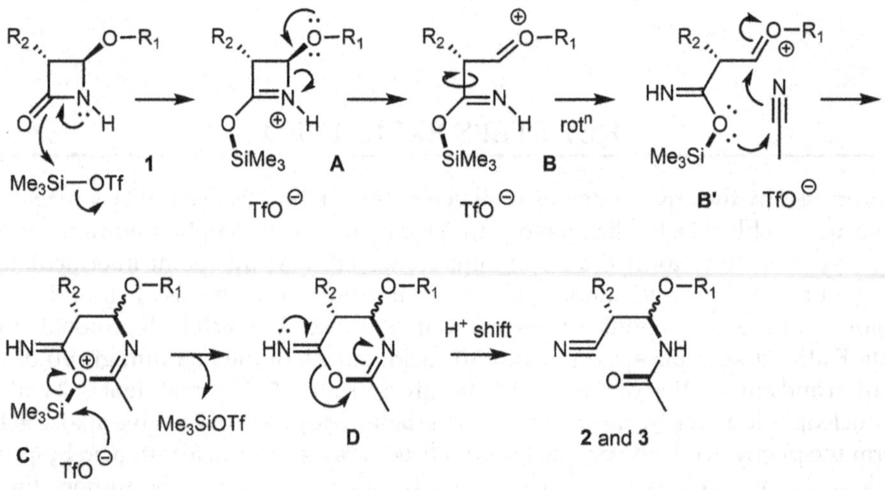

KEY STEPS EXPLAINED

The mechanism proposed by the authors for the transformation of **1** to the β-amido nitriles **2** and **3** begins with reaction of **1** with trimethylsilyl triflate to form the iminium triflate **A** which ring opens to give the oxonium ion **B**. Upon bond rotation to **B′**, reaction between **B′** and acetonitrile solvent leads to a [4 + 2] cycloaddition which generates intermediate **C**. At this point, triflate ion attacks the trimethylsilyl group on **C** to regenerate the TMSOTf catalyst and form **D**. Ring opening of **D** followed by an intramolecular proton shift generates an isomeric mixture of products **2** and **3**. It is worthwhile to note that this mechanism was postulated by the authors with the absence of supporting experimental and theoretical evidence. In fact, the authors claimed that based on their proposed mechanism, it is uncertain why the *anti* isomer **2** is preferred over the *syn* isomer. To help answer this question, we propose an additional mechanism (Scheme 2) which preserves the stereochemistry found in the original starting material and which may be operational at the same time as the mechanism proposed by the authors. Such a parallel mechanism leads stereoselectively to product **2**. We encourage future research on this problem to seek out further experimental and theoretical analysis to help support or reject either of these mechanisms. For example, an interesting experiment would be to start with

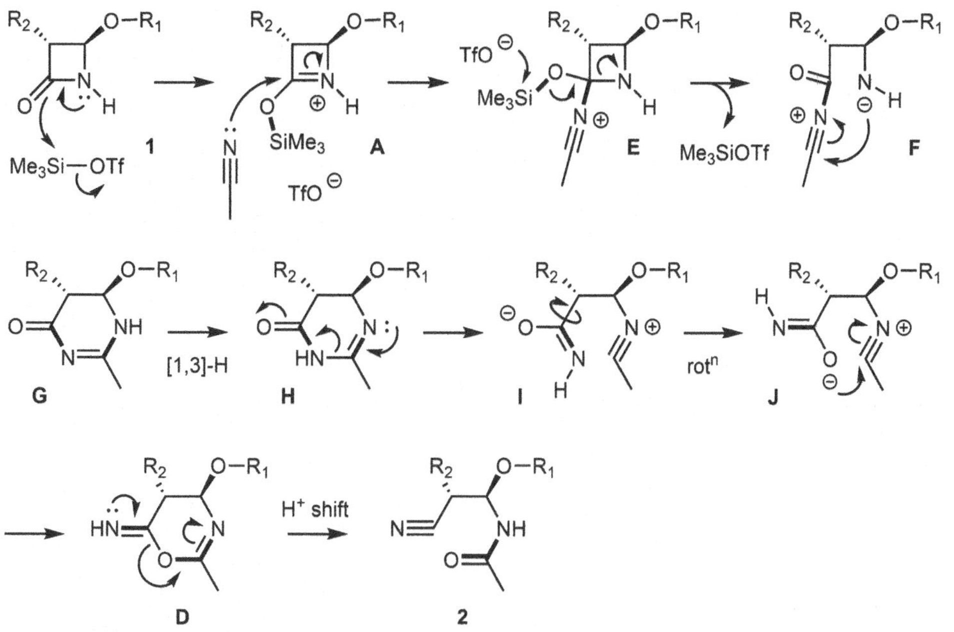

SCHEME 2

various enantiomers of **1** under the same reaction conditions. Identification of products and comparison of their yields would help shed more light on the mechanism.

Another possibility is to begin with a starting material that has the R_1 and R_2 groups tethered in a ring in a *syn* orientation and then to trace the stereochemistry of the carbon atom bearing the ether group to see if retention of configuration at that center occurs.

ADDITIONAL RESOURCES

We recommend an article by the same authors which seeks to apply this transformation to sulfur analogs of **1**.[2]

References

1. Kita Y, Shibata N, Yoshida N, Kawano N, Matsumoto K. An unprecedented cleavage of the β-lactam ring: stereoselective synthesis of chiral β-amido cyanides. *J Org Chem*. 1994;59:938–939. https://doi.org/10.1021/jo00084a002.
2. Kita Y, Shibata N, Kawano N, Yoshida N, Matsumoto K, Takebe Y. An unprecedented cleavage of the β-lactam ring: a novel synthesis of acyclic N, O- and N,S-acetals. *J Chem Soc Perkin Trans 1*. 1996;1996:2321–2330. https://doi.org/10.1039/P19960002321.

Question 168: A New Route to 1,3-Disubstituted Naphthalenes

In a 1994 study of the synthesis of 1,3-disubstituted naphthalenes, American and New Zealand chemists reacted aldol **3** with lithium 4-methylpiperazide to give the naphthalene product **4** in 30% yield (Scheme 1).[1] Devise a mechanism to explain the course of this transformation.

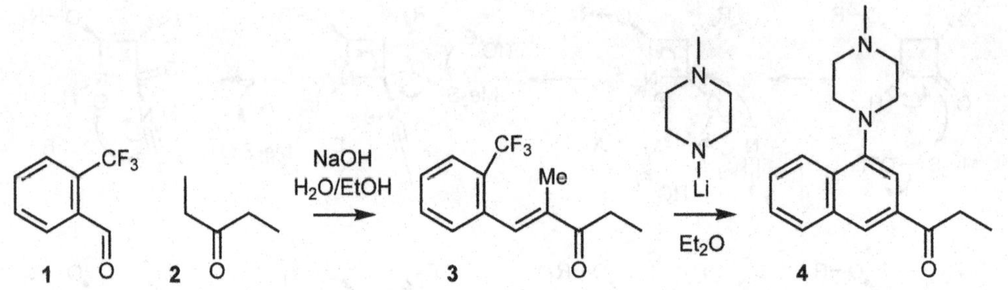

SCHEME 1

SOLUTION

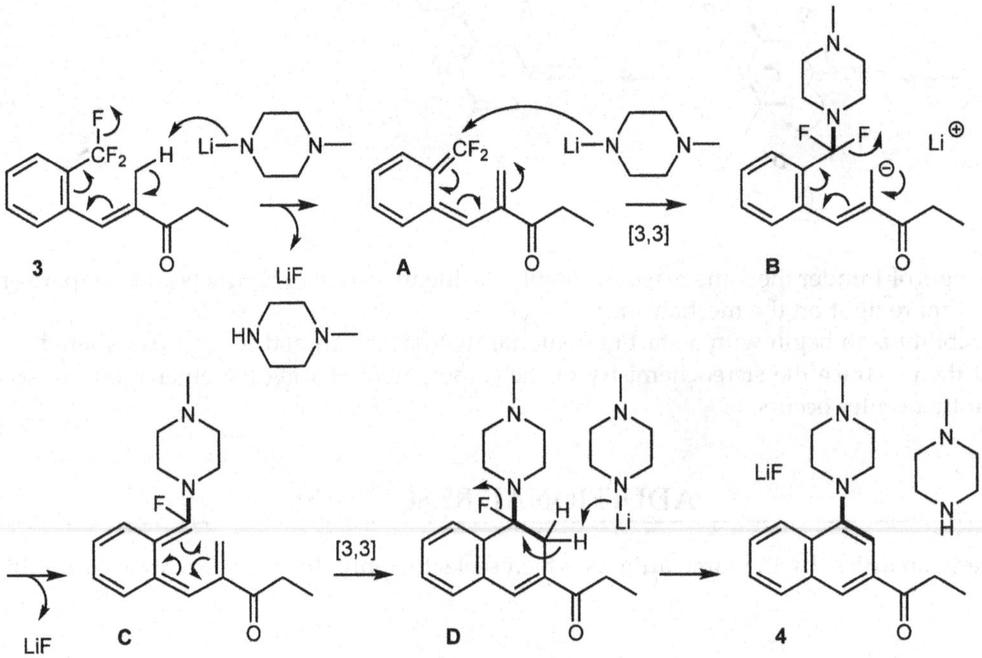

BALANCED CHEMICAL EQUATION

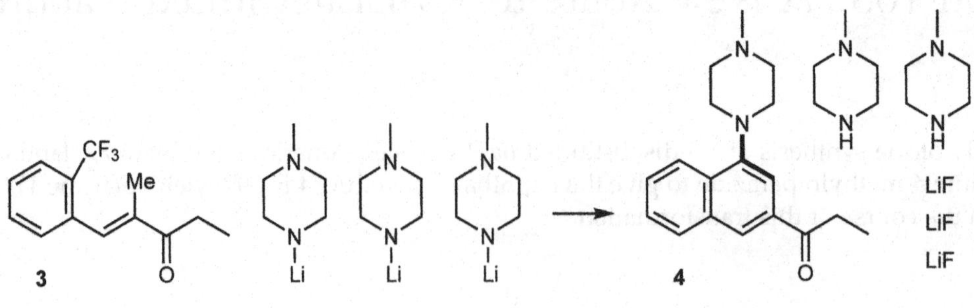

KEY STEPS EXPLAINED

The mechanism proposed by the authors for the transformation of **3** to **4** begins with lithium 4-methylpiperazide-mediated deprotonation of the methyl group of **3** which facilitates electron delocalization and elimination of fluoride ion giving the triene intermediate **A**. At this point, intermediate **A** undergoes a nucleophilically induced [3,3]-sigmatropic rearrangement to the rearomatized carbanion intermediate **B** via addition of a second equivalent of lithium 4-methylpiperazide. Electron reshuffling in **B** results in the elimination of a second fluoride ion to generate **C** which undergoes a [3,3]-sigmatropic rearrangement with rearomatization to form **D**. Lastly, lithium 4-methylpiperazide facilitates deprotonation of **D** which allows for elimination of a third fluoride ion to generate the aromatized naphthalene product **4**. Although there was no further experimental and theoretical evidence to support this proposed mechanism, the authors did discount the possible electrocyclization of **A** to **E** (which can lead to **4**) based on literature precedent and on the observation that the presence of **E** was not detected by GC-MS analysis regardless of the reaction progress (Scheme 2).

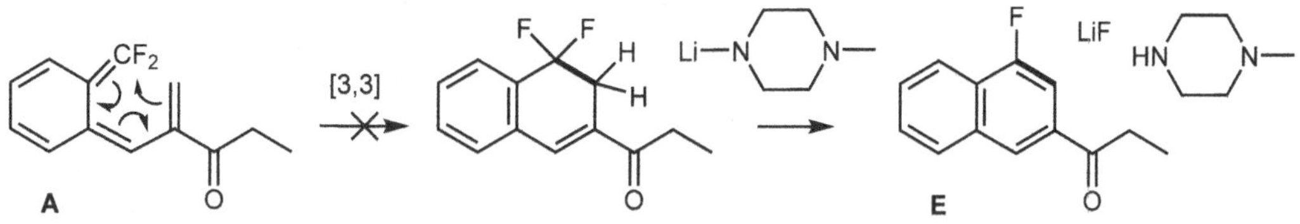

SCHEME 2

ADDITIONAL RESOURCES

For further work by these authors which explores several applications of this transformation with mechanistic discussion, see Refs. 2–7.

References

1. Strekowski L, Wydra RL, Kiselyov AS, Baird JH, Burritt A, Coxon JM. A novel synthetic route to 1,3-disubstituted naphthalenes. *Synth Commun.* 1994;24(2):257–266. https://doi.org/10.1080/00397919408013825.
2. Kiselyov AS, Dominguez C. A novel synthesis of 3,4-disubstituted cinnolines from o-trifluorophenyl hydrazones. *Tetrahedron Lett.* 1999;40:5111–5114. https://doi.org/10.1016/S0040-4039(99)00949-1.
3. Smith L, Kiselyov AS. A novel and highly efficient synthesis of the aza analogs of tacrine. *Tetrahedron Lett.* 1998;40:5643–5646. https://doi.org/10.1016/S0040-4039(99)01122-3.
4. Kiselyov AS. A facile one-pot synthesis of polysubstituted naphthalenes. *Tetrahedron Lett.* 2001;42:3053–3056. https://doi.org/10.1016/S0040-4039(01)00373-2.
5. Kiselyov AS. A novel synthesis of polysubstituted naphthalenes. *Tetrahedron.* 2001;57:5321–5326. https://doi.org/10.1016/S0040-4020(01)00450-1.
6. Kiselyov AS, Piatnitski EL, Doody J. Synthesis of polysubstituted 4-fluoroquinolinones. *Org Lett.* 2004;6(22):4061–4063. https://doi.org/10.1021/ol048257f.
7. Kiselyov AS. A convenient procedure for the synthesis of fused fluoro isoquinolines. *Tetrahedron.* 2006;62:543–548. https://doi.org/10.1016/j.tet.2005.10.025.

Question 169: An Efficient Annulation Route to 6-Substituted Indoles

In a 1992 synthetic study of the antitumor antibiotic CC-1065, Japanese chemists reacted the *N*-Cbz protected hydroxylamine **1** with methyl propiolate in nitromethane and in the presence of Hünig's base and formed the 6-substituted indole product **2** in 89% yield (Scheme 1).[1] Provide a mechanism for this transformation.

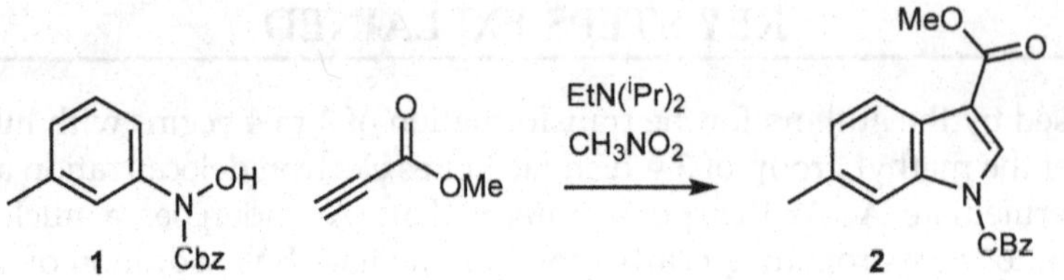

SCHEME 1

SOLUTION

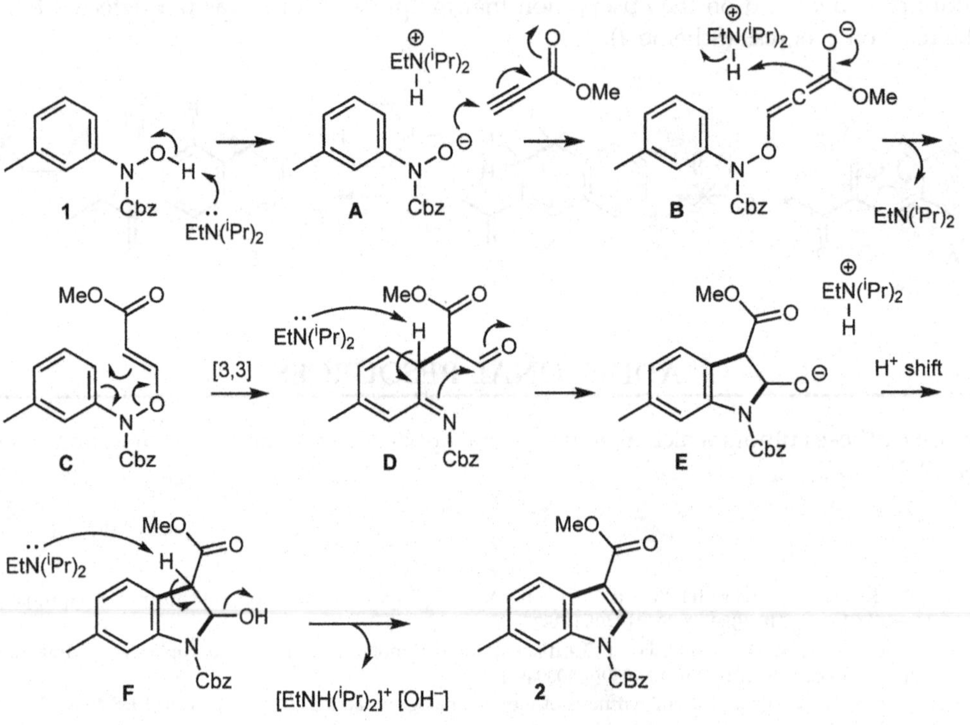

BALANCED CHEMICAL EQUATION

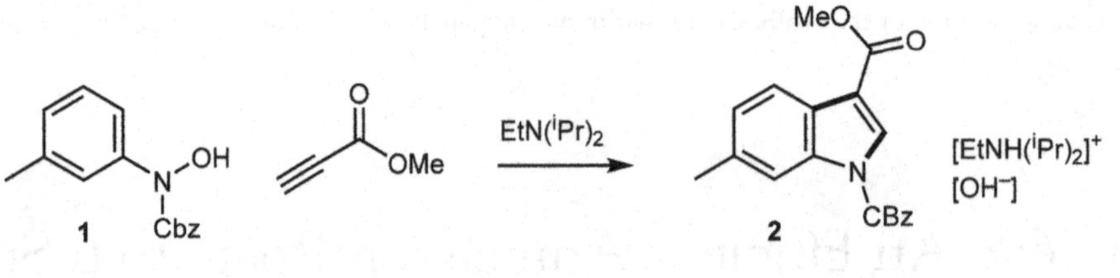

KEY STEPS EXPLAINED

The mechanism proposed by the authors for the transformation of **1** and methyl propiolate to the 6-substituted indole product **2** begins with deprotonation of the hydroxylamine group of **1** via Hünig's base to form an alkoxide **A** which attacks methyl propiolate in a Michael addition to generate the allene **B**. This intermediate abstracts a proton from Hünig's base to give **C** which undergoes a [3,3]-sigmatropic Cope rearrangement to form **D**. Deprotonation of **D** establishes aromatization of the aryl group which drives cyclization to indole **E** which is protonated to **F**. Dehydration of **F** via Hünig's base leads to the final 6-substituted indole product **2**. Although the authors do not provide further experimental and theoretical evidence to support their proposed mechanism, they do argue that structure **C** undergoes Cope rearrangement with the unsubstituted side of the aryl group likely because of unfavorable steric interaction between the ester group and the Ar-Me group if the rearrangement were to occur on the other side of the substrate.

ADDITIONAL RESOURCES

We recommend a research article and two review papers that cover interesting synthetic applications of the Cope rearrangement and other molecular rearrangements.[2-4]

References

1. Toyota M, Fukumoto K. Tandem Michael addition–[3,3]sigmatropic rearrangement processes. Part 2. Construction of cyclopropa[3,4]pyrrolo[3,2-e]indol-4-one (CPI) unit of antitumour antibiotic CC-1065. *J Chem Soc Perkin Trans 1*. 1992;1992:547–552. https://doi.org/10.1039/P19920000547.
2. Zhou L, Li Z, Zou Y, et al. Tandem nucleophilic addition/oxy-2-azonia-Cope rearrangement for the formation of homoallylic amides and lactams: total synthesis and structural verification of motuporamine G. *J Am Chem Soc*. 2012;134:20009–20012. https://doi.org/10.1021/ja310002m.
3. Paquette LA. Recent applications of anionic oxy-Cope rearrangements. *Tetrahedron*. 1997;53(41):13971–14020. https://doi.org/10.1016/S0040-4020(97)00679-0.
4. Overman LE. Molecular rearrangements in the construction of complex molecules. *Tetrahedron*. 2009;65:6432–6446. https://doi.org/10.1016/j.tet.2009.05.067.

Question 170: Failure of a Rearrangement: From a Useful Compound to a Useless Product

In a 1994 study of pyrethroid insecticides, Bedekar et al. reacted methyl permethrate **1**, a mixture of *cis* and *trans* isomers, under pyrolysis (260–270°C) and isolated methyl *o*-toluate in 78% yield (Scheme 1).[1] Suggest a mechanism for this transformation.

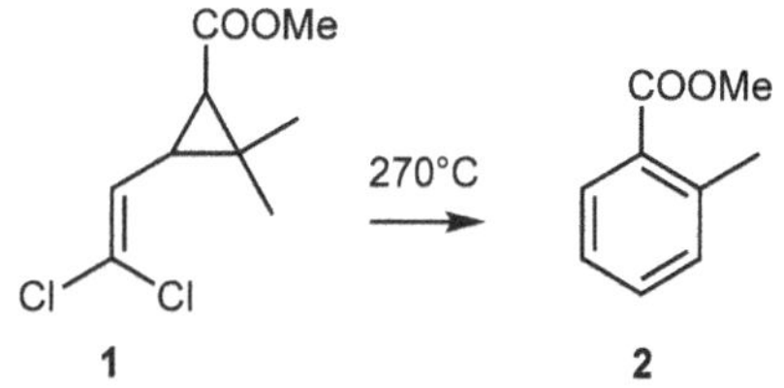

SCHEME 1

SOLUTION

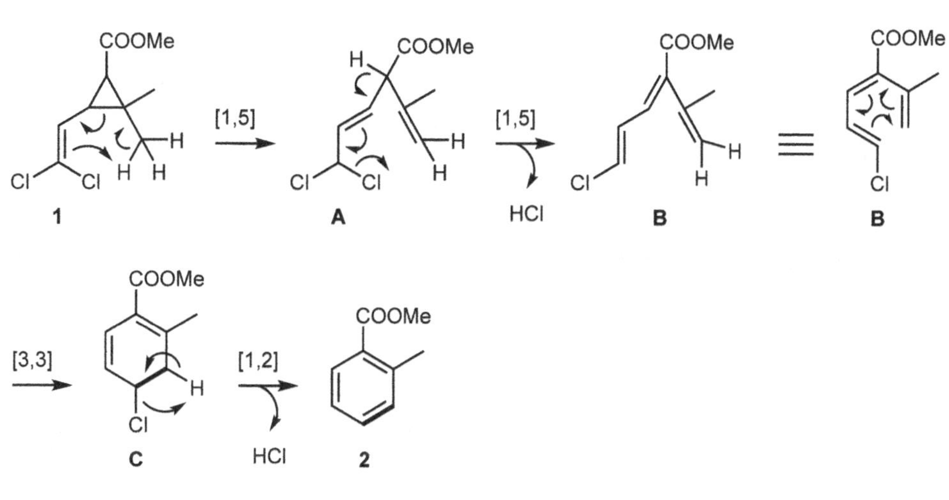

KEY STEPS EXPLAINED

The mechanism proposed by the authors to explain the pyrolysis of **1** to **2** begins with a [1,5]-ene rearrangement which forms diene **A** followed by a second [1,5]-ene rearrangement with elimination of HCl to give the triene **B**. This intermediate undergoes a [3,3]-sigmatropic rearrangement to **C** which is followed by a final rearomatization-driven elimination of HCl to generate product **2**. It is important to note that this transformation constituted an unexpected result for the authors. What they initially intended to carry out was the radical-based vinylcyclopropane-cyclopentene rearrangement shown in Scheme 2. Nevertheless, the authors did not provide additional experimental or theoretical evidence for their proposed mechanism.

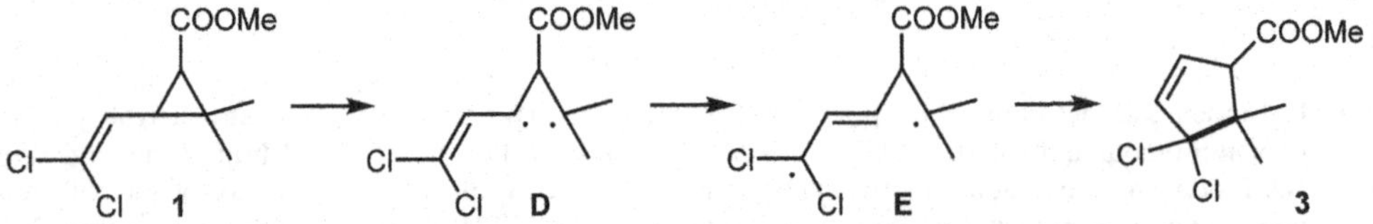

SCHEME 2

ADDITIONAL RESOURCES

We direct the reader to an excellent review article which covers many transformations of similar carbocyclic aromatic systems.[2] We also recommend an article which describes an unexpected aromatization during an N-alkylation reaction.[3]

References

1. Bedekar AV, Nair KB, Soman R. *J Chem Res (S)*. 1994;1994:52.
2. Williams AC. The synthesis of carbocyclic aromatic systems. *Contemp Org Synth*. 1996;3:535–567. https://doi.org/10.1039/CO9960300535.
3. Lopez-Cara LC, Camacho ME, Carrion MD, Gallo MA, Espinosa A, Entrena A. An unexpected aromatization during the N-alkylation reaction of 3,4-dihydro-1H-pyrazole derivatives: insight into the reaction mechanism. *Tetrahedron Lett*. 2006;47:6239–6242. https://doi.org/10.1016/j.tetlet.2006.06.141.

Question 171: Synthesis of Trisubstituted Isoxazoles

In 1993 Japanese chemists investigating synthetic approaches to isoxazole ring containing products reacted dihydropyran **2** (the product of the condensation of 1-ethoxyethene with **1**) with 2 or 3 equivalents of ethanolic hydroxylamine hydrochloride under reflux to form the isoxazole product **3** in 99% yield (Scheme 1).[1] Provide a mechanism for this transformation.

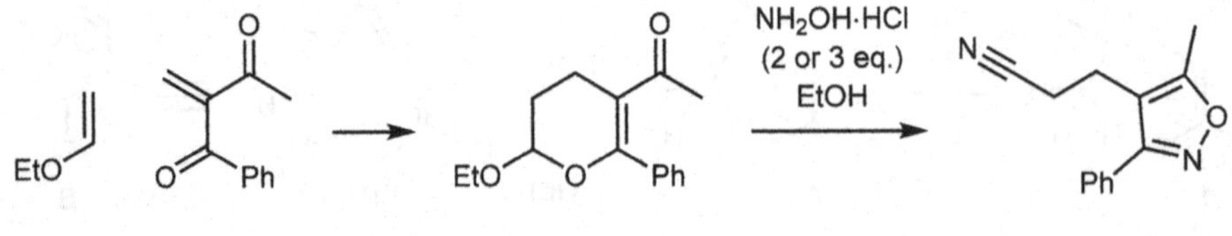

SCHEME 1

SOLUTION: 1 TO 2 AND 2 TO 3

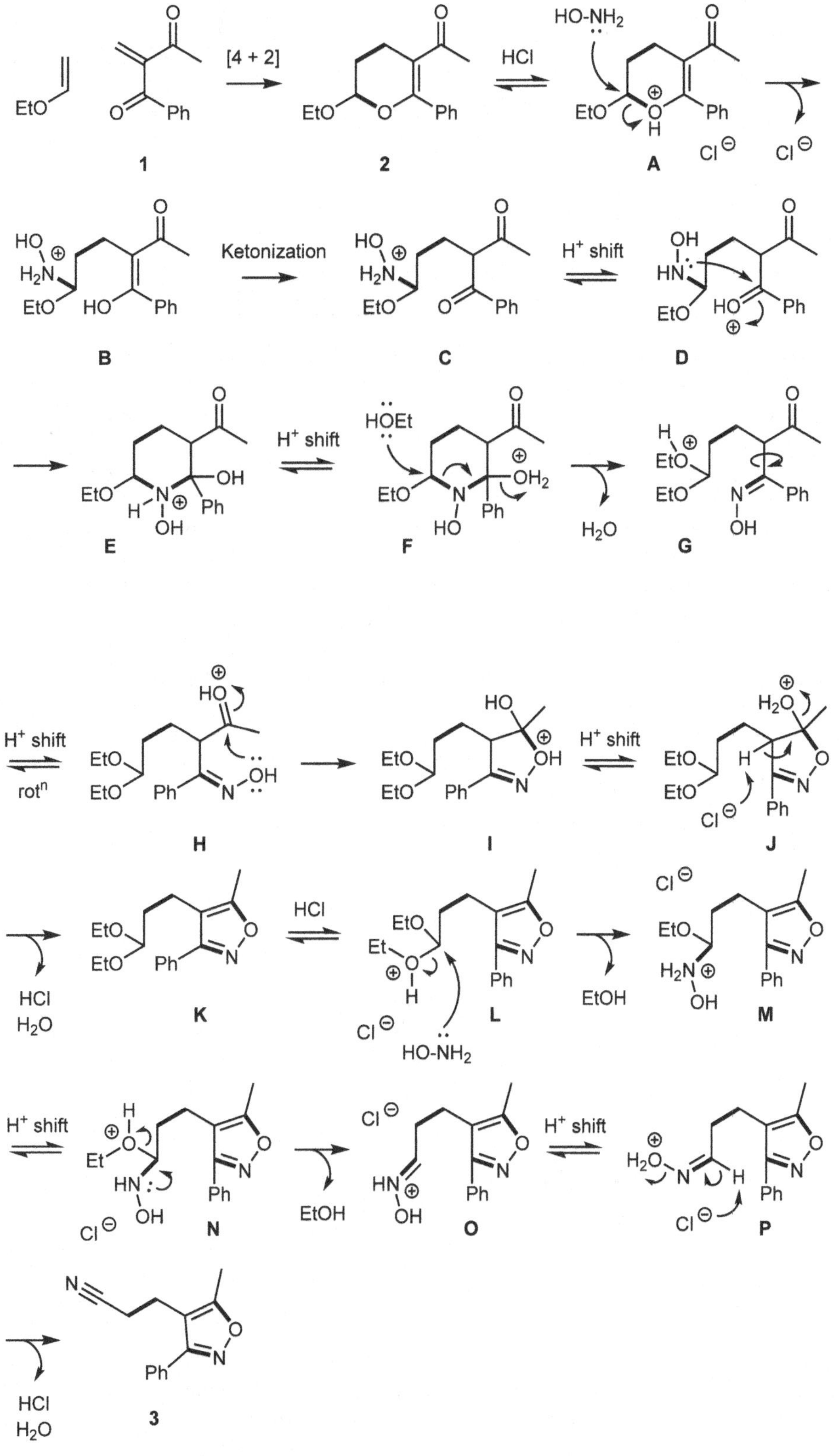

BALANCED CHEMICAL EQUATION

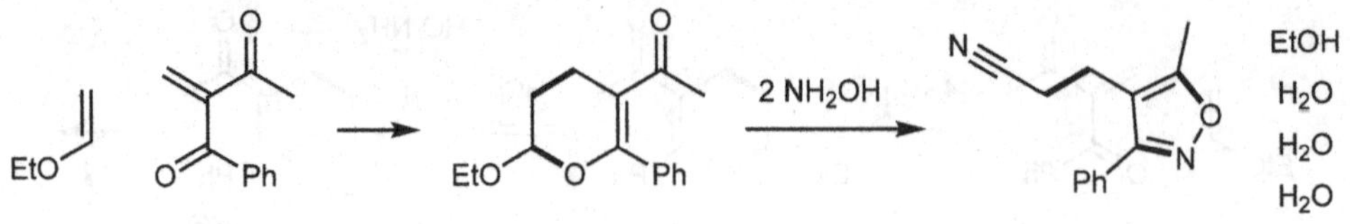

KEY STEPS EXPLAINED

The mechanism proposed by the authors begins with an initial [4+2] Diels-Alder cycloaddition of 1-ethoxyethene and **1** to form product **2**. The treatment of this product with two equivalents of hydroxylamine is explained by initial protonation of the ring oxygen atom via the acidic medium to form **A** followed by nucleophilic addition of hydroxylamine at the acetal position to cause ring opening to **B**. Ketonization of the enol group of **B** followed by proton transfer to form **D** via **C** facilitates ring closure to **E**. An additional proton transfer to **F** allows for the nucleophilic addition of ethanol from the reaction medium to the same acetal carbon attacked previously by hydroxylamine which causes ring cleavage and elimination of water to form **G**. After bond rotation and proton transfer to **H**, ring cyclization generates **I** which undergoes dehydration to **K** via **J**. At this point, an additional equivalent of hydroxylamine nucleophilically attacks the acetal position to eliminate ethanol and form **M**. After a series of proton shifts, this intermediate eliminates another equivalent of ethanol and water to generate the final isoxazole product **3**.

Although it is possible to imagine the addition of hydroxylamine to either the vinyl or carbonyl positions of product **2**, the authors performed two experiments in a subsequent paper which show evidence in support of the proposed hydroxylamine attack at the acetal position (Scheme 2).[2] First, the authors carried out the same reaction with **4** which has both vinyl and carbonyl positions available for hydroxylamine attack. Nevertheless, the expected products of such attacks, namely, **5** and **6**, were not observed. The reaction simply produced unreacted **4**. In the second experiment, reactant **7** had a vinyl position, a carbonyl position, and an acetal group available for hydroxylamine attack. Nevertheless, the isolated product **8** simply contained a newly formed cyano group at the carbon which was originally part of the acetal group. Both of these experiments provide evidentiary support for the proposed hydroxylamine attack onto the acetal group of **A**.

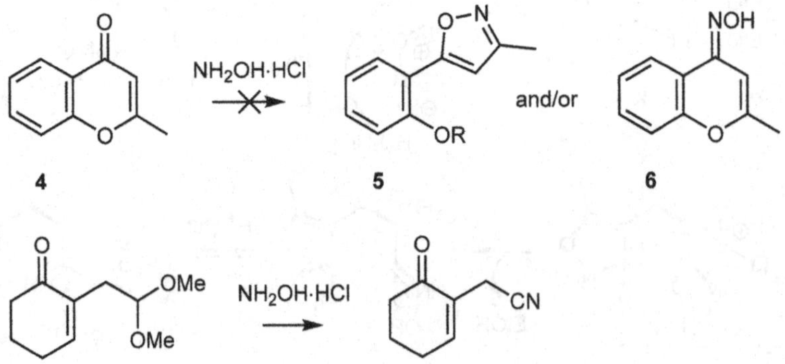

SCHEME 2

The authors provided further experimental support for their proposed mechanism when they were able to isolate the reactive intermediate **K** when only one equivalent of hydroxylamine was reacted with **2**. When **K** was reacted on its own with an additional equivalent of hydroxylamine, product **3** was isolated. This experiment proved that intermediate **K** was indeed part of the mechanism for this transformation. Furthermore, it is interesting to note that in their original article, the authors believed hydrogen bonding between the hydroxylamine group and the carbonyl group in

C and **D** played a role in the mechanism. Unfortunately this assumption was discounted in the subsequent paper on account of consideration for the high temperature required for the reaction under which conditions hydrogen bonding would be less likely. The authors required a high temperature to mitigate the insolubility of hydroxylamine in ethanol. Nevertheless, despite the experimental evidence, the authors did not perform a theoretical analysis to further support their proposed mechanism.

ADDITIONAL RESOURCES

We recommend further work by the same authors which include further support for this mechanism[3] and an example of a similar transformation with an interesting mechanism.[4] For an example of a similar synthesis of isoxazoles which highlights a **K**-like intermediate and hydrogen bonding, see the work of Doleschall et al.[5]

References

1. Yamauchi M, Akiyama S, Watanabe T, Okamura K, Date T. A novel synthesis of 4-cyanoethylisoxazoles. *J Chem Soc Chem Commun.* 1993;1993:17–18. https://doi.org/10.1039/C39930000017.
2. Yamauchi M, Akiyama S, Watanabe T, Okamura K, Date T. A facile conversion of ethoxydihydropyrans to 4-cyanoethylisoxazoles. *J Heterocyclic Chem.* 1996;33:383–387. https://doi.org/10.1002/jhet.5570330229.
3. Yamauchi M. Facile conversion of acetals to nitriles. *Chem Pharm Bull.* 1993;41(11):2042–2043. https://doi.org/10.1248/cpb.41.2042.
4. Katayama S, Hiramatsu H, Aoe K, Yamauchi M. Synthesis of bicyclo[4.1.0]hept-2-enes (trinorcarenes) by photochemical reaction of bicyclo[2.2.2] oct-5-en-2-ones. *J Chem Soc Perkin Trans 1.* 1997;1997:561–576. https://doi.org/10.1039/A601900E.
5. Doleschall G, Seres P, Parkanyi L, Toth G, Almasy A, Bihatsi-Karsai E. Novel rearrangement of an isoxazole to a pyrrole. X-Ray molecular structure of diethyl 5-(ethoxycarbonylmethyl)pyrrole-2,4-dicarboxylate. *J Chem Soc Perkin Trans 1.* 1986;1986:927–932. https://doi.org/10.1039/P19860000927.

Question 172: 2,3-Dihydrobenzofurans From 1,4-Benzoquinones

In 1993 Chilean chemists formed product **3** by condensing the benzoquinone **1** with the diene **2** (Scheme 1).[1] Treatment of **3** with HCl formed the dihydrobenzofuran **4**. In a similar study, French chemists reacted benzoquinone **5** with the dimethylhydrazine **6** and BF₃•(OEt)₂ Lewis acid catalyst to achieve product **7**. When **7** was treated with copper(II) acetate in buffered aqueous THF the 2,3-dihydrobenzofuran product **8** was formed in excellent yield (Scheme 1).[2] Suggest mechanisms to explain both sequences of reactions.

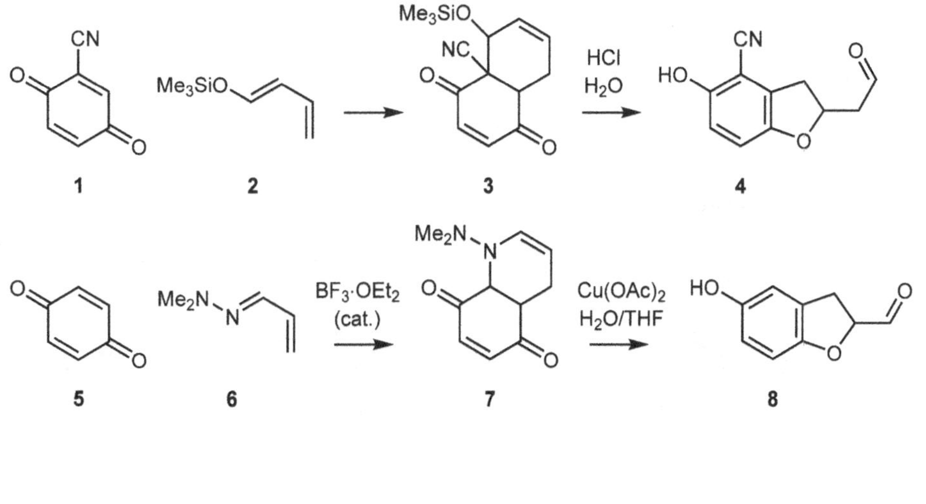

SCHEME 1

SOLUTION: 1 AND 2 TO 3 AND 3 TO 4

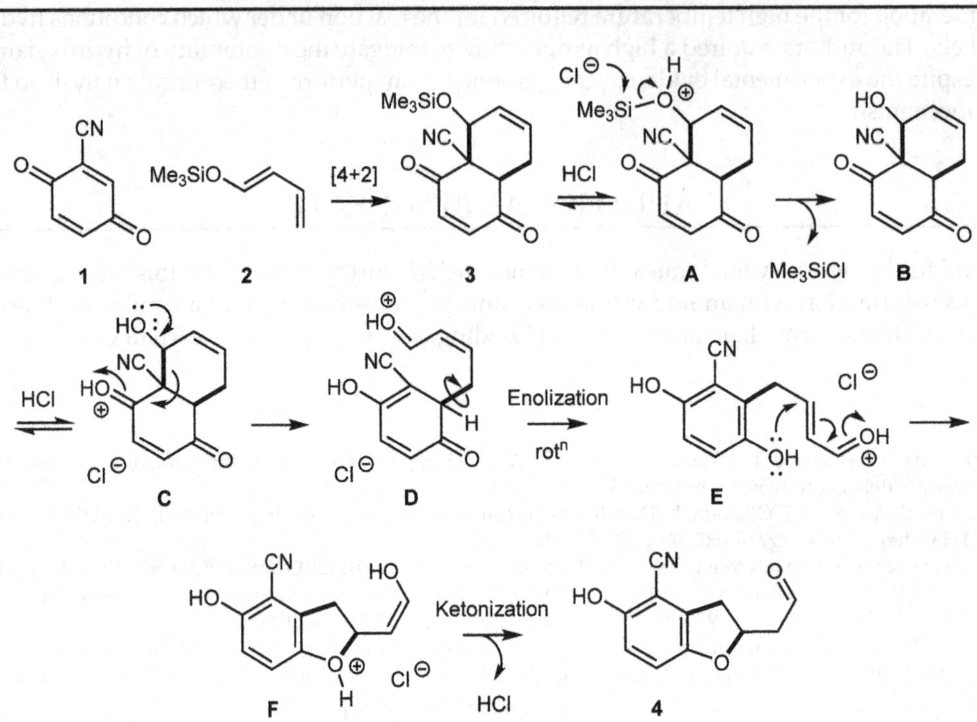

SOLUTION: 5 AND 6 TO 7

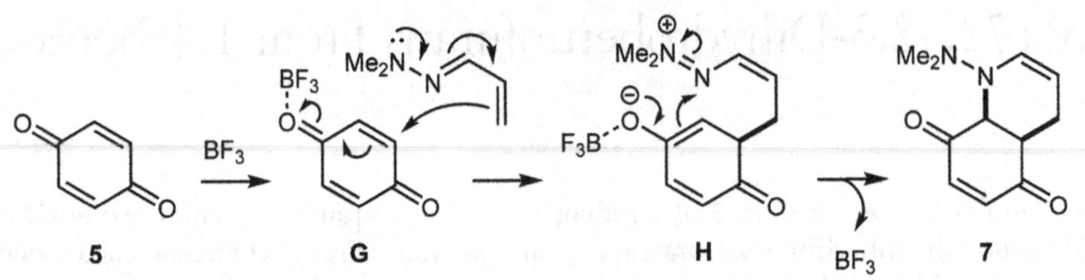

SOLUTION: 7 TO 8

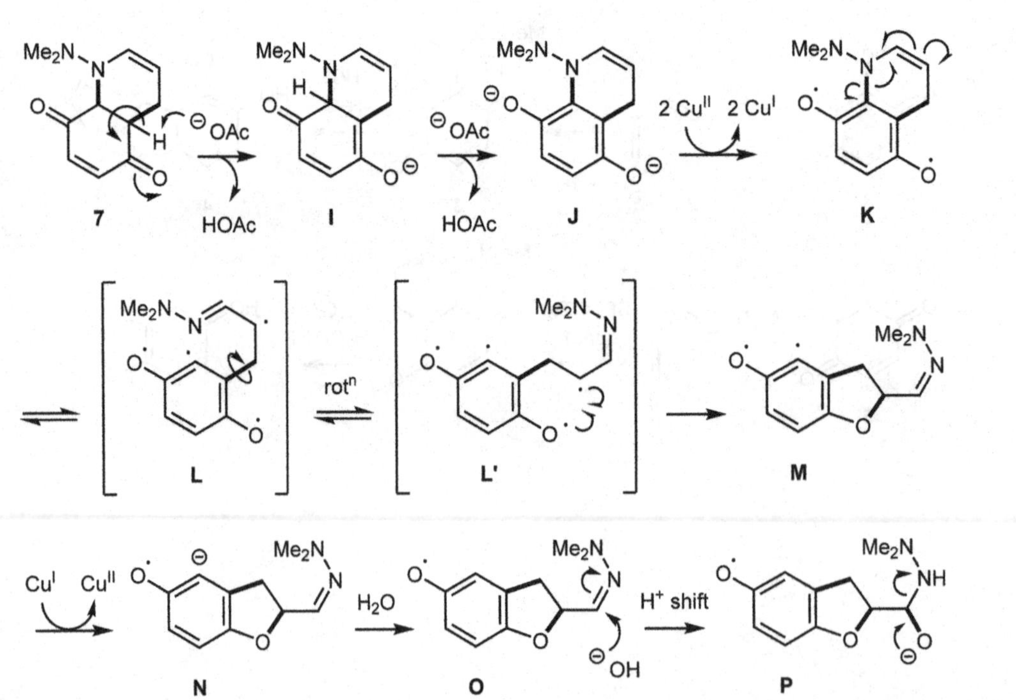

BALANCED CHEMICAL EQUATION

KEY STEPS EXPLAINED

The mechanism proposed by Santos et al. to explain the transformation of **1** and **2** to **3** and eventually to **4** begins with a [4+2] cycloaddition of **1** and **2** to form **3** which is protonated by acid trimethylsilyloxy group to give **A** and chloride anion. Hydrolysis of the trimethylsilyl group via chloride attack leads to intermediate **B** which undergoes further protonation on the carbonyl group attached to the benzoquinone structure to generate the oxonium ion **C**. In this form, **C** undergoes ring cleavage to form the carbocation intermediate **D** which after enolization and bond rotation gives intermediate **E**. Cyclization of this intermediate followed by deprotonation and ketonization gives the final dihydrobenzofuran **4**. In order to help support their proposed mechanism, Santos et al. performed kinetics and spectrophotometric analysis to determine the relevant rate constants between each of the postulated mechanistic steps. We encourage the reader to visit the original article for more details on these experiments. It is interesting to note that the authors also suggested the possibility of direct formation of product **4** from the ring-opened carbocationic intermediate **D** through a 5-*exo-trig* cyclization. We also note that the authors did not perform a theoretical analysis to support their mechanism.

Conversely, the transformation of **5** and **6** to **7** begins with complex formation between the boron trifluoride Lewis acid catalyst and the carbonyl oxygen atom of benzoquinone to form **G**. This enables the dimethylhydrazine reagent to attack **G** in a Michael fashion to form intermediate **H** which cyclizes with expulsion of boron trifluoride to product **7**. With **7** in hand, it is possible to undertake the transformation to the 2,3-dihydrobenzofuran product **8** by starting with two acetate mediated deprotonations of **7** that result in formation of dioxyanion **J**. This intermediate then reacts with two equivalents of Cu(II), which functions as an electron transfer catalyst, to facilitate the formation of diradical intermediate **K**. What follows is a homolytic cleavage of the central C—N bond in **K** to allow for cyclization to the dihydrofuran ring and formation of the diradical intermediate **M** which may gain an electron from Cu(I) to form radical anion **N** which is protonated by water to form the oxygen radical **O** and hydroxide anion. At this point, hydroxide ion attack onto the imino group causes hydrolysis to the aldehyde group to form the oxygen radical **R** which gains an electron from Cu(I) and a proton from water to generate the final product **8** and regenerate the Cu(II) electron transfer catalyst. It is important to note that Nebois et al. did not discuss any aspect of this transformation in any meaningful mechanistic sense. As such the proposed mechanism for the transformation of **5** and **6** to **7** and then to **8** represents our best efforts to answer the question in the absence of experimental and theoretical evidence.

ADDITIONAL RESOURCES

For further work by Valderrama et al. and Nebois et al., see Refs. 3–6. For analogous transformations with interesting mechanisms, see Refs. 7, 8.

References

1. Santos JG, Robert P, Valderrama JA. Kinetic studies on the rearrangement of Diels–Alder adducts of activated benzoquinones with (E)-1-trimethylsiloxybuta-1,3-diene. *J Chem Soc Perkin Trans 2*. 1993;1993:1841–1845. https://doi.org/10.1039/P29930001841.

2. Nebois P, Fillion H, Benameur L, Fenet B, Luche J-L. Synthesis and NMR structural study of furoquinolines and naphthofurans from quinones and a 1-azadiene. *Tetrahedron*. 1993;49(43):9767–9774. https://doi.org/10.1016/S0040-4020(01)80179-4.

3. Cassis R, Tapia R, Valderrama JA. Studies on quinones. XIII. Synthesis and reactivity in acid medium of cyclic *O,N*-ketals derived from acylbenzoquinones and enamines. *J Heterocyclic Chem*. 1984;21(3):869–872. https://doi.org/10.1002/jhet.5570210346.

4. Nebois P, Bouaziz Z, Fillion H, et al. The Diels-Alder cycloaddition, an intriguing problem in organic sonochemistry. *Ultrason Sonochem*. 1996;3 (1):7–13. https://doi.org/10.1016/1350-4177(95)00039-9.

5. Nebois P, Fillion H. Diels-Alder reactions of benzo[*b*]furan-4,5-diones and benzo[*b*]furan-4,7-diones. *Heterocycles*. 1999;50(2):1137–1156. https://doi.org/10.3987/REV-98-SR(H)7.

6. Pautet F, Nebois P, Bouaziz Z, Fillion H. Cycloadditions of α,β-unsaturated *N,N*-dimethylhydrazones. A Diels-Alder strategy for the building of aza-hetero rings. *Heterocycles*. 2001;54(2):1095–1138. https://doi.org/10.3987/REV-00-SR(I)5.

7. Eshavarren AM. Lewis acid catalyzed reactions of α,β-unsaturated *N,N*-dimethylhydrazones with 1,4-benzoquinone. Formation indoles by a novel oxidative rearrangement. *J Org Chem*. 1990;55(14):4255–4260. https://doi.org/10.1021/jo00301a009.

8. Perez JM, Lopez-Alvarado P, Avendano C, Menendez JC. Hetero Diels–Alder reactions of 1-acetylamino- and 1-dimethylamino-1-azadienes with benzoquinones. *Tetrahedron*. 2000;56:1561–1567. https://doi.org/10.1016/S0040-4020(00)00058-2.

Question 173: Easy Construction of a Tricyclic Indole Related to the Mitomycins

During studies on mytomycins in 1979, Japanese chemists treated **1** with NaH in dry THF and achieved an epimeric mixture of carbinolamines **2** which when treated with glacial acetic acid at room temperature for 5 min underwent dehydration to generate the tricyclic indole product **3** (Scheme 1).[1] Provide a mechanism for the overall sequence **1** to **2** and **2** to **3**.

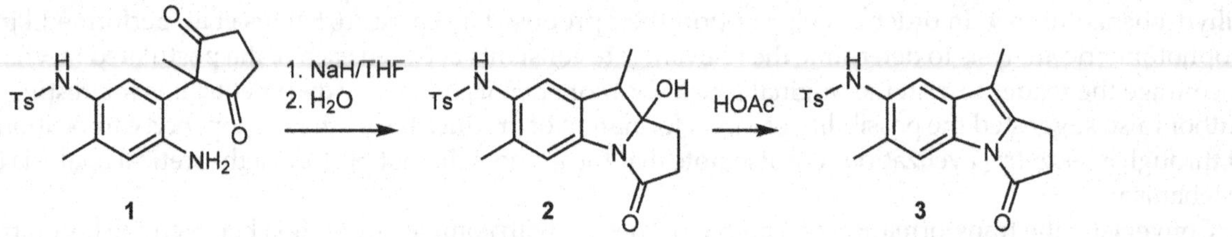

SCHEME 1

SOLUTION: 1 TO 2

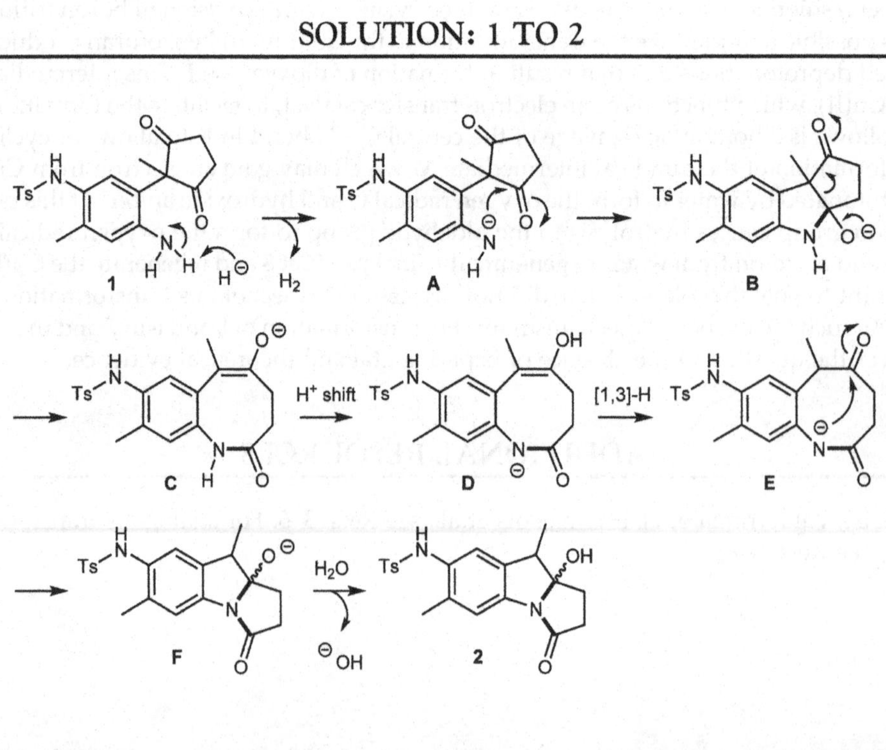

SOLUTION: 2 TO 3

KEY STEPS EXPLAINED

The mechanism proposed by the authors for the transformation of **1** to **2** begins with deprotonation of the amino group by hydride ion to give hydrogen gas and amide **A** which cyclizes to the tricyclic alkoxide **B**. Ring opening of **B** via ketonization establishes the [6.4.0] bicyclic amide **E** after two proton shifts. This amide undergoes recyclization to form the rearranged epimeric tricyclic alkoxides **F** which are protonated by water to form the epimeric mixture of carbinolamine products **2**. Afterward, protonation of the hydroxyl group of **2** via acetic acid results in the formation of **G** and acetate anion which facilitates the subsequent deprotonation and dehydration sequence to generate the fused indole product **3**. We note that the overall target bond mapping indicates that the ring construction strategy is [(5 +0)+(4+1)]. Furthermore, we also note that the authors did not provide additional experimental and theoretical evidence to support their postulated mechanism.

ADDITIONAL RESOURCES

We refer the reader to several articles describing the mechanistic pathways of transformations involving interesting fused bicyclic and tricyclic indole containing compounds.[2–5]

References

1. Ohnuma T, Sekine Y, Ban Y. Synthetic studies on mitomycins. 2. A synthesis of 9a-hydroxy-5,8-dideoxomitosane skeleton through a novel retro-aldol type of ring-opening reaction. *Tetrahedron Lett*. 1979;20(27):2537–2540. https://doi.org/10.1016/S0040-4039(01)86342-5.
2. Clayton KA, Black DS, Harper JB. Mechanisms of cyclisation of indolo oxime ethers. Part 2: Formation of ethyl 6,8-dimethoxypyrazolo[4,5,1-*hi*]indole-5-carboxylates. *Tetrahedron*. 2008;64:3183–3189. https://doi.org/10.1016/j.tet.2008.01.100.
3. Yokosaka T, Nemoto T, Hamada Y. An acid-promoted novel skeletal rearrangement initiated by intramolecular *ipso*-Friedel–Crafts-type addition to 3-alkylidene indolenium cations. *Chem Commun*. 2012;48:5431–5433. https://doi.org/10.1039/c2cc31699d.
4. Ozuduru G, Schubach T, Boysen MMK. Enantioselective cyclopropanation of indoles: construction of all-carbon quaternary stereocenters. *Org Lett*. 2012;14(19):4990–4993. https://doi.org/10.1021/ol302388t.
5. Park I-K, Park J, Cho C-G. Intramolecular Fischer indole synthesis and its combination with an aromatic [3,3]-sigmatropic rearrangement for the preparation of tricyclic benzo[*cd*]indoles. *Angew Chem Int Ed*. 2012;51:2496–2499. https://doi.org/10.1002/anie.201108970.

Question 174: Cycloaddition to a Benzothiopyrylium Salt

In 1996 Japanese chemists studied the synthetic sequence shown in Scheme 1.[1] They observed that treatment of cycloadduct **3** with triethylamine in ethanol formed a mixture of products **4** (51% yield) and **5** (28% yield). This mixture was also obtained with other base/solvent systems and careful NMR analysis revealed the presence of an unstable intermediate. Also, when oxygen gas was bubbled through the reaction, the yield of **4** decreased to 7% and the yield of **5** increased to 73%. Considering these observations, suggest mechanisms for the formation of **4** and **5**.

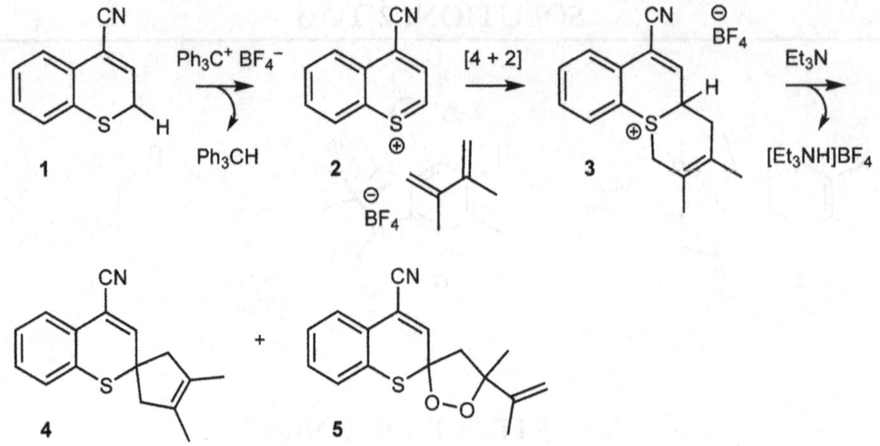

SCHEME 1

SOLUTION: 3 TO 4

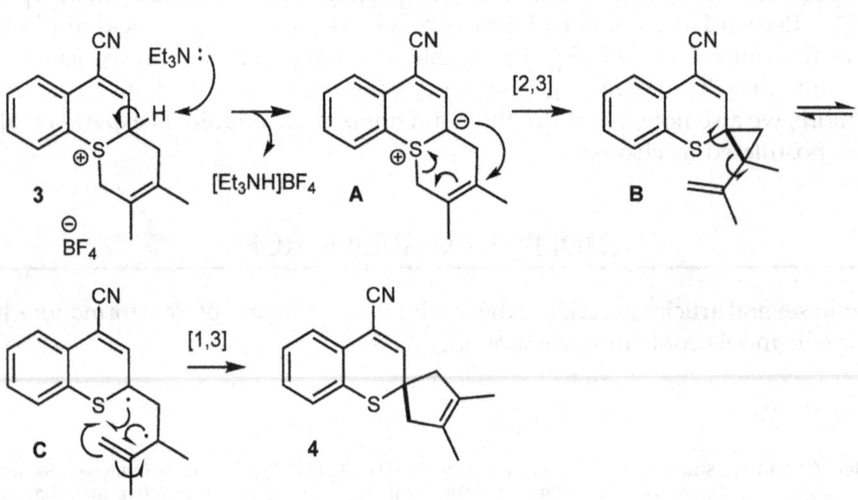

SOLUTION: 3 TO 5

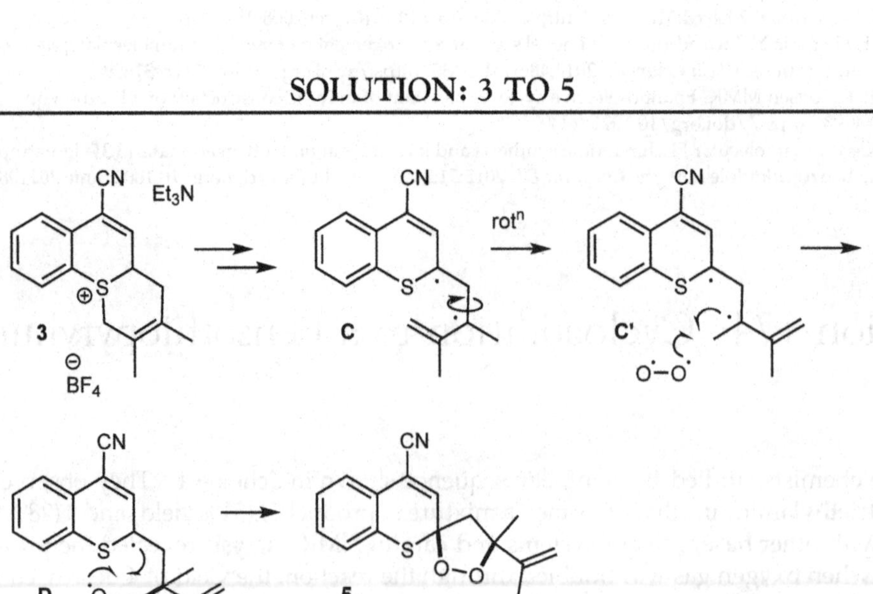

KEY STEPS EXPLAINED

The mechanisms proposed by the authors for the transformation of **3** to a mixture of **4** and **5** both start with deprotonation of the acidic proton adjacent to sulfur in **3** via triethylamine base to form ylide **A**. This ylide intermediate then undergoes a [2,3] sigmatropic rearrangement to generate the spirocyclopropane intermediate **B** for which the authors were able to observe ¹H NMR signals. This unstable intermediate fragments to the diradical **C** which can either proceed with a 5-*endo-trig* cyclization to give the [6,5]-*spiro* product **4** or react with oxygen (ground state triplet state) to form hydroperoxide radical **D** which then cyclizes to the [6,5]-*spiro* peroxide product **6**. Experimentally, the authors provided evidence for the existence of an oxygen labile intermediate such as **C** by showing that the product ratio of **4** to **5** decreases from 51:28 under standard room temperature conditions to 7:73 under bubbling oxygen conditions. Furthermore, the authors surmised that formation of diradical **C** may be facilitated by stabilization via captodative substituents and allyl resonance. Lastly, the authors did not perform a theoretical analysis to help support their postulated mechanisms.

ADDITIONAL RESOURCES

We recommend three articles which cover interesting transformations with a mechanistic discussion that involve sulfur-containing *spiro* structures such as products **4** and **5**.[2–4]

References

1. Shimizu H, Miyazaki S, Kataoka T. Polar cycloaddition of 1-benzothiopyrylium salts with conjugated dienes and some transformations of the cycloadducts. *J Chem Soc Perkin Trans 1*. 1996;1996:2227–2235. https://doi.org/10.1039/P19960002227.
2. Sun X-Y, Tian X-Y, Li Z-W, Peng X-S, Wong HNC. Total synthesis of plakortide E and biomimetic synthesis of plakortone B. *Chem A Eur J*. 2011;17:5874–5880. https://doi.org/10.1002/chem.201003309.
3. Ohsugi S, Nishide K, Node M. A novel tandem [4⁺ +2] cycloaddition–elimination reaction: 2-alkenyl-4,4-dimethyl-1,3-oxathianes as synthetic equivalents for α,β-unsaturated thioaldehydes. *Tetrahedron*. 2003;59:1859–1871. https://doi.org/10.1016/S0040-4020(03)00179-0.
4. Appel TR, Yehia NAM, Baumeister U, et al. Electrophilic cyclisation of bis(4-methoxybenzylthio)acetylene—competition between Ar₂-6 and Ar₁-5 routes, yielding 1*H*-2-benzothiopyrans or spiro derivatives of cyclohexadienone. *Eur J Org Chem*. 2003;2003:47–53. https://doi.org/10.1002/1099-0690(200301)2003:1<47::AID-EJOC47>3.0.CO;2-B.

Question 175: Benzotriazole From 1,2,4-Benzotriazine N-Oxides

In 1982 Jordanian chemists observed that aqueous NaOH treatment of either the 1- or 2-oxides of 3-methyl-1,2,4-benzotriazine **1** or **2** will give benzotriazole **3** in low yield (Scheme 1).[1] Show mechanisms for both possible transformations.

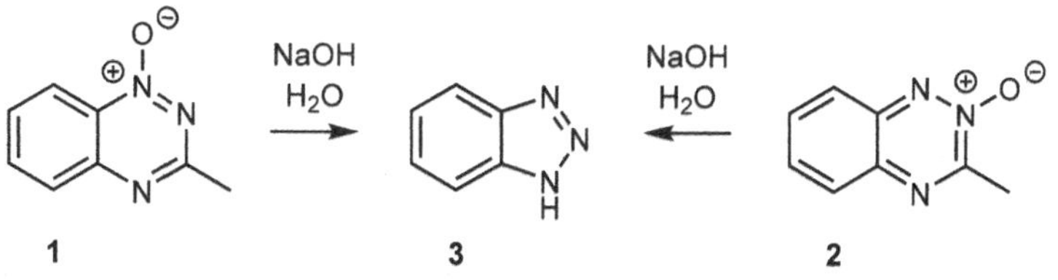

SCHEME 1

SOLUTION: 1 TO 3

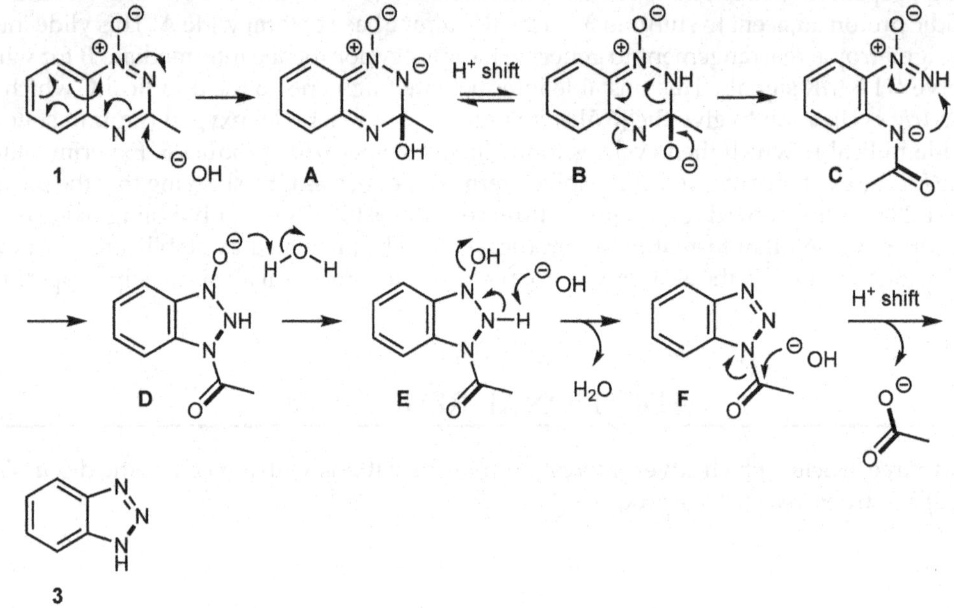

SOLUTION: 2 TO 3

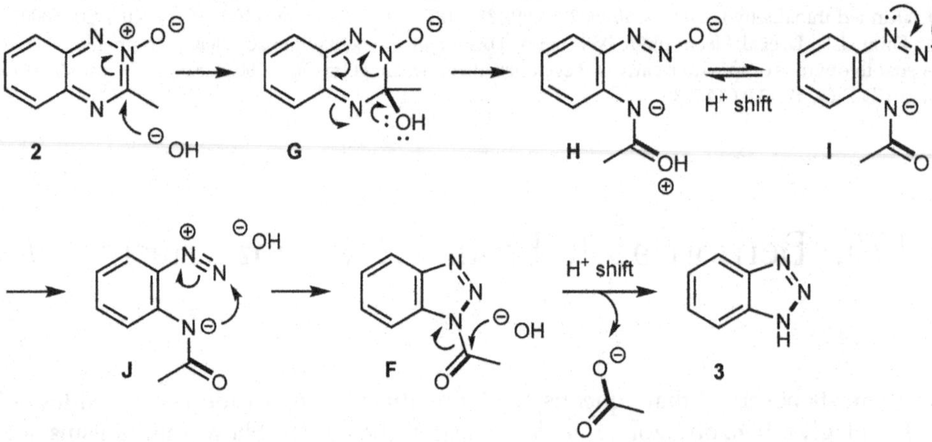

KEY STEPS EXPLAINED

The mechanism proposed by the authors for the transformation of 3-methyl-1,2,4-benzotriazine 1-oxide **1** to the benzotriazole product **3** begins with nucleophilic attack by hydroxide anion at C_3 to form **A**. A proton transfer enables the formation of **B** which through ketonization causes ring opening to **C** followed by recyclization to **D**, dehydration to **F**, and hydroxide-mediated deacylation to the final product **3**. A similar transformation occurs starting from the 3-methyl-1,2,4-benzotriazine 2-oxide **2** with the formation of the initial intermediates **G** and **H**. In this case, however, proton transfer from **H** to **I** enables elimination of hydroxide to form the benzodiazonium zwitterion **J** which undergoes recyclization to the common intermediate **F**. Deacylation of **F** via hydroxide anion attack, as seen with the previous mechanism, leads to the formation of benzotriazole **3**. It is important to note that the authors did not provide further experimental and theoretical evidence to support their postulated mechanism.

Interestingly, an oxidation number analysis on key atoms for each reaction shows that both transformations are internal redox reactions (see Scheme 2).

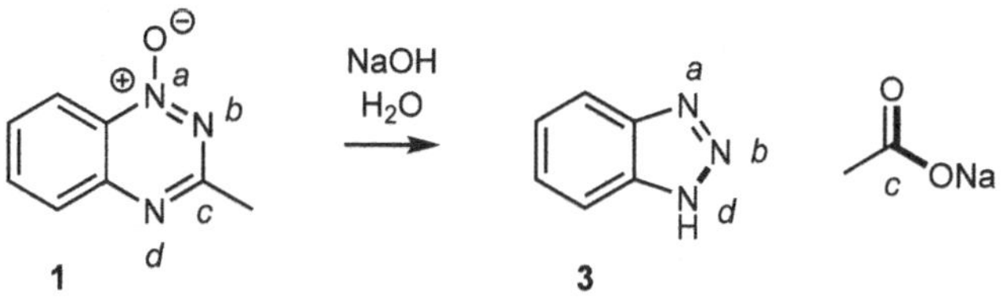

N_a (reduction): +1 to −1 (change of −2)
N_b (oxidation): −1 to 0 (change of +1)
C_c (null): +3 to +3 (no change)
N_d (oxidation): −3 to −2 (change of +1)

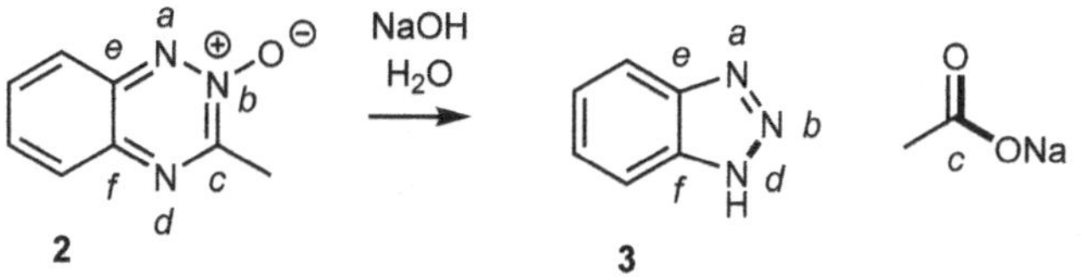

N_a (oxidation): −2 to −1 (change of +1)
N_b (null): 0 to 0 (no change)
C_c (null): +3 to +3 (no change)
N_d (oxidation): −3 to −2 (change of +1)
C_e (reduction): +2 to +1 (change of −1)
C_f (reduction): +2 to +1 (change of −1)

SCHEME 2

ADDITIONAL RESOURCES

We recommend two articles and one review which describe synthetic approaches to 1,2,4-tiazine N-oxide derivatives and which also include interesting mechanistic analyses.[2–4]

References

1. Atallah RH, Nazer MZ. Oxides of 3-methyl-1,2,4-benzotriazine. *Tetrahedron*. 1982;38(12):1793–1796. https://doi.org/10.1016/0040-4020(82)80252-4.
2. Katritzky AR, Wang J, Karodia N, Li J. Facile synthesis of benzotriazines and indoles by ring-scissions of α-benzotriazol-1-yl hydrazones. *Synth Commun*. 1997;27(22):3963–3976. https://doi.org/10.1080/00397919708005918.
3. Guo H, Liu J, Wang X, Huang G. Copper-catalyzed domino reaction of 2-haloanilines with hydrazides: a new route for the synthesis of benzo[e][1,2,4]triazine derivatives. *Synlett*. 2012;23:903–906. https://doi.org/10.1055/s-0031-1290615.
4. Kozhevnikov DN, Rusinov VL, Chupakhin ON. 1,2,4-Triazine N-oxides. *Adv Heterocycl Chem*. 2002;82:261–305. https://doi.org/10.1016/S0065-2725(02)82029-3.

Question 176: A Furan → Furan Transformation

In a 1970 study of the reactivity of furans with morpholine, Japanese chemists discovered that treatment of **1** with morpholine at room temperature resulted in the exothermic formation of **2** after addition of water (Scheme 1).[1] Suggest a mechanism for this transformation.

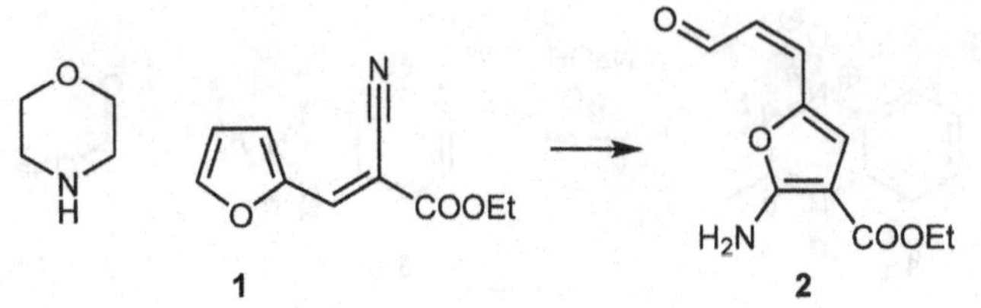

SCHEME 1

SOLUTION

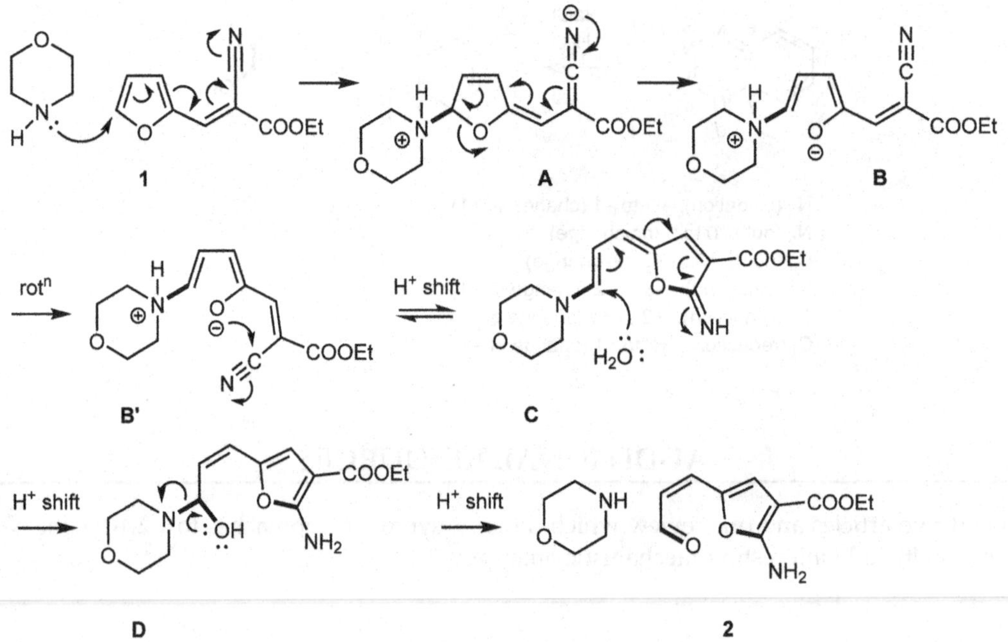

KEY STEPS EXPLAINED

The mechanism proposed by the authors begins with nucleophilic attack by morpholine onto the unsubstituted 2-position of the furan ring of **1** to form intermediate **A** which rearranges to the oxide intermediate **B** after ring opening of the furan ring. With bond rotation, the oxide group of **B'** attacks the nitrile group to form a new ring after proton shift thus giving **C**. This structure is then attacked by water molecule at the carbon atom bearing the morpholine group causing an electron cascade sequence that restores aromaticity in the newly formed furan ring while providing an amino group after proton transfer to give **E**. Elimination of morpholine group in **D** generates the aldehyde group giving the final product **2**. We note that reforming of morpholine suggests the possibility for its recycling in this reaction to minimize waste. We also note that the authors observed a change of color to dark red during the transformation even though the color of the isolated product **2** was yellowish-green. We attribute this color change to the presence of zwitterions **A** and **B** which have an extensive conjugated π-bond system as a notable feature of this morpholine-induced rearrangement. Nevertheless, we also acknowledge that the authors did not carry out further experimental or theoretical analysis to provide evidence for their postulated mechanism.

ADDITIONAL RESOURCES

We recommend an article describing a mechanistic analysis by Slovak chemists of a morpholine-induced ring-opening transformation with an interesting point of divergence.[2] Lastly, we direct the interested reader to a review article which covers transformations of 2-aminofurans and 3-aminofurans.[3]

References

1. Yasuda H, Hayashi T, Midorikawa H. Base-catalyzed reactions of α-cyano-β-furylacrylic esters. *J Org Chem.* 1970;35(4):1234–1235. https://doi.org/10.1021/jo00829a102.
2. Safar P, Povazanec F, Pronayova N, et al. Dichotomy in the ring-opening reaction of 5-[(2-furyl)methylidene]-2,2-dimethyl-1,3-dioxane-4,6-dione with cyclic secondary amines. *Collect Czechoslov Chem Commun.* 2000;65:1911–1938. https://doi.org/10.1135/cccc20001911.
3. Ramsden CA, Milata V. 2-Aminofurans and 3-aminofurans. *Adv Heterocycl Chem.* 2006;92:1–54. https://doi.org/10.1016/S0065-2725(06)92001-7.

Question 177: Ring Expansion of Both Rings of Penicillin Sulfoxides

In 1981 Israeli chemists attempted to synthesize sulfilimine (also known as sulfimide) products by undertaking the reaction of the penicillin sulfoxides **1** with two equivalents of ethoxycarbonyl isocyanate in THF under heat.[1] The products obtained in 31%–54% yield were the unexpected imidazo[5,1-*c*][1,4]thiazines **2** (Scheme 1). Suggest a mechanism for this transformation.

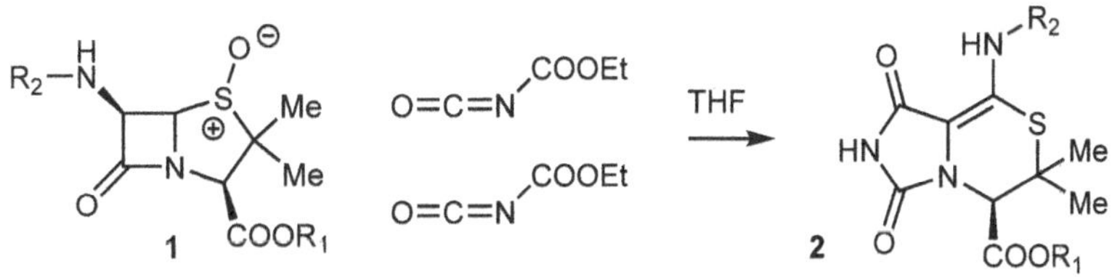

SCHEME 1

SOLUTION

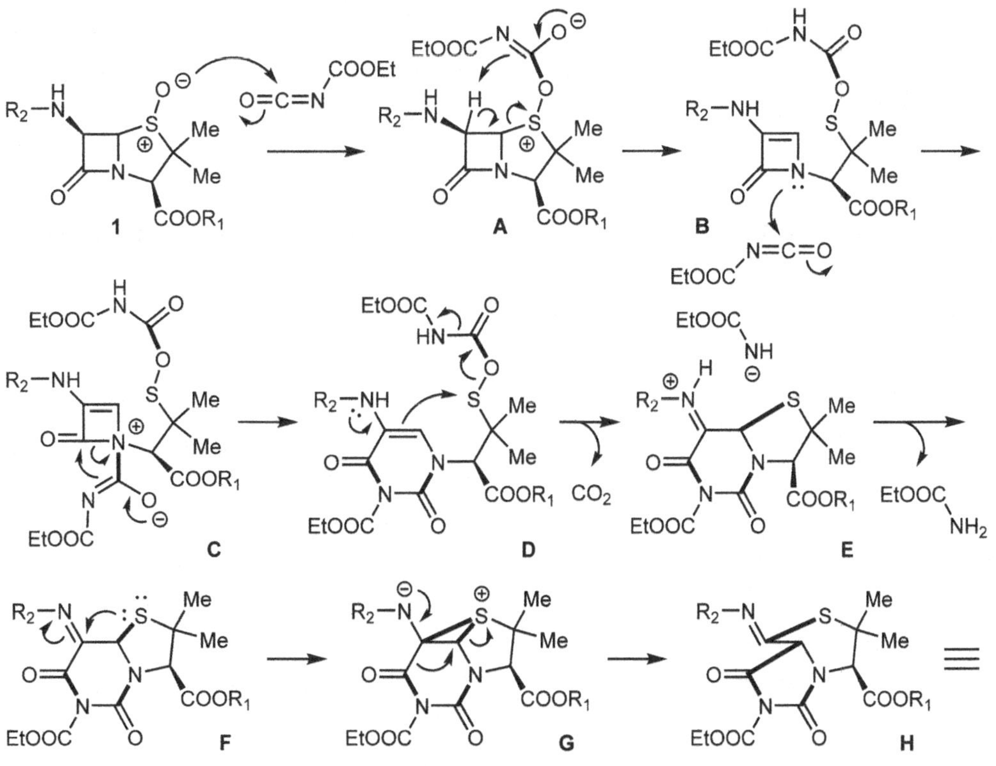

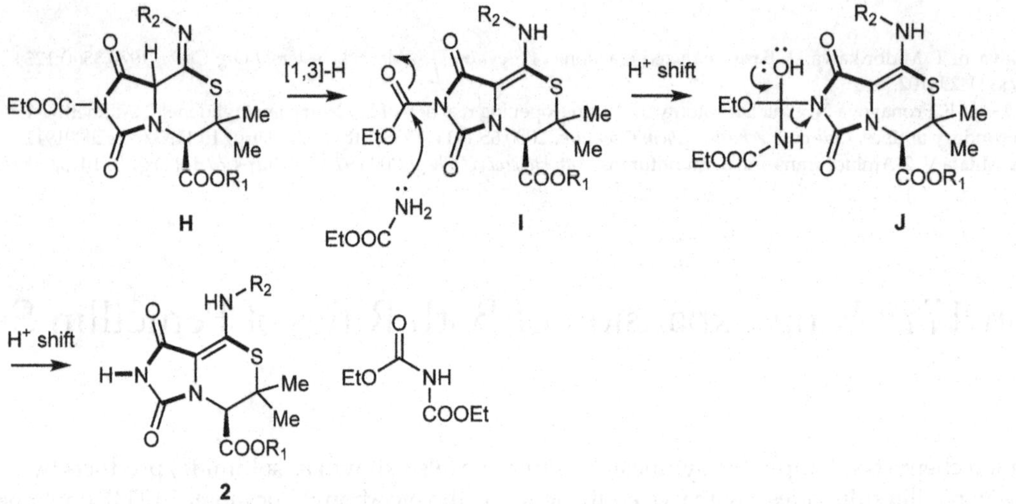

BALANCED CHEMICAL EQUATION

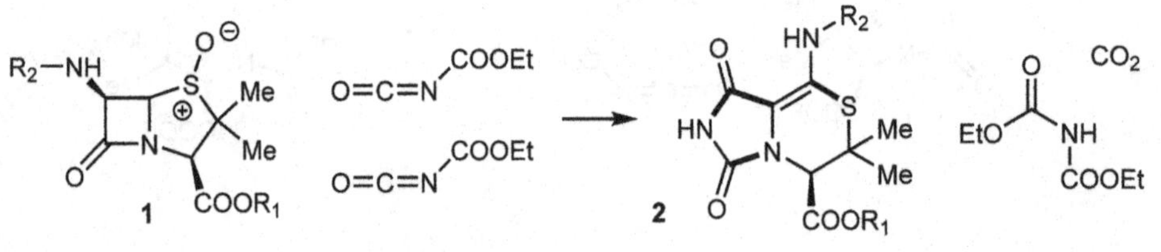

KEY STEPS EXPLAINED

We begin by noting that the authors originally intended to form the sulfilimine (sulfimide) products **4** according to the strategy illustrated in Scheme 2.[2] Nevertheless, the transformation of the penicillin sulfoxides **1** and ethoxycarbonyl isocyanate led to the unexpected formation of imidazo[5,1-*c*][1,4]thiazines **2**. The mechanism postulated by the authors to explain this transformation starts with acylation of the sulfoxide group of **1** via reaction with the first equivalent of ethoxycarbonyl isocyanate to generate intermediate **A**. This structure rearranges to the azet-2-one **B** by means of keto-nization, deprotonation, and C_5—S bond cleavage. The unsaturated β-lactam intermediate **B** then reacts with a second

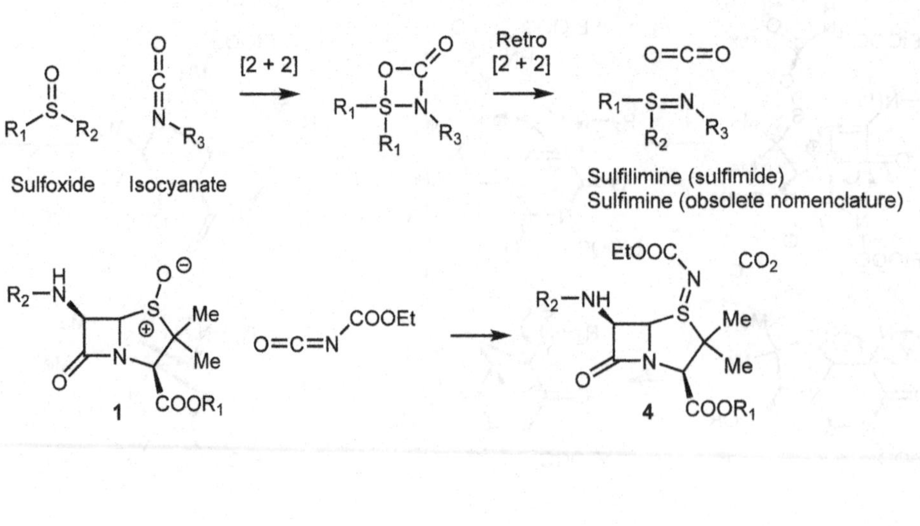

SCHEME 2

equivalent of ethoxycarbonyl isocyanate to form the ring expanded intermediate **D** via **C**. Intermediate **D** cyclizes to a [4.3.0] bicyclic intermediate **F** after decarboxylation and elimination of ethyl carbamate via **E**. Cyclization of **F** to a tricyclic zwitterionic episulfonium intermediate **G** facilitates subsequent rearrangement to the [4.3.0] tricyclic intermediate **H** which after a [1,3]-H shift and elimination of bis(carboxyethyl)amine via ethyl carbamate generate the final thiazine products **2**. The authors stipulate the intervention of the episulfonium intermediate **G** based on literature precedent from the rearrangement of a similar penicillin sulfoxide.[3] Unfortunately the authors do not provide further experimental or theoretical evidence to support their postulated mechanism.

As such, we offer an alternative mechanistic proposal which avoids the need for the sterically hindered high energy tricyclic episulfonium intermediate **G** (Scheme 3). Here, the **A** to **H** transformation occurs by ring cleavage of **A** to form the zwitterionic iminium ion **K** which contains an isocyanate group (possibly detectable by FT-IR spectroscopy). An intramolecular deprotonation of **K** leads to **L** which rearranges to form the azirinium ion **M** after decarboxylation and elimination of ethyl carbamate anion. The ethyl carbamate then attacks the isocyanate group of **M** causing a cascade sequence which results in the formation of a fused [4.1.0] aziridine ring system in **N**. This intermediate may undergo ring cleavage and rearrangement to form the energetically and sterically stable intermediate **P** via **O**. Alkylation of **P** via a second equivalent of ethoxycarbonyl isocyanate leads to intermediate **R** which ring closes to form the bicyclic intermediate **S**. This structure then undergoes a proton shift and elimination of ethyl carbamate to generate the familiar intermediate **H** which proceeds to form product **3** after elimination of bis(carboxyethyl)amine by-product as seen previously.

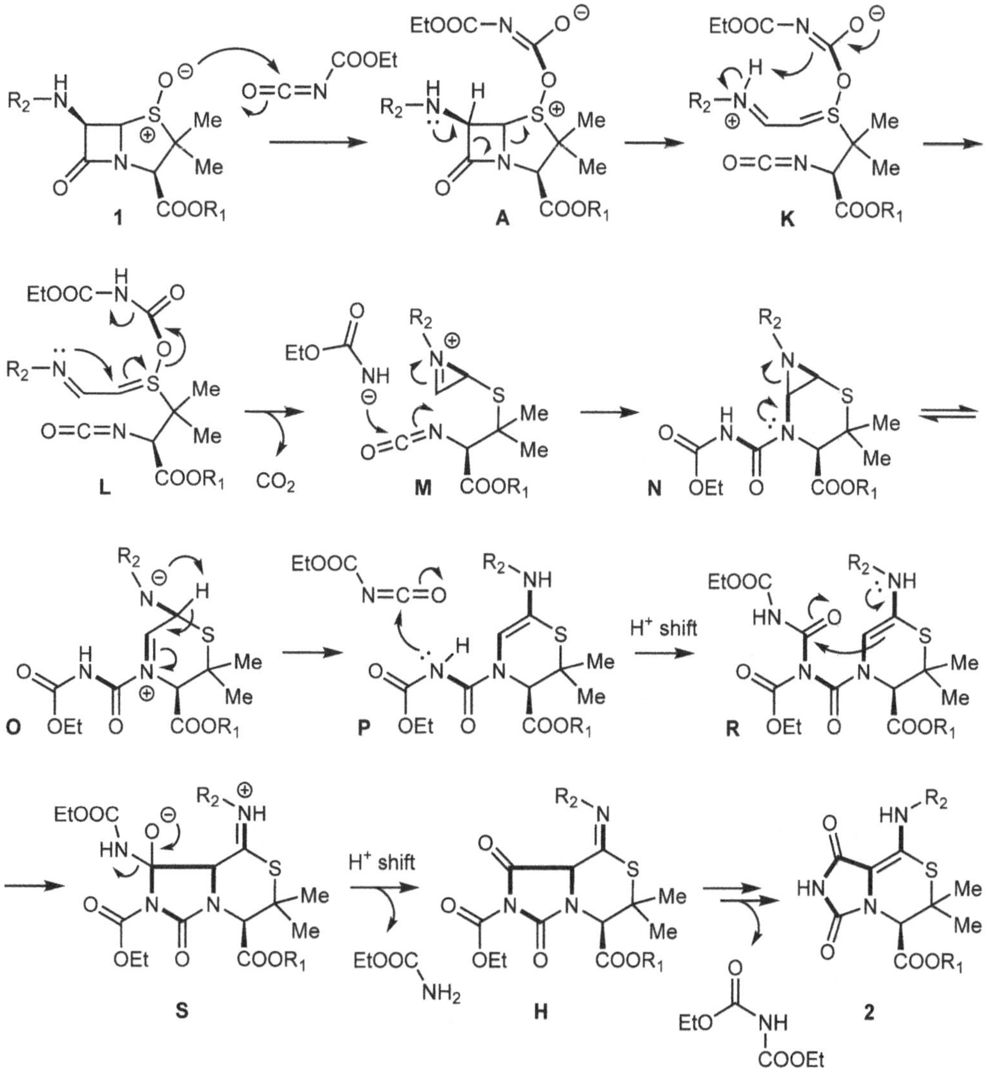

SCHEME 3

We note that one important difference between the two proposed mechanisms is the target bond mapping in the product structure. Whereas in the authors' mechanism the mapping is $[(6+0)+(2+2+1)]$, in the alternative mechanism it is $[(6+0)+(2+1+1+1)]$. Furthermore, the fused C—N bond is a target bond made in the alternative mechanism and not in the author's mechanism. We believe that a future reinvestigation of this transformation ought to undertake a theoretical analysis to help distinguish between the two mechanistic proposals in terms of energy differences on a reaction coordinate diagram.

ADDITIONAL RESOURCES

We direct the reader to two interesting analyses of transformations involving ring expansions which also involve theoretical analyses.[4,5]

References

1. Nudelman A, Haran TE, Shakked Z. Rearrangements of penicillin sulfoxides. 2. Spectral data and X-ray crystallography of the novel imidazo[5,1-c] [1,4]thiazine ring system. *J Org Chem*. 1981;46(15):3026–3029. https://doi.org/10.1021/jo00328a007.
2. Neidlein R, Heukelbach E. Zur Bildung von A-Acyl-sulfiminen. *Arch Pharm*. 1966;299:64–66. https://doi.org/10.1002/ardp.19662990110.
3. Chou TS, Spitzer WA, Dorman DE, et al. New rearrangement of penicillin sulfoxide. *J Org Chem*. 1978;43(20):3835–3837. https://doi.org/10.1021/jo00414a009.
4. Sakai T, Ito S, Furuta H, Kawahara Y, Mori Y. Mechanism of the regio- and diastereoselective ring expansion reaction using trimethylsilyldiazomethane. *Org Lett*. 2012;14(17):4564–4567. https://doi.org/10.1021/ol302032w.
5. Campomanes P, Menendez MI, Sordo TL. Theoretical study of the mechanism of the formation of 3-unsubstituted 4,4-disubstituted β-lactams by silver-induced ring expansion of alkoxycyclopropylamines: a new synthetic route to 4-alkoxycarbonyl-4-alkyl-2-azetidinones. *J Org Chem*. 2003;68:6685–6689. https://doi.org/10.1021/jo034803r.

Question 178: A Retro-Pictet-Spengler Reaction

In a 1997 article exploring enantioselective Pictet-Spengler reactions, Cox et al. observed that stirring of the *cis* and/ or *trans* forms of **1** (R═Me) in acidic mixture at room temperature gave the *trans* form of **1** (R═Me) in 96% yield whereas heating of **1** (R═SO$_2$Ph) in acidic mixture led to products **2** and **3** (Scheme 1).[1] Suggest mechanisms to account for these observations.

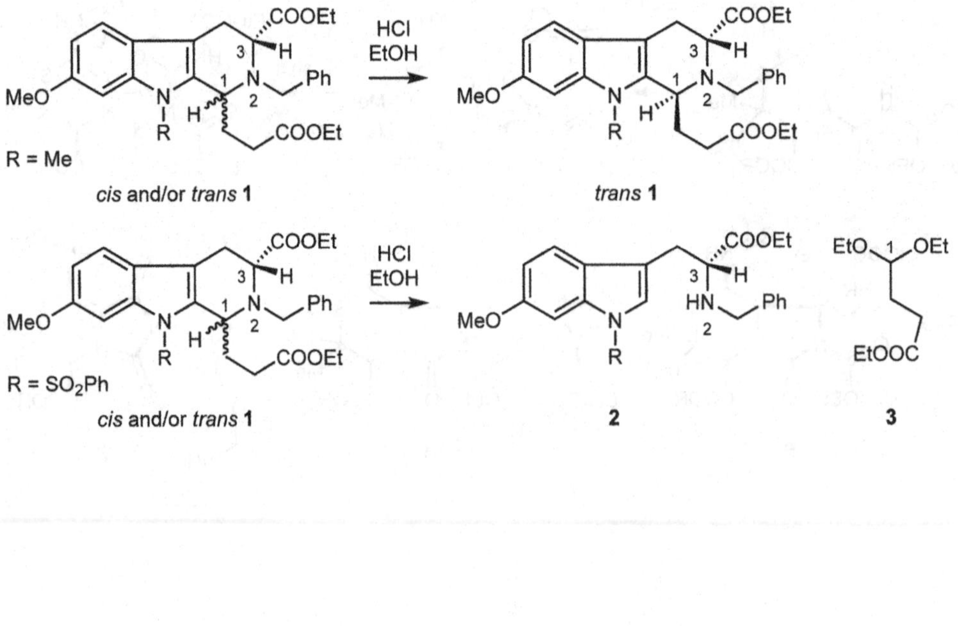

SCHEME 1

SOLUTION: EPIMERIZATION MECHANISM 1 (C₁—N₂ BOND SCISSION)

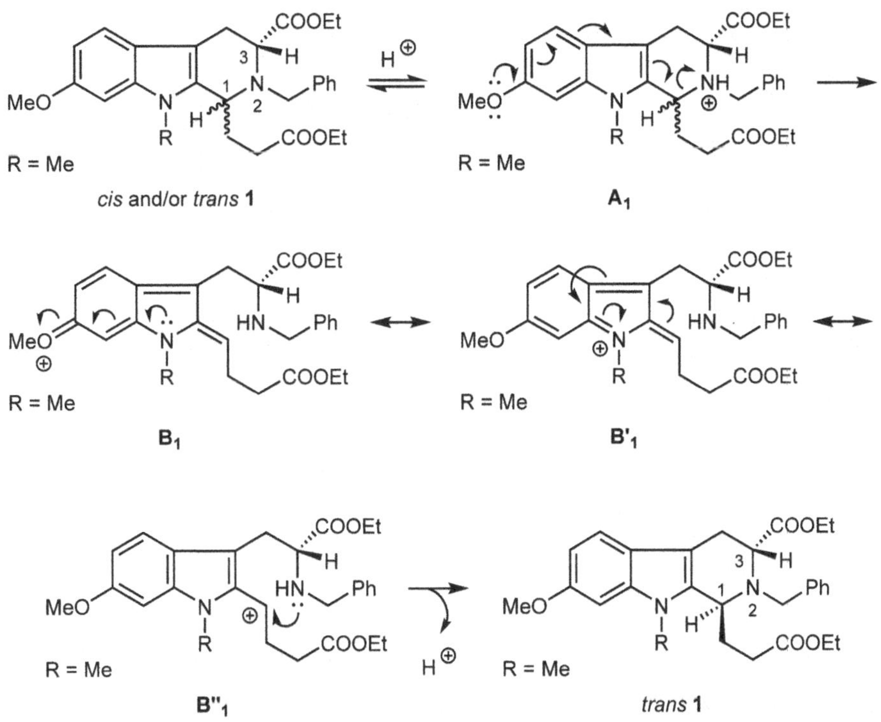

cis and/or trans **1**   **A₁**

B₁   **B'₁**

B"₁   trans **1**

SOLUTION: EPIMERIZATION MECHANISM 2 (OLEFIN PROTONATION)

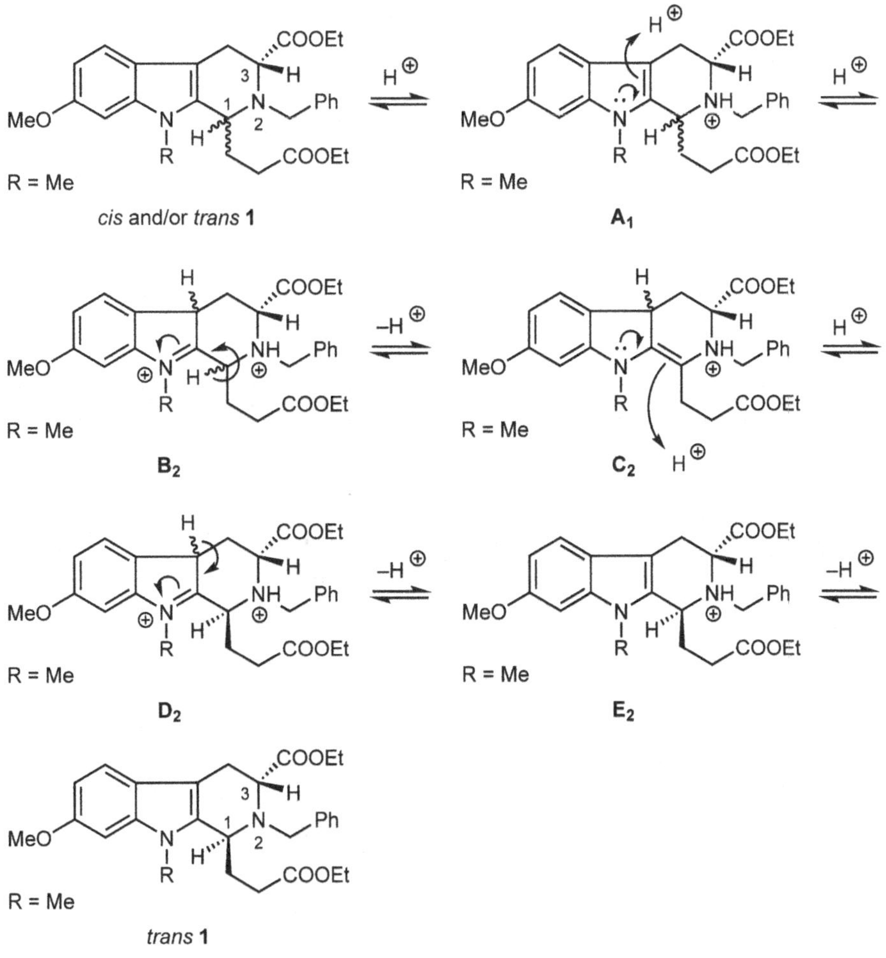

cis and/or trans **1**   **A₁**

B₂   **C₂**

D₂   **E₂**

trans **1**

SOLUTION: EPIMERIZATION MECHANISM 3 (IMINIUM ION FORMATION)

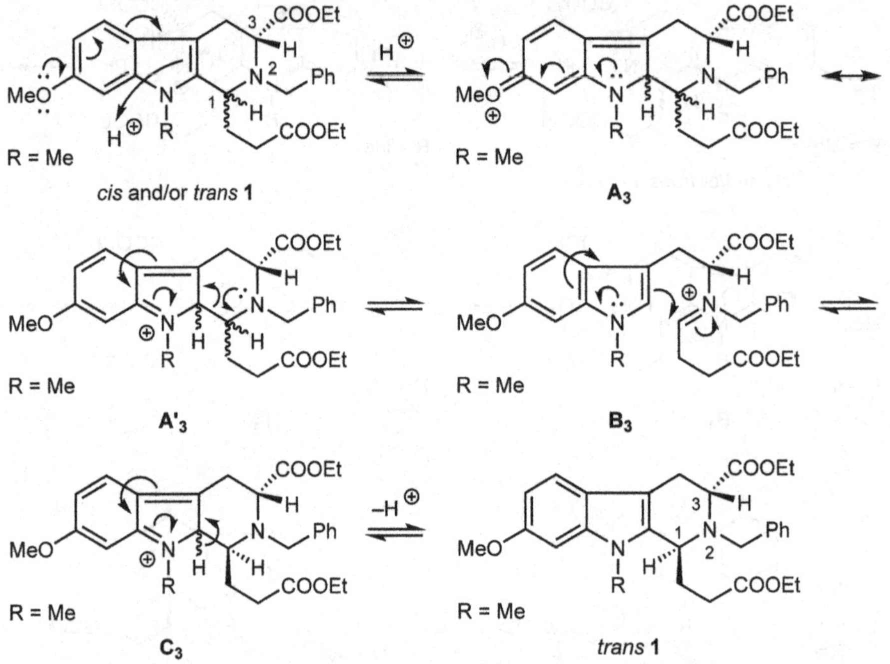
R = Me
cis and/or trans 1
A3
A'3
B3
C3
trans 1

SOLUTION: DEGRADATION MECHANISM 1 (RETRO-PICTET-SPENGLER) PATH A

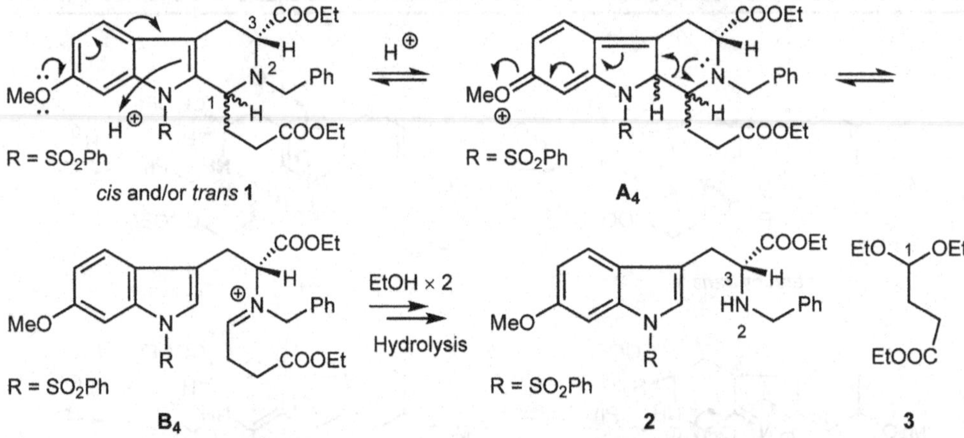
R = SO2Ph
cis and/or trans 1
A4
EtOH × 2
Hydrolysis
B4
2
3

SOLUTION: DEGRADATION MECHANISM 1 (RETRO-PICTET-SPENGLER) PATH B

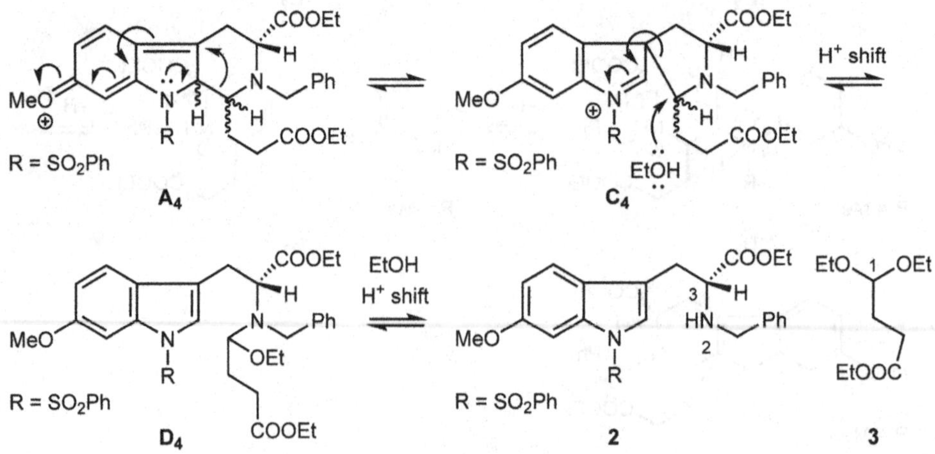
R = SO2Ph
A4
H+ shift
C4
EtOH
D4
EtOH
H+ shift
2
3

KEY STEPS EXPLAINED

Cox et al. considered three different mechanisms for the epimerization of the *cis* form of diester **1** into its *trans* epimer. First, thanks to earlier research on a similar chemical system,[2] the authors entertained Mechanism 1. This mechanism begins with protonation of N_2 to form A_1 which then undergoes electron delocalization from the aryl methoxy group to facilitate cleavage of the C_1—N_2 bond to form the carbocationic intermediate B_1 which is stabilized through resonance in two additional forms. A nucleophilic attack by N_2 of the free amine group then results in recyclization and formation of the more stable *trans* epimer of **1**.

Similarly, Mechanism 2 also begins with protonation of N_2 to form A_1 but diverges at this point with further protonation of the indole olefinic bond to form the iminium ion B_2. Deprotonation of this intermediate at C_1 results in the formation of the olefinic intermediate C_2 which is protonated to D_2 thus establishing the more stable *trans* configuration at C_1. Two further deprotonation steps result in the formation of the *trans* epimer of **1**.

Lastly, the third epimerization mechanism considered by the authors begins with protonation of the indole olefin to form A_3 which is stabilized through resonance in one additional form. Imine formation via N_2 causes bond cleavage between the indole carbon atom and C_1 with rearomatization to generate the imine B_3. This intermediate then undergoes cyclization to form C_3 while also establishing the *trans* configuration at C_1. A last deprotonation step restores aromaticity to form the *trans* epimer of **1**.

These three mechanisms align with the experimental observation that an acidic medium is required for the epimerization to occur. In fact, in nonpolar aprotic media, the authors did not observe a reaction and instead saw only decomposition by-products. The addition of acids such as trifluoroacetic acid (TFA) allowed the epimerization to occur. It was also shown from equilibration studies that the *trans* epimer of **1** was thermodynamically more stable than the *cis* epimer as *cis-trans* mixtures were converted into *trans* product in acidic media in excellent yield. Furthermore, under nonpolar aprotic conditions, analysis of the residues leftover after by-product removal showed the presence of only the *trans* diastereomer even though a mixture of *cis* and *trans* starting material was used. After conducting a theoretical calculation, the authors concluded that the theoretical *anti* intermediate precursor to the *trans* diastereomer was 2.1 kcal/mol less in energy than the analogous *syn* intermediate which was the precursor to the *cis* diastereomer (see Scheme 3 in Ref. 1). These effects, according to the authors, were all governed by steric interactions primarily between the indole nitrogen substituent and the substituents at C_1. Various combinations of substituents were shown to influence the *cis-trans* ratio in the small fraction of isolated product. The addition of TFA, however, shifted all ratios toward the *trans* epimer which at no instance reacted by itself to form the *cis* epimer (i.e., retro-Pictet-Spengler process).

As such, the authors performed a deuterium labeling experiment where CF_3COOH was replaced with CF_3COOD as the acid catalyst. According to Mechanism 2, the expected *trans* epimeric product should have deuterium incorporation at C_1 thanks to the transformation of C_2 to D_2. When the authors analyzed the product of this experiment using ^{1}H NMR and mass spectrometry, they noticed that there was no evidence of deuterium incorporation anywhere in the *trans* product, least of all at C_1. This experiment therefore unequivocally proved that Mechanism 2 was not operational for the *cis-trans* epimerization.

Further experimental evidence was collected when the substituent on the indole nitrogen atom was changed from a methyl/alkyl group to the electron-withdrawing sulfonamide group SO_2Ph. Under these conditions, the authors did not observe the epimerization reaction take place and instead observed the decomposition products **2** and **3**. Thus they argued that because the sulfonamide group worked to withdraw electron density from the indole ring, the likely carbocationic intermediate B_1 (which is pivotal for the epimerization Mechanism 1) could not be efficiently stabilized by electronic means as with the case of a methyl substituted indole nitrogen group. While this piece of evidence both supports Mechanism 1 (i.e., inefficient stabilization of proposed carbocationic intermediate) and does not support Mechanism 3 (i.e., no detection of *trans* epimer of **1**), the authors felt they needed more direct evidence to contradict the operation of Mechanism 3.

They therefore ran the reaction (R=Me) under heat and in proton-free solution which consisted of dry distilled chloroform free from EtOH and HCl with added 4.7 equivalents of anhydrous $ZnCl_2$. Under these conditions, the *cis* to *trans* epimerization was successful and had a yield of 92%.[1] This outcome represented indirect evidence in support of Mechanism 1 (i.e., the authors thought that the Lewis acid $ZnCl_2$ promoted the formation of the carbocationic intermediate B_1). It also represented direct evidence which contradicts the operation of Mechanism 3 in the sense that there are no sources of protons in the reaction mixture to allow for the initial protonation of **1** to A_3. The role of $ZnCl_2$ was confirmed by the control experiment where the absence of $ZnCl_2$ led to no reaction.

Interestingly, the authors also carried out the reaction in the presence of sodium borohydride ($NaBH_4$, i.e., reducing conditions). The results of this experiment are shown in Scheme 2. This experiment showed that product **4** must arise from sodium borohydride reduction of a carbocationic intermediate such as B'_1 (Mechanism 1) rather than borane-mediated cleavage of C_1—N_2 bond (Mechanism 3). If Mechanism 3 were operational, it would be expected that both

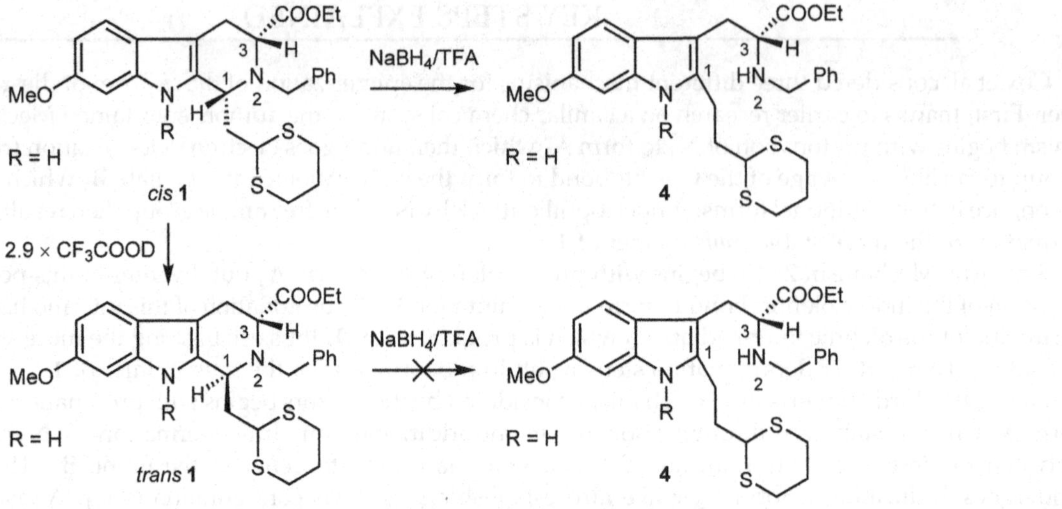

SCHEME 2

cis and trans epimers of **1** react under NaBH₄/TFA to form the cleavage product **4**. Although the authors did not undertake a theoretical analysis of any of the mechanisms they analyzed, because all their experimental evidence supported Mechanism 1 and none supported Mechanisms 2 and 3, it was concluded that Mechanism 1 was operational for the epimerization process.

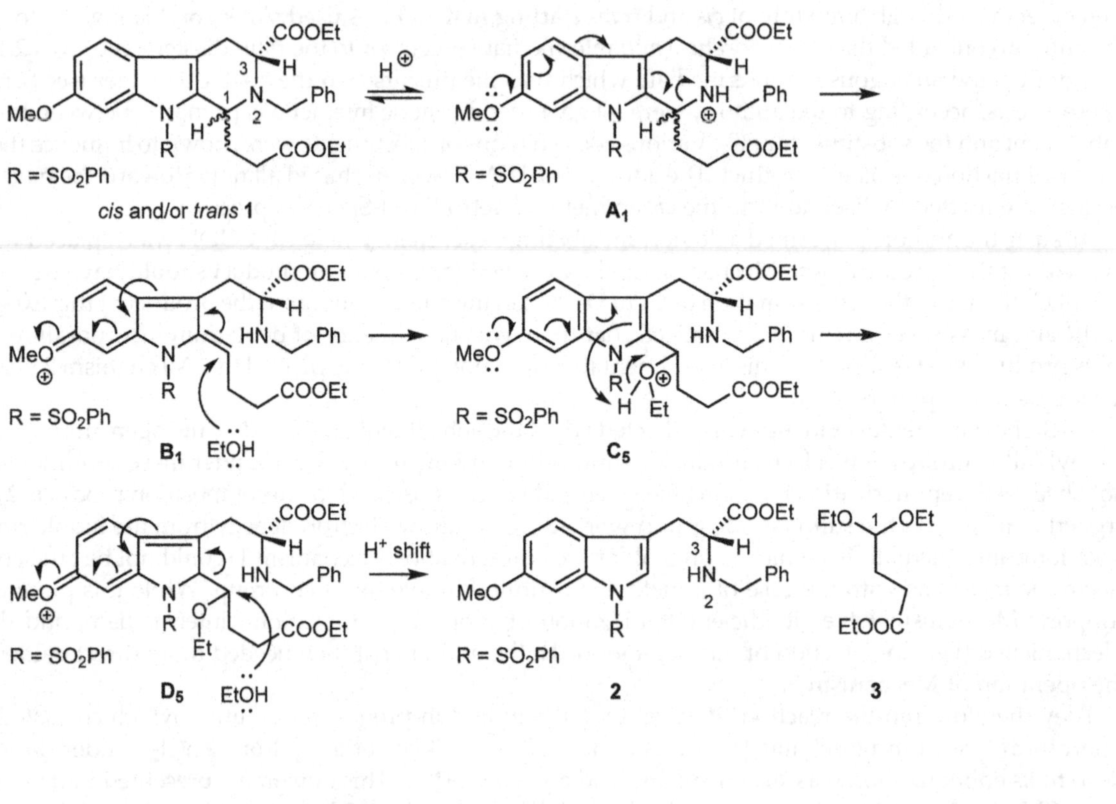

SCHEME 3

Consequently, by a process of elimination, the authors proposed Mechanism 3 to explain the fragmentation of **1** to decomposition products **2** and **3**. This is essentially a retro-Pictet-Spengler process which begins with protonation of the indole olefin to form **A₄** followed by imine formation and rearomatization to **B₄** followed by two hydrolysis steps by the solvent (ethanol) that cause elimination of **3** to form **2**. The transformation of **A₄** to **2** was also presented as following two possible paths which cannot be judged in the absence of further experimental and theoretical evidence. Furthermore, Mechanism 1 was discounted because no *trans* epimer of **1** was detected among products **2** and **3** likely due to the electron-withdrawing effect of the sulfonamide group on the indole nitrogen atom. Unfortunately the authors did not perform a deuterium labeling study under these conditions to verify if an alternative mechanism analogous to Mechanism 2 should also be discounted (Scheme 3). According to the author's mechanism, using CF₃COOD would allow for deuterium incorporation at the nonaryl carbon atom *alpha* to the indole nitrogen confirming the initial protonation of **1** to **A₄**. Such a label would be absent in product **2** if the alternative mechanism were operational. In either case, a theoretical study would also work to help differentiate between the two possible mechanisms for the degradation of **1** to **2** and **3**.

ADDITIONAL RESOURCES

For further work by Cook et al. involving synthetic approaches to indole-containing alkaloids along with mechanistic discussion, see Refs. 2, 3. For a review article describing work aimed at the formation of chiral heterocycles by iminium ion cyclization (including the Pictet-Spengler reaction), see the work of Royer et al.[4]

References

1. Cox ED, Hamaker LK, Li J, et al. Enantiospecific formation of *trans* 1,3-disubstituted tetrahydro-β-carbolines by the Pictet-Spengler reaction and conversion of *cis* diastereomers into their *trans* counterparts by scission of the C-1/N-2 bond. *J Org Chem.* 1997;62(1):44–61. https://doi.org/10.1021/jo951170a.
2. Zhang L-H, Gupta A, Cook JM. Reinvestigation of the mechanism of the acid-catalyzed epimerization of reserpine to isoreserpine. *J Org Chem.* 1989;54(19):4708–4712. https://doi.org/10.1021/jo00280a052.
3. Yu J, Wang T, Liu X, et al. General approach for the synthesis of sarpagine indole alkaloids. Enantiospecific total synthesis of (+)-vellosimine, (+)-normacusine B, (-)-alkaloid Q3, (-)-panarine, (+)-N_a-methylvellosimine, and (+)-N_a-methyl-16-epipericyclivine. *J Org Chem.* 2003;68(20):7565–7581. https://doi.org/10.1021/jo030006h.
4. Royer J, Bonin M, Micouin L. Chiral heterocycles by iminium ion cyclization. *Chem Rev.* 2004;104:2311–2352. https://doi.org/10.1021/cr020083x.

Question 179: An Olefin to α-Hydroxy Ketone Transformation

In 1982 chemists from Scotland observed that the olefin **1** can be oxidized using excess monoperphthalic acid in ether/dichloromethane at 4°C in the dark to a 74% yield of a 6:1 mixture of epimeric ketols **2** (*trans*) and **3** (*cis*), respectively (Scheme 1).[1] Suggest a mechanistic explanation for this outcome.

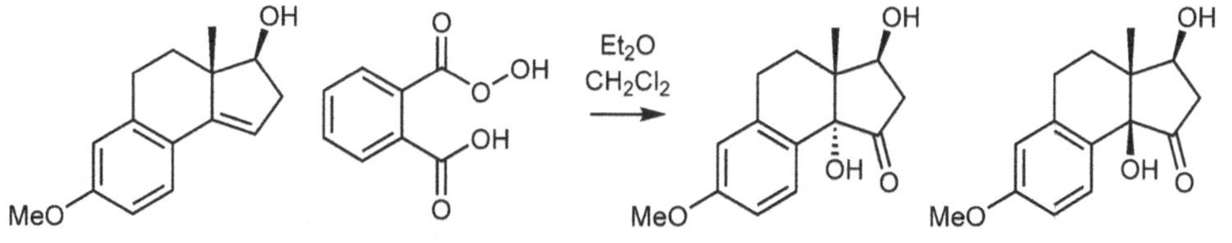

SCHEME 1

SOLUTION

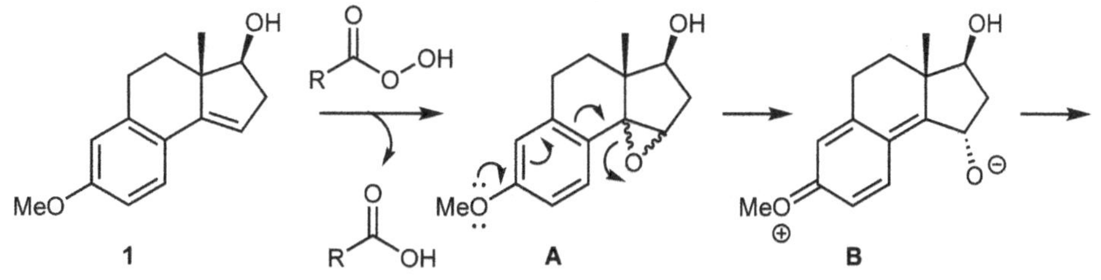

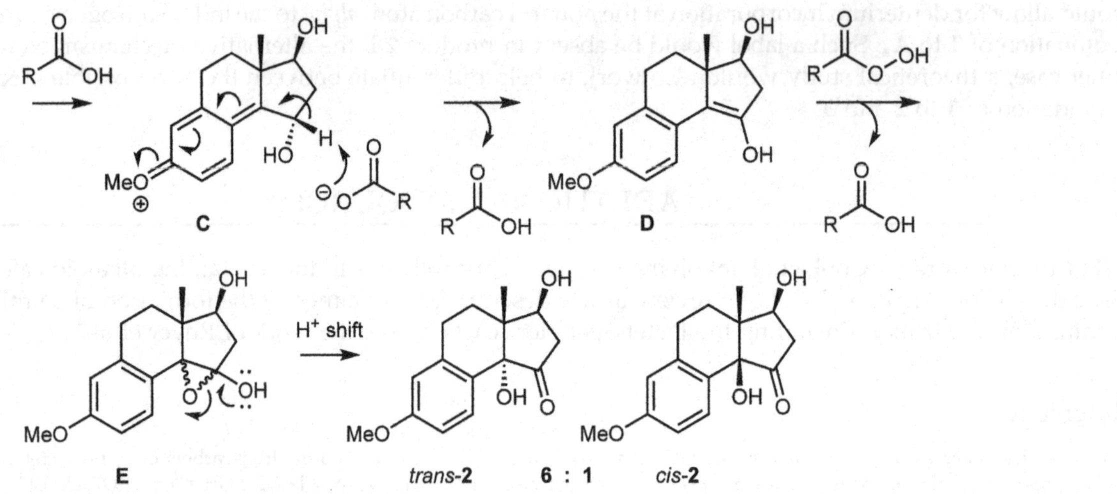

BALANCED CHEMICAL EQUATION

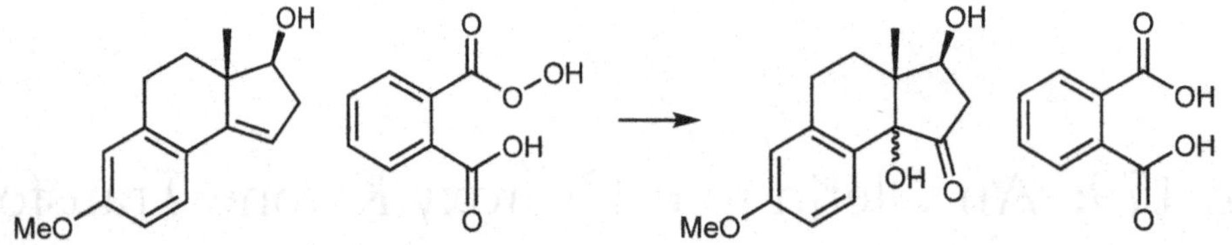

KEY STEPS EXPLAINED

The mechanism proposed by the authors begins with epoxidation of the olefinic bond in the five-member ring of **1** using one equivalent of peroxyacid reagent to give epoxide **A**. Next, electron delocalization from the aryl methoxy group causes ring opening of the newly formed epoxide to form a zwitterionic intermediate **B**. This species undergoes protonation to form **C** followed by deprotonation to form the enol **D** after rearomatization of the aryl ring. Intermediate **D** then reacts with a second equivalent of peroxyacid reagent to form epoxide **E** which has its hemiketal group undergo fragmentation with concomitant epoxide ring opening to generate product **2** after a final proton shift. The authors argue that both epoxidation events are more likely to occur on the less sterically hindered α-styrene face in **1** and **D** which they believe explains the six-fold stereoselectivity of the transformation in favor of the *trans* product. Nevertheless, the authors do not provide additional experimental and theoretical evidence to help support their postulated mechanism.

ADDITIONAL RESOURCES

For interesting transformations of compounds involving the α-hydroxy ketone functionality, see Refs. 2, 3.

References

1. Cox PJ, Howie RA, Nowicki AW, Turner AB. Ketol formation by peracid oxidation of a tricyclic styrene. *J Chem Soc Perkin Trans 1*. 1982;1982:657–664. https://doi.org/10.1039/P19820000657.
2. Palmisano G, Danieli B, Lesma G, Mauro M. α-Hydroxy ketone rearrangement as a key step en route to the calebassinine skeleton. *J Chem Soc Chem Commun*. 1986;1986:1564–1565. https://doi.org/10.1039/C39860001564.
3. Frearson MJ, Brown DM. Skeletal rearrangement of α-hydroxy-ketones upon electron impact. *J Chem Soc C*. 1968;1968:2909–2912. https://doi.org/10.1039/J39680002909.

Question 180: A Benzofuran From a Cyclopropachromone

During studies on the reactivity of cyclopropyl epoxides, Irish chemists reacted the cyclopropachromone **1** in DMSO with a solution of dimethylsulfonium methylide in DMSO/THF at room temperature for 1.5h followed by aqueous quenching to generate the benzofuran product **2** in 30% yield (Scheme 1).[1] Provide a mechanism for this transformation.

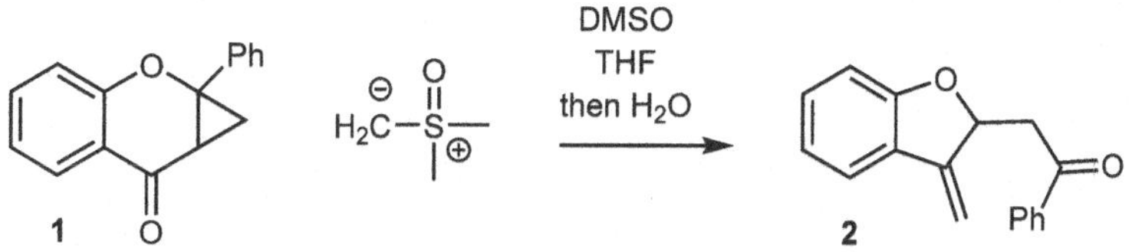

SCHEME 1

SOLUTION

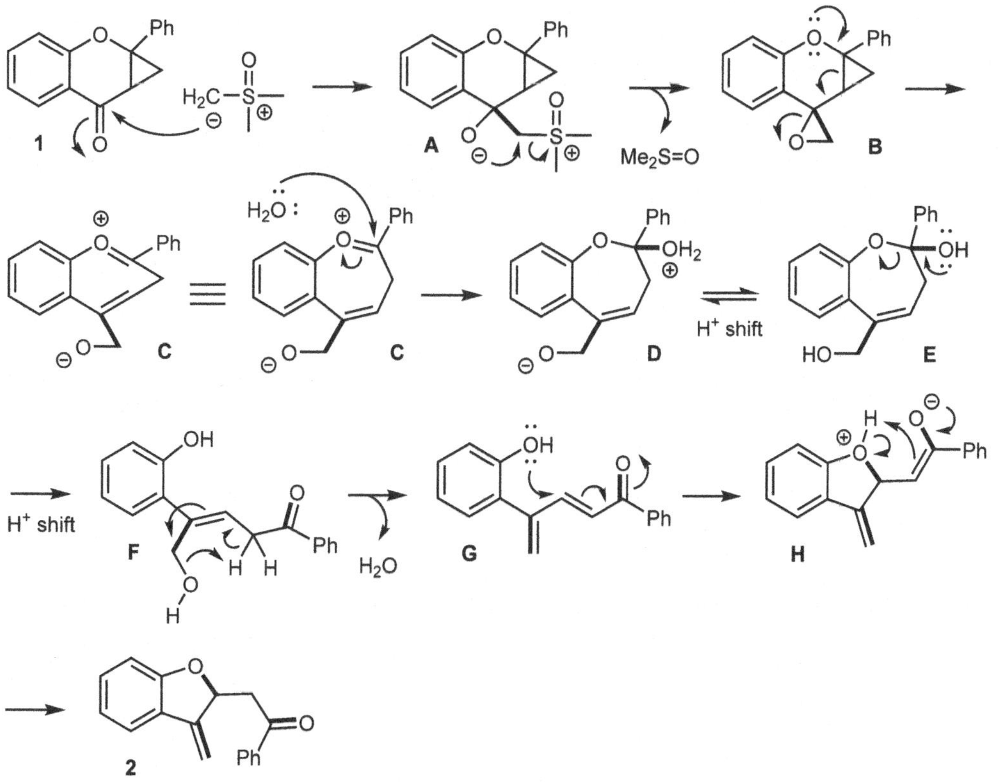

BALANCED CHEMICAL EQUATION

KEY STEPS EXPLAINED

The mechanism proposed by the authors for this transformation begins with nucleophilic addition of dimethylsulfonium methylide to the chroman-4-one ketone group of **1** to give sulfonium alkoxide **A**. This intermediate then ring closes to epoxide **B** after elimination of dimethyl sulfoxide. The steps following this sequence likely take place in the aqueous work-up phase of the reaction. Essentially, intermediate **B** undergoes ring opening to form the seven-membered ring oxonium alkoxide **C** which is then attacked by a water molecule at the electrophilic carbonyl group to give hemiketal **E** after proton shift via **D**. Intermediate **E** then undergoes ring opening to generate the β,γ-unsaturated ketone **F** which after elimination of water gives the conjugated dienone intermediate **G**. This intermediate undergoes cyclization via the phenolic hydroxyl group in a Michael fashion at C$_\beta$ to generate the final benzofuran product **2** which contains an exocyclic methylene group. We note that the authors did not provide further experimental or theoretical data to support their postulated mechanism.

ADDITIONAL RESOURCES

For two examples of interesting mechanisms involving, on one hand, Michael-Michael-ring-closure, and on the other hand, a very similar transformation to the one explored in this problem but with sulfur chemistry, see Refs. 2, 3. Lastly, we also recommend a review article highlighting the use of dimethylsulfonium methylide in synthetic chemistry.[4]

References

1. Bennett P, Donnelly JA, Fox MJ. The molecular rearrangements of cyclopropyl epoxides generated from various flavonoid systems. *J Chem Soc Perkin Trans 1*. 1979;1979:2990–2994. https://doi.org/10.1039/P19790002990.
2. Posner GH, Mallamo JP, Black AY. Tandem Michael-Michael-ring closure (MIMIRC) reactions: one-pot steroid total synthesis-(±)-9,11-dehydroestrones. *Tetrahedron*. 1981;37(23):3921–3926. https://doi.org/10.1016/S0040-4020(01)93265-X.
3. Makosza M, Sypniewski M. Reaction of sulfonium salts of formaldehyde dithioacetals with aromatic aldehydes and rearrangements of the produced thioalkyl oxiranes. *Tetrahedron*. 1995;51(38):10593–10600. https://doi.org/10.1016/0040-4020(95)00632-I.
4. Gololobov YG, Nesmeyanov AN, Lysenko VP, Boldeskul IE. Twenty-five years of dimethylsulfonium methylide (Corey's reagent). *Tetrahedron*. 1987;43(12):2609–2651. https://doi.org/10.1016/S0040-4020(01)86869-1.

Question 181: A Pyridocarbazole to Pyridazinocarbazole Rearrangement

In 1978 Manchester chemists tried to reduce the ketone group of the tetrahydrocarbazole derivative **1** under Wolff-Kishner conditions which unexpectedly gave the pyridazinocarbazole product **2** in 62% yield (Scheme 1).[1] Devise a mechanism for this transformation.

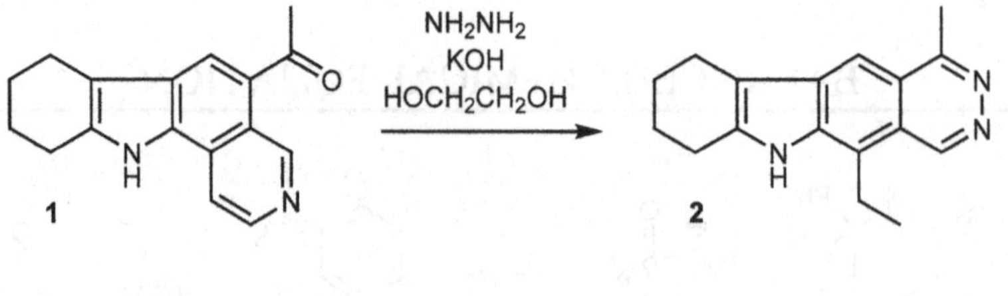

SCHEME 1

SOLUTION

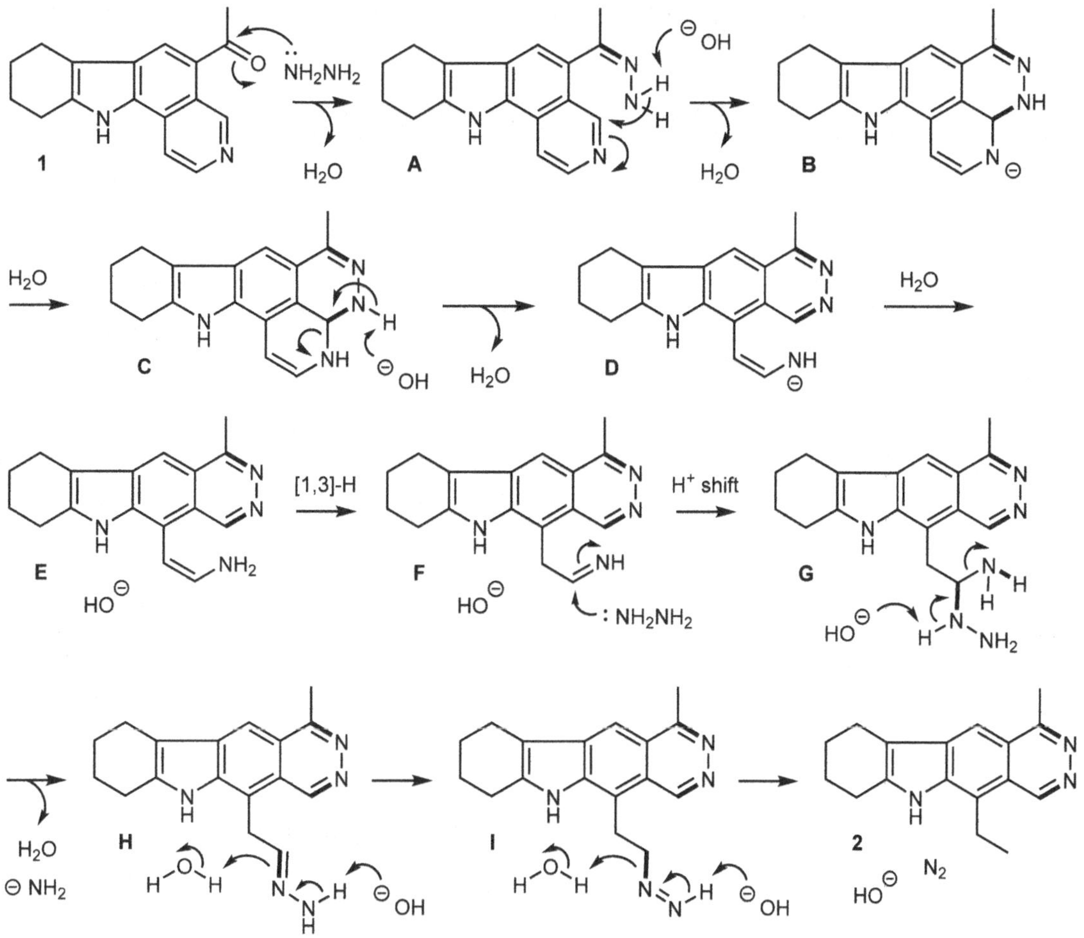

BALANCED CHEMICAL EQUATION

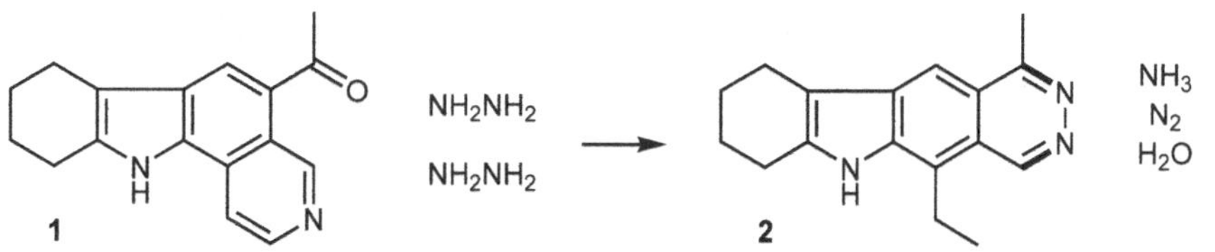

KEY STEPS EXPLAINED

We note first that the original intention of the authors was to carry out the Wolff-Kishner reduction of the ketone group of **1** to a methylene group as shown in Scheme 2.

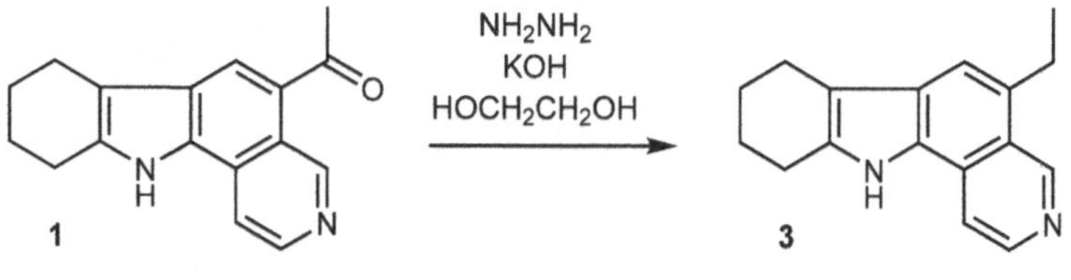

SCHEME 2

Instead, Baradarani and Joule obtained the unexpected product **2**. Following this experimental observation, the mechanism they proposed begins with the normal formation of hydrazone **A** using one equivalent of hydrazine. Base removal of a terminal hydrogen atom from the hydrazone moiety then leads to cyclization which generates intermediate **C**. A second base removal of the remaining hydrogen atom causes cleavage of the pyridine ring to form enamine **E** which then undergoes a [1,3]-H shift yielding imine **F**. Imine **F** is then captured by another equivalent of hydrazine to give intermediate **G**. The remaining steps leading to product **2** involve successive deprotonations by base causing elimination of ammonia and nitrogen gas as by-products. The authors, however, did not provide any further experimental or theoretical data to help support their postulated mechanism.

ADDITIONAL RESOURCES

For further work by Joule et al., see Ref. 2.

References

1. Baradarani MM, Joule JA. Synthesis of pyridazo[4,5-b]carbazoles. *J Chem Soc Chem Commun.* 1978;1978:309–310. https://doi.org/10.1039/C39780000309.
2. Ashcroft WR, Beal MG, Joule JA. Efficient syntheses of 'ellipticine quinone' and the other three isomeric 5*H*-pyrido[x,y-b]carbazole-5,11(6*H*)-diones. *J Chem Soc Chem Commun.* 1981;1981:994–995. https://doi.org/10.1039/C39810000994.

Question 182: Substituent Group Effect During Sulfide → Sulfoxide Oxidation

During a 1979 study involving the oxidation of sulfides, chemists from Iowa State University observed the reaction sequences depicted in Scheme 1.[1] Suggest a mechanism for the transformation of **1** to the *p*-nitrobenzaldehyde product **2** under the given reaction conditions.

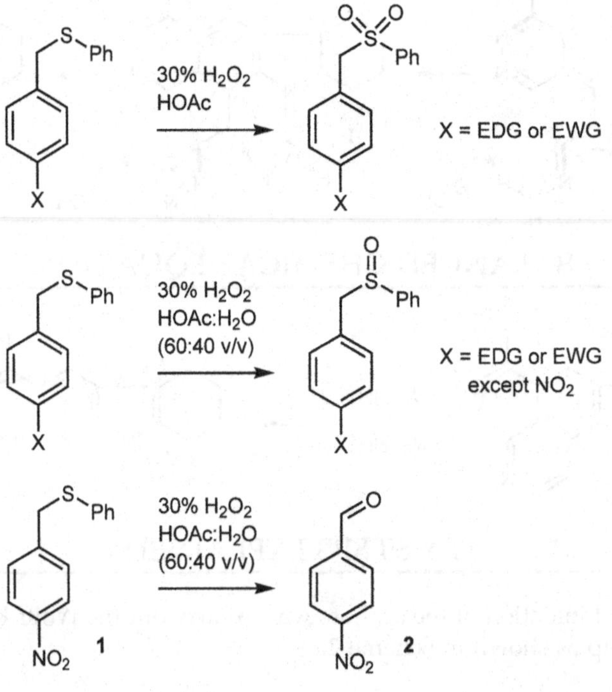

SCHEME 1

SOLUTION

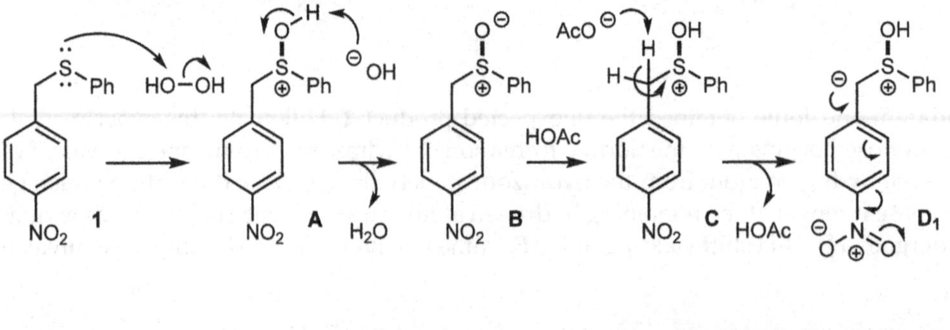

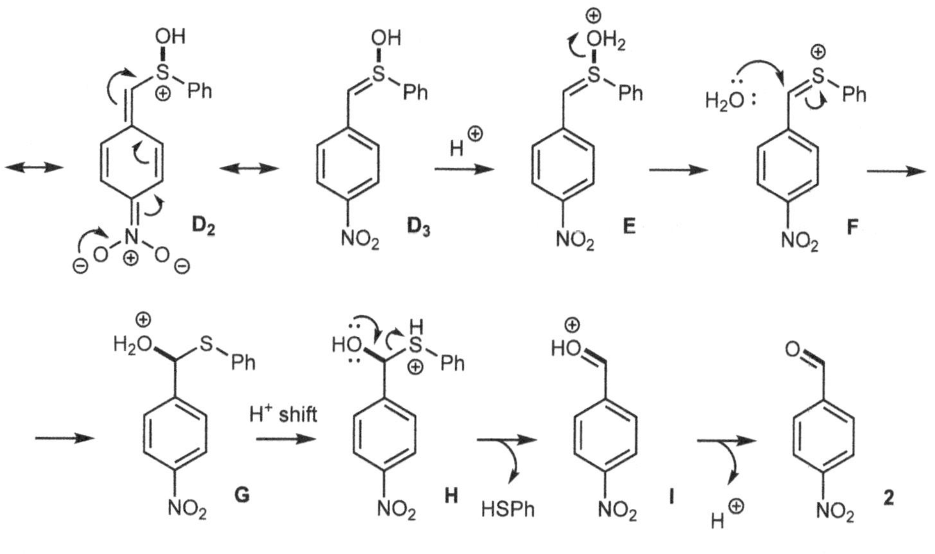

BALANCED CHEMICAL EQUATION

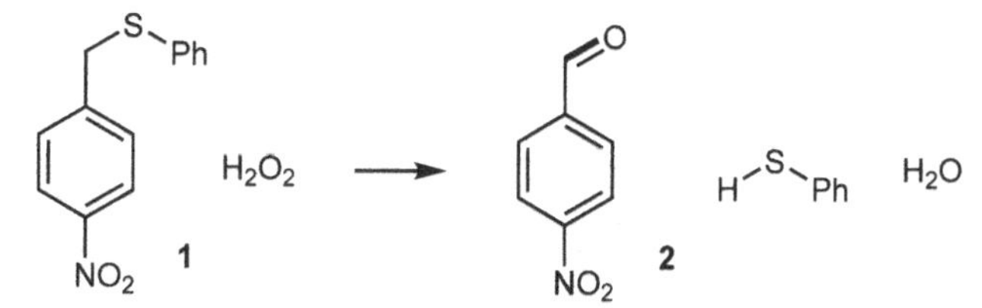

KEY STEPS EXPLAINED

The mechanism proposed by the authors to explain the transformation of the sulfide **1** into the benzaldehyde **2** begins with hydrogen peroxide-mediated oxidation of **1** to the sulfoxide intermediate **B** via **A**. The sulfoxide oxygen atom of **B** is subsequently protonated in the acidic medium to form **C** which is then deprotonated at the benzylic position to form the zwitterionic intermediate **D$_1$** which in turn is resonance stabilized by the nitro group in the *para* position via **D$_2$** and **D$_3$** resonance structures. The authors reasoned that this stabilization is why the nitro-substituted sulfides promote the Pummerer reaction as opposed to the formation of the sulfone products seen in Scheme 1 where the *para* substituent is not a nitro group. Therefore, to sum up the mechanism, from structure **D$_3$**, protonation of the hydroxyl group by the acidic medium forms **E** which eliminates water to form the sulfonium ion **F** which is then attacked by water at the benzyl carbon position to generate intermediate **G**. A final proton shift which forms **H** promotes the elimination of benzenethiol to form **I** which loses a proton to generate the final *p*-nitrobenzaldehyde product **2**. It is important to note that the authors did not perform further experimental and theoretical work to help support their postulated mechanism. For example, by-product identification of benzenethiol would produce one such supporting piece of evidence. We also note that in the transformation of sulfoxide **B** to product **2**, the sulfur atom is reduced from 0 to −2 (net change of −2) and the benzylic carbon atom is oxidized from −1 to +1 (net change of +2).

ADDITIONAL RESOURCES

For an interesting article on substituent effects in the context of the photocleavage mechanism for benzyl-sulfur bonds, this time involving a *meta* effect, see the work of Fleming and Jensen.[2]

References

1. Russell GA, Pecoraro JM. Pummerer reaction of para-substituted benzylic sulfoxides. *J Org Chem*. 1979;44(22):3990–3991. https://doi.org/10.1021/jo01336a060.
2. Fleming SA, Jensen AW. Substituent effects on the photocleavage of benzyl-sulfur bonds. Observation of the "*meta* effect" *J Org Chem*. 1996;61(20):7040–7044. https://doi.org/10.1021/jo9606923.

Question 183: Ring Contraction of a Benzodiazepine

In 1976 researchers from Michigan reacted the 1,4-benzodiazepine **1** with 2,4-pentanedione and formed, among other products, 11% of the quinoline product **2** (Scheme 1).[1] Provide two possible mechanisms for this transformation.

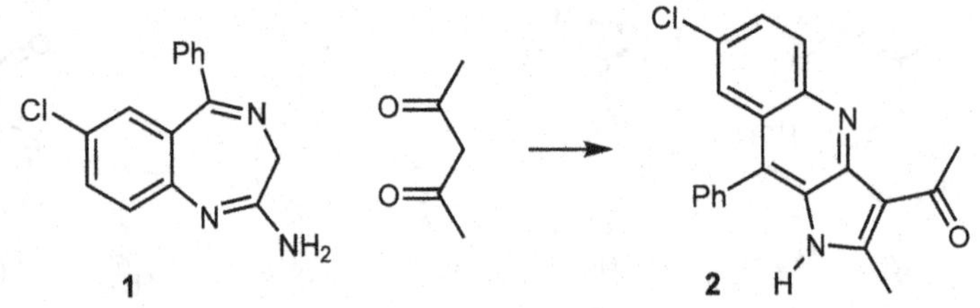

SCHEME 1

SOLUTION: MECHANISM 1A (AUTHORS' MECHANISM)

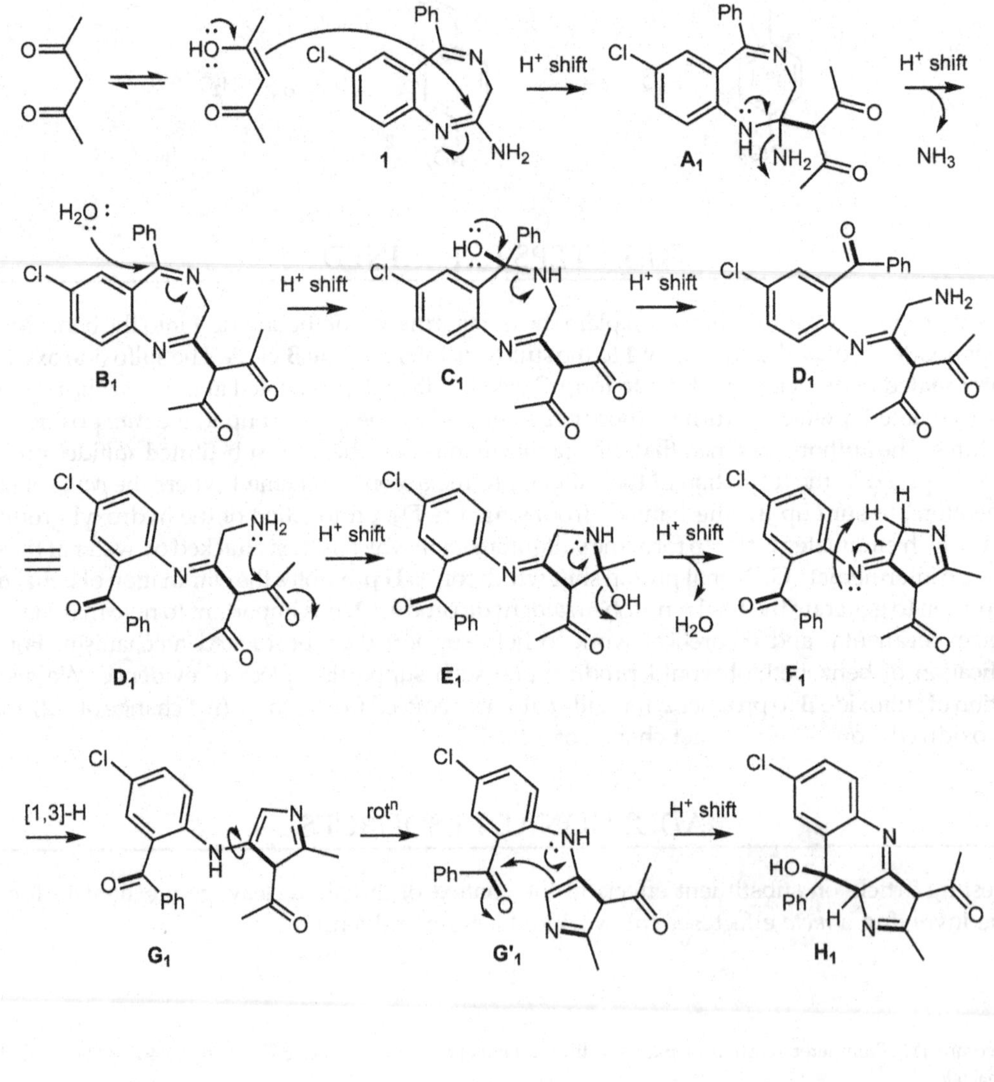

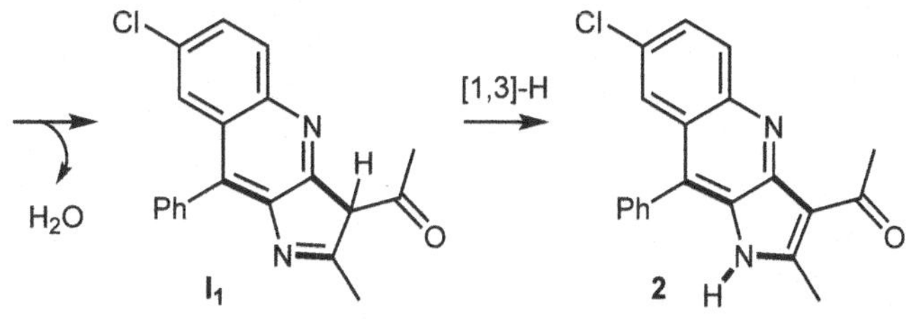

SOLUTION: MECHANISM 1B (AUTHORS' MECHANISM)

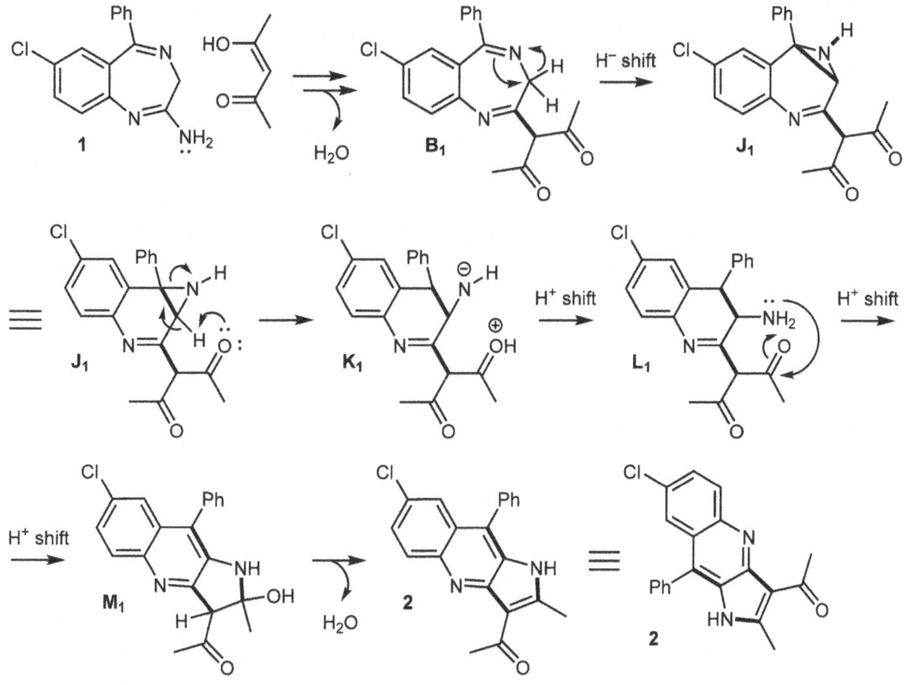

SOLUTION: MECHANISM 2 (ALTERNATIVE MECHANISM)

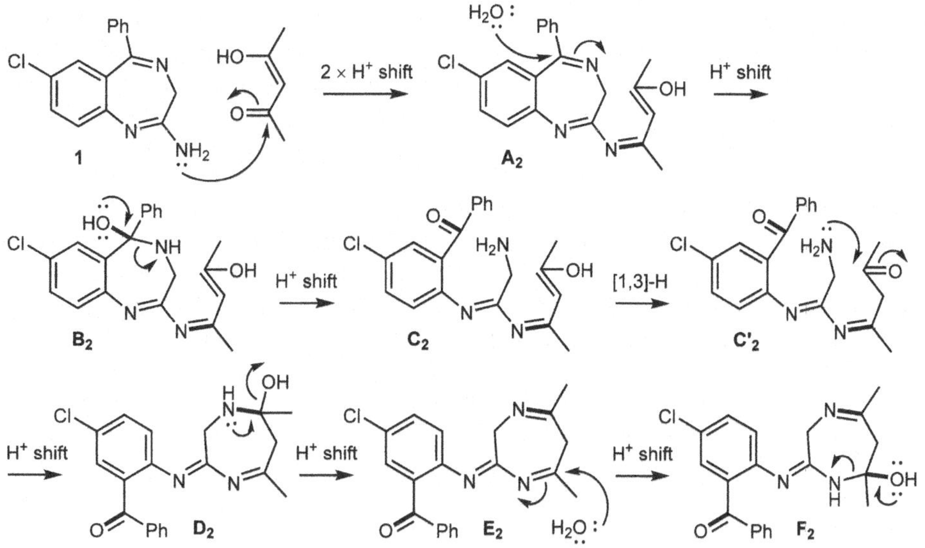

H+ shift   Enolization   H+ shift

G₂   **H₂**   **I₂**

H+ shift   [1,3]-H   rotⁿ

NH₃

J₂   **K₂**   **K₂**

H+ shift   H₂O   [1,3]-H

L₂   **M₂**   **2**

BALANCED CHEMICAL EQUATION

1   **2**   NH₃   H₂O

KEY STEPS EXPLAINED

The two mechanisms proposed by the authors to explain the transformation of benzodiazepine **1** and 2,4-pentanedione into the quinoline product **2** are represented as Mechanism 1A and Mechanism 1B. These mechanisms are the same until structure B_1 where a divergence occurs to form either the open-chain amino ketone intermediate D_1 (Mechanism 1A) or the aziridine intermediate J_1 (Mechanism 1B). To arrive at B_1, the authors proposed an initial alkylation of 2,4-pentanedione where a bond is made between C_h and C_d (see labels assigned in the Balanced Chemical Equation scheme given earlier) to give intermediate A_1 which then eliminates ammonia to give the imine intermediate B_1. At this point, it is possible to have water adding to C_a to cause the seven-membered ring to open and generate intermediate D_1 via C_1. The free N_b amino group of D_1 then attacks C_g in a nucleophilic manner to create a five-membered ring with water elimination to form intermediate F_1. After a [1,3]-H shift of F_1 to G_1 and rotation about the N_e—C_d bond, intermediate G_1 cyclizes to form a six-membered ring at C_a leading to H_1. Product **2** is obtained after elimination of water followed by a final [1,3]-H shift. The target bond mapping of the product structure according to Mechanism 1A indicates that the ring construction strategy is [(6+0)+(3+2)].

Moreover, the authors proposed that intermediate B_1 can also undergo a hydride shift to form the aziridine intermediate J_1 which then has one of the ketone oxygen atoms abstract the C_h hydrogen atom from the aziridine ring causing it to open and form L_1 after a hydrogen shift (Mechanism 1B). The resulting free N_b amino group of L_1 now participates in recyclization at C_g to give intermediate N_1 which after a final dehydration generates the quinoline product **2**. We note that the same ring frame target bond map for the product structure is obtained as for mechanism 1A.

Although the authors claim to have isolated the enol intermediate B_1, they do not provide further experimental and especially theoretical evidence to support either of their postulated mechanisms. Therefore, for contrast, we present an alternative Mechanism 2 and suggest that in a future mechanistic analysis of this transformation, further evidence should be produced to adequately judge the likelihood of all three mechanisms in terms of steric and energetic factors.

Thus Mechanism 2 begins with imination occurring between N_f of **1** and C_i of 2,4-pentanedione to generate **A_2**. Addition of water at the C_a atom of **A_2** then causes the seven-membered ring to open and form **C_2**. The free N_b amino group of **C_2** then cyclizes at C_g to form another seven-membered ring which dehydrates to form intermediate **E_2**. The liberated water molecule then adds again at C_i to cause a second ring opening of the seven-membered ring generating **G_2**. Enolization of **G_2** followed by recyclization to form a five-membered ring leads to **I_2** where a bond is formed between C_h and C_d. Furthermore, sequential ammonia elimination, a [1,3]-H shift, and bond rotation promote the cyclization of intermediate **K_2** to form a six-membered ring giving **L_2** where a bond is made between C_a and C_c. Product **2** is then obtained after dehydration and a final [1,3]-H shift. We note once again that the same ring frame target bond map for the product structure is expected as for Mechanisms 1A and 1B except that one of the carbonyl groups is also highlighted indicating that it was made by water addition at C_i.

ADDITIONAL RESOURCES

For an example of a very similar transformation involving an electrocyclization-oxidation process, see the work of Boisse et al.[2] For an example of an interesting transformation involving aziridine chemistry, see Ref. 3. Lastly, we also recommend an article describing a thermal ring contraction involving radical chemistry.[4]

References

1. Szmuszkovicz J, Baczynskyj L, Chidester CC, Duchamp DJ. Reaction of 2-amino-7-chloro-5-phenyl-3H-[1,4]benzodiazepine with 1,3-dicarbonyl compounds. *J Org Chem*. 1976;41(10):1743–1747. https://doi.org/10.1021/jo00872a016.
2. Boisse T, Gautret P, Rigo B, Goossens L, Henichart J-P, Gavara L. A new synthesis of pyrrolo[3,2-b]quinolines by a tandem electrocyclizations-oxidation process. *Tetrahedron*. 2008;64:7266–7272. https://doi.org/10.1016/j.tet.2008.05.071.
3. Baktharaman S, Afagh N, Vandersteen A, Yudin AK. Unprotected vinyl aziridines: facile synthesis and cascade transformations. *Org Lett*. 2010;12 (2):240–243. https://doi.org/10.1021/ol902550q.
4. Crawford LA, McNab H, Mount AR, Wharton SI. Thermal ring contraction of dibenz[b,f]azepin-5-yl radicals: new routes to pyrrolo[3,2,1-jk]carbazoles. *J Org Chem*. 2008;73:6642–6646. https://doi.org/10.1021/jo800637u.

Question 184: An Indane-1,3-dione Synthesis

During a 1970 synthetic study of 1,3-indandiones, Mosher and Meier reacted 2,4-pentanedione **1** with 3-nitrophthalic anhydride **2** in the presence of pyridine and piperidine and formed 76% of the substituted nitroindane-1,3-dione **3** after aqueous workup (Scheme 1).[1] Provide a mechanism for this transformation.

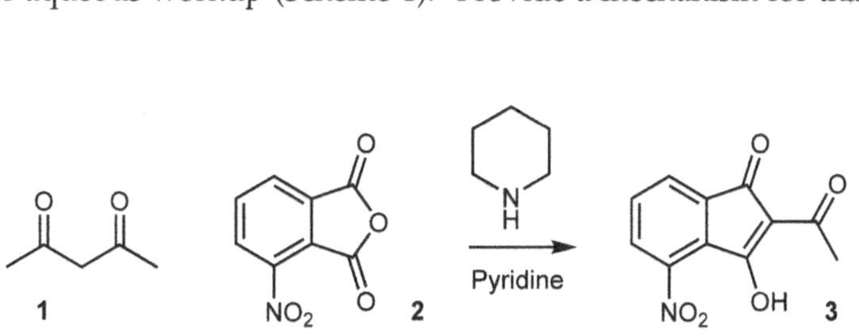

SCHEME 1

SOLUTION: MECHANISM 1 (MOSHER AND MEIER)

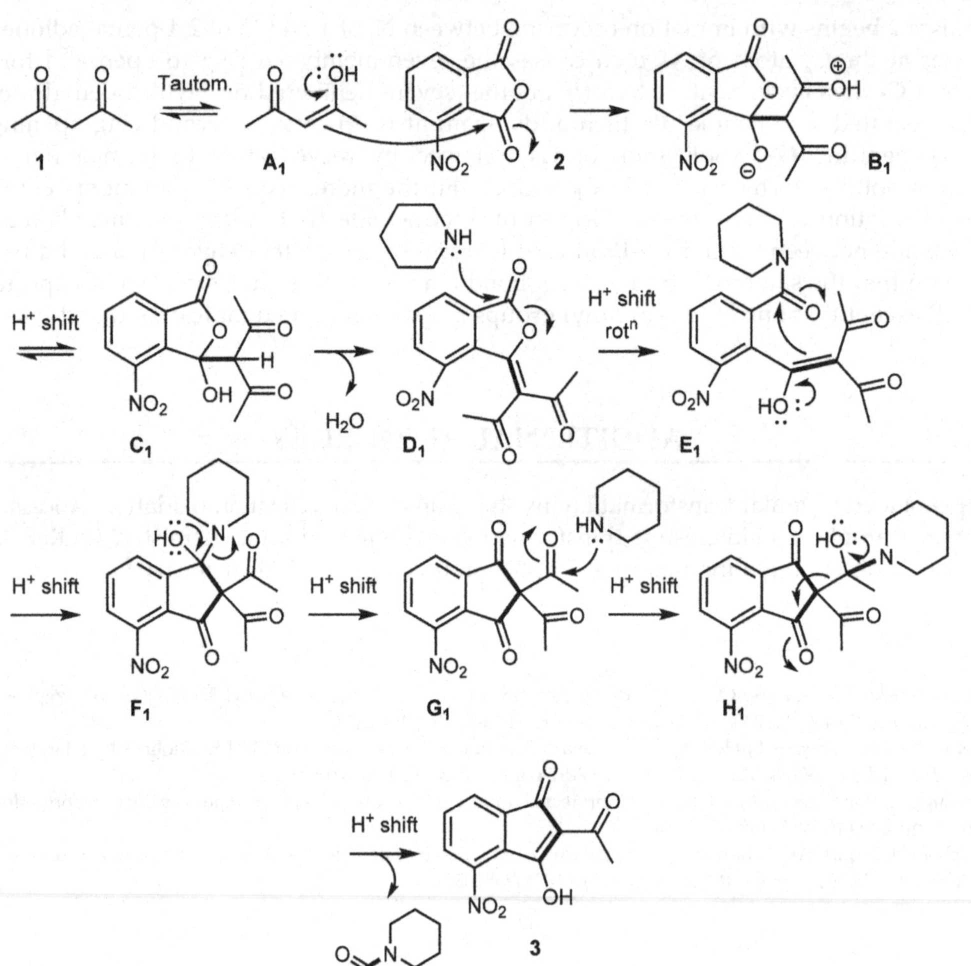

SOLUTION: MECHANISM 2 (ALTERNATIVE)

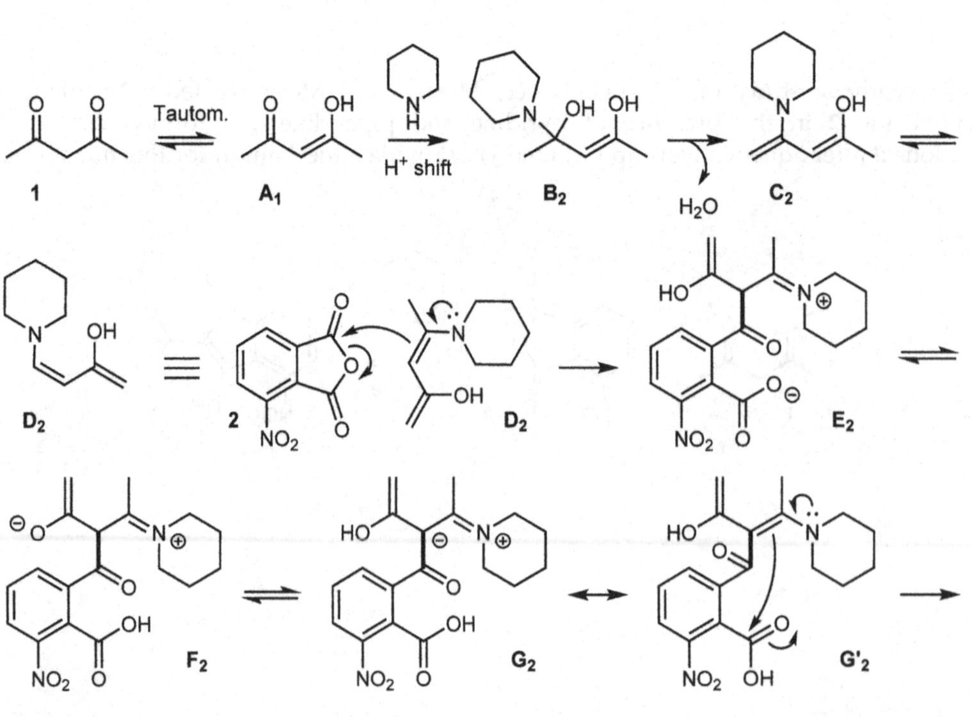

BALANCED CHEMICAL EQUATION

KEY STEPS EXPLAINED

We start by emphasizing that a number of mechanisms can be written down to explain the transformation of **1** and **2** into **3**. Mosher and Meier, for instance, postulated the initial formation of the enol of **1** (structure A_1) which adds to **2** to form the anhydride intermediate D_1 after proton shift and dehydration. Next, piperidine nucleophilically attacks at the lactone carbonyl group of D_1 to form E_1 which cyclizes to the carbocycle F_1. The piperidinyl moiety of F_1 is then eliminated as a piperidine molecule after ketonization thus forming G_1. Piperidine then returns to attack one of the exocyclic ketone groups of G_1 to form H_1 after a proton shift. A final ketonization of the hydroxyl group of H_1 causes the elimination of 1-(piperidin-1-yl)ethan-1-one by-product with formation of the substituted nitroindane-1,3-dione product **3**.

By contrast, the alternative mechanism models a Stork enamine synthesis strategy. Here, the enol A_1 reacts with piperidine to produce the enamine D_2. Intermediate D_2 then nucleophilically attacks the anhydride **2** causing it to ring open to zwitterion G_2 after a series of proton transfers. Recyclization of G_2 forms the carbocycle H_2 which eliminates hydroxide anion through ketonization to form the nitroindane diketone I_2 which is then attacked by the free hydroxide ion at the imine position to form J_2. Elimination of 1-(piperidin-1-yl)ethan-1-one by-product followed by two proton shifts will generate the final product **3**.

It is important to note that the main difference between the two mechanisms concerns what species nucleophilically opens the anhydride ring. Another difference concerns the stepwise 1,3-migration of a piperidinyl group in the authors' mechanism (from F_1 to H_1) versus the migration of a hydroxyl group in the alternative mechanism (from H_2 to J_2, respectively). From the perspective of steric factors, it is possible to argue that the smaller hydroxyl group is more prone to migration as compared to the bulkier piperidinyl group. With respect to the formations of D_1 and D_2 in the two mechanisms, Mosher and Meier cite literature precedent to explain their suggested formation of D_1 based on the known condensations of phthalic anhydride with an active methylene compound. Nevertheless, they do not consider the pivotal work of Stork[2] which encompasses the known reactivity of piperidine with anhydrides such as **1** to form the enamines C_2 and D_2 as the first steps in the transformation before enamine alkylation can occur to form E_2.

Interestingly, although the authors did not isolate their postulated intermediate D_1, in their actual experimental procedure, the pyridine salt of product **3** was isolated first and then the free ketone product was obtained in 76% overall yield after treating the salt with hydrochloric acid. Nevertheless, Mosher and Meier did isolate D_1 using a much less efficient method for the synthesis of **3** (Scheme 2). For example, compare the yield of 26% for **3** achieved using the less efficient method with the yield of 76% using the strategy shown in Scheme 1. This low conversion of D_1 to **3** suggests that D_1 is less likely to be an intermediate in the mechanistic pathway as suggested by the authors. We believe the higher yield can be better explained by the alternative Stork enamine mechanism. In fact using a literature search we were able to find two articles by Japanese chemists who carried out the reaction between 2,4-pentanedione and piperidine and were able to isolate the enamine product C_2/D_2 in its keto form which was later used in alkylation reactions.[3,4] We believe this literature constitutes evidence to support the alternative mechanism.

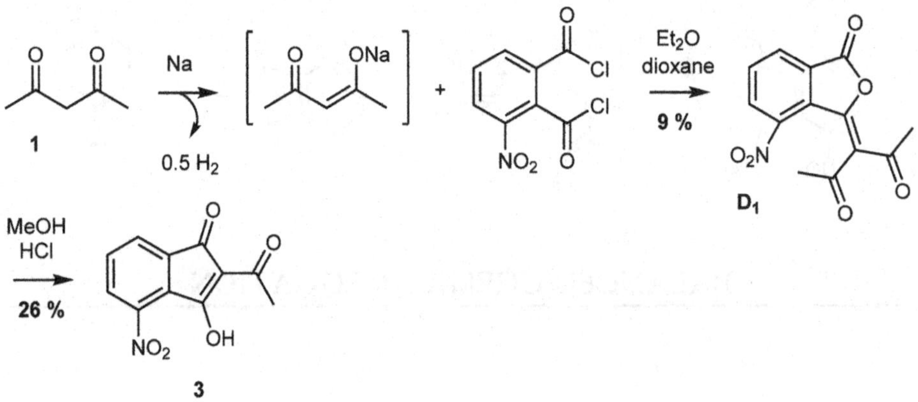

SCHEME 2

ADDITIONAL RESOURCES

For an example of an interesting transformation involving enamine chemistry in a similar context as the one encountered here, see Ref. 5. For a recent review by Stork outlining interesting highlights and sidelights across the span of his academic career, see Ref. 6.

References

1. Mosher WA, Meier WE. Benzene-ring-substituted 2-acetyl-1,3-indandiones. *J Org Chem*. 1970;35(9):2924–2926. https://doi.org/10.1021/jo00834a015.
2. Stork G, Dowd S. A new method for the alkylation of ketones and aldehydes: the C-alkylation of the magnesium salts of *N*-substituted imines. *J Am Chem Soc*. 1963;85(14):2178–2180. https://doi.org/10.1021/ja00897a040.
3. Tsuge O, Inaba A. Studies of enamines. II. The reaction of 4-(1-piperidyl)- and 4-(1-pyrrolidinyl)-3-penten-2-ones with aryl isocyanates. *Bull Chem Soc Jpn*. 1973;46(1):286–290. https://doi.org/10.1246/bcsj.46.286.
4. Tsuge O, Inaba A. Studies of enamines. III. The reaction of 4-(1-piperidyl)- and 4-(1-pyrrolidinyl)-3-penten-2-ones with aryl isothiocyanates. *Bull Chem Soc Jpn*. 1973;46(7):2221–2225. https://doi.org/10.1246/bcsj.46.2221.
5. Mamedov VA, Hafizova EA, Zamaletdinova AI, et al. Sequential substitution/ring cleavage/addition reaction of 1-(cyclohex-1-enyl)-piperidine and -pyrrolidine with chloropyruvates for the efficient synthesis of substituted 4,5,6,7-tetrahydro-1*H*-indole derivatives. *Tetrahedron*. 2015;71:9143–9153. https://doi.org/10.1016/j.tet.2015.10.004.
6. Stork G. Chemical reminiscences. *Tetrahedron*. 2011;67:9754–9764. https://doi.org/10.1016/j.tet.2011.10.007.

Question 185: An Azepine to Cyclohexadienone Ring Contraction

In the course of an investigation of the reactions of 3*H*-azepines, American chemists heated the 3*H*-azepine **1** under reflux in glacial acetic acid for 4 h and obtained a complex mixture of products which contained 12% of the cyclohexadienone **2** (Scheme 1).[1] Explain this transformation in mechanistic terms.

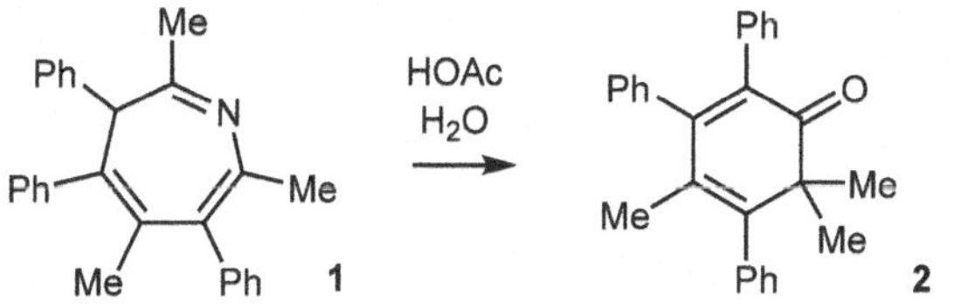

SCHEME 1

SOLUTION

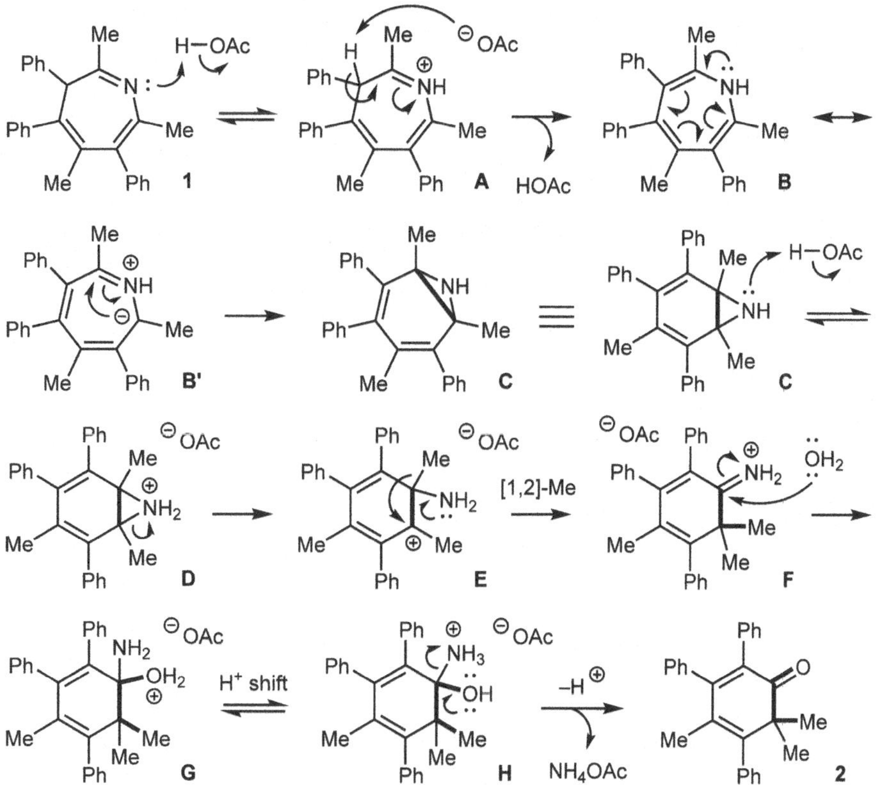

BALANCED CHEMICAL EQUATION

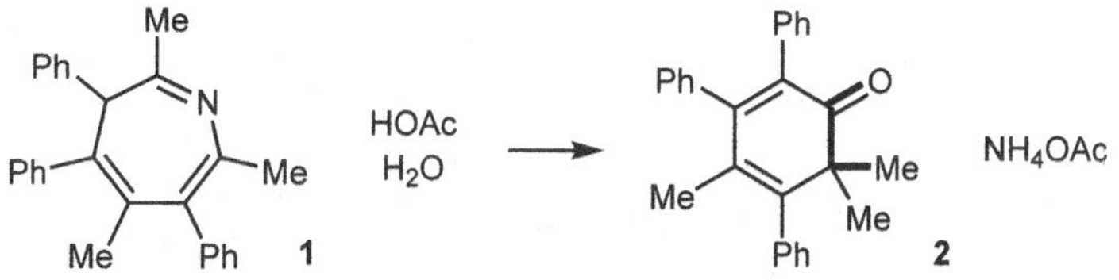

KEY STEPS EXPLAINED

The mechanism proposed by Anderson et al. to explain the ring contraction of **1** to **2** begins with rearrangement of the seven-membered 3*H*-azepine **1** to a [4.1.0] bicyclic intermediate **C** under acid catalysis conditions. Next, protonation of the aziridine ring of **C** leads to ring opening to form carbocation **E** which then undergoes a [1,2]-methyl shift to form the iminium ion **F**. Addition of water to this intermediate followed by elimination of ammonia (via ammonium acetate) and ketonization forms the final product **2**. It is important to note that since intermediate **C** is asymmetric, the aziridine ring could also open the other way leading to an isomeric ketone product **3** as shown in Scheme 2.

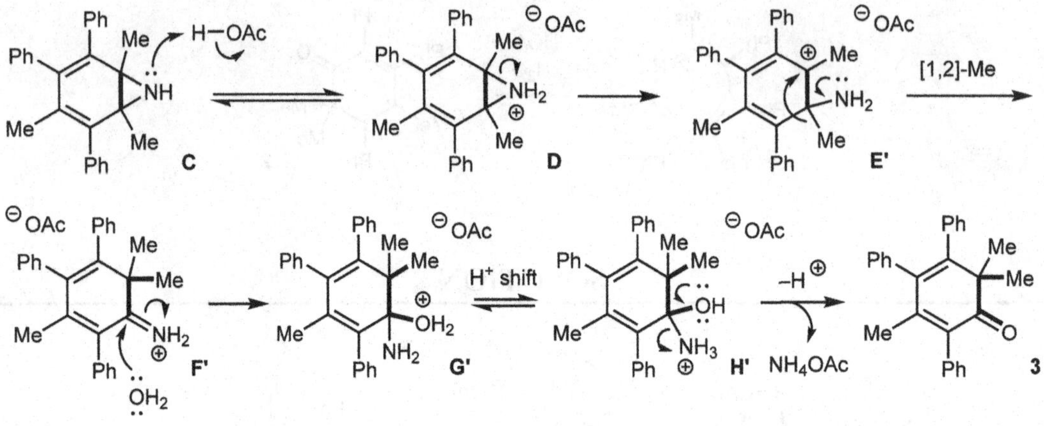

SCHEME 2

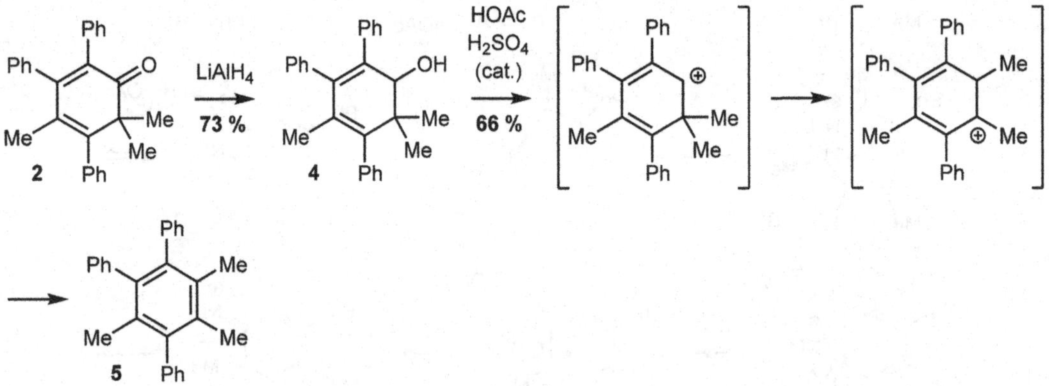

SCHEME 3

The authors concluded that the structure of the ketone product was in fact **2** and not **3** on the basis of results from a two-step synthesis shown in Scheme 3. One can thus compare the experimental results with the expected outcomes from both ketone isomers. The collected ketone product from the azepine reaction was isolated and then subjected to a lithium aluminum hydride reduction followed by acid-catalyzed dehydration. Compounds **4** and **5** were isolated and characterized at each stage. By inference they arose from a ketone with structure **2** as shown. Scheme 4 highlights the results of the same sequence of reactions if the ketone had structure **3**. In the dehydration step, instead of methyl group migration to again give ketone **2**, the allyl cation intermediate was expected to be deprotonated to form the quinoid product **6** by analogy with the results of another previously reported experiment done on an analog ketone precursor, namely, **7**, having all methyl groups.[2] Nevertheless, further experimental or theoretical analyses to help support this postulated mechanism have not been done.

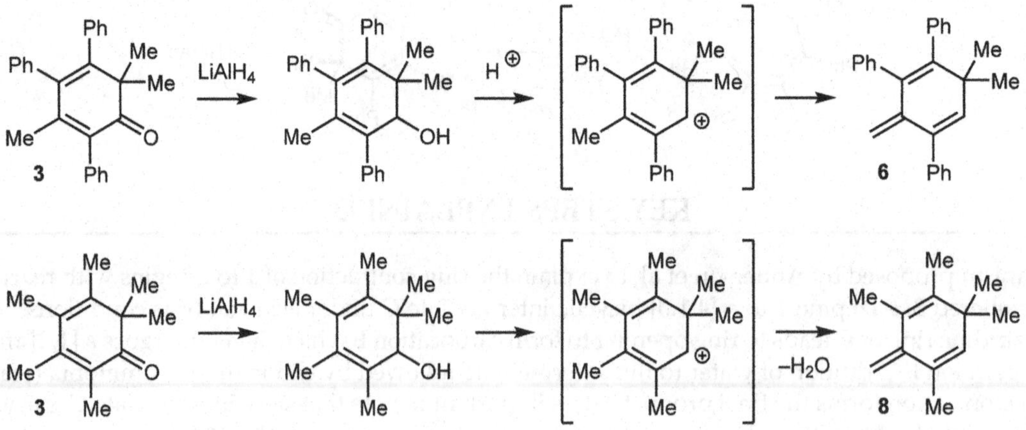

SCHEME 4

ADDITIONAL RESOURCES

For further examples of ring contraction transformations of a very similar nature to the one encountered here, see Refs. 3, 4. For examples of interesting transformations involving mechanistic analysis in the context of 1,2-methyl shifts, see Refs. 5, 6.

References

1. Anderson DJ, Hassner A, Tang DY. Reactions of 3*H*-azepines derived from cyclopentadienones and 1-azirines. *J Org Chem*. 1974;39(21):3076–3080. https://doi.org/10.1021/jo00935a003.
2. Waring AJ, Hart H. Conversion of hexasubstituted benzenes to cyclohexadienones. *J Am Chem Soc*. 1964;86(7):1454–1456. https://doi.org/10.1021/ja01061a051.
3. Satake K, Takaoka K, Hashimoto M, Okamoto H, Kimura M, Morosawa S. A new ring contraction rearrangement of 2,5- and 3,6-di-*tert*-butyl-3*H*-azepines to pyridine derivatives. *Chem Lett*. 1996;1996(12):1129–1130.
4. Dardonville C, Jimeno ML, Alkorta I, Elguero J. The behavior of 5H-dibenz[b,f]azepine dissolved in sulfuric acid. *ARKIVOC*. 2004;2004 (2):206–212.
5. Hart H, Huang I, Lavrik P. Acid-catalyzed rearrangement of two cyclohexadienone monoepoxides. *J Org Chem*. 1974;39(7):999–1005. https://doi.org/10.1021/jo00921a030.
6. Vitullo VP, Logue EA. Cyclohexadienyl cations. 6. Methyl group isotope effects in the dienone-phenol rearrangement. *J Am Chem Soc*. 1976;98 (19):5906–5909. https://doi.org/10.1021/ja00435a026.

Question 186: Xanthopterin From Pterin 8-Oxide

In 1973 American chemists investigating the anticancer potential of certain naturally occurring pteridines stirred the pterin 8-oxide **1** in a 1:1 mixture of trifluoroacetic acid-trifluoroacetic anhydride at 50°C for 1h. Following solvent evaporation and hydrolysis of the solid residue with ammonium hydroxide/sodium hydroxide, acidification provided the xanthopterin product **2** in quantitative yield (Scheme 1).[1] Suggest a mechanism for this transformation.

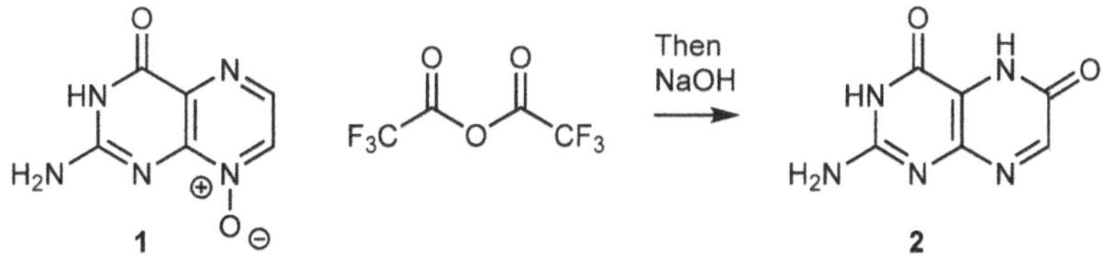

SCHEME 1

SOLUTION

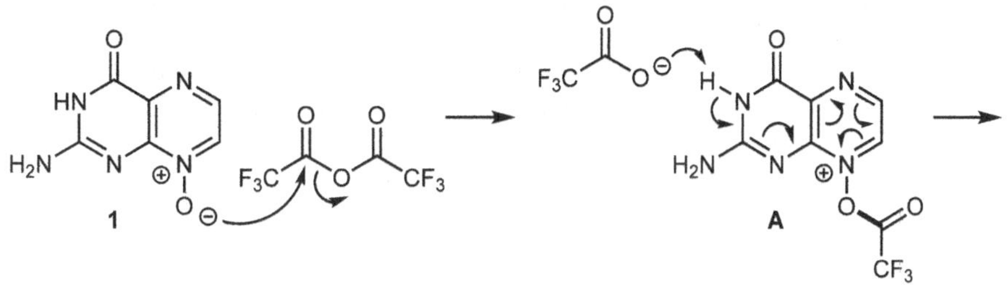

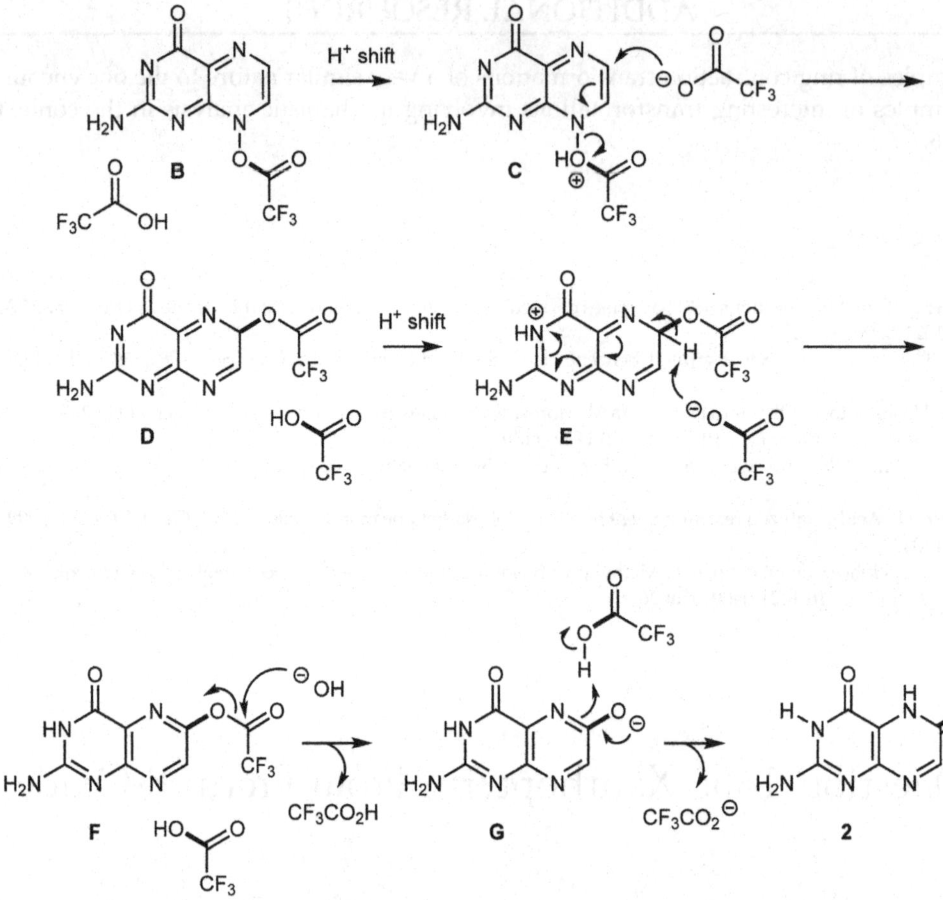

BALANCED CHEMICAL EQUATION

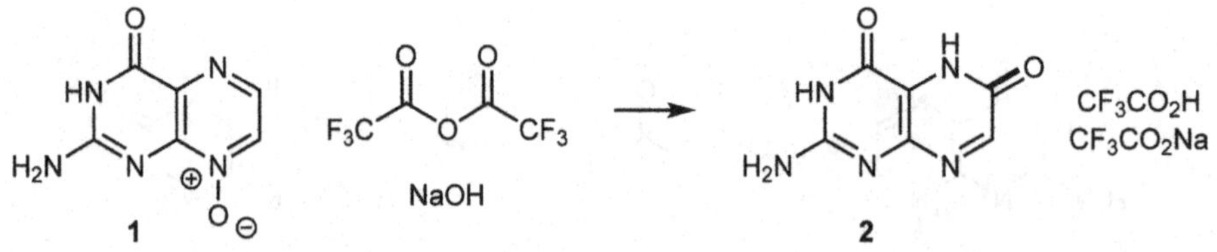

KEY STEPS EXPLAINED

The mechanism proposed by Taylor et al. to explain the transformation of **1** to **2** begins with *O*-acetylation of the *N*-oxide group of 2-amino-8-oxy-3*H*-pteridin-4-one **1** to give intermediate **B** via **A**. Trifluoroacetate then adds to C_6 liberating trifluoroacetic acid to form the quinoid intermediate **D** (via **C**) which rearomatizes to intermediate **F** after deprotonation via **E**. Hydroxide attack on the trifluoroacetate group of **F** leads to enolate **G** which finally ketonizes to product **2**. The authors did not provide any further experimental or theoretical analysis to help support their proposed mechanism.

Furthermore, examination of the oxidation state changes for this reaction shows that the *N*-oxide nitrogen atom is reduced from −1 to −3 for a net change of −2 and C_6 is oxidized from +1 to +3 for a compensating net change of +2.

ADDITIONAL RESOURCES

We direct the reader to further work by Taylor et al. on very similar transformations involving both pterins and discussion of mechanism.[2,3]

References

1. Taylor EC, Jacobi PA. Pteridines. XXX. A facile synthesis of xanthopterin. *J Am Chem Soc.* 1973;95(13):4455–4456. https://doi.org/10.1021/ja00794a071.
2. Taylor EC, Jacobi PA. Pteridines. XXXIV. Synthesis of 8-hydroxy-7(8H)-pteridinones (pteridine hydroxamic acids). *J Org Chem.* 1975;40(16):2332–2336. https://doi.org/10.1021/jo00904a015.
3. Taylor EC, Abdulla RF, Tanaka K, Jacobi PA. Pteridines. XXXVI. Syntheses of xanthopterin and isoxanthopterin. Application of *N*-oxide chemistry to highly functionalized pyrazines and pteridines. *J Org Chem.* 1975;40(16):2341–2347. https://doi.org/10.1021/jo00904a017.

Question 187: Fused Dihydro-1,4-dithiins From Chromanones

During a 1991 synthetic investigation for the preparation of *gem*-difluoro compounds, Hungarian chemists reacted the 4-chromanone **1** with 1,2-ethanedithiol and 1,3-dibromo-5,5-dimethylhydantoin (DBH) to form the dihydro-1,4-dithiin derivative **2** in 78% yield (Scheme 1).[1] Suggest a mechanism for this transformation.

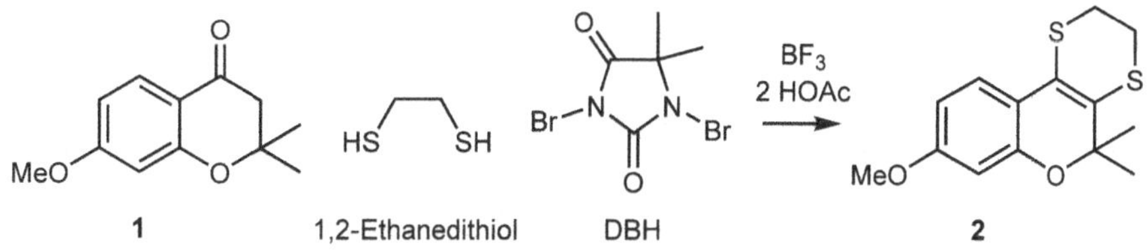

SCHEME 1

SOLUTION

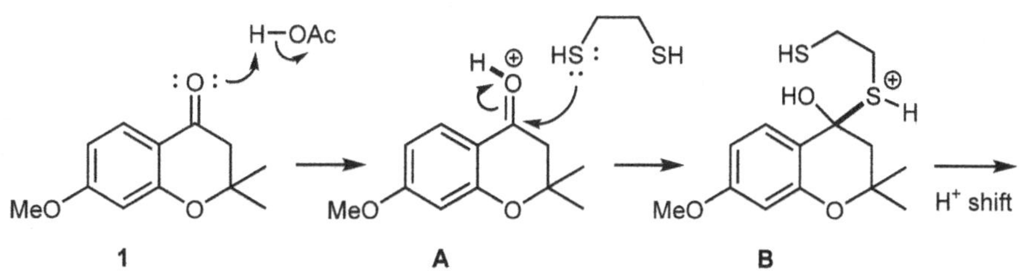

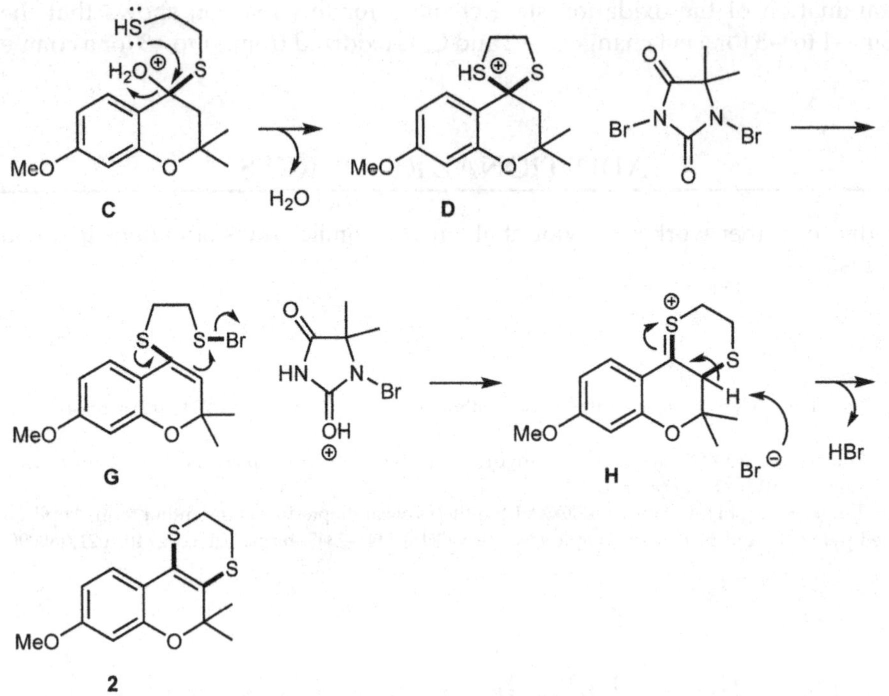

BALANCED CHEMICAL EQUATION

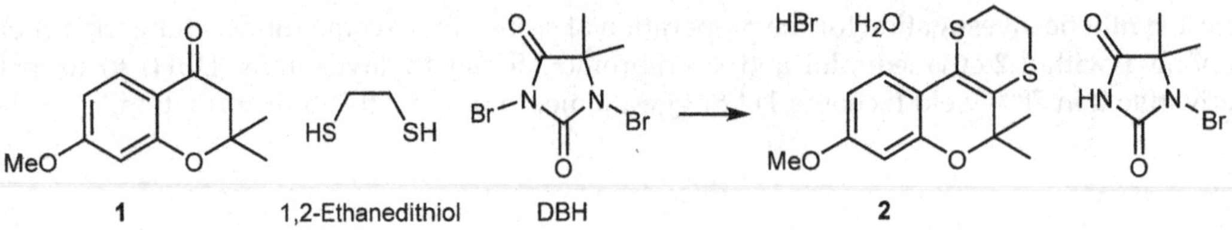

KEY STEPS EXPLAINED

The mechanism proposed by the authors for the transformation of **1** to **2** begins with acid-catalyzed addition of ethane-1,2-dithiol to 7-methoxy-2,2-dimethyl-chroman-4-one **1** to form, eventually the thioketal **E** after loss of water and cyclization. Next, the sulfur atom of **E** abstracts a bromine atom from protonated 1,3-dibromo-5,5-dimethyl-imidazolidine-2,4-dione to give bromosulfonium ion **F**. This intermediate rearranges to the sulfonium bromide **H** which then undergoes deprotonation to generate product **2**. Interestingly, the intended target of the transformation with the addition of excess HF and pyridine was the *gem*-difluorinated product **3** (Scheme 2). The authors were not able to isolate this product when no HF/pyridine was added and when a large excess of HF/pyridine was added, suggesting that the sequence of **E** to **2** occurs rapidly. Nevertheless, the authors did not carry out any further experimental or theoretical analyses to help support their postulated mechanism.

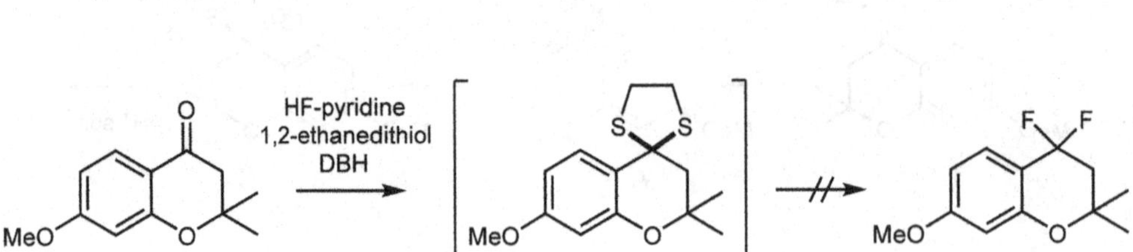

SCHEME 2

ADDITIONAL RESOURCES

We highly recommend Refs. 2–4 for examples of interesting mechanisms which involve very similar chemical systems to the one explored in this problem. For ways to prepare *gem*-difluoromethylene compounds, see the work of Tozer and Herpin.[5] For synthetic applications of *N*-halo compounds such as DBH, see Ref. 6.

References

1. Jeko J, Timar T, Jaszberenyi JC. Synthesis of benzopyran derivatives. XVIII. Gem-difluorination vs. 1,3-dithiolane-dihydro-1,4-dithiin rearrangement. The role of benzylic carbons. *J Org Chem*. 1991;56(24):6748–6751. https://doi.org/10.1021/jo00024a010.
2. Arote ND, Telvekar VN, Akamanchi KG. Rapid method for the ring expansion of 1,3-dithiolanes and 1,3-dithianes with *tert*-butyl hypochlorite. *Synlett*. 2005;2005(19):2935–2938. https://doi.org/10.1055/s-2005-921906.
3. Firouzabadi H, Iranpoor N, Garzan A, Shaterian HR, Ebrahimzadeh F. Facile ring-expansion substitution reactions of 1,3-dithiolanes and 1,3-dithianes initiated by electrophilic reagents to produce monohalo-, -cyano-, -azido- and -thiocyanato-1,4-dithiins and -1,4-dithiepins. *Eur J Org Chem*. 2005;2005:416–428. https://doi.org/10.1002/ejoc.200400502.
4. Murru S, Kavala V, Singh CB, Patel BK. A one-pot synthesis of 1,4-dithiins and 1,4-benzodithiins from ketones using the recyclable reagent 1,1'-(ethane-1,2-diyl)dipyridinium bistribromide (EDPBT). *Tetrahedron Lett*. 2007;48:1007–1011. https://doi.org/10.1016/j.tetlet.2006.12.003.
5. Tozer MJ, Herpin TP. Methods for the synthesis of *gem*-difluoromethylene compounds. *Tetrahedron*. 1996;52(26):8619–8683. https://doi.org/10.1016/0040-4020(96)00311-0.
6. Veisi H, Ghorbani-Vaghei R, Zolfigol MA. Recent progress in the use of *N*-halo compounds in organic synthesis. *Org Prep Proced Int*. 2011;43:489–540. https://doi.org/10.1080/00304948.2011.629553.

Question 188: The "Additive Pummerer Reaction"

In the early 1990s, English chemists reacted vinylic sulfoxides such as **1** with electrophiles such as trifluoroacetic anhydride (TFAA) or triflic anhydride and sodium acetate in acetic anhydride to give products **3** or **4**, respectively (Scheme 1).[1, 2] Suggest mechanisms for these transformations which the authors referred to as examples of the "additive Pummerer reaction."[1,2]

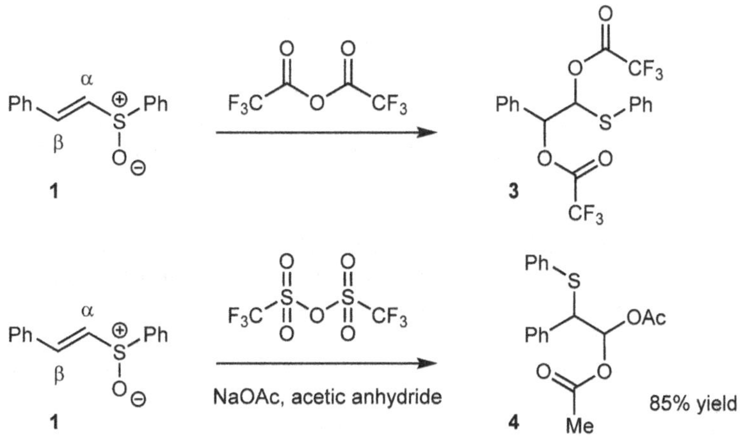

SCHEME 1

SOLUTION: 1 TO 3

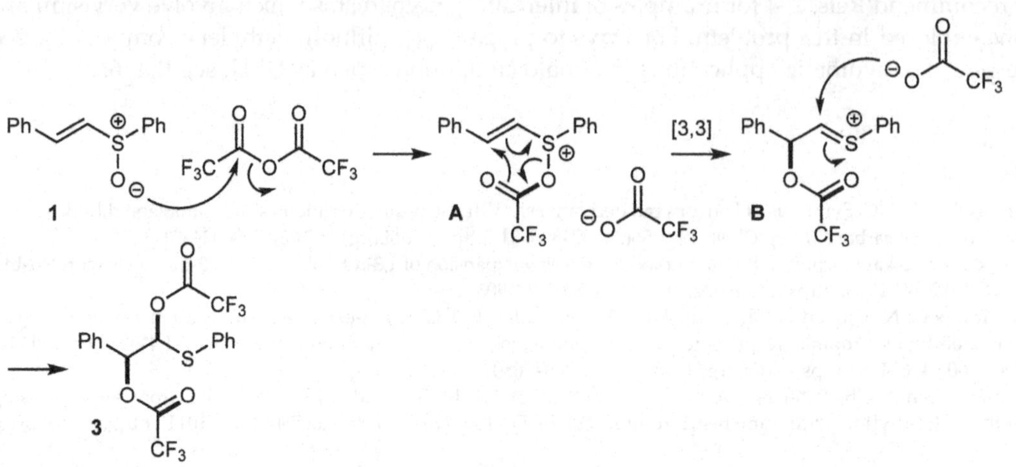

SOLUTION: 1 TO 4

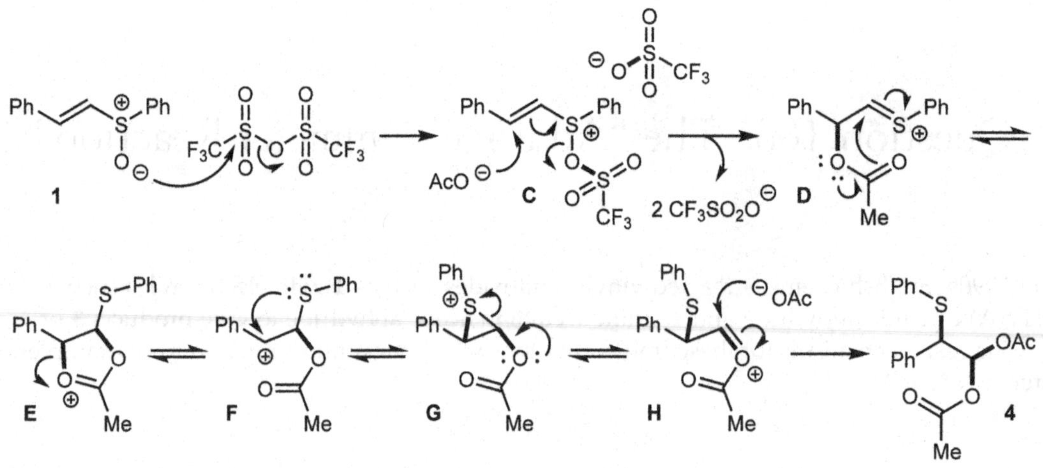

KEY STEPS EXPLAINED

The mechanism proposed by the authors for the transformation of **1** to **2** begins with nucleophilic attack by the sulfoxide group of **1** onto trifluoroacetic anhydride to form the sulfonium ion **A** after elimination of trifluoroacetate. Intermediate **A** then undergoes a [3,3]-sigmatropic rearrangement to form intermediate **B** which is attacked by acetate anion to generate product **3**.

The authors named this reaction an "additive Pummerer" reaction according to the general mechanistic pathway shown in Scheme 2 which may be compared with the traditional and vinylogous Pummerer reactions. For example, whereas the additive version of the reaction involves addition of two nucleophiles to the substrate, the traditional and vinylogous versions involve addition of one nucleophile.

With respect to the transformation of **1** to **4**, the authors proposed once again an initial nucleophilic attack by the sulfoxide group of **1** onto the triflic anhydride reagent to form sulfonium ion **C** after elimination of triflate ion. Intermediate **C** is then attacked by acetate anion to eliminate sodium triflate and form the anchimerically stabilized sulfonium ion intermediate **D**. This structure undergoes a [5 + 0] cyclization to generate the oxonium ion intermediate **E** which then rearranges to a thiiranium ion **G** via carbocation **F**. Rearrangement of **G** yields the O-acetyloxonium ion **H** (after what effectively constitutes the migration of the phenylsulfenyl group to the β position of the substrate). A final nucleophilic attack by a second acetate anion onto **H** gives the final product **4**. Experimentally, the authors observed that the reaction was not successful when the substituents at the β position of **1** were alkyl group and hydrogen atom,

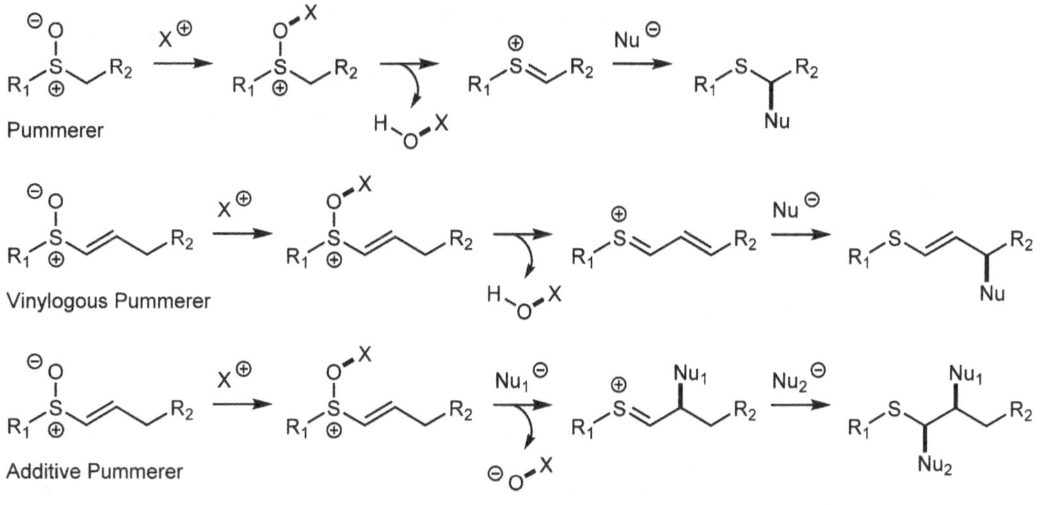
Pummerer
Vinylogous Pummerer
Additive Pummerer

SCHEME 2

respectively. The authors accounted for this by arguing that alkyl groups at the β position of **1** would help destabilize carbocation **F** thus restricting the formation of **G**. Another pivotal observation was that the reaction produced a low yield of **4** when the aryl group at the β position contained a strongly electron-donating substituent. In this case the authors argued that the carbocation **F** would be too stable and once again restrict the formation of **G**. Nevertheless, despite these postulated arguments about energetics, the authors did not perform a much needed theoretical analysis to help confirm these hypotheses.

Lastly, it is interesting to note that the authors regard the additive Pummerer reaction as proceeding through synthon **2** (Fig. 1). We may verify this claim by applying oxidation number analyses to the balanced chemical equations for both reactions (Scheme 3).

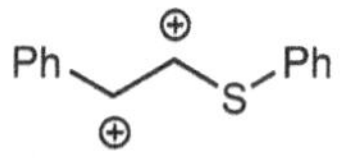

FIG. 1 Structure for synthon **2** showing two electrophilic sites.

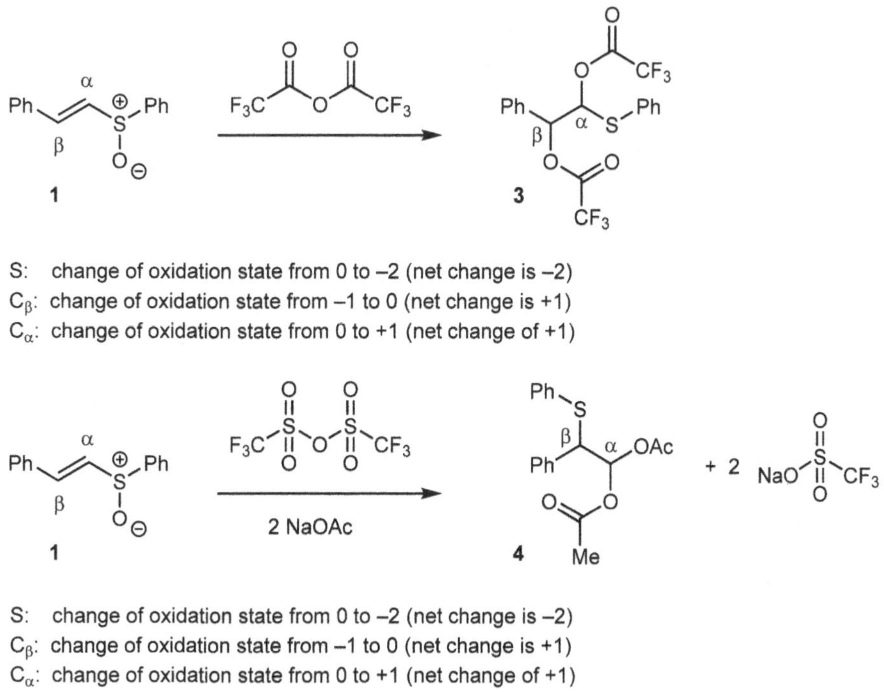
S: change of oxidation state from 0 to –2 (net change is –2)
Cβ: change of oxidation state from –1 to 0 (net change is +1)
Cα: change of oxidation state from 0 to +1 (net change of +1)
2 NaOAc
S: change of oxidation state from 0 to –2 (net change is –2)
Cβ: change of oxidation state from –1 to 0 (net change is +1)
Cα: change of oxidation state from 0 to +1 (net change of +1)

SCHEME 3

ADDITIONAL RESOURCES

For readers interested in chemical structures of the types **F** and **G**, we highly recommend an article by Andres et al.[3] For those interested in other chemical transformations where triflic anhydride plays an important role, see Ref. 4. For an example of a nonoxidative Pummerer reaction, see Ref. 5. Lastly, for a review of synthetic applications of the Pummerer reaction, see the work of Padwa et al.[6]

References

1. Craig D, Daniels K, MacKenzie AR. Additive Pummerer reactions of vinylic sulphoxides. Synthesis of 2-(phenylsulfenyl) aldehydes and primary alcohols. *Tetrahedron Lett*. 1991;32(47):6973–6976. https://doi.org/10.1016/0040-4039(91)80458-I.
2. Craig D, Daniels K, MacKenzie AR. Additive Pummerer reactions of vinylic sulphoxides. Synthesis of γ-hydroxy-α,β-unsaturated esters, α-hydroxyketones, and 2-phenylsulfenyl aldehydes and primary alcohols. *Tetrahedron*. 1993;49(48):11263–11304. https://doi.org/10.1016/S0040-4020(01)81812-3.
3. Andres DF, Laurent EG, Marquet BS, Benotmane H, Bansadat A. Anodic functionalization of vinyl sulfides. Formal access to *gem* or vicinal aryl thioether dications. *Tetrahedron*. 1995;51(9):2605–2618. https://doi.org/10.1016/0040-4020(95)00009-W.
4. Baraznenok IL, Nenajdenko VG, Balenkova ES. Chemical transformations induced by triflic anhydride. *Tetrahedron*. 2000;56:3077–3119. https://doi.org/10.1016/S0040-4020(00)00093-4.
5. Crucianelli M, Bravo P, Arnone A, Corradi E, Meille SV, Zanda M. The "non-oxidative" Pummerer reaction: Conclusive evidence for S_N2-type stereoselectivity, mechanistic insight, and synthesis of enantiopure L-α-trifluoromethylthreoninate and D-α-trifluoromethyl-*allo*-threoninate. *J Org Chem*. 2000;65:2965–2971. https://doi.org/10.1021/jo991534p.
6. Padwa A, Gunn Jr. DE, Osterhout MH. Application of the Pummerer reaction toward the synthesis of complex carbocycles and heterocycles. *Synthesis*. 1997;1997(12):1353–1377. https://doi.org/10.1055/s-1997-1384.

Question 189: An Illustration of the Problem of Artifacts in Natural Product Chemistry

In 1981 during the course of separation experiments at Penn State, Manikumar and Shamma found that simple chromatography of oxyberberine **1** on silica using chloroform as eluent gave magallanesine **2** in excellent yield showing that **2** is likely not a natural product (Scheme 1).[1] Suggest a mechanism for this transformation.

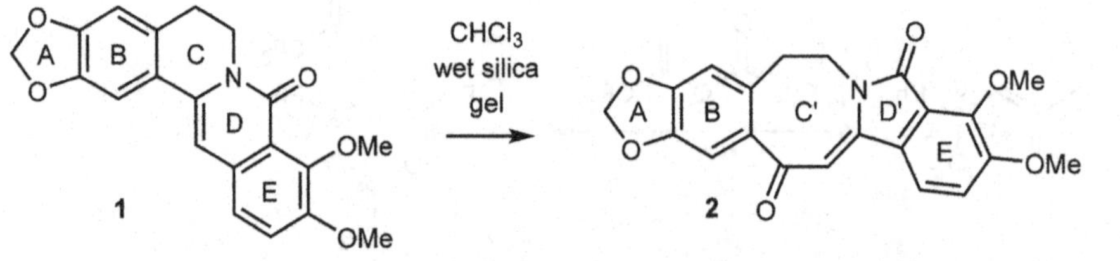

SCHEME 1

SOLUTION

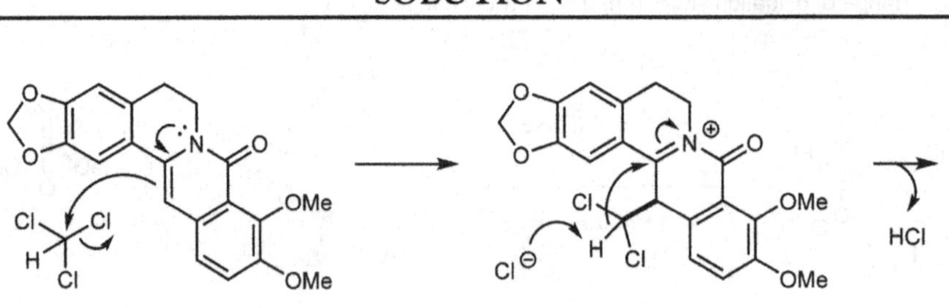

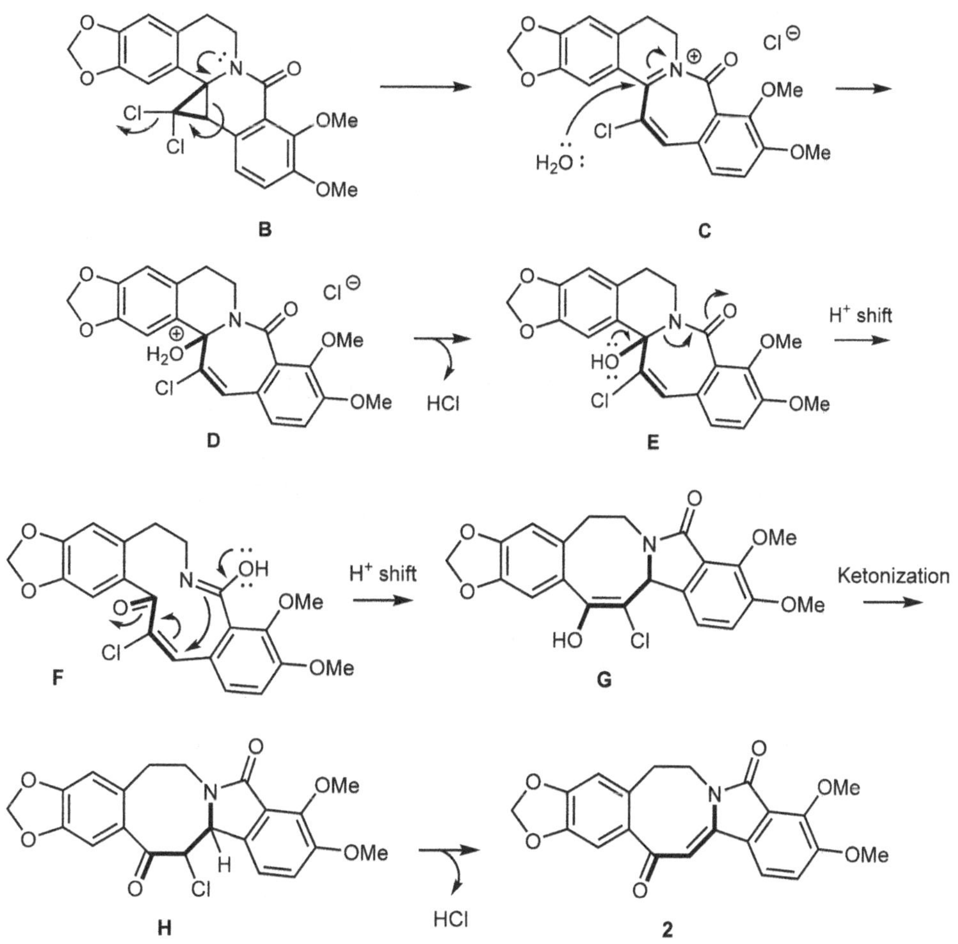

KEY STEPS EXPLAINED

The mechanism proposed by Manikumar and Shamma for the **1** to **2** transformation begins with essentially the addition of dichlorocarbene to the olefinic bond in ring D to give the cyclopropane adduct **B** after nucleophilic addition to **1** via **A**. Afterward, intermediate **B** ring expands to a seven-membered ring iminium ion intermediate **C** which is then hydrated to form **D** which, in turn, loses HCl to form the unstable chlorohydrin intermediate **E**. Fragmentation of the 7-membered ring of **E** yields the 11-membered ring intermediate **F** which then undergoes an intramolecular condensation to form the bicyclic intermediate **G**. Ketonization of **G** to **H** followed by elimination of hydrochloric acid generates the final product **2**. The result of this transformation is thus the rearrangement of the [4.4.0] bicyclic structure of rings C and D in **1** to the [6.3.0] bicyclic structure of rings C′ and D′ in **2**. The final target bond mapping may be encoded as $[(6+1+1)_{C'} + (5+0)_{D'}]$. Moreover, the authors carried out a two-step synthesis shown in Scheme 2 to lend support to their mechanistic hypothesis.

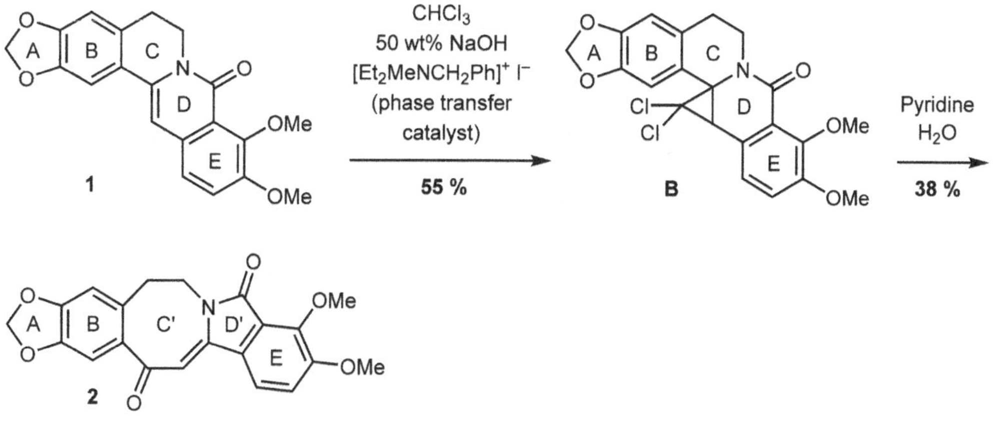

SCHEME 2

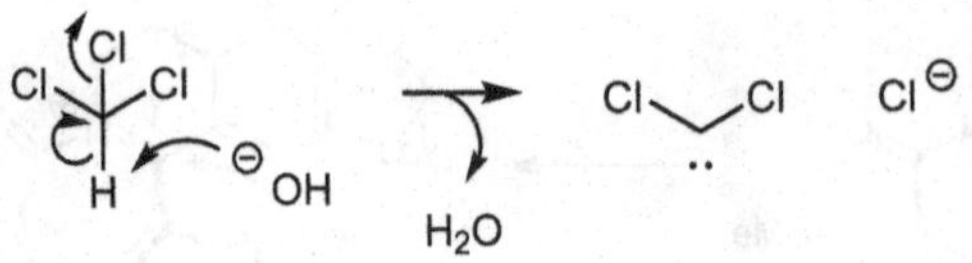

SCHEME 3

The product for the first step of this experiment is intermediate **B** which originates from insertion of dichlorocarbene, generated from chloroform and sodium hydroxide (Scheme 3), into the C=C bond of ring D. Nevertheless, the authors did not carry out a theoretical analysis to help add further support for their postulated mechanism.

ADDITIONAL RESOURCES

We refer the reader to three examples of interesting mechanistic transformations involving alkaloid chemistry of a similar nature to the solution presented here.[2–4] We also recommend an article which describes a synthetic approach to magallanesine **2** via the [1,2]-Meisenheimer rearrangement and Heck cyclization.[5]

References

1. Manikumar G, Shamma M. Addition of dichlorocarbene to oxyberberine and berberine. *J Org Chem.* 1981;46(2):386–389. https://doi.org/10.1021/jo00315a031.
2. Wildman WC, Bailey DT. Novel alkaloids containing the [2]benzoyrano[3,4-c]indole nucleus. *J Org Chem.* 1968;33(10):3749–3753. https://doi.org/10.1021/jo01274a014.
3. Dumbacher JP, Spande TF, Daly JW. Batrachotoxin alkaloids from passerine birds: a second toxic bird genus (*Ifrita kowaldi*) from New Guinea. *Proc Natl Acad Sci.* 2000;97(24):12970–12975. https://doi.org/10.1073/pnas.200346897.
4. Rochfort SJ, Moore S, Craft C, Martin NH, Van Wagoner RM, Wright JLC. Further studies on the chemistry of the Flustra alkaloids from the bryozoan *Flustra foliacea. J Nat Prod.* 2009;72:1773–1781. https://doi.org/10.1021/np900282j.
5. Yoneda R, Sakamoto Y, Oketo Y, Harusawa S, Kurihara T. An efficient synthesis of magallanesine using [1,2]-Meisenheimer rearrangement and Heck cyclization. *Tetrahedron.* 1996;52(46):14563–14576. https://doi.org/10.1016/0040-4020(96)00900-3.

Question 190: Vinylogy, and a Stereochemical Puzzle

During the course of 4 years, Canadian, Japanese, and British chemists examined the mechanism of the transformation of 1,1,1-trichloro-2-penten-4-one **1** to 5-chloro-*trans*-2-*cis*-4-pentadienoic acid **2** after aqueous sodium hydroxide followed by acid work-up treatment (Scheme 1).[1] Elucidate the reaction mechanism to explain the observed stereochemical outcome.

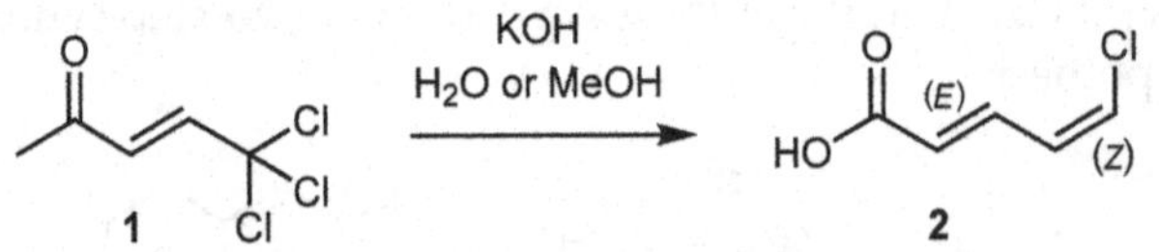

SCHEME 1

SOLUTION: MECHANISM 1 (KIEHLMANN ET AL.)

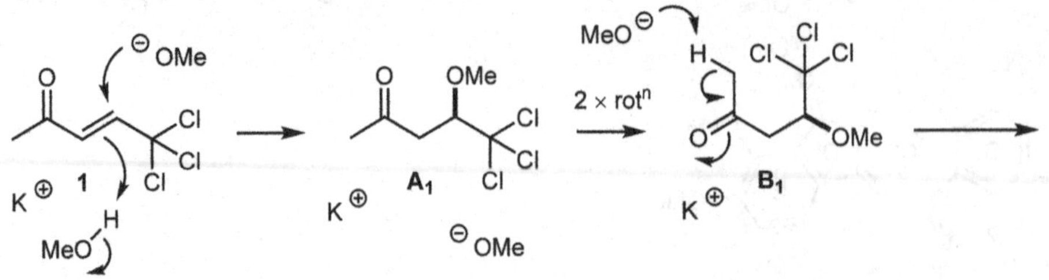

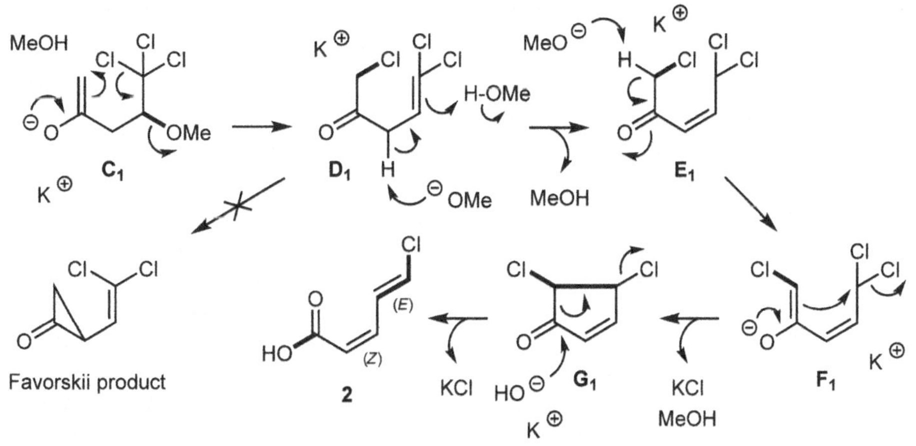

BALANCED CHEMICAL EQUATION: MECHANISM 1

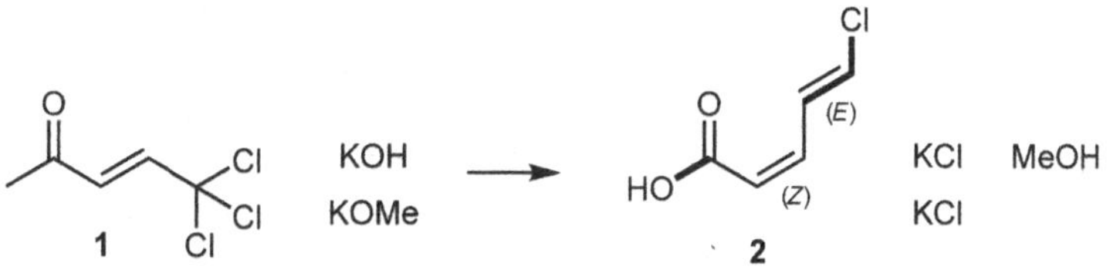

SOLUTION: MECHANISM 2 (ACEVADO ET AL.)

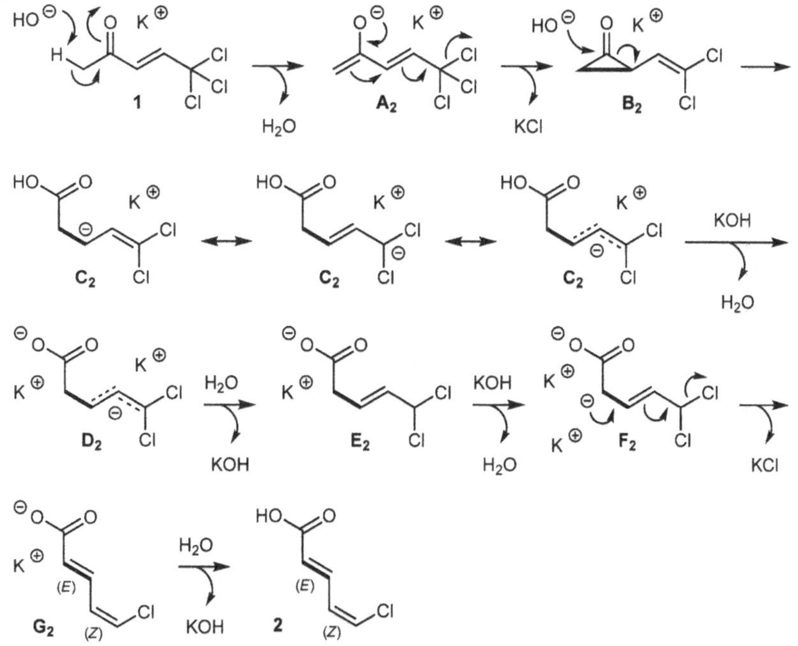

BALANCED CHEMICAL EQUATION: MECHANISM 2

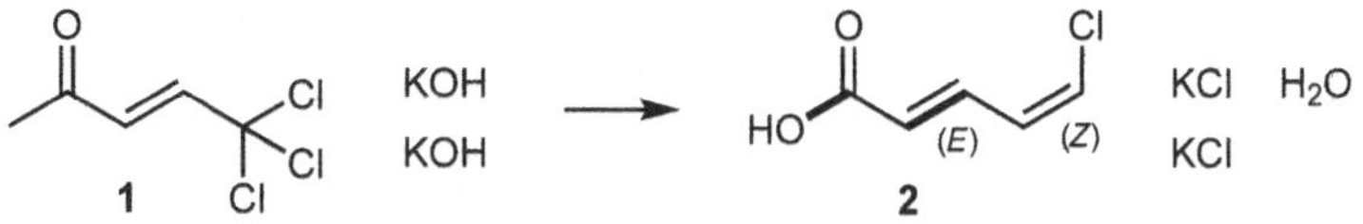

KEY STEPS EXPLAINED

The mechanism proposed by Kiehlmann et al. for the formation of the chloro-pentadienoic acid **2** from trichloro-pentadienone **1** begins with base-catalyzed addition of methoxide to the α,β-unsaturated C=C bond of **1** to generate intermediate A_1. After two bond rotations that culminate in B_1, base-catalyzed enolization of B_1 forms C_1 which sets up the migration of chlorine to give D_1 after ketonization. Intermediate D_1 undergoes a prototropic shift to form the more stable α,β-unsaturated ketone E_1 which undergoes base-catalyzed enolization to F_1 followed by ring closure to the substituted 2-cyclopentenone G_1. Ring cleavage of G_1 via attack of hydroxide anion at the carbonyl group with elimination of chloride ion generates product **2**. According to this mechanism, the implied stereochemistry in **2** is expected to be 5-chloro-penta-2Z,4E-dienoic acid, that is, 5-chloro-*cis*-2-*trans*-4-pentadienoic acid. This stereochemical outcome was not discussed by Kiehlmann et al. and would seem to contradict the desired stereochemical outcome of 5-chloro-*trans*-2-*cis*-4-pentadienoic acid. Nevertheless, the authors cited several pieces of mostly experimental evidence in favor of mechanism 1.

First, Kiehlmann et al. observed that the transformation requires the presence of two geminal α-hydrogen atoms in the olefinic starting material **1** because reactions of 1,1,1-trichloro-2-alken-4-ones lacking at least two geminal α-hydrogen atoms did not produce the expected dienoic acid products **2**. When the reaction was tested with an equimolar amount of base, instead of fivefold excess used, and **1** (along with other derivatives of **1**) the authors obtained mixtures of products that supported the structure of intermediate A_1 and its corresponding analogs (Scheme 2). This happened because some of the analogs had alkyl groups in place of the required two geminal α-hydrogen atoms. This prevents the proposed enolization of the A_1 intermediates. Specifically, enolization of B_1 and E_1 could not occur. Furthermore, in the course of several experiments with equimolar amounts of **1** or A_1 and KOH in methanol, the authors were able to isolate mixtures of A_1 and E_1 thus further supporting these proposed intermediates. Although the authors could not isolate any cyclopentenone G_1 intermediates, they argued that the likely driving forces for the last step of the mechanism were relief of ring strain coupled with resonance stabilization of the highly conjugated product **2**. It was also remarked that no Favorskii rearrangement products were isolated, a result which the authors attributed to the developing negative charge on the unsubstituted α-carbon atom of intermediate D_1 which is part of an allylic system and thus is unavailable to form the cyclopropane intermediate required for the Favorskii rearrangement.[1]

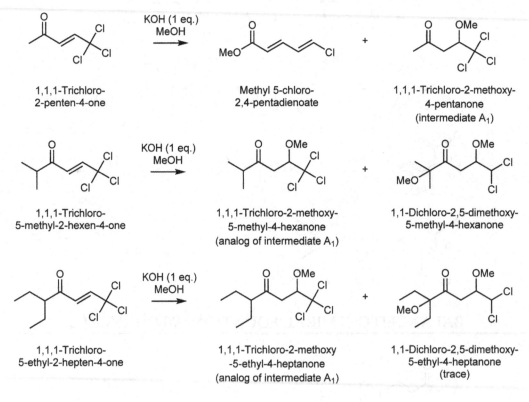

1,1,1-Trichloro-
2-penten-4-one

Methyl 5-chloro-
2,4-pentadienoate

1,1,1-Trichloro-2-methoxy-
4-pentanone
(intermediate A_1)

1,1,1-Trichloro-
5-methyl-2-hexen-4-one

1,1,1-Trichloro-2-methoxy-
5-methyl-4-hexanone
(analog of intermediate A_1)

1,1-Dichloro-2,5-dimethoxy-
5-methyl-4-hexanone

1,1,1-Trichloro-
5-ethyl-2-hepten-4-one

1,1,1-Trichloro-2-methoxy
-5-ethyl-4-heptanone
(analog of intermediate A_1)

1,1-Dichloro-2,5-dimethoxy-
5-ethyl-4-heptanone
(trace)

SCHEME 2

Following the article of Kiehlmann et al. in 1972, two separate groups subsequently proposed an alternative mechanism (mechanism 2) for the transformation of **1** to **2**.[2,3] Essentially, instead of beginning with methoxide incorporation into **1**, mechanism 2 starts with hydroxide-mediated enolization of **1** to oxide **A_2** which ketonizes to form the cyclopropanone **B_2** after elimination of chloride ion. At this point hydroxide anion attacks the cyclopropanone to cause ring opening and formation of the carbanion intermediate **C_2** which can be stabilized through resonance until base-mediated deprotonation gives the carboxylate carbanion intermediate **D_2**. At this point, protonation of **D_2** may lead to carboxylate **E_2** which undergoes deprotonation to the carboxylate carbanion intermediate **F_2**. This intermediate rearranges to the α,β-unsaturated carboxylate **G_2** after elimination of chloride ion while also establishing the required 2E,4Z-stereochemical outcome. Lastly, protonation of **G_2** forms the final product **2** with the desired stereochemistry of 5-chloro-*trans*-2-*cis*-4-pentadienoic acid.

The basis for proposing this alternative mechanism by Acevado et al. (which resembles the earlier proposal by Takeda and Tsuboi)[2] stems from ^{14}C-labeling experiments.[3] These experiments support the conclusion that the mechanism proceeds via a cyclopropanone intermediate **B_2** that later fragments to a delocalized carbanion intermediate **C_2** (see Schemes 3 and 4). It should be noted that the formation of cyclopropanone intermediate **B_2** arises from a non-Favorskii precursor (compare intermediate **A_2** versus intermediate **D_1** in mechanism 1).

As such, mechanism 1 which involves 1,5-chlorine migration from **C_1** to **D_1** can be ruled out. If we apply the same kinds of ^{14}C labels on mechanism 1 we end up with the following results: (a) if the ^{14}C label is on the carbonyl group in **1**, then it ends up on the carboxylic acid group of product **2** which is in agreement with mechanism 2 (Scheme 5); (b) if the ^{14}C label is on the trichloromethyl group in **1**, then it ends up on the C_4 carbon atom in product **2**, not on the C_5 carbon atom as observed according to mechanism 2 (Scheme 6).

Further experiments undertaken by Acevado et al. were the following: (a) when the reaction was carried out in deuterated water, product **2** was completely deuterated; (b) when the reaction was carried out in the presence of bromide ions, no bromine was incorporated in product **2**; and (c) when the reaction was carried out in the presence of cyclohexene, no products incorporating cyclohexene were observed. These results show that all the C—H bonds in **1** are potentially exchangeable for C—D bonds. Moreover, lack of bromine incorporation suggests that the rates of all steps in mechanism 2 are faster than reaction of any carbanion intermediate with bromine. The lack of cyclohexene incorporation suggests that carbene intermediates are not involved.

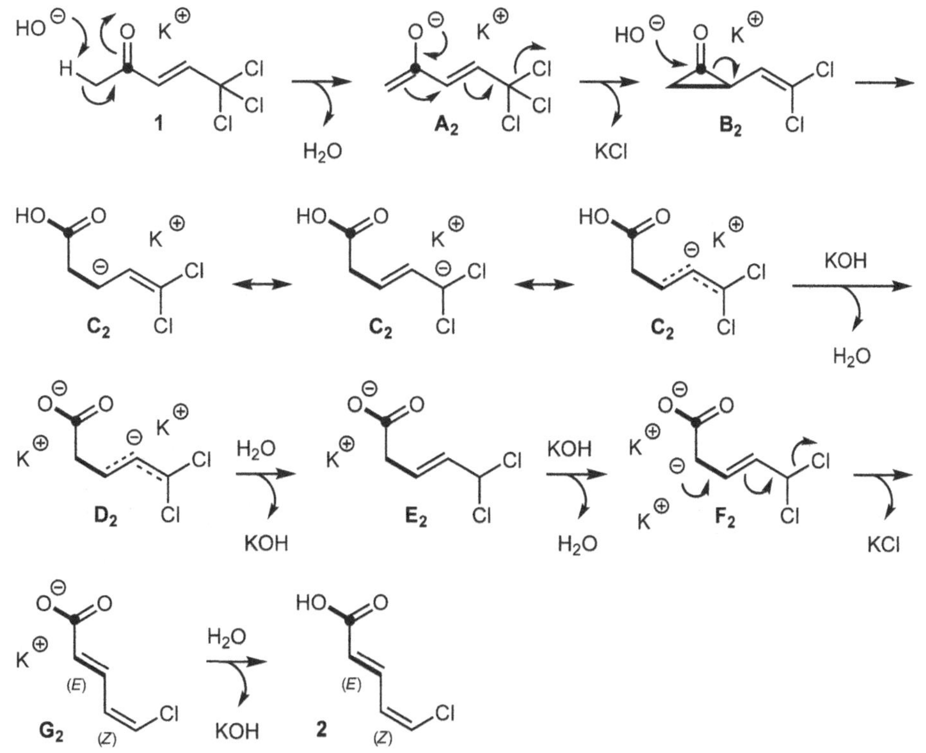

SCHEME 3

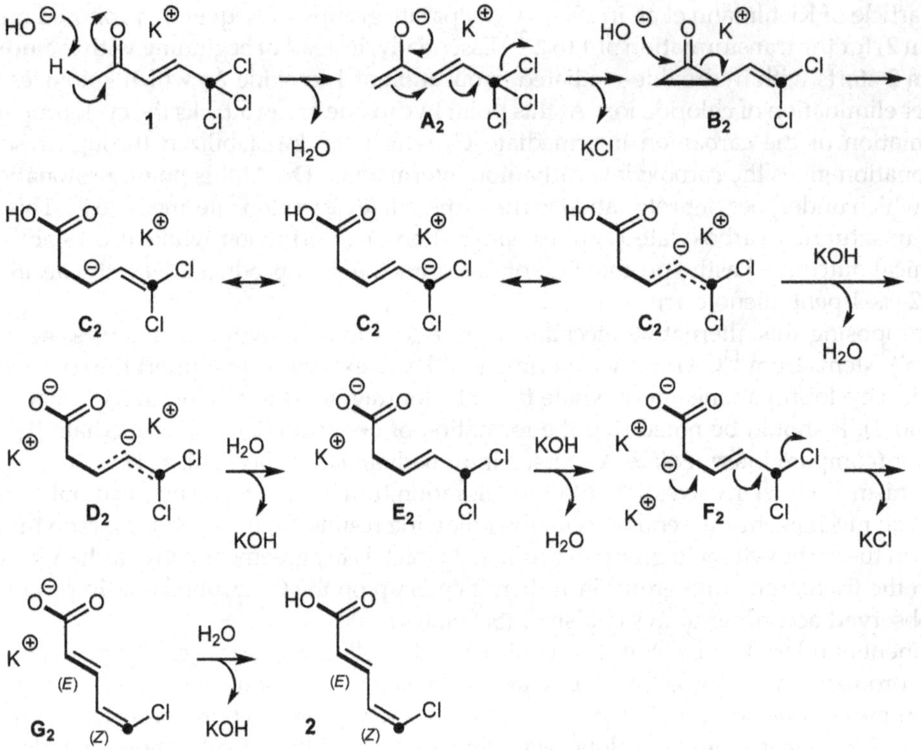

SCHEME 4

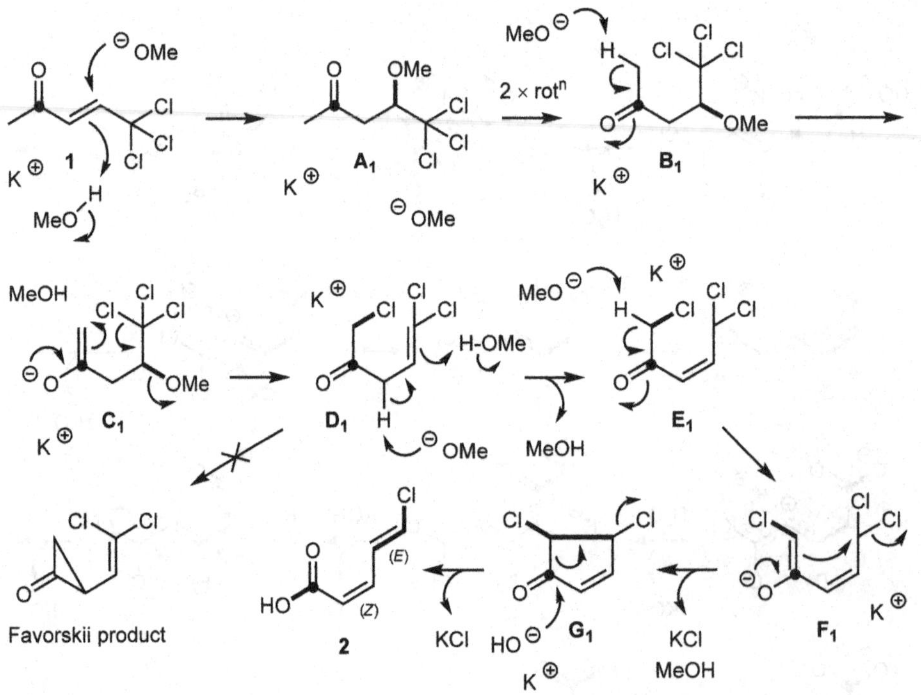

SCHEME 5

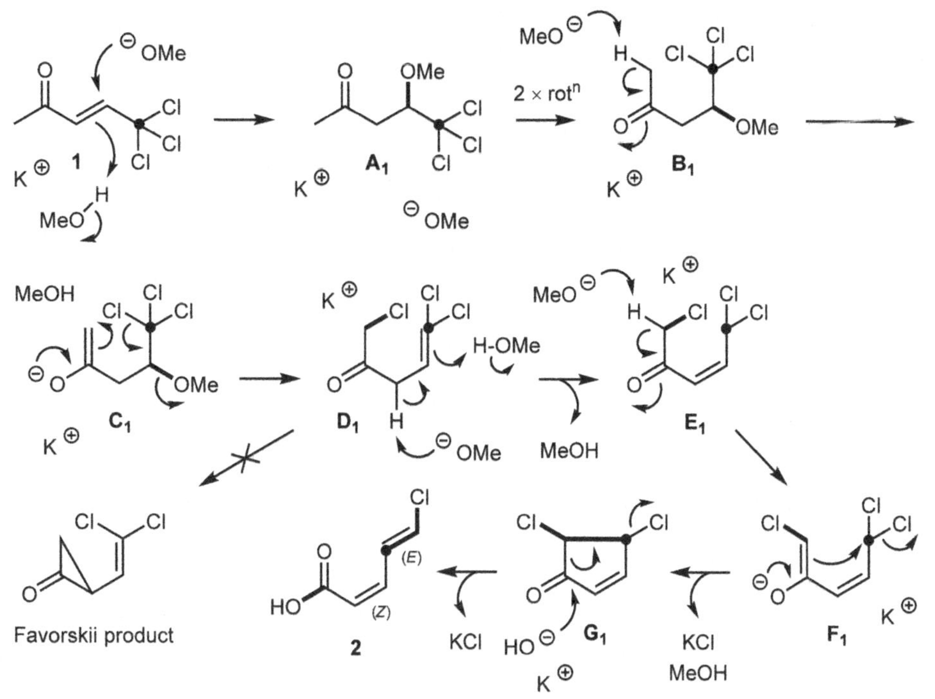

SCHEME 6

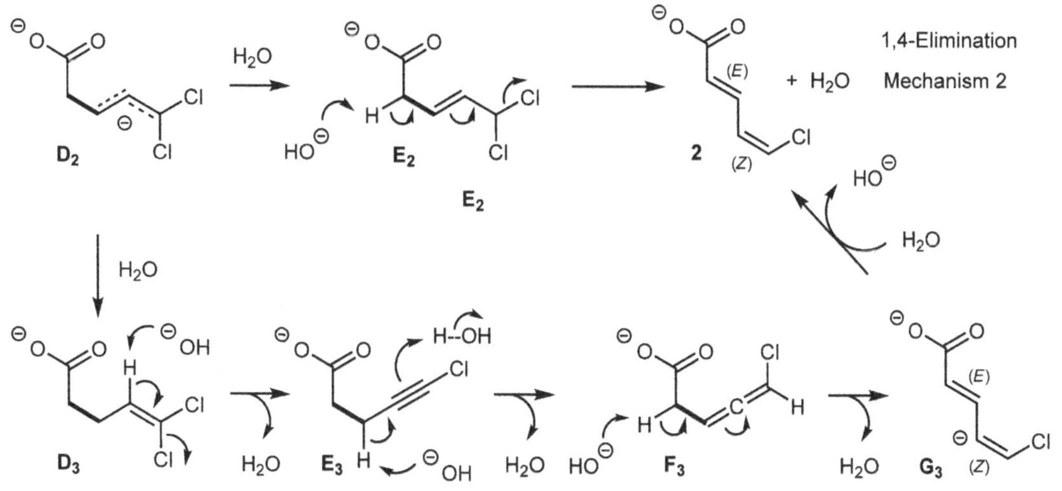

SCHEME 7

In addition, the authors also suggested an alternative pathway to the 1,4-elimination process from carbanion D_2 in mechanism 2 to explain the stereochemistry observed in the product (see Scheme 7). Essentially, the delocalized carbanion intermediate D_2 from mechanism 2 is protonated at C_3 to give intermediate D_3 which then eliminates chloride ion yielding chloroalkyne E_3. Further deprotonation forms allenyl chloride F_3 and another deprotonation leads to vinyl carbanion G_3 which finally protonates at C_4 to give product **2**. The authors stated that "vinyl carbanions are known to possess stereochemical stability in *cis*-forms" which would lead to the observed $2E,4Z$-structure shown in product **2**.[4,5]

In conclusion, there is stronger experimental evidence in favor of mechanism 2 as the operative mechanism in the transformation of **1** to **2**. The authors who proposed mechanism 1 did so due to results of product studies on a limited series of structurally related substrates and literature precedent. By comparison to isotope labeling studies, these experiments constitute weaker indirect evidence. Also, Kiehlmann et al. failed to account for the stereoselectivity of this reaction. Lastly, all authors had omitted theoretical analyses of their proposed mechanisms which may have shed much needed light on this interesting transformation. We also remark on the fact that the two proposed mechanisms have slightly different balanced chemical equations.

ADDITIONAL RESOURCES

For two examples of interesting discussions of reaction mechanisms involving carbocation chemistry in the context of conjugated carbonyl systems such as the system encountered in this problem, see Refs. 6, 7.

References

1. Kiehlmann E, Menon BC, Wells JI. Novel rearrangement during alkaline hydrolysis of 1,1,1-trichloro-2-alken-4-ones. *Can J Chem.* 1972;50:2561–2567. https://doi.org/10.1139/v72-412.
2. Takeda A, Tsuboi S. The allylic rearrangement. III. A Favorskii-type rearrangement of the vinylogs of α-chloroacetones. *J Org Chem.* 1973;38 (9):1709–1713. https://doi.org/10.1021/jo00949a020.
3. Acevado S, Bowden K, Henry MP. The mechanism of the alkaline rearrangement of 1,1,1-trichloro-2-penten-4-one. *Tetrahedron Lett.* 1976; (52):4837–4838. https://doi.org/10.1016/S0040-4039(00)78925-8.
4. Bowden K, Price MJ. Addition to unsaturated carbonyl compounds. Part I. The addition of hydrogen halides to propiolic acid. *J Chem Soc B.* 1970;1970:1466–1472. https://doi.org/10.1039/J29700001466.
5. Hunter DH, Cram DJ. Electrophilic substitution at saturated carbon. XXIII. Stereochemical stability of allylic and vinyl anions. *J Am Chem Soc.* 1964;86(24):5478–54901. https://doi.org/10.1021/ja01078a020.
6. Pirkle WH, McKendry LH. Photochemical reactions of 2-pyrone and thermal reactions of the 2-pyrone photoproducts. *J Am Chem Soc.* 1969;91 (5):1179–1186. https://doi.org/10.1021/ja01033a025.
7. Wang G, Zou Y, Li Z, Wang Q, Goeke A. Unexpected cycloisomerizations of nonclassical carbocation intermediates in gold(I)-catalyzed homo-Rautenstrauch cyclizations. *J Org Chem.* 2011;76:5825–5831. https://doi.org/10.1021/jo200416d.

Question 191: A Simple, High Yielding Route to a Cage Compound

During the course of two studies in the 1970s, Spanish chemists investigating synthetic routes to the caged product **3** heated a solution of triketone **1** in dimethyl ether (DME) containing a catalytic amount of *p*-TsOH under reflux for 48 h to form **2** quantitatively (Scheme 1).[1,2] The reaction of **1** with *t*-BuOK/*t*-BuOH in THF at room temperature for 16 h gave **3** in 78% yield (Scheme 1). Provide mechanisms to explain these transformations.

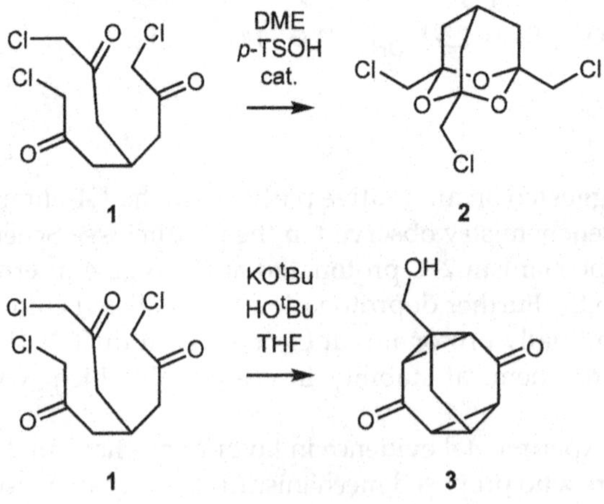

SCHEME 1

SOLUTION: 1 TO 2

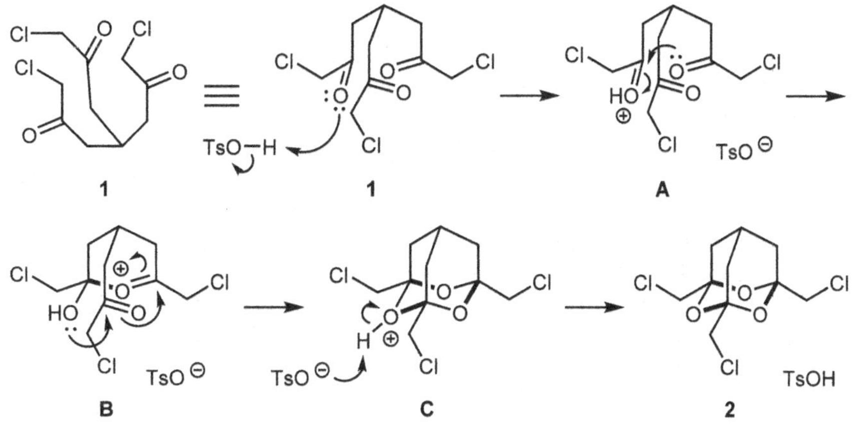

SOLUTION: 1 TO 3

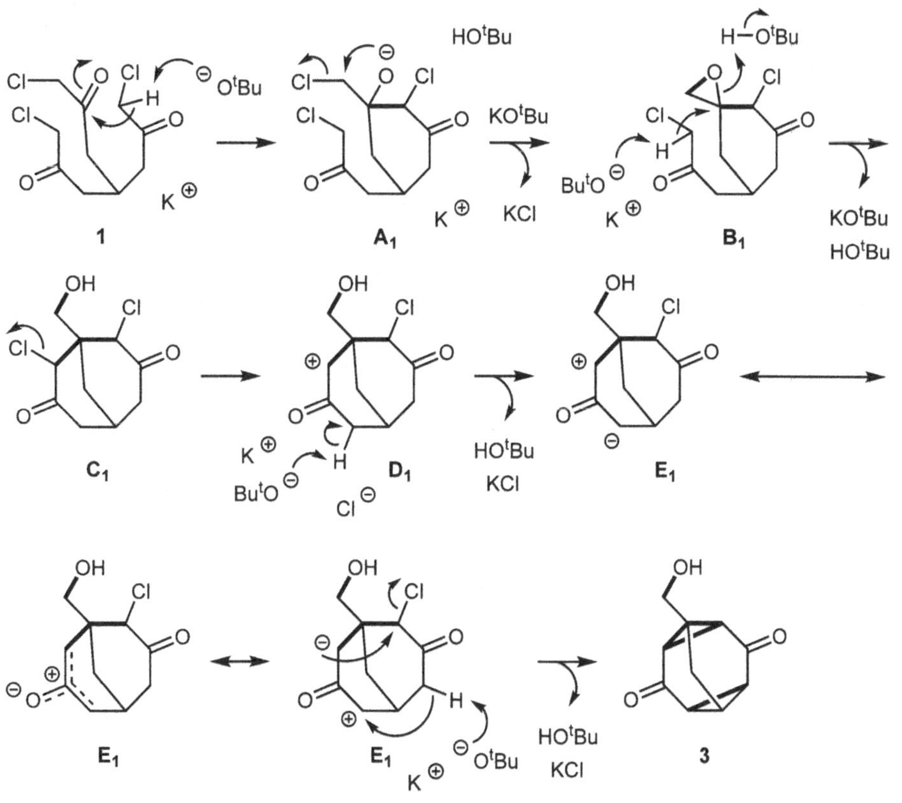

KEY STEPS EXPLAINED

The mechanism for the cyclization of **1** to **2** begins with protonation of one of the ketone groups of **1** to form **A** followed by intramolecular aldol cyclization to **B** which undergoes another tandem intramolecular cyclization sequence to **C**. This intermediate forms **2** after deprotonation and regeneration of the acid catalyst.

By contrast, the mechanism proposed by the authors for the transformation of **1** to **3** begins with deprotonation of one of the hydrogen atoms α to both the chloride and carbonyl groups to form a stable anion. It is worthwhile to note that according to the authors, citing the literature,[3] the C—H bond adjacent to both the chloride and carbonyl groups is significantly more acidic (by 2 pK$_A$ units) than the C—H methylene bonds adjacent to only the carbonyl groups. Given the soft basic reaction conditions of t-BuOK/t-BuOH, the authors argue that the more acidic hydrogen atom will be abstracted. Moreover, because of the high nucleophilicity of the base, the authors argue that deprotonation would out-compete attack on the carbonyl group by the base. The mechanism thus proceeds with reaction of the stable anion in an intramolecular manner with a carbonyl group from another side chain (six atoms apart) to form alkoxide A_1. This alk-oxide may attack the neighboring carbon atom to eliminate chloride anion and form the epoxide B_1. Another depro-tonation of the C—H adjacent to chloride and carbonyl groups of the third side chain of the substrate may lead to cyclization and ring opening of the epoxide to form the bicyclic primary alcohol C_1 following protonation. At this point the authors contend that dehydrochlorination of the bicyclic carbinol intermediate C_1 via loss of chloride (i.e., structure D_1) and deprotonation via excess base may lead to the oxyallyl dipolar ion E_1. This intermediate exists in two other resonance forms, the last of which may facilitate an anomalous Favorskii rearrangement initiated by a deprotonation step. This leads directly to the caged product **3** after elimination of chloride. Interestingly, the authors do not discuss the possibility of forming product **3** from C_1 via the strained intermediate F_1 (see Scheme 2). Presumably this possibility may be supported or ruled out after conducting a theoretical analysis to calculate the energy required to transform C_1 into F_1 as compared to forming D_1 and subsequently E_1.

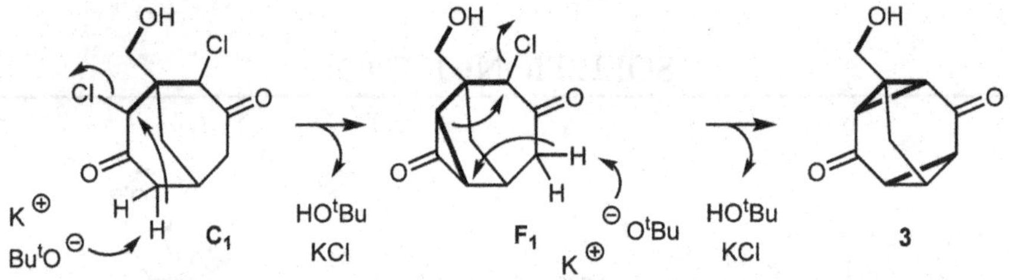

SCHEME 2

Another interesting point not discussed by the authors concerns the formation of the epoxide group in interme-diate B_1. For example, there are two C—Cl groups adjacent to the oxide group and they are both susceptible to the formation of an epoxide. With this consideration, a slightly different alternative mechanism may be proposed which makes use of the fused epoxide B_2 and the fused oxetanes D_2 and E_2 intermediates (Scheme 3). The differentiator

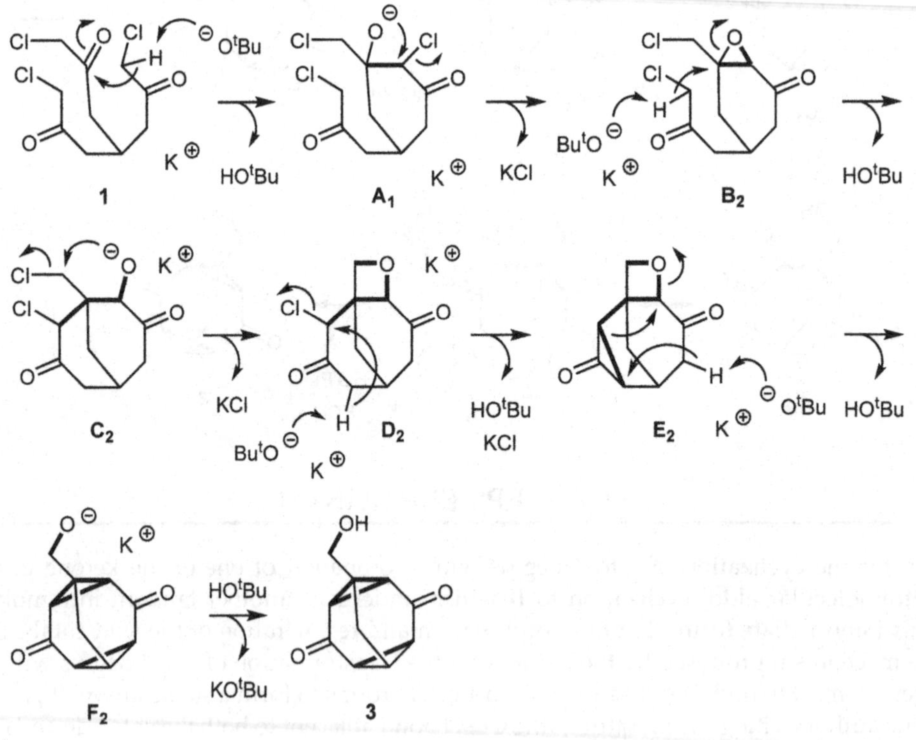

SCHEME 3

between this mechanism and the one illustrated in the solution earlier pertains to establishing the energy barriers to formation of intermediates B_2 and D_2 as compared to B_1 and D_1. A theoretical study can help shed light on these contrasting proposals.

ADDITIONAL RESOURCES

For three examples of very interesting transformations of caged compounds and compounds with a similar carbon skeleton to the system encountered here, see Refs. 4–6.

References

1. Herranz E, Serratosa F. Reaction of methane(tri-chloroacetone) under Favorskii conditions: a direct entry to the triasterane structure. *Tetrahedron Lett.* 1975;16(38):3335–3336. https://doi.org/10.1016/S0040-4039(00)91442-4.
2. Herranz E, Serratosa F. Protonation of methane(tri-α-diazoacetone) in acid softening solvents: acid and base-induced intramolecular cyclizations OM methane(tri-chloroacetone) to trioxaadamantane and triasterane derivatives. *Tetrahedron.* 1977;33(9):995–998. https://doi.org/10.1016/0040-4020(77)80214-7.
3. House HO, Fischer Jr. WF, Gall M, McLaughlin TE, Peet NP. Chemistry of carbanions. XX. Comparison of .alpha.-chloro enolate anions and. alpha.-diazo ketones. *J Org Chem.* 1971;36(22):3429–3437. https://doi.org/10.1021/jo00821a034.
4. James B, Rath NP, Suresh E, Nair MS. Formation of novel polycyclic cage compounds through 'uncaging' of readily accessible higher cage compounds. *Tetrahedron Lett.* 2006;47:5775–5779. https://doi.org/10.1016/j.tetlet.2006.06.005.
5. Lledo A, Benet-Buchholz J, Sole A, Olivella S, Verdaguer X, Riera A. Photochemical rearrangements of norbornadiene Pauson-Khand cycloadducts. *Angew Chem Int Ed.* 2007;46:5943–5946. https://doi.org/10.1002/anie.200701658.
6. Kokubo K, Koizumi T, Yamaguchi H, Oshima T. Lewis acid-catalyzed successive skeletal rearrangement of cyclobutene-fused diphenylhomoquinone. *Tetrahedron Lett.* 2001;42:5025–5028. https://doi.org/10.1016/S0040-4039(01)00908-X.

Question 192: Not All Ketals Hydrolyze Easily in the Expected Manner

In the course of a 1977 hydrolysis of the 9-ethylenedioxybicyclo[3.3.1]nonane-3,7-dione **1** under acidic conditions, Swiss chemists formed either the indanone **2** (14%) or a mixture of **2** (13%) and the bis(ethylenedioxy)bicycle[3.3.1] nonanone product **3** (13%) depending on reagents used (Scheme 1).[1] Provide a mechanistic explanation for these results.

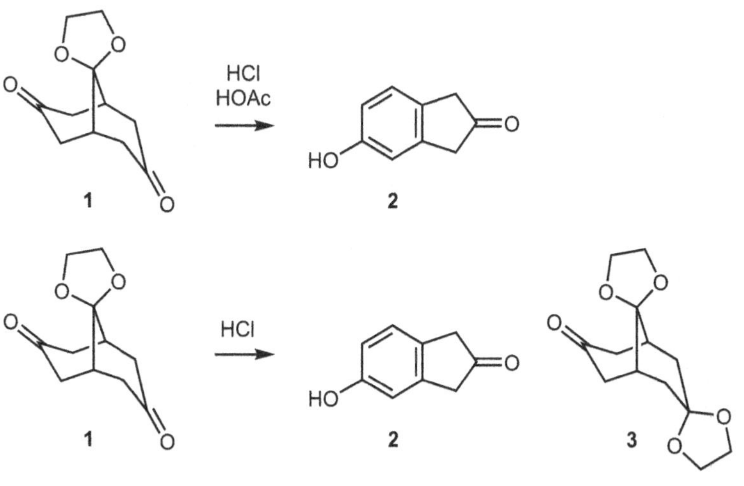

SCHEME 1

SOLUTION: 1 TO 2

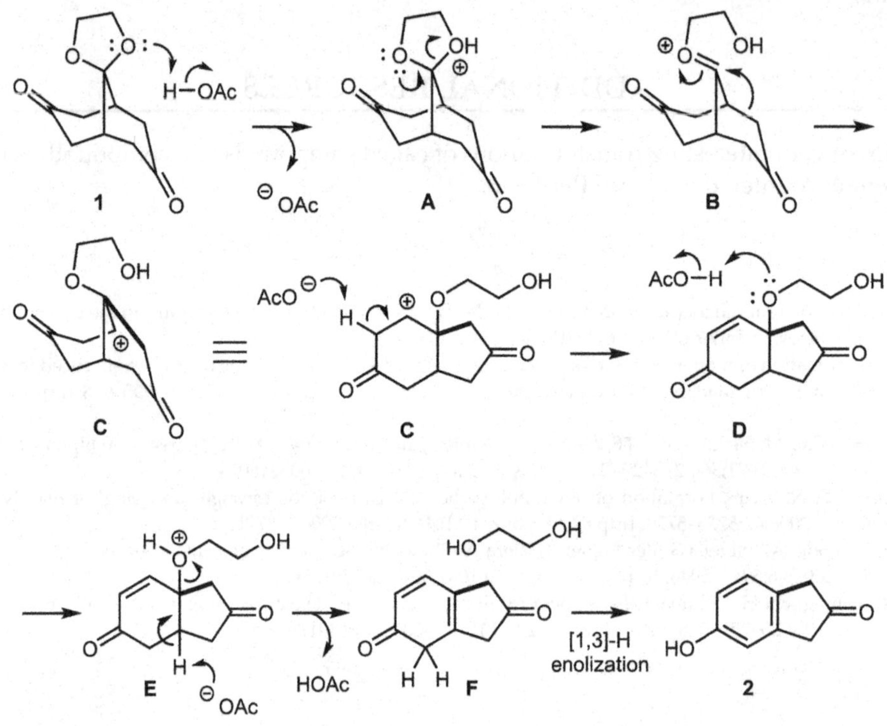

SOLUTION: 1 TO 2 AND 3

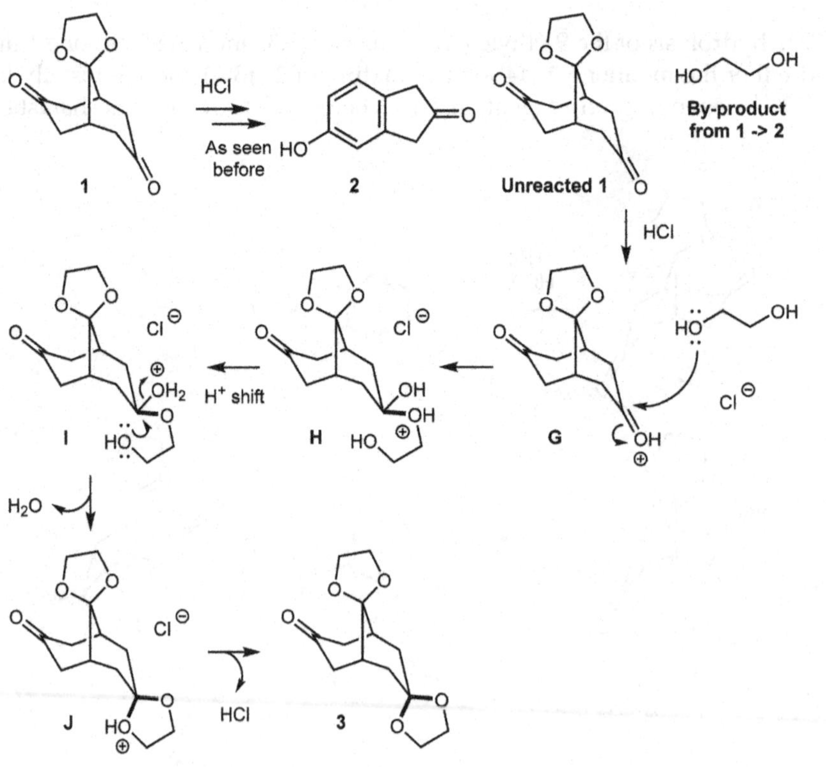

KEY STEPS EXPLAINED

The authors rationalize their proposed mechanism by contending that acid-promoted acetal ring-cleavage in **1** via oxonium intermediate **A** would lead to intermediate **B** which undergoes a 1,2-Wagner-Meerwein shift to form intermediate **C**. We note that this rearrangement changes the original [3.3.1] ring framework to a [4.3.0] framework. Deprotonation of **C** via the acetate counterion forms **D** which may eliminate ethylene glycol after a protonation-deprotonation sequence via **E** and **F** to form the indanone product **2** after enolization and aromatization.

Under the more strongly acidic conditions of using just hydrochloric acid, the authors rationalize that once enough of **2** along with ethylene glycol by-product accumulate in solution (according to the same mechanism presented before but substituting chlorine for acetate counterion), a side reaction between unreacted **1** and ethylene glycol leads to the formation of bis(ethylenedioxy)bicycle[3.3.1]nonanone product **3**. We know that there is likely to be a large amount of unreacted starting material **1** since the yield of **2** is only 14% and 13%, respectively, in the two reactions illustrated before. As such, with unreacted **1** and ethylene glycol in hand, intermediate **H** forms via acid-catalyzed addition of ethylene glycol to the ketone group of **1**. A proton shift then forms **I** which undergoes ring closure to **J** with elimination of water followed by deprotonation to generate product **3**.

Lastly, the authors attribute the lack of hydrolysis of **2** to sufficient ring strain in intermediate **B** preventing the addition of water to the carbocation position in order to cause the substitution of ethylene glycol for a hydroxyl group. Nevertheless, the transformation of **1** to **2** and subsequently to **3** proves experimentally that addition of water to **B** is not favorable. However, absent a theoretical analysis highlighting the calculated energies of **B**, **C** and the relevant transition states surrounding this step, it is not possible to say with certainty that ring strain alone is responsible for why **B** preferentially adopts a Wagner-Meerwein rearrangement to form **C**.

ADDITIONAL RESOURCES

For additional examples of interesting transformations in similar chemical systems, see Refs. 2, 3.

References

1. Chapleo CB, Dreiding AS. 145. Formation of 5-hydroxy-indan-2-one from attempts to hydrolyze 9-ethylenedioxy-bicyclo[3.3.1]nonane-3,7-dione. *Helv Chim Acta*. 1977;60(4):1448–1451. https://doi.org/10.1002/hlca.19770600437.
2. Song S, Lee J, Kim H. Short synthesis of γ-hydroxy octalone utilizing an unusual decarboethoxylation. *Bull Kor Chem Soc*. 1993;14(4):435. https://doi.org/10.1002/chin.199412163.
3. Oda M, Miyazaki H, Kayama Y, Kitahara Y. A new synthesis and thermal rearrangement of 3,4-benzobicyclo[4.2.0]octa-3,7-diene-2,5-dione. *Chem Lett*. 1975;4(6):627–630. https://doi.org/10.1246/cl.1975.627.

Question 193: From a Tricycle to a Ring Expanded Bicycle

During a 1977 study of the total synthesis of gibberellic acid, American chemists heated lactone **1** under reflux with potassium *t*-butoxide in *t*-butanol and achieved compound **2** in 30% yield after acidification and product separation (Scheme 1).[1] Explain this result in mechanistic terms.

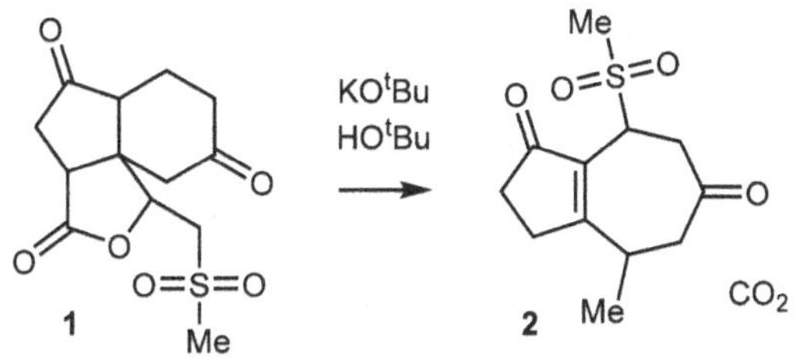

SCHEME 1

SOLUTION

KEY STEPS EXPLAINED

We first note that the structure of lactone **1** as drawn by the authors in their paper is missing an oxygen atom in the ring. The mechanism proposed by the authors begins with base-promoted deprotonation and ring opening of the lactone functionality of **1** to form the carboxylate intermediate **A** which may be further deprotonated by a second equivalent of base to form the oxonium carboxylate intermediate **B**. Ketonization of **B** then leads to formation of a cyclopropane ring to generate **C** which after a protonation-deprotonation sequence via **D** forms the ring-expanded intermediate **E**. Ketonization of **E** to **F** followed by deprotonation to **G** may set up the elimination of methanesulfonate group to form the diketodiene intermediate **H**. This intermediate then undergoes two prototropic rearrangements to form **J** via **I** (which the authors believe to be facile transformations under the strongly basic reaction conditions). With **J** in hand, the authors claim that acidification of the reaction medium facilitates the nucleophilic addition of methanesulfonate at the α,β-unsaturated ketone group to form **K** followed by decarboxylation to **L** and a final ketonization-protonation sequence to form the desired product **2**. Despite this reasoning, the authors did not provide further experimental and theoretical evidence to support their postulated mechanism. In particular, a discussion about why the decarboxylation is expected at the end of the transformation as compared to, for example, the very first step of the mechanism, is not available. As such, we present the mechanism the authors provided with the goal of ensuring that as many mechanistic steps as possible have a clearly identified electron sink that would facilitate the various deprotonation-protonation sequences required to form **2**. For completeness we show an alternative mechanism where early decarboxylation takes place as shown in Scheme 2.

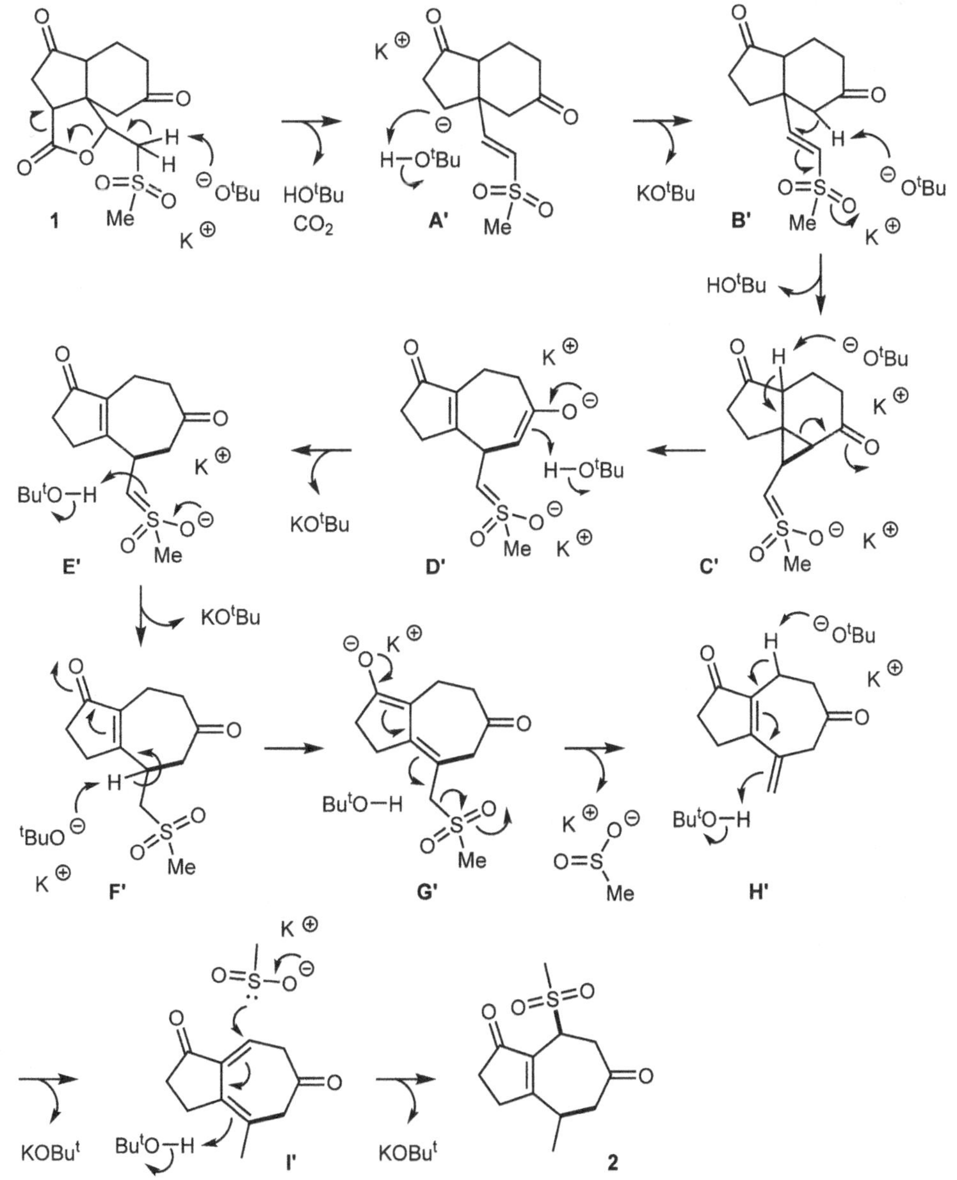

SCHEME 2

ADDITIONAL RESOURCES

We recommend two articles which involve interesting mechanistic discussions involving methanesulfonate group participation.[2,3] We also refer the reader to a review article which covers the synthetic utility of acylvinyl and vinylogous groups which includes a few interesting transformations from the standpoint of reaction mechanism.[4]

References

1. Fayos J, Clardy J, Dolby LJ, Farnham T. The chemistry of γ-oxo sulfones. 1. A novel rearrangement and a method for the β-alkylation of α,β-unsaturated ketones. *J Org Chem*. 1977;42(8):1349–1352. https://doi.org/10.1021/jo00428a017.
2. Alcaide B, Almendros P, Aragoncillo C, Redondo MC. Stereoselective synthesis of 1,2,3-trisubstituted 1,3-dienes through novel [3,3]-sigmatropic rearrangements in α-allenic methanesulfonates: application to the preparation of fused tricyclic systems by tandem rearrangement/Diels-Alder reaction. *Eur J Org Chem*. 2005;2005:98–106. https://doi.org/10.1002/ejoc.200400527.
3. Harger MJP, Smith A. Migration of the amino group in the base-induced rearrangements of *N*-(aminophosphinoyl)-*O*-sulphonylhydroxylamines. *J Chem Soc Perkin Trans 1*. 1986;1986:2169–2172. https://doi.org/10.1039/P19860002169.
4. Chinchilla R, Najera C. Acylvinyl and vinylogous synthons. *Chem Rev*. 2000;100(6):1891–1928. https://doi.org/10.1021/cr9900174.

Question 194: A Flavone From a Chromanone

In 1979 Irish chemists studying the reactivity of chroman-4-ones noticed that different products were obtained when the temperature and acid-base medium of the reaction conditions were changed. Their major observations are illustrated in Scheme 1.[1] Provide an explanation for these results in the context of a mechanistic analysis.

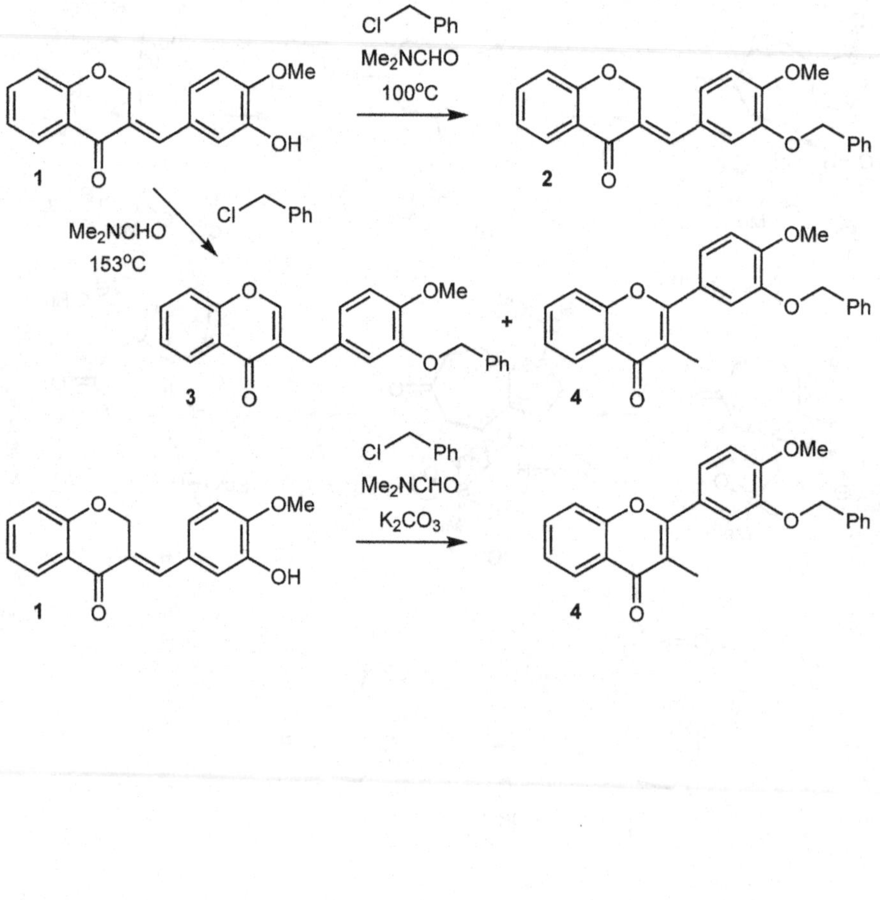

SCHEME 1

SOLUTION: 1 TO 2

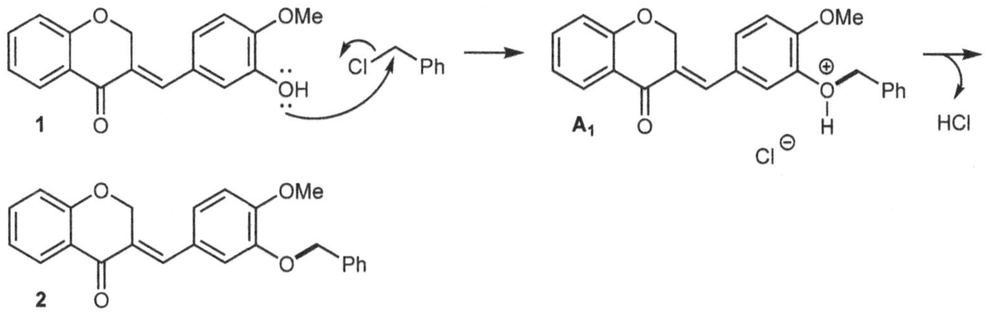

SOLUTION: 2 TO 3 (AT ELEVATED TEMPERATURE)

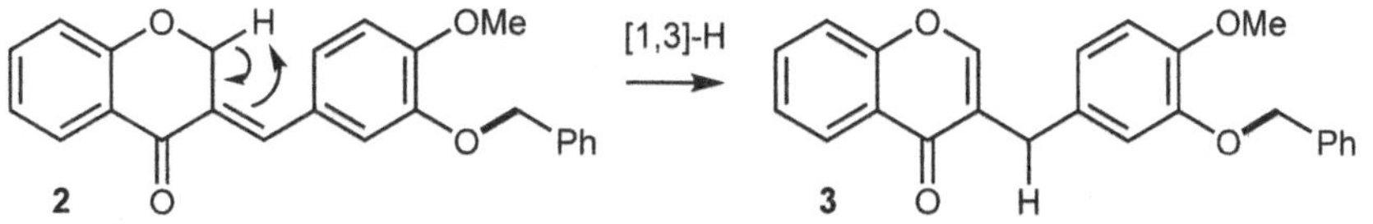

SOLUTION: 2 TO 4 (AT ELEVATED TEMPERATURE NO ADDED BASE)

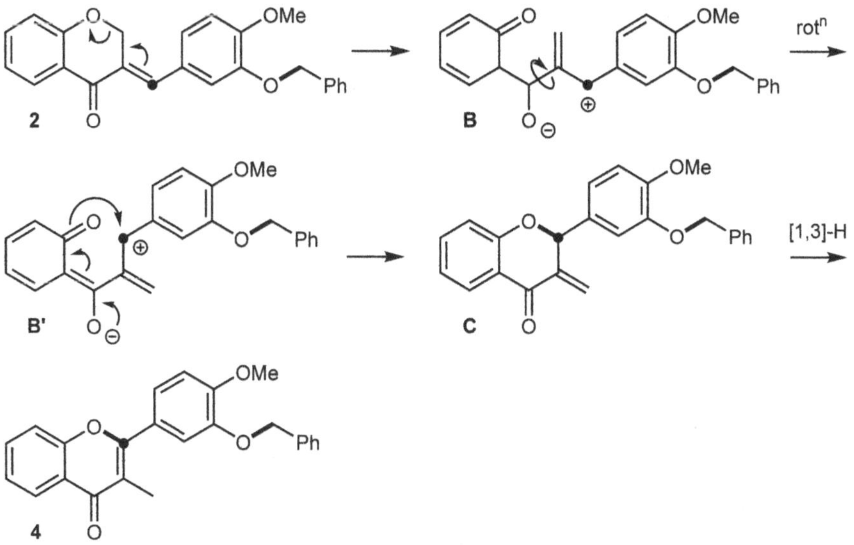

SOLUTION: 1 TO 4 (WITH ADDED BASE)

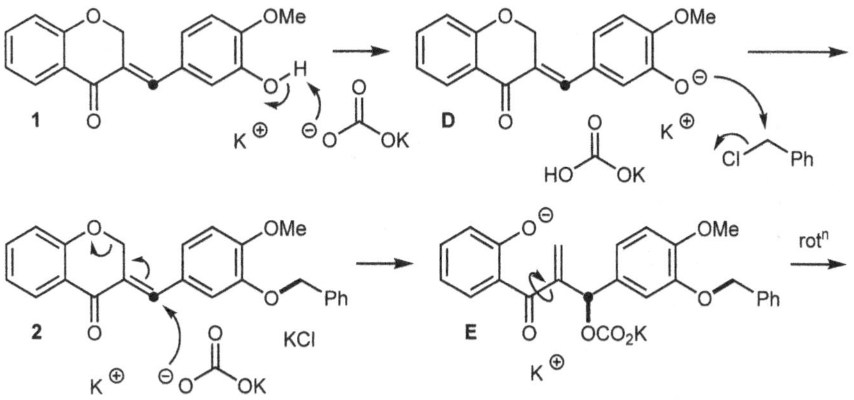

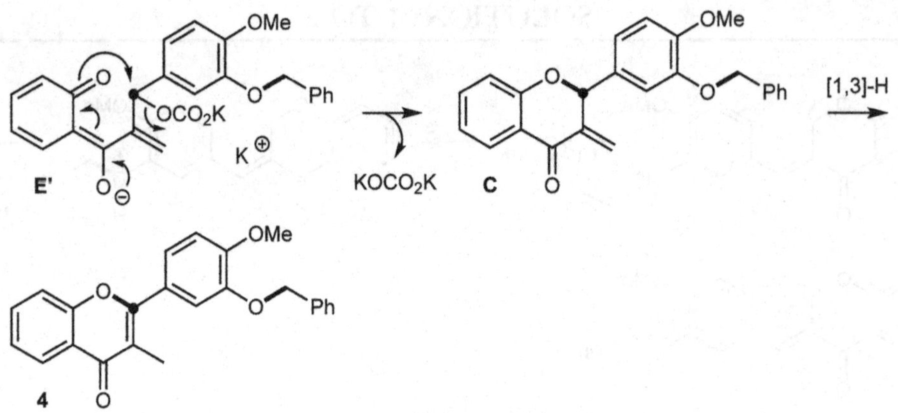

KEY STEPS EXPLAINED

During the course of these synthetic experiments, the authors proved the structures of products **3** and **4** by check syntheses as shown in Scheme 2. In addition, the transformation of **1** to **2** represents a simple uncatalyzed etherification reaction. Under elevated temperature, product **2** further undergoes either a [1,3]-protic shift to form product **3** or ring opening to form the zwitterion **B** which after bond rotation to **B′** recyclizes to form intermediate **C** which proceeds to product **4** after a final [1,3]-protic shift.

With respect to the transformation of **1** to **4** when potassium carbonate base is added to the reaction mixture, the proposed mechanism begins with a base-catalyzed etherification that forms **2**. At this point, a second equivalent of base acts as a nucleophile and attacks **2** to cause a cascade ring opening transformation that leads to enolate **E**. After bond rotation to **E′** and recyclization to **C** with regeneration of base, product **4** is formed after a [1,3]-protic shift.

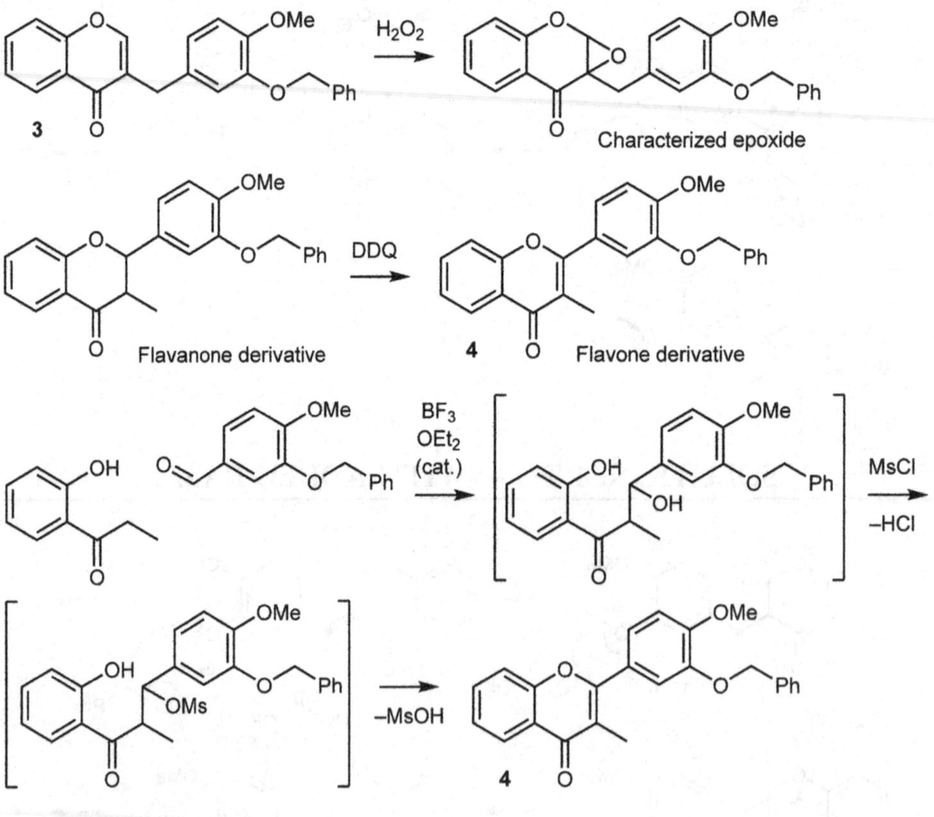

SCHEME 2

The authors presented evidence for the proposed rearrangement mechanism for the **1** to **4** transformation based on results obtained from a ^{14}C labeling experiment as indicated by the carbon atom labeled with a dot in the mechanisms shown earlier. Furthermore, a simplified labeled flavone starting material was made by condensing chroman-4-one with ^{14}C labeled benzaldehyde to form a product which once subjected to base-catalyzed rearrangement conditions generated an analog of **4** (Scheme 3).

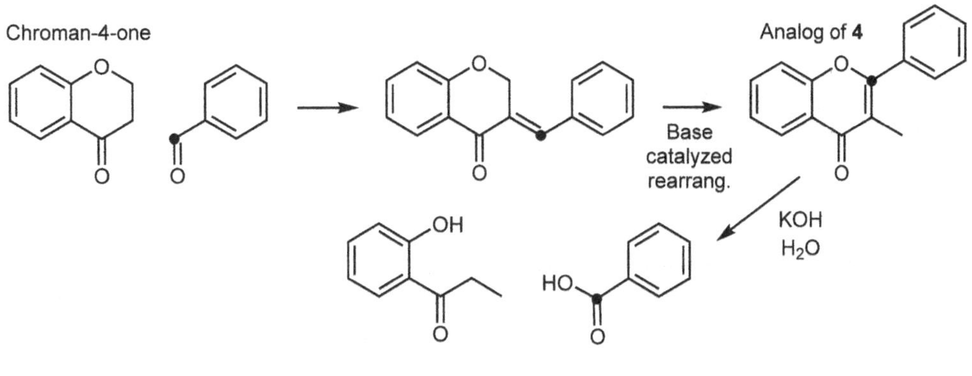

SCHEME 3

Degradation of this analog of **4** in aqueous potassium hydroxide solution resulted in the production of unlabeled 1-(2-hydroxy-phenyl)-propan-1-one and ^{14}C labeled benzoic acid in the carboxyl group (see proposed mechanism in Scheme 4). We note that the authors omitted the hydroxyl group in the unlabeled product in the scheme as drawn in their paper. We also note that a theoretical analysis is absent from the analysis of the authors.

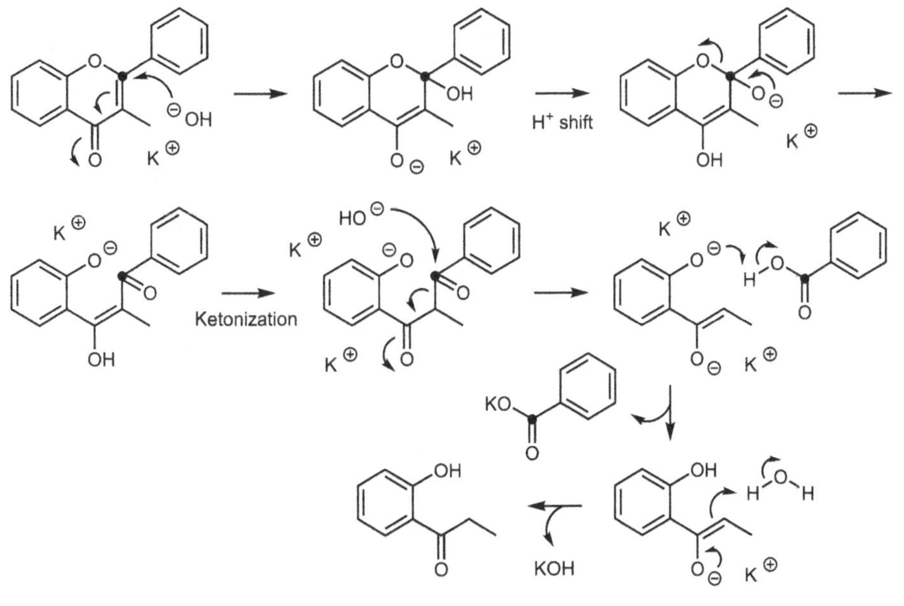

SCHEME 4

ADDITIONAL RESOURCES

For another example of a flavanone transformation, see the work of Kinoshita et al.[2] For a wonderful example of a temperature-dependent transformation with two alternative mechanisms operating at different temperatures, see the work of Xiong et al.[3]

References

1. Mulvagh D, Meegan MJ, Donnelly D. Rearrangements of 3-benzylidenechroman-4-ones. *J Chem Res (S)*. 1979;1979:137.
2. Kinoshita T, Ichinose K, Sankawa U. One-step conversion of flavanones into isoflavones: a new facile biomimetic synthesis of isoflavones. *Tetrahedron Lett*. 1990;31(50):7355–7356. https://doi.org/10.1016/S0040-4039(00)88565-2.
3. Xiong Y, Schaus SE, Porco Jr JA. Metal-catalyzed cascade rearrangements of 3-alkynyl flavone ethers. *Org Lett*. 2013;15(8):1962–1965. https://doi.org/10.1021/ol400631b.

Question 195: Thiophene Ylide Rearrangement

In the course of the thermolysis of the ylide **1** in anisole containing a catalytic amount of BF$_3$•OEt$_2$, Scottish chemists achieved a mixture of products the major component of which was the thieno-furan carboxylate **2** (Scheme 1).[1] Provide a mechanism for this transformation.

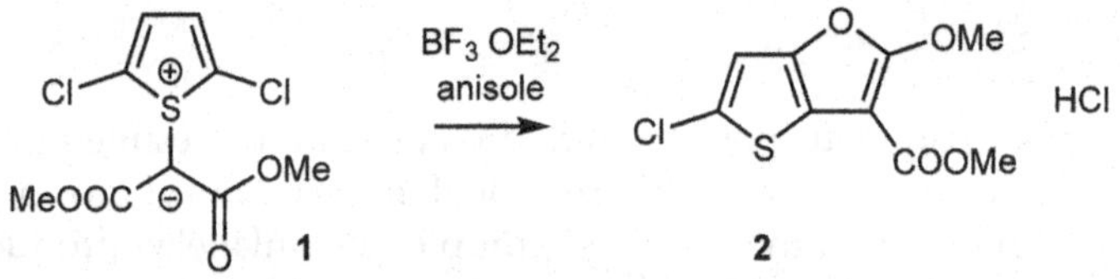

SCHEME 1

SOLUTION

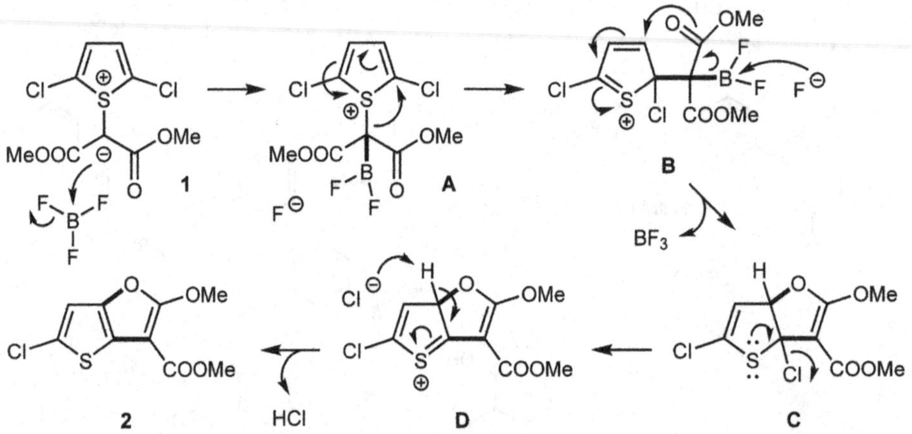

KEY STEPS EXPLAINED

The mechanism proposed by the authors for the eliminative rearrangement of **1** into **2** begins with addition of ylide **1** to the Lewis acid catalyst boron trifluoride to give intermediate **A**. This intermediate undergoes an intramolecular malonate substituent transfer from the sulfur atom to the neighboring carbon of the thiophene ring to form intermediate **B**. Intermediate **B** is subsequently attacked by fluoride anion to regenerate the Lewis acid catalyst and simultaneously facilitate ring cyclization to the furan group in intermediate **C** which upon loss of hydrogen chloride generates the final product **2**.

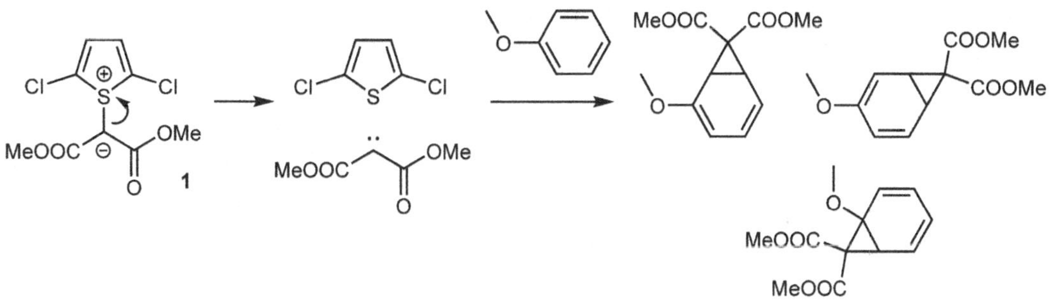

SCHEME 2

Furthermore, when the authors carried out the reaction of the same ylide **1** in the presence of metal catalysts such as copper(II)acetylacetonate, malonation of the solvent anisole occurred via a carbene insertion reaction as shown in Scheme 2. Since no malonation of anisole took place when boron trifluoride was used this was interpreted as evidence that carbene intermediates were not involved in that mechanism.

ADDITIONAL RESOURCES

We recommend an example of an interesting transformation involving thiophene derivatives in a very similar chemical system to the one presented here.[2] For a discussion of synthetic approaches to thieno[3,2-*b*]furan derivatives, see the work of Hergue et al.[3] Lastly, we recommend an excellent review article which discusses the intramolecular generation and rearrangement of oxonium ylides.[4]

References

1. Gillespie RJ, Murray-Rust J, Murray-Rust P, Porter AEA. On the mechanism of the catalysed and uncatalysed thermolysis of 2,5-dichlorothiophenium bismethoxycarbonylmethylide: X-ray crystal and molecular structure of methyl 5-chloro-2-methoxythieno[3,2-*b*]furan-3-carboxylate. *J Chem Soc Chem Commun*. 1979;1979:366–367. https://doi.org/10.1039/C39790000366.
2. Jenks WS, Heying MJ, Stoffregen SA, Rockafellow EM. Reaction of dicarbomethoxycarbene with thiophene derivatives. *J Org Chem*. 2009;74:2765–2770. https://doi.org/10.1021/jo802823s.
3. Hergue N, Mallet C, Touvron J, Allain M, Leriche P, Frere P. Facile synthesis of 3-substituted thieno[3,2-*b*]furan derivatives. *Tetrahedron Lett*. 2008;49:2425–2428. https://doi.org/10.1016/j.tetlet.2008.02.058.
4. Murphy GK, Stewart C, West FG. Intramolecular generation and rearrangement of oxonium ylides: methodology studies and their application in synthesis. *Tetrahedron*. 2013;69:2667–2686. https://doi.org/10.1016/j.tet.2013.01.051.

Question 196: A Stepwise $2\pi + 2\pi$ Intermolecular Cycloaddition

In 1992 American chemists reacted 1,2-dimethylcyclohexene with the ethylene glycol acetal of acrolein **1** in methylene chloride in the presence of 25 mol% of BF$_3$•OEt$_2$ at −78 to −10°C for 2h to form the cycloadduct **2** in 70% yield (Scheme 1).[1] Suggest a stepwise mechanism for this formal $2\pi + 2\pi$ intermolecular cycloaddition.

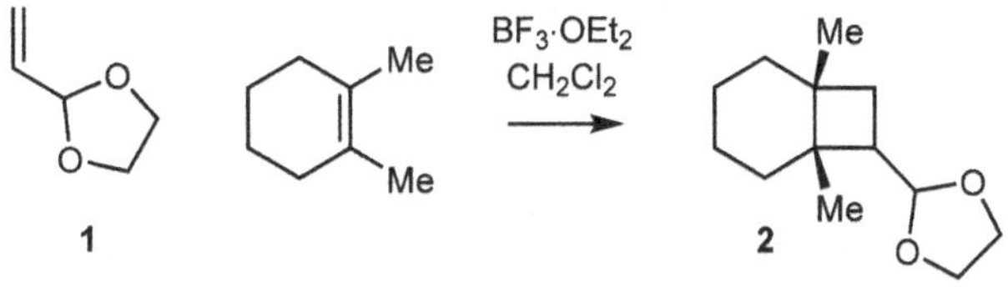

SCHEME 1

SOLUTION

1   A   B

2   D   C

KEY STEPS EXPLAINED

The stepwise mechanism proposed by the authors for the transformation of **1** to **2** begins with the reaction between **1** and the Lewis acid catalyst boron trifluoride to produce oxonium intermediate **A**. Intermediate **A** then undergoes ring opening to form the oxonium ion **B** which reacts with 1,2-dimethylcyclohexene in an overall [2+2] fashion to give oxonium intermediate **D** which is attacked by fluoride anion to help form the final product **2** after regeneration of the boron trifluoride catalyst.

ADDITIONAL RESOURCES

We highly recommend two articles describing interesting cationic [2+2] cycloadditions[2,3] as well as a review article discussing cationic cycloaddition reactions in a more general context.[4] Lastly, we also include an example of a transformation involving Lewis acid catalysis which has an interesting mechanistic discussion involving cationic cyclization routes.[5]

References

1. Gassman PG, Lottes AC. Cyclobutane formation in the $2\pi + 2\pi$ cycloaddition of allyl and related cations to unactivated olefins. Evidence for the second step in the proposed mechanism of the ionic Diels-Alder reaction. *Tetrahedron Lett*. 1992;33(2):157–160. https://doi.org/10.1016/0040-4039(92)88038-7.

2. Deng J, Hsung RP, Ko C. Gassman's cationic [2 + 2] cycloadditions using temporary tethers. *Org Lett*. 2012;14(21):5562–5565. https://doi.org/10.1021/ol3026796.

3. Kurdyumov AV, Hsung RP. An unusual cationic [2 + 2] cycloaddition in a divergent total synthesis of hongoquercin A and rhododaurichromanic acid A. *J Am Chem Soc*. 2006;128:6272–6273. https://doi.org/10.1021/ja054872i.

4. Harmata M, Rashatasakhon P. Cycloaddition reactions of vinyl oxocarbenium ions. *Tetrahedron*. 2003;59:2371–2395. https://doi.org/10.1016/S0040-4020(03)00253-9.

5. Engler TA, Ali MH, Takusagawa F. Studies on the synthesis of acanthodoral and nanaimoal: evaluation of cationic cyclization routes. *J Org Chem*. 1996;61(24):8456–8463. https://doi.org/10.1021/jo9610568.

Question 197: Rearrangement of an Aryl Propargyl Ether

During a 1992 study of the Claisen rearrangement of benzopyran-4-one derivatives, Indian chemists heated the propargyl ether **1** in refluxing dimethylaniline for 5h and formed the unexpected angular tricyclic ketone **2** (Scheme 1).[1] Suggest a mechanism for this transformation.

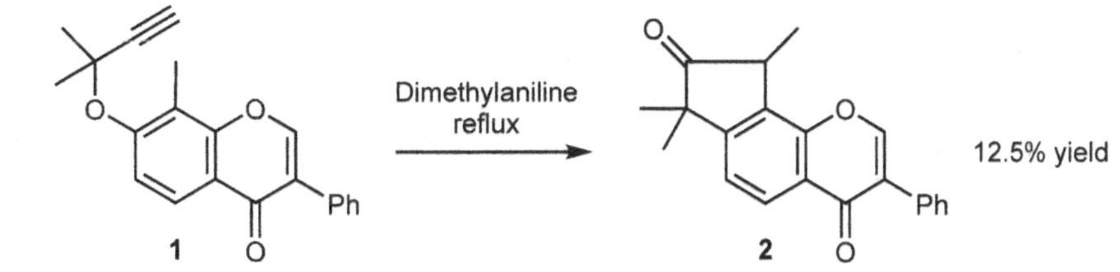

SCHEME 1

SOLUTION

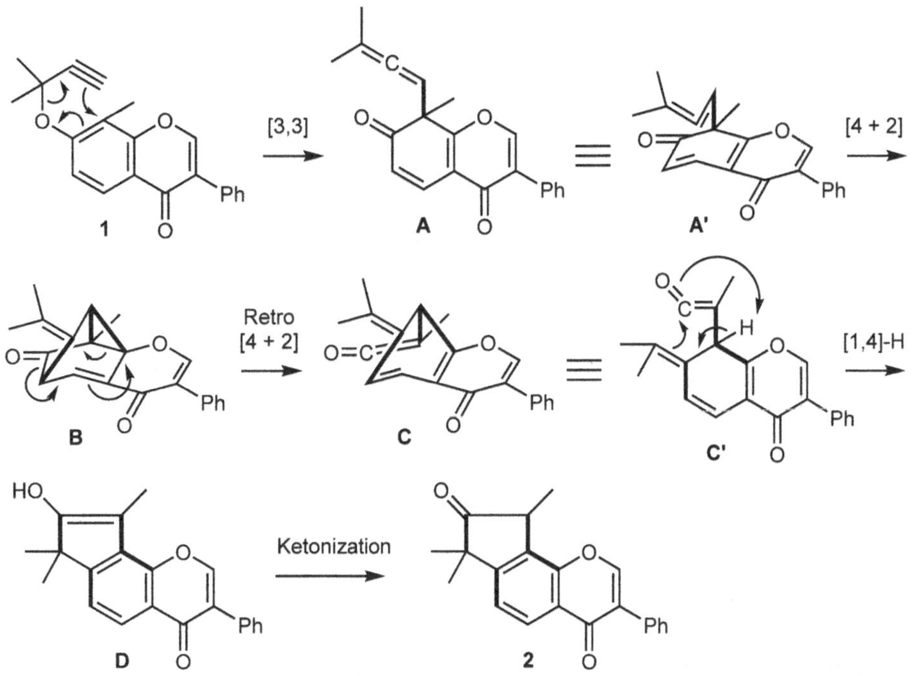

KEY STEPS EXPLAINED

The mechanism proposed by the authors to explain the unexpected transformation of the propargyl ether **1** to the angular tricyclic ketone **2** begins with an abnormal [3,3] Claisen rearrangement that forms allene **A** which then undergoes a [4+2] cycloaddition to generate the tricyclic intermediate **B**. This intermediate then undergoes a retro [4+2] transformation to form the ketene **C/C'** which after a [1,4]-H shift to the enol **D** ketonizes to form the final product **2**. We note that the ring construction strategy for this transformation is [(4+2)+(3+2)]. It is interesting, for comparison, to provide the mechanism of the transformation the authors expected to occur with regard to the reaction conditions (see Scheme 2).

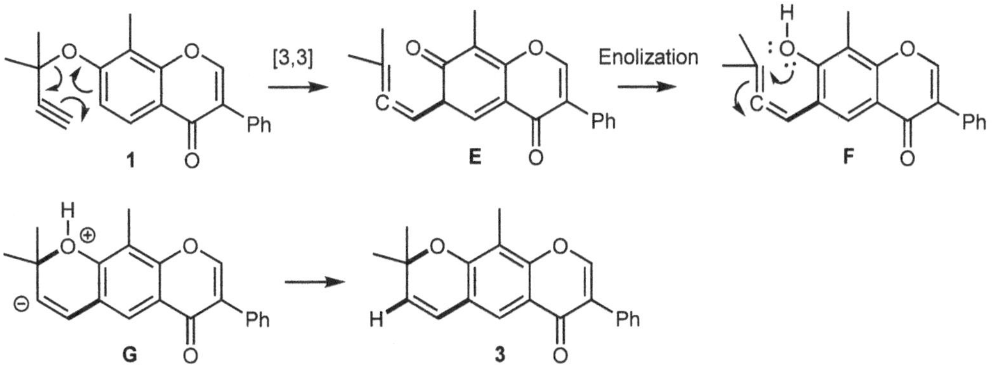

SCHEME 2

The expected mechanism which leads to the formation of the linear tricyclic product **3** begins with a Claisen rearrangement which occurs in the opposite (less hindered) direction to form the keto allene intermediate **E**. This intermediate enolizes to allene **F** which then cyclizes to product **3** after a prototropic shift. The overall transformation is a [3+3] ring construction mapping. For future study we highly recommend undertaking a theoretical analysis to help explain why the unexpected Claisen rearrangement from **1** to **A** is favored compared to the expected Claisen rearrangement leading from **1** to **E**.

ADDITIONAL RESOURCES

For the reader interested in examples of other unusual Claisen rearrangements, electrocyclizations, and hydride shifts, we recommend Refs. 2–4, respectively.

References

1. Joshi SC, Trivedi KN. A study of Claisen rearrangements of (1,1-dimethyl-3-prop-2-ynyloxy)-[4H]-1-benzopyran-4-one derivatives. *Tetrahedron.* 1992;48(3):563–570. https://doi.org/10.1016/S0040-4020(01)89017-7.
2. Al-Maharik N, Botting NP. Synthesis of lupiwighteone via a para-Claisen–Cope rearrangement. *Tetrahedron.* 2003;59:4177–4181. https://doi.org/10.1016/S0040-4020(03)00579-9.
3. Nicolaou KC, Sasmal PK, Xu H. Biomimetically inspired total synthesis and structure activity relationships of 1-O-methyllateriflorone. 6π electrocyclizations in organic synthesis. *J Am Chem Soc.* 2004;126:5493–5501. https://doi.org/10.1021/ja040037+.
4. Gabbutt CD, Heron BM, Instone AC, et al. Observations on the synthesis of photochromic naphthopyrans. *Eur J Org Chem.* 2003;2003(7):1220–1230. https://doi.org/10.1002/ejoc.200390176.

Question 198: The Best Laid Plans......

In 1992 American chemists attempted to synthesize the tetracycle **2** in two steps by treating indole **1** with methyl chloroformate and a nonnucleophilic base such as Hünig's base followed by an acid-catalyzed rearrangement (Scheme 1).[1] Although the first reaction was successful, the latter acid-catalyzed rearrangement produced the carbazole **3** instead of the expected product **2**. Show by means of mechanisms the authors' reasoning about how tetracycle **2** could be formed. Explain the observed outcome of carbazole **3**. Is the observed reaction in the second step a true rearrangement?

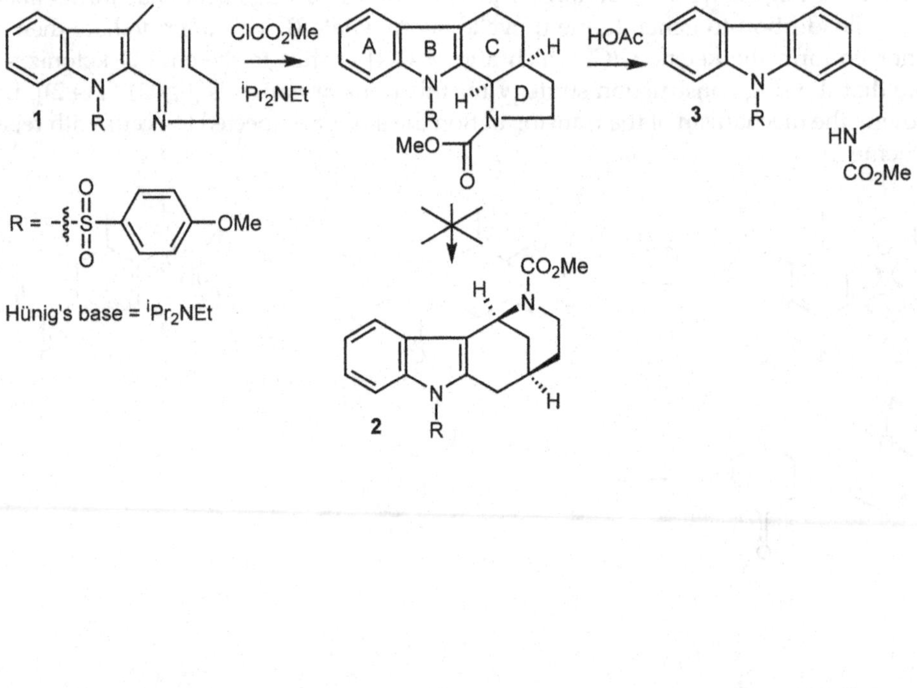

SCHEME 1

SOLUTION: 1 TO 2 (EXPECTED RESULT) STEP 1

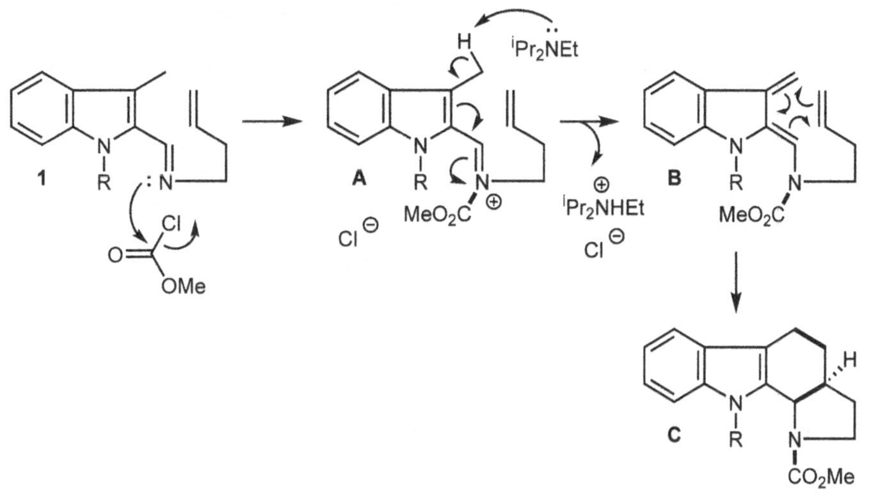

SOLUTION: 1 TO 2 (EXPECTED RESULT) STEP 2

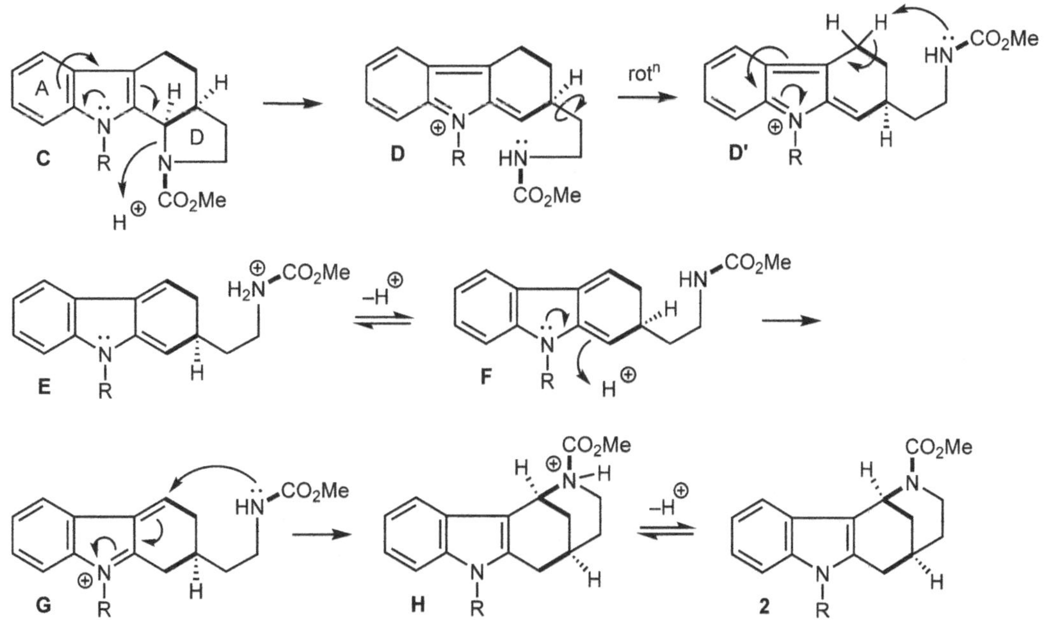

SOLUTION: 1 TO 3 (OBSERVED RESULT)

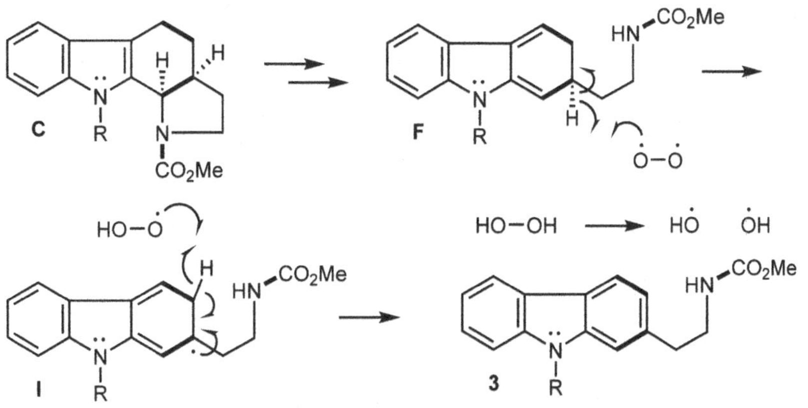

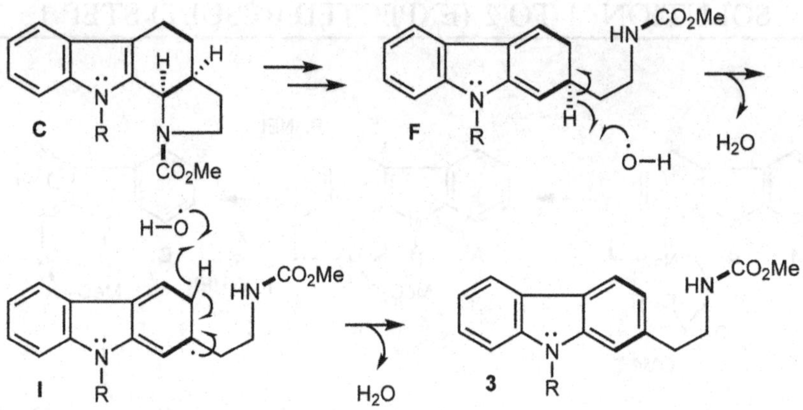

ALTERNATIVE PATHWAY FROM INTERMEDIATE D′

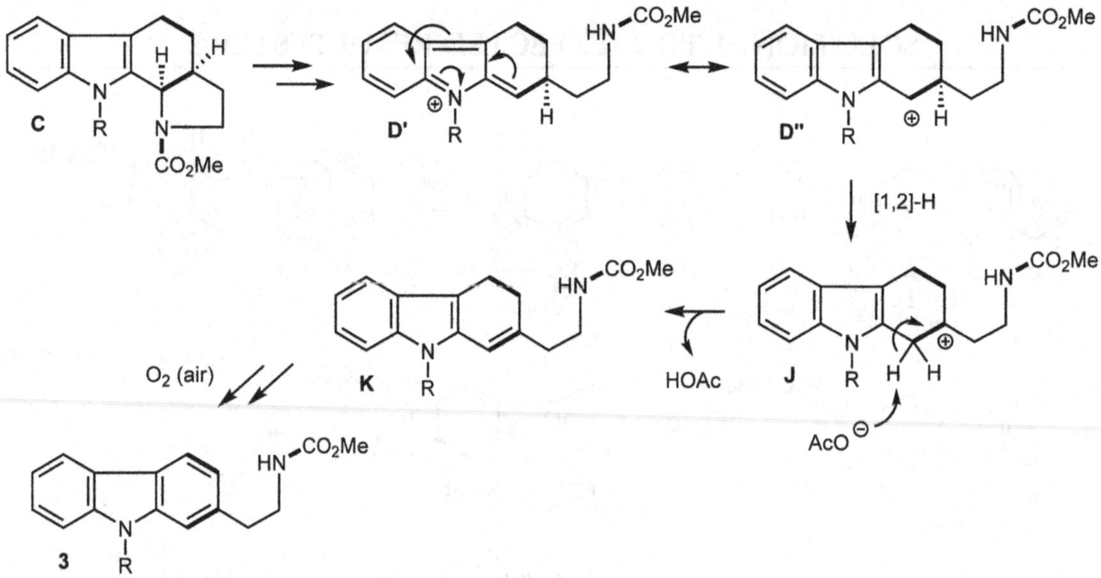

BALANCED CHEMICAL EQUATIONS

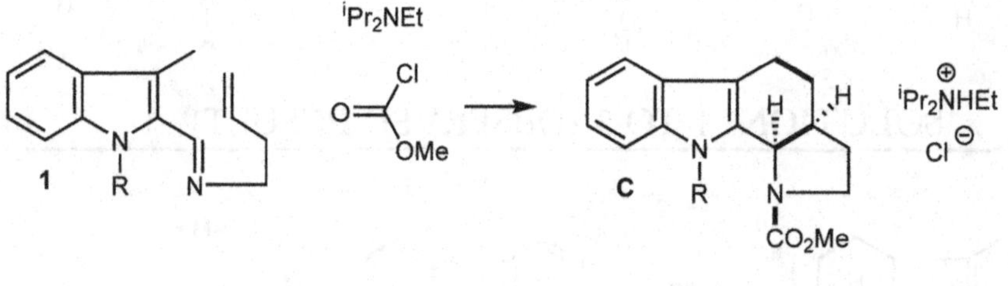

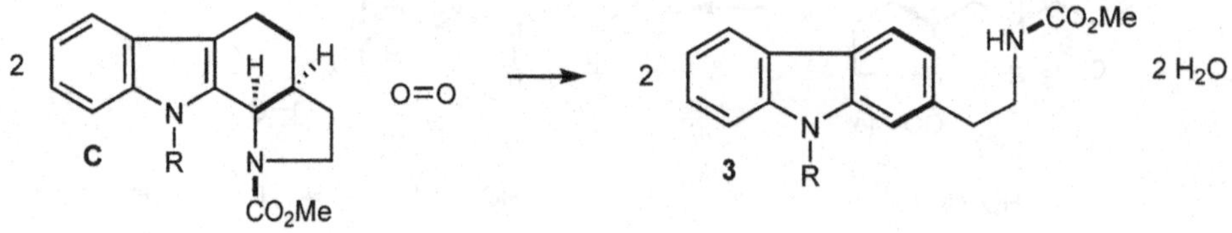

KEY STEPS EXPLAINED

The mechanism of the first step in the synthesis sequence begins with *N*-acylation of indole **1** via nucleophilic attack by the imine nitrogen atom onto the methyl chloroformate reagent to form iminium chloride **A**. Hünig's base then deprotonates **A** forming triene **B** which then undergoes a 6π electrocyclization forming the recovered product **C**. In this cyclization two rings are produced via a $[(4+2)+(5+0)]$ strategy. In the critical second step involving acid-catalyzed rearrangement leading presumably to product **2**, Magnus et al. conjectured that intermediate **C** undergoes fragmentation of ring D at the C—N carbamate bond shown with concomitant protonation of the carbamate nitrogen atom leading to iminium ion **D**. This process was envisaged to be induced by electron flow from the lone pair of electrons at the indole nitrogen atom. The resulting untethered chain with a terminal amino group can then undergo C—C bond rotation and intramolecularly deprotonate a methylene proton from ring C leading to ammonium ion **E** bearing a diene moiety. After deprotonation at the ammonium group, presumably by acetate ion, to diene **F** followed by another indole-induced reprotonation, iminium intermediate **G** is formed. Finally, the bicyclic [3.3.1] bridged structure in product **2** is formed when the carbamate nitrogen atom cyclizes onto the previous methylene carbon atom of ring C. Two key factors working against this mechanism are the weak basicity of the *N*-carbamate group in abstracting a proton on going from iminium ion **D'** to ammonium ion **E**, and the entropic constraint in carrying out the final cyclization from **G** to **H**.

The authors' observation of product **3** instead of product **2** suggested that their conjectured mechanistic proposal did not go completely to plan. The outcome may be rationalized as follows. The same sequence of steps is followed from intermediate **C** to diene intermediate **F**. However, the transformation of diene **F** to carbazole product **2** involves an energetically facile oxidative aromatization of ring C. Alternatively, intermediate **C** leads to iminium ion **D'** as before, however, since it has a carbocation ionic resonance form **D''** it can undergo a [1,2]-prototropic shift leading to carbocation **J** which then is deprotonated by acetate ion to diene **K**. This diene then aromatizes to product **3** via a similar oxidation process. In the absence of any specified reagents we must assume that both of these processes occur via air autoxidation. When the authors carried out the acid-catalyzed reaction in the presence of zinc dust, intermediate **C** was reduced to product **4**, in unspecified yield, as shown in Scheme 2. The authors interpreted this result as confirmation of the intermediacy of iminium ion **D/D'** and diene **K**, and that an oxidative process must have occurred in the absence of zinc. It should be noted that McKillop's original posing of this problem omitted this important oxidative step. Therefore, the second step of the synthesis is an acid-catalyzed rearrangement of **C** to **F** or **C** to **K** followed by autoxidation of **F** or **K** to **3**. For completeness we show possible reductive pathways to product **4** from **C** that go through either intermediates **D/D'/D''** or **K** (see Scheme 3).

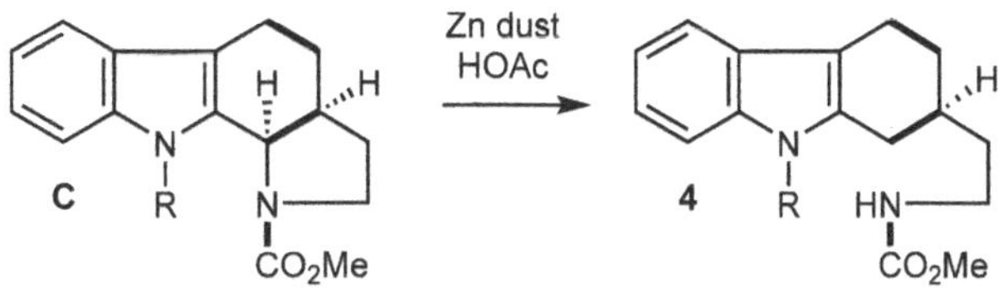

SCHEME 2

Prior work done by the same authors[2,3] demonstrated that they were able to carry out aromatization of ring C in analogs of intermediate **C** without cleaving ring D using DDQ (2,3-dichloro-5,6-dicyano-1,4-quinone) as oxidant. In this case, no rearrangement takes place. It should be noted that though the authors used excess DDQ, stoichiometrically only two equivalents of DDQ are required. Scheme 4 shows an example synthesis sequence. Furthermore, analogs of the intended tetracyclic product **2** have been successfully prepared by Bonjoch et al.[4] by other ring construction strategies.

ADDITIONAL RESOURCES

For a further example of unexpected transformations involving indolo-2,3-quinodimethanes, see Ref. 5. Lastly, for an elegant synthesis of the indole ring functionality, see the recent work of Gordon W. Gribble.[6]

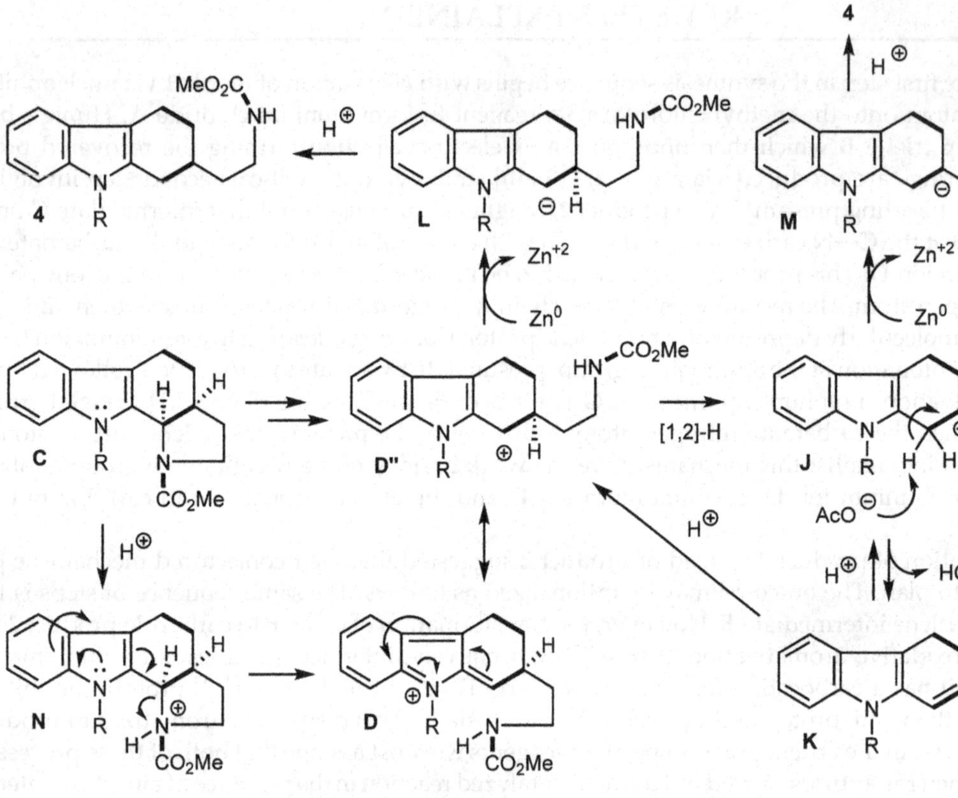

SCHEME 3

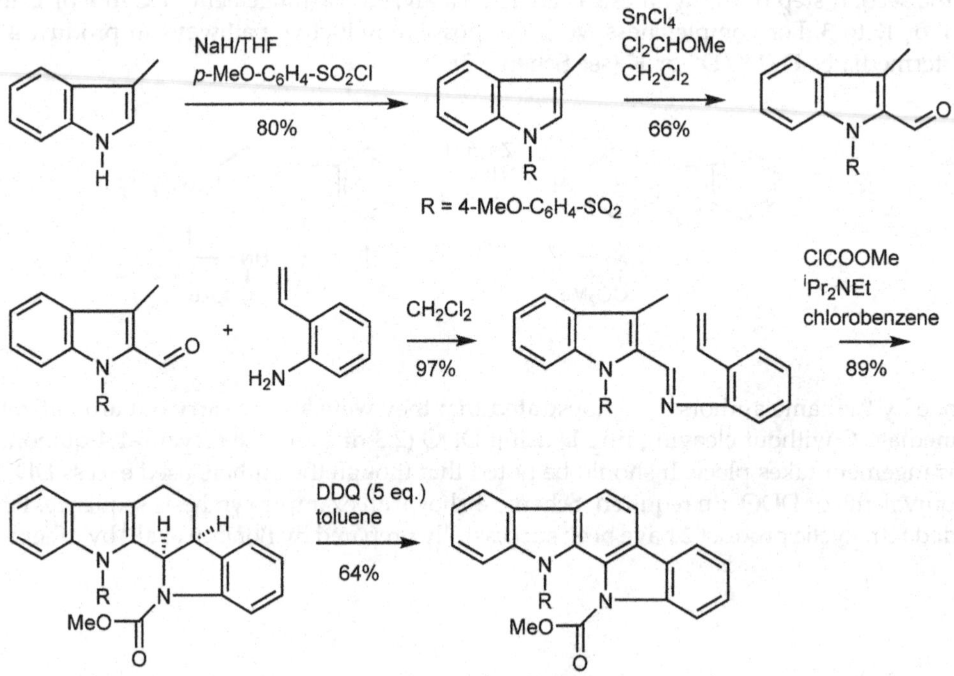

R = 4-MeO-C₆H₄-SO₂

SCHEME 4

References

1. Magnus P, Sear NL, Kim CS, Vicker N. Studies on the synthesis of *Strychnos* alkaloids. A new entry into the azocino[4.3-*b*]indole core structure and related studies. *J Org Chem.* 1992;57(1):70–78. https://doi.org/10.1021/jo00027a016.
2. Magnus PD, Exon C, Sear NL. Indole-2,3-quinodimethanes: synthesis of indolocarbazoles for the synthesis of the fused dimeric indole alkaloid staurosporinone. *Tetrahedron.* 1983;39(22):3725–3729. https://doi.org/10.1016/S0040-4020(01)88612-9.
3. Exon C, Gallagher T, Magnus P. Synthesis of elusive 1,4-dihydrocarbazoles *via* intramolecular trapping of an indole-2,3-quinodimethane. *J Chem Soc Chem Commun.* 1982;1982:613–614. https://doi.org/10.1039/C39820000613.
4. Bonjoch J, Quirante J, Solé D, Castells J, Galceran M, Bosch J. 8-Aryl-2-azabicyclo[3.3.1]nonan-7-ones. Synthesis and retro-Michael ring opening. *Tetrahedron.* 1991;47(25):4417–4428. https://doi.org/10.1016/S0040-4020(01)87110-6.
5. Perumal PT, Nagarajan R. An unusual hydrogen addition of indolo-2,3-quinodimethanes to dimethylindoles in the presence of 1,3-azoles. *J Chem Sci.* 2006;118(2):195–198. https://doi.org/10.1007/BF02708473.
6. Gribble GW. *Indole Ring Synthesis: From Natural Products to Drug Discovery.* New York, NY: John Wiley & Sons, Inc.; 2016. https://doi.org/10.1002/9781118695692

Question 199: A New Synthesis of Substituted Ninhydrins

Over the years there has been much interest in developing synthetic approaches to ninhydrin **3** on account of its ability to reveal latent fingerprints on paper caused by its reactivity with amino acids released by eccrine sweat glands. In 1991 Joullié et al. discussed a synthesis of **3** by the sequence depicted in Scheme 1.[1,2] Suggest a mechanism for the transformation of **2** to **3**.

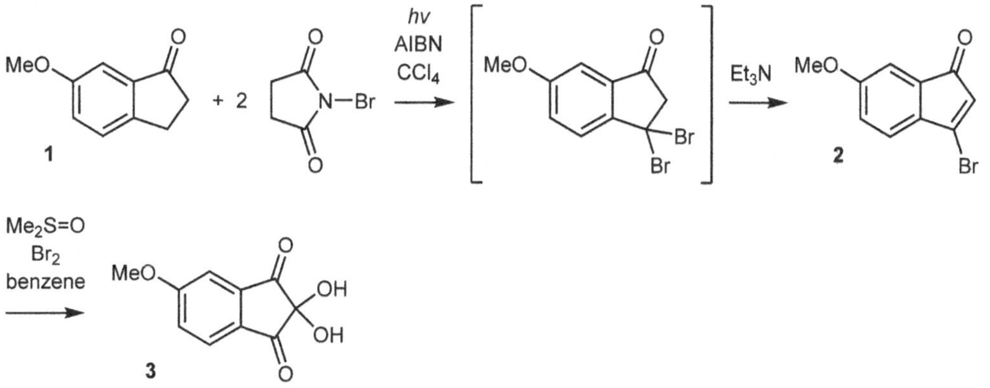

SCHEME 1

SOLUTION

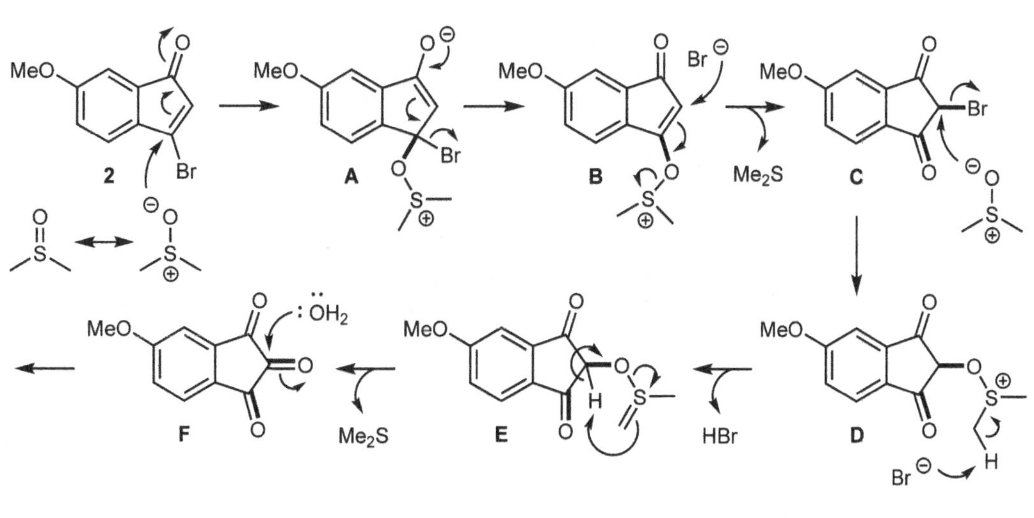

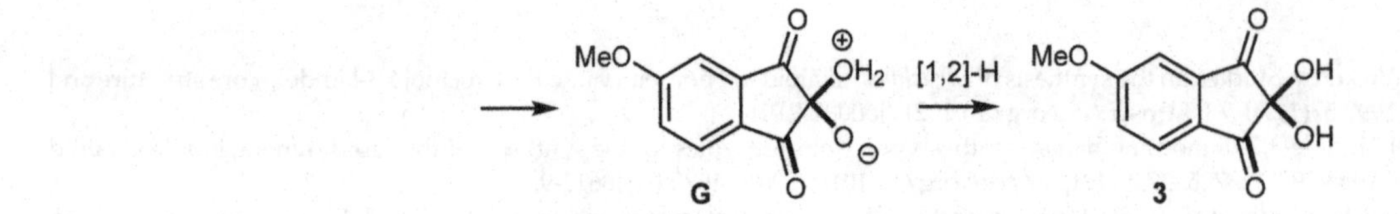

BALANCED CHEMICAL EQUATION

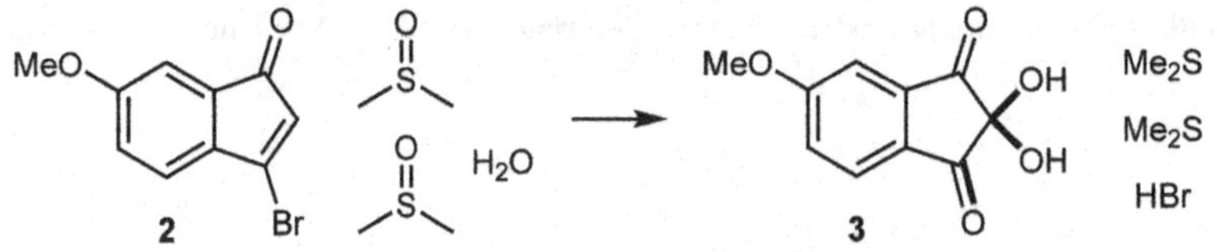

KEY STEPS EXPLAINED

The mechanism postulated by Joullié et al. for the transformation of **2** to the substituted ninhydrin product **3** begins with nucleophilic attack by the ylide of dimethyl sulfoxide onto the carbon bearing the bromide atom in **2** to generate the zwitterion intermediate **A**. Ketonization of **A** with elimination of bromide anion leads to the sulfonium intermediate **B** which may ketonize after the nucleophilic attack of bromide which also helps eliminate dimethyl sulfide by-product to form intermediate **C**. At this point, a nucleophilic substitution via a second equivalent of the ylide of dimethyl sulfoxide with elimination of bromide anion generates **D** which eliminates hydrogen bromide by-product to form **E**. Intermediate **E** undergoes fragmentation and ketonization with production of dimethyl sulfide to form **F** which after addition of water and a hydride shift leads to the final ninhydrin product **3**. Although the authors did not provide any experimental and theoretical evidence for their proposed mechanism, subsequent work[3] had pointed out that the sequence from **C** to **F** represented an instance of the Kornblum oxidation.[4] We would like to point out that the mechanism of this oxidation was also postulated in the absence of supporting experimental and theoretical evidence.

Nevertheless, since the earlier mechanism is entirely conjectured, we would like to offer an alternative mechanism which makes use of the bromine reagent (Scheme 2). The balanced chemical equation for this mechanism is provided in Scheme 3.

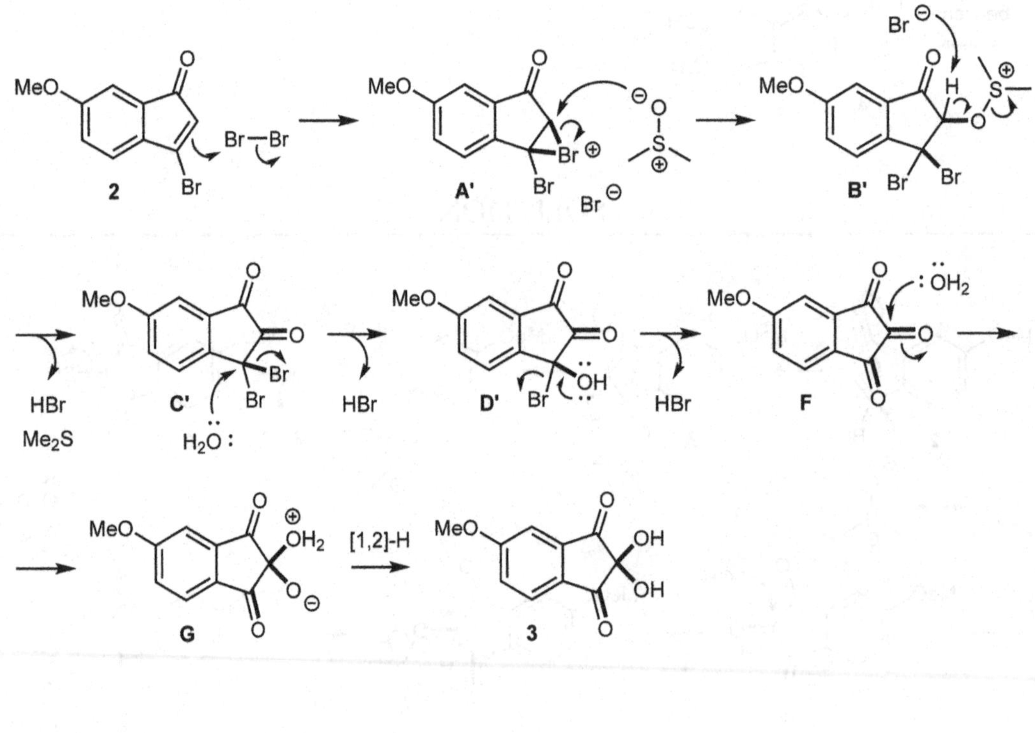

SCHEME 2

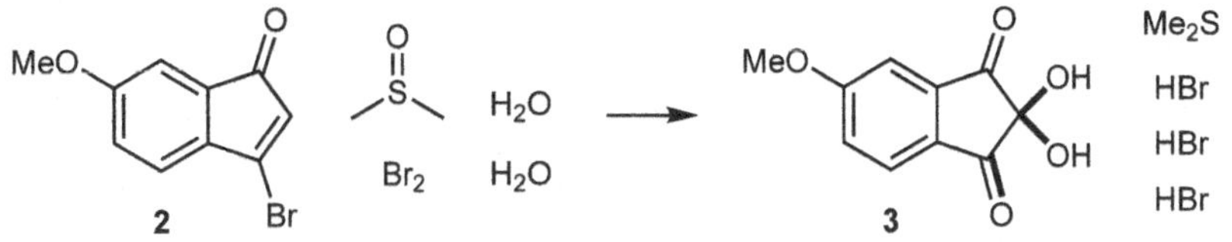

SCHEME 3

The first step of the alternative mechanism constitutes addition of bromine across the olefinic bond to give bromonium ion **A′** which is then attacked by the oxygen atom of dimethyl sulfoxide ylide to form **B′**. Bromide anion then abstracts the hydrogen atom to create a carbonyl group and eliminate dimethyl sulfide to form **C′**. Note that the sulfur atom is reduced from an oxidation state of 0 to -2 (net loss of -2). The remaining bromine atoms in intermediate **C′** are then successively substituted with a hydroxyl group from water which then ketonizes to form **F** during the aqueous workup. Addition of a water molecule then gives the ninhydrin product **3**. To compensate for the reduction taking place on the sulfur atom and on each of the bromine atoms in Br_2 (0 to -1, net loss of $2 \times (-1) = -2$), the carbon atom bearing a bromine substituent in **2** is oxidized from +1 to +2 (net gain of +1) and the other olefinic carbon atom in **2** is also oxidized from -1 to +2 (net gain of +3). The overall net gain of +4 for the oxidation component is balanced by the overall net loss of -4 for the reduction component.

We propose this alternative mechanism which makes use of the bromine reagent because the authors mentioned in their experimental procedure that: "Caution should be used when adding an oxidizing agent to dimethyl sulfoxide. The addition of bromine to the solution of 3-bromo-6-methoxy-inden-1-one in benzene and dimethyl sulfoxide must be carried out dropwise and with cooling."[2] Without further experimental and theoretical evidence, it is not possible to determine whether bromine plays a mechanistically significant role in this reaction since the comment about the dropwise addition of bromine to a solution containing dimethyl sulfoxide can also be explained by the side reaction that occurs between dimethyl sulfoxide and bromine (Scheme 4).

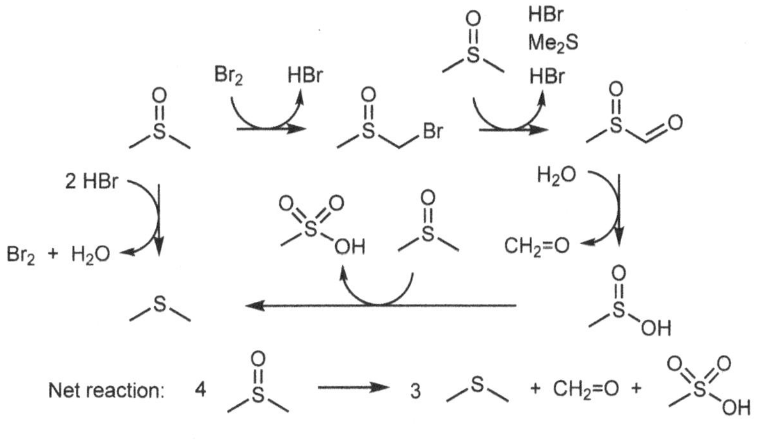

SCHEME 4

In this side reaction, a possible α-bromination initially forms bromomethyl methyl sulfoxide and hydrobromic acid which reacts further in a catalytic redox cycle to eventually reduce the initial dimethyl sulfoxide to dimethyl sulfide making it less able to act as the intended oxidant for this transformation (see Refs. 5, 6). This is what explains the need for the drop-wise addition of bromine.

Lastly, although the two mechanisms proposed cannot be confirmed in the absence of further evidence, it is possible that they may both participate in the formation of **3** by having the drop-wise addition of bromine open up an alternative mechanistic pathway. It is also interesting to note the difference in stoichiometry with regard to dimethyl sulfoxide, which acts as both a solvent and a reagent, in the context of the two postulated mechanisms.

ADDITIONAL RESOURCES

It is interesting to note that two Japanese authors have also synthesized compound **F** by two successive treatments of **1** with N-bromosuccinimide (NBS) in dimethyl sulfoxide (DMSO) as shown in Scheme 5.[7] Here, there were two oxidation steps involved with the use of two equivalents NBS and DMSO each, respectively. For additional information on the synthesis and application of ninhydrin derivatives, see the two review articles in Refs. 8, 9.

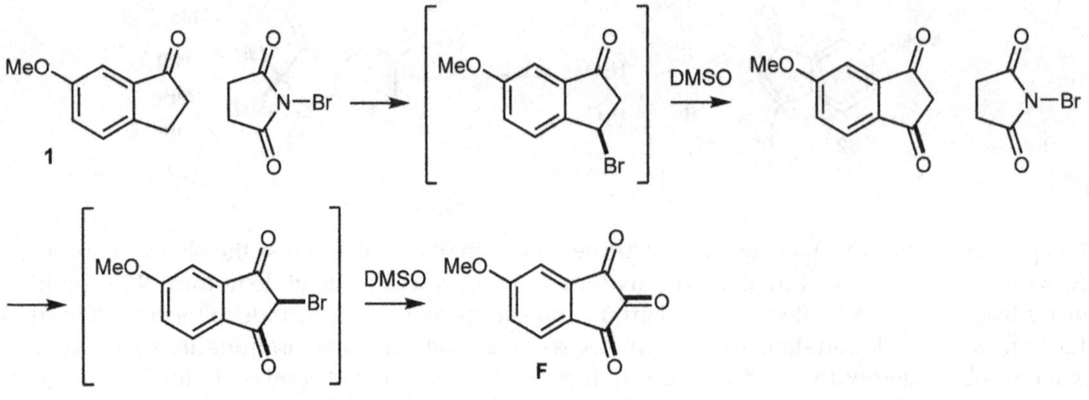

SCHEME 5

References

1. Joullié MM, Thompson TR, Nemeroff NH. Ninhydrin and ninhydrin analogs. Syntheses and applications. *Tetrahedron*. 1991;47(42):8791–8830. https://doi.org/10.1016/S0040-4020(01)80997-2.

2. Heffner RJ, Joullié MM. Synthetic routes to ninhydrins. Preparation of ninhydrin, 5-methoxyninhydrin, and 5-(methylthio)ninhydrin. *Synth Commun*. 1991;21:2231–2256. https://doi.org/10.1080/00397919108055457.

3. Marminon C, Nacereddine A, Bouaziz Z, Nebois P, Jose J, le Borgne M. Microwave-assisted oxidation of indan-1-ones into ninhydrins. *Tetrahedron Lett*. 2015;56(14):1840–1842. https://doi.org/10.1016/j.tetlet.2015.02.086.

4. Wang Z, ed. *Comprehensive Organic Name Reactions and Reagents*. New York, NY: John Wiley & Sons, Inc.; 2010:1672–1674. https://doi.org/10.1002/9780470638859.conrr373

5. Iriuchijima S, Tsuchihashi G. Synthesis of α-bromosulfoxides. *Synthesis*. 1970;1970(11):588–589. https://doi.org/10.1055/s-1970-21646.

6. Aida T, Akasaka T, Furukawa N, Oae S. Catalytic reduction of sulfoxide by bromine-hydrogen bromide system. *Bull Chem Soc Jpn*. 1976;49(4):1117–1121. https://doi.org/10.1246/bcsj.49.1117.

7. Tatsugi J, Izawa Y. A convenient one-pot synthesis of indane-1,2,3-triones by oxidation of indan-1-ones with *N*-bromosuccinimide-dimethyl sulfoxide reagent. *Synth Commun*. 1998;28(5):859–864. https://doi.org/10.1080/00032719808006484.

8. Hark RR, Hauze DB, Petrovskaia O, Jouillé MM. Synthetic studies of novel ninhydrin analogs. *Can J Chem*. 2001;79(11):1632–1654. https://doi.org/10.1139/cjc-79-11-1632.

9. Hansen DB, Jouillé MM. The development of novel ninhydrin analogues. *Chem Soc Rev*. 2005;34:408–417. https://doi.org/10.1039/b315496n.

Question 200: Thermal Elimination Reactions Often Fail

When Katritzky et al. subjected dihydropyridine **1** to pyrolysis at 180°C in 1982, the products obtained were, unexpectedly, 2,4,6-triphenylpyridine **2** and *m*-chlorostyrene **3** (Scheme 1).[1] Provide a mechanism for this transformation.

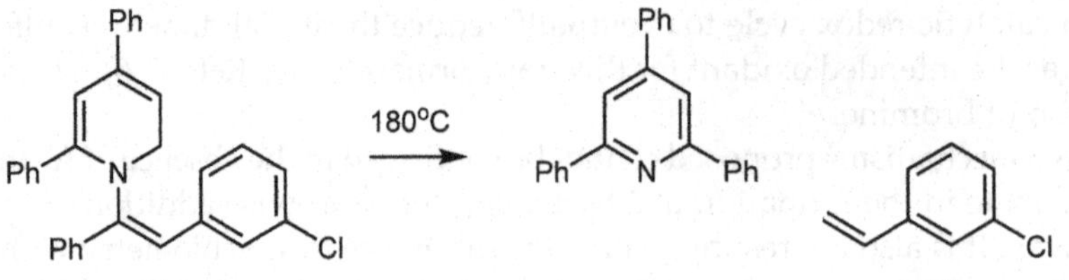

SCHEME 1

SOLUTION

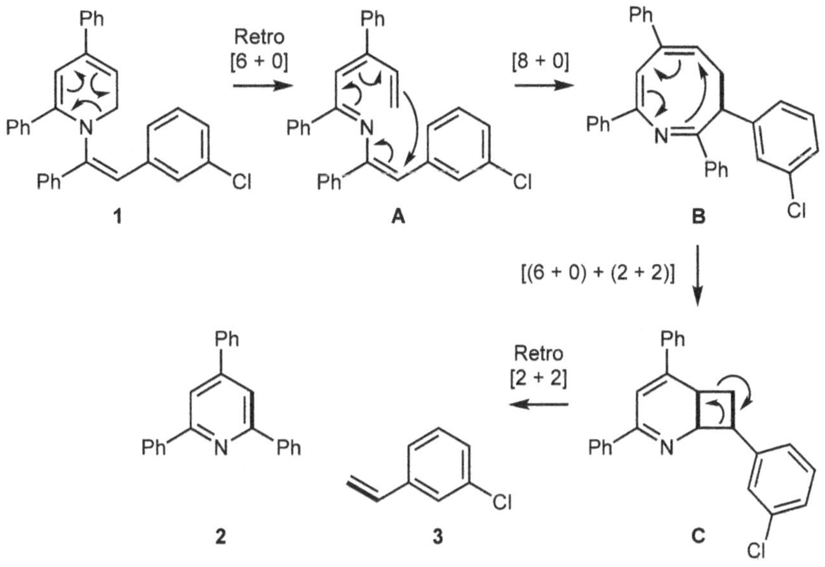

KEY STEPS EXPLAINED

The mechanism proposed by the authors involves the following steps: (1) a retro $[6+0]$ electrocyclic $4\pi + 2\sigma$ reaction generating intermediate **A**, (2) cyclization of **A** via an electrocyclic 8π reaction giving intermediate **B**, (3) **B** undergoes an electrocyclic 6π reaction to give the [4.2.0] bicyclic intermediate **C**, and lastly (4) **C** undergoes a retro $[2+2]$ degradation to generate products **2** and **3**. This mechanism is entirely conjectured and further experimental and especially theoretical evidence is needed to support it. We also highlight that the intended products of this reaction were **4** and **5** (Scheme 2).

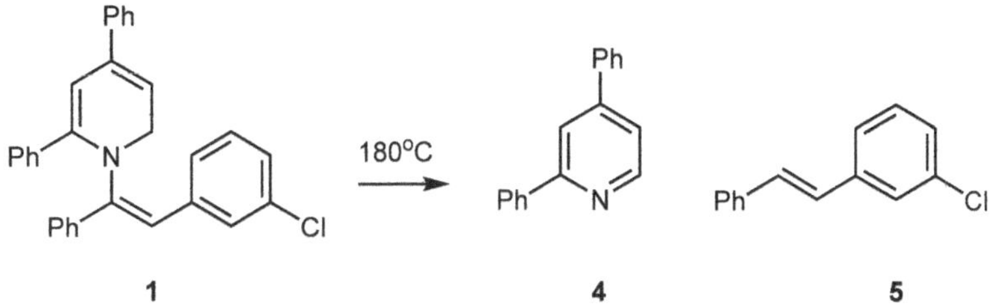

SCHEME 2

ADDITIONAL RESOURCES

We recommend an article which highlights a more complex mechanism for the synthesis of triarylpyridines.[2] We also include an article by Japanese chemists which tackles a nickel-catalyzed $[2+2+2]$ cycloaddition involving two alkynes and an imine.[3]

References

1. Katritzky AR, Chermprapai A, Patel RC, Tarraga-Tomas A. Pyridinium ylides derived from pyryliums and amines and a novel rearrangement of 1-vinyl-1,2-dihydropyridines. *J Org Chem*. 1982;47:492–497. https://doi.org/10.1021/jo00342a024.
2. Kumar A, Koul S, Razdan TK, Kapoor KK. A new and convenient one-pot solid supported synthesis of 2,4,6-triarylpyridines. *Tetrahedron*. 2006;47:837–842. https://doi.org/10.1016/j.tetlet.2005.11.043.
3. Ogoshi S, Ikeda H, Kurosawa H. Nickel-catalyzed $[2+2+2]$ cycloaddition of two alkynes and an imine. *Pure Appl Chem*. 2008;80(5):1115–1125. https://doi.org/10.1351/pac200880051115.

Chapter 8: Solutions 201 – 250

Question 201: The Wrong Choice of Reaction Conditions? How Not to Prepare an Acid Chloride

In 1979 American chemists heated the indole derivative **1** under reflux in thionyl chloride for 5h which was followed by distillation of the excess thionyl chloride and treatment of the remaining residue with methanol to form the unexpected tricyclic product **2** in 63% yield (Scheme 1).[1] Suggest a mechaonism for this transformation.

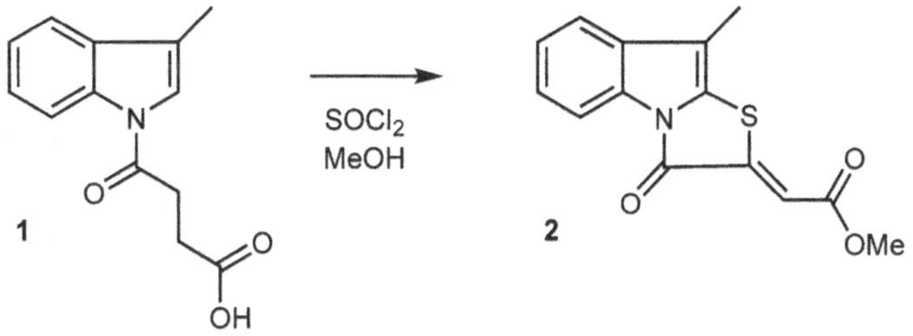

SCHEME 1

SOLUTION

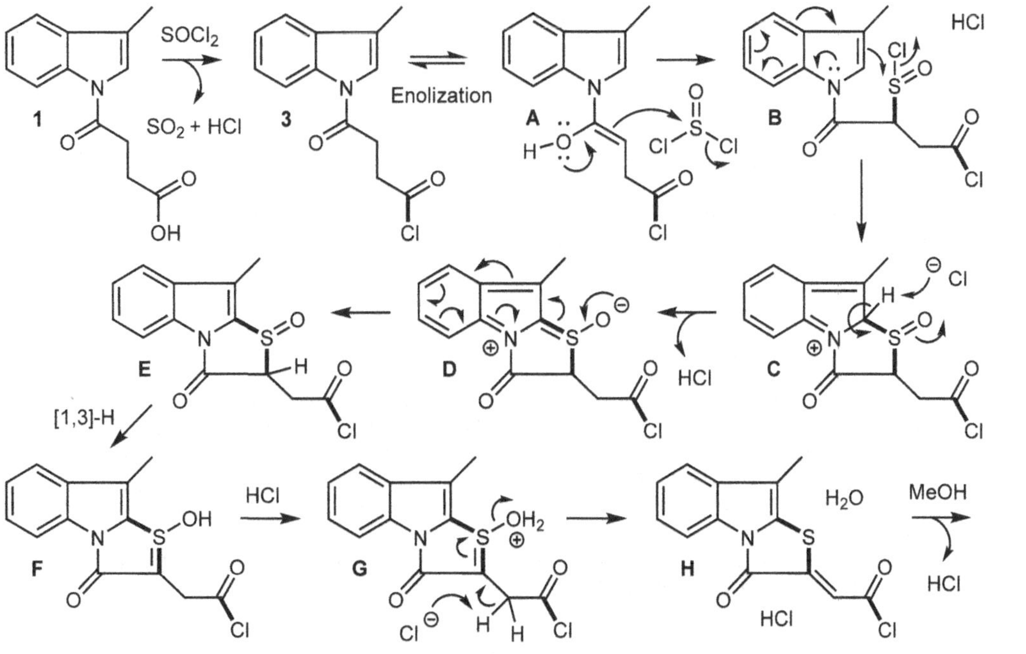

Strategies and Solutions to Advanced Organic Reaction Mechanisms
https://doi.org/10.1016/B978-0-12-812823-7.00317-7

BALANCED CHEMICAL EQUATION

KEY STEPS EXPLAINED

It is worth mentioning that the authors did not expect that their reaction would produce **2**. Instead, their anticipated synthetic sequence was the reaction of **1** with thionyl chloride to form the expected acid chloride **3** followed by reaction with methanol to form the corresponding ester as we see in the last step of the mechanism presented before. Nevertheless, the transformation did not stop with the acid chloride **3**. Instead, **3** turned out to be an intermediate which eventually led to **2**. To explain this progression, the authors proposed initial formation of the desired acid chloride **3** from the reaction of **1** with thionyl chloride. Next, enolization of **3** to form **A** sets up an electrophilic addition of thionyl chloride to form the sulfinyl chloride intermediate **B** which cyclizes to the sulfone **E** after chloride elimination, Pummerer rearrangement and electronic rearrangement. Sulfone **E** undergoes hydrolysis in an acid-catalyzed process to form the cyclic sulfide intermediate **H** which finally reacts with methanol to form the corresponding ester product **2**. This mechanism is purely conjectured and further experimental and theoretical analysis are needed to help support it.

ADDITIONAL RESOURCES

We recommend three articles highlighting some unusual applications of thionyl chloride in synthesis in the context of similar chemical systems to the one encountered in this solution.[2-4]

References

1. Showalter HDH, Shipchandler MT, Mitscher LA, Hagaman EW. Facile entry into the thiazolo[3,2-*a*]indol-3(2*H*)-one system via an unusual reaction with thionyl chloride. *J Org Chem.* 1979;44(22):3994–3996. https://doi.org/10.1021/jo01336a062.
2. Brown D, Griffiths D. Reaction of an α-methylpyrrole with thionyl chloride. An unusual synthesis of pyrrole α-thiocarboxamides. *Synth Commun.* 1983;13(11):913–918. https://doi.org/10.1080/00397918308059545.
3. Beattie JF, Hales NJ. Unusual oxidation in thionyl chloride: novel synthesis of methyl 3-alkoxy-1,4-dioxo-1,2,3,4-tetrahydroisoquinoline-3-carboxylates. *J Chem Soc Perkin Trans 1.* 1992;1992:751–752. https://doi.org/10.1039/P19920000751.
4. Katritzky AR, Fedoseyenko D, Kim MS, Steel PJ. Chiral 1,2,4-triazoles: stereoselective acylation and chlorination. *Tetrahedron Asymmetry.* 2010;21:51–57. https://doi.org/10.1016/j.tetasy.2009.12.007.

Question 202: Quantitative Yield Isomerization of a Xylenol Derivative...

In 1979 a Canadian chemist investigating a catechol to phenol transformation observed that the reaction of 2-acet-oxy-4-methylthio-3,4-xylenol **1** in benzene containing excess triethylamine and methyl isothiocyanate under heat and reflux conditions for 18 h produced xylenol **2** in quantitative yield (Scheme 1).[1] It was later shown that methyl isothiocyanate did not participate in the transformation since the results were the same as in its absence. Propose a mechanism to explain these results.

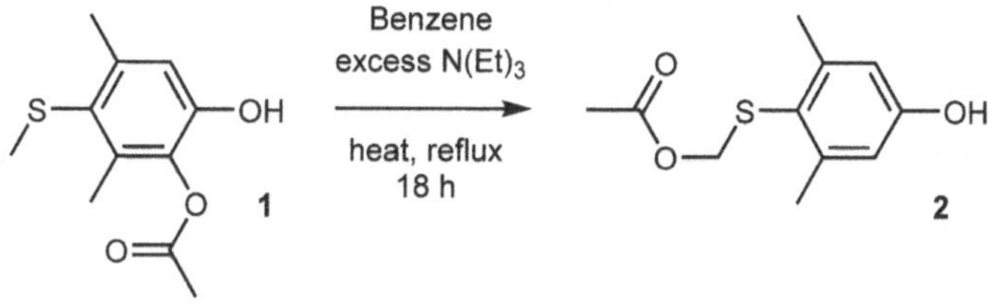

SCHEME 1

SOLUTION

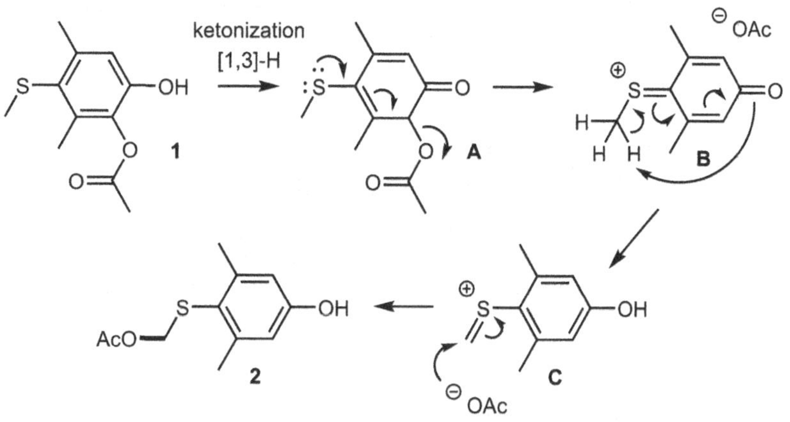

KEY STEPS EXPLAINED

The original intention was to synthesize the *N*-methylthiocarbamate of **1** with methylisothiocyanate. Nevertheless, the mechanism proposed by King for the transformation of **1** to **2** begins with ketonization of **1** (presumably via base catalysis) to the 2,4-dienone intermediate **A** which eliminates acetate anion forming sulfonium intermediate **B** via electron pair donation from the methylthio group. Intermediate **B** enolizes to **C** which is driven by rearomatization of the benzene ring. Nucleophilic attack by acetate anion generates product **2**. The author bases his proposed mechanism on the experimental observation that the 2,5-dienone **3** also rearranges to product **2** (Scheme 2). Nevertheless, King does not provide any additional experimental or theoretical evidence to support this mechanism. As a consequence, we

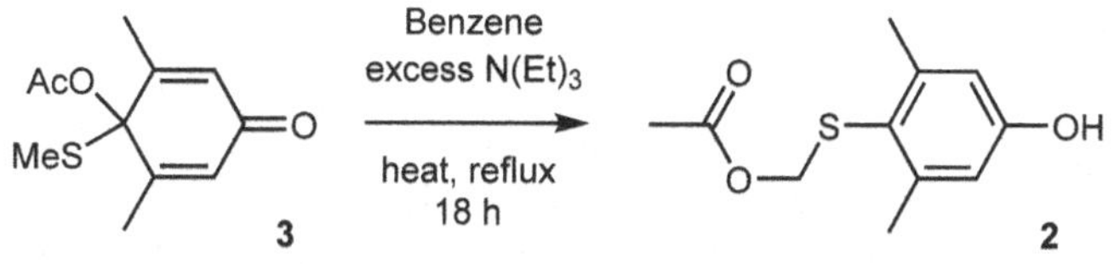

SCHEME 2

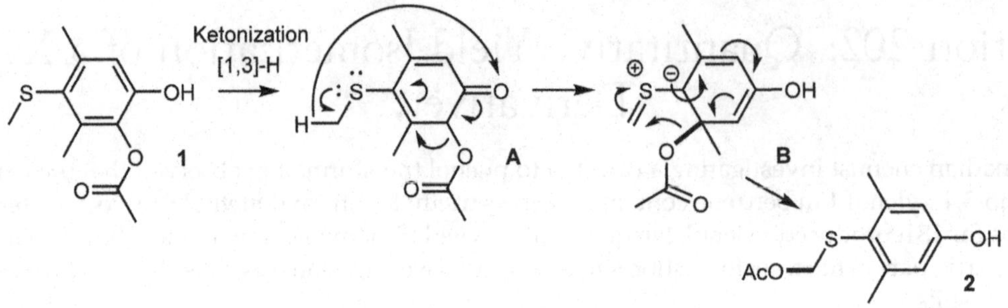

SCHEME 3

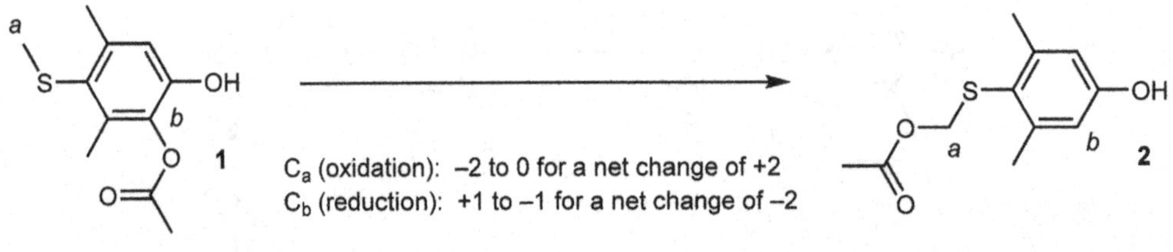

C_a (oxidation): -2 to 0 for a net change of $+2$
C_b (reduction): $+1$ to -1 for a net change of -2

SCHEME 4

present an alternative mechanism which involves intramolecular migration of the acetoxyl group two times in order to achieve product **2** (Scheme 3). For a future study, a theoretical investigation comparing the energetics of these two mechanisms would be very informative. Furthermore, we note that this thermal rearrangement is also a redox reaction (Scheme 4).

ADDITIONAL RESOURCES

We highly recommend an article discussing the mechanism of an interesting catechol transformation,[2] an article describing the utility of catechol in the synthesis of macrocyclic rods,[3] and lastly a research paper examining the mechanism of dioxygenase-catalyzed ring expansion of substituted 2,4-cyclohexadienones.[4]

References

1. King RR. Acetoxyl group migration in 2-acetoxy-4-(methylthio)-3,5-xylenol. A novel catechol to phenol transformation. *J Org Chem.* 1979;44 (23):4194–4195. https://doi.org/10.1021/jo01337a041.
2. Xin M, Bugg TDH. Evidence from mechanistic probes for distinct hydroperoxide rearrangement mechanisms in the intradiol and extradiol catechol dioxygenases. *J Am Chem Soc.* 2008;130:10422–10430. https://doi.org/10.1021/ja8029569.
3. Weibel N, Mishchenko A, Wandlowski T, Neuburger M, Leroux Y, Mayor M. Catechol-based macrocyclic rods: en route to redox-active molecular switches. *Eur J Org Chem.* 2009;2009:6140–6150. https://doi.org/10.1002/ejoc.200900751.
4. Xin M, Bugg TDH. Biomimetic formation of 2-tropolones by dioxygenase-catalysed ring expansion of substituted 2,4-cyclohexadienones. *Chembiochem.* 2010;11:272–276. https://doi.org/10.1002/cbic.200900631.

Question 203: ...and an Alternative Route to the Starting Material for Problem 202

A Canadian chemist working in 1978 on dienone intermediates relevant to the Pummerer rearrangement reacted the 4-methylsulphinyl-3,5-xylenol **1** in a 2:1 mixture of acetic anhydride-acetic acid solution at 100°C to achieve products **2** (major product) and **A** (minor product) after an hour (Scheme 1).[1] Reaction of **A** in heated benzene under reflux produced **2** in 50% yield. Propose mechanisms to explain the **1** to **A + 2** and the **A** to **2** transformations.

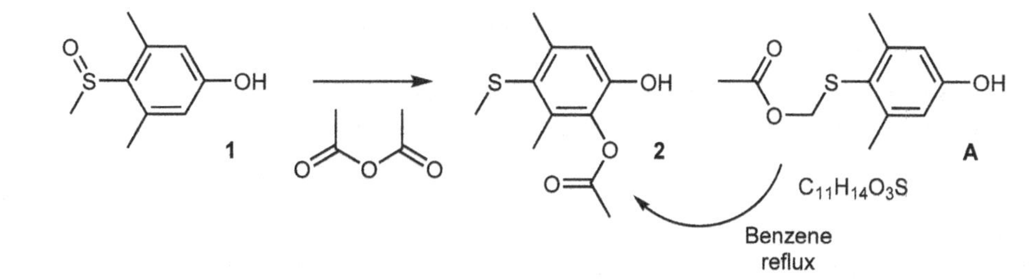

SCHEME 1

SOLUTION: 1 TO 2

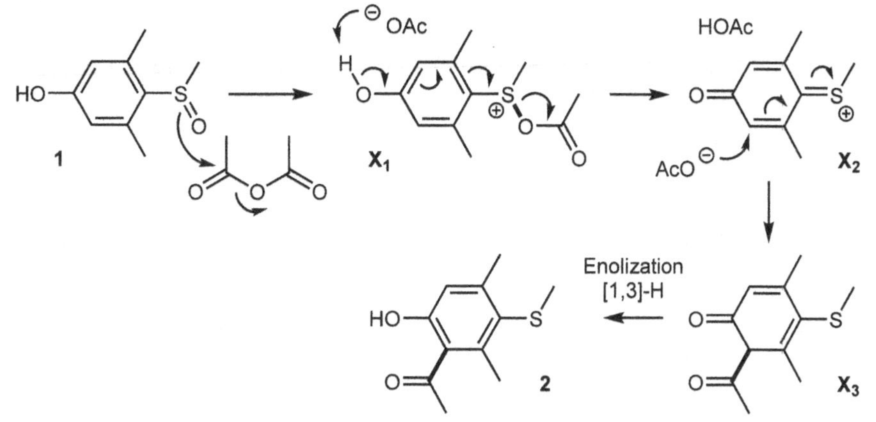

SOLUTION: 1 TO A

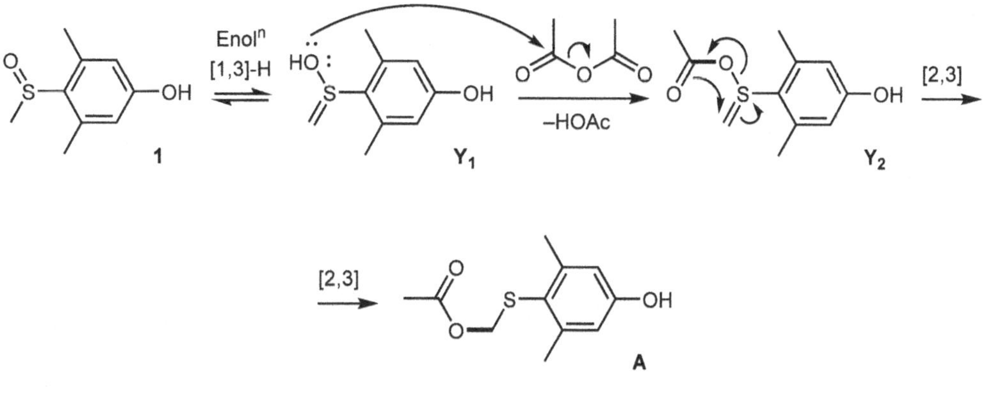

SOLUTION: A TO 2

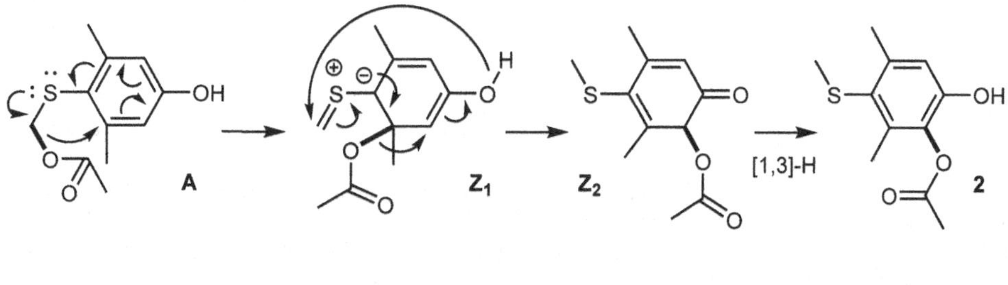

KEY STEPS EXPLAINED

The mechanism proposed by King for the **1** to **2** transformation (Pummerer rearrangement) begins with acetylation of the sulfoxide oxygen atom of **1** via the acetic anhydride to give intermediate X_1. Subsequent deprotonation of X_1 via acetate anion forms the quinoid intermediate X_2 after elimination of acetate. The newly formed acetate anion then adds *ortho* to the ketone group of X_2 in a nucleophilic manner to form X_3 which through enolization driven by the rearomatization of the benzene ring forms product **2**.

Concerning the transformation of **1** to **A**, the authors suggested an initial enolization of the sulfoxide group of **1** to form intermediate Y_1 which reacts with acetic anhydride to cause acetylation to intermediate Y_2. A [2,3]-sigmatropic rearrangement of Y_2 then furnishes product **A**.

Lastly, with respect to the transformation of **A** to **2**, the suggested mechanism starts with migration of the acetate group of **A** to the *ortho* position of the ring relative to the sulfur group following electrocyclic rearrangement to intermediate Z_1. A subsequent migration of the acetate group to the *meta* position following electrocyclic rearrangement provides intermediate Z_2 which undergoes enolization driven by rearomatization as seen previously to form product **2**.

A closer examination of all three reactions shows that they are also redox reactions in addition to being thermal rearrangement reactions. See Scheme 2 for the important changes in oxidation number with respect to the key atoms involved. In conclusion we note that the author did not provide any further experimental or theoretical evidence to support the mechanisms outlined earlier.

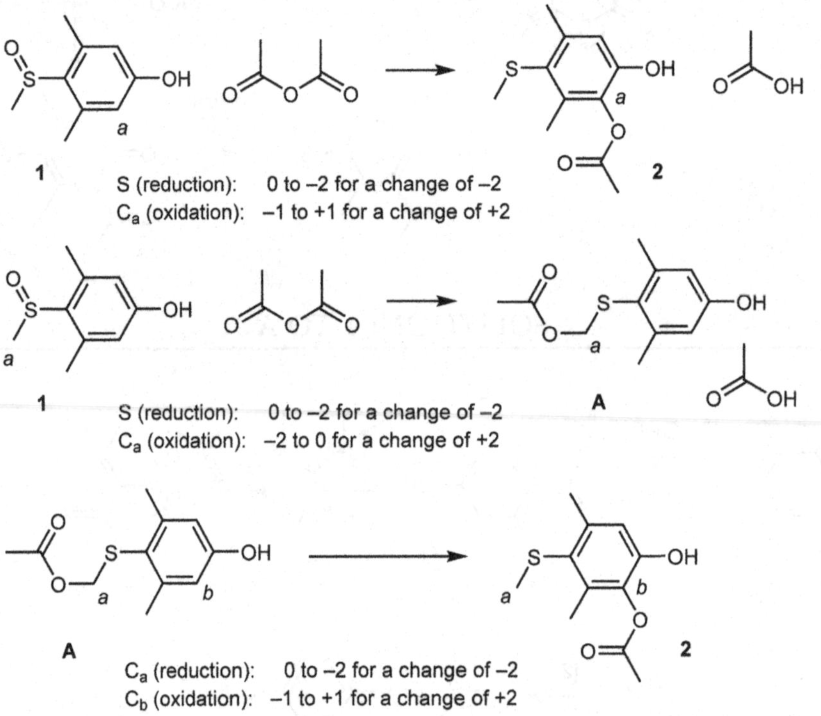

SCHEME 2

ADDITIONAL RESOURCES

For two interesting articles focusing on Pummerer-type rearrangements in similar chemical systems as the one discussed here, see Refs. 2, 3. For a review article discussing the role of Pummerer-type reactions in a modern context, see the work of Feldman.[4]

References

1. King RR. Dienone intermediates in the Pummerer rearrangement of 4-methylsulphinyl-3,5-xylenol. *J Org Chem*. 1978;43(19):3784–3785. https://doi.org/10.1021/jo00413a037.
2. Kita Y, Tekeda Y, Matsugi M, et al. Isolation of the quinone mono *O,S*-acetal intermediates of the aromatic Pummerer-type rearrangement of *p*-sulfinylphenols with 1-ethoxyvinyl esters. *Angew Chem Int Ed Engl*. 1997;36(13/14):1529–1531. https://doi.org/10.1002/anie.199715291.
3. Matsugi M, Murata K, Anilkumar G, Nambu H, Kita Y. Regioselective nucleophilic addition of methoxybenzene derivatives to the β-carbon of *p*-benzoquinone mono *O,S*-acetal. *Chem Pharm Bull*. 2001;49(12):1658–1659. https://doi.org/10.1248/cpb.49.1658.
4. Feldman KS. Modern Pummerer-type reactions. *Tetrahedron*. 2006;62:5003–5034. https://doi.org/10.1016/j.tet.2006.03.004.

Question 204: Hydride-Induced Rearrangements With Indole Alkaloid Intermediates

Japanese chemists working on the synthesis of *Aspidosperma*-type alkaloids in 1979 reacted compound **1** with phosphorus oxychloride followed by sodium borohydride to form the pentacyclic product **2** in 50% yield (Scheme 1).[1] Suggest a mechanism for this transformation.

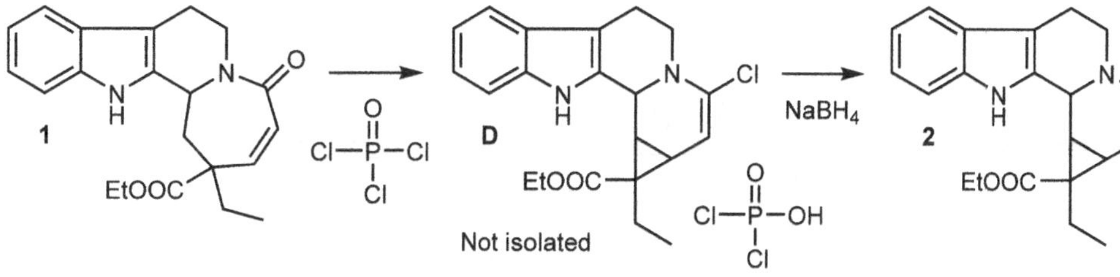

SCHEME 1

SOLUTION

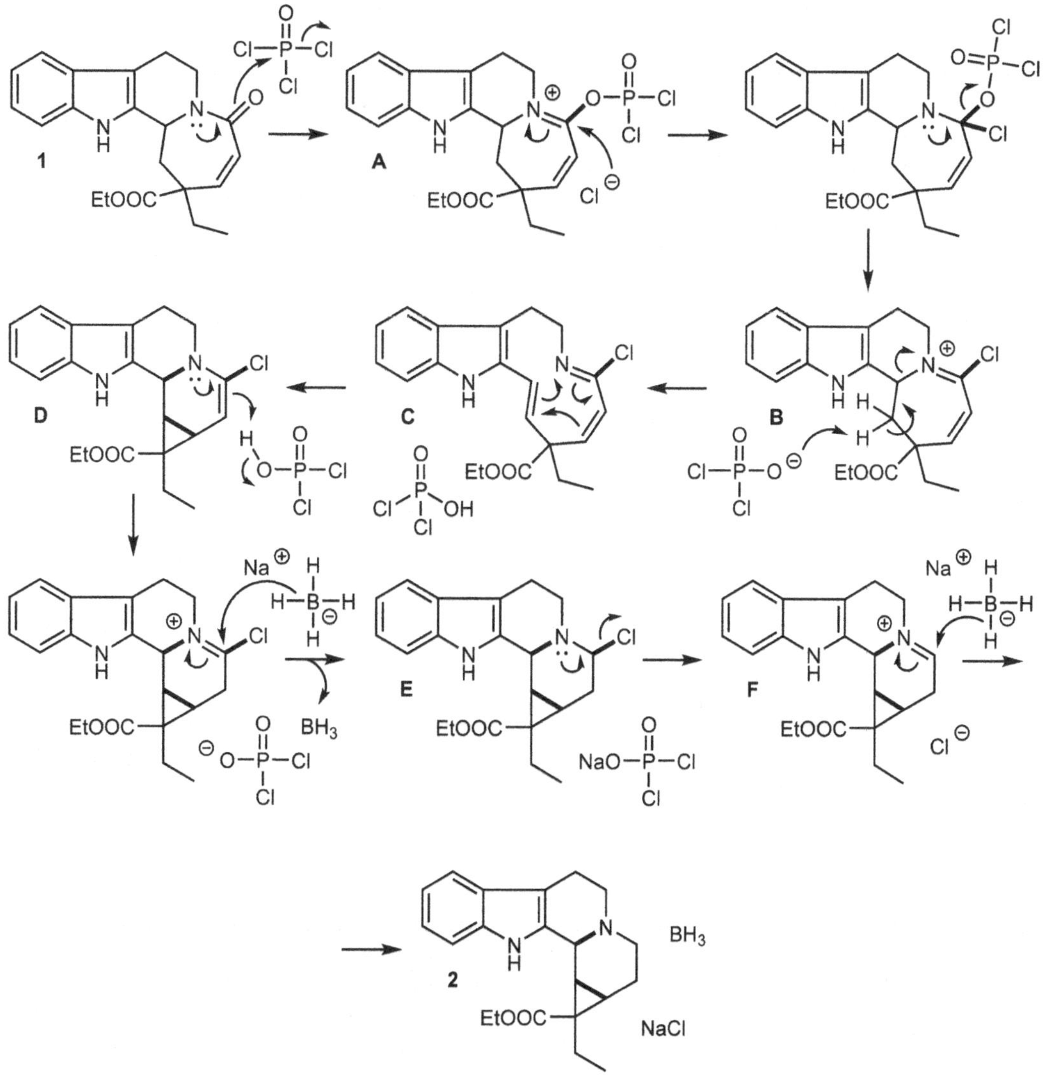

BALANCED CHEMICAL EQUATION

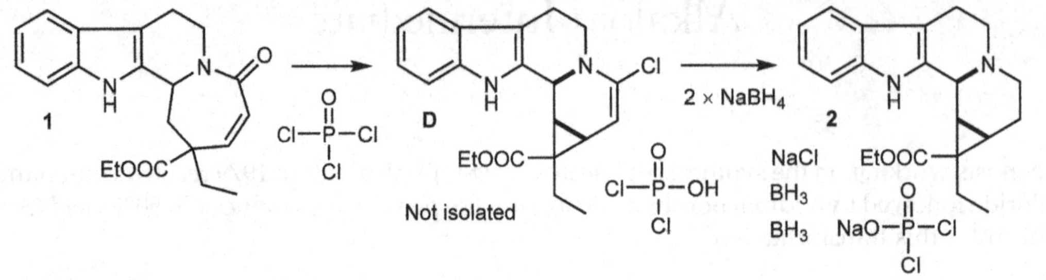

WORK-UP REACTION

$$2BH_3 + 6H_2O \rightarrow 2B(OH)_3 + 6H_2$$

KEY STEPS EXPLAINED

The mechanism proposed by the authors begins with nucleophilic attack by the oxygen atom of the amide group of **1** onto phosphorus oxychloride with substitution of chloride to form intermediate **A**. Nucleophilic attack by chloride onto **A** followed by iminium formation aids in the elimination of dichlorophosphate anion to form **B**. This anion deprotonates **B** to cause ring expansion to **C** which undergoes an electrocyclic 6π cyclization to the fused [4.1.0] bicyclic intermediate **D**. This intermediate then forms an iminium with protonation and hydride attack via the sodium borohydride that gives intermediate **E** after expulsion of borane and sodium dichlorophosphate by-products. A second iminium formation and elimination of chloride anion generates **F** which is attacked by a second equivalent of hydride via sodium borohydride reducing agent to form the final product **2** following expulsion of a second equivalent of borane and sodium chloride by-products. It is important to note that the authors did not provide any further experimental or theoretical evidence to support their postulated mechanism. Furthermore, the authors originally intended to simply reduce the amide group of **1** without affecting the carbon skeleton or the alkene C=C bond. Treatment with lithium aluminum hydride did not achieve this goal, as both the amide and the C=C double bond were reduced (see Scheme 2), and treatment with phosphorus oxychloride and sodium borohydride led to the rearranged product **2** (Scheme 1).

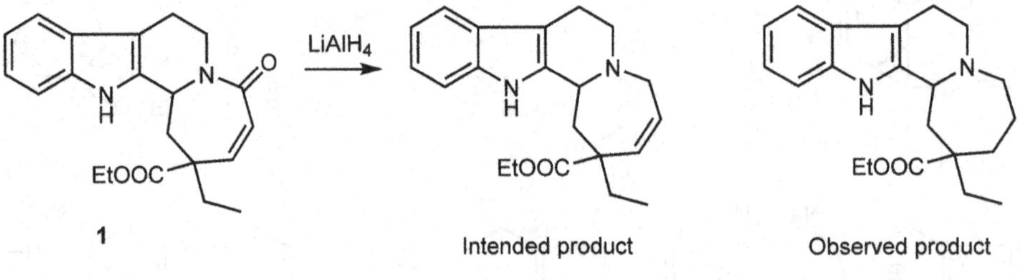

SCHEME 2

ADDITIONAL RESOURCES

We highly recommend a review article discussing the mechanisms of several very similar alkaloid transformations as the one covered in this solution.[2]

References

1. Takano S, Hatakeyama S, Ogasawara K. Synthesis of the nontryptamine moiety of the *Aspidosperma*-type indole alkaloids via cleavage of a cyclic α-diketone monothioketal. An efficient synthesis of (±)-quebrachamine and a formal synthesis of (±)-tabersonine. *J Am Chem Soc.* 1979;101 (21):6414–6420. https://doi.org/10.1021/ja00515a042.
2. Hajicek J. A review on the recent developments in syntheses of the post-secodine indole alkaloids. Part II: Modified alkaloid types. *Collect Czechoslov Chem Commun.* 2007;72(7):821–898. https://doi.org/10.1135/cccc20070821.

Question 205: An Unusual "Hydrolysis" Product of 2-Nitrosopyridine

In 1986 American chemists studied the unusual hydrolysis of 2-nitrosopyridine **1** to product **2** (Scheme 1).[1] Propose a mechanism for this transformation.

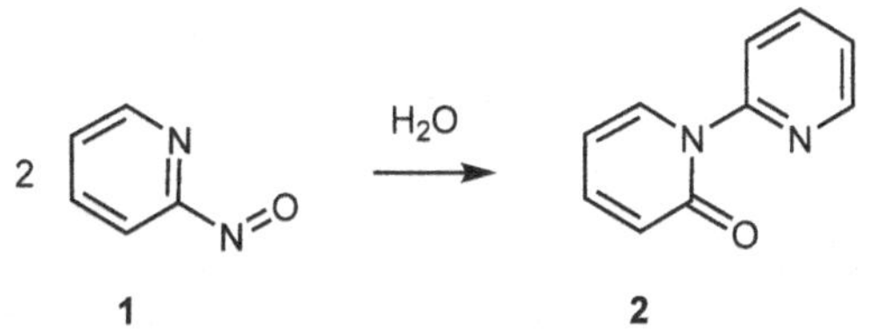

SCHEME 1

SOLUTION

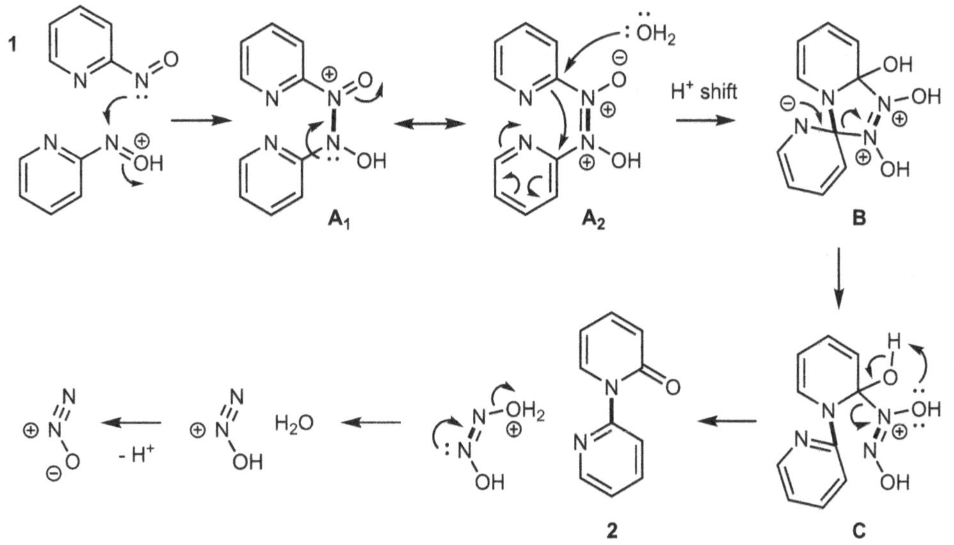

BALANCED CHEMICAL EQUATION

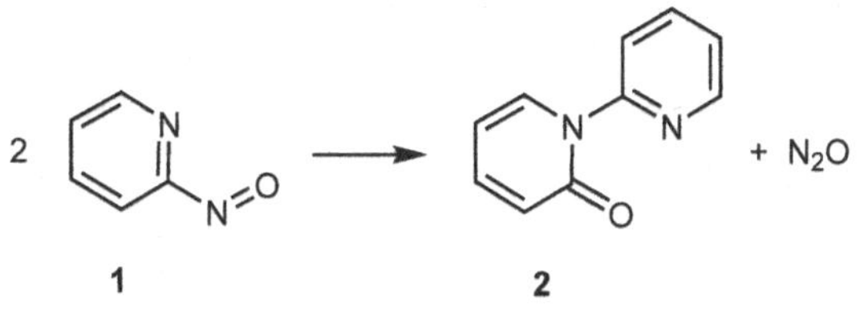

KEY STEPS EXPLAINED

The authors' proposed mechanism of **1** to **2** under acidic conditions begins with azodioxy dimerization of **1** and its protonated form yielding intermediate A_1/A_2 existing in two resonance forms which is subsequently attacked by water at the *ipso* C_2 position forming tricyclic *spiro* intermediate **B** after a proton shift. Rearomatization of the lower pyridine ring causes cleavage of the *spiro* C—N bond to form intermediate **C** which eliminates protonated 1,2-dihydroxydiazene by-product to form the final pyridone product **2**. It is worth mentioning that according to the literature, the 1,2-dihydroxydiazene is expected to fragment further to nitrous oxide and water.[2,3] It is interesting that experimentally the authors observed that: (a) no conversion of **1** to **2** took place when 25% dioxane/aqueous phosphate buffer was used, (b) the solution changed color slowly from pale green to light yellow as the reaction progressed, (c) the rate of formation of **2** in water increased with time, and (d) addition of one drop of acid to a solution of 25% dioxane/water drastically reduced the reaction time leading to complete conversion of **1** to **2**. The authors accounted for these observations by pointing out that intermediates such as **A** and **B** are likely responsible for the observed pale green color of the solution during the beginning stages of the reaction. The addition of one drop of acid and the increased rate of reaction with time suggested that the transformation was acid catalyzed. In particular, when examining the addition of water at the *ipso* C_2 position of A_2, the authors noted that this position is likely more electrophilic via protonation of the *N*-oxide group thereby creating an attached group bearing two positively charged nitrogen atoms which is expected to impart a net electron-withdrawing effect on the *ipso* carbon atom. Moreover, the authors ran a crossover experiment (Scheme 2) to prove that an alternative mechanism which involves initial hydrolysis of 2-nitrosopyridine **1** to 2-pyridone **3** followed by coupling of **3** and **1** to form **2** (Scheme 3) did not occur. Since the expected mixed pyridylpyridone products **4'** and **5'** were not observed, the alternative mechanism was ruled out. Hence, product **2** could only be formed by the reaction of two equivalents of **1**. In the crossover experiment 1*H*-pyridin-2-one **3** does

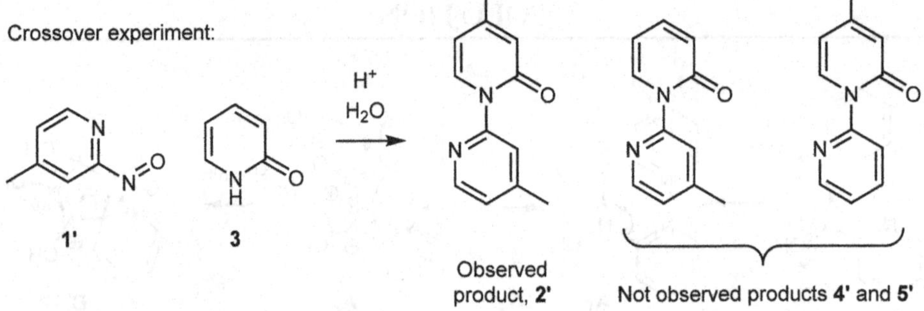

SCHEME 2

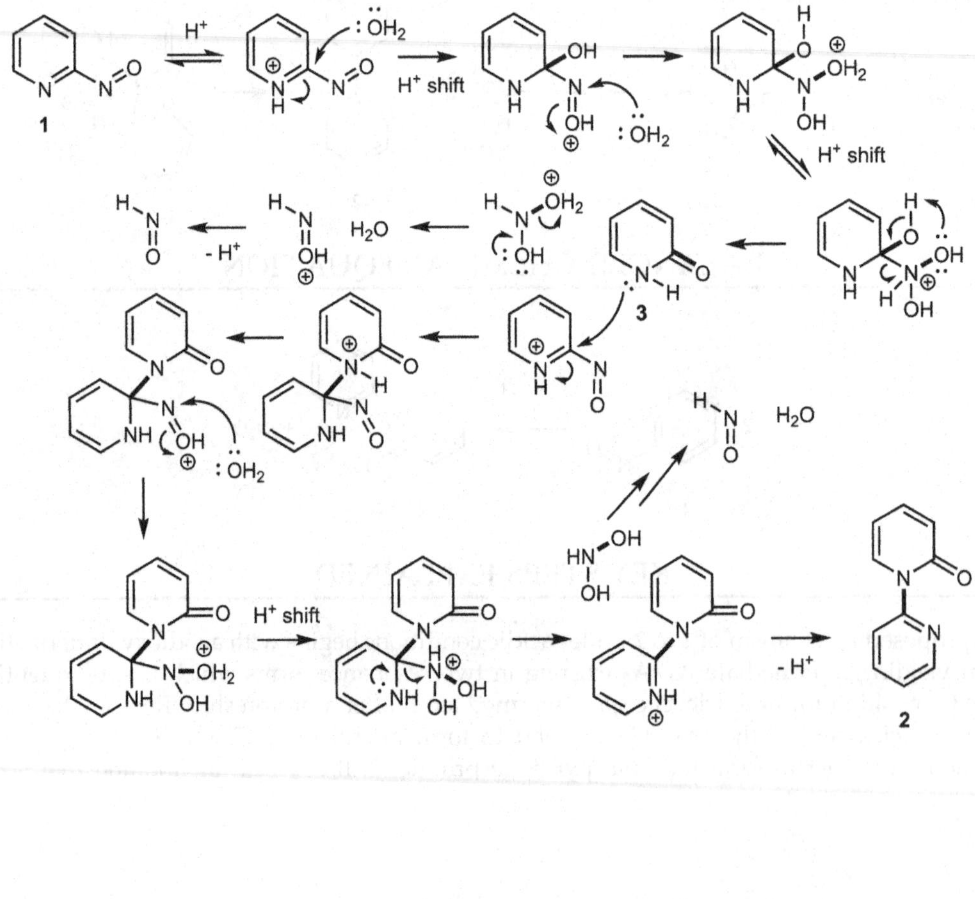

SCHEME 3

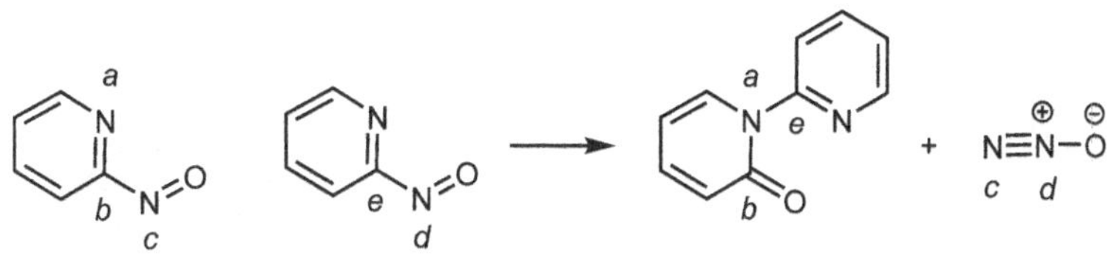

SCHEME 4

not participate but acts as a spectator. Furthermore, the authors' speculation at the outset that some of the starting 2-nitrosopyridine could be hydrolyzed while the rest remained intact is a low probability scenario given the described experimental protocol which involved simply dissolving the 2-nitrosopyridine in water (neutral conditions) and stirring for 8 h at room temperature. A partitioning of the substrate via two mechanistic pathways is therefore not expected. The alternative mechanism also leads to different expected by-products as shown in Scheme 3. An oxidation number analysis for the balanced chemical equation of the solution mechanism shows that the transformation is an internal redox reaction (see Scheme 4).

Recent X-ray crystallographic and variable temperature ^{1}H NMR work[4] on the solid- and solution-state structures of 2-nitrosopyridine confirmed that it dimerizes to the Z-form in solution where it exists as torsional conformers as shown in Scheme 5 and that solutions of the dimer in organic solvents are pale green in color. From dynamic NMR line shape analysis it was determined that at equilibrium the dimer is 4.6 kcal/mol lower in energy than the two monomers and the energy barrier for dimer dissociation is 70.4 kcal/mol. This implies that the energy barrier for the reverse dimerization reaction is $70.4 - 4.6 = 65.8$ kcal/mol. These findings therefore confirm the first step in the mechanism proposed by the authors. However, we offer an alternative mechanism that circumvents the formation of the expectedly high-energy *spiro* tricyclic intermediate **B** but is consistent with both the intramolecular constraint indicated by the result of the crossover experiment and the observation that the reaction is acid catalyzed. We suggest that the geometry of the Z-dimer having significantly twisted pyridine rings is not favorable for the cyclization that is required to form intermediate **B**. Scheme 6 shows our alternative mechanism under neutral conditions where the first step is formation of Z-dimer $\mathbf{D_1(Z)}$ which then undergoes a 1,2-pyridine shift onto the adjacent oxygen atom of the N-oxide group via the resonance form $\mathbf{D_2(Z)}$ (reminiscent of a 1,2-Meisenheimer or Martynoff rearrangement of N-oxides) yielding azooxy intermediate **E** which then rearranges again to azooxonium ion **F**. The final step is fragmentation of **F** to product **2** with liberation of nitrous oxide. Note that this final step generates nitrous oxide directly without having to go through 1,2-dihydroxydiazene as in the authors' proposed mechanism. The observed rate acceleration under acidic conditions is likely because the dimerization of **1** in the first step can be acid catalyzed.

In order to resolve which mechanism is operative, we recommend for future work a theoretical investigation of the energy reaction coordinate profiles and an experimental investigation of the reaction using ^{18}O labeled water. The authors' proposed mechanism predicts that the ^{18}O label should end up in product **2** since it originates from water attack at the *ipso* C_2 carbon atom of $\mathbf{A_2}$; whereas, the alternative mechanism predicts no ^{18}O label incorporation in the product since the oxygen atom in **2** originates from one of the nitroso oxygen atoms in **1**. Of particular interest from a computational point of view are estimations of the energy barriers for the formation of the tricyclic *spiro* intermediate **B** proposed by the authors and for the isomerization and rearrangement steps proposed in the alternative mechanism.

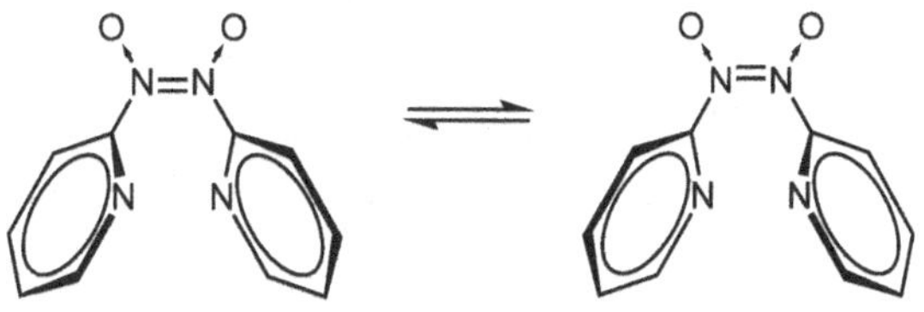

SCHEME 5

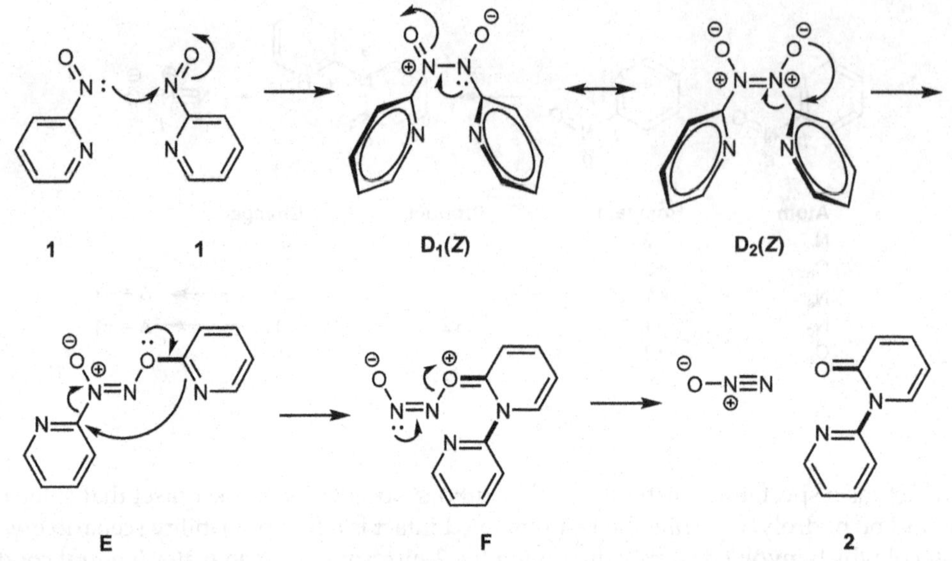

SCHEME 6

ADDITIONAL RESOURCES

For further interesting examples of unexpected hydrolysis reactions, see Refs. 5, 6.

References

1. Taylor EC, Harrison KA, Rampal JB. Unusual "hydrolysis" of 2-nitrosopyridines: formation of 1-(2-pyridyl)-2(1H)-pyridones. *J Org Chem.* 1986;51:101–102. https://doi.org/10.1021/jo00351a023.
2. Olah GA, Salem G, Staral JS, Ho T-L. Preparative carbocation chemistry. 13. Preparation of carbocations from hydrocarbons via hydrogen abstraction with nitrosonium hexafluorophosphate and sodium nitrite-trifluoromethanesulfonic acid. *J Org Chem.* 1978;43(1):173–175. https://doi.org/10.1021/jo00395a045.
3. Harteck P. Die darstellung von HNO bzw. [HNO]n. *Ber Dtsch Chem Ges A/B.* 1933;66(3):423–426. https://doi.org/10.1002/cber.19330660325.
4. Gowenlock BG, Maidment MJ, Orrell KG, et al. The solid- and solution-state structures of 2-nitrosopyridine and its 3- and 4-methyl derivatives. *J Chem Soc Perkin Trans 2.* 2000;2280–2286. https://doi.org/10.1039/b004270f.
5. Vasiliev AN, Kayukov YS, Nasakin OE, et al. Synthesis of alkyl 5,6-dialkyl-2-amino-3-cyanopyridine-4-carboxylates. *Chem Heterocycl Compd.* 2001;37(3):309–314. https://doi.org/10.1023/A:1017503015471.
6. Makarov AY, Shakirov MM, Shuvaev KV, Bagryanskaya IY, Gatilov YV, Zibarev AV. 1,2,4,3,5-Benzotrithiadiazepine and its unexpected hydrolysis to unusual 7H,14H-dibenzo[d,i][1,2,6,7,3,8]tetrathiadiazecine. *Chem Commun.* 2001;2001:1774–1775. https://doi.org/10.1039/B105001J.

Question 206: A Quinolizine to Indolizine Transformation

In 1971 German chemists synthesized the indolizine **2** from quinolizine **1** according to the synthetic sequence depicted in Scheme 1.[1] Explain these transformations in mechanistic terms.

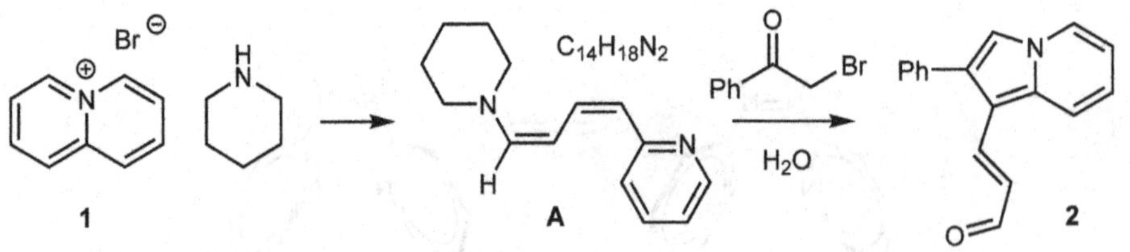

SCHEME 1

SOLUTION: 1 TO A

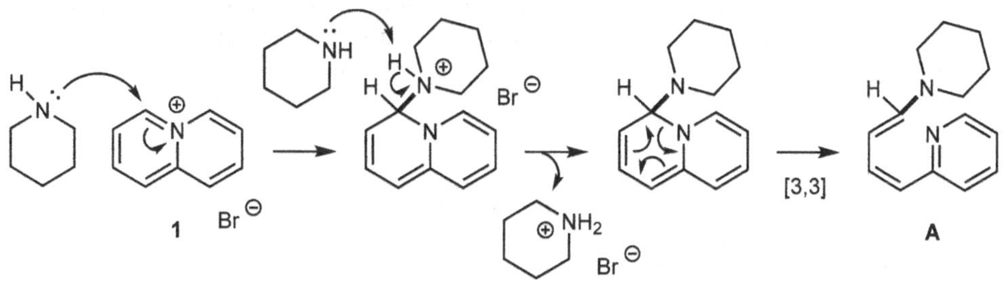

SOLUTION: A TO 2

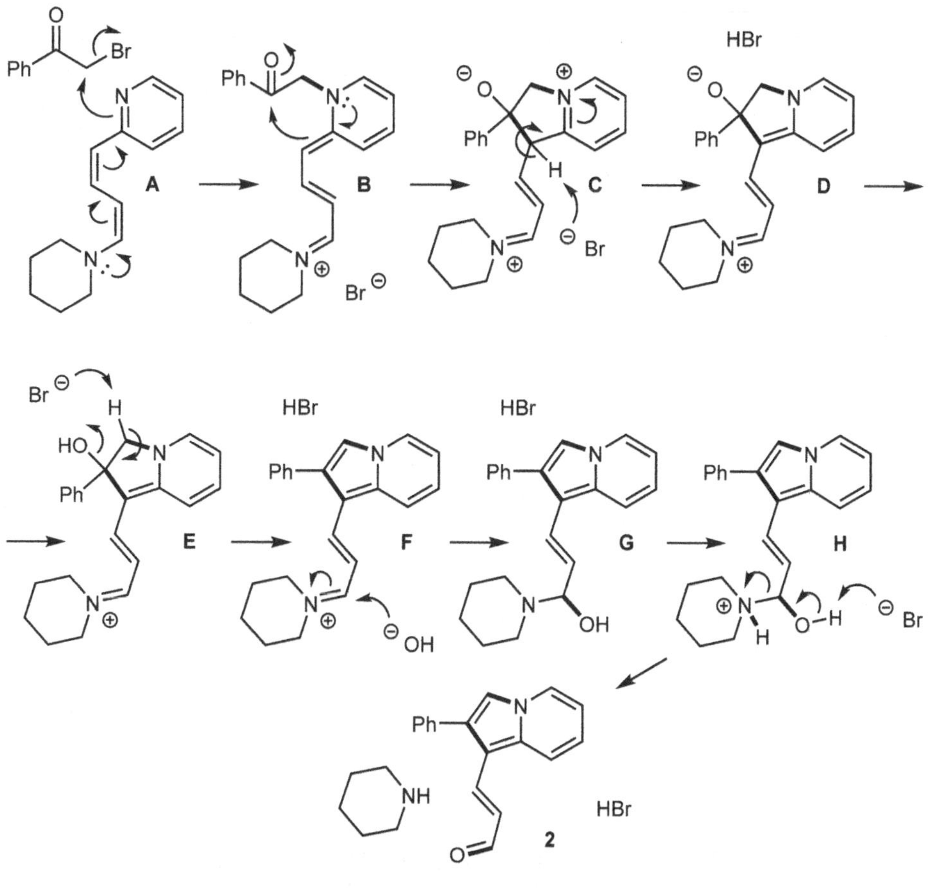

KEY STEPS EXPLAINED

To explain the formation of compound **A**, the authors suggested a mechanism where one equivalent of piperidine acting as a nucleophile attacks the quinolizine **1** starting material. Deprotonation of the resulting product by a second equivalent of piperidine acting as a base causes elimination of piperidinium bromide by-product and formation of 4H-quinolizine intermediate. This intermediate undergoes an electrocyclic [3,3] sigmatropic rearrangement to generate product **A**. From then on, the reaction of **A** with phenacyl bromide and water begins with an electron "push" from the nitrogen atom of the terminal piperidinyl group of **A** to facilitate a nucleophilic attack onto phenacyl bromide to install a new C—N linkage and generate intermediate **B**. Rearomatization of the pyridine ring and cyclization at the carbonyl group completes the [3+2] framework for the formation of **2** after several prototropic rearrangements and a final hydrolysis step of the piperidinyl group to an aldehyde. The authors do not present any additional experimental and theoretical evidence to support their postulated mechanism.

ADDITIONAL RESOURCES

We highly recommend two articles examining the synthesis of indolizines from the standpoint of new transformations involving interesting mechanistic pathways.[2,3] Next, we found two chapters on aromatic quinolizines and on heterocycles containing a ring-junction nitrogen atom to be highly informative.[4,5] Lastly, for more advanced transformations employing very similar chemical systems, see Refs. 6, 7.

References

1. Morler D, Krohnke F. Eine neue ringoffnung an chinoliziniumsalzen und deren folgereaktionen. *Justus Liebigs Ann Chem.* 1971;744(1):65–80. https://doi.org/10.1002/jlac.19717440110.
2. Shang Y, Zhang M, Yu S, Ju K, Wang C, He X. New route synthesis of indolizines via 1,3-dipolar cycloaddition of pyridiniums and alkynes. *Tetrahedron Lett.* 2009;50:6981–6984. https://doi.org/10.1016/j.tetlet.2009.09.143.
3. Sashida H, Kato M, Tsuchiya T. Thermal rearrangements of cyclic amine ylides. VIII. Intramolecular cyclization of 2-ethylpyridine *N*-ylides into indolizines and cycl[3.2.2]azines. *Chem Pharm Bull.* 1988;36(10):3826–3832. https://doi.org/10.1248/cpb.36.3826.
4. Jones G. Aromatic quinolizines. In: Katritzky AR, ed. *Advances in Heterocyclic Chemistry.* New York, NY: Academic Press, Inc.; 1982:1–62. vol. 31. https://doi.org/10.1016/S0065-2725(08)60395-5
5. Vaquero JJ, Alvarez-Builla J. Heterocycles containing a ring-junction nitrogen. In: Alvarez-Builla J, Vaquero JJ, Barluenga J, eds. *Modern Heterocyclic Chemistry.* 1st ed. Weinheim: Wiley-VCH Verlag GmbH & Co. KGaA; 2011:1989–2070. https://doi.org/10.1002/9783527637737.ch22.
6. Gupta CM, Rizvi RK, Kumar S, Anand N, Murthy MRN, Venkatesan K. Cycloaddition reactions: reaction of 3-methyl-2-phenyl-pyrrocoline with dimethyl acteylenedicarboxylate. *Indian J Chem Sect B.* 1981;20B:735–741.
7. Timari G, Hajos G, Messmer A. Synthesis, alkylation, and ring opening of two differently fused pyridoquinazolones. *J Heterocyclic Chem.* 1990;27(7):2005–2009. https://doi.org/10.1002/jhet.5570270730.

Question 207: S$_N$HetAr Reactions Often Proceed With Complications

In 1969 Dutch chemists reacted 2-bromo-6-phenoxypyridine **1** with potassium amide in liquid ammonia to form the three isomeric products **2**, **3**, and **4** (Scheme 1).[1] Show mechanisms for the formation of each of these products.

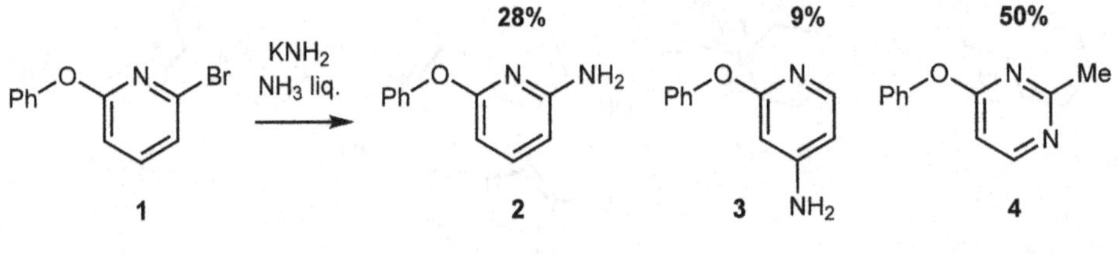

SCHEME 1

SOLUTION: 1 TO 2 (*ipso* SUBSTITUTION)

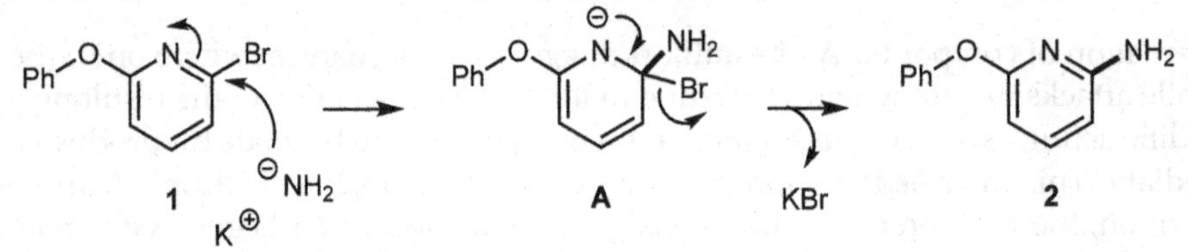

SOLUTION: 1 TO 3 (*tele* SUBSTITUTION)

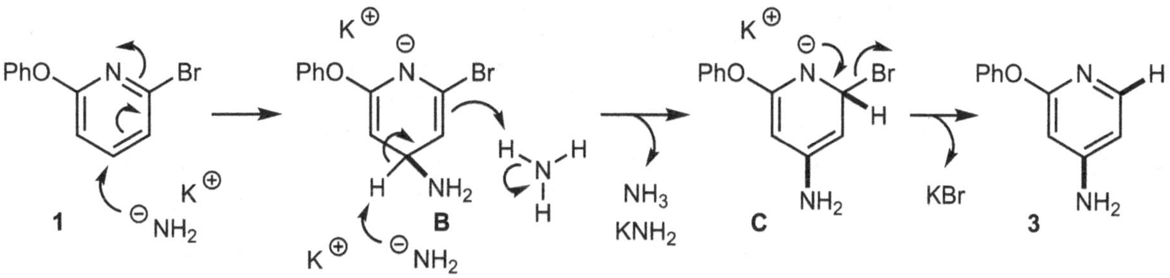

SOLUTION: 1 TO 4 (ANRORC MECHANISM)

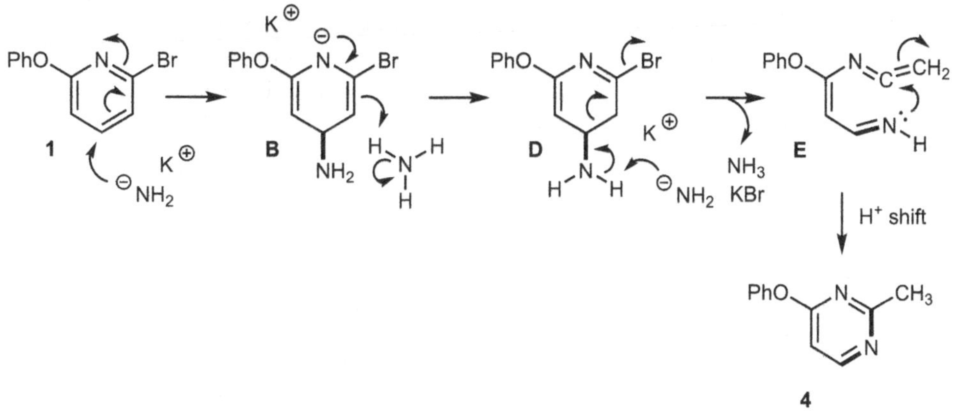

BALANCED CHEMICAL EQUATIONS

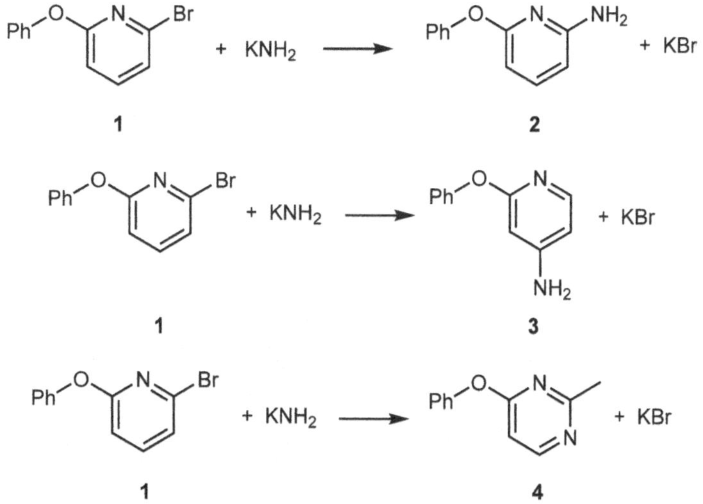

KEY STEPS EXPLAINED

The transformation of **1** to **2** constitutes according to the authors a simple addition-elimination-type mechanism, also referred to as S_NHetAr. In this case, amide anion adds in a nucleophilic fashion to the *ipso* 2-bromo position to give anion **A** which eliminates bromide anion in the form of potassium bromide by-product to give product **2**. In the case of the *tele* substitution, the authors described the mechanism of this transformation in a subsequent paper.[2] Essentially, since the reaction conditions are strongly basic, amide anion is expected to attack in a nucleophilic manner at the *tele* position of **1** to generate anion **B** which undergoes a prototropic rearrangement to generate intermediate **C**.

Here, amide anion acts as the proton acceptor and ammonia acts as the proton donor. Simple E2 elimination of bromide anion in the form of potassium bromide by-product generates the *tele* substitution product **3**. Lastly, the **1** to **4** transformation involves an S_N(ANRORC) mechanism[3-5] (substitution nucleophilic addition of the nucleophile, ring opening, and ring closure) and is a different kind of substitution reaction compared to the previous two conventional cases. Essentially amide anion adds at the *tele* position to form anion **B** as seen previously. Nevertheless, instead of a prototropic rearrangement as was the case for the *tele* substitution, we instead have protonation via ammonia to form **D** followed by deprotonation of the amino group with concomitant ring opening and elimination of bromide anion (potassium bromide by-product) to form the ketene imine **E**. This intermediate undergoes ring closure at the central carbon position of the ketene imine to give the final product **4** which has an *exo* methyl group at the *ipso* position and a newly formed pyrimidine ring. The transformation from **1** to **4** constitutes a [5+1] ring construction strategy for the synthesis of pyrimidine rings. It is important to note that the authors do not provide much in terms of experimental and theoretical evidence to support these postulated mechanisms, especially when it comes to explaining the interesting product distribution given two entirely distinct pathways and two mechanisms which diverge at intermediate **B**. However, experimental evidence for the S_N(ANRORC) mechanism has been demonstrated using ^{15}N and ^{14}C isotopic labeling experiments in later work.[3] Schemes 2 and 3 show possible outcomes applied to the **1** to **4** transformation discussed in this problem using variously labeled reactants. For example, using ^{15}N labeled potassium amide in ammonia solution would result in ^{15}N incorporation into the pyrimidine ring (see Scheme 2). If the C_3 carbon atom of the starting pyridine substrate is labeled as ^{14}C, then the mechanism predicts that the label ends up on the newly formed methyl substituent on the pyrimidine ring (see Scheme 3).

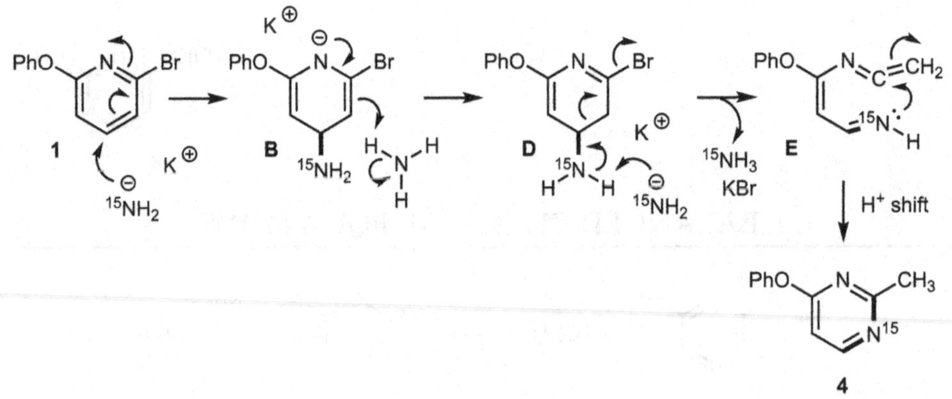

SCHEME 2

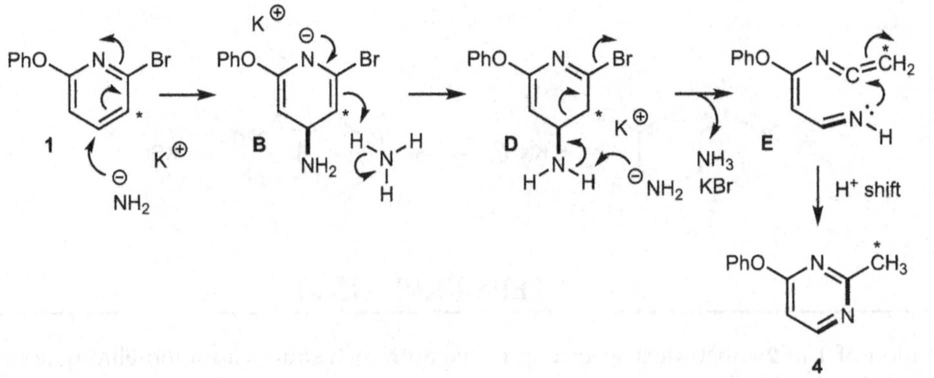

SCHEME 3

ADDITIONAL RESOURCES

We direct the reader to further work by the same authors concerning mechanistic analysis within a very similar chemical system.[6] In addition, we also recommend two review articles by the same authors which are filled with mechanistic illustrations of very similar heterocyclic ring transformations.[7,8]

References

1. Streef JW, den Hertog HJ. Action of potassium amide on 6-substituted derivatives of 2-bromopyridine in liquid ammonia. *Recl Trav Chim Pays-Bas*. 1969;88:1391–1412. https://doi.org/10.1002/recl.19690881205.
2. Streef JW, den Hertog HJ, van der Plas HC. Reactivity of some 3-substituted derivatives of 2,6-dihalogenopyridines towards potassium amide in liquid ammonia. *J Heterocyclic Chem*. 1985;22(4):985–991. https://doi.org/10.1002/jhet.5570220411.
3. Van der Plas HC. The S$_N$(ANRORC) mechanism: a new mechanism for nucleophilic substitution. *Acc Chem Res*. 1978;11(12):462–468. https://doi.org/10.1021/ar50132a005.
4. Van der Plas HC. S$_N$(ANRORC) reactions in azines containing an "outside" leaving group. *Adv Heterocycl Chem*. 1999;74:9–86. https://doi.org/10.1016/S0065-2725(08)60809-0.
5. Van der Plas HC. S$_N$(ANRORC) reactions in azaheterocycles containing an "inside" leaving group. *Adv Heterocycl Chem*. 1999;74:87–151. https://doi.org/10.1016/S0065-2725(08)60810-7.
6. Streef JW, van der Plas HC, Wei YY, Declercq JP, van Meerasche M. A new rearrangement in the methylation of 2-(diethylamino)-3-(ethoxycarbonyl)-5-phenylazepinyl anion. *Heterocycles*. 1987;26(3):685–688. https://doi.org/10.3987/R-1987-03-0685.
7. Bird CW, Cheeseman GWH, van der Plas HC, Streef JW. Ring transformations. In: Bird CW, Cheeseman GWH, eds. *Aromatic and Heteroaromatic Chemistry*. London: The Chemical Society; 1977:163–259. vol. 5. https://doi.org/10.1039/9781847555731-00163.
8. Bird CW, Cheeseman GWH, van der Plas HC, Streef JW. Ring transformations. In: Bird CW, Cheeseman GWH, eds. *Aromatic and Heteroaromatic Chemistry*. London: The Chemical Society; 1976:146–226. vol. 4. https://doi.org/10.1039/9781847555724-00146.

Question 208: Rearrangement During Recrystallization

While studying the reactivity of activated cyclopropanes and enamines, American chemists carried out the synthetic sequence in Scheme 1.[1] Explain these results in mechanistic terms.

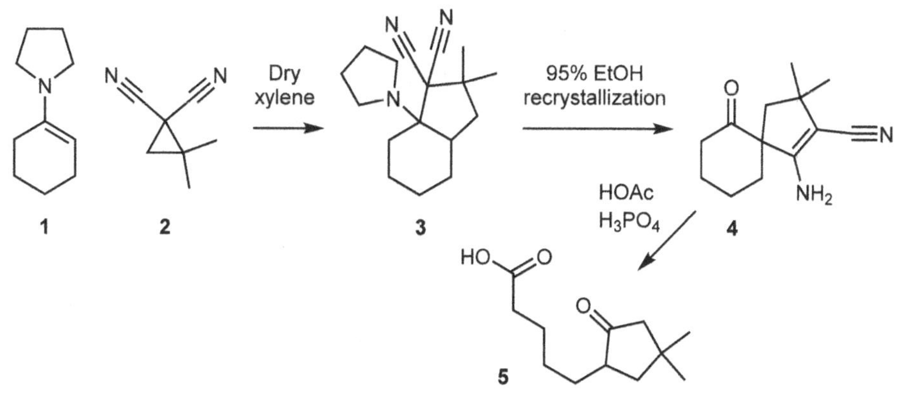

SCHEME 1

SOLUTION: 1 AND 2 TO 3

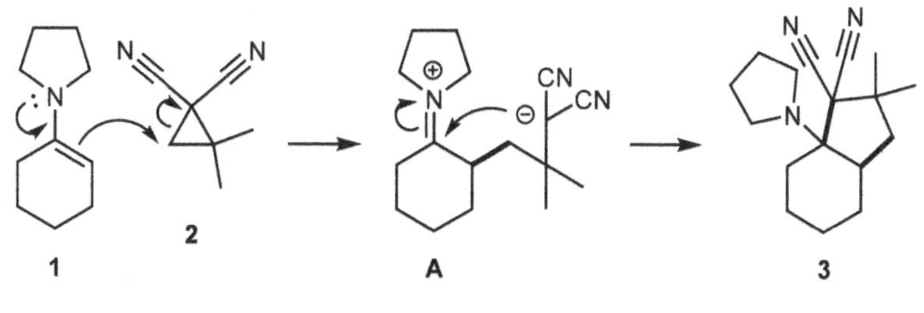

SOLUTION: 3 TO 4

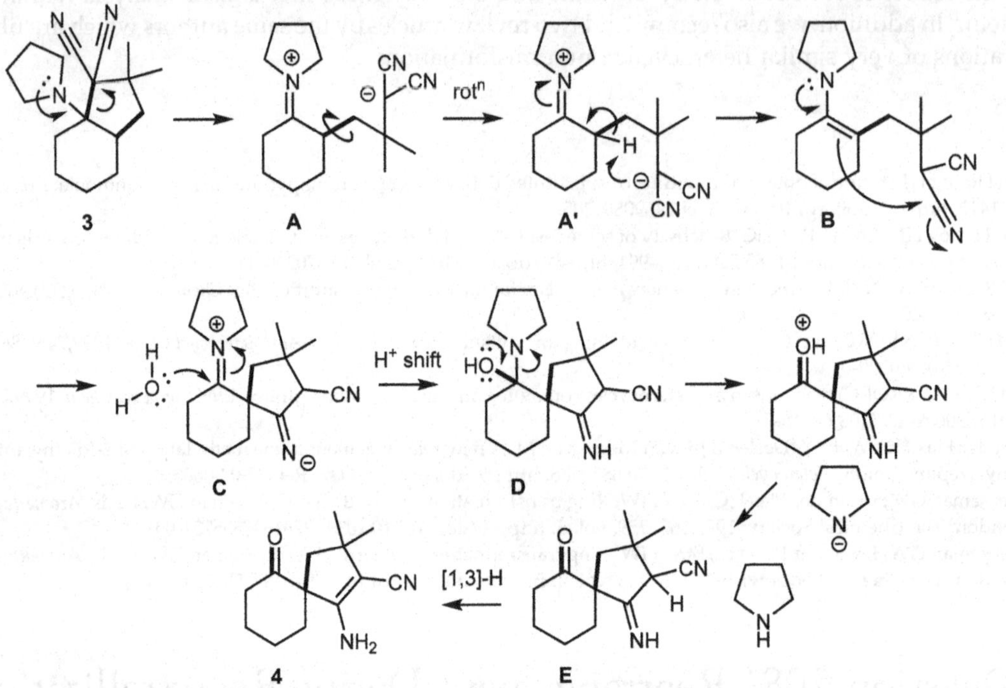

SOLUTION: 4 TO 5

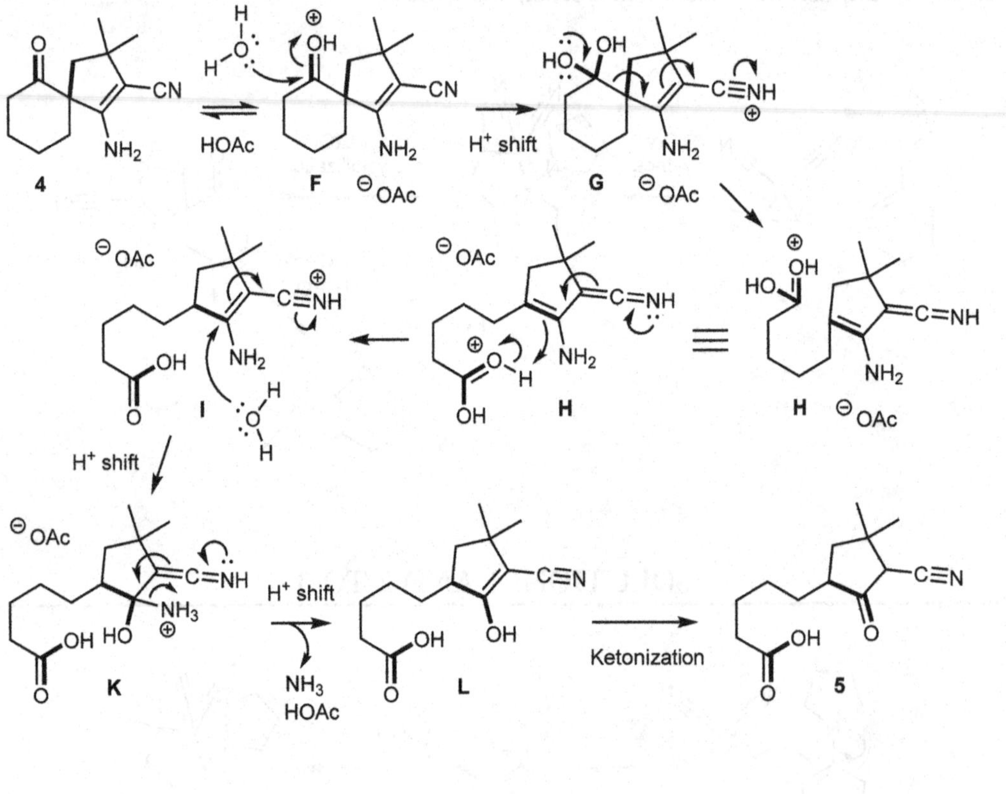

BALANCED CHEMICAL EQUATION (4 TO 5)

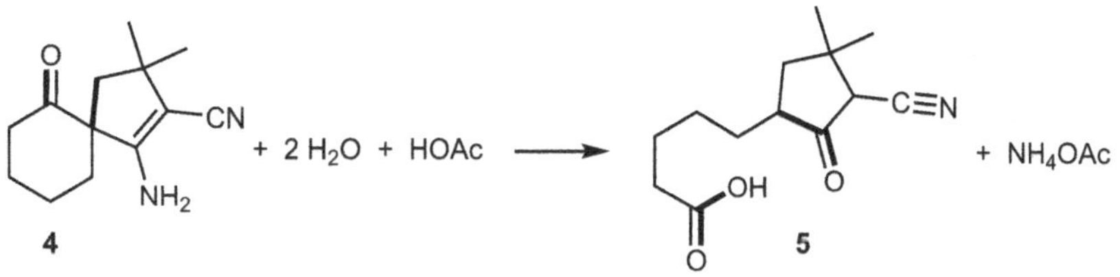

KEY STEPS EXPLAINED

To explain the reaction between **1** and **2**, the authors proposed a mechanism in which the pyrrolidine group of **1** facilitates an S_N2 attack by the double bond onto the unsubstituted cyclopropane carbon of **2** to form the zwitterion iminium intermediate **A** which cyclizes to **3**. It is noteworthy that the authors had to rule out an alternative mechanism for this transformation based on experimental results (Scheme 2). This alternative mechanism involves initial heterolytic cleavage of the cyclopropane ring between the quaternary centers of **2** followed by nucleophilic addition of the generated 1,3-dipolar zwitterion **A'** onto the C=C bond of **1** in a [3+2] manner. The mechanism was excluded because it would have produced an unobserved product **3'** rather than the observed product **3**. Note that the positive charge in **A'** is centered on a tertiary carbocation and the negative charge is centered on the carbon bearing the two electron-withdrawing nitrile groups. Essentially the difference between the two mechanisms lies in which bond is cleaved in the cyclopropane ring. In the operative mechanism the bond cleaved is the one between the dicyanocarbon atom and the unsubstituted carbon atom.

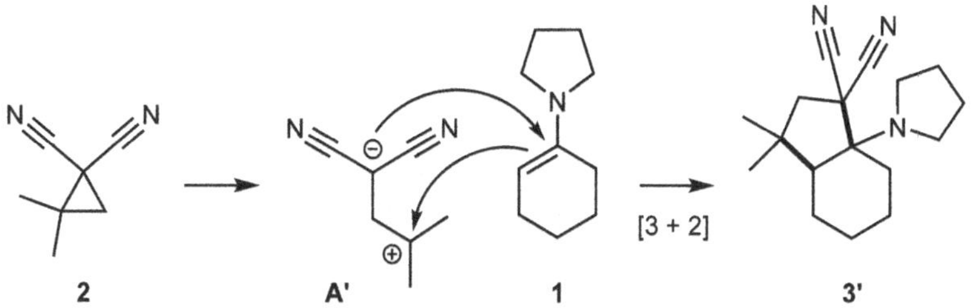

SCHEME 2

Once product **3** is formed, recrystallization begins with pyrrolidine-assisted ring opening back to intermediate **A** followed by bond rotation to **A'** which undergoes intramolecular proton transfer to generate enamine **B**. This enamine cyclizes to a [6,5]-*spiro* zwitterionic intermediate **C** which reacts with water to eliminate the pyrrolidine group and form the [6,5]-*spiro* keto-imine intermediate **E** which tautomerizes to the keto-amine product **4**.

The final acid-catalyzed transformation of **4** to **5** begins with water-mediated cleavage of the cyclohexanone ring leading to vinylideneamine intermediate **H** which isomerizes to **I**. Nucleophilic addition of a second equivalent of water which after two proton shifts and a nitrile-assisted expulsion of ammonia generates intermediate **L** which ketonizes to the final product **5**. The authors did not present any experimental or theoretical evidence to help support the mechanisms postulated for the **3** to **4** and the **4** to **5** transformations, respectively.

ADDITIONAL RESOURCES

Firstly, we would highly recommend a recent article describing essentially the same transformation as **1** and **2** to **3** only the context is palladium catalysis with a focus on mechanism.[2] Secondly, we refer the reader to a review article describing recyclization reactions of compounds containing three-, four-, five-, and six-membered rings.[3]

References

1. Berkowitz WF, Grenetz SC. Cycloaddition of an enamine to an activated cyclopropane. *J Org Chem.* 1976;41(1):10–13. https://doi.org/10.1021/jo00863a002.
2. Trost BM, Morris PJ, Sprague SJ. Palladium-catalyzed diastereo- and enantioselective formal [3 + 2]-cycloadditions of substituted vinylcyclopropanes. *J Am Chem Soc.* 2012;134:17823–17831. https://doi.org/10.1021/ja309003x.
3. Litvinov VP. Recyclisation of carbo- and heterocyclic compounds involving malononitrile and its derivatives. *Russ Chem Rev.* 1999;68(1):39–53. https://doi.org/10.1070/RC1999v068n01ABEH000273.

Question 209: Another Attempt to Reduce a Ketone Goes Wrong

A 1976 attempt to reduce the amino ketone **1** at 150–160°C for 15 min in a constant boiling liquid salt trimethylammonium formate led to the isolation of two unexpected products, the major one of which was product **2** in 58% yield (Scheme 1).[1] Suggest a mechanism for the formation of **2**.

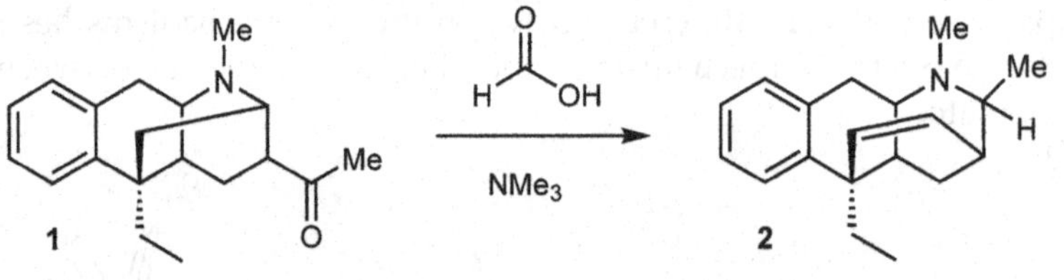

SCHEME 1

SOLUTION

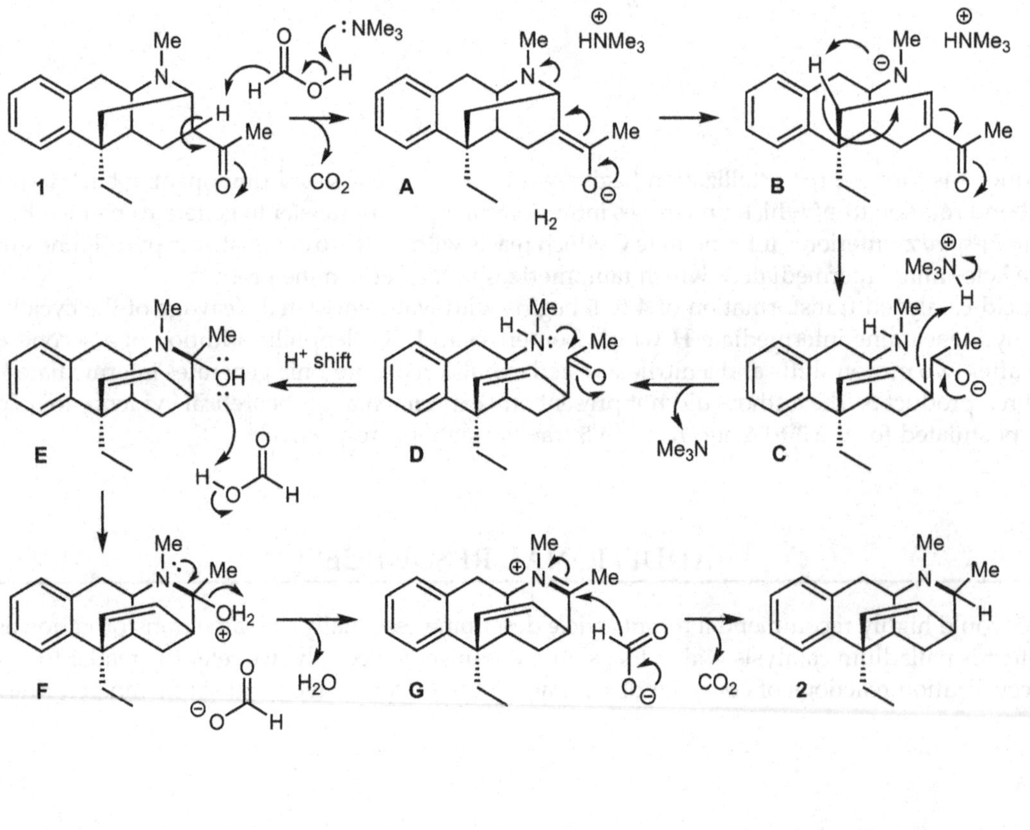

BALANCED CHEMICAL EQUATION

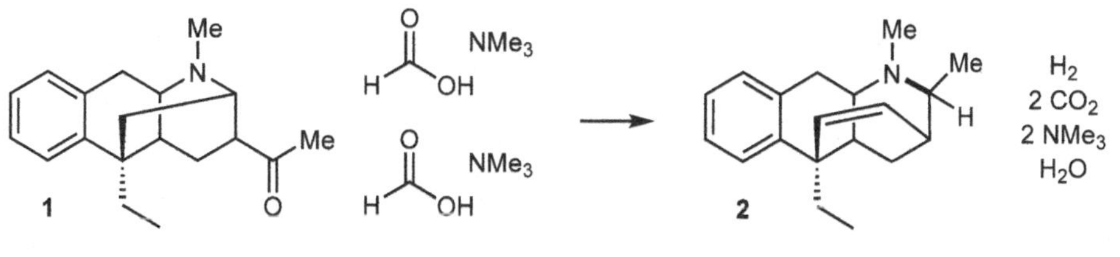

KEY STEPS EXPLAINED

The mechanism for the conversion of **1** to **2** begins with decomposition of formate to carbon dioxide and hydride which abstracts a proton from **1** to form enolate **A** and hydrogen gas. Enolate **A** then undergoes a retro-Michael reaction to form amide **B** which abstracts a proton from the methylene group shown in an intramolecular fashion to generate enolate **C** which ketonizes to intermediate **D** which ring closes to intermediate **E**. Protonation of the hydroxyl group of **E** facilitates the elimination of water to form the iminium ion **G** which is reduced to **2** following a second formate decomposition to carbon dioxide and hydride. It is noteworthy that the authors also isolated product **3** in trace amounts. To explain this transformation they proposed a pathway consisting of initial retro-Mannich reaction followed by reduction of iminium to form product **3** (Scheme 2). It is important to highlight that both mechanisms are speculative since the authors do not present any experimental or theoretical evidence in support of the postulated intermediates. We would add that a theoretical analysis would greatly help to explain why a retro-Michael reaction is favored to a retro-Mannich reaction as the initial step of this transformation.

Lastly, the original transformation the authors intended to perform appears in Scheme 3.

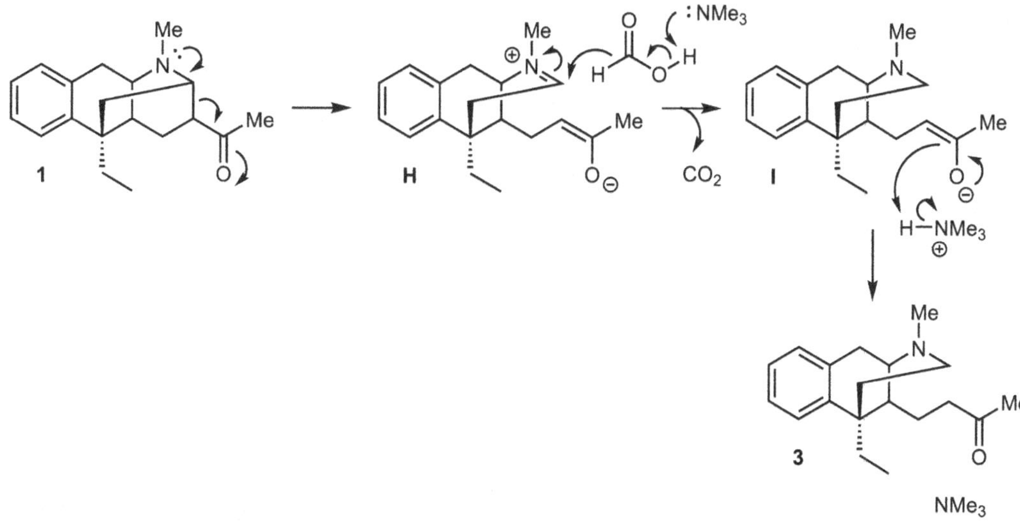

SCHEME 2

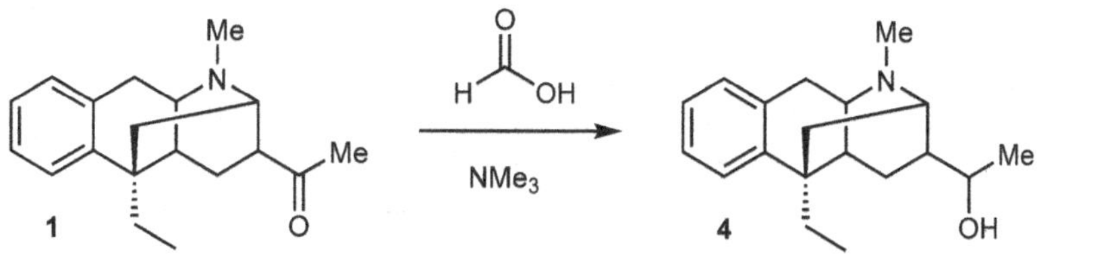

SCHEME 3

ADDITIONAL RESOURCES

We recommend several articles highlighting interesting mechanistic transformations involving retro-Michael rearrangements.[2-4] For interesting retro-Mannich transformations, see Refs. 5–7. Lastly, for an interesting ketone reduction which involves a detailed theoretical analysis, see the work of Yang.[8]

References

1. Michne WF. A 2,6-methano-3-benzazocine related to the thebaine Diels-Alder adduct derivatives. *J Org Chem.* 1976;41(5):894–896. https://doi.org/10.1021/jo00867a037.
2. Clarke DS, Gabbutt CD, Hepworth JD, Heron BM. Synthesis of 3-alkenyl-2-arylchromones and 2,3-dialkenylchromones via acid-catalysed retro-Michael ring opening of 3-acylchroman-4-ones. *Tetrahedron Lett.* 2005;46:5515–5519. https://doi.org/10.1016/j.tetlet.2005.06.058.
3. Roman LU, Morales NR, Hernandez JD, et al. Generation of the new quirogane skeleton by a vinylogous retro-Michael type rearrangement of longipinene derivatives. *Tetrahedron.* 2001;57(34):7269–7275. https://doi.org/10.1016/S0040-4020(01)00718-9.
4. Naidu BN. Tandem retro-Michael addition–Claisen rearrangement–intramolecular cyclization: one-pot synthesis of densely functionalized ethyl dihydropyrimidine-4-carboxylates from simple building blocks. *Synlett.* 2008;2008(4):547–550. https://doi.org/10.1055/s-2008-1032088.
5. Funk P, Motyka K, Soural M, et al. Study of 2-aminoquinolin-4(1H)-one under Mannich and retro-Mannich reaction. *PLoS ONE.* 2017;12(5):1–15. https://doi.org/10.1371/journal.pone.0175364.
6. Chen P, Carroll PJ, Sieburth SM. 3-4'-Bipiperidines via sequential [4 + 4]-[3,3]-retro-Mannich reactions. *Org Lett.* 2009;11(20):4540–4543. https://doi.org/10.1021/ol901743p.
7. Cramer N, Juretschke J, Laschat S, Baro A, Frey W. Acid-promoted retro-Mannich reaction of *N*-protected tropenones to 2-substituted pyrroles. *Eur J Org Chem.* 2004;2004(7):1397–1400. https://doi.org/10.1002/ejoc.200300804.
8. Yang X. Unexpected direct reduction mechanism for hydrogenation of ketones catalyzed by iron PNP pincer complexes. *Inorg Chem.* 2011;50:12836–12843. https://doi.org/10.1021/ic2020176.

Question 210: A More Complex Benzocyclobutane to Isochroman-3-one Rearrangement (c.f. Problem 138)

In 1986 Japanese chemists heated the benzocyclobutene derivative **1** in degassed *o*-dichlorobenzene at 180°C for 2h to form the spirocyclic product **2** in 88% yield (Scheme 1).[1] Provide a mechanism for this transformation.

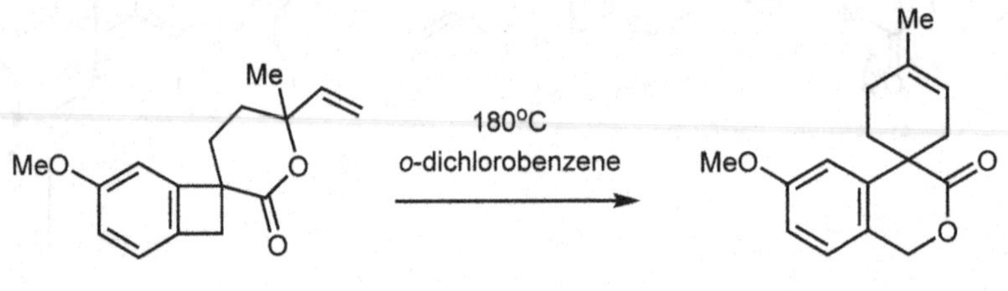

SCHEME 1

SOLUTION

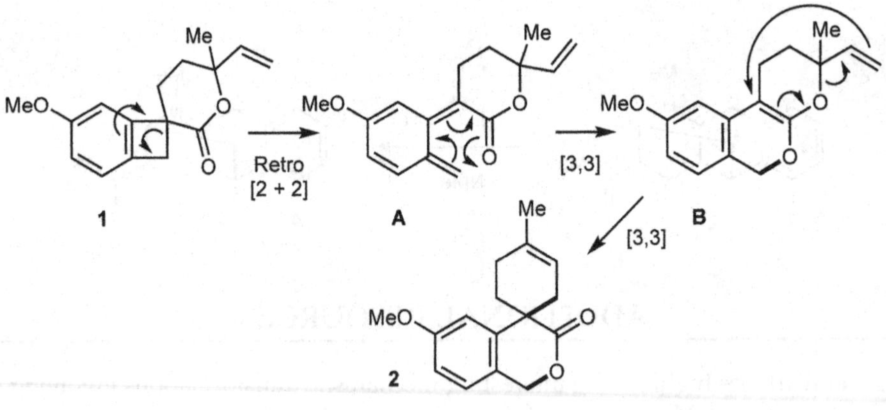

KEY STEPS EXPLAINED

The transformation of **1** to **2** may be regarded as a thermal 100% atom economical reaction which involves an initial retro [2+2] ring opening of **1** to the *o*-quinodimethane intermediate **A** followed by two successive [3,3]-sigmatropic rearrangements to generate the [6,6]-*spiro* product **2**. We note that the ring construction strategy here may be described as *spiro*-[(6+0)+(6+0)]. Interestingly, when the authors carried out this reaction with a substrate containing a *spiro* five-membered ring instead of a *spiro* six-membered ring, two products were obtained (Scheme 2). Product **4** was the analog of **2** and product **5** was a mixture of [6,3]-*spiro* diastereomers.

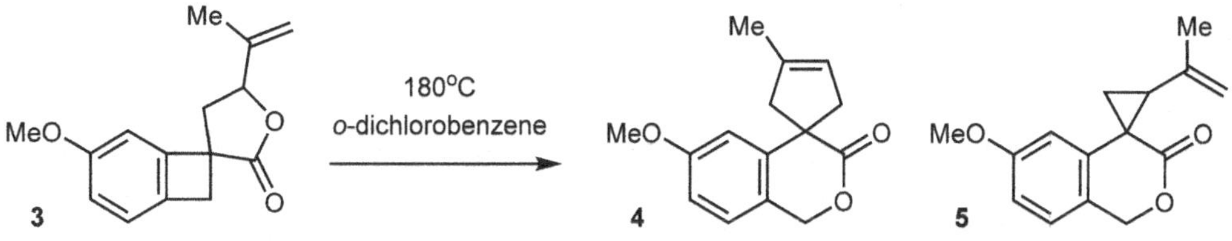

SCHEME 2

To explain this transformation, the authors invoked a diradical intermediate **C** after the initial retro [2+2] fragmentation and [3,3]-sigmatropic rearrangement (Scheme 3). **C** can thus cyclize in two ways: product **5** can come from the C_2 form and product **4** can come from the C_3 form (Scheme 4). Moreover, experimental observation of [6,3]-*spiro* product **5** suggested a diradical intermediate. If product **4** was observed by analogy with the **1** to **2** transformation, its outcome could be rationalized via an electrocyclic mechanism without invoking a radical intermediate.

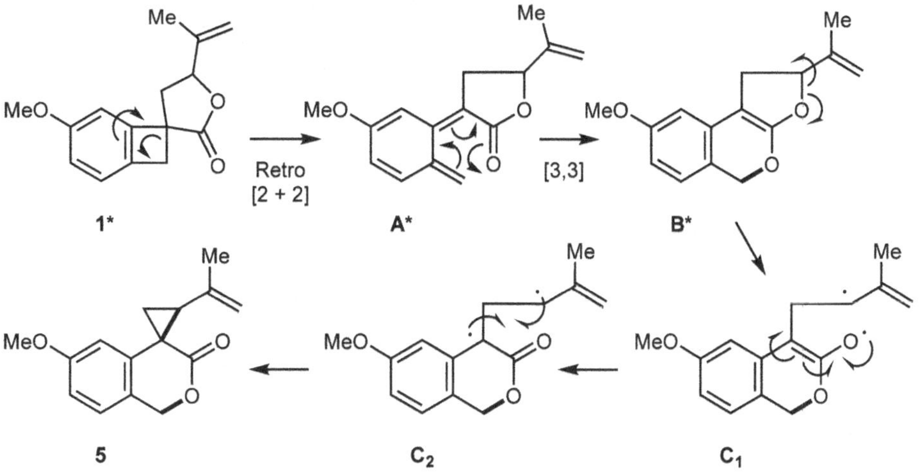

SCHEME 3

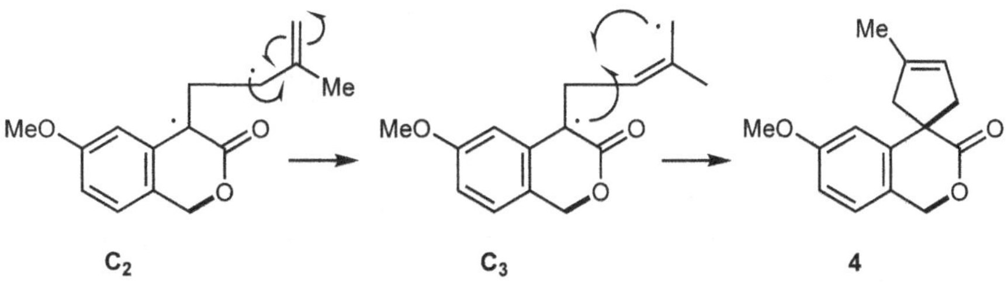

SCHEME 4

ADDITIONAL RESOURCES

We recommend an excellent review of cycloaddition reactions of *ortho*-quinodimethanes derived from benzocyclobutenes in synthesis.[2] We also refer the reader to two articles describing the mechanisms of two very interesting [3+3] cycloaddition transformations.[3,4]

References

1. Shishido K, Hiroya K, Fukomoto K, Kametani T. Tandem electrocyclic-sigmatropic reaction of benzocyclobutenes. 2. A new route to isochroman-3-one-4-spiro-1'-cycloalk-3'-enes. *Tetrahedron Lett.* 1986;27(8):971–974. https://doi.org/10.1016/S0040-4039(00)84151-9.
2. Michellys P-Y, Pellissier H, Santelli M. Cycloadditions of *ortho*-quinodimethanes derived from benzocyclobutenes in organic synthesis. A review. *Org Prep Proced Int.* 1996;28(5):545–608. https://doi.org/10.1080/00304949609458572.
3. Sklenicka HM, Hsung RP, McLaughlin MJ, Wei L-I, Gerasyuto AI, Brennessel WB. Stereoselective formal [3 + 3] cycloaddition approach to *cis*-1-azadecalins and synthesis of (-)-4*a*,8*a-diepi*-pumiliotoxin C. Evidence for the first highly stereoselective 6π-electron electrocyclic ring closures of 1-azatrienes. *J Am Chem Soc.* 2002;124:10435–10442. https://doi.org/10.1021/ja020698b.
4. Shen HC, Wang J, Cole KP, et al. A formal [3 + 3] cycloaddition reaction. Improved reactivity using α,β-unsaturated iminium salts and evidence for reversibility of 6π-electron electrocyclic ring closure of 1-oxatrienes. *J Org Chem.* 2003;68:1729–1735. https://doi.org/10.1021/jo020688t.

Question 211: Indoles by Solvometalation Ring Closure

During synthetic studies to mitosenes, American chemists have shown that reaction of **1** with mercuric acetate in THF containing sodium bicarbonate leads to the indole derivate **2** in over 50% yield (Scheme 1).[1] Propose a mechanism for this transformation.

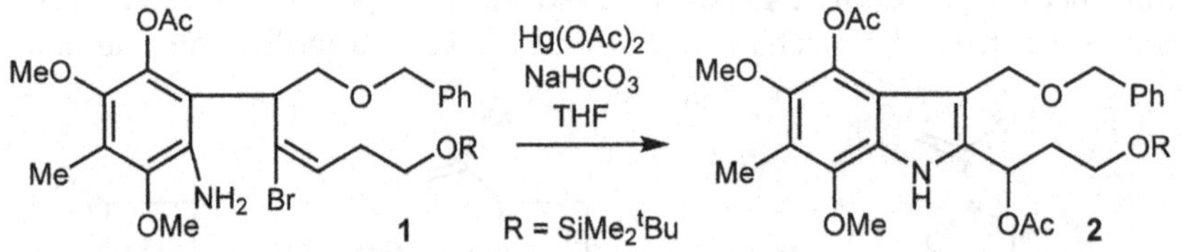

SCHEME 1

SOLUTION

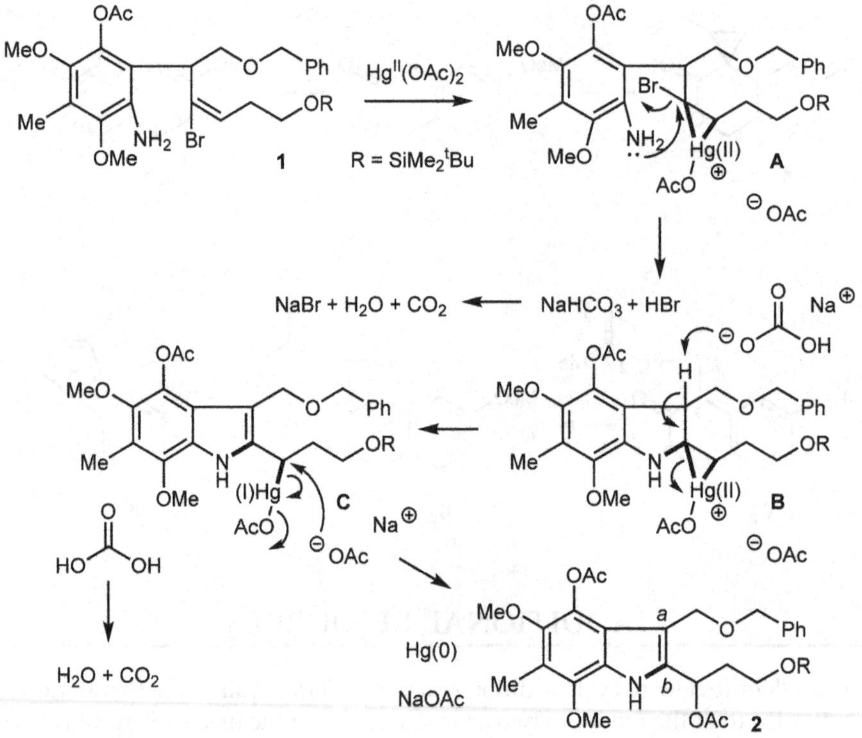

BALANCED CHEMICAL EQUATION

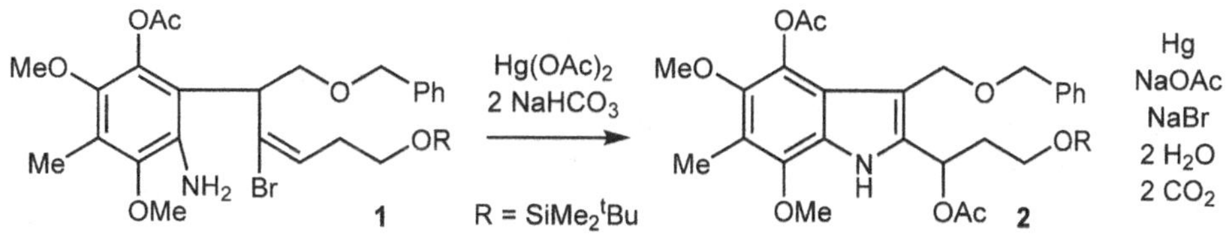

KEY STEPS EXPLAINED

The mechanism proposed by Danishefsky and Regan begins with addition of mercury(II)acetate across the olefinic bond of **1** to obtain intermediate **A** which then ring closes via nucleophilic attack by the neighboring amino group to form intermediate **B** with expulsion of hydrogen bromide. The hydrogen bromide by-product then reacts with one equivalent of sodium bicarbonate to form sodium bromide, water, and carbon dioxide. Another equivalent of sodium bicarbonate deprotonates **B** in order to establish the indole functional group with reduction of Hg(+2) to Hg(+1) thereby forming intermediate **C** and carbonic acid by-product. The carbonic acid will then easily decompose to water and carbon dioxide. Lastly, acetate anion attacks **C** in a nucleophilic substitution which displaces Hg(0), forms the final indole derivative product **2** along with sodium acetate by-product. We note that this is a redox reaction where mercury is reduced from +2 to 0 which is compensated by the oxidation of C_a and C_b each from −1 to 0. Lastly, we highlight the fact that the authors admit that they have no further experimental or theoretical evidence to support their postulated mechanism.

ADDITIONAL RESOURCES

We refer the reader to three articles which discuss the mechanistic role of mercury(II)-mediated rearrangements of three distinct transformations.[2–4]

References

1. Danishefsky S, Regan J. A mercury mediated route to the mitosenes. *Tetrahedron Lett.* 1981;22(40):3919–3922. https://doi.org/10.1016/S0040-4039(01)82026-8.
2. Bach RD, Klix RC. A mercury-mediated acyl migration in a Pinacol-type rearrangement. Model studies toward the synthesis of Fredericamycin A. *J Org Chem.* 1986;51(5):749–752. https://doi.org/10.1021/jo00355a036.
3. Majumdar KC, Das U. Synthesis of 1-alkoxy-1,2,3,4-tetrahydrocarbazoles by mercury(II) mediated heterocyclization of 2-cyclohex-2′-enyl-N-alkylanilines. *Can J Chem.* 1996;74:1592–1596. https://doi.org/10.1139/v96-175.
4. Ghorai S, Bhattacharjya A. Mercury(II) chloride-mediated cyclization-rearrangement of o-propargylglycolaldehyde dithioacetals to 3-pyranone dithioketals: an expeditious access to 3-pyranones. *Org Lett.* 2005;7(2):207–210. https://doi.org/10.1021/ol047893a.

Question 212: An Unusual—But Inefficient—Synthesis of Methyl 1-Hydroxynaphthalene-2-carboxylate

During a 1983 study, Japanese chemists treated homophthalic anhydride **1** with methyl propiolate in toluene at 150°C for 24 h and realized methyl 1-hydroxynaphthalene-2-carboxylate **2** in 19% yield (Scheme 1).[1] Provide at least two possible mechanisms for this transformation.

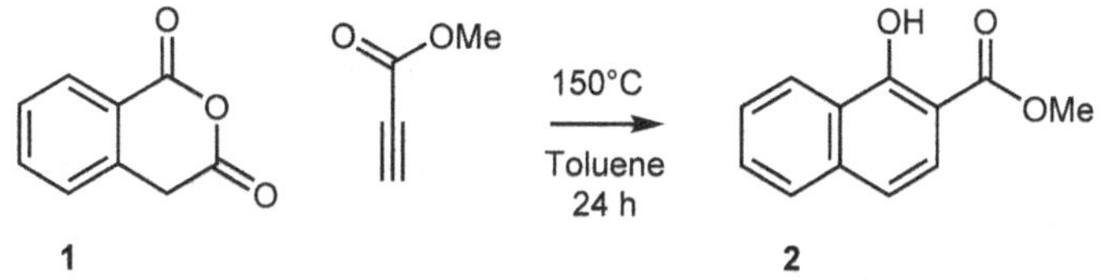

SCHEME 1

SOLUTION: MECHANISM A

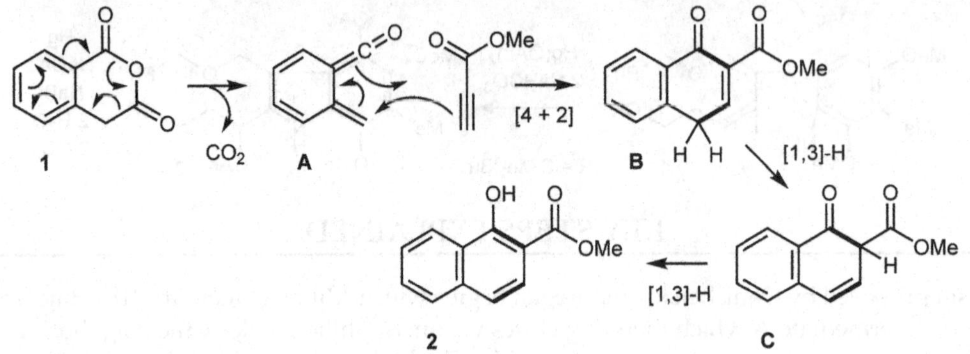

SOLUTION: MECHANISM B

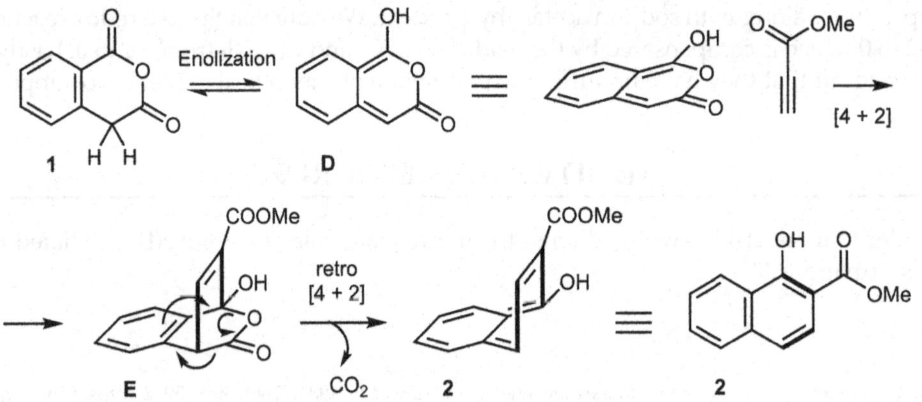

SOLUTION: MECHANISM C

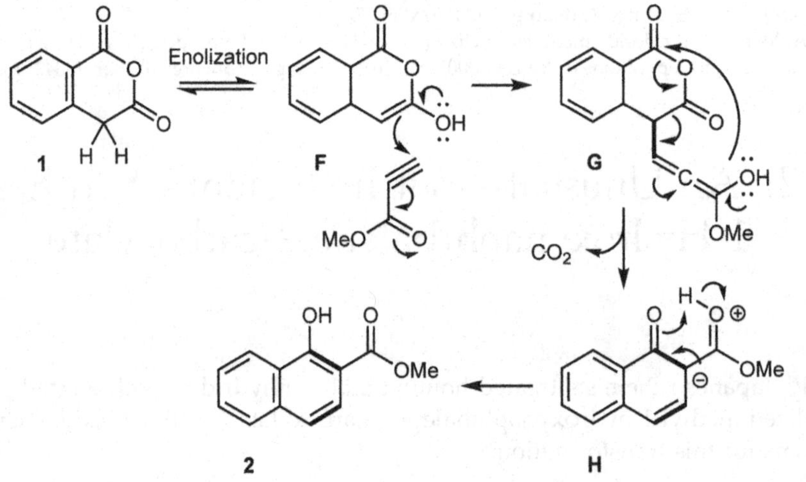

KEY STEPS EXPLAINED

We present here three possible mechanisms for the conversion of homophthalic anhydride **1** to methyl 1-hydroxy-naphthalene-2-carboxylate **2**. In mechanism A we have an initial rearrangement of **1** with elimination of carbon dioxide by-product to form the ketene intermediate **A** which undergoes a [4+2] Diels-Alder cycloaddition with methyl propiolate to give intermediate **B**. This intermediate undergoes two successive [1,3]-protic shifts to generate the aromatized product **2**. This may also occur in one step via a [1,5]-protic shift.

With respect to this mechanism, the authors ruled it out on the basis of several observations. First, since ketene **A** is formed during the course of this mechanism, it would be expected that its [2+2] cycloaddition product (i.e., benzocyclobuten-1(2H)-one, Scheme 2) should have been observed. According to the authors, this product was not formed. Second, the yield of product **2** from **1** and methyl propiolate is 65%; whereas, the yield of product **2** from benzocyclobuten-1(2H)-one and methyl propiolate is only 8%. If mechanism A were operational, one would certainly not expect such a large difference in the product yields under the two contrasting reaction conditions.

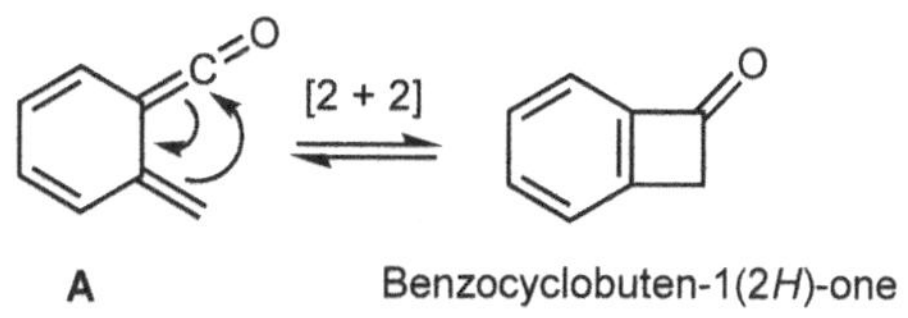

A   Benzocyclobuten-1(2H)-one

SCHEME 2

Given this evidence, the authors next entertained the possibility of mechanism B. In this mechanism, the first step is tautomerization of **1** to the enol intermediate **D** which undergoes a [4+2] Diels-Alder cycloaddition with methyl propiolate to give the [2.2.2] bicyclic intermediate **E**. This structure can subsequently proceed with a retro [4+2] degradation to give product **2** and carbon dioxide by-product.

With respect to mechanism B, the authors favored it based on the experimental evidence that the reaction of homophthalic anhydride **1** with 2 equivalents of 1-phenyl-pyrrole-2,5-dione leads to an isolable adduct **3** according to the analogous mechanism shown in Scheme 3.

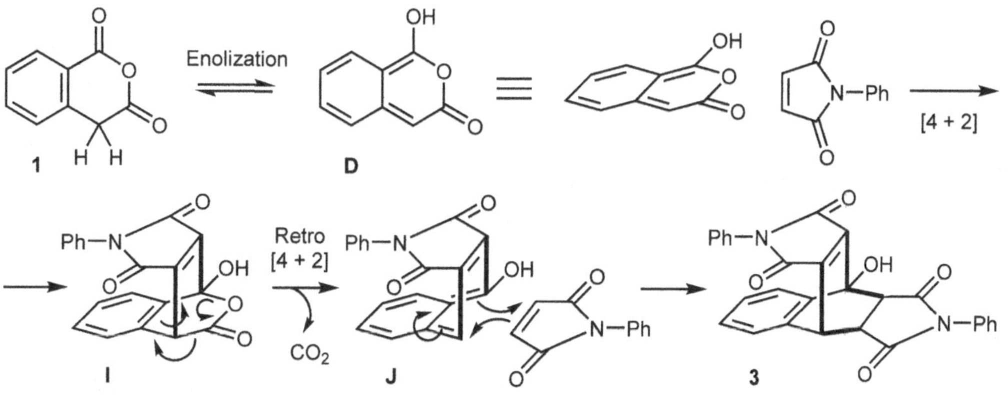

SCHEME 3

The last explanation provided for the conversion of **1** to **2** constitutes mechanism C. In this mechanism, initial tautomerization of **1** (at the alternate carbonyl group when compared to the first step of mechanism B) forms intermediate **F**. This intermediate undergoes nucleophilic addition to methyl propiolate in a Michael 1,4-addition fashion to give the allenol intermediate **G**. Cyclization of **G** to **H** with concomitant extrusion of carbon dioxide by-product followed by a final proton transfer leads to product **2**.

We note that although the authors could not definitively rule out mechanism C, they did show that when homophthalic anhydride **1** was reacted with 2-bromo-8-hydroxy-[1,4]naphthoquinone, the isolated product was product **4** and not, based on mechanism C, the expected product **5** (Scheme 4). The distinction here was that Michael addition occurred at the C_3 position of the anthraquinone and not at C_2 which is fully occupied via bonding to the bromo group.

ADDITIONAL RESOURCES

We refer the reader to two articles by the same authors exploring very similar transformations where several mechanistic possibilities are discussed.[2,3] We also recommend an excellent review of the role of cyclic anhydrides in formal cycloadditions and multicomponent reactions.[4]

Mechanism B expectation:

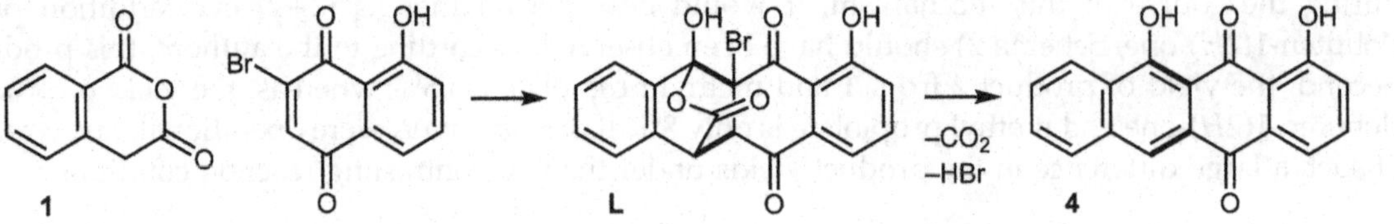

Mechanism C expectation:

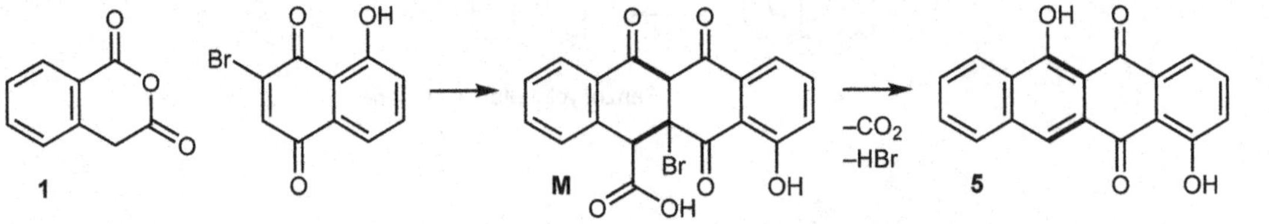

SCHEME 4

References

1. Tamura Y, Wada A, Sasho M, Kita Y. Linearly condensed *peri*-hydroxy aromatic compounds derived from the cycloaddition reaction of homophthalic anhydrides with dienophiles. *Chem Pharm Bull.* 1983;31(8):2691–2697. https://doi.org/10.1248/cpb.31.2691.
2. Iio K, Okajima A, Takeda Y, et al. Synthesis of *p*-phenylthio-*peri*-hydroxy polyaromatic compounds by strong-base-induced [4+2] cycloaddition of 4-(phenylthio)homophthalic anhydrides with phenylsulfinyl-dienophiles. *ARKIVOC.* 2003;(8):144–162.
3. Kita Y, Fujioka H. Syntheses of anthracyclines and fredericamycin A via strong base-induced cycloaddition reaction of homophthalic anhydrides. In: Krohn K, ed. *Topics in Current Chemistry.* Heidelberg, Berlin: Springer; 2007:299–319. vol. 282. https://doi.org/10.1007/128_2007_10.
4. Gonzalez-Lopez M, Shaw JT. Cyclic anhydrides in formal cycloadditions and multicomponent reactions. *Chem Rev.* 2009;109(1):164–189. https://doi.org/10.1021/cr8002714.

Question 213: Synthesis of α-Arylalkanoic Acids From Acetophenones

In 1981 Japanese chemists carried out the synthetic sequence depicted in Scheme 1 which begins with anisole **1** and ends up with methyl α-(4-methoxyphenyl)propionate **2**.[1] Provide mechanisms for the transformations: **B** to **C**, **C** to **D**, and **D** to **2**.

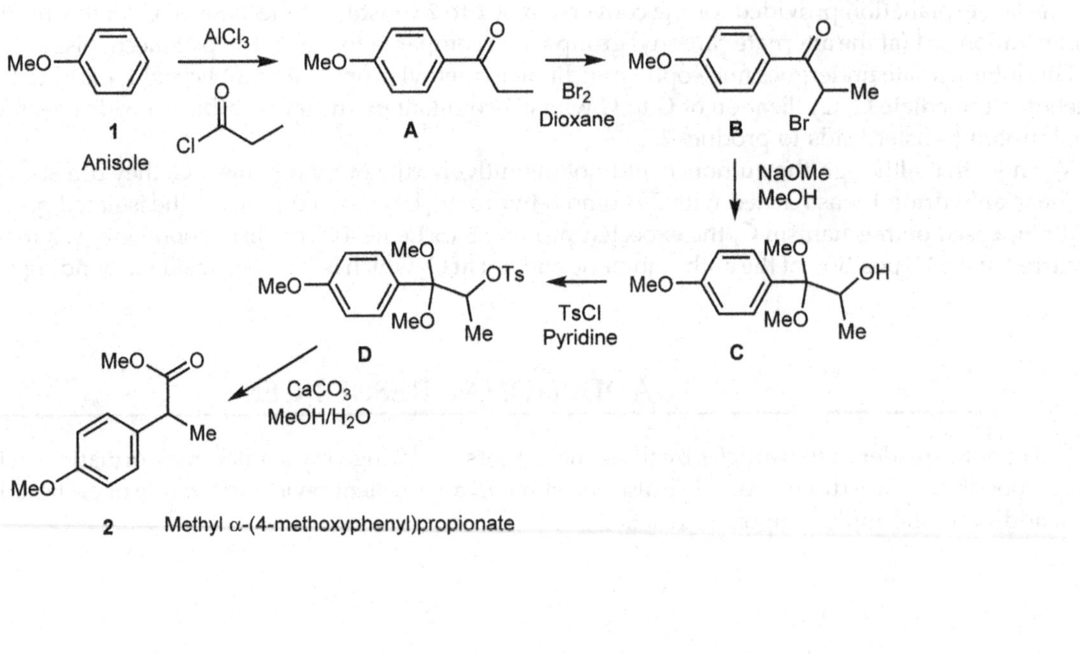

SCHEME 1

SOLUTION: B TO C

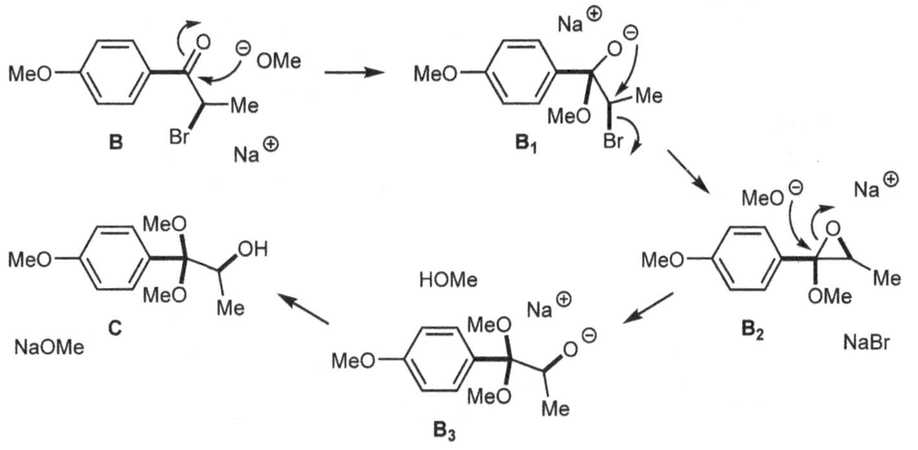

SOLUTION: C TO D

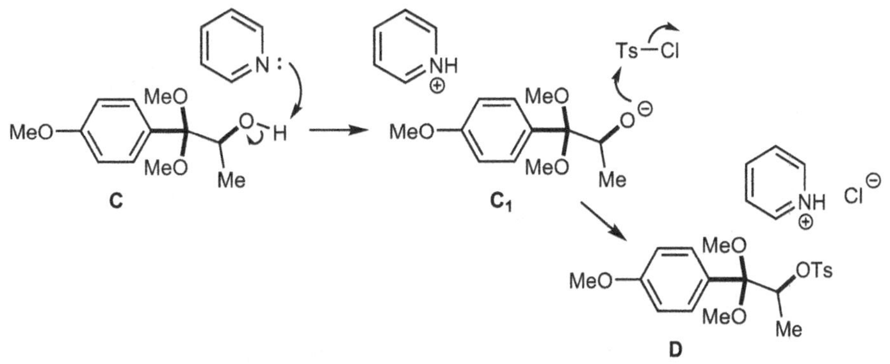

SOLUTION: D TO 2

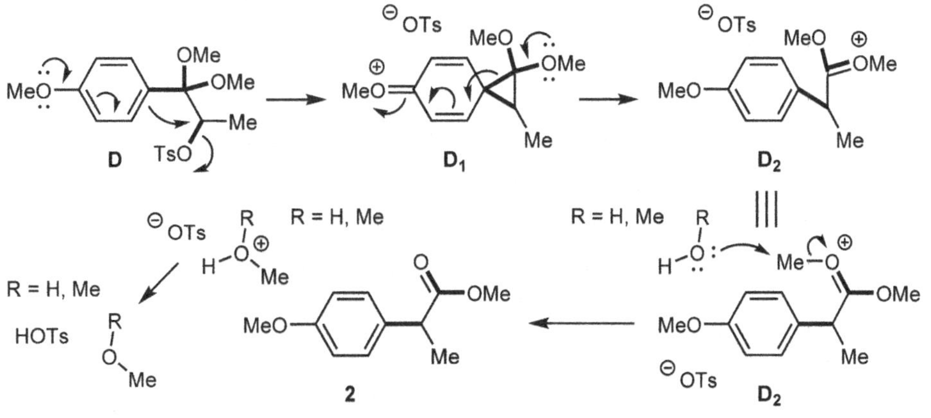

KEY STEPS EXPLAINED

The mechanism for the conversion of **B** to **C** begins with nucleophilic attack by methoxide anion onto the carbonyl group of **B** to form intermediate **B₁** which undergoes epoxide formation with concomitant elimination of bromide anion to form the epoxide intermediate **B₂**. Methoxide anion attack at the more hindered (but more electrophilic) carbon atom of the epoxide group of **B₂** results in formation of **B₃** which after protonation by the solvent (methanol) generates the product **C**. The conversion of the hydroxyl group of **C** to its tosylated form in **D** proceeds via a standard base (pyridine) catalyzed mechanism. Finally, the conversion of **D** to the final methyl α-(4-methoxyphenyl)propionate product **2** proceeds via initial formation of a phenonium cation intermediate **D₁** (which is greatly facilitated by electron-donating groups located at the *para* position of the aromatic ring as the authors observed experimentally). Ring opening of the phenonium cation intermediate **D₁** leads to intermediate **D₂** which is demethylated to form the ester product **2**. Overall this transformation constitutes a 1,2-migration of the aryl group where the by-product according to the scheme is either methanol or dimethylether.

It is important to highlight that the authors also isolated the unrearranged side product **3** when **D** did not contain a good electron-donating group at the *para* position of the aryl group. This pathway may occur via the mechanism shown in Scheme 2.

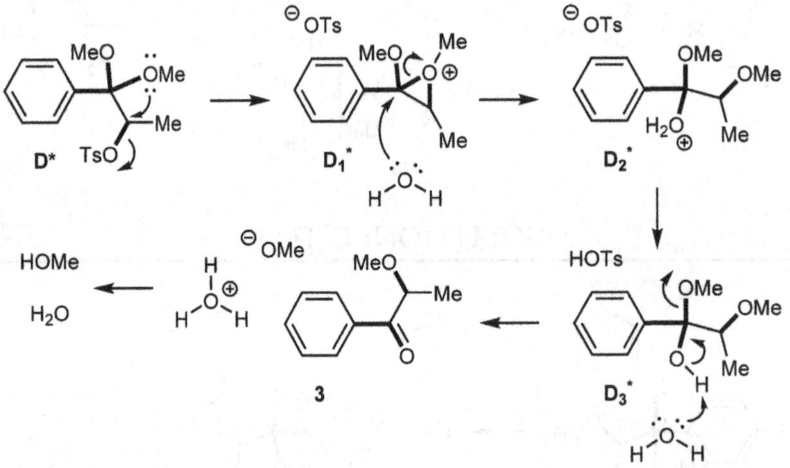

SCHEME 2

ADDITIONAL RESOURCES

We direct the reader to two articles which describe similar transformations to those encountered here, but nevertheless the context is iodine chemistry.[2,3] We also highly recommend a review by Nicolaou et al. which discusses cascade reactions in total synthesis.[4]

References

1. Tsuchihashi G, Kitajima K, Mitamura S. A new method for the synthesis of α-arylalkanoic acids by the use of 1,2-rearrangement of the aryl group. *Tetrahedron Lett.* 1981;22(43):4305–4308. https://doi.org/10.1016/S0040-4039(01)82941-5.
2. Yamauchi T, Hattori K, Nakao K, Tamaki K. A facile and efficient preparative method of methyl 2-arylpropanoates by treatment of propiophenones and their derivatives with iodine or iodine chlorides. *J Org Chem.* 1988;53(20):4859–4862. https://doi.org/10.1021/jo00255a037.
3. Elinson MN, Feducovich SK, Dorofeev AS, Vereshchagin AN, Nikishin GI. Indirect electrochemical oxidation of aryl alkyl ketones mediated by NaI-NaOH system: facile and effective way to α-hydroxyketals. *Tetrahedron.* 2000;56(51):9999–10003. https://doi.org/10.1016/S0040-4020(00)00951-0.
4. Nicolaou KC, Edmonds DJ, Bulger PG. Cascade reactions in total synthesis. *Angew Chem Int Ed.* 2006;45:7134–7186. https://doi.org/10.1002/anie.200601872.

Question 214: Triazene-Triazole-Triazole Interconversions

During a 1981 study of the synthesis of triazole compounds, Canadian chemists reacted triazene **1** under recrystallization conditions from absolute ethanol and formed the triazole product **2** in 80% yield. Heating of either **1** or **2** under reflux for an hour resulted in smooth isomerization to the triazole **3** in 99% yield (Scheme 1).[1] Provide mechanisms to explain these transformations.

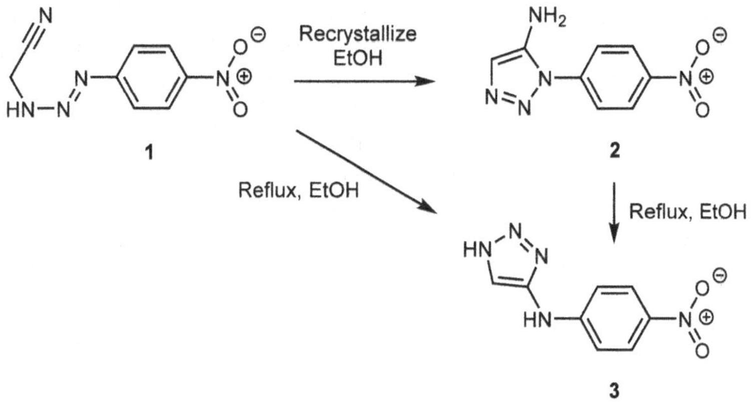

SCHEME 1

SOLUTION: 1 TO 2 TO 3

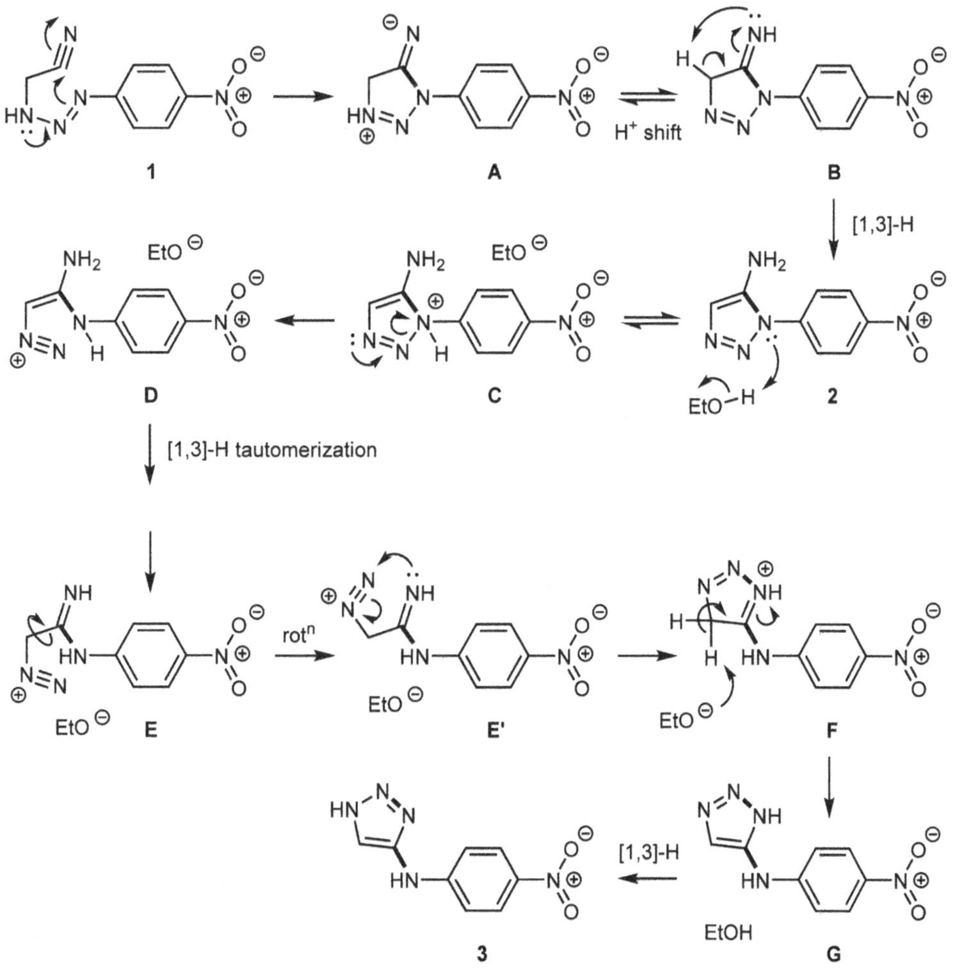

KEY STEPS EXPLAINED

The mechanism proposed by Baines et al. for the transformation of **1** to **2** begins with initial cyclization of **1** via intramolecular nucleophilic attack at the cyano group to form intermediate **A** which undergoes a prototropic rearrangement to **B**. This intermediate can then undergo tautomerization to the more stable aromatic 5-aminotriazole product **2**. Since the authors showed experimentally that treatment of either **1** or **2** under reflux in ethanol leads to **3**, whereas recrystallization of **1** from ethanol leads only to **2**, it was proven that **2** is an intermediate along the path from **1** to **3**. Therefore from **2** to **3** (a transformation often called the Dimroth rearrangement) the mechanism continues with protonation by the solvent at the N_1 position to form **C** followed by ring opening to form the diazonium intermediate **D**. This structure must tautomerize to the iminium diazonium intermediate **E** in order to have bond rotation around C_4—C_5 giving **E'** which is now able to undergo recyclization to **F**, ethoxide promoted deprotonation to **G** and a final [1,3] hydride shift to the final triazole Dimroth isomer product **3**. For future reference, we would encourage researchers to undertake a theoretical study of this mechanism in order to help support this postulated pathway.

ADDITIONAL RESOURCES

We direct the reader to two recent works describing unexpected Dimroth rearrangements encountered during synthetic attempts to substituted pyridines.[2,3]

References

1. Baines KM, Rourke TW, Vaughan K, Hooper DL. 5-(arylamino)-1,2,3-triazoles and 5-amino-1-aryl-1,2,3-triazoles from 3-(cyanomethyl)triazenes. *J Org Chem*. 1981;46(5):856–859. https://doi.org/10.1021/jo00318a006.
2. Lauria A, Patella C, Abbate I, Martorana A, Almerico AM. An unexpected Dimroth rearrangement leading to annelated thieno [3,2-*d*][1,2,3] triazolo[1,5-*a*]pyrimidines with potent antitumor activity. *Eur J Med Chem*. 2013;65:381–388. https://doi.org/10.1016/j.ejmech.2013.05.012.
3. Baiazitov RY, Sydorenko N, Ren H, Moon Y-C. Unexpected observation of the Dimroth rearrangement in the ribosylation of 4-aminopyrimidines. *J Org Chem*. 2017;82(11):5881–5889. https://doi.org/10.1021/acs.joc.7b00780.

Question 215: Reissert Compounds as Precursors to Novel Phthalides

In 1981 American chemists treated the Reissert compound **1** with sodium hydroxide in aqueous methanol and realized phthalideisoquinoline **2** in 58% yield (Scheme 1).[1] Suggest a mechanism for this transformation.

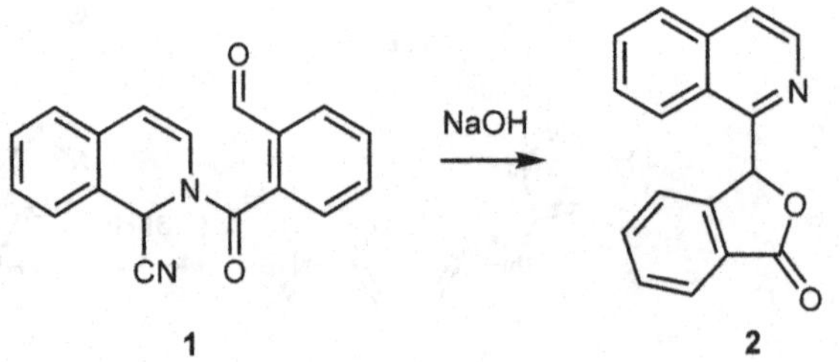

SCHEME 1

SOLUTION

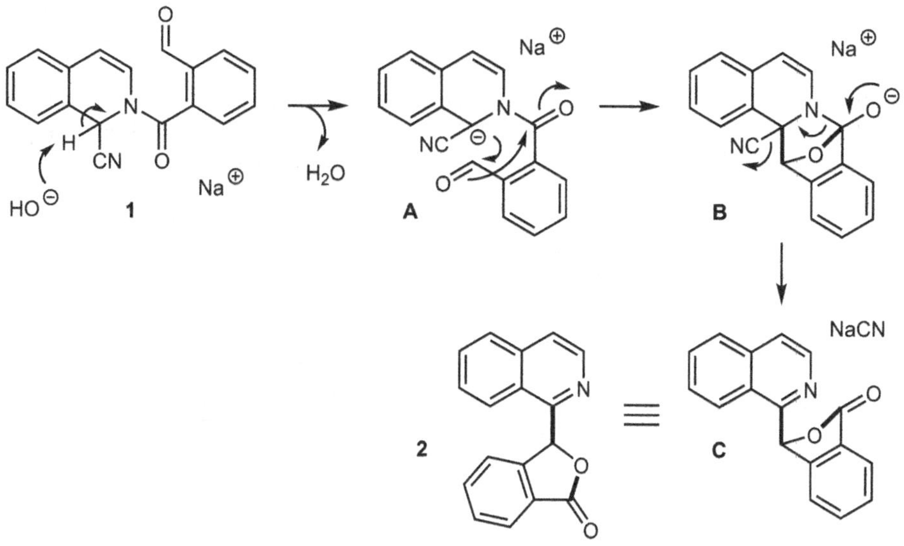

BALANCED CHEMICAL EQUATION

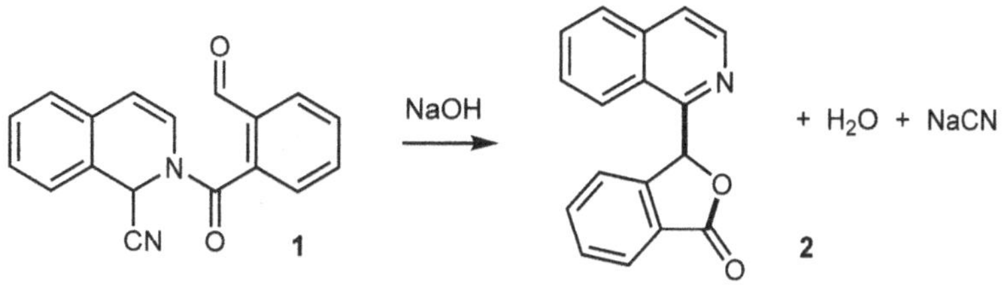

KEY STEPS EXPLAINED

The mechanism proposed by Tyrell and McEwen begins with deprotonation at the carbon atom bearing the cyano group to form carbanion **A** which then cyclizes to the [2.2.1] bicyclic alkoxide intermediate **B**. This intermediate can ketonize with elimination of cyanide anion to generate the phthalide product **2**. Unfortunately, the authors did not conduct a theoretical study to help shed light on the interesting **A** to **B** mechanistic step on which this entire transformation rests.

ADDITIONAL RESOURCES

We direct the reader to two research articles discussing the mechanistic aspects of preparing Reissert compounds and utilizing them in synthesis.[2,3] We also recommend an article by the same authors on whose work the problem is based which describes a Stevens-type rearrangement of open-chain Reissert analogs.[4]

References

1. Tyrell III JA, McEwen WE. Intramolecular reactions of Reissert compounds. *J Org Chem*. 1981;46(12):2476–2479. https://doi.org/10.1021/jo00325a010.
2. Hahn J-T, Kant J, Popp FD, Chhabra SR, Uff BC. Reissert compound studies. LXV [1]. Preparation of Reissert compounds derived from α,β-unsaturated acid chlorides. *J Heterocyclic Chem*. 1992;29(5):1165–1176. https://doi.org/10.1002/jhet.5570290521.
3. Gibson HW, Brumfield KK, Grisle RA, Hermann CKF. Synthesis of heterocyclic monomers via Reissert chemistry. *J Polym Sci A Polym Chem*. 2010;48(17):3856–3867. https://doi.org/10.1002/pola.24172.
4. Stamegna AP, McEwen WE. Stevens-type rearrangement of open-chain analogues of Reissert compounds. Evidence for participation by radical intermediates. *J Org Chem*. 1981;46:1653–1655. https://doi.org/10.1021/jo00321a026.

Question 216: Di-*t*-butylacetylene Does Not Cycloadd to 2-Pyrone

During a 1972 synthetic investigation of Diels-Alder reactions by American chemists, it was shown that reaction between 2-pyrone and di-*t*-butylacetylene did not produce the expected 1,2-di-*t*-butylbenzene product. Instead, the only isolated product was *trans*-cinnamic acid (Scheme 1).[1] Suggest a mechanism for this transformation.

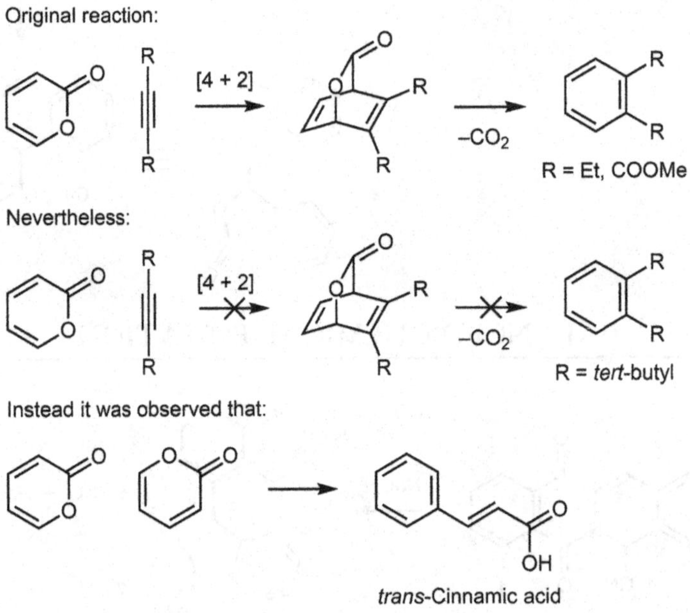

Original reaction:

Nevertheless:

Instead it was observed that:

trans-Cinnamic acid

SCHEME 1

SOLUTION

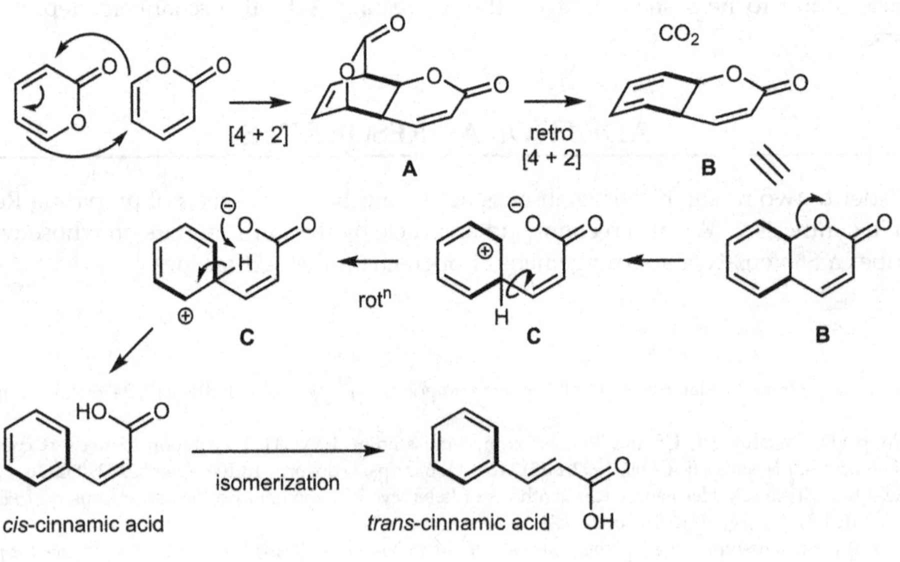

CO$_2$

cis-cinnamic acid   isomerization   *trans*-cinnamic acid

KEY STEPS EXPLAINED

The mechanism proposed by White and Seyferth for the transformation of 2-pyrone and di-*t*-butylacetylene to *trans*-cinnamic acid begins with self-dimerization of 2-pyrone in a standard [4+2] hetero-Diels-Alder cycloaddition leading to a [2.2.2] bicyclic intermediate **A**. Afterward, carbon dioxide is liberated in a retro-[4+2] manner to obtain intermediate **B**. Due to the drastic reaction conditions employed, the lactone thermally ring opens to a zwitterion intermediate **C** which has both a carbocation and a carboxylate group. A final aromatization and isomerization lead to the *trans*-cinnamic acid product observed. The authors argue that the likely reason for why the reaction between 2-pyrone and di-*t*-butylacetylene does not produce the intended 1,2-di-*t*-butylbenzene is because of steric hindrance between the bulky *tert*-butyl groups on the acetylene reagent and 2-pyrone. When the R group on the acetylene reagent is an ethyl or methoxycarbonyl group, the reaction proceeds to the doubly substituted benzene product as intended. Lastly, the isomerization from *cis*-cinnamic acid to its *trans* isomer is likely to proceed via electron delocalization to the carbonyl group to form a carbocation which is stabilized by resonance from the neighboring phenyl group, especially at elevated temperature. Bond rotation and electron delocalization can then afford the *trans*-cinnamic isomer. For a future study we recommend a theoretical analysis to shed light on the energetic demands of this interesting transformation and to help support the postulated mechanism.

ADDITIONAL RESOURCES

We would like to first recommend an article focusing on a theoretical study of the mechanism of the cycloaddition of acetylene to α-pyrone.[2] Next, we also recommend two recent articles and a review discussing the hetero-Diels-Alder and retro-Diels-Alder reactions with an emphasis on mechanism where unexpected results were also present in some cases.[3–5]

References

1. White DL, Seyferth D. The Diels-Alder dimerization of 2-pyrone. *J Org Chem.* 1972;37(22):3545–3546. https://doi.org/10.1021/jo00795a036.
2. Goldstein E, Kallel A, Beauchamp PS. Theoretical study of the mechanism of the cycloaddition of acetylene to α-pyrone: asynchronism and substituent effects. *J Mol Struct (THEOCHEM).* 1987;151:297–305. https://doi.org/10.1016/0166-1280(87)85065-0.
3. Domingo LR, Picher MT, Saez JA. Toward an understanding of the unexpected regioselective hetero-Diels-Alder reactions of asymmetric tetrazines with electron-rich ethylenes: a DFT study. *J Org Chem.* 2009;74:2726–2735. https://doi.org/10.1021/jo802822u.
4. Zhang J, Wu J-L. Tandem intramolecular Diels–Alder/retro-Diels–Alder cycloaddition of 2*H*-chromen-2-one as dienes with the expulsion of CO$_2$. *Chin Chem Lett.* 2016;27(9):1537–1540. https://doi.org/10.1016/j.cclet.2016.03.034.
5. Foster RAA, Willis MC. Tandem inverse-electron-demand hetero-/retro-Diels–Alder reactions for aromatic nitrogen heterocycle synthesis. *Chem Soc Rev.* 2013;42:63–76. https://doi.org/10.1039/c2cs35316d.

Question 217: A Failed Thorpe-Dieckmann Cyclization: "Obvious" Reactions Are Not Always Well Behaved...

In a 1980 synthetic study of benzophenanthridine alkaloids, American chemists treated the nitrile ester **1** with sodium hydride in THF in the hopes that a Thorpe-Dieckmann cyclization would occur. Nevertheless, this did not occur and the naphthopyran derivative **2** was isolated instead in 44% yield (Scheme 1).[1] Propose a mechanism for this transformation.

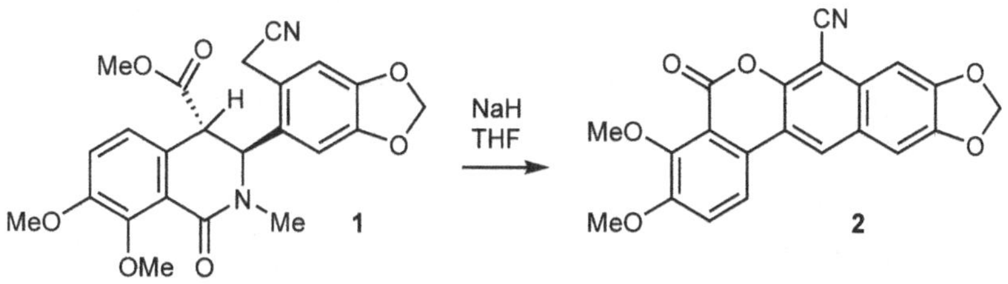

SCHEME 1

SOLUTION

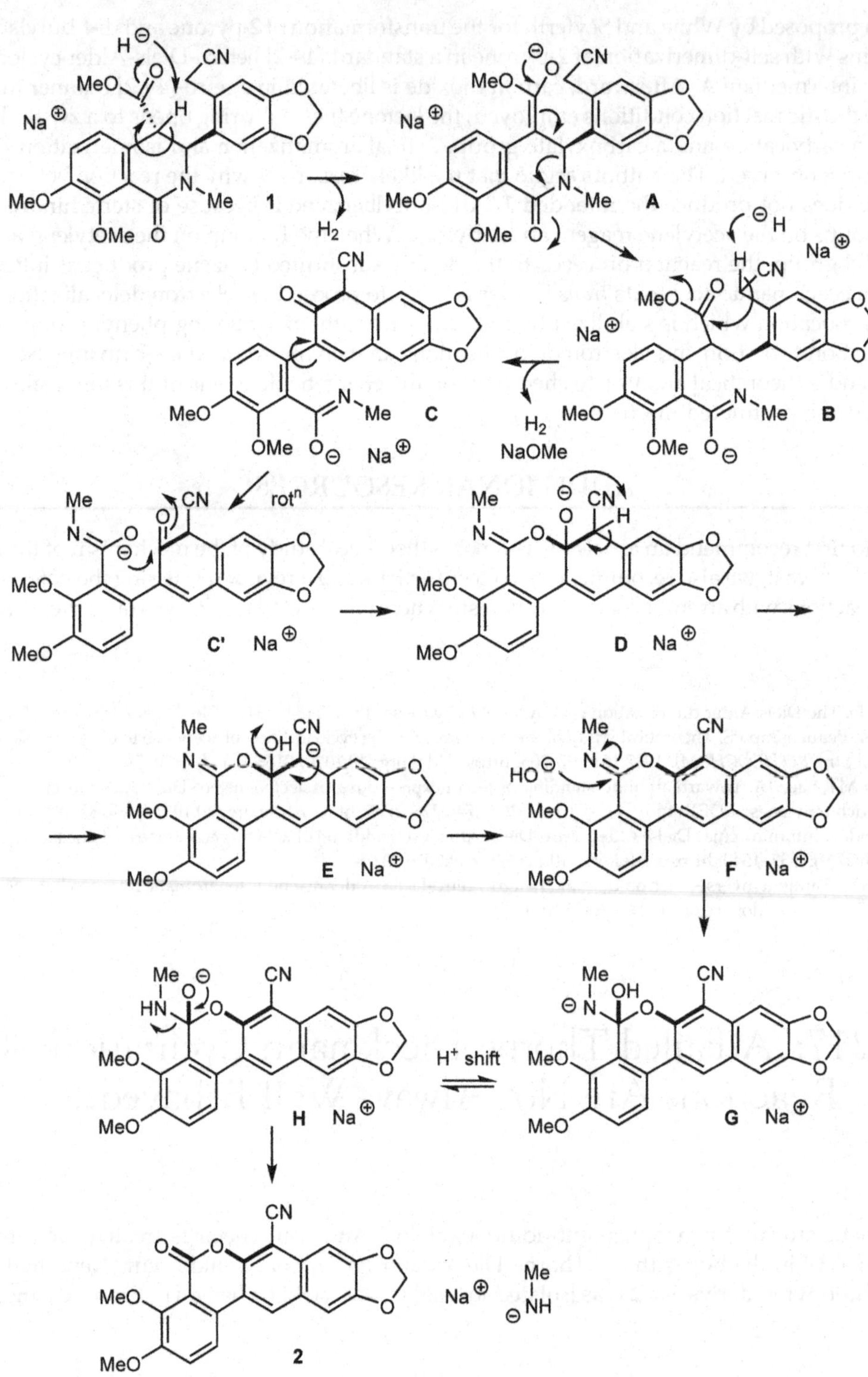

BALANCED CHEMICAL EQUATION

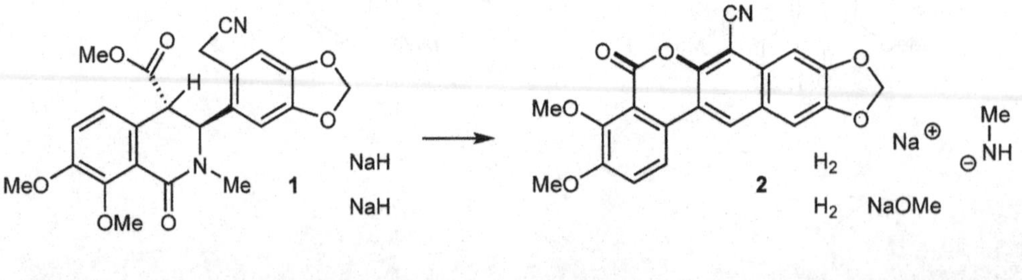

KEY STEPS EXPLAINED

The original intention of the authors was to carry out the simple Thorpe-Dieckmann cyclization shown in Scheme 2.

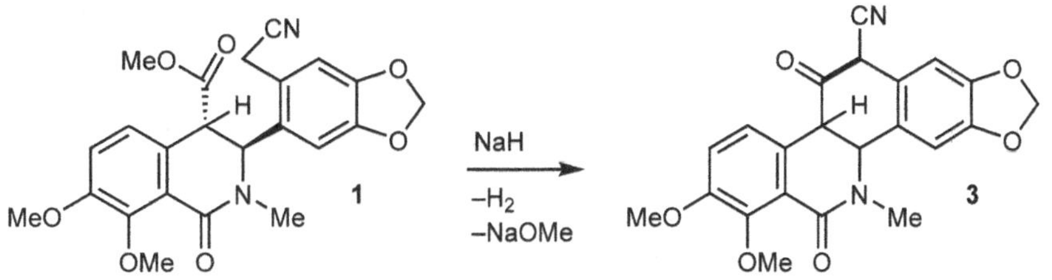

SCHEME 2

Instead, the transformation followed a different pathway to product **2**. The authors suggested that this pathway begins with initial deprotonation at the *alpha* position to the ester group rather than at the methylene group *ortho* to the cyano group as intended by Scheme 2. Ester enolate **A** then fragments to amide enolate **B**. The second equivalent of base now abstracts a proton from the methylene *ortho* to the cyano group resulting in ring closure to intermediate **C**. The aromatic amide enolate group of intermediate **C** then rotates so as to effect ring closure to intermediate **D**. Finally, in the last steps, a series of proton transfers and final elimination of methylamide produce the observed product **2**. It is important to note that the authors do not present any further experimental or theoretical analysis to help support this postulated mechanism.

ADDITIONAL RESOURCES

We refer the reader to an article where a Thorpe-Dieckmann cyclization was itself the unexpected result.[2] Furthermore, we also recommend an article discussing the mechanism of a Thorpe-Ziegler cyclization where unexpected results were observed.[3] Lastly, we refer the reader to a review describing cyclizations of *N*-acyliminium ions similar to intermediate **C'** in the current solution.[4]

References

1. Cushman M, Dikshit DK. Formation of the 5-benzo[d]naphtho[2,3-b]pyran system during an attempted benzophenanthridine synthesis. *J Org Chem*. 1980;45(25):5064–5067. https://doi.org/10.1021/jo01313a010.
2. Kovacs L. Unexpected Thorpe reaction of an α-alkoxynitrile. *Molecules*. 2000;5(2):127–131. https://doi.org/10.3390/50200127.
3. Kim YM, Kim KH, Park S, Kim JN. Synthesis of 3-aminoindole derivatives: Combination of Thorpe–Ziegler cyclization and unexpected allylindium-mediated decyanation. *Tetrahedron Lett*. 2011;52(12):1378–1382. https://doi.org/10.1016/j.tetlet.2011.01.085.
4. Maryanoff BE, Zhang H-C, Cohen JH, Turchi IJ, Maryanoff CA. Cyclizations of *N*-acyliminium ions. *Chem Rev*. 2004;104:1431–1628. https://doi.org/10.1021/y0306182.

Question 218: ...and Can Lead to Remarkable Rearrangements: A Failed Thorpe-Ziegler Cyclization

During an attempted Thorpe-Ziegler cyclization of 2-(3-cyanopropylthio)pyridine-3-carbonitrile **1** to 5-amino-2,3-dihydrothiepino[2,3-b]pyridine-4-carbonitrile **2** by Japanese chemists in 1995, a different product, the thieno[2,3-h][1,6] naphthyridine **3** was instead isolated in 82% yield (Scheme 1).[1] Provide a mechanism for this unexpected result.

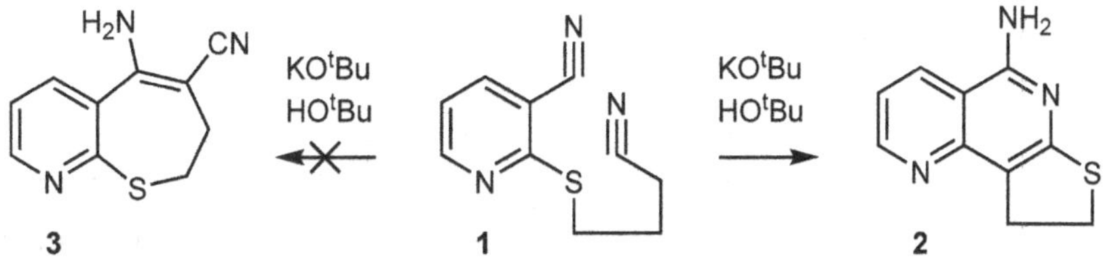

SCHEME 1

SOLUTION

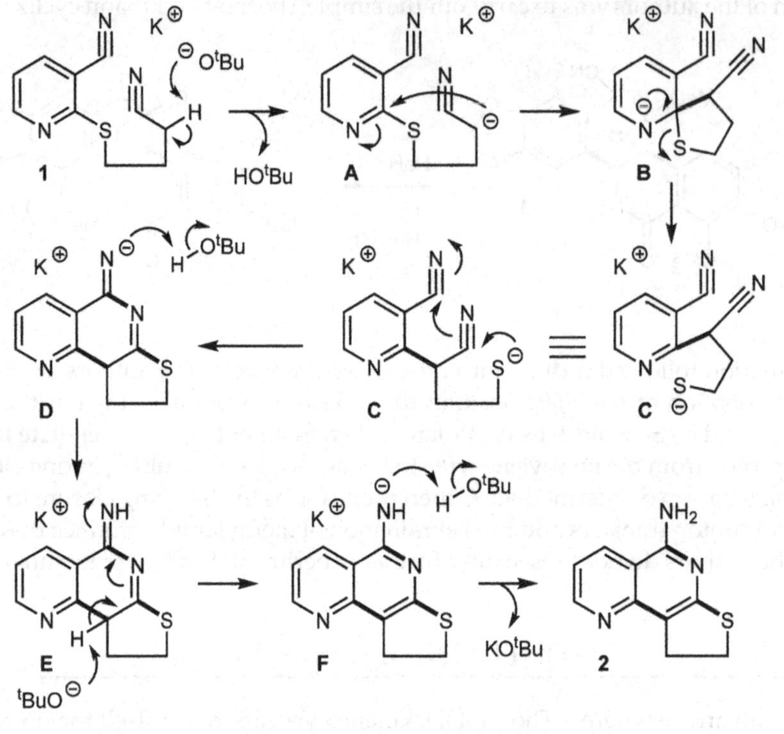

KEY STEPS EXPLAINED

The mechanism proposed by the authors for the unexpected formation of **3** from **1** begins with deprotonation of **1** to form carbanion **A** which cyclizes at the 2-position of the pyridine ring creating a *spiro* intermediate **B**. This structure can then undergo ring opening to the thiol anion **C** which recyclizes to intermediate **D**. The net result is a [(3+3)+(5+0)] ring construction transformation rather than the intended simple [7+0] cyclization resulting from initial cyclization of carbanion **A** at the 3-cyano position (Scheme 2).

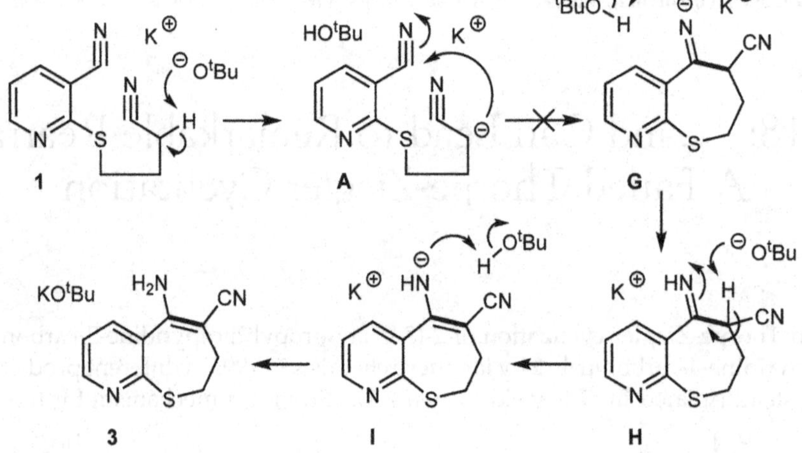

SCHEME 2

ADDITIONAL RESOURCES

We refer the reader to three interesting review articles covering the synthesis and properties of pyridines and naphthyridines.[2–4]

References

1. Sasaki K, Rouf AS, Kashino S, Hirota T. Polycyclic *N*-heterocyclic compounds. 46 synthesis of thieno[2,3-*h*][1,6]naphthyridines from 2-(3-cyano-propylthio)pyridine-3-carbonitrile. *Heterocycles*. 1995;41(6):1307–1318. https://doi.org/10.3987/COM-95-7064.
2. Yurovskaya MA, Mit'kin OD, Zaitseva FV. Functionalization of pyridines. 2. Synthesis of acylpyridines, pyridinecarboxylic acids, and their derivatives. Review. *Chem Heterocycl Compd*. 1998;34(8):871–899. https://doi.org/10.1007/BF02311322.
3. Litvinov VP, Roman SV, Dyachenko VD. Naphthyridines. Structure, physicochemical properties and general methods of synthesis. *Russ Chem Rev*. 2000;69(3):201–220. https://doi.org/10.1070/RC2000v069nABEH000553.
4. Litvinov VP, Krivokolysko SG, Dyachenko VD. Synthesis and properties of 3-cyanopyridine-2(1*H*)-chalcogenones. Review. *Chem Heterocycl Compd*. 1999;35(5):509–540. https://doi.org/10.1007/BF02324634.

Question 219: A Remarkably Stable Tertiary Alcohol by Solvolysis of a Primary Tosylate

In 1999 British chemists synthesized a remarkably stable tertiary alcohol fortesol **3** by treating nopol tosylate **1** with acetic acid and aqueous work-up according to Scheme 1.[1] The authors discussed several possible mechanisms for this transformation. Outline these mechanisms and discuss the operational likelihood of each.

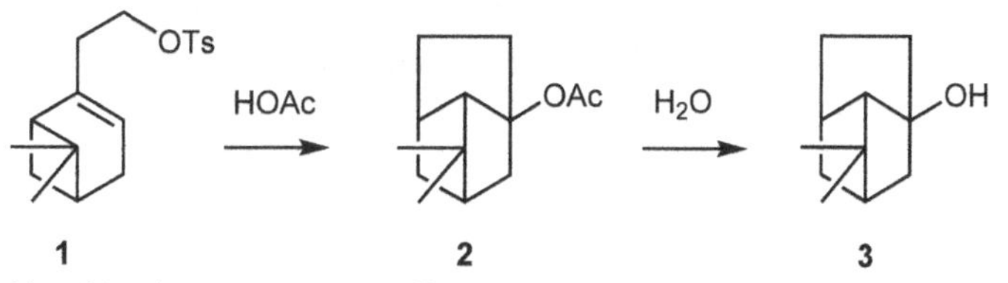

1
Nopol tosylate

2
Fortesol acetate

3
Fortesol

SCHEME 1

SOLUTION: MECHANISM A—BORNYL PATH (HELICOPTER VIEW)

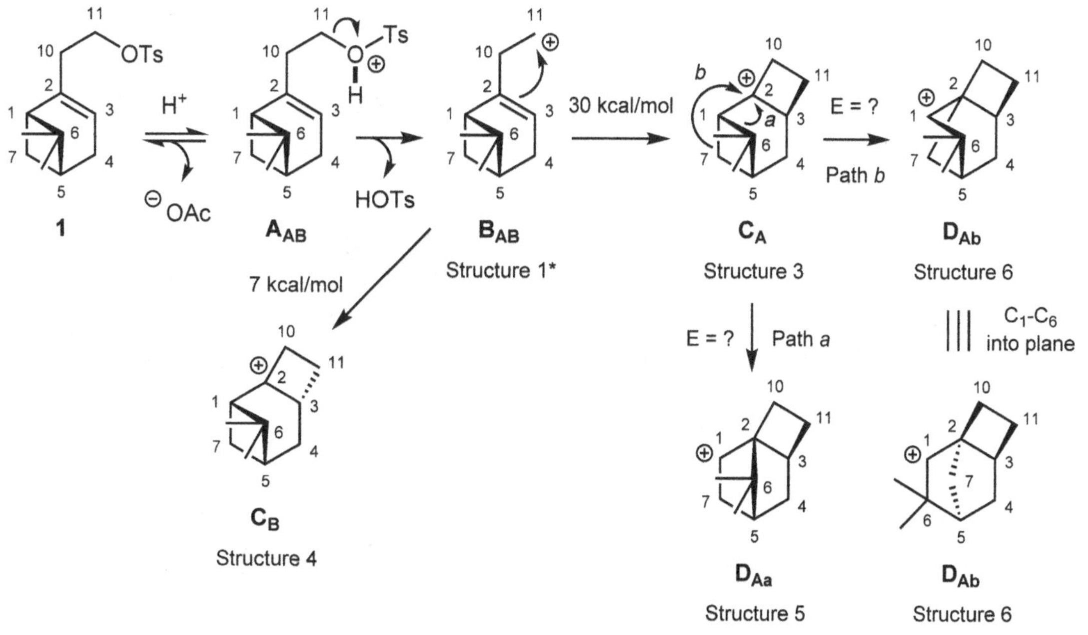

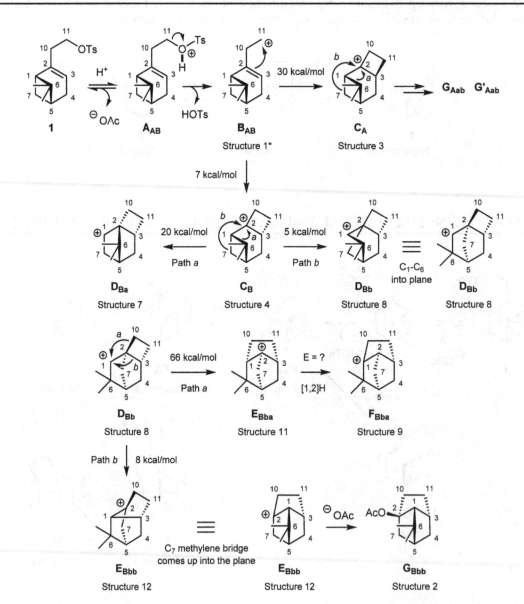

SOLUTION: MECHANISM B—FENCHYL PATH (HELICOPTER VIEW)

KEY STEPS EXPLAINED

To further emphasize the importance of visual perspective, we present the two postulated mechanistic pathways in a 3D-like view to help highlight important geometrical and stereochemical considerations (see Schemes 2 and 3 for mechanisms A and B, respectively). We note that the energy values depicted for each step, where available, were obtained from the work of the original authors who utilized the MOPAC PM3 method for all their theoretical calculations.[2]

Without further ado, our mechanistic analysis begins with recognizing that the first steps for both mechanisms constitute acid-promoted elimination of toluene-*p*-sulfonate group from the starting nopol tosylate **1** to form the common cationic intermediate $\mathbf{B_{AB}}$ (structure 1* in the authors' paper).[1] Cyclization of the nopyl side chain of $\mathbf{B_{AB}}$ then leads to formation of a second cyclobutyl ring which may either be oriented toward the *gem*-dimethyl bridge ($\mathbf{C_A}$) or away from the *gem*-dimethyl bridge ($\mathbf{C_B}$). According to the authors, this junction represents the main divergence point between the two proposed mechanisms with structure $\mathbf{C_A}$ giving the *exo* orientation (*syn*) between the newly formed cyclobutyl ring and the *gem*-dimethyl bridge and structure $\mathbf{C_B}$ giving the *endo* orientation (*anti*). At this initial junction, theoretical calculations indicated that formation of the *exo* product $\mathbf{C_A}$ requires an energy barrier of 30 kcal/mol whereas formation of the *endo* product $\mathbf{C_B}$ requires 7 kcal/mol, a difference arising mainly from steric factors and which favors mechanism B at least kinetically. Following mechanism A, we proceed from $\mathbf{C_A}$ with either a shift by the methylene bridge to form $\mathbf{D_{Ab}}$ (structure 6, along path *b*) or a *gem*-dimethyl bridge shift to form $\mathbf{D_{Aa}}$ (structure 5, along path *a*). Although theoretical calculations are not provided

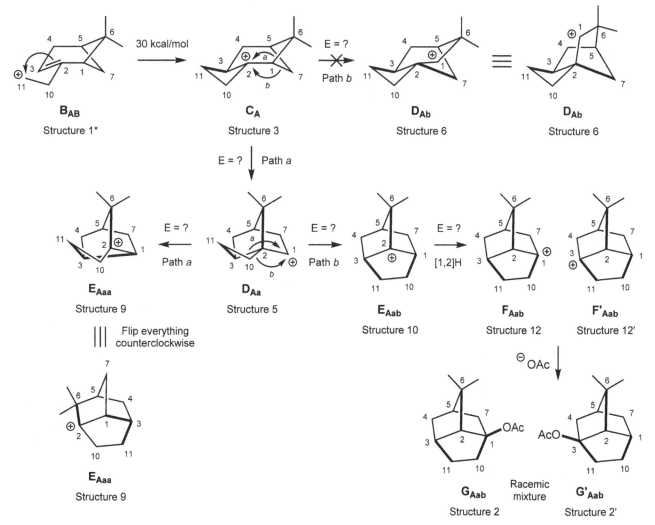

SCHEME 2 3D-like view of mechanism A.

for this junction, the authors believe that C_2 and C_7 are too far apart to form a C—C bond to give $\mathbf{D_{Ab}}$ (an expected cationic species easily stabilized by the neighboring *gem*-dimethyl group), even though bond distances are not provided. Instead, they argue that despite steric constraints, the strained bornyl intermediate $\mathbf{D_{Aa}}$ is formed instead.

Mechanism A then proceeds either through rearrangement to form the symmetrical bridgehead cationic species $\mathbf{E_{Aab}}$ (structure 10, along path *b*) or the cationic species $\mathbf{E_{Aaa}}$ (structure 9, along path *a*). Due to its symmetrical nature, $\mathbf{E_{Aab}}$ may form either $\mathbf{F_{Aab}}$ or $\mathbf{F'_{Aab}}$ after a [1,2]-hydrogen shift which then reacts with acetate anion to form a racemic mixture of the fortesol acetate product 2. As such, according to mechanism A, the final product, being a racemic mixture, should not be optically active. Nevertheless, the final product obtained by the authors was optically active thus throwing further doubt on the plausibility of mechanism A.

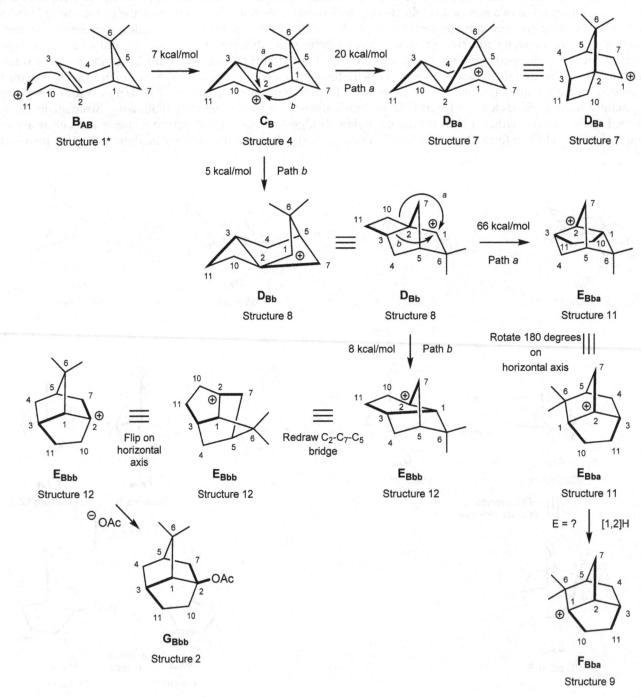

SCHEME 3   3D-like view of mechanism **B**.

As such, the authors considered mechanism B which begins with formation of the less sterically hindered *endo* cationic intermediate C_B (structure 4) from the common intermediate B_{AB}. Once again, structure C_B may either react via a methylene bridge shift to form the fenchyl D_{Bb} (path *b*, structure 8) or via a *gem*-dimethyl shift to form D_{Ba} (path *a*, structure 7). In this case, the authors once again invoke the unrealizable bond formation between C_2 and C_6 as being responsible for preventing the expected *gem*-dimethyl bridge shift (which was operative in mechanism A) from taking place to form D_{Ba}. Instead, an unexpected methylene bridge shift occurs via path *b* to form D_{Bb} which has the methylene cation stabilized inductively by the neighboring *gem*-dimethyl group (the exact opposite result compared to mechanism A). Indeed, theoretical calculations performed by the authors supported this line of reasoning even though the bond distance between C_2 and C_6 was absent from the paper. From D_{Bb}, the mechanism proceeds with another possible junction to form either the energetically unfavorable bridgehead cationic species E_{Bba} (structure 11, along path *a*) or the cationic intermediate E_{Bbb} (structure 12, along path *b*) which was shown to be much more energetically favorable by theoretical calculations. Reaction of E_{Bbb} with acetate anion provides the sole enantiomer 2 thus leading to the optically active product species observed experimentally by the authors.

To eliminate the effects of ion pairs which may lead to observed optical activity, the authors conducted a deuterium labeling experiment which provided the sole product 4 (in accordance with mechanism B) as opposed to the product mixture 4 and 5 which would have been expected if mechanism A were operational (Scheme 4). Since different carbon skeletal rearrangements occur by the different pathways, another possible labeling experiment is the use of ^{13}C labeling as shown in Scheme 5. We also include a depiction of *exo* and *endo* denomination to facilitate this understanding (see Fig. 1). Lastly, we note that the authors also suggested a third mechanism C which involves only electron delocalization steps as opposed to concrete step-wise transformations as seen in mechanisms A and B. This mechanism was discounted because it also involved a symmetrical bridgehead cationic species like E_{Aab} that leads to a racemic mixture of products. We omit it here because we believe it does not provide value to the solution.

Finally, although the authors performed sufficient experimental and theoretical analysis to allow them to conclude with confidence that mechanism B is operational, we do criticize their lack of transparency with regard to providing full details with respect to their theoretical analysis, especially with regard to the complete lack of theoretical calculations for the steps involved in mechanism A. We believe that such an analysis along with a discussion of bond lengths and angles would greatly facilitate one's understanding of the geometrical and stereochemical considerations relevant to this solution. Lastly, we advise the interested reader to consider the carbon skeletal rearrangements occurring in both mechanisms A and B and how the same structures (e.g., E_{Aaa} versus F_{Bba}) contain different carbon atom connections. A future investigation of this transformation would also preferably focus on structures 6, 7, and 9 which the authors did not spend sufficient time discussing.

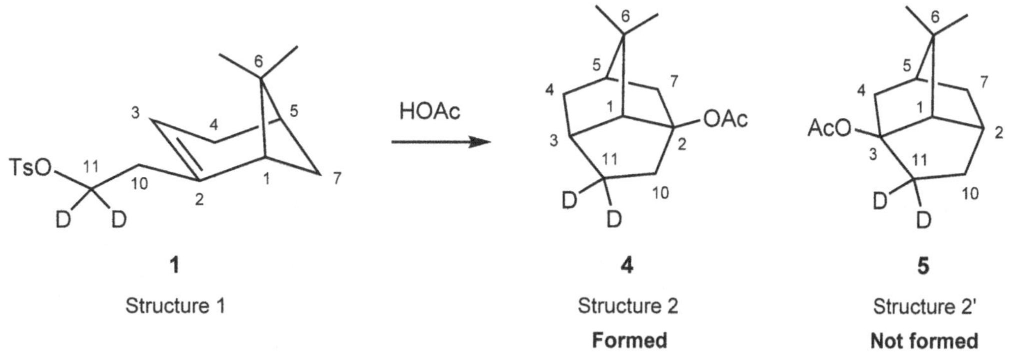

SCHEME 4

ADDITIONAL RESOURCES

For an interesting rearrangement involving the same nopol system involved in this solution, see the work of Abraham et al.[3] Also, for a discussion of the literature surrounding such chemical systems, see a review article by Grayson.[4]

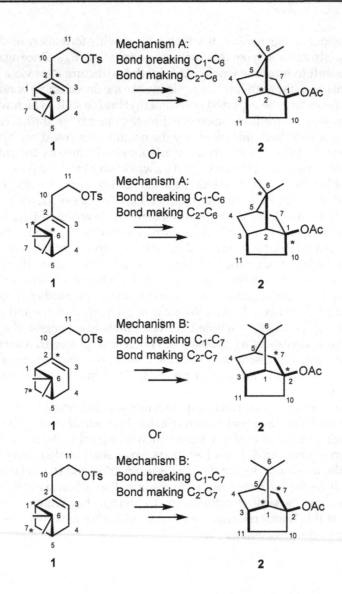

SCHEME 5

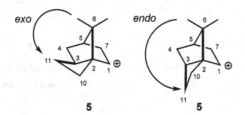

exo = group is oriented toward the bridgehead
endo = group is oriented away from the bridgehead

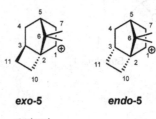

strained

FIG. 1   3D-like and helicopter representations of exo-**5** and endo-**5**.

References

1. Al-Qallaf FAH, Fortes AG, Johnstone RAW, Thompson I, Whittaker D. The mechanism of formation of 8,8-dimethyl[4.2.1.03,7]nonan-6-yl acetate (fortesyl acetate) during acetolysis of nopyl toluene-*p*-sulfonate. *J Chem Soc Perkin Trans 2.* 1999;1999:789–794. https://doi.org/10.1039/A809268K.
2. Stewart JJP. Optimization of parameters for semiempirical methods I. Method. *J Comput Chem.* 1989;10(2):209–220. https://doi.org/10.1002/jcc.540100208.
3. Abraham RJ, Jones-Parry R, Giddings RM, Guy J, Whittaker D. 3α-Acetoxy-6,6-dimethylbicyclo[3.1.1]heptane-2-spiro-1'-cyclopropane from nopylamine deamination. *J Chem Soc Perkin Trans 1.* 1998;1998(4):643–646. https://doi.org/10.1039/A707911G.
4. Grayson DH. Monoterpenoids. *Nat Prod Rep.* 1997;1997(5):477–522. https://doi.org/10.1039/NP9971400477.

Question 220: "Anionic Activation" for the Preparation of Fluoroheterocycles

Using a novel approach of "anionic activation" of trifluoromethyl groups, Russian chemists performed in 1994 the three transformations depicted in Scheme 1.[1] Provide mechanisms for these transformations.

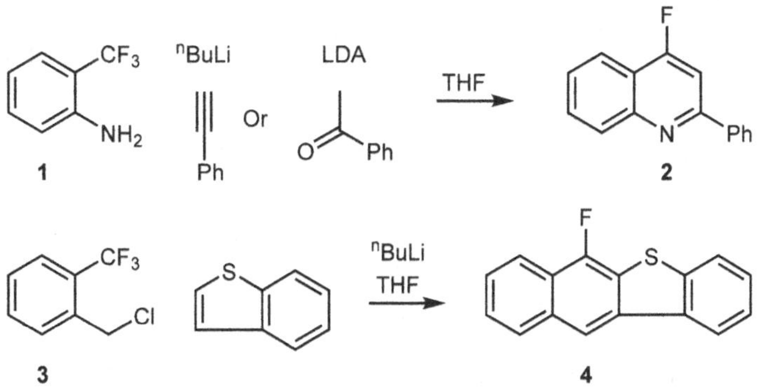

SCHEME 1

SOLUTION: 1 TO 2 VIA PHENYLACETYLENE

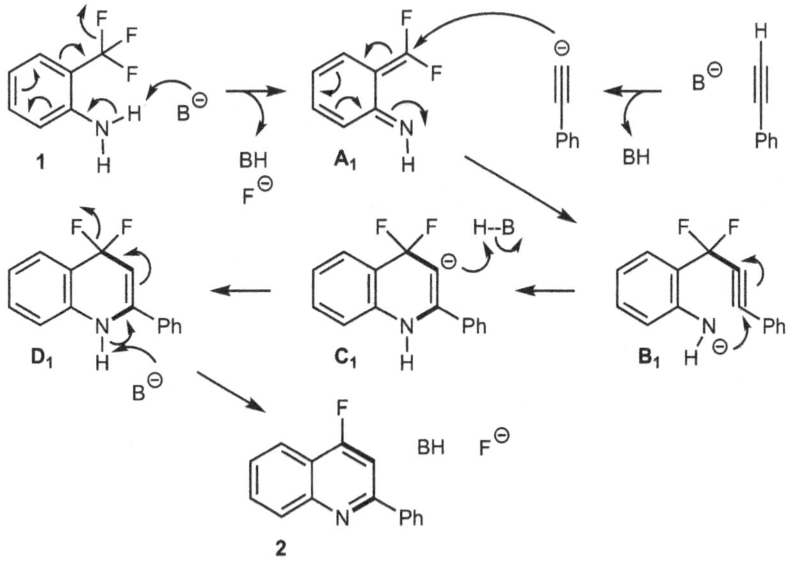

BALANCED CHEMICAL EQUATION: 1 TO 2 VIA PHENYLACETYLENE

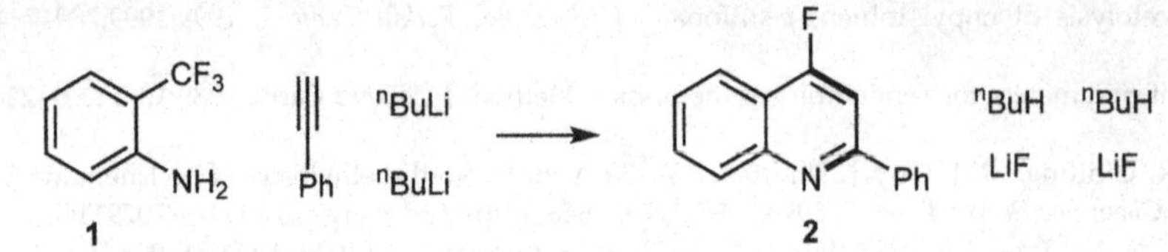

SOLUTION: 1 TO 2 VIA ACETOPHENONE

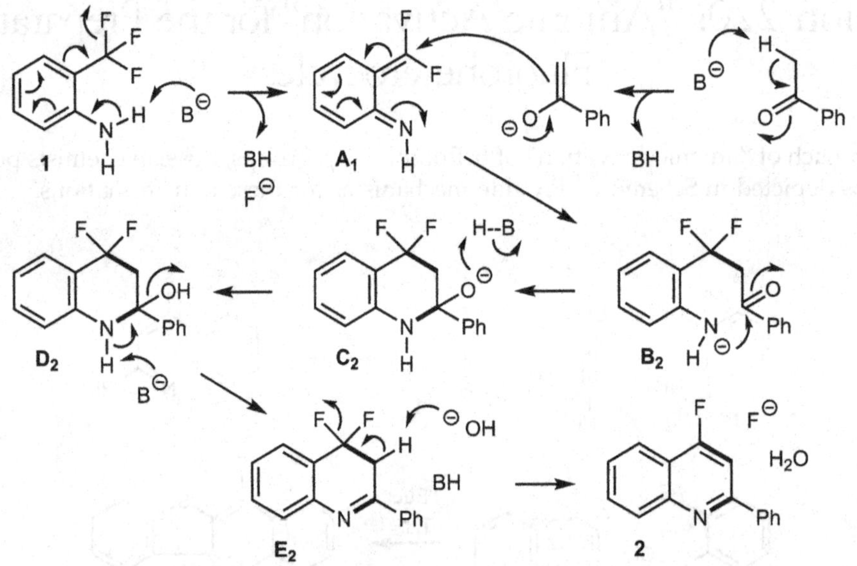

BALANCED CHEMICAL EQUATION: 1 TO 2 VIA ACETOPHENONE

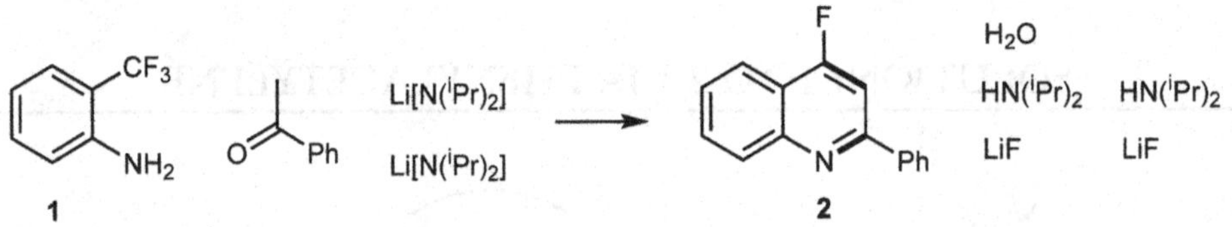

SOLUTION: 3 TO 4

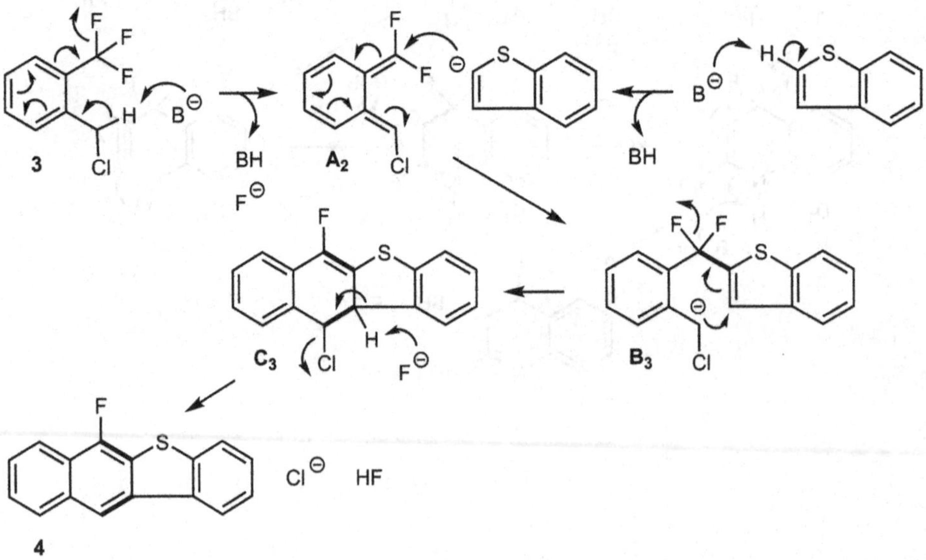

BALANCED CHEMICAL EQUATION: 3 TO 4

KEY STEPS EXPLAINED

All three transformations performed by the authors follow the same mechanistic pattern: (i) initial base promoted generation of orthoquinomethide intermediates (A_1 or A_2) from precursors **1** or **3**, (ii) capture of A_1 or A_2 by carbanions (phenylacetylide, acetophenone enolate, or benzothiophene carbanion), (iii) cyclization to quinoline products (**2**) or fused aromatic products (**4**). The ring construction strategy employed in all three examples is [4+2] cycloaddition. The quinoline syntheses may be contrasted with the following named reactions: (a) [3+3] Combes, (b) [3+2+1] Doebner, (c) [4+2] Friedländer, and (d) [3+3] Skraup. The synthetic advantage of the method presented by the authors is the facile construction of fluorine-substituted quinolines where the fluorine appears in the fused pyridine ring. Lastly, the authors also considered two other possible mechanisms for the transformation of **3** to **4** but had to rule them out because they conflicted with the experimental observation that product **4** was isolated in high yield without the presence of any isomers.

ADDITIONAL RESOURCES

We direct the reader to several literature examples involving mechanistic discussion of very similar fluorine and quinoline containing heterocyclic systems.[2–6]

References

1. Kiselyov AS, Strekowski L. An unconventional synthetic approach to fluoro heteroaromatic compounds by a novel transformation of an anionically activated trifluoromethyl group. *Tetrahedron Lett.* 1994;35(41):7597–7600. https://doi.org/10.1016/S0040-4039(00)78352-3.
2. Torres JC, Garden SJ, Pinto AC, da Silva FSQ, Boechat N. A synthesis of 3-fluoroindoles and 3,3-difluoroindolines by reduction of 3,3-difluoro-2-oxindoles using a borane tetrahydrofuran complex. *Tetrahedron.* 1999;55:1881–1892. https://doi.org/10.1016/S0040-4020(98)01229-0.
3. Kiselyov AS. Unexpected behavior of imines derived from trifluoromethylaryl ketones under basic conditions: convenient synthesis of 2-arylbenzimidazoles and 2-arylbenzoxazoles. *Tetrahedron Lett.* 1999;40:4119–4122. https://doi.org/10.1016/S0040-4039(99)00632-2.
4. Erian AW. Recent trends in the chemistry of fluorinated five and six-membered heterocycles. *J Heterocyclic Chem.* 2001;38(4):793–808. https://doi.org/10.1002/jhet.5570380401.
5. Marco-Contelles J, Perez-Mayoral E, Samadi A, Carreiras MC, Soriano E. Recent advances in the Friedländer reaction. *Chem Rev.* 2009;109:2652–2671. https://doi.org/10.1021/cr800482c.
6. Wojciechowski K. Aza-*ortho*-xylylenes in organic synthesis. *Eur J Org Chem.* 2001;2001:3587–3605. https://doi.org/10.1002/1099-0690(200110)2001:19<3587::AID-EJOC3587>3.0.CO;2-5.

Question 221: Synthesis of a 5-(2-Quinolyl)pyrimidine

During a 1995 investigation by American chemists, the synthesis of the 5-(2-quinolyl)pyrimidine **D** was carried out by means of the synthetic sequence depicted in Scheme 1.[1] Provide a mechanism for the transformation of intermediate **B** to intermediate **C**.

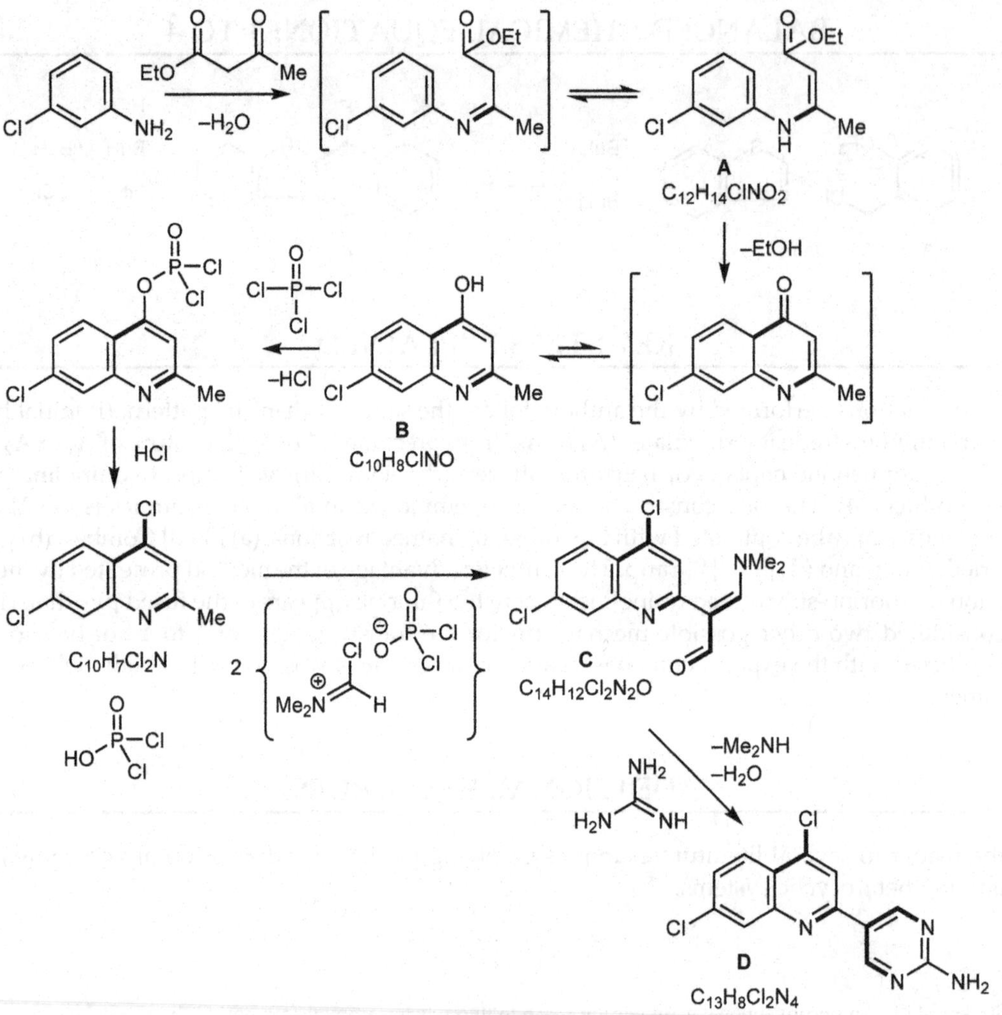

SCHEME 1

SOLUTION: B TO C

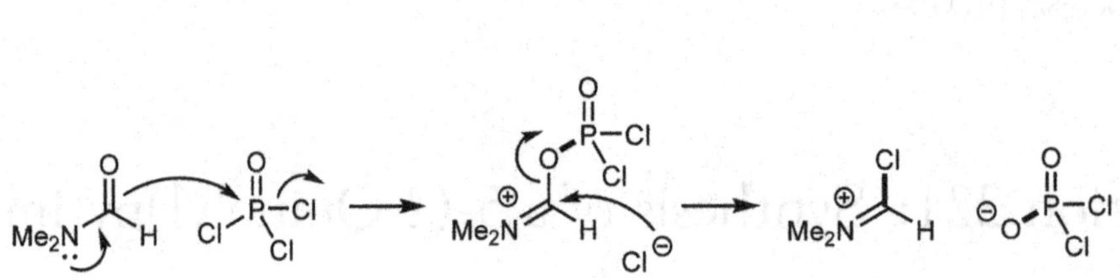

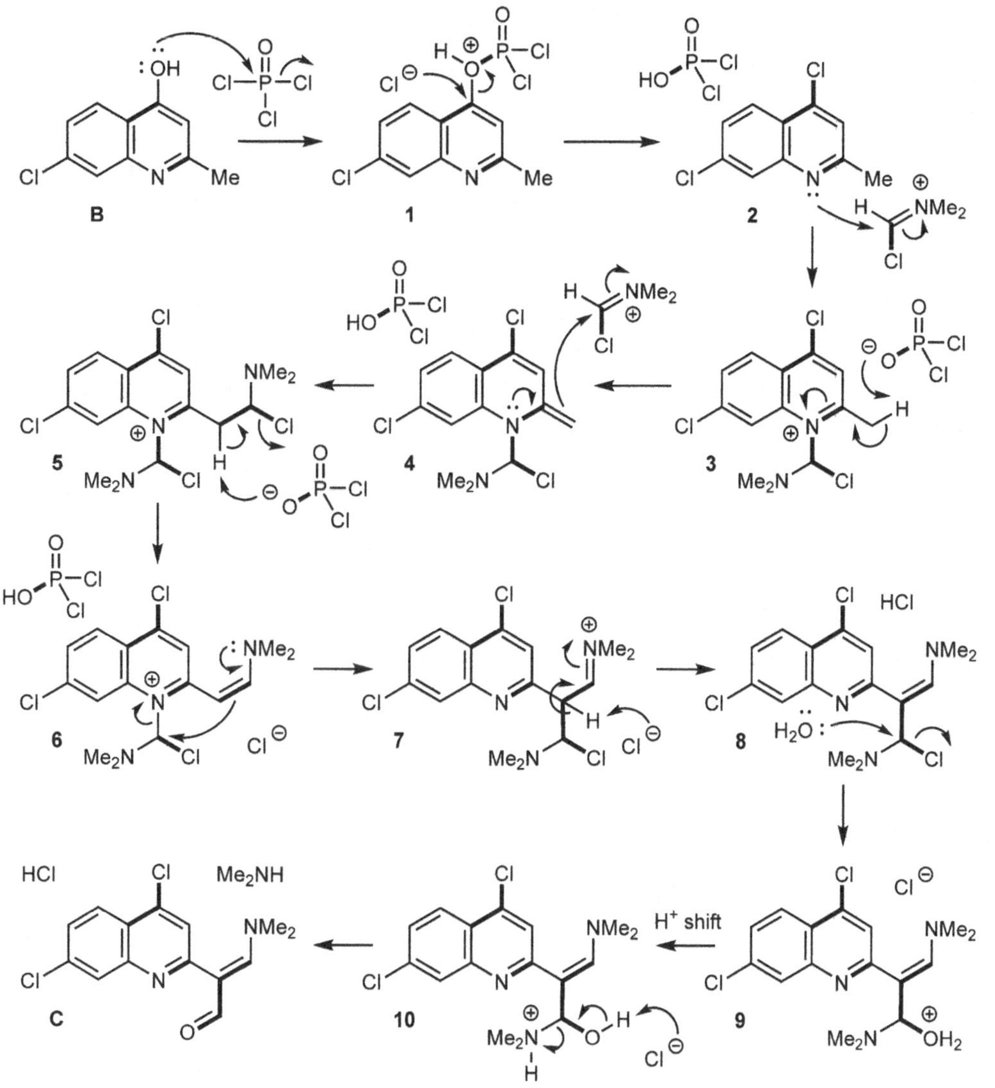

BALANCED CHEMICAL EQUATION: B TO C

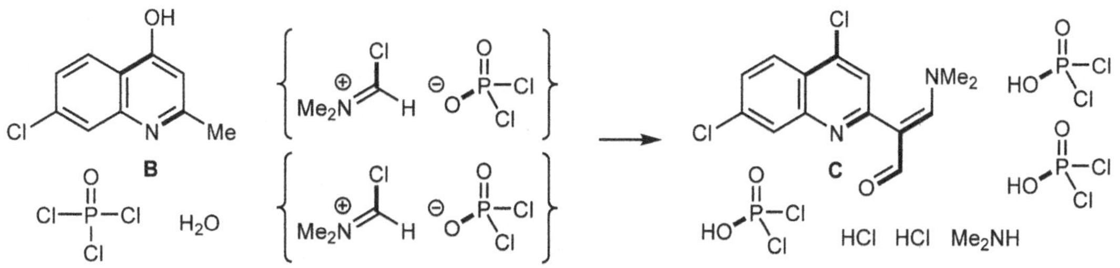

KEY STEPS EXPLAINED

The mechanism for the transformation of **B** to **C** starts with a substitution of the phenolic hydroxyl group for chlorine using phosphorus oxychloride thus leading from **B** to intermediate **1** to the substitution intermediate **2**. The authors believe that the reactivity at C_2 of **B** is enhanced by conjugation with the ring nitrogen atom. Next, the two equivalents of the Vilsmeier-Haack reagent are used to construct intermediate **6**, which then rearranges to **7**. This is then dechlorinated via hydrolysis to intermediate **9** at which point one of the dimethylamino groups is cleaved generating an aldehyde group as part of an α,β-unsaturated electrophilic moiety in product **C** that can be captured by guanidine in a [3+3] cyclization to give the final product **D**. In the final product target bond mapping, we see that in this synthesis sequence the quinoline and the pyrimidine ring systems are made by [3+3] and [3+1+1+1] ring construction strategies, respectively. Nevertheless, the mechanism of this transformation is by no means established as the authors point out, and therefore we would recommend future experimental and theoretical investigation in order to help support the postulated mechanism.

ADDITIONAL RESOURCES

We direct the reader to two mechanistic examples in the literature exploring transformations of a similar nature to that encountered in this problem.[2,3]

References

1. De D, Mague JT, Byers LD, Krogstad DJ. Synthesis of (*E*)-2-(4,7-dichloroquinolin-2-yl)-3-dimethylamino-2-propene-1-al and its use as a synthetic intermediate. *Tetrahedron Lett*. 1995;36(2):205–208. https://doi.org/10.1016/0040-4039(94)02248-A.
2. Dominguez E, Ibeas E, Marigorta EM, Palacios JK, SanMartin R. A convenient one-pot preparative method for 4,5-diarylisoxazoles involving amine exchange reactions. *J Org Chem*. 1996;61(16):5435–5439. https://doi.org/10.1021/jo960024h.
3. Shaker RM. Synthesis and characterization of some new 4,4'-(1,4-phenylene)dipyrimidine and 6,6'-(1,4-phenylene)-di(pyridin-2(1*H*)-one) derivatives. *Heteroat Chem*. 2005;16(6):507–512. https://doi.org/10.1002/hc.20150.

Question 222: "Obvious" and "Nonobvious" Pathways to a Highly Substituted Pyridine and Aniline

In 1995 Spanish chemists reported that the condensation of (*E*)-4-phenylbut-3-en-2-one **1** and malononitrile **2** in boiling sodium methoxide/methanol gave a mixture of the pyridine **3** and the aniline **4** (Scheme 1).[1] In light of observations that relative stereochemistry and amounts of **1** and **2** played a role in the transformation, obvious and nonobvious pathways were proposed. Show both the obvious and nonobvious mechanisms for transforming **1** and **2** to products **3** and **4**.

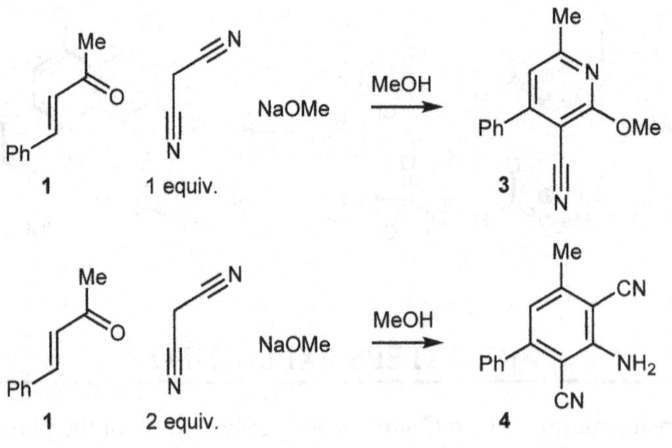

SCHEME 1

SOLUTION: FORMATION OF 3

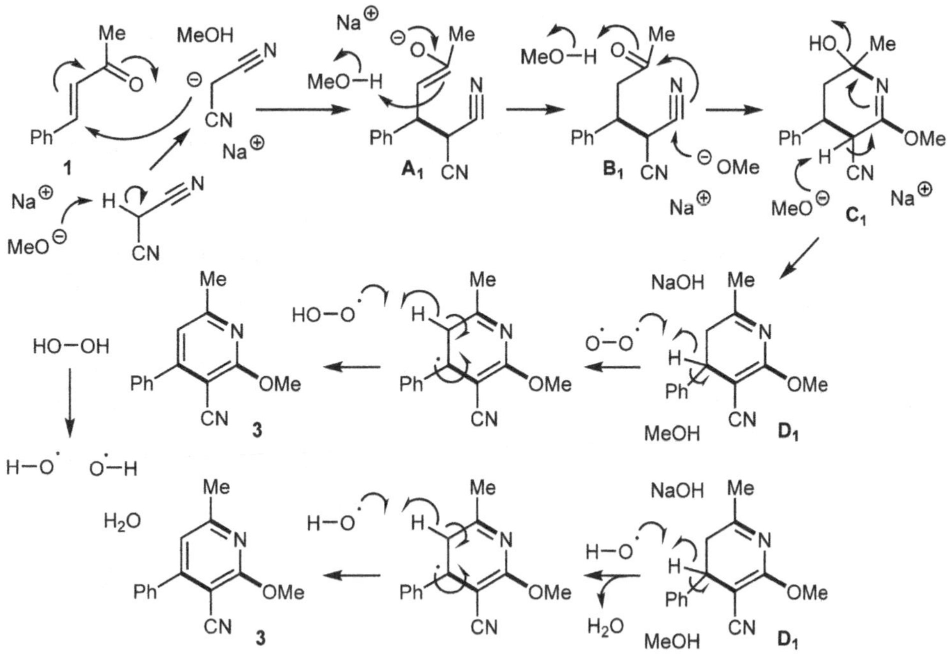

BALANCED CHEMICAL EQUATION FOR FORMATION OF 3

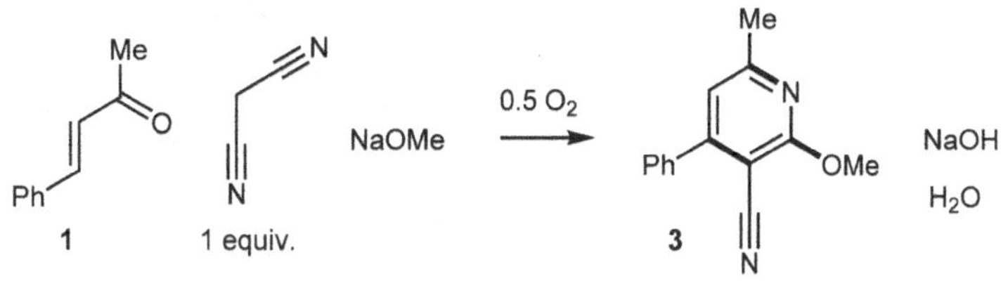

SOLUTION: FORMATION OF 4 OBVIOUS PATHWAY

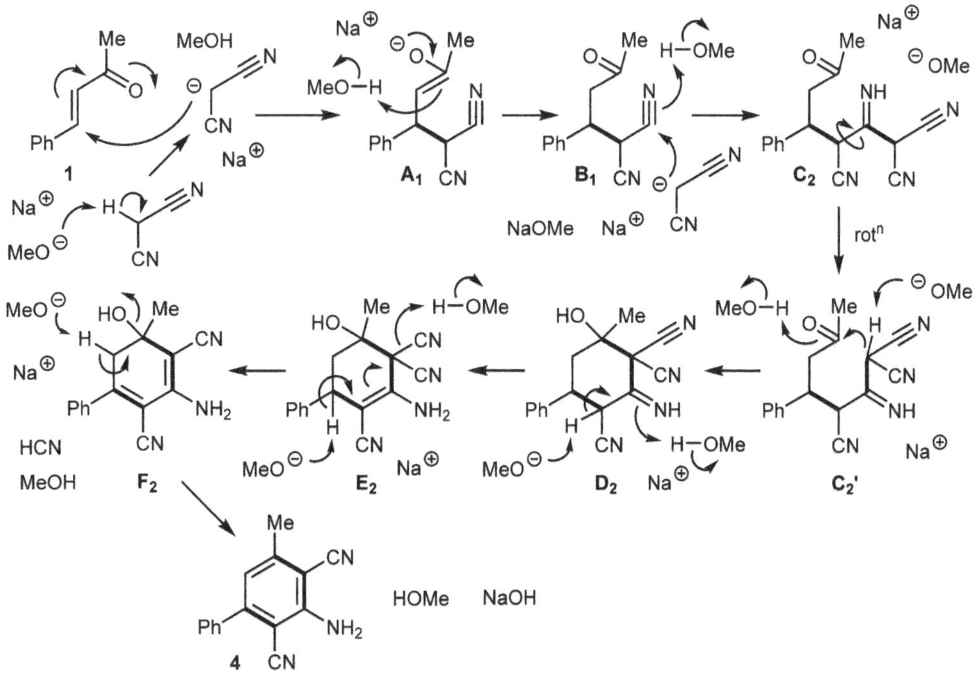

BALANCED CHEMICAL EQUATION FOR FORMATION OF 4 OBVIOUS PATHWAY

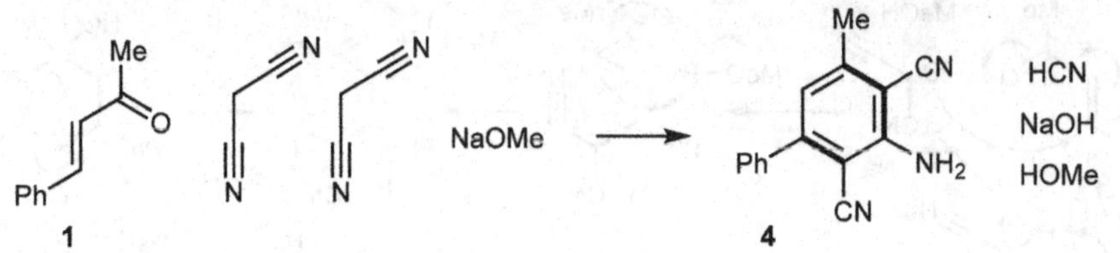

SOLUTION: FORMATION OF 4 NONOBVIOUS PATHWAY

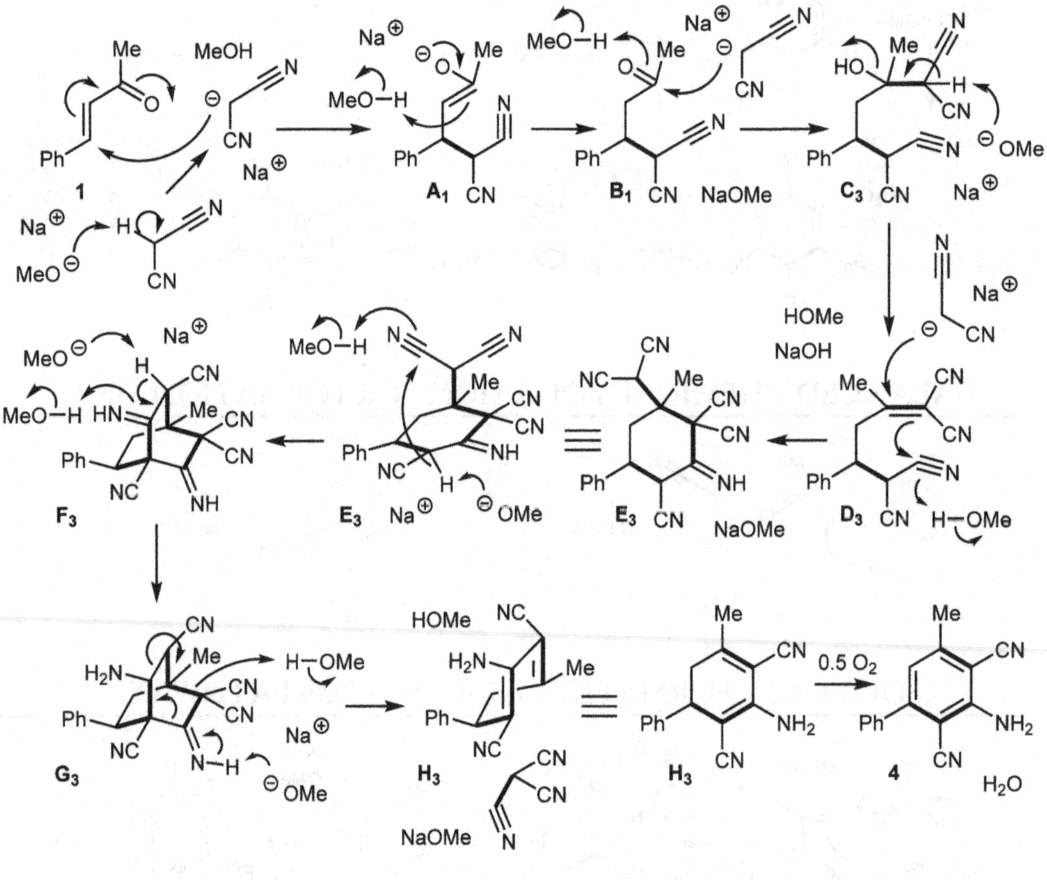

BALANCED CHEMICAL EQUATION FOR FORMATION OF 4 NONOBVIOUS PATHWAY

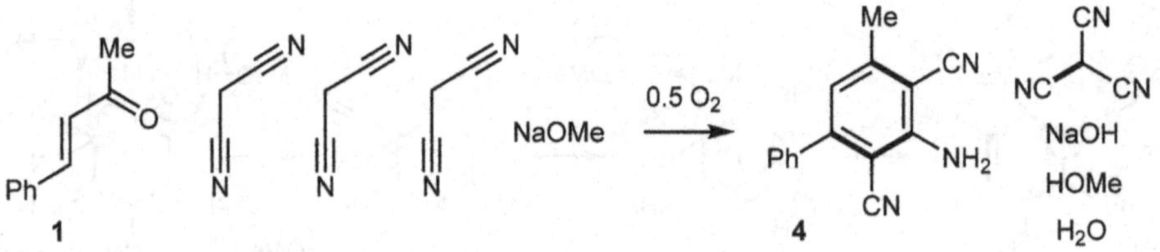

KEY STEPS EXPLAINED

The mechanism proposed by the authors for the formation of **3** begins with generation of the anion of malononitrile **2** via sodium methoxide deprotonation. This anion can then attack the carbon atom *beta* to the carbonyl group in a Michael addition to form intermediate A_1 which may ketonize to B_1. At this point methoxide anion can attack the nitrile group causing cyclization to the six-membered ring intermediate C_1 which can be deprotonated to form intermediate D_1. Air oxidation of this intermediate helps establish the aromaticity in the pyridine ring giving product 3. We note that one equivalent of malononitrile is needed for this [3 + 3] ring construction strategy from 1 to 3.

With regard to the "obvious" pathway to 4, the authors propose the same beginning Michael addition sequence to intermediate B_1 where one equivalent of malononitrile is consumed. Afterward, a second Michael addition of the anion of a second equivalent of malononitrile occurs at the nitrile carbon of B_1 which after protonation forms intermediate C_2. Bond rotation to C_2' sets up a Knoevenagel condensation to D_2 which after proton shift to E_2 and elimination of hydrogen cyanide and sodium hydroxide generates the final product 4 by an overall [3 + 2 + 1] ring construction strategy. In this case, air oxidation is not required.

After several experiments, the authors proposed a "nonobvious" mechanistic route to product 4 which differs significantly from the "obvious" route. Here, a different [3 + 2 + 1] ring construction mapping for 4 is obtained by utilizing three equivalents of malononitrile 2. Essentially, the mechanism begins with the same Michael addition sequence to B_1, only this time we have a Knoevenagel condensation of B_1 with the anion of malononitrile to generate the intermediate C_3 (i.e., attack at the carbonyl group instead of at the nitrile group as represented in the "obvious" route). Dehydration of C_3 leads to D_3 which may undergo a Thorpe cyclization via addition of a third equivalent of the anion of malononitrile to D_3 and thus form E_3. Intermediate E_3 undergoes a second Thorpe cyclization to F_3 which after a prototropic rearrangement to G_3 and base-mediated elimination of 1,1,1-tricyanomethane by-product to H_3 undergoes a final air oxidation to product 4.

In order to support this alternative "nonobvious" pathway, the authors presented several pieces of experimental evidence. For example, when the reaction was carried out under milder basic conditions (sodium bicarbonate in methanol at room temperature versus boiling sodium methoxide in methanol), the authors observed: (a) isolation and spectroscopic characterization of intermediate G_3; (b) confirmation of the structure of G_3 by carrying out its retro-Diels-Alder thermal degradation to styrene and compound 5 (see Scheme 2); (c) isolation, spectroscopic characterization, and X-ray analysis of penultimate intermediate H_3; and (d) demonstration that product 4 could be obtained under the same reaction conditions from either isolated G_3 or isolated H_3.

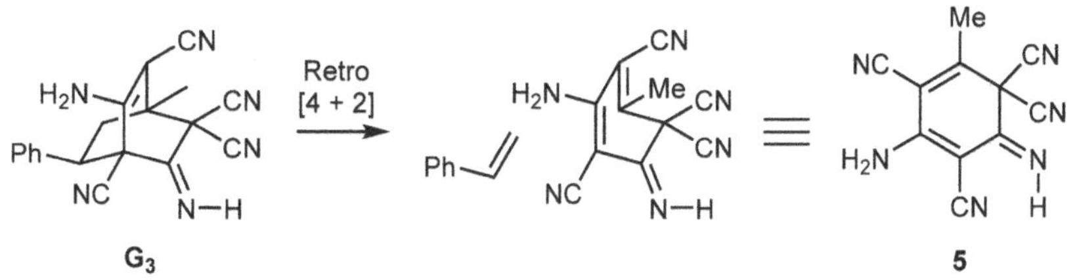

SCHEME 2

When the authors carried out a reaction between (E)-3-phenylpropenal and excess malononitrile under boiling sodium methoxide-methanol conditions and analyzed the product mixture by GC-MS they were able to identify 16 products including the nonmethyl analog of intermediate G_3 and its thermal decomposition product vinyl benzene, and addition products via dicyanocarbene that could only have been derived from decomposition of 1,1,1-tricyanomethane anion. The authors could not isolate 1,1,1-tricyanomethane from the reaction mixture as definitive proof of this by-product since it was not stable under the strongly basic reaction conditions.

Though the authors could not rule out that the "obvious" reaction pathway was not also occurring in parallel with the "nonobvious" one, they correctly warn chemists that their work "shows how far reaction pathways can be from the simple and elegant mechanistic rationalizations most of us propose and accept without question."[1]

From a synthesis strategy point of view the obvious mechanistic route can be categorized as a three-component coupling reaction to make substituted nitrile and amino substituted benzenes involving α,β-unsaturated ketones and two equivalents of malononitrile. In contrast, the nonobvious route is also a three-component coupling leading to the same product involving α,β-unsaturated ketones and three equivalents of malononitrile where one of the malononitriles is used sacrificially and ends up as 1,1,1-tricyanomethane. For future study we would highly recommend a theoretical analysis comparing the energetics of both the "obvious" and the "nonobvious" mechanistic pathways for the 1 to 4 transformation.

ADDITIONAL RESOURCES

For examples of very closely related transformations involving the same mechanistic considerations as encountered in this problem, see Refs. 2–5.

References

1. Victory P, Alvarez-Larena A, Germain G, Kessels R, Piniella JF, Vidal-Ferran A. A non-obvious reaction pathway in the formation of 2-amino-benzene-1,3-dicarbonitriles from α,β-unsaturated ketones or aldehydes. *Tetrahedron*. 1995;51(1):235–242. https://doi.org/10.1016/0040-4020(94)00937-P.

2. Agarwal N, Saxena AS, Farhanullah, Goel A, Ram VJ. A diversity oriented synthesis of highly functionalized unsymmetrical biaryls through carbanion induced ring transformation of 2H-pyran-2-ones. *Tetrahedron*. 2002;58(43):8793–8798. https://doi.org/10.1016/S0040-4020(02)01066-9.

3. Han Y-F, Xia M. Multicomponent synthesis of cyclic frameworks on Knoevenagel-initiated domino reactions. *Curr Org Chem*. 2010;14(4):379–413. https://doi.org/10.1021/ja973118x.

4. Jiang B, Wang X, Shi F, Tu S-J, Li G. New multicomponent cyclization: domino synthesis of pentasubstituted pyridines under solvent-free conditions. *Org Biomol Chem*. 2011;9:4025–4028. https://doi.org/10.1039/c0ob01258k.

5. Yi C, Blum C, Liu S-X, et al. An efficient and facile synthesis of highly substituted 2,6-dicyanoanilines. *J Org Chem*. 2008;73:3596–3599. https://doi.org/10.1021/jo800260b.

Question 223: A 1-Isoquinolone Synthesis

In 1979 Japanese chemists reacted methyl 2-formylbenzoate **1** with methyl isocyanoacetate in the presence of sodium hydride at 30–40°C to form methyl 1-oxo-l,2-dihydroisoquinoline-3-carboxylate **2** in 42% yield (Scheme 1).[1] Suggest a mechanism for this transformation.

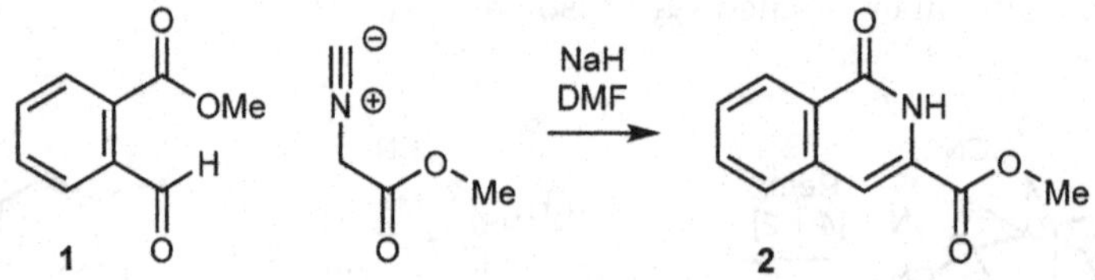

SCHEME 1

SOLUTION

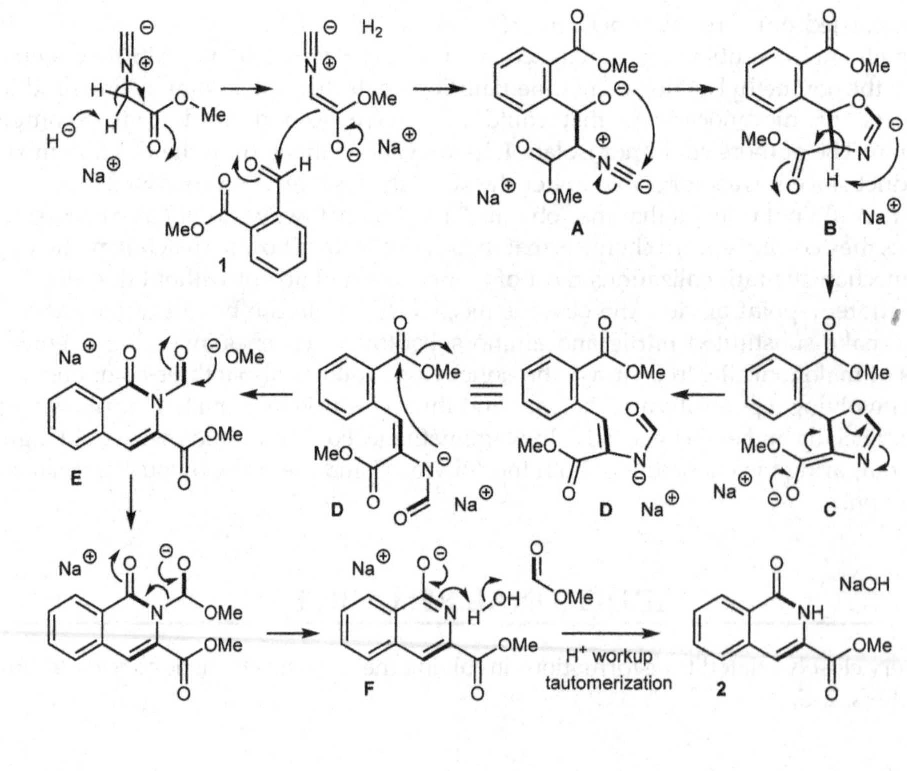

BALANCED CHEMICAL EQUATION

KEY STEPS EXPLAINED

The mechanism proposed by Nunami et al. begins with deprotonation of the isocyanide via sodium hydride to form a carbanion which attacks the aldehyde group of **1** to generate intermediate **A**. This structure cyclizes to the oxazole carbanion **B** which undergoes a proton transfer to form the oxazole carbanion **C**. This structure in turn undergoes ring opening to form the nitrogen anion **D** which cyclizes to the 1,2-dihydro-isoquinoline intermediate **E** with simultaneous elimination of methoxide anion. Elimination of methyl formate by-product via methoxide attack on the formyl group of **E** results in intermediate **F** which tautomerizes to the final product **2** upon aqueous work-up. In this synthesis the terminal carbon atom of the isocyanide reagent ends up as the carbonyl ester carbon of methyl formate. We note that the ring construction strategy employed in making product **2** is [4 + 2]. For future study we would recommend experimental and theoretical analyses to help support this suggested mechanism.

ADDITIONAL RESOURCES

For a mechanistic discussion featuring intermediates such as structure **B**, see the work of Schollköpf et al.[2] For an example of the mechanistic significance of amide substitution on the synthesis of iminohydantoins, see Ref. 3. We also recommend two excellent reviews covering the reactivities of isocyanoacetate derivatives and isocyanides, respectively.[4,5]

References

1. Nunami K, Suzuki M, Yoneda N. Synthesis of heterocyclic compounds using isocyano compounds. 5. One-step synthesis of 1-oxo-1,2-dihydroisoquinoline-3-carboxylic acid derivatives. *J Org Chem*. 1979;44(11):1887–1888. https://doi.org/10.1021/jo01325a036.
2. Schollköpf U, Gerhart F, Schröder R, Hoppe D. Synthesen mit α-metallierten Isocyaniden, XVI β-Substituierte α-Formylamino-acrylsäureäthylester aus α-metallierten Isocyanessigestern und Carbonylverbindungen (Formylaminomethylenierung von Carbonylverbindungen). *Justus Liebigs Ann Chem*. 1973;766(1):116–129. https://doi.org/10.1002/jlac.19727660113.
3. Garcia-Valverde M, Marcaccini S, Gonzalez-Ortega A, Rodriguez FJ, Rojo J, Torroba T. Complementary regioselectivity in the synthesis of iminohydantoins: remarkable effect of amide substitution on the cyclization. *Org Biomol Chem*. 2013;11:721–725. https://doi.org/10.1039/c2ob27098f.
4. Gulevich AV, Zhdanko AG, Orru RVA, Nenajdenko VG. Isocyanoacetate derivatives: synthesis, reactivity, and application. *Chem Rev*. 2010;110:5235–5331. https://doi.org/10.1021/cr900411f.
5. Giustiniano M, Basso A, Mercalli V, et al. To each his own: isonitriles for all flavors. Functionalized isocyanides as valuable tools in organic synthesis. *Chem Soc Rev*. 2017;46:1295–1357. https://doi.org/10.1039/C6CS00444J.

Question 224: Selective Cleavage of the Mycinose Sugar From the Macrolide Antibiotic Tylosin

In 1979 Pfizer chemists showed that mild acid treatment of the macrolide antibiotic tylosin **1** resulted in selective hydrolysis of the mycinose sugar **3** plus formation of the pyrrole **2** (Scheme 1).[1] Explain the transformation of **1** to **2** in mechanistic terms.

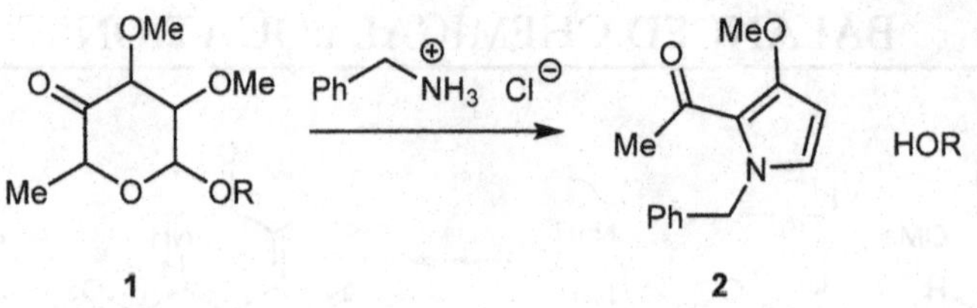

SCHEME 1

SOLUTION

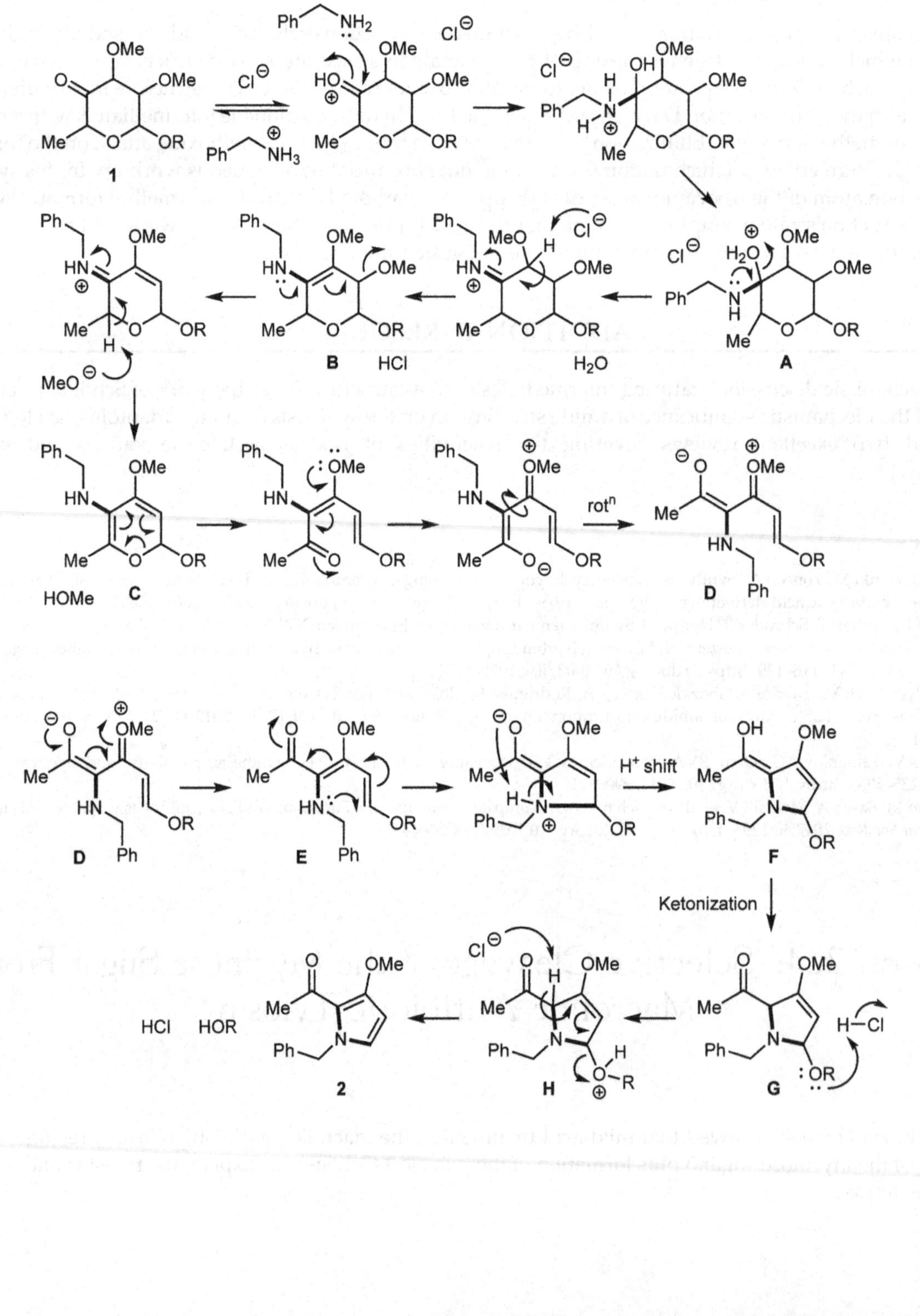

BALANCED CHEMICAL EQUATION

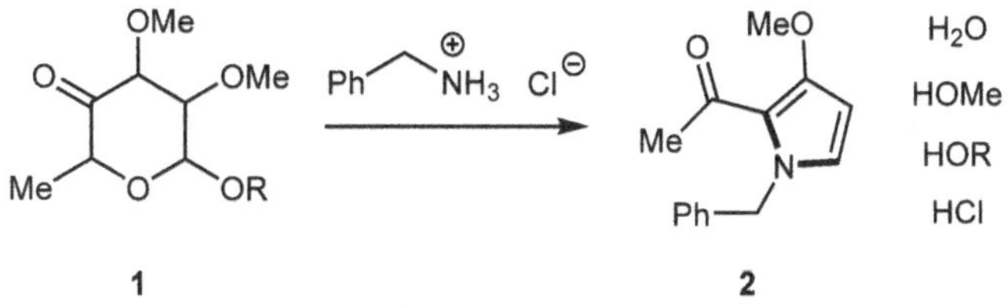

1
Tylosin macrolide   R = protected macrolide

2

KEY STEPS EXPLAINED

The mechanism proposed by Nagel and Vincent begins with amine attack on the carbonyl to form **A** which eliminates water and undergoes a deprotonation to form the endocyclic enamine **B**. This structure eliminates methoxide anion to give an exocyclic enamine which isomerizes via methoxide deprotonation to the endocyclic enamine **C** with production of methanol by-product. Intermediate **C** undergoes a [3,3]-sigmatropic rearrangement with ring opening followed by bond rotation, possibly through a zwitterionic intermediate, to form intermediate **D** which ketonizes to intermediate **E**. This intermediate cyclizes to a zwitterion which after a proton shift forms the enol **F** which ketonizes to **G** which undergoes acid-catalyzed hydrolysis of the macrolide chain with concomitant aromatization and formation of the pyrrole **2** and hydrochloric acid. We note that this degradative sequence to pyrrole **2** constitutes a [4+1] ring construction strategy. Lastly, we also note that the mechanism presented here represents a speculative suggestion by the authors who did not provide further experimental or theoretical evidence to support the postulated steps and intermediates involved along the way.

ADDITIONAL RESOURCES

For an interesting mechanistic discussion of 2,3-unsaturated sugars which also contains experimental and theoretical analysis, see the work of Bert Fraser-Reid.[2]

References

1. Nagel AA, Vincent LA. Selective cleavage of the mycinose sugar from the macrolide antibiotic tylosin: a unique glycosidic scission. *J Org Chem.* 1979;44(12):2050–2052. https://doi.org/10.1021/jo01326a043.
2. Fraser-Reid B. Some progeny of 2,3-unsaturated sugars—they little resemble grandfather glucose: twenty years later. *Acc Chem Res.* 1996;29(2):57–66. https://doi.org/10.1021/ar950104s.

Question 225: Unexpected Formation of an Enamide

During a 1979 attempted synthesis of the spiro[indane-2,2'-pyrrolidine] **3** from the reaction of pyrrolidine derivative **1** with trifluoroacetic anhydride, British chemists had unexpectedly isolated the isomeric enamines **2** (Scheme 1).[1] Explain these results in mechanistic terms.

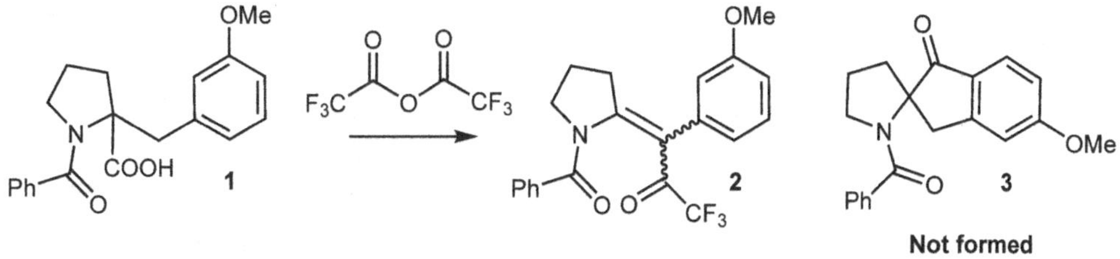

Not formed

SCHEME 1

SOLUTION

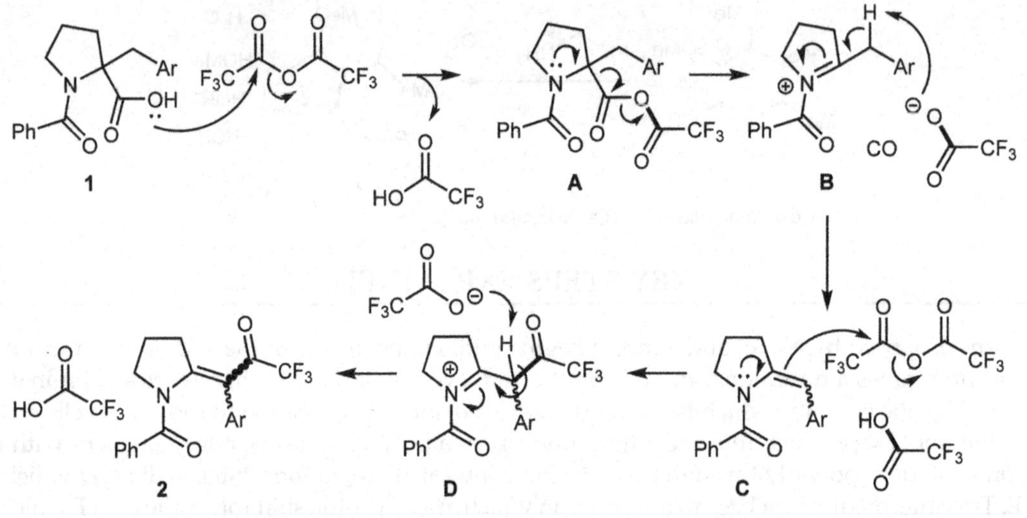

BALANCED CHEMICAL EQUATION

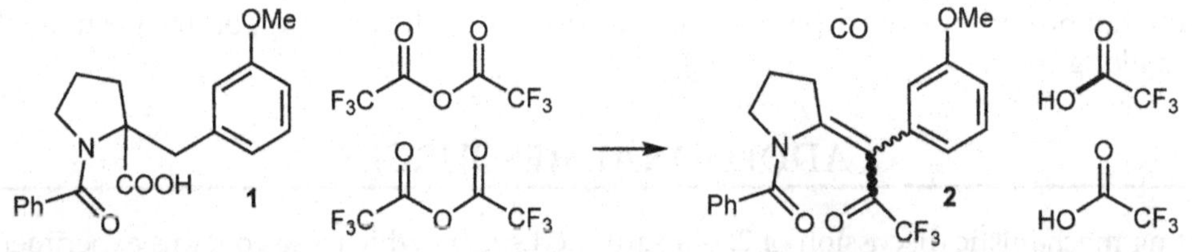

KEY STEPS EXPLAINED

The original intention of the authors was to synthesize the spiro[indane-2,2'-pyrrolidine] **3** from **1** according to the mechanism presented in Scheme 2. The likely reasons why this pathway did not occur are the increased ring strain in creating a [5,5]-*spiro* structure and the conformational difficulties in causing the aromatic group to cyclize onto the carboxyl group as indicated by the **A** to **E** transformation. Instead intermediate **A** fragments to the iminium ion **B** which then deprotonates to give the exocyclic olefin **C**. Alkylation of **C** with a second equivalent of trifluoroacetic anhydride leads to iminium ion **D** which then deprotonates to give the observed product **2**.

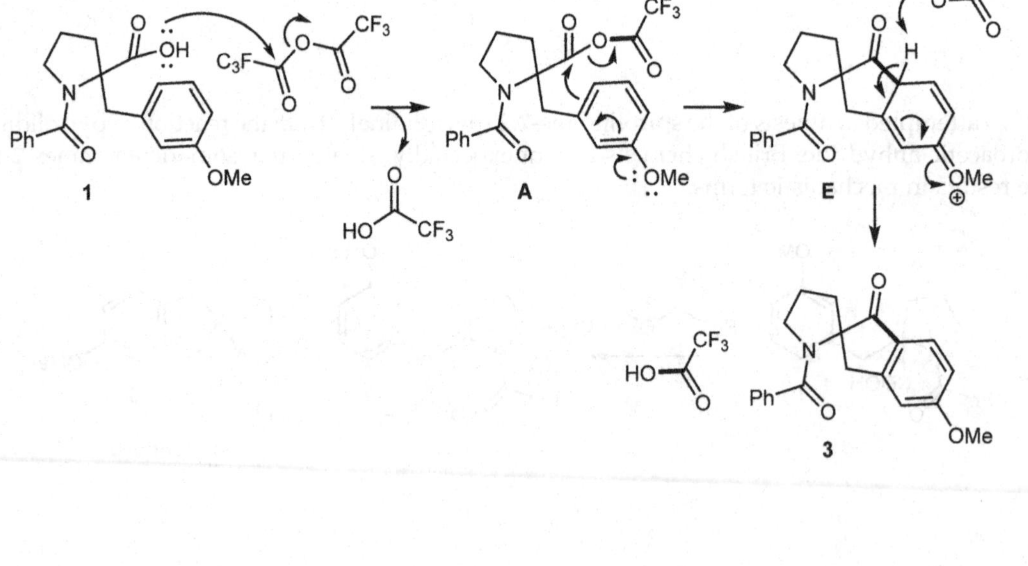

SCHEME 2

ADDITIONAL RESOURCES

We direct the reader to further work by Crooks et al. with regard to synthetic approaches to similar *spiro* compounds and pyrroles, respectively.[2,3] For an example of photocyclization process involving postulated *spiro* intermediates such as compound **3**, see the work of Nishio et al.[4]

References

1. Crooks PA, Rosenberg HE. Synthesis of 5-hydroxy- and 5,6-dihydroxy-derivatives of spiro[indane-2,2′-pyrrolidine], rigid analogues of tyramine and dopamine respectively. *J Chem Soc Perkin Trans 1*. 1979;1979:2719–2726. https://doi.org/10.1039/P19790002719.
2. Şommerville R, Rosenberg HE, Crooks PA. Synthesis and pharmacological evaluation of aromatic dihydroxylated spiro[indan-1,3′-pyrrolidine] and spiro[indan-2,2′-pyrrolidine] derivatives. *J Pharm Sci*. 1985;74(5):553–555. https://doi.org/10.1002/jps.2600740512.
3. Crooks PA, DeSimone F, Ramundo E. Synthesis of 2,3,3a,8a-tetrahydroindeno[2,1-*b*]pyrrole derivatives. *J Heterocyclic Chem*. 1982;19(6):1433–1436. https://doi.org/10.1002/jhet.5570190635.
4. Nishio T, Koyama H, Sasaki D, Sakamoto M. A novel intramolecular photocyclization of *N*-(2-bromoalkanoyl) derivatives of 2-acylanilines via 1,8-hydrogen abstraction. *Helv Chim Acta*. 2005;88(5):996–1003. https://doi.org/10.1002/hlca.200590095

Question 226: Highly Functionalized Furans From 3-Bromochromone

In 1979 an American chemist described the reaction of 3-bromochromone **1** with acetylacetone at room temperature in chloroform under basic conditions which forms the highly functionalized furan product **2** (Scheme 1).[1] Provide a mechanism for this transformation.

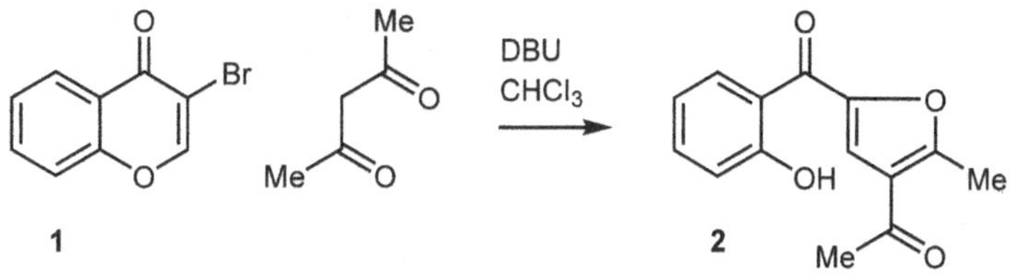

SCHEME 1

SOLUTION

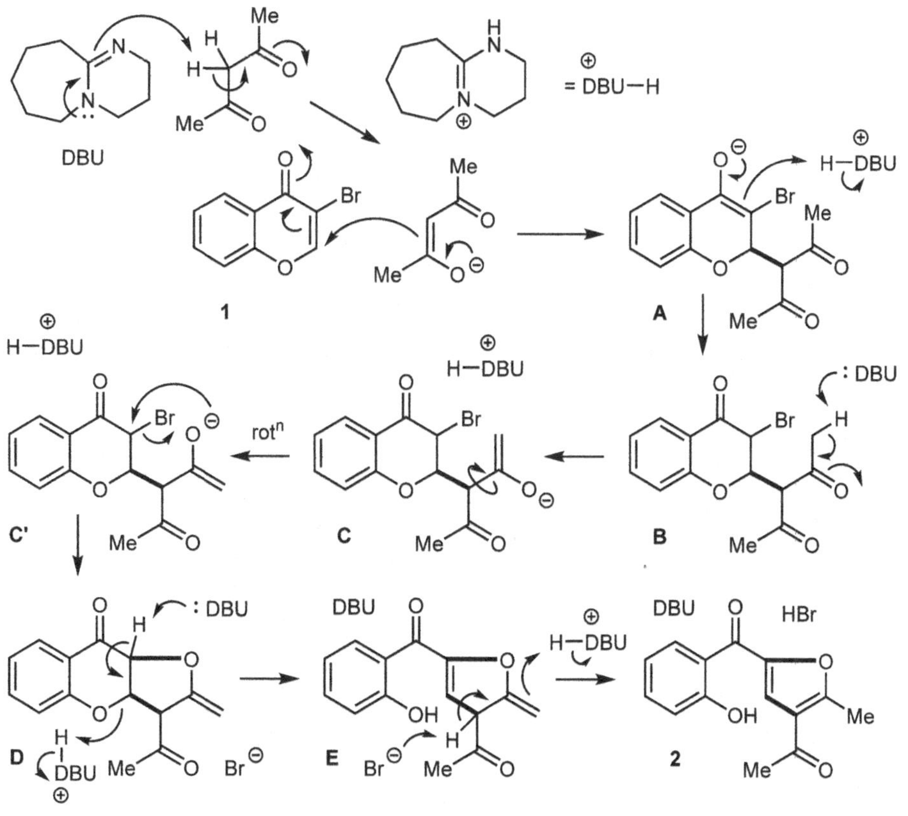

KEY STEPS EXPLAINED

The mechanism proposed by Gammill for the transformation of **1** to **2** begins with a Michael 1,4-addition of a 1,3-diketoenolate generated from the deprotonation of acetylacetone via the base DBU (1,8-diazabicyclo[5.4.0]-undec-7-ene) onto the 3-bromochromone **1**. The result of this sequence is the enolate intermediate **A** which ketonizes to **B** after protonation by the conjugate acid of DBU. **B** is then deprotonated by the regenerated DBU to form the enolate **C** which after bond rotation to **C'** cyclizes to intermediate **D** after S$_N$2 type attack by the oxide group with displacement of bromide anion. Base-promoted ring opening of the chromene-4-one ring of **D** helps establish the phenol group of **E** after protonation. A final deprotonation-protonation sequence with regard to **E** helps establish the aromaticity of the furan ring and leads directly to the highly functionalized furan product **2** and hydrobromic acid by-product. Although this postulated mechanism would suggest that the DBU base acts in a catalytic manner because it is regenerated at the end of the transformation, the authors showed experimentally that two equivalents of DBU are necessary for complete conversion. We rationalize this observation by noting that the **C'** to **2** sequence requires the presence of two equivalents of DBU which are both regenerated at the end of the transformation. Overall one equivalent of DBU is consumed and ends up as the hydrobromide salt upon neutralization of hydrobromic acid by-product. Furthermore, we would recommend for future study, a theoretical analysis to help provide supporting evidence for this mechanism. Lastly, we note that the ring construction strategy employed to make the furan ring of **2** is a [3 + 2] cycloaddition.

ADDITIONAL RESOURCES

We refer the reader to an article describing an oxidative ring closure process in the context of the synthesis of annulated furans.[2] We also recommend an excellent review of synthetic approaches toward halogen-containing chromones.[3]

References

1. Gammill RB. Reaction of 3-bromo-4*H*-1-benzopyran-4-one with β-diketones and β-keto esters to give functionalized furans. *J Org Chem*. 1979;44 (22):3988–3990. https://doi.org/10.1021/jo01336a059.
2. Sperry JB, Whitehead CR, Ghiviriga I, Walczak RM, Wright DL. Electrooxidative coupling of furans and silyl enol ethers: application to the synthesis of annulated furans. *J Org Chem*. 2004;69:3726–3734. https://doi.org/10.1021/jo049889i.
3. Sosnovskikh VY. Synthesis and reactions of halogen-containing chromones. *Russ Chem Rev*. 2003;72(6):489–516. https://doi.org/10.1070/RC2003v072n06ABEH000770.

Question 227: A One-Pot Benzene → Naphthalene Transformation

A 1985 study by Japanese chemists has revealed a new one-pot method for the synthesis of naphthalene derivatives **2** from benzhydrols **1** and maleic anhydride (Scheme 1).[1] Suggest a mechanism for this transformation.

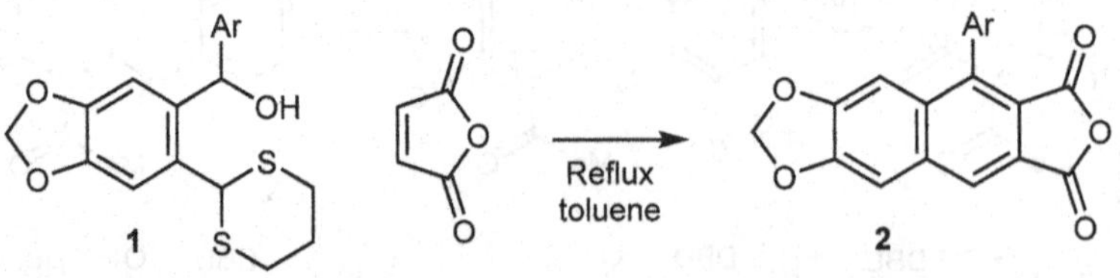

SCHEME 1

SOLUTION

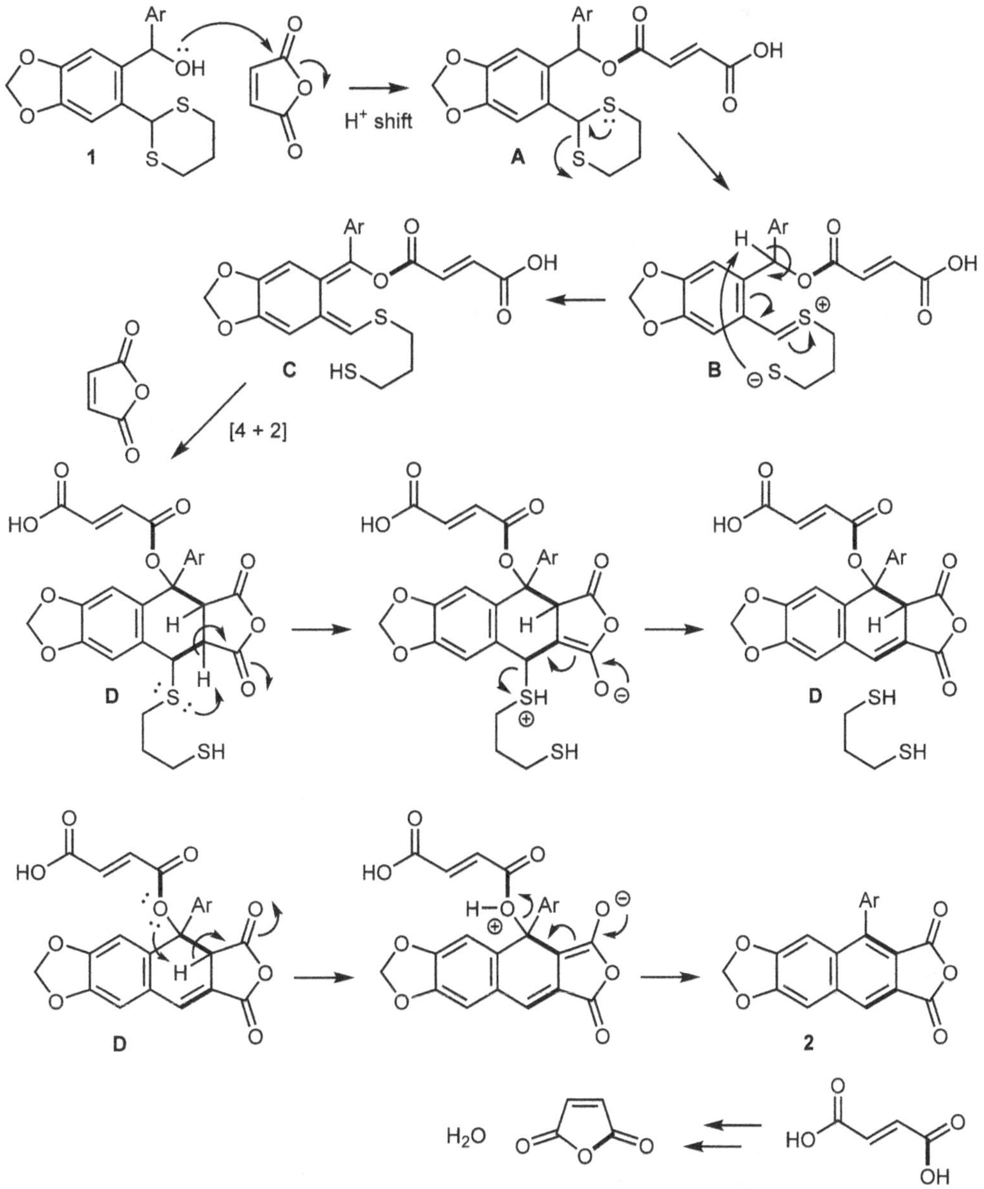

KEY STEPS EXPLAINED

The mechanism suggested by Takano et al. utilizes two equivalents of maleic anhydride, one of which is used as a sacrificial reagent. Here, we start with ring opening of the anhydride by the hydroxyl group on the benzhydrol **1** and proton transfer to form intermediate **A** which undergoes ring opening of the 1,3-dithiane ring to form intermediate **B** which after proton transfer gives the unaromatized intermediate **C**. This structure then undergoes a [4+2] Diels-Alder cycloaddition with a second molecule of maleic anhydride to form intermediate **D**. This intermediate eliminates 1,3-propanethiol and maleic anhydride together with water all driven by rearomatization of its central aryl substituted ring to form the naphthalene product **2**.

As an interesting aside, we would like to suggest an alternative mechanism which does not involve sacrificing an equivalent of maleic anhydride (Scheme 2). In this mechanism the first step constitutes ring opening of the 1,3-dithiane followed by proton transfer to generate the sulfoxide **G** which undergoes a [4+2] Diels-Alder cycloaddition with maleic anhydride to form the sulfoxide **H**. The intermediate **H** then fragments to give intermediate **I** and 3-hydroxysulfanyl-propane-1-thiol, which fragments in a homolytic manner to form hydroxyl radical and the radical of propane-1,3-dithiol. These two radicals may each abstract a proton from **I** in order to form the aromatized naphthalene product **2**.

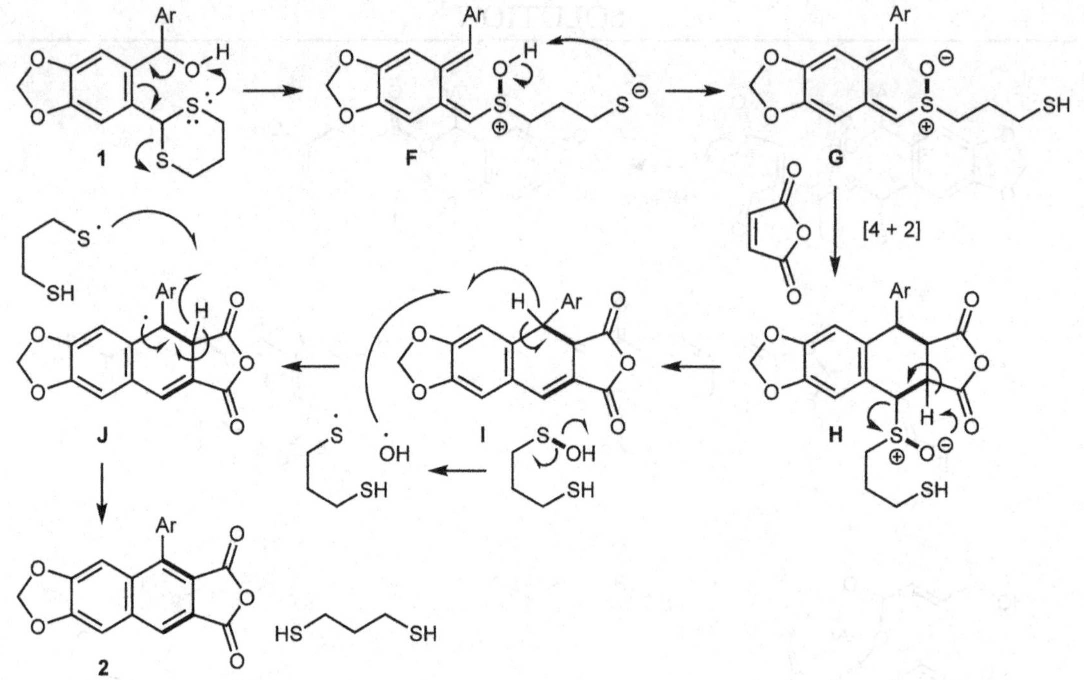

SCHEME 2

Although both mechanisms lack experimental and theoretical evidence, we believe that the second mechanism is more plausible under the reaction conditions of refluxing toluene with no other added reagents in a hydrophobic environment. Moreover, the authors' mechanism has steric constraints in intermediates **B, D,** and **E** that have to be overcome whereas the second mechanism has far fewer steric constraints. The presence of radicals may be experimentally checked using radical quenchers.

ADDITIONAL RESOURCES

We direct the reader to two examples of the synthesis of 1-arylnaphthalene lignans which involve a mechanistic discussion.[2,3]

References

1. Takano S, Otaki S, Ogasawara K. A new route to 1-phenylnaphthalenes by cycloaddition: a simple and selective synthesis of some naphthalene lignan lactones. *Tetrahedron Lett.* 1985;26(13):1659–1660. https://doi.org/10.1016/S0040-4039(00)98577-0.
2. Padwa A, Cochran JE, Kappe CO. Tandem Pummerer-Diels-Alder reaction sequence. A novel cascade process for the preparation of 1-arylnaphthalene lignans. *J Org Chem.* 1996;61(11):3706–3714. https://doi.org/10.1021/jo960295s.
3. Sato Y, Tamura T, Kinbara A, Mori M. Synthesis of biaryls via palladium-catalyzed [2+2+2] cocyclization of arynes and diynes: application to the synthesis of arylnaphthalene lignans. *Adv Synth Catal.* 2007;349:647–661. https://doi.org/10.1002/adsc.200600587.

Question 228: Vicarious Nucleophilic Substitution Routes From Simple to Complex Phenols

During a 1994 study of vicarious nucleophilic substitution reactions to synthesize complex phenols, American chemists have recorded the following three transformations illustrated in Scheme 1.[1] Suggest mechanisms for these three transformations.

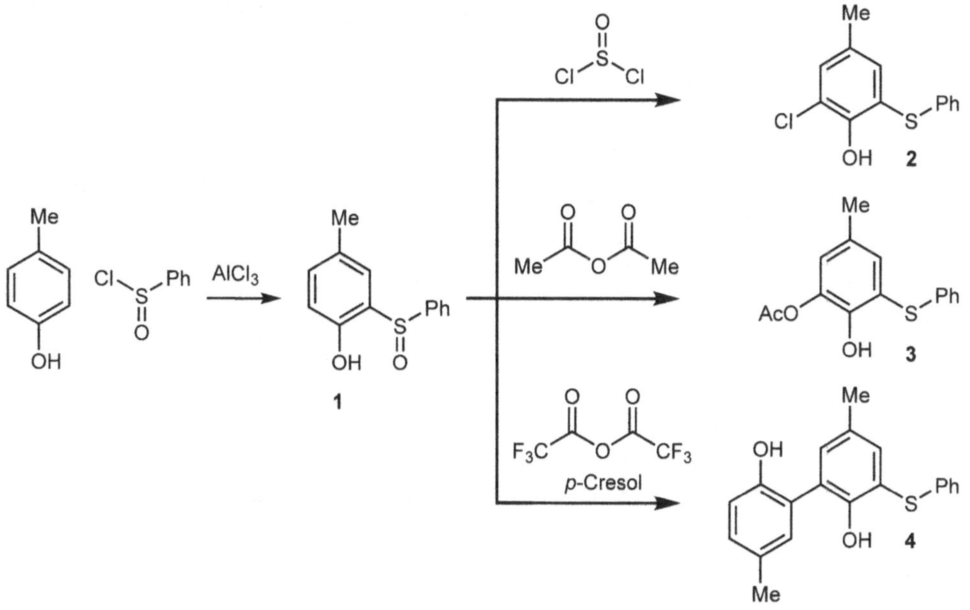

SCHEME 1

SOLUTION: FORMATION OF 1

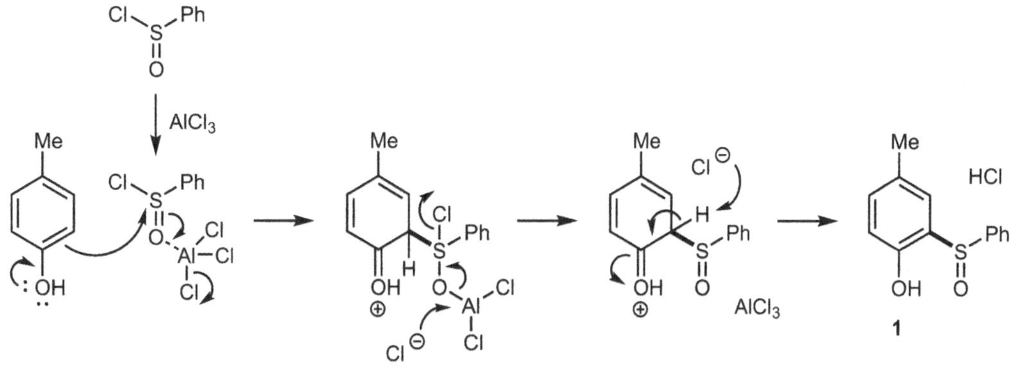

SOLUTION: FORMATION OF 2 (MECHANISM VERSION 1)

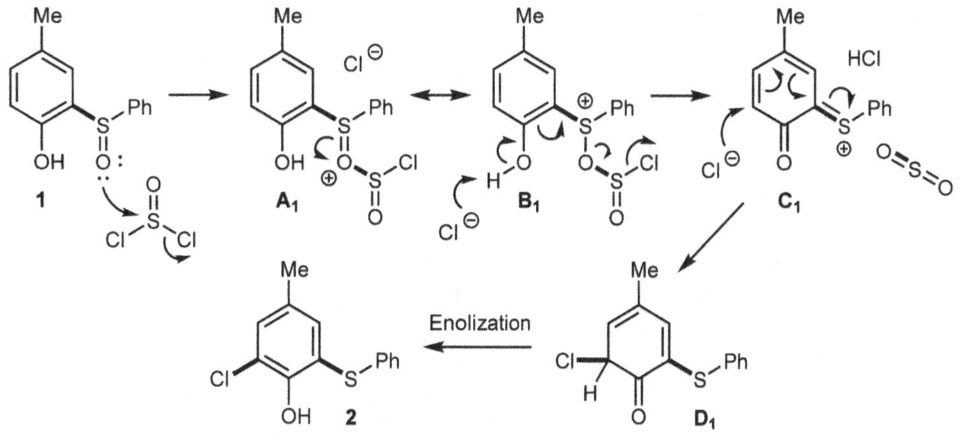

SOLUTION: FORMATION OF 3 (MECHANISM VERSION 2)

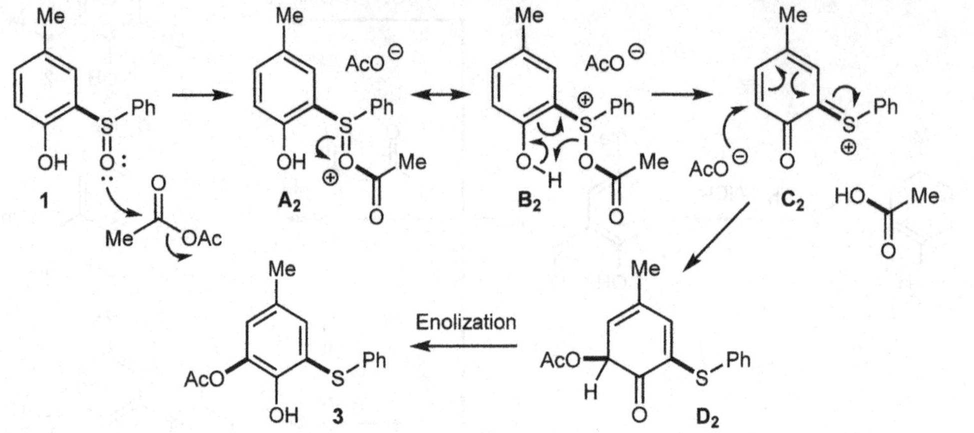

SOLUTION: FORMATION OF 4 (MECHANISM VERSION 3)

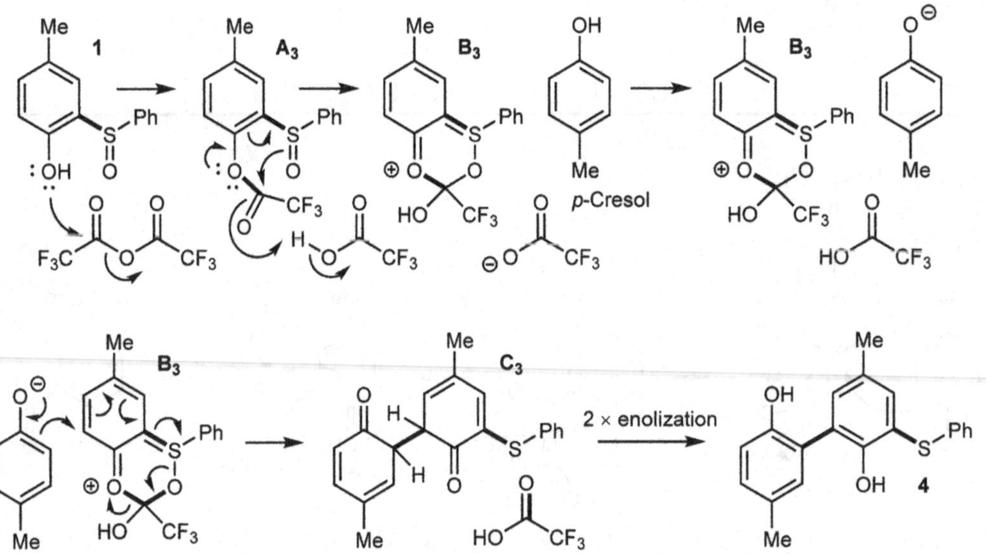

KEY STEPS EXPLAINED

The mechanism proposed by Jung et al. begins with an initial formation of the sulfone **1** via a standard Lewis acid catalyzed Friedel-Crafts acylation reaction where substitution takes place in the *ortho* position of the aromatic ring. With **1** in hand, for each of the transformations (**1** to **2**, **1** to **3**, and **1** to **4**) we believe there exist three versions of the mechanism which all produce the same results.

We begin by highlighting version 1 of the mechanism, which closely resembles what Jung et al. proposed, to explain the transformation of **1** to **2**. We thus have the nucleophilic sulfonic oxygen attacking the electrophilic center of thionyl chloride to eliminate chloride anion and form the oxonium cation intermediate A_1 which then forms the sulfonium cation intermediate B_1 via electron delocalization. Intermolecular deprotonation of the phenolic hydroxide via the chloride anion leads to ketonization and formation of the sulfonium cation C_1 after elimination of SO_2 and chloride anion. The chloride anion then attacks the *ortho* position of C_1 to form the substituted intermediate D_1 which undergoes enolization to form the substituted phenol product **2**.

A similar version of this mechanism (version 2) is presented to explain the transformation of **1** to **3**. This is actually the version proposed by Jung et al. Once again, it is assumed that the nucleophilic sulfonic oxygen attacks the electrophilic center of acetic anhydride to generate A_2 which forms B_2 as discussed before, only in this case the deprotonation is an intramolecular process with concomitant elimination of acetic acid. The rest of the mechanism follows the same nucleophilic *ortho* attack and enolization sequence seen before to generate product **3**.

The last version of the **A** to **C** sequence (version 3) is offered to explain the transformation of **1** to **4**. Here we assume that the nucleophilic phenolic oxygen attacks the triflic anhydride (as opposed to the sulfonic oxygen) to form **A₃** which cyclizes to **B₃**. Intermediate **B₃**, following *ortho* attack by the *p*-cresol via its phenoxide group, eliminates trifluoroacetic acid to form **C₃** which undergoes double enolization to the final product **4**.

It is important to note that in the absence of experimental or theoretical analysis, it is not possible to rule out any of the postulated mechanisms and that at this point we should assume that any version may be operative for any of the three transformations. From the authors' perspective, it is thought that a putative α-ketosulfonium ion intermediate **A₄**, known to participate in Pummerer rearrangement reactions, helps explain why the sulfonium oxygen atom is more nucleophilic than the phenolic oxygen atom (Scheme 2). In the absence of further evidence, we cannot endorse one explanation over the others. Moreover, the phrase "vicarious nucleophilic substitution" was used to describe these reactions since the substitution occurred in an unusual or unexpected location.

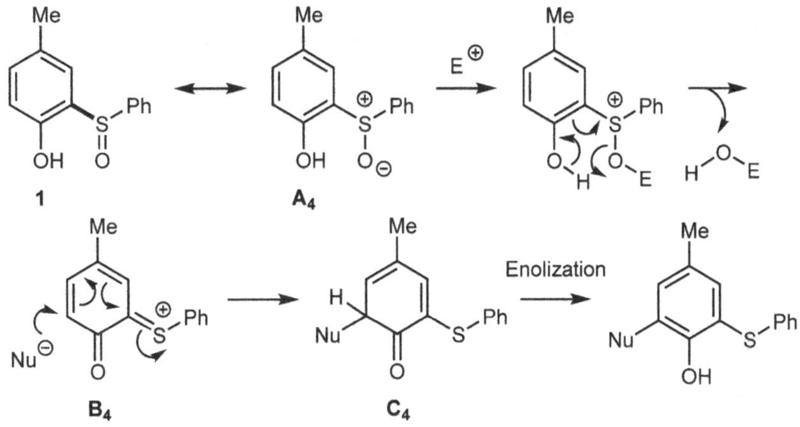

SCHEME 2

A closer examination of the transformation reveals that oxidation number changes take place at the sulfur atom and the *ortho* carbon atom that is substituted. The sulfur atom is reduced from 0 in sulfone **1** to −2 in the sulfide products, and the substituted *ortho* carbon atom in **1** is oxidized from −1 to +1 in the case of chloride or acetate substitution. In the third case where *p*-cresol is the nucleophile both *ortho* carbon atoms in the original sulfone substrate and in the *p*-cresol each undergo a net oxidation from −1 to 0. Since there are two of them this balances the net reduction of −2 found in the sulfur atom.

ADDITIONAL RESOURCES

For other work by Jung et al. involving mechanistic discussion of Pummerer-type rearrangements, see Ref. 2. For other interesting examples of the Pummerer rearrangement and its usefulness in synthesis, see Refs. 3–5.

References

1. Jung ME, Kim C, Bussche L. Vicarious nucleophilic aromatic substitution via trapping of an α-ketosulfonium ion generated by Pummerer-type rearrangement of 2-(phenylsulfinyl)phenols: preparation of biaryls. *J Org Chem*. 1994;59(12):3248–3249. https://doi.org/10.1021/jo00091a003.
2. Jung ME, Starkey LS. The total synthesis of (*S*,*S*)-isodityrosine. *Tetrahedron*. 1997;53(26):8815–8824. https://doi.org/10.1016/S0040-4020(97)90393-8.
3. Padwa A, Kuethe JT. Additive and vinylogous Pummerer reactions of amido sulfoxides and their use in the preparation of nitrogen containing heterocycles. *J Org Chem*. 1998;63(13):4256–4268. https://doi.org/10.1021/jo972093h.
4. Feldman KS, Skoumbourdis AP. Extending Pummerer reaction chemistry. Synthesis of (±)-dibromophakellstatin by oxidative cyclization of an imidazole derivative. *Org Lett*. 2005;7(5):929–931. https://doi.org/10.1021/ol0500113.
5. Feldman KS, Vidulova DB, Karatjas AG. Extending Pummerer reaction chemistry. Development of a strategy for the regio- and stereoselective oxidative cyclization of 3-(ω-nucleophile)-tethered indoles. *J Org Chem*. 2005;70(16):6429–6440. https://doi.org/10.1021/jo050896w.

Question 229: Lewis Acid-Catalyzed Rearrangement of Humulene 8,9-Epoxide

Scottish chemists investigating the rearrangement of humulene-8,9-epoxide **1** showed that under Lewis acid conditions, the bicyclic alcohol product **2** could be isolated as the major product following initial isomerization (Scheme 1).[1] Propose a mechanism for this transformation.

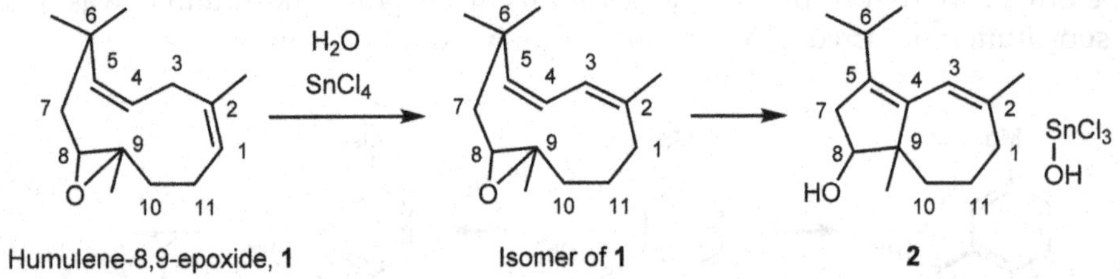

Humulene-8,9-epoxide, **1**   Isomer of **1**   **2**

SCHEME 1

SOLUTION

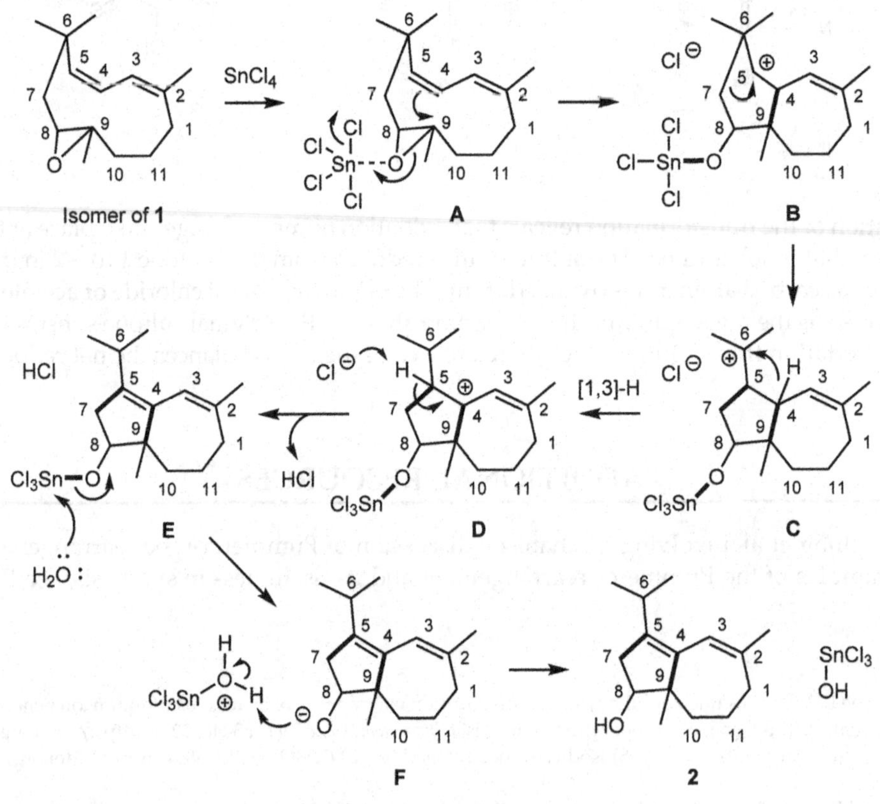

Isomer of **1**   A   B

E   D   C

F   2

KEY STEPS EXPLAINED

The mechanism proposed by Bryson et al. begins with isomerization of the C_1—C_2 double bond of **1** to the isomer which has a new C_2—C_3 double bond. This initial step was supported by means of experimental and molecular modeling techniques based on other literature.[1] Following this isomerization, we have coordination of the Lewis acid tin(IV)chloride with the oxygen atom of the epoxide group of the isomer of **1** to form intermediate **A**. This intermediate can then rearrange by means of C_4—C_5 double bond participation to form the bicyclic secondary carbocation intermediate **B** which may easily rearrange to the more stable tertiary carbocation intermediate **C**. A [1,3]-hydride shift would then lead to carbocation **D** which would undergo deprotonation via a chloride anion to generate the diene **E** which after hydrolysis of the tin oxide moiety results in formation of product **2**. We also note that the overall ring construction strategy for the formation of the bicyclic [5.3.0] ring system is [(7+0)+(3+2)]. A theoretical analysis for a future study is highly recommended to add supporting evidence for this postulated mechanism.

ADDITIONAL RESOURCES

For an interesting rearrangement of a humulene-like skeletal structure, see the work of Enev and Tsankova.[2] We also recommend several other articles highlighting transformations of very similar chemical systems.[3–6]

References

1. Bryson I, Roberts JS, Sattar A. Rearrangement of humulene-8,9-epoxide. *Tetrahedron Lett.* 1980;21:201–204. https://doi.org/10.1016/S0040-4039(00)71413-4.
2. Enev V, Tsankova E. Biomimetic germacrene-humulene rearrangement. *Tetrahedron Lett.* 1988;29(15):1829–1832. https://doi.org/10.1016/S0040-4039(00)82056-0.
3. Dauben WG. Transformations of sesqui- and diterpenes. *J Agric Food Chem.* 1974;22(2):156–162. https://doi.org/10.1021/jf60192a011.
4. Totobenazara J, Haroun H, Remond J, et al. Tandem Payne/Meinwald versus Meinwald rearrangements on the α-hydroxy- or α-silyloxy-spiro epoxide skeleton. *Org Biomol Chem.* 2012;10:502–505. https://doi.org/10.1039/c1ob06776a.
5. Pulido FJ, Barbero A, Castreno P. Seven-membered ring formation from cyclopropanated oxo- and epoxyallylsilanes. *J Org Chem.* 2011;76:5850–5855. https://doi.org/10.1021/jo200705u.
6. Haib AE, Benharref A, Parres-Maynadie S, Manoury E, Urrutigoity M, Gouygou M. Lewis acid- and Bronsted acid-catalyzed stereoselective rearrangement of epoxides derived from himachalenes: access to new chiral polycyclic structures. *Tetrahedron Asymmetry.* 2011;22:101–108. https://doi.org/10.1016/j.tetasy.2010.12.013.

Question 230: From a Dihydrofuran to an Indole-3-acetate

In 1994 Japanese chemists reacted *N*-2-iodophenylcarbamate with 2,5-dihydro-2,5-dimethoxyfuran in DMF in the presence of Hünig's base, diisopropylethylamine (DIPEA), and a catalytic amount of palladium(II) acetate at 80°C for 10h to form **1** as a mixture of diastereomers (Scheme 1).[1] Treatment of **1** at room temperature in DCM (dichloromethane) containing 6% v/v TFA (trifluoroacetic acid) gave **2** in 65% overall yield. Illustrate the reaction sequence and provide a mechanistic explanation for the transformation of **1** to **2**.

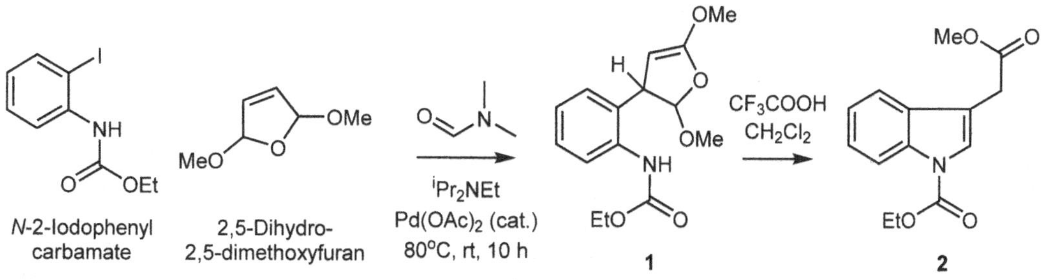

SCHEME 1

SOLUTION: FORMATION OF 1

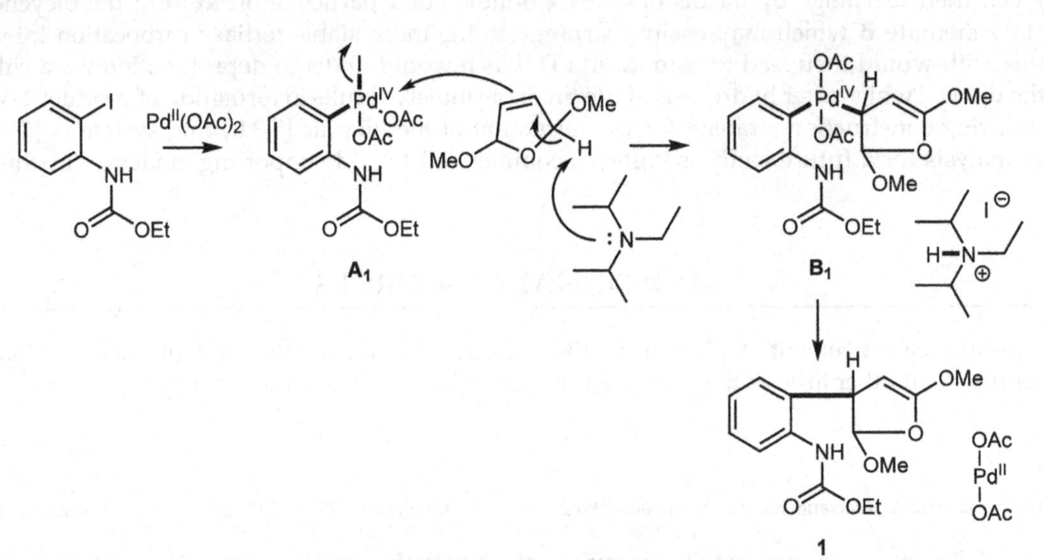

SOLUTION: FORMATION OF 2

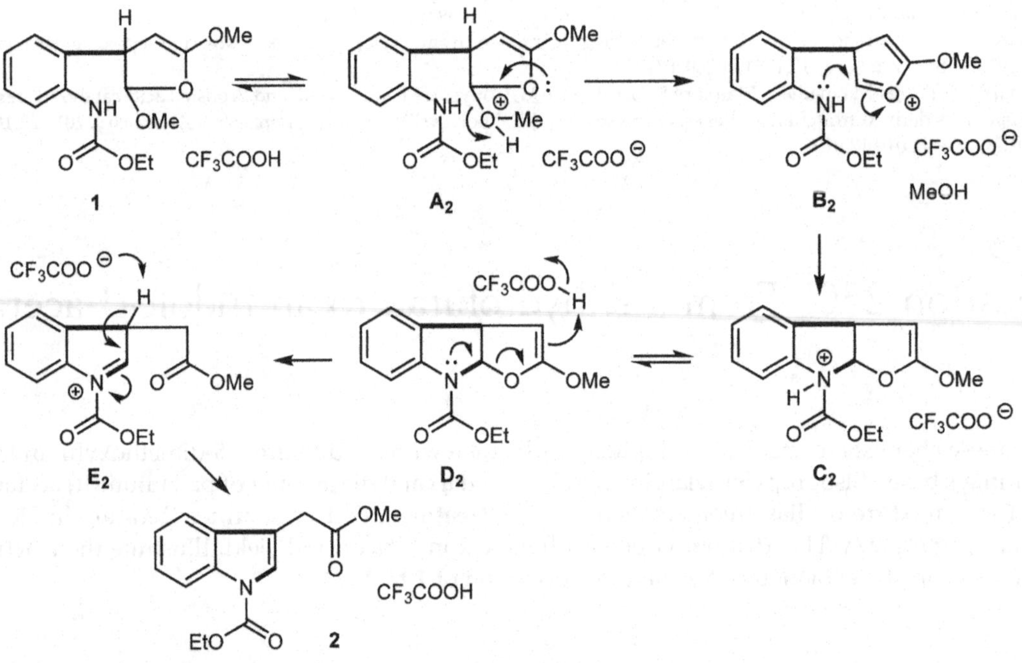

KEY STEPS EXPLAINED

The mechanism proposed by Samizu and Ogasawara begins with a Heck coupling between N-2-iodophenylcarbamate and 2,5-dihydro-2,5-dimethoxyfuran which is catalyzed by palladium(II) acetate. Therefore we start with oxidative addition of palladium to N-2-iodophenylcarbamate where palladium is oxidized from +2 to +4 in intermediate A_1. Next we have base promoted nucleophilic addition of 2,5-dihydro-2,5-dimethoxyfuran to A_1 via Hünig's base to give B_1 after elimination of iodide anion. Lastly, B_1 undergoes reductive elimination where the two organic moieties are coupled to give product 1 while the palladium is reduced from +4 to +2, regenerating it to its original oxidation state in the catalyst.

Concerning the acid treatment of 1 to form the indole-3-acetate 2, the authors presume that initial protonation of methoxy to form A_2 leads to furan-promoted elimination of methanol to form the oxonium intermediate B_2 which may ring close to C_2 which is deprotonated to the aminoacetal D_2. Iminium formation in D_2 would promote ring opening and ketonization with regard to the furan ring leading to the iminium E_2 which would undergo a last deprotonation to generate the final indole-3-acetate product **2**.

Since the authors do not provide any experimental or theoretical analysis to help support their postulated mechanism for the transformation of **1** to **2**, we would like to pose a slightly different alternative mechanism here (Scheme 2). In this alternative mechanism, we begin with initial protonation of the furan oxygen atom followed by simultaneous pyrrolidine ring closure and dihydrofuran ring cleavage to yield intermediate B_3. Ketonization of B_3 and acid-catalyzed elimination of methanol yield the final indole product **2**.

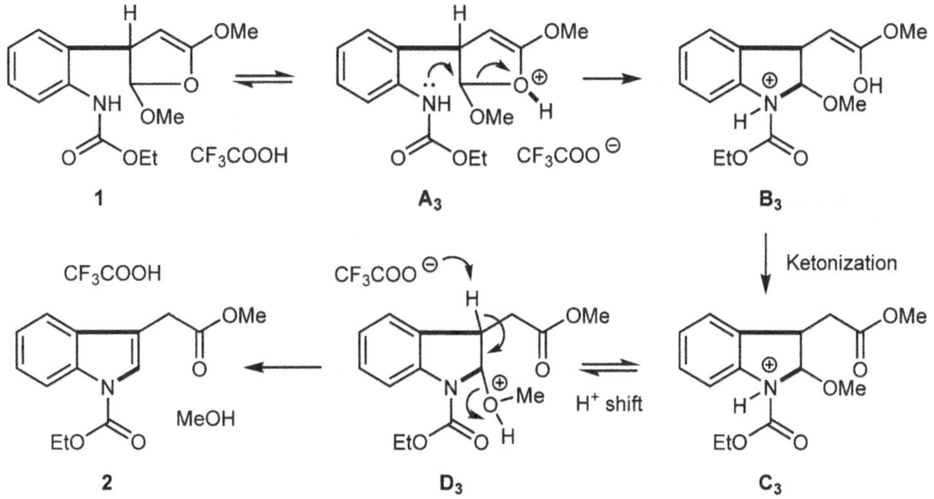

SCHEME 2

ADDITIONAL RESOURCES

We recommend an interesting transformation involving furan ring opening and indole ring closure reported recently by Pilipenko et al.[2] Cacchi et al. have also reported an interesting palladium-catalyzed synthesis of 3-allylindoles which involves a mechanistic discussion.[3] Lastly, we recommend a review of recent developments in indole ring synthesis.[4]

References

1. Samizu K, Ogasawara K. An expedited route to indole-3-acetic acid derivatives. *Synlett.* 1994;1994(7):499–500. https://doi.org/10.1055/s-1994-22903.
2. Pilipenko AS, Mel'chin VV, Trushkov IV, Cheshkov DA, Butin AV. Furan ring opening–indole ring closure: recyclization of 2-(2-aminophenyl) furans into 2-(2-oxoalkyl)indoles. *Tetrahedron.* 2012;68:619–627. https://doi.org/10.1016/j.tet.2011.10.114.
3. Cacchi S, Fabrizi G, Pace P. Palladium-catalyzed cyclization of *o*-alkynyltrifluoroacetanilides with allyl esters. A regioselective synthesis of 3-allylindoles. *J Org Chem.* 1998;63(4):1001–1011. https://doi.org/10.1021/jo971237p.
4. Gribble GW. Recent developments in indole ring synthesis—methodology and applications. *J Chem Soc Perkin Trans 1.* 2000;2000:1045–1075. https://doi.org/10.1039/A909834H.

Question 231: Reaction of Fervenulin 4-Oxide With DMAD: The Role of Solvent

In a 1979 study, Japanese chemists showed that solvent plays a role in the reaction of fervenulin 4-oxide **1** with DMAD since products **2**, **3**, and **4** can be isolated when different solvents are used (Scheme 1).[1] Suggest mechanisms to explain the role of the solvent in these transformations.

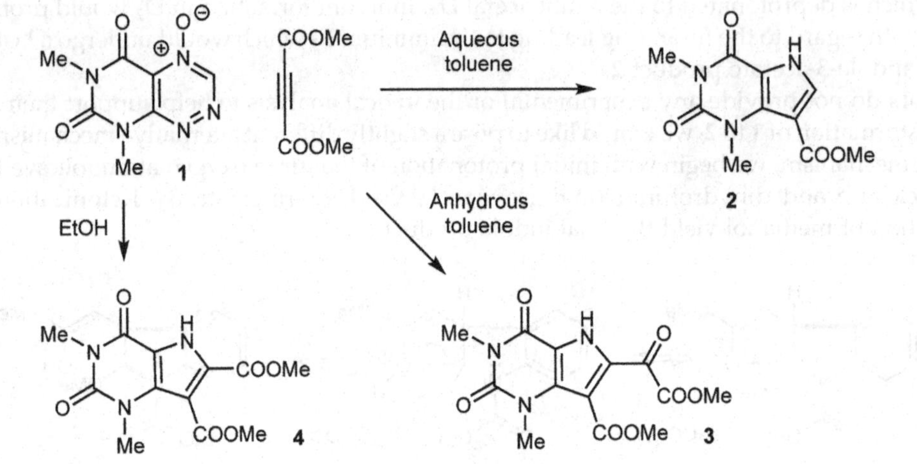

SCHEME 1

SOLUTION: 1 → 2

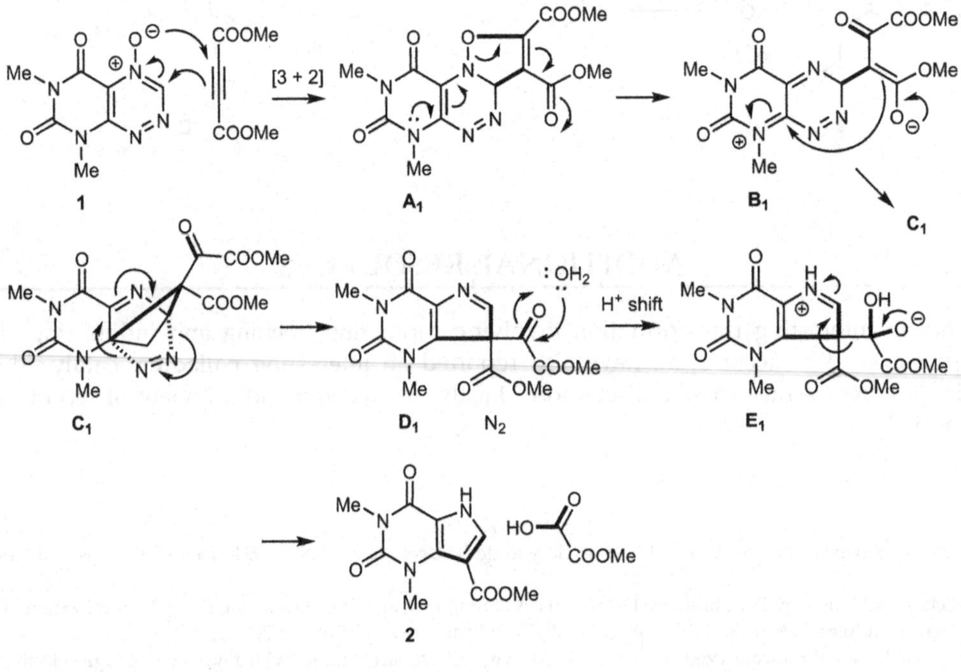

SOLUTION: 1 → 3

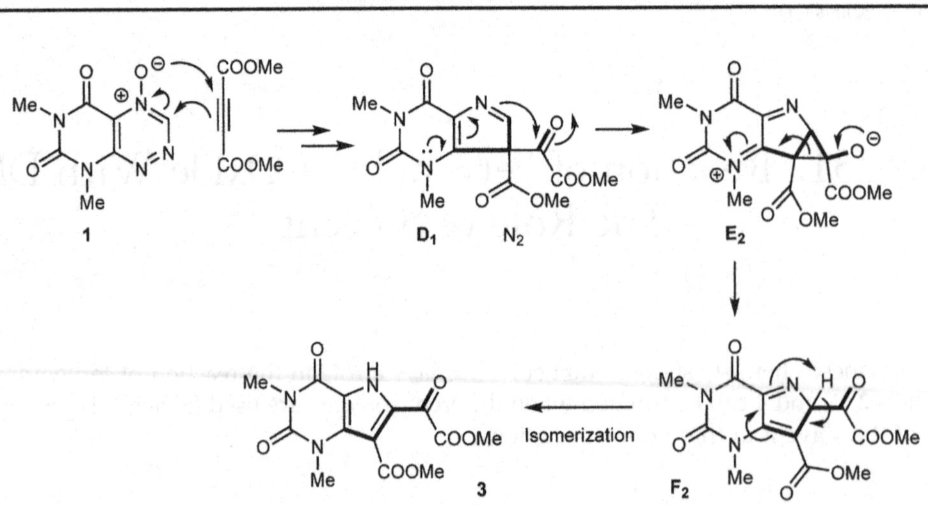

SOLUTION: 1 → 4

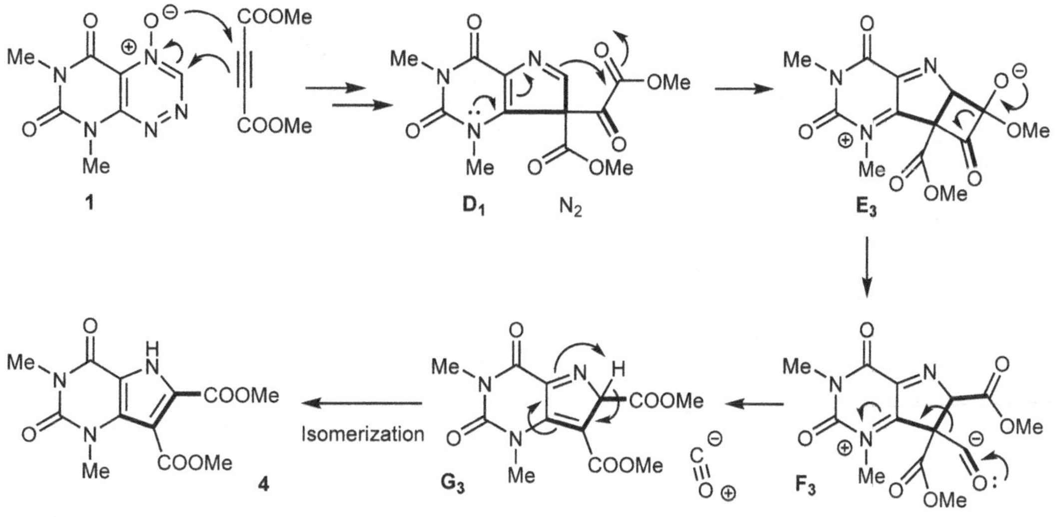

BALANCED CHEMICAL EQUATIONS

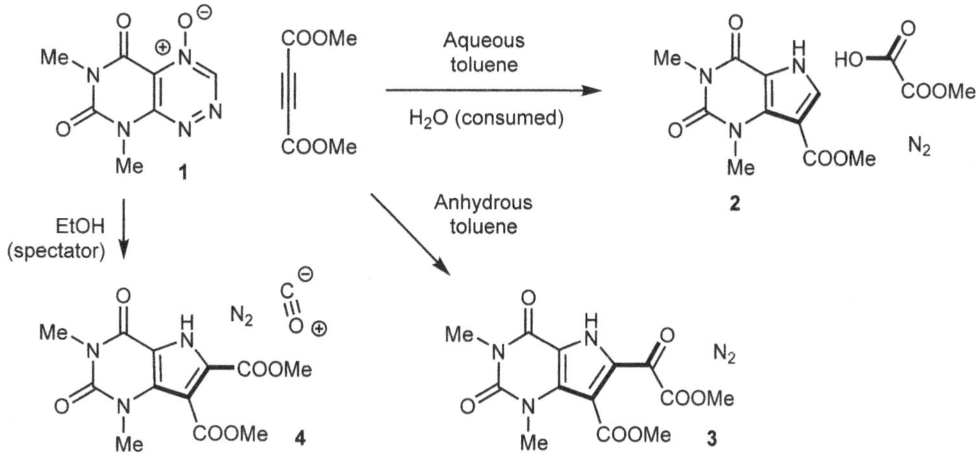

KEY STEPS EXPLAINED

The authors presented the mechanism for the transformations of **1** to **2**, **3** and **4** as all beginning with a [3+2] cycloaddition reaction between fervenulin 4-oxide **1** and DMAD to form the tricyclic isoxazoline intermediate **A₁** which rearranges via isoxazoline ring cleavage to the 1,8-dipolar intermediate **B₁** after electron delocalization. Intermediate **B₁** undergoes intramolecular cyclization to a tricyclic intermediate **C₁** which undergoes extrusion of nitrogen gas via a retro [4+2] process to generate intermediate **D₁**. At this point the mechanistic pathways for the three transformations diverge along three pathways which are solvent dependent. Along the first path, if water is present in the reaction mixture, it may add in a nucleophilic manner to the exocyclic ketone group of **D₁** thus initiating a cascade sequence that eliminates monomethyl oxalate as a by-product giving the fused pyrrole product **2**. If water is not present in the reaction mixture, intermediate **D₁** rearranges to **F₂** via the tricyclic cyclopropane ring containing intermediate **E₂**. Isomerization of **F₂** via the prototropic rearrangement depicted leads to the fused pyrrole product **3**. Alternatively, in the polar protic solvent ethanol, hydrogen bonding is thought to help stabilize the rearrangement of **D₁** to the tricyclic cyclobutane ring containing intermediate **F₃** (by making the β carbonyl group of **D₁** more electrophilic than the α carbonyl group, see Scheme 2). Intermediate **F₃** may eliminate carbon monoxide by-product to form **G₃**. Once again isomerization of **G₃** via a prototropic rearrangement leads to the fused pyrrole product **4**.

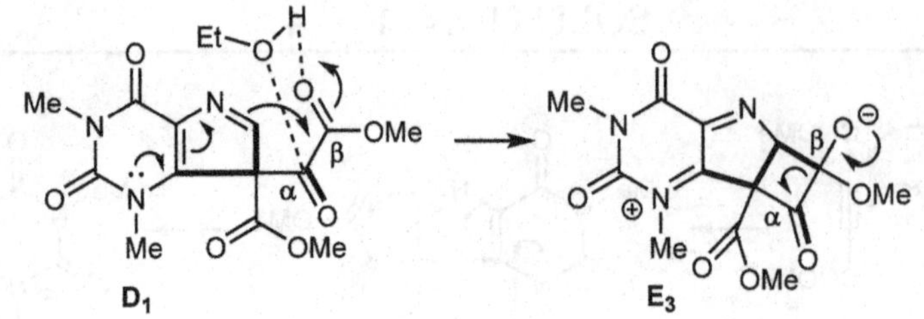

SCHEME 2

With regard to the isomerization of F_2 to **3**, the authors show that if the hydrogen atom undergoing prototropic rearrangement were an alkyl group, decarbonylation would occur to form product **5** (Scheme 3). This experiment lends support for the postulated prototropic isomerization. Moreover, the three transformations together support the idea of solvent participation in this transformation. Lastly, we believe a future theoretical study would greatly enhance one's understanding of the energetics which determine the favorable mechanistic pathway insofar as solvent participation is concerned.

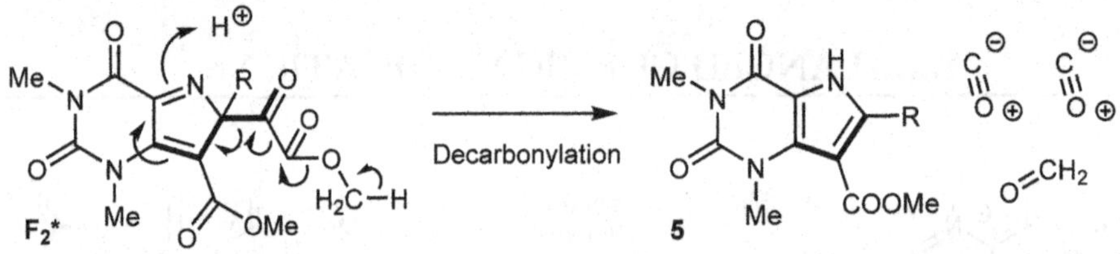

SCHEME 3

ADDITIONAL RESOURCES

We refer the reader to an interesting synthesis of furo[3,2-d]pyrimidines via radical cyclization from the work of Majumdar and Mondal.[2]

References

1. Senga K, Ichiba M, Nishigaki S. New synthesis of pyrrolo[3,2-d]pyrimidines (9-deazapurines) by the 1,3-dipolar cycloaddition reaction of ferve-nulin 4-oxides with acetylenic esters. *J Org Chem*. 1979;44(22):3830–3834. https://doi.org/10.1021/jo01336a018.
2. Majumdar KC, Mondal S. An expedient approach for the synthesis of pyrrolo[3,2-d]pyrimidines (9-deazaxanthines) and furo[3,2-d]pyrimidine via radical cyclization. *Tetrahedron*. 2009;65:9604–9608. https://doi.org/10.1016/j.tet.2009.09.059.

Question 232: Rearrangement of 4-Quinazolinylhydrazines

During studies of heterocyclic rearrangements, chemists at Zeneca Pharmaceuticals observed that acidic and basic treatment of the trimethoxyquinazolin hydrazine **1** produced the hydrolyzed quinazolin-4-one product **3** whereas at neutral pH (pH 4–7.7), **1** rearranged smoothly to the isomer **2** (Scheme 1).[1] Suggest mechanisms for these transformations.

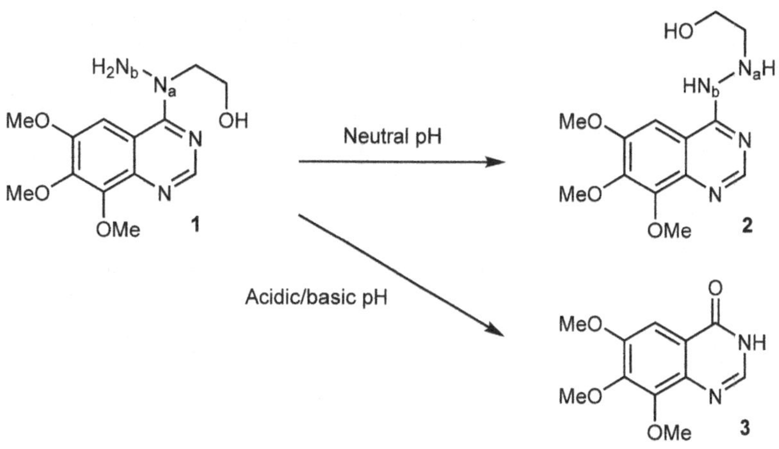

SCHEME 1

SOLUTION: REARRANGEMENT UNDER NEUTRAL pH

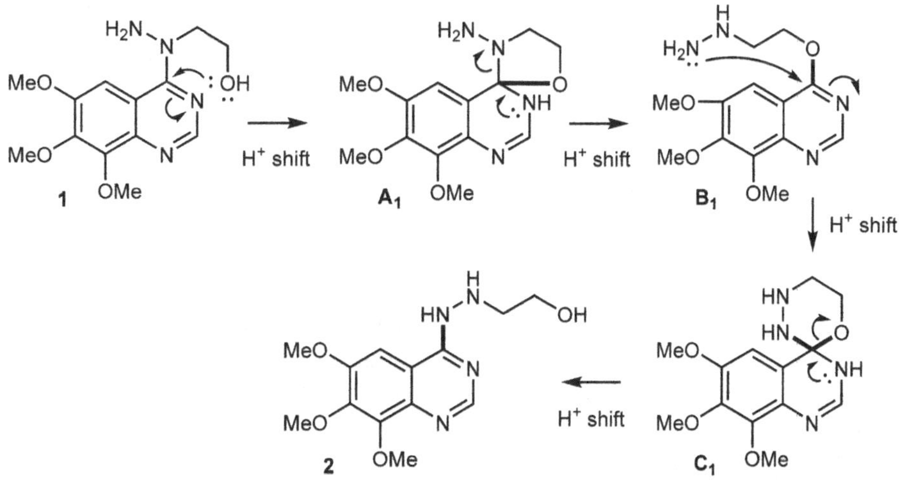

SOLUTION: HYDROLYSIS UNDER ACIDIC pH

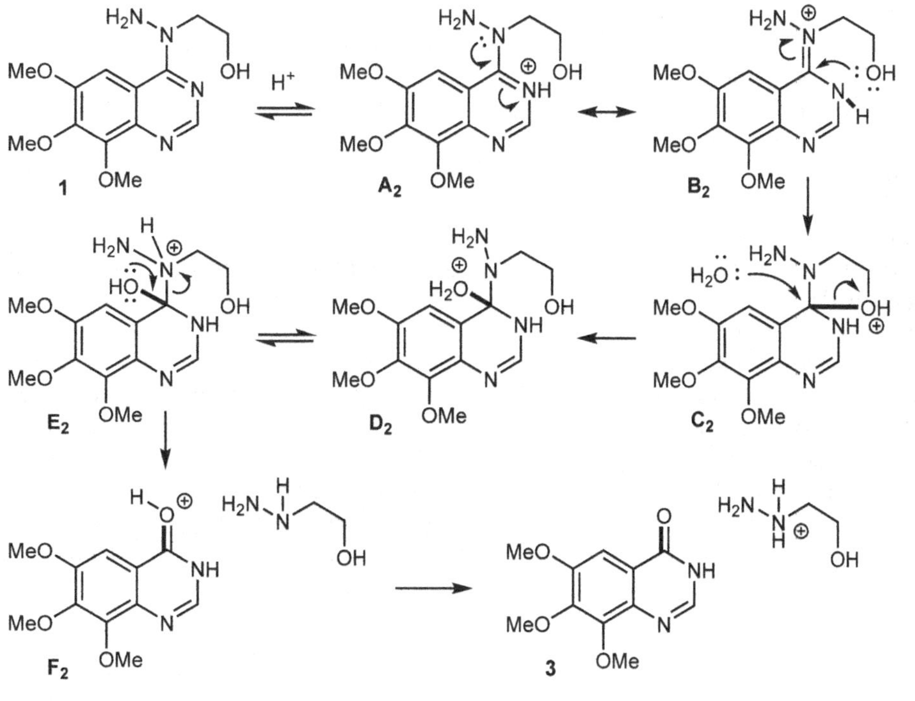

SOLUTION: HYDROLYSIS UNDER BASIC pH

KEY STEPS EXPLAINED

The transformation of **1** to **2** under neutral conditions entails a double Smiles rearrangement via the [5,6]-*spiro* and [6,6]-*spiro* intermediates A_1 and C_1 to form product **2**. Essentially the transformation is one that maintains the quinazoline moiety intact except that its attachment to the 2-hydrazino-ethanol moiety changes from a C—N_a attachment to a C—N_b attachment. Concerning the acid- and base-catalyzed processes, there are several possible mechanisms one could write in order to arrive at the 3*H*-quinazolin-4-one product **3**. Here we present simplified mechanisms with the least number of steps.

Therefore, under acidic conditions, an oxygen protonated [5,6]-*spiro* intermediate C_2 is formed (analogous to A_1 intermediate). Nucleophilic addition of water leads to C—O cleavage forming D_2, which after a proton shift and ketonization with elimination of 2-hydrazineylethan-1-ol generates the 3*H*-quinazolin-4-one ring in product **3**.

Under basic conditions on the other hand, a nitrogen deprotonated [5,6]-*spiro* intermediate B_3 is formed (analogous to A_1 intermediate). The five-membered ring undergoes C—O bond cleavage, as under acid conditions, except that hydroxide ion attacks the *spiro* carbon atom yielding alkoxide intermediate D_3. Proton transfer between the two oxygen atoms leads to a different alkoxide that both eliminates the hydrazine moiety and creates a carbonyl group to generate the 3*H*-quinazolin-4-one ring in product **3**.

To back up their proposed mechanisms, the authors show in their paper a complex pH-rate profile for the disappearance of starting material **1** which is composed of four distinct regions: (a) pH < 2, concave down region indicating acid catalysis with saturation; (b) 4 < pH < 8, concave down region indicating base catalysis with saturation; (c) 8 < pH < 10, a plateau region indicating an uncatalyzed process; and (d) pH > 10, concave up region indicating a second base catalyzed process.

The general shape of the pH-rate profile is reminiscent of that for the hydrolysis of aspirin except for the region of discontinuity.[2] Good pedagogical discussions of the aspirin example may be found in the *Journal of Chemical Education*.[3,4] Also, these papers describe how to parse and interpret regions of pH-rate profiles and link those regions to rate laws and reaction mechanisms. A key generalization highlighted was that a concave down region is associated with a change of rate determining step in the same mechanism and a concave up region indicates that a change of mechanism has taken place. Another useful critical review discusses literature misinterpretations of pH-rate profiles.[5]

In the present example, there are significant problems with the authors' presented pH profile that hamper its proper analysis. Firstly, the authors do not explain the discontinuity in the pH profile in the region 2 < pH < 4. Secondly, the authors did not mention if the observed rate constants recorded in the region 4 < pH < 8 were corrected for the rearrangement reaction, that is, in this pH region the rate constant contribution for the rearrangement reaction needs to be subtracted from the total in order to obtain the hydrolysis reaction contribution. This is needed in order to properly link this pH region with the other two regions at low and high pH so that all regions represent the same reaction, namely, hydrolysis of **1** to **3**. Thirdly, the authors did not specify if they performed product studies in *each* pH region to determine the reaction products. We are only told that product **2** is formed in the region $4 < pH < 8$ and that product **3** is formed under acidic and basic conditions, which we interpret must correspond to regions $pH < 2$ and $pH > 10$. The other regions of the pH profile are unaccounted for with respect to product studies. The authors mention that there is a build-up of a nonisolable intermediate on the reaction path between **1** and **3**, as evidenced by time-resolved UV absorption changes, in the pH regions $pH < 2$ and $pH > 10$. The observation of an intermediate that accumulates suggests that its rate of formation is faster than its rate of decomposition. The authors do not mention any unique characteristics of the UV absorption spectrum of this accumulating intermediate in comparison to that of starting material **1**. It may not be the case that the authors' pH-rate profile actually represents the rate of disappearance of **1**, but may in fact be the rate of disappearance of the accumulating intermediate as a function of pH.

Based on shape of the present pH-rate profile and the few details provided by the authors we have the following tentative interpretations which rest on the assumption that the pH-rate profile represents the rate of disappearance of **1** as a function of pH: (a) region $pH < 2$ is an acid-catalyzed mechanism that leads to hydrolysis product **3** via an intermediate that accumulates; (b) region $4 < pH < 8$ is a combination of general base-catalyzed mechanisms that lead to hydrolysis product **3** and rearranged product **4**; (c) region $8 < pH < 10$ is a combination of uncatalyzed mechanisms that lead to products **3** and **4**; and (d) region $pH > 10$ is a base-catalyzed mechanism that leads to product **3** via the same intermediate that accumulates in the acidic region $pH < 2$.

Given the mechanisms presented here we postulate that the accumulated intermediate observed under acidic and basic regions is the [5,6]-*spiro* intermediate having structures C_2 (acid-catalyzed mechanism) and B_3 or C_3 (base-catalyzed mechanism). The authors, however, suggested that the accumulated intermediate **4** has the structure shown in Fig. 1. Following, we illustrate how **4** could be formed from **1** under acidic (Scheme 2) and basic (Scheme 3) conditions, respectively.

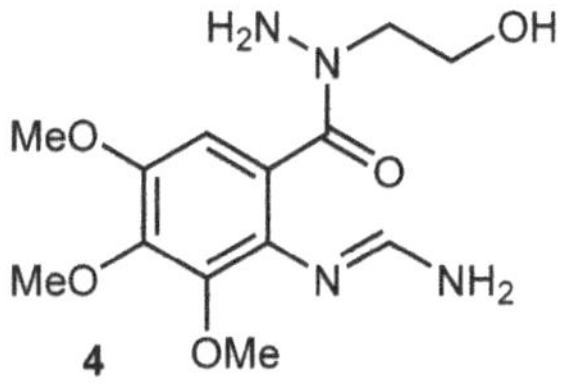

FIG. 1   Structure of intermediate **4**.

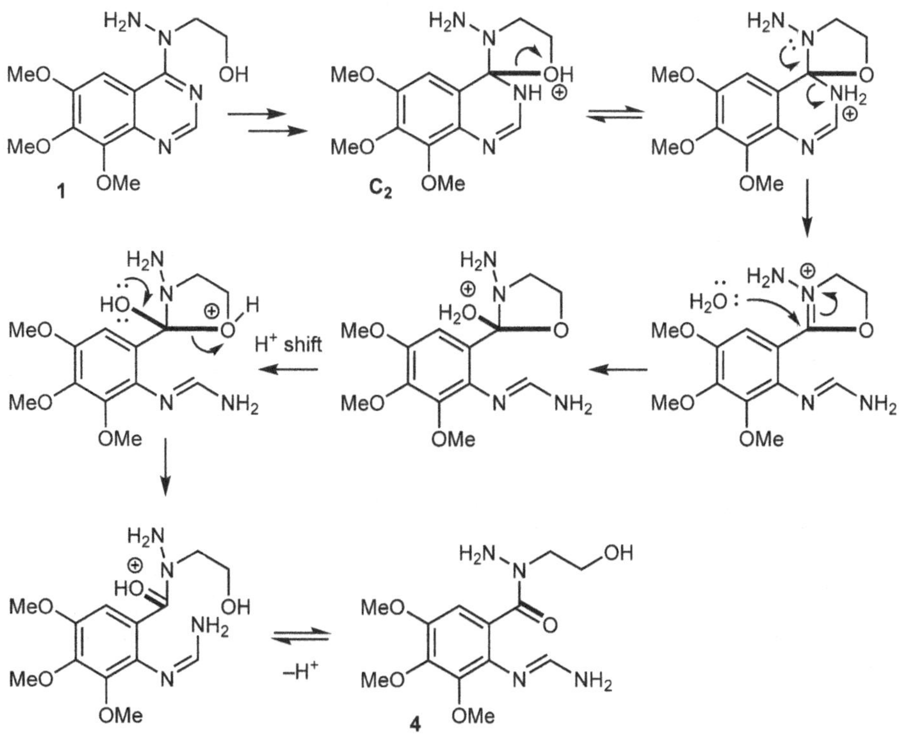

SCHEME 2

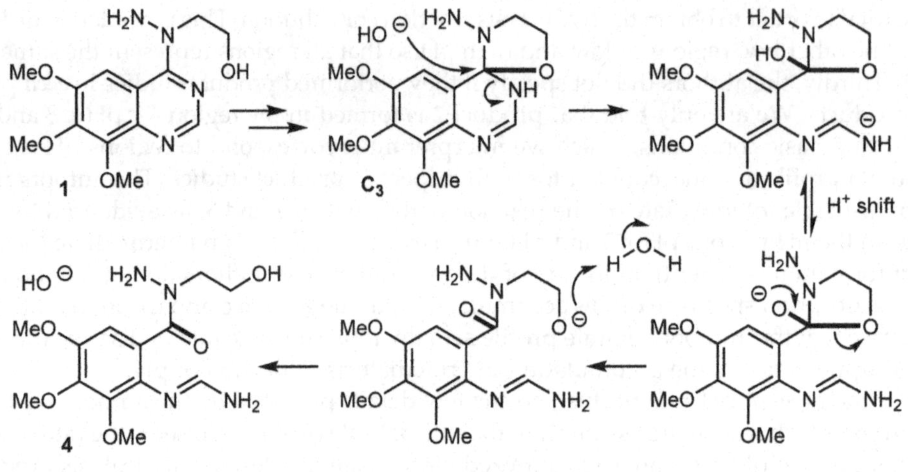

SCHEME 3

It is unlikely that a structure such as **4** would go on further to form product **3**. Such a process would require reforming the pyrimidine ring found in structure **1**. In our mechanisms presented here we see that the pyrimidine ring is not cleaved and that all of the action takes place at carbon C_4 (*spiro* carbon atom in the [5,6]-*spiro* intermediate). It is expected that **4** would be hydrolyzed further under either acidic or basic conditions to different products as shown in Scheme 4. Hence, we believe that careful product studies should be undertaken in each pH region to clarify the problem of resolving the mechanistic interpretations of this interesting reaction. This entails improving the quality of the pH-rate profile to cover *all* pH regions including (a) correlating the rate of disappearance of **1** with the rate of appearance of accumulating intermediate $C_2/B_2/B_3$, (b) correcting the observed rate constants determined in the neutral pH region for the rearrangement reaction by determining the rates of accumulation of products **2** and **3** independently.

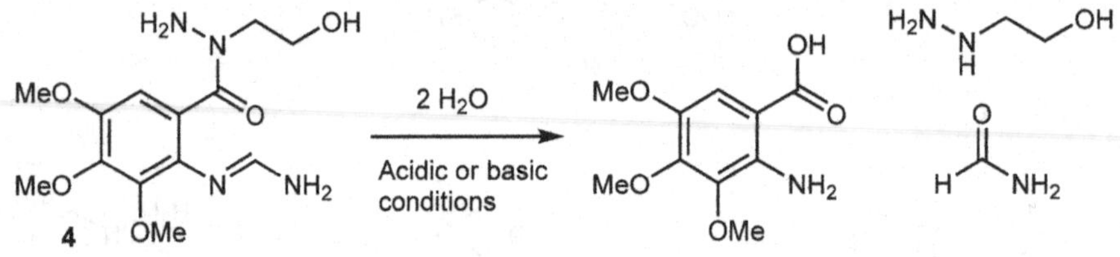

SCHEME 4

In summary, given the limited data available, we believe the most reasonable general kinetic scheme that describes the hydrolysis of **1** is given in Scheme 5.

Since the substrate **1** has three ionizable groups, it can exist in four different forms depending on the pH of the medium. Each of these forms can undergo hydrolysis via specific acid catalysis, specific base catalysis, and uncatalyzed processes. In principle, the overall pH-rate profile will have $3 \times 4 = 12$ terms and the function should exhibit three inflection points corresponding to the three equilibrium constants. Depending on the experimental results, a number of the rate constants in the kinetic model may drop out indicating that some of the arms of the kinetic model may not be applicable or may be negligible.

ADDITIONAL RESOURCES

A literature search of prior work by the same authors shows that they investigated pH-dependent hydrolyses of other heterocyclic systems including 3-amino-8-*n*-propyl-s-triazolo[4,3-*a*]pyrazine **5**,[6] 2-amino-6-methyl-4-*n*-propyl-s-triazolo-[1,5-*a*]pyrimidin-5(4*H*)-one **6**,[7] and 1,5-diamino-1*H*-s-triazolo-[1,5-*c*]quinazolinium bromide **7**,[8] the structures of which are shown in Fig. 2.

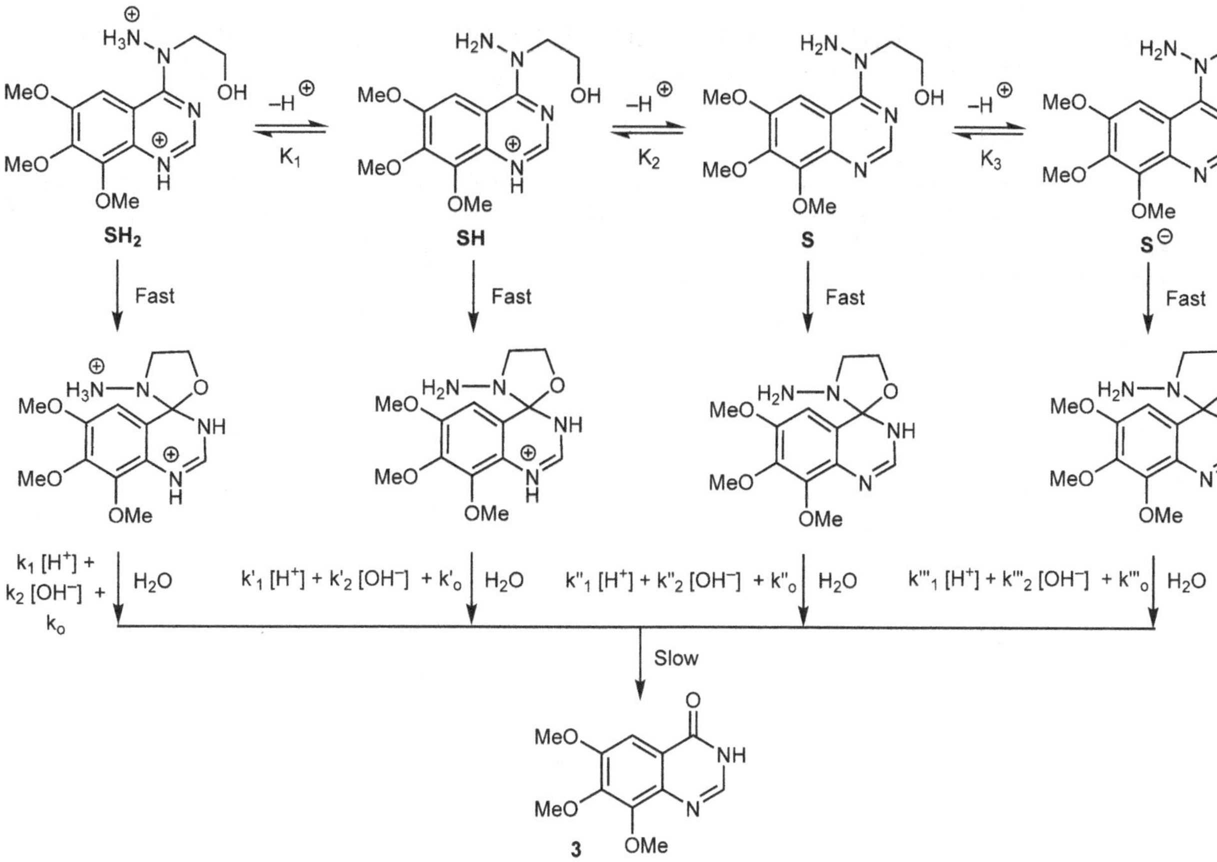

SCHEME 5

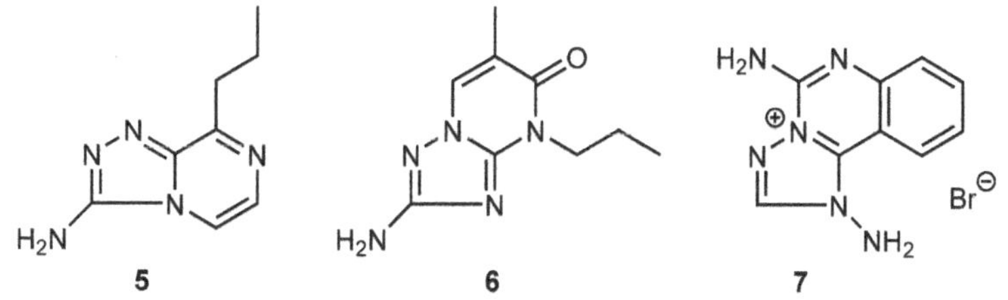

FIG. 2 Structures of heterocyclic compounds **5**, **6**, and **7**.

References

1. Barker AC, Barlow JJ, Copeland RJ, Nicholson S, Taylor PJ. Covalent hydrates as intermediates in heterocyclic rearrangements. Part 4. The hydrolysis and double Smiles rearrangement of N-(2-hydroxyethyl)-N-(6,7,8-trimethoxyquinazolin-4-yl)hydrazine. *J Chem Res (S)*. 1994;1994:248–249.
2. Edwards LJ. The hydrolysis of aspirin. A determination of the thermodynamic dissociation constant and a study of the reaction kinetics by ultraviolet spectrophotometry. *Trans Faraday Soc*. 1950;46:723–735. https://doi.org/10.1039/TF9504600723.
3. Loudon GM. Mechanistic interpretation of pH-rate profiles. *J Chem Educ*. 1991;68(12):973–984. https://doi.org/10.1021/ed068p973.
4. Marrs PS. Class projects in physical organic chemistry: the hydrolysis of aspirin. *J Chem Educ*. 2004;81(6):870–873. https://doi.org/10.1021/ed081p870.
5. van der Houwen OAGL, de Loos MR, Beijnen JH, Bult A, Underberg WJM. Systematic interpretation of pH-degradation profiles. A critical review. *Int J Pharm*. 1997;155(2):137–152. https://doi.org/10.1016/S0378-5173(97)00156-7.
6. Nicholson S, Stacey GJ, Taylor PJ. Covalent hydrates as transient species in heterocyclic rearrangements. Part I. The ring fission of some s-triazolopyrazines. *J Chem Soc Perkin Trans 2*. 1972;1972:4–11. https://doi.org/10.1039/P29720000004.
7. Dukes M, Nicholson S, Stacey GJ. Covalent hydrates as intermediates in heterocyclic rearrangements. Part II. The hydrolysis of 2-amino-6-methyl-4-n-propyl-s-triazolo-[1,5-a]pyrimidin-5 (4H)-one. *J Chem Soc Perkin Trans 2*. 1972;1972:1695–1697. https://doi.org/10.1039/P29720001695.
8. Bowie RA, Edwards PN, Nicholson S, Taylor PJ, Thomson DA. Covalent hydrates as intermediates in heterocyclic rearrangements. Part 3. The alkali-catalysed transformations of 1,5-diamino-1H-s-triazolo[1,5-c]quinazolinium bromide. *J Chem Soc Perkin Trans 2*. 1979;1979:1708–1714. https://doi.org/10.1039/P29790001708.

Question 233: A Failed Approach to the Oxetan-3-one System

During a 1984 attempted preparation of the oxetane-3-one **5** based on a conventional one-pot method starting from the 1,3-dithiane **1** (see the **1** to **2** to **3** synthetic sequence), Thai chemists have unexpectedly isolated the ring expanded products **6** and **7** instead according to the reactions depicted in Scheme 1.[1] Explain both the expected and unexpected results in mechanistic terms.

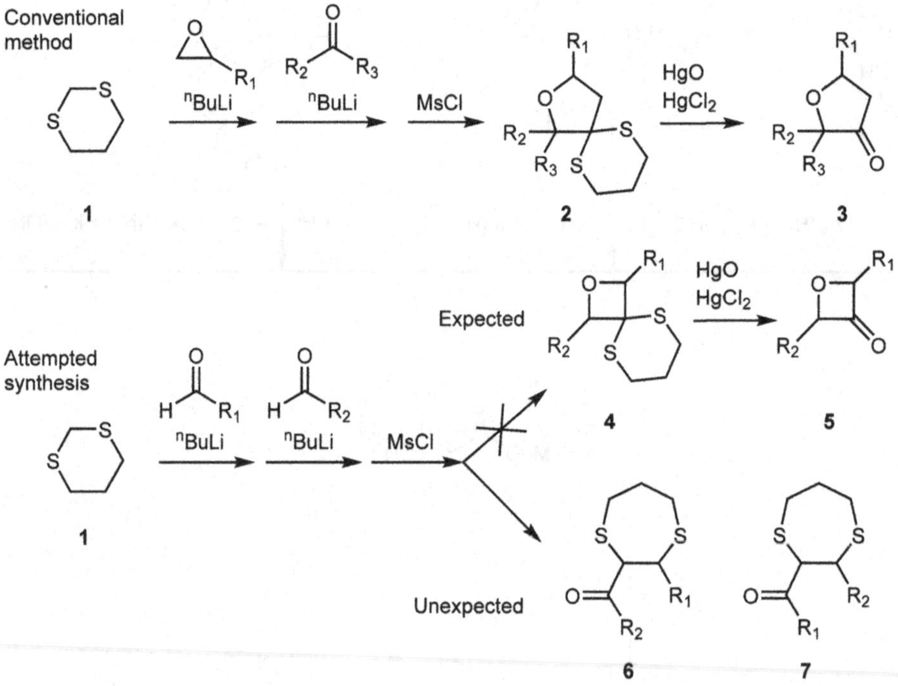

SCHEME 1

We shall begin with the mechanism for the conventional method of transforming 1,3-dithiane **1** to product **2** (Scheme 2). We see therefore that the mechanism begins with deprotonation of 1,3-dithiane by the n-butyllithium base resulting in a carbanion that attacks the less sterically hindered carbon atom of the epoxide to give intermediate **A**. A second deprotonation occurs on the dithiane moiety generating a second carbanion that attacks the ketone to give intermediate **C**. Mesylation leads to either intermediate **D$_1$** or **D$_2$**. Nevertheless, both of these intermediates are able to cyclize to the common [5,6]-*spiro* product **2**. The net result is a [3 + 2] cycloaddition strategy. When the epoxide and ketone materials are replaced by two aldehydes, a very similar mechanistic sequence follows which results in the formation of intermediates **H$_1$** and **H$_2$** (Scheme 3).

Unfortunately intermediates **H$_1$** and **H$_2$** as the authors observed do not react further to form the expected product **4** (Scheme 4). Presumably this is because of increased ring strain in creating a four-membered ring. Instead, **H$_1$** and **H$_2$** cyclize to give sulfonium ion intermediates **I$_1$** and **I$_2$**, which rearrange to epoxides **J$_1$** and **J$_2$**. These epoxides then ring open to give sulfonium alkoxide species **K$_1$** and **K$_2$** which undergo a final 1,2-hydride shift to yield the isomeric products **6** and **7**, respectively (Schemes 5 and 6).

Interestingly, the authors showed one synthetic example using different aldehydes, acetaldehyde and benzaldehyde, in a sequence that leads to *one* product corresponding to the structure of product **7** where R$_1$=Ph and R$_2$=Me, or the structure of product **6** where R$_1$=Me and R$_2$=Ph. No explanation was given as to why a single geometric isomer was produced though two possible mesylated intermediates **H$_1$** and **H$_2$** can be generated from the dilithiated salt **G**. A theoretical study would lend more credibility to the postulated intermediates along these mechanistic pathways.

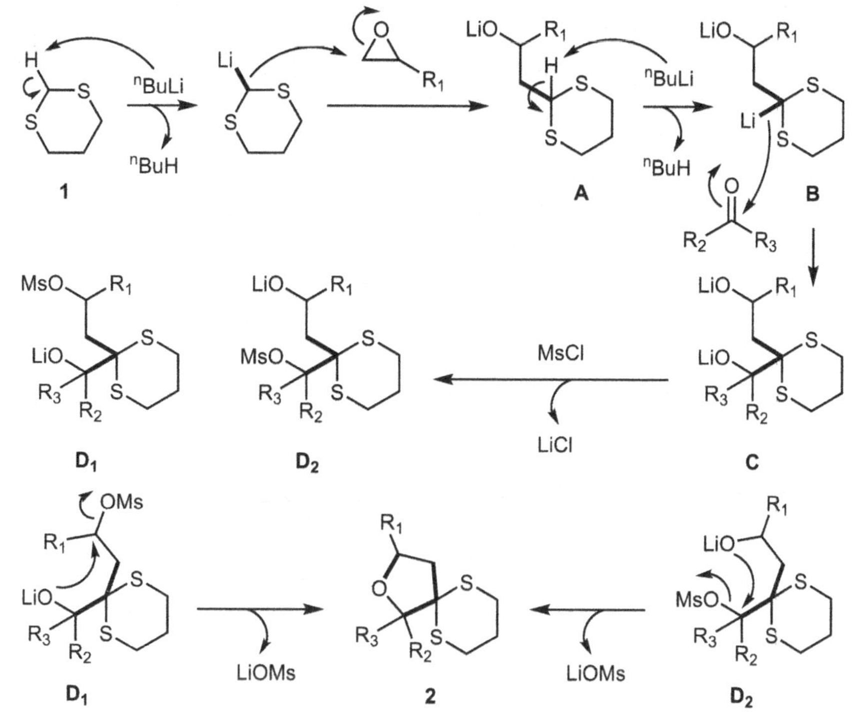

SCHEME 2

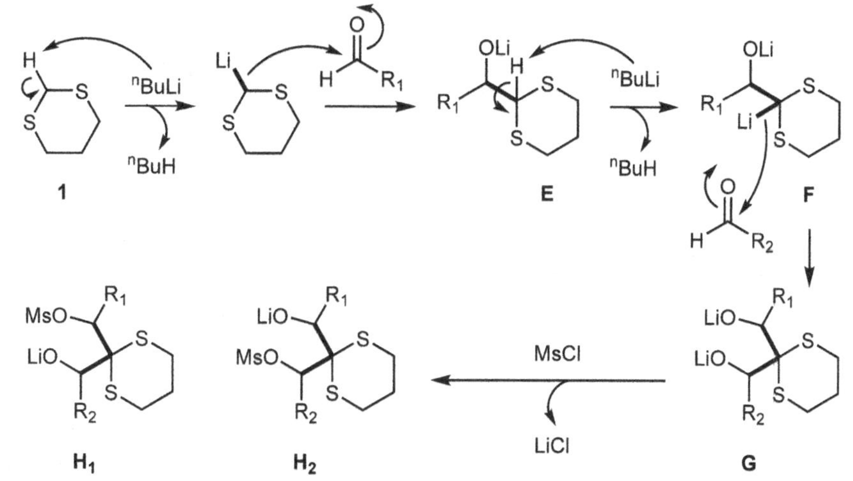

SCHEME 3

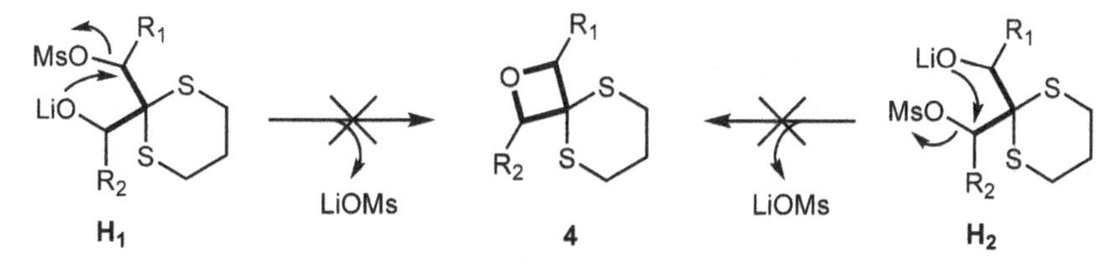

SCHEME 4

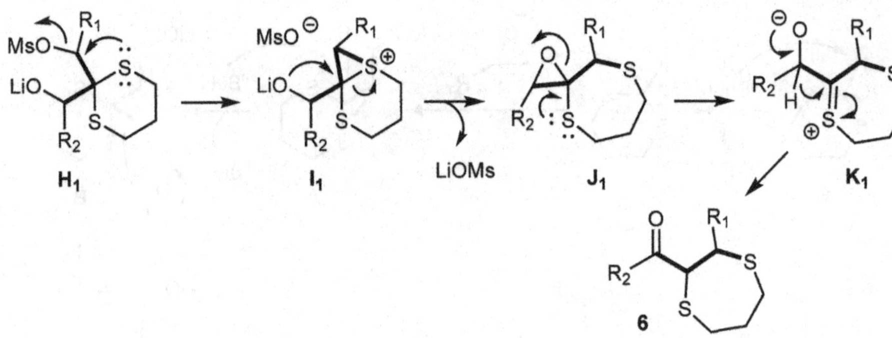

SCHEME 5

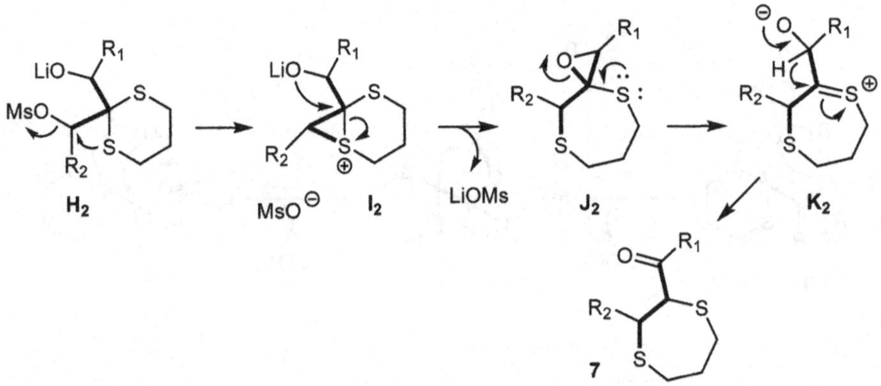

SCHEME 6

BALANCED CHEMICAL EQUATIONS

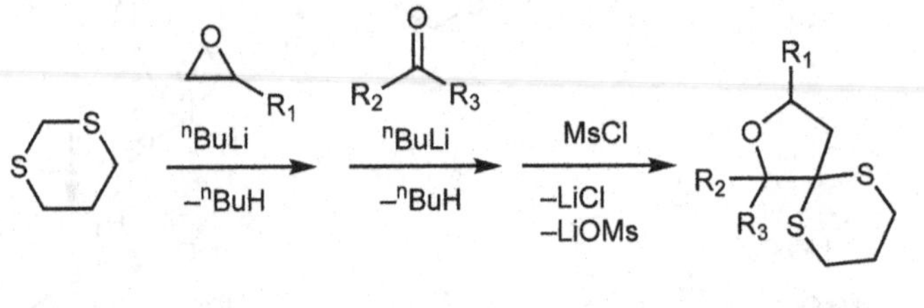

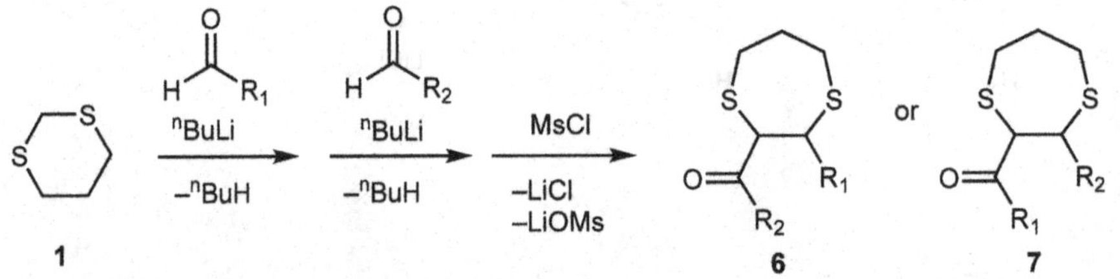

ADDITIONAL RESOURCES

For an interesting paper outlining the possible geometric configurations of aryl sulfimides, see Ref. 2. For a mechanistic discussion about a ring expansion transformation in the context of 1,3-dithiane chemistry, see the work of Ranu and Jana.[3]

References

1. Tarnchompoo B, Thebtaranonth Y. A condensed synthesis of dihydro-3(2H)-furanone. *Tetrahedron Lett.* 1984;25(48):5567–5570. https://doi.org/10.1016/S0040-4039(01)81628-2.

2. Bailer J, Claus PK, Vierhapper FW. Configurationally and conformationally homogeneous cyclic n-aryl sulfimides—IV: Synthesis, configuration and stereochemistry of the rearrangement of 1,3-di-thiane-1-N-p-chlorophenylimides. *Tetrahedron*. 1980;36(7):901–911. https://doi.org/10.1016/0040-4020(80)80041-X.
3. Ranu BC, Jana U. A new redundant rearrangement of aromatic ring fused cyclic α-hydroxydithiane derivatives. Synthesis of aromatic ring fused cyclic 1,2-diketones with one-carbon ring expansion. *J Org Chem*. 1999;64:6380–6386. https://doi.org/10.1021/jo990634s.

Question 234: Tricyclics From Furfural

During the 1990s, Japanese chemists working on synthesizing tricyclic molecules achieved the synthesis of **6** from furfural according to Scheme 1.[1] Explain this synthetic sequence in mechanistic terms.

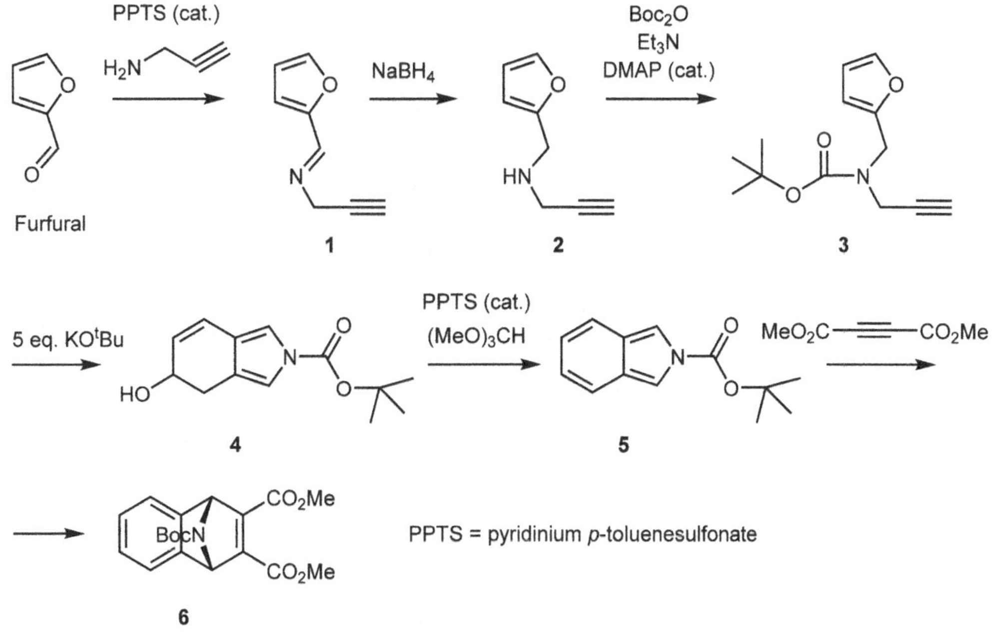

SCHEME 1

SOLUTION: FURFURAL TO 2

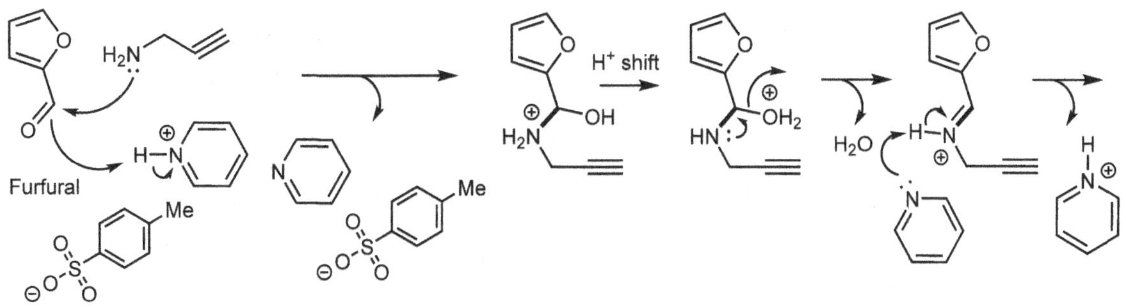

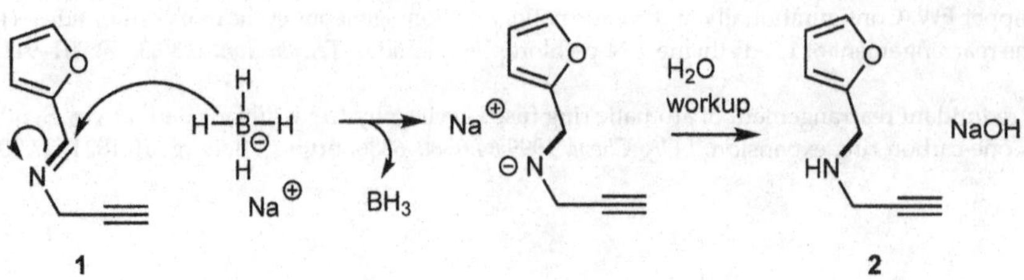

SOLUTION: 2 TO 3

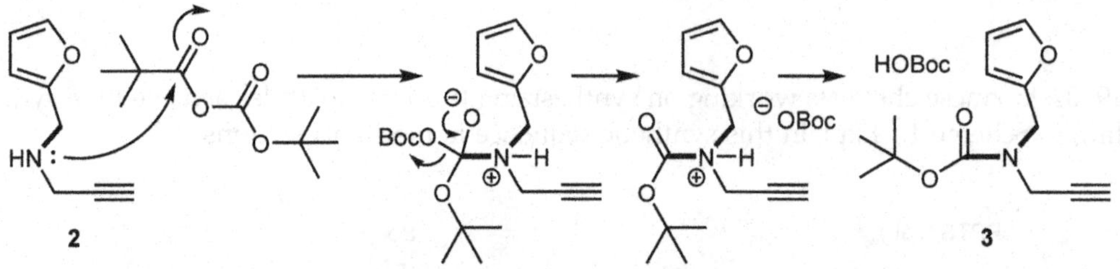

SOLUTION: 3 TO 4

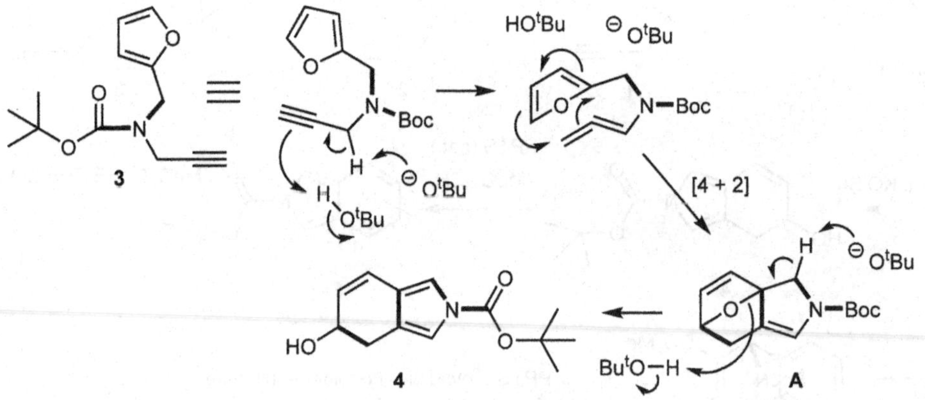

SOLUTION: 4 TO 5

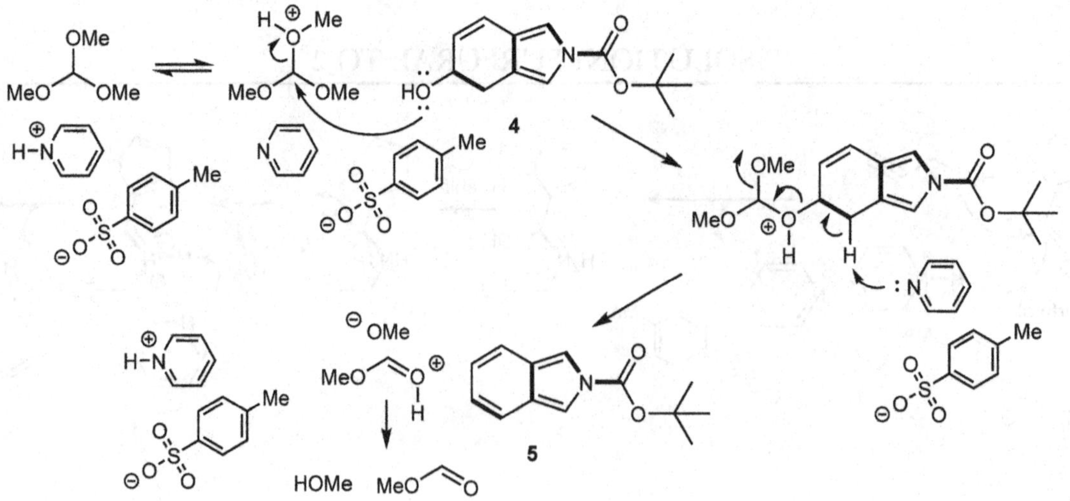

SOLUTION: 5 TO 6

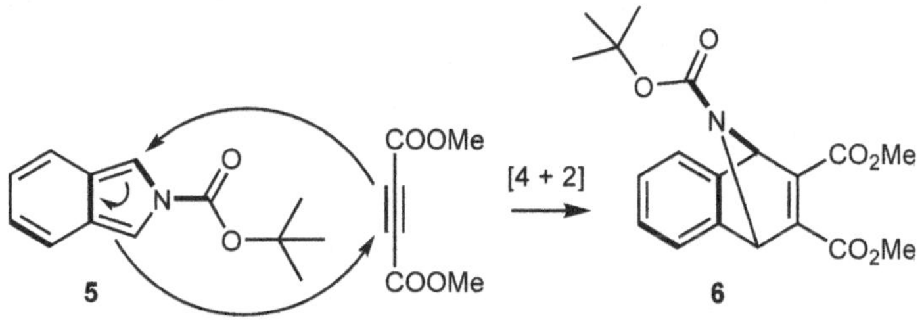

KEY STEPS EXPLAINED

The synthetic plan performed by Lee et al. begins with imination of furfural by propargylamine which proceeds via a catalytic amount of pyridinium *p*-toluenesulfonate (PPTS) and generates the imine **1** after elimination of water. A subsequent reduction using sodium borohydride transforms the imino group of **1** into an alkyl amino group to give **2**. The third step constitutes an amide bond formation where an N-Boc protecting group is installed onto the alkyl amino group and thus product **3** is formed. In the fourth synthetic step, arguably the most interesting of the entire synthesis, we observe the transformation of a terminal alkyne functional group to an allene which can then undergo a Diels-Alder [4+2] cyclization to give intermediate **A**. This *tert*-butoxide base-mediated transformation results in an overall [(4+2)+(5+0)] ring construction strategy for this step. Reaction of *tert*-butoxide base with **A** causes ring opening of **A** to generate product **4**. Moreover, the fifth step in the synthesis constitutes a dehydration assisted by PPTS acid catalyst and trimethylorthoformate to give product **5**. Finally, a second [4+2] Diels-Alder cyclization between **5** and dimethyl acetylenedicarboxylate (DMAD) leads to the final tricyclic product **6**.

In terms of experimental evidence, the authors reported that use of unprotected **2** or use of methyl substituted propargylamine did not lead to the desired products but only to decomposition.[1] Thus the authors rationalize the importance of the Boc protecting group in two ways. First, the electron-withdrawing nature of Boc is believed to suppress the electron-donating properties of the lone electron pair on nitrogen thus preventing a potential further prototropic isomerization of the allene group to propyn-1-ylamine as shown in Scheme 2.[1] Second, the bulky nature of the Boc group is thought to present an entropic effect that might "(enable) the allene group to adopt the correct conformation to react with (the) furan" group in the [4+2] cycloaddition leading to intermediate **A**.[1] The dehydration of **4** to **5** could not be effected directly without the use of auxiliary reagents.[1] Lastly, the authors were able to isolate the transient isoindole **5** by reacting it with various reactive dienophiles.[1] Certain other properties of [4+2] cycloadditions were also exploited by the authors with respect to the last synthetic step, that is, the transformation of **5** to **6**. In view of these experimental results which support the postulated mechanisms, we would only comment that a theoretical analysis would make an even stronger case for the proposed solution.

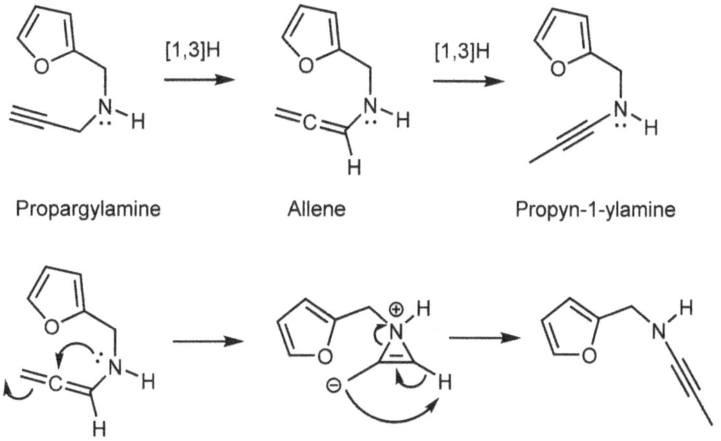

Propargylamine   Allene   Propyn-1-ylamine

SCHEME 2

ADDITIONAL RESOURCES

For an excellent review article outlining the synthetic applications of Diels-Alder reactions involving furan chemistry, see the work of Kappe et al.[2] For two related research papers containing mechanistic discussion about very similar transformations, see Refs. 3, 4.

References

1. Lee M, Moritomo H, Kanematsu K. A new route to the isoindole nucleus via furan-pyrrole ring-exchange. *J Chem Soc Chem Commun.* 1994;1994:1535. https://doi.org/10.1039/C39940001535.
2. Kappe CO, Murphree SS, Padwa A. Synthetic applications of furan Diels-Alder chemistry. *Tetrahedron.* 1997;53(42):14179–14233. https://doi.org/10.1016/S0040-4020(97)00747-3.
3. Wu H-J, Yen C-H, Chuang C-T. Intramolecular Diels-Alder reaction of furans with allenyl ethers followed by sulfur and silicon atom-containing group rearrangement. *J Org Chem.* 1998;63:5064–5070. https://doi.org/10.1021/jo980240l.
4. Wu H-J, Lin C-F, Wang Z, Lin H-C. Intramolecular Diels–Alder reaction of 2-diphenylphosphinyl-5-(propargyloxymethyl)furans followed by nucleophilic 1,2-rearrangement of the phosphinyl group. *Tetrahedron Lett.* 2007;48:6192–6194. https://doi.org/10.1016/j.tetlet.2007.06.163.

Further Reading

5. Lee M, Moritomo H, Kanematsu K. Construction of three types of fused isoindoles via furan-pyrrole ring exchange reaction. *Tetrahedron.* 1996;52(24):8169–8180. https://doi.org/10.1016/0040-4020(96)00386-9.

Question 235: An Ylide-Based Synthesis of 4-Phenylisocoumarin

A 1984 investigation of the reactions of 2-acylbenzoates with dimethyloxosulphonium methylide found that reactions of 2-benzoylbenzoate **1** with one equivalent of methylide formed 4-phenylisocumarin **2** (52% yield) whereas use of two equivalents of methylide gave the stable ylide **3** in 96% yield (Scheme 1).[1] Suggest mechanisms for these two transformations.

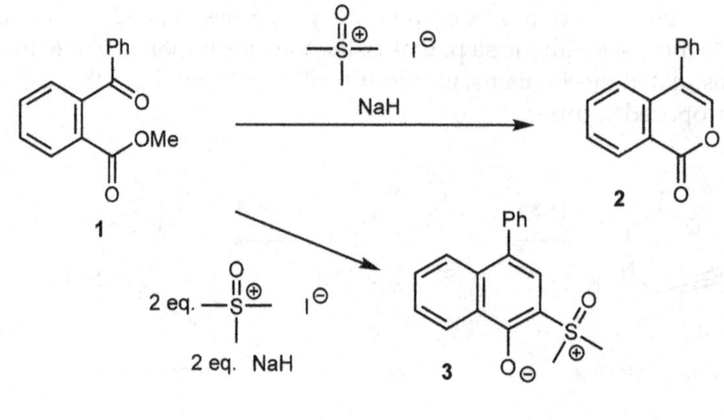

SCHEME 1

685

SOLUTION: 1 → 2

SOLUTION: 2 → 3

BALANCED CHEMICAL EQUATIONS

KEY STEPS EXPLAINED

The mechanism for the transformation of 2-benzoylbenzoate **1** to 4-phenylisocumarin **2** begins with generation of oxo-sulfonium ylide which then attacks the ketone carbonyl group of **1** to form A_1 which undergoes cyclization to the lactone B_1. A deprotonation via methoxide anion then causes ring opening of the lactone to form intermediate C_1 which recy-clizes to form 4-phenylisocumarin product **2** and dimethyl sulfoxide by-product. If another equivalent of oxosulfonium ylide is generated, it further reacts with **2** to open up the isocoumarin ring and generate the enolate A_2 which ketonizes to the aldehyde B_2. Deprotonation of B_2 via methoxide anion leads to the enolate C_2 which then cyclizes to the six-membered carbocycle D_2 after protonation via methanol. Methoxide deprotonation enables elimination of hydroxide anion to generate intermediate E_2 which undergoes dehydration to form the aromatized ylide product **3**.

In *lieu* of any theoretical or experimental analysis to help support any of the proposed intermediates, the authors did perform an experiment in which **2** was treated with one equivalent of methylide in order to form the stable ylide **3** thus confirming the expectation that **2** is a stable intermediate along the mechanistic pathway from **1** to **3**.

ADDITIONAL RESOURCES

For a discussion of an interesting rearrangement occurring during the synthesis of a stable ylide using methylide as a reagent, see the work of Sanders et al.[2] For an example of an usual reaction of dimethylsulfoxonium methylide with esters, see the work of Leggio et al.[3]

References

1. Beautement K, Clough JM. Reactions of 2-acylbenzoates with dimethyloxosulphonium methylide: a novel route to isocumarins. *Tetrahedron Lett.* 1984;25(28):3025–3028. https://doi.org/10.1016/S0040-4039(01)81355-1.
2. Sanders WJ, Zhang X, Wagner R. Rearrangements encountered in the attempted syntheses of pyridoazepinone carboxylic acids. *Org Lett.* 2004;6 (24):4527–4530. https://doi.org/10.1021/ol0481378.
3. Leggio A, Marco RD, Perri F, Spinella M, Liguori A. Unusual reactivity of dimethylsulfonium methylide with esters. *Eur J Org Chem.* 2012;2012 (1):114–118. https://doi.org/10.1002/ejoc.201101031.

Question 236: Model Studies for Mitomycin A Synthesis Lead to a New Preparation of Pyrroloindoles

In 1989 French chemists working on the synthesis of mitomycin A showed that treatment of (*E*,*E*)-hexa-2,4-dienal **1** and nitrosobenzene in absolute ethanol at room temperature gave a mixture containing the pyrroloindole analog **2**, the betaine **3** and the pyrrole **4** (Scheme 1).[1] Suggest mechanisms for all of these transformations.

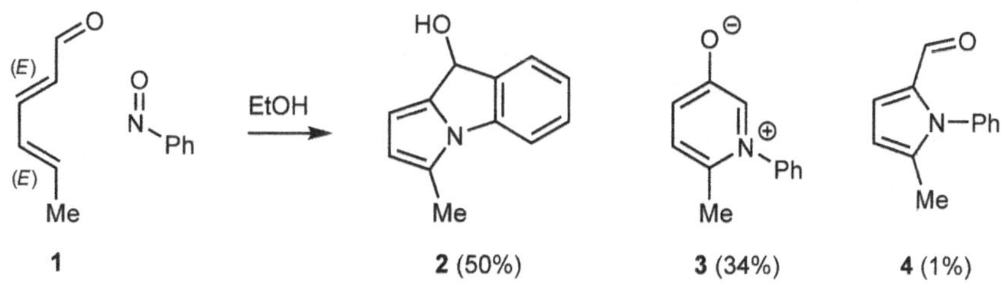

SCHEME 1

SOLUTION: 1 → 2

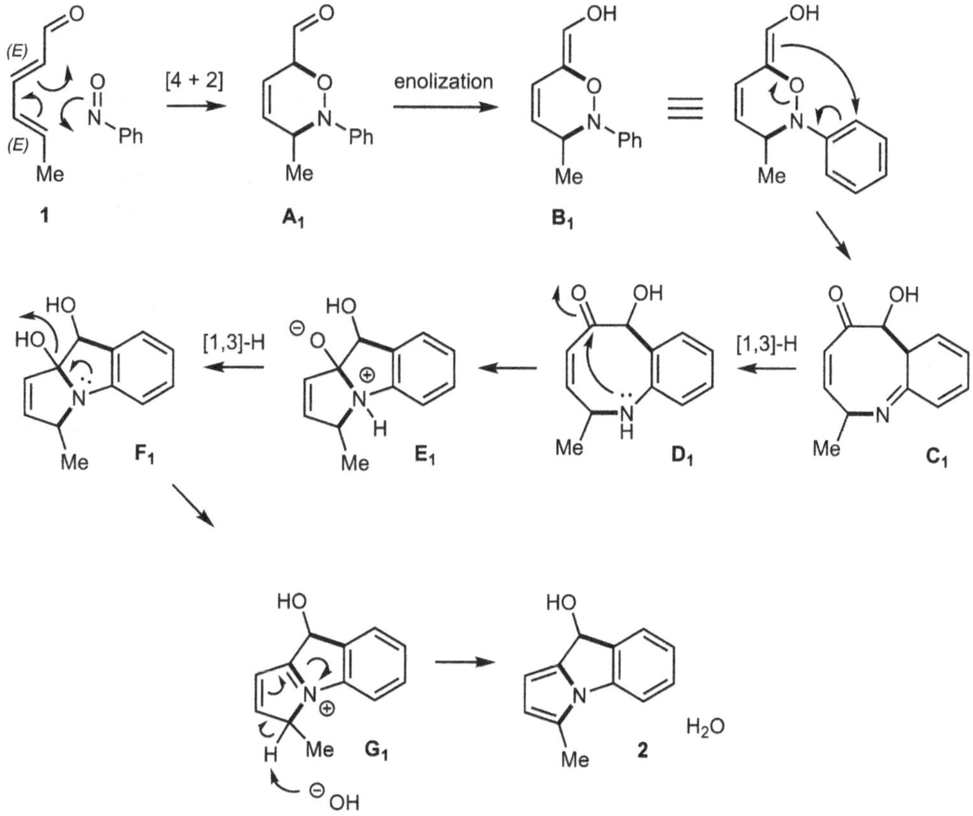

SOLUTION: 1 → 3

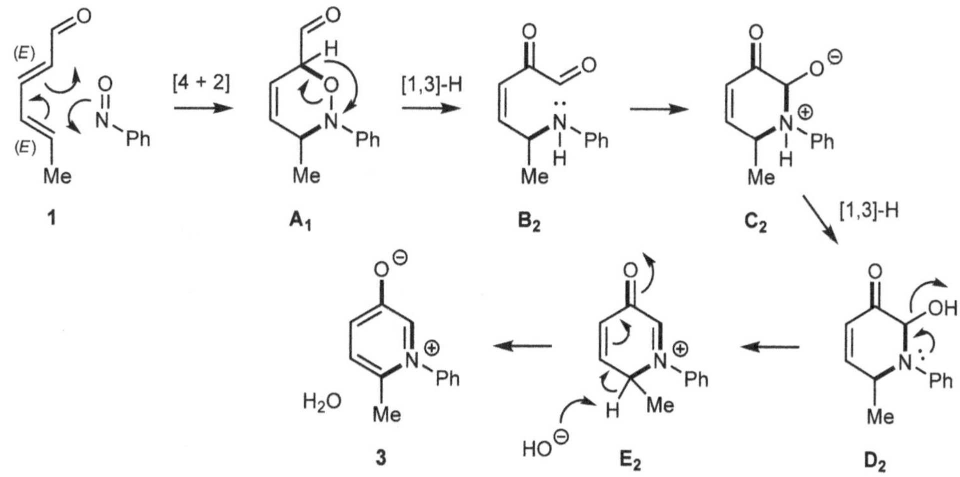

SOLUTION: 1 → 4

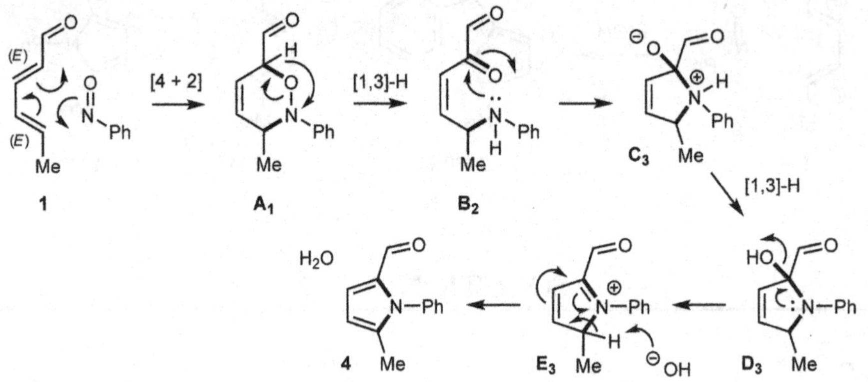

KEY STEPS EXPLAINED

For the transformation of (E,E)-hexa-2,4-dienal **1** and nitrosobenzene into the pyrroloindole analog **2**, the mechanism begins with a [4+2] hetero-Diels-Alder cycloaddition to form intermediate **A₁** followed by enolization of the aldehyde group to form intermediate **B₁**. Cyclization of **B₁** at the phenyl group forms the eight-membered ring in **C₁** which undergoes a [1,3] hydride shift to rearomatize the phenyl ring and lead to **D₁**. This intermediate can then cyclize in a transannular manner to form **E₁** which has the fused bicyclic [3.3.0] ring system. A second [1,3] hydride shift forms **F₁** which eliminates hydroxide anion to form the iminium ion intermediate **G₁**. A final deprotonation leads to the pyrroloindole analog **2** plus water as a by-product.

The transformation of **1** and nitrosobenzene to the betaine **3** follows the same initial hetero-Diels-Alder cycloaddition to **A₁** but instead of an enolization to **B₁** we find a [1,3] hydride shift which results in fragmentation of the six-membered ring to generate intermediate **B₂**. This intermediate recyclizes at the terminal aldehyde group to give a new six-membered ring intermediate **C₂** which undergoes a [1,3] hydride shift, hydroxide ion elimination, and deprotonation sequence to form the betaine product **3** and water as a by-product.

Finally, the transformation of **1** and nitrosobenzene to the pyrrole **4** follows the same mechanistic pathway to intermediate **B₂** (hetero-Diels-Alder cycloaddition plus [1,3] hydride shift), but now instead of having recyclization at the terminal aldehyde group to give **C₂**, we have recyclization at the ketone group to give the five-membered ring intermediate **C₃**. This intermediate too undergoes a similar [1,3] hydride shift, hydroxide ion elimination, and deprotonation sequence to form the pyrrole **4** plus water as a by-product.

For the purpose of visualization, we include a trajectory of the mechanistic fate of **1** and nitrosobenzene in Scheme 2. Therein we can clearly see a divergence point (two possible mechanistic paths) at intermediate **A₁**, thus distinguishing the **1** to **2** transformation from the other two, and at intermediate **B₂**, which differentiates transformations **1** to **3** and **1** to **4**.

Although the reported yields indicate the favorability of each mechanistic pathway (i.e., enolization is more favorable than [1,3] hydride shift in **A₁**, and recyclization with terminal aldehyde is much more favorable than with internal ketone in **B₂**), an important follow-up would be to conduct a theoretical analysis to confirm the energetics surrounding these divergences.

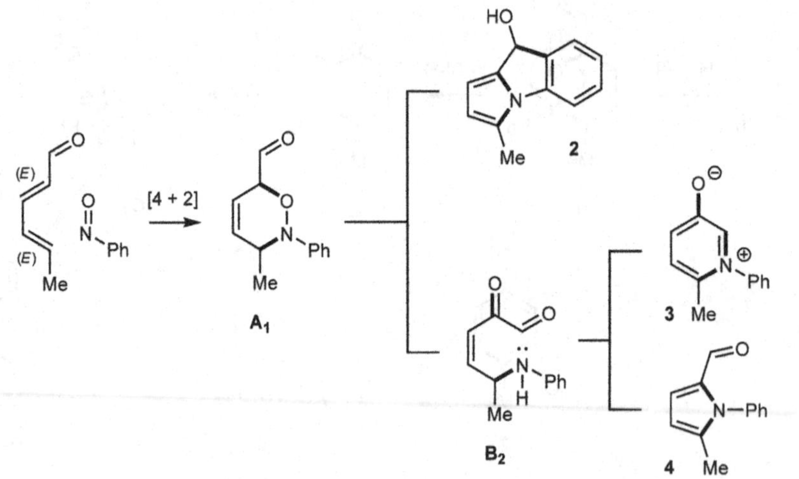

SCHEME 2

ADDITIONAL RESOURCES

We direct the reader to an interesting mechanistic investigation of various thermal cyclization mechanistic pathways available to a chemical system very similar to the one encountered here.[2] In addition, we recommend an excellent review by the same authors which covers cascade rearrangements in the context of hetero-Diels-Alder cycloadditions with nitroso dienophiles.[3]

References

1. Defoin A, Geffroy G, Nouen DL, Spileers D, Streith J. 131. Cascade reactions. A simple one-pot synthesis of the mitomycin skeleton. *Helv Chim Acta*. 1989;72(6):1199–1215. https://doi.org/10.1002/hlca.19890720604.
2. McNab H, Reed D, Tipping ID, Tyas RG. Thermal cyclisation reactions of methyl 2-(pyrrol-1-yl)cinnamate and methyl 3-(1-phenylpyrrol-2-yl) propenoate. *ARKIVOC*. 2007;2007:85–95.
3. Streith J, Defoin A. Azasugar syntheses and multistep cascade rearrangements via hetero Diels-Alder cycloadditions with nitroso dienophiles. *Synlett*. 1996;1996(3):189–200. https://doi.org/10.1055/s-1996-5366.

Question 237: Formation of N-Cyanofluoren-9-imine From 9-Dinitromethylenefluorene

In 1979 Israeli chemists studying the reactions of nitroolefins with nucleophiles reacted 9-dinitromethylenefluorene (DNF) **1** with a solution of sodium azide in DMSO at ambient temperature. Their reaction released a gaseous by-product and the *N*-cyanofluoren-9-imine **2** (Scheme 1).[1] Suggest a mechanism for this transformation.

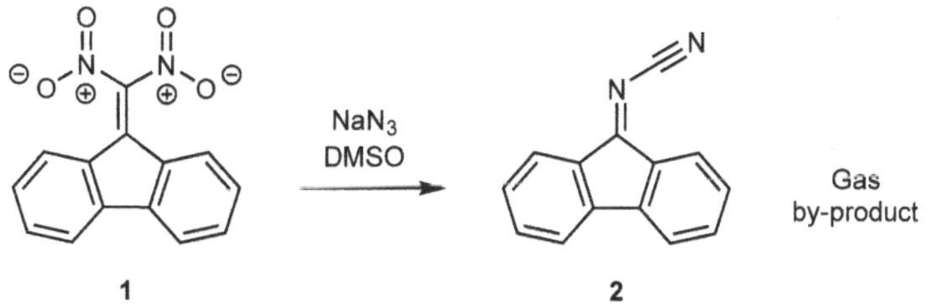

SCHEME 1

SOLUTION

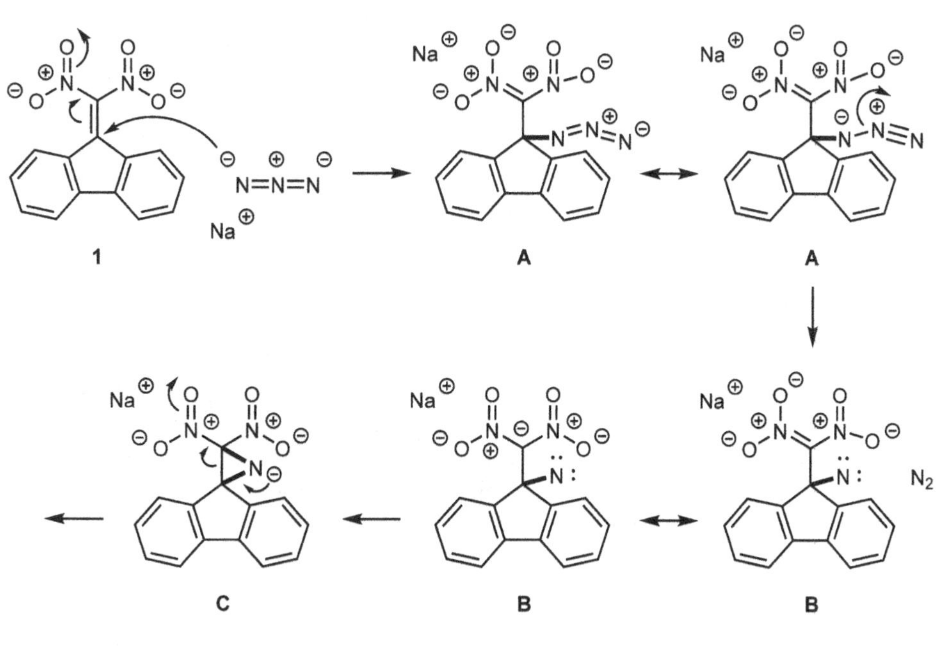

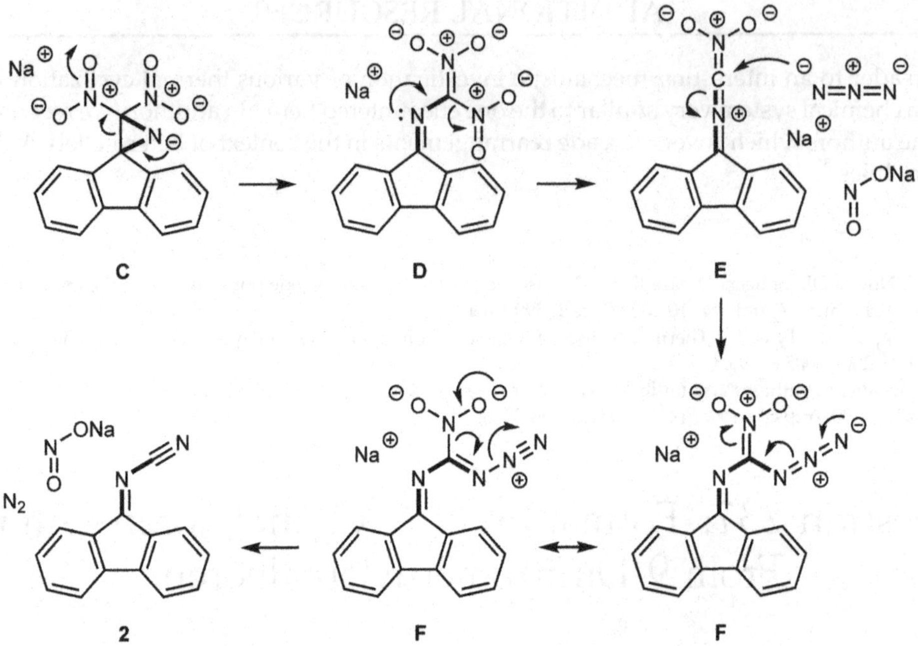

BALANCED CHEMICAL EQUATION

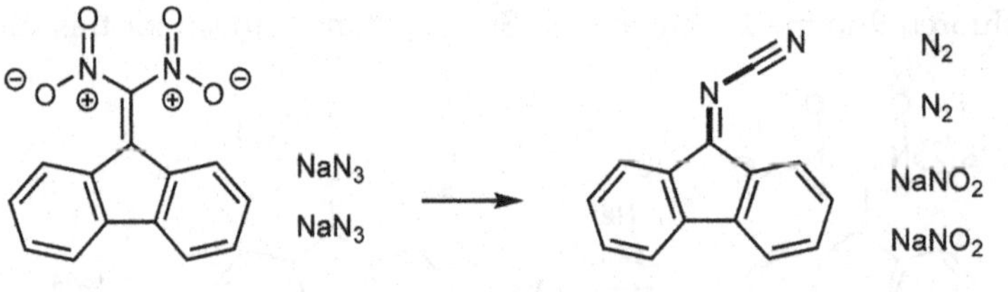

KEY STEPS EXPLAINED

The mechanism proposed by Hoz and Speizman begins with azide anion nucleophilic attack at the C_9 position of the fluorene moiety of **1** followed by nitrogen gas elimination to the carbene intermediate **B** which rearranges to the aziridine **C**. Intermediate **C** can undergo facile rearrangement to the imine intermediate **D** which can eliminate sodium nitrite to form a resonance stabilized nitrile ylide **E**. This ylide is attacked by another equivalent of azide anion at the C_1 position (especially since **E** is also activated by the electron-withdrawing nature of the nitro group) to form intermediate **F** which can easily eliminate sodium nitrite and nitrogen gas to form the final N-cyanofluoren-9-imine product **2**.

Concerning the mechanism of this transformation, the authors obtained the following supporting experimental evidence: (1) a positive test for nitrite using the iodide-starch paper test method; (2) the structure of **2** was corroborated by use of elemental analysis, mass spectrometry, IR, and ^1H NMR spectroscopy; and (3) vigorous gas evolution corresponding to production of two equivalents of nitrogen gas.

Furthermore, the authors do not provide any evidence to support the postulated aziridine intermediate **C**. They nevertheless believe that the transformation of **A** to **D** could proceed without the intermediacy of the "highly energetic" nitrene **B**. We include both structures for illustrative purposes of completeness noting that this question can be resolved by performing a computational study that maps the energy landscape of the intermediates involved along the entire mechanistic pathway. An experimental possibility would be to add an exclusive nitrene trap to the reaction mixture (e.g., imines, nitriles, nitroso compounds, thioketones, or ketones) and analyze the product mixture for the resultant trapped product.[2-4] For example, one may conduct the reaction in acetonitrile solvent to see if the nitrene could be trapped as shown in Scheme 2. This may be difficult to carry out since such an intermolecular trap would have to intercept a putative nitrene faster than it intramolecularly rearranges. Hence, to be successful a strict kinetic condition is required which may or may not be fulfilled.

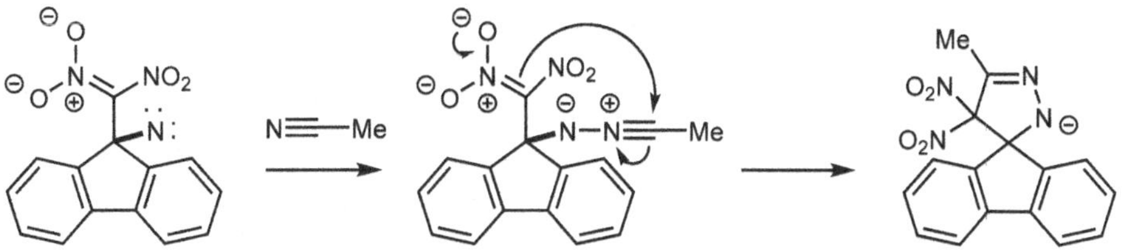

SCHEME 2

ADDITIONAL RESOURCES

We would like to refer the reader first to the work of Lindley et al. which describes a competitive cyclization between singlet and triplet nitrenes.[5] We would also like to refer the reader to three interesting transformations involving aziridines in the context of mechanistic analysis.[6-8]

References

1. Hoz S, Speizman D. An unusual reaction of azide with nitroolefins. *Tetrahedron Lett.* 1979;20(50):4855–4856. https://doi.org/10.1016/S0040-4039(01)86731-9.
2. Granier M, Baceiredo A, Gruetzmacher H, Pritzkow H, Bertand G. Direct evidence for a nitrile imine–imidoylnitrene rearrangement: X-ray crystal structure of an unusual nitrene complex. *Angew Chem Int Ed.* 1990;29(6):659–661. https://doi.org/10.1002/anie.199006591.
3. Tomioka H. Versatile reactions undergone by carbenes and nitrenes in noble gas matrices at cryogenic temperatures. *Bull Chem Soc Jpn.* 1998;71(7):1501–1524. https://doi.org/10.1246/bcsj.71.1501.
4. Murata S, Abe S, Tomioka H. Photochemical reactions of mesityl azide with tetracyanoethylene: competitive trapping of singlet nitrene and didehydroazepine. *J Org Chem.* 1997;62(10):3055–3061. https://doi.org/10.1021/jo9622901.
5. Lindley JM, McRobbie IM, Meth-Cohn O, Suschitzky H. Competitive cyclisations of singlet and triplet nitrenes. Part 5. Mechanism of cyclisation of 2-nitrenobiphenyls and related systems. *J Chem Soc Perkin Trans 1.* 1977;1977:2194–2204. https://doi.org/10.1039/P19770002194.
6. Phung C, Tantillo DJ, Hein JE, Pinhas AR. The mechanism of the reaction between an aziridine and carbon dioxide with no added catalyst. *J Phys Org Chem.* 2017;https://doi.org/10.1002/poc.3735.
7. Stamm H. Nucleophilic ring opening of aziridines. *J Prakt Chem.* 1999;341(4):319–331. https://doi.org/10.1002/(SICI)1521-3897(199905)341:4<319::AID-PRAC319>3.0.CO;2-9.
8. Jarvis AN, McLaren AB, Osborn HMI, Sweeney J. Preparation and ring-opening reactions of *N*-diphenylphosphinyl vinyl aziridines. *Beilstein J Org Chem.* 2013;9:852–859. https://doi.org/10.3762/bjoc.9.98.

Question 238: Methylphenylacetic Acids From Butenolides

During studies of the aromatization of aliphatic compounds, Italian chemists observed that treatment of butenolides **1** and **3** with molten pyridine hydrochloride gave the methylphenylacetic acids **2** and **4** (Scheme 1).[1] Provide mechanisms for both transformations.

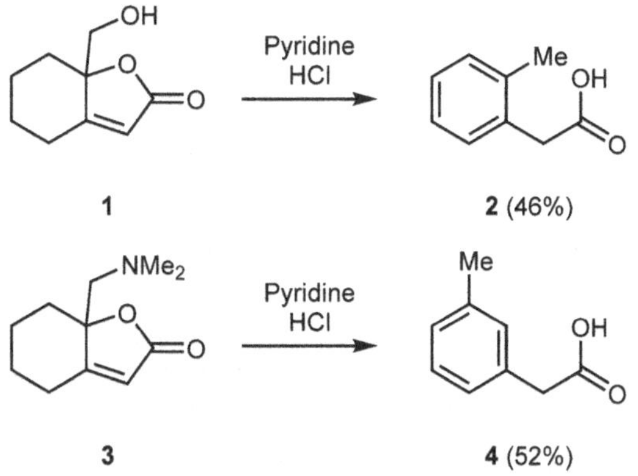

SCHEME 1

SOLUTION: 1 → 2

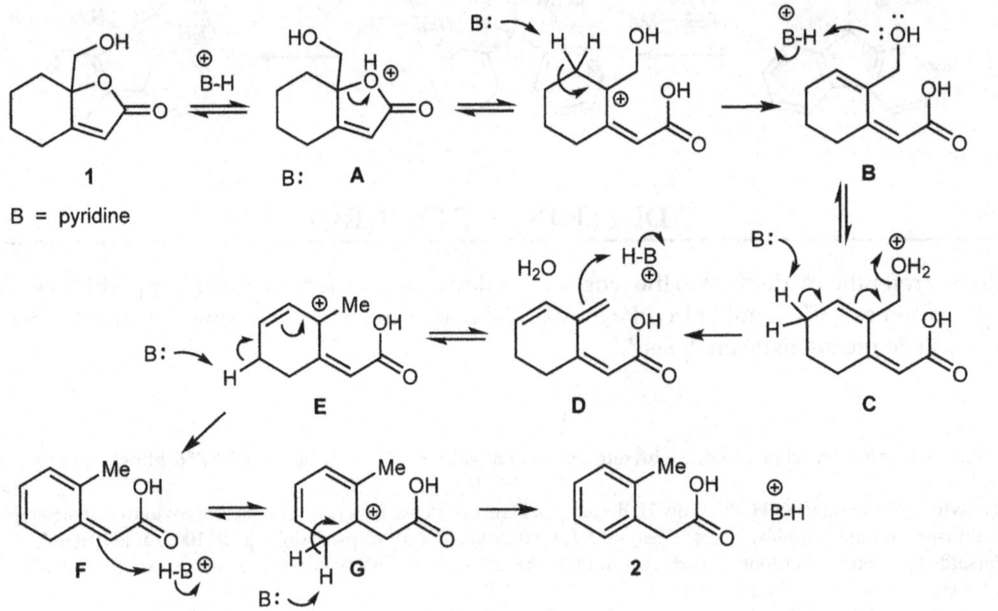

1
B = pyridine
A
B
C
D
E
F
G
2

SOLUTION: 3 → 4

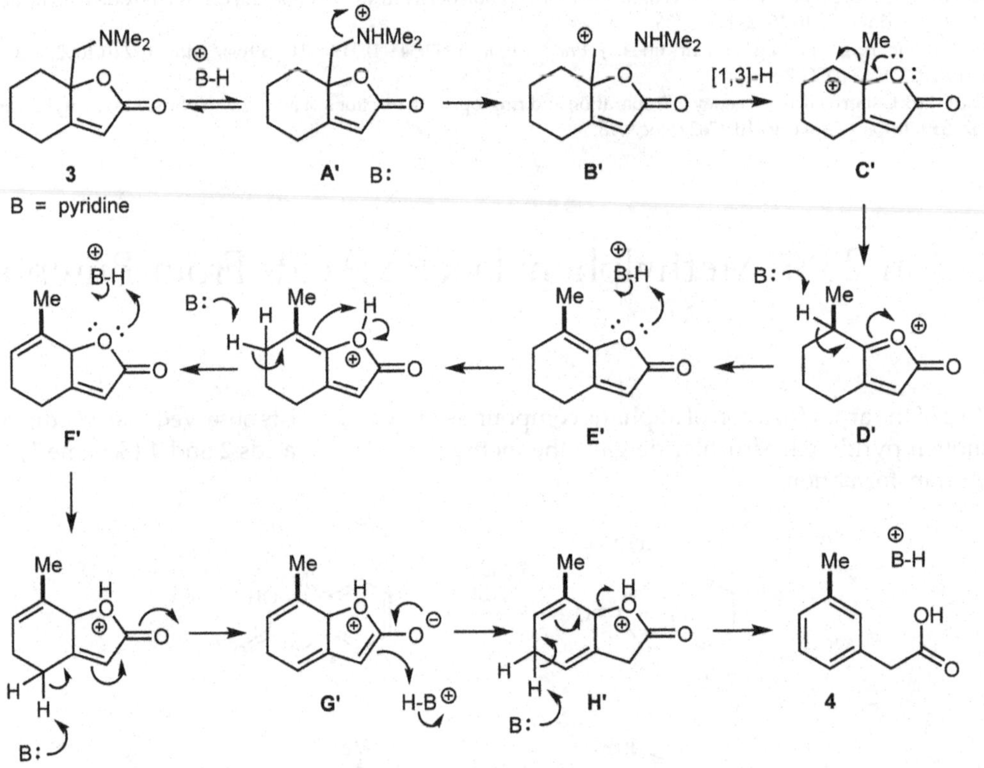

3
B = pyridine
A'
B'
[1,3]-H
C'
D'
E'
F'
G'
H'
4

KEY STEPS EXPLAINED

The mechanisms for both transformations involve lactone ring opening, aromatization of the cyclohexane ring, and elimination of either water or dimethylamine. The authors in their paper highlighted the key carbocation intermediates in both mechanisms including the 1,2-methyl migration that takes place in the **3** to **4** transformation. However, for both reactions they did not show the mechanism of aromatization. Here, we have presented the full mechanistic pathways for the two reactions.

Concerning the **1** to **2** transformation, the first step is protonation of the lactone ring oxygen atom to give **A** which ring opens via the pyridine base and deprotonates to the intermediate **B**. Next, protonation of the hydroxymethyl group gives **C** after which water is eliminated to give **D** again assisted by pyridine base. The remaining steps to aromatize the ring involve a sequence of protonation and deprotonation processes until product **2** is reached. Analogously, for the **3** to **4** transformation, the mechanism begins with protonation of the N,N-dimethyl group which cleaves by an S_N1 process to give carbocation **B'**. After a [1,3] protic shift to **C'**, the methyl group undergoes a 1,2-migration to give the oxonium intermediate **D'**. As previously seen, the remaining steps lead to aromatization of the ring including ring opening of the lactone ultimately forming the final product **4**. More experimental and computational work is needed in order to supplement these postulated mechanisms.

ADDITIONAL RESOURCES

We recommend a series of articles highlighting experimental, computational, and mechanistic analysis of metal-catalyzed syntheses of butenolides.[2-4] Lastly, Ref. 5 contains an interesting mechanism involving a starting butenolide.

References

1. Baiocchi L, Giannangeli M. Aromatization of aliphatic compounds. o- and m-tolylacetic acids. *Tetrahedron Lett.* 1979;20(46):4499–4500. https://doi.org/10.1016/S0040-4039(01)86631-4.
2. Gronnier C, Kramer S, Odabachian Y, Gagosz F. Cu(I)-catalyzed oxidative cyclization of alkynyl oxiranes and oxetanes. *J Am Chem Soc.* 2012;134:828–831. https://doi.org/10.1021/ja209866a.
3. Xie X, Li Y, Fox JM. Selective syntheses of $\Delta^{\alpha,\beta}$ and $\Delta^{\beta,\gamma}$ butenolides from allylic cyclopropenecarboxylates via tandem ring expansion/[3,3]-sigmatropic rearrangements. *Org Lett.* 2013;15(7):1500–1503. https://doi.org/10.1021/ol400264a.
4. Hadfield MS, Haeller JL, Lee A-L, Macgregor SA, O'Neill JAT, Watson AM. Computational studies on the mechanism of the gold(I)-catalysed rearrangement of cyclopropenes. *Org Biomol Chem.* 2012;10:4433–4440. https://doi.org/10.1039/c2ob25183c.
5. Dai M, Zhang X, Khim S-K, Schultz AG. Ammonia-promoted fragmentation of 2-alkyl- and 2,4-dialkyl-3-iodo-1-oxocyclohexan-2,4-carbolactones. *J Org Chem.* 2005;70:384–387. https://doi.org/10.1021/jo048408s.

Question 239: Base-Induced Quinoline Rearrangements

In 1979 a study of phosphodiesterase inhibitors revealed that treatment of compound **1** with 1.1 equivalents of sodium methoxide in refluxing methanol for 15–30 min formed compound **2** in almost quantitative yield. Treatment of either **1** or **2** under excess sodium methoxide and the same reaction conditions furnished the quinoline derivative **3** in 92% yield (Scheme 1).[1] Provide a mechanism for this transformation.

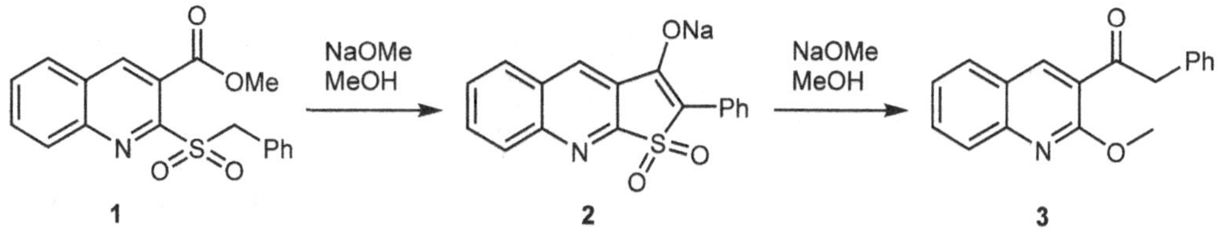

SCHEME 1

SOLUTION

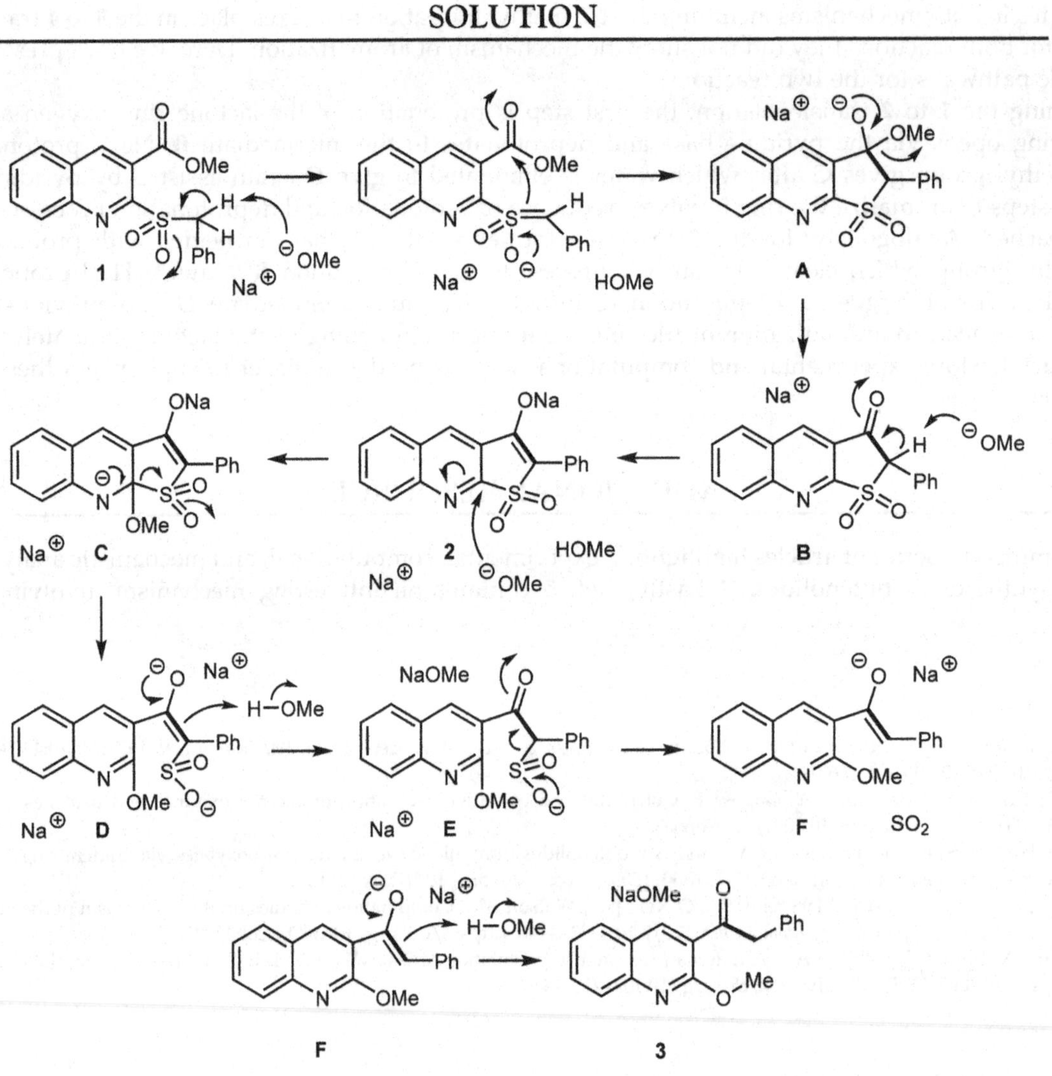

BALANCED CHEMICAL EQUATION

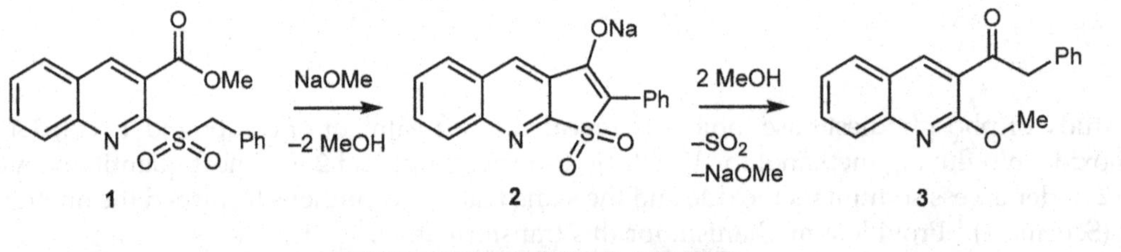

KEY STEPS EXPLAINED

The mechanism proposed by Harrison and Rice begins with deprotonation of the benzyl proton of **1** by methoxide anion to form a carbanion which undergoes intramolecular cyclization to the tetrahedral intermediate **A**. At this point, ketonization with elimination of methoxide anion forms **B** which then enolizes to the sodium enolate salt **2**. This salt can be isolated or left in solution to continue to react further. The nitrogen anion intermediate **C** is formed when methoxide ion attacks the *ortho* carbon atom next to the quinoline nitrogen atom of **2** further enabling a cascade sequence which involves fragmentation of the five-membered ring, ketonization, simultaneous sulfur dioxide elimination and reenolization, and final ketonization to form the quinoline derivative **3**.

We would like to mention that the authors obtained product **3** in two ways. In the first case, the enolate salt **2** was isolated and characterized and then taken further to the final product. In the second case, the enolate salt was generated in situ without isolation in a one-pot operation leading to the same final product. These experiments helped establish the fact that **2** is a well-defined intermediate along the reaction pathway. Moreover, we would also recommend a theoretical analysis which we think would help explain why intermediate **2** is isolable and also shed much needed light on the energetics of breaking the five-membered sulfone ring.

ADDITIONAL RESOURCES

We direct the reader to four examples of transformations whose mechanisms involve ring-breakage and nitrogen anion formation events similar to those illustrated in this problem.[2–5]

References

1. Harrison Jr EA, Rice KC. Novel alkaline ring cleavage of 2-phenyl-3-hydroxythieno[2,3-*b*] quinoline-1,1-dioxide, a potent inhibitor of cyclic adenosine 5'-monophosphate phosphodiesterase. *J Org Chem.* 1979;44(17):2977–2979. https://doi.org/10.1021/jo01331a005.
2. Kafka S, Kosmrlj J, Klasek A, Pevec A. Rearrangement of furo[2,3-*c*]quinoline-2,4(3a*H*,5*H*)-diones to furo[3,4-*c*]quinoline-3,4(1*H*,5*H*)-diones. *Tetrahedron Lett.* 2008;49:90–93. https://doi.org/10.1016/j.tetlet.2007.11.010.
3. Haddadin MJ, El-Nachef C, Kisserwani H, Chaaban Y, Kurth MJ, Fettinger JC. 2-Benzyliden-2*H*-thieto[3,2-*b*]quinoline: a new heterocycle and its rearrangement to 2-phenylthieno[3,2-*b*]quinoline. *Tetrahedron Lett.* 2010;51:6687–6689. https://doi.org/10.1016/j.tetlet.2010.10.039.
4. Donati C, Janowski WK, Prager RH, Taylor MR, Vilkins LM. Base-catalysed rearrangement of isoxazolinyl heterocycles: synthesis of annelated pyrimidines. *Aust J Chem.* 1989;42:2161–2169. https://doi.org/10.1071/CH9892161.
5. James KJ, Grundon MF. Quinoline alkaloids. Part 17. Mechanism of base-catalysed rearrangement of hydroxyisopropyldihydrofuroquinolones and of dihydrodimethylpyranoquinolones. *J Chem Soc Perkin Trans 1.* 1979;1979:1467–1471. https://doi.org/10.1039/P19790001467.

Question 240: The Angucyclines: Rearrangement of Angular to Linear Tetracycles

In 1970 Japanese scientists investigating the structure of the angular tetracyclic antibiotic aquayamycin **1** noticed that heating **1** at 220°C for a short period of time readily produced the linear tetracycle **2** (Scheme 1).[1] Suggest a mechanism for this transformation.

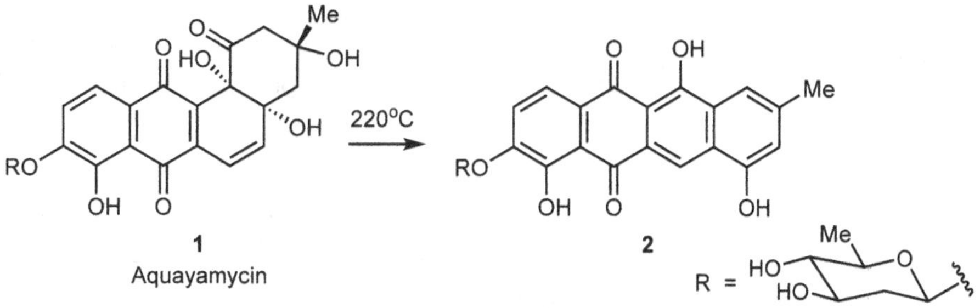

SCHEME 1

SOLUTION

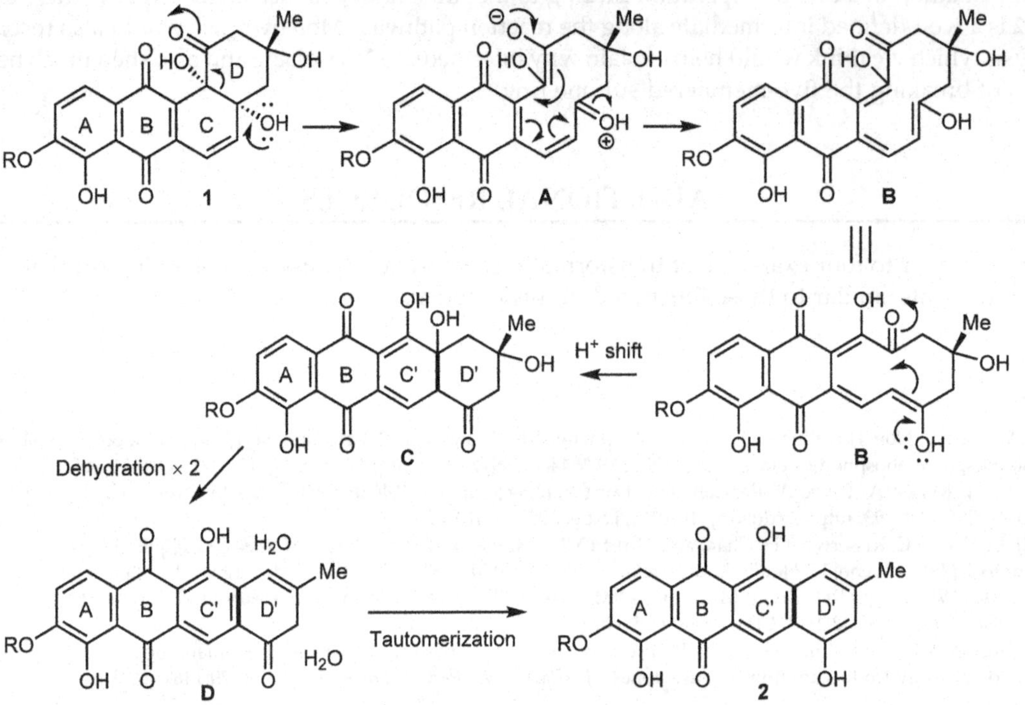

KEY STEPS EXPLAINED

The mechanism for the heat-induced rearrangement of **1** to **2** begins with ring opening of rings C and D in **1** to form intermediate **A** which then experiences electron delocalization that results in simultaneous ketonization and enolization to generate intermediate **B**. Hydroxyl-directed ring closure results in the formation of two new rings, that is, C′ and D′, thus giving rise to the linear tetracyclic intermediate **C**. After two dehydration events which lead to the aromatization of ring C′, intermediate **C** becomes **D** which undergoes a final tautomerization promoted by the aromatization of ring D′ to generate final product **2**. The authors note that the same transformation from **1** to **2** was also achieved under basic conditions using barium hydroxide.

Certain spectroscopic evidence obtained by the authors supported the structural assignment of product **2**, namely: (1) the UV spectrum of **2** was similar to that of 1,6-dihydroxynaphthacenequinone; (2) the IR spectrum of **2** showed a band at $1615\,cm^{-1}$ characteristic of a chelated (hydrogen bonded) C=O quinone group and no band at $1725\,cm^{-1}$ as was found in the starting material **1**; and (3) the ^{1}H NMR spectrum of **2** showed one phenolic OH group, two chelated (hydrogen bonded) phenolic groups, a 0.98 ppm downfield shift in the methyl protons compared to **1**, and three new aromatic singlets appeared instead of four methylene protons and two vicinal olefinic protons as were observed in the spectrum of **1**. We believe that further theoretical analysis would highly complement the postulated reaction mechanism.

ADDITIONAL RESOURCES

We would highly recommend the work of Krohn and Rohr for those interested in learning more about the synthesis and mechanistic aspects of transformations involving angucyclines.[2] For an interesting article highlighting possible transformations of an ABC tricyclic analog of aquayamycin from a mechanistic perspective, see Ref. 3.

References

1. Sezaki M, Kondo S, Maeda K, Umezawa H. The structure of aquayamycin. *Tetrahedron.* 1970;26(22):5171–5190. https://doi.org/10.1016/S0040-4020(01)98726-5.

2. Krohn K, Rohr J. Angucyclines: total syntheses, new structures, and biosynthetic studies of an emerging new class of antibiotics. In: Rohr J, ed. *Bioorganic Chemistry Deoxysugars, Polyketides and Related Classes: Synthesis, Biosynthesis, Enzymes*. Heidelberg, Berlin: Springer; 1997:127–195. Topics in Current Chemistry; vol. 188. https://doi.org/10.1007/BFb0119236.
3. Lebrasseur N, Fan G-J, Oxoby M, Looney MA, Quideau S. λ³-Iodane-mediated arenol dearomatization. Synthesis of five-membered ring-containing analogues of the aquayamycin ABC tricyclic unit and novel access to the apoptosis inducer menadione. *Tetrahedron*. 2005;61:1551–1562. https://doi.org/10.1016/j.tet.2004.11.072.

Question 241: 1,2-Dihydropyridines From 2,3-Dihydro-4-pyridones

In 1992 North Carolina scientists applied a Vilsmeier reagent in order to effect the transformation of the 2,3-dihydro-4-pyridone **1** into the 1,2-dihydropyridine **2** (Scheme 1).[1] Suggest a mechanism for this transformation.

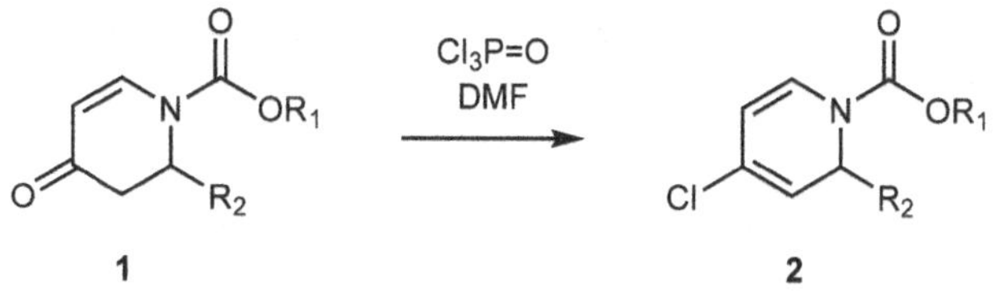

SCHEME 1

SOLUTION

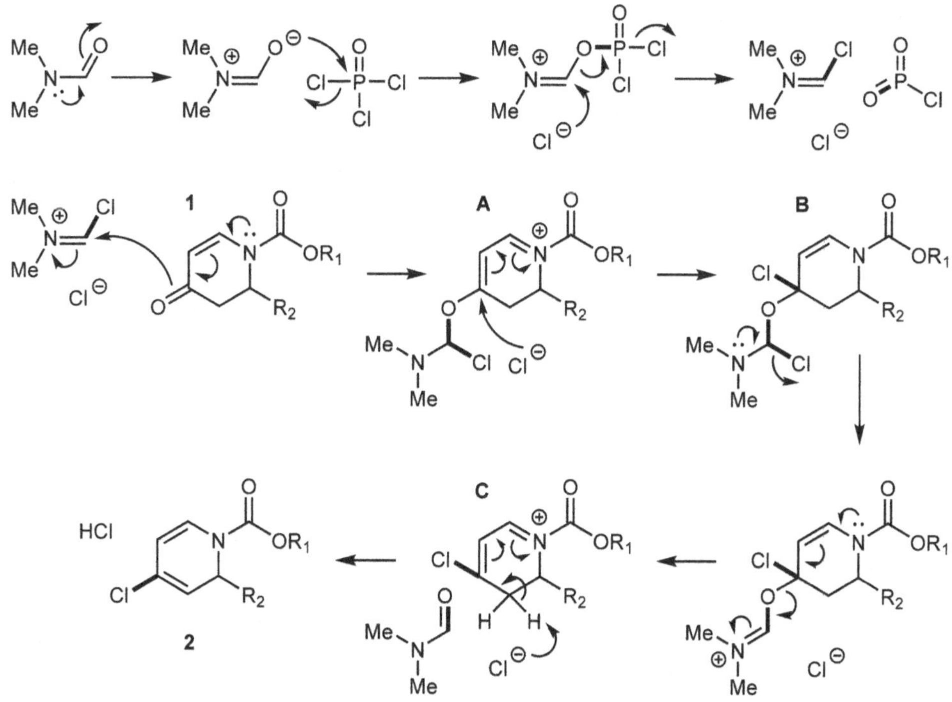

KEY STEPS EXPLAINED

The mechanism proposed by the authors begins with generation of the Vilsmeier-Haack reagent, namely, dimethylformamide attack on phosphorus oxychloride to generate an iminium ion intermediate which then partitions to the reagent via chloride ion attack. Once the Vilsmeier-Haack reagent is available, it reacts with the 2,3-dihydro-4-pyridone **1** in a nucleophilic manner to form the iminium chloride intermediate **A** which then reacts with chloride anion to form the neutral intermediate **B** which undergoes fragmentation to the chloroiminium intermediate **C** after elimination of dimethylformamide. Deprotonation by chloride anion generates the 1,2-dihydropyridine **2**.

We note the interesting observation that in this reaction, dimethylformamide acts both as a solvent and as a reagent. Its regeneration at the end of the reaction suggests that it can be recovered for further cycles of the same transformation.

In terms of experimental evidence, the authors carried out the same reaction in the absence of DMF and recovered only the starting material. Further experimental and theoretical analysis to support the postulated intermediates in this mechanism awaits future investigation.

ADDITIONAL RESOURCES

For an example of a mechanistic analysis examining two postulated mechanisms involving the synthesis of dihydropyridones, see the work of Song et al.[2] For a review article covering synthetic approaches to pyridine and dihydropyridine derivatives, see Ref. 3. We would also like to include an article highlighting the importance of computational analysis for providing evidence in support of a postulated reaction mechanism.[4]

References

1. Al-awar RS, Joseph SP, Comins DL. Conversion of N-acyl-2,3-dihydro-4-pyridones to 4-chloro-1,2-dihydropyridines using the Vilsmeier reagent. *Tetrahedron Lett.* 1992;33(50):7635–7638. https://doi.org/10.1016/0040-4039(93)88003-2.
2. Song D, Rostami A, West FG. Domino electrocyclization/azide-capture/Schmidt rearrangement of dienones: one-step synthesis of dihydropyridones from simple building blocks. *J Am Chem Soc.* 2007;129:12019–12022. https://doi.org/10.1021/ja071041z.
3. Bull JA, Mousseau JJ, Pelletier G, Charette AB. Synthesis of pyridine and dihydropyridine derivatives by regio- and stereoselective addition to N-activated pyridines. *Chem Rev.* 2012;112:2642–2713. https://doi.org/10.1021/cr200251d.
4. Paton RS, Steinhardt SE, Vanderwal CD, Houk KN. Unraveling the mechanism of cascade reactions of Zincke aldehydes. *J Am Chem Soc.* 2011;133:3895–3905. https://doi.org/10.1021/ja107988b.

Question 242: The Vitamin K to Vitamin K Oxide Transformation

The mechanism of the oxidation of vitamin K to vitamin K oxide has received much attention over the years. In 1997 Korean chemists utilized ^{18}O labeling in the form of $^{18}O_2$ in order to differentiate between two postulated mechanisms (Scheme 1).[1,2] Provide the two possible mechanisms for this transformation and explain how ^{18}O labeling can be used to differentiate between them.

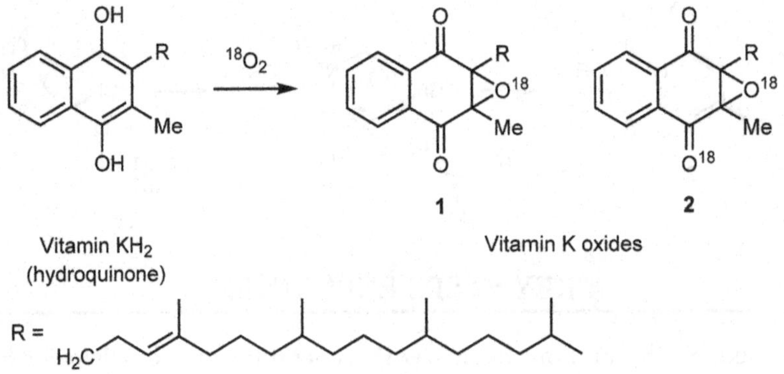

Vitamin KH$_2$
(hydroquinone)

Vitamin K oxides

$$R =$$

SCHEME 1

SOLUTION: MECHANISM 1

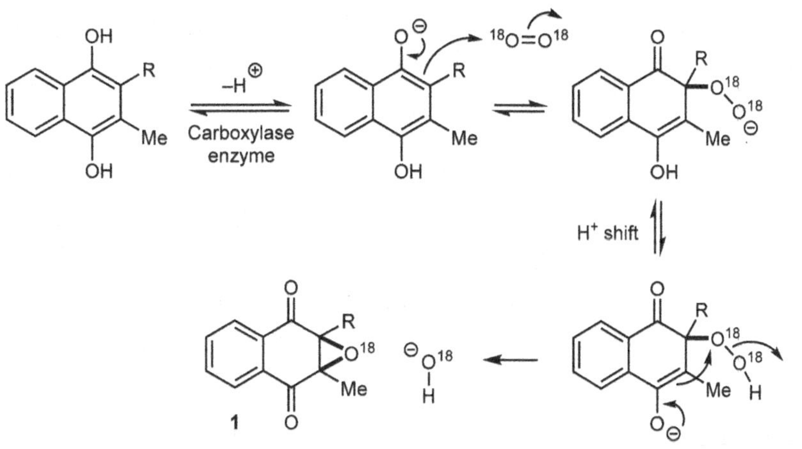

SOLUTION: MECHANISM 2

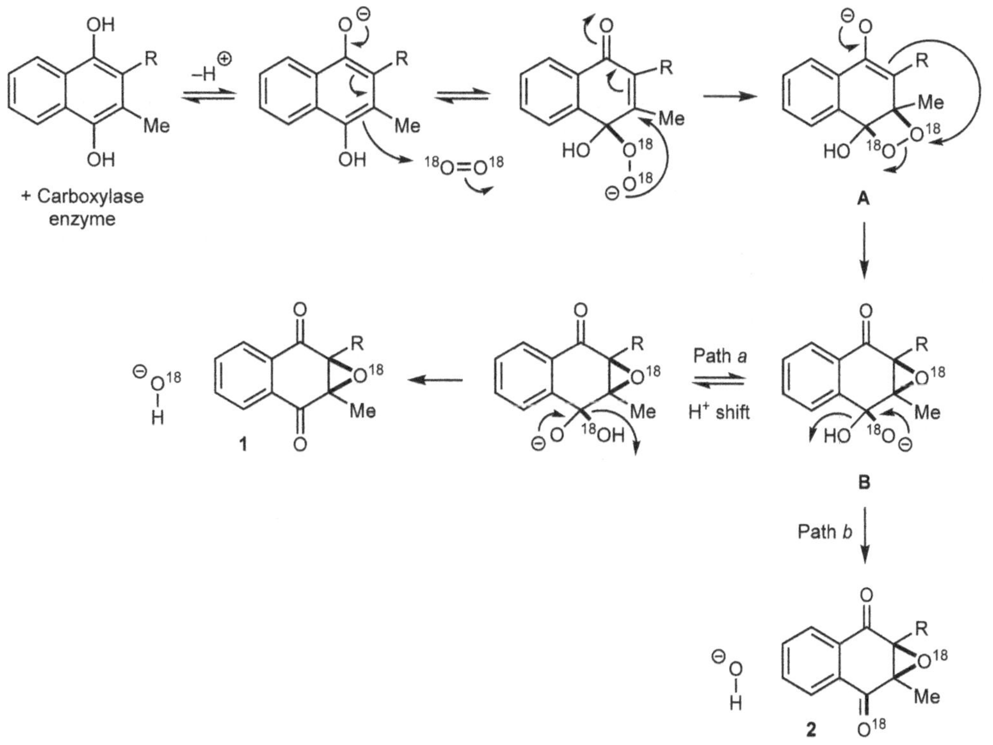

KEY STEPS EXPLAINED

As reported by the authors, both postulated mechanisms begin with deprotonation of the hydroquinone form of vitamin K (pK$_A$ = 9.3)[1] in the presence of carboxylase enzyme to form a phenoxide which reacts with labeled dioxygen. In mechanism 1, dioxygen adds *ortho* to the carboxy group enabling a facile epoxide closure directly to 1 after elimination of ^{18}OH anion. By contrast, in mechanism 2, the dioxygen species adds to the *para* position relative to the carboxy anion which then enables formation of the dioxetane intermediate A. This intermediate can rearrange to the epoxide intermediate B which then has two possible pathways to the formation of vitamin K oxide. The first, path *a*, involves a proton shift and closure to 1 with elimination of ^{18}OH anion as seen before in mechanism 1. The second, path *b*, involves direct ketonization to 2 with expulsion of unlabeled OH anion. Concerning the products of this transformation, the authors observed that the epoxide oxygen atom was 100% labeled, whereas only 34% labeling was observed for the quinone oxygen atom. Therefore, mechanism 1, which predicts only the formation of product 1 with the single oxygen label found at the epoxide position can be ruled out in favor of mechanism 2 which predicts the formation of both products 1 and 2.

For a further study of this interesting transformation and a mechanistic investigation involving oxygen labeling experiments in the context of a very similar chemical system, see Refs. 3, 4, respectively.

References

1. Ham SW, Lee G-H. Mechanism of oxygenation of naphthoquinone derivative. *Tetrahedron Lett.* 1998;39(23):4087–4090. https://doi.org/10.1016/S0040-4039(98)00714-X.
2. Ham SW, Yoo JS. Mechanism of oxygenation of vitamin K hydroquinone. *Chem Commun.* 1997;1997:929–930. https://doi.org/10.1039/A701091E.
3. Li S, Furie BC, Furie B, Walsh CT. The propeptide of the vitamin K-dependent carboxylase substrate accelerates formation of the γ-glutamyl carbanion intermediate. *Biochemistry.* 1997;36(21):6384–6390. https://doi.org/10.1021/bi962816b.
4. Rafanan Jr ER, Hutchinson CR, Shen B. Triple hydroxylation of tetracenomycin A2 to tetracenomycin C involving two molecules of O₂ and one molecule of H₂O. *Org Lett.* 2000;2(20):3225–3227. https://doi.org/10.1021/ol0002267.

Question 243: Rearrangement During Intramolecular Cyclization to the Indole 4-Position

During studies of regular Friedel-Crafts intramolecular cyclizations, Merck chemists encountered the unusual rearrangement of **1** to the 4-sulfur-substituted tricyclic indole **2** (Scheme 1).[1] Suggest a mechanism for this transformation.

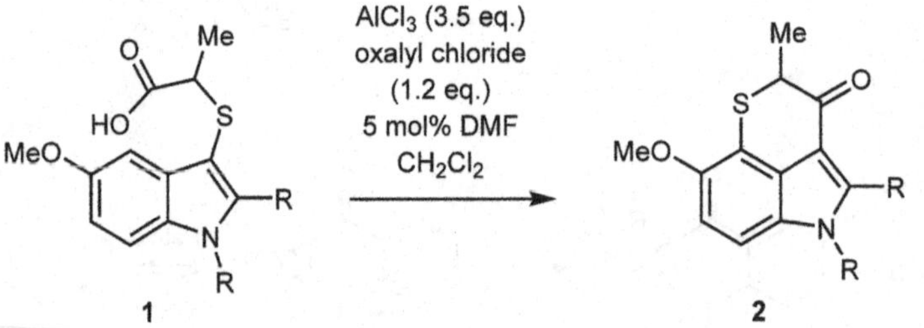

SCHEME 1

SOLUTION

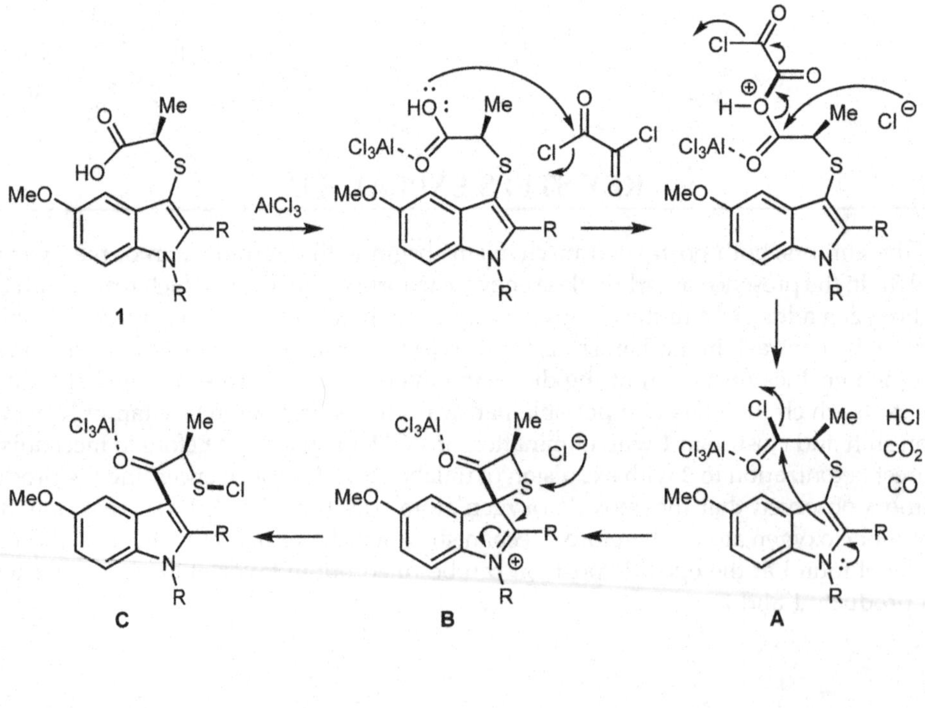

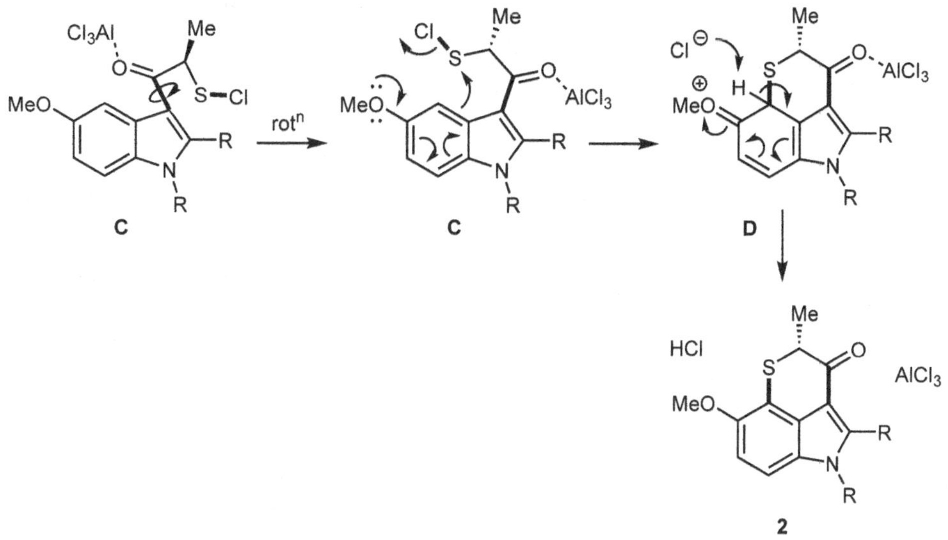

KEY STEPS EXPLAINED

We would first like to highlight the fact that the authors expected this reaction to produce the standard Friedel-Crafts product **3** instead of the rearranged isomer **2** (Scheme 2).

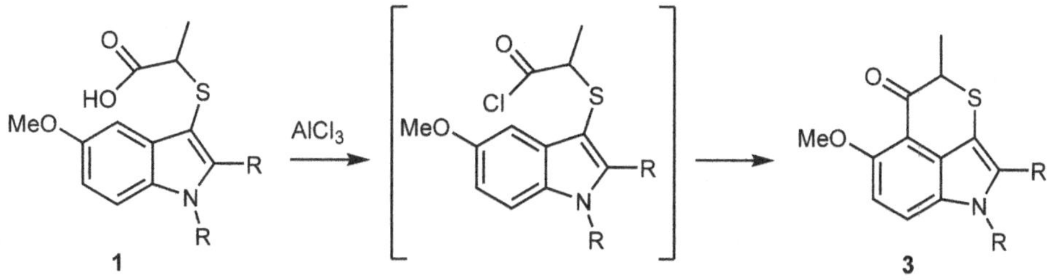

SCHEME 2

Therefore the rearrangement of **1** to **2** begins with the expected formation of the acid chloride **A** after reaction of **1** with oxalyl chloride and elimination of CO, CO_2, and HCl as reaction by-products. Next, instead of the expected Friedel-Crafts acylation taking place *ortho* to the methoxy group on the aromatic ring, acid chloride **A** cyclizes to *spiro* intermediate **B** and ring opens to intermediate **C** after chloride anion attack on sulfur. We thus have the original C—S bond being cleaved in order to allow for bond rotation around the newly formed C—C bond which maintains the stereochemical configuration of the methyl group. Regular Friedel-Crafts alkylation then connects the sulfur atom and the *ortho* carbon atom with respect to the methoxy group on the aromatic ring to form intermediate **D** which is rearomatized to the final product **2**.

Furthermore, as far as experimental work is concerned, the authors demonstrated that they had an isomeric product **2** instead of the expected product **3** when they examined a derivative of **2** (**2′**) that was different in spectral (^{1}H NMR, ^{13}C NMR, and MS) and chromatographic properties (TLC and HPLC) when compared with an authentic sample of an analogous derivative of **3** (**3′**) that was prepared by the same synthetic sequence (Scheme 3). Moreover, the authors demonstrated retention of configuration at the carbon atom bearing the methyl group by subjecting enantiopure starting material to the same Friedel-Crafts conditions thus showing that the bonds emanating from that stereocarbon atom are not broken during the course of this transformation.

Lastly, concerning the postulated *spiro* intermediate **B**, the authors cite literature precedent for such an intermediate from the work of Ungemach and Cook.[2]

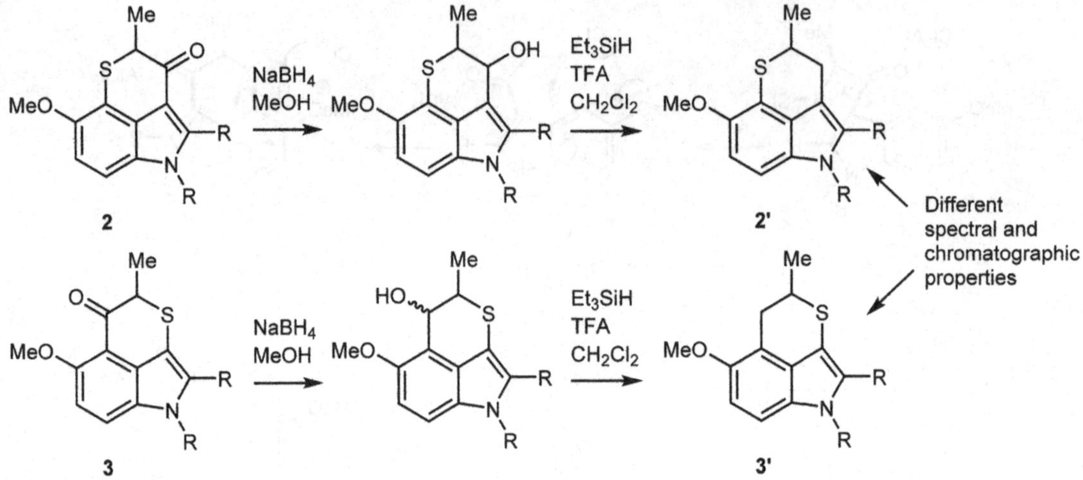

SCHEME 3

ADDITIONAL RESOURCES

We direct the reader to three interesting heterocyclic transformations involving mechanistic discussion in a similar context to that encountered in the present solution.[3-5]

References

1. Chung JYL, Reamer RA, Reider PJ. Friedel-Crafts cyclization of 2-(3-indolythio)propionic acids. An unusual rearrangement leading to 4-sulfur-substituted tricyclic indoles. *Tetrahedron Lett.* 1992;33(33):4717–4720. https://doi.org/10.1016/S0040-4039(00)61267-4.
2. Ungemach F, Cook JM. The spiroindolenine intermediate, a review. *Heterocycles.* 1978;9(8):1089–1119. https://doi.org/10.3987/R-1978-08-1089.
3. Sporn MB, Liby KT, Yore MM, Fu L, Lopchuk JM, Gribble GW. New synthetic triterpenoids: potent agents for prevention and treatment of tissue injury caused by inflammatory and oxidative stress. *J Nat Prod.* 2011;74:537–545. https://doi.org/10.1021/np100826q.
4. Kupai K, Banoczi G, Hornyanszky G, Kolonits P, Novak L. A convenient method for the preparation of cyclohepta[b]indole derivatives. *Cent Eur J Chem.* 2012;10(1):91–95. https://doi.org/10.2478/s11532-011-0117-4.
5. Tomioka Y, Ohkubo K, Maruoka H. A novel synthesis of indole derivatives by the reaction of N-arylhydroxamic acids with malononitrile. *J Heterocyclic Chem.* 2007;44(2):419–424. https://doi.org/10.1002/jhet.5570440222.

Question 244: 4-Quinolone Antibacterials: A New Synthesis

In 1992 chemists at Pfizer working on synthesizing antibacterials developed a novel short synthesis of 4-quinolone analogs illustrated by the three-step transformation of **1** to **4** (Scheme 1).[1] Elucidate the mechanism of this overall transformation and show the one-step procedure that the authors later invented on the basis of the mechanistic insight they gained from their investigation.

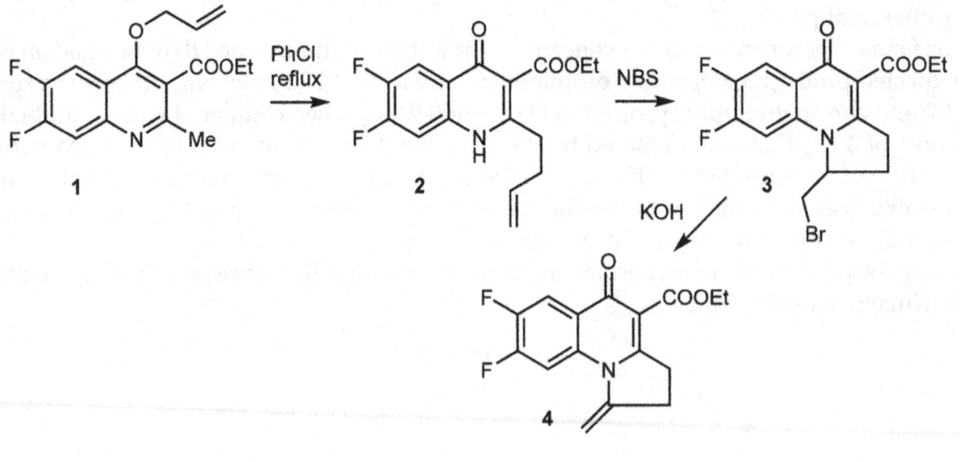

SCHEME 1

SOLUTION: 1 TO 2

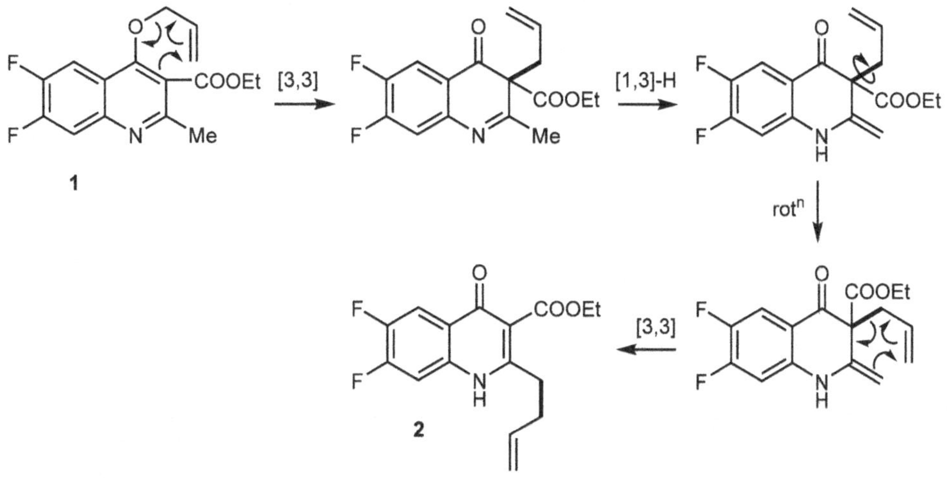

SOLUTION: 2 TO 3

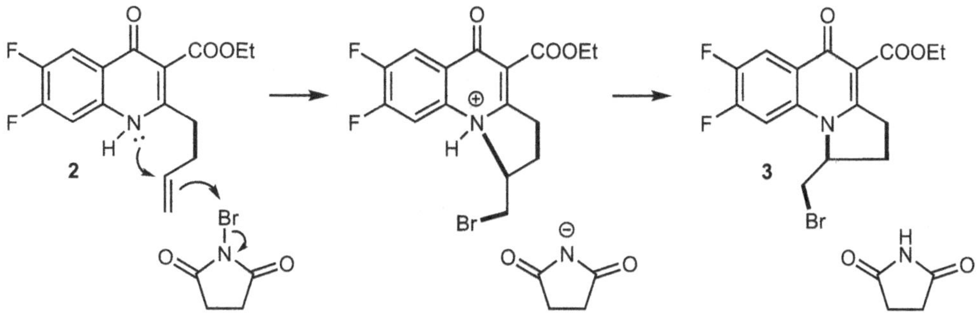

SOLUTION: 3 TO 4

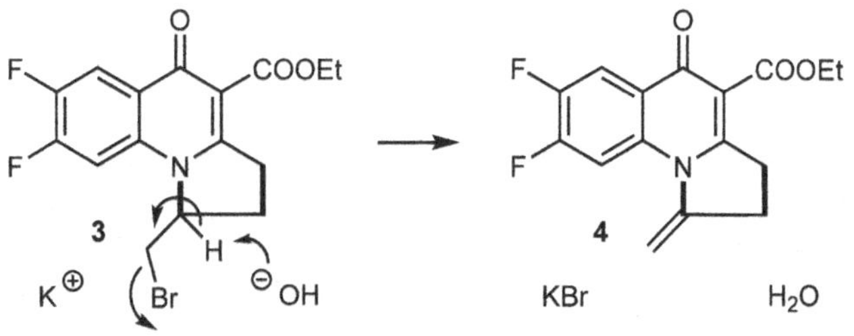

KEY STEPS EXPLAINED

The **1** to **4** transformation begins with a tandem Claisen thermal rearrangement where the allyl group bonded to oxygen migrates first to the *ortho* carbon atom of the quinoline, then, after a [1,3]-H shift between the nitrogen atom and the methyl group, to the generated exocyclic carbon atom that was once methyl in the starting material in order to form **2**. The second synthetic step constitutes an N-bromosuccinimide reaction which is also a 5-*exo-trig* cyclization leading to the bromo product **3**. Finally, a base-promoted elimination of potassium bromide is utilized to generate the exocyclic C=C bond in the final 4-quinolone product **4**.

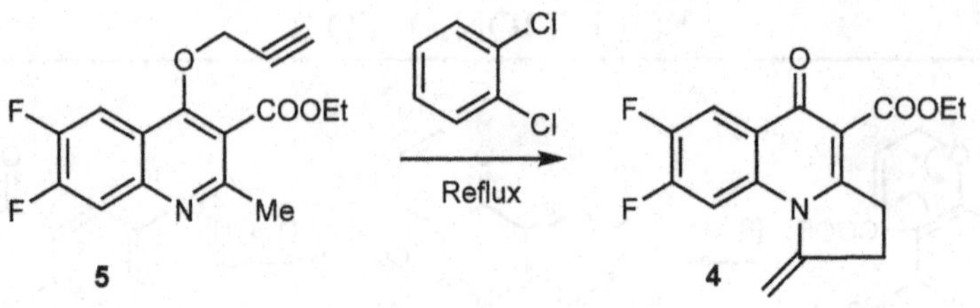

SCHEME 2

Moreover, the more direct one-step route that the authors developed was to start off with simple thermolysis of 4-(propargyloxy)-quinoline **5** in refluxing *o*-dichlorobenzene (Scheme 2).

ADDITIONAL RESOURCES

We refer the interested reader to three articles which include mechanistic discussion of similar transformations involving ring forming reactions via cyclization of terminal alkynes, in some cases catalyzed by iodine.[2–4]

References

1. Newhouse BJ, Bordner J, Augeri DJ, Litts CS, Kleinman EF. Novel [3+2] and [3+3] 4-quinolone annulations by tandem Claisen-Cope-amidoalkylation reaction. *J Org Chem.* 1992;57(25):6991–6995. https://doi.org/10.1021/jo00051a061.
2. Verma AK, Aggarwal T, Rustagi V, Larock RC. Iodine-catalyzed and solvent-controlled selective electrophilic cyclization and oxidative esterification of *ortho*-alkynyl aldehydes. *Chem Commun.* 2010;46:4064–4066. https://doi.org/10.1039/b927185f.
3. Verma AK, Rustagi V, Aggarwal T, Singh AP. Iodine-mediated solvent-controlled selective electrophilic cyclization and oxidative esterification of *o*-alkynyl aldehydes: an easy access to pyranoquinolines, pyranoquinolinones, and isocumarins. *J Org Chem.* 2010;75:7691–7703. https://doi.org/10.1021/jo101526b.
4. Ali S, Zhu H-T, Xia X-F, et al. Electrophile-driven regioselective synthesis of functionalized quinolines. *Org Lett.* 2011;13(10):2598–2601. https://doi.org/10.1021/ol2007154.

Question 245: An Efficient Synthesis of Fused 1,3-Dithiol-2-ones

French chemists have demonstrated in 1992 an elegant synthesis of fused 1,3-dithiol-2-ones. For example, they treated 4-benzoyloxy-1-bromo-2-butyne **1** with potassium *O*-methyl xanthate in methanol to form the expected xanthate ester **2** in excellent yield followed by reflux of **2** in chlorobenzene for 1 h in the presence of dimethyl fumarate to form the fused 1,3-dithiol-2-one **3** in quantitative yield (Scheme 1).[1] Explain the reaction sequence and provide a mechanism for these transformations.

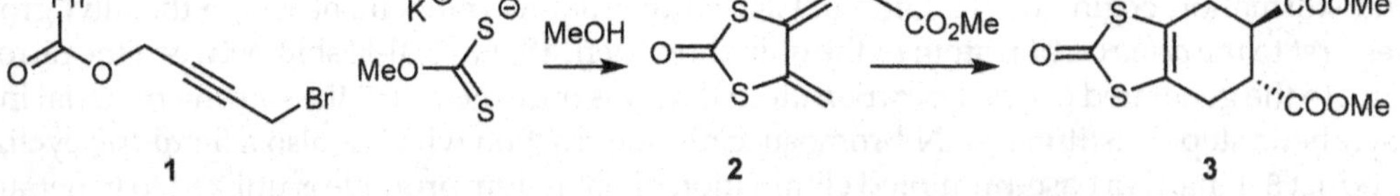

SCHEME 1

SOLUTION

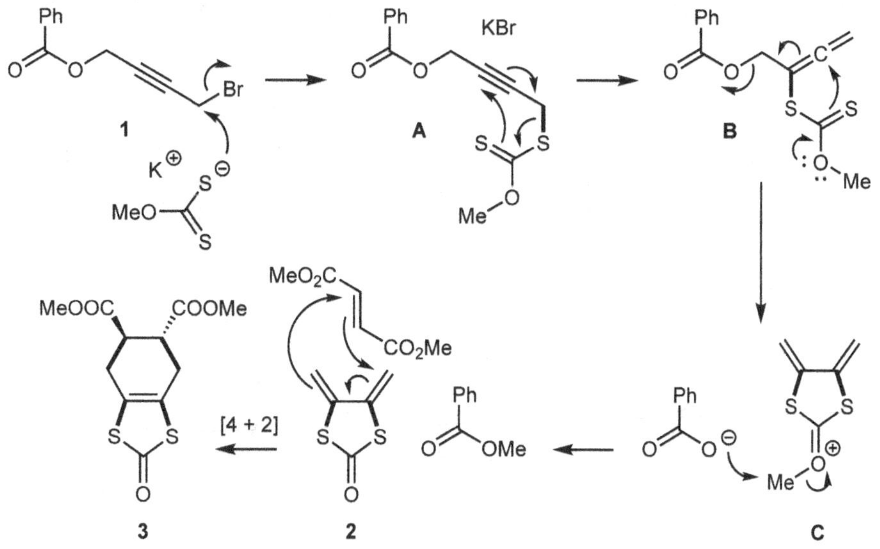

KEY STEPS EXPLAINED

The mechanism proposed by the authors begins with xanthate nucleophilic attack onto the 4-benzoyloxy-1-bromo-2-butyne **1** to form intermediate **A** with expulsion of KBr followed by thermal rearrangement of **A** to the allene intermediate **B** which cyclizes to an oxonium intermediate **C** after elimination of benzoate. The benzoate then reacts with **C** to form the expected xanthate ester **2** and methyl benzoate by-product. Following this, **2** reacts with dimethylfumarate in a Diels-Alder [4+2] cycloaddition to give final fused 1,3-dithiol-2-one product **3**. We note that when **2** reacts with dimethyl maleate, we would have the *cis* analog of **3** formed instead of the *trans* product. This two-step sequence thus results in a [4.3.0] bicyclic framework made by a [(3+2)+(4+2)] strategy. In order to help support this proposed mechanism, we recommend further theoretical, synthetic, and spectroscopic work in order to try to confirm the presence of intermediates **A** through **C**.

ADDITIONAL RESOURCES

We refer the reader to further mechanistic work by Zard with respect to the transformations of xanthates.[2,3] We also include two references which highlight reactions of similar sulfur containing starting materials with alkynes and allenes.[4,5]

References

1. Boivin J, Tailhan C, Zard SZ. An unusual access to 4,5-bis(alkylidene)-1,3-dithiolan-2-ones: a new class of cisoid dienes. *Tetrahedron Lett.* 1992;33 (51):7853–7856. https://doi.org/10.1016/S0040-4039(00)74761-7.
2. Denieul M-P, Quiclet-Sire B, Zard SZ. A synthetically useful source of propargyl radicals. *Tetrahedron Lett.* 1996;37(31):5495–5498. https://doi.org/10.1016/0040-4039(96)01146-X.
3. Zard SZ. On the trail of xanthates: some new chemistry from an old functional group. *Angew Chem Int Ed Engl.* 1997;36(7):672–685. https://doi.org/10.1002/anie.199706721.
4. Gorgues A, Hudhomme P, Salle M. Highly functionalized tetrathiafulvalenes: riding along the synthetic trail from electrophilic alkynes. *Chem Rev.* 2004;104:5151–5184. https://doi.org/10.1021/cr0306485.
5. Banert K. New functionalized allenes: synthesis using sigmatropic rearrangements and unusual reactivity. *Liebigs Ann Recl.* 1997;1997 (10):2005–2018. https://doi.org/10.1002/jlac.199719971003.

Question 246: The "Double Functional Group Transformation": Terminally Unsaturated Nitriles From 1-Nitrocycloalkenes

Taiwanese chemists have demonstrated in 1993 an example of the "double functional group transformation" by treating the 1-nitrocycloalkene **1** first with trimethylsilylmethylmagnesium chloride in THF at −20°C and then, in situ, with PCl₃ at 67°C to form the terminally unsaturated nitrile **2** in 33% yield (Scheme 1).[1] Suggest a mechanism for this transformation.

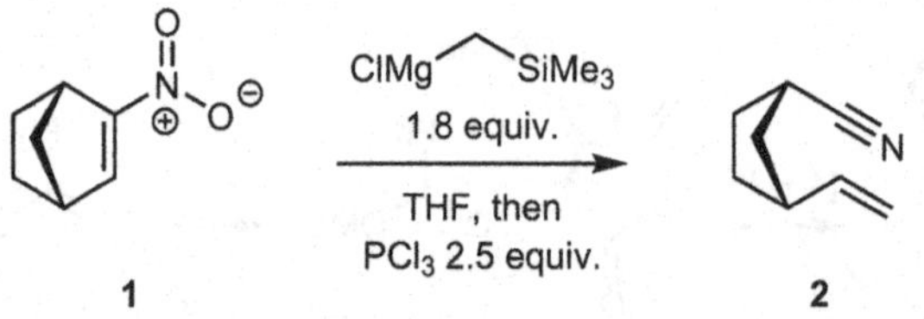

SCHEME 1

SOLUTION

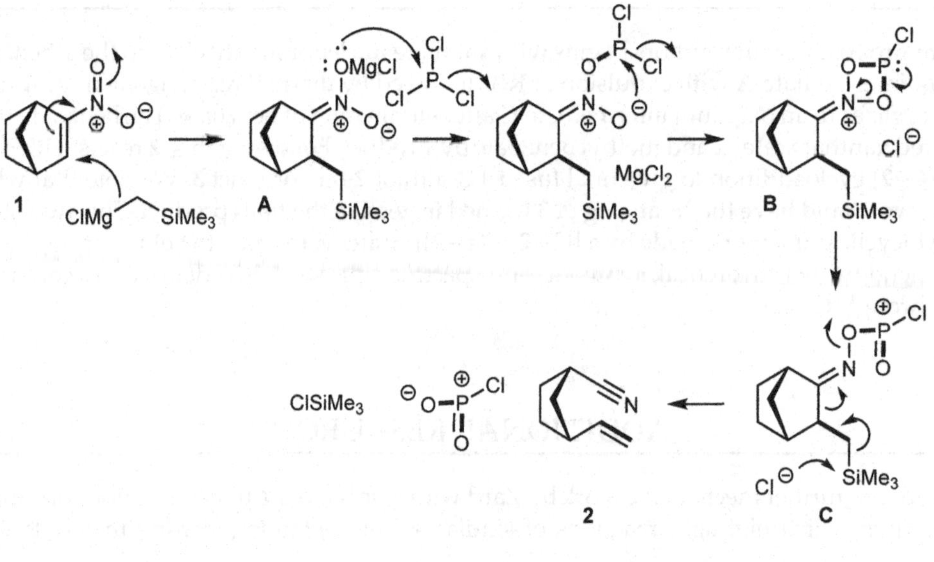

BALANCED CHEMICAL EQUATION

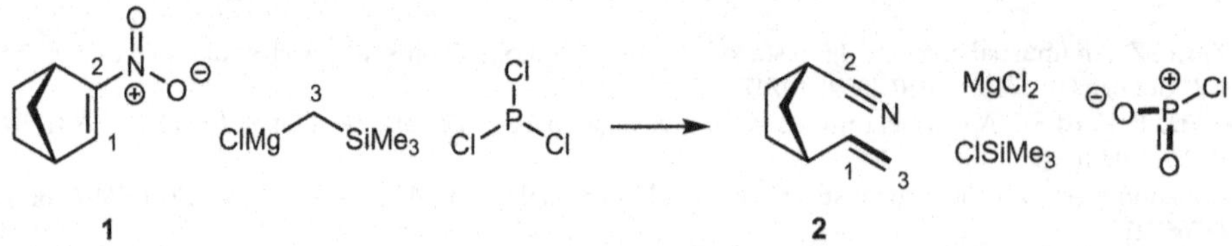

KEY STEPS EXPLAINED

The authors described this reaction as a "one flask double functional group transformation" since the nitro group becomes a nitrile group and the ring olefinic group opens to become an exocyclic vinyl group. Therefore the proposed mechanism begins with a Michael 1,4-addition of the silyl Grignard reagent to the 1-nitrocycloalkene **1** to give intermediate **A**. This intermediate then attacks the phosphorus atom of phosphorus trichloride eliminating chloride anion in the form of magnesium dichloride by-product. What follows is a cyclization to a four-membered ring intermediate **B** which ring opens to give **C** which undergoes a silicon-directed second-order Beckmann fragmentation to yield the terminally unsaturated nitrile product **2**. Use of the one flask technique also circumvented the need to isolate reaction intermediates along the way which ultimately reduced waste production.

A closer examination of the reaction shows that key atoms undergo oxidation number changes, for example: C_2 changes from +1 to +3 for a net gain of +2; N changes from +3 to −3 for a net loss of −6; P changes from +3 to +5 for a net gain of +2; and C_3 changes from −4 to −2, for a net gain of +2. C_1 does not change its oxidation state. We see that the overall net gain of $(+2)+(+2)+(+2)=+6$ is compensated by an overall net loss of −6 indicating that the reaction is a six-electron redox reaction.

We would also like to note that the authors gave the stereochemistry of the product as shown in Fig. 1 which is opposite to what is expected from the reaction mechanism.

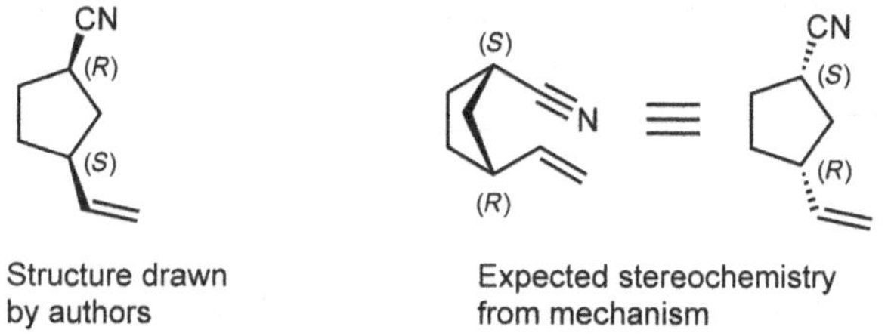

FIG. 1 Comparison of product structure as drawn by authors and what is expected from the reaction mechanism.

Lastly, this mechanism can be described as a tandem sequence of attack and counterattack steps: step 1 involves attack of the Grignard reagent on the substrate, step 2 involves counterattack of intermediate **A** onto phosphorus trichloride, step 3 involves internal attack at the phosphorus atom, step 4 involves counterattack ring opening in intermediate **B**, and step 5 involves attack of chloride ion on intermediate **B** leading to the product structure. Analogous "one-flask" reactions involving double functional group transformations using counterattack reagents by the same authors are presented in Scheme 2.[2,3]

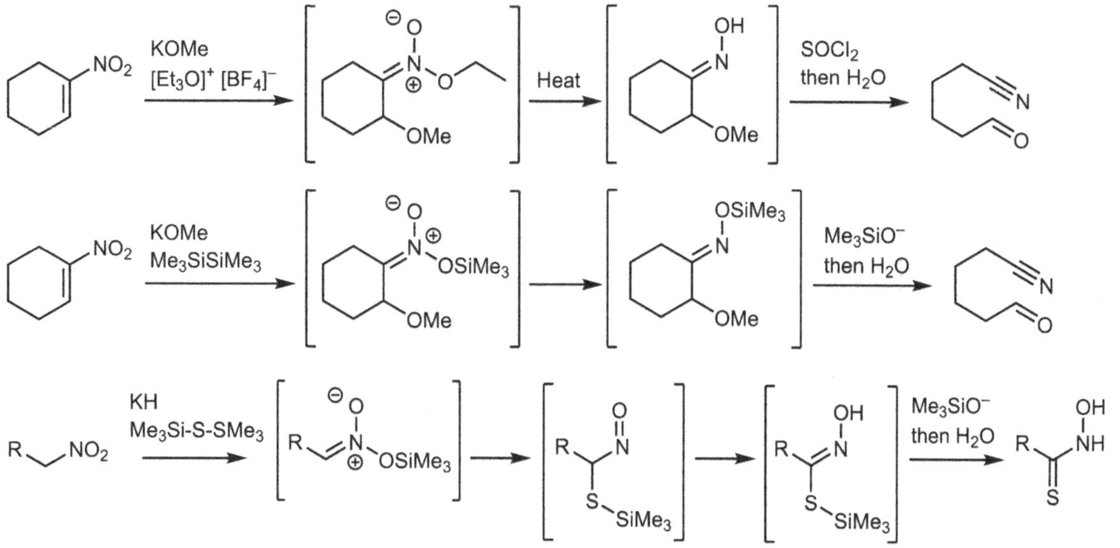

SCHEME 2

ADDITIONAL RESOURCES

For further examples of similar mechanistic analyses involving nitration and ring opening transformations, see Refs. 4, 5.

References

1. Tso H-H, Gilbert BA, Hwu JR. Novel double functional group transformation: 'one-flask' conversion of 1-nitrocycloalkenes to terminally unsaturated nitriles. *J Chem Soc Chem Commun.* 1993;1993:669–670. https://doi.org/10.1039/C39930000669.
2. Hwu JR, Tsay SC, King KY, Horng DN. Silicon reagents in chemical transformations: the concept of 'counterattack reagent'. *Pure Appl Chem.* 1999;71(3):445–451. https://doi.org/10.1351/pac199971030445.
3. Hwu JR, Lin CF, Tsay SC. Hexamethyldisilathiane in novel chemical transformations: concept of "counterattack reagent" *Phosphorus Sulfur Silicon Relat Elem.* 2005;180(5–6):1389–1393. https://doi.org/10.1080/10426500590912691.
4. Mezzetti A, Nitti P, Pitacco G, Valentin E. Carbocyclization reactions of 1-nitrocycloalkenes and cross-conjugated dienamines. *Tetrahedron.* 1985;41 (7):1415–1422. https://doi.org/10.1016/S0040-4020(01)96544-5.
5. Patil GS, Nagendrappa G. Nitration of cyclic vinylsilanes with acetyl nitrate: effect of silyl moiety and ring size. *Chem Commun.* 1999;1999:1079–1080. https://doi.org/10.1039/A902388G.

Question 247: Lewis Acid-Catalyzed Condensation of Indole With 1,3-Cydohexanedione

When Banerji et al. added $BF_3 \cdot OEt_2$ to a solution of indole and 1,3-cyclohexanedione they obtained a mixture of **1** and **2** in 55% and 20% yield, respectively (Scheme 1).[1] Propose mechanisms for these transformations.

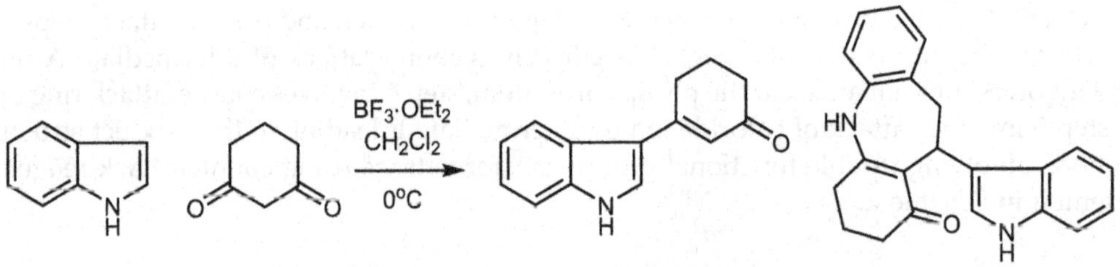

1 (55%)   **2** (20%)

SCHEME 1

SOLUTION: FORMATION OF 1

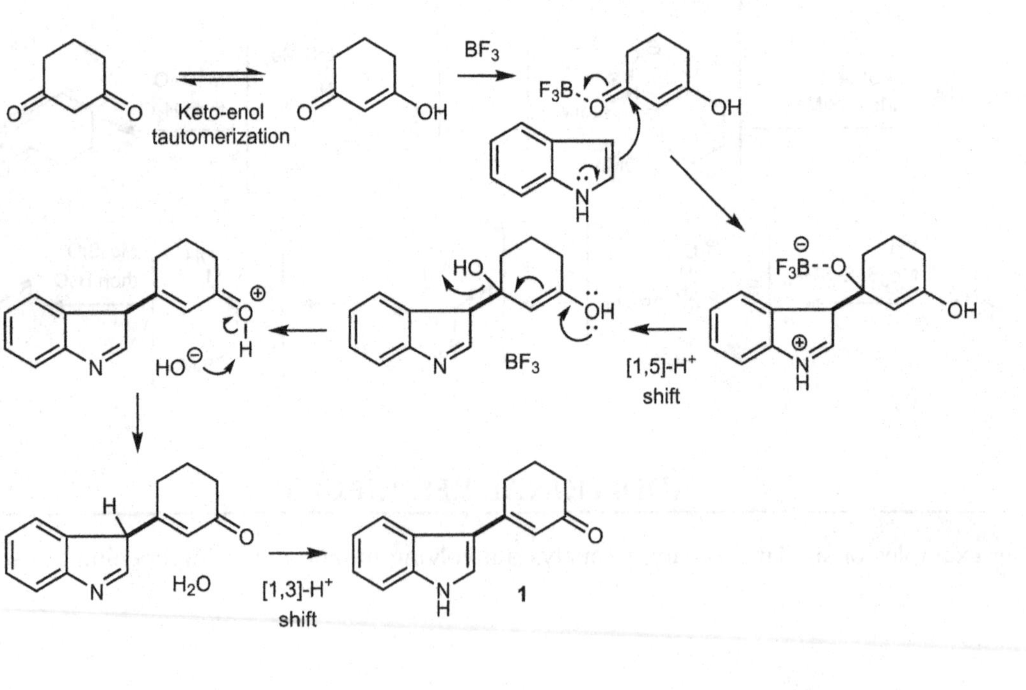

SOLUTION: FORMATION OF 2

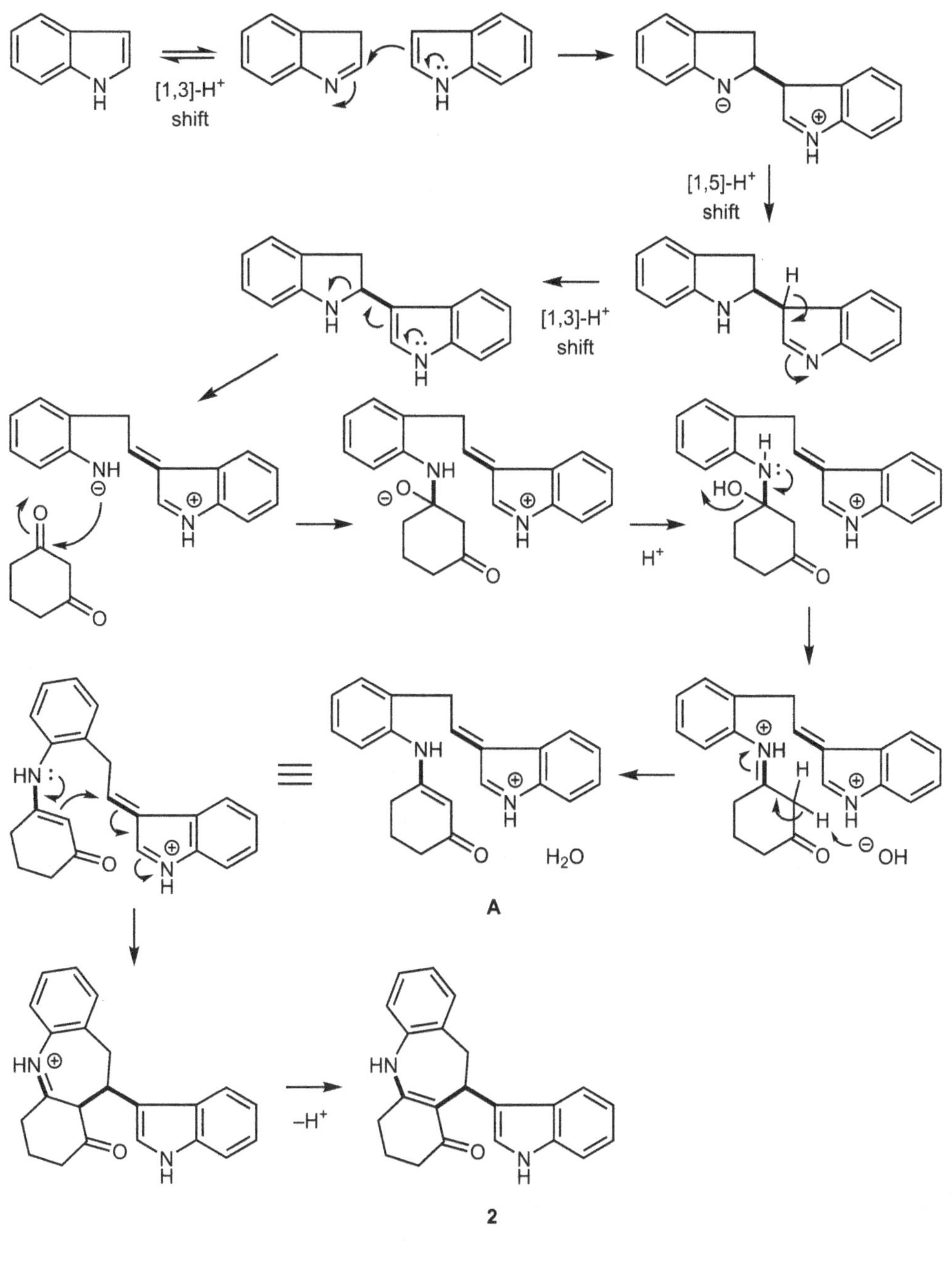

A

2

KEY STEPS EXPLAINED

During the course of these transformations it is important to note that boron trifluoride acts as a Lewis acid catalyst which can coordinate to carbonyl groups and facilitate hydrogen shifts. Concerning the formation of product **1**, we first begin with facile keto-enol tautomerization of cyclohexane-1,3-dione, whose central carbon is very acidic. The enol form, along with boron trifluoride coordination to its carbonyl group, facilitates nucleophilic attack by indole which leads to product **1** after loss of water and two prototropic rearrangements.

Pertaining to the formation of **2**, we first begin with dimerization of indole to form 2,3-dihydro-1*H*,1'*H*-[2,3']biindolyl which has one of its indole rings open up to form a free amino anion which can attack cyclohexane-1,3-dione to form after loss of water intermediate **A**. This intermediate can readily close to the seven-membered ring product **2** in two ways. The authors chose the closure taking place via the enamine keto form of intermediate **A**; however, the same ring closure can also take place via the imine enol form as shown in Scheme 2. The ring construction mapping for

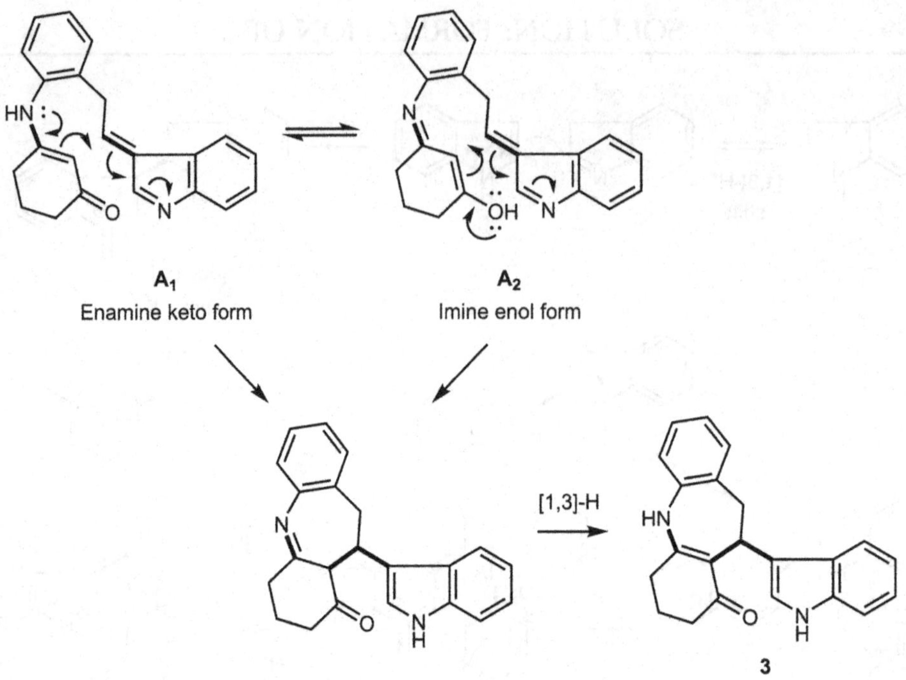

SCHEME 2

formation of product **2** is [5 + 2]. Though the authors characterized the structure of product **2** by X-ray crystallography, it is important to note that further experimental and computation work is required to produce evidence that supports the steps of these two mechanisms postulated by the authors.

ADDITIONAL RESOURCES

We refer the interested reader to the work of Sadak et al. who have explored very similar mechanistic transformations as those encountered in this problem.[2]

References

1. Banerji J, Saha M, Chakrabarti R, et al. Electrophilic substitution of indole. Part 12. Synthesis and structure of 2,3,4,5,10,11-hexahydro-11-(indol-3-yl)-dibenz[b,f]azepin-1-one. *J Chem Res (S)*. 1993;8:320–321.
2. Sadak AE, Arslan T, Celebioglu N, Saracoglu N. New 3-vinylation products of indole and investigation of its Diels–Alder reactivity: synthesis of unusual Morita–Baylis–Hillman-type products. *Tetrahedron*. 2010;66:3214–3221. https://doi.org/10.1016/j.tet.2010.02.081.

Question 248: Serendipitous Preparation of a Pyrrole Precursor to Porphyrins

In 1996 Curran and Keaney unexpectedly achieved a one-pot synthesis of ethyl 5-methylpyrrole-2-carboxylate **3** by reacting 1,4-dichloro-2-butyne **1** with diethyl acetamidomalonate **2** in excess sodium ethoxide and hot ethanol (Scheme 1).[1] Propose a mechanism for this transformation.

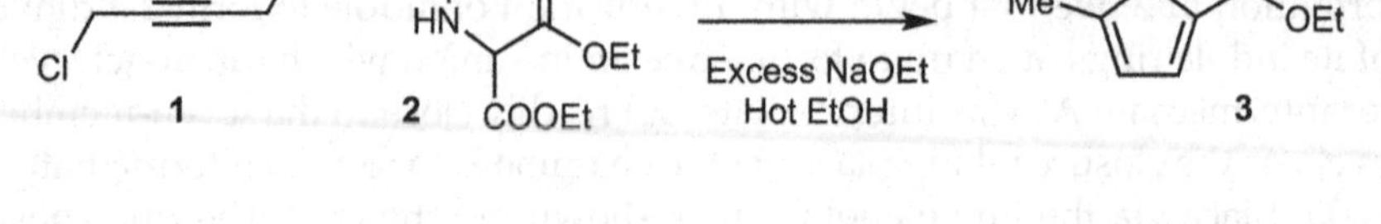

SCHEME 1

SOLUTION

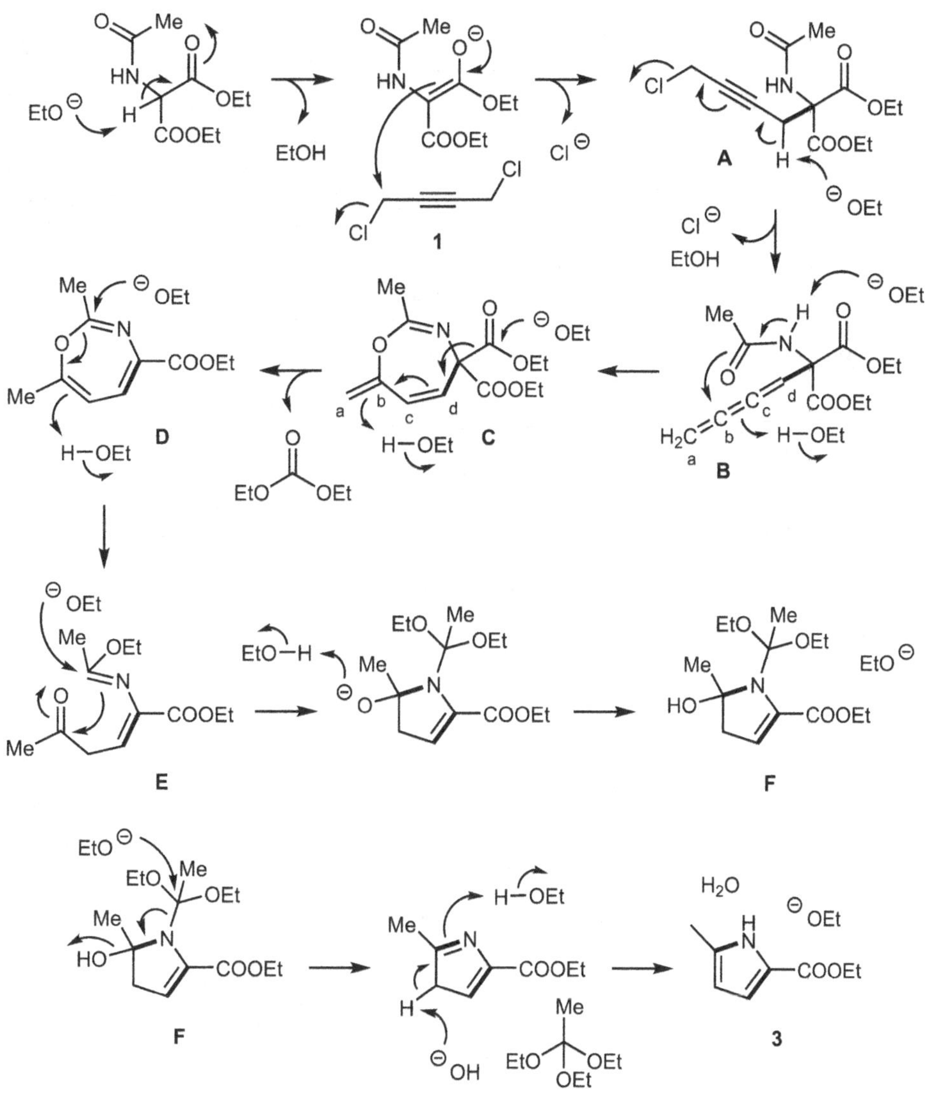

BALANCED CHEMICAL EQUATION

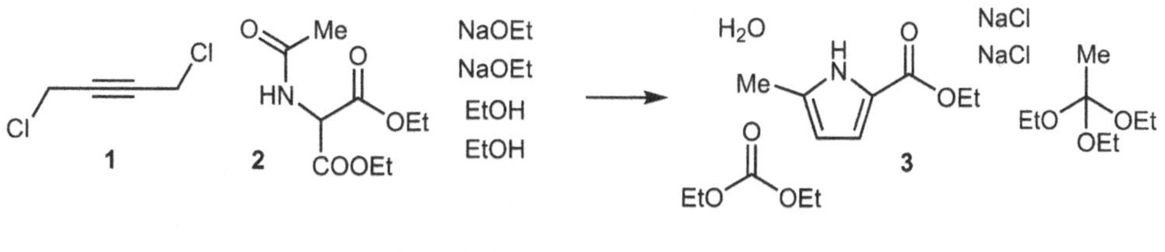

KEY STEPS EXPLAINED

The mechanism for the conversion of **1** and **2** to **3** begins with C-alkylation of the sodium salt of **2** with 1,4-dichloro-but-2-yne **1** to form the propargyl chloride intermediate **A** which undergoes base-catalyzed 1,4-elimination of chloride anion to form the butatriene intermediate **B**. This intermediate then undergoes base-catalyzed nucleophilic attack of the amide carbonyl group onto the central double bond of the butatriene group to form the oxazepine intermediate **C** which eliminates diethyl carbonate to produce the isomeric 1,3-oxazepine monoester **D**. Nucleophilic attack by ethoxide anion onto **D** allows for fragmentation which leads to ketone **E**. This intermediate undergoes another ethoxide anion attack to promote a *5-exo-trig* cyclization to the corresponding pyrrolidine **F** which after elimination of 1,1,1-triethoxyethane and base-catalyzed isomerization forms the final pyrrole product **3**.

With respect to the postulated butatriene intermediate **B**, the authors reference studies examining both the known instability of butatriene compounds and their tendency to undergo addition reactions at the central double bond labeled as $C_b{=}C_c$.[2] Moreover, research by Williams and Kwast from 1988 includes a postulated formation and enolate addition to a butatriene intermediate during the synthesis of bicyclic piperazine-2,5-diones.[3] In addition, the authors cite literature precedent for the base-catalyzed fragmentation of 1,3-oxazepines like **D** to form ketones like **E** which further react to form pyrrolidines like **F**.[4] Experimentally, the authors show that both **1** and **2** are required in equimolar amounts and that aside from heat, an amount greater than two equivalents of base is required. None of the postulated intermediates are synthesized or isolated nor do the authors conduct a theoretical analysis to help support this mechanism from the perspective of geometry and energy calculations.

Based on recent work by Spanish chemists on palladium-catalyzed cyclization reactions of α-cumenols[5] to substituted furans (see Scheme 2) we suggest, by analogy, the possibility of an alternative pathway for the earlier reaction that does not involve oxazepine intermediate **C**, but instead proceeds via a *5-endo-dig* cyclization as shown in Scheme 3.

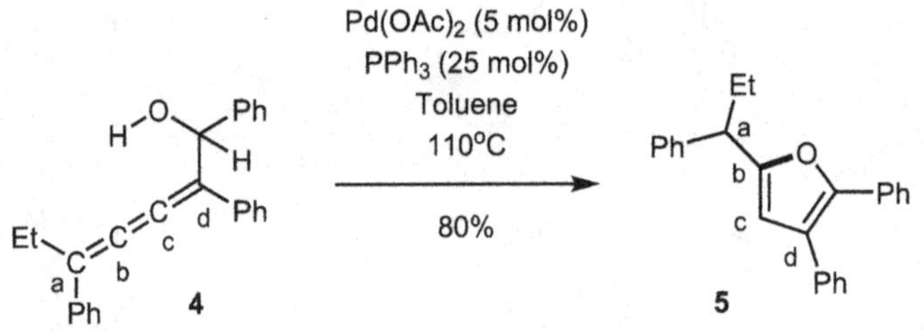

SCHEME 2

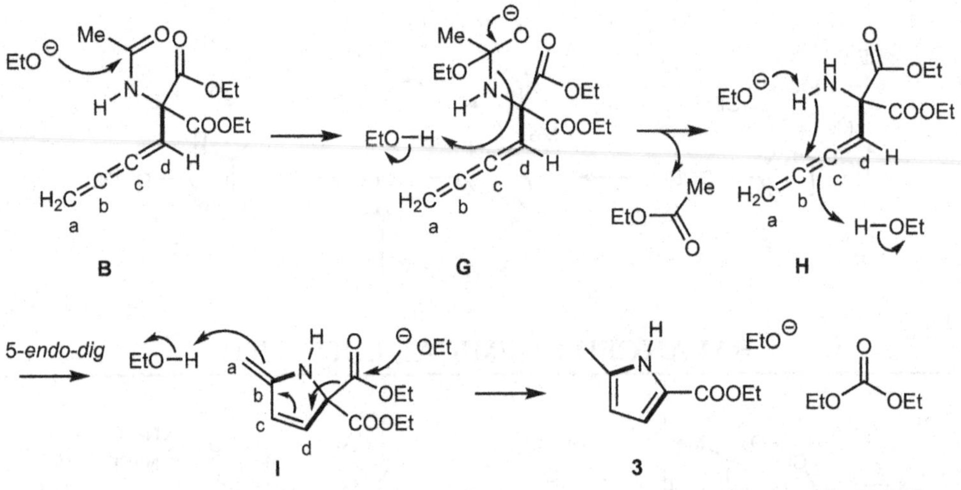

SCHEME 3

Under the harsh alkaline reaction conditions, ethanolysis of the amide group of butatriene intermediate **B** leads to **H** and the elimination of ethyl acetate by-product. Then, base-catalyzed *5-endo-dig* cyclization to a 5-methylene-1,5-dihydro-pyrrole intermediate **I** followed by elimination of diethyl carbonate results in the formation of pyrrole product **3**. The feasibility of the key cyclization step from **H** to **I** and therefore the likelihood of generating pyrrole product **3** depends on deformation of the $C_a{=}C_b{=}C_c$ bonds of the cumulene moiety under the alkaline conditions employed. In the furan example, though this deformation would be expected upon complexation of palladium(II) salts onto these bonds, the geometry of such a deformation was not shown by the authors.[5] Based on the Cartesian coordinates they supplied of the carbon atoms in the model system used in their calculation for the uncomplexed and complexed forms, we were able to determine that the $C_aC_bC_c$ and $C_bC_cC_d$ angles were deformed upon palladium complexation by 10.3 degrees and 7.8 degrees, respectively (see Scheme 4).

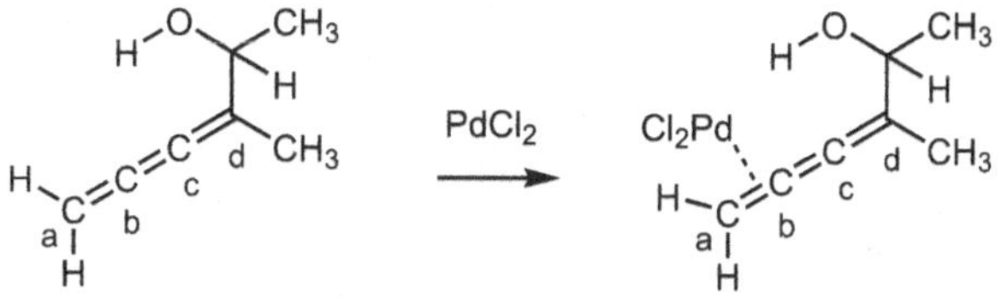

6

Angle $C_aC_bC_c$ = 179.9 degrees

Angle $C_bC_cC_d$ = 179.1 degrees

7

Angle $C_aC_bC_c$ = 169.6 degrees

Angle $C_bC_cC_d$ = 171.3 degrees

Deformation

10.3 degrees

7.8 degrees

SCHEME 4

Theoretical calculations on the furan example at 298 K showed that π-bond complexation has the effect of lowering the initial state by about 27 kcal mol^{-1} and that the 5-*endo-dig* cyclization has the lowest energy barrier at about 17 kcal mol^{-1} compared to other possible modes of cyclization. Analogous computational calculations mapping out the energetics of the two pathways leading to pyrrole **3** would be helpful in identifying the likelihood of formation of oxazepine intermediates versus direct cyclization of butatriene intermediate **B**. A similar 5-*endo-dig* cyclization involving a cumulenyl alkoxide intermediate leading to a furan in the presence of *tert*-butoxide base was also invoked for the mechanism given in Problem #48 by Katritzky and Li[6] (see Scheme 5).

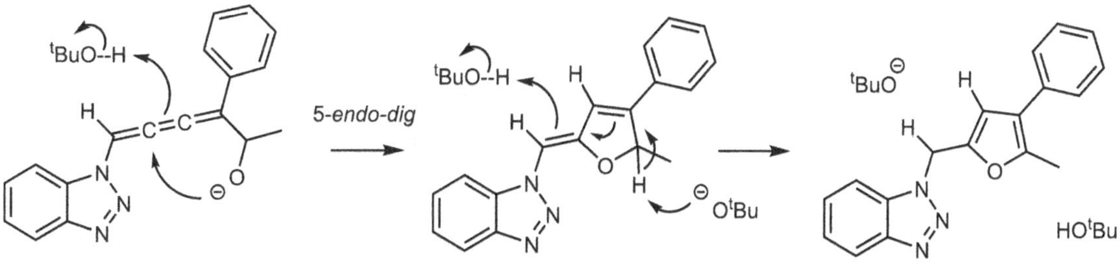

SCHEME 5

Furthermore, if such a cyclization is operative according to our conjectured alternative mechanism this would result in a different overall balanced chemical equation for the transformation. Comparing the net balanced chemical equation shown in Scheme 6 with the previous one indicates that ethyl acetate is formed as a by-product instead of 1,1,1-triethoxyethane. Again, an obvious way to distinguish these mechanisms is to identify reaction by-products, albeit under the reaction conditions it can be seen that 1,1,1-triethoxyethane could result from attack of ethyl acetate with two equivalents of ethoxide or ethanol. Also, ethanol solvent does not appear in the second balanced equation unlike previously where two equivalents were required. However, this distinction will be difficult to resolve since the reaction is carried out in ethanol solvent which is used in excess compared to the molar concentrations of reactants **1** and **2**.

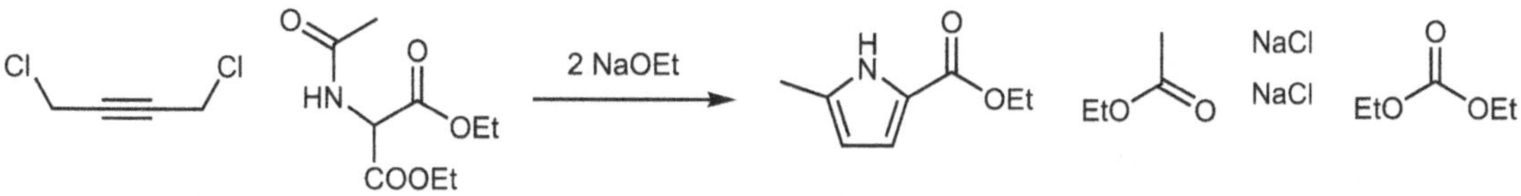

SCHEME 6

ADDITIONAL RESOURCES

For other interesting mechanisms involving the formation of pyrroles, see Refs. 7–9.

References

1. Curran TP, Keaney M. A novel pyrrole synthesis: one-pot preparation of 5-methylpyrrole-2-carboxylate. *J Org Chem*. 1996;61(25):9068–9069. https://doi.org/10.1021/jo961322h.
2. Fischer H. Cumulenes. In: Patai S, ed. *The Chemistry of Alkenes*. London: Interscience; 1964:1100.
3. Williams RM, Kwast A. Versatile new approach to the synthesis of monosubstituted and bicyclic piperazine-2,5-diones: unusual in situ generation and enolate addition to a cumulene. *J Org Chem*. 1988;53(24):5785–5787. https://doi.org/10.1021/jo00259a036.

4. Pedersen CL, Buchardt O. Seven-membered heterocyclic rings. III. Thermolysis and hydrolysis of 2,4,5,7-tetraphenyl-1,3-oxazepine. Pyrrole formation. *Acta Chem Scand*. 1973;27:271–275. https://doi.org/10.3891/acta.chem.scand.27-0271.

5. Alcaide B, Almendros B, Cembellín S, Fernández I, del Campo TM. Metal-catalyzed cyclization reactions of 2,3,4-trien-1-ols: a joint experimental-computational study. *Chem A Eur J*. 2016;22:11667–11676. https://doi.org/10.1002/chem.201601838.

6. Katritzky AR, Li J. A novel furan ring construction and syntheses of 4- and 4,5-substituted 2-(α-heterocycloalkyl)furans. *J Org Chem*. 1995;60:638–643. https://doi.org/10.1021/jo00108a028.

7. Knight DW, Redfern AL, Gilmore J. An approach to 2,3-dihydropyrroles and β-iodopyrroles based on 5-*endo-dig* cyclisations. *J Chem Soc Perkin Trans 1*. 2002;2002:622–628. https://doi.org/10.1039/b110751h.

8. Mullen GB, Georgiev VS. Synthesis of α- and β-nicotyrines. Use of phenyl vinyl sulfoxide as a masked equivalent of acetylene dipolarophile. *J Org Chem*. 1989;54(10):2476–2478. https://doi.org/10.1021/jo00271a052.

9. Sheradsky T. The rearrangement of *o*-vinyloximes a new synthesis of substituted pyrroles. *Tetrahedron Lett*. 1970;11(1):25–26. https://doi.org/10.1016/S0040-4039(01)87556-0.

Question 249: A Chromone Ring Contraction

In 1979 Curran et al. synthesized 5-nitro-2,3-benzofurandione-(Z)-2-oxime **2** by reacting chromone-3-carbaldehyde **1** with red fuming nitric acid in concentrated sulfuric acid at ice bath temperature (Scheme 1).[1] Devise a mechanistic explanation for the formation of the oxime.

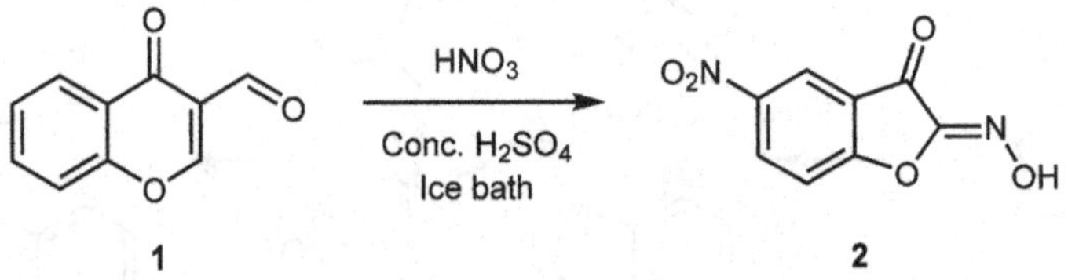

SCHEME 1

SOLUTION

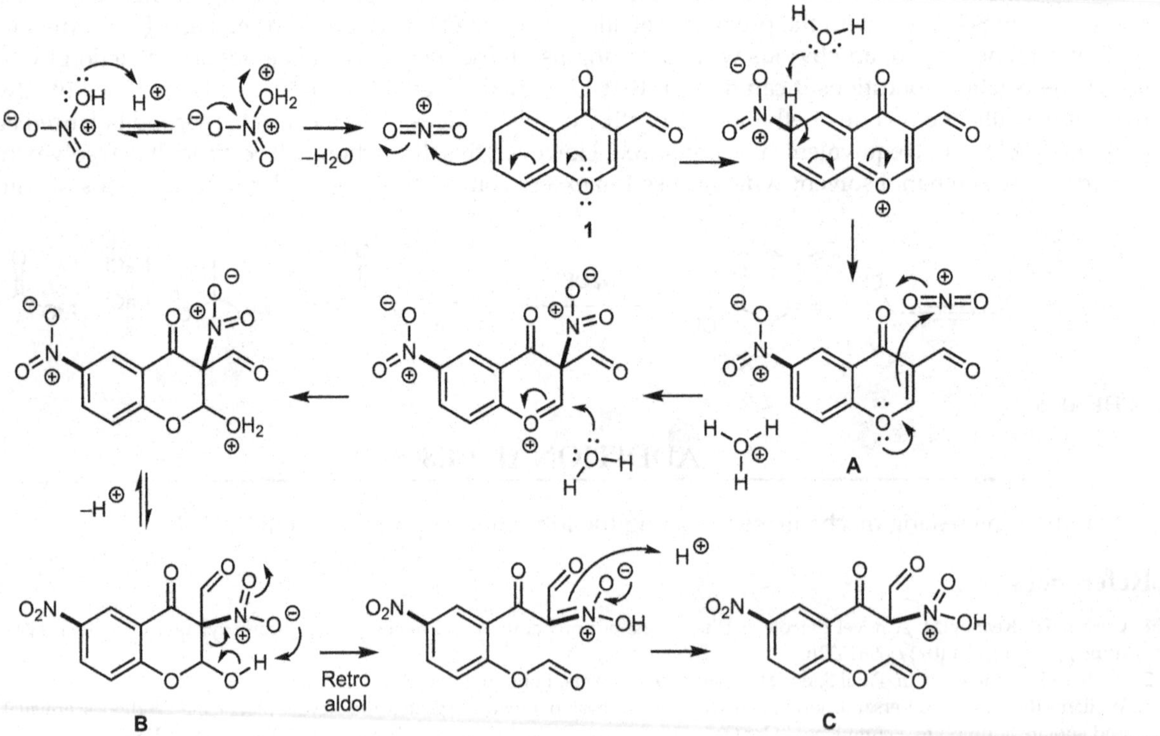

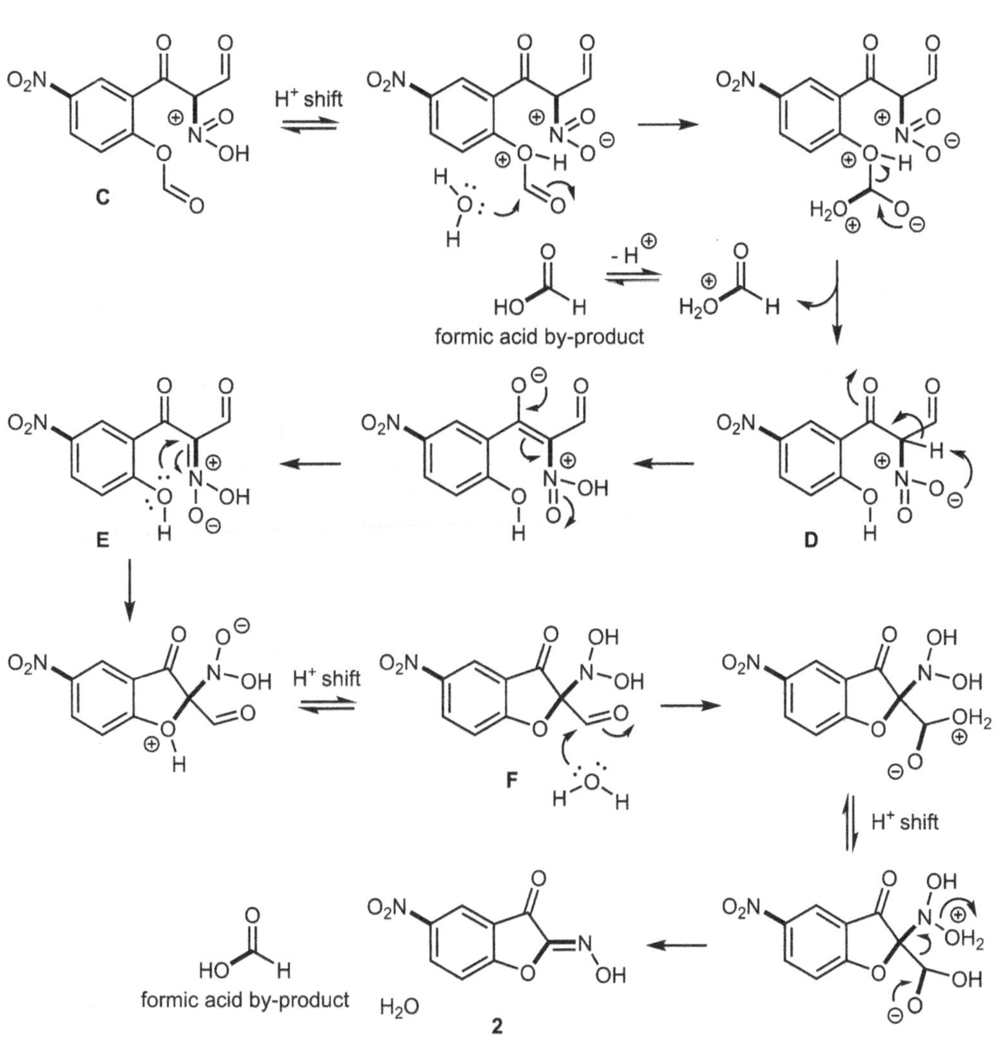

BALANCED CHEMICAL EQUATION

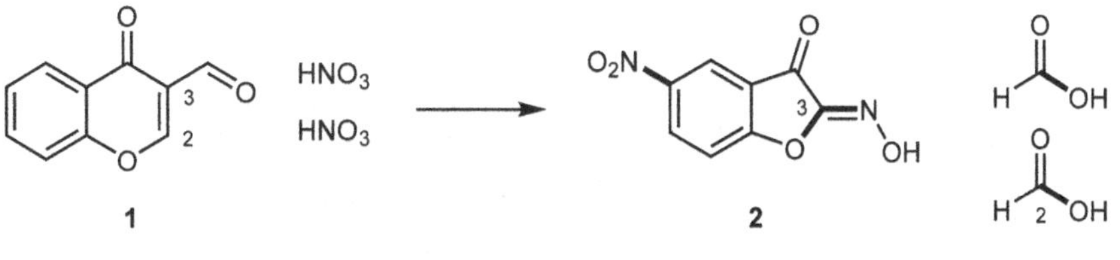

KEY STEPS EXPLAINED

The mechanism proposed here closely resembles that of Curran et al. in that they depict a much shorter mechanism for the carboxylic acid analog of the starting chromone **1** (which ultimately forms the same product and one equivalent of CO_2 and formic acid each, respectively).[1] Nevertheless, we believe that this pathway begins with generation of nitronium ion from nitric acid followed by aromatic nitration of **1** to yield intermediate **A**. Afterward, a second equivalent of nitronium ion is generated and incorporated between the ketone and aldehyde groups of **A** leading to intermediate **B** whose chromone ring undergoes retro-aldol fragmentation to form intermediate **C**. At this stage, **C** undergoes a prototropic rearrangement (possibly acid catalyzed) followed by fragmentation of the formyl ester moiety with elimination of formic acid by-product and formation of **D** which after a prototropic shift transforms into the *aci*-nitro group containing intermediate **E**. This intermediate can easily cyclize to the benzofuranone **F** by having its phenolic hydroxyl group attack the carbon atom bonded to the *aci*-nitro group. A final fragmentation with loss of another equivalent of formic acid by-product (possibly acid catalyzed once again) leads to the formation of the final oxime product **2**.

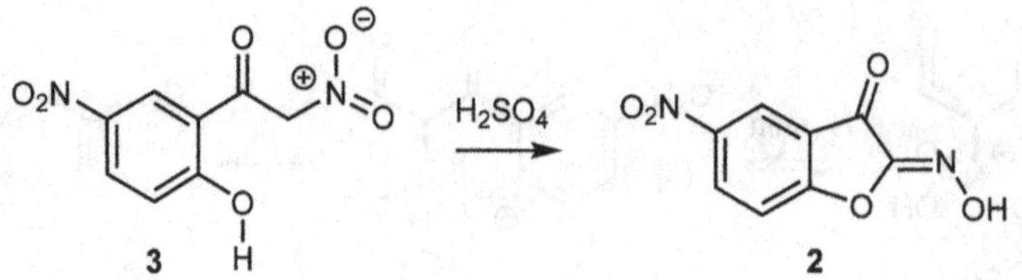

SCHEME 2

We note that although the authors do not present a complete mechanism nor supply a theoretical analysis in support of their mechanism, they do present experimental supporting evidence. For example, Curran et al. have synthesized the conjectured nitroketone **3** intermediate and subjected it to sulfuric acid treatment forming the same oxime product **2** in good yield (Scheme 2). We note that **3** constitutes an intermediate along the analogous mechanism which starts with the carboxylic acid analog of **1** and which involves a late-stage decarboxylation in place of formic acid elimination.

Finally, a closer examination of the mechanism and final balanced chemical equation shows that oxidation number changes occur in key atoms indicating that the reaction is also a combined fragmentation-redox reaction. The nitrogen atom in nitric acid, for instance, has a +5 oxidation state and ends up as −1 in the oxime functional group thereby indicating that a six-electron reduction has taken place. Offsetting this is the change of oxidation number of C_2 in the chromone substrate from 0 to +2 in formic acid by-product, and of C_3 in the chromone substrate from 0 to +3 in the product. Lastly, the aldehyde moiety in the substrate changes from +1 to +2 in the formic acid by-product. In total, a net counterbalancing oxidative change of $(+2)+(+3)+(+1)=+6$ takes place during the course of this transformation.

ADDITIONAL RESOURCES

We direct the reader to several articles highlighting interesting rearrangement mechanisms involving ring contractions and chemical systems similar to the ones encountered in this problem.[2–4] We also recommend a review article by Chandra Ghosh.[5]

References

1. Curran WV, Lovell FM, Perkinson NA. A novel conversion of chromone-3-carboxaldehyde to 5-nitro-2,3-benzofurandione-(Z)-2-oxime. *Tetrahedron Lett.* 1979;20(24):2221–2224. https://doi.org/10.1016/S0040-4039(01)93681-0.
2. Korotaev VY, Barkov AY, Slepukhin PA, Kodess MI, Sosnovskikh VY. Unexpected spontaneous ring-contraction rearrangement of trifluoromethylated 1,2-oxazine N-oxides to 1-pyrroline N-oxides. *Mendeleev Commun.* 2011;21:277–279. https://doi.org/10.1016/j.mencom.2011.09.016.
3. Miao M, Cao J, Zhang J, Huang X, Wu L. PdCl$_2$-catalyzed oxidative cycloisomerization of 3-cyclopropylideneprop-2-en-1-ones. *Org Lett.* 2012;14(11):2718–2721. https://doi.org/10.1021/ol300927n.
4. Mamedov VA, Saifina DF, Rizvanov IK, Gubaidullin AT. A versatile one-step method for the synthesis of benzimidazoles from quinoxalinones and arylenediamines via a novel rearrangement. *Tetrahedron Lett.* 2008;49:4644–4647. https://doi.org/10.1016/j.tetlet.2008.05.060.
5. Ghosh CK. Chemistry of 4-oxo-4H-[1]benzopyran-3-carboxaldehyde. *J Heterocyclic Chem.* 1983;20:1437–1445. https://doi.org/10.1002/jhet.5570200601.

Question 250: A Simple Conversion of Hydroquinone to a Benzofuran, but Which Mechanism?

In 1979 Renfrew et al. examined the mechanism of a hydroquinone to benzofuran transformation (Scheme 1).[1] Out of several possible mechanisms, they argued in favor of one which they considered the likeliest. Explain their reasoning.

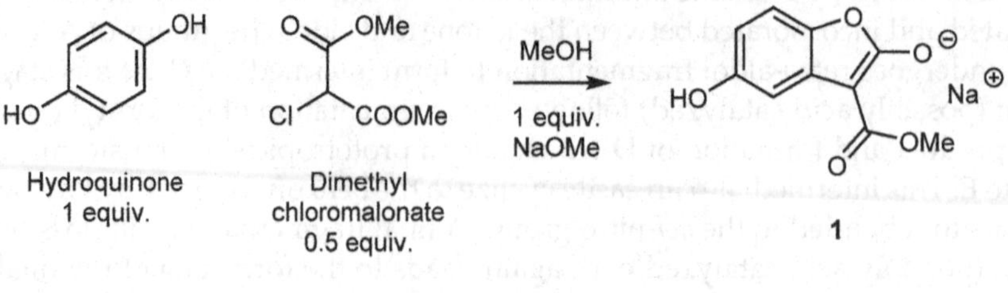

SCHEME 1

SOLUTION: MECHANISM 1 (C-ALKYLATION-LACTONIZATION)

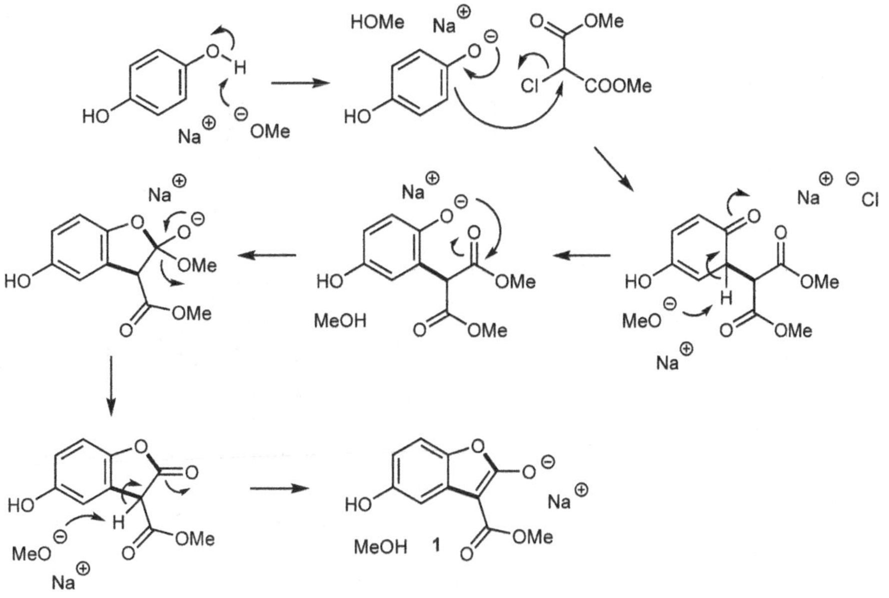

SOLUTION: MECHANISM 2 (O-ALKYLATION-MICHAEL-LACTONIZATION)

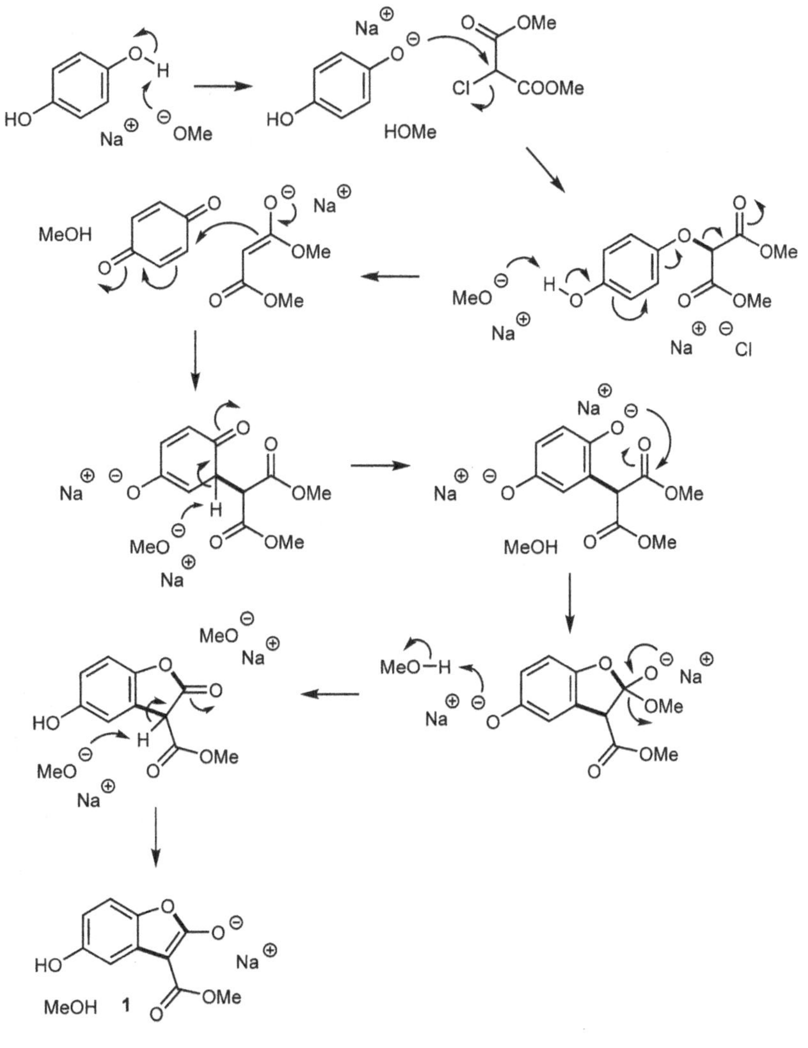

BALANCED CHEMICAL EQUATION

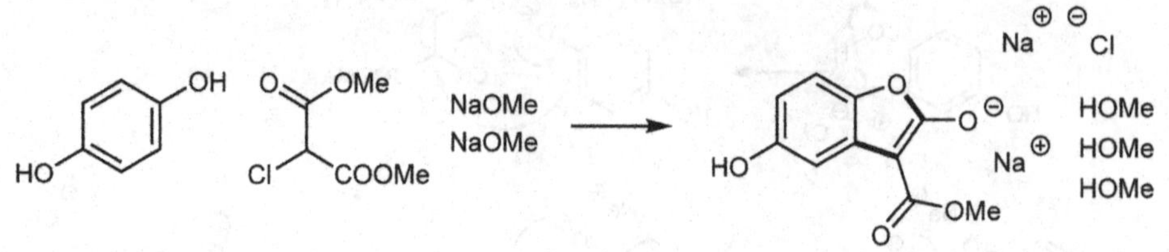

KEY STEPS EXPLAINED

The two mechanisms presented differ in the position of alkylation of the hydroquinone with the dimethyl sodio-malonate. In mechanism 1, alkylation occurs at the carbon ring atom *ortho* to the hydroxyl group deprotonated earlier. Intramolecular cyclization forms the lactone functional group which becomes the conjugated lactone product **1**. In mechanism 2, alkylation occurs at the oxygen atom which leads to base-catalyzed fragmentation followed by 1,4-addition to the nascent 1,4-benzoquinone followed by the regular intramolecular cyclization and enolization seen in mechanism 1 to produce the final product **1**.

The authors rationalize a preference for mechanism 2, the *O*-alkylation mechanism, using a literature precedent as well as experimental by-product identification. They say that the literature examples of observed base-catalyzed hydroquinone alkylations are all of the *O*-alkylation variety. Experimentally they explain that a variety of mono-substituted hydroquinones form only *O*-alkylated products and that only in certain cases, for example, where the starting hydroquinone lacks —*I* (inductive) and —*M* (mesomeric, i.e., resonance stabilized) groups in addition to a few other cases, does one observe the *C*-alkylated product. Nevertheless, a closer examination of the main reaction of this problem revealed to the authors the generation of the three by-products **2**, **3**, and **4** (Scheme 2).

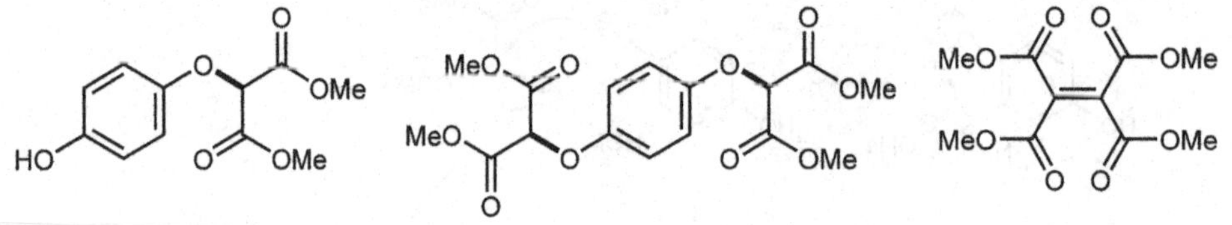

2 (5.3%) **3** (1.9%) **4** (3.2%)

SCHEME 2

These by-products represent the mono-*O*-alkylated intermediate **2**, the di-*O*-alkylated product **3**, and by-product **4**, which is the product of the self-reaction of two molecules of dimethyl sodiomalonate. By-product **2** is the first intermediate generated in mechanism 2. By-product **3** is generated from **2** upon reaction with base and another molecule of dimethyl chloromalonate. According to the pathway in mechanism 1 both of these by-products cannot form. By-product **4** can arise via two possible pathways. Scheme 3 shows a mechanism involving a base-mediated tandem addition-elimination sequence.

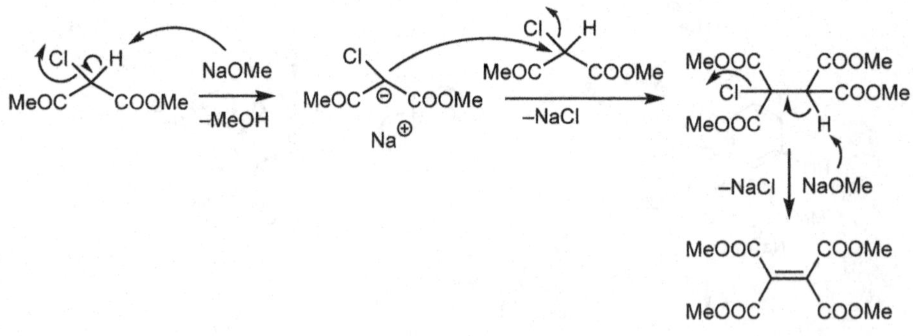

4 (3.2%)

SCHEME 3

Another way of generating by-product **4** is via a redox reaction from the liberated benzoquinone shown in mechanism 2. Scheme 4 shows a mechanism where the benzoquinone and dimethyl malonyl carbanion intermediates generated in step 3 of mechanism 2 are, respectively, reduced back to hydroquinone and oxidized to dicarboxymethyl carbene which can then dimerize to the olefin. This possibility was not discussed by the authors. The occurrence of by-products **3** and **4** is more likely in the presence of excess dimethyl chloromalonate. In any event, the key take home lesson is that it is the experimental observation and identification of minor products that offers clues as to which mechanism is more likely to be operative in the reaction.

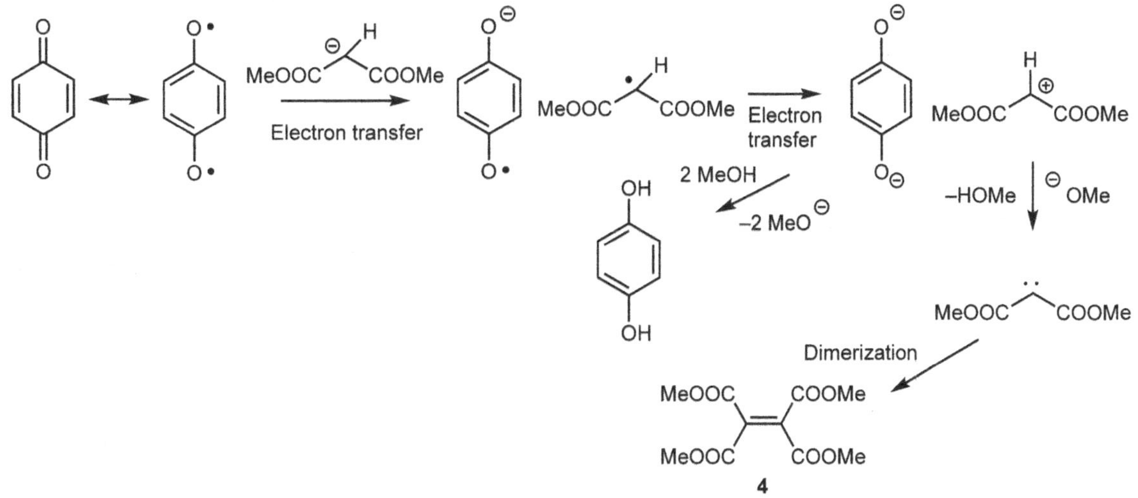

SCHEME 4

ADDITIONAL RESOURCES

We direct the interested reader to a modern approach to the synthesis of benzofurans using similar chemistry as the current problem.[2] An electrochemical example[3] as well as an interesting mechanism example[4] involving hydroquinone is also available. Lastly, we present a theoretical study of the comparison of *O*- and *C*-alkylation of phenoxide ion given various solvent effects.[5]

References

1. Renfrew AH, Bostock SB, Ecob CM. The reaction of hydroquinone and monosubstituted hydroquinones with dimethyl chloromalonate. *J Chem Soc Perkin Trans 1*. 1979;2382–2386. https://doi.org/10.1039/P19790002382.
2. Kokubo K, Harada K, Mochizuki E, Oshima T. A new approach to benzofuran synthesis: Lewis acid mediated cycloaddition of benzoquinones with stilbene oxides. *Tetrahedron Lett*. 2010;51:955–958. https://doi.org/10.1016/j.tetlet.2009.12.047.
3. Nematollahi D, Amani A, Tammari E. Electrosynthesis of symmetric and highly conjugated benzofuran via a unique ECECCC electrochemical mechanism: evidence for predominance of electrochemical oxidation versus intramolecular cyclization. *J Org Chem*. 2007;72:3646–3651. https://doi.org/10.1021/jo062468b.
4. Urban S, Capon RJ. Marine sesquiterpene quinones and hydroquinones: acid-catalyzed rearrangements and stereochemical investigations. *Aust J Chem*. 1994;47(6):1023–1029. https://doi.org/10.1071/CH9941023.
5. Breslow R, Groves K, Mayer MU. Antihydrophobic cosolvent effects for alkylation reactions in water solution, particularly oxygen versus carbon alkylations of phenoxide ions. *J Am Chem Soc*. 2002;124:3622–3635. https://doi.org/10.1021/ja012293h.

Chapter 9: Solutions 251 – 300

Question 251: A Remarkable Loss of One Carbon Atom in the Indole Alkaloid Field

After reacting anhydrovinblastine N_b-oxide **1** with trifluoroacetic anhydride in methylene chloride and treatment with water and THF, French chemists obtained pure 5′-noranhydrovinblastine **2** in 27% yield along with several other products (Scheme 1).[1] Provide a reasonable mechanism for this transformation.

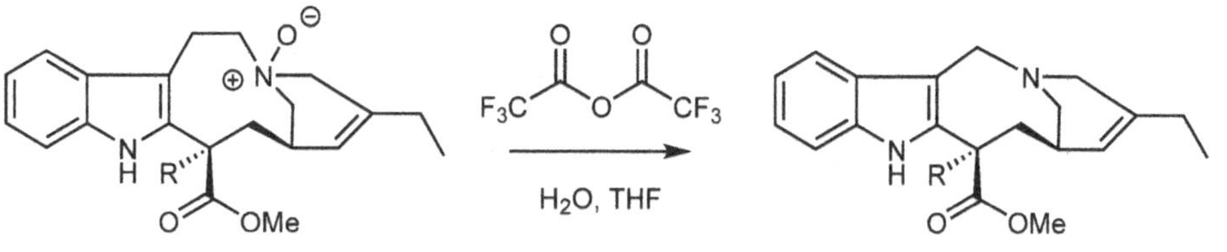

SCHEME 1

SOLUTION

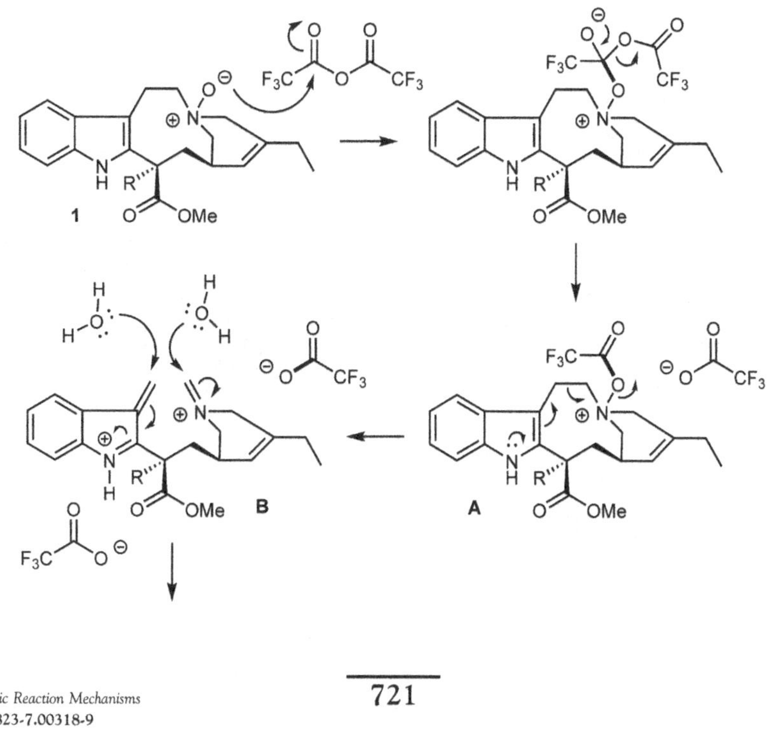

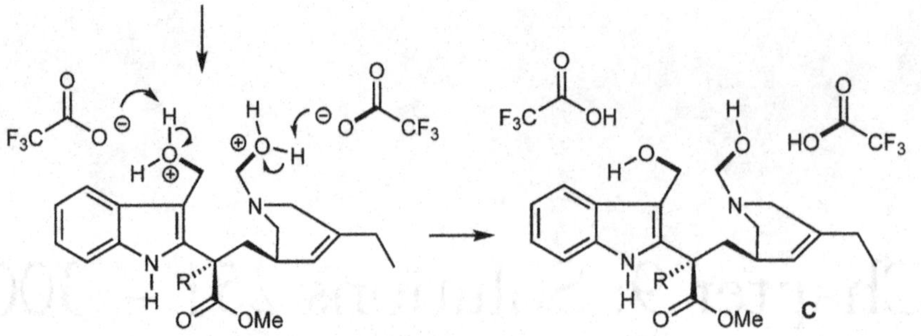

SOLUTION—POSSIBILITY 1 (WITH INTERVENTION OF RETRO-MANNICH PROCESS)

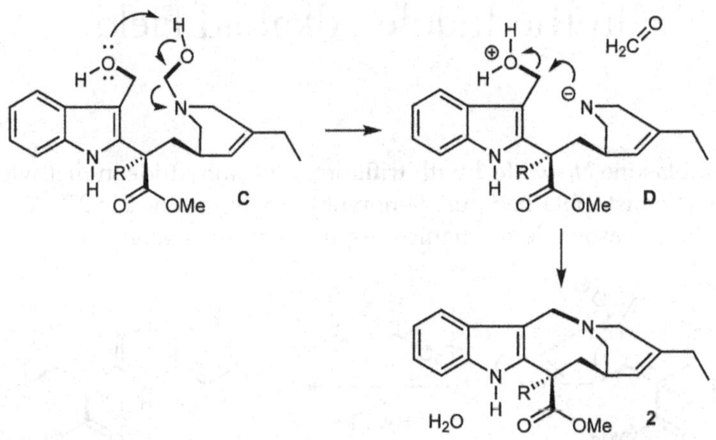

SOLUTION—POSSIBILITY 2 (WITHOUT INTERVENTION OF RETRO-MANNICH PROCESS)

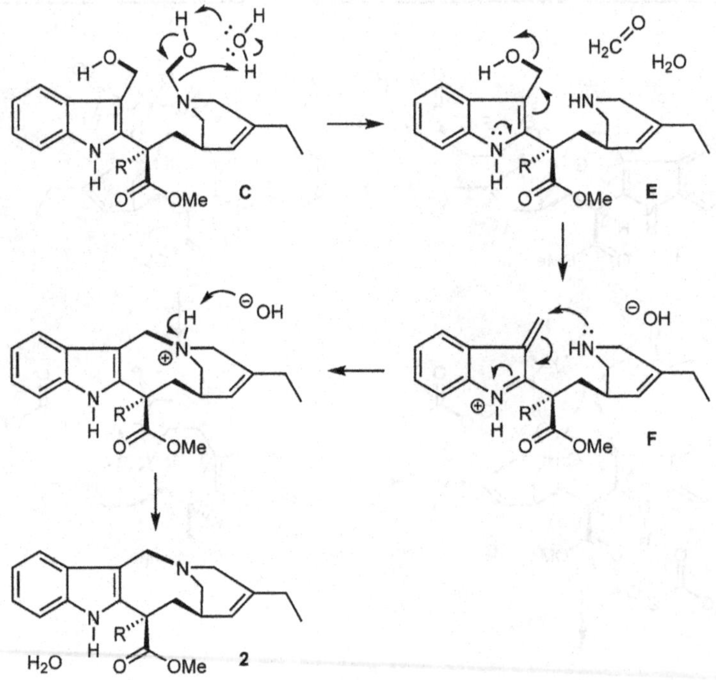

BALANCED CHEMICAL EQUATION

KEY STEPS EXPLAINED

The authors describe the transformation as starting with the reaction between anhydrovinblastine N_b-oxide **1** and trifluoroacetic anhydride in order to set up the fragmentation of **A** by means of electron shuffle and loss of trifluoroacetate anion to form **B**. At this point two equivalents of a nucleophile (in this example, water) add to the bisimmonium salt **B** to form **C** and upon proton shuffle lose two equivalents of trifluoroacetic acid. Now the authors propose two possible avenues for the formation of **2** from **C**. The first (not depicted in their paper) involves the *retro*-Mannich process in which a geometrically neighboring hydroxyl group abstracts a proton in an intramolecular fashion from the other hydroxyl group to initiate loss of formaldehyde molecule and to set up the final intramolecular cyclization step which leads to **2** with loss of water. The second possibility, which is actually depicted in the original paper, involves an intermolecular process making use of a molecule of water to effect the loss of formaldehyde to give **E** which can lose hydroxide anion in order to set up the final intramolecular cyclization step which leads to **2**. Unfortunately, without the use of theoretical analysis (absent from the original work) and geometry optimization capable of calculating the theoretical energy barriers for structures **D**, **E**, and **F**, it is not possible to determine which of the two possible pathways from **C** to **2** is the likeliest.

ADDITIONAL RESOURCES

For more of the work of Langlois and Potier, we recommend Refs. 2, 3 which contain discussions that are very relevant to the current problem. Finally, we refer the interested reader to Ref. 4 which encompasses a mechanistic analysis of a very similar catharanthine analog.

References

1. Mangeney P, Andriamialisoa RZ, Lallemand J-Y, Langlois N, Langlois Y, Potier P. 5'-Nor anhydrovinblastine prototype of a new class of vinblastine derivatives. *Tetrahedron.* 1979;35(18):2175–2179. https://doi.org/10.1016/0040-4020(79)87036-2.
2. Langlois N, Potier P. Antitumour alkaloids of the vinblastine-type: air oxidation of anhydrovinblastine. *J Chem Soc Chem Commun.* 1979;(13):582–584. https://doi.org/10.1039/C39790000582.
3. Langlois N, Andriamialisoa RZ, Langlois Y. Oxo-15 dihydro-15,20S catharanthine, nouvel intermediaire dans l'hemisynthese de l'anhydrovinblastine. *Tetrahedron.* 1981;37(10):1951–1954. https://doi.org/10.1016/S0040-4020(01)97945-1.
4. Sundberg RJ, Hong J, Smith SQ, Sabat M, Tabakovic I. Synthesis and oxidative fragmentation of catharanthine analogs. Comparison to the fragmentation—coupling of catharanthine and vindoline. *Tetrahedron.* 1998;54(23):6259–6292. https://doi.org/10.1016/S0040-4020(98)00289-0.

Question 252: A "Nonobvious" Cycloaddition Reaction

The exothermic reaction of diene **1** with dimethyl acetylenedicarboxylate in benzene at room temperature was found by Belgian chemists to produce dimethyl 4-*N,N*-dimethylamino-5-methylphthalate **2** in 70% yield (Scheme 1).[1] Provide a mechanism for this transformation.

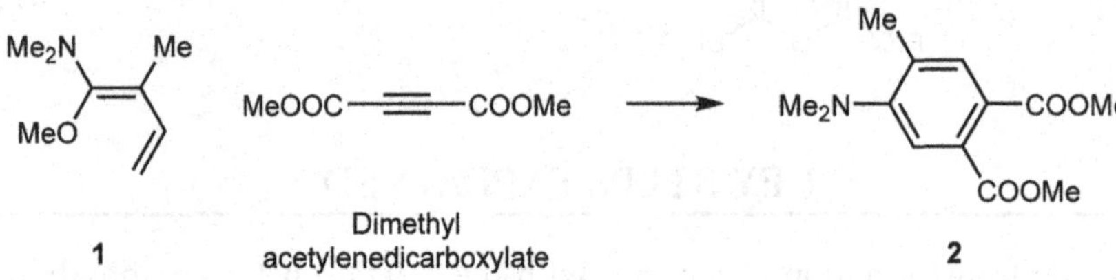

SCHEME 1

SOLUTION

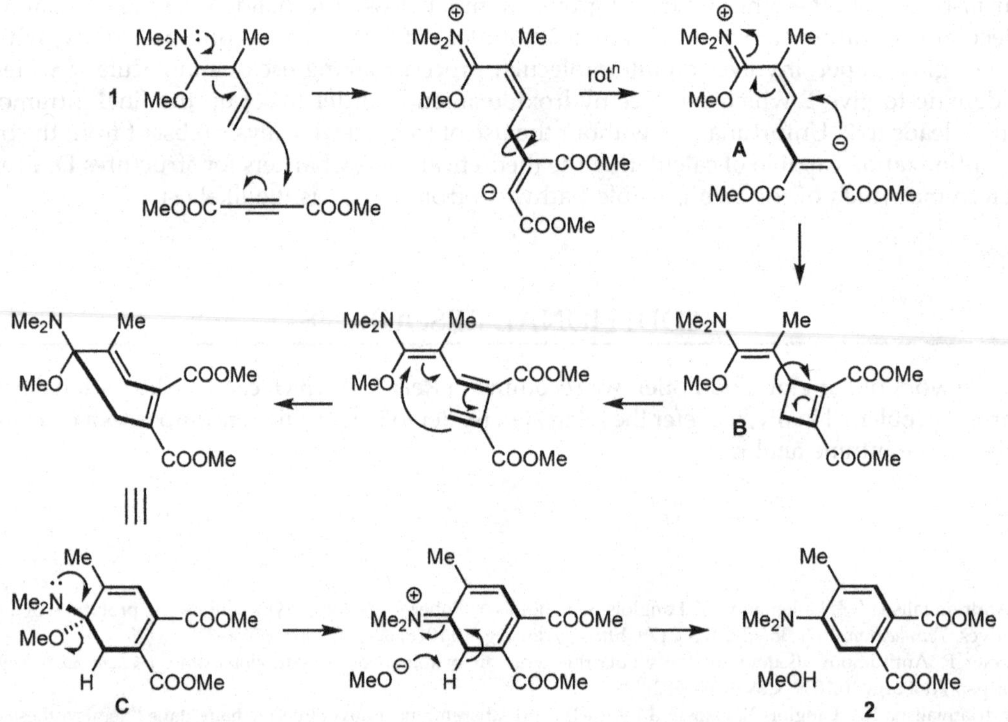

BALANCED CHEMICAL EQUATION

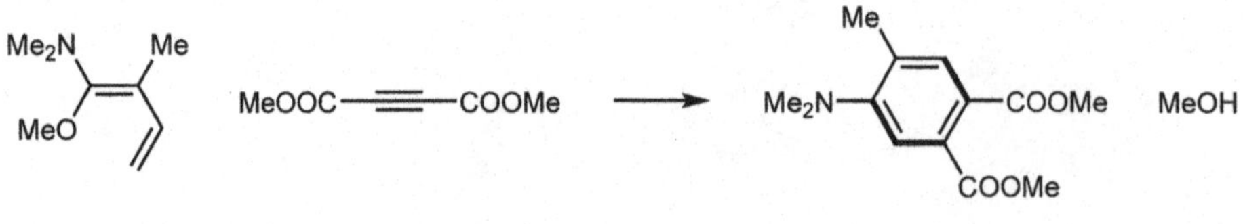

KEY STEPS EXPLAINED

Gillard et al. explain the transformation of diene **1** with dimethyl acetylenedicarboxylate in benzene at room temperature as following an ionic pathway as opposed to a Diels-Alder or even [2+2] cycloaddition coupling pathway. We know, for instance, that the [2+2] pathway from **1** directly to **B** is forbidden under the Woodward Hoffmann rules of cycloadditions because it does not occur under thermal conditions (as relevant here) but only under photochemical conditions. The authors further explain that in cases where the methoxide group of **1** is replaced with a leaving group such as Cl⁻ or I⁻, a Diels-Alder reaction is more likely to occur via a vinylketeniminium halide intermediate. Interestingly, when this is done the resulting product has a benzene ring with the same groups as in structure 2 but in different positions as shown in Scheme 2. Lastly, replacing this group with other more electron-donating groups and reacting the dienes with substituted alkenes has opened up new synthetic opportunities according to the authors. Nevertheless, the mechanistic pathway relevant for this problem begins by coupling of **1** with dimethyl acetylenedicarboxylate followed by bond rotation to form **A** which ring closes to **B** which fragments to achieve π-conjugation and subsequently ring closes to **C** which upon loss of methanol molecule leads to the final product **2**.

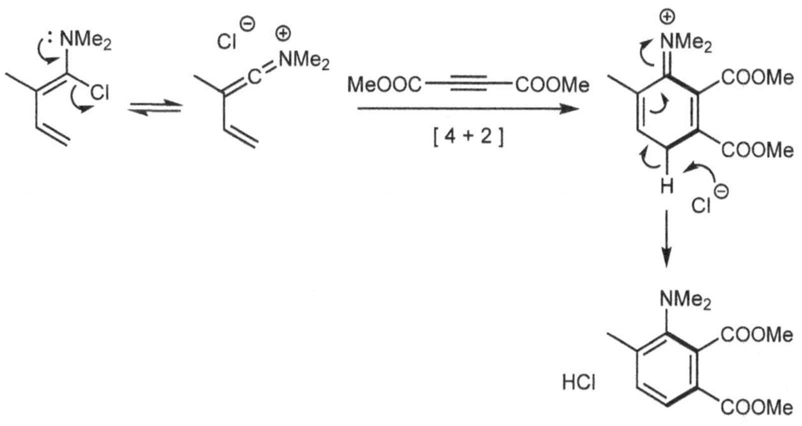

SCHEME 2 Diels-Alder adduct resulting when methoxide is replaced by a halide group in the starting diene.

ADDITIONAL RESOURCES

We direct the reader to two works which highlight interesting mechanistic outlines of similar coupling reactions involving cyclic oxonitriles[2] and push-pull alkenes[3] as well as their synthetic utility.

References

1. Gillard M, T'Kint C, Sonveaux E, Ghosez L. Diels-Alder reactions of "pull-push" activated isoprenes. *J Am Chem Soc*. 1979;101(19):5837–5839. https://doi.org/10.1021/ja00513a069.
2. Fleming FF, Iyer PS. Cyclic oxonitriles: synergistic juxtaposition of ketone and nitrile functionalities. *Synthesis*. 2006;(6):893–913. https://doi.org/10.1055/s-2006-926363.
3. Ye G, Chatterjee S, Li M, et al. Push–pull alkenes from cyclic ketene-*N,N'*-acetals: a wide span of double bond lengths and twist angles. *Tetrahedron*. 2010;66(16):2919–2927. https://doi.org/10.1016/j.tet.2010.02.071.

Question 253: More Radical Cascades and a Formal [2 + 2 + 2] Cycloaddition of a Dienyne

In 1996 a Spanish chemist described the reaction of dienyne **1** with triphenyltin hydride and triethylborane in toluene at room temperature for 3–6 h. The result consisted of the three products **2**, **3**, and **4** all obtained diastereomerically pure (Scheme 1).[1] Show mechanisms for the formation of **2**, **3**, and **4**.

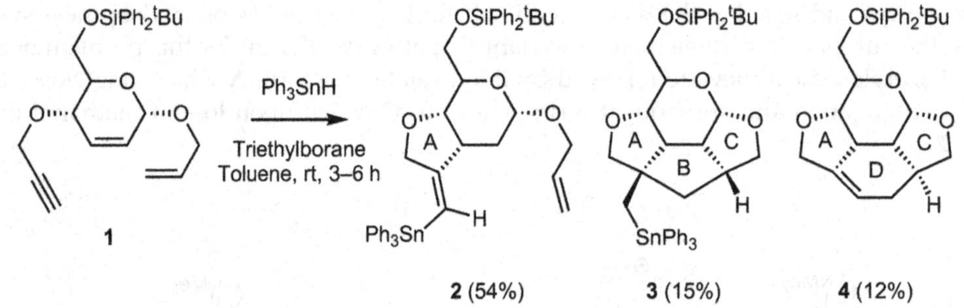

2 (54%)   **3 (15%)**   **4 (12%)**

SCHEME 1

SOLUTION

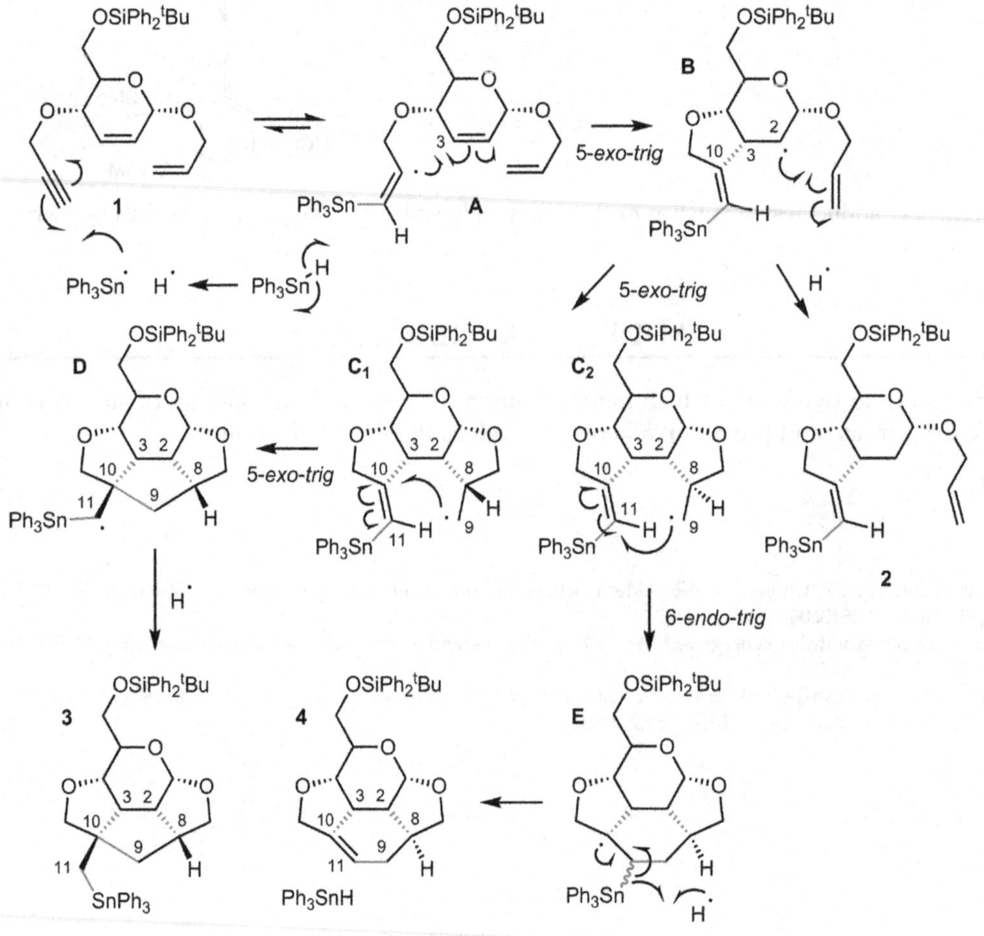

KEY STEPS EXPLAINED

In his original work, Marco-Contelles explains the formation of **2**, **3**, and **4** from the dienyne **1** as a series of three tandem radical cyclizations. For example, after homolytic fragmentation of triphenyltin hydride into its respective radical species, triphenyltin radical may react with the terminal end of the acetylene group of **1** to form **A** which can proceed with a 5-*exo-trig* radical cyclization (nomenclature for ring closure according to Baldwin's rules, see Fig. 1)[2] to produce **B** by forming a C—C bond between C_{10} and C_3 (sugar numbering). The newly formed radical at C_2 can either react with hydride radical to form **2** right away (which explains the disproportionately high yield of **2**) or it may react with the terminal alkene group again in a 5-*exo-trig* manner to form two radical intermediates C_1 and C_2 which differ in the stereochemistry of the methylene radical group at C_9. Interestingly, the radical intermediate species C_1 may cyclize in a 5-*exo-trig* manner to form **D** which can react with hydride radical to form **3**. Alternatively, C_2 may close in a 6-*endo-trig* manner to form a racemic species **E** which can eliminate triphenyltin radical to give the product **4**. It is also possible to label the products formed by ring formation nomenclature as such: **2** is [5+0], **3** is $[(5+0)_A +(2+2+1)_B +(5+0)_C]$, and **4** is $[(5+0)_A +(2+2+2)_D +(5+0)_C]$. Lastly, the "[2+2+2]" designation in the title of the author's paper refers to the number of π bonds involved in the formation of the cycloaddition products and not to ring construction connectivity as in our nomenclature. The only common [2+2+2] designation between the two nomenclatures applies to the central ring D formed in product **4**.

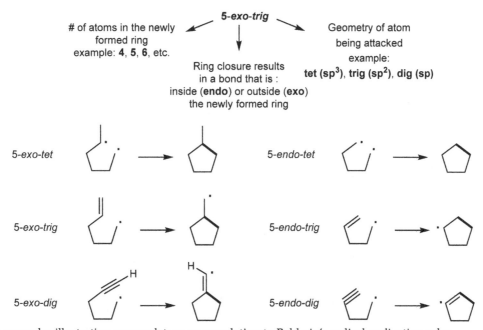

FIG. 1 Various examples illustrating nomenclature usage relating to Baldwin's radical cyclization rules.

ADDITIONAL RESOURCES

It is interesting to note that products **3** and **4** are formed exclusively via the respective 5-*exo-trig* and 6-*endo-trig* cyclizations from C_1 and C_2, respectively. The yields of these products are also much smaller. The author attributes this consequence to the unfavorable nature of a radical attack on a tri-substituted vinyl alkene group and the consequences involved in the nature of ring-forming reactions. For a recent review of this topic which includes a discussion of Baldwin's rules, see the work of Alabugin and Gilmore.[3] For other examples of radical-mediated cyclizations that can take different mechanistic paths, see Refs. 4–6.

References

1. Marco-Contelles J. Synthesis of polycyclic molecules via cascade radical carbocyclizations of dienynes: the first SnPh₃ radical-mediated [2 + 2 + 2] formal cycloaddition of dodeca-1,6-dien-11-ynes. *Chem Commun.* 1996;2629–2630. https://doi.org/10.1039/CC9960002629.
2. Baldwin JE, Thomas RC, Kruse LI, Silberman L. Rules for ring closure: ring formation by conjugate addition of oxygen nucleophiles. *J Org Chem.* 1977;42(24):3846–3852. https://doi.org/10.1021/jo00444a011.
3. Alabugin IV, Gilmore K. Finding the right path: Baldwin "rules for ring closure" and stereoelectronic control of cyclizations. *Chem Commun.* 2013;49:11246–11250. https://doi.org/10.1039/c3cc43872d.
4. Spino C, Barriault N. Radical cyclization of polyenes initiated by attack of trialkyltin or germanium radical on an ynone. *J Org Chem.* 1999;64:5292–5298. https://doi.org/10.1021/jo9905524.
5. Kelly DR, Picton MR. Catalytic tin radical mediated tricyclisations. Part 2. *J Chem Soc Perkin Trans 1.* 2000;2000:1571–1586. https://doi.org/10.1039/b000662i.
6. Fensterbank L, Mainetti E, Devin P, Malacria M. New advances in radical cascades based on the use of vinyl radicals. *Synlett.* 2000;2000 (9):1342–1344. https://doi.org/10.1055/s-2000-7139.

Question 254: LTA-Induced Acetoxylation and Rearrangement of a Phenol

During a 1980 investigation of the reactivity of tetrahydroisoquinolines, Hara et al. showed that oxidation of phenol **1** with lead tetraacetate (LTA) in acetic acid gave phenol **2** in 60% yield (Scheme 1).[1] Devise a mechanism for this transformation.

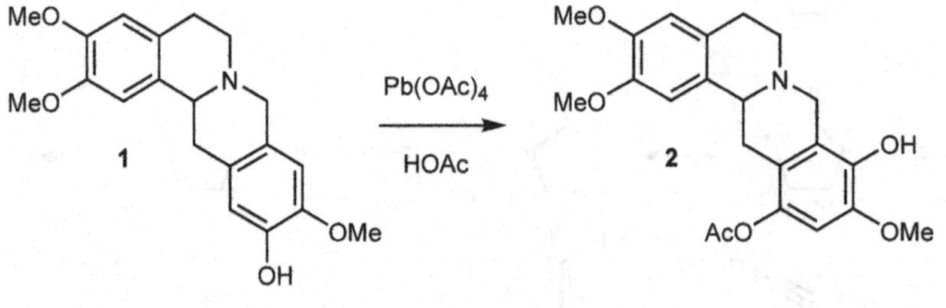

SCHEME 1

KEY STEPS EXPLAINED

In their work, Hara et al. present the mechanism for the transformation of **1** to **2** (see Solution below) as beginning with a nucleophilic reaction between **1** and lead tetraacetate to give **A** with loss of acetate which is protonated to acetic acid to give a neutral species. Acetic acid can then attack the same carbon atom where lead triacetate had attached to displace lead (II) diacetate and acetate to facilitate the formation of **B**. Structure **B** thus undergoes retro-Mannich fragmentation to form iminium ion **C** after rotation about the aryl group to allow for facile *ortho* recyclization to a ketone structure which after tautomerization gives the final rearranged phenol product **2**. It is worth noting that **2** is not the only product formed in this transformation. The authors also reported the formation of the phenol **3** as a product. We thus present a mechanism for the formation of **3**. In this mechanism, the branching point occurs at structure **A** whereby instead of attacking the carbon bearing the lead triacetate group, acetate deprotonates **A** and subsequently attacks on the other side of the ring to cause aromatization and displacement of lead (II) diacetate and acetate to help form **3**. In their paper, the authors discuss other experimental results that help support the proposed mechanism.

SOLUTION

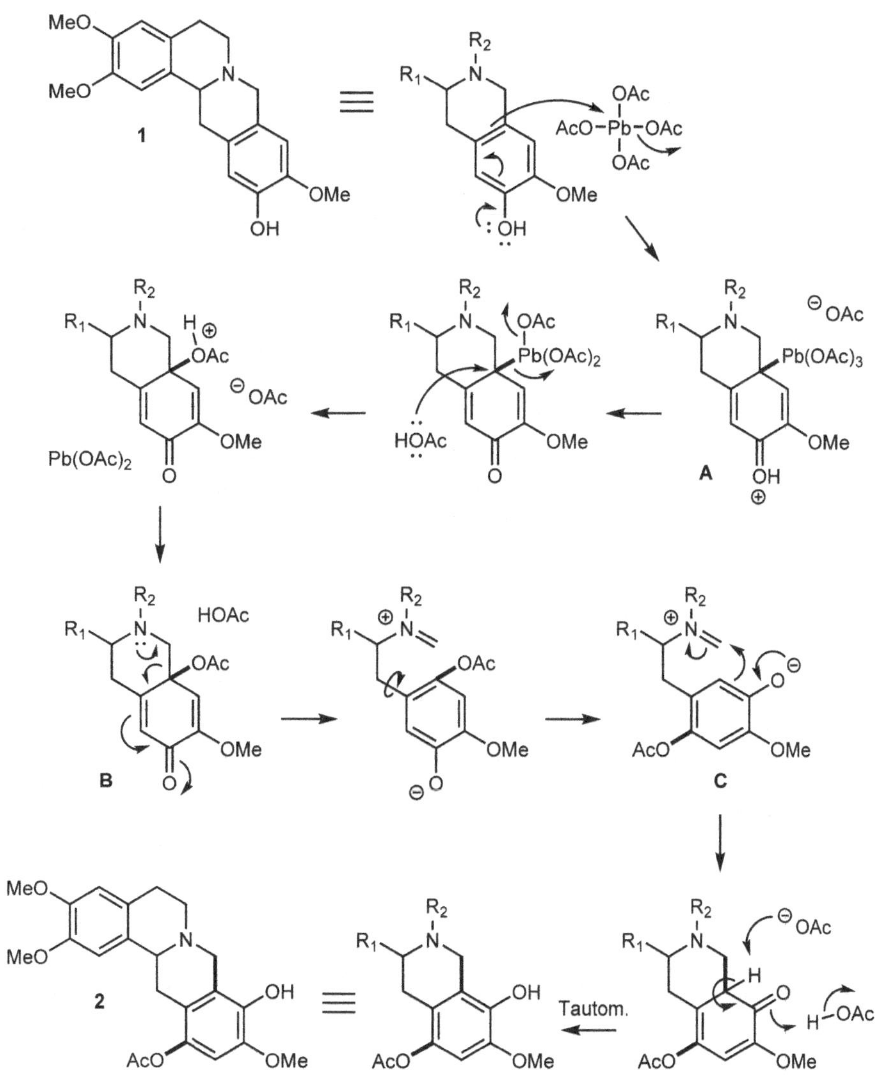

BALANCED CHEMICAL EQUATION

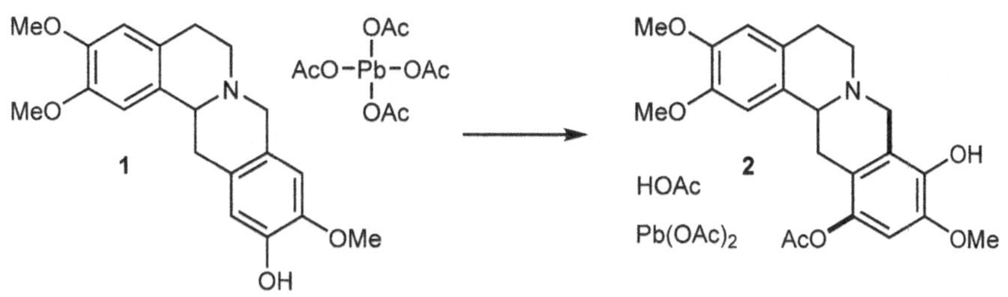

MECHANISM FOR 3

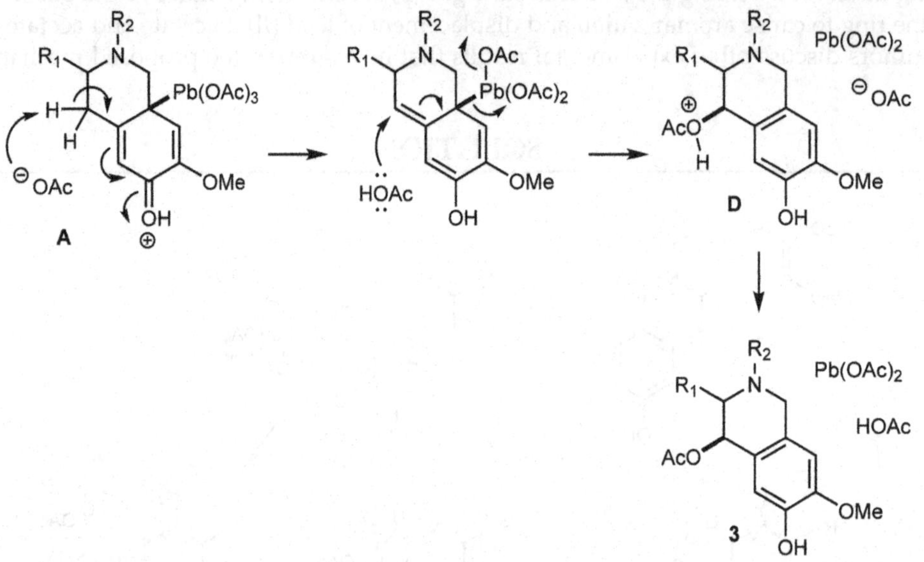

ADDITIONAL RESOURCES

For a similar transformation involving ring fragmentation, bond rotation, and recyclization steps in the context of carbohydrate chemistry, see Ref. 2. For an example of the mechanistic pathways involved in LTA-mediated C—C bond oxidation reactions, see Ref. 3.

References

1. Hara H, Hosaka M, Hoshino O, Umezawa B. Studies on tetrahydroisoquinolines. Part 15. Lead tetra-acetate oxidation of four hydroxytetrahy-droprotoberberines, (±)-govanine, (±)-discretine, (±)-corytencine, and (±)-10-hydroxy-2,3,11-trimethoxytetrahydroprotoberberine. *J Chem Soc Perkin Trans 1*. 1980;1169–1175. https://doi.org/10.1039/P19800001169.
2. Munyololo M, Gammon DW, Mohrholz I. Extending the scope of the Ferrier reaction: fragmentation-rearrangement reactions of selectively substituted 1,2-cyclopropanated glucose derivatives. *Carbohydr Res*. 2012;351:49–55. https://doi.org/10.1016/j.carres.2012.01.006.
3. Trost BM, Fleming I, Ley SV. *Comprehensive Organic Synthesis: Selectivity, Strategy and Efficiency in Modern Organic Chemistry. Oxidation*; vol. 7:. Oxford: Pergamon Press; 1991;710.

Question 255: How Cocaine Decomposes at 550°C

In 1989 American scientists showed that (−)-cocaine 1 under flash vacuum pyrolysis decomposes to benzoic acid, N-methylpyrrole, and methyl 3-butenoate (Scheme 1).[1] Devise a mechanism for this transformation.

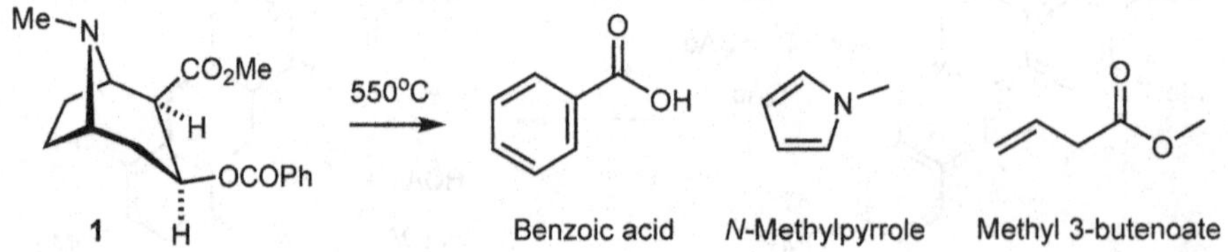

SCHEME 1

SOLUTION

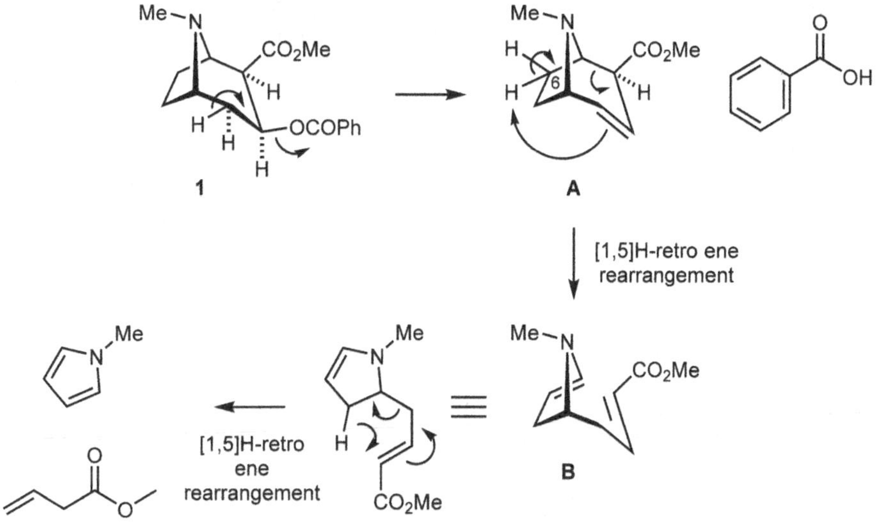

KEY STEPS EXPLAINED

Under very high dilution conditions, meant to limit bimolecular reactions, Sisti et al. have used flash vacuum pyrolysis to demonstrate the decomposition pathway of (−)-cocaine. This pathway begins with E2 elimination of benzoic acid to form **A** which undergoes a [1,5]H-retro ene rearrangement. This rearrangement makes use of the double bond abstracting the *endo* γ-hydrogen atom on C_6 of **A** on account of structural and geometrical constraints. The product **B** can undergo another [1,5]H-retro ene rearrangement to produce the final N-methylpyrrole and methyl 3-butenoate products. The authors argue that energetic driving forces for this decomposition consist of the favorable formation of the aromatic pyrrole compound and the favorable TΔS term. Also, the authors highlight the importance of the ester group for the facilitation of the retro-ene rearrangement. Nevertheless, a theoretical investigation would be beneficial to support the proposed mechanism.

Interestingly, a more recent article describes the presence of an alternative mechanism occurring with differences in reaction temperature.[2] For example, the author points out that indeed above 500°C, a cis-elimination of benzoic acid (such as is presented here) occurs, whereas in the 200–500°C reaction temperature range, a trans-elimination of benzoic acid making use of the adjacent ester group (1,7-H rearrangement followed by 1,5-H rearrangement) results in an alkaloid analog of **A** (Scheme 2). A further theoretical analysis would also complement this work.

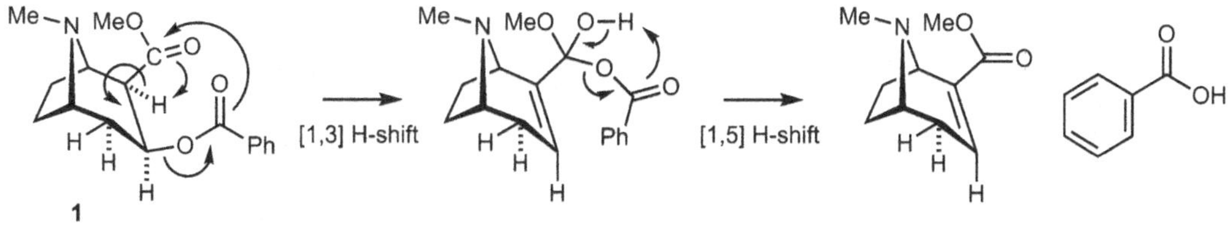

SCHEME 2

ADDITIONAL RESOURCES

For two recent publications outlining theoretical analyses of enzymatic degradations of cocaine, see Refs. 3, 4.

References

1. Sisti NJ, Fowler FW, Fowler JS. The flash vacuum thermolysis of (-)-cocaine. *Tetrahedron Lett.* 1989;30(44):5977–5980. https://doi.org/10.1016/S0040-4039(01)93832-8.

2. Novak M. Temperature-dependent benzoic acid elimination mechanisms in pyrolysis of (-)-cocaine. *Quim Nova.* 2011;34(4):573–576. https://doi.org/10.1590/S0100-40422011000400004.

3. Liu J, Zhao X, Yang W, Zhan CG. Reaction mechanism for cocaine esterase-catalyzed hydrolyses of (+)- and (-)-cocaine: unexpected common rate-determining step. *J Phys Chem B.* 2011;115:5017–5025. https://doi.org/10.1021/jp200975v.

4. Liu J, Zhan CG. Reaction pathway and free energy profile for cocaine hydrolase-catalyzed hydrolysis of (-)-cocaine. *J Chem Theory Comput.* 2012;8:1426–1435. https://doi.org/10.1021/ct200810d.

Question 256: A Highly Efficient Anilide to Benzimidazole Transformation

In 1963 chemists Meth-Cohn and Suschitzky reacted anilides **1** with a mixture of formic acid and hydrogen peroxide which formed benzimidazole **2** in 85%–95% yield (Scheme 1).[1] Suggest a mechanism for this transformation.

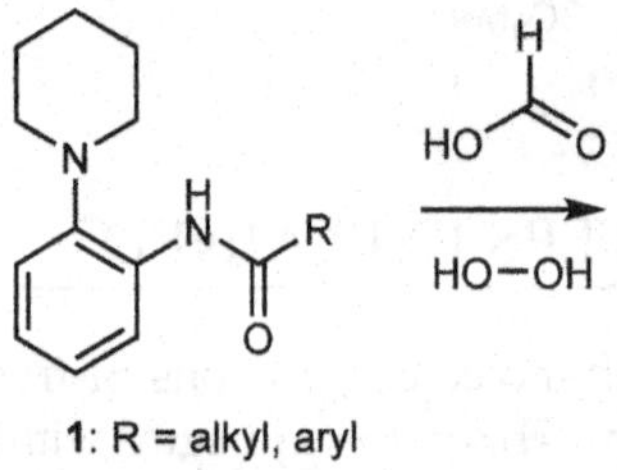

1: R = alkyl, aryl   **2**

SCHEME 1

SOLUTION

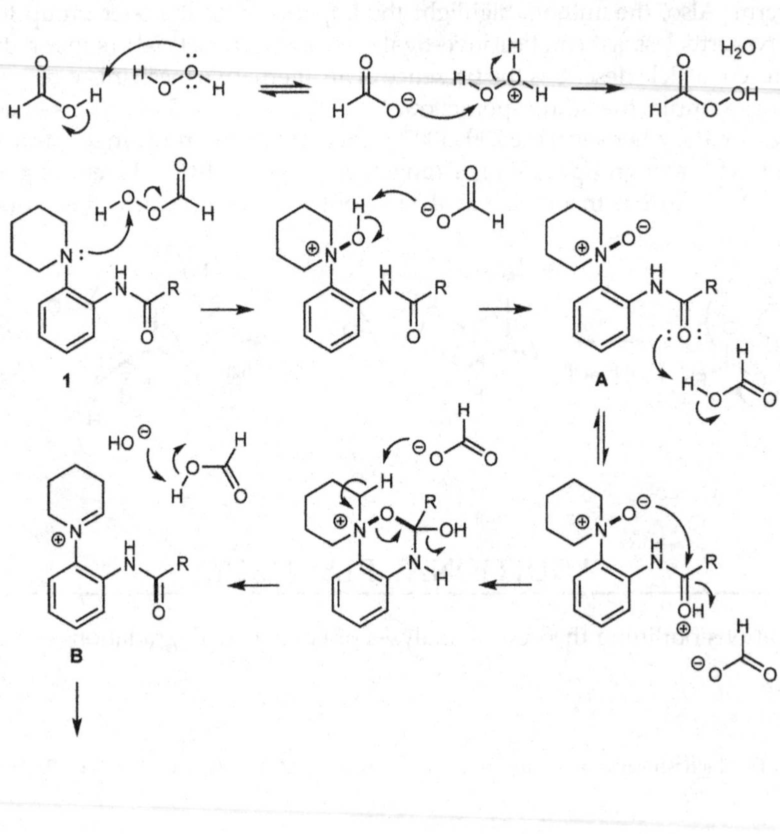

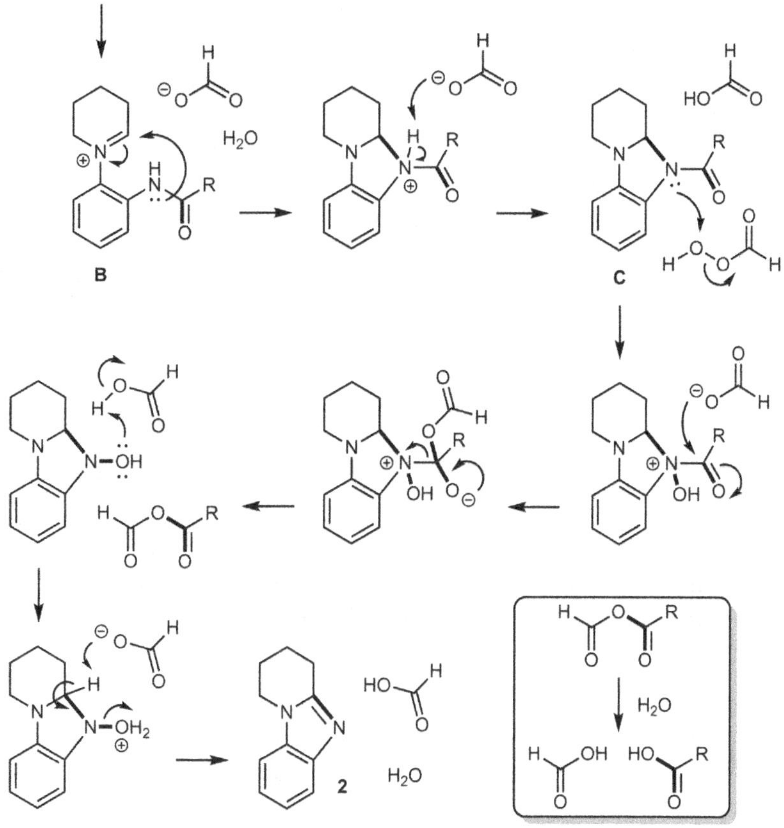

BALANCED CHEMICAL EQUATION

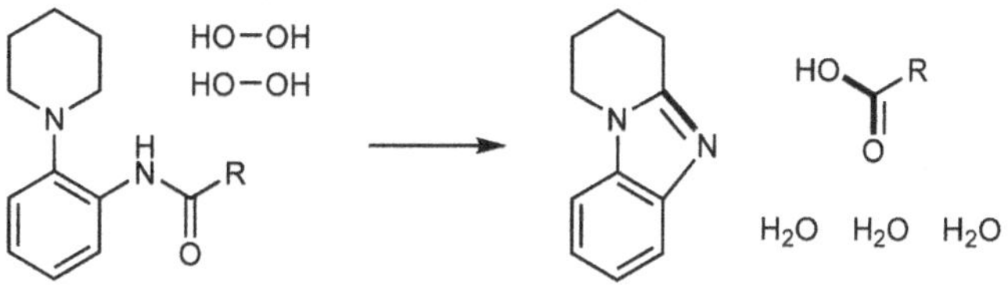

KEY STEPS EXPLAINED

In their original work, Meth-Cohn and Suschitzky explain the transformation of anilides **1** to benzimidazole **2** as beginning with the oxidation of the tertiary amine to an *N*-oxide group using peroxyformic acid (made from the reaction of formic acid with hydrogen peroxide) to form **A** which then undergoes protonation in order to set up an acid-catalyzed acylation of the *N*-oxide to give **B**. After loss of water and fragmentation **B** is ready for cyclization to the benzimidazole precursor **C**. In order to proceed to the final product, **C** has its tertiary amine oxidized with peroxyformic acid (as seen previously) losing RCO$_2$COH (which in turn is hydrolyzed to RCOOH and formic acid) in the process. The product is finally hydrolyzed to the final benzimidazole product **2**. This mechanism which has an intramolecular acylation was confirmed experimentally by the authors using two test syntheses, one of which is depicted in Scheme 2.

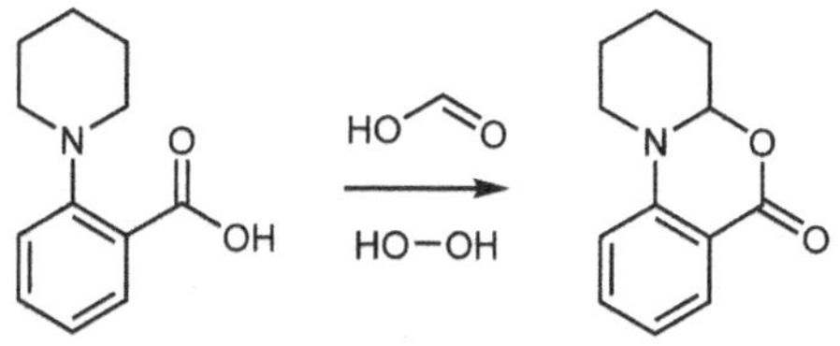

SCHEME 2

ADDITIONAL RESOURCES

For further work of Meth-Cohn and Suschitzky highlighting mechanistic analysis of analogs very similar to the ones in this problem, we recommend Ref. 2. Ref. 3 showcases an interesting mechanistic analysis which, despite lacking an N-oxide participating group, does feature a **B**-like intermediate.

References

1. Meth-Cohn O, Suschitzky H. 893. Syntheses of heterocyclic compounds. Part IV. Oxidative cyclisation of aromatic amines and their N-acyl derivatives. *J Chem Soc.* 1963;4666–4669. https://doi.org/10.1039/JR9630004666.
2. Fielden R, Meth-Cohn O, Suschitzky H. Thermal and photolytic cyclisation, rearrangement, and denitration reactions of o-nitro-t-anilines. *Tetrahedron Lett.* 1970;11(15):1229–1234. https://doi.org/10.1016/S0040-4039(01)91595-3.
3. Rees CW, Tsoi SC. A new redox-denitration reaction of aromatic nitro compounds. *Chem Commun.* 2000;415–416. https://doi.org/10.1039/a910290f.

Question 257: The Baeyer-Drewson Synthesis of Indigo

The classical synthesis of indigo **1**, known as the Baeyer-Drewson synthesis, involves reaction of o-nitrobenzaldehyde with acetone in the presence of aqueous sodium hydroxide (Scheme 1).[1] Suggest a mechanism for the transformation.

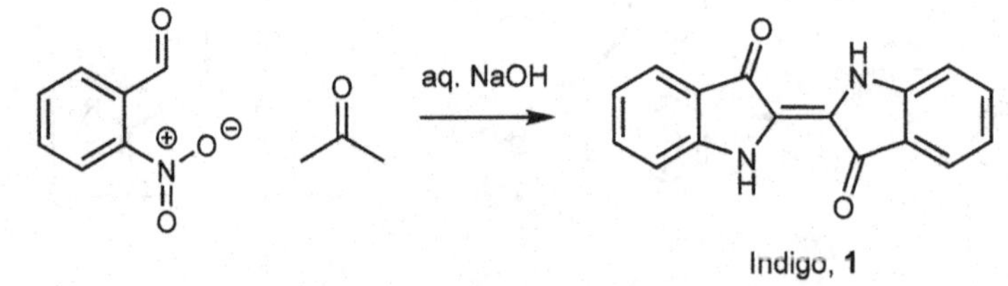

Indigo, **1**

SCHEME 1

SOLUTION

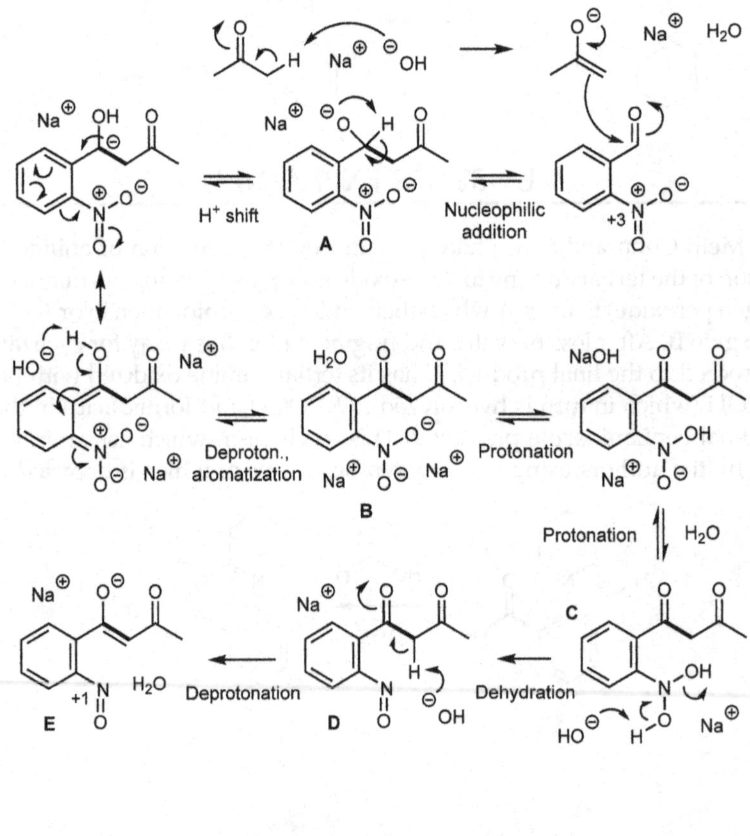

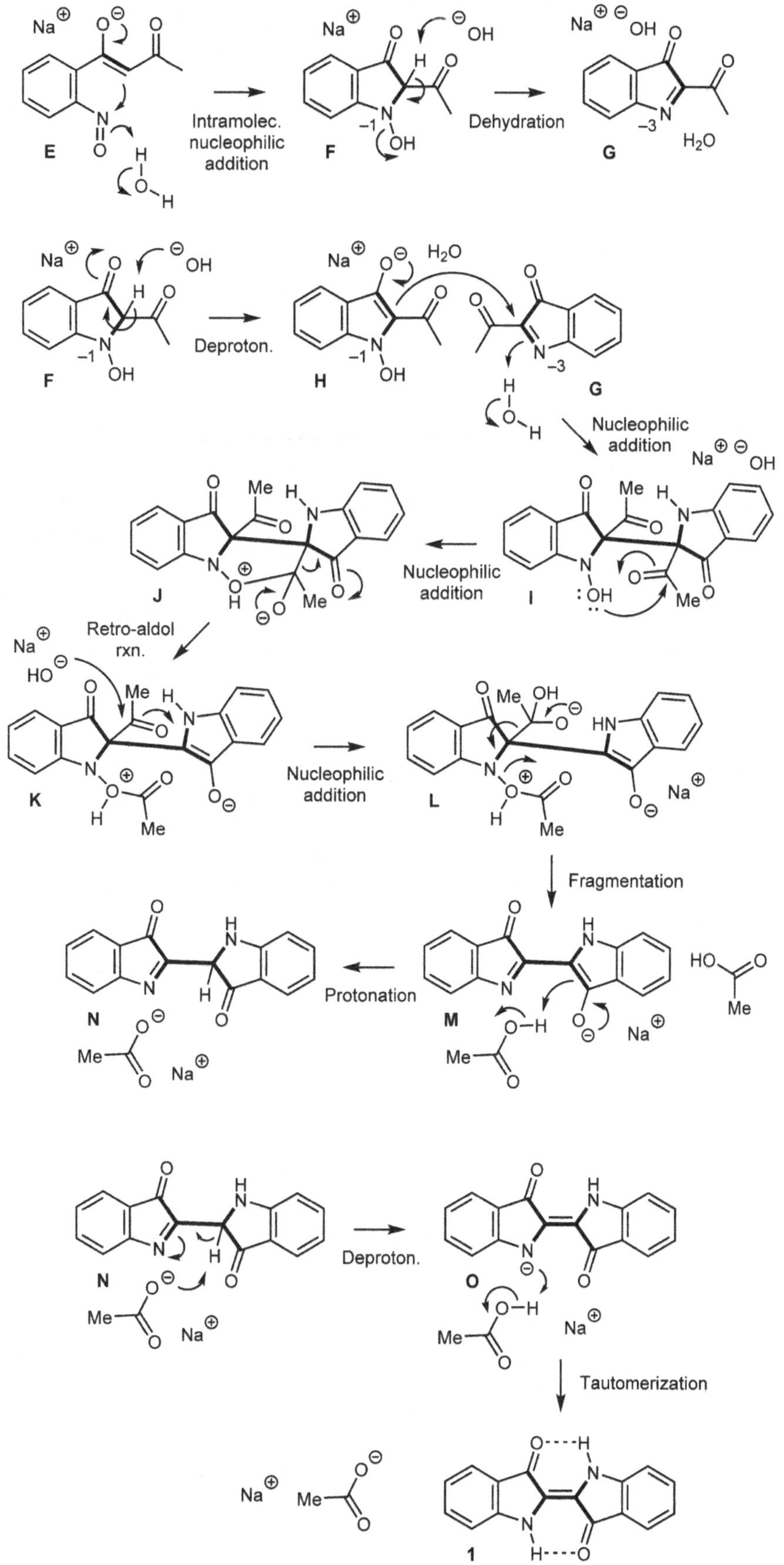
Intramolec. nucleophilic addition
Dehydration
Deproton.
Nucleophilic addition
Nucleophilic addition
Retro-aldol rxn.
Nucleophilic addition
Fragmentation
Protonation
Deproton.
Tautomerization
E
F
G
F
H
G
I
J
K
L
M
N
N
O
1

BALANCED CHEMICAL EQUATION

KEY STEPS EXPLAINED

In his original article relating the story of the synthesis of indigo, a paper we consider of the highest encountered educational value, Ranganathan presents the detailed mechanism outlined in the solution earlier. He explains that the synthesis of indigo is unique because of the stepped change in oxidation number of the nitrogen atom which goes from +3 at the start of the transformation (nitro group of o-nitrobenzaldehyde) to −3 at its end (amino groups of indigo product) via nitroso (+1 state) and hydroxylamine (−1 state) intermediates. He also described the need for an alkali medium to help produce needed nucleophilic species. Hydrogen transfers are also facilitated by the acid-base equilibria established in this reaction.

As such, the transformation begins with formation of the enolate anion of acetone using alkali medium. This anion attacks the starting o-nitrobenzaldehyde via a nucleophilic addition which then undergoes a proton shift and reorganization of electrons to set up a deprotonation and aromatization sequence to form **B**. This intermediate is protonated twice to form **C** which then undergoes base-mediated dehydration to form **D** which in turn deprotonates to an enolate **E** which undergoes an intramolecular nucleophilic addition to form the key indolinone species **F**. This species can undergo either dehydration to form **G** or deprotonation to form **H**. The two species **G** and **H** can react in a nucleophilic addition fashion to form the indigo precursor **I**. From this point forward, a sequence of steps involving nucleophilic addition, retro-aldol cleavage, nucleophilic addition, fragmentation, and tautomerization leads to the final indigo product **1**. It is noteworthy that, as Ranganathan explains, the oxidation number of nitrogen goes from +3 to −3 during the course of this transformation. The major changes seem to occur during dehydration and fragmentation steps. We note that the proposed mechanism by Ranganathan, represented here, does not have experimental or theoretical evidence in its support. We thus recommend such analyses for future research projects.

ADDITIONAL RESOURCES

We direct the interested reader to an article describing a very similar mechanism involving indolinone products and intermediates which might actually provide some precedent for some of the mechanistic steps presented here.[2] For recent examples of novel approaches to the synthesis of indigo, including arguably greener chemistry approaches utilizing enzymes, see the work of Rebelo et al.[3] Lastly, Ref. 4 provides a review of recent significant literature on indigo and related products.

References

1. Ranganathan S. Jeans and means 3. The story of indigo. *Resonance.* 1996;1(8):22–27. https://doi.org/10.1007/BF02837019.
2. Azadi-Ardakani M, Alkhader MA, Lippiatt JH, Patel DI, Smalley RK, Higson S. 2,2-Disubstituted-1,2-dihydro-3*H*-indol-3-ones by base- and thermal-induced cyclisations of *o*-azidophenyl s-alkyl ketones and *o*-azidobenzoyl esters. *J Chem Soc Perkin Trans 1.* 1986;1107–1111. https://doi.org/10.1039/P19860001107.
3. Rebelo SLH, Linhares M, Simoes MMQ, et al. Indigo dye production by enzymatic mimicking based on an iron(III)porphyrin. *J Catal.* 2014;315:33–40. https://doi.org/10.1016/j.jcat.2014.04.012.
4. Cooksey CJ. An annotated bibliography of recent significant publications on indigo and related compounds. *Biotech Histochem.* 2012;87(7):439–463. https://doi.org/10.3109/10520295.2012.698308.

Question 258: The Marschalk Reaction

The Marschalk reaction was discovered in 1936[1] and has been described in a later review to be "by far the most important reaction for anthracyclinone synthesis using anthraquinones as starting materials," see Scheme 1.[2] Devise a reasonable mechanism for this transformation.

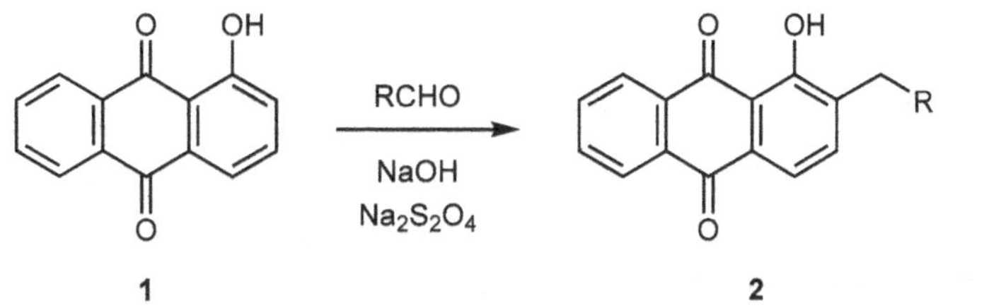

SCHEME 1

SOLUTION

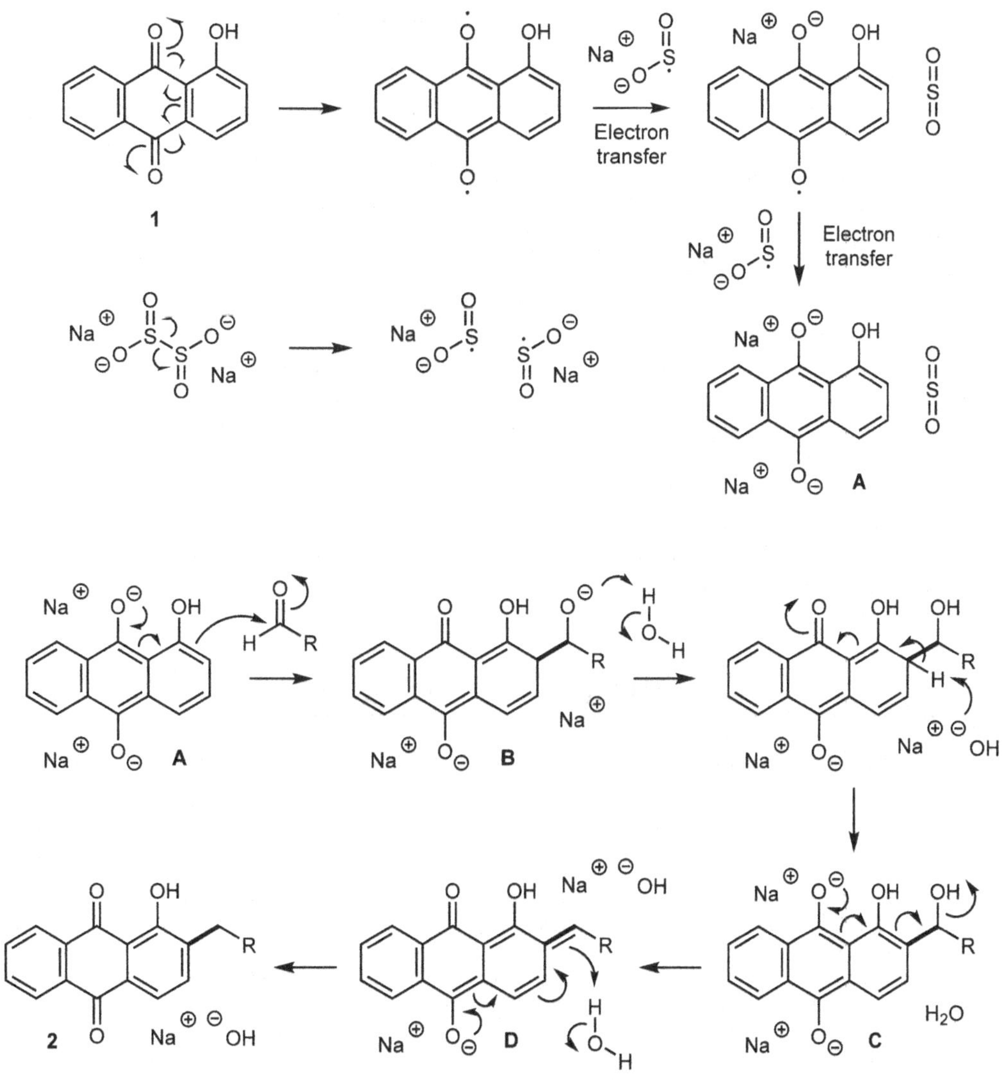

BALANCED CHEMICAL EQUATION

KEY STEPS EXPLAINED

In his review of Marschalk's work, Krohn writes that there are four important observations with regard to the Marschalk reaction, for example: (1) generally "the reduction to the corresponding hydroquinones is necessary prior to the reaction with aldehydes," (2) that "formaldehyde is the most reactive," and (3) that "the alkyl side chain is *always* introduced *ortho* to a free phenolic hydroxy group."[2] Nevertheless, Krohn and the authors that have dealt with the Marschalk reaction mechanism since its discovery have not presented a full mechanism accounting for all reagents and reaction steps. Given their leads and in accordance with their observations, we have sought to present the full story here. In so doing, we must recognize that our mechanism lacks any new experimental evidence to justify the new additions and we have not completed a theoretical analysis to indicate that our proposal is likely to have the lowest energy pathway. Therefore we offer the present solution as our best attempt at a mechanism given all the experimental evidence available in the literature.

The Marschalk reaction is recognizable as a redox reaction involving sodium dithionite that is a known strong reducing agent. Dithionite ion gets oxidized to two equivalents of sulfur dioxide and the anthraquinone and aldehyde starting materials get reduced to the final product. Essentially, the mechanism begins with homolytic fragmentation of sodium dithionite into two equivalents of $[SO_2]^-$ radical anion. This process is expected under aqueous conditions, since the sulfur—sulfur bond is known to be rather weak and $[SO_2]^-$ radical has been observed in solutions of sodium dithionite by electron spin resonance spectroscopy (ESR).[3] In fact, it is estimated that the enthalpy of reaction for the dissociation of dithionite anion into two SO_2 radical anions in the gas phase is $-271 \, kJ/mol$ (or $-64.8 \, kcal/mol$) indicating that this is an exothermic process.[4] The standard reduction potential for the half reaction $2 \, SO_2 + 2 \, e \rightarrow S_2O_4^{-2}$ is estimated to be $-0.52 \, V$.[5] At the same time, the structure of anthraquinone **1** can be transformed into a hydroquinone diradical oxygen species, which upon interaction with sulfur dioxide radical anion results in electron transfer occurring twice in order to produce two equivalents of sulfur dioxide and the dianion structure **A**. We note that structure **A** represents the reduced hydroquinone dianion that may now react with aldehyde to form **B** which is isomerized to the hydroxyalkylated anthrahydroquinone anion **C** which eliminates hydroxyl group to form the quinone methide-type structure **D** which abstracts a proton to yield the final anthraclinone product **2**. The evidence in support of this pathway is documented in Ref. 2. Our contribution consists of elucidating the initial reduction of **1** to **A** and completing the missing pieces along the **A** to **2** pathway. Key pieces of evidence in support of this mechanism are the observation of sulfur dioxide by-product and the stoichiometry of the reaction.

ADDITIONAL RESOURCES

For synthetic examples of the usefulness of the Marschalk reaction, we direct the reader to Refs. 6, 7.

References

1. Marschalk C, Koenig F, Ourousoff N. New methods of introducing side chains into the anthraquinone nucleus. *Bull Soc Chim Fr*. 1936;3:1545–1575.
2. Krohn K. Synthesis of anthracyclinones by electrophilic and nucleophilic addition to anthraquinones. *Tetrahedron*. 1990;46(2):291–318. https://doi.org/10.1016/S0040-4020(01)85414-4.
3. Janzen EG. Electron spin resonance study of the SO_2^- formation in the thermal decomposition of sodium dithionite, sodium and potassium metabisulfite, and sodium hydrogen sulfite. *J Phys Chem*. 1972;76(2):157–162. https://doi.org/10.1021/j100646a002.
4. Steudel R, Steiger T. Sulfur compounds. Part 162. Geometries and energies of the radical anion SO_2^- and of three isomeric structures of the dithionite anion $S_2O_4^{2-}$. *J Mol Struct (THEOCHEM)*. 1993;284:55–59. https://doi.org/10.1016/0166-1280(93)87179-H.

5. Neta P, Huie RE. Free-radical chemistry of sulfite. *Environ Health Perspect*. 1985;64:209–217. https://doi.org/10.2307/3430011.
6. Jin HS, Zhao LM. A contribution to the study of the modified Marschalk reaction: hydroxymethylation of 6,8-*O*-dimethyl emodin. *Chin Chem Lett*. 2010;21:568–571. https://doi.org/10.1016/j.cclet.2010.01.013.
7. Zhao LM, Ma FY, Jin HS, Ma J, Wang H, Fu CZ. Facile installation of a hydroxyalkyl group into hydroxyanthraquinones and aminoanthraquinones through the modified Marschalk reaction. *Eur J Org Chem*. 2013;(31):7193–7199. https://doi.org/10.1002/ejoc.201300891.

Question 259: A 1,3-Cyclohexanedione to 2-Cyclohexenone Conversion

In 1992 Japanese chemists reacted 2,2-disubstituted 1,3-cyclohexanediones **1** with dimethyl methanephosphonate in tetrahydrofuran (THF) in the presence of lithium di-isopropyl amide (LDA) to achieve moderate to very good yields of 3-substituted 2-cyclohexenones **2** (Scheme 1).[1] Provide a mechanism for this transformation.

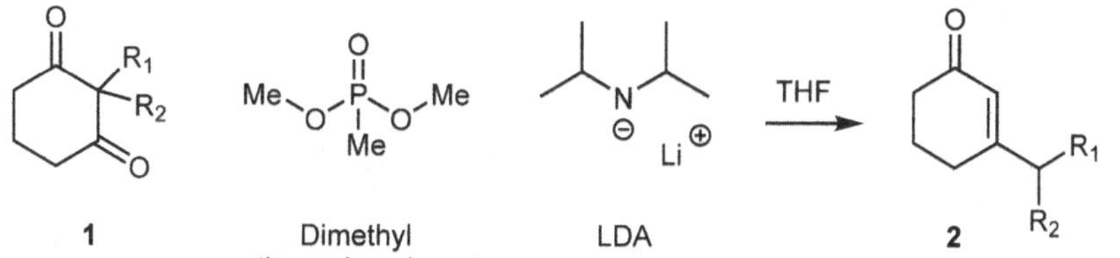

SCHEME 1

SOLUTION

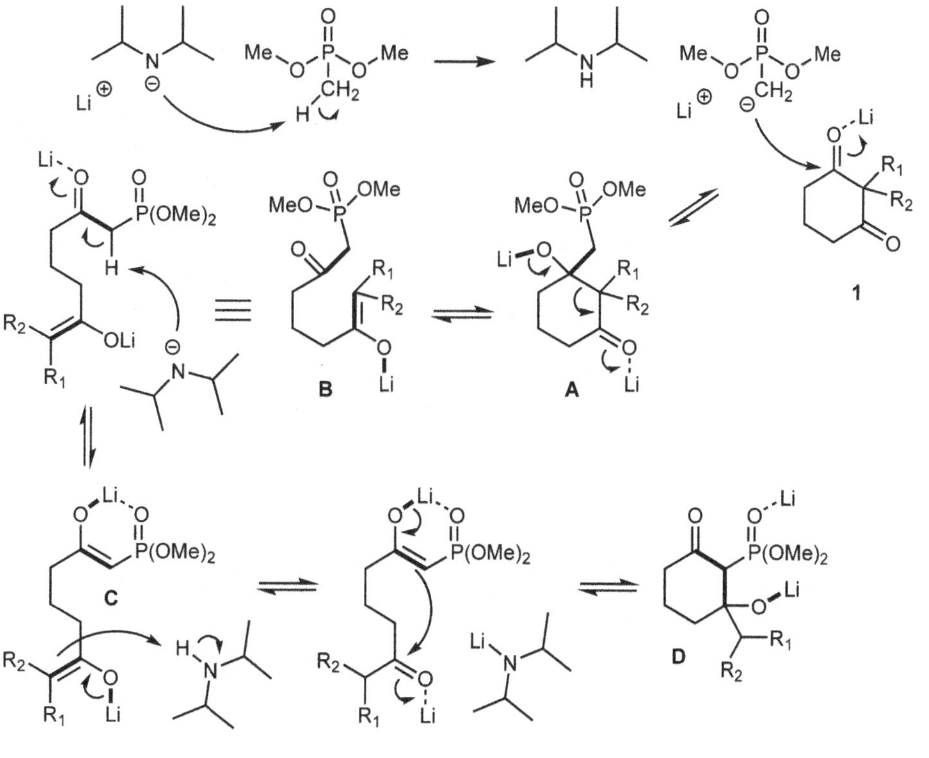

BALANCED CHEMICAL EQUATION

KEY STEPS EXPLAINED

In their work, Furuta et al. explain the transformation of **1** to **2** as following initial formation of a phosphonate anion from LDA base attack on dimethyl methanephosphonate starting material. This anion then attacks a ketone group on **1**, a process facilitated by coordination between oxygen and lithium metal, in order to form a new C—C bond to give **A**. The structure **A** can then coordinate with lithium in order to undergo retro-aldol cleavage (ring opening) to form **B** which after prototropic rearrangement can undergo an intramolecular Horner-Wadsworth-Emmons condensation to eliminate lithium dimethyl phosphate and complete the conversion to **2**. It is noteworthy that the authors report several other experimental results, namely, that the use of n-butyl lithium instead of LDA dramatically reduces the yield of the reaction, use of trimethyl silyl chloride (TMSCl) as an additive dramatically increases the yield, and use of ether as a solvent over THF decreases the yield. Therefore both base and solvent effects seem to be important vis-à-vis this transformation. One can explain these results by noting that n-BuLi is a strong nonbulky base and is thus likely to reduce the effectiveness of the proton rearrangement necessary for the conversion of **B** to **D**. Also, TMSCl works well to stabilize negatively charged oxygen atoms through silylation, and this mechanism is ripe with such opportunities. Lastly, the authors note an interesting observation in the sense that LDA is observed to be ineffective in converting five-membered ring derivatives of **1** into their corresponding counterparts **2**. The authors explain this result by referring to the geometrical conformations of six-membered and five-membered cycloalkane rings, respectively, in the sense that for the case of six-membered rings, lithium works well to coordinate between the oxy- and phosphate groups of structure **A** in order to facilitate the retro-aldol cleavage to **B**. This is not the case for five-membered analogs, where, due to geometrical constraints, the carbonyl group cannot assist in the retro-aldol cleavage and where a Li—O bond, which has more covalent character, is difficult to break. Therefore larger metal atoms such as sodium (Na) and potassium (K) are needed since their bonds with oxygen have a more pronounced ionic character and are thus more easily broken.[1]

ADDITIONAL RESOURCES

We direct the reader to a doctoral dissertation on the mechanism and stereochemistry of phosphonate anion condensations for further examples of the role of phosphonate anions in organic transformations.[2] Since the work by Furuta et al. contained no theoretical analysis, we recommend a recent work looking at the roles of chiral catalysts in the epoxidation of cyclic enones with respect to theoretical investigation of possible reaction mechanisms.[3]

References

1. Furuta T, Oshima E, Yamamoto Y. A new method for construction of cyclic enones via phosphonate anions. *Heteroat Chem*. 1992;3(5/6):471–478. https://doi.org/10.1002/hc.520030504.
2. Mandanas BY. *Studies on the Mechanism and Stereochemistry of Phosphonate Anion Condensations* [dissertation]. Ames: Iowa State University; 1971.
3. Lv PL, Zhu RX, Zhang DJ, Duan CG, Liu CB. Theoretical investigation on the chiral diamine-catalyzed epoxidation of cyclic enones: mechanism and effects of cocatalyst. *J Phys Chem A*. 2012;116:1251–1260. https://doi.org/10.1021/jp207914h.

Question 260: Reaction of a Steroidal Olefin With Br_2/AgOAc

Treating the steroid derivative **1** with bromine and silver acetate in a mixture of chloroform and pyridine at low temperature followed by quenching with aqueous acid was found to give a mixture of compounds, one of which was **2** (Scheme 1).[1,2] Propose a mechanism for this transformation.

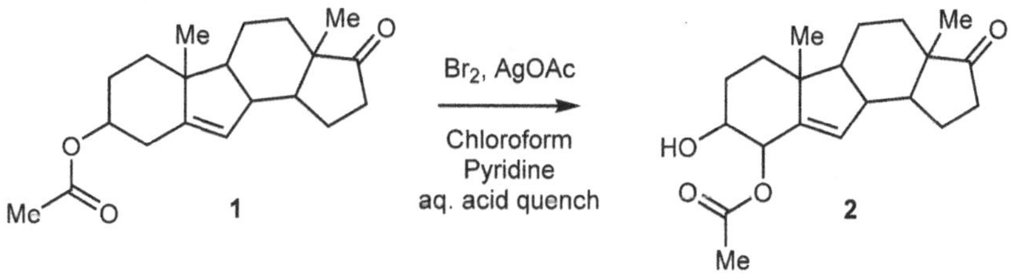

SCHEME 1

KEY STEPS EXPLAINED

The solution to this problem (see Scheme below) is assembled with reference to the work of Paul B. Reese et al.[1,2] Essentially the mechanism consists of initial dibromination of **1** using bromine and deprotonation via pyridine to eliminate hydrogen bromide which results in **A**. At this stage the carbonyl oxygen atom of the neighboring acetoxy group can form a five-membered ring and eliminate bromine anion (which is facilitated by coordination with silver metal ion). Introduction of water molecule, prototropic rearrangement, and ring excision followed by final quench leads to **2**.

SOLUTION

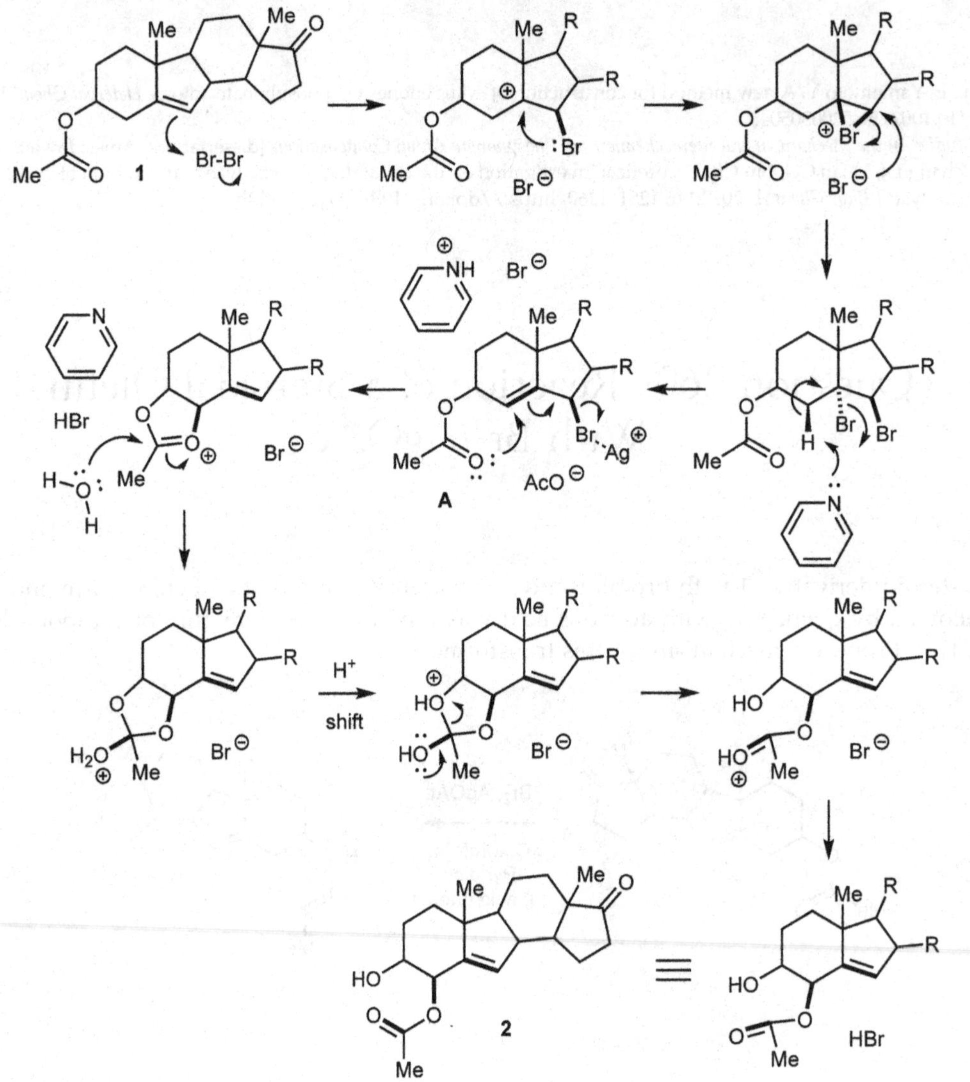

BALANCED CHEMICAL EQUATION

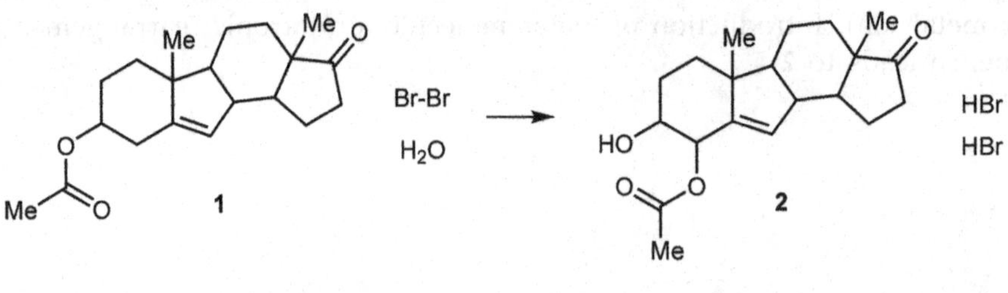

ADDITIONAL RESOURCES

We direct the reader to further work by Paul B. Reese where a similar mechanism is described, this time involving the use of metal complex catalysts and chlorine.[3] Other literature examples showcase several interesting mechanism problems involving similar steroids and structures.[4,5]

References

1. Hanson JR, Reese PB, Wadsworth HJ. Allylic acetoxylation of Δ^5-steroids at C-4. *J Chem Soc Perkin Trans 1*. 1984;2941–2944. https://doi.org/10.1039/P19840002941.
2. Ruddock PLD, Reese PB. The effect of 4β and 19 ester functionalities on some electrophilic addition reactions of Δ^5-steroids. *Steroids*. 1999;64(12):812–819. https://doi.org/10.1016/S0039-128X(99)00062-8.
3. Ruddock PL, Williams DJ, Reese PB. The reactions of palladium(II), thallium(III) and lead(IV) trifluoroacetates with 3-acetoxyandrost-5-en-17-one: crystal structure of the first trifluoroacetate bridged 5,6,7-π-allyl steroid palladium dimer. *Steroids*. 2004;69:193–199. https://doi.org/10.1016/j.steroids.2004.01.001.
4. Akhrem AA, Reshetova IG, Titov YA. Fluorinated steroids. *Russ Chem Rev*. 1965;34(12):926–942. https://doi.org/10.1070/RC1965v034n12ABEH001576.
5. Numazawa M, Yamada K. Synthesis of 19-oxygenated derivatives of the competitive inhibitor of aromatase, 5-androstene-4,17-dione. *Steroids*. 1999;64(5):320–327. https://doi.org/10.1016/S0039-128X(98)00113-5.

Question 261: A Most Unusual Synthesis of Tropones From Phenols

During studies of intramolecular oxidative coupling reactions of the **1** to **2** variety, American chemists Kende et al. reacted *p*-(nitrobutyl)phenol **3** with KOH, $K_3Fe(CN)_6$ and acidification in order to form the tropone **4** in 80% yield (Scheme 1).[1] Show mechanisms for both transformations.

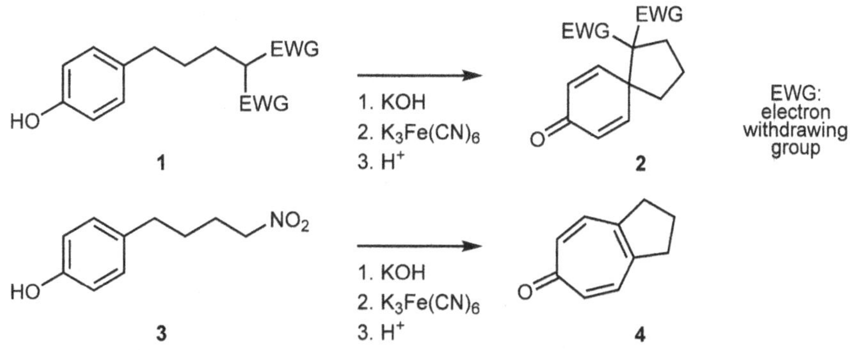

SCHEME 1

SOLUTION: 1 TO 2

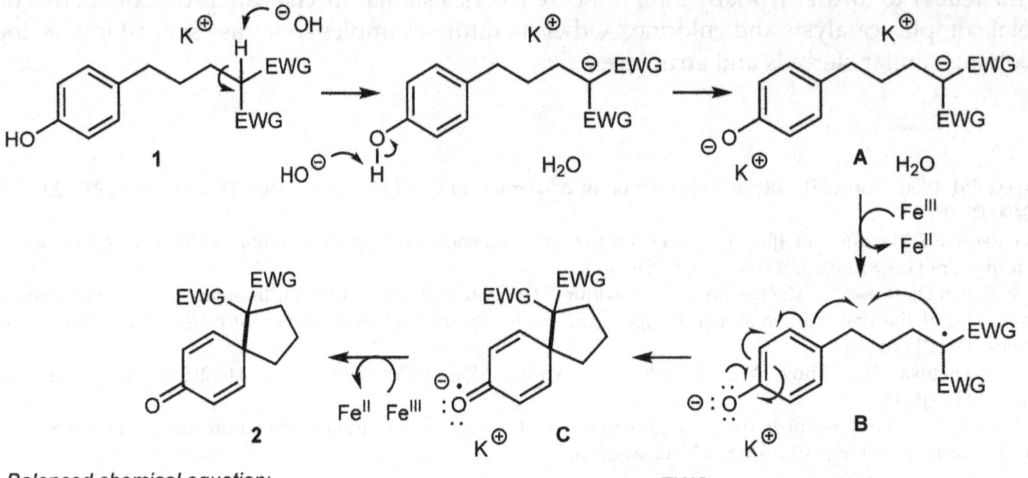

Balanced chemical equation:

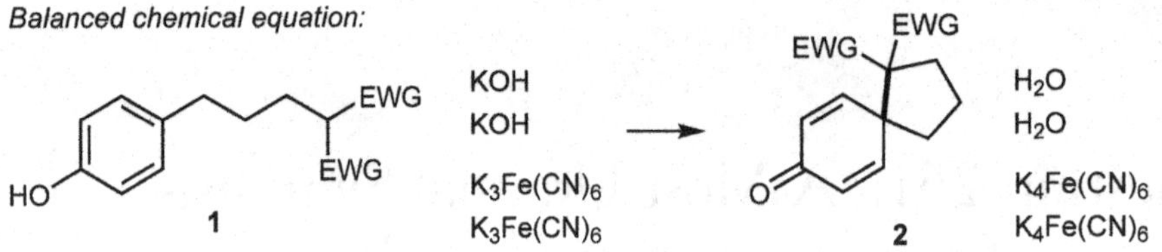

SOLUTION: 3 TO G

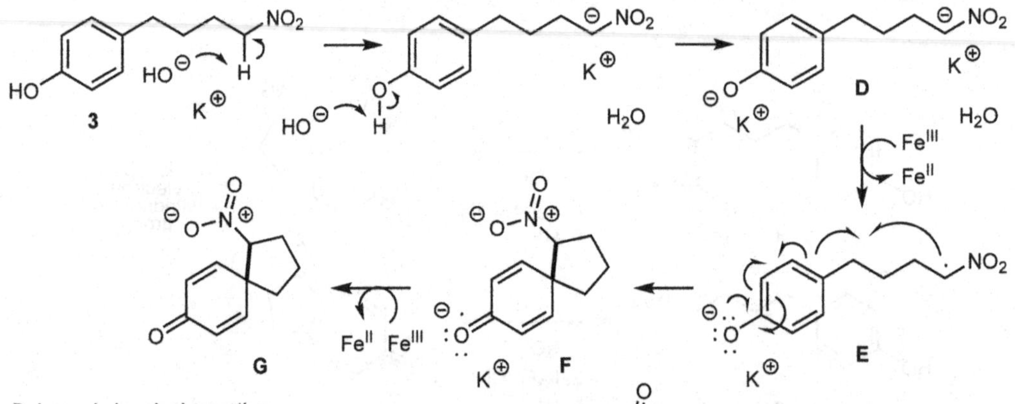

Balanced chemical equation:

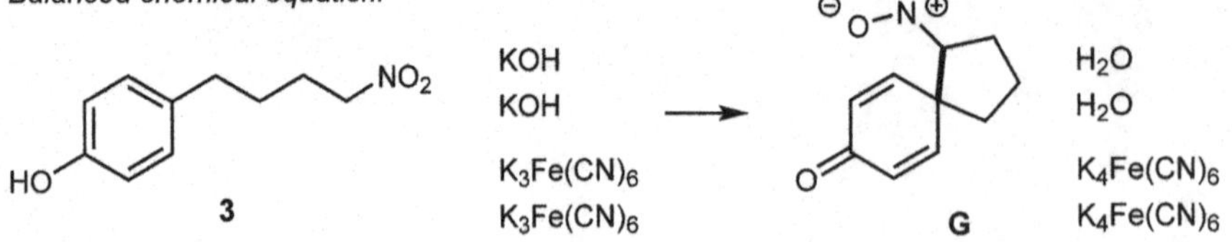

SOLUTION: G TO 4

BALANCED CHEMICAL EQUATION: G TO 4

KEY STEPS EXPLAINED

Through extensive research spanning several years, Kende et al. proposed the two mechanisms illustrated earlier for the transformations of **1** to **2** and **3** to **4**.[1–4] Essentially both **1** and **3** undergo double deprotonation via two equivalents of KOH base in order to form the dianions **A** and **D**, respectively. Next, one electron transfer from **A** (and **D**) to the oxidant Fe^{III} results in the formation of the radical **B** (and **E**) plus the reduced Fe^{II} by-product. The radical **B** (and **D**) can now add to the phenoxide ring to form a radical anion intermediate **C** (and **F**) which after a rapid electron transfer process to the oxidant Fe^{III} forms **2** (and **G**). We can see that stoichiometrically this sequence utilizes two equivalents of KOH and $K_3Fe(CN)_6$, respectively, and forms two equivalents of water and $K_4Fe(CN)_6$, respectively. From **G** to **4** the authors propose first a proton shift to a presumed nitronic acid intermediate **H** followed by an "ene"-type rearrangement to the strained cyclopropane intermediate **I** (for literature precedent see Ref. 5) followed by an acid-catalyzed keto-enol tautomerization to the enol **J** followed by a facile fragmentation (with relief of ring strain) to the bicyclic intermediate **K** which after loss of HNO_2 forms the final tropone product **2**.

Experimentally, the authors note that compound **3** does not cyclize at pH values below 11 which means that considering the pK_A values of phenol ($pK_A = 10.1$) and that of nitroalkane ($pK_A = 8.8$), it must be the case that the starting steps involve the formation of the dianions **A** and **D**, respectively.[1] Furthermore, intermediate **G** was isolated in quantitative yield when **3** was treated with 2.1 equivalents of KOH and extracted directly with $CHCl_3$ after 10 min thus lending experimental confirmation of **2** and **G**. Next, direct treatment of **G** with acid produced tropone **4** in quantitative yield while under basic conditions, the rearrangement was much slower. In addition, kinetic studies at fixed base concentration showed that the rate of cyclization for **3** was proportional to the concentration of $K_3Fe(CN)_6$ oxidant in solution which lends further support for the proposed one-electron transfer steps. The last experimental results of the authors constitute the analogous C_{11} system, the results of which are presented in Scheme 2. Here all structures are isolated under the given conditions. The cyclic nitronates ester **7** represents the kinetic product of the treatment of **6** (analogous with **2** and **G**) with dilute base for a few minutes whereas reaction of either **6** or **7** under more basic conditions for several hours produced the tropone product **8** in quantitative yield. These results not only confirm the versatility of this transformation but also shed light on the outcomes under basic reaction conditions.

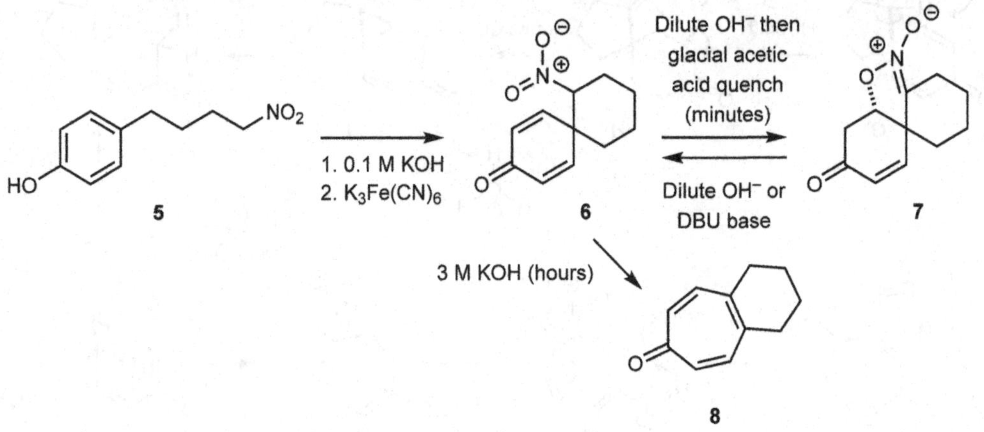

SCHEME 2

Although it is fairly clear that the sequence from **1** to **2** and **3** to **G** is well established experimentally, we would like to propose two alternative mechanisms for the **G** to **4** transformation. The first alternative involves an acid-catalyzed process which avoids an "ene"-rearrangement and maintains the highly strained cyclopropane intermediate **I** (Scheme 3). The second alternative avoids both the "ene"-rearrangement and intermediate **I** (Scheme 4).

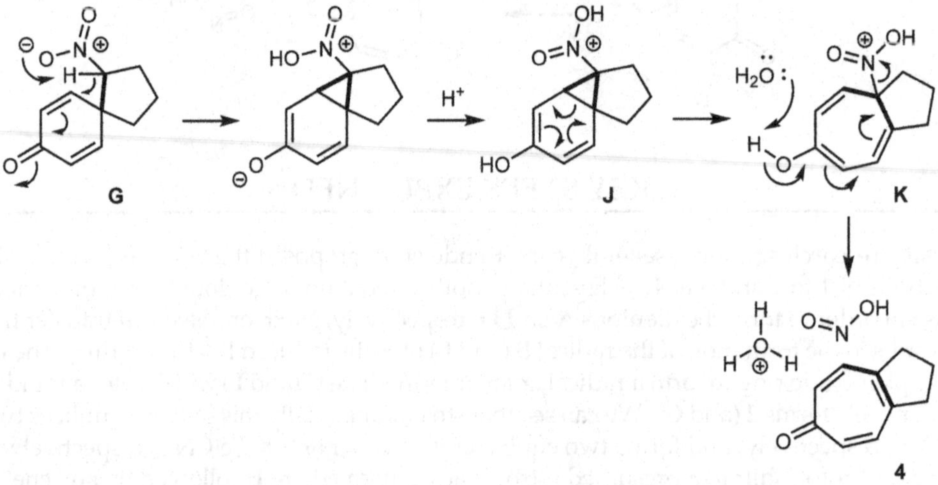

SCHEME 3

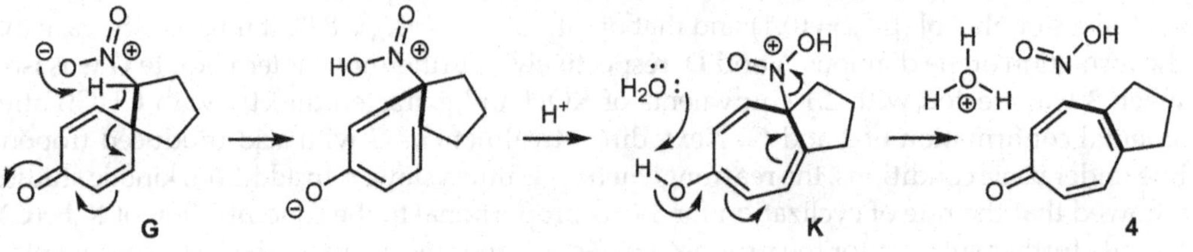

SCHEME 4

We note that although these two alternative mechanisms are shorter than the pathway proposed by Kende et al., it is possible to argue that the initial deprotonation would create electron density that would be partially weakened by the electron-withdrawing nature of the nitro group thus preventing direct attack of the quinoid ring either to form **I** or **K** directly. This criticism is what the introduction of the "ene"-rearrangement stipulated by Kende et al. resolves. Nevertheless, we believe that a future theoretical study comparing the energetics of breaking a quinoid ring as in Scheme 4 versus a highly strained tricyclic cyclopropane intermediate **I** (Solution: **G** to **4** and Scheme 3) would shed much needed light on the likelihood of the three proposed pathways for the **G** to **4** transformation.

ADDITIONAL RESOURCES

For further literature examples which involve radical anion cyclizations and proposed highly strained cyclopropane intermediates such as **I** which fragment into ring-expanded products such as **K**, see Refs. 6, 7.

References

1. Kende AS, Koch K. Intramolecular radical cyclization of phenolic nitronates: facile synthesis of annelated tropone and tropolone derivatives. *Tetrahedron Lett.* 1986;27(50):6051–6054. https://doi.org/10.1016/S0040-4039(00)85396-4.
2. Leboff A, Carbonnelle A-C, Alazard J-P, Kende AS. Intramolecular radical coupling of a phenolic enolate: oxidative fragmentation of the spirodiketone intermediate. *Tetrahedron Lett.* 1987;28(36):4163–4164. https://doi.org/10.1016/S0040-4039(00)95567-9.
3. Kende AS, Koch K, Smith CA. Intramolecular radical cyclization of phenolic enolates. *J Am Chem Soc.* 1988;110(7):2210–2218. https://doi.org/10.1021/ja00215a034.
4. Thebtaranonth C, Thebtaranonth Y. Developments in cyclisation reactions. *Tetrahedron.* 1990;46(5):1385–1489. https://doi.org/10.1016/S0040-4020(01)81956-6.
5. Iwata C, Yamada M, Shinoo Y, Kobayashi K, Okada H. Intramolecular cyclization of phenolic α-diazoketones. Novel synthesis of the spiro[4.5] decane carbon framework. *J Chem Soc Chem Commun.* 1977;1977:888–889. https://doi.org/10.1039/C39770000888.
6. Varin M, Chiaroni A, Lallemand J-Y, Iorga B, Guillou C. A new access to dihydrotropones through ring expansion of spirocyclohexadienones: synthesis and mechanism. *J Org Chem.* 2007;72:6421–6426. https://doi.org/10.1021/jo070594p.
7. Hong S-K, Kim H, Seo Y, Lee SH, Cha JK, Kim YG. Total synthesis of pareitropone via radical anion coupling. *Org Lett.* 2010;12(17):3954–3956. https://doi.org/10.1021/ol1017849.

Question 262: Side-Chain Manipulation With a Purine Derivative: Unexpected Formation of a Thietane

In 1990 American chemists Press et al. discovered that the reaction of the chlorohydrin derivative of 6-mercaptopurine with two equivalents of sodium methoxide in methanol at room temperature for 24 h did not produce the expected epoxide nor the expected methoxy-containing derivative. Instead, a 32% yield of the substituted thietane product **1** was obtained (Scheme 1).[1] Suggest a mechanism for this transformation.

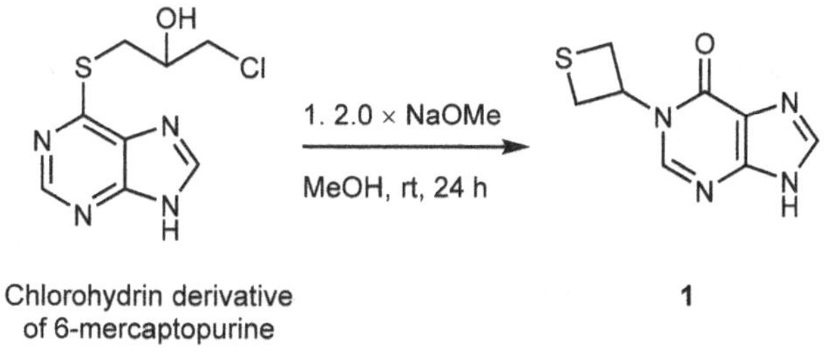

Chlorohydrin derivative
of 6-mercaptopurine

1. 2.0 × NaOMe

MeOH, rt, 24 h

1

SCHEME 1

SOLUTION

Chlorohydrin derivative
of 6-mercaptopurine

KEY STEPS EXPLAINED

The mechanism proposed by Press et al. begins with deprotonation of the alcohol by methoxide base followed by epoxide formation to **A** which after bond rotation (possibly through protonation and deprotonation of the purine N_1 nitrogen atom) allows for alkylation of the N_1 nitrogen and ring formation in a 5-*exo-trig* manner to give intermediate **B** which has the open chain alkoxide. This alkoxide can then carry out an intramolecular attack at the activated C_6 position of the purine ring to form the bicyclic intermediate **C** which can open to an isomeric mercaptide anion **D** with an open chain alkyl thiolate group that can then attack the oxygen-bearing methylene group to form irreversibly a stable hypoxanthine system and thietane ring product **1**.

Experimentally the authors point out that under stronger base conditions, product **1** is formed in substantially lower yield and that reexposure of **1** to the reaction conditions gives **1** unchanged. In an interesting discussion comparing intermediates **B** and **D**, the authors note that although intermediate **B** either does not form an oxetane ring (or the ring forms and reopens), the analogous mercaptide intermediate **D** closes to the thietane ring product **1** irreversibly. The authors attribute this thietane ring preference to the enhanced nucleophilicity of sulfur over oxygen and the longer length of the S—C bond which presumably allows for a better trajectory with regard to the postulated S_N2 process (i.e., **D** to **1** versus **B** to the oxetane analog of **1**). This irreversible formation of the thietane ring is what drives the reaction to completion following a series of facile equilibria.

We also note that a future theoretical analysis would highlight the energy barriers for the various proposed structures which comprise this interesting mechanism, especially the energy barrier between structures **B**, **C**, and **D** which we expect to be rather small.

ADDITIONAL RESOURCES

For further work by Press et al. in the context of examining the base and solvent effects that may contribute to this interesting transformation, see Refs. 2, 3.

References

1. Press JB, Hajos ZG, Sawyers RA. A remarkable thietane formation from a 6-mercaptopurine derivative. *Tetrahedron Lett.* 1990;31(10):1373–1376. https://doi.org/10.1016/S0040-4039(00)88809-7.
2. Barton DL, Press JB, Hajos ZG, Sawyers RA. Preparation of positive inotropes using glycidyl derivatives: influence of metal ions and solvent on stereochemical outcome. *Tetrahedron Asymmetry.* 1992;3(9):1189–1196. https://doi.org/10.1016/S0957-4166(00)82104-7.
3. Press JB, McNally JJ, Hajos ZG, Sawyers RA. Synthesis of *N*-thietan-3-yl-α-oxo nitrogen heterocycles from imino thioethers. A novel transformation. *J Org Chem.* 1992;57(23):6335–6339. https://doi.org/10.1021/jo00049a052.

Question 263: An Isoxazoline to Pyridine N-Oxide Transformation

American chemists have shown in 1990 that treatment of 5-cyanomethylisoxazolines **1** with catalytic amounts of the base DBU in boiling xylene gives good to excellent yields of the 6-substituted-2-aminopyridine *N*-oxides **2** (Scheme 1).[1] Devise a mechanism for this transformation.

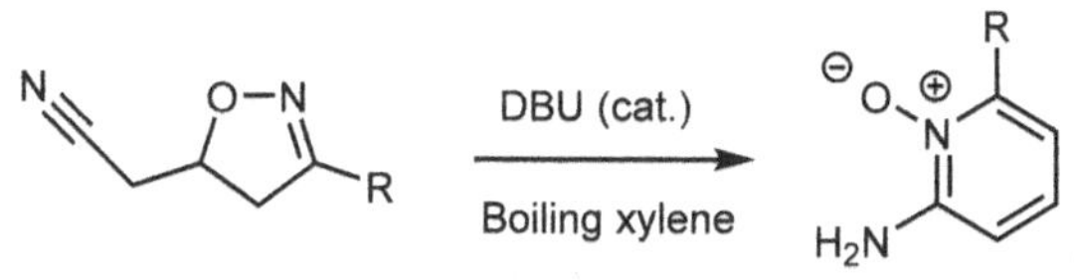

SCHEME 1

SOLUTION

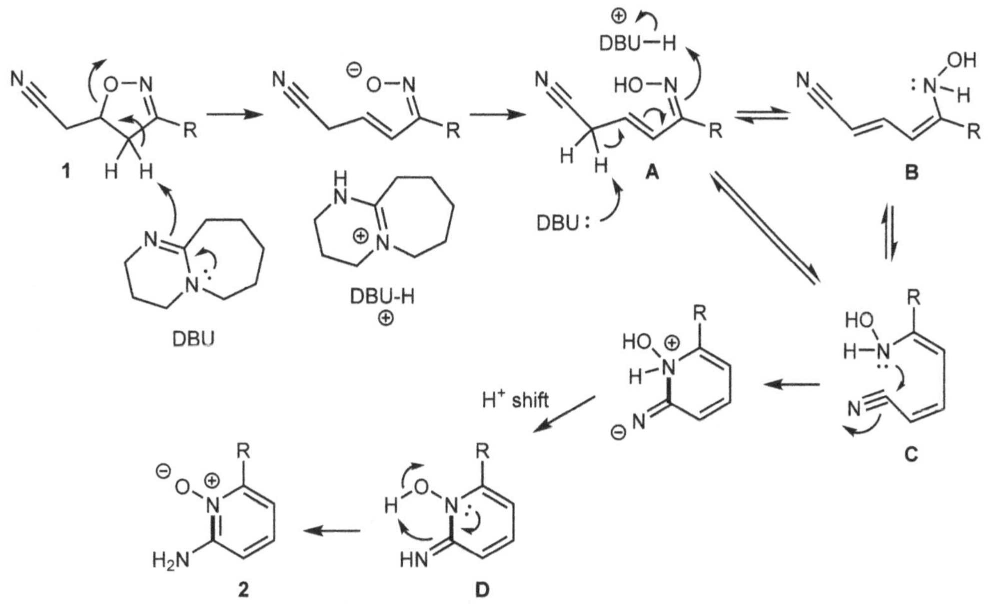

KEY STEPS EXPLAINED

The transformation of 5-cyanomethylisoxazolines **1** to the 6-substituted-2-aminopyridine *N*-oxides **2** begins with a base-catalyzed deprotonation of **1** to form the α,β-enoxime **A** which after a second base-catalyzed deprotonation gives the *E*-isomer of the vinyl-ene-hydroxylamine intermediate **B** and/or the *Z*-isomer **C**. The authors note that it is very likely that under the reaction conditions **B** can isomerize to **C** by the reverse pathway. The important takeaway is that only the *Z*-isomer **C** will spontaneously cyclize to the six-member ring intermediate **D** due to geometrical proximity factors. After a prototropic rearrangement **D** will form the final 6-substituted-2-aminopyridine *N*-oxide product **2**.

In terms of experimental evidence the authors were unable to isolate any of the putative open-chain intermediates **A**, **B**, and **C**. Nevertheless, they point out that the research of Gewald and Hain who synthesized 2-amino pyridine *N*-oxides by reacting α-ylidene malonitrile with various nitrile oxides lends support for intermediate **C**.[2] An interesting question would also be the energetic difference between intermediates **B** and **C**.

For an interesting example of a rearrangement of a compound similar to **1**, see the work of Lopes et al.[3]

References

1. Chucholowski AW, Uhlendorf S. Base catalyzed rearrangement of 5-cyanomethyl-2-isoxazolines: novel pathway for the formation of 2-aminopyridine N-oxides. *Tetrahedron Lett.* 1990;31(14):1949–1952. https://doi.org/10.1016/S0040-4039(00)88886-3.
2. Gewald K, Hain U. Reaktion von nitriloxiden mit ylidenmalononitrilen. *Z Chem.* 1986;26(12):434–435. https://doi.org/10.1002/zfch.19860261205.
3. Lopes SMM, Nunes CM, Melo TMVDP. 4-isoxazolines and pyrroles from allenoates. *Tetrahedron.* 2010;66:6078–6084. https://doi.org/10.1016/j.tet.2010.06.010.

Question 264: A Simple Synthesis of the Lignan Carpanone

The synthesis of carpanone can be carried out by means of four simple reactions (Scheme 1).[1] Devise a mechanistic pathway for this synthetic sequence.

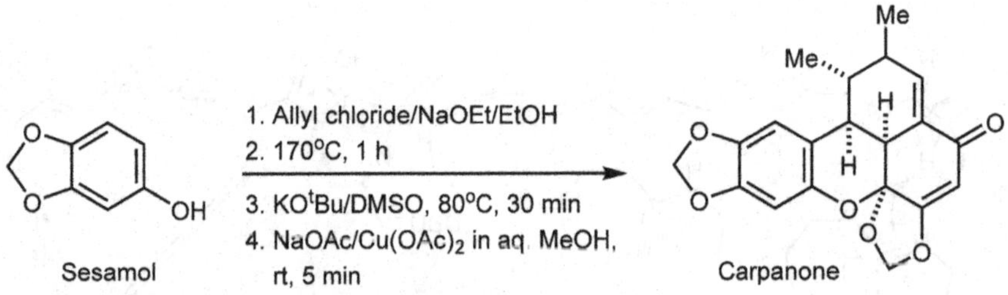

1. Allyl chloride/NaOEt/EtOH
2. 170°C, 1 h
3. KO^{t}Bu/DMSO, 80°C, 30 min
4. NaOAc/Cu(OAc)$_2$ in aq. MeOH, rt, 5 min

SCHEME 1

SOLUTION

A

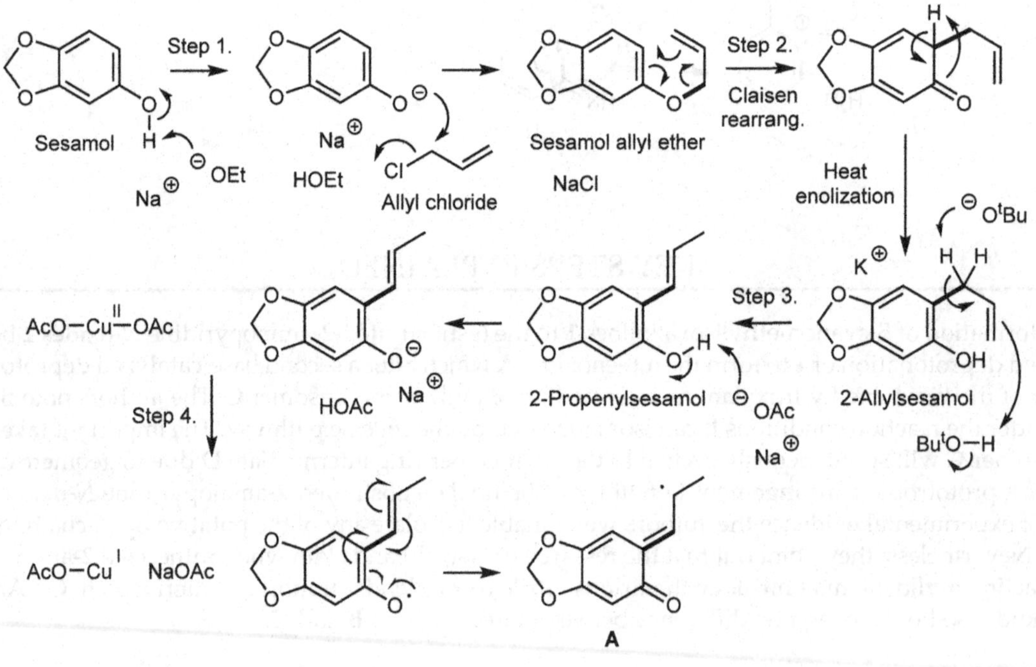

Carpanone

KEY STEPS EXPLAINED

For simplicity we show the entire mechanistic pathway of all four synthetic steps together when in reality these are separate transformations. Interestingly, this experiment can be carried out with little purification of intermediates from one synthetic step to the next given the reaction conditions.[1] Therefore the synthesis of carpanone begins with a Williamson ether synthesis of sesamol allyl ether from sesamol and allyl chloride. This is followed by a heat-initiated Claisen rearrangement and prototropic rearrangement to 2-allylsesamol. This intermediate then isomerizes under basic conditions to 2-propenylsesamol. Finally, 2-propenylsesamol undergoes deprotonation and one electron transfer from Cu(OAc)$_2$ to form oxy radical which rearranges to form the more stable alkyl radical **A**. Two equivalents of **A** then react in an oxidative radical dimerization to form intermediate **B** which undergoes a facile hetero-Diels-Alder [4 + 2] cycloaddition to produce the final carpanone product. Aside from **A** and **B**, all intermediates along this route are isolated and purified.

ADDITIONAL RESOURCES

For supporting evidence for the proposed mechanism, especially with regard to the fact that the dimerization of **A** results in the *trans* dimer **B** (with respect to the two methyl groups) and not the *cis* isomer, see the work of Matsumoto and Kuroda.[2] For a mechanistic analysis of the interaction of novel catalysts with 2-propenylsesamol for the synthesis of carpanone, see Ref. 3. For a review of the use of the Diels-Alder reaction in synthesis in the context of similar cyclizations as the one presented "earlier", see Ref. 4.

References

1. Sloop JC. Microscale synthesis of the natural products carpanone and piperine. *J Chem Educ.* 1995;72(2):A25–A27. https://doi.org/10.1021/ed072pA25.
2. Matsumoto M, Kuroda K. Transition metal(II) Schiff's base complexes catalyzed oxidation of *trans*-2-(1-propenyl)-4,5-methylenedioxyphenol to carpanone by molecular oxygen. *Tetrahedron Lett.* 1981;22(44):4437–4440. https://doi.org/10.1016/S0040-4039(01)82977-4.
3. Constantin MA, Conrad J, Merisor E, Koschorreck K, Urlacher VB, Beifuss U. Oxidative dimerization of (*E*)- and (*Z*)-2-propenylsesamol with O$_2$ in the presence and absence of laccases and other catalysts: selective formation of carpanones and benzopyrans under different reaction conditions. *J Org Chem.* 2012;77(10):4528–4543. https://doi.org/10.1021/jo300263k.
4. Nicolaou KC, Snyder SA, Montagnon T, Vassilikogiannakis G. The Diels-Alder reaction in total synthesis. *Angew Chem Int Ed.* 2002;41(10):1668–1698. https://doi.org/10.1002/1521-3773(20020517)41:10<1668::AID-ANIE1668>3.0.CO;2-Z.

Question 265: Cyclopentaquinolines by Tandem Reactions

In 1991 American chemists irradiated a mixture of *p*-fluorophenyl isocyanide and 5-iodo-1-pentyne in *t*-butylbenzene containing hexamethylditin under nitrogen in a pyrex flask at 150°C for several hours until all of the alkyne was consumed (followed by ^{1}H NMR) and after standard work-up obtained a 56% yield of an 85/15 ratio mixture of cyclopentaquinolines **1** and **2**, respectively (Scheme 1).[1] Show mechanisms for the formation of **1** and **2**.

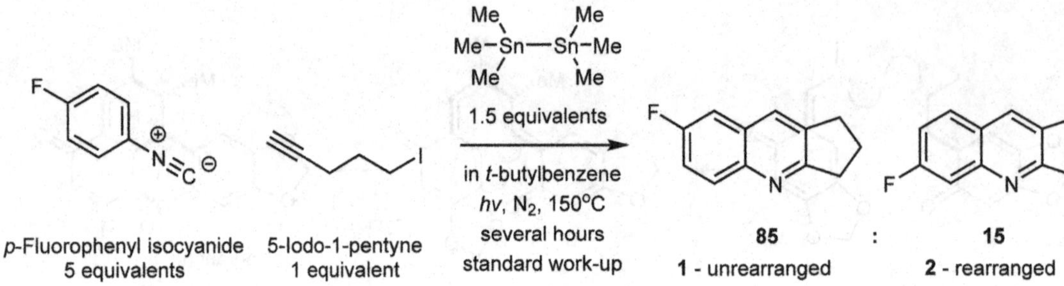

SCHEME 1

SOLUTION

The mechanism proposed by Curran and Liu begins with radical generation which the authors believe can occur either by photolytic cleavage of hexamethylditin followed by iodine abstraction (possibility 1) or directly by C-I photolytic cleavage followed by iodine capture by hexamethylditin (possibility 2) to generate radical intermediate **A** (Scheme 2).[2]

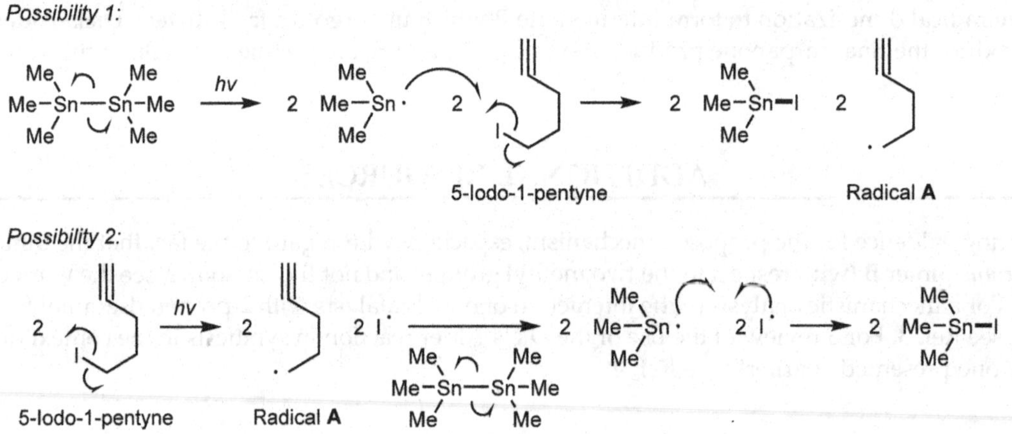

SCHEME 2

Once alkyl radical **A** is generated it reacts with the nitrile group of *p*-fluorophenyl isocyanide[3] to form presumably the more stable nonlinear[4] imidoyl radical *E*-**B** which exists in equilibrium with its *Z*-isomer *Z*-**B**. Radical intermediate *E*-**B** can then cyclize with the alkyne group in predominantly a 5-*exo* manner to form the productive vinyl radical *Z*-**C** (Scheme 3).[5]

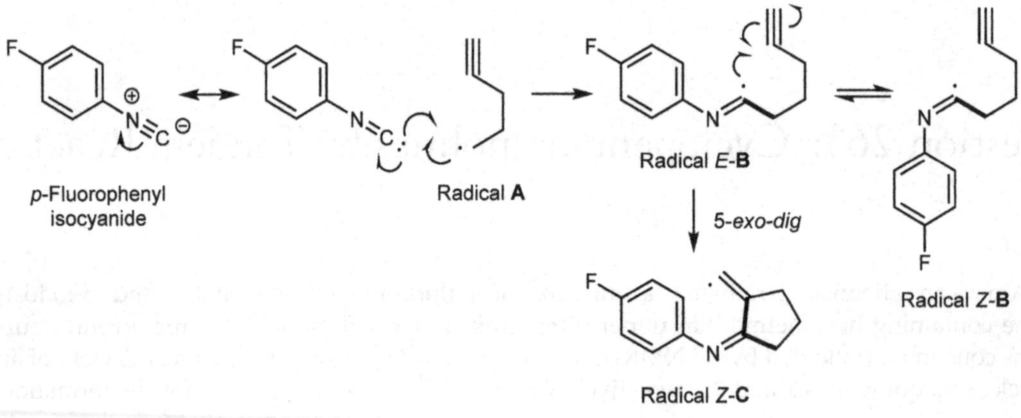

SCHEME 3

The authors note that they are not aware of any 5-*exo* ring openings of vinyl radicals and therefore they believe that the *E*-**B** to *Z*-**C** transformation is likely irreversible. Despite this fact, we think a theoretical analysis should be performed to confirm the energetics. The authors also state that they do not know whether *E*-**B** is formed kinetically and then it cyclizes to *Z*-**C** faster than it isomerizes to *Z*-**B** or whether the two radicals are in rapid equilibrium and cyclization occurs predominantly through *E*-**B** according to Curtin-Hammett kinetics.[6] Relevant here is the fact that related disubstituted vinyl radicals have low inversion energy barriers which means that radical isomerization occurs much faster than cyclization for these species whereas for imidoyl radicals, since the electronegative nitrogen atom pyramidalizes the radical, the expected inversion barriers should be higher.[1] Nevertheless, according to the authors, it is not clear whether the inversion barriers for imidoyl radicals are sufficiently high to ensure that cyclization of *E*-**B** to *Z*-**C** outcompetes its isomerization to *Z*-**B**. Moreover, we note that even if isomerization to *Z*-**B** outcompetes the cyclization to *Z*-**C**, cyclization of the resulting *Z*-**B** imidoyl radical with the alkyne group would produce the *E* isomer *E*-**C** which must decompose by nonproductive pathways since subsequent cyclization steps to the aromatic ring are geometrically forbidden.[1] Interestingly, according to Blum and Roberts, theoretical calculations of simple imidoyl radicals predict that the *E* isomer is more stable than the *Z* isomer by 2.5 kcal/mol and that the barrier to interconversion is about 10 kcal/mol.[4] These findings support the proposed mechanism. Moreover, it has been shown that in contrast to imidoyl radicals, β-stannyl vinyl radicals cyclize only through the *Z* isomer.[7] Lastly, further study of the nature of equilibria and reaction rates for stereoisomeric sp^2-hybridized radicals, say the authors, will lead to a better understanding of the **A** to *Z*-**C** transformation. We would also add a theoretical analysis to this.

From radical intermediate *Z*-**C**, the mechanism then proceeds with cyclization of the vinyl radical group of *Z*-**C** onto the aromatic ring. This cyclization can occur at two different positions (*ortho* and *ipso*) on the aromatic ring as shown in Scheme 4 to produce the *ortho* radical **D** and *ipso* radical **E**, respectively.[5]

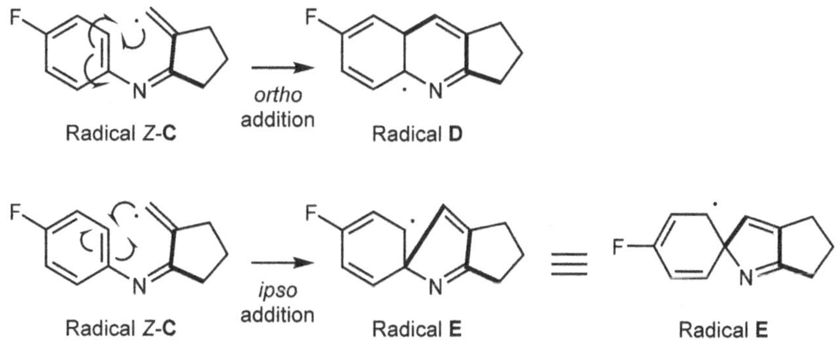

SCHEME 4

From the *ortho* radical **D**, the only further step required to form the final unrearranged cyclopentaquinoline **1** is an oxidation step, which will be covered later. The fate of the *ipso* radical **E** (which is expected to lead to the rearranged cyclopentaquinoline **2**) is more important at this stage. Indeed, the authors present two possible pathways for radical **E**. These consist of either a ring opening step followed by a recyclization step or formation of a strained intermediate followed by ring expansion to relieve the accumulated ring strain (Scheme 5).

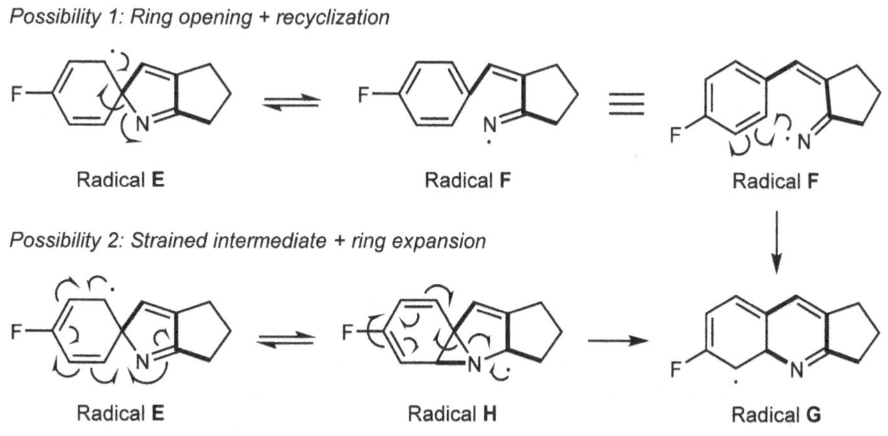

SCHEME 5

With respect to this sequence, the authors first explain that they regard the reverse cyclization of *ipso* radical **E** back to radical Z-**C** to be unlikely. As such, the first possible pathway radical **E** might take in the authors' view consists of ring opening via cleavage of the C—N bond to form the iminyl radical **F** which may then recyclize at the *ortho* position of the aromatic ring to form the rearranged radical **G** (i.e., the expected precursor to the rearranged cyclopentaquinoline **2** product). The authors indicate that although they expect radical **F** to undergo reverse cyclization back to radical **E** (based on work which showed the easy reversibility of aminyl radical cyclizations),[8] it is not known whether this actually occurs. A theoretical study would certainly shed more light on the likelihood of this step being an equilibrium step.

Alternatively, a proposed 3-*exo* closure of radical **E** (unknown whether this step is reversible or not) would form the strained radical **H** which can ring expand via 3-*exo* cleavage of the intraannular bond to form the same rearranged radical **G**. The authors explain that such mechanistic pathways are well documented for simple β multiply bonded alkyl radicals even though there are no known analogous examples for allyl and dienyl radicals.[9] We would also like to add that strained intermediates and products of a similar structure as radical **H** have been reported elsewhere.[10, 11]

Lastly, the authors have performed an interesting experiment to show that if intermediate radical **F** is formed, it has to cyclize to **G** under the reaction conditions (Scheme 6). This experiment is quite clever since the aromatic group is already built into the starting material thus avoiding the imidoyl radical **B**. Here, after bromine abstraction, the vinyl radical derived from **3** would likely react with the nitrile group to form the cyclopentane ring and a radical on the nitrogen atom thus providing a radical **F** type structure which exclusively leads to rearranged cyclopentaquinoline product **4** as well as reduced starting material. No unrearranged cyclopentaquinoline products are formed thus indicating that the radical **F** intermediate is, according to the authors, "a viable but not obligatory intermediate" in the transformation of **E** to **G**.[1]

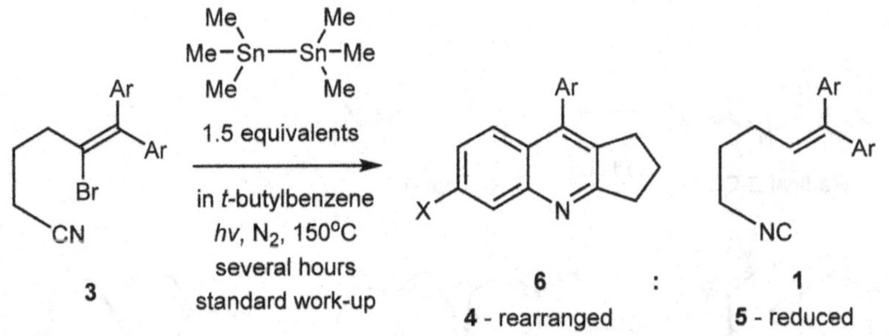

SCHEME 6

Supposing that radicals **D** and **G** constitute the precursors to final products **1** and **2**, respectively, the final step of this mechanism must involve radical oxidation followed by aromatization (Scheme 7).

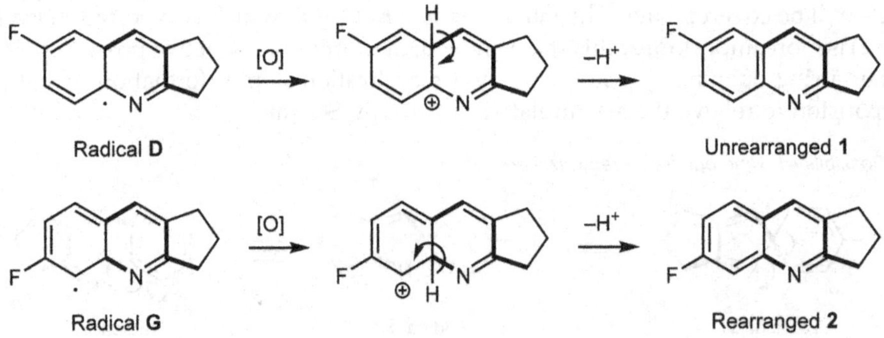

SCHEME 7

Concerning these last steps, the authors indicate that the oxidant needed to effect these transformations is not immediately obvious from the reaction conditions. One possibility, they explain, could be trace amounts of O_2 gas which is not rigorously removed by freeze and thaw techniques (even though the reactions are degassed).[1] Nevertheless, the authors do not believe that there is sufficient O_2 in solution for carrying out stoichiometric oxidations. Another possibility consists of iodine as the oxidant. The authors discard this idea because if iodine were the oxidant, a chain would result and stoichiometric quantities of tin reagent would not be required.[12] A third possibility is that the trimethyltin iodide formed at the beginning of this mechanism participates as the oxidant at its end. Here, trimethyltin iodide would react with radicals **D** or **G** to form **1** or **2**, respectively, plus trimethyltin radical and hydrogen iodide as a reaction product from the aromatization (i.e., deprotonation) step. The newly formed HI may react with the hexamethylditin reagent (thus consuming it) to regenerate trimethyltin iodide plus trimethyltin hydride (Scheme 8).

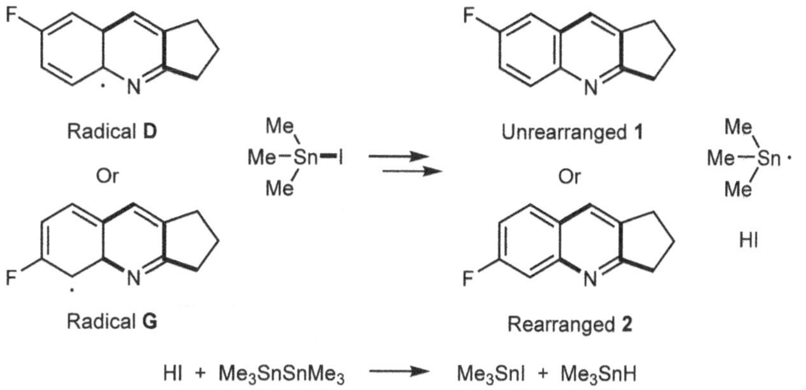

Radical **D**
Or
Radical **G**

$Me-Sn-I$ (with two Me groups)

Unrearranged **1**
Or
Rearranged **2**

$Me-Sn\cdot$ (with two Me groups)

HI

$HI + Me_3SnSnMe_3 \longrightarrow Me_3SnI + Me_3SnH$

SCHEME 8

Once again the authors do not believe that this chain is efficient because the high reaction temperature and long reaction time are not good factors for efficient chains and also there is no evidence that trimethyltin hydride accumulates as a reaction product.[1]

The last possibility for oxidation of **D** and **G** to **1** and **2** considered by the authors consists of radical disproportionation (Scheme 9). Here, two equivalents of radical species react in a redox fashion (with one equivalent acting as the oxidant and the other as the reductant) in order to form one equivalent of desired quinoline product and one equivalent of undesired dihydroquinoline by-product. The authors argue that since the yield of quinoline from these transformations often exceeds 50%, one must conclude that some of the dihydroquinoline by-product must be air oxidized to quinoline during the reaction and/or work-up stage. Nevertheless, when reactions were conducted in C_6D_6 and subject to NMR analysis prior to air exposure, the authors did not detect any dihydroquinoline by-product. Once again a theoretical analysis would help shed light on the probable pathway for the final oxidation step.

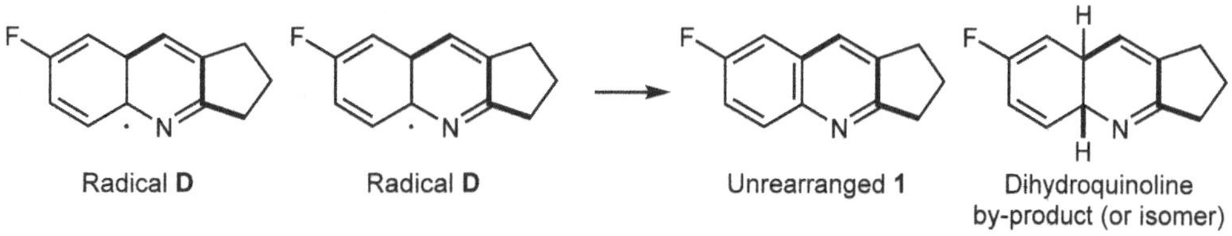

Radical **D**   Radical **D**   Unrearranged **1**   Dihydroquinoline by-product (or isomer)

SCHEME 9

EXPERIMENTAL RESULTS AND VERIFICATION

Putting everything together, the authors explain that this mechanism consists of the following sequence of steps: (1) radical addition to an isonitrile, (2) cyclization of the resulting imidoyl radical to the alkyne group (monitored by ^{1}H NMR), (3) addition of the resulting vinyl radical to the aromatic ring in either the *ipso* or *ortho* position, (4) oxidation and rearomatization. The major products are always the unrearranged cyclopentaquinolines. The concentration of the reaction components is important as higher product yields are achieved at lower concentrations. The reactions are also highly sensitive to temperatures above 80°C which are required for appreciable product yields. Moreover, excess isonitrile starting material is needed, perhaps due to its susceptibility to thermal decomposition during reaction, as the authors believe given that they were able to detect polar high molecular weight by-product. Furthermore, two interesting experimental observations were made with regard to the starting materials (Scheme 10).

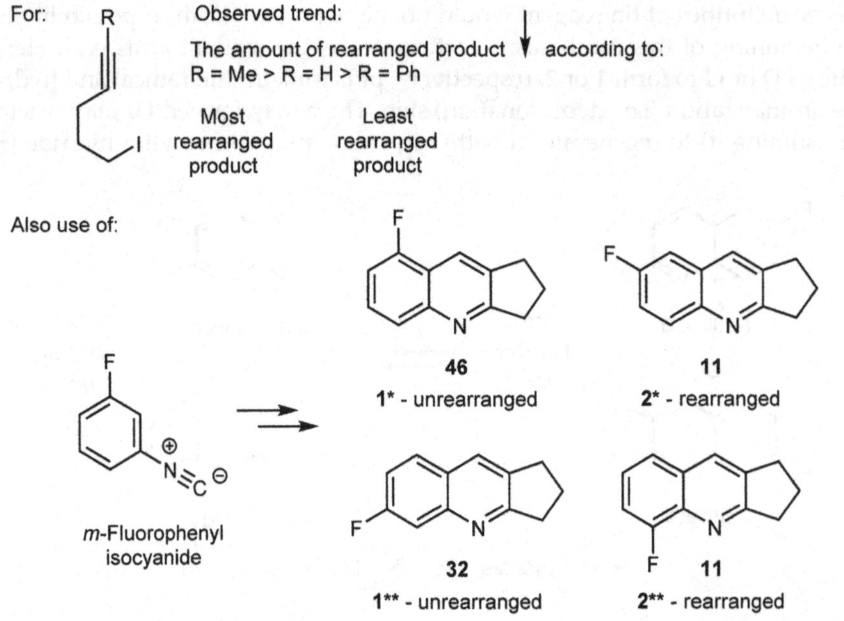

SCHEME 10

To explain the first observation in terms of the trend surrounding the R group on the starting 5-iodo-1-pentyne, one can invoke the formation of radical Z-**C** which for steric reasons is probably less likely to form the *ipso* radical **E** when R=Ph. Interestingly, when R=Me the reaction produces the most rearranged product. This is likely due to a lower energy barrier with regard to the formation of vinyl radical Z-**C** but without a theoretical investigation this reasoning must be regarded as speculative at best. More interestingly, however, the use of *m*-fluorophenyl isocyanide as starting material produces a distribution of four products (two unrearranged and two rearranged), a result which we can use to double check the mechanism proposed by the authors (Schemes 11 and 12). Indeed, the mechanism does lead to all four products.

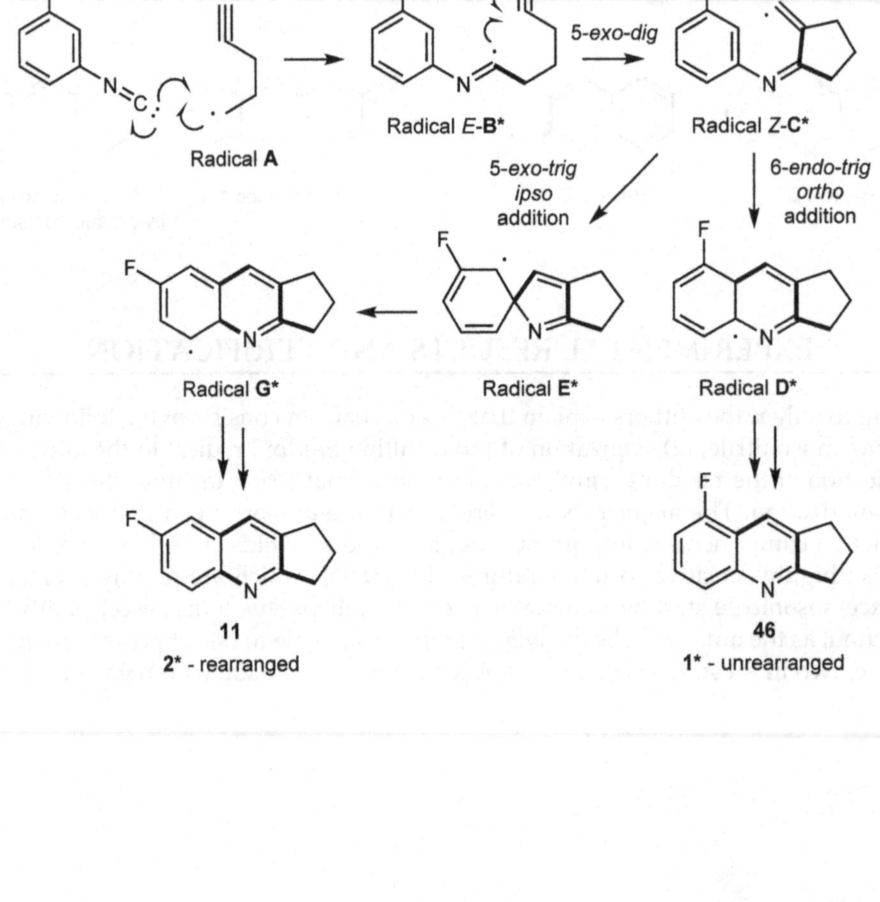

SCHEME 11

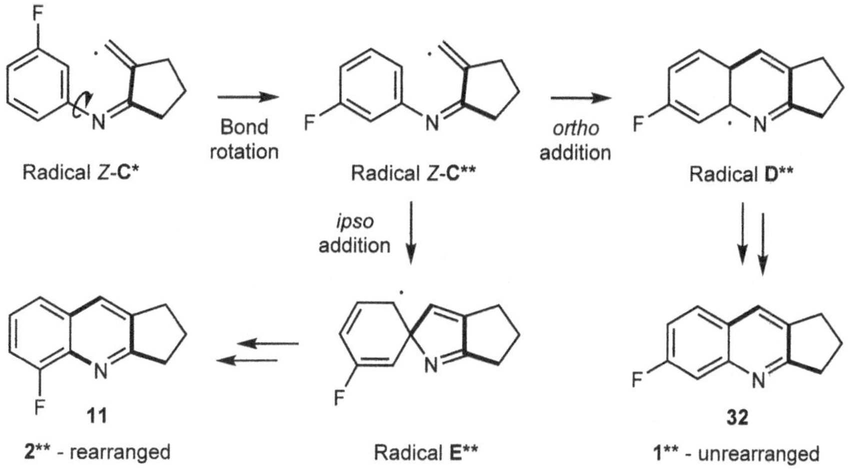

SCHEME 12

ADDITIONAL RESOURCES

We direct the reader to further work by Curran et al. both with respect to the synthetic applications of this transformation[13] as well as other interesting mechanistic analyses.[14–20] For review articles by Curran and other works involving interesting transformations in the context of radical chemistry, see Refs. 21–23.

References

1. Curran DP, Liu H. 4 + 1 radical annulations with isonitriles: a simple route to cyclopenta-fused quinolines. *J Am Chem Soc*. 1991;113(6):2127–2132. https://doi.org/10.1021/ja00006a033.
2. Curran DP, Chen MH, Kim D. Atom transfer cyclization reactions of hex-5-ynyl iodides: synthetic and mechanistic studies. *J Am Chem Soc*. 1989;111(16):6265–6276. https://doi.org/10.1021/ja00198a043.
3. Meier M, Ruechardt C. A free radical chain mechanism for the isonitrile-nitrile rearrangement in solution and its inhibition. *Tetrahedron Lett*. 1983;24(43):4671–4674. https://doi.org/10.1016/S0040-4039(00)86223-1.
4. Blum PM, Roberts BP. An electron spin resonance study of radical addition to alkyl isocyanides. *J Chem Soc Perkin Trans 2*. 1978;1313–1319. https://doi.org/10.1039/P29780001313.
5. Leardini R, Pedulli GF, Tundo A, Zanardi G. Aromatic annelation by reaction of arylimidoyl radicals with alkynes: a new synthesis of quinolines. *J Chem Soc Chem Commun*. 1984;1984:1320–1321. https://doi.org/10.1039/C39840001320.
6. Seeman JI. Effect of conformational change on reactivity in organic chemistry. Evaluations, applications, and extensions of Curtin-Hammett Winstein-Holness kinetics. *Chem Rev*. 1983;83(2):83–134. https://doi.org/10.1021/cr00054a001.
7. Stork G, Mook RJ. Vinyl radical cyclizations mediated by the addition of stannyl radicals to triple bonds. *J Am Chem Soc*. 1987;109(9):2829–2831. https://doi.org/10.1021/ja00243a049.
8. Newcomb M, Deeb TM, Marquardt DJ. N-hydroxypyridine-2-thione carbamates. IV. A comparison of 5-exo cyclizations of an aminyl radical and an aminium cation radical. *Tetrahedron*. 1990;46(7):2317–2328. https://doi.org/10.1016/S0040-4020(01)82012-3.
9. Dowd P, Choi S-C. Novel free radical ring-expansion reactions. *Tetrahedron*. 1989;45(1):77–90. https://doi.org/10.1016/0040-4020(89)80035-3.
10. McBurney RT, Walton JC. Interplay of ortho- with spiro-cyclisation during iminyl radical closures onto arenes and heteroarenes. *Beilstein J Org Chem*. 2013;9:1083–1092. https://doi.org/10.3762/bjoc.9.120.
11. Kaga A, Tnay YL, Chiba S. Synthesis of fasicularin. *Org Lett*. 2016;18(14):3506–3508. https://doi.org/10.1021/acs.orglett.6b01669.
12. Kolt RJ, Wayner DDM, Griller D. Using electron-transfer reactions to propagate radical chain processes. *J Org Chem*. 1989;54(17):4259–4260. https://doi.org/10.1021/jo00278a059.
13. Curran DP, Liu H, Josien H, Ko S-B. Tandem radical reactions of isonitriles with 2-pyridonyl and other aryl radicals: scope and limitations, and a first generation synthesis of (±)-camptothecin. *Tetrahedron*. 1996;52(35):11385–11404. https://doi.org/10.1016/0040-4020(96)00633-3.
14. Curran DP, Seong CM. Radical annulation reactions of allyl iodomalononitriles. *Tetrahedron*. 1992;48(11):2175–2190. https://doi.org/10.1016/S0040-4020(01)88882-7.
15. de Turiso FG-L, Curran DP. Radical cyclization approach to spirocyclohexadienones. *Org Lett*. 2005;7(1):151–154. https://doi.org/10.1021/ol0477226.

16. Zhang H, Jeon KO, Hay EB, Geib SJ, Curran DP, LaPorte MG. Radical [3 + 2]-annulation of divinylcyclopropanes: rapid synthesis of complex meloscine analogs. *Org Lett.* 2014;16(1):94–97. https://doi.org/10.1021/ol403078e.

17. Haney BP, Curran DP. Round trip radical reactions from acyclic precursors to tricyclo[5.3.1.02,6]undecanes. A new cascade radical cyclization approach to (±)-isogymnomitrene and (±)-gymnomitrene. *J Org Chem.* 2000;65(7):2007–2013. https://doi.org/10.1021/jo9914871.

18. Bruch A, Froehlich R, Grimme S, Studer A, Curran DP. One product, two pathways: initially divergent radical reactions reconverge to form a single product in high yield. *J Am Chem Soc.* 2011;133:16270–16276. https://doi.org/10.1021/ja2070347.

19. Hay EB, Zhang H, Curran DP. Rearrangement reactions of 1,1-divinyl-2-phenylcyclopropanes. *J Am Chem Soc.* 2015;137(1):322–327. https://doi.org/10.1021/ja510608u.

20. Curran DP, Sisko J, Balog A, Sonoda N, Nagahara K, Ryu I. Carbonylative radical cyclization approaches to tri- and tetraquinanes: sequential formation of three, four and five carbon-carbon bonds. *J Chem Soc Perkin Trans 1.* 1998;1591–1594. https://doi.org/10.1039/A800352A.

21. Ryu I, Sonoda N, Curran DP. Tandem radical reactions of carbon monoxide, isonitriles, and other reagent equivalents of the geminal radical acceptor/radical precursor synthon. *Chem Rev.* 1996;96(1):177–194. https://doi.org/10.1021/cr9400626.

22. Studer A, Curran DP. The electron is a catalyst. *Nat Chem.* 2014;6:765–773. https://doi.org/10.1038/nchem.2031.

23. Studer A, Curran DP. Catalysis of radical reactions: a radical chemistry perspective. *Angew Chem Int Ed.* 2016;55(1):58–102. https://doi.org/10.1002/anie.201505090.

Question 266: Epimerization of Herqueinone

In 1972 UK chemists treated herqueinone **1** with anhydrous potassium carbonate in anhydrous acetone under reflux and heat for 1 h to form the C_4 epimer **2** (Scheme 1).[1] Suggest a mechanism for this transformation.

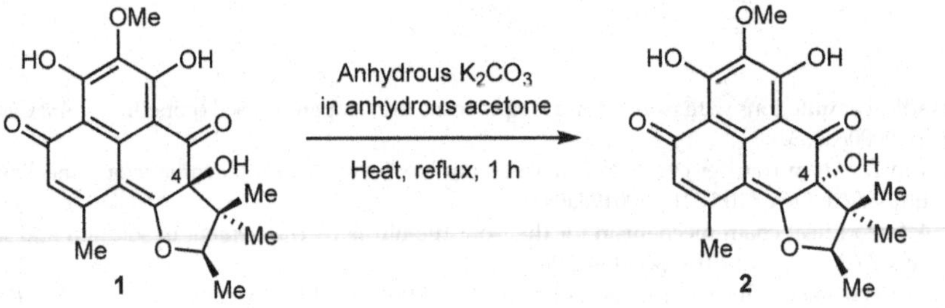

SCHEME 1

SOLUTION: BROOKS MECHANISM

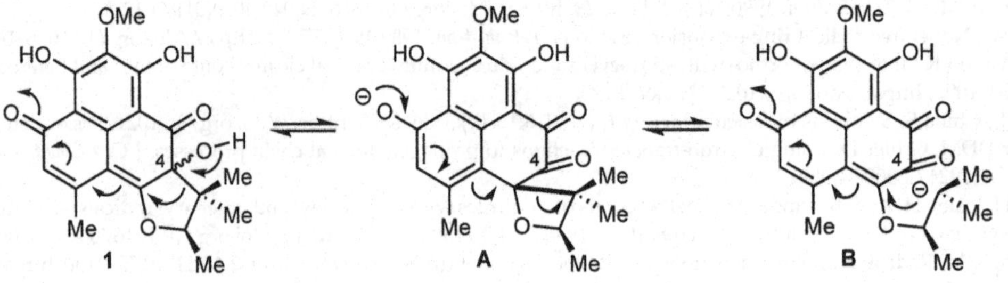

SOLUTION: PROPOSED MECHANISM

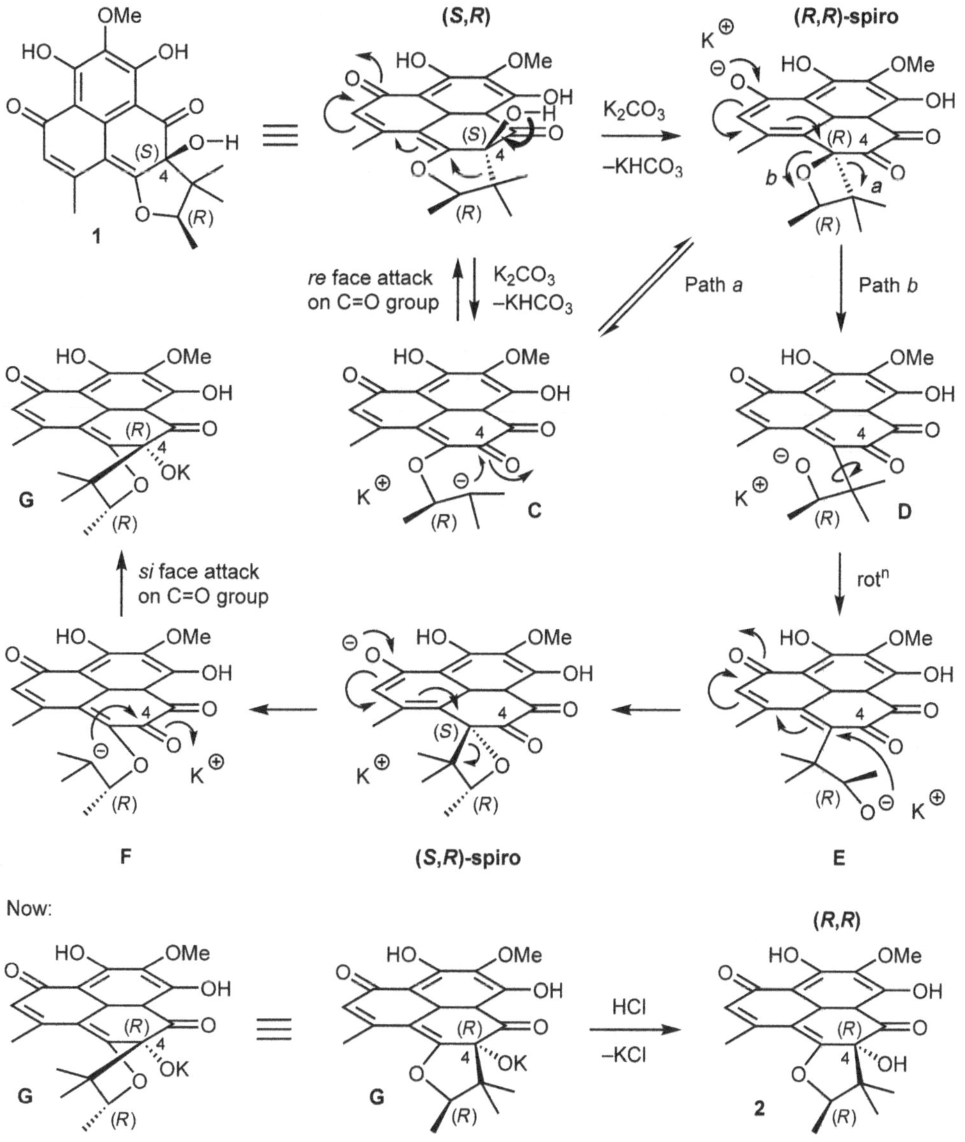

KEY STEPS EXPLAINED

The mechanism proposed by us here is merely an expansion over that proposed by Brooks and Morrison. Given the postulated spirointermediate **A** arising from base-mediated ketonization of **1** (for which there exists an electron sink) and the carbanion **B**, we sought to depict the missing geometrical and stereochemical considerations that lead from **1** to **2**. Thus, we first note that herqueinone **1** has an S-configuration at C_4 and an R-configuration at the furan carbon bearing the single methyl group. Base-mediated ketonization therefore helps form the (R,R)-spiro intermediate which may ketonize back to break the spiroring by one of two possible pathways: path (a) cleavage of C—C bond to form the carbanion **C**, or path (b) cleavage of C—O bond to form the carboxy anion **D**. We note that carbanion **C** closes back to **1** by re facial nucleophilic attack onto the ketone group at C_4. Indeed, base-mediated ketonization of **1** can also lead directly to **C**. From **D**, bond rotation allows for closure to an (S,R)-spiro intermediate which ring opens to the carbanion **F** which after si facial nucleophilic attack onto the ketone group at C_4 produces the (R,R)-furan intermediate **G** which upon acidic work-up forms the final herqueinone epimer **2**. A simplified depiction of this process is available in Fig. 1. For future research we would highly recommend a theoretical analysis of this mechanism to help elucidate whether all steps are equilibrium steps or whether some are irreversible based on energy calculations and geometrical optimization.

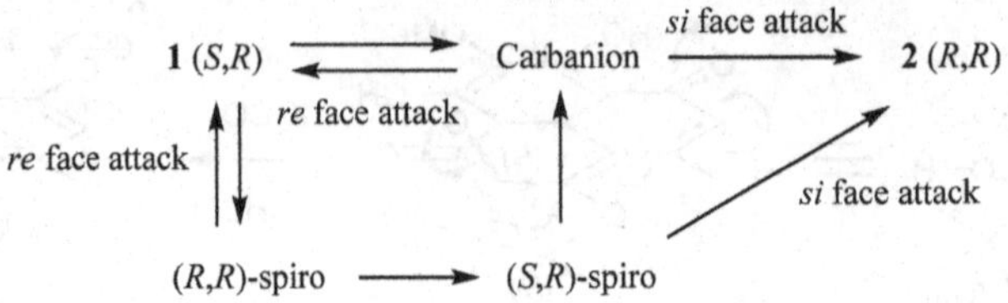

FIG. 1

ADDITIONAL RESOURCES

We refer the interested reader to research highlighting the possible transformations of herqueinone[2] and its biosynthetic pathways[3,4] for further mechanistic insight.

References

1. Brooks JS, Morrison GA. Naturally occurring compounds related to phenalenone. Part III. The structure of herqueinone and norherqueinone and their relationships with isoherqueinone and isonorherqueinone. *J Chem Soc Perkin Trans 1*. 1972;1972:421–437. https://doi.org/10.1039/P19720000421.
2. Quick A, Thomas R, Williams DJ. X-ray crystal structure and absolute configuration of the fungal phenalenone herqueinone. *J Chem Soc Chem Commun*. 1980;1980:1051–1053. https://doi.org/10.1039/C39800001051.
3. Elsebai MF, Saleem M, Tejesvi MV, et al. Fungal phenalenones: chemistry, biology, biosynthesis and phylogeny. *Nat Prod Rep*. 2014;31:628–645. https://doi.org/10.1039/c3np70088g.
4. Gao S-S, Garcia-Borras M, Barber JS, et al. Enzyme-catalyzed intramolecular enantioselective hydroalkoxylation. *J Am Chem Soc*. 2017;139:3639–3642. https://doi.org/10.1021/jacs.7b01089.

Question 267: A New Method for Amide Bond Formation

In 1996 Japanese chemists used 2,2-dichloro-5-(2-phenylethyl)-4-trimethylsilyl-3-furanone **1** (DPTF) as a water-resistant activating reagent for an amide reaction as illustrated in Scheme 1.[1] Devise a mechanism for this transformation and explain the formation of the 5-(2-phenylethyl)-4-trimethylsilylfuran-2,3-dione by-product.

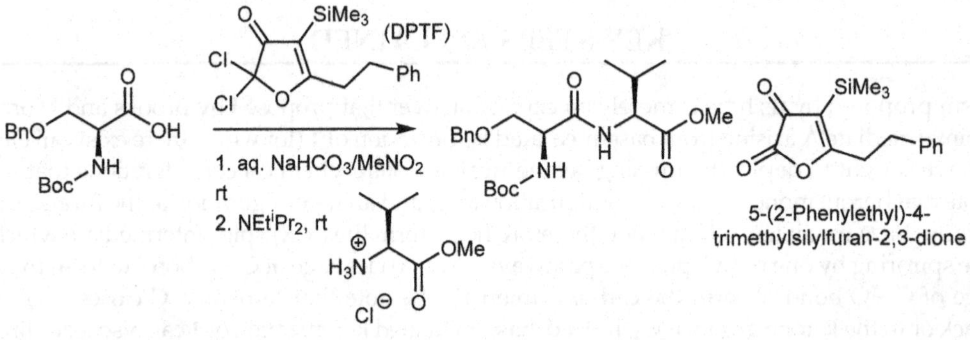

SCHEME 1

SOLUTION

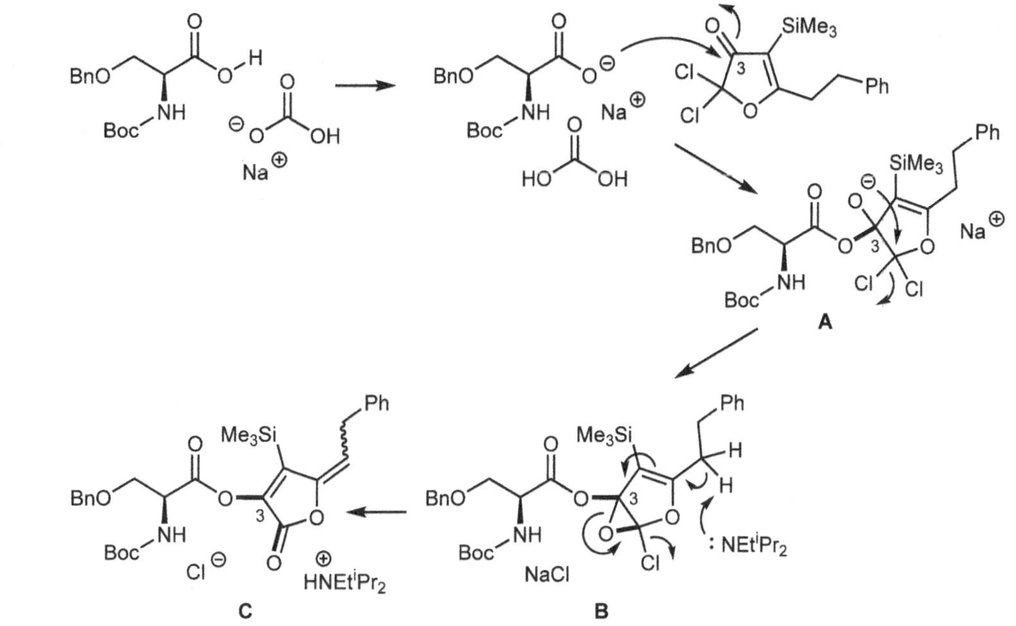

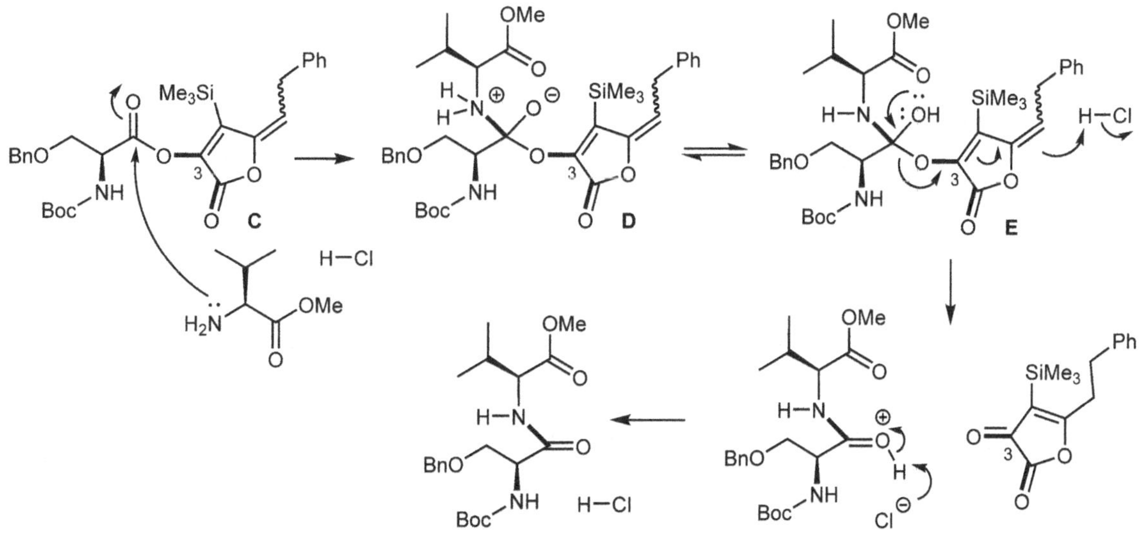

KEY STEPS EXPLAINED

The mechanism proposed by Murakami et al. begins with deprotonation of the starting doubly protected amino acid to form a carboxylate anion which selectively attacks in a nucleophilic fashion the C_3 position of DPTF to form the carboxy anionic intermediate **A**. The authors believe that the C_3 position of DPTF is attacked as such based on previous research outlining the synthesis of DPTF, wherein it was observed that the reaction of

DPTF with methanol under basic conditions led to selective nucleophilic attack at the C_3 position of DPTF.[2] Next, closure of **A** to an epoxyfuran **B** sets up a deprotonation on the side chain of **B** which leads to an electron shift that opens the epoxy ring forming **C** after elimination of HCl. The authors actually isolate **C**-like intermediates in excellent yields for representative peptide reactions. It is also noteworthy that intermediate **C** is the activated carboxylic acid which usually constitutes the first step in most amide-forming reactions. Luckily for the authors, this intermediate also proved to be resistant to aqueous conditions. The next step in the sequence follows reaction between the amino end of the second amino acid and **C** to form complex **D** which, after intramolecular proton transfer, cleaves to the final peptide chain product and the 5-(2-phenylethyl)-4-trimethylsilylfuran-2,3-dione by-product.

ADDITIONAL RESOURCES

We refer the reader to two review articles highlighting the development of coupling reagents for amide bond-forming reactions along with mechanistic illustrations.[3,4]

References

1. Murakami M, Hayashi M, Tamura N, Hoshino Y, Ito Y. A new water-compatible dehydrating agent DPTF. *Tetrahedron Lett.* 1996;37 (42):7541–7544. https://doi.org/10.1016/0040-4039(96)01712-1.
2. Murakami M, Hayashi M, Ito Y. Preparation of 2,2-dichloro-3(2H)-furanone and its reactions with heteronucleophiles. *J Org Chem.* 1994;59 (25):7910–7914. https://doi.org/10.1021/jo00104a058.
3. Han S-Y, Kim Y-A. Recent development of peptide coupling reagents in organic synthesis. *Tetrahedron.* 2004;60:2447–2467. https://doi.org/10.1016/j.tet.2004.01.020.
4. Valeur E, Bradley M. Amide bond formation: beyond the myth of coupling reagents. *Chem Soc Rev.* 2009;38:606–631. https://doi.org/10.1039/b701677h.

Question 268: A Benzothiazole From Oxidation of Mammalian Red Hair With Hydrogen Peroxide

Pheomelanins are sulfur-containing pigments thought to play a role in the susceptibility of red-haired fair-skinned individuals to sun burns and skin cancer.[1] In a 1996 report, Italian scientists oxidized the known pheomelanin precursor **1** with the enzyme tyrosinase followed by degradation with 1% hydrogen peroxide in 1 M NaOH to obtain **2** in 25% yield (Scheme 1).[2] Suggest a mechanism for the transformation of **1** to **2**.

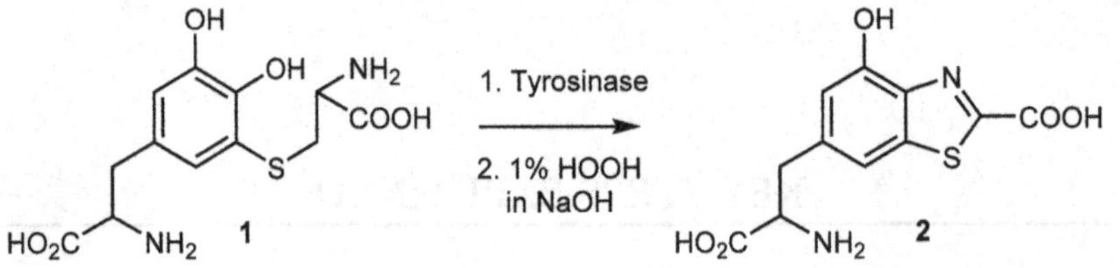

SCHEME 1

SOLUTION

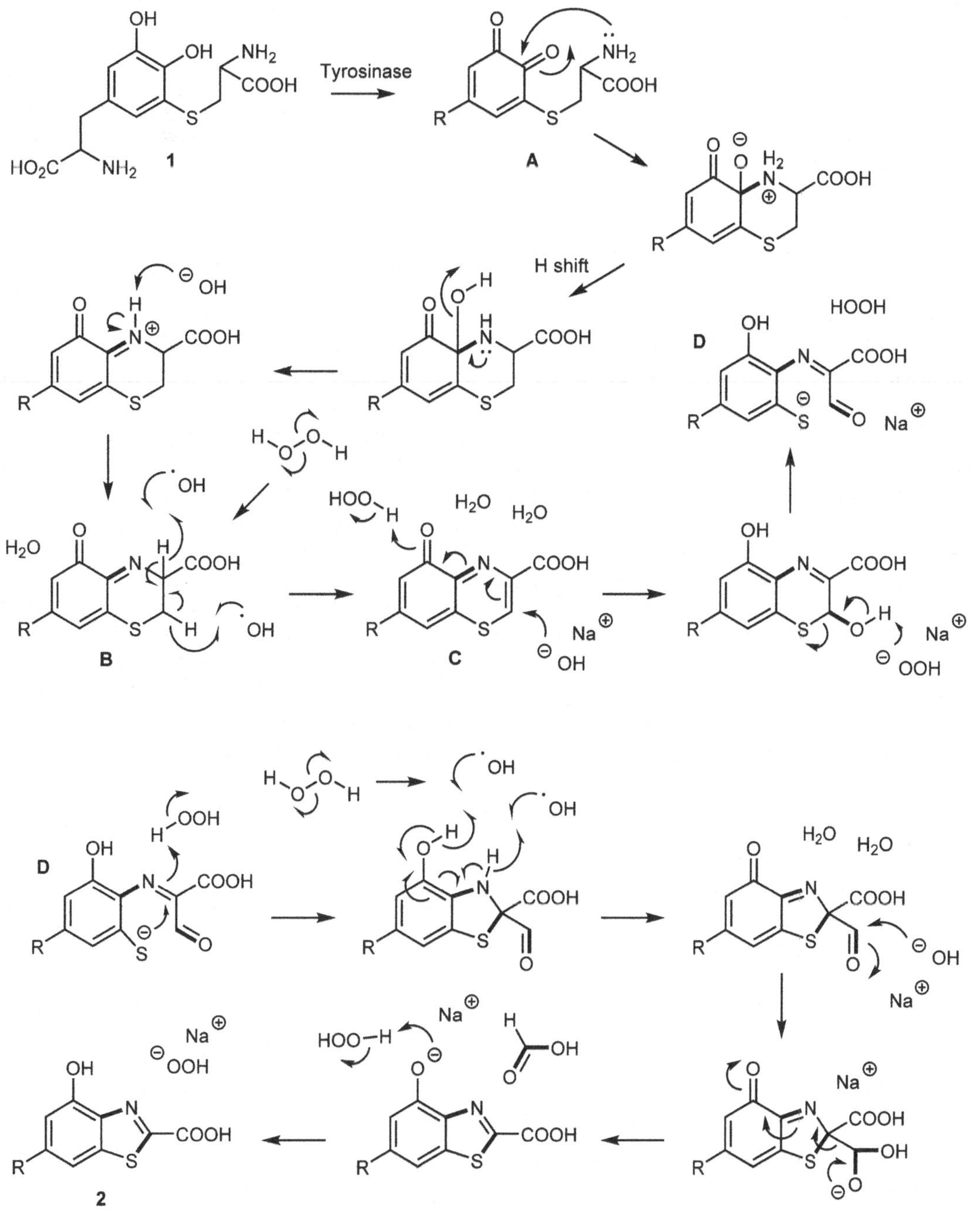

*Tyrosinase oxidation of **1** would likely involve a cofactor such as NAD+ or FAD+ or NADP+.

BALANCED CHEMICAL EQUATION

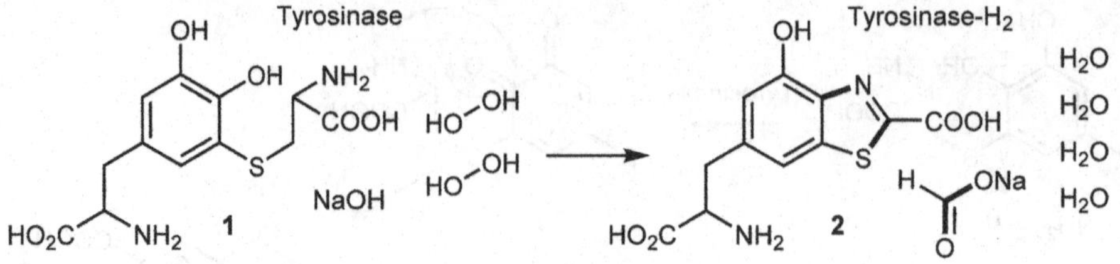

KEY STEPS EXPLAINED

Essentially the mechanism proposed by Napolitano et al. begins with enzymatic oxidation of **1** by tyrosinase (with a cofactor) to the diketone which reacts with the amine to form the imine **B** after loss of water. Radical oxidation involving hydrogen peroxide facilitates formation of **C** which is attacked by hydroxide anion and after fragmentation produces **D**. Compound **D** then recyclizes to a five-member ring by sulfide attack on the imine. This cyclization also affords an aldehyde group. A second radical-type oxidation with hydrogen peroxide forms a ketone which provides the electron sink for hydroxide attack and loss of formic acid to form the final product **2**. The authors also explain that other mechanisms are possible for the degradation pathway on account of the difficulty to obtain experimental evidence in support of the proposed mechanism. For example, in a recent report, using hydrogen peroxide, $K_3Fe(CN)_6$, and $ZnSO_4$, the same authors proposed a very similar mechanism in which zinc coordination and a subsequent hydrogen peroxide oxidation of the aldehyde group to a carboxylic acid group ($CHO \rightarrow COOH$) facilitates a final decarboxylation to produce **2**.[3] Here, the by-product is not formic acid but rather carbon dioxide gas. Since sodium does not have the same coordination properties as zinc, we cannot endorse this mechanism for the current reaction conditions relevant to this problem. Nevertheless, an oxidation and final decarboxylation, in the absence of experimental or theoretical evidence cannot be ruled out. For future study, we recommend that other chemists conduct theoretical and experimental analyses to support a proposed mechanism, especially the identification and detection of by-products (formic acid versus carbon dioxide) and detailed accounting of stoichiometric amounts of reagents used. For illustrative purposes, we also include the ending of the alternative decarboxylation mechanism below (Scheme 2).

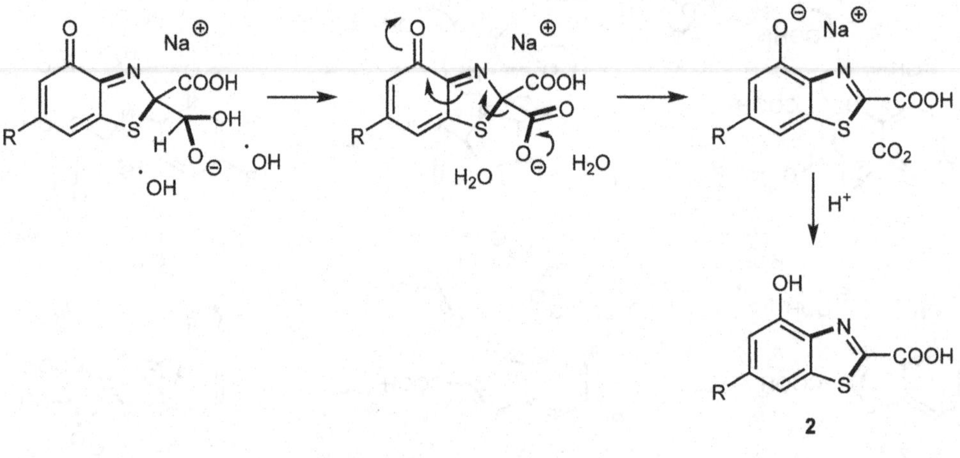

SCHEME 2

ADDITIONAL RESOURCES

For mechanistic proposals involving dimerizations and other metal ions by the same authors, see Refs. 4, 5, respectively.

References

1. Nasti TH, Timares L. MC1R, eumelanin and pheomelanin: their role in determining the susceptibility to skin cancer. *Photochem Photobiol.* 2015;91 (1):188–200. https://doi.org/10.1111/php.12335.
2. Napolitano A, Vincensi MR, d'Ishchia M, Prota G. A new benzothiazole derivative by degradation of pheomelanins with alkaline hydrogen peroxide. *Tetrahedron Lett.* 1996;37(37):6799–6802. https://doi.org/10.1016/S0040-4039(96)01483-9.

3. Greco G, Panzella L, Napolitano A, d'Ishchia M. Biologically inspired one-pot access routes to 4-hydroxybenzothiazole amino acids, red hair-specific markers of UV susceptibility and skin cancer risk. *Tetrahedron Lett.* 2009;50:3095–3097. https://doi.org/10.1016/j.tetlet.2009.04.041.
4. Greco G, Panzella L, Verotta L, d'Ishchia M, Napolitano A. Uncovering the structure of human red hair pheomelanin: benzothiazolylthiazino-dihydroisoquinolines as key building blocks. *J Nat Prod.* 2011;74:675–682. https://doi.org/10.1021/np100740n.
5. Donato PD, Napolitano A, Prota G. Metal ions as potential regulatory factors in the biosynthesis of red hair pigments: a new benzothiazole intermediate in the iron or copper assisted oxidation of 5-S-cysteinyldopa. *Biochim Biophys Acta.* 2002;1571(2):157–166. https://doi.org/10.1016/S0304-4165(02)00212-X.

Question 269: Cyclopentenones From 1,3-Cyclopentanediones

In 1996 German chemists investigating β-dicarbonyl cleavage reactions of triketones reacted **1** (R=Me) with two equivalents of sodium hydroxide in water at room temperature and achieved an 85%–95% yield of a mixture of two products (Scheme 1).[1] Product **2** was expected but product **3** was unexpected. It was also found that the yield of product **3** increased with increasing size of the R group. Explain these findings using mechanistic reasoning.

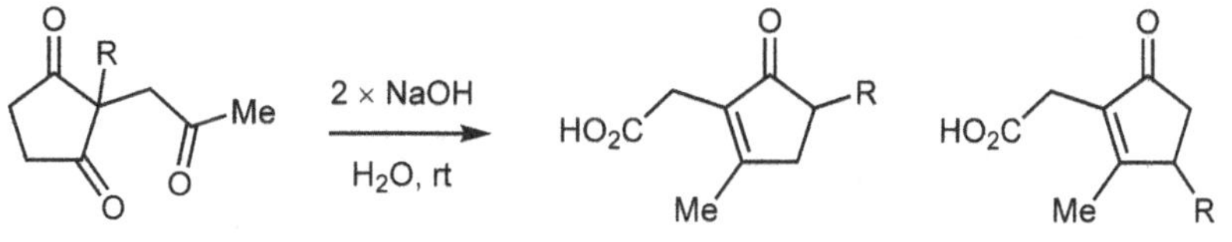

SCHEME 1

SOLUTION: 1 → 2

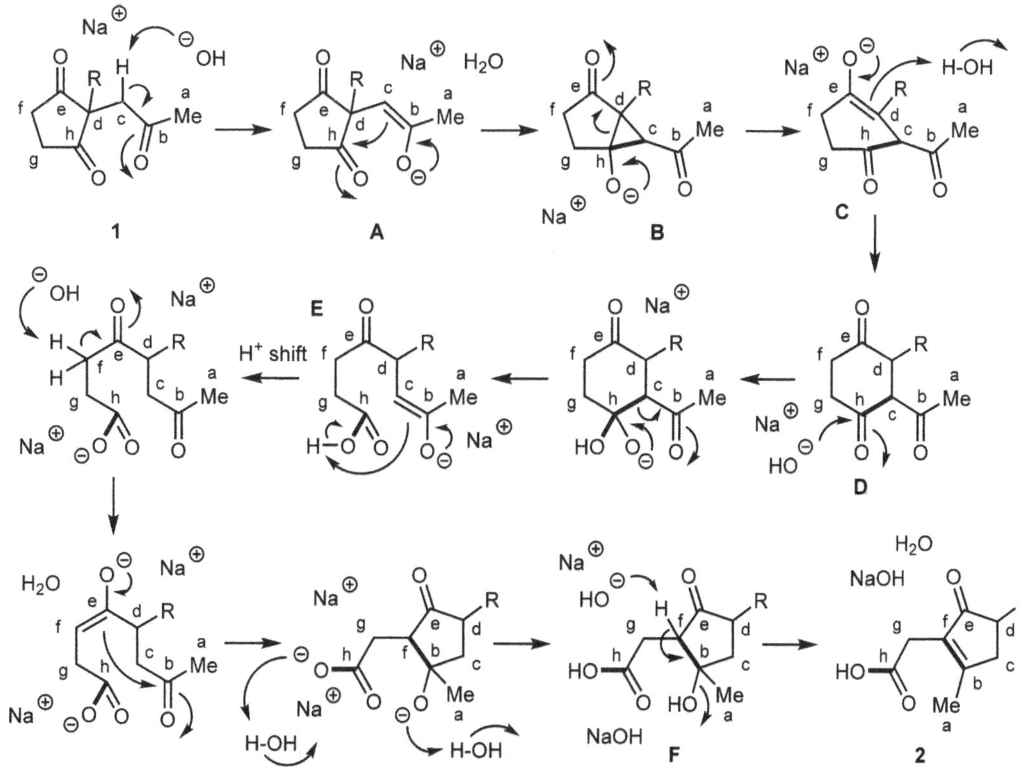

SOLUTION: C → 3

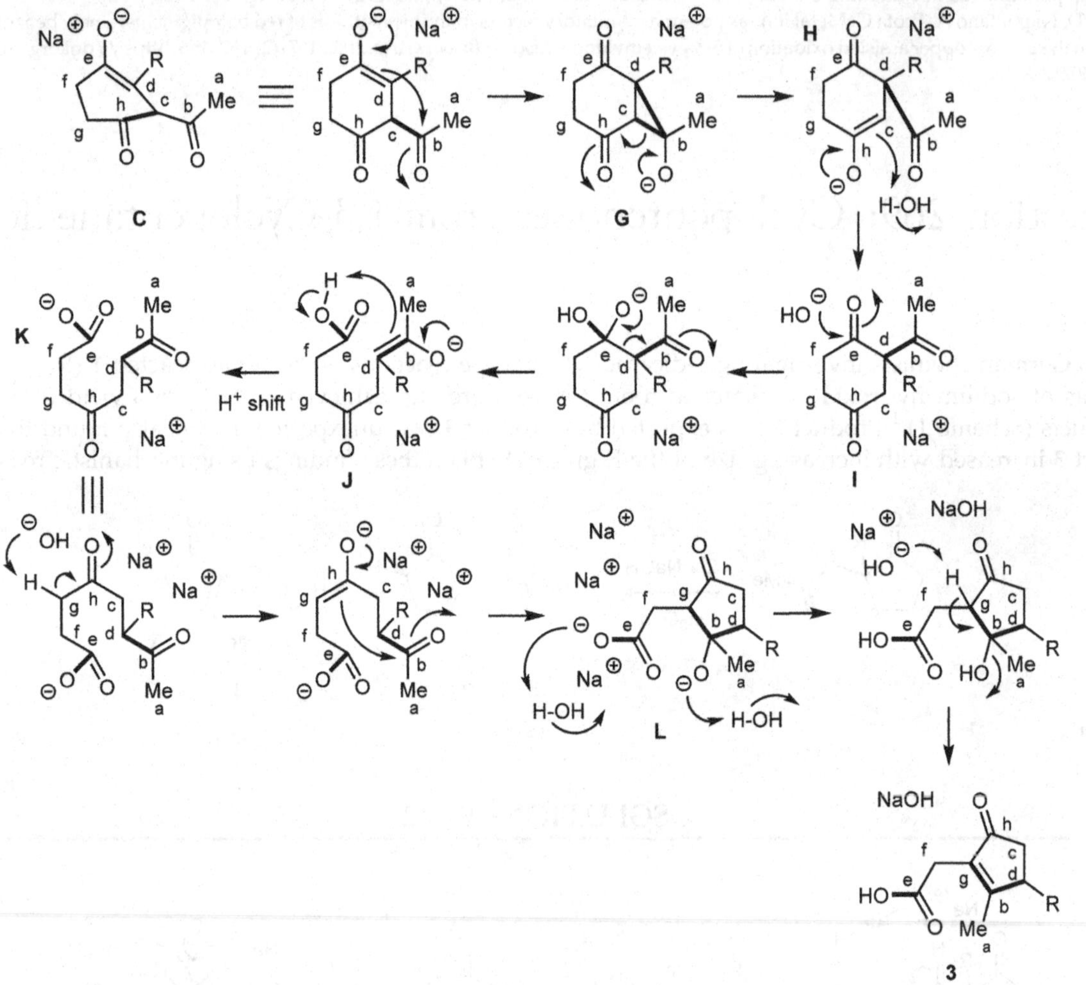

KEY STEPS EXPLAINED

According to Schick et al., the mechanism for the transformation of 1,3-cyclopentanedione **1** to the cyclopentenone **2** begins with sodium hydroxide deprotonation of **1** to form the enolate **A** at C_c.[1] This negative charge buildup at C_c can then be delocalized in a nucleophilic attack between C_c and C_h to form the diacylcyclopropanolate structure **B** which immediately fragments and leads to the common ring-enlarged intermediate **C**. From this point onward, protonation leads to the 2-acetylcyclohexan-1,4-dione **D** which is also a 1,3-diketone. As such, **D** is attacked by hydroxide anion at C_h in order to produce the ring-opened species **E** which after hydrogen shift and enolate formation at C_f can cyclize once more in an aldol-type reaction to the cyclopentanone species **F** which upon alkali-promoted loss of H_2O leads to the final cyclopentenone product **2**. As experimental evidence in support of this mechanism, the authors isolated intermediate **D** in 70% yield after the reaction of **1** (R=Me) with one equivalent of NaOH was quenched after 2 min by acidification. This intermediate was also identified using ^{1}H NMR spectroscopy.

In order to explain the **1** to **3** transformation, it is important to recognize that structure **C** is also an enolate which can close to allow for a nucleophilic attack between C_d and C_b to form the 2,3-diacylcyclopropanolate species **G**. This structure can easily open to break the C_b—C_c bond and to form an enolate which after protonation forms the 2-acetylcyclohexane-1,4-dione **I**. The authors note that the **C** to **I** transformation illustrates a 1,2-acyl transposition (acyl migration) from C_c to C_d. As seen before, the carbonyl group at C_e in **I** is easily attackable by a hydroxide anion because it is part of a 1,3-diketone system which allows for ring opening to the carboxylate **J** which may follow a similar sequence of steps as seen previously (i.e., aldol condensation and loss of H_2O) to lead to the final rearranged cyclopentenone product **3**. The authors note that this remarkable rearrangement is illustrated by recognizing the connectivity in the carbon skeleton of the starting reactant **1** (i.e., ab-cd-efgh) and that of **3** (i.e., ab-dc-hgfe).[1]

Another piece of experimental evidence in support of this mechanism offered by the authors consists of the fact that **2** can be prepared exclusively (without the presence of **3**) by treating **D** with aqueous sodium hydroxide. The reason for this, the authors point out, is the fact that the protonation of **C** to form **D** is likely irreversible since the C_c—H bond of **D** is more acidic than the newly formed C_d—H bond of **D** because C_c—H is positioned between two electron-withdrawing ketone groups and also because an alkyl R group can impart an inductive effect onto C_d—H to make it less acidic. Therefore the reverse reaction, that is, deprotonation of **D** back to **C**, is not likely to occur at C_d.

Lastly, although the authors do not mention it, the observation that the yield of **3** increases with increasing size of the R group can be rationalized by comparing the protonation of **C** (*en route* to **2**) with the enolate closure of **C** (*en route* to **3**). This essentially entails a steric effect. An increased size of the R group would increasingly restrict the protonation step (i.e., make the C_e=C_d bond less accessible) in favor of the enolate closure. Inductive considerations among alkyl groups are not expected to play a significant role in explaining this observation. Nevertheless, a theoretical analysis and an energy reaction coordinate diagram would shed more light on this issue and this remarkable rearrangement.

ADDITIONAL RESOURCES

We direct the reader to several papers outlining interesting carbon skeletal rearrangements and the experiments used to elucidate their mechanisms.[2–6] We would also like to highlight the rearrangement of cinenic acid to geronic acid which highlights the importance of structural perspective in mechanistic analysis (discussed in detail in Chapter 3).[7] The fascinating aspect of this mechanism is that the pairwise relationship between the starting material and product seems to influence one's perception of which group is migrating, either a methyl group or a carboxylic acid group (see Scheme 2).

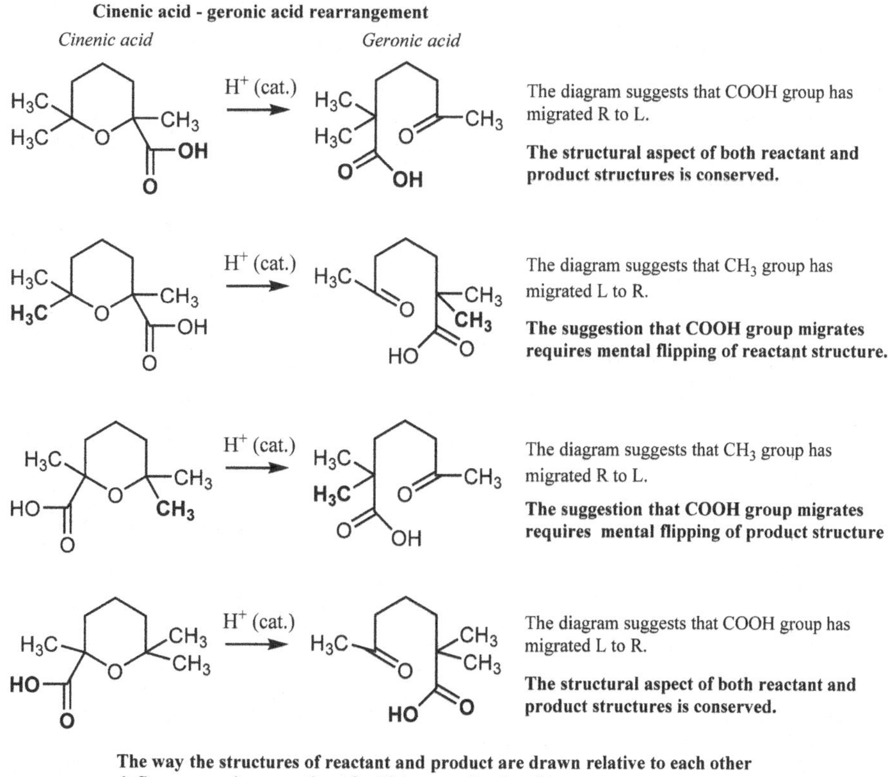

Cinenic acid - geronic acid rearrangement

Cinenic acid → *Geronic acid*

The diagram suggests that COOH group has migrated R to L.

The structural aspect of both reactant and product structures is conserved.

The diagram suggests that CH₃ group has migrated L to R.

The suggestion that COOH group migrates requires mental flipping of reactant structure.

The diagram suggests that CH₃ group has migrated R to L.

The suggestion that COOH group migrates requires mental flipping of product structure

The diagram suggests that COOH group has migrated L to R.

The structural aspect of both reactant and product structures is conserved.

The way the structures of reactant and product are drawn relative to each other influences one's perception of which group is migrating.

SCHEME 2

References

1. Schick H, Roatsch B, Schramm S, Gilsing H-D, Ramm M, Gruendemann E. Conversion of 2-alkyl-2-(2-oxopropyl)cyclopentane-1,3-diones into 2,3,5- and 2,3,4-trisubstituted cyclopent-2-enones by intramolecular aldolizations to 2,3-diacylcyclopropanolates followed by remarkable skeletal rearrangements. *J Org Chem.* 1996;61(17):5788–5792. https://doi.org/10.1021/jo960189q.
2. Enev VS, Tsankova ET. Lewis acid catalyzed rearrangement of 7,11-epoxyisogermacrone. Formation of a new carbon skeleton. *Tetrahedron.* 1991;47(32):6399–6406. https://doi.org/10.1016/S0040-4020(01)86568-6.
3. Madhushaw RJ, Lo C-Y, Hwang C-W, et al. Rutheniun-catalyzed cycloisomerization of o-(ethynyl)phenylalkenes to diene derivatives via skeletal rearrangement. *J Am Chem Soc.* 2004;126(47):15560–15565. https://doi.org/10.1021/ja045516n.
4. Fuerstner A, Hannen P. Platinum- and gold-catalyzed rearrangement reactions of propargyl acetates: total syntheses of (−)-α-cubebene, (−)-cubebol, sesquicarene and related terpenes. *Chem A Eur J.* 2006;12(11):3006–3019. https://doi.org/10.1002/chem.200501299.
5. Zhao F, Wang C, Liu L, Zhang W-X, Xi Z. Skeletal rearrangement of all-carbon spiro skeletons mediated by a Lewis acid. *Chem Commun.* 2009;2009:6569–6571. https://doi.org/10.1039/b914619a.
6. Valeev RF, Selezneva NK, Starikova ZA, Pankrat'ev EY, Miftakhov MS. Chiral building blocks from *R*-(-)-carvone: *N*-bromosuccinimide-mediated addition-skeletal rearrangement of (-)-*cis*-carveol. *Mendeleev Commun.* 2010;20:77–79. https://doi.org/10.1016/j.mencom.2010.03.004.
7. (a)Meinwald J. The acid-catalyzed rearrangement of cinenic acid. *J Am Chem Soc.* 1955;77(6):1617–1620. https://doi.org/10.1021/ja01611a063; (b)Meinwald J, Cornwall CC. The acid-catalyzed rearrangement of cinenic acid. II. Geronic acid from 6-hydroxy-2,2,6-trimethylcyclohexanone. *J Am Chem Soc.* 1955;77(22):5991–5992. https://doi.org/10.1021/ja01627a058; (c)Meinwald J, Hwang HC. The acid-catalyzed rearrangement of cinenic acid. III. Structure and synthesis of the lactonic product. *J Am Chem Soc.* 1957;79(11):2910–2912. https://doi.org/10.1021/ja01568a059; (d)Meinwald J, Ouderkirk JT. The acid-catalyzed rearrangement of cinenic acid. IV. Synthesis and rearrangement of 6-carboxy-6-ethyl-2,2-dimethyltetrahydropyran. *J Am Chem Soc.* 1960;82(2):480–483. https://doi.org/10.1021/ja01487a059; (e)Meinwald J, Hwang HC, Christman D, Wolf AP. The acid-catalyzed rearrangement of cinenic acid. V. Evidence for a decarbonylation-recarbonylation mechanism. *J Am Chem Soc.* 1960;82(2):483–486. https://doi.org/10.1021/ja01487a060.

Question 270: A New Synthesis of 1- and 2-Chloronaphthalenes by an Annulation Process

For a number of years Japanese chemists have worked to elucidate the different reactivity patterns of dihalocyclopropylmethanols as precursors to naphthalene. One such difference arises from the synthesis of 1-chloronaphthalenes **2** by reaction of **1** with either one equivalent of Lewis acid in dichloromethane (DCM) or by dissolving **1** in trifluoroacetic acid (TFA) as compared to the synthesis of 2-chloronaphthalenes **4** by reacting **3** in TFA solvent (Scheme 1).[1,2] Show mechanisms for the formation of **2** and **4**.

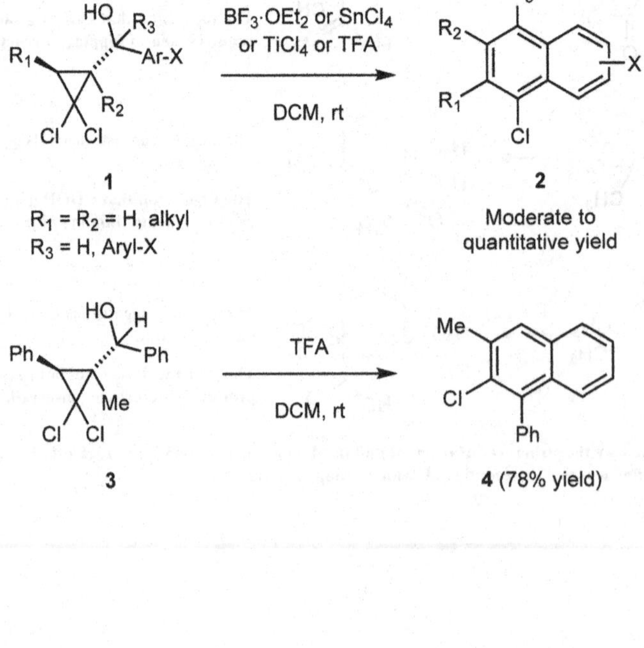

SCHEME 1

SOLUTION: 1 → 2

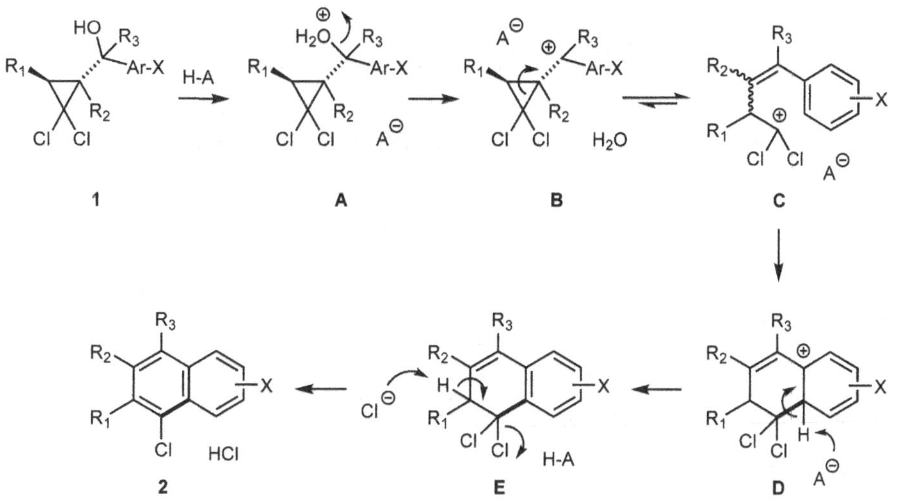

SOLUTION: 3 → 4

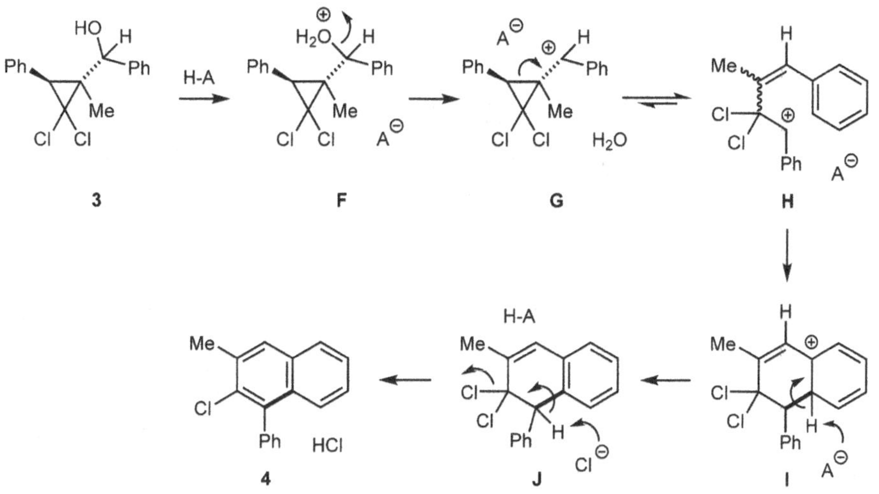

KEY STEPS EXPLAINED

The authors note that the difference in reactivity between the **1** to **2** transformation and the **3** to **4** transformation has to do with differential cleavage of the cyclopropane ring to relieve ring strain and form a stable cation which in this case can take two possible routes with breakage of either the Cl_2C—CR_2 bond in intermediate **B** or the R_1C—CR_2 bond in intermediate **G**. The overall mechanism for this benzannulation proceeds with protonation of the hydroxy group, E_1 elimination of water molecule (from **A** and **F**) to form a benzyl cation (**B** and **G**) which undergoes skeletal rearrangement into a homoallyl cation intermediate (**C** and **H**) through regioselective bond cleavage of a specific bond of the cyclopropane ring as illustrated in the solutions. An intramolecular Friedel-Crafts reaction with the aryl group temporarily disrupts the aromaticity of this group but restores it with a deprotonation in the next step and finally the elimination of HCl affords the final halogenonaphthalene products **2** and **4**. The authors note that while the *E* alkene forms of **C** and **H** are likely to be formed predominantly, they are expected to isomerize quickly to the *Z* forms via equilibration from **B** and **G**. Lastly, since cation **H** is more stable than **C** (due to electronic delocalization availability from the phenyl group), in an overall system, product **4** is expected to predominate over product **2**. This is why the authors regard the regioselective cleavage of cyclopropane rings as a highly tunable transformation which is of great value for synthesis strategy. For future study we recommend carrying out a kinetic and theoretical analysis to study the effects of different substituents attached to the aryl group as well as variations with respect to the three R groups present in **1**. Hammett analyses would be particularly useful in shedding light on the effect of electron-withdrawing and electron-donating aryl substituents on the transformation **1 → 2**.

ADDITIONAL RESOURCES

We direct the reader to further mechanistic work on these interesting transformations by the same authors.[3,4] Lastly, we also highlight the use of this transformation in achieving the synthesis of chiral pesticides which was also carried out by Tanabe's group.[5]

References

1. Tanabe Y, Seko S, Nishii Y, Yoshida T, Utsumi N, Suzukamo G. Novel method for the synthesis of α- and β-halogenonaphthalenes by regioselective benzannulation of aryl(*gem*-dihalogenocyclopropyl)methanols: application to the total synthesis of the lignan lactones, justicidin E and taiwanin C[1]. *J Chem Soc Perkin Trans 1*. 1996;1996:2157–2165. https://doi.org/10.1039/P19960002157.
2. Seko S, Tanabe Y, Suzukamo G. A novel method for the synthesis of α- and β-halonaphthalenes via regioselective ring cleavage of aryl(*gem*-dihalogenocyclopropyl)methanols and its application to total synthesis of lignan lactones, justicidin E and taiwanin C. *Tetrahedron Lett*. 1990;31(47):6883–6886. https://doi.org/10.1016/S0040-4039(00)97197-1.
3. Nishii Y, Yoshida T, Asano H, et al. Regiocontrolled benzannulation of diaryl(*gem*-dichlorocyclopropyl)methanols for the synthesis of unsymmetrically substituted α-arylnaphthalenes: application to total synthesis of natural lignan lactones. *J Org Chem*. 2005;70:2667–2678. https://doi.org/10.1021/jo047751u.
4. Nishii Y, Tanabe Y. Sequential and regioselective Friedel–Crafts reactions of *gem*-dihalogenocyclopropanecarbonyl chlorides with benzenes for the synthesis of 4-aryl-1-naphthol derivatives. *J Chem Soc Perkin Trans 1*. 1997;1997:477–486. https://doi.org/10.1039/A605030A.
5. Yasukochi H, Atago T, Tanaka A, et al. Practical, general, and systematic method for optical resolution of *gem*-dihalo- and monohalocyclopropanecarboxylic acids utilizing chiral 1,1-binaphtholmonomethyl ethers: application to the synthesis of three chiral pesticides. *Org Biomol Chem*. 2008;6:540–547. https://doi.org/10.1039/b714614k.

Question 271: From the Diterpene Carnosol to the Benzodiazepine Agonist Miltirone

During a 1996 synthesis of miltirone **4**, Spanish chemists synthesized intermediate **A** through methylation of the abundant diterpene carnosol **1** (Scheme 1).[1] This intermediate was further reacted with boron tribromide to form **2a** and **2b**. The ester **2b** was then treated with KO^tBu base in DMSO in order to form the acid **3** which upon treatment with boron tribromide followed by air oxidation formed miltirone **4**. Show mechanisms for the **1** to **A** transformation and elucidate the structure of intermediate **A**. Show mechanisms for the **A** to **2a** and **2b** transformations and devise a mechanism for the **2b** to **3** to **4** transformation.

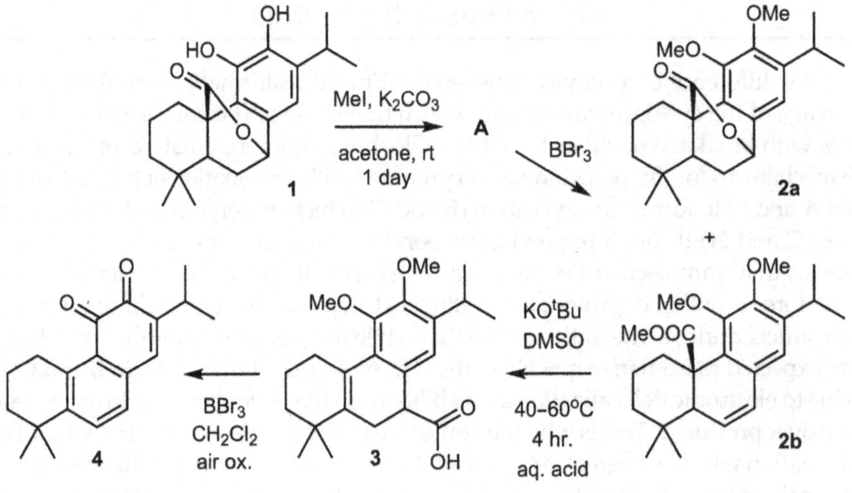

SCHEME 1

SOLUTION: 1 → A

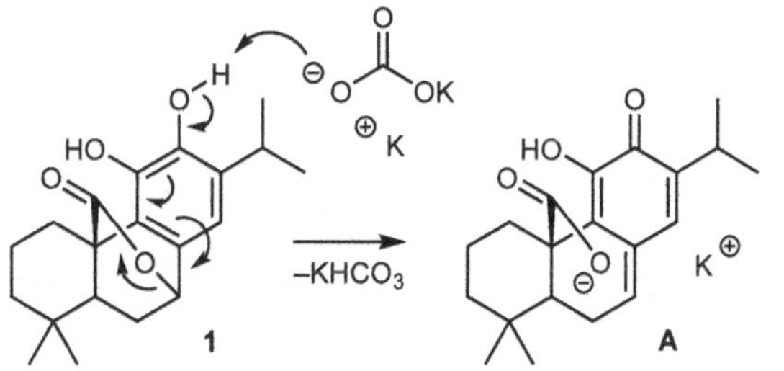

SOLUTION: A → 2A

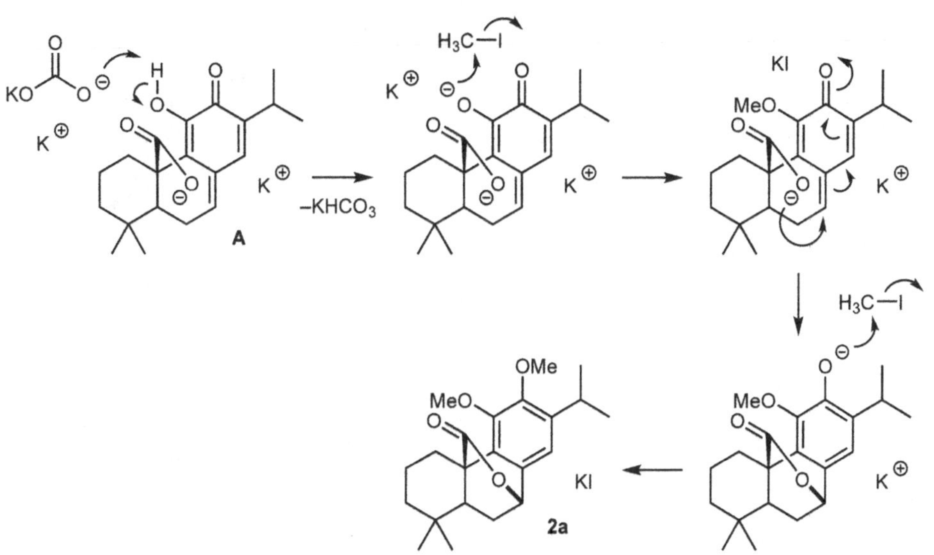

SOLUTION: A → 2B

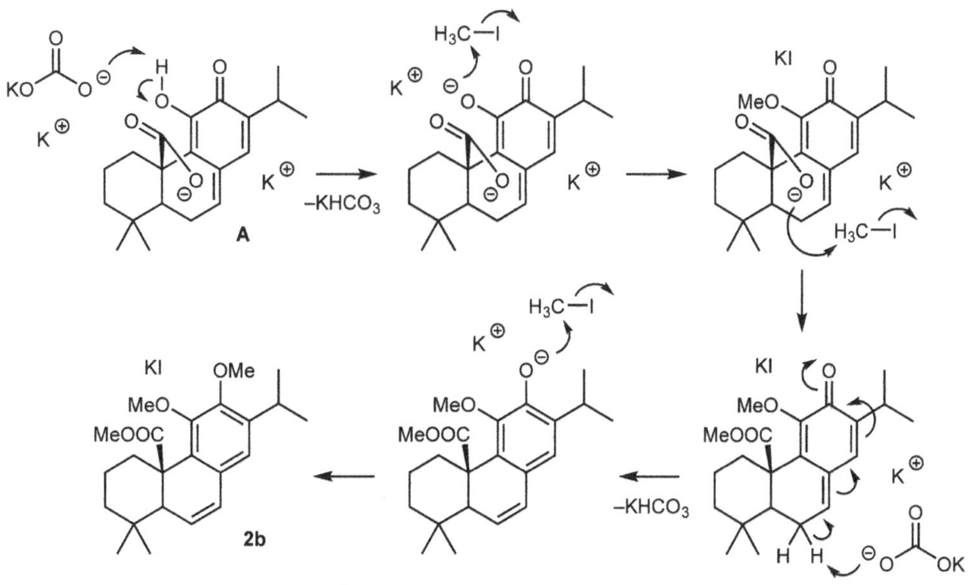

BALANCED CHEMICAL EQUATION: 1 → 2A

SOLUTION: 2B → 3

SOLUTION: 3 → 4

BALANCED CHEMICAL EQUATION: 1 → 2B

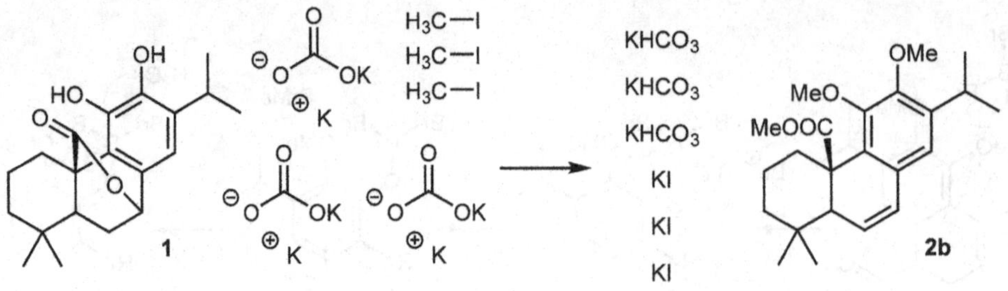

BALANCED CHEMICAL EQUATION: 2B → 3

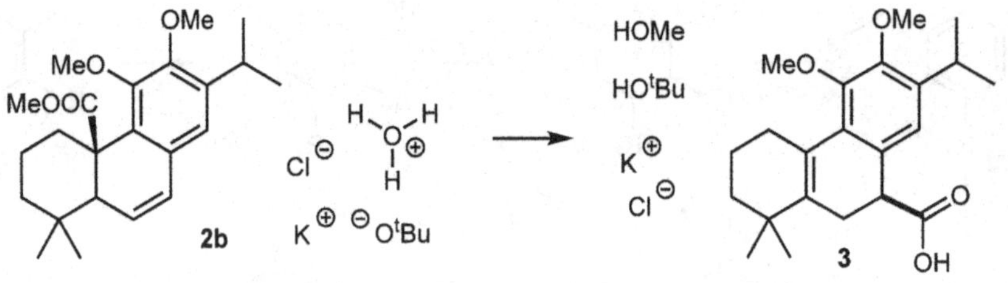

BALANCED CHEMICAL EQUATION: 3 → 4

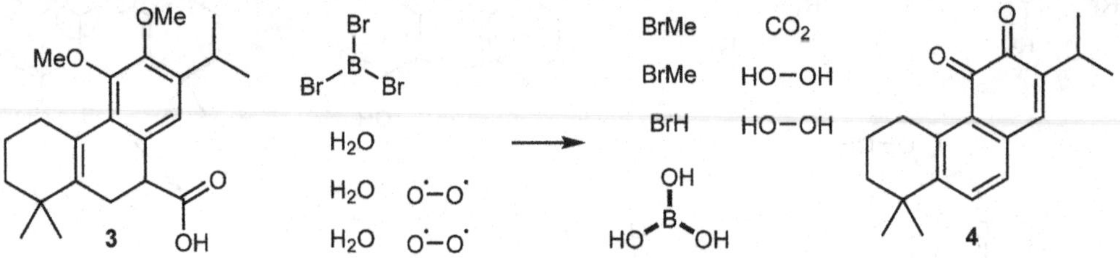

KEY STEPS EXPLAINED

The original work of Luis et al. describes the transformation of the diterpene carnosol **1** into several products according to the synthesis shown in Scheme 1.[1] At the very start, **1** is reacted with potassium carbonate and methyl iodide in acetone at room temperature for 1 day. The reaction produces products **2a** and **2b**. The authors describe the role of potassium carbonate in deprotonating **1** to form the common intermediate **A** (whose structure in the paper is drawn incorrectly and so we refer the reader to the present solution for the correct structure).

En route to **2a**, intermediate **A** is further deprotonated by another equivalent of potassium carbonate base to form a phenoxide which then extracts a methyl group from methyl iodide in solution to give a methoxy group. The quinoid ring then rearomatizes to a benzoid ring by transannular reattachment of the carboxylate group to re-form a bicyclic lactone and generate a second phenoxide which is methylated as before to produce the final product **2a**. This entire sequence utilizes two equivalents each of potassium carbonate and methyl iodide.

En route to **2b**, once again, intermediate **A** undergoes first a deprotonation to form a phenoxide group followed by a methylation to form a methoxy group. Nevertheless in this case, the carboxylate group is methylated to an ester group instead of reattaching itself to the ring. A further deprotonation in the lower proton of the central ring helps to rearomatize the quinoid ring to a benzoid ring forming a new phenoxide group in the process which is then methylated to produce **2b**. We note the difference in stoichiometric amounts of base and methyl iodide (3 and 3) for the **1** to **2b** transformation as compared to the **1** to **2a** transformation (2 and 2).

From **2b** to **3** we have first deprotonation with potassium *t*-butoxide base in order to form a negative charge which can delocalize into the neighboring π bond to set up a transannular carbocyclization to a quaternary intermediate which breaks apart the bicyclic structure to an ester which is racemized under the strongly basic conditions and with the addition of quenching aqueous acid, after which water addition causes ester hydrolysis which helps form the final carboxylic acid product **3**.

Finally, from **3** to **4** we have first demethylation of methoxy groups of **3** to hydroxy groups using boron tribromide with the expulsion of two equivalents of methyl bromide and one equivalent of hydrogen bromide as well as the by-product boric acid. The dihydroxy structure **B** then undergoes air oxidation (O_2 diradical process) to a diketone which enables an acid-catalyzed decarboxylation and subsequent deprotonation to reform the dihydroxy structure after loss of carbon dioxide gas and a final air oxidation to the miltirone **4**. It is important to note that when the authors used weaker bases than potassium *t*-butoxide they were unsuccessful in transforming **2b** to **3**, which highlights the relative stability of the structures that have to undergo deprotonation at key steps along this transformation. The authors also showed the importance of air oxidation by running an experiment devoid of any oxygen which did not produce any trace of **4**. Nevertheless, the importance of stoichiometry and by-product identification, even of the relatively trivial production of carbon dioxide gas, never mind the less trivial differentiation between the **A** to **2a** and **A** to **2b** transformations are not mentioned by the authors. Also absent is a theoretical analysis showing a reaction coordinate diagram of these processes. With the addition of errors present in some of the structures and schemes of the original paper, we recommend a new study of the mechanistic aspects of this interesting transformation in order to help confirm the mechanisms postulated herein.

ADDITIONAL RESOURCES

In an earlier paper, Luis et al. had proposed the possible participation of a perepoxide intermediate in what seems to be an analogous transformation to that of **2b** to **3**, even though in that research **3** is not the target product.[2] We include this reference merely for an interesting perspective on the difference between considering the mechanistic implications of air oxidation via the oxygen gas diradical triplet species shown above versus the $^-$O—O—H anion proposed by the authors in that paper *vis-à-vis* synthesis strategy. For a review of diterpenoids, we refer the reader to Ref. 3.

References

1. Luis JG, Andres LS, Fletcher WQ, Lahlou EH, Perales A. Rearrangement of methyl 11,12-di-*O*-methyl-6,7-didehydrocarnosate in basic medium. Easy hemisynthesis of miltirone. *J Chem Soc Perkin Trans 1*. 1996;1996:2207–2211. https://doi.org/10.1039/P19960002207.
2. Luis JG, Andres LS, Fletcher WQ. Chemical evidence for the participation of a perepoxide intermediate in the reaction of singlet oxygen with mono-olefins in relationship with the biogenetic pathway to highly oxidized abietane diterpenes. *Tetrahedron Lett*. 1994;35(1):179–182. https://doi.org/10.1016/0040-4039(94)88195-2.
3. Hanson JR. Diterpenoids. *Nat Prod Rep*. 1996;13:59–71. https://doi.org/10.1039/NP9961300059.

Question 272: Synthesis of Chiral Phthalimidine Derivatives

The synthesis of *N*-substituted isoindolin-1-ones can be carried out by condensation of *o*-phthalaldehyde either with primary aliphatic amines such as α-methylbenzylamine to produce **1** or with α-amino acids such as L-valine to produce **2** (Scheme 1).[1] Show mechanisms for both transformations and explain why using α-amino acids rather than primary aliphatic amines leads to a much better product yield.

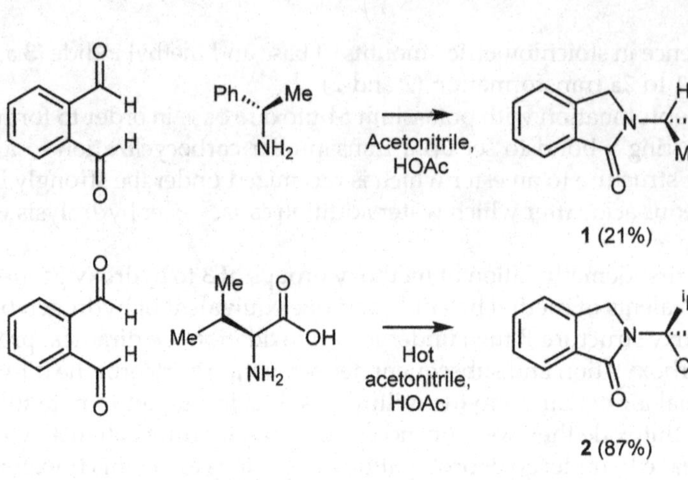

SCHEME 1

SOLUTION: SYNTHESIS OF 1

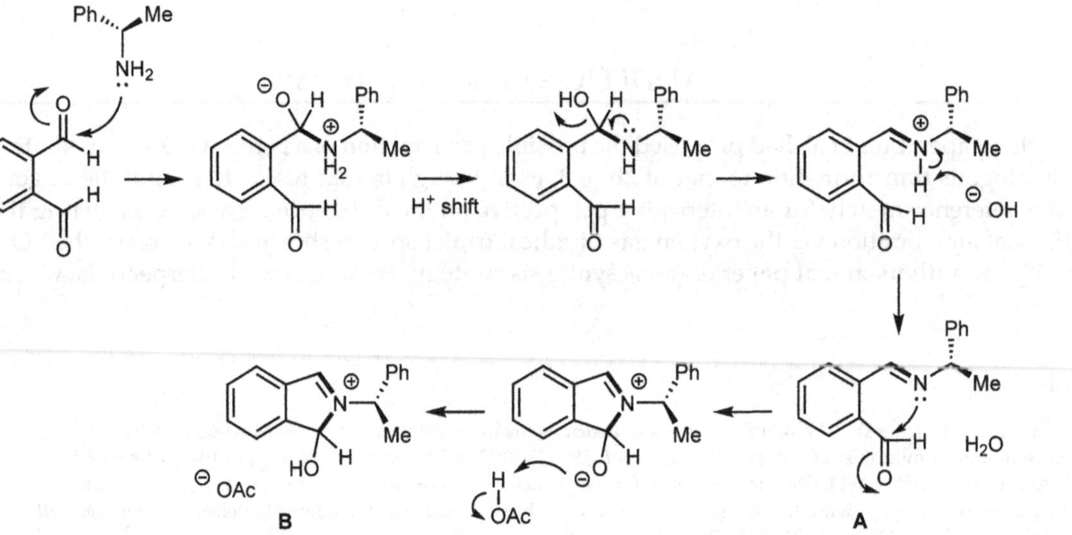

SOLUTION: SYNTHESIS OF 1: PATHWAY A

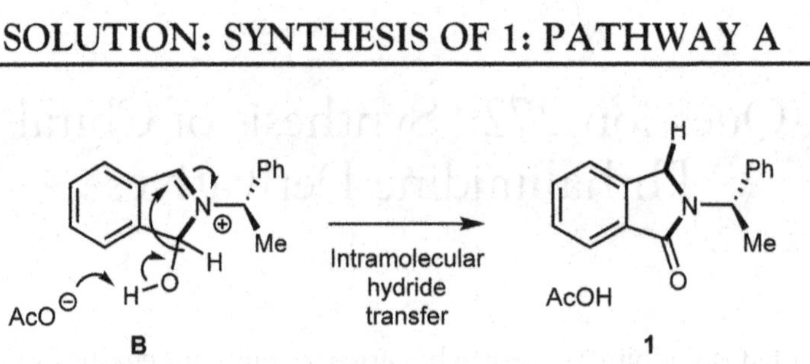

SOLUTION: SYNTHESIS OF 1: PATHWAY B

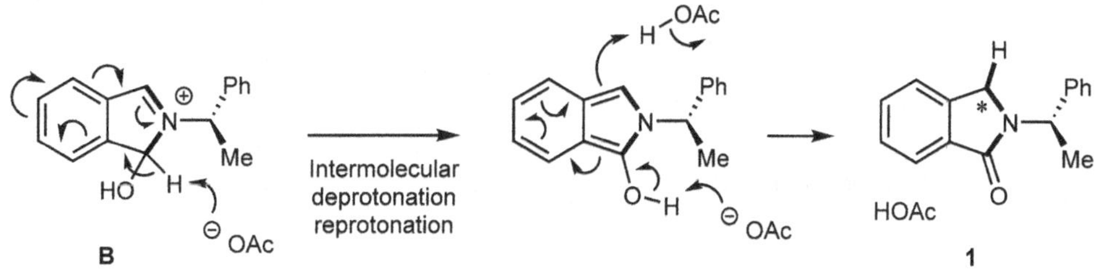

B → Intermolecular deprotonation reprotonation → 1

SOLUTION: SYNTHESIS OF 2

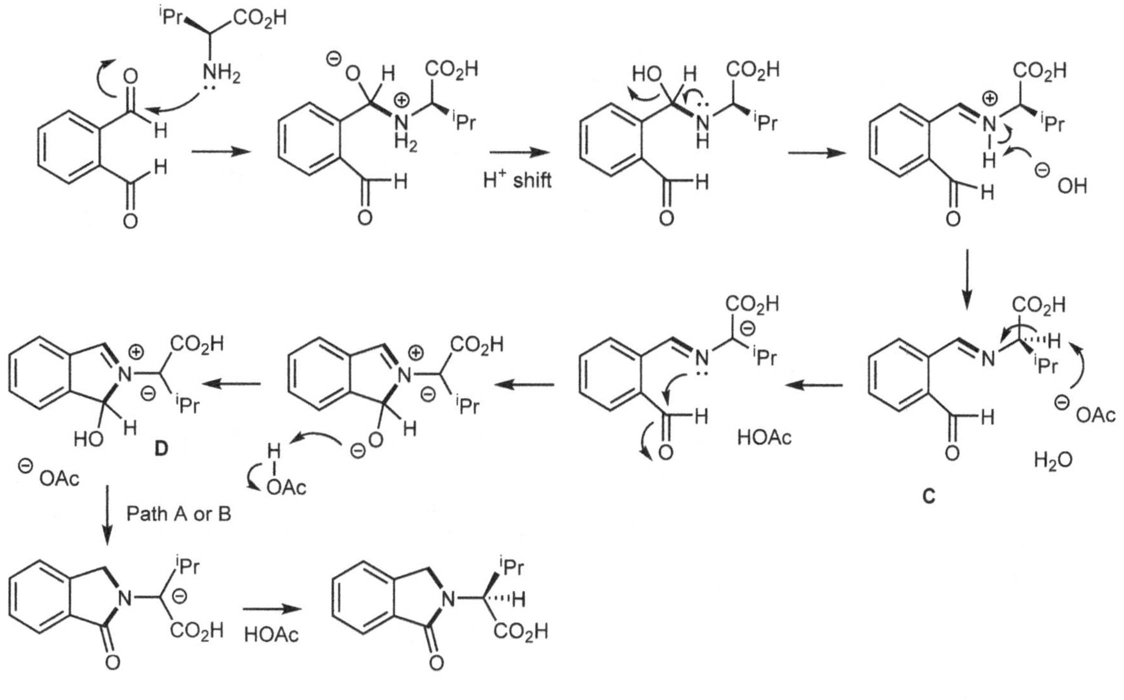

H⁺ shift

C

D

Path A or B

2

BALANCED CHEMICAL EQUATION FOR THE SYNTHESIS OF 1

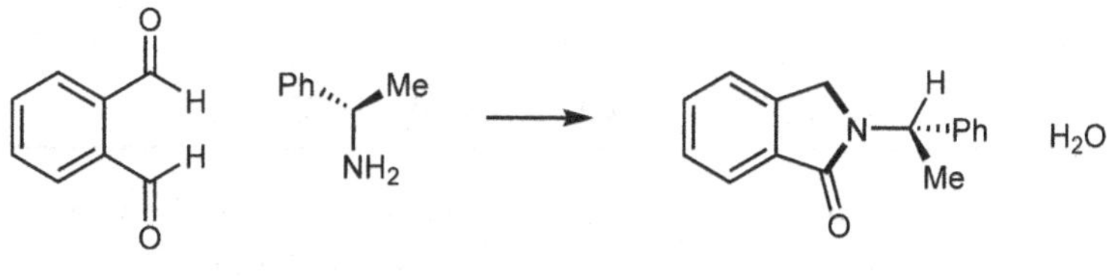

1 (21%)

KEY STEPS EXPLAINED

This transformation starts with imination of an aldehyde group of *o*-phthalaldehyde by amine with loss of water followed by cyclization of the resulting imine to a hydroxyiminium intermediate with structure **B** (or **D** depending on the choice of amine). From **B** (or **D**) the mechanism may follow either pathway A, that is, an intramolecular [1,3]-hydride shift, or pathway B, that is, an intermolecular deprotonation-reprotonation sequence via an isoindolinol intermediate. To differentiate these two paths, the authors used perdeuterated acetic acid as the quenching reagent and noticed deuterium incorporation at C* indicating that pathway B is operational (in pathway A deuterium incorporation at C* is not expected). This experiment directly supports pathway B, nevertheless we recommend an alternative experiment (Scheme 2).

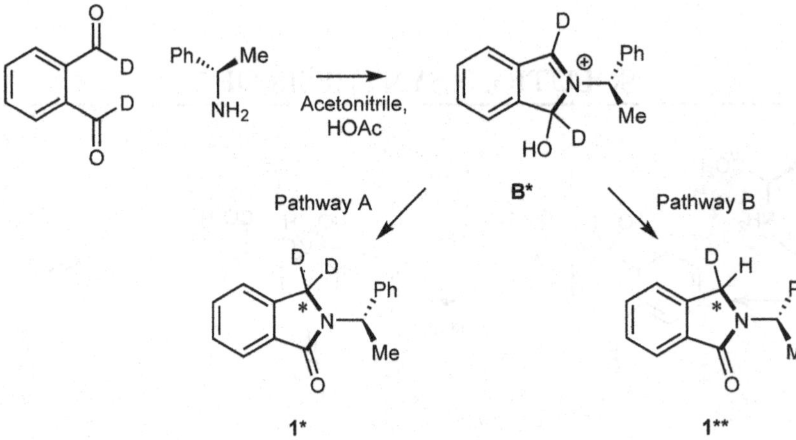

SCHEME 2

Here, instead of using labeled solvent, we begin with doubly labeled *o*-phthalaldehyde which enables the production of two distinctly labeled products depending on which mechanistic pathway the transformation follows. This alternative labeling experiment also eliminates the question of H/D exchange/loss to the solvent. Lastly, we believe that a theoretical analysis would shed important light on the energy difference between pathways A and B and therefore should be considered for a future study.

Next, although the authors did not explain the stark difference in yields between the use of primary aliphatic amines and α-amino acids, in the solution for the synthesis of **2** we show the possible formation of an ylide **D** which has thermodynamic stability over structure **B** (seen before) and which can further be stabilized electronically and inductively by the neighboring carboxylic acid group. An interesting experiment which would further support the proposed mechanism would be to use an amine similar to L-valine but instead of the aliphatic isopropyl group, have a second carboxylic acid group (or a strong electron withdrawing group) in order to further stabilize the ylide **D**, which might perhaps also allow for its isolation, which would be the ultimate proof of the proposed mechanism. We should note that we do not have any evidence that would indicate ylide **D** has to be present along the mechanistic pathway for the synthesis of **2**. We merely suggest this as a possible explanation for the stark difference in yields.

ADDITIONAL RESOURCES

We direct the reader to the work of Augner et al. who also present a similar mechanism for the same transformation (i.e., *o*-phthalaldehyde condensation with amine) but following pathway B and avoiding the imine formation before cyclization sequence.[2] Instead they show cyclization followed by imine formation with loss of water (Scheme 3).

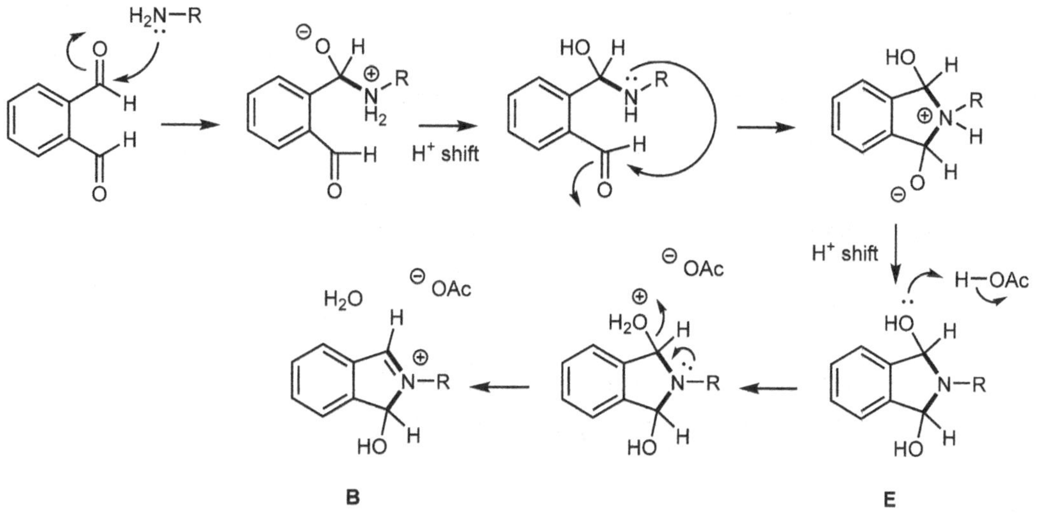

SCHEME 3

This mechanistic explanation appears in other articles as well.[3–5] We note that it is possible for compound **E** to be an intermediate in the mechanism instead of compound **A**, for which there is also literature precedent,[6] because it is impossible to determine the order of mechanistic steps by experimental means. Nevertheless, a theoretical analysis could shed light on the likelihood of each possibility. Such a theoretical study has been conducted by Alajarin et al. in order to investigate the possibility of a still fourth possible mechanistic pathway for this transformation (Scheme 4).

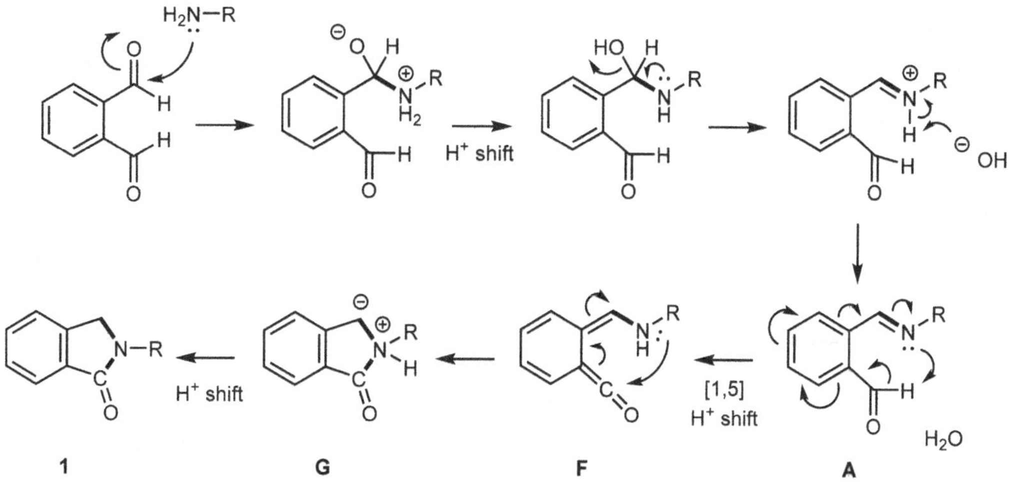

SCHEME 4

It is interesting to note that the labeling experiment we proposed in Scheme 2 would lead exclusively to product **1***
for the mechanism shown in Scheme 4 whereas for that in Scheme 3, due to the nature of the exchangeability of either aldehyde hydrogen in **E** to help form **B**, it is possible to end up with some percentage of product containing no deuterium, which would differentiate this mechanism from the ones proposed in the solutions as well as from that in Scheme 4. A further study looking at the results of labeling experiments with the support of theoretical analysis would add much needed depth to the mechanistic investigation of this interesting transformation.

References

1. Grigg R, Gunaratne HQN, Sridharan V. A simple one-step synthesis of *N*-substituted isoindolin-1-ones. Diastereofacially selective protonation of an intermediate isoindolinol. *J Chem Soc Chem Commun.* 1985;1985:1183–1184. https://doi.org/10.1039/C39850001183.
2. Augner D, Gerbino DC, Slavov N, Neudoerfl J-M, Schmalz H-G. *N*-Capping of primary amines with 2-acylbenzaldehydes to give isoindolinones. *Org Lett.* 2011;13(19):5374–5377. https://doi.org/10.1021/ol202271k.

3. Alajarin M, Sanchez-Andrada P, Lopez-Leonardo C, Alvarez A. On the mechanism of phthalimidine formation via o-phthalaldehyde monoimines. New [1,5]-H sigmatropic rearrangements in molecules with the 5-aza-2,4-pentadienal skeleton. *J Org Chem.* 2005;70(19):7617–7623. https://doi.org/10.1021/jo0508494.
4. Poli G, Baffoni SC, Giambastiani G, Reginato G. A new asymmetric approach toward 5-substituted pyrrolidin-2-one derivatives. *Tetrahedron.* 1998;54:10403–10418. https://doi.org/10.1016/S0040-4020(98)00494-3.
5. Wan J, Wu B, Pan Y. Novel one-step synthesis of 2-carbonyl/thiocarbonyl isoindolinones and mechanistic disclosure on the rearrangement reaction of o-phthalaldehyde with amide/thioamide analogs. *Tetrahedron.* 2007;63:9338–9344. https://doi.org/10.1016/j.tet.2007.07.009.
6. Aubert T, Farnier M, Guilard R. Nouvelle voie de synthèse d'isoindolones et d'isoquinoléines par condensation d'iminophosphoranes avec l'ortho-phtalaldéhyde: réactions, mécanismes et étude structurale. *Can J Chem.* 1990;68(6):842–851. https://doi.org/10.1139/v90-133.

Question 273: Remarkable Rearrangement of a Camphor Derivative

In working toward the synthesis of enantiopure cyclopentane derivatives for use in terpene and steroid synthesis, Canadian chemists Ferguson et al. have uncovered a remarkable rearrangement of *endo*-3-bromo-4-methylcamphor **3** to the tribromo derivative **4** (Scheme 1).[1] Devise a mechanism for this transformation which involves a combination of four Wagner-Meerwein rearrangements.

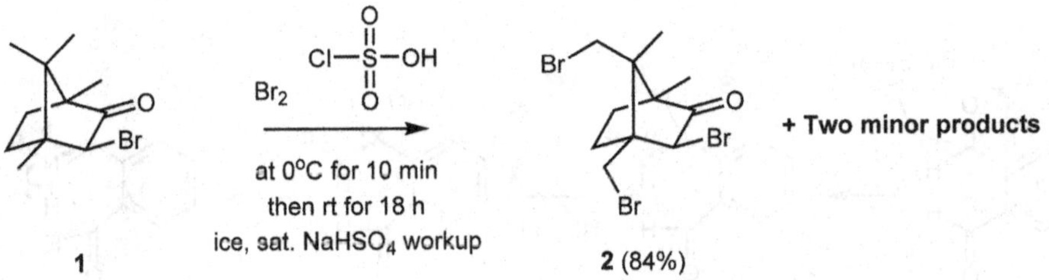

SCHEME 1

SOLUTION

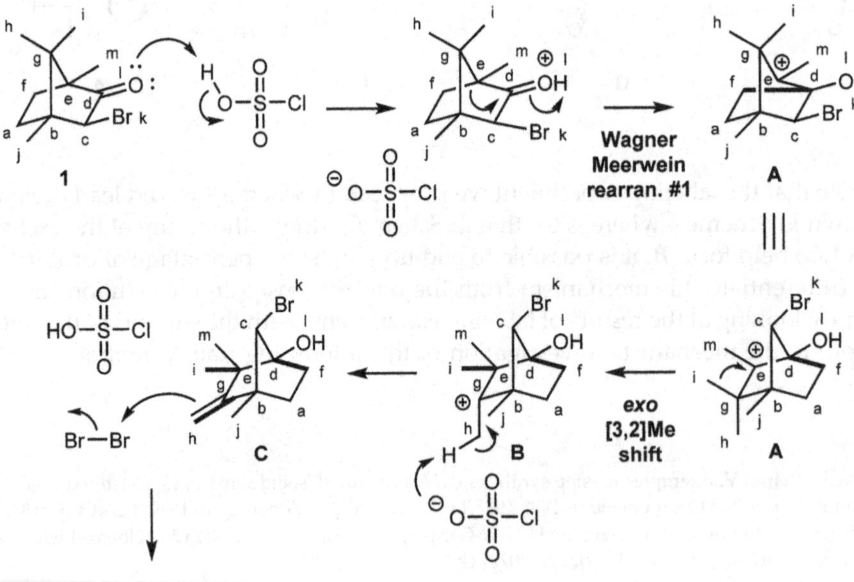

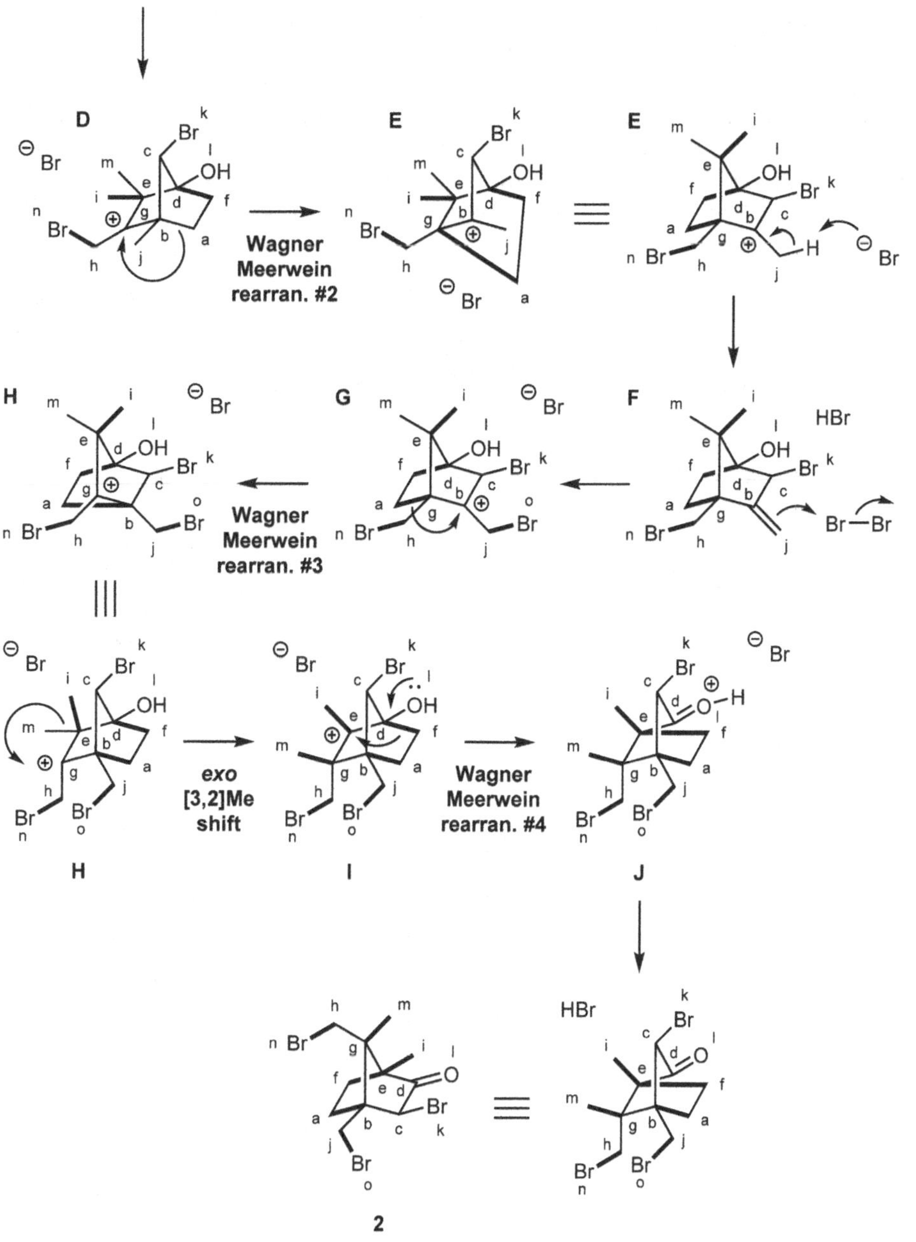

BALANCED CHEMICAL EQUATION

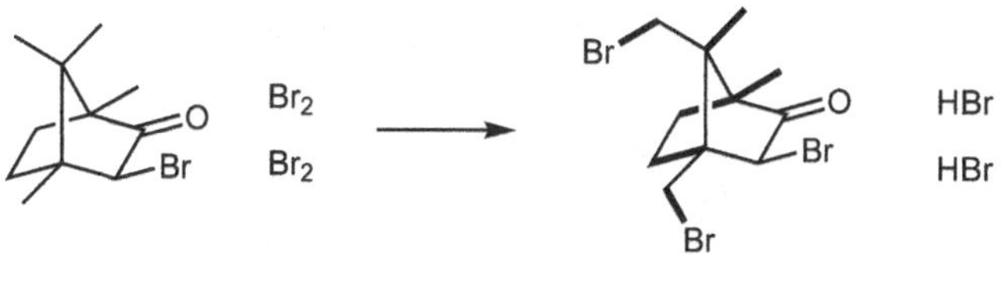

1   **2** (84%)

KEY STEPS EXPLAINED

To explain this remarkable rearrangement, Money et al. devised a series of deuterium labeling experiments in order to arrive at the mechanism illustrated here.[1,2] Essentially, the mechanism begins with protonation of the carbonyl group of **1** followed by a first Wagner-Meerwein rearrangement to form carbocation **A** followed by an *exo*-3,2-methyl shift to form carbocation **B** followed by a deprotonation to give the exocyclic olefin **C**. This newly formed alkene group can then be brominated using Br_2 to form carbocation **D** which undergoes a second Wagner-Meerwein rearrangement to form carbocation **E** which after elimination of HBr forms the exocyclic olefin **F** which is again brominated by Br_2 to form carbocation **G** which undergoes a third Wagner-Meerwein rearrangement to form carbocation **H**. This structure once again undergoes an *exo*-3,2-methyl shift to form carbocation **I** which is followed by a fourth Wagner-Meerwein rearrangement and deprotonation of alcohol to form the final tribrominated ketone product **2**. In addition, two minor products—**3** and **4** were also isolated and we present mechanisms for these in Schemes 2 and 3, respectively.

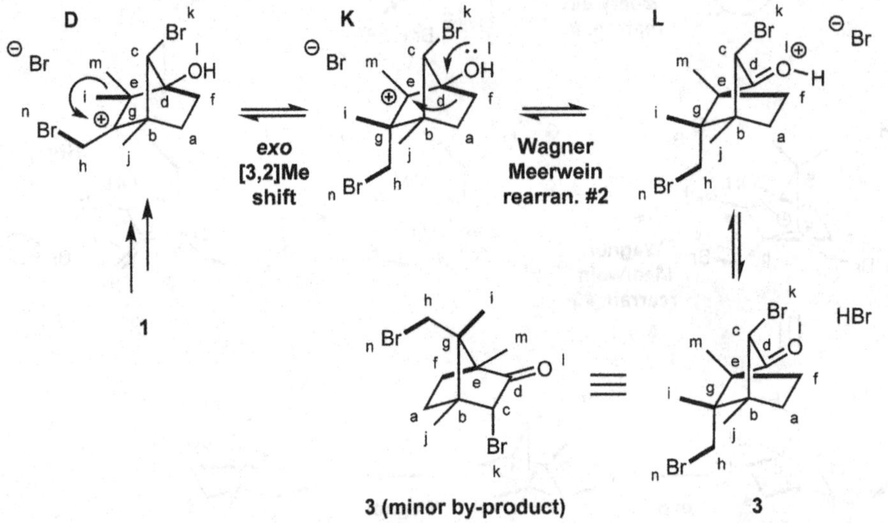

SCHEME 2

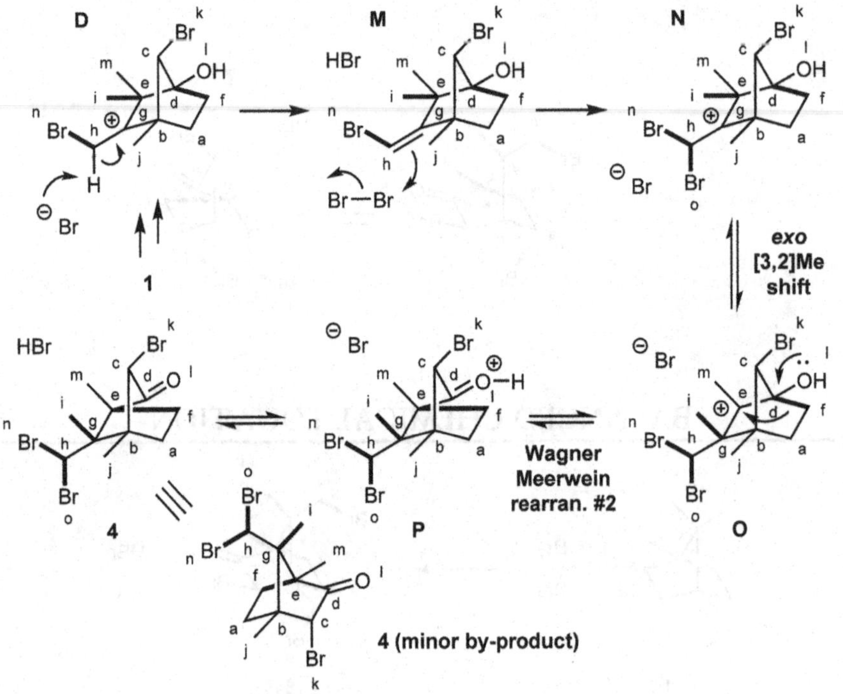

SCHEME 3

We also note the interesting carbon skeleton rearrangement that has occurred from **1** to **2** (namely, abcdef/hgi/em/bj for **1** transformed into abcdef/hgm/ei/bj with bromine attachments at C_j and C_h). The two by-products did not exhibit any carbon skeleton rearrangement. The authors also observed that treatment of pure **3** with Br_2 leads to **2** indicating that the production of this by-product is a reversible process involving equilibrium mechanistic steps. It is also noteworthy that both minor by-products are formed from the common carbocation intermediate **D**.

Furthermore, one noteworthy observation obtained from the extensive deuterium labeling experiments performed by the authors includes the observed possible loss of deuterium label from C_h and C_j (which is consistent with the two *exo*-methylene structures **C** and **F**). Essentially the authors began with doubly deuterated compound **1-d$_2$** and obtained compound **2-d$_2$** under the same reaction conditions as going from compound **1** to compound **2** (Scheme 4). We note that the methyl groups labeled "m" and "i" essentially trade places. We also include Scheme 5 as a useful summary of the synthetic isotopic labeling experiments the authors carried out. We once again note the interchange between methyl groups "m" and "i" as well as the fact that the boxed structures represent the same compound (even though the methyl groups have traded places).

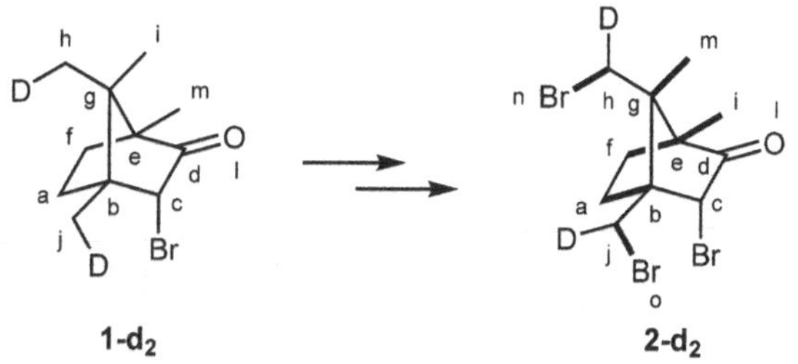

SCHEME 4

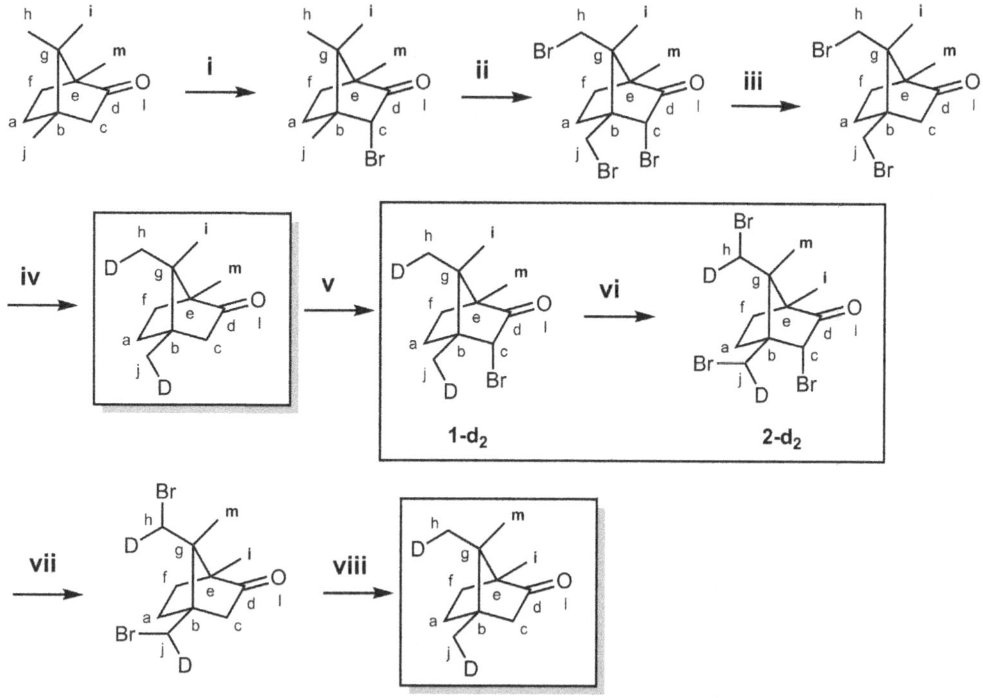

(i) Br$_2$/HOAc (96%)
(ii) Br$_2$/ClSO$_3$H (84%)
(iii) Zn/HOAc/Et$_2$O (81%)
(iv) Bu$_3$SnD/AIBN/benzene (58%)
(v) Br$_2$/HOAc (96%)
(vi) Br$_2$/ClSO$_3$H (84%)
(vii) Zn/HOAc/Et$_2$O (81%)
(viii) Bu$_3$SnH/AIBN/benzene (64%)

SCHEME 5

For an extensive discussion of this interesting camphor transformation, we highly recommend a review article by Thomas Money.[3]

References

1. Ferguson CG, Money T, Pontillo J, Whitelaw PDM, Wong MKC. A remarkable multiple rearrangement process in the bromination of *endo*-3-bromo-4-methylcamphor: intermediates for triterpenoid synthesis. *Tetrahedron*. 1996;52(47):14661–14672. https://doi.org/10.1016/0040-4020(96)00944-1.
2. Clase AJ, Li DLF, Lo L, Money T. A simple synthetic route to 4-methylcamphor: mechanistic aspects. *Can J Chem*. 1990;68(10):1829–1836. https://doi.org/10.1139/v90-285.
3. Money T. Camphor: a chiral starting material in natural product synthesis. *Nat Prod Rep*. 1985;2:253–289. https://doi.org/10.1039/NP9850200253.

Question 274: A Failed Attempt to Prepare Benzylidenethiophthalide by the "Obvious" Method

In 1996, Belgian scientists attempted the synthesis of benzylidenethiophthalide **1** by condensing 2-thiophthalide **2** with benzaldehyde in THF at 0°C using potassium *t*-butoxide as base and with the work-up conditions: addition of water, HCl acidification, and chloroform extraction. Instead of the expected **1**, the authors isolated the stilbene acid **3** in 52% yield (Scheme 1).[1] Propose a mechanism for the **2** to **3** transformation.

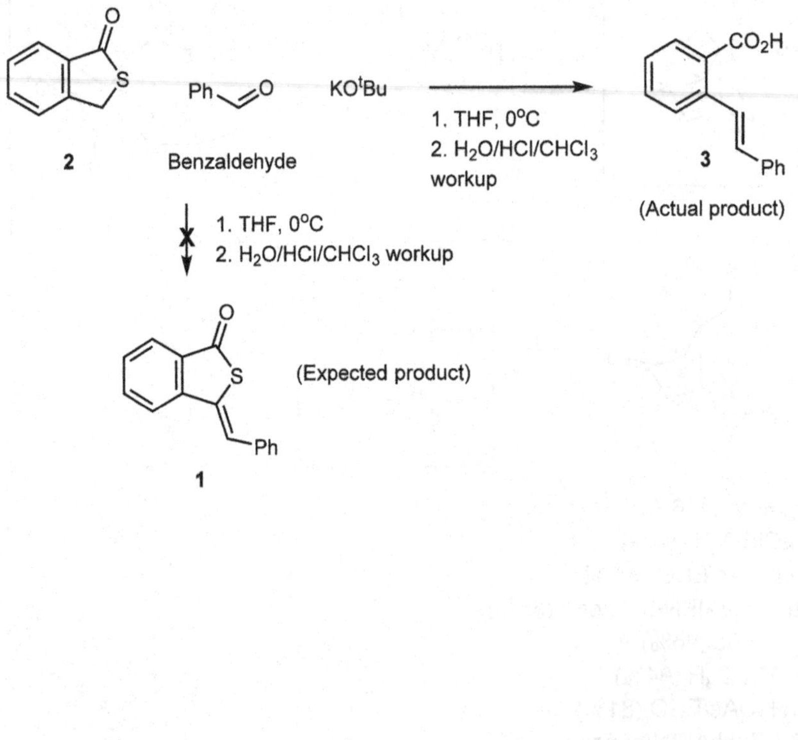

SCHEME 1

SOLUTION

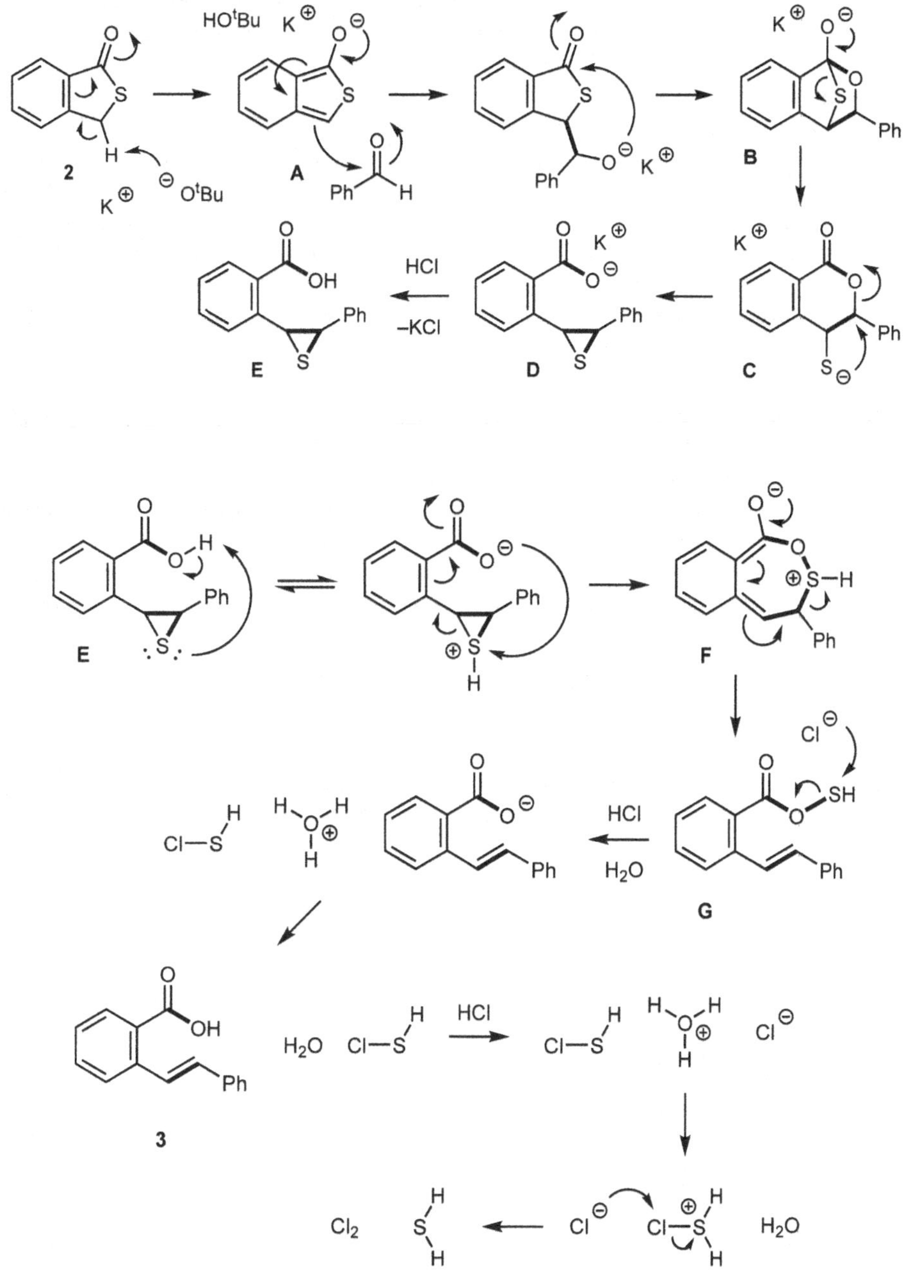

BALANCED CHEMICAL EQUATION

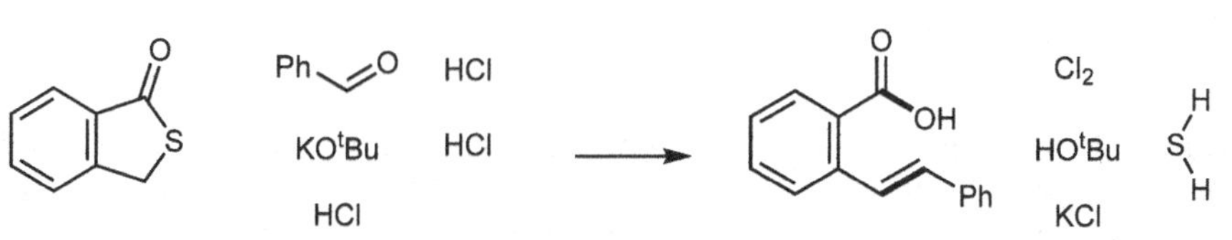

KEY STEPS EXPLAINED

The proposed mechanism begins with deprotonation of **2** to intermediate **A** followed by acylation with benzaldehyde and a cyclization to [2.2.1] bicyclic intermediate **B**. This structure then undergoes ring opening to the thiolate lactone **C** followed by a tandem cyclization and ring opening to thiirane carboxylate and an acidic quench to intermediate **E**. From this point forward, the authors do not show in their paper the mechanistic pathway of how sulfur is eliminated from the thiirane, indicating that elemental sulfur is likely eliminated to get from **E** to **3** in one step. This process is expected to occur, as the authors claim, during a mild acidic work-up process. Based on literature references involving similar transformations with loss of elemental sulfur,[2–5] we are skeptical of the claim that elemental sulfur can be eliminated during a mild acidic work-up. Indeed, evidence shows that the extrusion of elemental sulfur is an energetically demanding process which requires significant thermal energy (at least 100°C) which is clearly not available during the work-up procedure. We therefore propose an intramolecular route for the elimination of sulfur from **E** shown in part 2 of the solution which might be more likely under the mild acidic work-up conditions. This mechanism begins with an internal proton transfer from the carboxylic acid to the thiirane followed by ring opening of thiiranium ion to form a seven-membered ring intermediate **F** which opens to restore the aromaticity of the benzene ring to form **G**. Here, the chloride anion from the mildly acidic work-up environment can attack the —SH group bound to the carboxyl group to produce HSCl and **3** (after protonation). The HSCl by-product can further decompose to H_2S and chlorine gases as by-products of this mechanism. Unfortunately the experimental description given by the authors did not state any observations such as gas evolution, precipitation, color changes, or detection of bad smells emanating from the reaction vessel which would be expected if hydrogen sulfide and chlorine gases are produced so we cannot say for certain that this second part of the mechanism is operational. Further experimental work is needed to identify these by-products in order to provide evidence for this mechanism. Until then, the explanation of how sulfur is eliminated from **E** remains unconfirmed.

ADDITIONAL RESOURCES

It is also possible to envision the intermolecular addition of chloride anion without the participation of the neighboring carboxylic acid group. Nevertheless, we highly recommend Refs. 2–5 for understanding the possible ways and underlying energetics of eliminating sulfur from thiirane-containing compounds.

References

1. Paulussen H, Adriaensens P, Vanderzande D, Gelan J. Unexpected rearrangements in the synthesis of arylidene- or alkylidene-2-thiophthalides. *Tetrahedron*. 1996;52(36):11867–11878. https://doi.org/10.1016/0040-4020(96)00678-3.
2. Steudel Y, Steudel R, Wong MW. The thermal decomposition of thiirane: a mechanistic study by ab initio MO theory. *Chem A Eur J*. 2002;8 (1):217–228. https://doi.org/10.1002/1521-3765(20020104)8:1<217::AID-CHEM217>3.0.CO;2-0.
3. Chin WS, Ek BW, Mok CY, Huang HH. Thermal decomposition of thiirane and 2-methylthiirane: an experimental and theoretical study. *J Chem Soc Perkin Trans 2*. 1994;1994:883–889. https://doi.org/10.1039/P29940000883.
4. Chew W, Harpp DN. Thermal decomposition of 3′,3′-dichlorospiro[fluorene-9,2′-thiirane]: kinetics and mechanism. *J Org Chem*. 1993;58 (16):4405–4410. https://doi.org/10.1021/jo00068a040.
5. Lutz E, Biellmann JF. Mechanism of sulfur extrusion from thiirane. *Tetrahedron Lett*. 1985;26(23):2789–2792. https://doi.org/10.1016/S0040-4039 (00)94913-X.

Question 275: Base-Catalyzed Reactions of Highly Hindered Phenols Used as Antioxidants

In 1977 American chemists synthesized the highly hindered phenol **A** by reacting 3,5-di-*t*-butyl-4-hydroxybenzyl chloride **1** with the anion of dimethyl sulfoxide **2** (Scheme 1).[1] The spectroscopic characteristics of **A** include a chemical formula of $C_{44}H_{66}O_3$ as well as strong IR (KBr) bands at 3618, 1655, and $1640\,cm^{-1}$ with UV spectrum (MeOH) maxima at 231 (ε 23,000) and 272 nm (ε 5200). The ^1H NMR spectrum ($CDCl_3$) consisted of six singlets at 1.10 (9H), 1.38 (18H), 2.90 (2H), 5.01 (1H, exchangeable), 6.56 (1H), and 6.81 ppm (2H) downfield from internal TMS. The proton decoupled

^{13}C NMR spectrum ($CDCl_3$) showed signals at 29.4 (6C), 30.2 (12C), 34.1 (4C), 34.6 (2C), 46.2 (1C), 47.2 (2C), 126.4 (4C), 127.7 (2C), 135.1 (4C), 145.2 (2C), 147.0 (2C), 152.5 (2C), and 186.4 (1C) ppm. Furthermore, the presence of 2,6-di-*t*-butylphenol **3** as an impurity proved to be pivotal for the formation of **A**. Devise the structure of **A** and provide a mechanism explaining its formation.

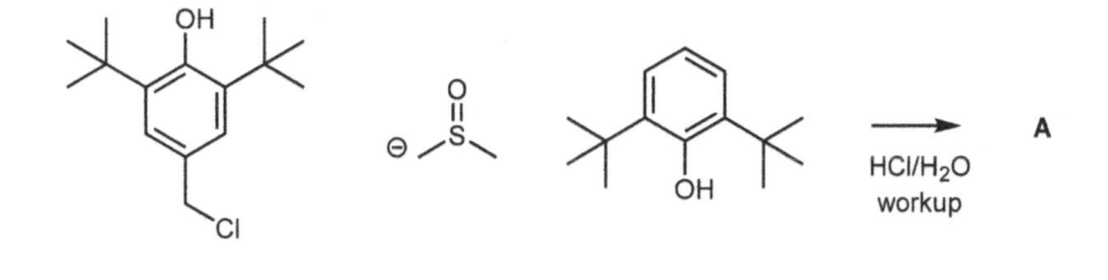
OH
Cl
O
S
OH
HCl/H$_2$O
workup
A

3,5-Di-*t*-butyl-4-hydroxybenzyl chloride (**1**) **2** 2,6-Di-*t*-butylphenol (**2**) $C_{44}H_{66}O_3$

SCHEME 1

SOLUTION

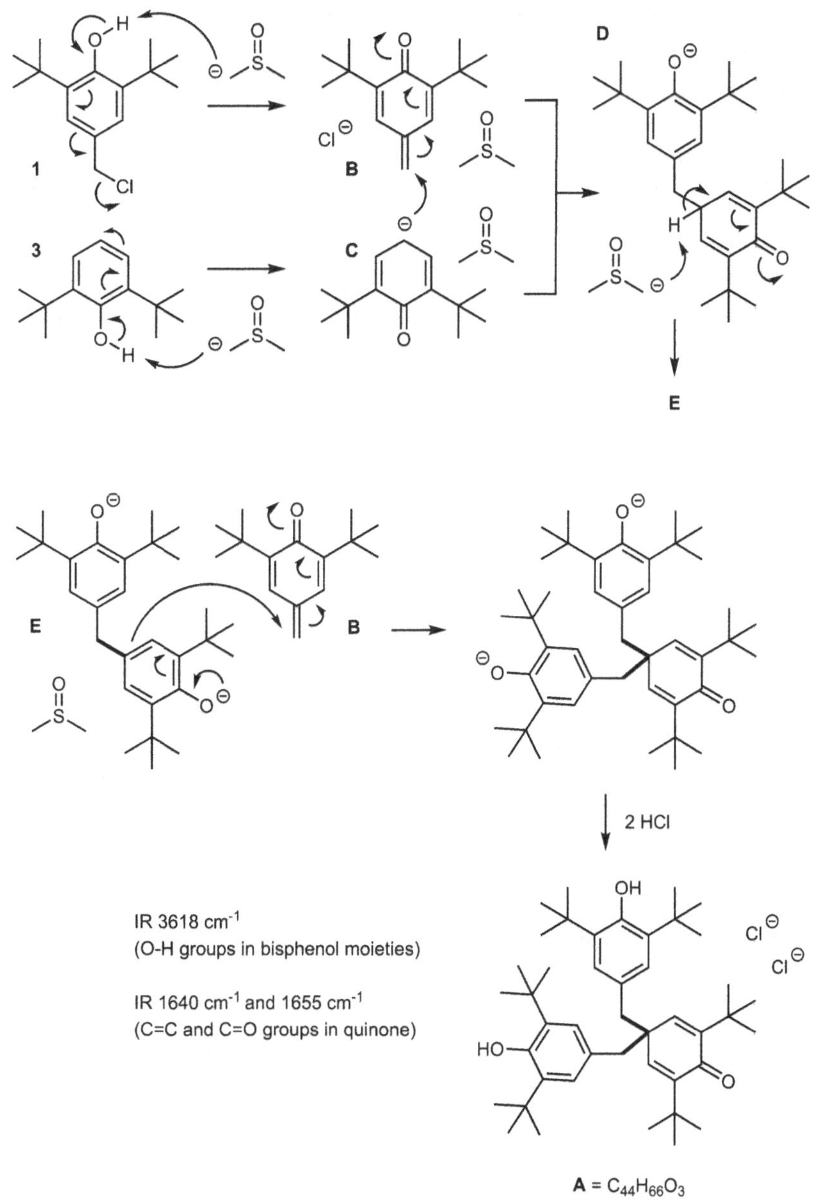
1
3
B
C
D
E
2 HCl
IR 3618 cm^{-1}
(O-H groups in bisphenol moieties)
IR 1640 cm^{-1} and 1655 cm^{-1}
(C=C and C=O groups in quinone)
A = $C_{44}H_{66}O_3$

BALANCED CHEMICAL EQUATION

A = $C_{44}H_{66}O_3$

KEY STEPS EXPLAINED

The mechanism suggested by Chasar and Westfahl begins with deprotonation of the phenyl of **1** by the dimsyl anion to produce a quinoid intermediate **B**. On an adjacent path, there is a subsequent deprotonation of the impurity **3** by dimsyl anion to form the quinone anion **C** which attacks the methylene group of **B** to form the intermediate **D** which is itself deprotonated by dimsyl anion to form the bis-phenoxide intermediate **E** which attacks another equivalent of **B** to form the hindered bis-phenolic species **A** after acidic quench with HCl in water. We also refer the reader to the balanced chemical equation for this transformation because it tells a more compact story as compared to the convergent mechanism presented earlier. As far as spectroscopic evidence for **A**, we have the IR bands at 1640 and 1655 cm^{-1} which correspond to the quinone carbonyl group and the band at 3618 cm^{-1} which corresponds to the hydroxyl O—H groups of the bisphenol moieties. The ^{1}H NMR spectrum provided differentiates only 33 hydrogen atoms indicating symmetry in the molecule as is shown in the structure of **A**. There are also 12 expected unsaturation elements to **A** based on the molecular formula provided which are consistent with the presence of 3 rings and 9 double bonds in the structure of **A**. Furthermore, the authors showed experimentally that without **3** as an impurity, no **A** is produced, whereas when **E** is synthesized on its own (phenolic form) and reacted with base and **B** (from **1**), the final product **A** was formed regardless of the base chosen. This fact provides experimental evidence for the presence of intermediate **E** as well as the importance of **3**.

ADDITIONAL RESOURCES

A similar transformation making use of a **B**-like intermediate, this time with the participation of an amino group attached to the benzene ring, is available for study in the work of Schmidt and Brunetti.[2]

References

1. Chasar DW, Westfahl JC. 2,6-Di-*tert*-butyl-4,4-bis(3,5-di-*tert*-butyl-4-hydroxybenzyl)-2,5-cyclohexadienone. A new reaction product of a hindered phenol. *J Org Chem*. 1977;42(12):2177–2179. https://doi.org/10.1021/jo00432a039.
2. Schmidt A, Brunetti H. 55. *p*-Hydroxybenzylierung von carbanionen mit chinonmethid-liefernden verbindungen. *Helv Chim Acta*. 1976;59(2):522–532. https://doi.org/10.1002/hlca.19760590218.

Question 276: Oxidative Rearrangement of an Aconitine Derivative

In 1963 Canadian chemists treated the oxonine **1** with one equivalent of periodic acid in order to form **2**, a product which showed no IR absorptions resembling those of aldehydes and ketones. When **2** was heated in dilute aqueous base with air passing through the solution, product **3** was obtained (Scheme 1).[1] Provide a mechanism for the transformation of **1** to **3**.

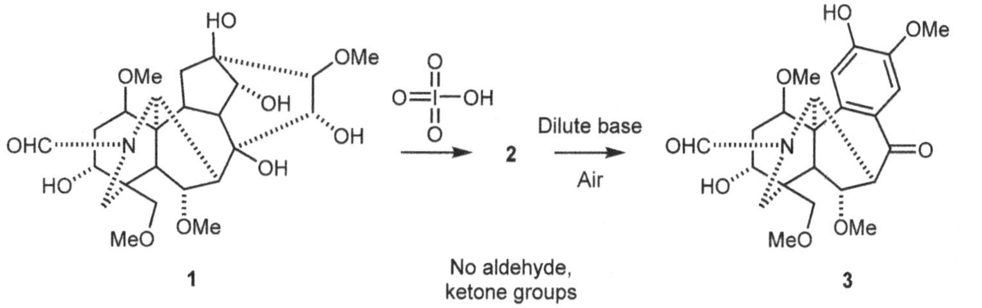

SCHEME 1

SOLUTION: 1 TO 2

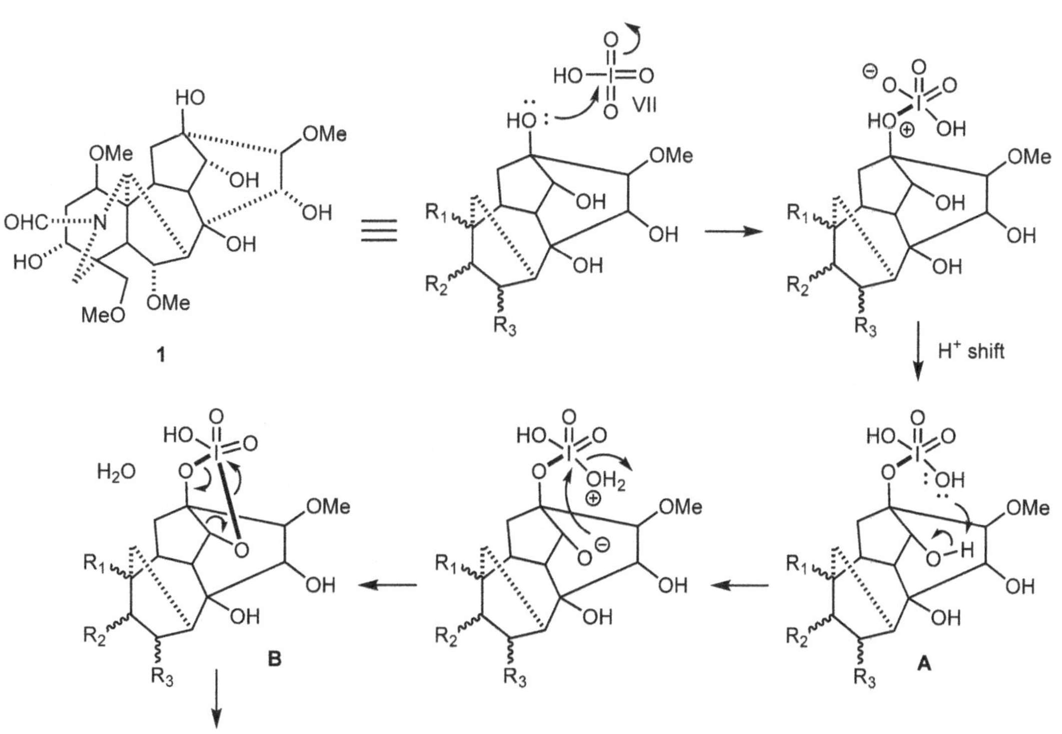

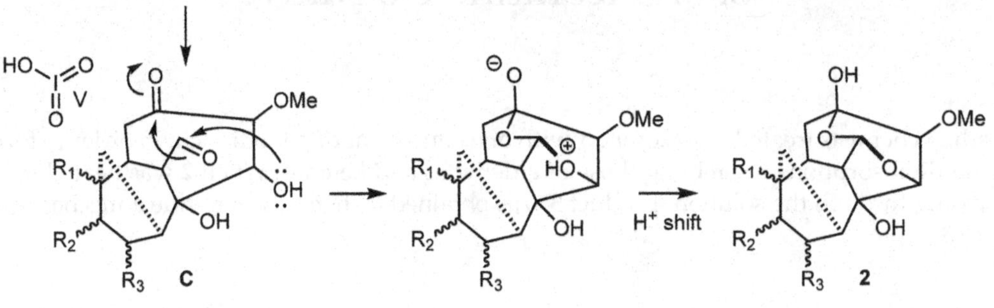

BALANCED REDOX CHEMICAL EQUATION FOR 1 TO 2

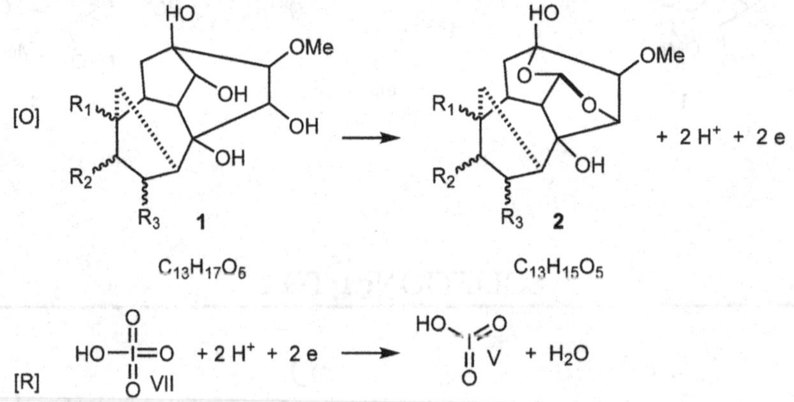

$C_{13}H_{17}O_5$ $C_{13}H_{15}O_5$

SOLUTION: 2 TO 3

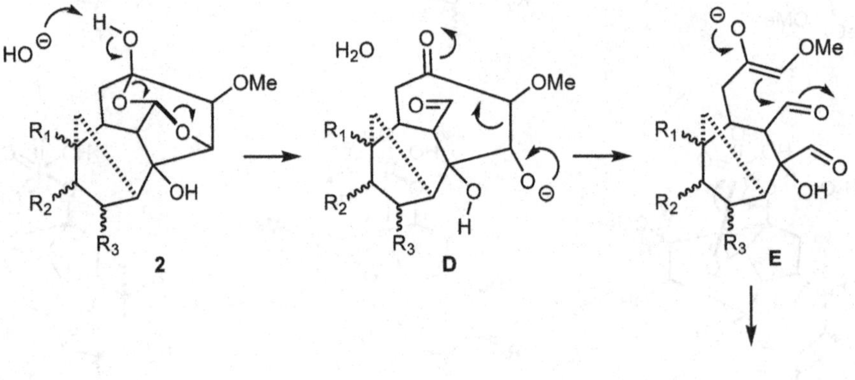

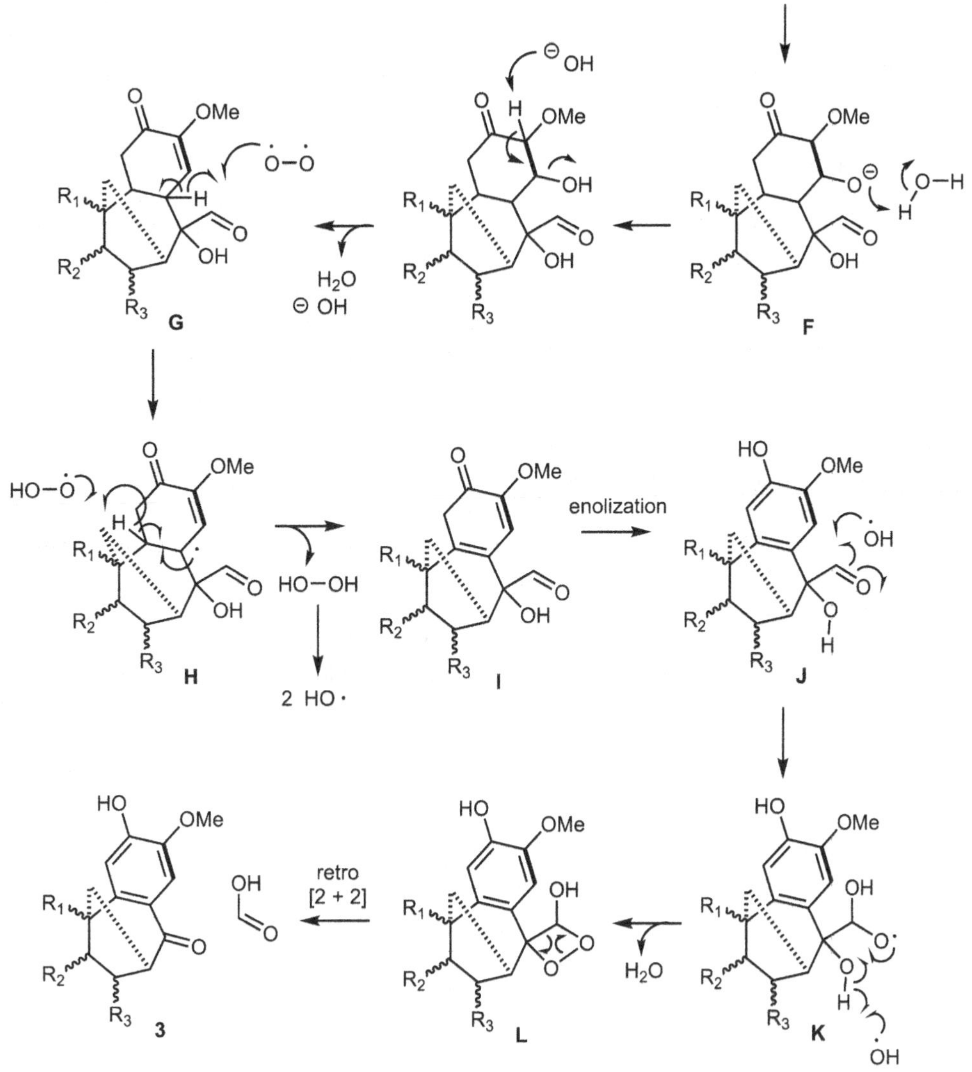

BALANCED CHEMICAL EQUATION FOR 2 TO 3

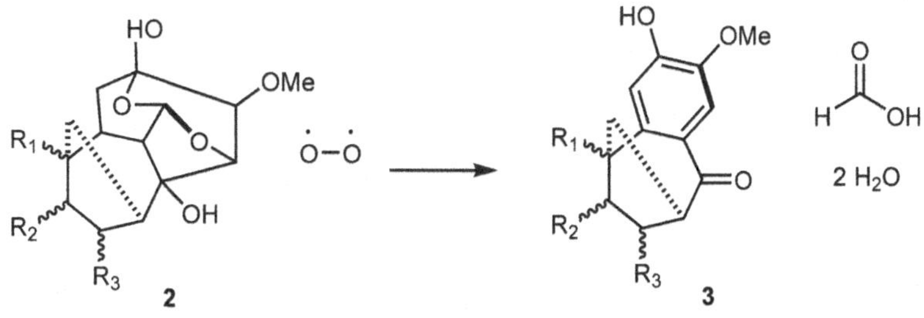

KEY STEPS EXPLAINED

In their original work, Canadian chemists Wiesner et al. described the mechanism for this transformation as beginning with periodic acid oxidation of **1** which leads to structures **A** followed by **B** followed by the expected ketone aldehyde cleavage product **C**. Nevertheless, the authors explain that **C** cyclizes to the acetal form **2** which disguises both the ketone and the aldehyde groups. This initial oxidation constitutes a redox reaction whose balanced chemical equation is provided earlier. Next, under dilute base conditions, we have dealdolization of **2** to form an oxy anion **D** which can close to the ring-fragmented dialdehyde **E** which ring closes to the six-membered ring alkoxide intermediate **F**. This intermediate protonates and then undergoes base-catalyzed dehydration to cyclohexenone **G** which undergoes radical-induced air oxidation to **I**. Intermediate **I** ketonizes to phenol **J** which contains a hydroxyformyl group. At this point the authors maintain that autoxidation (oxygen diradical process) of **J** leads directly to **3**. Nevertheless, they do not provide a mechanism and instead cite a literature precedent involving oxidative cleavage of hydroxydiphenyla-cetaldehyde to benzophenone and formic acid in the presence of silver(I)oxide generated in situ from silver nitrate under alkaline conditions. Unfortunately, that reference also did not contain a mechanism for that transformation.[2] Based on known autoxidation mechanisms of aldehydes via acyl radicals,[3] we propose here a tentative mechanism for the transformation of **J** to **3** which involves successive attack of the hydroxyformyl group by two hydroxyl radicals generated from hydrogen peroxide by-product arising from the preceding transformation of **G** to **I**. Addition of the first hydroxyl radical yields acyl radical intermediate **K**. A second hydroxyl radical abstracts a hydrogen atom from the tertiary hydroxyl group resulting in formation of 1,2-dioxetane intermediate **L** which then undergoes a retro [2 + 2] thermal cleavage to yield product **3** and formic acid. The 1,2-dioxetane intermediate could also fragment in a stepwise radical process to the same products. In terms of experimental evidence, Wiesner et al. provide only IR spectroscopic evidence in favor of structure **2**. They also report a very low yield (approx. 9.5%) for **3** which indicates the possibility of several other mechanistic pathways in which **2** may undergo fragmentation. The elucidation of such possible pathways is left for the reader as an exercise with the only hint being contained in Refs. 2, 4. Another useful exercise would be to depict compounds **1**, **2**, and **3** in their boat-like geometrical conformations to highlight another perspective on this interesting transformation. For the sake of completeness, a proposed mechanism for the hydroxydiphenylacetalde-hyde oxidative fragmentation is shown in Scheme 2 which is significantly different from the autoxidation mechanism given earlier. In this mechanism, two equivalents of silver(I) ion are used to oxidize the aldehyde to a carboxylic acid group yielding benzilic acid. Then, base-catalyzed decarboxylation leads to benzhydrol which is then further oxidized to benzophenone with two more equivalents of silver(I) ion.

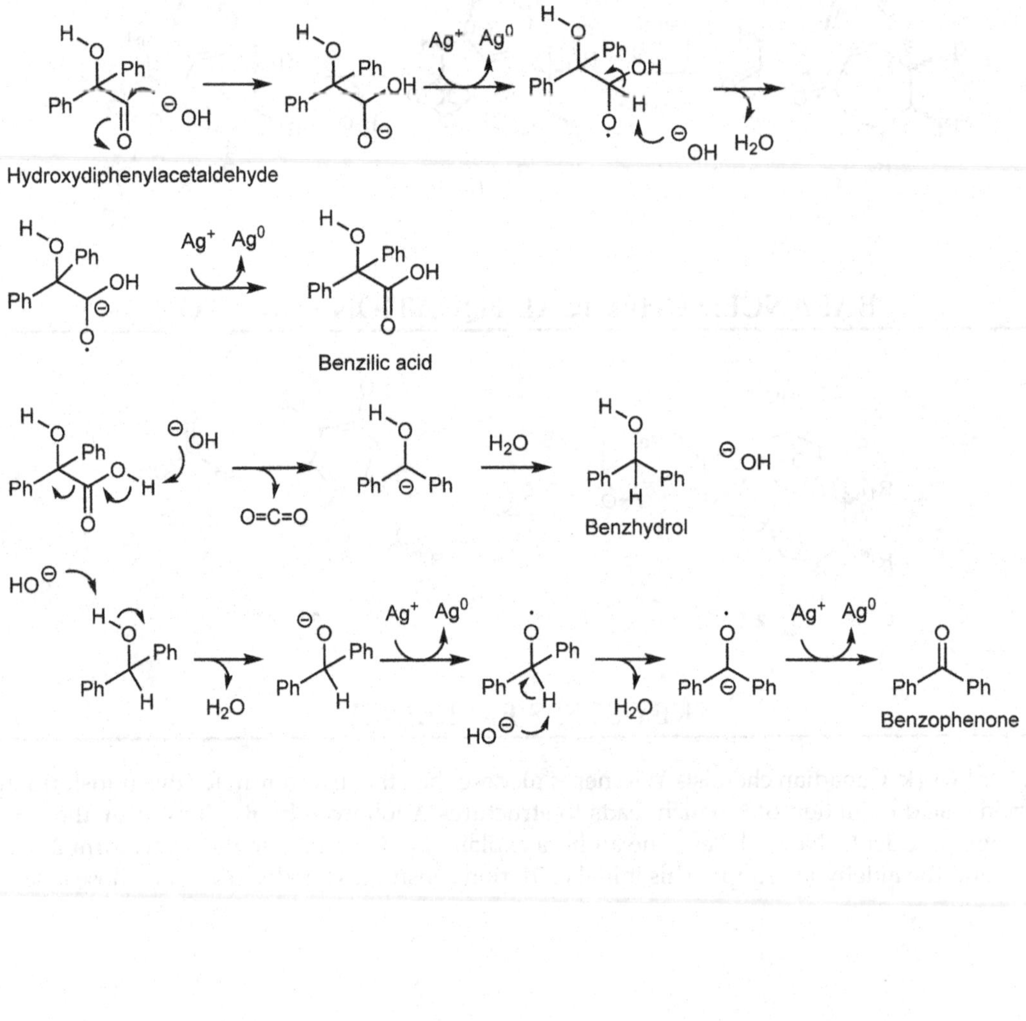

SCHEME 2

BALANCED CHEMICAL EQUATION FOR HYDROXYDIPHENYLACETALDEHYDE TO BENZOPHENONE TRANSFORMATION

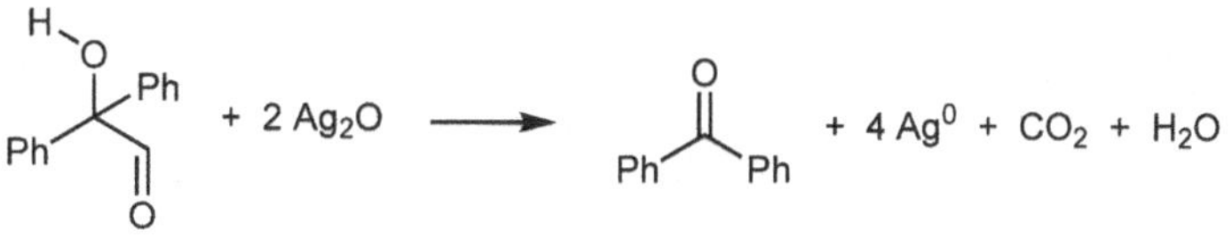

ADDITIONAL RESOURCES

Given the low yield of product **3** and the possibility for further mechanistic pathways involving Aconitine, we recommend the mechanistic analyses of Wang et al. and Ayer and Deshpande.[5,6]

References

1. Wiesner K, Goetz M, Simmons DL, Fowler LR. The structure of aconitine. *Collect Czechoslov Chem Commun.* 1963;28:2462–2478. https://doi.org/10.1135/cccc19632462.
2. Danilow S. Isomerisation der oxy-aldehyde, I.: umwandlung von diphenyl-glykolaldehyd in benzoin. *Ber Dtsch Chem Ges.* 1927;60:2390–2401. https://doi.org/10.1002/cber.19270601026.
3. Chudasama V, Fitzmaurice RJ, Caddick S. Hydroacylation of α,β-unsaturated esters via aerobic C–H activation. *Nat Chem.* 2010;2:592–596. https://doi.org/10.1038/nchem.685.
4. Patai S. *The Carbonyl Group.* vol. 1. Chicester: John Wiley & Sons Ltd; 1966. https://doi.org/10.1002/9780470771051.
5. Wang FP, Chen QH, Li ZB, Li BG. Novel products from oxidation of the norditerpenoid alkaloid pseudaconine with HIO₄. *Chem Pharm Bull.* 2001;49(6):689–694. https://doi.org/10.1248/cpb.49.689.
6. Ayer WA, Deshpande PD. Rearrangements in the diterpenoid series. I. The solvolysis of methyl 15β-tosyloxy-13-isopropyl-17-noratis-13-en-l8-oate. *Can J Chem.* 1973;51:77–86. https://doi.org/10.1139/v73-011.

Question 277: Decomposition of Akuammicine

When scientists Edwards and Smith applied heat (100°C) in methanol to the alkaloid akuammicine **1** for 3 h, decomposition had occurred to produce the betaine **2** in 70% yield (Scheme 1).[1] When the temperature was raised to 140°C, the decomposition process was completed within 2 h and the isolated products were 3-ethylpyride **3** and 3-hydroxycarbazole **4**. Provide mechanisms for these transformations.

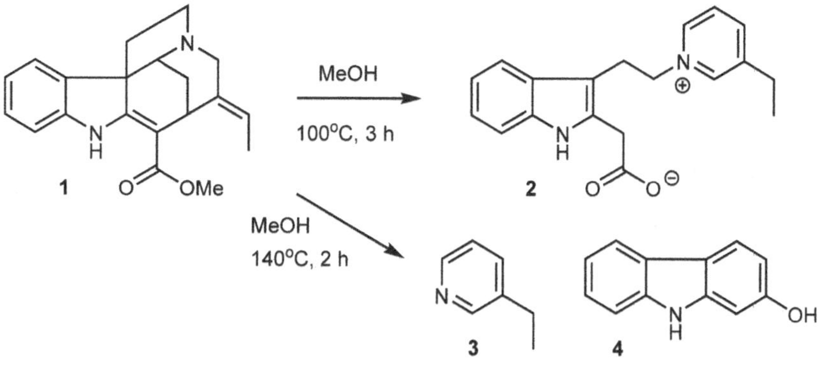

SCHEME 1

SOLUTION: DECOMPOSITION AT 100°C

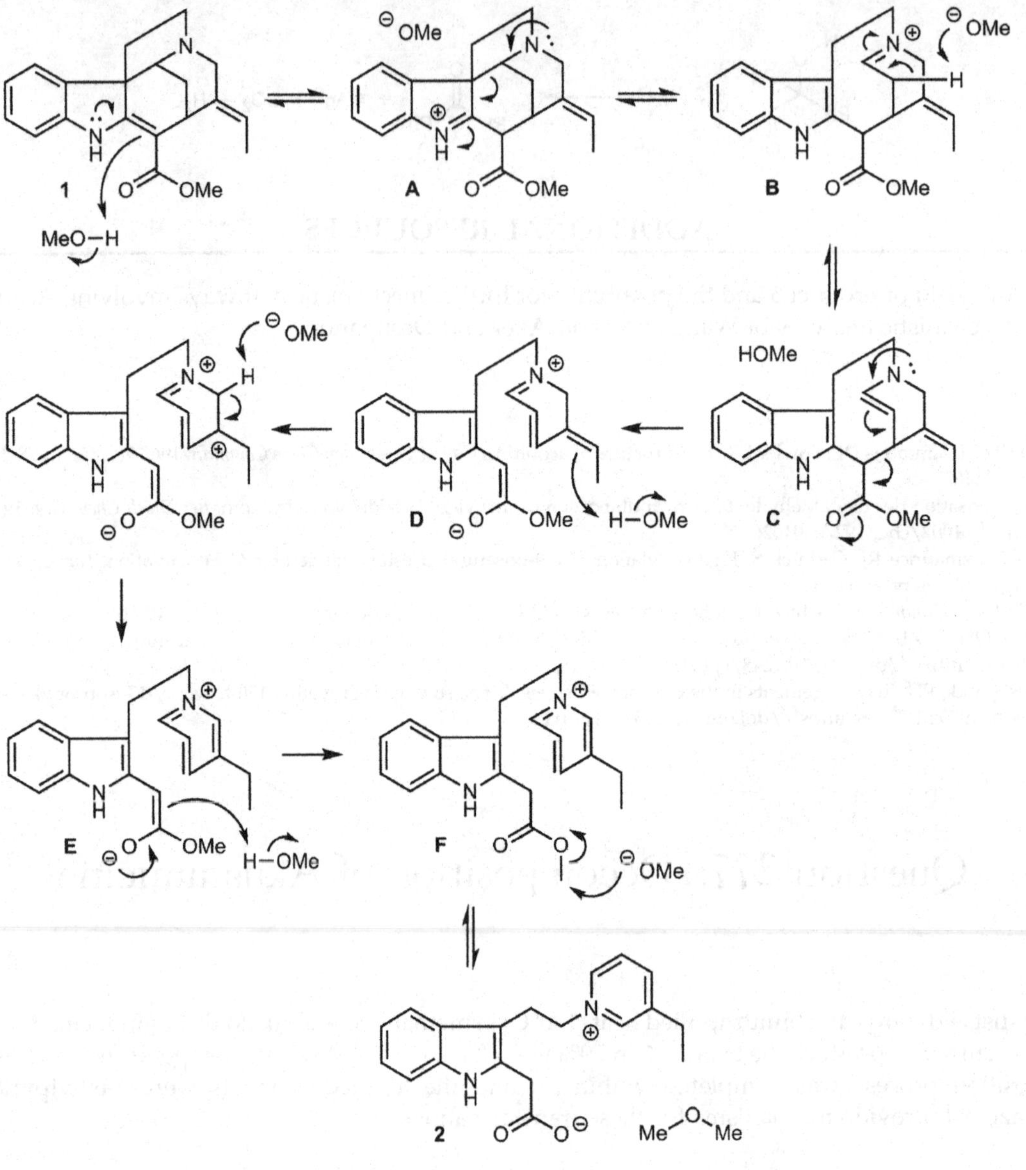

SOLUTION: DECOMPOSITION AT 140°C

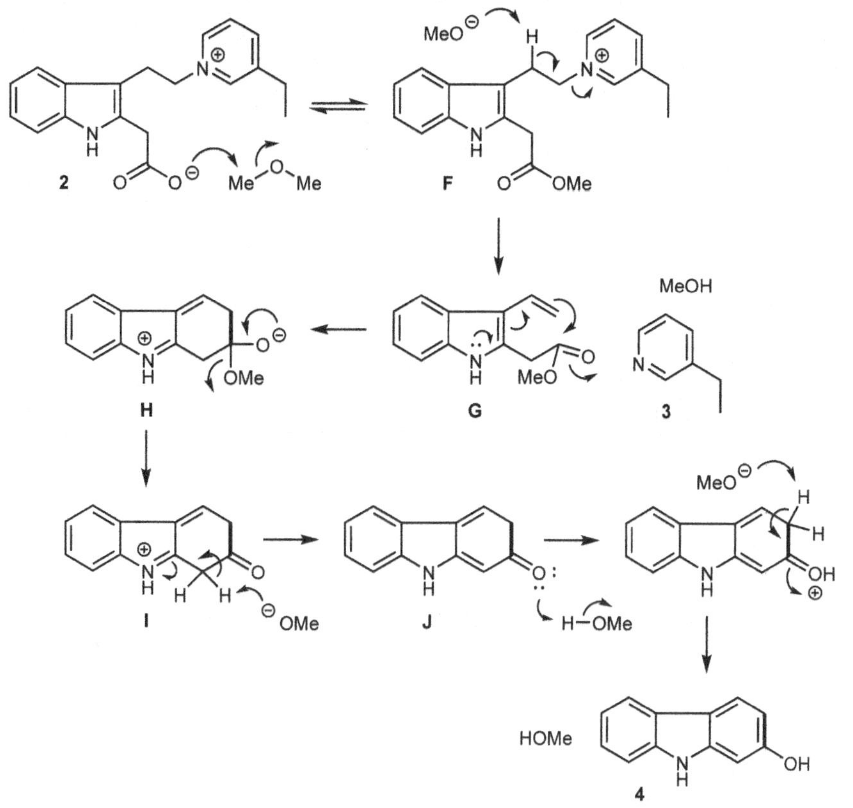

BALANCED CHEMICAL EQUATION: DECOMPOSITION AT 100°C

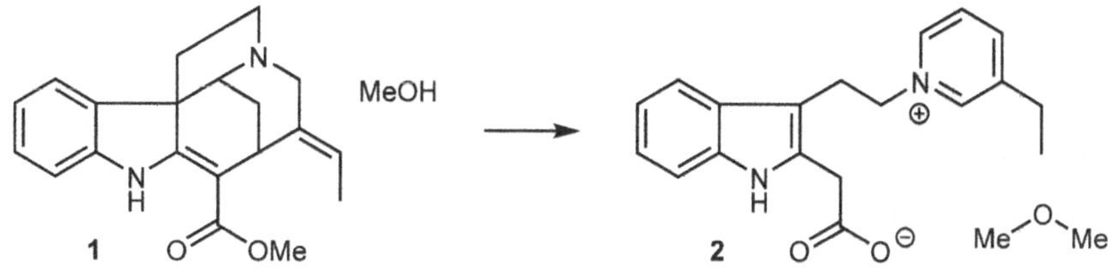

BALANCED CHEMICAL EQUATION: DECOMPOSITION AT 140°C

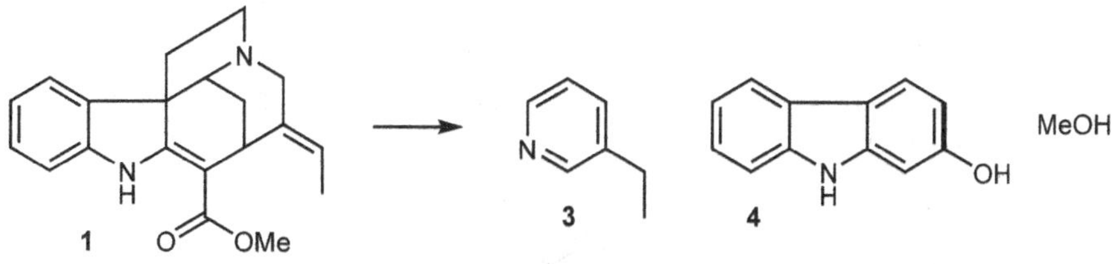

KEY STEPS EXPLAINED

At 100°C, the authors explained that akuammicine **1** undergoes protonation under equilibrium conditions to form structure **A** and methoxide anion which are also in equilibrium with the ring-expanded structure **B** which forms, after deprotonation, the α,β-unsaturated trialkyl amine **C**. This intermediate can undergo irreversible amine-induced ring opening to the enamine **D** which can isomerize after a 1,3-hydrogen shift to structure **E** containing the pyridine ring. Keto-enol tautomerization forms **F** which can be demethylated using methoxide (note that water or hydroxide is not present in solution at 100°C) to form the betaine **2** along with dimethyl ether by-product. When the temperature is increased to 140°C, **2** can revert to **F** since this is an equilibrium step to regenerate methoxide anion which can then carry out a deprotonation to form the exocyclic alkene **G** and expel 3-ethylpyridine **3** product. Compound **G** can then ring close onto the ester group to eventually expel methoxide anion and after a series of prototropic rearrangements, help form the aromatized 3-hydroxycarbazole product **4**. The identification of these decomposition products as a result of one prevailing mechanism provides important experimental evidence in support of the proposed mechanism.

ADDITIONAL RESOURCES

For an example of mechanistic analysis involving flow thermolysis rearrangements of indole alkaloids such as akuammicine and strictamine, see the work of Hugel et al.[2] For an example of a radical process involving a decomposition reaction in the context of a similar starting material, see the earlier work of Hugel and Levy.[3]

References

1. Edwards PN, Smith GF. 287. Akuamma alkaloids. Part IV. The decomposition of akuammicine in methanol. *J Chem Soc.* 1961;1961:1458–1462. https://doi.org/10.1039/JR9610001458.
2. Hugel G, Royer D, Le Men-Olivier L, Richard B, Jacquier M-J, Levy J. Flow thermolysis rearrangements in the indole alkaloid series: strictamine and akuammicine derivatives. The absolute configurations of ngouniensine and *epi*-ngouniensine. *J Org Chem.* 1997;62(3):578–583. https://doi.org/10.1021/jo961816e.
3. Hugel G, Levy J. Radical-ion chemistry of natural indolenines: 19-20-dehydrotubifoline. *Tetrahedron Lett.* 1993;34(4):633–634. https://doi.org/10.1016/S0040-4039(00)61638-6.

Question 278: Extending the Favorskii Reaction

During studies aimed at fine-tuning the Favorskii rearrangement for synthetic purposes, Japanese chemists treated chloroketone **1** and benzaldehyde with 14% ethanolic KOH at room temperature for 3h and formed lactone **2** as the major product along with the unsaturated acid **3** and the alcohol **4** as minor products. On the other hand, treatment with 14% aqueous KOH produced only alcohol **4** (Scheme 1).[1] Show mechanisms for the formation of **2**, **3**, and **4**.

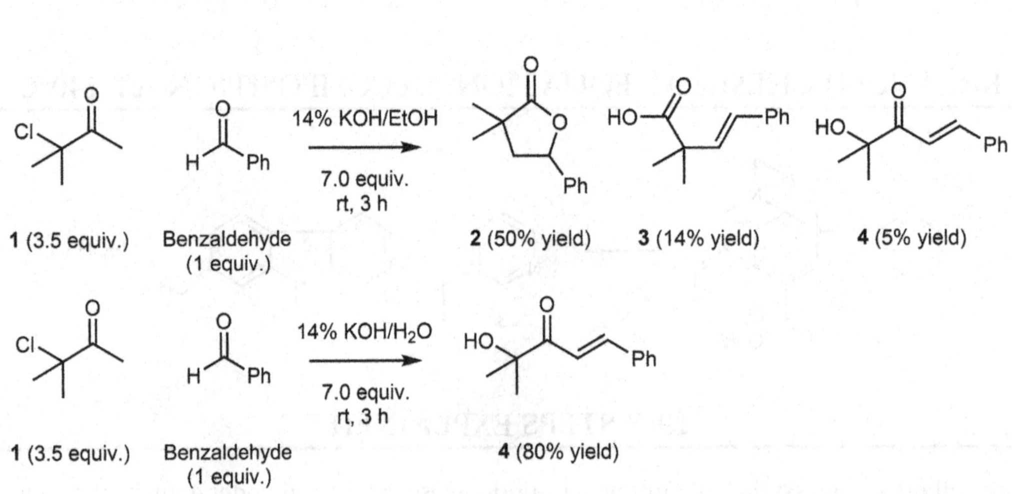

SCHEME 1

SOLUTION: 1 → 2

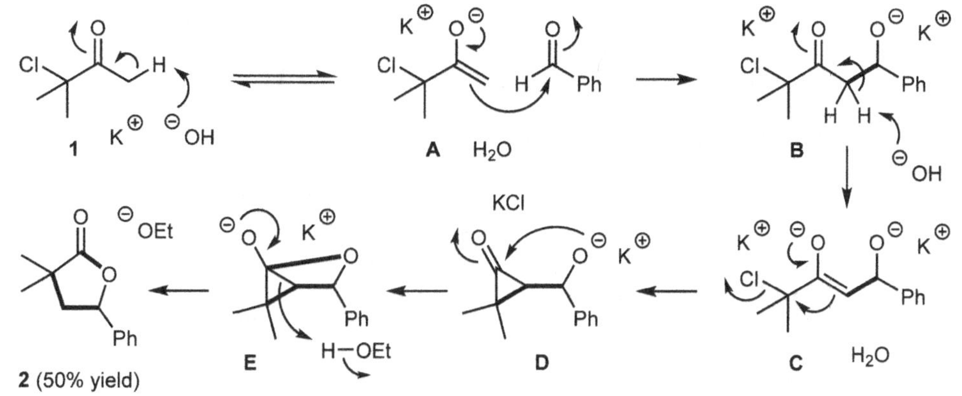

2 (50% yield)

SOLUTION: 1 → 3

3 (14% yield)

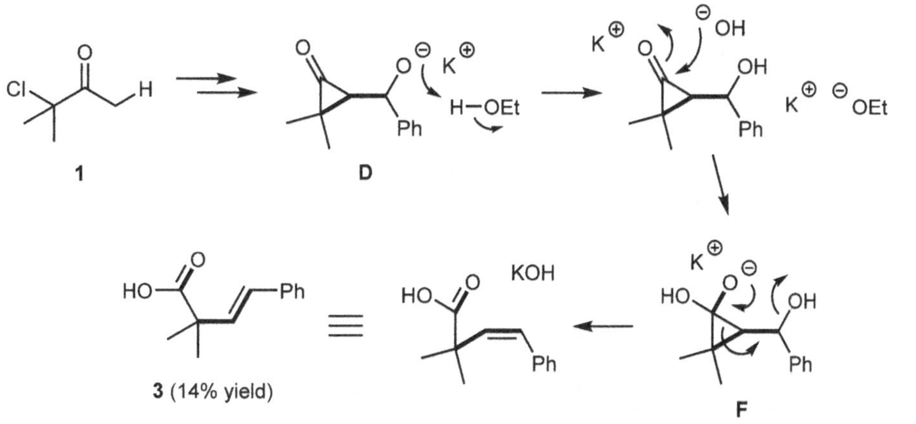

SOLUTION: 1 → 4

4 (5% yield)

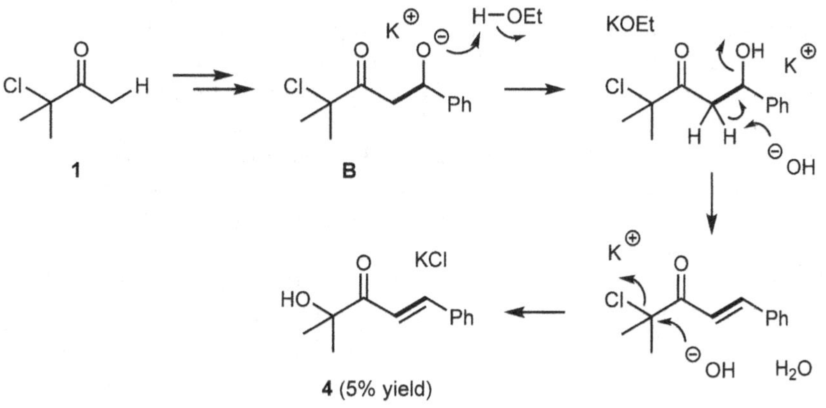

BALANCED CHEMICAL EQUATION: 1 → 2

1 (3.5 equiv.)   Benzaldehyde   2 (50% yield)
                  (1 equiv.)

BALANCED CHEMICAL EQUATION: 1 → 3

1 (3.5 equiv.)   Benzaldehyde   3 (14% yield)
                  (1 equiv.)

BALANCED CHEMICAL EQUATION: 1 → 4

1 (3.5 equiv.)   Benzaldehyde   4 (5% yield)
                  (1 equiv.)

KEY STEPS EXPLAINED

The mechanism proposed by Sakai et al. begins with chloroketone **1** undergoing an aldol reaction with benzaldehyde to form the key intermediate **B**. This compound is then deprotonated and undergoes Favorskii cyclization to form the cyclopropanone intermediate **D** which may ring close to structure **E** and ring expand subsequently to form product **2** after protonation. Moreover, at the junction of intermediate **D** it is also possible to have protonation followed by hydroxide anion attack on the cyclopropanone carbonyl followed by ring opening and loss of hydroxide anion to form product **3**. Lastly, from intermediate **B**, it is possible to have protonation followed by dehydration and lastly an S_N2 attack of hydroxide anion to replace the chloride group and form product **4**. The authors explain the difference in products obtained under aqueous KOH and ethanolic KOH conditions as having to do with the possible hydrolysis of **1** at the very beginning of the reaction (i.e., S_N2 replacement of Cl group by OH group). This reaction under the harsher aqueous KOH conditions would compete with the initial aldol reaction thus removing the possibility of easily forming the key cyclopropanone intermediate **D** which leads to **2** and **3**. If hydrolysis occurs first, the only available pathway would be toward the formation of **4** (see Scheme 2). The authors also explicitly state that they have obtained no evidence to support intermediate **D**. Since all three products have essentially the same balanced chemical equations in terms of inputs and by-product outputs, we recommend carrying out a theoretical analysis in order to help shed light on the energetic likelihood of forming intermediate **D**. Another experimental possibility is to synthesize intermediate **D** and treat it with 14% ethanolic KOH and determine whether the products formed are **2** and **3** and whether their ratio more or less corresponds to that reported by Sakai et al. In such an experiment, product **4** would not be expected to be formed at all. We also note that product **4** is the major product when the reaction is carried out in aqueous solution, whereas it is a minor product when the solvent is ethanol. This constitutes evidence that hydrolysis of **1** occurs first before reaction with benzaldehyde in aqueous solvent.

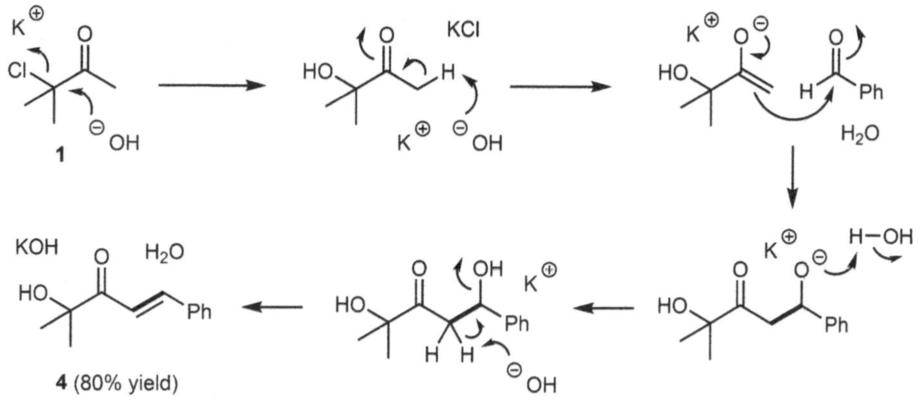

SCHEME 2

ADDITIONAL RESOURCES

For interesting examples of Favorskii rearrangements in the context of bridged polycyclic compounds and alkylated steroids, see Refs. 2, 3, respectively. For an interesting mechanistic analysis highlighting a competition between a Darzens, Favorskii, and Gabriel reaction, see the work of Mamedov et al.[4] Lastly, for an example of a radical process involving the Favorskii rearrangement, see the work of Givens et al.[5]

References

1. Sakai T, Yamawaki A, Katayama T, Okada H, Utaka M, Takeda A. Novel type of the Favorskii rearrangement combined with aldol reaction leading to γ-butyrolactone derivatives. *Bull Chem Soc Jpn.* 1987;60:1067–1069. https://doi.org/10.1246/bcsj.60.1067.
2. Chenier PJ. The Favorskii rearrangement in bridged polycyclic compounds. *J Chem Educ.* 1978;55(5):286–291. https://doi.org/10.1021/ed055p286.
3. Logan RT, Roy RG, Woods GF. Alkylated steroids. Part 4. An unusual Favorskii rearrangement. *J Chem Soc Perkin Trans 1.* 1981;1981:2631–2636. https://doi.org/10.1039/P19810002631.
4. Mamedov VA, Tsuboi S, Mustakimova LV, et al. 1,4-dioxins from methyl phenylchloropyruvate. Competition of the Darzens, Favorskii, and Gabriel reactions. *Chem Heterocycl Compd.* 2000;36(8):911–922. https://doi.org/10.1007/BF02256975.
5. Givens RS, Heger D, Hellrung B, et al. The photo-Favorskii reaction of *p*-hydroxyphenacyl compounds is initiated by water-assisted, adiabatic extrusion of a triplet biradical. *J Am Chem Soc.* 2008;130:3307–3309. https://doi.org/10.1021/ja7109579.

Question 279: Dihydroxylation/Base Treatment of the Westphalen Ketone

When Merck scientists attempted the dihydroxylation of the Westphalen diketone **1** with OsO$_4$ they formed a 9,10-dihydroxy derivative which under warm methanolic KOH conditions produced **2**. When **2** was reacted with methanolic KOH under more drastic conditions, the aromatized product **3** was formed (Scheme 1).[1] Devise mechanisms for the formation of **2** and **3**.

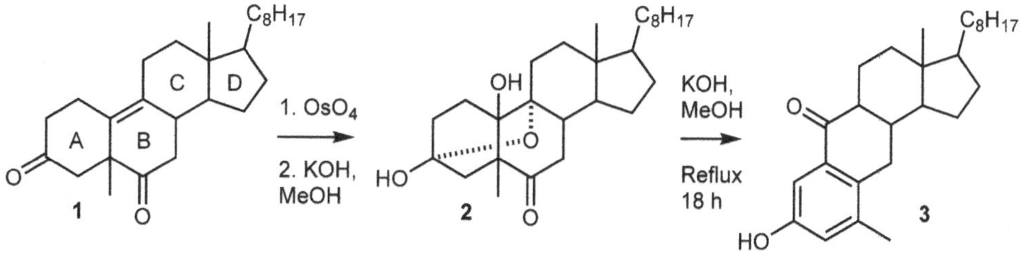

SCHEME 1

SOLUTION: 1 → A

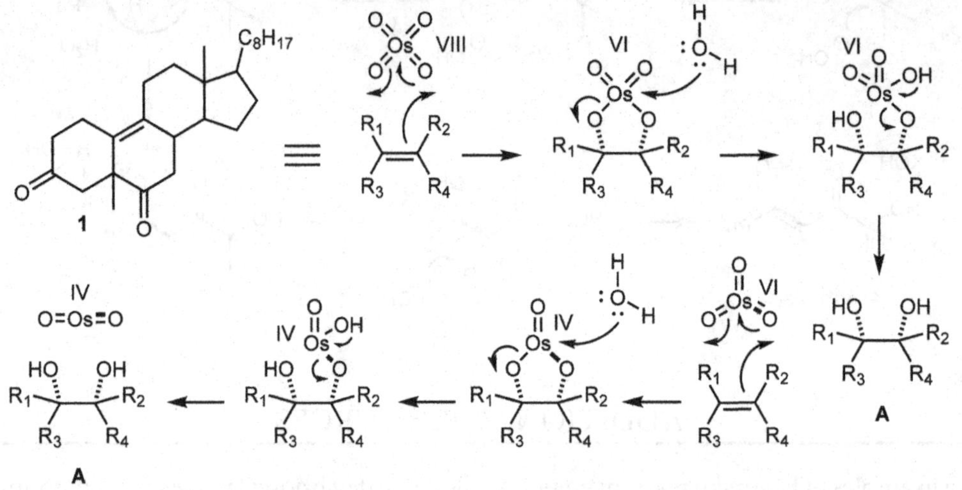

BALANCED CHEMICAL EQUATION: 1 → A

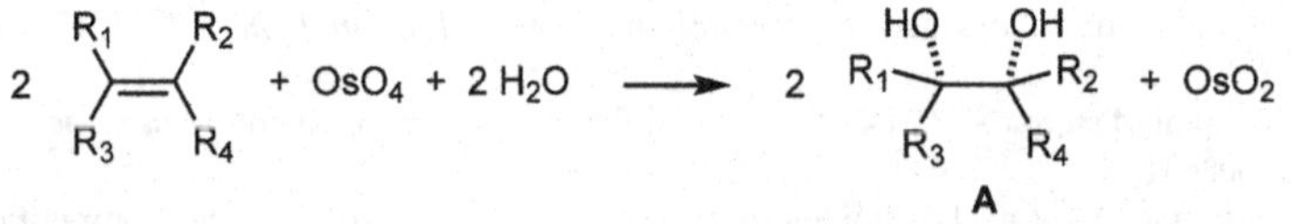

SOLUTION: A → 2

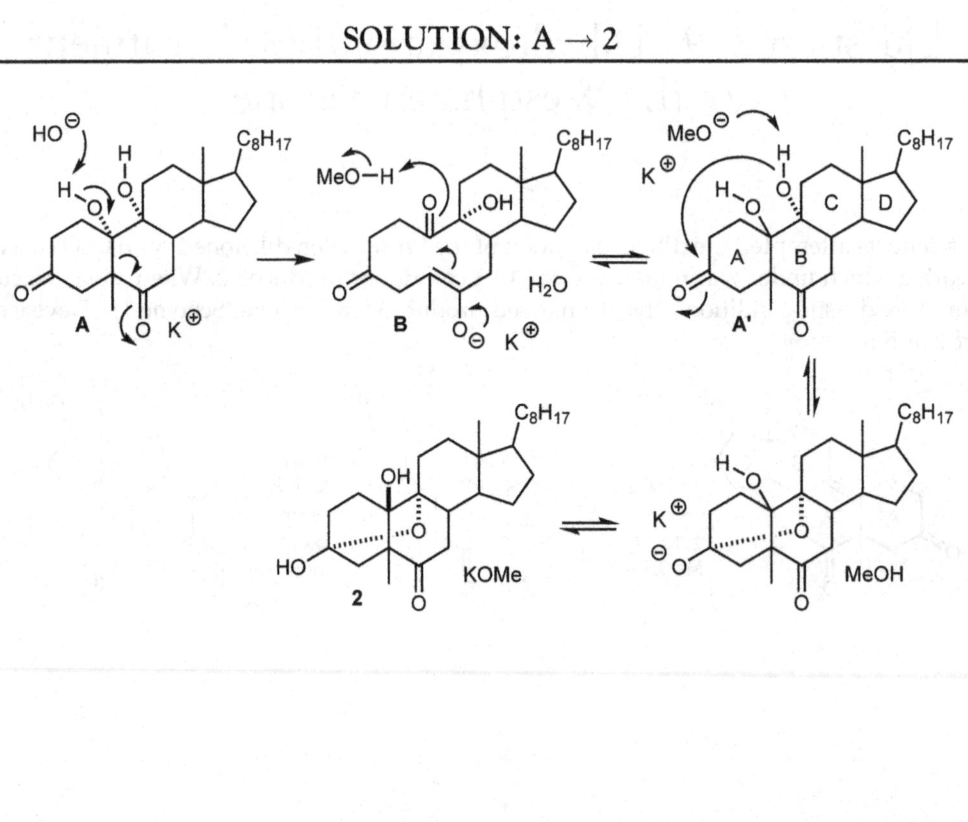

SOLUTION: B → 3

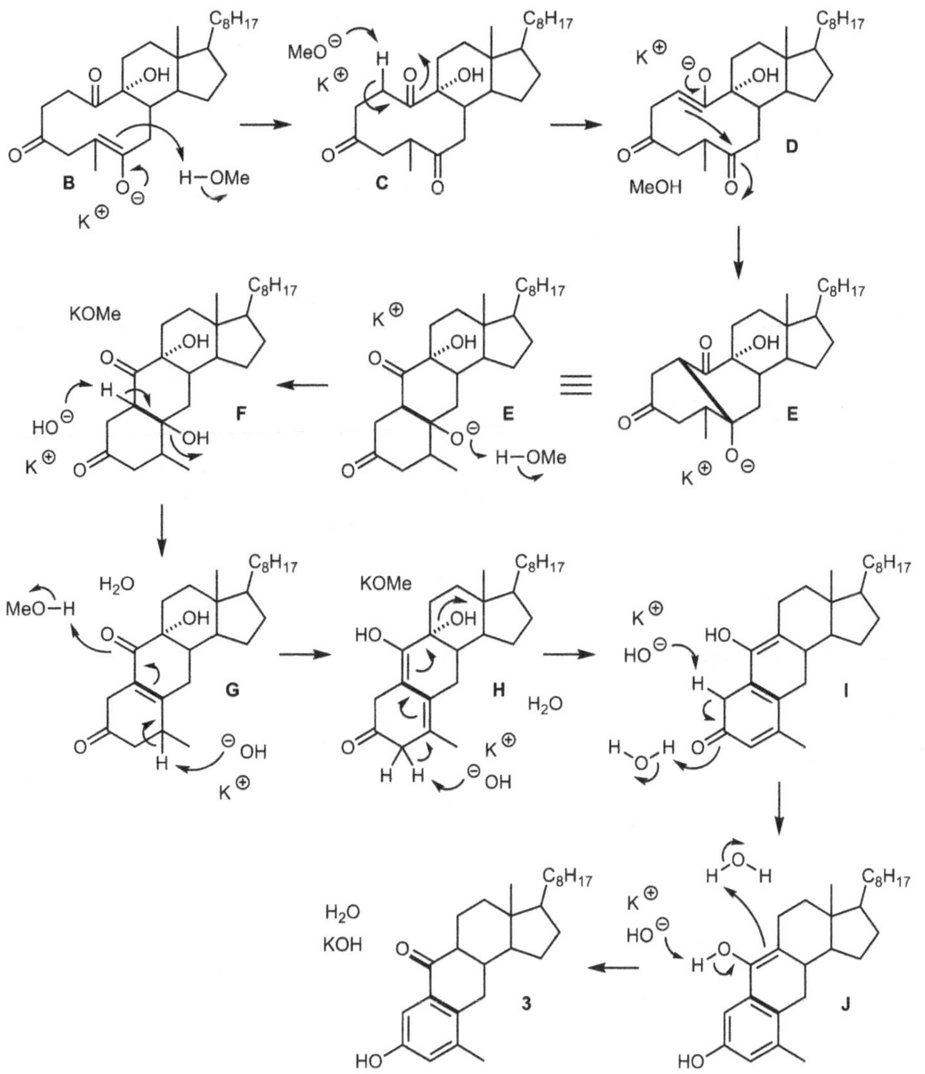

BALANCED CHEMICAL EQUATION: A → 2

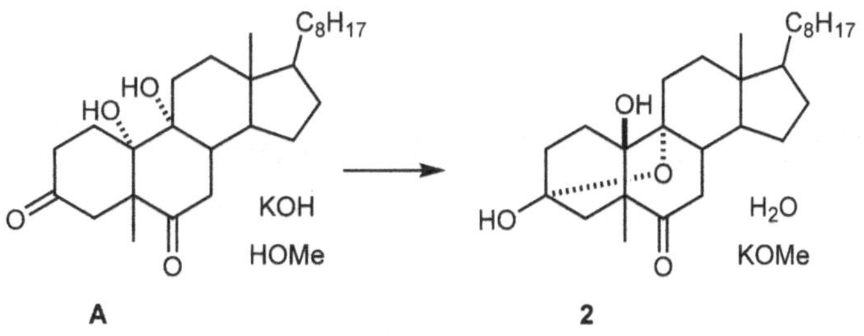

BALANCED CHEMICAL EQUATION: 2 → 3

KEY STEPS EXPLAINED

The transformation of the Westphalen diketone **1** into **2** and eventually **3** begins with a dihydroxylation reaction utilizing the oxidizing reagent OsO_4 which installs a diol across the alkene double bond of ultimately two equivalents of **1** to form two equivalents of **A** (A/B *trans*) together with the reduced OsO_2 by-product. Mild treatment of **A** with warm methanolic KOH results in base-catalyzed retroaldol ring cleavage (Grob fragmentation) and subsequent isomerization to the more stable *trans* diol isomer **A'** (A/B *cis*) which can undergo a base-catalyzed transannular cyclization to **2**. The authors note that **A** has a constrained B-ring boat conformation and therefore isomerization to the more stable **A'** is required for the formation of **2**. Continuous treatment of **2** with methanolic KOH under more drastic reflux conditions for several hours can revert back to the enolate **B** using the reverse transformations described earlier. Enolate **B** can then ketonize by abstracting a proton from methanol to form **C** which can be deprotonated to form the enolate **D** which can undergo transannular cyclization to give a new [6.6.0] bicyclic linkage and form **E**. After protonation and base-catalyzed dehydration of **E** we obtain **G**. Base-catalyzed enolization of **G** sets up a base-catalyzed elimination of hydroxide to give **I** which can be aromatized by a base-catalyzed enolization to **J** which may ketonize to the final product **3**. Furthermore, the authors were successful in intercepting intermediate **A'** as an acetate which supports its existence. For future study, we recommend carrying out a theoretical analysis which will shed much needed light on this interesting mechanism.

ADDITIONAL RESOURCES

For an interesting mechanistic transformation involving an alkaloid ring system very similar to the one involved in this problem, see the work of Edwards and Paryzek.[2] For a discussion of a rearrangement known as the Westphalen rearrangement, see Ref. 3.

References

1. Slates HL, Wendler NL. Retroaldol transformations in the Westphalen diol series. *Experientia.* 1961;17(4):161. https://doi.org/10.1007/BF02160356.
2. Edwards OE, Paryzek Z. Lanostane-to-cucurbitane transformation. *Can J Chem.* 1983;61:1973–1980. https://doi.org/10.1139/v83-341.
3. Hanson J, Hitchcock PB, Kiran I. The Westphalen rearrangement of a tricyclic steroid. *J Chem Res Synop.* 2003;2003:222–224. https://doi.org/10.3184/030823403103173552.

Question 280: Quantitative Conversion of a Vinylogous Thioamide Into a Thiophene

While studying the synthesis of 2,5-disubstituted thiophenes at Shell in 1969, Smutny had proposed a novel approach (Scheme 1).[1] Suggest a mechanism for the formation of the ethyl 2-methylthiothiophene-5-carboxylate product.

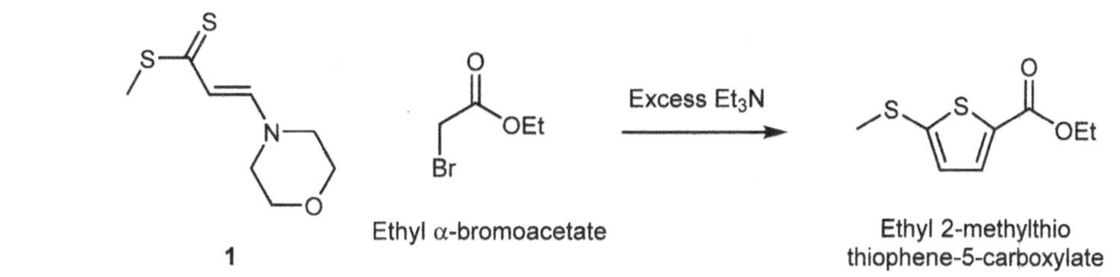

SCHEME 1

SOLUTION: POSSIBILITY 1: SULFONIUM ION INTERMEDIATE

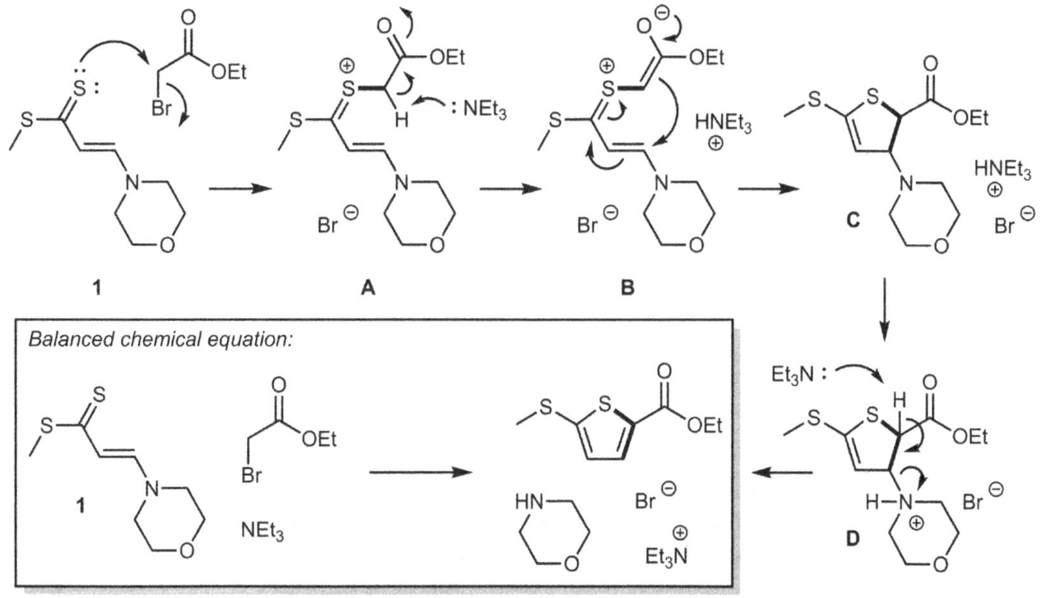

SOLUTION: POSSIBILITY 2: IMINIUM ION INTERMEDIATE

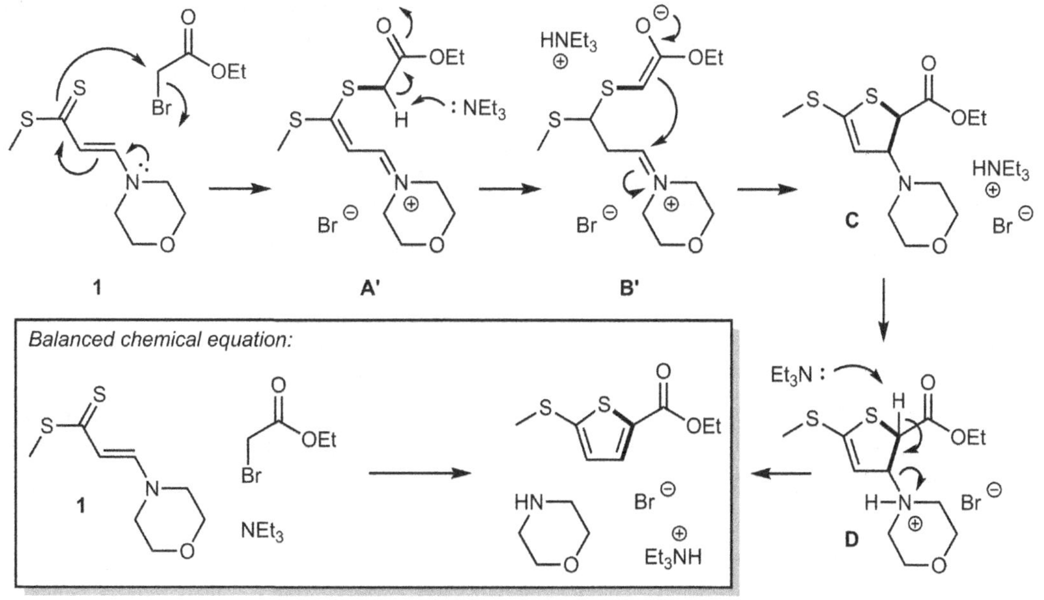

SOLUTION: POSSIBILITY 3

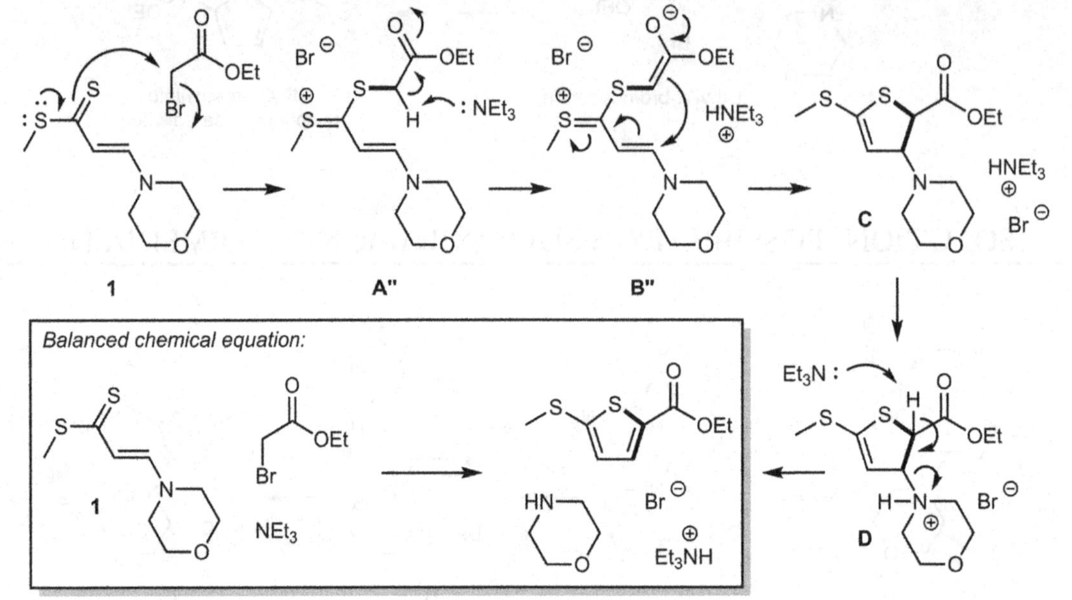

1 **A"** **B"** **C**

Balanced chemical equation:

D

KEY STEPS EXPLAINED

To explain the formation of ethyl 2-methylthiothiophene-5-carboxylate from **1** and ethyl-α-bromoacetate, Smutny proposed the mechanism outlined in Possibility 1 via the sulfonium ion intermediate. Essentially, thioester **1** attacks the ethyl bromoacetate to replace bromide in order to form a sulfonium ion intermediate which upon enolate formation can cyclize in a [4 + 1] manner to a dihydrothiophene intermediate using one equivalent of triethylamine as base. A thiophene ring is then formed with subsequent elimination of morpholine by-product using a second equivalent of triethylamine. We note that the alternative mechanisms proposed in Possibility 2 and 3, respectively, utilize slightly different processes for achieving the same common intermediate **C**. In Possibility 2 we have the morpholine group playing a role in establishing an iminium ion intermediate while in Possibility 3 we have the thioalkyl group participating to form another sulfonium ion intermediate. Due to the close similarities of these three mechanistic proposals, we cannot say with confidence, in the absence of a detailed theoretical investigation showing the relative energetics of intermediates **A** versus **A'** versus **A"** and of **B** versus **B'** versus **B"**, which mechanism is likeliest to occur.

ADDITIONAL RESOURCES

For an interesting example highlighting the mechanism of a similar transformation by the same authors, see the work of Kalish et al.[2]

References

1. Smutny EJ. Synthesis of 2,5-disubstituted thiophenes. *J Am Chem Soc.* 1969;91(1):208–209. https://doi.org/10.1021/ja01029a048.
2. Kalish R, Smith AE, Smutny EJ. Synthesis of 2*H*-thiopyrans from 3-dialkylaminodithioacrylates and maleic anhydride. *Tetrahedron Lett.* 1971;12 (24):2241–2244. https://doi.org/10.1016/S0040-4039(01)96830-3.

Question 281: Rearrangements During Synthetic Studies on Carba-Sugars

Throughout synthetic studies of carba-sugar glycosidase inhibitors, which can be used in chemotherapy as antidiabetics or antiviral agents, for example, chemists Shing and Wan from Hong Kong carried out a synthesis of valiolamine in 1996.[1] During their work they encountered a number of unexpected rearrangements as are illustrated in Scheme 1. Devise mechanisms to account for these transformations.

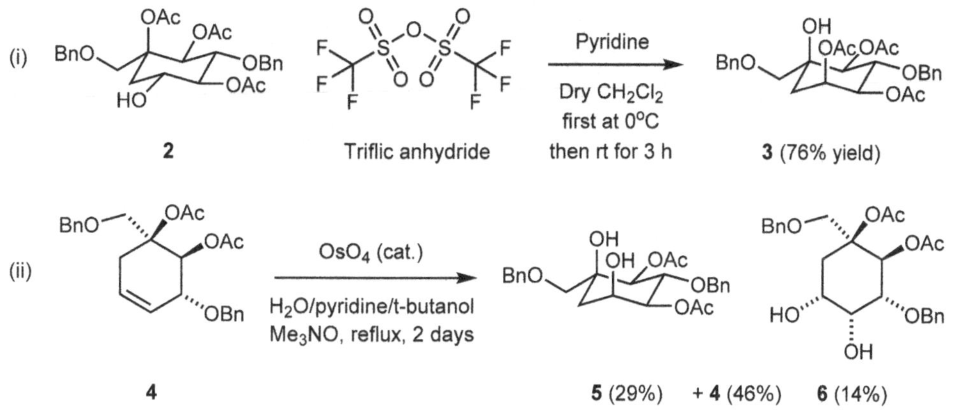

SCHEME 1

SOLUTION: (I) 2 → 3 WITH TRIFLIC ANHYDRIDE

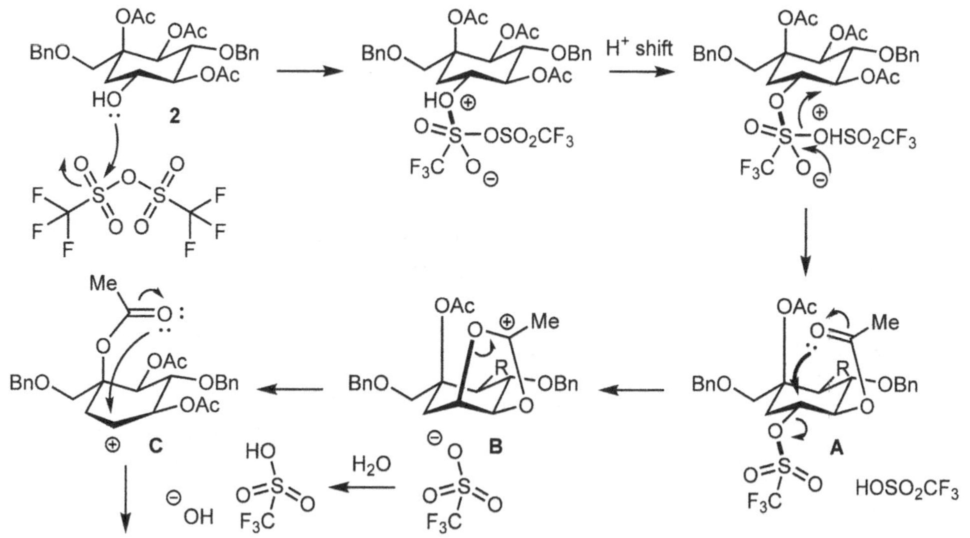

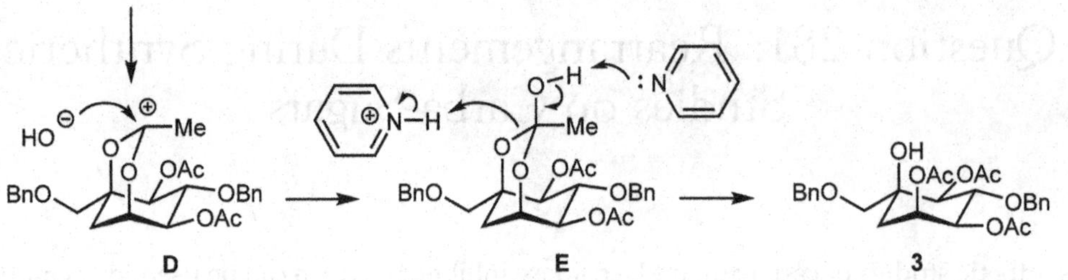

SOLUTION: (II) DIHYDROXYLATION MECHANISM

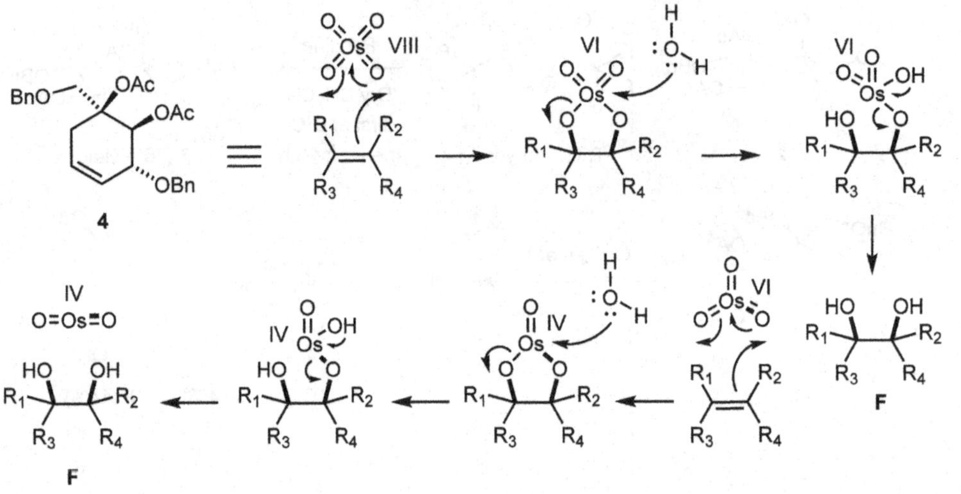

BALANCED CHEMICAL EQUATION: DIHYDROXYLATION OF 4

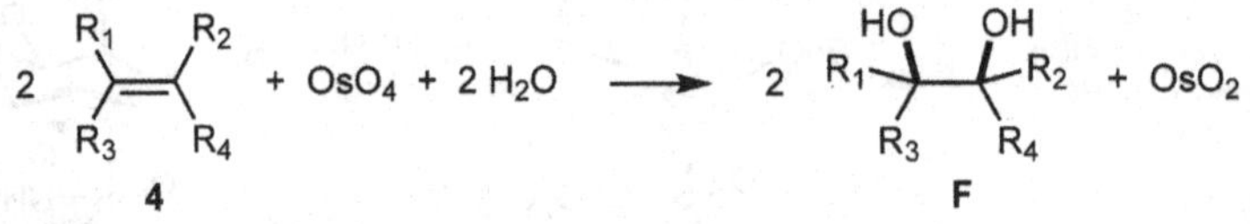

SOLUTION: (II) FORMATION OF 6

If dihydroxylation occurs from the bottom face we have:

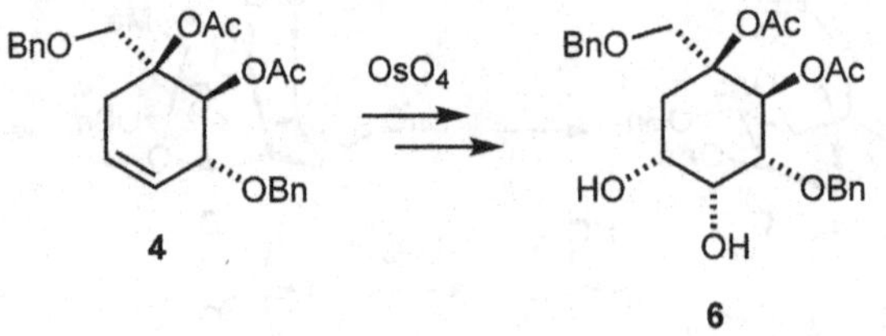

SOLUTION: (II) FORMATION OF 5

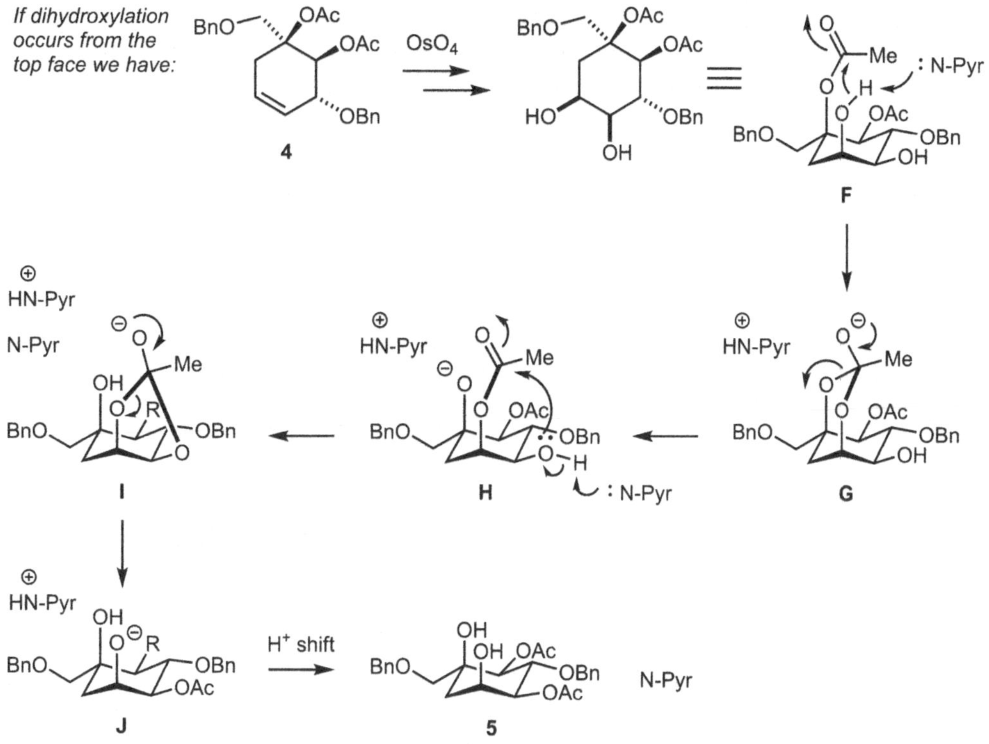

If dihydroxylation occurs from the top face we have:

KEY STEPS EXPLAINED

The mechanism proposed by Shing and Wan for the transformation of **2** to **3** begins with alcoholic attack of **2** on the triflic anhydride followed by a proton shift and elimination of trifluoromethanesulfonic acid in order to form the triflate intermediate **A**. At this point, rather than have the OAc group on C_5 attack the C_1—O—triflate group directly in order to cause ionization, it is proposed that the neighboring OAc group on C_2 participates in displacing the triflate to form the cationic species **B** which can easily produce the cation **C** which is attackable by the OAc group on C_5 finally which after hydroxide anion attack, ketonization and protonation gives **3**. The reason why the authors propose the participation of the neighboring C_2—OAc group is because in the analogous experiments where the C_2—OAc group on **2** is replaced with C_2—OH and when there is a C_1—O—SO_2—O—C_2 cyclic sulfate utilized there is no observed rearranged product **3**. It is thought therefore that the secondary OAc group on C_2 plays a pivotal role in establishing the ionization necessary for the tertiary C_5—OAc to attack the C_1-position.

In the case of the **4** to **5** and **6** transformation, we first begin with dihydroxylation which installs a diol across the alkene double bond of **4** whereby both OH groups are oriented *syn*. If the dihydroxylation occurs from the bottom face of the ring, we end up with a *cis* relationship between the newly installed OH groups and the OBn and CH_2OBn groups thereby providing product **6**. If the dihydroxylation occurs from the top face of the ring, we end up with structure **F** which is deprotonated by pyridine. This deprotonation allows for C_1—O attack on C_5—OAc to form the quaternary intermediate **G** which undergoes ketonization to structure **H** where the acetyl group has migrated to C_1 (the more stable secondary axial position as compared to the tertiary axial position it used to occupy before). At this point C_5—O is protonated and C_2—OH is deprotonated again by use of pyridine base to allow for attack of C_2—O onto C_1—OAc to form the quaternary intermediate **I** which upon ketonization and hydrogen shift gives the final rearranged product **5**. We note that once again the acetyl group has migrated to a more stable position (i.e., from secondary axial to secondary equatorial position) in **3**. Furthermore, the authors note that this transformation is subject to base effect whereby increasing concentrations of pyridine caused an increased yield of **5**. Lastly, it was noted experimentally that if any of the C_5—OAc, C_2—OH, and C_3—OH groups of **4** were *trans* with respect to each other, there was no rearranged product **5** observed. These experimental results form a strong case in support of the proposed mechanisms for both transformations.

ADDITIONAL RESOURCES

For an extensive review of internal displacements caused by neighboring group participation in carbohydrates, see the work of Goodman.[2] For a review of carba-sugars in the context of synthesis and geometrical aspects, see Ref. 3. For further mechanistic work by Shing on a very similar system in the context of epoxide chemistry, see Ref. 4. Lastly, for an interesting mechanistic example involving nitrogen chemistry in the context of carbohydrate chemistry, see Ref. 5.

References

1. Shing TKM, Wan LH. Facile syntheses of valiolamine and its diastereomers from (-)-quinic acid.[1] Nucleophilic substitution reactions of 5-(hydroxymethyl)cyclohexane-1,2,3,4,5-pentol. *J Org Chem*. 1996;61:8468–8479. https://doi.org/10.1021/jo9607828.
2. Goodman L. Neighboring-group participation in sugars. *Adv Carbohydr Chem*. 1967;22:109–175. https://doi.org/10.1016/S0096-5332(08)60152-6.
3. Arjona O, Gomez AM, Lopez JC, Plumet J. Synthesis and conformational and biological aspects of carbasugars. *Chem Rev*. 2007;107(5):1919–2036. https://doi.org/10.1021/cr0203701.
4. Shing TKM, Tam EKW. Enantiospecific syntheses of (+)-crotepoxide, (+)-boesenoxide, (+)-β-senepoxide, (+)-pipoxide acetate, (-)-iso-crotepoxide, (-)-senepoxide, and (-)-tingtanoxide from (-)-quinic acid. *J Org Chem*. 1998;63(5):1547–1554. https://doi.org/10.1021/jo970907o.
5. Sudau A, Muench W, Bats JW, Nubbemeyer U. Planar chirality: cycloaddition and transannular reactions of optically active azoninones that contain (E)-olefins. *Chem A Eur J*. 2001;7(3):611–621. https://doi.org/10.1002/1521-3765(20010202)7:3<611::AID-CHEM611>3.0.CO;2-9.

Question 282: A Pyrimidine to Pyrazole Transformation

In 1988 Japanese scientists Matsuura et al. had accomplished a pyrimidine to pyrazole ring transformation by treating 5-acylaminouracil **1** with 5% aq. NaOH in ethanol at reflux in order to form pyrazole **2** in generally good yield (Scheme 1).[1] Devise a mechanism to account for this transformation.

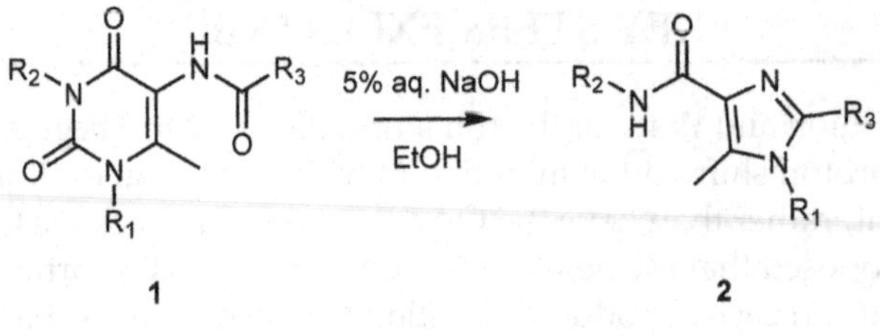

SCHEME 1

SOLUTION

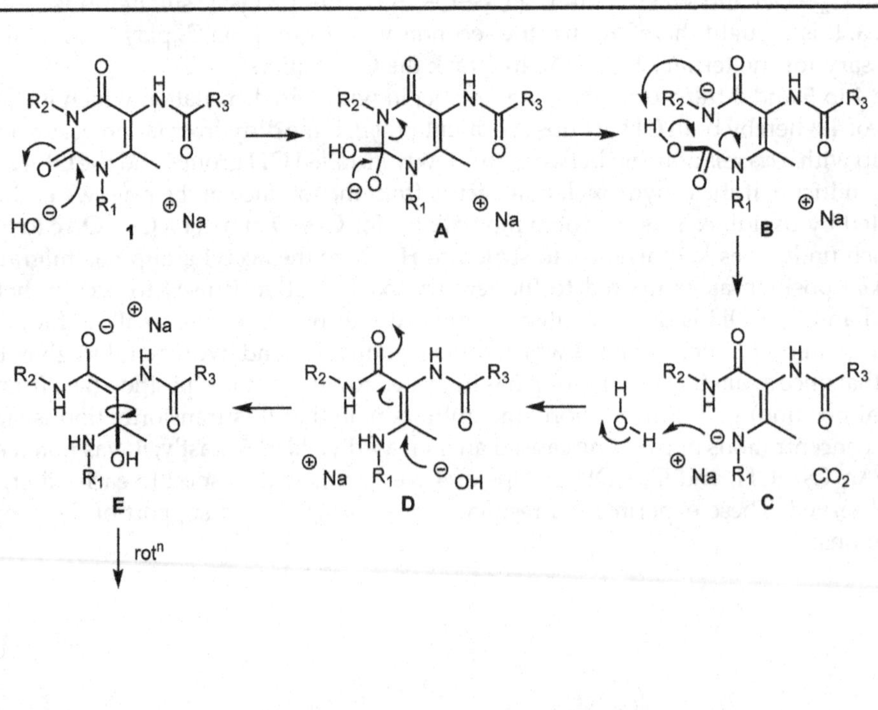

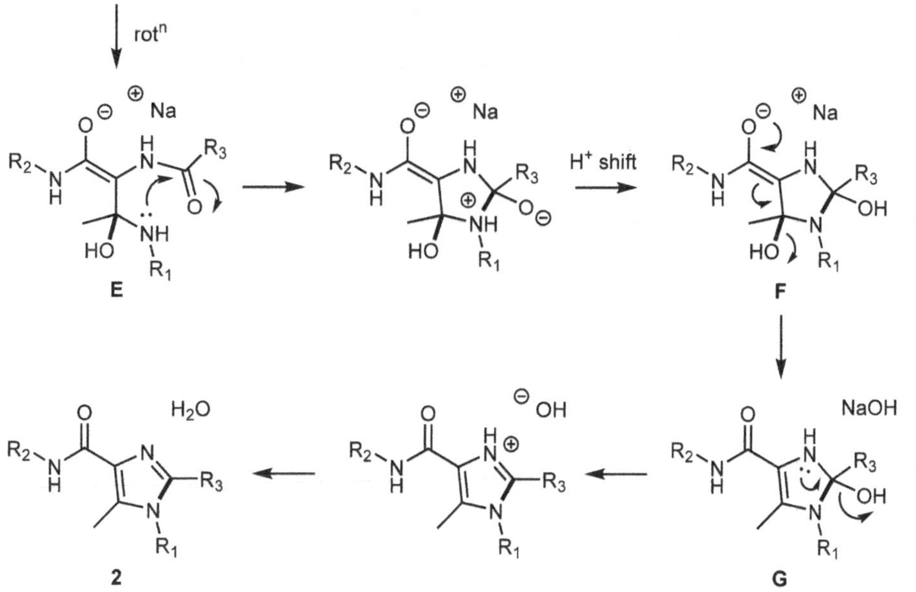

BALANCED CHEMICAL EQUATION

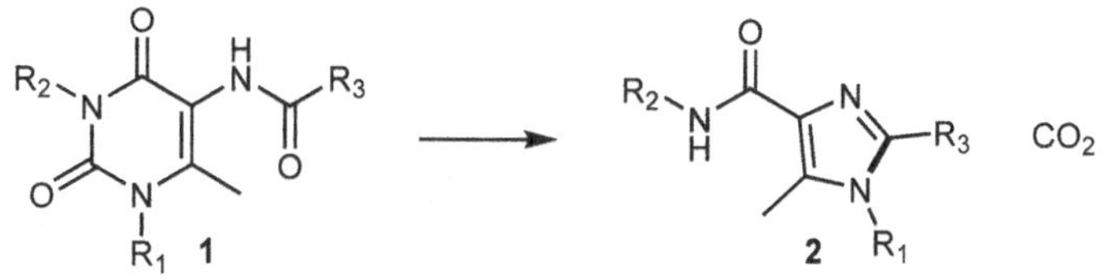

KEY STEPS EXPLAINED

The mechanism proposed by the authors begins with hydroxide anion attack onto the carbonyl group of the pyrimidine ring in order to form a quaternary intermediate **A** which can close to break the pyrimidine ring forming the anion **B** which may extract a proton from the carboxylic acid to cause a decarboxylation to the anion **C** which protonates via water to intermediate **D**. A nucleophilic attack by hydroxide anion forms the tertiary alcohol **E** which upon bond rotation facilitates a nucleophilic attack by the newly proximate amine group onto the amide carbonyl in order to form the five-member ring intermediate **F** after a proton shift. **F** then ketonizes with the elimination of hydroxide to form **G** which after dehydration forms the final pyrazole product **2**.

ADDITIONAL RESOURCES

We direct the reader to further work by the same authors for interesting mechanistic examples of pyrimidine to imidazole ring transformations.[2–4]

References

1. Ueda T, Matsuura I, Murakami N, Nagai S, Sakakibara J, Goto M. A novel ring transformation of 5-acylaminouracils and 5-acylaminopyrimidin-4(3H)-ones into imidazoles. *Tetrahedron Lett.* 1988;29(36):4607–4610. https://doi.org/10.1016/S0040-4039(00)80560-2.
2. Matsuura I, Ueda T, Murakami N, Nagai S, Sakakibara J. Synthesis of 1H-imidazoles by the simple ring transformation of 5-acylaminouracils and 5-acylaminopyrimidin-4(3H)-ones. *J Chem Soc Perkin Trans 1.* 1991;2821–2826. https://doi.org/10.1039/P19910002821.

3. Matsuura I, Ueda T, Nagai S, et al. Oxidative transformation of 5-aminouracil into imidazolone by thallium(III) nitrate trihydrate in methanol. *J Chem Soc Chem Commun.* 1992;1992:1474–1475. https://doi.org/10.1039/C39920001474.
4. Matsuura I, Ueda T, Murakami N, Nagai S, Nagatsu A, Sakakibara J. Synthesis of imidazoles by the oxidative transformation of 5-aminopyrimidinones. *J Chem Soc Perkin Trans 1.* 1993;1993:965–968. https://doi.org/10.1039/P19930000965.

Question 283: Attempted Diastereocontrol in Synthesis of a Homoallylic Alcohol

For a key component of the synthesis of amphotericin B, Brueckner attempted the transformation of sulfone **2** to the diene **1** using 3 equivalents of allyllithium followed by addition of HMPA.[1] The products of this reaction were **1** (43% yield) mixed with **3** (31% yield, Scheme 1). Provide mechanisms for the formation of these two products.

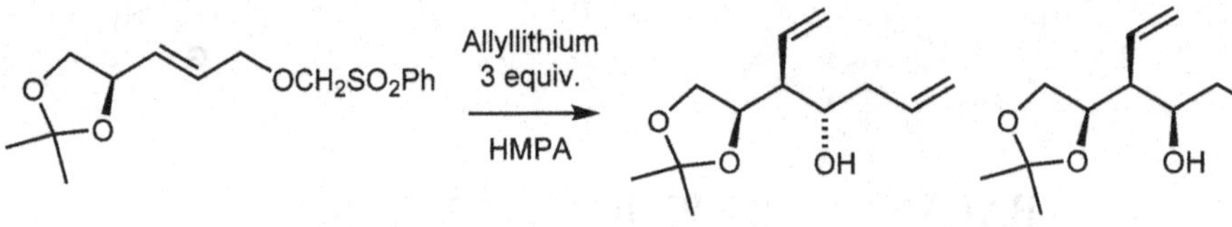

SCHEME 1

SOLUTION

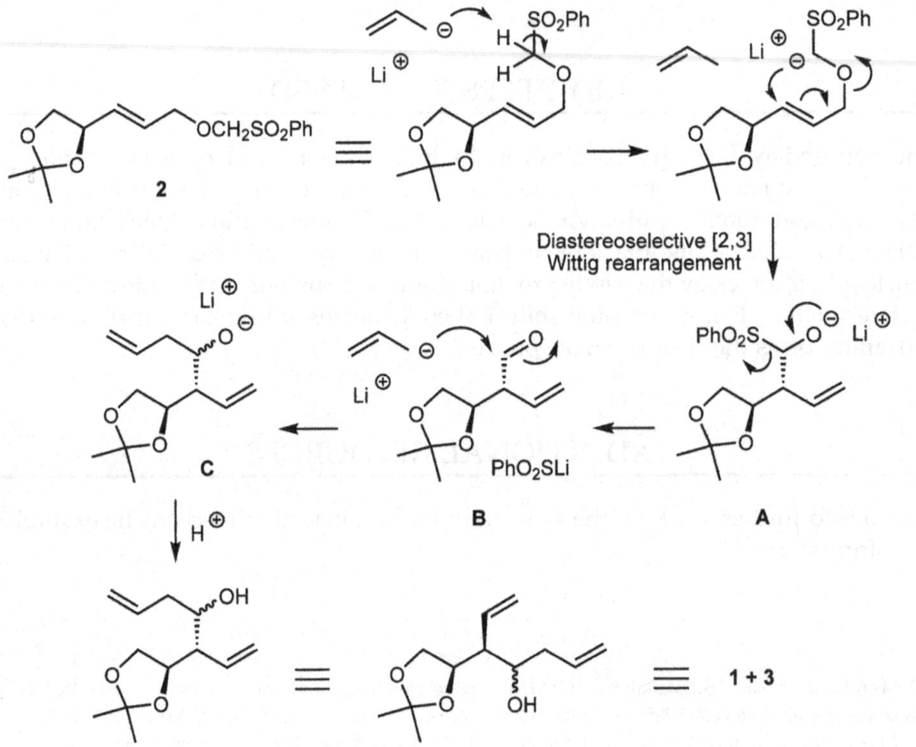

KEY STEPS EXPLAINED

The transformation of **2** to the epimeric mixture of **1** and **3** begins with lithiation of **2** to form an anion at the carbon α to the sulfonyl group and the ether group. This anion, upon the addition of hexamethylphosphoramide (HMPA), undergoes a stereospecific diastereoselective [2,3]-Wittig rearrangement to form the intermediate **A**. This intermediate can then eliminate a phenylsulfinate group to form the aldehyde **B** which can undergo a nonstereospecific addition reaction with the excess allyl lithium on the aldehyde group to give a mixture of epimers with the required stereochemistry for aqueous quench to the mixture of **1** and **3**.

As Brucckner explained in the original and in a second article,[1,2] the key to this transformation has to do with the nature of the transition state involved in the [2,3]-Wittig rearrangement step. The geometry of this transition state follows from the theory of Houk which proposes that the allylic σ-bond is oriented antiperiplanar with respect to the trajectory of the approaching nucleophile in order to both minimize steric hindrance and to maximize stabilization by overlap with properly disposed σ and σ* orbitals in the transition state.[3] Accordingly, the trans orientation is the outcome for **A** as a result of an endo-type transition state for this Wittig rearrangement (Fig. 1).[2]

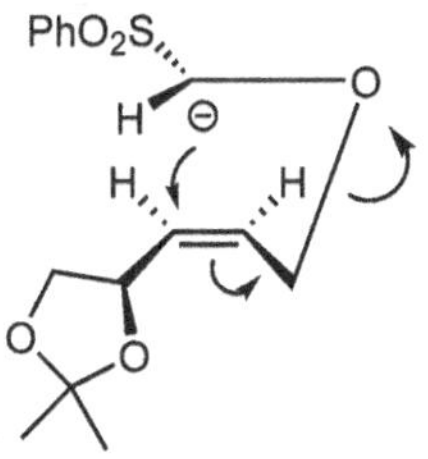

FIG. 1

ADDITIONAL RESOURCES

The [2,3]-Wittig rearrangement has appeared in other synthetic and mechanistic research.[4] Additionally, Brueckner has also continued to exploit the Wittig-type reactions in the context of stereocontrolled synthesis.[5]

References

1. Brueckner R. Stereocontrolled synthesis of a C_{14}-C_{20} building block for amphotericin B using a novel [2,3] Wittig rearrangement. *Tetrahedron Lett.* 1988;29(45):5747–5750. https://doi.org/10.1016/S0040-4039(00)82180-2.
2. Brueckner R. Asymmetric induction in the [2,3] Wittig rearrangement the stereoselective synthesis of unsaturated alcohols with three contiguous stereogenic centers. *Chem Ber.* 1989;122(1):193–198. https://doi.org/10.1002/cber.19891220130.
3. Houk KN, Paddon-Row MN, Rondan NG, et al. Theory and modeling of stereoselective organic reactions. *Science.* 1986;231(4742):1108–1117. https://doi.org/10.1126/science.3945819.
4. Katritzky AR, Wu H, Xie L. Benzotriazole-mediated [2,3]-Wittig rearrangement. General and stereocontrolled syntheses of homoallyl alcohols and β,γ-unsaturated ketones. *J Org Chem.* 1996;61(12):4035–4039. https://doi.org/10.1021/jo960019d.
5. Berkenbusch T, Brueckner R. Stereocontrolled synthesis of the C^{21}-C^{38} fragment of the unnatural enantiomer of the antibiotic nystatin A_1. *Chem A Eur J.* 2004;10:1545–1557. https://doi.org/10.1002/chem.200305540.

Question 284: Arsonium Ylides for Cyclopropane Synthesis

To carry out a novel synthesis of highly substituted cyclopropanes, Moorhoff proceeded to couple 2*H*-pyran-5-carboxylates such as **1** with an arsonium ylide **2** in THF at 0°C to produce two *trans*-diastereomers **3** in 64% yield (Scheme 1).[1] Explain the mechanism of this transformation.

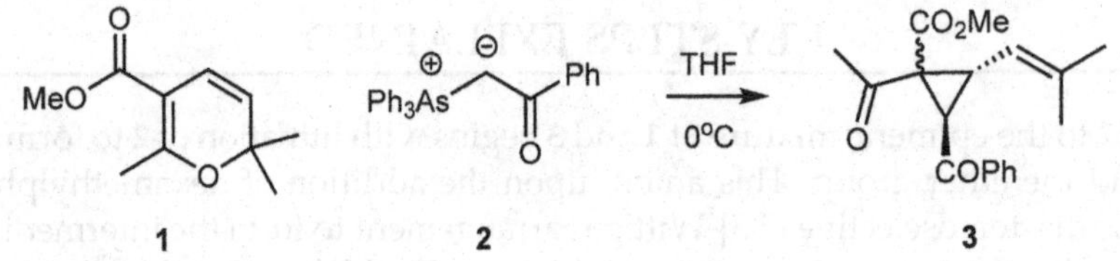

SCHEME 1

SOLUTION

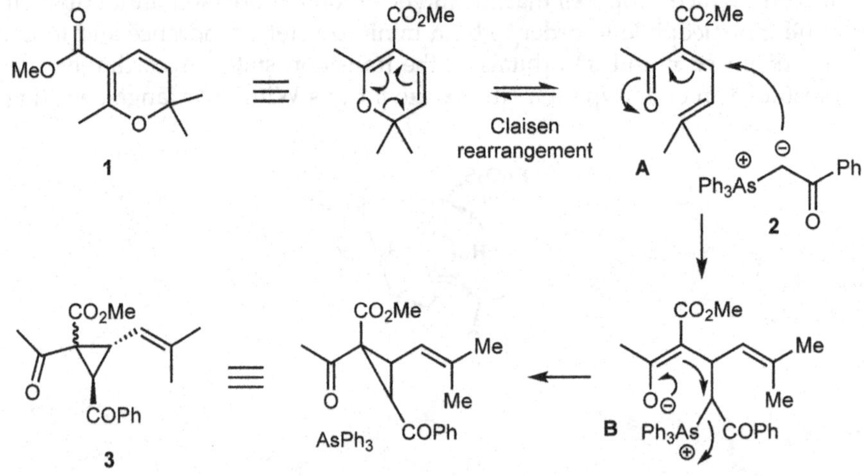

ALTERNATE POSSIBLE PATHWAY

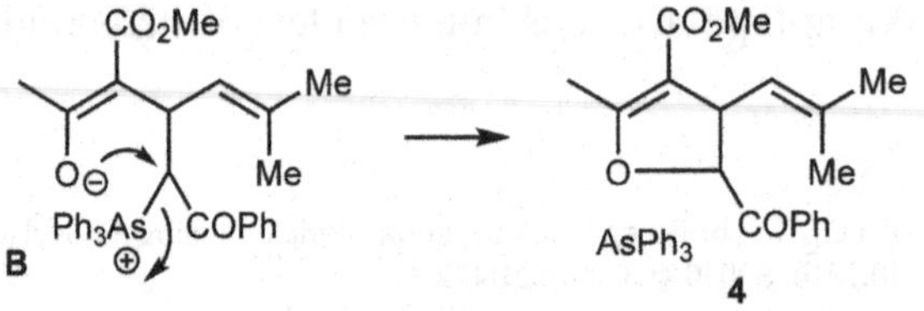

KEY STEPS EXPLAINED

According to Dr. Moorhoff, 2H-pyran-5-carboxylates **1** are interesting because they can undergo Claisen-type reversible electrocyclic ring-opening rearrangements to **A** which can be attacked by nucleophiles. In this problem the nucleophile is an arsonium ylide **2** and after attacking **A** produces **B**. Note that due to the presence of the carboxylate group at the beta position, structure **B** can be resonance stabilized. From this structure it is possible to form the preferred cyclopropane diastereomers **3** via *trans* elimination of the bulky triphenylarsine group. In some cases by an alternate pathway one can also form the substituted vinyldihydrofurans **4**.

ADDITIONAL RESOURCES

In a subsequent article, Dr. Moorhoff continued his work with 2H-pyran-5-carboxylates and arsonium ylides in order to show new synthetic applications of this procedure.[2] The various uses of arsonium ylides in organic chemistry have also been reviewed in the subsequent years.[3] Recently, similar transformations have been realized using pyridinium ylides undergoing similar mechanistic pathways.[4,5]

References

1. Moorhoff CM. Novel reactions of arsonium ylides and substituted 2*H*-pyran-5-carboxylates, a new preparation for functionalised vinylcyclopropanecarboxylates and dihydrofurans. *Tetrahedron Lett.* 1996;37(52):9349–9352. https://doi.org/10.1016/S0040-4039(97)82961-9.

2. Moorhoff CM, Winkler D. Transformation of substituted 2*H*-pyran-5-carboxylates into 3R*-vinyl-1,2R*-cyclopropanedicarboxylates. *New J Chem.* 1998;22:1485–1492. https://doi.org/10.1039/A804237C.

3. He HS, Chung CWY, But TYS, Toy PH. Arsonium ylides in organic synthesis. *Tetrahedron.* 2005;61:1385–1405. https://doi.org/10.1016/j.tet.2004.11.031.

4. Wang QF, Hou H, Hui L, Yan CG. Diastereoselective synthesis of trans-2,3-dihydrofurans with pyridinium ylide assisted tandem reaction. *J Org Chem.* 2009;74:7403–7406. https://doi.org/10.1021/jo901379h.

5. Chuang CP, Chen KP. *N*-phenacylpyridinium bromides in the one-pot synthesis of 2,3-dihydrofurans. *Tetrahedron.* 2012;68(5):1401–1406. https://doi.org/10.1016/j.tet.2011.12.035.

Question 285: Pyridoacridines From 4-Quinolones

In 1997 Spanish chemists Ajana et al. described a rapid method for the synthesis of pyridoacridines **4** from 4-quinolones **1** (Scheme 1).[1] Show mechanisms for all transformations.

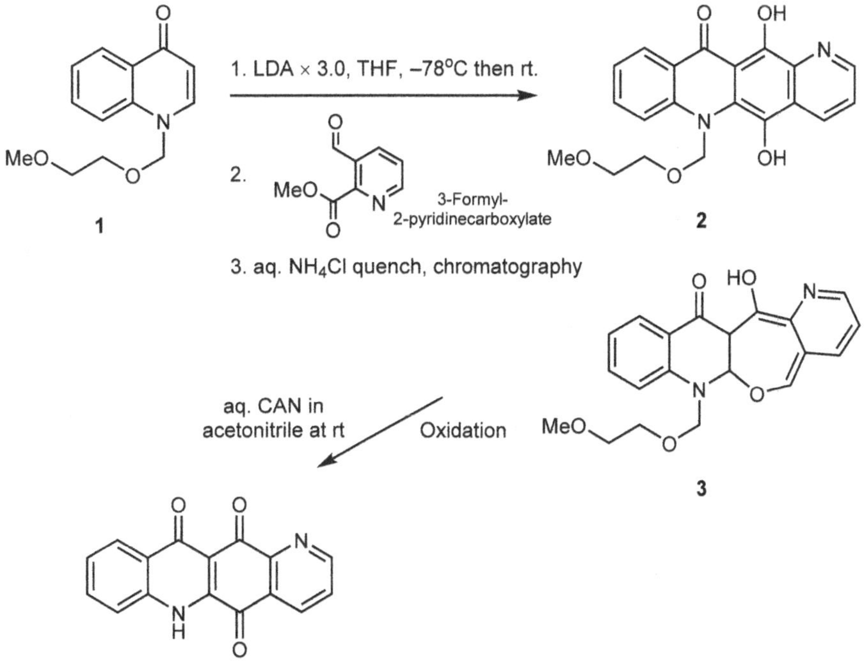

SCHEME 1

We begin the analysis by recognizing the initial treatment of **1** with three equivalents of lithium diisopropylamide (LDA). The authors noted this amount of LDA is needed in order to suppress the tendency of the compounds toward dimerization.[1] They also mentioned that the interaction of the LDA with **1** represents an instance of directed *ortho* metalation (DoM). Although they do not present the mechanistic details, Ajana et al. reference the work of Victor Snieckus. In his Chemical Reviews article, Snieckus explained that directed *ortho* metalation occurs when a directing metalation group (DMG) on a conjugated ring system interacts with a strong metal-containing base to direct deprotonation at the *ortho* position relative to the DMG. As seen in Part 1 of the solution, the first step involves coordination of an accumulated (RLi)$_n$ aggregate to the heteroatom containing the DMG, which in this case is the moderately directing —NR$_2$ group present on **1** in order to form the **A** aggregate. This aggregate can then undergo an energetically favorable deprotonation to give the coordinated *ortho*-lithiated species **B**. At this stage of the DoM process, electrophilic substitution can occur with the 3-formyl-2-pyridinecarboxylate reactant thus forming intermediate **C**. At this point, on the basis of arguments presented by Snieckus for a similar chemical transformation,[2] the authors propose a second DoM process which takes advantage of the newly established —CH₂OLi directing metalation group, essentially repeating the same mechanistic steps seen earlier to establish a doubly lithiated species **E**.[1] This, it is argued, occurs due to the inherent efficiency of the tandem DoM process and also in order to effect the subsequent cyclization step.[2]

SOLUTION: 1 TO 2

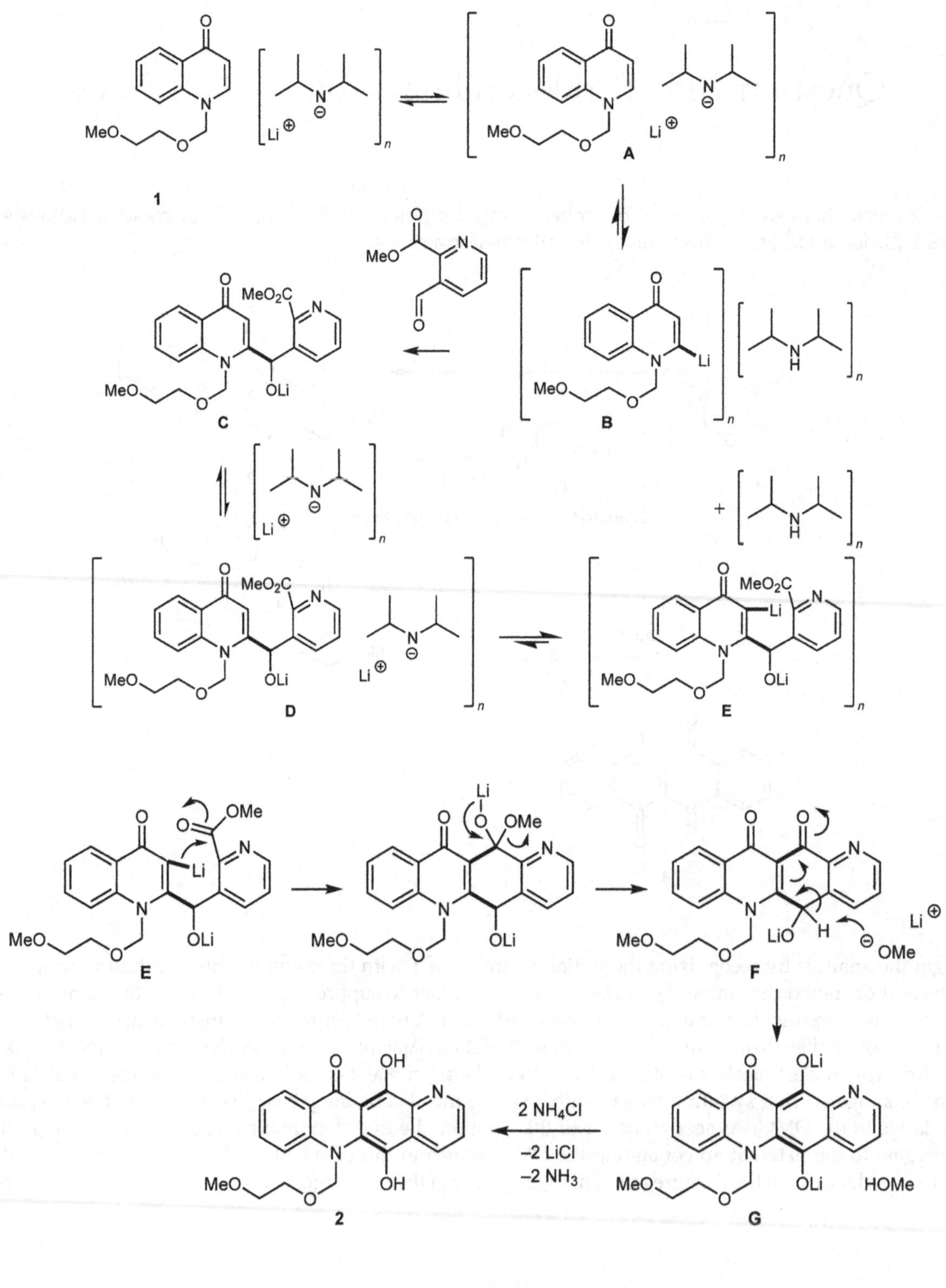

In part 2 of the solution we can see that **E** may undergo intramolecular acylation and deprotonation to arrive at **G** after loss of methanol. During work-up, 2 equivalents of ammonium chloride are used for protonation to arrive at **2**. To explain the formation of **3**, we have to revisit structure **F**. From **F** it is possible to form the epoxide **H** which after electrocyclic ring opening and ammonium chloride quench produces **3**.

BALANCED CHEMICAL EQUATION: 1 → 2

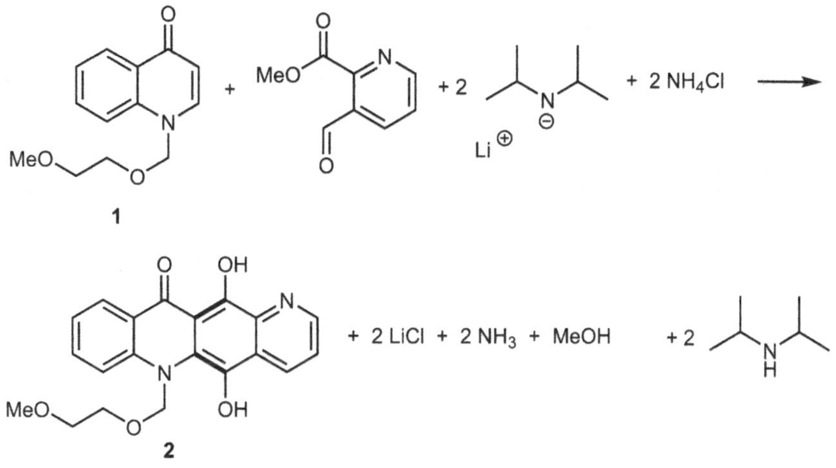

SOLUTION: F TO 3

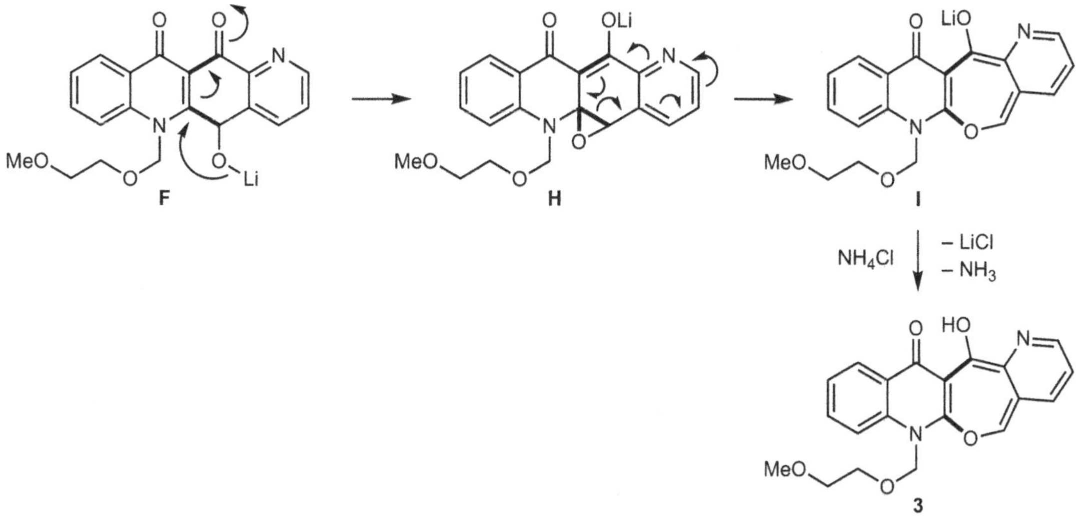

Finally, we are told that aqueous CAN (cerium ammonium nitrate) oxidation of **2** and **3** in acetonitrile at room temperature produces **4**. The mechanisms for this transformation starting from **2** and **3** are shown as follows.

BALANCED CHEMICAL EQUATION: 1 → 3

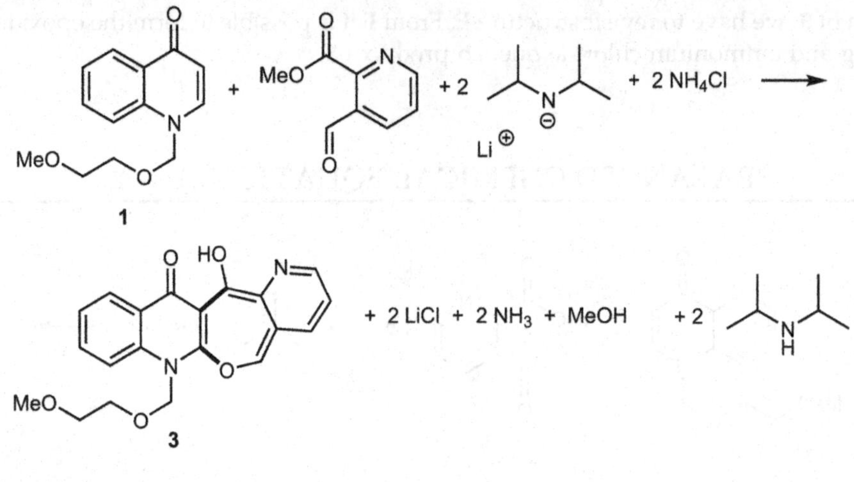

SOLUTION: CAN OXIDATION OF 2 → 4

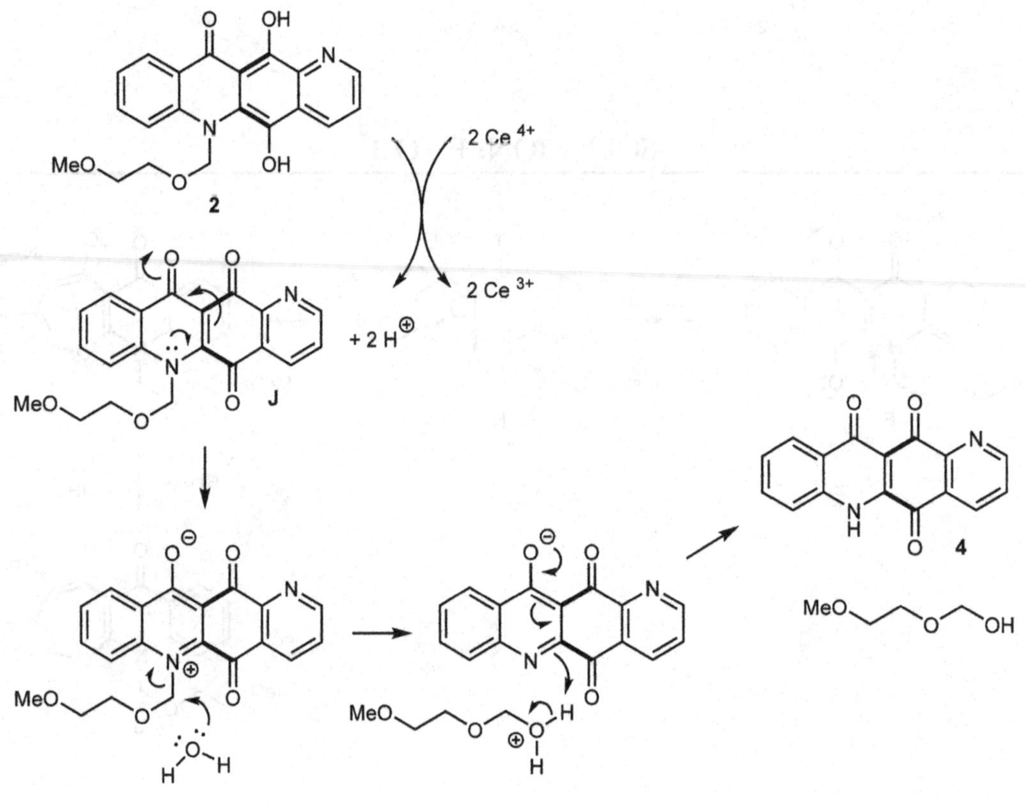

BALANCED CHEMICAL EQUATION: 2 → 4

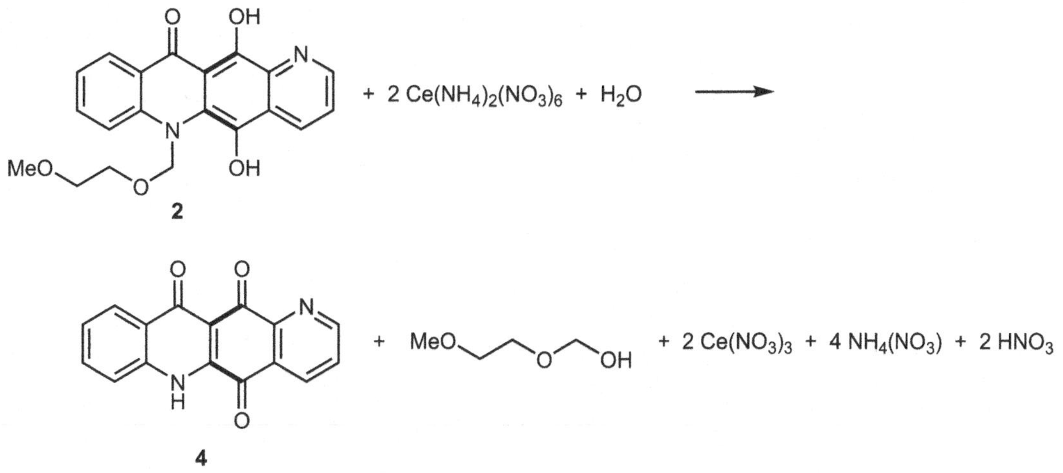

+ 2 Ce(NH4)2(NO3)6 + H2O
MeO
2
4
+ MeO OH + 2 Ce(NO3)3 + 4 NH4(NO3) + 2 HNO3

SOLUTION: CAN OXIDATION OF 3 → 4

3
Ce 4+
Ce 3+
+ H⊕
4
H2O quench
Ce 3+
Ce 4+

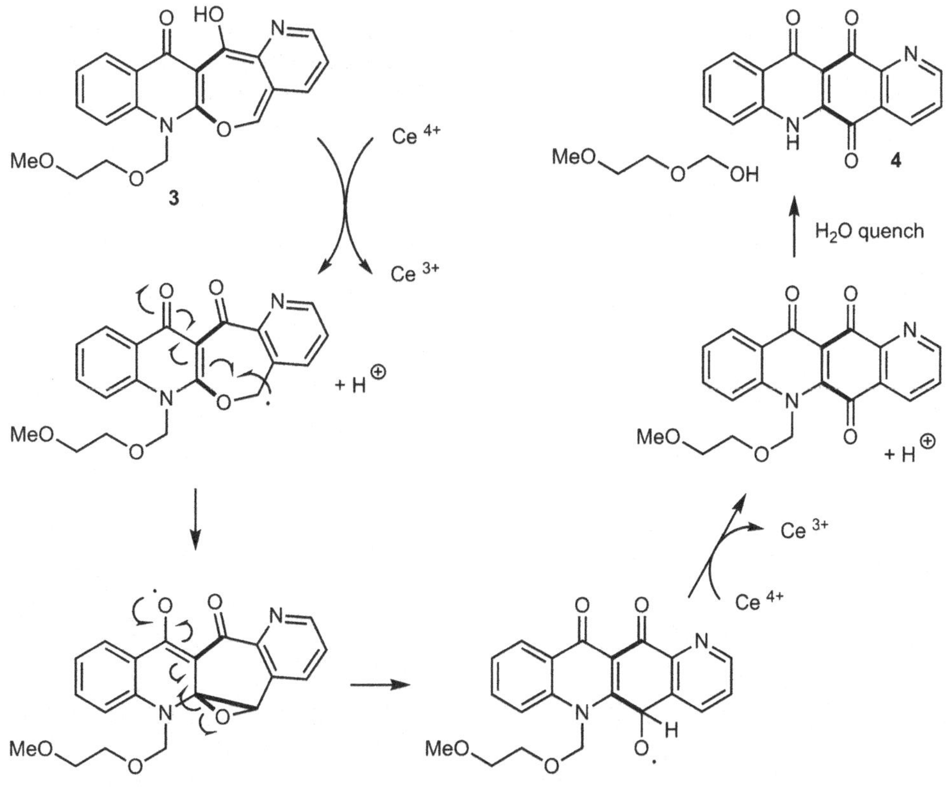

BALANCED CHEMICAL EQUATION: 3 → 4

$+ \ 2 \ Ce(NH_4)_2(NO_3)_6 \ + \ H_2O \longrightarrow$

3

$+ \ 2 \ Ce(NO_3)_3 \ + \ 4 \ NH_4(NO_3) \ + \ 2 \ HNO_3$

4

ADDITIONAL RESOURCES

For those interested in the mechanism behind deprotometalation reactions, we recommend a recent review from *Dalton Transactions*.[3] For an example which involves chemistry similar to that found in this problem, we refer the reader to a recent article which described the synthesis of indolin-3-ones and tetrahydro-4-quinolones.[4]

References

1. Ajana W, Lopez-Calahorra F, Joule JA, Alvarez M. Synthesis of pyrido[2,3-b]acridine-5,11,12-triones. *Tetrahedron Lett.* 1997;53(1):341–356. https://doi.org/10.1016/S0040-4020(96)00990-8.
2. Watanabe M, Snieckus V. Tandem directed metalation reactions. Short syntheses of polycyclic aromatic hydrocarbons and ellipticine alkaloids. *J Am Chem Soc.* 1980;102(4):1457–1460. https://doi.org/10.1021/ja00524a059.
3. Harford PJ, Peel AJ, Chevallier F, et al. New avenues in the directed deprotometallation of aromatics: recent advances in directed cupration. *Dalton Trans.* 2014;43:14181–14203. https://doi.org/10.1039/c4dt01130a.
4. Shimizu M, Takao Y, Katsurayama H, Mizota I. Synthesis of indolin-3-ones and tetrahydro-4-quinolones from α-imino esters. *Asian J Org Chem.* 2013;2:130–134. https://doi.org/10.1002/ajoc.201200174.

Further Reading

5. Snieckus V. Directed ortho metalation. Tertiary amide and *o*-carbamate directors in synthetic strategies for polysubstituted aromatics. *Chem Rev.* 1990;90(6):879–933. https://doi.org/10.1021/cr00104a001.

Question 286: Imidoyl Radicals in a New Quinoline Synthesis

In order to carry out a new annulation synthesis of quinolines, Nanni et al. decided to use the approach of hydrogen abstraction from imines.[1] Imine **1** was reacted with phenylacetylene and di-isopropyl peroxydicarbonate in benzene at 60°C to produce a mixture of quinolines **2** and **3** in 65% yield (ratio of 2:3 = 4.4, Scheme 1). Show mechanisms for both transformations.

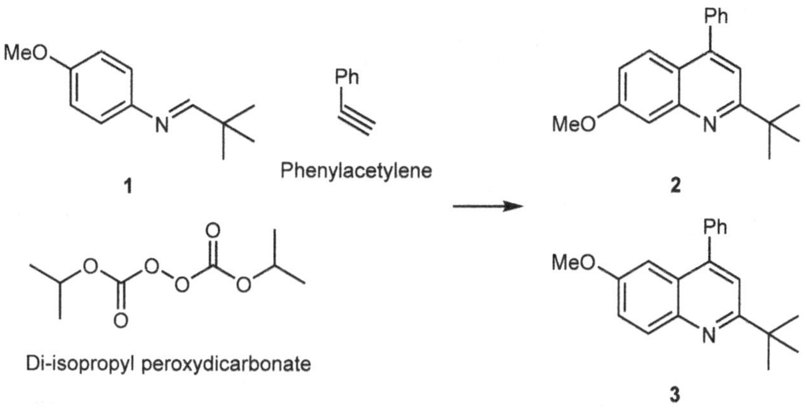

SCHEME 1

SOLUTION—1 TO B

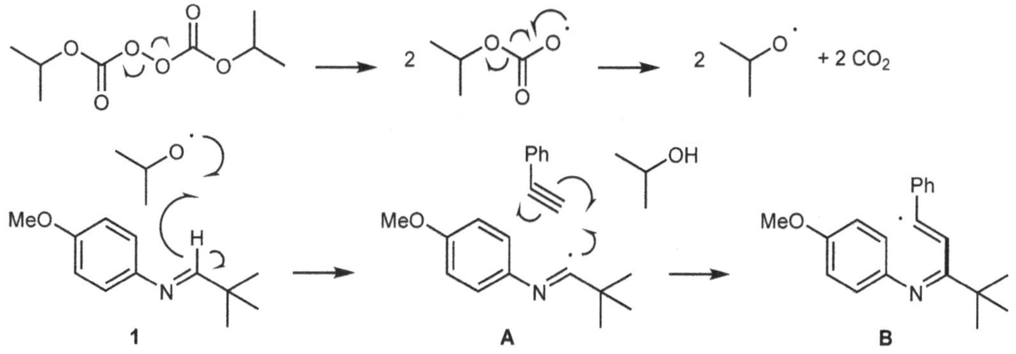

SOLUTION—B TO 2

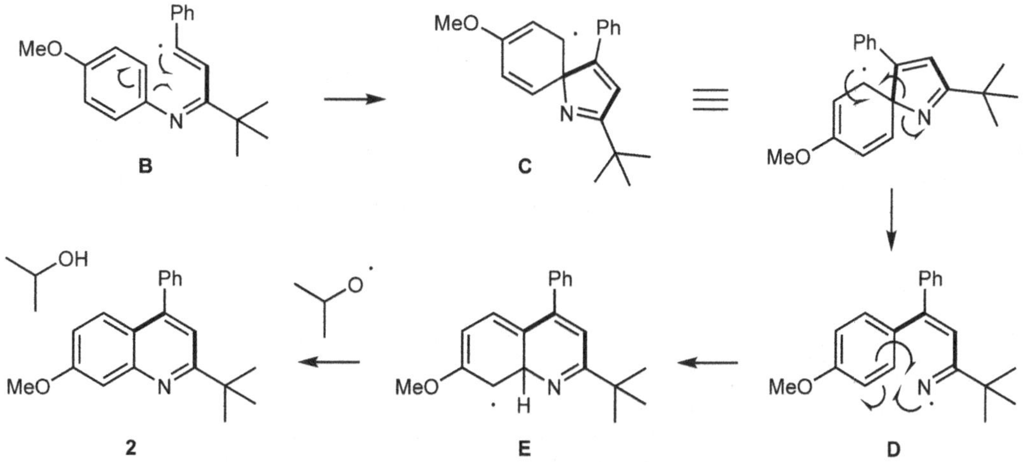

SOLUTION—B TO 3

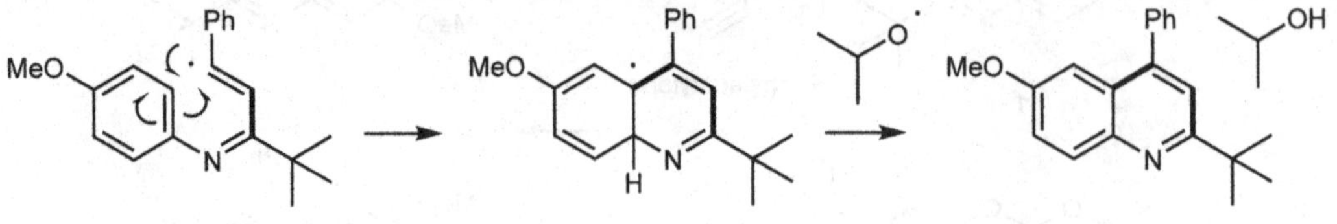

B   F   3

KEY STEPS EXPLAINED

During their study of the synthetic use of isonitriles as a source of imidoyl radicals, Nanni et al. described the synthesis of quinolines **2** and **3** starting from a radical-induced reaction of imine **1** with phenylacetylene.[1] The initiation part of this transformation begins with the homolytic cleavage and decarboxylation of di-isopropyl peroxydicarbonate to form two equivalents of 2-propane oxide radical. One equivalent of this radical species abstracts a proton from **1** to generate the imidoyl radical **A** which reacts with phenylacetylene to form a vinyl radical **B**. The species **B** can cyclize either in a 6-*exo* fashion (in order to lead to the quinoline product **3** after a final proton abstraction and rearomatization), or it may cyclize via a 5-*endo* route to produce the *spiro* radical intermediate **C**.[1] Upon rearomatization and homolytically induced cleavage of the five-membered ring of **C**, we are left with iminyl **D** which can subsequently cyclize in the normal fashion and after radical-induced proton abstraction and rearomatization, **D** will lead to the formation of the final quinoline product **2**.

The authors describe the initial proton abstraction from the imine **1** as a novel homolytic α-fragmentation.[1] Furthermore, they demonstrate the importance of structure **D** by experimentally isolating minute amounts of the side product **4** (Scheme 2). The isolation of this product in varying yields depending on the ability of the R group (β to the nitrogen atom) to stabilize a radical species upon β-fragmentation provides experimental support for structure **D**. With regard to the rest of the mechanism, the authors do leave open the possibility for alternative rearrangement pathways, especially concerning the often debated rearrangement of the azaspirocyclohexadienyl species **C**.

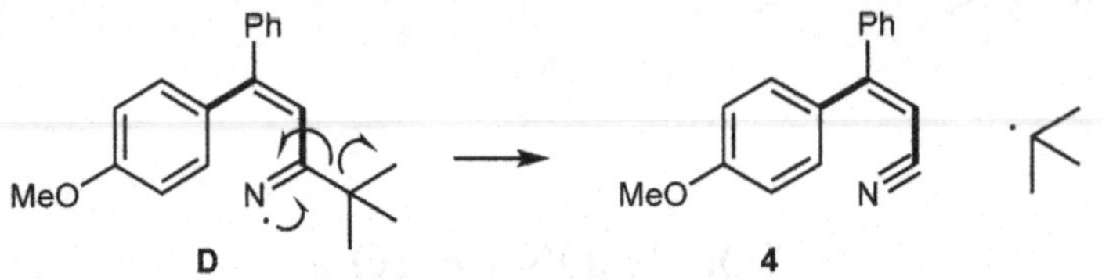

D   4

SCHEME 2

ADDITIONAL RESOURCES

In a previous publication, Nanni et al. have gone into a larger scope analysis of the mechanism, mechanism by-products, and side products of this reaction including other possible rearrangements of **C**.[2] We highly recommend this reference. Additional examples of the mechanistic roles of iminyl radicals (such as **D**) in synthesis which also involve the work of Nanni et al. have appeared in the literature over the years.[3–5]

References

1. Nanni D, Pareschi P, Tundo A. Isonitriles as source and fate of imidoyl radicals: a novel homolytic α-fragmentation. *Tetrahedron Lett*. 1996;37 (52):9337–9340. https://doi.org/10.1016/S0040-4039(97)82958-9.
2. Nanni D, Pareschi P, Rizzoli C, Sgarabotto P, Tundo A. Radical annulations and cyclisations with isonitriles: the fate of the intermediate imidoyl and cyclohexadienyl radicals. *Tetrahedron*. 1995;51(33):9045–9062. https://doi.org/10.1016/0040-4020(95)00348-C.
3. Camaggi CM, Leardini R, Nanni D, Zanardi G. Radical annulations with nitriles: novel cascade reactions of cyano-substituted alkyl and sulfanyl radicals with isonitriles. *Tetrahedron*. 1998;54(21):5587–5598. https://doi.org/10.1016/S0040-4020(98)00230-0.
4. Leardini R, Nanni D, Pareschi P, Tundo A, Zanardi G. α-(arylthio)imidoyl radicals: [3 + 2] radical annulation of aryl isothiocyanates with 2-cyano-substituted aryl radicals. *J Org Chem*. 1997;62(24):8394–8399. https://doi.org/10.1021/jo971128a.
5. Leardini R, Nanni D, Zanardi G. Radical addition to isonitriles: a route to polyfunctionalized alkenes through a novel three-component radical cascade reaction. *J Org Chem*. 2000;65(9):2763–2772. https://doi.org/10.1021/jo991871y.

Question 287: Electrolytic Fluorinative Ring Expansion Reactions

In 1996 Japanese chemists Hara et al. had described an electrochemically induced fluorinative ring expansion reaction involving the formation of difluorocycloalkyl esters **2** from the electrolysis of cycloalkylideneacetates **1** using $Et_3N \cdot 5HF$ at $-20°C$ (Scheme 1).[1] Suggest a mechanism for this transformation.

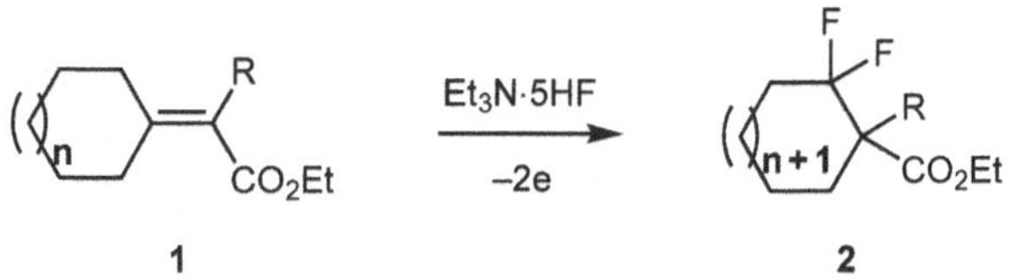

SCHEME 1

SOLUTION

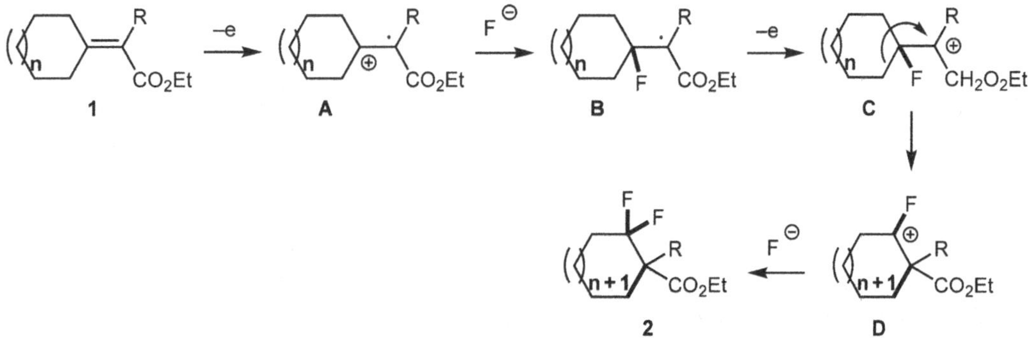

KEY STEPS EXPLAINED

The authors describe the mechanism as beginning with a one-electron oxidation of the starting cycloalkylideneacetate **1** to form the radical cation species **A**.[1] Fluorination of **A** by $Et_3N \cdot 5HF$ produces the fluoroalkyl radical **B** which can undergo one-electron oxidation once more to form an unstable carbocation **C** which can rearrange into the more stable one —CH_2— group ring-expansion intermediate **D**. Fluorination of **D** gives the final difluorocycloalkyl ester **2**. The authors also present a mechanistic pathway consisting of deprotonation, oxidation, and fluorination of **A** in order to form a mono-fluorinated by-product which competes with the formation of the target product **2** (Scheme 2).[1] Furthermore, the authors rationalize that this unusual ring expansion transformation is driven by the formation of the carbocation intermediate **C** which is destabilized inductively by the ester electron-withdrawing group. This intermediate is therefore aided in its isomerization to the more stable intermediate **D**. A methyl group, for instance, inhibits this transformation. Certain experimental conditions and their consequences are also briefly touched upon in the original article.

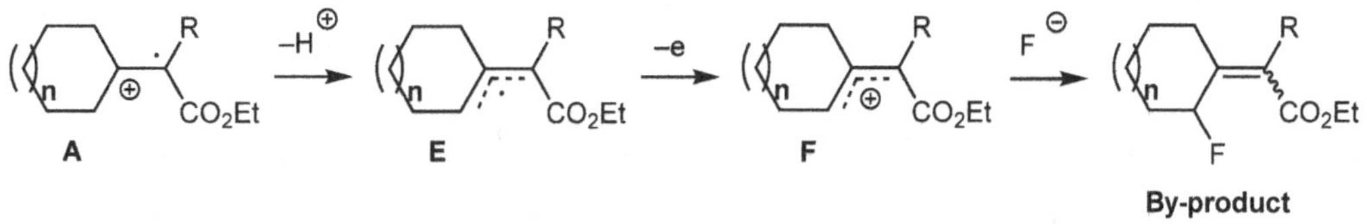

SCHEME 2

ADDITIONAL RESOURCES

Over the years, several reports describing oxidative fluorination transformations[2,3] appeared in the literature including one detailed mechanistic analysis of the electrochemical fluorination of chlorobenzene.[4] We would also like to highlight two other noteworthy examples, one where fluorination of cyclic alkenes results in ring contraction[5] and the other example having to do with the ring expansion of cyclic ethers.[6]

References

1. Hara S, Chen SQ, Hoshio T, Fukuhara T, Yoneda N. Electrochemically induced fluorinative ring expansion of cycloalkylideneacetates. *Tetrahedron Lett.* 1996;37(47):8511–8514. https://doi.org/10.1016/0040-4039(96)01953-3.
2. Meurs JHH, Eilenberg W. Oxidative fluorination in amine-HF mixtures. *Tetrahedron.* 1991;47(4–5):705–714. https://doi.org/10.1016/S0040-4020(01)87060-5.
3. Chen SQ, Hatakeyama T, Fukuhara T, Hara S, Yoneda N. Electrochemical fluorination of aliphatic aldehydes and cyclic ketones using Et$_3$N-5HF electrolyte. *Electrochim Acta.* 1997;42(13–14):1951–1960. https://doi.org/10.1016/S0013-4686(97)85466-7.
4. Momota K, Horio H, Kato K, Morita M, Matsuda Y. Electrochemical fluorination of aromatic compounds in liquid R$_4$NF.mHF—Part IV. Fluorination of chlorobenzene. *Electrochim Acta.* 1995;40(2):233–240. https://doi.org/10.1016/0013-4686(94)E0183-Z.
5. Hara S, Nakahigashi J, Ishi-i K, Fukuhara T, Yoneda N. Fluorinative ring-contraction of cyclic alkenes with *p*-iodotoluene difluoride. *Tetrahedron Lett.* 1998;39(17):2589–2592. https://doi.org/10.1016/S0040-4039(98)00276-7.
6. Inagaki T, Nakamura Y, Sawaguchi M, Yoneda N, Ayuba S, Hara S. Fluorinative ring-expansion of cyclic ethers using *p*-iodotoluene difluoride. Stereoselective synthesis of fluoro cyclic ethers. *Tetrahedron Lett.* 2003;44(21):4117–4119. https://doi.org/10.1016/S0040-4039(03)00841-4.

Question 288: An Isoxazole to 1,3-Oxazine Transformation

During a 1986 reexamination of a Vilsmeier reaction, Anderson showed that reacting 3-phenyl-5-isoxazolone with POCl$_3$/DMF at room temperature then for 30 min at 60°C gave the isoxazolone **1** in 35% yield upon standard neutralization with 5% NaHCO$_3$ solution (Scheme 1).[1] Under longer reaction times or higher temperatures (>80°C), the product was the 1,3-oxazinone **2** isolated in 71% yield, and **1** was shown to be a precursor to **2**. Suggest mechanisms to account for the formation of **1** and **2**.

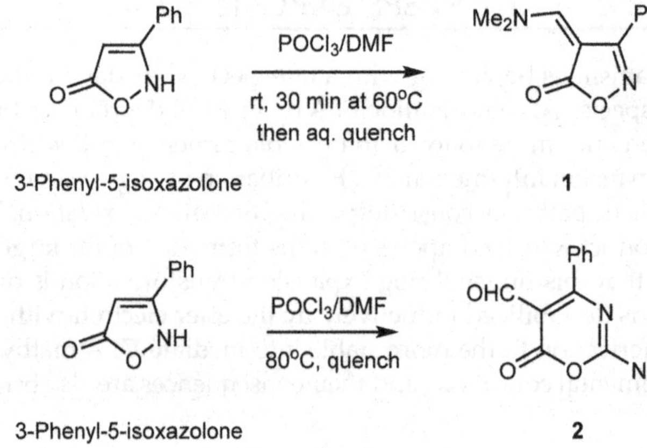

SCHEME 1

KEY STEPS EXPLAINED

The 3-phenyl-5-isoxazolone to **1** transformation begins with the formation of the Vilsmeier reagent **A** which reacts with 3-phenyl-5-isoxazolone in order to lose HCl and to form **1**. The authors stipulate that this transformation occurs during milder conditions with faster reaction times. Under prolonged reaction times with harsher conditions, given an excess of the Vilsmeier reagent **A**, **1** can undergo further reactivity in order to add a second equivalent of **A**. This in turn can lead to ring opening which can immediately close back into a six-membered oxazinone structure **B** which upon aqueous quench, HCl loss, and NHMe₂ loss can form the resultant aldehyde **2**. In their paper, the authors go on to explain the possible formation of a third product from **1**, which competes with the ring expansion to **2**. To save space, we did not include the details of this possibility here and instead refer the reader to Ref. 1.

SOLUTION: 3-PHENYL-5-ISOXAZOLONE → 1

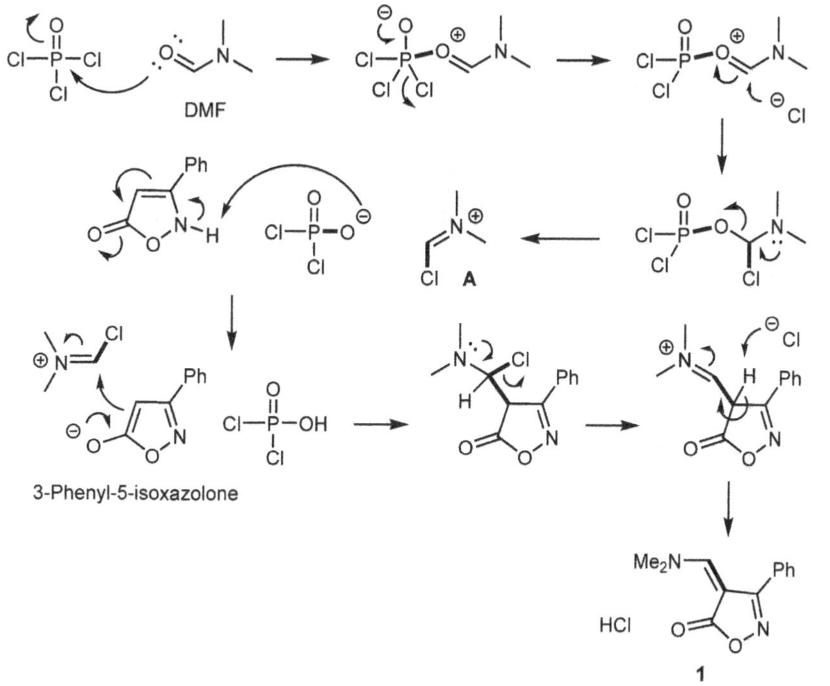

SOLUTION: 1 → 2

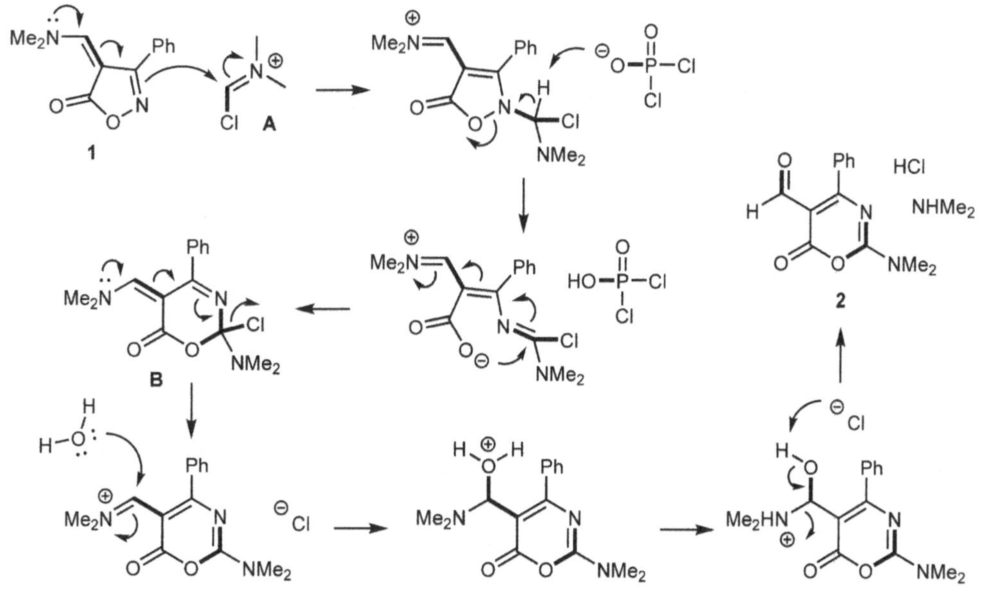

ADDITIONAL RESOURCES

In previous years there have appeared examples in the literature involving mechanistic analysis of chemistry involving similar transformations with Vilsmeier reagents that have led to both ring expansion[2,3] and ring maintenance.[4] A short review expanding upon the mechanistic considerations related to nitrogen-containing heterocycles is also recommended.[5]

References

1. Anderson DJ. A reinvestigation of the Vilsmeier reaction of 3-phenyl-5-isoxazolinone. Isolation of 1,3-oxazin-6-ones. J Org Chem. 1986;51:945–947. https://doi.org/10.1021/jo00356a039.
2. Beccalli EM, Marchesini A. The Vilsmeier-Haack reaction of isoxazolin-5-ones. Synthesis and reactivity of 2-(dialkylamino)-1,3-oxazin-6-ones. J Org Chem. 1987;52(15):3426–3434. https://doi.org/10.1021/jo00391a048.
3. Li KL, Du ZB, Guo CC, Chen QY. Regioselective syntheses of 2- and 4-formylpyrido[2,1-b]benzoxazoles. J Org Chem. 2009;74(9):3286–3292. https://doi.org/10.1021/jo900267c.
4. Becker C, Roschupkina G, Rybalova T, Gatilov Y, Reznikov V. Transformations of 2,2-dimethyl-2,4-dihydro-3H-pyrrol-3-on-1-oxide derivatives in the Vilsmeier–Haack reaction conditions. Tetrahedron. 2008;64:9191–9196. https://doi.org/10.1016/j.tet.2008.07.070.
5. Banert K. New functionalized allenes: synthesis using sigmatropic rearrangements and unusual reactivity. Liebigs Ann Recl. 1997;1997(10):2005–2018. https://doi.org/10.1002/jlac.199719971003.

Question 289: Synthesis of Folate Antimetabolites: A Furan to Pyrrole Transformation

During studies for the synthesis of furo[2,3-d]-pyrimidines from o-aminonitriles 1 and guanidine free base (Scheme 1), Taylor et al. obtained an unexpected result (Scheme 1).[1] Devise a mechanism to account for this result.

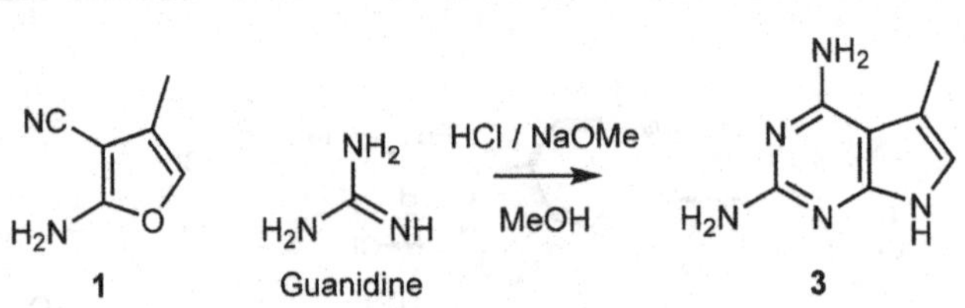

SCHEME 1

SOLUTION: MECHANISM A

SOLUTION: MECHANISM B

BALANCED CHEMICAL EQUATION (APPLIES TO BOTH MECHANISM A AND B)

KEY STEPS EXPLAINED

The authors encountered this interesting transformation during studies of the synthesis of furo[2,3-*d*]-pyrimidines which are valuable compounds for cancer therapy.[1] It is evident from the proposed mechanisms that *o*-aminonitriles **1** react with guanidine free base to produce nitro derivatives of the furo[2,3-*d*]-pyrimidines upon loss of water. Taylor et al. have proposed two possible pathways for this transformation.

In mechanism A, guanidine reacts as a nucleophile and adds to **1** resulting in electron density being delocalized onto the cyano group. After a hydrogen shift, the electron density can return to the ring to reestablish the cyano group and lead to ring opening giving the enol **B**. Within this form, **B** can ring close on the cyano side to form, upon prototropic rearrangements, the pyrimidine structure **C**. Keto-enol tautomerization can then establish the aldehyde which can ring close and lose water in order to form the final product **3**.

Mechanism B begins with guanidine reacting as a base to deprotonate **1** and to generate electron density which is delocalized onto the cyano group of **1**. Regeneration of the cyano group results in ring opening and a prototropic rearrangement produces an interesting dinitrile species **E**. The authors propose that at this point guanidine comes in as a nucleophile to add to **E** and, following several prototropic rearrangements, eventually establishes the pyrimidine **F**. As seen before, **F** can undergo a ring-closing condensation, losing water, in order to form the final product **3**.

Although the authors do not present any experimental and computational evidence in support of either mechanism, in a subsequent paper, Jun, an original coauthor, provided a literature based argument in favor of mechanism A.[2] Therein, Jun argued that the literature supports mechanism A for all furan derivatives whereas mechanism B is encountered only for certain thiophene derivatives. We have examined these references and concluded that they did not establish any hard evidence for either mechanism in the context of the compounds and chemistry encountered in this problem. For example, the literature examples lack the pyrimidine ring functionality, the dinitrile **E** or similar structures, and finally estimates of pK_A and step energy barriers are absent from all discussions. In one case relevant to thiophene derivatives (ref. 5b) in Ref. 2, the main reactant was a symmetrical nitro furan derivative where the thio group and the amine group simply swap positions without loss of any components to form the nitro furan derivative. Without pK_A and energy values, spectroscopic evidence or the presence of similar compounds undergoing similar transformations, it is impossible to construct an objective argument in support of either mechanism. We therefore recommend for future study, a serious mechanistic analysis to establish whether guanidine reacts primarily as a nucleophile or as a base. The experimenter can engage the use of low temperature FT-IR spectroscopy which can help identify the presence of dinitrile species **E** as required by mechanism B. Alternatively, one can independently synthesize **E** and then investigate whether its reaction with guanidine leads to product **3**.

We would however like to present a brief speculative argument in favor of mechanism A. For this we turn the reader's attention to the structure of **E** in mechanism B. This species is highly prone to deprotonation at two positions, the β carbon relative to the aldehyde and the γ carbon relative to the aldehyde (i.e., β to the two cyano groups). Of the two, the more acidic position is likely the γ carbon (β to the two cyano groups). This is because the cyano groups contribute both inductively and electronically to withdraw electron density from the γ carbon and therefore they can stabilize any negative charge that might develop there. When compared to the amine group of **1**, this position is in our judgment much more prone to deprotonation. Deprotonation at this position can only lead to decomposition and will not produce **3**. Since guanidine is a strong base (pK$_A$ of conjugate acid is 13.6), speculatively we would not expect guanidine to react as a nucleophile with **E** as depicted in mechanism B. Moreover, we view it as a mistake to expect that guanidine will react as a base with **1** and as a nucleophile with **E**. Nevertheless, lacking definitive pK$_A$ values, we cannot validate this argument and must state it only as our current thinking on the issue. For this reason, we speculatively favor mechanism A in this problem.

ADDITIONAL RESOURCES

For more examples of the work of Prof. E.C. Taylor with regard to synthetic achievements, we recommend Ref. 3.

References

1. Taylor EC, Patel HH, Jun JG. A one-step ring transformation/ring annulation approach to pyrrolo[2,3-*d*]pyrimidines. A new synthesis of the potent inhibitor TNP-351. *J Org Chem*. 1995;60:6684–6687. https://doi.org/10.1021/jo00126a017.
2. Jun JG. Regioselective synthesis of 2-amino-3-cyano-furan derivatives and its guanidine cyclization reaction. *Bull Korean Chem Soc*. 1996;17 (8):676–678.
3. Taylor EC, Liu B. A new and efficient synthesis of pyrrolo[2,3-*d*]pyrimidine anticancer agents: alimta (LY231514, MTA), homo-alimta, TNP-351, and some aryl 5-substituted pyrrolo[2,3-*d*]pyrimidines. *J Org Chem*. 2003;68:9938–9947. https://doi.org/10.1021/jo030248h.

Question 290: Rearrangement of a Silylacetylenic Ketone

In 1988 during synthetic studies of bicyclic keto silanes, American chemists Kende et al. noticed an unexpected rearrangement when the cycloalkanone **1** bearing an ω-silylacetylenic chain β to the carbonyl group was heated neat at 300°C for 2h to produce the bicyclic enol silyl ether **2** in 60%–65% yield (Scheme 1).[1] Provide a mechanism for this transformation.

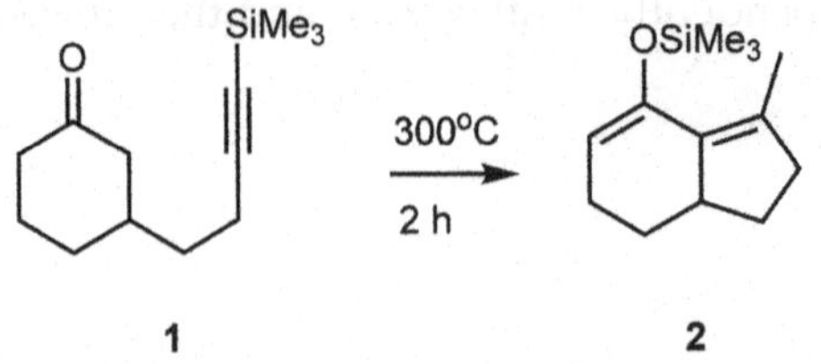

SCHEME 1

SOLUTION

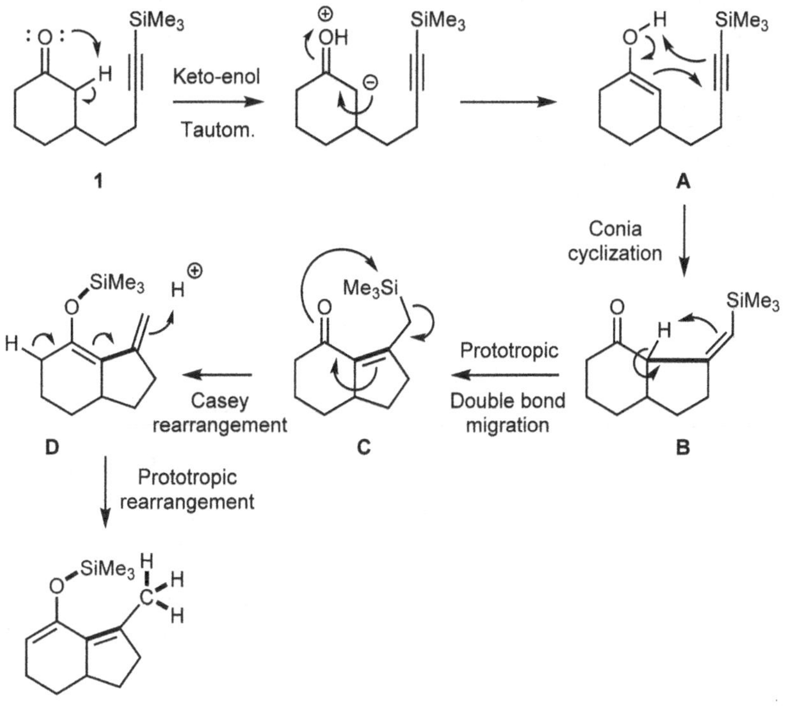

KEY STEPS EXPLAINED

In the work of Kende et al.,[1] the mechanism for this rearrangement sequence begins with the keto-enol tautomerization of cycloalkanone **1** to the enol **A**. At this point, due to the proximity of the ω-silylacetylenic chain (which was β to the carbonyl group), a so-called Conia cyclization can occur forming **B** which has the vinylsilane functionality.[2] A prototropic double bond migration produces the conjugated enone **C** which can undergo Casey rearrangement of silicon to oxygen giving **D**.[3] A final prototropic rearrangement gives the final bicyclic enol silyl ether **2**.

ADDITIONAL RESOURCES

Over the years, examples of novel syntheses of functionalized allylsilanes[4], radical-mediated intramolecular cyclization of silylacetylenic α-iodo ketones,[5] and double bond migrations[6] have been reported. These examples work well to complement some of the mechanistic steps encountered in this problem.

References

1. Kende AS, Hebeisen P, Newbold RC. Synthesis of bicyclic keto silanes by tandem rearrangement of silylacetylenic ketones. *J Am Chem Soc*. 1988;110:3315–3317. https://doi.org/10.1021/ja00218a060.
2. Conia JM, Le Perchec P. The thermal cyclisation of unsaturated carbonyl compounds. *Synthesis*. 1975;1975(1):1–19. https://doi.org/10.1055/s-1975-23652.
3. Casey CP, Jones CR, Tukada H. Interconversion of .gamma.-silyl .alpha.,.beta.-unsaturated carbonyl compounds and siloxybutadienes by 1,5-shifts of silicon between carbon and oxygen. *J Org Chem*. 1981;46(10):2089–2092. https://doi.org/10.1021/jo00323a022.
4. Li WDZ, Yang JH. A novel synthesis of functionalized allylsilanes. *Org Lett*. 2004;6(11):1849–1852. https://doi.org/10.1021/ol049311v.
5. Sha CK, Jean TS, Wang DC. Intramolecular radical cyclization of silylacetylenic or olefinic α-iodo ketones: application to the total synthesis of (±)-modhephene. *Tetrahedron Lett*. 1990;31(26):3745–3748. https://doi.org/10.1016/S0040-4039(00)97460-4.
6. Kobychev VB. Double bond migration mechanism in allyl systems involving the hydroxide ion. 2. Polarizable continuum model (PCM). *J Struct Chem*. 2004;45(1):20–27. https://doi.org/10.1023/B:JORY.0000041497.26313.a8.

Question 291: And More Silyl Rearrangements: A Brook-Retro-Brook Sequence

In 1996 American chemists prepared the epoxide **1** and treated it with *t*-butyllithium in THF/HMPA at −78°C in the hopes that a newly formed benzylic carbanion would undergo 4-*exo*-epoxide ring opening and a [1,2]-silyl shift (Brook rearrangement) to generate oxetane **2** with complete stereocontrol (Scheme 1).[1] After successful formation of the anion and treatment with 1 mL 1 M HCl, the aldehyde **3** was obtained pure in 40% yield (Scheme 2). There was no oxetane produced. Explain the mechanism of this transformation **1 → 3**.

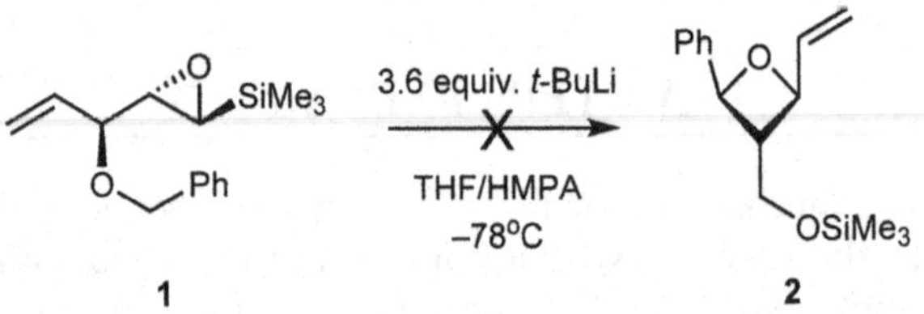

SCHEME 1

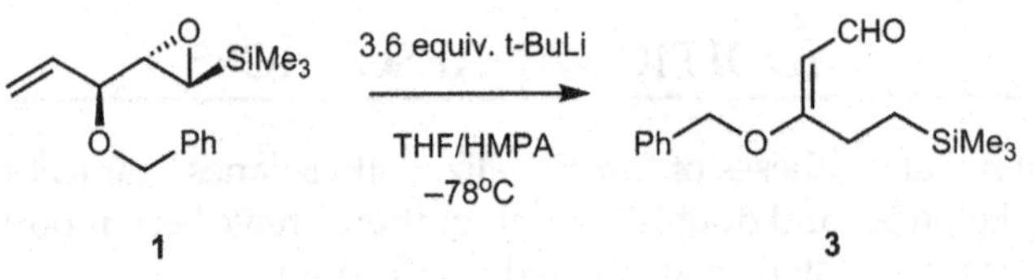

SCHEME 2

SOLUTION

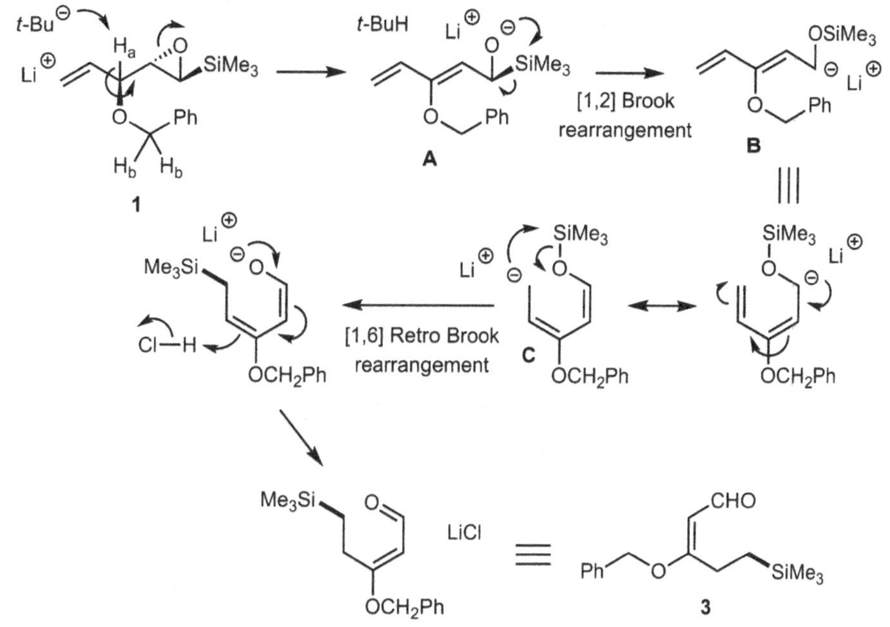

A

B

C

[1,2] Brook rearrangement

[1,6] Retro Brook rearrangement

3

BALANCED CHEMICAL EQUATION

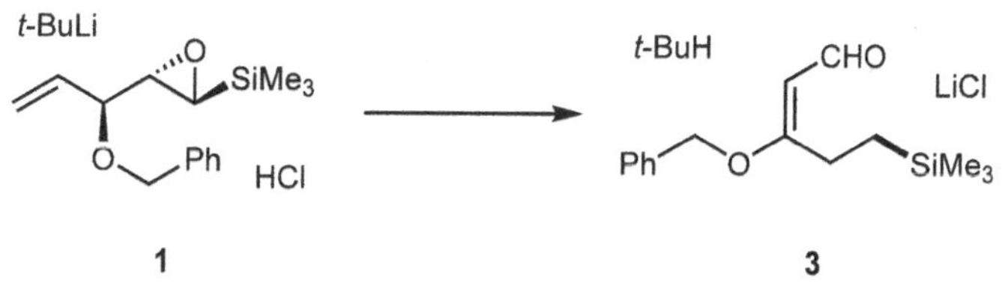

1

3

KEY STEPS EXPLAINED

In their original work, Jung and Nichols attempted to transform **1** into **2** using *t*-BuLi in THF/HMPA at −78°C.[1] The intent was to deprotonate at the benzylic position (H_b) and to have the anion formed attack the epoxide to conduct a 4-*exo*-ring opening to effect a standard [1,2]-silyl shift (Brook rearrangement) and form **2** with complete stereocontrol.[2] Deprotonation did not occur at the benzylic position as intended but did occur at the vinylic position (H_a) with ring opening and Brook rearrangement forming **B**. Structure **B** is strongly resonance stabilized and in its **C** form undergoes a reverse, or a retro-Brook rearrangement, that is, [1,6]-silyl shift. The resulting structure can isomerize to the thermodynamically stable conjugated aldehyde **3** upon acid work-up. This example is claimed to be the first reported case of an intramolecular [1,6]-retro-Brook rearrangement.

ADDITIONAL RESOURCES

Over the years there has been immense interest in the Brook rearrangement. A recent review chapter captures the essence of much of this work and thus is highly recommended.[3]

References

1. Jung ME, Nichols CJ. A novel tandem [1,2]-Brook/retro-[1,6]-Brook rearrangement of a 1-(trimethylsilyl)-2,4-pentadien-1-ol anion. *J Org Chem.* 1996;61:9065–9067. https://doi.org/10.1021/jo961265s.
2. Rojas CM. Brook rearrangement. In: Li JJ, ed. *Name Reactions for Homologations Part II.* Hoboken, NJ: John Wiley & Sons, Inc.; 2009:406–437. https://doi.org/10.1002/9780470487044.ch1
3. Sasaki M, Takeda K. Brook rearrangement. In: Rojas CM, ed. *Molecular Rearrangements in Organic Synthesis.* Hoboken, NJ: John Wiley & Sons, Inc.; 2015. https://doi.org/10.1002/9781118939901.ch6

Question 292: Two Syntheses of Dehydrorotenone

Dehydrorotenone **1** can be synthesized in two ways using DCC (dicyclohexylcarbodiimide) as a coupling agent: (i) addition of derrisic acid **2** to a solution of DCC/Et₃N followed by reaction of the intermediate with sodium propanoate in ethanol;[1] and (ii) condensation of tubaic acid **3** with the pyrrolidine enamine of **4** in the presence of DCC (Scheme 1).[2] Elucidate these syntheses and explain the mechanistic role of DCC.

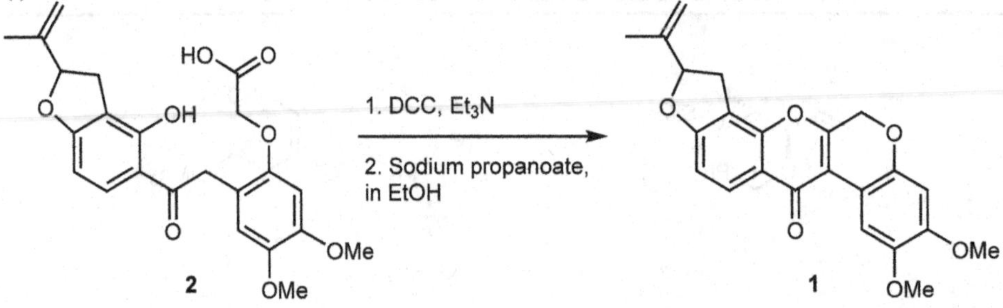

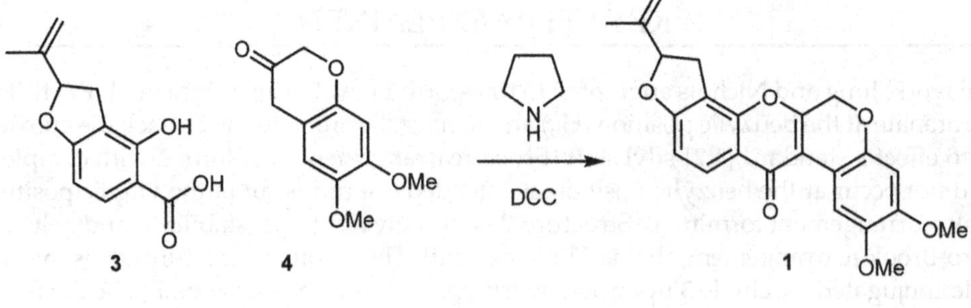

SCHEME 1

SOLUTION: (I) FROM 2[1]

DCC = dicyclohexylcarbodiimide =
2
A
B
C
rotn
EtOH
Na⊕
EtO⊖
OMe
Cy

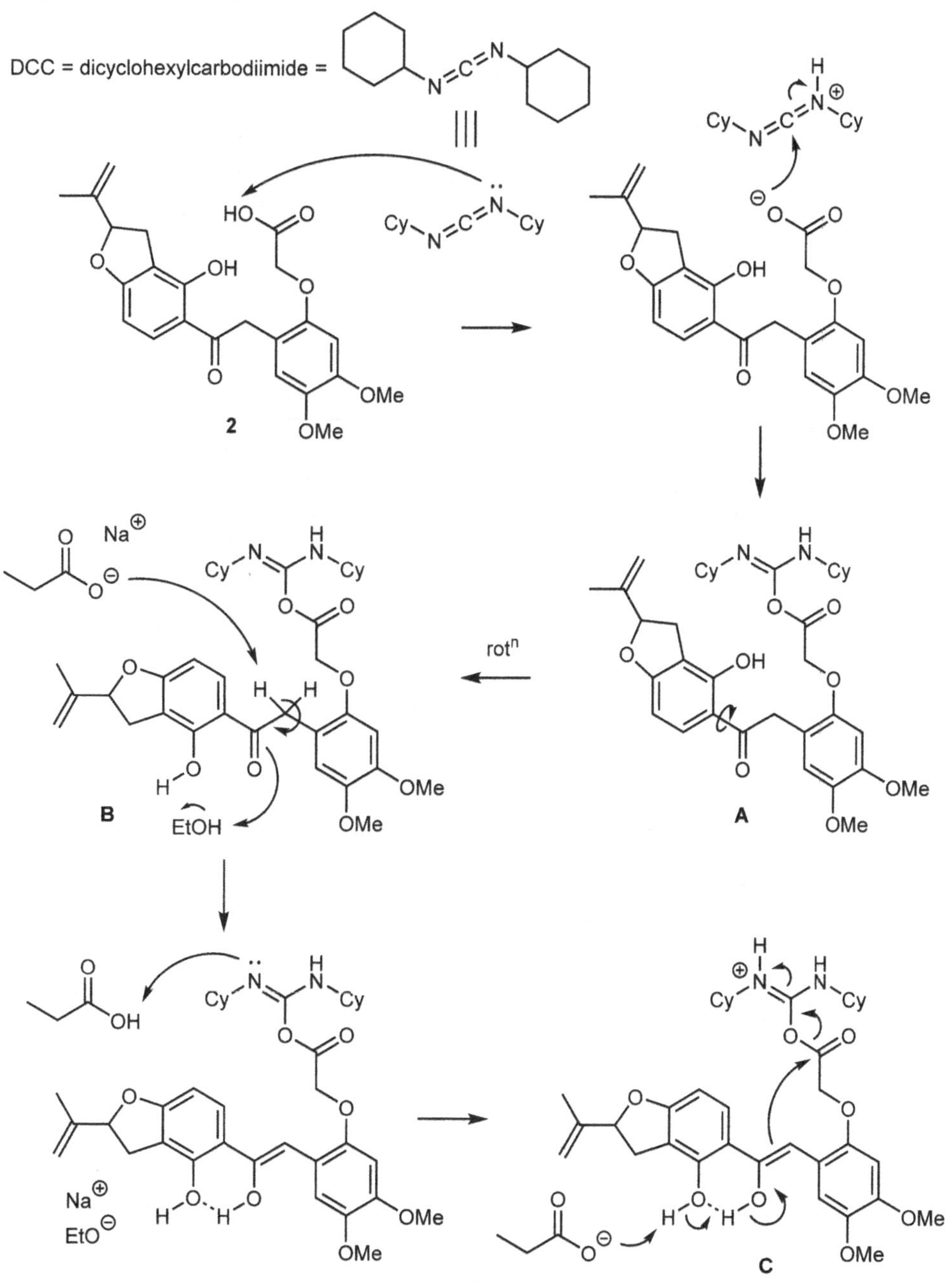

BALANCED CHEMICAL EQUATION FOR (I)

SOLUTION: (II) FROM 3 AND 4[2] MECHANISM A

H-shift
OMe
OMe
4
F
H_2O
HO

OH
3
Cy
DCC = dicyclohexylcarbodiimide =
HN
G

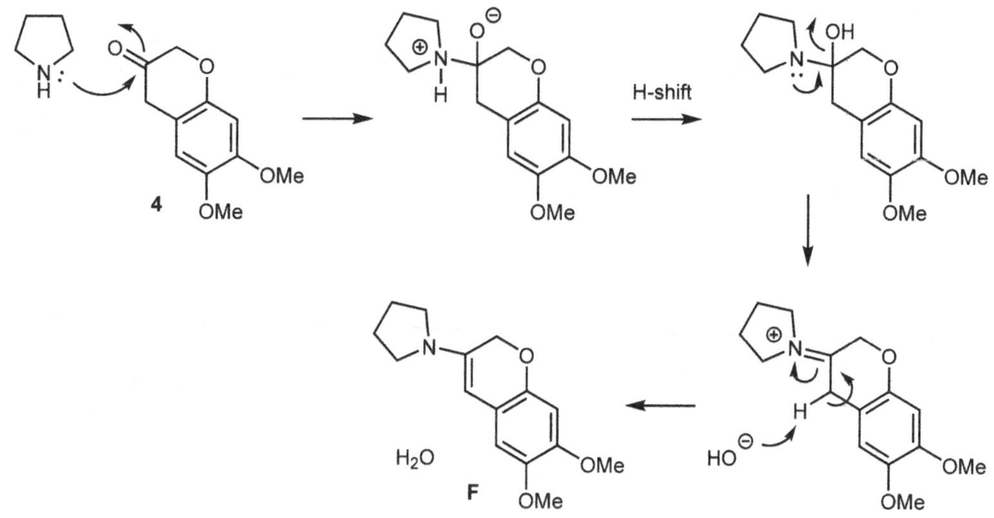

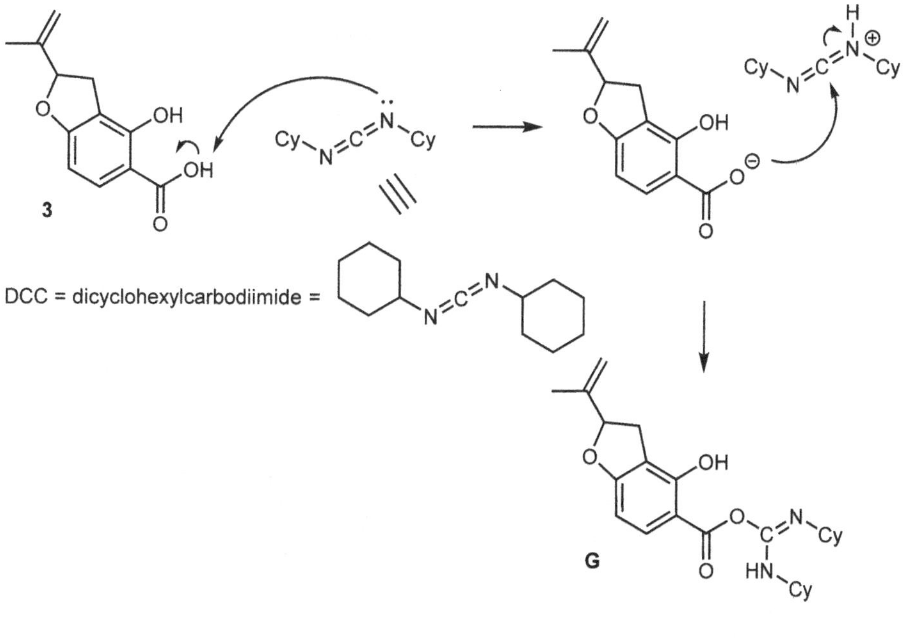

BALANCED CHEMICAL EQUATION FOR MECHANISM A

SOLUTION: (II) FROM 3 AND 4[2] MECHANISM B

4
H-shift
F
H2O
HO
3
DCC = dicyclohexylcarbodiimide =
G
H-shift
H
I
Cy
NH2

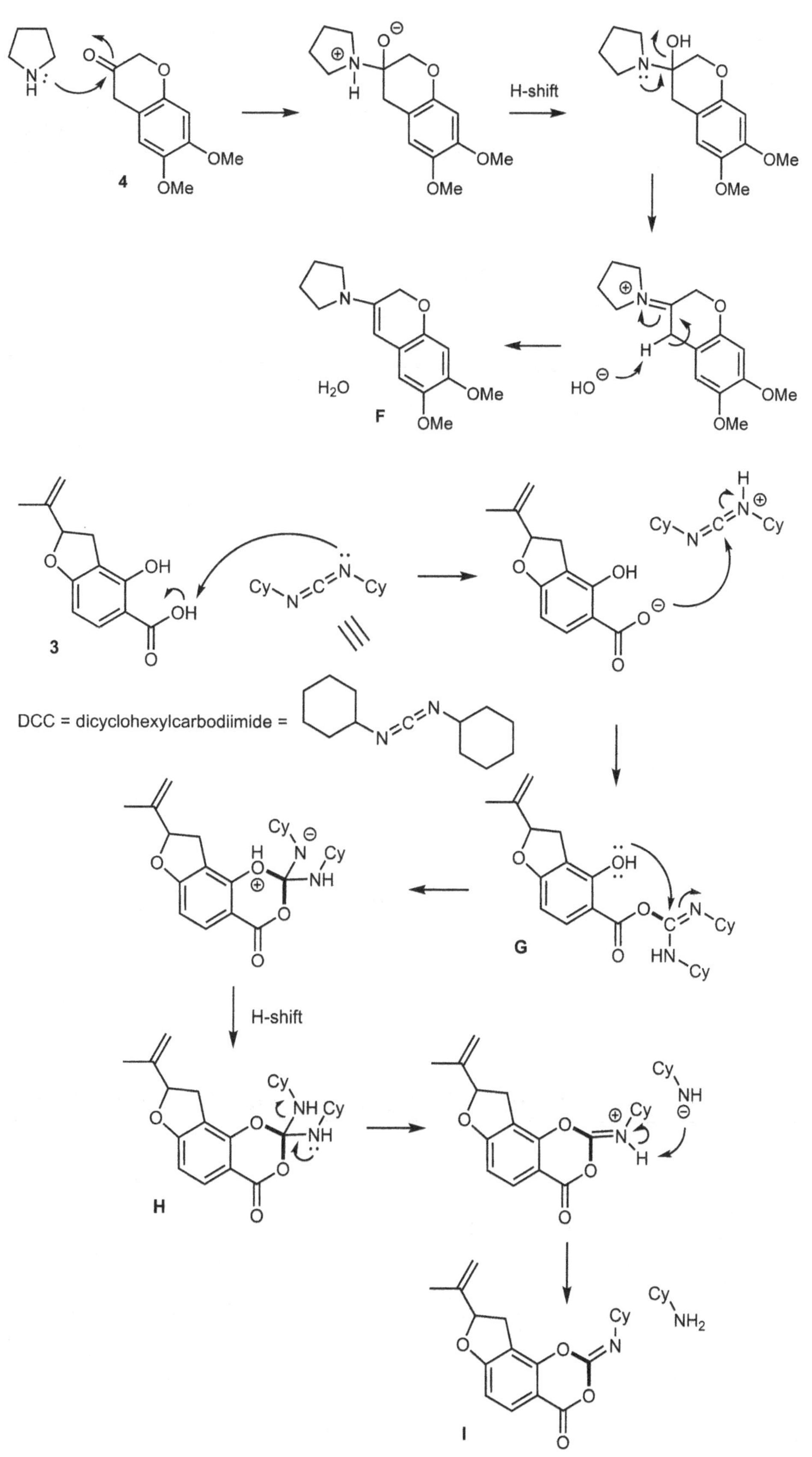

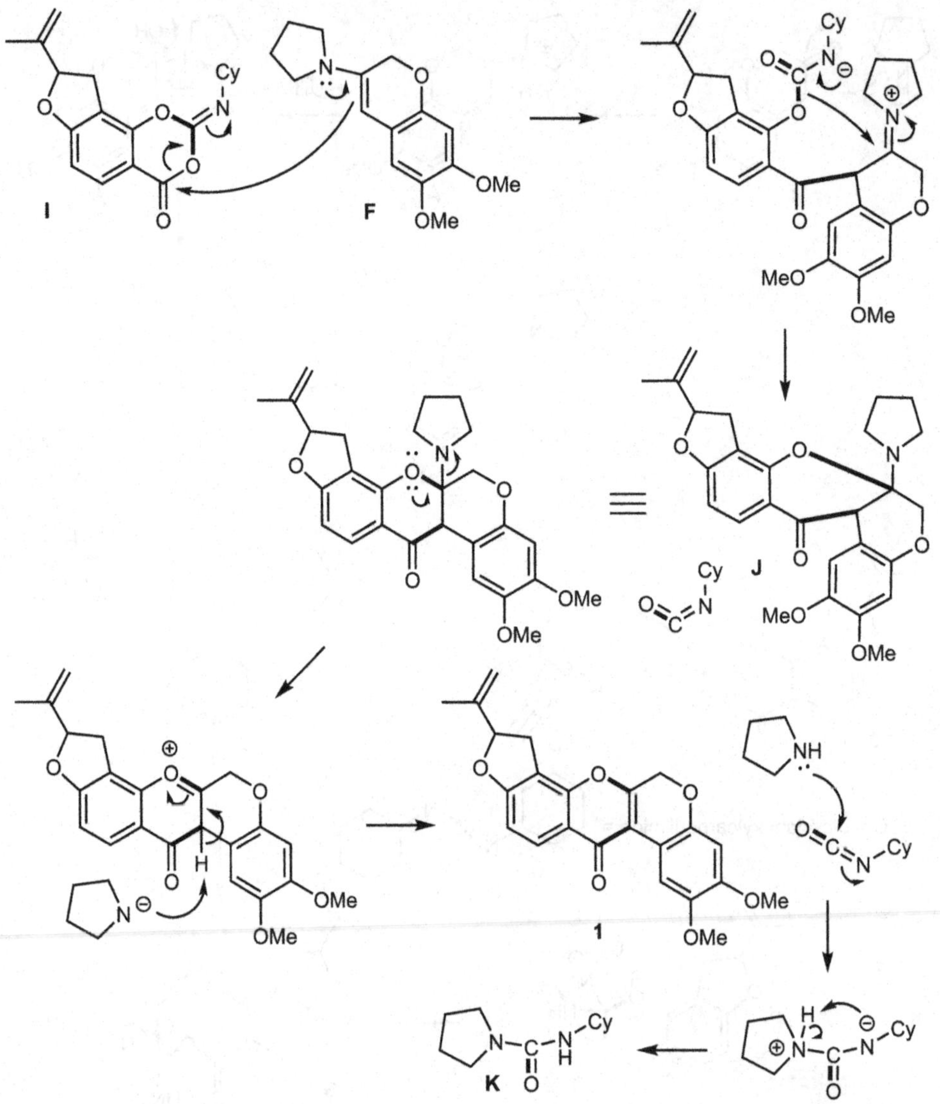

BALANCED CHEMICAL EQUATION FOR MECHANISM B

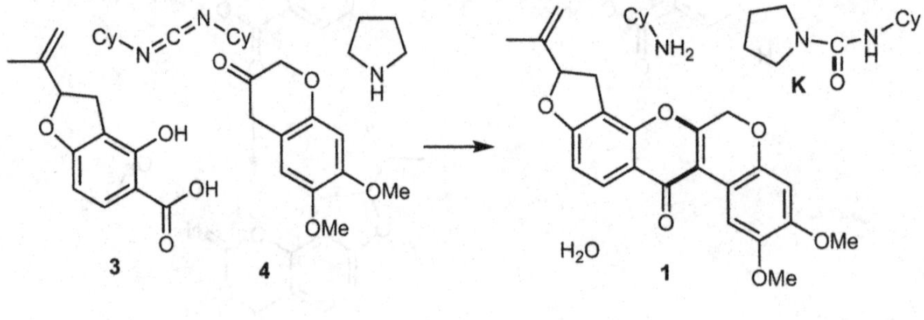

KEY STEPS EXPLAINED

In his synthesis of dehydrorotenone **1**, Miyano uses DCC (dicyclohexylcarbodiimide) as a dehydrating reagent whose role is to couple to carboxylic acid functionalities.[1,2] In his first synthesis,[1] using derrisic acid **2**, the sequence begins with DCC coupling onto the carboxylic acid end of **2** to give **A**. This structure, upon rotation and keto-enol tautomerization can lead to **C** which rearranges through a 4-membered transition state to **D** eliminating dicyclohexylurea as a by-product. Upon rotation, hydroxyl-mediated ring closure and subsequent dehydration **D** transforms into the final dehydrorotenone **1**. We note that before bond rotation takes place, the internal H-bonded structure forming the six-membered ring must break.

In another synthetic approach to **1**, Miyano utilizes tubaic acid **3** with the pyrrolidine enamine of **4** in the presence of DCC for a condensation reaction.[2] In this case, as seen in other syntheses, it is possible to envision one mechanism (mechanism A), where the pyrrolidine enamine of **4** (**F**) attacks the DCC coupled form of **3** (**G**) to give a structure which can ring close and eliminate pyrrolidine along with dicyclohexylurea as by-products. Nevertheless, Miyano recognized the odor of isocyanate in the reaction mixture and went on to characterize the morpholine derivate of structure **K** in a similar experiment thereby leading him to suggest mechanism B as a mechanism which is better supported by the experimental evidence.[2] Here, instead of nucleophilic enamine attack onto **G**, we have **G** undergoing intramolecular displacement of cyclohexanamine to produce **I**. Via nucleophilic enamine attack, **F** can come in and attack **I** at the carbonyl center to ring open the compound. This can then ring close to form **J** and eliminate an isocyanate by-product. After expulsion of pyrrolidine, **1** is formed with pyrrolidine and the isocyanate by-products formed reacting to produce by-product **K**, a variant of which was isolated and identified by Miyano. On comparing the balanced chemical equations from mechanisms A and B, we find that in mechanism A DCC is transformed to DCU (dicyclohexylurea) and pyrrolidine is regenerated which is consistent with it acting as a catalyst. In mechanism B, conversely, DCC reacts with pyrrolidine via cyclohexylisocyanate to form cyclohexylamine and mixed urea **K** (pyrrolidine-1-carboxylic acid cyclohexylamide). In this case there is a net consumption of pyrrolidine and so pyrrolidine is not a catalyst. This is an example where the mechanisms can be experimentally distinguished based on identifying the kinds of by-products made.

ADDITIONAL RESOURCES

For those interested in further examples of mechanisms involving ring transformations in rotenoids, we highly recommend Refs. 3, 4. For a synthetic example of Miyano's other works with rotenoids in the context of total synthesis, see Ref. 5.

References

1. Miyano M. Rotenoids. XXI.[1] Cyclization of derrisic acid to dehydrorotenone. *J Am Chem Soc.* 1965;87(17):3962–3964. https://doi.org/10.1021/ja01095a031.
2. Miyano M. Rotenoids. XX.[1] Total synthesis of rotenone. *J Am Chem Soc.* 1965;87(17):3958–3962. https://doi.org/10.1021/ja01095a030.
3. Crombie L, Godin PJ, Whiting DA, Siddalingaiah KS. Some chemistry of the B/C-ring system of rotenoids. *J Chem Soc.* 1961;2876–2889. https://doi.org/10.1039/JR9610002876.
4. Sakakibara J, Nagai S, Akiyama T, Ueda T, Oda N, Kidouchi K. Studies on the chemical transformations of rotenoids. III[1]. Ring conversions of methyl rotenononate and β-rotenonone. *Heterocycles.* 1988;27(2):423–435. https://doi.org/10.3987/COM-87-4382.
5. Miyano M. Rotenoids. XXII. Total synthesis of isomillettone. *J Org Chem.* 1970;35(1):246–249. https://doi.org/10.1021/jo00826a054.

Question 293: Radical-Induced Decarboxylation of a Lactone

During an investigation of antibiotic frenolicin-B **1** to evaluate its predisposition toward hydrolysis and radical-induced decomposition, the major racemic pyranonaphthoquinone product **2** was isolated (Scheme 1).[1] Suggest two mechanisms for the radical-induced degradation of **1 → 2**.

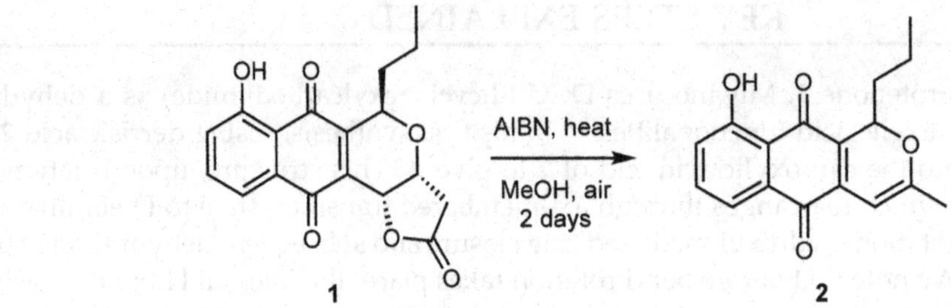

SCHEME 1

SOLUTION: PATHWAY A

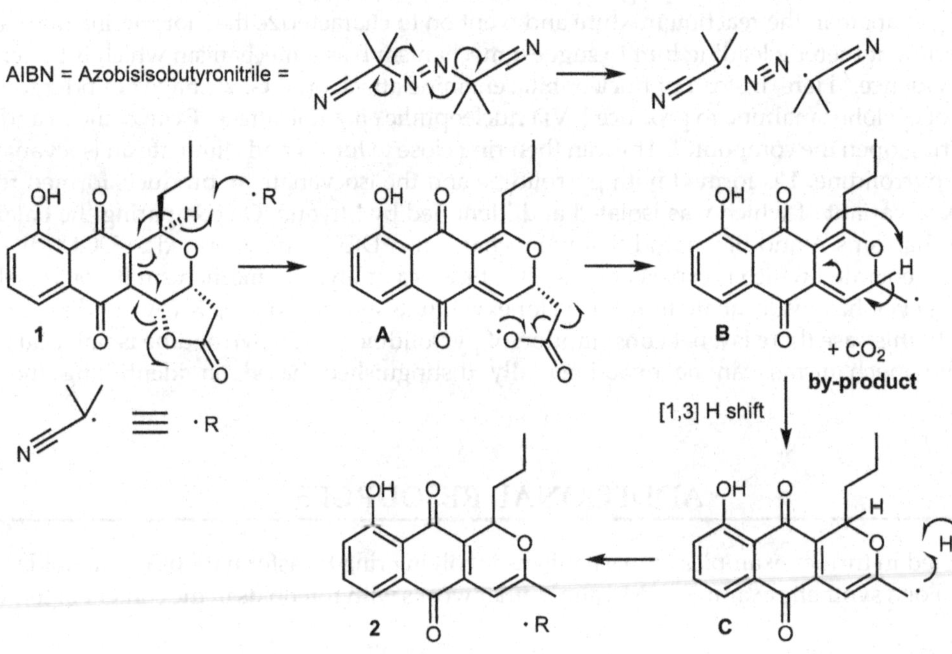

PATHWAY B

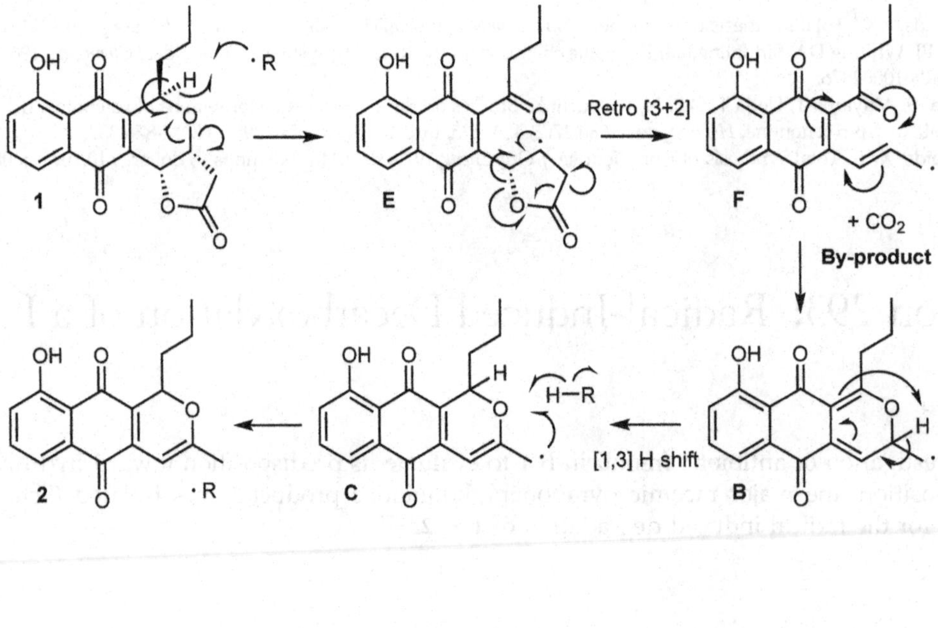

KEY STEPS EXPLAINED

The authors of this work have suggested two possible radical-induced mechanisms for the **1** → **2** transformation.[1] They both begin with radical-induced proton abstraction at the same position of **1**. They then diverge with lactone ring opening shown in pathway A and pyran ring opening shown in pathway B. Decarboxylation, [1,3] H-shift, and one-electron shuffling lead to the final product **2** in both mechanisms with the caveat that the pyran ring must be reformed in pathway B. According to the authors, PM3 calculations suggested that the activation enthalpy for lactone ring opening (pathway A) is 46 kcal/mol whereas that of pyran ring opening (pathway B) is 35 kcal/mol. Further experimental work suggested by the reviewer for their manuscript pointed the authors in the direction of undertaking a control experiment where **2** was subjected to the reaction conditions with and without AIBN in d_4-methanol at reflux for 1 day. This deuterium labeling experiment resulted in no H-D exchange at the position of the initial radical-induced hydrogen abstraction thus, according to the authors, further supporting pathway B. We note that this experiment does not suggest that pathway B is favored over pathway A since the chemical environments of **1** and **2** at the position of initial radical initiation are very similar chemically. Moreover, the authors did not mention that in pathway A, the thermodynamically stable naphthoquinone moiety found in **1** is disrupted in structure **A**. This disruption would likely account for the large energy difference between **A** and **E** which the authors mistakenly attributed to the activation energy of lactone ring opening versus pyran ring opening. Furthermore, the original paper does not present reaction coordinate diagrams depicting the energetics of both proposed mechanistic pathways. Such diagrams are needed in order to make a statement about which ring opening steps are rate limiting in order to support one of the proposed mechanisms over the other. Since both the experimental and computational arguments for this mechanistic analysis are weak in our view, we cannot endorse one mechanistic pathway over the other. This problem requires revisiting by more careful experimental and computational chemists.

ADDITIONAL RESOURCES

For further examples of radical-induced mechanisms involving decarboxylation and other possible mechanistic pathways, the reader is referred to Refs. 2, 3.

References

1. Van Arnum SD, Stepsus N, Carpenter BK. An unexpected oxidative decarboxylation reaction of frenolicin-B. *Tetrahedron Lett.* 1997;38(3):305–308. https://doi.org/10.1016/S0040-4039(96)02339-8.
2. Steffen LK, Glass RS, Sabahi M, et al. OH radical induced decarboxylation of amino acids. Decarboxylation vs bond formation in radical intermediates. *J Am Chem Soc.* 1991;113(6):2141–2145. https://doi.org/10.1021/ja00006a035.
3. Kondo K, Kurihara M, Miyata N, Suzuki T, Toyoda M. Mechanistic studies of catechins as antioxidants against radical oxidation. *Arch Biochem Biophys.* 1999;362(1):79–86. https://doi.org/10.1006/abbi.1998.1015.

Question 294: Aryl Azide Thermolysis: "A Series of Rather Involved Rearrangement Reactions"

Indian chemists in 1996 reported on the thermolysis of a highly substituted aryl azide **1** (chlorobenzene, 130°C, 4h) to make an unusual product **2** in 19% yield (Scheme 1).[1] Suggest a mechanism for this transformation.

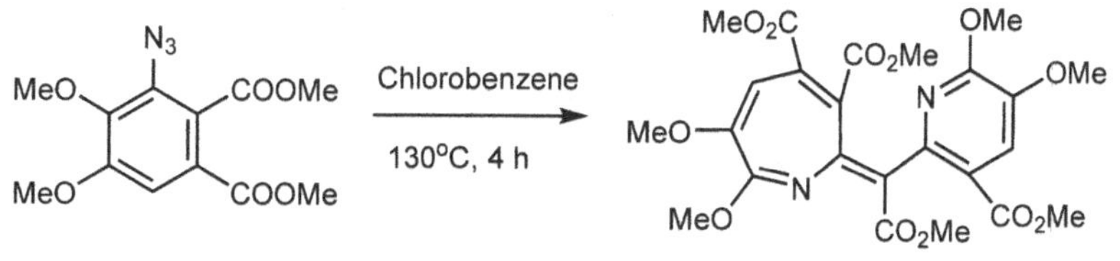

SCHEME 1

SOLUTION

BALANCED CHEMICAL EQUATION

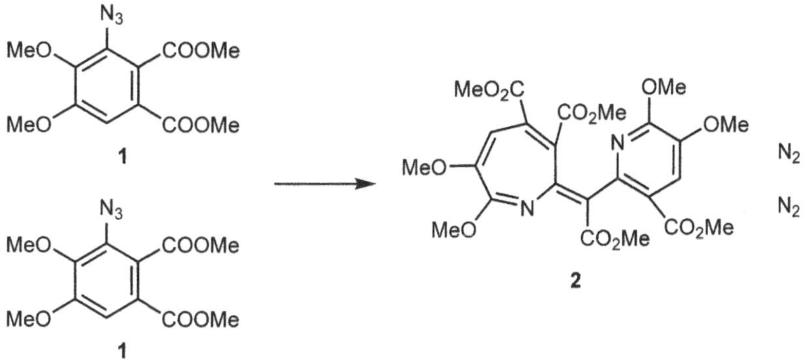

KEY STEPS EXPLAINED

In their 1996 paper, Eswaran et al. did not provide a mechanistic explanation of the thermolysis of aryl azide **1**.[1] Nevertheless, years later, a collaboration between Eswaran and Platz revealed a mechanistic explanation using spectroscopic data and density functional theory methods.[2] Our mechanism here is largely a representation from this latest literature example of synthesis and mechanistic chemists coming together to better understand the mechanics of an interesting transformation.

Therefore, as explained in their paper, the mechanism of **1** → **2** begins with structure **1** which is an aryl azide. Aryl azides can eject N_2 gas to form persistent singlet nitrenes like structure **A**. Such nitrene species can undergo ring expansion by means of nitrene insertion leading to ring expanded azepine structures (cyclic ketenimines) like **C** upon π-electron reshuffling. As this compound accumulates, a fraction of it is likely to convert by ring contraction to pyridine **D** which also contains a carbene functionality. This carbene can then attack the azepine **C** leading to carbene insertion and π-electron reshuffling to form **2**.

Interestingly, in their latest report, the authors discovered the formation of two side products, namely, **3** and **4** which occur via a rare intramolecular nitrene insertion into an adjacent C—H methoxy bond from **A** thus leading to an oxazolidine ring (Scheme 2). After nucleophilic attack on **C** the mechanism will lead to **3** (Scheme 3). The ylide **D** formed previously can interact with a neighboring carbonyl ester (Scheme 5). After demethylation (by chlorobenzene acting as a reagent in this case), a protonation step produces the side product **4**. The balanced chemical equations for these two transformations are represented in Schemes 4 and 6, respectively. For more information about these transformations and the experimental evidence to support their mechanisms, we highly recommend Ref. 2. As an aside, we emphasize this reference as an example of chemical understanding realized through the collaboration between synthetic and mechanistic specialists.

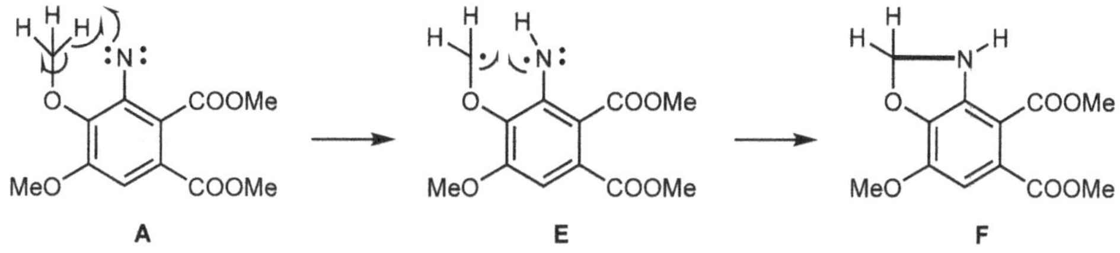

SCHEME 2

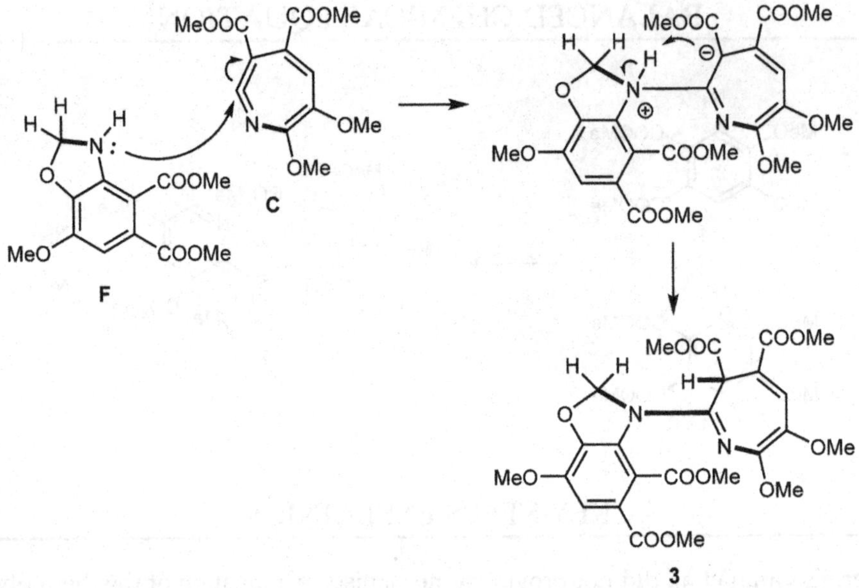

SCHEME 3

BALANCED CHEMICAL EQUATION

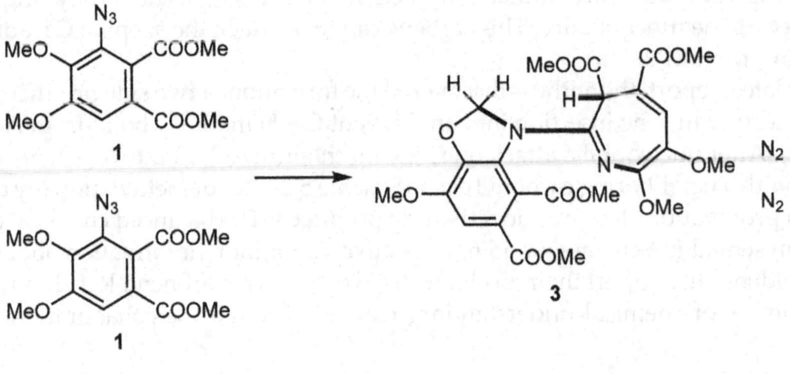

SCHEME 4

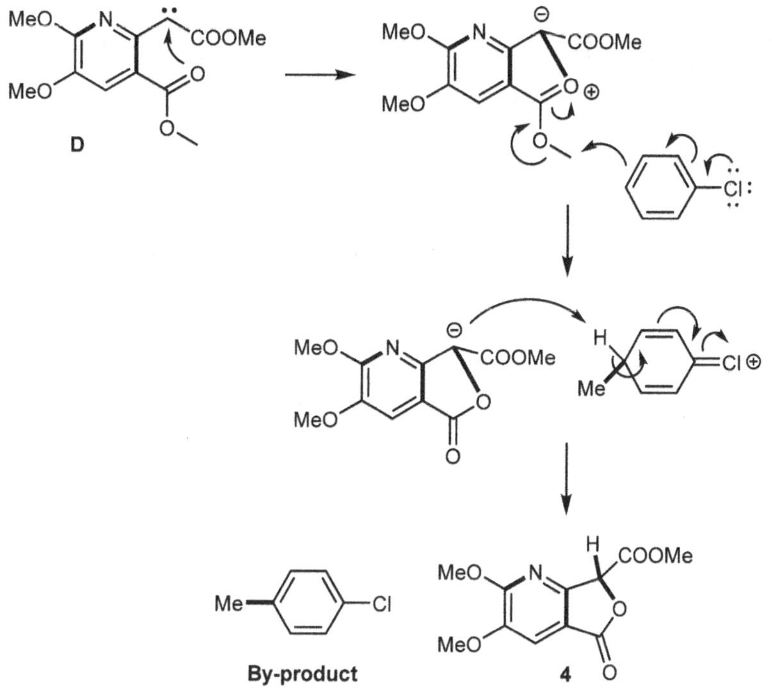

SCHEME 5

BALANCED CHEMICAL EQUATION

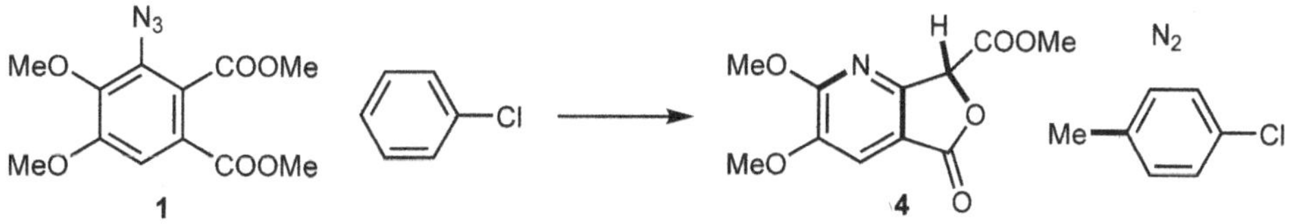

SCHEME 6

ADDITIONAL RESOURCES

In addition to these two fascinating works, readers interested in other examples of the thermolysis of substituted aryl azides are referred to Refs. 3, 4.

References

1. Eswaran SV, Neela HY, Ramakumar S, Viswamitra MA. The unusual formation of methyl α-(5,6-dimethoxycarbonyl-2,3-dimethoxyazepin-7-ylidene)-α-(5-methoxycarbonyl-2,3-dimethoxypyrid-6-yl)acetate during the pyrolysis of "azido-*meta*-hemipinate": first example of a reaction involving a concomitant ring expansion and ring extrusion. *J Heterocyclic Chem*. 1996;33(4):1333–1337. https://doi.org/10.1002/jhet.5570330454.
2. Kaur D, Luk HL, Coldren W, et al. Concomitant nitrene and carbene insertion accompanying ring expansion: spectroscopic, X-ray, and computational studies. *J Org Chem*. 2014;79:1199–1205. https://doi.org/10.1021/jo402621w.
3. Patel DI, Smalley RK. Thermolysis of aryl azides in phenyl isocyanate. *J Chem Soc Perkin Trans 1*. 1984;2587–2590. https://doi.org/10.1039/P19840002587.
4. Ohba Y, Kubo S, Nakai M, Nagai A, Yoshimoto M. Synthesis of the 3*H*-azepines utilizing the thermolysis of substituted aryl azides. *Bull Chem Soc Jpn*. 1986;59:2317–2320. https://doi.org/10.1246/bcsj.59.2317.

Question 295: An Efficient Route to Hexakis(trifluoromethyl) cyclopentadiene

In 1995 British scientists described an efficient synthesis of hexakis(trifluoromethyl)cyclopentadiene **3** in 74% yield (Scheme 1).[1] Provide a mechanism for this transformation.

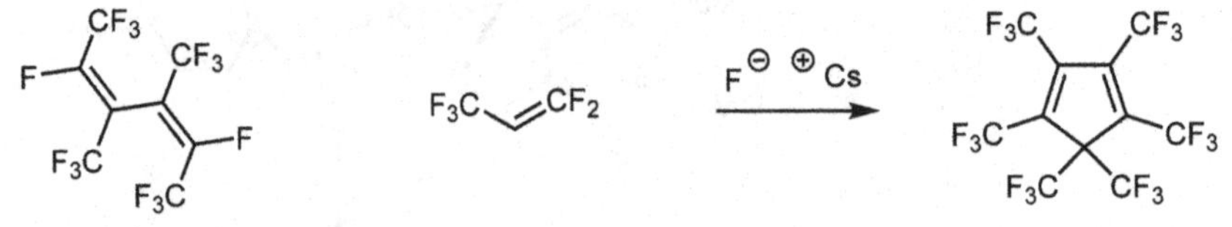

1 equiv.

Perfluoro-3,4-dimethyl hexa-2,4-diene, **1**

2 equiv.

Pentafluoropropene, **2**

4 equiv.

3

SCHEME 1

SOLUTION

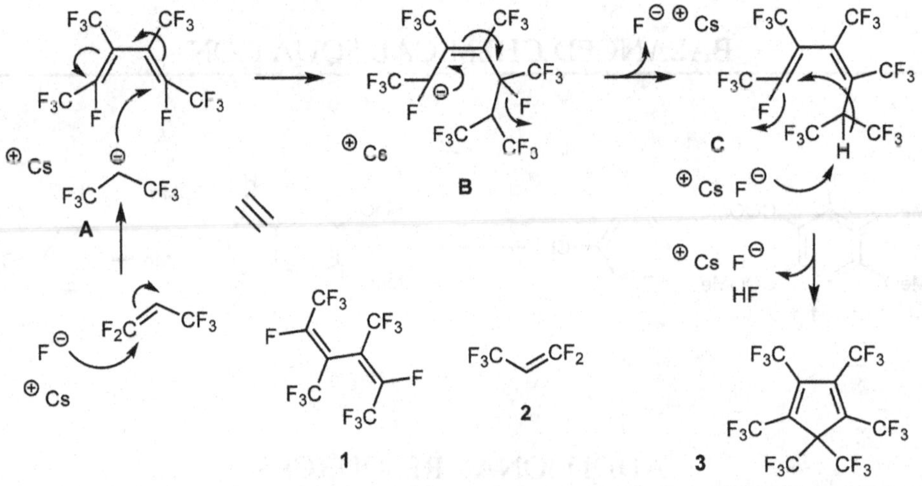

BALANCED CHEMICAL EQUATION

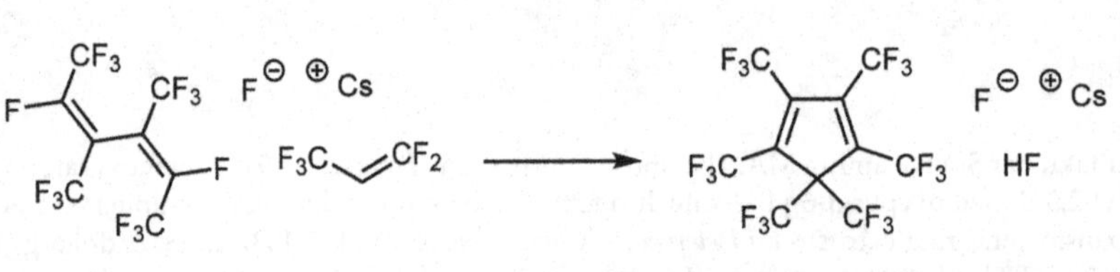

We note that when compared to the actual amounts of reagents used, the balanced chemical equation alone suggests that equimolar amounts are sufficient to carry out this transformation. It is possible the authors used extra pentafluoropropene and cesium fluoride to push this reaction to completion and achieve a higher yield.

KEY STEPS EXPLAINED

The mechanistic pathway envisioned by the authors begins with attack of fluoride anion (from excess cesium fluoride) onto pentafluoropropene **2** to generate a vinylic anion **A** which can attack diene **1** to generate another anion **B**.[1] This anion readily eliminates fluoride anion to produce **C**. Fluoride ion, which also acts as a base, can deprotonate the hydrogen atom at the allylic position of **C** forming a new anion which can cyclize to give the final cyclopentadiene product **3**.

ADDITIONAL RESOURCES

Chambers et al. have released a newer finding which includes the same example as this problem in addition to the case of derived cyclopentadienide salts.[2] Moreover, Dr. Chambers published work on several other chemical systems with similar ring constructions where fluoride chemistry plays an important mechanistic role.[3,4] In a review several years ago, Chambers explained the progression of his work over the past decade.[5]

References

1. Chambers RD, Mullins SJ, Roche AJ, Vaughan JFS. Direct syntheses of pentakis(trifluoromethyl)cyclopentadienide salts and related dienes. *J Chem Soc Chem Commun.* 1995;841–842. https://doi.org/10.1039/C39950000841.
2. Chambers RD, Gray WK, Vaughan JFS, et al. Reactions involving fluoride ion. Part 41.[1,2] Synthesis of hexakis(trifluoromethyl)cyclopentadiene and derived cyclopentadienide salts. *J Chem Soc Perkin Trans 1.* 1997;135–145. https://doi.org/10.1039/A604087J.
3. Chambers RD, Gray WK, Mullins SJ, Korn SR. Reactions involving fluoride ion. Part 42.[1] Heterocyclic compounds from perfluoro-3,4-dimethyl-hexa-2,4-diene. *J Chem Soc Perkin Trans 1.* 1997;1457–1463. https://doi.org/10.1039/A608584I.
4. Chambers RD, Nishimura S, Sandford G. Reactions involving fluoride ion Part 43[1] Oligomerisations of hexafluoro-1,3-butadiene and -but-2-yne. *J Fluor Chem.* 1997;91:63–68. https://doi.org/10.1016/s0022-1139(98)00213-9.
5. Chambers RD. Footsteps of a fluorine chemist. *J Fluor Chem.* 2010;131:665–675. https://doi.org/10.1016/j.jfluchem.2010.02.009.

Question 296: Exploitation of the Boulton-Katritzky Rearrangement: Synthesis of 4,4′-Diamino-3,3′-bifurazan

An improved one-pot synthesis of 4,4′-diamino-3,3′-bifurazan **2** from the furoxan **1** (overall yield 18%) appeared several years ago (Scheme 1).[1] Compared to a previous multistep synthesis (overall yield 3%), this much shorter and simpler method marked a remarkable synthetic achievement by authors Sheremetev and Mantseva. Provide a mechanism for this transformation.

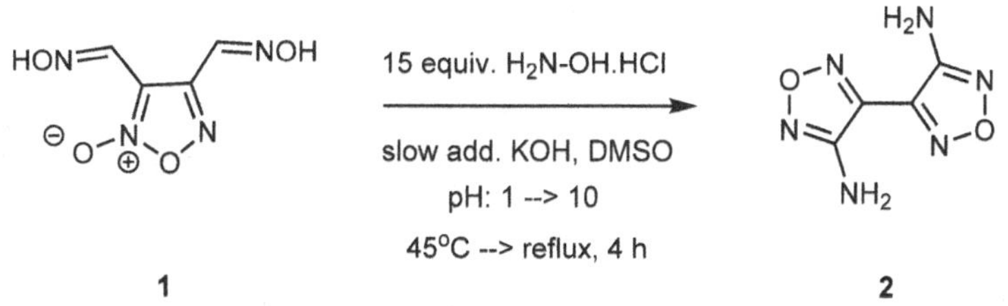

SCHEME 1

SOLUTION

We start by examining the structure of **1** and the reaction conditions. As drawn, the orientation of **1** has the hydroximino groups pointing toward the furoxan ring. In the original reference it is not specified whether this orientation follows from the synthesis of **1**.[1] We cannot assume it does. Hydroxylamine hydrochloride salt is used in great excess (hydroxylamine free base is to be liberated from its salt at elevated temperature), and it is expected that any added base will neutralize the hydrochloride salt in a highly exothermic reaction. This is important because only the orientation of the structure having the oxime groups pointing toward the furoxan ring allows the Boulton-Katritzky rearrangement to occur later on. Thanks to Dr. Sheremetev, we have also learned that the EZ isomerization of oximes requires acidic media, which never occurs in this reaction. We believe that the lack of isomerization pathways may contribute to the low reported yield of the final product as there are no productive mechanistic pathways for the other possible stereo-isomers of **1** (EE, EZ, ZE). All steps in the mechanism should be base catalyzed. The base, potassium hydroxide, is added slowly to prevent a sudden highly exothermic reaction with hydroxylamine hydrochloride. At elevated temperature, direct addition of base to the mixture decomposes the hydroxylamine and, according to Dr. Sheremetev, will boil the reaction medium and eject it from the flask. Lastly, placing the reduced furazan compound **A** derived from furoxan **1** directly into basic solution leads to very poor yields of **2**. Thus the starting furoxan **1**, excess hydroxylamine hydrochloride, and slow addition of potassium hydroxide are all necessary steps for the success of this transformation.

Accordingly, Dr. Sheremetev suggested to us three mechanistic possibilities for the initial transformation of **1**. These possibilities are (i) reduction followed by rearrangement (Scheme 2A and B), (ii) rearrangement followed by reduction (Scheme 3), and (iii) rearrangement followed by substitution (Scheme 4).

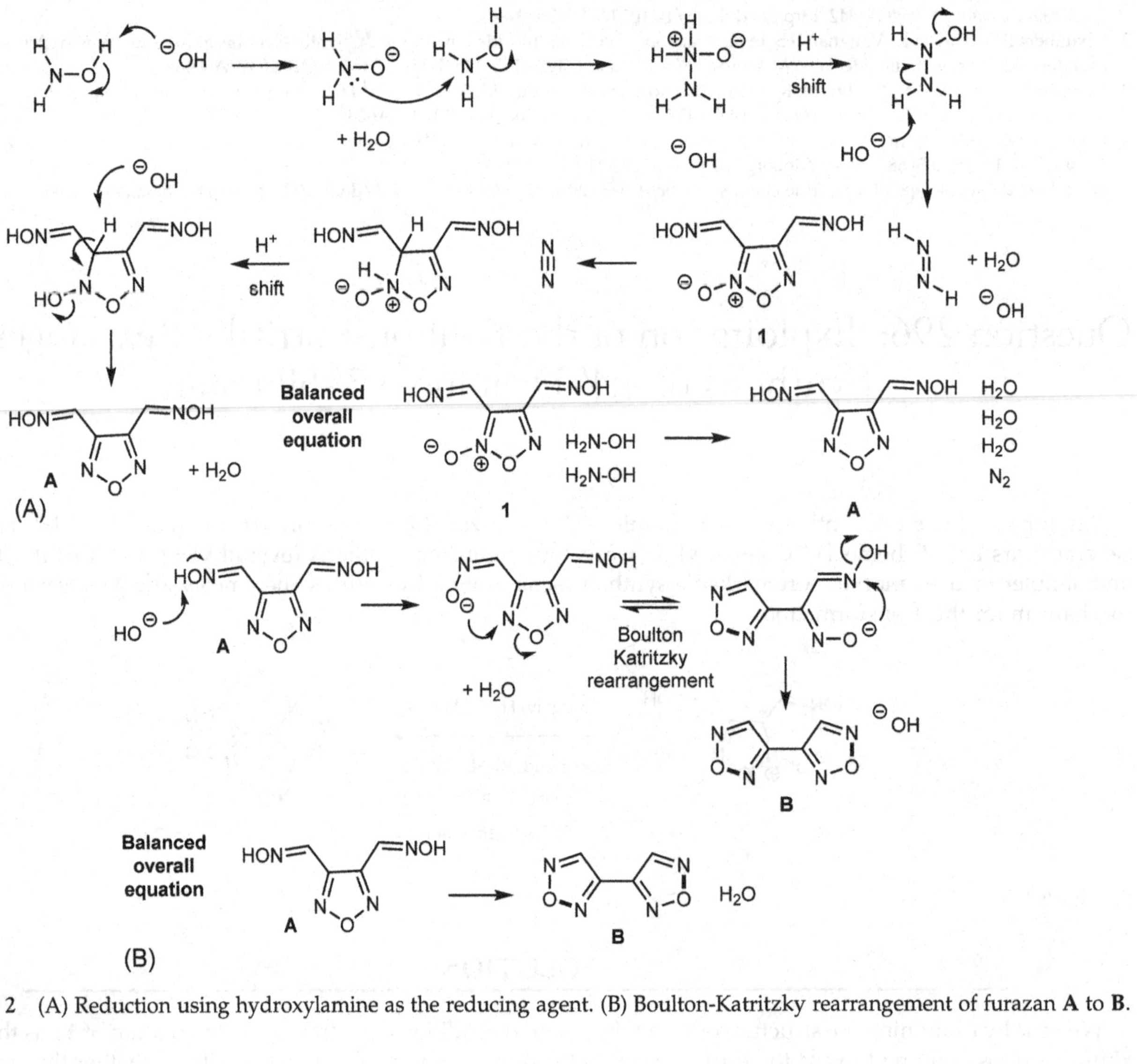

SCHEME 2 (A) Reduction using hydroxylamine as the reducing agent. (B) Boulton-Katritzky rearrangement of furazan **A** to **B**.

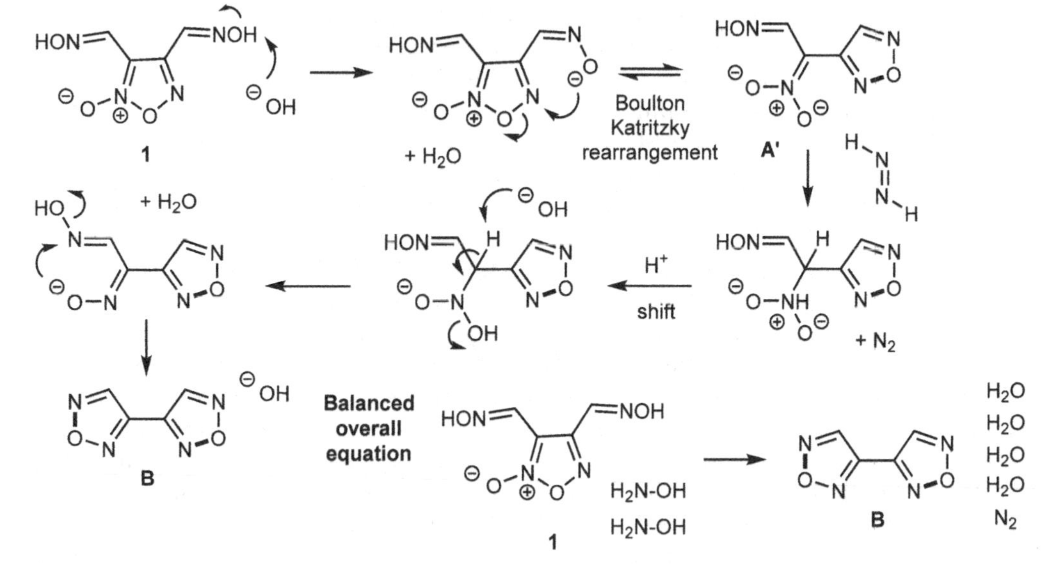

SCHEME 3 Rearrangement followed by reduction.

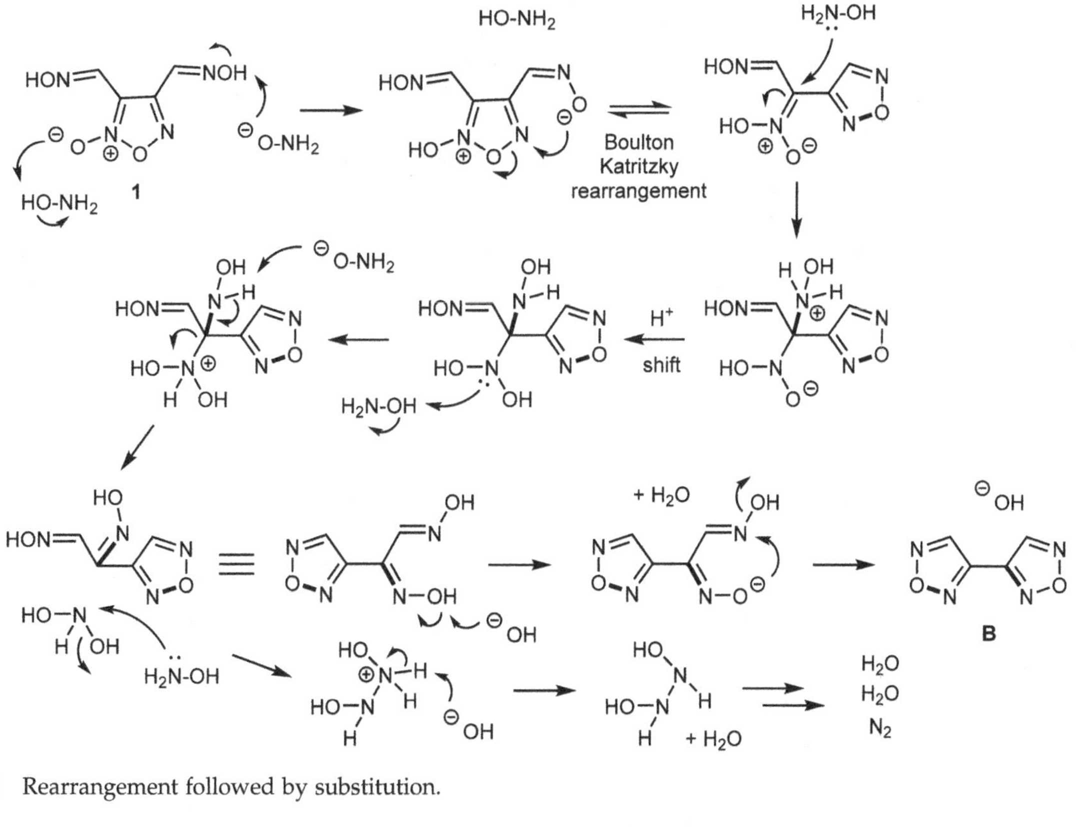

SCHEME 4 Rearrangement followed by substitution.

It is worth noting that this proposed mechanism produces nitrogen gas as a final by-product. In the original reference this by-product is not mentioned but in our recent correspondence, Sheremetev offered us the suggestion. Such a by-product would be easy to identify experimentally in order to confirm the suggested mechanistic pathway. We also note the production of diimide (N_2H_2) which hydrogenates **1**. After a proton shift and dehydration, the reduced furazan **A** is formed. Although there is no experimental evidence for diimide mentioned, we suggest it based on literature examples regarding similar reactions with hydroxylamine and investigated thermal decomposition pathways of hydroxylamine.[2,3] It is also known that diimide is an excellent hydrogenating agent.[4]

For case (ii) we begin with rearrangement as the first step followed by reduction (Scheme 3). We note that in this sequence the mechanism does not involve the furazan structure **A** as in case (i) but rather the structure **A'** which undergoes the usual reduction sequence using diimide hydrogenation. The balanced chemical equation and target bonds formed are identical to case (i). It is interesting to note that the experiment starting from the reduced furazan **A** when treated with base significantly reduces the yield of **2**. This result was communicated to us by Sheremetev and might be explained by the lack of oxime isomerization pathways for possible stereoisomers of **A** in basic solution. And so it does not necessarily reject the possibility of **A**. We recommend a theoretical calculation to shed light upon the energetics and reaction coordinate diagrams of cases (i) and (ii). Lastly, case (iii) involves rearrangement followed by substitution using hydroxylamine (Scheme 4). The balanced chemical equation for this case is identical to cases (i) and (ii). The target bonds formed are different however. Experimentally one can differentiate this mechanism from the prior two by using ^{18}O or ^{15}N labeling of hydroxylamine and observe any ^{18}O or ^{15}N incorporation into **B**. Theoretical analysis would also be recommended.

With structure **B** resulting from one of the three mechanistic cases outlined earlier, the base-catalyzed pathway to **2** is represented in Scheme 5. This transformation begins with hydroxide deprotonation to form a nitrile and open up a furazan ring. Hydroxylamine attack on the nitrile group followed by two prototropic shifts and a ring-closing step allows for the incorporation of an amine group in one of the original furazan rings with regeneration of hydroxide anion. The same sequence can then repeat itself for the other furazan ring to generate the final 4,4'-diamino-3,3'-bifurazan **2**. The balanced chemical equation for this sequence is represented in Scheme 6. Finally, the overall balanced chemical equations involving cases (i) and (ii), on one hand, and case (iii), on the other, are illustrated in Scheme 7.

Lastly, Sheremetev and Mantseva explain in the original paper that two side products are also recovered upon acidic workup, namely: 3-aminofurazancarboxylic acid (**3**, 10%) and furazandicarboxylic acid (**4**, 7%).[1] We will complete the story by showing how these side products are formed. To do this we must identify points of possible divergence in our overall mechanism for **1** → **2**. We identify these points as structures **C** → **3** (Scheme 8) and **B** → **4** (Scheme 9). Both of these pathways are base catalyzed. In the first instance (formation of **3**), we note the production of CN^- by-product and in the second instance (formation of **4**), we note the production of NH_3 by-product. These by-products along with N_2 gas must be experimentally identified for these mechanisms to be supported. This concludes our mechanistic analysis of this problem.

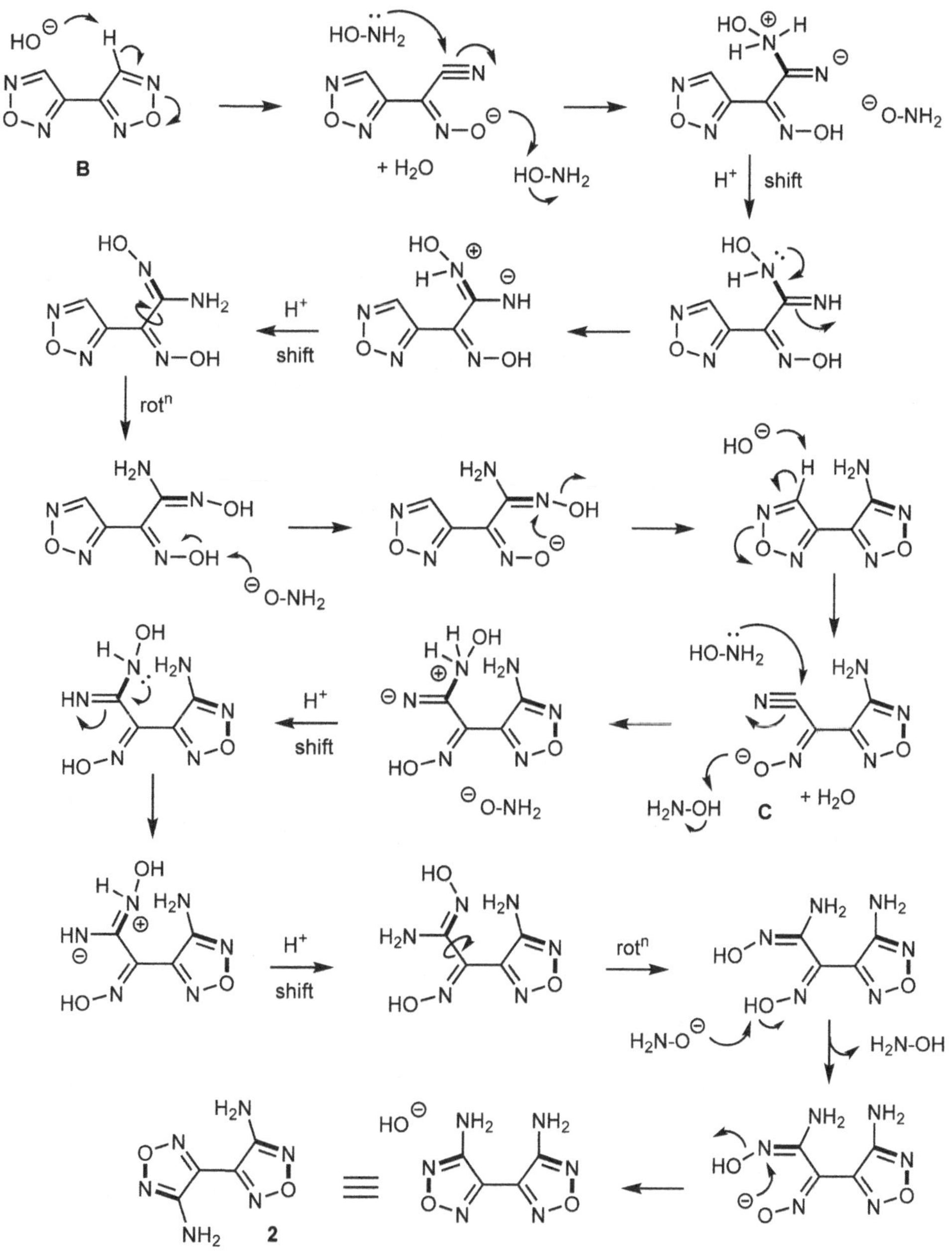

SCHEME 5   Mechanistic pathway for **B** to final product **2**.

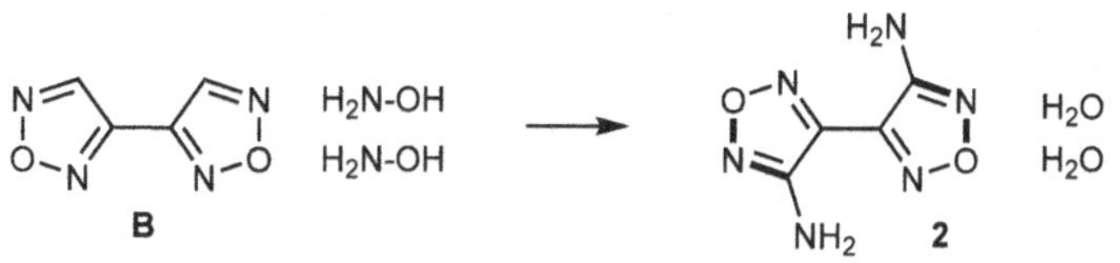

SCHEME 6   Balanced chemical equation for the sequence: **B** to **2**.

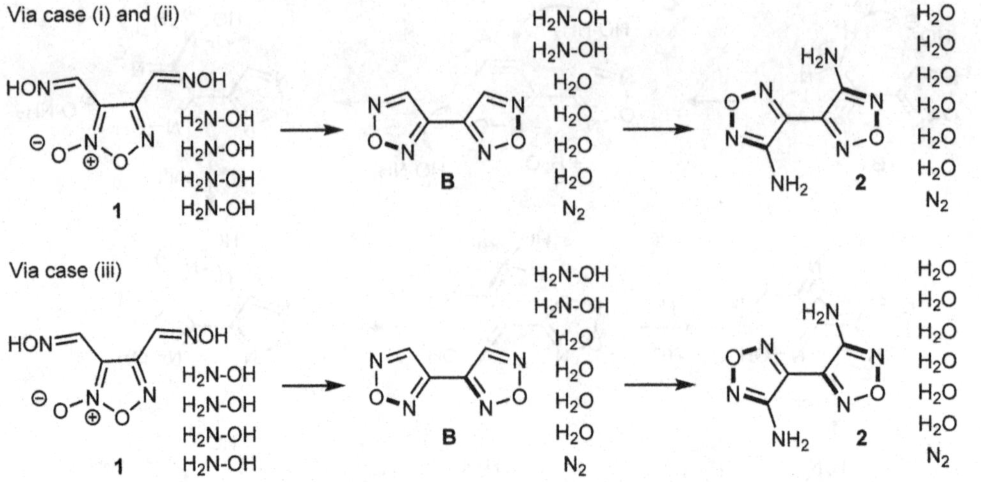

852

Via case (i) and (ii)

Via case (iii)

SCHEME 7   Balanced overall chemical equations for the sequences: **1** to **2** via cases (i) and (ii), and via case (iii).

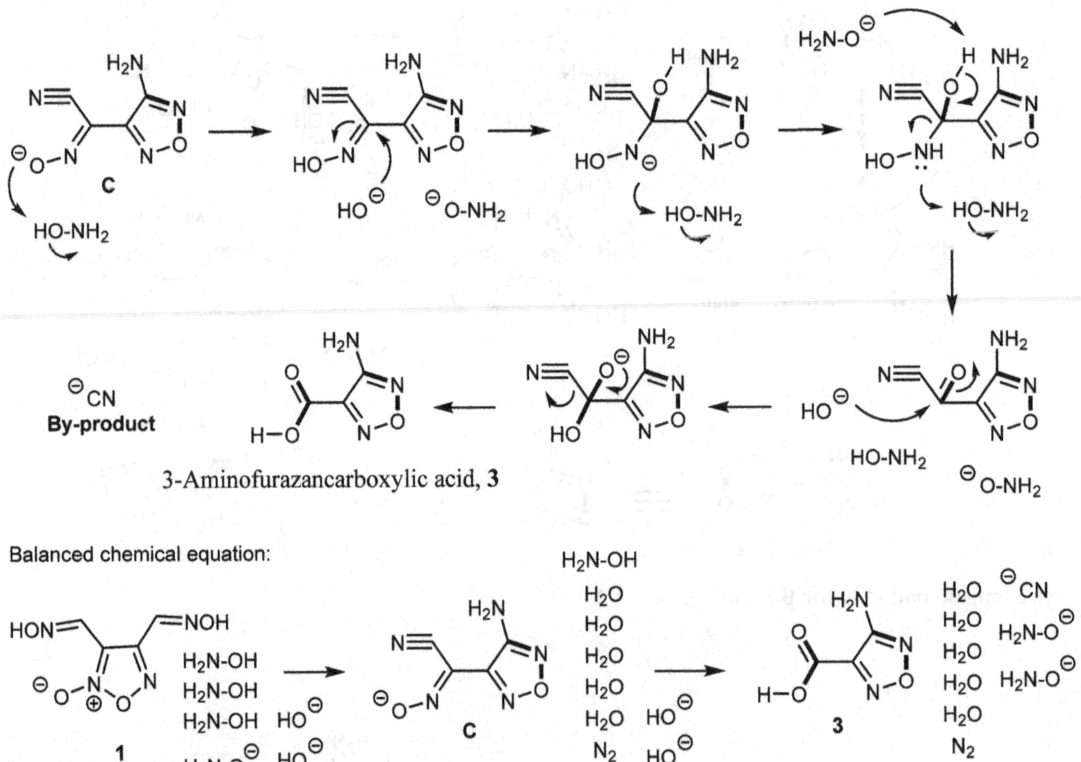

3-Aminofurazancarboxylic acid, **3**

Balanced chemical equation:

SCHEME 8   Formation of side-product **3** along with balanced chemical equation.

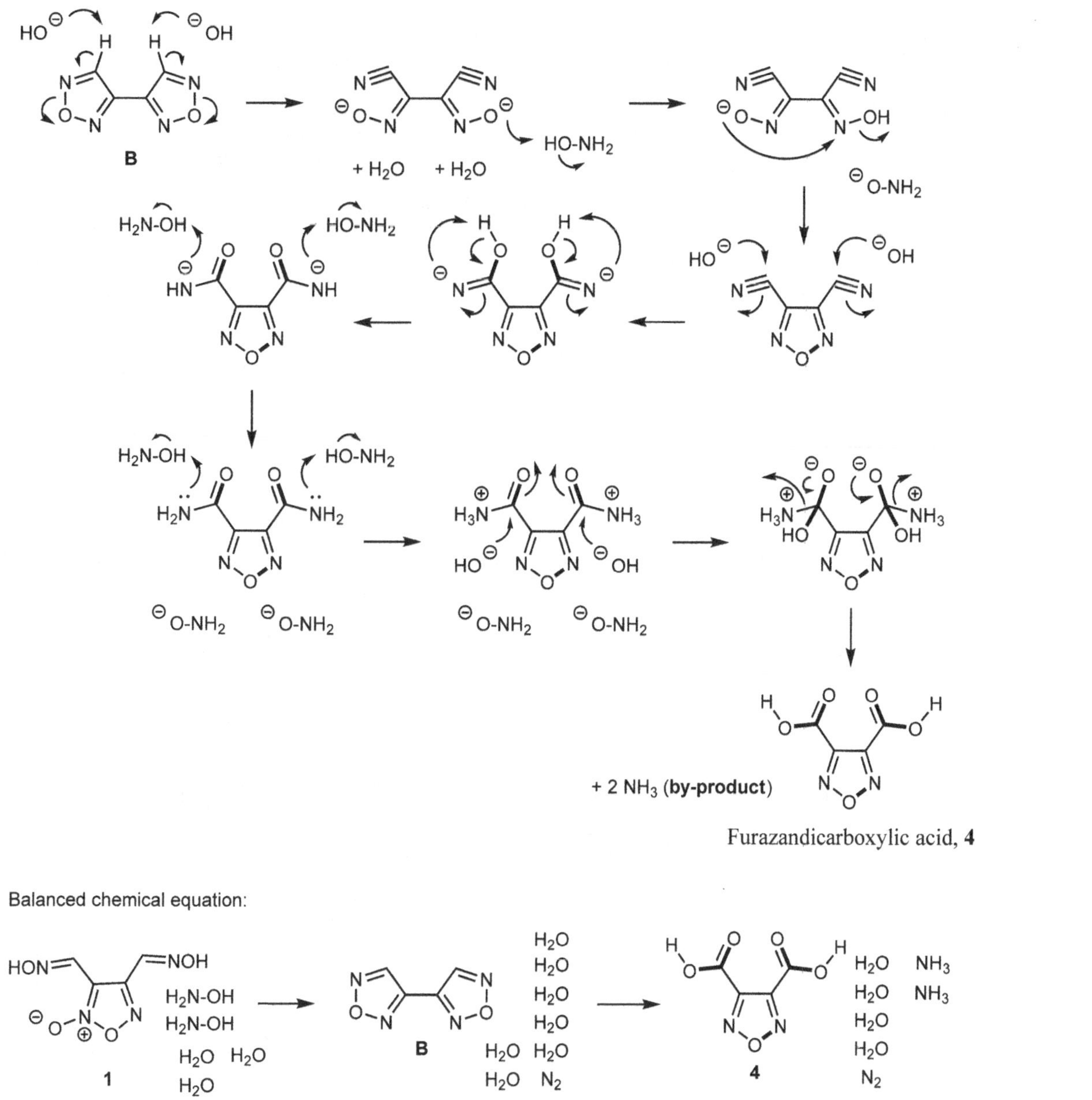

+ 2 NH$_3$ (by-product)

Furazandicarboxylic acid, **4**

Balanced chemical equation:

SCHEME 9 Formation of side-product **4** along with balanced chemical equation.

ADDITIONAL RESOURCES

Over the past 20 years, Dr. Sheremetev has published extensively on the type of chemistry described in this problem.[5–11] We would particularly like to emphasize Refs. 5, 8, 9, and which contain synthesis examples using very similar chemistry with mechanisms explained. Refs. 6, 7 are review articles on furazans and furoxans which are also very useful. An interesting and more recent example of the synthesis of aminofurazans also appeared in 2014 with a mechanism provided.[10] And for an example of an unexpected result involving several mechanisms, we recommend Ref. 11 by Sheremetev et al. Finally, for the role of hydroxylamine in two different mechanisms which depend on whether one R group on the amine is either H or Me, we highly recommend a recent article in *Tetrahedron Letters*.[12] For a synthetic example of the usefulness of compound **2**, we recommend Ref. 13.

References

1. Sheremetev AB, Mantseva EV. One-pot synthesis of 4,4′-diamino-3,3′-bifurazan. *Mendeleev Commun.* 1996;6(6):246–247. https://doi.org/10.1070/MC1996v006n06ABEH000745.

2. Gangadhar A, Rao TC, Subbarao R, Lakshminarayana G. Hydrogenation of unsaturated fatty acid methyl esters with diimide from hydroxylamine-ethyl acetate. *J Am Oil Chem Soc.* 1989;66(10):1507–1508. https://doi.org/10.1007/BF02661982.

3. Wei C, Saraf SR, Rogers WJ, Mannan MS. Thermal runaway reaction hazards and mechanisms of hydroxylamine with acid/base contaminants. *Thermochim Acta.* 2004;421:1–9. https://doi.org/10.1016/j.tca.2004.02.012.

4. Pasto DJ, Taylor RT. Reduction with diimide. *Org React.* 1991;40(2):91–155. https://doi.org/10.1002/0471264180.or040.02.

5. Sheremetev AB. Novel synthesis of 4-aminofurazan-3-acetic acid. *Mendeleev Commun.* 1998;8(4):135–136. https://doi.org/10.1070/MC1998v008n04ABEH000984.

6. Sheremetev AB. The chemistry of furazans fused to six- and seven-membered heterocycles with one heteroatom. *Russ Chem Rev.* 1999;68(2):137–148. https://doi.org/10.1070/RC1999v068n02ABEH000449.

7. Sheremetev AB, Makhova NN, Friedrichsen W. Monocyclic furazans and furoxans. *Adv Heterocycl Chem.* 2001;78:65–188. https://doi.org/10.1016/S0065-2725(01)78003-8.

8. Sheremetev AB. One-pot synthesis of 3-amino-4-aryl- and 3-amino-4-hetarylfurazans. *Russ Chem Bull Int Ed.* 2005;54(4):1057–1059. https://doi.org/10.1007/s11172-005-0359-4.

9. Sheremetev AB, Shamshina YL, Dmitriev DE. Synthesis of 3-alkyl-4-aminofurazans. *Russ Chem Bull Int Ed.* 2005;54(4):1032–1037. https://doi.org/10.1007/s11172-005-0352-y.

10. Sheremetev AB, Zabusov SG, Tukhbatshin TR, Palysaeva NV, Suponitsky KY. Synthesis of 4-acyl-3-aminofurazans from 3,4-diacylfuroxans. *Chem Heterocycl Compd.* 2014;50(8):1154–1165. https://doi.org/10.1007/s10593-014-1576-2.

11. Yudin IL, Palysaeva NV, Averkiev BB, Sheremetev AB. Unexpected formation of (trinitromethyl)pyrazines. *Mendeleev Commun.* 2015;25:193–195. https://doi.org/10.1016/j.mencom.2015.05.011.

12. Piccionello AP, Pace A, Buscemi S, Vivona N, Giorgi G. Synthesis of fluorinated 1,2,4-oxadiazin-6-ones through ANRORC rearrangement of 1,2,4-oxadiazoles. *Tetrahedron Lett.* 2009;50:1472–1474. https://doi.org/10.1016/j.tetlet.2009.01.071.

13. Epishina MA, Kulikov AS, Makhova NN. Synthesis of macrocyclic systems from 4,4′-diamino-3,3′-bi-1,2,5-oxadiazole and 3(4)-amino-4(3)-(4-amino-1,2,5-oxadiazol-3-yl)-1,2,5-oxadiazole 2-oxides. *Russ Chem Bull Int Ed.* 2008;57(3):644–651. https://doi.org/10.1007/s11172-008-0101-0.

Question 297: Rearrangement During Hydrolysis of a Cyclohexadienone

During a study for the synthesis of diterpenoid products found in certain plant species, Australian chemists attempted the hydrolysis of **1** in both acidic and basic solutions and found it to be unsuccessful.[1] Instead, under a basic mixture followed by acidic work-up, extraction, and chromatography over silica, the products **2**, **3**, and **4** were formed in 11.7%, 63.7%, and 24.7% yields, respectively (Scheme 1). Suggest mechanisms for their formation.

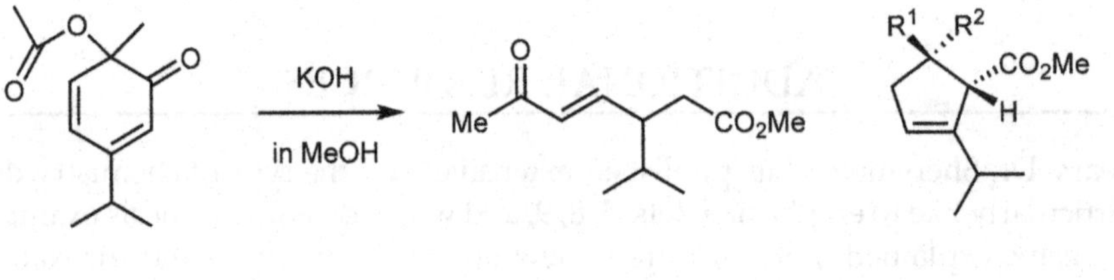

SCHEME 1

SOLUTION: 1 → 2

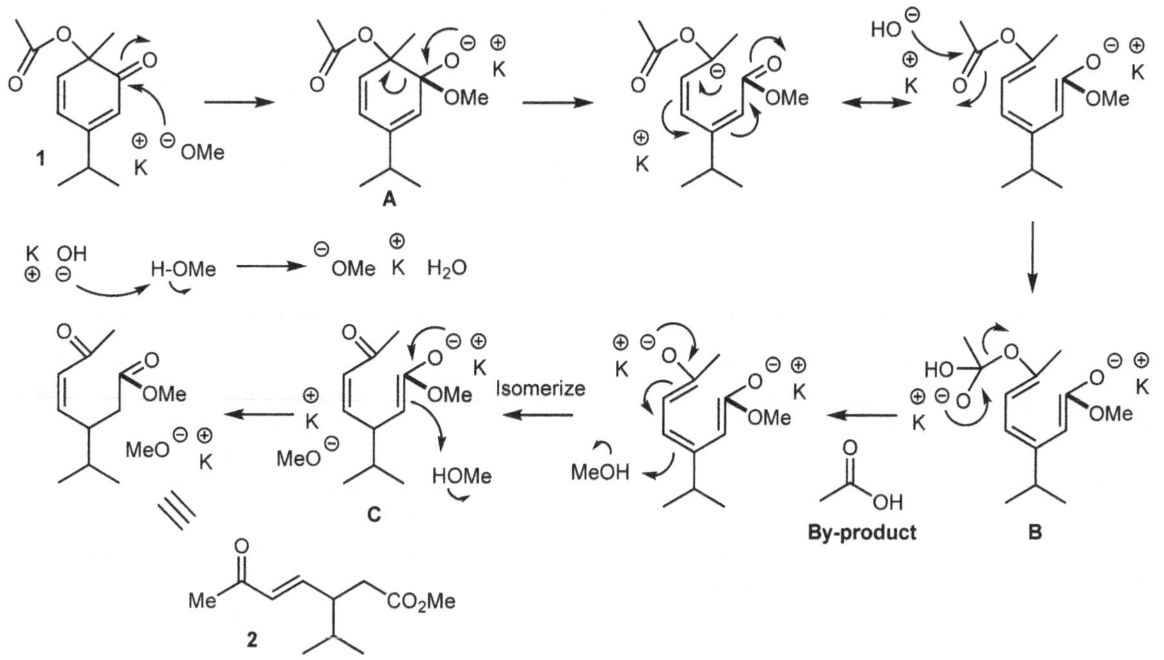

BALANCED CHEMICAL EQUATION: 1 → 2

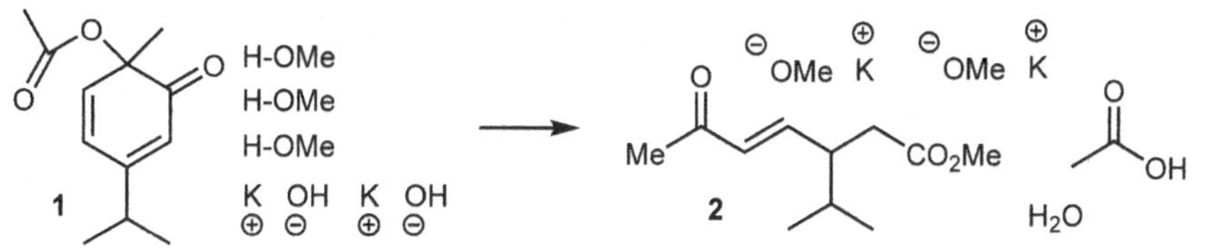

SOLUTION: 2 → 3, 4

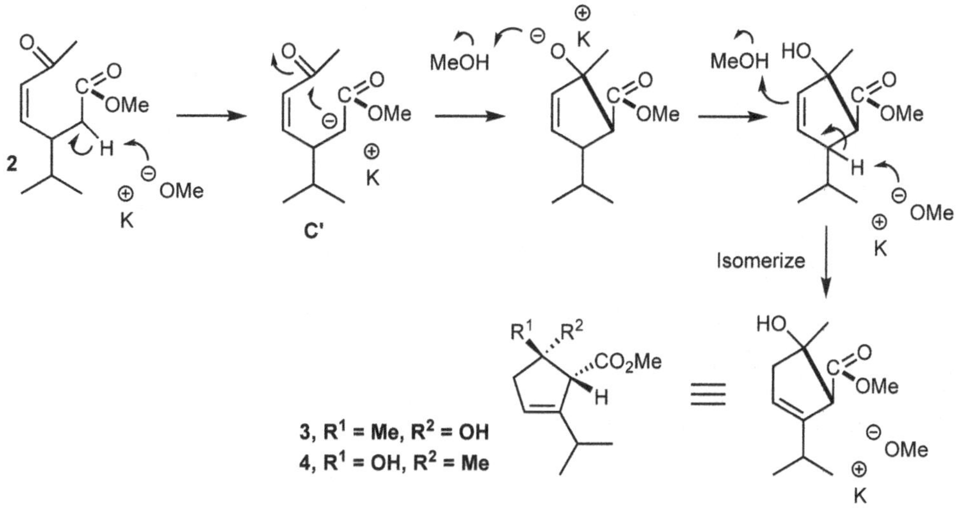

3, R¹ = Me, R² = OH
4, R¹ = OH, R² = Me

BALANCED CHEMICAL EQUATION: 1 → 3, 4

3, R^1 = Me, R^2 = OH
4, R^1 = OH, R^2 = Me

KEY STEPS EXPLAINED

The authors of this work described the mechanism for the **1** → **2** transformation as occurring first with attack of methoxide ion on the ring carbonyl to make an oxyanion intermediate, which decomposes via ring opening to a carbanion intermediate. This structure is resonance stabilized by charge delocalization into the newly formed ester group. After base-mediated hydrolysis of the acetate group of structure **B**, isomerization and keto-enol tautomerization follow to form the open chain product **2**. This product along with structure **C** (the structure preceding it) can also undergo anionic cyclization leading to a diastereomeric mixture of **3** and **4**. The authors went on to discuss reasons why the *cis* isomer of **2** as well as other anions of this structure do not form. The main reason is that in a *cis* conformation, compound **2** would likely cyclize either by anionic attack of either carbonyl groups or by aldol condensation thereby removing the presence of the cis isomer through cyclization. Other anions of **2**, though possible, become constrained geometrically from attacking either of the carbonyl groups to enable cyclization. The yield of compound **2** is therefore low due to the possibility of forming these anions, the majority of which can lead to products **3** and **4**.

ADDITIONAL RESOURCES

Two reviews of chemistry undertaken in similar systems for use in synthesis are provided.[2,3] It is interesting to note that under photochemical excitation, the transformation of **1** to **2** is likely to proceed via a ketene intermediate, for which there are numerous examples in the literature.[4–8] During our research, we have been unable to find an example of a thermally generated ketene in a similar system. This has led us to hypothesize that in the absence of photochemical excitation, a thermal two-electron transfer mechanism would not involve the generation of a ketene intermediate.

References

1. Carman RM, Van Dongen JMAM. A novel anionic rearrangement of a cyclohexadienone. A serendipitous synthesis of methyl (E)-3-isopropyl-6-oxohept-4-enoate. *Aust J Chem*. 1986;39:2171–2175. https://doi.org/10.1071/CH9862171.
2. Magdziak D, Meek SJ, Pettus TRR. Cyclohexadienone ketals and quinols: four building blocks potentially useful for enantioselective synthesis. *Chem Rev*. 2004;104:1383–1429. https://doi.org/10.1021/cr0306900.
3. Quideau S, Pouysegu L. Synthetic uses of orthoquinone monoketals and their orthoquinol variants. A review. *Org Prep Proced Int*. 1999;31(6):617–680. https://doi.org/10.1080/00304949909355348.
4. Snider BB, Shi Z. Synthesis of antitumor cyclic peroxy ketals related to Chondrillin and Xestins A and B. *J Org Chem*. 1990;55(22):5669–5671. https://doi.org/10.1021/jo00309a003.
5. Quinkert G, Heim N, Glenneberg J, et al. Total synthesis of the enantiomerically pure lichen macrolide (+)-aspicilin. *Angew Chem Int Ed Engl*. 1987;26:362–364. https://doi.org/10.1002/anie.198703621.
6. Quinkert G, Scherer S, Reichert D, et al. Stereoselective ring opening of electronically excited cyclohexa-2,4-dienones: cause and effect. *Helv Chim Acta*. 1997;80:1683–1772. https://doi.org/10.1002/hlca.19970800602.
7. Tidwell TT. *Ketenes II*. Hoboken, NJ: John Wiley & Sons, Inc.; 2006.
8. Henry CH, Bolien D, Ibanescu B, Bloodworth S, Harrowven DC, Zhang X, Craven A, Sneddon HF, Whitby RJ. Generation and trapping of ketenes in flow. *Eur J Org Chem*. 2015;1491–1499. https://doi.org/10.1002/ejoc.201403603.

Question 298: Less Common Sigmatropic Rearrangements: [3,4], [3,5], or Even [2,3]?

In 1997 American chemists investigated the thionation of strained 1,4-diketones with the expectation that low-temperature oxidation conditions would produce fused 1,2-dithiins,[1] an unusual antiaromatic ring system present in several natural products. During their investigation it was found that treatment of the *endo, exo*-diketone **1** with bis(tricyclohexyltin) sulfide gave a mixture of **2** (45%) and **3** (20%) while the *endo, endo*-diketone **2** formed only **3** (75%) (Scheme 1). Suggest mechanisms for the formation of **2** and **3**.

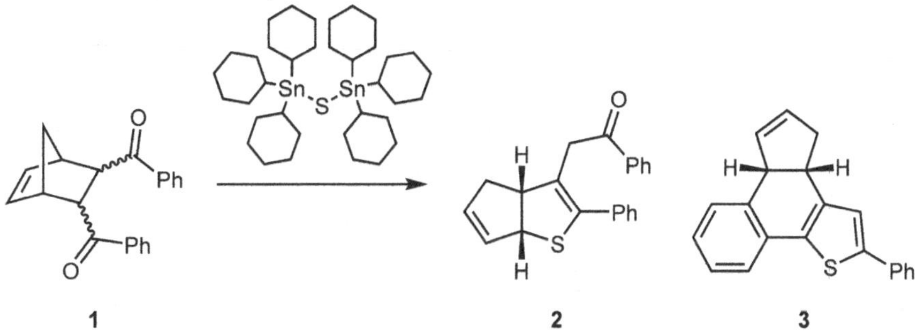

SCHEME 1

SOLUTION: THIONATION MECHANISM

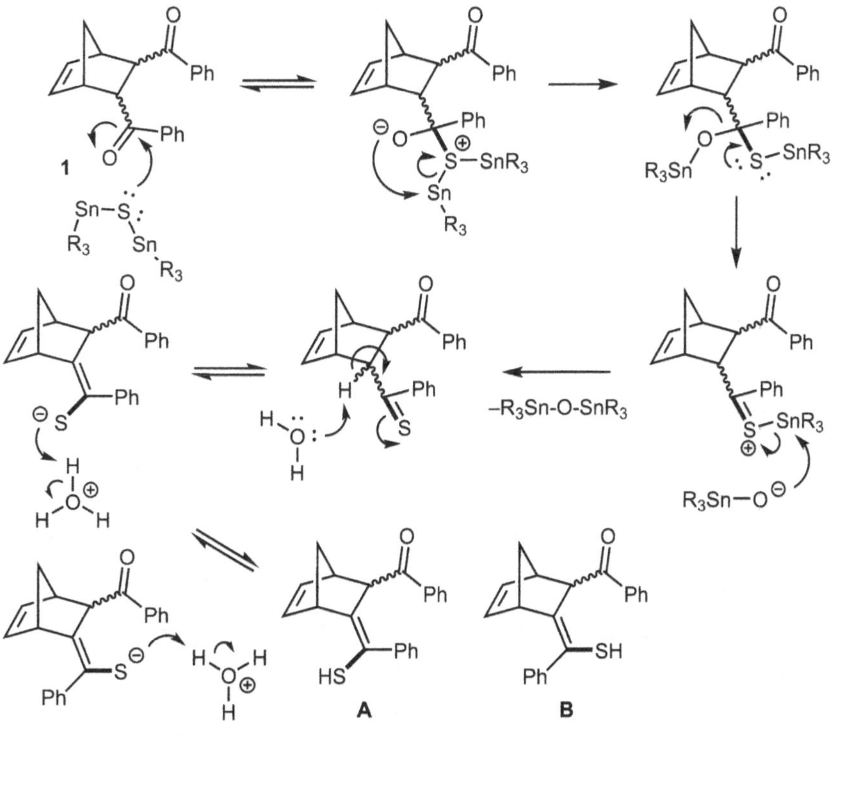

SOLUTION: A → 2

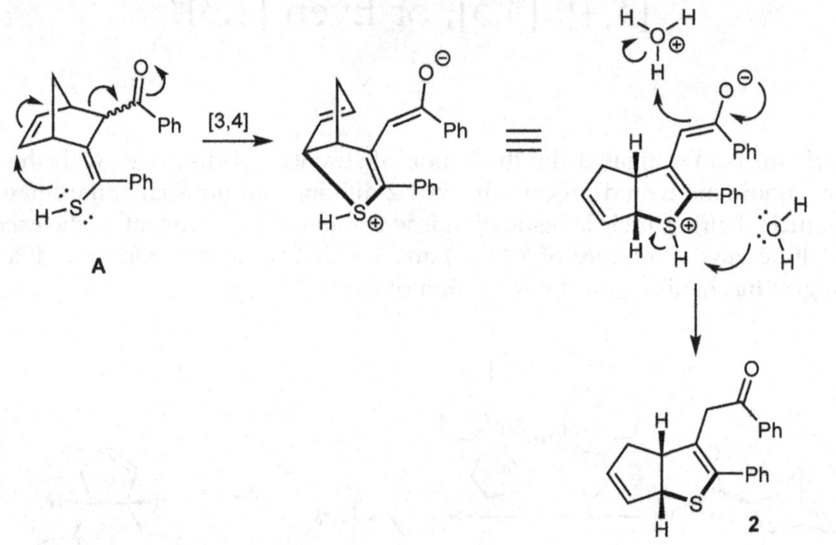

BALANCED CHEMICAL EQUATION: 1 → 2

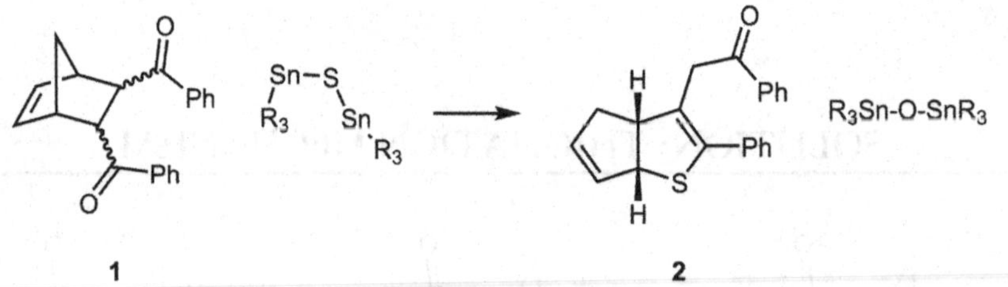

1   2

SOLUTION: B → 3

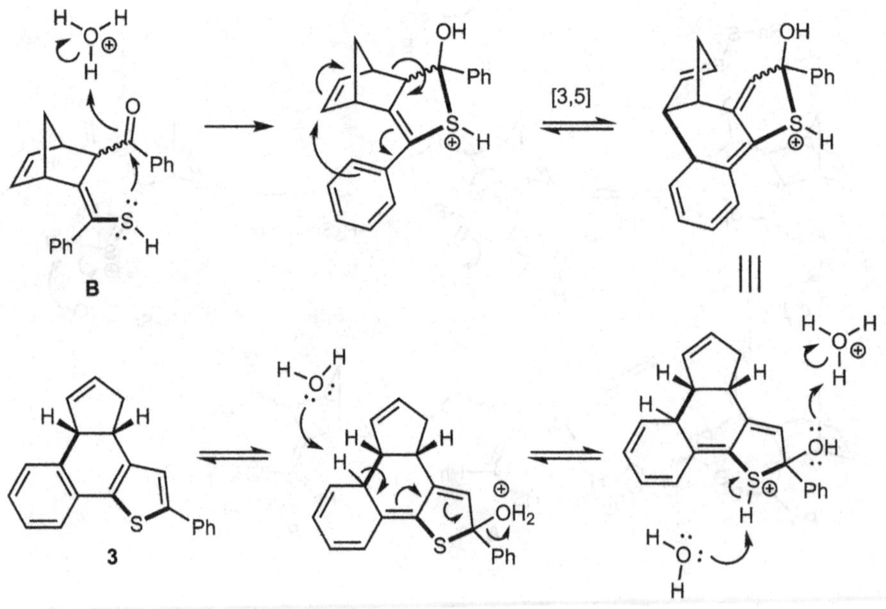

BALANCED CHEMICAL EQUATION: 1 → 3

1   **3**

KEY STEPS EXPLAINED

The unexpected [3,4] and [3,5] rearrangements observed by the authors can be explained by considering the enethiols **A** and **B** formed at the end of the thionation reaction.[1] When an initial mixture of the *endo, exo*-diketone **1** undergoes thionation with bis(tricyclohexyltin) sulfide, it is assumed that both enethiols **A** and **B** are formed whereas when the *endo, endo*-diketone **1** reacts, only the enethiol **B** is formed. This result, the authors stipulate, occurs in order for the enethiol **B** generated to minimize ring strain from the neighboring aroyl group. This strain exists between the enethiol and phenyl groups. Once both enethiols are formed, their orientation can assist the [3,4] and [3,5]-sigmatropic rearrangements leading to products **2** and **3**, respectively.[2, 3] In terms of energetics, the authors believed that expulsion of the energy in the norbornene ring system and formation of the thiophene ring system were the driving force behind these unusual rearrangements. They also proposed a possible [2,3]-sigmatropic rearrangement that can take a **B**-like structure to product **3**.

ADDITIONAL RESOURCES

For further examples of [3,4][4] and [3,5]-sigmatropic rearrangements,[5,6] we recommend the following reports. An unusual [9,9]-sigmatropic rearrangement had also been reported.[7]

References

1. Kim DSHL, Freeman F. Enethiol assisted [3,4] and [3,5] sigmatropic rearrangements during thionation of 2,3-diaroylbicyclo[2.2.1]hepta-5-enes with boron sulfide. *Tetrahedron Lett.* 1997;38:799–802. https://doi.org/10.1016/S0040-4039(96)02485-9.
2. Battye RJ, Jones DW, Tucker HP. Positional selectivity in the rearrangement of 7-formyl and 7-vinyl norcaradienes; evidence for 3,5-sigmatropy. *J Chem Soc Chem Commun.* 1988;495–496. https://doi.org/10.1039/C39880000495.
3. Battye PJ, Jones DW. Evidence for thermal 3,5-sigmatropy of 7-vinylnorcaradienes. *J Chem Soc Chem Commun.* 1986;1807–1808. https://doi.org/10.1039/C39860001807.
4. Erden I, Xu FP, Cao WG. Sigmatropic shifts in allene oxide rearrangements: first general route to [3,4] shifts in aliphatic systems. *Angew Chem Int Ed Engl.* 1997;36:1516–1518. https://doi.org/10.1002/anie.199715161.
5. Risitano F, Grassi G, Foti F, Filocamo F. 5-exo-trigonal cyclization and [3,5] rearrangement of N-aryl benzamidoximes by reaction with nitrile oxides. *Tetrahedron.* 1997;53:1089–1098. https://doi.org/10.1016/S0040-4020(96)01032-0.
6. Jacobi PA, Buddhu SC, Fry D, Rajeswari S. Studies on the synthesis of phytochrome and related tetrapyrroles. Dihydropyrromethenones by photochemical rearrangement of N-pyrrolo enamides. *J Org Chem.* 1997;62:2894–2906. https://doi.org/10.1021/jo970288j.
7. Park KH, Kang JS. [9,9]-Sigmatropic shift in a benzidine-type rearrangement. *J Org Chem.* 1997;62:3794–3795. https://doi.org/10.1021/n9703966.

Question 299: Failure to Construct an Oxetane by S$_N$' Rearrangement

During the design phase of a synthesis aimed at providing access to the ring system found in 2,7-dioxatricyclo [4.2.1.0^{3,8}]nonane, scientists discovered that treatment of a key subunit **1** with base produced no reaction while treatment with a catalytic amount of acid resulted in the formation of 4-methylacetophenone (Scheme 1).[1] This transformation also occurred in the storage of **1**. Suggest a mechanism to account for the formation of 4-methylacetophenone from **1**.

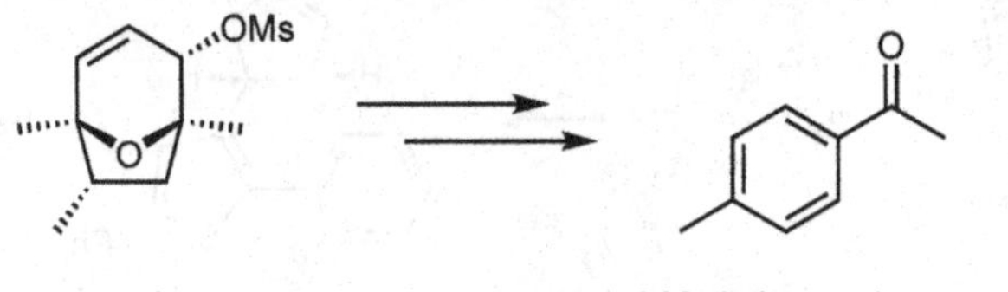

1   4-Methylacetophenone

SCHEME 1

SOLUTION

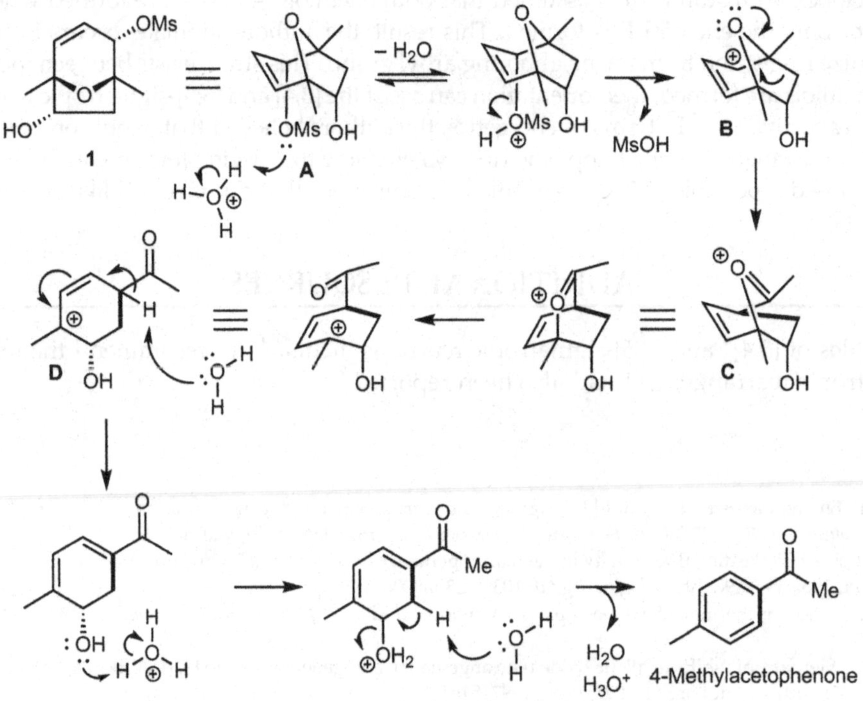

BALANCED CHEMICAL EQUATION

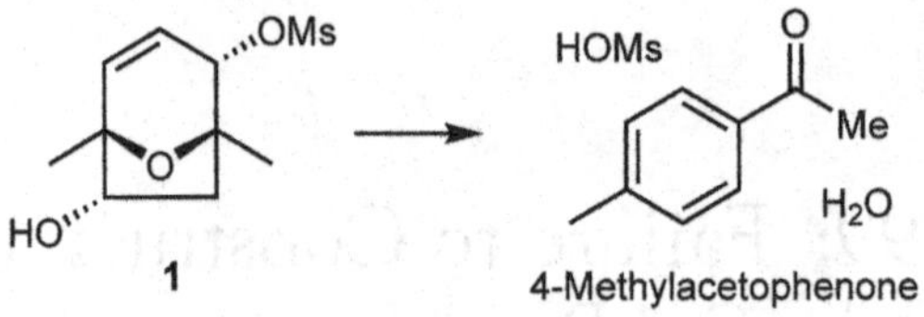

1   4-Methylacetophenone

KEY STEPS EXPLAINED

The authors of this work were initially interested in transforming the subunit **1** into the 4-membered oxetane ring system under base or thermal conditions with an expected allylic displacement strategy (Scheme 2).[1] The unexpected rearrangement they observed occurred under thermal heating with catalytic acidic conditions and was depicted as a mechanistic scheme of interest while the focus shifted to new approaches for the synthesis of the oxetane ring. As for the acid-catalyzed mechanism depicted earlier, the key to understanding this mechanism is being able to visualize the structures in boat-type conformations and to maintain the structural aspect of the illustrations through the very end, as is obvious from structures **1** to **A** and **C** to **D**. The rearrangement portion of this mechanism occurs between structures **B** and **C** with the formation of the oxonium ion and 1,2-C—C bond migration.

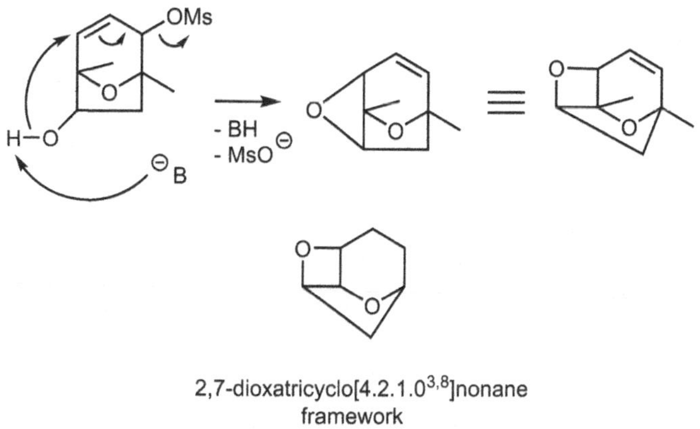

2,7-dioxatricyclo[4.2.1.0^{3,8}]nonane
framework

SCHEME 2

ADDITIONAL RESOURCES

In a recent review of the important role of oxetanes in drug discovery and synthesis, we direct the reader to the work of Burkhard et al.[2]

References

1. Marshall KA, Mapp AK, Heathcock CH. Synthesis of a 2,7-dioxatricyclo[4.2.1.0^{3,8}]nonane: a model study for possible application in a synthesis of dictyoxetane. *J Org Chem*. 1996;61:9135–9145. https://doi.org/10.1021/jo961680k.
2. Burkhard JA, Wuitschik G, Rogers-Evans M, Müller K, Carreira EM. Oxetanes as versatile elements in drug discovery and synthesis. *Angew Chem Int Ed*. 2010;49:9052–9067. https://doi.org/10.1002/anie.200907155.

Question 300: Acid-Catalyzed Isomerization of a Tetraspiroketone

In 1987 German chemists synthesized the tetraspiroketone **1** and sought to investigate whether it underwent acid-catalyzed rearrangement preferentially or exclusively to the bispropellanone **2** or to the pentacyclic ketone **3** (Scheme 1).[1] Devise mechanisms for both possible transformations.

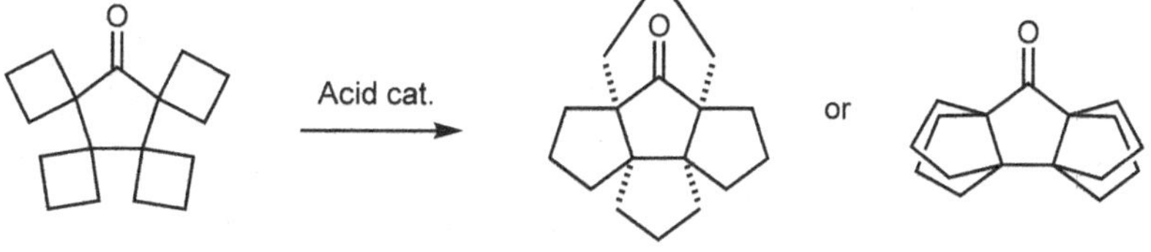

SCHEME 1

SOLUTION: 1 → 2

SOLUTION: 1 → 3

KEY STEPS EXPLAINED

In their original work, Fitjer and Quabeck wanted to demonstrate the preferential acid-catalyzed rearrangement pathway for the starting tetraspiroketone **1**. Their two proposed pathways began with acid-catalyzed formation of an oxonium ion followed by subsequent dual 1,2-carbon bond shifts to the common structure **A**, a β-hydroxy carbenium ion. At this point, the two proposed mechanisms diverge wherein the (**1** → **2**) pathway continues along the expected 1,2-carbon bond shift trajectory with structure **B** until arriving at the final bispropellanone product **2** whereas the (**1** → **3**) pathway encounters two 1,3-carbon bond shifts leading from structure **C** to the final pentacyclic ketone product **3**. Although theoretical calculations could not conclusively rule out either of these pathways, experimentally, under a solution of *p*-toluenesulfonic acid in benzene at 80°C, it was shown that the **1** → **3** pathway dominates preferentially. The authors attributed this result to nonbonding interactions that are "minimized by a slight folding of the central ring, which cause the peripheral rings to be crossed pairwise."[1] Observing the target bonds formed, it becomes easier to observe how these interactions might occur.

ADDITIONAL RESOURCES

We direct the reader to further work by Dr. Lutz Fitjer for examples of similar carbon-walk type transformations.[2–4] For an example of an oxygen-walk type of transformation, we recommend the work of Bruice et al.[5]

References

1. Fitjer L, Quabeck U. Pentacyclo[11.3.0.0^{1,5}.0^{5,9}.0^{9,13}]hexadecane ([4.5]coronane). *Angew Chem Int Ed*. 1987;26:1023–1025. https://doi.org/10.1002/anie.198710231.
2. Wehle D, Fitjer L. Heptacyclo[19.3.0.0^{1,5}.0^{5,9}.0^{9,13}.0^{13,17}.0^{17,21}]-tetracosane([6.5]coronane). *Angew Chem Int Ed*. 1987;26:130–132. https://doi.org/10.1002/anie.198701301.
3. Fitjer L, Steeneck C, Gaini-Rahimi S, et al. New rotane family: synthesis, structure, conformation, and dynamics of [3.4]-, [4.4]-, [5.4]-, and [6.4] rotane. *J Am Chem Soc*. 1998;120(2):317–328. https://doi.org/10.1021/ja973118x.
4. Fitjer L, Kanschik A, Gerke R. A new approach to helical primary structures of four-membered rings: (P)- and (M)-tetraspiro[3.0.0.0.3.2.2.2]hexadecane. *Tetrahedron*. 2004;60:1205–1213. https://doi.org/10.1016/j.tet.2003.11.081.
5. Bruice PY, Kasperek GJ, Bruice TC, Yagi H, Jerina DM. Oxygen walk as a complementary observation to the NIH shift. *J Am Chem Soc*. 1973;95(5):1673–1674. https://doi.org/10.1021/ja00786a060.

Index

Note: Page numbers followed by *f* indicate figures, *t* indicate tables, and *s* indicate schemes.

Printed in the United States
By Bookmasters